MOVING THE EARTH

THE WORKBOOK OF EXCAVATION

MOVING
THE EARTH

THE WORKBOOK OF EXCAVATION

BY

Herbert L. Nichols, Jr.

ILLUSTRATIONS BY HELEN SCHWAGERMAN
AND JOSEPH A. ROMEO

•THIRD EDITION•

McGraw-Hill Publishing Company

New York   St. Louis   San Francisco   Auckland
Bogotá   Hamburg   London   Madrid   Milan   Mexico
Montreal   New Delhi   Panama   Paris   São Paulo
Singapore   Sydney   Tokyo   Toronto

ISBN 0-07-046483-9

Library of Congress Catalog Card Number 76-17135
International Standard Book Number 0-911040-12-9
Manufactured in the U.S.A.
Composition by Lettick Incorporated, Bridgeport, Connecticut 06606

For more information about other McGraw-Hill materials, call 1-800-2-MCGRAW *in the United States. In other countries, call your nearest McGraw-Hill office.*

ACKNOWLEDGMENTS

Acknowledgment is gratefully made of the assistance supplied by thousands of manufacturers, associations, equipment dealers, contractors, supervisors, operators, engineers, architects, job inspectors, and public officials, who contributed generously of their time and knowledge to make this book possible.

The author sincerely regrets that he has not known the names of most of these helpers and well-wishers, and that the names of so many others are lost among the mountains of paper involved in the preparation of this book, so that it is not possible to thank them personally.

Apologies are offered to those whose contributions have been omitted or wrongly credited.

Particular thanks are now extended to my family, employees, and associates who assumed most of the burdens of my work so that I could have time to write, and to the people, companies and references listed below, each of whom has supplied a definite part of the information included in the book.

Many contributors have moved, changed names, gone out of business, or died during the 24 years since publication of the first edition of MOVING THE EARTH.

When present address and status are unknown, the last known name and address are given, followed by an asterisk to indicate that they may not be current. The publisher will appreciate any recent information about these contributors.

AASHTO (American Association of State Highway and Transportation Officials), National Press Bldg., Washington, D.C. 20004.

Accubeam Division, Constructors Supply Co., 15629 Clanton Circle, Sante Fe Springs, Calif. 90670.

Acker Drill Co., Inc., Box 830, Scranton, Pa. 18501.

Aeroquip Corporation, 300 So. East Ave., Jackson, Mich. 49203.

Ahlberg Bearing Co., Chicago, Ill. 60632.*

Allied Steel & Tractor Products, Inc., 5800 Harper Road, Solon, Ohio 44139.

American Brakeblok Div., American Brake Shoe Co., 530 Fifth Ave., New York, N.Y. 10036.

American Concrete Paving Association, 1112 W. 22nd St., Oak Brook, Ill. 60523.

American Road Builders Association, 525 School Street S.W., Washington, D.C. 20024.

American Cyanamid Company, 1937 W. Main St., Stamford, Conn. 06902.

American Steel Foundries, 1001 Prudential Plaza, Chicago, Ill. 60601.

Ametek/U.S. Gauge Division, 911 Clymer Ave., Sellersville, Pa. 18960.

Amorossi, Ricky, Brookfield, Conn. 06804.

ACKNOWLEDGMENTS

Amsco Division of Abex Corp., 387 E. 14th St., Chicago Heights, Ill. 60411.

Armco Steel Corp., Metal Products Div., (Armco Drainage & Metal Products, Inc.), 1001 Grove St., Middletown, Ohio 45042.

Arps Corporation, New Holstein, Wis. 53061.

Asphalt Equipment Company, Inc., 13333 U.S. Highway 24 West, Fort Wayne, Ind. 46804.

Asphalt Institute, The, College Park, Md. 20740.

Associated Equipment Distributors (AED), 615 W. 22nd St., Oak Brook, Ill. 60521.

Associated General Contractors of America, 1957 E Street N.W., Washington, D.C. 20006.

ATECO (American Tractor Equipment Corp.), 9131 San Leandro St., Oakland, Calif. 94603.

Athey Products Corp., Box 669, Raleigh, N.C. 27602.

Automatic Grade Light of America (AGL), Div. of Blount & George, Inc., Jacksonville, Ark. 72076.

Balderson, Inc., 430 Lincoln, Wamego, Ks. 66547.

Baldwin-Lima-Hamilton, *see* Clark Equipment Co., Lima Div.

Barber-Greene Co., 400 N. Highland Ave., Aurora, Ill. 60507.

Barco Manufacturing Co., Barrington, Illinois.*

BASIC PROCEDURES OF DIAMOND AND SHOT CORE DRILLING, Acker Drill Company, Scranton, Pa. 18501.

Beebe Bros. Inc., 2724 6th Ave. South, Seattle, Wash. 98124.

Be-Ge Division, Box 67, Gilroy, Calif. 95020.

Bendix Products Div., Bendix Aviation Corp., South Bend, Ind. 46620.*

Bethlehem Steel Company, Bethlehem, Pa. 18016.

Bid-Well Corporation, Subsidiary of CMI Corp., Canton, S.D. 57013.

BLASTERS' HANDBOOK, 15th Edition, E. I. duPont de Nemours & Co. (Inc.), Wilmington, Delaware 19898.

Bomag (Canada Limited), 1300 Aerowood Drive, Mississauga, Ontario.

Boston Lockport Block Co., 100 Condor, Boston, Mass. 02128.

William Bros Boiler & Mfg. Co., Minneapolis, Minn. 55414.*

Buckeye Div., Gar Wood Industries, Findlay, Ohio 45840.

Bucyrus-Erie Company, Box 56, S. Milwaukee, Wis. 53170.

Buffalo Bomag, Koehring Road Division, Springfield, Ohio 45501.

Bush Hog - Div. of Allied Products Corp., P.O. Box 1039, Selman, Ala. 36701.

California, State Board of Water Resources, Sacramento, Calif. 95814.

California, State Dept. of Public Works, Sacramento, Calif. 95814.

Calweld - Division of Smith International, Santa Fe Springs, Calif. 90670.

Canadian Construction Association, 151 O'Connor Street, Ottawa, Canada K2P 1T2.

J. I. Case, A Tenneco Company, 700 State St., Racine, Wis. 53404.

Caterpillar Tractor Company, 100 N.E. Adams Street, Peoria, Ill. 61602.

Celanese Fibers Marketing Co., 1211 Avenue of the Americas, New York, N.Y. 10036.

Central Scientific Company, 260 South Kestner Ave., Chicago, Ill. 60623.

Chain Systems, Division of R. K. Carter & Co., Box 838, Church Street Station, New York, N.Y. 10008.

Challenge-Cook Bros., Inc., 15421 East Gale Ave., Industry, Calif. 91745.

Charles Machine Works, Inc., West Ditch Witch Road, Perry, Okla. 73077.

Chevron Asphalt Co., 555 Market St., San Francisco, Calif. 94120.

Chicago Pneumatic Tool Company, 6 E. 44th St., New York, N.Y. 10017.

Clark Equipment Company, Austin Western Division, 601 N. Farnsworth Ave., Aurora, Ill. 60507.

Clark Equipment Company, Construction Machinery Division, Pipestone Road, Benton Harbor, Mich. 49022.

Clark Equipment Company, Lima Division, 1046 So. Main St., Lima, Ohio 45802.

Clark Equipment Company, Melroe Division, Gwinner, N.D. 58040.

Cleveland Trencher Co., Div. of American Hoist & Derrick Co., 63 S. Robert St., St. Paul, Minn. 55107.

CMI Champion, 5600 E. 39th Ave., Denver, Colo. 80207.

CMI Corp., 1-40 Morgan Rd., Oklahoma City, Okla. 73101.

COMPILATION OF RENTAL RATES FOR CONSTRUCTION EQUIPMENT, 26th Edition, Associated Equipment Distributors, Oakbrook, Ill. 60521.

State of Connecticut, State Department of Transportation, Hartford, Conn. 06115.

CONSTRUCTION COST CONTROL. American Society of Civil Engineers, New York, N.Y. 10017.

Construction Industry Mfrs. Assn., CIMA, Chicago, Ill. 60603.

CONSTRUCTION METHODS & EQUIPMENT, 1221 Avenue of the Americas, New York, N.Y. 10020.

Construction Safety Association of Ontario, 74 Victoria St., Toronto, Canada M5C 2A5.

Construction Supply Company, P.O. Box 10113, Palo Alto, Calif. 94303.

Construction Technology Inc., (CONTECH), 607 Madisen Plaza Bldg., 90 Madisen St., Denver, Colo. 80206.

Contractor's Pump Bureau, c/o Associated General Contractors of America, 1957 E Street N.W., Washington, D.C. 20006.

Conveyor Equipment Manufacturers Assn., 1 Thomas Circle, Washington, D.C. 20005.*

Crisafulli Pump Co., Inc., Box 1051, Glendive, Mont. 59330.

Cubic Industrial Corp., 4285 Ponderosa Ave., San Diego, Calif. 92138.

Cummins Engine Co., Inc., 1000 Fifth Street, Columbus, Ind. 47201.

Dart Truck Company, P.O. Box 321 - 1301 Chouteau Trafficway, Kansas City, Mo. 64141.

DEEP ROTARY DRILLING by Robert L. Hazlett. The Ohio Oil Co., Findlay, Ohio 45840.

Deere & Company, John Deere Road, Moline, Ill. 61265.

Delaware Water Supply News, Water Supply, Gas & Electricity Dept., New York, N.Y. 10007.

Dempster Brothers, Box 3127, Knoxville, Tenn. 37917.

DESIGN, Volume One, by Elwyn E. Seelye. John Wiley and Sons, Inc., New York, N.Y. 10016.

Detroit Diesel Allison, Div. General Motors, 13400 West Outer Drive, Detroit, Mich. 48228.

Deutz Diesel Corporation, 7585 Ponce de Leon Circle, Atlanta, Ga. 30340.

H. S. Dhillon, Perini Corp., Framingham, Mass. 01701.

Diamond Ironworks, Div. of Goodman Mfg. Co., Halsted St., Chicago, Ill. 60609.

Digmor Equipment & Engineering Co., Inc., 6308 Woodman Ave., Van Nuys, Calif. 91401.

Dixie Dredge Corp., Miami, Fla. 33150.*

Dodge Div., Chrysler Corp., Detroit, Mich. 48231.

Dorr Oliver Inc., 77 Havemeyer Lane, Stamford, Conn. 06904.

Drott Mfg. Company, Box 1087, Wausau, Wis. 54401.

Duncan, Archie, Norwalk, Conn. 06856.

E. I. du Pont de Nemours & Co. (Inc.), 1007 Market St., Wilmington, Del. 19898.

Eagle Crusher Company, Inc., Rt. 2, Box 72M, Galion, Ohio 44833.

Eagle Iron Works, 203 Holcomb Ave., Des Moines, Iowa 50313.

Eimco Division, Envirotech Corporation, 669 W. 2nd South, Salt Lake City, Utah 84110.

Ellicott Machine Corp., 1613 Bush St., Baltimore, Md. 21230.

Emaco, Inc., 111 Van Riper Ave., Elmwood Park, N.J. 07407

ENGINEERING NEWS-RECORD, 1221 Avenue of the Americas, New York, N.Y. 10020.

Ensign-Bickford Company, 660 Hopmeadow St., Simsbury, Conn. 06070.

Equipment Guide-Book Company, 3980 Fabian Way, Palo Alto, Calif. 94303.

Equipment Supply Div., 3980 Fabian Way, Palo Alto, Calif. 94303.

Erie-Strayer Company, 400 Rudolph Ave., Erie, Pa. 16502.

Essick Mfg. Company, 1950 Santa Fe Ave., Los Angeles, Calif. 90021.

E. D. Etnyre & Company, 200 Jefferson Street, Oregon, Ill. 61061.

Euclid, Inc., Subsidiary of White Motor Corp., 22221 St. Clair Ave., Cleveland, Ohio 44117.

Al Evans Winches, Inc., Gloucester, Va. 23061.*

Eversman Mfg. Company, Curtis at Fifth St., Denver, Colo. 80204.

EXCAVATING CONTRACTOR, 21590 Greenfield Road, Oak Park, Mich. 48237.

FAMOUS SUBWAYS AND TUNNELS OF THE WORLD by Edward and Muriel White, Random House, New York 10022.

Farnum, H. E., Salt Lake City, Utah 84104.

Fiat-Allis Construction Machinery Inc., 245 E. North Ave., Carol Stream, Ill. 60187.

Firestone Tire & Rubber Company, 1200 Firestone Parkway, Akron, Ohio 44317.

Fleco Corporation, Box 2370, Jacksonville, Fla. 32203.

Ford Motor Company, Ford Tractor Operations, 2500 E. Maple Road, Troy, Mich. 48084.

L. B. Foster Company, Seven Parkway Center, Pittsburg, Penna. 15220.

Four Wheel Drive Auto Co., Clintonville, Wisc. 54929.*

Fruehauf Div., Fruehauf Corp., 10900 Harper Ave., Detroit, Mich. 48213.

Fullam, Harry, Armonk, N.Y. 10504.

Fuller Form, Inc., 24 E. Pioneer St., Phoenix, Ariz. 85040.

Galion Manufacturing Co., Division of Dresser Industries, Galion, Ohio 44833.

Gar Wood Division - Sargent Industries, 32500 Van Born Road, Wayne, Mich. 48184.

Gardner-Denver Company, 1863 Gardner Expressway, Quincy, Ill. 62301.

Geophysical Specialties Co., 3110 Shores Blvd., Wayzata, Minn.

Gleason, William P. F., Port Chester, N.Y. 10573.

Goff, Patrick, North Salem, N.Y. 10560.

Goodman Equipment Corp., 4834 S. Halsted St., Chicago, Ill. 60609.

Goodyear Tire & Rubber Company, 1144 E. Market St., Akron, Ohio 44316.

Grove Mfg. Company, Div. of Walter Kidde & Co., Rte. 16, Box 21, Shady Grove, Pa. 17256.

Gruendler Crusher & Pulverizer Co., 2915 South Market St., St. Louis, Mo. 63106.

George Haiss Mfg. Co., New York, N.Y. 10001.*

HANDBOOK OF CULVERT AND DRAINAGE PRACTICE, Armco Drainage Products Assn., Middletown, Ohio 45042.

HANDBOOK OF HEAVY CONSTRUCTION, 2nd Edition, edited by Havers & Stubbs. McGraw-Hill. 1971.

Harnischfeger Corporation, 4400 West National Ave., Milwaukee, Wis. 53246.

Harpold, Ed, Huntington Beach, Calif. 92647.

ACKNOWLEDGMENTS

Heil Company, 3000 W. Montana St., Milwaukee, Wis. 54201.

Hein-Werner Corporation, 1400 National Ave., Waukesha, Wis. 53186.

Hewitt-Robins, Div. of Litton Industries, 270 Passaic Ave., Passaic, N.J. 07055.

Highway Equipment Co., Inc., 616 D. Ave. N.W., Cedar Rapids, Iowa 52405.

Highway Research Board, *See* Transportation Research Bd.

Homelite, a Division of Textron Inc., P.O. Box 7047, Charlotte, N.C. 28217

Frank G. Hough Co. *See* International Harvester Co.

Huber Corporation, A division of A-T-O Inc., Box 501, Marion, Ohio 43302.

Hughes Tool Company, 5425 Polk Ave., Houston, Texas 77023.

Huron Mfg. Corporation, Box 1398, Huron, S.D. 57350.

Hyster Company, Box 289, Kewanee, Ill. 61443.

State of Illinois, Division of Highways, Springfield, Ill. 62706.

Ingersoll-Rand Company, Phillipsburg, N.J. 08865.

Insley Mfg. Corporation, 801 N. Olney St., Indianapolis, Ind. 46201.

International Harvester Company, 401 N. Michigan Ave., Chicago, Ill. 60611.

International Harvester Company, Construction Equipment Div., 10400 West North Ave., Melrose Park, Ill. 60160.

International Union of Operating Engineers, 1125 17th St. N.W., Washington, D.C. 20036.

Intrafix, Inc., 205 N. Elizabeth, Wichita, Kan. 67203.

Iowa Manufacturing Company, 916-16th St. N.E., Cedar Rapids, Iowa 52402.

ITT Marlow, Box 200, Midland Park, N.J. 07432.

Jackson Vibrators, Inc., 1905 Bernice Rd., Lansing, Ill. 60438.

Jacques Power Saw Company, Denison, Texas 75020.*

Jaeger Machine Company, 550 W. Spring St., Columbus, Ohio 43216.

Jeffrey Mfg. Company, Columbus Ohio 43216.*

Joy Manufacturing Company, Montgomeryville Industrial Center, Montgomeryville, Pa. 18936.

Kalba, Budd, White Plains, N.Y. 10604

Keller, Norman W., M.D., Greenwich, Ct. 06830

Keuffel & Esser Company, 20 Whippany St., Morristown, N.J. 07960

King-Seely Corp., Ann Arbor, Mich. 48107*

Koehring Company, 780 N. Water St., Milwaukee, Wis. 53201.

Koehring Road Division, 1210 Kenton St., Springfield, Ohio 73701.

Koehring Company, Lorain Division, 1374 E. 28th St., Lorain, Ohio 44053.

Koehring, MKT Division, Dover, N.J. 07801.

Kolman Division of Athey Products Corp., 5100 W. 12th St., Sioux Falls, S.D. 57101.

Krupp International, Fried Krupp, 350 Executive Blvd., Elmsford, N.Y. 10523.

La Bolsa Tile Co., Huntington Beach, Calif. 92647.*

Laser Alignment Inc., 6330 28th St., Grand Rapids, Mich. 49506.

Laserplane Corp., 5475 Kellenburger, Dayton, Ohio 45424.

Thomas Laughlin Company, Portland, Maine.*

Legge, Ernest C., Greenwich, Conn. 06830.

"Led" Ballast Inc., Boulder, Colo. 80302.*

LeRoi Div. - Dresser Industries Inc., Box 90, Sidney, Ohio 45365.

LeTourneau - Westinghouse, *See* WABCO

Liebherr - America, 4100 Chestnut Ave., Newport News, Va. 23605.

Linde Division, Union Carbide Co., 270 Park Ave., New York, N.Y. 10017.

Link-Belt Speeder Division, FMC Corp., 1201 Sixth Street S.W., Cedar Rapids, Iowa 52406.

MacDonald, Ralph, Armonk, N.Y. 10504.

Mack Trucks, Inc., Allentown, Pa. 18105.

Marathon LeTourneau Company, Sub. of Marathon Mfg. Co., Mobberly Ave., Longview, Texas 75601.

Marcincuk, Joseph, Byram, Conn. 10573.

Marion Power Shovel Company, Inc., 617 West Center St., Marion, Ohio 43302.

Marmon - Herrington Co., Indianapolis, Ind. 46204.*

Marvin Landplane Co., Box 209, Woodland, Calif. 95695.

Master Vibrator Co., Dayton, Ohio 45403.*

Mattson, George, Denver, Colo. 80209.*

McCaffrey - Ruddock Tagline Corp., Ruddock Bldg., Los Angeles, Calif. 90058.

Metal Forms Corp., 3334 N. Booth St., Milwaukee, Wis. 53212.

Micro-Grade Laser Systems Inc., 2352 Charleston Road, Mountain View, Calif. 94043.

MINING ENGINEERING, 345 E. 47th St., New York, N.Y. 10017.

Miskin Scraper Works, Inc., Ucon, Idaho 83454.

Mitts & Merrill Inc., 1939 S. Water St., Saginaw, Mich. 48601.

Monroe Standard, Inc., Galion, Ohio 44833.*

Morbark Industries, Winn, Mich. 48896.

Morris Machine Works, 31 E. Genesee St., Baldwinsville, N.Y. 13027.

Morse Chain Division of Borg Warner, S. Aurora St., Ithaca, N.Y. 14850.

M-R-S Manufacturing Company, Kearny Park, Box 199, Flora, Miss. 39071.

ACKNOWLEDGMENTS

Mud Cat Division, National Car Rental System, 5501 Green Valley Drive, Minneapolis, Minn. 55437.

National School of Heavy Equipment Operation, Charlotte, N.C. 28208.

State of New York, Department of Transportation, Poughkeepsie, N.Y. 14302.

Walter Nold Company, 24 Birch Rd., Natick, Mass. 01760.

Nordberg - Division of Rex Chainbelt Inc., Box 383, Milwaukee, Wis. 53201.

Northwest Engineering Company, 135 South La-Salle St., Chicago, Ill. 60603.

Occupational Safety and Health Administration (OSHA), Washington, D.C. 20210.

The Ohio Oil Company, Findlay, Ohio 45840

O & K Orenstein & Koppel, Clifton, New Jersey 07013

The Oliver Corp., 400 W. Madison St., Chicago, Ill. 60606.*

Omark Industries, Inc., 9701 S.E. McLaughlin Blvd., Portland, Ore. 97222.

Ottawa Steel Div., Ottawa, Kansas 66067.*

Owen Bucket Co., 6031 Breakwater Ave., Cleveland, Ohio 44102.

Oxy - Catalyst Inc., East Biddle St., West Chester, Pa. 19380.

Pacific Car & Foundry Company (CARCO), 1400 North 4th St., Renton, Wash. 98055.

Page Engineering Company, Clearing Post Office, Chicago, Ill. 60638.

Paris Mfg. Company, Paris, Ill. 61944.*

Parsons Company, Newton, Iowa 50208.*

Peabody Coal Company, 301 Olive St., St. Louis, Mo. 63102.

Peckham Industries, 686 Canal St., Stamford, Conn. 06902.

Peerless Chain Company, 1416 E. 8th St., Winona, Minn. 55987.

Pennsylvania, State of, Department of Reclamation, Harrisburg, Pa. 17120.

Perini Corporation, Framingham, Mass. 01701.*

Petersen Engineering Company, Santa Clara, Calif. 95052.*

Pettibone Corporation, 9501 West Devon Ave., Rosemont, Ill. 60018.

Pit and Quarry Publications, 105 W. Adams St., Chicago, Ill. 60603.

Poclain, 3401 Tidewater Trail, Fredericksburg, Va. 22401.

Portland Cement Association, Old Orchard Rd., Skokie, Ill. 60076.

Power Shovel & Crane Association, PSCA, Bureau of CIMA, 111 E. Wisconsin Ave., Milwaukee, Wis. 53202.

PRACTICAL TUNNEL DRIVING by Richardson and Mayo, McGraw-Hill Book Co., New York, N.Y. 10036.

PROCEDURE FOR ESTIMATING COSTS OF TUNNEL CONSTRUCTION (Bulletin No. 78, Appendix C), State Printing Plant, Sacramento, Calif. 95814.

Pullman - Standard Div. Pullman, Inc., 200 S. Michigan Ave., Chicago, Ill. 60604.

Racine Federated Industries Corp., 1637 Goold St., Racine, Wis. 53404.

Ramsey Winch Company, Box 15829, Tulsa, Okla. 74115.

RayGo Inc., 13500 Country Road 6, Minneapolis Industrial Park, Minneapolis, Minn. 55441.

Re-Bo Mfg. Co., Inc., 357 Park Ave., Newark, N.J. 07107.

RENTAL RATES ON CONTRACTORS' EQUIPMENT. Canadian Construction Association, Ottawa, Canada K2P IT2.

Rexnord Inc., P.O. Box 2022, Milwaukee, Wis. 53201.

Ridge Construction Co., Riversville Road, Greenwich, Conn. 06830.

Rivinius, Inc., RR2, Box 125, Eureka, Ill. 61530.

Robbins Company, The, 650 South Orcas St., Seattle, Wash. 98108.

ROCK TUNNELING WITH STEEL SUPPORTS by Proctor, White and Terzogh. Youngstown, Ohio, Commercial Shearing and Stamping Co.

Rogers Brothers Corporation, Albion, Pa. 16401.

Rolatape Corp., 4221 Redwood Ave., Marina Del Rey, Los Angeles, Cal. 90066.

Rome Industries, Cedartown, Georgia 30125.

Rosco Manufacturing Co., 3118 Snelling Ave., Minneapolis, Minn. 55046.

ROTARY DRILLING HANDBOOK, Fourth Edition, by J. E. Brantly. Palmer Publications, New York 10036.

Roth, Albert Jr., West Lane, Greenwich, Conn. 06830.

Rubber Manufacturers Assn., 444 Madison Ave., New York, N.Y. 10022.*

Ruhr Industries, Inc., Philadelphia, Pa. 19101.*

Ruth Company, The, Denver, Colo. 80202*

Salem Tool Company, 767 S. Ellsworth Ave., Salem, Ohio 44460.

Sanborn, Leonard F., Kingston, N.H. 03848.

Sauerman Bros., Inc., 620 S. 28th Avenue, Bellwood, Ill. 60104.

Schilt, Vernon, Greenwich, Conn. 06830.

Schramm, Inc., 901 E. Virginia Ave., West Chester, Pa. 19380.

Seaman Co., (A Div. of Stowell Industries, Inc.), Box 8331, Milwaukee, Wis. 53225.

Shunk Manufacturing Co., A Subsidiary of Chromalloy American Corp., 1460 Auto Ave., Bucyrus, Ohio 44820.

Smith Engineering Works, Div. of Barber-Greene, 532 E. Capitol Dr., Milwaukee, Wis. 53212.

Spencer Turbine Company, 486 New Park Ave., Hartford, Ct. 06110.

ACKNOWLEDGMENTS

Spencer, White & Prentis, Inc., 10 E. 40th St., New York, N.Y. 10016.

Sperry - Vickers, Salem Valve Division, 1441 So. Ellsworth St., Salem, Ohio 44460.

Sperry - Vickers Tulsa Division, 7217 East Pine St., Tulsa, Okla. 74115.

Spicer Transmission Div. Dana Corp. (Spicer Manufacturing), Box 986, Toledo, Ohio 43696.

Stardrill - Keystone, Beaver Falls, Pa. 15010.

Albert Stauffer Machine Shops, Honey Brook, Pa. 19344.*

Stevenson, Marion, Greenwich, Conn. 06830.

Stevenson, William S., Greenwich, Conn. 06830.

Stewart - Warner Corp., Alemite and Instrument Division, 1826 Diversey Parkway, Chicago, Ill. 60614.

Stiehl, Walter Alan, Somerville, Mass. 02145.

Talbert Trailers, Box 38, Rensselaer, Ind. 47978.

Tampo Manufacturing Co., Box 7248, San Antonio, Texas 78285.

Taylor, John C., Armonk, N.Y. 10504.

Taylor, S. G. Chain Co., Inc., 3 - 141st St., Hammond, Ind. 46325.

TEREX Division, General Motors Corporation, Hudson, Ohio 44236.

Thew Shovel Co., See Koehring Company, Lorain Division.

Thor Power Tool Company, 175 N. State St., Aurora, Ill. 60507.

Timken Company, The, Rock Bit Division, 333 East Fillmore St., Colorado Springs, Colo. 80901.

Timken - Detroit Axle Co., Detroit, Mich. 48209.*

Transport Trailers, A Div. of Barnard & Leas Mfg. Co. Inc., 1200 Twelfth St. S.W., Cedar Rapids, Iowa 52406.

Transportation Research Board, National Academy of Sciences, 2101 Constitution Avenue, N.W., Washington, D.C. 20418.

Triumph Machinery Company, Box 110, Hackettstown, N.J. 07840.

Troeber, Otto B.

Tube-Lok Products Division, Portland Wire & Iron Works, 4644 S.E. 17th Ave., Portland, Ore. 97202.

Tulsa Products, see Sperry - Vickers - Tulsa

Twin Disc Inc., 1330 Racine St., Racine, Wis. 53403.

W. S. Tyler Screening Division, A Subsidiary of Combustion Engineering Inc., 8200 Tyler Boulevard, Mentor, Ohio 44060.

Unit Crane & Shovel Corp., 1915 S. Moorland Road, New Berlin, Wis. 53151.

U.S. Department of Agriculture, Washington, D.C. 20250.

U.S. Army Engineers, Washington, D.C. 20315.

U.S. Coast & Geodetic Survey, Washington, D.C.

U.S. Department of Defense, Washington, D.C. 20305.

U.S. Forest Service, Washington, D.C. 20251.

U.S. Department of Transportation, Washington, D.C. 20590.

Universal Engineering Corp., 625 C Avenue N.W., Cedar Rapids. Iowa 52405.

Vermeer Mfg. Company, 3902 New Sharon Road, Pella, Iowa 50219.

Vibro-Plus Products, Inc., Stanhope, N.J. 07874.

Vickers, Inc., see Sperry-Vickers.

Vulcan Iron Works, Inc., 2725 North Australian Ave., West Palm Beach, Fla. 33407.

WABCO Construction and Mining Equipment Group — an American-Standard Company, 2301 N. Adams St., Peoria, Ill. 61601.

Wacker Corporation, 3808 W. Elm St., Milwaukee, Wis. 53209.

Walter Motor Truck Co., School Road, Voorheesville, N.Y. 12186.

Ware Machine Works, Inc., E. Main St., Ware, Mass. 69624.

Warner & Swasey Company, 31700 Solon Road, Solon, Ohio 44139.

Warren Group, Div. of Warren Tool, Box 68, Hiram, Ohio 44234.

Waukesha Motor Company, 100 W. St. Paul Ave., Waukesha, Wis. 53186.

Wegener, J. Peter, Stamford, Conn. 06902.

Wegener, Maj-Greth, Greenwich, Conn. 06830.

Wellman Engineering Company, Cleveland, Ohio 44113.*

Westchester, County of, Department of Health, County Office Bldg., White Plains, N.Y. 10601.

Western Rock Bit Mfg. Co., Salt Lake City, Utah 84101.*

David White Instruments, a division of Realist, Inc., Menomonee Falls, Wis. 53051.

White, Edward E., Larchmont, N.Y. 10538.

Wickwire Spencer Steel Div., Colorado Fuel & Iron Corp., New York, N.Y. 10022.*

Hugh B. Williams Mfg. Company, Dallas, Texas 75226.*

Williams Patent Crusher & Pulverizer Co., 2701 N. Broadway, St. Louis, Mo. 63102.

Worthington-CEI, Inc., Box 431, Holyoke, Mass. 01040.

York Modern Corp., Unadilla, N.Y. 13849.

Young Corp., Box 3522, Seattle, Wash. 98124.

Carl Zeiss, 7082 Oberkochen, West Germany.

Zygmont, George, Banksville, N.Y. 10506.

Jacket photograph courtesy of Koehring Company.

* Last known address

x

PREFACE

TO THE THIRD EDITION

This second revision of "Moving the Earth" adds to it most of the machines and many of the techniques developed during the fourteen years since the second edition was published.

At the same time, many obsolete machines have had their references shortened, or eliminated entirely. Others, however, have been kept in full. In general, units that are still widely used, or that demonstrate mechanisms of particular interest, have been kept. Occasionally, descriptions of machines of marginal interest have been either retained or dropped for convenience in page layout.

Old machines may be found in illustrations for newer text, when new pictures of equal usefulness were not found.

Makes and models of machines were often chosen for description on the basis of clarity and appropriateness of photographs, drawings and/or written material made available by manufacturers. Inclusion of particular makes or models, and omission of similar ones, therefore does not imply that those discussed or illustrated are preferred or considered superior to machines not listed or less fully described.

Herbert L. Nichols, Jr.
GREENWICH, CONN.
June, 1976

CONTENTS

Part One—The Work

CHAPTER ONE • LAND CLEARING

CONTENTS

CONTENTS

CONTENTS

CHAPTER FIVE
DITCHING AND DEWATERING

DITCHING

CONTENTS

CONTENTS

CONTENTS

CONTENTS

CHAPTER NINE
BLASTING AND TUNNELING

CONTENTS

CONTENTS

CHAPTER ELEVEN • COSTS

CONTENTS

Part Two—The Machines

CHAPTER TWELVE
BASIC INFORMATION

CONTENTS

CONTENTS

CHAPTER THIRTEEN
REVOLVING SHOVELS

CONTENTS

CHAPTER FOURTEEN
CONVEYOR MACHINERY

CONTENTS

CONTENTS

CHAPTER SIXTEEN
TRACTOR LOADERS

CONTENTS

CONTENTS

CHAPTER NINETEEN • GRADING AND COMPACTING MACHINERY

GRADERS

CONTENTS

CHAPTER TWENTY
COMPRESSORS AND DRILLS

CONTENTS

CHAPTER TWENTY-ONE
MISCELLANEOUS EQUIPMENT

CONTENTS

INTRODUCTION

"Moving the Earth" has been prepared to supply the first working coverage of the whole excavation industry.

The past twenty-five years have brought a total revolution in the machinery, the methods, and volume of excavation. The rate of development and diversification of machines is steadily increasing, and the techniques of efficient use become more complicated daily. The annual income from their work is now in billions of dollars.

But the literature on the subject has failed to keep up with its advances, and today a very large proportion of the highly specialized knowledge existing about these machines and their proper use has not been put in writing in such a form that it can be used for reference.

The result is that a tragic loss is incurred every year through unwise selection of equipment, damage to machinery through ignorance of its functions and weak points, and waste of time, material, and money in learning by trial and error. While the industry has proved that it is big and dynamic enough to absorb such losses, it would unquestionably be in an even better position today if they had been reduced.

"Moving the Earth" has been written primarily to fill the needs of those closest to the actual earth moving: the small contractor, the foreman, and the operator. Since their work is basic to all excavation and to the planning and direction of even the largest projects, the know-how for them should also answer most of the needs of the engineer, the architect, the superintendent, and the student.

Some sections of the book have been arranged to meet the needs of the property owner and home builder, and to assist company and public officials whose duties include planning or supervision of earth moving work.

Knowledge of the contractors' viewpoint on problems of site selection, cellar excavation and drainage, backfill, landscaping, topsoiling, planting, and lawn making can result in more satisfactory results at lower cost. The owner or responsible official can learn from these pages what can be done, and how. He can use this information in making his plans, in explaining to his architect and contractor exactly what he wants to do, and in defending his ideas against criticism.

The preparation of "Moving the Earth" has proved a more ambitious project than was apparent at the beginning. Earthmoving is not one industry, but many. It includes hundreds of highly specialized skills, which often have no relation to each other except in their common aim to move dirt or rock. Many of these are important and complex enough to deserve individual reference books.

To drive a tunnel deep underground is a totally different job than to make a fill across a swamp, but the contractor may lump them as a single operation, the waste from one becoming the building material for the other.

The operator of a baby bulldozer grading around a house has nothing in common with the man at the controls of a gigantic stripping shovel, except that they are both classified as operating engineers.

On the other hand, one size and make of bulldozer may be found backfilling a desert pipeline, uprooting jungle stumps in the path of a highway, pushing a scraper on an irrigation land leveling job, spreading tailings on a mine dump, or placing rock on the face of a dam. One busy hoe shovel may stack logs, pull stumps, pile topsoil, dig a cellar, load trucks, cut ditches, and lay pipe, all on one small job.

It has been difficult to arrange a comprehensive working guide to an industry with so many varied and separate occupations, which cannot be kept separate because of the way they overlap and interlock. The approach that has been chosen is to divide the book into two distinct sections.

Part One discusses the jobs the excavating contractor is called upon to do, including the preliminary ones of land clearing and rough surveying. The first ten chapters are concerned with the work itself; the basic ways to do it, the problems that arise, and the applications of different types of equipment to the work. Chapter Eleven discusses financing, bookkeeping, estimating, and insurance in regard to the requirements of the excavating contractor.

Part Two is focussed on the machines themselves. Every important type of excavating, hauling and grading equipment is discussed and illustrated. Treatment includes mechanical description of the parts and assemblies that go into the machines, the underlying principles of construction, the ways in which special work requirements are met, and suggestions on adjustment and maintenance.

The operation of each machine is explained. In many cases instruction is sufficiently detailed to enable a man to learn from the book alone, although this is less safe and satisfactory than personal instruction. However, because of the complexity of the skills involved, and the secondary importance of the units in general excava-

tion, operation of some machines (oil well drills, for example) has been described for the spectator rather than the operator. Even then, the fundamentals of the work methods are carefully explained.

Special operating techniques, and uses of a machine in various classes of work, will be found both under the main discussion of the unit in Part Two and in appropriate places in Part One. Reference to particular jobs may also be found under several headings. This separation and distribution of subject matter has been necessary to present information in its proper context, to avoid duplication, and to keep an already over-large book from growing out of bounds.

For example, there is a chapter devoted entirely to cellars, but further reference to the subject will be found under drainage and landscaping. Crawler tractors and bulldozers have a chapter of their own, but because of almost universal application to earthmoving work, are mentioned in at least nineteen of the other twenty chapters.

In order to provide quick and easy access to all the information on any subject, the subheadings are included in the Table of Contents, and abundant listings and cross references are supplied in an unusually complete index.

Terminology is a difficult problem in this industry, and may well have discouraged attempts of others to write about it. Some words, such as bulldozer and catskinner, have about the same meaning anywhere. Others, such as scraper and ditcher, may be used in reference to several completely different types of equipment. One machine may have many names. An outstanding example is the hoe shovel, which answers to this name, and to hoe, backhoe, pull shovel, drag shovel, ditching shovel, ditcher, and trencher.

This confusion exists partly because of the extremely rapid growth of the business and change in its equipment and methods, and partly because of the complete absence of standard reference material.

All (or almost all) terms used in the text that are peculiar to the construction industry have been defined in the glossary. The definitions chosen are those that seemed most simple, reasonable, or descriptive. Undoubtedly, many interesting and useful terms have been omitted, and in some instances there may be more appropriate meanings than those given. The glossary may therefore arouse more controversy than the text, but it at least represents a first step toward desirable standardization.

Most of the descriptions of equipment are based on specific current or recent models. In general, one representative machine has been selected for careful description, and others in the same class described according to their differences. In no case has an attempt been made to include the whole line of equipment of any manufacturer, nor to judge the respective merits of different makes. Because of changes constantly being made, some information may not apply to current models at the time of reading.

Machines have been chosen for description because of my personal experience in operating them, availability of descriptive material and illustrations, or by chance. Such choice does not imply any recommendation of such makes or models over others listed or not listed.

Equipment is usually identified and indexed by manufacturer's name and often by model number as well. This is not only a courtesy to those who so liberally supplied illustrations and information, but also serves to make the discussion clearer and more interesting.

All sections dealing with either the construction or use of particular machines, and with specialized work methods have been submitted to manufacturers and other contractors for criticism and correction. If any errors have survived this checking, or mistakes have been made in final arrangement of copy, the publisher will appreciate being notified.

"Moving the Earth" is a book about machines and their work. Therefore, comparatively little is said about the men who direct and operate the equipment, or about the hand labor needed to supplement it.

However, I am keenly aware that without the men the machines are nothing. Before there were wheelbarrows, men built roads and dams, quarried and transported rock, and drove tunnels. But without men, the finest machines can do nothing but rust. Without skillful operation, few pieces of equipment can justify their cost. Without maintenance, none can work for long. Without competent direction, all their work is waste. Lack of repeated mention of the men in and behind the machines therefore does not indicate lack of respect and appreciation.

Herbert L. Nichols, Jr.
GREENWICH, CONN.
January, 1955

PART ONE

THE WORK

CHAPTER ONE

LAND CLEARING

MACHINES AND MEN

Clearing of vegetation is usually necessary and almost always desirable as a preliminary to moving or shaping ground. Any growth makes dirt or rock difficult to handle, and its decay will cause settlement of fills.

But most clearing of growth heavier than grass or weeds is done almost as an end in itself, for agricultural purposes. It makes possible replacing woods and brushlands with pasture, crops, or tree farms.

Clearing is preferably a machine job. It may be done by a wide variety of standard excavators, particularly by bulldozers, front loaders, and backhoes. But if the job is large and/or difficult, it will probably pay to buy or rent one or more of the specialized clearing machines or attachments discussed in Chapter 21.

However, hand labor may be used in addition to or instead of machine work. A small piece of equipment may be able to do the work of a much larger one if occasional oversize trees are cut, or stumps are blasted ahead of it.

Clearing may be done in areas that are too small or too remote to justify bringing in equipment big enough to handle the vegetation. Ground may be too rough or too soft for the machines to work efficiently. Tree trunks are often salable if felled and cut into proper lengths by hand.

Land clearing equipment types, including chain saws, are described in Chapter 21.

DISPOSAL

Cut or uprooted vegetation must be processed or removed as part of most clearing jobs. Possible ways include burial, allowing time to decay, burning, shredding, or chipping; removal from the area, and various combinations of these methods.

Disposal methods will be discussed below, and throughout the chapter. They are considered first because of the extent to which they affect techniques of clearing.

Nitrogen. Disposal of vegetation by any method other than burning is likely to be complicated by absorption of nitrogen by decay processes. This element is essential for life of every kind. Although abundant as a free gas in the air, its quantity in fixed or usable form in the soil is limited.

The microorganisms that cause the decay of vegetation require more nitrogen than is present in wood. They often obtain it from the soil around it, and if decay is rapid, they may so impoverish the soil that nothing will grow on it. This situation is temporary, as the organisms die when the wood is used up, and their nitrogen returns to the soil.

Nitrogen deficiency must always be considered when mixing vegetation into soil. The effect is strongest but brief—perhaps only a week or three—when the material is soft and green and the soil is warm and damp. More durable parts, like wood chips, may not cause such complete takeup of nitrogen, but the lesser deficiency may continue for years.

Sometimes the effect is desirable, as in use of a layer of wood chips to discourage the growth of grass, when planting seedling trees with roots that extend below the zone of impoverishment. But leaving piles of chips may cause barren spots lasting for tens of years.

Heavy layers of wood chips rich in tannin —from oaks, for example—have an acidifying effect on soil, and interfere with growth of many plants.

The problem of nitrogen deficiency is usually not a strong objection to disposal of vegetation by any reasonable method such as surface decay, or mixing into soil or plowing it under. Nor does it prohibit the use of wood chips for soil cover. But it must be considered in assessing both the immediate and possible long term effects of the work.

The deficiency can be largely corrected by addition of suitable amounts of nitrogen-rich fertilizer throughout the period of decay, or by planting legumes that can obtain nitrogen from the air. The end result is usually substantial enrichment of the soil, as the presence of abundant nitrogen aids the conversion of vegetation into soil humus.

Surface Decay. Cut or uprooted vegetation is sometimes left on the ground to rot. Soft material such as grass and non-woody plants may disappear in a few weeks, but trees will make the area unusable for years. Prevention of re-growth is made difficult.

Disadvantages of disposal by surface decay are reduced, and may even by eliminated, if woody material is reduced to small pieces as part of the clearing operation. This may be done by using a shredder or a heavy rotary mower for clearing, or a chipper to grind up pieces after cutting.

This method is suitable for construction only if clearing is done long in advance of removal of topsoil.

Burial. In agricultural work, burial is often the preferred means of disposal if equipment of the proper type is available,

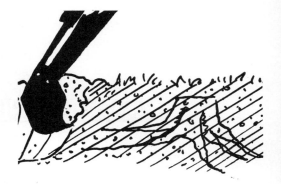

Fig. 1-1. A buried stump may cause trouble

and is big enough to handle the growth density and trunk sizes involved, and if the soil is soft enough to permit it.

Grass, weeds, brush and sometimes saplings may be buried intact by a brush-breaker plow, or slashed, chopped, and partially or wholly buried by a heavy disk harrow. Rolling choppers can disintegrate and partially bury medium size trees, including trunks.

Burial of this type is used in agricultural rather than construction clearing, and only when the ground can be left undisturbed (except for planting of a cover crop) until a large part of the material has decayed. This process may take weeks or years, depending on vegetation type and maturity, and weather.

In construction, about the only permissible burial is undisturbed, low-cut stumps under deep fills. When allowed, this exposes the fill and its supported structures (usually road pavement) to eventual settlement as the wood decays. This danger is often outweighed by advantages in prevention of sliding of fill down a slope, and in economy.

Loose stumps are often buried, but the operation is likely to be expensive and unsatisfactory. Attached roots make them enormously bulky; cutting roots back to the buttresses is likely to ruin saw chains by contact with clinging dirt. They are still awkward after cutting back.

Fill including stumps will almost always settle badly. Spaces left in and under irregularities will gradually fill with soil, allowing the surface to sink. Rotting of the wood will cause slow, long term settlement. Both effects are at a minimum if burial is in permanently wet mud that will flow around them immediately, and preserve them against decay.

In any soil, one stump can be packed in more solidly than a number of them. A common practice, when there are just a few big stumps, is to dig each one out with a backhoe, dig the hole deeper, put the stump back in it, and backfill. This shortcut is often not on the plans, and may cause surprise and dismay later, when the surface settles, or a small machine tries to dig a trench through the area.

Burning. Where burial is not practical, burning is usually the most efficient method of disposal, and does the least long range damage to the environment. Most of the discussions in this chapter are based on the presumption that fire will be used.

However, since open burning is prohibited or severely restricted in an increasing number of states and localities, it is worth while to examine the relative advantages and disadvantages of fire compared with other methods of disposal.

It will be assumed that burning is handled by reasonably experienced crews, with proper regard for safety, and confinement of the fire to the vegetation being cleared.

Piles of brush or trees usually contain 1/5 to 1/10 solid matter, the rest being air space. These solids average at least half water, and most of their dry weight is cellulose, lignin, and other burnables. The percentage of ash that is left after efficient burning is only a few per cent of the dry matter (exact figures are difficult to obtain). A good fire will therefore reduce the vegetation to a small fraction of a per cent of the original bulk.

The ash residue is too fine to be good fill material, but its quantity is so insignificant that it can usually be incorporated in other soils, or pushed aside, without difficulty.

It may be reasonably held that efficient burning results in the total removal of cut and uprooted vegetation.

Soil under a hot fire is rendered unfit for supporting growth for one to three years, but can be restored by plowing or ripping, and fertilizing.

A fire in dry material, with plenty of air, will burn mostly with hot clean flames, which produce carbon dioxide and water vapor, with few pollutants. Green wood and leaves, wet or dirty piles, and most weakly burning fires will give off large amounts of smoke, containing variable quantities of methanol, methane, acetic acid, tars and oils, and carbon monoxide.

Such fires, if upwind from inhabited areas, may create an extreme local nuisance, but its duration is very short. It is doubtful if the pollutants they put into the atmosphere equal those that would have been discharged during the natural lifetime of the plants themselves if they had not been destroyed. The burning concentrates them into a few hours or days.

The pollution problem that faces the world does not arise from country areas, nor from any activities (including open fires) normally conducted in them. It is a problem of cities, factories, and internal combustion engines.

It is therefore quite unreasonable that burning should come under total or almost total bans, while the real offenders are usually let off with moderate percentage reductions of their offensiveness.

Regulations against burning are costly. Substitute means of disposal, which are discussed in following sections, are more expensive under most circumstances, require vastly increased consumption of fuel, and may create environmental problems of long duration. The extra cost in highway con-

Fig. 1-2. Brush chipper

struction alone is probably already in the tens of millions of dollars a year.

Clearing by hand in cold weather may be practical only in the presence of hot fires, both for their emotional uplift and their actual prevention of acute discomfort, including frostbite. Even without need for heat, there is little satisfaction in clearing brushy land, if the debris must be left to litter the ground, making it dangerous for men and animals; or heaped into unsightly piles that take years to rot. And loss of unfarmed and unmowed fields to brush is a serious and increasing ecological problem.

The cost of buying or renting shredding equipment, and the noise, fueling problem, and danger to inexperienced operators, put these machines out of reach of most people who wish to do their own clearing.

Chipping. Brush, saplings, and even big trees may be fed into machines that reduce them to chips of small and fairly regular size, by action of a rotating toothed drum. These chips may be scattered or piled in the work area, or fed through a chute into dump trucks.

A small machine can be towed behind a pickup truck, and often maneuvered on the job by hand. It is hand fed with bundles of brush, and with saplings up to 3 or 4 inches diameter.

At the opposite end of the scale are monsters which can gulp down entire trees, with trunks up to 20 inches in diameter, without need to even trim the branches.

The chips that are produced may be a definite asset, a problem, or have no importance.

A few modern paper mills are able to digest wood chips that include bark and twigs, and will pay good prices for them. If such a mill is within economical hauling range of a clearing job, chipping pays off both in money and in utilization of the wood. In other areas, chips might be sold for processing into pressed wood, or charcoal and distillation products.

There is also a possibility that chips can be used on the job, either because they are really useful or because it is the best way to get rid of them. They can hold soil on slopes while vegetation becomes established, and add organic material to poor soils. Some applications are discussed in Chapter 7, together with possible problems.

Chips from light to medium thickness of vegetation may be left scattered on the ground, to be incorporated with the topsoil when it is pushed off or cultivated.

If the growth is heavy, the chips are likely to make the soil critically short of nitrogen, and difficult to work, and to accumulate in spots as pockets of almost un-diluted wood.

Chipping is usually not practical for up-rooted stumps, unless they are very small in proportion to machine capacity. Their bulk and shape make it difficult or impos-sible for the guards to pass them or the drums to grip them, and the dirt and rocks stuck to them damage cutters and make the chips unsalable.

Chips made from stumps in the ground are always contaminated with dirt.

Piling chips, either by keeping a dis-charge chute in one direction, or by dump-ing them from trucks off the work area, should be permitted only when they are to be reclaimed later. Such piles are likely to remain for many years before decaying enough to permit growth of vegetation.

Chipping machines are expensive, con-sume large amounts of fuel, are extremely noisy, and may be dangerous to personnel. Their use in mass clearing is justified when there is good use for the chips, and in areas of high fire danger, or where smoke cannot be tolerated.

In addition, chippers are valuable in low volume or selective clearing and trimming, where the cutting would otherwise have to be hauled away.

Removal from Area. Removal of cleared vegetation from the work area may be anything from a sound and profitable operation to a financial and ecological disaster.

Where lumber or paper mills are within economical hauling distance, it may be pos-sible to sell cut trees profitably. In some cases, the user of the wood may be glad to cut and remove the part of the vegetation that he can use, and pay for the privilege.

The usability and value of trees varies greatly, both in quality to be found on the job, and the processing equipment that will handle them. Some users are very narrowly restricted as to species, size, straightness, and soundness. Others will take (usually at

a lower price) almost anything that is recog-nizable as wood. In any big job, it may be worth while to invest considerable time in investigating possible outlets.

Firewood is another possibility. Very high prices are often paid for wood cut in two-foot lengths and split to cross sections averaging 30 square inches or less. Lower but still interesting prices may be paid by cordwood dealers for cut (and perhaps trimmed) trees which they process them-selves. But this market is largely limited to the vicinity of cities.

If the vegetation must be removed, and nobody wants it, the cheapest disposal is to just push it off the right of way, or out of the construction area, and hope to forget it. Fortunately this practice is usually not al-lowed. Even if it is, it may have disastrous effects on high-priced surveyors' reference points.

If the contract requirement is off-site dis-posal at a distance, there may be several possibilities. Both bulk and problems can be greatly reduced, although possibly at considerable cost, by chipping the vegeta-tion and hauling out the chips. Otherwise, trees should be trimmed into lengths suit-able for dump trucks or trailers, or flat beds of either type, with all angles in trunks or branches cut to make them lie flat. Non dumpers require a log-handling crane at the disposal site. Brush may be chipped and loaded, or loaded whole.

Fig. 1-3A. Stacked firewood (cordwood)

Courtesy of Fleco Corp.

Fig. 1-3B. Clearing with a rake blade

Except with medium to large tree trunks, or chips, these loads are likely to be mostly air. Haul cost per pound will be proportionately high.

Unchipped vegetation that is hauled off the job, and is not to be burnt, may be dumped off a cliff, dumped in piles over a wide area, or stacked in high piles with a log grapple or clamshell.

The result is almost always an environmental nightmare. Bulk is enormous in relation to the amount of clearing done, appearance is generally a first class eyesore, and the dumps may be dangerous or impossible to cross for many animals, and for men. They may serve as inaccessible infection points for plant insects and diseases.

Depending on the size and variety of vegetation involved, and climatic conditions, these unfavorable conditions may persist for five to twenty or more years.

SPROUTING

Many species of trees, and most kinds of brush, will sprout vigorously from stumps that are left in the ground with their root systems intact. Some varieties can sprout even from fragments of cut or mashed roots.

Sprouts can convert a cleared area into a dense mass of brush in one season. They are not only a nuisance in themselves, but they keep the root systems alive, strong, and resistant to removal.

When the old vegetation has been cut off flush with the ground, the new growth may be controlled by repeated mowing. Most of the root systems will be killed this way eventually, but it may (or may not) take years.

The most efficient control of sprouting is application of a weed killer (herbicide) such as 2-4-D. This may be mixed with fuel oil and sprayed or painted on the stumps immediately after cutting, or any time until the sprouts are big enough to get in the way. The kill may be close to 100 per cent if the work is done with reasonable care.

After sprouting has developed strongly, the same chemical may be mixed with water and sprayed onto the leaves, and is usually highly effective. With either method, spot spraying may be needed after a while to get plants that were missed.

Application of weed killer before clearing is usually too expensive in chemicals and labor to the practical, except in small areas.

Herbicides must be handled and applied with care. They can drift long distances from from a sprayer, to kill or injure crops or ornamental plants. Excessive applications may soak into the ground, killing desirable microorganisms. Runoff into streams will damage food supplies for fish.

BRUSH CLEARING

Dozer. Dozers and loaders are basic machines for clearing, both with regular blades or buckets, and with special attachments. They work best when the ground is firm enough for support, and where they are not hampered by holes, gulleys, sharp ridges, and rock.

Uneven surfaces make it hard to keep the blade in contact with the ground, and lead to burial rather than removal of vegetation in hollows. However, there are few places where a dozer cannot aid hand clearing crews, by clearing areas where it can work, moving logs and cut brush, cutting roads for supply trucks, or firebreaks.

Dozers have a particular advantage over hand crews where briars and vines are abundant, as these are very tedious to cut but can be readily stripped off by the blade, provided the operator does not take too long a pass and get caught in the tangle.

Unless he is completely protected by a cab with windows, an operator should have hand clippers to cut his way out if necessary.

Brush and small trees may be removed by a bulldozer walking with its blade in light contact with the ground. It will uproot or break off a number of the stems, and bend the rest over so that by a return trip in the opposite direction, it can take out a number more. If the distance is short, it is best to doze the whole patch in one direction, then across or backward.

Individual small trees are first knocked over then pushed out with another pass in the same direction.

Results will vary with the type of vegetation and the condition of the soil. Hardbaked soils will cause a high percentage of broken stems, while wet or sandy conditions favor uprooting, which is more satisfactory for most purposes. The work can be speeded by having a laborer cut out or pick up individual bushes that would otherwise require another pass by the dozer.

If the job requires removal of light stumps and roots, they may be overturned in one pass and pushed out in the next. It may be necessary to dig several inches into the soil to get a grip on them, then backblade the soil into the holes.

Brush heaps may be largely freed of dry loose dirt by rolling them over with the blade and shaking the blade up and down. If this is ineffective, rolling them over backward or pushing them from the side may be tried. A dozer with a blade which can be easily tilted down on either end is very good at this work, as one corner can be used for taking out roots, and pushing piles without taking a bladeful of dirt along with it, and the blade returned to flat position to skim off surface brush.

Rake Blade. Rake blades, which are made for the larger bulldozers and loaders, add to clearing efficiency under most conditions. They allow working below ground level, to take out roots as well as surface material, usually without bringing the soil along with them, if it is dry or sandy.

However, they may be somewhat specialized. A blade with teeth close-set enough to handle brush may bend a tine if it collides with something solid, while one strong enough for impact is apt to have too wide spacing for brush.

With or without rakes, any mechanical

loosening and removal of brush that is to be burned should be done when the soil is dry for best results. If it is wet, it lumps and sticks.

If the loosened material is allowed to dry, much of the dirt can be shaken out while piling, by rolling and shaking.

Work Patterns. In agricultural clearing, vegetation is frequently pushed into straight windrows, where it is burned after drying, or is allowed to decay. The width of cleared

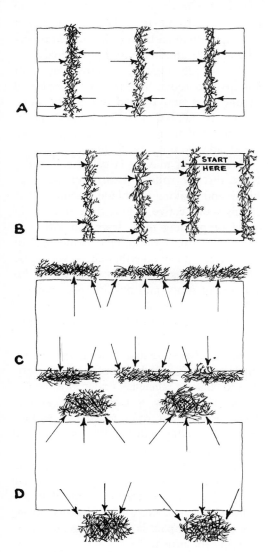

Fig. 1-4. Piling brush with dozer

aisles will depend on dozer power and the heaviness of the growth. Single pushes may range from fifty to over two hundred feet.

The simplest method of windrowing, Figure 1-4 (A), is to push from two directions toward a center line. This has the disadvantage of burying some vegetation without loosening it, which will increase difficulties in burning or moving the pile, and require further clearing after burning.

The pattern in (B) avoids this difficulty by building the windrows on cleared land. However, any windrow or large pile of brush, particularly if it includes trees, may be very difficult to rehandle. Brush, roots, trees, stumps, and dirt are crushed into a mat which often cannot be moved again by the dozer that built it. Except in highly inflammable growths, such as fat pine and palmetto, the mat may be difficult or impossible to burn, even after long drying.

If it becomes necessary to remove a tangled heap, a larger dozer, or a clamshell, dragline, or hoe shovel may have to be used. Taking it apart by hand might be more costly than the original clearing.

When piling coarse brush or trees, it is often advisable to have them cut into lengths of ten to twenty feet, after uprooting. This will enable the dozer to handle larger loads and to place them more accurately. Piles can be more readily separated or moved if necessary.

When a right of way is cleared, and immediate burning is not required, the brush may be stacked in windrows, as in (C) or in piles as in (D), outside of the work area.

Burning. In general, however, it is the best practice to burn machine cleared vegetation at the same time that it is piled. A hot fire, including heavy wood, is prepared and brush piles pushed up on it. A new fire is made when the push gets too long.

Best results are obtained if the vegetation is uprooted and allowed to dry at least a few days before burning. This may be done by backing the dozer into the woods from the cleared edge, and uprooting small patches, or individual trees, pushing them clear of the ground, and then leaving them.

The trash dries more rapidly scattered on the ground than in piles. Dirt will tend to dry and break away from stumps, and to sift out of roots and stems.

When burning, the brush nearest the fire is put on it first.

Fires fed by a dozer tend to get choked up with dirt. In general, matted light brush is more difficult to clear and to burn than heavy brush or small trees, as it tends to slip under the blade, or to bring too much dirt with it.

Dozer Protection. When a dozer is clearing dense undergrowth, there is the danger that it will fall into some hole, natural or artificial, whose presence is concealed by the brush. This may be guarded against by scouting the area on foot, and by moving forward in a succession of short pushes overlapping each other on the side, as in Figure 1-5A. This enables the operator to watch from one side, without getting branches in his face, and to observe the nature of the ground. In addition, it avoids tangling the dozer in branches and vines.

Any dozer used for clearing work should be thoroughly protected with crankcase and radiator guards; the latter including screen with holes not over one quarter of an inch, and accessible for removal of leaves and trash. The engine needs side guards. The operator should carry hand tools to cut himself out of tangles.

Minimum operator protection is a strong overhead structure. Most new clearing units have complete, extra-strong cabs, often heated and air conditioned.

Accidents have been caused by branches moving throttle and clutch controls.

Other Machines. There are a number of types of equipment that are used for chopping or shredding brush, which may bury it, or leave it on the surface to rot or to be re-

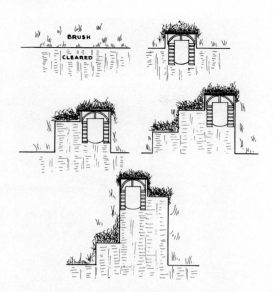

Fig. 1-5A. Clearing thick brush

moved by other equipment or laborers.

A big rotary mower mounted on the rear of a wheel tractor is highly effective up to its thickness-of-stem limit, which may be 1½ to 3 inches. Vertical rotary shredders may handle double that size. Brush is mostly chopped or shattered into small pieces, but root systems are seldom disturbed. There is no suppression of re-growth.

The sickle bar or hay cutter, in heavy duty models, will cut brush up to ¾-inch, but with considerable wear and breakage. It does not chop the stems, and is little used in clearing.

Either a rotary or a sickle bar can be used to suppress re-growth by repeated mowings.

A big moldboard plow, preferably a brush breaker model, can put brush and saplings underground, cover them neatly with dirt, and leave them to rot. The area is usually harrowed lightly and planted to grass or a cover crop immediately.

A big heavy disc harrow, with discs 24 inches in diameter or larger, chops brush and buries a large part of it. Big pieces may be loosened, chewed, and pushed around without burying.

Courtesy of Rome Industries

Fig. 1-5B. Clearing with a disc harrow

Both the plow and the harrow tend to create ridges and troughs in the ground surface, because they move loosened dirt to the side.

A rolling chopper knocks down, tears, and mashes both brush and trees, and cuts near-the-surface roots. A small portion of its cutting is buried.

Carrier. When the final cleanup of any kind of brush clearing is done by hand, it is helpful to furnish laborers with stick carriers, which may consist of a piece of heavy canvas, six or more feet long, and two to three wide, with handles on the ends, as in Figure 1-6. This is laid on the ground, and sticks and branches piled across it. The handles may then be picked up and several armfuls carried at a time with minimum effort.

TREE REMOVAL

Mechanized Logging. In some large scale operations, logging may be almost completely mechanized, with felling, trimming, bucking, and transport (or piling and burning) done by highly specialized machines, some of which are described briefly in Chapter 21.

However, most clearing-for-excavation projects must rely on more standard machines, and/or hand labor.

Cutting or Uprooting. Big equipment can handle small trees in the same manner as brush. But big trees, or any trees too large to be walked down by the equipment on the job, may require special kinds of work.

Cutting for possible use as lumber is discussed in the next section.

Pushing a tree over with a dozer or pulling it down with a cable follows the general methods described for stumps later in this chapter. However, a machine can generally uproot a much larger tree than a stump, because of greater leverage from a higher push or pull point, and help from the weight

1-10

of the tree which tends to tear out its roots as it leans.

But see page 1-15 for special overhead dangers.

Handling Trunks. Tree trunks, even in sapling sizes, may be tricky and dangerous to push around. One may ride up over the top of even a big dozer blade during ordinary pushing. It may be put under tension by pushing while an end is blocked, which may result in its whipping with great force against the cab, or into other machines or men in the area. An operator must be vigilant in avoiding such a situation.

When felled trees are pushed into heaps for burning, the piles contain so much air space that they may be very fire resistant. Heavy branches and crooked trunks increase this difficulty.

Branches may be cut off and trunks cut into pieces to make more compact piles. Or the pieces may be placed on an existing fire by a clamshell.

Subcontracts. If trees are to be removed which are of no value on the job, an attempt should be made to sell them. To the contractor desiring to confine himself to dirt work, the best arrangement is to get the customer, whether sawmill, firewood dealer, or whatever, to buy the trees on the stump and cut and remove them. A danger is that the logger may fail to do the work in the time specified, and so force the contractor to do it himself at the last moment. In making such an arrangement, the disposal of the scrap wood and brush and the height of the stumps should be specified.

A sawmill operator is interested only in large sound trunks, whereas a pulp or firewood man can use bulky branches also. The mill will ordinarily pay the best prices, but do the least work toward cleanup of the tract, unless it has an arrangement with pulp or firewood users to take its tops and limbs.

No one wants the rotten trees, crooked

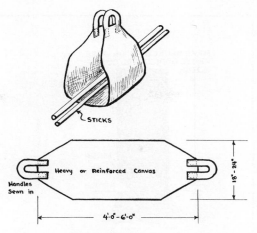

Fig. 1-6. Trash carrier

branches, and brush, but the lumbermen may agree to burn them, if this is a part of local logging practice; or if the contractor accepts a complete cleanup job as part or full payment for the wood.

Cooperative clearing arrangements may be made in which the logger is assisted by the contractor's tractors or trucks.

Stump Height. Stump height may be determined by local law or lumbering custom. From a clearing standpoint, high stumps are more easily removed than low ones, and are especially desirable when the machinery is undersize for the job, or depends primarily on winches. Low stumps are more difficult to cut, particularly where the trunk flares out widely at the bottom, but do not impede machines as much, and can often be filled over and left.

Cutting. If the trees are valuable and lumbermen will not clear them out in time, the contractor may cut and stack them for future sale. This, as a logging proposition, is somewhat out of the field of this book and will be considered very briefly.

Practically all wood cutting is now done with chain saws operated by one man. Their construction and operation are described in Chapter 21.

There are a number of sizes and models. Thick logs require long blades and in-

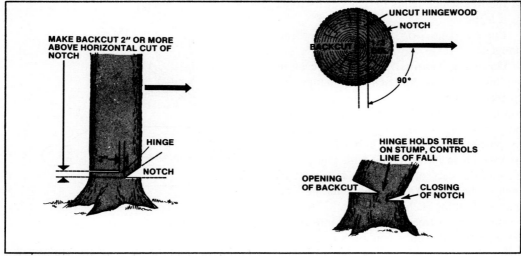

Courtesy of Homelite

Fig. 1-7A. Felling a tree

creased power. For any log thickness or blade length, increase in engine power means faster cutting, but also more weight to handle.

Most experts recommend use of a bar long enough to cut the average size tree on the job in one cut.

Undergrowth should be roughly cleared around the tree before cutting it, to reduce danger to men and tangles with fallen trees.

A tree should be notched by a V-cut up to one-third of trunk thickness, on the side toward which it is expected to fall. The bottom cut of the notch should be made first and should be horizontal, the other sloped down to meet it.

The tree is then cut through from the other side with a level cut 2 inches or more above the floor of the notch.

If the saw bar is longer than the tree diameter, or a hand crosscut is used, the line of cut is parallel to the back of the notch. With a short bar, procedures shown in Chapter 21 are used to obtain the same effect.

The cut is theoretically finished when the strip of uncut wood (the hingewood) has been reduced to 1/10 to 1/20 of trunk di-

ameter. The tree should now fall toward the notched side in a direction at right angles to the length of the hingewood, if its balance has been judged directly. If the hingewood tapers, direction will tend to shift toward the thicker side.

The hinge may be crushed as the tree starts to lean, and may narrow the edge of the cut, binding or even crushing the saw. It is necessary to be alert to pull the bar out quickly. It can be put back in if the fall does not occur.

Direction of Fall. Natural direction of fall of a tree is influenced by a number of factors. Location of its center of gravity, as affected by lean, twist, and limb location, is

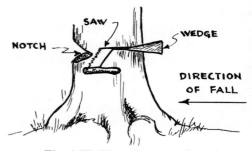

Fig. 1-7B. Use of saw and wedge

primary. Direction and velocity of wind may be important.

When practical, cutting should be done so as to take advantage of natural factors, to drop the tree in the direction that it tends to fall. This direction can be made positive, or altered moderately, by appropriate notching and cutting.

For greater changes in direction, including tipping the tree oppositely from its normal inclination, wedging or line pull is needed.

Wedging. If there is the slightest question about the direction the tree will fall, a wedge of wood, plastic, or magnesium should be driven lightly into the back of the cut, to prevent wrong-way leaning. This protects the saw against binding, and may influence direction of fall.

A large tree that has a moderate tendency to fall in a wrong direction can be forced into the right one by wedging. The wedge or wedges may have to be iron or steel, as other materials may not stand heavy driving. Saw manufacturers oppose any use of hard metal.

The moving saw chain must never touch a hard wedge, as it would suffer immediate and severe damage. The wedge therefore cannot be inserted until the cut is deep

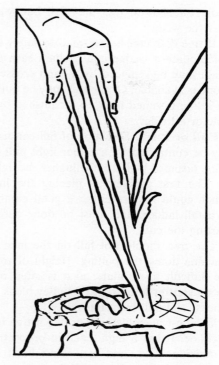

Fig. 1-8B. Making a wood wedge

enough to provide ample working space for the saw, and the operator must exercise extreme care.

Wedging tends to overturn the tree to the side directly opposite. But the fall will be affected by other factors, such as unbalance to the side, taper in the hingewood, or wind pressure. These factors should be allowed for in placing the wedge.

Wedging may be started with a soft wedge, to keep the cut open until it is deep enough to use the hard one.

One wedge may not be sufficient to tilt the tree. A similar wedge may be driven just above it, Figure 1-8 (A), or a thicker wedge driven beside it. Since a thin entering edge is not needed now, a made-on-the-job wood wedge, (B), may be used.

The hingewood may be cut thinner to make wedging easier. But if this strip is made too thin or cut through, or if it is weak because the wood is brittle or decayed, a

Fig. 1-8A. Double wedging

badly unbalanced tree may fall backward over the wedge.

If tree diameter is too small to safely accommodate a wedge and a chain saw, the chain may be replaced by a hand crosscut, if one is available. Or the tree may be pulled by a line fastened high up, or pushed by a pole or a loader.

Pull or Push. Direction of fall can usually be controlled by a rather light pull on a line fastened high up, the higher the lighter. The first problem is placing the line, which could require either a good climber or a tall ladder, and must be done before starting the cut.

The tree should not fall on the man or machine doing the pulling. Height of trees are difficult to estimate, so it is safer, and uses a shorter line, to use a pulley block or blocks, and pull from the side.

The best machine for pushing a tree is a backhoe. It has a high reach in proportion to machine size, and can stand back far enough not to interfere with the cutting.

Pull or push is usually light at first, and is increased as the cut deepens.

DANGERS

Tree cutting is dangerous work, because of the nature of the tools used, and the sometimes unpredictable behavior of trees and their parts.

Direction. In spite of best judgment and efforts, a tree may fall in an unexpected direction, even backward across wedges. And a fall may occur much sooner than expected, because of concealed weakness in the base, or a puff of wind.

No one should be in a cutting area except the man or men doing the work, at the foot of the tree. They must be alert to move quickly. The critical area should be clear of brush and litter, as tripping over it might have fatal consequences.

A chain saw that is not being used, or any other valuable equipment, should be

Fig. 1-9A. Pulling from the side

placed behind a tree or other protection during tree felling.

Overhead Breaks. Some trees come apart while being cut. Blows of a hammer on a wedge, rough contact of a pusher bucket, or even the vibration of a saw, may loosen dead limbs so that they come crashing down around the base. If the trunk is decayed it may break as it starts down, with the lower part leaning and falling conventionally, but leaving the upper parts in the air to come straight down.

This will not happen when a tree is live and healthy throughout, but dead branches and decay are not necessarily visible from the ground. It is more common in tall trees than in short ones, and the pieces fall harder.

Hard hats are not sufficient protection against the weight of pieces that might fall. The top must be watched closely during work, and men must not be ashamed to run instantly if anything breaks loose. The danger area is usually small and close to the trunk, although this varies with limb spread.

Fig. 1-9B. Trees may come apart

But of course, do not take the direction the tree is expected to go.

Leaners. A tree may be held from falling by other trees. This may be by comparatively light contact of branches while the cut tree is nearly vertical, or because of fall into a crotch or across heavy limbs.

Such a tree may be brought down by moving the lower end away from the direction of fall. A small or medium tree can be moved by prying the butt up with a pole, and swinging the pole. This can be repeated as many times as necessary. Medium to large trees may be lifted or pulled by a machine, or pulled by a block and fall.

If the trunk cannot be moved, it must be cut in the air. Because of the strain of its position, only shallow cuts or notches can be made in the upper side, with the main cut from below. Pinching of the saw and erratic movements of the trunk can be expected, and can be dangerous.

As each piece is cut and knocked or pried out of the bottom, the remainder should slide down to rest on the ground, usually at a steeper angle.

Cutting the supporting tree is too dangerous to be attempted. It is usually under great extra tension, and is likely to snap and fall suddenly during work. The leaning tree would then be apt to drop on the men doing the cutting.

Trimming. When the tree is down, the branches should be cut off nearly flush with the trunk, before any other trees are dropped across it to make a tangle. Light branches and any heavier wood which is

Fig. 1-10A. Moving hung-up tree with pole

1-15

Fig. 1-10B. Never cut the support tree

to be wasted, should be piled, burned, chopped up, or taken away by methods described for brush.

Removal. The trunks may be dragged out of the woods by tractors, or cut (bucked) into lengths where they fall. Saw logs for small mills may be from eight to sixteen feet long, the size being largely determined by the use for the sawed wood, the capacity of the mill, and of the trucks that carry the logs. Piling, which may often be made from thinner trunks than saw logs, is left full length. Cordwood is usually in four foot sections, and split to a

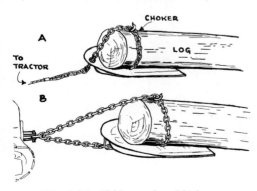

Fig. 1-11. Skid-pan log hitches

size that one man can handle. Pulpwood varies in length in different localities and is usually peeled but not split.

Dragging (Skidding). Crawler tractors are the standard equipment for dragging long logs, but with good ground conditions, they may be replaced by rubber tired machines with either four wheel or two wheel drive.

Crawlers and four wheel drives usually carry loaders or dozers. Their efficiency may be increased by mounting winches, and by use of log carriers and various rigs too specialized for description here.

A log is usually pulled by means of a chain or cable fastened around its butt, choker fashion, and attached to the tractor drawbar or winch. The most important consideration in arranging this is to get the butt off the ground, or riding on it very lightly, as digging in will take greatly increased power and will rip up the trail. A short line, particularly to the top of a winch, is helpful unless the log has a greater diameter than the height of the drawpoint.

The log may also be pulled onto a stoneboat, or other sled, and the line passed through the eye, or two lines used as in Figure 1-11.

If the tractor is sufficiently powerful, several logs may be pulled at a time by attaching them individually to different lines. If only one line is available, they may be fastened with one choker, as in Figure 1-12, which should be fastened well back, as such piles often come apart while being towed.

Two-wheel drive tractors can drag logs or bundles of logs on dry ground. Loads must be small, but they move briskly. If the tractor has a hydraulic lift drawbar which can be chained to the log to lift its butt off the ground, its efficiency is more than doubled, as the weight on the driving wheels is increased, and friction greatly reduced.

Bucking. Bucking is the operation of cutting tree trunks or heavy limbs into short lengths. It consists of making a series of vertical cuts, which under favorable conditions may be directly from top to bottom, at right angles to the grain.

The whole trunk is usually measured off first, and cut locations marked by chipping with an axe or scratching with a saw. Location of limbs, bends, or defects often makes it necessary to adjust the spacing.

The principal problems in bucking are keeping the saw from being pinched in closing cuts, and being damaged by cutting dirt. These are discussed in Chapter 21, under the subheadings **Stresses** and **Ground Damage.**

Completion of a cut may be prevented by

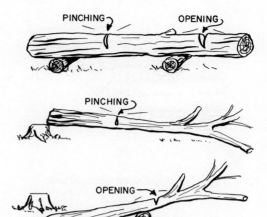

Fig. 1-13A. Pinching and opening cuts

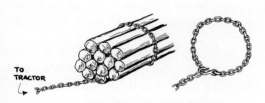

Fig. 1-12. Fastening poles for towing

obstacles such as stumps, brush, rocks, or dirt. Sometimes the obstacle can be moved, sometimes the trunk can be slid, rolled, or lifted free, or cuts may be made at other places until the trunk can be moved. The most frequent trouble is the tree partially burying itself in soft earth, so that cuts cannot be carried all the way through. This may be prevented by dropping the tree on logs placed to support it, or overcome by trenching the dirt for the saw, or jacking up the trunk.

Moving Short Wood. When trucks can get in the woods, cordwood and pulpwood are usually cut to size and trucked out. As wood is much lighter than dirt, a dump truck can carry several times its body rating, if the pile stays on. Figure 1-13B

shows a method of placing planks, poles, or thin split logs vertically along the body sides to permit high piling. These are held in place by the piled logs. If the road is rough, it is wise to pass a chain from the body over each row of logs and to tighten it with a load binder.

If the wood is cut short, and rain or unforeseen mud conditions make trucking impractical, it may be dragged out by tractors. If a stoneboat is available, logs may be piled on it, and the tractor line threaded through the eyehole over the pile and

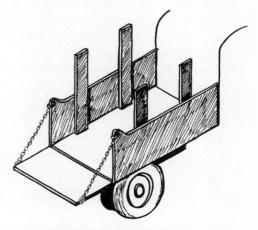

Fig. 1-13B. Sideboards to hold cordwood

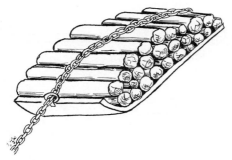

Fig. 1-14. Cordwood on stoneboat

anchored on the back. The eyehole should be beveled so that a chain or cable can slide freely through it, as the tractor pull will then hold the logs to the stoneboat. See Figure 1-14. If no boat is available, the logs may be piled on the line which is then looped around it as a choker. Parts of the piles may be chained by snaking the line under them, without repiling.

Storage. Wood should be stored outside of the work area where it will be accessible both during and after the digging. Poles and logs may be very useful in shoring up banks, making corduroy roads, getting machinery out of the mud, and other purposes. A buyer might be found for the wood at any time. Stored logs or cordwood should be stacked so as to be off the ground and well ventilated. This makes it easier to remove them later and delays damage from rot and borers. Cordwood is usually stacked in easily measured units.

Personnel. If possible, experienced log-gers should be employed for lumbering. They will be able to do it much more efficiently than equally energetic and resourceful men not used to the work. The difference may be as much as 10 to 1.

STUMPS

A stump is the base of a tree trunk and its attached root system. The trunk part may be anywhere from a few inches to many feet in diameter. It usually flares out near ground level into root buttresses, which connect it to the major roots. Its top may be flush with the ground or several feet high.

Roots form a network near the ground surface. A few species of trees have a taproot, a strong root that extends more or less straight down from the center of the trunk. It makes stump removal much more difficult.

Stumps are a major problem in most clearing that involves trees. They can sometimes be cut low and left under deep fills, but usually must be removed.

Stumps may be broken out by uprooting the whole tree, then disposed of as part of the tree or separately. But it is more usual to cut and remove the tree, and then take out the stump.

Stumps may be pushed out by powerful dozers, dug out by somewhat less powerful dozers, rippers, or hoes; pulled out with cables or chains, or blasted. Blasting may be combined with other methods. It is oc-

Fig. 1-15. Two passes to push out a stump

casionally practical to burn stumps in the ground.

Once out of the ground, stumps present a problem of disposal. As massive pieces of green wood caked with dirt, they are difficult to burn. Nevertheless, this is usually best way to get rid of them. They are so bulky and irregular in shape that they are hard to bury, and they are an eyesore in piles. Since they rot, they cannot be used in fills under structures.

PUSHING AND DIGGING OUT

Pushing. Crawler tractors, with dozer blade or loader bucket, or a special narrow stumping blade, are standard for uprooting by pushing. Stumps should be cut high — at least 36 inches — for good leverage.

The blade or bucket is lowered to contact the stump a few inches below the top, and the machine is moved forward in low gear. If the stump yields, forward push is continued until the trunk leans so far that effective contact is lost, or until the roots bulge in front of the tracks.

The tractor is then backed, and the edge forced under the upturned roots. Lifting while moving forward slowly should roll the stump out of the ground, breaking all roots except those on the far side.

Further pushing to get out the far roots may drop or mire the tractor in the stump hole, or may roll up a too-big ball of roots and soil. It may be better to finish freeing the stump by going to its other side.

A number of stumps may be overturned in one direction, then the tractor turned to finish them up from the other.

The same method is followed to uproot a standing tree, except that a high push point is used for greater leverage, and that the capacity of the machine relative to trunk diameter is improved.

Before pushing a tree with any type of machine, the operator should look to make sure that it is alive or at least sound. If a rotten tree is pushed near the base it may

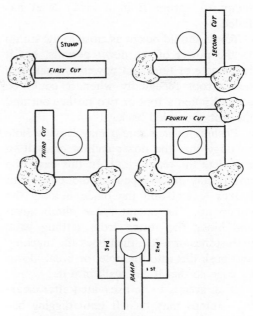

Fig. 1-16. Digging out stump

break high up and a top section fall on the dozer. Large dead branches are sometimes dropped with equally disastrous consequences. A dozer to be used extensively for tree pushing should carry overhead guards for the operator.

A tree may bend or split without affecting the roots, in which case the push should be applied lower on the trunk, or from a different direction.

Digging Out. If it does not yield at all to pushing, it must be dug out. This is done by trenching around it with the dozer to cut the roots. If done systematically, this may follow the pattern in Figure 1-16, but it is frequently unnecessary to go through the whole procedure. Each time the dozer cuts a big root, it may turn and push the stump to see if it is loosened. The operator will often be able to tell when it has been softened up by the way it shakes as the roots break. Many operators do not bother with cut number 4, but it helps with very heavy stumps, particularly when cut low. The ramp need be built only after an attempt to uproot from a lower level has failed.

Roots should be cut as close to the stump as the power of the dozer permits, but it is a waste of time and power to buck at a heavy root repeatedly when it could be easily broken a foot or two farther out and the stub crumpled back.

Pushing over a stump may leave a hole so large that the dozer cannot cross it to complete the tearing out. In this case, the dozer may be stopped at the edge with brakes locked and the blade holding the stump up. The operator can climb down and block the stump from settling back with stones or a log, then back the machine and push dirt into the hole, or break down its edge so that it can walk into it.

If an area is to be excavated after clearing, stumps may be left until digging has undermined them and cut many of their roots when they can be easily removed.

Shovel Dozer. A shovel dozer can remove stumps in the same manner as a bulldozer, or make use of the hydraulic control bucket in special techniques. A stump of small to medium size may be dug by tilting the bucket floor downward from thirty to sixty degrees and forcing it into the ground close to the stump, as in Figure 1-17, using both down pressure and forward motion. With the machine pushing forward, the bucket is then flattened and may be driven under the stump, cutting and tearing the roots, then rolled back, lifting the stump from the ground. If it falls off, the bucket is dropped to contact it and to roll it out of the ground. If it stays in the bucket, it can be carried to a pile or loaded directly on a truck.

The high lift gives the dozer shovel excellent leverage for pushing over trees.

Bucket teeth are desirable for stumping as they aid penetration, get a better grip on the stump, and are useful for knocking dirt off the ball, and for raking out roots.

Methods of handling stumps with a

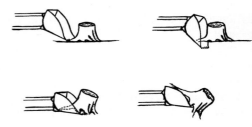

Fig. 1-17. Stumping with shovel dozer

dozer bucket are described in Chapter 16.

Selection of Machinery. Big machines, and special machines, greatly reduce time, effort, and breakage in clearing work, and should be used whenever a job is large enough to justify their purchases or hire. Stumps that can be knocked right out of the ground may be removed at the rate of one a minute or better; moderately resistant ones may take two to five minutes; and those which are definitely oversize may take an hour or more. It is easy to see the time that can be saved by applying overwhelming power.

A good clearing team may be made from a heavy tractor with a stumper, assisted by a smaller one with a shovel dozer. If these machines work together closely, the stumper can devote its entire time to breaking out the big ones, while the dozer takes out small stumps, knocks down brush, finishes off loosened stumps, piles and removes them, and smooths off the ground.

Revolving Shovels. The backhoe is probably the best stumper among the shovel attachments. Usually, the operator tries to take the stump by a direct pull first. If it resists, it can be weakened by chopping roots on the far side and by trenching at the sides. Except in the case of very large stumps, digging and removal can be done from one position. It may be necessary to put blocks against the tracks to prevent the shovel from dragging toward the stump when power is applied.

A dragline is used in the same manner

but is less efficient at chopping roots, and does not have as strong a pull. However, it can often take out a number of stumps without changing position, and is able to backfill the holes and grade the area at the same time.

A dipper stick is used in somewhat the same manner described for a dozer shovel. It has more penetration and can trench to cut roots more readily but is less maneuverable.

The hoe and the dipper are more effective than dozers in rocky ground and among interlocked stumps as they can apply their power in smaller spaces. It is often necessary to devote time to digging out rocks before the stump can be attacked. If the roots are strongly entrenched in bedrock or over-size boulders, the rock may have to be blasted before the stump can be pulled.

Rippers and Pans. Rippers may be used to cut stump roots and to pull out stumps directly. With two teeth (or one tooth mounted at the side), the ripper can cut roots close to the trunk on all sides. The tractor is then backed with raised teeth to catch the far side of the stump, then forward to pull it over. Another back pass enables the operator to get a tooth under the stump, and lift and roll it out.

Scrapers (non-elevator models) can take out small and low stumps, but this involves complicated maneuvering, and danger of getting the stump jammed in the bowl.

Close Quarters. Stumps are often so located that uprooting them would damage pavement or buildings. Such a stump may be removed in chip form, with little disturbance, by a carbide-toothed wheel driven by a tractor or truck power takeoff. Cutting is usually carried 4 to 6 inches below ground level.

If such a cutter is not available, hand work may be required. Dirt and stones are dug away with shovel and trowel to expose the roots, which are cut with a sharp grub axe (mattock), with possible help from chisels or other implements.

After the main roots are cut, the smaller laterals are liable to break so that the trunk will come out with very little heaving of the soil. If a taproot is present, it may be cut with long chisels while pulling. Any tendency to heave will show the presence of heavy roots which have not been cut, so that further digging and chopping may be required. This same technique may be used to weaken trees or stumps that are too resistant for the pulling power available.

Men accustomed to grubbing out stumps by hand can do the job in a fraction of the time required by ordinary laborers.

Tree Killing. Under most conditions dead stumps are easier to remove than live ones because of the disappearance of the hair roots which bind them to the earth, and the weakening of the larger roots. Soft woods in well drained soils may show perceptible weakening in a few months, while rot resistant stumps in saturated ground may remain firm for many years. In any case, dead stumps contain lighter wood and hold less dirt than live ones.

It may therefore be advantageous to kill trees well in advance of removal in cases where plans are made early enough. This may be done by cutting, girdling, poisoning, burning or drowning. Cutting and girdling are ineffective with trees that sprout from the stumps, unless the sprouts are cut back several times, or poison is inserted under the stump bark. Poisoning may be done without cutting by nicking the tree in late summer or fall and putting sodium arsenite or weed killing chemicals in the cuts. The same methods are not equally effective in different localities so advice should be obtained from local tree experts.

Burning peat soils in the dry season will kill trees and loosen the stumps, but because of smoke and smell nuisance, and danger of spreading, it should be done only under carefully controlled conditions.

If it is possible to dam a stream so as to flood a wooded area for several months during the growing season, some or all of the trees may be killed. Unfortunately this is usually possible only in swamps, and swamp trees are more resistant to drowning than those in dry locations.

PULLING STUMPS

Stumps may be pulled out of the ground by a cable to a power source. This method, although widely used, is less popular than in the past. Bigger machinery and special attachments have made it possible to push or dig out most stumps encountered on construction jobs, without taking the extra time to rig lines.

However, such equipment is not always available, it may cause too much damage, and it cannot work efficiently on soft ground. Pulling will probably continue to be an important method of stump removal.

The pulling line may be a chain, cable, or rope, and the power may be direct pull by a machine or animal, winding in of cable on a winch, either machine or hand powered, or a combination of these methods with pulley blocks.

The stump line is generally a choker type which tightens its grip as the pull increases. In smaller sizes, chain is preferred because it is easier to carry, safer to handle, and more resistant to abuse. However, it is much heavier than cable for the same strength, and in large sizes is too weighty to be practical.

Line pulling is preferred when the ground is too rough or soft to allow machinery to get at stumps directly, and when available force needs to be increased by multiple lines.

Chains. A more detailed description of chain and fittings will be found in Chapter 21. A standard tow or logging chain is composed of short straight links, carries a round hook on one end and a grabhook on the other. The round hook may be fastened to the chain by a ring, or a ring may be used instead of this hook.

Either the round hook or the ring can be used in chokers. The hook is easier to attach and to detach, but may fall away from the chain when it is slack. The ring may be used by passing the grabhook through it and pulling from the grabhook end; or for stumping, the chain near the ring may be pulled through it to form a loop that is dropped over the stump.

The grabhook fits over any individual chain link, and will not slide along the chain. It is used to adjust the length of chain by increasing or decreasing the amount of double line, by moving it toward or away from the choker end, or passing the chain behind a tractor drawbar pin, and preventing it from being pulled out again by attaching the grabhook to the slack side, making it too large to be pulled through the space. In this case the surplus chain is slacked, and if it is long, must be hung on some part of the tractor. See Figure 1-18.

Grabhooks are used to anchor a chain to a tree that is to be saved. The lack of sliding pressure makes it possible to protect the bark by pads and sticks placed on the side receiving the pull. The grabhook may also be used to make a ring which can be used to make up a choker.

Figure 1-19 shows three ways of fastening a line to a stump. In each case, the stump is shown to be grooved by an ax at the back. This cut is quickly made and will prevent the chain from squeezing off during the pull, and will delay its slipping off as the stump leans. (A) is the easiest and most usual method, pulling at the center; (B) is a side pull, a little harder to arrange, but which puts less of a kink in the line, so that it can be used with cable as well as chain, and gives the advantage of a twist on the stump; and (C) the overhead method which requires an inverted T notch. This gives the greatest leverage but is more likely to slip off than the others.

Care should be exercised not to put loads on a chain that is twisted or kinked as it will be broken or damaged. It can be readily checked for straightness as the links which are in one plane should lie in an almost straight line.

Alloy steel chains weigh only about a third as much as standard chains in proportion to strength. If a crew is careful enough not to lose chains, and is conscientious

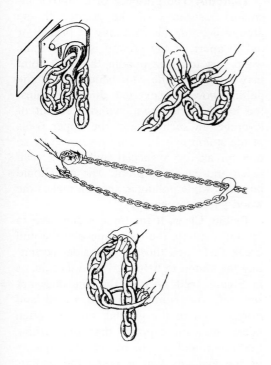

Fig. 1-18. Grab hook uses, and stump choker

enough not to abuse them by kinking or gross overloading, alloy chains will amply repay their much higher cost in reduced labor and fatigue, and by greater efficiency.

As an example, one ⅜″ alloy chain, weighing 1.6 pounds per foot, is thirty percent stronger than the same make of ⅝″ ordinary chain, weighing 4.1 pounds per foot.

It is recommended that the alloy chain be dipped in bright red paint so that it can be easily recognized, and recovered readily if mislaid.

Broken Chains. A broken chain is best repaired by having new links forged into it by a blacksmith. However, good field repairs may be made with a variety of patent repair links, or by shackles with removable pins. Links that have been stretched thin may not admit the proper size repair piece, and may have to be cut off with a chisel or hacksaw, or opened by supporting on a block with a hole in it, and spreading with a punch and hammer.

A broken chain may be used temporarily by tying the broken ends into a square knot, then tying the end links together with wire to prevent coming apart. If there is not enough chain, a half knot should be used and the ends fastened with a bolt. A bolt may be used without any knot, but will pull apart more easily than a link. Two grabhooks connected by a ring can be used for temporary repair and for shortening.

If the chain is too weak for the job, an attempt should be made to double it. It may be looped around the load and fastened by its two end hooks or rings to the drawbar. If it is looped around the drawpin, links may be bent or crushed.

Cables. Only the method shown in Figure 1-19A (B) should be used in pulling a stump with a cable choker, as the sharp bends involved in the others will cause early breakage of the cable.

If a double cable line is used to reduce strain, or to shorten the rope, it should not be bent around sharp angles. A stump is generally round and smooth enough not to cut a cable wrapped around it, and the end hooks or loops can be attached to the drawbar. If the load is angular, it is better to fasten a snatch block to it with a chain or sling choker, and to run the long cable through the block pulley.

If a double cable is so wrapped around

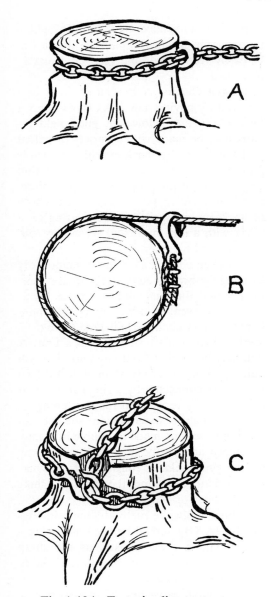

Fig. 1-19A. Fastening line to stump

lateral roots, the hook is placed to grip one of these, power is applied, and the root torn out. This process is repeated until the stump is sufficiently weakened to be taken out on one of the root pulls, or by direct pull on the butt.

The root hook may also be laid on top of a stump, with the teeth in a notch on the back. A pull on this gives excellent leverage, but the edge of the stump is liable to tear off.

Taproots. The presence of a taproot increases resistance of the stump. If the ground is hard, this root may be broken or pulled apart. If the ground is soft, or the wood very tough or pliable, the pivot point may crush and the root bend so that the pulling power is exerted directly against the length of the root, without benefit of leverage. In such a case, the upper roots of the stump may be torn up sufficiently so that an ax, or a special long chisel, can reach and cut the taproot. The cut should be made while pulling as tension makes the wood part more easily.

Pulling Clear. If the force is sufficient to uproot a stump, the roots opposite the pull break first, then those at the side, permitting the stump to be pulled onto its side, as in Figure 1-20 (A). If the line does not slip off, the stump may be rolled and dragged out of the ground, but this often takes much more power than overturning the stump, and may be beyond the capacity of the machine that is doing the pulling.

If the stump will not come all the way, the line may be slacked and a log placed or chained against the stump, as in (B). This log will provide a new fulcrum and aid the breaking out. Or the line may be taken off and the tractor moved to pull in the opposite direction, which should free it without difficulty. If a number of stumps are being pulled, all of them may be overturned one way, before pulling the tough ones in the opposite direction.

Half uprooted stumps are easily

the load that it cannot slide around it, great care must be taken to adjust it so that both ends share the strain equally, unless a single line is strong enough to take the entire pull alone.

Root Hook. A root hook may be used when a stump is too big to pull directly Enough soil is dug away to expose the

knocked out by dozers, and may be left for them to save the trouble of re-rigging.

Resistance. A stump's resistance varies in different directions. If on a slope, downhill pull is most effective. Otherwise it should be pulled toward its strongest roots, as these are easier to bend than to pull apart, and can be dealt with more easily when the rest of the stump is loosened.

The most obvious variable in stump resistance is its height. Greater height means greater leverage and easier pulling. Limiting factors are difficulty of high cutting, of fastening heavy chains at a height, and of the trunk breaking under pull.

A buried stump is the hardest of all to pull and usually must be dug out. On filled land, two separate systems of lateral roots may be found, one under the old ground level and the other near the surface, in which case it may be necessary to cut the trunk below the upper roots, in the same way as a taproot.

Fig. 1-19B. Root hook

A stump which yields to pull but will not break loose, can often be uprooted by moving it as far as possible, slacking off to allow it to settle back, and pulling again, repeating this process a number of times. This is most effective if done slowly and smoothly, whether with winch or traction. This method is very effective with trees, as the trunk will bend with a whipping motion that exaggerates the force of both the pull and the snap back.

Chopping the roots on the side opposite

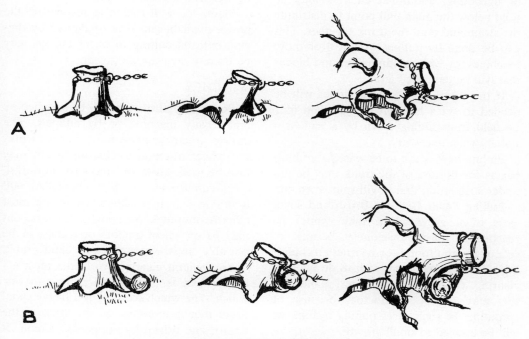

Fig. 1-20. Stump pulling

the pull, while they are under maximum tension, weakens the resistance. A moderate amount of digging will generally expose the main lateral roots.

When a stump has been split by blasting, the pieces are most easily pulled away from the center, rather than across it.

Uprooting Trees. If trees are so large that their stumps will be difficult to remove, it may be advisable to pull the trees over rather than to cut them down. This gives the opportunity to fasten lines as high as desired and to make use of the weight of the tree. As soon as the tree is pulled toward the tractor, its center of gravity shifts to that side and aids greatly at breaking out the roots. If a large log is chained to its base, on the pull side, the force of the tree's fall will be more effective at breaking roots on that side. The log will also serve to prevent the trunk from digging into the ground where it would be difficult to cut.

If the tree tends to break or split instead of uprooting, additional chokers may be used below the main pull point to distribute the strain and bind the trunk together. This can be done by pulling with two or more machines, or with multiple lines and blocks that will be described later.

If the trunk is smooth, a ladder will be needed to get a high grip. The chain may be held from sliding down by a nail or a notch, when necessary.

Pulling trees is apt to be wasteful of lumber as the bottom of the trunk may be put under such strain that it will split when cut.

Pulling Small Growth. Brush and small trees often grow where they cannot be reached by pusher machinery, because of soft or rough ground, or nearness to buildings. A landowner may wish to do his own clearing without hiring a dozer. Hand cutting may not be satisfactory because of sprouting. In such cases pulling techniques will be applied to small growth.

An automobile has sufficient power for pulling some brush and small, stiff-trunked trees, but the work doesn't do it any good. Trucks and farm tractors usually put more power on the job and are less likely to be damaged by the exertion.

If the stems are stiff, fastening may be made high for leverage. If they are flexible, height does not matter, and the greater strength of the base may make it the best place.

Chains tend to slide along smooth stems and they often can be made to grip by wrapping once or twice around before fastening. Light chain with small links holds much better than coarser types. A round hook or ring should be used to make a choker. If stems are close together, it is often possible to pull several at a time by putting a single choker around the group. It will slide up until it can pull them all tight together and then should hold.

Brush tongs get a good grip on small trees and flexible plants, and are easy to attach and to remove, but their weight may outweigh these advantages.

Plants too well rooted to respond to the power available may be weakened by digging out and cutting roots, or pulleys may be used to step up the power.

WINCHES

Power to pull stumps may be supplied by almost any machine or by animals. The crawler tractor is preferred for heavy work, but wheel tractors, trucks, and oxen may also be used. However, the most powerful, most convenient and best controlled pull is obtained from winches. These are most efficient mounted on crawler tractors, but may be on wheel tractors or trucks, or be portable units operated by a hand crank.

For general clearing work, the most effective tool is a dozer carrying a power winch. The winch consists of a heavy spool drum that is mounted on the back of the tractor and driven by the power takeoff. It is controlled by the tractor main clutch and the power takeoff engagement lever. In addition, it may have a transmission, giving rotation of the drum in either direction, and in large machines permitting several speeds of rotation. A jaw clutch or neutral gear is used to disconnect the drum from the drive shaft, to allow it to turn freely when the cable is being removed. A brake is provided to slow or lock the drum when necessary.

The winch may hold two hundred or more feet of cable of a size proportionate to its power. Additional cable can be carried on a separate spool and connected to the winch cable by a choker device when needed.

In small sizes, the winch cable generally is fastened at the working end to a short piece of chain equipped with a round hook. Larger cables may be fastened directly, or through a swivel or single link, to a round hook, or a wide face cable grip hook. The cable is generally underwound on the drum, that is, leads from the work to the lower part of the drum. This gives better stability under heavy load than overwinding.

Stump Pulling. To winch out a stump, the tractor should be placed facing directly away from it, and both brakes locked on. The winch jaw clutch should be released, its brake set to drag very slightly, and the cable pulled to the stump by hand. If the brake is not used, the drum may continue to spin after being pulled, and unwind and snarl the cable.

If the winch will not freewheel, or the cable is very heavy, the drum is turned backwards by the engine to pay it out. It is convenient to have two men, one to operate the winch and the other to pull the cable. If no helper is available, the operator can stand near the winch while it turns, stripping the cable and coiling it on the ground until he thinks he has enough. He then stops the winch and drags the cable to the stump. The cable must then be whipped up and down and twists

worked out to avoid kinking when pulled.

The winch cable may be put around the stump directly, hooked to a choker chain or cable, or may be run through one or more snatch blocks.

Power is applied to the winch and the cable is reeled in, care being taken to see that it feeds onto the drum properly. The stump may come out or the tractor may be dragged backward. If the latter, the tractor may be anchored by a chain from the blade or front pull hook to a tree. Resistance to pull may also be increased by backing it against a log or bank, or by trying to pull the stump by tractor pull and allowing the tracks to spin until they have built mounds behind them. If the anchoring or blocking is effective, the stump will come out — if nothing breaks, slips, or stalls.

The drum carries a number of layers of cable so that it has a greater spool diameter full than empty. It therefore reels in cable more slowly and powerfully on a bare drum than on a full one. On a bare drum, logging winches will give 50 to 100 percent more pull than the tractor itself; on a full drum, the same pull as the tractor or somewhat less. But with torque converter drive, strongest pull may be with a full drum. See page 21-15.

Jammed Cables. Using a nearly bare drum not only gives the greatest pull but reduces damage to the cable. If a long cable is wound smoothly onto a drum under moderate tension, and a heavy pull applied when it has built up several layers, the last wrap may squeeze between the wraps below, as in Figure 1-21 (A). This scrapes and wears the cable and jams it so that it will not spool off again. The best way to free it is to turn the drum until the catch is in the position shown in (B), and jerking it, or anchoring the end and driving the tractor away. Or, in the same position on the drum, the cable may be given a couple of wraps around the drawbar, and the winch turned backward as in (C).

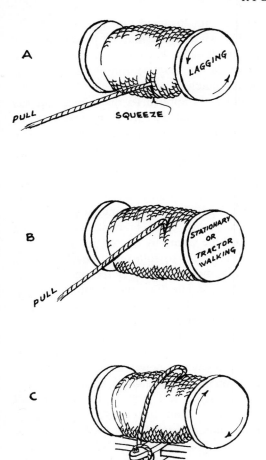

Fig. 1-21. Freeing winch cable

If the cable is wound unevenly onto the drum, with the wraps crossing each other at random, it cannot cut down between lower layers readily, but may put severe kinks in sections of cable that cross under it, and this cross wrapping may not entirely prevent it from squeezing in and sticking.

In spite of these difficulties, a long cable is desirable for general work. If reasonable effort is made to spool it in evenly while working, it will usually be rough enough to prevent excessive sticking, without too much bending or crushing.

Two-Part Line. Where the distance to the stump is less than half the cable length, a two-part line may be used by attaching a pulley to the stump and by running the line from the winch around the pulley and back to the drawbar. The useful strength of the cable and the pull between the tractor and the stump are doubled.

The tractor may have to be backed against a heavy log or an outside anchor used in the manner to be described below. The tractor should not be anchored by the front pull hook while using a double line anchored on the drawbar, unless the manufacturer will state that it is strong enough to take the strain.

Rocking. The winch and tractor pulls differ in quality and it may happen that the pull of the tracks will do jobs that the winch will not. Use of the tractor drive helps in "rocking" stumps or trees out. The line is left slightly slack and the tractor moved forward in low. As the line tightens, the stump may lean a few inches, then stop. When the tracks start to spin and the clutch is released, the weight of the stump, combined with the spring in the roots and in the line will pull the tractor back. The clutch is immediately re-engaged and held until the tracks spin or the engine lugs down again.

If the stump is within the tractor's power range, repeating this maneuver should gradually break it out. A long cable has more elasticity than a short one, or a chain, and will be more effective at rocking.

This procedure should not be allowed to degenerate into yanking, where the tractor is given a long enough slack run to be brought up with a jerk when it tightens. This will break more tractors and cables than it will pull stumps.

Cable Breakage. Cable or chain breakage is a serious danger to both operator and helpers. A cable particularly stretches under strain, and if it breaks suddenly, may whip with great force. The danger to the operator is greatest if the break is fairly near him. The cable used should be the

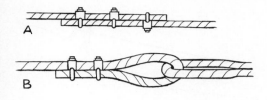

Fig. 1-22. Temporary cable repair

best quality, in the largest size recommended by the winch manufacturer, if the tractor is to be anchored or used for rocking; and it should be inspected frequently for weak spots.

Another danger inherent in the use of cable is cutting and tearing of the hands and clothing on broken wires. Preformed cable gives minimum trouble of this kind and should be used when possible. Leather palmed gloves are good protection for the hands.

Either hemp center or wire center cable may be used, according to preference or manuafcturer's recommendation. Wire center is about 10 per cent stronger, size for size, is stiffer, and is not as easily deformed by crushing. It is more difficult to handle, and when kinked or crushed is much harder to straighten. Standard 6 x 19 constructions are usually recommended.

It is good practice to work a winch at less than its maximum capacity, and to avoid anchoring the tractor unless absolutely necessary. Moderate loads give long life to cables and winch parts, and avoid severe catching on the drum. If the work is heavy, strain can be reduced by the use of pulleys and multiple lines.

Broken cables can be repaired by splicing, but the length of cable used in the splice, and the labor involved, may be too great to justify this method for the short cables ordinarily used in land clearing.

A rough repair may be made by trimming back the broken ends, overlapping them as in Figure 1-22, and fastening them with two or three cable clamps, for sizes up to half inch or five eighths, or with

three or more for larger sizes. Or two interlocked loops may be made, as in (B), fastened with clamps, or any type of loop fastening. Cables repaired in this manner are weakened but may last a long time. The patch will not go through pulleys and is inconvenient in other ways.

Winches on Wheel Tractors. If a winch is mounted on a wheel tractor or truck, it is usually necessary to anchor it for heavy pulls. The anchor chain or cable should be attached to the winch frame, or to a heavy member as near to it as possible, to reduce strain on the tractor.

An important consideration in the use of these winches is the fact that the cable will tend to take a straight line between the work and the anchor. In making a high pull, as in Figure 1-23 (A), the tightening cable may lift the tractor and turn it over sideward. In (B) the downward pull may blow the tires, unless the axle housing is blocked up, as in (C).

If a wheel tractor is not anchored, a rear winch must be underwound, and care must be taken that the machine does not overturn through rising on the front, a danger which is particularly serious if the tractor is driven to move the load.

Truck Winches. A truck winch is usually of the horizontal drum type. It may be mounted in the front bumper, on a flatbed body, or between the body and the cab.

Rotation may be for underwinding, overwinding, or both. Power is from the power takeoff, controlled by the truck engine clutch pedal.

The principal handicap of a truck winch is the difficulty of maneuvering it into position for a straight pull. One or more pulleys may be required to obtain a proper direction of pull and a straight line onto the winch. The truck should have all the wheels blocked, or be anchored by a line from a frame member near the winch.

A gypsy spool or capstan winch, Figure 1-24, may be mounted vertically on the for-

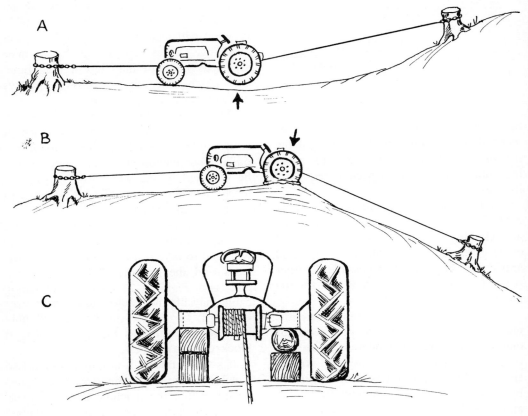

Fig. 1-23. Vertical effects of winch pull

ward end of a flat body, chiefly for dragging loads onto the truck. A hemp rope is looped around it two or three times, with one end attached to the work and the other end held by the operator. If the operator leaves it slack, the spool will turn inside the rope; if he pulls it tight, the working end of the rope will be pulled with great force. The slippage on the spool absorbs shocks that would break the rope and enables it to do very heavy pulling, under exact control. However, the gypsy is not ordinarily used for stumping.

Hand Winches. Hand winches are turned by a hand crank, operating through one or more sets of reduction gears. Under most conditions, it is not possible to make a full turn of the handle because it strikes obstructions, or passes through awkward po-

sitions. A large part of the work of winching consists in removing and replacing this handle, and if much work is to be done, a ratchet handle should be purchased, or made up by adapting one from a heavy socket set.

The winch is usually equipped with a friction brake and a pawl that can be en-

Fig. 1-24. Capstan winch

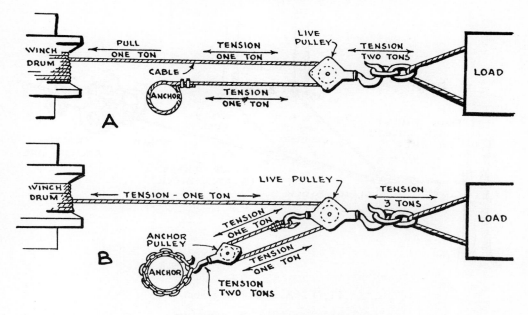

Fig. 1-25. Stump pulling layouts

gaged to prevent it from turning backward when the handle is released.

Operation of these devices is tedious because of the number of crank turns which must be made to reel in the cable; and exhausting because of the force which must be applied to the handle to develop the rated pull of the winch. It is important that it be thoroughly lubricated.

Hand winches can be used in places inaccessible to power equipment, are comparatively inexpensive and are surprisingly powerful. Sometimes they can take out tougher stumps than a power winch of the same pull because the line can be left taut and tightened gradually or from time to time as the stump yields. Their weight, with cable, may be from 75 to 300 pounds, so that carrying one of them any distance is at least a two-man job. It can often be transported in a loader bucket.

Hand winches are sometimes mounted on a truck, in which case they serve largely as a spool to carry cable, most of the pulling being done by the power of the truck. If the job is too heavy for the truck, it may be anchored or blocked and the work done with the winch handle.

If not mounted on a truck or other carrier, the winch should have a V-shaped towbar, or a subframe by which it can be anchored. Planks should be provided to build up a base in line with the pull, as if this is high, the winch will be lifted off the ground and will not be steady enough to allow turning the handle.

MULTIPLE LINES

Snatch Blocks. If pulling stumps takes the full power of the tractor or winch, it may be advisable to use snatch blocks to obtain greater power at slower speed. These devices, also known as blocks and as pulleys, are pulleys set in frames that are provided with one or two round hooks or rings, usually on swivel connections. For most field work, single pulley wheels, with a latch arrangement permitting insertion of a cable at the side, are best, as cables usually carry attachments too large for threading; a tedious job even when possible.

These blocks can be obtained in sizes to

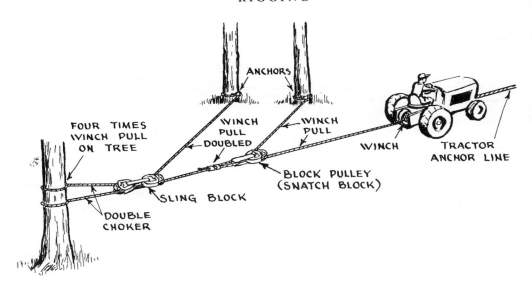

Fig. 1-26A. Use of sling block

match any cable or strain. In large sizes they are very heavy, and several men, or a loader, or light winch or other lines, may be used to move them.

Figure 1-25 shows several riggings using pulleys. If the lines are approximately parallel, and the pulley bushings lubricated, each additional line will add about 90 per cent to the single line pull. This puts no extra strain on the winch cable, but the chokers holding the blocks must take the combined pull of all the lines fastened to them.

The advantage obtained from the use of a block is decreased when the lines are not parallel, bcoming zero when the angle between the lines is 114°. Still wider angles result in loss of power.

The number of lines that can be put on one stump may be determined by the number of blocks on hand, the space available for fastening them, the strength of the available anchorage, or the amount of cable. Light machines may use six or eight blocks on a heavy stump.

Rigging is simplified by the use of series and sling blocks, as in Figure 1-26A, (A). The tractor line passes around a block pulley to an anchor, doubling the pull at

the block frame. This is attached to a heavier line which passes around another pulley to an anchor. The second pulley is pulled with almost four times the power of the winch.

The sling block makes possible use of a double choker on the tree being pulled. The choker cable is approximately centered in the sling pulley, and both ends are hooked around the tree. Such a double choker can be of lighter and more flexible cable than a single choker.

Rigging. Multiple blocks require care in rigging and pulling. Two anchors are better than one as they spread the cables over a wider space where they are less apt to interfere with each other. Each block is best fastened to a separate choker, but one may be fastened to each end of a chain passed behind the stump if it is strong enough to take a double pull; or one to both ends of a chain given one turn around the stump. It is good practice to notch the stump for each chain used so that the chains and blocks will not slide into each other as it yields.

Rigging is done with the lines slack. When they are pulled tight an inspection

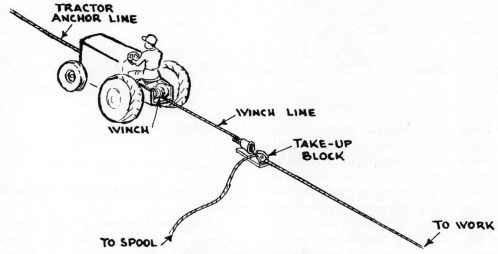

Fig. 1-26B. Take-up block

should be made to make sure that no pulley latches have fallen open, as a pull on an open block will bend it and cut the cable; that no pulleys are jammed with debris, or liable to pull into each other, and that no chain hooks have become disengaged. As the line is wound in, all blocks should be watched to make certain that they do not collide. Lines should not be allowed to drag on each other.

If the winch does not carry sufficient line for the distance or the number of lines involved, extra line can be added by the use of a take-up block, as in Figure 1-26B. A standard practice is to use the winch line from the tractor to near the first snatch block, and the extra line for reeving. The take-up cannot be pulled through a snatch block, and is liable to cause trouble if included in the multiple lines.

The extra line is often carried on a spool supported on a pipe axle and brackets. This should have a drag brake of some kind to prevent spinning when paying out. The brake might consist only of a log leaned against the face or side, or a man's gloved hand.

Anchors. It is often a question of whether the stump or the anchor will yield.

Anchor lines should be as low as possible and stump lines high. It may be best to pull the largest stumps first, using several smaller ones for anchorage if necessary. In a clean clearing job, there is always one last stump for which there is no anchor, and if it is small, it may be pulled out directly; or in any case it wil respond to less elaborate artificial anchors than a large one. On the other hand, a large stump will be a dependable anchor, and will prevent the need of frequent re-rigging when anchors pull out.

The final stump may be pulled by use of a living tree as an anchor. A choker should not be used under any circumstances on a tree which is to be preserved; padding and blocks should be used with a grabhook loop.

If no anchor is available, one may be made, ground conditions permitting, by digging a T shaped trench, two or more feet in depth, as shown in Figure 1-27. A log is placed in the crossbar, the cable anchored to it and led up through the sloping trench toward the work. Load and local conditions will determine the depth of cut and size of log. In medium soil, a standard railroad tie two feet down should hold a horizontal pull of five tons.

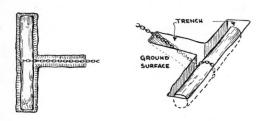

Fig. 1-27. Artificial anchor

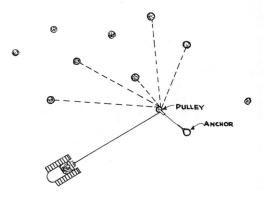

Fig. 1-28. Use of guide pulley

Advantages of Blocks. Stump pulling with a winch and blocks takes more time and care than direct winch pull, but results are generally more satisfactory. Jerks and jars which are destructive to machinery and cables are largely eliminated. Lighter cable may be used and a sufficient number of lines will reduce the tension on any one so much that squeezing and crushing on the drum will not occur.

A snatch block may be used to advantage with an anchored winch to avoid shifting its direction for each stump. Figure 1-28 shows a plot in which a number of scattered stumps are to be pulled. A snatch block is anchored to one of them and the line led through it. The line can then be attached to stumps straight in line with the winch, or to any on the opposite side from the anchor, as the pulley will lead the line almost straight in, at the price of a small friction loss. The pulley is best placed at a moderate distance from the winch so that the cable can feed evenly onto the drum, instead of tending to pile up in the center. The line from pulley to anchor should be short to keep shifting of the pulley to a minimum.

STUMP BLASTING

Blasting may be used to break up and remove stumps when machinery heavy enough for the work is not available. Explosives may do the complete job, or just weaken them for removal by equipment.

The standard explosive is 40% dynamite, but only because it is most convenient. Slower dynamites and even black powder may be more effective. Charges are gener-ally too small to justify using ammonium nitrate.

Underground Charges. A hole, or several holes can be made under the stump by punching with a crowbar, or paving breaker, or drilling with an auger. Easiest penetration is usually obtained close to the trunk, between heavy root buttresses. Several tries may be necessary. If no hole can be made deep enough exploding about a tenth of a stick of dynamite in a shallow one may soften up the resistance.

A heavy enough charge should be placed to blow a crater in the ground large enough to include the stump and its major roots. A pound of dynamite should move about a yard of dirt, but the charges should be varied according to results obtained.

If the stump has a taproot, a substantial part of the charge should be in direct contact with it in order to shatter it.

If one charge is placed, it may be exploded by either fuse or electric caps. If more than one, electricity should be used so that they will go off together. If the charge is too small, it may merely create an underground chamber; if too large, it may carry the stump or its fragments into forbidden territory. The desired result is to split it, and lift it clear out of the ground, without excessive after-travel.

Wood Drilling. A stump may also be blasted apart with a smaller charge by drilling directly into the wood. A one and one quarter inch drill will make a hole

Fig. 1-29A. A stump-blasting hazard

large enough for standard dynamite sticks. If smaller drills are used, the dynamite may be obtained in thinner sticks, or unwrapped and stuffed into the holes. Hand extension bit drills are generally used, but if a quantity of stumps are to be blasted this way, an electric drill powered by a portable generator will be a big time saver.

A stump blasted in this manner is usually shattered but not blown out of the ground. The pieces are generally easier to pull or dig out than the whole stump, but not always.

Mudcapping. If it is impossible to get under the stump and no drill is available, it may be broken by placing charges between the root buttresses and packing mud over them. The mud turns the force of the explosion inward and lessens the noise and concussion. The mud must be free of stones and pebbles as they fly like bullets. A stump blasted in this manner is easier to dig out than when whole, but may be more difficult to pull.

This type of stump may also be loosened by cutting the main roots with blasts several feet out from the trunk. The dynamite is removed from its wrapping, packed closely against or around the root, and covered with mud or damp earth.

If the first blast does not loosen the stump sufficiently, another can be placed.

In circumstances where the charges cannot be properly placed because of rock, lack of tools, or danger of damage to nearby property, a series of blasts may be needed. If the dirt has blown away from the roots, their effectiveness is greatly reduced and an ax may do better work than more powder. If the roots are entangled with bedrock or massive boulders, the rock may have to be drilled and blasted.

A standing tree may be blasted in the same manner as a stump, but a much heavier charge is required to lift it enough to break the roots. However, a moderate off-center blast may cause the tree to fall, in which case its own weight will snap a large number of roots. Trees felled by blasting are likely to be badly split and useless for lumber.

Precautions. Stump blasting is dangerous work at best because of unpredictable conditions underground. Particularly, a rock may be held just over the charge by roots in a position that will cause it to take off like a mortar shell. Split pieces of wood will also fly long distances. Mud or earth packs over charges should be free of stones or pebbles, and personnel should move a long way back from the blast, unless good shelter is available.

All stumps should be accounted for after a blast, as they are sometimes blown up in trees, where they stay until dislodged by wind or another blast, with serious results to persons underneath.

If the blasting is to be done near buildings, logs or saplings chained together should be piled on the ground on the side of the stump toward the building, to stop stones and fragments from flying. Regular blasting mats are safer, but if machinery capable of handling them is on the job, it generally can pull the stumps without the necessity of using explosives in close quarters.

Applications. Blasting ahead of stump moving machines has several uses. It reduces the power requirement, which is good when the stump is big enough to argue with the machine, and saves time in any case. A factor which is often more important is that blasting the stump loosens the mass of earth held by the roots, so that much or all of it will fall off during handling. This earth is a large part of the weight of many stumps and increases the difficulty of moving and burning them. Blasting is ineffective at removing the dirt once the stump is out of the ground.

Stump blasting is generally ineffective in sandy or powdery soils, as the force of the explosion flows around the roots with minimum breaking effect. These soils also fall off the stump rather readily without blasting.

The relative size of holes left by blasting and by pulling varies with tree species and soil conditions.

BURNING STUMPS

In the Ground. Dead, dry stumps can sometimes be burnt without taking out of the ground, but the process is usually slow and laborious.

A standard tool for stump burning is a large kerosene blowtorch, called a flame gun. It may be blocked to flame against a stump. One man can operate several,

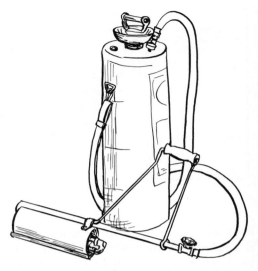

Fig. 1-29B. Flame gun

or do other jobs while taking care of one. When the wood starts to burn, the torch may be moved to another stump and brought back if the fire dies down. Green wood will require continuous heat for many hours.

The torch will operate most effectively if directed into a cavity with an opening in the far end so that a draft can move through it. If the flame is aimed into a dead end hollow, very high temperatures will be attained, but because of lack of oxygen the wood will distill rather than burn and will be destroyed slowly. If the flame is used against the outside of the stump, it should be directed upward to draw a current of fresh air between the flame and the wood.

Dry stumps may also be burned by starting a wood fire alongside them, and keeping it supplied with logs and snags placed to almost touch the stump. The draft and reflection of heat will keep both surfaces burning, but the loose wood must be moved in rather frequently. This method may remove only the top and outside of the stump, leaving a conical core.

Care should be taken to avoid spreading stump fires. Roots may burn underground to start surface fires at a distance. Soils rich in humus, such as swamp peat or forest loam, may burn unless saturated with water, and are very difficult to extinguish.

In a Pile. Because of difficulties in burial, and scenic and environmental damage in piling, the best way to get rid of a pile of stumps is to burn them. However, it is becoming increasingly difficult to get a permit for the work.

The practical side is also difficult. A stump may cling to more than its own weight of dirt and rock, most of which should be knocked off to make it burnable. This may be done by allowing the dirt to dry, then kicking them around with a dozer or other machine, or by picking up with a grapple and clamshell and dropping.

Even if clean, green stumps are difficult to ignite. If they must be piled before burning, the base of the pile should include a substantial quantity of old tires, preferably topped by a layer of logs.

But the surest way is to burn them at the same time they are piled. Build a hot fire by the methods described in the next section, supply it with enough heavy wood to keep it hot for several hours; then roll, push, or lower stumps onto it, in sufficient quantity to make a thick mass that can sustain fire by itself.

A grapple or clamshell is the best machine to feed a stump fire, even if they are pushed to it by dozers. It can place them properly, stack them high, and avoid pushing dirt with them.

Stump fires are very hot and may burn for weeks. There is usually little smoke, and few sparks.

BURNING BRUSH

A great deal of time and effort are wasted in ineffective attempts to burn brush, and for this reason proper procedure will be discussed in some detail.

Even green, wet brush and logs will burn vigorously once properly started, but considerable heat is required to boil off the sap and water, and to ignite the wood. This heat may be obtained originally from a carefully built fire, or by use of inflammable chemicals.

Building a Fire. The fire should be on level ground, or on a hump. If built in a hollow or against a rock or stump, inward flow of air will be hindered, and brush added to the top of the fire will be held up away from the heat. All inflammable material should be cleared or burnt away from around the site, particularly downwind. Fire fighting tools should be available.

Figure 1-30 illustrates two ways of starting the fire—andirons and tepee. The

Fig. 1-30. Starting a fire

"andirons" consist of a pair of small logs, or rocks, or ridges of dirt. Twigs and sticks, preferably dry, are laid across the andirons. These should be laid in one direction so that they will lie close together, but should not fit together so well as to prevent air and heat going between them. No leaves or grass should be included.

This pile may be ignited by burning paper, grass, or leaves under it. The material must be dry, and must not be packed tightly, as this reduces the oxygen supply and the heat of the flame.

A self-feeding starter may be made by tearing a section of 10 to 30 pages of newspaper into a strip that will fit easily between the base logs, lying flat. Crumple the top sheet, and light it.

As it burns, the heat will cause the next sheet to curl up and burn. The process repeats for every sheet, keeping a brisk fire going for long enough to ignite dry logs. No kindling is needed, but of course small dry wood starts faster than thick green pieces, for which the process may have to be repeated.

When the cross sticks start to burn, more and heavier sticks are added, then partly trimmed branches, and finally, when a good bed of embers and strong flames are present, untrimmed bushes and branches. It is a good plan to put a few logs or snags on at this time to give the fire staying power.

The tepee is similar in principle. The sticks are piled on end around the kindling. As heavier pieces are added, the tepee is crushed, but if it is burning well this will not matter.

A danger in transition from the hand tended fire to the roughly piled one is that the untrimmed brush may include so much air space that the heat cannot cross it effectively. The fire may burn a dome shaped hole over itself, then die down. In such a case, sticks should be poked into the fire itself to build it up, and the brush over it should be compacted by rearrangement or piling on of heavy sticks. This is tiresome work and may fail. It is better to tend the fire longer before piling on loose material, to be sure it will not have to be worked over afterward.

Artificial Helps. Old tires provide excellent material for starting a fire. Trimmed brush can be piled on them as soon as they are burning.

A dying fire may be pepped up by use of kerosene, fuel oil, gasoline, or similar fluids. To be effective, these must be applied at the base of the pile. Because of its explosive qualities, gasoline should be applied only as a stream from a blowtorch or similar pressure device with a fine nozzle, and only when it burns as it is ejected. If it does not burn, it may accumulate in sufficient quantity to cause an explosion.

Putting inflammable fluids on the heap itself may produce a fine flame, but it will have little kindling effect as the evaporating fuel will absorb the heat that radiates downward.

Flame guns produce a hot flame up to 20 inches long. They are effective kindlers when directed into the base of a pile.

A brush burner is a portable unit that combines a heavy fan to direct a strong wind into a pile being burned, with a mist of fuel oil for kindling. Few piles can resist one for long.

If the fire dies down in spite of nursing, it may be best to build a new fire nearby, with greater care to avoid air spaces and coarse green wood early in its life.

Transferring Embers. Once a good fire is burning on the job, its embers may be shoveled out and used for starting other fires. This should be done rather frequently as a long brush carry adds greatly to labor costs.

Four or five shovels of hot embers may be laid on the ground in a pile, and fine brush, or dry twigs and wood, piled on it. Or the embers may be sifted down through

Fig. 1-31. Brush burner

piled brush. The embers give a sustained heat and consume little oxygen, so that a strong new fire starts quite quickly.

Feeding. It usually takes at least two men cutting and dragging brush to keep one fire burning briskly. If it is allowed to burn down, it is good practice to put the unburned ends in the center hot spot, before piling on more brush.

When a dozer is used, ample supplies of fuel can be brought to the fire, and it is usually well packed by the pressure of the blade and the weight of the machine if it climbs, up on the pile.

The principal problem of dozer feeding is dirt. This tends to block the fire from spreading into new material, and to smother parts already burning. Every effort must be made to reduce the amount of dirt by rolling and jostling piles, holding the blade high enough not to dig in, and giving the vegetation and mud a chance to dry before bringing it in.

A hot fire will burn through quite a lot of soil, but it will seldom burn clean. After it cools, the remaining stems and stumps can be sifted out by the dozer and used in building the next fire.

Good results in fire feeding are obtained only if most of the new material is placed on top of the flames.

Burning Piles. If the brush is piled a long time before being burnt, dropping a match in it on a hot day may accomplish its complete removal. If it has been piled only a few hours or a few days, a fire may be built on the windward side against it but not under it. This fire may be caused to spread into the heap by keeping it buried under compact brush, so that the fire is fed and the heat reflected into the pile. If the brush has leaves, it is good practice to cover any place where flames show through. A strong fire cannot be smothered with hand piled brush.

Brush piles may be pushed on top of fires by a dozer, placed by a clamshell, or rolled on by a number of men using long poles.

If brush is being cut in an area presenting unusual fire hazard, or the cutting is in small, scattered areas, it may be desirable to truck it to a central burning place. A continuous fire may be maintained with incoming loads dozed or hand piled onto it, or the brush may be piled to dry and burnt off occasionally.

Brush up to a few inches in diameter can be reduced to chips by a chopping machine after which it can either be left on the ground or easily trucked to a dump.

Clamshell. One excellent combination for heavy clearing and burning is a large dozer, preferably with a rake blade; and a clamshell shovel. The dozer uproots and pushes in brush and trees, and the clamshell picks them up, shakes dirt out, and places them on top of the fire.

The clamshell can also maintain a fire, moving in unburned ends, and can bury it under dirt at the end of the working day.

A clamshell is also often the best tool for high stacking of vegetation that is to be left to rot, and for burning old piles that need rehandling.

Banking Fires. If the job is not extensive enough to justify the employment of a night man to watch the fires, and any inflammable material is nearby, they should be buried under a few inches of clean dirt at the end of the work. Humus or rich topsoil should not be used. The soil cover will prevent sparks from blowing, preserve a hot bed for use in the morning, and, if the cover is not removed, may make a fair grade of charcoal.

FIRE CONTROL

Any contractor burning brush in an area subject to brush or forest fires is subject

Fig. 1-32. A good fire tender

to heavy responsibility if one of his fires spreads. Also, in the presence of extensive forest fires from any cause, the contractor may be required by authorities to use his men and equipment to control them. At such a time there might not be experienced fire fighting personnel available to direct his work. A brief outline of fire fighting techniques is therefore considered appropriate.

Hand Tools. Where the material burning is largely grass and associated weeds, or thin brush, fire can be beaten out. Household brooms, occasionally dipped in water if possible, are very effective. Shovels or leafy bushes or branches can be used with good effect. Each blow should be directed so that flying sparks are knocked toward the burned area.

The fire may also be starved by scraping away the vegetation just beyond the flames. This may be done with shovels, hoes, rakes, grub axes, or almost any piece of metal. A special type of fire fighting tool, shaped like a heavy rake and fitted with sickle bar teeth instead of tines, is quite effective. Bushes may be cut with axes, machetes, bush hooks, or pruners. ·

Extinguishers. Back pack fire extinguishers, which consist of a water tank carried like a knapsack, a flexible hose, a hand pump, and a nozzle are important pieces of equipment. If the grass is low or thin, spraying in the path of the fire may stop it. If the fire is strong and moving rapidly, the water may be most effectively used for putting out smoldering spots behind the beaters. Addition of a wetting agent—a small quantity of almost any detergent will do if regular compounds are not available—increases the effectiveness of the water by enabling it to soak through vegetable litter and punky wood.

Pumps. If streams or ponds are available, the contractor's pumps, particularly the light centrifugal type, are very valuable. A welder or machinist can usually

make adapters quite quickly that will permit fire hose to be attached to the pump outlet. The high pressures used in regular fire pumps will probably not be developed, but sufficient pressure will be available for wetting down firebreaks, or making direct attacks on anything short of a crown fire.

Sprayers. Tree spraying outfits make good fire fighters. These usually consist of a tank holding from two to five hundred gallons, a high pressure pump driven by a small gasoline engine, or by the power takeoff of a towing tractor, or a carrying truck; a reel of hose, and a nozzle. Those having an engine are generally mounted on a wagon chassis that can be towed by

Fig. 1-33. Back Pack extinguisher

almost any motor vehicle. If the pump is tractor driven, adaptation to most wheel tractors can be made quickly. The handiest models are those mounted on a motor truck.

Such equipment can generally be rented or borrowed in almost any area. The volume of water delivered through the nozzle is small, but pressure is high and results are usually excellent.

Dozers. A bulldozer can put out a grass fire by starting behind the fire and straddling its line as in Figure 1-34. It may be able to scrape off the grass without cutting much into the ground. If this is not practical, it can skim off the sod until the load is heavy, then swing it into the burnt area, or raise the blade and spread the sod over the next few feet of flames, smothering them. An angle dozer can side cast the sod into the burnt area, and a hand beater, or extinguisher, should follow to put out any spots that are missed.

Method of Attack. Windblown fires should not be attacked directly at the front as this procedure is both dangerous and ineffective. A new fire running before a wind will assume a shape similar to that in Figure 1-34. A direct attack on the front means fighting flames several feet deep. If these should be put out, fire blowing up the sides could rekindle them in a few seconds. A crippled man or machine ahead of the fire could not escape being burned.

Pinching off the sides is both effective and reasonably safe. The fire is extinguished starting at the back so that the heat and smoke are blown away from the workers. Provided a constant watch is kept behind them for rekindled spots, the fire cannot repossess the extinguished area. When the front is reached, it is attacked from directly behind as well as on the sides.

If the fire is too strong for the force fighting it, the front will continue to advance, but the work on the flanks will limit its

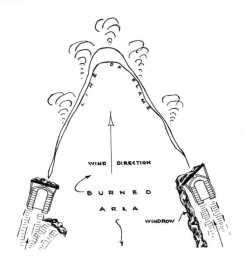

Fig. 1-34. Fire-fighting with dozers

width and make easier the task of stopping it with firebreaks or backfires, or after a shift in wind direction. It can sometimes be turned by concentrating on one flank.

Firebreaks. A firebreak is any strip bare enough of inflammable vegetation to delay or stop the spread of fire across it. Roads, open water, plowed fields, close cut lawns, and even footpaths may be used. In addition, breaks may be prepared in anticipation of fire along the crest of hills or mountains, at property lines, or at the edge of the areas being cleared.

Advantage should be taken of any existing breaks when deciding where to place one to stop a fire already burning. A short line is preferable, and valuable property or highly inflammable areas should be protected. The break should be far enough from the fire to allow time to finish it and to start backfires; it should be in vegetation least apt to make a spark-producing or a high fire, and on terrain favorable to operation of machinery. A compromise among these features must usually be made.

A bulldozer may be walked along the line of the break, alternately cutting and filling, so as to mix the vegetation with dirt. Hand workers with cutting or digging tools follow to cut out any spots where fire might cross. If the brush is heavy, the dozer may turn to push heaps of it out of the path. An angledozer or a heavy grader might be able to make a single clean cut in each direction, turning the sod and brush out from the center.

In grass or light brush, a plow or heavy disc harrow might do a better or faster job than a dozer. The plow makes a rather narrow strip with each trip, and is subject to jamming with brush but does not have to go back over its work. A harrow may require several trips and might not be effective.

The hand tools listed earlier may be used to build a complete break, or to work one over after the machinery has passed.

Backfires. A strong fire adds to the force of the wind which is moving it, somewhat as a blowtorch builds up its fuel pressure. The combined force may be enough to project a sheet of flame many feet in front of the burning line, and to shower sparks for long distances ahead. For this reason, the fire may cross a break of any practical width and make the area too hot for fire fighters.

The principal use of the firebreak is to provide a line from which backfires can be started. Since the break is made on the downwind side of the fire, a new fire started on that edge burns upwind. The backfire should be made in a continuous strip along the break so that it will not be able to turn and blow back toward it. It will increase in strength as it progresses, but will be steadily farther away from the protected side. When it meets the main fire, there is liable to be a spectacular flareup and heavy production of sparks. If the backfire has been started in time, this should be far enough away from the break so that few sparks will cross it, and those can be extinguished by men patrolling the break. If no shift in wind occurs, the sides of the fire can then be put out by the crew working from

behind, aided by the firebreak crewmen.

If the break is made in a forest where the flames might crown (burn in the tops), the trees on each side of the break should be bulldozed or cut so as to fall away from the center.

Since a change in wind direction may occur at any time, care should be taken not to start backfires prematurely, and to keep men and machines in positions where they can get away if the fire turns toward them. The burnt-over area, ponds or wet swamps, plowed land or well grazed pastures, are suitable retreats.

Men on the fire lines must be kept provided with food, water, and tools, and relieved for rest periods. Machinery must have fuel but may be skimped on other maintenance in sufficiently dangerous situations.

Backfiring, and possibly other phases of fire fighting, may be regulated or prohibited by local laws.

Re-Kindling. After the spreading of a fire has been checked, it must be patrolled until all danger of its making a fresh start has passed. A grass fire in a clean field may be safe to leave within an hour, while wooded areas containing dead or fallen trees, or rich dry soil, may be dangerous until after several soaking rains.

Dead stumps may burn a long time and are difficult to extinguish unless ample supplies of water are available. Fires burning under and between logs on the ground can often be put out by moving the logs apart, or can be caused to burn out more quickly by piling additional wood on them.

The worst hazards are standing dead or hollow trees, called snags by the lumbermen. If close to the line, snags may set fire to the unburned area by falling into it. They frequently produce sparks that may drift long distances. Even thorough soaking may not extinguish them, and it may be necessary to cut them down or maintain an expensive patrol for days or weeks.

Cutting a burning tree is a tricky and dangerous job as the cutters are in constant danger of being hit by falling pieces, and temperatures at the base may be too high for them or their tools. This job is best left to experienced fire fighting crews.

Snags may be pushed over by bulldozers but the tops are apt to fall on the machine. A cab with maximum-strength overhead protection is needed for operator safety.

The best time to check a burned area for hot spots is immediately after a rain, or a heavy dew, as the moisture near the fires will steam.

Underground Fires. Underground fires, such as occur in rich forest soils and dried out swamps, constitute a special problem. When fire gets in them, often by smoldering down a dead root, they will burn hot and persistently. Plain water has little effect on such a fire unless applied in such quantities that the area is flooded. Smaller quantities do not penetrate the deeper burning zones, which have sufficient heat to evaporate quantities of water from surrounding peat, and then spread through the dried material.

Special nozzles consisting of pipes long enough to reach the bottom of the fire are helpful. The lower end is plugged and a fairly large hole is drilled in the plug to wash humus out of the way as the pipe is pushed down, and smaller holes in the side spread a soaking spray. The use of wetting agents will substantially reduce the amount of water required, and may make the difference between success and failure where the water supply is limited.

Such a fire may be confined by trenching down to inorganic or saturated soil. The digging may be quite difficult because of roots, and a backhoe or dragline shovel might have to be used.

Peat fires spread very slowly unless they ignite surface vegetation or litter which set fire to the soil at new points. If equipment is not immediately available to extinguish or ditch the fire, leaves and inflammable trash should be removed for ten or more feet around it, to prevent rapid spreading while arrangements are made to put it out.

BOULDERS AND BUILDINGS

Boulders. An area may be so strewn with loose or partially buried boulders that work is difficult, and the removal of these rocks may properly be considered clearing.

If large enough machinery and suitable disposal points are available, the rocks may be turned or dug out and pushed away. If they are too large for easy handling and disposal, they should be broken up.

Breaking may be done by blockhole or mudcap blasting, backhoe mounted demolition hammers, drilling followed by plug and feathers splitting, or a muscle-operated sledgehammer.

A plug and feathers set consists of a pair of half-cylinders (feathers) with outer surfaces fitting in a drilled hole, with inner faces shaped for driving a thin steel wedge (plug) between them by air or hand hammer, or by hydraulic pressure.

The very gradual taper of wedge and cylinder halves causes blows or pressure on the plug to be converted into a tremendous sideward pressure, which can split large boulders, and break off chunks of bedrock.

Under many circumstances, however, a contractor may prefer to get rid of the rocks by digging and pushing. The dozer is the standard tool for this work. Efficiency can be increased by use of a tilting blade, a dozer shovel bucket, a stumper, or a heavy duty rake blade.

A dozer can move quite a large rock on firm ground, perhaps several times its own weight. If the stone is too large for direct pushing, it can be pushed first on one side, then on the other, as in Figure 1-36. If it is rounded, it can be rolled by lifting the blade while pushing. If the blade does not have enough lift to roll it over, it can hold it in a partially rolled position, with locked brakes, while the stone is blocked up. The blade may then be lowered and the push and lift repeated.

Partly buried rocks may be pushed or dug out in somewhat the same manner as stumps. The resistance they offer is usually more rigid and brittle than that of stumps, and if a rock will shake in the first few direct blows of the blade or bucket, it should come out. It is sometimes very difficult to get a grip on smooth sloping surfaces, so that an excessive amount of digging must be done just to get a hold.

When a grip is obtained with a dozer blade, the rock may be raised and pushed. The engine clutch should be slipped, or the converter-equipped engine throttled so as to supply just enough forward pressure to keep the blade in contact with the rock while it lifts it vertically, and, when it is

high enough, rolls it out. The rock may slip back into its hole at any time, and it is good to have a helper throw stones or logs under it so the blade can be dropped and a fresh grip obtained. If no helper is available, the operator can lock the brakes to hold pressure against the stone and do the hole filling himself.

If a big stone is rolled out without blocking, it may leave such a large hole that the tractor may be damaged if it falls in it. The danger is more serious than with stumps as rocks leave sharper edged and harder holes.

A rock should be pushed from all angles before digging it out as it may be susceptible to pressure from only one direction. If it is to be dug out, a bowl-shaped crater of considerable size is excavated, working on three sides, if a good grip is available at the top, or all around it if the top is

Courtesy of Emaco, Inc.

Fig. 1-35. Boulder split by plug and feathers, hydraulic-powered

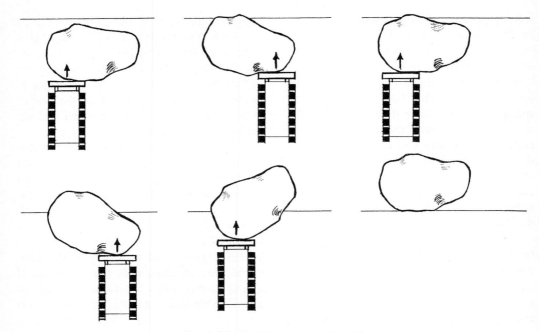

Fig. 1.36. Pushing oversize boulder

smooth. When it is finally loosened, it may be found that it is so heavy that the dozer cannot get it up out of the hole.

It probably can be pushed out by following the procedure outlined in Figure 1-37. The dozer builds a ramp out into the crater, shaped so that the machine will be pitched downward when its blade meets the rock. With gravity assisting, it should be able to push the stone a short distance up the opposite slope. The ramp is then built out, and another push made until the stone is out of the hole.

Loose boulders may be pushed out of the work area and scattered, piled, or arranged in walls; or they may be buried, being either used as a fill or wasted in holes. Holes may be dug to bury them.

Where many boulders are pushed into a hole they afford unsafe footing for a bulldozer and may pile up above the desired grade. A moderate amount of dirt, either scraped off the bank or trucked to the spot, will allow the dozer to fill in the holes and stabilize the rocks so that it

can walk across them in pushing other boulders to their resting place.

If the area is to be finished to a grade, it pays to be liberal in supplying covering soil, for if the layer is thin the dozer work-

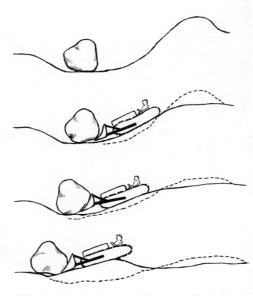

Fig. 1-37. Getting boulder out of a hole

ing on it may hook into freshly buried rocks and turn them into high positions. They can seldom be put back in place because of soil and other rocks getting under them, and it may be necessary to knock their tops off with hammers or explosives, or dig them out and rebury them. Digging a boulder out from among others is very difficult, and is likely to turn them up.

Rocks may be trucked away from the job. They may be loaded by lifting in or on a loader or dipper shovel bucket, by clamping in a clamshell or grapple, or between a hydraulic hoe bucket and the stick.

Or a crane may lift them by tongs, chains, or cable slings. Light chains grip well, but are easily damaged.

Ordinary dump truck bodies may be severely battered by oversize rocks. The floor may be protected by an extra sheet of steel, a layer of planks, or by just a few inches of dirt.

Stone Walls. Stone walls built to dispose of boulders removed from farm land are very common in some sections of the country, and may include rocks large enough to present a problem to machinery. The big base stones are often partly or completely buried, interlocked, and bound in place by tree roots. The smaller stones may be valuable for use in masonry, and may be removed by hand before or during the wrecking of the wall.

A dozer of sufficient size can walk right through the wall and scatter it around, but an undersize machine may have to start at a gateway, or find a weak spot to break through and widen the hole by worrying the rocks out one at a time. If the wall cannot be broken from one side it should be tried from the other.

Foundations. Old foundations and other masonry structures usually yield readily to heavy machinery. High walls should be pulled down as they might fall on a machine pushing them.

If a foundation is too strong for avail-

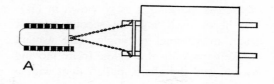

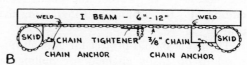

Fig. 1-38. Bracing tow skids

able machinery, it may be weakened by blasting along the lines where it meets the floor and other walls, by mudcapping or drilling. Demolishing very heavy or extensive structures, however, is a house wrecking job out of the field of this book.

Small Buildings. Moving buildings properly is also a highly specialized trade, but an excavating contractor may be called upon to move small structures of minor value out of the work area, or to drag his work buildings around on a job.

The easiest method is to jack the building up, or to lift it one end at a time with a shovel or crane, and pull a pair of substantial skid logs under it. These should be beveled at the front so as not to dig in, and notched on the top for the sills of the building. They should be spiked or bolted to the sills. If the building needs additional rigidity, cross logs may be used above the skids and the walls may be braced with diagonal planks.

The skids should be rigidly fastened to each other. If the building is to be pulled by one machine, through a double chain as in Figure 1-38 (A), which is the usual method, the bracing between the front of the skids takes a tremendous inward pressure. Ordinary log or timber braces may not hold, unless very expertly installed, or the pull is light.

It is usually worth while to make a

steel cross brace such as is shown in (B). The beam itself may be second hand I beam, or channel or angle. The welded brackets prevent the skids from moving inward, and the chain, pulled tight with a load binder, prevents them from moving out. Notches in the bottom of the skids are necessary to prevent the chain from cutting into the ground and increasing the draft.

The skids should project far enough forward to be easily chained for pulling or lifting. They should be high enough to carry the sills or cross logs over any irregularities in the ground. If this is not practical, rollers, consisting of short logs, may be put under the front of the skids. As the building goes over these, they must be watched so that they will not turn up and injure the structure. When they are left at the back, they may be picked up and carried to the front.

Rollers are also used when the tractor is not powerful enough to pull the skids on the ground.

CHAPTER TWO

SURVEYS AND MEASUREMENTS

GENERAL CONSIDERATIONS

Surveying is a profession in itself and contractors and their employees seldom have time to master it. However, it is possible for a layman to run levels, to re-establish lines and locations obliterated by construction, and do rough layout work.

If a job involves as much as a day of work for a surveying crew, it is usually economical to hire professionals. They work more rapidly and efficiently than amateurs, and are less liable to make costly mistakes. Unfortunately, it is frequently not possible to obtain the services of engineers exactly when needed, and there are many jobs which are too small, or too simple, to justify calling them in.

Also, it is sometimes desirable for the owner or contractor to make a rough survey of a project to determine the amount of work to be done, and possible layouts, before bringing in surveyors to provide detailed information. A man can usually obtain a much clearer idea of the problems involved by running his own levels than by reading the findings of another.

The methods outlined in this chapter will in some cases be those used by surveyors, but will often be shortcuts and substitutes which can be used by amateurs with reasonably satisfactory results, and which generally are easier to learn, but less accurate, than professional methods.

More detailed information about survey-ing may be found in textbooks on plane surveying. Work with a survey crew is the soundest training in field methods.

TELESCOPIC LEVELS

The basic surveyors' tool is a telescopic level mounted on a turntable which in turn is usually based on a tripod. The entire unit is often referred to as an instrument. There is a great variety, but most of them may be classified under three headings—level, convertible level, and transit. The difference is partly that in the first the telescope is always used in a horizontal position; in the second the telescope may be lifted out of its frame and reset so as to pivot vertically; and in the transit it is permanently mounted so as to swivel vertically as well as horizontally. However, these general distinctions are not always true in regard to particular models.

Automatic leveling and laser instruments will be discussesd later in this chapter.

Builders' Level. Figure 2-1 shows a type of builders' level which is convenient for general contractors' use. The telescope is held rigidly in a U Frame that rotates on a vertical spindle, which is perpendicular to the line of sight of the telescope. A spirit level with a graduated glass is mounted on the frame.

The leveling head on which the spindle rotates is shown in Figure 2-2 fitted with a horizontal circle marked in degrees, in contact with a pointer fastened to the

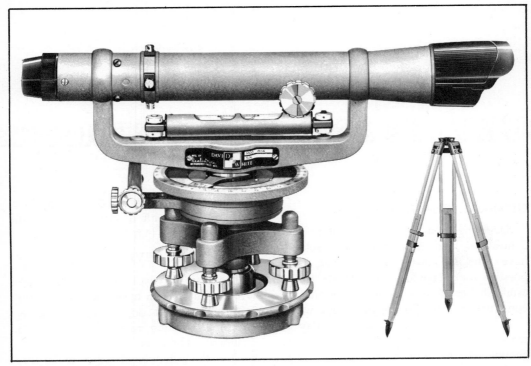

Courtesy of David White Instruments

Fig. 2-1. Builders' level and tripod

spindle. Many levels do not have this circle, but it is essential for the location work to be described.

Courtesy of David White Instruments

Fig. 2-2. Horizontal circle

Vernier. The pointer may be expanded into a vernier such as shown in Figure 2-3. This is a device for reading fractions of a scale, which in the example is calibrated to 30′ (half degree) divisions. The length of twenty-nine divisions on the circle is divided into thirty divisions on the vernier. Each vernier space therefore represents 29/30 of a space on the circle, and is 1/30 shorter.

In the illustration, if an angle on the main scale is being read from left to right, the zero, or center of the vernier, shows a reading slightly higher than 2° 30′. Reading the vernier to the right, it will be found that the tenth division line matches exactly with a line on the circle scale. This indicates that the zero mark was 10/30, or 1/3, of the way from 2½° mark to the 3° mark, as the difference is cancelled out in the course of subtracting the 1/30 difference 10 times.

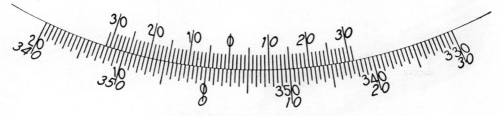

Fig. 2-3. Vernier

One third of 30′ is 10′. The angle is therefore 2° 30′ plus 10′, or 2° 40′.

If the angle were being read from right to left, the main scale would read 357° and a fraction. Reading the vernier to the left, the twentieth division is found to correspond with a line on the circle. The angle is therefore 357° plus 20/30 of 30′, or 357° 20′.

The telescope may be locked against swinging by means of a thumbscrew for convenience in reading the scale, or holding it in a certain direction. Another thumbscrew (tangent screw) will then move it slowly for fine adjustments.

Telescope. The length and power of the telescope and the length of the spirit level, determine the range of the instrument and, to a considerable degree, its accuracy. Telescopes range from ten to eighteen inches in length, and from ten to thirty-five power in magnification. Spirit levels may be three to ten inches long.

The telescope is focused by means of a knob on the side or top, and possibly by a turning eyepiece also.

The field of view of the telescope is divided into quarters by the cross hairs, shown in Figure 2-4 (A), which are held in a frame or diaphragm inside the telescope. Provision is usually made to make these visible or invisible by focusing the eyepiece. The horizontal hair is used for taking levels. If it is correctly placed in the telescope, and the telescope is properly leveled, it indicates the slice of the field of view which is level with the observer's eye. The vertical hair is used to sight a given point or line, and indicates the exact center of the field of view for determining horizontal angles.

Stadia Hairs. Stadia hairs (B) may be fitted into the same frame. These are horizontal and are located above and below the center hair. The distance between the stadia hairs is fixed at a ratio with the telescope, usually 1 to 100, so that if a measuring rod is sighted through the scope, the inches or feet seen between the stadia hairs may be multiplied by 100 to give the distance of the rod from the instrument.

Amateurs are apt to confuse one or the other stadia hair with the cross hair in taking levels, with resultant serious error. If this trouble persists, additional hairs may be installed, as in (C) in the form of a letter X, which should make the center hair easy to distinguish.

Base. The leveling head is mounted on the turntable or base by means of a center pin on which it can both tip and rotate, and four leveling screws. These screws are threaded into the leveling head and rest on the leveling plate or turntable. They are expanded into knurled wheels for convenience in turning with the fingers, and have expanded feet which do not turn with the screw, and which protect the plate.

The turntable base has internal threads by means of which it can be screwed on

Courtesy of David White Instruments

Fig. 2-4. Sighting hairs

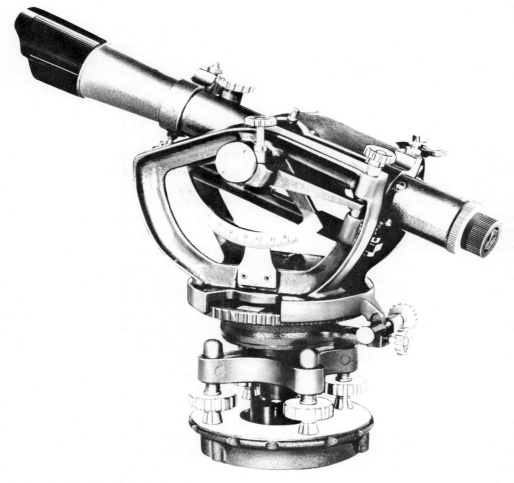

Courtesy of David White Instruments

Fig. 2-5A Level transit

to the tripod head. It may have a hook on the bottom, at the center, from which a plumb bob (pointed-tip weight) may be hung by a chain and string. The base may be made to slide a limited distance horizontally, relative to the tripod head, for convenience in centering the unit directly over a mark.

Tripod. A tripod consists of three metal or wood legs, hinged together by a top plate which is threaded for the instrument. These threads should be protected by a cap whenever the instrument is not mounted.

The legs may be one piece, or two pieces sliding on each other and locked by a screw clamp.

The base may be rested directly on a flat rock or stump, if it is not possible to set up the tripod, but this is not recommended.

Transit. A transit has a vertical swivel and support yoke mounted on the turntable, which permits the telescope to be tilted up and down, in addition to its horizontal rotation. Angle from the horizontal is shown on a vertical scale with a vernier. When

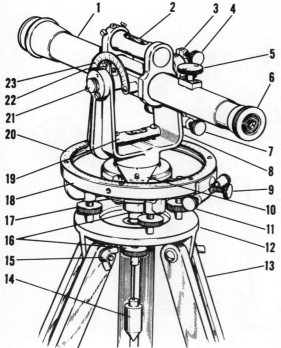

1. Telescope
2. Telescope Bubble Assy
3. Vertical Clamp
4. Vertical Clamp Screw
5. Focusing Screw
6. Eyepiece Cap
7. Vertical Tangent Screw
8. Telescope Support
9. Horizontal Clamp Screw
10. Horizontal Circle
    Vernier Plate
11. Horizontal Tangent Screw
12. Tripod Head and Base Plate
13. Tripod Leg
14. Plumb Bob
15. Tripod Wing Nut
16. Center Screw
17. Leveling Screw
18. Leveling Head
19. Support Level Tube
20. Horizontal Circle Scale
21. Telescope Trunion
22. Vertical Arc Pointer
23. Vertical Arc Scale

Courtesy of David White Instruments

Fig. 2-5B. Parts of level transit

used as a level, the reading on this scale should be zero.

A level transit, Figure 2-5A and B, can be tilted about 45°, either up or down.

Full transits have an extended support yoke that permits the telescope to turn in a full vertical circle within it. This makes it possible to make a back sight (180° turn) without changing the setting of the horizontal circle.

Compass. Compasses are standard equipment in transits, and can usually be obtained for other type instruments that have a horizontal scale. They are not necessary for the work to be described in this chapter, although it is often convenient to know the general directions of lines.

Surveys are generally based on the true north, from which the compass north varies rather widely. Part of this variation may be obtained approximately from the map, Fig-ure 2-6A, or exactly from local sources.

If you are in an area of west magnetic declination, the compass needle will point west of the true north by the amount shown on the map.

Another source of error is the magnetic attraction of magnets, iron, and iron ore for the compass needle. It is also affected by the time of day. No confidence should be placed in a compass reading taken near machinery or electrical apparatus. Metal objects in the observer's pockets may cause errors.

Setting Up. The first step in using the instrument is to set up the tripod. The top should be as level as possible, and the legs pushed into the ground firmly. On a slope, two legs should be downhill. The protecting cap is removed and the instrument screwed on. The telescope frame should be unlocked so that it is free to rotate. The telescope

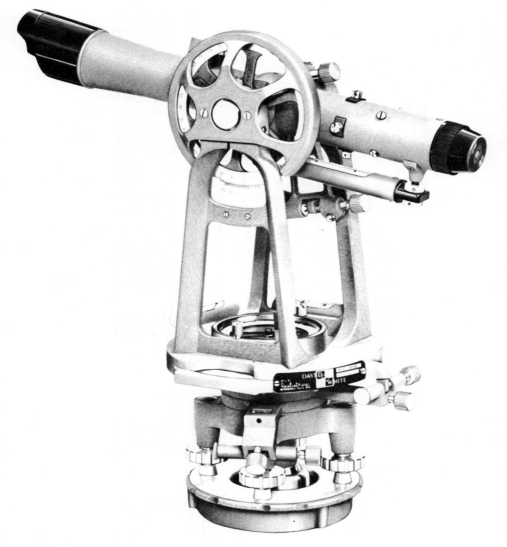

Courtesy of David White Instruments

Fig. 2-5C. Transit (Full transit)

can then be held in one hand and the base screwed on the tripod with the other.

Leveling. The instrument must now be leveled by means of the four screws. The telescope is turned so that it is over two of them, and those screws adjusted until the bubble in the level is exactly in the center of the scale. The screws are turned at the same time in opposite directions, so that one pushes the leveling head up while the other makes space for it to come down, as it pivots on its center pin.

The bubble moves in the same direction as the left thumb, as indicated in Figure 2-6B. If the two screws are turned exactly the same amount, the tension on them will remain constant. If the screw toward which the bubble is moving is turned farther, it

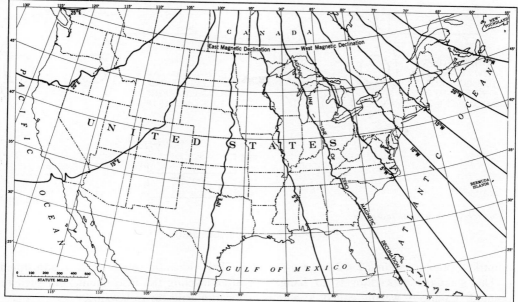

Courtesy of U. S. Coast and Geodetic Survey

Fig. 2-6A. Magnetic declination map

will jam both screws. If the screw behind the bubble is turned the most, the tension will be reduced and the screws may lose contact with the turntable.

If both screws are turned to the left (clockwise), or one is turned left while the other is stationary, they will be jammed; while if one or both are turned to the right (counter clockwise) they will both lose contact.

The screws should be kept in light contact with the plate during the adjustment and tightened somewhat as it is finished.

When the bubble is approximately centered, the telescope should be swung ninety degrees so as to be over the other pair of screws, which are used to center the bubble in the same manner. This adjustment will disturb the first one, so the telescope must be swung into its original position and leveled, this time more exactly. It should then be checked in the second position, and adjustment in the two positions made alternately until it does not move during

the swing through this quarter circle arc.

The telescope should now be swung through the other three quarters of the circle. If the adjusting screws are tight, and the tripod has not been disturbed, the bubble should not move. If it does move, and the table cannot be leveled so that it will not move when swung, the spirit level is probably out of adjustment.

Re-Leveling. During use, the spirit level should be checked occasionally and the instrument re-leveled if necessary. The tripod may settle into the ground, particularly if on some unstable base such as ice, wet clay, or oil road top. Jars from focusing the telescope, or the wind, or other causes, may disturb it. Sometimes it is necessary to put small boards under the legs to avoid settlement.

Centering. Exact placement of the instrument is usually not important in setting grades, but it is essential in most other work. It involves locating the vertical axis on which the telescope swings directly over

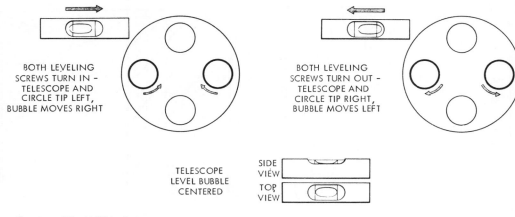

BOTH LEVELING
SCREWS TURN IN –
TELESCOPE AND
CIRCLE TIP LEFT,
BUBBLE MOVES RIGHT

BOTH LEVELING
SCREWS TURN OUT –
TELESCOPE AND
CIRCLE TIP RIGHT,
BUBBLE MOVES LEFT

TELESCOPE
LEVEL BUBBLE
CENTERED

SIDE VIEW

TOP VIEW

Courtesy of David White Instruments

Fig. 2-6B. Leveling screw action

a marker, often a nail in a 2 x 2 stake.

The standard procedure is to hang a pointed, balanced weight called a plumb bob from the center of the instrument by a light chain and/or string. The tripod is maneuvered until the point of the bob is just over the center of the nail head.

This location may need readjustment after leveling the instrument, and such re-location calls for re-leveling.

The tripod head in Figure 2-7A allows horizontal shifting a distance of two inches in any direction, without disturbing its level. This device saves time and trouble.

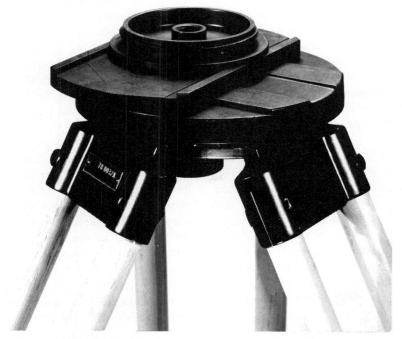

Courtesy of Keuffel & Esser Co.

Fig. 2-7A. Shiftable tripod head

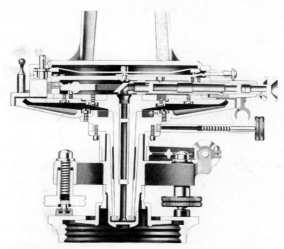

Courtesy of Keuffel & Esser Co.

Fig. 2-7B. Optical plummet

The optical plummet, Fig. 2-7B, replaces the plumb bob with a low magnification telescope in the rotating base, which, by means of a prism, looks down the vertical axis. A small bullseye and cross line make possible exact alignment with a marker.

SELF-LEVELING LEVEL

The self-leveling or automatic level uses a three-screw leveling base with a circular level vial, for approximate leveling of the instrument.

Fine leveling is done by a gravity-con-

Courtesy of Keuffel & Esser

Fig. 2-8A. Self-leveling level

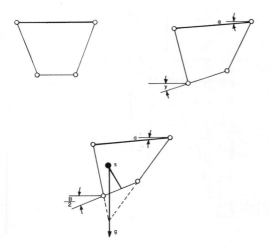

Fig. 2-8B. Trapezoid leveling effect

trolled device (pendulum or compensator) inside the telescope.

In principle, a telescope may be made self-leveling by being made part of a pendulum (hanging weight), so that gravity would keep it horizontal regardless of the tilt of the weight's supports. However, problems of size, weight, vibration, and sway make such construction impractical.

Instead, the pendulum or gravity device is built into the optics of the telescope. This may be done in a number of different ways. In the Zeiss Ni 2 instrument in Figure 2-8 A to D, there is a prism or mirror supported by wires, so that it forms the smallest side of a trapezoid, or irregular 4-sided figure.

If a hinged trapezoid of this type is tilted, as in Figure 2-8B, the lower side will assume a greater change of angle with the horizontal than the upper. The extent of the difference is determined by the arrangement of the fixing points of the wires that form the sides.

The difference in tilt may also be increased by raising the center of gravity of the lower member, as in the bottom drawing, or decreased by lowering it.

When the upper surface of the lower member is either a prism or a mirror included in the path of light through the telescope, the proportions can be arranged so that the tilt of the body of the instrument is offset optically by the tilt of the refracting or reflecting unit. That is, the line of sight through the instrument will be level, even if the instrument itself is not.

Such a mechanism must not be affected by temperature, and must include a dampening device to prevent swinging and vibrating in response to wind action or ground movement. There may be means to keep the leveling bubble in the field of view, to assure that the instrument does not go off-level beyond the somewhat limited range of the compensator.

If kept within this range, the field of view is almost as automatically level as the surface of still water.

There are very small inaccuracies which may appear in automatic levels, which may be detected, limited, or counteracted by procedures too complicated to merit decription here. But these are generally smaller than errors in standard levels.

LEVELING ROD

The surveying instrument's best companion piece is the leveling or target rod. This is a measuring stick, marked in feet,

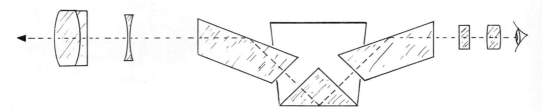

Fig. 2-8C. Line of sight in self-leveling level

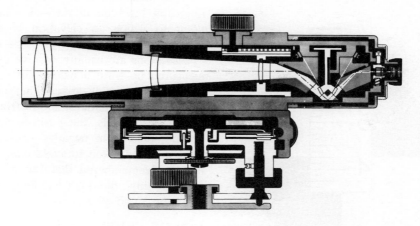

Courtesy of Zeiss

Fig. 2-8D. Cross section of self-leveling level

tenths, and hundredths of feet; or in feet, inches, and eighths of inches. It may be eight to fifteen feet long, and usually is in two or three pieces which slide on each other, or, occasionally, are hinged or pegged together. The sliding type must be fully closed or fully open to be accurate.

Long rods are very desirable in hilly country.

Spaces may be marked by fine lines similar to those on a ruler, in which case it is called a New York rod. A Philadelphia rod uses the division lines as units of measurement in themselves. See Figure 2-9. A rod in the decimal scale has the tenths of feet each divided into ten equal

sections, alternating black and white. If inches are shown, each is divided into eight equal bars of alternating color.

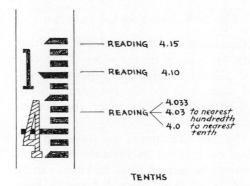

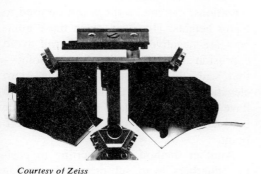

Courtesy of Zeiss

Fig. 2-8E. Trapezoid suspension

Fig. 2-9. Philadelphia rods

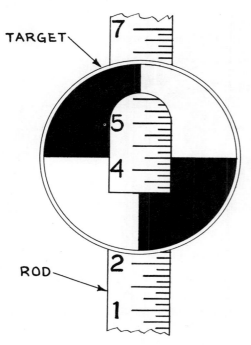

Fig. 2-10. New York rod and target

The target, 2-10, is a metal disc that slides on a track on the sides of the rod. It is painted in quadrants, alternately red and white, with the division lines horizontal and vertical, or in other conspicuous patterns.

The rod man moves the target up and down on the rod in response to signals from the instrument man. Readings can be taken in this way when distance or haze prevents reading figures on the rod.

The target may include a vernier, in which case it can be used for precise work requiring reading of fractions of the smallest divisions of the rod scale.

The rod is used for measuring the distance from the instrument's line of sight down to a point. If the point is almost as high as the instrument, this distance will be short; if it is much lower, the distance will be long.

The elevation of a point is the distance which it is above some standard level. This may be mean sea level—halfway between high and low tide marks—or some local point to which an elevation is assigned arbitrarily.

Elevations are usually positive numbers, measured up from a base point or plane. Rod readings are negative, being measured down from the plane of the instrument.

In taking levels, the positive elevations are obtained from negative rod readings. Care must be taken to avoid confusion. It must be remembered that for any instrument setting, the high readings are low elevations, and vice versa.

USE OF LEVEL

Use of the instrument as a level depends upon the fact that its cross hair indicates a horizontal plane, level with the observer's eye in all directions. By use of the rod, the amount by which a point is lower than this plane can be measured, and the relative elevation of any number of points within range can be calculated from rod readings. Points above the cross hair cannot be measured, except on vertical walls, without moving the instrument higher and resetting it.

Levels are most accurate over short distances. If the instrument must be used when out of adjustment, readings should be taken as nearly as possible at equal distances.

Converting Readings to Elevations. If the instrument is to be set up only once on a job, and no record is to be made of observations, the rod readings can be used to figure heights. However, since these numbers are negative, the beginner will avoid confusion by calling the lowest point—that with the highest rod reading—zero elevation. The other points will then each have an elevation equal to the difference between its reading and that of the zero point.

Another method of converting to positive numbers is to subtract each reading from some number larger than the highest

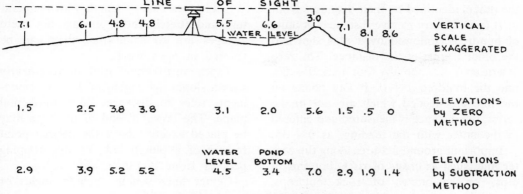

Fig. 2-11. Rod readings

reading. Figure 2-11 shows a series of readings taken to determine in which of two locations a drainage ditch should be dug, with figuring done by both the zero and subtraction (from plus 10) methods.

Bench Marks. If the elevations are to be recorded for future use, it is necessary to have some fixed reference point which will not be disturbed and which can be readily identified. A knob on firm bedrock, a nail projecting from a tree trunk, a mark on a building, or a stake hammered flush with the ground may be used. Such a point is called a benchmark, and is abbreviated as BM. A reading is taken the first time the instrument is used, and again each time it is set up. These readings will probably all be different, but in each case the elevation of the instrument may be found by adding the rod reading to the benchmark elevation.

Since the benchmark is the most permanent point observed, it is good practice to assign an elevation to it, and to calculate all other elevations from that. An assigned number should be large enough so that no elevation less than zero will be found on the job, as working with minus figures may cause confusion and error.

If levels have been taken previously in the area, engineers' benchmarks may be found, in which case it is wise to use them. If possible, the elevation assigned to them in the previous survey should be used to facilitate comparison between the two sets of levels.

Even if engineers' benchmarks cannot be used directly in surveying the job, it may be advantageous to run a level to one, and note its elevation in relation to the contractor's own benchmark, so that the two systems can be compared if necessary.

If the job is a type that will involve frequent checks of levels, as on a road where stakes may be knocked out by machinery, it is a good plan to set up benchmarks so that one will be visible from each point where the instrument will be used. This saves time in taking grades on a few stakes and eliminates common errors in moving the instrument, or in taking an elevation from the wrong line stake, or a stake which has been disturbed.

All benchmarks should be figured very carefully, and re-checked at least once.

Recording Readings. Another requirement in recording observations is to identify the spot at which each reading is taken. This is usually done by taking readings at set intervals, such as ten, or fifty, or one hundred feet. These distances should be marked by stakes, pegs, small rock cairns, or in other ways. The first stake or mark of the series is called the zero stake, and the others identified by their distances from it. It is customary to give distances in units of hundreds, followed by a plus sign and the other figures of the distance. The zero is written $0 + 0$, the fifty foot mark $0 + 50$, and the hundred $1 + 0$. If any points on the line are needed which are not in the series, the distance is measured and entered in the notes with the reading, as $0 + 35$.

Important ground features along the center line, such as crests of rises, bottoms of dips, or beginnings of rock outcrops, should be taken in addition to the stake readings.

Elevations may be taken from the ground, from the top of the stake, or more rarely, from a mark on a stake. If taken from the ground, it should be stamped or cut flat. Such readings are not as accurate as those taken from the top of a stake, and may be very difficult to check back, but they can be used directly in preparing profiles and figuring cut and fill. If the top of the stake is used, it is necessary to measure the stake height.

Tapes. Measuring is usually done with a steel tape, often called a chain by surveyors. Fifty foot and hundred foot lengths are standard, and will suffice for most purposes. They should have a non-rusting finish as it is often difficult to dry and oil them immediately after wet work. Care should be taken not to kink a tape, or to bend it sharply, as such abuse may break it.

If the numbers become illegible, they can be fixed for rough work by measuring and marking the feet, and perhaps some fine divisions, with paint. A broken tape can be repaired by means of a splint and two rivets.

Cloth tapes stretch readily, and are not accurate enough for even rough use. Metallic tapes, composed of cloth with interwoven wires, are variable in quality and resistance to stretching. If used, they should be checked occasionally by a good steel tape.

Steel tapes change length with temperature and stretch under tension, but these changes are so small that they can be ignored in open work.

Tapes must be held level, or very nearly so, on slopes, as engineers' land measurements refer to distances on a horizontal plane. The downhill end of the tape may be placed exactly above the desired point by use of a plumb bob, or by dropping pebbles from the tape end.

Center lines usually include angles or curves. If the former, measurements must be made to and from the angle point, rather than by a shortcut. Gradual curves may be measured in a series of chords (straight lines beginning and ending in the curve). Sharper curves may require a reduction in the length of the chords, as from a hundred to fifty, or twenty-five or even ten feet. The difference in length between the chord and the arc of the curve may be readily found by laying the tape along the curve from one chord point to another; or measuring a distance along it in very short chords, then measuring the distance between the two points directly. If no significant difference is found, the chords are not too long.

Tapes are best suited to two-man use. However, the loop on the zero end can be anchored in dirt with a screwdriver, and to stakes with a pushpin or thumbtack, and measurements made by one man.

Ground measurements may also be made with the rod, with a short rule, a stick of known length, or for very rough work, by pacing.

Stadia. If the instrument is equipped with stadia hairs, it may be used to measure distance as well as elevation. If the stadia ratio is the usual 1 to 100, and the rod is marked in feet, tenths and hundredths of feet, each tenth visible between the stadia hairs indicates a distance of ten feet from the center of the telescope to the rod. Six tenths would mean a distance of sixty feet, a foot would mean a hundred feet. This distance may be noted at the same time as the cross hair reading.

If the rod is marked in feet, inches, and eighths of inches, each inch indicates a distance of eight and a third feet, each foot a hundred feet.

If a distance is to be measured off, the rod is held at increasing distances from the instrument in response to signals, until the proper number of markings show between the stadia hairs.

If the rod is partially hidden by brush, a reading may be made between the center hair and either stadia hair, and multiplied by two, with only slight loss of accuracy.

The rod should be held perpendicular to the line of sight. If the telescope is level, the rod should be vertical; looking downhill it should lean away from the instrument; and looking uphill lean toward it. If it is not at the proper angle the reading will be too large. The correct angle can be found by pivoting it slowly toward and away from the instrument, until the minimum reading is obtained.

Turning Points. If elevations are to be taken for any points above the cross hair, the instrument must be picked up and reset at a higher elevation. It must be located so that it can take a reading on at least one point that was taken from the old setting. This point (turning or transfer point) is preferably one of the higher elevations (low readings) taken, and should lie between the two instrument locations. It is best taken from the top of a firm stake, or a knob or a well marked spot on rock or hard ground, so that the rod set on it will be at exactly the same height at the second reading as at the first. Accuracy in reading at the turning point is very important as any error made will persist through the rest of the survey. Amateurs are advised to use two turning points with each move, as mistakes in reading or

in arithmetic should then show up immediately.

The new instrument elevation (abbreviated H.I. for Height of Instrument) is found by subtracting the smaller reading from the larger one for each turning point, and, in an uphill move, adding the result to the first elevation of the instrument.

Recording and Figuring. Figure 2-12 shows some of this work. (A) shows the slope, the location of benchmarks and stakes, and the two instrument positions used. (B) is an informal set of notes of rod readings and calculated elevations. (C) is a profile drawn on cross section paper from the notes in (B). It is made by drawing a base line, assigning it an elevation lower than those of the stakes, and making each square represent a certain distance. In this diagram, each square represents one foot vertically and ten feet horizontally. This vertical exaggeration is necessary to have a large enough scale without making the drawing impossibly long.

The profile is useful in giving a picture of the slope, in determining gradients of roads or ditches, and in figuring the cut and fill necessary to convert the present grade to the new one.

The dotted line is the subgrade for a proposed road. It will be seen that the depth of cut or fill on this line may be approximately determined by measurement with a ruler; the elevation of any point on the road, in relation to the benchmark, can be found in the same manner.

Moving Downhill. If, at the original or at any later location of the instrument, points to be taken are so low that the rod is below the cross hair, the instrument must be moved downhill. A turning point (or points) is chosen with the lowest possible elevation (highest reading), the instrument moved, and new readings taken. The low reading is subtracted from the high one and the result subtracted from the earlier instrument elevation.

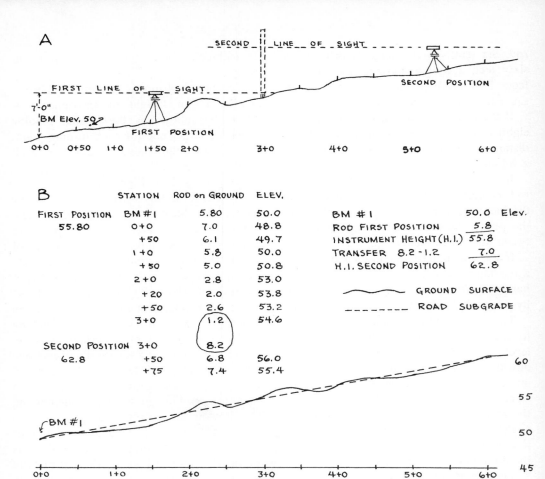

Fig. 2-12. Recording and figuring

If only one or two points slightly below the cross hair must be taken, the rod may be rested on a stake, and the height of the stake added to the reading; or a ruler may be used at either the top or the bottom of the rod to extend it.

Check Runs. When all the necessary points have been taken, the accuracy of the work may be checked by taking levels back to the starting point. This is usually a faster operation than the outward trip, as it is only necessary to take transfer points and benchmarks. If frequent benchmarks have not been placed, it is advisable to use the same turning points, or to take readings on a few of the grade points, so that if an error is present, it may be local-

ized. It is not necessary or desirable to set up the instrument on the same points for the return trip.

The two elevations found for each point should agree, but a difference, varying with the care with which the work is done, generally exists. Benchmark runs should be held to within a few hundredths of a foot, even in rough work where a difference of several inches on a grade point might be allowable. If any considerable amount of cut or fill is needed, even benchmarks may be left as approximations, until skill or time is available for a more careful run. Any discrepancies found in the check run should be listed in the notes.

If benchmarks are set at the beginning

2-12

and end of a run, and check properly on the return trip, it will not be necessary to back check any later run on which these two elevations show correctly. However, if benchmarks have been set by other parties in some previous survey, they should be checked the first time they are used, as they may be wrong or their description misunderstood.

Grade Stakes. Centerline road stakes are set by instrument and measurements from a prepared base line. Shoulder, slope and other side stakes are set from the centerline.

Grading information, that is, the cut or fill necessary at each stake, may be determined in several ways. The preferred method is to take the ground level elevation at each stake by instrument and rod to the nearest hundredth of a foot or eighth of an inch, figure the difference from the desired elevation, and mark the difference on the stake.

Ground is often so rough that some dirt has to be patted down flat at the foot of the stake to provide a recognizable base for the rod. It may be advisable to put a crayon mark on the stake at ground level, in case anything should change it. This mark, usually a horizontal line, will be needed when grades are marked on it.

Many surveyors prefer to use the top of the stake rather than the ground level. In this way they have a firm base for the rod, and they do not have to be concerned with the possibility that dirt might be kicked away or added. However, this usually requires measuring the stake height to give ground level for yardage calculations.

A horizontal crayon line on the stake may be used instead of the top or ground.

The elevations observed at the stakes are subtracted from those required by the road plans. Plus numbers indicate that fill is needed; minus numbers that the ground must be cut or removed. The symbols written on the stakes are "F" for fill and "C" for cut. Each stake must show location.

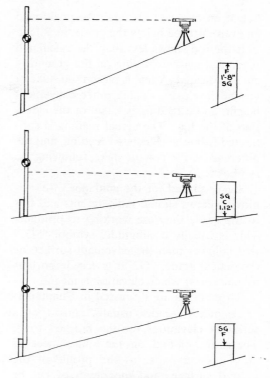

Fig. 2-13A. Setting grades

Another method is to supply the surveying crew with the required grade elevation for each stake. These are subtracted from the H.I. (height of instrument) to show the reading on a rod held with its bottom at correct grade at that location.

If the rod can be held at an elevation so that the reading is seen at the instrument cross hair, the stake is marked with a horizontal line at the base of the rod. This line is marked SG or G. The indication is that ground level must be raised to this line.

If the top of the stake is below the grade, the rod is placed on its top. The calculated reading is subtracted from the actual reading, and the difference written on the stake, following the symbol F. A crayon line is made at the top of the stake, and/or an arrow is drawn from the figures to the top.

Such a measurement for fill can also be made from a line drawn at any convenient height on the stake. It might be positioned an even distance below the grade, as 2 feet.

If the rod shows less than the calculated reading when it is resting on the ground, a cut is indicated. A line is put on the stake at ground level, or at some other convenient height, and a reading is taken of the elevation of the line. The actual reading is subtracted from the calculated reading, and the difference written on the stake, following the symbol C.

Stakes placed for the guidance of earthmoving crews should indicate cuts and fills to the grade they are working to produce. This is usually a subgrade, symbol "SG," that is lower than the pavement surface or theoretical grade, "G." It is very important to make clear which elevation is meant.

Road stakes are discussed in Chapter 8.

Engineers' grades usually consist of a series of elevations for the finished road. These are plotted on the same sheet of cross section paper as the profile of the ground surface, and the depth of cut or fill determined by measuring the distance between the two lines. These figures, if used directly, will not be accurate for most subgrade work, as the thickness of the pavement or gravel and of any special subgrade material must be subtracted to obtain the rough grade elevations.

A misunderstanding as to whether figures on grade stakes are for finish grade or subgrade can be very expensive. Use of subgrade figures for preparing subgrades is usually most satisfactory.

The contractor may obtain from the engineer a list or profile showing subgrade elevation at each station, and information as to the location and elevation of benchmarks. This, combined with sufficient field references to show the center line, will enable him to replace stakes which have been knocked out, and to find the depth of cut or fill required, by comparing the ground elevation with that required for the road.

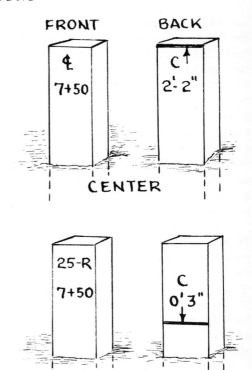

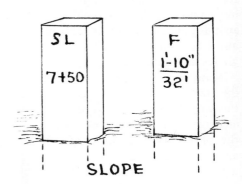

Fig. 2-13B. Stake markings

LOCATIONS

Turning Angles. When an instrument is used to turn angles—that is, to measure the horizontal angle between two lines or directions—the axis of revolution of the telescope must be exactly above the intersection of the lines, which may be marked

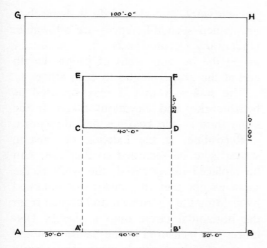

Fig. 2-14. Locating house site

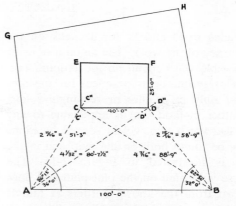

ALL ANGLES AND DISTANCES SCALED FROM PLOT PLAN

Fig. 2-15. Locating house site in irregular plot

by a nail in a stake driven flush with the ground, a cross chiseled in rock, or by markings on concrete or metal plugs.

When the instrument is set up over a point, a plumb bob should be hung at its center of rotation from a hook or through a hole usually provided. The point of the plumb bob should be just above the mark, and an amateur may have to move the tripod repeatedly before it is placed right.

If the instrument has a shifting base so that it can slide on the tripod, setting up over a point is greatly simplified.

A range pole is a convenient accessory in line and angle work. It is a pole seven or eight feet long, equipped with a metal point. It is painted alternately red and white in bands one foot wide. It is lighter than a rod and because of its conspicuous pattern is more readily seen at a distance.

This pole is set on one of the lines in question. The instrument is swung so that the vertical cross hair is on the pole. The rotation is locked, and the hair lined exactly on the pole by turning the horizontal tangent screw.

The reading on the horizontal circle and on the vernier is recorded.

The pole is placed on the other line and sighted in the same way. The difference between the two readings is the angle between the lines.

Line and angle work may be done to stake out on the ground locations described on a blueprint or map; or to make a record of ground features or locations on paper so that they may be used in figuring, or replaced or relocated if necessary.

Staking out is best left to engineers if possible, as accurate work involves trigonometry and skillful use of the instrument, and inaccurate work may result in very expensive mistakes. However, in emergency, or when results need be only approximate, the contractor can do it himself.

Staking from a Map. An example of staking out from a map is shown in Figure 2-14. A building, 25' x 40', is to be erected at the location shown on a plot 100' square. Lot corner stakes are at A and B.

The instrument is set up on A and sighted at B. The distances AA' and A'B' are measured along the line of sight, and stakes driven at A' and B'. The instrument is then set up on A', sighted on B, and turned 90°. The distances A'C and CE are measured and stakes put at C and E.

The transit is now set on C, sighted at A', and turned 90°, the distance CD measured, and a stake placed at D. F is located from the instrument set at E in the same manner. The instrument is now set on F, sighted at E, and turned 90°, measuring the distance to D and B′ for a check on the accuracy of the work. The amount of error allowable will depend on job requirement.

This technique is practical for the amateur only on square or rectangular lots. Another method that is applicable to any lot for which two widely spaced locations may be found both on the map and in the lot, is illustrated in Figure 2-15. The house is located on the plot plan and lines drawn from the known corners A and B to the near corners of the building, as shown. These lines are measured and converted to feet according to the scale, and the angles they make with the line AB and with each other are measured with a protractor. Figures are written on the plan.

The instrument is set up on A and a bearing taken on B. A 36° angle is turned, the line AD measured, and the stake D placed. An additional angle of 26° 15′ is turned, AC measured, and C marked.

· The instrument is now set up at B, sighted on A, turned 32°, and BC measured. The end of this line should be the stake previously driven at C, but if it is not, a second stake C is placed. The instrument is turned an additional 22° 30′, and BD measured in the same manner. If the same locations are found for C and D from both A and B, and the line CD is the required length, the work thus far is correct. If serious disagreement is found, the work must be rechecked.

The instrument may next be set at C, sighted on D, and turned 90° left. The distance CE is measured and stake E driven. F is located by setting the instrument at D in the same manner. This part of the work may be checked by setting up on E, sighting C, turning 90°, and measuring EF.

The accuracy of the location of the house in the lot will depend on exactness of the measurement on the map, and the ability to read horizontal angles correctly. Amateurs may be off several feet

in such work, and should do it themselves only when such differences are allowable. Under any circumstances, it is necessary to get the building walls of proper length and at the proper angles to each other.

The stakes A and B may be used as benchmarks, and elevations taken at the same time as the bearings and directions.

Recording. If the location of existing stakes is to be recorded so that they can be replaced if destroyed, the work is the same except that the angles are obtained by sighting the instrument and copied from the horizontal circle onto a sketch. Distances are measured in the field and noted on the sketch, which is most conveniently made on cross section paper, roughly to scale. This sketch is used in the same manner as the map in the previous discussion in replacing the stakes. Results are generally much better as the field figures are more accurate than those obtained with ruler and protractor from the map.

If field observations are to be entered on a map, the baseline or points should be related to features shown on the map, as corner stakes, points measured on a line between diagonal corners, or measured along a boundary. When the baseline is correctly drawn, angles and distances can be marked in with protractor and ruler.

Without Instruments. Simple location work can also be done without instruments. Figure 2-16 shows the same square building plot. Lines are drawn on the print or tracing prolonging each side of the house to the plot boundaries, from where the distance to the corners is measured. These distances are then measured off on the ground and stakes set.

The distances of the house corners from the boundary lines may be scaled from the map and measured on the ground in directions found by sighting between pairs of boundary stakes.

Sighting may be done by placing a thin straight stake, as at L and another at Q.

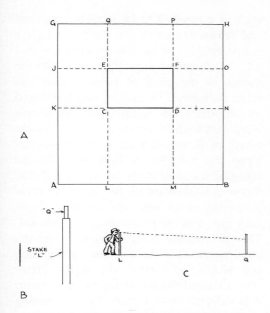

Fig. 2-16. Staking without instruments

have been set for a building in a plot without definite boundaries, and the contractor wishes to be able to reset them if necessary, there are several ways in which markers can be set without instruments.

In Figure 2-17 the house wall lines are shown continued in straight lines out of the digging area. These lines may be established by putting sighting poles on the corner stakes and finding a distant position from which two of these are in line—that is, one partly or completely hides the other. This sighting should be done with one eye and a pole held vertically in line with them. A stake is driven to mark the position of this third pole. The distance from this to the nearest corner stake is measured. This process is repeated for each pair of stakes at the foundation. In the figure, the reference stakes are indicated by Xs and the sight lines by dotted lines. A sketch should be made showing distances.

Any missing stake may be found by sighting from one marker to the other one on the same sight line, and measuring from the nearest marker. Even if the sketch is not available, the point may be found by the intersection of two lines of sight, as described under instrument work.

A man may stand behind the stake at L in such a position that, when he looks with one eye, the stake at Q is centered on L and just above it, as in Figure (B). Another man, carrying a third stake, measures the distance QE, keeping on the line LQ in response to directions from the observer. The measuring is best done by pinning the tape to Q. The stake is set at E so as to be directly in line between stakes L and Q. The distance EC is then measured and stake C set in the same way. CL is measured for a check.

Stakes F and D may be placed according to sighting from M to P, and measurements similar to the method used for E and C. The four corners of the building are thus located, and in a regular plot such as this no more work would be needed.

However, as a precaution against error, or in an irregular plot, or one with poorly defined boundaries, it is wise to prolong the other sides of the house into the lines JO and KN, and to sight and measure the corners again from J and K.

Reference Stakes. If the corner stakes

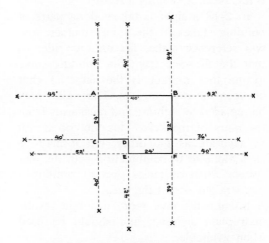

Fig. 2-17. Cross reference stakes

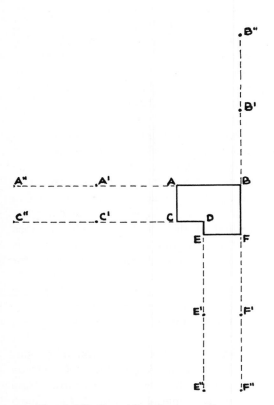

Fig. 2-18. One-side reference stakes

such as a pond, is to be roughly measured and drawn into an existing map, a base line is first established and two points measured off. A number of pegs are driven into the shores of the pond at points which will serve to indicate its outline, and numbered in rotation, as in Figure 2-19. An instrument with stadia hairs is set up at A, a sight taken on B, or on a more distant marker along the baseline. A sight may also be taken on a corner of the house for a check. Sights are taken on all the stakes in rotation, starting at one, the angle read for each one, and the stadia distance recorded. This information is sufficient to locate the pond by drawing the baseline on the map, and plotting distances and angles. However, to avoid the possibility of gross error, it is safer to set up at B, take a bearing on A, and record the angle and distance of some or all of the points observed from A.

The area of a pond so plotted can be easily obtained by counting squares on cross section paper, or by the use of a planimeter, which is a small instrument used for measuring areas on paper.

If each reference stake is set the same distance out from the nearest stake, there is less need of keeping a record.

In 2-18 a sight is taken along pairs of building stakes in the same manner, and two reference stakes set on one side, so that the distance from the building stake to the first marker is the same as that between the two markers. The stakes may be replaced by sighting and measuring from the pairs of markers. This method is not as accurate as the other, but it may be used alone or in combination with the first system when obstacles prevent running a line straight across the area.

Instruments give more accurate results than plain sighting, and should be used when available.

Locating a Pond. If an irregular shape,

Grids. If it is necessary to map an area, locating buildings, drainageways, trees, or other features, or to take elevations over a large area in order to prepare grading or drainage plans, a grid should be laid out. This consists of pegs or stakes at set intervals. They should be in straight lines, crossing each other at right angles. These lines, intersecting at the pegs, generally divide the area into squares. The interval may be five to twenty feet or more.

The grid may be laid out in a number of ways. A baseline should be laid out along an edge of the area. The instrument, preferably a transit, is set up at a corner of the proposed grid and sighted along the baseline. Pegs are set every ten feet, or at any other desired interval, measured from the instrument, to the end of the grid. Tape measurement is preferable.

The instrument is now turned ninety

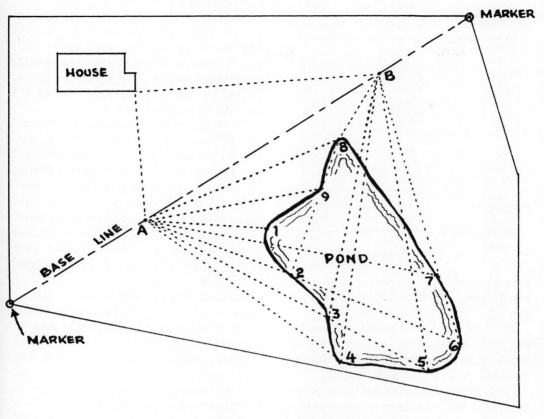

Fig. 2-19. Locating by stadia

degrees, and pegs set at the same interval along the line of sight to the end of the grid. The instrument is set up at the end, a backsight taken, and a ninety degree turn made. Pegs are set at the same intervals along this third line.

The interior pegs may be placed by the use of a long tape from opposite pegs, or the instrument may be set up over each peg in either the first or third lines, sighted at the corresponding peg in the other line, and pegs set according to its vertical hair and measurement.

Obstacles may make it impossible to set all the pegs by any of these systems. Usually, if as many pegs as possible are placed the rest can be filled in by sighting along lines of pegs, with reasonable accuracy.

The grid should now be copied on cross section paper with a point representing each peg. Any landscape features may be readily sketched in by estimating or measuring the distance from the nearest peg, and noting the place of the peg in the grid.

Elevations are now taken on each peg, preferably doing them a complete line at a time to avoid confusion. The rod reading may be written just above each point. Readings should also be taken on high and low spots, drain channels, and anything else of interest, and noted in the correct place on the paper.

When the instrument work is finished, the readings are preferably converted to

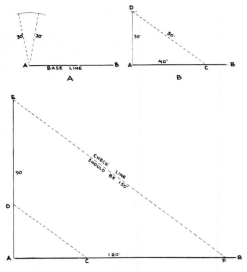

Fig. 2-20A. Right angles

gle. The tape is moved back and forth in an arc that is marked on the ground.

The tape is then pinned to the baseline at C, forty feet from the end, and an arc of fifty feet radius described, crossing the first arc. A stake is driven at the point where these arcs intersect. The line AE may be located by sighting along stakes A and D, and will be perpendicular to AB.

These figures have been given for the use of a fifty foot tape, but any measurements may be used as long as the 3-4-5 relationship is preserved. Larger triangles will give greater accuracy. If the grid is large in proportion to the triangle, a diagonal should be measured from E to a point on the baseline either ¾ or 1⅓ as far from A as the distance AE, and any necessary correction made if the diagonal is not in the proper proportion.

A very rough grid may be made by sighting along the sides of a building to obtain the right angles, and spotting in the pegs by eye and measurement.

Obstructions. Buildings, vegetation, and rough ground interfere seriously with primitive instrument techniques and make it more economical to hire an engineer.

Large permanent obstructions require layout of additional lines and angles to work around them.

Brush clearing for sight lines is laborious and sometimes quite destructive. It is handled by setting up the instrument, pointing it in the desired direction, and directing the cutters so that their work will be kept close to the line of sight.

In heavy undergrowth a mistake in turning an angle may waste hours of cutting work.

Contour Lines. A grid is frequently used to make a contour or topographic map of an area, as in Figure 2-20B.

A contour line is a line joining points of equal elevation on a surface, or the representation of such a line on a map. In mapping, contours are drawn at set intervals

positive numbers that can be penciled below the points, and the rod reading crossed out.

This grid sheet can be used for reference for any locations or grading estimates which may be required, and in drawing contours, profiles, and cross sections.

Grids Without Instruments. If no instrument that will turn angles is available, a grid may be laid out with a tape, and elevations taken with a hand level. A baseline is decided upon and a tall stake set at each end. A tape, the longer the better, is pinned at one end and extended toward the other, and lined up by sighting across it from one stake to another. The intervals are measured and the tape moved on and lined up again.

The right angle may be laid out by referring back to the ancient engineering knowledge that if the sides of a triangle are in the proportion of 3 to 4 to 5, the angle between 3 and 4 is a right angle. The process is illustrated in Figure 2-20.

First the baseline is laid out, measured off, and pegs set. The tape is pinned at A, one end of the baseline, and thirty feet measured off at approximately a right an-

whose size depend on the roughness of the ground and the scale of the map. In nearly flat areas contour interval may be one foot, while in steep mountains 100 foot intervals are usual. Every fifth line is drawn more heavily than the others, and its elevation is printed in it.

A contour interval of two feet is used in the illustration. The profile shown below the map is plotted along the dotted line. Note that grid points alone would not have shown the depression crossing it.

Since few of the grid points are exactly at a contour elevation, it is necessary to estimate the locations of the lines as they pass between higher and lower points, a process that is called interpolation. The 88 contour is placed half way between point A-3, elevation 87.4, and B-3, elevation 88.6, as it is the same distance above one as it is below the other. The same line passes between B-1, and B-2, elevations 84.1 and 88.1 respectively. In this case the 88 contour is placed very close to B-2.

In irregular land it is very important that the grid observations include the crests of ridges and the bottoms of gullies, as the contours cannot give an accurate picture of the ground if they are omitted. The grid interval must also be close enough to show all important ground features.

Placing of contours therefore cannot be an entirely mechanical operation. In the first place, the man making the field notes must understand topography sufficiently to take extra readings where necessary to show up special forms. The valley at the top of this map would not have been shown at all by readings on the grid points only.

The man drawing the contours should have a feeling for landscape forms, to avoid misinterpreting the data from which he works.

Topographic Map. A topographic map is usually a contour map on which both natural and artificial features are indicated. The U.S. Geological Survey, Washington, 25,

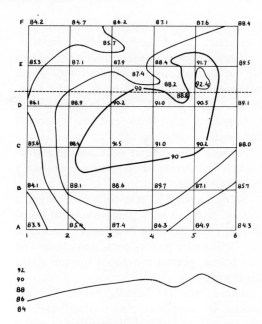

Fig. 2-20B. Contour map

D.C., has small-scale maps covering most areas of the country that show contours, vegetation, roads, trails, lakes, streams, and much other information. These maps are sold by mail and in bookstores at nominal prices. They are handy for many purposes and essential for some jobs.

Profiles and cross sections may be taken from a contour map by laying a ruler across it, measuring the interval between contour lines, and posting distances and elevations to cross section paper. The proposed new grades can then be drawn in, and the differences calculated.

The topographic map has a wide variety of uses, including locating highways, haul roads, borrow pits, and dump areas; studies of drainage and stream flow, and estimation of quantities in cuts and fills.

INSTRUMENT ADJUSTMENTS

Surveying instruments are delicate and are easily put out of adjustment by failure of parts, careless handling, or accidents. It is often not possible to have them checked or repaired locally; the return to the factory may mean loss of use for weeks or months.

It is therefore desirable that a person using an instrument be familiar with some adjustments that can be made in the field, without special skill. These include setting of telescope spirit level, and the horizontal and vertical cross hairs.

If these are properly set, and the instrument is inaccurate, shop service is probably necessary.

Spirit Level. The telescope spirit level is usually fastened by a pair of vertical bolts. A single nut holds it in a fixed position at one end, and a pair of nuts, one above and one below, permit moving it up and down on the other.

These nuts are usually round and are turned by a special pin, a small nail, or the smooth end of a drill, inserted in radial holes.

Both nuts are turned down, or clockwise, to move the bubble away from the adjusting bolt, and counter clockwise to bring it closer. They must first be unlocked by turning one away from the other, and should be locked again as soon as adjustment is made.

This level can be checked each time the instrument is set up. When the turntable is level, the telescope should be able to swing in a full circle without changing the position of the bubble. If no turntable screw adjustment will permit this, the level is presumed to be at fault.

To adjust, the turntable is leveled as accurately as possible and the bubble centered. The telescope is swung a half circle, causing the bubble to shift. The bubble is brought one quarter of the way back to center by the adjusting nuts, and the rest of the way by using the turntable leveling screws.

The telescope is then swung to its original position, the bubble moved one quarter way to center by adjustment, and centered by the leveling screws. This process is repeated until swinging the telescope does not affect the bubble.

Cross Hair. If the horizontal cross hair is not exactly centered, all readings on the rod will be too high or too low. Readings taken at about equal distances will agree. Greatest errors will be found on long sights.

A reasonably accurate check and adjustment of this hair can be made with the help of a still pond. Two stakes are driven flush with the water surface, about a hundred feet apart. The instrument is set in line with them, ten feet beyond one.

A rod is set on the near stake and a reading taken. This is assumed to be accurate, because the distance is too short for a perceptible error. The target is locked to the rod at this reading, or a note made of it.

The rod is set on the far stake. If the hair is correctly adjusted, the reading should be the same. If it is not, the hair should be raised or lowered until it agrees.

This is done by turning set screws at the top and bottom of the frame which holds the hair. The screws are unlocked by twisting one or the other a quarter or half turn, after which both are turned in the same direction.

Loosening the bottom and tightening the top moves the cross hair down, and turning them oppositely raises it. When the adjustment is finished, they are locked by turning gently against each other. Another reading should be taken to make sure that this does not disturb the adjustment.

The mechanism is fragile and may be damaged by forceful turning of the screws.

The side screws should not be turned during this adjustment.

Small errors are usually present in this method as it is hard to get the stake tops exactly at water level, and if the pond is overflowing, it will slope downward from inlet to outlet. Wind will raise the water against the bank toward which it blows.

If no pond is available, or a more accurate adjustment is required, two stakes

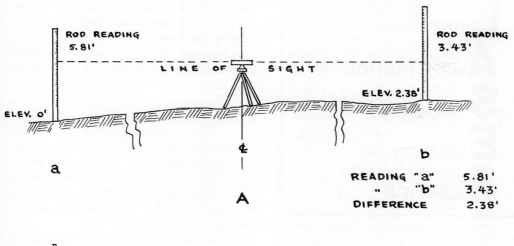

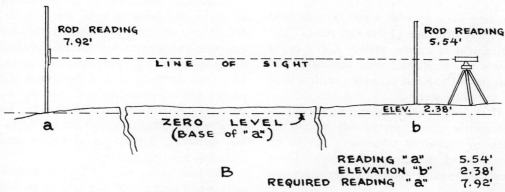

Fig. 2-21. Checking the cross hair

should be driven firmly into the ground, two to four hundred feet apart, and at almost the same level. The instrument is set up halfway between them as in Figure 2-21 (A), and leveled carefully.

A leveling rod is held on each of the outer stakes, and an exact reading taken according to the cross hair. These readings will be accurate with reference to each other, as any error in cross hair height is cancelled out in observations taken at equal distance.

The stake standing on lower ground is assigned an elevation of zero, and has the higher rod reading. The elevation of stake (b) is the difference between the two readings.

The instrument is now set up in line with the two stakes as in (B), about ten feet beyond stake (b). A reading is taken of (b). Then a reading is taken at (a), which should equal the elevation of stake (b) plus the reading there.

If it does not, adjustment is made in the manner described above.

Vertical Hair. The vertical cross or direction hair may be checked by driving three stakes exactly in line at two hundred foot intervals. The line may be determined by sighting with the vertical hair, and the distances measured by stadia or tape. The stakes should be about on the same level.

The instrument is set up over the center stake, leveled carefully, and turned so that

LIGHT
AMPLIFICATION BY
STIMULATED
EMISSION OF
RADIATION

Fig. 2-22. Meaning of "laser".

the hair lines up with a rod or pole held vertically over one of the end stakes. The instrument is turned exactly 180°, or, if it is a transit, flipped over vertically. The hair should now line up with a vertical stick on the other stake.

If it does not, the cross hair should be moved one quarter of the way toward the stick by means of adjustment screws on the sides of the cross hair frame. These work in the same manner as the upper and lower screws.

After the one quarter adjustment, the telescope is turned until hair and stick coincide. A 180° angle is again measured off

and the rod on the first stake sighted. The hair is adjusted to move toward it one quarter of any distance, then centered on it by moving the scope.

Additional half circle turns, and adjustments, are made until the hair will coincide with both sticks, 180° apart.

LASER

The word "laser" stands for "light amplification by stimulated emission of radiation". The type most used in construction is a tube filled with a mixture of helium and neon gases, which is stimulated by electric current.

Stimulation causes the gas atoms to emit energy in the form of light, which is reflected back and forth in the tube between a complete and a partial mirror This process produces light with only one frequency, with such intensity that it penetrates the partial mirror, emerging as a continuous, coherent, narrow beam of red light. Its sides remain almost parallel for considerable distances, instead of diverging and widening like ordinary light.

The laser beam may be kept as narrow as one-half inch at 500 feet, and might still be narrow and intense enough for useful work at 3000 feet.

The laser tube is mounted in an enclo-

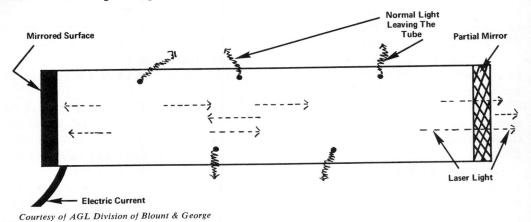

Courtesy of AGL Division of Blount & George

Fig. 2-23. Inside a laser tube

sure which may be mounted on a tripod, in a manhole, in a pipe, or in other ways. The design of the instrument and mountings provides threaded adjustments of great delicacy and accuracy, which permit precise regulation of direction (line) and grade.

There are usually two adjustments for vertical alignment. One levels the instrument. The other adjusts the laser tube inside the case for per cent of grade.

This per cent or gradient is zero for level work. In pipe laying, it is set for whatever plus or minus gradient is required. The setting is indicated by a dial gauge, or by a digital readout.

Power. The power source is usually a rechargeable 12-volt truck or automobile battery. It is important to use correct polarity, as the laser will not operate if connected to the wrong terminals. Most models will blow a fuse, some suffer serious damage. Others have a built-in electronic safeguard. guard.

For this reason, one wire (+ plus) and the matching battery terminal should be marked conspicuously with red plastic tape, or some other bright material.

Laser beams used in construction are usually "tail light red" of sufficient intensity to cause possible eye irritation and damage if looked at directly. Their effect is quite similar to that of light from arc welding.

Fig. 2-24. Accubeam gradient readout

However, the laser beam is harmless and is usually invisible when viewed from the side or the rear. It should be visible and harmless when seen in or through a plastic target. OSHA regulations currently require workers exposed to it to wear goggles designed specifically for protection against the wave lengths and intensity encountered.

Significance. The laser supplies a visible line-of-sight indicator that can be fixed, or set to move in a pattern, without an operator. It can be used by anyone in its path, by means of simple equipment, to determine grade and/or direction, with a high degree of accuracy. It makes otherwise imaginary reference lines real to the workmen.

The laser beam is sometimes compared to a taut, no sag stringline supported at one end only.

A transit or other sight instrument is needed in job layout, and perhaps each time a laser is set up. In addition, there should be a program of frequent checks with a transit to check work being done according to laser guidance.

Maintaining Accuracy. Once set up accurately, a laser remains accurate unless disturbed. It must be, and usually is, protected against violent disturbance. But a person or object might brush against it or strike it, ground may yield or a support may move, so that it goes off line or grade without the change being noticed.

Devices are available that will shut a laser off if it is disturbed. It can then be turned back on, but must be checked before further use.

With or without such a protective attachment, it is wise to re-check the setting of a laser frequently during precise work, and occasionally in rough work.

PIPE LAYING

For guidance in laying pipe, a laser is set in a manhole or at an angle point in the line, and the beam is directed along the inside of the pipe. Depending on the system being

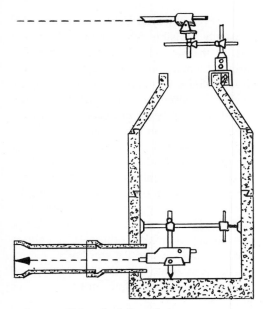

Courtesy of Micro-Grade Laser Systems

Fig. 2-25. Transit and laser

used, the beam may be centered in the pipe, or aligned just above the center of the invert (inside bottom).

The laser location is usually determined by setting a transit directly above the angle point, which (for purposes of this description) is assumed to be approximately centered in a newly built manhole.

The transit may be mounted on a rod based at the bottom of the excavation and supported by an adjustable cross frame, by rods based on a clamp or clamps on the manhole edges, or on a batter board. It is located exactly over the angle or location point by reference to base line and hub points.

Measurement is made from the transit down to the grade at which the laser is to be set. A measuring rod support is used directly, being checked with a carpenters' level to assure that it is vertical. Otherwise, a steel tape and plumb bob are hung from the center of the transit.

The laser is set inside the manhole, at the level of the pipe center or other measure-

ment level. There are a number of ways of setting up, of which two will be described.

Vertical Rod Mounting. In this system, both the transit and the laser are mounted on the same vertical column, which is composed of calibrated sections with lockable male-female joints. Cross section is square.

This column may be supported toward the top by an expandable cross brace inside a manhole, or by a batter board. At the bottom, it may be held by a tripod, or by forcing its lower end into soft dirt, or to a support on a steel plate. On fresh concrete, it may rest on a steel pin driven down through the concrete.

The elevation of a point on the rod may be determined by setting a batter board alongside it, or by direct measurement with a transit or other instrument.

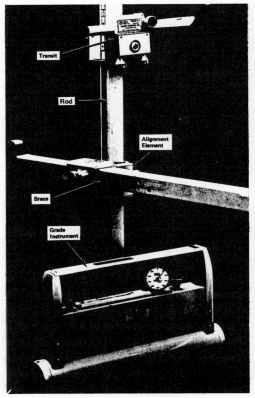

Courtesy of AGL Division of Blount & George

Fig. 2-26. Vertical rod mounting

By reference to rod calibrations, and/or by measuring with a tape, the laser is clamped to the rod at the desired level, which may be at the pipe center, just above the invert, or at any convenient point between. It faces in the direction in which the pipe is to be laid.

The laser grade indicator is set at the correct slope or gradient for the pipe, up or down, then the laser is connected to its battery and turned on.

The transit is clamped to the upper part of the column, so that it can look in the direction of pipe laying.

When the ditch has been dug 15 feet or more toward the next manhole, the laser beam should be shining along its centerline, or at least very close to it. A stake or target is placed approximately in the center of the ditch to make the beam visible. The transit is rotated on the column by adjusting a knob, until its vertical crosshair is centered on the beam.

The transit is now pointed straight down at the laser, and aligned with a reference mark on the barrel by adjusting a different knob than the one mentioned above. The transit is alternately aligned with the beam

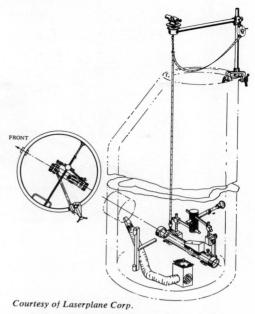

Courtesy of Laserplane Corp.

Fig. 2-28. Vertical measurement

and with the mark, until it remains exactly centered on both. The vertical line of sight (vertical hair) through the transit now includes the laser beam.

The transit is now aligned with the marker on the next forward manhole by rotating the vertical rod (to which both transit and

Courtesy of AGL Division of Blount & George

Fig. 2-27. Lining up the beam

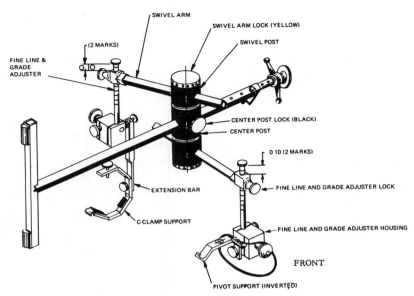

Courtesy of Lase Alignment

Fig. 2-29. Speed-Set T-Bar Platform

laser are clamped) relative to its supporting crossbar, with a fine adjustment. Both instruments, being clamped to the rod, remain in alignment with each other.

Therefore, when the marker is bisected by the transit's vertical crosshair, the laser is projected accurately along the line for the pipe.

The laser is now rechecked for level and for correct grade setting, and another check of the alignment of the two instruments is made. The level vial should be checked intermittently during the work.

Separate Mounting A transit may be mounted on a clamp frame at the top of a manhole, while the laser is secured separately to a crossbar at the bottom.

The transit is set first, exactly over the centerline or angle point of the pipe. In the model shown, a plumb bob is hung on a specially calibrated transit tape, which is fed over a flat hook hanging from the center of the transit.

The reading of this tape at the hook shows the total distance from the centerline of the transit telescope to the tip of the

plumb bob, plus one-half the diameter of the laser projector.

This laser mounting is called a Speed-Set T-Bar Platform. It is assembled at the surface and lowered into the manhole.

The support member is a square bar with a cross piece at one end and a threaded adjustment at the other.

Courtesy of Micro-Grade Laser Systems

Fig. 2-30. In-pipe mounting

A short, thick cylindrical center post slides on the bar, or is locked to it by turning a knob. The upper and lower sections of the post swivel horizontally on its center. They support sliding bars that separately carry the front and rear support or clamps for the laser unit.

The front or pivot support is held in inverted position while alignment is being set. It is then rotated downward to position to support the instrument. This, and the rear of C-clamp support, have vertically adjustable connections to the bars that carry them.

A plumb bob is hung by the transit tape at the exact reading for the laser. The T-bar assembly is manipulated by sliding and swiveling movements, until the target cross on the inverted pivot support or clamp is at the point of the bob.

The pivot clamp is then turned to supporting or bottom position, and the laser is installed in the two clamps. It is leveled, set to correct gradient, and turned on.

Direction is checked by sighting the transit at a trench line marker, then down at the beam in the trench, as described earlier.

In-Pipe Mounting. Many lasers are equipped with mountings that permit insertion in a pipe, or if the pipe is very small, just outside one end and looking through.

If the pipe has been carefully set to grade and alignment, and the laser is properly mounted, the laser can project the correct centerline or flow line from one pipe through the rest of a straight run.

However, alignment should be checked on a transit-set pole or target 15 to 25 feet ahead.

Refraction. All light beams, including lasers, may be bent out of a straight line if they pass through air of varying density and/or temperature.

If sun-warmed pipe is set in a cold trench, or cold pipe is put in a warm trench, or if hot sealing compound is used, a mixture of masses of air of different temperatures may occur inside a pipe, causing the laser beam to vary from a straight line.

This situation can be (and must be) corrected by using a blower to direct a steady

Courtesy of Laser Alignment

Fig. 2-31. Target

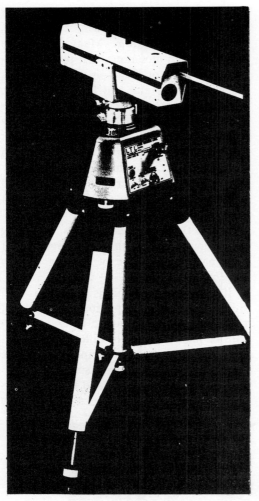

Fig. 2-32. Tracking level

current of one-temperature air through the pipe. Blowers using the same 12-volt current as the laser are usually supplied with them, and their use is a routine precaution that often avoids costly confusion and mistakes.

Targets. Targets for use in pipes are usually translucent plastic, or metal with a plastic window. They may be pre-marked with centering or other information lines, or such indicators may be put on with erasable markers on the job.

One size of target may fit several or all sizes of pipe, or there may be a separate size for each pipe. Some are snap-fitted into the pipe, others have fairly elaborate holding and extension devices, and may include a level vial to check the installation.

Ordinarily, a pipe target is inserted in each pipe at the end away from the laser, and the pipe is shifted until the beam is centered on indicators, usually cross lines, with or without bull's-eye circles. If the laser is set correctly, and the target properly used, accuracy to within ⅛ or 1/16 inch is readily obtained. The target is removed as soon as the pipe has been secured 'in place, and put in the next pipe when it is ready.

Other targets may be range poles or surveyors' rods, with a surface on which the beam can be seen readily even in brightly lighted areas. There are also electronic sensors, that can register the beam even when it cannot be seen.

Changes. From any one setting, an inside-pipe laser can be used for only one straight section. If the direction or gradient change, the instrument must be re-set at the angle point betwen the two lines.

If such changes are frequent, it may be better to put the laser on the surface, as described in the next section. It can then be used for gradient only, with direction determined by transit.

A curved pipe must be set from surface instruments.

Ditch Depth. The laser may be used to regulate ditch depth in between setting pipe sections. Readings may be taken on the bucket of a hoe, the shoe of a ditcher, or a rod held on the ditch floor close to the excavator.

This continual guidance greatly reduces the need for hand grading while laying the pipe.

LASERS ABOVE GROUND

Potential. The potential of lasers in construction work above ground is very great.

A general simplification and improvement in handling a variety of leveling and grading problems may be expected.

Here we will deal with two applications: grade reference within a specific construction area, and guidance of excavating machines.

Tracking Level. The Laser Tracking Level is mounted on a substantial tripod, leveled, and then rotated slowly in a horizontal plane by an electric motor. When set for automatic operation its beam passes over the whole area of a construction site at frequent intervals, perhaps twice a minute.

Each work crew that needs grade checks has a stadia rod, and a beacon and rechargeable battery that can be clamped to it. The beacon is preferably a few inches below the level of the laser, while the rod is being used.

When the beacon is turned on, it sends a signal that is received by a tracking sensor in the laser head. The sensor causes the rotation of the head to stop when the beam reaches the beacon, causing the beam to "lock on" to the rod, on which it makes a visible spot of light.

The beam will then remain locked on to the rod, while readings are taken at one or many locations, until the beacon is turned off.

The laser level is known, or can be quickly determined by reference to bench marks. Subtracting the rod reading of the center of the light spot from the laser elevation gives the elevation of the base of the rod.

When the beacon is turned off, the laser, after a 15-second delay, resumes its rotation and continues it until stopped by the turning on of a beacon, the same or another. The delay guards against unlocking because of accidental interruption of the line of sight by a vehicle or person.

It is of course very important that the laser beam elevation be accurate. The instrument is set up in somewhat the same manner as a self-leveling engineer's level,

Courtesy of AGL Division of Blount & George

Fig. 2-33. Rod and beacon

by adjustment of three leveling screws. It is turned on, and allowed a two minute warm-up period before being placed in automatic rotation.

This rotation may shut off automatically if the instrument is disturbed.

A rod and beacon are set on a bench mark, and a reading taken. If the elevation is a fractional number, it can be rounded off for greater convenience by lifting or lowering the laser slightly, using a fine vertical adjustment on top of the motor head.

If the locked-on beam misses the rod, the

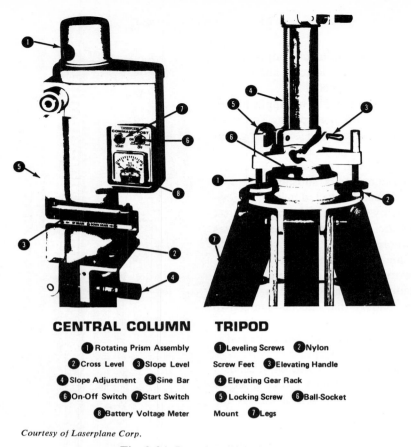

CENTRAL COLUMN

1 Rotating Prism Assembly
2 Cross Level **3** Slope Level
4 Slope Adjustment **5** Sine Bar
6 On-Off Switch **7** Start Switch
8 Battery Voltage Meter

TRIPOD

1 Leveling Screws **2** Nylon
Screw Feet **3** Elevating Handle
4 Elevating Gear Rack
5 Locking Screw **6** Ball-Socket
Mount **7** Legs

Courtesy of Laserplane Corp.

Fig. 2-34. Rotating prism laser

beacon may be leaned to one side or the other to bring it into register.

It is good practice to then set up one or two other bench marks in the work area, to encourage quick and easy re-checking of level during work.

The laser may have a distance range adjustment, which should be correct for job conditions. There may be two positions, one for 15 to 100 feet, the other for 75 to 700 feet.

The automatic tracking level does not need an operator when beacons are used. If no beacons are available, it may be used by having an operator swing it manually, in the same manner as a non-tracking instrument.

Oscillation. When work area are typically long and narrow, as in trenching or highway jobs, the laser head may oscillate from side to side, in an arc of 30 or more degrees, instead of rotating.

This oscillation may be quite rapid, and is used more often in automatic control of depth of excavation than as a reference for work crews. Such instruments may be used similarly to the rotating prism models described below.

Non-Tracking. A non-tracking laser is used in somewhat the manner as a builders' level or a transit. An operator directs it to a rod held on the spot being checked, and elevation is read from the spot of light made by the beam.

Its advantage is that the spot of light can be seen by the rod man, and used for reference without communicating with the instrument man.

Low visibility of the beam may be a

problem on bright days. It may be located by moving a piece of reflective material in its probable path.

Non-tracking levels may be supplemented by a telescope on the rotating head, to assist the operator in locating the light spot, and in reading rod or other markings associated with it.

Rotating Prism. In the Laserplane system, the laser is mounted so that its beam is vertical, shining down on a rotating 90° prism, which deflects it into a horizontal plane.

This plane reference can be tilted when a slope angle is desired.

The prism makes a full circle either five or ten times a second, depending on the model.

This unit may be used in conjunction with an electronic surveyor's rod. The latest type, Figure 2-35, consists of a rod, a detector-head and a signal box.

The rod man slides the detector up and down the rod until a steady beep' and a green light indicates that it is exactly on grade.

Profiling. There is also an automatic system, in which a hydraulically activated mast is mounted on a vehicle that tows a measuring wheel. A receiver on the mast senses the energy plane of the laser, and raises and lowers the mast to keep itself locked into it.

The vertical movement of the mast and the distance measured by the wheel are automatically plotted onto a chart, which presents a profile of each line that is traversed.

Automatic Grade Control. The basic application of the rotating prism is the direct, automatic control of cutting depth (or filling height) of grading machines.

The machine to be controlled must have hydraulic lift for its cutting edge or parts. A vertical mast is erected on a frame resting on the edge, and is kept vertical by a pendulum-controlled valve and hydraulic

Courtesy of Laserplane Corp.

Fig. 2-35. Detector-head

mini-cylinders. It has a telscopic adjustment for height.

The mast carries a receiver with a number of cells sensitive to the laser light, which are wired to a solenoid valve that controls the hoist for the cutting edge. This valve opens almost instantaneously when activated by the receiver cells.

Courtesy of Laserplane Corp.

Fig. 2-36. Automatic survey vehicle

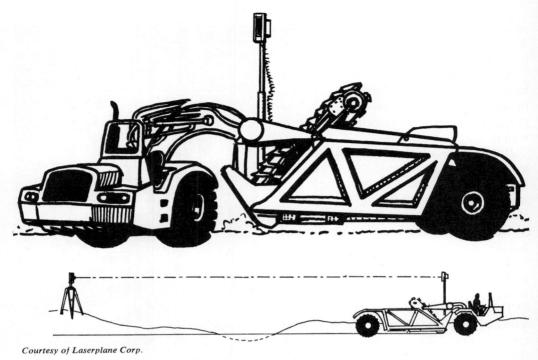

Courtesy of Laserplane Corp.

Fig. 2-37. Laser-controlled scraper

Center cells are neutral or on grade indicators, and do not activate the valve. Those just above or just below center will open the valve for 50 ms (milliseconds) or 1/20 second. The top and bottom ones open it for 200 ms, or 1/5 second.

There may be additional cells to activate a control that will swivel the mast to keep the receiver facing the laser.

The mast height is adjusted so that the center of the receiver is in the plane of the laser beam when the cutting edge is exactly on grade.

The machine is then driven and steered in the normal manner. If the blade drops slightly below grade, the hoist valve will be opened for 1/20 second to raise it. This action will be repeated five times a second until it is on grade.

If the drop is enough to activate the top cells, they will open the valve for 1/5 second. If the demand for lift is repeated when the beam returns, the valve will stay open continuously until correction shifts the work to the 1/20-second adjustment, or to neutral.

If the blade moves above grade, similar action will put the valve in DOWN position. Lights keep the operator informed of the actions of the system.

A refinement for fast-moving machines (scraper or dozer) is a proportional current system, responding to the 10-sweeps-per-second laser. In this, the opening of the valve is proportional to the amount of correction needed.

Repetition of automatic adjustments five or ten times a second results in running a very accurate grade, usually within 0.02 or 0.04 of specification. Accuracy is greatest close in, and declines within these limits to a distance of 1,000 feet, the longest recommended.

With increasing distance, the beam widens so as to lose its pin point accuracy, and problems with diffraction become more

Fig. 2-38A. Cubitape

sense loads, and may therefore tend to make a machine dig into more material than it can move.

When the automatic overloads the machine, the operator can override it, and raise the blade. He can then return for additional manually controlled passes, finally returning it to automatic for trimming.

At present control is practical only in a flat plane. A change of grade requires re-

likely. Errors caused by curvature of the earth's surface are given in Figure 2-39.

Applications. At present this system is well adapted to automatic control of ditchers, landplanes, scrapers, dozers, tube and cable plows, and graders.

The grader requires an additional cross-slope sensor. The laser controls the lift cylinder for the leading edge, while the slope regulator keeps the trailing edge in proper relationship.

Automatic control is largely limited to finishing operations, as it is not designed to

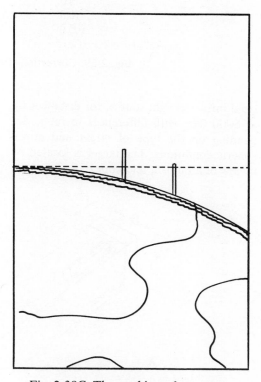

Fig. 2-38C. The earth's surface curves

setting the laser. There is no provision for working vertical curves. However, development work is in progress in these areas.

DISTANCE METERS

Several battery-powered instruments are available that can read the distance to a target by measuring reflection time of multiple-phase signals.

The short range Cubitape uses a modu-

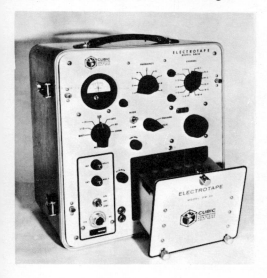

Fig. 2-38B. Electrotape

DISTANCE		SUBTRACT FROM READING	
Feet	Miles	Feet	Inches
100	0.019		0.002
500	0.095	0.006	0.070
1000	0.189	0.023	0.279
1500	0.284	0.053	0.633
2500	0.473	0.148	1.774
5000	0.947	0.592	7.096
5280	1.000	0.660	7.920
	2.0	2.64	31.68
	5.0	16.5	
	10.0	66.0	
	100.0	6666.0	

Courtesy of Walter A. Stiehl

Fig. 2-39. Corrections for curve of earth's surface

lated infra-red light source, for distances up to 6000 feet, with differences in range depending on the type of target and atmospheric conditions. The target is located by means of a built-in telescope.

The distance is automatically displayed in digits for feet (or meters) and their decimal fractions.

The Electrotape uses a microwave beam for distances of 100 feet to 30 miles. Two

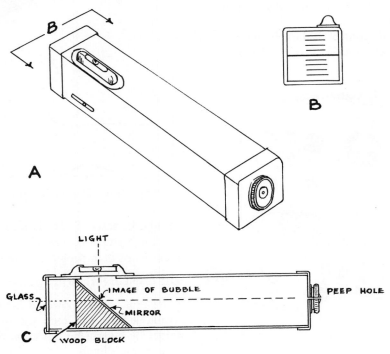

Fig. 2-40. Hand level

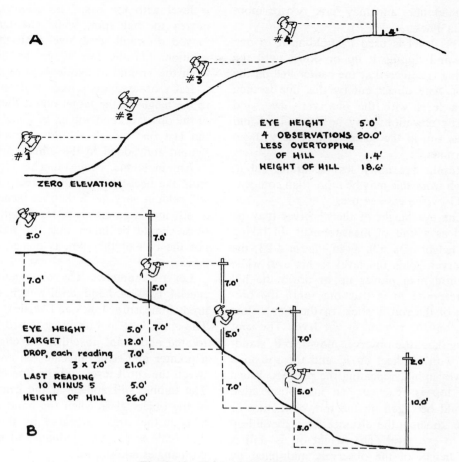

A

EYE HEIGHT 5.0'
4 OBSERVATIONS 20.0'
LESS OVERTOPPING
OF HILL 1.4'
HEIGHT OF HILL 18.6'

ZERO ELEVATION

EYE HEIGHT 5.0'
TARGET 12.0'
DROP, each reading 7.0'
3 x 7.0' 21.0'
LAST READING
10 MINUS 5 5.0'
HEIGHT OF HILL 26.0'

B

Fig. 2-41. Using a hand level

identical instruments are used, with built-in short wave communication between them for the operators. The reading are taken first by one unit, then by the other, as an accuracy check.

These instruments measure direct line-of-sight (slope) distances. If there is a difference in elevation between the points, the horizontal distance must be calculated.

Correction for the curve of the earth's surface is needed if elevations are required. An approximate formula, good up to distances of a few hundred miles, is:

$$\text{Correction} = \frac{(\text{Distance in miles})^2}{8,000}$$

Figure 2-39 is a table for various distances. The correction is subtracted from the apparent elevation indicated by the instrument. It will be noted that the curvature is very slight within ordinary survey distances, but must be considered in long sights.

MINOR INSTRUMENTS

Hand Level. Rough levels may also be run with hand levels, such as the one shown in Figure 2-41. This consists of a sighting tube, in the top of which is a small spirit level parallel with the line of sight. A slanted mirror reflects the spirit level so that it is seen vertically beside the field of view. The object glass is marked with a center line, and may have two or more stadia lines.

This level is used by holding it to one eye, and tipping it up or down until the bubble is centered at the center line on the glass. Any object cut by this line is then on a level with the observer's eye, and nearby elevations may be determined and levels run in the same manner as with an instrument.

Results are much less accurate, but in rough work this may be more than compensated by the ease of use.

The eye height of the observer may be used as a unit of measurement. In taking the height of a hill, as in Figure 2-23, the observer holds the level to his eye, while the rod man moves up or down the hill in response to instructions, until the bottom of the rod, a stick, or the man's shoe rests on the ground at eye level. The spot is marked, the observer moves to it, standing with his heel on it, and the rod man moves uphill, repeating the process. When the top is reached, the last observation should be taken on the rod, or a ruler or tape, to show the distance from the hilltop up to eye level. The height of the hill is the height of the observer, multiplied by the number of observations, less the rod reading on the last observation.

In working downhill, (B) a target fastened at the top of the rod, or a mark on a long stick, is sighted and moved until it is level with the eye. The observer then moves to that spot, while the target is moved downhill until level with the new position. On the last sight, the distance from the ground to eye level is measured.

Each observation covers a drop equal to the height of the target minus the height of the observer, and will all be equal except the last one, which should be separately figured and added to the others.

An individual's eye height measured from the heel when he is in erect position, will seldom vary more than an inch, which is not too large an error for rough work. Care should be taken that the heel, and not the ball of the foot, is placed on the mark.

Level-Clinometer. The clinometer is a special type of hand level which can be used to measure slopes and vertical angles. The spirit level is hinged so that it can be rotated about 45° in either direction, and a pointer and scale indicate the angle between the spirit level and the line of sight. The bubble will appear at the center line of the object glass when the hand level is held at the angle indicated by the scale. The scale is usually graduated to indicate both angles and slopes.

The angle of a slope may be measured by setting a target at eye height at one end and sighting it from the other. With the center line on the target, the spirit level may be adjusted until the bubble is beside

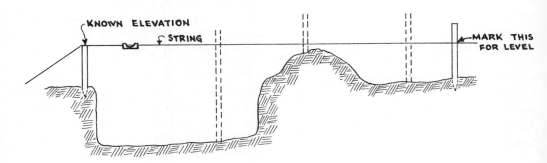

Fig. 2-42. Using a string level

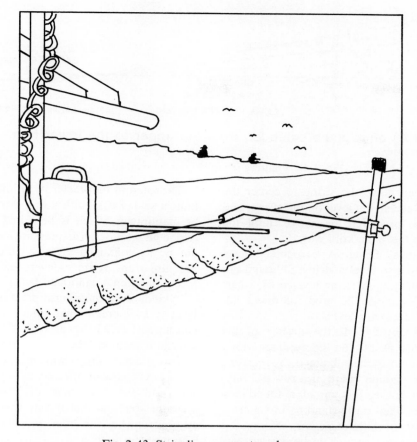

Fig. 2-43. Stringline, support, and sensor

the line. The pointer on the scale will then indicate the slope of the hill.

It may also be used as an inclinometer by placing it on the slope to be measured, and setting the spirit level until the bubble is centered. It is usually good practice to lay a board on the ground surface and take its slope to eliminate effect of small irregularities.

String Level. A string level is convenient to use over short distances. It is a spirit level fitted with prongs by which it can be hung from a string stretched tightly between two marks. Elevations may be taken from the end of the string, or by measuring down from any part of it, as illustrated in Figure 2-42.

The string used should be strong enough to take sufficient pull to remove all sag. If it is at all slack it will give false readings, showing slope at the ends of a level stretch, and level somewhere near the middle of an inclined string.

Most string levels use flexible prongs which are easily bent by light pressure. It is therefore best to check such a level before every use, and occasionally during a job. This is done by leveling a string according to it, slacking the string and reversing the direction of the level on it, and tightening the string to the same marks. If the reading is the same, the level is all right. If it disagrees, bend a prong sufficiently to move the bubble one quarter way to the

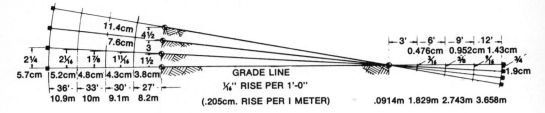

(%) on a plane extended from the upper to the lower hub.

Courtesy of CMI Corp.

Fig. 2-44. Figuring string heights on banked curve

-center, then move the string to center the bubble. Reverse the level on the string, and repeat the procedure as above, until the reading is the same both ways.

Carpenters' Level. A carpenters' level may be used to level a string, although not as conveniently. A string leveled by a carpenters' level may be used for direct adjustment of a string level.

Leveling by Eye. In the absence of any instruments at all, an approximate level may sometimes be obtained by sighting along a horizontal board or a row of bricks of a new house, the top of a foundation wall, or marked posts standing in water.

Altimeter. Altimeters or barometers which are small enough to be carried, and sensitive enough to react to small changes in elevation, may be used for taking preliminary levels.

The aneroid barometer or altimeter contains a sealed case with a thin flexible wall or diaphragm, which is bent in by an increase in atmospheric pressure, and bent out by its contained air when the outside pressure drops. It moves levers and gears to turn a hand on a calibrated face.

"Standard" atmospheric pressure at sea level is 14.7 pounds per square inch, and can support 29.92 (say 30.0) inches of mercury in a vacuum tube.

This pressure drops with increasing altitude, but it is also affected by wind pressure and the movement of storms and pressure areas, so that it may vary considerably from hour to hour and day to day. The problem is to separate changes due to altitude from those caused by weather.

This may be done by making the observations in such a short time that no impor-

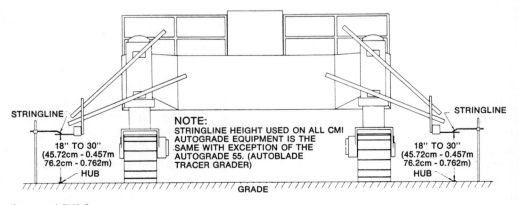

Courtesy of CMI Corp.

Fig. 2-45. Stringline heights

tant change will take place; by checking the instrument at the starting place, or some other spot of known altitude, at the end of a run, or by the use of two instruments, one of which is kept at a fixed point, and its reading recorded every hour, or oftener, while the other is used on the job, and the time of each reading noted.

In the second case, the change found may be used to correct the readings on a somewhat arbitrary basis, the most recent being the most affected. When the two instruments are used, each recording is corrected according to the barometer reading at the time it was taken.

The most convenient altimeters are mountaineers' pocket models, the size of a large watch. Engineers' models, which are rather rare, have a glass for magnifying the scale, for fine reading.

An airplane altimeter of the sensitive type, having two or more hands, may often be purchased reasonably at an airplane instrument repair shop, where it can also be checked for accuracy. These are easily read but are somewhat more bulky to carry. Single hand airplane altimeters are slightly smaller and usually cheaper, but the scale and lack of sensitivity make them unsuitable for any but the roughest work.

Altimeters usually carry two scales on the dial—an altitude reading, calibrated in feet, and a barometer index. The hands or the dial may be turned to permit correction of the altitude reading, as required by changes in local pressure.

Airplane altimeter hands turn clockwise with increased altitude, many pocket models counter-clockwise.

An altimeter is set for correct or assumed altitude when work with it is started. As it is carried up or down hill, the hands will point to higher or lower altitudes on the scale, and notes may be made of the reading wherever desired. It is advisable to tap the instrument before each reading.

Altimeters provide the quickest and easiest means of finding heights and depths in rugged and overgrown country. They are not accurate enough to be used in setting grades, except in very experienced hands.

Water Hose. A garden hose will provide an accurate level for distances it can reach. The ends are turned up so that it can be filled with water, then they are moved until the water is even with their openings. They are then at the same level and can be used for checking with little chance of error.

Specially designed liquid-filled hoses may be purchased for use in this type of leveling.

STRINGLINE

A stringline is a string or wire that indicates location, grade or both location and grade. Its use in construction is widespread and very old. A familiar example is string stretched between batter boards for reference in laying foundations and pipes.

A stringline may serve as a visual guide for a machine operator to prevent him from running off line between widely spaced stakes. If the machine is equipped with a pointer or a suspended plumb bob on the frame of its grading blade, ditching wheel or other grade-making tool, it can be adjusted so that following the string accurately with it assures working to proper alignment and grade.

Lately, the principal and increasing use of stringlines has been in automatic control of grading, excavating, and paving machines. These controls, which are described in Chapter 19, are usually made up of a tracer or wand that stays in contact with the string, and a sensor that reacts to any change in tracer position by actuating hydraulic valves to restore proper relationship with the string.

A machine may need two stringlines, one at each side. However, if it has an automatic cross-slope control, only one line is required.

Accuracy in locating such stringlines is

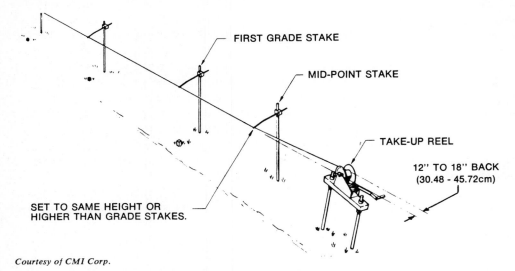

FIRST GRADE STAKE

MID-POINT STAKE

TAKE-UP REEL

12" TO 18" BACK
(30.48 - 45.72cm)

SET TO SAME HEIGHT OR
HIGHER THAN GRADE STAKES.

Courtesy of CMI Corp.

Fig. 2-46. Stringline and takeup reel

of the utmost importance, as line and grade produced in the automatic operation are absolutely dependent on them.

There are various ways of setting up a stringline. The following description follows a booklet, STRINGLINE MANUAL, published by the CMI Corporation.

Apparatus. The string holder is made up of a metal stake (standard length 42 inches), a sliding bracket that can be secured to the stake with a set screw, and a horizontal rod that slides in the bracket, and can be locked in it with another screw. One end of the rod is notched to hold the string.

The string is usually a special yellow polyethylene selected for strength, stretch characteristics, visibility, and rot resistance. Regular strong white cord may be substituted in an emergency.

The string is carried between jobs on storage reels. When in position, it is tightened by means of take-up reels, one at each end.

An engineers' steel rule is used to measure the distance from string down to the top of the gradestake (tack line). This distance has been previously determined, and is the same for each gradestake in a section.

A plumb bob may or may not be used for horizontal alignment.

Location. The stringline is placed outside the strip or lane being processed. In highway work, this usually means in the shoulder, two or more feet beyond the edge of the lane. It is preferably 18 to 30 inches above grade, although some machines can use higher and lower settings. It should be at least 5 inches above the ground.

It may be possible to use one setting for up to three processes—subgrade finishing, laying selected sub-base, and paving. Clearance from lane edge must be sufficient to accommodate windrowing and removing waste material, passage of machine tracks if outside the lane, and minimum tracing arm lengths.

If a paver is to be supplied by side dump trucks, the line should be on the side not used by them.

Stakes are driven about one foot outside of the line proposed for the string, alongside marking hubs or gradestakes.

Hubs (Gradestakes). Hubs are usually 2 x 2 wood stakes, placed by surveyors along the line of the string, parallel to the roadway centerline. They are driven to grade (bluetops) or to some fixed distance, the same for each stake, above grade. They are usually spaced 50 feet apart for uncomplicated work, and 25 feet for super elevations, vertical curves, and other special situations.

Reference grade is level with the centerline on straightaways. On super elevated (banked) curves, the reference is to the tipped plane of the cross section, so that it will be lower than the near edge on the inside of a curve, and higher than it on the outside. Figure 3-44.

A tack is driven into the top of each hub for precise location of horizontal alignment.

Setting Up. A string stake is driven alongside each hub, about one foot to the outside. It must be vertical, and driven deep enough for good stability. The rod should extend directly toward the stake, and can be slid inward or outward until its slotted end is directly over the tack. The bracket is raised or lowered until the rod is ¼ inch above the selected stringline height. Adjustments are locked with setscrews.

A steel engineer's rule is used to measure the height of the rod, and to check its location over the tack.

Take-up reels are then placed, 25 feet before the first stake in the section, and 25 feet past the last stake. These should be offset 12 to 18 inches outside the line of the string, with the crank to the outside. The reel is first located on the ground, stakes are driven through holes in its base,

and it is then raised to a convenient cranking height, and clamped.

A metal mid-point stake is driven in line with the string stakes, half way between the last (or the first) one and the reel. The rod may be at stringline grade, or higher.

The string itself is reeled off a storage reel (or two reels if there are two lines) held on a bar on a truck. It is anchored by winding about 25 feet onto one of the take-up reels, and is paid out as the truck drives to a few feet beyond the other end. The string is cut and secured to that reel.

The string is pulled taut by hand and by turning the reels. Then it is inserted into each rod notch. It should slip into them easily, but require moderate force to be pulled out. Slot openings can be widened with a screwdriver, or narrowed with pliers. There should be no visible sag between stakes.

Using an engineer's rule, adjust the rod brackets so that the string is at its exact height above the tacks, and with a plumb bob to make sure it is in line. Sight down the completed line, and re-check any break in a smooth flow.

Visual inspection should be repeated just before use, and occasionally as work progresses.

Any breaks in the string can be repaired by relaxing the tension, tying the broken ends together with a square knot, and re-tightening.

MEASURES OF PRODUCTION

IN-PLACE MEASUREMENTS

Distance. Surface distances may be measured approximately by pacing, stadia, or car speedometer; or more exactly by tape (chaining), the distance meters mentioned earlier, or by measuring wheel.

Distances may also be read off job plans

or maps. But a possibility of an error in a map, or in interpreting it or its scale, must be guarded against.

A measuring wheel is a convenient and sufficiently accurate means for measuring both short and long distances. Small wheels are hand-rolled along the ground, while large ones may be either hand-rolled or towed slowly by a vehicle. A counter registers the distance covered in feet. Readings down to two inches may be obtained.

Bank. A bank of earth or other material may be measured before and during the time it is dug away. This is usually done to establish a basis for payment to the owner of the land, or from the job to which it is being hauled.

A grid is established before digging starts. Ten or even five foot intervals are used between stakes for close measurement of rough ground, and wider spacing for more even ground and less exact work.

Bench marks and reference points are established well outside of the digging area.

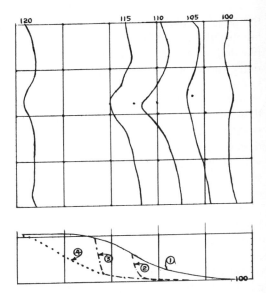

Fig. 2-48. Measuring a bank

Then the elevation of the ground is taken at each stake. Elevations are also taken of humps or hollows lying between the stakes. A contour map is usually made, and ground elevations are plotted on cross section paper. Vertical scale may be exaggerated. This original surface, called grade, is a solid line, as in Figure 2-48. Any desired profile or section that does not follow the stake lines can be made up from the contour map.

Elevation lines may also be plotted by laser and a measuring wheel.

When the amount that has been excavated is to be measured, the bank may be evened up a bit, and any stakes that have been disturbed by the digging are reset if work is to be exact, and new ground elevations are taken. If only a rough measurement for a progress report is required, the stakes may left out, and their former locations sighted in approximaately by stadia. The new elevations are plotted on the same paper as the originals, and are usually indicated by dotted lines, as illustrated.

Excavated material has been removed from between the solid and the dotted lines on each of the profiles. This area is meas-

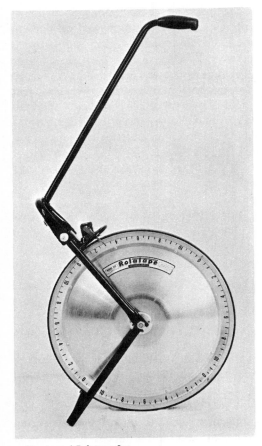

Courtesy of Rolatape, Inc.

Fig. 2-47. Measuring wheel

ured by counting squares, dividing into rectangles and triangles, or by mechanical methods. If the horizontal scale is 1 inch to 25 feet, and the vertical scale 1 inch to 5 feet, a square inch indicates 125 square feet or about 13.9 yards. It is best to keep the figuring in feet until the end, for greater accuracy.

The vertical area is determined for each profile, then areas are added and the total divided by their number to obtain an average. This average is multiplied by the length of the excavation at right angles to the profiles. This gives the amount excavated in cubic feet. Dividing it by 27 reduces it to cubic yards, bank measure.

The same procedure is followed each time the bank is measured. On all calculations after the first it is usual to determine both the amount removed since the previous measurement, and the total amount removed since the job began.

Pile. An experienced estimator can often judge the volume of a pile quite accurately by inspection, without measurement. However, appearances are often deceptive, and most people are safer if they make at least approximate measurements as a check.

Piles can often be conveniently measured by calculating the volume of regular masses of similar outline, and making plus or minus adjustments for differences.

A pile of clean, dry sand may have a conical shape, or be a ridge with a triangular cross section, ending in half cones. Measurments should be taken to determine base size and height.

The area of the circular base of a cone is found approximately from the circumference by the formula:

$$\text{Area} = \frac{\text{Circumference}^2}{12.6}$$

and from half the diameter by

$$\text{Area} = 3.14 \times \text{Radius}^2$$

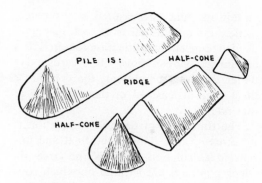

Fig. 2-49. Pile measurement

The volume of a cone is the height times one third the base area.

The long part of the pile is figured by the formula:

$$\text{Volume} = \frac{\text{height} \times \text{width} \times \text{length}}{2}$$

A long pile will have the volume of the center section, plus the volume of one cone, as each of the ends is a half cone.

When the pile shape is flattened, different, or irregular, profiles and cross sections are taken in the same manner described for banks.

PRODUCTION MEASUREMENTS

In the construction industry the word "production" is assigned the same meaning as "output," that is defined in the dictionary as "the quantity or amount produced, as in a given time." Perhaps a better definition for these words in construction is "useful work accomplished" as in the early stages of a job the chief product is likely to be destruction and upheaval.

Production may be figured in at least three ways. First is on the basis of job requirements. If a time schedule allows 200 working days to move 3,000,000 yards of earth, the contractor's earthmoving machines must move or "produce" 15,000 yards a day. Second, the production of a certain machine is measured or estimated, to determine the number of such machines needed to meet the production required. If a certain scraper can move 1,000 yards a day under this job's conditions, the contractor must keep at least 15 of them working at that rate.

The third way of figuring production is in terms of cost. This is the final and important calculation, as it is the basis on which contracts are let, and on which contractors ride in Cadillacs or go broke. But cost calculations are unlikely to be accurate or useful until job conditions and equipment are known. See Chapter 11.

Earthmoving quantities may be measured in terms of bulk of material, weight of material, area to be worked, or lineal feet or yards processed.

Cubic Measure. Most earthmoving is computed in cubic yards. A cubic yard is a cube 3 feet long, 3 feet wide, and 3 feet high. It is about 6 inches greater in each dimension than the space occupied by a card table.

"Cubic yard" may be stated and written in full, but is usually shortened to "yard" or abbreviated to cu. yd. or cy. The full designation is used in discussions where the type of yard might be in doubt, as where there is a possibility of confusion with square yards or linear yards. Abbreviations are used in formulas, figuring, and texts where terms are frequently repeated.

Cubic yards may refer to bank, loose, or compacted measure. This will be discussed below under Swell and Shrinkage.

Many dimensions in field measurements and contract plans are in feet, so that if they are multiplied together to obtain bulk, results are obtained in cubic feet.

These are usually changed to cubic yards by dividing by 27 (there are 27 cubic feet in one cubic yard). Of course it is also possible to divide the original linear measurements by 3 to convert them into yards, and then multiply, but this may lead to working

in fractions, decimals, and mixed numbers.

Most countries use metric measures. See pages A-5 and A-32. A meter is 39.37 inches, 3.2808 feet, or 1.0936 yards. A cubic meter is 35.315 cubic feet or 1.308 cubic yards, a bit less than $1\frac{1}{3}$ yards.

Measurements are sometimes given in feet and inches. A cubic inch is too small a unit to be used for dirt (it takes 1,728 of them to make a cubic foot, 46,656 of them to make a cubic yard) but it is used in measurements of engine and compressor displacement and other mechanical descriptions.

It is difficult to multiply feet and inches by feet and inches, so the inches are expressed as either regular or decimal fractions of feet. Regular fractions (2/3, 1/12, etc.) are more accurate, decimals are more convenient.

Weight. In highway jobs, cellar excavation, and site preparation the estimator is chiefly interested in the bulk and digging characteristics of the material to be moved. Its weight has importance chiefly in figuring loading and grade ability of haulers, and is only occasionally used as the basis for estimating. See Figure 3-5.

In mines and quarries weight is likely to be a more important factor than bulk. Products such as coal, crushed stone, cement, ores, and ore concentrates are usually sold by weight, and therefore are measured by weight as they are dug. There are exceptions, however. Some open pit mines compute their operations on a yardage basis. Others figure ore by the ton and overburden by the yard.

The standard weight measurement is the short ton of 2,000 pounds. There is also a long ton of 2,240 pounds that is used in some mines. It should always be specified as a long ton, and the unmodified word ton should mean short ton only.

Other measures are the hundredweight of 100 pounds, and the British stone that weighs 14 pounds. In the metric system a kilogram, often called a kilo, is 2.2046 pounds, say 2.2 for short. A metric ton is 1,000 kilograms, or 2,204.6 pounds, about ten per cent more weight than in a short ton, and a little less than a long ton.

Area. In many operations bulk and weight do not provide a suitable basis for measuring the amount of work. For example, with any given land clearing condition, the time and cost of the work vary with the area. Clearing is usually measured by the acre, which is 43,560 square feet or 4,840 square yards. Many estimators do their figuring in acres of 40,000 square feet to simplify arithmetic, and make allowance afterward for the difference of almost 10 per cent.

In other situations both area and bulk are important. Pavements are usually figured by the square yard, because a substantial part of the work is done in finishing the subgrade and the pavement surface or surfaces. On the other hand the cost of material in a 10 inch thick concrete slab will obviously be much greater than in an 8 inch slab. The estimator must therefore figure both area and bulk.

Grading land where cuts and fills are shallow is likely to show costs in relation to area rather than bulk of soil to be moved. But when cuts and fills are deep, bulk becomes more important than area. Exact grade requirements increase grading costs in proportion to area. The estimator must keep track of these factors.

Topsoiling and seeding are figured mostly on the basis of area, although of course the depth of topsoil and the heaviness of seeding are important in determining the rate per square yard.

In the metric system the square yard is replaced by the square meter. This contains 10.62 square feet or 1.196 (say 1.2) square yards. The next larger measure is the "are" that contains 100 square meters, 119.6 square yards, or .0247 acres. Then there is the hectare, that contains 10,000 square

meters, 11,960 square yards, or 2.451 (say 2½ acres. The largest measurement is the square kilometer, 1,000,000 square meters, 245.1 acres, or .3861 (say 2/5) square miles.

Lineal Measurement. Work such as ditching, installing pipe, tile, and fencing is measured by the lineal or running foot, yard, or mile. There are 3 feet in a yard, and 5,280 feet or 1,760 yards in a mile. Metric measures are the meter, 39.37 inches, 3.28 feet, or 1.09 yards; and the kilometer, that is 1,000 meters, 3,281 feet, 1,094 yards, or .6212 (slightly more than 3/5) miles.

From both a time and cost viewpoint it is of course necessary to know also the width and depth of the ditch, the soil and other special conditions, the size and type of pipe, and the type and quality of fencing.

Drilling is usually measured by the lineal foot, with the size of hole specified.

TYPES OF PRODUCTION

A construction machine may work in an intermittent cycle, in a continuous flow, or in ways that are intermediate between the two types. Figure 2-50 shows the category in which various types of equipment fall.

Intermittent Cycle. This group includes the most important machines used in primary excavation. They each have a bucket, bowl, or body that is loaded, moved, dumped, and returned to the loading point. One complete set of operations is called a work cycle.

For example, a revolving shovel digs in the bank, swings the bucket over a truck body, dumps it, swings back to the bank and positions the bucket to dig. A scraper loads in the cut, travels to the fill, dumps, turns, returns to the cut, turns, and gets in position to load. In each case, the set of operations is a cycle.

Production rate depends on the size and efficiency of the earthmoving container, whether bucket, bowl, blade or belt, and the time that it takes to go through a com-

Cycle	Intermediate	Continuous Flow
Revolving shovel, all rigs	Rock drill	Conveyor belt
Shovel dozer	Grader	Belt loader
Bulldozer	Roller	Bucket loader
Scraper	Ripper	Wheel Ditcher
Pusher	Plow	Ladder ditcher
Truck	Hopper	Wheel excavator
Pile driver		Paddle loader
Cable excavator		Rock crusher
Concrete mixer		Gravel washer
Pug mill		Screen
		Aggregate spreader
		Compressor
		Hydraulic dredge
		Hydraulic monitor

Fig. 2-50. Cycle classification

plete cycle. The cycle time in turn, depends on the rate at which the container is loaded, moved, dumped, and returned to the loading point.

The distance the load must be moved may be a few feet in the swing of the shovel, or a number of miles in truck haulage. Distance is often the determining factor in the production cycle.

The size of the container is rated by the manufacturer, usually for level and for heaped loads. Container efficiency is the amount of actual load in relation to rated load, and will be discussed later.

The probable production of a machine can be calculated by multiplying its actual capacity by the number of times it can go through its cycle in a given period. Actual production may be found by measurement in the bank, the haulers, or the fill of material removed during a measured time, and/or by measurement of individual loads and timing of cycles.

Continuous Flow. Continuous flow is found chiefly in equipment using belts, pumps, and/or pipes. It includes machines such as wheel ditchers that dig by means of individual buckets, but which have a number of buckets digging at the same time. The individual bucket cycles overlap and production is continuous.

Production of a belt machine is found by averaging a number of measurements of the cross section of the load on the belt, and multiplying this figure by the speed of the belt in feet per minute. The cross section is usually measured in square feet, so the result is divided by 27 to obtain loose cubic yards.

Ditchers, bucket conveyers, and other machines using numerous small buckets may be checked by multiplying bucket capacity by the number of buckets per minute, or by measuring the load going through a discharge belt or chute.

Measurements may also be made in the bank, in haulers, or in the fill.

In dredging, the volume of water handled is computed by multiplying a cross section of the stream in the pipe times the rate of flow. This production in cubic feet or yards of water per minute is then multiplied by the percentage of solids in the water —usually 10 to 20 per cent—to obtain useful production.

Each belt or pipe might be considered to be a container with a measurable capacity. However, this capacity is related to haul distance rather than rate of production. A conveyer belt carrying a load with a cross section of 2 square feet and traveling 500 feet a minute will deliver 1,000 cubic feet a minute, regardless of whether it is a 10 foot feeder belt or a half-mile-long hauler. The only production difference is the extra time —practically 5 minutes—that it will take to reload the longer belt after unloading it at the end of a shift.

The cost of belt construction and maintenance varies almost directly with the length, if width, load, and support requirements remain the same.

Intermediate. This class of machines requires individual methods of study to determine production.

A grader is a continuous flow machine in sidecasting until it reaches the end of a work area and must turn or reverse. If the area is very short, or it is pushing material bulldozer fashion, it has a cycle. Most grader work is measured by the area processed, either as square yards or as linear feet of roadway of a specified width. Production may be figured on a basis of width processed times speed.

A drill will cut continuously to the end of its stroke, but steels are changed and/or new holes are started frequently. Measurement is by speed in feet per minute or per hour, less non-drilling time.

A hopper often serves to convert intermittent cycle loads to continuous flow, as in a dragline feeding a conveyor belt. Its production rate is usually either that of the excavator or the belt, as it can be adjusted to almost any rate of feed within its capacity

A special application is in speeding up truck loading. A 2 yard excavator might load sand into a hopper at the rate of 3 yards a minute, and the hopper might be able to fill a 15 yard truck in a minute. It serves to convert a rapidly repeated 2 yard shovel cycle into a slower 15 yard truck cycle.

Non-Cycle Time. There is almost always a wide gap between production that should be obtained on a basis of cycle speed and capacity, and what is actually produced. Checking the cause of these delays is a drawn out matter that can usually be best done by studying lost time reports made out by operators or checkers. Some however, are frequent enough to be included in brief time studies.

For a shovel there are cleanup and move-up operations, and waits for trucks. Trucks wait to get into loading position, are delayed by traffic, and get stuck. Scrapers wait for pushers, have traffic problems, and sometimes get stuck.

If a breakdown occurs, it is useful to make a note of its nature, and of how long how many machines are put out of action or slowed down.

SWELL AND SHRINKAGE

Swell. When soil or rock is dug or blasted out of its original position it breaks up into particles or chunks, that lie loosely on each other. This rearrangement creates spaces or voids, and adds to its bulk. This increase from bank yards to loose yards is called swell.

Swell is expressed as a percentage of the bulk in the bank. If one bank yard puffs up to 1¼ loose yards, the swell is 1/4 x 100 per cent, or 25 per cent.

When converting bank yards to loose yards, the measurement is increased by the percentage of swell. In the above example, 10 bank yards could be converted as 10 plus 10 × .25, or as 10 × 1.25, giving 12.5 loose yards.

Swell Factor. The swell factor is the percentage of bank yards in loose yards. The contractor uses it to make a conversion on paper of the loose yards he is hauling back to the bank yards for which he may be getting paid. The swell factor is found by dividing the bulk of a bank yard by the bulk of a loose yard, per the formula:

$$\text{Swell Factor} = \frac{\text{one}}{\text{one plus per cent of swell}}$$

If the swell is 25 per cent, this works out:

Swell Factor =

$$\frac{1}{1 \text{ plus } .25} = \frac{1}{1.25} = .8 \text{ or } 80\%$$

The percentage of voids is found by subtracting the swell factor from 1.00.

The tables provided in Figure 2-51 show average swell, swell factor, and voids for various classes of soil. It also shows the shrinkage factor, which is the reduction in bulk to compacted yards in a fill.

Shrinkage. When soil placed in a fill is thoroughly compacted by rolling, it will shrink, the amount of shrinkage depending on its character, its structure in the bank, the thickness of fill layers, and the weight

TABLE I

	SWELL	VOIDS
Clean sand or gravel	5 to 15%	4.75 to 13 %
Top soil	10 to 25%	9 to 20 %
Sandy, clayey loam	10 to 35%	9 to 26 %
Good common earth	20 to 45%	17.7 to 31 %
Clay with sand or gravel	25 to 55%	20 to 35.5%
Clay- friable and light	30 to 60%	23 to 37.5%
Clay- dry, lumpy and tough, with rock	35 to 70%	26 to 41 %
Shale and soft rock	40 to 85%	28.5 to 46 %
Hard rock - well to poorly blasted	50 to 100%	33.3 to 50 %

The loose or aerated part of the load uses space as a percentage of the full heaped pile and is shown above in the second column as "voids".

TABLE II

As a means of simplifying the consideration of these factors, they have been reduced to a representative four as follows:

Sand	10 % Voids
Common earth	20% Voids
Clay	30% Voids
Shot rock	40% Voids

TABLE III

Excavation, or place yards may be judged then to represent:

In sand	90 % of the heaped maximum capacity
In common earth	80% of the heaped maximum capacity
In clay	70% of the heaped maximum capacity
In rock	60% of the heaped maximum capacity

TABLE IV

Loose yards will represent:

In sand	111 % of bank yards
In common earth	125% of bank yards
In clay	143% of bank yards
In rock	167% of bank yards

Fig. 2-51. Soil swell and shrinkage

and type of roller. Blasted rock may retain some swell, while ordinary loam may be reduced to 80 or 90 per cent of bank volume. Measurement in the fill is described as compacted yards, or yards after shrinkage.

$$\text{Shrinkage factor} = \frac{\text{volume in fill}}{\text{volume in bank}} \text{ or}$$

$$\frac{\text{volume in fill}}{\text{volume in loose yards} \times \text{swell factor}}$$

Shrinkage per cent = 1 minus shrinkage factor

Shrinkage may also occur in undisturbed soil on which fills are placed, particularly if the 50 tons plus super rollers are used. Such shrinkage may be calculated by the following formula, if more exact information is lacking:

$$\text{Settlement} = \frac{\text{depth of compaction} \times \text{shrinkage per cent}}{2}$$

This is just an averaging of maximum compaction at the top and zero compaction just below where the effect is felt. This depth would be about 48 inches in a clay loam soil under a 50 ton roller. Such soil may have a shrinkage factor of .2 so by this formula:

$$\text{Settlement, in.} = \frac{48 \times .2}{2} = \frac{9.6}{2} = 4.8$$

In wet conditions, or on steep slopes, displacement of subgrade material to the side may reduce its useful bulk much more than its compaction does.

Exceptional Soils. Soils composed of volcanic ash or pumice, and soils built up by streams on flats in arid regions, may show very peculiar behavior. Particularly, they may show excessive compaction, sometimes to 40 per cent of bank volume. In some cases, they even show shrinkage during digging, so that loose yards are heavier than bank yards.

Such soils are found only in limited areas where contractors may be aware of their possibilities. However, they constitute a hazard for the estimator, and any indication of their existence on a job should be carefully checked.

One western state highway department has adopted a policy of paying for highway borrow by weight rather than by volume to avoid unpleasant surprises for either the budget or the contractor when such materials appear in borrow pits. The necessity of measurement before starting work is avoided also. Portable truck scales that can be set up in a few minutes are used for measurement of pay quantities.

CONTAINERS

Container is used here as a broad term to include transporting boxes such as truck bodies, digging and carrying buckets or bowls as in shovels and scrapers, and digging and pushing blades on dozers and graders. But it does not include belts or pipes.

Measurement. Most buckets and bodies are rated by the manufacturer as to carrying capacity in loose yards. This rating may be water level (the yardage of liquid which could be carried if it didn't leak out), line of plate or struck measure, which is water level plus any space between parts of the rim which project above its low point, or heaped.

Shovel dipper buckets and highway truck bodies are normally rated at water level, carrying scrapers and off the road trucks at both struck and heaped measure, and clamshells at water level, line of plate and heaped.

Body capacity should be indicated by a printed plate. It is usually in yards, but may be in feet. It ordinarily does not include sideboards, or other removable extensions which increase its volume. The extra load between added boards may be found by the proportion between the heights of the body wall and of the board. For example, if a straight-sided body with 4-foot sides

holds 8 yards, a one-foot sideboard added on each side will increase capacity by one-fourth, or two yards.

If there is no plate, size can be determined by measurement of the inside length, width, and height, usually in feet and fractions of feet. These are multiplied together, to obtain cubic feet that are divided by 27 to get cubic yards.

Heaps. The heap on top of a load can be measured only approximately because of its shape. It is usually calculated on the basis of an assumed even slope of the material. The slopes from opposite sides are continued until they meet. In general, the volume of a heap is determined by the slope of the material and the area of the body. A steep slope and a wide body give it maximum yardage.

According to S.A.E. (Society of Automotive Engineers) standards, a heap is figured on a basis of slopes of 3 to 1 (3 horizontal feet to each foot of rise) inward from the top of the body walls. This covers most conditions, but some materials will stack up much higher, particularly if the trip is short and smooth, as in a dragline bucket, while a few will slide or flow to flatter slopes as the container is moved or shaken.

Truck Bodies. Bodies such as are mounted on highway trucks have a rectangular floor, and vertical body walls, with provision for increasing the height of sidewalls by adding wood or metal extensions. The rear wall (tailgate) is hinged to dump the load when the body is raised at the front. Most of such bodies are seven feet wide inside. Some standard heights and capacities are shown in Figure 2-52.

Bodies of off-the-road trucks are difficult to measure, as they may slope outward from bottom to top and the sloped fixed tailboard ends at a much lower level than the body sides. Ratings generally assume a load struck off by a plank resting on the two sides and moved from front to rear. In the absence of indications to the contrary, manufacturers' ratings for struck load may be accepted.

The more important rating for off-the-road trucks is capacity in tons. This is usually one-and-one-half times the rating in yards. The trucks are designed to carry a certain payload, and the body is designed to carry that load in material that weighs 3,000 pounds to the loose yard. Lighter materials can be heaped, heavier materials should not fill the body completely. Bodies can be purchased in special sizes for regular handling of material of very different weights.

Capacity Tests. When a capacity test is to be run the bank is trimmed and measured and a number of approximately equal loads are taken and counted. The bank is then trimmed off and measured again, and the difference calculated in yards. Then:

$$\text{Truck load} = \frac{\text{volume removed from bank}}{\text{number of truck loads taken}}$$

Such tests should be run on the largest practical volume for best accuracy. Minimum amount would be about 200 yards. Counting of excavator bucket loads during the test will supply container and output data for it also.

Scraper Bowl. Scraper bowls are irregular in shape and in height of rim. Manufacturers rating on struck load may be accepted if known. If not, inside measurements are taken and capacity is worked out, making allowance for curves and irregularities.

Scrapers are usually rated for heaps with 1 to 1 slopes, three times as steep as the S.A.E. standard. Dirt may stand a little more steeply above a scraper because of the way it is crowded up from the bottom, but there is no justification for assuming that it can stand on such a slope, either at the moment of loading or during the shaking and vibration of a haul. Also, such slopes are never carried up to peaks.

A good rule of thumb is to take the scraper's maximum heap at one-half the

Length, Inside, in Feet	Height Sides in Inches	Capacity Capacity Cubic Yards
9	15–1/2	3
	20–7/8	4
	25–3/4	5
9–1/2	14–3/8	3
	19–1/2	4
	24–1/2	5
10	18–3/4	4
	23	5
10–1/2	20–1/8	4–1/2
	22–1/4	5
11	20–7/8	5
	25	6
12	23	6
	31	8
	38–3/4	10
14	33–3/8	10

Fig. 2-52. Capacity of standard truck bodies

manufacturer's rating. That is, a scraper rated at 20 yards struck and 28 yards heaped can be assumed to have a heaped capacity of up to 24 loose yards, unless there are definite indications to the contrary. If swell factor were .80, the heaped load would contain 19.2 bank yards. Extensive checking by the Bureau of Public Roads indicates that scraper loads in bank yards seldom equal their rated struck capacity, so this works out well enough.

Push loading compacts material, so for the same heap a scraper carries fewer bank yards if it is loaded from the top.

Shovel Bucket. Dipper buckets are rated at struck measure, in yards and fractions of yards. Size can be checked fairly accurately by measuring and multiplying length, width, and height, as in the case of the truck body. Taper, curves, and extension of the lip can be used to increase the result to the nearest quarter yard. Heaps, when digging conditions permit, range between 1/8 and 1/4 of struck capacity.

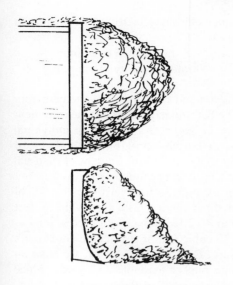

Fig. 2-53. Bulldozer blade load

Hoe buckets are rated and measured in the same manner. Heaps are usually small or absent. Width is an important factor in loading efficiency. Deep narrow buckets may average loads less than 1/3 of rated capacity.

The open front of the dragline bucket makes it difficult to measure accurately, as the capacity is determined largely by the steepness of slope of the load. A chunky mud that can bulge over the teeth and pile steeply over the top may have almost as much in the heap as there is in the struck load.

Clamshell buckets may be rated water level with the end plates, struck measure along the tops of the sides, or heaped. Another rating that may be important is that of deck area—the length and breadth of the space on a level floor occupied by the bucket when fully open. The same space is required between braces in a trench. See Figure 13-91.

Loader Bucket. Front end loader buckets are rated at struck capacity across the sidewalls. If the bucket can be tipped back 20 degrees or more at ground level, and digging conditions and tractor stability permit, it can carry a very large heap, perhaps 50 per cent of struck capacity. A bucket that will not tip back seldom can take 1/8 more than struck capacity. However, in hard digging few buckets can pick up even a struck load, and tractor stability may not permit carrying an overload. On the average these buckets are underloaded by loose measurement more often than they are heaped.

Dozer Blade. The capacity of a bulldozer blade cannot be computed exactly. The blade forms only one of the six sides of a shapeless pay load. Since the load must slide or roll along the ground, friction is a very important limiting factor on load size. Materials with low internal friction or light weight, and downhill pushing favor big loads. Pushing through a slot or between windrows, or with the blade almost touching that of another dozer, increase load by reducing side spillage.

A good average load may be said to be 20 per cent more than would be measured by the full width and height of the blade, with a forward slope of 1 to 1. The actual load would be higher in the center and skimped in at the corners, averaging out as above. See Figure 2-53.

The blade on a 25 ton bulldozer might be 12 feet wide and 4½ high. The profile of the theoretical load would be a right triangle 4½ by 4½, with an area of 10.25 square feet. Multiplying this by the 12 foot blade width, we have 123 cubic feet. Adding 20 per cent for center bulges, this would be 148 cubic feet or 5.5 loose cubic yards. Different manufacturers rate blades of this size at 5 to 7½ loose yards.

Under very good conditions the load might average 6 inches higher than the blade top, and extend forward on a 1½ to 1 slope. Then the profile would be 5 feet x 7½ feet or 19 square feet. Multiplying by 12 foot width, we get 230 cubic feet or about 8½ yards loose.

Capacity of a blade is greatly affected by

grade. Downhill the load tends to slide along with minimum pressure, uphill its friction rises and it spills more off the sides. Possible load, both from standpoint of power to push it and ability to keep it in front of the blade, increases about 4 per cent for each per cent of down slope. That is, a dozer might push double its normal load down a 25 per cent grade.

Uphill the load falls off at first almost as rapidly, but the decline slows with steeper grades. A half load might be pushed up a 20 per cent grade, and a quarter load up 100 per cent.

When a job is in progress dozer blade capacity can be checked by counting the number of full passes required to dig a known yardage of a bank, or by counting loads pushed into a pile, and measuring the pile.

The first result will be in bank yards, the second in loose yards.

Container Efficiency. A machine may go through many production cycles with less than its rated load. The proportion between the container's capacity and its actual load in loose yards is called the efficiency factor of the bucket, bowl, or body.

An excavator may not be able to load its bucket or bowl fully because of hard digging, inadequate power, improper design, dull teeth, traction, heavy material, operator's haste or carelessness or combinations of these and other unfavorable conditions.

A truck may be run partly empty to enable it to climb steep grades, to reduce strain on rough ground, to enable it to cross soft ground, to increase haul speed, or because of poor mechanical or tire conditions.

When material is coarse in proportion to bucket size, chunks may be partly supported by the sides and each other, leaving excessive voids and reducing the actual load below that indicated by the capacity of the container and the bulk of the load. The container efficiency factor (CEF) formula is:

$$CEF = \frac{\text{material in container}}{\text{rated capacity of container}}$$

The amount of material in the container may be determined by careful measurement and/or weighing of a number of individual loads, or by measuring either the bank or the fill to determine the amount of material moved in a certain number of cycles.

For example, a half yard backhoe might dig a ditch section 2½ feet wide, 6 feet deep and 12 feet long in 25 cycles, removing 180 cubic feet or 6.67 cubic yards bank measure. Dividing 6.67 by 25, we have an average bucket load of .267 yards. Since the rated capacity of the bucket is .5 yards, we have:

$$CEF \text{ (bank yards)} = \frac{.267}{.5} = .534$$

The efficiency factor in loose yards would be greater by the percentage of swell. In clay loam with 20 per cent swell (see below) the loose yards would be 6.67 x 1.20, or 8.0. Dividing by 25 cycles, we would have an average load of 3.2 loose yards. Then:

$$CEF \text{ (loose yards)} = \frac{.32}{.5} = .64$$

It is important to always specify whether any container efficiency factor is for loose yards or bank yards. One may be readily converted to the other, the loose yards figure almost always being the larger, as it is increased by the percentage of swell.

OUTPUT

Work Cycles. The work cycle may be timed as a whole in figuring output for existing conditions. For accuracy it is necessary to take the average of a large number of passes as there may be a considerable variation among them.

If a study is made of a cycle, either to find a way to speed it up, or to use its time intervals as a basis for figuring production under different conditions, it can be broken down into individual operations which are timed separately.

A study of bulldozer and of scraper operation may include some or all of the divisions listed below. Each one should be timed. Digging and traveling distances should be measured. A record should be made of all grades, as these machines are much less efficient going up hill than down.

Bulldozer	Scraper
Dig	Load
Shift into second	Shift
Transport	Transport
Dump	Shift
Shift to reverse	Spread
(raise blade)	Shift
Return	Deadhead to turn
Shift to low	Turn
(lower blade)	Return to digging area
	Deadhead to turn
	Turn
	Get to loading position
	Shift into low
	Wait for pusher

Machine Efficiency. Non-working time such as delays for moving the machine, minor repair adjustment, rest, getting instructions, or looking at grade stakes is not averaged into cycle time. It is considered separately as the efficiency factor of the machine. It may work out between 70% and 85% for short periods with expert operators on a properly run job, with machinery in good condition and weather favorable.

A convenient way of making rule-of-thumb allowance for efficiency of about 83% is to consider that an hour contains only fifty working minutes. On this basis, a machine with a thirty second cycle can-not be expected to perform it more than one hundred times an hour.

If two machines are interdependent, the efficiency factor is usually lower so far as the job is concerned. If two eighty per cent efficient machines are down at the same time every time the efficiency factor remains at 80%. If they are down separately each time the factor drops to 60%.

As the period under study becomes longer efficiency drops steadily as weather, work sequence stoppages, and major overhauls must be considered. Time losses can be further increased by inefficient management, bad pit layout, poor morale of employees, or other unfavorable conditions.

Output Formulas. A formula which can be used for figuring production of any machine with a regular cycle is:

OUTPUT, yards per hour =
$$\frac{Q \times K \times E \times 60 \times f}{Cm}$$

Where: Q = capacity, either struck or heaped
K = efficiency factor of bucket or body
E = efficiency factor of machine
60 = sixty minutes in an hour
f = soil conversion factor
Cm = cycle time in minutes

If the result is to be in bank yards, f has a value obtained from observed swell, or from Table 3, Figure 2-29.

If the result is to be in loose yards, f equals one, and can be dropped from the equation.

The factor K is not required when full, solid loads can be taken.

If efficiency is approximated at $\frac{5}{6}$, 50 minutes are used instead of 60, for the hour.

Under these conditions, the following simplified formula can be used:

OUTPUT, loose yards per hour = $\dfrac{Q \times 50}{Cm}$

If a 45 minute hour is used, this formula is:

OUTPUT, loose yards per hour $= \dfrac{Q \times 45}{Cm}$

When timing machines whose cycle is less than a minute, it is more convenient to figure it in seconds. This is done by multiplying the number of minutes by 60, and using Cs—cycle time in seconds—instead of Cm. With these substitutions, the formula will be:

OUTPUT, yards per hour $=$

$$\dfrac{Q \times K \times E \times 3600 \times f}{Cs}$$

Or simplified for a 50 minute hour to:

OUTPUT, loose yards per hour $=$

$$\dfrac{Q \times 3000}{Cs}$$

For a 45 minute hour we would have:

OUTPUT, loose yards per hour $=$

$$\dfrac{Q \times 2700}{Cs}$$

Additional data on output will be found in Chapter 11, the Appendix, and included in discussions of particular machines.

CHAPTER THREE

SOIL AND MUD

SOIL

SOIL AND ROCK

Soil is loose surface material. Rock is the hard crust of the earth, which underlies and often projects through the soil cover. There is no clear distinction between soil and rock. Geologically, all soils are considered to be rock formations. In ordinary usage, rock is something hard, firm, and stable.

A contractors' definition is that rock is any material which cannot be dug or loosened by available machinery, but this distinction from soil may depend more on the power, size, and digging efficiency of the machinery, than on the material itself.

Material to be excavated can also be roughly divided into three classes, rock, hard digging, and easy digging.

Rock is anything that needs blasting or ripping for efficient digging by most machines. Hard digging is compacted, cemented, or rocky dirt, clay, soft shale, and rotten rock, that can be dug directly by heavy machinery, or loosened readily by rippers. Easy to medium digging is any soft or fine, firm or loose deposit.

Soil. Soil is composed of particles of various sizes and chemical composition. It can be analyzed as to sizes by sifting a dried and weighed sample through a set of testing sieves, such as are shown in Figure 3-1, and weighing the material retained on each screen.

If further analysis is required for particles passing the smallest (200 mesh) sieve,

it is done by hydrometer. This process is based on the fact that the speed of settlement of such particles is proportional to their size.

Figure 3-2 indicates the size particles which are included in the common soil classifications. There are several scales, in which the boundaries between different classes may vary. The differences among them are not important to the average contractor.

Fine grained soils are known as heavy, and sandy ones as light. Heavy soils may or may not permit circulation of ground water, light ones almost always do. Heavy soils are more readily softened by water. They are often called plastic soils, even when not in a plastic condition.

A plastic soil is one which can be rolled, as between the hands, into strings 1/8 " in diameter without falling apart. Plasticity is a function of soil character and of moisture content. The minimum amount of water in terms of per cent of oven-dry weight of the soil which will make it plastic is defined as the plastic limit of the soil. If no amount of water will allow it to roll into strings it is called non-plastic, with the symbol NP.

The liquid limit is minimum moisture content, in terms of per cent of oven-dry weight, which will cause the soil to flow if jarred slightly.

The plasticity index is the difference between the plastic limit and the liquid limit;

Courtesy of W. S. Tyler Co.

Fig. 3-1. Testing sieves

that is to say, the range of moisture content in which the soil is plastic.

Soils or their particles may also be classified as to grain shape or hardness, and mineral and organic content. These factors will affect their resistance to weather, stability under load, wear on digging parts, and internal friction.

Most soils are inorganic, and are made up of products of decay and breaking up of rock. Organic soils and organic material in soils, are largely humus which is formed by decay of vegetation and has no definite particle size. Organic materials may also consist of lime from shells or from limestone originally formed from shells, and animal bones and excrements.

Rock. Geologically, rocks are classified as to the way in which they were made. Those which solidified out of a molten state are called igneous, and are subdivided into volcanics cooled at the surface, and plutonic hardened deep underground.

Sedimentary rocks are built up of soil or plant or animal remains and have been hardened by pressure, time, and depositing of natural cements.

Metamorphic rocks were originally ig-

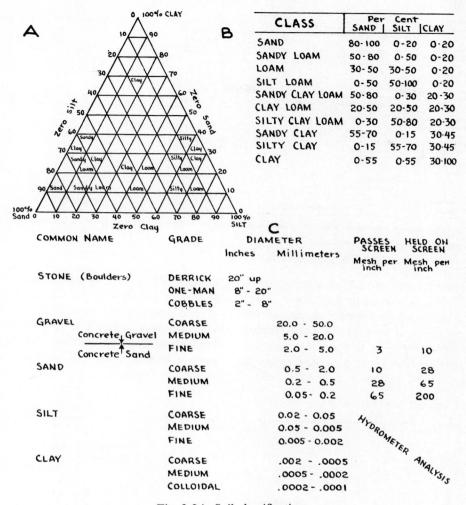

B

CLASS	Per Cent		
	SAND	SILT	CLAY
SAND	80-100	0-20	0-20
SANDY LOAM	50-80	0-50	0-20
LOAM	30-50	30-50	0-20
SILT LOAM	0-50	50-100	0-20
SANDY CLAY LOAM	50-80	0-30	20-30
CLAY LOAM	20-50	20-50	20-30
SILTY CLAY LOAM	0-30	50-80	20-30
SANDY CLAY	55-70	0-15	30-45
SILTY CLAY	0-15	55-70	30-45
CLAY	0-55	0-55	30-100

C

COMMON NAME	GRADE	DIAMETER		PASSES SCREEN	HELD ON SCREEN
		Inches	Millimeters	Mesh per inch	Mesh per inch
STONE (Boulders)	DERRICK	20" up			
	ONE-MAN	8" - 20"			
	COBBLES	2" - 8"			
GRAVEL	COARSE		20.0 - 50.0		
Concrete, Gravel	MEDIUM		5.0 - 20.0		
Concrete Sand	FINE		2.0 - 5.0	3	10
SAND	COARSE		0.5 - 2.0	10	28
	MEDIUM		0.2 - 0.5	28	65
	FINE		0.05 - 0.2	65	200
SILT	COARSE		0.02 - 0.05		
	MEDIUM		0.05 - 0.005		
	FINE		0.005 - 0.002		
CLAY	COARSE		.002 - .0005		
	MEDIUM		.0005 - .0002		
	COLLOIDAL		.0002 - .0001		

HYDROMETER ANALYSIS

Fig. 3-2A. Soil classification

neous or sedimentary, but have been altered by extreme heat and pressure.

Figure 3-3 contains tables classifying rocks as to type and hardness. The latter quality is quite variable, even in one formation, and may be made up of different factors, as resistance to penetration, abrasion, or crushing.

Digging Resistance. The resistance which must be overcome to dig a formation will be made up largely of hardness, coarseness, friction, adhesion, cohesion, and weight.

In digging, hardness is resistance to pen-

etration. It is increased by close packing of soil, or filling of voids with finer particles, or lime or other natural cements. Clay soils are hard when dry, and soft when wet.

Cobbles, boulders, or hard lumps increase the power requirement for penetration. They are most troublesome when they are oversize for the machine, or packed so firmly in place that they cannot slide or rotate away from the cutting edge.

As the digging edge penetrates, friction absorbs an increasing proportion of its force. It is affected by particle size and hardness, by the amount of moisture, and

General classification	Granular materials (35% or less passing No. 200)							Silt-clay materials (More than 35% passing No. 200)			
	A-1		A-3	A-2				A-4	A-5	A-6	A-7 A-7-5, A-7-6
Group classification	A-1-a	A-1-b		A-2-4	A-2-5	A-2-6	A-2-7				
Sieve analysis, percent passing:											
No. 10	50 max.	—	—	—	—	—	—	—	—	—	—
No. 40	30 max.	50 max.	51 min.	—	—	—	—	—	—	—	—
No. 200	15 max.	25 max.	10 max.	35 max.	35 max.	35 max.	35 max.	36 min.	36 min.	36 min.	36 min.
Characteristics of fraction passing No. 40:											
Liquid limit	—		—	40 max.	41 min.	40 max.	41 min.	40 max.	41 min.	40 max.	41 min.
Plasticity index	6 max.		N.P.	10 max.	10 max.	11 min.	11 min.	10 max.	10 max.	11 min.	11 min.*
Usual types of significant constituent materials	Stone fragments, gravel and sand		Fine sand	Silty or clayey gravel and sand				Silty soils		Clayey soils	
General rating as subgrade	Excellent to good							Fair to poor			

*Plasticity index of A-7-5 subgroup is equal to or less than L.L. minus 30. Plasticity index of A-7-6 subgroup is greater than L.L. minus 30.

Courtesy of Portland Cement Association

Fig. 3-2B. Classification of soils and soil-aggregate mixtures

the presence or absence of natural lubricants such as humus or soft clay.

Adhesion is the sticking of soil to the digging parts. It may increase the friction load substantially in wet work.

Cohesion is resistance to tearing apart. Firm or hard materials may split readily along bedding or cleavage planes so that they can be dug rather easily from the proper direction. Relatively soft clay banks may be very difficult to dig because of strong and uniform cohesion. A tough formation lacking planes of weakness is described as "tight."

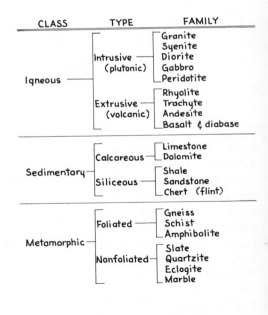

ROCK	WEIGHT lbs. per cu. ft.	PER CENT OF WEAR	HARDNESS	TOUGHNESS
Granite	167	4.3	18.3	11
Syenite	171	3.3	18.3	15
Diorite	179	3.0	18.2	17
Gabbro	185	3.0	17.7	14
Peridotite	182	4.0	14.2	11
Rhyolite	159	3.7	18.3	19
Trachyte	170	2.9	18.1	24
Andesite	166	3.9	17.0	18
Basalt	177	3.0	17.1	18
Diabase	186	2.4	18.0	22
Limestone	165	5.0	14.1	9
Dolomite	170	5.5	14.9	9
Sandstone	164	6.2	14.4	10
Chert	159	9.4	18.2	12
Gneiss	172	4.9	17.4	10
Schist	180	4.7	16.6	13
Amphibolite	188	2.8	17.5	19
Quartzite	169	3.2	18.8	18
Eclogite	194	2.4	18.4	22
Marble	173	5.7	13.1	6

CLASS	TYPE	FAMILY
Igneous	Intrusive (plutonic)	Granite, Syenite, Diorite, Gabbro, Peridotite
	Extrusive (volcanic)	Rhyolite, Trachyte, Andesite, Basalt & diabase
Sedimentary	Calcareous	Limestone, Dolomite
	Siliceous	Shale, Sandstone, Chert (flint)
Metamorphic	Foliated	Gneiss, Schist, Amphibolite
	Nonfoliated	Slate, Quartzite, Eclogite, Marble

Fig. 3-3. Rock types

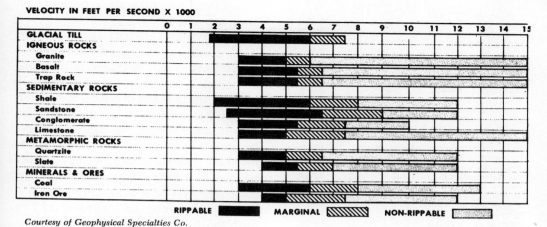

Fig. 3-4. Ripability and seismic vibration

Ripability. This is a measure of the ease or difficulty with which a rock can be broken by heavy rippers into pieces that can be economically moved by other equipment, usually scrapers.

At least three factors are involved: resistance to breakage of the rock material itself, the extent to which it is weakened by bedding layers (lamination) or by joint cracks or fault movement, and the degree to which the rock has been softened and weakened by weathering.

Many rocks are readily ripable at the surface, but become increasingly resistant with depth because of less exposure to weathering. A generally ripable rock may contain seams or boulders that are difficult or impossible to rip.

Some rocks are non-ripable in their natural condition, but can be ripped after shaking up by light blasting.

In general the igneous rocks such as basalt, granite, diorite, felsite, and lava are not ripable unless they are greatly weakened by weathering. They do not respond well to light blasting unless they have a closely jointed structure.

Massive metamorphic rocks such as gneiss, marble, and quartzite are usually non-ripable, but slate and thinly bedded schist may be ripped.

Some rocks have an absolute resistance to ripping. That is, they cannot be shattered by any amount of pressure that can be exerted by a ripper tooth. More often, ripability is determined by how much force can be applied. Limitations are the strength of teeth and shanks, the power availalbe, and the economics of applying the power.

There are also rocks that will break or tear rather readily, but remain in such large pieces that they cannot be handled economically. Oversize is sometimes broken by walking the ripper-tractor on chunks, by a crane or drop ball, or by blasting.

Ripping is usually not economical unless the product is fine enough for efficient scraper loading, or can be made so by secondary work.

The ripability of rock can often but not always be determined in advance by study of its characteristics, particularly toughness, shear, and bedding planes or cracks. Tests with a ripper are not conclusive unless made below the zone of weathering.

Seismic methods may be used. Ripable rocks seem to carry vibration differently from non-ripable formations, as indicated in Figure 3-4.

Weight. Figure 3-5 gives the approximate weights per cubic yard of various materials.

Weight may limit the amount that can

WEIGHTS OF MATERIALS

(ONE YARD — WEIGHTS IN POUNDS)

MATERIAL	1 YARD 27 CU. FT.
ASHES — PILED DRY	945
BRICK BATS	1485
CEMENT — PORTLAND	2538
CINDERS	1485
CLINKER — PORTLAND CEMENT	2295
CLAY — DRY, IN LUMPS	1701
CLAY — COMPACT, NATURAL BED	2943
COAL — ANTHRACITE	1512
COAL — BITUMINOUS, R. OF M. PILED	1485
COAL — BITUMINOUS SLACK, PILED	1350
CONCRETE — READY TO POUR	3996
DOLOMITE — CRUSHED FINE	2565
DOLOMITE — BROKEN LUMP	2565
EARTH — LOAMY, DRY, LOOSE	2025
EARTH — DRY, PACKED	2565
EARTH — WET (MUD)	2970
GYPSUM — CRUSHED TO 3"	2565
GYPSUM — CALCINED	1620
GRAVEL — DRY, LOOSE	2970
GRAVEL — DRY, PACKED	3051
GRAVEL — WET, PACKED	3240
IRON ORE — 60% IRON	8100
IRON ORE — 50% IRON	6750
IRON ORE — 40% IRON	5400
LIMESTONE — RUN OF CRUSHER	2565
LIMESTONE — FINES OUT	2700
LIMESTONE — 1 ½" OR 2" GRADE	2295
LIMESTONE — ABOVE 2" GRADE	2160
PHOSPHATE ROCK	2160
SAND — DRY, LOOSE	2565
SAND — WET, PACKED	3240
SHALE — BROKEN	2295
SLAG — BLAST FURNACE, BROKEN	3726
SLAG — OPEN HEARTH, CRUSHED	2835
SLAG — GRANULATED, DRY	1025
SLAG — GRANULATED, WET	1566
SULPHUR — BROKEN	1620

Fig. 3-5. Weights of materials

be dug or carried in a bucket or body, and the speed with which the load can be hoisted or transported. It is a critical factor in selection of dragline and clamshell bucket sizes, and in regard to the length and angle of the boom that carries them.

TRACTION

Soils differ greatly in their ability to support and permit movement of vehicles. An important characteristic is the amount of friction that exists between the ground surface and the drive tires or tracks of a machine on it.

Tractive Efficiency. This is a measure of the proportion of the weight resting on drive wheels or tracks that can be converted into movement of a machine. Soil charac-

teristics and condition are very important in determining it.

It is difficult to estimate and classify the tractive efficiency of ground surfaces, as wide variations may be caused by the shape of particles and their size gradation. In addition, the presence of certain compounds that do not show up in a soil specification may increase traction by improving packing qualities and even cementing particles together. Other substances may reduce traction by lubricating grains so that they move freely on each other and on tire surfaces. For example, salt improves packing qualities, lime acts as cement, and soapstone (serpentine, mostly $H_4Mg_3Si_2O$) acts as a lubricant.

The amount of water present is an important variable. For any soil there is a certain moisture content that gives best traction. In sandy or gravel soils less water is likely to allow them to become loose and unsatisfactory, while more water has little effect. In heavy soils too little water (once the soil is compacted) may not have any harmful effect except causing dust, but too much water will make them first slippery and then soft. A soaking shower on the dusty loam surface of a heavily traveled haul road may make it temporarily almost as slippery as ice.

Figure 3-6 gives a general guide to traction afforded by various surfaces, expressed as a decimal fraction of the weight on drive wheels or tracks. The variations between low and high values are sufficiently wide to take care of most differences. If the higher values are used, the possibility of having to deal with lower ones should be kept in mind.

This table assigns much lower values to loose and wet soils than most references do.

Weight Distribution. The ability of a machine to propel itself on slippery footing is affected by its weight distribution.

For example, if a machine having a weight of 10 tons on the drive wheels can exert a

Standard Tables	Type of Surface	Tractive Efficiency: Wheel Factor* Dry	Wheel Factor* Wet Surface	Track Factor (Grousers) Dry	Track Factor (Grousers) Wet Surface
	Smooth blacktop	.8 – 1.0	.6 – .9	–	–
.88 – 1.0	Rough concrete	.9 – 1.0	.8 – 1.0	.3 – .6	.3 – .6
	Hard smooth clay	.6 – 1.0	.1 – .3	.4 – .7	.2 – .4
.40 – .58	Hard clay loam	.5 – .8	.15 – .4	.6 – .9	.4 – .9
	Firm sandy loam	.4 – .8	.25 – .8	.6 – 1.0	.6 – 1.0
	Spongy clay loam	.4 – .6	.15 – .3	.7 – 1.0	.6 – .9
.40 – .44	Rutted clay loam	.3 – .5	.15 – .3	.7 – 1.0	.6 – .9
.20 – .35	Rutted sandy loam	.3 – .4	.2 – .5	.7 – 1.0	.7 – 1.0
.36	Gravel road, firm	.5 – .8	.3 – .9	.7 – .9	.7 – .9
	Gravel, not compacted	.3 – .5	.4 – .6	.5 – .9	.6 – 1.0
	Gravel, loose	.2 – .4	.3 – .5	.4 – .7	.5 – .8
.20 to .35	Sand, loose	.1 – .2	.1 – .4	.3 – .5	.4 – .7
.20	Snow, packed	.1 – .4	.0 – .3	.2 – .6	.2 – .6
.12	Ice, roughened	.1 – .3	.0 – .2	.1 – .4	.0 – .3
	Ice, smooth	.0 – .1	.0 – .0	.0 – .1	.0 – .1

*May be increased by extra wide or extra soft tires.

Fig. 3-6. Tractive efficiency of surfaces

drawbar pull of only 4 tons because of wheel slippage, the tractive efficiency of the ground is 40 per cent, or .40. It does not matter whether the whole machine weighs 10 tons or 50 tons, or whether its rim pull is 5 tons or 20, as for this particular calculation we use only weight holding the drive wheels (or tracks) against the ground. However, the total weight of the machine and its load make up the resistance that the drive wheels must move.

Increasing the weight on slipping drive wheels increases drawbar pull in direct proportion, up to the maximum that can be produced by the engine and gears. Increasing weight on non-drive wheels increases resistance. Shift of existing weight to the drive wheels increases potential traction, while shifting weight from drive to non-drive wheels reduces traction. Neither shift affects resistance.

While resistance to movement is in proportion to the weight of the whole machine, ability to move the machine, if power is adequate, depends on the weight on the drive units. All-wheel drive trucks and most crawlers keep all their weight on the drivers. In other machines weight distribution is important to performance when traction is limited.

Manufacturers usually provide detailed weight specifications for haulers, showing the total weight and its distribution, both loaded and empty.

Material	Bearing Power, pounds per square inch
Solid rock	350
Shattered rock	70
Dry clay	55
Medium dry clay	25
Soft clay	12
Cemented gravel	110
Compact sand	100
Clean dry sand	25
Quick sand	5

Fig. 3-7A. Load bearing capacity

If this information is not available the hauler should be weighed, one axle at a time, both empty and loaded. Different load distributions, to the front or rear of the body, can also be checked on scales.

A highway truck may be taken to scales. Off-the-road trucks and scrapers might have to have scales brought to them, which might prove difficult and expensive.

If there is no specific information on weight distribution and scales are not available, it may be assumed that a rear drive dump truck carries about 50 per cent of its empty weight or 70 per cent of its loaded weight on the drive wheels. A front pull scraper, wagon, or rocker usually has 50 to 60 per cent of its empty weight or 40 to 50 per cent of its loaded weight on the drivers.

On climbs, vehicle weight shifts toward rear drives and away from front drives by about one and one-half per cent for each per cent of grade. This factor increases the already considerable traction advantage of the rear drive truck in slippery climbs.

Driving torque of a vehicle also affects its weight distribution. As will be explained in Chapter 15, a vehicle shifts some of its weight to the rear when it is driven forward, and to the front when in reverse.

Tire Treads. Tire treads are an important variable in traction. There are a number of tread designs, all of which eventually wear smooth. For most situations high lugs or cleats give best traction, but smoother surfaces are better on dry sand and ice. When tire chains are used their effectiveness is reduced by high tread blocks.

Since a deep cleated tire gives best traction on soft loam and a smooth tire on loose sand, there are likely to be intermediate surfaces on which they are equally effective, or perhaps ineffective. Sometimes a well worn tire will give as good or better traction than the new one ordered to replace it.

FLOTATION

Flotation means the weight supporting ability of a tire, crawler track, or platform on soft ground. This ability, or lack of it, is the result of a relationship between the weight, the area of contact, and the load bearing ability of the ground.

The weight divided by the contact area in square inches gives the downward pressure in pounds per square inch (psi). If this pressure is greater than the load carrying ability of the ground, the machine will sink until it finds enough contact area to support it. Sinking increases rolling resistance. If it is severe it may prevent the machine from moving under its own power.

Tire Pressure. When a tire carries enough weight so that it tends to sink into the ground, its behavior depends largely on the relationship between its inflation pressure and the load bearing ability of the ground.

Ground with a bearing strength of 50 pounds per square inch (psi) will allow a loaded tire with 75 pounds air pressure to sink into it until the same area of tire is in contact with the ground as there would be if the tire had 50 pounds pressure and flexed to spread out on the surface.

It follows that if a larger tire were used,

that could carry the load at 50 pounds pressure, it would not sink in at all. It does not always work out just this way, but there is no question but that bigger and softer tires greatly reduce problems with vehicles sinking in soft ground.

The formula for finding the area in square inches over which a tire will contact the ground is:

$$\text{Contact area} = \frac{.9 \times \text{tire load}}{\text{inflation pressure}}$$

On a truck with a gross weight of 60,000 pounds, of which 70 per cent is on four drive wheels equipped with 14.00 x 24 tires, 20 ply, with 75 pounds pressure, we would have a load on each tire of ¼ of 60,000 x .7, or 10,500 pounds. Then, by our formula:

Contact Area =
$$\frac{.9 \times 10,500}{75} = \frac{9,450}{75} = 126$$

If the tires were 16.00 x 24, 16 ply, 45 pounds pressure, we would have:

Contact Area =
$$\frac{.9 \times 10,500}{45} = \frac{9,450}{45} = 210$$

Since these loads are within the rated capacity of the tires, they will show only normal flexing, and any additional bearing surface needed will be obtained by sinking into the ground.

To find whether the ground will support the load we use the formula:

$$\text{Bearing Area} = \frac{.9 \times \text{tire load}}{\text{ground bearing capacity}}$$

If the ground has a bearing capacity of 50 pounds per square inch, this problem works out:

Bearing Area =
$$\frac{.9 \times 10,500}{50} = \frac{9,450}{50} = 189$$

We find that the necessary ground area to bear this load is 189 square inches. The

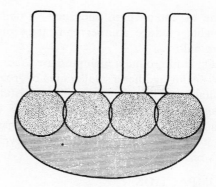

Fig. 3-7B. Pressure bulb

harder tire has only 126 inches of ground contact, so it will sink until an additional 54 square inches of its surface are in contact. On the other hand, the tire with 45 pounds pressure spreads its load over 210 square inches, and should not sink at all.

Figure 3-7A shows the bearing capacities that may be expected of various types of soils. The figures should not be trusted very far, as there are many unknowns in this field.

Other Factors. For repeated crossings, ground should have two or more times the minimum bearing capacity needed for a static load as above, as impact shock during travel greatly increases effective weight. Once ground begins to yield, truck wheels dropping into holes may deliver blows several hundred per cent greater than static weight.

Bearing qualities cannot be measured entirely in square inches. A country road of native soil may easily support a 200-pound man balancing on a one-inch metal cube, but break down rapidly under 30-ton trucks having only one-third as much ground pressure per square inch.

This may be explained by the formation of a bulb of pressure that is built up underground by the sum of individual surface points or areas of pressure. Figure 3-7B shows a cross section of soil under a rubber tired roller, where each tire produces its own zone of compaction, and their combined weights produce another and larger pressure bulb.

On hard ground a high pressure tire does less flexing and has a smaller area of ground contact, and therefore less rolling resistance than a low pressure tire. However, if the ground is soft enough for the tire to sink in at all, the low pressure tire will develop less resistance.

COMPACTION

The term compaction refers to the act of artificially increasing the density of soil. It involves the pressing of soil particles together into closer contact, and expelling air or water from spaces between them.

When the same process occurs in nature, as a result of wetting, drying, freezing, thawing, ground water movement, and weight of higher soil layers, it is called settlement.

The density of soil is measured in terms of its volume-weight, which may be expressed as pounds of wet soil or dry soil per cubic foot, or as porosity in per cent of total volume. A high porosity indicates a low density.

The purpose of compaction is to stabilize soil, particularly in built up fills, embankments, and dams, so that it will show minimum change in volume or shape under influences of weather and time, and under the weight of structures, pavement, and traffic.

Compaction is also useful or necessary to the work of building a fill. Loose soil causes haulers to use excessive power, break drive train parts, and get stuck. Rain that would wet only the surface of a compacted fill might sink into a loose one far enough to create several feet of impassable mud that would stop the work.

Background. The desirability of compacting road fills has been recognized since ancient times. Early methods included driving sheep or cattle back and forth on the fill, and towing weighted wooden rollers with horses.

Up to a few years ago, compaction by haulers and by casual rolling, together with natural settlement, could usually be relied upon to stabilize embankments so that they would retain their shape and support the loads that would be placed on them. Such settlement takes six months, preferably including a winter or a wet season, to a year or two. Some states may still retain a specification of a waiting period between completion of a fill and placing a pavement on it.

Where time permits, this may be good practice, as it will reduce or eliminate damage from poor compaction. However, fills with a high clay content that have been very thoroughly compacted may absorb excessive amounts of rain, and will swell unless protected by pavement.

Reliance on natural settlement and/or simple rule of thumb compaction methods is becoming impractical, because of the increasing loads carried by the modern highway. There is constant pressure to revise State and Federal laws to permit heavier axle and gross loads, in spite of the fact that the bearing capacity of ordinary soils has already been reached and exceeded in many areas.

There has been great progress in the science of soil compaction in the last fifteen years. Laboratory studies have solved many problems of soil behavior, and manufacturers have designed a wide range of equipment to produce maximum compaction under different conditions. However, much of this progress is in the nature of a rear guard action, as even the best possible compaction may not be sufficient to make up for gross overloading.

Equipment. Compaction equipment is described in Chapter 19.

Soil may be compacted by pressure,, kneading, vibration, impact or by combinations. The steel wheel and sheepsfoot rollers supply pressure, pneumatic tired rollers pressure with some kneading, wobble-wheel rollers kneading and pressure, and vibratory rollers supply both pressure and vibration. There are also vibrating plates that supply negligible pressure.

The weight of a smooth steel roller is applied along a straight line across the direction of travel. A high spot may carry the full weight of the roll, while a low spot may be bridged over and receive no compaction. With adequate weight, this causes soil to be squeezed from high to low spots, equalizing irregularities in spreading and producing a smooth surface. But high spots that are too hard to yield will support the roll and prevent any compaction along side or between them.

Steel wheel rollers give good results on all types of soil except clean sands, in layers from 4 to 12 inches deep depending on soil type and roller weight. Clay soil layers should be limited to 4 to 6 inches, to avoid possible compaction of the top of the layer only.

Sheepsfoot rollers compact mostly with the soles of their feet, from the bottom up. As the soil is compacted the roller rises and walks out of the ground. They do best on fine grained soils of the plastic groups, and are least efficient in sandy and gravelly types. Excessive weight may have to be avoided, as the feet may shear soil and damage its structure.

Rubber tired rollers are suitable for use in any type of soil, but weight and tire pressure must be proper for the soil type. Results are affected by shape of tires and their air pressure and by total wheel or axle load, not by tire pressure only as is often supposed. Very heavy units, of 50 tons or more, may be effective at compressing rock fills.

Vibration is most effective in sand or gravel soils, but may increase the effectiveness of a roller in any soil. It is particularly effective at bringing excess moisture to the surface.

Trench fills and other small areas may be compacted by impact of air, hydraulic or gasoline hammers, or by vibrators. Gravel fills may be compacted by puddling—adding water until the soil is semi-liquid—then allowing it to dry and settle. An immersion (concrete) vibrator will speed drying of puddled fill, by bringing its water to the surface.

Moisture Content. The most critical factor in compaction of a soil is its moisture content, since it can be most thoroughly and conveniently compacted only if it contains just the right amount of water. This quantity is called the optimum moisture content. It must be sufficient to provide a lubricant to allow soil grains to slide on each other as they are pushed together, and not enough to form an incompressible cushion between any of them.

Figure 3-7C shows the relationship between moisture content and compaction for several classes of soil.

A soil that contains too much moisture is likely to become rubbery under a roller, pushing in waves ahead of and beside it, and springing back into its original position when it has passed. This is a very common condition in highway work. Problems arising from such excess moisture will be discussed later in this article.

Soil that is too dry may become loose or powdery under pressure, or may be firm but not as dense as it should be. It is standard practice to add water to each such layer by means of sprinkler trucks or trailers. Extra payment may be made for watering.

Stability. If a road or runway fill is not compacted it is likely to shrink and settle, injuring or destroying pavements and any other structures on it. Poor compaction allows the same behavior to a smaller extent.

Indications are beginning to appear that an embankment that is over-compacted may swell, producing similar unfavorable results. It appears that for some soils compaction should be stopped short of the maximum obtainable. What soils and what conditions

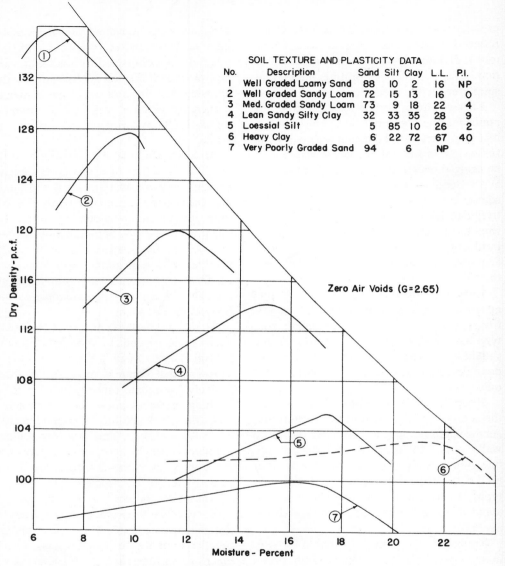

SOIL TEXTURE AND PLASTICITY DATA

No.	Description	Sand	Silt	Clay	L.L.	P.I.
1	Well Graded Loamy Sand	88	10	2	16	NP
2	Well Graded Sandy Loam	72	15	13	16	0
3	Med. Graded Sandy Loam	73	9	18	22	4
4	Lean Sandy Silty Clay	32	33	35	28	9
5	Loessial Silt	5	85	10	26	2
6	Heavy Clay	6	22	72	67	40
7	Very Poorly Graded Sand	94		6	NP	

Moisture-density relationships for seven soils compacted according to AASHO Method AASHO T99 (in part after "Public Roads").

Courtesy of Highway Research Board

Fig. 3-7C. Moisture-density relationships

are involved are now the subject of study.

Clay soils provide poor embankment material at best, as they will usually absorb moisture in wet seasons and swell. High, but not excessive, compaction reduces this tendency.

Variable compaction may cause the greatest damage. A whole road may settle a few inches with little noticeable harm. But subsidence of a narrow strip of trench fill across a firm embankment will ruin the road surface, as will also subsidence of a poorly compacted embankment on each side of a well compacted trench fill.

Unpaved road shoulders are more exposed to moisture than the pavement subgrade. If they absorb it, shoulder surfaces may rise noticeably above the pavement, interfering with drainage.

Layers. Many compaction specifications stress the thickness of layers to be rolled. The best a field crew can hope to do is come somewhere near the standard. There are some grading foremen who can keep exact control over spreading thickness and areas of a fleet of scrapers, and of the work patterns of rollers. But it is more usual for this work to be somewhat hit-or-miss. The important thing is that the layers should not be thicker than the compaction equipment on the job can handle.

Rock Fills. Layer specifications reach a point of absurdity in rock fills. Often they are subject to the same maximum thickness regulation of 6 to 12 inches as the soil. In practice, the thickness of rock lifts is often determined by the maximum size rock that the shovel can load.

A rock fill is usually stable if the rock is angular and is well mixed as to size. Such conditions are characteristic of both blasted and ripped rock.

Steel wheel and sheepsfoot rollers are useless on coarse, hard rock and will be damaged. Very heavy (50 tons and more) rubber tired rollers, and big hauling equipment provide about as much compaction as is possible or necessary.

The behavior of rock fills deserves more study than reports at hand indicate that it is getting.

Earth Dams. Discussion so far has been about problems of highway and similar fills. Dams do not ordinarily carry structures, but they must resist penetration by water, changes of shape, and any tendency to slump.

Fill used in dams is carefully selected for quality, with heavy soils used in the core and lighter ones on the slopes. Methods of compaction are about the same as for high-

ways, but methods and results are very carefully specified and checked.

Specifications. There are four basic ways in which compaction may be specified. They are:

Method only.
Method and result.
Suggested method and result.
Result only (Performance specification).

Specifying the method only is usually the most fair to the contractor. He is told that a certain number of passes with a specified type and weight of roller, on layers of a particular thickness, constitute acceptable compaction. He knows exactly what he has to do, and the equipment that he will need to do it.

Unfortunately, most such specifications are archaic, mentioning only steel wheel and/or sheepsfoot rollers, and giving neither the State nor the contractor the opportunity to take advantage of more recent and specialized compactors.

"Method only" relieves the contractor of responsibility for soil moisture content or other conditions beyond his control. If brought up to date with a wider range of usable equipment, and selection by the State of proper types for the soils in the contract area, it would provide a good contract basis.

In "Method and result" the contractor is told what equipment to use and how, and also the result in density that he must obtain. If the specified equipment will not produce the desired result, he is in trouble. He will need a special arrangement to either change the equipment or to get by with less compactions.

The usefulness of "Suggested method and result" may depend on the wording and intent of the specification. If the result must be obtained, there is no important difference from specifying result only. However, if the contractor is not held absolutely liable for the result if he follows the suggested method, this is a reasonable approach.

A performance specification (result only) allows the contractor to choose his own method, but requires a certain density. The State will have established by laboratory tests that this density is possible, but it may develop that it cannot be obtained in the field because of moisture or other conditions, or can be obtained only by far greater effort than was contemplated by either the State or the contractor. In many jobs it works out very well, but for reasons to be considered below it may be unfair to the contractor unless softened by irregular practices on the part of State engineers and inspectors. It should include a provision for re-negotiation.

Required density is usually specified as a percentage of maximum density produced in a particular test, such as 95% Modified Proctor.

Laboratory Tests. Tests are required to determine the optimum moisture content of a soil, the extent to which it can be compacted, and whether it is being sufficiently compacted on the job.

The Standard Proctor or AASHO (American Association of State Highway Officials) Standard Test T99-57, is made as follows:

A sample of soil is moistened and is then compacted in a standard mold 4 inches in diameter with a volume of 1/30 cu. ft. The soil is placed in 3 layers of approximately equal thickness, and each layer is subjected to 25 blows of a rammer with a striking face of 2″ diameter and a weight of 5½ lbs., falling freely a distance of 12″. This produces 12,400 ft. lb. of energy per cu. ft. of soil.

The sample, which contains a known volume, is then weighed and dried at 105° Centigrade for twenty minutes. After that, it is weighed again. The moisture content is computed by the difference between the wet and dry weight. The dry weight is recorded as well as the moisture content. The moisture content is recorded in percentage of weight of the entire dry sample.

By plotting the results of a series of these tests, using the same soil but with different moisture factors, a curve similar to that of Figure 3-7C will be produced. This curve shows the resulting dry weights obtained in a series of tests on a single sample compacted under a uniform method with varying amounts of moisture.

A slight variation exists between the

Fig. 3-7D. Compacting edge of a fill

Proctor and the AASHO Standard Test in that the free falling weight is given a slight initial impetus with the hand in the Proctor Test.

There are a number of modifications of this basic test, so any reference to an AASHO Modified Test should specify which one. In general, they employ a heavier hammer or plunger with a longer stroke than in the basic test. The use of modifications is becoming more general because of the higher densities required in construction.

Additional material on soil classificaton and compaction will be found in the Appendix. More detailed information will be found in bulletins published by the Transportation Research Board, the American Road Builders Association, and various manufacturers of compaction equipment.

Field Tests. The standard method of testing actual density of the embankment is to remove a measured sample for laboratory comparison. A sample may be of bulk to leave a hole 8 inches wide and 10 deep. Measurement is made by filling the hole with with a measured quantity of sand or water. The water is prevented from soaking away by lining the hole with a rubber balloon.

Sand may blow in the wind, or it may be shaken down by vibration of traffic. In either case density will appear to be less than it is. Balloons break. Both sand and water are heavy to carry. Measurements may be affected by the skill and judgment of the person taking the samples.

The Proctor needle pentrometer measures the resistance of soil to penetration by a heavy needle, by means of a spring and gauge in the push handle. It is used to compare the density of the embankment with a laboratory sample. It is very useful to check on how things are going, but should be reinforced with regular volume tests. Results in gravel are likely to be misleading.

The most satisfactory and accurate, but also the most expensive, method is nuclear testing. A device using radioactive material directs a beam of gamma or neutron rays into the ground, and counts those that are reflected back into the instrument. This count is compared with a laboratory sample. The higher the count, the greater the density.

This type of instrument is non-destructive, does not interfere with grading operations, does not depend on individual judgment, permits a large number of tests in a short time, and can be used for stony or frozen ground. On the other hand, the instruments are now both expensive and fragile, and an operator might be exposed to radiation if he handled one very carelessly.

A piezometer is a pipe gauge that is placed in the fill as it is built. It will indicate any movement of soil or soil water that would be likely to threaten to damage the embankment.

Too Much Moisture. The material taken from road cuts often contains too much moisture to meet compaction specifications in the fills.

In dry summer weather layers of fill may be stirred up with a disc harrow, moved from side to side by graders, or sprayed into the air by a rotary tiller with the hood up. When moisture is sufficiently reduced, the surface is leveled off and the layer is compacted. If it rains during this operation the result will be mud.

If some fill is wet and some is dry, the two types may be built up in alternating layers, that may or may not be mixed together.

In a wet season surface drying is impractical, and no dry fill may be found in any of the cuts. Under such conditions it may not be possible to meet density specifications.

This is a very serious problem that is almost completely overlooked in the literature on compaction. It is generally passed off with a paragraph or two on correction by substituting other material or by kiln drying.

Most highway designs succeed in coming

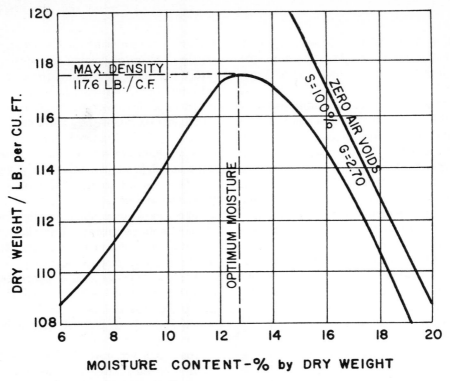

Courtesy of Essick Manufacturing Company

Fig. 3-7E. Moisture-density curve

somewhere near a balance of cut and fill, and interstate highway standards in rough country may involve moving of hundreds of thousands or even millions of yards within a few miles. Very often, if the highway cuts are wet possible borrow is wet also. Even if better material is available, wasting the excavated material and mining an equal amount of borrow would be likely to produce unacceptable landscape scars.

Kiln drying of such quantities would involve setting up huge and expensive plants, drastic changes in digging and hauling methods, and lengthening of hauls. It is questionable whether such operations could be included in a highway budget, or even whether improved pavement life would justify the cost.

The most critical point for the contractor is that in many states he is required by the terms of his contract to produce a certain density, that can be achieved only at or near optimum moisture content. If he cannot get the soil down to this content, it is unlikely that he can obtain the density.

Theoretically the contractor should get no pay for his embankment work, and would probably go bankrupt. Actually, this does not seem to occur. Field men in public works departments are more realistic than the design engineers, and are inclined to pass the work as long as sincere and intelligent efforts have been made to reach the specification. However, this puts both sides in the wrong if their actions become the subject of an official investigation.

Even when the work is accepted, the contractor may be caused serious extra expense in producing required density. He may have to change from one type of rolling equipment to another, taking severe losses on the original units. He may have to limit his earthmoving to certain seasons, perhaps late summer and early autumn, when bank moisture and/or rainfall are at a minimum, or when there are enough fair days for drying soil in the sun. These risks should be weighed when setting a price on the work.

The eventual solution of this problem will probably lie in the development of efficient mobile soil drying equipment. Such a unit will probably consist of a rotary tiller that will mix pulverized soil with air heated and dried by an oil or kerosene burner. A contract could provide for extra payment for use of soil driers when required.

MUD

Before proceeding with the details of various types of excavation, it is in order to consider some of the general problems. Mud is one of the most important of these.

NATURE OF MUD

Water Content. Mud is soil saturated with water to such an extent that it loses its structure and takes on some of the properties of a liquid. Even the driest soils contain some water in very thin films, and moderate additional amounts may give added firmness by acting as a binder. But when the quantity of contained water is sufficient to build up water films around the grains thick enough to serve as a lubricant so that they can move freely on each other, the soil becomes mud.

Particle Size. The quantity of water necessary to turn mineral soils into mud varies with the size, shape, and arrangement of the particles. Small grains have much less volume in proportion to the thickness of the water film they hold than have large ones, and therefore they form more fluid muds. Sharp angular grains have projections which penetrate the film and interlock, and large grains and pebbles develop high enough contact pressures to cut through the film. If there are enough fine particles in a mixed soil to prevent the coarse ones from touching, the mud will have the qualities of the fines.

A fine-textured (heavy) soil such as clay will also remain saturated much longer than a coarse one, as the spaces between the grains are so small that water moves through them very slowly.

Humus. Humus, or peat, which is decayed organic material, absorbs water somewhat as a sponge does, in large quantities, and holds it stubbornly against evaporation and drainage. When saturated, nearly pure humus, as found in some swamps and peat beds, resembles a jelly, fibrous or smooth in texture, and black or brown in color. It is the slipperiest and most treacherous of the muds. It dries very slowly, with shrinkage of 50 per cent or more, to a light, fluffy soil. When mixed with inorganic soils, as in topsoil and mucks, it greatly reduces their load bearing qualities and makes them muddier under wet conditions.

Quicksand. Quicksand is usually fine sand or silt through which water is moving upward with enough pressure to prevent the grains from settling into firm contact with each other. It provides practically no support for machinery unless its weight is distributed over a large area by platforms.

Making Mud. When undisturbed, inorganic soil is usually quite closely packed, with its grains fitted together closely, and often lightly cemented by mineral deposits. When it is dug up or pushed around, the grains are shaken away from each other into a loose structure. In this condition it can quickly absorb a large quantity of water and become a very soft mud. As it dries, the grains settle together so that less water is absorbed with each subsequent wetting. If it is compacted by rolling, tamping, or vibration before being soaked, it may become even more water resistant than in its original state.

When a firm, dry soil is covered with water, it gradually absorbs some of it and expands in volume, but it never becomes as soft as if disturbed before wetting.

If a firm, fine textured soil has a film of rainwater on its surface, and is passed over by a vehicle tire, the water will be forced between the surface particles and the resulting mud will be wiped off and pushed aside, leaving a new surface exposed to the next raindrops and wheel passage. Repetitions of this result in a slippery road, ruts, and mudholes.

Frost. When soil freezes, the expansion of ice crystals between the particles pushes them out of place. When the soil thaws, it is likely to become a slippery, structureless mud, often resembling toothpaste in consistency. It ordinarily firms up fairly quickly, particularly if vibrated by a heavy rain, but may persist for several months when upward seepage of water prevents settlement. In its extreme condition it will not support loads, and is made more dangerous by its occurrence in places that are normally firm, and under sod which bridges and hides it. Such places can be detected by sounding with a crowbar, and should be avoided or treated with the same precautions as soft swamps.

In northern winters, frost may stabilize a swamp so that it can be worked easily. Ice and frozen earth are liable to be variable in thickness and treacherous because of snow cover and the heat of decay of organic material, but any traveled route will gain in stability as long as freezing continues.

Freezing provides a pavement-like support for machinery, and may stabilize unfrozen mud as well. Ice lenses at the frost line absorb capillary moisture from below. If upward flow is restricted by coarse soil or an impervious formation, the mud will be dried and made firmer.

Mud from thawing is apt to render dirt road surfaces slippery, particularly on sunny slopes. Early mornings such roads are often frozen hard, and work must be done, then, in cloudy weather or at night. Tire chains are useful but may not be adequate.

Fabric Barrier. The Celanese Corporation has recently developed a strong, virtually rot-proof, nonwoven fabric of polypropylene and nylon fibers that may be used as a barrier to prevent loss of aggregate through mixing with mud in roads and drains, at the time of construction or afterward.

This Mirafi 140 fabric performs three functions in soil stabilization: separation, filtration, and reinforcement.

In building a road across a swamp, the strip is first cleared of all hard or sharp objects or bumps. Roots, sod, and sometimes soft brush are left undisturbed.

The fabric is unrolled on the surface, with overlapping at any joints. Aggregate, usually bank gravel, is truck-dumped and spread, preferably by light equipment.

There should be a foot or more of aggregate between the fabric and truck tires, to protect it. Any excess can be shaved off after compaction, which is preferably done with a light vibratory roller.

Sand. Clean sand is as troublesome as mud to two-wheel drive vehicles. It can be stabilized with pneumatic tired rollers and plenty of water, but the surface will loosen up as soon as it dries. Tires spin and dig down into it rapidly with a jerking motion that is very damaging to the drive mechanism. All-wheel drive vehicles ordinarily have less difficulty with it, but it consumes considerable power. The general problem is one of getting traction without digging in, but there is no danger of simply sinking as in mud. Partially deflating the tires may help; smooth tires will do better than those with tread as they will not dig down as readily. Mats of brush, wire, grass, or a thin layer of dirt may suffice to give traction.

Tracked vehicles can travel on sand without difficulty, but if equipped with grousers, care must be taken about pulling heavy loads that may cause tracks to spin, as they will then hang up quickly. The silica which makes up most sand is very wearing to the track parts, particularly when particles are angular.

Work Delays. Mud is an impediment to work in many ways. Deep mud causes equipment to bog down and to become useless until pulled out, and a film of mud may render firm footing dangerously slippery. Mud sticks to shovel buckets and truck bodies instead of dumping, and builds up in chains and tracks until they jam. It holds objects lying on it by powerful suction so that they become difficult or impossible to lift. When frozen it can lock together and immobilize the most powerful machines. And it is much heavier than the same amount of dry soil.

The most severe mud conditions are encountered in swamps because of permanently saturated conditions and the usually high organic content of the soil, but because the contractor expects this and is prepared, fewer difficulties arise than when mud appears on drier jobs.

EQUIPMENT FOR MUD

Mud trouble can be reduced by using proper equipment. In general, crawlers are preferable to wheels; tracks should be the longest and widest obtainable; tires should be big, soft, and cleated; and units should be the smallest that will do the work. All-wheel drive is desirable for trucks.

The ability of a machine to stay on top of soft ground is affected by its ground pressure, which is usually measured in terms of pounds weight on each square inch of ground contact; shear, which is the load on the edge of the track or tire; and total weight.

Ground pressure is the most important factor in loose soils such as sand or dust. Shear is most important when a soft soil is protected by a harder crust or sod. It is increased when the machine is tipped, and

when it pushes a load. Total weight affects deep mud which may creep or flow from under the machine.

Grousers, cleats and tire chains cut and churn up their footing but are necessary to get a grip on slippery surfaces.

Some wheel tractors can be fitted with temporary metal and rubber tracks which enable them to work in fairly soft places.

Crawlers may have oak planks or four-by-fours bolted to their pads so as to project several feet on the outside. These will stand up well if kept on soft ground and give excellent flotation. They cannot be used with a dozer that has outside push arms.

Special vehicles, such as the military weasel and the swamp buggies used for exploration by oil crews, are very useful in supplying fuel and other essentials to machinery working in swamps.

TEMPORARY ROADS

Wheeled Equipment. Wheeled equipment is best kept out of swamps unless they are frozen, dried out artificially or by drought, or have roads built into them. The minimum road is a strip in which the soft spots are stuffed with brush or bridged with planks. If poles are used, they should be closely fastened so that they cannot work apart and let wheels down between. Whenever possible surface poles should be at right angles to the direction of travel.

Pole Tracks. Poles may also be used as tracks to be straddled by dual wheels. They should be straight, free of stubs or sharp projections which might cut tires, and should be large enough so that they cannot pass between the tires toward the hub, and small enough so that they will not slip sideways out of the groove between the tires. The poles should be overlapped at their ends so that the wheels will not be left without support while passing from one to another.

Front wheels may be placed on skids or

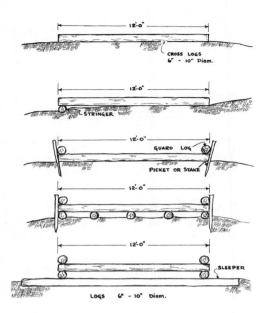

Fig. 3-8. Corduroy road cross sections

The upper surface of cross logs should be smoothed by removal of stubs and bumps, and perhaps planing down with adze, ax, or ripsaw.

Guard logs are desirable to prevent vehicles from sliding off, to reinforce the road structure, and to retain any surfacing which may be added. Picket stakes may be placed to hold guards in position, and to bind cross logs to the outer stringers. Edges may be bound together by heavy wire, by cable, spikes, or lag bolts. Such fastenings are important when the road surface is well above the ground, where the mud will not act to hold it in place.

Drainage ditches may be dug on one or both sides to remove standing water. These should be kept from three to ten feet away so that mud will not flow into them from under the logs. In very thin mud their use is inadvisable.

When a drainage channel or small stream must be crossed, the stringers can be increased in number and strength and set on sleepers on each side of the channel so as to serve as a log bridge. If clearance is doubtful, metal pipe may be used to carry the water under the stringers.

Corduroy roads are quick and fairly easy to construct if logs are on the site, and usually are very strong. However, their surfaces are extremely rough and it is advisable to cover them with gravel or other surfacing if they are to have much use. This saves damage to both machinery and road.

A corduroy road made of oak, cypress, or other strong and rot resistant wood may have a very long life. Some soft woods, such as poplar, will rot out in two or three years. If mixed species must be used, the inferior ones will give their best service as stringers or sleepers.

Saplings or brush may be wired into tight bundles and used instead of logs for light corduroy.

Plank Roads. Plank road constructions are shown in Figure 3-9. These are usually

runners and chained down. If the mud is not very deep, they may be left to make their own way.

This procedure is for short emergency moves only.

Corduroy. A corduroy road can be built to support heavy machinery on very soft ground. It consists of logs or half logs laid across the traveled way, touching each other. They may be laid directly on the ground, on one or more stringers running lengthwise, or on both stringers and longer cross logs, called sleepers. Several constructions are shown in Figure 3-8.

A minimum width of 12 feet is recommended for one way traffic, although it might be possible to get by with ten. Curves should be two to four feet wider. Cross logs may be extended beyond the road edge for additional stability.

Logs six to ten inches in diameter are generally strong enough for the work, without being excessively heavy. Thinner ones may be used as stringers. Heavier sizes may be split for cross logs, or partly buried for stringers.

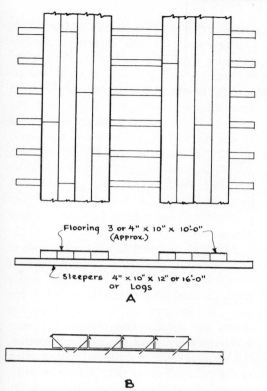

Fig. 3-9. Plank road

Heavy vehicles are seven, eight, or more feet wide, and cannot safely be used on a single row of standard ties. Several systems are used to obtain greater width, but none of them is entirely satisfactory. A center stringer, Figure 3-10A (A), of two logs securely fastened together permits using two rows of ties. Alternating side ties on a single center stringer makes a surface too sketchy for normal use, but will serve as good support for plank treads. The slightly staggered construction shown in (B) provides a wider surface than a straight row.

Stone Walls. Sometimes a stone wall extending into a swamp may be profitably used as a base for a road. It may be wrecked and scattered over the desired width by hand labor or by a bulldozer. The dozer can advance a limited distance, pushing and twisting the top stones to each side, and walking on them until the going gets too rough. It should then back out, some fill should be dumped, and the dozer push it into the hollows so that it can advance farther. The large stones, and the

more expensive than corduroy, but are easier to lay and provide a smoother surface. Rain or mud may make them very slippery so that sanding will be required.

Nails should be driven diagonally from the sides of the tread planks, as in (B), to prevent them from working up and puncturing tires. Even with this precaution, frequent inspections for nails and splinters should be made. If the wood is hard, nails should be soaped or greased before driving. Splitting may be reduced by making thin pilot holes with an electric drill.

Railroad Ties. Used railroad ties are often available in large quantities at cordwood prices, or less. They can be used for roads and platforms, as well as for supports, blocks, and braces. They are usually of hardwood, 8½ feet long, and about 7″ by 9″. Longer ones, used under switches, are occasionally available.

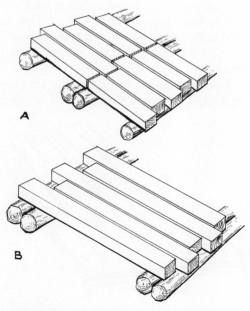

Fig. 3-10A. Railroad ties for truck road

age of the structure, assure stability and large amounts of fill are saved.

Surface Protection. A wet soil with good load bearing qualities which might be readily churned into mud, may be protected by a few inches of gravel, spread before traffic uses it. Two feet of broken rock with a skin of gravel may carry heavy traffic for a while over almost any mud. Brush mats or pole corduroy under the rock give added stability. Between these extremes is every type of condition and cure.

Fill for mud coverage should be as dry as possible and possess good packing qualities. It should be put on in sufficient thickness so that truck tires will not reach through to mix it with the mud. It is strengthened by compacting with a roller, jeep, or empty truck before carrying loaded trucks. A layer of fine textured, hard packing soil topped with gravel is often satisfactory, but clean broken rock in coarse sizes up to one half the fill thickness, is longest lasting.

Flat stones should be placed on edge, whenever possible, so that they will not shift under load. Natural drainage must not be blocked. Iron or steel culvert pipe is most satisfactory on soft bottoms, as it will keep its position and resist separation and breakage.

Swamp Surfaces. Most swamps are covered with vegetation that gives them some surface stability. Wet swamp sod in which a man's feet will sink slightly, ordinarily will support a light crawler machine moving steadily across it in a straight line or gradual curves. Crossing should be tried with caution, particularly if the unit has narrow tracks, or high grousers which tend to cut the sod. Sharp turns, if necessary, should be made in the firmest or best matted sections, or extra support should be provided. Bushes and close growing saplings which can be walked down by the machine, without cutting, provide excellent natural support.

Although a shovel can be safely walked on quite soft ground, it cannot stand or work on it. If standing, its weight slowly displaces mud beneath it, and as that creeps away, the sod or crust left without support shears and breaks. This process is greatly accelerated by working, as the vibration, the variable load, and the twisting reaction to the swing all cause the machine to settle, particularly at the end where the loads are picked up.

PLATFORMS

Platforms, also known as pontoons or mats, are wood supports for machinery working on soft ground. They are most often used under draglines or other revolving shovels.

Their primary function is to spread the weight of the machine over a wide area. They also serve to provide a relatively clean, dry place from which to work, and will act as temporary bridges over ditches and holes.

Construction is not standardized. They are generally made by the contractor who uses them according to local methods.

In general, mats for half yard shovels are twelve to fourteen feet wide, and for three quarter yard, fourteen to eighteen feet. The length should be about half the overall length of the crawlers, except that they should be shorter if necessary to fit

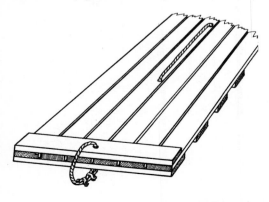

Fig. 3-10B. Open platform for light shovels

into the truck body that will have to carry them. If they are made for two sizes of shovel, they must be strong enough for the large one, and short so as not to be too heavy for the smaller machine.

Oak is quite generally used because of its strength and resistance to decay, and elm because of its toughness. These hardwoods have the disadvantage that they cannot be nailed readily, so it is necessary to drill and bolt. Spruce, although soft, is tough and resists splitting, can be nailed, and is lighter than the hardwoods.

Green lumber can be used. It is tougher and heavier than after curing.

Figure 3-10B shows a sample of very light construction. It is easily handled because it has a minimum amount of material and the slots in its surface help to free it from suction.

It has the disadvantage of allowing very soft muds to work up to the surface, keeping it muddy and slippery and reducing its carrying power. It will also break up rather readily if used over boulders, humps, or heavy roots, or if one side rests on a ridge which does not extend under the shovel tracks. It will not support trucks or other heavy wheel machinery.

The loops shown at each end may be of any size cable from ⅜″ up, fastened with two clamps. A chain is hooked to both of these and to the shovel bucket when the mats are moved. Three eighths chain is satisfactory for ordinary service with light shovels.

It is a good plan to make up a special chain about two feet longer than the platforms are wide. Round hooks should be used on the ends, preferably the lock-on type. The center should have a ring and a special hook large enough to hook on the bucket, or on the drag chain links next to the bucket.

A mat may be lifted and dragged by one side loop, but this is not as convenient as using two.

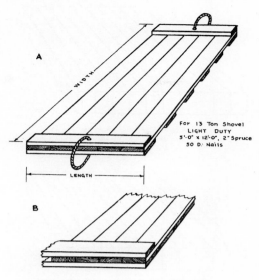

Fig. 3-10C. Butt joint platform

The center cable can be used for picking the mat up with bucket teeth, without chaining. This is difficult with a dragline, but fairly easy with a pull shovel or a dipper stick.

The side rails, above the deck, prevent the shovel from sliding off the side. They get in the way occasionally but on the whole are well worth having.

When nails are used, they should be long enough to hammer through the wood. The projecting point should be hammered over flat.

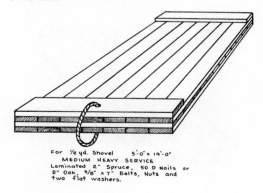

Fig. 3-10D. Laminated platform

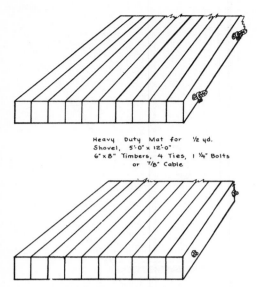

Heavy Duty Mat for ½ yd.
Shovel, 5'-0" x 12'-0"
6"x 8" Timbers, 4 Ties, 1 ¼" Bolts
or ⅞" Cable

Fig. 3-10E. Timber platform, heavy duty

Bolts should have washers on both ends. Heads should be countersunk into the deck.

The platforms in Figure 3-10C are of the same type except that the deck boards are laid together without slots, and additional tie boards are used. This is suitable for service on mud with a soft surface and is somewhat stronger, although subject to the same general limitations as the first type.

The laminated platform in Figure 3-10D is suitable for any but the roughest service and will support light trucks. It is heavier to handle.

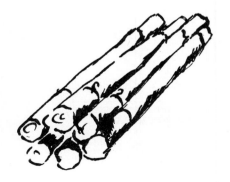

Fig. 3-10F. Rolled-up pole platform

The heavy duty type in Figure 3-10E will take any abuse a half yard shovel can give it, for a while. It is suitable for use on rough or frozen ground, and among stumps and boulders, but it is too heavy to be convenient for ordinary soft ground.

These designs can be varied almost any way as long as they are large enough to support the shovel, light enough to be handled, and strong enough to stay together. Sometimes single timbers are used with a ring at one end. Saplings may be tied into narrow mats which can be rolled up when placed in a truck.

Handling. Platforms are usually trucked as near to the job as possible, then dragged or carried by a loader or the shovel.

To walk a shovel across an open swamp on platforms, pick up two or more of them and lay them on the mud, as shown in Figure 3-11 (A). Move the shovel to the front of these, then pick up the remaining platforms and lay them ahead of the shovel as in (B). The minimum number to be used is three—two to stand on and one to move. Four are safer and easier to use, and the maximum is the number the shovel can reach when they are laid in a line behind it. Most draglines use three or four on small jobs, and four to six on large ones.

The shovel then walks to the front of the platforms, (C), and picks up those behind it, starting at the rearmost, and lays them in a line in front of itself. In (D), the platforms are laid in a curve to avoid a rock. (E) to (G) show patterns for laying them in sharp curves. These should be avoided where possible, as the turning of the shovel has a destructive grinding action, and overlapping the platforms causes severe strains, particularly if they are equipped with side rails.

Shovel Rigs. Techniques of working from platforms vary with the type of shovel attachment being used. The most efficient swamp worker is the dragline. Its long reach enables it to keep well away

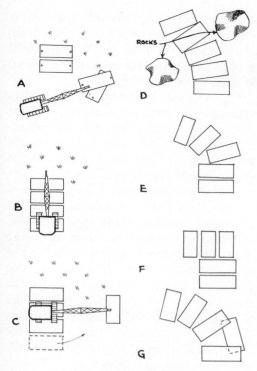

Fig. 3-11. Platforms

from the hole it is digging, and to pile spoil far enough away to reduce slumping back into the pit or against the shovel. The sliding action of the bucket during digging and hoisting reduces trouble with suction.

A dragline or a hoe stands at the back of its line of platforms and digs at the rear or sides. When its work from one position is finished, it walks ahead one or more platform widths, then picks up the platforms behind it, and swings and places them in the front; repeating this process as often as a move is required.

Hydraulic full revolving hoes are competitive with draglines. They can dig much harder material and to greater depths, and have more precise control in both digging and dumping. They are much better able to get themselves out of trouble, and to work without platforms. But their reach is much shorter in proportion to weight and bucket size. In swamp work, long reach is highly

desirable, and weight increases problems, cost, and risks.

Cable hoes are not as well suited to mud excavation. Digging radius is short, and dumping radius is worse. It is difficult to keep thin mud in the bucket. But like the hydraulic machine, it can dig hard ground, and save labor by picking up platforms by sliding its teeth under cross chains.

The tractor mounted hoe cannot work on platforms unless another machine places them. Digging and dumping reach are poor. But if the tractor also carries a loader, with skillful operation it can work in (and get out of) very soft areas without supports.

A clamshell has almost as much reach as a dragline, and can work without dragging debris, such as roots, stumps, and boulders, up against the platforms and itself. However, it has the severe handicap of pulling the bucket straight up, which in sticky mud requires overcoming suction more resistant than the weight of the bucket and load. Digging is best done off the rear or sides of the platforms, but the ground ahead can be graded before placing the platforms.

Dipper sticks are seldom seen in swamps. They can do shallow digging and uprooting of stumps behind and at the sides of the platforms, but any serious digging must be done at the front, and there is limited to fairly narrow ditching in firm mud. The spoil must be dumped far enough to the side so that platforms can be placed across the ditch. Even with extra wide platforms, the weight is liable to cause the sides of the ditch to cave or flow in, and the footing for the platforms is not secure.

The spaces between platforms will vary with the nature of the footing. For very soft conditions, they should be placed in contact with chains fastening them together, or even laid in two layers, as in Figure 3-12 (A).

For ordinarily soft or suspicious ground, the widest spacing allowed should be that

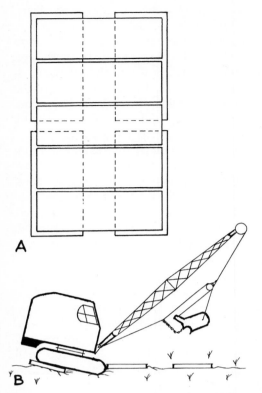

Fig. 3-12. Overlapping and spacing platforms

changes most of the forward thrust of the shovel into a downward one against the platform, and the plank anchor should hold it in place. One of these should be placed in front of each track.

On a solid platform, wood wedges chained to the platform, as shown in Figure 3-13 (A), are effective. For side protection, plank rails, permanently bolted to the side edges, or held there with pins, as shown in (B), are usually adequate.

Chaining to Platforms. For extreme conditions, the shovel may be chained to the platforms. There are many ways to do this, of which one sample is shown in Figure 3-14. The platforms have a cable loop at each corner, by some of which the two platforms are chained together. To the outer corners of the truck frames are welded brackets holding rings or hooks. Chains are fastened to these and to the corresponding outer corner loops of the platforms, and are drawn up taut with chain tighteners in the rear. The tightness is important, as the momentum of a sliding shovel will break chains or tear out loops which would hold against a direct pull.

All accessories, such as blocks, rails, chains, and tighteners that are to be used

which will enable the shovel to reach the next platform before its center of gravity reaches the edge of the one it is on, so that the shovel will not tip forward, as in (B), so much that the track will push the next platform instead of climbing on it.

Blocking. Normally, a shovel can be worked on platforms without blocking. There is always a risk, however, that a dragline, or particularly a pull shovel, will unexpectedly hook into something solid and drag itself off the platform before the operator realizes what has happened. If the platforms are slippery with mud or ice, this danger is greatly increased, and the machine is also likely to slide sideways in reaction from swinging.

If the platform is the spaced plank type, an effective block can be made of a wood wedge fastened to a stub of plank that fits between the platform planks. The wedge

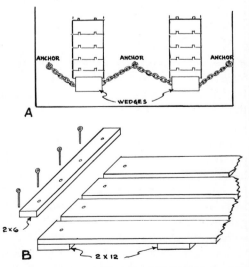

Fig. 3-13. Blocks and side rails

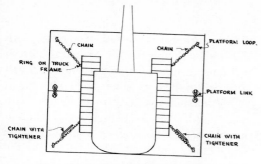

Fig. 3-14. Chaining to platforms

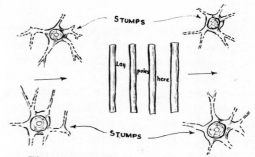

Fig. 3-15. Support by roots and poles

in mud, should be kept painted some brilliant color (not green). They are very easily lost and dug under, particularly if some emergency interrupts the routine of their use, and the bright color will greatly increase chances of salvage.

Platforms can be used efficiently only in places where the shovel has room to swing. If trees prevent swinging they must be cut, or other methods used to support the shovel.

Suction. Lifting platforms or other objects from mud is greatly impeded by atmospheric pressure (14.7 pounds per square inch), which is felt as suction.

A platform 4½ x 12 feet carries an atmospheric pressure of over fifty tons. When air can flow under it freely, the pressure is upward as well as downward, and is not felt in lifting it. If air is sealed from the bottom by mud, then the lift of the shovel merely serves to reduce the weight of atmosphere and platform on the mud beneath, causing mud at the edges which carries the full atmospheric load to squeeze in under it. A thin mud will flow quickly, a thick tenacious one may stretch but not flow and keep the air locked away from the bottom.

If the suction is too strong to break by a direct lift on the whole platform, the pull should be concentrated on one side or one corner. If the mud seal is broken at one spot, it should shear off from the whole under surface. The platform might also be lifted and dragged at the same time. Suction presents comparatively little resistance to

sliding, and this motion greatly increases the shearing effect on the mud seal. Unfortunately, a dragline is apt to make a pile of mud and debris which would prevent the platform from sliding forward, and the lattice boom is apt to twist or collapse if subjected to heavy side strain.

A method of releasing is to catch the forward edge with the bucket teeth and hoist while pulling back enough to keep hooked.

Root Mats. Trees growing closely in swamps usually form a mat of interlocked roots that will support crawler equipment, but will give it a rough trip. Openings may occur where the roots are lacking, which may be crossed by means of green saplings laid across the path of the machine, as indicated in Figure 3-15. If it is necessary to remove trees to give the shovel space to get through, they should be cut as close to the ground as possible, as the shovel might sink while going over the stump and get hung up on it. Logs laid on each side of the stumps across the line of travel are protection against this.

In a swamp studded with stumps, or rock, platforms break up very readily. The spaced plank type, which is generally preferred because of light weight and reduced suction, will break if used across a stump or heavy buttress root, and the heavier types will strain and splinter. Some operators prefer to pull the stumps before walking over the spot, thus exchanging the platform-breaking obstacles for a wet and uncertain footing.

It is dangerous to move a machine onto platforms which are placed with one side on firm ground and the other side in loose mud. The mud side may sink enough to cause the shovel to slide to that end and tip over. The soft side may be braced with platforms or logs, or the firm side can be ripped up, before placing the platforms.

Poles. A shovel may be walked and worked on fairly soft footing by the use of saplings instead of platforms. These should be of firm wood, preferably green, two to ten inches in diameter, and long enough to project two or more feet beyond the tracks on each side. They should be laid across the path of the shovel, widely spaced on fairly safe ground, a foot or two apart on soft. An extra precaution is to place outrigger poles under their ends, parallel to and outside of the machine's path. When the shovel has passed over the poles, they may be retrieved and used again, but the mortality rate is usually high, particularly among soft woods.

When the shovel is working, it will often be found that a pole or two under the front, or where the lifting of the load is done, will suffice to support it. With worse conditions, more cross poles and outrigger poles should be used.

If long poles are not available, short ones may be used, centered under each track, but they are not nearly as satisfactory. There is a danger that they might tip under the shovel and jam into the machinery.

WET DIGGING

In any extensive wet digging, draining or surface pumping is liable to leave a sheet of at least a few inches of water on the bottom. This is a convenience in establishing a flat bottom grade, but, unless special precautions are taken, large quantities of water will be dug out along with the soil.

This water may run directly back into the hole, in which case the time and fuel used in lifting it is wasted. If it mixes with soil, either in the bucket or in the pile, it will tend to liquefy it so that the spoil will not stand in the high, steep sided piles which afford maximum dragline production with each stand. Piles with high water content are subject to flowing or slumping while being built, and may move back into the excavation afterward.

The spoil may come out as soft mud even if no separate water is included, but mixing of water will make it more sloppy.

Compartments. Shallow bottom water may be largely kept out of the bucket by compartment digging, as illustrated in Figure 3-16. A ridge is left between the digging and the water until bottom grade is reached. The ridge is then dug out and water allowed to flow in. Some mud will be washed onto the digging surface by inflowing water and by the removal of the ridge. Compartments may be large or small.

This method gives water-free digging for the bulk of the excavation, and can be used, with diminishing efficiency, in increasing depths of water, but not when the whole digging area is under water, as in deepening an undrained pond.

If little surface water is present, but ground water drains into the excavation rapidly enough to be a nuisance, digging can be in two or more compartments. First, the whole area is dug in layers until it gets sloppy, after which the digging is concentrated in one spot until most of the water flows into the hole made. An adjoining area, separated by a ridge, is then dug deeper and the ridge cut.

The water will now flow into the deeper hole, leaving the first one nearly dry and ready to be deepened in its turn. This alternation of digging spots can be continued to the bottom of the cut. If the area is too large to be dug in this manner, the water can be concentrated in gouges dug behind the regular layer cut.

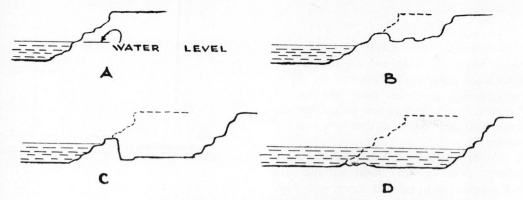

Fig. 3-16. Compartment digging

Spoil. If enough dry or stiff dirt can be dug to build a dike along the excavation edge of the proposed spoil pile, as in Figure 3-17, it will prevent mud behind it from flowing into the hole. There is usually ample space behind the pile so that the mud which moves in that direction makes place for more in the pile, and increases the amount of digging that can be done from one stand. The effectiveness of the dike will of course depend on the quantity and type of dirt available for it.

If the dirt is being loaded in trucks, an effort should be made to get dry material on the bottom to reduce trouble with the load sticking in the body.

Cable Excavators. Most excavating in swamps and mud is done by revolving shovels with dragline attachments. However, when the dug material is to be moved a great distance, or the amount is very large, cable excavators may be used. Since they can be anchored on one side of the swamp and operated from the other, they eliminate many of the difficulties and dangers of swamp digging. However, the work involved in setting them up and moving them, and the need for other equipment to remove the spoil from the tower foot, discourages their use except for large or long term operations, or where the digging cannot be done by any other available equipment.

Bulldozers. Bulldozers are not suited to mud excavation since they cannot work

efficiently on artificial supports. However, they can skim shallow layers of mud off hardpan and dig cautiously in muds compact enough to give them some traction. In skimming work, it is usually best to start at one side of the mud area and make a pass removing the mud cleanly. Each successive pass should be to full depth and cut just enough into the side of the muck to fill the blade, without allowing it to slop off the other side of the blade into the cleared area. The mud should be pushed far back as the dozer cannot climb up on it to make high piles, and there is danger of its flowing back into the hole.

Wide gauge dozers are preferred, as they usually have wider track shoes for greater flotation, their weight is spread over a larger area, and they turn more readily on poor footing.

Cleats. Smooth or semi-grouser tracks are unsatisfactory as they have little traction on wet slippery surfaces. In general, tracks

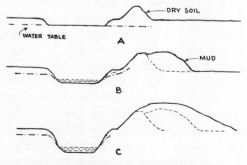

Fig. 3-17. Dike method of piling mud

3-27

with new and high grousers are better than those which are worn and rounded. However, such grousers on loose soil will dig in very rapidly, and many muds will pack in between them to make a new smooth surface even with their edges so that they become ineffective.

The best solution for straight mud work where rocks or very hard subsoils are not involved is to use flat shoes, with high grousers bolted on every fourth to sixth one. This should put two cleats on each track under the machine at all times, and will give a good grip which will not clog.

A similar effect can be obtained by building up some of the cleats on a set of standard track shoes, or replacing some shoes of a worn set with new ones.

A machine with cleats of variable lengths is likely to be almost incapable of working or traveling on hard ground.

Dragline Road Cut. A permanent road should not be built over unstable mud unless absolutely necessary. Such muds, particularly if rich in organic matter, will gradually be compressed and displaced by the weight of the road, allowing it to sink unevenly. The mud should therefore be removed down to firm bottom and replaced with clean fill if possible. If it is too deep, or otherwise too difficult to move, measures must be taken to stabilize it.

Shallow mud may be removed by a dragline working just ahead of the fill, and piling it to the side. The rate at which the mud pushes back into the hole determines how far ahead the dragline can be operated. A foot or two of mud liquefied by the mixing action of the shovel bucket is usually pushed out by the weight of the fill, but sliding in of the banks must be avoided.

Surface water should be diverted and sufficient pumps used to keep the hole fairly dry. This enables the dragline operator to see to clean the mud out thoroughly without unnecessary digging of firm soil, and will prevent the fill from being turned

SPECIFICATIONS FOR SINGLE-LINE COLUMN LOADS

CARTRIDGES PER HOLE	DEPTH TO TOP OF CHARGE	DEPTH OF DITCH	TOP WIDTH OF DITCH	DISTANCE BETWEEN HOLES	POUNDS PER 100'
½	6-8"	1½-2'	4-5'	12"	25
1	6-12"	2½-3'	6'	15"	40
2	6-12"	3-3½'	8'	18"	67
3	6-12"	4-4½'	10'	21"	86
4	6-12"	5-5½'	13'	24"	100
5	6-12"	6-6½'	16'	24"	125

CROSS-SECTION LOADING METHOD

CARTRIDGES PER HOLE	DISTANCE BETWEEN HOLES	DISTANCE BETWEEN ROWS	DEPTH OF DITCH
1	15"	30"	2½-3'
2	18"	36"	3-3½'
3	21"	42"	4-4½'
4	24"	48"	5-5½'
5	24"	48"	6-6½'

SPECIFICATIONS FOR POST-HOLE LOADING

No. of sticks (1¼ x 8") per hole	6	10	20	30	50	100
No. of lbs. per hole	3	5	10	15	25	50
Distance between holes-ft.	3	3½	4	4½	5	6
Depth of ditch "A"-ft.	4	5	6	7	8½	12
Bottom width of ditch "A"-ft.	4	5	6	7	8½	12
Top width of ditch "3A"-ft.	12	15	18	21	25½	36
Depth of load (⅔ of "A")-ft.	2⅔	3⅓	4	4⅔	5⅔	8
Diameter of post-hole-in.	4	4	6	6	8	8
Dynamite per 100 ft.-lbs.	100	142½	250	333	500	833
Material moved per 100ft.-cu.yds.	118	185	266	363	533	1,067

Fig. 3-18. Tables for mud blasting

to mud as it slides down the front of the slope. If the fill is too wide for the dragline to clear, it may be built out in sections.

MUD BLASTING FOR FILL SETTLEMENT

If the mud is too deep for available draglines to handle, or if it slumps into the excavation before fill can be placed, or if there is no adequate equipment to keep the water out, blasting may be resorted to.

Mud is easily blasted out of a limited area and spreads in a thin film over the landscape, leaving no heaps to dispose of. If a nitroglycerin or other sensitive dynamite is used, the concussion from one explosion may detonate other charges of dynamite in the mud nearby, without the necessity of using additional caps. The process, called propagation, greatly simplifies mud blasting.

There are three principal techniques: One is to blast a ditch along the right of way then dump fill in it. Another is to

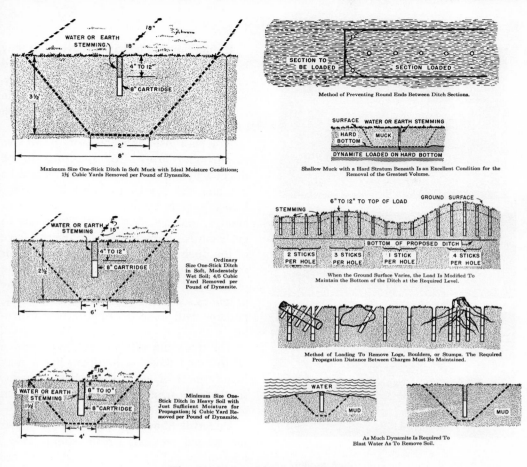

Fig. 3-19. Loading for ditch blast

build the road, or a section of it, on top of the swamp and blast the muck from under it. The third, toe-shooting, consists of making a big heap of fill and blasting the mud ahead and to the sides from under that.

Ditch Shooting. The tables in Figure 3-18 give average loading requirements for various types of open ditch blasts. Figure 3-19 shows how to load for rather narrow ditches, which are used chiefly for drainage, or to make possible placing of an earth dike to cut off movement of water or mud. It will be noted that the nature of the mud is very important in determining the size ditch which will be obtained from a given charge.

It should also be remembered that water counts as soil in loading calculations.

Blasting techniques for wider ditches, which are used to remove muck to secure a firmer base for road fills, are indicated in

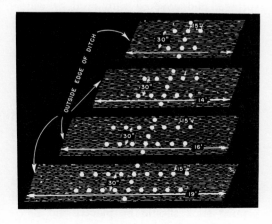

Fig. 3-20. Cross-section loading

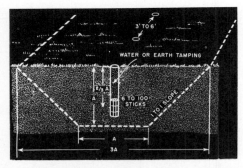

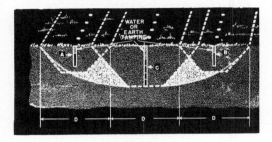

Fig. 3-21. Post hole and relief loading

Figures 3-20 and 3-21, which show the cross section and relief methods.

The charges should be 50% or 60% straight dynamite, at least 4″ below the surface, and covered with mud or water. The blasting cap should be in a stick of dynamite at or near the center of the group to be blasted, and with the loads and distances shown in the illustrations, should detonate them all.

However, if the structure of the muck changes between holes, it may shear, and direct the force of the concussion upward or to the side, and charges beyond that point will not explode. In this case, one of the remaining charges should be lifted and primed with a cap, or an additional primed stick should be placed and detonated.

If many reprimings are necessary, it may be better to put a cap in each hole and explode them simultaneously by electricity. As a great many caps are generally used in a single blast of this type, a powerful jolt of electricity is needed.

It is not usually necessary to remove stumps and brush from an area to be ditched by explosives, but trees should be cut and the trunks removed if possible. Extra charges should be placed under stumps, and an effort should be made to get them to throw toward the nearest edge of the ditch. If the stumps are so heavy or so numerous that they prevent accurate ditch work, they may be removed first by draglines, winches, or other machinery, perhaps with some blasting to loosen them up.

The dynamite will probably not remove all the mud down to hardpan, but it may liquefy most or all of that which remains, so that the fresh fill will sink through it to firm bottom.

Underfill Shooting. Figure 3-22 shows the second method. Brush, stumps, logs, and other debris are removed from the right of way, and any heavy sod or brush broken up by machinery or light blasting. Fill is placed in an amount calculated to reach from firm bottom up to the desired road grade. Charges are placed in the mud in casings driven through the fill. Shooting is by Primacord, or by caps in each hole, rather than by propagation.

The mud explosion is confined by the mass of fill above, and should blow out the sides, driving out most of the mud and liquefying the remainder so that the fill will sink to hardpan.

If the fill is clay or other compact material that might tend to bridge over the blasted cavity instead of settling in it; or if the weight is not considered to be sufficient to resist the explosion, deep ditches may be blasted on each side of the fill after it has been placed and the charge

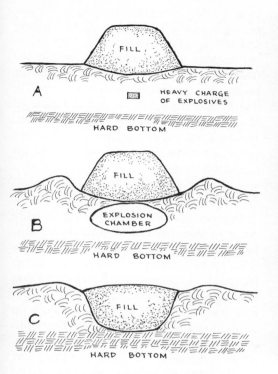

Fig. 3-22. Bottom shooting

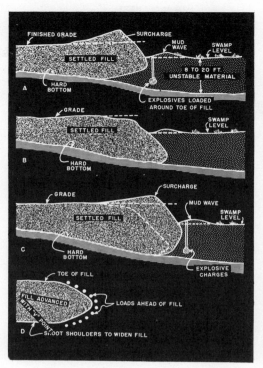

Fig. 3-23. Toe shooting

set off under it afterward. Blasting these relief ditches makes it much easier for the explosion to drive the mud from under the fill, but it also puts a thick layer of muck on top of it, which must be bladed off.

The charges under the fill must be so placed that there will be no danger of them going off by propagation. 30 or 40 per cent gelatin is recommended.

Toe Shooting. Toe shooting is illustrated in Figure 3-23. This is the second method —fill, then blast—done in such short sections that the muck will be driven out to the front as well as to the sides. It has the advantage of allowing the blaster to have frequent checks on the efficiency of his work, so that techniques of loading or filling may be altered readily. A disadvantage is that the muck blown ahead sometimes piles up in a large hill or wave of mud, making continually larger charges necessary to move it. This tendency may be re-

duced by making the fill with a V-shaped front so that most of the material will be thrown to the sides. The mud wave may be blasted away in separate operations, but part of it will then have to be cleaned off the fill.

Toe shooting is an excellent method to widen fills, the extra material being piled on the sides and a single line of charges set off in the mud underneath.

Placing Explosives. Shallow placement of explosives in mud may be made by pushing them down with a stick, or first punching a hole with a crowbar, then using the stick. For deeper work the apparatus shown in Figure 3-24 is effective. A plain piece of iron pipe of the necessary length is used for a casing, and a round wood pole or a plugged iron pipe making a loose fit serves as a core. The core is set in an iron handle larger than the casing and long enough for a good two-handed grip.

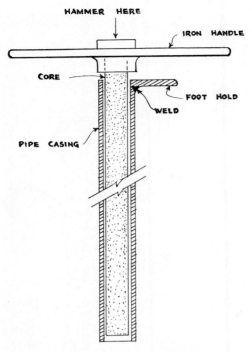

HAMMER HERE

IRON HANDLE

CORE

FOOT HOLD

WELD

PIPE CASING

Fig. 3-24. Cartridge-placing tool

The core is placed in the casing and the two pushed or hammered down into the mud by means of the handle. The core is withdrawn, the casing being held down if necessary by a foot on a bump welded on the side. The charge may then be dropped down in the casing, or pushed down by a long wood tamping stick, the same diameter as the core. This stick is used to hold the charge down while the casing is pulled out. Difficulties experienced will increase with the size and length of the pipe, and it may be necessary to use chain jacks or hoists to pull the casing.

Space permits only this brief summary of mud blasting techniques. Further information may be obtained in other parts of this volume, and from various books and bulletins dealing with explosives, but it is advisable to have an experienced mud blaster on any important job.

Results. A fault of all mud removal techniques mentioned is that they require placing a considerable depth of fill in a single layer so that it cannot be properly compacted; or the fill is shaken up by blasting and sinking after being placed. This is liable to cause uneven settlement, and in addition there is the danger of pockets or layers of muck getting included in the fill, due to miscalculation, or partial failure of blasts, or rapid slumping after excavation. However, the stability of the fill will always be greater than if it were simply laid on top of the swamp.

STABILIZATION

Dewatering. One method offers stability equal to that obtained in normal, dry fills, but at high cost. The right of way can be dried up by draining, or well point or sump pumping, as described under Ditching and Drainage, then excavated to firm bottom with machinery. Fill may then be built up to road grade in thin, thoroughly compacted layers and with properly sloped sides. When it is complete, pumping may be discontinued and the muck allowed to settle back against the slopes.

Removal of muck in any way is expensive work, except in the rare cases where it can be sold locally for humus. The expense increases very rapidly with depth, and a point will be reached where it is good practice to stabilize the mud rather than remove it. This is particularly true if the area involved is wide, or conditions do not permit side casting or blasting.

Sand Drains. Muck can be stabilized if enough water can be squeezed out of it to convert it from a semi-liquid into a solid. The weight of a heavy road fill has a squeezing effect, but the mud and water flow together and very little compaction is obtained for a long time.

Vertical sand drains may be used to dry up the mud so that it will support a load. They consist of columns of sand extending from hard bottom to the top of the mud, and connected at the top to each other, and to an outlet by a sheet of sand or gravel, or other drainage systems. They are discussed in Chapter 5.

Chemicals. If digging is done in a water bearing sand or gravel, very soft conditions and extensive caving of banks may be experienced. This difficulty may be avoided by drying the area with well points, or by other methods described in Chapter 5, but it may be more economical to prevent water from moving through the soil. This may be done by means of cement grout, discussed in Chapters 6 and 9, or by other chemical treatment. The grouting material displaces water, and then hardens or gels into a mass that prevents more water from getting between the soil particles.

One chemical treatment involves pumping a solution of silicate of soda into the ground through drill holes, then pumping in a solution of calcium chloride that causes the silicate to form a dense hard mass.

American Cyanamid's AM-9 chemical grout is a thin water solution of two acrylic chemicals and one catalyst, that is mixed with another catalyst solution just before injection into the ground.

A chemical reaction causes the mixture to form a permanent, water-impermeable gel, at a time that is controllable between 3 seconds and several hours.

These characteristics are particularly useful in sewer rehabilitation work.

Freezing. Any soil in which pipes may be sunk can be stabilized by freezing, but the presence of salt, or certain other chemicals, may make it difficult.

It is accomplished by sinking a number of metal pipes, rather closely spaced, in the area to be stabilized. Tubing containing a refrigerant, usually ammonia, is placed in these pipes and connected to heavy-duty refrigerating apparatus. The soil water will freeze most rapidly if it is stagnant, but even a steady flow can be checked.

This is an expensive job, mostly confined to very fine grained mud, or deep work which does not respond to well points.

FILLS

Compaction. In fresh fills the most serious mud difficulties are due to improper compaction. Here it is sufficient to say excellent compaction may be obtained if a fill is made in layers six to eighteen inches in depth, and each layer is thoroughly rolled or tamped in all parts. If rollers are not available, trucks may be used, first empty and then loaded. Such a fill usually will be incapable of absorbing enough water to turn to deep mud. The extra expense and nuisance of making a fill in this manner, when specifications do not call for it, may be regarded as an insurance premium against the loss of having to stop a job for days or weeks because of a soft dump. The expense may be not only the extra dozing required to spread, and the roller or machine time to compact, but the fact that from 10 to 40 per cent more material might be required to fill the same hole.

Under many conditions, such as stumps and boulders in fill, rough or liquid original grade, poor access or insufficient machinery, deep single layer fills are unavoidable. If the top is properly compacted by rollers or traffic it may suffice to shed rain water, and to resist water soaking up from below. Areas receiving no compaction except from grading of the fill by a light-footed dozer, are liable to soak water up like a sponge, and be a hopeless quagmire for a while, and also tend to wet and soften surface compacted areas.

Clay. Clay fills are liable to become soft and slippery during rain. The wet surface may be bulldozed off and replaced with dry material, or the dry material may be dumped over the mud in sufficient quantity to carry traffic over it, and be removed later. If rain is anticipated, the original fill may be left low to leave room to cover it.

If fill is coming from two or more sources, it is sometimes possible to keep the clay in lower parts of the fill and better material on the top.

Area Dumping. If trucked soil is too wet to be stabilized and will not support trucks after being spread, and if the ground (or lower layer of the fill) is passable to trucks, area dumping may be resorted to. A calculation is made of the amount of fill needed in a given area, and the amount of each incoming load. Allowance must be made for shrinkage of the soil upon drying and after compaction, and the number of loads required in such an area may be found by dividing the cubic yards required by the corrected yardage of a truckload.

The trucks may then dump their loads in piles which can be left to dry, then be spread and compacted. The easiest way to figure pile spacing is from center to center.

If more material is needed than can be left by dumping piles, higher bodied trucks may be hired, or a bulldozer employed to push the newest piles into and over the older ones.

Fill dumped in this manner is seldom in just the right quantity in the right place, but it can be shifted around during the grading process.

Buried Mud. If it is impossible to truck in the area to be filled, the mud must be covered as it is dumped with a layer of dry fill, gravel, or rock sufficient to support the trucks. Sometimes the mud will support a bulldozer which can roughly grade the soft fill. The dry surfacing is advanced close behind the face of the mud fill, trucks dump the mud at its edge, and the dozer pushes it over the face, digging down sufficiently to allow placement of more surfacing. See Figure 3-25.

Fig. 3-25. Two-layer fill

In such mud work, it is advisable to have at least two dozers, so that the one working in mud can be promptly rescued if stuck. Under extreme conditions two dozers may be attached to each other, back to back, by a chain or cable. As one of them pushes mud toward the face, the other backs up toward it, keeping the line slack. When the pusher backs up, the other one pulls it to firm footing. The helper cat need not have a dozer, and preferably should be larger than the one doing the work. Wide tracks are a big asset in mud.

STUCK MACHINERY

The next problem to consider is how to rescue machinery that has sunk so far into mud that it cannot move under its own power. The easiest and most attractive method is to stretch a chain or cable to some power source and pull it out. Often there is no such power available, but if there is it should be used with caution. If the stuck machine has not sunk, but has simply lost traction on a slippery surface,

Fig. 3-26. Dumping on a side slope

power can usually be applied without damage. However, if it is in badly, use of moderate power may be useless, and too much power may pull it apart.

TRUCKS

Dump trucks are probably the most frequently stuck type of equipment. If one is loaded, it may be possible and desirable to dump the load, after which the truck may pull out under its own power. But if the truck is tipped sideways, as with one pair of rear wheels bogged down, the other resting on the surface, the raising of the load preparatory to dumping increases the sideward strain, and may overturn the truck or tear the body off its base. A winch cable attached to the front top of the body on the high side, and pulling uphill, may permit dumping of a tipped truck, but even with skillful operation, the strain on the body is considerable. Unloading by hand, or partial unloading by power shovel, hoe, or front loader may be necessary.

A shovel is a highly effective rescuer of bogged trucks. A dipper bucket may be placed under the rear of the frame (or, with some risk of damage, under the rear of the body) and lift and push at the same time. A front loader can do the same, but might walk into the same mud hole; and its wider bucket may be blocked by tires. A dozer has the same problems.

A hydraulic hoe can get most of the load out of the body, then push and lift with the back of its bucket.

Digging Out. If the truck cannot be emptied, or is empty and is still stuck, and no powerful equipment is at hand, the next procedure is to dig out the wheels in the direction toward which it is hoped to move. Two-wheel drive trucks have best traction going forward, and in many situations, such as sinking in a soft shoulder, an attempt to back out will cause the front wheels to get in worse difficulties. Any digging helps, but the best procedure is to go to the bottom of all tires, make a ramp up to the surface with a length of three or more feet to every foot of depth, and put a board or boards on this slope, with the lower end against or under the tire. If boards are not available, matted brush,

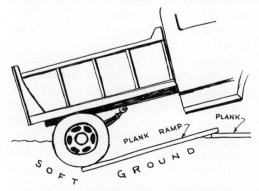

Fig. 3-27. Digging out of a mud hole

Fig. 3-28. Too much power

stones, gravel, or anything but mud may be used. If the axles or frame are resting on the ground, an attempt should be made to free them, but this is often not possible.

Even if this digging does not enable the truck to get itself out, it makes it much easier for a light machine to pull it, and it greatly reduces the danger of damage if pulled by a large machine.

Breakage. Applying brute force to pulling a deeply bogged truck may result in getting it out minus its rear axle assembly and wheels, a partial victory that brings little satisfaction. In most trucks the rear axle is attached to the frame only by the spring shackles and propeller shaft.

Lesser and more common damages are tearing off of bumpers and bending of front axles. Many trucks have no satisfactory pull points on the front end. The bumper should be used only at the fastening to the frame. If it is necessary to use the axle, the line should be attached as close to a spring as possible, and care should be used to prevent it from catching any part of the steering when tightened, or when the wheels are turned.

Pull Line. The most generally effective device for extricating bogged machinery is a winch. It may be mounted on and powered by a tractor or a truck, or be a portable affair. A power winch may have 200 or more feet of cable which enables it to reach a long way from firm ground, or on shorter pulls, to multiply its power many times by means of pulleys and anchors. The procedure for rigging and operating the

winch is the same as that described under stump pulling in Chapter 1. The use of multiple lines is often advisable, even when the machine may be de-bogged by direct pull, as the slower speed is less liable to cause damage. Hand winches are slow and laborious to wind in, but are powerful and can be used in places inaccessible to larger machines.

Where possible, it is best to pull the truck straight out with the truck wheels driving. If it is necessary to pull at an angle, the truck should be steered toward the pull.

A cable or chain may be stretched from the stuck truck to another truck or a tractor, with or without pulleys, and a traction pull used. Care should be exercised not to use too short a line and get the assisting machine stuck also.

Push. A dozer or front loader tractor can get behind it to lift and shove at the same time, but a man should be stationed to put poles in front of it as it moves so that it will not get stuck in the hole the truck leaves.

Body Hoist. A dump body usually extends a short distance behind its hinge, so that when the body is raised the rear edge of it goes down a few inches. Advantage may be taken of this feature by chaining the axle tightly to the frame, perhaps lifting it first by chaining it to the front of the body and hoisting; placing a plank on the ground under the rear of the body, and fitting a stout log between plank and body. If the body can be raised, the rear edge will push down against the log to lift the frame, and through the chain, the axle and wheels. Axle or wheels can then be blocked up, the body lowered, and the log shimmed up to meet the body at its new height, and the process repeated until the wheels are high enough so that a support or ramp can be put under them. See Figure 3-29.

Jacking. If a heavy jack is available, the axle may be chained to the frame, the jack placed on a plank or other support, and frame and wheels raised and blocked in the same manner as when the body lift is used. If it is not possible to chain the axle to the frame first raise the frame with jack or body hoist, then jack the wheels up with a light jack based on a board and placed under the wheel rim or hub.

Wheel Tractors. A wheel tractor in mud presents similar problems, except there is no load to be removed, and it takes more digging to get to the bottom of its big rear tires.

A tractor may have a differential lock, that can be engaged to prevent one wheel from spinning independently of the other. This should be engaged at the first suspicion of bogging down.

Letting some air out of drive tires may be helpful, but if overdone may cause the tire to slip on the rim, tearing out the valve.

If the stuck tractor has a loader, the bucket may be used to push it out of trouble, preferably backwards. The procedure is similar to that described for crawler tractors on page 3-40.

If the tractor carries a hoe, it can lift the rear wheels clear of the ground by forcing the bucket down. It can then push it forward, or swing it to the side.

A tractor equipped with both a loader and a hoe can, with skillful operation, work in extremely soft ground.

If a truck or wheel tractor is stuck more because of slippery surface than sinking, tire chains, short chains threaded through wheel openings, or rope or rags tied around

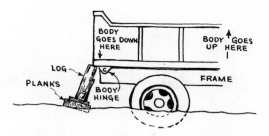

Fig. 3-29. Jacking out with body hoist

the tires will help to give traction.

Wheeled vehicles do best in mud or sand with large tires, or standard tires with reduced pressure. Truck tires break down rapidly if under-inflated, but tractor tires usually have thin flexible sidewalls and operate at such low speeds that they can run soft for considerable periods before being damaged. However, they may slip on the rim and tear out their valve stem if too soft.

Churning. A very important item in the sticking of vehicles is to know when to stop struggling. Often a minor tow job is changed into a major project by a driver continuing to spin his wheels and buck back and forth until he has sunk his vehicle completely. In rescue work, the strongest measures available should be applied early, as every attempt that fails is likely to make the job more difficult, and danger of damage to the sunk vehicle more severe.

CRAWLER TRACTORS

Crawler machines do not bog down as readily as wheeled ones, but can do an even more thorough job of it.

A crawler tractor may sink in mud too soft to support it, or it may dig itself in while pulling a load, or both. If the tracks are allowed to spin, the grousers act like buckets on a ditcher, digging soil from underneath and piling it behind. On soft ground they can work down rapidly this way, until the frame parts of the tractor are resting on the ground, or on a stump or

Fig. 3-30. Walking out on cross poles

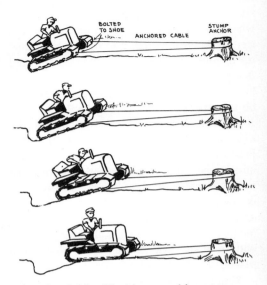

Fig. 3-31. Climbing a cable ramp

other object its normal clearance would have taken it over. When the weight of the tractor rests thus on the frame, the tracks churn helplessly in air or loose mud.

Pulling Out. If outside power is available, the machine may be pulled out by a line attached to its drawbar, front pull hook, or other hold. It should be pulled straight forward or backward, if possible, with its own power being used also. The drawbar, or a dozer blade, can take almost any pull, but use of a front pull hook may pull the engine out of some models.

Poling Out. If no outside power is available, and the stuck cat has no dozer, winch, or other helpful equipment, the first thing to try is digging a shallow ditch in front of (or behind) the tracks, a foot or two wider than the tractor. In this put a green sapling or strong board as long as the ditch, pressing or wedging it tightly against the tracks, which should then be turned slowly so as to pull the stick underneath. When it is well under, press in another stick, and perhaps more. Their effect will be to lift the tractor and restore the weight to the tracks, provide a wide base for support and traction, and to cut off or uproot obstacles under the tractor. They are almost certain to get the machine out if it will pull them under.

If the tractor has flat shoes or grousers that will not grip the size of pole available, and no bolt-on cleats can be obtained, or if there is an aversion to digging ditches, planks or heavy angle irons may be drilled and bolted to track shoes on both sides. The effect is then positive, but it is very necessary to unbolt and remove them as they come up at the other end. Poles or logs may be fastened to the shoes by loops of cable and used in the same manner. Short sticks should not be used unless absolutely necessary, as they do not afford nearly as much lift or traction as long ones, and they are liable to turn underneath and jam things.

These stratagems are equally effective at moving the tractor forward or backward, but reverse gear often has less power than low, and backward movement under difficult conditions is a severe strain on the tracks.

Cable Ramps. Another system is to fasten a cable to each track, or the two ends of a single cable to the tracks, perhaps by passing it from the outside through a hole in the shoe, and catching it inside with a loop and clamps. These cables should be stretched parallel or nearly so straight ahead of the machine to anchors of some sort. When the tracks are turned to move forward, they will advance on top of the cables, which will prevent them from spinning and provide a tightrope ramp on which they can climb. This technique should be used with caution in reverse as the strain on the track may break it. See Figure 3-31.

Chains may be bolted to the track and used instead of cables, but they are much heavier for the same strength and are seldom long enough for the job.

The tractor may also be jacked up on planks, in the same manner as a truck, and the hole filled in or bridged.

Winching Out. If the tractor is equipped with a towing winch, the cable may be fastened to an anchor behind it, and the machine will come out of the mud as the cable winds in. This should be done with caution if the presence of a stump or large stone underneath is suspected, as it might force a track off, or do some other damage. Risk can be reduced by putting blocks behind the tracks so that it will move up as well as back, or by anchoring the cable at a height, as in a large tree.

If the tractor is too badly bogged to turn, the only available anchor is not directly in line, and the winch is of a type that the

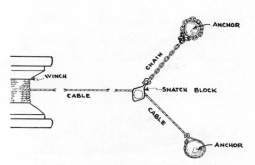

Fig. 3-32. Changing angle of pull

cable runs off the spool on an angle pull, run the cable out until it is slack then hold it with a crowbar or stick so that it reels on the side of the spool opposite to the direction of the anchor. When the pull begins, it will be off center and will have a tendency to turn the tractor in line with the anchor. If it does not do this, it will wind onto the drum in a spiral, making one or more loops before reaching the edge on the anchor side, moving the tractor a short distance. The line can then be slacked and the procedure repeated.

If the angle between tractor direction and anchor is too great, a second anchor must be used and a chain from that adjusted to hold the cable in line. Use of a snatch or pulley block, as shown in Figure 3-32, will prevent damage to the cable from the chain hook. If alignment is satisfactory but the winch does not have enough power to de-bog the machine, a pulley may be chained to the anchor and the winch line passed around that and back to the tractor drawbar. This will nearly double the pull.

Dozer Down Pressure. If the tractor is equipped with a hydraulic bulldozer, but no winch, the blade should be raised; planks, poles or other floats placed, dug or driven under it, and down pressure applied to the blade. If the hydraulic system is in good condition, this will raise the front of the tracks clear of the mud. Poles may then

Fig. 3-33. Pulling dozer shovel out with bucket

be forced under the tracks, or the holes filled with rocks, and the tractor can walk out after raising the blade. If the dozer will not raise the tractor, and there is enough oil in the system, try letting the machine cool off, as a worn pump will develop better pressure on thick, cold oil. If it still will not work, or if there is no time to wait, proceed as suggested for a plain tractor.

Cable Dozer. If a tractor equipped with a cable bulldozer is stuck, and plenty of pulleys and extra cable are available, the blade might be supported by a chain, and the control drum rigged to pull it out. The heavy pull of multiple lines should be through the drawbar or blade.

Dozer Shovel. Dozer shovels equipped with flat or semi-grouser shoes get stuck easily in wet holes. The bucket can be used to push it out, preferably backward. Place it on the ground in fully dumped position, apply down pressure, then rotate the floor forward to a flat position, usually turning the tracks slowly backward at the same time.

Poles may be placed under the bucket and behind the tracks, for extra bearing in very soft ground.

A forward pull may be obtained by placing the bucket flat on the ground, applying enough down pressure to raise the front slightly, and then dumping it slowly. It is not as effective as in reverse.

The bucket can also be used for short distance self-moving of a machine that cannot travel because of clutch or transmission failure.

The front of the tracks can be raised off the ground by using down pressure with the bucket floor vertical, and supported by poles or blocks. The rear can be raised by shoring up the front and piling the bucket with heavy material.

Hanging Up. Tractors are frequently hung up on stumps or rocks in the absence of any mud at all, usually through digging in the tracks while moving a heavy load. The first warning is liable to be a failure of the tractor to keep its direction or to steer. The tractor should be stopped immediately and the situation checked over. If the stump (or rock) is under the crankcase guard or other smooth surface, the tractor may escape by backing or turning, or by climbing a rock or a stick placed under one track. If the obstacle has wedged into a hole or raised section, large rocks or logs under the tracks, or even jacking, may be required to get free.

A crawler tractor may also be hung up by walking over a large flat stone or short log which flips up and catches in the chassis or between the tracks. This situation deserves careful examination before working, as applying too much power in the wrong direction may break a track or do other damage. Movement in the right direction, particularly turning one track only, may free it; or climbing up on blocks or using another power source to pull the rock out the same way it went in. Sterner measures are removing the track, or turning the machine on its side in order to break up the stone, or to chop the log.

Overturning. If a crawler tractor is ly-

ing on its side, it may be rolled back on
its tracks by pull on a line fastened to the
highest substantial structure near the center of the up side. If there is no spot above
the center of gravity which will take the pull
the line should be run across the up side
and what is usually the top of the tractor
until a hold is found. Logs or other blocks
under the line should be used to protect
the tractor parts against crushing. An improvement on this is to use a pulley in a
tree, or in a heavy tripod, so that the
line will lift as well as pull.

If the machine is upside down two pulls
may be necessary, one to get it on its side,
the other to right it. This is easiest downhill, but blocks must be arranged to prevent
it from rolling farther than planned. Power
should be applied slowly to avoid bending
or crushing of parts.

If no power is available, hand jacks
should be obtained together with planks
and blocks. A jack placed on a plank
should be used to raise some portion of the
tractor, blocks placed to hold it at that
height, the jack released, blocks placed under the jack, and the process repeated. One
jack will do it, but two are easier. It is
good procedure to start jacking the part of
the machine which will move the longest
distance in resuming upright position, and
work the jacks in as space opens up. If the
cat is dozer equipped, the dozer frame and
blade will safely take the strain of jacking.
The dozer may sometimes be moved advantageously, using the starter for power.

The type of jack shown in Figure 3-34
is particularly useful in machine salvage.

SHOVELS

Planks. Power shovel tracks usually have
flat shoes, so that it is difficult to get them
to grip poles or planks and drag them under, without excessive trenching, or bolting or cabling them on. Before working a
dragline or other shovel in risky places, it
is a good plan to get a plank 2″ × 12″,

Fig. 3-34. Equipment salvage jack

3″ x 12″ or heavier, two feet longer than
the shovel is wide, and drill holes in it
to match the holes in the track shoes so it
may be readily bolted, and carry it for
use when necessary as described under
tractor rescue. Strain on plank and shovel
will be reduced by trenching deeply to the
under slope of the idler if possible. If the

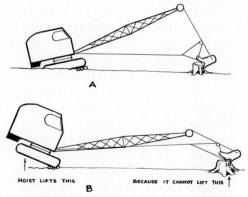

Fig. 3-35. Raising by an anchored bucket

shovel can climb up on this plank, saplings or other helps can be placed in front of it to assist it to firm footing.

Hoisting Bucket. If a dragline is down on one end, logs or planks can be forced or dug under the high end, and the low end raised by hooking the bucket or hoist cable to a stump or other anchor, as in Figure 3-35 (A) and (B). Pulling in the hoist cable, or raising the boom, should tip the shovel forward, lifting the rear sufficiently to allow shoring up with logs.

If the anchor is close, a low boom gives best leverage. If it is distant, a high boom is better. However, a boom angle of over 45° may be dangerous, because if the hoist line breaks or comes off the anchor, the reaction might throw the boom back on the cab.

The same procedure can be used if one track is bogged and the other one is free.

If the shovel is down all around, trenches can be dug to permit bracing under one end. This is used as a pivot to raise the other end, which is then shored up. The boom is swung to an anchor on the opposite side and the hoisting and blocking repeated.

Two anchors may be chained together for greater strength, or to secure a better direction of pull.

If no anchors are located in the proper direction, and it is not practical to make one, the bucket can be used as counterweight. The dump cable is shortened to permit holding a load far out. A bucketful of the heaviest dirt available is dug, the boom lowered so that the bucket can be held just above the ground. Rocks or weights are piled on the bucket until the shovel tips. If the rear of the tracks is not high enough, it can be brought up by raising the boom.

In any blocking work a loose track is a nuisance as it will hang down and impede placing of supports. The slack can be readily taken out by placing a jack on the track frame and raising the upper section.

Drum Line. A shovel can sometimes pull itself out of trouble by a line attached to one of its own drums, or to a drag bucket. These lines are too fast for a smooth, steady pull, and the drums are so high that there is a tendency to pull the front of the shovel down. Better results will therefore be obtained if the line is passed through a pulley chained to the anchor, and back to the bottom of the dead axle. If several pulleys are available, and all additional lines run to the dead axle, a very powerful, well directed pull will be obtained.

Jamming Chains. A special precaution to be observed with shovels is to keep the drive chains free from material of any kind. Stretching or jamming these with mud, gravel, or debris is not only a severe strain on the chain, but by increasing friction within the chain and with the sprockets absorbs a large part of the drive power needed at the tracks.

Straddling. It is risky to work a shovel that is straddling a stump or boulder, unless the ground is known to be entirely firm. The rhythm of load, swing, and rebound causes the tracks to work their way down into soft earth, particularly at the load end. If there is a thin layer of sod or hard soil overlying unstable material, the top will give an appearance of solidity, but if it breaks through, will let the machine sink rapidly. Ordinarily, slight sinking can be ignored or checked by poles under the tracks, but if it causes the shovel under parts to rest on a stump, it is a more ticklish situation.

The under parts of a shovel, between the dead axles, are more vulnerable to damage than any part under a tractor. In mud there is often nothing to do but to get out any way and hope for the best, but if possible, any projection reaching up into this vulnerable area should be carefully watched during rescue work.

The shovel may be freed from an obstruction by anchoring the bucket and hoisting, or by walking it up on logs, or using a plank bolted to the tracks, or even by using a long crosscut saw to cut off a stump while still under it.

Counterbalancing. If a shovel sinks so far on one side that it is in danger of overturning, the boom should be swung to the high side and lowered as much as practicable, the bucket extended, and weighted or anchored, then raised to serve as a counterbalance. This should prevent further tipping and may make it possible to shore it up.

Overturning. If a shovel is overturned or so steeply tipped that it cannot be operated, the swing lock should be set, a line should be attached to the top of the A-frame and to any available anchor on the high side, and drawn taut to prevent further settling. If this line can be attached to a power source of sufficient strength, the shovel may be pulled upright, but this should be done slowly, with careful attention to any tendency toward bending the A-frame or other parts.

If no adequate power is available, a platform must be laid or constructed on the ground, and the shovel raised with jacks and blocks. The best and cheapest blocks are old railroad ties, but any sort of beams or heavy planking may be used. The actual raising of a shovel is so intricate, so dependent on the position and construction of the machine, and so liable to result in severe damage if done improperly that it cannot profitably be described here. The best method is to hire people who specialize in rigging and machine moving, work with them, and remember their methods.

If a shovel starts to tip during a swing, the bucket and load should be dropped, if possible.

Counterweight. If a shovel is heavily counterweighted to carry a long boom, and the boom is removed to install another attachment, the shovel has a tendency to turn over backwards. Removing counter

Fig. 3-36. A basic shovel is tail heavy

weight is usually too much work to be practical. The shovel may be walked slowly to its other attachment, with a high heavy sawhorse or some other support dragged or pushed along under its tail. Or a line may be rigged from its A-frame to a tractor or heavy truck moving ahead, and kept taut, a device which can be used not only to steady it but to pull it upright if it does "sit down." Another system is to bring the attachment to the shovel.

Backhoe. A cable backhoe is generally similar to a dragline in working its way out of mud. Its shorter reach makes it less likely that it can get its bucket teeth in anything solid enough to be helpful, but it has double the pull in proportion to weight if it does.

If an anchor is near ground level, pulling on it tends to force the near side of the tracks down, so they must be given some kind of support in advance.

In case of a sudden loss of grip, there is serious danger of the boom and jackboom falling back on the cab.

HYDRAULIC BACKHOES

Hydraulic backhoes sink in mud just as readily as cable rigs, but they are much better able to get themselves out.

If any solid anchor is within reach, it can be gripped with the bucket teeth, the bucket crowded in to pull the shovel toward the anchor, and down pressure used to raise the near end of the tracks as they move.

If there is no anchor, or ground firm

enough to act as one, poles or other supports may be placed close to the tracks (or wheels). The back of the bucket is placed on them, and down pressure exerted to raise the machine high enough to put supports under it.

The bucket may also be placed against a bank, the ground, or artificial helps, and forced outward to move the shovel back.

If the carrier is a tractor or a truck, the working end of the machine can be moved sideward by forcing the bucket down to raise it, and then swinging. This may also be done with a crawler if the end opposite the bucket is not deeply buried.

FREEZING DOWN

Frozen Mud. Mud may cause serious difficulty when it freezes, as it sets like concrete and anything in it is likely to stay there. Crawler machines are particularly vulnerable, for if they are muddy, tracks, track wheels, and chassis are liable to be combined into one immovable unit. If they are clean, they are still apt to freeze to the ground.

Precautions to be taken are to clean all mud from the tracks and wheels at the end of the shift, and to park the machine on rock, metal, or wood, with as much of the track length as possible in the air. Mud should be at least scraped off with a trowel, screwdriver, or stick, but it is safer to wash it with a hose also.

Frozen mud is hard, heat resistant, and tough, particularly when very cold. However, the bond between mud and other materials may be a film of brittle ice, which will resist pull or twist but shatter at a quick blow.

Breaking Loose. If the tracks are frozen down, but the track wheels are free, the machine can often be broken loose by rocking it forward and backward with clutch and gears. Jerky clutch action is much more effective than smooth, but also much harder on the machine. Moving forward tends to loosen the rear pad, backward the front. The hardest part of the job is getting the first crack in the frost, after that the additional movement makes the breaking away progressively easier. It is sometimes advantageous to apply power to one track at a time, using the steering clutch to disconnect the other.

If there is frozen mud holding the wheels, it is unlikely that the machine can break loose with its own power, and the mud must be chipped or melted away. Chipping is very tedious and often impossible because of lack of space to work. Heat is usually more practical, but if the weather is extremely cold, very large amounts are required.

Melting Out. A blowtorch, the larger the better, is the usual tool for this work. The flame is moved back and forth over the mud surface, and the thawed material scraped off. The wheels, tracks, and other metal parts will conduct heat to the mud effectively, so that chunks can be undermined and broken off; but care must be taken not to heat the metal enough to take out its temper. If the temperature is very low, a small section should be worked at a time; if it is near or above freezing, the flame may be moved back and forth over the whole area.

Hot water is effective if available in large quantities, as it will wash off the mud particles as they are thawed, exposing new surfaces to the heat. However, it may freeze on other sections of the machine while draining off. A portable heater may be used to blow a current of warm air on it, preferably under a tarpaulin. A flexible tube may be used to conduct hot engine exhaust.

More drastic measures are to erect a tent, or tarpaulin shelter, over the machine and thaw it with a stove; or to drag and push it with locked tracks into a heated building. But often the most practical method is to leave it until a thaw occurs, clean it off, and make good resolutions.

Wheeled Equipment. Rubber tired vehicles are not as apt to freeze down or together, as the flexibility of the tires prevents the ice from holding them effectively, and the rotating parts are ordinarily not as close to the fixed parts. If such freezing does occur, the same means may be used to thaw it, except that the rubber must be protected from heat.

The parts most vulnerable to freezing are the brakes, and here water will do as badly as mud. If the vehicle is used in very wet or slushy conditions, then allowed to stand in freezing temperatures, a film of ice will bind the brake linings to the drums. Driving back and forth on a dry stretch of road, even if only a few yards long, with the brakes applied, for a few minutes before parking will usually dry them out again enough to prevent this trouble.

Frozen brakes may be loosened by rocking the car back and forth with its own power, or can be thawed with hot water or flame.

JUMPING THE TRACK

A weakness of tracked vehicles is the possibility of going off the track. When this happens, it may mean that the track is in a heap alongside the machine, the track wheels are resting on the ground; but more often it means that the track rails are not engaging the wheel flanges properly, being displaced to either side, and contacting the inner surface of the track shoe. Operation in this condition soon leads to the complete separation of machine and track.

Jumping or running off the track occurs most often during sharp turns on uneven ground, and is likely to indicate that tracks are too loose or have a broken link, or track and wheel flanges are worn, or that wheels are out of line. It is usually accompanied by a snapping or grinding noise, and if suspected, the machine should be stopped immediately and inspected, as

every inch of movement makes it harder to get back on the track.

If the track is off the truck rollers only, it will usually swing back into place if the track is raised off the ground, by running the bull wheel or idler onto a log or other lift, or by jacking. If the bull wheel is on hard ground, forcing down a hydraulic dozer blade will often raise the track sufficiently.

The principle involved in getting a track back on a bull wheel or idler is similar to that of installing a tight fan belt on a car. It cannot be pulled or pried enough to stretch it over the flange of the pulley, but if it is held in one end of its place in the pulley, and the pulley turned, the wedging action will stretch the belt over the flange, and the part of the belt already in the groove will draw the rest of it in.

Similarly, with the track it is a problem of getting part of the track in line with the flange, and turning the wheel to draw it on. If the track is partially on the wheel, simply turning it in the proper direction does the job. If it is off the flange completely, it may have to be pried into line with a crowbar, jack, or chain, and the track adjustment will usually have to be loosened as well.

If the track is off the upper part of the bull wheel, but still engaged with even one tooth at the bottom, and with the truck rollers, the machine should be moved forward slowly. The sprocket teeth will mesh properly with track appearing from beneath the rear truck roller, and will carry the wrongly meshed section overhead into the slack upper section, where it will be straightened out by the support roller. If the upper part of the bull wheel is correctly engaged, and the lower section off, the track will work into place if the machine is reversed. However, it might be necessary to pry with a crowbar to prevent the track from jumping off the rear roller.

If the track is entirely off the bull wheel

flange, but still meshed with the truck rollers, the machine should be moved forward. The bull wheel will roll onto track held correctly by the truck rollers, and, perhaps with the aid of vigorous prying, should mesh with it and pass this correct meshing up around itself. If the track is also off one or more truck rollers, the tractor should be backed so that the bull teeth can mesh with track held correctly by the support rollers. This may require more prying, or a pull in the correct side direction from another power source, or if the disabled machine is a shovel, from a line to its boom.

Should the track be off at the idler, the above methods are still good, with directions of travel reversed. If the track is tight, it might be loosened to facilitate crossing the wheel flanges. If the track is off the support roller, it usually can be replaced by lifting with a crowbar.

If the track is off the bull wheel, truck rollers, and idler but is still on the support rollers, the whole track and wheel assembly should be raised off the ground, and the bull wheel rotated backward. The track can be engaged with this at the top by use of a crowbar, and will be pulled into engagement with the rest of the bull wheel, then the truck rollers and finally the idler. Caution should be used to prevent the track from coming off completely in one place while being meshed in another.

If none of these stratagems work, or if the track is entirely off the wheels, the track adjustment should be loosened and the track "broken" by removing a lock and driving out one of the hinge pins, using the bull wheel as a brace. On most makes, there is only one master pin which can be used. The side of the machine should be raised off the ground, the track placed correctly under the wheels, the machine lowered, and the track wrapped around the bull wheel and idler. The ends should be pulled or forced together at the bull wheel or between it and the top of the idler, by chain tighteners, block and fall, winch, jacks, elbow grease, or any other means available, and the pin inserted and locked. This is a difficult and laborious job, even on a small machine, except for experienced personnel, and on a large one may require the use of other machinery to handle the track.

It is sometimes easier to replace a track, after opening it, by walking the machine off it onto a plank or beam, aligning the track behind it, and walking it back, than to shift the track around under the machine.

This type of work has been rendered much more difficult by changes in track design and fabrication. A single master pin may be difficult to identify and to get at. When found, it may be so tight that it cannot be driven out until the links are heated with a torch, and the links may be bent by the heavy hammering, during either removal or replacement.

It is common practice among service men to cut a pair of links with a torch; then take the whole track to a press to be reassembled. A minor incident may be turned into an expensive and time consuming project in this way.

Split master links make this drastic step unnecessary. They permit opening a track by removing a shoe, then taking out four bolts. Opening the track, and closing it, are greatly simplified.

CHAPTER FOUR

CELLARS

Cellar excavations may be roughly classified as the dig-and-pile and the dig-and-haul-away types, which will be referred to in this discussion as residential and commercial, respectively, as the larger part of them fall into these categories.

Backfilling around the foundation is discussed in Chapter 7. It must be emphasized here that a wet fill can crumple in a foundation and even move footings, and that this danger is especially severe if the masonry has not had time to cure.

PRELIMINARY WORK

Tree Protection. In residential excavation, any clearing that may be required is likely to be of the selective type. Large trees, or trees of desirable species, may determine the location of the house, and they must be guarded against damage and burial. It may be advisable to wrap the trunks of such trees in cloth, and protect them by a collar of vertical boards, as in Figure 4-1. If their bases may be temporarily buried, the original ground line should be marked on the bark with paint, so that the fill may be removed accurately, and burial or overcutting avoided.

Topsoil. Topsoil is usually present and if it will be needed for landscaping it should be saved. This involves taking it off the area to be excavated, and preferably from the areas where spoil is to be piled. This stripping may be a substantial part of the cost of the excavation but is required in most localities.

In places where there is no well defined topsoil, or the topsoil makes up a third or more of the spoil and the subsoil will mix well with it, stripping may not be needed.

A method of stripping topsoil which is often most economical in the long run is to remove it completely from all areas to be involved in the digging. Figure 4-2 (A) shows one method of doing this by placing it in compact piles well away from two corners of the proposed house. This will usually keep it out of the way of digging, piling, and ditching, and leaves it in a position for straight line spreading. But it should not be put in corners where a dozer will be unable to get behind it.

(B) shows a more usual method. The digging and piling area is stripped, the topsoil being pushed into two piles, so placed as to be just beyond the spaces for the piles of spoil. If too small allowance is made for the spoil, the topsoil may have to be moved back further, or may get buried by the fill and partly lost. In any case it will probably interfere with backfilling.

In (C) the topsoil is pushed to the sides and fill piled to the front and rear.

Topsoil stripping is discussed in Chapter 10. It is customarily done by a dozer, which does the cleanest work. Hoe shovels, and to a less extent other shovel rigs, will remove topsoil rapidly, and if the soil is heavy and wet, may be preferred because they do not compact it and cause it to cake. If large areas are involved scrapers may be used.

If the topsoil is thin and will be required

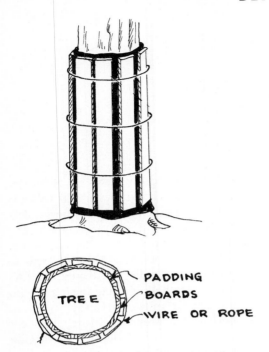

Fig. 4-1. Protection for tree trunk

for finishing off, it may be deliberately mixed with some fill while stripping to increase its bulk. If the subsoil has a loose texture, little or no harm will be done to fertility, and regrading will be simplified.

Artificial Obstacles. Serious digging difficulties may be caused by obstructions placed by the builder. Batter boards are used for reference in exact placement of foundation walls. They will increase excavation cost by restricting the movement of machinery, particularly if the house is to be irregular in shape. Piles of building material may also be much in the way.

Corner stakes, with back reference

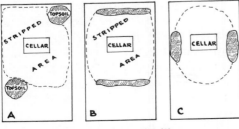

Fig. 4-2. Topsoil piling

points such as are described in Chapter 2, are adequate location marks for the excavator, and other preparations for building should be postponed until digging is complete.

Depth. The depth of the excavation depends on the first floor height relative to original grade, and the depth of substructure below it. Substructure might include floor thickness, rafters, sills, cellar headroom, and thickness of the cellar floor and gravel or crushed rock under it. If the full area is to be dug to the bottom of the footings their depth must be considered also.

On sloping ground, excavation depth will vary at different points.

Factors in determining first floor level, including a way to probe for rock, are discussed in Chapter 7.

Rock. Digging in ledge rock or large boulders may increase excavation cost three to five times, or more. Its presence may result in raising the house, or substituting a slab or crawlway for a cellar.

If it is to be removed, dirt removal should be completed and the rock surface cleaned before drilling.

Most cellar jobs are so located that mats must be used to cover blasts, and particular care taken to safeguard passersby and traffic, and to avoid damage to property.

DIGGING

The three standard machines for digging small cellars are the shovel dozer, the bulldozer, and the hoe. The bulldozer is now much less used than the shovel dozer, a reversal of the situation of a few years ago.

However, bulldozer methods are described first, as its work is basic for both machines.

Bulldozer. A bulldozer can dig a cellar very deep, if there is no heavy rock or mud, but it is at its best if the excavation is shallow. This is because it must dig away a considerable amount of the bank

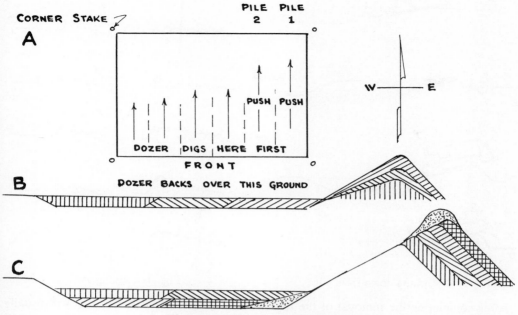

Fig. 4-3. Cellar digging sequence

to ramp itself in and out, and the whole weight of the machine must come up out of the hole with every pass. It can push much larger loads on a level or a moderate upgrade than on the steep rises from a deep cellar.

Digging techniques vary with the operator and the locality. Several methods will be described, but they are presented only as samples and should be followed only where they give satisfactory results.

It is good practice to leave a ramp that will allow trucks to back into the floor of the excavation, for convenience in delivering foundation materials to the point of use.

Example. The first case to be considered is that of a cellar, 20x30 feet, four feet deep, dug in a large, level, treeless lot, from which topsoil has been stripped. The dimensions given are those of the outside of the foundation walls, and an additional eighteen inches on each side should be allowed for wide footings and working space for the masons, so that the dimensions of the hole should be 23x33 feet. Stakes are

set six inches outside the digging line, as in Figure 4-3 (A) to avoid accidents to them. Any temporary guide pegs the operator needs are set along the edge, or just outside it.

OPEN FRONT METHOD

Top Layers. The bulldozer is first worked along the short dimension, inside the stake lines. The blade may be dropped at the front or south line for a fairly deep bite, and when filled it is lifted to ride the load over the undug ground until the north line is crossed. The dirt may be dropped at the line or pushed a few feet back. The dozer then backs to the south line and takes another bite in a strip adjoining or overlapping the first. It may work the whole width of the front line, as in (A), or only a section of it, before digging the area over which the soil has been pushed. The back edge of the cut is worked north by successive bites until the rear line of the excavation is reached, approximately as shown in the cross section (B).

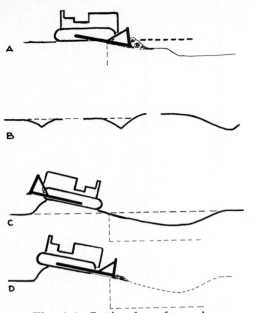

Fig. 4-4. Cutting down from edge

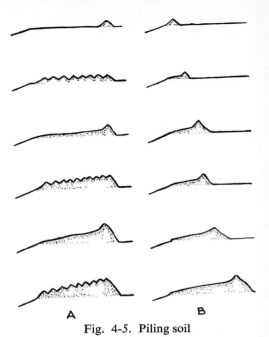

Fig. 4-5. Piling soil

After completing the removal of the top layer, the dozer may cut out and pile a second layer in the same manner, as in (C). This deeper cut will not extend to either digging line, as the slope down and the slope up are kept inside these lines. The east and west edges are also tapered in to provide slopes the dozer can climb.

This digging can be done about as well by making the first cuts against the north line and working successive cuts back toward the south line. However, the machine should not start a cut at the front line and make it shallow enough to continue to the back line, as the long move to the front is partly wasted unless a full blade is obtained there.

If the soil is so hard that the blade cannot be filled in a short distance, the dozer may be worked over a short digging area only, for several passes, after which the pile of loosened dirt may be pushed to the spoil pile. If a ripper or subsoil plow is available, it might be profitably used to break up layers of dirt in advance of digging.

Ramping Down. A hydraulic dozer, which is preferred for cellar work because of its ability to cut hard soils, will usually not cut a steep ramp down without special handling. Figure 4-4 (A) shows a dozer cutting down from a line in soft soil. If digging is good, the blade will penetrate rapidly until maximum depth is reached. It will make a level cut until the machine's center of gravity moves over the cut. It will then fall forward, and the blade will resume cutting until full penetration is reached, when it will level off again. Such a series of steps makes an unsatisfactory ramp.

One way to make a smooth ramp, shown in (B), is to start it well back from the edge with a gradual curve that is made steeper at the digging line. Cutting is then regulated so that the full depth of penetration will be reached as the center of gravity crosses the steepened part of the curve. Several passes are made in digging this ramp as indicated. This curve may be made steadily steeper as depth increases because the tractor itself is at a downward slope.

If the cut cannot be made far enough back into the bank for ramping in, the procedure shown in (C) may be followed. Dirt is pushed out of the excavation into a steep pile, the dozer is backed up on this, and

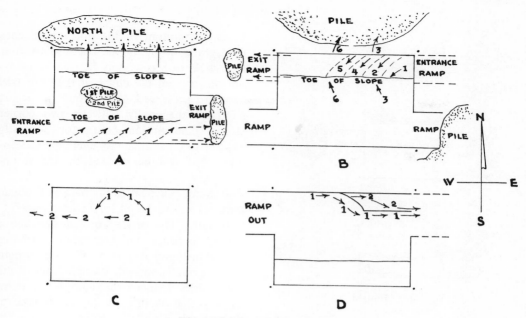

Fig. 4-6. Cutting lower layers

is thus pitched down a steep enough angle to cut down sharply.

Piling. The area to be occupied by the pile should be calculated or guessed at, and the first piles placed at its far side. Successive loads may be dropped toward the excavation, then a load carried over these, pushing the tops off some of the heaps, and dropped at the back. The pile is thus built up in a series of wedges, with their thin ends towards the excavation, as in Figure 4-5 (A).

Or the first load may be dropped at the near side of the intended pile, and the next pushed through it with a raised blade so that the approximate up slope is established from the beginning, and successive loads dumped off the back as in the (B) series. In either case, the pile may show a tendency to build toward the hole, and may have to be dug into severely to cut it back to proper distance.

The up slope of the pile should be made according to the power and traction of the dozer, the judgment of the operator, and may be between 1 on 5 and 1 on 2½. The easier gradients make it possible to push larger loads, but they must be moved farther and require more ground space.

Lower Layers. After the digging has reached the half depth, the dozer should be moved out onto the west bank and headed east along the south line of the excavation. An entrance ramp should be started several feet back of the west line, as in Figure 4-6 (A), and a steep ramp cut down toward the two foot level. As the blade fills, the machine is swung toward the center of the hole and the dirt left on its floor. The dozer returns to the south line, makes another cut, again swinging the spoil out into the center, pushing it somewhat farther. It thus cuts out the full width ramp by which the dozer entered the pit from the south, and occasionally shifts to pushing the dirt obtained from the ramp up on the north pile. As the east edge is approached, dirt may be pushed up on its bank, cutting a ramp, instead of to the north.

The north slope may then be cut away in the same manner, the spoil first being pushed up the undestroyed section of the ramp, as in (B), and finally onto the west

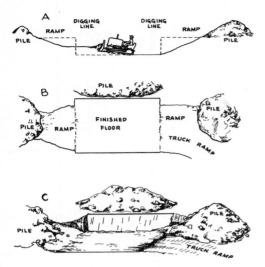

Fig. 4-7. Finishing

bank. These cuts result in vertical walls along the two long sides of the hole.

The inside ramps may also be removed by oblique passes from near the center (C), steering the dozer so that the blade is parallel with the bank by the time it reaches it. The slope ahead of the dozer can be gouged away in this manner, with the spoil being edged out into the open and pushed up the west bank. With the west section of ramp removed, the dozer can be turned to cut away the east part as in (D).

Finishing. After removal of the north and south slopes, the floor of the cut may be deepened by pushing to east and west. These ramps up to original grade should have their lower ends at the excavation line, and will therefore be cut back deeply into the bank, as in Figure 4-7 (A). Since a steep ramp means less extra excavation, the slopes should be made as steep as practical. They may be cut all the way through to begin with, for an easy gradient, and steepened as the hole deepens. The amount of dirt removed for ramps can be slightly reduced by narrowing them as they go up, as in (B). The ramps can be partly filled by the last loads pushed, as in (C), except where space is left for supply trucks.

As the bottom of the hole approaches final grade, it should be checked. A four foot stick, or a rule, may be used to measure down from the edges, if they are well enough preserved, or from a string stretched diagonally between two corner stakes. The mason contractor generally expects to do some hand leveling and trimming, but will be pleased to find it unnecessary.

It is not practical to dig narrow footwall trenches below the pit level with a dozer.

Results. The procedure outlined above should produce an excavation such as shown in Figure 4-7 (C) with two straight walls correctly spaced. Ramping out of the short sides reduces the amount of extra digging that is one of the drawbacks of dozer work. The whole front is left free for access and for piling building material.

However, the spoil may not be properly placed for backfill and grading. In such a location, a house would usually have the fill spread all around it, with particular attention to building up the front. Here, fill for the front yard would have to be obtained from the sides, which, in turn, might need to be partly replenished from the back. This involves extra pushing, and the presence of a few trees might make distributing it a major problem.

Time and Cost. This cellar involves about 120 yards of excavation. It might take a 40 horsepower bulldozer (a good size for the job) from one to three hours to strip the topsoil, and two to six hours to do the digging.

The machine might rate between $10 and $15 per hour, plus $10 to $40 for transportation to the job. Because of lost time between jobs, additional time or a full day might be charged.

Big savings can be effected if two or more jobs can be done in the area at the same time. It is often possible to dig one cellar, and backfill and grade for another, on the same trip.

Rock (even if not removed), water, irregularities, or obstacles will increase the cost of the work.

OTHER DOZER WORK

Four Pile Method. Another pattern for digging this same cellar is shown in Figure 4-8. The soil is pushed in four directions. The east and west ends, for a distance of perhaps eight feet from the line, are pushed up on the east and west piles, respectively, with ramps cut out of the bank and partly refilled as in the previous example. The section between these two cuts is pushed onto piles to the north and south, with ramps cut out of the bank beyond the digging line, each pile being supplied mainly from soil on its side of the center.

By this plan, the whole surface of the excavation can be worked down as a unit, the dozer always pushing dirt to the nearest pile. The four directions of push may be taken in rotation, or varied according to the operator's inclination.

Advantages of this method are efficiency in that pushing distance is kept to a minimum, and adaptability to grading plans. The four spoil piles are shown to be about equal, but their relative size may be changed without varying the method. Access and space for material is not as good as with the open front method. A larger amount of dirt is dug out for ramps and must be pushed back later.

Grading. If the topsoil has been pushed well back, it is possible to rough grade the fill at the same time that it is pushed out of the cellar, or immediately afterward. This has the advantage of making all four sides accessible to the builders.

Ordinarily, it is easier to spread several small piles than one big one. For this reason, it is wise to stop excavating occasionally and to spread the piles which have been built up.

The grade should be kept at least a foot above its final level, to provide excess soil for backfilling around the foundation.

If the digging is wet, the spoil may be too sloppy to spread immediately.

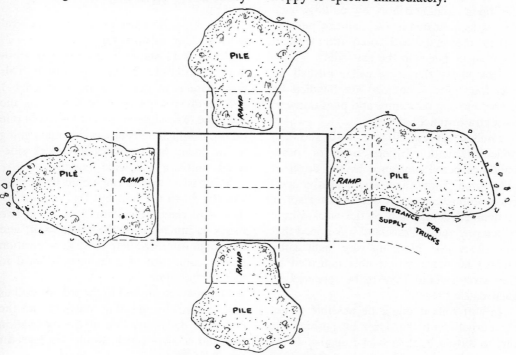

Fig. 4-8. Four-pile excavation

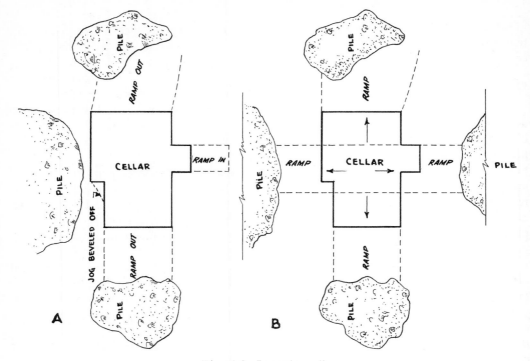

Fig. 4-9. Irregular cellar

Irregular Cellar. Figure 4-9 shows a cellar of irregular shape. This may be dug by the open front method by cutting back the jog as indicated in (A), and by pushing dirt for the small south room into the main cellar to be piled on the east side.

(B) shows the same cellar pushed up into four piles. The jogs are handled by simply moving the ramps and piles outward the extra distance.

Limited Access. Figure 4-10 shows a difficult situation where trees or other buildings permit entrance by the dozer at only three points, and where all the spoil must be pushed out at one of these spots. A detailed description of the work would be lengthy and tedious, and it is hoped that the successive diagrams are sufficiently clear. The dozer movements indicated by the arrows would have to be repeated on each cutting level.

In soft soils it might be possible to cut the corners with the dozer by pushing the dirt to loosen it, then backdragging. How-

ever, it is usually cheaper to dig them by hand and to throw the dirt out into the dozer path.

This type of excavation takes several times as long as open digging.

Figure 4-11 shows a succession of steps in excavation of a cellar in a hillside. This floor base is at grade on the east (right) and requires a cut of eight feet along the west line. The slope of this hill is about one on three, which a dozer can negotiate going up or down, but cannot safely work sideward, unless it is a wide model.

The easiest way to dig would be to push straight down the hill, but the yardage removed to ramp down would be about two thirds as much as for the cellar itself; and the spoil would be left in such a position that it could not be conveniently used to backfill the ramp.

Dirt may be pushed to the side as well as the bottom by pushing down from the center of the upper line of the excavation to build a step which would be level or

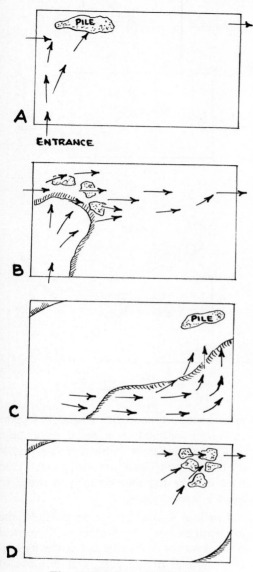

Fig. 4-10. One-exit digging

the dozer must do the whole job, the back wall cuts can be continued into slot ramps, which should be at least a foot or two wider than the blade to avoid getting the machine jammed against boulders or in slides.

SHOVEL DOZER

Digging. A dozer shovel or front loader, with a standard full width hydraulic dump bucket, may dig an open type of cellar in the same manner as a bulldozer. It should be expected to do a quicker job than a dozer of the same power, because of better penetration in hard soil and ability to push larger loads. An exception would be wet slippery soils where the smooth tracks of the shovel would not grip.

The ability of the dozer shovel to cut straight ahead, then back and turn with the load, makes it possible to reduce the amount of excavation outside the digging lines for ramps. If the cellar is dug by the three pile or open front method, the procedure shown in Figure 4-12 may be fol-

tipped oppositely to the hill slope. The dozer may then be operated on this step, pushing to the north and south, along the west line. The step will broaden as it is deepened, and will allow the dozer space to push downhill, first diagonally and then directly, in addition to the back cut parallel to the rear wall. By leaving the two rear corners for hand cutting, the excavation can be completed with small ramp cuts. If

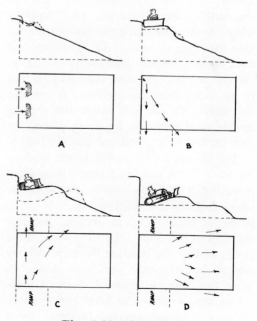

Fig. 4-11. Side hill cellar

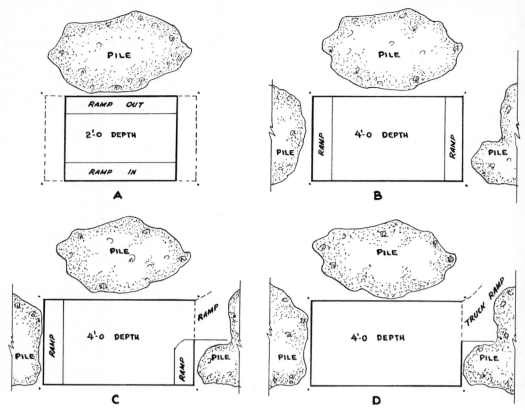

Fig. 4-12. Three-pile system

lowed. The first layer is not cut quite to the ends of the excavation, (A), and the ramps built in reaching the bottom are inside the digging lines, (B). They are cut back to the steepest practicable grades on the last pushes, and then one ramp, a foot or so wider than the machine, is cut into the bank. All soil left inside the digging line is then removed by being picked up and carried or pushed out the slot ramp.

The machine can carry a larger load up the ramp when moving forward than when reversing. This is because it is nose heavy when loaded, a condition which is made worse in backing up an incline, both by the shift of the center of gravity toward the front, and by the reaction from the driving torque which pushes the front down. In ascending a grade forward, both of these forces tend to raise the front and improve stability. These effects become more pronounced as the incline becomes steeper.

Turning around in a small excavation may be difficult, or impossible, so that it is often better to back out with a small load than to take the time to turn in order to carry a larger one.

Cutting Walls. So far as possible, the excavation should be finished with vertical walls. Long ones may be cut neatly by digging parallel to the edge, but because of limited space, much finishing must be done by working straight toward the digging line.

Until the line is reached, bank digging procedures in Chapters 10 and 16 may be followed.

A vertical face may be cut by allowing bank resistance to move the tractor back as the bucket rises. This may be managed by slipping the clutch with mechanical drive, or slowing the engine that has a torque converter. If the wall is high, the machine must be moved forward again after the bucket lip gets above the push arm hinges.

This back and forth motion compensates for the curved path of the rising bucket. The same effect may be achieved by careful regulation of the bucket tilt, as it makes a shallower cut when rolled back than when flat.

Any jogs or irregularities can be cut by digging into the wall from the excavation without making additional ramps. It is easiest to take these out in layers as the floor of the hole is worked down, but they can also be dug after completion of the main work.

Soil carried out of the hole may be spread or distributed to nearby low spots much more easily than by a dozer, and can also be readily loaded into trucks.

Crawler loaders are usually preferred for cellar work, but it is done by four wheel drive machines also.

Digging Under Buildings. A special feature of most dozer shovels is their ability to dig cellars under existing buildings. Particular attention must be paid to bracing the structure over the hole through which entrance is effected, and across any interior pillars which are to be moved. Digging should be done cautiously to avoid damaging the building through collision with rocks, beams, or soil pipes in direct contact with it.

If the building is not large enough so that the machine can dig a turning place inside it, it will be necessary to use hand labor to cut back to some of the walls.

Ventilation is very important, and usually requires at least the use of a powerful electric fan to keep the air reasonably free of exhaust fumes.

A machine that is to do much work indoors must be diesel, with a scrubber (fume reducer) on the exhaust.

Hydraulic shovels equipped with telescopic booms can dig under a building from outside, either directly or by removing hand-dug dirt.

BACK HOE SHOVEL

The hoe (backhoe or dragshovel) is the dozer's principal competitor for small cellar work. There are three types: full revolving cable operated, full revolving hydraulic, and tractor mounted part swing hydraulic. Cable models are handicapped by lack of down pressure, and no control of digging angle. Tractor mountings get in difficulties because of restricted swing, and often limited reach. But all of them can dig a cellar efficiently.

Any hoe is capable of shallow digging, but compares most favorably with the dozers when the hole is to be over six feet deep, or when unfavorable bottom conditions, such as water, mud, boulders, or ledge are encountered. It is able to take care of any necessary ditching without change of attachments.

It is recommended that the digging lines be set a few inches outside of the required excavation, although in even textured soil the back hoe can do a very exact job. In addition to the corner stakes, intermediate guide pegs should be set at short intervals along the digging lines, as the operator cannot sight along these lines without getting down from the shovel, and the finished wall is established with the first cut.

Figure 4-13 shows the twenty by thirty cellar with the depth increased to eight feet.

Lining Up. Accurate lining up of the machine is essential for a clean job. If the cut is to begin along the south line, the shovel is placed as in (A) with the bucket about three quarters extended and resting a few inches beyond the southwest corner. The boom and the tracks are parallel with the south digging line. Lining it up in this manner is greatly simplified by marking the width of the bucket (including side cutters, if used), centered, on the bottom of both dead axles, with paint, or better, stubs of welding rod. Sighting across the outside pair of these marks from the rear, the outer edge of the bucket should be exactly in line

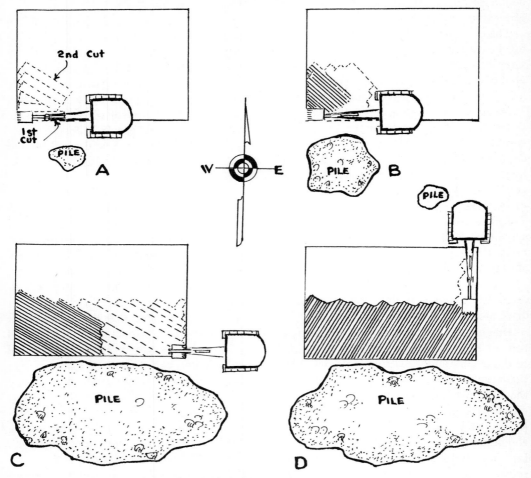

Fig. 4-13. Cellar digging with a hoe

with them, and all three points should be on the digging line.

Digging. A ditch is now dug to bottom grade with its left edge on the digging line, and the spoil is dumped to the south. The far end of the ditch will curve rather sharply inward. When the shovel has dug as much of the south wall as it can from its position, it reaches to the center of the west line and digs a trench back from there. The triangle included in these ditches is dug in layers to bottom grade, which may be found by a measuring stick, and as near to straight down from the west line as possible.

The shovel is then backed up a few feet to position (B). It can now cut the west end of the ditch to be almost vertical because of the more extended position of the bucket. The south wall ditch is then extended as near the shovel as possible, and material between it and the center cut down in layers to the bottom. The center line will be irregular.

The spoil pile will tend to build up too sharply at the edge of the hole unless pushed back. This pushing may be done by regulating the outward swing of the bucket during the dump so that it strikes the pile at a spot where its momentum will push a considerable quantity of dirt outward, without stopping its own motion. Knocking dirt back should be started early, before the pile gets high. The hoist clutch must be engaged during this operation. The quantity of dirt that can be put in a pile is greatly increased this way.

Digging is continued in the same manner with careful attention to a clean, level bottom until the east end is reached. The shovel can probably cut this to a nearly vertical wall immediately in front of it, but will leave a ragged edge, as (C) further north. The shovel is then turned and walked into the unexcavated north section. When its center is a half bucket width inside the east line, as in (D), it stops and shaves the end of the excavation, then trenches to the north edge. It next straddles the north line, and is lined up in the same manner as before, with the bucket resting in the hole in the northeast corner.

The north section is excavated in the same manner as described for the south, and completes the excavation. The west edge may be cleaned up, if necessary, by turning the shovel to walk parallel to the edge, so that the bucket can dig straight up. The shovel should not be put in this position, however, unless the soil is firm and is known to have good load bearing qualities, as a crawler machine is vulnerable to cave-ins or slumping under one track.

This edge may also be trimmed from the north and south banks.

The completed excavation and spoil piles are shown in Figure 4-14. It will be noticed that the piles are somewhat offset from the hole, so that the south pile can easily be used for fill on the east end, and the north pile on the west end. Both ends are left open for access and storage.

The north cut could have been made in the same direction as the south cut, if the east ditch were shorter, but the fill would then have been concentrated toward the east end.

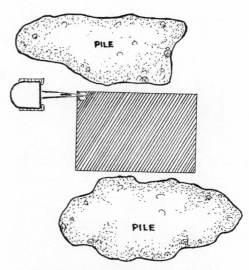

Fig. 4-14. Finished excavation

If the spoil showed a tendency to slump into low piles, or the cellar were deeper, a cable hoe might not be able to stack the spoil in these two piles, unless it occasionally left the digging and walked behind them and dragged them back, thus making room for more.

A more finished hole could be made by starting the digging with a ditch along the west edge, dug from the south. The spoil pile would largely block the access to that side, unless the soil were piled in the cellar area for rehandling. If access were not important, this ditch could be widened toward the center, reducing the amount to be piled to south and north. Existence of such a ditch would make it necessary to work the north section toward the east.

Loading Trucks. A hoe can load the spoil in trucks instead of dumping it on the ground. Where the piles will be so large that they will have to be dragged back, a truck or trucks may be used to take part of it away, the shovel continuing to dump on the piles when no truck is in loading position. If grading plans have been prepared requiring use of the fill away from the foundation, it may be cheaper to truck

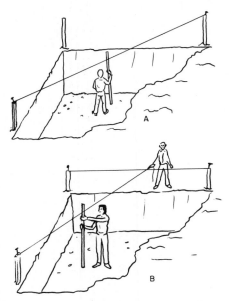

Fig. 4-15. Checking bottom grade

ment or a string level. A stick should be cut nine feet long.

If the operator is checking the grades alone, he may fasten a taut string between two stakes so that it will go over the spot in question and measure the distance from the floor to the string with the stick. Spots which cannot be crossed by the string, or measured directly from the height of the wall, may be checked from a known spot by eye, or with a hand or carpenter's level. See Figure 4-15 (A).

If two men are doing the work, a string may be stretched between stakes on one side. Another string fastened to another stake across the excavation may be held in any desired position on the first string, as in (B), while the other man holds the stick.

Hand, transit, or laser levels may be used. A long rod is required when the instrument is set outside the hole.

A shovel cannot cut a perfect floor to a pit because of the projection of the teeth and the tooth bases. The smoothest grade is obtained when the bottom of the bucket is used for finishing rather than the teeth.

Irregular Edge. Figure 4-16 shows a cellar of the same irregular shape as that in Figure 4-9. The principal considerations in doing complicated excavations with a hoe are to avoid digging it into a trap; to avoid surrounding it or blocking it from other work by piles of spoil; and to work either parallel or at right angles to outside edges.

There are several ways in which this cellar can be dug. The north side can be dug from the east end as in (A) and (B). When the jog is reached it is finished off with a vertical cut, the shovel backed away, and brought back in position to cut along the inner line. If the start is made at the west, the cut is brought a little beyond the jog, and position then shifted to dig along the outer line.

The machine may dig the south side by entering from the west and starting at

it than to push it later with a dozer. However, enough of a pile should be left by the hole for backfilling between the foundation and the edge of the excavation.

The pull shovel can dig footing trenches below the floor level where it is working parallel to the edge, as along the south, east, and north walls in Figure 14-13. The hydraulic hoe can dig them anywhere, but the parallel position is easiest.

Checking Grade. Cutting the bottom to proper grade is more difficult with a shovel than with a dozer, as the shovel operator looks down at the grade rather than along it; has more difficulty climbing down to check it; and cannot move the machine back to grade over mistakes.

It is very helpful to the operator to have a man to check his work, although he can manage alone if necessary. The corner stakes, and preferably some other stakes, may be marked at a certain height, as nine feet above the bottom. In a level field the marks would all be a foot above the ground; in a sloping one the highest stake should be marked a little above the ground, and the others with the aid of an instru-

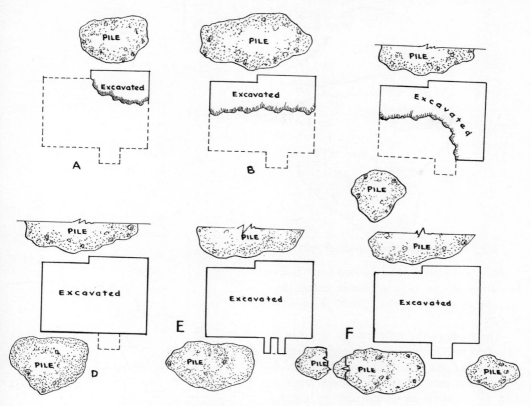

Fig. 4-16. Cutting jogs with a hoe

the southeast corner. Excavation is carried back to the west line first ditching the edge then digging out the center. Care is taken to begin the spoil pile well west of the south room. The shovel is then moved off to the south and up to the room. It is first lined up to the west side of this room and makes a cut from the main excavation to the south line of the room. It is then moved so as to cut the east wall of the room, then digs out the rest of the room.

Another way to do this would be in the same manner as the north wall, treating the room as a double jog. However, this would involve extra digging because the bucket needs considerable width in which to cut down.

If the excavation site is a hillside, the work should be managed so that shovel tracks will head up- or downhill, not across. If the grade is steep, the shovel should dig from downhill to avoid danger of being pulled into the hole if the bucket hooks into something solid.

If work must be done from the upper side the stability of the ground should be checked, and both tracks must be securely blocked against sliding.

OTHER SHOVEL RIGS

Dipper Stick. Dipper stick shovels are seldom used in residential cellar excavations, but they can do a good job. For satisfactory results, the ground should be firm at bottom grade, and the spoil should build into steep-sided piles.

A 20x30 cellar, six feet deep, can be dug as shown in Figure 4-17 (A). A somewhat larger excavation would be dug according to the diagrams, Figure 4-18. In each case, a ramp must be dug outside the excavation line with a slope, usually 3½ and 2½ on 1, which the shovel can climb when the job is finished.

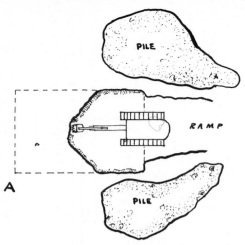

Fig. 4-17. Large dipper in small cellar

The spoil piles can be pushed back from the edge of the hole to some extent by crowding against them with a closed bucket.

The walls of the hole tend to slope in at the bottom, and to be somewhat jagged because of the different angles at which the bucket cuts them. Both these features can be reduced or eliminated by careful digging, but extra time will be consumed.

Clamshell. Clamshells are not ordinarily used in this type of excavation because they do not move as many yards an hour as competitive types. However, they turn out as accurate a job as a hoe, and for small deep excavations, will do the work cheaper than a bulldozer. Digging is done from the top so that no ramps are required.

An edge may be cut with the tracks parallel with it, and the tagline chains attached to one jaw, or with the tracks at right angles to the digging line, and the tagline on both jaws. Either of these arrangements will permit cutting straight sided trenches along the outer lines. The center is best cut in layers, or in sections behind completed edges.

A medium or heavy duty bucket should be used. The spoil may be placed in isolated piles, in windrows, or in trucks as desired.

MULTIPLE CELLARS

In many residential developments, small

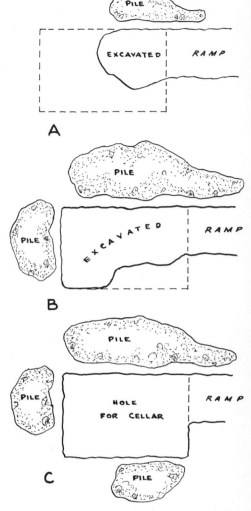

Fig. 4-18. Two-pass dipper excavation

houses are built close together in straight rows. Under such circumstances it may be economical to dig a wide trench straight through the block, and backfill between the houses after the foundations are built.

In Figure 4-19 the digging is done by a dragline or a large dipper shovel which piles the spoil on both sides. Materials for foundations are trucked and piled on the floor of the trench. After the foundations are up, the piles are bulldozed around and between them, and the surplus used to build up the grade throughout the area.

Use of part of the spoil for raising the

ground level decreases the depth of digging necessary.

Another method of line digging is to use a dipper or tractor shovel, load all spoil into trucks, sell or dispose of part of it, and use as much as is necessary for backfilling spaces between other houses in the same development.

A third method is to haul away all the spoil for use elsewhere, and obtain backfill by dozer cutting on the uphill side of the houses.

It is also possible to dig such a trench with dozers, working across it and piling soil on one or both sides. Or scrapers, working the long way of the cut, can haul to any desired location.

The first method requires at least partial backfilling as soon as the foundations are up, to provide access to building material. The weight of the fill and of the large dozers commonly used, are likely to break in uncured foundations, particularly if they are not braced by first floor beams.

A large dragline might be able to pile all the spoil on one side to allow access to the other.

The other methods give immediate access to the front and back, and allow space for piling materials. Backfilling can be postponed until the building is completed. Deeper excavation is required as the general grade is not raised by the dug material.

Relative cost will depend partly on the value of the spoil removed.

HAUL-AWAY DIGGING

Trucking Spoil. In large commercial excavations of the haul-away type, one of the most important considerations is arranging for the disposal of the spoil, except when it is to be used as fill on the same project. It may be possible to sell it profitably, or it might be necessary to pay someone for the privilege of dumping it.

Disposal arrangements may not only determine the price to be charged for the dig-

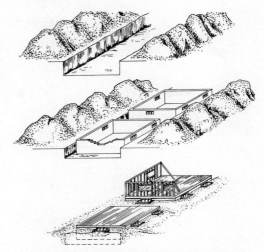

Fig. 4-19. Multiple cellar cut

ging, but also the time of starting the work, and the number and type of excavating and hauling units to be used.

The distance to the dump may be much farther for the trucks than for a car because of restrictions on trucking on residential streets. An inspection should be made of the dumping site, and any price quoted for fill should contain provision for additional charges for dumping delays.

Permits. An excavation contract should specify that the owner, or general contractor, should be responsible for obtaining all permits necessary for the work. If the job is stopped because of failure to have such permits in order, the excavator should have the privilege of charging for the tied-up equipment on an hourly basis.

Machinery. In a medium to large excavation, a crawler loader or a dipper stick is usually preferred, unless the bottom is too wet or sandy for trucks. The dipper needs less space, does not tear up the floor, is better in rock, and can tolerate rough or steep footing.

Increasing size and decreasing depth favor the loader. It can replace the shovel on most jobs.

A hydraulic hoe of sufficient size does the neatest work. If the bottom is bad it can

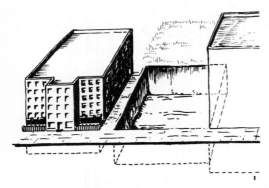

Fig. 4-20. Haulaway cellar job

load on top, but production is greatest with haulers below it.

Ramps. In most cases, the dipper shovel cuts a ramp down, inside the digging lines, which must be of such grade and material that loaded trucks can climb it. The grade may be between one on five and one on twelve, depending on the power of the trucks and the loads placed on them. The slope is made as gentle as the length and depth of the hole permit, for larger loads and fewer breakdowns.

If the plot is sloped, the ramp should be cut in from the lowest point on the edge to which trucks have access.

Earth ramps are generally removed immediately upon completion of the excavation they serviced. It is usually necessary to bring in a backhoe or a clamshell for this job, as the yardage is too great for hand labor.

Timber ramps afford less tractive resistance and better footing than earth, and so can be built with a steeper slope. They can be left in place during construction of the foundation for convenience in moving building material. However, timber work is so expensive that these ramps are largely limited to use in excavations that are very deep in proportion to their size.

If space allows, the ramp may be dug outside of the cellar area. This involves extra digging and backfill, but may be justified if a dipper is the only shovel available

and spoil can be piled near by, or when other work on or near the premises will produce enough spoil for backfill. Refilling cannot be done until the foundation is built and first floor timbers placed, and it should be carefully compacted so that it cannot soak up enough water to become a liquid mud and exert hydraulic pressure against the wall.

Pit Floors. The shovel may dig the floor of the pit exactly to grade, or may dig portions of it below grade, to allow room for disposal of spoil from hand dug trenches or removal of the ramp.

EXAMPLE

Note for the Third Edition:

It is now common practice to use larger machines, and often different machines, than are discussed in this Example. However, the problems and the basic approaches are the same, so the original description has been kept, with few changes.

Now, a contractor would probably prefer to use a dipper shovel rated at one to 2½ yards, or a 2 to 4 yard crawler loader. His hoe would be hydraulic, one yard or larger.

Figure 4-20 shows a basement layout for a business building. The excavation is to be 90′ by 120′ for an 88′ by 116′ building, eighteen feet deep, in a level plot measuring 100′ by 200′. The structure will be against the sidewalk line, seventeen feet back from the curb, will touch an existing store building on the east, a 12-foot driveway backed by another building on the west, and a proposed parking lot. The footings of the east building are below the eighteen foot level, those on the west are twelve feet below the surface.

The site has been cleared, but two large stumps have been left. The topsoil is of such poor quality that it will not be separated. The contractor owns a three quarter yard shovel with dipper, pull shovel, and clamshell attachments. This machine is somewhat undersize for the job but may be used unless a premium is placed on speed.

The dipper stick is used because of digging and loading efficiency. The ramp is located at the street as there is no access to the other sides except through the driveway, which is too narrow for heavy trucks and will be undermined by the digging. The ramp is next to the driveway, as if the other corner were used the store building might be damaged by collision or vibration.

Moving In. Arrangements are made to prohibit parking in front of the work area before machinery is brought in.

The shovel is unloaded from the trailer onto planks laid on the street to protect the pavement, or directly onto the sidewalk which is not protected because the trucking will destroy it anyhow. Digging is commenced in the sidewalk, or at its rear edge, and sloped down at about a one on five or 20 percent grade, in a cut thirty to forty feet wide. Trucks are first loaded when standing in the street, parallel to the curb. As the digging progresses, they are backed across the sidewalk and down the ramp, as in Figure 4-21.

It is necessary to have one man, preferably two, assigned to prevention of tangles between traffic and trucks. These should be policemen, or contractor's employees authorized by the police to do this work.

Trucks should be backed into both sides of the ramp, as in (B), and faced directly away from the shovel while being loaded. At least one truck should always be in loading position.

Trimming. As the cut progresses, the foreman checks the left edge for accuracy. Because of the angle at which the bucket works, it cannot make flat cuts on the wall, and one or more laborers should trim the face, either working from the top or from the ramp surface beside the boom. Checking may be done by stretching a string along the digging line at the top, and lower-

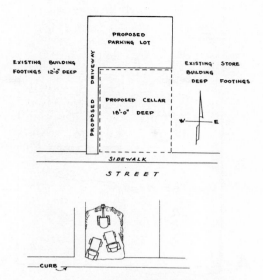

Fig. 4-21. Starting large cellar

ing another string to which a plumb bob is attached.

Bank Height. The shovel would be able to take the full eighteen foot depth in one cut but this would not be good procedure. The shovel cannot cut a straight face higher than the level at which the bucket teeth start to turn away from the bank, and the face above might overhang or break back beyond the digging line. Hand trimming is more difficult on a high face, and caving is more likely and more serious than from a lower one.

The bottom is more likely to be muddy, or to contain rock outcrops than higher levels, and it is economical to remove as much soil as possible under good conditions, before tackling the difficulties. In addition, trucks hauling from an upper level have an easier climb to the street.

Under average conditions, this job would be dug in two layers or benches of about nine feet each.

Dozer. A bulldozer or front loader may be used to advantage to dress up the ramp, smooth the pit floor, and be ready to push overloaded or weak trucks up the ramp.

A crawler loader is preferred, as it does

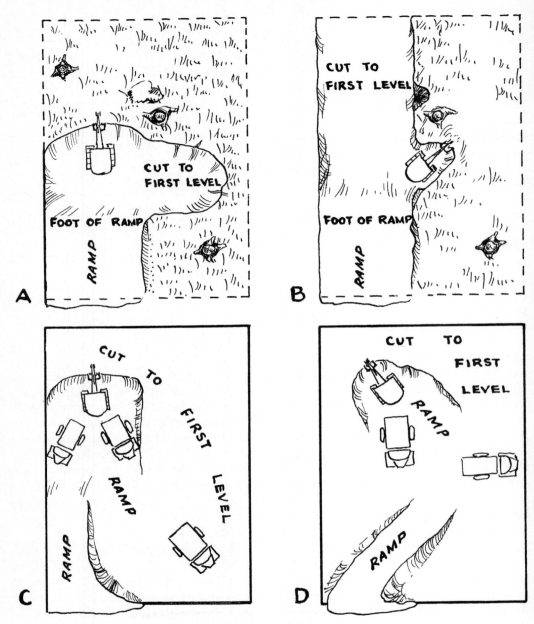

Fig. 4-22. Cutting first floor and ramp

better trimming, and can help with loading in an emergency. A big one is most useful, but also most in the way.

Trucks. Trucks with six to eight yard capacity are well matched to this size shovel, but anything from six to fourteen yards can be used. Small ones can maneuver more easily and work well on certain types of soft ground. Large ones cause less traffic congestion, and under favorable conditions will move soil at lower cost. If the pit is wet or sandy, all wheel drive machines are preferable. Trucks must be in good condition to carry capacity loads up the ramp.

First Cut. In making the ramp, the shovel has worked straight ahead. Upon reaching the floor level of the first cut, it may continue to the back wall, or it may first make a side cut, as in Figure 4-22 (A), so that trucks can turn around in the pit. If it works straight through to the back wall, it is then walked to the foot of the ramp again to take another slice toward the back, as in (B), as double spotting of trucks is easier working away from the ramp than toward it.

Once the pit floor is widened enough to allow trucks to turn, the digging may be extended in any direction so long as the shovel may be easily reached by trucks. Sooner or later the bank to the east of the ramp is dug away, and usually part of the ramp itself, leaving it wide enough for one truck only. Since there will eventually be an 18 foot drop off its side, it is wise to leave a generous width, and, if the stability of the soil is doubtful, to shore it up as well.

Stumps. When stumps are encountered, the shovel should dig around them before tackling them directly. The depth of this cut is such that they can be undermined enough so that their own weight will help to break them out. Roots can be cut easily at a distance of a few feet from the trunk, and the stubs splintered back. Many operators will waste much time, strain their machines, and break cables by direct attacks on stumps which would shortly fall out in the ordinary course of digging.

When the stump is loose, it should be knocked around with the bucket to loosen the dirt, then placed in a truck. The tailboard should be folded down or removed unless the stump is to be lifted off. With skill and luck, the shovel operator may be able to balance the stump on the bucket, then tip it off onto the truck. It may be picked up also by a chain. If it is too heavy for the shovel to lift, the dirt should be dug out of it by hand, and all possible wood removed from it by sawing and chopping. It may be blasted apart if necessary.

Barricades. As digging approaches the sidewalk, barricades must be erected to keep spectators from climbing or falling into the pit, and from using any part of the sidewalk likely to cave in. These barricades may be solid wood fences, or may be perforated so that sidewalk engineers may watch the work. The contractor may be able to build up local goodwill by encouraging spectators.

Furnishing adequate windows or peep holes reduces the dangerous practice of spectators standing in the truck driveways to watch the work.

Finishing First Cut. The floor grade of this first drop is approximate, and a foot up or a foot down is not important as long as it is easily passable to trucks. However, the walls must be cut to whatever finish the job calls for. A good operator can cut a straight wall and an almost square corner with the bucket, but hand finishing is neater and saves machine time.

Bottom Cut. When the first level is complete, ramp cutting is resumed until the bottom is reached, as in 4-22 (C). Trucks will drive down the upper section, turn on the upper level, and back down to the shovel. It is not necessary that the ramp continue in the same direction, but this is the most economical method where the pit is long enough. Any turn must be made very wide for the convenience of the trucks.

Cutting of the lower level proceeds in much the same way as the upper, except that the strip alongside the ramp is left until the last for any bracing value that it may have; and the floor grade must be carefully watched. This is usually checked with a transit or a builder's level. If foot wall trenches below the floor level are required, they may be dug by hand immediately after trimming of the wall is complete, and the spoil moved to the shovel by bulldozer or wheelbarrow, or spread on the pit floor.

A somewhat gentler ramp gradient could

be obtained by using a diagonal or a zigzag ramp, as in (D).

SHOVEL TEAMS

Two Shovels. This job is big enough to justify the use of two shovels of the three quarter size or larger. The second shovel might ramp down from the sidewalk along the east side, and cut through to meet the first one at the center. After this, one ramp might be used as an entrance and the other as an exit; or one might be cut away. Or the second shovel might be brought after the first one reached the upper level, and using the same ramp assist it on that level or ramp down to the bottom.

Traffic. An external factor which may limit the number and size of shovels in such an excavation is traffic congestion on the street. This may create a bottleneck that would leave a line of empty trucks parked waiting in the street with the shovels half idle for lack of trucks to load. In congested areas, traffic may be one of the principal problems of the digging.

Ramp Removal. When the digging is complete except for removal of the ramp, a hoe or a clamshell must be employed. This ramp may contain three to five hundred yards, a sufficient amount to make the use of the faster digging pull shovel (hoe) better than the clam. If the hoe has an effective downward reach of sixteen feet, it will leave a bit of the foundation of the ramp for hand labor; but the clam, in taking the whole ramp out, is apt to require a larger amount of hand labor assistance while working. Another factor may be that the shovel at this job with a crane (clamshell) boom, might pick up extra work lowering materials into the excavation. The hoe rig is awkward to handle and to transport when detached, so that it might be more economical to move the shovel to the yard to change over and bring it back than to pick up the hoe equipment, bring it to the pit, and take away the shovel front.

The clam rig could be loaded on a truck by a chain hoist or a tractor loader, and moved with little or no blocking.

It might also be good business for the contractor to hire a hoe or clam and move his dipper on to another job.

Whatever machine is used, it will probably stand on the ramp as it tears it up, as the driveway is too narrow for swing space, and for safety at such height.

Teaming Dipper and Hoe. If two shovels are to be used for the whole excavation, it may be that the larger one of them would be a dipper and the other a hoe, although under the ideal conditions considered so far, it is not likely. In such a case, the dipper stick ramps down on the building side, unless danger of vibration damage is unusually severe. The strength of the foundation wall might be checked, and permission obtained to brace it from inside if the weight of trucks on the ramp seem to threaten it.

The hoe shovel is assigned to cutting the north and west walls of the pit because of its ability to make a smooth straight cut without hand trimming. It would preferably start on the north side, cutting from east to west, keeping the line in the manner described earlier, and digging out as much of the center as could be conveniently reached, as in Figure 4-23 (A). The edge ditch should be made as deep as it could reach, but the rest of the digging only nine feet. The center digging is discontinued in the last few feet of the north line as the shovel is then backed up against the building beyond the driveway, and turns to get in digging position on the west line. Whether this corner could be cut square would depend on the length and tail swing of the shovel, but in general, it could not. In any case, it could not be squared to full depth.

The west line is ditched back to the sidewalk, with some additional material moved from the center as in (C). The hoe can

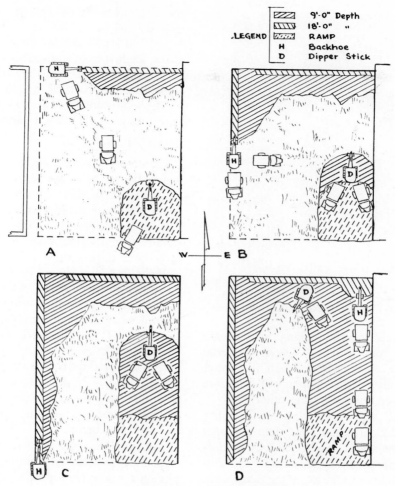

Fig. 4-23A. Teaming dipper and hoe

then work a wide cut back from any convenient starting place, taking care that its efforts, combined with those of the dipper stick below, do not cut off its exit.

The trucks carrying the spoil from the hoe may be loaded sideward to the shovel for safety, or from the back for convenience. If the body sides are very high, loading will be inconvenient and spillage excessive. This difficulty may be reduced by loading directly behind the shovel, so that it will have to walk over the spilled material, which will raise it so that loading will be easier. This spillage needs some manipu-

lation to make a smooth ramp, particularly if the soil contains boulders.

Another method of loading trucks easily would be to start the cut at the sidewalk, making a ramp down for trucks wide enough so that a truck could be backed against one part of the face to be loaded, while the hoe digs beside it.

When the combined efforts of the shovels have removed enough of the top cut so that there is room for both of them on its floor, the hoe shovel may be moved down to do the digging to final grade, while the dipper completes the upper cut, as in (D).

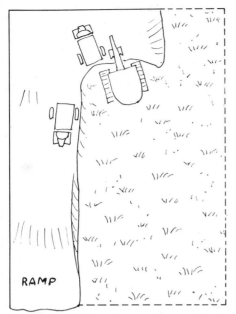

RAMP

Fig. 4-23B. Full depth excavation with a hoe

Both shovels will be working on the same floor, but one will be digging material above, the other below. When the upper layer is finished, the dipper will leave the job to be completed by the other.

The hoe will move from ten to thirty percent less dirt each hour than a dipper of the same size, a loss which may be only partly compensated by the straight wall cuts and the ability to take away the ramp without calling in another shovel or rig. However, if certain difficulties develop, the pull shovel output will be unaffected, while that of the dipper will be sharply reduced, and the presence of the hoe is insurance against undue loss of time from such causes.

HYDRAULIC HOE

This cellar might also be dug entirely by one full-revolving hydraulic hoe, bucket size 1½ yards or larger. Such a machine may take an 18-foot depth in one layer, and cut footwall trenches also.

If ground conditions permit, trucks are loaded on the pit floor, for maximum out-

put. The hoe would cut the ramp first, backing away from the street.

This 90-foot width is best taken in three strips, the first from front to back, the others from back to front. Walls are trimmed and trenches cut as parts of the main digging.

The final step would be to move around onto the ramp, then dig it backing up.

GROUND WATER

The most common difficulty is ground water. It may be in the form of springs or underground streams, or a nearly stagnant water table with capillary water moistening the soil for several feet above it. Wet soils usually turn to mud when loaded or disturbed and impede or bog down trucks.

If the first level should have a firm floor, but water be encountered in the next layer, trucks would not be able to operate on the bottom without expensive aids, so that removal of this bottom layer with a dipper stick would be impractical. The hoe would not be bothered unless there were sufficient water to hide the bottom, in which case it would have to be pumped out. Special dangers connected with such pumping will be discussed below.

Information about underground conditions may be obtained from test borings or pits on the site; from people who have dug cellars or ditches in the neighborhood, and from geologists. Such data may predict with reasonable accuracy the depth at which mud, water, loose sand, or rock might be expected, and digging plans made accordingly.

Special conditions might require taking off the ground in three cuts, or in one. The pattern should be such that the maximum amount of dirt would be dug by dipper sticks, on floors which permit trucking. When thin cuts are made the dipper can load trucks standing on the upper level, the extra dumping height slows the digging and in some materials the bank would not be stable enough to support trucks.

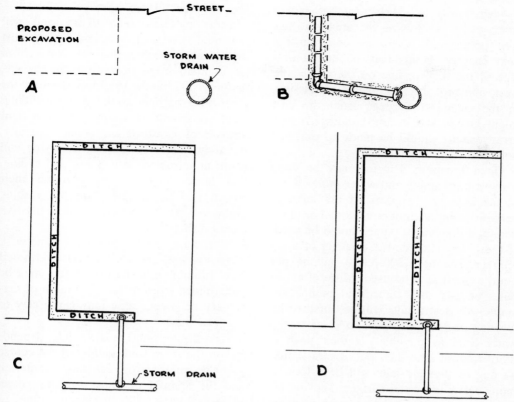

Fig. 4-24. Drainage

Drainage. Mud can be dried by draining or pumping the water. If the storm water drain in the street is sufficiently low, arrangements should be made to connect with it before excavating. A ditch is dug from the pipe line in the street to a spot several feet inside the excavation area, and a pipe with sealed joints laid, opening into the storm drain. At the cellar end, a vertical pipe of tile or concrete sections with unmortared joints or a perforated pipe is erected, as in Figure 4-24 (A) and (B). Sand or clean gravel is placed around the vertical pipe as the trench is backfilled, or a wooden barrier is placed to prevent backfill from closing the hole around it.

Each floor made during the digging should be sloped to drain to this pipe, which can be opened at any level.

This installation will also serve to remove some ground water from the site, before excavation.

A general lowering of the water table may be obtained by ditching on the three open sides, as in (C), or ditching the center also, as in (D). The edge ditches make the digging easier but the interior trenches complicate it. Heavy wood mats are required wherever shovels or trucks cross them, and these are expensive to build and a nuisance to handle.

If the storm water drain is not low enough to be useful, similar ditches may be dug and connected with a piped or open sump from which water can be pumped to a catch basin in the street.

An overloaded storm drain may push water into an otherwise dry excavation, unless a check or shutoff valve is provided.

Well Points. A satisfactory but expensive way of predraining the area is to use well points, which are discussed in the next chapter. Points may be driven outside of the digging line on the north and west, and probably, by special permission, in the sidewalk. Seepage from the east might be blocked by the building. If not, arrangements should be made to put well points in its basement.

Open Pumping. Digging may be done without predraining and water pumped out of the hole as it appears. If the water is very dirty, and quantities are small or modderate, a diaphragm pump should be used. If the inflow exceeds the capacity of a diaphragm, about 1,500-3,000 gallons per hour, several may be used. More often, in holes of this size, centrifugal pumps are employed. Best results will be obtained by locating centrifugal pumps as close to the water level as possible, as their push is more efficient than their pull. Holes should be dug so that the inlet will be a foot or more below the water surface. Sucking air in shallow water may be reduced by floating a piece of board over the inlet, where it will block the formation of whirlpools which would conduct air down to the inlet center, or by arranging the hose so that it rises vertically out of the water.

Pumping may be done on a 24-hour day basis, or only during or just before digging operations. If pumps are to be shut down overnight and holidays in very wet holes, it may be wise to take them up each time, or to put them on floats for protection against unexpected rises in water level. Other equipment should be moved up to a safe level when work is shut down at the end of the day.

Caving. Caving of banks and undermining of adjacent structures must be guarded against, particularly in connection with pumping. Caving banks involve hazards to men and equipment, and to adjoining struc-

tures, and increase the amount of excavation and backfill necessary.

Some materials, such as dry sand, will not stand in vertical walls, and digging must be figured to include natural slopes from the foundation line outward to the surface, or provision made to drive sheeting, or erect other barriers, to hold it from sliding. Sands or sandy soils containing the right amount of moisture will stand vertically, but they cannot be trusted, as drying will result in surface disintegration and sliding, and heavy rainfall may increase their weight and undermine them by washing grains out at the bottom so that massive caving will follow.

Silts, clays, and loams usually stand well, if not too wet, but if resting on a saturated layer draining into the excavation, may be undermined so as to fall. Vibration of machinery or street traffic may cause clay to creep or flow.

Gravel may stand or may slide, depending on the shape and grading of the coarse particles, presence of cementing material, and the amount of fines. Angular gravel of several sizes, with just enough fines to stick it together, will stand wet or dry unless subject to excessive water flow, or wave action. Very clean gravel, particularly if it includes a large proportion of cobbles and rounded pebbles, may slide in much the manner of dry sand.

Causes of Caving. Danger of caving continues for days, or sometimes weeks after the cut is made. In its natural state the soil is in both static and dynamic balance—static because of inertia and the manner in which its particles are fitted and stuck together, and dynamic because the weight overlying soil or structures exerts a sideward as well as a downward thrust, which is met by equal counter thrusts from surrounding soil on the sides and below.

When a cut is made, the soil pressure toward it is balanced only by the soil inertia. This may hold it permanently in place, or the pressure may deform the soil and cause breaking apart and rearrangement of its particles, gradually weakening it until it falls. The effect may be likened to the collapse of a building under the weight of snow on its roof, which may occur hours or days after the storm and even after part of the snow is gone.

Ground water is very effective in both holding and bringing down banks. While in very thin films it serves as a glue or binder. In contact with clay minerals it forms a lubricant, making it easier for particles to change position in response to pressure to such an extent that certain plastic clays will flow slowly. In larger quantities, water will seep or flow through the soil, carrying fine particles with it and cutting minute channels that weaken the structure. The flow of water is much slower through soil than through an open ditch, and it exerts pressure proportional to the restriction of flow.

If the water is allowed to stand in the excavation at its natural level it will cease to carry particles out of the bank, and will exert a back pressure against the bank that will tend to hold it in place. However, this will not prevent the part of the bank above water from creeping under soil pressure or absorption of capillary water, and wave action set up by wind or dropping of stones or clods will cut into the bank at water level and undermine the top.

In general, where unstable soils or abundant ground water is expected, open excavation should not be done until preparations have been made to build walls immediately after its completion; and if construction is delayed, it is better not to keep it pumped dry.

Side Effects of Dewatering. Often the most serious aspect of removing water from an excavation is the effect on adjoining property. Water makes up a substantial part of the bulk of some soils, and its removal, even if it does not carry particles,

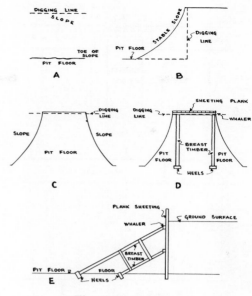

Fig. 4-25. Bracing

sometimes causes shrinkage, with settlement of the surface and overlying structures. Damage to structures may also be caused by creeping of plastic soils from beneath them into the pit.

SHORING

Wall Bracing. Movements of soil into a pit can almost be stopped and water intake reduced by the use of timber bulkheads or sheet piling. These are required by law in many cities, and are often good, although expensive, insurance against costly repairs and underpinning.

Installing such bulkheads is a highly technical operation, involving knowledge of soil behavior, engineering calculations, and skilled personnel. There is sufficient space available in this volume for only a brief sketch of general methods.

Bracing Stable Soil. Figure 4-25 illustrates installation of thorough bracing in an excavation where a short section of face will stand for a while without support. A long section is cut back by the shovel to a slope which is expected to be stable, (A) and (B). Then a short section, perhaps ten

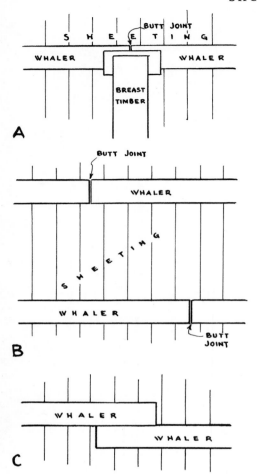

Fig. 4-26. Detail of bracing

Beams or plank mats called heels are placed on firm, undisturbed soil in the pit floor, sloping down toward the wall. These are used as abutments to take the thrust of the breast timbers that extend from the heels to the whalers. These should be 10″ x 10″ or larger. Each whaler must have two or more breast timbers, spaced five or more feet apart. If the heels are firm, the spacing of breast timbers can be increased by using heavier whalers.

While this bracing is being installed, an adjoining section of the wall is trimmed. This is braced in the same manner and the work continued in successive sections.

The sections may be tied together in several ways. The breast timbers may be placed against the whalers where they are butted together, as in Figure 4-26 (A), with or without the plate shown. The joints in different whalers may be staggered, as in (B), or may be overlapped, as in (C).

Nailing is kept to a minimum to avoid damage to the lumber. The bracing is dismantled after the foundation is placed and the material removed for re-use. The sheeting is usually pulled by a crane or pile driver equipped with a special clamp for gripping the top of the planks.

feet, is cut and trimmed to final shape, (C). Sheeting plank, 12″ by 3″ or heavier, is placed vertically against the dirt wall. This plank should be long enough to reach a foot or two below the bottom of the pit, and one or two feet above the ground surface. Bottom penetration may be obtained by ditching, or by driving the planks down with an air hammer fitted with a special head for the thickness of plank used.

Horizontal timbers, called whalers, are placed along the face of the sheeting, being temporarily supported on cleats nailed to the planks. The whalers should be 6″ x 6″ or larger, and should not be more than five feet apart vertically.

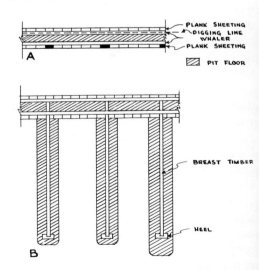

Fig. 4-27. Bracing, trench method

Unstable Face. If the soil is so unstable that it cannot be trusted to stand even in short sections, the sequence shown in Figure 4-27 may be followed. A trench is dug with the outer edge at the digging line. This is braced with sheeting, whalers, and sheeting jacks in the manner described in Chapter 5, except that planks are left out of the sheeting on the inner wall at regular intervals.

Additional trenches are now dug into the excavation at right angles to the edge ditch. Heels are placed in them at or below floor level, and breast timbers run from the heels to the whalers on the out wall, through the spaces in the inner sheeting. The dirt between the breast timbers is now dug out, usually with a clamshell with laborers assisting, and the sheeting jacks and inner wall bracing removed.

Steel sheet piling (see Chapter 19) may be driven along the digging line without the ditch. Breast timber ditches are dug in the same manner as described. A ditch is then dug on the inner side of the steel piling, and a whaler and breast timbers placed. Digging is then carried down to the level of the next whaler, which is placed and braced.

Steel piling does not require as close spacing of the whalers as wood sheeting. A single whaler near the top is often sufficient, and in some cases it is not braced at all.

Cofferdams. When dry excavation is carried a considerable distance below the water table without dewatering the area, the heavy walls constructed to keep out soil and water are called cofferdams. The bracing structures already described can be included in this term.

Cofferdams consisting of a single row of interlocked steel piling, with interior bracing, have been used for depths up to 60 feet, although ordinary practice limits them to 40. They may be installed by hammering the piling in undisturbed ground until it reaches bedrock, or to sufficient depth below the excavation floor to be considered safe. All the sections should be placed and driven to moderate depth before any of them are driven all the way, to make sure that all joints interlock properly.

Excavation is likely to be done by clamshells. Bracing is placed against the inside of the wall as it becomes exposed.

If the soil is very porous, great difficulty may be experienced getting the water down the first few feet, as the joints between sections leak quite freely until forced together by water pressure. More or larger pumps may be used at this stage of the job than at any later time. It may be necessary to trench outside the wall to place a clay seal part way down, or to partially seal the soil with cement grout.

Porous soil under the bottom of the wall may permit excessive quantities of water and sand to boil up in the bottom of the excavation as final grade is approached. If the bottom is in clay, but porous soil is immediately below, the job may proceed easily, and then suffer from a sudden and disastrous blow-up of the bottom.

The double wall cofferdam is a more elaborate structure, which provides means to combat the problems encountered, and to work at greater depths. Two rows of steel interlock piling are driven. The space between them may be excavated wet, and the bottom concreted to form a good seal against a rock bottom and to protect the ends of the piles from being bent in a blow-in by water pressure. The walls may or may not be cross braced to each other. They are filled to the top with clay or other soil.

An area protected by a cofferdam may be dug wet, in which case the structure serves to prevent soil from slumping into it.

Caissons. A caisson is a structure which serves to keep soil and water out of an excavation, and forms part of the perma-

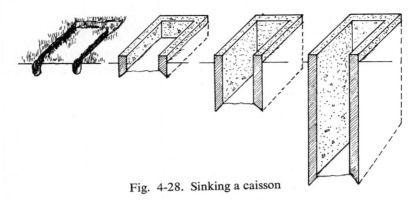

Fig. 4-28. Sinking a caisson

nent structure for which the excavation is made.

A simple type of open top caisson, and stages in its growth, are shown in Figure 4-28. A hollow square, ring, or other shape is made of reinforced concrete, with the bottom tapered to an inside edge. If the work starts on dry ground, it may be built in a shallow excavation where it is to be used. If the start is under water, it is made elsewhere with walls high enough to keep out water when it is lowered into place. Transportation is usually by barge.

The caisson is lowered by digging inside to undermine it, and building the top to provide more weight, and to keep it above ground or water as it descends. Most of the digging is done underwater, and it is a very ticklish job to do it accurately enough so that the caisson will sink straight. When it comes to the bottom, investigation must be made to determine whether it is on bedrock or boulders. If the rock surface slopes, concrete must be pumped underneath to give it firm bearing on the low side.

The pneumatic caisson has an air tight cap over the bottom, with sufficient air pressure maintained under it to keep water out. Air locks and chambers are provided for entrance and exit of men and material. Much of the digging is done by hand, and in deep work at high pressures men may be limited to less than an hour of work at a stretch, with long periods spent in entering and leaving the high pressure work

chamber. Depths up to 100 or 110 feet can be reached.

ROCK

Bedrock. If bedrock is encountered that is too hard for the shovel to tear apart it must be blasted. Generally it is best to complete the earth excavation first, to reveal the full extent and as much of the grain and quality of the rock as possible, before going to work on it.

Sometimes, however, drilling and blasting are started as soon as the rock is found, and the shovel doing the earth excavation can be utilized for handling logs or blasting mats. This may save shovel time, as under city conditions it is not often practical to blast rock fast enough to keep a shovel busy, and a shovel whose only duties are handling mats and removing blasted rock is likely to be idle most of the time. On the other hand, earth hauling trucks will be stopped while the shovel places mats and during blasting.

Hoe shovels and small dozers are good machines for cleaning the bulk of earth off ledges, but there is almost always need for hand work also.

Procedures for the rock blasting and removal are outlined in Chapter 9. However, it should be emphasized that blasting near streets and buildings is a much more dangerous and specialized job than the same work in a quarry or a country highway cut. Elaborate precautions must be taken to prevent material from flying, and large blasts, or small blasts following each other quickly at regular intervals, must be avoided because of danger of concussion and vibration damage to nearby buildings. Jobs must be inspected in advance by the insurance company in order to set a rate in line with the risks.

Boulders. Boulders in the soil slow excavation by creating digging resistance and complications, and often by difficulties of disposal after they are dug out.

The hydraulic backhoe usually does the best job in proportion to its size. The comparatively narrow, toothed bucket with wrist action can work around and under large objects, and has great prying force in its closing action.

It is competent at picking them up, but only if they are small enough to be held in or on the bucket, or clamped between it and the stick. Larger ones may be held against the wall of the excavation, or a specially dug slope, and pulled to the top.

The front loader is clumsier at digging out boulders, but does much better at removing the big ones. The wide deep bucket can pick many up directly, or by crowding against a bank while lifting and curling the bucket. Extra large ones that would fall out can sometimes be chained, as in Figure 16-9A, or just pushed out of the way to wait for equipment to break them.

Boulders may be loaded into a truck by a hoe or a loader, or more gently by a clamshell, grapple, or crane. For crane work, the rock must be firmly gripped by chains or slings.

Chains should be of the lightest size that will lift the weight, as a thin chain grips rock much more closely than a thick one. Undersize chains break frequently, and spares and repair links and hooks should be kept on hand.

Alloy chains are expensive but are small and light in proportion to strength.

Small cables grip rock well but wear and

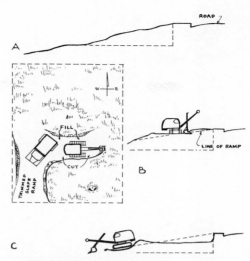

Fig. 4-29. First cut in a down slope

fray rapidly, so that sharp ends of broken wires make them dangerous to handle.

Slings may be made of several strands of light cable or chain, and combine the grip of small sizes with the strength of large ones.

Boulders may be broken by blasting but in city areas mud capping is not permissible. Splitting may also be done with sledge hammers, air hammers, or drills and plug and feather sets.

HILLSIDE SITES

Downslope. So far we have considered excavation in a level plot. As the cellar depth is calculated from street or sidewalk level, a downward pitch to the rear would decrease the amount of excavation, and an upward one would increase it.

If the lot slopes down to the north, as in Figure 4-29, the natural grade can be cut to the proper slope for a ramp by a bulldozer, and the material removed used to build a flat shelf at the first cutting level on which the shovel and trucks can start work. If insufficient dirt is cut in making the ramp, the shovel can dig into the hill and sidecast below, to build it up to the desired size.

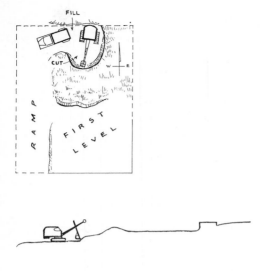

Fig. 4-30. Second cut

Excavation is carried back to the side of the ramp and to the south and west digging lines, in any convenient manner, while the bulldozer shapes the bottom level, making a flat space as before.

When the shovel starts work at the bottom, Figure 4-29, the excavation and ramp removal are carried out in the manner described earlier.

If two shovels are used, one can work on each level. The upper one should work across to the east side and finish it first, so that the one on the lower level can work in without cutting it off.

It is unlikely that a back hoe would be used on such a job, except in removing the ramp, unless mud conditions are encountered. Sometimes soft footing can be economically handled by surfacing the truck road with gravel, crushed stone, or dry fill.

Bottom Access. It may be possible to arrange for the movement of machinery and trucks into the lower end of the lot, as in Figure 4-30. A bulldozer may then cut a

truck road and turn around into one side of the lot. Loaded trucks will now move downhill and maximum loads can be carried.

It may be difficult for the empty trucks to turn on the slope and to back uphill, particularly in sloppy going. If the shovel first digs a wide shelf as in Figure 4-31, the trucks can turn on it, and another roadway can be graded later for exit so that no uphill backing will be necessary.

Cut and Fill Digging. Figure 4-32 shows the same sloping lot with a retaining wall built along its back lines. The spoil from the cellar is to be used to fill up to this wall for parking area.

A dipper shovel and trucks may still be effectively used for the digging, but the short haul makes possible the use of other machines.

However the soil is moved, it should be

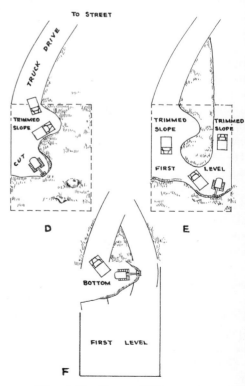

Fig. 4-31. Slope with rear access

spread in thin layers and thoroughly compacted by rolling in open spaces, and tamping where rollers cannot reach. This will prevent serious mud difficulties during the work, possible damage to the retaining wall from pressure or fluid mud after heavy rain, and excessive settling of the finished fill.

The average length of push is about one hundred and twenty feet, very slightly downhill. This is within the economical range of medium to large bulldozers, or small, self-loading scrapers, but help will be needed from a loader or hand labor to cut out the south corners. The equipment should be small enough to leave by the driveway when the job is done.

A dozer first cuts a shelf, level or sloping opposite to the hill at the top, just below the sidewalk. This is done by digging along the edge line until a bladeful is obtained, then turning downhill, lifting the blade at the same time, so that the fill is built higher than the cut to allow for compaction. Pans (scrapers) can be used to cut down this shelf as soon as it is a few feet wider than they are, but a dozer will be needed to keep the walls trimmed back to a vertical. The dozer can also cut much further into the corners than the pan by the process of gouging and then swinging out.

The dozer shovel can square the corners by working against one side, parallel to it, and digging into the bank until the other side of the corner is reached. The spoil is picked up, moved back, and dumped in the path of the pans. Best work can be done if the corners are kept cut down within a few feet of the level on which the pans are working.

The scrapers may be kept moving in a rotary path, as in Figure 4-33 digging at the south end, and dumping and spreading along the retaining wall. As the fill rises, it will enlarge to the south. At the same time, a bulldozer can be working down the center section in the soil to be moved the

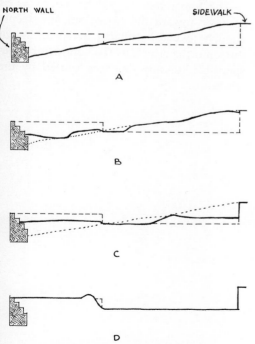

Fig. 4-32. Filling against retaining wall

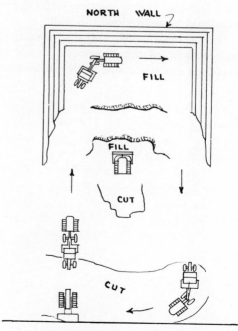

Fig. 4-33. Scraper digging

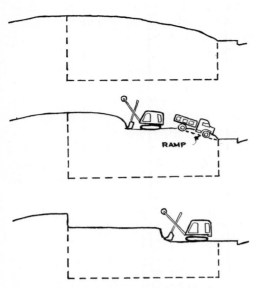

Fig. 4-34. Cellar digging in a hill

shortest distance. This dozer may also take care of trimming the fill for the pans and pushing it into the corners.

A self-powered tamping roller should be kept moving over the fill in both pan and dozer sections. Hand, gasoline, or air tampers should be used along the wall.

The parking lot fill, and that needed in the rear part of the driveway, cannot be placed on the south side until the founda-

tion of the building is in place. Material needed for this can be piled on the edge, ready to be pushed in place.

The sidewalk edge of the pit might also be cut by a clamshell standing on the sidewalk. The dirt could be either loaded into trucks or cast out into the pit in reach of the pans.

If additional fill is needed in the parking area, it should not be trucked in until the building foundation has set long enough to give support to the driveway.

Hill Removal. It often happens that the building site slopes up from the street, sometimes very abruptly. The hill must be removed, in layers if it is high enough, before digging down from the street.

Figure 4-34 shows one such situation. The first cut starts above street level so a ramp is dug up to it. When the top has been removed, digging is started at the street level, and the underground cuts taken afterward. Two or more shovels can work on the job, usually on different levels.

The upper cuts should be sloped so as to drain toward the street, but not steeply enough to cause gullying and washing of dirt onto the street, as the contractor is responsible for any damages caused by the work.

CHAPTER FIVE

DITCHING AND DEWATERING

DITCHING

Drainage by ditching is a very ancient type of excavation, and even drainage tunnels were built in prehistoric times. The purpose of this work was generally reclamation of land for agriculture.

Modern advances consist largely in the use of machinery for ditching, some improved types of pipe, and use of pumps to dewater areas that cannot be readily drained by gravity flow.

Most dry ditching is done by hoes or wheel, ladder, or drag ditchers. In soft swamps, draglines may be best. Clamshells are used for deep and tricky work. Shallow ditches in dry ground may be dug by hoes, ditchers, dipper shovels, draglines, graders, and dozers of straight, angle, and shovel varieties.

DITCHING WITH A HOE

Types. There are three types of hoe (backhoe or pull shovel) used in ditching. There is the old standby, the cable operated machine on a full revolving base, either crawler or wheel mounted. More recently, we have the full revolving hydraulic hoe, with similar undercarriages. And there are smaller hydraulic hoes carried on the back of tractors (or rarely, trucks), whose arc of swing is 200 degrees or less.

Design and operation of these various machines are discussed in Chapter 13.

The hydraulic models are superior in most applications. They have complete control of most bucket motions, with power for up and down, and pull and push. The "wrist action" bucket can be adjusted in angle during the digging pass for precise control of cutting, for tremendous breakout force on obstructions, for non-spill carrying of bucket loads to the dumping point, and for selection of place and rate of dumping.

Some crawler mounted hydraulics have separate speed and direction control of the two tracks, a refinement that helps maneuvering in restricted space, and reduces cutting up of the ground.

The primary discussion in this section, however, is based on the cable hoe, chiefly because it has more problems. Differences in hydraulic operation are indicated where necessary.

Width. Ditching with any machine is easiest and neatest if the trench is the same width as the bucket cut. This width is that of the bucket itself, plus any side cutters it may carry. The bucket should be wider at the front than the back, to prevent the sides from binding in the cut, and to simplify dumping.

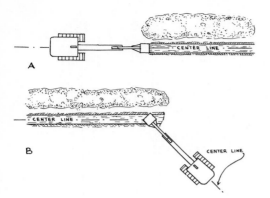

Fig. 5-1. Lining up a hoe

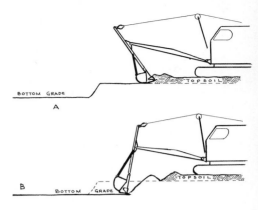

Fig. 5-2. Separating topsoil

Standard bucket widths, with or without side cutters, are 18 to 42 inches. Small machines may be as narrow as 12 or even 6 inches, and big ones 5 feet wide or more.

Direction of Work. A hoe should start at, and dig away from, any obstacle it cannot cross, such as a building. If there are two such obstacles, separate starts are made at each, and an extra-work connection is made between the two trenches.

If the centerline is on a grade, working in the uphill direction makes digging more difficult, by reducing digging force and increasing the tendency of the bucket to pull the machine into the ditch.

While digging downhill is easier, the working end of the trench may fill with water if the ground is wet or the job stands unfinished during a rain. Underwater work is sloppy, inaccurate, and often unstable.

Starting. The machine is placed so it is centered on the centerline of the ditch, with the tracks or wheels parallel to it, and the bucket extended to almost its full reach, and resting on the starting point, as in Figure 5-1 (A).

Actual digging procedures with the different types of hoe are described in Chapter 13.

Briefly, the soil is taken out in layers down to the required depth. The starting point may be squared off with a vertical face

from top to base. The bottom is smoothed off and checked for depth as it is made.

When the desired depth is obtained along the space the shovel can reach, it is walked away from the ditch from two to twelve feet, and a section of that length excavated. Short moves are made in connection with deep ditching, cutting the bottom to an exact grade, or cutting curves; longer moves are feasible for rough shallow work.

Curves. Curves are dug as a succession of short, straight ditches but a skilled operator can bevel the edges to produce a smooth curve. The machine stands with its center a little outside of the center line, and digging is done in the outer half of the bucket reach. Moves are short.

Angles. Many kinds of pipe require laying in straight lines and angles rather than curves, and trenches in which they are to be placed are dug accordingly. Angles are made by digging slightly past the angle point, then shifting the shovel to straddle the new center line, as in (B).

Spoil Piles. Spoil from the ditch is usually piled on one side, far enough back to allow a footpath or working space between it and the ditch. If a large volume of dirt is being moved, the pile must be pushed back by the bucket as it is built, and, in addition, it may be necessary to allow the spoil to come to the edge. Piling on

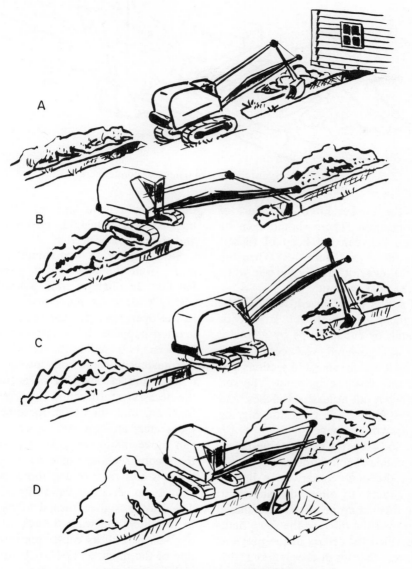

Fig. 5-3A. Connecting trench sections

both sides is usually avoided because of backfilling work. It does serve to block off the ditch so that people are less likely to walk into it absentmindedly or in the dark, although it is not adequate barricading.

Topsoil. If topsoil is to be saved and put back on top of the other fill, it may be piled on the opposite side of the ditch from the deeper digging. If the volume of

the spoil is not large, the topsoil may be placed on the same side as the fill, but farther back, so that when a dozer backfills, the topsoil will be next to the blade and will reach the ditch after the fill.

Topsoil is salvaged during the digging by scraping it off first, and bringing the bucket as near the shovel as possible on the last bite. The body of the ditch is then

Fig. 5-3B. Hydraulics make narrower connections

dug, with the bucket lifted out short of its closest position. There should then be a separation between the heap of topsoil and that of fill, as in Figure 5-2. When the shovel backs away, it can dig the pile while stripping the next section.

Sod. If sod is to be saved, it should be removed ahead of the shovel. It may be dug by hand, or cut in strips by a tractor drawn or self-powered sod cutter. The strips left by it may be sliced in sections and piled at a safe distance by hand. The sod should be taken out at least six inches, and preferably a foot, back from the digging side lines to avoid damage.

Guides. Sod removal serves as an excellent and unmistakable indication of the location of the ditch, otherwise a line of pegs low enough to allow the shovel to walk over them may be used. The shovel is lined up over the ditch in the same manner as described for cellars in the previous chapter, except that for ditches marked with centerlines, the dead axles should be marked to indicate bucket center, not sides.

If the bottom is to follow ground contour, the bucket stick may be given a paint mark that will be even with the surface when it is straight down at the proper depth. The distance may also be marked on a stick to be used for checking.

A bottom gradient that is independent of surface levels is usually set and checked by instrument from reference points. Any-thing from a hand level to a laser beam may be used, depending on the conditions and the accuracy needed.

Side Digging. A hoe should be worked away from the end of the ditch that is blocked. In ditching from a house to the street, it starts at the house and finishes in the open space of the street. However, it often happens that a ditch must be dug between two buildings, or under other circumstances where both ends are blocked.

The simplest method of accomplishing the necessary turnaround is to dig the ditch from one end, then from the other, having them meet at some spot where the shovel can move off to the side. The digging of the second section should be stopped while there is still comfortable room to turn the shovel and get it out, as in Figure 5-3 (A). The shovel is then turned at right angles to the ditch and walked back into the undug space, with its center pin in the center line of the ditch, as in (C). It then digs as close to its tracks as possible on both sides and backs away, connecting the ditch sections by digging first on a slant and then at right angles to the trench line.

This method calls for an excavation that is very wide if it is a cable hoe, and usually two or more ditch widths with a hydraulic. Sometimes such a connection can be made where extra width is needed for a manhole, pumphouse, or side ditch.

Tractor Mounting. The smaller hydraulics mounted on the back of wheel tractors are usually the most economical and efficient hoes for work around buildings.

They have the advantages of light weight and rubber tires, so are less likely to damage lawns and paths than the heavy crawlers, either while working or while getting in and out. Accidental damage to trees and structures caused by operator mistakes are likely to be small.

Another asset is that their small buckets are usually narrower than those of larger machines. For the usual pipe or wire, they take out less soil to put back. However, in this respect they are not nearly as economical as small drag trenchers.

Their fast cycle, and the prying power of the wrist action bucket, often enables them to do work and show production that seems greatly out of proportion to their size.

The tractors usually carry front loaders also, enabling them to backfill their ditches, carry pipe, and do miscellaneous work.

In spite of usually having two wheel drive, these machines are seldom stuck. The weight of the hoe on the driving wheels supplies excellent traction, and if that fails, downward and outward pressure on the bucket will lift the rear wheels and push the tractor right out of a mud hole. Or the bucket can pull it out backward, if that is what is needed.

Overlapping. On firm soil, ditch sections can be connected by overlapping.

In Figure 5-4(A) the machine, in digging the first section of the ditch has piled the spoil well back from the edge. It then can start at the opposite end of the ditch and cut until it is connected with the first section, which it straddles for the last part of the digging (B). Logs or beams are then placed across the ditch, as in (C), and may be cut down into its sides. Railroad ties are excellent for this job. The shovel is then turned so that the track next to the spoil pile will walk across the ditch on the beams. This turn may be sharp to increase

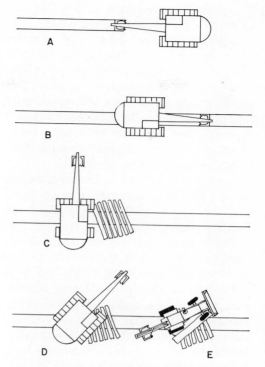

Fig. 5-4. Overlap method

the bridging action of the track itself and reduce weight on the beams; or it may be gradual to reduce danger of caving.

If the ditch line includes an angle, the crossing should be made there as it enables the shovel to walk across with less turning.

For wheeled mounting, a wide plank or planks should be laid across the beams, and the machine must be watched to make sure it stays on them.

Wide Ditches. When a ditch is to be more than one bucket width, one or both edges will be slightly uneven because the bucket will move inward, toward the center pin of the shovel. Usually one side is made straight by lining the shovel to that side, and the hacking done on the other side. If neatness is important, the ridges can be smoothed by drawing the bucket in while lightly swinging against the edge.

The full width of the ditch should be taken off in layers if it is to be dug from one position, rather than cutting one side to depth then starting on the other.

A ditch with two straight sides may be made by lining the hoe up to cut one straight side, and completing digging that can be reached from that position. The shovel is then moved back and maneuvered into position to cut the other edge straight, repeating this operation with each backward move. This works best when the ditch is two or more bucket widths.

If sloping beds of shale are encountered, digging should be arranged, if possible, so that the bucket teeth will cut along the bedding planes, as in Figure 5-5.

Shale dug in this manner at moderate depths is apt to come up in sheets so that the ditch will be widened irregularly.

Production. The rate of ditching depends on a number of variables, including depth and width of the ditch, bucket size and efficiency, cycle time of the hoe, digging qualities of the soil, obstacles and hazards both below and above ground, presence of rock, accuracy of grade required, and need of separating top soil.

A ditch that is shallow, with soil piled on the edge, offers the fastest digging cycle but the bucket is not apt to fill well. Deeper digging slows the cycle, means more soil to move, but allows better filling of the bucket. A narrow bucket does not fill as well as a wide one at any depth. A ditch that is wider than the bucket takes much more time.

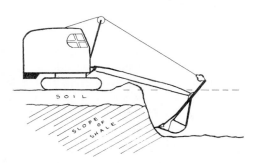

Fig. 5-5. Best angle to dig shale

Boulders and heavy roots slow the digging. Presence of buried pipes or conduits that must not be broken cause serious delays, particularly if their exact location is not known. Buildings or trees that interfere with maneuvering cut down production, as does lack of space to pile spoil.

It takes much longer to clean an irregular rock surface for blasting than to dig a clean trench to grade. Then there is the extra expense of drilling, blasting, and redigging.

Need to keep an accurate grade makes an operator work more slowly, and occasional stops are needed to check gradient or depth. A smooth bottom finish is produced readily by a wrist action bucket, but with some difficulty by a rigid one.

Stripping sod and topsoil separately will slow the digging from 5 to 30 per cent.

When a trench needs to be braced during or immediately after the digging, production will be determined by the rate at which bracing is set, which is almost always much slower than the digging.

An example of calculating digging rate is:

Assume that a half yard hoe with a 36 inch wide bucket, including side cutters, is digging a ditch 3 feet wide and 6 feet deep in common earth, with no special complications.

This ditch has a width of one yard and a depth of two, so its cross section is 2 square yards. There will be 2 cubic yards removed for each lineal yard of digging, or ⅔ yard per lineal foot.

This machine may have a cycle of 13 seconds, and an efficiency hour of 45 minutes. It should complete 206 cycles an hour.

The soil has a swell factor of .8 (25 per cent swell). The bucket averages only 4/5 of a load in loose yards, that is, its efficiency factor is .8. Multiplying the swell factor by the efficiency factor by the half yard capacity of the bucket, we have:

.8 × .8 × .5 = .32 bank yards per cycle

Multiplying the 206 cycles per hour by the .32 bucket load, we have a production

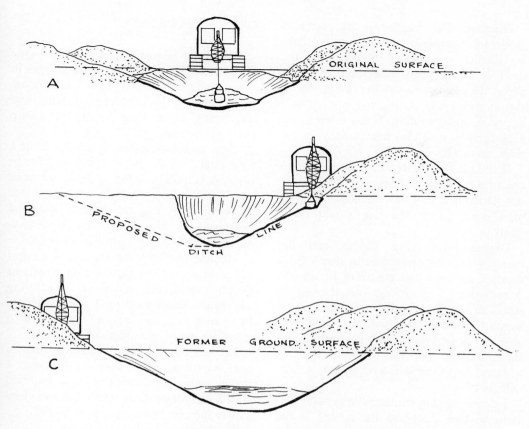

Fig. 5-6. Wide dragline ditch

of 65.92, say 66 yards an hour. Since there are two cubic yards to each running yard of ditch, the ditching rate is 33 yards or 99 feet (say 100) feet per hour.

A 30 inch wide trench with a 30 inch bucket would come out about the same, as what was gained in handling smaller volume would be lost in poorer bucket efficiency. However, if the ditch were 12 feet deep, either bucket would probably fill well.

OTHER SHOVEL RIGS

Clamshell. A clamshell ditches best when on the center line. If the ditch is narrow, the tagline chains are fastened to one jaw, or for a very wide cut, to both jaws. A ditch of intermediate width is made with the chains in the one-jaw position, and the soil is taken out in layers.

Connections between ditch sections are made by attaching the tagline chains to both jaws after completion of the main ditching, and digging the connection from the side. Whole ditches may be done from the side in this manner, but it is harder to keep on the correct line. The side position is desirable in deepening an existing ditch, or in digging beside a wall.

Smooth curves may be dug either by frequent readjustments of the position of the shovel, in the same manner as with a backhoe, or by having a man on the ground twist the bucket into proper position by pushing it by hand or with a stick as it is about to touch the ground.

Dragline. The dragline is the preferred shovel for ditching in swamps, and for making ditches with sloped banks when the spoil is to be piled alongside. It works along the centerline of the ditch, as in Figure 5-6 (A), cutting the bottom and slopes in one operation. If the ditch is too wide for this, two cuts are made from the sides, as in (B) and (C).

If the fill is to be trucked away, a dragline or a backhoe may be used in this manner. Draglines may have difficulty digging hard earth which the hoe would move easily.

Dipper Stick. The dipper can dig trenches four to seven feet in depth from the top, or wide trenches from the inside. A neat ditch may be dug from the top by straddling it, if the soil is very firm, or if support platforms are used. Trenching may also be done beside and parallel to the shovel's path, but this involves quite a wide cut in proportion to depth and is difficult to trim.

Interior digging conforms in general patterns to that discussed for cellars in the previous chapter. Part swing shovels can dig narrower slots than conventional models, as they do not need space for tail swing, but they cannot load trucks behind them.

Comparisons. The hoe is the best machine for ditches of moderate depth and width where boulders or stumps may be encountered. It will break up heavily fractured hard rock, and soft or thin bedded shale, and dig very hard soils if the bucket teeth are long and sharp. It can dig out large boulders by widening the trench as much as necessary, and dragging them up the slope toward itself. The ditch can be easily curved around boulders too large to lift or pull.

The clam is a slower machine but is able to dig to any depth desired, and can work close to obstructions, except those that are overhead.

OTHER DITCHERS

Ditching Machines. Continuous-type ditching machines offer great advantages in areas where hard bedrock and boulders are rare. They dig by continuous picking and sidecasting, rather than the dig and dump cycle of the hoe.

These machines excavate rapidly; make a neat ditch, usually with a curved bottom which is helpful in lining up pipe; can work with less headroom and do not need space to swing. They can dig certain classes of homogeneous soft rock which a shovel cannot, and seldom tear up banks in shale.

Medium and large machines have a number of small buckets mounted on wheels or double chains, that dump on side-casting conveyors. Small units have a single chain, fitted with teeth that cut soil and drag it to a surface auger that sidecasts it.

Buckets may be much narrower than in hoes, and drag chains may cut slots only 4 inches wide. Some machines may be fitted with a carbide-toothed wheel that can chew slowly through boulders and bedrock.

Ditchers can be equipped with shoes or reels to lay tile or flexible conduits immediately behind the digging so that shoring is not necessary.

A hoe may sometimes be mounted on the other end of the carrier.

Design, operation, and applications are discussed in Chapter 14.

Graders and Dozers. Graders can make shallow ditches with sloped sides rapidly and neatly by road building processes described in Chapters 8 and 19.

If there is no use for the excavated soil, it is usually spread and blended beside the edges.

The bulldozer can dig a wide, shallow trench from the side, as shown in Figure 5-7 (A). The volume of excavation required increases very rapidly with depth because of space needed for ramps.

When the practical limit for side excava-

Fig. 5-7. Bulldozer ditching

tion is reached, the dozer can work in the ditch, pushing dirt into heaps, which it then pushes to the side, as in (B) and (C).

An angling dozer can excavate by side casting in the same manner as a grader, but it may be harder to keep lined up.

ROCK

Stripping. Ditches frequently encounter rock that is too resistant to be dug by the available equipment. Occasionally the line of work may be shifted, but it is usually necessary to blast.

Dirt and rotten rock are removed by conventional methods. Spoil should be piled far enough back to allow space for the drilling equipment, and for the shovel when it returns.

After machinery has removed the soil, the rock surface should be cleaned by hand. If the trench walls are liable to crumble and slide from drilling vibration, they should be shored up, even if depth is shallow.

Opencut blasting is described in Chapter 9. Trench work differs chiefly in the restricted working space, and in the fact that all shots are tight. Loading must be fifty to one hundred percent heavier than on wide faces.

Drilling. Jackhammers can be used with the operator standing on the rock, or on the surface beside the ditch. Crawler drills are usually on the bank. Special ditching drills may be suspended over the work by cranes.

Several drilling patterns for three or four foot widths are shown in Figure 5-8. In each of these blasting is done back from an edge or face of rock exposed by digging, or by previous blasting.

The distance or length of ditch that can be blasted in a single shot depends on the near presence of buildings, whether it is permissible to overbreak the sides, and whether delay caps are used.

The holes next to the face can throw their burden along the line of the ditch. Any holes behind them, shot at the same time, will tend to expend more of their

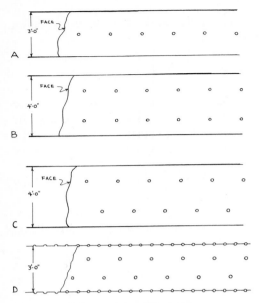

Fig. 5-8. Rock drilling patterns

energy to the side, so that they will need heavier loading. As a result, they tend to break rock outside the digging lines (overbreak) and produce finer pieces.

In (D) there are extra edge holes, called relief holes, which are lightly loaded or empty. Their use makes a smoother edge, reducing both overbreak and underbreak.

Delay Caps. After the blaster has decided on the area to be blasted, the number of holes included in it may be shot at one time; or a much larger number may be fired with delay caps. The first series will be at the face, with the other two groups following in succession.

Millisecond delay caps give the best results, as they provide thorough breakage with minimum concussion. They are arranged so that the wave of explosion travels back from the face, with such short intervals between rows that each is partially confined by the force of the previous blast. This condition is favorable to good fragmentation and reduced overbreak.

Damages. Nearby buildings, or more distant installations containing delicate apparatus, may dictate the size and type of shots. The conservative procedure is to fire one row at a time, and muffle and restrain the explosion with dirt and mats. Millisecond delays may permit much larger shots because the explosion is spread over enough time to reduce its sharpness.

Standard delays permit firing a succession of small blasts with one jolt, but there is danger that they will set a periodic vibration in some building or object which will damage it more than a heavy single explosion.

Deep shots, or those well buried, produce less damage than shallow ones. If the explosion is so confined that it acts about equally in all directions, concussion is at a minimum. If any part of the force can escape rather readily, its blowout will produce a reaction or back-kick against the solid rock, which will shake the surrounding area severely.

Air waves, which are serious offenders in breaking windows, can be prevented by thorough muffling.

In any blasting near buildings, roads, or people, elaborate safety precautions must be taken. An absolute minimum is confining the explosion so that no fragments can fly, giving ample advance notice of the blast, and blocking all roads and paths into the danger area. Woven steel mats are the only safe cover for close blasting.

The contractor's liability insurance company usually makes an inspection of the job before any blasting is done. This may include condition of nearby buildings.

A discussion of blasting damage starts on Page 9-39.

Burial. When the rock surface is well underground, the ditch may be refilled with dirt after the rock is drilled and loaded. This confines the explosion, prevents rock from scattering, muffles the noise, and protects the ditch walls from caving. It provides a safe working area for the machine, usually a backhoe or clamshell, which will re-dig the trench after the blast.

If the dirt fill is shallow, or rocky, it may be necessary to use mats or logs in addition.

Wiring must be thoroughly protected before filling on top of loaded rock. Fine dirt is hand shoveled over the connecting lines. If the rest of the fill is fine, a thin cushion is enough. If there are heavy or sharp rocks which cannot be kept out, a deep protection is required.

The lead wires may be run along the rock surface beyond the area to be filled, or led up to the surface inside air hose or other tubing.

Two caps may be placed in each hole with duplicate wiring, or Primacord can be used.

Backfill may be pushed in from the side by a dozer, or trucked from an excavator digging another section of the ditch.

Removal. Blasted rock may be dug from the trench by machine, by hand, or by both methods.

If the ditch is narrow, with hard, irregular walls, projecting stubs may jam a bucket so frequently that hand work may be cheaper. A bucket or container is frequently loaded by hand and hoisted by a machine.

Large rocks are lifted with slings or tongs.

In general, it is good policy to blast sufficient width to insure working space for a clamshell or hoe bucket.

PAVEMENT

Street Pavements. Hard pavements are generally broken up in advance of the digging. If the street is to be repaved after the trench work is completed, the whole pavement may be smashed and removed by a large dipper stick or dozer shovel; or torn up with a heavy ripper and dug away by smaller machines.

If the pavement is to be preserved so far as possible, it should be broken along the line of the ditch by pavement breakers. Wide edge chisels are used for blacktop,

narrow or pointed tools for concrete, and weight-dropping machines for either. The two edges are wholly or partially cut. The strip to be removed may be broken up by the tools or by the excavator.

Reinforcement in concrete greatly increases the cost of this work. Thin bars may be cut by the chisels, heavier ones may require a cutting torch or heavy bolt cutters. The work involved in exposing the reinforcement may be greater than that in cutting it. The broken pieces of pavement may be removed by hand, or may be left for the ditching shovel to take out with the dirt.

Many street pavements are underlain by an intricate network of pipes and conduits. Great care may be required in breaking the pavement, and then in digging the trench. Cooperation of utility companies is essential.

Temporary Cover. In populated areas, and particularly in roads and paths, it is often necessary to cover an open ditch between work periods, partially or completely, to reduce danger and nuisance from its presence.

Wide trenches such as subway cuts, which will be work sites for some time, are usually roofed with massive steel and timber structures. For lesser holes, heavy steel sheet is a safe and economical cover.

Such sheets or plates are usually ordinary mild steel, one or 1¼ inches thick, in squares or rectangles measuring 4 to 8 feet on a side. One-inch weighs 41 pounds per square foot. On a 4-foot span it will carry almost any load that is permitted on a highway.

Sheets may be supplied in standard sizes by a steel mill, or be cut from larger units (or bank vault scrap) with torches. Any edges that might cut tires should be removed by grinding.

Each sheet should have a hole, or preferably two, one inch or more its smallest way, to allow use of hooks or other gripping devices in lifting, moving, and setting them.

Even if the standard handling method is a sling with hooks or clamps for the edges, holes may be needed when edges are not accessible.

Plates can be handled by any machine that can carry a line to lift their moderate weight. This includes front loaders, cranes, and almost all shovel or hoe rigs. They may be pushed by dozers, and adjusted by crowbars.

A rigid sheet used as a bridge or cover over a hole must be firmly and evenly supported along two opposite sides. If one corner is either low or high, the plate will rock under moving loads, disturbing its bedding and perhaps turning up a corner enough to damage tires or to get shoved sideward.

Sheets may be simply laid on top of existing pavement or ground, with edges higher than the adjoining surface. Such edges can be sloped to the surface by a wedge of blacktop, gravel, or other material.

Placement will be neater and more secure if the surface is cut away where the plate is to rest, to lower it flush with its surroundings. Either blacktopping or lowering gives protection against shifts in position.

Side shifting enough to expose any part of the hole is dangerous and must be prevented. The open hole is unsafe in itself, and a heavy load on an unsupported edge might flip up the whole plate.

CAVING OF BANKS

Many soils will not stand in vertical walls, so the sides of the ditches must be either sloped back or braced. A few soils will not stand on even a moderate slope, and these will require very heavy bracing.

Ground water is the most important single factor in collapse of edges of cuts. It acts as a lubricant that enables the soil particles to move on each other readily, and exerts pressure that moves the particles toward the ditch. Sand faces may fall from this cause, or from the drying action of water draining and evaporating, so that the bond between the grains is weakened.

If water is allowed to stand in a ditch, danger of caving from flow of ground water into it and from other dynamic forces acting in the soil is reduced. However, wave action set up by dropping of stones or chunks of dirt may cut into the walls and undermine them.

It may happen that the upper part of a trench wall will be firm dry material, but that a shallow layer at the bottom is waterlogged and unstable. The sides will stand for only a short time before becoming undermined by movement of the lower layer.

In general, caving of ditch sides is more apt to happen minutes, hours, or days after the digging than immediately. The exceptions are usually loose sandy or semi-liquid muds that flow into the excavation, rather than cave or slide into it.

Stabilizing. Sloping of sides for stability is a technique chiefly used for permanent open ditches, which will be discussed later in this chapter. It is seldom used for trenches for burial of pipes because of the large amount of extra digging, the space required, or the area of pavement, lawn, or other surface disturbed.

Vertical trench walls may be stabilized by bracing, draining, freezing, or chemicals. Bracing is the most common technique, and may be required by law.

Bracing Structures. Figure 5-9 (A) shows a light system of bracing or shoring used where danger of caving is slight. Planks are placed vertically in the trench at five to fifteen foot intervals, and pressed against the dirt by means of push-type turnbuckles, called sheeting jacks. Bracing timbers are inserted and the jacks removed. The planks are usually two or three inches thick, the cross braces six by six or larger. Wide ditches require heavier cross braces than narrow ones.

A heavier type of shoring is shown in (B). The sides of the ditch are lined solidly

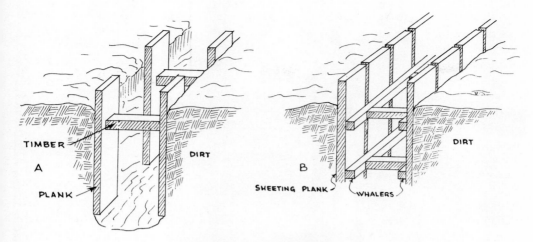

Fig. 5-9. Bracing trench sides

with vertical planks, called sheeting planks, held from falling inward by horizontal beams, known as whalers, which are braced to each other across the ditch by timbers. These timbers are sprung into place by forcing the whalers apart by sheeting jacks.

The weight and spacing of the planks and timbers will be determined by the depth and width of the ditch, and the instability of the soil. It is possible for a usually safe soil to be dangerously soft locally, due to disturbance of underground drainage, leakage of water mains, or other causes, so an ample margin of safety should be allowed.

Figure 5-10A shows a photograph of a shored trench. Complete structural diagrams will be found in the Appendix.

Delayed Bracing. The method of placing the braces is determined by the promptness with which the banks are expected to collapse. If they are so stable that the necessity of bracing is questionable, the ditch may be dug by shovel or ditcher, closely followed by the carpenters who build the bracing in place as a completed structure, as close against the sides as is practicable.

Immediate Bracing. If the banks are not

trustworthy, or the contract calls for immediate bracing, the ditch is made enough wider than the bucket so that it can work between the whalers. The ditch is dug to a depth of about two feet, full width, and the top pair of whalers placed, and cross braces set with such spaces that the bucket can get down between them. Planks are set vertically touching each other outside the whalers.

The shovel, preferably a clamshell, now digs a foot or two below the whaler. Men with handtools dig the dirt out from under the vertical planks, allowing them to settle, and also remove dirt under the crossbeams which the buckets cannot reach. This dirt is piled in the middle of the ditch and is taken out by the bucket when the laborers are out of range. The shovel then digs deeper and is followed by handtool work. At a depth of two to five feet below the top whaler, another pair of whalers is set inside the planks and braced across the ditch. Alternate excavation by shovel and handtools, undermining and dropping of side planks outside the whalers, and installation of additional beams is continued until bottom grade is reached. Ordinarily, the whalers and crossbeams are either heavier

Fig. 5-10A. Braced trench

or more closely spaced with increasing depth as the potential pressure increases.

If the ditch is deeper than the length of available planks, those started at the surface should be of variable length. As each one drops below the top whaler, another plank is placed on top of it to follow it down. Mixed lengths make this possible without weakening the structure by having a row of these joints occur together.

The whaler beams are also of different lengths so that both members of a pair do not end together. The joint between any two can therefore be braced against a solid beam on the other side.

Two or three inch sheeting, and six by six whalers and crossbeams, spaced five to eight feet apart, are strong enough for moderate depths in most soils. If more protection is needed, heavier wood may be used, additional planks can be driven outside the sheeting, or inserted inside by a complicated process of removing and replacing whalers. Steel sheet piling is much stronger than wood.

Movable Bracing. When the work which is to be done in the ditch can be completed in short sections so that the ditch can be backfilled a few yards behind a hoe shovel, a portable bracing structure can be used.

It may be made up of steel or wood, and should be equipped with a tow bar or chain at the front bottom, which can be gripped by the bucket teeth. It is lowered into the first section dug, and the pipe laying or other work done inside it while another section is dug. The shovel drags it along in the ditch whenever sufficient digging or pipe laying has been completed to justify moving it.

Such a device can result in tremendous savings. However, it cannot be used on many jobs because of the necessity of checking the work, or having it inspected. Also, if the sides should close in on it, it might be very difficult to free up for moving.

Backfilling should be done as soon as possible, as allowing the sides to cave may damage the pipe, or shift it out of line.

Flowing Banks. If the sides are so unstable that they cave or flow immediately upon being cut, the sheeting planks must be driven down by air hammers or pile drivers, and the dirt dug from between them afterward. Penetration and control of direction are usually best if the planks are driven only a short distance below the digging. However, mud may flow so readily that the sheeting must be down several feet below excavation level to prevent it from swelling on the bottom. All graduations between this condition and stable banks may be encountered in a short distance.

Washout Failures. Only wet ground or loose sand exerts very heavy thrusts against the shoring. Water draining down the sides of a braced trench may erode them, so the sheeting moves outward, thus loosening the cross beams or jacks and allowing them to fall, after which the sheeting can be pushed in by any movement of the banks.

Stabilization. Soft ground may be stabilized by chemical treatment, well point pumping, drainage, or freezing, by procedures described elsewhere in this book.

STREAM CROSSING

Pipelines must often be built across both large and small streams.

If a wide stream or a pond has a soft bottom and is not used by ships or barges, the pipe may be laid directly on the river bottom. However, it is often necessary to protect it by burying it.

Excavation may be done by a grapple dredge if it is available, by a clamshell, dragline, or hoe on a barge made up for the purpose, by a cable excavator on the bank, or from a temporary jetty.

Digging a straight trench from a dredge or barge takes experience. Anchors and winch lines must be so arranged that the barge can both pull itself across the stream, and keep or regain position against pressure of the current. Alignment is checked by surveying instruments that may be on the barge, on the shore, or in both places.

Many streams have sufficient current to fill in a trench as fast as it is dug. At low water periods it is usually possible to block off a substantial part of the stream channel with a fill or jetty, trench in it or just downstream from it, lay the pipe, and remove the fill. The operation is then repeated in another section. Erosion at the ends of the fill may be severe.

The jetty may be built from one bank to the stream center, and then from the other bank to connect, or a dragline may build itself an island that it moves by digging at the rear and filling the front.

The pipe may be assembled in sections as trench space becomes available, or the whole crossing may be put together at one time, dragged and floated across the water, anchored in approximate position, and sunk into the trench as sections of it are completed. Floating and sinking of the pipe are regulated by the amount of air or water in it, and/or by floats and weights.

Work of this type is subject to disastrous damage if the stream floods. If there is a little warning, equipment can be pulled back from shore-based fills, but a dragline on its own island is likely to be lost. A wire rope connection should be kept to insure getting back the crew if the water rises too suddenly to permit rescue by boat, and to help locate and salvage the machine if it capsizes. Floating pipe may be pulled back to shore.

Because of flood danger, all possible preparations are made in advance, and the crossing pushed as rapidly as possible once it is started. The urgency depends largely on the past behavior of the stream at that time of year, but few streams are ever entirely secure against flooding.

Small Stream. It is seldom practical to ditch directly across a brook without taking precautions to keep it out of the trench. This may be done by digging a sump hole upstream, and using a pump or pumps to move the water across the ditch line and back into the stream.

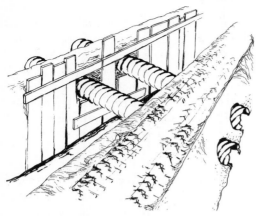

Fig. 5-10B. Ditching under a small stream

A less expensive method is to confine the stream flow to pipes, and work under them, as in Figure 5-10B. Sufficient pipe to accommodate the expected flow is laid in the stream across the ditch line. Sectional corrugated pipe should have seams filled with mastic to prevent drip leaks.

Dams are then built across the brook above and below the ditch area, confining the flow to the pipes. There may have to be two sets of dams, a first pair of light temporary ones toward the pipe ends to permit drying out areas where the regular dams can be built of selected soil, carefully tamped around the pipes. Bentonite might be added to the soil to improve water resistance. One dam may be made wide enough to serve as a road for machinery.

The ditch is then dug in the regular manner, under and on each side of the pipes. Water oozing into the ditch can be pumped out, diverted by well pumping, or blocked off by grouting.

When the pipe has been laid in the trench and the path or road across the stream is no longer needed, the dams are dug away and the trench backfilled, and the pipes are lifted out for re-use elsewhere.

Diversion pipe must be long enough to cross the ditch, go under the dams, and project on each side. Lengths of 20 to 40 feet are needed, depending on dam height and the need for a driveway.

If stream flow increases beyond pipe capacity it will overflow the dams, fill the trench, and stop work. After the flow subsides the dams are repaired, and the trench is pumped out. Redigging to remove washed-in soil may be necessary.

PERMANENT DITCHES

When a ditch is to be left open permanently, its sides usually must be protected by masonry or rot-resistant sheet piling, or sloped back far enough so that they will not slump, cave, or wash into the bottom. If a large volume of water may flow through the trench at any time, the bottom should also be protected against erosion, unless the gradient is so flat, or the water so burdened with silt, that cutting will not occur.

Ditches with low gradients, or which carry dirty water, must be cleaned out periodically by a dragline shovel or other excavator. Masonry, riprap, and particularly vertical stone walls interfere with machine digging and are liable to be damaged. This should be borne in mind in designing any artificial protection.

Sloped Banks. The most satisfactory bank protection for a country ditch is a stable slope and a good cover of vegetation. This can be reinforced on the outside of bends and other places subject to strong current action, by placing large boulders, walls, riprap, or light piling holding wire or brush mats.

Stable slopes vary in steepness with the character of the soil. Loess may stand indefinitely in vertical cliffs, while certain types of clay may slump if the slope is one on six. Generally, it may be said that slopes of one or two and a half or less are advisable if the soil contains much clay or silt; if there is movement of ground water through it toward the ditch; if there is considerable drainage of surface water running down its face, or if it is in layers that dip toward the ditch.

Vegetation. If trees are to be planted or allowed to grow, the slope can be steeper than if it is to be kept in grass. Trees have greater holding power, and maintenance of grass may require the use of mowing machines, whose ability to work on side slopes may determine the grade. However, trees will interfere with access for cleaning.

If the trench is partly or wholly in barren soil, topsoil may have to be spread on the banks to encourage vegetation, although some plants can generally be found that will grow well on the subsoil if encouraged with lime or fertilizer.

Banks of permanently wet ditches may be strengthened by laying willow poles or logs up and down the bank, two to four feet apart. They should be settled well into the soil for their full length, with the lower end in water or bottom mud, and staked or wired in place.

These poles should grow tops and a continuous root mattress.

Bottom scour may be largely prevented by keeping a gentle gradient (see **Page 5-20**). If part of the trench is too steep, check dams may be built.

This localizes the fall of the water at the erosion-resistant aprons of the dams. Good results can sometimes be obtained with plank dams, or even heavily anchored brush mats.

Spoil Arrangements. The spoil piles from a ditch are apt to interfere with local surface drainage. If the land is flat, spoil may be piled on both sides, but frequent breaks should be made in the windrows so that water will not pond behind them. If the ditch cuts across a slope, these breaks need be made in the upper pile only. Until the ditch slope is protected by vegetation; gullies may form at these spots unless the area is protected by pipe, flumes or stone.

The "W" ditch, Figure 5-11, is a double ditch, separated by sufficient space to provide disposal area for the spoil. This eliminates any blocking of drainage on either side, and allows maintenance of field grade to the ditch edge.

If the depth of the ditch is determined by the flow capacity required, two can be more shallow than one. This construction is only slightly more expensive than the single drainageway, although it usually takes more land out of production.

If depth is determined by a flow gradient, a W-ditch will about double the amount of excavation required. If the depth is considerable, the additional spoil is likely to damage an excessive area.

Whenever possible, the spoil piles from permanent ditches should be rounded off so as to permit easy access to the ditch and to make them less prominent in the landscape. This should be done before they are overgrown by trees.

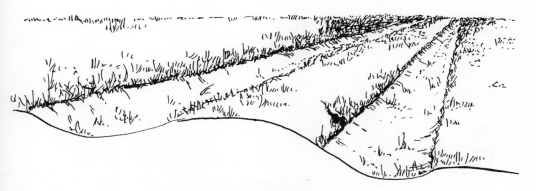

Fig. 5-11. "W" ditch

PIPE AND CONDUIT TRENCHES

Most trenches are dug to bury pipes or conduits. Conduits, and pipes for gas and water supply, run at more or less fixed depth below the surface. Sewers, storm drains, and other gravity flow pipes must maintain a minimum gradient from source to outlet or booster pump, and will have a variable depth below an irregular surface.

Fixed Depth. In cold-winter areas, water pipes are laid below the frost line in the ground. Conduit and wires are laid only deep enough to protect them against accident. In either case, depth may be increased under sharp ridges to provide smooth vertical curves.

Depth is usually specified in a contract, or chosen according to local custom. Items to consider in arriving at it include, in street work, possibility of repaving with a thicker pavement with very high loads on the subgrade during construction, or of lowering the street surface. In fields, the most important menace .is the moldboard plow, which penetrates eight or ten inches. There is a chance that a subsoil plow, with a penetration of eighteen inches to two feet, might be used, and in addition land even on gentle slopes gradually washes away, and the surface may be lowered several inches during the life of the conduit. Depths of two to four feet are usual, and are figured from the surface of the ground regardless of slope.

Since the bottom of such a ditch follows the surface at a fixed distance, depth measurements during digging are easily made. A rule or marked pole is set vertically at the side, or a board is placed across the ditch and its distance from the bottom measured.

If the depth requirement is only approximate, a mark on the stick of a backhoe, the cable of a clamshell, or the depth gauge of the ditcher may be sufficient guide.

Gravity Systems. Digging accurately for a sewer or other gravity system requires close supervision. A number of methods are used to keep on grade, of which a few examples will be given.

The grade can be checked from inside the ditch with a transit level. The beginning of the ditch is dug to proper depth, which may be measured from the surface or dictated by a connection with a pipe or manhole. This starting point is taken as a bench mark (see Chapter 2), and usually as zero station. Other stations are measured off as digging progresses, and their elevation taken with the instrument. Each successive station should be higher or lower than the starting point, according to the gradient. In the example in Figure 5-12, the drop is one foot in sixty feet, and the readings are taken at ten foot intervals.

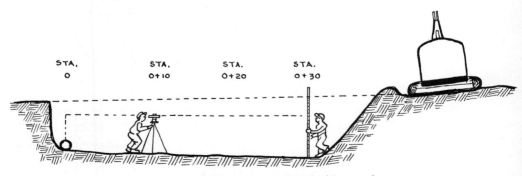

Fig. 5-12. Measuring gradient inside trench

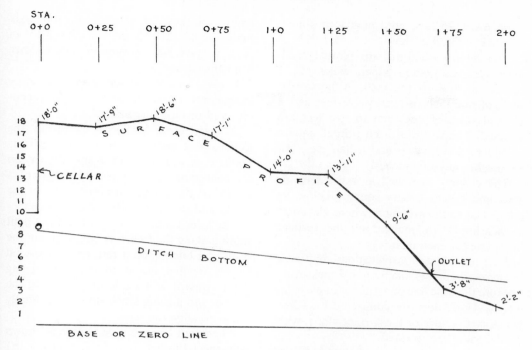

Fig. 5-13. Calculating gradient

If the ditch is correct, each rod reading should be two inches higher (the lower the ground, the higher the reading) than the previous one. If distances are measured with a tape, it must be horizontal to get an exact result.

Readings are taken as close to the digging machine as possible.

If the ditch makes an angular turn, the instrument is set at the angle point. If the ditch curves, inside work is not practical.

This way of setting gradient is liable to suffer from cumulative errors in either measurement or rod reading, unless it can be checked against surface features occasionally.

Measuring from Surface. If depth is to be measured from the surface, a profile of the ground on the center line, or on a side or offset line of the ditch, is taken. This consists in marking it off with station stakes at regular intervals, and taking the elevation of each. If the elevation of the ditch bottom is given in the plans, the amount of

cut at any station may be found by subtracting the bottom elevation from the surface elevation. Stick or rule measurements may be made at these points during the digging, and any intermediate levels that are needed can be found by a hand, transit, or string level used inside the ditch.

Plotting Profiles. If there are no plans, or if they merely specify a gradient, the surface profile should be drawn to scale on a sheet or strip of cross section paper.

In Figure 5-13, a cellar has been dug and it is desired to lay a pipe on a slope of one foot in fifty feet, to take water from a tile drain laid around the outside of the footings. An instrument is set up and the elevation of the cellar floor taken. This is arbitrarily assigned a value, say 10, and is used as a bench mark for the rest of the work. Another bench mark on a tree or some surface spot not affected by building work should also be taken for future reference.

A profile is then taken along a down

slope, every 25 feet, until points are found well below the floor level.

The figures obtained are plotted on a piece of cross section paper, which might be ten squares to the inch. A horizontal scale of one inch to twenty-five feet and a vertical scale of one inch to five feet are selected. The width of each printed square will then indicate two and a half feet, but its height only six inches.

The cellar is sketched in and a base or zero line drawn twenty squares below its floor. The stations (points where elevation is measured) are marked on the vertical lines, one for each inch.

Each of the station elevations may now be marked on the diagram by measuring up from the base line one square for each six inches. These dots are connected by a line which is a picture of the surface slope, with its steepness exaggerated.

The ditch may now be drawn in. A distance of 1½ feet below the cellar floor is marked 3 squares down, for the tile. At the last station, 2+0 (200 feet), measure 8 squares (4 feet) down from the tile.

A straight line drawn between these points represents a drop of 4 feet in 200, which is the one foot in 50 that is wanted.

Measurements on this sketch will now give the length of pipe needed, the distance on the ground to the outlet, the elevation of any point on the ditch bottom, and the depth of the ditch anywhere.

A larger scale in which each square would represent a smaller distance will give more accurate readings.

Finish Levels. When the digging of a ditch section is finished, boards may be placed on edge across the ditch at 10 to 25 foot intervals, and staked or otherwise firmly fastened in undisturbed soil, in positions such that a tight string may be run over the ditch and adjusted by instrument readings to be parallel with the final grade of the bottom. Extra strips of wood may be nailed on, or notches cut into original boards, if they are too low or too high. Finishing of the bottom, and placement of pipe, is governed by measurements down from this string.

Laser. Lasers provide means for very exact control over line and grade in trenching, and in pipe placement.

A laser beam projector is set in the trench, in a manhole, or at some starting point. It is usually located and aligned on the centerline of the pipe, by means of a transit on the surface, a target stake in the ditch, and its own grade-indicating unit.

The laser's thin beam of intense red light shining along the trench is picked up on the back of a bucket, on a rod, or on a special target. Its first use is to regulate digging depth, then it is a guide in smoothing the bottom and placing any bedding, and finally serves to align the pipe.

Alignment of each pipe section is shown by a special target sheet fitted over (or in) the free end, which is usually farthest from the projector. The first pipe must be lined up at both ends; for the others the free end is sufficient.

When the laser beam is centered in the bullseye of the target on the pipe, that pipe is accurately on line and grade.

Lasers are discussed in Chapter 2.

BACKFILLING

General Methods. Trenches dug for laying of pipes or conduits must be backfilled when the installation is complete. The dirt taken out is pushed or pulled back in. This job can be handled by most earth moving machines, but the bulldozer is the standard tool for the purpose.

If the backfill need not be compacted from the bottom up, it may be pushed into the trench in the ways shown in Figure 5-14. The bulldozer operates at right angles to the trench, taking as large a slice of the pile as it can handle comfortably. Dirt which drifts across the blade is left in windrows that are pushed in a separate

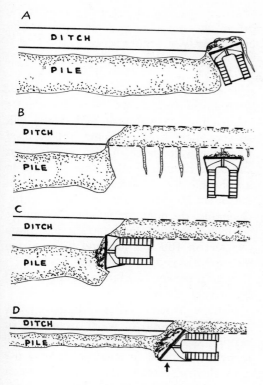

Fig. 5-14. Backfilling

series of passes, as in (B). Any remaining soil is pushed parallel with the ditch into the main pile as in (C), or, when the end is reached, distributed along the ditch.

There will usually be too much soil because of increase in volume of disturbed soil, and space occupied by the pipe laid. The excess may be mounded up over the trench and partly compacted by use of a roller, or driving the dozer or a truck along it. Full natural settlement may take as much as a year, and is liable to leave low and high spots to be graded in.

If the trench is small, it may be refilled by running an angle dozer or a grader through the pile, with the blade set to side cast it into the ditch as in (D).

Heavy backfill may be done by a dipper shovel walking through the length of the pile, digging it and dumping in the ditch. A backhoe or a dragline can work from across the ditch, pulling the soil into it. A dragline's efficiency will be greatly increased by fastening a heavy plank or other block across the mouth of the bucket so that it will not fill. Shovels are often used to move the bulk of the backfill with a bulldozer doing the final cleanup.

When pavement along the sides of the ditch is to be preserved, the best tool is a rubber tired loader or dozer, but light or medium crawler machines with semi-grousers or flat shoes may be used.

Special dragline type backfillers are often used on cross-country trenches.

Compacted Backfill. If a pavement is to be laid over the refilled trench immediately, the backfill must be carefully compacted from the bottom up. This may be done by dozing or hand shoveling fill slowly, while men in the trench compact it with hand or pneumatic tampers. A mechanical tamper may work from the side or straddle. The top layer may be compacted by use of a trench roller with a large wheel that will fit inside the ditch, or by running any heavy machine back and forth along it.

Open textured soils may be effectively compacted by puddling. Enough water is added to the fill to make it into mud, which, upon drying, will shrink considerably. Heavy soils take a long time to dry so are not as readily handled this way.

Machinery should not be run along wet trench fills as it is almost sure to get stuck in them.

Imported Backfill. Drainage trenches may be filled with better-draining, more porous material. Spoil removed in the original digging is trucked away or used in grading, and the gravel trucked in for refilling. This may be dumped in piles in and alongside the ditch, and pushed into it by a dozer or grader or hand shoveled. If considerable work of this type is to be done, a backfilling machine may be profitably used. This carries a hopper that is moved parallel with the ditch by rubber tired driving wheels.

Trucks dump into the hopper from which a belt carries the soil to the ditch and dumps it. The backfiller can push the truck which is dumping into it, so that the truck driver can concentrate his attention on lifting the body at proper speed.

DEWATERING

DRAINAGE

Both the surface and the subsurface water may be removed by a seasonal drop in ground water level; drainage through ditches, pipes, or siphons; by pumping; by walling off or diverting the source of water, and very often by combinations of two or more of these methods.

The purpose of dewatering may be to promote growth of crops; to dry out swamps or other objectionable wet areas; to stabilize slopes, foundations, and road subgrades, or to facilitate excavation for any purpose.

All of these objects except the last are accomplished chiefly by drainage—that is causing the unwanted water to flow away through artificial and natural channels or conduits. Pumps may be used to remove water from a sump or low point of a drainage system.

Gradients. The slope or gradient of a drain will depend on the work it has to do. In tidal marshes and other practically flat swamps, ditches with zero gradient may serve to lower the water level substantially. In general, water will flow through a flat ditch, but it is easily choked by sediment, growth of weeds, and dirt falling from banks, as the water flowing through it will have little or no ability to clean it. Too steep a ditch gradient may cause erosion of the bottom, undermining of banks through stream action, and damage from depositing of mud below the discharge point.

The slope must be adjusted first to the necessities of the situation, and second to the relation between the amount of water to be carried and the nature of the soil. A bottom gradient between one foot drop to 1000 feet and of two feet to 100 feet is desirable under most conditions encountered.

Drainage pipes should not be flat as costs in cleaning out sediment and debris will be very high. Low gradients can be used when the water is clean, the pipe is short and large enough to allow men to work in it conveniently, or there is a sharp fall at the outlet so that water will flow rapidly. Generally, the minimum gradient should be six inches to 1000 feet, and the maximum two feet in 100 feet for land tile, and ten feet in 100 for tight joint pipe.

Special erosion resistant pipe may be laid on steeper slopes.

Surface Water. Surface drainage may consist of disposal of water from rain or melting snow, or lowering the water level in ponds, ditches, or swamps. It may use open channels, conduits, or both. The water is usually led to a natural stream or body of water.

Such drainage may be accomplished by deepening, enlarging, or straightening and protecting existing streambeds, by digging artificial channels, or installing underground pipes or tunnels.

There is no definite separation between surface and subsurface drainage as they operate on different parts of the same water mass.

Tunneling. If it is not practical to ditch to install a drain or diversion pipe, boring or tunneling may be used. Diamond drills will put small holes into rock hillsides at any angle desired, for long distances. Augers will drill soft rock up to a hundred feet or more. Where such machines are not available, or where a larger drain is needed, tunneling may be done with explosives and hand digging.

Siphons. If a drainage line is to be used only occasionally, the expense of ditching

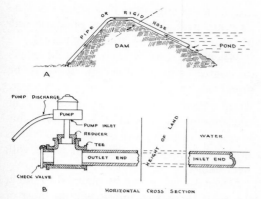

Fig. 5-15. Siphon and priming pump

or tunneling may be avoided by use of a siphon. This is an airtight pipe or stiff hose across the water barrier, with one end in the water to be drained, the other at a lower level, Figure 5-15(A). Maximum rise above water level is about 25 feet. Lower lifts work better.

When the siphon pipe has been filled with water, that which is between the high point and the lower or discharge end moves down the pipe by gravity, while atmospheric pressure, acting through the pond water, pushes the shorter and lighter column of water after it. This water is in turn renewed from the pond so that movement continues until the water level drops sufficiently to outbalance the suction, or to allow air to enter the pipe; air enters through leaks or through the discharge end, or water rises around the outlet to the same level as the intake.

The rate of flow will depend chiefly on the drop between the top of the water being drained, and the point where the water loses contact with the outlet end of the siphon. As a pond is drained and its level drops, the flow will become slower.

Siphons in which water moves slowly are likely to be stopped by air entering the top of the outlet. This may be prevented by putting the outlet in a box or small pool so that the opening will be under water. A slow current may also allow an air lock to

form from the accumulation of air or other gases escaping from the water in the pipe, or leaks from the outside.

Very small siphons may be started by mouth suction and a medium size by inverting it so that the ends are higher than the middle, filling it with water and holding the ends closed while placing it in position. Or a tee connection in the top may be used to pour water in, keeping the ends plugged until the pipe is full and the tee tightly plugged.

The most satisfactory way to start a large siphon is with a suction pump. A way to connect it is shown in (B). A tee is placed on the outlet end, with the side opening reduced to fit the inlet hose of a small diaphragm or centrifugal pump. Means are provided to prevent air from entering through the lower opening of the tee, by means of a check valve, a screw plug, or a piece of plywood with mud on it. With this stop in place, the pump is started and the air sucked out of the siphon so that water from the pond is drawn through it into the pump. The stop on the main pipe then opens or is removed and the pump is shut off.

Channels. Channels may consist of natural watercouses; watercourses which have been enlarged, straightened, or paved; or artificial ditches.

A stream bed may be dredged to lower its level, to increase its depth or capacity, to keep it from changing its course, or to change its course.

Level may be lowered to drain surface water from a swamp or pond, or to provide better underdrainage for land in its vicinity.

Depth is usually increased to assist navigation, or to provide for more rapid runoff of flood water. Widening and straightening increase capacity, often at the expense of depth.

Streams normally wander in their courses, cutting away banks in some places

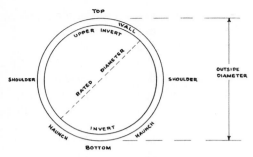

Fig. 5-16. Pipe details

and building them in others. When valuable property or structures are threatened by these changes, the channel may be artificially shaped to direct the force of the water away from them. This may involve turning the water back to its original direction, or forcing it to flow in a new one.

Dredging of small streams is generally done from the banks by draglines or clam-shells, and of large ones by floating dredges. The material dug may be piled on the banks, or removed by trucks or barges.

When the spoil is used to build banks to control stream direction, it must be protected by paving, masonry, rock, logs, wired brush, sod, or other material. The best emergency protection for a bank that is being washed away is drilled boulders fastened together in groups of three with steel cable.

River dredging may be planned to direct the river current so that it will do most of the excavating in the new channel.

Drainage channels are often paved to protect them from erosion or slumping, to prevent changing of course, and to increase capacity by reducing friction.

Irrigation canal pavements may be used for any of these purposes and to prevent

INSIDE PIPE DIA. (INCHES)	CONCRETE SEWER PIPE						CONCRETE CULVERT PIPE						CORRUGATED METAL PIPE				
	PLAIN A.S.T.M. SPEC. C-14-41.			REINFORCED A.S.T.M. SPEC. C-75-41.			REINFD. STAND. STRENGTH A.S.T.M. SPEC. C-76-41.			REINFD. EXTRA STRENGTH A.S.T.M. SPEC. C-76-41.			16 GAUGE	14 GAUGE	12 GAUGE	10 GAUGE	8 GAUGE
	SHELL THICK-NESS (INCHES)	WT. PER LIN.FT. (lbs.)	ULT. STRENGTH 3 EDGE BEARING LBS. PER LIN. FT.	SHELL THICK-NESS (INCHES)	WT. PER LIN. FT. (lbs)	ULT. STRENGTH 3 EDGE BEARING LBS. PER LIN. FT.	SHELL THICK-NESS (INCHES)	WT. PER LIN. FT. (lbs.)	ULT. STRENGTH 3 EDGE BEARING LBS. PER LIN. FT.	SHELL THICK-NESS (INCHES)	WT. PER LIN. FT. (lbs.)	ULT. STRENGTH 3 EDGE BEARING LBS. PER LIN. FT.	WT. PER LIN. FT. (lbs.)	WT. PER LIN. FT. (lbs.)	WT. PER LIN. FT. (lbs.)	WT. PER LIN. FT. (lbs.)	WT. PER LIN. FT. (lbs.)
4	9/16		1000														
6	5/8	25	1100														
8	3/4	35	1300										7.6	9.3			
10	7/8	48	1400										9.3	11.4			
12	1	60	1500	2 §	90	1800	2	90	2250				10.8	13.3	18.5		
15	1 1/4	90	1750	2 1/4 §	125	2000	2 1/4	125	2625				13.3	16.4	22.7		
18	1 1/2	120	2000	2 1/2 §	160	2200	2 1/2	160	3000				15.8	19.5	27.0		
21	1 3/4	190	2200	2 3/4 §	205	2400							18.3	22.5	31.2	39.7	
24	2 1/8	225	2400	2 5/8 *	225	2400	3	260	3000	3	320	4000	21.0	26.0	35.9	45.7	
30				3 "	315	2700	3 1/2	370	3375	3 1/2	470	5000		31.7	43.9	55.9	
36				3 3/8 "	450	3000	4	520	4050	4	600	6000		37.9	52.4	66.7	81.1
42				3 3/4 "	560	3200	4 1/2	680	4725	4 1/2	750	7000		44.4	61.5	78.3	95.1
48				4 1/4 "	720	3400	5	850	5400	5	1000	8000		50.5	70.0	89.1	108.5
54				4 5/8 "	880	3700	5 1/2	1050	5850	5 1/2	1050	9000		57.8	80.1	102.0	123.6
60				5 "	1060	4000	6	1280	6000	6	1280	9000			88.2	112.3	136.4
66				5 3/8 "	1250	4250	6 1/2	1480	6300	6 1/2	1480	9500			96.6	123.1	149.5
72				5 3/4 "	1560	4500	7	1835	6600	7	1835	9900			105.1	133.9	162.6
84				8 §	2000		8	2300		8	2300					156.6	190.3

* Conc. 3500 p.s.i.   § Conc. 3000 p.s.i.   † Conc. 4500 p.s.i.

Ultimate strength given for reinforced concrete pipe is A.S.T.M. "first crack" strength.
Standard laying length - 4 ft.
Weights per lin. ft. furnished by Universal Concrete Pipe Co.

Furnished in any length in multiples of 2 ft. Data furnished for Armco Pipe by Shelt Co. Elmira, N.Y.

Reprinted with permission from Elwyn E. Seelye, DESIGN, 1945 Edition, John Wiley & Sons, Inc.

Fig. 5-17. Pipe classes and properties

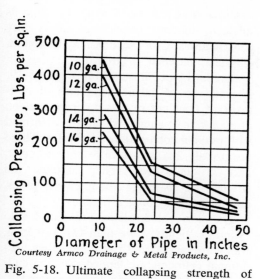

Courtesy Armco Drainage & Metal Products, Inc.

Fig. 5-18. Ultimate collapsing strength of corrugated pipe from exterior hydrostatic pressure

water from leaking out of the canal into surrounding soil.

Check Dams. When the slope of a channel or gutter is so steep as to make erosion likely, it can be divided into a series of easy gradients, and separated by check dams over which the water falls steeply.

It is important that each dam have a center spillway large enough to prevent water from overtopping the edges and eroding the earth alongside. An apron is also necessary to prevent undermining.

Where elaborate structures are not practical, crude ones made out of brush and logs or loose stones may serve the purpose.

PIPE

Drain and culvert pipe is made in sizes with inside diameters ranging from three inches to fifteen feet. Materials include concrete, tile (vitrified clay), transite (cement-asbestos), and corrugated iron or steel. Figure 5-16 indicates the names of various parts of a pipe cross section.

The tables in Figure 5-17 show some properties of concrete and corrugated pipe. Strength of the metal pipe is not shown

here because this is in part a function of the support it has on the sides.

Ordinary vitrified clay pipe is comparable in strength to plain concrete sewer pipe, and extra strength clay to reinforced concrete culvert pipe. Transite is available in a number of classes that vary from the strength of plain concrete to twice that of reinforced concrete. Both the tile and the transite are brittle and fragile to handle.

Figure 5-18 shows the approximate strength of corrugated pipe.

Concrete. Concrete pipe may be plain or reinforced, the joints may be butt, bell, slip joint, or gasketed. Size range is from four inches in inside diameter and up. Lengths are two, three, four, and eight feet.

Butt (open) joints are used for land tile.

Bell joints are resistant to chipping, will hold the pipe against slipping downhill, and, if the joints are open, will reduce flow or seepage of water along the outside of the pipe, but are difficult to lay.

Slip joints are easier to handle and to lay because of the uniform outside diameter.

Pipes over 12 inches inside diameter are usually reinforced, and this construction is required on most jobs because of its additional strength.

Concrete may be attacked by water carrying certain alkali salts or other chemicals. It is subject to erosion from fast flowing water carrying abrasive material, and may scale or disintegrate slowly from weathering. Structural difficulties may arise from the comparative weakness of its joints. However, under a wide range of conditions, it is long lived enough to be considered a permanent installation.

Tile. Tile may be porous or glazed, and is chiefly made in small and medium diameters, and in one to four foot lengths. The porous type usually has butt joints and is called land tile. Standard glazed or vitrified tile has bell joints and is called sewer tile.

Tile is lighter than concrete and has ex-

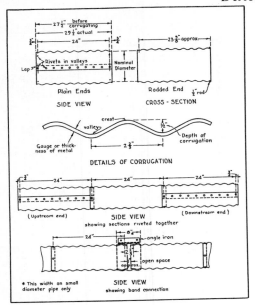

Fig. 5-19. Corrugated pipe detail

cellent bearing strength and resistance to weathering and corrosive chemicals. Its glazing resists erosion. It is fragile, and must be handled with care. In small sizes it is cheap and easy to lay except on unstable ground.

Orangeburg. In small sizes, tile may be replaced by Orangeburg pipe formed from asphalt-impregnated paper. Standard pieces are 10 feet long, with butt ends that may be linked by sliding collars. Bottoms may be solid to carry water, or perforated to take it in.

This pipe is light, easy to lay, and long lasting. It is not as strong as tile against crushing, but it is not brittle, and it is seldom broken in handling.

Transite. Transite, or cement asbestos pipe, is a light composition with very high pressure and crushing resistance. Diameters range from four inches to 36 inches. Sections are 13 feet long with gasketed collar joints. They must be handled carefully to avoid breakage.

Transite is used for water and sewer lines, but is too expensive for most drains.

Corrugated Metal. Corrugated pipe is made in standard, helical, and heavy duty constructions. Cross section may be round, elliptical, flattened, or arched.

Standard pipe, illustrated in Figure 5-19, is made up of galvanized plates of rust-resistant (but not rustproof) iron or steel, 16 gauge to 8 gauge, which are deformed with parallel corrugations or ripples. These are usually 2⅔ inches from crest to crest, and ½ inch deep. They increase the strength of 16 gauge about eleven times, and of 8 gauge about 3½ times.

The corrugated plates are rolled into cylinders slightly more than two feet long, which are lapped and riveted together. Additional cylinders are lapped over the ends and riveted to obtain the desired length.

Inside diameters of standard riveted pipe range from six inches to eight feet. Lengths may be made up in any multiple of two feet, but transportation problems usually limit single pieces to 20 or 24 feet. Ends may be beveled or skewed.

Features of arched pipe are lower clearance, greater bottom capacity, and less tendency to settle in soft ground.

Pipe sections are fastened together on the job by band collars. These may be one-piece or two-piece. One-piece bands are usually fastened by compression bolts only. Two-piece may be riveted or bolted to the sections. Because of allowance for overlap, each pipe section is 1½ inches longer than its nominal length. Each joint adds the width of one corrugation.

Under normal conditions, corrugated pipe gives long service, but its life may be shortened by chemicals or electro-chemical action, and by erosion of the bottom.

Corrosion can be checked by an asphalt coating, or for more severe conditions, by asbestos bonding. A mat of asbestos fibers is pressed into the molten zinc coating as a final step in the galvanizing process. The exposed fibers are saturated with asphalt. Erosion is reduced or eliminated by a paved invert. This is a wear-resisting asphalt pavement on the bottom which fills the valleys and covers the crests of the corrugations ⅛th inch deep. It can be applied to either galvanized or coated pipes.

Corrugated pipe is very much lighter than concrete or tile; it is not as readily damaged by carelessness or abuse; it is easily placed, connected, extended, or removed for salvage, and resists movements of fill which would pull short jointed pipes apart. Its internal flow resistance is higher than in other types. Its corrugations tend to keep it from moving in the fill, and discourage seepage or overflow following the outside. It will bridge low or weak spots in its supports.

For subdrainage it can be drilled with ⁵⁄₁₆th or ⅜th inch holes through the haunches.

Helical (twisted) pipe is long, narrow corrugated sheets bonded to each other by folding the edges. It is made in small 6″ to 21″ sizes, and 18 and 16 gauge metal. It is designed primarily for subdrainage.

Armco multi-plate pipe, Figure 5-20, is made of plate from 12 gauge to 1 gauge (⁹⁄₃₂″). Corrugations are 6 by 2 inches. Inside diameters range from 60 inches to 180 inches.

This pipe is shipped in the form of curved and drilled plates that are bolted together in location, as shown. It may also be made up into part circles or arches.

Other Types. Wood pipe is made of wood staves running the length of the section, cut to fit each other closely and bound together by wire loops. It should be full of water to keep the wood expanded and the seams watertight. It is light, easy to handle, and not particularly fragile. If its base is uneven, or if it dries out, it may leak. It is not commonly used.

Oil or grease drums may have the ends cut out and the cylinders tack welded together. Such conduit is easy to handle, but

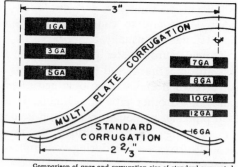

Comparison of gage and corrugation size of standard corrugated sheets and Multi Plate plates.

Courtesy Armco Drainage & Metal Products, Inc.

Fig. 5-20. Multi Plate pipe construction

compressive strength and resistance to corrosion are poor.

It may be strengthened somewhat by stretching it vertically, in the manner to be described for corrugated pipe. Struts may be left in permanently, but are likely to catch debris and to cause clogging. Such installations are usually temporary.

Threaded iron water pipes or well casings are sometimes used for small drains or culverts when they can be obtained second hand, or no other pipe is available. If the used pipe can be bought cheaply, it is ideal for draining small quantities of water through drives or fills during construction, if the regular drainage system is delayed. It will not pull apart under any natural stress, resists bending, can be cleaned out with a plumbers' snake, and is easily salvaged by pulling out from the end, or digging out.

BRIDGES

Where a road or other continuous embankment crosses a stream or drainageway, it is usually carried over it on a bridge or a culvert.

These structures may be distinguished from each other on a basis of width of opening. The critical width, or span, at which a bridge becomes a culvert varies from five to 20 feet in different localities.

Log. Figure 5-21 shows a type of log bridge suitable for carrying a pioneer or haul road, or a driveway, across a small stream.

Sill logs are set into the bank parallel with the stream, far enough back from it to be secure against being undermined. They may be braced by bolting to stumps or driven piles, or by cables stretched to anchors behind them. This anchoring is very important, as streams often flood sufficiently to float wood bridges.

Logs strong enough to carry expected loads are then placed close together across the stream, resting on the sills and preferably being fastened to them and to each other by lag bolts in drilled holes. Butts and tops should be alternated, so that log taper will not make one side wider and higher than the other. These logs are called stringers.

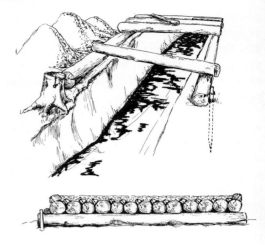

Fig. 5-21. Log bridge

Next, straight grained logs of smaller size should be split in quarters or other wedge shaped fractions and placed, split sides down, between the tops of the big logs. They may be cut into as short sections as necessary to fit snugly, and fill the cracks between the logs. Sections of round poles may be used the same way.

The split pieces should each be fastened to one log with spikes or lag bolts. Fastening can be on either side, but must not be on both, as that would permit movements of the logs to split them or pull the fastenings loose.

Quarter logs or poles are then fastened to the outside logs, to serve as curbings to prevent vehicles from running off, and/or to retain surfacing. The wood surface may be left exposed for light or occasional use, or covered with gravel or dirt to make it smoother and reduce danger of tire damage if nails work up.

Decks may also be made of planks or split logs nailed at right angles to the stringers.

If the span is long, or the loads heavy, a stone filled log or timber crib may be used as a center support. Cribbing may also be used at one or both banks if they are too low or soft to give proper support.

The strength of wood of various species, and in different conditions, varies so widely that individual judgment must be exercised in selecting the logs. If the bridge is to be used over a period of years, resistance to rot may be more important than initial strength.

Green wood is strong but lacks rigidity, and tends to give too much bounce to a long bridge. It will usually bend and splinter before breaking.

If only one side of the stream is accessible to machinery, the logs may be pulled across from that side by the use of a cable through a snatch block anchored on the opposite bank. The block should be placed high, if possible, to prevent the end from digging into the bank.

Concrete. Concrete bridges consist in general of two abutments supporting a slab. The slab usually includes guardrails, and supporting ribs or stringers which may be flat or arched. The abutments are usually continued into wing walls to direct the stream through the opening and to protect the embankment against sliding or erosion.

Even small structures are quite heavy and require that the abutments rest on solid footings. The flow of the stream should not be restricted, as it might then scour out the material against the abutments and undermine them. Abutments must be strong enough to resist the horizontal thrust of the fill behind them.

Reinforcement should be used throughout the structure, and is particularly important in the slab and its ribs.

The forms for the slab must be supported on a temporary wood or steel bridge of considerable rigidity.

Bridges should be engineered for the site and conditions. Construction should not be attempted without experienced supervision.

CULVERT DESIGN

A culvert may be made of almost any structural material. Reinforced concrete or corrugated metal pipe, and poured reinforced concrete are standard for highways and railroads. Tile and plain concrete may be used for light service. Log and timber construction are usual in pioneer and military roads.

Water passages (barrels), may be round, arched, rectangular, or in special shapes. More than one may be used.

Capacity. A culvert serves to carry the water from a drainage area or watershed of a certain size. This water includes surface runoff of rain and melted snow and ice, and whatever ground water comes to the surface within the area.

The size of culvert opening should be determined by the amount of rain which is likely to fall in the watershed within a certain period, and the character and slope of the ground so far as it affects the percentage of water that will run off, and the speed of its flow.

Additional factors to consider are the opening required by normal stream flow before it rains, the extent to which the opening may be restricted by silting, the velocity of water in the culvert, the extent to which water not passed through it can pond against the embankment before overtopping it or damaging property behind it by flooding; and the probable damage from overtopping.

Runoff. Rate of runoff is determined by intensity of rainfall, the size and shape of the watershed, and the slope, plant cover, and the permeability of the soil.

Rainfall is measured in inches, and its intensity in inches per hour, although the period of measurement may be less than an hour. For example, a rainfall of three inches might fall at the rate of six inches an hour for thirty minutes. In calculating runoff, an adjusted or equivalent rate can be used which makes allowance for variations in rate and duration.

Each watershed has a period of concentration, at the end of which the runoff is assumed to be at a maximum. This is the time required for water to flow from the farthest point in the shed to the culvert. If rainfall is continuous, and ground conditions are unchanging, the runoff at the culvert will increase from the beginning of the rain until it includes water from the whole area, after which it will continue at the same rate.

This period will be longer for long narrow watersheds than for square or round ones of similar area.

The assumptions involved are not strictly accurate as runoff increases as the ground becomes saturated, as water penetrating the soil emerges at lower levels, and the rate of flow is more rapid as the volume in channels becomes larger. However, there are so many variables that exact results cannot be obtained, and the average culvert is not important enough to justify an individual study of its drainage area.

The intensity of rainfall will determine the amount of water that will fall on an acre. The ground, slope, and vegetation will regulate how much of that water will flow off, and the speed of its flow. The number of acres in the watershed will determine the total amount of water delivered to the culvert. The period of concentration will determine the length of rainfall necessary to bring the area to the point of full discharge.

There are a number of formulas used in runoff calculations. These may give the volume of water to be expected, or the area in square feet of the culvert or bridge opening required. Information can also be obtained from performance of existing culverts or bridges, and observed heights of flood water.

The value of results obtained varies with the care with which field studies are made and with a number of factors that are difficult to work out. However, for the contractor who wishes a general guide to cul-

vert size requirements the simplest method is the best.

Figure 5-22 contains two maps showing adjusted rainfall rates in inches per hour for average requirements, and for any installation where overflow or backing up is particularly undesirable. The table supplies the number of square feet of culvert openings required to drain various areas on the basis of one inch of rain per hour.

To determine the size of a culvert, the drainage area is measured or estimated. Topographic or airplane maps are particularly useful for this purpose. The number of acres, or the next higher figure, is selected in the left hand column. The figure opposite this acreage, in the vertical column whose description best fits the area in question, is taken and multiplied by the rainfall rate shown for the locality by the appropriate map.

This will give the culvert area in square feet. To obtain the diameter of the proper size round pipe, use the formula

$$\text{Diameter} = 2\sqrt{\frac{\text{area}}{\pi}}$$

(twice the square root of the area divided by 3.14).

The indicated size should be increased if full culvert capacity may not be available, or any local conditions (such as abnormally intense rainfall, or extremely steep and non-absorbent slopes) indicate the need.

Even generously designed culverts may be inadequate for exceptional storms as it is seldom economically practical to provide for them.

Sidewalls or Headwalls. Sidewalls serve to hold embankments from falling into inlet or outlet channels; to direct water into and away from the passage or barrel to reduce turbulence and prevent undercutting of the embankment; to support the ends of the culvert, and to hold pipe sections against separating inside the fill.

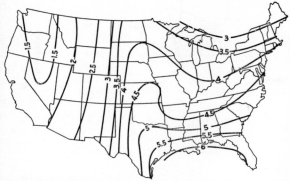

Equivalent rainfall rates in inches per hour for *average* design conditions.

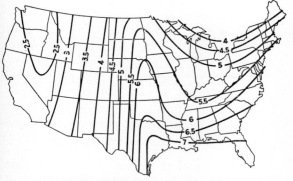

Equivalent rainfall rates in inches per hour for *unusual* design conditions.

Waterway Areas (Sq.Ft.) Required to Drain Different Acreages, *M*, for Equivalent Rainfall Rate of 1 In. per Hour

M acres	Flat areas not affected by accumulated snow. Length several times width	Rolling farm land. Length of watershed three or four times the width	Rough, hilly watersheds having moderate slopes	Steep, barren watersheds having abrupt slopes
2	0.08	0.14	0.28	0.42
4	0.14	0.24	0.47	0.71
6	0.19	0.32	0.64	0.96
8	0.24	0.40	0.79	1.19
10	0.28	0.47	0.94	1.41
15	0.38	0.63	1.27	1.91
20	0.48	0.79	1.58	2.36
25	0.56	0.93	1.86	2.80
30	0.64	1.07	2.14	3.21
35	0.72	1.20	2.40	3.60
40	0.80	1.33	2.65	3.98
45	0.87	1.45	2.89	4.34
50	0.94	1.57	3.14	4.70
60	1.08	1.80	3.59	5.39
70	1.21	2.02	4.03	6.05
80	1.34	2.23	4.46	6.69
90	1.46	2.43	4.87	7.31
100	1.58	2.63	5.27	7.91
150	2.14	3.57	7.14	10.7
200	2.66	4.43	8.87	13.3
250	3.14	5.24	10.5	15.7
300	3.60	6.00	12.0	18.0
350	4.04	6.74	13.5	20.2
400	4.47	7.45	14.9	22.4
450	4.89	8.14	16.3	24.4
500	5.29	8.80	17.6	26.4
600	6.06	10.1	20.2	30.3
700	6.81	11.3	22.7	34.0
800	7.52	12.5	25.1	37.6
900	8.22	13.7	27.4	41.1
1000	8.89	14.8	29.6	44.5
1200	10.2	17.0	34.0	51.0
1400	11.5	19.1	38.1	57.2
1600	12.7	21.1	42.2	63.3
1800	13.8	23.0	46.0	69.1
2000	15.0	24.9	49.8	74.8
2500	17.7	29.5	59.0	88.4
3000	20.3	33.8	67.6	101.4
3500	22.8	37.9	75.8	113.8
4000	25.2	41.9	83.9	125.8
4500	27.5	45.8	91.6	137.5
5000	29.7	49.5	99.1	148.7

10

Fig. 5-22. Culvert capacity maps and table.

The wall requirement is reduced by lengthening the pipe, as Figure 5-25, top. Pipe resting on the original grade, or projecting clear of the bank as in Figure 5-23, does not usually need a wall at the outlet.

A sidewall is usually of reinforced concrete but may be of stone, wood, or metal. It may be of heavy construction and firmly founded to resist movement in any direction; or it may be light and superficial so

Fig. 5-23. Projecting culvert pipe

Fig. 5-24. Metal headwall

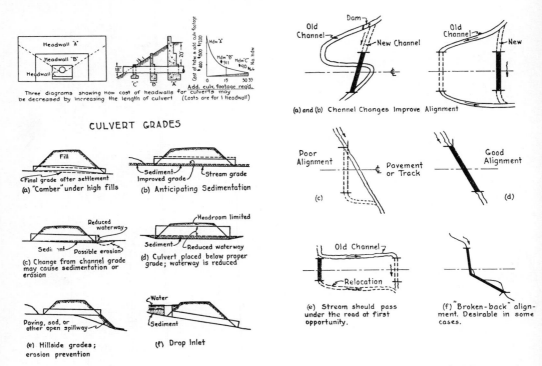

CULVERT GRADES

(a) "Camber" under high fills

(b) Anticipating Sedimentation

(c) Change from channel grade may cause sedimentation or erosion

(d) Culvert placed below proper grade; waterway is reduced

(e) Hillside grades; erosion prevention

(f) Drop Inlet

Fig. 5-25. Culvert headwalls and grades

(a) and (b) Channel Changes Improve Alignment

(c)

(d)

(e) Stream should pass under the road at first opportunity.

(f) "Broken-back" alignment. Desirable in some cases.

Fig. 5-26. Culvert alignment

that any settlement will affect it to the same extent as the pipe.

It is most convenient to place wall footings before laying the end pipes.

Metal headwalls, Figure 5-24, are fastened to corrugated pipe by standard couplings. They can be removed and re-used if the culvert is lengthened.

Wingwalls. Wingwalls are extensions of the headwall which serve to direct the stream, and may buttress the headwall against outward thrust. Several designs are shown in Figure 5-27.

Alignment. Culvert barrels should be straight under most circumstances.

It is usually desirable to use the original channel for the culvert passage and to have the culvert cross the road at right angles to the center line.

These two objectives are often in conflict.

The original channel may be undesirable

if it is crooked, crosses at a sharp angle or skew, shows rock ridges, is made up of soft mud, or has a strong flow of water. In such cases, it may be more economical to dig a trench nearby, lay the pipe, and then divert the stream into it.

Right angle alignment may be ignored if the natural channel is diagonal, and can be easily prepared for the culvert; or when excessive trenching is required to bring the stream straight across.

Referring to Figure 5-26 (C), it will be seen that a slight change in stream alignment under the road can lead to a considerable amount of digging on the side, that the new stream channel will be out of balance, and it may require mats or revetments to protect the outside banks of the curves.

On the other hand, changes may involve comparatively little excavating and produce more satisfactory channels than the original.

Fig. 5-27. Concrete head and wing walls

RECOMMENDED GAGES FOR STANDARD CORRUGATED PIPE UNDER EMBANKMENTS (Revised 1939)
For Highway Culverts, Municipal Drains, and Railroad Culverts Not Under Track
With Proper Backfilling and Tamping, Using Firm Material

Diam. Inches	Area Sq. Ft.	Fills Up to 15 Ft.	15 to 20 Ft. Fill	20 to 25 Ft. Fill	25 to 30 Ft. Fill	30 to 35 Ft. Fill	35 to 40 Ft. Fill	40 to 45 Ft. Fill	45 to 50 Ft. Fill	50 to 60 Ft. Fill	60 to 70 Ft. Fill	70 to 80 Ft. Fill	80 to 100 Ft. Fill
8	.35	16	16	16	16	16	16	16	16	16	16	16	14
10	.55	16	16	16	16	16	16	16	16	16	16	14	14
12	.79	16	16	16	16	16	16	16	16	16	14	14	14
15	1.23	16	16	16	16	16	16	16	16	14	14	12	12
18	1.77	16	16	16	16	16	14	14	14	14	12	12	12
21	2.41	16	16	16	16	16	14	14	14	12	12	12	10
24	3.14	14	14	14	14	14	14	14	12	12	12	10	10
30	4.91	14	14	14	14	12	12	12	10	10	10	8*	8*
36	7.07	12	12	12	12	10	10	10	8	8	8*	8*	8*
42	9.62	12	12	12	10	10	8	8	8	8	8*	8*	8*
48	12.57	12	12	12	10	10	8	8	8	8	8*	8*	8*
54	15.90	12	12	10	10	10	8	8	8*	8*	8*	8*	
60	19.64	10	10	10	8	8	8						
66	23.76	10	10	8	8								
72	28.27	10	8	8									
78	33.18	8	8	8									
84	38.49	8	8			Use Multi Plate Culverts							
90	44.18	8											
96	50.27	8											

The gages in the first column are minimum requirements for highway embankment conditions and agree with the recommendations of the A.A.S.H.O. specification.

Culverts below the heavy line should be strutted during installation.

Culverts with the mark (*) should be trenched one diameter.

The minimum height of cover should be as follows: for highways with unpaved surfaces, one-half diameter, with 12" minimum; for highways with flexible and rigid pavements, 12" minimum.

Gages heavier than those given here should be used where conditions are unusually severe.

Fig. 5-28. Recommended gauges, corrugated pipe

If good alignment between stream and culvert cannot be obtained on both sides, the upstream side should be favored. When the capacity of the culvert is heavily taxed by a storm, it is advantageous to get the water into the culvert smoothly.

Gradient. It is desirable to lay the culvert on the floor of the natural channel, on the original ground surface, or in a smoothly dug ditch. This gives firmer support than fresh fill. Inequalities in channel or ground are smoothed by cutting off ridges and tamping fill into hollows.

The passage should have at least a one half percent slope, and two to four percent is preferable. It should not be over eight or ten percent because of probable erosion of the bottom of the lining. The gradient of the waterway down from the discharge end should be as great as that above the inlet, for a long enough distance so that it will not silt up.

If the culvert is on fill or a foundation which is expected to settle, it may be laid in a vertical curve, known as camber, or to include a vertical angle. In each case, a slight gradient is used at the inlet end and a steeper one toward the outlet. Center settlement will tend to straighten out the passage.

Disjointing. If a fill settles unevenly, or part of it moves laterally, any culvert which is in it will be put under tension. Such forces are not often sufficient to break pipe but they are apt to pull apart the joints.

Metal pipe is very resistant to such disruptive forces and can be further strengthened by special joint fastenings. Short-section rigid pipe may be braced at each end by a heavy headwall, founded on underlying stable material, or may be cabled together.

Depth. Fills over a culvert, to a depth of four to seven feet, serve to protect it by spreading out the weight of vehicles on the surface. Deeper fills have diminishing protective effect and impose the load of their own weight, which at great depths may be sufficient to crush the pipe.

Figures 5-28 and 5-29 contain tables showing the approximate required strength for highway culvert or sewer pipe under various depths of fill. It should be noted that in table (A) the required strength listed must be multiplied by the diameter of the pipe in feet. This is because the increase in crushing strength with increased size of pipe only partly compensates for its greater surface area, against which pressure is exerted. In short, although large pipe is stronger in itself than small pipe, it is weaker in regard to burial loads.

The tables also indicate the importance of pipe bedding in determining strength. Corrugated pipe requires good compaction of fill at the sides and proper bedding.

It is questionable whether the formed bed shown for the ordinary and first class installations is often perfect enough to give the calculated support. However, careful tamping of the fill under the curve of the pipe will give similar results.

DRAINAGE & SEWERAGE - LOADING ON PIPES

TABLE A - REQUIRED ULTIMATE STRENGTH, LBS. PER LIN. FT., OF PIPE FOR VARIOUS DEPTHS UNDER H-20 HIGHWAY LOADING.

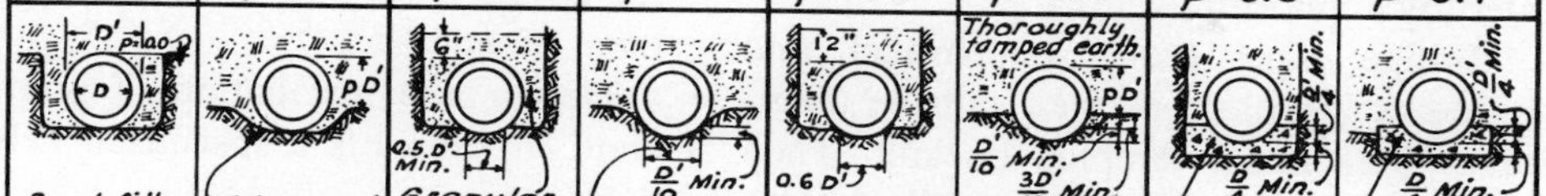

DEPTH OF COVER OVER PIPE (feet)	BEDDING CONDITIONS							
	IMPERMISSIBLE		ORDINARY		FIRST CLASS		CONCRETE CRADLE	
	p=0.0	p=0.7	p=0.0	p=0.7	p=0.0	p=0.7	p=0.0	p=0.7
	Backfill untamped.	Not shaped to fit pipe.	Granular materials Shovel placed and tamped.	Accurately shaped to fit pipe.	Backfill Carefully tamped in thin layers.	Accurately shaped to fit pipe.	2000# Concrete.	2000# Concrete.
2	2430 D	2490 D	2280 D	2320 D	2250 D	2260 D	2190 D	2190 D
3	2060 D	2190 D	1840 D	1900 D	1770 D	1820 D	1700 D	1690 D
4	1880 D	2040 D	1580 D	1670 D	1500 D	1560 D	1400 D	1400 D
5	1760 D	2180 D	1390 D	1540 D	1290 D	1480 D	1160 D	1240 D
6	1820 D	2220 D	1380 D	1600 D	1260 D	1440 D	1100 D	1160 D
7	1900 D	2380 D	1390 D	1660 D	1240 D	14500	1070 D	1140 D
8	2050 D	2550 D	1460 D	1740 D	1290 D	1510 D	1090 D	1160 D
9	2200 D	3030 D	1530 D	1990 D	1340 D	1720 D	1110 D	1290 D
10	2350 D	3240 D	1610 D	2130 D	1400 D	1820 D	1150 D	1320 D
12	2650 D	3780 D	1770 D	2430 D	1520 D	2040 D	1210 D	1450 D
14	3020 D	4340 D	2000 D	2830 D	1710 D	2360 D	1350 D	1660 D
16	3390 D	5010 D	2210 D	3170 D	1890 D	2650 D	1500 D	1860 D
18	3780 D	5780 D	2470 D	3650 D	2100 D	3030 D	1640 D	2090 D
20	4170 D	6300 D	2700 D	3970 D	2290 D	3400 D	1790 D	2270 D

For depth greater than 20 ft., use Table B below. D = inside diameter in feet.

EXAMPLE: Assume 24"ϕ pipe, depth 10 ft., ordinary bedding p=0.7
Table A, above: Ultimate strength per lin. ft. 2130 D = 2130 × 2 = 4260.
From page 5-21 select vitrified clay extra strength pipe or Class 2, Transite pipe.

TABLE B - MAX. HEIGHT IN FEET OF FILL OVER PIPE (NO LIVE LOAD).*

ULT. STRENGTH 3 EDGE BEARING LBS/L.F.	BEDDING CONDITIONS							
	IMPERMISSIBLE		ORDINARY		FIRST CLASS		CONCRETE CRADLE	
	p=0.0	p=0.7	p=0.0	p=0.7	p=0.0	p=0.7	p=0.0	p=0.7
1200 D	6.0	4.5	9.0	7.0	11.0	8.0	14.0	11.5
1500 D	7.5	5.5	11.5	8.5	13.5	10.0	18.0	14.0
1800 D	9.0	6.5	14.0	10.0	16.5	11.5	21.5	17.0
2000 D	10.0	7.0	15.5	11.0	18.0	13.0	23.5	18.5
3000 D	14.5	10.0	23.0	16.0	27.0	19.0	35.5	27.0
4000 D	20.0	13.0	30.5	20.5	36.5	24.5	47.5	36.0

NOTES FOR TABLES A & B :- In rock excavations an earth cushion less than 8" is classed as IMPERMISSIBLE bedding. ORDINARY & FIRST CLASS beddings require earth cushions of 8" minimum.
 The values shown in the columns headed p=0.7 should be used for projection ratios from 0.3 to 0.9 and those shown in columns headed p=0.0 should be used for installations where the pipe is installed in a trench dug to a depth at least equal to the outside diameter of the pipe.
 There is a difference of opinion in regard to the practicability of dishing the bed, as called for under ordinary and first class conditions.
 * From American Highway Practice by Laurence I. Hewes.
Reprinted with permission from Elwyne. Seelye, DESIGN, 1945 Edition, John Wiley & Sons, Inc.

Fig. 5-29. Loading on pipes

Restricted Height. If there is not sufficient space between the channel bed and the embankment surface to install a round pipe of adequate size, two or more round pipes, or low clearance pipe with a flatter cross section or a pipe arch may be used.

Poured concrete structures may use a flat rectangle or two or more openings separated by supporting walls.

Multiple pipes should be parallel, and spaced at least half of their outside diameters, to facilitate tamping backfill. They should not be used for streams that carry coarse debris which might choke the openings.

Poured Culverts. Culvert barrels may be made of monolithic (one piece) structure of concrete poured into forms at the final location. This operation allows a wide latitude in sizes and shapes of barrels; allows a fairly exact proportioning of strength to load, and avoids problems sometimes met in procuring, transporting, and handling large pipe.

Poured culverts should always be reinforced. Even in casual construction when reinforcing bars are not available, the use of scrap metal or cable will both strengthen the structure and reduce the amount of concrete required.

A reinforced concrete barrel has great tensile strength and is unlikely to pull apart due to stresses in a settling fill. It can be made one piece with the sidewalls and wingwalls.

Relative costs of pipe and poured concrete vary with the locality, conditions, and in the case of concrete pipe particularly, with the distance from the manufacturer. In general, as the structure gets larger, pouring becomes more economical than concrete pipe. Remote areas and very high fills favor metal construction.

Small culverts carrying light fills usually have a square or rectangular barrel. When either the embankment or the stream bed is shallow in relation to the volume of water, two or more barrels or cells are placed side by side.

With increased depth of fill, or length of span, arch construction becomes more desirable.

Cost may be reduced if the same forms can be used for two or more structures. Knock-down steel forms can sometimes be rented for less than the cost of building them of wood.

Design and installation of poured concrete structures is a specalized field.

LAYING CORRUGATED PIPE

Handling. Corrugated pipe can be made up in any multiple of two feet. Lengths of eight to twenty feet are usually carried in stock, and longer or shorter ones are obtainable on special order. Shorter pieces can also be made by cutting with a torch.

Small and medium sizes are usually unloaded and placed by hand as in Figure 5-30, and medium to large ones lowered with a rolling hitch or with a crane. The crane may use a hitch around each end or around the middle with the help of a man to keep it balanced. Soft rope should be used so as not to scratch the galvanizing.

Pipe should not be dragged around on abrasive ground, nor scratched or banged against anything, as such abuse will damage the galvanizing and shorten the life of the metal.

It is lowered into trenches in the same manner as it is unloaded.

Foundation. The base should be shaped to fit the lower part of the pipe as closely as possible, by cutting the ground to shape, or building up with well tamped fill. The work can be checked by placing and removing the pipe, and noting whether it was in full contact.

If the floor is rock, it should be cut from six inches to three feet below pipe grade, the depth depending on the height of fill to be placed, and backfilled with earth or pea gravel. If it is mud, space should be

Fig. 5-30. Handling corrugated pipe

allowed for enough pea gravel to stabilize the surface. Saplings or wire lath might be laid under the gravel to provide extra stability.

If one end of the culvert is to rest on fill and the other in a cut, the fill under it should be thoroughly tamped to avoid unequal settlement. If the cut is rock, its edge should be beveled to distribute any settlement strains over a longer piece of pipe.

Placement. Each section should be placed with the longitudinal seams at the sides. The cross joints should have the external part of the overlap upstream so that if the joint is not tight, seepage will tend to move into the pipe instead of out of it.

If the pipe has a metal date tag, or an instruction tag, it should be on the upstream end of the pipe.

Joints. Sections are usually fastened together by a one piece band. This is placed under the end of the first piece, and the second laid so that the band will overlap each by the same number of corrugations. The coupler is then drawn tight by turning down the nuts with a wrench.

This joint resists any force tending to pull it apart, being about as strong as the pipe itself. If the pipe will be subject to very severe stress, a two piece coupler riveted to one or both sections may be used. The pipes are placed so that the collar lines up, and it is fastened with the bolts and nuts.

If the pipe is large enough to work inside (36″ or more ordinarily, but less if a small worker is available), a one piece band may be installed in the regular way, and holes then drilled through the band and the pipes, and bolts used to strengthen the joint.

If water tightness is important, an asphalt sealer may be placed inside the band before installing, or a special coupling may be used.

Strutting. Large corrugated pipe may be strengthened against heavy loads by strutting. This consists of deforming it so that it is higher than it is wide. It is usually done by putting beams and soft compression caps along the top and bottom, pushing them apart with heavy jacks, and placing pre-cut struts to hold them in position.

Joints are made and fill placed and compacted in the usual manner.

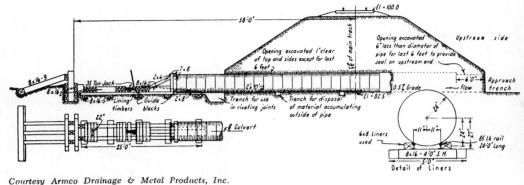

Courtesy Armco Drainage & Metal Products, Inc.

Fig. 5-31. Jacking pipe through fill

Struts are removed after a settlement period of three to six months, except that if the pipe shows a tendency to bend beside the head beam, or floods are expected, they are taken out immediately.

The pipe will then tend to resume its round shape and, in doing so, will press heavily against the fill on each side, further compacting it. This extra compaction, which is roughly proportional to the load carried by the pipe, gives additional support against flattening.

Paved or coated pipe is strutted at the factory before treatment. It is elongated with beams and jacks, which are removed after the sides are tied together with cross rods or wires. Paving or coating is then added.

These pipes are usually tamped in until the wires do not have tension, but not until they are perceptibly slack. Struts are cut when they have served their purpose.

Side Assembly. In rush jobs, where traffic is stopped or detoured inconveniently during work, the pipe sections may be assembled nearby, and carried, pushed, or dragged into position. If pushed by a bulldozer, the front should be held slightly above the ground by a crane, other machine, or a crew of men.

Jacking. Corrugated pipe can be pushed through an ordinary fill without trenching. Figure 5-31 shows a diagram of this process. An approach trench is dug to line the pipe up, and a heavy backstop built. A trough is made across the trench at the toe of the fill for access to the pipe. A wood frame is placed against the pipes to take the push of the jacks.

A man can work inside a large pipe, digging from the face to reduce pressure needed to push it. Dirt may be loosened ahead of smaller pipes by augers or water jets.

Trench drills such as those described in Chapter 20 make jacking unnecessary under many conditions, and can be used for jacking when it is necessary.

LAYING CONCRETE PIPE

Handling. The standard length of concrete pipe is four feet, except for land tile. The pieces may be unloaded by rolling each pipe so that it will fall from the truck endwise onto soft ground or a couple of old tires. Bell-jointed pipe should be dropped on the straight end.

Pipe up to two feet may often be rolled and pried into place by laborers with bar and poles. However, in any size it is more convenient to use some sort of hoist.

Pipe can be handled by any lifting device with power to pick it up, and enough reach and maneuverability to place it easily. A crane is most suitable for placing it in a ditch, and a dozer shovel for laying it in the open.

A pipe hook, Figure 5-32, is very useful. It permits holding the pipe at the free end, and avoids difficulties in balancing it on a chain or sling, and in disturbing the pipe after it is laid by withdrawing the chain. (A) is alloy bar, 2″ x 4″, for pipe up to 60″. In (B) 4″ iron pipe will handle 24″ concrete safely.

Care must be taken not to let the hooked end of the pipe rest on anything, as contact might tip it so that it would slide off.

Eight foot lengths may have a two inch hole through the wall at the center. These can be picked up by means of an eyebolt inserted in the hole and caught inside with a nut and washer.

Foundation. The base is shaped and tamped in much the same manner as for metal pipe.

If pipe is laid in running water, leveling the bottom may be difficult. Weighted planks or troughs of preservative-treated wood may be placed on the bottom and carefully leveled and blocked up. Straight-sided pipe can be set on this and only minor adjustment should be required.

If the load of fill or traffic is to be heavy, or the foundation is partly on fill and partly on rock, or is unstable, a concrete bed may be used. A stiff mixture of concrete, as wide as the trench, or in forms a foot or two wider than the outside diameter of the pipe, and one-quarter of the pipe diameter in depth, is placed on a well-compacted earth base. It is roughly hollowed for the pipe which is settled into it by rocking, or slight raising and lowering.

Bell joint pipe requires cross grooves in the bed to accommodate the oversize end.

Placement. Pipe is usually laid with the bell or female ends upstream. If this is the case, placement should start at the downstream end. The first pipe can be lowered level, but the others should have the male or free end slightly lower so that it can be guided into place without scraping on the bottom. This tip is arranged by inserting the hook only part way into the pipe,

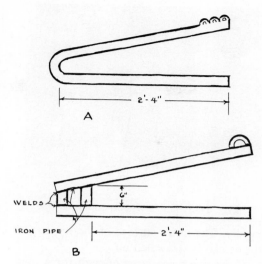

Fig. 5-32. Pipe hooks

ing the hook only part way into the pipe, or by pushing down on the free end.

If a section is not held in proper position by the bed, it should be chocked securely with stones or blocks until several more sections are laid, or the culvert is completed. This makes it possible to make any necessary readjustments with less work than if fill is tamped in immediately.

It is difficult to get each joint tight without considerable practice. However, it is often possible to lay several loosely, and then push them together from an end. This may be done by a small dozer in the trench, or by a cable threaded through the culvert to a crossbeam.

Joints. Except in informal or temporary work, joints should be cemented. This is particularly important if water may pond above the outlet so that it will go through under pressure, which might force it out through open joints and cause softening and channeling of the embankment.

Small pipe is cemented by wetting the ends to be joined, and troweling a rich mortar on the upper half of the male side and the lower half of the female. It is desirable to rotate the free pipe slightly, after it is in position, to spread the cement

Fig. 5-33. Cable ties on concrete pipe

evenly. The outer surface of the upper two thirds can be troweled off.

If the pipes are large enough for a man to work inside, the whole culvert may be placed dry. Oakum is then hand tamped into the joint cracks, and cement or bituminous mortar applied from the inside.

Ties. If foundation conditions are such that the fill may spread and pull the pipe apart at the joints, the culvert may be held together by heavy, deep based headwalls, or by tie lines.

Tie lines may consist of three rods, cables, or chains, hooked around the end pipes. Turnbuckles or loadbinders are used to tighten them. They may be internal, or external as shown in Figure 5-33.

The inside ties will reduce culvert capacity slightly, and may cause jamming of debris and complete stopping. However, they are accessible for inspection and tightening. Outside installations are difficult to service.

A loose cable is sometimes left inside a small diameter culvert for use if it becomes plugged with silt. Pulling the cable back and forth will make a small hole which can be enlarged by forcing water through.

OTHER FORMS OF CONSTRUCTION

Wood Culverts. Culverts may be constructed of wood when they are for tempo-rary use, or when time or expense prohibit obtaining more permanent materials.

Construction may be to almost any desired strength. Life expectancy will vary with the type and size of wood, preservative applied, and moisture conditions. In general, the parts which are permanently wet will have a much longer life than those which are exposed to air.

Several designs are shown in Figure 5-34.

Casual Placement. There are many situations in which it may be unnecessary or impossible to place culverts with the care required for permanent installations. These would include light traffic driveways and farm lanes; pioneer or access roads to be used for only a short period; and urgent construction in which it is necessary to get traffic through without delay, even at the cost of possible repair or reconstruction later.

Good standards should be approached as closely as possible, however.

If heavy traffic will ride directly on the pipe, or very close over it, a strong construction should be used. If silting and trash are not a problem, several small pipes will be better than one large as they are more resistant to concentrated loads, and they can be provided with an adequate depth of cover more readily.

If the foundation is unstable so that a part of the culvert will sink, oversize pipe may be used to provide adequate capacity after settlement and silting. If silting can be prevented, a badly sagging pipe may act as an inverted siphon.

The pipe should be long enough not to require large headwalls, unless they can be easily built with big stones or logs available on the site. Wingwalls, where required, can be made of rocks, of saplings hammered in as piling, or of brush mats.

On wet bottoms, pea gravel or crushed stone should be used under the haunches. Ordinary earth can be used as soon as

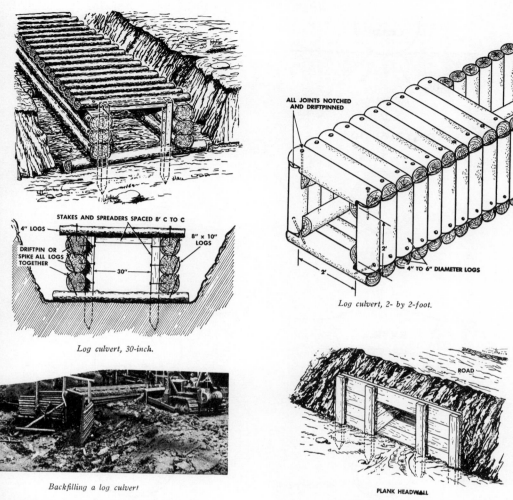

STAKES AND SPREADERS SPACED 8' C TO C

4" LOGS

DRIFTPIN OR
SPIKE ALL LOGS
TOGETHER

8" x 10"
LOGS

30"

Log culvert, 30-inch.

ALL JOINTS NOTCHED
AND DRIFTPINNED

2'

2'

4" TO 6" DIAMETER LOGS

Log culvert, 2- by 2-foot.

Backfilling a log culvert

ROAD

PLANK HEADWALL

Courtesy U. S. Army Engineers

Fig. 5-34. Wood culverts

the gravel has been built up above water level, but it will not consolidate under water.

BACKFILL

Proper placing and compacting of backfill affects both the strength of the pipe and the load it has to carry.

Load Distribution. If a round pipe lies on a hard flat surface, and is subjected to load, the entire pressure falls on the line of bottom contact. If the surface is curved, the area of contact is greatly increased, and

the load per square inch reduced correspondingly. As the haunches curve upward, the amount of vertical support to each square inch of surface decreases until it is zero at the widest point.

Corrugated pipe is flexible and requires horizontal support as well. A normal load tends to flatten the top and spread the sides. If the sides are held in firmly, the arch form of the load-carrying section is retained and strength kept at a maximum.

No part of the foundation or backfill touching a flexible pipe should be rigid, as

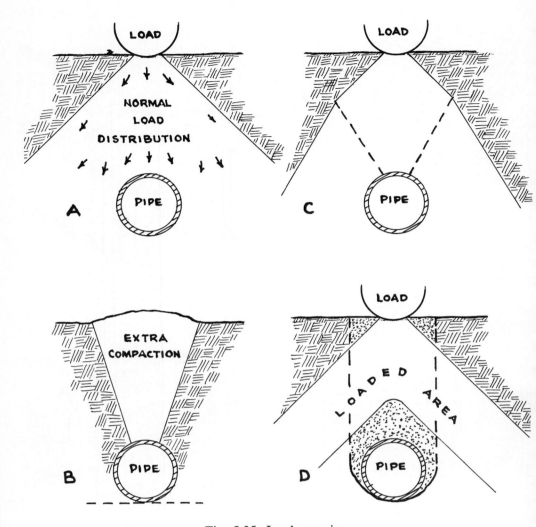

Fig. 5-35. Load over pipe

the whole pipe should be able to change cross section shape as it deflects under load, and any rigid support will cause excessive strains, particularly at the edges of the contact.

Rigid pipe receives only nominal support from the side fill.

Surface loads on soil masses are ordinarily distributed over increasing areas on lower levels, as in Figure 5-35 (A), so that pressure per square inch diminishes rapidly. If there is a difference in bearing power of the soils within the affected cone,

the more rigid soil may carry most or all of the load.

In (B) backfill has been placed loosely over an exposed or projected pipe. Settlement under traffic, or from the effects of weather, will be a fraction or percentage of its depth, so that the thinner fill over the pipe will not sink as far and will project as a surface ridge. This ridge may receive heavier loads, and be more thoroughly compacted, than the soil on each side.

If the surface is then bladed smooth, as in (C), the more compacted soil will transmit most of its loads straight down to the pipe, rather than allowing them to spread out through the softer fill to each side.

Loose backfill may have similar behavior in a wide trench. In a narrow one, however, loads will be carried by the firmer undisturbed walls, as in (D). This arching effect will relieve the culvert of part of the weight of the upper soil, and of traffic.

If a trench is loosely filled, settlement will result in a surface trough. Vehicles bumping across this will cause heavy impact loads. If the pipe is near the surface, it may be damaged or destroyed. If deep, the loads will be largely carried by the walls.

A tightly tamped backfill should distribute loads evenly over the whole area and subject the culvert to normal loads for its depth, as in (A).

Whatever load is imposed on the immediate vicinity of the pipe will be shared by it and by the fill on each side. If the fill is tightly tamped, it will bear a larger part of the burden, relieving the top of the pipe. Vertical pressure on the side fill is partly converted into horizontal pressure against the sides of the pipe, providing support for the top load.

Tamping. Fill must be tamped under the pipe haunches. It should be free of lumps, stones and trash, and should contain enough moisture to pack, but not enough to make it rubbery. It is placed with a hand shovel in thin layers.

The tamper should have a narrow edge to enable it to get well under the pipe, and, if the trench is narrow, may require a curved handle.

Filling and tamping are done evenly on both sides of the trench, to avoid shifting the pipe to the side. It may be necessary to wedge it in place temporarily with rocks or other blocking. Such material may be left and buried if the pipe is rigid, but removed if it is flexible.

Tamping blows should not be so vigorous as to wedge the pipe up out of position.

When sufficient fill has been placed so that its surface is out from under the pipe, mechanical tampers can be used. Fill should be compacted to the full width of the trench, or, if the pipe is on the surface (projection condition), for one pipe diameter on each side. Layers should be four to six inches deep.

When the pipe is nearly or wholly covered, layers of six to twelve inches may be placed, and tamping continued, until the trench is filled to grade.

Side Fill. When an embankment is to be built up on each side of the pipe and above it, much of the compaction can be done by machinery first moving parallel with the pipe, then across it.

A roller working parallel must be kept far enough away so as not to exert a horizontal thrust that will move the pipe. Fill and compaction should be kept even on both sides. Since there is often only one roller available, and it may not be able to get across, the other side may be compressed with a truck or by tamping.

Soil between the rolled strips and the pipe, and between the rolled strips and over the pipe, must be thoroughly tamped.

Successive layers can be rolled closer to the pipe center line. It is good practice to postpone crossing it until the fill is as deep as the outside diameter of the pipe, to avoid pushing it out of line.

Material is pushed to the pipe by a dozer or grader.

Loosened Backfill. A compacted embankment may be built up one diameter above the pipe, then ditched over it. This trench is filled in loosely, and layers of compacted fill built to the top of the embankment in the regular manner.

The soft fill over the pipe causes the load to be transferred to the solid sidewalls.

If any trench containing pipe is backfilled by a dozer, care should be taken not to drop rocks on the pipe.

FORDS AND DIPS

Farm or pioneer roads sometimes cross a shallow stream on its bottom so that vehicles must drive in the water. This type of crossing is called a ford. It may be satisfactory for light or occasional traffic, but it is subject to interruption by high water and ice, and may develop bad bottom conditions which would be difficult to remedy.

Crusher rock, in mixed sizes from ½" to 2½" makes a good patching and paving material. If bank gravel is used, thorough raking will allow the water to take away excessive fines.

In arid regions, many watercourses are dry most of the time, but will occasionally carry such large volumes of water that adequate bridges would be very expensive. In such cases, the road may run across the channel at its natural grade, with no provision made for passing water under it. The section of road in the channel is usually heavily built, reinforced concrete slabs up to two feet in thickness sometimes being used. This slab should be sloped on its upstream side, and may have a cut off or curtain wall extending below its main mass.

Occasionally a culvert may be placed under or beside the dip to pass small water flows, or a culvert structure may include a spillway for flood water.

Second-class roads may cross such a stream bed on graded local material that must usually be worked over after each flow of water. Roads may also run considerable distances in stream beds, as this may be the only route which is practicable without heavy blasting and grading.

SOIL MOISTURE

Water Table. Subsurface water exists in three states or zones. The lowest of the series is hydrostatic or free ground water. Its upper surface is known as the water table, or ground water level. It follows the contour of the land in a general way, but tends to be farther under the surface in hills and pervious soils than in hollows and heavy soils. If it rises to or above the surface, it makes swamps, ponds, or springs.

The actions of this water are controlled by gravity, causing it to seek lower levels by the resistance of the soil to its movement, and by fresh supplies of water reaching it from the surface.

The water table may be static, or fluctuate only slightly, or it may shift up and down widely in response to season or rainfall.

Soil which is saturated with ground water is usually unstable under load, will turn to mud if disturbed, and does not permit the growth of roots of most plants.

If a hole is dug below the water table, it should fill with water.

Capillary Zone. The capillary zone lies above the water table. It may be a few inches deep in coarse sand, and eight feet or more in heavy soils. It contains a substantial quantity of water that is held above the gravity surface by capillary attraction and other forces tending to attract and hold it in the finer soil spaces.

The amount of contained water diminishes from the bottom to the top of this zone.

Capillary movement in coarse soils is rapid, in fine ones quite slow.

Raising or lowering the water table may raise or lower the capillary table.

Medium and fine soils in this zone usually contain too much water for stability, and may be subject to frost heaving. In climates where rainfall exceeds evaporation, this zone offers the best conditions for root growth. In arid regions, the water may deposit alkali in the soil and render it unfit for cultivation.

Upper Zone. The upper or hygroscopic zone contains water which is in very thin films on the particles, or is in chemical or physical combination with them. Some of this water is hygroscopic—absorbed from the atmosphere—and is greatest in amount when humidity is high.

These small quantities of water often give the soil maximum stability, by acting as a cement or binder. Much of the water is too firmly attached to be removed by plant roots, or any method but oven baking.

This zone may also contain varying quantities of rain water, moving downward by gravity or capillarity, or adhering to soil particles. This is available to plants and may be found in sufficient quantity to make the ground unstable.

SUBSURFACE DRAINAGE

Purpose. Subsurface drainage lowers the water table. Deep drains, or those in porous soil, will lower the capillary surface also.

Soil must be drained when its water content makes it incapable of supporting roads or other structures on it, or causes frost heaving.

Playgrounds, golf courses, and other recreation areas may require draining to dry up spots which remain wet and soft long after rains.

Farmland drainage may serve to eliminate wet spots that cannot be worked as early as the surrounding land; to speed up the drying and the warming of soil in the spring; to encourage plants to form deep root systems, with resulting increase in vigor and drought resistance; and to leach out harmful substances which may accumulate in the soil.

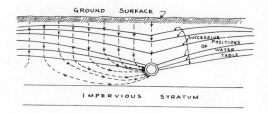

Fig. 5-37. Ground water movement

Methods. Ground water level may be artificially lowered by open channels or ditches, or by buried pipe or porous material. Such pipe is generally referred to as tile, even though it might be made of other materials.

In soils that will stand on steep slopes, ditches are the most economical construction down to a depth of a few feet. When wide cuts must be made to produce stable slopes, or when greater depth is required, the open ditch may involve so much excavation as to be more costly than tile.

Ditches, together with any space required for spoil, or for protection of machinery against falling in, may occupy rather wide strips of land. If in farms, they cut up the fields and add to the expense of planting and cultivation. They are hazardous when near roads. In any location, they will require occasional culverts or bridges for crossings.

Ditches usually require maintenance. This may include removing silt and cave-

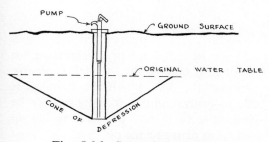

Fig. 5-36. Cone of depression

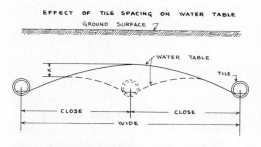

Fig. 5-38. Effect of tile spacing

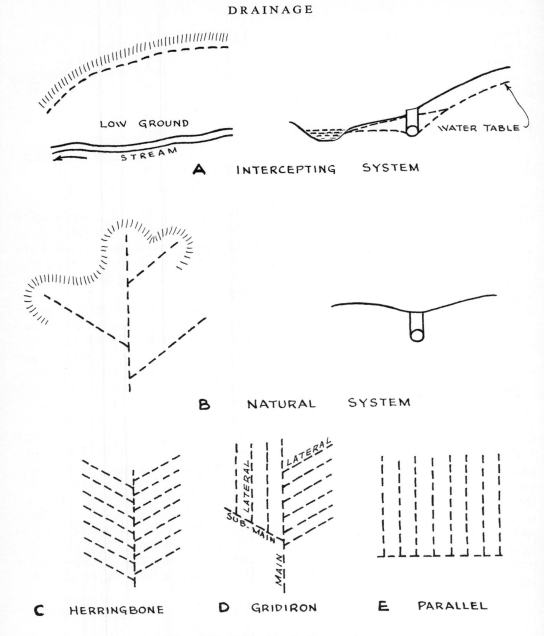

Fig. 5-39. Subsurface drainage patterns

ins, repairing erosion damage, and cleaning out vegetation. Neglect may result in general deterioration, with eventual stoppage, or expensive re-digging and clearing.

Buried drains do not cut up the fields, nor offer hazards along roadsides. However, if one becomes plugged, as a result of poor design, improper installation, or accident, it may be difficult and expensive to locate the difficulty. If the stoppage is due to general silting, it will probably be cheaper to lay a new line than to dig up and clean or repair the old one.

Choice of the type of drainage will de-

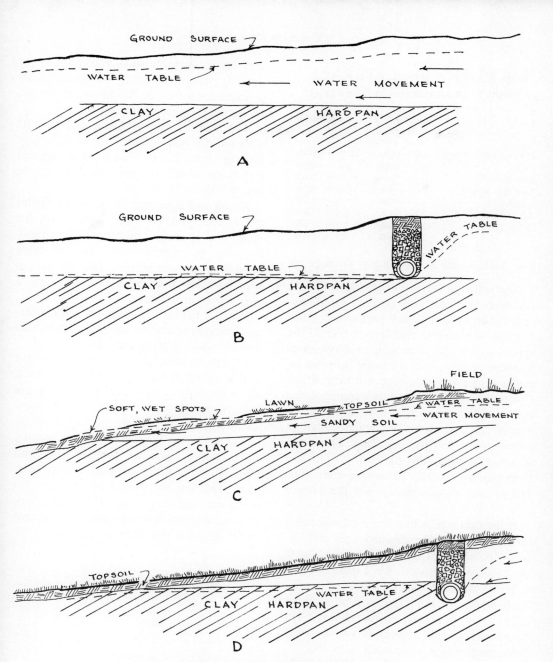

Fig. 5-40. Intercepting drains

pend on local conditions and on individual judgment.

Water Slope. The porosity and bedding of the soil largely determine the depth and spacing and to some extent the size of drains required for a given project.

A pool of surface water will assume a slight but measurable slope from its inlet down to an outlet or drain. If the pool is choked with weeds and brush, water may be removed more rapidly than it can flow through the obstructions, so the level at the drain may be several inches below other parts of the pond as long as flow continues.

LOCATION OF LATERALS

AVERAGE AGRICULTURAL PRACTICE

Soil	Depth in inches	Spacing in feet
Irrigated land	60 - 100	125 - 1,000
Peat	42 - 60	75 - 200
Sand	42 - 48	100 - 400
Sandy loam	36 - 48	100 - 250
Loam	36 - 48	40 - 100
Silt loam	36 - 48	35 - 80
Clay loam	30 - 48	30 - 66
Clay	30 - 48	20 - 40

Fig. 5-41. Depth and spacing of drainage tile

The water table may be considered to be the surface of an underground pond, obstructed in its flow by soil particles. If these particles are coarse and loosely fitted, the spaces will be large enough to allow some freedom of flow, and the slope up from an outlet of the water surface will be gradual. If the soil is fine grained and compact, the spaces will be so small that flow will be almost stopped and the gradient down to a drain point will be very steep. This slope is called the hydraulic gradient.

Slopes will usually be steeper after rains and in wet seasons than when the surface is dry.

If the drainage is to a single pipe opening in uniform soil, the drained area will assume the shape of an inverted cone, called the cone of depression. See Figure 5-36.

If the drainage is to a ditch or porous horizontal pipe, the shape will be a trough of roughly triangular cross section. The water surface and movement are shown in 5-37, and effects of spacing in 5-38.

Drainage Layout. Figure 5-39 shows the standard patterns used for subdrainage. Where practical, the intercepting or curtain drain is the most economical. Figure 5-40 shows two types of condition where it should be used.

The natural system involves use of the natural drainageways for ditch lines, and involves minimum excavation. Difficulties may be unfavorable surface conditions for ditching; irregular pattern which duplicates in some spots and is inadequate in others; or excessively crooked lines.

The herringbone, gridiron, and parallel systems are best suited to level or evenly sloping land. The choice will depend largely on which will most readily provide best gradients in the lines.

Depth and Spacing. The table in Figure 5-41 gives general recommendations for depth and spacing. They cannot be strictly followed in every case because of wide variations in conditions. Tile should at least be below the frost line and danger of crushing by machinery.

When a field is first tiled, the widest permissible spaces may be used, and additional laterals added later if they are required.

French Drains. These drains, also called rubble or blind drains, consist of a rock fill in the bottom of a trench, as in Figure 5-42, with finer material over it, to prevent dirt from working down. The usual practice is to put the large rocks in the bottom.

The more elaborate ones in (A) and (B) will serve the same purposes as open joint pipe, but the others are not suitable for water carrying sediment, as the lack of concentrated flow will allow the spaces to fill up until the drain is blocked.

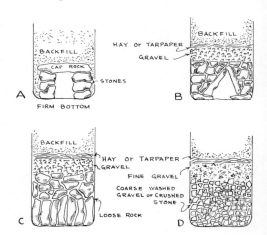

Fig. 5-42. French drains

Clogging of porous material may be prevented by wrapping it in the Mirafi fabric mentioned on page 3-18. A strip of the cloth is unrolled along the top of the completed ditch, then its center is pushed to the bottom. The porous material is placed to the desired height, the fabric is folded over the top, and backfilling is completed.

This cloth sifts most dirt particles out of water entering the drain.

French drains are used chiefly where supplies of suitable material are abundant.

Moles. Certain types of stiff plastic soils may be drained by opening pipe-like channels in the soil. This is done by attaching a mole, which is a metal piece shaped like an elongated egg, to the heel of a subsoil plow, as in Figure 5-43. The plow is set to penetrate to the desired depth and the mole is dragged through the ground. It pushes the soil aside and compacts it. Under favorable conditions these tubes will stay open for five years or more, and may open permanent channels. In other conditions, they may close immediately or within a few months.

Drainage is usually to a stream, ditch, or hole. The mole is dropped in this and pulled straight into the bank and lifted out at the upper end of the run.

A tile or metal pipe screened with

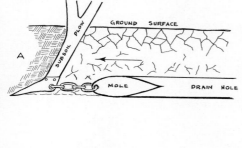
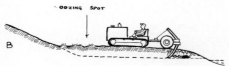

Fig. 5-43. Mole drainage

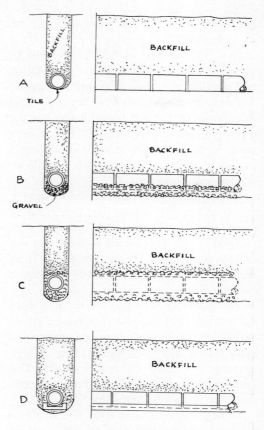

Fig. 5-44. Tile drains

coarse mesh wire should be placed in the outlet to protect it against erosion and plugging by entrance of small animals.

TILE DRAINS

The simplest type of drainage by underground tiling is shown in Figure 5-44 (A). A trench is excavated, the bottom smoothed to the desired gradient, butt joint land tile may be laid with ends touching, or with spaces up to ¼″, and the trench backfilled.

Water from the affected area drains into the tile through the joints, and to a less extent through the walls and flows inside the tile to the outlet.

Under favorable conditions, such a drain may function for a very long time. However, dirt falling in through the joint and entering with the water may fill it up, or plug low spots left by subsidence of the ditch bottom.

In (B) the tile is laid on a bed of gravel. This provides a firmer foundation and provides a porous space into which dirt can drop through joints from the pipe. This storage space for silt may fill so that the pipe will ultimately fill also; but it may serve to trap all dirt brought in during a period of adjustment, after which little or no dirt will move.

In (C) the pipe is surrounded by gravel, which preferably should be a mixture of pieces ranging from coarse sand to ½″ crushed stone. This serves to filter dirt out of incoming water, keeps loose dirt from reaching the pipe joints, and provides a good bedding.

Tar paper or hay can be used in connection with any of these techniques. It can be laid over the pipe where it prevents dirt from falling in, particularly when a large opening is caused by misalignment. The joints may be wrapped individually, or covered by a continuous strip.

It may also be placed over a gravel topping to prevent soil from working down into it. The under surface of the dirt often becomes so well stabilized that it will not cause trouble after hay has rotted out.

When the tile is laid on a curve, the wide spaces at the outside of the joints should be covered with pieces of broken tile.

Connections of branch lines may be made with sewer tile "Y's" or tees, or by junction pits which may be made large enough to serve as line cleanouts. A "Y" provides a smoother flow and larger water capacity than a tee of the same size.

Cradling. If the ground is muck, or otherwise unstable, the tile should be supported by boards, as in (D). Cleated boards supporting the haunches are preferable to flat boards because of better support and more permanent alignment.

Corrugated metal pipe may be used instead of tile and cradles.

Laying Land Tile. A large part of the land or drain tile used is in farmland. The work is usually on a fairly large scale, on regular grades and with adequate space. Costs must generally be kept to a minimum.

Ditching machines are particularly adapted to a rapid sequence of operations.

Small machines, with buckets as narrow as six or eight inches, may be used for depths up to four feet. These involve minimum excavation and assure lining up of tile. As the maximum depth is approached, it becomes more difficult to place tile accurately, and very difficult to remove stones or earth that may fall from the sides. It is usually not possible to use a tile laying shoe. It may be inconvenient or impossible to place gravel or tar paper with the tile.

Wider buckets will eliminate these difficulties, but will increase the amount of excavation and backfill.

The tile supply is laid on the field, parallel with the ditch line, just far enough to clear the ditcher, on the side away from the intended spoil pile. Pieces are placed

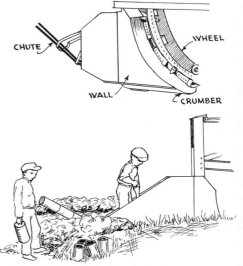

Fig. 5-45. Tile laying shoe, fed from top

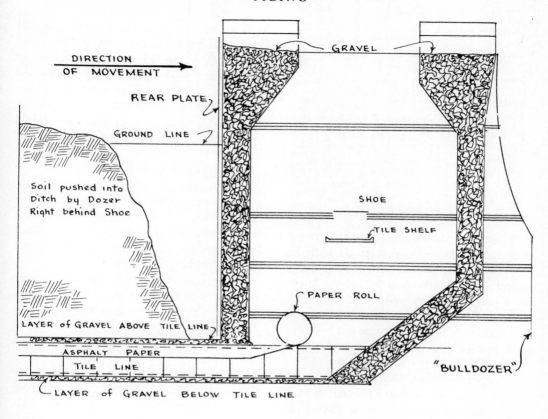

DIRECTION OF MOVEMENT

REAR PLATE

GROUND LINE

GRAVEL

Soil pushed into Ditch by Dozer Right behind Shoe

SHOE

TILE SHELF

PAPER ROLL

LAYER of GRAVEL ABOVE TILE LINE

ASPHALT PAPER

TILE LINE

"BULLDOZER"

LAYER of GRAVEL BELOW TILE LINE

Courtesy La Bolsa Tile Co.

Fig. 5-46. Tile shoe, man inside

end to end to give the correct number, with a few extra placed at frequent intervals to make up for broken or imperfect tiles.

The tile should be placed on the ditch bottom immediately behind the ditcher to minimize the danger of "losing" the ditch through caving of the sides. The first tile should be plugged with a stone or half brick to protect the line against entrance of dirt or animals. Pieces are usually picked up and placed with an L-shaped rod of light iron. A curve bottom ditch will tend to center them, but they must be checked for alignment anyway.

Tar paper, if used, should be in a narrow continuous strip in a roll, laid over the tile.

If the ditch is wide enough to work in, the tile may be laid in the same manner or

by hand. In the latter case, a picker may be used to supply tiles to the ditch man.

Gravel is sometimes laid under or over the tile by a dump truck with a small opening in the rear gate, similar to that used for supplying automatic sand spreaders. It straddles the ditch. The gravel may pour by gravity, or may be raked or shoveled down the body floor by the man controlling the gate opening.

It is important to smooth off a bottom layer of gravel before placing tile.

Tile Boxes. If the ground is not firm enough to stand until the pipe and accessories are laid down, a tile-laying box or shoe towed by the ditcher must be used. A number of varieties are available, many of them of only local distribution.

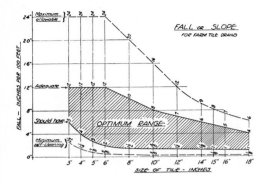

Fig. 5-47. Tile slope chart

A tile box should be slightly narrower than the bucket side cutters. It includes the bulldozer or crumber attachment which smooths the ditch bottom behind the buckets, a pair of parallel walls which will slide between the ditch walls, and a chute on which tile may be placed to be fed by gravity or manual control into the ditch bottom.

Figure 5-45 shows a simple type of box that is operated from above.

A more elaborate box in which a man can work, and which permits placing of tile, bottom and top gravel, and tar paper, is shown diagrammatically in Figure 5-46. This is suitable for the greater depths required in irrigated fields.

Two gravel hoppers are mounted on the box, front and rear. A roll of tar paper is mounted on an axle across the inside of the box. The tile layer sits near the bottom, his back to the "bulldozer," a cleaning blade which follows the wheel. The tiles rest on a shelf in front of him, and are replaced as he uses them by a man standing on the box, who picks them up with a rod.

As the ditcher moves forward, the blade smooths and shapes the bottom of the ditch. The front hopper spreads a strip of gravel on the bottom. The thickness of this strip is regulated by raising or lowering the hopper spout. The tile is laid on the gravel, with the ends touching. The tar paper rolls off its spool to cover the pipe and the rear hopper deposits gravel on it.

The bottom of the rear plate may be curved to smooth over the top of the gravel, or may be set to ride several inches above it. A dozer works immediately behind the machine backfilling. This is necessary, for if the ditch is allowed to stand open an appreciable time, one of the walls might move horizontally and slide the tile out of line.

The gravel may be piled beside the digging line, outside of the tile string, and placed in the hoppers by a small tractor front loader. A hydraulic control clamshell bucket is more efficient than the regular loader bucket, as it picks up the gravel without pushing it around.

A tile box is best adapted to a wheel ditcher, as it does not disturb its digging balance. Operation on a ladder ditcher is possible, but is more difficult as it tends to pull the bottom of the ladder backward and upward.

Backfilling. In agricultural work, it is customary to backfill with the soil dug from the ditch after placing the pipe and whatever porous material is required. However, if it is to act as an intercepting or curtain drain, imported porous material—such as gravel, sand, or corncobs—may be used to near plow depth. A top layer of native soil should be used to prevent surface water from washing in with its probable burden of silt and trash.

Whatever type of material is used next to the tile, it may be advisable to place it carefully by hand until there is no danger of the pipe sections rolling out of alignment. Compacting it immediately over the pipe (blinding) may reduce silting.

Gradients (Fall or Slope). Figure 5-47 indicates graphically the most desirable gradients for land tile of various sizes. It will be noted that up to 6″ diameter a maximum of one percent is desirable, over two percent is not permissible, and that larger sizes require flatter slopes.

Steeper slopes may give sufficient velocity

AREA IN ACRES DRAINED BY TILE WITH GIVEN DIAMETER AND SLOPE

PERCOLATION = 3/8 IN 24 HOURS — COEFFICIENT OF ROUGHNESS, n = .013

TILE DIAMETER IN INCHES	FALL OR SLOPE — FEET PER 100 FOOT STATION				
	.10	.20	.30	.50	1.00
4	3	4	5	7	10
5	6	9	10	13	19
6	10	14	18	22	31
7	15	22	27	34	48
8	22	32	39	50	71
10	41	58	71	92	130
12	67	96	118	153	213
14	105	149	182	235	333
15	127	179	218	283	400
16	149	211	260	336	473
18	209	295	360	467	660
20	280	393	483	625	880
22	360	510	626	800	1,140
24	453	643	794	1,020	1,440

Fig. 5-48. Area-tile table

to create eddies that will erode material below the joint, with possible undermining of the tile. This may result in stoppage, in blowing out to the surface, or in washing out of sections of line.

When steeper overall slopes are required by the topography, tiling may be done in a series of benches or levels, connected by inclines of sewer tile with cemented joints. Flatter slopes increase danger of silting.

Tile Size. Figure 5-48 shows the number of acres that will be drained by tile of various sizes, at slopes up to one percent. Some of the sizes listed may not be generally available, and these figures are for average conditions.

It is good practice to use tile of larger than minimum diameter for a line so that effectiveness will not be lost as readily by silting or misalignment.

Surface Inlets. A tile drainage system may include one or more places where surface water can drain into it. The flow of water should be calculated and the tile increased in size to accommodate it.

Such inlets require careful design. They must be arranged so that dirt cannot be washed in, so that animals cannot enter, and so that hydraulic pressure cannot be exerted on the tile underneath the opening.

For example, if a six-inch inlet to a six-inch line is used, water ponding over it due to excessive rainfall may put hydraulic pressure on the underground pipe, causing too fast a flow and probable erosion.

If the inlet is choked down to three or four inches, or provisions are made for surface overflow, this difficulty should not occur.

Outlets. The tile line or system may terminate in a drainage ditch, a large drain system, a river, lake or sea, or in a sump. If a sump, an automatic electric pump should be used to keep the water level below the bottom of the tile to insure free drainage.

A projecting metal pipe may be used, as in Figure 5-49, or a masonry spillway, 5-50, to avoid bank erosion and undermining. Since outfall lines often lie under surface channels, it may be necessary to make provision for surface flow also, as shown.

The outlet should be protected against entrance of small animals by coarse mesh wire, a grating, Figure 5-51 or an automatic gate, as in Figure 5-52.

A drainage system that ends in salt water between high and low tides, or in a waterway subject to flooding, should be fitted with a check valve or gate that will let the drain water out, but not allow the sea or flood water in. To make this effective, the last few feet of tile should be glazed and a concrete headwall used.

Other Pipe. Light weight, perforated, corrugated metal pipe may be used for tiling. It has four parallel rows of 5/16 or 3/8 holes, and is laid so that they are in the undersloops or haunches. Joints have collars to preserve alignment.

This pipe is too expensive for ordinary agricultural work, but is widely used for road subdrains because of its resistance to crushing and its dependable alignment.

Perforated asphaltic pipe is even lighter

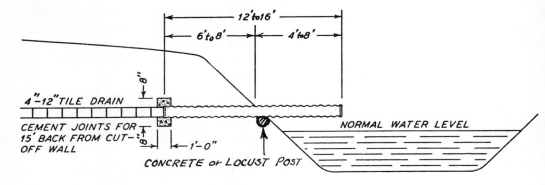

Fig. 5-49. Projecting drainage outlet

and is cheaper, but it is less strong. Both types are made in long pieces and are easy to lay, except where caving soil requires use of a tile-laying box.

Plugging. Tiles may become partially plugged with mud entering with ground water, particularly when no protecting gravel and paper are used. Flooded streams may also cause plugging, either by backing up the tile or causing the water to stagnate in them.

It is sometimes possible to flush tiles clean by utilizing a surface inlet, or opening the upper end and putting a large volume of clean water through. The flow should be started slowly, as a rush of water might move enough mud to form a solid block,

which would necessitate abandonment of the line.

It usually is cheaper to lay new tile than to dig up and clean an old one, unless the stoppage is in a small area and can be located accurately.

SEPTIC DRAINS

Some drains have no open discharge but depend on the porosity of the soil through which they run, or in which they end. A common type is the septic field for disposal of domestic waste, a set of standards for which is shown in Figures 5-53 and 5-54. Septic tank overflow is carried in a sealed pipe to the head of the field, which consists of a number of lines of porous land

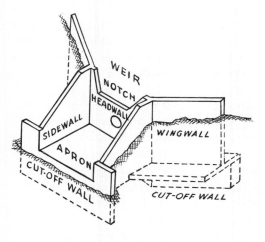

Fig. 5-50. Outlet for tile and for surface drainage

Fig. 5-51. Protected pipe outlet

tile with a maximum slope of one inch to 16 feet, with open joints on a base of graded washed gravel that is brought up the sides flush with the top. Building paper or hay is placed over this to prevent dirt fill from sifting into the spaces. Fluids in the pipe leak out at the joints and percolate through the gravel and subsoil down to the ground water. The gravel serves to widen the area over which the sewage can come in contact with the soil.

Purification is affected by microorganisms in the septic tank and in the soil.

For successful operation, a septic field must distribute the sewage evenly over an area large enough to absorb it completely. If the system is overburdened, or if some part of it carries too large a share of the total load, the fluids may force a channel and flow out on the surface.

Pervious soils absorb the sewage much more rapidly than tight ones, and can handle much larger quantities per foot of tile and square foot of trench.

The size of a system is calculated on the basis of volume of flow and the rate of absorption. For this set of specifications, it is figured there will be 100 gallons per person per day in residences. Day schools have a rate of 15 to 35, and restaurants and business buildings 10.

A percolation test is made on the soil. Holes a foot square are dug to the bottom level of the proposed trenches, and filled with water to a depth of two feet. The water is allowed to drain until it is six inches deep. The time required to fall to five inches is recorded.

This process is repeated until the drop from six to five inches takes the same time on two successive tries. This time is matched to the same or the next higher figure in column 1, Figure 5-54. The center column gives the absorption rate, the right hand one the number of square feet of trench bottom needed.

The tank should have a capacity of not

Fig. 5-52. Automatic outlet gate

less than 100 gallons per person. Any watertight construction may be permitted. Concrete block, special concrete slabs, poured concrete, and prefabricated tanks are used.

Where a garbage destructor is used, 50% should be added to all capacities.

As with most permanent installations, it is good policy to be generous in figuring. The load might increase, or the soil might lose absorption efficiency.

The field should not be heavily shaded, and cannot be crossed by vehicles. It must be at least 25 feet from any pond or stream, and 50 to 100 feet from any drilled well or reservoir. It must not drain toward a surface well or spring used for household water, at any distance.

If the native soil is not suitable, or the water table is high, it may be necessary to install a curtain drain to divert water, or to haul in hundreds of yards of sand to construct a filter bed in which to lay the tile field. This can be expensive enough to be a determining factor in selecting a lot.

Local regulations should be consulted before arranging for sewage disposal, as methods vary in different areas and exact conformance may be required.

TYPICAL DETAILS FOR SMALL SEWAGE DISPOSAL SYSTEMS

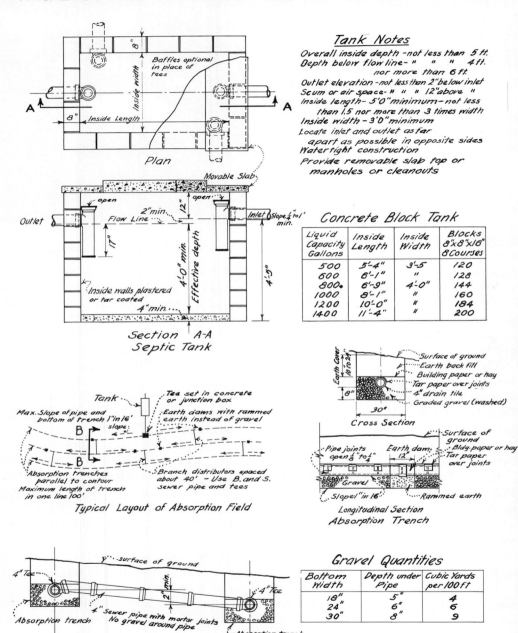

Tank Notes

Overall inside depth –not less than 5 ft.
Depth below flow line– " " " 4 ft.
nor more than 6 ft.
Outlet elevation –not less than 2" below inlet
Scum or air space– " " " 12" above "
Inside length– 5'0" minimum – not less
than 1.5 nor more than 3 times width
Inside width – 3'0" minimum
Locate inlet and outlet as far
apart as possible in opposite sides
Water tight construction
Provide removable slab top or
manholes or cleanouts

Plan

Section A-A
Septic Tank

Concrete Block Tank

Liquid Capacity Gallons	Inside Length	Inside Width	Blocks 8"x8"x16" 8 Courses
500	5'-4"	3'-5	120
600	6'-1"	"	128
800	6'-9"	4'-0"	144
1000	8'-1"	"	160
1200	10'-0"	"	184
1400	11'-4"	"	200

Cross Section

Longitudinal Section
Absorption Trench

Typical Layout of Absorption Field

Gravel Quantities

Bottom Width	Depth under Pipe	Cubic Yards per 100 ft
18"	5"	4
24"	6"	6
30"	8"	9

Section B-B
Branch Distributor

Rev. S.D. 23 Febr. 1951

Fig. 5-53. Details for small sewage disposal systems

TIME FOR 1" FALL Minutes	TYPE OF SOIL	ALLOWABLE ABSORPTION RATE Gals. per Sq. Ft. per Day	BOTTOM AREA REQUIRED PER 100 GALS. OF FLOW * Square Feet
5 or less	Sand, gravelly loam	2.5	40 *
8	Light loam	2.0	50
10		1.7	60
12	Loam	1.5	67
15		1.3	76
22	Heavy Loam	1.0	100
30		0.8	125 **
60	Hardpan	0.6	165 **
60	Heavy Clay	None (Use other means of disposal)	

* 100 sq.ft. minimum per system. To estimate length of trench, multiply by 2/3 for 18 in. trench, by 1/2 for 24 in. trench, and by 2/5 for 30 in. wide trench.

** NOTE: To be used only by special permission.

Reproduced with permission of Westchester County (N. Y.) Department of Health

Fig. 5-54. Septic field data

Septic systems are used only where public sewers are not accessible, and are usually abandoned if a sewer is laid.

A drain may also terminate in a dry well. The well consists of a hole in the earth, filled with loose rock, rubble, or clean gravel. The pipe discharges into it, and the fluid soaks out of the sides or bottom, or becomes part of the general circulation of ground water. This system depends on pervious soil in contact with the dry well.

VERTICAL DRAINS

Vertical drains of sand or porous soil can be used to move water up or down. They are often made by drilling or blasting holes in an impervious layer lying over a pervious one. Figure 5-55 (A) shows a section of ground in which opening up the hardpan would allow water to drain into underlying gravel, and (B) shows a pond which is able to exist above the water table in a pervious sand, because silt and organic matter deposited from the water has sealed all spaces in a thin layer on its bottom and sides. Breaking this layer by any means will cause the pond to drain unless sufficient silt is stirred up to seal it again.

Vertical drains, through which water rises because of being displaced by the weight of fill, are used to stabilize deep layers of saturated peat. Such soils may contain 50 percent or more of water, and

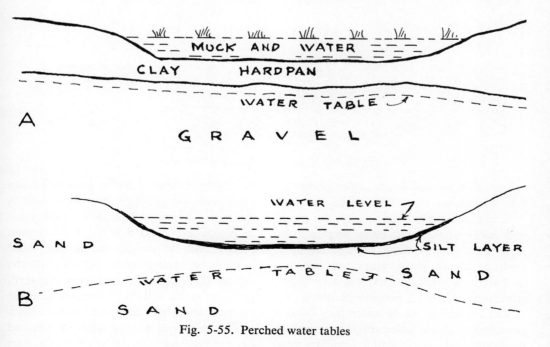

Fig. 5-55. Perched water tables

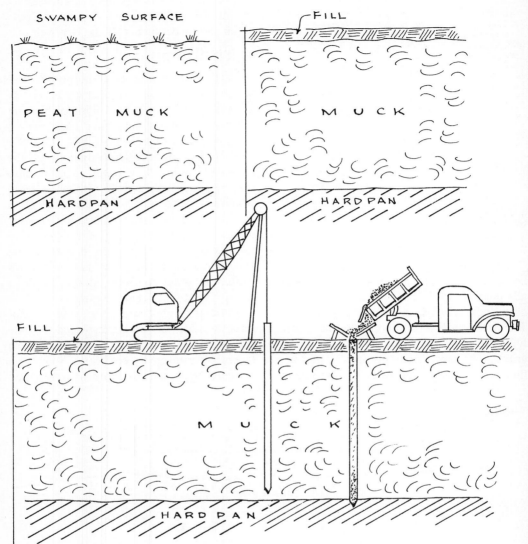

Fig. 5-56. Vertical sand drain installation

behave like viscous fluids under pressure. Fills placed on them sink, owing in part to water being squeezed out of the mud, and partly to the mud being displaced to each side. This settling of the fill may continue a great many years, and become so uneven as to make pavements or other structures on the fill become unusable.

The squeezing out of water may be accelerated, and the sideward movement of mud practically eliminated, by making vertical holes in the mud, filling them with clean sand, and connecting their tops with a drainage system, as indicated by diagrams in Figure 5-56 and 5-57. The fill is then placed and its weight causes the water to enter the sand columns and rise into the drains. If the columns are properly spaced, sufficient water will be removed to stabilize the mud enough to carry the intended load.

This technique is comparatively new and has not yet been reduced to a formula. Part

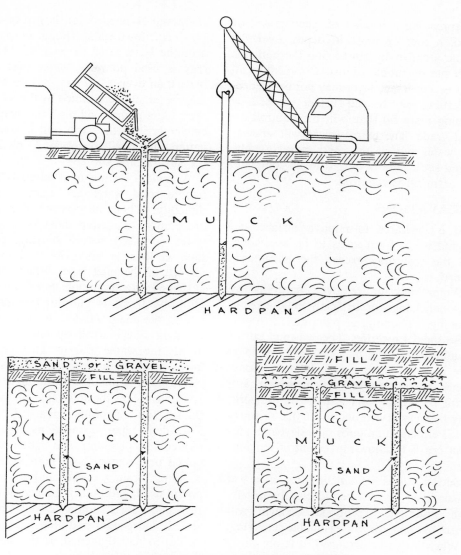

Fig. 5-57. Vertical sand drain installation (continued)

of the fill may be placed on the swamp, then the holes made by sinking a hollow walled pipe by jetting with water and compressed air, or by driving a hollow pile with a detachable head. In either case, when the tube has reached the bottom of the soft layer, it is filled with sand and pulled, leaving the sand in the hole. Hole diameters of 16 to 24 inches are commonly used, with spacing varying between 8 and 22 feet.

If the tube is pulled by conventional methods, the sand may stick to it and be raised sufficiently to allow mud to enter crevices in it and interrupt the drainage. This may be avoided by attaching an airtight head to the tube after the sand has been placed, and pumping compressed air between the head and the sand. This raises the tube but exerts an equal downward force against the sand and holds it in place and together.

The tops of the columns may be drained

by spreading a blanket of clean gravel or sand, a foot or more in depth, over the whole area; or by connecting the columns with tile or rubble drains. If the fill is all gravel, a drainage layer may not be needed.

Settling may be still further speeded by placing more fill than will be required for final grade. The extra weight will squeeze the water out of the mud more quickly. When settlement is judged to be complete, the extra fill is removed.

EXCAVATIONS

It is desirable to remove surface and ground water from areas to be excavated, but the cost may exceed the advantages gained.

Water may be removed naturally by seasonal change, or artificially by diversion, draining, siphoning, or pumping.

Seasonal Lowering. The seasonal fall of ground water level may be quite considerable in areas having dry summers. Some places which are so wet in winter and spring as to be very expensive to work, may become dry to depths of five to thirty feet. Permanent swamps may develop sufficient crusts to allow movement of machinery.

Such changes are not uniform, as a wet season may keep water levels abnormally high while an exceptional drought will cause extreme or unseasonable lowering.

When it can be arranged, it is obviously good practice to undertake wet excavation in a dry period, as any reduction in either mud or unwanted water will reduce costs. The economy is greatest in work in marshy areas and in shallow excavations, but may be noticeable even in deep work.

PUMPING

Dewatering of excavations commonly requires use of pumps.

If the water is small in volume or contains a heavy load of mud or other solids, a diaphragm pump is preferred. A centrifu-

gal pump is needed for larger quantities.

A centrifugal pump should be placed as near the water level as possible. More energy is used pumping water over a high bank than over a low one, but the total lift may be largely determined by the job. However, these pumps will push the water more efficiently than they will pull it, and much better output will be obtained by keeping the suction line short.

A standard centrifugal pump is built to handle solid water and pumps air with difficulty. When the pump and inlet hose are empty, the pump must be filled with water (primed) before starting, and will then work for a while slowly drawing the air out of the inlet pipe. When the water is sucked into the pump its efficiency rises abruptly. If air is permitted to enter the intake, the pump will lose its grip and have to again slowly work up enough vacuum to lift the water into it. This process is fairly quick in new pumps, but in worn ones is slow if the lift is at all high.

Air leaks on the inlet side of the pump should be carefully guarded against. Threaded pipe connections should be treated with pipe dope and tightened firmly; bolted ones should have both faces of the coupling clean, a gasket and sealing compound used, and bolts tightened down well. Inspection ports should be well seated and sealed, and gauges and plug fittings should be doped and tightened. If the shaft from the engine goes through a packing box, the packing should be tight enough so that there will be no drip when the pump is stopped.

Even small air leaks may prevent a pump from picking up water, and large ones developing while running may cause it to lose its prime. A pump may also fail to work because of a clogged inlet pipe. It is very desirable to keep a vacuum gauge on the inlet side, as it will indicate the conditions of the pump. Low vacuum indicates air leaks, lack of sufficient water in the pump, or worn or broken impeller and wear plates. Abnormally high vacuum indicates plugged intake line.

If the vacuum is low, but no leaks are found, wet clay should be plastered over all joints and possible cracks or fissures. This makes a good temporary seal for small leaks. The pump casing should be checked for cracks and the packing tightened.

Inlet Protection. The low end of the inlet pipe should be fitted with a screen which will prevent entry of any object large enough to plug the pipe or damage the pump. Water containing leaves or other fibrous matter will clog such a screen readily, and may make necessary the placing of an outer screen of quarter or half inch mesh, or some similar wire. This outer screen is best located far enough from the inlet so that water going through it will not have force enough to hold rubbish against it to block it. See Figure 5-58.

Another inlet problem is that of the pipe and screen sinking into a muddy bottom and becoming blocked, or being buried by soil washed over it. Placing the inlet in a wooden box will prevent sinking and make it easier for the pump to suck up washings.

In long inlet lines, or large ones, it is good to have a foot valve in the bottom to hold water in the pipe while the pump is not running, unless the pump is tight and the inlet under water at all times. This

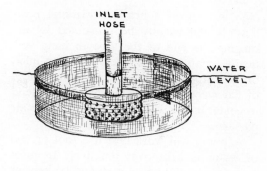

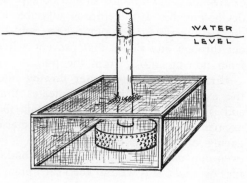

Fig. 5-58. Trash protection for inlet

will save time pumping out air each time the pump is started. Unfortunately, foot valves are subject to jamming and sticking, and may need frequent attention to keep them working.

A whirlpool may form over the suction end of the inlet pipe which will allow air to enter the pipe through several feet of water. This is most apt to happen if the

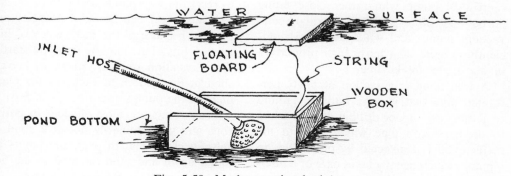

Fig. 5-59. Mud protection for inlet

pipe is lying in a nearly horizontal position. It can usually be prevented by arranging the end of the pipe to hang vertically or attaching a shield over the inlet, or by throwing a square or round piece of flat wood in the water, which will tend to center in the whirlpool and block the air passage.

A pump will work best if a foot or two of water is kept over the inlet. In most excavations, the bottoms should be kept as dry as possible. It therefore is advisable to dig a sump pit for the pump hose, and to cut through any ridges that prevent water from flowing into it from the pit.

Dirty Water. If the water flowing into the excavation is dirty, it indicates that soil is being brought in from outside the excavation. Continued pumping may cause caving of banks due to undermining; or even may cause sinking of adjoining buildings or roads. It is wise to keep such pumping to the minimum required for the work and to finish the job as rapidly as possible, even at extra expense. It may be necessary to dry the area by well points, or to block the water off by grout, chemicals, or freezing.

Contractors' liability and property damage insurance ordinarily does not cover damage to structures by undermining, even in the "comprehensive" policies. A special endorsement is necessary, and inspection of the job is usually required.

Well Points. A well point pump is a centrifugal pump with rather close fitting parts, and often with an auxiliary air-vacuum pump, and which can work efficiently in spite of a fairly high proportion of air in the intake lines.

A well point is a section of finely perforated pipe that is sunk into the ground by jetting, driving, or drilling. It is attached to ordinary iron pipe which rises to the surface and is connected by other lines to a pump, which usually takes care of a number of points.

When the pump is running, the ground water in contact with the well point is drawn through the holes or slits into the pipe and pumped away. The holes are so small that only very fine particles of earth will pass through them. The continued suction gradually removes all such particles from the area immediately around the pipe, leaving the coarser ones. This makes a porous screen with an outside area several times larger than that of the point, and improves its gathering efficiency.

Each well point will remove ground water from a cone of depression around it, the slope of which depends largely on the porosity of the soil.

If well points are placed in a line so that their cones overlap, a continuous band of soil can be dewatered, as in Figure 5-60.

A ditch could be dug in this band without encountering ground water, regardless of otherwise saturated conditions throughout the area.

The well points may also be set in a square pattern to dry up the ground for a cellar or similar excavation. It is sometimes possible to dewater such an area by using points as a curtain drain where the source and depth of the water are known.

In order to eliminate mud difficulties, the points should be placed deep enough so that the excavation will not reach capillary water standing above the artificially lowered water table.

In a deep excavation, it will be necessary to reset the well points and pumps on successively lower benches as the digging progresses, because of the inefficiency of high suction lifts. This may be done by starting the excavation oversize so as to leave a shelf at the bottom of each cut for placement of the pumps and lines, as in 5-61.

Well points are most efficient in porous soils and will ordinarily not give good results in clay soils. In peat, the points are jetted down and sand dumped in the hole around them to increase contact area.

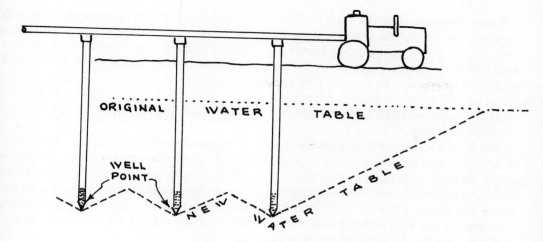

Fig. 5-60. Well points

Proper use of well points involves considerable work in placing, connecting, and moving points, and pumping is usually on a twenty-four hour a day basis for the duration of the job. In addition, considerable experience is desirable in order to avoid wasted time and possible failure to keep the job dry. It is generally advisable to subcontract this work to specialists.

Deep Hole Pumping. An excavation area may be predrained by sinking a number of shafts, lined with timber or pipe, and pumping from the bottom. The pumps used are usually small with electric motors. The shafts are more widely spaced than well points and can be used to much greater depths. Drilling and lining is expensive.

Deep well pumps, of the piston or the jet type used for water supply, may be used if equipped with good sand filters.

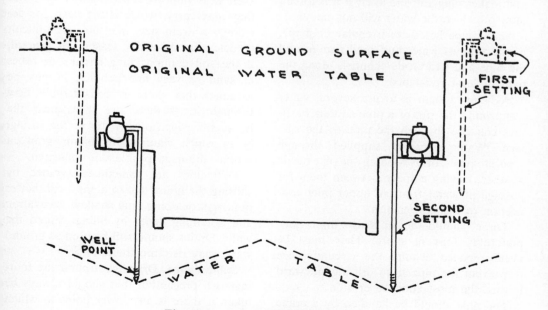

Fig. 5-61. Deep well point pumping

Sump Pumping. Shallow layers of soil may be dried by digging a deep hole in the area and keeping it pumped out. Effectiveness and promptness of drying may be improved by a system of ditches draining into the sump. These may run along the outside edges of the site and into the interior in any convenient pattern.

This is an excellent and inexpensive way to dewater a swamp before digging a pond, unless the flow of water into the area requires an excessive amount of pumping during the drying process.

Jetting. Jetting with high pressure water, or less commonly, compressed air, is used in making deep narrow holes for setting piles, installing vertical drains, obtaining soil samples, and for various other purposes.

Pressures required range from a few pounds for penetrating loose fine deposits to several hundred for tough clays.

A single pipe with a nozzle or reduction in size at the tip may be used in probing for rock or other obstructions. The tip reduction increases the velocity of the water and makes plugging less likely if it is forced into soil which the water will not cut.

Single pipe holes are irregular in shape, as the exhaust water and spoil rise around the pipe and will erode channels along the path of least resistance.

A better system is to use several water jets around the rim of a pipe so that washings can rise through the pipe to the surface. Water may be supplied through separate pipes, or by welding one pipe inside another, leaving a space between them for passage of water from an upper inlet connection to the bottom jets.

There should be at least three jets, preferably four or more. They must be evenly spaced around the circumference to prevent the pipe from drifting sideward toward the most effective erosion.

The pipe should be handled by a crane or some other type of hoist.

The lower end is sometimes fitted with teeth, and is lifted and dropped to loosen hard materials. The nozzles must be well protected against contact with hard dirt if this method is used.

CELLAR DRAINAGE

Excavating contractors are often consulted about the feasibility of having a cellar under a house. The problem may be one of the cost of dealing with rock on the site, or a fear of water conditions which would make the basement wet and unusable.

If proper procedures are followed, a cellar can be kept dry in any location where water does not spill in the windows or over the top of its wall. The cost ranges from the sometimes nominal expense of installing subdrains, up to a dollar or more a square foot (1975 prices) for complete waterproofing of floor and walls.

Soils and Locations. The tight soils such as clay or the various varieties of hardpan tend to become saturated in wet seasons, even near hilltops. The quantity of water they may carry, which is the basis for deciding on drain size, may be very difficult to determine in a dry season. In general, if the soil contains long streaks or tubes of sand or very fine pebbles, it may be assumed that there is considerable flow through it. If there are spots near the house site which ooze water in the spring, or in which water-loving plants grow, a serious drainage problem is indicated.

Difficulties are sometimes avoided by shifting the house site to a spot with better drainage, or doing only shallow excavation and obtaining depth by filling around the walls. Drains should still be used as ground water may rise into the fill.

Subdrainage. Drainage around the footings is a precaution that should always be taken if there is any lower point to which water can be led. A porous soil such as sand or gravel can seldom hold enough water to wet a cellar, but it may be part of a waterlogged lowland or a gradual slope up from one, or have water held in it by layers or sills of clay.

The standard cellar subdrainage consists of a line of land tile laid completely around the outside of the footing, and preferably a foot to eighteen inches below cellar floor level. It should be laid in a fine crushed stone, protected with tar paper or hay, and backfilled promptly. Such tile has a downward pitch of one-half to one percent from a point opposite the outlet.

The outlet may be land tile, but because of the danger of entrance of plant roots, glazed sewer tile with cemented joints is better. It should slope down away from the house at one to five percent grade to a disposal point. This may be a storm water drain under the street, a stream, or lower ground.

A storm water drain complication is that water entering the system at higher levels may back up through the tile and saturate the ground around its walls temporarily.

When there is no storm water drain, or connection to it is considered unwise, a discharge point on the same property should be sought. It is often easy to get permission to lay pipe through a neighbor's yard, but impossible to get a formal easement to keep it there.

A pipe having an open discharge should always be kept covered with coarse screening to prevent animals from entering it. If the pipe is large, a flap gate can be used, but these are not satisfactory in small sizes.

In many situations an open flow of water from a pipe is objectionable. In such cases the drain may lead into a dry well, or into tile laid out in the same manner as a septic field. An overflow exit may be provided, or the water may be left to work that out for itself.

Standard practice calls for four inch tile around the foundation. This is often too small, and its inadequacy is the cause of endless trouble. Even a small house can block a considerable area of horizontal movement of ground water, which will try to enter it unless it is drained off. After a heavy rain, a previously unnoticed seepage vein or group of small channels may carry more water than a small tile can hope to accommodate, and the foundation wall may cut into a number of them. Six inches is a safer minimum size.

The outlet drain can be the same size if the slope is steeper, or the next larger if it is the same.

If the house is on a slope most although not necessarily all of the water will be against the upper wall or walls, and may require eight to twelve inch pipe. If there is a long slope above the house, surface water may constitute a serious problem that is best solved by leading it through gratings and vertical pipes to the footing tile, which may then be as large as twenty inches. The size needed can be figured in the same way shown earlier for culverts, plus a liberal allowance for underground flow.

The floor should be laid on four to eight inches of crushed stone or gravel. This should be connected to the outside subdrain by a tile through or under the footing. If less gravel is used, one or two lines of tile might be laid under the floor. The gravel will serve to catch any vertical seepage of water, and will insulate and strengthen the floor also. The tile serves only for drainage.

Gutter leaders can be tied directly into the footing tile, emptied into nearby dry wells which will ultimately drain into it, or provided with dry wells or outlets at a distance.

If they dump directly into the tile, it must be large enough to carry easily all the water that will enter it from the ground and from the gutters. If the gutter water tries to enter a tile which is already full, it will accumulate in the leader and may build up to a head of fifteen or more feet. The resulting pressure inside the tile will force water out of the joints and cause erosion and misalignment that may result in entire failure.

On the other hand, if tile size is ample, the swift current of gutter water will tend to carry away dirt which may work into the tile from the ground.

If dry wells are used, it is best to place them well away from the house.

Where porous fill is available, it should be used for backfill against the foundation to prevent water pockets from forming against the wall. Surface water can be kept out of it by sloping up toward the foundation, and by placing a capping of topsoil.

Porous breather pipes or a hollow tile outer wall may be carried from the tile to the surface, against the foundation. Air chilled by contact with the ground tends to flow down the drain to the outlet, and when it can be replaced by warm air pulled down from the surface the resulting circulation warms the soil and the wall, and reduces the problem of condensation inside the cellar.

A house which has been built without subdrains or with inadequate ones can often be protected by installing a deep curtain drain along the uphill slope. This may be built to cut off underground water only, or to take care of surface water as well.

If a cellar is to be built in a hole blasted out of solid rock, it is good practice to provide a sump hole, about three feet deep and square, in a corner.

If water difficulties develop, an automatic pump can be quickly installed to remove it from the cellar itself, or from the floor base where it might otherwise build up enough pressure to cause heaving and cracking.

The sump can be protected with a man-hole cover when not in use. A 36 inch sewer tile open at the bottom makes an excellent lining.

This is cheap insurance against the possibility of needing an expensive and damaging ditch blasting operation, or an elaborate waterproofing job.

Waterproofing. If there is no downhill point or storm water pipe to which subsurface water can be drained, and location and soil type indicate the probability of ground water, the cellar should be waterproofed. This can be done more thoroughly and economically during construction than afterward.

A minimum treatment which should be given to the underground part of any house wall, except possibly in arid regions, is a coat of waterproof paint. This will seldom suffice to keep out water that is trying to force its way in, but will cut off the capillary water which is a large factor in wall dampness. Mixing waterproofing into the surface coat of the floor is also helpful.

A complete waterproofing job starts when the footing is complete, and before the walls are built. A concrete sub-floor, perhaps an inch thick, is laid and smoothed off flush with the tops of the footings. When this has set, three or four ply membrane roofing, consisting of alternate layers of roofing mastic and tar paper, or of similar materials, is laid over the whole floor area, so that it projects beyond the footing on all sides. The walls are then built of poured concrete or blocks, and membrane roofing tied in with the projecting ends of the floor covering and carried up the walls to grade line.

Water exerts considerable pressure so that larger than standard blocks should be used in the lower part of the wall, and they should be filled with cement and reinforced as well. At grade line a transition to smaller blocks may be made with the step on the outside. The waterproofing membrane is curved in on this step, and caught under a stucco coat which goes up to the sill line. Any pipes going through the wall are surrounded by flashing.

Water pressure is even more critical under the floor. A reinforced concrete slab is poured over the waterproofing, at least four inches thick, and a surface smoothing layer of one inch may be added. Water standing up to five feet above the floor level on the outside may require a seven inch slab, plus topping. The thickness required is somewhat affected by the span between walls, or between weight-bearing partitions or columns. It is essential that the waterproofing extend under the chimney, and its weight is an effective hold-down for the slab in its immediate area.

Another method is to pour a one piece reinforced concrete floor and walls with waterproofing compound added to the entire mix. This may or may not be placed over membrane waterproofing on the floor area, but does not require that the membrane be extended up the sides. The whole job should be done at one time as leaks may develop at joints between pours made on successive days.

A floor and a foundation constructed in either manner should be permanently waterproofed, regardless of water conditions outside them. There is even some danger that if a house is not built on it, it might be floated out of the ground in a flood.

A cellar usually has an inside drainage system to take care of leaks, condensation, overflowed or broken plumbing, and such.

This may open into the outlet of the outside subdrain, a storm water drain, or be caught in a sump which is emptied by an automatic pump.

Any outlet to either a tile system or a storm drain should have a check valve or a screw or clamp cap which can be used to prevent water from backing through it into the cellar.

Waterproofing a finished cellar involves multiple coats of waterproofing material on the inside of the walls, and laying membrane roofing on the floor and pouring a new slab on top of it. If there is not enough room for this, the old floor has to be broken up. The furnace should be treated in the same way as the walls, and any inside passages below the danger level thoroughly sealed off.

Condensation. Before undertaking any considerable expense to stop leaks into a cellar, it should be found out definitely whether they exist. Condensation may make substantial amounts of water appear on walls and floor. If the trouble occurs in hot weather, it is probably condensation; if in the wet season or during heavy storms, it is most likely leakage. If a piece of cardboard is secured against a suspected spot, it will get wet on the wall side if it is leakage, and on the room side if it is condensation.

Condensation may be checked by coating the wall with cement plaster or some other coarse absorbent material, or by running an electric dehumidifier in the room.

CHAPTER SIX

PONDS

The discussion in this chapter will be limited to ponds varying from a few square yards to a few hundred acres in area, which may be built according to the judgment of the landowner or the contractor, rather than according to detailed specifications.

Such ponds may serve as small reservoirs for domestic and industrial use, or to provide water for fire fighting, for animals, and for fishing and other recreation. They may also be useful in swamp reclamation, ground water replenishment, and flood control.

They may be made by damming streams, digging holes for streams to fill, digging below the water table, or by combinations of these techniques. Dry hollows may sometimes be converted into ponds by diverting streams, tapping springs, or lifting underground water by means of windmills or other pumps.

SWAMP RECLAMATION

Soil Conditions. Swamps which are wet all year are logical places to dig ponds. The spoil taken out of the excavation can be used to build up the area around it so that the section worked is changed from a bog into open water and dry land.

Swamps commonly have a top layer of soft peat or muck soil, which may be of any thickness from a few inches to a hundred feet or more. This organic material is easy to dig but provides very treacherous support for machinery. Below the muck, any type of soil or rock may be found.

The reader is referred to Chapter 3 for techniques in handling the mud that is one of the usual problems of swamp work.

Mud may be reduced or eliminated by working in a dry season, by diverting, draining, or pumping out the water before or during the work, or drying up the area by sump or well point pumping. These techniques are discussed below and in Chapter 5.

It is very advantageous to get rid of as much water as is reasonably possible. Water prevents the operator from seeing the bottom he is cutting, with resultant wasted passes and gouging. It reduces the digging effectiveness of the bucket so that some soils which can easily be dug when dry cannot be penetrated when under a foot or two of water. Even with skillful operation, water will mix with soil in the bucket, making sloppy spoil piles that reduce the amount of digging at a stand, and which sometimes will flow back into the excavation or cut off the shovel's exit.

Excavators. At the time that this chapter was first written, the dragline was almost the only excavator that was widely used in swamp excavation.

Hydraulic full-revolving hoes are now used also in this work. They have advantages in being able to dig hard ground, in handling platforms and stumps with little or no hand assistance, and in precise control

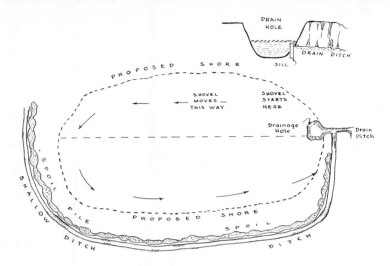

Fig. 6-1. Digging plan for swamp pond

of digging and dumping. They are handicapped by having substantially less reach in
proportion to their weight and cost.

Pond digging procedures are about the
same with the two types of excavator. The
descriptions in this chapter, based almost
entirely on draglines, have therefore not
been changed.

Operation is discussed in Chapter 13.

Digging Plan. Figure 6-1 shows a general plan for digging a pond in a swamp
and using the spoil to build up the unexcavated parts. A drainage hole is dug at
the downstream end, and the water level
lowered by a ditch drain or by pumping.
A sill may be placed at the entrance to the
ditch or the sump hole to hold water back
a few inches above the floor of the proposed pond.

If the swamp is fairly dry, and the digging is fast and continuous, the removal of
water may be unnecessary as the expanding
excavation may keep the water at a low
enough level so that it will not cause
trouble. Surface water may be diverted
around the excavation by shallow ditches
or dikes as shown, or be allowed to flow
into the hole.

If no obstructions prevent, the pond is
dug from the center toward both sides,
with the dragline walking along the longest

dimension, which is usually parallel with
the direction of water flow, as in Figures
6-1 and 6-2. The machine keeps back far
enough from the center line so that it
can reach it with an easy cast. It usually
works on platforms or other artificial supports, but if the swamp has been drained
enough in advance to be firm, or has
gravel soil, or a heavy mat of bushes, these
may not be necessary.

The bottom is kept on grade by digging
just enough to let the water back over it.
If there is not enough water to cover the
enlarging bottom, the grade may be checked
in the same manner as in a cellar excavation.

The length of a pond of this type can
be increased indefinitely without change of
method. The width, however, is limited by
the reach of the dragline and the depth of
the hole. The reach determines the width
of the strip in which it can dig and pile, and
the height of the piles; the depth governs
the part of that width which must be reserved for piling spoil.

If a wider pond is required than the
machine can dig in one round trip, as
illustrated, it must go behind the piles,
drag or swing them away from the excavation, and then widen the hole.

Size and Depth. Calculation of the size

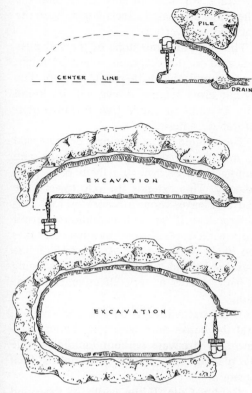

Fig. 6-2. Making the excavation

the water at the edge, and sloping up away from the pond for drainage. In a limited area that is to be reclaimed, an increase in water surface reduces the area of the banks and the amount of fill needed for them. Fewer yards need be moved for a large shallow pond than a small deep one of the same capacity; although a larger proportion of the yardage may have to be moved more than once.

Cut and Pile Relationships. Figures 6-3 and 6-4 show some of the relationships between the cut and the spoil piles. The diagrams show a machine with a 40' boom, digging beside and behind itself, and have been simplified by assuming a constant dumping height of twelve feet; an increase in volume of the spoil of 20 percent, nearly vertical slopes in the cuts, a one on one-and-a-half slope on the piles, and soft soil permitting deep digging. Figures on the

and depth of the pond should involve a number of factors. A large shallow pond gives the most for the investment, at first appearance. A deep pond is desirable in that it can be fed by seepage from lower levels, loses a smaller percentage of its water by evaporation, does not lose area by silting as readily, discourages growth of bottom weeds, and is more suitable for fishing and swimming. Against these advantages are increased cost and a possible drowning hazard.

Deep ponds may often be obtained from shallow excavations, or without excavation, by building of dams and dikes; but for the present we will consider results obtained by excavation only.

The pond should be dug to a clean bottom, if possible, and should yield enough spoil to build banks a foot or more above

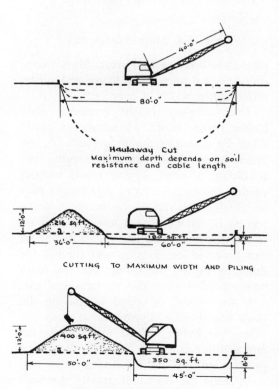

Fig. 6-3. Dragline cuts

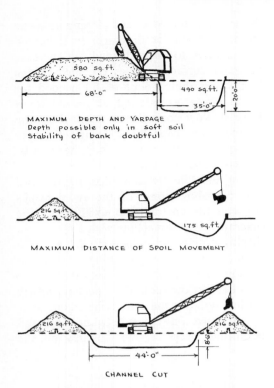

MAXIMUM DEPTH AND YARDAGE
Depth possible only in soft soil
Stability of bank doubtful

MAXIMUM DISTANCE OF SPOIL MOVEMENT

CHANNEL CUT

Fig. 6-4. More dragline cuts

piles show cross section areas. Muddy conditions cause piles to be lower and wider, reducing the width of cuts.

The dumping height increases with higher boom angles, an advantage partly offset by a shorter dumping reach. A low boom is preferable for deep digging and to obtain a good digging reach without casting the bucket. Extra pile height may be obtained by raising the boom to put a top on it, but keeping it low for the bulk of the digging. A dragline with a live boom can lower it for digging, and raise it for dumping during the swing; but the extra power needed limits this to occasional use.

If the shovel is tipped by having the track toward the digging lower than the other, the boom will be low for digging and high for dumping. This is a dangerous practice on any but the firmest soils, and as it shifts the center of gravity downhill so that lifting capacity is reduced, and the

low track will tend to sink more than the other. The combined effects might cause the machine to overturn.

The swell and the slope of the soil piles indicated in the diagrams might or might not be applicable to a particular job. The regular cross section shown will be found only in sand or other free flowing, non-saturated material.

The volume of the piles may be increased by moving the dragline behind them, and pulling them or swinging them back periodically. This extends the pile toward the rear and leaves room for more on top. This is practical if ground conditions permit work without platforms, but otherwise takes too much walking time. A second dragline working behind the windrow and swinging it back makes a very effective combination.

If the excavation is narrow and shallow, and the banks narrow, a long boom dragline may excavate without building a pile by placing each bucket load in its final position.

Double Cuts. If the width of the pond is to be greater than can be dug in the two cuts described, the dragline may make additional cuts, first removing the spoil windrow, then the ground under it.

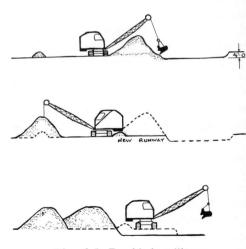

Fig. 6-5. Double handling

The windrow is usually moved by standing behind it and digging with a short dump cable, so that the loaded bucket can be picked up without pulling in. It is swung to dump as far back as possible, and a level runway made to permit the dragline to work along the rim of the original cut, as in Figure 6-5.

If the pile is too large to allow the bucket to reach across it, part of it may be dug away, as in Figure 6-6, and the balance removed at the same time that fresh ground is dug.

The runway is normally made at the original grade, as vegetation or drying makes it more firm than the soil underneath. However, if the soil is hard to cut, the runway may be lowered to improve digging efficiency. If digging is easy, and the dug soil dry and firm enough to support the shovel, the ramp may be made higher, as in Figure 6-7, to increase the dumping height. The cross section of a pile increases about in proportion to the square

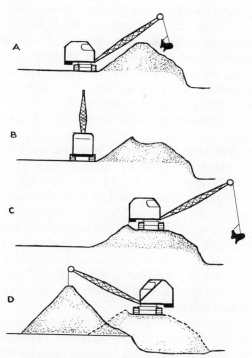

Fig. 6-7. Working from top of pile

of the height, so the advantage gained is important. The freshly moved dirt is left higher than the undisturbed part of the pile, as it may settle seriously under the dragline. If possible, the machine should be kept on the consolidated part.

If the pile has a wide top, it may simply be leveled by a bulldozer, or by the dragline raking and patting it as it travels along it. However, it should be remembered that the tops of piles are treacherous at best, the machine should be kept back from the edges, and the pile watched carefully for evidences of caving, particularly toward the digging.

As shown in the illustration, a dragline working on top of a pile is able to not only move the whole pile back but to dig the ground under it in one operation.

For even wider ponds, additional cuts may be made. Each additional slice involves moving all the material which has

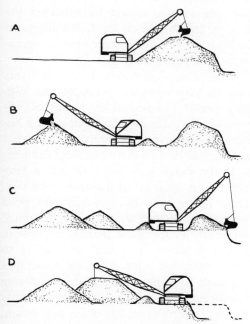

Fig. 6-6. Working back a heavy pile

been dug, with increasingly complicated patterns, and expense mounts rapidly. Usually more than two cuts are made in one direction only when the operation involves cleaning off vegetation and shallow digging. In deep work, trucking the spoil or removing it with slackline excavators or conveyors, is more economical when several rehandlings with a dragline are necessary.

Trucking. A flattened windrow, such as is shown in Figure 6-7, may be used as a truck road if it is dry and substantial enough, and is connected with dry land at one end. The dragline, or a hoe, may be started at the far end and worked toward the exit. It may dig to final grade, leave a shelf for further dragline work, or it may load the dry material in the trucks and cast the wet stuff from the bottom to build up a new windrow.

Abnormal delays may be experienced in the trucking. It is usually not practical to build turnarounds or two-lane roads, so the trucks must back in the full distance from shore, and no other truck can enter until the previous one has left, thus leaving the shovel idle each time. There is also danger of trucks going too close to the edge, or encountering soft spots, so that they get stuck or overturn.

If the windrow can be connected with dry land at both ends, scrapers may be used to remove the dry upper part.

Material may be trucked away economically if the pond is to be large in proportion to the size of the shovel, so that several shovel handlings would be required; if there is insufficient space in the digging area to pile the spoil; if fill is needed elsewhere on the project, or if the spoil can be sold. Under some circumstances, the entire cost of making the pond may be repaid by the sale of the spoil.

Selling Spoil. Materials that might be sold out of a swamp include the organic earths, such as topsoil, muck, and peat

(humus); and inorganic subsoils such as sand, gravel, clay, and loam. Frequently, the values of these materials are destroyed if they are mixed together, as the presence of organic matter makes subsoil undesirable or useless as a structural fill; topsoil or other organic earths are not salable if coarsely mixed with subsoil.

Swamp topsoil and mucks are generally too heavy in texture for use in lawns or gardens, but may improve greatly if left in piles for a year or two, or if mixed with sand or light loam. Nearly pure peats, however, are widely used to enrich soil, and command a fairly good market. These have a very high water content, and may shrink 50 percent or more in the pile.

If the spoil is to be used for the pond banks, or similar unloaded fills, organic and inorganic dirt is generally dug and handled together. If they are to be sold, the topsoil should be stripped and piled to the side, and the subsoil dug afterward in a separate series of operations. As the surface of the subsoil is often below water level, adequate drainage or a dependable pumping system should be provided before putting the shovel on it. The diagrams in Figure 6-8 show a succession of operations in selective digging.

Separation of materials in several layers may be very complex. Examples are digging gravel interrupted by seams of unwanted clay, digging clay seams lying between sandy layers, the decision as to what is desirable being a matter of local demand.

If the dragline is large, or has a long boom in proportion to the area to be dug, these layers may be removed and piled separately, or they may be stripped back separately in the same manner as topsoil. Often, however, a better solution is to truck away the least bulky or sometimes the most valuable of the materials during the digging. Trucking may be possible only during a short period late in the dry season, after a protracted period of artificial dry-

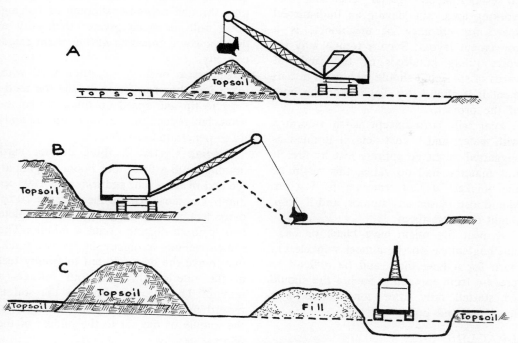

Fig. 6-8. Selective digging

ing by drainage or pumping, or only after construction of fills or timber roadways.

Selective digging should be done in the dry, so that the operator can see what he is doing. It will be discussed further in connection with borrow pits in Chapter 10.

Predraining. Whenever the soil is to be trucked out, or when very sloppy conditions are found in a swamp, it is advisable to investigate the possibility of drying up the area before work starts.

Almost any spot can be dewatered, at a price, by diversion of any streams or other surface inflow, and well point pumping.

Sump pumping may be more economical when surface water can be diverted, and when underground reservoirs are small, or when relatively impervious soils cause ground water movement to be slow. The soil must be fairly stable.

The sump is a deep hole with a bottom below the proposed digging. One side should be sloped gradually or terraced so that a

pump can be set up wherever desired. Outlet hose, or a flume or ditch, must be provided to lead the water out of the area being dried.

For best results, pumping should be started around the beginning of the dry season. If the swamp is underlain by porous gravel or sand, most of the water can be removed from small basins in a few days, and additional pumping will be required only occasionally. In large basins, continuous pumping might be needed during the job. If the soil is tight, less water will be removed at first, and seepage into the hole will be reduced very gradually.

This operation should cause the swamp to dry up, so that difficulties with mud will be reduced or eliminated. However, the effect may be largely lost if heavy rains saturate it again before work is started.

Both the speed and effectiveness of sump pumping can be greatly increased by drainage ditches leading into the sump. A horseshoe-shaped trench enclosing the working area, but leaving an undisturbed space for entrance of machinery, is a convenient layout. Such a trench may involve piling hundreds or thousands of yards of spoil, but should be a good investment if it allows dry digging for the bulk of the project.

Peat will burn except when saturated with water, and a peat deposit drained as suggested might be entirely lost by fire. If this material had no value, this might be the quickest way of removing it. The fire would also loosen any stumps, and if deep, might consume them.

A peat fire might have black or bad-smelling smoke (or be almost smokeless), burn for a long time, and be difficult to prevent from spreading. Such a fire should be started only after consultation with local fire and police officials.

DRAGLINE SIZE

Choosing the right size dragline for pond dredging in swamps involves consideration of a number of factors. Small machines are more easily supported on soft ground, are easier to salvage if they get in trouble, can work in restricted quarters, and usually have a faster digging cycle. However, they must handle material more often to move it the same distance; cannot dig as deep or pile as high or penetrate hard material or take out stumps as effectively as larger machines, and cost more for each yard of digging.

Small draglines may be equipped with long booms to match the reach of large machines, but this reduces their speed and may interfere with stability. Additional counterweight is advisable, and the inertia of this and the reduced leverage on the bucket load causes it to take more time to start and stop each swing, increasing the cycle time by one to five percent for each foot of boom added. If the swing clutches or engine power are barely adequate to manage a standard boom, loss of time through excessive slipping or lugging down will be much greater than with a high powered machine of the same rated capacity.

Undersize buckets are often used with extra long booms. These reduce the tendency to tip and speed up the cycle somewhat, but reduce the payload and the ability to penetrate hard soil.

Digging Cycles. A three quarter yard dragline with a forty-foot boom digging at the end of its reach, and swinging 180° to dump, can move part of the material 70 or more feet at each handling and complete two to three digging cycles a minute. The same machine, in digging up its own track, may move the dirt only ten to twenty feet at the rate of three to four buckets a minute. The average distance the soil is moved may be found by measuring from the middle of the cut to the middle of the pile.

If the spoil is to be moved back through several handlings by the dragline, the long move is desirable; but if it is being stockpiled, the short quick cycle is best. Most digging patterns include both long and short swings to get maximum work out of minimum walking.

If the ground is firm enough to allow the dragline to work without supports, and the spoil must be handled several times, it may be advantageous to make only the far part of the first cut, as in the "maximum distance" diagram in Figure 6-4, after which the dragline moves behind the pile and moves it back through another 180° swing. It then works from the spot where the first windrow had been, widening the cut somewhat, and building another pile. These piles may be moved back again to leave space for further digging.

This requires much more moving to do the same amount of digging than when the dragline digs as much as it can. It may also be found that even if the ground sup-

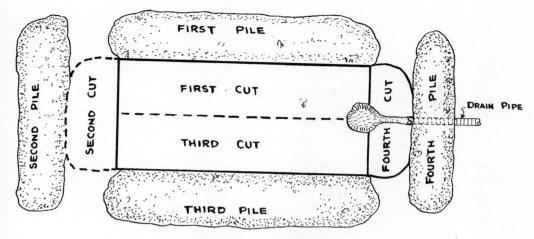

Fig. 6-9. Straight end cuts

ports the shovel before work starts, piling and removing such amounts of mud may soften it so that platforms must be used.

It is sometimes necessary to make breaks in spoil windrows to allow areas behind them to drain. These may be left open during the piling, but this is difficult when the windrow is as large as the shovel can build. They may also be made after piling by moving the machine behind the windrow and cutting a slot.

Building Up Ends. If fill is needed at the upper and lower ends of the pond, the dragline may dig and pile in a curve as it rounds the ends, as in Figure 6-2. Another method is to make straight cuts across the ends, as in Figure 6-9. Such work at the outlet will block the drainage ditch, or interfere with pumping, so that it should be the last spot to be dug.

Drainage may be maintained by laying pipe in the ditch before covering it. If the spoil is to be spread after the first cut, such pipe may be left permanently, being fitted with a cap or valve so that it may be blocked to fill the pond, or opened to drain it. If two or more handlings of the spoil are necessary, temporary pipe must be placed and then lifted as the digging progresses, for which purpose corrugated steel is most satisfactory because of ease of handling and resistance to rough treatment.

Islands. Both the quantity and difficulty of digging can often be greatly reduced by building islands. In cases where a shallow cut is made, and the balance of the depth obtained by building a dam, most of the spoil can be disposed of in islands. If cutting is deep, some spoil may still be piled on them and their bases do not have to be excavated.

Islands may be built at spots most convenient for disposal of spoil, or according to other requirements. A finished height of two or three feet above water is desirable. In humus, piles six feet or more above water level may be needed because of loss through shrinkage, smoothing the top, and slumping under water.

Grading should be done by the dragline immediately after piling, or at least before the pond is filled.

It is desirable to have some uninterrupted stretches of water in a pond. These may sometimes be located at low spots, where the digging is not deep enough to make rehandling burdensome, and the islands are located where disposal of material is more of a problem.

Under some circumstances, however, the open water will be located in formerly high ground, where the spoil can be trucked

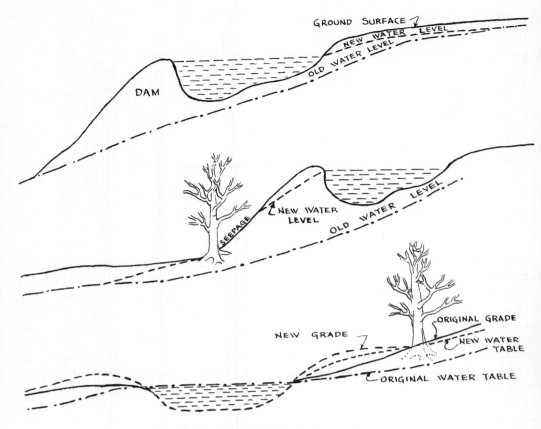

Fig. 6-10. Changes in level of ground water

away and the islands built in the soft areas.

Trees. Pond sites are usually not as open and regular as the examples discussed above. One of the commonest obstacles in the way of systematic work is a tree, or a group of trees. One tree can cause an increase in dragline time of three or four hundred percent for the digging in its immediate neighborhood, and two trees may make the digging impossible without the use of trucks.

This is one reason why most pond diggers recommend making a clean sweep of all trees within boom reach of the excavation area. Another factor is that if the water level is raised by construction of a dam, or if fill is placed around a tree, even with protection for the trunk, it is liable to die, and large and old trees are particularly

sensitive to any such changes around its roots.

Pond construction may cause injury or death to trees at some distance from the work. If a dam is built, the water level may be raised not only at the pond edge but to a decreasing amount for hundreds of feet back. Water level may be raised even when a dam is not built by ground water backing up behind impermeable fill from the pond, or as a result of raising the grade of the banks. In addition, seepage may drown out trees below a dam. See Figure 6-10.

Moderate lowering of the water table will only occasionally damage trees. Such lowering may be caused by digging into and draining a water bearing layer formerly having an outlet on higher ground, or, in the case of a pond without a dam, by digging back into a bank.

Methods of protecting trees during construction and from the effects of changes in ground and water level will be discussed in the chapter on landscaping.

Removal of small or worthless trees near the pond, and of trees which would shade a beach or float, is desirable from landscaping and recreational standpoints. A strip of grass on the pond edge that can be trimmed will make it easier to keep the shoreline free of bushes and water weeds.

However, none of these factors justify an indiscriminate destruction of trees in the pond area. A tree may be a landscape feature of as much value as the pond, and the cost of digging around rather than through it may be only a small fraction of the entire cost of the project. Also, the cost of cutting and removing a large tree and disposing of the stump, may be greater than the cost of operating the shovel for an extra day.

It is a good plan to consult a local tree expert, informing him of the scope of the excavation, the changes in water level, and the height of any fill to be placed near the tree, to get his opinion as to its chances of survival. This information may be given to the owner with an estimate for the job both with and without removing the tree for his decision.

Rock. Ledge rock and boulders cause less difficulty as they interfere only with the digging, not with the swinging. They may often be advantageously used for shoreline or islands, as a rock slope is usually more attractive than a mud or grass bank.

Rock to be blasted must be cleaned off and should be above water during the work if possible. Standard blasting techniques are used.

Bank Preservation. If the pond site includes a dry bank of suitable height, it may be advisable to leave it as one edge of the pond, digging away from it as if it were

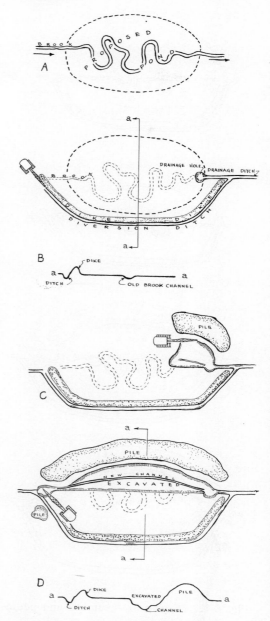

Fig. 6-11. Digging in stream bed—gravity drainage

the centerline in Figure 6-1 for the first cut, and disposing of all spoil on other edges. This technique can also be used to avoid tangling with trees or landscaped areas. If the pond is wide and without truck access, saving the bank may be too costly.

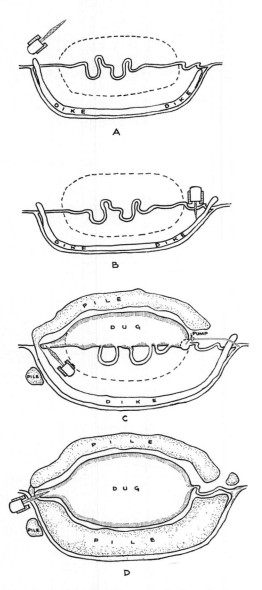

Fig. 6-12. Digging in streambed and pumping seepage

STREAM CONTROL

Digging Pattern. If a pond is to be excavated in the valley of a flowing stream, special precautions should be taken.

The diagrams in Figure 6-11 show procedures in excavating a pond similar to that in Figures 6-1 and 6-2, except that a brook runs through the center of the swamp.

A diversion ditch deep and wide enough to contain the brook is dug, starting downstream at a point lower than the proposed pond bottom, where it meets a ditch from a drainage hole, somewhat deeper than that described earlier. The diversion ditch is continued back to a meeting with the brook on the upstream side of the job. Spoil is piled on the pond side of the ditch to form a dike and to dam the original channel.

The dragline will have to walk across the brook. The banks may be dug away and platforms laid on them or in the water, or the stream may be partly filled with saplings or long logs to permit a crossing.

After diverting the water, one side of the pond is dug. A channel is cut below grade, midway between the center and side lines. At the upper end, the shovel cuts a gradual ramp up from the channel bottom, usually after crossing the dry stream bed. This ramp should be covered with a brush mat weighted with rocks, or logs or planks should be placed crosswise in the bottom to avoid excessive erosion when the dike is cut and water permitted to flow down it. Sometimes a hole is dug below grade at the foot of the incline to trap most of the dirt washed down, and the protections are omitted. The water should now follow the channel in the pond bottom, and flow over the sill into the drainage ditch, keeping away from the centerline where further digging is to be done.

If it is not practical to carry the diversion ditch to a point below the pond floor level, the pattern in Figure 6-12 may be followed. The diversion ditch is dug far enough back as to be behind the spoil pile. After the stream has been turned into it, the lower end of the stream channel is blocked off to prevent water from following it back into the excavation. A drainage hole and sump is dug, and excess water pumped into the stream below the block.

When the digging is finished, this block is removed and the dikes cut. The stream will now flow into the pond, fill it, and overflow into its old channel. Streambed erosion can be reduced by allowing the pond to fill by seepage or by controlled flow through a pipe or siphon before cutting through.

Still another method is to straighten out the stream channel so that it lies on one side of the centerline. The other side is dug and the pile extended well across the stream at the upper end, forcing it to find its way behind the pile and back to the stream below the pond.

Temporary Stream Diversion. It is sometimes possible to put a temporary dam across the stream well above the work area, and to divert it across a low ridge into another valley, or into a trench, flume, or pipe running along a hill slope in the same valley. A large pump may be used to raise the water into such diversions.

Before arranging to pump out a stream, or building a flume or pipe line to carry it, its volume of flow should be measured. This may be roughly done by placing a rectangular wood trough or flume 15 to 20 feet long in the stream bed, and packing around the upper end with mud so that all the water will enter it. Chips of wood may be dropped at the upper end, and the time they take to drift through checked with a stopwatch. Mud can be dropped in the water to find if the bottom or sides have a perceptibly slower current than the top.

The flow in cubic feet per second may be found by multiplying the depth of the water in feet or fractions of feet, by the inside width of the flume, and multiplying the product by the distance the wood chip traveled in one second.

If the stream has a fairly regular channel, the cross section area of the water in it may be calculated, and the speed of flow measured in the same manner.

Most streams are subject to considerable and sudden changes in volume, and pumps used should have extra capacity, unless it is possible to abandon the job during high water and return to it when the flow is reduced to a volume that the pumps can handle. Diversion channels, pipes, dikes, and dams should also be built to withstand high water.

Digging in Streams. If the stream cannot be diverted, the digging of each strip should start at the upstream end and move downstream. The dirt loosened but not picked up by the bucket will be washed downstream in considerable quantities which might entirely silt up any downstream excavation.

Riparian Rights. Laws relative to stream use and pond construction vary in different states and localities. Generally, owners of land on a stream below the job must give their permission before the stream can be diverted, even temporarily. They also can collect damages if mud from the excavation work chokes the stream, or is washed onto their property. Excavation permits are often required. It is well to have the law and the neighbors consulted before starting any important pond project.

Permanent Stream Diversion. A pond usually is kept in the best condition and appearance if a strong flow of water goes through it. However, if the pond is to be managed for the production of fish; or if the stream is likely to fill the pond with silt, it may be advisable to make a channel for the stream around the pond, keeping only a controlled flow from it into the pond through a pipe or ditch.

It is difficult to overestimate the power and destructiveness of even a small stream in flood, and it is at floodtime that the greatest damage can be done to a pond. It is therefore important to take every precaution to prevent the stream from breaking out of its prepared channel.

The diversion should start, if possible, in a stream section headed in the right direc-

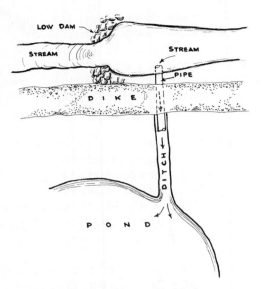

Fig. 6-13. Controlled inflow

tion; and often should be reinforced with heavy rocks or posts driven into the base of the bank on the outside of the curve. If the turn must be in the artificial channel, it should be a gradual one, protected with rocks, posts, or a well anchored timber bulkhead. A high earth dike should be built between the stream and the pond, planted with sod, bushes, or trees.

Figure 6-13 shows a safe arrangement. A low dam of concrete, masonry, or fitted rocks is placed across the stream to raise its level a foot or so. A pipe leading to the pond is placed below this water level, and a dike of earth or concrete placed over the pipe.

Water in excess of that which will pass through the pipe will flow over the dam into the channel. The pipe may open into a ditch, or continue into the pond. Flow may be reduced by means of a gate valve, or by partially obstructing it with a board, or by placing stones at its mouth.

SHORES

Spreading Piles. After a pond is dug, it is usually surrounded by spoil piles whose size and arrangement depend on the reach of the machine, the digging plan, the shape of the pond, and obstacles to digging or walking. These piles may be left to dry for ultimate sale or removal, but more often are knocked down and graded into banks and slopes. This can be done immediately, but if time is available and considerable yardage is involved, they may be left to dry and shrink. Light soils and some heavy ones should become firm enough so that the shovel can work without platforms, saving time and work. Peats and mucks are less likely to become firm, but lose substantially in bulk and weight.

The dragline is the preferred tool for spreading such piles. A bulldozer can be used if they are dry, and is frequently used for finishing after the dragline has knocked them down.

A dragline spreads piles by a combination of dragging down with lifting and swinging. First the machine approaches the pile closely and digs off the top. Each time the bucket is filled it is pulled closer to the shovel than necessary, sliding several times its capacity ahead of it. It is then lifted, swung, and dumped in a low spot, and the process repeated until the dirt piles against the tracks.

The shovel is then backed a few feet, and the digging and dragging continued, cutting to somewhere near final grade. The shovel continues to back and dig until the pile is exhausted, when it pulls down the lip in front of it, and walks up on the freshly graded area to work on the portion of the pile that was originally beyond its reach.

In Figure 6-14 the pile is shown to be on the edge of the pond excavation. The dragline digs this shore to its final slope, widening the pond in the process. It is good procedure to cut banks back to a slope which will be stable under water, as it reduces the accumulation of soft mud at the edge of the bottom from parts of the bank sliding and falling in.

If the spoil is in windrows, the shovel may be walked parallel to the pile, digging and pulling it down until it starts to fall against the tracks, then moving on to wreck another section, continuing until the end is reached. It then comes back, parallel with the windrow but further back, digging and dragging in the same manner. The ridge pulled against the tracks can be dug and spread behind the shovel. If the windrow is small, one trip may be sufficient.

Grading. It is sometimes possible to do the final grading while spreading the piles, but more often it is necessary to distribute them roughly to get an idea of the amount of material available, after which the area, or parts of it, are gone over for a light regrading. The finishing is done by raking and patting with the bucket.

There is probably no type of dragline work in which a skillful operator is as valuable as for spreading piles and grading banks. A long boom helps him considerably.

Unless the operator is an expert, a better

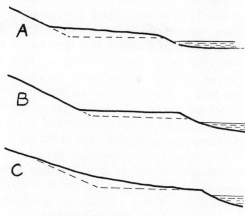

Fig. 6-15. Shore profiles

result will be obtained by finishing with a bulldozer or a grader, if the soil will support them. Also, it may be economical to make a quick rough grade with the dragline, and release it for other work while another machine finishes up.

When grading is complete, the area is usually disked and planted. Most peat soils need some time to drain, with the addition of lime, before they can support ordinary field vegetation. Testing outfits may be purchased at garden supply stores which may show the cause of trouble if things do not grow well.

Shore Drainage. The grades of the above water parts of pond banks are affected by disposal of material, securing proper drainage, the nature of the surrounding area, and the personal preference of the person in charge. Three main types may be distinguished and are shown in Figure 6-15. The shore may be low and an even slope carried up higher ground, as in (A), or the banks may be high and topped by an almost flat terrace, as in (B). A low shore and a concave slope, as in (C), is attractive and reduces the danger of water pockets at the junction of the fill and the original grade, but may stay too wet near the pond.

A difficult drainage and landscaping problem is presented by a swamp sloping

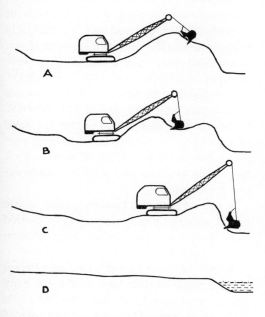

Fig. 6-14. Leveling piles with dragline

gently up away from the center, changing gradually from wet swamp to dry meadow. Even when tapered to a thin edge, the fill is liable to create a wet spot where it meets the meadow, either because a dike is formed holding surface water on the lowest part of the meadow, or because the whole fill is relatively impervious to water, stopping underground seepage and causing it to overflow at the top edge of the fill.

If possible, the excavation should yield a sufficient volume of spoil to carry it far enough into the meadow to be well above pond level. When sufficiently dried, the lower part of the meadow and the upper part of the fill should be deeply plowed and disked, then blended together with a bulldozer or grader.

If wet spots still appear, they can usually be relieved by mole drainage starting at the pond shore. If this is ineffective, tile or rubble drains may be required.

Shore Erosion. Freshly built banks will wash and gully badly in heavy rains unless protected. Drainage coming from undisturbed slopes across the fill is particularly destructive. This can often be diverted by shallow ditches made with a plow or by hand. These may be leveled after the banks are anchored by a firm sod.

Disking hay or straw into the surface of the ground increases its resistance to erosion and may supply ample grass and weed seeds. Unless applied with a nitrogen fertilizer, it may delay the growth of vegetation by temporarily absorbing this element from the soil. Such mixing in or scattering hay on the surface is helpful in holding soil that has been graded too late in the year for planting.

Beaches. If a pond is to be used for swimming, a beach is very desirable. It may also be of use for wading, picnicking, and getting small boats in and out of the water.

A maximum exposure to sunlight is desirable for at least part of a beach. This is best obtained by locating it on the north or east bank, so that midday or afternoon sun, or both, comes over the water. Reflection intensifies its heat and the slope of the beach is favorable to its reception. In most localities, more swimming is done in the afternoon than in the morning.

If the beach must be located so that its sunlight comes over the land, it may be necessary to cut a number of trees to obtain exposure. If the beach is large, enough trees should be spared to give shade over some part of it, or over a lawn area adjoining it. Sometimes a tall tree that is removed may be profitably replaced with one or more smaller ones to shade a smaller area.

If the pond is being dug, or can be emptied, the beach site can be graded. A gradual underwater slope is desirable for small children and non-swimming adults. Vigorous swimmers are likely to prefer a steep underwater slope, particularly if the water is usually cold. The dry section is usually gently sloped or flat.

A beach must be protected against runoff of water from surrounding land as this will wash away the sand, spread dirt on it, or do both. A grass-covered ridge immediately behind the sanded area will serve to divert water, and may also function as a very welcome windbreak and heat-conserver.

It is desirable to have the beach subgrade a cut rather than a fill, and of firm material. If this is the case, three inches of sand might suffice for a cover for swimming purposes, but not for building of sand castles. Six inches to a foot is a safer but more expensive depth.

If part of the pond bottom is sand or fine gravel, some of it may be pushed or carried to the proper location during the excavation work.

If the subgrade is soft mud, an attempt may be made to stabilize it with clean bank gravel, pea gravel, or fine crushed rock. A layer of lawn clippings or hay placed immediately before the sand may prevent mixing with the mud. This, of course, is not practical underwater.

An attempt should be made to extend the sand blanket to a depth of four or more feet below pond level, so that swimmers who are sensitive about walking on mud will be able to take off before they reach it.

Any clean sand that is suitable for concrete or plaster can be used for beach sand. Coarse grades are more attractive than fine, and light colors are better than dark. Where obtainable, white sand from ocean beaches or bars is most satisfactory, but it is apt to be much more expensive than pit or mason's grades. Sometimes the bulk of the beach is made with a cheap quality, and the surface dressed up periodically with a better grade.

A beach will usually require an additional two or three inches of sand after the first year or two, and occasional freshening up with smaller quantities afterward.

Fire Control. A pond is a valuable asset for fire fighting. Country fire apparatus and many city units have suction pumps so that they can get their water supply from ponds as easily as from hydrants. Even a small pond provides enough water to supply hoses for a considerable time. Many fire crews carry enough hose to utilize water a half mile or more from the blaze. An accessible pond may reduce fire insurance rates substantially.

Suction lines are short, so a rock fill or other firm surface should be provided to allow equipment to get close to the water in any weather. A deep hole should be dug near the shore. A wood or masonry wall to allow the suction hose to enter the water vertically instead of sloping down a bank adds to pumping efficiency. Any shallow bars separating the pumping hole from the bulk of the pond should be ditched.

Such a deep spot also serves well as a location for a diving board and an entrance ladder.

SPECIAL PROBLEMS

Clearing. Land clearing adds materially to the cost of reclaiming swamps, sometimes being more expensive than the earth moving. Trees, and usually brush, should be removed or burned in advance of digging. Because of soft footing the cutting is usually done by hand, but tree trunks may be dragged out by tractors or winches.

Stumps. The stumps should be cut high in the area to be excavated, and very low where the spoil is to be piled. Height gives leverage which helps in digging them out, and affords a grip for chains for handling them, but makes disposal more difficult.

High stumps in the area to be filled will cause major difficulties during grading by hanging up machines, or tipping or breaking platforms, and by requiring excessive depth of fill.

When a large stump is dug, a slow process of cutting roots, overturning, and dragging out may be required. They are sometimes too heavy to pick up, and must be dragged out of the way, cleaned and reduced by hand, or split by blasting.

Lighter stumps, which can be hoisted, usually cannot be held in a bucket and must be gripped with tongs, or chained.

Stumps should not be allowed to drag along the ground while being swung as they will put a serious twisting strain on the boom.

Swamp stumps seldom have tap roots, and the lateral roots are very close to the surface, so that they tend to come out as rather thin sheets. These can be most conveniently picked up by a pair of tongs inserted somewhere in the root mat, and chained to the bucket. If the ground is soft, the butt of the stump may be turned down, and driven into the mud in the fill area by patting the roots with the bucket. It is sometimes possible to entirely dispose of large numbers of stumps in this manner.

If this cannot be done, and the dragline

is working alone, it may be necessary to put the stumps in the spoil pile where they cause trouble in re-handling. When the spoil is being spread, an effort should be made to rake out the stumps first and put them in the lowest spots, driving them in if possible in order to bury them. If they are too big to bury, they must be piled up to dry for eventual burning, or, more rarely, loaded in trucks and removed.

Digging in stumpy swamps is greatly simplified by the help of a crawler tractor, preferably equipped with a winch, which can stay on firm ground and pull the stumps away as the dragline digs them out. These stumps may be winched or bulldozed into low spots for burial under the spoil; scattered around to dry before piling for burning, or piled immediately by passing the winch line through a pulley held high by a tripod, or a stout tree.

The high pulley arrangement decreases the power needed to drag the stumps, but unchaining them on the pile is a messy and somewhat dangerous job. The tripod or tree is usually destroyed when the stumps are burnt.

Logging tongs are the preferred tool for gripping muddy stumps for winching. When a chain is used, notching the butt reduces the inclination to slide off.

Boulders. Large boulders also interfere with digging and are very likely to cut the drag cable if lodged in front of the tracks. It is sometimes possible to dig deep holes in which they can be buried, or to line them up along the edge of the pond where they should improve the appearance of the bank. However, they are more difficult to winch out than stumps, because of difficulty in getting a grip on them, and are an even greater nuisance in rehandling spoil. Often it is best to break them with dynamite, or air or hand tools, into pieces small enough to bury or mix with the spoil.

Hard Digging. A small dragline has great difficulty digging hard or rocky soil.

It will do its best if the soil is not covered by water; if the bucket teeth are sharp; the boom held at a low angle, and the shovel footing kept as low as possible.

Occasionally a swamp floor may be of cemented gravel or decomposed rock which can be broken up with a tractor drawn ripper. Such floors, and most hardpans, can be effectively dynamited if charges can be sunk deep enough in drilled or hand dug holes. It is sometimes sufficient to blast a small area in which the dragline will be able to cut to depth, as it may be able to maintain this depth through the undisturbed material around it. If the whole bottom needs to be blasted, it will probably be cheaper to use other machines. Sometimes a single heavy blast will soften clay throughout the whole area.

Backhoe. A hoe shovel has quite effective penetration, but is hampered in pond digging by the inability to pile spoil at a distance. This limits the amount it can dig and exposes it to the danger of getting caught in slumping piles. A very good working team is a pull shovel digging the hard soil and a dragline taking it away as fast as it is dumped. Best results are obtained if they work together, but because of the exact timing required to avoid accident, it is safer for the hoe to cut as much as it can pile and move on, with the dragline following at a discreet distance behind. At the end of the strip the hoe may turn and work back, building a new pile to be removed by the dragline.

Clamshell. A clamshell with a heavy bucket has good penetration, but works quite slowly, and is at a disadvantage in sticky soils because of suction holding the bucket down. If used, it may do the digging and the casting back and spreading; or the rehandling of loosened material may be left to a dragline.

Dipper Stick. A dipper stick can break up the bottom providing drainage is adequate and dependable. The machine can

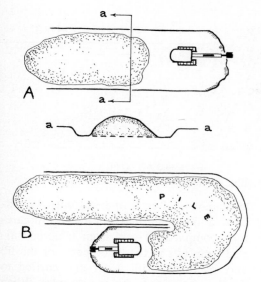

Fig. 6-16. Loosening hard bottom with a dipper shovel

operate on pontoons and ramp itself down to the required depth, and dig as wide a cut as it can reach, or is required, dumping the spoil in a ridge behind it. It may come out of the pond site elsewhere, or turn and emerge near the entrance point, as in Figure 6-16. Material broken up in this way can be easily dug by a dragline, but may be so soft as to be unsafe even on platforms, until it is well drained.

It rarely happens that a layer under a swamp is sufficiently firm to carry trucks, but in this case regular cellar digging techniques can be used.

Bulldozer digging in softer mud is done by methods described below for pond cleaning.

Water Level. The highest water level in a pond depends on the height of the overflow point, whether it is a stream bed, or an artificial spillway.

The best way to decide the new water level is to lay out a grid and take elevations throughout the area. The boundaries of a pond at any level can readily be sketched in, and the amount of cut required for desired depth, and the spoil to be disposed

of in the dam and the banks can be roughly calculated.

A less laborious method which is usually satisfactory is to select some spot that would make a good shore and use a transit or hand level to find the corresponding shoreline at other points. This is done by reading a rod set on the selected point; and moving the rod up and down any slope in question until the reading is the same, at which place it will be on the same level as the original point.

Readings can be taken on the dam site and on high, low, and normal points in the pond basis, and distances measured with a tape or by stadia.

DAMS

Digging may be done according to patterns outlined previously, with one or more cuts made across the bottom of the pond and piled for the dam. Or the digging may be done parallel with the dam and all the spoil used in its construction.

A dam should fulfill three requirements. It must be high enough in relation to the spillway so that water will never flow over unprotected parts; it should be stable enough not to break, slump, or move under any conditions, and it should not leak.

Usually earth dams for small ponds are given a freeboard, or height above water, of two feet. If the spillway is wide, wave action very weak, and the material thoroughly consolidated one foot may be enough, but under reverse conditions, three or more feet may be required. For further protection, an earth dam should be covered with a strong sod, or bushes and trees.

Small masonry dams are usually built so that part of the dam is overtopped by the water and serves as a spillway. As long as the masonry is strong, no harm should result, and the expense of an extra structure is saved.

No dam should be built to hold a depth of over six feet of water, or any consider-

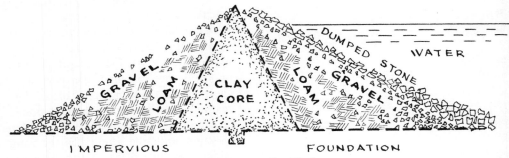

Fig. 6-17. Hydraulic fill dam

able volume of water which would flood an inhabited area if released, without competent engineering advice. In many localities, plans must be filed and permits obtained before building any dam.

Earth Dams. For stability, an earth dam should rest on a base of firm soil or rock without stratification dipping away from the pond. It should be well bonded to its base by removing vegetation (this is very important), and plowing or ditching parallel to the axis of the dam.

The dam should be at least six feet thick at water level, and slopes should not exceed one on two on the downstream face, nor one on three upstream. If the top is to be used as a roadway, it should be at least ten feet wide.

The soil used should be stable enough to hold itself up, to resist both the push and the softening effect of the water, and to carry any traffic or other loads on the top of the dam. It should also be fine grained and compact enough to give maximum resistance to movement of water through it.

Stability in the presence of water is best obtained by the use of broken rock or clean gravel, but these materials allow easy passage of water. Clay, and soils rich in clay, are best for sealing off water, but may be inclined to slump and flow when saturated.

Large Earth Dams. In large dams, a clay core may be used to stop the water, with loam or gravel faces to support the clay, as in Figure 6-17. In such dams, the type

and amount of each material must be calculated. The dam may be built up in carefully compacted layers from material carried from pits in trucks or pans; or the fill may be mixed with water and carried to the dam in pipes laid along its sides. When carried mechanically, the clay core is usually built up a step or two ahead of the faces. The hydraulic method mixes all the soil types together, but as they come out of the pipe, the coarse material is dropped first, at the edge, and successively finer particles as the water flows inward. At the center a pond forms and fine clay and silt particles are deposited to build the core.

Small Dams. These methods are not well adapted to small dams. The expense of setting up hydraulic equipment can be justified only by large scale operations. Mechanical transportation and spreading is handicapped by lack of width. Even small trucks, pans, and dozers cannot work readily in strips less than eight feet wide, nor in comfort on less than twelve feet. The three sections would accordingly produce a width greatly in excess of that needed. A narrower dam could be built by the use of undersize equipment or hand labor.

A reasonably satisfactory dam may be made of mixed soil dug out of the pond or obtained nearby. This may be piled wet by a dragline or built up in compacted layers in the same manner as a road fill. Dusty soil should be dampened. If much sand or gravel is included, it should be mixed with heavier soil, or placed on the downstream face as much as possible.

If the dam is built of dry, uncompacted material, it should be allowed several months and some soaking rains to settle it before using it to impound water.

If the soil is porous, the dam will leak unless sealed on the upstream side. It is not safe to wait for sediment in the pond to accomplish this as leakage may liquefy the soil and cause the dam to fail.

The upstream face may be covered with a blanket of clay, heavy soil, or a bentonite mixture.

Bentonite. Bentonite is a volcanic clay which absorbs large quantities of water, changing to a jelly that effectively seals soil against water seepage. It is used in many industrial processes and is available in most cities. The pellet size is more desirable for pond work than the powder forms, which tend to float on the water surface for long periods, and may be lost over the spillway.

A recommended practice is to mix one part of bentonite with four parts of sandy soil, or six or eight parts of heavy loam, and place a four inch layer of the mixture over the areas to be waterproofed. When more convenient, the pure material is spread over the ground and raked in. Satisfactory results are often obtained from more economical amounts, applied either in leaner mixtures or thinner layers.

Either bentonite or the mixture can be shoveled into a pond over leaks, and allowed to settle into them. This is best done when there is no overflow.

Usable Materials. Small stumps can be used in a dam if the fill is muddy so that it will form a close bond and fill cavities. It is good practice to cut roots back close to the butt. Boulders may be used in either wet or dry fills if the soil is carefully puddled or tamped around them, and they are not close to each other.

If the fill is rich in organic matter, con-siderable shrinkage must be allowed for in both height and thickness. Even after years of use, the dam may shrink still farther if the pond is dry for an extended period.

Cutoff Trench. If the soil on the dam site it porous or unsubstantial, a trench should be dug down to better material, approximately under the center line of the dam. This should be filled with clay, well tamped or puddled.

If a deep layer of peat is found at the dam site, it would be best to find another place for the dam. If this is impractical, the peat may be dug or blasted out, or compacted by sand hole vertical drainage. If the budget does not include funds for any of this work, the dam may be built on the peat, and access for machinery provided so that it can be built up later if it sinks. If bulges appear in the peat above or below the dam, they should be left as they serve to partly counterbalance the weight of the dam.

Settling and Cracking. In an all-earth dam, troubles to be guarded against are settling, cracking, slumping, seepage, erosion, and damage by burrowing animals. Settling is prevented by building on a firm base, using fill low in organic matter, and tamping or rolling it in thin layers if built dry. Cracking may occur in a dam with a high clay content when the pond level is low, and may be avoided by mixing in lighter soils. Such cracking rarely causes dam failure.

Slumping. Slumping may occur while building a dam with wet fill, and usually necessitates stopping work on the affected section until it has partially dried. Much more serious slumping may occur when water is impounded behind the dam before it is thoroughly consolidated. Wet fills that have not dried, or uncompacted dry fills which have not stood long enough to settle together, are apt to have this trouble. Seepage of pond water into the dam, softening it, and water pressure giving it a push, act together.

Water in a pond exerts pressure against its shores, that tends to balance inward pressure from the water they contain. If the pond is drained, removal of the water support may cause extensive slumping, which may be disastrous if it occurs in a dam.

A dam or causeway separating two ponds is particularly vulnerable if the lower pond is drained.

For this reason, it is important to face dams with coarse, self-sustaining material that will resist slumping.

Seepage. No earth dam is watertight as there is a slow movement of water even through clay. Water working its way from the pond through the dam is usually called seepage only when it is sufficient in quantity to show on the downstream side, where it may make wet spots on the dam face or marshy patches below it. Aside from the loss of water, such seepage may damage the dam by liquefying it until it slumps; or by making channels of increasing size by washing out particles of earth. Once definite channels are established the volume of flow may enable it to tunnel and destroy the dam.

The seepage appearing below the dam may damage it by undermining, but more often merely produces soft wet areas that may detract seriously from the value of the pond area.

Seepage may be largely prevented by cleaning and scarifying the subgrade, careful construction of the dam, using sufficient impervious material, compacting it well, and allowing it to set before raising the water level.

The surest and most expensive cure for seepage in an existing dam is trenching along it, with a hoe shovel or clamshell, to solid foundation, and building or pouring a masonry core. This may be quite thin if of dense masonry treated with waterproofing on the pond side. Since a leaking dam is liable to have extensive soft spots in its interior, such a ditch may be dug safely only if the pond has been drained for several months, or very heavy bracing is used.

Driving a single line of sheet piling, or tongue and groove sheeting down the center line of the dam, with grouting on the upstream side, is often effective.

The leaks may be stopped by laying a clay blanket on the pond side. The pond should be drained, if possible, to allow inspection. The leakage may be through the upstream side of the dam or in the pond bottom nearby. If the spots cannot be found, clay or heavy soil should be laid six inches to two feet thick on the whole face of the dam, and on the bottom, back about twice the height of the dam, and should be thoroughly tamped. The slope should be gentle, one on four to one on six, and the clay should be covered with gravel or cobbles where subjected to action of waves. If the leaks are found, dig them out about 2 feet deep, then tamp in clay or heavy earth patches. Bentonite mixture may be used instead of clay.

Impervious patches should never be applied on the downstream side, where the water is leaking out, if the leaks are low on the face. The water will generally work into or around the patch, soften it, and force it out. If the patch holds, the water held in the dam may liquefy parts of it, causing slumping and possibly complete failure.

The best treatment for the downstream slope of a leaking dam is to face it with gravel, with an underdrain below the bottom of the dam opening into the outlet brook, as in Figure 6-18. The first coating of gravel should be bank run to allow the passage of water, while holding back any soil particles carried with it. Over the bank gravel should be clean coarse gravel or crushed stone to correct any tendency toward sliding when saturated. If the area

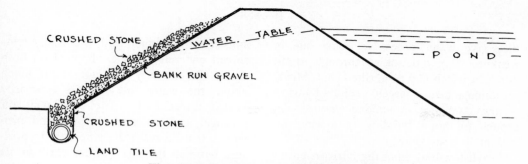

Fig. 6-18. Seepage apron, small earth dam

is to be planted with grass, stone should be covered with straw, hay, or cut weeds before placing topsoil.

This gravel blanket does not reduce loss of water, but it does stop damage to the dam and eliminates surface wet spots.

Seepage at the foot of the dam may be kept underground by tile and gravel, or stone drains, of the same type used in draining farm land.

If the dam is of pervious material the methods suggested later in this chapter for stopping seepage into porous soil may be of use.

Overtopping. If the water is allowed to flow over the top of an ordinary earth dam, it may cut a gully to the bottom of it, draining the pond, wrecking the dam, and perhaps causing flood damage below. Freshly built dams are much more subject to damage from overtopping than old established ones that have set and are covered with vegetation.

Overtopping is due to the dam settling or slumping below a safe height, or an inadequate or too high spillway allowing the pond level to rise too much.

If a dam starts to slump, the water should be drained if possible, the dam allowed to dry, and then be rebuilt with more or better material. If it is not possible to drain the pond and pumping or siphoning are not practical, the dam should be reinforced by putting first gravel, then a heavy fill of coarse rock on the downstream side. An attempt should be made to puddle or blanket the pond side, and the top should

be filled to grade. If it settles badly without slumping, the top should be built up, preferably with compacted fill. Sandbags, if obtainable, make an excellent temporary stop.

Sometimes a dam can be saved by partly draining the pond through a trench dug in firm ground nearby. Undisturbed soils can often carry a heavy flow of clean water without severe gullying, particularly if reinforced with roots, boulders, or brush mats.

Repair. When a gullied dam is fixed, the sides of the break should be smoothed and sloped sufficiently so that the fill can be tamped against all parts of them, but it should not be cut into a straight ditch. The bottom should be dried up if possible. Fill should be dumped on the edge and pushed or shoveled down gradually, while men at the bottom spread it in thin layers, tamping or tramping it thoroughly. If the break is large enough to allow machinery to work in it, it can do most of the spreading and compacting, but the bond with the walls must be done by hand. Dusting bentonite against the sides while filling should prevent seepage along them.

If it is not practical to dry up the bottom, fill should be dumped and kneaded until the water is absorbed into a stiff mud on which a layered fill may be built.

Burrowing Animals. Earth dams may be damaged by animals burrowing part or all the way through them. Muskrats make holes which run under water to well under the bank, where they rise above the water. Such tunnels will cause leaks only when they give water access to some line of weakness that did not go through to the pond, or which had been silted shut. Muskrat damage can be largely avoided by using a low dam not containing enough dry ground for home building, or a wide one without porous veins.

Crayfish will at times dig burrows all the way through a dam, creating a water channel large enough to enlarge by erosion, unless a fortunate cave-in should block it. This damage is most apt to occur in soft peat soil and it may sometimes be cured by injections of cement grout.

Burrowing animals may be discouraged by including quarter inch mesh wire in the underwater part of the upstream slope. This affords fairly good protection for a number of years. It is usually laid on the dam, and six inches to a foot of fill are spread on it.

Masonry Dams. Masonry dams may be used instead of earth fills. They are most suited to comparatively narrow sites with firm bedrock near the surface of bottom and sides. Reinforced concrete is the strongest construction, but field stone masonry is more attractive and may be less expensive in inaccessible spots.

Earth and decayed rock should be cleaned off the dam site, and the bedrock shaped or gouged in such a way that the dam will not be able to slide on it in any direction. Holes two or more feet in depth should be drilled in the rock, and reinforcing iron cemented into them so that it will project into the concrete or other masonry.

If the dam is to be more than a few feet high, it is advisable to have an engineer or a geologist check the ground as fractured rock can make a leaky and unstable foundation.

The dam should have a bottom thickness of at least two to three feet for every three feet of height.

Masonry Cores. A masonry core dam consists of a thinner wall, preferably reinforced concrete, with earth piled on both sides. The masonry does not extend much above the water line, and is ordinarily buried under earth. The core seals off seepage, and the sides support and protect it. It must resist the difference in pressure between the wet and dry earth on its two sides. Thickness is about ¼th of height.

The core should be founded on a firm, impermeable material, preferably rock. The original surface is ditched for footings. The sides are carried into the banks until they meet rock, or until they are far enough from the water to make seepage unlikely. Rock should be roughened to hold the masonry against shifting.

The core is built and allowed to cure before placing the earth fill. The upstream face should be painted with waterproofing. If its ends are not keyed into rock, they should be fitted with vertical metal baffles sealed to the concrete, and the fill near the baffles should be mixed with bentonite. Failure to take these precautions may lead to serious leakage around the core.

Fill should be placed on both sides of the wall at the same time to avoid unbalanced pressure. If the dam is high, the fill should be carefully compacted. If it is low, this is not necessary unless final grading is to be done immediately.

The masonry core dam is the safest and most satisfactory construction for ponds, but is too expensive for casual use.

Removable Wood Dams. If a small pond is built on a small but fast flowing stream subject to flood, there is not only the danger that earth or weak masonry dams will be washed out, but that, if the dam holds, the pond may fill completely with mud and debris in a single season, because the slowing and widening of the stream causes it to drop a part of its burden.

A removable wood dam may be used to advantage under such conditions. If the stream is narrow, ten feet or less, a heavy,

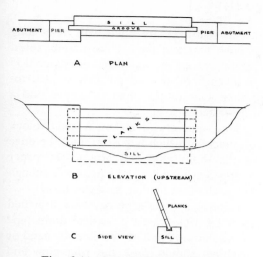

Fig. 6-19. Removable wood dam

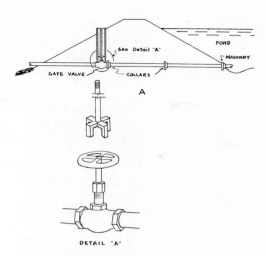

Fig. 6-20. Drain pipe and gate valve

well founded masonry wall is put on each bank, having slots to receive two to three inch plank, as in Figure 6-19. A masonry sill, similarly slotted, connects the piers on the stream bottom. Planks cut to the correct size and length are slid down the pier slots, resting on the sill and on each other, until the desired height is achieved. This structure will leak but will impede a brisk stream enough so that part of it will flow over the top board, and the desired water height may be maintained. If the stream shrinks, the leaks may be reduced by jamming a tarpaulin in the sill slot, upstream, and pulling it over the dam face and top, and tying weights on the downstream side. Or tongue and groove planks may be used to cut leakage, with the top plank fastened down and all the joints packed.

When a flood is expected, or pond use stopped for the season, the planks may be taken out and stored, allowing the stream a clear passage.

DRAINS

Gate Valve. When possible, means should be provided to drain a pond for repair, cleaning, and other purposes. The best, but most expensive, means is to place an iron pipe under the dam, connected with a gate valve, which may be located in the dam or at either end. Figure 6-20 shows an installation in which the valve is in the downstream face below frost line. To prevent burial and clogging, a vertical eight inch pipe placed over the valve wheel extends to the surface, where it is plugged or covered. The valve is opened or closed by removing this cover, and turning the valve wheel by means of a jaw on the bottom of a rod which can be turned from the top.

If the cover should be left off, and the vertical pipe filled with dirt and trash, it may be jetted out by the use of an engine-driven water pump, delivering water at pressure through a small pipe which is pushed down inside the casing, where it can break up and wash out the debris.

Elbow Drains. A much less expensive installation which can be used in climates where freezing is not expected is shown in Figure 6-21 (A). An iron drain pipe under the dam is fitted with an elbow on the downstream end into which a vertical pipe is threaded. Space is provided so that this pipe can be turned into a horizontal position.

If the open end of the pipe is higher than the water in the pond no water will move through it. If it is lower, the water will flow through it until the pond level is lowered to the same elevation. The pond

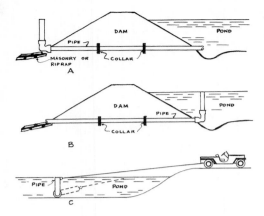

Fig. 6-21. Elbow drain

level can therefore be adjusted to any height desired by turning the pipe up or down.

In cold climates, the exposed pipe would be subject to breakage because of water freezing in it. This is not likely to occur if the movable pipe is placed in the pond as in (B) because of less severe freezing and inward pressure of pond ice. However, the water makes the pipe difficult to get at so that it must usually be moved by a line stretched to shore, as in (C). This will not pull it into a horizontal position and may have difficulty raising it from down position also.

A more satisfactory arrangement for underwater use is shown in Figure 6-22. The drain pipe is extended by means of a tee, a close nipple, another tee, a short pipe, and

a cap. The tees are fitted with pipes long enough to reach the surface of the water. These are set at an angle of about 45° from each other, and the tees welded together. One of the pipes is capped and a ring fastened to it.

This apparatus rests on a small block of concrete, which is cast around the edge of the drain pipe and around tar paper wrapped around the end pipe, but is enough below the tees so that they can turn.

Control is by a rope or cable stretched from the ring, past the vertical drain pipe to the shore. A pull on this line, by hand or machine, should raise the ring pipe and turn the drain pipe down. The drain pipe can be raised by pulling the line from the opposite bank.

With some risk of twisting the end off instead of turning it, the masonry block may be omitted and the outer tee replaced by a street ell welded to the inner tee.

The threads should be treated with waterproof grease or plumber's dope, and wrought iron fittings should be used if possible.

Metal pipe is expensive in large sizes and six inch is about the minimum for a pond drain, except for use in dry seasons only. Considerable expense may be saved by using concrete or tile pipe under the dam, connecting it near the end with metal pipe to the valve or other drain arrangement.

Vertical Tile. An overflow or trickle drain can also be made entirely with tile. A pipe is laid under the dam, ending on the upstream side in a concrete junction box, as in Figure 6-23. From this a tile pipe with joints sealed with soft mastic rises to the surface. One of the pipes may have to be chipped short to obtain the proper height. The pond height is limited by overflow into the pipe.

The pond is drained by pulling the top pipe out of its joint and removing the next section when the water has gone down sufficiently, repeating the process until the bot-

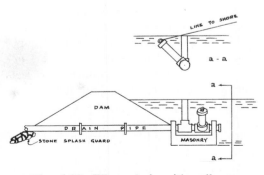

Fig. 6-22. Elbow drain with pull arm

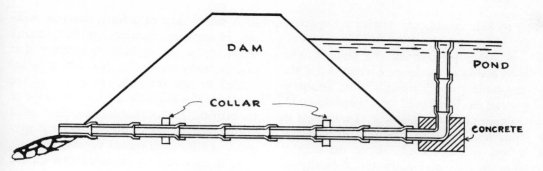

Fig. 6-23. Spillway drain pipe

tom is reached. Sometimes the whole pipe will pull out of the box and drain the pond all at once. At other times, a pipe may refuse to move and may have to be broken with a hammer or crowbar.

The overflow type of pond drain serves to some extent as a spillway, but a regular or emergency spillway also should be provided for flood conditions, and because of the possibility of the pipe becoming clogged.

Pipes reaching the surface of the water can be protected against external ice pressure by tying several sticks or boards to the outside.

Plugged Pipe. Plug drains are shown in Figure 6-24. A stopper is made by reinforcing a piece of half inch to one inch marine plywood with iron bands, passing a cable through it and placing it over the pond end of the pipe. If the pipe is rough or chipped, it may be smoothed over with cement grout and painted with a soft mastic.

As the water rises, its pressure should hold the wood firmly against the pipe. If any leakage occurs, the contact may be packed with clay and tied with burlap, or the pipe end buried in mud.

The pond is drained by sliding the wood off the pipe by means of the cable and a tractor or car on the bank. Sometimes wood and concrete will adhere so firmly that the pipe separates at the next joint and comes out with the plug, in which case it

can be reset after the water goes down. If the end pipe is set in masonry it will stay put.

The upper end of the cable can be fastened to a buoy or to an anchor in the bank.

A permanent plug such as that shown in (B) may be placed in the pipe and the pipe pulled with a cable. Hammered-in wood plugs are satisfactory for diameters

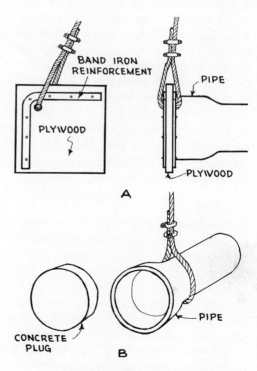

Fig. 6-24. Drain pipe plugs

6-27

up to four inches, and reinforced concrete for larger sizes.

A drain pipe should be straight with good access to the lower end, so that if the plug-pulling device doesn't work, an opening can be hammered or blasted through.

Drain pipes are a source of weakness to dams and must be carefully installed. It is best to place them before the dam is built, as this eliminates the difficulty of making a proper bond between fill and the wall of a ditch. Pipe joints should be watertight.

One or two collars of metal or masonry should be built out from the pipe, as indicated in the illustrations, and sealed to it by cement or welding. These will discourage seepage from following the outside of the pipe and cutting a channel along it. Clay, or soil mixed with bentonite, should be tamped or puddled around the pipe and the collars.

The first layer of fill should be spread rather evenly along the masonry pipe, as a full load in one spot might push it down enough to open the joints. If the ditch bottom is not firm, the pipe may be set on a reinforced concrete slab the width of the pipe and up to six inches thick.

A wood box, 3 feet square or larger, may be built of rot-resistant wood around the upper end of a drain pipe and topped with quarter or half-inch mesh screen, to keep fish in while lowering the pond level.

SPILLWAYS

Construction. Ponds which are made by excavation only, and do not raise the original water level, usually overflow through stabilized streams or channels that do not require any artificial protection against erosion. If an earth or masonry core dam is used, however, an artificial overflow channel, called a spillway, must be prepared.

A spillway may have a surface of any material that will resist the destructive action of the water which might flow across it. A steady flow calls for a structure, usually of stone or concrete, but occasionally wood, metal, or asphalt. A spillway that carries water rarely, as one which is intended to care for floods in excess of the capacity of a masonry or pipe spillway, or to provide for occasional overflow of a normally static pond, may be planted with grass or other well rooted vegetation.

Spillway size may be calculated on the basis of the area drained, type of land and vegetation, and rainfall records in the same manner as culverts. However, a greater margin for safety should be allowed.

It is good practice to keep the spillway and the dam separate if possible, as each is a source of weakness to the other. A recently constructed dam ordinarily lacks the stability necessary to support heavy masonry, and it is difficult to get a leakproof bond between dirt and stone. Any leaking through or around a spillway will be much more destructive to an earth dam than to a long established subgrade. On the other hand, practical and esthetic considerations frequently require placement of the spillway in the dam.

If the dam including a spillway has a masonry core, the two structures can be combined. However, the core must be widened or buttressed, or the spillway provided with additional foundations as firm as the core wall. If the spillway is supported by a thin core wall and dirt fill, and the fill settles, the spillway will be left supported only at the core, and may break, or may twist and break the core. A preferred method is to extend the core footings far enough to carry piers to support the spillway.

If the overflow is to be carried around the dam, standard practice for masonry structures may be followed. Two more or less parallel walls carrying the water race, which may be a curve or a series of steps, is a standard type of construction. The structure is strongest if of reinforced concrete well tied together, but stone and mortar make a more attractive appearance. The fill under the water race should be clean sand or gravel with good bottom drainage if the ground freezes in winter.

Settlement. If the spillway is to be part of a newly made dam, it may be based on the fill material, or may have footings in the native soil underneath the dam. In the first case, any settlement is liable to tilt or break the spillway and to settle away from it leaving channels for leakage. In the second case, the masonry will stand firm while the dam settles under it and away from it. If the structure includes a core wall long enough to tie into the earth on each side, such settlement may not be serious.

Grouting. Leakage under a masonry spillway surface, resulting from dirt settling away from it, may be stopped by drilling holes in the masonry and pouring or pumping a cement and water grout into them.

A grout injector may be an air pressure tank or a pump. The tank is provided with an agitator to prevent separation. It is partly filled with grout and tightly closed. Compressed air is piped into the top of the tank, forcing the grout out through a pipe or hose in the bottom. The tank is opened and a fresh batch of grout poured in as often as necessary. Air should not be allowed to enter the outlet hose.

Special pumps may be purchased, or a fluid grease dispenser or a tractor grease gun used. Pumping can be continuous, with extra grout added as necessary.

The holes are drilled or punched to a depth where the leaks are suspected. The grouting tube may be fitted with a rubber collar to fit the holes and held in place by hand, if low pressures are used. For high pressures, a threaded iron pipe is cemented into the hole some days before and the grout pipe coupled to it.

The grout forced underground may penetrate and seal the leaks, it may be washed away by water or may escape to the surface of the ground. If possible, the pond level should be lowered to stop the water flow during grouting. The whole area should be watched for the appearance of grout, particularly at the leakage points.

A very thin grout made with forty-five gallons of water to a sack of cement is good for sealing fine porous soil, but will escape readily through small channels. The thickest grout used, four and a half gallons of water to a sack of cement, will escape only through large openings, but does not seal fine passages effectively. Sand mixtures are not recommended for amateur use because of the tendency to separate, but sawdust or fine shavings may be mixed with grout used from pressure containers if the grout is otherwise washed out by water which cannot be stopped.

If grout is applied at a pressure of more than a few pounds, care should be taken that it does not lift or break the spillway, or even split bedrock beneath. A tractor grease gun can develop pressure of thousands of pounds per square inch, and will break up strong masonry with little effort.

All grouting equipment should be thoroughly cleaned immediately after finishing the job, or for any shutdowns of more than a few minutes.

Detailed information on the use of grout for stopping leakage and for other purposes may be obtained by writing to the Portland Cement Association, Old Orchard Road, Skokie, Illinois 60076.

Wood Spillway. Trouble from settling under a spillway may be avoided by putting in a temporary structure upon completion of the pond, and removing or destroying it after complete settlement, then building the permanent spillway. Tongue and groove plank made into a box is a satisfactory construction. The dam surface on which the wood rests should be coated with bentonite, clay, or other heavy soil, and puddled until semi-fluid. The spillway should be stirred around or vibrated when set, and mud packed in along the sides.

A wood spillway may give satisfactory service for a great many years under favorable conditions.

Horizontal Pipe. Concrete, tile, or corrugated steel pipe of large size may be used, either as described under drains or laid horizontally through the dam at water level, with the same precautions against seepage.

WATER SUPPLY AND LOSSES

Water Supply. The ability of a pond to remain nearly full of water through a dry season is to a large extent the measure of its usefulness, except in semi-arid sections where it is considered a success if it retains any water at all.

A pond level is kept up by water entering it through rainfall, surface wash, springs and seepage, and streams. It is lowered by evaporation, outflow, leaks, and seepage through sides and floor.

Once a pond is built, little can be done to add water to it except by pumping water from a well, by windmills or engine-driven pumps, or more rarely, diverting water into it. It is therefore important to locate and build it in such a manner as to take full advantage of sources of water.

Ponds dug in swamps may depend primarily on the water table existing before work is started. If possible, fluctuations of this should be watched for a year or two. A hole may be dug by hand in the wet season until the bottom fills with water. If the water dries up the holes should be deepened. The water table can be followed down and its changes observed in this way.

A dug pond may cut into active springs or extensive seepage areas which had previously been draining below the site, so that the pond may keep a higher level than the ground water did. On the other hand, the swamp water might overlie a layer of clay or hardpan, which, when cut, would allow all the water to drain down into unsaturated porous soil, in which case it might be difficult to keep water in the pond.

The best way to estimate the water supply is to measure the drainage area. Figure 6-25 indicates approximate requirements throughout the country.

Seepage into Porous Soil. Outgoing seepage can be greatly reduced and sometimes stopped altogether by keeping mud in suspension in the pond water for some time. The water in seeping out of the pond takes the suspended particles with it and lodges them in the fine passages through which it travels, thus clogging them up. This process operating naturally over a period of years makes possible the existence of rain fed ponds and swamps on sand dunes and gravel banks, high above the water table.

Digging in a pond will keep it muddy, as will driving livestock around in it several times a day. Fine grained silt, powdered clay, or pellet bentonite may be scattered on the water with hand shovels, preferably when there is no overflow.

If the water is leaking through channels too large to be plugged by sediment, a layer of clay or a soil-bentonite mixture several inches in thickness should be spread over any outcrops of porous veins. If this fails to hold, the pond should be pumped dry and any leakage holes appearing in the clay should be dug out and filled with the blanketing material to a depth of a foot or more.

If the porous vein is comparatively thin and close to the surface, it may be sealed by injections of cement grout in the same manner suggested for spillways.

If leakage is along sod or brush which was not removed before placing fill for the dam, it may be stopped by chopping and mixing. A tamper such as is used in breaking street pavements can drive a narrow tool several feet underground, and repeated blows struck close together will mix the vegetation into the dirt so thoroughly that

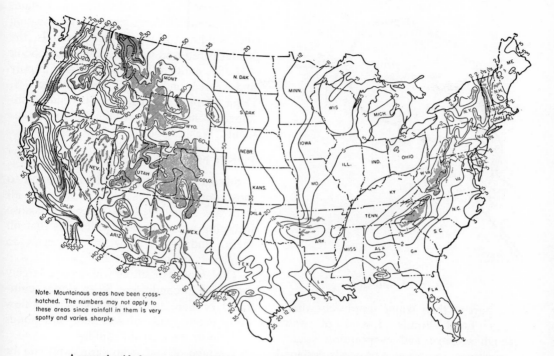

Note. Mountainous areas have been cross-hatched. The numbers may not apply to these areas since rainfall in them is very spotty and varies sharply.

A general guide for use in estimating the approximate size of drainage area required for a desired storage capacity in either excavated or impounding reservoirs. The numbers on the chart show the number of acres of drainage area required for 1 acre-foot of water impounded.

Courtesy U. S. Department of Agriculture

Fig. 6-25. Drainage area map

it will no longer provide water channels. Additional fill can be added to the surface if necessary, and a wide face tamping tool used for compaction.

Seepage along the old ground surface may cease when the vegetation rots, but this cure cannot be depended upon.

Seepage cannot be stopped entirely but will fall to a very small amount in a well sealed pond, particularly if the water level is not high enough to create a strong pressure toward a nearby low spot.

Movement of ground water is often nearly horizontal so that much of the loss from a pond is through the banks rather than the bottom. This is one factor in the excessive shrinking of some small ponds during dry spells.

Evaporation. Evaporation acts constantly to remove surface water. It varies with heat, humidity, and exposure to sunlight and wind, and may lower the level of a stagnant pond from five to fifteen feet during a summer. This loss is most pronounced in desert regions.

The rate of evaporation is higher on small ponds than on large, and on shallow ponds as compared with deep ones. A number of factors are involved: The banks heat more readily than the water surface; capillary attraction draws water several feet up on the banks, thus increasing the surface exposed to evaporation, and a large body of water warms more slowly than a small one.

This loss from the water surface may be reduced by shading it with trees, but it is a question whether the trees do not use as much water as they save. If they are set well back from the edge, they may find a

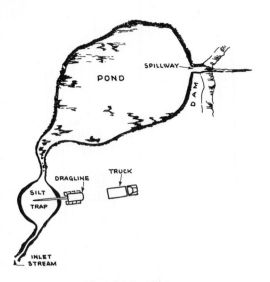

Fig. 6-26. Silt trap

large part of their water supply elsewhere.

Dry Land Ponds. Losses of water through seepage and evaporation assume their greatest importance in ponds designed to fill with surface run-off in the winter or spring, and to hold this water through a dry summer, even though the water table drops many feet below their bottoms.

Such a pond should be so located that the drainage from a large area will flow into it; not only so that it will fill even in years of subnormal rainfall, but so that it will get the fullest advantage from any freak rains that might fall in the summer. But it should not be placed in the channel of a stream having enough force to fill the pond with sediment during flood time, or to require an unreasonably expensive spillway.

Such a pond may generally be dug in the dry season without any interference from ground water. Any dry land excavator or set of excavators down to small tractor drawn scrapers and scoops may be used. Techniques are similar to those used in borrow pits and cellars, except that banks must be sloped, not more steeply than one on one, and it is usual to place a large part of the spoil so as to build up a dam.

For detailed discussion of the locating and building of such dry land ponds, the reader is referred to Agriculture Handbook AH 387, entitled "Ponds for Water Supply and Recreation" (1971), issued by the U.S. Department of Agriculture. This can be obtained from the Superintendent of Documents, Washington, D.C. 20402.

POND MAINTENANCE

Silting. Silting is a problem common to most ponds and reservoirs. Lakes of all sizes are short lived geologically, because incoming water deposits sediments that fill them, and water flowing out tends to deepen its channel.

The amount of silting will depend largely on the local conditions. Steep slopes, cultivated or bare land, and fast stream flow bring heavy loads of sediment into ponds and cause them to fill rapidly.

Wastage of soil from farm land can be greatly reduced by contour plowing, terracing, and planting steep slopes to permanent grass or trees, with beneficial results to the land, the stream, and the ponds.

If it is not possible to alter watershed conditions, silt traps may be constructed. These may consist of small ponds built above the main one, or a very deep hole on the upstream end of the pond. Such traps should be so located that a dragline shovel and trucks can reach them for periodic cleaning.

Mud deposits found in ponds and lakes are made up of soil brought in by water or slumping from the banks; dust, leaves, pollen, and other debris falling from the air, and remains of plants and animals living in the pond. A combination of these sources usually produces a soft black mud which dirties and shallows the water. Near inlets and steep banks it may be chiefly silt or sand, and away from shores it is largely organic.

Removal. A hydraulic dredge removes

such a deposit without draining the pond, but its use is often not practical. There must be enough work to justify transporting and launching it, enough water inflow or return flow for its needs, and adequate disposal areas.

Removal by machinery usually requires draining or pumping out of the water to avoid distributing disturbed mud throughout the pond.

After draining, the mud deposit will often be found to be so soft that it will not support machinery safely even on platforms. Given time, it will drain and compact so as to be fairly firm, in which condition it will not only support platforms but will stay in a dragline bucket. This hardening process, which may reduce its bulk as much as 80 percent, can be greatly accelerated by hand ditching into the subsoil for more thorough drainage. The ditching, however, is a sloppy job, and will be very discouraging at first because of mud flowing or slumping into the ditch and blocking it. The first digging should be very shallow and can be gradually deepened as the banks drain.

If the pond is narrow and accessible enough so that all parts can be reached by a dragline on the banks, or if the mud overlies firm material that will support a dozer which can push the mud to a dragline, it may not be absolutely necessary to let the mud dry. If it is too thin to be picked up in the bucket, digging the ground under it several times may suffice to get enough of it out. In any event, such undercutting will eventually lower it so much that it will no longer be a nuisance.

It is seldom practical to just skim even dry mud off the old bottom. At least several inches of native soil are ordinarily dug with it, and this opportunity is often taken to deepen the pond substantially. In some cleaning methods, it is necessary to take enough subsoil to build firm piles.

The cleaning process differs from the original digging in the usually shallower cuts, the peculiar nature of the mud, the fact that trees and landscaping on the banks often must not be disturbed, and the undesirability of reducing the pond area by piling spoil inside it.

Bottom mud is generally useless for agriculture when freshly dug, but makes excellent topsoil after curing in piles for a year or two. Mixing with sandy subsoil speeds curing and improves its quality. It is often necessary to add lime to correct acidity.

Dragline. If a dragline can do the necessary cleaning from the banks, the problems are chiefly avoiding or cutting trees, and providing either places to pile the spoil or means of access for trucks to haul it away.

If the width is too great for the boom length, an unassisted dragline must work from the pond bottom, usually on platforms. From there it may pile spoil on the banks to be leveled off later; against the banks, to make a new shore for a smaller pond; load it in trucks on the bank, or build one or more windrows in the pond to be trucked out later.

Trucking windrows must contain enough inorganic soil so that they will become firm as they dry, and must be high enough so that capillary water will not keep them soft. This height will vary from about three feet for a sandy mixture to seven to ten for silt or clay. Lower piles, or any piles containing a lot of humus, may require a surfacing of better soil or gravel before they will support trucks. The height of the roadway will be substantially lower than that of the top of the original windrow.

The dragline may roughly level the piles as it builds them, or this work may be left to a bulldozer. It may be advisable to use the lightest dozer that can do the work, as unexpected soft spots may be found, due either to slower drying of sections of heavy soil or excessive amounts of humus in spots.

Fig. 6-27. Trucking out piled mud

Since cuts are usually shallow near shore, and trees may interfere with maneuverability, it may not be practical to build the piles large enough to make a good land connection, in which case extra fill might be trucked in to bridge the gap.

When the windrows have dried and been leveled off for a roadway, the dragline can walk out to the end of one, possibly with the precaution of using platforms or poles, and dig it back from the end, loading trucks backed to it from the shore. It may just dig the piled material, or go down into the pond, either to deepen it or to obtain fill. Sections of roadway may be left to form islands.

Use of this method involves deepening the pond six inches to two feet or more. The double handling, the trucking, and the volume of material to be removed may make it prohibitively expensive for large areas, although it results in a pond which is better than new.

Dozer. If the bottom is firm enough to

Fig. 6-28. Pushing mud up slot ramp

support a dozer, and if the mud is thick enough so that a good load will stay in front of the blade, a dozer may provide the fastest and cheapest cleaning job.

The mud can be skimmed off gravel subsoils with little mixing. On softer footings, or wet soils which churn to mud readily, several inches may have to be taken with the mud. In any case, the digging down need not be as deep as with dragline work.

Disposal of the spoil may be a critical problem. It is liable to be too sloppy to pile up high enough for a bank, and to contain too much organic matter to make a satisfactory shore.

It can often be pushed out. The average pond edge is too steep for a dozer to climb with a load, so a ramp or ramps must be cut in it. If the shore is a dam, with low ground beyond, very liquid muds can be trapped in the ramp entrance and pushed through. Because of light friction, a dozer may push five to ten times its normal yardage on each trip through the slot, but a part of the volume will be water.

The ramp is apt to soften and break down, particularly at the bottom. Also, disposal areas at its head may fill up rapidly. For this reason, a number of ramps are liable to be required, and backfilling these later may be a major project.

The shovel dozer with grousers bolted on every fourth or fifth shoe on each track is the preferred tool for this work. The widely spaced cleats do not clog with mud. In cutting through heavy deposits, a shovel dozer is more adept at side casting than a standard dozer, and can backdrag material out of bad spots. When this is done by filling the bucket, it may be necessary to float it while backing to better ground, as lifting it tends to make the front of the machine sink in. This machine can often unstick itself by using the bucket dump as a pushing or pulling device.

The second choice is a wide gauge bulldozer. It usually has wide shoes that reduce its tendency to sink, and the width gives extra leverage for turning with loads on slippery footing. If grousers are worn down, a few of them can be built up to provide non-clogging traction.

When the bottom is reliably hard, a large dozer may be used. It is desirable because of greater production in both the volume of mud moved and area left cleaned by a single pass. It can also back into a deeper layer of soft mud without getting hung up than smaller machines with less

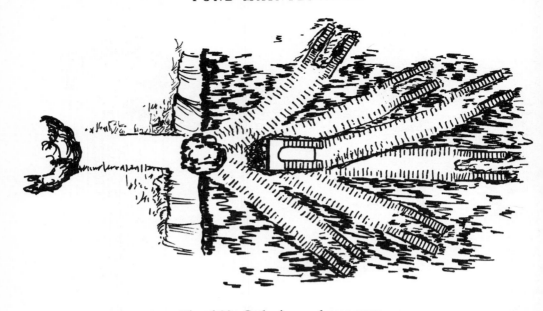

Fig. 6-29. Gathering mud near ramp

clearance.

On soft or doubtful bottoms, lighter machines are much less apt to get stuck and are easier to rescue if they do.

Saturated clay, silt, or very fine sand may look and act firm when work starts, but soften under the weight and vibration of machinery. This change will be caused much more quickly by heavy than by light units. However, such soil will often continue to give adequate support to a dozer as long as it keeps moving, even after becoming too soft for comfortable walking. No machinery should be left standing for any length of time on it, particularly if unattended.

When a dozer is used for swamp digging, means should be provided for prompt rescue in case the bottom proves too soft, or careless operation gets it stuck. If the dozer does not have a winch, a hand or machine winch, or equipment capable of exerting a heavy pull, should be on the bank with sufficient cable or chain to reach any part of the area. Cut green saplings and hand shovels should also be available.

Drain holes in the flywheel and steering clutch housings should be plugged to prevent the entrance of water and mud. Plugs should be taken out periodically to drain any oil that might leak into them.

Fully sealed rollers which are greased on a twice a year schedule require no special attention. Other types may require greasing every two to four hours to prevent mud from working past the seals. Sand in the mud may make it very abrasive so that track wear may be several times as rapid as normal.

Under average conditions, dozer work in a pond bottom offers considerable danger of getting bogged down, and conditions are often found to be so sloppy that little effective work is done. But when it works, it's fine.

Ramps. A bulldozer ordinarily cuts a ramp by digging from the side, parallel to the bank. The last few yards on the pond edge may be pushed out into the pond and to the side, and brought back up with the mud.

A shovel dozer may make the main cut in the same way, but is more likely to cut

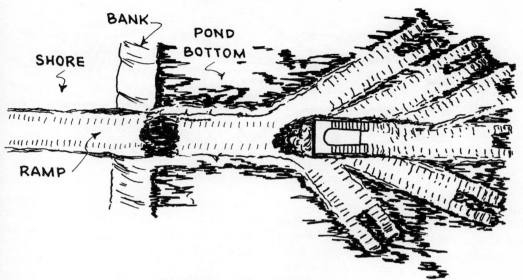

Fig. 6-30. Cleaning far section first

it lengthwise from the top, facing the pond and carrying or dragging the spoil back up it. It will not need to push as much out into the pond.

The ramp should be at the easiest possible gradient to facilitate pushing large loads and to minimize churning under the tracks.

When the ramp is roughed out, the dozer is backed into the pond mud until a good load is ahead of the blade or bucket. This is then pushed through the ramp to the disposal point, or parked in the ramp to be moved along with an additional load or loads.

The floor of the ramp will usually soften from absorbing water out of the mud being moved over it, and it will be worn down continuously by the push of the tracks and cuts by the blade. These effects are liable to be most severe in the pond at the foot where the dozer turns upward for its climb. A deep hole may be gouged here which will usually fill with a very thin mud. This ordinarily does not bother the machine any more than the same quantity of water, but will eventually reach the fan or other non-submersible parts, and the ramp will have to be abandoned or its foot relocated.

Such a hole may be convenient in freezing weather as the tractor may be placed in it overnight, so that the tracks will be under water and the mud on them will not freeze. This will save a long and messy job of putting it up on blocks and of cleaning and hosing it at the end of the day. It is of course not practical unless the bottom is entirely safe.

The cleared space may be widened by other cuts fanning out from the ramp. This uniform expansion of area is not particularly efficient from the pushing standpoint, as mud tends to spread on each push over ground cleaned by previous passes, and both mud and subsoil become increasingly sloppy from reworking. However, it keeps the dozer close to dry land while bottom conditions are observed. This pattern is shown in Figure 6-29.

If the original strip is worked back to the limit of the area to be served by the ramp, then widened at the far end, windrows will form at its sides that will allow moving larger loads. The cleared space will not be subject to being crossed by other loads of mud. However, the risk of getting stuck is greater, and if the ramp

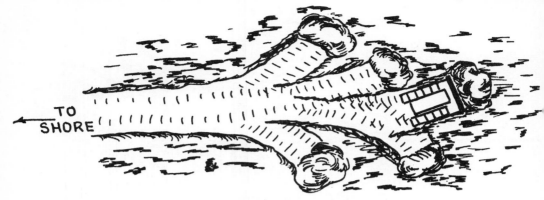

Fig. 6-31. Breaking a path

breaks down and becomes unusable, the material near it, which should have been the shortest and easiest push, will have to be moved to another ramp. Figure 6-30.

The long slot is made in shallow mud by backing away from the ramp. In deeper mud it may be necessary to work away from the ramp, herringbone fashion, before starting to push in, as in 6-31.

Loads brought in at a sharp angle to the ramp are generally dropped temporarily in front of it to avoid the excessive churning of swinging a load.

Dozer and Dragline. Ramp difficulties, or lack of nearby disposal areas, may make dozer cleaning impractical even when the bottom conditions are favorable. In such cases, the dozer may push the mud so that it can be reached by a dragline standing on the shore or the pond bottom, which can pile it on the bank or load it into trucks, as in Figure 6-32.

This method can be rather widely applied and is usually more economical than doing the whole job with the dragline.

Pump. Cleaning by machinery usually mixes some of the mud with so much water that it becomes too thin to be picked up or pushed, but can often be pumped.

A diaphragm pump will handle heavier mud than a centrifugal, but the volume moved is much smaller.

If a water source is nearby, clean water can be pumped into a hose line and the mud stirred up, thinned, and driven to the

mud pump or gravity outlet by a stream directed from a nozzle. If patience and man power are sufficient, whole ponds can be cleaned in this manner.

After removal from the pond, the mud may be allowed to flow away from the work area, or to accumulate in natural depressions; it may be held in a settling basin from which it can be dug after it has dried; or it may be placed directly in tight-bodied trucks. The very thin muds which pump most easily are usually the hardest to dispose of. The contractor is liable for mud damage downstream or on adjoining property.

Water Plants. Vegetation in ponds is of many kinds, few of which are desirable.

The blue-green filamentous (stringy) algae, which grow in surface masses that may have considerable depth, are often a major nuisance. They may be dense enough to make swimming and even boating impossible. These algae grow vigorously (bloom) in the Spring, and at irregular intervals throughout the warm months.

In large lakes, these plants can flourish only if the water is over-enriched by drainage carrying sewage, detergent, fertilizer, or other nutrients. In small and shallow ponds, however, they often grow very well without such assistance.

Fortunately, such algae are very susceptible to poisoning by tiny amounts of copper sulphate, and die within a few days of contact with it. This chemical may be ob-

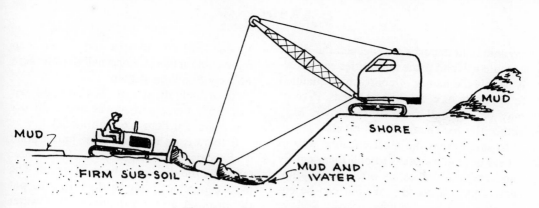

Fig. 6-32. Dozer-dragline team

tained in the form of coarse blue crystals from large hardware stores, or from dealers in commercial or agricultural chemicals.

The easiest way to apply it is to put it in a burlap or other loose weave bag, then tow it behind a boat, or pull it by hand while wading or swimming.

Two to three pounds to a million gallons of water, applied two or three times a year, should keep a pond clean. If it is allowed to become heavily overgrown, applications of two to three times that amount, at two week intervals, may be required.

Pond area may be roughly calculated from dimensions found by pacing, or by stadia as in Figure 2-19. Soundings will indicate the depth. Area in square feet times average depth will give cubic feet of water. One cubic foot equals eight gallons.

Copper sulphate cuts down the food supply of fish by reducing the vegetable food that is the basis of all animal life in the pond. Usually a balance can be preserved in which the plants are reduced enough to be unobjectionable but sufficient food is left for a large number of fish.

In the proportions recommended, or even in much heavier doses, this chemical is harmless to swimmers and to fish. However, the abrupt killing of very heavy plant growth may suffocate fish by absorbing the water's oxygen into decaying material.

This danger can be avoided by using light, repeated doses, or by treating only part of the pond at a time.

There are a number of chemicals and compounds used to control algae in swimming pools, which can also be effective in ponds. However, their price is usually higher, and application may be more difficult without special equipment.

Many water weeds that resist copper sulphate may be killed by sodium arsenite. However, it has been found that this very poisonous chemical tends to accumulate in mud bottoms. Its use is prohibited in many areas, and even where permitted, it should be applied by an expert.

Pellets of clay containing 2,4-D or similar materials may be broadcast over the water surface. The pellets break up and release the poison gradually at the plant roots.

Emergent weeds such as cattails and water-lilies can be killed by a 1.0 percent spray solution of 2,4-D, or kept from spreading by about $\frac{1}{10}$ that strength.

Duckweed (lemna minor), a very simple plant consisting of small, flat, floating ovals with roots dangling several inches below, is a problem in many shallow ponds. It often covers the surface completely, giving the appearance of a pale green lawn.

Most of the growth can be killed by chemicals, including sprayed fuel oil spiked with 2,4-D, but it may grow back almost immediately. It may be necessary to remove organic mud deposits and deepen the pond.

State fish and game authorities can usually supply the names of contractors qualified to eradicate or control pond weeds. Permits may be required from State and/or local authorities for chemical treatments.

In small areas the most effective way to get rid of pond weeds is to pull them out by hand and with rakes, as often as they appear. This is a muddy job, but can be enjoyable as a group activity.

HYDRAULIC DREDGES

Hydraulic dredges are very efficient excavators in wet land. However, their use in digging small ponds is limited by the expense of bringing them in and setting them up, interference of brush and stumps with cutters and suction lines, lack of sufficient water, and problems in spoil disposal.

Pond cleaning work does not usually involve handling stumps and brush, and most water weeds can be removed by hand or simple equipment. The pond provides some water. The competitive position of the dredge is improved by the great difficulties that land machines have in working pond bottoms too wide to reach from the shore.

This discussion will be limited to pond cleaning. However, the same problems arise in digging a new pond where water is available and vegetation can be handled by other equipment.

Dredge construction, work characteristics, and operation are described in Chapter 14.

Water Supply. A six inch hydraulic dredge will pump water out of a pond at a rate between 800 and 1,200 gallons a minute, and will move 1,000 to 2,000 gallons for each yard of soil. An acre of water contains about 325,000 gallons for each foot of depth (this is a measurement called an acre-foot), or enough water for 4½ to 6½ hours of operation. Because of the slope of banks, each successive foot of water will contain fewer gallons than the one above it.

A pond one acre in area and 5 feet deep might contain two and one-half to three acre feet, say a million gallons.

An 8 inch dredge needs about 60 per cent more water and moves about 60 per cent more soil than a 6 inch model.

If there is no natural inflow into a pond it is necessary to return most of the water taken by the dredge to be used again. The most economical way to do this is to put the disposal area upstream, so that water will drain back. If this is not practical, a pump is placed so that it can return the water through a hose or pipe. A heavy duty 4 inch pump can usually handle enough water to keep a six inch dredge busy.

Return water is usually dirty, so that part of the dredge capacity is wasted rehandling the fines that did not settle out. The amount of circulating soil particles is greatest when the soil is fine textured, and when settling ponds or areas are small.

The fine organic mud that forms a large part of the deposits on pond bottoms may stay suspended in water for long periods.

The experimental use of cyclones for separating soil particles from dredge water and of polymers for clotting organic slime which may reduce this problem, are discussed in Chapter 14.

Dirt in return water increases when the suction strainer of a return pump is too low, or the water is allowed to run on the ground and erode it. It is reduced by having a large settlement pond or area, and/or very sluggish open flow of return water. It tends to increase as the settlement pond fills up with use.

Re-using dredge water has the important advantage of reducing or eliminating the problem of fouling streams or other property below the fill with muddy water.

Any natural flow into a pond reduces the water problem. A pond with no inflow in the summer or dry season may have an ample supply pouring into it in the wet season. If the dredge is operated on a single

Courtesy of Dixie-Dredge Corp.

Fig. 6-33. Loading a small dredge for transportation

shift it could operate on a steady inflow of half or one-third of its output.

Disposal Area. Desirable features for a fill disposal area include: location close to the pond, drainage back to the pond, good conditions for separation of soil and water, need for fill, and absence of trees and brush.

A six inch dredge may be expected to pump spoil from 800 to 1,500 feet, depending on its coarseness and weight, the percentage carried in the water, and the alignment and gradient of the pipe. Maximum distance is obtained with light load, straight pipe, and low lift. A downhill line could be much longer.

Production is reduced by long lines and uphill flow, as these conditions reduce both the volume of liquid and the percentage of solids that can be carried. For maximum output the spoil should be discharged close to the dredge, or down hill from it.

If it necessary to re-use the water, and two or more dump areas are available, the loss of production from uphill pumping to get gravity flow back to the pond must be balanced against the expense of a return pump at the same or a lower level.

A close location reduces the cost of providing and handling pipe.

Soil settles out most rapidly and completely when water is still. A discharge or settlement pond is usually kept more or less agitated by flow from the pipe. Increasing its area and depth reduces rate of water movement and allows more particles to settle out. The overflow or the sump for pumping should be as far from the flowing water as possible. The size and depth of the pond diminish as it fills during work.

The cost of pond cleaning may be best justified when good use can be made of the material removed. The value of swamp land may be greatly increased by building it up to a higher level. Rocky or stumpy fields may be filled over to smooth surfaces.

Fill obtained from cleaning a pond should

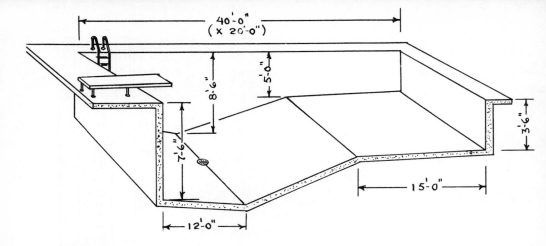

Fig. 6-34. Dimensions, family swimming pool

include both a layer of the original bottom formation, and rich black mud built up by the pond water. It is usually of rich fill or topsoil quality, and may be salable after drying.

Mud is held in the area to be filled by putting a dike around it. This may be built by a dozer on firm ground or a dragline where it is soft. A wood dam with movable boards, as in Figure 6-19, will provide for overflow and height regulation.

Both the owner and the contractor may be liable for any damages caused on neighboring property or downstream by mud or too much water.

Filling Around Trees. Many swampy areas that should be filled are covered with vegetation ranging from brush to big trees. Undesirable brush usually survives partial burial, but trees are very likely to die. The biggest and most valuable specimens have a smaller chance of survival than younger ones.

Trees that will die anyhow should be cut before the fill is brought in. This will allow burial of stumps and unwanted logs. Such logs should be cut in short pieces to avoid later interference with trenching or cellar digging.

If a tree is cut after ground is filled there will be an unsightly stump. This will be extremely difficult to remove because of the depth of the roots under the new surface.

Brush should be cut and burned, or at least knocked down flat, before filling. Afterward it will be difficult to take out and will make grading very difficult.

Wood may decay very slowly under a hydraulic fill.

SWIMMING POOLS

A concrete-lined swimming pool is a special type of artificial pond which is enjoying increasing popularity. It is tremendously more expensive in relation of area and volume of water. On the average, it requires more upkeep. However, it is much more flexible in location and is more limited in space requirement. Many persons who have an aversion to mud or water snakes or even just nature will carefully avoid a natural or dug pond, but swim happily in a pool.

A pool should be made of reinforced concrete. Concrete block construction is risky, although it is often successful in well drained soils in frost-free regions. It sometimes ends up with a reinforced concrete wall inside it.

The standard type of pool has vertical walls joined at right angles, and a floor slope which allows diving at one end and wading at the other. A usual dimension for

6-42

family use is 20 x 40 feet with a maximum depth of nine feet. An excavation is made in about the same manner as for a cellar except that more care is taken in finding firm and uniform footings.

A less common but somewhat more economical pool may be made by digging a hole with irregular shape and sloping banks, setting reinforcing rods on the bottom and sides, and spraying on concrete to make a structure conforming with the irregularities of the excavation. This requires very firm and uniform soil for proper support. Whether the less conspicuous but also less conventional appearance is good or bad is a matter of personal taste.

A pool must be stronger and better supported than a house foundation, because of the variable load of water it must carry. When it is full, it weighs heavily; when empty, it is light enough to float. In fact, ground water and tide have been known to float pools out of position.

A gravity drain should be provided if possible, for convenience in emptying.

Plastic pool liners are offered in such variety of design, material, and quality that it is difficult to report on them. The plastic is usually a liner that is placed so as to fit the inside of an excavation, with or without walls or floor. It should be installed by factory trained men, as it is a very extensive piece of material, it usually has little strength, and one tear or puncture might destroy its usefulness.

Some plastic is tough and some is tender, some stays good for years and some deteriorates rapidly, some can be repaired and some cannot. The reputation of the manufacturer and his local representative, and the type of guarantee that they will supply, are very important.

The purchase of a moderately priced plastic lining may serve to correct otherwise hopeless leaks in a badly built masonry pool, or to provide recreation for one to five seasons until funds are available for a more permanent lining. It may be that there are or soon will be stronger and longer lasting linings than those that have been reviewed for this article.

It is almost necessary to equip a pool with a pump and filtering apparatus to remove dust, pollen, and other materials that fall into the water. The attractiveness and the cleanliness of the water are improved, and the frequency of needing to clean the bottom and walls are reduced.

If the capital budget is small, a pool for family use only can be kept usable by frequent addition of new clean water through a garden hose, and draining and scrubbing as often as necessary.

Maintenance includes supplying small amounts of chemicals to kill algae, removal of bottom debris with a suction hose, and occasional draining, cleaning, and painting.

For freezing weather, a pool should be kept full, as its ice will brace the walls against pressure from freezing ground water, and will provide a small but convenient skating rink.

CHAPTER SEVEN

LANDSCAPING AND AGRICULTURAL GRADING

HOME LANDSCAPING

Landscaping may include the processes of cutting, filling, or grading to change ground contours; retaining or placing adequate topsoil; preserving, moving, or adding vegetation, and planning and installing walls, drives, and game courts.

An important purpose is to produce a pleasing appearance. This may be an end in itself but is usually secondary to the use of the land.

Landscaping is often the final step in jobs which involve earth moving. It is required in connection with highways, particularly of the parkway or thruway type; to improve the appearance of home or business buildings not surrounded closely by other buildings and paved areas; to beautify parks, and to provide them with suitable recreation areas.

Plans should take into account proper drainage, which may include subdrainage.

Landscaping is often done under the personal direction of the landowner or his representative, but may be finished to grade stakes or left largely to the contractor's judgment.

A large part of the annual landscaping bill is for work around homes. Much of this is done during house construction or immediately after its completion, in connection with backfilling around the foundation, disposing of dirt dug for the cellar or footings, and restoring surface drainage.

Such landscaping may include construction of terraces, retaining walls, and driveways, moving or planting of trees and shrubs, and making lawns.

The excavating contractor may perform the entire job or only the heavier parts.

CHOOSING THE SITE

House Elevation. The type of grading close to the house is determined by its elevation relative to the land. The wood sills or trim should be at least four inches above the finished grade of the topsoil. In general, exposure of more than a foot or two of foundation causes a house to look too high for current styles. The ground should slope down away from the house enough to prevent surface water from standing against the wall.

A house may be set high enough so that dirt from the cellar excavation can be used entirely in backfilling and grading up to it. If the floor level is determined in reference to the original grade, the bulk of the piles must be "lost" on the grounds, or trucked away. If one side of the house is level with or cut into an up slope, massive

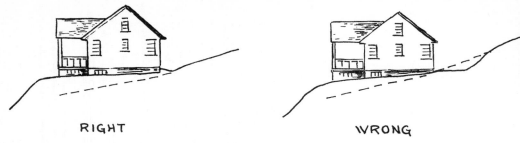

RIGHT WRONG

Fig. 7-1. Elevation of house on slope

excavation will be required to give outward drainage. This is costly and will produce an artificial appearance. See Figure 7-1.

Grading is also affected by the extent and type of cellar excavation. A deep, full cellar produces large quantities of fill, while digging for footings and a floor slab may yield little or none. When the house is to have a cellar, is to set low, and is to be built on a plot having a good grade, it will probably be economical to haul away all dirt not required for backfill around the foundation.

Desired depth of the foundation below ground line may be obtained by digging full depth and removing spoil; by putting the cellar floor at the original surface and filling; or by an intermediate method. In general, the most economical way is to cut just enough to provide the necessary amount of fill to build the ground up to the house.

Rock and Water. The presence of rock or water near the surface may make a plot a poor investment, and in any case is important in deciding whether to have a cellar, and the depth to place its floor.

Shallow rock can be found with a probe made of four or five feet of $\frac{5}{16}''$ stainless steel rod, with a sharp point at one end and a handle at the other. This can be pushed down into any but the hardest soils.

However, it will not tell whether resistance is a cobble or ledge. A long sharp crowbar or prybar can be sunk by repeated dropping and turning. If it is stopped by an obstruction, lack of vibration as it strikes indicates a small stone, vibration only near

the hole a boulder, and a general jarring, a formation of bedrock.

Vegetation will tell a lot about water conditions. Bush willows and bog or bunch grass must have it wet in the spring at least. Such water-loving plants on a flat indicate swampy conditions. On a slope they show a spring or seepage, and may warn of ledge rock as well.

If rock or a high water table is found on the site or surface drainage is poor, it is often good practice to reduce the depth of excavation and truck in fill.

No fill should interfere with drainage from adjoining property. If the land must be raised, drains should be placed under or around any dam that is formed.

A septic field on low or impervious ground may have to be placed in a filter bed (pervious fill) which may be quite costly.

Hill or Valley. Choice between a hill or valley site may be largely a matter of personal preference. Some people like to look down, most like to look around, but a few enjoy a closed-in feeling. Often, however, these preferences will not be strong enough to outweigh other considerations.

A hill top is almost always well drained, so that the wet cellar difficulties discussed in Chapter 5 will not arise. On the other hand, it is much more likely to have rock close to the surface, so that the expense of cellar digging may be three to six times greater than for dirt excavation.

Ground drainage can be too good. A person wanting to enjoy lawns and gardens will have difficulty with them in dry weather if they are on a heap of sand or gravel. Top-

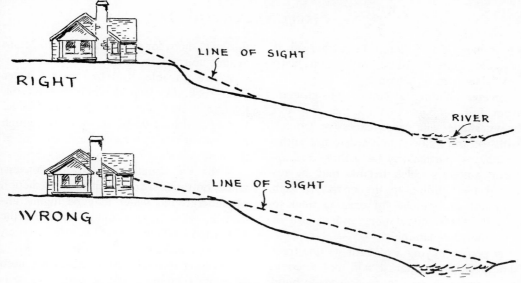

Fig. 7-2. View down slope

soil is likely to be poor, thin, and stony.

A top-of-the-world house gets whatever breeze exists in hot weather, but may get rather too much wind at other times. In icy weather it may be hard to get home up a slippery driveway, but it would take very bad conditions to make it impossible to get out in the morning. Allergic trouble with pollen and molds is usually somewhat reduced.

Building on low ground risks water trouble in the cellar, if any, and the possibility of serious flooding from streams or drains. It limits view to the immediate surroundings, provides a higher average temperature but increases danger of frost damage (cold air flows downhill), reduces effect of cooling breezes in the summer and even more cooling gales in the winter, and usually provides rich and moist soil for lawn and garden.

If at all damp, a low site is dangerous to the health of arthritis and asthma victims.

Slopes may offer any combination of features of high and low land. Special factors to consider are that if the land slopes down to the south it will be warm (or hot), and if down to the north it will be cold. Western exposure offers sunsets, which are much more popular than sunrises. Steep slopes

make landscaping difficult and expensive, although the final result may be worth it.

Steep driveways are a perpetual nuisance. Elderly people may be prevented from walking and visiting by the necessity of climbing a slope. Severe wet cellar difficulties are rare but not unknown, and some water trouble is common.

View. A house on high ground may be largely deprived of the enjoyment of a fine view by being set too low or too far back from a slope, or by careless grading or planting.

The majority of houses are now the one-story type. When two-story construction is used, the ground floor, particularly the living room and terraces, is the level from which scenery is most often enjoyed. Scenic potentialities should therefore be worked out in reference to a person seated on the ground floor.

A common error is building or failing to remove a high spot which, although lower than the house, blocks the view of nearby down slopes and hollows. See Figure 7-2.

There is often conflict between trees and view which must be decided on a basis of individual preference. In general, ordinary young trees may be quite readily sacrificed while old trees or fine specimens of younger ones should be preserved if possible. Drastic pruning will often serve the same purpose as removal.

Shade. Shading a house and grounds from full sunlight is desirable, but too heavy shade will cause excessive trouble with rot and mildew, and create unhealthy conditions, particularly for asthma and arthritis sufferers. Such trouble may be reduced by building in the open, by high trimming of branches of existing trees to permit full air circulation, and by use of discretion in planting.

A person buying a plot for its fine trees should be sure that it will not become necessary to remove them in order to build a house.

Noise. If noise from a highway or railroad is of critical importance in determining house location, it should be remembered that it travels chiefly upward, partly because of reflection from the pavement or roadbed. Even hundreds of feet up a hillside will not reduce it substantially if the source remains within sight.

If the river in Figure 7-2 were a noisy highway, the construction which is wrong from a scenic standpoint would become right when noise only is considered. An earth bank is a more effective sound deflector than a hedge or other planting.

Water Well Drilling. A substantial portion of both home and industrial building is in areas not reached by water mains. Most farms depend on ground water for domestic use, and many use it for irrigation also. Factories, theaters, and other large users of water may find that they need a supply in addition to city water. Under such circumstances, the only method of getting a dependable supply of safe water may be to drill for it.

In sandy or gravelly soils surface water outcrops, such as ponds and springs, give a rather good indication of the level and abundance of subsurface water. However, a well should go substantially deeper than this level, both for purity and for protection against unusual dry spells.

Where possible, it is best to get water from rock, or deep down in sandy soil. Danger of contamination is then negligible. Casing is driven down at least far enough to keep surface water and loose soil out of the hole.

Wells are usually located for convenience, on the first try at least, as prediction of underground water may be highly uncertain. This is particularly so when the soil is too shallow to provide safe supplies, and water must be obtained from rock.

Divining rods of various kinds are used in many sections to locate water. In tests these "dip sticks" have shown a somewhat better record than random drilling, but the difference can usually be accounted for by the good judgment of the experienced man who carries it.

The best place for a well for a residence is just outside the foundation line, so that it can be included in a small extension of the cellar or connected by a short pipe, but can still be reached vertically from outside for pulling underground equipment and servicing the underground part of the pump. It is usually drilled and lined (cased) before the cellar is dug.

Placing the well away from the house involves building a rather costly separate pump house, which may offer a landscaping problem and which will have to be connected to the house by water and electric lines. It does have the advantage of freeing the house from the noise of the pump and automatic switch, and the possible nuisance of water from leaks.

A well under the house is very convenient, and lately has become permissible because of improvements in pump design. The flexible plastic pipe and jet pumps, now most commonly used in drilled wells, make it possible to service them in spite of limited headroom.

Distance between sewage septic fields

and wells may be subject to local regulations. Under ordinary circumstances, there is no conflict between having them in the same place if the well is deep, but there is a slight chance that the casing might crack or become disjointed and allow leakage into the water. For this reason, prudence dictates that the well top should be higher than the field, and at least 50 feet away from it.

The truck-mounted spudding or well drill has been the standard tool for the drilling, but rotaries and down-hole units have largely replaced it. Flow in the well is measured by pumping or bailing.

A hand pump may be put on a well for use during building construction.

A flow of four gallons per minute is considered adequate for a small residence, but double this is desirable to assure a generous supply. A small water flow can be partly compensated by a large storage tank.

SHAPING THE LAND

Backfilling. In general, it is most satisfactory to backfill around a foundation before the framing of the house is started. This removes the piles of fill that form an obstacle and a hazard during construction, and provides space for entrance and piling of materials.

Backfill against fresh masonry must be done carefully. A heavy dozer should keep farther away from the wall than the diameter of the largest stone found in the fill, to avoid accidental punching of holes. It should not walk on fresh backfill parallel to the wall because if it sinks on the side toward the house it will exert a heavy thrust, and be almost impossible to get out without causing damage.

Foundation backfill is seldom tamped when it is placed, but failure to compact it offers the danger of the loose dirt soaking up enough water during a heavy rain to crush the wall by hydraulic pressure. Good underdrainage around the footings, a

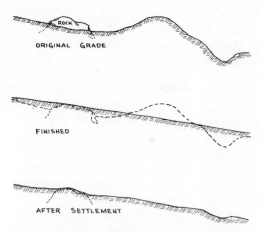

Fig. 7-3. Irregular settlement after grading

proper surface slope away from the house, and compaction of the surface make such a disaster unlikely. Placing floor beams strengthens the foundation.

A foundation of concrete block is subject to damage even after curing. Unless the fill is wet the weight of the dozer is unlikely to cause damage, but a stone may still be punched through it.

A shovel dozer is the preferred tool for backfilling and grading around a house. Its ability to back and turn with loads, to cross graded ground with a load without excessive damage, and to place dirt exactly where it is needed, enable it to accomplish much more work than a bulldozer of the same size. However, it cannot grade quite as close to a wall because of the overhang of the back of the bucket in dumped position, and the fact that the bucket is little, if any, wider than the tracks.

Grading. Grading may be mostly or entirely a problem of disposing of surplus fill to the best advantage. At other times it will consist of arranging for proper drainage, removing objectionable humps or filling gullies; disposing of stone walls or boulders; reshaping to obtain a desirable view or to avoid an undesirable one, or rearranging contours for better appearance. These operations may produce a surplus of soil, or may require bringing in hundreds or even thousands of yards.

Soil in trenches and fills should be thoroughly compacted before the fine grading is done. Unfortunately, it is not common practice to attend to this on small jobs, with the result that an originally pleasing appearance degenerates badly in a year or two. Effects are bad when a level or evenly sloping lawn settles into humps and hollows, and are worse when game courts, stone walls, or paved drives are involved. See Figure 7-3.

Trench backfill can be compacted by hand, with air, gasoline, or mechanical hammers, or with electric vibrators. If ample time will elapse before grading, ditches can be loosely filled then puddled by flooding with water. Full shrinkage will not occur until they have dried out, a process which takes a few days with porous soils and weeks with heavy ones. While wet, a puddled ditch is a dangerous trap for machinery.

Septic fields and tanks can be easily damaged by machinery or trucks.

Fills should be compacted by rollers or trucks. If trucks are used, each fill layer (preferably not higher than ten inches) should be thoroughly rolled first empty and then loaded. Running a loaded truck on loose fill puts a severe strain on its power train.

If an area to be filled is cut to an even grade or the high spots broken up first, results of settling will be less damaging than if fill is placed over an irregular surface.

A medium textured fill is more satisfactory for most purposes than either very porous or very heavy soils.

Lawns should not be perfectly flat for any appreciable distance. The maximum slope which it is convenient to mow is about 1 on 6 for long grades, and 1 on 3 for short terraces that are hand cut. Steeper grades may be left in long grass, planted with vines, shrubs, or fixed as rock gardens.

Fig. 7-4. Stone walls may be very solid

Old Walls. In New England and many other sections of the country, utilizing or disposing of old stone walls is a common problem in landscaping. They often contain huge stones which are so buried and bound that they offer a problem to any but the largest machinery. For this reason, and because of the beauty of many of them, it is advisable to leave them in place when possible.

A dozer can move a wall but can seldom rebuild it properly. A shovel or a shovel dozer can dig out such a wall and roughly rebuild it with the bucket, or with the help of chains and a good rock man, reassemble it better than new. The chain work is slow and expensive.

If the wall is to be removed an attempt should be made to sell it. Weathered field stone in small sizes is often in demand. Boulders can occasionally be used in deep fills, stream bank riprap, or breakwater construction. Prices obtained for large stone seldom more than repay the expense of handling.

If there is no market for stone, an attempt should be made to bury it. The bulk can be roughly calculated by measuring the length and the average height and width of

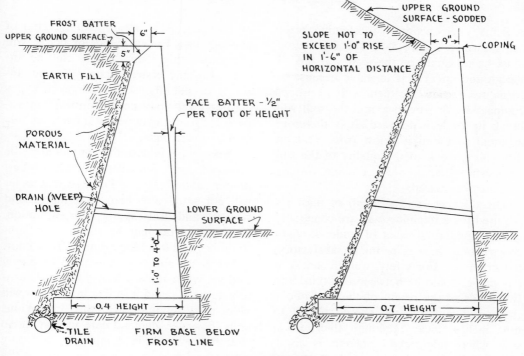

Fig. 7-5. Retaining wall sections

the wall, including the underground part. If no gully or other natural disposal point is available, a hole or holes should be dug to contain somewhat more than the calculated yardage, allowing for a foot or more of fill over the top.

Excavation is done in the same manner as for a cellar. Topsoil should be stripped off the area that is to be dug and regraded. The hole should be deep rather than wide, and might well be dug by a hoe rather than a dozer, if one is on the job. A hoe may dig a trench close along the wall, followed immediately by a dozer pushing the stones into it and regrading.

A hoe is often more efficient than a dozer at breaking up a stubborn wall, as it can work out one stone at a time. However, it cannot transport the stone readily.

The rocks can be trucked away if burial is impractical because of shallow soil, trees, or landscaping. A shovel dozer, a dipper stick, or a big clamshell can break up the wall and load it. Trucks should preferably

have bodies built to carry rock, or be so old and beat-up that damage will not matter.

Loading a wall is slow work. Even small stones may be hard to dig out when in groups, and big ones are hard to get securely in the bucket. Production in yards-per-hour may be pitifully low.

Retaining Walls. Masonry walls are frequently used to separate different ground levels. They may be required where the slope is too steep for earth, or used largely for the sake of appearance. In the first case, the wall may make up only part of the required rise, and an earth slope is continued from its top. Such walls must be strong and well founded if they are to give good service. They are subject to very heavy pressure from the dirt behind them, particularly if it slopes up from the top of the wall, and if it becomes saturated. Freezing will cause a push against the top of the wall and disruptive forces inside it. Tree roots can act to lift and overturn it.

Some cross-sections of retaining walls are shown in Figure 7-5. The foundation must be adequate or the wall will fail. If the ground under it is unstable because of its nature, recent placement without proper compaction, or frost heaving, the wall will break up or lean outward. It is therefore essential to found it below frost level on firm soil or rock. If the quality of the soil is questionable, a wide concrete footing slab may be poured.

The thickness and strength of masonry required for a retaining wall are commonly underestimated. Results of under-strength construction are sometimes satisfactory, but often not. For safety, wall thickness at any point should be between one-third and two-thirds of the height of the wall above that point, and the top should be six to nine inches thick. Minimum thickness is safe when reinforced concrete is used, when height is moderate, and the retained soil well drained and stable.

Maximum thickness is required when a steep slope rises from the rear of the wall, and when the ground is very unstable. Other considerations are the strength of the masonry. Reinforced concrete is the strongest used. Plain poured concrete is considered stronger than concrete block, brick, or mortared stone. Dry stone walls have little resistance against thrust and should be kept low.

The push from dirt behind the wall can be minimized by keeping it well drained. A layer of gravel or other porous material should be placed along the rear face of the wall. A tile drain should be placed beneath the foundation, and there should also be "weep" holes through the wall itself.

Ground expands when the water in it freezes, and the surface slab formed in this manner can exert a considerable thrust. A slope or batter at the rear corner will deflect this pressure upward so the slab will slide on the wall instead of pushing it.

A vertical wall often has an appearance of overhanging. A backward lean or batter of one-half inch for each foot of height will counteract this. Such batter can be increased to any desired slope with some increase in stability. A face slope in a dry masonry wall may permit outward movement for some years before it becomes vertical or overhanging.

Drainage. It is desirable that all areas be provided with sufficient surface slopes, proper subdrainage, or both, so that water will not stand anywhere, and the ground will dry and firm rapidly after saturation. Particular care may be required to subdrain any soil touching cellar walls or floor.

When the soil is porous sand or gravel and the water table is low, drainage is usually automatic and mistakes in gradient will show only briefly during rains. Impervious soils, however, demand care in shaping so that they will drain completely, not only when the job is completed but after settlement of fills.

If pervious fill is placed on a relatively impervious native soil, the lower surface should be shaped to drain, to avoid trapping underground water in pockets. If the native soil to be buried is pervious it need not be graded for drainage, regardless of the type of fill, although shaping to avoid uneven depth of fill is still advisable.

Where areas are large, rain water flowing on the surface may constitute a serious nuisance even if it does not erode the ground. At a price, such water can be caught in catch basins, then removed through underground pipes. Because of the expense of such an installation it is best to have it designed by someone familiar with the work. If this is not possible, pipe size should be figured in the same manner as culvert capacity, according to the maps and tables in Chapter 5. Sometimes undersize pipe is used for economy, on the basis that occasional overflow along the surface under extreme conditions will do no great harm. However, eight inch pipe is the smallest that should be used with such catch basins.

If land tile is used, it will also function as a subdrain. However, care must be taken not to allow more surface water to enter it than it can easily handle, as the hydraulic pressure resulting from water standing in or over the inlets may force channels outside the tile, which will undermine or misalign them, with resultant impairment or destruction. If the important problem is surface water, concrete pipe or sewer tile with mortared joints is preferable.

All inlets should be protected with gratings firmly set in masonry. Lack of these may permit entrance of large objects or masses of material which will plug the drain. Gratings are usually larger in area than their pipe, to allow for partial clogging with leaves. The vertical or steeply sloped pipes up to the catch basin should have tight joints.

If backfill is not tamped in the trench made for the drain, it may settle and leave the grating standing up above the sod. This is unsightly, makes it vulnerable to breakage, and interferes with reception of water.

If a garage is below ground level, a catchbasin in the driveway just outside it is necessary. This drain MUST be adequate, as its failure in a heavy storm will flood the garage and perhaps the cellar.

Gutter drainage to dry wells or tile is discussed in Chapter 5 under Cellar Drainage.

Subdrainage. Land tile subdrains may be installed under lawns and gardens to correct saturated or oozing conditions, to speed up drying after rain, or to provide better growth conditions for plants. Because of expense, they are generally limited to "show" sections or game courts.

The techniques to be described for agricultural drainage are employed, except that the interval between tile lines is often much smaller, spacing as close as ten feet sometimes being used under tennis courts.

Subdrains may be tied in with the tiling around the cellar and with catch basin systems for surface runoff. They should drain to low areas when possible, as opening into storm water drains expose them to damage from backed-up flood water.

FINISHING OFF

Topsoil. Topsoil which has been salvaged in advance of digging may be spread as soon as the fill is graded off, or left piled until the house is finished. Immediate spreading provides a cleaner appearance which is of particular value to houses built for sale, but the topsoil is liable to become mixed with various sorts of waste, and to be severely packed by supply trucks.

Two-ton "toy" dozers or small compact loaders, are good spreaders, as they are so light that they leave average topsoil in condition to be finished off by hand, where heavier machines compress it so that machine tillage is required. It also can maneuver among trees, retaining walls, and other obstacles with less danger of damage and far less loss of time than larger dozers. It can work over the average septic tank or field, and can often do a complicated job more quickly than its big brothers.

An excellent team for residential grading is a shovel dozer as big as the traffic can bear for the long and heavy pushes and carries, and a two-ton to distribute and smooth off the material.

A light wheel tractor with a front bucket or rear grader blade can do light grading.

Freshly spread topsoil or undisturbed field sod which is to be reworked into lawn is often loosened up with a rotary tiller. This machine leaves it soft and easy to work, and if the topsoil is thin will increase its usefulness by mixing in some subsoil.

It is not possible to state a general rule for the amount of topsoil needed around a house. Good topsoil has three important characteristics: it contains humus which absorbs water and doles it out to plants in dry weather, it contains a supply of available fertilizer, and it has a grain size and arrangement that is favorable to plants.

A lawn made with poor or too thin topsoil may be persuaded to grow vigorously by proper fertilizing. However, it will tend to burn out during dry spells unless it is shaded. It will dry out more readily if it is over gravel or sand subsoil than over heavier soil. The minimum topsoil depth for a lawn under most conditions is two inches, and four inches is safer. However, benefits are obtained from greater depths, and it is common practice to use up whatever piles are around. If the original soil was thin, or had been lost, more must be trucked in.

Gardens, flower beds, and shrubs like to have about eight inches, and depths up to two feet are recommended for some species.

If peat (humus) is obtainable locally at a low price, it may be spread on subsoil and mixed in by hand or machinery. With the addition of lime and fertilizer it may serve well as topsoil and might be much cheaper.

Converting Field to Lawn. The original lawn made around a new house may be partly or wholly at the grade of an existing field supporting a growth of mixed grass, wild flowers, and weeds. The standard way to make a lawn is to use a rotary tiller or a plow and a harrow to turn in the existing growth, and pulverize the soil for planting of grass seed. It is usually good practice to bury the old vegetation several weeks before planting. Decay of vegetable residues often temporarily deprives the soil of nitrogen to such an extent that the new crop cannot obtain enough for a start. Also, sprouts from roots of undesirable plants can be readily destroyed as they appear on bare ground, and can be reduced or eliminated before planting.

A less expensive method, which is usually satisfactory, is to pull out any brush, then mow the field repeatedly. It may be reduced in one operation from field length

to lawn length, then kept short, but better results are apt to be obtained from starting with a high cut as if for hay, and progressively trimming shorter.

The effect of the cutting is to kill or place at a disadvantage plants that prefer to grow tall, and to encourage those which are not damaged by mowing. Most fields contain enough lawn-type grass and clover to take over the whole area within a year when encouraged by repeated cutting.

A similar effect may be obtained by moderate driving and parking of cars and trucks on the field.

Field surface may be too rough for a lawn. Large inequalities may have to be cut or filled. Smaller ones can be rolled down, filled with dirt or both.

A power lawnmower having heavy steel or rubber rolls will tend to flatten ground. Its effect is quite marked when inequalities are small and choppy and the soil is soft, and becomes negligible as the ground bakes in the summer.

A steel wheel roller, weighing from three to five tons, of the type used in driveway construction and blacktop patching is a very effective lawn flattener, but must be used when soil is in the right condition. If it is too soft, it will make ridges, and probably get stuck. Heavy wet soils may pack so hard as to discourage growth. If the ground is too hard, it will be ineffective. Heavier rollers are sometimes used, but the expense of transporting them and the danger of sticking them is greater.

Hollows may also be filled with topsoil. This may be applied in layers a quarter to half an inch thick so that grass will grow up through them, or the fill made to the level of the surroundings in one lift, then new seed planted where necessary. Sometimes the seed is mixed with the topsoil prior to placing it.

Fills of more than an inch or two on bare ground and any fill over grass tend to settle noticeably so that the work will require doing over, although with less material, the following year. Over-filling enough to compensate for this settlement requires expert judgment. Firm tamping is sure to reduce this difficulty and may eliminate it, but might make it difficult for existing grass to push up through the new soil.

Planting Grass. Soil should be loosened and smoothed before planting. If the area is too large for hand digging and raking, a rotary tiller of appropriate size is the preferred tool. A plow and a disc harrow, or a harrow only, followed by dragging with a plank will often do a good job. Fertilizer, lime, manure, peat moss, or other soil-enriching materials are best mixed in by hand or machinery. If the area is large, a spike tooth harrow can be used for both smoothing over the ground and covering the seed.

After the machinery is finished, the ground is hand raked. This looks simple but is not. A certain knack is required to get a smooth surface.

Soil acidity can be tested with litmus paper and color scales that are obtainable at garden supply and drug stores. Most lawn grasses like a pH 6. If the soil is pH 5, about 75 pounds of ground limestone should be added to each 1000 square feet. A pH of 4 will call for 100 to 200 pounds for the same area.

Slaked or hydrated lime has a higher calcium content, 100 pounds of it being equivalent to about 135 pounds of ground limestone. Quicklime is still more concentrated, but is inconvenient and dangerous to handle.

It is good practice to add some fertilizer, but the amount varies widely with circumstances and individual judgment. If vegetation has been mixed into the soil within the last two or three weeks, either on the spot or at the source from which it is being hauled, some nitrogen fertilizer must be added, to make up for that borrowed by the decay processes. In general, an addition of a moderate amount of general fertilizer is cheap insurance for the work and expense invested in preparing and planting the lawn.

Any of the lawn seed mixtures are good, and the differences can be explained by the store selling them. A mixture should almost always be used instead of a single kind, as each variety has its own special and often obscure likes and dislikes in soil conditions. If several kinds are planted, there is a good chance one of them will like the circumstances, and its vigorous growth will compensate for any sulking by other varieties.

Seed scattered on the ground surface in very early spring may be worked underground by freezing and thawing so that no effort is required to cover. Seed lying on the surface will sprout and take root successfully during a long wet spell, but hot sun may kill it, or birds eat it. Raking after seeding reduces these dangers.

The grass will come up faster and better if the ground is rolled lightly after seeding. This may be done with a muscle-powered lawn roller, a hand tamper, a power lawn mower of the roller type, or the smallest size of steel wheel gasoline roller. If the job is a rough one, driving a car or empty truck back and forth on it will give good results.

Erosion. A sloping lawn is vulnerable to severe damage from flowing water from the time the topsoil is spread until the grass has made a good root growth. The probability of such damage should be figured into cost estimates by both the owner and the contractor.

Danger of erosion can be reduced by mixing straw or lawn clippings into the surface. This necessitates a heavy addition of nitrogen fertilizer, and makes it harder to cover the seed.

If a seeded surface is sprayed with asphalt emulsion, it will be held against ordinary erosion, without preventing growth of the seedlings.

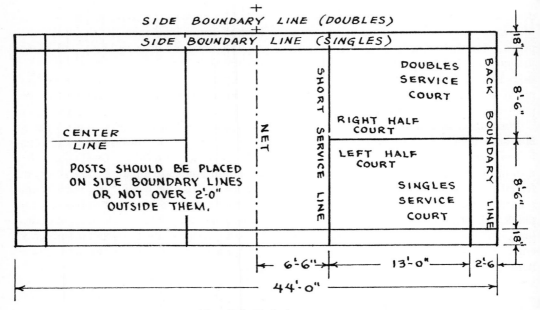

Fig. 7-6. Badminton court

It is sometimes possible to divert drainage into other areas. One section may be fixed up and seeded first, and water routed through the part which is only rough graded. After grass is firmly established, drainage may be shifted to go over it, so the rough area can be smoothed and planted.

Occasionally, it is possible to install temporary pipes or flumes to carry rain water while grass is getting a grip.

Sod. Drainageways and steep slopes can be protected with sod. This is cut out of existing lawns or mowed fields by means of hand tools or an engine-driven sod cutter, laid on freshly loosened and smoothed topsoil and tamped into firm contact. It is sometimes fastened in place by driving pegs or thin stakes, or by pegging chickenwire firmly over the whole area.

Sod may be cut in strips 12 or 15 inches wide and 6 to 10 feet long or in squares or rectangles of any convenient size. A depth of 1½ inches usually suffices to get practically all the roots.

It is essential that newly placed sod be thoroughly watered and tamped to estab-

lish its contact with the ground. It should be watered as necessary until it shows that it can take care of itself.

Home Games. If a house plot has sufficient area, the owner may wish some kind of a play court included in the landscaping plans. Tennis, badminton, and croquet are among the games in most common demand.

Tennis and badminton courts may have the standard dimensions shown in Figure 7-6 and 7-7 or may be smaller or of some different layout, to conform to restricted space, local custom, or individual preference.

Grass blends in best with general landscaping, but to give satisfactory results it must be well fertilized, protected from beetle grubs and burrowing animals, and frequently cut and rolled. Even with the best of care, it will wear bare in spots which have too frequent use.

Clay, and various special materials which give a similar surface, provide the preferred topping for tennis courts. However, they require excessive maintenance in weeding, scraping, lining, rolling, and add-

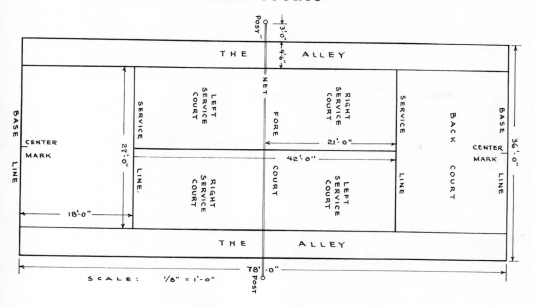

Fig. 7-7. Tennis court

ng of fresh material. Many of them cannot be played for some time after a rain.
This last characteristic will vary with the
type and condition of surface and efficiency
of subdrainage. Clay courts need more
maintenance if idle than if well played.

Blacktop has become the most popular
court surface. It requires a good, highway-
quality base, subdrainage, and extreme care
in shaping and finishing.

The playing surface should be nearly
level. If the topsoil is porous, and subdrainage is good, it may be possible to have it
exactly so. More often it is better to have
a slight slope, of ¼ to ½ percent to a side
or corner. Grades should be set by instrument. Particular care must be taken to roll
or tamp any fresh fills to avoid an uneven
surface after settlement.

Even when surface drainage is provided,
subdrainage is needed under clay or grass
to provide earlier usability after a rain.
However, too good drainage, such as
through a coarse gravel fill, might cause
burning out of lawn in dry summers unless
about eight inches of topsoil rich in humus
is placed.

Croquet is less rigid in its requirements.
Two layouts are shown. The first, Figure
7-8 is the professional type, which is supposed to be on practically level ground. The
other arrangement can be varied in distances to suit the available space and preferences of the players, and can be played
on sloping or irregular ground. In each
case, grass is the preferred surfacing.

A satisfactory court layout may be endangered by pulling the wickets to mow
the grass. Replacements are seldom in exactly the same place, unless markers are
used. The best markers are 2 x 2 redwood
or locust stakes about 18 inches long,
sharpened at the bottom and drilled at the
top with sockets which provide a snug fit
for the wicket wire. These are hammered
flush with the ground, and serve both as
guides in placing wickets and as braces to
hold them upright. Spools can be used in
the same manner, but are less satisfactory
and usually have to be replaced every year.

TREE PROTECTION AND REMOVAL

Trees are liable to destruction or damage
from various causes during construction

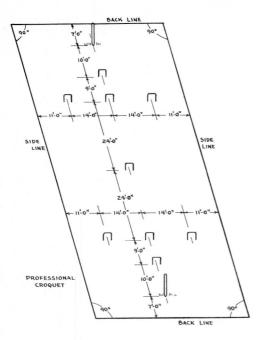

Fig. 7-8. Professional croquet court

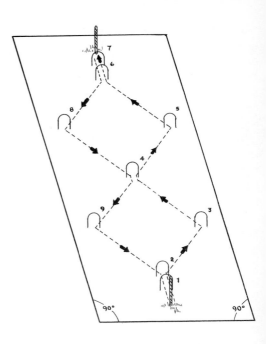

Fig. 7-9. Croquet court, home style

work. Trunks or branches may be broken or scraped by accidental contact with machines; roots may be dug away by ditching or lowering of grade, lessening the tree's ability to obtain food and water and rendering it more vulnerable to uprooting by wind; its trunk may die because of dirt piled around it, or its roots may be drowned or suffocated by placing of fill.

In general, the larger and more valuable trees are less subject to fatal damage from collisions, although the scars they do get heal more slowly; but they are much more likely to die from root cutting or suffocating than younger and more adaptable specimens.

Bark Damage. Trees can be partly or wholly protected from collisions by wrapping with burlap or other cloth, and tying thin wood strips around the trunks and any particularly exposed branches. If used for anchors in pulling out machinery, trunks must have very heavy padding and thick wood pieces between the bark and the chain, and the chain loop should be fas-

tened with a grab hook, bolt, or knot that will not slide. The choker effect obtained from round hooks or rings can readily crush bark and wood all around a tree so that it will be fatally injured.

If a tree is girdled by removing even a narrow ring of bark around the whole trunk, it is supposed to die. However, if the sapwood is not injured and the damage is kept shaded so that the wood will not dry out, young and vigorous trees may repair the cut by growing several inches of callus and new bark across the injury from top and bottom.

If the gap is wide, a skillful worker may be able to graft strips of bark across the injury. If circulation is established through these, they will serve to keep the tree alive and they will widen out so that the damage may heal over entirely.

Scars on trunks or branches should be promptly covered with black tree paint. This protects the tree from fungus and reduces the owner's annoyance about the damage.

Ditching. Ditching on one side of a tree ordinarily does not injure it severely. However, it is best to keep the cut as far from the trunk as possible, thus reducing the number of roots lost, minimizing the danger of tearing the trunk, and making the digging easier.

If a hoe or dozer digs within two trunk diameters of the tree, the roots should be uncovered, and then cut by hand to avoid danger of splitting the trunk while tearing them up.

A close cut weakens the tree's resistance against a wind that tends to tip it away from the ditch. If uprooting in that direction would cause it to fall on a building or across a highway, a tree expert should be consulted about the advisability of providing cable support, Figure 7-10, or removing some of the upper branches.

Burial. A tree's reaction to having its trunk buried varies with its species, health, and nature of the dirt. Burial is fatal to the majority if the fill is deep enough or of such a nature that it will smother the bark and support organisms which will destroy it.

The fill also changes the air and water content of the topsoil and subsoil around the roots. Such changes may damage or kill the roots directly, or indirectly by changing the nature of the soil population.

The best defense a tree can muster is to put out new roots near the surface. The willow does this automatically, but the majority of temperate zone trees do it with difficulty or not at all.

Trunk damage can be avoided by building a stone wall around the tree on the original ground, at a sufficient distance to allow free air circulation. See Figure 7-11. The space inside is called the well. Sometimes the fill is made and part of it dug away by hand to make space for the wall. Or the wall may be built first, and the dirt placed around it.

The first method is expensive and offers

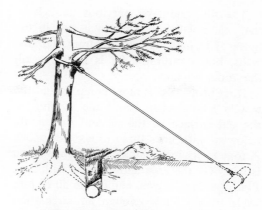

Fig. 7-10. Temporary tree brace

some danger of damaging the tree with the digging tools. The second is subject to the danger of knocking over the wall while placing fill, or accidentally spilling dirt over it that will fill up the well.

In general, the most satisfactory technique is to build the wall first, and fill the well with easily removed material such as stones, wood scrap, or crumpled newspaper. Such items will prevent any appreciable amount of dirt from entering the hole, and are easily taken out when grading is complete.

Sometimes pebbles or crushed stone collars are used to avoid unsightly or dangerous holes. These will usually allow sufficient air circulation when new but are likely to plug up with dirt.

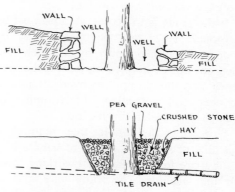

Fig. 7-11. Tree trunk protection in fill

The fill should be pervious enough so that water will not stand in it and in the holes. Tile may be laid to drain the tree wells, but a saturated fill is liable to kill the roots anyhow.

Root protection is more complex and the results less certain. Land tile is laid on the old surface of the ground or slightly below it, with lines three to six feet apart. A four to six inch blanket of crushed stone is laid over the area and covered with hay. Pipe openings at each end of the fill or into wells should allow enough air circulation to preserve favorable conditions long enough to enable the tree to adjust to the changing conditions. Wire mesh must be placed across openings to keep animals out.

If it is not economically feasible to take these precautions, the fill should be made of clean bank gravel or coarse sand. Trees may survive heavy additions of such open textured material.

Removal. Landscaping work may involve removal of trees. If they are to be destroyed the job resembles the land clearing described in another chapter, except that interference from buildings, wires, valuable trees, and other obstacles is much more common.

If the ground is to be filled, trees may be cut as nearly flush with the ground as possible. This may also be done if the grade is not to be changed and the presence of the stump is considered less objectionable than the cost of removing it.

If the grade is to be cut, stumps must be uprooted. Trees may be pushed over by a dozer or pulled down by a winch and then cut up, or may be sawed or chopped down and the stumps removed afterward. A deep soil cut allows undermining of stumps and easier digging out.

Uprooting the whole tree requires a machine of sufficient power to handle it, a clear space for it to fall, and working room for the machine. The trunk may split or be placed under such tension that it will be difficult to saw. Direction of fall can be rather closely controlled.

When a tree is cut it may be harder to control, the trunk should be intact and easier to saw, and the stump will be more difficult to remove, particularly if cut low.

A dozer, hoe, or winch can take out large stumps, if space permits. But the surest, neatest method is to hire a stump-chipping wheel for the job.

Such a machine can remove a stump of any size in a comparatively short time, reducing it to chips down below ground level. It must be operated with caution if rocks are present.

Damage to Property. Property may be damaged by falling trees. It is often wise to measure them before felling to see if they have clearance. This can be most conveniently done on a sunny day by setting a vertical stake, measuring it, its shadow, and the tree's shadow. See Figure 7-12. The proportion between the stick's shadow length and its height are used to figure the height of the tree from the length of its shadow. For example, if a four-foot stick casts a three-foot shadow and the tree shadow is sixty feet, the tree height is eighty feet and it will reach that distance when it is down.

A "false shadow" may be cast by a projecting branch so as to exaggerate the tree height. True readings can be taken only from a shadow of the top.

It is often necessary to take trees down in sections, starting at the top. This is a job for specialists.

When a tree or stump is uprooted near a curbing or other masonry, it is apt to pull it up or break it. This damage may be avoided by hand digging and cutting the roots on that side and pulling away from it, but it is often cheaper to break and then repair the structure.

Direction of Fall. Unless a tree is badly out of balance its direction of fall may be controlled by expert sawing and wedging, or by a line pulling it in the right direction.

The line should be fastened high, particularly if its power is weak. Hand pull is not effective for large trees.

Trucks or cars can be used for line pull, but the line may raise their wheels until they have a little traction. They may also stall at the critical moment. A way to use them is to put maximum tension on the line while the tree is still sound and block the wheels to prevent rolling back.

A taut line decreases the effort necessary to saw or cut on the far side of the tree, but makes splitting of the trunk more likely.

A tree pulled by a line will start its fall in that direction, but will usually fall rapidly enough to slacken the line, after which it might veer to either side.

Any machine used for pulling over a tree should be out of its reach. The line may be passed through a pulley block or around a grooved stump to permit the machine to be off to the side.

PLANTING

Selection of Trees and Shrubs. A new house standing entirely in the open often presents a bare and unattractive appearance which can be relieved by proper planting. Selection of suitable sizes and varieties may offer a knotty problem about which the grading contractor may be asked to offer suggestions.

If enough bushes and trees are placed to completely relieve the bare appearance, the house is liable to disappear into a jungle within a few years. This danger can be lessened by selecting plants which are slow growing or which respond well to suppressive pruning. Even if these precautions are taken, the householder should resign himself to the periodic necessity of cutting, transplanting, or selling some of his prized vegetation in order to prevent crowding and to preserve view and healthful conditions.

Cost of planting increases spectacularly with the size of the units. As an example, in

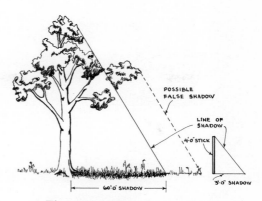

Fig. 7-12. Measuring tree height

one locality white pines five feet high can be purchased growing in fields for a dollar or two each, or in nurseries for five or ten. Anyone with a strong back, a sharp shovel, and the use of a truck or a station wagon can move them himself.

In the same area, white pines thirty feet high cost over fifty dollars in the ground, and transplanting prices range upward from five hundred dollars.

Occasionally a nursery or farm with a surplus of large size trees will sell them at bargain rates. Because of the high cost of handling them, they are still very expensive by the time they are set in place.

In most areas, a wide variety of trees and shrubs can be obtained and grown, and individual preference and price govern selection. Attention should be paid to the local prevalence of diseases or insects which may mar or kill the plant. Poplars and willows, which are desirable because of very rapid growth, should not be planted near tile drains as they have a disagreeable habit of filling them with roots.

If desirable plants are growing elsewhere on the property, or are available in the vicinity, the contractor may use his men and machinery in transplanting, by methods to be outlined later. However, he should not guarantee that they will survive as there are many factors of soil type, drainage, and pests that may kill them even

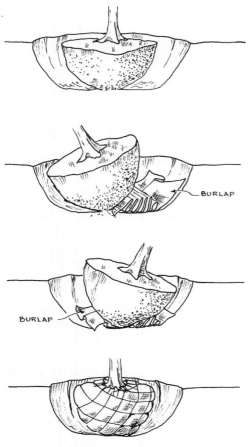

Fig. 7-13. Balling a tree

The very fine root hairs are vital to the life of a tree and they are quickly destroyed by drying. Either sun or wind can kill them in a few minutes, or sometimes even seconds of exposure. Transplanting is best done on cloudy days with little wind, or in rain or fog.

Somewhat larger trees are handled as in Figure 7-13. A trench is dug around the tree, about four to six inches out for each inch of trunk diameter. Roots are cut off flush with the inner wall of the trench. Below the roots, which are usually shallow, the ball is undercut until the tree can be pushed over. A piece of burlap is folded accordion-fashion and tucked under the raised side of the ball. The tree is then leaned the opposite way, the burlap is pulled through under the ball and used to wrap it. The cloth is tied in place with heavy twine or soft rope and is planted with the tree.

If the tree is too heavy to lift, it is dug around and undercut in the same manner. It is pulled over by a winch or block and fall. A square wood platform, with a ring or other fastening for a tow line, is placed under the raised side and pleated burlap over that.

The ball is then rocked back onto the platform and wrapped with the burlap. One side of the trench wall is beveled off, skids are placed, and a tow line fastened to the platform and the ball. They may be dragged to a nearby destination, or winched up on a truck or trailer for a longer move. Fast transportation should be avoided because of the "burning" effect of wind on the leaves.

Care must be taken to protect the bark against injury from attaching lines. The trunk or branch should be wrapped with burlap or other cloth, and flat sticks placed between that and the line. The line loop should not be a choker—that is, it should not tighten up under tension.

Branches which interfere with digging

if he handles them skillfully and which may not be obvious even to a specialist.

Transplanting. Transplanting trees is a specialty best left to nurserymen. However, they may not be available when required so that the grading contractor may have to do the work to get them out of the way.

Very small trees may be dug around with a spade and pried out. If the ball of earth attached to the roots hangs together, the tree may be carried to the planting space with one hand under the ball, the other on the stem. If the soil tends to fall away, it may be tied with burlap or shaken off and the roots covered with damp material until planted.

are tied up, or in toward the trunk, with soft rope.

In the new location dirt should be tamped firmly around the roots or ball and the ground thoroughly soaked. The dirt should come to the same height on the trunk as before, as the above-ground bark may rot if buried. If in doubt about the proper level, it is better to place the dirt too low rather than too high.

The transplant may require bracing to keep it upright. Lines, from three to six in number, are anchored to stakes driven firmly into the ground. They may be rope, wire, or cable. Chafing of the tree bark is prevented by wrapping it and passing the line through rubber hose at the point of contact.

If a tree is transplanted from a shady to a sunny spot, or suddenly exposed to sunlight by the removal of surrounding trees, the trunk and larger limbs should be wrapped to prevent sunburn or sunscald.

Trees and shrubs make most vigorous growth in a deep, rich topsoil. It is good practice to dig planting holes oversize to provide space for putting extra topsoil. This may be enriched by mixing with a little manure or fertilizer, but not enough to injure the roots.

If the planting is in a field where there is competition with other plants, the ground around it may be closely covered with scrap wood or stones. These will not interfere with the tree and will discourage other growth.

Transplanting by Machinery. Trees and shrubs of small to moderate size may often be dug efficiently by machinery.

A bulldozer can cut trenches around the tree and push it and the remaining block of earth onto a skid or stone boat, or along the ground to its destination. A shovel dozer can get the bucket floor under the block and lift and carry the tree. The trunk usually has to be tied to the top of the bucket, or held so that it will not fall for-

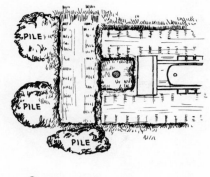

Fig. 7-14. Transplanting tree with shovel dozer

ward. See Figure 7-14 for this procedure.

Dipper shovels can often take out good sized trees unharmed by either trenching or direct digging. Draglines and hoes are limited to trees short or flexible enough not to be broken by the overhang of the bucket, and some scarring of the trunk is to be expected.

When conditions are favorable, trees can be machine dug at a fraction of the cost of hand work. Results may or may not be as satisfactory. Risk of damage is greater, except by the trenching method with a shovel dozer.

Planting holes can be dug rapidly by a clamshell or hoe, and less readily by other machines. If the dug soil is poor, it should be removed and replaced with topsoil.

Heeling In. It is often necessary to dig up small trees and shrubs before their new location is ready for them. They should

Fig. 7-15. Heeling in

then be planted temporarily, a process known as heeling in.

A trench is dug large enough to hold the roots and the plants are placed in it side by side or in groups. Enough dirt is shoveled or pushed over them to cover the roots, and the fill and nearby ground are thoroughly soaked. The stems may be upright, or leaned over sharply as in Figure 7-15. If the heeling in is to be for only a few days a shady spot is best. If for much longer, there should be little more shade than the species can tolerate in normal growth. The trench should not hold water for more than a few hours after a rain as the roots may die from lack of air.

Large trees with wrapped balls of dirt can be left standing on the surface of the ground if kept watered and not subjected to much sun and wind.

Delay between digging out and solid replanting always harms a tree. Heeling in and other precautions serve to keep this damage to a minimum, but should never be regarded as good substitutes for prompt and direct installation in a permanent location.

Transplants. A tree which has been transplanted before and has had a chance to recuperate can be moved again with less risk and damage. This is largely because the long roots cut with the first transplanting are replaced by new growth close to the tree, which are included in the ball or root mass that is moved with it the next time.

When small trees are purchased from a nursery for planting in fields where they will have to struggle against the native vegetation, transplants are usually a better investment than seedlings of the same size. Their much higher survival rate more than compensates for their higher cost. On the other hand, if they are to be set in beds where they will receive care, the seedlings may be the better buy.

Another factor is that of height. While a small plant usually sustains less shock and damage from transplanting than a large one, there may be a critical factor of overtopping. If grass and weeds can grow over a little tree so as to cut off its sunlight, in addition to competing with its roots for nourishment, it starts out with two strikes against it. If it is tall enough to overtop its immediate neighbors, its chances are much better, so that the tall seedling may make a better showing than the short transplant at about the same price. However, extra height which is not enough to get it to sunlight will not be of great value.

An exception to the above discussion will be found in trees such as the beech and dogwood which thrive in shade. However, these are not suitable for open field planting.

Pruning. Some roots are always lost in planting from cutting off, scarring, or drying. The remaining roots may not be adequate to supply water to the tree, and their ability to do so is further lessened by the disturbance of capillary water movement by the planting process. Shortage of sap may cause the tree to die, or may weaken it so that growth will be poor and it will be very vulnerable to damage from disease or weather.

If enough branches are cut to restore a

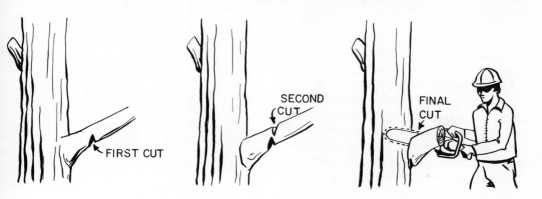

Fig. 7-16. Cutting large limb

balance between the top and the diminished root capacity, damage will be less likely. The type of pruning depends on the variety of plant, and its extent is determined by the amount of root loss, the amount of watering it will receive, and whether it is in shade or sunlight.

Some plants will go into shock if pruned drastically. Others will be unfavorably affected as to ability to produce blossoms or fruit for the next several years, or will lose the pruned branches because of inability to start new buds in the stubs. Because of the great number of plants and regional differences in their behavior, exact information cannot be given here.

A young tree which is meant to grow into a tree shape should almost never have its central or top shoot cut. A new top will grow from one or more side shoots, but the trunk will develop a permanent kink or a weak fork below the cut.

Deciduous (broad leaf) trees and shrubs are usually pruned at least in part by cutting back some or all of the branches. A very drastic pruning that will cause death by shock to some species involves cutting all limbs back to bare stubs. Less severe pruning may consist of trimming the ends which project too far, or removing whole limbs which are crowded or undesirable.

The conifers—pines, spruces, hemlocks, etc.—usually die if they are cut back to bare stubs, and many of them cannot stand any general pruning. Leafage can be re-duced by taking out entire branches in places where others will fill in the gaps, or by removing all the lower branches for several feet above the ground.

Care must be taken in removing heavy limbs not to allow them to tear the bark below the cut. The three steps shown in Figure 7-16 should be followed. The final cut should be as near to flush with the trunk as possible, and the wound should be covered promptly with tree paint.

Watering. Watering a freshly transplanted tree or shrub serves not only to supply it in such quantity that even injured roots can take up enough, but also settles the ground so that it will be better able to supply them with natural soil moisture afterward.

In dry seasons watering should be continued for days or weeks, depending on soil character and moisture, and the degree of adaption the tree makes to its new environment. Watering may have to be resumed if extreme dry spells occur within a year or two of the planting.

The tree should be set in a slight hollow, or a dike should be built around it to retain water. The hollow is preferable as it will catch rain also.

Subdivision Planting. In real estate developments, street grading often precedes the selling of the lots by some years. The appearance and value of the development can usually be enhanced by planting of shade trees immediately after road work is completed, so that they can have maximum time to grow.

Unfortunately, trees planted in fields or freshly graded land usually do not prosper unless cared for, and neglect may cause the loss of most or all of them. However, there are planting procedures which will enable them to fend for themselves quite well.

The trees should not be taken from dense woods or heavy shade, as sudden exposure to full sunlight is likely to cause burning and cracking of the bark. They should have adequate roots with wrapped balls of dirt. If they are not wrapped, they must be moved promptly from their original location to the new one, with roots protected from sun and wind.

The tree locations are indicated by stakes or other markers, set by instrument, measurement, or eye.

A tractor backhoe is used to dig the holes. Under most conditions, a small size will be adequate. The hole is dug oversize in both depth and area for the tree to be planted, to allow for errors in calculation and for placing topsoil. The dirt is loaded into a truck and hauled away.

If the trees to be planted do not have balls of dirt, the holes are refilled with good topsoil. This can be brought by truck, and perhaps picked out by the excavator, which can place it directly in the holes. The soil in the truck bottom which cannot be scraped up may be dumped at a hole, and hand shoveled in or left in the truck when it goes for another load.

If the trees are balled, sufficient topsoil should be placed in or beside the hole to set the tree properly.

Dirt is tamped around the tree and water supplied in the conventional manner. If there is any question about the quality of the topsoil, a moderate amount of manure or general fertilizer should be mixed in the surface before watering.

The ground for several feet around the tree should then be covered with material

Fig. 7-17. Tree protection against sod

that will prevent grass and weeds from competing with the tree for nourishment. When available, flat stones laid so as to overlap each other as in Figure 7-17 are the most effective cover. Scrap shingles or other flat pieces of wood are a good second choice. If neither are available any mulch or ground cover such as wood chips, hay, or leaves may be used. These last materials need to be placed in a layer four to six inches thick, and may require renewal after a few months or a year.

It is a good precaution to guy trees against blowing over, but it may not be necessary if they are small or have heavy roots.

Trees planted in this manner are supplied with adequate nourishment, and are protected against the competition of vigorous sod.

DRIVEWAYS

Most home landscaping involves planning of a driveway. It may be a straight connection to a street a few feet away, or a long roadway involving considerable problems.

Short straight drives can be as narrow as eight feet for use by passenger cars only, but twelve feet is more comfortable. A long drive, one in a slot between walls, or any drive to be used by trucks, should be at least twelve feet wide.

Curves should be one to three feet wider than straightaways, the sharper turns requiring the greater width.

The entrance from the street should be thirty feet wide at the curb, in order to permit turning into it from the near side of the street. An effort should be made to avoid entering the road through a deep cut, between large trees, or in or very near a curve, as any of these features add to the danger of accident.

Sidehills. A long driveway of the farm or estate type may have to cross a hill slope. It is notched into it in the same way as a pioneer road. However, since it is a permanent improvement which should have a pleasant appearance, special procedures are followed.

Such a drive serves not only as an automobile route but also as a drainageway. Unless diversion ditches are made above it, whatever water flows down the slope will land in the driveway cut, and flow down it or across it. Unless an ample channel is provided, the drive may often resemble a stream bed.

A driveway crossing a hillside below a long slope should be 14 to 16 feet in width, including drive, shoulder, and gutter, but not the slopes. The gutter should be at least three feet wide and deep enough to carry all the water. It can be relieved at frequent intervals by diagonal cross drains to the lower slope. The drive cross-section may be crowned or sloped oppositely to the hill, but should never slope with it. See Figures 7-18 and 8-1.

The cheapest gutter is sod and it may hold on quite steep slopes if well established. Temporary diversion ditches can be made with a plow to keep much of the water off until it is well established. Stone, concrete, and blacktop are water resistant, but are subject to frost heaving unless on a stone or gravel base.

The slopes of the cut and fill should be topsoiled and seeded. They are often too steep for mowing.

Garage Level. Driveways offer minimum trouble in use and maintenance if they are nearly level. Unnecessary expense and inconvenience may be caused by placing the garage so that grades are created or exaggerated.

If the driveway is long, the garage should be at about the grade of the ground around its entrance after landscaping. If the drive is short, the garage should be at about street level.

It is unusual for a garage to be higher than the grade around it, but it is very common practice to place it under the house at cellar level. In this case the driveway often must enter it through a deep cut bordered by steep slopes or retaining walls, and usually descends more or less steeply as well.

A descending drive must level off several feet outside of the doors, and must not do it so abruptly as to cause a car's bumpers to scrape when entering or leaving. Extra width is needed between walls. One or more grating drains should be provided outside the doors, with plenty of capacity. Drainage from a long drive or from the lawn should be diverted to other drains or

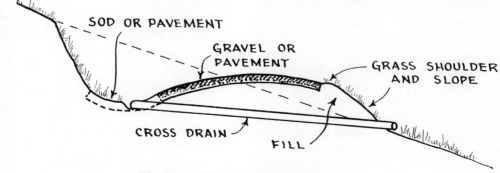

SOD OR PAVEMENT

GRAVEL OR PAVEMENT

GRASS SHOULDER AND SLOPE

CROSS DRAIN

FILL

Fig. 7-18. Hill slope driveway section

7-23

channels. Failure to observe both of these precautions may result in a flooded garage or cellar. It is also necessary to keep the area clear of leaves and trash that might block the drain.

If the drive is short, its slope will be determined by the difference in level between the garage and the street. Driveways as steep as 30 percent grade are in use, but they are both difficult and dangerous.

Any steep drive requires care in designing the vertical curves at each end so that the center of the car will not scrape on a convex curve, nor its bumpers hit on a concave one. There should be a parking place which is moderately level if possible.

In climates where freezing weather occurs, steep grades may become dangerous or impassable. If there should be a choice, a grade up to the garage is to be preferred to one leading down to it. In the former case, it is usually possible to reach the street, and the importance of getting somewhere is usually greater than that of putting the car in the garage rather than leaving it at the curb. Also, drifting snow may entirely fill a cut to a low garage, from which it will probably have to be dug rather than plowed.

Snow Melting. Heating pipes can be installed under driveways, walks, and outside stairways to prevent snow and ice from resting on them. Cost of such an installation is now between one dollar and three dollars a square foot of surface, and under many circumstances it is a worthwhile investment. Operating expense is said to be only a fraction of the cost of shoveling or plowing.

The preferred method is to lay wrought iron pipe or copper tubing in the pavement slab or immediately below it, and circulate an antifreeze solution heated by a heat exchanger in the house steam or hot water boiler. It can be turned on and off by hand, or by automatic controls operated by the weight of the snow, or by its interference with a light beam reflected off a polished surface to an electric eye.

The system must have ample capacity, or will occasionally do more harm than good. If it does not quite keep up with the snowfall, at the end of the storm it may leave the area covered by a layer of slush, which might then be frozen by an extreme drop in temperature and kept frozen until the weather warmed slightly.

Electrically heated wires, available in hardware stores, may be placed on ice for emergency melting. Covering with cloth or paper increases effectiveness.

Turnarounds. A driveway which does not include a turning place requires that a car be backed out of it or into it. This is entirely impractical on long or curving drives, and is a nuisance and a danger in any case. Wherever lot size permits, a turnaround should be provided.

The best way to lay one out is to have the people who are to use it make some trial turns, add a few feet to the space they require to allow for carelessness or a bigger car, and build the drive accordingly. If there is no opportunity to practice, any of the layouts shown in Figure 7-19 should prove satisfactory.

Allowance should always be made for car overhang. This may be two feet front, up to four feet rear, and eight inches at the sides. It is desirable to have a curbing that will keep the wheels about a foot away from vertical walls to protect the car from scraping at the side.

Extra space may be provided in a turnaround for parking, or to supply peace of mind to uncertain drivers.

Surfacing. Four inches of good bank gravel, crushed rock, shell, or similar materials should be stable enough for a house drive on well drained soil. Under average conditions six inches is safer, and when the ground is soft and wet, eight inches to a foot or more may be required.

A stone fill underneath can be used to

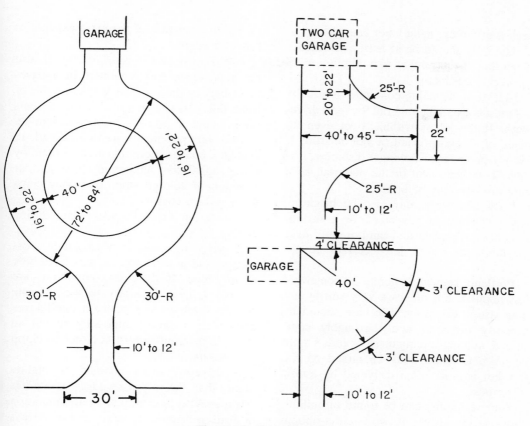

Fig. 7-19. Turnarounds

reduce gravel requirement. Any flat stones near the surface should be set on edge so that they will not rock and disturb the top dressing.

If the driveway is long, it may pay to try to get by with a minimum depth and add more material to any soft spots as they develop. However, it is often necessary to dig away the softened gravel, as it mixes with mud underneath. If the driveway is short or the budget liberal it is good practice to put down a safe depth in the first place.

These materials may be used for both the bulk and the surface of the drive, may be given a surface treatment, or serve only as a base course.

If used alone, varying amounts of difficulty may be found with loose stones, gullies, dust, tracking small particles into houses and cars, muddy surfaces, ruts or mudholes, or scattering on the grass, depending on the kind and quality of material used and the circumstances.

Calcium chloride, either scattered on the surface or mixed in, will prevent the drive from becoming dusty, and will help to hold it against washing and scattering. It should not be used in the immediate vicinity of the house where it might do damage if tracked in.

A thin layer of loose pebbles or fine crushed rock makes an attractive surface for light and slow-moving traffic, but scatters under fast traffic. Snowplows are likely to move a large part of it to the lawn.

Pavement. Any of the road pavements or surfaces discussed in the next chapter may be used in a driveway, to provide a long-lasting hard surface. Principles of placement are similar, but because of smaller scale work, more hand labor is involved.

The base should be at least a few inches wider than the pavement, to minimize cracking under edge loads.

Blacktop (asphaltic concrete) is usually the most convenient material for short driveways. If the job is within 10 or even 20 miles of a mixing plant, a hot mixture is usually preferred. For longer distances, it is difficult to keep heat in the material, so a cold mixture may be better.

Long driveways may also be blacktop, but a penetration pavement, or a surface treatment of asphalt and sand or chips over oil-stabilized gravel, have certain advantages.

Portland cement concrete is somewhat less popular, but makes a very sturdy and long lasting driveway. The base should be properly prepared and thoroughly compacted, and reinforcing mesh or rods should be used. Concrete may be mixed on the job, but is more often delivered by transit mix truck.

Some economy can be found in building short straight drives of two parallel ribbons of pavement instead of a full width slab. The saving is not proportionate to the reduced area because of the extra expense of forms or edging.

Asphalt dissolves in gasoline and oil, and the appearance and usefulness of a pavement may be damaged by drips or spills of these fluids. It can be protected by a coating of a coal tar preparation such as Jennite.

Driveway pavements are often tinted. In asphaltic construction, this may be done by using asphalt-like binders in color, and/or selection of colored stone or plastic chips for a surface coat.

Concrete is tinted by adding dye during mixing.

A paved driveway is often flat in cross section, and in level areas it may have little or no slope. If these two conditions are combined, it is necessary to use extreme care in finishing to produce a puddle-free surface. It is almost impossible to put an invisible patch afterward in either blacktop or concrete.

A driveway surface should be slightly higher than the soil at its edges. Lawn has a tendency to rise because of accumulation of dead or trapped material, and driveways often sink. Either change can lead to an unsatisfactory canal or pond appearance.

AGRICULTURAL GRADING

Need for Grading. It is often necessary or desirable to regrade land in order to use it for farming. In arid regions, land is leveled to permit even distribution of irrigation water. In semi-arid climates, sloping land may be terraced to hold rainfall behind dikes so that it will soak into the ground instead of flowing off.

Where the rainfall is adequate or excessive, terracing may be necessary to reduce washing of soil from cultivated slopes. Under any conditions of climate or soil, leveling may be desirable to allow use of large or high speed machinery. Alone or in conjunction with underdrainage it may increase yields by eliminating burning out of crops on ridges and drowning in hollows.

Agricultural grading differs from other types of earth moving in the large areas to be treated in proportion to the money available, the flexibility in engineering requirements to suit conditions and cost factors, and in problems relating to the handling of topsoil.

Cuts and fills are typically shallow, vertical movement of soil is slight, and horizontal movement is relatively great.

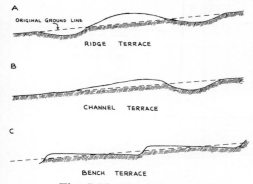

Fig. 7-20. Terrace types

TERRACING

Terracing land is the grading process of interrupting slopes with ridges, channels, or benches, or combinations of them, in order to slow or stop the flow of rainwater, and to prevent harmful soil erosion.

Terracing may serve to hold water on the slope so that it will soak into the ground; allow water to flow off it while keeping the loss of soil to a minimum, or to reduce slopes so as to make them more readily workable or irrigable.

Terrace Types. Three principal types of terrace are used. Each is constructed along level or contour lines. The ridge terrace, Figure 7-20 (A), is a ridge built of soil obtained from both sides. The channel terrace, (B), is a ridge constructed of dirt from the upper side only, and the channel formed by this excavation is an essential part of the structure. The bench terrace, (D), has a stair structure with steep risers separating relatively flat cultivated areas.

Ridge and channel terraces are usually built with sufficiently gentle slopes to allow farm machinery to work along or across them. Best results are obtained if farming operations are done parallel with their center lines.

Ridge Terraces. The ridge or absorptive-type terrace is used primarily to conserve water in regions of deficient rainfall. Each ridge serves as a dam for a pond, which is deepest in the excavated area immediately

above it. Water may also be impounded in the trough formed below this ridge by borrow of material. See Figure 7-21.

A larger area and quantity of water can be held on slight gradients .than on steep ones, by any one size of ridge. Not only is more water retained per yard of dirt used in the ridge, but its distribution over the land is more uniform.

Too great a depth of water may drown out crops immediately above the ridge.

It is ordinarily not economical to construct terraces for water conservation alone on slopes over 3 percent, and structures for reducing soil erosion are more often of the channel or intermediate types.

Overflow channels may be provided to carry off rain in excess of that for which the system was designed. These should be protected like channel terrace spillways.

Channel Terraces. Channel or drainage-type terraces are essentially shallow diversion ditches which catch water flowing down a hill and lead it off to drainageways that have been protected against erosion.

The channel depends on the ridge of excavated material for much of its capacity. Its grade is flat, or nearly so, so that only extremely fine soil particles can be carried by the water it discharges.

Bench Terraces. The principal application of bench terraces in the United States is in connection with irrigation. If the original slope of the land is greater than that of the graded fields, each field will consti-

Courtesy U. S. Department of Agriculture

Fig. 7-21. Ridge terraces after rain

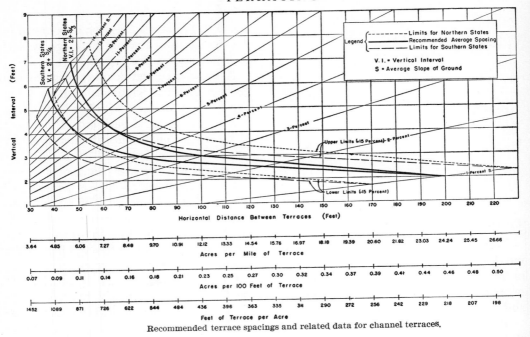

Recommended terrace spacings and related data for channel terraces.

Courtesy U. S. Department of Agriculture

Fig. 7-22. Channel terrace data

tute a terrace, separated from fields above and below by comparatively steep slopes.

Benches may also be made in steep cultivated land by leaving narrow contour strips in grass or other permanent vegetation. Soil washing from the wider strips between will be caught by the grass and will tend to build up the low side of the cultivated piece, while its top is lowered by erosion. This process, often accelerated by plowing so as to throw dirt downhill, will ultimately result in gentle slopes separated by steep banks.

This work is ordinarily done by farmers without assistance from contractors.

Surveying. A terrace system must be carefully surveyed and planned before construction starts.

The interval between terraces may be taken from the chart in Figure 7-22, or better, determined after conference with soil conservation specialists.

Stakes are placed from top to bottom of the field at the selected intervals. From each

of these a level line is run the full width of the field or area to be processed. These lines, known as contours or contour lines, will bend toward the high side of the slope in hollows, and away from it on ridges.

Each level line may be found by setting a level transit or laser at a point, then measuring its height above the ground. A marker or laser receptor is set at that height on a rod.

At various distances, the rod is moved

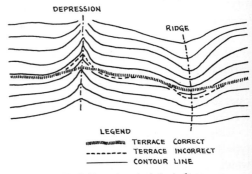

Courtesy U. S. Department of Agriculture

Fig. 7-23. Terrace line correction

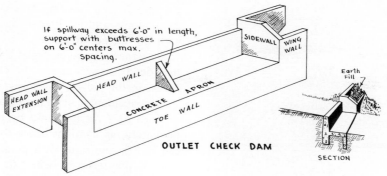

If spillway exceeds 6'-0" in length, support with buttresses on 6'-0" centers max. spacing.

SIDEWALL WING WALL

HEAD WALL

APRON

HEAD WALL EXTENSION

CONCRETE

TOE WALL

OUTLET CHECK DAM

Earth Fill

SECTION

Courtesy U. S. Department of Agriculture

Fig. 7-24. Check dam

up and down the slope until the marker is at instrument height when the rod is on the ground. A marking stake is placed.

A series of such readings indicates a level line. Re-checking is advisable.

Each line indicates the location of a terrace. However, sharp angles and extremely irregular lines are not desirable and can often be reduced or eliminated by minor adjustments, as in Figure 7-23. Farming is simplified when adjoining terraces can be made parallel.

If an angle in a gully is eliminated by moving the line downhill, the terrace ridge will have to be built higher above ground level to preserve its grade, and water will be ponded behind it.

If the line is moved uphill to cut off a bend on a ridge, the channel will have to be dug deeper to allow flow of water through it. However, such a ridge may be used as a divide or drainage head, in which case little water will be present.

Stakes are ordinarily used only for location guides, but may be marked with grades where the terrace is to be higher or the channel deeper than standard.

It is desirable that the top of the terrace system be also the top of the drainage area. The top terrace should serve the same width of ground as those below it. If a larger area must be served because of flow from higher fields, the channel capacity

must be increased proportionally, or some other type of intercepting drain used.

Grades. Ridge terraces usually have a level grade.

Channel terraces may be level for the section most distant from the outlet, and slope increasingly to about a ½ percent grade at the outlet.

Drainage is normally from ridges toward hollows.

Short terraces require less maintenance than long ones.

Outlets. The discharge from a terrace should be into a waterway that is capable of carrying the water directly down the slope, without eroding. Shallow depressions carrying a permanent sod are often satisfactory. These should have enough drop from side to center to insure gathering of all water discharged from the channels, but should not concentrate enough flow at the center to cause erosive velocities.

The strength of the sod, as affected by fertilizing and grazing, and the condition of the soil will determine the maximum safe gradient for a meadow outlet. This is ordinarily six percent or less.

Sod should extend several feet above flow lines on the sides of the waterway. Steeper or narrower channels may be protected by ungrazed and uncut growth of grass, weeds, bushes, or trees.

For extreme conditions, a channel pro-

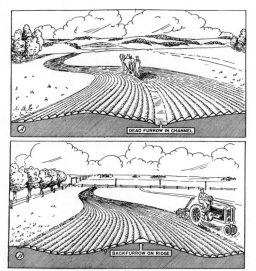

Courtesy U. S. Department of Agriculture

Fig. 7-25. Plowing to preserve terrace

tected with check dams (Figure 7-24), pavement, or other artificial structures may be required. Steep banks require sod or artificial flumes for the terrace discharges.

Permission of owners of land below the farm must be obtained if the terrace system alters the path or concentration of water on their properties.

Outlets MUST be completed and protected before terraces are built.

Construction. Terrace construction is primarily a matter of sidecasting. The work is commonly done with graders of either the powered or towed types. Bulldozers working at right angles to the terrace are also effective, particularly in the channel type.

Belt loaders, of both standard and special terracing models, give excellent results.

If the channel or cut depth is greater than that of the topsoil, a barren strip will be left which will yield poor crops. This damage may be reduced by overcutting the channel so as to leave it below grade, and blading some topsoil from above to cover it.

Maintenance. Terraces will serve their

purpose best if plowing and cultivating is done along contour lines—that is, parallel with the terrace lines.

A terrace can be enlarged by plowing so that dead furrows are in the channels and lands on the ridges, as in Figure 7-25.

Any accidental blocking of a channel or damage to a ridge should be repaired immediately, as it might cause the terrace to fail in a heavy rain, with possible destruction of the terraces below it.

GULLIES

Characteristics. So far as this discussion is concerned, a gully is a drainage channel that has become so deepened or enlarged that its banks are unstable, and tend to extend destructively into surrounding land.

Control of gullies is largely an agricultural problem, but may also be required to protect highways or structures.

Gullies are a sign of the beginning of a new cycle of erosion which tends to dissect smooth slopes or high levels of ground into table-lands separated by steep walled channels or canyons. Unless controlled, it will eventually narrow such tables into peaks. Geologically, they are small examples of the type of stream erosion which carves rising land into mountains.

The new erosion cycle may be started by land rising so that steepening channels add to the velocity of flowing water; by lowering the outlet of a stream with the same result, or by reducing the resistance to erosion of the land.

Gullies are caused most frequently by the destruction of the vegetation which protects the land surface, although they may also be started by lowering of outlets, due to highway or stream channel work, or land slips.

The majority of gullies contain intermittent streams which flow only during or immediately after rains, or in wet seasons. Permanent streams are less often affected as their beds are unsuitable for agriculture. These are most apt to be raised and choked

Courtesy U. S. Department of Agriculture

Fig. 7-26. Active gully

by silt deposits resulting from bad farming.

Growth. When a slope is covered with vegetation, whether sod, bushes, or forest, rain water tends to move downward as a flowing sheet, and is only gradually gathered in definite drainageways. Its eroding action on the ground is slight as it is held from contact with the dirt and its velocity is lowered by stems and roots that form a protective mat.

If the vegetation is removed by plowing, disking, close grazing, or fire, the water comes in direct contact with the soil and tends to remove the surface particles. This effect is usually rather uniformly distributed at first, and is called sheet erosion. It can be reduced to slight proportions by proper farming, including contour plowing and cultivating, and terracing or return to sod when necessary.

Erosion is most active where the amount or velocity of water is greatest, or the soil is least resistant. Such places tend to wash out more than the surrounding area, and then, being lower as a result, will catch the runoff from a larger area, increasing the quantity and velocity of water, and its

eroding effect. The deepening of the channel therefore tends to build up forces which will make it deepen more rapidly.

In its early stages such a gully may be destroyed by plowing or harrowing, so that it is choked by clods and some of its water is diverted elsewhere. However, unless close growing vegetation is planted, or weeds allowed to grow, new storms will re-form the channel or create new ones nearby, and they may eventually become too deep to be choked by plowing or even to be crossed by a plow.

Once a gully is formed it enlarges by three separate processes. One of these is channel erosion—the scouring action of the water deepening the bottom. This is accompanied by the falling in of the sides as they are undermined.

The upper ends (heads) of gullies advance into the land by waterfall erosion, both along the main drainage line and branches which are acquired. Subsoil is often less resistant to erosion than topsoil. Water pouring into the gully will cut it into steep banks, undermining the topsoil and causing it to fall. The impact of the

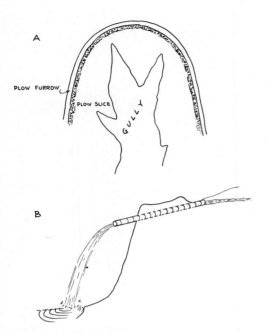

Fig. 7-27. Diversion of water from gully head

waterfall on the bottom gouges holes which accelerate channel erosion.

Waterfall erosion usually produces a gully with a U cross-section. It becomes less active as it approaches the head of drainage and the quantity of water is reduced.

If the subsoil is equal or superior to the topsoil in resistance, waterfalls will not develop, but the gully can progress by extensions of channel erosion.

The third major factor in extending gullies is sloughing off of soil due to alternate freezing and thawing, or following saturation by heavy rains. This process is most active on southern and eastern slopes, and will eat through a field with little regard to slopes or drainage lines.

Continued progress of either waterfall or sloughing erosion depends on sufficiently active channel erosion to remove the loosened dirt.

Once well established, a gully will continue to enlarge even if the surrounding land is planted in erosion resistant vegetation, as the head and side slopes will undermine the surface.

A gully may advance downstream by channel erosion. More often, it deposits debris in a delta fan at or near in its mouth, so that the land is built up. This process is destructive also as it buries topsoil under subsoil.

Damage. The damage from gullies includes actual destruction of farm land, cutting up fields so that they cannot be worked economically, lowering the water table so that crops dry up, undermining buildings, roads, and bridges, burial of lower lands under barren subsoil, and choking of streams with silt.

An individual gully can do damage amounting to many thousands of dollars, and the national loss from them is in the hundreds of millions. Their control is therefore of great importance.

CONTROL

Control measures after gullying has started may include diverting water to other drainageways, planting, breaking down walls, building check dams, and proper use of the affected land.

Diversion. Water entering the gully can sometimes be diverted by plowing an arc around its head, as in Figure 7-27 (A). The slice should be turned toward the gully to make a dam to back up the furrow.

More often it is necessary to build dams or to dig ditches. Dams are safer, as a new ditch or even a plow furrow may start a new gully, unless watched and controlled.

Where diversion is not practical, waterfall erosion may be checked by conducting the water into an overhanging pipe or flume, as in (B).

If water can be permanently diverted from a gully, the walls will eventually break down into stable slopes, which can be covered and held by vegetation.

Planting. Gully growth may be checked by planting. Small ones may be held by pasture plants or vines; larger ones require shrubs and trees.

The plants should be of types that can grow well in poor soil, have extensive root systems, and will not be injured by partial burial. If the ground is damp most of the year, willows are probably the most efficient, particularly as they will grow from poles or logs secured horizontally in the floor across the direction of flow, thus providing mechanical control while roots and stems are sprouting.

Black locust grows vigorously in poor soil, has a widespread root system, and yields a crop of fence post material. Many of the pines do well in barren soil but their roots are not as strong.

Vines such as kudzu and honeysuckle are used successfully.

Whatever plants are chosen, enough soil should be loosened to give them a good grip. Whenever possible, fertilizer, manure, or topsoil pockets should be provided to give them a good start. Animals should be fenced out of the area.

It is usually not practical to plant very steep walls. When vegetation is growing well on the bottom, soil caving from the walls will be held so that the slopes will gradually become less abrupt, and will allow growth of self-seeded plants.

It is often necessary to divert water or to reduce its velocity in the channel before planting can be successful.

Breaking Down Walls. The healing of a gully scar can be greatly accelerated by breaking down the walls into slopes gentle enough to support vegetation. When practical, it is desirable to have the slopes such that farm machinery can cross them in any direction.

Very small gullies can be broken down with plows and other farm implements. Somewhat larger ones can be reduced by graders or angle dozers working parallel with the edge. Big ones require dozers pushing the bank more or less straight into the

gully, or a dragline pulling it from below.

Gullies may be of such large size— depths of fifty feet or more—that it is not economically practical to grade them in even with the largest machinery.

The new slopes produced are planted in somewhat the same manner as road banks.

Check Dams. In order to obtain permanent control, it is usually necessary to slow, stop, or reverse the process of channel erosion. An actively cutting channel will steepen and undermine its banks and the new grading or planting they support.

A channel interrupted by dams or other obstacles will tend to silt up to a higher gradient, reducing the height of its walls and encouraging plant growth.

If flow is only occasional, close plantings of bushes in the channel and on its banks may be sufficient. These are often planted in lines across the gully, and protected against washing out by wire stretched across deeply driven posts.

Sod, combinations of cut brush and wire, or stone or logs, can be used in building check dams. Some constructions are shown in Figure 7-28.

In general, any structure which water cannot get under or around, which is tight enough to prevent dirt from going through it, and is strong enough not to be washed out, will serve as a check dam.

IRRIGATION SYSTEMS

The design of an irrigation system is an engineering problem out of the field of this book. It is in order, however, to mention briefly some of the factors.

Wells. Water may be obtained from wells, occasionally by natural flow but more often by pumping. Such water, if used in the immediate vicinity, involves minimum piping and distribution difficulties. The most serious problem is the likelihood of using too much water, so that wells must be deepened or new ones drilled to keep in touch with the falling water table. In many

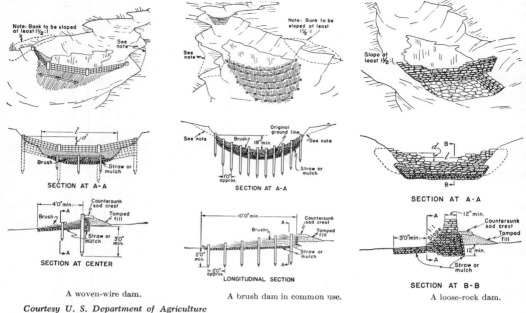

A woven-wire dam.

A brush dam in common use.

A loose-rock dam.

Courtesy U. S. Department of Agriculture

Fig. 7-28. Gully check dams

localities such over-pumping will eventually reach salty or alkaline water unfit for agriculture.

River water may be pumped up to the land to be benefited, but more often flows onto the land from a higher point in the river. A ditch or canal is cut from the stream, and run horizontally or with a very slight gradient along the side of the valley. The land between the canal and the river is irrigated by gravity flow from the canal into ditches, and from them onto the fields.

Most rivers have a wide variation in level so that such a canal might be subject to being left dry at times and flooded out at others. A more constant level can be obtained by damming the stream below the canal entrance to keep the water high enough to enter it, and by providing gates and a protective dike to prevent flooding. A dam and control gates above the inlet will permit regulation of the river level and inflow.

In dry climates streams may flow only during the winter and spring or in occasional floods. Water from such a source may be useful for crops which will mature in the wet season or as a supplement to water obtained from wells, but the cost of handling it may not be justified.

Reservoirs. Such streams may be utilized by building one or more impounding reservoirs upstream. These will permit storing the heavy flow of some months to be released in drier periods. Large reservoirs of this type may be ideal sites for hydroelectric plants, as the irrigation water can be used to turn turbines on its way through the dam.

Most of the large irrigation systems of the United States depend on storage reservoirs.

Lakes may be used as sources of water. It may be pumped up onto adjoining land or conducted through siphons or canals to lower land. Occasionally, a lake may be tapped by means of a tunnel and used for stream regulation in the same way as an artificial reservoir.

Sediment. Direct river flow into an irrigation canal carries varying quantities of sediment with the water. The flat gradients and slow motion of the water in the system will cause extensive silting, which will ultimately reduce water capacity, may cause continuous trouble in operation of gates and other devices, and will clog pipes. The design should take this quality of the water into account. Very dirty water requires oversize waterways, steepened gradients, particularly in pipes; cleanout traps or flushing devices for pipes, and access to all ditches for cleaning by machinery or by hand.

Upstream reservoirs catch most, or sometimes all, of the nonsoluble sediment. Any residue may be still further reduced by settling basins.

Dirty water has certain advantages. In most localities it is good for the land, carrying topsoil and minerals which help to replace natural losses. It also tends to seal leaks in canals and ditches so that wastage by seepage is reduced.

Its disadvantage is the greatly increased cleaning and maintenance work required, which is so important a problem that most irrigation engineers will secure clean water whenever possible.

Canals. The main canal may be an unlined ditch or a ditch lined with special impervious soils or a concrete or other artificial lining. If it is carried well below natural grade at any point, it is customary to use a lined tunnel or concrete pipe at such a section.

The unlined ditch is the most economical to construct, and may be used where soils are impervious, where the water carries enough sediment to seal leaks, or where the supply of water is so large that seepage losses are less important than the expense of preventing them.

A ditch may be lined, either in sections or as a whole, by placing a layer of clay or impervious silt on the bottom and sides.

Courtesy U. S. Department of Agriculture

Fig. 7-29. Concrete lining in irrigation canal

This may be placed in a dry ditch, or fine clay may be added to the irrigation water until it has coated the canal.

A ditch must have bank slopes which will not cave or slump into the water. Vegetation may be used to hold dirt banks and will permit steeper slopes, but its consumption of water is often excessive. Banks should be protected from erosion by surface water.

Concrete lining is expensive construction, and if properly done is most satisfactory. Such leaks as occur will generally be localized and comparatively easy to find and repair.

Damage from leaking canals is not confined to the water loss and the breaking down of the canal structure. Farm land near the leaks may be rendered unusable by excessive water and resulting alkali deposits, unless subdrainage is provided.

Pipes. Either ditches or pipes may be used to distribute the water from the canal. Ditches are more economical to construct, but cut up the land, are expensive to main-

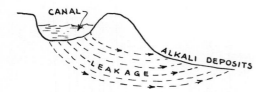

Fig. 7-30. Damage to land from canal seepage

tain, form breeding spots for insect pests, are a hazard for children, and may allow substantial water losses by seepage and evaporation.

Pipes may be concrete, glazed tile, iron or steel, or composition. The first two sometimes separate at the joints when filled with cold water, either after lying empty or being used to carry warm water. Iron pipes may corrode rapidly due to alkalies in either the soil or the water.

Various methods have been developed of joining concrete and of treating iron pipes, which reduce the difficulties mentioned. On the whole, pipes require less maintenance and waste much less water than ditches. However, they must be laid on steeper grades to prevent silting. If they become blocked, they are very expensive to clean.

Large farms or ranches may control an entire irrigation system, but water is usually distributed to a number of users by a water

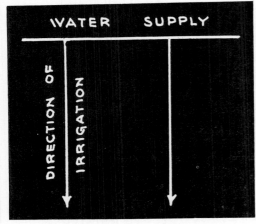

Fig. 7-31. Direction of irrigation

district or other government agency, or by a water company. One water gate or weir is usually provided for each land unit, and distribution within that unit is cared for by the farmer himself.

This may be handled by running a pipe line along one edge of the field, with standpipes and valves at ten to twenty foot intervals. If the field is long several distribution lines may be used, connected by a main.

Ditches may be used instead of pipes but they require more maintenance and may interfere with tillage.

The "direction of irrigation" is the direction the water flows on the surface of the field from the distribution pipe or ditch. It is generally at right angles to the pipe.

LAND LEVELING

Slope Patterns. Land leveling may be divided into six classes, according to the result obtained. These are:

1. Spot grading.
2. General downward slope away from water supply—for sprinklers.
3. Uniform grade in direction of irrigation.
4. Uniform grade in direction of irrigation and at right angles to it.
5. Uniform grade in direction of irrigation and exact level at right angles to it.
6. Exact level.

Spot grading consists in removing humps or filling hollows, without establishing a uniform grade in any direction. It is sometimes done in advance of better leveling for irrigation, and is of general use to make possible faster tillage and more even production.

If water distribution is to be by means of sprinklers, perfectly uniform slopes are not required. For water distribution, it is only necessary that the land have a general slope down from the source of water. In climates where deep freezing of soil occurs, the slope should be uniform enough to make possible drainage of sprinkler pipe laid at a fairly regular depth.

When the water reaches the individual plants by flowing on the surface of the ground, it is necessary to have an almost uniform slope in the direction of irrigation. The steepness of slope may be determined by the character of the soil, the crop to be planted, the original grade, and the rate of water use.

Economies may be affected on many plots by leveling only in the direction of irrigation, and following the original profile at right angles to it. This type of job is used chiefly in orchards which can readily be cultivated into ridges that will regulate water drifting across the field.

Choice between the fourth and fifth method will depend largely upon economies in working over the natural grade. In very large fields, the two-way slope will facilitate movement of water through the cross distribution pipes. The cross grade should be so slight that even light ridging will prevent sideward drift of the water.

Entirely level plots are usually limited to rice fields, and alfalfa and other crops which can tolerate flooding.

Flow and Absorption. The rate of water flow and absorption should balance so that water will reach the lower end of the slope in sufficient quantity for the crop, without flooding or running off. In practice, this balance may be difficult to achieve, and provision is made for draining off excess water when necessary.

Increase of gradient will accelerate the flow and decrease the rate of penetration. Light sand or gravel soils absorb water rapidly and require steeper grades than clay, which may be almost waterproof unless freshly loosened.

The maximum gradient would be below that which will cause the soil to wash and gully during irrigation or heavy rains. The minimum is flat.

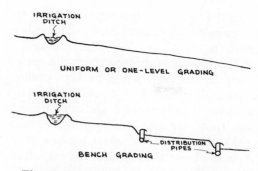

Fig. 7-32. One-level and bench grading

If the maximum practical gradient is not sufficient to move the water the length of the field, additional distribution lines can be installed. Very porous soils require pipes and sprinklers.

Figuring Gradients. Earth moving should be kept to a minimum to save money and to conserve topsoil. If conditions justify the use of any one of several slopes, the one will be chosen which conforms most closely to the natural topography.

If the steepest possible gradient is so much flatter than the original grade that excessive earth moving will be required, because of deep cuts at the top and high fills at the bottom, the field may be divided into two or more levels or benches. These will have the desired slope and will be separated by steep banks. A separate water line is required for each bench.

The high corner or end of the field must be below the level of water in the ditch or its head in the standpipe.

The new gradient must usually be placed at a level, or levels, where cut and fill will balance, as cost may be greatly increased by bringing in borrow or dumping surplus soil. After the surrounding areas are largely under cultivation, borrow or disposal might not be possible. Occasionally, special circumstances, such as nearby channel or road construction, might make transfer of material profitable.

Topsoil is not a problem in many arid valleys where soil is fertile to a considerable depth. However, when topsoil is thin and rests on layers of soil which are infertile or hard to work, or when the surface soil differs sharply in character from that underlying it, the cuts should be kept shallow.

Stakes. Stakes set in a square grid at one hundred foot intervals, and on high and low spots, are used in measuring the original surface and for marking the new grade. One or more bench marks are set outside the grading area.

The new grades are marked on the stakes in any convenient manner. On fills, time may be saved by tying strips of cloth on grade lines to enable operators to see them without getting off their machines. Where soils are loose, two stakes may be used to advantage—one hammered down to ground level, the other left to project two or more feet. Cut and fill are figured from the top of the lower stake and marked on the upper one.

Clearing. Much of the growth in arid regions is of light and brittle character which breaks up during grading and mixes with the soil, and has value as a binder and a source of plant food that is more important than the slight difficulties it causes during finishing. This group includes the various kinds of sage, tumbleweed, and many of the smaller cacti.

However, larger shrubs and trees such as mesquite, greasewood, and acacia, the presence of which often indicates good soil, are tough and deeply rooted and their removal requires heavy machinery. In thick stands of these plants clearing is more expensive than grading.

Such growth can usually be piled and burned immediately after removal. The leaves and sapwood are resinous and the dry soil sifts out of the piles so that they burn readily. Heavy trunks are more difficult to ignite, and if only a few are present it may be easier to haul them to a dump.

Clearing methods and machinery are discussed in Chapters 1 and 21.

Savings may be effected by cutting the trees flush with the ground wherever the fill will be deep enough to permit tilling over them. However, this is not recommended as it will prohibit the future use of pan breakers or other deep tillage tools, and will add greatly to the expense of installing underdrains if they should become necessary. The same objection applies to burying logs in the fills.

Wind Damage. Clearing and grading should not be started until irrigation water is available. The native vegetation, even when very sparse, has some power to break the wind and hold the soil. The weathered ground surface usually has a crust which resists wind scour. These natural protections are destroyed by the work, and unless water can be put on and a holding crop started immediately, the best part of the soil may be blown away or piled into dunes that may be more costly to level than the original surface.

Wind damage during the work may often be avoided by choosing a season in which windstorms are infrequent. If this is not possible, the final leveling and planing should follow immediately behind the rough grade, as a perfectly smooth surface is much more resistant to scour and dune formation than one having ridges or tracks of machinery on it.

If such a planed surface becomes roughened by wind, it should be replaned before the next storm, and kept flattened until a crop can be grown.

Machinery. Dozers are used to clear, to take the tops off ridges and dunes, to bevel steep slopes, and to fill in pits; for cut and fill work on short pushes, and for pusher work with pans.

A drag leveler can be used to smooth out rough spots wherever it is possible to walk the tractor over them. It can transport soil long distances, although its efficiency diminishes rapidly over two hundred feet. As compared with the dozer, it has the advantage of making a wider cut, with little tendency toward scalloping, and has a greater transporting capacity and speed. As compared with pans, it has greater stability against overturning, smooths a wider area with each pass, cuts down and fills more quickly, and can be dumped promptly if the tractor gets stuck. It has a smaller transporting capacity and generally will not make as smooth a grade.

On any large area, the bulk of the dirt moving is most efficiently done with scrapers. Because of the width of most of these grading jobs, and the small slope of the land, these can be used in almost any pattern preferred by the foreman or operators.

Grading. One technique is to produce a rough finish grade in the high corner of the field and to expand this grade as continuously as possible. Where necessary, spot grading operations are done beyond this area in order to secure fill or to dispose of surplus.

Economies may be effected by loading the pans toward adjoining depressions so that the soil pushed along in their efforts to load will fill them. If the soil is very loose, this may be more important than loading in the direction of the dump, but it is often possible to do both.

Fills are usually made in thin layers in order to get maximum compaction from hauling equipment, as rollers are seldom used. Tamping rollers will produce a more permanent grade where fills of more than a foot or two are required, but close competition for the work may not permit the necessary increase in price.

Rough graded sections may be settled by flooding before doing finish work.

It is often possible to keep the short range equipment, dozers and drag scrapers, working all the way through the job if the area contains a number of adjacent humps and hollows. The drag level can also do the light finishing more economically than the pans.

Finishing Off. As soon as any considerable section is rough finished, grades are re-checked and additional stakes may be placed. These can be on fifty foot centers both ways, or fifty feet one way and a hundred the other, or at the intersection of diagonals across the original squares. Any changes of grade necessary at this stage can usually be made by the drag level.

If a small error has been made in balancing cut and fill, the gradient in the lower part of a field or bench may be increased or decreased slightly to change the quantity needed to balance. Large errors may require lifting or lowering the whole field, or one of its benches, or making arrangements for disposal or borrow outside the plot.

When the entire field, or a large section of it, has been brought as near the finish grade as is practical, all stakes are removed and the job finished with a land plane. This will flatten ridges and hollows around stakes, plane off spill windrows, piles, and track marks, and even off local inaccuracies at hitting the grade.

Planing also serves as a maintenance operation and is sometimes repeated after each harvest.

Distribution pipes are usually laid immediately after completion of leveling. Ditches should not be dug until the pipe is on the job, as drifting soil can fill them very rapidly. It is usual to lay and cement the pipe, to partially cover it by hand, and to allow it to cure. This fill is removed by hand where standpipes are installed. The ditch is then backfilled and graded over by machinery.

IRRIGATION DRAINAGE

Alkali. All soils contain some salts, both in soluble and insoluble forms, and these are necessary for plant growth. In rainy climates the soluble salts tend to be leached out of the soil and carried away in underground water about as rapidly as they are made soluble by plant action and weathering, or added as fertilizer.

In dry climates where there is not sufficient rain to leach them effectively, they accumulate in the soil, accounting for the great richness and productivity of the land. However, in flat low areas and some other places the salts, then known as alkali, may be concentrated so heavily that they kill plants instead of aiding them. Alkali may also appear as a surface crust where ground water comes to the surface and evaporates.

Underground Pools. Where soaking rains are rare, natural underground drainage tends to be poorly developed or non-existent. If such an area is irrigated, water absorbed in excess of that required by the crops will accumulate in a stagnant underground pond whose top may rise close to the surface.

This water will dissolve minerals on its way down and while lying underground, and usually becomes so alkaline that it injures or kills plants which absorb it. If it does not become loaded enough to do this, it still may injure plants by drowning their lower roots. Also, when the water table is near enough to the surface so that capillary attraction will lift it to the surface —a short distance for light soils, a long one for heavy—its evaporation will form an alkali crust.

When such a stagnant or semi-stagnant pool forms, the land above it usually becomes unfit for crops; and even if irrigation is stopped and the water slowly drains away, the alkali deposits in the soil may render it unusable. Artificial leaching would re-establish the underground water.

Drainage. The area can usually be put back in production by the installation of an adequate system of drains. These will serve to lower the water table below the trouble line, or to give the water enough flow toward the drains so that it will not stay in the soil long enough to become alkaline. Such drains are preferably deep, six to seven and a half feet being usual, and spaced from 75' to 800'. Close spacing is for impervious soils, wide for pervious ones. However, for sub-drainage purposes, the porosity of the soil cannot be judged from casual inspection, or even by analysis of samples. Heavy impervious clays often respond readily to tiling, because they are filled with fissures, either open or sand filled, which conduct the water. Many really tight soils will not require drainage because of their refusal to absorb the irrigation water.

Some irrigated lands are composed of alternating layers of heavy and porous soil, which are in the form of lenses tapering to nothing on each end, so that natural drainage must move through both types of soil. Ditching cuts and drains the porous lenses.

The tile lines which do most of the work of draining are called laterals, and the larger pipe into which they empty is called the base line. Four or five inch laterals and six or eight inch base lines are usual. Sizes are ordinarily much smaller in proportion to acreage than in non-irrigated fields, but layouts are similar.

If the problem is water leaking from an adjoining canal, an intercepting drain should be placed parallel to the canal, at a distance of fifty to seventy feet. Water may leak under it if it is too close or too shallow.

It is best practice to lay all drainage tile on gravel, and under tar paper and gravel, as described in Chapter 5, as the effective life of the system will be many times that of plain tile. Since tiling is generally done in saturated land, a tile box should be used to avoid danger of cave-ins.

Leaching. After a field is subdrained, alkali can be leached out of it by repeated soaking with irrigation water. This dissolves the chemicals and removes them through the tile lines.

CHAPTER EIGHT

ROADS

Roads are of many kinds, from cart tracks to superhighways. The importance of the simpler types is often overlooked. They are essential in their own places, and they show principles that are basic to the more elaborate highways.

ROAD TYPES

Pioneer Roads. Pioneer roads are access roads built along the route of a highway, pipeline, or other heavy construction project to allow the movement of equipment to and between different sections of the job. If such a road is required, it should be the first work undertaken; and any delays in cutting it through will slow the starting of the job and may keep men and equipment idle.

It is best to locate it sufficiently to the side so that it will not be blocked or cut off by the main work, and if it must cross the construction strip, it should do so where it is close to subgrade.

The importance of the pioneer road decreases as sections of the main road become passable for trucks, but it often retains at least emergency or detour value until the job is finished.

If it is to be used only for moving in equipment, it may be narrow, crooked, and steep for the sake of economy or haste. Specifications written, and the route surveyed or walked through for it, serve as guides rather than instructions, and the job supervisors are usually given wide latitude in altering them for the sake of speed or economy.

Pioneer roads are most often needed in mountainous and timber country where severe obstacles hinder cross-country travel. Where fill is available, trees are cut flush and the stumps buried; otherwise they are uprooted and the holes graded in. Topsoil is handled as fill.

Rock is avoided as much as possible in the layout of the road, and when found is often buried instead of blasted. If an excessive amount of rock must be moved, it may be economical to place the pioneer road in the route of the highway, as the cost of the separate blasting may outweigh the advantage of the independent road.

Grades follow the land contour as closely as possible. The maximum grade will depend on the use. Shovels, tractors, and lightly loaded trucks should be able to negotiate grades up to thirty percent, but serious delays can be caused by stalling of weak units, or as a result of skidding. Ten to fifteen percent grades are more practical.

Curves should be wide enough to enable the longest units to get around them somehow, and the machines in steady use should be able to make them without backing. Attention should be paid to the lane width needed, so that inside rear wheels will not run off the road. Width requirement increases with length of wheelbase and sharpness of turn.

The road width is determined by its intended traffic, construction problems, and haste. It is desirable that it be two lanes wide, but this is often not practical. On steep slopes, two one way roads may be constructed, one above the other.

Two way traffic on one lane will require turnouts at 100 to 500 foot intervals. It is best to make these of two lane roads the length of two vehicles, but dead end turnoffs may be easier to build and will serve the same purpose. A vehicle may turn into one, and back out on the road again when opposing traffic has passed.

Small streams are best bridged with corrugated metal culvert pipe and fill. Occasionally, bottoms may be hard enough to permit easy fording.

Fords are the most economical means of crossing larger streams. A soft bottom can sometimes be made safe by a rock fill. Its downstream edge should contain heavy boulders. Its surface can be crushed rock or clean gravel. It may be placed over a culvert pipe that will handle normal stream flow but not high water.

If a ford is not practical, multiple culvert pipe, log or timber bridges or trestles, or prefabricated steel bridges may be used.

Roads built for use in a dry season may be so constructed that they will be washed out when the rains come if the contractor believes they will have served their purpose by then.

The bulldozer, or angle dozer, is usually the primary tool for cutting a pioneer road. Methods are described in a later section.

In sidehill cuts, the road surface should slope down to the bank or inner side, and may have a berm (ridge) along the outer edge. This shape allows for fill settlement, reduces washing of the fill slope, and decreases danger of sliding off the road.

Drainage from the road surface and the hill slope is carried along the inside bank to culverts, or to outward dipping sections of road reinforced with rock or blacktop. Overhangs or sluices must be provided to carry the water across the fill.

One of the constructions used by the U. S. Forest Service is shown in Figure 8-1.

Access and Farm Roads. A pioneer road is an access road for each otherwise isolated piece of the job it services. However, the term access road usually means a road by which a whole job is connected to a highway system, and is generally used in connection with pits and dams.

The quality of construction is variable. If the project is small, or to be quickly finished, and no substantial amount of raw material is to be trucked in, or products to be taken out, rough pioneer construction may suffice. More often, it must be built as a haul road. Occasionally, a first-class highway will be required.

Farm roads are usually graded native soil, two lanes wide, with gravel, dirt, or other low cost surfacing.

Haul and Logging Roads. There is no sharp distinction between these two types. Both must carry heavily loaded trucks at a good speed, and are ordinarily located according to a favorable terrain, rather than property lines. The logging road is likely to be longer, to climb to much greater elevations, and, under modern lumbering practice, to be permanent. The haul road will carry a much greater traffic for a limited time, and then often will be abandoned.

As compared with the pioneer and access types, these roads differ in that grades are limited. Ten percent is the usual maximum for the logging road, and in haul roads grade is sometimes kept as low as three percent of climb in the direction of load movement. Culverts and bridge capacities are designed according to the period of use, and the comparative expense of large openings, or repairing washouts over smaller ones.

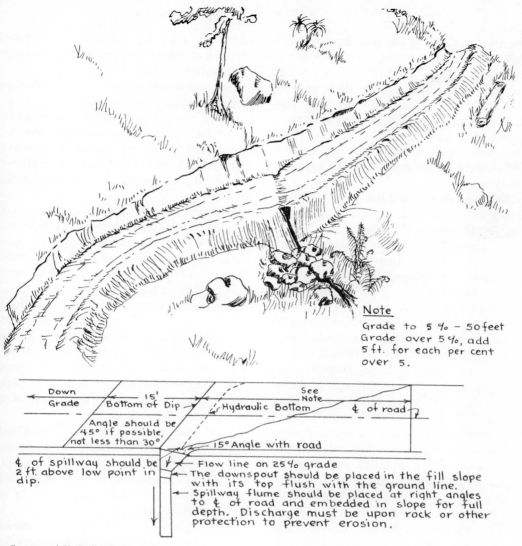

Note
Grade to 5% – 50 feet
Grade over 5%, add
5 ft. for each per cent
over 5.

Courtesy of U. S. Forest Service

Fig. 8-1. Hillside roadway

The long climbs needed on log roads in mountainous country are best ascended at even grades, which can only be attained by careful survey of possible routes. Where the direct distance along a valley wall is too short to provide the ascent at the required grade, the road may be run back into spur valleys instead of crossing them on trestles, or may ascend the slope in a series of switchbacks, or hairpin turns. The turns require a wide space, which, for economy, should be placed where the grade is flatter than ordinary, or where excavation will require minimum blasting.

These factors limit the route rather closely to that originally surveyed, although occasionally, if the contractor runs into unexpected difficulties, he can have the road shifted to avoid them.

The haul road seldom has long ascents and descents, but switch-backs and side wanders must often be used to get them out of a deep pit or over a massive ridge.

Logging roads are surfaced with local

material where possible, from cuts or borrow pits along the road. Any fairly hard and porous material, such as gravel, desintegrated granite, or broken shale can be used, as traffic is ordinarily not long sustained. Haul roads may be oiled to control dust and speed traffic.

Trouble with snow or ice is minimized by locating on the north or east slopes of valleys.

Development Roads. Roads built for real estate subdivision vary in quality from the crudest pioneer type to city streets. Differences depend on the type of development, local regulations, value of land, capital available for improvement, terrain, and other factors.

Rural subdivisions are seldom regulated, but those in and near cities may have to have roads built to high standards. However, the developer may be allowed latitude in locating roads, or shifting them to avoid obstacles, to run cuts through banks of desirable fill or gravel, to change lot lines, or to obtain a more attractive appearance.

Subdivision roads may be financed partly by sale of topsoil, gravel, fill, and other surplus material. Construction costs may be reduced and swamp land "reclaimed" by using such areas as dumps for quarry waste or other clean, solid fill.

Fills made over swamps are subject to severe settlement. For a good quality road, mud should be removed for use in landscaping lots, so that road fill may rest on a firm bottom.

City Streets. City streets are built to exact specifications, often under circumstances which do not allow maximum output from either machines or men.

All operations are likely to be impeded by traffic, which will probably require working the job in sections limited to a few blocks, and frequently to half the street width. Provisions must often be made to pass traffic on intersecting streets through the work. In addition to direct interference

with work schedules, congestion will probably delay trucks and machines entering and leaving the job.

Removal of old pavement is usually the first construction step. Asphalt paving, on a gravel or crushed rock base, can be dug by most front loaders, hoes, or dipper shovels. Occasionally, it is hard enough to require preliminary breaking with a ripper or scarifier, or direct loading with a large size excavator.

A shovel can dig close to manholes, but care should be taken not to hook into them, or into a widened masonry base, as these are easily broken or crushed. Pavement chunks sliding up on the manhole cover may be thrown into the bucket by hand.

Concrete pavements and bases are tough, particularly if reinforced. They may be bonded to the manholes or their bases so as to require breaking away by air hammers, ahead of shovel digging. They break out in big slabs which are difficult to pick up in the bucket, and to dump out of a small or medium truck.

Soil beneath the pavement is removed with it to required depth. It may be native soil, or rock, dirt, or even garbage fill. It may be honeycombed with pipes and conduits that may belong to the city, or to various utility companies.

If the grade is to be lowered, some of the pipes may have to be dug in deeper. In any case, extensive repairs, enlargements, or relocations of piping is liable to be done between the removal of the old pavement and the laying of the new. This will involve a lot of ditching and probably considerable delay.

The paving contractor should see to it that backfill is thoroughly tamped in all ditches.

After the completion of underground work, all manholes, catch basin gratings, and other street openings are fixed to line up with the new pavement surface. Checking should be done carefully by instrument.

The subgrade is graded and compacted according to specifications. Because of interference with manholes, and the need for working in short sections, a large amount of handwork will probably be required.

Highways. Highways make up the bulk of the excavating contractors' road work. Modern standards of width, grade, and alignment require heavy cuts and fills in rough or rolling land, and grading and compaction of subgrades involve heavy work on any terrain.

Highways are built to exact specifications, although the engineer in charge is generally allowed some latitude in interpreting them.

Contracts may be let on a basis of a fixed price for a job; a fixed price plus specified extras, such as allowance for overhaul, rock blasting, slides, or other difficulties whose extent cannot be conveniently estimated in advance; or on a price per yard basis. Less frequently, they are constructed on a cost plus or equipment rental arrangement.

Highway jobs may be resurfacing or paving of existing roads, widening and straightening of roads, building a new road in the approximate location of the present one, building a new road which will run along or cross the old one only occasionally, or a totally new road crossing undeveloped country. There are of course no definite lines of distinction among these types.

A requirement of most highway construction is providing for continuance of traffic along any roads running along or crossing the job. This may be a controlling factor in job sequence.

Airports. An airport runway is essentially a very wide, short, straight road. It is usually located on the flattest land available, but deep fills are often required.

Banks of cuts must be graded back to very gentle slopes to avoid choppy air currents.

Borrow is frequently obtained from the glide areas at the ends of the runway. It is standard practice to cut away any ridges which might be hit by a plane climbing slowly off either end of the runway.

The runway may have a level centerline, crowned up from the sides slightly for drainage, or have a flat cross section and a longitudinal slope. In either case, drainage slopes are very slight, and the surface must be exactly on grade to avoid puddles.

Taxiways and plane parking areas are roads surfaced to an ample width to carry the wheels of a plane running on the ground. Additional areas on each side must be cleared and lowered to allow clearance for the wings.

Airport subgrades and pavement may have to exceed standards for heavy truck highways if maximum size planes are to be carried.

Road Markers. A construction or mining road should be plainly marked as such, to prevent accidental entrance by motorists. Cars and big machinery do not mix well, and also drivers who are lost or confused may get in the way of a blast or run off a cliff.

Warning signs should be placed on highways at least 400 feet on each side of a haul road crossing. If either road is a busy one, the intersection should be protected by a flag man or a traffic light.

One way haul roads should be marked plainly and frequently with direction signs. There should be a sign wherever any vehicle could enter, and additional signs along the roadways to warn drivers going the wrong way. If the road is paved, arrows should be painted on the pavement, pointing in the direction of travel. A conventional dash stripe line down the center could have arrow heads painted on some of the dash lines.

Signs at entrances to one way sections are not enough, as they may be destroyed by accident or vandalism, or obscured.

Failure to provide sufficient notice and warning of traffic direction is the cause of

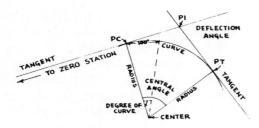

Fig. 8-2. Curve and tangents

many head-on crashes on divided State and Federal highways. There the blame is put on the wrong-way driver, but in a private construction road the contractor is likely to be held responsible.

General safety precautions are discussed in Chapter Eleven.

HIGHWAY LAYOUT

Highways are planned and staked out with consideration for horizontal alignment, vertical alignment, and cross section.

Horizontal Alignment. Horizontal alignment is the route as it would appear on a map, with detail enough to enable field engineers and contractors to lay out and build the road exactly as it was planned. It is figured in terms of the location of the center line.

Curves. Curves are laid out as arcs of circles. Each point on such an arc is equally distant from the center of the circle that would be formed by continuing the arc on the same curve.

A curve is described or defined by its degree of curve or by the length of its radius.

Its degree of curve is the number of degrees in the angle at the center that is made by drawing lines from the center to points on the curve that are 100 feet apart. A sharp curve will have a higher degree of curve than a gradual one.

Curves may also be defined according to the length of the radius, that is the distance from any point on the curve to the center. The radius of a one degree curve is 5,730 feet. A two degree curve will have half that

radius, or 2,865 feet. A short radius means a sharp curve.

Either measurement may be converted into the other by using one of the formulas:

$$\text{Radius} = \frac{5730}{\text{Degree of Curve}}$$

$$\text{Degree of Curve} = \frac{5730}{\text{Radius}}$$

If a man were in a hurry, or trying to work the problem out in his head, he could change the 5,730 to 6,000 and be less than 5 per cent off.

A highway curve may be compound. A compound curve includes two or more arcs having different degrees of curve, and may include some short straight lines also.

Tangents. In highway work and in many other surveys, straight lines are called tangents. This is because they are tangent to the connecting curves. This one word makes clear that a straight line forms a smooth continuation of an adjoining curve.

Tangents vary in length from a few inches up to many miles.

Whenever two tangents are joined by a curve lines are drawn on the plan continuing them until they cross each other outside the curve. The point where they meet is called the Point of Intersection, or PI for short. The tangent distance is measured between the point where the tangent meets the curve (called PC at the beginning of the curve, or PT at its end) to the PI. The side or deflection (def.) angles formed by the intersection of the extended tangents equals the central angle of the simple curve between the tangents.

Base Line. In making a road survey, the engineers first lay out a base line that follows the general route of the road, but that may be partly or wholly outside of the right of way. This line is often made before the exact location of the highway has been decided.

The base line is very carefully surveyed and marked. Some points on it, called hubs, are more important to the engineers than any of the road line stakes. They may be conspicuously marked, or concealed with leaves or rocks.

No contractor or employee should ever destroy, move, or otherwise interfere with any stake or marker on the job, regardless of whether it seems to fit in with the markers he recognizes.

Centerline. The centerline is the basic location reference for the highway itself. It is the center of the pavement in a single road, or the center of the median division of a dual highway whose two roadways are at a fixed distance from each other.

The engineers set the centerline according to angles and distances from base line points. Measurements are made along it with a steel tape (an operation often called chaining) and stakes are set at 100 foot intervals.

All distances are measured along the centerline, and structures and stakes are located in reference to it. It is also the basis for grade calculations for single roadways.

There are a number of construction lines that run parallel or almost parallel to the centerline. These include pavement, shoulder, gutter, and slope edges. They are usually measured off from the center line, at right angles to it on straight stretches, and along radial lines on curves.

Profile. The profile of a road is the vertical alignment of the centerline or of a theoretical grade line. It is a representation of its rise and fall, without indication of whether its route is straight or curved.

Two profiles are prepared, one of the existing ground surface, the other of the proposed pavement surface. Both of these are drawn on one sheet or roll of cross section paper marked off in squares of 1/10 inch, with inch squares indicated by heavier lines.

The usual scale is 100 feet to the inch horizontally and 10 feet to the inch vertically. The exaggeration of the vertical scale

is necessary because the ups and downs are usually quite small compared to the horizontal distances, and would be hard to measure accurately on a small scale.

The road profile is made of a series of straight lines or grades connected by curves. These vertical curves are usually arcs of parabolas, not circles. Plus grades go up as they go away from zero station, and minus ones go down.

The ground profile is prepared from topographic maps, often made by stereoscopic photography from the air. Profiles may be prepared for several possible routes, and highway profiles sketched along them. A ground survey is made along the selected route to serve as the basis for final plans.

A rough estimate of the volume of cuts and fills can be made from the profile, but accurate determination usually requires cross sections showing side slope of the ground, slopes proposed for highway cuts and fills, and other details.

Cross Sections. There are two types of highway cross section. The plans usually include a set of "typical road sections" that show the details of pavement width and thickness, shoulder and gutter width, crown or side slope, and other construction information. These typical sections serve as guides in staking out and building the road.

An ordinary cross section is a profile taken at right angles to the center line. It is at least long enough to include the full width that will be graded. It is usually taken with a transit, but for rough calculations a hand level may be satisfactory.

The number of cross sections taken depends chiefly on the irregularity of the ground. In hilly country they are taken at each 100 foot station, and at additional points where the ground surface changes. On perfectly flat land only one or two might be taken on a whole project.

The cross section of the ground is drawn on cross-section paper with the vertical and horizontal scale the same, or the vertical

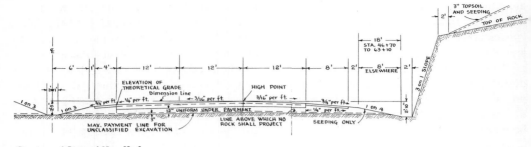

Courtesy of State of New York

Fig. 8-3. Typical cross section

scale exaggerated. Then the proper typical road section is selected and its subgrade line is drawn in, on the same scale and in proper location. Wherever present grade is above proposed subgrade, material must be "cut" or dug and removed. Where present grade is lower than subgrade, material must be added or filled in.

Such cross sections provide data to figure the cut and fill for the road.

STAKES

Stakes are used to guide the contractor and his employees in following the engineer's plans. They also assist inspectors in checking up on the contractor's performance.

The first working stakes on the job may be the centerline, showing depth of cut and height of fill, and slope stakes that show the outer limits of the area to be cleared, grubbed, and graded, and usually the cut and fill information also.

When heavy cuts and fills are required, most of the work may be done with guidance of only slope stakes, both the originals and others that are set up or down the slopes as work progresses.

When the working levels approach the subgrade, additional stakes are needed. Centerline stakes will be restored, and lines of shoulders and gutters may be staked.

Finishing may be done with blue tops, that are stakes driven until the tops are at the grade desired, usually subgrade, and/or string on shoulder stakes.

Centerline. The centerline is usually

staked at 100 foot intervals in preliminary work, and sometimes as closely as every 25 feet in narrow, winding roads, or in finishing operations.

These stakes are called stations. The first one, the zero station, is at the beginning of the road or other project. The distance in feet from zero is marked as a double figure.

Stakes at 100 foot intervals are called full stations, others are called plus stations. Station numbers are made up of the distance from zero, with the hundreds divided from the last two figures by a plus sign. For example, a full station 500 feet from zero is 5 + 00, and the part or plus station at 545 feet is 5 + 45. A distance of 3,456.2 feet from zero is station 34 + 56.2

If changes in plan should cause the project to be extended to the other side of the zero point, minus stations would be used. A stake 180 feet beyond zero would be station — 1 + 80.

Location measurements refer to distances on a horizontal plane, unless specified otherwise. As a result, 100 foot stations will appear to be more than 100 feet apart if measurement is made along steeply sloping ground. On a 1 on 3 slope (33⅓ per cent grade) the surface distance would be about 105.4 feet. But if the stakes were extended upward or downward, any horizontal line between them would be 100 feet long.

Centerline stakes are in the middle of a single roadway, and usually in the median of a double one. They are marked L, C, or /c. They show depth of cut or height of

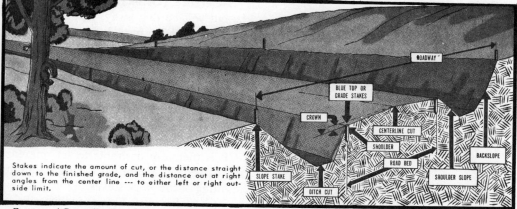

Stakes indicate the amount of cut, or the distance straight down to the finished grade, and the distance out at right angles from the center line --- to either left or right outside limit.

Courtesy of Caterpillar Tractor Co.

Fig. 8-4A. Stakes for cut

fill needed, and usually carry information about location of culverts, structures, and other features.

Slope. Slope stakes are set where the outer slopes of the cuts or fills meet the original grade, usually at 100 foot intervals, and also at other points where ground slope changes or special structures affect the slope. They are always at points of no cut and no fill.

Slope stakes are usually set with a transit or dumpy level and a 100 foot steel tape. They provide the first markers the work crews need, as they show the outer limits of the area to be cleared.

Cut slope stakes may be slanted slightly away from the centerline, while fill slope stakes may slant toward it.

Each stake should show the cut or fill necessary to make the ground level with the centerline at that point, and the distance to the centerline. It should also show the steepness of the slope, but it often does not. If this is the same for the whole job, the grading foreman can carry the information in his head.

Stakes indicate the amount of fill, or the distance straight up to the finished grade, and the distance out at right angles from the center line --- to either left or right outside limit.

Courtesy of Caterpillar Tractor Co.

Fig. 8-4B. Stakes for fill

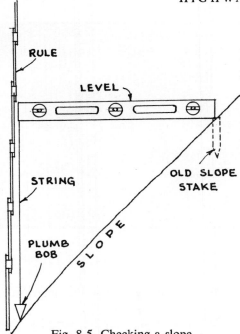

Fig. 8-5. Checking a slope

Slope and other side stakes are usually marked with the station number, the distance from the centerline, and the direction of the centerline. Direction is indicated by the letter R for right or L for left. Such directions for plus stations are read looking from the zero stake. In some localities 25-R means 25 feet right of the centerline, in others that the centerline is 25 feet to the right of the stake.

A slope that is not known can be figured by subtracting half the width of the road (including a gutter, if there is one) from the distance from slope stake to center, and dividing the remainder by the cut or fill measurement.

For example, if the distance from slope stake to center were 28 feet and the half width of the road were 24 feet, the width of the slope would be 4 feet. If its height were 2 feet, the slope would be 1 on 2.

As a cut deepens or a fill is built, it is usually necessary to check the slope with new stakes. These may be set from the original with a string or carpenters' level, a rule, and a plumb bob.

Reference (Offset). Stakes on areas to be cut will be dug away, and those inside fill lines will be buried. In shallow cuts, stakes can be left temporarily in islands; and in shallow fills long stakes may be used which will project from the top, unless they are knocked over. Slope stakes are liable to be undercut or buried. Any stakes are apt to be moved by accidents, particularly if the ground is stony or frozen.

It is therefore desirable to set reference stakes well outside the work lines to simplify resetting of the work stakes.

Such stakes may be set on one side or on both sides. They are marked with the station and the distance from the centerline, and may be identified by lettering such as "OFF" or "REF."

If the road strip is narrow or of moderate length, reference stakes on both sides will permit replacement of the working stakes simply by measuring between the offsets.

Where trees or heavy rocks are near the road, nails may be driven into trees, or marks chiseled on rocks, on opposite sides of the road, and a tape stretched between them. The reading at the center line and ends and the station are noted. With these notes it should be possible to find the center again quickly and accurately.

When a few center line points can be found from side references, it is often possible to sight in the rest of the missing stakes by eye with reasonable accuracy.

Grades may be marked on offset stakes, or separate system of bench marks may be used.

Surveyors often set a line of offset stakes instead of a centerline, leaving the center and other stakes to the contractor's crew.

Grade. Grade stakes show the distance that the ground surface is above or below a desired elevation or grade. Vertical distances to grade are marked on the stakes in feet, inches, and eighths of inches, or in feet, tenths, and hundredths of feet. Figures are preceded by the letter "C" for cut if the ground is high and must be cut or dug away, or by "F" if it is low and must be filled.

Cuts and fills may be figured from the base of the stake (ground level), from its top, or a line drawn on it. Any basis except ground level is confusing to operators and may cause serious mistakes. However, ground level should be marked in case soil falls away or is added without disturbing the stake.

If the fill is less than the height of the stake, the grade may be marked directly on it with crayon. It is an excellent practice to tie a rag around the mark to make it readily visible to the operator.

Shallow cuts may be marked temporarily with rags a specified distance, as one or two feet, above grade, so that operators will not have to dismount to read the figures.

A great number of rags can be made of one old sheet by tearing it in narrow strips. If none is available, unsterilized one-inch bandage can be bought quite cheaply for the purpose. These cloths are easily dyed.

Original centerline stakes are usually marked to show finish grade, that is, the surface of the pavement, since it is the line that forms the basis for engineers calculations. The letter "G" indicates that reference is to finish grade.

Subgrade is the surface of the native soil after cutting, filling, grading, and compaction. It is lower than finish grade by the thickness of the pavement, and any pavement base and/or sub-base that may be required. The combined thickness of these layers may be almost nothing, if the surfacing is to be oil or cement stabilization of native soil; or three or more feet for very heavy construction.

Since the cut and fill operations are directed toward reaching subgrade, much confusion will be avoided by making all such stake markings refer to subgrade, using the symbol "SG."

A misunderstanding as to whether figures on grade stakes are for grade or subgrade can be very expensive.

Centerline stakes are quickly lost in most heavy grading, and the work is checked mostly by slope stakes. However, as sections are brought near the correct subgrade the center stakes are replaced, and additional lines of markers are used to show the crown or cross slope of the road, and shoulders and gutters.

Stakes at the edge of any grading area should be set back about six inches to a foot, so that they can serve as a guide without interfering with the work.

It is frequently necessary to remove spilled dirt or level around a stake by hand, so that the operator can read it, and see whether his grade is high or low in reference to it.

Blue Tops. The final or fine grading operation is often guided by blue tops. These are usually 2 x 2 grade stakes driven down until tops are at subgrade. The tops are often colored with blue crayon to make them more visible. Any that are driven below the surface are marked by a light stake alongside.

An expert grader operator can work over blue tops without disturbing them. However, it is necessary for a man on the ground to remove spill piles that hide them, and to expose them if they become buried. Even with this precaution, varying numbers of

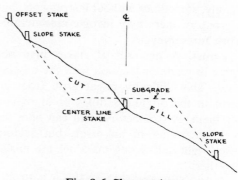

Fig. 8-6. Slope stakes

these stakes are caught by the blade, or rolled over by tires, so that they have to be reset.

To set blue tops, a telescopic level of any type is set up, and its height (HI) figured from a bench mark. The correct rod reading for grade is figured for each stake location from the center or the theoretical grade profile, with allowance for crown or banking where necessary. The rod man starts the stake and holds the rod on top of it, and is told by the instrument man how far high he is. The rod man drives the stake the approximate distance, another reading is taken, and the process repeated until the top is with a few hundredths of feet of grade.

It is common practice to set blue tops 50 feet apart along the length of the roadway, and at 12 to 15 foot spacing across it. The 50 foot spacing sometimes produces a low wave-like effect, as a grader operator may get the grade perfectly at the stakes, but have a tendency to run consistently high or low between them. This may be prevented by reducing spacing to 25 feet.

A grader equipped with automatic blade control can grade widths up to 40 feet from a single line of blue tops, resulting in substantial savings in grading and staking time.

Care. Operators should be very careful when working around stakes as they are valuable, both as guides to correct work and in relation to replacement cost. In general, an occasional stake can be replaced readily, sometimes without instruments, but a group of them may involve considerable work for surveyors.

Errors. A new set of stakes may not agree in grade or location with the missing ones. This difficulty might arise from an error in the original settings or in replacing them. A satisfactory road can often be built according to an error, but seldom when right and wrong markings are mixed together.

Stakes are accepted as correct until dis-

crepancies are noted. If any stake appears to be out of line, or badly off grade, it may have been moved or disturbed; it may be a base line or other marker, or a mistake may have been made in placing or marking it.

When possible, the surveying crew should be recalled to check it. If this is not practical, the foreman may be called upon to use his own judgment about whether it should be re-measured. It should not be disturbed, however, unless absolutely necessary, as the suspected stake may be right and others wrong.

GRAVEL ROADS

Surfaces of bank gravel and other low cost materials are so frequently required for haul, access, and other work roads that a brief discussion of them is in order.

Bank Gravel. Bank gravel is a natural mixture of pebbles and sand. For road building purposes it should contain some fines that will act as a binder. Most deposits contain cobblestones and boulders.

Specifications for road gravel vary greatly. The following spread includes most of them:

Sieve	% Passing Each Sieve
2 inch	80-100
1 inch	60-100
1/4 inch	40- 85
10 mesh	15- 70
200 mesh (Fines)	5- 25

The Ohio Specifications I 22.02 have been widely adopted through the country, and are as follows:

	% Passing Each Sieve	
Sieve	"A" Gravel	"B" Gravel
3 inch	100	100
2 inch	90-100	90-100
1 inch	70-100	70-100
#10	25- 75	25- 75
#200	0- 10	5- 15

Gravel characteristics and tests are described in Chapter 10.

In general, gravels with over ten percent fines are not suitable for roads that will be subjected to freezing. Less than five percent may lead to loosening up in hot, dry weather. However, an increase in the percentage of coarse particles will lessen the softness caused by too much binder. Variations in particle shape and material will also affect results considerably. Increase of depth may make up for weakness.

There is no consistent difference between the parts of gravel banks which are above and below the water table. Water levels were usually different at the time the material was deposited. However, there is very often a difference in color due to above-water oxidation of certain pigments.

Engineers frequently write ideal specifications for gravel which is not obtainable, and contracts are let to use practical grades on a price or availability basis.

Screened Gravel. Specifications may call for screening gravel to be used in the top course or in the full road depth. Maximum size stones may be limited to one, two, or four inch diameters.

Screening is desirable to obtain a smooth, easily worked surface, but it often involves wasting of an excessive amount of stone which could be worked into the road. The resulting loss of strength may affect the road stability, particularly in crossing soft or wet ground.

In general, most oversize stone can be eliminated during the spreading and grading processes at less expense than pit screening, except in patching work.

Crusher Gravel. Bank gravel which is short of pebbles and long on stones, may be run through a crusher to reduce the oversize to pebbles. The result may be superior to run-of-bank of similar size distribution because of the angular shape of the crushed pieces.

Blasted rock which is run through a crusher, without separation of the product, will often produce a material similar in size, distribution, and performance to the best of bank gravels.

Crusher gravel is usually more expensive than run-of-bank because of the extra processing.

Similar Materials. Any hard material which is broken into particles of the gravel size range may be used in its place. The breakage may be from blasting, rooting or digging, burning, or the effect of heat and cold. Such materials include shale, soft limestone, fine blasted rock of any kind, scoria, red dog, slag, disintegrated granite, cinders, and shell.

Exposures of shale rock are frequently soft enough to be dug by a small shovel without blasting. The broken shale has the appearance of excellent road material, but breaks down readily into mud. Some very expensive road failures have been caused by allowing traffic to use a shale subgrade, then putting a concrete pavement over it after its usefulness had been destroyed.

Soft limestone is the "coral" of the island military bases. It is often dug from the solid by loosening with heavy rippers, or hydraulic dozer blades fitted with teeth. It is easy to drill, but blasting may require as heavy loading as hard rock.

Such limestone is used as it comes from the pit. It should be rolled promptly after placing, as rain can make a soggy mess of it when loose. After compaction wetting sets it into a hard surface that requires less maintenance than gravel.

Rock from tunnels (muck) is well suited for road fills as the tight, heavily loaded shots cause fine fragmentation.

Scoria comes from clay beds that have been cooked by the underground burning of adjacent coal seams. It resembles broken brick. Red dog is a similar material that is produced by the burning of piles of waste bituminous coal with a high clay content. Both of these substances may break down into mud under traffic unless protected by some other surfacing.

Disintegrated granite is the standard low cost road material in many parts of the southwest. It is a rough coarse sand with excellent compaction and drainage characteristics.

Slag is a by-product of hot refining of metals, that may be poured molten onto dumps where it hardens into rock, or may be cooled and broken up by a water spray.

Cinders are of two kinds—refuse from steam power plants burning lump coal, and aerated rock blown from volcanoes.

Those from power plants are light, easily worked, and free draining. However, they pound into mud quickly under traffic, and are useful for light duty footings and emergency surfaces only. They are becoming rare because of power plants changing over to powdered coal and other fuels.

Weight of volcanic cinders and ash ranges from 60 to 120 pounds a cubic foot, compacted. The difference is chiefly in entrained air, the lighter qualities being so full of bubbles that they have little strength.

Most volcanic cinders from 90 pounds up make good road material, but care may be needed selecting them in the pit. They are too resilient for use under rigid pavements, but have good frost resistance because of their air content.

Shells are dredged in enormous quantities from bars along the Gulf of Mexico, and serve locally for the principal low cost road base and surface.

Preparing Subgrade. The subgrade should be finished as accurately as possible. Ridges or hummocks of subsoil which extend up into the gravel weaken it. If the subgrade is clay or silt, it is good practice to place a blanket of clean, coarse sand to interrupt capillary flow and add to road stability.

The subgrade should be compacted if it is practical to do so. However, temporary gravel roads are often put across wet spots that are not workable. Rock fill, or extra

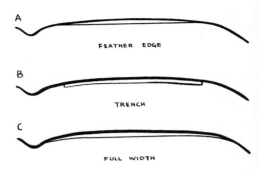

Fig. 8-7. Cross sections of gravel roads

depth of gravel is used to make up for lack of subgrade preparation.

Cross Sections. Three cross sections in common use are shown in Figure 8-7. The feather-edge construction in (A) calls for a flat subgrade. Its advantage is ease of construction. Disadvantages include poor drainage of water out of the center gravel, deficient strength at the edges, and the necessity of blading fill from gutters or shoulders into the road during maintenance.

The trench section (B) provides center drainage and strength to the outer edge of the gravel. However, frequent bleeder drains through the shoulders may be needed to prevent water from ponding in the edges, soft shoulders may be a hazard in wet weather, and maintenance work will put dirt over the gravel.

The shoulders and gutters may be shaped before laying gravel, or the gravel may be laid and gutters then cut to obtain shoulder material.

The full width surfacing in (C) is the best construction, and is to be recommended wherever the price of gravel is not a controlling factor. It saves the trouble and expense of edging, provides hard shoulders and good drainage throughout the surface, and minimizes maintenance difficulties.

Placing Gravel. On good subgrades, gravel may be very thin, but it is the best practice to use six to eight inches com-

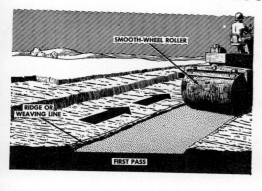

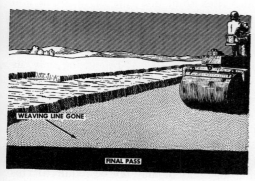
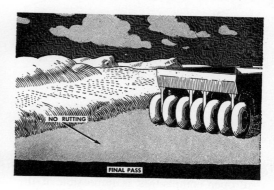

Courtesy of War Department

Fig. 8-8. Subgrade compaction

pacted depth, and to spread it in two layers. On soft ground, the depth may be twelve inches or more. The greater part of deep gravel is usually in the bottom layer.

The best gravel available should be in the top layer. It should not contain many stones larger than one inch, or at the most, two inches in diameter. It should be coarse enough to resist the action of tire suction, water, and wind, and should have enough binder to hold it in dry weather, but not enough to make it sloppy when wet or thawing.

In the bottom, stones up to two thirds of the layer thickness can be tolerated. Clean sand without stone may serve if the top layer is thick and well bound enough to hold it together.

Gravel is ordinarily trucked in and spread by a dozer or grader. Occasionally, hauling and spreading can be done by scrapers.

Oversize stone may be bladed to the side, or picked out by hand. Small loose stone may also be taken off, or it may be left to be pushed back into the gravel by the roller.

Oversize stones that remain in the gravel after spreading and smoothing may be pulled out by a rake grader blade or a spike tooth harrow.

Hand picked stones may be thrown directly into trucks, or placed alongside the road for later removal. Second handling is easier if they are piled rather than scattered along the edge.

Compaction. Each layer should be thoroughly compacted by pneumatic tired or steel wheel rollers, or by traffic. A heavy steel wheel roller will work back into the gravel all small stone pulled out by spreading work, and it gives a well finished appearance.

The edges should be rolled first, and

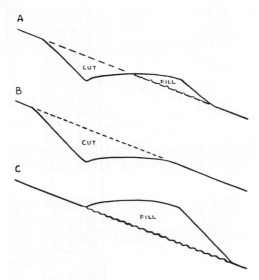

Fig. 8-9. Sidehill cut and fill cross sections

strips should be overlapped. This preserves the crown of the road, which should be at least 4 inches for a 20 foot road.

Proper compaction is impossible if the gravel is entirely dry, and difficult if it is too wet. However, dry gravel can be watered, and wet gravel will usually drain quite quickly. Full compaction on a gravel surface is not as important as in subgrades for pavements.

Dust Control. Drying out with resultant dust nuisance and aggravation of washboarding may be prevented by use of calcium chloride. This is spread by hand shovels or machinery on the surface, and absorbs air moisture that soaks into and dampens the surface.

Recommended application is 1½ pounds per square yard at the beginning of the dust season, and 1/2 pound a month or two later. However, satisfactory results may be obtained with much lighter applications where summers are not entirely dry.

Lignum sulfonate, a byproduct of paper manufacture, is used for the same purpose. It is sold in 50 per cent solution form in drums, diluted to 10 per cent and spread by watering trucks.

Gravel road maintenance is discussed in Chapter 19. For haul use the surface should be kept free of loose stone and coarse gravel pieces, as these may cut tire mileage by as much as 50 per cent.

Quantities of gravel needed for various areas and depths are given in the Appendix.

SIDEHILL CUTS

In hilly or mountainous country, roads are largely notched into slopes so that the land rises from one side of the road and dips away on the other. Such a road may be constructed by digging on the high side and using the spoil to build up the low side, as in Figure 8-9 (A); by cutting only, as in (B); or, less commonly, by building a shelf of fill as in (C).

Difficulties of design, excavation, draining, and stabilizing increase rapidly as hill slopes become steeper.

Stripping. Removal of topsoil, stumps, and logs may or may not be required. This matter will be decided by the job specifications, or by the judgment of the engineer or contractor.

In general, stripping of topsoil becomes both more difficult and less important as the slope increases, as deep cuts in steep hills increase the proportion of subsoil in the dirt moved.

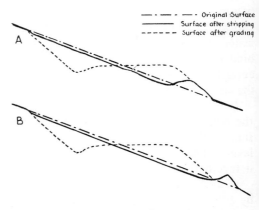

Fig. 8-10. Topsoil stripping

When stripping is required, the topsoil can most economically be pushed straight downhill by dozers to form the toe of the fill, as in Figure 8-10(A), or a windrow below it, as in (B). Such a windrow may be moved by scrapers to a stockpile, or left to be pushed or pulled back up the slope to cover the completed fill.

If the hill is too steep to back up, the dozer may be equipped with a towing winch, and the line anchored above the work so that it can pull itself up the slope. It may also be helped by winch or direct pull from a tractor above it, or by a line around a pulley anchored above it to a tractor on its own level or on a lower one.

Loose stumps can be used in pioneer road fills but are unsuitable for highways. Their use in intermediate classes of roads will depend on job conditions, the estimated useful life of the road, and the rate of decay of the stumps.

Logs placed at the toe of the fill are useful in catching rolling boulders and checking slides. These, or more elaborate precautions, are often required to protect roads or structures below. They may be held by their own weight, by resting against stumps, or by cables and anchors. They may be used as temporary expedients in most work.

Stumps left intact in steep fill areas may serve to prevent the completed fill from sliding downhill as a mass. Specifications often permit leaving them if they will be covered two or three feet deep.

If a sidehill is cleared and stripped, the areas to be filled should be plowed or roughened across the slope to reduce the danger of slides.

Dozer Digging. If the side slope is gentle, the road shelf may be cut by pushing downhill. Steeper slopes may be started in the same manner and finished by working along the roadline, as in Figure 8-11.

In general, when the upper bank becomes so steep that the dozer cannot back up it without assistance, it is more economi-

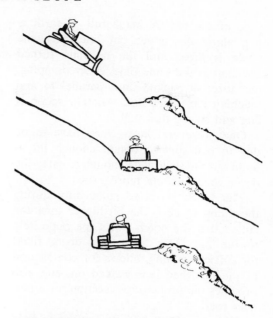

Fig. 8-11. Starting sidehill cut with dozer

cal to work from the side. However, if the line of cut is interrupted by rock ribs, which are not to be blasted until the softer parts of the road are made, a dozer with a helper cable may be used to cut benches in each section, at least long enough to permit it to start a sidecasting cut.

Pushing from above, where practical, is faster than sidecasting.

Sidecasting. The standard method of notching a steep sidehill is sidecasting with a dozer. A wide track, close coupled dozer with a blade that can be tilted to cut low on the uphill side is most efficient. An angling blade, set with the uphill side low, and angled to cast down the hill, is useful, particularly in light soils and shallow cuts. The advanced position of the blade may make it difficult to turn with heavy loads.

A non-tilting bulldozer blade, or a shovel dozer bucket, requires special care in preserving a counter slope.

Work is started near the upper slope stakes at a spot naturally or artifically level enough to permit the dozer to work parallel with the road center line, at the upper edge of the cut. A blade full of earth is dug along the upper cut line, then the blade is lifted and the machine turned downhill at the same time. After dumping, the dozer is backed until parallel to and touching the upper line. Another scoop is dug and swung downhill.

One or several layers may have to be dug in one spot to obtain enough fill to build out the shelf wide enough to carry the dozer. Steeper slope, more passes.

The blade is raised sufficiently during the dump to keep the fill higher than the cut so that the notch will slope oppositely to the hill. This keeps the dozer tilted for efficient cutting, allows for compaction of the fill when it is walked on, and also provides the proper cross section for a pioneer road.

When the shelf is wide enough to hold the dozer, further procedures are varied to suit the slope, the soil, the machine, and the operator's preference. The cut can be lengthened to the end of the slope, then cut in successive layers to grade and width, or it may be developed to full size in a single cut.

Layer cutting involves more rehandling of the dirt, as the loads dropped from the first cut are moved again as it is deepened. However, it is easier for a dozer to make shallow cuts, and the angle blade sidecasts most effectively, and puts minimum strain on the tractor when the cut is light.

Deep single cuts make it difficult to trim the bank and may have to be avoided for that reason.

Rock Slopes. If the slope is composed of material that the dozer cannot dig, or that it can dig only with difficulty great enough to reduce production and increase repair costs substantially, the material should be softened ahead of the dozer.

Hard clay and soft rock on moderate slopes may be loosened with a tractor mounted ripper. But if the rock is hard or the slope is steep, drilling and blasting will probably be necessary for the pioneer cut.

Very steep or broken slopes call for hand drills. Holes may be drilled along the top line of the cut or horizontally at the first floor level. It usually pays to drill closely and load heavily, as ribs or poorly fractured rock will delay the dozer operation out of proportion to any saving in costs.

When the rock is patchily covered with loose overburden it may be necessary to dig holes before drilling. At other times only the exposed rock is drilled, and secondary work done on the parts that are uncovered during dozer work.

If there is a considerable amount of pioneering in rock slopes, best results may be obtained by use of light self propelled drills on crawler mountings. They can reach and work in very difficult places, and can tow their own compressors except under extreme conditions.

After blasting, dozer sidecasting proceeds in the same manner as in naturally loose soil.

Once the pioneer bench is established, the character of the rock will determine whether rippers or drills and explosives should be used to loosen it.

Lower levels may utilize dozer sidecasting, shovel sidecasting, scraper hauling, or shovel and truck hauling, depending on the job plan. If the material is used elsewhere it is of course desirable to take it away immediately, rather than sidecast it first, and then re-dig and haul it.

Belt Loaders. Once a cut of sufficient width has been made between two areas that are wide and level enough for turning, a belt loader (see Chapter 14) is sometimes used for widening and deepening the cut by sidecasting if the soil is suitable. This machine may work in only one direction. It may be followed on each trip by a dozer grading off the spoil.

Dipper Shovel. The dipper shovel can be used instead of a dozer for notching a slope. It can usually do the rough work in one trip, as in Figure 8-12, but if the bank must be trimmed or the cut is very deep, it may be done in layers.

When the width of the cut will allow it, it is good practice to keep the shovel on its floor rather than with one track on the fill. For narrow roads and deep cuts, a small shovel with a short rear overhang is desirable. The cut should be kept sloped into the bank to keep the weight off the edge.

The fill is kept higher than the cut, particularly if used for footing. Poles or platforms can be used for extra support under both tracks, or under the outside track only.

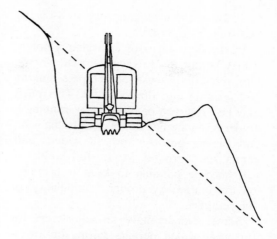

Fig. 8-12. Sidehill cut with shovel

When the ground is soft or wet, the slope very steep, soil layers slope with the hillside, or smooth bedrock is just under the cut, the smallest shovel which can handle the digging should be used. The weight of a large machine, together with the vibration of its work, may cause a slide.

Shovel spoil can most conveniently be sidecast, but can be loaded into trucks backed up to it. If the road is long and narrow, trucking out all the spoil will be very slow work.

Rock exposures along the road line should be blasted as a shovel cannot be readily moved up and down steep slopes to bypass them.

Use of a shovel is indicated when soil is too soft or rocky for effective dozer work, when cuts are deep, and when spoil is to be used at a distance.

The work is ordinarily left rough to be finished off by a dozer or grader.

Side Cuts. When the notch is to be largely or entirely a cut and the spoil is to be used nearby on the job, dozer sidecasting is used only until the shelf is of ample width to hold the machinery. The material is then pushed or carried along the shelf to the fill area.

Big dozers can be used for pushes up to two hundred feet on the level, and farther

downhill with fair efficiency. When the cut is too narrow to allow machines to pass each other, their production can be stepped up, at some additional cost, by using two or more dozers in relays. One, working from the back of the cut, will push a load part way to the fill and spread it a bit in dumping it. The dozer below it will back over the heap and push it to the end of its beat.

Scrapers. The possibility of using scrapers should be considered. Their use on short runs is discussed later under Intermediate Hauls.

Tractor-drawn, full trailer scrapers are difficult to back, so in a narrow road they require an additional road to bring them back from the fill. This may have to go back to the beginning of the hillside, or enter it at some intermediate point. In either case, the scraper travel is apt to be much longer than that of the bulldozer.

A semi-trailer scraper can be backed up a narrow road, but only by a skillful operator. Travel distance would be the same as for a dozer, and production should be very much greater in proportion to power. However, only one scraper can be used conveniently until turnouts can be provided.

If some spoil is being sidecast, and some

hauled away, a dozer can work on widening and serve as a pusher.

In the first stages of enlarging a notch, it may be difficult to keep the road sloping into the hill because of scrapers sinking and gouging into the loose fill. This pitch may be preserved or restored by running a grader or an angling dozer close to the wall, and casting out. As the cut widens and enters solid ground for its full width, it will become possible to keep it trimmed on the bottom by proper manipulation of the scrapers.

Some scrapers have an adjustment for tire wear which permits raising or lowering one of the rear wheel sets. Such a machine may be set to cut low on the inside during the pioneer stage of the cut.

When a steep hill contains boulders, stumps, or ledge, sidecasting is to be preferred to hauling, and dozers will probably be both safer and more economical than scrapers if short or medium hauls are required.

Pioneering is done by dozers. If the cuts are in material they can handle, scrapers take over the job. In very rugged terrain crawler tractors with standard or undersize scrapers are to be preferred, as they can work on the steepest grades and need minimum turn space. After grades are reduced to 20 per cent self powered scrapers can take over, if the haul is long enough to justify their use.

For rubber tire jobs turnarounds should be kept nearly level, and machines should be driven directly up and directly down the grades. Overhung scrapers are in their most vulnerable position in regard to overturning when turning downhill on a side slope.

Compaction. When a wide road is notched into a hillside by cut and fill methods, it may be difficult or impossible to compact the fill if it is sidecast.

If compaction is required, two pioneer notches may be made, Figure 8-13, at the top and the bottom of the cut. Scrapers are then used to cut the top down and build the bottom up. Compaction of the fill can be handled by rollers following the scrapers, until sufficient width is obtained to permit them to pass the scrapers on the fill, after which they can operate in both directions.

BANK SLOPES

Angle. The angle at which bank slopes will stand in cuts and on fills, is an important factor in the cost, and sometimes in the feasibility, of sidehill construction. It is also a limiting factor in the depth of through cuts.

There are two approaches to determining how steep a cut slope may be left. One is the behavior of the same or similar material in cuts and on natural slopes, the other is soil analysis and calculations. They are frequently used in conjunction, a tentative slope being determined by field observation and then checked by engineering research.

Natural slopes are seldom steeper than those that can be used in the same material

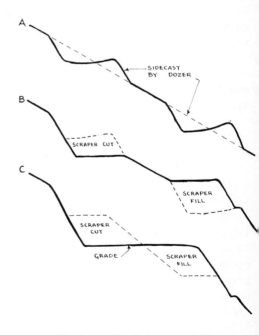

Fig. 8-13. Parallel cut and fill

in a highway cut. The exceptions usually involve ground water problems or the binding effect of vegetation.

However, natural slopes may not be nearly as steep as the soil qualities permit. In general, a hill whose foot is being vigorously eroded by a fast flowing stream will approach maximum steepness, and one rising above meadows will tend to have a flatter slope.

Old cuts give a much more accurate indication. However, before dependence can be placed on them, it would have to be ascertained that the material is actually the same, that it is subject to the same weather conditions (freezing and thawing loosen faces more actively on the north than on the south walls of canyons), to the same dip of strata, as in Figure 8-14, and that ground water conditions are similar.

It is desirable to cut back to entirely safe slopes, but this may not be possible. In notching along the side of a mountain the cut wall must be substantially steeper than the natural slope to avoid excavating tremendous yardages. Also, a fill slope must be steeper than that of the sidehill on which it rests if it is to support a road.

Some soils, such as loess in the Midwest and lightly cemented gravels in the Southwest, will stand for long periods with almost vertical faces. In general, slopes can be steeper in arid climates than in wet ones.

There are varieties of clay that will stand steeply when first cut, but under influence of surface freezing and thawing, ground water pressure, and vibration will slump to a 10 per cent (1 on 10) grade unless stabilized by topsoiling and planting.

In general, cut slopes range from vertical or even overhanging in rock and 5 on 1 in the most stable soils down to about 1 on 6. It is the engineer's responsibility to decide where in this wide range the requirements of stability, safety, and the highway budget can best be compromised.

Slides. The most serious problem associated with deep highway cuts is that of landslides. These may occur during the work, or at any time after completion. Dangers include loss of life or injury among those building or using the road, destruction of excavators, trucks, cars, and other equipment, and loss of use of the highway for long periods.

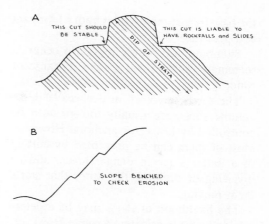

Fig. 8-14. Slope stability

The likelihood of slides increases with height and steepness of banks, but they are caused by internal conditions.

In rock the cause is usually a dip of seams or joint structure that provides an inclined slide for the cut layers, as in the right side of Figure 8-14(A). This structure, when well lubricated by ground water, may produce anything from a series of minor rockfalls to a 100,000 yard road block.

Much the same effect is produced in unconsolidated soils that have sloping layers of pervious and impervious material, or where a slightly pervious soil mass rests on a sloping base of clay or rock.

The existence of such a condition may be revealed by preliminary borings, or be shown by a line of springs as the top of the impervious layer is excavated.

In this case a slide is most likely to occur after heavy rains, when the loose soil is heavy with water, and water movement on the base formation provides both a lubricant and pressure.

Slumping and sliding may also occur in seemingly uniform soil masses, because of water seepage or saturation zones.

The forces involved in deepseated large volume slides are usually too great to be controlled by braces or anchors. However, most of them can be prevented by cutting to a flatter slope in danger areas, and/or diverting or draining the water that starts them moving.

The likelihood of slides may be a determining factor in selecting a route. However, there are many areas in the world where mountain highways must be constructed for long distances along steep soil slopes, and where money is not available for engineering investigations or drainage works.

Under such conditions a pioneer road may be cut by dozer sidecasting down slopes and pushing fills across canyons. The road is widened and improved by cutting down its floor.

When slides occur the dozers simply cut new road shelves across them, repeating as often as necessary. When fills wash out they are replaced, cutting new slots on the slopes above the road cut where necessary.

In the course of a few years it is usually found that the larger part of the route is reasonably stable, and that other sections stabilize with repeated working. The true problem spots that remain can then receive detailed engineering investigation and corrective measures, at a fraction of the cost of similar work for the entire route.

Checking. A technique is being developed for testing the stability of slopes with the Seismitron, described briefly at the end of Chapter Nine. The instrument's probe or receiver may be placed on the slope surface, but results are more accurate if it is placed in a drilled hole in the bank. It picks up tiny sounds of ground movement called microseisms, and amplifies them so that they can be heard in earphones.

An increasing or high frequency, say over 25 or 30 microseisms a minute, indicates danger of a slide. A low or decreasing rate is an assurance of stability. It has been found that slopes that failed when wet had given warning of possible failure while they were dry.

Clay, mud, or fine sand may not produce warning noises that can be recognized. Most other formations do.

Stabilization. When the bank is high it may be necessary to drill long holes into the toe to prevent water from causing the face to slump. Since this trouble usually occurs in soil that is firm enough to leave in high banks, but not hard enough to resist percolating water, augers may be the preferred drilling tool. Perforated metal pipe is inserted in the holes while drilling or immediately afterward.

Long slopes may be benched, as in (B), to break the flow of surface water. Each bench has a reverse slope so that it acts as a diversion ditch, with water flowing along the back. A gentle grade spills the water toward one or both ends of the cut.

Benches may also serve to catch falling rock. Their effectiveness for this is increased by a berm of dirt along the outer edge. This construction is particularly useful in banks of cemented gravel, 1 on 1 or steeper, from which surface cobbles are released.

Slopes can be stabilized by growth of vegetation. Most types will provide surface protection, and types with deep or interlocking roots may hold against some internal pressure as well.

Artificial protections include supporting walls, drainage systems to intercept or remove ground water, and fences to catch rolling pieces.

Walls may be of masonry, interlocked concrete, or metal bins. Strength of the last two constructions depends on their being filled with coarse, pervious fill. Any of these must rest on a solid footing that can resist both weight and thrust.

Logs can be used for temporary retaining walls and to catch boulders rolling during work.

Drainage. Freshly worked embankments should be protected against surface water flowing from adjoining ground. In cuts, a diversion ditch may be dug a few feet back from the upper edge. Unless its gradient is gentle, its bottom may need protection to prevent it from developing into a gully that would damage land below it, and eventually break out through the bank.

Such protection may include establishment of a strong sod, construction of a series of check dams, paving with resistant materials, diversion of some of the natural drainage at higher points, or use of discharge flumes down the slope.

If the slope is threatened by softening or washing by ground water, subdrainage may be required also. Land tile may be laid under the surface channel if its floor is impervious enough not to allow excessive surface water to enter the tile. Underdrainage may be required in the gutter at the foot of the slope, and in or behind wet spots in the slope to catch the seepage.

In areas of rapid runoff a highway may be protected along its entire uphill side by a system of diversion ditches that channel all drainage into culverts or across dips. For economic reasons this type of work is limited to diggable soils on slopes that are accessible to machinery.

Fills usually have less drainage across them, but because they are not as well bonded together, they are more subject to surface erosion than cuts. Water may flow onto them from the road and from slopes above the road. They can be protected by berms along the outer edge of the road shoulders, which will prevent water from going down the side of the fill, except at points protected by pipes, flumes, or pavement.

Fills which are built on sidehills have a tendency to slide along the old surface, unless it is well roughened. Leaving of stumps and boulders, roughening by plowing, or placing of subdrains to stop seepage of water along the joint, are common methods of reducing this danger.

Any soil, whether original bank or fill, which rests on smooth steep rock slopes is liable to slide. The most important step in preventing slippage is diverting ground water moving down the surface of the rock.

Grading. Steep side slopes should be finish-graded as they are made, as it may be difficult and dangerous to work them afterward. But if it becomes necessary, a wide track dozer may work a long slope in strips, from the top down or diagonally.

Horizontal trimming by use of graders or dozers on steep side slopes may be made safer by cabling to another machine moving parallel to it on the top of the bank. Two cables are used attached to the front and rear of the lower machine.

It is not safe to operate unsupported heavy equipment along slopes which contain rocks, soft spots, or frozen ground.

Topsoiling. The best protection for a dirt slope is a good cover of vegetation. Grass, weeds, bushes, and trees are all effective controllers of erosion. The type selected will depend on the locality, soil, and season.

On most jobs, it is necessary to place a layer of topsoil over the fill or exposed earth in order to get a good growth. Occasionally plants will grow well enough on raw earth, or with the aid of some lime or fertilizer.

Deep topsoil is favorable to growth but it may discourage plants from rooting into the subsoil, and absorb too much water so that it will slide off during rains. For this reason, and for economy, topsoiling of steep slopes is usually limited to a depth of two to four inches.

The fill surface should be roughened so as to bond with the topsoil. A sheepsfoot or tamping roller is one of the best tools for accomplishing this. If the slope cannot be

Fig. 8-15. Finishing a slope with a grader

worked, the roller may be operated by a dragline at the top. The drag cable is used to pull the roller up and to let it down, and the walking of the shovel moves it along the slope.

Topsoil may be pushed up a slope from stockpiles at the bottom, pushed down it from piles trucked to the top, or distributed over the surface by a clamshell working from either top or bottom, and the resulting piles shoveled or raked out by hand.

The Gradall is an excellent tool for final shaping of the subsoil and spreading topsoil on any area it can reach.

Freshly spread topsoil gullies readily and needs protection on slopes. A thin coat of sprayed asphalt emulsion will carry off rain, and still allow grass to grow through it.

A layer of hay or straw may be mixed into topsoil by a tamping roller to hold it. The hay should be well cured, as rapid decay would make its useful life too short. It is apt to absorb so much nitrogen from the soil as to interfere with growth of seedlings. Use of barn straw that contains some manure, or adding nitrogen fertilizer, cures this difficulty.

Some hay and straw contains enough grass and weed seeds to establish a good cover. Other types are deficient and require that the ground be seeded. Seed can be mixed with water and sprayed onto slopes.

On small areas, topsoil may be held by adding straw, and holding it with chicken wire firmly pegged down. Horizontal wood slats are sometimes used. Placing and tamping cut sod in drainageways, in horizontal strips on slopes, or on the whole surface, is very effective but the cost is high.

Rock Faces. Rock cuts can be left with very steep or vertical faces, and occasionally are allowed to overhang. Such faces usually cause a hazard of rock falls to the pavement, but the expense of cutting rock back to completely safe slopes can seldom be justified.

Some rock formations tend to break up into gravel or small stones at the face because of temperature changes, and will at times subject the road to an almost continuous bombardment. Such faces should be cut back sufficiently to permit a wall or fence to be put beside the road, with space behind to catch falling stones.

More massive cliffs may present the danger of occasional falls of larger rocks or of whole sections. These may be checked in the danger season by a man with a bar, supported by a rope held at the top. Loose pieces can be pried out.

Long expansion bolts, similar to those used to secure tunnel roofs, can be placed to fasten a whole slope into a solid and safe unit. They are particularly efficient in shale beds parallel to the slope.

Vegetation tends to break up rock faces, so artificial planting should not be attempted.

THROUGH CUTS

A through cut has a high wall on each side so that little or no material can be excavated by sidecasting.

If it is on a sidehill, one edge will be higher than the other. The part of the cut which is above the low wall is actually a sidehill cut, and may be handled as one or as a through cut.

Through cuts are seldom used in building pioneer roads, except where borrow is needed to cross a ravine. When roads are

Fig. 8-16. Slope too steep to work

narrow, and the sharpness of curves is not an important consideration, sidehill work is faster and less expensive.

SCRAPERS

Scrapers, which are described in Chapter 17, are the standard excavator for alternating cuts and fills, where the soil is soft or fine enough for them to work, or can be made so by rippers and explosives, and where the haul is too long for dozers.

Preparation. The first requirement is to smooth over the cut and the fill areas so that scrapers can work them. This is usually a dozer job. The ground is cleared of vegetation and boulders, holes and gullies broken in and ramped over, sharp ridges beveled off, side slopes notched, and turning places graded off.

It is not absolutely necessary to prepare the whole area in order to have the scrapers move in. Their work can start on the high part of the cut and the low part of the fill, while the dozers are clearing and smoothing the balance of the area.

If the cut has a high side, it is cut to a passable driveway by straight pushing or sidecasting. The bottom of this cut is sloped oppositely to the hill.

If the hill is high in the center of the cut, the hump is graded off sufficiently to afford good footing for scrapers.

It is sometimes economical to make small fills in areas which are to be lowered, and small cuts under future fills in order to smooth out working areas quickly.

When a dozer is not available, a scraper can smooth moderately rough ground by driving through it with the knife held low enough to cut off the bumps and high spots. If the tailgate is held near dumped position, it will act as a dozer blade.

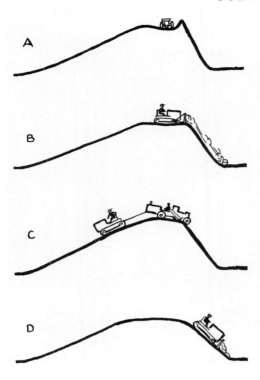

Fig. 8-17. Breaking down steep slope

Scraper work on side slopes is simplified by first cutting a shelf with a dozer. If no dozer is available, the scraper can be taken uphill to the start of the cut, the blade dropped, and the scraper turned to dig along the upper cut line. The turn will cause the edge to cut deepest on the uphill side, and if done repeatedly, will level the digging area, or slope it oppositely from the hill.

Cutting Ridgetops. If the slope up from the fill is too steep for the scraper to climb, it may be broken down into a ramp by dozers, or the cut made with shovels.

If the slope away from the fill is too steep for scrapers, as in Figure 8-16, the top can be lowered by the combined work of scrapers and dozers, as shown in Figure 8-17. Full trailer scrapers will dig across the cut as they turn, as in (A). Semi-trailers can be backed up to the edge, as

in (C), and if a snatch tractor is available, can be backed over it. Digging is then done straight toward the fill.

The undug lip left by the first method is pushed over the edge by a dozer, as in (B). This filling, and the cutting into the slope, will extend the floor and allow scrapers to work farther back.

Eventually the bank will be lowered sufficiently to make it practical to break it down with dozers (D), so that scrapers can go through to dump on the far side, or turn to continue hauling in the original direction.

Shaping. The outer edges of the slope should be determined before starting work so that steep banks may be cut to final grade from the first. They are taken down in a series of steps. If the slope is one on three, and the pans are taking six inch slices each new cut should be eighteen inches further from the bank than the previous one.

The slope should be checked frequently by engineers for correctness, and trimmed off by a grader working on the floor of the cut, as it may become very difficult to reshape when the floor has been cut too far down.

The floor of the cut should slope down toward the edges. This slope may be originally established by an angle dozer or a

1 WORK IN LAYERS ACROSS AREA TO BE SLOPED
2 WORK FROM TOP OF SLOPE DOWN
3 AFTER EACH LAYER IS CUT, START NEXT CUT AGAINST SLOPE, ALLOWING BLADE TO CUT TO GRADE
4 STEPS WILL REMAIN TO BE FINISHED BY OTHER METHODS OR EQUIPMENT

Courtesy of U.S. Army Engineers

Fig. 8-18. Cutting slopes with scraper

grader, after which it will tend to perpetuate itself, as the weight of the machine will be greater on the down side so that it will tend to cut low there. If the slope becomes too great the upper part may be readily planed off.

Machinery may not be available to shape the original surface, or the crown may be lost because of oppositely sloping strata, or by careless operation. A scraper can cut a crown by taking advantage of the fact that the oscillating tractor part does not affect the side-to-side tilt of the knife, which is determined almost entirely by the rear axle. A gouge taken heading up a slope can be used to tilt the rear axle so that the knife can cut on the uphill side when turned along the slope. See Figure' 8-20.

Whenever possible, the cut should be arranged for digging downhill and toward the fill. The first factor is usually more important. The grade of the cut is most important when it is or can be made steep, and when power and traction are small in proportion to the size load desired.

To facilitate rapid movement and easy loading, it is important to keep the pit from getting too rough or ridged.

Hard Digging. Scrapers can penetrate

Courtesy of U.S. Army Engineers

Fig. 8-19.   Cross section of scraper cut

fairly hard soils, as the cutting edge is sharp and held at an effective angle. The machine cannot be overbalanced by suction, as the knife is carried between the axles, but a plastic soil may pull the edge a few inches deeper than intended by flexing the tires or causing them to sink.

There are hardpans and rocky soils which the knife will not cut and many others which can be dug only by the ex-

If no bulldozer is available, scrapers can pioneer side-hill cuts. At the point where the cut should be started, turn sharply toward the row of stakes above you.

When tractor has completed a full turn, drop the cutting edge to start loading. As the scraper turns, the hillside corner of the blade will cut into the high side of the hill.

In this manner a bench will be started. Repeat this procedure whenever loading next to the bank. Result will be cuts that always slope toward the high side. This prevents scraper from drifting.

Fig. 8-20.   Scraper pioneering a sidehill cut

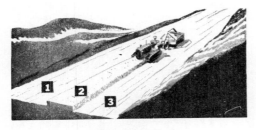

Courtesy of U.S. Army Engineers

Fig. 8-21. Straddle loading

penditure of so much power in penetrating that little force is left for the loading. In such cases, the use of rippers ahead of the pans is advisable. Single or widely spaced teeth give best results, as coarsely broken ground is usually easier to load than that which is reduced to very fine pieces or pulverized.

In hard digging, a straddle loading sequence is often helpful. Parallel cuts are made, leaving a ridge between which is narrower than the bowl. The ridge is then taken out on a third pass, and it will be found that the digging resistance is more nearly in proportion to the shallow cuts at each side than to the deep one under the ridge. See Figure 17-33.

This method should be used with caution near the edges of cuts as it may destroy the crown.

If bedrock needing blasting is found in the cut, a dozer might strip the overlying soil and push it out to be picked up by the scrapers. A scraper is more vulnerable to damage from contact with rock, and its loading will be slowed by any effort made by the operator to avoid such damage.

Trimming Banks. Successive cuts are set back from the edge to provide proper slope. Scrapers will not cut vertical walls but will leave very steep faces.

Slopes between 1 on 1 and 1 on 4 are usual in soil cuts. If the scraper takes a six inch slice, a 1 on 2 slope would require each pass to be a foot inside the last. If 1 on 4, it would be spaced two feet.

The steps are best trimmed to a smooth slope by a grader working on the floor. The excess material is cut and slides to the bottom to be removed by the scrapers.

If trimming is done with a scraper, one rear wheel should be on the bottom, the other on the slope. If it is steep, the tailgate should be carried well forward so that loosened dirt will slide downhill rather than enter the bowl.

The cut should not be deepened so far between trimmings that the grader cannot reach all the steps. This is particularly important when the slope is so steep that it cannot be worked by machinery later.

As the cut deepens, new slope stakes are placed. They are often set from the originals with a string level, rule, and plumb bob. If driven in flush and marked with light sticks, a good grader operator can trim the bank without knocking them out.

Finishing Subgrade. The bulk of a deep cut can be made without staking except for the slopes. As it approaches bottom, however, grade stakes should be set, and digging done with sufficient care to avoid overcutting and resulting need for patch fills.

Good scraper operators can hit a grade within a fraction of an inch if the soil is smooth, but it is often better economy to have them run a rough grade and go on to other work while a grader finishes up. The grader will probably be required to cut and shape gutters, in any event.

The road, or the shoulders, are sometimes overcut to allow space for spoil from ditches. If this is not done, ditch cuttings may be windrowed on the road for later removal.

Any areas that are cut and refilled in this manner must be thoroughly compacted. A few patches may be rolled by trucks or scrapers, but regular rolling equipment should be brought in for any extensive areas.

Failure to compact may lead to local settlement and pavement failure.

Courtesy of Caterpillar Tractor Co.

Fig. 8-22. Scraper and pusher

Selected Base. A layer of porous soil with high bearing strength is usually placed between the native soil and the pavement, in both cuts and fills. If it is obtained from nearby pits it will probably be economical to bring it in and spread it with scrapers.

The pit wall or floor may be shaped for direct scraper digging. If this is not possible, hauling may be done with either scrapers or trucks loaded by shovels or other equipment. The scrapers offer the advantage of doing their own spreading.

Self-loading scrapers are good in this work, where quantities are relatively small and pit shape is irregular, as they do not have to be part of a balanced spread which may be difficult to organize.

PUSHER

The pusher is a separate tractor that pushes the scraper while it is loading. It is almost always required for efficient full loading of single engine self-powered scrapers, and is usually desirable with two engine scrapers, and with crawler drawn scrapers that are oversize for the tractor or that are in hard digging.

Types. Pushers are usually crawlers with tractor weights of 20 tons or more.

There are also four wheel drive pushers with similar or greater total weight. These have the advantage of much greater speed. They can often get a scraper through the cut much more rapidly and with less scraper strain and wheel spin than a crawler of

similar weight can, and they can make fast moves from one scraper to the next, and between jobs. However, they do not have nearly as much push at stall speeds, they may lose their speed advantage by requiring shallower and longer cuts. Tires lose traction badly on wet slippery surfaces.

Crawler tractors with torque converters have maximum pushing power and traction, and have some speed flexibility.

Several thousand dollars may be saved by buying a pusher with a fixed push plate instead of a center-reinforced dozer blade. However, the fixed plate will not permit the tractor to do cleanup or backripping work in the pit, and it is likely to find little use outside of it. It may cause difficulty by not meeting the scraper at an efficient angle, or by losing contact on rough ground.

A good push plate needs little maintenance, but even the most heavily reinforced dozer blades are likely to cave in when used for scraper pushing.

There are also small push plates that mount on the C-frame of a dozer or angle dozer. This installation is more expensive than the fixed plate, but permits accurate lining up with scrapers, can be used for pushing big boulders, and can be easily replaced with a dozer blade.

Either type of plate is much less likely to damage scraper rear tires than a dozer blade is.

Loading Effect. Soil and slope conditions being equal, loading time and the size of load in a particular pan are determined by the power applied to it, regardless of whether that power comes from one, two, or three tractors.

The rule of thumb on the subject is that a pound of push puts a pound of dirt a minute in the scraper. This "push" includes the scraper's effort.

Tandem Pushers. It follows that size of scraper loads and speed of loading can both be increased by increasing pusher power. Additional power may be obtained by using bigger tractors, using two or more tractors together, or in both ways.

The standard arrangement for pushing with two or more tractors is to line them up behind the scraper. The front dozer pushes the scraper, it is pushed by the machine behind, and sometimes that is pushed by still another. However, two pushers are much more usual than three.

These tractors must have push brackets at the rear, fastened to the side frames so as to put the thrust directly against the dozer arms.

Tandem pushing involves extra delays in getting the machines in contact with each other and loading, but it usually more than makes up for this in extra speed and depth of slice in the cut.

The effects of the extra power will be discussed later in this section under the heading "Super Pushing."

Snatch Tractors. A few contractors use a pull tractor instead of or in addition to pusher tractor(s) for loading scrapers. Coupling may be by means of a short tow cable, or preferably by a coupling that locks automatically on contact and can be opened by the puller operator from his seat.

The tow cable requires a full stop and the services of a ground man; the automatic coupler a high degree of operating skill.

Snatch tractors show good results with experienced crews, but most contractors are better off if they stick to pushing.

Scraper-Pushers. Scrapers may be

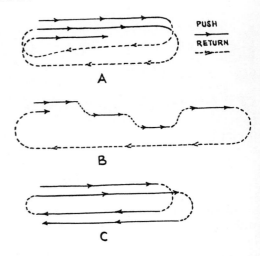

Fig. 8-23. Pusher patterns

equipped with pusher plates or blades so that they can push each other to help in loading. This arrangement is usually not as efficient as using a separate pusher. A scraper cannot deliver nearly the push of a tractor of the same price. There is a chain loading effect that requires a long cut or better than average supervision.

There are a few jobs where scraper-pushers have worked out well. They may save or postpone a big investment in a crawler tractor, or fill in time while a pusher is out of service.

Grader. A heavy grader equipped with a dozer blade may make a good pusher, and may be used for this purpose in an emergency. Under most conditions, however, it is much more valuable as a first rate finishing and maintenance machine than as a second rate pusher.

Super Pushing. Advantages in speed of loading and size of loading are obtained by "super pushing"—using extra large pushers, or two or even three standard units in tandem. Whether it is economical to do so must be carefully figured out on each job.

An important factor that must be taken into account is the ability of the scrapers to stand up in such heavy service.

Many large scrapers that are now in service, or in process of manufacture, have been designed for use in medium to easy digging, and have shown high repair costs and short life when used in hard, rough soil. These units are likely to have the same serious difficulties when pushed by too much power, even when the ground conditions are good.

Also, scrapers that stand up well under all conditions with ordinary pusher power may break up under a combination of super pushing and rough or hard ground.

Loading on Grades. The effect of gravity is to pull a load going downhill to the extent of about 20 pounds per ton of weight per per cent of grade. An uphill load will be held back to the same extent.

This will cause a pusher weighing 25 tons to gain or lose 1,000 pounds in net drawbar pull on a 2 per cent grade, while a 50 ton loaded scraper will be affected doubly.

The difference between a 2 per cent up grade and 2 per cent down grade will be 5,000 pounds of pull, about one seventh of the pusher's power. Using the rule of thumb that a pound of push produces a pound of dirt a minute in the scraper, direction on this grade would mean a difference of three tons or two yards of earth in the load, or a proportionate amount of time in obtaining the load.

It is therefore profitable to keep the loading down grade even on a gentle slope, whenever job conditions permit.

Pushing. The pusher is driven up behind the scraper in low or sometimes in second gear, and contact with its bumper is made as smoothly as possible. The scraper unit, in low gear, applies as much power as it can without spinning the wheels or drawing away from the pusher.

The pusher moves as fast as it can without making the scraper jackknife or either unit twist sideward. Twisting or off-center pushing is likely to cause a dozer blade edge to rip into a tire.

Most pushing is done in low gear because it is important to keep going right through the pass. Second gear may be used to advantage when the pusher is overpowered for the size of the scraper or the load it is to carry. There is an extra hazard of damage from making contact too roughly. Cushion dozer blades may be used to reduce the shock.

Patterns. The simplest pusher loading pattern is called back-track or shuttle loading. See Figure 8-23(A). The scraper drops its bowl at the beginning of the cut, and the pusher makes contact. The scraper pulls and the pusher pushes until the desired load is obtained. The pan bowl is raised, informing the pusher operator that he is not needed, if he has not already learned this from watching the load. The scraper tractor then shifts into higher gear and departs.

The pusher returns to the beginning of the cut. It may make this move in reverse, or by turning and using forward speeds. The length of the cut and the relative speeds the machine can make in the highest usable reverse and forward speeds, determine which should be used.

If the next scraper gets in loading position before return of the pusher, it will start to load itself, as it can pick up part of a load readily, and will thereby decrease the distance the pusher must come back in order to get behind it.

Another pattern, shown in (B), is suitable only for long cuts. After loading the first scraper the pusher waits for the next empty one to come up along side it, pushes that until it is full; pushes the next from its stopping point, and so on until the end of the cut is reached. The pusher is then turned and run back to the beginning of the cut to start another series.

A third system, which is useful where there are more scrapers than the pusher can readily handle, or where the dirt can be moved in two directions, is outlined in (C). Each scraper is loaded moving in the opposite direction from the previous one, so that the pusher need only turn around to be in position for the next push, instead of having to move back to a starting place.

A pusher should have as many scrapers as it can conveniently handle, but it is difficult to maintain a proper proportion because of changes in the length of haul. Two pans might keep a pusher busy on a very short haul, where a dozen might not work it steadily on a long run.

Where there are more units than a pusher can service so that one or more are waiting, and it is not possible to shift any to longer runs, it may be wise to have the more powerful tractors or those with the shortest runs to load without assistance, so that all can be kept moving, although with a smaller average yardage.

Time and Distance. A scraper loading pass may take from 20 seconds to 2½ minutes. It is best to keep it down to a minute or less.

Speed of both machines in the cut is determined by that of the pusher. Low gear speed of most gearshift crawler tractors is about 1½ miles an hour, say from 2.2 to 2.5 feet per second. In heavy pushing speed may be reduced below a foot per second by track slippage. Torque converter tractors in low gear may go up to 4 miles an hour, or 6 feet per second when loads are light, but in heavy pushing may be slower than clutch type units.

Length of the pass should be kept down between 90 and 125 feet, as long runs are wasteful of both pusher and scraper time. The pusher usually has to go back the same distance it worked forward, and the scraper should shift up and start its haul.

Rubber tired pushers may have low gear speeds of 2½ miles an hour or more, and usually have the speed flexibility of the torque converter as well. The higher pushing speed synchronizes better with the low gear speed of scrapers. Very fast travel gears are available for the return trip.

The higher speed of the four wheel drive pushers makes it practical to use longer runs in the cut, perhaps up to 200 feet. But the extra length in the pass largely cancels out the speed advantage.

Quickest loads, both in time and distance, are obtained with easy digging, downhill loading, and small loads in proportion to total digging power.

When job study figures are not available, an estimator may assume that an average push is one minute long. This push represents about half the pusher cycle, as it will spend about as much time getting to and contacting each scraper as it does pushing it.

This gives a rule of thumb pusher cycle time of two minutes. With good operation and supervision it should be much shorter, but under field conditions it is just as likely to be longer.

Scrapers Serviced. The number of scrapers that can be serviced by a pusher depends on the relationship between the length of the pusher and scraper cycles. The formula is:

$$\text{Scrapers per pusher} = \frac{\text{Scraper cycle}}{\text{Pusher cycle}}$$

If a scraper cycle were five minutes (300 seconds) and the pusher cycle two minutes (120 seconds), the pusher could take care of 2½ scrapers. This could be managed on a big job by using two pushers and five scrapers, but on a small one the contractor would run either three or two.

Three scrapers would mean waiting time for scrapers; two of them would not keep the pusher busy.

If pushing could be speeded up by more efficient patterns, better operating skills, or a faster machine, so that the pusher cycle would be reduced to 100 seconds, one pusher could take care of the three scrapers.

The pusher cycle might be shortened to 90 seconds by keeping the scraper in the cut only 45 seconds and sending it out with a lighter load. Scraper cycle would be short-

Number of scrapers	3	2	3	3	5	5
Number of pushers	1	1	1	1	2	2
Push time, seconds	60	60	60	45	60	45
Scraper cycle time, seconds	300	300	300	275	300	275
Scraper waiting time for pusher	60	—	—	—	—	—
Pusher cycle time, seconds	120	120	100	90	120	90
Size of load, cubic yards	18	18	18	16	18	16
Loads per 45 minute hour	22.5	18	27	29.5	45	49
Yards per 45 minute hour	405	324	486	472	800	784
Scraper cost @ $25 per hour	$75	$50	$75	$75	$125	$125
Pusher cost @ $20 per hour	20	20	20	20	40	40
Equipment cost per hour, total	95	70	95	95	165	165
Cost per cubic yard, cents	23.4	21.6	19.6	20.1	20.6	21.0

Fig. 8-24. Costs in scraper and pusher cycles

ened by 15 seconds in the cut and by about 10 seconds in the haul and dump, by faster acceleration and faster spreading. On this basis one pusher could take care of three scrapers.

Figure 8-24 shows how these various arrangements would work out on a basis of a 45 minute hour, an hourly cost of $25 for a scraper and $20 for a pusher, an 18 yard load in 60 seconds, and a 16 yard load in 45 seconds.

Other cost calculations on scraper and pusher costs will be found in Chapter 17.

Other Work for Pushers. In ordinary backtrack work the pusher uses half its travel distance and about a third of its time moving from loaded scrapers to empty ones, usually in reverse. In addition, it will spend a variable but often considerable amount of its time waiting for scrapers. The thrifty contractor will wish to make profitable use of this non-productive time.

Backup time of dozer pushers may be used either to smooth the surface of the cut by backdragging, or to loosen the ground with backripper teeth hinged to the blade.

This ground smoothing requires no extra equipment on a dozer pusher, and uses only a little extra time to vary return paths. It is quite effective in loose or sandy soils, but results may be poor on hard or stony ground.

Grading that must be done in forward gears reduces the time available for pushing scrapers, and the effort to combine duties is apt to lead to inefficiency in both assignments.

Ripping. If the ground is too hard for good scraper loading it may pay to rip it. This work can often be done by the pusher.

Back rippers are economical to buy, and permit doing most of the ripping while the pusher is backing from one scraper to the next. This usually does not work out so that the whole area can be loosened, but if the scrapers are able to load the material anyhow, whatever work the teeth do is so much to the good.

Use of the teeth may cost $1.50 to $3.00 an hour, for ripping 200 to 500 yards. The pusher may not use any extra time for this work. On the other hand, the operator may take his ripping so seriously that he will let scrapers wait while he finishes a strip. This can result in considerable loss of production.

Soils too hard for back rippers, and many rock formations, can be ripped for scraper loading by a rear mounted ripper with hydraulic down pressure.

Methods of use and rate of production of

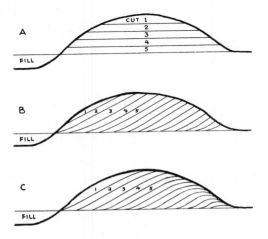

Fig. 8-25. Taking off a hill

these tools are discussed in Chapter 21. In most soils a big tractor should be able to keep three teeth in the ground, and break soil much faster than an ordinary fleet of scrapers could remove it. In frost or rock one tooth would be used, and output might be so reduced that one scraper could handle it.

A pusher can do heavy ripping only if it has enough time between scrapers to prepare enough soil for them. If it does not have this much time the ripping will be incomplete or the scrapers will have to wait, or both. Each case must be judged on its own merits, but under such conditions it is usually more efficient to use one tractor for ripping and another for pushing.

If this arrangement leaves the ripper with idle time, it can be used as a tanden pusher when it is available, with the regular pusher carrying the whole load during ripping periods.

Economics of Ripping. The Bureau of Public Roads has found no difference in loading time or in size of loads between ripped and unripped jobs. This may be taken to mean that ripping usually serves only to cancel out the disadvantage of working in hard soil, and that such soils when ripped have no better loading quali-

ties than ordinary soils have in their natural condition.

Many soils become so loose when they are broken up that they are harder to load, or provide smaller loads, than when in their solid state. Unless loading results are clearly and definitely favorable, ripping should not be done.

Ripping rock for scrapers pays only when a fairly fine and uniform breakage is obtained. Scrapers load badly in coarse rock, and unless specially reinforced, may sustain severe damage from repeated contacts with big pieces and unbroken ledge.

WORK PATTERNS

Scraper work patterns should be arranged to allow for as many of the following as possible:

1. Digging downhill.
2. Digging in the direction of the work.
3. Utilization of pushed soil.
4. Efficient turns with minimum deadheading.
5. Start cuts at high points, and fills at low ones.

Direction of Digging. A favorable grade increases the speed and the effectiveness of loading (see page 8-31), and reduces wear. The advantage becomes more marked as the downgrades get steeper.

Figure 8-25 shows three ways to make a deep scraper cut. (A) is inefficient because the downgrade is used in transporting where little power is required and does not assist the digging. (B) takes full advantage of the downgrade but may create an inconveniently sharp angle at the beginning of the cut.

In (C) the digging is started on the upgrade, just before the crest. The power requirement for the first few yards is small as resistance increases with load. The machine is rounded into the downgrade for the bulk of the load. This keeps the crest cut down without sacrificing much of the advantage of the slope.

Digging in the direction of the work is desirable. A loaded pan moves slower, wears more, uses more fuel, and may be less stable on turns than an empty one. If the load is picked up heading toward the fill, it is able to take the shortest path to the dump, and to make the turns and the longer run beween them empty.

However, there is often sufficient reason for digging away from the dump. Digging downhill is more important than direction. Occasionally a pusher can be best utilized if scrapers are loading in both directions, which, in a single cut and fill, would require about half the units load before turning to go to the fill.

Pushed Dirt. The scraper knife usually pushes some dirt ahead of it, the amount increasing with the size of the load. Loose material such as sand may be moved in considerable quantities. This is left in low piles when the bowl is lifted.

This dirt can be utilized to build up the fill where it meets the cut, by allowing the bowl to drag slightly until the fill is reached. However, dragging may cause a loss of speed which outweighs the importance of the dirt moved.

Care should be taken not to cut below grade at the junction with the fill unless it is necessary to make a ramp.

Turns. The time consumed in making a U turn with a scraper may vary from five to sixty seconds or more, depending on space available, ground conditions, type of machine, traffic, and operator. Fifteen seconds is a fair average.

Time consumed deadheading from the working area to the turn and back, may be considered part of either the turn or the haul, but it is better practice to consider it a separate part of the cycle.

On short hauls, turns and deadhead time have an important effect on production. As hauls become longer, their significance decreases.

There are four major patterns of scraper

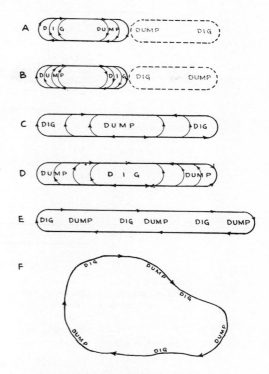

Fig. 8-26. Scraper patterns

operation which are shown in Figure 8-26. In the first, (A) and (B), there are two turns to each dig-dump cycle, in the second, (C) and (D), one, and in (E) one half. (F), with no U turns, is only practical when a very wide area, such as a field or runway, is being graded.

When both cut and fill are wide enough for easy turns, the (A) and (B) layouts may be most efficient, particularly when work areas are long and tractor speeds low. The advantage is that the scraper can turn to start a new cycle immediately after digging or dumping. The length of haul can therefore be figured between the centers of mass of the cut and the fill, as the longer and shorter runs will average out.

The diagram and arithmetic in Figure 8-27 indicate the advantage of operating one cut with one fill under low speed conditions, and combining them when travel speeds are higher.

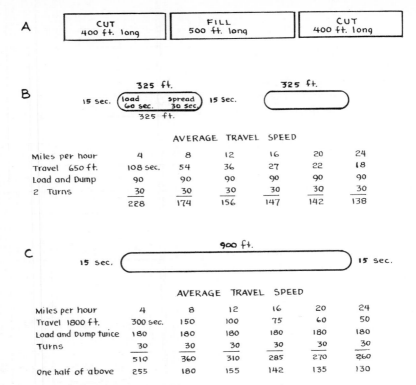

A — CUT 400 ft. long | FILL 500 ft. long | CUT 400 ft. long

B — 325 ft. | 325 ft.
15 sec. (load 60 sec. / spread 30 sec.) 15 sec.
325 ft.

AVERAGE TRAVEL SPEED

Miles per hour	4	8	12	16	20	24
Travel 650 ft.	108 sec.	54	36	27	22	18
Load and Dump	90	90	90	90	90	90
2 Turns	30	30	30	30	30	30
	228	174	156	147	142	138

C — 900 ft.
15 sec. 15 sec.

AVERAGE TRAVEL SPEED

Miles per hour	4	8	12	16	20	24
Travel 1800 ft.	300 sec.	150	100	75	60	50
Load and Dump twice	180	180	180	180	180	180
Turns	30	30	30	30	30	30
	510	360	310	285	270	260
One half of above	255	180	155	142	135	130

Fig. 8-27. Turns may save time

If turns cannot be made immediately at the end of the work, the time required to travel the average distance from the ends of spreading runs, and from the beginning of cuts to their turns, must be added to the cycles, as in Figure 8-28.

Through travel highway patterns, Figure 8-26 (E), have their greatest use where the graded area is too narrow for turns, and cuts and fills are rather short and closely spaced. Their efficiency depends on the extra time required for through travel, compared with that used for turns and deadheading.

Turn patterns are of course subject to the overall plan for the job which may include some very complex factors of distribution, work sequence, and separating of different types of soil.

These examples are somewhat oversimplified for demonstration purposes.

Turnarounds. The location of the turnaround in a narrow one-way cut is affected by the difficulty of making it. For efficiency, it should be slightly across the hill crest from the fill, so that the scraper can be straightened out to load just before it crosses the crest. However, the digging will work the crest back and destroy the turnaround quite soon. It may therefore be wise to locate it well back from the crest.

Whenever possible, a turning place should be wide enough for the machines using it to get around without backing. Space requirements vary greatly in different sizes and types of scrapers. Time may be saved and accidents reduced, by providing more space than the minimum requirement, particularly for sharp-turning models.

A big scraper will need 30 to 50 feet to make a U-turn (180 degrees) on good ground, up to 80 feet on bad footing. If it is not possible to grade enough space for a full turn where it is needed, the machine must back one or more times, or go on until space is available.

The bottom of a cut for a four lane undivided highway will usually allow somewhat more than tight turn space for a scraper, two lane cuts may not permit a non stop turn. Wider roads and the upper parts of any deep cut will allow ample space.

Deadheading. If the cut and/or the fill are too narrow for turns, or there are traffic difficulties, the scrapers will go past the work areas to a turnaround. The extra travel is called deadheading, and is usually for only short distances.

At the cut the trip to the turnaround will be made in one direction at full travel speed, except for any necessary slowing for traffic, while the return will be at the speed of the turn or only slightly better, because of lack of space for acceleration. Beyond the fill the turn will be approached at dumping speed, the return from it will be a part of the empty return and will be at full acceleration if traffic allows.

Deadheading distances are often increased unnecessarily by careless dumping or loading that leaves the fill or cut too rough for a turn, or by placing too many grade stakes to permit turning in the work area.

Intermediate Hauls. Some hauls are too short for normal scraper use and too long for dozers.

A crawler dozer's output falls rapidly as length of haul increases. It becomes uneconomical somewhere between 100 and 200 feet on level ground, although it may be used for much longer distances. Because of complications of turning and distances required to load and to spread, scraper use generally starts at two to four hundred feet. These figures are for large machines. Small bulldozers lose effectiveness at lesser distances, and small scrapers are operated on much shorter runs.

Because of their higher speed, rubber-tired dozers may be used for substantially longer pushes than crawlers. They are less

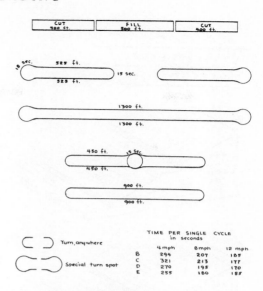

Fig. 8-28. Fixed turn patterns

efficient in a short run, as except in the easiest digging, the loading part of their cycle is longer.

Scrapers can often be used more efficiently on very short hauls than is generally believed. They can waste considerable time deadheading to turns, and carry undersize loads because of short digging runs, and still move much more dirt than a dozer.

When the fill is too short for proper spreading, part of the load can be carried around the turn and dropped on the way back to the cut.

Semi-trailer scrapers can be used shuttle fashion. The load is spread dumped, and the machine backed into the cut for another. In each pass some of the spoil will be moved, dozer fashion, in front of the knife, and dropped at the beginning of the fill.

Two-wheel drag scrapers, such as are used for land leveling, can be used shuttle fashion and in easy digging will move about double the load of a dozer blade. Large models are very wide so as to be inconvenient to transport and unable to work

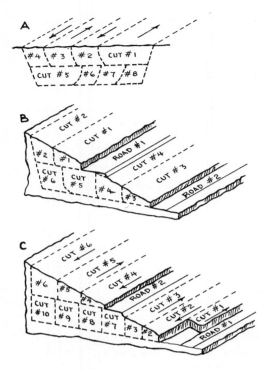

Fig. 8-29. Sequence of shovel cuts

with other machines in ordinary roadways.

Haul Speed. Scrapers have advertised top speeds from 20 to 30 and even 40 miles per hour. However, the average construction job does not permit much use of these high speeds, and even when they are used, the average speed straightaway hauling is very much lower.

Top speed is limited by the space needed to attain it. A loaded scraper or other hauler in its highest gear has little surplus power for acceleration, after taking care of the rolling resistance of big tires and dirt roads. It might take such a machine a half mile of level road to reach top speed in its top gear, and this distance is longer than the average scraper haul.

A favorable grade will encourage rapid acceleration, but it is likely to make high speed dangerous. And for every favorable grade in a round trip there must be an unfavorable one to balance it.

Scrapers are rough riding vehicles. While the big tires absorb bumps, no way has been found to fit them with shock absorbers, and bouncing will prohibit full speed unless exceptional attention is paid to maintaining the haul road.

Scrapers with overhung, two-wheel tractors are likely to develop a rhythmic galloping effect even on smooth roads, unless equipped with a cushion hitch.

Hauler speed is discussed in some detail in Chapter 12 and later in this chapter. When there is no time to work out level ground performance in detail, figure 40 per cent of top speed in the highest gear to be used for hauls up to 1,500 feet, 50 per cent to 2,500 feet, and 60 per cent from there on.

SHOVEL CUTS

Through cuts are made by dipper shovels and trucks when the original surface cannot be readily leveled for scraper operation, when the ground is rocky or wet, when the fill is too soft or too narrow for surface dumping, and when the haul is too long for scrapers.

In general, shovels do best in banks that are about as high as their shipper shafts. Soft, sliding banks of sand or gravel provide best digging when they are very high, but there may be danger from slides.

If the cut is considerably deeper than the favorable bank height, it may be taken in two or more layers or benches. On through cuts with a fairly level cross section, as in Figure 8-29 (A), the top is removed first. On sidehills there is an option of taking the top first, as in (B), or cutting the toe, then the top, then the floor of the upper cut, as in (C).

It is first necessary to build a haul road between the start of the shovel cut and the fill. Most of this may be already provided by highways or construction grading. It should be wide enough for two trucks, although for short or small jobs, one lane with turnouts may be adequate. Slope up to the shovel should not be over fifteen percent, although in special cases, grades up to 35 percent have been used.

If no natural turnaround exists at the start of the excavation, and the grade is easy, trucks may be backed in at first. As excavation progresses, the pit floor will provide turning space.

The roads in (B) and (C) are usually cut by dozers. They allow a rotary movement of one-way traffic past the shovel on its first cut on each level. They eliminate the necessity of turning at the shovel in cramped quarters. However, they may be too expensive or inconvenient to build.

When there is no through road, the shovel starts each level by taking as wide a cut as it can reach, as in Figure 8-30 (A), to allow space for two-way traffic, turning, and spotting two trucks at a time. Subsequent cuts are made about half as wide, as in (B), to facilitate loading and truck movement.

If the cut floor is soft, it is the best practice to use a dragline working from the top. However, if none is available, or the digging is too hard for it, the dipper shovel may work from supporting platforms and have a gravel, stone, or other road built behind it for the trucks. If it is working on the bottom level, the road can be left to facilitate surfacing work.

A busy shovel should have at least occasional help from a dozer, which can level the pit floor, clean up spilled dirt, get boulders out of the way, and assist stalled trucks. Even if the shovel can handle the operation without assistance, it will produce more if it needs only to dig.

A shovel cut should be started on the low side of the grade, and worked uphill so that it will drain. The floor should be shaped carefully to avoid excessive working over

LOADER CUTS

Cuts made with front end loaders, of

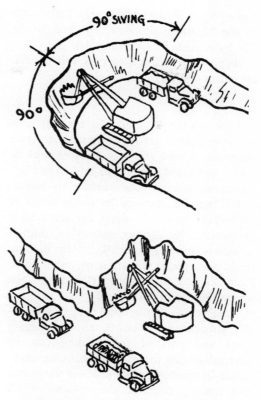

Fig. 8-30. Through and side cuts

either the crawler or wheel-mounted type, are similar to shovel cuts in many basic features.

They are moderately deep, usually between the height of the push arm hinges and the maximum lift of the bucket edge. They should be sloped to drain, to avoid difficulties with rain or ground water.

The loader is not as efficient as the shovel if the ground is muddy, soft, or loose; or if the bank is extremely hard. While it can work in narrow places, production may be greatly reduced by lack of generous space in which to turn.

The loader can keep its working area smoothed and free of rocks. However, if there are enough trucks to keep it busy, it is often better to send in another machine for cleanup.

If the distance is 1200 feet or less, large

rubber tired loaders may carry their loads to the fill more economically than trucks can haul them. See Chapter 16.

Belt and wheel loader cuts will be discussed in Chapter 14.

FILLS

Fills are made to bring a road or area up to a desired grade, to elevate it above water or drifting snow, to bury stumps or rocks, or to add strength to ground too unstable to support road surface or traffic.

Fill may be obtained by the removal of high spots or banks along the same road or project, by digging gutters or ditches alongside or near the fill, or by hauling from necessary excavation on other jobs, from commercial pits, or borrow pits opened just to obtain the fill.

Nearby cuts on the same project are usually the cheapest source, as the digging costs and part of the hauling can be charged against the excavation. Also, excavation in adjacent hills will lower the grade, and thereby decrease the volume of fill needed to carry the road across hollows.

Roads in hilly country are often engineered to balance the cuts and fills, so that all the material cut out of high spots is just enough to build up all the low spots. However, where the road crosses ridges of hard rock close to the surface, good borrow is available nearer fills, or snow removal problems are severe, it may be advisable to keep cuts to a minimum and haul in dirt.

Where very heavy hill cuts must be made without a corresponding need for fill, the surplus may be wasted in dumps off the road. This may be preferred to raising road levels to absorb the fill, because of the economy of a waste dump as compared with a compacted highway fill. Also, high fills may require purchase of extra land width to avoid need for steep and dangerous side slopes.

Cuts and fills on a road are sometimes so far apart that combining them would cost more than wasting spoil from the cut nearby, and getting the fill from borrow pits.

Types of Fill. Any type of mineral earth or rock can be used as road fill, but clay and silt are generally .undesirable. They soften when wet, frequently with changes in volume, and may act as a wick to bring ground water to the surface. Humus is avoided, particularly in its pure state, because of lack of bearing strength and excessive water absorption. Topsoil, a mixture of mineral soil and humus, may or may not be permissible, depending on its qualities and its location in the fill.

Sand and loose, clean gravel have excellent bearing power but afford poor traction, are hard to compact, and must be held in by other materials.

The most desirable fills are mixtures of two or more simple types. Varying proportions of clay, silt, sand, gravel, and stones are found in loams, boulder clay, and glacial till. Sand and gravel are most desirable when mixed with enough clay or silt to bind them together. Various soil mixtures are described in Chapter 3.

Light soils with a high percentage of sand or gravel are desirable when work must be done in rainy places or seasons. They absorb and drain off large quantities of water, and do not get slippery easily.

Moisture Content. The water content of soils largely determines their behavior on a fill. Each soil has a best (optimum) water content which favors compaction. Less water will allow the grains too free motion in relation to each other, and more will permit soil to bend or creep away from pressure.

A soil which contains too much moisture will develop a rubbery quality. It will move away from the roller, and when its weight has passed, spring back into nearly its original position.

A loose soil may hold too much moisture for best compaction and still appear fairly dry. When the grains are squeezed together, water films between them are displaced and tend to work up toward the surface, rendering it wet. This condition may be cumulative through a number of layers of fill.

Some compaction is accomplished by rolling a rubbery soil, and the operation warms the ground and brings moisture to the top so that drying is speeded up.

The problem caused by soil that is too wet for specified compaction is discussed in Chapter 3.

If the soil is too dry, it is watered by sprinkler-equipped tank trucks or trailers while being spread and rolled.

Swell and Compaction. Undisturbed soil has generally been in the same position for long periods. The particles are well settled against each other, leaving little space. Natural cements may bind them together.

When such a soil is dug or disturbed, it breaks up into chunks or grains which are thrown against each other in a disorderly arrangement, leaving air spaces or voids between. This increases the bulk of the soil, and increases its ability to absorb and conduct water. Such a loosened soil will turn to a very soft mud if soaked.

The process of soaking and then drying will settle the grains together somewhat, reducing the voids. Repeated wetting and drying will cause it to shrink to about its original bulk. Freezing and thawing will accelerate this settlement, as will also the weight of traffic or additional fill.

Compaction by Hauling Units. Considerable packing down of fill can be done by hauling and grading equipment. Ground pressure under loaded scraper tires may be thirty to forty pounds per square inch, and the kneading effect of these tires, and/or the vibration of crawler tracks are quite effective.

However, compaction tends to decrease with distance from the cut, as all the fill material must pass over the near portion, and only a small fraction over the far end.

In addition, it is difficult, sometimes impossible, to get the operators to vary their routes enough to give systematic rolling to the full width. Routes may have to be shifted by stationing one or more men along the way to tell or signal the operators where to go, or by the use of movable obstacles.

It is usually inadvisable to have a heavily loaded unit break a new path in soft fill, as the power requirement and strain on the machine are excessive. Trail breaking should be done on the empty return trip, and loaded units then turned into those tracks.

Excessive rutting may be avoided by having a whole strip rolled by empties before using it for loads.

If enough units are hauling to make collisions likely, the two directions of traffic should be separated, and their routes alternated as necessary. If two-way traffic can use the same route safely, it can be gradually shifted to the side.

If a road fill is dumped loosely and not rolled, it may settle unevenly over a period of from six months to two years. Any surface placed on it during this period will be warped or broken.

It is generally not convenient, and often not possible, to allow a long period for subgrade settlement. Also, it is very difficult to estimate the compaction a fill receives from traffic, or from rains, while building. This makes it impractical to finish grade with a loose fill and obtain the desired surface after settlement.

These difficulties can be avoided by compacting the fill as it is placed. Rollers of various types are used on thin layers of fill to squeeze the grains into even closer contact than they had in the bank. They are aided by the weight of grading and hauling units. Loam soils may be reduced to ninety percent of their bank volume by thorough compaction.

A properly compacted fill should not shrink on exposure to time and weather, so that it is theoretically possible to put a permanent surface on it immediately. In addition, it has the highest bearing power possible to its particular soil type, so that wheels and tracks will not sink into it much, and speed and capacity of hauling equipment on it is increased.

A compacted fill will not absorb rain water readily, so that the fill should remain hard enough to work even after heavy rains. Whether the surface will become greasy depends on the clay content and the possible presence of a layer of dry uncompacted dust before the rain.

Rollers. Rollers may be smooth steel wheel, tamping (sheeps-foot), or grid; or rubber tired. Some have rubber on two drive wheels, and a steel roll, rough or smooth. Rollers may depend entirely on weight and shape for effectiveness, or may have vibrators also. See Chapter 19.

The smooth steel wheel models are just known as rollers. They are usually self-powered, and may have either two or three rolls. Weights range from 1½ to over 20 tons. They are primarily finishing machines, used more often on surfaces than on subgrades. Tops of thick layers may be better compacted than bottoms.

These machines have little traction, particularly with tandem construction, and are not suited to rough ground. Grid or segmented drive rolls overcome this difficulty.

Towed tamping or sheepsfoot rollers are steel drums 4 or 5 feet long and 40 to 60 inches in diameter, fitted with projecting lugs (feet or legs) 7 or more inches long. There may be three lugs to every two square feet of drum surface.

Feet may have expanded soles (which may kick up soft dirt), or taper from a wide base to a flat end. Either way, they penetrate soft fill until weight is carried on the sole, where compaction begins. As the ground is compressed on successive passes, the feet do not sink as far and start to "walk out" of the ground.

Drums may be filled with sand or water. Sole pressures range from 250 to 750 pounds per square inch.

The drums are mounted in box frames fitted with drawbars. Up to three may be mounted side by side, and two pairs may follow each other. Working speed is 2 to 3 miles per hour. Power requirements are high, particularly in soft fill.

These units are being replaced by self-powered units with one or two tamping drums, and a pair of rubber tired drive wheels. This is in line with a general move away from towed equipment.

Pneumatic tired rollers are ballast boxes supported by wheels with smooth tread tires. The wheels may roll straight, vibrate, or move up and down or wobble as they revolve. They compact by a combination of weight and kneading action of the soft tire walls. Weights vary up to more than 80 tons. They can compact single fill layers as deep as 24 inches.

Fill Bases. It is desirable that a fill be firmly bonded to the surface on which it rests to prevent formation of saturated zones, water channels, and possible sliding downslopes. This is usually accomplished by removing vegetation and topsoil, and plowing ridges across any slopes.

Methods of removing humus and other muds from the location to be filled, and of stabilizing such muds when removal is impractical, have been discussed in Chapter 3.

When the area to be filled is wet, rough, or otherwise impassable to machinery, the first layer is built by dump trucks and dozers to a height at least sufficient to carry the hauling units over the soft spots or obstacles. After a usable floor is estab-

If the surface is uneven but passable, low spots may be built up first with compacted layers, or high spots removed, before the main fill is placed.

Rock Separation. Handling and compaction of fill material is rendered difficult by the presence of loose stone.

Fig. 8-31. Screening oversize rock from fill

Rocks of even small sizes interfere with grading. If their diameter is greater than that of the fill layer, they will project from the top. If two or more rocks are in contact, they are liable to prevent even distribution and compaction of fill under their adjoining edges.

For this reason, the size and number of rocks present in thin or layered fills are often limited. This may be done by using selected borrow, or by putting bouldery material through a grizzly.

The arrangement shown in Figure 8-31 represents a minimum of equipment for screening. A truck on a high level dumps on a sloping grizzly, dirt falls between the bars into a truck parked below, boulders roll to the side.

Oversize material may be allowed to roll directly into trucks, be loaded from beside the grizzly, dozed away from it to a stockpile, or, if the grizzly can be located on the edge of an abandoned pit, allowed to accumulate.

One or two men are needed to free oversize stones stuck between the bars and to coordinate the trucks. If the stones are a substantial part of the bulk of the soil, smaller trucks may be used under the grizzly than on top of it.

If sticking of stones or sliding off of chunks of earth is much of a problem, a vibrating grizzly, or a standard grizzly with a vibrator bolted to it, may be desirable. The flat slope illustrated is suitable only for loose soil and large openings.

Rock Fill. Various results are obtained from all-rock fills. If the largest pieces are smaller than the depth of the fill, and sizes are mixed, including a good proportion of fines, a solid fill with a good surface may be obtained by pushing piles off an edge with a bulldozer. Large pieces tend to move ahead and over the bank, while smaller ones drift under the blade to form a topping.

If there are not enough small pieces to provide a working surface, finer material should be brought to fill surface holes and even off the top.

Rocks too large to fit in the fill can be rolled ahead of it until a hole is found or is made to bury them, or they are reduced by splitting or blasting.

Rock fills are generally almost incompressible, exceptions being when rock is soft or fissured, and very heavy weights are used. However, they are apt to be subject to only minor and local settlement, where fines are shaken or washed into spaces between rocks below them.

Rock is desirable fill material for the bottom layer in crossing water or mud, as it is not softened by contact with water and spreads surface loads over large areas of the base. In such locations it may settle due to displacement or compression of the ground under it.

The volume of fill is greater than the unbroken rock in the bank. The difference will vary with the quality of rock, type of fragmentation, and amount and kind of compaction. 50 per cent is a rule of thumb average that can be used except where there are indications to the contrary.

If the rock must be used in the fill, it is best placed at the bottom. Unfortunately, rock is ordinarily the last material to be taken from a cut as soil is stripped prior to blasting.

SCRAPER FILLS

The standard method of building a fill with scrapers is to start with an area sufficiently leveled to allow the scrapers to travel on it, and to build it up in thin layers, starting at the outside edges, or at low spots.

Spreading depth may vary from two inches to the maximum lift of the bowl— eight inches to a foot. Thin layers favor compaction, particularly if the scrapers are depended on for the rolling and facilitate smooth building up of the grade. Chunky, sticky, or rocky fill will not spread thin, or can be made to do so only by very slow travel during the dump.

Thick spreading is liable to flow out of the bowl more smoothly; can be done at higher speed, and reduces the dump time. However, it tends to make a rough fill which will require slower travel speeds, or smoothing work with a dozer or grader.

Edges. If a fill is high, the edges may be troublesome and dangerous unless carefully made. The problems involved are keeping it at the correct toe alignment, proper slope, at full density or compaction, and not rolling any machinery off it.

These problems are affected by the nature of the fill and by its height and slope.

Loose fills of sand, clean gravel, or too-dry dirt tend to cave under the weight of machinery close to the edges. Finer grained fills may have excellent bearing power if well compacted and not too wet. However, while being compacted, they tend to squeeze outward, and an allowance for this creeping must be made when placing the first fill so that it will not move out past the toe stakes.

The behavior of the fill on edges may be anticipated by making soil analysis or by consulting with contractors or machinery operators who have worked with the same formation.

Except for allowance made for creeping under load, or spillage from above, which seldom should be more than a foot or two, the fill is started at the toe line and built up of layers, usually not over six or eight inches, loose. Each layer should be rolled with a tamping roller that is allowed to project slightly beyond the edge. For this purpose, two or more rollers should be fastened in a single yoke so that their width will be substantially greater than that of the tow tractor, which should not have to walk on the edge. This is particularly important with high banks and wheel tractors.

If watering is required for proper compaction, application may be somewhat heavier at the edge to allow for side evaporation. However, it should not be sufficient to make it soft or muddy.

The fill should slope up at the edges in order to incline the center of gravity of the machinery toward the center and minimize the danger of caving. If the fill is narrow it will have a trough shape, and if wide, it will be flat with raised sides.

This slope is most easily started by a grader or an angle dozer working over the first layer or two left by the scrapers. Once made, it will tend to preserve itself as the tilt will tend to make the inside wheels of the scraper sink deeper. If it becomes too steep, it is readily reduced by filling toward the center.

If the job is shut down during any period when rain is expected, it may be wise to build the fill up to a crown in order to allow it to drain. This involves resloping the edges on resumption of work, and, if the work is done under exact compaction specifications, may cause confusion in the treatment of the tapered layers required.

Another treatment is to preserve the trough but so grade it that all water will flow to selected low spots. Here ditches are dug through the raised edges, and troughs of metal or wood placed to lead the water down the slope. This is readily done in hilly country where most of the road is on definite gradients, but not in level country.

Such drain ditches may be made wide and gentle, so that they can be dug, backfilled, and compacted by machinery, or may be hand dug, refilled, and tamped.

If the trough shape is left without precautions, a center gully may be scoured by a heavy rain, a pond formed in low spots, and damage done to edges by overtopping and concentrated runoff.

Scraper distance from the edge is determined by depth of spread and slope. If a slope is one on two, spreads are six inches deep, and compaction is one third, each pass will be eight inches inside the previous one.

If the edge is not firm enough to support scrapers at the proper distance to dump loads, they should be spread farther back and the dirt cast out to the edge by a grader or angling dozer.

Additional slope stakes should be set as a high fill is built up to maintain the correct width.

DRAINAGE

Drainage is an important factor in the construction of most roads. Ground water must be kept far enough below the surface so as not to damage it, or weaken the subgrade directly or by supplying capillary water. Water falling on the surface of the road must be conducted off it, and run-off or streams crossing the road must be provided for.

Frost Heaving. Capillary action in silt soils is largely responsible for frost heaving of pavements. This may occur when the ground freezes below the pavement level to the top of a silty layer in contact with ground water. The water in the top forms an ice crust and capillary water feeding from below adds to it from the bottom. An ice lens of considerable thickness may form in this manner, pushing the pavement up as long as the temperature provides a balance between the heat liberated by the ris-

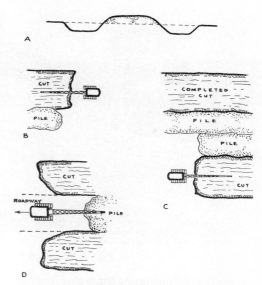

Fig. 8-32. Road fill from gutters

ing and freezing water, and the loss of heat through the pavement.

Deeper freezing may plug the circulation and stop the heaving, or cause formation of additional lenses at greater depths. Thawing will allow the lens to melt and liberate excessive water, and the broken pavement above it may collapse.

Capillary water can be controlled by using a coarse, pervious fill under the surface to a depth sufficient to provide necessary bearing power and to get below frost line; or by lowering ground water level and intercepting surface water which might enter the subgrade.

Raising the Grade. In swamps and lowlands, the only practical method of getting the road well above ground water is to build a high fill. If the base course can be made entirely of rock, it will break any contact between the water and the balance of the fill. Clean gravel or coarse sand may serve the same purpose.

Proper quality of fill can reduce the required height substantially. However, it is often more economical to make a higher fill of inferior material obtained from roadside ditches, as in Figure 8-32 (A), and it

Fig. 8-33. Borrowing road fill from sides

is often possible to lower the water level by the same operation. Draglines are generally used, but dipper dredges may be preferred when much of the land is under water, or it is intersected by numerous channels.

Either machine may work along the ditch lines, piling spoil toward the center, as in (B). The dragline will work away from the cut, as shown, but the dredge will float in it.

If a dragline has a sufficiently long boom, it can travel on the road centerline, and dig both ditches and pile the spoil in one pass, as in (D).

Road fill may also be obtained by ditching in dry flatlands where the road is to be raised above floods or snowdrifts. In such circumstances elevating graders may be used as shown in Figure 8-33, or dozers or scrapers.

Tiling. In sloping land, it is usually more economical to lower the water table by drainage. The standard method, Figure 8-34, is to put shallow ditches to carry surface water through cuts (A), and, if necessary, to place porous tile or other drains (B) two to three feet deeper in loam soils. Silt or clay deposits may require a drain depth of as much as seven feet, but in such a case better results may be obtained by a

normal drain depth and by the use of a layer of pervious material under the road.

The design of subsurface drains must be carefully adapted to the requirements of the particular job. The ground may drain naturally so that no work is necessary. There may be a saturated condition that could be relieved by providing a drain through an impervious barrier (C), or by cutting off the source of water (D). There may be springs or seepage rising under the road which would require center or lateral drains (E) and (F). Such drains may also be required to take off water soaking through a porous road surface.

When the ground is generally dry and firm, but has local springs or seepage, the wet areas should be dug well below the intended drain, and backfilled with stones and clean gravel, topped with sand. The drain itself may be any type of pipe, laid at the lowest convenient level, and opening into side drains, a catchbasin, or a gutter.

The rock fill directs the water toward the pipe and reduces or eliminates softening of adjacent areas.

JOB STUDY

Road construction may involve clearing vegetation; stripping and storing of topsoil;

excavating soil and rock to cut natural levels to road grades; hauling the spoil to road fills or waste dumps, building culverts, bridges, and drainage systems; raising low areas to road grade by fill obtained from highway cuts or borrow pits, and finishing, topsoiling, and seeding of slopes; and cleaning up the work area.

Usually, this work must be accomplished within a time limit. It is desirable to get the maximum number of machines and men on the job as soon as possible after the start, but it is more important to keep them efficiently employed once they are there.

Sequences. When time permits, it is often desirable to perform complete operations in sequence. If an entire work area is cleared, it will usually be easier to arrange dirt moving sequences than if the excavators have to be limited to a few small sections. Culvert construction should be completed before fills are raised high enough to go over them, unless they are to be installed by ditching the completed subgrade.

Liberal areas of rock should be cleaned before drilling starts. Pioneer bulldozer work should be well advanced before scrapers operate.

If the schedule is close, delay in one operation will delay others that have to wait for it, which may be more costly in machine and man time. These secondary delays are much more serious when maximum amount of equipment is crammed into a job than when a few units are doing it over a longer period.

Basic Factors. Basic factors to be considered in figuring grading for a road may include:

1. Clearing costs.
2. Topsoil stripping, storage, reclamation, spreading, and planting.
3. Amount and type of soil excavation in cuts or borrow pits.
4. Amount and type of rock excavation.

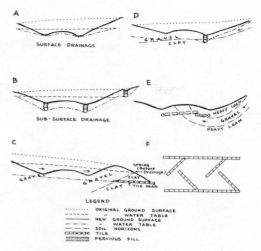

Fig. 8-34.   Road subdrainage

5. Availability of suitable borrow and cost of purchase.
6. Haul road construction and maintenance, and length of hauls.
7. Quality of fill required, and processing required of material from cuts and pits.
8. Fill compaction, shrinkage, and disposal of surplus.
9. Slope finishing and protection.
10. Ground water conditions and drainage requirements.
11. Structures such as bridges, culverts, and retaining walls.
12. Possession or availability of proper machinery, with necessary parts and supplies. Extra costs of using second-choice or beat-up equipment.
13. Availability of construction supplies such as pipe, forms, etc.
14. Labor supply.
15. Weather—rain, snow, ice, dust, frozen ground, frozen equipment, mud.
16. Time of completion of related structures such as bridges, being built under separate contract.

In highway work, the amount, kind, and location of cut, borrow, and fill, and the length of haul, may be specified. Haul may be described as "normal" or free, up to

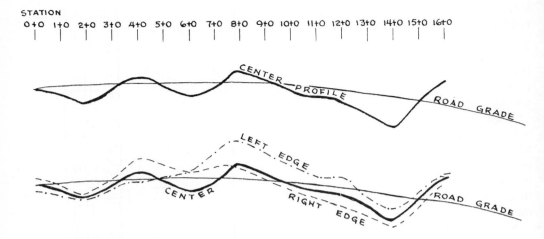

STATION

0+0 1+0 2+0 3+0 4+0 5+0 6+0 7+0 8+0 9+0 10+0 11+0 12+0 13+0 14+0 15+0 16+0

Fig. 8-35. Center and side profiles

a certain distance, which may be 300 to 1000 feet, and longer hauls called "overhaul." Excavation may be described as "unclassified," or divided into rock yards and dirt yards.

In less formal jobs, these factors may be indicated only approximately, or may be figured by the contractor from grade or route plans.

Casual Estimating. Where cuts and fills are shallow, and side slopes lacking or moderate, grading can often be estimated fairly accurately by inspection of centerline stakes. The exact yardage is sometimes not of primary importance, as stripping topsoil and working over a piece of ground represents an amount of machine time that may be only moderately increased by the cuts and fills.

Several errors must be watched for, however. Cuts and fills on the stakes may be figured from the top of the stake, from ground level, or from a line on the stake. The grade indicated may be subgrade, in which case it is taken at face value, or finish grade when the depth of base courses and of surfacing must be added to the cuts and subtracted from the fills. The width to be figured on is not only the road and shoul-

ders, but also gutters and slopes. The depth of topsoil to be stripped is subtracted from the cuts, added to the fills, and is considered separately as an important cost factor.

When cuts or fills are deep, side slopes exist, topsoil need not be stripped, or when the job is large, yardages should be carefully calculated. If this is not done on the plans, the contractor can do it for himself.

YARDAGE CALCULATION

Center Profile. The minimum staking for a road is the centerline. When this is done, a profile is taken, showing the elevation of the ground at each stake. These elevations are plotted on cross section paper, usually with the vertical scale ten times the horizontal, and the points connected by a line. A profile for the road is then sketched in according to the standards of grade and vertical curve required, or from some previously formed plan. This line should represent the subgrade before the addition of any imported material.

Distances measured from the road line to the ground line will indicate the depths of cut and fill required to establish the road grade. If topsoil is to be stripped, its

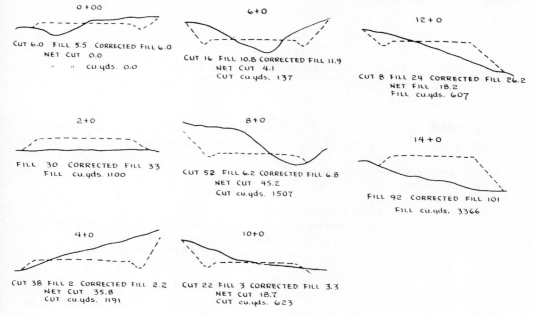

Fig. 8-36. Cross sections

depth should be added to the fills and subtracted from the cuts.

If the ground does not slope across the line of the road, this type of profile, shown in Figure 8-35 (A), should give a reasonably accurate picture of the relative volume of cuts and fills, and the distances they are to be moved. However, to obtain yardages, cross sections usually must be calculated, as described below.

Side Profiles. If the road is laid out on side hills, side stakes and slope stakes may be set. The side stakes may be at the edge of the pavement, at the outer edge of the shoulder, or the far side of the gutter, if any. In general, the shoulder or the gutter locations are preferable. Slope stakes are placed where the intended cut in a bank reaches its top, or at the outer, base edge of a proposed fill. These are not placed until cross sections are calculated.

If the side stake elevations are plotted in the same manner as the centerline, two additional profiles can be drawn, as in (B). These will give additional information about the bulk of material to be moved, but since they often do not include cuts for gutters, and cannot show the volume which

must be dug or filled for side slopes outside the road lines, they are not an adequate basis for careful calculation.

Cross Sections. A cross section is a profile taken at right angles to the line of the road. It is at least long enough to include the full width that will be graded. Such profiles are sometimes taken with hand or string levels. They may be taken at each 100' station, plus points where the ground surface changes, or, in smooth terrain, less frequently.

This cross profile is also drawn on cross section paper, preferably on the same vertical scale as the center profile. Horizontal scale may be the same as vertical, or at any convenient proportion to it. The cross section of the road subgrade is drawn in.

A number of such cross sections are shown in Figure 8-36, together with the cut and fill for each.

Wherever the ground line is above the road line, there will be a cut; and where the road line is higher than the ground line, there will be a fill. If topsoil is to be stripped and saved, it may be well to lower the ground line by the depth of the topsoil to save confusion.

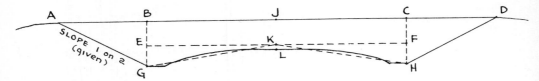

Fig. 8-37. Figuring a cross section

Figure 8-37 shows a sample cross section with the arithmetic involved in computing its area. The road and gutter surface have been simplified in the calculation, as this reduces the work without introducing too large an error for rough figuring. The problem is also simplified by a horizontal ground surface.

The cut is divided into slope triangles; and a road section, which in turn is divided by the line EF into a rectangle and two triangles. The data given by the engineer is labeled "given," and that measured off the diagram as "scaled." The areas of the triangles and the rectangle are readily computed, their measurements in feet being used for convenience and the result converted into square yards by dividing by nine.

The areas of a succession of cross sections are obtained in this manner and averaged by adding together and dividing by the number of sections added. The result is multiplied by the length in yards of the area in which the sections were taken, giving the number of cubic yards of excavation required. Figures from in-place measurements are in bank yards.

Where the ground slopes irregularly the ground surface is simplified by drawing straight lines, and the cut and fill areas are divided into triangles.

The road and gutter cuts could be figured by averaging the width and average depth at each cross section, then multiplying their product by the length of the sectioned area. The slope sections cannot be averaged as their areas vary with the square of the depth of cut, and use of average depth would indicate a much smaller yardage than that actually required.

The most convenient way to measure the areas of cut and fill is by counting squares and fractions of squares. If a lot of work is to be done, areas can be measured by means of a planimeter.

Fill Shrinkage. When fills are rolled to the compaction required in modern highways, the material is often compressed into a smaller space than it occupied in the bank. This shrinkage should be allowed for in figuring cross sections. Loam soils often shrink 10 percent, clean sand 5 percent or less, and blasted rock, not mixed with other dirt, will show a minus shrinkage, or swell. Compaction by hauling equipment without rolling is variable and will seldom cause shrinkage.

The examples in Figure 8-36 use a shrinkage factor of 10 percent, but the figure selected should depend on job conditions.

Net Cut or Fill. On side hills, one station is likely to include both cut and fill. The smaller amount is subtracted from the larger, giving net cut or net corrected fill.

Converting to Cubic Yards. The net square yards of the cross section are converted into cubic yards by multiplying by the length of the road it represents. If sections are taken at 100 foot intervals, each will represent a piece 100 feet long, that is, halfway to the next section, on each side. If a special section is taken 40 feet from a 100 foot station, it will cover 20 feet on one side and 30 on the other—a total of 50 feet. The adjoining sections will be reduced proportionately.

When the 100 foot interval is used, it represents $33\frac{1}{3}$ yards. It is easier to multiply the section square yards by 100, then divide by 3, than to multiply by $33\frac{1}{3}$.

The net cut and net fill figures, when converted to cubic yards, are used in making a mass profile. The gross cut figures are converted to cubic yards, in the same manner, to determine the total excavation, exclusive of topsoil.

Topsoil volume is figured by multiplying the length of the road, the average width to be stripped as indicated by the cross sections, and the average depth.

Cubic yards of net cut are added together and compared with the total of net fill yards, to determine whether extra fill will have to be obtained from pits; or whether fill will have to be wasted outside the road area.

Mass Profile. A mass profile is prepared by drawing on cross section paper a straight line to indicate the road grade, dividing it into stations, and posting cubic yards of net cut above it and net corrected fill below

it, on any convenient scale. It is sometimes helpful to draw in blocks representing the fill at each station as in Figure 8-38.

A curved line, the mass profile, is drawn connecting the station points. The amount of net cut or net fill at any point along the road can now be scaled off, as well as the haul distance between cuts and fills.

The haul distance is measured between the centers of mass, or centers of gravity, of the cut and fill. The longer and shorter hauls should average out.

Mass Diagram. Many engineers prefer to use the mass diagram shown in (B). A straight base or zero line is drawn on cross section paper, and marked off for road stations, and plus and minus yardages in the same manner as for the mass profile.

Points are plotted for cumulative or total yardages, starting at zero station. Points are placed half an interval farther to the right than the station they represent, as the full yardage figured for each station is not accumulated until the end of that station block.

At station 1 + 50, the minus yards of fill for station 1 + 0 is entered. At station 2 + 50, the total of the fill for stations 1 + 0 and 2 + 0 is posted, and at 3 + 50, the total fills for stations 1, 2 and 3.

When a cut is reached, at 4 + 0, the cut yardage is subtracted from the accumulated fill so that the line turns up. This line, called the mass curve, crosses the baseline when accumulated cut equals the accumulated fill, and continued cuts raise it above that line, until a fill is reached and pulls it down.

In short, wherever accumulated fill, starting at zero station, exceeds accumulated cut, the mass curve will be below the base. If there is an excess of cut, it will be above.

The mass curve line does not show total yardages of either cut or fill.

The points of loops which are farthest from the base line indicate changes from cut to fill, or fill to cut. They also represent

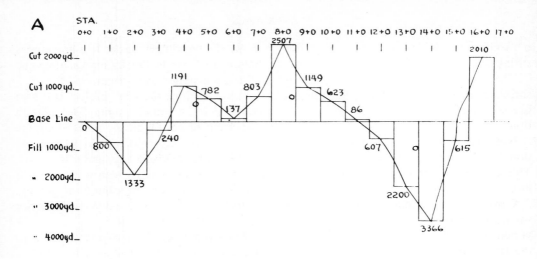

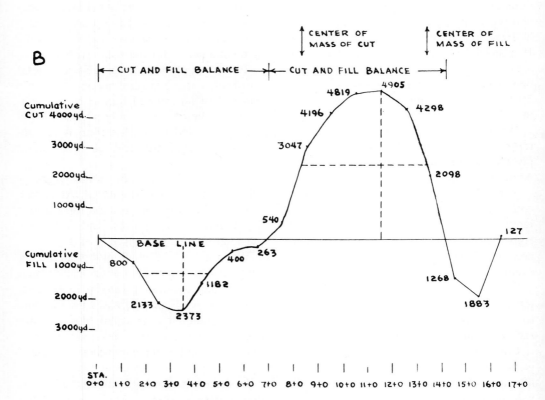

Fig. 8-38. Mass profile and mass diagram

the total net yardage to be moved from cut to fill along the road line, but disregard sidecast material.

Any horizontal line drawn on the diagram is called a balance line. The yardages between any two places at which it intersects the curve have a balance of cut and fill. The baseline will often serve as a balance line, as in the illustration.

The centers of mass of a cut or fill can

be found by drawing a vertical line from the outermost point of a loop to the balance line. A horizontal line is drawn through the center point of this vertical. Its points of intersection with the sides of the loop are approximately at the center of gravity of the cut and the fill for that balance line.

A single balance line may be used for the whole road. Any part of the mass curve which extends beyond the last balance point to the first or last station of the road, will represent yardage to be borrowed, if it is below the base; or to be wasted, if it is above.

A loop above the balance line indicates fill movement to the right in the diagram, and below it to the left.

Any number of balance lines can be used so long as they end in points on the mass curve, and do not overlap. The vertical distance between two balance lines represents borrow or waste in the part of the curve connecting them.

The mass diagram is a very flexible and useful aid in studying yardage distribution. However, it is so confusing to persons not accustomed to this type of computation that the average contractor working out such a problem may do better to use a mass profile.

PAVING

PURPOSE OF PAVEMENT

Pavement is a surfacing for traveled areas, which is intended to provide a long-lasting, smooth, clean, supporting surface; to spread loads sufficiently so that base material can support them; and to protect the base against damage by traffic and weather.

DESIGN

There are two basic types of pavement; rigid and flexible.

Both are mixtures of a binder and a mineral aggregate. The rigid binder is Portland cement; flexible binder is usually asphalt. Aggregate may be sand, natural gravel of selected sizes, crushed stone; alone or in mixtures of various proportions.

A rigid pavement may form a bridge over soft or sunken spots in its base, but if the span and the load exceed its strength, it will break and sink. A flexible pavement will settle in all such areas naturally, with loss of surface smoothness but with less structural damage.

In both constructions, resistance to damage from excessively heavy loads may depend more upon the strength of the supporting subgrade than the strength of the pavement.

Width. Design width of a pavement depends on two factors: the number of traffic lanes required, and the width of the lanes.

Traffic counts are out of the field of this book. However, it seems to be standard practice to underestimate future traffic, with costly results in congestion and inadequately utilized rights of way.

Traffic lanes in cities and built up areas are usually 10 to 12 feet wide, and curbside parking lanes are 7 or 8 feet. For express highways, city or country, 12-foot lanes are standard, with 13 or wider used occasionally. Maximum number of lanes may be 3 or 4, with additional ones in interchanges.

Some light traffic rural roads are still built with two 9-foot lanes, with narrow shoulders.

Many paving machines can lay two lanes at a time; some can do three or more. Very wide construction at one pass is more common on airports than highways.

SUBGRADE

The construction of subgrades by excavation and fill was discussed earlier in this

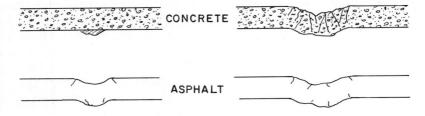

Fig. 8-39. Effect of hollows on rigid and flexible pavements

chapter. That work is followed by finishing or dressing operations, just before placement of pavement.

Comments below apply equally to sub-base or selected material courses placed between subgrade and pavement. But base courses of stabilized material will be considered separately.

It is important that the surface lying directly under pavement be finished as exactly to grade as practical, and that it be evenly compacted. The pavement material is costly, so that none of it should be used for fill to correct low spots. Full pavement thickness is necessary for it to have its design strength (and for the contractor to get paid), and this will be skimped if the subgrade is high.

If compaction does not accompany and follow final finishing, filled areas will be loose and will not give proper support to the pavement. A settlement of ¼ or even ⅛ of an inch in a one-inch deep patch-fill is enough to leave a concrete slab unsupported, and to spoil the smooth surface of an asphaltic pavement.

The base must be firm enough to support the trucks carrying paving materials, without any rutting, or the paver must be loaded from the side.

Trimming (Fine Grading). The subgrade or sub-base should be carefully finished at the time of its construction, with grading to blue tops, stringlines, or other exact references. However, there is often a lapse of time before the paving crews ar-

Fig. 8-40. The subgrade may need dampening

rive, so that the surface may be marred by vehicle use, settlement, washing, or blowing. Even if it looks intact, it is sound policy to re-check old work, and re-finish if necessary, as part of the paving operation.

Final trimming is easiest and most accurate if the subgrade was originally left a little high, as then all or almost all of the corrections will be cuts, with less problem of compaction. However, any substantial amount of extra material at this time may be a nuisance.

Even if there are no fills, a thin layer of the surface is loosened by regrading, and must be compacted. This is usually done by either a steel wheel or a rubber tired roller, working closely with the grading equipment. There may be a tolerance of ¼ inch for errors in grade.

If the soil is wet, fine grading must be postponed. If it is too dry, watering is required. It should be damp when concrete is placed on it, so that it will not absorb water from the mix.

Fine grading is done just ahead of the paver, but far enough ahead to avoid delaying it.

A finish subcourse of trucked-in granular material may be run through a paver to distribute it to a uniform depth.

Grading Equipment. Final grading may be done by motor graders, by an automatic grademaker, or both.

A grader carrying a blade near its center has been the standard grade finishing tool for many years. It may be used in its natural state, depending entirely on operator skill for accuracy of results, or it may be equipped with automatic blade control (ABC). With this device, the operator may set the control at the desired cross slope, and then need to keep exact grade only at the leading edge of the blade.

The leading edge may also be put under automatic control of sensors responding to a stringline, a guide wheel, or a laser beam. The grader may do the complete finish-

ing job, or just preliminary work in advance of a grademaker.

The automatic grademaker processes the full width of the lane or lanes being paved, in a single pass. It is customarily guided by either one or two strings carefully set at edges of the work strip.

Final check may be mostly by eye, backed up with an instrument for questionable spots. If concrete forms are used, a scratch templet will mark any high spots, and show up low ones by failing to touch.

ASPHALT

Asphalt is the essential ingredient in practically all flexible pavements being built at this time. It is a black or brownish-black amorphous solid which is hard or even brittle when cold. When heated, it gradually softens and then liquefies, without any definite melting point.

Most of our asphalt is obtained as the end product of the refining of petroleum. It also occurs naturally in lakes or seeps in many parts of the world.

Most road asphalt is used in one of three forms: asphalt cement, liquid asphalt (cutbacks or oil), or emulsion. Each of these is in turn divided into a number of categories, largely on the basis of hardness or viscosity, and of rate of setting or hardening.

It is necessary that asphalt be quite fluid while it is being applied to or mixed with aggregate, and that it become thick and firm within a reasonable time after application. Liquefying of asphalt cement is done by heat alone, of cutbacks by both heat and solvents, and of emulsions by moderate heat and homogenization with water.

Description presents a number of problems, which have been approached in a number of ways. The result is a variety of descriptions and specifications, some of which mean almost (but not quite) the same thing. They can be very confusing, but some familiarity with them is necessary for anyone working with asphalt.

Fig. 8-41. Automatic grademaker

Classification. Until recently, hard asphalt (asphalt cement) was classified by its resistance to penetration. The standard test, which may still be made, involves penetration of a sample by a needle carrying a weight of 100 grams. Temperature is held at 77° F., and a time of 5 seconds is allowed.

The distance the needle penetrates is measured in hundredths of a centimeter (cm/100). The result is the penetration number. Hard asphalt has a low number, soft asphalt a high one. Commercial as-

phalts have penetration grades from the hardest, 40-50 to the softest, 200-300. These grades are now superseded by viscosity numbers.

Penetration tests cannot be applied to liquids. They are therefore rated by viscosity; that is, thickness or resistance to flow. A sample at a specified temperature is caused to flow by gravity or otherwise through a measured aperture, and the rate of flow is measured. These tests are applied to the cements also.

Several types of apparatus are used for this test. Their details are not of primary interest to the user of asphalt, but he should be familiar with their names. Figure 8-42A shows their approximate relationship to each other. Even more approximately, their relationship may be expressed as:

1 poise = 100 centipoises = 1 stoke = 100 centistokes = 50 Saybolt Furol, seconds = 500 Saybolt Universal, seconds = 12½ Engler degrees

A poise is a measure of absolute viscosity, widely used in physics. A stoke has

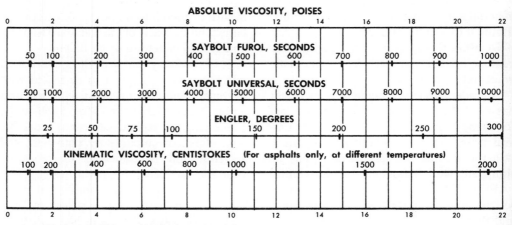

Fig. 8-42A. Measurements of asphalt viscosity

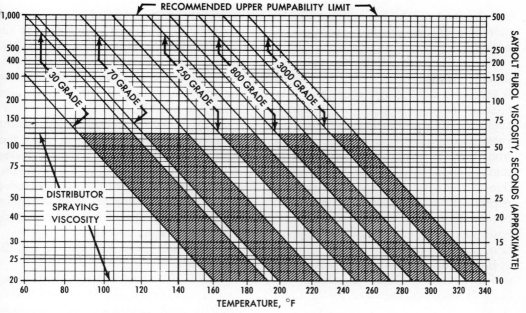

Fig. 8-42B. Temperature-viscosity in handling liquid asphalts

pecific reference to asphalt, and Saybolt efers to a particular measuring device.

The viscosity of various asphalts, when ound in a table, may not be comparable, as different grades are tested at different temperatures. And they are ordinarily used at much higher temperature, and therefore ower viscosity, than in tests.

For spraying with a distributor, recommended viscosity is between 20 and 120 centistokes.

Cement. Asphalt cement is usually the primary, untreated material as it comes rom the refinery. It may be classified by either the penetration or viscosity standards.

Penetration grades range from 40-50 (the hardest) to 200-300.

Viscosity is measured at 140° F. (60° C.). The grade classification is the number of poises, minus two zeros, preceded by the initials AC (for asphalt cement). The four recognized grades are AC-5, AC-10, AC-20 and AC-40.

This scale runs oppositely to penetration, so that larger numbers indicate harder asphalt. AC-5 is about 120 penetration, and AC-20 is equivalent to penetration 60-70.

Since it is a solid at ordinary temperatures, asphalt cement is usually made quite hot in order to mix with aggregate to make blacktop (asphalt concrete), or to use directly on road in penetration macadams or seal coats. Temperature of application is usually between 275 and 400° F.

Flash point of the thinner grades is 350 and 425° F., and of the heavier ones, 450. Fire and explosion danger is therefore much less than with the cutbacks, but constant vigilance is needed to avoid severe burns from the material and from the apparatus.

Liquid Asphalt (Cutback or Oil). The temperature at which asphalt becomes fluid enough for use can be reduced substantially by mixing with a thin, more or less volatile petroleum liquid, which may be called a cutback or a solvent. The cutback asphalts are classified on the basis of the time required to harden or cure, and their viscosity.

For rapid cure (RC) the solvent is usually naphtha, for medium cure (MC) it is kerosene, and for slow cure (SC) it is a light oil. Other solvents with similar characteristics are sometimes used. Some thickening occurs quickly, as the solution cools from tank to road temperature, but full setting and hardening depends on evaporation of the solvent. The rate is primarily dependent on the volatility of the solvent, but is affected by temperature, humidity, absorption by aggregate, depth of application, and other factors.

Each of these rate-of-cure designations is divided into five classes or grades on the basis of viscosity, indicated by the initials

of its rate of cure (RC, MC, or SC) followed by a hyphen and a full kinematic viscosity number in centistokes, 30, 70, 250, 800, or 3,000. This number represents the lowest viscosity the class may have, and it maximum is twice that number. Measurement is at 140° F. (60° C.).

There is another classification system discarded officially a number of years ago, which is found in many reference books and is still used on jobs to a considerable extent. In this, there are six viscosity grades, with zero representing the thinnest (equivalent to the new 30 grade), and numbers up to 5 for increasing viscosity.

Figure 8-43 shows both the old and the new grades.

An RC-30 cutback contains about 55 per cent asphalt cement and 45 solvent. In RC 3000 the asphalt content is increased to 80 per cent, with a corresponding reduction in solvent.

Temperature of application varies with the material and job conditions. Figure 8-42A gives the possible range of use. A 30 grade might be put down as cool as 90° F but 180 is more usual. A 3000 grade might be used as hot as 340, and becomes too cool for distributor pumping at about 240.

Figure 8-44A shows typical or average temperatures actually used with cements, cutbacks, and emulsions, in both blacktop mixes and surface applications.

Since flash points in RC liquids are 100 to 150° F., there is always severe fire danger in handling and spraying them, in addition to danger of direct burns from the fluid and the equipment.

Gloves and heavy protective clothing are required. Smoking must be absolutely prohibited for the crew while on the job, and spraying must not be done near any open flame or sparks.

Emulsion. Emulsified asphalt is a stable mixture of asphalt cement, water, and an emulsifier. It is increasing in importance because of the rising cost of the petroleum

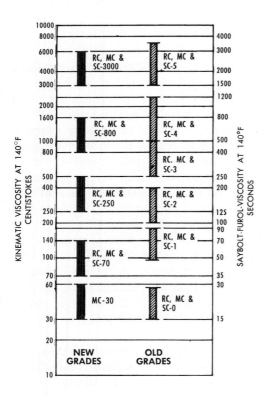

Fig. 8-43. Asphalt grades

Type and Grade of Asphalt	Pugmill Mixture Temperatures[1]		Spraying Temperatures[5]	
	Dense-Graded Mixes	Open-Graded Mixes	Road Mixes	Surface Treatments
Asphalt Cements				
AC-2.5	235-280	180-250	–	270+
AC-5	250-295	180-250	–	280+
AC-10	250-315	180-250	–	280+
AC-20	265-330	180-250	–	295+
AC-40	270-340	180-250	–	300+
AR-10	225-275	180-250	–	275+
AR-20	275-325	180-250	–	285+
AR-40	275-325	180-250	–	290+
AR-80	275-325	180-250	–	295+
AR-160	300-350	180-250	–	–
200-300 pen.	235-305	180-250	–	265+
120-150 pen.	245-310	180-250	–	270+
85-100 pen.	250-325	180-250	–	280+
60-70 pen.	265-335	180-250	–	295+
40-50 pen.	270-350	180-250	–	300+
Cutback Asphalts (RC, MC, SC)[2]				
30 (MC only)	–	–	–	85+
70	–	–	65+	120+
250	135-175[3]	–	105+	165+
800	165-210[3]	–	135+	200+
3000	180-240[3]	–	–	230+
Emulsified Asphalts				
RS-1	–		–	70-140
RS-2	–		–	125-175
MS-1	50-160[4]		70-160	–
MS-2	50-160[4]		70-160	–
MS-2h	50-160[4]		70-160	–
SS-1	50-160[4]		70-160	–
SS-1h	50-160[4]		70-160	–
CRS-1	–		–	70-140
CRS-2	–		–	125-175
CMS-2	50-160[4]		70-160	–
CMS-2h	50-160[4]		70-160	–
CSS-1	50-160[4]		70-160	–
CSS-1h	50-160[4]		70-160	–

NOTES:

Temperatures for asphalt cements and cutback asphalts are guides only.

[1] Temperature of mixture immediately after discharge from the pugmill rather than temperature of asphalt cement or cutback asphalt.

[2] Application temperatures may, in some cases, be above the flash point of the material. Caution must therefore be exercised to prevent fire or an explosion.

[3] Rapid Curing (RC) grades are not recommended for hot pugmill mixing.

[4] Temperature of the emulsified asphalt in the pugmill mixture.

[5] The maximum temperature (asphalt cement and cutback asphalt) shall be below that at which fogging occurs.

Courtesy of The Asphalt Institute

Fig. 8-44A. Typical asphalt use temperatures, Fahrenheit

...uids used in cutbacks. The non-polluting ...ature of evaporating water, and lower and ...ss dangerous application temperatures, ...e other advantages of emulsion.

Asphalt does not ordinarily mix with water. But if it is broken up into very fine particles or globules, from 1/25,000 to 1/2,500 inches (1 to 10 microns) in diam-

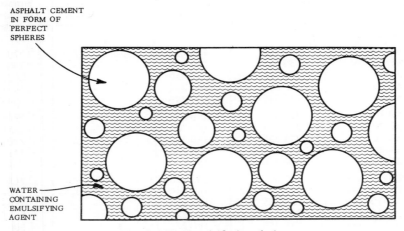

ASPHALT CEMENT
IN FORM OF
PERFECT
SPHERES

WATER
CONTAINING
EMULSIFYING
AGENT

Fig. 8-44B. Emulsified asphalt

eter, it will mix with water in the form of an emulsion, and will remain mixed unless the particles re-combine with each other to increase in size.

Such combining is prevented by presence of an emulsifying agent, usually soap, which is dissolved in the water, and may be said to form a protective film around each globule. There may also be a stabilizing agent to keep the films intact. These chemicals must be able to prevent spontaneous breakdown of the emulsion during storage and handling.

Salts and electrical charges on the surface of aggregate particles tend to destroy the films, allowing the globules to break down and the asphalt to coagulate and to stick to the stone, while the water drains off or evaporates. The rate at which this process takes place is determined largely by the type of emulsifiers and stabilizers used, and partly by type of aggregate, temperature, and other conditions.

Standard emulsions are the anionic type, in which the water-soap solution has a positive charge, and the asphalt globules are negative. Cationic emulsions use different emulsifiers that produce a negative fluid and positive globules.

Either type can be used for satisfactory results most of the time. But if an aggregate carries a definite surface charge (limestone is usually positive), the emulsion with oppositely charged asphalt will give the best results. The two types are not compatible and must be kept separated in storage and handling.

Figure 8-44C shows an exaggerated picture of the change in wetting effect of asphalt on a piece of aggregate caused by electrical charges. The poor results in the bottom picture can also result from dusty aggregate.

A sensitive standard emulsion that separates readily on contact with aggregate is called RS, for rapid-set. If the cationic type, it is called CRS. Viscosity is designated by a following hyphen and a numeral, as RS-1 for thin or RS-2 for thicker.

The RS type may be called penetration grade, because of original use in penetration macadam.

More stable emulsions are called MS, for medium set, or mixing grade, because of their use in replacing asphalt cements in blacktop mixes. The most stable type, SS or slow-set, is practically immune to globule breakdown, and depends on evaporation of the water to release the asphalt.

Recommended spraying temperatures are moderate, from 70 to 175 degrees Fahrenheit. When used at less than 100 degrees the liquid may be called cold emulsion.

Blown Asphalt. A thick asphalt residue may have distillation stopped while it is still liquid. It is placed in a converter, where it is held at high temperature while air is blown through it, until the desired properties are obtained.

Blown asphalt may be very stiff, and is less affected by temperature than asphalt cement, that is, it takes greater heat to melt it, and deeper cold to make it brittle. It is little used in pavement, except as crack and joint fillers for concrete. It is valuable for roofing, car undercoating and waterproof paint. Road asphalt would have a tendency to creep and drip in these applications.

Cracking. Refinery residual material, which includes the asphalt fraction of petroleum, may be subjected to extreme heat and manipulation to break some of it down into the lighter and more saleable hydrocarbons. This process is called cracking.

Asphalt obtained from the remainder after this processing is called cracked asphalt. It weathers and hardens much more rapidly than first-run asphalt, and its use is forbidden by many specifications.

Badly cracked asphalt has a dull surface and a high specific gravity. Borderline cases may be detected by dissolving a specimen in a solvent, then placing a drop on filter paper. A stain of uniform color indicates that it is uncracked; a dark center with a lighter ring around it indicates cracking. Unfortunately, good asphalts from certain crudes show up as cracked on this test, so it might have to be followed by other investigation.

Durability. Asphalt must be somewhat plastic to serve its purpose in pavement. Deterioration in service is largely a process of losing plasticity. The rate of this loss is affected by many factors, most of them arising from exposure to air and weather changes. They are often lumped under the heading of weathering.

Oxidation and volatilization are the prin-

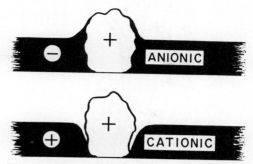

Fig. 8-44C. Anionic and catonic emulsion, with positive aggregate

cipal weather effects. Oxygen in the air attacks asphalt chemically, and lighter hydrocarbons evaporate. Both processes are much more rapid at high temperatures than at low ones, and both cause hardening of the asphalt. They are surface actions, and at any given temperature will proceed at a rate that is directly proportional to the amount of surface exposed.

There is also an age hardening that is independent of these factors. This proceeds rapidly for a few hours after cooling, then becomes quite slow and decreases with time, so that after a year the rate may not be measureable. It seems to be due to rearrangement of molecules in the gel structure. The action is reversible, so that in the laboratory asphalt can be brought back to its original penetration rating by melting or pounding.

When asphalt hardens to a 30 penetration it is likely to become brittle and crack; at 20 penetration it is almost sure to. Thin surface coats will crumble, thicker pavements will tend to break away at edges, ravel, and form potholes.

These failures can be postponed by using a fairly soft asphalt cement in the original work. An AC-10 asphalt has a much greater margin of safety than an AC-20.

Asphalt dissolves readily in most petroleum-based fluids, and its pavement surfaces can be severely damaged by gasoline or oil spills. Parking lots and driveways are often protected by a coating of coal tar.

Fig. 8-45. Tool heating pail

Spillage on highways is usually spread so thinly that it rarely does damage, and it may even prolong pavement life by replacing volatiles lost by asphalt during aging.

Stickiness. At low temperatures, basic asphalt (asphalt cement) is solid or semi-solid, and clean to the touch. If warmed above 70° to 120° F., depending on the grade, it becomes very sticky, so that it will adhere to anything it touches until it is dissolved off, or heated further.

At temperatures of 200 to 250° F. or hotter it becomes a thick oil, which will adhere to non-porous objects only as a film which can be wiped off. But if the object is cold or cool, it will chill the asphalt so that it will stick.

Cutback asphalts show the same range of characteristics, but at temperatures that vary with the type and amount of solvent present.

As a result of these characteristics, asphalt can be manipulated without difficulty

(but with caution) as long as both it and the tools are hot, but will stick vigorously to everything as it cools. It flows like oil in distributing systems at proper heat, but can plug them completely if chilled.

Hand tools are kept hot by placing in a fire bucket when not in use. The one in Figure 8-45 is a large pail with a few sticks to act as a wick, or some ashes, and some burning kerosene (or #1 fuel or diesel oil) in the bottom, mounted on a 2-wheeled cart. Air supply to the fuel is poor, so it burns at a low temperature, but it is more than hot enough to prevent tools from sticking to the asphalt.

Other objects that are too large or too inconvenient to heat, such as wheelbarrows, tailgate doors, or forms, may be protected by occasional spraying or painting with kerosene. A 2-gallon pump sprayer with a mist nozzle, such as might be used for insecticide in a garden or small orchard, is easier to use and more effective than a can and brush.

Hot asphalt will of course burn the skin; and will also stick to it. Warm asphalt just sticks. It can be removed with kerosene or gasoline, but it is easier if it is softened first by rubbing with lubricating oil. Removal of asphalt from fabrics is very difficult by any method.

BLACKTOP (ASPHALT CONCRETE)

The principal use of asphalt in first class highways is in the form of some hot-mix combination of asphalt cement, crushed stone, sand, fines, and a small amount of air. The mixture is known throughout the industry as blacktop, the name that will be used in this article. Its official designations are asphalt concrete, asphaltic concrete, and bituminous concrete. These terms are too long for casual use. Although they are used in all specifications and bids, they are seldom heard in plants or on jobs. However, the abbreviation A.C. is coming into use.

Blacktop is usually manufactured at a

Fig. 8-46. Blacktop plant

central location in a hot-mix plant, delivered to the job in dump trucks, spread by a paving machine, and rolled.

There are also hot mixes which are spread cold, and cold mixes. Either hot or cold blacktop may be made on the job in a portable pugmill. But the dominant type is central plant hot-mix hot-spread.

A large amount of asphalt concrete is mixed on road surfaces, but this is ordinarily called road mix or oil cake, rather than blacktop

The manufacture of blacktop is discussed in Chapter 21.

Aggregate. Asphaltic concrete depends on its aggregate for strength. The asphalt functions chiefly as a binder that holds it

together and protects it from the weather.

The stone in the aggregate must be tough and hard enough to withstand rolling and carry traffic without crumbling or breaking. This property is tested by putting a measured sample and some steel balls in a slowly rotating drum, for 500 revolutions. The product is shaken over a 12-mesh sieve.

The part that goes through is measured, and its percentage of the original charge is called the Los Angeles abrasion value, or the per cent wear. Stone for mixes must have a value of 50 per cent or less, while cover stone for surface treatments must not be over 27 to 35 per cent.

Soundness is a quality of weather resis-

Mix[1] Type	2½ in.	1½ in.	1 in.	¾ in.	½ in.	⅜ in.	#4	#8	#16	#30	#50	#100	#200	Percent[2] Asphalt
I a'	100	35-70		0-15				0-5					0-3	3.0-4.5
II a						100	40-85	5-20					0-4	4.0-5.0
II b					100	70-100	20-40	5-20					0-4	4.0-5.0
II c				100	70-100	45-75	20-40	5-20					0-4	3.0-6.0
II d			100	70-100		35-60	15-35	5-20					0-4	3.0-6.0
II e		100	70-100	50-80		25-60	10-30	5-20					0-4	3.0-6.0
III a						100	75-100	35-55	20-35	10-22	6-16	4-12	2-8	3.0-6.0
III b					100	75-100	60-85	35-55	20-35	10-22	6-16	4-12	2-8	3.0-6.0
III c				100	75-100	60-85	30-50	20-35		5-20	3-12	2-8	0-4	3.0-6.0
III d			100	75-100		45-70	30-50	20-35		5-20	3-12	2-8	0-4	3.0-6.0
III e		100	75-100	60-85		40-65	30-50	20-35		5-20	3-12	2-8	0-4	3.0-6.0
IV a						100	80-100	55-75	35-50	18-29	13-23	8-16	4-10	3.5-7.0
IV b					100	80-100	70-90	50-70	35-50	18-29	13-23	8-16	4-10	3.5-7.0
IV c				100	80-100	60-80	48-65	35-50		19-30	13-23	7-15	0-8	3.5-7.0
IV d			100	80-100	70-90	55-75	45-62	35-50		19-30	13-23	7-15	0-8	3.5-7.0
V a						100	85-100	65-80	50-65	37-52	25-40	18-30	10-20	4.0-7.5
V b					100	85-100	65-80	50-65	37-52	25-40	18-30	10-20	3-10	4.0-7.5
VI a						100	85-100	65-78	50-70	35-60	25-48	15-30	6-12	4.5-8.5
VI b					100	85-100	65-80	47-68		30-55	20-40	10-25	3-8	4.5-8.5
VII a						100	85-100	80-95	70-89	55-80	30-60	10-35	4-14	7.0-11.0
VIII a							100	95-100	85-98	70-95	40-75	20-40	8-16	7.5-12.0

[1] See article 4.0.7.
[2] By weight as a percentage of total mix.

Courtesy of The Asphalt Institute

Fig. 8-47A. Blacktop mix compositions

tance. An unsound stone (shale, for example) will absorb water, changing in bulk, and disintegrating if it freezes. A test is made by alternately soaking a sample in a sodium or magnesium sulfate solution and drying it in an oven.

Internal friction is a desirable quality of resistance of movement of stones past each other. Slag, which is rough and jagged, is high in this property, basalt and hard limestone are good, and rounded, uncrushed gravel is poor.

The degree of affinity for asphalt is called the surface property of stone. Aggregates with a high affinity to asphalt are easily coated by it, and tend to retain it against rubbing or wetting. They are called hydrophobic (water-hating), and are usually basalt, limestone or dolomite.

Aggregates with poor affinity for asphalt are called hydrophilic (water-loving). They include quartzite and other rocks high in silicon. Asphalt films do not form on them readily, and are easily stripped off. The condition may be aggravated by smooth surfaces, as in water-rounded gravel.

Cleanliness and purity are important. Value as aggregate is lessened or lost if there are important quantities of shale, vegetation, humus, other soft particles, clay lumps, or clay coatings. Such contaminants can often be found by visual inspection, and can be confirmed by analysis of washings.

Widely used aggregates include crushed basalt (trap rock), limestone, dolomite, slag, gravel, and sand.

Aggregate Size. Figure 8-47, mix compositions, gives the sizes of U.S. Standard square sieves for asphalt specifications. The following terms are generally used in regard to sizing:

Coarse aggregate: any material held on the #10 sieve.

Fine aggregate: any material passing the #10 sieve.

Mineral filler: fines, passing the 200 sieve.

Graded aggregate: a mixture of sizes from coarse to fine, the largest being many times larger than the smallest.

Close-cut aggregate: graded aggregate in which the largest size only slightly bigger than the smallest. May be called "one-size".

Top size: the smallest sieve which will pass all the material.

Size number	Nominal size square openings (1)	4	3½	3	2½	2	1½	1	¾	½	⅜	No. 4	No. 8	No. 16	No. 50	No. 100
						Amounts finer than each laboratory sieve (square openings), percentage by weight										
1	3½ to 1½	100	90 to 100		25 to 60		0 to 15		0 to 5							
2	2½ to 1½			100	90 to 100	35 to 70	0 to 15		0 to 5							
24	2½ to ¾			100	90 to 100		25 to 60		0 to 10	0 to 5						
3	2 to 1				100	90 to 100	35 to 70	0 to 15		0 to 5						
357	2 to No. 4				100	95 to 100		35 to 70	0 to 15	10 to 30		0 to 5				
4	1½ to ¾					100	90 to 100	20 to 55	0 to 15	0 to 5						
467	1½ to No. 4					100	95 to 100		35 to 70		10 to 30	0 to 5				
5	1 to ½						100	90 to 100	20 to 55	0 to 10	0 to 5					
56	1 to ⅜						100	90 to 100	40 to 75	15 to 35	0 to 15	0 to 5				
57	1 to No. 4						100	95 to 100		25 to 60		0 to 10	0 to 5			
6	¾ to ⅜							100	90 to 100	20 to 55	0 to 15	0 to 5				
67	¾ to No. 4							100	90 to 100		20 to 55	0 to 10	0 to 5			
68	¾ to No. 8							100	90 to 100		30 to 65	5 to 25	0 to 10	0 to 5		
7	½ to No. 4								100	90 to 100	40 to 70	0 to 15	0 to 5			
78	½ to No. 8								100	90 to 100	40 to 75	5 to 25	0 to 10	0 to 5		
8	⅜ to No. 8									100	85 to 100	10 to 30	0 to 10	0 to 5		
89	⅜ to No. 16									100	90 to 100	20 to 55	5 to 30	0 to 10	0 to 5	
9	No. 4 to No. 16										100	85 to 100	10 to 40	0 to 10	0 to 5	
10	No. 4 to 0 (2)										100	85 to 100				10 to 30

(1) In inches, except where otherwise indicated. Numbered sieves are those of the United States Standard Sieve Series.
(2) Screenings.

Reprinted from AASHO Designation M43 - Standard Sizes of Coarse Aggregate For Highway Construction

Courtesy of American Association of State Highway Officials

Fig. 8-47B. Coarse aggregate sizes

Most blacktop mixes are of the graded type, in which smaller pieces fill spaces between larger ones, in several size ranges. Exceptions are usually for two layer paving, with a coarse bottom layer, followed by a finer surface layer.

The largest stone ordinarily used in blacktop is 2 or 2½ inches, and surfacing mixes may be limited to ¾-inch or smaller. Large stone is economical of asphalt (smaller average surface area for bulk), and offers good stability. But the material is usually laid in layers one to three inches thick, and pieces should be smaller than layer thickness. Large sizes may also cause difficulties in mixing and in spreading.

Mixture Design. The proportions of a mix should be such that the aggregate would have good stability and load bearing capacity by itself, and that the proportion of asphalt be enough to coat all surfaces for effective binding and sealing, but not enough to separate the particles and cause instability.

Instability is avoided by using a proportion of asphalt and fines that does not quite fill spaces in the aggregate, so that air bubbles make up from two to six per cent of the bulk of the mixture. This assures enough resiliency so that aggregate can always rest on aggregate when under pressure.

More air than six per cent does not interfere with stability, but it increases exposure of the asphalt to oxidation and weathering.

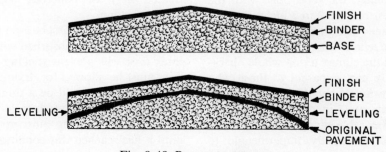

Fig. 8-48. Pavement courses

LAYING BLACKTOP

Courses. An asphalt concrete pavement may have one, two, three, or more layers or courses.

The **base course** may be subgrade or sub-base stabilized by mixing with asphalt, by methods described later under ROAD MIX, or by plant mixing. Portland cement may be used instead of asphalt. Sometimes such bases are of high enough quality to be considered a pavement course. Thickness may be 3 to 10 inches.

If all courses above subgrade are asphalt mixes, it is a full depth asphalt pavement.

A **leveling course** is required only in resurfacing projects. Its purpose is to smooth out irregularities in the old pavement, so that they will not interfere with the smoothness of the new. It may be limited to hand patching, grader spreading, or short paver runs over rough stretches, or be continuous throughout the job. It often includes changes in crown or in banking, and widening. If placed on the sides of a road to reduce its crown, it may be called wedging. Thickness is "as required".

A **binder course** is the layer below the surface course. It sometimes is the leveling course, and is often called by that name, even if there is patch-leveling beneath it.

Binder course mix is designed primarily for stability, as it is not directly exposed to weather or traffic. Aggregate is almost always larger than in the surface layer, with limited gradation to produce a coarse surface after rolling, for good bond with the top. Thickness may be 2 to 3 inches.

The **surface course,** which is sometimes the only course and therefore the complete pavement, is the climax of the whole operation; the only part apparent to those who use the highway. It should be smooth yet skid resistant, quiet in contact with tires, cohesive and durable enough to resist traffic stresses, and sufficiently impermeable to water and air to resist weathering. Thick-

ness is usually 1 to 2 inches, but under special circumstances, such as scarcity of skid-resistant aggregate, might be as thin as ½ inch.

Prime Coat. A prime coat is an application of thin bituminous material to a porous base, before putting on a surface. It serves to stabilize the base, and bind it to the paving material.

Priming is discussed a little later, under the headings **Road Oiling** and **Prime Coat.**

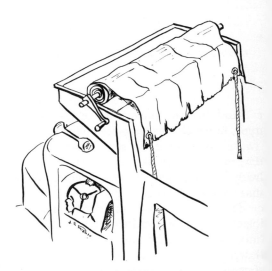

Fig. 8-49. Rolled up tarp cover

Tack Coat. The adhesion of fresh blacktop to old surfaces is often doubtful, but can be made sure by spraying or painting the surface with a tack coat (often called priming coat) beforehand. The material treated may be concrete, asphalt, iron, brick, or stone.

The surface must first be swept or flushed. Thin dust might be absorbed without harm, thick dust will prevent sticking. If flushed, time must be allowed for drying.

The asphalt should be a thin liquid that will dry or cure to produce a firm, tacky surface. The blacktop must not be placed until it has reached this condition, or it will act more as a lubricant than an adhesive.

Materials used include: RC-70 and 250, RS-1 and 2, and RT-7 and 8. The thinnest covering application is used, which on pavements means up to 0.1 gallons per square yard. On joints and curbs, any handy thin asphalt may be put on with a brush, but coating should be as thin as possible.

Tack coat is often applied by dipping a broom into the liquid, and shaking it to leave numerous small blobs on the surface.

Hauling. Blacktop is hauled from the mix plant to the job in dump trucks, which may be medium or large size. They are usually of standard construction, but are (or certainly should be) equipped with a heavy tarpaulin to cover the whole surface of the load. Otherwise, the cooling effect of the wind is likely to form a crust during even a short run, and this may be a lumpy nuisance during spreading.

A full load of 10 to 20 tons of blacktop may stay in workable condition for several hours in a covered truck on a warm day, so that the distance-limiting factor is often the cost of hauling, rather than hardening of the load.

A double bottom improves heat retention. It is usually made by placing an extra sheet of steel over the original floor. If it has been battered, no spacers may be needed.

Bodies specially designed for hauling hot mix may have double bottoms and sides, with ducts to carry engine exhaust through them for a positive heating effect.

HAND WORK

Blacktop must be spread and leveled by hand on jobs that are too small, either in value or in usable space, to justify bringing in a spreading machine. Also, much machine time can be saved and quality of finish improved by doing hand work around a machine in areas that it cannot reach without excessive manipulation.

Paving machines are often accompanied by one or more laborers, who spread shovels of blacktop over spots that have been

Fig. 8-50. Hand tools

gouged (center) or skipped (edges) by the screed.

Tools and Equipment. The basic hand tools for blacktop are the rake, the lute, the shovel, and the wheelbarrow.

A blacktop rake is an oversize, overweight edition of a garden rake. It is usually 16 inches wide, with strong 3-inch teeth. The 6½-foot handle is wood, except that the lower 18 inches is steel, to avoid damage in heating pots. Weight is about 6 pounds.

Use of a rake is forbidden on many State jobs, because of possible segregation caused by action of the teeth. No perceptible damage seems to be done to driveways or patches, however.

The rake is used for pulling and pushing down piles of loose blacktop, filling in hollows, and general smoothing and leveling.

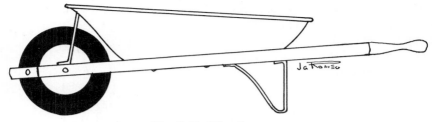

Fig. 8-51. Wheelbarrow

Pushing and smoothing are usually done with it upside down; that is, with the teeth pointing up.

The lute is an aluminum tool shaped like rake, with the teeth replaced by a solid strip. It may have two straight edges, or one straight and one serrated. The handle may be one piece, up to 7 feet long, or telescoping. Width may be 24, 30, or 36 inches.

Shovels may be the garden or digging type, with a curved back and a rounded point, or flat with turned up sides and a square front. The first is better for penetration, the second for cleaning up and spreading.

Wheelbarrows are usually shallow, all-metal garden types with a single wheel.

Special shoes or sandals with thick wooden soles may be worn to reduce discomfort while walking or standing on fresh, hot mix surfaces.

The truck that brings the blacktop should have one or more small sliding gates in the tailgate, and possibly one in the rear section of the side of the body. A small gate allows neat and efficient loading of a wheelbarrow from a partially dumped body.

A heating pail is needed. This was described earlier.

A garden-type sprayer (a 2-gallon size is convenient) is used to apply kerosene to any cool surfaces that might be in contact with hot asphalt. This may prevent sticking, and will often remove crusts after they have formed.

A hand tamper (stomper) is necessary on jobs too small for a roller, and is handy on any job including small or irregular pav-

ing areas. It consists of a flat heavy metal plate, 8 to 12 inches square, with a vertical center handle.

Where new blacktop begins at an old solid pavement, the connection should be both hand-tamped and over-filled. Roller pressure into the bottom angle is poor, and the roll will shove loose material away from the joint.

Dumping. It is considered bad practice to dump a truck load of blacktop on a surface where it is to be spread by hand. In the first place it is likely to segregate, with larger stones sliding down the outside of the growing pile and making a bottom layer of different texture. More importantly, it is extra labor to dig into and spread a big pile.

These objections are reduced but not eliminated if the load is spread-dumped instead of pile-dumped.

The preferred method is to station a wheelbarrow under a sliding gate in the tailgate, raise the body enough to get good pressure of the load to the rear, open the slide to fill the wheelbarrow, then close it until the next load. If there are two slides, two loads may be taken at the same time.

The wheelbarrow is taken to the edge of the paving area, or close to the last blacktop spread, and either pile or spread dumped. Usually, there is a man to spread it while the barrow operator goes back for another.

If there is a small wheel-mounted loader on the job, it may dig out of the dump body, and carry bucket loads to where they are needed. It thus serves as a motorized wheelbarrow. It may do some spreading and smoothing with its bucket. It has much greater space requirements than the real wheelbarrow.

Spreading. Each small pile left by a wheelbarrow or a loader may be spread with a rake or a lute, or partly spread and partly moved to other areas by a shovel. In all cases, the object is to reduce the loose backtop to an almost level surface slightly higher than the finish after-rolling level desired.

The material (unless too coarse for a hand job) is usually smooth and semi-flowing, with the large pieces carried with the small. It is readily pulled, pushed, or patted into the approximate grade desired, but the surface will be rough in a small way. Except in cold weather (when such work is seldom done) there is liberal time to shape it, before it must be rolled or tamped.

A light touch is needed. The rake or lute must be partly or wholly supported for leveling, as otherwise it will move too much material and gouge hollows which must be refilled.

Hot mix will keep the rake or lute hot enough not to stick. But it should be put in the fire pail whenever it is not in use.

Patching. Patching is the operation of filling small holes, wide cracks, and ruts with a blacktop mixture. Procedures are affected by whether the patch is the whole job, or is in preparation for surface treatment or paving; whether the patching material is hot or cold; and by weather conditions.

If a patch is to be unprotected, it must be stuck or bonded to the old pavement. If that is dry, clean, and warm, the patch may hold if it is put down hot (or, if cold-patch, while fresh enough to be sticky) and compacted immediately.

For less favorable circumstances, a thin tack coat of liquid cutback or emulsion asphalt should be applied first by spray or brush, at least around the edges. Very cold pavement should be heated with a blow-torch before priming.

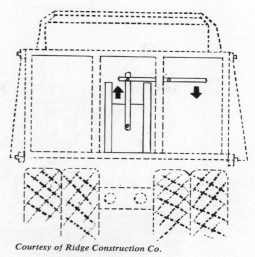

Courtesy of Ridge Construction Co.

Fig. 8-52. Blacktop door

Patch material should be mounded over the hole — the deeper the hole, the higher the mound. It should be compacted as soon as possible, preferably with a roller, but with a tamper or even a sledgehammer if necessary. The compactor must be wet with water or kerosene to avoid picking up pieces.

If compaction is poor or lacking, and traffic will cross the patch soon, its surface should be covered with sand, dust, or any loose material to avoid picking up by tires. In cold weather work, the usual ignoring of this precaution leads to entire patches being disintegrated and scattered before they can harden.

Bonding is less important, or even unnecessary, when the road is to have a surface treatment strong enough to hold the patching material down.

In patching before a seal coat, the mix should be fine grained (sheet) to reduce absorption of oil. A curing time of at least a week is recommended. With a coarse or fresh patch surface, so much liquid asphalt may soak in that the aggregate would not be properly bound.

MACHINE WORK

Paving Machine. The modern blacktop spreader, known variously as a paving ma-

Fig. 8-53. Paving machine

chine, paver, or finisher, is a highly sophisticated piece of equipment, capable of producing smooth pavement under a variety of conditions.

Its construction is described in Chapter 21. Basically, it consists of a crawler or wheel tractor with a front receiving hopper, feeders to transfer material through the tractor to spreading augers in the rear, and a towed screed unit which shapes, smooths, and partially compacts the paving mixture.

The tractor has push arms with rollers at their tips, with which it can push a truck by its tires, while it is backed up against the hopper to dump into it.

The paving machine leaves a ribbon of hot blacktop that should be uniform in surface grade and edges, with a dense, smooth, pressed-down surface. The next and final step is compacting with rollers.

Compaction. There are many different specifications, practices, and beliefs in regard to rolling of blacktop, of which only a sample can be given here.

Blacktop immediately behind the machine is often too hot and soft to be rolled. Time to set up for a roller varies between zero and 10 minutes. It is affected by the type of mix (some can be rolled at 250° F.,

others not above 200), its temperature in the machine, the weight of the roller, the temperature of the air and the subgrade, and presence or absence of sunshine and wind. The test is whether it will support the roller.

On the job, readiness for rolling is a matter of judgment, and is usually estimated in distance behind the paver, rather than in minutes.

There are usually two stages, breakdown rolling and finish rolling, both customarily done with steel wheels. There are often substantial advantages to intermediate compaction with a rubber tired roller. Some authorities are now recommending the rubber unit for breakdown rolling also.

Sometimes a big paver will put down a stable mix at such low temperature that it will support a roller immediately, so adequately that there is no true breakdown phase in the rolling.

A roller must have a good roll-wetting system, as hot asphalt tends to stick to a dry roll, spoiling both the pavement surface and the smoothness of the roll This is most important during the breakdown, but should not be neglected for finishing.

Breakdown. The breakdown roller typi-

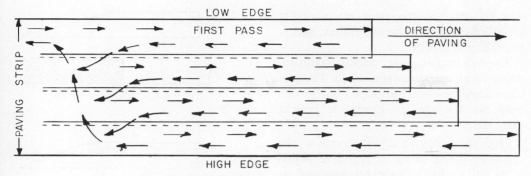

Fig. 8-54A. Rolling sequence, blacktop

cally works on soft, hot material which tends to move under it and even ahead of it, and to show marks at roll edges.

It is usually recommended that the drive rolls be kept toward the paver. This is because the steering roll, being turned only by friction with the blacktop, tends to shove it ahead more than a drive roll, which tends to climb up on it. On the other hand, the steering roll is lighter and less likely to squeeze material out from underneath, and it is less affected by any jerking from imperfections in the reversing mechanism.

When working uphill, the drive roll should be downhill.

Three-wheel rollers may be preferred for breakdown work, as the difference in width, diameter, and weight between drive and steering rolls favors compaction, and steering backward along an edge is easier. They must be backed toward the paver in order to roll the edge first.

If the strip is a half road, with a longitudinal joint with a previously laid strip at the center line, the first roller pass is on this joint. The edge of the roll or rolls is supported by a narrow strip of the completed pavement, and overlaps onto the fresh material.

The standard rolling pattern after pressing the joint (or to start, if there is no joint) is to start on the low edge. If the mix is set enough for rolling, the roll can be run along the edge without squeezing it sideward noticeably.

The edge firms up a little faster than the interior.

Any flowing to either side or ahead of the rolls means that additional cooling must be allowed, in order to avoid making the surface uneven. Other possible symptoms of over-softness are cracks forming at roller edges, or blisters appearing behind.

If too much cooling is allowed, the mix will be stiff and stubborn, and full compaction will not be obtained.

In an ordinary pattern of rolling the roller moves straight back toward the paver and then straight forward through the fresher and softer areas, and does its turning on cooler and firmer surface However, gradual changes of direction can be made on return trips.

Each pass is of a different length, so that the roller never stops a second pass on the same cross line, as this would tend to create or intensify a hollow resulting from extra compaction during change of direction. The first pass is usually the shortest in a set, and the last the longest, but an opposite procedure may be followed.

The roller must never be left standing on any surface that is not completely compacted and cooled, as it would sink and create a hollow.

Number of passes in breakdown rolling vary, but coverage of every part of the surface twice may be enough. Marks of the roll edges are usually still visible.

Finish. On a small job, finish rolling

may be done by the same machine. On larger jobs there will be two rollers, with a possibility of a third.

The finishing machine, usually a tandem, rolls until all marks are eliminated. The surface should be closely inspected during this work, as it is the last chance to correct irregularities.

High spots may be taken off with a rake or shovel. Large pieces of aggregate pulled to the surface by this should be removed, as they would give the patch a different appearance. Low spots should be loosened with the rake, spread with hot mix, and rolled.

Tolerance for surface irregularities is usually ⅛ inch in ten feet. But a good paving gang is likely to feel disgraced unless its work is much smoother than that.

A rubber tired roller may be used between the breakdown and finishing machines. Its weight is accompanied by a kneading action that more nearly duplicates natural traffic load, and which may reduce settlement in use after completion. The finish roller then adds little to compaction, but its smooth rolls improve the surface.

Joints. A joint is any surface line where hot blacktop is joined to blacktop that has cooled or hardened, or to concrete or any other material.

When a paving machine lays blacktop in two or more parallel strips, the joining lines are called longitudinal joints. The screed is usually overlapped about 3 inches on the older surface, both to heat and soften it, and to make sure that the new material fills right to its edge. Strikeoff is at a slightly higher level than the old. The extra height, called fluff, allows for roller compaction. This may amount to ⅛ to ¼ inch per inch of layer thickness, depending on the amount of compaction achieved under the screed. Bringing up extra fines with a lute before compaction may make the joint tighter.

When work is stopped temporarily, the paver is moved ahead to allow the rollers

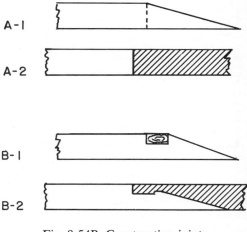

Fig. 8-54B. Construction joints

to compact the end of the lane into a ramp down to the base level. This ramp is cut back to where it is full depth, preferably with a pavement breaker, and primed, when fresh material is to be placed against it.

Another procedure is to sink a plank or board at the end of the full-depth blacktop, with a downward ramp extending beyond it. This board is rolled in flush with the pavement surface, then the roller is taken across it to park on the base layer.

To resume work, the roller is put back on the new pavement, and the board is lifted and set aside. The slot left between pavement end and ramp provides a lock-type joint. It should be primed, or hot blacktop may be piled temporarily on top, to soften the first material and make a tighter joint.

These two methods are illustrated in Figure 8-54B.

A roller should not be moved onto an unpaved or dirty surface, where its wet rolls could pick up dirt, leaves, or trash. Boards may be placed on the base (or on shoulders), or a thin layer of mix may be spread ahead on the base to protect them.

When blacktop is placed against concrete at grade, as along a gutter or at the ends of a bridge slab, it should be left a little high, perhaps ¼ inch after rolling, to allow for

Courtesy of Young Corp.

Fig. 8-55. Pulverizing old road surface

further compaction under traffic. The concrete should be painted or sprayed with a tack coat in all areas of contact.

The same precaution should be observed in joining with firm pavements of asphalt, brick, and other material.

MIXED-IN-PLACE ASPHALT CONCRETE (ROAD MIX)

Asphalt and aggregate can be mixed on the road where they are used. The resulting paving material is usually not as good quality as that mixed in a plant, where conditions can be exactly controlled. However, road mixing is much more economical where long distance transportation is a problem, and usually provides a satisfactory result. The method can be used for either surface or base courses.

Mixing. Mixing may be done by graders, which blade the materials back and forth across the road until they are mixed; traveling plants that can pick up a windrow of loose material, mix it with asphalt, and leave it in windrow form; and rotary mixers that pulverize material lying in place on the road, mix it with asphalt, and spread it out flat again.

For the grader and the traveling plant, there should be a windrow of uniform height,

width and slope, so that its volume can be calculated, and the proper quantity of asphalt used. Provided good penetration can be obtained, the quantity processed by the rotary can be figured by the width and the depth of the cut it takes.

The asphalt is usually rapid or medium cure cutback in number 250 or 800 grade, or MS-2 emulsion with added water. Requirement is usually 0.4 to 0.6 gallons per square yard, for each inch of depth. Layers may be 4 to 8 inches deep, with 6 to 8 most usual.

The aggregate may be carefully selected (and sometimes blended) material brought to the job, an old road surface originally made of imported material, or native soil.

Special machines, such as grid rollers, or crusher-grinder units for mounting on graders or loaders, may be used to pulverize old bituminous pavements to fineness suitable for mixing.

Aeration. Road mixed aggregate is usually more or less damp. This is an aid to mixing, but the presence of more than 2 per cent moisture, or of any considerable amount of cutback solvent, will make compaction unsatisfactory .

These excess fluids are encouraged to evaporate by aerating the mix. This may be done by blading it back and forth with graders, working it with disk harrows, or re-pulverizing it with a rotary with the back cover up The number of passes needed will vary widely with original conditions, temperature, sun, and wind.

When proper condition for compaction is reached, the material is spread and carefully finished with graders, and rolled with rubber-tired and/or steel wheel rollers.

Off-Road-Mix. Road mix is often mixed on an unused road, or any hard, flat area, and hauled to its place of use. This may be done when the material is used for patching, or when traffic blockage on the main road must be kept to an absolute minimum.

SOIL-ASPHALT STABILIZATION

Asphalt stabilization of soil follows the general practices used for road mix. However, the product is intended to be protected by a wearing surface of different type, so that the only problem is to achieve stability. Poor quality aggregate may be used.

General requirement is that the soil be one that can be pulverized for efficient mixing, and that it contain not more than 45 per cent of combined silt and clay.

Courtesy of Seaman Co.

Fig. 8-56. Rotary mixer

Rapid-curing cutback is used with sandy soils with a plasticity index of 5 or less. Medium-cure is used for loams with an index of 6 to 10, and slow-cure for heavier soils.

Mixing, aeration, and compaction follow road mix practices, except that deep layers may be used, requiring heavier rollers or sheepsfoot rollers. After a sheepsfoot has walked itself out to one-inch indentations, the top is bladed, regraded, and then compacted by a rubber tired or wheel roller.

The completed stabilization is allowed to cure for about a week, then prime-coated, and surface-treated or blacktopped.

ASPHALT SURFACE TREATMENTS

Asphalt surface treatments are applications of asphalt and (usually) aggregate on existing road surfaces, new or old. The purpose may be to preserve a base course, to check deterioration in an old road, to provide protection against skidding, or to change color or appearance.

Single or multiple courses (layers) may be applied. A newly completed flexible base may be capped with up to four surface courses, with a total thickness of two inches or more. An old road being freshened may have a single coat of asphalt cutback and sand less than ¼ inch thick. Almost every possible variation between these extremes, and even outside them, may be used.

All these surface treatments require liquid asphalt, a distributor to spread it, and (usually) a thin layer or coating of aggregate (small stone or sand) to cover it.

Liquid asphalts (cutbacks) were discussed earlier in this chapter.

Aggregate. Aggregate used for covering or blotting surface treatments of liquid asphalt must be of sound, clean material with good resistance to abrasion, that forms a good bond with the asphalt. Size should be small enough to permit the pieces to sink to ⅔ to ¾ of their height into the asphalt, so that they can be held firmly.

In many areas, the preferred material is crushed basalt (trap rock), slag, or hard limestone, with largest size ¼, ⅜, or ½ inch, and smallest not less than half the maximum. Even gradation is preferred to mixed sizes. Cubical pieces give best results, thin flats are most difficult.

Some highway departments avoid stone, and specify medium to coarse sand, either natural or artificial. Others may alternate stone and sand treatments in successive treatments of a road, several years apart.

Aggregate must be available in sufficient quantity, and with proper equipment, to provide prompt cover for all the asphalt laid down by the distributor. If the aggregate supply is at a distance, schedules should provide for loaded trucks waiting in line to spread. This over supply of equipment is usually less costly than holding up the job for brief road delays.

On a big job, a single mechanical spreader will usually take care of all the trucks. On small work, there may be only one or two trucks with tailgate spreaders, which with a long haul would cause a bottleneck.

It is often possible to pile aggregate in advance in the immediate neighborhood of the job, and load it into the truck or trucks with a rubber tired loader.

Such a loader can be used to advantage in carrying smaller quantities of aggregate for hand covering of spots skipped or skimped by the spreading trucks.

Aggregate must not be dusty, as a dust film prevents adhesion to the asphalt. It must not be wet, unless the asphalt is emulsified, but slight dampness is usually permissible. Adhesion is sometimes improved by a light coating of fuel oil, applied as a spray at a loading belt. Or it may be heated, then lightly coated with about one per cent of asphalt in a pugmill, but this adds unreasonably to costs.

Distributor. Most liquid asphalt applied directly to roads is handled by a pressure distributor. This consists basically of a heated tank mounted on a truck or trailer chassis, with a pump that forces the fluid in the tank into a spray bar mounted across the rear of the truck.

These machines are rather complex. They are discussed in Chapter 21.

The distributor is required to cover a road surface with an even sheet of liquid asphalt. The application (amount per square

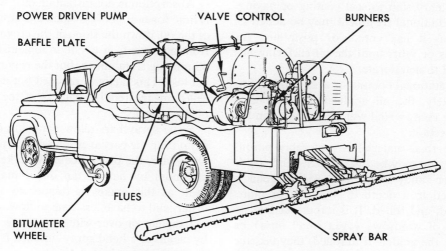

POWER DRIVEN PUMP   VALVE CONTROL   BURNERS

BAFFLE PLATE

FLUES

BITUMETER WHEEL

SPRAY BAR

Courtesy of The Asphalt Institute

Fig. 8-57. Asphalt distributor

yard) is regulated by pumped volume and by truck speed. Opening a regulating valve increases volume. At any valve setting, increasing truck speed will reduce the application.

The depth of the asphalt layer left on the road is limited by the rate of application, and by its viscosity. These should be coordinated to avoid runoff to the sides, which is waste. Viscosity of asphalt depends first on its makeup, and second on its temperature. The hotter, the thinner.

Desired depth of application (asphalt table) varies with the type being used, and with aggregate size. Using too thin a fluid (cutback or emulsion) or overheating the right fluid, will make it impossible to hold that depth on the pavement. A lesser depth is economical of asphalt, but produces an inferior pavement surface, sometimes hopelessly inferior.

Road Oiling. Road oiling, widely known as MCO treatment, is liberal application of a thin cutback asphalt to an absorbent road surface. It stabilizes a surface layer, often an inch or more in depth, and makes it dust free and water resistant.

Oiling may be done to create an all-weather road for light traffic, and/or as a preliminary step to seal coating or paving. Any additional treatment may be added as soon as it has cured, or postponed for months or years until the oil surfacing has started to deteriorate.

If additional operations are to follow immediately, the oil application is called a **prime coat,** which will be discussed in the next section.

The road material is preferably porous gravel, but may be tight gravel, sand, sand-clay, or sandy loam. The original road is scarified to a depth of 4 inches or more, shaped, and rolled. It is important that no ridges of windrows (even small ones) be left by the grader. If unrolled, they become dry and dusty, and will not absorb oil; if rolled they are still likely to remain as low ridges of over-fine and over-dry material. They tend to cause unpenetrated or poorly penetrated strips, which weaken the road.

For best penetration, the road should be damp enough to hold together without becoming dusty, but it must not be wet. If dusty, it should be lightly sprinkled a while before oiling, as oil tends to run over dust without sinking in.

The "oil" is usually MC-30 (MCO) or some other thin liquid asphalt or distillate. It is sprayed at its maximum permissible temperature, 175 to 200° F. (this is above its flash point, so smoking and other fire dangers must be avoided), in the maximum amount the road will absorb. The rule of thumb is one gallon per square yard. This may sometimes be exceeded, but more often must be reduced. Any starting rate may have to be reduced if there is a substantial amount of runoff of unabsorbed fluid. A too-tight soil might take only 0.2 gallons.

Most oiling of this type is done on crowned two-lane roads, 18 to 22 feet in width, with a single pass of a pressure distributor. If the liquid is discharged uniformly along the width of the spray bar, the center or top of the road will almost always retain less oil than the sides.

Absorption is not instantaneous, so there is time for more or less sideward movement of the oil, down the slope of the crown. Loss from any point on a side is replenished by oil moving from above, but the center has a net loss. This factor is reduced but not cancelled by the relative flatness of the center of the crown.

Some spray bars allow compensation by permitting the partial closing of side nozzles. Otherwise, the difference may be kept to a minimum by having the oil at maximum heat, as this makes it thinner to soak in faster, and to have less time to flow.

Irregular or over-width areas are treated by means of a hand spray supplied from the distributor's pressure system, or if very small, with a can with a spout.

Puddles may be left in depressions caused by grading errors, or ruts from distributor tires. These should be blotted by filling with road material or sand, or fine crushed rock.

The road may be left to cure by itself, with traffic kept off it for 24 hours. But if it is to receive no further treatment, it is better to compact it immediately, with a steel wheel or rubber tired roller with wet rolls.

The oiled road surface should be smooth, firm, and dust free. It is waterproof or at least water resistant, so that it protects the base from soaking from above. It will not stand up under heavy traffic or no traffic, however, and even under light or medium traffic will start to deteriorate in one to three years. It can be protected and built up with a seal coat, then or any time earlier.

The whole oil job must be done in a single application, as it seals the surface against absorbing any substantial amount from a second treatment. The only exception is that an oiled road that has gone badly off grade may be sacrified, shaped, and re-oiled. It is likely to absorb much less fluid the second time than the first, and to show increase in strength.

Prime Coat. Priming is usually a road oiling job, done with a smaller quantity of liquid asphalt, as a preliminary to seal coating or blacktopping. The oil may be the same MC-30, or a heavier grade up to 250, or SC-70 or 250. Thicker grades are used on porous material, thinner on tighter surfaces.

Rate of application varies from 0.2 to 0.6 gallons per square yard.

Runoff is guarded against, and puddles blotted, in the same manner as in regular oiling.

A seal coat will bond best, and retain the highest percentage of its aggregate, if the road is not rolled after oiling. However, many engineers specify this rolling, perhaps on the ground that it reduces total surface aggregate requirement.

Absorption time may be a few minutes up to 24 hours, and another 24 hours curing time may be allowed before coating or paving.

Seal Coat. A seal coat is usually a thin layer of liquid asphalt covered (blotted) by a layer of aggregate. It is used as a surfacing for new macadam or oiled gravel, and for rejuvenation and protection for old asphalt pavements that have weathered and hardened so that they are checking, cracking, chipping, and/or breaking up.

The aging asphalt may be softened and toughened by absorption of part of the new material, and it is given protection by the coat of tough new material with embedded aggregate.

The liquid needs a reasonably high viscosity to prevent running off. In warm weather, which is the preferred condition for this work, it should be a thick cutback such as RC-800, RC-3000, or MC-3000, or asphalt cement A-5. In cool weather, RC-250 would be preferable. These asphalts will not bond with wet or damp surfaces.

Emulsified asphalts RS-1 and RS-2 are less sensitive to temperature and moisture. They can be used in either warm or cool weather down 40° F. But freezing ruins them, breaking down the oil globules so that the mixture separates and clots.

If the aggregate is sand, 30 pounds to

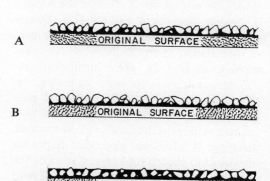

Fig. 8-58. Stone embedment

Size of aggre-gate	Pounds of aggre-gate per sq yd†,§	Gallons of asphalt per sq yd†,¶	Hot weather		Cool weather	
			Hard aggregate	Absorbent aggregate	Hard aggregate	Absorbent aggregate
¾–⅜	40–55	0.28–0.35	120–150 RC3000, RS2	RC3000, RS2	RC800, RS2‡	RC800, RS2‡
¾–No. 8	30–45	0.23–0.30	200–300 RC800, RS2	RC800, RS2	RC800 RS2‡	RC800 RS2‡
½–No. 4	25–35	0.20–0.25	200–300 RC250,800 RS1, 2	RC250,800 RS1, 2	RC250,800 RS1‡ RS2‡	RC250,800 RS1‡ RS2‡
½–No. 8	25–35	0.20–0.25	RC250,800 RS1, 2	RC250,800 RS1, 2	RC250,800 RS1‡ RS2‡	RC250,800 RS1‡ RS2‡
⅜–No. 4	20–25	0.20–0.25	RC250,800 RS1, 2	RC250,800 RS1, 2	RC250,800 RS1‡ RS2‡	RC250,800 RS1‡ RS2‡
⅜–No. 8	20–25	0.20–0.25	RC250,800 RS1, 2	RC250,800 RS1, 2	RC250,800 RS1‡ RS2‡	RC250,800 RS1‡ RS2‡
¼–No. 8	15–20	0.15–0.20	RC250,800 RS1, 2	RC250,800 RS1, 2	RC250,800 RS1‡ RS2‡	RC250,800 RS1‡ RS2‡
Sand	10–15	0.10–0.15	RC250,800 RS1, 2	RC250,800 RS1, 2	RC250,800 RS1‡	RC250,800 RS1‡ SS1‡

* These quantities and types of materials may be varied according to local conditions and experience.
† The lower application rates of asphalt shown in the table should be used for aggregate having gradings on the fine side of the limits specified. The higher application rates should be used for aggregate having gradings on the coarse side of the limits specified.
‡ Caution should be exercised when using this material under poor drying conditions.
§ The weight of aggregate shown in the table is based on aggregate with a specific gravity of 2.65. In case the specific gravity of the aggregate used is less than 2.55 or more than 2.75 the amount shown in the table above should be multiplied by the ratio which the bulk specific gravity of the aggregate used bears to 2.65.
¶ Under certain conditions, the heavier grades of MC liquid asphalts may be used in cool weather.

Fig. 8-59. Quantities, single surface treatments and seal coats

the square yard provides good cover. If it is stone, the rate of application of the asphalt is largely dependent on its size. Stones should be embedded to about 60 per cent of their depth, as in Figure 8-58(A). Less asphalt, as in (B), gives insufficient anchorage and leads to early loosening, although it may be good for anti-skid purposes. The too-high asphalt level in (C) reduces skid resistance and is likely to cause bleeding. If additional aggregate is put on afterward to absorb the surplus, most of the pieces will be weakly attached and will tend to ravel off.

There is an absorption factor also. A deeply weathered surface will absorb a measurable quantity of fluid, a sounder one probably will not. The aggregate itself might or might not be absorbent. Fresh blacktop patches absorb greedily.

As indicated in Figure 8-59, quantity recommended ranges from 0.10 up to 0.35 gallons per square yard.

Fog Seal. A fog seal is asphalt only, usually slow emulsion mixed 1:1 to 1:3 with water. Its primary purpose is rejuvenating and sealing an old surface by soaking in and filming over.

Application is at the rate of 0.1 to 0.2 gallons per square yard. It is usually left exposed during curing, without aggregate cover. It contributes little or no strength to the pavement.

Slurry Seal. When pavement is badly cracked and weathered, regular seal coat liquids may not be satisfactory because of variable amounts of absorption.

A slurry coat is emulsion combined with graded fine aggregates to a creamy, flowing consistency. It is spread with mechanical or hand squeegees. No additional aggregate is added. Traffic is kept off for the 2 to 4 hours needed for setting.

This process fills cracks and shallow holes, and leaves a rejuvenating film on intact surfaces.

Mixing may be done in a blacktop plant, or mix-on-the-job trucks. Spreading may be done by a squeegee box towed by the truck. The road may be dampened by a water mist spray just ahead of the box.

Anti-Skid Coat. An anti-skid surface is a seal coat, with minimum quantity of asphalt, and an angular aggregate preferably composed of stone that breaks to sharp angles and rough surfaces.

This may be applied only at certain points where accidents have been caused by skidding, or used for long stretches as a combined pavement rejuvenator and accident preventative.

Multiple Surface Treatment. Two or three seal coats may be applied to a pavement or base in succession, to build up a substantial surface. The procedure is sometimes called "inverted penetration" because the liquid is applied at the bottom of each layer, and works up.

The bottom layer has the coarsest aggregate, and size is decreased with each successive course.

Each course is completed, with asphalt, aggregate, rolling, and brooming, before the next is added.

Color Coat. A color coat is applied in the same way as a seal or anti-skid coat. The aggregate, however, has some distinctive color, usually to distinguish it from other areas. A black road may have near-white shoulders. Red aggregate may be used on ornamental driveways, or on freeway ramps.

There are also pavements formed in depth from plastic aggregates of brilliant colors mixed with matching resins in pugmills, that are laid by conventional blacktop methods.

Spray Joints. The asphalt for seal coats is applied by pressure distributor (see Chapter 21), at a rate carefully worked out in advance to conform to aggregate size and other conditions. A distributor carries a limited load, which may be renewed a number of times in doing one stretch of road.

A distributor is not a precision instrument, and cannot be turned off and on within a fraction of an inch. As a result, it can-

Courtesy of Highway Equipment Co.

Fig. 8-60 Applying slurry seal

not start off at an even line where the job begins. Also, finishing one load and then starting with another may cause an overlap with double liquid, or a gap with none. The overlap will cause immediate bleeding. If it is blotted with extra aggregate it will make a bump, otherwise it will be sticky at first, then slippery, and look badly. A gap means no seal coat.

This problem is kept to a minimum by skill in operation.

First, the spray is cut off immediately when the first sign of an empty tank appears, usually irregularity or puffing in the output. Using the last few gallons produces a long strip full of skipped spots, due to spreading a variable air-asphalt mixture.

Aggregate is spread to within a few inches of the end of the solid coverage.

With the next load, the truck is settled into its exact spreading speed, and the operator turns on the spray bar just as it crosses the end line of the last load. He often gauges this so exactly that no hand work is needed.

If the new asphalt overlaps the old, making it too thick, the surplus is distributed by a laborer with a squeegee or a lute. If there is too much to be blended in and "lost" in a yard or two, the surplus is pulled off the side, for later removal.

A narrow or partial skip may be filled in with the squeegee, drawing from and slightly thinning the nearby layer of asphalt. A larger area is covered by using a hand pour pot, which may have been filled and left by the previous distributor, for use in case of need.

This working over of the spray joint is done very rapidly, so as not to unreasonably delay the aggregate spreader.

An almost perfect joint may be made by putting a strip or two of building paper across the bare road at the starting point, and another a short distance short of the point at which the distributor should empty enough to be shut off. The paper should be weighted enough to prevent flapping, or blowing away.

Fig. 8-61. Starting on paper

The distributor is run across the first paper at proper speed, and the bar is turned on as it crosses it. It is turned off over the second paper, and the truck may be stopped abruptly so that the paper will catch drip from the nozzles.

When the paper is removed, a sharp and (hopefully) accurate edge is left. The next load is started on paper (usually the same) lined up on the edge of the first application, and held out of it by aggregate.

This excellent method is seldom used because of confidence in the distributor operator, the small nuisance of obtaining the paper, and the often big nuisance of disposing of it.

The driver should have a guide to keep him in correct alignment. A good edge to the road, or the strip built or graded for treatment, is usually enough. If the edge is doubtful, or is being altered, a string and pointer arrangement, along one side, is easy to follow, but considerable work to set up. Sometimes pegs or pegs and flags, are placed in doubtful areas only.

Distributors and their metering arrangements are described in Chapter 21.

Spreading Aggregate. Aggregate may be crushed stone up to one inch in size, but ⅜ and less is more usual. Fine natural gravel with sand and fines removed (native grits), or sand, may be used instead of crushed stone.

Fig. 8-62A. Aggregate spreader

This material is spread behind the machine while backing over the fresh-sprayed asphalt, in sufficient quantity to cover or "blot" it, so that the aggregate spreader and the roller will not touch any asphalt. Rate of spreading must be closely watched, and adjustments made as needed.

At least three types of power spreader are available. The most sophisticated is a self-powered machine equipped with a hopper to receive aggregate from a dump truck, and means to push the truck while it is dumping. The aggregate is power-fed to gravity openings or spinners.

Another type of spreader is a hopper and distributing mechanism on wheels, that is temporarily attached to the aggregate truck as it backs and dumps.

A power spreader may be firmly and semi-permanently attached to the truck itself. Feed may be by an auger or chain conveyor in the bottom slot of a special V-shaped body, or by dumping through a restricted tailgate. Distribution to the pavement is usually by a spinning, ridged, horizontal wheel.

A gravity spreader is usually a dump truck tailgate provided with precise control over the bottom opening. It may be flared

Courtesy of Ridge Construction Co.

Fig. 8-62B. Using a tailgate spreader

8-81

Courtesy of Ridge Construction Co.

Fig. 8-63. Using a hand spray

at the bottom to a 9-foot width, or wider. There is a full width slot and closure along the bottom, with positions from closed to fully open controlled by a manual lever. The operator may have a seat on the back of the gate, from which he can watch the flow of material, and move the lever as required.

Aggregate moves through the slot by gravity. With a full load, the truck body is started out level, then gradually raised toward dump position, just enough to keep a good supply over the slot.

Full dumping may be prevented by trees or wires, or it may be ineffective going down hill. Remainders can usually be slid back by jerky acceleration of the truck.

If foreign material (leaves, branches, lumps) restricts flow, the operator can open the gate much wider for a moment, pass the obstruction, then return it to the original setting. The extra thick ridge or mound of aggregate is distributed along the pavement during touching up by hand, chiefly with lutes.

As the truck is backed during spreading, it rides on the stone it spreads. The driver steers by watching his mirrors. He signals the operator with his horn, the operator signals the driver by extending his hand past the side of the truck.

Some spreaders have split gates, so that it is possible to open one side in covering a narrow strip. In single-gate models, the spread can be narrowed by blocking part of the slot with a metal plate or piece of wood.

In lack of spreading equipment, aggregate may be distributed by hand shoveling out of the truck. This is much more work, and with average skill provides poorer coverage with greatly increased quantity of material.

It is important that aggregate be spread promptly ("immediately" is the standard instruction), so that it will sink into the asphalt while it is still liquid, with maximum stickiness. Delay results in increasing waste of aggregate that fails to stick, and therefore becames a traffic nuisance and cleanup expense, instead of a part of the pavement.

Touching Up. Neither spraying of asphalt nor spreading of aggregate are precision operations, except in some big jobs. The machines themselves may be difficult to steer accurately, and they are subject to variations in performance, and the surfaces to be covered may vary in width and shape.

On most jobs it is therefore necessary to do some hand work.

If asphalt is too heavy because of overlapping at an end of a strip, or on a parallel one, the excess material should be distributed with a lute, to avoid a local concentration. If there is much too much of it to blend in, it may be necessary to pull some of it off the side, for burial or some other disposal.

Small bare spots or strips can be covered by pushing or pulling asphalt off adjoining areas. Medium or large spots are filled by pouring from a hand can, which can be filled from a connection for that purpose which is built into the distributor.

When a paving area is irregular, a large part or all of it may be asphalted with a hand spray, which is supplied through flexible hose from the distributor. A strip may be hand sprayed on a straight road, to reduce problems of spray bar width, or to avoid an extra trip by the distributor.

Aggregate is often too thin in some spots, and too thick or in piles in others. Unless the thin areas are so large that the spreader should make a second trip, it is sufficient to spread some more stone (or sand) lightly with hand shovels.

Thick spots and piles may be shoveled or raked only if so thick that the work will not pull up any embedded pieces on the bottom, as this is likely to loosen them permanently. They usually should be left until the asphalt has set.

Any moving of aggregate that must be done before the surface has set, should be done with a lute or the back of a rake, not with a broom. Hard edges move only the loose surface, while broom fibers tend to reach down and loosen embedded pieces.

There may be a narrow strip of bare asphalt on the edge of the pavement, or on the shoulder or in the gutter. The pavement strip must be covered with aggregate. Spill beyond its edge must usually be covered also, because exposed asphalt may stay sticky a long time, and be a hazard to shoes and clothing. It is often disastrously attractive to children.

Non-pavement asphalt may be blotted and covered by pulling in sand or dirt from the shoulder or gutter.

Finishing. After aggregate is spread and tidied, it must be firmly embedded in the asphalt by rolling. The rollers may be steel wheel, rubber tired, or both.

After a few passes, the area should be broomed lightly. Embedded pieces should not move, loose ones will shift around to cover bare spots. The area is then re-rolled.

It is sometimes necessary to add aggregate in spots (by hand) or over considerable areas (by truck spreader) if there is too little blotting material or too much liquid. The final step is always rolling.

On the following day, the surface may be dragged with a plastic fiber broom. This does not loosen well-embedded pieces, but moves loose ones along, helping to cover any skimped spots or bleed areas.

PENETRATION MACADAM

A macadam pavement consists of one or more layers of crushed rock of fairly uniform size, compacted with a binder. At present, the binder is almost always asphalt.

Base courses may be 3 to 4 inches thick after compaction; surface courses 2½ to 3 inches. Stone size should be about half an inch less than layer thickness.

The aggregate is spread on a graded and compacted base by a paver, truck spreader, or any convenient means. It is rolled by a heavy steel wheel, vibratory, or rubber tired roller. Compaction is from the edges toward the center, with wide overlaps, usually one-half of roller width. It is desirable to have the shoulders already in place, with precisely cut inner edges, to hold the stone against spreading.

Rolling is continued until a smooth, compact surface has been obtained, with thorough locking of the aggregate pieces to each other.

Hot asphalt cement, AC-5 or AC-20, a RC-800 or RC-3000 cutback, or RS-1, RS-2, CRS-1, or CRS-2 emulsion is applied by a distributor at the rate of about 1½ gallons per square yard, for a 3-inch base course. The rule of thumb is to use ½ gallon of asphalt per square yard for each inch of stone depth.

Immediately after the asphalt is applied, it is covered by a fine aggregate, usually ⅜ crushed rock, to fill surface voids and keep the roller wheels from contact with the asphalt. Rolling is resumed, and continued until the surface is firmly bound, and there is no movement under the roller.

The asphalt is then allowed time to set, before applying the next course. Traffic can usually be permitted after a few hours, but one to several days should be allowed before applying another course, as some curing is desirable.

The next step is to broom off all loose stone, to provide solid contact of the next course. This may be another layer of macadam, or a seal coat.

A second course of macadam is made in the same manner as the first — spread, roll, spray, cover, and roll again.

The surface course is covered with a seal coat, usually a heavy application of asphalt, 0.3 or 0.4 gallons to the square yard, plus immediate blotting with aggregate, and rolling as described earlier. The road may be rolled again after a 48-hour curing period.

Penetration macadam is considered to be a heavy duty pavement, largely because of the coarse aggregate used.

More information on asphalt paving will be found in Chapter 21.

PORTLAND CEMENT

Portland cement is the essential ingredient in rigid pavements, and in most structural concrete. It is a fine dry powder which, when combined with water, forms a paste that sets into a stone-like substance. This is moderately strong in itself, and has the quality of binding sand and stones into rigid masses with great strength.

Cement and water are mixed with clean sand to produce mortar, which is used chiefly as a binder and filler between bricks, stones, and blocks; and with sand and graded sizes of crushed rock or similar material to make concrete.

Portland cement is by far the most important of a class of materials called hydraulic cements, because they not only produce a rock-like solid when combined with water, but will also set under water.

Figure 8-64 shows the composition range of portland cement. It is ordinarily made from a mixture of limestone and clay or shale, with additions and modifications to obtain the desired proportions. Blast furnace slag, or pozzolan materials such as volcanic ash, fly ash, or diatomaceous earth, may provide part of the raw material.

The ingredients are crushed and ground to powder, mixed, then burned or calcined

	per cent
Lime, CaO	60 - 67
Silica, SiO_2	18 - 25
Alumina, Al_2O_3	3 - 8
Iron oxide, Fe_2O_3	0.5 - 6
Magnesia, MgO	0.0 - 5
Sulfur trioxide, SO_3	1.0 - 3

Fig. 8-64. Composition of portland cement

Type of portland cement		Compound composition, per cent*				Fineness, sq.cm. per g.**
ASTM	CSA	C_3S	C_2S	C_3A	C_4AF	
I	Normal	50	24	11	8	1800
II	Moderate	42	33	5	13	1800
III	High-Early-Strength	60	13	9	8	2600
IV	Low-Heat	26	50	5	12	1900
V	Sulfate-Resisting	40	40	4	9	1900

*The compound compositions shown are typical. Deviations from these values do not indicate unsatisfactory performance. For specification limits see ASTM C150 or CSA A5.

**Fineness as determined by Wagner turbidimeter test.

Courtesy of Portland Cement Association

Fig. 8-65. Portland cement compositions

in a rotary kiln at temperatures up to 3000° F. The resulting clinkers are ground with a small amount of added gypsum to a very fine powder, averaging about 10 microns (a micron is 1/1000 of a millimeter) in diameter. Approximately 96 per cent passes through a 200 mesh sieve, and 90 per cent through a 325 mesh.

Standard Specifications. The American Society for Testing and Materials (ASTM) has defined five types of portland cement, designated by Roman numerals. They are:

Type I. "Ordinary" cement for general use.

Type II: Moderate resistance to action of sulfates, and somewhat reduced liberation of heat while setting.

Type III: High-early-strength, known in the field as high early or quick setting and in Britain as rapid hardening, is more finely ground than other types, and develops good strength within 24 hours.

Type IV: Low-heat cement, used in massive structures where cooling during setting is a problem.

Type V: High sulfate resistance.

In pavements, almost all work is done with Type I, except that III may be used at intersections, and under other conditions where early opening to traffic is essential.

Special Portland Cements. There are a number of cements that are prepared for special purposes, that are not important enough for a type designation.

White portland cement is made in the regular way, but of selected materials low in iron and manganese oxide, so that its color is white instead of gray. It is used to afford a color contrast in itself, or as a carrier for clear colors which may be mixed into it, or applied as stain or paint. It is often used in traffic dividers and curbs, and in acceleration-deceleration lanes at highway interchanges.

Air entraining chemicals may be added in very small amounts, to enable concrete to hold microscopic air bubbles. These help concrete to resist weathering, and the action of salts. If they are in the cement, the letter "A" is added to its type number. They may also be added separately, when the concrete is mixed.

Waterproofed portland cement is made by adding stearate of aluminum or calcium to the clinker before grinding. Again, the same result can be obtained by adding chemicals during mixing, or even by reducing the quantity of water used. Waterproofing is usually not completely effective.

Masonry cements are mixtures of portland cement, air-entraining chemicals, and supplemental materials that improve workability, plasticity, and water retention in cement-sand mortar for masonry.

Up to 12 per cent of plasticizing agents may be added to the cement during manufacture, to produce plastic cement. This variety has superior adhesion and water resistance. It is used in making mortar, plaster, and stucco, and is particularly valuable in patching work.

Natural Cement. Natural cement is made in the same general manner as portland cement, but the raw material is usually a single formation of "cement rock" that contains the necessary ingredients in about the right proportions. Other materials may or may not be added during processing.

Cement rock may be used as the principal ingredient in portland cement, with other substances added to bring it to a standard formula.

CONCRETE INGREDIENTS

Concrete is usually made of cement, water, fine aggregate, and coarse aggregate. In pavements, and where either exposure to weather or ease of placement is required, air is included as an ingredient.

In addition, small amounts of other materials, called additives, may be used to regulate setting time, freeze resistance, color, and other characteristics.

Water. Water combines chemically with cement to produce a hard solid, a process called hydration. The reaction proceeds rapidly at first, then more slowly. Its speed varies with the cement used, temperature, chemicals in the mix, and other factors. For practical purposes, hydration is assumed to be complete in 28 days, although the concrete may be used for some time before that. There may be a small increase in strength afterward.

Concrete can be made with almost any water, including seawater, acid mine drainage, and dilute sewage. However, it is best to use drinking quality water when possible. In the first place, specifications may require it.

There are also particular problems. Special precautions have to be taken to protect steel reinforcement from corrosion by mixed-in seawater. Salts of zinc, copper, or lead may retard setting and reduce strength. Varied effects result from other impurities, and an expensive analysis might be required for proper use.

Various sulfate compounds disolved in ground water may attack concrete after it has set, by chemical replacement and crystalliation. Damage is most likely near effluent from industrial plants, in installations where alternate wetting and drying move the solutions through the concrete by capil-

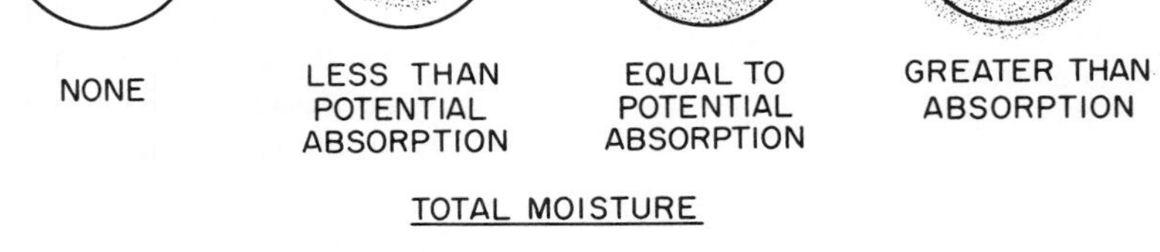

STATE

Courtesy of Portland Cement Association

Fig. 8-66. Moisture conditions of aggregates

lary action. Special sulfate-resistant cements are used where exposure is severe.

Sand (Fine Aggregate). For concrete, sand consists of natural or crushed particles of rock ranging in size from ¼ inch (longest dimension) downward to somewhere around 1/50 of an inch. There are a number of different specifications in use. Most of them call for 10 to 30 per cent of the sand to pass a No. 50 sieve, with 3 per cent or less passing a No. 100. One limits the No. 50 to 5 per cent, and the No. 100 to zero.

Sand particles are preferably irregular, with points and edges. Rounded sand from desert dunes or sometimes from ocean beaches may not make a good bond with the cement paste.

Sand usually has a high proportion (up to 100 per cent) of hard silica particles, as this mineral is most resistant to the abrasion that produces sand. Occasionally, there is an excess of particles of mica, which is soft and laminated, so that it weakens concrete. Contamination from dirt must be guarded against.

Stone (Coarse Aggregate). Coarse aggregate is usually crushed rock, but it may be natural river or beach gravel, slag from steel mills, or even oyster shells.

Basalt (trap rock) and limestone are

preferred rock types for concrete aggregate. Shale should never be used, as it is soft, absorbs water, and crumbles.

Specifications for coarse aggregate are more variable than for the fine. Maximum size is fixed, but latitude may be allowed in gradation of sizes below it. The smallest particles are those that are held on a No. 4 (quarter-inch) sieve. Pieces that go through are sand.

Large maximum size pieces are economical of cement, but make placing of concrete difficult, and cause presence of voids (honeycombing) in thin pieces and against forms or reinforcement. In general, the largest pieces should be less than one-third of the thickness of the slab, and less than three-fourths of the spacing of reinforcing bars from each other and from surfaces of the concrete.

Cleanliness. It is important that aggregates be substantially free of organic substances, silt, clay, coal and low-strength particles or pieces. The effect of these may be anything from negligible to disastrous, depending on the nature of the material and the form in which it is present.

For example, a small quantity of fines passing a No. 200 sieve, evenly distributed in the fine aggregate, might not affect the mix except in a slight increase in water requirement. But if it were in the form of a film on coarse aggregate surfaces, it would weaken the vital bond between cement paste and aggregate. If it occurred as clay balls hard enough to stay intact into the finished concrete, each would be a point of no-strength and probable disruption in the slab.

Moisture. Aggregate may carry absorbed moisture, surface moisture, or both. This water will affect the water proportion in the mix, either by absorbing or yielding water during mixing, transport and setting.

The four moisture-states of aggregate are shown in Figure 8-66. The allowance for this moisture in mix design varies with the

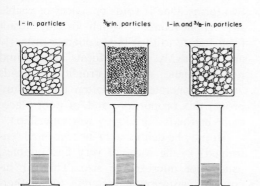

Fig. 8-67. Mixed aggregate sizes reduce voids

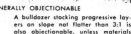

PREFERABLE
A crane or other equipment should stockpile material in separate batches, each no larger than a truckload, so that it remains where placed and does not run down slopes.

OBJECTIONABLE
Do not use methods that permit the aggregate to roll down the slope as it is added to the pile, or permit hauling equipment to operate over the same level repeatedly.

LIMITED ACCEPTABILITY—GENERALLY OBJECTIONABLE
Generally, a pile should not be built radially in horizontal layers by a bulldozer working with materials as dropped from a conveyor belt. A rock ladder may be needed in this setup.

A bulldozer stocking progressive layers on slope not flatter than 3:1 is also objectionable, unless materials strongly resist breakage.

FINISHED COARSE AGGREGATE STORAGE

FINE AGGREGATE STORAGE

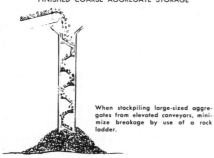

When stockpiling large-sized aggregates from elevated conveyors, minimize breakage by use of a rock ladder.

CORRECT
Chimney should surround material falling from end of conveyor, to prevent wind from separating fine and coarse materials. Openings should be provided as required to discharge materials at various elevations on the pile.

INCORRECT
Do not allow free fall of material from high end of conveyor, which would permit wind to separate fine from coarse material.

Courtesy of Portland Cement Association

Fig. 8-68. Stockpiling of coarse aggregate

aggregate, as some lots are much more absorbent than others.

Gradation. Any mass of broken pieces has spaces between the pieces or particles, called voids. If the pieces are of uniform size, the relative volume of voids will be about the same, whether the pieces are large or small. But if two or more sizes are mixed together, the result is a graded aggregate in which the voids will be substantially reduced by small pieces filling spaces between larger ones, as in Figure 8-67.

In concrete, voids are filled by cement paste. This is the most expensive ingredient, so reduction in voids decreases cost. In addition, the graded aggregates tend to produce a stronger concrete.

When gradation of aggregate is wrong for the specifications, it can be corrected (at a price) by crushing to reduce oversize fractions, and/or screening into separate sizes, then recombining or wasting portions that do not fit the specifications.

Segregation (Separation). Coarse aggregate, blended to produce a desired gra-

dation in size, tends to separate or segregate during handling, so that parts of a pile or bin will be coarser or finer than others.

Unless this tendency is controlled or corrected, successive batches of concrete might have very different composition.

This problem is discussed in Chapter 10. Figure 8-68 gives some specific suggestions for concrete aggregates.

Air. Air may be present in concrete in two forms. One is voids or spaces of visible size (honeycomb), which usually result from failure to distribute concrete properly and thoroughly, against forms and around reinforcement and obstructions. Voids weaken concrete. They often indicate improper mix proportions and/or poor placement methods (or sloppy workmanship).

Air may be deliberately incorporated (entrained) in the form of very fine bubbles, usually of microscopic size. This air increases resistance of concrete to weathering, freezing and thawing disruption, and attack by de-icing chemicals. It causes some reduction in strength.

Entrained air makes concrete more workable. This makes it possible to use stiffer (and therefore stronger) mixes, regaining some or all of the strength lost to the bubbles.

Protection from entrained air is partly its ability to yield internally to expansion, relieving strains which would otherwise tend to rupture the concrete.

Rotary mixers incorporate air into concrete with every turn, but most of it escapes unless chemicals (entraining agents) capable of subdividing and retaining it are incorporated in the cement at the mill, or added to the batch in the mixer.

The desired quantities of air in concrete, using various sizes of aggregate, vary between 3 and 9 per cent. The percentage of air is controlled by adjusting the quantity of entraining agent.

Other Additives (Admixtures). A considerable variety of chemicals may be added to concrete mixes to change some characteristic. Accelerators for faster setting, and retarders for slowing it, are discussed below under **Workability Time.**

Water reducers increase the slump of concrete for a given water content, making it possible to reduce water while retaining a desired slump. Some of these admixtures are compounded so as to also act as retarders, and/or air entraining agents.

CONSISTENCY

The consistency of wet mix for concrete is important both in regard to its workability, and its relationship to the strength of the cured concrete.

Slump. The softness or liquidity of a mix is measured by the slump test. The testing cone, Figure 8-70, is a sheet metal fabrication in the form of a truncated cone (frustrum of a cone) with an open base 8 inches in diameter, an open top 4 inches in diameter, and a height of 12 inches.

This cone is placed on a hard, level surface. The freshly mixed concrete to be tested

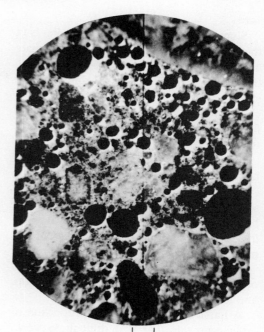

|—| 0.01 in.

Courtesy of Portland Cement Association

Fig. 8-69. Air-entrained concrete

is packed to fill it completely, then the cone is lifted clear. The concrete will slump, bulging out at the sides and sinking at the top. Measure its height in inches, subtract from 12, and you have the slump rating of the batch being tested.

A very stiff mix might have a slump of one inch or less, a fluid or wet one 4 inches, and an exceptionally fluid one 6 inches.

Courtesy of Central Scientific Co.

Fig. 8-70. Slump cone

Slump is affected by all the ingredients in the concrete. If large pieces of aggregate are removed to get complete packing into the cone, a higher-than-actual slump will be shown. The test is used primarily as a check on the water-cement ratio, but is useful for this only if the other factors are kept the same.

A stiff, low-slump mix is desirable for high strength, minimum shrinkage during curing, and good water resistance. It is essential for slip-forming. A soft, fluid, or high-slump concrete is used for efficiency in placement, filling of small places, and preventing voids.

Rich, Lean, and Harsh. A rich mix is one with a high cement-to-aggregate ratio, and lean one the reverse.

A harsh mix has so high a proportion of large aggregate that it is difficult to work.

PROPORTIONS

Characteristics of concrete are largely dependent on the proportions of its ingredients. These may be expressed in ratios, percentages, or units. For large quantities, measurements (except air) are by weight; in small jobs they may be by volume.

Water-Cement. The water-cement ratio is measured by weight, and expressed as a decimal. A ratio of 0.50 means a half pound of water to each pound of cement. If expressed in volume, this would be about 1.53 cubic feet of water to one cubic foot of cement. In units, it would be 5.6 gallons of water to a 94-pound bag of cement.

A ratio of 0.25 is usually enough to complete the chemical hardening (hydration) of cement, and provide maximum paste strength after setting. However, such a dry mix is not workable. The stiffest practical ratio, 0.40, is used when the concrete is exposed in use to sea water and sulfates. Pavements in mild climates should have 0.53, but if exposed to freezing and thawing, 0.49.

Massive concrete may have a ratio of 0.58, or higher under special conditions.

Entrained air reduces the water ratio needed for workability. This relationship, expressed now in pounds of water per cubic yard of concrete, is shown in Figure 8-71.

Cement-Aggregate. Because of variations in aggregate, this proportion, and that between fine and coarse aggregates, are usually set in a laboratory or in the field, using samples of the actual sand and stone. The batch is often 20 pounds of mixed concrete; figures are worked out on basis of both weight and volume.

Trials are based on tables giving suggested quantities for various water-cement ratios and maximum size of aggregate.

The trial batch is mixed dry, then water from a measured container mixed in, until the desired slump is obtained. If it is reached before using all the water, the balance is withheld and a note made of the quantity. If more water is needed for the slump, it may or may not be provided.

The rule of thumb is to use as much coarse aggregate as is practical without making the mix harsh and difficult to work.

In general, aggregate makes up from 60 to 80 per cent of the mix, by weight or by volume.

Aggregate, Fine-Coarse. Sand may make up from 30 to 60 per cent of the total aggregate. Maximum proportion of sand is used when the coarse aggregate is small, the difference between ⅜ and 1½ maximum size being more than 2 to 1 in stiff mixtures, and 1.6 to 1 in sloppy ones.

With other conditions equal, up to 8 per cent more sand may be used if its particles are coarse than if they are fine.

A mixture that has too little sand in proportion to the stone tends to be harsh and soupy, and to segregate into mortar and stone if not carefully handled. Too much sand is uneconomical, because of high cement and water requirements, and the concrete will be subject to shrinkage.

There is sometimes a substantial differ-

APPROXIMATE MIXING WATER REQUIREMENTS FOR DIFFERENT SLUMPS
AND MAXIMUM SIZES OF AGGREGATES*

Maximum size of aggregate, in.	Air-entrained concrete				Non-air-entrained concrete			
	Recommended average total air content, per cent **	Slump, in.			Approximate amount of entrapped air, per cent '	Slump, in.		
		1 to 2	3 to 4	5 to 6		1 to 2	3 to 4	5 to 6
		Water, lb. per cu.yd. of concrete†				Water, lb. per cu.yd. of concrete†		
3/8	7.5	310	340	360	3.0	350	385	410
1/2	7.5	300	325	340	2.5	335	365	385
3/4	6.0	275	300	315	2.0	310	340	360
1	6.0	260	285	300	1.5	300	325	340
1½	5.0	240	265	285	1.0	275	300	315
2	5.0	225	250	265	0.5	260	285	300
3	4.0	210	235	—	0.3	240	265	—
6	3.0	185	200	—	0.2	210	235	—

*Adapted from Recommended Practice for Selecting Proportions for Concrete (ACI 613-54).
**Plus or minus 1 per cent.
†These quantities of mixing water are for use in computing cement factors for trial batches. They are maxima for reasonably well-shaped angular coarse aggregates graded within limits of accepted specifications.

Courtesy of Portland Cement Association

Fig. 8-71. Water requirement

ence in cost between aggregates of different sizes, which may be allowed for in designing the mix.

Strength. Strength of concrete is usually measured as compressive strength—pressure in pounds per square inch (psi) required to crush it. For various mixes and preparation methods, this ranges from 1000 to 12,000, with most pavements apparently requiring 4000 to 7000 pounds.

Pavement concrete is more often rated by bending strength, or modulus of rupture, which should be between 600 and 750 psi.

These tests are made after 28 days of curing, unless another time period is mentioned in the rating.

CONCRETE PROCESSING

Mixers. Most concrete is mixed in batches in rotary drums, which are fitted with longitudinal, internal blades. There are also fixed, open-top drums with blades or paddles revolving inside, pugmill fashion. Measured dry ingredients are dumped in, while water is added more gradually.

Mixers may be in stationary plants with a wide size range, mounted on trucks (transit mixers) and activated just before delivery, or in paving machines. Any type may have a loading skip or hopper, and there may be a swinging and/or extensible discharge chute.

There are batch plants that simply measure the dry ingredients and discharge them into trucks or transit mixers, and others (shrink mixers) that mix them partially. This mixing reduces bulk, as smaller pieces or particles work into the spaces among the larger ones.

There are various standards for mixing time, which necessarily allow for different mixers and ingredients. In general, 40 seconds is an absolute minimum, 50 or 60 seconds is usual, and 75 may be required in the absence of tests. A transit mix may be 100 revolutions at a speed of 8 to 12 revolutions per minute.

Plant mixer capacity may vary from 2 up to 12 yards or more. Transit mixers may be 5 to 16 yards, with 7 and 9 quite usual. It is important not to overload, as this reduces fall from the blades and prevents proper mixing action.

A tilting mixer is tipped like a dump truck, so that concrete flows out of the discharge port at the back. Its discharge is aided by continued revolving of the blades.

Non-tilting mixers discharge by reversing

Courtesy of CMI Corp.

Fig. 8-72A. Dumping in paving lane

rotation, or through gates in fixed drums.

Transport. Concrete that is prepared in a central mix plant must be trucked to the job. Trucks are of two basic types; agitating and non-agitating. Either type may be side dump or rear dump.

An agitator truck may be a standard transit mix unit, with the drum turning slowly, perhaps 2 to 6 revolutions per minute. Its load as an agitator may be ⅓ larger than as a mixer. Or it may be a special unit, with a cylindrical body open at the top, and a set of blades or paddles turning inside it.

Agitation assures that the concrete will not separate, but may damage it by over-mixing.

Non-agitating trucks are gaining in popularity. The preferred body is a dumper in which floor and walls curve smoothly into each other, to avoid angles and corners that invite separation and are difficult to clean. The tailgate in rear dumps is replaced by a narrower gate with tapered sides.

Ordinary flat-body, wide-gate dump bodies are often used because of convenience. They are satisfactory for short to medium hauls on smooth roads, if they are inspected

and cleaned as necessary after each dump.

Successful use of non-agitating transport depends on smooth roads that keep jiggling to a minimum, and a concrete mixture that does not separate readily.

Rear-dump or side-dump depends on paving equipment and job arrangements. In general, side dumpers are more convenient, as they can haul alongside the paving strip, and dump into it or into a hopper on it. Rear dumping is satisfactory if the trucks can use the paving lane and back up to the paver, or if the paver has a receiving hopper that can be carried alongside. Sometimes there is space for rear dumpers to turn and back up to the paving lane from the side.

Workability Time. Hydration may start in a small way when cement is mixed with aggregate that is even slightly moist. However, for most purposes, the adding of water to the mix is taken as the starting point.

The rate of setting of a standard concrete mixture is fastest when it is warm and slowest when it is cold. Small effects depend on cement-water ratio, aggregate type, method of mixing employed, and other factors.

Courtesy of Portland Cement Association

Fig. 8-72B. Dumping in side-loading hopper

As a rule of thumb, time from mixing in the water to stiffening sufficient to be noticeable and possibly awkward, may be as little as 45 minutes in piles or in a non-agitating truck or one hour in an agitating truck. A median or normal time is 1½ hours. Maximum time with all conditions favoring slow setting may be as long as 3 or 4 hours.

Time may be altered by additives. Calcium chloride, dissolved in its weight of water, is used in weights up to 2 per cent of the mix, and may make it set up to 15 per cent faster. Its effect at close to freezing temperatures may be greater, and it offers a little protection against freezing also. Any mixture including this salt must be handled more rapidly.

A wide range of chemicals can be used as retarders. They include gypsum, some of which is in standard cement anyhow, sucrose (cane/beet sugar), starch, cellulose,

APPROXIMATE RELATIVE STRENGTH OF CONCRETE
AS AFFECTED BY TYPE OF CEMENT

Type of portland cement		Compressive strength, per cent of strength of Type I or Normal portland cement concrete			
ASTM	CSA	1 day	7 days	28 days	3 mos.
I	Normal	100	100	100	100
II	Moderate	75	85	90	100
III	High-Early-Strength	190	120	110	100
IV	Low-Heat	55	55	75	100
V	Sulfate-Resisting	65	75	85	100

Courtesy of Portland Cement Association

Fig. 8-72C. Comparison of setting times

and many ordinarily unfamiliar substances. One is a sulfonated lignin from wood pulp, which may also act as an air entraining agent.

It is said that a 0.05 per cent admixture of sucrose (which is a tiny amount) will delay initial setting by 4 hours, improve workability, and increase ultimate strength. A 0.20 per cent admixture may delay initial hardening by several days, so that this additive must be used with restraint. Effect on both long and short term compressive strength is good, with added strength up to 8 per cent.

Most concrete is mixed and used without either accelerators or retarders. However, it is good to keep in mind that they can be added when needed.

If delays in transportation or placement cause concrete to partially set so that it is too stiff to be workable, it may either be discarded, or softened by mixing (usually by hand) with more water. The added water weakens ultimate compressive strength to at least the same value as if the water had been added in the original mixing.

In concrete pavements, specification strength is critical and the remixing may not be allowed. In many other applications, however, such as foundations for light structures where wall thickness is in excess of required strength, it may be a legitimate salvage operation.

Transfer. Mixed concrete must be transferred between the mixer and the transports and/or final location. Care must be taken at any transfer to avoid separation of aggregate.

The preferred mechanism is a round-bottomed, smooth-walled chute. This should not have a slope flatter than 1 on 3, and except for very stiff mixes, should not be steeper than 1 on 2.

Concrete should not be allowed to flow, particularly if it is a thin (high-slump) mix, as it then tends to separate into a thin liquid going ahead, followed by cement-sand mortar, with coarse aggregate tending to lag behind. A steep slope, with motion sliding rather than flowing, reduces or eliminates this effect.

The chute should end in a vertical pipe, preferably a movable and flexible one (elephant trunk), the outlet of which can be kept below the level of the concrete being discharged.

Transfer will be discussed again under **Placement.**

PAVEMENT DESIGN

Portland cement concrete pavement is basically a rigid slab extending the width and length of the roadway, but usually interrupted by joints. Such joints allow limited horizontal movement to take care of contraction, and sometimes expansion, but should be reinforced or interlocked against vertical movement.

Courtesy of Portland Cement Association

Fig. 8-73. Chute end should be lower

Concrete forms a strong bridge across small spots in its base, which may settle or become soft. However, for reasons of economy, the slab is not made strong enough to sustain loads over wide settlements, nor on a generally weak base.

A well-drained, well-compacted, and accurately finished subgrade with good bearing strength is necessary for proper pavement performance, and is assumed in the design of the slab. In addition, on poor ground, or for very heavy traffic loads, a base (often called sub-base) of selected granular material should be placed on the subgrade. It may be stabilized with cement or asphalt.

Slab Dimensions. Concrete pavement is usually divided by longitudinal joints into single-lane widths. These may be 10 to 13 feet wide, with 12 feet fairly standard for roads carrying heavy traffic.

Transverse or cross joints to allow for contraction may be 12 to 21 feet apart in unreinforced pavement, 30 to 100 feet in reinforced standard construction, and lacking in continuously reinforced concrete (CRC).

Slab thickness ranges from 6 to 10 inches for highways, and upward from 12 inches for airport runways. Differences are related to weight and intensity of traffic, strength of the base, and experiences of the engineer.

Reinforcement. Steel reinforcement, officially called distributed steel, is commonly placed in rigid highway pavement. It is sometimes omitted because of arguments about its usefulness.

City streets often use unreinforced concrete, because it is easier to cut for utility openings.

Reinforcement may be welded wire mesh, or mats fabricated from reinforcing bars. The mesh has stronger steel, 70,000 pounds per square inch (psi) yield point, as compared with 60,000 for bar. Mesh may be more efficient because smaller diameter permits closer spacing and more even distribution, but it is more expensive.

Most bars and some mesh wires are deformed. This means that they are crimped to provide an irregular or embossed surface that resists any tendency to slide within the concrete.

It is not economic to use steel heavy enough to add much to the flexure or bending strength of the pavement slab. Its primary uses are to increase tensile strength so that most shrinkage cracking will be limited to cut joints, and to hold the slab together if it does crack.

The holding effect is important. It keeps irregularities tied together like a lock joint, preserves the bridging action of the slab, keeps out foreign matter, prevents development of a bump at the break, and reduces raveling and enlargement of the crack.

To keep a crack closed, steel must be strong enough to overcome friction between the slab and subgrade from the crack to the nearest contraction joint.

An unreinforced slab would make equal openings at the crack and the joint. Reinforcing drags the slab across the subgrade as it shrinks, keeping it in tight contact.

Steel is customarily placed about ⅓ of the thickness of the slab from the top. In a 9-inch slab, it would be 3 inches below the surface. This is a small measurement compared to the length and width of the sheets. It is therefore very important to handle reinforcing carefully so that it will not be bent, as this might cause parts to reach the surface. A 2-inch layer of concrete over the steel is usually the minimum allowed.

Cross reinforcing is usually much lighter than the longitudinal, and may even be omitted altogether, as the lane-width spacing of lengthwise joints make contraction cracks parallel to them unlikely. The cross wires or bars serve mostly for spacing. But if they are extended across a longitudinal joint, more weight is needed.

The longitudinal steel usually has a cross section about 0.06 per cent of the cross section of the slab in which it is placed.

(a) Expansion Joint   (b) Contraction Joints

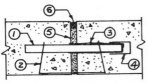

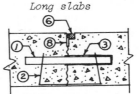

Long slabs   Short slabs

1.	Dowel	4.	Expansion cap	7.	Stop strip
2.	Basket	5.	Filler board	8.	Groove
3.	Coating	6.	Seal	9.	Interlocking aggregate

(c) Between-Lane Joints

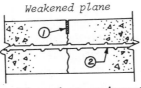

Weakened plane   Keyed construction

1.	Groove with sealant or insert	3.	Threaded insert tie (or bent re-bar)
2.	Deformed steel tie	4.	Molded key and keyway

Courtesy of Transportation Research Board

Fig. 8-74. Joint types

Joints. Four types of joints are used in concrete pavements. Three of them, contraction, longitudinal, and expansion, are planned. The construction joint just happens at the end of a work period.

The construction joint is usually rigid, unless deliberately made as one of the others. The three not only serve the purposes indicated by their names, but also relieve stresses caused by slab warping.

The top of the slab expands and contracts with temperature and moisture changes more often and to a greater extent than the bottom. Either drying (hot sun, low humidity) or chilling will make it contract. The bottom is usually damp all the time, except in deserts, and it takes much longer for cooling to reach it.

When the top contracts more than the bottom, the slab tends to curl up at the sides and ends. Actual movement may be infinitesimal, but the strain may be enough to cause cracks unless relieved. Slight movement at the joint is sufficient to prevent damage.

Contraction Joint. Concrete normally shrinks during curing, and whenever it is chilled. Since it has little tensile strength, this contraction causes cracks which, if not controlled, will be irregular and unsightly. It is therefore almost standard practice (the exception is CRCP, continuously reinforced concrete pavement) to create straight lines of weakness in the slab, or regular or patterned intervals, at which it will crack when shrinkage occurs.

This is usually done by sawing with a carbide or diamond toothed wheel. Two cuts are made. The first, as narrow as saw construction permits, is made to about ¼ of the depth of the slab, and provides the breakage line. The second is wider (this dimension varies with design), and is about one inch deep. It provides a reservoir or socket for sealing material. The two cuts may be made at the same time by separate blades mounted on a single frame, or the second may be made later.

Timing of the first cut is critical. It may be 4 to 6 hours after finishing, depending

Courtesy of Rexnord Inc.

Fig. 8-75A. Keyed joint and reinforcing

on job conditions. The concrete must have set enough so that the saw will not tear out pieces of aggregate. Minor raveling may be allowable, or may even be accepted as evidence of proper timing. But concrete must not be allowed to shrink enough to crack elsewhere before sawing.

Weakening may also be provided by insertion of a plastic strip or a metal bar just below the surface, or by forming grooves in the wet concrete. These non-saw methods include a possibility of disturbing the finishing of the surface across the joint line, creating a permanent bump.

The two sides of the joint are held in vertical alignment by interlocking effect of pieces of aggregate in irregular breakage. Plain concrete depends entirely on this feature for stability. Slabs are made short, 12 to 21 feet, so that gaps will not widen enough to weaken this support. A gap of only 0.04 to 0.035 is damaging.

Joints in reinforced pavement are also supported by dowels. These are usually round bars, about one inch or ⅛ of pavement thickness in diameter and 18 inches long. However, they are available from ½-inch diameter and 10 inches long, up to 1½ by 20 inches. The whole dowel, or sometimes half of it, is made slippery with plastic coating, paint, or lubricant, so that the concrete can slide on it when it contracts. Spacing is about 12 inches on centers. The dowels must be exactly in line with slab movement—that is parallel with the slab surface and with the centerline—as any distortion will create strains in the concrete. They are usually halfway down in the slab.

Contraction joints sometimes fill with soil or sand while widely open, and then cannot re-close. Such material usually pumps up from the base, but may sift in from the top. It contributes to excessive expansion pressure and blow-ups. It is rare if there is a first class subgrade, and a good joint sealer is used.

Longitudinal Joint. Any concrete road slab more than 16 feet wide tends to develop an off-center longitudinal crack, un-

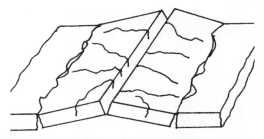

Courtesy of Transportation Research Board

Fig. 8-75B. Joint blowup (buckling)

less a joint is provided to relieve warping stresses. Such a joint may be made by weakening the centerline of the slab by a plastic insert or by sawing. It may then be called a warping joint. Or it may be a construction joint between slabs laid at different times.

An important secondary use for longitudinal joints is service as traffic lane markers that do not wear off. They also serve as absolutely accurate guides for paint-spray crews putting on more conspicuous marking.

The two sides are almost always locked against vertical movement. If both slabs are poured at the same time, the reinforcing is usually continuous across all lanes, so the joint is tied together by the transverse (cross) reinforcing. In roads where there is no transverse reinforcing, bars 30 inches long are usually laid across the joint.

If the lanes are laid separately, locking may be by means of a keyway, a continuous projection or groove on the side of the first slab, into which concrete of the second slab will flow. This may be made by additions to side forms, or a specially shaped slipform. Means of putting reinforcing across either keyed or unkeyed joints between separately built lanes will be discussed under **Construction Joints.**

Longitudinal joints ordinarily do not have sufficient opening or other motion to require sealers.

Expansion Joint. An expansion joint is usually an even-sided slot all the way through the slab, with a compressible filler. It is seldom found in modern highway construction except at the ends of bridges and other long, rigid structures.

Under most conditions, concrete slabs are at their maximum length when formed, as shrinking during hydration is greater than later expansion from heat. However, there are at least two causes for occasional over-

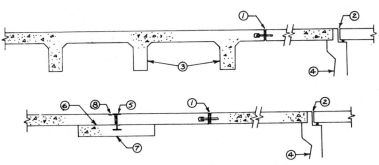

1. One or more conventional expansion joints
2. Bridge joint
3. Reinforced transverse beams
4. Bridge abutment
5. Wide flange beam or proprietary device
6. Friction reducer (plastic, asphalt, etc.)
7. Sleeper slab
8. Seal

Courtesy of Portland Cement Association

Fig. 8-75C. Bridge approach joints

expansion that results in severe pavement damage by upward buckling.

Contraction joints may fill with incompressible material when they are open during cold weather. This may be silt or sand pumped up from wet subgrade as the slab sections flex under heavy loads, or de-icing sand penetrating from the top. This filling of joints makes the pavement longer.

Concrete swells somewhat when it absorbs water. A wet Spring may expand the slabs against each other, or against sand in the joints, so that on an extremely hot day there is no space left. The resulting pressure usually is dissipated to the side or held by the strength of the pavement, but it occasionally causes upward buckling at a joint, which may be high enough to stop traffic.

Expansion joints have often been found ineffective at preventing this damage. They lose effectiveness when the filler (a one-inch board, or asphalt-impregnated fiber) loses resiliency, sand enters during cold weather, or the presence of the joint causes adjoining contraction joints to open too far, and fill with sand.

When used, an expansion joint is held in vertical alignment by dowels. Half of the dowel is lubricated and slides in a cap, which provides space in which it can move during expansion. A seal across the top must be carefully maintained.

At a bridge approach, several expansion joints may be close together, special sliding slab may be used, or the end of the roadway may be anchored by transverse beams made by continuing the slab concrete and reinforcement down into cross ditches. See Figure 8-75B. Or the problem may be avoided (except for hot-day bumps) by putting a strip of asphaltic concrete, (flexible pavement) 20 or more feet long, where highway and bridge deck meet.

Construction Joint. A construction joint occurs where old and new concrete work meet. The time lapse between the two pours may be anywhere from overnight to a period of years.

They are longitudinal when adjoining pavement lanes are made separately, or when a concrete shoulder is placed alongside a concrete pavement. They are often as long as the paving project, and forming them is an integral part of the paving operation.

The trickiest detail is extending reinforcement through the joint. This may be done by bending bars to fit inside the form while concrete is placed, then pulling them straight after the concrete is firm and before it is hard.

Long bolts may be set in the first concrete with threaded ends against the forms. The threads are chipped free and pieces with matching sockets screwed in.

A slipform paver may have a machine following behind the form to shoot rods or bolts into the side of freshly exposed concrete with an air gun. Since there is no interference with forms, these can be left projecting.

A transverse construction joint is usually made after the last load of concrete on the shift arrives. Side forms or heavy braced boards are set across the work strip, and concrete is poured and carefully finished right up to them.

Reinforcement is a major problem here also, as it is necessary to overlap the bars in the completed section with those that will be placed for the next pour. Boards may be drilled so that they can be slid along the bars into place, leaving the ends projecting behind them. Or bars may be bent back into the concrete, and found and straightened the next day, after removing the form.

The exposed concrete may or may not be painted with a cement and water grout before starting the new pour. The joining is hand finished for smoothness, but sometimes requires grinding afterward.

An important problem in transverse construction joints is the quality of the concrete. Finishing often tends to segregate certain lighter and thinner portions of the mix, which are pushed ahead and concentrated in the last few feet. The first few batches of concrete mixed or transported are the most likely to have errors in proportioning. Sticking of cement paste to clean dry mixer and truck body interiors may reduce cement proportion left in the mix.

An inspector may require wasting of the last few yards of concrete if they appear substandard, and adjusting the first few batches of the next day with extra cement and water, or with a reduction in coarse aggregate, to compensate for losses.

Another problem is that hand finishing of one side, and hand starting of the other, may provide too little consolidation.

Continuous Reinforcement. Continuously reinforced concrete pavement, CRCP, is a design in which the pavement is built without transverse contraction or expansion joints, except at the ends, where it meets other types of construction. Longitudinal reinforcement is continuous from end to end.

Contraction causes random cracking, which starts within a few days of placement, and may continue for 3 or 4 years. Close spacing of 3 to 10 feet is considered desirable, as it results in very narrow cracks, which ordinarily do not require sealing or other surface treatment. In general, increasing the strength of reinforcement results in closer spacing. Ordinarily, cracks are in the top and often the bottom of the slab, but do not penetrate to the steel.

The reinforcing may be deformed bars, bar mats, or wire fabric. Smooth wire is not used. Steel percentage ranges from 0.5 to 0.7 per cent of the slab. Placement is at mid-depth or slightly above. Overlaps between pieces are 16 inches or more, and are staggered or skewed to prevent more than one-third occurring in any cross section.

Emphasis is on longitudinal strength, and cross reinforcement is usually weaker and

may be omitted. Longitudinal joints between lanes may be tied together by reinforcement, or supported by dowels.

Pavement ends tend to slide on the subgrade, sometimes for a length up to 500 feet. This results in wide contraction openings, up to 2 inches, at joints with bridge slabs or other pavements. Movement may be reduced by cross trenches, at the end of the paving strip, filled with concrete as the slab is poured, as in Figure 8-75B. Sometimes several joints are installed.

Lack of joints permits use of a slab an inch or more thinner than standard design. This saving is at least partly offset by more stringent base requirements, preferably with stabilization by cement, lime, or asphalt.

PAVING WITH FORMS

Fixed forms for concrete pavement are being superseded by slip-form methods. However, forms are still used for a considerable amount of straight paving, and continue to be essential for complicated, interrupted, and variable-width pavements.

A description of their use is therefore justified in its own right, and will also form a background for description of slip-forming.

Base Preparation. The base is prepared, graded, stabilized, and finished according to methods described earlier in this chapter, or by other approved procedures. The quality of this work is important for setting forms as they must be absolutely accurate in grade to act as rails to carry the finishing machinery. (It is even more important in slipforming, where the base must carry the finishing equipment directly.)

Any irregularity or softness in the base must be compensated by a tedious extra amount of hand tamping, adjustment, and blocking of forms.

Finish grading of the paving strip must be extended to the sides to include the strip on which the forms will be set.

The base must be damp when the con

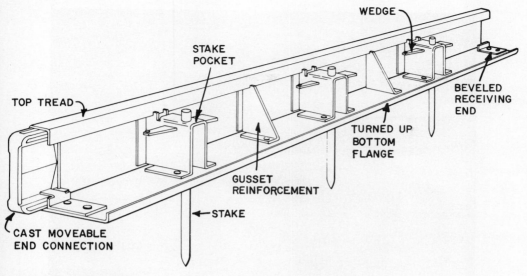

WEDGE

STAKE
POCKET

TOP TREAD

BEVELED
RECEIVING
END

TURNED UP
BOTTOM
FLANGE

GUSSET
REINFORCEMENT

STAKE

CAST MOVEABLE
END CONNECTION

Courtesy of Metal Forms Corp.

Fig. 8-76. Form for concrete pavement

crete is placed, so that it will not absorb water from the mix.

Forms. A typical paving form, Figure 8-76, is a complex steel channel section, with a face the height of proposed pavement thickness, a wide flat base for stability, an upper rail to carry equipment, reinforcing gussets, and provision for fastening to the ground with spikes.

Wedges are driven and locked between spikes and forms for fine adjustments of alignment.

Length of sections is 10 feet or more, steel thickness 3/16 to 5/16, and weight per foot 12 to 45 pounds.

The face of the channel holds the concrete, the rail-like top provides the strikeoff level for the surface, and usually is a track for finishing machinery. Forms that are twisted, bent, dented, or encrusted should not be used.

Forms are usually removed about 24 hours after the concrete is placed, if standard cement is used. This period allows time for sufficient hardening to reduce chance of breaking edges. Care must still be exercised in separating them from the con-

crete. If it is necessary to use wedges, they should be wood or plastic, never metal, and driving should be gentle.

Forms should be painted lightly with oil or any of several special compounds before use, to reduce or prevent sticking.

Forms must be inspected and thoroughly cleaned before each re-use. Cleaning is absolutely necessary on the face and the top. Other parts should be cleaned also, to keep down weight, allow inspection for defects, and avoid a messy appearance.

Drum Pavers. Until rather recently, most concrete for pavements was mixed on the spot, by machines similar to that shown in Figure 8-77. Now, more mixing is done in central plants.

The drum paver, or concrete paving machine, is usually crawler mounted, and operates on the subgrade. A wide, shallow hopper is hinged to the front. Dry cement and aggregate, proportioned at a batch plant, are dumped into this by a truck that has a number of compartments, each holding one batch.

The loaded hopper is raised to dump its load into a revolving drum, with capacity

Courtesy of Rexnord Inc.

Fig. 8-77. Dual drum paver

of 1½ yards or so. A measured quantity of water is added, usually through a hose from a tank truck alongside.

After a specified mixing time, the drum discharges into a bucket that can be moved inward and outward on a horizontal swing-ing boom. The boom and bucket are manipulated to deposit the concrete in the desired area. To avoid separation, concrete is deposited against forms, or against previously placed concrete, to approximately proper depth.

Courtesy of Portland Cement Association

Fig. 8-78A. Spreader, and side dump truck

Courtesy of Portland Cement Association

Fig. 8-78B. Auger spreader

The drum may mix enough to fill the bucket several times.

The hopper can be refilled by the truck immediately after discharging into the mixer. But if there is only one drum, it cannot accept material from the hopper until it has discharged all the mixed batch.

In dual and triple drum mixers, only part of the mixing is done in the first drum, which can accept another load while the original batch is being processed by the other(s). In this way, mixing is in process almost continuously.

Spreaders. But even this is not fast enough for modern road jobs, so the mixer is usually replaced by a distributing machine, which accepts mixed concrete in full truck loads, and spreads it between the forms. There are a number of types.

Hopper spreaders, with capacities up to 3 yards, are loaded by side dump trucks, and move back and forth across the paving strip, spreading to a pre-selected depth.

A horizontal auger, Figure 8-78B, is a basic design for spreading hauled-in con-

crete. With side dump trucks and a side hopper, it may be a continuous screw. If the hopper is in the center, for rear dump trucks, the two sides of the auger are pitched oppositely.

The auger is positioned and speed-regulated to leave the surface of the wet concrete (often called mud) lumpy and a little higher than the desired surface. It is followed by a strike-off bar or plate (often called the front screed) that pushes and cuts it down to almost final level, perhaps 3/16 inch high. It should push a roll of concrete 4 to 6 inches in diameter ahead of it.

This roll should be fairly uniform in diameter. It creates a back pressure against the screed, and a high or thick spot tends to force extra concrete to flow under the screed, creating a ridge.

Then there is a rear screed, a wide plate that has either side to side or vibratory motion as it advances. It should move a smaller roll of surplus ahead of it. Its effect is to make the concrete flow into exact contact

Courtesy of Bidwell Corp.

Fig. 8-79. Paving bridge

with its lower surface. Both screeds are about ⅛ inch higher in front than in the rear.

Spreading and smoothing may also be accomplished by means of auger and roller combinations working crosswise on the slab, supported by a bridge or cross frame between the forms. These structures may have a span well over 100 feet.

This roll should be fairly uniform in diameter. It creates a back pressure against the screed, and a high or thick spot tends to force extra concrete to flow under the screed, creating a ridge.

Vibration. Vibration is important in the consolidation of practically all concrete pavements. The method and frequency to be used are, however, not well agreed upon.

Courtesy of CMI Corp.

Fig. 8-80A. Placing reinforcement

The effect of the vibration is to cause the concrete to become temporarily more liquid, and to supply it with an impetus to flow freely into restricted spaces. Properly vibrated concrete should pack smoothly against forms, and under and around reinforcement, without voids. It lends to desirable low slump concrete the fluidity otherwise obtained only with high slump, lower strength mixtures.

However, if vibration is continued too long, or is the wrong type, it will cause the concrete to segregate. The tendency is for liquid to separate at the top, and coarse aggregate to sink, with bad effects on weather resistance and strength.

Vibration may be applied through surface plates or screeds, by cylinders immersed in the concrete, or by both. Immersion causes interference with reinforcement, and is less widely used.

Reinforcement. Concrete pavement is usually reinforced with steel mesh or mats (distributed steel) described earlier. The sections may be placed alongside the forms ahead of the paving operation, and positioned in the concrete manually, or by a manually supplied placing machine. They are usually the full width of the strip being laid, running across any longitudinal joints (lane divisions) that are included.

The standard method of installation is to spread and strike off the concrete roughly to about ⅔ of the form height, then place the reinforcing on this surface. Additional concrete is then spread to finish grade.

Sections of wire mesh may be overlapped by one cross wire plus one inch, and reinforcing bars (rebars) in mats by one foot. Standards vary among states, both in overlap and in whether the pieces are fastened or even welded together.

The steel does not cross contraction joint lines, but is stopped 6 to 12 inches short of them on both sides. The lengths in which the sections are supplied should divide evenly into the between-joints length of slabs, after allowing for overlap, so that no cutting is required.

Reinforcement may also be placed in a full-depth poured slab and vibrated down to correct position.

Courtesy of Rexnord Inc.

Fig. 8-80B. Continuous reinforcement on chairs

Courtesy of Rexnord Inc.

Fig. 8-81. Dowels on chairs

Dowels. Reinforcement in contraction joints is in the form of dowels, which were described earlier. They allow the slabs to move longitudinally in relation to each other, but resist any vertical movement.

Dowels are mid-depth in the slab. They may be supported in correct position by racks called baskets or chairs, placed on the subgrade before pouring. Or they may be vibrated by machine down into the full depth slab after striking off and before finishing.

With either method, it is absolutely essential that an accurate location system be used, so that the cut for the joint will be made over the center line of the dowels.

Finishing. The operations just described leave a smooth top on the slab. They are followed by further operations to correct any deficiencies, and provide desired finish.

Finishing may include a float device, usually moving in strokes forward and backward as a carriage moves it from side to side, or a diagonal tube finisher. This is a very long aluminum tube about 8 inches in diameter, which is moved back and forth by hydraulic controls until finish grade is satisfactory.

Final finish may be provided by a sheet of burlap dragged longitudinally, a long-handled, stiff-bristled brush pulled across, or other means.

Depth of brush pattern may be varied by type and spacing of bristles, and the amount the concrete has set. The ridges between bristle grooves should not slump or flow together (concrete too green), nor ravel into sand or crumbs (set too much).

Much deeper, sharper grooves may be cut by sets of diamond or carbide saws, after the concrete has set so that no pieces of aggregate will be loosened, or at any time after that during the life of the pavement.

Curing. Water must not be allowed to evaporate from the surface until hydration has progressed far enough to lock the necessary amount into the concrete. Either the hydration process, or the means taken to prevent loss of water, may be referred to as curing. The period for regular cement mixtures may be 3 to 10 days, with the first few hours the most important.

Curing procedures are usually closely regulated by specifications. In general, the protection should be furnished immediately

Courtesy of Rexnord Inc.

Fig. 8-82. Final finishing

after finishing. This is very urgent under hot, dry, and sunny conditions, and somewhat less so when fog or drizzle indicate that the evaporation rate is very low.

At present, the two methods most used are waterproof membrane or polyethylene sheets. Older methods included paper sheets, and cover with hay or dirt.

The membrane is a chemical that is sprayed mechanically onto the concrete to form a thin film that stops evaporation. The substance itself is transparent, but it is usually mixed with a white dye that shows accuracy of application, and keeps the concrete cooler by reflecting the sun's rays. The membrane wears off naturally when traffic uses the road.

The sheets are heavy gauge white polyethylene, somewhat wider than the pavement, and very long. They are rolled for storage and transportation, and unrolled on the concrete behind the finishers, either by hand or by machine. The edges are weighted with loose dirt, stones, or scrap to hold them from blowing.

After curing is complete, the hold-down material is removed, and the sheets are re-rolled to move ahead to another section.

Several miles of sheets may be needed in high speed paving.

The sheets provide protection against heavy rain, which might damage bare or sprayed concrete very soon after finishing.

SLIP-FORMING

In slip-form paving, the forms are attached to the paving machine and move with it, leaving the slab unsupported. These forms may be 16 to 48 feet long, and pavers may move at 6 to 12 feet a minute.

The slip-form paver performs a number of different operations in succession. These may include some or all of the following: receiving the truck-delivered concrete, spreading it to full width, compacting, eliminating voids, striking off and leveling the surface, installing reinforcements, surface finishing, and placing membrane or polyethylene sheets for moisture retention during curing.

These operations may be done by one machine, but more usually by several units. All or almost all of them may operate within the limited space provided by the length of the sliding forms.

It is customary to pave at least two lanes

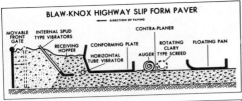

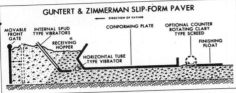

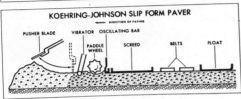

Courtesy of American Road Builders Association

Fig. 8-83. Slip form paver designs

at one time, and sometimes three lanes and occasionally four are laid as a single strip. One reason is that a small amount of edge slumping is not serious at the edge of a road, but could be a nuisance along a longitudinal joint between lanes.

Advance Preparation. The subgrade must be prepared to sufficient width to accommodate the tracks of the paver outside of the paving lanes. This may require 18 inches to 2 feet extra on each side.

Subgrade accuracy is vital to smoothness of finished pavement, as the paver operates directly from it. If the paver does not have automatic grade controls, any irregularity in the base is reflected directly in the pavement surface. If there are automatic controls, irregularities activate them, and problems of overcompensation or undercompensation may arise.

As a result, it is more and more common practice to prepare the base courses with automatic grading equipment, which reduces trouble with irregular bases to an absolute minimum. If paving follows immediately after base preparation, the two processes

may be controlled by the same stringline, assuring coordination of the two levels.

Stringlines and automatic grading equipment are discussed in Chapter 2, also in Chapter 19.

If concrete-carrying trucks are to use the roadway, it must be firm enough to support them without any rutting. If they will use a side lane, the stringline must be on the other side of the paver, or across the truck lane with a delicate and often unsatisfactory overhead sensing structure.

Automatic controls are first set by hand for pavement depth and height and distance from the edge of the string, and thereafter are self regulating except when deliberately overriden by the operator.

Placement. The wet concrete may be supplied into a hopper by side dump trucks, or onto the subgrade by rear dumps. Spreader boxes, attached temporarily to the truck while dumping, prevent segregation that might result from uncontrolled dumping on the subgrade. They also help the paver by measuring out the amount it needs, saving it from oversupply and starvation.

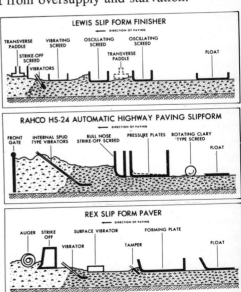

Courtesy of American Road Builders Association

Fig. 8-84. More paver designs

Courtesy of Portland Cement Association

Fig. 8-85. Slipformed edge

A substantial advantage of using the sub-grade for trucking is that there is room for two or more side by side, when paving covers two lanes or more. This usually allows for smoother traffic flow and ability to dump more trucks in a given time.

Figures 8-83 and 8-84 are sketches of six types of slip-form pavers, to give an idea of the variety of methods that can be employed.

The hopper or receiving front usually has internal vibrators that help to spread the mix evenly from side to side, and reduce resistance to going under the front screed or conforming plate. The next sets of processes vary greatly, but there is general agreement on finishing with a float.

Reinforcing. Reinforcing is customarily placed by a two-layer method. A slip-form paver with forms set 2 or 3 inches in from the edges puts down ⅔ of the pavement depth. Reinforcing and dowels are placed, by hand or machine. A second paver with forms set for full width then spreads concrete over and beside the first strip to form a finished slab.

In continuously reinforced pavement, the reinforcing bars may be laid on the base ahead of the paver, carried over its top, and forced down to proper level in the concrete ahead of the final finishers.

SOIL AND CEMENT MIXTURES

Types. There are three general types of soil and cement mixtures: compacted soil-cement, cement modified soil, and plastic soil-cement. The first of these is by far the most important, and will be discussed in some detail in a following section.

In all of these, portland cement is mixed with native (in place) soil or with borrow from nearby sources, to produce a low-cost, consolidated material. For reasons of economy, cement content is usually held to the minimum to produce a specific result.

Cement-Modified Soil. This is the most economical and casual of these cement mixtures. The cement used may be only one or 2 per cent of the soil volume.

The purpose is usually to improve a soil that is unfit for a pavement base or subgrade. The modified product may be caked or slightly hardened, but the principal object is to reduce plasticity, water-holding and volume-change capacities, and increase its load bearing strength.

Soft spots encountered in underlying soil in building secondary roads are sometimes stabilized by mixing in small quantities of cement.

Modification is applicable to a wide range of soils, including expansive clays.

Mixing is usually done on the road, in the same manner as soil-cement, but less attention is paid to compaction.

Plastic Soil-Cement. This is a mixture in which sufficient water is included to make it soft, like plastering mortar. It does not require compaction, and therefore can be used in places inaccessible to road-building equipment.

Plastic soil-cement is used to line or pave ditches, slopes, canal banks and other places that are subject to erosion.

It may also be made by mixing high early cement into the natural material in mudholes, for a temporary emergency patch. Hand shovels and rakes are the usual tools.

COMPACTED SOIL-CEMENT

Compacted soil-cement is usually referred to just as soil-cement, as the other two kinds of soil-cement mixtures have comparatively little use.

This soil cement is a mix of pulverized soil and carefully calculated amounts of portland cement and water, compacted to a high density. The resulting material is a rigid slab, with moderate compressive strength, which is resistant to the disintegrating effects of wetting and drying and freezing and thawing.

Soil. Practically all sub-standard soils can be improved for use as structural material by mixing with portland cement. However, many of them require excessive quantities of cement or are difficult to work, and therefore are seldom processed in this manner.

Stabilization is most efficient with sandy or gravelly soils with 10 to 35 per cent silt and clay, with not over 45 per cent pieces larger than ¼ inch.

Sandy soils with few fines, or none, require somewhat more cement, and may

Normal Range of Cement Requirements for B- and C-Horizon Soils†

AASHO SOIL GROUP	Percent by vol.	Percent by wt.
A-1-a	5-7	3-5
A-1-b	7-9	5-8
A-2-4		
A-2-5	7-10	5-9
A-2-6		
A-2-7		
A-3	8-12	7-11
A-4	8-12	7-12
A-5	8-12	8-13
A-6	10-14	9-15
A-7	10-14	10-16

†A-horizon soils (topsoils) may contain organic or other material detrimental to cement reaction and may require higher cement factors. For dark grey to grey A-horizon soils, increase the above cement contents 4 percentage points; for black A-horizon soils, 6 percentage points.

Average Cement Requirements of Miscellaneous Materials

MATERIAL	Percent by vol.	Percent by wt.
Caliche	8	7
Chat	8	7
Chert	9	8
Cinders	8	8
Limestone screenings	7	5
Marl	11	11
Red dog	9	8
Scoria containing plus No. 4 material	12	11
Scoria (minus No. 4 material only)	8	7
Shale or disintegrated shale	11	10
Shell soils	8	7
Slag (air-cooled)	9	7
Slag (water-cooled)	10	12

Courtesy of Portland Cement Association

Fig. 8-86. Cement requirements for soils

create traction problems for rubber tired processing equipment.

Silty and clayey soils make satisfactory soil cement, but clay may have a high cement requirement, and may be unusable if it cannot be pulverized. Both season and weather are important when working with these soils.

Figure 8-86 gives average cement requirements, by both volume and weight, for various soil types and miscellaneous materials.

Stones over an inch or two in diameter are considered highly undesirable. Old blacktop can be included in the mix if it can be broken into fine enough pieces.

Organic material in the soil usually has a very unfavorable effect on soil-cement, so the use of topsoil, or soil contaminated with topsoil, should be avoided whenever possible. However, the effect depends partly on other soil qualities, and satisfactory results have been obtained with organic content as high as 3 per cent, without unreasonable increase in cement.

Mixture Design. Presumably because of its appeal as a low-cost road material, soil-cement is not mixed to obtain maximum strength or durability, but to use the minimum amount of cement that will enable it to pass two standard laboratory tests.

The wet-dry test involves samples containing varying proportions of cement that have cured in high humidity for 7 days. They are weighed, submerged in tap water at room temperature for 5 hours, then placed in an oven at 160° Fahrenheit (71° C.) for 42 hours. They are then given two firm strokes on all sides with a wire brush to remove material loosened by the wetting and drying. Re-weighing and subtracting the new weight from the old indicates the amount of disintegration that occurred during the cycle.

This process is repeated 12 times. Passing grade ranges from 14 per cent loss for sandy or gravelly soils down to 7 per cent for clayey soil.

In the freeze-thaw tests, 7-day specimens are placed on moist blotters, refrigerated at −10° F. (−23° C.) for 24 hours. They are then thawed in a moist atmosphere at 70° F. (21° C.) for 23 hours. They are brushed, and half-loose scales are pried off with an ice pick. After 12 cycles, the specimens are oven-dried and weighed. Permissible, or perhape desirable, loss is the same as in the wet-dry test.

Compressive strength, in spite of its great importance in the pavement, is not necessarily tested directly, but is assumed from these two. The 7-day strength ranges from 500 to 800 pounds per square inch. After long curing, it may be 1200 to 3000 psi.

Additional tests are firm pressure with a dull ice pick (penetration should not be over ⅛ inch), and tapping two test cylinders together. Both of these indicate strength developed at the time of testing.

Water. Water requirement is figured out very carefully for optimum compaction, without regard to the quantity needed for hydration of the cement. Air-dry, pulverized soil is weighed in a laboratory, then blended with the proper amount of cement. Water is mixed in, in small measured quantities, and compaction tested after each addition. A moisture-density curve is plotted from these results.

Exact proportions cannot be obtained in the field. It is not practical to determine the moisture content of the soil on the road in every section nor for every hour of work. Nor is it possible to know just how much moisture will be lost during processing, as the rate changes with humidity, wind, and intensity of sunlight.

Moisture content of soil is averaged from a number of spot-check tests, and subtracted from the optimum moisture. That is, if the soil should have 10 per cent water by weight, and contains 4, then 6 per cent must be added. An arbitrary assumption of combined absorption by the cement, and loss from evaporation, is made, perhaps 2 per cent. The result is that this soil-cement mixture needs 8 per cent of its dry weight in additional water.

The weight of material going through the mixer per minute is calculated and corrected to dry weight. Assuming that this worked out to 9,000 pounds, the water requirement would be .08 x 9,000, 720 pounds. At 8.33 pounds per gallon, this would be 86.4 gallons a minute.

This water is supplied to on-the-road mixers by tank truck. The mixer should have a small tank to keep it working during truck changes.

When weather is dry, additional water may be added almost constantly to the surface during processing, as a light or fog spray from a distributor truck. The amount of such addition should just compensate for

Fig. 8-87A. Distributing bags of cement

evaporation, and depends on judgment of the foreman or inspector.

Pavement Design. Soil-cement is used in two principal ways. In one, it is the structural part of a pavement, but is protected against surface abrasion by a bituminous coating or pavement. In the other, it is a base (called subbase) for a concrete pavement.

Methods of construction are almost identical. The pavement use will be considered first.

Soil-cement usually has a compacted depth of 6 to 8 inches. On firm subgrades and light traffic conditions, this may be reduced to 5 or even 4 inches, but then its advantage over granular bases becomes doubtful.

Twelve or more inches may be placed under special circumstances, but 8 inches is the most that should be put down in one layer.

To develop its full strength, the correct proportions of cement, water, and soil must be present, and compaction must be 95 to 100 per cent or (more) of maximum density.

Fig. 8-87B. Preliminary mixing of cement

Fig. 8-87C. Mechanical cement spreader

Courtesy of Portland Cement Association

Fig. 8-88. Single shaft mixer

Preparation. If the original road surface is to provide the soil for the new pavement, it must be carefully shaped to final grade. Very little displacement of material occurs during mixing, and any inaccuracies in original grade will cause extra work at the end, and probably result in too-thin and too-thick areas in the pavement.

If the roadbed contains oversize stone, or is so hard that the mixing machines might not reach design depth, it should be scarified or ripped until loose enough to work, and until all stones have been pulled to the surface and removed. Grading, or re-grading, should follow this operation.

Any soft spots that develop during this work should be repaired, as proper compaction cannot be obtained on a yielding base. It may be necessary to dig out and replace wet material. It might be possible to stabilize limited areas by mixing in high early cement. Permanent drainage should be installed where indicated.

Clayey soils that are in dry hard lumps usually are easier to pulverize if water is added a day or so in advance. If the lumps are wet, they should be partly dried by scarifying or disking repeatedly to expose them to air and sun.

If the mix soil is to be hauled in, the road must be shaped and made firm, but scarifying is not required unless to remove projecting stone.

Cement. On a small job, cement may be placed on the road in bags in a regular pattern. A marked tape or string can be used as a guide in distributing them properly. The bags are opened by hand and spread in windrows across the work strip. These windrows are then spread lengthwise by a small tractor pulling a spike tooth harrow or some other drag. Several trips back and forth should distribute the cement in a fairly even layer.

But in most of this work, the cement is brought in trucks, and distributed by a mechanical spreader. Ordinary dump trucks, fitted with a body cover to prevent blowing, can be used. There are bulk cement trucks that discharge by means of an auger, and agricultural lime trucks which might be able to both haul and spread.

The spreader should be towed rather than pushed to avoid picking up of cement by the truck tires. It should be kept at least half full, to avoid changes in rate of flow.

The rate of cement delivery is rather critical. There must be enough to keep the mixer working, but cement is vulnerable to loss by wind and spoilage by rain, so to spread it far in advance is risky.

Mixing. Two types of in-place mixers are used, single shaft and multiple shaft. A single-shaft machine is shown in Figure 8-89. As the machine moves forward, rapidly revolving tines strike downward at the cement and earth, pulverizing and mixing with repeated blows as the mixture is carried around inside a covering hood. Water may be added by a spray bar in the hood, or it may have been applied to the soil ahead of the mixer.

This machine might accomplish the whole mixing operation in one pass, but two or more are usually made to be sure.

Multiple axle machines have either two

Courtesy of Portland Cement Association

Fig. 8-89. Placing pre-mixed soil cement

or three axles, each equipped with tines. The first set pulverizes and mixes, water is then added by a spray bar supplied by a tank truck, the second and possibly the third set complete the job. These machines are often called single-pass.

If the rear plate of a mixer hood is lifted, the material is ejected a considerable distance to the rear. On a sunny day this, as a preliminary operation, will produce quick drying of too-wet soil.

If the soil is trucked in, it is dumped, then bladed by a grader or shaped by a towed proportioner into a symmetrical windrow for the width being worked, with cement usually added on the top. Another type of mixer picks up the windrow and passes it through a revolving drum. The first few paddles blend the dry material, then water is sprayed in, and mixing completed by the remaining paddles.

The material is discharged to the rear, and spread by an attached strike-off blade, by a following grader, or by both.

The width that can be processed in one pass is limited by the width of the in-place mixer, or the capacity of the windrow machine. This is usually around 10 feet, or half the width of a two lane road.

If traffic must be maintained, a quite long section of one lane is completed, traffic turned onto it, and the other lane worked. This method causes several problems, the most serious being the center joint between the two strips.

When traffic can be detoured, the two strips are worked alternately, leap-frog fashion. First 300 to 500 feet are done on one side, then 600 to 1000 on the other, so that the second lane can be placed before the first one has set.

Pre-Mix. If mixing is done at a central plant, the blended and moistened material is brought by trucks and dumped into aggregate spreaders. These should not be allowed to run empty, but should be stopped during truck changes. If the subgrade is dry, it must be moistened before the mix is placed on it.

Compaction. A variety of rollers may be used for compaction of soil cement. The sheepsfoot or tamping roller, either towed or self-powered, is favored for all except the most granular soils. Weight of ballast is adjusted so that there will be initial penetration to near the bottom, followed by walking out to the surface in a few passes. Unit pressure may be as low as 75 psi for friable soils, and as high as 300 for heavy clay.

If the feet will not penetrate and pressure cannot be increased, the material may be loosened by scarifying or dry running through a mixer.

Pneumatic tired rollers are favored for coarse, loose mixes. Models that permit varying tire pressure during work are the most efficient. Three-wheel steel rollers in the 12-ton class may also be used, for the whole job or just for final smooth-up rolling.

Grid rollers, segmented rollers, and various types of vibratory rollers all have their backers for soil-cement compaction.

The most important factor is that compaction be started immediately after spreading, and continued steadily until completion. Hydration of cement starts as soon as it is wet, and best strength will be obtained with least work if all processing is finished quickly.

Finishing. For various reasons, the compacted surface will not be entirely smooth, and will usually need some reshaping with a motor grader. This involves trimming high spots, ridges, and other irregularities, and either using the material to fill depressions, or wasting it at the road edges.

Differences from trimming a gravel or dirt road include effects of partial setting or drying of the surface; and the possible presence of compaction planes. These are smooth or dense surfaces, usually tire prints, close to final grade. They are found chiefly in heavy soils. Material bladed onto them may not form a good bond, so that loosening and spalling will occur in the finished pavement.

Compaction planes are eliminated or made harmless by scratching the surface with a spike tooth harrow, weeder, or steel broom pulled by a small tractor. Final grading may then be done in a conventional manner, except that the surface must be kept damp during the work.

If compaction planes are at or very close

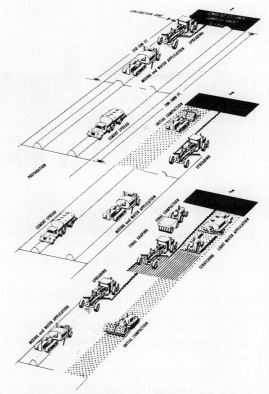

Courtesy of Portland Cement Association

Fig. 8-90. Windrow-processing sequence

to final grade, or if the surface is partly set, it may be economical to waste any cuttings at the side, rather than to try to use them in the road.

Final grading is followed immediately by moistening, and compaction with a pneumatic tired roller.

If another layer of soil-cement is to be built on top of this one, exact surface grade

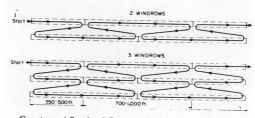

Courtesy of Portland Cement Association

Fig. 8-91. Another processing sequence

Courtesy of Portland Cement Association

Fig. 8-92. Multiple shaft mixer, and compactor

is not critical, and compaction planes need not be roughened.

Joints. Joints occur at the end of each day's construction, and between parallel work strips.

A vertical joint at the end of a day is made by cutting straight down into the mixed and compacted material, either immediately after stopping work, or before starting the next morning. The toe of a grader blade, a special disk attachment on a blade, or hand grub axes or shovels may be used.

When work is started again, the cut surface is brushed clean of loose material, dampened, and new mix placed against it. Compaction may be by a roller or grader operated across the road, or by a hand vibrator or tamper. Excess material should be placed at the fresh side, to be pressed into the joint by regular longitudinal rolling. Remaining excess, if any, can be bladed off when finishing.

A longitudinal joint between two work strips is made by overlapping the mixer a few inches, if the material is still fresh and

Courtesy of Portland Cement Association

Fig. 8-93. Cutting a joint

Courtesy of Portland Cement Association

Fig. 8-94. Scratching out compaction planes

soft, or only partially hardened. Test holes may be made with a pick to locate the firm edge of the original strip, and stakes or string placed as a guide for the operator.

Fully hardened soil cement is shaved back to solid material with a grader blade or disc, with hand-tool assistance as necessary, cleaned of loose material and moistened, after which the new mix is placed against it with some excess.

Curing. The freshly finished soil-cement surface must be protected from drying. The standard method is to cover it with a thin bituminous coating, as soon as possible after completion.

The surface should be damp, and free of any loose material. Cutbacks RC-250, MC-250, and emulsion RS-2 are suitable curing materials, applied at the rate of 0.15 to 0.30 per square yard.

Sanding is required if traffic is to use the surface before the asphalt has set. The soil cement can then safely carry loads that are not greater than the compaction equipment used, but low speeds are advisable for a few days.

Asphalt is the most convenient curing material, but anything that will hold moisture will do. Plastic or waterproof paper sheets, wet cloth, straw, or sand, or membrane spray can be used.

Surface. A soil cement surface is ordinarily too friable and absorbent to stand up under exposure to traffic. It is therefore customary to apply a regular bituminous surface on top of the curing coat. This may be placed immediately, or delayed a month or more to allow shrinkage cracks to develop.

For light traffic, a single seal coat may be sufficient. A double seal coat, with a total thickness of ¾ inch is suitable for general use, except in snow areas. There 1½ inches of plant mix blacktop is recommended, for assurance against peeling by snow plows.

A single seal coat may be applied immediately, and a second coat or blacktop put on several months later, preferably in cool weather, to bridge over shrinkage cracks.

These cracks are inevitable in soil-cement, as the hydration process causes shrinkage. However, breaks are so irregular that the edges support each other, so structural weakening is slight. The principal damage is to appearance of the surface, which can be minimized by delaying the top coat until after they have formed.

SOIL-CEMENT BASE

Soil-cement base for concrete pavement is constructed in the manner just described, except that curing is likely to be done with some non-bituminous material.

Getting correct surface grade is critical, and may be somewhat difficult. If a gravel or other base is found to be off-grade just before concrete is laid, correction is made rather easily by removing material or bringing it in. With hardened soil cement, this is not practical, so great care must be taken to get it right.

CHAPTER NINE

BLASTING AND TUNNELING

For the purposes of this chapter, rock is defined as material which requires loosening by explosives or rippers in order to be dug economically by machinery.

PURPOSES OF ROCK EXCAVATION

Surface excavation of rock is done chiefly for the following purposes:

1. Stripping—the removal and wasting of any type of rock or dirt in order to uncover valuable layers.

2. Cutting—removal primarily to lower the surface. In road and airport construction the spoil is generally used for fill elsewhere on the project. In ditching, it is often used for backfill after installation of pipes.

3. Quarrying or mining—excavation of rock which has value in itself, either before or after processing. A rough distinction can be made between these two in that quarries are ordinarily concerned with the physical characteristics of the stone, and mines with its chemical composition. However, the terms will be used interchangeably in this discussion.

One excavation can involve all three classifications, as in a heavy road cut where some material is wasted, some is used for road fill, and the best rock is crushed and used for aggregate.

Blasting may be divided into a primary operation in which rock is loosened from its original position in bulk, and secondary work which consists of reducing oversize fragments, and breaking back ridges and spurs. The latter is done in the same manner as other light blasting, such as breaking boulders and chipping out ledges.

Rock work may also be classified as to the type and fineness of breakage required. Quarrying of building or dimension stone involves loosening large solid pieces from the parent rock, while blasting for fill or crushed rock requires pieces small enough to fit in the shovel bucket, the fill layer, or the crusher.

Stripping. In most stripping work, the spoil has no value so that the cheapest way of handling it is the best way. It is often possible to dump it in excavated areas from which pay material has been removed.

It is common practice to shoot and dig overburdens over a hundred feet deep in a single layer, and the use of the largest shovels and draglines is required for such work.

Drilling may be done horizontally from a face, as in Figure 9-1 (A), vertically from the top, as in (B), or in combination, as in (C).

Horizontal drilling has its best use when the mineral deposit is immediately under soft rock. Auger type drills with extensions six to ten feet long, and diameters of about five inches, are used. These have a tendency to drift downward, and since distances of thirty to seventy-five feet are

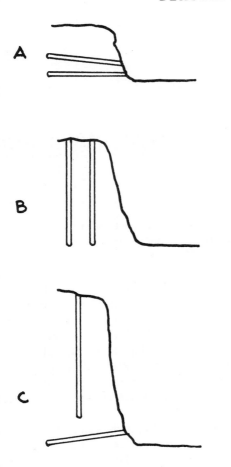

Fig. 9-1. Horizontal and vertical drill holes

commonly drilled, it is necessary to start them several feet above the deposit, or to start them at a slight upward slant. Spacing may be ten to thirty feet.

For harder bottom rock, crawler-mounted (track) air drills may be used. The hole is usually started several feet above the bottom. Lower holes are slanted downward. Higher ones, if drilled, may be horizontal or slant upward.

If thick higher layers of the overburden are hard rock such as sandstone or limestone, holes are drilled vertically, as in (B) or (C), using rotary or downhole drills with bits from 4 to 12 inches in diameter.

Burdens (distance from face) from 25 to 35 feet, and spacings in rows from 25 to 35

feet are common in high faces. Track drills are used generally to depths of 50 feet, and occasionally to 100.

Scrapers are sometimes used to remove most of the loose soil before drilling. This saves a considerable drilling footage, makes casings unnecessary, and, on low and medium faces, simplifies the use of track drills.

If the pit area is to be regraded when work is completed, the scrapers can be used to fill the hollows between the piles left by the shovels, or to place topsoil over regraded areas.

Because of the tremendous size and power of the excavators used in large pits, the blasting need only shake up, crack, and loosen the overburden without producing fragmentation comparable to that required in a cut or quarry. Wide spaced holes and light charges can therefore be used successfully.

Highway Cuts. Rock cuts for highways may be of the through type, Figure 9-2 (A), and the sidehill (B). Material from a sidehill cut may be thrown down to make a fill, as in (C).

The area to be cut should first be cleared and stripped of loose soil, and preferably of rotten rock. This may be done with dozers, scrapers, or shovels, depending on the conditions and the equipment available.

If the rock is soft, its upper surface may be loosened with rippers and removed, along with any dirt pockets it may contain. If it is hard and irregular, extensive cleaning by hand and with small equipment may be necessary. It is desirable to remove all loose dirt, particularly if the rock is to be used as crusher aggregate for road topping.

When water and disposal areas are available, hydraulic cleaning with contractors' pumps and fire hose or with special equipment may be used.

If cleaning is not practical over the whole area, the spots to be drilled can be cleaned individually. The top layer may then be drilled, shot, and removed for fill, and any clean rock required can be obtained from lower levels.

If the cut is twenty feet or less in depth it may be taken in a single layer, but depths of 12 to 15 feet are generally considered most satisfactory for track drills, and digging by one to 2½ yard shovels, or medium-large front loaders.

In a through cut, the full width is used as a face to provide maximum space for machinery. On a sidehill, the same technique or one or more bench faces parallel with the centerline may be used.

Degree of fragmentation required is determined largely by the depth of fill layers where the spoil is used.

Mining and Quarrying. Pit operations are largely conducted to obtain certain classes of rock or earth. The general aspects of this work will be discussed in the next chapter.

Rock excavation may follow highway techniques in exploiting comparatively narrow or irregular veins; or large scale stripping work may be necessary to make pay rock accessible to surface digging units.

Pits are often distinguished by the use of high and wide faces; or holes sunk be-

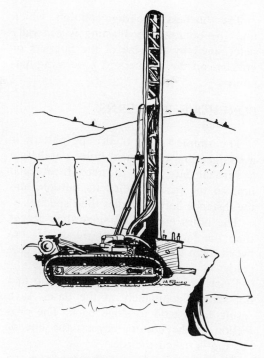

Fig. 9-2. Drilling near a high face

low surrounding grades, with access by ramps or inclined or vertical hoists.

It is advantageous to have the face wide enough so that several operations can be carried on in different sections with minimum interference.

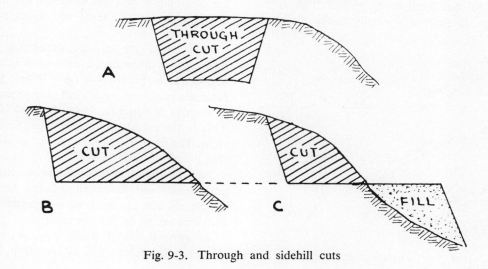

Fig. 9-3. Through and sidehill cuts

The fineness of fragmentation which must be obtained by blasting is generally determined by the size of the hoppers or grizzlies on the crushers or processing machines.

BOREHOLE PATTERNS

The simplest type of drilling pattern is a straight line of vertical holes parallel to a vertical face. The distance from each hole to the face is called its burden, and the distance between holes their spacing.

The holes are drilled somewhat deeper than the face so that any ridges left between them will not project above the new grade.

Blasts tend to overbreak at the top and not shatter completely at the base. As a result, faces tend to slope back. The projection of the bottom beyond the line of the top is called the toe.

The extra burden at the toe may be handled by bottom drilling, or heavier loading (more powerful explosive or tighter packing) in the bottoms of vertical holes.

Holes may be drilled at an angle so that they are parallel to the slope of the face. This angle may be from 5 to 30 degrees.

Angle or slope drilling keeps the burden at the toe from being greater than at the top, so that no specially heavy charge is

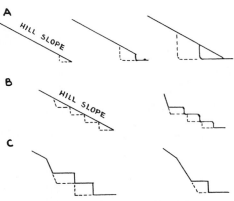

Fig. 9-4A. High face and bench cuts

required at the bottom. Improved bottom breakage reduces the need for drilling below the floor, so that the total length of hole is usually not greater than in vertical drilling.

Cleaning. Small and medium air drills are not suited for working through soil overburden. It tends to choke up their air passages and to fall in on top of the bit. If this falling occurs after penetration into rock, it may make pulling the steel difficult or even impossible. Soil drills readily only when frozen.

Problems of stuck steels are discussed in Chapter 20, pages 41C and D, and 51A.

Stripping is usually essential before use of these drills.

The whole area is usually machine stripped, and the actual spots to be drilled cleaned by hand. At other times the soil is cleaned off only in slots, and drilling is done in lines along them.

Face Height. A rock mass may be taken out in a single layer, or in a series of benches. Highway contractors usually prefer benches, as do most open pit miners, but quarry operators may take single slices 100 or even 200 feet high. This is partly because quarry rock is often sound enough to stand at great heights without much danger of collapse, and the length of fall may break up the top as a result of blasting only the bottom.

High faces are usually developed by pushing low or moderate ones back into a hillside, as in Figure 9-4 (A). Where low faces are preferred, such a cut may be made in a series of benches, as in (B) and (C). This is the safest method in most formations. The final slope should not be so steep that it cannot stand by itself.

Face height affects the method of drilling and the size and placement of holes. Accessibility of the top may be a determining factor in taking off the top layer.

In general, hand drills can be used efficiently for holes up to 10 feet deep, and

LOADS

they can be carried anywhere a man can go. Their comparatively light air hoses can be strung long distances, although at the cost of reduced pressure.

Track mounted drills work well from one to 50 feet deep, and may go to 100 or more. They can climb and work on slopes over 30 degreees. However, risks and delays are involved in working them on very steep or rough ground, and it may be more economical to take a first cut with hand drills.

Depths over 30 feet call for blast hole drills, of the rotary, down hole, or churn types. These heavy and expensive machines should be kept on safe, fairly even ground. They can dig through any depth of overburden readily, and can start their work from pioneer roads notched into soil slopes by dozers.

Hole Size. Hand drills will produce hole diameters of between one and 2 inches, wagon and track drills 1½ to 5 inches, and blast hole drills from 4 to 12 inches. Pneumatic drills of all sizes make tapering holes with steel bits, as the hole gets smaller toward the bottom as they wear. Carbide bits, and rotary and churn drills produce holes with little or no taper.

Bore Hole Loads. Figure 9-5 (A) shows the cubic foot capacity of holes of various diameters for each linear foot. For example, to find out how much bulk of explosive will be needed to fill a hole 3 inches in diameter and 9 feet deep, find the figure 3 in the first column. Next to it is the capacity of one linear foot, .05 (1/20) cubic feet. Multiplying this last figure by the 9 foot depth (below stemming) of the hole, we find the capacity is 0.45 or 9/20 cubic feet.

The rest of the columns indicate the weight of explosive of various densities that it will take to fill the hole completely. The actual load will usually be somewhat less except with free running materials, because of waste space around cartridges. The amount of difference will depend on the efficiency of tamping.

To find the amount of dynamite with a rated density of 164 sticks to the 50 pound case that can be tamped into this hole, follow the horizontal line in the table from the 3 inch line to the right to the fourth column, under the figures 47 and 164. We find that a 3 inch borehole will hold 2.3 pounds of this powder per lineal foot. Multiplying this

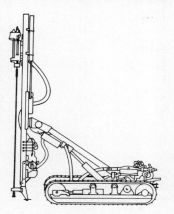

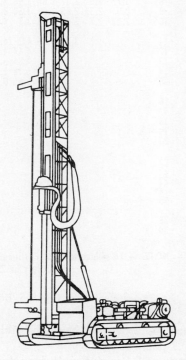

Fig. 9-4B. Track and blast hole drills

9-5

A

BOREHOLE CAPACITY

Specific gravity of explosive	.64	.75	.80	.96	1.12	1.28	1.44
Pounds per cubic foot	40	47	50	60	70	80	90
Number 1–1/4 x 8 cartridges in 50 lb. case	192	164	154	128	110	96	85

Hole Dia., Inches	No. Cu. Ft. Per Linear Ft.	Capacity, pounds per linear foot						
1–1/2	.012	.48	.6	.6	.7	.8	1.0	1.1
2	.022	.88	1.0	1.1	1.3	1.5	1.8	2.0
2–1/4	.027	1.08	1.3	1.4	1.6	1.9	2.2	2.4
2–1/2	.034	1.4	1.6	1.9	2.0	2.4	2.7	3.1
3	.05	2.0	2.3	2.5	3.0	3.5	4.0	4.5
3–1/2	.06	2.4	3.0	3.0	3.6	4.2	4.8	5.4
4	.09	3.6	4.1	4.5	5.4	6.3	7.2	8.1
4–1/2	.11	4.4	5.2	5.5	6.6	7.7	8.8	9.9
5	.14	5.6	6.4	7.0	8.4	9.8	11.2	12.6
5–1/2	.16	6.4	7.7	8.0	9.6	11.2	12.8	14.4
6	.20	8.0	9.2	10.0	12.0	14.0	16.0	18.0
7	.27	10.8	12.5	13.5	18.2	18.9	21.6	24.3
8	.35	14.0	16.4	18.5	21.0	24.5	28.0	31.5
9	.44	17.6	20.7	22.0	26.4	30.8	36.2	39.6
10	.54	21.6	25.6	27.0	32.4	37.8	43.2	48.6
12	.78	31.2	36.9	39.0	46.8	54.6	62.4	70.2

B

CUBIC YARDS DISPLACED PER FOOT OF BOREHOLE

Average Burden on Boreholes	SPACING OF BOREHOLES														
	6	7	8	9	10	11	12	13	14	15	16	17	18	19	20
6	1.33	1.55	1.77	2.0	2.22	2.44	2.65								
7	1.55	1.81	2.0	2.33	2.7	2.85	3.11								
8	1.77	2.0	2.37	2.65	2.96	3.26	3.55								
9	2.0	2.33	2.65	3.0	3.33	3.66	4.0								
10	2.22	2.7	2.96	3.33	3.7	4.1	4.44	4.81	5.18	5.55	5.92				
11			3.26	3.66	4.1	4.48	4.88	5.3	5.7	6.11	6.52				
12				4.0	4.44	4.88	5.33	5.77	6.22	6.66	7.11				
13					4.81	5.3	5.77	6.26	6.74	7.22	7.70				
14					5.18	5.7	6.22	6.74	7.26	7.77	8.30				
15					5.55	6.11	6.66	7.22	7.77	8.33	8.88	9.44	10.0	10.55	11.11
16							7.11	7.70	8.30	8.88	9.48	10.07	10.66	11.3	11.85
17							7.55	8.18	8.81	9.41	10.07	10.70	11.33	11.96	12.59
18							8.0	8.66	9.33	10.0	10.66	11.33	12.0	12.66	13.33
19								9.15	9.85	10.55	11.3	11.96	12.66	13.37	14.07
20								9.63	10.37	11.11	11.85	12.59	13.33	14.07	14.81
21										11.66	12.44	13.22	14.37	14.77	15.55
22										12.22	13.03	13.85	14.66	15.48	16.30
23										12.78	13.63	14.48	15.33	16.18	17.03
24										13.33	14.22	15.11	16.0	16.88	17.77
25										13.88	14.81	15.74	16.66	17.60	18.51
26										14.44	15.74	16.37	17.33	18.30	19.26
30										16.66	17.77	18.88	20.0	21.1	22.22

NOTE: – To reduce to tons: For limestone, shale, granite, etc., multiply by 2¼; for trap rock, multiply by 2½; for sandstone, multiply by 2⅛.

Fig. 9-5. Borehole capacity and burden

by the 9 foot length, we have a possible load of 20.7 pounds.

Untamped Cartridges. If cartridges are not slit and tamped, capacity is figured on the basis of the length and weight of the cartridges, and the number that can be placed side by side. Three of the standard 1¼ x 8 inch cartridges would fit in the hole if tied together. It would take 13 of these bundles to fill the hole within 4 inches of the stemming line. There would be 39 cartridges, each weighing 1/164 of 50 pounds, or about .3 pounds per stick. The total load by this method would be 11.7 pounds, just a little more than half of the possible load.

Burden. The explosive in each hole is supposed to break out a section of the rock between the line of holes and the face. Only experience with the particular rock and explosive will indicate exactly the amount and type required.

In a general way, however, it may be said that a pound of 40 per cent dynamite should break up and move two yards of soft rock, or one yard of medium hard rock, on an open face. In soft, layered, or rubbery rocks, 20 per cent dynamite might move more per pound; while in very hard rocks, even higher strength dynamites might have smaller production. In tight holes, at edges, and in corners heavier loading is required.

Figure 9-5 (B) is a table showing the yardage of rock to be moved per foot of hole for various burdens and spacings.

As an example of its use, consider that the powder that packed 2.3 pounds to the foot of 3 inch borehole would move 1½ yards of rock per pound, or 3.45 cubic yards to the foot of borehole.

According to the table, 3.33 yards of rock would be moved if spacing were 9 feet and burden 10, or the other way around. An 8 foot spacing and 12 foot burden would yield 3.55 yards.

Pull. If a hole fails to move its burden, it is said that it did not "pull." This usually occurs at the bottom of the hole, and most often in edge or corner holes where the rock is held back on two or three sides. Such failures may be due to too heavy a burden or too wide a spacing, to improper stemming of a shallow hole, use of the wrong explosive, explosive not reaching the

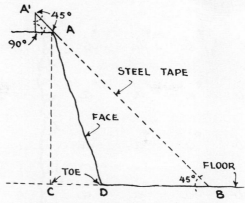

Fig. 9-6. Measuring face height

bottom of the hole, or a partial misfire. It is generally necessary to remove the blasted rock, then drill and shoot the bottom again.

Measurement. In order to drill and load holes accurately, it is necessary to know the height of the face and the amount of the toe. With low faces, or in casual operations, or in working upper lifts to temporary grades, depths may be estimated, although this is always risky.

Faces between ten and seventy feet may be measured by the device and method shown in Figure 9-6. A 45° right triangle is carefully made of two by twos or two by fours, with the sides of the right angle equal and from two to four feet long.

This is placed on the top edge of the face, as shown, and the bottom carefully leveled. A sight is taken along the line A'AB and the spot B marked on the quarry floor by an assistant. The distance AB is then measured with a steel tape. Multiplying its length by .71 (the sine of 45°), gives the height of the face and also the distance BC. The distance BD is then measured, and when subtracted from BC, gives CD, the projection of the toe.

This measurement may be repeated at various points along the face.

If the face is high, its top irregular, or considerable accuracy is necessary, it may be preferable to make a transit survey of the site, establish benchmarks and loca-

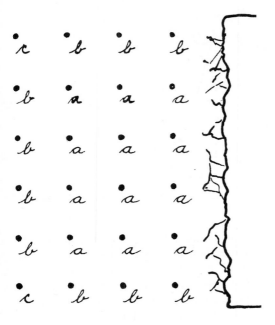

Fig. 9-7. Open and tight holes

tion points on all levels, and calculate from direct measurements from these points.

Each drill hole may be marked according to the cut to bottom grade, or by the drilling depth desired. For convenience in loading, the projection of the toe may also be noted on the marker.

Bottom Grade. The bottom grade should have a slope for drainage that may be away from the face or toward the sides, but not toward the face. If natural drainage is not possible, adequate pumps should be provided.

If blasting and excavating are accidentally carried below grade, hollows can be readily filled with fine shot rock. If the floor is too high, and the rock is soft, it may be possible to take it down with rippers and dozers. If it is too hard for machinery, a tedious job of shallow drilling and blasting is required.

Spacing. In general, large drill holes are increasingly prevalent in faces over 30 feet high, with proportionate increase in spacing and burden.

With most rock types, a point will be reached where enlarging and spreading the holes will result in poor fragmentation in the center of the blocks.

Faces fifty feet or more high are usually shot with a single row of holes at a time. Low faces use additional rows, fired simultaneously or in short-interval succession. In any height, holes in the same row are now fired in sequence.

Each blast should supply enough rock to keep the shovel busy for at least half a day, therefore, the lower faces must be shot back deeper than the high ones, particularly if the shovel is large and its working area narrow.

The entire group of holes may be drilled on a rectangular pattern, as in Figure 9-7, or they may be staggered to improve fragmentation.

Best results from multiple rows are obtained if there is a free cleavage plane at the new grade. If the bottom is very hard to pull, heavier loading may be required than when firing single rows; or burdens may have to be reduced progressively toward the back, resulting in higher costs.

Tight Holes. When blasted rock must be sheared away on two or more planes the shots are called tight. In Figure 9-7 the holes marked (a) are open, those marked (b) are tighter, having to shear off the back or side as well as the bottom, and the (c) holes must shear along back, side, and bottom. In general, the tighter the hole, the greater the likelihood that it will fail to pull. It is usually cheaper to take special precautions with tight holes in the original blast, than to do secondary drilling and blasting.

The tightest blasting found in open work is the start of a cut down from the surface. The first rock blasted can move upward only. If the whole set of drill holes are shot together, each of them will be very hard to pull. However, if the center holes are made oversize, loaded heavier, and fired first in a delay sequence arrangement that progresses toward the outer limits of the area, the adjoining holes can throw into the space left, then the next holes throw into their space.

Buffers. Blasted rock may be entirely dug away before the next blast, or varying quantities may be left against the face. Complete cleanup is required if the toe is to be accurately measured or drilled horizontally, and is considered good practice with high faces.

If the working space is narrow, and the face low, it may be desirable to leave some shot rock as a buffer or "blanket." This confines the force of the explosion, and may prevent blocking of the work area and aid fragmentation. However, the buffer should be small enough so that some scat-

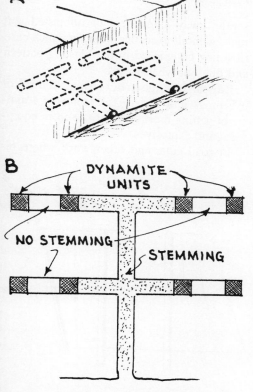

Fig. 9-8. Coyote holes

tering occurs, as this makes it easier to find and blast oversize rock before the shovel gets in the heap. Heavier loading is required.

Snake Holing. Sometimes a face can be most economically blasted from the bottom only. This may be when a cliff (face) is being cut into a steep hill with poor access to the crest, the rock has a vertical cleavage so that it will break away without dangerous overhang when the bottom is blasted out, and is brittle enough to shatter when it falls. Such an operation may be very economical of powder, as the breaking out of the crest and fragmentation upon hitting the floor are accomplished partly by gravity.

Coyote Holes. Coyote holes, illustrated in Figure 9-8, are used for heavy undermining blasts. They may be used alone to topple a cliff, or to break out heavy toe burdens in conjunction with well drill holes loaded from the top. Faces should be at least 60 to 80 feet high to justify their use.

Coyote holes are used when material is difficult to drill, a large yardage is required at one time, and when the rock fractures readily.

They consist of small tunnels, three or four feet in diameter, driven horizontally into the face at floor level, and one or two cross tunnels parallel to the face. Explosives are stacked in the cross tunnels and the entrance is securely blocked with stemming. Firing is best done by Primacord.

CONTROLLED BLASTING

Overbreak. Rock usually tends to overbreak at the top of the bank, and special drilling or loading may be required to avoid leaving a hard bottom rib at each blast junction.

It may be necessary to set back the first row of holes for the next blast for more than the normal burden, as their burden may be partly shattered, so that the shot is not confined as well as in the other holes.

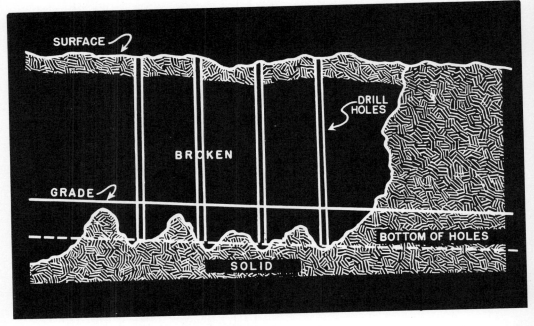

Fig. 9-9. Breakage lines

Drills ....y not be able to work close to the edge, or may not be able to penetrate shattered material.

In such conditions, the front row may not pull the bottom, unless it is drilled considerably deeper than the others, some horizontal bottom snake holes are drilled, or a denser or faster explosive is used.

However, shovels are frequently able to scrape away a few feet of unblasted soft rock, without excessive wear, in which case ribs may be ignored in the blasting pattern and dug out when found.

Edges. In large-area blasting, control of overbreak and underbreak is mostly a matter of getting most efficient use from equipment and explosives.

In road cuts, and other work where re-

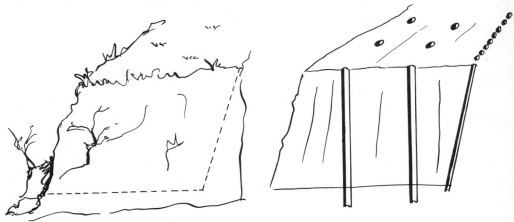

Fig. 9-10A. Preparing for a cushioned blast

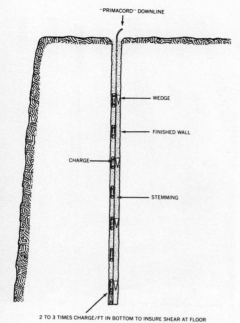

"PRIMACORD" DOWNLINE

WEDGE

FINISHED WALL

CHARGE

STEMMING

2 TO 3 TIMES CHARGE/FT IN BOTTOM TO INSURE SHEAR AT FLOOR

Courtesy of E. I. du Pont de Nemours & Co. (Inc.)

Fig. 9-10B. Charge placement for cushion blasting

maining rock is a permanent part of the finished work, accurate finishing may be required by the contract, either directly as a specification, or indirectly in the form of non-payment or penalties for excess removal.

In most rock formations, good to precise faces may be cut by line drilling, cushion blasting, pre-shearing, and various combination techniques. However, selection and refinement of method for a particular formation may be difficult.

For success, the rock must be reasonably sound and cohesive, and capable of standing at the slope to which it is trimmed. In starting a new project, it is highly advisable to obtain advice from your explosive company's representative, and from local sources.

It is essential that the drills and the drillers be able to keep the holes in accurate alignment, both to obtain the desired edge, and because the holes often show in the completed work.

Line Drilling. In line drilling, a single row of holes, usually 3-inch or less, is drilled along the edge, either from top to bottom, or in benches. Spacing is very close, from 2 to 4 hole diameters.

These holes are left empty. They create a line of weakness along which the rock should be sheared by the preliminary blast.

The back row of primary holes may be closer to the line than the normal between-row interval, and be more closely spaced and lightly loaded than other rows.

This method is best suited to formations with a minimum of bedding planes, joints, and other weaknesses which may affect breakage more than the line of holes. Weak-bedded rock may respond satisfactorily if its planes are nearly at right angles with the proposed slope, or exactly parallel to it.

Cushion Blasting. Drilling for cushion blasting is similar to that for line drilling, except that larger holes and wider spacings may be used.

The holes are loaded with light charges, with undersize and/or spaced cartridges, often strung out on detonating cord. All space not occupied by explosive is usually

Courtesy of Chicago Pneumatic Tool Company

Fig. 9-10C. Angle drilling

filled with stemming, although some blasters prefer to stem only the top.

Closer drilling or heavier loading, or both, are needed in cutting angles in the face. Extra unloaded relief holes may be put between the loaded ones.

Standard procedure is to blast and usually to excavate the main cut back to within a few feet of the final line, either before or after drilling the cushion holes, and fire this final row separately. Detonating cord or instantaneous caps are normally used. If shock and noise are problems, close-interval MS delay caps may be substituted.

Cushion blasting is more economical and accurate than line drilling because of the smaller number of larger holes, and greater depth can be obtained. The method is less sensitive to the quality of the rock.

Pre-Shearing. Pre-shearing resembles cushion blasting in drilling and loading, but firing is in advance of the main blast. Since the explosive force has nowhere to go (except up) it may be expected to produce finer and more even fragmentation between the holes, and a totally effective shear plane for the primary blast. The two blasts may be fired together, using delay caps.

EXPLOSIVES

General Properties. Explosives are chemical compounds that can decompose quickly and violently. The original solid or liquid chemicals are largely changed into gases, including steam, that have a much greater volume. Heat is generated by the change, and serves to expand the gases greatly.

Explosion by rapid burning is called deflagration, and by almost instantaneous decomposition is called detonation. High explosives detonate.

Properties to be considered in selecting an explosive include sensitivity, density, strength, velocity, water resistance, fumes, price, and availability.

Sensitivity is a measure of the ease with which a substance can be caused to explode

Courtesy of E. I. du Pont de Nemours & Co. (Inc.)

Fig. 9-10D. Cushion blasting result

and its capacity to maintain an explosion through the length of a borehole. It is also a measure of safety—the higher the sensitivity, the greater the risks of handling.

Nitroglycerin is so sensitive that it must be mixed with other substances before it can be used in commercial explosives. Compounds such as fulminate of mercury and lead azide that are used in detonators are so sensitive they will explode at a light hammer blow or moderate heat. At the other extreme, ammonium nitrate is so insensitive that few precautions and no permits are required for shipping and storing it.

Density is the volume of an explosive in proportion to its weight. It is measured in terms of pounds per cubic foot, or in number of 1¼ x 8 inch cartridges in a 50 pound case for wrapped dynamites. Such a cartridge contains about 9.78 cubic inches. A count of 100 sticks to the case is roughly equivalent to a density of 77 pounds to the cubic foot. A cubic foot of water weighs 62.4 pounds.

Strength is the energy content of an explosive in relation to its weight. In general, maximum explosive power can be obtained from a given borehole by using a high density, high strength explosive in it.

Velocity is a measure expressed in feet per second of the speed at which the burning or the detonation wave travels through an explosive. It varies from 1,000 to 3,000 feet per second for black powders to 23,000 feet per second for blasting gelatin.

Low velocity explosion has a heaving and separating effect, while high velocity crushes and shatters.

Water resistance is an important factor in wet rock, and varies not only with the character of the explosive, but the manner in which it is packed and wrapped. Manufacturers are increasingly able to put water resistance in the explosive rather than in the wrapper.

The gases resulting from explosions vary in toxic and irritating qualities. This is very important in underground work, particularly if ventilation is poor. Explosives are rated by the manufacturers according to fumes, as excellent, good, fair, and poor.

Explosives vary widely in the length of time they can be kept under various conditions before deterioration makes them dangerous or useless. Dynamite was formerly damaged by freezing, but this difficulty has been entirely overcome. Spoiling may be a serious factor if use is subject to delay, particularly in hot wet weather.

Different explosives vary widely in price. The most economical one for a certain use may be one with a high cost. It is important to select an explosive on the basis of all the related factors, rather than purchase cost alone.

In many areas, very few types of explosive may be available, and because of the complications of shipping, delivery of special orders may be delayed weeks or months. Under such conditions, use of the standard dynamite may be advisable even if it is not exactly suited to the job.

If special explosives are purchased from a contractor or a quarry, it may be necessary to handle the transaction through a dealer in explosives to comply with state laws.

Permissible dynamites are those approved by the U. S. Bureau of Mines for use in gassy and dusty coal mines. Their most important feature is minimum flame in the explosion.

Black Powder. The explosives which explode by burning, are called "low explosives." Black powder is the only commercially important member of this class, and is the oldest explosive known.

Black powder is ordinarily composed of sodium nitrate, sulfur, and charcoal, finely ground and combined in grains of various sizes. The grains are then coated with graphite or other glazing to make it free-running. A more expensive powder for special purposes uses potassium nitrate instead of the sodium compound.

Fine grain powders burn and explode faster than coarse grains. Somewhat more powder can be packed in a borehole by mixing two or more grain sizes.

Black powder can also be obtained in pellets, which are short cylinders of compressed powder with a center hole. They are wrapped into eight inch cartridges resembling dynamite in appearance. They are more convenient to use in small boreholes than the loose powder, and are somewhat safer to handle.

Black powder may be ignited or exploded by flame, heat, sparks, or concussion, and requires more careful handling than most dynamites. A special hazard is that powder spilled on the ground or on the magazine floor may ignite if stepped on or scuffed.

The blasting action will depend on the degree of confinement, the bulk, the grain size, and the closeness of packing. Unconfined powder will flash burn, without explosion; and poorly confined powder will waste much of its energy along the path of least resistance.

Black powder produces considerable smoke, and quite toxic fumes the quantities of which vary considerably in different blasting procedures.

Black powder can be used to advantage when large, firm pieces of rock are desired, or when the material being blasted is soft and resilient enough to absorb the shattering blow of high explosives.

It cannot be used underground where ventilation is poor, or where the air may contain inflammable gas which may be ignited by the flame from the powder, or in wet holes. It has been replaced by other explosives in most applications.

Dynamite. Dynamite is the best known and one of the most widely used commercial high explosives. The name includes several different chemical groups, wrapped and marketed in about the same manner as each other. Only the most general distinctions may be made among them, as research is steadily widening their applications.

The "straight" dynamites consist primarily of a mixture of nitroglycerin, sodium nitrate, and combustible absorbents such as wood pulp, wrapped in strong paper to make a cylindrical cartridge. Although a wide variety of sizes is available, the most popular are eight inches long and one and one eighth or one and one quarter inches in diameter.

The percentage of nitroglycerin by weight contained in the mixture is used to identify it, according to strength. From 15 to 60 percent may be used.

Strength does not increase in proportion to the percentage of nitroglycerin because the other ingredients also contribute gas and heat. For example, a sixty percent dynamite is about one and a half times as strong as a twenty percent.

Higher percentages are faster and more sensitive. Speed is desirable in hard rock and where the explosive is not confined, as in mud capping boulders. Sensitivity is necessary when blasting mud ditches by the propagation method.

Straight dynamites have fair water resistance. Their fumes are poor, however, and they are never recommended for underground work.

Any type of dynamite of the general purpose 40% strength will explode if subjected to sharp concussion, such as explosion of a blasting cap; from impact of a rifle bullet; from excessive heat, whether produced by fire, friction or impact, and from sparks.

When dynamite is burned—usually to destroy surplus or deteriorated stock—it is spread in a thin layer on straw or other combustible material, which is ignited. All personnel should keep at a safe distance from the fire. Dynamite will usually burn without incident, but there is always a chance that it may explode.

Spoiled dynamite may soak into its containers, and render them explosive. The cases and wrappings should therefore be burned with the same precautions as would be taken with dynamite.

The "ammonia" dynamites use ammonium nitrate as the principal explosive, in combination with some nitroglycerin. They do not catch fire as easily as straight dynamites, and are less sensitive to shock and friction. Water resistance is generally inferior, but fumes are less objectionable.

These are rated on a percentage strength basis, but the figures do not indicate anything of their chemical composition, but simply that performance is comparable to that of a straight dynamite of the same rating.

A third type of explosive used in commercial blasting is gelatin dynamite. This is based upon a jelly made by dissolving nitrocotton in nitroglycerin. Various other ingredients are added.

The gelatin dynamites are dense, plastic, cohesive, and practically waterproof. Fumes are excellent in all but the highest strengths, which vary up to ninety percent. A 100% gelatin is produced, but is not used in construction or mining.

Shot with a standard cap, and when not confined, ordinary gelatin dynamites will explode at a velocity of about 5,000 ft. per second. If confined, or shot with a straight dynamite primer, velocities of 13,000 to 22,000, depending on the strength, will be obtained. Certain types may also be obtained which will always detonate at the higher velocity.

Gelatin dynamites are relatively insensitive to shock, and often will not explode by propagation from adjacent holes. Their plasticity makes it easy to load them solidly in boreholes, and to pack tightly in cracks for mudcapping. The velocity of the higher grades and their high density recommend them for hard, tight blasting; and the waterproof qualities for any underwater work not requiring propagation.

Ditching, stumping, and agricultural dynamites are usually one of the standard strengths best suited for the purpose, with a special designation.

Blasts are initiated or set off by timing devices, by remote electrical controls, or by a combination of these methods.

AMMONIUM NITRATE

Ammonium nitrate (NH_4NO_3), called AN for short, is a nitrogen fertilizer that has largely replaced dynamite in medium and large borehole blasting.

At ordinary temperatures AN is generally a stable compound. If heated to 300 to 400 degrees Fahrenheit it will decompose without exploding into water and nitrous oxide, a brownish red gas with a pungent smell.

If subjected to great heat under confinement, or direct detonation of high explosives, AN decomposes explosively into water, nitrogen, and oxygen. The formula is:

$$2NH_4NO_3 = 2N_2 + 4H_2O + O_2$$

Ammonium nitrate is so insensitive that it is not rated as an explosive. It is called a blasting agent, and it can be shipped and stored free of special regulations and permits. However, it does burn, and since it supplies its own oxygen it cannot be smothered. The only way to put it out is to use plenty of water.

If burning AN is confined in rooms, or if it is piled in such bulk (over 100 tons to a pile) that heat and pressure can build up inside it, it may explode as destructively as dynamite.

AN should never be transported or stored with high explosives.

Additives. The sensitivity and explosive power of AN are greatly increased by mixing or blending with organic materials that absorb oxygen when they burn. Several disastrous explosions have occurred in AN fertilizer that was treated with a small amount of wax to prevent sticking.

For use as an explosive, additives such as lampblack, powdered coal, sawdust, and fuel oil have been tried. Fuel oil is very successful, of which more will be said below. Any of these provide material to combine with the surplus oxygen to produce additional heat and gas volume.

Prills. Most AN is prepared in prill form. Prills, sometimes mis-called pellets, are globular, porous particles obtained by spraying a 95 per cent solution of AN into a rising current of warm, dry air. They are usually coated with about 3 per cent by weight of kieselguhr (diatomaceous earth, about 80 per cent finely divided silica, called "guhr" for short) to prevent sticking. There are also uncoated prills, and prills coated with minute amounts of other materials. Prills vary in density, size, and size mixtures according to brand and specification.

Other processes produce flakes, grains, or pellets that are usually denser. These have not been as successful in direct blasting as the prills up until now.

History. AN has been used as an ingredient in dynamite and other explosives since 1867. In 1935 a canned AN mixed with a small quantity of fuel was introduced by Du Pont under the name "Nitramon." Because of its low sensitivity it could not be detonated by caps, and special primers of amatol, a TNT and AN compound, were provided.

Nitramon and similar products are safe and clean to handle, economical, excellent in wet holes and coyote holes, but are now declining in use because of competition of fertilizer grade AN.

Nitrex, a canned AN, was used in a 3,300,000 pound blast to remove Ripple Rock above Vancouver.

Fuel Oil. In 1955 it was discovered that #2 fuel oil is an almost ideal material to mix with AN to make a practical explosive. The mixture is called AN-FO for convenience.

Maximum explosive power is obtained when the mixture contains about 94 per cent AN and 6 per cent fuel oil. This is about a gallon of oil to a 100 pound bag of fertilizer. This is also the proportion of oil that the AN absorbs most readily.

Mixing may be done at the factory, at a fixed plant near the job, by mobile equipment or by hand at the borehole, or in the borehole. Relative cost depends largely on the size of the job.

Factory mix is the most thorough and the most expensive, and is the least trouble on the job. The mixed AN-FO, called nitro-carbonitrate by the ICC, is more dangerous to handle than the unmixed AN. Various mild precautions must be taken, and vehicles carrying it must be marked "DANGEROUS."

AN-FO should not be handled or transported in the original AN bags unless the change in contents is plainly marked.

The danger in handling AN-FO is increased by the fact that manufacturers are experimenting with various coatings, some of which make it much more sensitive.

AN-FO may be made dangerously sensitive, or even caused to explode spontaneously after long standing in stemmed boreholes, by contamination with unidentified, naturally occurring chemicals.

Local plants should mix as well as the factory, and mobile equipment almost as well. Hand mixing is much less effective, and mixing in the borehole is rather hit-or-miss. A thorough blending is important for highest blasting efficiency, but it may cost more than it is worth.

Poor mix, or poor detonation for other reasons, is likely to be indicated by "Kodachrome clouds" of yellow, orange, or brown smoke from the explosion. The color is caused by nitrous oxide mixing with steam and other gases.

Borehole mixing is done by pouring in a bag or two of AN, adding the correct quantity of fuel oil, then putting in more AN and more oil alternately until the full charge is placed.

Hand mixing outside the hole may be done by pouring 3 quarts to a gallon of fuel oil into each 80 pound bag of the fertilizer, or 2 to 2½ quarts into each 50 pound bag, moving the bag around to help the liquid to distribute itself evenly, and allowing it to stand for 5 to 20 minutes before pouring it into the hole.

Somewhat better results may be obtained by putting the AN in a mixing trough such as is used to hand mix concrete or plaster, spraying the fuel oil over it, and mixing with a hoe or shovel.

Unopened bags may have fuel oil injected into them under high pressure by the use of an engine driven pump and a sharp-pointed nozzle that can be pushed through the wrappings. This method, pioneered by Monsanto Chemical Company, is called the Needle Fuel Injector System. The apparatus costs from $300 to $400, and it is said that an experienced man can treat eight tons of piled bags in an hour.

Mechanical mixing may be done with

engine driven cement mixers. Oil should be added as a spray rather than a solid stream.

A great deal of specialized equipment has been designed, both in the factory and in the field, for mixing quantities of prills with fuel oil, and discharging them rapidly into boreholes. Augers may move prills from a truck body into a stream of compressed air, in which they are mixed with oil and blown into the borehole through flexible hose.

Loading prills by compressed air may build up charges of static electricity in the borehole and its vicinity. Such charges create danger of premature discharge of electric caps, and even possible hazards with fuse caps. Minimum precautions are grounding of the pneumatic loader and use of semi-conducting hose. Caps should be of types least sensitive to stray currents.

This problem, and a number of important precautions, are discussed in a U.S. Bureau of Mines Report, IC 8179, titled "Safety Recommendations for Sensitized Ammonium Nitrate Blasting Agents" (196.3).

Priming. Most AN-FO cannot be detonated dependably by blasting caps or regular Primacord. Even when detonated it may not maintain full speed of explosion, or even any explosion, for the length of the hole.

It is therefore necessary to use primers or boosters that can be exploded by caps or Primacord, and that will produce sufficient explosion to detonate the AN-FO at high velocity; and to use enough of them in the hole so that the explosion wave will not have enough space between them to weaken.

Such primers may be made of one or more sticks of gelatin dynamite of 60 per cent or higher strengths, or of special "cast" explosives such as Pento-Mex or Procore. The cast boosters are somewhat more expensive than dynamite, but are safer to handle and are more powerful.

Figure 9-3A shows a cross section of a Procore booster. The sensitive explosive that is detonated by Primacord or a cap is completely surrounded and protected by another high explosive that cannot be detonated by a cap or any probable accident, but can be detonated by the core, and has sufficient strength to set off any AN-FO mixture.

Borehole diameter is the most important single factor affecting propagation of the explosion, as this dies out much more quickly in small holes than in large ones. Although AN-FO has been successfully used in holes as small as 2½ inches, variable results and cost of extra boosters limits its use in holes smaller than 4 inches. This minimum diameter may be expected to get smaller as research continues.

Poor confinement may also cause the detonation to slow or stop. A soft rock or mud seam between layers of hard rock may not confine the AN-FO sufficiently for it to carry the explosion across.

In general, there should be a primer at the bottom and at the top of the hole, and one at least every 20 feet. However, 50 grain Primacord and a single primer at the bottom may prove adequate.

Loading. Dry holes are loaded by placing the primers with the detonating cord or wire, then pouring the mixture or the two ingredients separately until the proper amount is placed. Stemming is then added, as described later under LOADING.

There is a tendency to load holes higher with AN-FO than with dynamite because

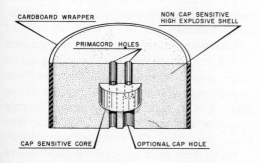

CARDBOARD WRAPPER   NON CAP SENSITIVE HIGH EXPLOSIVE SHELL
PRIMACORD HOLES
CAP SENSITIVE CORE   OPTIONAL CAP HOLE

Courtesy of Intermountain Research and Engineering Co.

Fig. 9-11. Procore booster

EXPLOSIVE	94/6 AN / FO	SLURRY BLASTING AGENT
POWDER FACTOR	1.5 lb./yd.	1.5 lb./yd.
SPACING	15 ft. x 15 ft.	21 ft. x 23 ft.
HOLE SIZE	10 in. x 40 ft.	10 in. x 40 ft.
POWDER COLUMN	18.5 ft.	22.8 ft.
POWDER COST / yd rock	S .0675	S .21
DRILLING COST / yd. rock	S .552	S .257
BROKEN ROCK COST / yd.	S .6195	S .467

Courtesy of H. E. Farnham

Fig. 9-12A. Drilling and slurry cost

of its lower cost, and to use less stemming. Overdoing this is likely to be a waste of the explosive, and to add to the hazard of high flying rocks.

Unprotected AN-FO cannot be expected to perform dependably in wet holes. It can be protected by putting in plastic bags. It may be bought this way, or the bags made up from sheet plastic with the help of a sealing machine.

Bagged AN-FO tends to float on any water in the hole, and must be forced down by the weight of explosive above it. Even with firm tamping, the bags will reduce the amount that can be placed because they will not conform perfectly to the walls. Bags may tear and allow water to ruin their contents, perhaps cutting off part of the blast.

Density. The density of AN ranges from 47 pounds per cubic foot for the prilled variety to 64 pounds for fine grained types. This compares with dynamite densities of 37 to 90 pounds.

In loading calculations the low density of prilled AN-FO compared to heavier dynamites is offset to a variable degree by complete filling of all hole space by the free running material. That is, if a dynamite with a 60 pound density filled only 80 per cent of the bore space in spite of slitting and tamping the cartridges, then its effective density in the hole would be .8 x 60, or 48 pounds per cubic foot, about the same as the prills.

The finer and denser types of ammonium

nitrate are not well suited for use as blasting agents with present techniques. They are more sensitive and less powerful than the prills, and are more difficult to mix with fuel oil. Some increase in density of prills may be obtained by mixing two or more sizes, but the heavier charge may still be offset by slower detonation.

SLURRY (WATER GEL)

Composition. Slurries, also known as water gels or dense blasting agents (DBA), are usually mixtures of a "sensitizer", an oxidizer, water, and thickener.

The sensitizer may be any of a number of reducing (oxygen-hungry) chemicals. This is usually the explosive TNT (trinitrotoluene), but may be (or include) finely divided aluminum, and/or other substances that may or may not be explosives themselves. They may be in sand size granules, very fine powder, or in other forms.

The oxidizer is an oxygen-rich chemical such as ammonium nitrate (AN) and/or sodium nitrate.

Characteristics. Consistency is regulated by the amount of thickener or jelling agent (often guar gum) that is used. It varies from that of pancake syrup to soft (flows if jiggled) jelly at room temperature. Stiffening occurs at low temperatures, but most formulas are resistant to damage by freezing.

Water resistance varies from good to excellent, unless water flow is sufficient to wash it away. Loading water gel in its

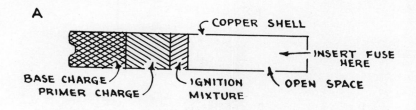

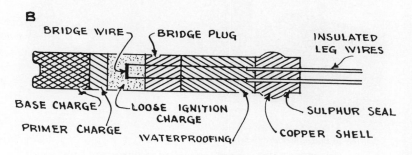

Fig. 9-12B. Detonating caps

sealed plastic packages is recommended for severe water conditions.

Packaged slurries are generally jelled in cylindrical shapes slightly smaller in diameter than the boreholes in which they are to be used. They are shipped in polyethylene bags, protected by burlap or by cardboard containers.

The bags are usually soft enough to allow slump to fill the borehole almost completely, and can be obtained in even softer, expandable types. However, it is normal to slit the top six inches of each bag, and drop it into the hole. Some water gels have unique gellant systems which are designed to permit them to be poured into the hole.

It doesn't mix readily with water, and its weight (specific gravity is 1.1 to 1.5) causes it to flow to the bottom, displacing any water and filling all borehole space. But if it is very thick, it may bridge over and retain water pockets.

For large scale use, slurries may be shipped and even mixed in tank trucks and pumped into holes. If the hose can be extended to the bottom of the hole, danger of bridging is eliminated.

Most slurries are insensitive, and have the same "B" rating as AN-FO for shipping and storage. They cannot be detonated by blasting caps or Primacord, and require special booster-primers.

Recently developed small and intermediate diameter grades of water gel are cap sensitive, so that they require shipment and storage as a class A explosive. This type is finding wide acceptance as a replacement for dynamite in bottom loads.

Applications. Slurries are less expensive than dynamite, but cost more than ammonium nitrate fuel oil mixtures (AN-FO). They are now used chiefly in open pit mines where rock is hard and/or holes are wet.

The high density of slurry (1.37 to 1.68) compared with AN-FO permits use of smaller diameter boreholes, or wider spacing, to obtain the same explosive power and fragmentation. The higher price of the slurry may be more than offset by reduced drilling cost. Figure 9-12A shows the results of a cost study made of a very hard rock formation by the Iron Ore Company of Canada.

In wet holes AN-FO is not practical, so the choice is between dynamites and slurries. Here the slurries have a smaller advantage in density, but have a price advantage as well.

Courtesy of Ensign-Bickford Company

Fig. 9-13A. Primacord is supplied in boxes and rolls

Some slurries have been employed successfully in boreholes down to one and one half inch diameter, using a special small diameter booster, Procore-1C.

INITIATORS

Safety Fuse. The original timing device is a fuse made up of a black powder core, surrounded by layers of protective wrappings. Two speed ranges are available, with burning speeds of ninety seconds and one hundred twenty seconds to the yard. These speeds must be considered approximate, as they are affected by altitude, weather, storage conditions, and possible damage to the fuse.

Fuse is water resistant except at the cut ends, which are immediately spoiled by contact with moisture. It should not be used unless it can be shot the same day it is loaded.

It is manufactured in lengths of fifty feet or more, and wound in coils or on spools to be cut to the desired length on the job. As short an interval as possible should be allowed between cutting and using.

Squibs. Electric squibs are devices for igniting charges of black powder, and may be used instead of fuses. They consist of copper or aluminum tubes with powder,

an electric firing element, and wires sealed into them. They are imbedded in the powder charges, and when sufficient current passes through them, they take fire and ignite the charge.

Blasting Caps. Figure 9-12B shows construction of blasting caps for both fuse and electric firing. Fuses are inserted in the hollow shell of the cap, and fastened in by crimping the metal with special tools or machines. The fuse should be cut off square, preferably by a razor blade or other very sharp edge, which will not pinch the wrappings together.

The electric firing device consists of a very thin wire lying in a highly combustible mixture. Passage of electricity causes the wire to become white hot and ignite the mixture, which explodes the primer and high explosive.

Delay firing may be obtained by use of delay electric blasting caps which have a slow burning composition between the ignition charge and the primer charge. Time interval averages about one second in "standard" caps, and 0.025 to 0.1 seconds in the millisecond type.

Wires may be obtained in almost any desired length, and should be long enough to reach the wires from adjacent charges, or to connect with the lead wire to the electrical source. They may be copper or iron, and are protected by a plastic insulation. The ends of the wires are fastened together into a bridge or shunt, to prevent premature firing through contact with stray electric currents.

Instantaneous and delay caps may be used on the same round. If they are in the same series, it is well to increase firing current by ⅓. Caps made by different manufacturers should never be fired together. They are almost certain to have different current requirements, so that one brand of cap will fire and break the circuit before the bridge wires in the other brand are heated enough to fire.

Caps contain explosives which are more sensitive than dynamite, and they must be handled with great care. Heat, friction, and shock must be avoided. In the original package, electric caps are usually cushioned in their own folded wires, and are often protected by a cardboard or paper wrapper in addition.

But the greatest danger of accidental firing of electrical caps comes from electricity. Even with their wires shorted by a soldered shunt in the original package, a very powerful nearby current might detonate them unexpectedly.

After the shunt is opened to connect to other wires, there is danger from any stray electrical current, even from radios.

The content of a cap is small, but one can blow off fingers and toes, and flying particles of the copper case may cause injury to personnel within a radius of thirty feet. The most serious danger in an accidental explosion of caps, however, is that of setting off primers or nearby heavy explosives.

Caps should be buried before exploding for test purposes.

Primacord. Primacord is a detonating (exploding) fuse, made up of a core of an insensitive high explosive, pentaerythrite-tetranitrate, that is called PETN for short, surrounded by a protective wrapping. Primacord is detonated by means of a blasting cap. The explosion travels along it at a rate of about 21,000 feet a second, and detonates any cap-sensitive explosives with which it is in contact.

It is produced in a number of types that are classified according to explosive content in grains per foot, and/or the type of protective covering around the explosive core.

Standard types usually have 50 to 60 grains per foot. Lighter grades, including "E-Cord" with 25 grains, are used chiefly for secondary and very shallow blasting. Heavier ones, with 100 to 400 grains, are used for continuous column initiation of ammonium nitrate fuel oil mixtures, or for cutting into short lengths for use as primers.

Wrappings are rated on the basis of strength and water resistance. All are water resistant, but plastic coated ones are essentially waterproof except at cut ends. The plastic is also resistant to oil, an important point when using an AN-FO mixture.

E-Cord, and Primacord with 45 or more grains will fire when wet with water or oil if it is initiated from a dry spot. Wet Primacord can be initiated only with a very powerful primer, such as 80 per cent gelatin dynamite or a special booster. It will not maintain detonation through a knot connection.

Cut ends of Primacord will pick up some moisture from capillary attraction, usually only a few inches in. If the cord is lower than the water level it may in time become soaked all the way through.

Textile reinforced Primacords are usually used in ordinary down holes, wire reinforcement in rough and jagged ones.

The individual lines in the holes, called branch or down lines, are usually connected to the blasting cap by a trunk line of Primacord. Fastening is done by simple knot connections. It is important that the lines be at right angles to each other. When very

Fig. 13-D—Trunkline spliced with square knot.

Fig. 13-E—Double wrap half-hitch knot connection between a reinforced trunkline and downline.

Fig. 13-F—Trunkline clove hitched over a downline.

Fig. 13-G—Special clove hitched trunkline to prevent slippage from subsidence of the charge.

Courtesy of E. I. du Pont de Nemours & Co. (Inc.)

Fig. 9-13B. Primacord connections

stiff cord is used it may be necessary or convenient to use plastic connectors instead of knots.

DuPont MS Connectors may be inserted in the trunkline for delay firing. They have four intervals, from 0.005 to 0.025 seconds. Two or more can be combined in series. They are non-electric, with delays similar to those in caps.

Electric caps may be used to combine quiet surface wiring with Primacord downlines. Connection is made by taping a cap (or preferably two caps) tightly to the Primacord. This may be a short piece, called a tail, which is tied to the downline with a square knot just before firing. The caps are wired into a conventional blasting circuit.

There are also non-electric delay caps, Primadets, that are made up attached to a light detonating cord, Primaline. They are strong enough to explode ANFO and most slurries directly, without a primer.

The cap is placed near the bottom of the hole, with or without a primer, before pouring in the main charge of explosive. The Primaline is knotted to the Primacord trunkline. It does not have sufficient strength to detonate ANFO or slurry, so timing is regulated by the delay feature in the cap. It is not used with dynamite or other explosives that are sensitive enough to be set off by the cord.

On all large blasts it is customary to arrange Primacord so that each hole may be reached by the explosion from two directions. Large blast holes, or holes of any size with deck loading, may have two cords strung on opposite sides of the hole to insure firing.

Primacord has become the standard method of setting off large blasts, because of its exceptional safety. As an explosive it is quite inert, and is less likely to be detonated by accident than the main charge of explosives. Particularly, it cannot be exploded by stray electrical currents, a serious hazard for electrical hookups. An entire blast can be prepared with detonating cord, and a fuse or electric cap attached at the last minute.

Primacord is manufactured by the Ensign-Bickford Company, Simsbury, Conn. Upon request, they will send a booklet containing much more detailed information than can be included in these pages.

Low Energy Detonating Cord. The noise made by the explosion of Primacord trunk lines above ground may be objectionable. This noise can be considerably reduced by covering it to depths from a few inches up to a foot with sand, dirt, drill cuttings, or other stone-free material, or almost eliminated by using low energy cord.

LEDC, a low energy cord that is relatively noise-free, has been developed jointly by the Ensign Bickford Company and E. I. Du Pont de Nemours Co. It contains two grains of PETN per foot, protected in a tiny lead tube covered by wrapping of cotton and plastic. The noise made by exploding 150 feet of it is equal to that caused by one blasting cap or two inches of reinforced Primacord.

Special connectors are available to fasten LEDC to Primacord and to itself. When the explosion travels from LEDC to LEDC or to Primacord a booster charge is needed. This may include a 0, 10, 15, or 25 millisecond delay unit.

The preferred method of firing is to fasten an electric blasting cap to one of the down lines of Primacord, and to fasten the LEDC to this down line with a non-explosive connector called a trunk line adapter. The LEDC is connected to other down lines by booster delay caps.

PRIMING

Primers. A primer is a stick of dynamite that contains a blasting cap; or is any other heavy explosive which has been fitted with a device for setting it off.

Since primers combine the power of the dynamite with most of the sensitivity of the cap, they must be handled with greater care than any other units of explosives.

They are ordinarily prepared at the borehole immediately before being placed, but may be made in some central place and delivered to the loaders as required.

The essentials of a good primer are that the cap must be powerful enough to produce detonation, there must be intimate contact between cap and explosive, they must be fastened together so that they will not separate while being placed, the cap should be shielded from shock or friction, and the wires or fuse should not be kinked or strained.

Black Powder. Black powder may be primed by placing a fuse in the hole and

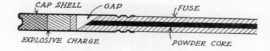

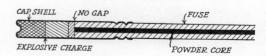

Courtesy of E. I. du Pont de Nemours & Company (Inc.)

Fig. 9-13E. Insertion of fuse in cap

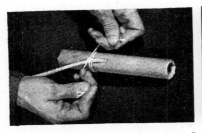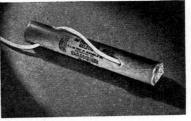

Courtesy of E. I. du Pont de Nemours & Co. (Inc.)

Fig. 9-14. Priming dynamite with fuse cap

pouring the powder around it. This method may be improved by tying a knot in the end of the fuse to anchor it, and making several slits into the core above the knot where they will be in contact with the powder. A paper cartridge may be prepared to hold powder closely around the knot and slits.

If the powder is to be exploded by an electric squib, a similar cartridge is made up to enclose the squib with some powder.

Blasting caps may also be used to explode black powder.

Fuse Caps. Preparation of dynamite and fuse cap primers includes two jobs—attaching the fuse to the cap and the cap to the powder. The fuse ends MUST be dry.

The fuse should be cut squarely with a

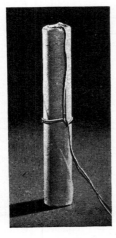

Courtesy of E. I. du Pont de Nemours & Co. (Inc.)

Fig. 9-15A. Priming small dynamite stick

clean, sharp blade, preferably a razor blade in a suitable mounting, and pushed into the cap until it is seated against the explosive compound. The copper shell is then crimped firmly onto the fuse with a hand or a bench crimper.

If the fuse is cut on a bevel, it may fail to make proper contact, Figure 9-13, top, or the end of the casing may curl over. and prevent contact. A good contact is shown at the bottom.

The interior of the cap should be clean. Any foreign matter in it should be tapped out or removed with a straw or toothpick. Blowing into it may dampen it and cause it to fail. If the cap is suspected of being damp from any cause, it should not be used.

Figure 9-14 shows two ways to place and fasten a fuse cap in dynamite. In each case a hole or slit is made in the cartridge and the cap inserted. The primer can be held together by lacing the fuse through another hole, or by tying it with string.

In shallow holes, and in blockholing or mudcapping, it is practical to simply insert the cap in the cartridge end, without fastening, as the primer need not go out of reach. Friction will hold it in place against a moderate pull, but not against yanking

Electric Cap Primer. Figure 9-15A illustrates the most common method of priming a small diameter stick of dynamite. The cap is pushed into the navel-like end of the wrapper and the wires caught in a half hitch around the center. If the dynamite is hard,

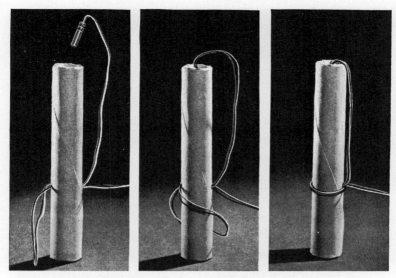

Courtesy of E. I. du Pont de Nemours & Company (Inc.)

Fig. 9-15B. Alternative method of priming small dynamite stick

a hole may be made in the end with a wood peg to make it easier to insert the cap securely so that it will not slip out during loading.

Fig. 9-15B shows steps in a method which involves placing the cap in the same manner, but passing a loop of the wires through a cross hole punched in the center of the stick. This eliminates a slight danger of damaging the wires in the half hitch, but is somewhat slower to make up.

Large diameter sticks are best primed by the sequence shown in Fig. 9-15C. The cap is inserted in the top, and a loop of wire

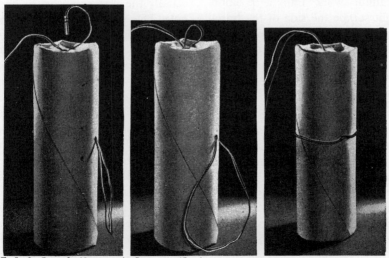

Courtesy of E. I. du Pont de Nemours & Company (Inc.)

Fig. 9-15C. Priming a large stick of dynamite

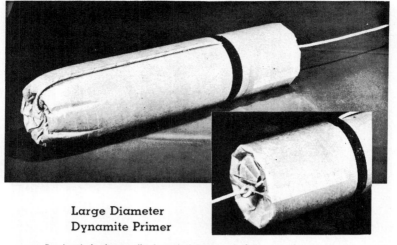

**Large Diameter
Dynamite Primer**

Punch a hole diagonally through the top end of the cartridge, tie the Primacord
on to the lowering tape allowing 8 to 12 inches plus the length of the cartridge
for Primacord length below the knot to complete the assembly. Thread this length
through the diagonal hole, then down the side of the cartridge and insert the end
of the Primacord into a hole punched in the bottom of the cartridge. For cartridge
reinforcement, wrap electrician's tape around the cartridge as shown.

Courtesy of Ensign-Bickford Co.

Fig. 9-15D. Primacord and a large dynamite stick

is pushed through a slanting hole and is caught around the middle.

In each case the cap is entirely inside the dynamite, cannot work into a position where it might scrape the side of the borehole, and its direction of explosion, away from the wires, is directed into the dynamite.

Primers for ANFO and slurries may be made up of a cartridge of high velocity gelatin dynamite and an initiator. But there are special cast or extruded primers made up of even faster explosives such as PETN, with a ready made hole for insertion of a cap or detonating cord.

The material is usually insensitive to almost any kind of rough handling, but can be detonated by high-strength caps or 50 grain cord, and will in turn set off ANFO and most slurries. Handling it does not cause headaches.

Detonating Cord. To make up a primer, Primacord is fastened to a large stick of dynamite by being threaded through and secured with tape and to a small one by being tied tightly. The resulting primer is usually the first explosive to be put in the hole. This puts the detonating cord in contact with the entire charge.

Placing Primers. If one primer is to be used in a borehole, it is best to place it at the top, or one cartridge down from the top. This keeps to a minimum the danger of damage to fuse or wires while placing and tamping the charge. Use of one stick above the primed one cushions the primer against jars from overzealous tamping, and from contact with abrasive stemming material.

If the hole is long, or the charge heavy, it may be considered a good precaution to use two or more primers. The second one is liable to be in the bottom and any additional ones spaced throughout the column.

With all types of delay firing the primer must be in or near the bottom. This largely avoids the danger of throwing the primer out in the muck pile, where it would be likely to detonate during digging, with a possibility of disastrous results.

Spaced charges are more likely to require additional priming than solid ones. Deck charges need a primer in each level.

Correct placing of primers, and even correct direction of the cap in the primer, are of importance under some circumstances and make little difference in others. Because of the speed and destructiveness of blasts, exact analysis of the mechanism and results are difficult.

Speed of Explosion. There are four classes of speed concerned in the firing of explosives. There is the slow burning of a fuse, explosive burning of confined black powder, extremely rapid but somewhat variable detonation of high explosives, and almost instantaneous passage of electricity.

Black powder is little used, but still serves to illustrate the effect of point of ignition on explosive performance in a blast hole. Its slow-explosive burning speed may be 1000 to 3000 feet per second, so that it would take $\frac{1}{5}$th to $\frac{1}{15}$th of a second for a 200 foot borehole to fire. If ignited at the top, the upper rock might be moved a considerable distance before the bottom fired. In one way, this would act to lighten the bottom burden and help it to pull, but it might also serve to "uncork" the borehole and allow the bottom of the charge to blow upward rather than horizontally. On the other hand, the force of the upper part of the explosion, reacting against a heavy burden, might press down on the unexploded charge with great force and seal it.

If the column were set off from the bottom by electricity, the toe would be well kicked out before the explosion reached the top. If several caps were placed at intervals in the column and fired together, the whole burden would be moved out at approximately the same time.

The same action is found in high explosives, although the rapid detonation makes

it of less effect. A two hundred foot column of a 40 percent dynamite with a velocity of 10,000 feet per second, fired from the top, would explode at the bottom $\frac{1}{50}$th of a second afterward. If Primacord were used to the full depth, the detonation would take only $\frac{1}{100}$th of a second. If electric caps were used at top and bottom, the time at the bottom would be about $\frac{1}{5,000,000}$th of a second after the top, but the lag to the center would be $\frac{1}{100}$th of a second.

At first glance, these small fractions of a second might not seem significant. In many cases they are not, but they sometimes have an important effect on both the performance and the concussion of a blast.

If long borehole blasts do not give the desired effect in fragmentation, throw, or

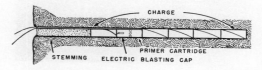

Courtesy of E. I. du Pont de Nemours & Company (Inc.)

Fig. 9-16. Loading with instantaneous cap primer

in any other way, it may be advisable to change the location, number, or type of primers to try for better results.

Cameras can be obtained that will take pictures of the various stages of the explosion for study.

Precautions. Precautions to be observed in regard to a primer in placing and leaving it in the hole include: placing it in such a manner that it is not subject to shock or jar, and is not penetrated by rock splinters or other sharp objects; that it is not to be wet for a longer time than the powder, the cap, the fuse, or the wiring can stand; that the fuse and wires lead to the top without kinks, and are held so as not to be damaged by placing and tamping of additional charges and stemming.

Courtesy of E. I. du Pont de Nemours & Company (Inc.)

Fig. 9-17. 700 ton blast recorded by camera

Water Resistance. This is the resistance of an explosive to penetration by moisture, and/or its ability to explode when wet or damp. It is important when holes are wet, or when there is a long time between loading and firing.

Gelatin dynamite and slurry are good to excellent. Other dynamites are good with intact wrappings, fair to poor if cartridges are torn or punctured.

Ammonium nitrate has very poor resistance, but can be kept dry in plastic bags.

Electric caps are waterproof. Fuse and Primacord are practically waterproof except at cut ends, where fuse has no resistance. Cut Primacord absorbs water slowly. Damp sections are difficult to detonate.

HANDLING EXPLOSIVES

Transporting. Large quantities of explosives should be transported in special vehicles marked in accordance with state laws. Smaller quantities may be carried in an ordinary car or truck, with any required warning signs made so that they can be removed when not in use.

Caps and explosives should be carried in different trips or vehicles, unless quantities are small, in which case they may be carried in one vehicle if kept well separated and if permitted by law.

ICC regulations are accepted by most states for intra-state transportation, but some have more restrictive laws.

Storage. Different classes of explosives should be kept in separate magazines. These should be far enough apart so that an explosion in one would not affect the other; and should be surrounded so far as possible by earth barricades or higher ground so that the force of an explosion would be deflected upward.

Magazine areas should be as far as practicable from roads, railroads, or structures, should be posted with warning signs and fenced if possible. A list of minimum distances will be found in the Appendix.

Magazines should be constructed of cohesive fire resistant material, such as sheet iron, or soft material such as brick which will tear or crush rather than separate into flying fragments. Ventilation and protection from grass fires and from excessive heat should be provided. Doors should be heavy and provided with strong locks.

Portable magazines to hold a few cases of powder or boxes of caps are most easily made from large metal tool or packing boxes fastened with padlocks. When properly marked, these are legal in most states, although many laws and regulations recommend more complicated units. When a portable magazine is to be left on a job, it should be chained and locked to a tree or other anchor.

Handling. Dynamite may cause severe headaches. This is especially apt to occur if it is unwrapped and handled with bare hands. Different brands and strengths differ in headache producing qualities, and individual reaction is highly variable.

Persons handling explosives should not smoke and preferably should not carry matches. A complete list of safety precautions recommended by the manufacturers will be found in each box of dynamite. A few basic rules may be emphasized here:

1. Do not expose any explosive to flame, heat, sparks or electrical current, shock or friction, and do not load or handle them during a thunderstorm.
2. Do not use iron tools.
3. Do not keep larger quantities of explosives on hand than are needed.
4. Keep caps and powder as widely separated and as thoroughly protected from each other as possible.
5. Keep records of explosives on the job, and those used in each shot, and make sure no unexploded material is left lying around the job, or concealed where children might find them.

A complete list of the "Don'ts" issued by the Institute of Makers of Explosives is included in the Appendix.

Boxes. Dynamite is usually packed in 50 or 60 pound boxes, although 25 pounds may be available. The wooden box, which was inseparable from blasting operations for many years, is no longer used.

The standard box is now made of fiberboard, with a full lift-up cover whose overlap provides double sides for the container. Sealing is by means of tape.

The box is lined by a polyethylene bag, which is an effective moisture and chemical barrier, and can be readily opened and reclosed in the field.

A small possibility of damage to both the dynamite and the liner, with resultant contamination of the fiberboard with explosive, is the reason that these boxes should not be burnt, except with the same precautions as with dynamite itself.

LOADING

Holes may be loaded in a number of ways. These may be classified as solid, string, spaced, deck, and spring.

Solid. In solid loading as much explosive is crammed into the hole as possible.

Free running explosive is poured into the hole, or blown in pneumatically from bulk

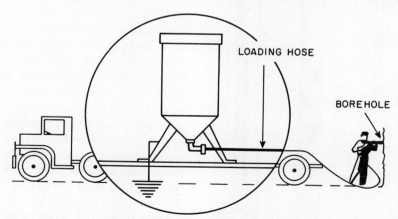

Fig. 15-E—Mounted Du Pont Airloaders should be grounded to the ground with cable or wire. DO NOT use chain.

Courtesy of E. I. du Pont de Nemours & Co. (Inc.)

Fig. 9-17A. Grounding a loading hopper

carriers, or from portable hoppers. The air tends to build up static electricity, so the unit must be grounded, and other precautions taken.

Water gels are usually loaded in their plastic containers, which may be intact or slit, but may be poured in. On large scale operations, it is economical to use a truck pumper, which may mix the ingredients also.

Cartridges smaller than the hole are slit up the sides in two to four places, so they will spread when tamped.

Slitting is better than unwrapping, because of reduced danger from spilled powder or of headache from skin contact, and the wrapper ends prevent powder from sticking to the tamper.

Some cartridged explosives have special perforated wrappers that do not need slitting, as they tear and unwind under heavy end pressure.

Tamping. Tamping is a process of com-

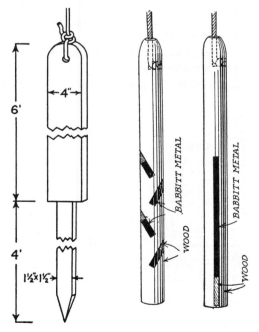

Fig. 9-17B. Tamping and cutting blocks

pacting explosives in boreholes by comparatively light blows, and/or pressure, of a stick or weight. This tool must not have exposed metal of any kind.

For best compaction, tamp each cartridge or layer separately with a firm, pressing stroke. Sharp blows are less effective, and should be avoided.

A tamping stick should be of round wood, with slightly smaller size than the smallest part of the hole, with a straight cut across the working end. If the hole is too deep for the use of a single tamping stick, several sticks should be drilled lengthwise and strung together with a cord. When the cord is slack, the stick will fold and can easily be fed into the hole. If any stick is held, and the cord tightened above it, the joints below the pull will be made rigid.

If the lower end of the stick wears to a taper, it should be cut back. The taper may punch holes in the tops of cartridges that would not be filled by pressure on the next one placed, it may grind some of the powder against the sides, and it may stick in cartridges and pick them up.

Large blast holes made by rotaries are usually tamped with a block on the end of a rope. The block should be of hard wood to resist abrasion, be slightly smaller than the bore of the hole, and have a flat end.

If weight is needed for heavy tamping, or working in wet holes that would float wood, the block may be drilled and weighted with lead or other heavy metal plugs, which should be covered with wood.

This type of block is not adapted to ramming down cartridges which have stuck in the hole above the bottom, as it may cause excessive side friction. A special block with a chisel point stake that will break up the stuck cartridges, is better.

These blocks are shown in Figure 9-10.

Deep Holes. It is a good plan to check a deep large bore hole before loading it, by inspecting it with a flashlight or sunlight reflected from a mirror, or sounding it with

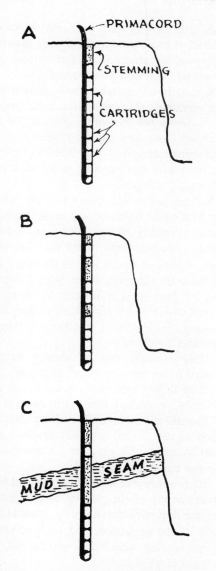

Fig. 9-17C. Loading deep holes

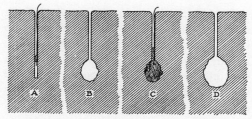

Courtesy of É. I. du Pont de Nemours & Company (Inc.)

Fig. 9-17D. Springing a hole

a tamping block, to make sure it is not obstructed. The block can be used to knock obstructing pieces or scale to the bottom.

Cartridges may be dropped into shallow holes or deep smooth ones. If the hole is deep and rough, and there is a possibility that they may stick part way down, they should be lowered. All explosives manufacturers will provide cartridges equipped with means to attach a lowering rope. A band of special Scotch Tape and a readily disengaged hook are used.

The impact of the cartridges on the bottom and the weight of the column above, frequently compress charges well enough so that tamping is not necessary.

If the hole is ragged or partially caved so that it is not practical to load it with cartridges, a free running explosive may be poured down it. If it blocks the hole and starts to build over an obstruction, it should be poked down with a long jointed pole or a dislodging block.

If such a hole is wet, it may be necessary to use a water-resistant slurry or water gel, either placed or poured.

A tamping block used for unwrapped dynamite should be kept clean by resting it on a box or some sacks when it is not in use.

String. If the borehole is wet enough so that slit or unwrapped dynamite would be spoiled, or if solid loading would make too heavy a charge, cartridges somewhat smaller than the borehole, but not small enough to fit side by side, may be dropped in one after the other without tamping, or after having tamped the bottom cartridge or two.

This is the easiest way to load, and is satisfactory for small or occasional blasts. However, it is inefficient. More rock must be drilled than is necessary to hold the charge that is used. Part of the strength of the explosive is wasted in the air cushion around it.

Spaced. Spacers may be used to string cartridges out along a hole that is not to be fully loaded. These may be square, round, or hollow pieces of wood, tile, lean concrete, or rolled cardboard. They are usually made up ahead of time, in lengths of eight to ten inches. There should be sufficient air space around them to allow for detonating cord or wires, without squeezing or rubbing.

Spacers may be alternated with cartridges or pairs of cartridges in the parts of the borehole that are not to be fully loaded. The primer cartridge should have at least one additional stick in contact with it.

Decking. In large boreholes, charges which are to be strung out are usually separated by solid plugs of sand or other stemming material, and each section of the charge primed separately, unless fired with Primacord or other detonating fuse.

Figure 9-11 illustrates well drill holes showing a solid column load (A), a deck load (B), designed to blast a heavy toe and relatively thin top burden, and a deck load (C) which avoids danger of blowing out a mud seam.

Springing. If the force of a blast is to be concentrated at the bottom or back of a hole, it may be necessary to make an enlarged chamber to hold extra explosive. This is done by exploding a small quantity of dynamite—one to six cartridges for a two inch hole—in the bottom. The hole may be left open, or lightly stemmed with dry sand, or with water. Quick acting dynamite is used. The charge must not be large enough to blow out the face.

Springing makes a cavity by crushing the surrounding rock and blowing it out of the hole. Two or more blasts of increasing size may be required to make a large enough chamber. See Figure 9-12.

The explosion creates considerable heat which is slow to dissipate. Unless the hole fills with water, several hours should elapse between a spring shot and further loading.

This time may be reduced, and the cavity cleared of loose sand and chips by lowering drill steel or air hose to the bottom and blowing it out with compressed air.

Sprung holes are most easily loaded with free running explosive that can be poured in. If this is not available, slit, cut, or unwrapped cartridges can be spread into the chamber by tamping. The chamber will not fill, even with free running powders, unless thoroughly tamped.

Sprung holes are not advisable where the burden is light, or the rock tends to fracture readily along joints or bedding planes. The springing blasts may loosen a slab of rock so that it will be thrown a long distance by the main blast. Such loosening can be kept to a minimum by using the fastest dynamite obtainable. Springing is becoming obsolete, as it is laborious, dangerous and inefficient.

Holes should not be sprung next to loaded holes because of propagation.

If springing blocks the hole, it can generally be re-opened with a drill or steel bar.

Stemming. Stemming is inert material such as dirt, sand, or finely crushed rock that is used to fill parts of a borehole that do not contain explosive.

Its primary use is in filling vertical holes from the top of the powder to the surface. Its use improves breakage by confining the force of the explosion, and adds to safety by preventing accidental igniting of the charge before it is fired.

Stemming is also used to space out charges as in Figure 9-11.

The minimum depth of top stemming should be about a foot in a 1½ inch hole, and 12 feet in a 12 inch hole. Deeper stemming is used where the top will be shattered simply by having its base blown out from under it. About 2/3 of a very high face may be broken in this way, with explosive in only the bottom third.

Carrying the load too high in a hole is at best a waste of powder, and at worst can mean excessive and damaging rock throw and noise. On the other hand, too little explosive and too much stemming will give poor top breakage.

Drill cuttings are often the best source of stemming, particularly from rotary drills or air drills equipped with dust collectors. It is important that stemming contain no sharp pieces or sizable stones that might cut wires or fly bullet fashion. Moist material usually is more effective than loose dry sand.

Mines and quarries that have rock crushing plants and that use large diameter holes often use fine crushed rock (screenings) for stemming. This may be hauled in Dumpcrete trucks that can dump directly in the hole through chutes. Ordinary dump trucks may leave a pile of screenings near each hole or group of holes, for distribution by hand shovel or wheelbarrow.

LIGHTING FUSES

The length of fuse determines the time which will elapse between lighting it, and the explosion. Regular sequences of firing can be obtained by varying the lengths of fuse in different holes, and lighting them at the same time.

Fuse does not light readily with a match because of the small area of powder exposed, and the likelihood of wax from the

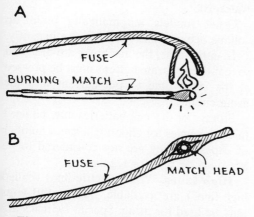

Fig. 9-18A. Lighting a fuse with a match

coverings being spread over the end while it is being cut.

If only a few fuses are to be lit, good results can be obtained by splitting each fuse with a knife or razor blade, as in Figure 9-18A, bending the fuse so the opening is down. It may then be lit with a match. Care should be taken to keep the fingers out of a line with the end of the fuse, as it will spit out a jet of flame.

The split fuse can be ignited more readily by having the opening horizontal or upward, placing a broken-off match head between the halves, and squeezing them together, as in (B). The match head gives a much hotter flame than the stick.

It is possible to buy a number of devices which simplify the lighting of fuses. The match lighter is a short paper tube which fits over the end of the un-slit fuse and is coated on one end with a compound similar to that of a safety match head. This is readily ignited with a match or the edge of a match box and subjects the fuse end to intense heat.

This lighter throws a jet of flame which resembles that caused by ignition of the fuse. To tell whether the fuse is actually burning it is necessary to observe a moment later whether a thin stream of smoke is issuing from it. If it is not smoking the lighter should be removed and another applied.

The pull wire lighter is of similar material, but clamps on the fuse and is ignited by pulling a wire.

The lead spitter is a coil of thin lead tubing containing black powder. A piece is cut off, lit with a match, and the resulting hot flame used to then light the fuses. The lead melts back as the powder burns.

The hot wire is similar to a fireworks sparkler. It burns slowly with a very hot flame, and is the safest and most dependable of the devices listed.

A burning fuse will light black powder by contact. When used to explode dynamite,

Fig. 13-A—Single row of holes connected by detonating cord and detonated by electric blasting caps or caps and fuse attached to detonating cord at A. Note detonating cord loop is tied to trunkline with right angle connection.

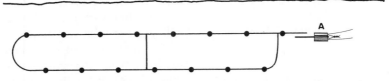

Fig. 13-B—Double row of holes connected by detonating cord. Note detonating cord from back line is tied to trunkline with right angle connection. Detonated by electric blasting caps or caps and fuse attached at point A.

Courtesy of E. I. du Pont de Nemours & Co. (Inc.)

Fig. 9-18B. Detonating cord layouts

it must be connected to an explosive cap.

ELECTRICAL FIRING

Electrical firing requires a complete circuit from the power source through all the caps and back to the power source. One cap, or several hundred, may be used.

A cap may detonate the charge through a primer in the borehole, or initiate a detonating cord hooked up with a number of holes.

Source of Current. One standard source of electrical energy for blasting is a blasting machine. This may be a generator which delivers high voltage when a handle is pushed or twisted vigorously.

These devices are rated according to the number of caps in series which they can fire at one time. Rating is usually conservative. Their efficiency should be tested occasionally, particularly if they are not in steady use, as they may deteriorate rapidly in damp storage. Testing is done by means of a special rheostat which sets up a resistance in the line; and from one to four blasting caps. If the machine will overcome the rheostat resistance of its rated capacity, and fire the caps in addition, it is in good condition.

Newer type blasting machines operate on the capacitor or condenser principle. Current is supplied by flashlight-type dry cells, or by a 12-volt rechargeable nickel-cadmium unit. This modest current is built up to high voltage in condensers, then discharged into the blasting circuit.

The high voltage may be built up when required, by pressing a CHARGE button or switch, then used by depressing a FIRE button, after a buildup time of 5 to 30 seconds.

The most sophisticated of these machines at present is the DuPont SS-1101. It has solid state circuits, a 2000 volt input and is waterproof.

Other models automatically build up voltage between uses, and can fire instantly.

In either case, the machines are designed so that two separate buttons must be pressed at the same time to put current into the firing circuit.

Storage or dry cell batteries may be used in emergencies for shooting a small number of caps, but they are not considered to be adequate or safe for regular use.

High Lines. Where electric lines (called high lines) are available, 220 to 440 volts may be used for firing. Special switches are made for connecting into such lines. They

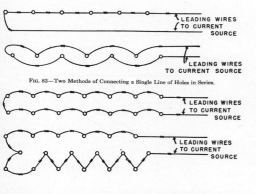

Fig. 83—Two Methods of Connecting a Single Line of Holes in Series.

Courtesy of E. I. du Pont de Nemours & Company Inc.)

Fig. 9-19. Series wiring

automatically shunt or short out the firing lines until the moment the switch is pulled.

Series. There are three basic types of circuit—series, parallel, and parallel series.

When the caps are arranged in series, Figure 9-19, the current must have enough force, or voltage, to overcome in succession the resistance of the lead wire, the caps and their wires, and the return lead wire, in addition to the variable resistance offered by connections between wires.

The voltage required can be calculated by Ohm's law.

This basic law states that the current, in amperes, in an electrical circuit will be equal to the potential, or voltage, of the power supply divided by the resistance, in ohms, of the circuit. That is, if sufficient current is supplied at 110 volts to a circuit with ten ohms resistance, the flow of current will be eleven amperes. If the voltage is six, the flow will be only .6 amperes.

A single cap requires a current of about .5 amperes. A series of caps takes 1.5 amperes, with sufficient voltage to overcome all resistances in the circuit.

The tables in Figure 9-20 indicate the resistance of caps and wires commonly used. The supply of current should be well over the calculated need, however. Minute differences in the bridge wires in caps may vary their resistance, so that a weak current might burn some of them through and break the circuit before all are exploded. Be specially liberal if the series includes both regular and delay caps.

Series circuits are easy to lay out, to hook up, and to test.

Caps made by different manufacturers must not be used in a series, because of variation in current requirement for detonation.

Parallel. In a parallel circuit, Figure 9-21, the current does not go through the caps one after another but through all of them at the same time. A poor connection on a cap wire affects only that cap. The voltage requirement is lower than when the same number of caps are shot in series, but more amperage is needed.

Two caps in series have twice the resistance of one cap. Two in parallel have only half the resistance of one, as less potential is required to force the current through a large conductor than a small one. But where 1.5 amperes was sufficient current

AWG GAUGE No.	OHMS PER 1,000 FEET	
	Copper	Aluminum
0	0.100	0.161
2	0.159	0.256
4	0.251	0.408
6	0.403	0.648
8	0.641	1.03
10	1.020	1.64
12	1.620	2.61
14	2.580	4.14
16	4.10	6.59
18	6.51	10.50
20	10.40	16.70
22	16.70	26.50

TABLE XIV
Resistance of Electric Blasting Caps in Ohms per Cap

LENGTH OF WIRE IN FEET	COPPER WIRE		IRON WIRE	
	Instantaneous Caps	Regular and "MS" Delay Caps	Instantaneous Caps	Regular and "MS" Delay Caps
4	0.94	1.50	1.71	2.19
6	1.00	1.56	2.11	2.59
7	1.04	1.60	2.31	2.79
8	1.07	1.63	2.51	3.00
9	1.10	1.66	2.72	3.20
10	1.13	1.69	2.92	3.40
12	1.20	1.76	3.32	3.80
16	1.32	1.89	4.13	4.61
20	1.45	2.01	4.94	5.42
24	1.58	2.14	5.75	6.23
30	1.41	1.98	6.96	7.44
40	1.62	2.18	8.98	9.46
50	1.82	2.38	11.00	11.48
60	2.02	2.58	....	....
80	2.43	2.99	....	....
100	2.83	3.39	....	....
150	3.84	4.40	....	....
200	4.85	5.41	....	....
250	5.86	6.42	....	....

Courtesy of E. I. du Pont de Nemours & Company (Inc.)

Fig. 9-20. Resistance of caps and wires

to shoot a whole series of caps, .5 amps is required for each cap placed in parallel.

Parallel wiring is therefore preferred where a source with low voltage and high amperage, such as a storage battery, is to be used. It is not suitable for blasting machines and the results with dry cell batteries are doubtful. High lines are equally efficient with either arrangement.

The most common simple parallel hookup is the second one shown, and it is not recognized as such by many who use it. It is the most convenient way to fire a small irregular group of blasts.

In figuring a parallel circuit, the resistance on one cap is divided by the total number of caps, and in a large blast, may be so small a figure that it can be ignored. The resistance of the two bus wires, between the leads and the last cap, is approximately one half the resistance of the same length of the same wire used for a lead. Lead wire resistance is the same as in a series circuit.

The lowered resistance of the bus wires is due to the fact that some of the current is diverted at each cap. Full current is present at the beginning, and zero current at the end, so that it averages out to about one half current for the full length of these wires.

Parallel Series. This layout, Figure 9-22, makes it possible to shoot large numbers of caps without requiring excessive voltage or amperage.

There is some disagreement about the balancing of the size of the different series. Technically, each series should have the same number of caps. Many blasters, however, claim better results when the series differ from each other by a set amount. This is said to be particularly advantageous when firing an excessive number of caps with a blasting machine.

When the series are equal, and juice put in the line, all caps are equally heated and should detonate simultaneously. If any

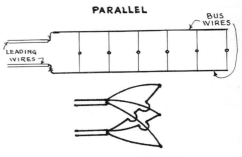

Fig. 9-21. Parallel hookup

series gets current but less than its share, due to a poor connection or other defects, it may not fire. However, as the other series fire, current will cease to go through them, and will all go through the remaining one, unless the wires are broken, until it fires also.

If the series have different numbers of caps, at first the current will flow most strongly through the series with the fewest caps and least resistance, and having fired that, will concentrate on the next longer string, and so through the whole group. If the current is weak, there may be a brief but definite time interval between the series, so that short strings should be at the face. If it is strong, explosions may be simultaneous.

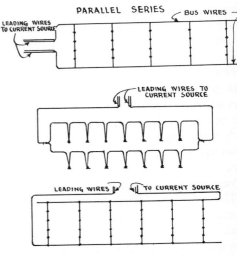

Fig. 9-22. Parallel series

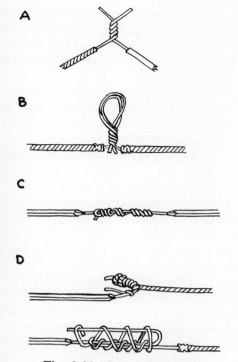

Fig. 9-23. Connecting wires

As the current from a blasting machine flows very briefly, it is possible that in a graded pattern only the shorter series would fire.

In most blasting patterns, the equal series hookup will be very much simpler than the graded series, which is generally considered obsolete.

Making Connections. It is most convenient to have the cap wires of such length that they meet each other with a moderate amount of slack between the holes. If they are short, an extra piece of wire must be spliced in at each connection; if they just reach, insulation or primers may be strained while splicing; and if much too long, they will need cutting, or will make a tangle of loose wire that may lead to mistakes in connecting, and accidents.

While connections are being made, the power or far end of the lead wires should be fastened together to ground out any induced current. The electrical source should be locked or at least removed from the immediate vicinity of the wires.

In a series circuit, the current runs from one lead wire through all the caps and their wires, to the other lead wire. The insulation on each lead is stripped back with a knife or plucked off with pliers for an inch or two, and the wire is bent into a tight loop.

The cap wires are pulled apart at their soldered connection, and each wire connected to one from an adjoining cap. When all the caps are connected in this manner, the two end wires are connected to the lead. Figures 9-23, (A) to (C), shows connections commonly used between caps, and (D) two cap-wire-to-lead-wire hookups.

It is usually possible to arrange the circuit so that it is convenient to hook both end caps to the lead. However, in a one row straight line layout, an extra wire must be used to connect the last cap to the lead.

If cap wires are long, such connecting wires are usually made up of scrap wire left from preceding blasts. It is good practice to extend the leading wires some distance with a lighter and less expensive wire, to reduce damage to the ends. If not enough surplus wire is available, connecting wire may be brought on the job on a spool and cut as needed.

The use of scrap wire involves a risk of misfires due to breaks inside the insulation, which is not justified by the small value of the wire saved.

When the last cap is connected, the whole series should be rechecked, to make sure that no hole has been omitted, that no loose ends of wire are lying around, and

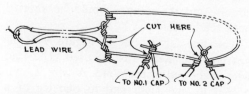

Fig. 9-24. Protecting caps against stray currents

that connections are tight. It is good practice to squeeze each connection with pliers at this time.

Bare connections may be propped up on sticks or rocks where necessary to keep them out of water, or from contacting wet ground. If any connections are unavoidably wet, they may be tightly taped or smeared with water resistant grease.

If the shot is only of a few caps in a limited area, and the electrical source is of ample power, precautions against bare wet joints are necessary only if the water has a high mineral content.

Electrical Hazards. Electric caps are supplied with a connection between the ends of the wires which is called a shunt. This prevents the accidental building up of opposite electric charges in the two wires, which might pass enough current through the cap to explode it.

Such charges may be caused by the near presence of electrical machinery or transmission lines, a radio transmitter, stray currents in the ground, thunderstorms, or static electricity from dust storms or escaping steam. Such currents in lead wires may often be detected by inserting a No. 47 radio pilot lamp in the circuit instead of the cap. If it glows, conditions are unsafe.

The best precaution to take in blasting near an electrical hazard is to use fuse caps and Primacord. However, certain precautions can be taken which will reduce the danger of using electric caps.

Lead or other wires should not run parallel to electric lines.

The shunts should be left on cap wires until they are connected into the blasting circuit, and the circuit should be shorted until ready to fire.

The two-cap layout in Figure 9-24 illustrates the method to be used. The lead wire is shorted at each end. A wire is connected to one lead, and to one leg of a cap. A connection is made from the other leg or the shunt, to one leg of the next cap

which has its other leg connected to the return lead. Making cuts where indicated will include the caps in a firing circuit which is closed until the leads are separated at the battery end.

Even with these precautions, however, blasting should be discontinued if there is a thunderstorm within five miles, or other severe hazards exist.

Crackling in a portable battery-type radio (not FM), left at high volume, provides warning of approach of a thunderstorm.

Testing. A circuit tester should be used for checking before attempting to blast. This device consists of a galvanometer and a silver chloride dry cell, which produces a current too weak to fire a standard blasting cap. The lead or cap wires are fastened to its terminals, and the action of the indicator needle shows the condition of the circuit.

If the circuit is good, the indicator needle will move an amount inversely proportional to the resistance offered by the caps and wire used. If the needle does not move, there is an open break. If it moves only slightly, a loose connection, or a break with

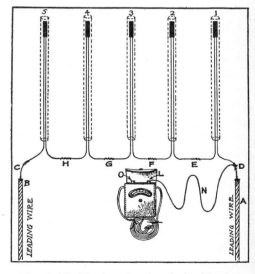

Fig. 9-25. Testing for break in blasting circuit

wires just touching, is indicated. If the needle moves farther than it should, a short or a ground is present.

A single test may be made from the power end of the lead wires when wiring is complete, or each hole may be tested before hooking into the circuit.

Any trouble in the system can be spotted by making a series of tests with a long connecting wire. In Figure 9-25, the connecting wire, N, is fastened to one lead and to a tester post. The other post, or a wire securely connected to it, is touched in succession to connections E, F, G, H, and C. The bad reading will show up whenever the difficulty is inside the circuit being tested. As an example, if normal readings are shown at E and F, but an abnormal one at G, the trouble is in number 3 cap or its wiring.

After the blasting circuit tests properly, an additional test is made at the power end of the leading wires.

Warning. Warning must be given of intention to blast. The type of warning may be determined by either law or custom. For large blasts, particularly in pits employing numerous men and machines, blasting should be done at specified times, as at twelve and at quitting time, and should be preceded by definite and well understood signals, such as horns, sirens, whistles, or yelling, long enough in advance to notify everyone and give them time to prepare.

A usual blast signal is a call of "Fire" or "Fire in the hole," repeated two or three times at intervals of ten to thirty seconds. Signals should be arranged by which men detailed to watch entrances to the area can stop the blast if necessary.

It is the responsibility of the blasting crew to make sure that all personnel is out of the way, machines are protected, and that no visitors or trespassers are where they can get hurt.

The area, and particularly any roads or paths leading into it, should be marked with warning signs. If any public roads are within the danger area, traffic should be stopped at a safe distance.

Firing. During the signaling, the wires are connected to the blasting machine or switch, or if a battery is used, one wire is connected to a post. To fire, the blasting machine handle is slammed to the bottom, or the switch is closed.

There is a wide divergence of opinion as to what constitutes a safe distance from which to fire a blast. Some experienced men will take shelter behind a nearby rock or tree, whereas others consider 500 feet a bare minimum. No one should stand in front of a face, at any distance.

Proper barricading may be as safe as distance and more convenient. Full protection requires some sort of roof or overhang. A very safe spot is in the tucked-in bucket of a big dipper shovel which is turned away from the blast. The shovel itself may be protected by wood lagging on the rear of the cab.

The return to the blast should be slow for several reasons. The fumes, which dissipate in a few minutes in the open, are toxic and may cause severe headaches and nausea. If more than one hole has been shot, one of them may fire late. Rocks are occasionally thrown so high that they take a long time returning. Rocks or debris may be lodged precariously in trees.

LIGHT BLASTING

Light blasting, Figure 9-26A, includes loosening up of shallow or small outcrops of rock and breaking boulders. It may constitute the entire job, be done in connection with dirt excavation, or follow heavy blasting which has failed to cut to grade or slope lines, or has left chunks too large to load.

Chip Blasting. Shallow rock outcrops are most conveniently broken up by drilling and blasting. Unless the rock breaks readily along planes more or less parallel with the

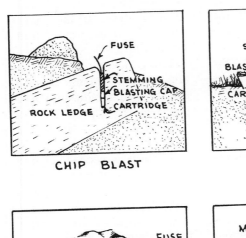

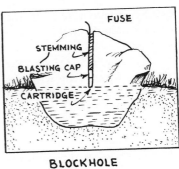

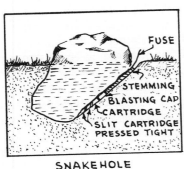

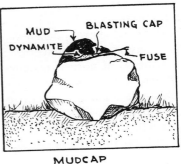

Fig. 9-26A. Spot blasts

surface desired, it will be necessary either to drill much deeper than grade or to space the holes closely. It is often good practice to blast each row before drilling the next.

Loading may be light, or very heavy, but in general it is necessary to use more powder per yard of solid rock than in heavier work.

Laminated or jointed formations may be shaken apart by light charges.

Fragments may be thrown long distances, and mats used to confine them are more subject to damage than with deeper blasts.

The amount and direction of throw can often be controlled to a large extent by drilling and loading procedures. A vertical hole causes scattering in all directions. A sloped hole tends to leave the lower slope in place and to throw the upper one away from it. Throw is reduced by increasing the number of holes, reducing the charges, or drilling deeper than required by the break-

age line. These two last place the powder deeper where more of its power is applied to breaking and less to scattering.

When breaking must be done exactly to a line, holes are drilled closely along the line and a variable number left without charges.

Blockholing. Boulders and oversize pieces of blasted rock may be broken by drilling a hole slightly more than halfway through, and exploding a small charge of dynamite in the hole.

Fragments may be thrown for long distances, so that protection should be provided for the blaster and other personnel. High velocity explosive, or large charge, will produce finest fragmentation.

Chip blasting may be called blockholing also.

Mudcapping. Ledges may be chipped and boulders broken by mudcapping instead of drilling. Heavy charges of dyn

DIAMETER OF BOULDER IN FEET	APPROXIMATE NUMBER OF CARTRIDGES, 1¼ x 8" — IN AVERAGE HARD STONE — REQUIRED FOR:		
	Mudcapping	Snakeholing	Blockholing
1½	2	1	¼
2	3	1	¼
3	4	1½	½
4	7	4	¾
5	12	6	1

Fig. 9-26B. Charges for boulder blasting

nite, preferably of the highest velocity type that is obtainable, are laid on the surface of the rock, primed, and covered with a few inches of mud. The explosion acts as a giant hammer blow and should split or crush the stone.

Knowledge of the grain and jointing of rock is important in successful mudcapping. The charge is placed in the same place which, in handbreaking, would be hit with a hammer or opened with a wedge. In general, hammer-like crushing is most effective on loose boulders, and splitting on ledges.

It is a common error to suppose that the force of black powder is chiefly exerted upward, and that of dynamite downward. In each case the explosion acts equally in all directions, but when it acts slowly it can find and follow paths of least resistance, where the quicker acting dynamites deliver such a rapid blow that they will crush objects under them, even when not confined. However, a study of the table in Figure 9-26B, showing quantities of dynamite used for blockholing and mudcapping, will show the waste involved in open explosions.

The mud pack over the charge is usually two to six inches thick. It serves to confine the explosion slightly, increasing the force exerted on the rock and reducing noise and air-borne concussion. Mud is to be preferred to any other substance. It is much more effective at confining the explosion than dry or damp dirt or sand, as it packs and sticks together better. It should be free of stones or pebbles that would create a hazard by flying long distances.

Charges can be fired on bare rock, but they are less efficient and even noisier.

Mudcapping is wasteful of powder, excessively noisy, and less certain in effect than drill hole blasting. However, it causes less rock scatter than other methods of shallow blasting, and does not require the presence of a compressor.

Snakeholing. Boulders are most readily broken if they are lying on the surface of the ground. If partly buried, the earth or other rock around them provide a support and a vibration-absorbing cushion that may prevent or reduce breakage.

Embedded boulders may resist machinery which can handle them readily once they are loosened up.

Snakeholing consists of making a hole beside or under a boulder, and firing a charge sufficient to roll it out of the ground, and preferably to break it also. Any further breakage required can then be accomplished by mudcapping.

Snakeholing is more laborious than mudcapping, but is more economical of powder and is much less noisy.

MECHANICAL BREAKAGE

Drop Ball. Balls of hard steel, familiarly known as drop balls or skullcrackers, may be carried by a crane, and used for breaking loose rock. Weights range from 1,000 to 8,000 pounds new. They wear down in use, and are replaced whenever they become too light for the work they are expected to do.

The ball is carried on the hoist line of a revolving crane. It is positioned high over the rock to be broken, the brake is released and the ball falls almost freely to strike the rock. The brake must be put on as it hits, to avoid spinning out of cable. Long booms held high increase striking power, but may reduce accuracy, particularly on windy days.

Expert operators may obtain greater force by a swing-and-drop technique.

Skullcrackers eliminate the danger and nuisance of secondary blasting, but they cause problems of their own. Rock chips fly from 50 to 200 feet, and they make the area around the operation one of continuous danger. The operator's station on the crane must be protected with wire mesh or bullet proof glass, and signals must be arranged to stop the work when the area must be entered by others.

Some rock breaks very well on impact, and some does not. If breakage is good, the rock can be readily reached by the crane, and other operations are not unreasonably delayed by it, this is a good method.

Some quarries or pits keep a drop ball on a crane at all times, and may use it continuously or only occasionally. Gravel pits may push boulders aside until there are too many around, and then put a ball on one of their draglines or clamshells for a day or two. Rock breaking may be done on days when the pit is shut down, to avoid interference.

Plug and Feathers. Rock may be broken by first drilling, then inserting a device in the hole that can be caused to expand until it breaks the rock. The Egyptians used dry wood plugs, which they supplied with water until they swelled and broke the rock. Now we use wedges or jacks to obtain the same result.

The plug and feathers is a three piece wedge set with a very gradual taper. The outer pieces, called feathers, are placed in the hole, and the plug is forced between them by blows from a sledgehammer or air hammer, or by hydraulic pressure.

Pressure may be developed by an attached hand jack, or by an engine driven hydraulic pump. Expansion force of 80 to over 400 tons may be developed by the wedging. Two or more sets may be used in adjacent holes. It is probable that any rock strong enough not to crumble, and that has a free face to move toward, can be split in this way. Solid rock faces are taken in shallow cuts.

In general, these devices are used where explosives are particularly dangerous or annoying, are prohibited, or the job is very small.

Backhoe. A hydraulic or air hammer mounted on a hydraulic backhoe in place of the bucket, is an effective boulder-breaker and is convenient to use.

ROCK MECHANICS

Rock is broken or fractured by an explosion in three ways—compression, shear, and tension. Compression is obtained by the direct hammer-like blow of the explosion against an unyielding rock mass. An explosive that is so deeply buried that it cannot break out to the surface breaks by compression only. This is the least effective way to use it.

Shear is movement of pieces or blocks of rock along lines of weakness. Tension is produced by reflection of the explosion back from an unconfined surface or face of the rock.

In a hard rock, maximum effectiveness of explosives is in tension. Tensile strength

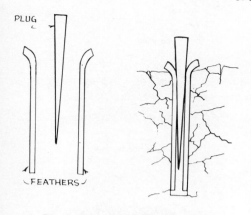

Fig. 9-26C. Plug and feathers

is only about 1/10 of shear strength, and shear strength is only 1/10 of compressive strength. This means that a blast that can break out to a free face at an efficient distance may produce 100 times the rock breakage of one that is completely confined.

The importance of tension has been demonstrated by setting off a tightly confined explosive far enough below a horizontal surface so that the explosion could not break out, but near enough so that surface rock was shattered in a cone shaped crater, separated from the small explosion chamber by solid rock.

It is very fortunate that rock breakage is not produced mostly by the outward movement of gases from an explosion, as in this case rock throw would be many times greater than it is. In a well engineered and successful blast most of the rock moves less by explosive force while being broken than it does afterward by slumping under the pull of gravity.

The improved fragmentation obtained by using miillisecond delay caps is partly due to creation of a series of free faces from which waves of succeeding explosions reflect to produce tension in the rock.

DAMAGE

One of the contractor's problems in connection with blasting is the possibility of real or imaginary damage being done to structures in the vicinity.

An explosive, if properly used, will expend most of its energy in shattering the rock immediately around it. The remaining energy will set up waves or vibrations in the ground, and sound and concussion in the air.

Noise. The noise of an explosion may cause most or all of the neighborhood difficulties. Mudcaps, shallow blasts, overloaded holes, fractured rock, and other conditions that allow the explosion to break out into open air before expending its energies, are productive of complaints all out of proportion to the amount of explosive used.

In the first place, the noise attracts attention to the fact that blasting is going on. It causes the householder to concentrate on trying to feel the jar or shake of the blast, to look for cracks in plaster, and to speculate about other damages that might be done. In many cases, the sound of the blasting will annoy sensitive people so that they will invent or exaggerate physical effects. The contractor or quarry operator's first rule is therefore to blast as quietly as he can in any area where there is a possibility of complaint.

This means a first rule of NO MUDCAPPING. This technique is not only wasteful of explosive, but a sure way to lose the good will of the neighborhood and of the insurance adjuster.

Boulders and oversize blast fragments should be drilled before blasting. The noise is tremendously reduced, and it will usually be found that the saving on explosive and the better fragmentation obtained will more than outweigh the cost of the drilling. When there are only a few pieces, the nuisance of clearing the pit for blasting may be avoided by plug-and-feather splitting. In brittle rock, a crane with a skull-cracker steel ball may be the most economical solution.

In primary blasting, the noise, and particularly the shock quality of the noise, can be reduced by use of short period delays. Their more important effect on ground vibration will be discussed later.

Long period delays are productive of complaints. They divide up a shot so that the amount of explosive detonated at one time is greatly reduced, but one explosion frequently uncovers the next in the series, making it very noisy.

If a solid rock blast is a good one, the sound should be dull or muffled. Even a good blaster cannot always get this effect, however. If the face is fractured in places, lighter loading or greater burden should prevent noisy breakout. But this may cause poor fragmentation, with greatly increased need for secondary blasting. About the only general rule here is that the blaster should consider avoidance of noise one of his important objectives.

This anti-noise advice is particularly applicable to quarries and open pit mines that are within earshot of residences. The contractor moves from job to job, his blasting may be only one of many nuisances associated with an improvement, he may be finished and gone before people complain seriously. But the pit operator is tied down to one location for many years, and upon exhaustion of the material in which he is working, will probably wish to move to a similar deposit in the same area. Cities and villages are acquiring an increasing power to regulate industrial activity, by means of zoning and nuisance regulations, so that the reputation the pit has acquired over a period of years may actually determine whether it can move, expand, or even stay in operation.

Sound travels rather slowly. Its distribution is affected by winds, as shown in Figure 9-27, by reflection from hills, clouds, or atmospheric layers, and by temperature and humidity.

Concussion. Air borne concussion is responsible for a large share of the damage

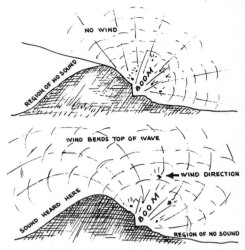

Fig. 9-27. Sound travel from blast

in bombing, and in accidental detonation of explosives, but is rarely a factor of importance in blast damage. It consists of one or more waves of highly compressed air moving outward from the explosion.

Sufficient explosive to cause concussion more than a few feet away should not be used in mudcapping. Even very heavy blasts in solid rock cause little or no concussion if they are laid out and loaded properly.

Any damage caused by concussion is usually obvious. Glass breakage in closed windows at right angles to the path of the waves is the most common result. In the absence of extensive glass breakage, it is very doubtful that any other parts of a structure could be damaged.

When blasting must be done very close to a few buildings of small or moderate value, the question may arise whether the shots should be kept small enough to avoid damage, or whether it would be economical to blast more freely and repair resulting damage. If the latter, the inconvenience to owners or users of the building is a factor in the amount of compensation.

If a building is to be endangered by blasting, windows should be opened or removed particularly those on the facing and far sides of the house. Store windows may be braced as a routine precaution. Careful check should be made of the condition of plaster and masonry, so that claims need not be paid on preexisting defects.

Rock Throw. Unexpected damage may be done by rock or other material thrown through the air by blasts. In general, shallow blasts, overloaded holes, shots in rock with irregular resistance, and blockholed boulders give the most trouble in proportion to the amount of powder used.

Thrown objects may cause injury or death, and their control is therefore of first importance. Property damage may or may not be severe, but at least claims filed on this ground are usually sincere.

Danger of damage from rock throw may be reduced by increasing the number of holes so that smaller charges may be used, by sloping holes to throw rock away from danger points; by reducing the quantity or strength of powder, and handling any resulting oversize fragments by blockholing under mats or by the use of larger machinery.

Covered Blasts. Throw can be closely controlled by working downward, using small blasts and covering them with mats or chained logs. If the cover is large and heavy in proportion of the strength of the explosion, it will prevent any scattering of fragments. If the charge is heavy enough to lift the cover, it will move somewhat less than the average distance of throw to be expected from an uncovered blast, and fragments with higher than usual velocity will be held in.

It is important that the cover extend several feet beyond the area being shot, particularly if the charge is heavy enough to lift the mat, as fragments might escape under its edges.

When a power shovel is used to remove the shot rock, it is advantageous to use a woven steel mat as it is easily handled with

chains, and provides a quicker and more secure cover than logs. The mat is lowered over the holes, or dragged in such a manner that it will not damage the wiring and cause misfires.

Logs are used when no mat is available, or when there is no machinery on the job which can handle one. They should be long enough to overlap the blast at both sides, and light enough to allow the crew to carry them by hand. Two chains should be laid on the ground first, the logs piled, and the chains fastened over them, preferably by wired square knots.

Chaining is important, as unfastened logs may be thrown farther than rocks.

Neither mats nor logs should be laid directly over mudcaps as they are liable to be thrown long distances, and severely damaged as well.

Blasting mats should be used wherever there is the slightest possibility of fragments reaching people or property. Even a scattering of sand or fine pebbles on their property will make people nervous and resentful, and is an indication that loading should be reduced or technique changed to a safer one.

Ground Waves. The vibration or wave motion set up in rock and soil by a blast constitutes the principal source of both actual and imagined damage to buildings. There are a number of varieties of waves, traveling both deep underground and along the surface. The latter are of primary importance to the blaster and to his neighbors.

The study and description of the movement of particles of earth or rock in an earth wave from a blast is a highly technical subject. There are push waves, that momentarily increase the density of the ground, in the same manner as the concussion wave acts on the air.

Particles are also shaken from side to side (shake or shear waves) and moved elliptically as well. These waves all start off together at a blast, but have different speeds, so they tend to spread out with distance.

On the surface, the ground is forced into waves similar in shape to those caused by wind on water, except that their height is in thousandths of an inch, and the distance from crest to crest is between 100 and 1000 feet. Waves of these dimensions cannot be seen, but they may be felt in intensities as small as 1/100 of that required to do damage.

A diagram of the spreading of ground waves outward from a blast is shown in Figure 9-28. The amplitude (height) of the waves is greatly exaggerated.

It takes a really huge wave of this type, such as might be found in a major earthquake close to its source, to damage any material by shifting its particles. The trouble in structures arises when the foundation moves or changes shape with the earth in which it is embedded, but the inertia of the rest of the building causes it to try to stay in its original place, with resulting stresses set up between foundation and upper structure. The effect is similar to that caused by a heavy wind that pushes and slightly deforms the house, but cannot affect the foundation, and its effect on the structure is seldom more serious.

Surface earth waves in soil ordinarily have a frequency of from 4 to 20 cycles a second. 10 cycles are used as a standard in calculating blast results. The body waves traveling deep underground may have a frequency between 20 and 90 in rock, and travel 8000 to 26,000 feet a second.

Loose soil moves further (makes a bigger wave) than rock, but it also absorbs and damps out the wave in a shorter distance. The wave height or amplitude, and therefore the possibility of damage, is about 10 times greater in normal overburden 50 feet or less in depth, than in rock. Very deep overburden may shake 30 times as much as rock.

Displacement. The Bureau of Mines has conducted extensive investigations into the effect of earthborne waves on buildings. It is interesting to note that they were forced to resort to machine-induced vibration, as they were unable to find any quarry or tunnel job that produced sufficient jarring in its vicinity to make investigation possible.

The finding is that displacements between 100 and 250 thousandths of an inch (.100 to .250) were usually required to crack or loosen plaster. Occasionally minor cracks were caused by motion of .050. In tests in buildings in the neighborhood of commercial blasting, many of them where complaints had originated, they found vibration to be only .01 to .001 of an inch. These observations were taken at a cycle frequency of 10.

Energy Ratio. Insurance companies have conducted tests that also included the part played by acceleration of earth movement in building damage. Acceleration is the rate at which a particle changes from a state of rest to the maximum motion imparted by the wave, and since inertia and resulting lag in motion of parts of a structure are the chief source of damage, this acceleration has great importance.

They use a term, Energy Ratio, abbreviated E.R., that is obtained by dividing the square of the acceleration by the square of the frequency of the waves. When E.R. is 3, old buildings that have been pre-stressed by uneven settlement or warping may show slight damage, but sound buildings will be unaffected. At 6, there is a strong probability of damage to residential structures. An E.R. of 3 corresponds very closely to the displacement of .050 inches displacement used by the Bureau of Mines, so the two separate approaches lead to the same result. However, there are serious discrepancies between these somewhat theoretical conclusions, and damages found near actual operations.

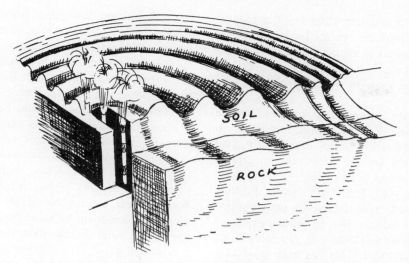

Fig. 9-28. Ground waves

Characteristics of Damage. In most cases, there is an observable difference between building damage caused solely by vibration, and that resulting only from structural defects and/or settlement that change the shape of the building. Most of the distinction is in the pattern of wall cracks.

Cracks due to change of shape of the building usually indicate the area and type of failure by their location and width, and often reveal sliding motion.

Such cracks may be wider at one end than the other, indicating spreading displacement at the wide end.

Much of the damage caused by blasting vibration is a result of various parts of a building reacting at different rates, which are affected by the material of which the parts are made, their thickness, and shape of assemblies; and by their supports.

A chimney reacts differently from the wall on each side of it, as does also a window or door frame. A masonry floor vibrates out of phase with an adjoining floor of wood, as does plaster with its supporting woodwork.

Cracks tend to form along vibration boundaries, and also to radiate out from corners or irregularities. Cracks in walls tend to close when vibration stops, but those in unconfined floors may become and remain wide.

Cracks caused by vibration are usually fairly uniform in width from end to end, with little or no displacement of the sides relative to each other, unless the shaking has been severe enough to cause structural damage.

Vibration cracks tend to fan out diagonally from corners of window and door frames. Glass often cracks diagonally inward.

Either type of stress frequently causes walls to crack along their junctions with a chimney. Settlement or tilting of the chimney only, or the house only, usually results in an opened or spread crack on one side, and a closed or even pressure-squeezed crack on the other. Vibration more often produces equal-width cracks on both sides.

A distinction can sometimes be made on an age basis. New cracks show clean surfaces of material; older ones become progressively grimier, the rate depending on conditions. Dust, spot deposits, and light-bleaching of color pigments are indicators. Recent cracks should be clean and unfaded.

Reasons for Complaints. Studies indicate that much greater vibrations are produced in house structures by slamming doors, running, and often by street traffic, than by even severe blasting. It would appear that at ordinary distances the ground and air vibrations set up by heavy blasts are so weak as to be incapable of affecting any structure. Yet complaints and claims for damage pour in on every blasting job. Why?

There are a number of reasons. One is that the ability of rock, soil, and water to transmit vibration varies much more than is indicated by the relatively superficial testing that has been done.

Bulletin 442, U.S. Government Printing Office, 1942, now deservedly out of print, had a table of vibration amplitude expected from various size blasts at distances from 100 to 6000 feet. This has been used as a guide to safe practices by contractors, often with the approval of insurance inspectors.

It would appear from this table that 600 pounds of explosive would not produce sufficient ground waves to damage a house on average overburden 100 feet away. (!!). But there are records of high five figure awards paid for damages done to a village two miles away from an underwater blast of this size, indicating a difference of over 10,000 per cent between theory and fact.

Contractors frequently blast much more heavily than is indicated by their records and statements, particularly when a job gets behind schedule. Mistakes in loading can occur. Variation in the strength and quality of explosives can be a factor.

Most of the checking of blast damage to date has been done by representatives of mining and insurance interests who are more interested in disproving it than in impartial study. Some of the instruments used for measurement leave much to be desired.

There are also psychological reasons for exaggeration of blast damage. Bomb damage received very extensive publicity during the last war and made people over conscious of the dangers of explosives. There is also fear and resentment of the unusual, that makes blast vibration appear more significant than that from a truck.

Still another factor is coincidence. Residences, and plastered buildings generally show cracks and changes in shape progressively throughout their life. Each crack must have a time of appearance. There is no reason why a crack should not appear at about the same time as a blast, even if there is no cause and effect relationship.

Many people are so unobservant that they can live in a house with cracked plaster and sagging beams for years, not noticing until the rumble or jar of an explosion makes them look for possible damage. Cracks and defects then appear to be a direct effect.

Pre-stressing. One of the principle defenses advanced by defendants in blasting damage suits is criticism of the condition of the structure before the blast. If it is in a condition of stress due to unequal settlement, warping or shrinking of timbers, or overloading, it will change in shape and its plaster will crack.

If a blast vibration is within "safe" limits, an over-stressed condition may cause cracking from the blast. The theory is that if the blast had not been set off, the same cracks might have developed shortly from natural causes.

In general, the poorer the quality of construction, the greater the probability that stresses will develop, plaster crack, and mis-alignment occur. But this is not always so.

The Bureau of Mines has prepared a list of 40 reasons for cracking of wall and ceiling plaster as a result of defects in construction. These are of interest not only in respect to blast damage, but as warnings of mistakes or economies to avoid when building. They are:

1. Building a house on a fill.
2. Failure to make the footings wide enough.

3. Failure to carry the footings below the frost line.
4. Width of footings not made proportional to the loads they carry.
5. The posts in the basement not provided with separate footings.
6. Failure to provide a base raised above the basement floor line for the setting of wooden posts.
7. Not enough cement used in the concrete.
8. Dirty sand or gravel used in the concrete.
9. Failure to protect beams and sills from rotting through dampness.
10. Setting floor joists one end on masonry and other end on wood.
11. Wooden beams used to support masonry over openings.
12. Mortar, plaster, or concrete work allowed to freeze before setting.
13. Braces omitted in wooden walls.
14. Sheathing omitted in wooden walls (excepting in "black-plastered" construction).
15. Drainage water from roof not carried away from foundations.
16. Floor joists too light.
17. Floor joists not bridged.
18. Supporting posts too small.
19. Cross beams too light.
20. Subflooring omitted.
21. Wooden walls not framed so as to equalize shrinkage.
22. Poor materials used in plaster.
23. Plaster applied too thinly.
24. Lath placed too closely together.
25. Lath run behind studs at corners.
26. Metal reinforcement omitted in plaster at corners.
27. Metal reinforcement omitted where wooden walls join masonry.
28. Metal lath omitted on wide expanses of ceiling.
29. Plaster applied directly on masonry.
30. Plaster applied on lath that is too dry.
31. Too much cement in the stucco.
32. Stucco not kept wet until set.
33. Subsoil drainage not carried away from walls.
34. First coat of plaster not properly keyed to backing.
35. Floor joists placed too far apart.
36. Wood beams spanned too long between posts.
37. Failure to use double joists under unsupported partitions.
38. Too few nails used.
39. Rafters too light or too far apart.
40. Failure to erect trusses over wide wooden openings.

The pre-stress argument is unpleasantly reminiscent of the whitewash given the Donora smog by a group of doctors. They said in effect that it was nothing to fuss about, as only people with a previous history of respiratory disease had died.

It would be unjust to allow reckless blasters to evade payment of damages on these grounds, or to make property owners go without recompense because their building standards fall short of those set up by the Bureau of Mines. However, blasters should not be compelled to subsidize substandard construction. It is likely that most cases where pre-stressing is actually proved should be subject to compromise settlements.

Water Supply. Blasting sometimes causes springs and even deep wells to go dry. The vibration causes underground movements that may close water passages or open new ones. However, explosives are probably responsible for only a fraction of the difficulties for which they are blamed.

Underground water circulation is under constant change. Old seepage veins become plugged with mineral deposits, new ones are opened by solution and erosion. Changes in rainfall pattern, in conversion

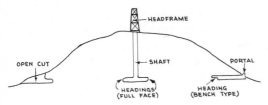

Fig. 9-29A. Tunnel layout

of forest land to farms, or back again, may alter the quantity and location of underground water over a wide area. Over-pumping will lower the water table.

A new well may tap into an underground reservoir of limited size, which once pumped out will not refill. Such a well may show a very high yield on its first test, but decline markedly after long use, when it comes to depend on circulating water only.

Keeping Out of Trouble. Under all ordinary circumstances, blasting should be kept light enough not to damage buildings. The job should be figured on a basis of conservative blasting, and the work done the same way.

Short period delays are a real friend to the man who wants heavy blasts, but is surrounded by structures. Up to 70 percent of the charge of an instantaneous blast can be used with EACH of ten or twelve short delay periods, without increasing the vibration. Or to look at it another way, the same loading can be used as for an instantaneous shot, and 10 periods used to cut the damage potential by four-fifths.

Even with ultra-conservative procedures, inspections should be made of nearby buildings before blasting. If property is valuable, vibration-testing devices will be supplied by the insurance company or by a blasting consultant to measure the disturbance caused. Such instruments should be used to check the next blast in any building or area from which complaints are received.

As detailed before, noise should be kept to a minimum. If there are few people in the area, it should be possible to notify them before blasts, so that they will not feel it necessary to be tense all day waiting for an explosion. Another method, applicable to heavily populated areas also, is to set a definite time or times each day for shooting, and stick to it.

If a claim is made and is justified it should of course be paid. But if it is clearly unjustified, it probably should not be paid even if apparently too small to be worth arguing about. One paid claim is likely to bring in a dozen or a hundred others, and the contractor might find himself replastering and decorating a whole town before he knew what hit him. Payment of any claim makes any other much harder to defend in front of a jury.

Of course, a contractor should protect himself with insurance, and usually does. But in the long run the premiums he pays are based on what the company pays out for damages, so their interests are identical.

DIGGING UNDERGROUND

TUNNEL WORK

Tunnels are underground passageways of any size, and may be natural (as in limestone caverns) or made by animals or men. Those discussed in this section are man made. They serve a variety of purposes, including mining, water supply and drainage, laying sewer and other pipes, railroad and vehicular shortcuts or water crossings, and air raid protection.

Rock tunnels are driven through solid material that usually requires blasting and may support itself permanently, or at least long enough to allow setting up of bracing after digging out a short section. Soft ground tunnels involve digging or pushing aside soil, and the roof (called the crown) and the walls may require support before removing the soil. Mixed-face involve going through both types of ground, either together or in different sections.

Men have driven tunnels since prehistoric times. They usually worked in rock, because the difficulty of digging it was more than compensated by its ability to hold itself up. Cutting was done with hand tools, or by heating the face with wood fires, then throwing cold water or cold water and vinegar on it to cause sections to crack off.

The vinegar technique, with little or no ventilation, must have been really rough on the men who did the work. A rough approximation of the atmosphere might be obtained by building a good blaze in a fireplace, shutting off the chimney damper, then putting out the fire with vinegar.

Layout and Problems. The methods used to drive a tunnel vary tremendously with the nature and water content of the material to be penetrated, depth and size required, surface conditions along the route, time allowed, and background of the men doing the job. There is space in this section to indicate only a few of the problems most often encountered.

The diagrams in Figures 9-29A and B show the layout of a simple tunnel job, and the names for some of its parts. If it is driven more or less horizontally into a hillside, the opening is called the portal. The working face, where the digging is done, is the heading. Vertical access tunnels descending from the surface to the main tunnel level are known as shafts.

In the tunnel itself, the floor is the invert, and the roof is the crown. The spring line is the meeting of the vertical side wall with the curve of the roof arch. A supporting shelf cut at this line is the hitch. A small pioneer or accessory tunnel is called a drift. Standard cross sections are rectangular, round, and horseshoe.

There are a great many special problems connected with even a simple tunnel project. To the open-cut man, one of the most impressive is lack of space. Many tunnels have been driven with cross sections as small as 4 x 4 feet—not even big enough to stand in. Twenty to thirty foot diameter tunnels are big, yet they provide a floor width that would be considered skimpy for a haul road on top.

Equipment to be managed at and near a tunnel heading may include a drill jumbo (a movable frame almost as big as the tunnel, carrying a battery of drills), a machine for loading muck (the below-ground name for spoil) and rail cars or rubber-tired trucks to remove it; the same or other cars or trucks to bring drill steel, bits, explosives and other supplies to the heading, a locomotive to push and pull cars, and a switching or passing device to permit hauling units to get past each other, although there is often room for only a single width of track or roadway.

There will be high pressure air pipes to supply the drills, and often large low pres-

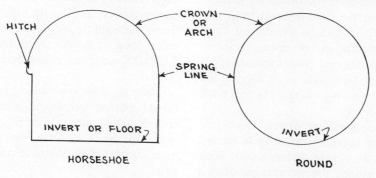

Fig. 9-29B. Tunnel cross sections

sure ducts for ventilation. Overhead wires or ground cables carry electricity for light, power, and blasting juice. Water under pressure may be supplied for wet drilling. A system of drainage, pumping, or both may have to handle tremendous volumes of water.

In addition to the regular equipment there may be need for a diamond drill to make test and grouting holes, grouting equipment to seal off leaks and solidify wet ground, and/or a movable buffer to confine rock throw from blasts.

If the tunnel is to be timbered for support, or lined for support or for permanent use, the crews and materials for this work may follow the digging closely, and in any event will have to work in and over the single entrance way.

If driving is from a shaft, its bottom is another crowded point. Haulage equipment may be lifted to the top to dump, or dump into containers at the bottom. Supplies must be unloaded from the elevator cages and reloaded for hauling to the face. Crews of men leaving work, and their supervisors and inspectors wait here for transportation to the surface. Pumps, compressors, and even drill and repair shops may be located in skimpy quarters excavated near the shaft.

Sequences are very exacting. The tunnel cycle (the succession of drilling, shooting, and mucking) must keep the largest possible number of men and machines usefully employed, and the time interval from any operation to its repetition should not vary. Whenever possible, two or more operations should be performed simultaneously, as drilling the top of a face while digging the bottom, and installing lining a few feet back at the same time.

When two headings driven from one shaft are close together, one may be drilled while the other is shot and mucked. With increasing distance the advantages of this arrangement are reduced.

Most tunnel crews are the universal type, and perform all operations in the cycle. This saves the contractor from paying a crew waiting time because of a delay in a prior operation.

Speed. Under favorable conditions, tunneling may progress very rapidly. The Owens River Gorge Power Tunnels in the Los Angeles water system were driven as fast as 104 feet in a day, and 2442 feet in the best 31 day period.

On a start-to-finish basis, the fastest tunnel on record is the six mile Carlton drainage tunnel in Cripple Creek, Colorado, which was completed in under 2 years. Both of these were about ten feet in diameter, and unlined.

The longest tunnel in the world is the 85-mile Delaware Aqueduct in the New York City water system. This was driven from 26 shafts, with no single section longer than 5 miles. The longest tunnel driven from just two headings is the 13-mile Alvah B. Adams in Colorado.

Gold mines at Kimberley, South Africa, hold the depth record at 9000 feet. These tunnels must be air conditioned, as otherwise the heat would make it impossible to work in them.

The 12-mile Simplon tunnel in the Alps is 7000 feet beneath the surface at one point. Temperatures up to 131 degrees were encountered in drilling it.

A record for maximum excavation in a single tunnel project was established in twin power tunnels at Niagara Falls in Canada. These are each 51 feet in diameter and 5½ miles long. Together they required over 5,000,000 yards of excavation. See Figure 9-36.

There is such constant improvement in tunnel driving techniques, and increase in confidence to undertake bigger projects, that some or all of these records may have been surpassed by the time this book is in print.

Plant. The plant at a tunnel may include

the tower, hoist, and hopper; compressors, low pressure ventilation system, water pumps, electric transformers or generators, change rooms with showers and lockers, provision for emergency treatment of injuries, a blacksmith, forging, and bit dressing shop; welding and repair equipment, and telephone or radio communication systems.

Compressors are usually at the surface. They are usually of the two-stage type, and have an aftercooler as well as an intercooler, to avoid transporting any heat of compression into the heading, which is often too hot already.

Alternating current is used. When possible, it is purchased from a utility. It is usually stepped down to 220 or 110 volts at the entrance, but on some jobs is taken in at several thousand volts, in armored parkway three-wire cable. Dry transformers (oil filled ones are a fire hazard underground) are set about a thousand feet back from the faces, and advanced in long jumps as progress warrants. This system avoids the power loss and voltage drop associated with long distance transmission of low voltage current.

There may be three electric circuits in the tunnel, a 220 or 440 volt for power, a 110 volt for light, and a high voltage line for firing explosives. Some operators standardize on 220 for both lighting and power. 220 bulbs are sometimes a nuisance to get in this country, but they have the advantage of being useless in an ordinary lighting circuit, so they are seldom pilfered.

No drill dust can be tolerated. It may be suppressed by detergent or foam, or drowned in wet drills with water supplied through pipes from outside the tunnel.

Surveying. Tunnel sections meet each other far from their portals or shafts, sometimes after curves, with uncanny precision. Differences usually vary between a small fraction of an inch up to several inches. These are too small to be noticed on the

walls, but are measured at the surveyed center line (axis).

An underground direction is obtained by establishing a base line at the surface, running close to the line of the tunnel. This is very carefully done, and it is marked at frequent intervals by permanent monuments, with exact points pricked into copper bolts embedded in concrete.

Two plumb bobs weighing twenty to thirty pounds each are suspended close to the bottom of the shaft by piano wire from the surface. They are as far apart as shaft width permits. Vibration and tendency to swing may be dampened by hanging them in pails of water. Very careful observations are taken of the wires at the surface, relative to the proposed tunnel center line. Direction is identical with that of the same wires at the bottom.

Careful observations are taken of the bottom part of wires, using a very accurate instrument and special sighting devices. Readings are taken over and over again, and the results averaged. The tunnel line is then established in the correct direction, by reference to surface readings.

This work must be done at a time when men and equipment are not working, as ventilating currents and vibration can disturb the wires.

The line is extended through the tunnel by laser beam and/or transit, and marked on spads (markers) driven into holes drilled in the roof.

Exploration. Tunnels are seldom driven blind. Preliminary drilling is done along the route to determine the type of rock, the amount of water to be expected, and the danger of mud slides. Test holes are drilled from the surface, usually with diamond drills that can bring up cores for inspection.

Diamond drilling may also be done from the heading, where dangerous conditions are expected. This precaution has often revealed the presence of such quantities of water or unstable soil ahead, that disaster

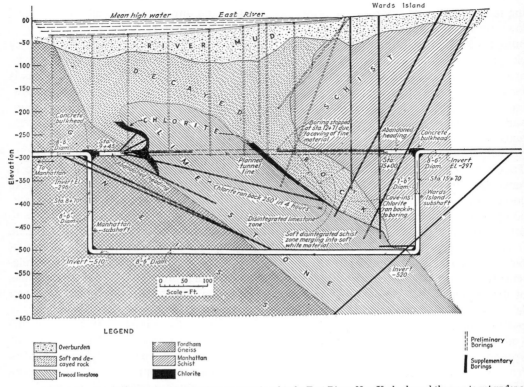

Extensive core drilling on Wards Island sewer tunnel under the East River, New York, showed the way to get under a bad fault that stopped driving on original upper level tunnel. Preliminary borings failed to reveal true conditions.

By permission from "Practical Tunnel Driving" by Richardson & Mayo, Copyright 1941, McGraw-Hill Book Company, Inc.

Fig. 9-30. Underground exploration and tunnel detour

might have resulted had it been broken into by a full-face blast.

Figure 9-30 shows extensive core drilling that was done for a sewer tunnel under the East River, New York City, in order to find a way to avoid a dangerous seam of decayed rock.

Dangers. Underground work is naturally very dangerous, and it is greatly to the credit of tunnel men and labor departments that there are so few accidents.

The most evident danger is that of collapse. Most soils and many rock formations will slump rather quickly into any hole cut under them. In any given material, this tendency increases markedly with depth. Below 500 feet even apparently firm rock may creep, and break off slabs with ex-

plosive violence. There is always danger of loose pieces falling.

Caving and breaking off are combatted with compressed air, timbers, steel and concrete linings, and holding bolts.

If the soil will not stand at all without support, bracing must be installed ahead of the digging; or the heading protected by a movable shield.

Water, with or without accompanying soil, may break into a tunnel in such volume as to flood it completely within minutes. Escape of men may be difficult, machinery is apt to be abandoned, and an expensive and tedious job of sealing off the water and pumping out the tunnel is often required before work can be resumed.

Fire must be carefully guarded against,

particularly on jobs using compressed air and/or timbers.

Air conditions are difficult to keep healthy. Drills produce rock dust, and most air powered machines have foul, oil-charged exhausts. Explosives produce fumes. Some clay and rock formations give off unpleasant or poisonous vapors. The increasing use of internal combustion engines underground makes tremendous demands on ventilation systems that try to clear out dangerous or irritating exhaust gases.

Silicosis, a lung disease caused by breathing dust from rock drills over long periods, was formerly one of the greatest health dangers of tunnel work. It now can be almost entirely avoided by wet drilling and good ventilation.

The Delaware Aqueduct used 10,000 feet of air a minute at each heading. A standard estimate is 1500 feet per drill.

Shallow rocks are cool, but in deeper work an increase of 1° F. for every fifty feet of depth can be expected.

Noise is deafening, particularly during drilling, as the sound echoes back and forth in the confined space. Diesel engines are adding to the uproar.

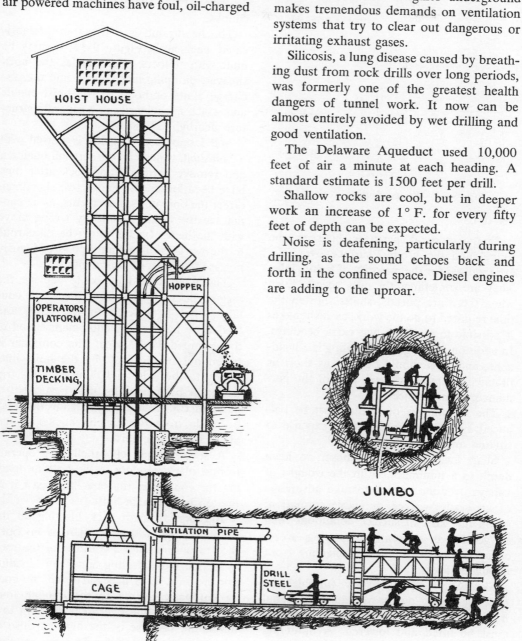

Courtesy of Delaware Water Supply News

Fig. 9-31. Shaft and heading equipment

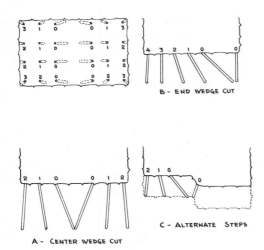

B - END WEDGE CUT

C - ALTERNATE STEPS

A - CENTER WEDGE CUT

Fig. 9-32. Wedge drilling in shaft

SHAFTS

Shafts—vertical passages between the tunnel and the ground surface over it—are required for the majority of tunnel jobs. They are sometimes the only access. Even when there are portals, shafts shorten the time required to do the work, as each makes it possible to work on two extra headings. In addition, underground hauling is a headache, and runs should be kept as short as practical. Some shafts are part of the permanent tunnel project.

The advantages of shafts must be balanced against the considerable expense of sinking and equipping them.

Shaft location may be chosen to keep depth to a minimum, as in the troughs of valleys over the tunnel; to take advantage of an easily worked or stable formation; or on the basis of surface conditions such as good access, nearby dumping areas, cheap land, or distance or shielding from populated areas.

Size. Shaft size is highly variable, depending largely on the volume of material it must handle and the size of objects that must be lifted and lowered. A minimum size, about 11 by 13 feet inside the lining, accommodates a single hoist and supply

elevator and a ladderway. The ladderway includes a ladder, electric and high pressure air lines, water pump discharge pipe, and ventilator ducts, all of which must be protected from swinging loads or falling chunks.

The headframe is a tower of prefabricated steel, as in Figure 9-31, or may be built with timbers. This carries the hoist sheaves, dumping mechanism, and the discharge chute or hopper. The hoist engine and winch is ordinarily in a separate structure nearby.

Soil Excavation. Digging is started with a clamshell, which can dig soft soil unaided, and remove hard soil and rock after they have been loosened. One or two signalmen direct the operator's movements, as he cannot see the bottom, and any wrong move with the heavy bucket might be disastrous to the workers. The clamshell is ordinarily not used below a 25 foot depth.

The next stage may be to replace the digging bucket with a light bucket or container that is lowered to the floor, and loaded by hand or by equipment suited to the cramped work space. The container is raised out of the shaft by the hoist line, swung to the side, and dumped by a trip device or by hand. This may be used to a depth of 100 feet, or a direct transition may be made from the digging bucket to use of the headframe hoist.

A special small clamshell may operate from a platform close to the bottom, loading the containers that are lifted past it to the top by the main hoist.

Blasting. In shaft rock blasting all the holes are tight—that is, there is no open face to permit sideward throw of the rock—so that close drilling and heavy loading are the rule. It is necessary that the rock be cut back cleanly to the digging lines and important that overbreak be kept to a minimum, because of the high expense of removing muck, and the frequent requirement of filling all spaces outside the lining.

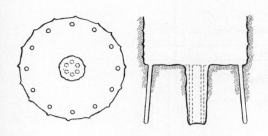

Fig. 9-33. Burn cut

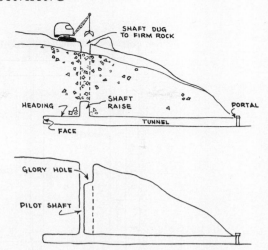

Fig. 9-34. Raise and glory hole shaft excavation

Figure 9-32 shows typical drilling patterns for shaft and tunnel work. A set of two or more converging angle holes (wedge holes) are drilled, and other sets of straight or slightly angled holes next to them, until the rim is reached. The wedge holes are heavily loaded, so that they crush and kick out the rock between them, making an opening into which the rock around can move sideward when the next ring of holes is fired. These in turn make space for the next set. Firing is best done by short period delays.

In (B) the floor is lowered on only one side in each shot. This permits drilling to be resumed on one side while muck is being loaded from the other.

Figure 9-33 shows a burn hole shot, with the center holes parallel instead of angled.

The blast is fired from the top, after all men and equipment are out of the shaft, except that in very deep work some equipment might be merely raised far enough to be out of immediate danger.

After the explosion the bottom will be full of fumes, which would take a long time to dissipate naturally. These may be blown out by lowering the tool air lines with the ends open, or extending low pressure ventilating ducts to the bottom. (These have to be dismantled or pulled back a considerable distance to avoid damage from the blast.) A suction line (foul air duct) is more effective at cleaning the air than a blow or pressure line, as the fumes tend to settle.

Some shafts are large enough to provide space to load the muck by machinery, but in many of them it is tossed, rolled, or hand shoveled into buckets or skips, that are removed by the hoist when filled. The best fragmentation for this type of loading is usually one-man stone, that is, pieces that one man can handle conveniently.

Drilling can be resumed as soon as part of the bottom is cleared. Six or seven foot steels giving a five foot penetration are often used, but longer or shorter ones may be better in particular circumstances. Hand and wagon drills are standard, although special jumbos have given good results.

Working Up. When a large shaft is required as part of the finished job, but is not needed for the early tunnel work, it may be more economically cut from below. With this method the blasted rock falls to the tunnel floor, and is removed through the portal.

The glory hole method is to sink a small pilot shaft to the tunnel, then to dig the large shaft from above, blasting or pushing the muck into the small shaft so that it will fall to the tunnel. See Figure 9-34.

Shaft Lining. In soft ground the shaft is protected from caving by setting sheeting

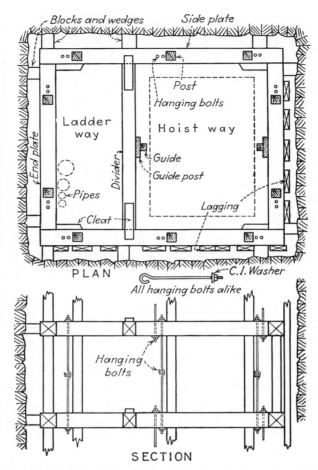

By permission from "Practical Tunnel Driving" by Richardson & Mayo, Copyright 1941, McGraw-Hill Book Company, Inc.

Fig. 9-35. Timbering for small shafts

planks or sheet piling, held by whalers, in much the same way described earlier for ditches and cellars. The whalers are interlocked at the corners to hold each other in position, and additional divider beams may be run across between the ladderway and the hoist way. See Figure 9-35.

If the soft soil is too deep to be held by sheeting driven from the top, successive layers can be driven from inside the shaft, inside the upper ones. If the sheeting is driven with an outward batter, shaft size will be preserved. If driven straight, the inside diameter of the shaft will decrease, so that the top would have to be oversize

to allow it to be full size at the bottom.

Shaft lining or timbering is required for the hoist and to a less extent for the utilities even when the soil or rock is self-supporting. When not needed for wall support, timbering may follow twenty or more feet behind the digging, to avoid interference with the work, and damage to the lower section in blasts.

The lining must be supported vertically to prevent it from slipping down. Each new set is fastened by hanging bolts to that above, and every hundred feet or so horizontal notches or shelves are cut into the walls to provide fresh support.

Fig. 9-36. Tunneling with big equipment

Timbered shafts are usually rectangular, while metal or concrete linings call for circular cross section. Steel ribs are made up, curved to the proper arc, and divided into two or three pieces that are lowered endways and then supported by hanging bolts until fastened into a full circle. The actual lining or lagging may be sheet piling or similar material, or the ribs may be built as liner plates, with curved flanges which butt against those above and below to make a continuous sheet.

A continuous lining or lagging is used in soil that might squeeze between ribs or timbers, or in rock that scales or breaks off so that falling pieces would endanger workmen. Unlagged walls may often be kept intact by spraying with concrete or a bituminous mixture.

Drainage. Most shafts are wet. If there is only a little water it can be bailed into the bucket and hoisted with the muck. More often it is removed by a pump with a discharge line reaching to the surface or, if the height is great, to one or more pumps that help push the water out of the shaft. All pumps used in deep shaft work should be able to develop very high discharge pressures, so that a good lift can be obtained between boosters.

If water conditions are severe, the area may be predrained by sinking 4 to 12 inch holes with rotary drills, and pumping from them. Depth is too great for ordinary well point work from the surface, but in flowing ground well points may be sunk from the shaft bottom or sides, and the water re-handled by the regular pumps.

A deep wet shaft should have gutters and sumps at intervals, to catch water running down the sides. Pumping to the top from intermediate points may be more efficient than allowing it to get down to the bottom, and raising it from there.

Shaft Sinking Machines. A large and increasing percentage of new shafts are cut in a single or dual operation by gigantic drills, described in Chapter 20.

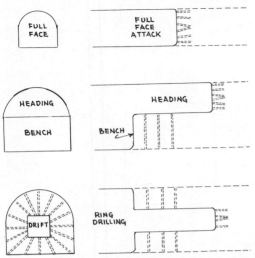

Fig. 9-37. Tunnel headings

HEADINGS

A heading is a digging face and its work area.

Conventional tunnel driving is discussed here. Tunneling machines (borers or moles) are described in Chapter 20.

When the shaft has reached the level of the proposed tunnel floor, two headings are started, one in each direction along the line of the tunnel. In addition, the foot of the shaft may be greatly expanded for storage and maneuver space, and one or more rooms may be built to house compressors, pumps, and other plant equipment.

At first only a single set of tunnel driving equipment may be used, as there will not be space enough for two, and greatest efficiency will be obtained by drilling at one face while mucking at the other. Room for two sets will be made very quickly, but alternate work is sometimes continued until the distance between headings is great, or sometimes for the whole job.

Drilling patterns may be similar to those described for shafts—wedge or burn holes, and successive rings breaking into the crushed-out area. The whole face is usually drilled and blasted in one operation (full-face attack), but a small tunnel (drift) may

be drilled full face, blasted, and cleaned out, then enlarged by radial drilling; or the top may be kept ahead of the bottom (bench-and-heading method). See Figures 9-36 and 9-37.

Pilot Tunnel. Shafts may be partly or wholly replaced by a small pilot tunnel, driven parallel and close to the main tunnel. Crosscuts are driven from this to the main tunnel wherever new headings are to be started. The main tunnel is opened up with a center drift, and enlargement started after it is cut through enough so that both tunnels can be used for traffic.

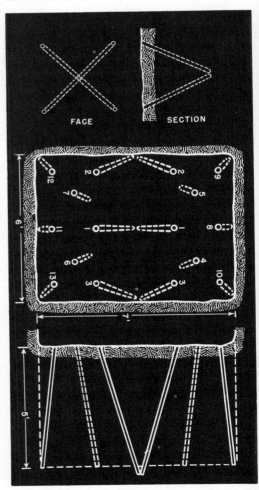

Fig. 9-38. Small tunnel drilling pattern

The extra tunnel may be used for ventilation, both during the work and afterward. It permits a great many operations to be performed at the same time, and may save considerable expense in sinking shafts. This method has been used chiefly for long railroad tunnels through mountains where depth was too great for shafts.

Drilling. The standard tool for small tunnel drilling has been the drifter, a medium weight hand drill with a hand or automatic feed, mounted on a vertical column or a horizontal bar of such length that it can be secured between the floor and roof, or between the sides, by screw-jack ends. Because of the weight of the columns, they become impractical for full face work in tunnels of greater cross section than 10 x 10 feet. It is now being replaced by hydraulic boom mountings.

The drifter permits the drill crew to resume work on the top of the face as soon as blast fumes have cleared away, with the drill men standing on the pile of muck until it is dug away. They can drill the bottom after it is cleared.

Larger tunnels were formerly done by the heading-and-bench method. This permits the use of drifters on short columns for the advance, and approximately vertical jackhammer or wagon drilling for the bench. Sometimes the heading is extended far ahead of the bench, and has its own hauling equipment that dumps over the bench face into other cars, or into a pile to be dug away.

Now the standard method is to use a drill carriage (jumbo) on which power feed drills can be mounted so as to reach all parts of the face at correct angle and to correct depth. Each drill usually does several holes. It can be positioned by hand, or by mechanical, air, or hydraulic controls. Such jumbos may be so constructed as to straddle hauling equipment, so that it need not interfere with removal of muck. They may also carry a cherry picker crane to

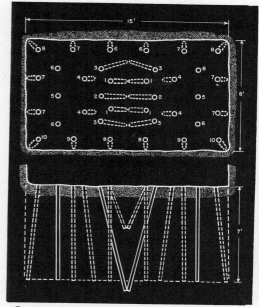

Courtesy of E. I. du Pont de Nemours & Co. (Inc.)

Fig. 9-39. Drilling pattern, large tunnel

pick up empty cars to switch loaded ones through. They are backed away from the face before each blast.

On very large tunnels jumbos may be used on both levels of heading-and-bench work.

Usual drilling depth is 10 to 12 feet, but in any case is seldom deeper than two thirds the smallest dimension of the tunnel.

Figures 9-38 and 9-39 show typical full face drilling patterns.

Bits. Recently tunnel drilling has been partly standardized to use steels threaded to carry detachable bits. These may be multi-use types that can be sharpened by grinding, or sharpened and reshaped by hot milling; one-use or throwaway bits that are discarded when dull; and carbide insert bits. In some mines carbide tipped steels are used, and in others the old fashioned steel with the business end forged into a bit is still doing business.

The carbide insert bit has caused a spectacular advance in speed and ease of hard-rock tunneling. Many tunnel men say,

"There is no such thing as hard rock any more." Carbide outwears steel at an average of about a hundred to one, and gives much more rapid hard rock penetration. The time of handling, transporting, and processing bits is reduced from a major to a minor problem.

Loading. Water resistant explosives with good fume characteristics are desirable in underground work. These qualities are found in gelatin dynamites.

When all holes in a face have been drilled, each is blown out with a high pressure air jet to remove loose cuttings and water. Cartridges are slit (unless the explosive is damaged by water and the hole is wet) and tamped firmly with a wooden pole. It is common practice to place the primer after the first cartridge, with the cap pointed toward the collar of the hole.

Stemming may be taken from the drill cuttings. It is most convenient to use if wrapped in paper bags of the same size as the cartridges. If this material is very high in silica its use as stemming might increase the silica in the air enough so that pre-wrapped blanks supplied by powder manufacturers might be preferred. There are also wood and rubber plugs that are very satisfactory.

It is good practice to place a wad of paper between the explosive and the stemming, so that the powder can be easily and safely located in case of a misfire.

There is danger of premature explosion from stray currents. A common precaution is to take down or "kill" all electric wiring within five hundred feet of the face before starting to load. Safety flashlights, of hand or cap models, or headlights from a battery locomotive can be used. It is sometimes a question whether the poor lighting obtained does not offer as much of a hazard as the electricity would.

Even the complete absence of electricity on the job would not guarantee a tunnel face against currents, as underground water is often highly mineralized and will conduct a charge for long distances. Metallic ores may be excellent conductors.

The precautions described earlier for blasting in the presence of electrical hazards should be followed.

Firing. Any wiring hookup can be used —series, parallel, or parallel series, depending on the preference of the blaster. If 440 volt electricity is available it is preferred for firing, although 220 or even 110 will do. Regular blasting machines are also used, but they should not be kept in the tunnel when not in use, because of possible damage from dampness.

All equipment is moved 500 to 1000 feet back from the face, as rocks caroming off the walls can travel long distances. Compressor pipe can be left fairly close to the blast, but ventilation conduit must be stripped way back.

Move-back requirements may be reduced by a portable metal buffer wheeled into place or set up on the jumbo before the blast.

Checking. It is important that a thorough check be made after the blast for misfires. Tunnel work brings a large number of men into close contact with the heading, and any accidental explosion during mucking or drilling would be disastrous. The best check is inspection by experienced men.

If an unexploded hole is found, and the wires are intact, they can be hooked up and fired. If the wires are missing, the stemming can be washed out by a water jet, and a new primer inserted and fired. Or a parallel hole, about two feet away, can be drilled, loaded, and fired. The muck must be inspected for unexploded cartridges.

MUCKING

Loading. In small tunnels blasted rock may be dug by hand, although the excellent mechanical loaders adapted to work in tight quarters that are now available, and the rising price of labor, are steadily reducing

Courtesy of The Robbins Company

Fig. 9-40. Bored tunnel face

the practice. Output for the loading gang is generally figured at about ½ to ⅔ yards per hour per man, although one man may load up to 2 yards under favorable conditions. The difference lies in work of loosening, handling cars, and other delays.

The swell or "growth" of rock in passing from the solid to the blasted state averages about 50%. In tunnels, mucking is usually calculated in terms of loose yards, in mines in number of tons loaded.

Slick sheets should be used in connection with hand loading. These are thin steel plate, ¼ or ⁵⁄₁₆ inch, in pieces about 4 x 6 feet, with holes punched for convenience in picking up for moving. They are laid to cover the tunnel floor for 10 to 25 feet back from the face before each shot. Large rocks are picked up and thrown into the cars individually, while the finer material is dug by shovels that slide easily along the metal surface.

Mechanical loaders include full revolving shovels with short booms and proportion-

Courtesy of The Robbins Company

Fig. 9-41. Tunnel borer

Courtesy of The Ruth Company

Fig. 9-42. Diesel-powered mine locomotive

ately larger buckets, that move either on crawlers or rails, and until very recently used only air or electric power. There are also railroad type shovels that use one track and load cars on another beside it, and may have a cherry picker for changing cars on the back.

Special tunnel-mucking machines are available in large variety. Most of them are rail-mounted, although crawlers are gaining in popularity. The bucket can be swung from side to side to reach the full floor area, and is filled by pushing into the pile.

It is then lifted, in some models over the machine to discharge into a car or conveyor belt behind; in others it loads a built-in conveyor that discharges to the rear. In either case, the car may be coupled to the mucker so that it is always in loading position.

BORING

A tunnel may be cut to full size in a single operation by boring with a tunneling machine, sometimes called a mole. These machines are described in Chapter 20.

A tunnel borer grinds, chips, or digs its way through formations, by rotary or oscillating motion of cutter teeth, and deposits the muck onto a conveyor belt for discharge into haulers at the rear.

In addition, it may and often must provide for placing steel or concrete linings around or behind itself. Such lining can be used to take the thrust of its crowding force.

These machines can usually be disassembled sufficiently to be brought down a large shaft, and assembled in a conventionally-dug tunnel section at its foot. However, whenever possible, they start work at a portal (outside entrance).

HAULING

Almost any type of hauling unit may be used in a tunnel, from a wheelbarrow to an off-the-road ten wheeler. It is a matter of tunnel size, speed of driving, ventilation, and preferences of the management.

The traditional system is small muck cars pulled along narrow gauge tracks by electric locomotives. The locomotives can take power from either batteries or high lines,

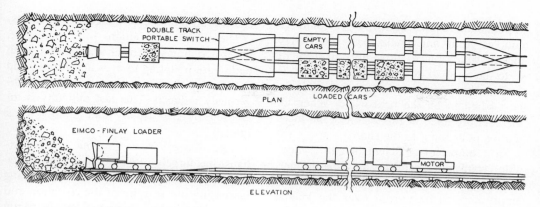

Fig. 9-43. Portable switch

and range in weight from 4 tons up. There is an increasing use of diesel locomotives with exhaust conditioners in well ventilated tunnels. See Figure 9-42.

Cars are usually side dump types, although many special constructions are found. The width is governed by the tunnel and the gauge of the track, and should be small enough to allow passing in the tunnel. Car width is generally about twice the track width.

The capacity of the car may be limited by switching arrangements. If they are pushed by hand, capacity is limited to one or two yards, as heavier cars will need to be pried along the tracks, rather than shoulder-pushed. The car must be low enough to go under the discharge of the mucking machine being used. If hand loaded, it must not be over four feet high.

The loaded muck cars are hauled to the shaft and run into hoisting cages, in which they are lifted to the top, where they are dumped by side tipping. There are also special cars that can be lifted directly, without entering a cage. Or they may be dumped at the bottom into a hoisting skip.

The perpetual problem in tunnel haulage, which becomes more acute as size decreases, is bypassing the empty cars (or trucks) going to the face around the full ones coming away from it. Empty cars may be switched to the side; or if they are small, be lifted or pushed off the track by hand, where there is space for only one track. Larger ones may be handled by a cherry picker. In either case the spotting arrangement shown in Figure 9-43 may be used.

The locomotive pulls a string of empties into the heading and stops to let the cherry picker take up the rearmost car and set it aside. The locomotive then backs far enough so that the car can be replaced on the track in front of it; then pushes that car up the loader. While it is being loaded, it backs so that another car can be picked off.

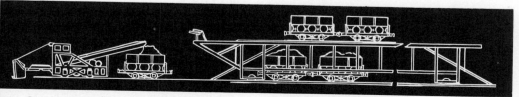

Fig. 9-44. Grasshopper overhead switch

When the car is loaded, the locomotive couples to it and backs past the cherry picker, which places the empty in front of it to be pushed to the face. While it is loaded, the rear empty is again set aside, to be pushed in on the next cycle. When all the cars are filled in this manner, the locomotive pulls them to the shaft.

In a tunnel of sufficient height, a movable framework called a Grasshopper, Figure 9-44, can be used. This allows the empty cars to be moved over the loaded ones, and can be pulled up to the face by the loader.

A conveyor belt may be set up so that a full train of cars can be backed under it, and loaded one by one from the front back.

Conveyor belts can also be set up to haul from the face to the shaft. No switching arrangements are required, but this unit cannot be used readily to bring supplies from the shaft to the face; considerable work is involved in dismantling or protecting it for a blast, and there is constant work adding sections to keep it in touch with the digging.

Diesel-powered trucks are increasing in underground popularity. They carry much bigger loads than mine cars, and if sufficient width is available to make passing possible, they get past each other with fewer complications than rail-mounted carriers. The shuttle types, such as the Dumptor, which are equally comfortable going backward or forward, are often better adapted to the work than those which have to be turned in the tunnel.

The use of internal-combustion engines fouls the air, so that very good ventilation is required.

Exhaust Gas. The exhaust from a gasoline engine contains carbon monoxide, an odorless but poisonous gas that soon makes any closed-in place deadly to life. Amounts of monoxide that are not sufficient to cause unconsciousness or death may temporarily damage judgment and reasoning power, causing an increase in danger of accidents.

Diesel exhaust contains little monoxide but it is rich in various chemicals that smell badly, are irritating to eyes and throat, and that fog up the air so that visibility is dangerously reduced. This last difficulty is increased by the usually bad lighting in a tunnel.

The danger from gasoline engine exhaust has largely prevented use of this type of power underground. Diesels are finding increasing use in spite of the irritation and danger they cause. Their presence is partly compensated by increasing the ventilation, but conditions do become very bad. They are often made worse by an astonishing lack of care in adjustment of the engines. Diesel trucks sometimes emerge from tunnels belching black smoke, presumably caused by defective or souped-up injectors that would justify arrest of the driver on an open highway.

Various types of scrubbers using water and chemicals to dissolve and neutralize gases, and secondary catalytic oxidizers that serve also as mufflers, are used to make internal combustion engines acceptable underground. These are described in Chapter 12.

Good ventilation and lots of it is a basic requirement, even when such devices are efficient. The most they can do is reduce the exhaust to carbon dioxide and water. Carbon dioxide is not poisonous or irritating, but in sufficient concentration it has a suffocating effect that can cause impairment of judgment, unconsciousness, and death.

WATER

Ground water is a problem in most tunnels, and may be the principal one in some. Many mining tunnels, some of them mile in length, are made solely to lower the water table. There may be seepage all along the line, adding up to a considerable volume to be drained or more often pumped way. Gushing springs may be exposed by any blast, or may open up from seepage points well behind the face. Underground lakes or rivers may be encountered that are capable of flooding the work in spite of continuous pumping. Veins of soft water soaked soil may be found in hard rock, that may break into and fill the tunnel.

The first necessity is to have adequate pump capacity. The tendency is to underestimate requirements, largely because pumps and lines are expensive, partly because even careful exploration from the top seldom reveals the full quantity and pressure of water that may be encountered.

If a tunnel runs uphill from a portal, drainage may be by natural flow through a ditch cut along the side. If an upgrade from a shaft, it can be drained to a pump inlet at the shaft foot. This arrangement is easy and inexpensive, but seldom satisfactory, because of repeated blocking of the ditch by rock falls from walls or from hauling equipment, resulting in water running over the floor, making it sloppy and often undermining the track or spoiling the road surface. The ditch also takes up more space than a pipe, and there has not yet been a tunnel with floor space to spare.

The conventional arrangement is to pump all water. A small centrifugal pump, usually air-driven, is kept near the face, and takes from a sump and discharges into a pipe running back toward the portal or shaft. Another sump is provided every 500 to 1500 feet back to collect local water for another centrifugal, usually electric-powered. Each pump may discharge into the sump behind it, which is kept down by another pump, usually of a larger size. Another arrangement is to have all pumps discharge through check valves into a common discharge line. A powerful electric pump of the piston or centrifugal jetting type is installed at the shaft bottom, and as many boosters as are required for the lift installed at intervals in niches in the shaft.

Pipe lines vary from 1½ to 10 inches in diameter.

The pump or pumps at the base of the shaft are sometimes placed in a sealed room, with power and control directly from the shaft top. In other cases the pumps are in the open, but are of the submersible type. These arrangements permit use of the units along with emergency pumps if the tunnel should be flooded.

Grouting. Water inflow can often be checked by grouting. This may be done by drilling deep into the rock in the direction of the supposed source of the water, sealing in pipes with cement, and then pumping in cement and water grout, either straight for seepage or mixed with sawdust or shavings for gushing flow. This may be done in advance of the tunnel driving in very wet areas, by fanning the grout holes out from the face and edges of the heading, as in Figure 9-45.

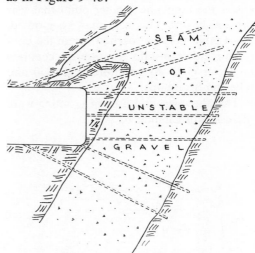

Fig. 9-45. Exploration and grouting holes

Grouting is also done through completed linings, either to check water or to fill in spaces between it and the wall. Grout pipes may be cemented into a concrete lining when it is poured.

Successful grouting of a wet seam sometimes merely diverts the water so that it

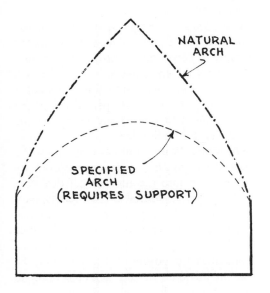

Fig. 9-46. Cutting crown for self-support

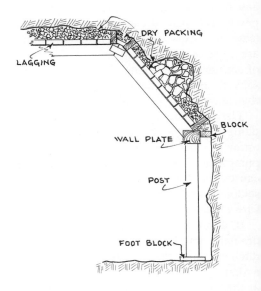

Fig. 9-48. Timber arch on posts

enters the tunnel at another point that was previously dry. This also may be grouted, but a point may be reached where the contractor either installs a complete concrete lining, or gives up the effort to seal off and relies on his pumps.

The above-ground uses of grouting were discussed in Chapter 6.

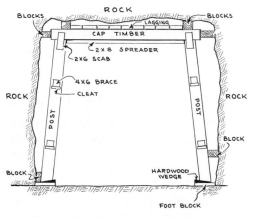

Fig. 9-47. Timber set, small tunnel

ROCK SUPPORT

Ground pressure in rock tunnels is difficult or impossible to estimate. This problem is dealt with in detail in "Rock Tunneling with Steel Supports." In firm formations there will be little or no pressure until depths over 500 feet are reached.

However there are soft, joined, or laminated formations that will scale off or fall from a flat or moderately curved roof, until a Gothic or pointed arch develops, after which it will be self supporting. If bracing is done only to support the roof, it is a question whether it will not be more economical to cut up to a stable roof line, and avoid placing of supports. See Figure 9-46.

In any roof problem, width is a very important factor, as wide spans will drop pieces or fall in much more readily than narrow ones.

Many rock tunnels are perfectly safe without any bracing. Many others get by without accidents. But very often it is necessary to place supports directly after the digging, or within a few days. Also, the

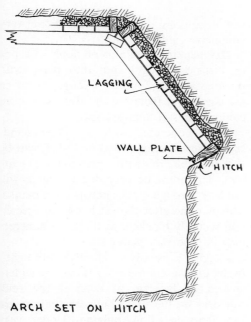

ARCH SET ON HITCH

Fig. 9-49. Timber arch on hitch

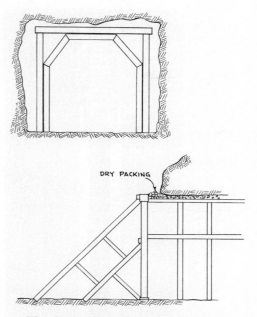

Fig. 9-50. Timber bracing at portal

majority of tunnels outside of mines are more or less permanent in nature, and except in very firm rock, will require lining to prevent deterioration and to reduce or eliminate maintenance.

Support or lining may be wood timber, steel ribs, plates or bolts, or concrete. Concrete is frequently placed inside one of the other types of support.

Timber is the oldest material used, and is found in ancient tunnels. Concrete was used to some extent by the Romans, and has become the standard for permanent installations. Steel liners and roof bolts are quite modern developments, and are rapidly replacing timbering.

Timbering. Figures 9-47 to 9-49 show some designs for timbering. The square-set framing is confined to small tunnels, and various forms of arch construction can be used in quite large ones. The arch may be supported on posts supported in the floor, or rest on a springline shelf (hitch) cut in the sidewalls. Support may vary between these methods with changes in ground, or in shape of the edges.

Posts should be fastened to the wall plate by dowels, lag bolts, or scabs (nailed-on pieces) so that they cannot fall if relieved of weight.

The weight of timbering varies with expected ground pressure. Sometimes it is merely a light roof to catch light rockfalls, at other times a high strength lining designed to resist squeeze from all directions, including the bottom.

Where timbering ends at a portal, or at an enlarged shaft base, it must be securely braced by diagonal beams, as in Figure 9-50, so that any compression developing in the tunnel will not squeeze it out.

Packing. In rock or soil that tends to push in, it is important not to leave any space between the lagging and the wall or roof, as any inward movement will increase the instability of the ground, and may cause it to exert tremendously more pressure than if it had been held in its original position.

An exception is found in swelling or squeezing ground that is allowed a limited space for movement.

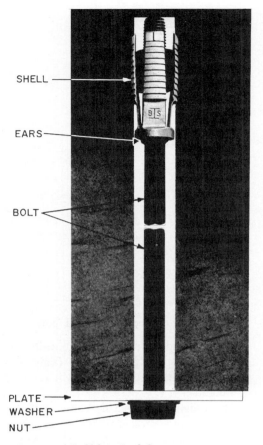

SHELL

EARS

BOLT

PLATE
WASHER
NUT

Courtesy of Bethlehem Steel Company

Fig. 9-51. Square head mine roof bolt with shell

Initial movement is prevented by packing the space between the lagging and the rock. The most economical system is to use a dry packing of fine muck, which is shoveled behind the planks as they are placed. At the crown it must be thrown in from the end and securely rammed—a tedious, disagreeable job that is seldom well done.

Large overbreaks may be filled with dry walls built of chunks that can be handled by one man. Packing of scrap wood is good only for temporary service, as it will ultimately decay and leave voids. Timbering for permanent or semi-permanent use should be treated with preservative, as the

wet conditions commonly found are ver conducive to decay.

Dry packing may also be done with pe or birds-eye gravel shot into place witl pneumatic guns, either through holes i the lagging, or from the end. Its use is mor common in soil than in rock.

Lean concrete, with a cement-sand small stone mix of 1:3:6 or 1:4:8, can b shot into the arch with a pneumatic placin; tool. This must be very dry so that it wil not leak through cracks between the planks This is done after the set has been erecte and securely blocked, as the fresh concret may impose very heavy stress.

Lagging is usually set closely (skin tight in the crown. On the walls it may be widel spaced or lacking, as even if the rocl squeezes in, the spans between timbers ar too short to allow bulging. Under ver heavy conditions, the timbers may be se skin tight, so that lagging is not needed.

If the tunnel is to be concreted, the lag ging may be placed inside the timbers, t provide a smoother outer form that save concrete yardage. The disadvantages ar that much more packing is needed and tha the fastening is under tension rather tha compression, so that heavy pressure ma make the lagging pop off the timbers. Th effect may be cumulative, as yielding o one fastening increases the strain on th next, so that a considerable length ma give way at one time.

Steel Ribs. Steel supports are no standard in tunnel work. They are easier t handle, and allow substantial saving in ex cavation. This is because for a give strength, they are only half as thick; anc the projections of ribs into a concrete lin ing are counted as reinforcing. In timbe construction, the outside line of the concrete is figured as the inside line of the timbers, and the concrete used to fill ou to the lagging is largely figured as waste. On small tunnels the saving in excavation may be 30 per cent and in concrete 50.

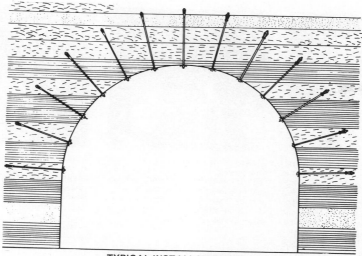

TYPICAL INSTALLATION OF
ROOF BOLTS FOR TUNNEL WORK
Courtesy of Bethlehem Steel Company

Fig. 9-52A. Roof bolts in a tunnel arch

However, steel liners are more vulnerable to blasting damage, and do not give warning of impending collapse under load by groaning, as timbers do.

The steel ribs are made in two pieces, occasionally more. They are brought in endways and set up individually. The lagging may be wood planks or steel liner plates. If the former, the ribs must be well strutted to each other to keep them in line.

As in the case of wood, steel lining may be only a roof or crown support based on shelves at the spring line in the side walls, or a complete tunnel enclosure.

Roof Bolts. It has been found in mining and tunneling operations that unsafe rock will often support itself safely over wide spans if it is reinforced with steel bolts.

In laminated (thin bedded) formations the effect is similar to that obtained in plywood and other layered wood constructions. Several weak and thin layers may be very strong when bonded together. In jointed and fissured rock, the bolts if used properly and in sufficient numbers, restore to the rock the massive strength it had before it separated into blocks and pieces.

Expansion bolts are used, rather similar to those that fasten wood framing to masonry. The type shown in Figure 9-51 is made in 5/8, 3/4, and 7/8 inch diameter. The 3/4 has a minimum breaking load of

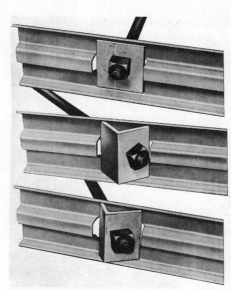

Courtesy of Bethlehem Steel Company

Fig. 9-52B. Mine roof ties, with bolts and washers

1 As with conventional roof bolts, holes are drilled in the roof to be bolted following proper placement pattern. A 1⅜-in. hole is used for ⅝-in., ¾-in., ⅞-in., and 1-in. bolts; a 1⅝-in. hole, for 1¼-in. bolts. Hole depth should be carefully controlled so that the protruding end of the bolt, when finally installed, will accommodate the bearing plate, hardened washer, and nut.

2 Capsule is inserted in the hole and pushed to the back of the hole with the bolt. The bolt end with the sealing ring or retaining washer is inserted first. The protruding end should be fitted with a cap nut or other device for driving.

3 As soon as the capsule is positioned at the back of the hole, the bolt is advanced and rotated through the capsule with an impact wrench or power rotating tool. This mixes the resin, stone aggregate, and hardening agent. *After the bolt reaches the back of the hole, mixing should be continued for a minimum of 15 seconds.*

4 Although it is not yet ready for a load, the bolt is securely held in the hole by the mixed resin. Installation equipment can now be moved to the next hole. Depending on rock and air temperature, a waiting period is required for the resin to harden prior to tensioning the bolt. At 70 F, a 30-minute curing time is suggested; at 60 F, a 60-minute curing time.

5 After waiting the proper time for curing, the bearing plate, hardened washer, and nut are attached. The bolt is then tensioned to a predetermined torque value. Use of hardened washers insures more accurate torque/tension relationships.

6 Tightening torque is dependent on the anchorage capacity of the installation. For this reason, a pull test should be made to determine the anchorage capacity.

Fig. 9-52C. Installation of resin-anchor roof bolts

22,500 pounds in regular strength, and 32,000 in high strength steel. Lengths are 2 to 8 feet, in 6″ steps.

The bolt threads into the plug of an expansion shell that fits into a 1⅜″ diameter hole. Ears on the bolt prevent it from sliding too far into the shell, so that tightening pulls the plug down into the shell, expanding it against the sides of the hole.

Roof ties, Figure 9-52, may be used to support the roof between bolts. Wire mesh can be used in addition to the ties, or instead of them, where the problem is separation of small pieces.

Flat or dished (reinforced) plates of 3/8 or 1/4″ thickness and 6″ diameter are used where ties are not needed. They are usually made with a 1⅜″ hole. A hard steel washer

1. Drill hole to required depth. Install bolts perpendicular to bearing surface. Pattern of bolt holes is predetermined according to conditions of particular strata.

1. Drill 1¼-in. diameter hole (for 1-in. slotted bolt) to a depth 3 in. less than the length of the bolt.

2. Insert assembly—square-head bolt, hardened steel washer, plate washer, and expansion shell and steel plug—into hole.

2. Insert steel wedge into slotted end of bolt. Place bolt through opening in both the hardened steel washer and the plate washer and drive unit into hole.

3. Tighten bolt head with power wrench. As bolt turns, plug is drawn down on threads. This action expands the serrated section of the shell.

3. Driving dolly must be used to protect threads. Wedge expands slotted end of bolt providing anchorage.

4. Shell is now anchored securely. Steel roof plate furnishes additional support. Steel ties, connecting a series of bolts, may be used in addition to steel plates.

4. Tighten nut with impact wrench. With nut drawn up snug, the steel anchor plate bears against the rock surface.

HEADED BOLT   SLOTTED BOLT

Courtesy of Bethlehem Steel Company

Fig. 9-52D. Installation of headed and slotted bolts

prevents the bolt head from pulling through.

The drilled hole must be at least as deep as the bolt is long, and may be deeper. The bolt is usually provided with a hard steel washer and assembled to the shell at the factory. The shell is pushed through a plate and as far as it will go into the hole, and the bolt head is tightened with an impact air wrench, usually to a torque of 150 to 200 pounds-feet.

The bolt head is held outside by the plate and washer, so the threaded plug is pulled

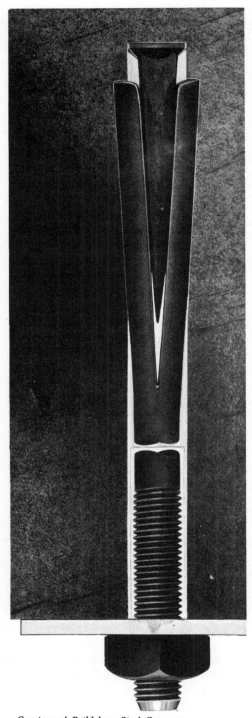

outward as it is tightened, squeezing the shell against and into the walls of the hole. The grip of the shell in hard shale, sandstone, or limestone is usually greater than the breaking load of the bolt. In soft shale grip is usually less than bolt strength. In rotten rock, grip may not be adequate.

Holding power can often be increased by injecting cement grout into the space around the bolt, after tightening.

A very strong grip is obtained by anchoring with resin. A capsule containing resin, fine aggregate, and a tube of hardener is placed in the hole, then broken and mixed by turning a bolt into it. After the mixture has set 30 to 60 minutes, a bearing plate and hard washer are tightened onto the bolt with a nut and a torque wrench.

The washers offer the advantage of producing greater bolt tension with the same effort, or the same tension with less effort. The more uniform tension provides greater security.

The one inch bolt in Figure 9-53 is intended for use in a 1¼ inch hole, which should be drilled to exact depth. A wedge is inserted in the slot at the unthreaded upper end of the bolt, which is inserted in the hole and driven to refusal with an air hammer. The end is spread by the wedge until it anchors firmly in the hole. A nut is then threaded on to the projecting thread and tightened. Similar washers are used.

If the roof is too low to permit inserting a bolt of the full length required, short bolts and extension pieces can be used.

Bolting requires from one-fifth to one tenth the steel required for ribs and lagging and under many conditions is equally strong. In addition, it saves the need of excavating space in which to set the steel structure, and reduces the amount of concrete required for permanent lining.

Elimination of all ribs and timbering makes a tunnel easier to work in, as there are fewer obstructions, and it provides for a smoother flow of ventilating air.

Courtesy of Bethlehem Steel Company

Fig. 9-53. Slotted mine roof bolt

Another important advantage is that the economy of the work causes it to be done on roofs that might be judged to be self supporting if bracing were time consuming and expensive. The bolts can also be installed right up to the face immediately after blasting, so that protection is available to the heading crew. As a result, their use in the rather wide range of conditions where they are applicable results in a marked decrease in roof-fall accidents.

Heavy wire mesh may be used to prevent falling of small fragments in between the bolts. In some instances gunnite is used to minimize air-slacking and spalling.

Rock anchor bolts, which are similar to the slotted mine roof bolts, are used along highway and railroad cuts to prevent rock falls and slides.

Concrete Lining. Installation of concrete lining is construction rather than excavation work (however necessary it may be to the excavation) and will be only briefly considered.

There are two general procedures—soft ground technique, in which it is placed immediately behind the digging, and is necessary to the driving of the tunnel, and hard ground. Under the second heading comes work in rock that is self supporting, and requires lining for permanence, scaling protection, or waterproofing; and unstable soil or rock that is adequately held in place by timber or steel.

The soft ground technique is to follow the heading closely, with some resulting interference between operations. Perhaps the most serious is maintaining a track for muck cars through the pouring operation, and across the freshly laid invert (floor). Steel beam bridges may be used to carry the track in this section.

The invert may be laid about 1½ inches low, protected with planking, and brought up to grade with a top dressing of cement mortar as a finishing operation after the tunnel is complete.

Traveling forms of various types are used. For fast schedules, it is essential to have telescoping forms that can be folded up and moved through other forms supporting more recently poured sections. On other jobs, forms are used that can be collapsed just enough to break away from the concrete surface so as to be moved ahead to the next section. In either case, the forms are carried on carriages, that may move on steel wheels and tracks, or on rubber tires, depending largely on the muck haulage method used.

Breakthroughs. It sometimes happens, in spite of precautions, that there will be a sudden rush of water or mud into the tunnel. This most often occurs at the face immediately after a blast. Sometimes the source is a limited underground pocket which will give no trouble after it once drains off. At other times a stream or large body of water will keep up a continuous flow. If the water is muddy, or the flow is partly or wholly mud, an unstable soil formation has been reached which may give increasing rather than diminishing trouble.

In any case the first step is to seal off the face with a bulkhead (wall) as quickly as possible. Timber, sandbags, or sandbags with timber may be used. Occasionally timber may be backed with concrete.

The bulkhead must not be used as a dam while being built. Pipes should be built into it large enough to take the water flow until the structure is complete. Otherwise water pressure will tend to destroy the bulkhead as it is being erected, and conditions will be very dangerous to personnel. With water discharged through pipes, the structure can be properly and strongly made and keyed into the tunnel rim. The water can then be controlled by valves on the pipes.

The bulkhead should also be fitted with pipes for grouting and concrete placement. After the water has been shut off, grout can be injected into the space between the bulkhead and the break, and will sometimes

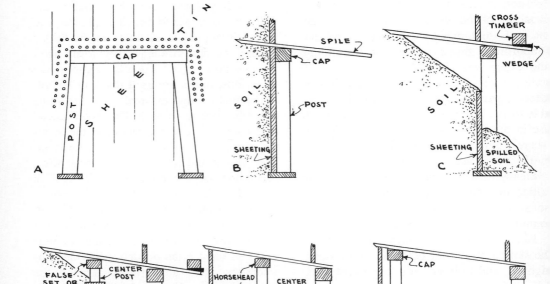

Fig. 9-54. Forepoling soft ground

work back along the water seam and stop or reduce the flow. Grouting may also be done through exploration holes drilled through the bulkhead and into the rock beyond. Over 90,000 bags of cement have been used to control one water pocket.

Further tunneling through such a spot is first in the form of drifts (small tunnels) each of which serves as a base for further grouting, until the ground is consolidated enough to drive the big tunnel.

SOFT GROUND TUNNELING

Soft ground is divided roughly into the following subclasses, description of which is abbreviated from "Practical Tunnel Driving" by Richardson & Mayo, McGraw-Hill 1941:

Running ground: Must be instantly sup-
ported. May be dry sand or gravel, quick-sand, silts and muds.

Soft ground: Roof must be instantly supported, but walls will stand vertically for a few minutes.

Firm ground: Roof will stay up unsupported for a few minutes, and the side walls and face for an hour.

Self-supporting ground: Will stand unsupported while the entire tunnel is driven a few feet ahead of the timbering.

The standard methods of driving through soft ground are forepoling with wood or steel, or working in a shield. Plenum method is keeping out soil and water with air pressure, with either forepoling or shield.

Forepoling. The use of plank forepoles was formerly the standard method of driving a tunnel through soft ground. While this technique has been largely replaced by steel liner and poling plates, it is still widely used on jobs too small to justify obtaining steel.

In forepoling, the tunnel is protected by timbering, and by breast boards set against the face. Planks are driven through slots cut in the breast board and supported cantilever fashion to make a temporary roof, under which dirt can be dug and permanent supports installed.

Figure 9-54 indicates the terminology of the parts and something of the method.

Starting from a shaft lined with plank sheeting, a bent (cap or roof timber and two post supports) is set and securely braced close to the sheeting. Close set holes are drilled through the sheeting in a double line just above the cap and a single line about 18 inches below it. Double vertical lines are drilled just outside the posts.

A set of light forepoles or spiles are made of 2 x 6 planks five to six feet long, sharpened to a chisel point on one end. A piece of sheeting is knocked out between the lines of drill holes above the cap, and a forepole is rested on the cap and driven through the hole, at an upward slant of about 2 inches per foot, for about half its length. Another bit of sheeting is cut or knocked out, and another forepole driven in the same manner, parallel to and touching the first. This process is repeated until the full width of the cap has been covered.

Spiles are then driven into the sides, flaring out about 2 inches per foot, to a penetration six or eight inches deeper than the roof pieces. These may be driven horizontally, or at an upward slant to keep contact with the roof.

A timber is now placed across the shaft, immediately above the free ends of the roof forepoles, which are then forced downward slightly by driving wedges under the timber. The poles are now supported on the cap and held down tightly by the timber and wedges at the rear, so that the front is supported cantilever fashion.

The sheeting is then broken out from the spiles down to the lower line of holes, and the ground allowed to run into the tunnel until it assumes its natural angle. The resulting slope will normally not extend back to the points of the forepoles, but will end at some intermediate position.

Next a horsehead, or false set, is placed under the poles about two feet beyond the sheeting. This consists of a cross piece under the spiles, and a center post set on a small supporting block in the dirt. The spiles are then driven to their full penetration, substituting the support of the horsehead for that of the cap rear timber.

Earth is then raked in until the points of the spiles are almost uncovered. A board the width of the cut and about 18 inches high is set vertically immediately under them. This serves as a breast board to keep more dirt from flowing in, and supports the spiles.

A cap timber is then set to line and grade, and is temporarily supported by a single center post. A "bridge" of 2 x 6 planks is fastened to the top of the cap but separated from it by 4 inch blocks.

The remainder of the side spiles are now driven. Some of these are tapered, and are used wide end forward, so as to reverse the upward slope of the roof spiles and the upper few wall spiles.

The forward cap is now supported by a pair of beams resting on short temporary posts and wedged down from a cross timber. The remainder of the sheeting below the first cap is now broken out from the top down, and the dirt pulled into the tunnel and hauled away. Additional breast boards are set under each other as space becomes available, and held in place by cleats nailed to the side spiles.

When the floor is cut to grade, side posts (legs) are set on below-grade blocks and wedged up until they take the weight of the cap. Wedges are driven between the posts and the side spiles to tighten them. Some-

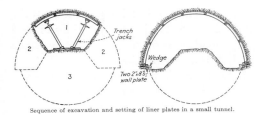

Sequence of excavation and setting of liner plates in a small tunnel.

By permission from "Practical Tunnel Driving" by Richardson & Mayo, Copyright 1941, McGraw-Hill Book Company, Inc.

Fig. 9-55. Setting liner plates

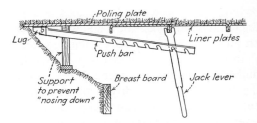

By permission from "Practical Tunnel Driving" by Richardson & Mayo, Copyright 1941, McGraw-Hill Book Company, Inc.

Fig. 9-56. Steel poling plate and jack

times a trench jack must be used to force the side spiles out while setting the legs.

The next set of roof spiles is entered through the bridge slot on top of the second cap, and is driven at the same upward angle. Space for side spiles outside the legs is obtained by knocking out the wedges as the spiles are placed for driving.

All spiles should be driven skin tight (touching throughout their length) except at the corners, where 1 inch boards (lacing) are tacked on. When necessary, cracks are stuffed with excelsior, salt hay, or other packing to prevent inward leakage of soil.

Each timbering set must be braced securely to that behind it, as any shifting will severely weaken the structure.

Spiles are usually driven with a sledgehammer or air hammer. Sometimes they are jacked in—a very tedious job—to avoid jarring the soil.

If the tunnel floor tends to get muddy it should be floored, for convenience of workmen, and to avoid possible shifting or settling of the foot blocks. Sometimes floor spiles are driven if the bottom tends to boil up, but compressed air is a better way to combat this and other difficulties with excessively soft ground.

This method is relatively easy to follow in many soils, but it takes an experienced crew to get through boulders, flowing mud, and other difficult conditions.

Forepoles may also be used with steel ribs instead of timber sets.

The standard soft ground tunneling hand

tool is a short-handled, round-pointed shovel, aided when necessary by a grub hoe (mattock), pickaxe, or crowbar, and often by paving breakers. Special grub hoes have one hammer face for use in driving wedges. In soft clay a curved two-handled draw knife can be used to advantage. It is pulled by two men or a power winch, and slices the clay off in strips.

Liner Plates. Corrugated steel liner plates, curved to match the tunnel rim, and supplied with drilled bolt holes in flanges or overlaps for fastening to each other, are increasingly used for soft ground tunneling. They are made in various sizes, with 16 x 36 inches in common use. A plate of this size made of $\frac{1}{8}$ inch metal weighs about 27 pounds, and if $\frac{1}{4}$ inch stock, 53 pounds. Short plates are available for fitting into the tunnel circumference.

Stiffening ribs are used when the tunnel is over 10 feet in diameter, and for heavy loads in any size opening. They are generally not used when the same strength can be supplied by a heavier gauge plate.

Liner plates are usually a temporary support to hold the tunnel until a concrete lining is installed, usually a matter of hours or a few days after the digging. They are sometimes "robbed" for re-use immediately before the concrete is placed. The safety of this practice depends on the character of the soil, which is a matter for engineers to pass on in each case.

Liner plates are placed from the top down. A small section is dug ahead, the

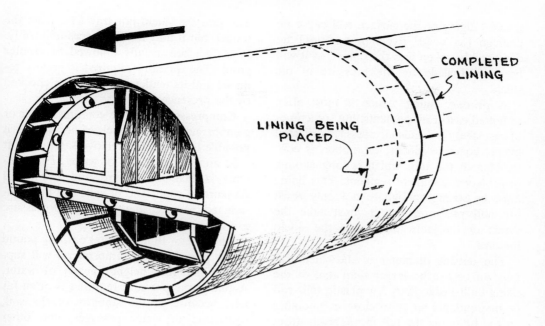

Fig. 9-57. Tunnel driving shield

center plate placed and braced with a post or jack; and then the sides dug away to place the adjoining plates. These are supported radially on cleated center blocks, as in Figure 9-55. When the spring line or base of the roof arch is reached, two 2 x 8 planks, called footing boards, are placed on each side. Wedges are driven between the two boards until they lift the arch of liner plates enough to take the weight off the jacks. The lower plates are then nailed to the boards to prevent slipping off them.

If the ground is too soft for this method, interlocked poling plates, Figure 9-56, can be placed outside and forward of the completed liner, and jacked forward from inside.

Shields. Shields have become the standard equipment for driving major tunnels in soft ground. A schematic view of one is shown in Figures 9-57 and 9-58. It resembles a tin can with an open back and controlled openings in the front. The front may be open, with grooves to allow setting breast board or plates if necessary, or

closed by a bulkhead with controlled ports. The back or tail is large enough to permit placing the tunnel lining inside it.

The shield is forced forward into the dirt by jacks based on tunnel lining. Doors in the front are opened to allow soil to flow in, or to be shoveled. In very soft ground

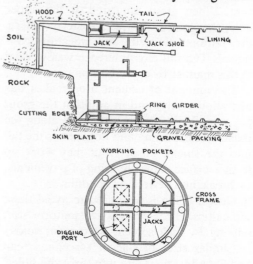

Fig. 9-58. Shield in mixed face heading

where bulging of the surface will cause no damage (as under rivers or swamps) no dirt need be taken into the tunnel, as it will be pushed aside by the pressure of the shield.

A primary lining, which is most often of bolted cast-iron segments, but sometimes of cast steel (for unusual stress), fabricated plates, concrete blocks, or timber, is constructed in the tail, which is long enough to protect a complete segment. This lining must be strong enough to not only resist full soil pressure, but it must take the thrust of the jacks that move the shield forward.

The outside diameter of the shield tail must of course be larger than that of the lining built inside it. A few plastic soils can be manipulated so as to close in smoothly on the lining as the tail moves away from it, but under most conditions the space must be filled. Failure to do so will leave the lining without proper side support, so that the arch will tend to sag.

Grout was originally the standard filler for this space. Grout plug holes were built into the liner pieces. When the tail cleared them, grout was forced into the bottom hole, with the next above used as an air vent. When grout appeared at the upper hole, the grout hose was transferred there, the bottom plugged, and injection continued. The full circumference was worked in this manner from the bottom up.

The amount of cement used makes this operation costly, and in addition the grout has a tendency to move forward along the outside of the lining and flow under the tail into the shield. Also, grout may work up to the surface, cause heaving of pavements, or break into sewers or conduits.

Gravel filling is now used to avoid these difficulties. Birds-eye gravel (uniform size, passing ¼ inch screen, 33 percent voids) or similar sizes of slag or screenings, can be blown by an air gun into the grout holes, also starting at the bottom. This will not leak into the shield nor travel far from the tunnel, but it may not fill spaces uniformly. It is therefore usually followed by regular grout. The quantity required is greatly reduced and its tendency to travel is checked by the presence of the gravel.

Compressed Air. The compressed air or plenum method of tunnel driving makes it possible to work with relative safety in soft mud and under bodies of water. The principle is that the inward and downward pressure of water and of soils can be counteracted by increasing the outward pressure exerted by air in the tunnel.

The rule of thumb is that each ½ pound of air pressure over atmospheric will support a one foot height (head) of water. Actually, the pressure required is often far less, because of the stability of the soil, restriction of water passages, and other factors.

The extra pressure is built up by low pressure compressors (converters) at the surface, and piped through a retaining bulkhead into the tunnel. Men and materials are passed through this bulkhead through one or more locks. Air in the tunnel may leak out through the soil as fast as it is supplied, or may be exhausted from the heading through a blowline.

A lock is a passageway between two airtight doors. In entering the tunnel the outer door is opened to admit men or materials. It is then closed, air pressure is raised to match that in the tunnel, and the inner door opened to complete the passage.

For exit, valves are opened to bring pressure in the lock up to that in the tunnel. The inner door is opened, the traffic moved into the lock, and the door is closed. Air is then allowed to escape from the lock until it is at atmospheric pressure. The outer door can then be opened.

This device permits maintenance of pressure in the tunnel, and limits traffic air loss to the relatively small amount in the lock at each use.

It is best practice to have at least two locks, one for men and one for materials. The man lock should be large enough for the whole crew, and must have valves by which pressure can be closely controlled so that it will drop gradually for minutes or hours while men leaving the tunnel are in it. This process of gradual reduction, called decompression, is necessary to prevent nitrogen dissolved in the blood from being suddenly liberated to cause a painful and sometimes fatal ailment called the bends.

There may also be a small emergency lock, high in the bulkhead so as to be the last place flooded. This is left open to the high pressure, so as to be ready for immediate use. One or more cross partitions may be placed in the crown to hold air pockets in case the tunnel should be flooded.

The materials lock should be long enough to accommodate hauling units of the size used. It may be small, so that one car at a time is pushed by hand in one end, and then pulled out the other, or it may be as much as eighty feet long, to accommodate a train and a locomotive. Lock construction is expensive, but a liberal size speeds work greatly.

Fire danger under high air pressure is severe. The extra supply of oxygen in close contact may cause even wet wood to burn vigorously. Smoking and other fire hazards must be avoided, and there should be a liberal supply of fire extinguishers, and fire hose connected to high pressure water.

Clay reacts most satisfactorily to compressed air, as it is so nearly impervious that it is well supported by the air, and seals it in. Primary bracing may not be required before placing the permanent lining.

On first exposure to compressed air, silt acts like clay, but it then tends to dry out and crumble off at the top, and to turn to mud and flow at the bottom. The higher the tunnel, the greater the differences between top and bottom behavior.

This is because the air pressure is the same on all parts of the tunnel rim, but the head of water that tends to force water into the tunnel, or resists its being forced out of the lining soil, is much less at the top than at the bottom. A partial cure for the difficulty is to excavate the upper or arch section first under low pressure, install liner plates or other support; then increase pressure and dig the bottom. Once a full lining is installed, the unbalanced condition becomes unimportant.

In sand the air penetrates several feet at the top, and leaves the bottom wet enough so that boards have to be stuffed with excelsior to stop sand runs. The best cure for this condition is to drive well points ahead of and below the face, and keep the lower sand dry until lining is placed.

Air will escape in any formation except tight clay, and will reach the surface by following porous veins, old wells, or even sewers. It is best conserved by getting the lining in immediately after the digging. Airtightness of the lining is not automatic, however. Grouting outside it (which is necessary for firmness also) and painting the inside of concrete with cement and water greatly reduce leaks.

Liner plates may be made airtight by spreading wet clay along the joints. Building paper can be used on wood lagging.

About 20 cubic feet of atmospheric air per minute are required for each square foot of face area, with an additional allowance for losses through the locks.

Blowouts. Sometimes the compressed air in the tunnel blows out the surface. This is particularly likely to occur in shallow tunneling in soft underwater mud. Any outward leak must be immediately plugged with any material on hand, valuable or otherwise. From the outside a blowout can be prevented or stopped by dumping enormous quantities of clay from barges.

Type of Rock

	Stratified or Schistose	Massive Moderately Jointed or Intact	Moderately Blocky and Seamy		Very Blocky and Seamy	Completely Crushed Rock or Unconsolidated Sediments	
	Dry	Dry	Dry	Wet	Dry	Dry	Wet
Heading advance in feet per 24 hr. day	42	36	40	15	38	23	8
Labor	78	90	81	217	36	142	407
Underground Equipment	26	31	28	74	29	48	139
Power	4	4.5	4	11.2	4.4	7.3	21
Pipe, track, etc.	21	21	21	21	21	21	21
Explosives	9.5	12	10	10	8.5	–	–
Drill bits & rods	1.5	1.5	1.50	1.5	1.5	–	–
Subtotal	140	160	145	335	149	218	588
Profit Contingencies	56	64	58	134	60	87	235
Grout	–	–	–	100	–	–	–
Miscellaneous	10	10	10	10	10	10	10
Total cost per Lineal Foot	206	234	213	579	219	315	833
Cost per cubic yard	45	51	46	126	48	68	181

Courtesy of State of California

Fig. 9-58A. Effect of rock type on cost
of 12 foot unlined tunnel

The blowout can be disastrous in itself, hurling men and equipment up into the water. The immediate drop in pressure allows water and mud to enter the tunnel, threatening those in it with drowning or suffocation. A job "lost" in this manner is expensive and tedious to resume, and sometimes driving can be more easily done on a different route.

COSTS OF TUNNELING

During the period 1956 to 1959, the Department of Water Resources of the State of California made an extensive study of cost and progress records of existing tunnels, in order to develop a standardized procedure for estimating tunnel costs.

The study included 99 tunnel projects constructed after the year 1930. Unlined bore sizes considered ranged from 9 to 24 feet, with headings up to 5 miles in length. The data developed refers only to tunnels driven from portals, and does not include shaft construction and hoisting, nor provision for internal hydraulic pressure.

The results of this study were published in 1959 as Appendix C, "Procedure for Estimating Costs of Tunnel Construction," of Bulletin No. 78, "Investigation of Alternative Aqueduct Systems to Serve Southern

DRY HEADING IN MODERATELY BLOCKY OR SEAMY ROCK / WET HEADING IN COMPETENT ROCK

	DRY HEADING IN MODERATELY BLOCKY OR SEAMY ROCK						WET HEADING IN COMPETENT ROCK					
Unlined Diameter	9	12	15	18	21	24	9	12	15	18	21	24
Cubic Yards per Lineal Foot	2.9	4.6	7.2	10.3	14.1	18.4	2.9	4.6	7.2	10.3	14.1	18.4
Heading advance, feet per 24 hour day	39	40	37	34	31	28	14	15	13	11	10	8
Labor (6 day week, 3 shift 24 hour day)	78	81	111.0	125.0	147.0	173.0	217.0	217.0	316.0	387.0	456.0	606.0
Underground Equipment	27	28	56.0	74.0	82.0	92.0	74.0	74.0	156.0	230.0	256.0	323.0
Power, one cent per horsepower hour	4.1	4.2	4.7	6.0	6.6	7.4	11.2	11.2	12.9	18.4	19.2	24.0
Pipe, track etc.	21	21	21.0	21.0	21.0	21.0	21.0	21.0	21.0	21.0	21.0	21.0
Explosives	6.9	10.1	13.0	17.1	21.4	26.3	6.4	10.1	13.0	17.1	21.4	26.3
Drill Bits & Rods	1.0	1.5	1.9	2.4	2.8	3.3	1.0	1.5	1.9	2.4	2.8	3.3
Subtotal	139	145	207	245	281	322	330.0	335.0	521.0	676.0	776.0	100.3
Overhead contingencies, and profit	56	58	83	98	112	129	132	134	208	270	310	401
Miscellaneous	5	9	15	21	29	38	5	9	15	21	29	38
Grout	–	–	–	–	–	–	100	100	100	100	100	100
Cost per lineal foot	200	213	305	364	422	489	567	574	844	1067	1215	1542
Cost per cubic yard	69	46	42	35	30	27	196	125	117	104	86	84

Courtesy of State of California

Fig. 9-58B. Effect of tunnel diameter on costs

California." This publication is a major contribution to the literature on tunneling, and provides the best available basis for preliminary cost estimates for tunnels.

The tables in this section are abbreviated from the much more complete coverage in the Bulletin. Because of the extent of the abbreviation, and the variable conditions of tunnel excavation, they should be used only as a very general guide.

Unit costs are given to reflect price levels of January, 1957. They can be adjusted to cost levels of any particular time by determining relative cost indices and applying them on the basis of 53 per cent of the total cost for labor, 28 per cent for equipment, and 13 per cent for materials.

The cost of operation of a dump near the tunnel portal is included in the excavation costs. It is assumed that there will be a 50 per cent swell in the tunnel muck, and that disposal areas are available nearby. A distant dump would involve additional haulage costs.

A more extensive selection of data from this report was published in the Engineering News Record, December 17, 1959.

The original bulletin is now out of print, so that it is no longer possible to get copies from the California State Printing Plant.

However, it is probably still available at many of the larger public libraries, and at libraries with special sections on mining and tunneling.

Unlined Diameter	9		12		15		18		21		24	
	Min.	Max.	Min.	Max.	Min.	Max.	Min.	Max.	Min.	Max.	Min.	Max.
Continuous steel horshoe rib without invert strut	5	35	10	70	15	100	25	165	35	250	60	320
Continuous steel horshoe rib with invert strut	10	50	15	85	20	140	35	230	45	350	110	510
Circular steel rib	55	55	75	120	100	250	120	450	210	740	320	1140
Concrete lining	45	65	70	90	100	130	140	175	185	225	230	285
Timber lagging	11	30	14	40	15	49	18	59	20	69	22	78

Courtesy of State of California

Fig. 9-58C. Cost of tunnel lining per lineal foot

Figure 9-58A shows the effect of ground conditions on the cost of excavating a tunnel with an unlined diameter of 12 feet, or a lined diameter of about 9.2 feet. The effect of tunnel diameter and water is shown in Figure 9-58B. Work is said to be wet if flow from a heading is more than 100 gallons per minute. The report assumes that flows less than this amount would not materially impede tunnel progress, and that such flows could readily be drained from the tunnel.

The cost of dewatering is not included in these figures. The formula for equipment cost (C) is:

$$C = \text{pipe cost} \times \text{length} + \text{pump cost}$$

Size of pipe is determined by volume of water, and varies from $4.00 to $7.50 a foot. Length is the distance from the wet location to the portal. Pumps cost from $3,000 to $18,000, depending on capacity. Power cost, at one cent a horsepower hour, is usually light.

The table in 9-58C gives the cost of several types of lining in a 12 foot tunnel bore. The price basis, as of 1957, is installed-in-place cost for steel of 25 cents a pound, concrete $35 a cubic yard, and grout at $3.50 a cubic foot.

The spread in prices is largely due to differences in weight and spacing of ribs,

or thickness of concrete, to meet different rock loads. The cost of timber lagging, based on a unit cost of $350 per 1,000 board feet in place, should be added to the cost of steel where it is needed.

The figures given here cover only actual tunneling costs, except for the overhead and profit items. In addition to this below-ground work, a variable amount of surface construction is required.

The following is a list of major items of this nature:
Access roads
    Construction
    Maintenance
Power supply
    Installation of lines
    Construction of generating plant if necessary
Surface buildings
    Change and washroom facilities
    Blacksmith shop
    Machine shop
    Compressor building
    Powder magazine
    Cap magazine
    Miscellaneous buildings
Construction camp (if needed)
Portal excavation
Water supply
Sewer system
It must be emphasized again that these

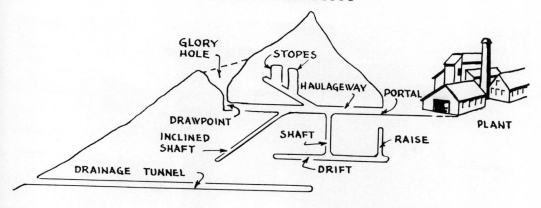

Fig. 9-59. Mine shafts and tunnels

very abbreviated figures may be used only as a general indication of relative costs.

Figures on cost per cubic yard have been added to the original report, because of their interest to above-ground contractors.

Tunneling machines, Figure 9-41 and described in Chapter 20, are having an important effect on methods and cost of solid rock tunneling.

MINING

A mine is an excavation made in order to obtain (recover) material that has valuable chemical or physical characteristics. If it is an open cut project, it may be called a pit or a quarry. Strictly speaking, a quarry is usually concerned chiefly with a material desired for its physical characteristics, as trap rock for road aggregate. However, if a quarry goes underground, it is called a mine.

The problem of mining is to get the highest possible percentage of the pay material out, at minimum expense. In some cases, the best system is confine excavation almost entirely to pay dirt, even if it requires a maze of small and irregular tunnels. In others it is more efficient to blast and remove a hundred feet of overburden so as to expose only a few feet of ore.

The first step in deciding upon an approach is to find how much ore or other pay material there is, exactly where it is located, the extent to which it is interrupted by other materials, and its physical and chemical characteristics. This information may be required not only to determine whether the deposit is worth mining and the method, but also the type and size of any processing plant required.

Mineral Deposits. Exploration is very complex, because of the number of factors that influence distribution of minerals. Sedimentary rocks are built in more or less horizontal layers, but may then be folded, twisted, or even turned upside down. Faults are breaks that extend across the layers, and are made by movement of whole blocks of the earth's crust. They may be a single clean sliding plane, or a width of hundreds of yards in which the rock is smashed up. Movement along a fault may be a fraction of an inch, so that the same formations are found on each side, or several miles, so that one section of a deposit may be at a great distance, or lost entirely. Movement may have taken place in the ancient past, or might occur from time to time during mining.

Most metallic minerals are associated with invasion of formations by molten rock from below. If fluid rock reaches the surface it becomes the lava, ash, and other usually valueless materials associated with volcanoes. If it stays far below, it hardens gradually into granite or other coarse grain

Fig. 9-60. Diesel mine hauler

rock, and while cooling may give off great quantities of minerals, in fluid or gaseous form, that penetrate the surrounding rock for miles. The weak or porous streaks through which they move and in which they are deposited, are the miners' pay veins. Parts of the main mass may become mineralized, often resulting in an extensive but low grade ore body with rather uniform composition.

One area may undergo several successive periods of mineralization, the later ones reworking, removing, or enriching some of the earlier deposits. When the area cools sufficiently ground water becomes active at dissolving, transporting, and redepositing material.

The result of these factors is that underground structure is often extremely complex, and while exploration can give a general picture of what to look for, only very extensive (and expensive) diamond drilling, or the actual removal of the ore, will give the complete story. Access to formations is often obtainable only by the

hardest and most costly type of digging—hard rock tunneling.

Exploration. There are a great many methods of exploring an area for valuable minerals. Until rather recently most deposits were found by surface inspection and sampling of the ground. Men on foot or horseback found pay outcrops, pieces of them below or downstream, or formations associated with them. Major finds are still made in this manner, but complex techniques have become more important.

A search may start with studies of geologic maps indicating more or less completely the rocks to be found in a region. Inspection and photographs from planes may reveal promising areas, which can then be scouted on foot. Test holes can be made with almost any tool from a pickaxe to a diamond drill. Radiation-sensitive instruments such as the Geiger counter are used to locate radioactive deposits. Local changes in gravity may indicate metallic ores.

A very interesting method is seismic

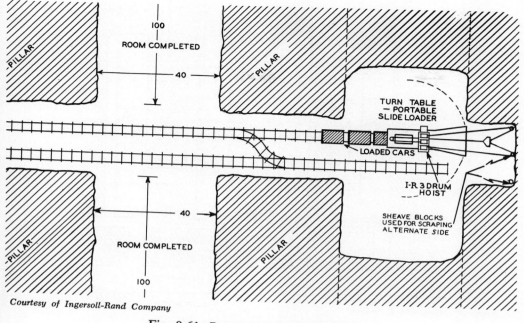

Courtesy of Ingersoll-Rand Company

Fig. 9-61. Room and pillar excavation

prospecting. A deep hole is made, usually with a diamond drill, and a heavy charge of explosives fired in it. Sensitive instru-

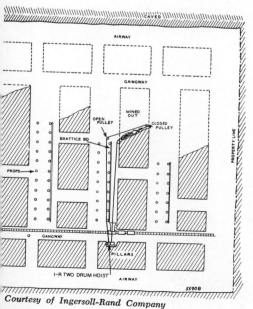

Courtesy of Ingersoll-Rand Company

Fig. 9-62. Pillar robbing with drag scraper

ments at selected spots in the area record the time and pattern of the resulting earth waves, which indicate the nature of the ground through which they pass. The information may be used directly in locating oil and some other deposits, or in working out underground structure to indicate the location of veins whose outcrops are confusing.

Following the Ore. When mining has started, with or without benefit of thorough exploration, the digging is kept in the ore whenever it does not make too complicated a pattern. In general, large well financed operations are more inclined to place their haulageways and shafts for long term efficiency, where the small operator keeps in pay rock as much as he possibly can, sometimes with most unfortunate effects on later operations.

Figure 9-59 shows some of the tunnels that might be included in a mine. The main route in from the portal is the haulageway. A drift is an approximately horizontal tunnel of small size, a stope is excavation

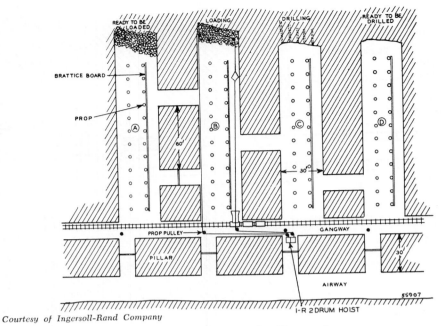

Courtesy of Ingersoll-Rand Company

Fig. 9-63. Four-heading cycle

of a room. A raise is a shaft worked from the bottom. Drainage tunnels, sometimes miles in length, are driven to save the cost and danger of heavy pumping inside wet highlands. The glory hole is a shaft enlarged from the top, with the muck descending by gravity to a floor from which it can be loaded by machinery. This loading area is a drawpoint.

Any of these except the glory hole may be in either ore or non-pay rock. They are contracted to minimum dimensions (which for a haulageway may still be quite large) when in country rock, and are expanded and supplemented by side drifts when in ore. If a vein is too narrow or too low for working space, excavation may include enough other rock to give head or side room.

The "arch" of the rock (the span of roof that can be allowed between walls or pillars, with or without timbering) and the height of the veins are important factors, as they determine the yardage (called tonnage underground) that can be taken out at one

stand. Larger volume and working space permits use of bigger and more efficient machinery.

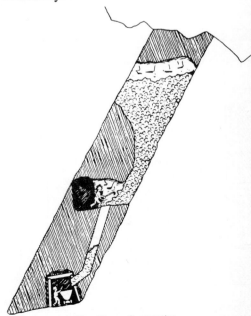

Courtesy of The Eimco Corporation

Fig. 9-64. Chute loading

Development. Development is that part of mining that prepares an ore body for removal. It is likely to include sinking shafts, driving tunnels and installing of chutes and transportation and drainage systems.

If development is done in the ore or other pay mineral, its expenses may be partly or wholly paid by the value of the product. In some procedures, as in room and pillar mining where the pillars are left, the development may be the entire mining operation.

Shafts. Shafts are similar to those used for other tunneling. They provide for entrance of men and supplies, removal of ore and waste, drainage, and ventilation. They vary in size upward from a drill hole with a casing eight inches or more in diameter that serves as a chute for concrete or fill, to openings large enough for four divisions —two for a pair of skips for bringing up ore and waste, one for an elevator for men, and one for pipes and ladders. Supplies may be lowered in man elevators, in ore skips, or by a separate system. There may be an additional system of ventilator shafts, that may be either dug or rotary-drilled.

Haulageways and Drifts. Tunnels that are made primarily for hauling ore or other materials from digging points to shafts or portals are called haulageways. Their size varies with that of the haulage units to be used in them. Old time man-and-wheelbarrow methods could get by with a width of four feet and a height of six and a half feet. Two way haulage with off the road trucks requires a width of about thirty feet and a height of twenty.

The majority of underground mines use track haulage although use of rubber tired trucks and shuttle cars is increasing. A few still use hand-push cars on 18 inch gauge tracks. Gauge is the spacing, center to center of the rails. Locomotive haulage may call for track gauges from 24 to 42 inches. Cars may project 18 inches beyond the track on each side. An additional space of at least 18 inches on one side is required so as not to crush men against the walls. There is usually a gutter or pipe for drainage water, piped high pressure air for power and low pressure air for ventilation, and electric wires or cable.

Haulageways that are to be used for a long time are given strong and permanent linings. Concrete without reinforcing is gaining popularity, because of its strength, comparative simplicity of placement, and the fact that when broken it is readily replaced or removed. However, steel lining, timber sets, roof bolting, or sometimes just shaping and scaling of a natural roof are all used under suitable conditions.

Drifts are tunnels made during exploration and development. They are usually smaller, shorter, or are expected to carry less material than a haulageway, but there is no clear distinction.

Extraction. Extraction is the operation of removing ore or other minerals that have already been made accessible by development work.

ROOM AND PILLAR EXCAVATION

Minerals in extensive beds of fairly uniform thickness may be mined by a method known as room and pillar. Coal, limestone, and salt are frequently dug in this way.

A number of parallel corridors (tunnels or drifts) are cut into the formation, and are connected by series of parallel corridors crossing at right angles. There must be at least two of the main corridors to provide for ventilation, and there may be as many as five in a set—two haulage, two ventilation, and one spare. The corridors are the rooms, and the blocks or walls between them are the pillars.

Figure 9-61 shows a layout during the development stage, as the rooms are being cut. The rooms and pillars are the same width. This could be as little as six feet with a weak roof, or as great as fifty feet under very favorable conditions. With such an equal width layout the development work

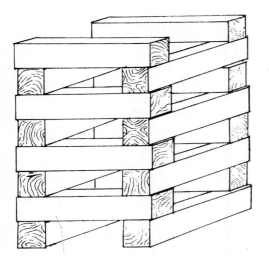

Fig. 9-65. Cribbing

line, or to the edge of the area to be worked. Then the pillars nearest to this edge are partly or completely removed, then the next ones toward the entrance, and so on. At some point the unsupported roof will collapse. It may stay up while a large number of pillars are weakened or entirely removed, or it may fall during removal of each pillar.

The miners have the tricky job of deciding how much of each pillar they dare to remove. The more they take out of each pillar the greater their percentage of recovery, but also the greater the danger of losing men and machinery in a collapse. The work is sometimes partly protected by temporary timbering, or by cribbing of cross piled timbers, Figure 9-65, that can be pulled down by ropes after the extraction is finished.

Danger may also be caused by failure of the roof to collapse. If a large area is deprived of support, and finally falls suddenly, it may cause a dangerous rush of air through the mine. In coal mines this air may be heavily charged with toxic and explosive gases.

Limitations. Room and pillar work of this type is usually limited to beds whose thickness is not much greater than the safe span of the roof between pillars. Tall pillars must be thicker than short ones to support the same weight, and in thick beds with narrow corridors they will contain too high a proportion of the mineral.

A mineral body must also be fairly regular for efficient room and pillar work. Interruption of the pattern by barren or soft ground, or sudden changes in pitch, increases costs and decreases recovery.

would remove 75 per cent of the mineral formation, if a clean cut were made to the floor and the roof.

Another method, Figure 9-63, is to leave the pillars as continuous walls except for occasional ventilation connections between the corridors. This is often done when the corridors will be backfilled to support the roof, and the pillars then dug out.

Leaving the Pillars. The pillars may be left permanently to support the roof. This may be done because the supply of the mineral is so large that it is more efficient to waste the pillars than to do the extra work of extracting them; when roof conditions make extraction particularly dangerous; or to prevent damage by subsidence of the ground above. Worked out mines of this type are beginning to be used for storage of records, and for underground factories.

Pillars left in this manner may be the full size left by the development, or they may be cut away (robbed) to the point where they are barely adequate to support the roof.

Pillar Extraction. Pillar material may be robbed until the roof caves. This is standard procedure in many bituminous coal regions. Development is carried on to the property

STOPING

The term "stoping" is used so differently in different mining areas that it has little specific meaning. In general, and in this discussion, it applies to any underground digging of valuable minerals that does not create a passageway. It covers most forms

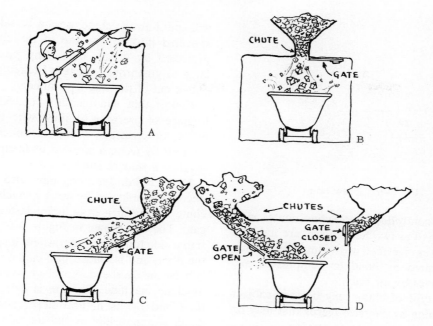

Fig. 9-66. Chutes

of ore extraction other than room and pillar and block caving.

A stope is the cavity in which stoping is being done or has been done.

Gravity. Almost everything is against economical work underground: restricted working space, tight blasting, need of roof and sometimes wall supports, cost of hoisting to the surface, darkness, limit on size of machinery, and drainage, pumping, and ventilation requirements.

There is one helpful force that the miner puts to his use very effectively—gravity. He employs it to help break rock, or sometimes to do the whole breaking job, and to convey the broken material through chutes to loading points.

Chutes. Underground operations in extraction or stoping are very complex, and almost any cutaway view and brief description is likely to be incomprehensible to the surface man. For this reason, the subject will be opened with an illustrated and over simplified description of procedures.

If ore lying above a development tunnel

is to be mined, it might theoretically be loosened by a bar or by blasting to fall directly into a mine car parked below, as in Figure 9-66 (A).

However, some of it would fall beside and between the cars, and it would be difficult to control the amount loaded. For this reason it would be better if the loosening were done somewhat higher, as in (B) and the broken ore fed down through a chute with a control gate.

But a vertical chute dumping chunks of rock out of the ceiling is dangerous, and the gate will be very difficult to work, as a great weight of rock can rest directly on it.

Therefore, the chute is made a sloping one coming in at the side, as in (C). And since a tunnel has two sides, two chutes can be cut into it at one point, as in (D). Several chutes or pairs of chutes may be spaced along the tunnel so that each can load a separate car of a parked train at one time. In this manner, a twelve-car train could be loaded by two chute locations with six stops or by three chutes with four stops.

Fig. 9-67. Draw point

Gate controlled car loading chutes may be as small as two feet square, are unlikely to be over five feet wide. Small sizes may be operated by hand or compressed air, larger ones by air only.

This type of chute may be very subject to jamming by oversize pieces that may be difficult and expensive to remove.

This problem is met by having a sorting and screening point between the stope and the chute, as in Figure 9-64. A tunnel is cut above the haulage drift, into which broken ore can flow by gravity, so that it will spill into the chute, or can be raked or shoveled into it. The top of the chute is protected by grizzly bars. Any piece too large to go through them is broken up, or pulled out of the way by a miner.

Chutes that will handle a considerable tonnage of ore may be lined with timber, metal, or concrete. Wear on concrete is reduced by using a stepped underslope instead of a smooth slide.

Draw Point. Jamming may also be reduced or eliminated by using a much larger chute ending in a draw point instead of a gate. The one shown in Figure 9-67 opens into a side tunnel or room where it spills on the floor, and is loaded into cars or onto a belt by an overhead shovel or some other mucking machine. The ore is held by the floor and by its natural slope, and feeds down automatically as the toe of the pile is dug away.

Another type of draw point or free flowing chute empties directly onto the tunnel floor, putting the ore in the working path of a drag scraper (slusher).

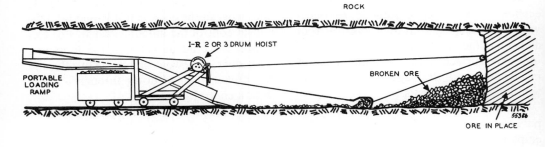

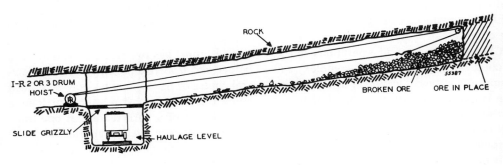

Fig. 9-68A. Drag scraper installations

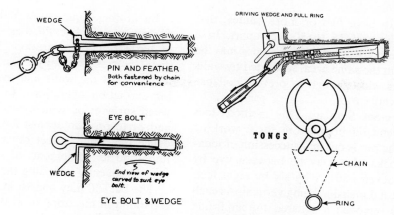

Fig. 9-68B. Drag scraper anchors

This machine, one version of which is shown in Figure 9-68A, is a bucket, usually bottomless, that is pulled to the digging area by a cable attached to its rear, and then pulled forward through its digging, transporting, and dumping cycle by the other end of the same line, attached to its front.

The line operates between a winch at or near the dumping point, and a pulley block anchored behind the digging area, several feet above the tunnel floor.

One scraper may service several draw points.

Shrinkage Stope. The shrinkage stope, Figure 9-68C, is for strong ore that is firm enough so that a wide ceiling may be left unsupported. The ore is broken by drilling and blasting overhead. The fallen pieces make the floor on which the miners stand as they drill.

Broken rock occupies from 50 to 100 per cent more space than solid rock, so that ore can and must be drawn off through the chute as the work progresses. The rate of drawing regulates the height of the working space. When all the ore has been blasted, continued drawing will remove all loosened material from the stope, leaving it empty unless it is backfilled to prevent later collapse that might result in damage to parts of the mine.

Shrinkage stopes are ordinarily large

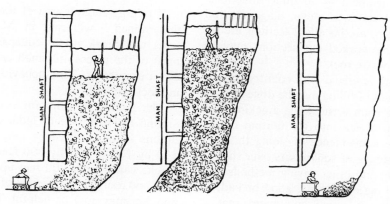

Fig. 9-68C. Shrinkage stope

enough to have two or more chutes, that may be spaced 25 to 50 feet apart. In a large ore body a number of stopes may be worked in the same area, separated by walls or pillars wide enough to prevent collapse of the whole roof.

Square-Set Stoping. This method does not depend on broken ore for a working floor. The ore is cut in a succession of identical blocks, that may be between five by five by seven and six by six by eight feet, the largest dimension being vertical. As each block is removed a squared timber frame is erected in the space and roofed with planks. The stability of the material will determine whether the blocks must be dug and shored one at a time, or if a number can be opened and then shored at one time.

This work may be started at the bottom of the ore body or of the section being worked, and the bottom level (or slice) is followed in sequence two or three sets behind by higher slices, so that the cross section of the front resembles the underside of stairs. The roof of each frame serves as the floor of the one above it. The line of advance may assume any one of a number of patterns, the straight wall from side to side of the ore or section, or a center advance followed by stepped back sides.

The strength of the timbering in a large stope depends largely on keeping the sets in exact alignment in all three directions. If any part starts to move, diagonals and bracing plates are installed, sections are rebuilt, or the worked out sections may be filled with waste rock.

If the ore is mixed with country rock, a large amount of filling may be done by separating ore from waste while mucking, and moving the waste into the bottom of the stope. The convenience of doing this is a great advantage of square sets over shrinkage stoping and other caving methods that require taking everything, good or bad, to mechanical separating equipment that is usually in the mill above ground.

Backfilling. Square sets and other types of stope may be backfilled as a regular part of the extraction operation. This may be done to strengthen any bracing structures and support the walls and overhangs during extraction operations, to provide elevated surfaces on which miners can work, and/or to provide against long term subsidence that might affect other parts of the mine and its equipment, or surface buildings (frequently including the mill).

The backfill may consist in part of waste separated in the stope or at the crusher, of tailings from the mill, or of granulated slag. When operated in connection with open pit or open stope mines, or when the proportion of waste is very high, this material may be all that is needed. Otherwise sand, gravel, or other suitable material is dug in the vicinity and hauled to the mine.

The waste fill is usually handled by a system of shafts and passages separate from that used for ore removal. Sand, boulderfree gravel and dry tailings may be dropped through chutes made from cased drill holes. Wet tailings may be pumped down similar bores, or through pipe rigged in the main shaftway. Cement may be mixed with backfill gravel to make a very lean concrete, in order to avoid settlement, and slumping into adjoining pillar extraction work.

BLOCK CAVING

If a large enough area of rock or ore is undermined it will cave in. Many formations will break up in collapsing so that most of the pieces are small enough to go down ore chutes. This behavior is utilized in block caving.

The minimum width for sure caving is around 100 to 150 feet. Under various conditions block dimensions may vary from 100 by 100 to 200 by 300 feet. Height of blocks varies from 200 feet up to perhaps ·400, and a substantial weight of overburden may (or may not) be helpful. The block is sometimes cut off from adjoining blocks,

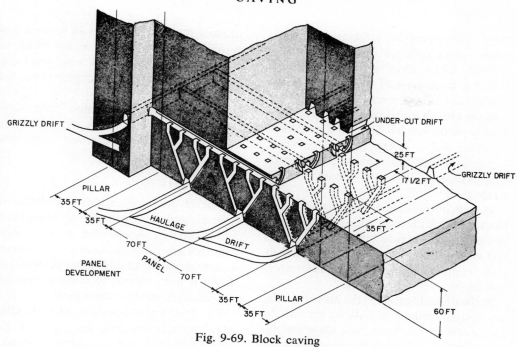

Fig. 9-69. Block caving

or from separating walls by shrinkage stoping or a network of raises or drifts.

Development work prior to caving is fairly complicated. Three levels of tunnels are needed: the undercutting level where the block is undermined, the grizzly level below it for sorting and feeding into chutes, and the bottom or haulage level. See Figure 9-69.

The levels are connected by raises. The upper set, from the grizzly to the undercut passages, are called finger raises. Their upper ends may be belled out into funnels. These terminate beside and above grizzly bar entrances to the control raises (chutes), so that oversize pieces may be broken or set aside.

When the development work has been completed for a block or a large enough part of one to start caving, the pillars separating the undercut drifts, and usually their ceilings as well, are drilled and blasted. The broken ore spreads and settles into the drifts, leaving the roof entirely unsupported, so that it breaks under its own weight and

crumbles into the raises. This process may start immediately, or after a delay of several days.

The number of control raises can be greatly reduced by using drag scrapers (slushers) to pull ore away from the bottoms of the finger raises to a few large raises.

A number of problems arise in block caving. First, there is the blocking of finger raises by oversize pieces, that must be broken by "bombs" of dynamite that are pushed up the raises on poles to the obstruction, and exploded. Then there is the arching of the ore over a raise or group of raises so that it will not feed, usually a tricky and sometimes a dangerous proposition to correct.

Another difficulty is that uneven breakage and settlement may put great pressure on certain sections, crushing even heavily reinforced control and haulage tunnels. If the blocks are separated by pillars (walls), and the barren capping rock is stronger than the ore, settling in the blocks may result in these walls supporting large areas of overburden, with resultant heavy pressure on passages under them. Another possibility is that a large mass of rock in one of the blocks may be more resistant to breakage than the ore, so that it rests more and more heavily on the ground beneath it as the ore crumbles and is drawn away. One block caving operation lost a haulageway for 18 months as a result of such a mass crushing it repeatedly.

In spite of these and other complications, block caving is the most economical method of mining large fairly uniform ore bodies that will break into small pieces under their own weight, and that are too far underground for open pit recovery.

Surface Subsidence. On the planning side, any mining method that allows the roof to cave as the ore is taken must allow for the effects on the surface. A thin even bed such as a six foot coal seam may let the surface down evenly with comparatively little damage to structures. But any substantial and irregular removal of underground material will result in subsidence pits. These may be hundreds of feet deep and thousands of feet wide, and are likely to destroy any road or structure within their reach.

The subsidence of a caving block may be nearly vertical at first, but the cleavage in the overburden is likely to fan out widely. The ultimate effect, in a wide variety of rock formations, is crumbling and sinking inside a slope line of 40 to 45 degrees outwardly from the bottom of the block or caving operation, with additional slip faults that might extend to a 38 degree slope. See Figure 9-70.

All too often ore treatment mills, shaft head structures, and towns are involved in such subsidence, so that very expensive relocation is required.

FORECASTING ROCKFALLS

One of the principal safety problems of underground work is the danger of rockfalls and rockbursts. Knowledge of where such incidents are likely to occur usually makes it possible to install additional supports, or if that is not possible, to remove men and equipment from the danger area.

The cracking and popping of rock, often heard by men working underground, has long been recognized as a warning of moving, unstable ground. It has now been discovered that sounds of the same type, so slight as to be heard only by very sensitive instruments, are made by ground that is under even moderate stress. Increasing frequency and intensity of the sounds, called microseisms, indicate an increasingly unstable condition that is likely to result in collapse.

Microseisms occur within the audio frequency range and, when amplified, sound like 'clicks' akin to the creaking of wooden stairways, floors and diving boards. Almost

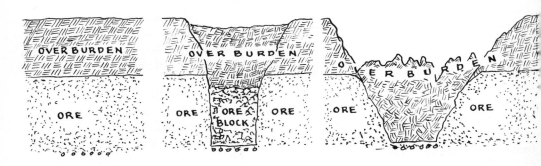

Fig. 9-70. Surface subsidence

any hard material, such as salt, rock, shales, sandstone, quartz, brick, glass, wood, steel, concrete, sand and sugar, will produce microseisms. Microseisms by their 'clicking' nature can be easily distinguished from sounds caused by foot shuffling, drill rigs, trucks, mucking cars, talking and other means. Microseismic apparatus can never be used when vibrations are so great as to overwhelm the equipment.

The Seismitron is an instrument based on the microseismic principle and has been in use since 1951. Its primary purpose is for the prediction of rockfalls in tunnels and mines as well as for the determination of stability of earth dams and slopes at the sides of mountains and highways. The apparatus is made by the Walter Nold Company of Natick, Massachusetts and is certified (PERMISSIBLE) for use in coal mines by the U.S. Bureau of Mines and by the Department of Mines and Technical Surveys for Canada.

Receiving sensors (geophones) are cylinders 1¼″ in diameter and nine inches long. They contain synthetically-grown crystals which, when stressed, generate low level electrical currents. A portable battery-powered amplifier increases the level to a condition that enables the microseisms to be heard by means of simple earphones.

The number of microseisms per minute which occur when the rock is at the failure point, or about to burst, is determined by monitoring the heading (of a tunnel, for instance) immediately after a blast. The microseismic rate of decay under such a state is the same as the rate of increase prior to a rockfall. This is the failure point.

Six ft. deep holes are drilled and spaced every 100 feet apart in the suspected area. Each of these 'stations' are monitored each week for microseismic activity. This is accomplished by placing the geophone in the hole and sealing the opening with waste. The microseisms heard are manually counted for a 15 minute period and the rate per

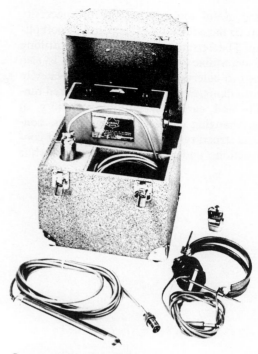

Courtesy of Walter Nold Company
Fig. 9-71. Seismitron

minute vs. dated time plotted on graph paper at a later time.

A history is thus built up for each station. A low, little-changing rate of microseismic activity is indicative of stable conditions. A continually increasing rate indicates suspect conditions especially when 120 to 180 microseisms occur per minute. A doubling of the rate within any 24 hour period of time indicates imminent collapse.

An excellent prediction of collapse, sometimes as much as 45 days ahead of time under ideal conditions, can be made by extrapolation of the plots on the graph to the rock failure point. The curvature becomes hyperbolic as failure approaches.

If the active rock, which is sometimes hidden by gunite, should become stabilized either naturally or by roofbolting or timbering, the decaying microseismic rate would immediately indicate the tendency toward safe conditions.

The instrument in this manner becomes a

very useful tool for judging the necessity for, or the effectivity of, expensive roofbolting. The Seismitron is used in earth drilling to determine depth of bedrock. It is also used to determine leaking running water in earth dams, evidences of life in blocked tunnels and earthquake rubble, stability of over-hanging cliffs, large ice masses, concrete structures during quiet periods of construction, underground foundations while underpinning is being replaced, communication through rock or pipe, robbing pillars of mineral-bearing ore, and so forth.

With the Seismitron, timbers about to break sound just like one would expect wood to sound like under these conditions. Rock about to fall sounds like rock about to fall. Earth about to slide sounds just like earth about to slide. Like the subject of physics; natural and normal.

CHAPTER TEN

PIT OPERATION

This chapter contains an outline of some of the principles of pit layout and operation. The term "pit" is intended to cover any open excavation made to obtain material of value, whether it be coal, mineral ore, quarry rock, gravel, or fill.

Because of the complexity of the subject, treatment will have to be brief, and many operations omitted entirely.

Techniques of drainage, road building, and blasting have been discussed in previous chapters.

Most pit operations are started with the removal of soil or rock lying over the deposit to be mined. The problems involved in stripping will therefore be considered first.

STRIPPING OVERBURDEN

Overburden may include topsoil, subsoil, sand, gravel, clay, shale, limestone, sandstone, and other sedimentary deposits. In some clay pits, the overburden is partly coal, and in many coal mines it is partly clay.

The depth of overburden that may be removed depends on its character and accessibility, on the value of the underlying formation, the comparative cost of underground mining, and the extent to which the spoil can be sold or utilized.

Need for Stripping. Stripping overburden may be a very large part of the cost of mining, and a number of factors should be considered before undertaking it.

First, is it necessary to strip it? It may be possible to mix it with the product, or to separate it at less expense during processing.

For example, much run-of-bank gravel does not contain enough fines to bind it for road use, and overlying soil, if permitted to mix with it by caving, may improve its quality.

If the pit has gravel or stone screening, separating, or washing equipment of adequate capacity, soil may be dug or blasted down with the pay dirt, and separated as part of the regular processing. In this case, thorough clearing is necessary as sod and brush clog screens and crushers.

Loose sand and gravel cannot be dug underground, but with other deposits, careful investigation should be made of the latest methods of underground mining, and a comparison with open cut costs made.

Utilization of Spoil. If the land must be stripped, the next question is possible profit or use to be obtained from the material. Near large cities, topsoil can often be sold at a higher price than the regular pit product. This may also be true of peat deposits.

Any substantial layer of clay, or fine earth, should be sampled and analyzed. Clays which are superficially similar in appearance are used in widely different products, such as fire, paving, and common brick; tile, pottery, portland cement, flux, mud for rotary drills, and specialized functions in chemical processing.

Good deposits of some of these earths are very rare and are valuable. Common clays may be in demand because of local shortage, or industry using them may be developed.

Limestone is often found in overburden, and it is extensively quarried for crusher rock, building stone, and cement manufacture.

If it is not possible to get good prices for any part of the spoil, it may still be possible to get enough for it as fill to repay some of the stripping cost. The stripper is in a good competitive position because he has to pay for digging, and often for dumping as well, and is better off selling for a fraction of the excavation cost than not selling at all.

The limiting factor is his additional cost in making the spoil salable or delivering it. If selective digging that will slow his operation, or finer fragmentation, is required, or if he must truck material which otherwise could be cast, the salvage may cost more than it is worth.

When the spoil is being removed in trucks or scrapers, and no market exists for it, and it is not practical to set up any plant to process it, it might be used for real estate improvement or for good will.

Swampland along railroads or highways can often be bought very cheaply, and can be converted into valuable industrial sites by filling and grading. Filled land near towns may be sold for residential purposes.

Pits are frequently unpopular with the neighborhood because of blasting, heavy trucking, or dust. Local authorities may impose restrictions which would make operation more expensive or impossible. Under any circumstances, cultivation of local good will is sound business practice.

Fill which is being hauled and wasted can be utilized to reclaim land for parks or parking areas, building road or airport fills, and blocking gullies. Topsoil can be

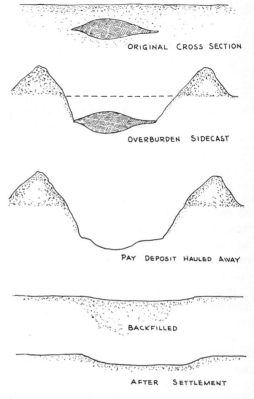

Fig. 10-1. Steps in strip-mining a lens

used to enrich farms, gardens, or lawns.

However, there is danger in such work of getting into much greater expense for extended hauling and grading than anticipated, so a careful study should be made in advance. The people or community benefited will often be willing to pay at least these extra costs.

As a general proposition, however, the pit operator must figure that his spoil from stripping will not be salable or of any use, and when it must be hauled away it may be a problem to find a disposal area. This is particularly likely in large operations.

Regrading. A number of states now require grading of spoil piles and breaking down of walls of worked-out pits. This work should be included in cost calculations.

Replacing of topsoil and/or replanting may be required by law, or by private contract between the pit operator and the land owner.

Pit floors, slopes, and spoil piles are usually quite barren, and are often very rocky as well. They may be entirely lacking in nitrogen, and their potash and phosphorous supplies are usually in insoluble forms that become available to plants only after long weathering.

Reclamation involves grading to smooth contours, planting with grass, hardy trees, or other vegetation; and protecting the graded land against erosion until the plants have taken hold.

Spreading topsoil makes immediate reclamation possible. Such topsoil my include subsoil that is fine in texture, and distinct from underlying barren layers.

A reclamation plan must usually be filed and bonded before work starts. If the pay seam is thin, restoration of original contours will probably be required. If it is thick, or underlies rough land, new topography may be permitted.

Removal of coal often makes ground water highly acid. Treatment of this condition within a project, and prevention of contamination of streams flowing from it, may be a major reclamation problem.

SIDECASTING

When the spoil cannot be utilized, the cheapest way to move it is by sidecasting—that is, to pile it alongside the cut within dumping range of the excavators. This disposal is possible when the pay formation has only a narrow exposure, or can be worked in narrow strips, so that the overburden can be economically moved across the pit, from the unopened to the worked-out area.

Sidecasting is generally not practical in working thick layers of quarry rock or mineral ore, because of need for wide working space below the face. Its best known application is in the strip mining of bituminous coal.

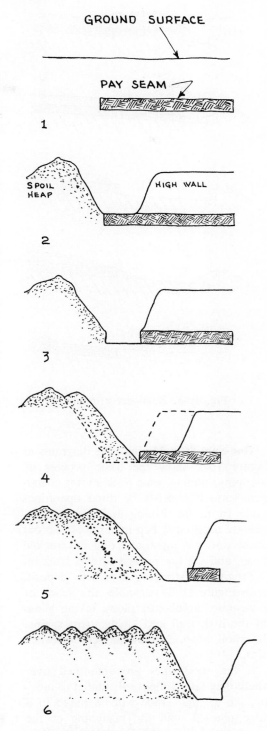

Fig. 10-2. Strip-mining a horizontal bed

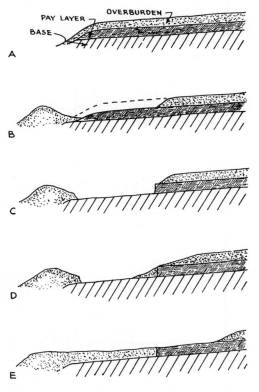

Fig. 10-3. Bulldozer stripping

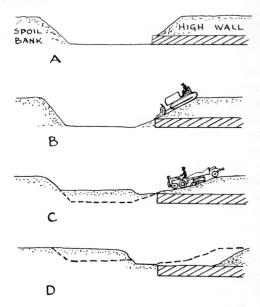

Fig. 10-4. Stripping shallow overburden with scrapers

One-Spot Strip Mining. The diagrams in Figure 10-1 show the basic process of stripping, mining, and backgrading a narrow lens. It consists of three operations aside from the drilling and blasting frequently necessary. Top material is cast out of the way, pay material is dug and trucked away, and the top pushed or cast back in.

Progressive stripping of a wide deposit, as in Figure 10-2, resembles the action of a gigantic moldboard plow, taking slices off the high wall and laying them against the spoil heap. Mining is done between trips.

If the spoil piles are smoothed over where they fall, without substantial re-moving, the original grade will be raised in a ridge parallel with the beginning of the work, and lowered at the last cut, these changes corresponding to the land and the dead furrow of the ploughed field.

Bulldozers. Where the deposit is shallow

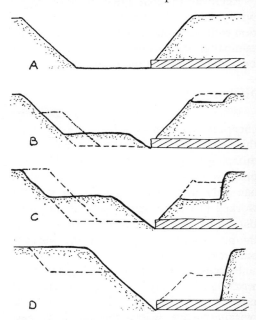

Fig. 10-5. Stripping deep overburden wtih scrapers

and the pit narrow, bulldozers make good stripping tools, particularly when the land slopes across the pit. Figure 10-3 shows a sequence of operations. This can be repeated in successive strips across the deposit if conditions remain the same.

Such shallow overburdens seldom require blasting but use of a rooter may speed the digging.

It will be noted that this machine backfills the face of the pay layer so that work must be done from the surface of the seam.

Bulldozers are also essential auxiliary tools in the heavier stripping work.

Scraper. Scrapers, or pans, are used in sidecasting, haul-away, and combined stripping. Their best application is to soils that can be dug easily with the power that is available, or that can be broken into fine fragments by ripping or blasting; and that are not deep enough to justify the use of shovels with enough reach to give the pit the width it requires.

Figures 10-4 and 10-5 illustrate two methods of scraper use. These are considered sidecasting since the spoil piles are placed immediately behind the pit, but otherwise the work is identical with haul-away with the same machines. A particular problem that becomes more important as depth increases, is maintenance of haul roads across the pit and up the spoil bank.

In 10-4 the high wall is smoothed off and the outcrop of the pay formation covered by bulldozers working down the slope. The pans then dig the high wall, working downhill, and build their fill on the full width of the empty pit, working against and parallel to the spoil bank.

In 10-5 the digging and spreading are both done the long way of the pit. The seam outcrop is crossed only occasionally by haul roads, the rest of the face being left exposed.

Lengthwise digging is well adapted to combined operations. If only part of the overburden were to be removed by the

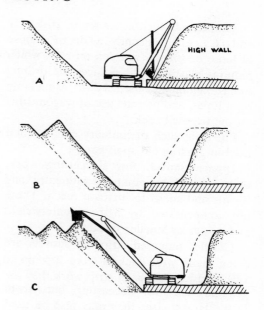

Fig. 10-6. Two-trip stripping with a dipper shovel

scrapers, and the balance by shovels, a narrower fill would be made, as indicated by dot-dash lines. The scraper work would thus serve to reduce the depth to be handled and the distance it would have to be thrown.

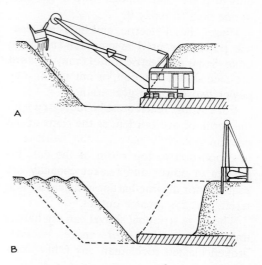

Fig. 10-7. One-trip stripping with a dipper shovel

Scraper cuts such as shown in 10-5 (B) and (C), are frequently made to dispose of loose or soft overburden on rock which requires blasting. This reduces the drilling footage, eliminates the need of casing large blast holes, and permits use of wagon drills on the exposed rock.

Fill from such preliminary cuts is often used in grading old spoil banks.

Dipper Shovel. For sidecasting work, dipper shovels are equipped with extra long booms and sticks to increase reach. These are compensated by extra counterweight and power or smaller buckets.

Crawler mounted stripping shovels are being made with buckets from 2½ up to 140 yard capacities. In strip work they are generally used for sidecasting, but except in the largest sizes they may also be used for loading trucks or trains that stand in the pit or on comparatively low walls.

Small and medium stripping shovels are generally more or less standard units that can be readily transported from job to job. Large ones are likely to be custom made, shipped to the job in pieces, and erected at considerable expense. Their high cost is only justified when enough work is available in one area to repay it.

Diesel, diesel-electric, and high-line electric power are used in medium and large units. Separate electric motors may be used for each shovel function, or one motor may power two or more gear trains.

For casting, reach must be increased with the depth of overburden, as the slope of the pile progressively narrows the work area in proportion to top width of the cut. Increases in power and bucket capacity are required for digging harder and coarser material, and for greater output.

The large stripping shovel can dig harder unblasted material, or more coarsely broken blasted rock, than any other excavator.

A shovel may make one, two, or more trips to clear a working area. It may be worked from the coal or pay seam, or from the floor left after its removal.

Figure 10-6 illustrates two-trip stripping by a shovel, and 10-7 one-trip by a bigger one. Maximum swing required is 180°.

The shovel is followed by one or more clean-up bulldozers. The toe of the high wall is scraped back, and the pay seam is cleaned of material left or dropped by the shovel, or which has slid or rolled from the pile. This debris is pushed into the spoil pile.

Coal may be injured by grousers of heavy bulldozers. Rubber tired dozers, or crawlers with the grousers trimmed back or covered by street shoes can be used to avoid damage.

Dragline. Stripping draglines are being manufactured to carry buckets up to 220 yards capacity, and booms over 300 feet long. Diesel and diesel electric power are standard up to 3½ yards, and are sometimes used in machines as large as 8 yards.

Crawler mountings are standard in small machines, and walking bases in large ones. The walker rests directly on the ground and moves by eccentric movement of shoes on each side. It is safer for use on high or loose banks because its ground pressure is low (5 to 12 pounds per square inch, as against 50 to 60 for comparable crawler machines) and it can change its travel direction without exerting any side thrust. The mechanism is described in Chapter 13.

Walkers are not adapted to use on pit floors because they can only walk away from the boom so that they may be trapped if unable to swing freely. In addition, they are usually wider than crawlers of the same weight.

The dragline strips by moving along the high wall, parallel with the pit, digging behind it and casting onto the spoil pile, as in Figure 10-8. Maximum swing is about ninety degrees. Standard practice is to dig within the radius of the boom point.

Large draglines can dig hard and coarse

Courtesy of the Marion Power Shovel Company

Fig. 10-8. Stripping with a dragline

formations but are somewhat less efficient in them than shovels.

Draglines can load part or all of the spoil into trucks or trains, on the wall or in the pit. They can dig selectively from the top down, handling different formations in succession from one stand, providing blasting has not disarranged and mixed them too badly.

This ability makes possible a rough division of the bank into select material to be hauled away, and waste to be sidecast.

Clean-up bulldozers follow immediately after the dragline and push their loads within reach of its bucket.

Stacker. Stackers are mobile elevating belts. The belt is carried in a boom which can usually be raised, lowered, and swung while operating, and may be adjustable in length as well.

The hopper is usually crawler mounted and self propelled. It is protected by a grizzly, so that pieces too large for the belt will be rejected onto the pit floor.

Figure 10-9 illustrates the use of one of these machines in conjunction with a standard-boom shovel. The shovel digs the bank and dumps in the hopper, from which the belt carries it to the spoil bank.

Dug soil can usually be moved more economically by a conveyor belt than by swinging a shovel heavy enough to carry a boom of the same range. However, the stacker is not as flexible and will not handle as coarse material as a stripping shovel or dragline.

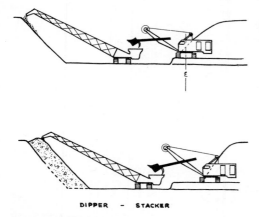

DIPPER - STACKER

Fig. 10-9. Dipper and stacker handling overburden

The stacker is also used in the pit for loading trucks up on the bank.

Wheel Excavator. The wheel excavator is the newest big machine to appear in the coal fields. It is a self propelled, crawler mounted unit that carries a cutting wheel on the end of a long boom that can be raised, lowered, and swung; it also has a

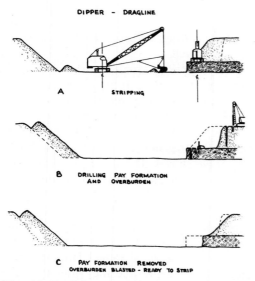

DIPPER - DRAGLINE

A   STRIPPING

B   DRILLING   PAY FORMATION
                AND   OVERBURDEN

C   PAY FORMATION   REMOVED
        OVERBURDEN BLASTED - READY TO STRIP

Fig. 10-10. Dipper and dragline handling overburden

stacker-like conveyor belt for discharging the spoil, that may or may not be separately controlled.

One of these machines is owned by the Peabody Coal Company and is used in their River King mine in southern Illinois. It is teamed with a 70 yard dipper shovel to handle overburden from 65 to 135 feet deep.

The wheel excavator takes the top third of the overburden, which is mostly soil. Where the cut is 90 feet the wheel takes off 30 feet. The bottom 40 feet contains hard rock that is drilled and blasted, this work being done from the shelf left by the wheel. The dipper digs this and piles it across the pit in the space from which coal has just been removed.

On the next pass, the wheel spoil is deposited above and behind the shovel piles, as in Figure 10-11A. This makes an excellent division of work, as the wheel cannot handle coarse rock, and its soft spoil is held from sliding back into the pit by the shovel's rock pile. On the high wall side, the wheel shelf provides a better surface for the drill than the original ground, and drill holes are much shorter. The shelf also catches bank slides that would otherwise go down on the coal.

The coal loading shovel follows the wheel, making a 45 foot cut. Overall pit width at the bottom, from high wall to the toe of the spoil bank, is 130 feet to provide space for these big machines to pass each other.

Cable Excavator. Overburden can sometimes be economically stripped with slack-line cableways. A long pit will require a mobile or portable tower and means for easily moving the tail tower.

A good practice is to locate the heavy head tower on undisturbed ground, and the tail tower on the old spoil, as in Figure 10-11B. The bucket and its mechanism are rigged to dump at the low end, and digging is done toward the head tower to avoid dragging spoil over the face.

Slacklines are best adapted to wide cuts. They may be assisted by rooters or blasting in hard deposits.

If access to the face is not required, a power drag scraper can be used to pull the overburden into the old pit.

Double Casting. If a space is required which is wider than that which can be stripped by available shovel and dragline equipment, and the spoil is too coarse for scrapers, a shovel may be used to swing the spoil out onto the pit floor, and a dragline to pile it on the bank. The illustration in Figure 10-10 is of a quarry requiring a wide working area below the face.

For best results, the two machines must work together as, if the dipper is allowed to build a substantial windrow, the dragline will need a longer reach to get to its far edge.

A dragline may also be operated on the previous spoil bank to take off the top of the pile being built by a shovel, to increase its disposal capacity.

The extra expense of double casting limits its use to special conditions.

Deep Stripping. Strip operations often run into areas where depth of overburden is too great for efficient sidecasting with available equipment. This may be handled by two trips of a shovel, by a shovel and wheel, or by a shovel and dragline. Such work is not double casting if most of the spoil is handled only once.

Two trips with excavators may be made for drilling and blasting convenience rather than to make up for a lack of reach. Taking

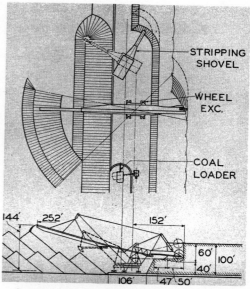

Courtesy of Bucyrus-Erie Company

Fig. 10-11A. Wheel and shovel team

a high bank in two levels may result in worthwhile savings.

The strata in the upper and lower lifts may differ in quality so that different drilling and blasting techniques are used. In some cases one will require blasting where the other does not.

Very large draglines may be able to dig solid shale, but not sandstone. In such formations only the sandstone layers would need to be drilled and blasted.

Grading Spoil. The spoil banks from light stripping operations can be leveled by bulldozers, scrapers, or graders. However, the huge machines used in heavy stripping

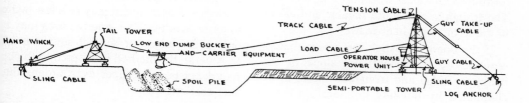

Fig. 10-11B. Overburden removed with drag scraper

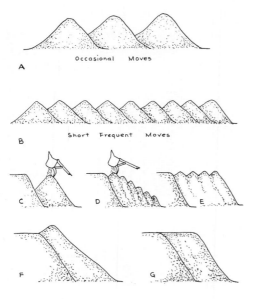

Fig. 10-12. Piling spoil

leave such large coarse piles that the biggest dozers cannot reduce them economically.

The problem is largely a new one, as prior to legislation of the subject, most of the stripped areas were left as permanent wastelands. Techniques are still in the experimental stage.

There are two ways of approach—one, to do original piling in such a manner that a minimum of grading will be required, and the other, to strip in a conventional manner, then work over the result.

If a dipper shovel is moved short distances frequently, the high peaks of its spoil ridge are greatly reduced as indicated in Figure 10-12 (A) and (B). The more even ridge top is the more readily broken down by a bulldozer working along its crest. The trough is also more regular and easier to grade out as a haul road.

If the boom and stick are longer than is required by the height and slope of the bank, the ridge and trough surface can be eliminated by building out level with the old bank, (D) and (E), instead of heaping toward the pit, as in (C). Accurate grad-

ing cannot be expected in such overhead dumping, but the bulk of the irregularities can be eliminated.

A dragline with extra boom length can build a flat top pile by dumping against the bank near the top at full reach, and building out the edge by shorter or longer swings as required. Approximately the same result can be obtained with a slightly shorter boom by throwing some spoil.

Another dragline procedure, requiring extra clearance at the pile foot, is to drag the top of the completed pile onto the slope to the pit, as in (F) and (G).

If the machine can be reeved so as to have a live boom, grading off spoil heaps may be made much easier as the reach can be varied by raising or lowering the boom.

When old piles too large for bulldozers must be leveled, a large dragline can be used to drag them down and spread them in the manner described in Chapter 6 for leveling spoil heaps from ponds.

Another system is to cut rough haul roads through the troughs, and build them up with spoil hauled from fresh stripping. Scrapers usually do this more economically than trucks as they can spread as well as dump. The troughs may be entirely filled, or just brought up far enough to enable dozers or other machines to take down the tops.

HAULAWAY STRIPPING

When sidecasting does not provide a wide enough work area for removal of a pay formation, or there is no place for sidecast spoil to be piled, the overburden is loaded and hauled to a dump.

Depth of overburden may be a few inches to more than 600 feet, and the area may be only a few acres or several square miles. In general, the method is adapted to thicker pay deposits and deeper overburdens than side cast stripping.

Mines in which stripping is chiefly done by sidecasting are called strip mines, while those whose overburden is hauled away are called open pits. Most sand and gravel is obtained from open pits.

Open pit overburden may be any material at all, from soft lake bottom mud to the hardest varieties of rock.

Stripping Area. The size of the area to be stripped for full recovery of a deposit is determined by the area of the pay material, its bottom depth, the thickness of the layer of waste, the slope of the natural ground surface, and the steepness of the safe slope of cuts.

As a practical matter, this ideal area may be limited by property lines, the presence of valuable buildings, by access or drainage requirements, by the money available for stripping, or by combinations of these factors.

Property lines may not be absolute barriers. Permission may often be obtained, at a price, to extend excavation across them. In the Mesabi range where open pit mines adjoin each other, different companies often exchange both overburden and ore, balancing such accounts over a period of years.

If the mineral is valuable enough, whole towns and mills worth millions of dollars may be moved or demolished to permit pit expansion.

Cut Slopes. It is seldom possible to limit stripping to the area immediately above the pay deposit, unless the overburden is very shallow or very firm. Cutting must be kept at a slope that will stand during the period of the work.

Even the firmest rock formations should not be left vertical for over 200 feet. The majority of rocks are not safe until slopes are reduced to 38 to 45 degrees. The only guide that is reasonably accurate is behavior of similar rock in other pits, as laboratory tests for shear and other characteristics are unlikely to include all the characteristics of large masses.

Unconsolidated soil has the same general slope requirements, except that a combina-

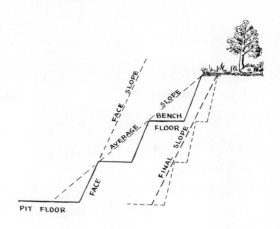

Fig. 10-12B. Benched slope

tion of fine textured soil and ground water may reduce possible slopes to 10 per cent or less. Water is the most important single factor in the stability of soil slopes.

While natural cliffs may be vertical for thousands of feet, they almost always have rock heaps at their feet that show falls that would have been disastrous in a busy pit.

In benched cuts there are three slopes: that of the individual faces, the average working slope, and the final or residual slope that is left when the work is completed. The average and final slopes are taken from crest to crest. See Figure 10-12B. The final slope is usually steeper than the working slope.

Rate of Removal. It is usually necessary to remove some overburden before the pay material can be dug. When money is short, extraction work may start before there is really room for it, with resulting inefficiency. A well financed project may remove much or all of the waste, perhaps at an expense of millions of dollars, before digging any ore.

It is usual for both stripping and extracting operations to be carried on through most of the life of a pit, either at the same time or alternately. In the Mesabi iron mines the same equipment that digs ore in the summer loads overburden in the winter.

Method	Haul, one way, feet	Cost per yard
Tractor—scraper	850	26.6
Self—propelled scraper	2,400	37.9
Shovel and truck	900	45.9
Walking dragline		11.8
Hydraulicking		7.3
Hydraulic dredge	5,000	8.2

Fig. 10-12C. Stripping costs

Gravel pits do their stripping in slack seasons, whenever they occur.

Some mines blast and load their own ore, but let contracts for stripping off waste, so that the two jobs are active side by side through the whole working year.

The relative rates of stripping and mining are an important factor in mine efficiency. When stripping is well ahead, the mine has plenty of working space, with wide bench floors. If mining is faster than stripping, benches narrow and space becomes crowded, reducing efficiency.

The size of the pay formation may change during mining. Additional reserves may be located, or the cutoff point between usable material and waste may be changed by variations in price, processing methods, or market. Any extension of mining will probably make it necessary to extend the stripping. However, abandonment of part of the project cannot un-strip the affected area, or recoup any part of its substantial cost.

Lag of stripping is often due to lack of funds, or to reluctance to invest them any sooner than is absolutely necessary.

Stripping Costs. Stripping costs are affected by the type and amount of material to be removed, the distance it must be hauled, grades on the haul route, and efficiency of the operation.

One mining company having diversified stripping operations reported the costs shown in Figure 10-12C, for the years 1953 to 1955. These figures include all loading and hauling costs including supervision, but not drilling and blasting.

Scrapers can move suitable material at 50 to 60 per cent of the cost of shovels and trucks, if conditions favor them. But in coarse and rough material their production goes down and costs go up rapidly. Costs may be as low as 20 cents a yard under very favorable conditions, and up to $1.75 or more in really rough work.

Hydraulicking and dredging are limited to soft overburden, abundant water supply, and dumps suitable for ponding hydraulic fill.

Rail haulage is likely to be cheaper than truck haulage if large volumes are to be moved more than two miles.

Small pits may be able to sell or use their overburden, medium sizes may have little difficulty finding places to dump it, and large operations may find considerable difficulty in finding adequate disposal areas. For this reason, large volume stripping may show a higher cost per yard than small volume, because of longer hauls needed.

Working at the Edge. A pit in a deep sand or gravel deposit usually expands rather slowly, and stripping work is done at intervals, often by the pit machinery in slack periods. Stripping may be postponed too long, until the face is pushed back against the overburden, as in Figure 10-13A (A).

Stripping may then be done by either pushing into the pit and loading from the bottom, or throwing back with a dragline, hoe, or clamshell. Consideration should be given to the question of caving of high banks. Bulldozers are most apt to cause collapse but may not be damaged. Revolving shovels are both heavy and easy to overturn, and should be kept well back from doubtful edges.

Once the edge is cleared, as in (B), the burden can be moved back farther by re-casting with the dragline, pushing with a dozer, or carrying in trucks or scrapers.

If there is no definite boundary to the pit area, pushing back the soil and leaving it, which is generally the easiest disposal at the time, will cause greater expense when it has to be moved again.

If the ground above the pit slopes up, it may be necessary to cut a platform parallel to the edge to support the dragline, as in (C) and (D). Because of the inefficiency of casting uphill, the platform should be used as a road for hauling the spoil.

It is vital to pit efficiency to keep over-burden stripped far enough ahead to be out of the way in rush periods. Failure to do this often results in spoiling or losing valuable material, and in inability to fill important orders.

Dump Location. A dump for waste from stripping should be as near the pit as possible to reduce haul costs, but it should not be within an area that might be dug away because of any conceivable extension of the stripping. These two considerations may be opposed, and decision between them may be difficult. It is a matter of regret that initial economies have often resulted in disproportionate later expense in re-digging and moving a dump pile.

For example, loading overburden might cost 40¢ a yard, hauling it one mile about 10¢, and additional miles 5¢ each. A single

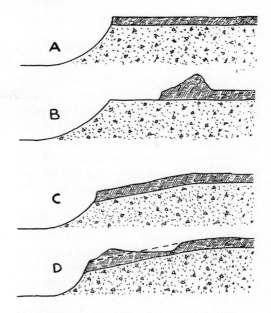

Fig. 10-13A. Stripping edge of expanding pit

move of two miles would cost 55¢, while moving it a mile, re-digging, and moving another mile would cost $1.00 a yard. Double handling is always expensive.

If it can be managed without substantial extra expense, different types of spoil should be placed in separate dumps in such a manner that they will be accessible for re-digging. Changed conditions may make previously worthless material valuable. Examples are the current reclaiming of mine tailings and slag heaps.

Haul Grades. Grade of haul roads is important to economical hauling. A level run from cut to dump is desirable for speed and economy. An adverse grade (upgrade in the direction of haul) will cut both the speed and the load carrying capacity. The extent of this loss depends on many variables. A rule of thumb is that production will be reduced about 5 per cent for each per cent of adverse grade.

The adverse grade will increase fuel consumption, tire wear, and maintenance costs. Wear on the truck engine and drive train are increased disproportionately on grades over 6 per cent.

A down grade in the direction of haul (favorable grade) is helpful to about 2 per cent, but steeper grades may reduce production about as much as an adverse grade. Downhill speed must be limited for safety reasons, and even empty trucks are slowed by upgrades.

Favorable grades over 2 per cent and adverse grades over 5 per cent call for special retarding devices in torque-converter equipped trucks.

Grades may change considerably during a stripping operation. The floor of the cut moves downward, but its edges move outward and often upward. Dumps may stay at the same level, but if space is restricted they usually build upward.

Haul Routes. Two way roads for heavy hauling should be from 4 to 4½ times as wide as the vehicles using them. That is, highway trucks should have 32 to 36 feet between gutters or banks, and 11 foot off the road haulers from 44 to 50 feet. Hauling can be done on much narrower roads when necessary, but liberal width pays whenever large volumes must be moved. Even wider roads are made for some mines.

A haul route that crosses a public road is subject to serious traffic delay. For example, an automatic traffic light that is set against pit traffic, but trips within ten seconds when a truck reaches it, will delay the hauler as much as an extra 1,000 feet at 20 miles an hour. A full stop sign will cause the same or greater delays, depending on the density and speed of highway traffic.

A signal man at the intersection reduces delays to a minimum if he is allowed to favor the pit traffic.

Hillside Dump. The easiest way to dispose of stripping waste in trucks is to dump it off a bank, that is high enough so that it grows outward quite slowly. Height may be anything from ten feet to several hundred. Such a dump may be started by flattening off a hilltop enough to give trucks space to turn, or by cutting a pioneer road along a slope and dumping from it.

Capacity can be figured in two ways. Annual or daily capacity depends chiefly on the length of the dumping face, and to a smaller extent on its height. Total capacity is the volume that can be dumped without spilling beyond the boundaries. It depends on the area than can be used, the height of the fill, and the slope of the dumped material.

Blasted rock dumped off a bank usually has an angle of repose of about 1 on 1, or 45 degrees, and is likely to stay at its original slope indefinitely unless the native soil beneath it moves outward.

Dumped soil tends to assume a somewhat flatter angle, which will depend on the size and shape of its particles. Wet soil flows, and it is important that no drainage from other areas flow into the dump. Rain falling on the turning area on the top may gully the slope and spread mud over a large area below it. This can be prevented by sloping the surface up toward the edge, keeping a berm or ridge of dirt at the edge, and providing another escape for the water.

Both rock and soil slopes offer a hazard of rounded and oversize pieces rolling far beyond the toe of the fill. In empty country this may not matter, and brush and trees check such objects naturally. But dirt or log barriers may have to be built to prevent rolling onto paths, roads, buildings, or other property.

The slope of dumped fill is likely to be between 27 and 35 degrees, with coarser material having the steeper slopes. For rough calculations, assume that it will be 1 on 2. If the waste is already being dumped or piled, the slope can be measured for more accurate figuring of areas and quantities.

Dump Operation. Trucks are backed square to the edge and dumped. Methods of keeping them from backing over the edge vary widely. Sometimes the driver is just supposed to stop in the right spot and at the right distance, with or without a spotter. Or a dump log may be place to both indicate the dump spot and to protect the truck against backing too far. The chief problem here is that a log heavy enough to stop 30 to 60 tons of loaded truck is difficult to move.

The simplest and best protection for trucks is an 18 inch to 2 foot ridge of dirt left at the edge by the grading dozer. If the edge is very soft a ridge may be built at a distance from the edge.

Off the road trucks with standard spill chutes instead of tail gates will dump their loads clear over a ridge at the edge, so that piling up does not occur at the top until it builds up from the bottom. Tail gate bodies may or may not spill part of their loads on the top.

It is good practice to maintain an upward slope from the truck entrance to the dumping edge. One-half per cent is sufficient on porous fills, and one and one-half per cent is sufficient for any material that is kept well graded.

Fills tend to settle under weight of traffic, and with time and weather. This sinking is most rapid toward the edges that are being built out. It is necessary to correct the resulting down slope with wedge fills to avoid down slopes, that are dangerous to trucks and may cause gullying.

A wedge fill is made by dumping on the surface at the back or thin edge of the settlement, and grading with a dozer to restore or increase the original up slope, as in Figure 10-13B. The dozer first pushes toward the edge to establish the slope, then parallel to it to smooth and compact it, and to leave an even windrow along the edge.

TOPSOIL

Topsoil is frequently the only material sold from a temporary pit. At other times,

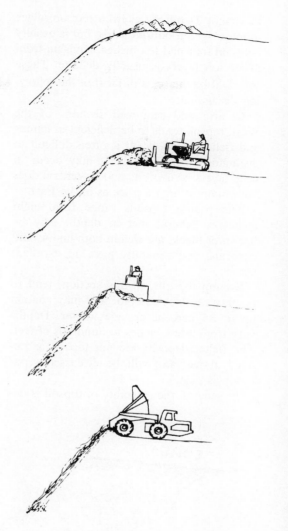

Fig. 10-13B. Building up a dump edge

it may be a highly profitable side line, or a costly stripping and wasting problem.

In this discussion, topsoil is defined as any layer or layers of soil containing sufficient humus (organic matter) and plant food to support a good growth of grass or other desirable vegetation. It is ordinarily on the surface but is occasionally buried by flood deposits or slides.

In the eastern states, topsoils are predominantly brown in color, with a humus content between 3 percent and 20 percent by weight. Depth varies from zero on ridges to many feet in bottom lands, but is usually between four and ten inches. Division from lower soil layers is usually definite. These soils will be the basis of most of the following discussion.

In arid and semi-arid sections of the West, topsoil tends to be deficient in humus and rich in minerals. It is often difficult to distinguish from subsoil. It may occur in deep layers or deposits and in general does not obtain as high a price as in the East.

The prairie topsoils range from eight inches to several feet in depth, may be brown or black, are rich in both humus and minerals, and generally have an excellent texture.

Swamp topsoils, in any section, tend to be gray to black in color and may contain up to 85 percent organic matter. Depths vary from a few inches to hundreds of feet. The richer deposits are not topsoil as defined above, and will be discussed separately as peat.

In general, the salability of topsoil is determined more by appearance and texture than by the ability to grow crops. The average topsoil buyer will seldom have soil tested, and tests are often not as reliable as good judgment.

Topsoils with high percentages of clay or silt will be heavy, slow draining, and inclined to pack into hard lumps if disturbed when wet. Increase of humus content will soften the lumps.

Sandy or gravelly topsoil is loose in texture, drains readily, tends to dry out, and can be worked when wet without caking. Most soils fall in an intermediate structure, with variable draining and lumping.

Tilth is the condition of the soil in regard to lump or particle size. It is affected by grain size, humus content, microorganisms, and the way in which it has been worked. For most crops, a loose structure made up of fine, soft lumps is desirable.

Appearance is largely a matter of color, and freedom from lumps, stones, subsoil, and trash. Dark soils are generally preferred, with strong differences of opinion

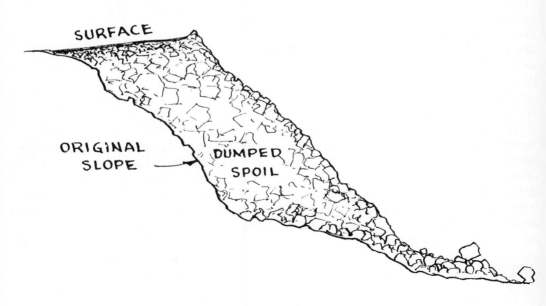

Fig. 10-13C. Cross section of spoil dump

between black and brown types. Red tinting is due to the presence of iron and has no significance in many localities. When entirely dry, topsoil loses its color and is practically indistinguishable from subsoil.

A plowed or cultivated field, or a topsoil pile will appear darker and richer when the observer is looking toward the sun than when he is looking away from it.

Testing. One test of topsoil is the observation of the type and condition of vegetation it supports before stripping. Allowance must be made for weather conditions and the type of crop. Old pastures or fields may become sod bound, or taken over by low growth, so that good topsoil will give a poor appearance.

The vigor of weed growth on piled topsoil is an excellent index to quality.

Laboratory or field tests can be made for humus content, grain size, acidity, and available plant food. Humus is measured by the ignition test to be described for peat, and grain size on screens used for testing sand and gravel.

The test for acid-alkaline balance is commonly made by pressing litmus paper against the damp soil, and comparing its new color with a chart. If the soil is dry, distilled water should be used to dampen it, as tap or pond water may give a false reading.

Acidity is expressed in terms of pH (percentage of free hydrogen ions). A reading of 7.0 is neutral, lower readings increasingly acid, and higher ones alkaline. Most plants will grow under quite a wide range of conditions. A slightly acid condition is desirable for most of them.

Excessive acidity is readily corrected by the addition of lime which can be spread on the field before plowing or disking. Soils are made more acid by mixing with humus, oak leaves, or aluminum sulfate.

Kits obtainable at garden supply stores can be used to measure available or soluble nitrogen, potash (potassium oxide), and phosphorus or phosphorus oxide. It should be remembered, however, that these chemicals are often taken up by plants, or leached out by rain, as fast as they become soluble. The real measure of prolonged fertility is the insoluble reserves that are gradually made available by soil organisms, plants, and weathering. Except for humus, which is rich in nitrogen, and often in the others, such reserves are difficult to measure.

Preparation for Stripping. The cost of properly preparing a field from which topsoil is to be sold is usually a small part of the total expense, and should increase the value of the soil so that it will either command a higher price or be more readily salable at a standard price.

Aside from clearing, field preparation may not be necessary if the soil is to be left piled long enough to rot the vegetation —usually four to six weeks of warm weather for sod—or if digging is to be done by a chain bucket loader. Plowed land is more easily and cheaply piled by a bulldozer than solid fields.

Removal of brush is described in Chapter 1. If no other excavation is to follow the topsoil stripping, medium and large trees should be left, as the amount of soil obtained from around their stumps is small, and the expense of removal and the loss of property value may be considerable.

The field should be plowed to the full depth of the topsoil, if possible, or at least deep enough to turn up most of the roots of the grass or crop. However, turning up of subsoil should be kept to a minimum, particularly if it is of a conspicuously different color. Therefore, if the topsoil depth is variable, or plow depth hard to control, it may be necessary to plow very lightly.

Plowing is usually necessary for sod and heavy crops. Disking may be enough to cut up soft light crops, and on shallow soils will avoid bringing up subsoil. Disking generally does not loosen soil deeply enough to help a bulldozer much, but may create an ideal condition for hoe or drag-line piling.

If a field is burned off, it can usually be disked without plowing. This practice is not recommended, as chopped up vegetation increases bulk and improves the quality and the appearance of the soil. However, occasionally it will be of advantage for economy in handling, or for immediate sale to customers wanting very clean soil.

After plowing, the field should be thoroughly disked so that the vegetation is chopped up and well mixed with the soil. It is then ready for stripping.

If soil is to be removed in a wet season, it may be necessary to leave some strips of sod intact to support trucks.

Noxious weeds can be reduced or eliminated by planting and turning under one or more vigorous, close growing cover crops. Buckwheat is particularly effective at smothering out.

Nitrogen Deficiency. When vegetation is turned under and mixed with soil, there is an immediate and rapid increase in the number of the microorganisms which cause it to decay. For their growth they need nitrogen. If the crop is a legume, such as clover or vetch, they will be able to obtain it as they break down the plant material, otherwise they obtain it, or as much as they can get, from the soil.

This is likely to result in temporary total exhaustion of available nitrogen. When the vegetation has decayed so that it no longer provides sufficient food, most of the organisms die and their nitrogen is largely returned to the soil.

During the interval, which is two or three weeks for fresh green material in warm weather, and longer when the material is dry or coarse, when soil moisture is deficient, or when the weather is cold, any crop which is planted will be starved for nitrogen and make little or no growth.

This effect is most severe when conditions favor rapid decay.

Buyers of topsoil containing large amounts of freshly mixed vegetation should therefore be cautioned to either allow sufficient time to elapse before planting grass seed, or to supply nitrogen fertilizer to give the plants a start. Otherwise considerable dissatisfaction may be expressed and future orders lost.

Packing Soil. The value of topsoil is reduced or lost if it is packed into lumps. Except for a few very light and friable soil types, varying degrees of damage will be done if it is worked wet, or trucked over except when thoroughly dry.

Wet working breaks down the soil particles into a structureless mud that dries into lumps and sheets. The probability of damage increases with the amount of contained water, and its severity decreases with increased proportions of sand or humus.

Packing under trucks or other heavy machinery may produce the same result by bringing water out of small spaces between particles so that it will make a mud. Trucking on rather dry soils will produce compression cakes which are usually softer than the mud lumps.

A rough test of condition may be made by rubbing a sample of soil between thumb and finger. If it smears, it is too wet, and if the particles remain separate, it is probably ready to work. The dirt turned by the plow or dozer blade should be watched for smearing, which indicates that soil is too wet.

Topsoil should not be trucked over, but it is often impossible to avoid doing it. The stripped areas may be too soft, or they may not offer enough space for maneuvering. If the latter, trucks may be routed to drive empty across the topsoil and run on the subsoil with loads.

It is usually better to completely ruin a narrow strip of soil by using it as a haul road than to damage a large area by allowing trucks to wander around on it.

Soil lumps are completely broken down

by freezing and thawing and usually disintegrate slowly in wet seasons. If still on the field, roots of a cover crop, or sometimes only a thorough rolling or disking, will reduce them.

If absolutely necessary, a hammer mill shredder of the type used for humus can be used to pulverize them.

Piling. The standard tool for piling topsoil is the front loader. The old standby, the bulldozer, does the same work but rather less efficiently, as it cannot make high piles without walking on them. But it may separate topsoil more cleanly.

When piling is done in advance of loading, the standard practice is to heap the soil in windrows (long piles). These may be run up and down the slope so as not to interfere with drainage; or across it to keep in uniform soil types, or for convenience in trucking. It may be necessary to make occasional breaks in windrows to prevent ponding of water above them. Piling should be started at the entrance to avoid trucking over unstripped areas.

Windrow size will vary with loading requirements and soil depth, and the size of the bulldozer. Small, closely spaced ridges are most easily piled but loading machines work best in large high piles. Large dozers and deep soil favor building big piles.

Building of the piles is described in Chapter 4. The width of the windrow should be figured in advance so that work will not be wasted by digging inside the pile area, then refilling.

If piles run up and down a slope, stripping should be done from the top down to correct any tendency to gouge on the lower side. If piles run across the slope, most or all of the soil should be moved downhill.

Under ordinary conditions, the top of the topsoil is piled first, so that the operator can concentrate on moving big loads without much concern about accurate work.

After a section has been rough stripped, it is gone over again, carefully cutting to the bottom of the good soil and working from the back of the cut. The windrows of topsoil left by the sides of the blade are then pushed in. If two dozers are working, the one with the more accurate blade control, or the smaller one, is used for the cleanup.

Careful separation of the topsoil from subsoil is a requirement of most stripping. Inclusion of even considerable amounts of loose fill in topsoil does not usually damage its usefulness, but it is a type of adulteration that is unpopular with the buyer. The damage to appearance and value is especially severe if the subsoil is a conspicuously different color, or texture, or is in the form of lumps or sheets. It is generally better to leave a thin sheet of topsoil on the field than to mix topsoil and fill.

However, if the stripping is done in order to get clean dirt or gravel for roads or other purposes, it may be better to concentrate on cleaning the surface, even if some fill mixes with the topsoil.

Topsoil and subsoil are separated chiefly on the basis of color. If the difference is prominent, the distinction is easy to make, but the results of a mistake are painfully obvious.

Light conditions may obscure the color difference. When the sun is very low, in the morning or evening, or high in a clear sky, subsoil and topsoil may look the same. Cloudy days and intermediate sun elevations give the clearest distinction.

Any shovel rig can be used for piling topsoil. The backhoe and dragline are best at it, and the toothed clamshell slowest. Shovels work rather slowly in shallow soil. The hoe is adept at salvaging soil along a wall or fence.

Shovels loosen and aerate the soil, and cause minimum damage in handling it when wet. Fill dug with the topsoil can be concealed by mixing in. Teeth prevent abso-

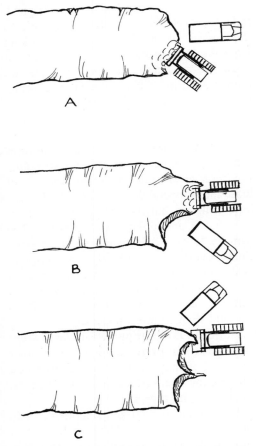

A

B

C

Fig. 10-14. Loading topsoil with shovel dozer

Loading from the Pile. Topsoil is comparatively light and normally has a low digging resistance. However, it tends to push ahead of narrow buckets, instead of entering them, and this factor, coupled with the small size of the usual pile, may reduce production below that of hard digging in a bank.

Difficulty in filling the bucket may be reduced by thorough chopping of sod and weeds before piling, building large piles, and building piles on undug areas so that the bottom of the bucket will work in firm soil.

The best topsoil loader (now obsolete) is the chain bucket loader. Its feed mechanism chops up sod and trash, and mixes, aerates, and puffs up the bulk of the soil. Although production rate is usually high, the soil is discharged in a thin stream, so that trash and rocks are readily removed.

It can dig directly from an unplowed field but output is higher in piles. It is seriously hindered and may be damaged by large rocks or roots.

Next in preference is the shovel dozer because of flexibility, ability to load from either pile or field, to clean up as it works, to pile when not loading, and high production in relation to purchase price. Trucks are backed into the end of the pile, in variable positions to keep them at an angle of about forty-five degrees to the digging, as shown in Figure 10-14. These wide buckets get heaping loads until the end of the pile is reached. The remnants can be pushed to the next pile.

The dipper shovel is widely used. It may have trouble filling the bucket but its principal difficulty is in cleaning up. If the pile is narrow, as in Figure 10-15 (A), it can walk down the center and scrape in the sides by swing dragging. If the pile is heavy, as in (B), it can dig the bulk from one side, easily cleaning as it goes, and come back through the small remnant. In either case, it is best to have a bulldozer

lute accurate work, and it is good practice to clean up afterward with a dozer.

Scrapers are used for piling topsoil chiefly when it has to be completely removed from a large work area. They generally make wide, low piles that are more readily rehandled by scrapers than by other excavators. They pack the soil heavily even when it is in good working condition.

If topsoil is stripped from one part of the job at the same time that it is spread on another, the scraper can combine the two operations very efficiently.

Scrapers drawn by fast wheel tractors may be used to dig topsoil and deliver it by road to local customers, who will appreciate having it spread.

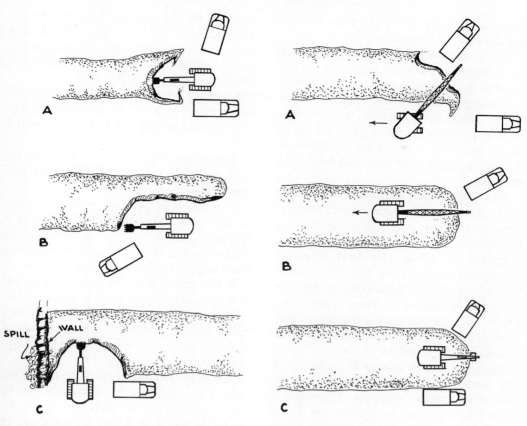

Fig. 10-15. Loading topsoil with dipper shovel

Fig. 10-16. Loading topsoil with dragline and hoe

work with it, at least part time, cleaning up. A light rubber-tired dozer is generally adequate.

When digging a windrow ending in a wall or property line, as in (C), the shovel should be started at that end and worked in, to minimize working soil over the end and losing it.

Draglines and clamshells can be worked from one side of the pile, as in Figure 10-16 (A), or preferably from the smoothed-off top, as in (B). The top of the pile is particularly suitable for hoes, as in (C), as they load more rapidly and easily if higher than the trucks. These rigs clean up the edges without extra work but can strip closer if aided by a dozer.

Any shovel cutting to a grade will occa-

sionally go below it so that fill will be dug. The mistake is immediately shown by the color of the bottom. The bucket should be dumped off to the side to be wasted, or on the pile, to break up and mix the slice of fill. Subsoil may be deliberately dug and mixed in this manner if the topsoil is rich and the price is low or the buyer in-different.

If the soil is uneven in quality, the shovel may mix it up during loading.

The buyer may not be able to judge the quality of the topsoil he is getting from observation of it in the truck, or after being dumped. The operator of a revolving shovel may put undesirable material in the back of the truck, good soil in the front and over the top. The top layer is all that is seen in transit and that which was in the bottom front after it is dumped.

If any doubt exists about the reliability of the supplier, the loads should be spread as soon as received.

Loading from the Field. Topsoil is often loaded direct from the field, with no systematic preliminary piling. The efficiency of the operation depends on the depth of the soil, the machinery used, and the output required.

Best results will be attained in deep soils, as increase in depth reduces the proportion of time spent in trimming the bottom and in moving into the digging.

Unless the soil is very hard or rocky, the bucket loader gives the best results; particularly as vegetation does not have to be disked ahead of it. Front loaders dig easily, but may roll up big masses of sod. Almost any loading machine can be used unless the soil is shallow or rocky.

High production cannot be expected of any machine, except possibly the bucket loader, in shallow stripping. However, if the trucks are scheduled so that the shovel dozer has free time between them, it can stockpile enough to enable it to fill the truck quickly.

Sometimes a bulldozer is used to boost output by feeding soil to the loader. This may be done on the whole job or just on thin spots. Dozer cleanup after digging is usually desirable.

Screening. Topsoil may be coarsely screened on a portable grizzly, which is placed on the truck body prior to loading, or may be up on legs so that the truck can back under it. The latter type requires large piles for convenient use. A truck with a permanently mounted screen can be used to haul to a nearby stockpile.

These are used either to market topsoil which is otherwise unsalable because of presence of rocks, roots, and trash; or to obtain a premium price. It is usually necessary to have a man to clean stuck material off the grizzly. Best results are obtained with dry or sandy soils. Heavy soils require a coarse mesh and must be put on a small quantity at a time, or too large a proportion may be rejected.

Reducing the slope of the screen will allow more material to go through, but will increase sticking and piling up on the surface.

Square openings from one half inch up to four inches are used, or similar bar spacings.

A much finer product, with less waste in rejections, can be obtained by using a revolving screen on the discharge of a bucket loader. A portable gravel screening plant might be used, with some changes.

If the topsoil is full of lumps, a hammer mill shredder may be used to reduce them.

If only occasional loads are required, hand screening will be more economical than the purchase or conversion of equipment.

Artificial Topsoil. Topsoil of fair to excellent quality can be manufactured by expert mixing of fill and peat (humus). Poor quality topsoil can be enriched, or dark rich soil made to go further, by similar mixing.

If the humus is pure (less than 20 percent ash when burned after drying), and air dry so as to contain less than 25 percent water, about one part of humus to two to four of fill is used. The fill should be of light texture, but not coarse or gravelly, and preferably taken from near the surface. Fine white sand is sometimes used instead of fill, to provide a mixture for topping golf greens.

The mixing is done by a bucket loader. The two materials may be fed to it from stockpiles by a dozer or shovel, or a combined pile may be made by dumping and leveling fill, and then dumping humus on top to make proper proportions.

Revolving shovels, and other excavators, can mix the materials by re-piling.

Peat is generally quite acid, and unless an acid soil is wanted, or the fill is quite alkaline, ground limestone should be added at the mixing paddles. Manure, or general purpose chemical fertilizer, such as 5-10-5, added at the same time, will improve the product.

Success in topsoil compounding depends on a supply of good cheap humus and suitable subsoil, proper fixing, and local acceptance of the product.

Storage. Topsoil is often left piled for long periods. Texture is generally improved by standing over the winter, and quality does not seem to be damaged much even by years of storage. In hot dry seasons, it may bake out to colorless dust, which is unpleasant to load and hard to sell, but a wet season will restore its original texture.

However, piled topsoil will support a luxuriant growth of grass and weeds, which quickly forms a sod that is very undesirable. The irregular shape of the piles makes cultivation with any ordinary tools almost impossible. Flame guns are tedious to use and ineffective, and chemical weed killers are selective and uncertain in action.

Such growth can be kept down, at a price, by cultivation with a bulldozer, as in Figure 10-17. The machine is run along the crown of the windrow, cutting off the top and allowing the dirt to flow down the sides. This uproots or buries a large part of the weeds. Later the bulldozer pushes the sides back up to the top, as in (C) and (D), destroying the rest of them. Most economical results are obtained by allowing a week or more to pass between the two operations. The job is done over as often as necessary.

This serves not only to keep the soil clean, but also to enrich it by the decay of the weeds.

Restoring Vegetation. When topsoil stripping is not followed by other work, areas are left denuded of vegetation and

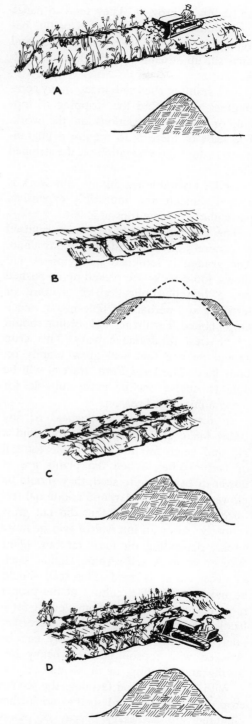

Fig. 10-17. Weed control on piled topsoil

the ability to grow it. They tend to cause a dust nuisance, and to erode badly, becoming a mass of unsightly gullies, and often silt up streams and block roads with the waste.

As a result of such nuisances, many communities now forbid the stripping of topsoil, or impose restrictions on the work. The least of these is to require a guarantee to restore the vegetation on the stripped areas.

Under favorable conditions, this work is neither difficult nor impossibly expensive, particularly if some topsoil is left.

The process is similar to that required to restore worn out and eroded farmland. The ground is loosened with rippers and plows, loose rocks are picked up or pushed off, a liberal application of manure or fertilizer and perhaps of lime made, and a crop planted. Fertilizing and seeding should be heaviest in drainage ways. This crop should be one that will grow readily on poor land. The local Farm Bureau will be able to advise one or more suitable for the conditions and season.

In many localities, buckwheat or soy beans make a good summer crop, and a rye and vetch mixture is suitable for fall or early spring. When the plants are in flower or beginning to seed, they should be plowed or disked under, and additional fertilizer supplied to spots that did not grow properly. After an interval of two or more weeks, depending on local custom, plant another crop. A self-seeding legume such as sweet clover is good. This seed should be inoculated with a culture of the proper nitrogen-fixing bacteria. Some patch fertilization and replanting may be necessary later, and any tendency to gully can be checked with topsoil patches, heavy fertilizing and planting, or brush mats.

With luck and good farming, the second crop may hold the land so that no further plantings are necessary. Decay of plants and nitrogen absorbed from the air will enrich the soil, so that the native vegetation of the area will soon be able to re-establish itself.

Reclamation probably gives best results on glacial till soils which will develop a new topsoil cover in a surprisingly short time. The author has taken good topsoil off land that was thoroughly stripped ten years before. In this case, however, three or more crops were turned under before sweet clover was left as a permanent cover.

Floors of deep pits will respond to the same methods but much more slowly. Deep subsoils rarely have proper tilth, or plant food in quickly available form for crops, so more "green manuring," or turning under of vegetation, is required.

Animal manure will give better results than chemical fertilizers, particularly on floors of deep excavations, largely because it contains many organisms essential to a healthy soil. However, in many localities it is difficult and expensive to obtain.

It is advantageous to plant as soon after stripping as possible. If an interval is to elapse between piling and loading, it is good practice to plant a cover crop between the piles to prevent the development of a dust nuisance and loss of topsoil remnants.

PEAT (HUMUS)

Characteristics. This is a light, soft, absorbent, organic substance formed by partial decay of vegetation, which accumulates in more or less pure form in swamps or shallow ponds. It is usually brown or black, and may be a structureless jelly, or fibrous, or lumpy with recognizable remnants of wood or leaves.

It may be mixed with varying quantities of soil which increase its weight and will make it lighter in color, particularly when dry. Water content is 50 percent or more by volume.

When pure from 70 to 85 percent of its dry weight is organic and the balance mineral ash.

In some countries it is used extensively as fuel, when partly or wholly dried.

If allowed to dry in stockpiles or de-watered deposits, it will take fire rather easily and burn slowly and persistently. It can be put out by complete soaking, or controlled by cutoff trenches. This problem is discussed in Chapter 1.

Peat is usually very acid. It may contain no immediately available plant food, but reserve supplies in insoluble form are good, particularly of nitrogen. It may be sterile, or able to support only limited growth, when first dug, but on exposure to weather it will gradually become fertile. It will make good topsoil when mixed with soil or sand, and lime and manure.

It is used to enrich lawns and gardens and occasionally farms. Golf courses use large quantities. It will soften heavy soils so that they will not bake hard; and make light soils absorbent so that they will not dry out readily. Use of large quantities at one application, particularly without thorough mixing, may unbalance the soil so as to make it temporarily unsuitable for some crops. Lime should be used liberally with it unless acid loving plants are to be grown. Manure, or to a less extent fertilizers, may prevent unbalance.

Humus may be tested for water content, for organic matter, and acidity. A quantity is weighed, baked dry at low heat, and weighed again. The difference is the water which it held.

The dry humus is then raised to a red heat, stirred occasionally, and kept at this temperature until it has been reduced to ash. The ash is weighed. The difference from dry weight is the organic content.

Confusion is caused by the use of humus and organic content in reference both to the total weight of dry plant or animal remnants, and the non-ash part of such remnants. As an example, a pure peat deposit may be said to be 100 percent humus or organic material; or 85 percent humus after allowing for the minerals contained in the woody fiber. Both usages are permissible.

Percentage tests are best performed in the laboratory but they can be done roughly in the field, or at home, with a heavy pot and kitchen or parcel post scales. The pot should be weighed separately, and all other weights taken in it. The largest quantity convenient should be processed to minimize errors in reading the small weight of the ash.

Burning humus may produce noxious fumes so good ventilation should be provided.

Tests for acidity are the same as used for topsoil. Extremely acid conditions are to be expected.

Digging. Peat is normally a water level or underwater deposit. It is often readily drained or pumped dry, and is soft and light to dig. However, it is difficult to recover, particularly in large deposits, as it is sometimes too unsubstantial to make safe footing even for crawler draglines on platforms, or to support ordinary haul roads.

Methods of digging and getting it out of the pit are discussed in Chapters 3 and 6. In general, the cable excavator mines it at lowest overall cost, and with fewest complications, but operations are often too small to justify the necessary investment.

Curing. A particular problem is the high water content. Half to three quarters of the water will drain out of a pile in a month or two of dry weather, reducing its bulk 50 per cent or more. The balance will stay in it unless baked out, and is a normal part of bulk or screened humus.

Before draining, the humus is too wet for handling as it becomes sloppier each time it is moved, and it will not respond to processing. Provision must therefore be made for curing between the pit and the plant or customer, unless the plant is equipped with drying apparatus.

A drag scraper or slackline may dump

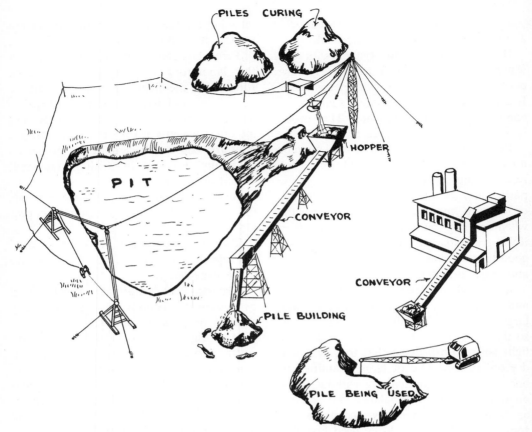

Fig. 10-18. Humus pit layout (scale distorted)

into a hopper, from which one or more movable conveyors carry it to piles, as in Figure 10-18. The conveyors may have to use cleated belts, or buckets, to prevent the peat from flowing back along them. Conveyors, loaded by dragline or clamshell, can reclaim the piles to the plant; or the material may be loaded into trucks for sale as "raw" humus as soon as it has drained.

Draglines may pile it up at the pit edge and return to load it.

Machines handling dry or partly dried humus can be equipped with oversize buckets, or oversize drum lagging, or be run in a higher gear than it can when handling heavier soils.

Processing. Humus is processed to break up lumps, to mix together different colors or grades, to mix in soil accidentally or deliberately added, and to increase bulk by fluffing and aerating.

A chain bucket loader will usually perform these functions, and will also load into trucks or storage at the same time. If lumps are soft, it can break them as well, and if they are hard but infrequent, they can be removed by a man stationed at the discharge. If the ground under the pile is too soft to support the loader, it can walk on saplings or planks placed behind the blade and ahead of the tracks; or it can stand on firm ground and be supplied by a shovel.

Humus loaded in this manner may or may not pass as a screened product.

When lumps are a serious problem, or the product must meet rigid specifications, a shredder, often a special type of high speed hammermill, is used. It is lighter and less costly than rock crushing models, but operates in the same way. Toothed drums may also be used.

In small operations, the shredder may be supplied with hand shovels, and the product hand shoveled into trucks or left in a windrow to be gathered by a loader.

A portable conveyor belt may take the humus from the discharge opening and load it in trucks, or on a pile. Larger portable mills may have a high hopper opening supplied by a hand loaded conveyor belt.

The shredder can also be fed mechanically by any kind of loader, without hand work.

Portable crushing and gravel plants are sometimes successfully adapted to processing humus.

Fixed plants should be on firm ground outside the pit, and may be supplied by cable excavator, conveyors, or trucks. They are capable of much higher production, and may turn out a finer or more uniform product. They can be equipped with drying and packaging machinery so that their product can be sold in stores.

In any calculation about humus it should be remembered that one yard in the wet deposit or pile means a half yard or less in the cured pile.

GRAVEL, SAND AND CLAY

Bank Gravel. Bank gravel is a useful and highly varied material. It consists chiefly of sand, pebbles, and cobbles, but may also contain clay, silt, and boulders, mixed in or in accompanying layers or pockets. The gravel proper is the pebbles and cobbles in sizes from ¼" to 2".

The specifications which gravel must meet to do certain jobs, and the proportions found in deposits, vary widely. The range of road gravel requirements is indicated in Chapter 8, pages 8-12 and 8-13.

Bank gravels consist mainly of deposits laid down by fast running streams, often of glacial origin, but they are also formed by waves on the seashore. The quality depends on the original stone, the proportion of sizes, and the angularity of the particles. Wave formed gravels are predominantly rounded, glacial ones subangular, and product of other streams variable.

Talus gravels, formed at the foot of cliffs by falling and sliding, may be coarse and angular, but are often weak stone.

If gravels are not sufficiently angular for their job, and contain oversize stones, they may be run through a crusher which will produce angular fragments.

Fines in bank gravel act as a cement or binder, holding it together when dry. Gravel without binder becomes too loose for road use in hot, dry weather.

Fines in excess of eight or ten percent may cause a gravel to become sloppy after repeated freezing and thawing when wet. Fines over fifteen percent may cause it to soften under prolonged soaking. Softening is made more likely by a high proportion of fine sand in the mixture, and less likely if thorough compaction precedes the freezing or soaking.

Any gravel will become sloppy if soaked when freshly dug, but if of good quality, should drain and firm quite quickly.

Gravels derived from continental glaciers are largely of hard rock. River and mountain glacier gravels are derived from upstream formations, and occasionally include too much shale or other soft rock for some purposes.

There are a number of tests for gravel, for field and laboratory use.

If a specimen is rolled between the fingers, it will separate into grains which, if inspected with a magnifying glass, will indicate something of the sharpness and assortment of the sand particles.

A sample, with stones over one quarter inch removed, can be shaken up with water in a glass jar, then allowed to stand. The pebbles will form a layer in the bottom, with coarse and then fine sand on top. Silt and clay will settle out more slowly, and may take an additional day to compact. The relative amounts of the different size particles can then be determined by inspection.

In the laboratory, gravel is dried, weighed, and put through a vibrating screen with many different meshes. The particles caught on each tray are weighed. Any lumps have to be broken up. This operation gives a classification of the specimen for size gradation.

Gravel can be tested for abrasion resistance by rolling in a cylinder with steel balls or other hard weights. Resistance to breaking up by freezing can be tested with cold, or with chemicals which duplicate its effect.

Clean bank gravel of proper sand-gravel proportions is frequently mixed directly with cement for concrete.

Sand. Most bank gravel deposits are more than half sand. In addition sand deposits occur in many areas where no gravel is found.

Ocean beaches are typically sand, and river deposits usually contain high proportions of it. If the river flows slowly, the sand may be mixed with silt and clay, which usually must be separated before use.

Most sand is largely particles of silicon dioxide, best known in the form of quartz. It is very hard and withstands the abrasion of water working which reduces other minerals occurring with it to fines. Calcium carbonate, mica, feldspar, gypsum, and many other minerals may also occur as sand.

Many sand banks are clean enough for use without processing, but in most cases it is safer to screen and wash before using in concrete.

Occurrence. Sand and gravel deposits occur in all parts of the world, and with special frequency on or near past or present shores, glaciers, and mountains. They may be thin, irregular deposits, or in heavy masses. In general, gravel is more variable than sand in size and type of particles, and thickness and shape of beds.

Running water needs higher velocity to carry large pieces than small, and in general, gravel is deposited nearer the source than sand, or at times of heavier stream flow. However, a stream which is building up a deposit, alternates bringing in materials and cutting parts of it away. Channels wander over the whole area. Oversize material beyond the capacity of the water to carry may be rolled long distances along the bottom. Clay and silt may be deposited in temporary pools and cut off and stagnant channels.

The result of these factors is that gravel, sand, and clay deposits are often extremely variable and uncertain. When this is the case, mining them requires constant good judgment in deciding which horizons should be combined and which separated; and what can be used and what must be wasted.

Processing. Sand and gravel may be processed to clean out dirt; to separate into different sizes; to combine different sizes and materials; to remove or crush oversize stones; and for combinations of these purposes.

In variable formations, the primary processing is selection at the bank as discussed under selective digging.

The processing plant proper may consist of a washer, a screen, a crusher, or multiples or combinations of these units, together with feed hopper, and transfer and discharge conveyors. These plants, available in both mobile and portable types, are described in Chapter 21.

By the use of units of proper size, any desired reduction, combination, or separation can be secured. It should be remembered, however, that no plant can produce a coarse product from fine particles. Deficiencies in gravel content must be made up by mixing in stone of proper size, or oversize up to the crusher capacity, in addition to the run-of-pit material.

Clay. Clay, like sand and gravel, may be found in massive deposits or in irregular layers and lenses. It is often interbedded or mixed with other materials in very complex ways.

Underwater clay may be soft enough to be dug with a small dragline, or quite hard. Dry clay grades from hard shovel digging to shales requiring heavy blasting.

Pit operators usually find it economical to loosen up dry clay with at least light blasting, to facilitate digging. Electric or gasoline driven augers are extensively used for drilling, and slow to standard velocity explosives for blasting.

When valuable clay is in narrow and confused beds it is often blasted, then separated by hand into piles which are loaded by machine.

LOADING OUT OF THE BANK

Most primary pit excavation is in formations deep enough to be loaded directly from the bank. The material may be in its natural state or loosened by blasting.

Bank Height. In free flowing material, such as loose dry sand, the only limit to bank height is that imposed by safety. This will be discussed below.

If a formation will stand in vertical or overhanging walls, and is dug from the bottom, the face should not be higher than the machine can reach, as it may be necessary to dislodge overhanging pieces with the bucket to avoid danger from falls. Half this height is usually more convenient and may allow greater production if the top of the bank does not keep falling as the lower face is cut.

For example, dipper shovels of 2½ yard capacity may do their fastest loading in banks in which the dipper teeth do not have to be lifted above 12 to 15 feet in the bank.

However, they can trim banks up to about 25 feet.

When working in rock, it is the height of the blasted rock heap that counts, not the height of the face. The amount of settlement depends largely on the proportion between the height of the face and the depth to which it is blasted.

A 200 foot face blasted back 20 feet may yield a muck pile only 15 or 20 feet high. A 20 foot face blasted back 200 feet might produce a 30 foot high rock pile.

The height of a rock face may be limited by the length of a drill feed. Changing steels or adding drill rod takes time. This can be saved by fixing face height somewhat lower than the length of the feed. Rotary drills with 50 foot masts are used extensively on 40 foot banks.

Low faces require frequent moves. Height of less than one half of shipper shaft height may make it difficult to fill the bucket.

Benching. Whenever a non-caving formation is too deep for convenient digging, it is removed in layers. In shovel work these are called benches. They may be anywhere from 6 to 200 feet high. In highway work 12 to 20 foot faces are common, being suited both to the mobile light drills favored by contractors, and to the 3/4 to 3½ yard shovels they use.

In very massive rock cuts in highways, and in large scale quarry and mine operations, heights of 30 to 60 feet are usual, with drilling by rotary or down-the-hole drills, and loading by 2½ to 8 yard shovels or the biggest front loaders.

The best place to start benching is at the top. This simple fact is often obscured by other considerations, so that hillside work may be started at the bottom or the middle, and the pattern straightened out later.

The width of a bench, from the edge to the toe of the unblasted rock, should be at least 50 feet. Greater widths are better.

Types of Machinery. Loading machinery used for pit excavation can be roughly divided into tractor loaders, which depend on traction on the pit floor for digging power; revolving shovels with dipper, clamshell, or skimmer front ends, which stand on the floor while working; revolving shovels, with dragline, hoe, or clamshell rigs, which load from the top of the bank; scrapers and bulldozers which work down the bank slope; and remote control cableway excavators.

Selection of machinery will depend on the location and digging characteristics of the formations, the volume of output required, the type and importance of other work that must be done by the same machines, the type of haulage or conveyor units, and the price tags.

Production Factors. Big machines are suited to hard and coarse formation and to high production requirements.

Practically all excavators are available in different sizes. Production usually does not increase in direct proportion to power and weight, as the more massive construction of heavier units may require lower speeds, and space may be lacking for convenient operation.

Manufacturers' data on output should not be accepted without careful study. Some firms deliberately underestimate production to avoid arguments; while others exaggerate it to make sales. Others base it on time-motion calculations, with little reference to field conditions.

Production ratings based on loose yards, or on a sixty minute hour, will be higher than for bank yards, or a fifty minute hour.

Also, there is room for honest difference of opinion about whether a formation is hard or soft, and conditions average or ideal.

A rough index to output can be obtained by timing a machine at work in various materials. A stopwatch should be used and the results written down. The cycle time is the elapsed time between a certain movement, as entering the bank, and the repetition of that movement. Average number of cycles per minute, from a number of observations, multiplied by the average bucket load in yards, will give the production rate in yards per minute in simple work such as sidecasting. Extra passes made to trim the bottom, or to break out or avoid boulders, may be averaged in or considered separately.

If the machine is loading, the loss of time in spotting trucks and trimming up their loads should be observed.

Data for calculating production are included in Chapter 3.

The figures obtained must be modified to allow for mechanical difficulties, maintenance, cigarette time, failure of trucks to keep up their schedule, blinding dust, and inspection of bank material. These are often lumped together as one-sixth of operating time, so that each hour is figured to have only fifty working minutes. If the calculation is to determine the number of days required to do a job, weather must be allowed for. This will include the time the job is shut down because of rain and resulting mud.

Big machines can almost always dig hard material better than small ones of the same type, but this factor is even harder to calculate. A rough index to penetration in material of even texture can be obtained by dividing the force which can be applied to the bucket by the width of the edge, or the combined width of the teeth. The extra resistance to the thicker teeth of the heavy bucket may be negligible in brittle formations and important in resilient ones.

In poorly blasted rock, or boulder-filled banks, the gain in penetration is much greater, as nearly the full power is often applied in succession to points of greatest resistance.

A wide bucket may be at a disadvantage because of inability to get between obstacles to attack them separately, or benefit from its capacity for large chunks.

In any digging, sharp cutting edges are essential to best work. In hard formations, teeth of proper spacing will give better results than straight edges.

Mobility is an important factor for machines which may dig for short periods from a number of different bank sections, or are used for loading from storage piles as well. Ability to do several types of work is liable to be useful, particularly in small pits.

It is good practice, although not always essential, to match the size of loading and hauling units. If large shovels are used with small trucks, time is wasted centering the bucket and material will be spilled off the sides. Truck tailgates may be jammed by oversize pieces and trucks damaged by impact. If the trucks are too large for the shovel, they must spend too much time being loaded; the shovel may be unable to fill them from one stand, and high body walls may hamper it.

Revolving shovels are usually teamed with trucks which will carry between five and ten bucket loads. Capacity is not as important for tractor-loaders but body walls should be low enough to permit easy placement.

Tractor Loader. Tractor loaders include crawler types, which can do quite hard digging and heavy bulldozing; four wheel drive rubber mounted units suited for medium hard banks, and two wheel drive loaders for soft or loose material and hard ground operation.

Crawler mounted loaders are easy enough to move around pits of moderate size, but the wheel mountings are superior in speed and cause less wear to themselves and to the roads while traveling.

Four wheel drive loaders may be obtained with standard buckets up to 18 yards in capacity. They can often replace an excavator-and-truck combination in filling hoppers, and in short to medium hauls.

Front loaders are described in Chapter

16. Almost all of them now have an open-front bucket that rotates forward to dump and backward into a cupped position for slicing upward in a bank, and for carrying a load.

Multi-purpose or 4-in-1 buckets are easier to dump into high trucks, and are invaluable in handling bulky objects. However, their greater weight is a disadvantage in ordinary digging and loading.

A good truck loading pattern for a front loader is shown in Figure 10-19. While these machines are flexible and can dig

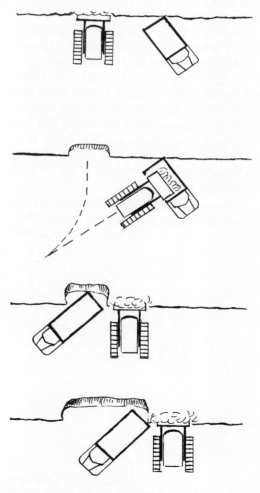

Fig. 10-19. Loading from bank with shovel dozer

Courtesy Frank G. Hough Co.

Fig. 10-20. Filling a truck

under very awkward conditions, best production is obtained if both angle of turn and walking distance are kept to a minimum. Best height for non-caving banks is somewhere between the height of the push arm hinges, and the maximum upward reach of the bucket edge.

Loaders retain a high degree of efficiency in very low banks, but are seriously exposed to slides or cave-ins in very high banks.

In spare time or in emergencies, either crawler or wheel mounted front loaders can do almost any bulldozer work except scraper pushing. They are used for tidying the pit, smoothing haul roads, shifting heavy machinery, carrying heavy or bulky objects, and rescuing or starting stuck trucks and other equipment.

Skimmer. All-cable skimmer shovels are found only occasionally in strip mines. They load thin layers of coal easily and accurately, making a clean separation at the bottom. But they cannot dig high banks, and are generally lacking in flexibility.

A dipper or loader front on a hydraulic shovel may be able to cut readily to flat grades. It carries a loader-type bucket that supplies great break-out force, and swings with the same ease as other revolving excavators.

Dipper Shovel. Dipper shovels are excellent machines for bank excavation. Although fastest loading is in soft material that will heap on the bucket, they can maintain good output in very hard or rough material. They are more costly in proportion to capacity than the tractor loaders, but have lower repair requirements as the tracks do not move during the digging cycle.

In the smallest sizes, and when mounted on wheels in any size, they have good mobility. Medium and large sizes are generally not moved around for less than a day's work.

They will dig from any graded floor that will support trucks. Best production in non-caving material is usually obtained when the bank is about as high as the shipper shaft (dipper stick hinge).

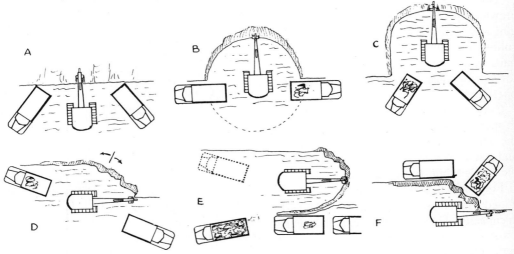

Fig. 10-21. Loading from bank with dipper shovel

10-32

A short arc of swing is important in getting maximum production. The bucket can usually be moved from break out position at the bank, to correct height and distance for dumping in a truck, during 30° to 45° of swing. Any longer swing required by truck position slows the digging cycle.

When a shovel is walked straight into a wide bank, the initial swing required is about sixty degrees. As the machine works in, the swing becomes longer, finally approaching 180°. See Figure 10-21, (A) to (C).

Except in the case of a through cut being made in just this width, it is obviously inefficient to penetrate so deeply into the bank on one path.

If the shovel is walked parallel to the bank, as in (D), trucks can be spotted ahead, at a slight angle to the bank, with a minimum swing of about 40° and a maximum of 140°. If trucks are also placed behind, the maximum swing can be reduced somewhat. As long as the shovel is kept slightly outside the toe line of the bank, as shown, it can do its own cleanup. However, production is increased if the shovel operator can dig roughly, depending on another machine to smooth out the floor.

If the shovel is kept deeper in the bank, as in (E), a ridge will be left near the toe line, reach to load across it will be longer and a dozer will be needed. However, production from each stand will be greater, which in a low bank may be an important factor.

When the pit floor is narrow, sandy, or wet, so that trucks must keep to beaten paths, a drive-through pattern, as in (E), can be used. The shovel again works along the toe of the bank, and the trucks run parallel to it, at a convenient loading distance. Only one truck can be spotted at a time, but it can be moved into position much more rapidly then when backed in.

If the pit floor is too soft for trucks, and

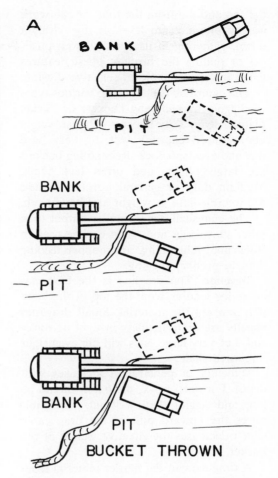

Fig. 10-22. Loading from bank with a drag-line

only a dipper shovel is available for loading, trucks may be put on top of a low bank. This works well only with a big shovel and low trucks.

In banks offering a danger of slides, the shovel should be worked straight in, as in (A), so as to be able to back out directly if partly buried. Cuts should be kept shallow by frequent moves to different parts of the face.

Clamshell. The clamshell is so versatile that it is difficult to set up patterns for it. It can stand at the foot of a fairly high bank and dig from the top, or stand on the top and dig from the foot, or can work at any intermediate level. It digs straight down, gathering in its load, without pushing or pulling the surplus. These features make it very valuable in selective digging.

The clamshell is adapted to various types of digging by changing buckets or bucket plates. Heavy duty buckets of great weight and reduced capacity will dig very hard dirt and even soft rock. Rehandling buckets are larger, light, and often lack teeth. Medium duty or general purpose models are intermediate in weight and have teeth.

The clamshell has a smaller output than other shovel rigs and is more often used in stock pile, rehandling, and hopper feeding than in primary digging.

Dragline. The dragline is the best machine for loading from the top of the bank if it can dig the material. Small draglines usually are quite helpless in tight or rocky soil, but very large ones will dig even tight unblasted shale.

Difficulty of penetration increases with depth. For deep work, the boom should be long and digging done well out. This minimizes the upward pull of the drag cable which decreases the effective weight of the bucket.

A dragline can dig harder material from a face than it can cut vertically. If a wide ditch is started by other machinery or by blasting, it can be continued back into the bank by a dragline. If hard, it will tend to narrow down and become shallow.

The most efficient arrangement for hard digging is that shown in Figure 10-22 (A). The machine works parallel to an existing cut and back from another one. The high wall is cut in the line of walk so that the bucket will not have to try to work down from the surface to keep a wider cut.

If the dragline is able to dig the material with little difficulty, it can cut up to a double width, as in (B).

A dragline's reach enables it to stand well back from treacherous banks so that it can usually make deep cuts safely.

The dragline loads best if it digs inside the boom point, at a medium depth, with a swing which takes no longer than the raising of the bucket, and the haul units are on the pit floor beside the bucket.

"Throwing" the bucket, that is, casting it so that it digs beyond the boom point, adds substantially to the area a dragline can reach from one stand or one line of work. But it is a practice that is best reserved for special situations, such as small cleanup jobs where access is difficult.

Throwing the bucket may add from ten to fifty percent to the cycle time, since it usually requires that the bucket be pulled in, swung out, then dragged in with a full load until it can be picked up. If digging is done close in, the bucket is simply dropped and raised as soon as it is filled.

The extra reach is rarely more important than the time consumed. Also, careless casting causes or increases damage to bucket and cables.

When the loading or piling area is on the top of the bank, digging becomes slower as the reach becomes deeper, because of the time required to reel in the additional hoist cable required. At usual hoist·line speeds, an extra second is needed for each two and a half to three feet of depth. However, if the swing is long and unobstructed, the time of raising the bucket may not affect the length of digging cycle. If trucks are in the pit, the bucket may be raised only a few feet, regardless of depth.

If the dragline is not overloaded, it should have power enough to perform simultaneously the three functions of raising and braking the bucket, and swinging, without lugging down the engine. If the bucket is lifted to dumping position before the swing is completed, it is the length of the swing which determines the loading speed. If the swing is delayed in order to raise the bucket, it is the hoist, and therefore digging depth, which regulates it.

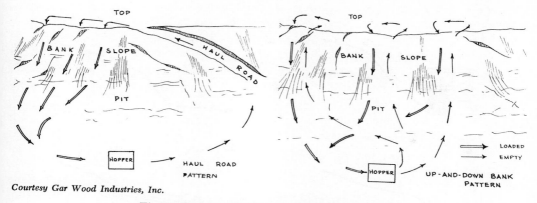

Courtesy Gar Wood Industries, Inc.

Fig. 10-23. Loading from bank with scrapers

Hoe. The hoe, also known as backhoe or pull shovel, is made in three distinct types. The original design is operated by cables, and is offered as an optional front on standard revolving shovels.

The more versatile hydraulic mechanism has largely replaced cables among the newer machines. Hydraulics may be comparatively small tractor attachments, used chiefly for ditching and miscellaneous work, or larger full-revolving units for both ditching and general excavation.

Any hoe has shorter reach than a dragline of comparable size, but will dig harder material and to a greater depth. The cable type loads more slowly than either a dragline or a hydraulic hoe under most conditions. It is fastest if the hauler is in the pit, or at least on a lower level.

If loading at its own level, trucks must be brought very close in. The dump is spread over a long strip so that it is most convenient to load the length of the body from the rear. However, there is the danger that a broken cable would allow the bucket to sweep forward and demolish the cab. For this reason, many operators will load only from the side.

Production can often be increased by bigger lagging on the hoist drum, or by reducing the number of parts in the hoist line. A dump bottom or hydraulically controlled bucket will also speed up loading.

The full-size hydraulic hoe, with down pressure on a wrist-action bucket, is less limited. It can dig harder material, handle oversize pieces more easily, and has little difficulty loading at its own level. But it can still benefit from having trucks spotted in the pit if it is deep, because hoist distance and time are shortened. It can then (and only then) compete with dippers and loaders in production rate.

Carrying Scraper. Scrapers are not ordinarily considered to be bank digging tools, but they may give lowest cost on combined digging and hauling.

Self-loading models, whether elevator, crawler-drawn, or two-engine, are most suitable for pit use, as they can work alone.

The bank is first shaped to a slope that may be between ten and twenty-five percent if the machines are to climb it, and steeper if they reach the top by a haul road. It is desirable that the top of the bank be flat or have only a gentle grade to reduce the danger of tipping while turning. The scrapers are loaded by driving straight down the slope, which should be long enough to give them plenty of space to load. Rippers and pushers should not be needed.

The scraper then hauls its load away to dump into a hopper, build a storage pile, or deliver it to a job. On its return, it is driven up the bank, or a haul road, turned on the top and again loaded coming down.

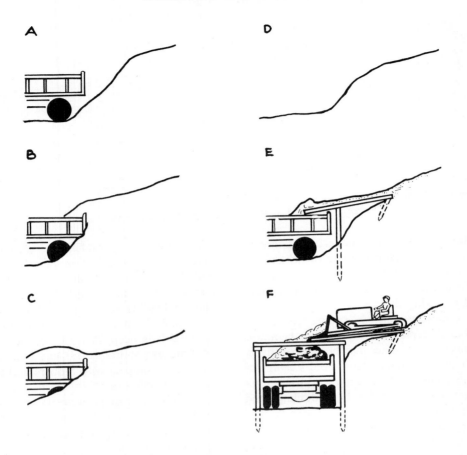

Fig. 10-24. Loading from bank with a bulldozer

The cycle is illustrated in Figure 10-23. Semi-trailer machines may be backed up the slope if turn space is lacking.

Once the bank is properly sloped, a single scraper may perform all the functions of digging, hauling, and storing without help from other machines or men. Such scrapers can also be used for digging in the pit floor, building haul roads, and grading.

Bulldozer. Bulldozers can load trucks from banks high enough to permit the machine to push into or over the body. For occasional loads, this may be done direct from the bank, as in Figure 10-24 (A) to (D). Considerable material is usually lost in building the bank out to the truck, and

repeated loading extends the bank out into the pit, requiring a longer push with each load. The truck may get stuck in the spill.

A retaining wall and platform, as in (E) and (F), will eliminate this difficulty. Many other constructions are used. The platform should have steel strips or rails to keep the blade from digging into the timbers. These should be spaced so that the tracks will not have to walk on them. If they are raised above the wood, they will cause a dirt cushion to be built up which will protect the wood from the grousers.

Dozer push loading is most effective in high banks with a slope steep enough to allow pushing of large loads and which will still allow the tractor to back up easily. Much steeper banks can be used in clay and hardpan than in loose sand or gravel.

Spoil can be pushed in the same manner into a hopper and conveyor, which may be a light homemade arrangement such as that in Figure 10-25(A), or a factory-built, high capacity belt loader such as that in (B). In either case, when the material within efficient range is exhausted, the hopper should be moved.

It is common practice to postpone moving for much too long. Dozer loading on level or slight grades is inefficient and should be avoided. No part of the push should ever be uphill.

Bulldozers are also used to push bank material within reach of excavators which are stopped by rock outcrops in the toe, and to keep high banks sloped to prevent undermining and caving.

In "glory hole" excavation, which is usually in rock, a tunnel is driven in from the toe and a connecting shaft run to the surface. Rock blasted from the sides of the shaft feeds by gravity to a conveyor, drag scraper, or cars, which haul it out of the tunnel. A bulldozer is not required until the pit has widened its slopes so that rock will no longer slide.

Cable Excavator. These machines are permanent or semi-permanent installations, which should have enough digging within reach to repay the investment. They usually serve as both digging and haulage equipment. See Chapter 14 for details.

The drag scraper is the easiest and safest means of digging a high bank which slides or caves. No equipment is needed near the toe, and only the light tail tower and anchors at the crest. If the far side of the crest is accessible, steepness can be reduced as digging progresses.

If the spoil is to be moved a considerable distance across the pit from the toe, a three drum slackline may be used instead of the drag scraper.

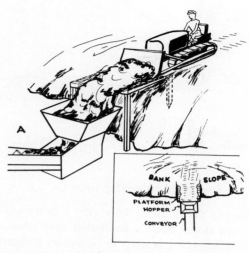

Fig. 10-25. Loading a hopper with a bulldozer

Dumping may be done into the processing plant itself, into a conveyor of any length feeding the plant or through a hopper into trucks.

Digging can also be done on the level and in a pit so that one setting may be used to convert a hill into a hole.

Ability to dig hard deposits increases with size in the same manner as in draglines.

Bank Slides. Most materials will rest temporarily at a steeper slope than their natural angle of repose. Some sand and gravel may stand in vertical or overhanging banks when freshly cut, but eventually fall or slide to slopes between one on one and one on two.

The danger of undercutting high, non-caving banks is obvious. It is less apparent when a bank caves and slides steadily when

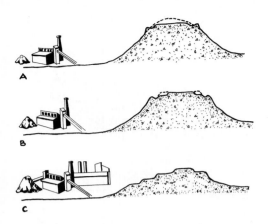

Fig. 10-26. Benching from the top

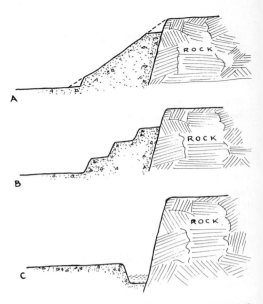

Fig. 10-27. Benching from the top and bottom

dug at the bottom, preserving a fairly uniform slope. Such a formation may gradually become too steep to be stable, without giving any indication of its condition. A change in moisture content, a blast, thunder, or the dropping of a pebble may start a slump, which reduces the slope of the face by lowering the crest and advancing the toe into the pit.

The danger from such slides increases rapidly with bank height and its steepness. In many cases, men have been killed and machinery buried in them.

Changes in moisture content affect both internal friction and weight, and either drying out or becoming soaked may create or intensify unbalanced conditions.

Aside from this danger, a high, sliding bank offers the best possible bottom loading conditions. Because of the constant supply of fresh, loose material digging is easy, and the excavator has to move forward only at long intervals.

Damp clay will usually stand vertically when cut, but will slump or fall eventually. Vibration from passing machinery or nearby drilling is liable to break down its structure so that it will flow. If the movement starts at the top, a dangerous collapse may be caused.

A face of clay or silt exposed to alternate freezing and thawing or internal water pressure may liquefy and flow out on the pit floor, eventually assuming a gradient as low as ten percent. This action is usually too slow to be dangerous, but should be allowed for in figuring clearances for haul roads or in parking machinery.

High, steep rock walls should be checked for fissures running parallel to the face, which would allow sections to fall off. These are particularly dangerous if filled with dirt that might absorb water and exert a push.

No high face of any kind should be undercut widely without adequate bracing.

The safest way to dig high steep banks in general is with a drag scraper. Sometimes a dragline with a very long light boom, and a backfiller, is used to pull the crest down to the excavators.

On lower slopes, and on firm material, a dozer can be used.

Benching. It is usually good practice to limit the height of shovel cuts by taking the materials in a series of layers or benches.

Two methods of benching a hill slope are described in Chapter 8. Pits are liable to take much larger areas and require many more benches.

Benching may be done from the top down, as in Figure 10-26, or from top and bottom, as in 10-27.

A boundary cut is frequently carried down below the pit floor when the higher parts are exhausted.

A large number of benches may be worked at the same time, or in rotation.

Each bench should be large enough to provide ample space for shovel and trucks. If it is accessible at each end, one way traffic can be maintained and the need for turnaround space avoided. However, narrow roads are often blocked by stalled vehicles, slides, rockfalls, or overbreaks.

If the bench is accessible from only one end, the shovel should work from that end so that the width of the new cut will be available to traffic.

When layers are taken from the top down, starting at a hillcrest, or a back line which will stand steeply, the working area may be made about as wide as desired. The excavating done on the top widens the area available for the next cut.

When cuts are worked up from the bottom, width is largely determined by slope gradient and face height. If the slope is 45°, a foot of height is required for each foot of bench width.

Top benching is preferred for steep slopes whenever immediate access can be had to the top.

LEVEL DIGGING

Material is frequently obtained by sinking the pit floor, or a part of it, in thin layers without developing a bank except at the boundaries of the excavation. The material may be piled before loading, in the same manner as topsoil, but it is usually more convenient to dig directly, with carrying scrapers or cable excavators.

Rooters may be used to loosen the ground for scrapers and for cable excavators except in wet digging.

The wheeled scraper is more flexible in digging, can vary the dumping spot readily, and by change of the number and size of the machines can excavate at almost any rate desired. The machine may also be used in other pit work or outside jobs when the cut is idle.

Under favorable conditions, the cable excavator can dig at less cost per yard. The fact that it is difficult to move is sometimes in its favor as it will be there when needed.

SURFACE WATER

Rain. Rain will usually stop excavating and hauling operations. In addition, it may soften stockpiles and turn pit floors and haul roads into swamps or ponds so that work may not be resumed for days or weeks.

Some pits are in such porous soil that any volume of water will soak away quickly, and neither mud nor standing water will delay work more than a few hours. Others are in such dry climates that it is better to run a small risk of water delay than to spend the necessary thought, time, and money on arranging for drainage. The majority, however, are so situated that at least routine precautions should be taken to keep them usable.

A first principle is to shape pit floors so that they will drain. In cutting into a hill, the floor should slope slightly upward toward the face so that water will flow away from this line of greatest activity. If this rear drainage is not practical, the slope may be made to the side or to channels or drains in the floor.

If the pit is sunken, drainage can sometimes be arranged into a deeper portion which will take it off the working floor and allow it to soak away gradually. This will often be a pond dug under the water table to supply the plant with water.

If pumping is necessary, it should be done from a sump that will hold a large volume of water, and in which the pump can be protected against being covered if heavy rain falls while the pit is shut down. Such a sump may serve as a storage reservoir for plant supply.

Runoff. A pit may be troubled by water running off surrounding areas, either during rains or in the form of permanent streams, which should be diverted around the working areas, or channeled through them in such a way that it will cause minimum interference.

The best practice is to dig diversion ditches to lead the flow in other directions. However, if the pit is expanding, these ditches will require relocating and may cost more than they are worth. This is particularly likely if the ground is steep or rocky so that ditching is difficult.

Also, the water may be needed in the pit for plant or dredge supply.

If the water flows only occasionally, it can be led through the pit in wide shallow channels which can be crossed by machinery and trucks at any point. If it flows often or continuously, it should be in a ditch and taken through haul roads in pipes or on rock paved fords.

GROUND WATER

Layout, machinery, and methods in a pit may be affected by the water table, or level of ground water.

This unseen water surface may be practically level, evenly sloped, or irregular. The water generally appears to be stagnant, but it is almost always in slow motion and will slope down from the source to the outlet. The angle of this slope is the hydraulic gradient which is determined by the resistance of the material to the passage of water, the pressure and volume of the supply, and the relative heights of inlet and outlet.

Porous materials, such as gravel and sand, have low gradients, and tight ones, such as silt or clay, steep ones.

The water table tends to follow the slope of the land, but at reduced grades so that it will be further from the surface in hilltops than in valleys.

Above the true water table is the so-called capillary table which is kept wet by water rising in the spaces between soil particles. The finer spaces cause greater rise in heavy soils than in porous ones. Capillary water gives comparatively little trouble in clean sand and gravel, but causes serious softening of other soils.

Capillary water may come up higher when ground is compressed, as under a haul road.

Underground water ranges in quality from fine spring water to solutions of salt or chemicals unfit for any use.

A shallow pit may be kept well above the water table, which is then of little interest except as a source of water for processing. A pit carried sideward into a hill may cut into the table and have drainage problems of varying severity. A downward cut will get into water eventually, except in arid climates, or very tight ground.

Surface water falling into a depressed pit as rain, or flowing into it from adjoining areas, must also be taken into account in coping with the ground water.

Permanent plants and all year haul roads should be above the highest water levels.

When it is necessary to use materials lying at or under the water table, they may be obtained by wet digging, digging in dry seasons only, draining, pumping, and by combinations of these methods.

Digging Under Water. Any machine which can dig from the top of a bank can dig under water to some extent. However, there are a number of special difficulties, including inability to see the work, weaker penetration because of decreased bucket weight and interference of water currents, loss of material carried out of the bucket by water, and sloppy condition of the spoil. Wet banks are also more liable to cave under a shovel.

Loose underwater material, such as sand,

s efficiently dug with a clamshell, as it is securely held in the bowl while being lifted. Draglines, or long boomed hydraulic hoes, have highest production. For large areas, hydraulic dredges are preferred.

Cable hoes have most load loss from water currents and often show poor peneration as well. Their reach is not adequate for most wet digging and they cannot dump sloppy spoil far enough away to keep from getting in it.

Cable excavators, in proper material, can dig large pits to a great depth from a single mast position. The drag scraper will bring in better loads of fine loose material, and the slackline will dig deeper and is faster, particularly on long carries.

If a lake or river borders a pit, or a large enough pond is dug in it, a hydraulic dredge can be trucked in, assembled, and floated. The largest dredges may be able to cut a hundred feet under water, but forty feet is the maximum depth usually recommended. The spoil is usually pumped through a pipe line directly into a processing plant, but storage piles can be built. It is also possible to put the plant on the same hull as the dredge and discharge processed material into barges or conveyors.

The dredge can also enlarge the pond by undermining banks so that they cave within reach of its suction. High banks that do not slide should be lowered by land machinery to avoid danger of damaging the dredge when they come down.

Unless the pond is very large, or has a large inflow of water, the dredge may depend on a prompt return of water pumped out with the spoil. If the product does not contain many fines, water may be returned direct to the pond through a pipe or sluice. Fines can be filtered by allowing it to drain back through gravel or sand, or by holding it a while in a settling basin. In either case a larger water supply may be required than for direct return.

When the dredge has cut the working area to the maximum depth allowed by its ladder, it may be possible to reach further supplies by lowering the pond level. This may be done by partial drainage, by diverting the waste water from the plant into other drainageways, or by combining diversion with pumping.

If the material is soft enough not to require cutting, suction pipes can be extended below the ladder to the desired depth, but recovery will probably not be complete.

Care should be taken not to locate sunken and wet pits where they will interfere with the orderly development of higher layers. It should be remembered that drainage through a bank may cause it to settle to a very gradual slope, which might run it much further back into the pit floor than was intended. This type of spreading is particularly apt to occur when the pond is used as a water source, and waste water returned by being allowed to soak into the ground.

Drainage. The costs of wet digging are generally higher than doing the same work dry. If the ground surface slopes down far enough in the vicinity of the pit, it may be more economical to undertake even a large drainage project than to dig wet or to pump.

Draining is particularly feasible when spoil from an open cut ditch is of the same type as that which is being mined in the pit.

Where possible, provision in the original work should be made to drain to the full depth of the outlet or of the deposit. This may involve a very wide top cut, stable side slopes, and trucking out of spoil. If the spoil cannot be used, trenching, installing of drain pipe, and backfilling may be more economical.

Sometimes it is practical to run a tunnel, or one or more drilled holes, from the low spot into the edge of the pit area where a connecting surface shaft can be sunk. Each level can then be connected in turn to drain into the shaft. Precautions must be taken against the entry of dirt or trash that might block the tunnels.

Sand and gravel will generally drain into either a shaft or trench with little further attention but other deposits may require trenching of various types. The problems are similar to those involved in digging and draining a large cellar, and are discussed in Chapters 4 and 5.

When a side hill cut gets into water, a curtain drain may be required at the toe of the bank to avoid damage to the floor from seepage or flowing water.

A high water table may be supported by an impervious layer separating it from well drained formations, as in Figure 10-28A. Such a perched water table can be drained by drilling or digging into the porous layer below.

Surface ponds on sand or gravel deposits will often drain if the silt layer on the bottom is opened. A few sticks of dynamite exploded on the pond floor, or a shallow dragline cut, may do the work.

Pumping. When drainage is not practical, water may be removed by pumping. This may be the preferred method if the water can be used in the processing plant, then discharged outside the pit.

Pit pumping follows the practices described in Chapters 5, 6, and 21. It usually consists of removing open water standing against or over the deposit being dug. A small sump is generally made by digging part of the hole deeper and the suction hose placed in that.

Success in pumping depends on the relation between the volume of water in the hole and the rate of inflow, and the capacity of the pump or pumps. Costs per gallon are usually smaller, and work can be started or resumed more promptly if the pumps are oversize and can handle many times the volume of inflow, so that a large part of their capacity can be used to lower the open water.

Extensive gravel layers may contain billions of gallons of water over areas of many square miles which will drain into the sump. If no rain falls, the rate of flow gradually declines as the continued drainage flattens the hydraulic gradient, but this situation requires handling so much water that expenses are usually too high.

If the gravel is of limited depth, much of the inflow might be sealed by cement grout forced down into the gravel at pit boundaries, or where flow channels are suspected. Pumping should be stopped before and during the grouting.

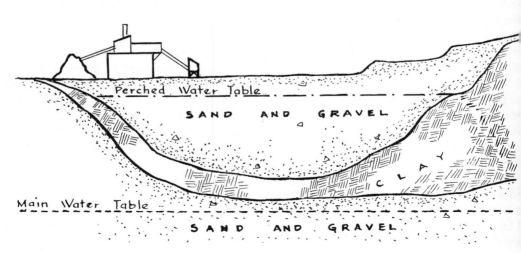

Fig. 10-28A. Perched water table

In gravel formations of limited size, and in tight materials such as clay and peat, the original rate of inflow usually declines rapidly, and the underground reservoir may be exhausted so that inflow will stop until it rains.

Each time a wet pit is enlarged by working, the pumping job of re-opening it becomes greater because of the increased pond volume.

Pumping should be done in dry seasons when the water is lowest and interruption from rain least likely. In general, it is better practice to pump out, dig a large volume as quickly as possible, and allow to refill than to maintain pumping and a slow digging rate over a long period. This is particularly true in the porous, quick filling formations.

Both surface and underground water are sometimes removed by pumping out of deep wells drilled in the pit floor or near its boundaries. This method is particularly suited to plant supply.

PIT PLANNING

Pits may be opened casually by digging in a roadside bank, or large sums may be spent on investigations, plans, road building, and site preparation, before work is started. Most of them start with small equipment and output and increase in scale if they prosper.

If a contractor can dig in his own land, or on a fixed price per yard basis elsewhere, and no access road is required, no investigation of the deposit may be necessary.

On the other hand, if a large scale operation is planned involving plant and other equipment bought specially, the building of haul roads, clearing and stripping, or if the deposit is of such value that every yard must be removed, the formation should be carefully investigated for extent, quality, accessibility, and water conditions. An option to buy or to develop should be secured before this investigation.

Zoning. In many areas, zoning regulations absolutely prohibit opening a pit as such. Frequently, however, if it can be shown that the land will be improved by the work, other types of permits can be obtained which will allow limited or complete operations.

Favorable conditions include deep road cuts through hills and the removal of underwater deposits to make a lake. Other projects include taking away ridges that block view or drainage, or leveling of land for residential or industrial development.

Sometimes a community will rezone an area for the purpose of preventing the opening or expanding of a pit. If the developer has purchased the land for the specific purpose of mining it, he has a fair case against the legality of such a move. If he has already started work, such zoning would probably not be enforceable against the area being dug, but would limit expansion.

It is probable that restriction of mining has been carried too far in many areas. Where acute shortages of sand and gravel exist, there are sometimes millions of yards of these materials made permanently inaccessible by reserving the land for houses which could just as well be built on the pit floor after digging.

On the other hand, there is no question but what a pit in a residential community is a dust, noise, and traffic nuisance, and is all too often an eyesore as well. A pit operator who takes care that no machinery is operated without mufflers, or on Sunday; that calcium chloride or oil is used freely to keep down dust; that the floors and banks are kept trimmed and reasonably neat, and that finished areas are promptly topsoiled and planted, will encounter minimum resistance to expansion or to opening other pits in the vicinity.

Wherever possible, operations should be screened from the public view by leaving or planting trees and shrubs, or by natural or artificial ridges covered with vegetation.

Permits. In areas which are not zoned against digging, it still may be necessary to get both state and local permits to operate, and to put up bonds to guarantee that the area will be smoothed and planted afterward.

Heavy fines may be imposed for any failure to abide by such regulations, and the operation may be closed down, perhaps permanently.

Investigation. The only certain way of finding exactly what is underground is to dig it out. However, inspection and mapping of surface indications, study of other pits or holes in the area, and talks with geologists and local old timers can provide at least an idea of where to look.

Fig. 10-28B. Be sure you have a permit

Next may come some test digging and/or boring. A tractor mounted hoe can dig inspection pits ten or more feet in depth. If the soil does not cave, quite deep holes can be sunk by a clamshell of sufficient bucket weight. But don't go down in them, unless protected by very heavy bracing. Get your information from what came out, and possibly by inspecting the walls with a light and a mirror.

Rotaries and downhole drills grind or pound soil and rock to chips, sand, and dust. Spudders (churns) and some rotaries complicate the problem further by mixing the spoil with water. The resulting mud may contain material eroded from the walls of the hole.

Interpretation of such samples requires knowledge and skill.

Augers give a fairly accurate picture of formations which they can penetrate, except that the material is loosened and mixed.

Core drills give excellent and reliable samples of rock, but may not pick up soft or loose material satisfactorily.

Any hole or shaft should eventually indicate the water level. In porous soils this takes a few minutes, in tight ones as much as a month. If a record of the seasonal rise and fall of the water is to be kept, the hole should be lined with perforated casing, unless it is in rock.

Market Analysis. The next step is to figure the extent and durability of the market for the products to be mined, and the likelihood of competition from similar projects, or from small workings with less overhead.

If the total consumption of the products within shipping range is small, investment in a big plant would be inadvisable unless new outlets could be opened by lower prices, superior quality, or building of consumer plants. If the demand is large, but is already adequately supplied, the question will be whether the new pit can supply better material or service, or cut prices, or create additional demand. The efficiency of existing pits and the extent of their reserves should be studied.

If the potential market is large, and the supply limited, the question of future competition will depend on the availability of similar material to others, and whether the intended plant will retain its relative efficiency long enough to repay its cost.

The amount of processing required to fit sand or gravel to specifications of highway departments and other wholesale users may be an important factor in costs.

Capital. A contractor already operating

in other lines may not need to make any extra investment in a pit until business is brisk enough to justify it. In general, however, the minimum capital required consists of down payments on machinery and land or digging rights, and money to carry payroll, operating expenses, and installments until there is income to take care of them.

If the pit must build up its market gradually, or if the demand is seasonal so that stock piles are accumulated throughout the year, to be sold during a short period, substantial capital will be needed to carry through the slack periods.

The necessity of selling on credit so that short to long periods elapse between loading the material and getting the money, will sometimes tie up more capital than all the other investments combined. This problem is discussed in the next chapter.

The risks of pit operation, as of any business venture, are greatly increased by a lack of surplus funds or borrowing capacity reserved for emergencies.

Selling Without Loading. A pit may be opened or operated without capital by the owner of the land or digging rights, by selling material "in the bank" to customers who will dig it themselves. Such arrangements are usually based on a price per yard, measured either in the bank or in the trucks. Occasionally, a certain portion of the deposit is marked off and sold, or the buyer may be allowed to dig all he needs for a certain job, for a lump sum.

A customer taking a substantial amount is usually expected to do his own stripping of overburden, keeping his section of the pit orderly and doing any required pushing back of topsoil after completion.

Sales of bank yards involve measuring the ground surface before and after their removal. This is usually done by surveyors, who work out a grid or a series of profiles. The original measurement may be fairly expensive. Later ones are much cheaper

if the bench and location marks have not been disturbed. This method is best adapted to large yardages.

Occasionally, such measurements are made at the fill site rather than the pit.

Sale by truck yards usually involves measuring the trucks and counting the loads. Unless the rate of excavation is very rapid, it is apt to be impractical to assign a man as a counter. It is common practice to measure the trucks first, agree as to how much of a heap is to be carried and how much of it paid for, and then depend on the buyer's records, or end of day checks with the shovel and truck operators. The seller is then exposed to being gypped, but he may lose less than he would spend measuring the bank or hiring a checker.

A bank yard has more material in it than a loose yard because of swell. A price of forty cents in the ground may be equivalent to fifty cents in an accurately measured and charged truck. If the same price can be obtained either way, the extra yield from the bank measurement should soon repay the cost of a survey.

There is sometimes a question as to the accuracy of measurement, but on the whole, sales by bank yards or pit section require minimum bookkeeping and fuss, and are productive of fewer misunderstandings than either loose yards or material-for-job arrangements.

It is often necessary to limit a buyer to a small area so that he will not wander around picking out pockets of especially good material, making unsightly holes, and leaving sub-standard remainders.

Working Space. A pit which sells direct from bank to customer may require only enough working space to back in a truck. However, it is usually desirable to have a flat area in which to park idle machinery, to pile topsoil or other good material not immediately salable, and to place boulders and stumps until they can be disposed of.

If a crushing, screening, or other proc-

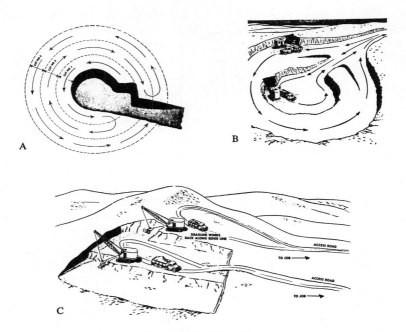

Fig. 10-29. Pit patterns

essing plant is used, space requirements are greatly increased. It is unusual to be able to sell products in the same proportions in which they are produced. There is usually a surplus of one or more grades that must be stored for future sale.

Since it is often easier to ship direct from the plants than to reclaim from storage, piles may grow when demand for their particular item is weak, and remain untouched when demand is good. This may result in steadily increasing storage requirements as the pit is enlarged.

If a pit is started in a small way and preserves a more or less level floor while being dug into a hillside, storage area may increase automatically with requirements. But if a big operation is to be started full scale, level land outside the pit area must be obtained or built up with waste overburden.

If the pit is to be sunk vertically, its area is not suitable for either plant or storage, The theoretical turnover of stock piles may

be such that the material would be reclaimed before it would have to be removed ahead of the digging, but this is liable not to work out in practice.

The plant is almost always located at or near the original ground level, and hauling products down to any of the cut benches for storage, and back out to sell, would be uneconomical. This type of pit will require storage space outside of the digging area which is best provided at the beginning of the work.

Excavating Patterns. Hill pits may be opened by a straight cut in or by benching. After reduction to the level of the surrounding land they are dug as sunken pits.

Subsurface workings, called sunken or dig-down pits, may be opened with shovels and ramps, or by dragline or hoe work from the top, in much the manner of a haul away cellar excavation. The circular pattern shown in Figure 10-29 (A), and (B), is also widely used, both for subsurface and slope with gentle gradients.

Draglines can take gentle slopes in a series of benches, as in (C). It is necessary to level a strip for walking, as accurate loading is difficult on a slant.

PROCESSING PLANTS

This heading includes screens, crushers, and washers with their feeding and discharge mechanisms. These units will be described in Chapter 21, and are discussed here only in their relations to pit layout and other operations.

Portable Plants. The simplest screening equipment is that described earlier in connection with topsoil. These pickup or skid grizzlies can be used wherever a shovel can work, but require ramps if they are to be used with tractor loaders. Their product is not well graded as narrow oversize pieces pass through readily.

Mobile plants having screening, usually crushing, and occasionally washing, equipment mounted on one or more wheeled trailers, require from a few minutes to several hours to move up to a bank and start work. Short moves in the pit require less down time than highway transportation as conveyors and other projecting parts need not be removed or folded in.

One of these units is usually able to eliminate primary hauling or to reduce it to a single truck shuttle, or a short conveyor. For direct loading from a low bank, particularly by a short-range excavator, it may be desirable to keep a tractor constantly on hand to move it up.

The use of portable units allows almost as much flexibility as in direct-from-the-bank selling. However, highway requirements keep their maximum size and weight below that required for many jobs; and the necessity of packing everything neatly in minimium space causes them to be harder to service than fixed plants of the same capacity.

They can often be used profitably for handling the more distant banks, or for filling special orders, in a pit that has a fixed plant. They may also be used for outside jobs or subsidiary pits. Hauling is a major cost factor in gravel and crushed rock, and the ability to open and process banks near the job may be valuable.

A mobile screener and crusher will reduce the risk involved in opening a new pit. Although not very readily disposed of, they are more or less standardized, and if found to be of the wrong size and type, can be sold or turned in, with a far smaller loss than a fixed plant.

Also, the use of one or more of these units may permit the pit to be developed until adequate space is dug for a fixed plant. If the pit is in a hillside, with steadily increasing bank height, getting the permanent plant well back in it will result in cheaper primary hauling over the whole job.

Fixed Plants. A fixed plant should be the right type for the material to be processed; it must be large enough for the job and within the capital budget. The first consideration is generally the most important, for if business is good, the plant can be expanded although at relatively higher cost; and exceeding a budget may be less damaging than getting the wrong equipment.

Plant manufacturers are ready to supply good engineering advice on every aspect of plant layout. It is a sound plan to get at least general recommendations from two or more companies, and to compare their findings with local practice. Even with these precautions, no person without a good working knowledge of the business should make a heavy investment in machinery from catalogs.

Plant Location. A permanent or semi-permanent plant should be as close to the digging as it can without being in danger from blasting and slides. If the pit is wide, or includes many sections, the plant should be near the center, or the side which produces the greatest yardage.

Dig-down pits may supply the plant by means of ramps and trucks, trucks with cable assists, vertical bucket conveyors, clamshells, or elevators. If supply is vertical, the plant should be located on the pit edge, or as near to it as firm footing can be found. This method is ordinarily used only in rock pits, but not all rock will support a factory on the edge of a cliff.

If the spoil is trucked up, it is best to locate the plant well back from the pit to allow for ramps, storage, and room for expansion.

Wherever a fixed plant is placed, the cost of hauling to it will increase as the digging progresses, whether laterally, downward, or both.

Taking down, moving, and setting up a permanent plant is usually a tedious and expensive job, particularly if it is a large size. Even if it is of prefabricated, knock-down construction; rust, wear, and patching may make it hard to handle and foundations are generally left behind. Many millwrights consider it best to salvage only the operating units, and to order or build new frames to carry them.

It is usually sound policy to charge the entire cost of such a plant against the material that can be handled at its original location. If the pit area is definitely limited by property boundaries, zoning restrictions, or change of ground, and the depth of the deposit is known, the yardage can be calculated. It is best to make a liberal allowance for occurrence of unexpected masses of unusable material.

HAULING

Pit hauling includes the movement of material from the bank to the plant or to storage, and between the plant and storage in both directions. It also involves delivery from these three locations to the job, although a variable amount of the product may be hauled from the pit in the customers' trucks.

The principal hauling units for pit use are trucks, including semi and full trailers, and conveyor belts. Railroad freight trains, of either narrow or full gauge, are used in big pits, the latter particularly in taking raw material from banks to a distant market. Digging units such as scrapers and cable excavators may also do a substantial amount of hauling.

Conveyor belts and cable excavators and, to a smaller extent, scrapers, are largely confined to work inside pits. Trucks are equally adapted to inside hauling and outside delivery. On very long hauls, heavy materials are more economically moved by standard gauge rail.

Conveyor belts may be considered either hauling units or part of the plant itself. They move and elevate material with minimum effort, but are usually difficult to set up and to relocate. They may be used instead of haul roads and trucks for delivery of a heavy volume of material to a single point many miles away.

Trucks are excellent flexible, general purpose units. They are available in a wide range of standard sizes and can be adapted to different size loaders or production schedules by varying the number on the run.

Cable excavators require a large yardage within reach to justify setting up. They can also be used in storage yards to pile, reclaim, and feed.

Scrapers, to operate as such, need ground they can dig and hoppers which they can drive across, or storage areas giving them room to maneuver. Banks which they cannot dig can be loaded into them. However, scrapers are more costly and are usually slower than dump trucks of the same size, so it is not good practice to use them steadily under shovels.

A scraper can dump beside a sunken hopper which is kept filled by a dozer.

Truck hauls may be kept short by adding conveyor belts to the plant. The new belt

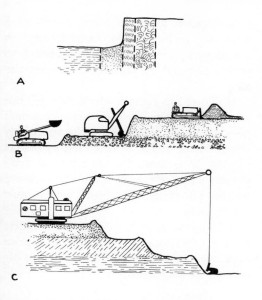

Fig. 10-30. Selective digging

will dump on the receiving end of the previous belt. Such installations may be quite long and are justified whenever considerable yardage will be handled.

Hoppers which are built so that the truck can drive straight across, instead of backing to dump, are more expensive to construct but will allow a faster truck cycle. Such hoppers can also be used for scrapers.

SELECTIVE DIGGING

Selective digging may be done to separate, at the face, two or more materials of value and to remove them; to remove one or more formations, leaving unwanted material; or to dig two or more materials so as to combine them.

Any or all of the spoil from these operations may be hauled away or sidecast.

Layers. If the different formations are in vertical sheets, as in Figure 10-30 (A), any machine which is accurate enough to work the narrowest vein can be used. If they lie horizontally, as in (B), any excavator can move them if they can be cut as separate banks. If they are horizontal, and two or more must be removed at once, the

excavator should be able to work from the top down. If divisions run in several directions, and separation must be exact, a clamshell, with assistance from hand labor, is probably required.

When horizontal layers are separated by a dragline, as in (C), it should have a boom at least twice as long as the bank is high. The boom angle should be low and the dump cable short to make possible picking up the bucket at a distance.

A clamshell can do the same work with a shorter boom as no allowance need be made for space to drag the load.

Inclined strata fall into any of the above classes. In general, it is bad practice to remove enough of any layer to leave the one above it overhanging.

Selective digging is quite commonly required in stripping overburden, and in gravel and clay pits. The operator may have the responsibility of choosing the section of bank most suitable to plant or customer requirements, and supplying deficiencies by mixing different sections or layers.

Mixing. A good way to mix at the bank is to build a stockpile by dumping the several materials on one spot. A conical pile will be built, with each bucket load separating and sliding down the sides. A succession of very thin layers will be made which, upon redigging to load, should mix together quite smoothly.

Such a pile will tend to concentrate round or coarse pieces at the bottom, but these will be remixed in handling by a machine working from the bottom.

Mixing is also done directly at the bank by digging from one formation, dumping on another, then loading the two together.

If the layers are horizontal, digging in slices from the bottom up will mix them.

Scrapers may bring to the top of the bank material to be mixed into it by loading.

Pockets. When an irregular deposit, with

sloped or vertical edges and numerous interruptions, is dug from the bottom, a cluttered pit is left. All parts of it are accessible to digging equipment and trucks, but there may be little room to maneuver and none for storage. If the digging is done from the top, the area may become a badlands, with little or no access and probably deficient drainage.

Such areas should be dug out, or smoothed down, at the first opportunity, particularly if any further work is to be done behind or under them. Many pits have strangled themselves out of business while they had ample reserves because leaving obstacles to orderly digging has forced haphazard development with increasing excavation and hauling costs.

Cutting a pit floor by pursuing a good vein across it is a bad practice which it is sometimes difficult to avoid. In general, it should not be done unless its value is sufficient to cover the cost of backfilling. If floor and walls of the cut are smoothed off, it may be filled with material to be stored which can be recovered when the whole floor is lowered. If no further digging is to be done in the floor, the slot can be filled with waste of a type which will not become too soft to carry pit traffic.

Cuts down from a floor, whether pockets or the start of new levels, should be near the outer edges of the pit.

Boulders. A common problem in pits which are dug without blasting is the occurrence of boulders too large for the loading or processing machinery. These are found in glacial and stream deposits, in disintegrated rock, and near steep slopes.

In pits selling only direct from the bank, there is no convenient way of utilizing boulders or of disposing of them. Blasting will reduce them to a size which can be loaded, but the market for coarse rock is so limited that they may have to be sold as second grade fill, or wasted. The pit operator will generally prefer to allow them to accumulate along the bank, or will have holes dug to bury them. Occasionally, an abandoned pit is close and deep enough to permit disposal by pushing them over the edge.

If allowed to remain where they fell out of the bank, or pushed into occasional piles, they will present obstacles to orderly development similar to those left by pocket digging. In general, the nuisance value increases with the size of the pieces, relative to the power of the dozer which must handle them.

It is occasionally possible to sell boulders for use in jetties or breakwaters, at a price high enough to justify hiring a machine big enough to load them.

STOCKPILING

Stockpiling is most efficiently done on hard, flat, clear areas. Dumping may be done on the flat, off piles, or from side banks. The location should be convenient to the face, the plant, and the market, the relative importance varying with the use of the material.

Trucks. If available space is very large compared with the bulk to be stored, trucks may dump piles against each other, as closely as possible, without further grading or heaping. Large trucks make high piles and place maximum yardage in an area.

This method takes a lot of space, forms a bank too low for efficient loading by many machines, and causes maximum danger of mixing the stored material with the floor.

If packing by trucks will not cause damage, such a piled area may be smoothed off by a dozer, and one or more additional layers added. Factors limiting the maximum height are the slope in from the edges and the more gradual grade for the truck ramp which steadily cut in on the area available at the bottom.

Figure 10-31 illustrates the building of a stockpile by backing trucks up on the

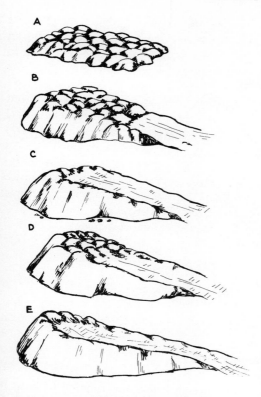

Fig. 10-31. Truck-and-dozer stockpile

dump and building it up in layers. Ramp grade should not be so steep as to strain the trucks or to prevent them from dumping cleanly.

At any time, the building of the top can be discontinued, and loads dumped off the end. The trucks are then usually driven up forward and turned on the pile.

These two methods can be alternated so that both height and area can be increased to the limits of the space available.

Figure 10-32 illustrates the building of a pile by the use of a spiral ramp. This is started as a narrow, back-up pile which is spread on the outside far enough to protect the road from caving, and well past the center on the inside. The road, steadily rising, is turned and comes back on the far side, parallel with the first section, but above it. Material is still dumped far enough to the outside to protect the road, and on the inside until its toe approaches the first part of the road.

The lower road is then widened to the outside to preserve its width as the fill encroaches on it from the next loop above. It is important to allow liberal space for a pile built by this method.

Digging from any part of this pile will undermine the road and make the top inaccessible until it is rebuilt.

If the material is somewhat too soft or loose to support trucks, the road may be strengthened by the use of wire mats, or

Fig. 10-32. Spiral stockpile

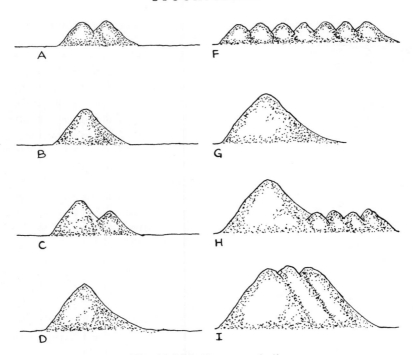

Fig. 10-33A. Dozer stockpile

small quantities of screenings, soil, or other binders, if their use will not spoil the value of the stockpile. However, it is usually easier and safer to use other piling methods with such material.

Trucks can build a stockpile by dumping off a high bank. This involves a minimum of dozer work. The special problems are loss and contamination because of mixing with the bank.

Reclaiming may be done by a loader at the foot of the pile, or a dragline or hoe at the top.

Trucks may be kept off a pile, either for safety or to avoid packing, by dumping on a level and piling by machine.

Heaping up may be done by dozer or by loader, on either tracks or wheels. The front loader is more efficient, as it can combine lift with push for higher, steeper piles with shorter moves and less power consumption. In discussions of jobs where either machine can do the work, "dozer" may refer to either of them.

If enough trucks are running to keep it busy, the machine may push each load up as it is dumped, as in Figure 10-33A, (A) to (D). The volume it can handle decreases as the height of the pile increases.

If the trucking is slow or irregular, dumping may be done closely over an area, and dozer heaping done now and then, as in (E) to (H).

If the stockpile can be laid out as a windrow, and the dumping kept close to it, the operation is almost as efficient as immediate heaping, allows the use of the dozer on other jobs, and makes hauling and piling work largely independent of each other.

The dozer heaps up stockpiles rapidly, is entirely flexible in placing them, or varying their size or shape, and can be used for a variety of other work. However, it must move its entire weight up the pile with each load, has constantly working tracks which may be subject to severe wear in sand or other abrasive piles, and may pack or crush soft materials so as to reduce their value seriously.

Choice of tracks or wheels depends on availability of equipment and the type of material. Wheels provide more compaction and cause less breakage into fines. They wear less in sand or gravel, but become ineffective under slippery conditions.

Loaders of either type can be used to reclaim the pile, loading it into trucks or carrying it to hoppers or to the area of use. If the carry becomes longer than 50 feet, the tires have an increasing advantage in speed and economy.

Clamshell. The clamshell is commonly used for building stock piles from shallow dumps, barges, and cars. A light rehandling bucket, with toothless lips, is used for loose material on a floor.

It can be used at the same time for loading trucks from either the dump or the stockpile. It often alternates hopper loading from the stockpile with the heaping work.

It is flexible in regard to placement and shape of piles, although not as much so as the dozer. It does not crush or pack.

Drag Scraper. If the storage location is fairly permanent, it may be more economical to heap up the truck dump with a drag scraper than with the trucks and bulldozer. The mast should be high enough to allow for piling the maximum amount to be stored, and the tail tower should be readily moved so that a wide area can be utilized.

The drag scraper may also be fixed to reclaim the material with a reversed bucket, and dump it into a hopper from which it is conveyed into the plant, dump bins, or directly into trucks.

If several classes of material are to be handled, both head and tail towers may be mobile.

A scraper installation does not have the flexibility of dozers and is not available for other work. However, it is very efficient in that only the weight of the bucket needs to be pulled up with the load, it does not pack

or crush the pile, and only the bucket is subject to abrasive wear.

Carrying Scraper. The rear dump scraper builds a pile in about the same manner as a truck. It can operate on a steeper ramp and in much tighter places, particularly if towed by a crawler.

The bottom dump scraper is most efficient at building a long pile, in the same manner as a highway fill. The sides are built up, starting at the outer edges, as steeply as the material will permit. The entrance ramp at one end, and the exit at the other, are started at gentle slopes and steepened as necessary. The pile may be started full length, or made short, and extended when additional capacity, or an easier ramp grade, is needed.

A dozer or a motor grader should be available for at least occasional trimming of the surface. If the material tends to get soggy when water soaked, or is unstable at the edges, compaction with sheepsfoot or rubber tired rollers might be advisable. Ordinarily, however, the scrapers themselves provide sufficient compaction for stockpiling purposes.

Scrapers may also make shallow piles for rehandling by drag scrapers.

Conveyors. Stackers and other conveyors of the boom type will build high piles with material dumped into a hopper. If these machines are wheel or track mounted, they can be towed away from the pile so that it will build into a windrow. Width of pile can be increased or separate piles made by pivoting the boom or the whole machine.

These are most easily fed by a drag scraper or revolving shovel, but a skid mounted ramp that will allow trucks to dump into the hopper can be made out of heavy timber.

Long conveyors, of either elevating or horizontal types, may be made so that a dumping device can be inserted at any spot desired. This makes possible the building of a windrow the full length of the conveyor,

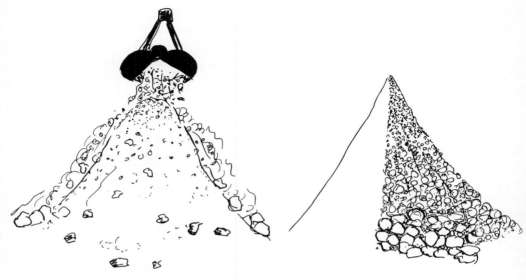

Fig. 10-33B. Segregation in a dry gravel pile

or making a series of separate piles of different grades. Lengthening the dump conveyor will increase the area which can be used.

SEGREGATION

Different-size pieces in a material being piled or disturbed have a tendency to separate from each other, so that a disproportionately large amount of coarse pieces will be found in one part of the pile, and finer ones in another. This process is called segregation or separation.

In many materials, and under many conditions, this is not important. Use in heavy road fill, for example, may not be affected by range of sizes. Sizes may be so uniform, or pieces so sticky, that separation does not occur. Method of loading out of the pile, or

later stages of processing, may provide re-mixing.

However, whenever size gradation is important to the use of the material, the system of handling must be checked to make sure that it will either not cause segregation, or will include adequate re-mixing.

Even if aggregate is to be re-screened before use, as in a mix plant, the screens must be supplied with a proper assortment of sizes.

Dry. In dry materials, a principal cause of separation is sliding and rolling down slopes. When a pile is built from the top material falls or settles onto the pile, picking up greater or less momentum from the downward movement.

Small particles develop little energy, and tend to come to rest almost immediately. Larger ones have enough momentum to slide or roll down the surface, the biggest tending to reach the ground, while intermediate sizes stop on the slope.

The sharpness of the separation increases as the range in sizes becomes greater, and when the material has low internal friction. Fine or sticky material tends to pile up to unstable slopes and then fall or slide in

Fig. 10-33C. A wide pile reduces segregation

masses, minimizing separation. For example, damp sand does not usually separate, dry sand or fine crushed rock separates a little, while coarse-and-fine rock and rounded river gravel may segregate almost completely.

When sliding in chutes, or subjected to random movements or strong vibration, large pieces tend to be wedged upward by small ones working under them, with a layering effect opposite to a top-built pile.

The segregation caused by building a pile of assorted-size pieces from the top, with a clamshell or a fixed-discharge conveyor, is shown in Figure 10-33B. The pile is always surrounded at ground level by a ring of the coarsest pieces. As the pile expands, its bottom layer is made up of these. Above it is a zone of somewhat smaller particles, with a fairly even gradation to mostly-fines at the top.

Such separation is almost never complete, as big pieces will get trapped in the top and small ones get to the bottom, but it is often sufficient to make the pile unusable until re-combined.

The clamshell operator can largely avoid layering by estimating pile area in advance, and building it in layers. Such successive layer must be enough smaller than the one below it to prevent sliding over the edges.

Re-Mixing. A dragline working from the top of a pile, or a long boom dipper shovel at the bottom, can make long shallow cuts from bottom to top, to get some of each layer in each bucket load. A clamshell should dig at various layers in turn, trying to provide a good average mix in the hopper or hauler.

Layers are more or less mixed when loading is done from the bottom by a loader. Undermining causes the finer upper part to fall and slide, mixing with the bottom.

A hopper gate or other fixed-in-place reclaiming unit under a pile must take what it gets, so it is essential to keep material properly unsorted above it.

Thin Liquids. Some processes, such as hydraulic dredging, move mixtures of water with heavier solids. These behave more or less like true fluids as long as they are in sufficiently rapid motion, but separate into liquid and solids when slowed or stopped.

If speed of flow is marginal, several zones may develop in a pipe or stream. Coarse pieces slide or roll along the bottom, somewhat finer pieces are partly pushed and partly carried, and still finer pieces remain in suspension and function as part of the liquid. The margins of the zones are moved up and down by turbulence.

With decreasing velocity, coarse pieces no longer move, medium sizes slide or roll sluggishly, until only fines are moving with the water. Buildup of the coarser parts is likely to plug the pipe.

On the other hand, if velocity is increased sufficiently, all pieces will be swept along as part of the fluid, although there is usually a higher percentage both of solids and of their coarse portion toward the bottom.

If such a high velocity fluid is released from confinement, as at the discharge end of a pipe, it will lose speed and force abruptly. Coarse pieces will be dropped immediately, finer ones carried somewhat further, and only very fine particles kept in suspension. This behavior is used to advantage in building hydraulic fill dams, Chapter 14.

In a washing plant, it means than an open-pipe discharge to a stockpile will build a coarse center and top, and finer edges. As it grows, it may tend to have a fine layer at the bottom, and coarser ones toward the top.

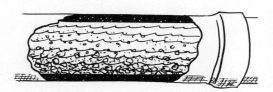

Fig. 10-33D. Layers in dredge pipe

However, if the material builds in steep slopes, this effect may be balanced or reversed by the gravity separation described for dry material.

Thick Liquids. A liquid such as a cement-water-aggregate mix, or asphaltic concrete, is usually thick or viscous enough so that there is little or no separation from or among pieces of aggregate, as long as the mixture is at rest.

However, if it is subjected to flowing motion or vibration, the liquid and the solids, and/or the fine and coarse aggregate, tend to separate. This is a problem principally with freshly mixed concrete, as the stickiness of asphalt limits aggregate movement.

Concrete tends to segregate into zones in response to several influences. Coarse aggregate is usually heavier than the fluid part, so will tend to sink. During flow it has higher friction, so that it hangs back and the fluid tends to flow out of and ahead of it.

Separation is much less of a problem in a stiff (low slump) mix than in a softer one. But it may have a special problem of its larger pieces falling and sliding down the slopes of a pile, even a small one, so that they may need to be hand shoveled back into it.

Vibrators are used to cause stiff concrete to flow into small spaces, and to pack smoothly against forms. They must be applied only long enough to accomplish their purpose, as continued vibration will cause even the stiffest mixtures to segregate.

In moving and distributing fresh concrete chutes should be steep enough so that it will slide rather than flow, it should not have a free fall, its initial placement should be under the surface of the concrete already placed and as close to its final position as possible, and vibration should be kept to the minimum necessary to move and consolidate it.

REFUSE DISPOSAL

A pit near a city may yield a much larger profit from being refilled than from the original digging. Disposal of raw garbage, incinerator ash, scrap, and industrial waste is a severe and increasing problem in many areas; and a worked out pit or pit section may provide an ideal dumping area.

The pit owner may charge the city or the garbage contractor on the basis of each load dumped, by the cubic yard of dumped and settled or compacted waste, on a monthly or yearly service basis, or at a flat price for use of a certain area. Leveling, compacting, covering, and/or burning may

Fig. 10-33E. Segregation in poured wet gravel

10-56

be done by either the refuse hauler or the pit owner.

Arrangements for refuse disposal in or near settled areas should include detailed coverage of the methods of operating the dump, as it can easily become a nuisance, and in any case it would be a subject to regulation, interference, and possible closing by state or local authorities.

There are a great many ways to dispose of refuse, ranging from open dumping of mixed garbage and scrap in fields, to filling of prepared holes with thorough compaction and immediate coverage with clean fill. If the dump is kept free of litter, and all garbage is promptly compacted and covered by dirt, it is a sanitary fill.

Segregation. A dump may be expected to handle a wide variety of material, from old automobiles and heavy steel scrap to semi-liquids. Both of these extremes must be avoided on the main area of a commercial dump, as they make it difficult or impossible to handle rubbish or garbage efficiently when they are mixed with it.

Burning. Some or all of the material on a dump is usually burned. This reduces its bulk substantially, and often makes it more manageable by destroying long or awkwardly shaped pieces. Such burning may be done in a series of separate fires on the top, by keeping the forward part of the dump itself continuously burning, or by both methods.

Surface fires of wood, dry paper, tires, and other flammable scrap seldom cause disagreeable smoke or odors. However, fires in mixed materials including garbage may be offensive to workers and drivers on the dump, and to nearby communities. Such fires may be started by a surface fire working down, by hot ashes or spontaneous combustion in the trash, or by deliberate setting. They may go out by themselves, or persist in spite of determined efforts to extinguish them.

Spreading. Refuse is dumped from trucks and spread by a bulldozer or shoveldozer, usually a crawler mounted model. Rubber tires are too apt to be damaged by sharp pieces of metal. Tracks may get badly snarled with wire, cable, or bedsprings, but can be freed by wire or bolt cutters, hacksaws or other tools with little or no damage. A dozer operator on a dump should develop a special knack to avoid track tangling.

The dump may be built out in a single high face, in a series of thin layers, or in compartments. A high face is the least work, makes the worst mess, gets the least compaction, and is likely to cause maximum difficulty with rats and insects.

A dozer crushes and compacts the surface of the trash as it moves over it. A bull clam or a rolled back loader bucket may be used to flatten it ahead of the machine. Compacting effect usually goes down one to 3 feet.

Compaction is desirable to increase the amount of refuse that can be put in the available space, to make a stable fill that will be able to support graded surfaces and buildings, and to reduce the number of rats.

Rats. Garbage dumps usually provide abundant food and shelter for large numbers of rats. They are a major nuisance. They spread disease. They kill or drive away many species of desirable birds and animals in adjoining wild areas. When a dump is closed thousands of starving rats are likely to desert it and pillage the neighborhood.

Rats may be controlled to a limited extent by poison, inoculation with disease, and encouragement of boys to use them as targets for .22 rifles. But the most effective way to suppress them is to compact the garbage so thoroughly that they cannot tunnel through it, and to cover it with dirt before they can feed on it. The same work eliminates breeding places for flies and mosquitoes.

Litter. Papers and other light trash arti-

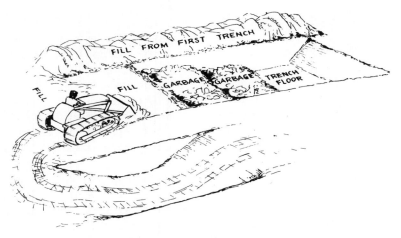

Fig. 10-34. Sanitary fill, dug cell type

cles blow around a dump and its neighborhood, producing an untidy appearance that may be one of the chief factors in local hostility to the operation.

Litter is controlled by enclosing the area with a high fence, putting wire or weights over papers as soon as they are dumped, and prompt burning or burial of all trash that might blow. It is often necessary to spend considerable time retrieving pieces by hand, that have escaped in spite of one or more of these control methods.

Smell. Objectionable odors around a dump may be caused by decay, burning, or chemicals, or various combinations of these sources.

Decay odors can be almost entirely prevented by prompt burial in sanitary fills. Separate burning of clean trash, and avoiding fires in the dump itself will eliminate burning odors. Dumping of chemicals may have to be controlled or prohibited.

SANITARY FILL

A sanitary fill is usually a garbage dump in which the rubbish is thoroughly compacted in thin layers and is promptly covered with clean fill, and that is therefore free of persistent bad odors, large rat populations, and severe litter nuisance.

Dug Cell Method. The cell method usu-

ally depends on obtaining clean fill from the dump area itself. Procedure is probably not exactly the same in any two dumps.

Figure 10-34 illustrates the parallel or double trench method. First a wide flat bottom trench is dug, usually by the dozer that spreads the garbage. The soil is piled to the side for use in final grading. Garbage is dumped at one end of the trench, where a ramp is provided to permit trucks to back in. Garbage is spread in one to six layers (two shown), depending on trench depth and thickness of layers. They are compacted by the back and forth and turning motions of the dozer.

Another trench is dug alongside the first one in waiting time between trucks. Fill obtained from it is pushed or carried across to cover the surface of the garbage, either the top layer only, as shown, or each layer as it is made. Thickness of dirt layers in between garbage lifts may be only 4 to 6 inches, but the final cover should be 18 to 36 inches.

A single trench may be used in progressive fashion, where fill for the day's garbage is obtained from digging another section for the next day. Or the trench may be made by sidecasting, and backfilled from the side to cover rubbish.

Trench cell methods are efficient only in

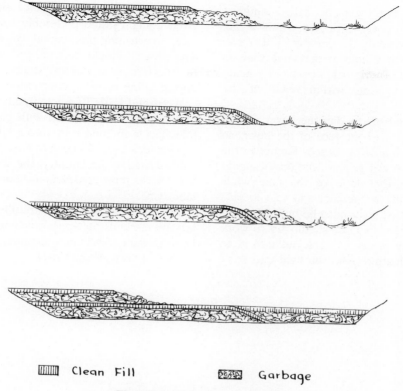

▨▨ Clean Fill   ▨▨ Garbage

Fig. 10-35. Built up sanitary fill

areas that have a good depth of firm dry soil, preferably with a sand or loam texture. Wet conditions would require using a dragline or hoe for excavation and a pump to keep the trench dry. Mud difficulties could be expected until the bottom layer of garbage could be placed. Pump failure or heavy rains might stop work, and float rubbish out of the area. These conditions involve extra expense and nuisance.

Hard, coarse, or boulder-filled ground such as makes up the floor of many old pits will make cell digging impractical.

Ground level can be raised only a few feet by this method, because of the inefficiency of dozers in deep trenching.

Dug cells are therefore poorly suited to land improvement, as they can be made only in land that is fairly good to begin with. A community may use its rubbish to better effect by filling swamps or burying rough, rocky, or stump filled areas.

Covered Fills. A sanitary dump for improvement of poor land may be made up of one or more layers of garbage covered by imported fill. The bottom layer should be deep enough to keep trucks and spreading equipment safely above water and obstructions. The dozer should make enough spreading passes to compact it thoroughly. The layer should then be sealed off with a 4 to 6 inch soil cover, and another garbage layer started.

Maximum density is obtained if layers are limited to a compacted depth of about two feet. High density permits putting more rubbish in the same space, and prevents future settlement. Some types of trash compact satisfactorily in lifts as high as 5 feet, but this is unusual.

A garbage fill can be called sanitary only if clean cover is put on within a few hours of spreading the garbage. Fill trucks or scrapers may haul at the same time as garbage trucks, or stockpiles may be made from time to time within reach of the dozer.

At the end of the day the cover should be extended over the face of the layer, so that no trash is left exposed. Except in the first layer on wet ground, the dozer should walk down the slope of the face while spreading fill, to compact and seal it.

When a first layer must be built in standing water, the operator should try to work large, heavy pieces of trash out into it, to minimize floating away of light pieces.

As with all fills, the surface should be given a slight grade preferably toward the face, to prevent ponding of water on it. The project should be arranged so as not to interfere with natural drainage either during filling or after completion.

Artificial Hills. If a community runs out of low places to be filled with garbage, it may decide to build it up into a hill.

This idea usually does not create popular enthusiasm. In flat country, the dump may become the most conspicuous feature of the landscape.

But as long as it is built carefully by sanitary fill methods, its appearance need never be distressing. And upon completion, it can become a very pleasant park.

CHAPTER ELEVEN

COSTS

Note: Many cost figures in this chapter are obsolete. Because of chaotic price conditions in 1975, newer figures might be out-dated before they were printed. A number of examples have therefore not been changed.

The reader should always check sample figures, even recent ones, against real current costs. But the principles of using the figures remains the same.

BOOKKEEPING

Adequate bookkeeping is a basic necessity both for intelligent estimating and profitable operation. Most earthmoving contractors start in business with some knowledge of how to get work done, but with little or no understanding of how to keep track of what they are doing.

Fortunately, it is not necessary for the contractor keep his books personally. Large organizations have their own bookkeeping departments with full time employees. A very small operator can hire an accountant or bookkeeper for part time work, even for one evening a week or a month, for a fraction of the money he will save for him. Even if the contractor, or his wife, can do his figuring, he will be wise to have a trained man check his books regularly.

A usual procedure for the small contractor is to hire a bookkeeper to make up a system, and to train him, his wife, or an employee in using it. Daily entries and rough work are done by the contractor, and the bookkeeper makes periodic inspections of the records, posts items to the proper accounts, balances the books, and calls

attention to mistakes and omissions. The frequency of his visits will depend on the volume of work, and upon the care and competence with which the contractor keeps his books.

Books should be kept on a double entry system, in which a record is made of two sides of each situation or transaction. For instance, a sale might result in the receipt of cash; therefore both the receipt and the sale are entered and the books are "in balance."

The checking account is usually the basis for the books and records. Entries are made on the stubs and/or in a separate book, to correspond with both bank deposits and checks written. The figures are reconciled with the bank statement monthly. In this manner each month is put into balance.

Balance Sheet. The balance sheet shows what the business owns and what it owes. An individual owner of a business that is not incorporated may include his non-business property and debts. It is better practice to keep them separated as much as he can.

A contractor's balance sheet might include the items in Figure 11-1.

Net worth is the amount left after subtracting total liabilities from total assets. It is listed as a liability in order to balance the two columns, and because it may be said that the business owes this amount to its owner or owners.

Day Book. Every contractor should keep

ASSETS		LIABILITIES	
Cash on hand and in bank	$ 3,640	Accounts payable	$ 4,430
Accounts receivable	3,830	Notes payable on equipment	6,880
Notes receivable	–	Notes payable, other	2,000
Work in progress	6,400	Interest, taxes, wages due	360
Depreciation reserve???	11,200	Advance payments not earned	100
Construction supplies	420	Mortgage on real estate	12,000
Machinery	59,860		
Land and buildings	18,700	TOTAL LIABILITIES	25,770
Processing plants	–	NET WORTH	78,280
TOTAL ASSETS	$104,050		$104,050

Fig. 11-1. Balance sheet

a daily record in a book of what he does. It should show jobs worked, man time, machine time, services provided, and materials used. Definite figures in feet, yards, tons, hours, and/or dollars are best. Such a record is easier to use than a collection of sales and job tickets, that are likely to get mixed up or lost. However, these tickets should be kept also, at least until payment for the work is received.

The day book may also serve as a diary for non-bookkeeping matters, such as important contacts with customers, promises of work and material to customers or from subcontractors or suppliers; important difficulties with weather, footing, breakdowns, or employees. It should record money spent, at least if it is in cash.

Such a daily record provides data for settling disputes about work done, payroll, and other matters, keeping track of work in progress and materials used, for obtaining adjustments in insurance rates, and backing up income tax returns.

Other Records. A contractor, like everyone else in business, must fill out forms for income tax for himself, and for withholding and social security for his employees. He will probably have to keep track of sales and use taxes, fuel taxes, compensation insurance, and perhaps truck mileage.

Other records will depend on his volume and variety of business, and how much he believes in paper work. Records can get too numerous and too detailed, but in the construction field they are usually too few and too carelessly kept.

DEFINITION OF COSTS

It is customary to divide contractors' costs into overhead and operating expenses. Overhead, often miscalled fixed cost, may be divided into overhead and job overhead.

Overhead. Overhead is made up of costs which do not vary immediately or directly with volume or type of work.

It may include the following items:

Drawing accounts, or living expenses, of owner or partners.

Management and supervision — salaries of executives, engineers, superintendents, and foremen.

Office rent, payroll, and supplies.

Interest paid on loans, or charged against capital investment.

Insurance for fire, theft, and liability if paid on the ownership of equipment and premises.

Ownership taxes on land, equipment, and other capital assets.

Depreciation.

Job Overhead. This heading may include any of the overhead items which are increased to take care of a particular job. When a contractor takes on a big project, his office and supervision force may be enlarged several hundred percent for its duration. This increase, arising from the one job, can justifiably be charged against it.

If job conditions require providing guaranteed pay, meals, rooms, or services to field employees, such expense may be labeled overhead, operating, or job overhead.

Job overhead may also include a proportion of home office overhead.

Operating Costs. This heading includes:

The field payroll of employees hired by the hour or day, or for the job.
Payroll taxes.
Liability and compensation insurance based on payroll, work, volume, or job conditions.
Machinery fuel, lubrication, maintenance, and repair.
Machinery rental, delivery, and changing rigs.
Expendable supplies.

Borderline Costs. It is often difficult to classify particular expenses, to decide just which account should carry them. As long as the contractor is consistent, he can list them very much as he wishes. However, following accepted practices makes it easier to keep bookkeepers, and to explain matters to banks or bonding companies when it is necessary to do so.

Personal Expenses. The contractor who runs his own business should keep books sufficiently to distinguish between business and personal expenses. However, he should bear in mind that these come out of the same pocket, and that his living costs are part of business overhead to the extent that it is up to the business to provide money to cover them.

It is common practice for owners to draw a fixed amount, and to consider this to be the only personal charge on the business. However, if personal expenses are in excess of the drawings, and the difference results in running up bills, these will ultimately have to be paid by the business, and might better be considered a charge against it from the first.

If personal expenses are not closely accounted for, a one-man business which is profitable in itself may go steadily downhill, without the proprietor ever understanding why.

RECEIVABLES

Importance. An important consideration for a contractor or a pit operator is the amount of capital required to carry customers' accounts. In most localities it is difficult or impossible to work on a cash basis. Even when the primary business is selling a commodity in great demand, as gravel in a gravel-scarce area, and operations are started successfully on a cash-for-each-load formula, good customers have a way of working away from it through a series of steps, such as pay after several loads, pay at the end of the day, at the end of the week, and at the end of the month, to a regular charge account, perhaps tying up thousands of dollars for long periods. Losses on jobs, or difficulty in collecting accounts, may change a well-heeled customer into a slow paying one.

Credit granted to one makes it triply difficult to refuse it to others.

It takes more backbone, or perhaps uncooperativeness, than is possessed by the average contractor to resist this technique of opening and increasing accounts. Also, it is often true that an enterprise cannot maintain a profitable volume except on credit, particularly if competition is severe.

The contractor who does small and medium size jobs for a number of different customers has no choice but to extend credit. Insistence on cash in advance or even on payments during work usually means the loss of too many jobs.

Home builders often pay promptly for cellar excavation, but very slowly for grading and landscaping. Big companies usually pay on schedule unless there are disputes about the extent or quality of the work.

Receivables not only tie up a large amount of working capital, but include a probability of bad debt losses. These can be minimized by good judgment in extending credit and skillful collection methods, but they cannot be entirely avoided.

A bank is usually willing to lend money on receivables. If the account has a good credit rating or local reputation, it may advance the full amount, less a discount which serves for an interest payment, on the understanding that any money received from that customer goes directly to the bank. Or a certain portion of the total amount of receivables may be lent on a regular interest bearing note.

The cost of such discounts or interest, and an allowance for uncollectible accounts, should be figured into the prices charged for material and services.

Offering discounts direct to the customer for cash or prompt payment is helpful in bringing quick money from good accounts, but is not very effective with those who are really hard up, and who constitute the major problem.

If a business is run partly with owned machinery, and partly with units rented from others, an over-large or doubtful account with a contractor may be tactfully collected by hiring the customer's machinery until he has worked it off. Sometimes an arrangement is made to pay the customer partly in cash to enable him to keep up with his payroll, and to apply the balance to the bill.

Liens. A contractor or subcontractor can usually file a lien for an unpaid bill against the property on which he did the work. Such a lien stays in force for a number of years, and must be paid when the property is sold, mortgaged, or re-mortgaged.

In most states it is necessary to file a lien quite soon after completion of the work. The contractor or supplier must be aware of the local time limitation, and not allow an unpaid account to run until it is too late.

An old account can sometimes be brought up to date for lien purposes by making one more shipment of material, or performing one more service for which a charge can be made, as it is the date of the last item that determines the last date for filing.

Filing a lien does not prevent a contractor from taking other collection action, such as a law suit or attachment of other assets.

Bonds. Government and government agencies can usually be depended on to pay their bills, although they are sometimes slow and they may dispute amounts. But a subcontractor may have to be careful that the general contractor does not collect without paying out.

In almost all public jobs, and in many large private ones, the general contractor must put up a bond. This usually means that a responsible insurance company guarantees that the contractor will complete the work and pay all suppliers and subcontractors. If he fails to pay any bill incurred on the job, the creditor can collect from the bonding company.

However, claims under bonds must be filed very promptly, often within 90 days of the date of the work, or protection is lost.

Most losses due to late filing of liens and claims under bonds are due to originally friendly relations between the parties, so that collection of the account is not handled in a business-like way.

Work in Progress. A contractor may tie up substantial amounts of cash and credit in jobs before he is able to even ask for payment. On small jobs he may have to wait until he is finished, on big one there is usually an arrangement by which he is paid in installments as the work progresses.

Installments may become due on completion and approval of parts or stages of the work, as a building contractor may receive a first payment when the foundation is completed, another when framing is done, and so on. The owner, in his turn, may receive installments on his mortgage loan at such times.

In highway and other large heavy construction projects a number of different stages may be worked at the same time. Rough grading and even clearing may be in progress on one part of the job while paving is being completed on another.

For this reason payments are made at regular intervals on the basis of measurements and estimates of the amount of work completed. Five or ten per cent of the amount due is usually held back until the end of the job. Payment may be made one to twenty days after the end of the work period, which is usually a month, but may be at shorter or longer intervals.

A schedule involving frequent and prompt payments reduces the contractor's need for working capital. However, he must have money on hand to keep him going if a payment is delayed by disagreements or other difficulties.

The extent of such delays is largely dependent on the policies of the owner, and his or its reputation may enable the contractor to make proper allowance for them in advance.

Cumulative Cost and Income. The graph in Figure 11-2 shows a simplified example of the drain on a contractor's resources during an installment payment contract. The job is assumed to use no more than the contractor's regular equipment, so that none need be purchased specially. Costs are actual expenditures, plus calculated machinery depreciation.

The cost curve shows the approximate amount spent at any time during the job. The stepped line indicates the total received in payments. The vertical distance between

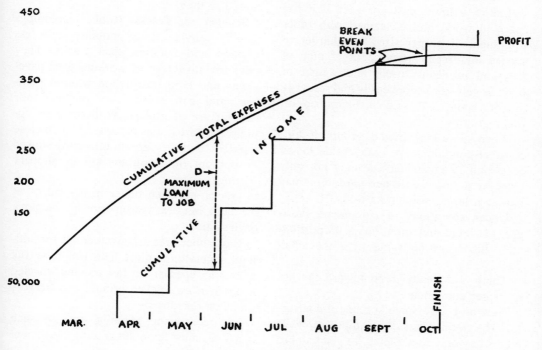

Fig. 11-2. Cash requirements of job

these lines first shows the amount "loaned" to the job, and later the profit.

The line, "Maximum Loan to Job," is the greatest distance, and indicates the minimum amount of cash and credit that will be required to carry the job under normal conditions. If the May payment were smaller or the June payment were delayed, the line would be longer, indicating need for more money.

MACHINERY PURCHASE

Purchase of a machine, whether new or used, involves consideration of the type and amount of work in hand and expected, price and availability of suitable models, as well as operator skills, work habits and personal preference.

Size. The arguments about machine size can appropriately be restated here. A big excavator is more costly to buy and to move, and requires more working space. It will dig more dirt in a given time, will handle harder and coarser formations, and will show a lower cost per yard if it has space to work, and is teamed with other equipment of proper size. It is harder to service and repair because of volume of fuel and lubricants used, and weight of parts. It gets stuck more easily and seriously in soft spots, but seldom hangs up on rough ground.

When space is restricted, ground is soft, or other conditions are unfavorable to the large unit, a small machine may not only work at a lower cost per yard, but may handle a larger volume as well.

Under conditions of equipment shortage, the large unit often has a proportionately higher resale value than the small one.

There is a steady trend toward the use of bigger equipment.

New or Used? Some successful contractors buy nothing but new equipment, while others buy only used pieces. In general, but not always, a new machine will have less mechanical trouble, and will receive better service from the dealer. It is more costly in purchase price, and in percentage of loss when sold. It has advertising or prestige value. It may be difficult or impossible to secure in the make, size, and model wanted within a reasonable time.

A purchaser of used equipment should have a good knowledge of mechanical condition and current values, and must be alert for liquidations and other forced sales where good values can be obtained. Considerable time may be required to find a particular make and model at a good price, and haste may make it necessary to pay too much. On the average, repairs will be more costly and service less satisfactory than on new units.

The expert buyer of used machinery is often able to sell his purchases at a profit, sometimes obtaining considerable work from them first. The average buyer, however, will seldom accomplish this, and is liable to be stuck with worthless machines now and then.

Rubber vs. Tracks. Rubber mountings usually provide more mobility and less traction and flotation than tracks. They offer the advantage of working over pavements and hard obstruction without damage, and can move over public roads without use of trailers. With some exceptions, they are not as maneuverable in close quarters, and they are more readily slowed or put out of action by soft or slippery footing.

Tracks are often better in the cut or at the bank, but tires are superior on the move or the haul.

Big rubber tires are given credit for adding to operator comfort. This would be true at crawler speeds, but fast moving wheeled equipment on uneven ground can be very rough on operators.

The contractor planning to use such machines must figure on grading equipment to keep haul routes smooth.

	New	Used
Shovel, complete with standard equipment	$32,400	$12,000
Diesel engine	1,540	–
Spare parts	160	300
Sales and use tax, 3%	1,023	369
Rail freight (44,000 lb. @ $1.60 per cwt)	704	–
Trailer haulage	40	90
Install center pin, overhaul engine	–	2,000
Hardface bucket teeth and lip	45	70
PURCHASE PRICE	$35,912	$14,629
Turn-in allowance on old shovel	6,000	–
Sale price of old shovel	–	4,000
NET PURCHASE COST	$29,912	$10,629
Finance charges, 18% of $22,000	3,960	–
Finance charges, 6% of $ 3,000	–	180
NET COST PLUS FINANCING	$33,872	$10,809

Fig. 11-3. Example of purchase cost, ¾ yard shovel

If rubber tired equipment is selected because of its ability to travel on roads without a trailer, the cost of licensing and insurance should be investigated. Some states license heavy equipment for a set fee of a few dollars. Others charge the same rates per pound or per horsepower as for trucks, which, on heavy equipment, may be a large sum.

Technicalities involved in obtaining permits to move over-width machines may be so tedious as to interfere with their use. This is a question not only of the law but of the attitude of local authorities toward its enforcement.

Cost. A contractor should figure the cost of an intended equipment purchase in two ways—total outlay of cash and credit involved in buying the machine and putting it to work, and the relationship between its cost of ownership and operation and the money it can earn.

The expenditure, particularly the cash down payment, is the most important figure to the contractor with limited capital, but may be merely a factor in considering long term costs for the large or well financed operator.

Care should be taken to include in the estimated cost all expenses involved. These may include list price, taxes, delivery to the freight station and then to the job or yard; extra front ends or other units to adapt the machine to different types of work; accessories such as cabs, lights, spare tires, parts, and special tools; repairs or alterations necessary immediately; and allied equipment required to get full use of the machine.

Some of these items are self-explanatory. Repairs are required only on used machines and include such items as replacement of worn tires or tracks, mechanical repairs, engine work, or complete overhaul.

Alterations may be changes made to adapt to overloads or special work, or may be necessary to correct mistakes or omissions of the manufacturer. They can include fishplating and other types of reinforcement, building up wearing surfaces with hard steel, and adding safety guards.

Backfillers, power, light	3	Hoists, air, electric, or steam	8
Medium or tractor	5	Gas	6
Heavy	6	Hand power	8
Breakers, pavement, pneumatic	3	Jacks, hydraulic	8
Buckets, cableway, clamshell,		Pile drivers, barge	8
Orange peel, scraper, and dragline	6	Steam, on skids	10
Concrete, elevator, and bail	5	Track	12
Bulldozers, tractor	4	Pile hammers, steam or air, heavy	10
Gradebuilders	8	Light	4
Compressors, gasoline, portable	6	Medium	5
Motor-truck unit	5	Pipe lines and fittings for floating	
Conveyors, belt, elevating, portable	3	dredges	10
Stationary	6	Pit and quarry plants	6
Bucket	6	Pumping units, gas or electric	
Cranes, crawler, gas, 2-1/2, 5 tons	5	Centrifugal or diaphragm	6
10, 15 tons	7	Highway contractor's	4
20 tons and over	12	Rollers, road, gas	10
Dragline	10	Scarifiers, attachments	4
Universal, gas, on 10 ton truck	6	Scrapers, blade, carryall	6
Crushers, rock, portable	8	Rotary	4
Stationary	10	Wheel	5
Drag lines, gasoline,		Shovels, electric or gasoline,	
1/2, 3/4 cubic yard	6	crawler or wheel, 1/2, 3/4 yard	5
1, 1-1/4, and 1-1/2 cubic yards	9	1 to 1-1/2 cubic yards	6
2 cubic yards and over	12	2 cubic yards and over	8
Dredges, clamshell	16	Tanks, gasoline, storage	6
Dipper	8	Water or air, storage, steel	10
Hydraulic	20	Tarpaulins and tents	3
Drill points, well	5	Tongs, chain	4
Drills, airdrifter	3	Tower excavators	12
Jackhammer	3	Towers, steel	6
Tunnel carriage	5	Tractors, gas or steam, 3-ton	4
Well	10	5-ton	6
Engines, gas	10	10-ton	8
Excavators, cableway complete	4	20-ton	10
Trench, gasoline, 7 to 12 foot depth	6	Trailers, dump, steel	10
18 foot depth	8	Washers, gravel	3
Wheel or ladder type	5	Wagons, dump, steel	6
Graders, blade, road, 7, 8 foot blade	4	Welding outfits, acetylene or	
9, 10 foot blade	5	electric	10
Over 10 foot blade	8	Winches, electric and	
Rooters, wheel	5	pneumatic	10

Fig. 11-4A. Depreciation periods from Bulletin F

Allied equipment might be a trailer to carry the machine, ramps for loading it, and different sizes of excavator or hauler to match its size.

It is also advisable to add up the interest or finance charges that will be incurred in making the purchase. As these are not actually part of the price they should be added on after the original cost figures are determined.

For example, if a contractor decides to replace an old shovel with either a new or a used 3/4 yard machine, he might study the comparative costs by setting up figures such as those in Figure 11-3.

DEPRECIATION

When a contractor buys a piece of equipment at a fair price, he does not "spend" the amount he pays. He invests it. He exchanges his money for something of equal value.

But the value of machinery starts to decrease as soon as it is delivered, because of use, wear, weathering, and passage of time. This decline in value represents the true spending of the invested money, and it is entered in the books and deducted from taxable income as an expense, depreciation.

It would be expensive and unsatisfactory to have a machine appraised every year to determine how much it had depreciated. It is also necessary to estimate in advance the rate of depreciation, as it is an important factor in establishing the price to be charged for the machine's work.

Depreciation is therefore calculated in advance according to various formulas. Each of them provides basis for balance sheets, profit and loss statements, and income tax. Annual depreciation is converted to an hourly figure for estimates and cost records.

More simply, hourly depreciation is the cost of a machine divided by the number of hours that it is expected to work.

Useful Life. Depreciation schedules must be based on the number of years the equipment is expected to be in service. Its useful years depend on the type of equipment, the class of work it will do, the care it receives, industry standards, and income tax decisions. At best the time selected represents only an informed guess.

"Bulletin F" was the Internal Revenue Service's guideline to equipment life for many years. Extracts are shown in Figure 11-4A. It has been retired as a guide, but may still be used as a reference and basis for discussion.

There is now a Class Life setup, with five possible life periods for the contractor, shown in Figure 11-4B. He may elect to use these figures, or ones 20 per cent higher or lower. He may also be able to make special arrangements based on his records, local experience, or other valid arguments.

Class	Description	Period, years
	General purpose trucks	
00.241	Light	4
00.242	Heavy	6
01.1	Agricultural machinery and equipment	10
10.0	Mining equipment	10
15.1	Contract construction other than marine	5

Fig. 11-4B. Guideline for depreciation periods

Schedules calling for more early depreciation than is allowed under declining balance method are likely to be disapproved.

Used equipment may be given the same depreciation period as if it were new, or any shorter period that appears to be reasonable.

Fast Writeoff. It is considered to be good practice to depreciate equipment at the fastest possible rate. This is called fast writeoff. It permits charging the largest proportion of costs against the machine when it is new and best able to carry the burden, and when it is doing the specific work for which it is bought.

Fast writeoff also keeps the book value of equipment down near its real value.

The most important advantage of fast writeoff is related to income tax. The faster the depreciation, the greater the deduction that can be made NOW, and the less to be left for an uncertain future. However, this expected advantage might turn out badly, as a contractor may waste his heavy depreciation on unprofitable years, and not have the deductions in later profitable periods.

Capital Gain. If a machine is sold for more than its depreciated value, the profit is a capital gain that is taxed at only half the rate of ordinary income, while depreciation is deductible at the full rate. A substantial tax saving may therefore result

	6 year life	10 year life
Purchase price of 3 yard shovel	$120,000	$120,000
Annual depreciation	20,000	12,000
Depreciation deducted during 4 years	80,000	48,000
Book value at time of sale after 4 years use	40,000	72,000
Sale price	60,000	60,000
Loss in sale for $60,000	–	12,000
Profit in sale for $60,000	20,000	–
Income taxes deducted, 52% of depreciation	41,600	24,960
Tax deduction on sale, 52% of loss	–	6,240
Tax paid on sale, 26% of profit	5,200	–
Total tax deductions on machine	36,400	31,200
Difference (saving by over–depreciation)	5,200	–

Fig. 11-5. Profit from over-depreciation

from a fast write-off that overdepreciates equipment.

For example, a contractor paying corporate income tax at the rate of 52 per cent might buy a shovel for $120,000 and sell it in four years for $60,000. At the time of purchase he had a choice of figuring its life at 6 years, the government's minimum allowance for a 3 yard shovel, or at a much more realistic figure of ten years. The resulting tax situation using straight line (same amount every year) depreciation is shown in Figure 11-5.

Capital gain treatment of the sale of over-depreciated equipment is likely to be taken out of the law some day, but while it is there it is poor business not to take advantage of it.

Salvage Value. This is the value of a machine after it is fully depreciated. It may be the actual sale price, or the value it might be assumed to have to the contractor when it is theoretically over age and worn out.

Salvage value varies greatly with the type of equipment, its condition, its scarcity and the local prosperity of the construction industry. Sometimes it is only a few dollars a ton for scrap, at other times, but more rarely, as much as 60 per cent of new cost.

It is usually estimated somewhere between 5 and 20 per cent of purchase price.

The declining balance depreciation method leaves a small salvage value automatically. With other methods, any estimated salvage is subtracted from purchase price before figuring depreciation.

Amounts allowed for salvage can be adjusted to simplify arithmetic. For example, if a machine with a five year life cost $16,346.93 and might be expected to bring $1,000 to $1,500 salvage, the salvage value could be taken as $1,346.93, leaving an even $15,000 to be depreciated.

Internal Revenue sometimes insists on deducting salvage value before figuring depreciation, and sometimes does not. It is better to depreciate the full purchase price when possible. This simplifies bookkeeping and makes an allowance for a probable increase in replacement cost, a matter that will be discussed below under **Price Increases.**

DEPRECIATION SCHEDULES

There are three official tax methods of computing depreciation for periods of three years or longer. These are known as straight line, sum-of-the-years-digits, and declining balance. For short periods only straight line is used.

There are also a number of unofficial but allowable-with-permission methods, of which we will describe hours of use and units of work.

Information in regard to taxes may become obsolete while it is being printed. It should always be checked before use.

Straight Line. Straight line is the simplest method, gives a uniform basis for figuring machine costs, and avoids complications in reserve for depreciation. The only thing it lacks is fast write-off.

The cost of the machine, less any salvage value, is divided by the number of years it is expected to be useful. The resulting figure is the annual depreciation. It is the same amount each year.

Declining Balance. This method is based on the total cost of the machine. The maximum depreciation rate is twice that allowed by the straight line method, but it is applied only to the value at the beginning of the year, which is the original cost less all depreciation that has been deducted.

For example, a $20,000 tractor with a five year useful life would depreciate 20 per cent or $4,000 each year under the straight line method. With declining balance, depreciation the first year would be 40 per cent of $20,000 or $8,000, the second year 40 per cent of $12,000, or $4,800, the third year 40 per cent of $7,200 or $2,880. At the end of the fifth year a salvage value of $1,555.20 would remain.

If the machine's life expectancy were eight years, the depreciation each year would be 25 per cent of the value at the beginning of the year.

Sum-of-the-Years-Digits. This method is based on cost less estimated salvage value. The number of years of useful life is taken as the first figure in a descending series, which for a five year period would be 5, 4, 3, 2, 1, and for eight years 8, 7, 6, 5, 4, 3, 2, 1. The series is added together, giving 15

for the five year period or 36 for eight years.

A fraction is made by placing the number of years of life from the beginning of the year over the total obtained by adding all the numbers in the series together. This is multiplied by the cost to give the depreciation for the year.

On a $20,000 machine with a five year life, first year depreciation will be 5/15 (1/3) times $20,000, or $6,666.67. In the second year, machine life from the beginning of the year is four years, so the fraction is 4/15, or $5,333.33.

The whole series of deductions would be 5/15, 4/15, 3/15, 2/15, and 1/15, totalling 15/15, or the entire cost.

On an eight year basis the first year depreciation would be 8/36 (2/9) of the cost, or $4,444.44. The next year would be 7/36, and so forth.

Choice of Method. Declining balance and sum-of-the-years-digits formulas are designed to put most of the depreciation at the beginning of the period. They provide the fast write-off that is liked by industry, and they conform most accurately to the actual loss of value of equipment in normal markets.

However, they cause problems of converting to an hourly basis for use in figuring job costs. Using a different rate each year would be difficult. If there were several machines of the same model but different years, the attempt to charge different prices for them would be confusing to the bookkeeper and aggravating to the customers, and would make accurate pricing of a job almost impossible.

The contractor should use straight line depreciation in figuring his hourly costs, regardless of what method he uses for income tax and annual reports.

Figure 11-6 shows graph lines for these three types of depreciation, and 11-7 gives annual depreciation figures per $1,000 of cost. This table can be applied for calculations on any price of machine, by multiplying by its cost divided by 1,000. That is, for a machine costing $15,500, the table figures are multiplied by 15½.

Hours of Use. A contractor may elect to depreciate his machinery on an hourly use basis, without regard to calendar time. He may buy a bulldozer for $25,000 and expect to use it 5,000 hours. He will charge $5.00 an hour against its jobs, and at the end of the year depreciate it by $5.00 times the number of hours it worked. If it were busy 600 hours, depreciation would be $3,000; if working time were 1,400, the year's depreciation would be $7,000.

Units of Work. A machine's production may be used as a basis for depreciation, if it can be measured accurately. A mine may buy a 2½ yard shovel for $100,000 in the expectation that it will work 20,000 hours and load 4,000,000 tons of rock and ore before it is scrapped. If business is good it may work 6,000 hours a year, if it is very poor the machine might be entirely idle.

Under such conditions annual depreciation would not be appropriate. Instead, the $100,000 value of the shovel might be divided by the 4,000,000 tons it is expected to handle, giving a depreciation figure of $0.025 (two and one-half cents) a ton. Each year it is depreciated on the basis of the number of tons loaded.

Tires. It is common practice to deduct the value of tires from the purchase price of equipment, charging them as operating expense and depreciating the balance as a capital investment.

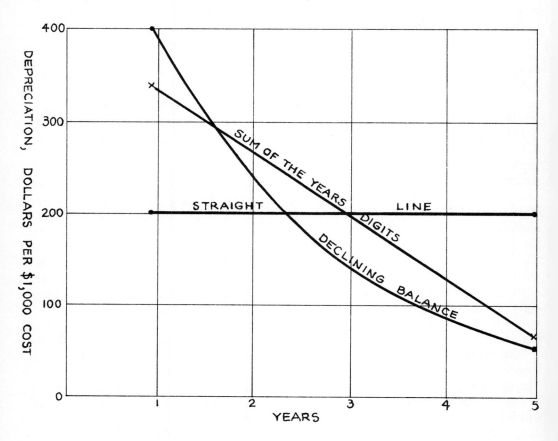

Fig. 11-6. Three depreciation methods

Depreciation Schedule

Term of Years	Year	Straight Line	Declining Balance	Sum-of-the-Years-Digits
2	1	$500.00	–	–
	2	500.00	–	–
3	1	333.33	$666.67	
	2	333.34	222.22	$500.00
	3	333.33	74.00	333.33
				166.67
4	1	250.00	500.00	400.00
	2	250.00	250.00	300.00
	3	250.00	125.00	200.00
	4	250.00	62.50	100.00
5	1	200.00	400.00	333.33
	2	200.00	240.00	266.67
	3	200.00	144.00	200.00
	4	200.00	86.40	133.33
	5	200.00	51.84	66.67
6	1	166.67	333.33	285.71
	2	166.67	222.22	238.09
	3	166.67	148.15	190.45
	4	166.67	98.77	142.83
	5	166.67	65.84	95.21
	6	166.67	43.90	47.59
8	1	125.00	250.00	222.22
	2	125.00	187.50	194.44
	3	125.00	140.62	166.66
	4	125.00	105.47	138.88
	5	125.00	79.10	111.10
	6	125.00	59.33	83.32
	7	125.00	44.50	55.54
	8	125.00	33.37	27.66
10	1	100.00	200.00	181.80
	2	100.00	160.00	163.62
	3	100.00	128.00	145.44
	4	100.00	102.40	127.56
	5	100.00	81.92	109.08
	6	100.00	65.54	90.90
	7	100.00	52.43	72.72
	8	100.00	41.94	54.54
	9	100.00	33.55	36.36
	10	100.00	26.84	18.18

Fig. 11-7. Annual depreciation on $1,000 cost

The advantages and disadvantages of this approach will be considered later. At present, it is subject to disapproval by the Internal Revenue Service unless the contractor can prove from his records that such tires usually last a year or less.

Repairs as Capital. Repairs are considered an operating expense as long as they do not add greatly to the value of a machine. But major overhauls, particularly if done late in the depreciation period, may be considered to be a capital expense that must be depreciated over several years.

For example, if a contractor spends $4,000 rebuilding a $10,000 machine in the last year of its depreciation schedule, he may have to list the expense as a capital investment, and set up a new schedule for it.

Fully Depreciated Equipment. If a machine is kept beyond the end of its depreciation period, no further depreciation is charged against it. However, the hourly price for it should remain the same. The part of its earnings that formerly paid for depreciation becomes profit, one of the "hidden profits" that help to keep contractors in business.

However, this extra profit may easily turn into a loss because of high repair costs and too much down time. It is not good business to run old machines unless they are in good condition.

Short Term Use. Many contractors buy machines for particular jobs, and sell them as soon as they finish. Others have a policy of turning in equipment after a certain amount of use, to reduce maintenance and job delays and to have the prestige value of up-to-date machines. Cost estimates are then based on the difference between purchase price and estimated sales price.

For example, a contractor may buy a fleet of $30,000 scrapers for use in two seasons of about 1,200 hours each, after which he will sell them for one-third of cost. Each of them will depreciate $20,000, and that cost per hour will be $8.33. This would compare with no-salvage depreciation of $3.00 for 10,000 hours or $6.00 an hour for 5,000 hours.

When periods of use are to be very short it may be cheaper and/or less risky to rent equipment. This will be discussed later in the chapter.

DEPRECIATION RESERVE

The contractor who intends to stay in business should set up a depreciation reserve, in which he can build up funds to replace his equipment as it wears out or becomes obsolete. This reserve may be a separate bank account, a fund maintained inside the regular account, or perhaps only a page in the ledger.

As depreciation is charged against a machine and deducted from income it should be paid into the reserve. If emergencies prevent saving the actual cash, the amount should at least be entered as a liability so that it will not be forgotten.

Need For Reserve. Machines wear out and must be replaced. Money is needed for the replacement, and it should be provided by the machines as they work. Otherwise the capital invested in them is consumed and destroyed.

A contractor who lives from one day to the next without much thought may figure that he is doing all right if he gets enough from his machinery to pay its out of pocket costs and his three square meals a day. But as equipment ages the repair bills get bigger and the down time longer, until it does not pay to run it.

If whatever money it made has been eaten up, the contractor may not even have the down payment on new equipment. Without adequate books he will find it hard to understand why he should finish a number of busy and apparently successful years without money to replace his machinery.

Inventory and Reserve. When a machine is purchased its value is listed as an asset under Equipment Inventory or some such

	Machinery Inventory	Depreciation	Depreciation Reserve
Purchase date	$16,000	$ –	$ –
End of first year	12,800	3,200	3,200
End of second year	9,600	3,200	6,400
End of third year	6,400	3,200	9,600
End of fourth year	3,200	3,200	12,800
End of fifth year	–	3,200	16,000
	$48,000	$16,000	$48,000

Fig. 11-8. Inventory depreciating into reserve, 5 year basis

heading. The depreciation is deducted from this each year, and added to the Depreciation Reserve. On a $16,000 machine with a five-year life, the straight line method would work out as in Figure 11-8.

Inadequate Depreciation. Most manufacturers recommend that construction equipment, aside from big shovels and special units, be depreciated on the basis of 10,000 hours use in five years. But as we will see later, most contractors are doing all right if they work 1,000 hours a year.

On the recommended basis a $20,000 bulldozer would depreciate $4,000 a year and $2.00 per hour, and its price on jobs set accordingly. But at the end of a year it might have worked only 1,100 hours because of weather, job delays, and repairs.

The jobs would owe the depreciation reserve $4,000 for the year, but the machine would have earned only $2,200 for this purpose. There would be a deficit in the reserve of $1,800, which would have to come out of profits, or if there were none, out of other funds.

If machine use had been more realistically figured at 1,000 hours a year for five years, depreciation would be $4,000 a year and $4.00 per hour. In 1,100 hours of work in a year the machine would have been able to provide the full $4,000 for the reserve, plus a $400 surplus for profit. The extra $2.00 an hour might make jobs harder to get, but if that is the actual depreciation,

the contractor must charge accordingly or lose money.

If experience shows that the machine will last 10 years at the 1,000 hours a year rate, its schedule could be set up for 10,000 hours in 10 years, and $2.00 an hour. On this basis the 1,100 hours of work would pay the $2,000 depreciation charge, with $200 left over. But if the machine had to be scrapped at the end of five years the reserve fund would be short $9,000 or $10,000.

It may seem to the reader that this is just a matter of juggling figures. But the figures are very real, and understanding them and arranging them properly may mean the difference between prosperity and bankruptcy.

If a contractor can really use a machine 10,000 hours he is justified in basing his estimates on long use. But if he is getting only 3,000, 5,000 or 6,000 hours out of his equipment now, basing costs on longer use is foolish and dangerous.

Price Increases. Contractors share with all other users of modern machinery the problem of price increases. The price of any equipment is likely to increase during its life, so that the replacement is more expensive than the original. This arises both from a general rise in prices, and from improvements in equipment that add to its cost.

A properly kept depreciation account for a single machine will seldom contain enough money to buy a new one if the original unit

DEPRECIATION

BEFORE PURCHASE

Cash on hand and in bank	$22,000	Accounts payable	$ 3,000
Accounts receivable	7,000		
Equipment inventory	43,000	NET WORTH	69,000
	$72,000		$72,000

AFTER PURCHASE FOR CASH

Cash on hand and in bank	$ 2,000	Accounts payable	$ 3,000
Accounts receivable	7,000		
Equipment inventory	63,000	NET WORTH	69,000
	$72,000		$72,000

AFTER PURCHASE BY INSTALLMENT FINANCING

Cash on hand and in bank	$17,000	Accounts payable	$ 3,000
Accounts receivable	7,000	Notes payable — principal	15,000
Equipment inventory	63,000	Notes payable — interest	1,800
			$19,800
		NET WORTH	67,200
	$87,000		$87,000

Fig. 11-9. Effect of equipment purchase on balance sheet

were scrapped at the end of its calculated life. Other funds or loans would have to be added to buy a replacement.

This difficulty may be partly or wholly overcome by figuring that the machine has no salvage value. Since it almost always has some salvage, even if only for junk steel at $10.00 a ton, and is often worth a substantial amount if in good condition, its value plus the depreciation reserve may provide fully for a replacement of the same type and size.

Another hedge against price inflation is to increase the depreciation charge against the machine whenever its replacement price is increased, so that it is the same as for a new model.

For example, if our $20,000 bulldozer were depreciated on a 5,000 hour basis, the hourly depreciation would be $4.00. If after two years the price of similar machines was increased by the manufacturer to $23,000 the depreciation charge against the old machine would be increased to $4.60. This is for internal bookkeeping only, and cannot be used on income tax returns.

This has the advantage of partially providing for replacement at an increased price, and of keeping prices uniform with new units that might be added.

Advance in prices to cover rise in replacement costs is particularly important for firms that obtain a substantial part of their income from renting out machinery.

Improvements may be made in equipment so that a new model is not strictly comparable to one two or three years old. However, the only point of importance in regard to upgrading the old model in price per hour is whether the changes produce an important increase in production. They often do not.

INVESTMENT

There are at least three different ways to consider an investment in equipment. They are: initial, total, and average annual investment.

Initial Investment. This is the net cost—the total of cash paid and debts incurred to buy a machine. It causes a shift in the balance sheet, adding to machinery inventory by reducing cash and/or increasing liabilities.

Figure 11-9 shows the effect of machinery purchase on a simplified balance sheet, both as a cash transaction and as a financed deal that includes notes for interest.

Total Investment. It is customary, although not particularly reasonable, to charge equipment with interest on any cash invested it. Also, property taxes and loss insurance are paid on the basis of machine value.

Since a certain part of the initial investment is amortized—that is, paid off in depreciation charges—each year, the investment on which such charges are figured is reduced in a series of steps. For the first year it will be the purchase price, the second year purchase price less one year's depreciation, and so on.

Such charges are most easily worked out for the whole life of a machine by using the total investment. This is found by adding the machinery inventory value for every year of the unit's life. In Figure 11-8 the total of the first column, $48,000, is the total investment for that machine.

Total investment (TI) may also be found by adding one to the depreciation period in years, DP,yrs, and multiplying by one-half the original cost. Stated as a formula, this is:

$$TI = (1 + DP,yrs) \times \frac{Cost}{2}$$

For a $16,000 machine used for five years, this would be:

$$TI = 6 \times \frac{\$16,000}{2}$$
$$= 6 \times \$8,000 = \$48,000$$

If interest were to be charged against the unit at 6 per cent, the total interest for its life would be 6 per cent of $48,000, or $2,880.

Average Annual Investment (AAI). This is a more realistic figure that averages the purchase cost over the years of machine life. It can be found by dividing the total investment by the number of years.

Average annual investment may also be obtained by multiplying the cost by the years of depreciation (DP,yrs) plus one, and dividing by twice the years of depreciation. Stated as a formula:

$$AAI = \frac{Cost \times (DP,yrs \text{ plus } 1)}{2 \times DP,yrs}$$

Figure 11-10 gives the average annual investment for $1,000 purchase cost for the most used depreciation periods up to 25 years. To use this, multiply the figure appearing after the number of years of life by the number of thousands of dollars in the machine cost.

INTEREST

Rates. Interest rates vary greatly with different type of loans, with the risk and bookkeeping involved for the lender, and with the general level of interest rates at the time the loan is made. They can be very confusing.

If a man borrows $100 and pays it back at the end of a year, plus $6.00 interest, his interest rate is 6 per cent. If he keeps the money for two years and pays only $6.00 interest at the end of that time, his rate is 3 per cent.

If the $100 is borrowed on a discount basis with the interest paid at the beginning, the borrower will receive $94 when the loan is made, and pay back $100 at the end of the year. Here his real rate is 6.38 per cent.

On an installment loan with an advertised rate of 6 per cent on the unpaid balance, the borrower will receive $100 and pay $106 in twelve equal monthly payments. The real interest rate is about 11 per cent, as the average indebtedness for the year is only $54.16.

True interest rates can be found by dividing the amount borrowed into the interest paid, and multiplying by a fraction made of one over the time in years, or 12 over the time in months.

If $5.00 in interest is paid on $100 borrowed for two years, we have:

$$\text{Rate} = \frac{5 \text{ (interest)}}{100 \text{ (loan)}} \times \frac{1}{2 \text{ years}}$$

$$= 1/40 = 2\tfrac{1}{2}\%$$

If the term of the loan were four months, then:

$$\text{Rate} = \frac{5 \text{ (interest)}}{100 \text{ (loan)}} \times \frac{12}{4 \text{ months}}$$

$$= 3/20 = 15\%$$

Interest on Equipment. As mentioned before, interest may be charged against a machine even if it is bought for cash. If the purchase is financed at a rate of more than 6 per cent the higher rate is charged, while if interest is less than 6 per cent or if none is paid, the charge is kept at 6 per cent. Interest is figured on a yearly basis on the average annual investment.

This custom does not conform to good accounting practice, and is in conflict with methods of treating money tied up in other ways.

Equipment Debt. There is good reason for charging interest on equipment purchase loans against the equipment, although a good case could also be made for charging it to general overhead. Carrying it as an equipment expense has the advantage of simplicity and of automatically identifying the source of the charge.

General Debt. A second case is charging bought-for-cash equipment with the same interest rate being paid on open loans for general purposes. It should be considered that money borrowed for general business use is a general overhead item, and responsibility for it is shared by field, shop, and office equipment, materials on hand, cash in the checking account, accounts receivable, and work in process.

The value of owned equipment is usually much greater than the amount of the general debt. If a machinery inventory of $100,000 were charged with 6 per cent interest because of a debt of $20,000 it would be paying $4,800 more than the cost of the interest.

If this equipment were charged with the actual interest, the $1,200 paid would be

Useful Life, Years	Average Annual Investment	Useful Life, Years	Average Annual Investment
1	$1,000.00	8	$ 562.50
2	750.00	9	555.55
3	666.67	10	550.00
4	625.00	12	541.67
5	600.00	15	533.33
6	583.33	20	525.00
7	571.42	25	520.00

Fig. 11-10. Average annual investment for $1,000 cost

11-18

applied to the whole $100,000, so that interest would be at the rate of 1.2 per cent. If the debt were carried by all the money consuming items mentioned above, which might easily total $150,000, the effective interest rate would be only .8 per cent.

No Debt. A third situation is assigning an interest charge of 6 per cent to equipment even though its owner pays no interest on any kind of debt. The idea is that the contractor could invest his money elsewhere if he didn't buy equipment. But the only investments that a contractor can make and still keep the money available in his business are savings accounts and short term bonds, that might pay 5 per cent or less.

It is of course both possible and likely that funds invested outside the business would fail to return 6 per cent interest, and might be partly or wholly lost.

Equipment is not a bond or mortgage that justifies itself by paying interest. It pays its way by production. To saddle it with interest charges is to ask it to produce a profit before it goes to work. In this highly competitive business, such an arbitrary increase in its cost basis may make the difference between getting a job and losing it.

It is the belief of the author that equipment should be charged only with interest costs incurred directly by its purchase. It should not be charged with interest that is not paid, and interest on general debt should not be charged to equipment at its full rate.

The contractor should include any interest that he believes that he owes to himself in his profit margin rather than in his equipment costs.

However, since assuming of interest charges is now a widespread practice in the industry, it will be taken into account in some of the cost calculations that follow.

Installment Interest. The interest rate on an installment contract is found by dividing the total debt into the total interest.

Total debt (TD) is a figure that is similar to total investment. It is found by adding

one to the number of monthly installments (I), multiplying by the loan, that is, the amount borrowed before interest, and dividing the result by 24. The formula is:

$$TD = (I \text{ plus } 1) \times \frac{Loan}{24}$$

The amount of interest is found by adding all the installments together, or multiplying their number by their amount, and subtracting the amount of the loan.

For example, $12,000 of a $16,000 purchase is financed in 36 monthly notes, 35 for $393.33 each and a final one of $393.45, totaling $14,160. Subtracting $12,000, we find that the interest totals $2,160. By the formula above, the total debt is $18,500. Dividing $18,500 into $2,160 we have .117, or an interest rate of 11.7 per cent.

Some contractors want to know the interest rate of finance charges on the whole purchase price. If the above machine had five year depreciation, its total investment would be $48,000. This would be divided into the $2,160 interest, showing a rate of 4.5 per cent on the machine.

If the machine is to be charged with 6 per cent interest on the non-financed part of the investment, the total debt is subtracted from the total investment, and the remainder multiplied by .06. In this example, we would have $48,000 minus $18,-500, or $29,500, multiplied by .06 to give an interest charge of $1,770. Added to the finance charge, this would give a total interest cost of $3,930 and an average rate of a little less than 8.2 per cent.

It is worthwhile for a contractor to understand interest rates. He will then have a clear picture of the extra expense involved in financing equipment, and he may be able to save substantial amounts by being able to detect mistakes or fraud in papers.

FINANCING

The cheapest way to finance the purchase of a piece of equipment is to borrow from a bank on a straight time note at their regular rate of interest, that used to be six per cent. Such a loan may be obtained by pledging collateral such as stocks, bonds, or accounts receivable. A substantial contractor may be able to obtain such a loan without putting up security.

Installment Plans. Most equipment financing is done on a straight line installment basis, with a down payment of 20 to 40 per cent (usually 25 per cent) and the balance plus interest paid in equal monthly installments over a period of one to five years, with eighteen month to three year terms the most common.

The finance or interest charge is usually 6 per cent per year on the original amount of the loan. That is, if $1,000 is borrowed to be repaid in 12 monthly installments, the interest is $60.00. If installments extend over three years, this charge is $180. It works out to an actual rate between 11 and 11.7 per cent, the higher cost being found on the longer terms.

Installment payments are secured by a chattel mortgage on the equipment, that is recorded in the town or city records. The borrower must be sure to have this cancelled by filing a release from the finance company or bank when he has completed payments.

When the value of the equipment and/or the contractor's ability to make the payments are questionable, the lender may ask for additional security, such as an endorser on the notes, or a mortgage on additional pieces of equipment that have no debt against them.

If loan installments are not paid on time an extra charge may be made for each one that is delayed. The lender also has the privilege of demanding immediate payment of the whole sum if even one payment is unreasonably delayed, and may seize and sell the equipment to collect. Machinery sold in such proceedings is not likely to bring its full value, and the contractor may find that he still owes a balance even after his equipment is lost.

Schedules may be made up to allow omitting payments in off seasons, usually three or four winter months. Such a provision will either make the other payments larger or stretch them over more years. Most contractors manage to make regular winter payments with surplus from working months, collection of accounts, or short term borrowing from banks.

Another financing plan, called PayD by its originator the C.I.T. Corp., provides for a down payment of only 10 per cent, then a graduated series of payments reduced annually over the life of the machine to follow the sum-of-the-years-digits depreciation formula. It is often called sum-of-digits financing. It is available only for new equipment and for terms of three to five years.

The finance or interest charge, that is added to the loan and divided proportionately among the payments, is 12.75 per cent of the loan for 3 years, or 21.25 per cent for 5 years.

Actual interest rate is about the same as in straight line financing. Advantages are the small down payment, and having the heaviest installments early in the machine's life when it is best able to earn the money to pay them. Total interest cost is less than for the same term in straight line payments because of the heavy early installments that reduce the debt rapidly.

OTHER OWNERSHIP COSTS

Property Tax. The contractor must pay a variety of taxes, including real estate, personal property, excise, and payroll levies. Here we are concerned with the personal property taxes payable on the assessed value of equipment.

This is entirely a local matter. In some states the local governments are permitted or required to tax machinery and other movable property in the same manner as real estate. This tax may range from 2 to

4½ per cent of the assessed value of the equipment. In other states or localities there are no property taxes whatever on construction equipment.

It is customary for estimating advice to suggest using the nationwide average tax of 1.5 to 2 per cent of value in figuring ownership costs. However, in this case average costs have little bearing on particular costs. A contractor must find out what taxes, if any, he will pay before he can use them in his figuring.

The tax is usually low in country districts and high in cities, but it varies with local financial policies. A high rate with a low assessment may mean a lower tax than a lower rate and full value assessment.

Assessments may or may not follow the depreciation schedule of the contractor. But it is a general practice to assess a machine for at least 20 per cent of its cost as long as it looks as if it might run.

Registration. Highway vehicles must have registration plates. The cost is moderate for cars, pickups, and jeeps, but may be very heavy for big trucks.

In most states this tax is on a basis of weight and/or capacity. In some there is an additional mileage charge. There is no close relationship to purchase price, so that it cannot be handled on a percentage basis.

Registration is an overhead expense, mileage an operating item. Both of them are added to other costs in setting a price on a truck's services, but this must be done on an individual basis.

Liability Insurance. Highway vehicles are not covered by a contractors general liability and property damage insurance. They have special coverage at much higher rates.

This is another ownership expense that is not related to purchase cost. Its amount is affected by vehicle weight, type of use, accident record of the owner, and miles driven.

Loss Insurance. The cost of insurance against fire, collision, upset, and theft is an ownership cost that is charged against each piece of equipment in proportion to its value.

The charge for insurance of this type is known in insurance circles as a judgment rate, as it is set for each locality or contractor according to the insurance companies' judgment of the risks involved.

The rate for fire, collision, and upset in a combined extended coverage policy is usually about one per cent of the actual value of the equipment, for the small contractor with a few machines used in miscellaneous work. Very large earthmoving or construction projects such as the St. Lawrence Seaway sections may be given a rate as low as one-half per cent. This is in spite of the fact that some of the machines work under very dangerous conditions, as the extreme risk positions are outbalanced by many behind the lines units working under safe and stodgy circumstances.

The highest rates for this coverage may be 1½ to 2 per cent. These are charged where the job conditions are more dangerous than average, or where the contractor is considered to be careless or reckless in management.

Theft insurance may be written into these policies as an extra coverage. With a $50.00 deductible clause it may be free in country districts where stealing is rare, up to one-half per cent of value in cities. One-quarter per cent is a usual charge. Companies may refuse to issue theft coverage at any price in certain cities or areas.

Premiums are usually charged on the basis of the contractor's valuation of his equipment, as long as he follows any reasonable and consistent system of depreciation. Each year, or at more frequent intervals, he sends the insurance company a list of his equipment, showing date of purchase, original cost, and present value. The premium is charged as a percentage of the total.

If a unit must be replaced because of

Purchase price - $25,000

Interest on investment, 6%	$1,500
Taxes, 4%	1,000
Maintenance	500
Night watchman	3,000
Annual cost	$6,000

Area, one acre, or 43,560 square feet

$$\text{Annual cost per square foot} = \frac{\$\,6,000}{43,560} = \$0.1376$$

Fig. 11-11. Cost of storage yard

insured loss, payment is made on the basis of the actual value of similar equipment in the locality at the time of loss. However, the company has the right (which it may not use) to refuse to pay more than the value of the machine stated in the policy schedule.

Therefore, if the value stated in the policy schedule, which should be the same as in the equipment inventory, is more than the actual value, the premium on the excess might be wasted money. If schedule value is less than real value, the equipment is not fully protected against complete loss. However, complete loss is rare except in very small units, and most payments under these policies are for repairs.

Storage. It is unusual for there to be any storage cost directly chargeable to a piece of equipment. Most contractors have at least one home lot, often near their repair shop. This has room for a number of pieces of equipment. The rest are kept out on jobs, where they must be to earn their keep. They usually can be left on or near a job until they are moved directly to the next one.

Ownership and maintenance of a storage yard is strictly a general overhead expense, as this facility is not expanded and reduced with purchase or sale of machines.

However, if a contractor wants to charge it against individual machines, he can do

so by finding the annual cost per square foot of the yard, and charging each machine according to the number of square feet i occupies when it is there.

For example, a piece of industrial land i the outskirts of a city may cost $25,000 a acre, including a graded and stabilized sur face. It might be assessed at full value, with a tax rate of 4 per cent. As this is not a income producing investment, 6 per cen interest may be charged against it. Cost pe square foot may be worked out as in Figur 11-11, to $0.1376.

A large scraper might occupy a space 5 feet long and 12 wide, or 600 square feet Allowing 400 more feet for maneuver space its requirement would be 1,000 square feet Annual cost would then be 1,000 × .1376 or $137.60. This machine would cost abou $50,000, and have an average annual in vestment of $30,000, so storage would b about 4/10 per cent of value. A shove of similar value would need less than hal as much space.

It is unusual to store large pieces o equipment indoors. If it is considered neces sary to do so because of vandalism, extrem cold, or other conditions, the cost may b as high as 5 per cent of investment.

Summary. Figure 11-12 shows the nor mal range in ownership costs or carryin charges, on a per year per $1,000 of averag annual investment basis.

There is a wide range, from .5 to 23. per cent. Most estimating advice recom mends using 10 to 13 per cent. Ten per cen is an easy figure to remember and to use but 8 per cent is likely to be more accurat if interest charges are limited to those actu ally paid.

The contractor or estimator should no rely on any general average of costs, bu should find what they really are for his ow situation.

EQUIPMENT WORK HOURS

Annual depreciation and other ownershi

	Minimum	Ordinary	Maximum
Interest on investment	$ 0.00	$ 45.00	$ 115.00
Loss insurance	5.00	15.00	25.00
Property taxes	0.00	20.00	45.00
Storage	0.00	0.00	50.00
Total	$ 5.00	$ 80.00	$ 235.00
Per cent of average annual investment	.5	8.0	23.5
Per cent of cost, with 5— year depreciation	.3	4.8	14.1

Fig. 11-12. Ownership costs per $1,000 average annual investment, without depreciation or highway use

costs are converted to an hourly figure as a basis for charging out equipment time.

At first glance this appears to be easy. It is only necessary to divide the annual costs by the hours worked per year, or the total work hours by the total costs.

For example, a machine whose fixed costs during a year are $3,600 and that worked 1,200 hours in that year, will show fixed cost per hour of $3.00. Or if its total costs for life are $18,000 and its total hours of work are 6,000, the figure is still $3.00.

But it is difficult to settle on the number of hours that equipment can be expected to work, as this is affected by a number of variable factors.

This discussion will be limited to the problems of contractors who work a single shift and are subject to delays in weather, getting jobs, and keeping equipment running, which includes most of them. It will also deal chiefly with the contractor's first line equipment that has work most of the time he has a job.

Maximum Use. Estimating advice from manufacturers usually recommends a basis of 2,000 hours a year, and a five year life.

But most construction work is done in eight hour days and 5 day weeks, with shutdowns for a minimum of 6 holidays. The maximum number of hours that can be worked in a year is 2,040 on this program. It is usual to lose an extra five days in special holidays or shutdowns, reducing the year to 2,000 work hours.

Bad Weather. Weather often makes outdoor heavy construction work impractical or impossible. One New England state highway department estimates that weather and ground conditions permit the following number of days per quarter:

January - March	35
April - June	55
July - September	60
October - December	55
	205

These figures are on the optimistic side for the area, and they are based on working Saturdays when necessary to make up for rained-out weekdays.

Another northeastern state highway department finds that workdays never average over 18 per month in any season, and that there are winter months in which no work is practical.

A five-year survey by the Bureau of Public Roads indicates that the nation-wide average of shut downs on highway jobs that are due to weather amount to about 1/5 of working time.

The Southeast and South Central states do not have to stop work for snow and ice, but they do have rain that may have equally bad results. Only in certain areas in the Southwest can the 2,000 hour figure be even closely approached on a permitted-by-weather basis.

Maximum working hours are affected by job conditions. Work in rock, gravel, or sand, or on surfaced haul roads, can continue under conditions that would make a job in loam or clay impossible. Pressure of a deadline can make it worthwhile to work under very unfavorable conditions, just in the hope of making some progress.

The type of equipment also affects lost time during weather. A dragline piling wet soil may not be affected by rain unless it is flooded out. Crawler equipment keeps going after rubber tired types give up. Vehicles may carry part loads where full ones would make them bog down.

In lack of specific information to the contrary, the estimator should allow for loss of 20 per cent of annual working time because of unfavorable weather.

No Work. Equipment can work only when there is a job for it. This means not only work in general, but for the specific machine under consideration.

Some contractors find little difficulty in keeping busy all working season or all year, others must get through frequent or prolonged periods of insufficient work or no work. The differences depend on construction activity in the area, the specialties of the contractor and the demand for such specialties, the aggressiveness and reputation of the contractor himself, and a factor of luck in bidding and in selling his services.

Even when a contractor has a job, he may not have it for all his equipment. He may even have to leave his own machinery idle and work with hired equipment at a job that is outside his regular field.

As a general average, a capable contractor may hope to keep his first line equipment busy on jobs about 80 per cent of the time that weather permits working.

Down Time. Even when weather is good and work is available, a machine may not be able to work because of need for repairs to itself or to another unit whose operation is necessary to it, or as a result of shortage of materials, strikes, or other causes. This non-working time on the job is called down time.

Bureau of Public Roads studies show that equipment down time on the job is likely to be between 20 and 65 per cent of working time, with age and condition of equipment and competence of management being the most important factors in the variation.

Most of this down time is considered to be working time (if the machine were rented, rental would be charged), but the owner must take its loss into consideration in figuring the work he gets out of the machine.

Such down time is in addition to the small delays that are taken into account by using a 45 or 50 minute work hour.

Work Hours Summary. A rule of thumb for the hours that heavy equipment will work is to assume a one shift, 2,000 hour year, take off 20 per cent for bad weather leaving 1,600 hours, take off another 20 per cent for lack of work, leaving 1,280 hours, and another 20 per cent (an absolute minimum) for lost time on the job, leaving a net working time of 1,024, or say 1,000 hours. This is the Rule of the Three Twenties.

Like all rules of thumb, this can be way off. But before he discards it, the estimator should study his own conditions carefully to see if they are really better, or quite possibly worse.

This rule does not apply to mines and pits, that may work three shifts on a seven day week, and have up to 8,600 scheduled machine work hours in a year. They do not ordinarily lose as high a proportion of this time.

A number of machine cost computations in this book use a 1,200 hour year as a basis. This is due partly to the fact that many contractors consider the lost time on the job to be working time, and partly to the longer-than-five-year life enjoyed by many machines. That is, the hourly costs come out nearly the same whether the machine is used 1,200 hours a year for 5 years or 1,000 hours for each of 6 years.

Auxiliary and Emergency Equipment. Most contractors own a certain amount of equipment for which they find little use. For example, a general contractor who seldom does rock work may keep a compressor and drill to have them immediately available if they are required.

Emergency equipment is kept to cope with unusually bad working conditions, or for rescue work. Part of a fleet of trucks may have all wheel drive, or a truck or tractor may carry a winch that is not used until something gets stuck. The trucks or tractors may be first line machines, but the extra cost features on them are not.

Other little-used equipment may include items that were bought for a special job and proved unsuitable for it and were also found to be unsalable, and are kept around in the hope that they can either be put to work or sold.

Such equipment may be used only a few hours a year. If a full year's depreciation is divided by these hours, the fixed cost will be so high that the machine cannot hope to earn it.

For example, if a contractor owns a compressor or pump that he uses only 50 hours a year, his depreciation per working hour is 20 times as much as that of another contractor who uses his 1,000 hours. The depreciation on such a unit must be charged to general overhead, as the machine cannot hope to earn it.

OPERATING EXPENSE

The expense of operating a piece of equipment is likely to include the following:

Fuel: both fuel and handling.

Lubrication: cost of oil, grease, lube equipment, and labor.

Fuel	Price in 1000 gallon lots	Federal and State Taxes
Premium gasoline		
Houston	$.355	$.09
New York	.376	.12
Seattle	.357	.13
Regular gasoline		
Houston	.320	.09
New York	.341	.12
Seattle	.322	.13
Kerosene		
Houston	.335	—
New York	.375	—
Seattle	.358	—
No. 1 diesel fuel		
Houston	.335	.105
New York	.379	.13
Seattle	.358	.13
No. 2 diesel fuel		
Houston	.325	.105
New York	.372	.13
Seattle	.339	.13
No. 2 heating fuel		
Houston	.325	—
New York	.365	—
Seattle	.339	—

Fig. 11-13. Fuel costs, 1975

Maintenance and repair: parts, supplies, shop equipment, and labor.

Labor: operator, oiler, helper, ground men, supervision.

Fuel. As a national average, fuel and lubricant take 3¢ of each dollar spent on highway construction, and 1/6 of non-labor equipment operating costs.

Fuel cost varies widely with the power, type, and condition of engines, the type and condition of equipment, type of work, and the grade of fuel.

Fuel consumption in relation to horsepower and load is discussed in the next chapter.

Fuel costs vary with the prices of crude oil, distance from the source, quantities delivered, seasonal demand, and the taxes imposed.

The delivery quantity may be very important. The contractor with a tank of 275 or 550 gallon capacity may have to pay up to 5¢ a gallon more than his big competitor who can take 2,000 or 3,000 gallons at a time. However, this difference can be reduced or eliminated if the small order can be filled on the same trip as others in the locality.

For example, a contractor who uses gasoline or #2 heating oil may be able to get it at a small markup from bulk prices, as the distributor can send out a full tank truck, including him with station, home, or industrial deliveries. But #2 diesel fuel would probably require a special trip and a service charge.

Current prices of standard types of fuel and the Federal tax schedule are given in Figure 11-13. In addition, there are state taxes of 5¢ to 8¢ a gallon that apply to fuel used in highway vehicles. Generally, any vehicle that is registered for highway use must be charged with the state tax, even if operation is off the roads.

Taxes may be paid by the distributor at the highest rate and passed on to the contractor, who then must report the amounts used at lower tax rates to obtain a refund. Or the fuel may be delivered tax free, and the user required to make monthly or quarterly statements of use, with payment of tax due. Payment by the distributor is usually most convenient.

These taxes are substantial enough so that it pays the contractor to keep careful

Lubricant	Quantity	Price per Gallon or Pound
Engine oil		
Low grade	55 gallon drum	1.23 (gal.)
Medium grade	55 gallon drum	1.28
	24 quart cans (case)	1.48
Premium & diesel	55 gallon drum	1.38
	24 quart cans (case)	1.58
Transmission oil		.2025 (lb.)
Grease		
Multipurpose	5 pound can	.43
	35 pound can	.345
	400 pound drum	.31
Track roll	5 pound can	.38
	35 pound can	.35
	400 pound drum	.27

Fig. 11-14. Lubricant prices, 1975

account of his use of fuel. A tally sheet must be kept at the pump or in the distribution truck, showing quantities, type of equipment, and class of use. The bookkeeper needs the information on these sheets when he makes up reports, either for tax payment or refunds.

Lubrication. Lubrication costs are made up of the amount paid for lubricant, ownership and operating cost of the dispensing equipment, and labor.

There is considerable variation in lubricant prices and applications, with resulting confusion to the purchaser. In general, the best quality and most suitable lubricant is the most economical regardless of its price per gallon or per pound, as the cost of labor in using it, and the expense of repairing wear and damage resulting from poor lubrication are vastly greater than the price differences.

The prices of some standard lubricants are shown in Figure 11-14.

Oil. Equipment manufacturers recommend that engine oil be changed at regular intervals, that may vary from 75 to 200 hours in different makes or models.

The time between changes may be shortened under dusty or extreme temperature conditions, or lengthened where work is light, air is dust-free, and/or a special type of filtering or reclaiming apparatus is used.

Crankcase capacities vary widely with size and design of engines. They may hold a quart of oil for every 3½ horsepower, or only a quart for 13 horsepower.

While oil consumption may be negligible in new engines, it may be as high as 1/20 of fuel consumption in engines that have badly worn piston rings and/or external leaks. However, no properly run job would tolerate oil loss of more than 1/50 of fuel use, as pumping oil into cylinders is accompanied by losses of fuel and power, and leaks are likely to allow dirt to get in.

Oil in transmissions, rear ends, and final drives is usually changed twice a year, the most important change being in the fall. Loss between changes is usually negligible, but may become severe because of failure of seals and gaskets, or cracks in housings. Any type of leak may allow dirt to enter, so prompt repair is important.

In general, an allowance of three times the reservoir capacity per year will take care of two changes and losses by leakage or accident.

Grease. Equipment varies tremendously in its requirement for grease. For example, a 20 ton crawler tractor may use from one to 5 pounds of grease in old fashioned track rollers every eight hours or less. A similar machine having positive seals may need lubricant in the rollers only at 1,000 hour intervals or when the rollers are rebuilt.

Here records are the only indication of what to expect. Even if they only indicate the pounds of grease bought and the total of equipment work hours in a year, they will at least provide an average requirement for the fleet.

Lube equipment varies in cost from $3.50 for a hand push gun to about $8,000 for a deluxe grease truck. Small equipment is carried in the tools account and is difficult to

separate, while big units are depreciated in the same manner as other equipment. Lacking information to the contrary, a 6,000 hour life may be assumed for them.

Lube Labor. The pay of the man or men who operate a grease truck or a stationary rack is definitely charged to lubrication. But an oiler on a shovel, in addition to taking care of oiling and greasing, is likely to assist the operator with other maintenance, repair, moves, and in many other ways. It is usual to carry his pay in the same account as that of the operators.

A great deal of lubrication is done by the operators themselves. They may be paid for a half hour overtime a day to take care of this and fueling, or may do it during the shift in pauses in the work.

A grease truck crew can take care of about 3 machines per man hour. This figure is an average of daily lubrications that may take five minutes or less, periodic thorough jobs where all points are reached and all reservoirs checked, and complete lubes including oil change.

Rule of Thumb. In view of all the variables and borderline costs, the estimator is justified in accepting and using the rule of thumb that costs of lubrication equal one-third of the cost of diesel fuel or one-quarter of the cost of gasoline. There will usually be some error as a result, but is likely to be less than that resulting from a superficial attempt to work out the actual figures.

The important thing about keeping track of these costs is to decide on a system and stick to it. A contractor who uses a different method each time he thinks of one will not be able to make comparisons between different jobs and different years.

Always keep in mind that the biggest lube expenses are the failures—the breakdowns that are caused by improper or neglected lubrication.

MAINTENANCE AND REPAIR

There is no definite line of division between maintenance and repair. It is usual to say that maintenance includes items such as cleaning, inspection, adjustment, routine replacements, and hard face or other build-up welding, while repair consists of fixing or replacing worn or broken parts.

Lubrication is often treated as a maintenance expense, and it is probably the most basic and important of all the maintenance operations.

Many contractors and most equipment rental firms divide repairs into two classes —major repairs, overhauls, and painting; and small repairs and maintenance. The first class may be called shop work, as it should be done in the repair shop even if it actually has to be done in the field, and the second class is called field repairs and maintenance.

In rental arrangements the shop repairs are usually done by the owners, the others by the lessee, although the contracts do not say so specifically.

A repair, whether in the shop or the field class, serves simply to fix or replace a defective or broken part, together with any associated parts that have caused the breakdown or have been affected by it. An overhaul involves thorough inspection and all necessary rebuilding of an entire unit.

For example, a transmission with a broken gear may be repaired by simply replacing the gear, but if it were overhauled it would be completely disassembled, all parts would be cleaned and checked, and any defective ones replaced.

Whatever classification is used, there is nothing that is more important to the contractor's success than careful maintenance and prompt repair, as it will save his equipment and his money.

Estimating Repairs. A contractor must have a fairly accurate idea of his future cost of maintaining and repairing a machine, before he can put a price on its use. If he has kept good records he can check on his own experience, and use it as a basis on which to allow for future expenses.

If there are no records, or if new equipment and/or new jobs are so different that old records do not apply, estimating must be done on the basis of reports of other people's records or ideas. These must be modified to suit particular conditions.

Most manufacturers and estimating texts recommend setting total non-tire repair cost during the life of the machine at 60 to 100 per cent of depreciation. However, most of these same sources set machine life at 10,-000 hours of use in five years. But we have seen that the contractor usually does not get over 5,000 or 6,000 hours machine time in five years. This leaves the question of whether these authorities really expect life to be 10,000 hours or five years.

There are so many variables in this field that experience records can be found to support almost any estimate. Costs are affected by the quality of the machine, the accessibility of its parts, standards of lubrication and maintenance, skill of mechanics, work conditions, hours and years of use, and quality of supervision and operation. There is also an important factor of luck.

The contractor who has just one important machine may be made or broken by different combinations of these factors. But possession of a number of machines will usually cause good and bad features of individual machines to average out, and a succession of jobs is likely to smooth out the ups and downs of work conditions.

Repair Factors. Figure 11-15 gives a table of repair factors that may be useful in determining probable repair costs over the life of a machine, in adjusting experience records to new conditions, or in explaining expenses that have already been incurred.

In using this table, the estimator selects the description under each heading that most nearly represents the conditions expected, and takes the figure that follows it. These figures are multiplied by each other to produce a combined repair factor, that is then multiplied by 1/10,000 of the purchase price of the piece of equipment.

Unless special conditions have an unusual effect on tire life, these factors may be used for the whole machine, including tires. When tires have exceptionally short or long life, these factors should apply only to the non-tire part of the equipment, and the factors in Figure 12-117 used to determine tire life.

For example, a contractor may buy a crawler mounted front loader for $40,000. It is a top quality machine, maintenance is

1. TYPE OF EQUIPMENT

Crane, revolving	.5
Compressor, air	.8
Truck, highway dump	.8
Dragline and clamshell	.9
Shovels, dipper and hoe	1.0
Truck, off the road	1.0
End loader, 4-wheel drive	1.0
Scraper, all types	1.1
Dozer, crawler	1.2
End loader, crawler	1.4
End loader, 2-wheel drive	1.6
Crawler and mounted ripper	2.5

2. TOTAL HOURS OF USE

1,000	.5
2,000	.5
3,000	.6
4,000	.7
5,000	.9
6,000	1.0
8,000	1.3
10,000	1.6
12,000	1.9
15,000	2.3
20,000	3.0

3. YEARS OF USEFUL LIFE

1	.6
2	.7
3	.8
4	.9
5	1.0
6	1.0
7	1.1
8	1.2
9	1.3
10	1.4
15	2.0

4. TEMPERATURE, FARENHEIT

Very hot	Over 100°	1.3
Hot	85 to 90°	1.1
Normal	32 to 84°	1.0
Cold	0 to 31°	1.2
Very cold	Under 0°	2.0

5. WORK CONDITIONS

Mostly standby	.4
Light	.8
Average	1.0
Heavy	1.4
Rough	2.0

6. MAINTENANCE

Excellent	.6
Good	.8
Average	1.0
Poor	1.5
None	3.0

7. TYPE OF SERVICE

Mine or large pit	.5
Small pit	.8
Contractor	1.0
Rental to others	1.4

8. OPERATORS

Exceptional	.8
Good	.9
Average	1.0
Rough	1.2
Cowboy	2.0

9. EXPERIENCE (LUCK?)

Excellent	.6
Good	.8
Average	1.0
Poor	1.5

10. EQUIPMENT QUALITY

Top	.8
Average	1.0
Poor	1.5

11. WORK PRESSURE

Leisurely	.9
Average	1.0
Desperate haste	1.5

To find hourly cost of repairs for a machine, select one factor from each group, omit any that are 1.0, multiply the others together, and multiply the product by 1/10,000 of the purchase price of the machine, minus tires if they are figured separately.

Fig. 11-15. Repair cost factors

expected to be good, work conditions heavy, temperature normal; experience, work pressure, and operation are average, and the machine is expected to be used a total of 6,000 hours in five years. We select the appropriate figure from each group of factors, obtaining:

1. Type of equipment 1.4
2. Total hours of use 1.0
3. Years of life 1.0
4. Temperature 1.0
5. Work conditions 1.4
6. Maintenance 8
7. Type of service 1.0
8. Operators 1.0
9. Experience 1.0
10. Equipment quality 8
11. Work pressure 1.0

Dropping the 1.0 factors because they do not affect the multiplication, we have:

$$1.4 \times 1.4 \times .8 \times .8 = 1.2544, \text{ say } 1.25$$

We multiply this by $4.00, which is 1/10,000 of the purchase price of $40,000.

Then:

Hourly repair cost $= 1.25 \times 4 = \$5.00$

These factors should be used only by persons experienced in heavy equipment use, as judgment is required in selecting the correct factors, and in deciding whether the results obtained are reasonable.

If equipment is not used in its proper jobs, in relation to its size and its design, repair costs will be affected. For example, a half yard shovel used in coarse blasted rock would be in the rough conditions classification, even if the bank were average digging for a 2½ yard machine.

A highway type dumper used in off the road work will suffer severely. Repairs are likely to be those of rough conditions, even if the job were average or light for an off the road hauler.

This $5.00 is an average of the hourly repair cost over the expected 6,000 hour life of the machine, and is important in its effect on the price to be charged for its work.

Hours of Use	Repair Cost Per Hour, Dollars					
	Whole Life			Last 1,000 Hours		
	Light	Normal	Heavy	Light	Normal	Heavy
1,000	.04	.05	.07	.04	.05	.07
2,000	.04	.05	.07	.04	.05	.07
3,000	.048	.06	.084	.064	.08	.112
4,000	.057	.072	.10	.088	.11	.154
5,000	.069	.086	.12	.112	.14	.196
6,000	.08	.10	.14	.136	.17	.238
7,000	.091	.114	.16	.16	.20	.280
8,000	.103	.129	.18	.184	.23	.322
9,000	.115	.144	.20	.208	.26	.364
10,000	.126	.158	.22	.232	.29	.406
12,000	.15	.187	.26	.280	.35	.49
15,000	.184	.23	.32	.352	.44	.616
20,000	.235	.306	.43	.472	.59	.826

Fig. 11-16. Revolving shovel repair costs for each $1,000 invested

End-of-Period Cost. Repair cost increases as equipment ages, and the increase is faster than is indicated by whole-life averages. It is desirable to estimate its actual rate at the end of possible life periods, to determine whether it is likely to be so high as to make it uneconomical to keep using the machine.

Average repair cost can be converted into the end-of-period rate by multiplying by the proper one of the following factors:

Hours of Use	Factor	Hours of Use	Factor
2,000	1.0	8,000	1.8
3,000	1.3	9,000	1.8
4,000	1.5	10,000	1.8
5,000	1.6	12,000	1.9
6,000	1.7	15,000	1.9
7,000	1.7	20,000	1.9

Repair Cost Table. Figure 11-16 is offered for those who prefer taking a quick approximation from a table to working out an answer with a set of factors. It shows hourly repair costs for a revolving shovel in light, average, and heavy work conditions on a $1,000 cost basis. Both average and end-of-period figures are included.

This table takes care of Groups 2 and 5 in the factor table. Its figures can be adjusted for any other of the groups simply by multiplying by a factor in that group.

Percentage of Depreciation. Repair costs are often calculated simply as a percentage of depreciation, varying from 60 to 100 per cent. Depreciation periods also vary, from 5,000 to 10,000 hours or more, so here the estimator can select from a wide range of possibilities.

The following are some of the costs per hour for $1,000 investment that can be obtained from a percentage of depreciation:

Per Cent	Depreciation Period, Hours	Repairs per Hour
100	10,000	$.10
100	6,000	.17
80	10,000	.08
80	6,000	.13
60	10,000	.06
60	6,000	.10

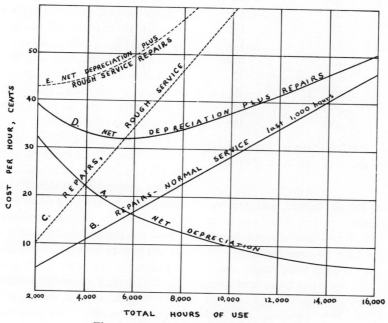

Fig. 11-17. Depreciation and repair

Percentage of Cost. The Associated General Contractors of America publish a table of ownership expense of construction equipment, compiled from reports from their members. This lists a combined item of "Overhauling, Major Repairs, Painting" that is 12, 15 or 20 per cent of the purchase price for most items of earth moving machinery.

Heavy repairs are considered as an ownership expense for the purpose of computing charges for renting equipment to others where the owner pays the major repairs and the user the smaller ones. They make up from 50 to 80 per cent of total repair and non-lubrication maintenance expenses.

Using these figures, the hourly cost for heavy repairs alone for each $1,000 investment would be:

Per Cent	Annual Use, Hours	Heavy Repairs per Hour
12	800	$.15
	1,000	.12
	1,200	.10
	2,000	.06
15	800	.19
	1,000	.15
	1,200	.125
	2,000	.075
20	800	.25
	1,000	.20
	1,200	.167
	2,000	.10

As we will see later, repair expense on rented equipment tends to be higher than normal.

Usefulness. Tables, factors, and percentages are all based on averages from large numbers of machines, and do not necessarily hold good for any one machine.

The increase in costs with longer use is also an average. Any machine, and even most fleets of equipment, will go through good periods of little expense, and bad periods of frequent breakdowns.

In the same manner, July should be a good month for earthmoving and January a bad one, but occasionally the reverse condition occurs. But the contractor still must figure on working the next July and losing time during the next January.

Rising Repair Costs. As a machine gets older its repair costs increase. Fuel and lubrication bills also increase unless held down by first class maintenance.

Depreciation cost per hour decreases steadily with longer use. This serves partly to offset rising repair costs. Figure 11-17 shows a curve, (A), for net depreciation, and lines (B) and (C) for normal and extra heavy repairs. Curves (D) and (E) give combined net depreciation and repair.

The low parts of the curves D and E show the most economical life period in hours. For average medium repair costs this is 5,000 to 6,000 hours, with little difference shown from 4,000 to 8,000 hours. For extra heavy repairs a 2,000 hour machine life appears to be less costly, with moderate increase to 4,000 hours and then a steep rise in expense.

Such very heavy repair costs with resulting short economical life can be combatted by using bigger and stronger equipment, reducing work pressure and loads, stepped up maintenance with frequent overhauls including bracing and substituting heavier components when they are available, and by various combinations of these methods.

Light Duty. The heavy repairs usually incurred by using a machine late in its useful life can sometimes be avoided by assigning the unit to light or standby duty. A bulldozer may be assigned to cleaning up loose dirt around a shovel or shovels, a tired wheel tractor may pull a water wagon, and an old shovel be kept in soft digging. Such assignments are more easily made in open pit mining than in general earthmoving, and they help to explain the large number of over-age machines that continue to render satisfactory service in pits.

Down Time. An item that does not appear on the bill at all is often the most expensive part of equipment repair. That is the lost time on the job while mechanics and parts are being obtained and the repair is being made. This cost is usually at its highest rate during the first few minutes or hours of the breakdown, while operating costs are still at a maximum and before the work program has been changed.

If the shutdown is a short one and only one machine is affected, for example, if a hose bursts in a self-loading scraper working alone, it causes loss of only the production of that one unit. The same difficulty in a loader might stop the loader, a string of trucks, a dozer, and compacting equipment also.

Conditions involved in down time are so variable that they cannot be put into graphs or tables. Losses are kept down by alert supervision, new equipment, and expert maintenance.

TIRES

Tires may represent an important part of the cost of new equipment, and they have several characteristics that make them difficult to fit into the same cost calculation as the rest of the machine.

Non-Capital Treatment. It is almost standard practice to deduct the cost of tires from the price of a new machine before setting up its depreciation account. If this deduction is made it should be on the basis of the actual cost of replacing such tires, but list price is sometimes used.

The chief reason for subtracting tire cost is to obtain the fastest possible writeoff. From 1/10 to 1/4 of the cost of rubber mounted equipment may be written off immediately in this way, if Internal Revenue permits. The excuse used is that tires are not physically part of the machine, and wear at a faster rate so that the ordinary long depreciation period does not apply.

A hauler may go through two or three sets of tires before it wears out itself. If the original tires are capitalized the owner may be still deducting depreciation on them years after they have been scrapped. He will have paid the cost of replacing them before having been able to get a tax reduction on all of their original cost.

Another reason for keeping tire accounts separate from the machines that carry them is they wear at different rates and are affected by different conditions. Mechanical parts may be little affected by differences between sandstone and shale, but they may cause a three to one difference in tire wear. Extreme cold may increase equipment repair costs and prolong tire life.

None of these factors entirely justify separation of tires from the rest of the machine in bookkeeping. The same arguments could be advanced for separate accounts for crawler tracks and rollers, blade and bowl edges, batteries, and even shovel buckets. And the double bookkeeping does exaggerate the cost of tires.

But to conform to general practice, and to make it easy to compare figures with other sources, tires will be separated from equipment before depreciation in examples in this book.

Figure 11-18 shows the hourly costs of a $50,000 scraper equipped with $10,000 worth of tires at fleet prices. They are computed in round numbers in two ways, first with the tires depreciated as part of the machine, second with the original tires deducted from the purchase price and put in operating cost. In this particular set of figures, separation doubles the apparent hourly cost of tires.

If earthmover tires have an average life of a year or less, the Internal Revenue Service agents will approve their deduction from capital investment. If average life is over a year they have the right to disapprove. In general, they are against any type of separation of a unit into various pieces for separate depreciation treatment.

If tires are depreciated with the machine, then:

Depreciation, 1/5,000 of $50,000	$10.00	
Interest, insurance, etc.	2.00	
Fixed expense per hour		12.00

2 new tires	$ 5,000	
2 recaps	2,500	
Tire purchase cost	$ 7,500	

Tire cost, 1/5,000 x $7,500	1.50	
Tire repair	.50	
Total tire cost per hour		2.00

Non-tire repairs and maintenance, 80% of $8	6.40	
Fuel and lubrication	2.40	8.80

Total hourly cost, without operator	22.80

If tire cost is separated before depreciation, then:

Depreciation, 1/5,000 of $40,000	$ 8.00	
Interest, insurance, etc.	2.00	
Fixed expense per hour		$10.00

Cost of original tires	$10,000
2 new tires	5,000
2 recaps	2,500
Tire purchase cost	$17,500

Tire cost, 1/5,000 x $17,500	3.50	
Tire repair, about 15%, say	.50	
Total tire cost per hour		4.00

Repairs and maintenance, 80% of $8.00	6.40	
Fuel and lubrication	2.40	8.80

Total hourly cost, without operator	$22.80

Fig. 11-18. Cost treatment of tires

Figure 12-111 gives the list prices of a number of sizes of earthmover tires. Contractors can usually get discounts of 25 to 33 per cent from these prices, and markdowns of as much as 50 per cent to large operators have been reported.

Operating Cost. The major operating expense of a tire is its replacement. The actual cost divided by the number of hours it operates gives this cost on an hourly basis. Since many tires reach an early and sudden end through accident or abuse, the life of a number of tires must be averaged to obtain a fair figure.

Tire maintenance and repair are assumed to cost about 15 per cent of replacement.

HOURLY WAGE RATES—Base rate + fringe benefits

Classification	Atlanta $	Boston $	Chicago $	Cleveland $	Los Angeles $	New Orleans $	New York $	St. Louis $	San Francisco $	Toronto $
COMMON LABOR										
Heavy const.	5.80	8.05	8.63	8.63	8.095	5.65	10.36	9.255	9.885	7.83
Building	5.80	8.05	8.63	9.92	8.095	6.21	10.36	9.255	9.735	7.68
SKILLED BUILDING										
Bricklayer	9.31	10.65	11.13	11.58	10.88	9.125	13.80	10.19	12.80	9.47
Carpenter	8.97	10.62	10.89	11.63		9.46	13.045	10.21	12.38	9.60
Struc. Ironworker ...	9.04	11.30	12.42	11.52	11.215	8.86	13.94	+10.275B	13.095	9.97
Plasterer	8.92	9.77	10.28	11.59	11.725	9.17	13.94	+10.275Ar	13.43	7.95
Electrician	11.27	11.60	12.10	11.51	11.725	8.06	+11.31Ar	10.085	13.43	9.97
Steamfitter	9.94	11.62	11.62	11.67	13.21	9.43	12.28	11.43	11.724	7.95
						9.99	12.552	11.55	12.325	10.35
					12.073					10.41
EQUIPMENT OPERATORS										
Hoist. 1 drum	8.72	11.03	11.75	11.04	10.63	8.97	13.35	9.69	11.96	10.11
Tractor, Incl. dozer .	8.37	10.83	10.45	11.19	10.63	8.97	12.335	10.24	12.78	9.46
Tractor scraper 15-16 cu. yd.	8.72	10.91	10.45	11.19	10.63	8.97	12.552	10.24	12.98	9.46
Power crane	8.72	11.03	11.75	11.54	10.73	8.97	14.332	10.24	13.26	10.11
Motor grader	8.37	10.91	11.75	11.19	10.73	8.97	12.552	10.24	12.33	10.11
Air Compressor	6.80	9.44	9.40	10.26	9.85	7.78	13.35	10.24	12.28	9.24
Air tool	5.97	8.40	8.75	9.57	8.485	7.78	10.36	9.255	12.48	9.24
TRUCK DRIVERS										
Pickup, under 1½ tons	5.90h	7.32	7.90	6.95	8.90m	6.27	10.21	8.87	10.64	7.89
Dump, under 4 cu. yd.	5.85j	7.47	8.15	7.10	9.14z	6.77	10.21	8.87	11.975	7.89
Dump, 4 cu. yd. & up.	6.10k	7.47	8.35	7.45	9.28s	6.82	10.21	8.87	14.84	8.00

Courtesy of Engineering News-Record

Fig. 11-19. Pay rates, 1975

LABOR

In spite of mechanization, labor accounts for a big part of every heavy construction dollar. No estimator can afford to overlook any of his labor costs.

Current pay rates on a national basis are shown in Figure 11-19.

Operator pay is often left out of equipment costs in estimating advice, because it varies so widely from place to place. The estimator must be very careful not to leave it out of his figures.

Almost every piece of construction equipment has an operator. Many shovels have an oiler too. Laborers are needed to handle supplies, spot trucks, direct traffic, pick up rocks, trim banks, scrape sticky soil out of bodies and buckets, and for many other tasks.

Supervision at the superintendent level is considered an overhead expense. Foremen may be charged to overhead also, but it is more usual to enter their pay as an operating expense. It may be charged against the job in a lump, or divided by the number of men supervised, and added in as part of operator payroll.

An operator is usually paid for a good many more hours than his machine works. He may get a full day's pay just for reporting, whether the machine runs or not. He is certainly kept on the payroll during short delays for adjustment and repair, and when standing by during various job delays. He may be paid for extra non-operating time in which he greases and services the equipment.

If an operator is paid on an annual salary basis, his wage should be divided by the number of hours the equipment works or is expected to work during the year to obtain an hourly rate.

Mine workers are usually paid on a portal to portal basis, that is, from the time they check in at the gate or the main building until they get back to the time cards. They receive full pay for time spent between the entrance and the place of work. This may make a substantial reduction in the time actually worked during a shift. To look at it another way, it means a higher per

11-35

hour cost for the time they are working.

Construction workers usually check in close to the work, and are usually expected to have equipment running and ready to go at the start of a shift.

To find the full cost of labor, it is necessary to add in payroll taxes, both for Social Security and unemployment, compensation insurance, and fringe benefits such as paid holidays and sick time, reserve for pensions special travel or subsistence allowances, and pay for non-working time on temporary job shut-downs.

These extras may increase base pay from 10 to 30 per cent or more.

Cost figuring in this book often uses already-obsolete hourly rates for operators and laborers, but any correction would itself be temporary. The reader should always substitute current rates.

Shift. A shift is the continuous (except for breaks for meals) time worked by one crew in one day. It is usually eight hours but it may be 7, 10, or even 12. The longer shifts usually include an overtime pay rate.

Multiple Shifts. Work may sometimes be speeded by working two or three shifts. Three shifts are commonly eight hours each, one crew taking over from another without any shutdown. The day shift is from 8:00 A.M. to 4:00 P.M., the "swing" until midnight, and the "graveyard" until the day gang takes over. Pay time is eight hours, but a "lunch" period, and time lost in the changing of the shifts, reduces work time to less than seven and a half.

Two shifts may be of either eight or ten hours each. The job is usually shut down after each shift, except for lubrication and repair crews.

Night work is less efficient than day because of the need for artificial light, and the lessened accuracy and usually lower mental and physical vigor of the men. Night grading is very destructive of stakes.

Multiple shifts may work at cross-purposes, or at least with insufficient understanding of what has been done. This difficulty is somewhat less when the new crew arrives before the other leaves. If there is no contact, the supervisors should meet in the idle period to discuss the work and coordinate their efforts.

There should be a system for rotating men among the shifts, but it should be admininstered intelligently. Night shifts are generally unpopular, but some individuals prefer them and they should be left in them. Swapping of shifts among equally qualified men should always be allowed. Rotation should be at rather long intervals to enable the men to adjust to changes in sleep and work hours. Two weeks is the shortest period which should be considered.

Overtime. Where the industry operates on five working days of eight hours each per week, additional hours worked on any of the five days, and any time worked Saturdays, are called overtime, and paid at 1½ or 2 times the regular hourly rate. Sunday and holiday work may be time and a half, double, or even triple time.

A contractor who must finish work before a contract deadline, or before bad weather, or who wishes to take advantage of a busy season, may ask his men to work overtime; hire additional men to work two or three shifts; or may buy or rent additional equipment to work one shift.

In general, it is profitable to work large machines overtime, as the extra wages are more than offset by the drop in hourly ownership costs caused by spreading them over a greater number of hours. Small machines may show either increased or reduced profits.

Overtime work with labor, supervisors, and other personnel ordinarily results in a loss, as the profit figured on a payroll is liable to be between 15 and 25 percent, as against the 50 percent increase in pay rate.

It should be remembered that payroll insurance premiums increase in direct proportion to the amount of pay, although some payroll taxes do not apply over a certain amount.

The small contractor whose machine operators are frequently able to work for extended periods without help or supervision, is more likely to work overtime than the large scale operator.

When work is being done on a fixed price contract, or at a fixed hourly rate, overtime costs must be carefully watched. If the work is on a basis of cost plus a fixed fee, overtime will merely require a larger investment to obtain the same profit. If payment is cost plus a percentage of cost, overtime, as well as any other extra expenses, will increase the contractor's profit.

Cost-plus contracts may require that a contractor obtain written permission before incurring overtime or special expenses.

WORKING TIME

Time can be used as a measurement in direct clock and calendar divisions, or on a basis of working times that are fractions or combinations of them.

Efficiency Hour. A second and a minute always have the same time meaning in construction that they do on a clock. But an hour may contain the regular 60 minutes, or may be an "efficiency hour" of 50 minutes or less. This special hour allows for lost time in a way that can be easily included in calculations.

For example, a machine may be able to move two yards of earth a minute in steady digging. This would represent an output of 120 yards in a full hour. But no machine can be counted on for absolutely steady work because of delays from such causes as need for adjustments and minor repairs, changing positions, lack of supporting equipment, cigarette time, digging obstacles, and so forth.

The actual production of the machine, averaged over many hours of work, may be only 90 yards an hour. This is 75 per cent of its maximum potential output, and is

called 75 per cent efficiency. It means that the machine is working to capacity only 3 out of every 4 minutes.

It is customary to express the reduced ability to produce by reducing the number of minutes in the hour, rather than by deducting a percentage from production. In this example the work hour would be 3/4 of 60, or 45 minutes. Multiplying 45 minutes by the maximum production of 2 yards a minute, we get an hourly production of 90 yards, which is where we started.

In the excavation industry it is usual to assume an average efficiency of 83 per cent, so that many calculations are based on a 50 minute hour. This efficiency is not unknown, but the average is very much lower. Most examples in this book will use a 45 minute hour.

To convert 45 minute hours to 50 minute (that is, to substitute 83 per cent efficiency for 75 per cent), number of cycles per hour and production are increased by 1/9 or 11 per cent and costs are decreased by 1/10 or 10 per cent.

In our example of 90 yards production in a 45 minute hour, production in a 50 minute hour would be 50/45 (10/9) of 90, or 100 yards. If delays multiplied until the average hour included only 20 minutes of useful work, a 20 minute hour would be used. Production would be either 20/60 (1/3) of 120, or 20/45 (4/9) of 90, or 40 yards an hour.

A basic approach is to multiply the production per uninterrupted minute by the number of useful minutes in the hour, as $50 \times 2 = 100$, $45 \times 2 = 90$, and $20 \times 2 = 40$.

Day and Week. A work day includes all the hours of work during a regular 24 hour period. This is usually 8 hours, but may be 8½, 10, or some other time. If there is one shift the shift and the day are the same time measurement.

In construction, a work week is usually 5 days, with overtime pay for working ad-

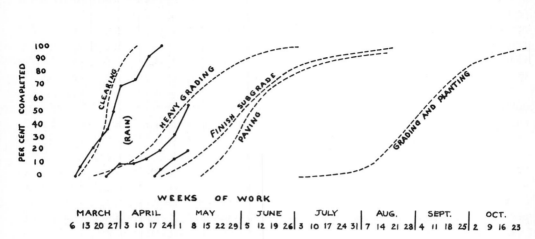

Fig. 11-20. Work schedule graph

ditional days. Weeks in which holidays occur are shorter. In mines six and seven day weeks are common.

If business is poor, work may be stretched out by shortening the work week to as little as one or 2 days. This practice is rare in construction, but is common in mines.

Year. A work year may be same as the calendar year, or it may be a fiscal year starting at any convenient time. It always includes the twelve calendar months to the same date the next year, but it may be rated in terms of fewer months because of work stoppages caused by weather. An 8 month work year usually means that it is not considered practical to work during 4 winter months. An eight-month year would correspond in efficiency to a 40-minute hour, but it is not often used as a basis for cost adjustment.

Job. The time that elapses between starting a job and finishing it is known as job time. The completion date may be stated in the contract, either with or without penalties for not meeting it, or may be set by the contractor's own schedules. It may be measured in hours, days, weeks, or years. The number of work days must be carefully

distinguished from the number of calendar days.

Job time gives an excellent check on progress. Daily or weekly plottings of accomplishment against percentage of time used will indicate whether the job is on schedule. For example, if laying pavement on a prepared base started on September 15th for completion on October 31st, and was 30 per cent complete on October 15th, the contractor should be out on the job asking some questions.

Progress Charts. A form that indicates the percentage of work intended and accomplished in each time period is shown in Figure 11-20. Time is shown on the horizontal scale, percentage on the vertical. Dash line curves are drawn in for the schedule, and a solid line is plotted in week by week according to progress made in the field, through the week of May 5 in this example.

The dash curves usually have somewhat the shape of a letter S, as work starts slowly, speeds up as men and equipment are adjusted to it, and slows again as the work force is reduced for finishing operations toward the end.

Taking the heavy grading work for an example, we can tell from the graph that it started a week late, but in one week was a little ahead of the two week schedule. But it then fell behind, as there was no work the next week because of rain and mud, and progress was poor the following week because of mud. After that progress was good, and another week at the present rate should reach or pass the scheduled output.

The next set of lines indicated that finish subgrade work started early and is running well ahead of schedule.

A highway job involves many other items, and if each is to be followed it will be necessary to use several graphs to avoid confusion from crossing lines. Colored pencils are always helpful in making graphs easy to understand.

A contractor seldom expects the work to correspond closely to his ideal curve, but it is important for him to know how far off it is, and why.

Progress sheets can also be made up to show various sections of a job separately, so that the contractor can tell at a glance how particular problem areas are shaping up.

EQUIPMENT RENTAL

There are many different types of equipment rental arrangements. The one that we will discuss here is the renting of a machine owned by a contractor, distributor, or rental agency to a contractor who will use it as his own during the period for which he makes payment.

Rental of equipment with fuel, maintenance, operator, and supervision will be considered later under CONTRACTS.

A contractor may decide to rent part or all of the machinery he will need on a job because of short period of use, availability, lack of confidence in future work, lack of capital, and/or other reasons.

Cost. Equipment rental is often casual in nature. A contractor who has a machine

that is idle or nearly so will rent it to another contractor who has need for it. Price is likely to be strongly affected by the amount of demand for the unit, its condition, and the financial position of the parties. Rental rates under these conditions may vary widely.

There are also concerns that make a substantial part of their income from renting out equipment, and who operate rather close to a fixed rate basis. Rates may be based on references on the next page, or some other formula. They are usually for the base machine, with separate rates for some buckets and equipment. Delivery; and operator, fuel, lubrication, or service, if available, are extras.

Rates are usually based on one shift of 8 hours per day, 40 hours per week, and 176 hours per month of a 30 consecutive day period. Contractors may find that it pays to work a 10 hour day or use other overtime arrangements to get full time on expensive rented equipment.

Overtime on the machine is paid at the same hourly rate as regular work time.

Time may be taken from the contractor's records, inspectors' reports, hour meters, engine revolution counters, or combinations of these methods.

Unless other arrangements are made, the rental period starts when the machine leaves the owner's yard, and does not end until it is back in the yard, or is taken to or by another contractor by arrangement with the owner.

Even if the machine does not work the full number of hours, or any hours at all, during the rental period, full charge will be made except under special conditions. Most firms renting equipment will make allowances for time lost through long breakdowns, excessive bad weather, or strikes or material shortages; but the conditions under which such allowances will be made should be clearly understood in advance.

A.E.D. The Associated Equipment Distributors (AED), publish every other year a "Rental Compilation". A condensation from some of their tables, and explanatory material, will be found in the Appendix. A warning in this Compilation reads:

"The rental rates and terms set forth in this compilation are for informational purposes only and not to suggest or to influence the rates or conditions of rental of any item of equipment, as this is a matter which must be determined by the lessee and the lessor of the equipment involved . . .

"For any distributor, or any other person, to enter into any agreement, understanding, combination or concerted action with one or more distributors, or with one or more other persons, to adhere to the rental rates shown in this Compilation, or to refrain from charging less than such rental rates, would be a violation of the Federal and State Anti-Trust Laws."

The Equipment Guide Book Company publishes the "Rental Rate Blue Book for Construction Equipment". This is also a compilation. It is more detailed in that it provides rates for specific make-and-model items of equipment, and has information on regional variations.

The two references differ widely on many items. However, either (or both) may be very useful as a quick guide to relative costs and values.

Addresses may be found in the Acknowledgments section, which is in the front of this book.

Canadian Rates. The Canadian Construction Association issues a compilation of rental rates in which yearly charges are shown as percentages of replacement cost. Depreciation varies from 9 to 100 per cent, and overhaul, major repairs, and painting are listed at 0 to 65 per cent, depending on the type and size of machine. General Overhead is figured at 10 per cent, made up of 4 per cent interest, 1 per cent storage, 1 per cent insurance and 4 per cent overhead.

Some sample Canadian figures are given in the Appendix. It is necessary to know the present prices of the equipment in order to change these percentages of cost figures into dollars.

Prices of new and used equipment may be obtained from a reference such as the Equipment Guide Company's "Green Guide."

Repairs. A definite understanding should be made about repairs in connection with every rental agreement. Policies of owners vary with local custom and the type and condition of equipment.

The owner should take care of overhauls, major repairs, cleaning, and painting. He should do this conscientiously enough so that the equipment is able to work through its rental period without major breakdowns. The renter is supposed to take care of small field repairs, all damage from abuse or accident, replacement of cables, cutting edges, and other fast wearing parts, and to return the machine in as good condition as he got it, except for normal wear.

Field repairs should be the responsibility of the owner if the breakdown is in parts known to be defective at the start of the rental period.

Some rental arrangements distinguish between non-tractor equipment, for which the owner assumes responsibility for wear and tear, and tractor and rubber tired scrapers and haulers, on which the owner expects the contractor to pay all expenses, including those resulting from ordinary wear and tear in normal use.

Misunderstandings often arise as to the responsibility for major repairs needed during the rental period, and the amount of wear that can be said to be normal.

Adjustment of these differences can mean a cost variation of several dollars an hour, so a clear understanding in advance is important.

The owner usually reserves the right to pull his equipment off a job where it is being abused.

Ownership vs Rental. The contractor who keeps his machinery from job to job and takes good care of it, operates at lower cost than if he rented his equipment. Rental prices include an allowance for greater than average major repairs because few people are as careful of rented equipment as they are of their own, and the owner's profit is of course added in.

However, for short jobs with no sure usefulness for the machinery after completion, renting is cheaper.

Contractors whose work is scattered over the country usually rent machinery at each job, instead of owning and moving it. This saves heavy transportation expense, reduces hostility to a "foreign" contractor, and makes it easier to hire and control local operators.

The following will serve as an example of figuring comparative costs on one job.

Assume that a job requires a two yard front loader (shovel) to work 600 hours during a 4-month period, and that the new price of such a machine is $40,000 including incidental expenses of purchase.

If the contractor buys the loader, and keeps it employed for 1000 hours a year for five years, then junks it, his average hourly costs will be:

Depreciation, 40,000/5,000	$8.00
Ownership	2.00
Repairs	4.00
Fuel and lubrication	1.00
Total non-labor cost per hour	$15.00

If he buys the shovel and sells it for $24,000 on completion of the job, his costs will be:

Depreciation 16,000/600	$26.67
Ownership	2.00
Repairs	1.20
Fuel and lubrication	.90
Total non-labor cost per hour	$30.77

If he should rent the machine at the rate of $2,000 per month, he may pay $8,000 plus a one-way delivery charge of $100.00. Then:

Rent, 8,100/600	$13.50
Loss insurance	.30
Repairs, 1/2 total	2.00
Fuel and lubrication	1.00
Total non-labor cost per hour	$16.80

If rental is on the Canadian basis, he will pay 6.5 per cent of $40,000 per month. Monthly rental is $2,600, total rent $10,400, plus delivery of $100. By the hour:

Rent, 10,500/600	$17.50
Loss insurance	.30
Repairs, 1/2 total	2.00
Fuel and lubrication	1.00
Total non-labor cost per hour	$20.80

ESTIMATING

A contractor is usually called upon to estimate the time, material, and expense involved in a piece of work. This estimate may involve careful calculation of all factors, may be made up from records of similar work, or the memory of them; or, in bidding on a small proposition, be only an informed guess.

An estimate may be used as a basis for making a fixed price bid, or simply to give the customer an idea of cost while performing the work on an hourly or cost plus basis.

The first requirement for most estimating is practical experience with the work involved. In large organizations, this experience may be only in handling cost, production, and time figures. In small firms, the figuring is often done by the same man who does or directs the work. He should be familiar not only with excavation in general, but with the specific type or types of work to be done.

Check List. Every estimator needs a check list of the items involved or possibly involved in the job he is figuring. For simple work or rough estimates he may keep it in his head, but it is better practice to have it in writing and to refer to it frequently.

The principal use of the check list is to remind the estimator of items he might forget. An experienced man might feel that he no longer has need for such artificial helps, but anybody can forget something.

Records of state highway departments indicate that careless mistakes are common even in multi-million dollar estimates produced by experienced men. Errors in arithmetic are the most common failing, and leaving out operations is the next. A contractor may estimate concrete at $55.00 a cubic yard and put it in a bid at $5.50. Or he may figure out to four decimal places what it costs him to drill, blast, and shovel load a rock ledge, and entirely forget the haul cost.

Check lists can also be overdone. A simple operation such as digging a cellar for a house may have 20 or more different items to consider. A bid on a section of interstate highway is likely to include hundreds of items, each of which is made up of many different purchases or processes. A check list to include everything would be so long that few men could stay awake while using it.

An estimator, whether he is a contractor or is hired by one, should work up his own check list for each type of work, refer to it, and add to it whenever necessary. It can be one of his most valuable assets.

Round Numbers. An estimate is an informed guess. No matter how solidly it is founded in experience and knowledge, it deals with future work in which unexpected conditions can upset the most careful calculations. It is also often the basis of a competitive bid that must be lower than that of any other qualified contractor in order to get the job.

Since the figures themselves may prove to be inaccurate, and because they may be changed in the bid to meet a price, it is usually pointless to work them out to several decimal places. Excessive detail adds greatly to the time and labor of making up a bid, and the estimator may become so lost in his complicated figures that he will overlook errors in arithmetic, or whole items that ought to be included.

A sense of proportion must be preserved. A per-yard cost of moving dirt might well be carried out into several decimals if there are a million yards to move. But as final figures are approached, pennies should be dropped, and dollars rounded off to the nearest ten, hundred, or thousand, depending on the size of the job. The rounding off should be indicated at the point where it is done to avoid confusion, by writing in a word such as "say" or "approximately" or an abbreviation such as "approx."

For example, an engineer's calculation may indicate that there are 27,685 yards of soil in a bank, and excavation cost is figured at 31¢ a yard. Cost of digging the whole bank would be:

$$27{,}685 \times .31 = 8{,}582.35, \text{ say } \$8{,}600$$

If the engineer had simplified his figure to approx. 27,700, the calculation would be:

$$27{,}700 \times .31 = 8{,}587.00, \text{ say } \$8{,}600$$

Slide Rule. Estimating work is often done by slide rule. This is a device that is so convenient that it is a necessity to engineers, but it is not absolutely accurate. The standard slide rule is set up to multiply figures by conversion to decimal logarithms, adding, and converting back to real figures. Inaccuracies are introduced both by the conversion in and out of decimals, and by limits on the fineness with which the rule can be read. For this reason, a slide rule is often called an "about stick." It gives results close enough for estimating purposes, but not accurate enough to justify keeping exactly to every figure it turns out, when rounding off the results would make work easier to do and to check.

Bulk and Hauling. The gross factors in estimating excavations are the quantity of material to be dug, its digging qualities, the distance it must be moved, haul conditions, and the manner of its use or disposal; all in relation to the equipment to be used.

Quantity, which is usually measured in bank yards, should include anything that must be dug, quarried, or moved in the course of the work. Material which is stored and reclaimed must be added in twice.

Digging qualities will include not only the hardness and coarseness of the bank, but water or sand conditions on the pit floor, danger of slides, etc. It will largely determine the type of excavators to be used, and whether blasting will be necessary or not.

The distance to be moved will decide whether it is more economical to push or to carry it, and the types of hauling unit to be used. In general, haulage is figured from the center of mass of the cut to the center of mass of the fill, but the length of the longest and shortest hauls must also be considered.

Haul calculations should include attention to the type of ground to be crossed, its probable carrying capacity and tractive resistance, grades to be climbed and the cost of making and keeping it passable.

Spoil can be dumped over a high bank more economically than it can be spread and compacted in a fill. Operations will be slowed unless there is space for equipment to maneuver and dump in, and unless the fill will support and give adequate traction to the hauling units.

Fill requirements can be greatly increased by a soft base that will compress or shift under its weight.

Digging Factors. The digging qualities of a soil are of great importance in estimating. If blasting is required, expenses are increased five times or more, with the extra costs per yard increasing if the quantities are small, or if precautions must be taken against damaging property.

Hard soil that can be barely dug without blasting will also prove expensive, in the terms of slower production, and increased breakage of equipment. It may require the purchase or rental of special or larger machines.

Wet digging requires working from above with hoes or draglines; may call for expensive drainage or pumping, and will cause mud difficulties at the dump. Operation on wet or muddy pit floors may require the use of tracked hauling units instead of rubber tired, with resultant drop in speed; or substitution of all wheel drive for conventional trucks.

Wet conditions may also make necessary the building of temporary roads of corduroy, crushed rock, or gravel.

If boulders are numerous, digging will be hard; there may be a problem of loading stone that is too large to dump through truck tail gates; and, in any event, time will be wasted in separating boulders from soil, either at the bank or on the fill, and in piling, blasting, or working around those which are too big or heavy to lift or haul.

Ledge outcrops in a pit floor complicate drainage and make haul roads rough.

Deep cuts which require benching involve more planning and systematic work and probable lost time than straight digging.

The amount of swell of the soil and rock when dug will affect the capacity of hauling units in bank yards, unless such steep grades must be climbed that weight is a more critical factor than bulk.

In scraper digging, the presence of a lubricating material in the soil may make possible bigger or quicker loads. Slippery surfaces, however, decrease production, particularly with rubber tired prime movers.

Fills. Trucked fill placed in thin layers requires more or larger dozers for spreading than when in high lifts. Even if no rollers are used, compaction and rain resistance will be improved because of better vertical distribution of the weight of the hauling units. If rollers are used, the thin layers will have more total surface to be treated, but compaction may be secured with lighter machines, or with fewer passes on each level.

When the spreading dozer is oversize for the quantity of dirt being trucked, the extra cost of thin spreading is negligible.

Scrapers naturally spread in thin layers, so depth requirements are more readily met.

Clean dry sand or gravel may not afford enough traction for trucks so that watering or adding thin layers of loam fill may be necessary.

Wet clay may require sandwiching with layers of sand or gravel to make a stable fill.

Specifications for compaction may be impractical, and compliance may be very costly in time and effort, or even impossible to accomplish.

Sequences. Excavation or grading projects often involve a sequence of two or more operations. Sufficient delay in one of them will slow or stop work on those which follow it. Increase in the number of operations makes the final ones more subject to delay. If each step in a series is followed closely by the next, through physical necessity, or haste, the possibility of continuing some work after a breakdown is reduced.

As an example, in laying subsurface drains, a ditch is dug, tile is laid in it, and the ditch is refilled. If the tile is laid and the ditch backfilled immediately behind the ditcher, it cannot even stop for fuel without making the tiling crew and the dozer idle. Any delay in the supplying or the placing of tile will shut down the dozer and, if the ditch is likely to cave, the ditcher as well.

If tile is supplied by truck as required, or a little ahead of use, truck breakdown will stop work quickly. On the other hand, if several hours' supply are laid out along the ditch line, work can continue while the truck is repaired or replaced.

If the ground will permit leaving open ditches, the ditcher may keep several hours ahead of the tiling crew, and the dozer work a long distance behind it so that only a major breakdown will stop more than the unit affected.

In shovel loading, the sequence is digging, hauling, and spreading. If the shovel stops, the job stops. If a truck stops, shovel and dozer work are usually slowed. If the dozer quits, work may shut down after a few loads, or continue some time, depending on dumping conditions.

If two shovels and two dozers are working, complete stoppage is much less likely, although there is an increase in the likelihood of slowdowns.

Slowing or stopping of a job increases the contractor's cost, especially when there is no other work to which machinery can be shifted for the time involved. Fixed expenses continue, and part or all of the payroll. The effect on contracts involving penalties for failure to finish on schedule may be even more serious.

Bottlenecks are another hazard of sequences. Any machine, or any operation, which is slower than those preceding and following it, will set the pace, or the lag, for the whole job, until the condition is corrected. This situation may arise through improper selection of a machine, delivery of the wrong size or type, mechanical or digging difficulties, man shortage, lack of skill, or mistakes in figuring.

In making an estimate, sequences should be studied carefully and allowance made for the probable delays.

Rush Jobs. Rush jobs usually involve very close sequences to such an extent that machines and men are so on top of each other that a great deal of time is wasted, even if no breakdowns or serious tieups occur. An extra charge should be made to cover this inefficiency.

Another type of rush which is frequently experienced is that a customer, often a home owner or building contractor, will demand that machinery be sent over immediately to backfill and grade around a house, dig ditches, or perform other work required to obtain a payment on a building mortgage, or to make the house look attractive to possible buyers on a week end.

If such a call is answered promptly without investigation, it will all too often be found that the site is not in workable condition. Perhaps the whole area is cluttered with piles of sand, gravel, bricks, and lumber; or the foundation has not been painted with waterproofing, nor the scaffolding removed; or neither the boss nor the plans can be found.

The contractor usually does not have the men or the information necessary to clean things up, or figure what to do. Using valuable supplies for fill, or working in the wrong place, does not lead to friendly customer relations, nor does charging time for the machinery while it is parked waiting for a chance to work.

It sometimes seems that the owner's resentment of mistakes or lost time arising from such negligence on his part, increases with the extent of his fault.

On one such job, a contractor was induced to take a backhoe off a job just long enough to ditch for a septic tank and field at a house whose owner was to be there at eight in the morning with the plans. He did not appear. The contractor wanted to get back to the other job, so he went to the Board of Health, got specifications for the work, and dug for a tank and 240 feet of field—all deep ditching in hardpan.

That night the owner was very angry as he was sure that he had told the contractor that he had received a special dispensation to use only 100 feet of field. He not only did not want to pay for the surplus ditch—he wanted damages for the field that was

dug up. Eventually, after much argument, he made partial payment.

Owner Delays. An extra amount may be allowed on an estimate for excessive job delays caused by inadequate or contradictory plans, expectation of changes during the work, and owner meddling with work methods.

Such an extra charge may be based on inspection of plans, on the owner's reputation, or both.

Some owners are poor credit risks, and work may have to be slowed or stopped during the job because of lack of money.

Public highway contracts may have a provision that excavation must be stopped immediately in any area where prehistorical or historical ruins or objects are encountered, until the objects are checked and possibly removed by experts. Such stoppages can interfere seriously with orderly work on a project.

Other contractors may have jobs in the construction area, installing or relocating utilities, that may cause confusion and delay.

Production. Most estimators are familiar with the output of the machines to be used on jobs that they figure. If they are not, production can be determined from field studies, taken from manufacturers' charts, or worked up on paper from references in Chapter Two, in discussions of various classes of equipment in Chapters 13 through 21, and from other sources.

Allowance must always be made for special conditions that will affect machine performance. These are usually on the bad side—water, mud, cramped working areas, high altitude, steep grades, and so forth. But there are also favorable possibilities, such as light, easily dug soil, rock with good fragmentation, or expert operators.

Cost of Production. The cost of owning and operating the job equipment must be known, so that its production can be converted into cost figures. If a shovel can

OFF THE ROAD REAR DUMP
15 ton 10 yd. 200 HP

OWNERSHIP

Depreciation period 5 years - Annual use 1,200 hours - Total use 6,000 hours - No Salvage

Price at factory (Stubbs)	$26,650
Extras	472
Freight, 30,000 lbs.	525
Sales tax, 0 to 4%, assume 2%	533
Purchase price, including $2,550 tire value	$28,180
Average annual investment, 5 year, .6 x 28,180	16,908

Hourly depreciation, 6,000 hours, no salvage   4.69

Annual carrying charges, per cent of average annual inventory

Interest or return on investment	0 to 11.3%	assume	4.0
Insurance	5 to 2.5%	assume	1.5
Taxes	0 to 4.5%	assume	2.0
Storage	0 to 5.4%	assume	.5

.08

Hourly carrying charges $\dfrac{.08 \times 16,908}{1,200}$ =   1.13

Total ownership charges per hour   5.82

OPERATING

	\ TYPE OF SERVICE			
	Light	Medium	Heavy	Severe
Tire replacement, 3600, 2400, 1800, & 500 hrs.*	.33	.67	1.00	3.60
Tire repairs, 15% of replacement	.05	.10	.15	.54
Mechanical repairs & maintenance	1.80	2.26	2.82	4.52
Diesel fuel, 200 HP x .02, .035, .05 & .07 @ 18¢	.72	1.26	1.80	2.52
Lube, 1/3 of fuel	.24	.42	.60	.77
	3.14	4.71	6.37	11.95
Ownership cost, above	5.82	5.82	5.82	5.82
Total hourly non-labor costs	8.96	10.53	12.19	17.77
Labor, operator	4.00	4.00	4.00	4.00
Total cost per hour	12.96	14.53	16.19	21.77
Cost per minute, 45 minute hour	.287	.323	.36	.483
Cost per minute, 50 minute hour	.26	.29	.33	.435
Cost per ton body capacity per hour	.86	.96	1.08	1.44
Cost per yard capacity per hour	1.30	1.45	1.62	2.18

*Each replacement set assumed to include retreads and to cost ¾ of original or $1,912

Fig. 11-21. Hourly cost calculation, 1962 prices

load an average of 100 yards an hour after allowance for average delays, and all costs including operator are $20.00 an hour, the loading cost is 20¢ a yard.

The time the machine will be on the job is found by dividing its production into the volume of work. This same shovel would take 1,000 hours to move 100,000 yards of dirt. This is a year's work, $20,000 worth. Total yards divided into total cost gives unit cost again.

It is important to figure all side expenses such as supervision, spotting, pit maintenance, and incidental labor into each part of a job. Figure 11-21 is a sample of figuring costs of owning and operating a piece of equipment, without side costs.

Materials and Supplies. A general contractor or a building contractor may win or lose by the accuracy of his estimation of job materials and their price. There are three classes of material to be considered: those that become part of the structure, those that are necessary or desirable for its construction but are not part of it (scaffolds, forms for concrete, temporary supports, etc.), and material of either class that is broken, spoiled, surplus, or otherwise wasted without being used.

Bidders on highway contracts must at least allow for the cost of culverts and pavement material. In addition, they may build expensive bridges, supply and install fences, guard rails, electric conduit, and light poles.

Firm price commitments must be obtained from suppliers before it is safe to submit a bid.

Overhead. When each part of a project has been figured, the costs are added together. Overhead expense must then be added. It is made up of the part of general overhead that will be devoted to the job, and any additional overhead costs that are incurred for it. This may be figured out separately for each bid, or an arbitrary percentage of the cost total may be used. Ten per cent is usual.

Profit. Profit is what the contractor gets out of his work and risk if he has estimated properly, gets the job, and does it successfully. It is usually figured as a percentage of total estimated cost.

The contractor must decide on this amount for himself. If he puts it too high and doesn't get the job there won't be any profit on this one. If he puts it too low and gets the job, he may wish he hadn't. Five to ten per cent are often used. A combined figure of fifteen per cent for overhead and profit is standard in many areas.

Jobs are sometimes bid on a no-profit basis to keep money turning over so that bills and installments can be paid, or to keep an organization together in hope of profitable jobs in the future. But it should be remembered that in this business, the man who breaks even is usually losing money after hidden and delayed costs are counted up.

Casual Estimating. Small jobs are often estimated on an average time basis. A hoe or a dozer may be able to dig small house cellars in three to six hours. A contractor specializing in this work may figure that the machine is usually tied up for about a day on such a job, and base his rates accordingly. Estimates may be given on a yardage basis without inspection, upon the customer's assurance that the site is clear and workable, and with the understanding that an additional charge could be made if it is not, or if rock or other special conditions require extra work.

Excavating contractors may avoid some of the pitfalls of estimating by working on fixed prices charged by the hour, day, week, or month for their machinery. When working for another contractor, or under supervision of an engineer, no estimating may be required.

Prices obtained are determined primarily by the size and type of machine, but are also affected by the duration of the work, whether it is hard or easy, probable amount of idle time on the job, and the amount of supervision required.

In small private jobs, an approximate estimate of total cost is usually required, even when working by the hour. While this should not include a guarantee of doing the work within the figure stated, it is good business to have it fairly accurate.

CONTRACTS

Small jobs may be done on the basis of verbal agreements, that may be quite specific and definite (or very vague). Big jobs should always have a written agreement, that is usually in the form of a contract.

The contract describes the work that is to be done and the price that is to be paid for it. This may be done in two paragraphs or in a hundred pages. There are usually drawings or plans, that may be one sheet or several hundred. Standard forms should be used when possible.

If good faith exists on both sides, it is usually easy to arrange a simple contract between contractors, or between a contractor and a man who is familiar with the work involved.

In making arrangements with persons of little or no knowledge of excavating procedures, the greatest care should be taken to explain both what will be done and what will not be done.

Payment. Payment basis may be a lump sum or fixed price for the whole job, unit prices that vary with quantities, cost plus, or combinations of these methods.

Any type of contract may call for either a single payment, or installments based on the contractor's investment, work, and/or accomplishment. Monthly payments based on a percentage of work completed are usual in large jobs.

Lump Sum. In a straight lump sum or fixed price contract, the owner agrees to pay an agreed price for a certain piece of work. This is a good arrangement where all the factors that will affect the job are

known, but it must be based on a thorough understanding of the nature and finish of the work by both the owner and the contractor.

In a fixed price contract the contractor is on his own as long as he keeps to the job specifications and time schedule. He can reduce his measurement, classification, timekeeping, and bookkeeping to what he needs himself. While the prudent owner will still have an inspector on the job, he has a minimum of measurement and timekeeping to do.

However, unless advance engineering work and site study are very complete, such contracts may result in disagreeable surprises for either party. The owner may find that he has paid blasting price for a volume of rock that is readily broken out by a shovel; or for removal of valuable material that could have been dug at a profit. The contractor may find himself digging rock where he looked for loam, or running pumps 24 hours a day where he thought he would be high and dry.

A fixed price job that is turning out disastrously for the contractor can sometimes be re-negotiated, but unless the provisions for possible change are written into the contract he is largely dependent on the good will and generosity of the owner for such relief.

However, the contractor can demand extra payment if the unfavorable conditions were known to and concealed by the owner, or where the owner withheld information that would have enabled the contractor to anticipate the difficulties.

Unit Prices. When quantities have not been determined exactly, or when they may be subject to considerable change during the job, parts of a contract or a whole contract may be let at unit rates.

For example, an owner might ask for bids on removing a hill of approximately 30,000 yards of dirt. He lets the job to a contractor who bids 60¢ a yard. The hill is measured before work is started, at intervals during the work, and after the job is complete. It is found that 37,000 yards have been moved. Payment to the contractor is .60 × 37,000, or $22,200.

If the contract had been let on a lump sum basis of the estimated yardage of 30,000 times .60, payment would be $18,-000. But the contractor would be likely to claim that the 30,000 figure was not honest, and disagreements and even lawsuits might be caused. On the unit basis the owner pays for just the volume that is moved, whether it be more or less than his estimate.

Unit prices reduce the requirement for careful pre-job investigations that can be very expensive if underground conditions are involved. On the other hand, measurement of quantities is difficult and sometimes inaccurate when cuts are shallow and the ground is irregular.

Quantities can also be measured by truck load or by measurement of fill.

Unit prices for earth and rock are usually based on cubic yards, trenches on linear feet or occasionally linear yards, and clearing on acres.

Unit price measurement is usually based on cubic yards for excavation, topsoil, and concrete; linear feet for trenches, curbs, pipes, and fences; acres for clearing; and square yards for rip rap and sodding.

Classified Excavation. On a big job that involves various digging conditions, excavation may be divided into a number of different classifications. These may be separated according to the type of work, as road cut, borrow, shallow trench, culvert, and deep trench. Or the classification may be according to difficulty of digging, as soil or rock, or dry or wet.

The most important classifications in regard to total money involved, and problems in estimating, bidding, and working, are earth and rock.

The practical distinction is that soil can be dug directly by shovels of normal size

for the job, while rock must be blasted or ripped before it is dug. The pay difference may be made on this basis, or according to geologic definitions.

It is fairly standard practice for the contractor to remove all or most of the soil over rock, then send for the owner to inspect and measure the rock for payment. Sometimes the two parties will agree on the amount of rock before excavation, depending on inspection of outcrops and depth of soil in test holes for the amounts.

Boulders are measured after they are freed from the bank, and before they are broken or loaded out.

In tunnels, and in some trenches and road cuts, payment for rock may be varied according to its position. Full price may be paid inside the bore, side, or slope lines, a lesser price for moderate overbreak, and nothing for excessive overcutting.

Numerous problems arise in connection with identifying and measuring rock. Many engineers and public works departments prefer to avoid them by letting excavation work on an unclassified basis. The contractor is given access to whatever boring and test hole data is available, allowed to look over the ground, and makes his own estimate or perhaps guess about how much rock he will find. This method diminishes risk for the owner and increases it for the contractor.

Excavation prices usually include hauling to the fill and compacting. In highway work in some areas the haul distance is limited to a few hundred yards, beyond which an item called overhaul or paid haul calls for additional payment.

Cost-Plus. If conditions are such that the contractor cannot readily tell the amount, kind, or conditions of excavation; if the amount of work to be done has not been determined; if it is not practical to clearly define the extent of the work and the condition in which it is to be left; or if the job is to be done a little at a time, as equipment or funds are available, the cost-plus or hourly basis will probably be the most satisfactory.

On a cost-plus arrangement the contractor will have all his costs in doing the work repaid, and will receive either a fixed fee or a percentage of his costs in addition. This type of contract is most often let in government or other work where haste prevents thorough investigation of the site, or plans are subject to change during operations.

The fixed fee basis is appropriate where the total amount of work can be estimated with fair accuracy, and the percentage where changes and extras can make up a substantial part of the job. The latter system is subject to grave abuses, as mistakes which add to the cost will increase the profit, so that inefficiency is rewarded.

A serious cost-plus difficulty is that it is apt to lead the customer to interfere with the contractor's policies and management on the job. This effort to lower costs is liable to be of the penny-wise pound-foolish variety, and increases expenses more often than it reduces them.

It is important for the contractor to include all indirect as well as direct costs in this type of bid.

A variation of cost-plus is a bid listing hourly or daily rates for all machines, services, and personnel to be employed on the job. The contractor figures his profit on each unit into the price charged for it.

Hourly Work. Working by the hour is almost the standard practice on small jobs in which the expense of investigations by the owner and estimating by the contractor are not justified by the money involved. It is also common in subcontracts and other arrangements between contractors.

A working hour may be considered to be the time that the machine is present on the job, the time it is present and ready to work, or only the time it is actively working, depending on the arrangements made.

In effect, the contractor who owns the equipment is renting it to his customer, but he usually retains the right to supervise, and pays all expenses, including the operator, fuel, lubricants, and repairs. Occasionally, the customer may furnish fuel or other items if he can do so more conveniently than the owner can.

If equipment is rented to a job without an operator it is a rental rather than a working agreement.

Pay for time during which the machine is stuck in mud is usually on the lessee, as it is a mishap caused by job conditions. However, if the fault lies with a disobedient or careless operator, or if the owner has warranted that the machine will not get stuck on that job, payment may be withheld.

The machine is not paid for time lost because of mechanical failure or absence of the operator. However, stops for adjustments, minor repairs, fueling, lubrication, or cigarettes, which average less than ten minutes an hour, may be considered working time if agreement is made to that effect.

Timing. Working or pay time may be taken from readings of electric hour meters, which register the time the engine is running; from mechanical counters which register engine revolutions in terms of hours of wide-open operation; from special checking by a foreman or timekeeper, from the lessee's job time sheets, or from the owner's pay roll records.

On operator work, it is good practice to check time daily and have the customer sign a ticket for it.

Timing by hour meter leads to owners' operators keeping the engine running, whether it is needed or not. Many jobs involve substantial amounts of waiting time, during which noise and wear would be reduced by stopping the engine, but this action by the operator would penalize his employer and possibly himself.

Hour meters should be checked frequently as they may become disconnected.

Unclassified Excavation Item

Stations	Total Usable Excavation	Embankment Fill Required	Deficiency	Excess	Balance
1 +00 − 20 +00	113,784	54,640		59,144	59,144
20 +00 − 40 +00	204,365	224,510	20,145		38,999
40 +00 − 60 +00	14,611	291,766	277,155		−238,156
60 +00 − 80 +00	310,444	15,193		295,251	57,095
80 +00 − 100 +00	40,043	52,607	12,564		44,531
100 +00 − 120 +00	16,633	189,312	172,679		−128,148
120 +00 − 140 +00	12,200	121,241	109,041		−237,189
140 +00 − 158 +60	382,988	145,807		237,181	− 8
	1,095,068	1,095,076			

Fig. 11-22. Calculated net yardage table

Engine revolution counters are more accurate, and seldom get out of order, but when used as a pay basis, offer the added disadvantage of placing a premium on running the engine at full throttle at all times. This may make it difficult to do precise or fine work and will cause excessive wear, waste, and noise.

Jokers. Many contracts are tricky and can be used to make the contractor do his work for part pay, or to take the responsibility for conditions beyond his control.

It is customary to make payments on account on jobs which take over a few weeks to complete, so that the contractor will not have to scratch for payroll and immediate expense money. In many types of work, it is usual for the owner to withhold a percentage, which may range from five to fifty percent of the value of work performed, as security for completion of the job and fulfillment of any guarantees.

Such contracts may leave the owner the option of not completing the job, thus withholding final payment indefinitely. For example, a contractor might bid on a development job of installing sewers, back-filling the ditches, laying gravel roads, and then blacktopping. This is one job, and 10 percent withheld from payments is not due until the blacktop is completed.

But the developer can sell his houses when he has the gravel in. He may lack the money to put on blacktop, or just figure it is a good idea not to do it, and hold on to the percentage. He may be able to do this under the contract, unless it specifies that each operation calls for a separate final settlement, or that all work may be performed in proper sequence, without specific authorization.

Another trick is to exact a guarantee of quality while specifying material or methods which may be sub-quality. If roads are to be constructed of bank-run gravel from a developer's pit, the contractor should not guarantee that they will be passable after a rain unless he is given the option of rejecting part or all of the material if it is substandard, and importing better gravel at additional cost.

On contracts which call for penalties for failure to complete by a certain date, the contractor should be protected against delays caused by the owner. These may include failure to complete prior operations, or to remove surplus material left from them; or not supplying plans, grade stakes, work permits, or access to property on time.

The contractor should also protect himself against shortage of materials, by means of delay-caused-by-circumstances-beyond-control clauses, option of substituting available for unavailable items, or both.

However, the greatest losses to contractors occur because of forgetting to leave a loophole for possible underground conditions. The big three are rock, mud, and flowing water, and one or all of them can crop up in most unexpected places.

Highway contracts may call for compaction of fills to specified density, that can only be reached if the soil contains just the right quantity of water. It may be impossible to reach this density if both the cuts and the weather are wet.

HIGHWAY CONTRACTS

Specifications. A state public works department or other agency responsible for highway construction usually has a book of specifications, available to engineers, contractors, and others at nominal cost.

Such a book constitutes the basic specifications, regulations and instructions for every highway contract. The estimator and contractor must be familiar with it in order to bid intelligently.

Usually, the book is revised only at long intervals, and it is not practical to make it so complete as to apply to every possible condition. It is therefore usual to make up a set of job specifications for each contract. These may serve to modernize or correct the material in the book, or to cover an unusual piece of work such as underpinning a structure or moving an aqueduct.

The job specifications may also give information about minimum and prevailing labor rates, the weights, qualities and perhaps sources of supply for materials, the provisions for closing roads or for maintaining traffic, and other matters pertaining to this one job.

The job specifications usually govern where there is any conflict with the book. However, both authorities leave a wide range of work to be "done to the satisfaction of the Engineer" and the Engineer should be consulted whenever there is doubt.

Plans. A complete set of plans is made up for the project. The originals are on tracing cloth, and blueprint copies are made for purchase or rental by prospective bidders. Reduced scale copies may be printed also, for easier reference. The scale is usually not corrected on the smaller copies. If the reduction is one half, the estimator must double each apparent dimension. Serious error may occur if this is neglected, so the full scale prints should be used for all important work.

There are many ways to draw up a set of plans, so the estimator must know how to hunt for what he wants. There may also be additional information not on the plans that is kept at the public works office for inspection by interested parties.

The first sheets usually show the proposed road in a plan—that is, in the form of a map drawn to scale—that indicates all construction, and usually all buildings to be removed. Property lines, existing roads, stone walls, landmarks and surveyors bench marks are shown.

If the project is a large one, these maps will be in several sections, and they may also be included twice—the first time on a small scale to give a general idea, and again later on a larger scale and in more detail.

The map or other sheets carry a profile showing both the present ground surface and the proposed road grade along the highway centerline. Where the road is to be below the present surface, there is a cut; where it is higher a fill is needed. Loca-

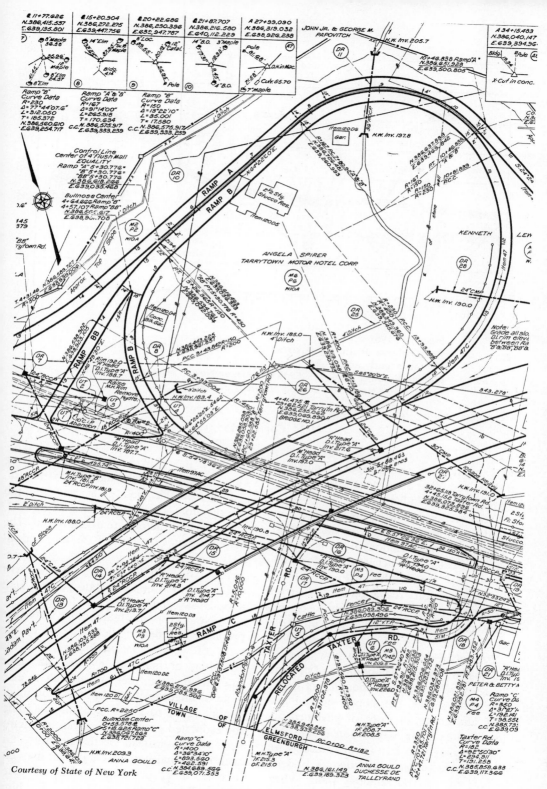

Courtesy of State of New York

Fig. 11-23. Detail plan of part of interchange

11-53

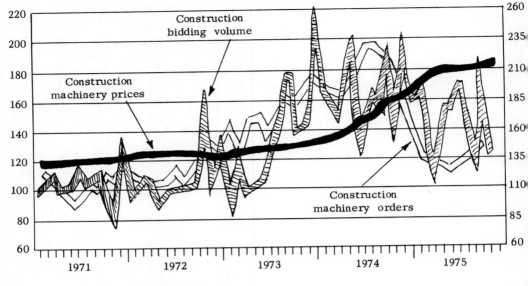

Indexes: 1967 = 100
New Orders: including exports, McGraw-Hill Economics Dept.,
     seasonally adjusted
Bidding Volume: source: CM&E Business News, seasonally adjusted
Machinery Prices: US Bureau of Labor Statistics

Courtesy of Construction Methods and Equipment

Fig. 11-24. Construction trends

tions of present and proposed culverts, bridges, and other cross structures or drainageways are also indicated.

Cross sections showing present and proposed levels may be given, or may not. If there are no cross sections there is usually a table of calculated net yardages for various sections, such as that in Figure 11-22.

If rock and soil excavation are to be paid at different prices there is likely to be another profile line showing the probable depth of soil.

Typical cross sections show how the road is adjusted to various conditions of cut and fill. These show width of shoulder, of gutter if any, fill slopes, cut slopes ion soil and rock, and crown or slope of pavement. A simplified example is shown in Figure 8-3.

There may also be typical drawings for standard structures such as culverts, bridges, tunnels, ramps, and retaining walls. Wherever a structure differs from the typical ex-

ample, it has individual drawings showing either the differences or the whole structure

Proposal. The job plans and the job specifications are made up into a Proposal on which contractors are asked to bid. The State estimates the cost of the job, and all bids for a greater amount are usually rejected. The estimate is published as part of the proposal in some states, and is kept a top level secret in others.

In New York and many other states the proposal includes an itemized list of the types of work to be performed, and the quantities of material involved. In some cases where quantities are difficult to estimate or might form the basis for too much argument, or to avoid weighting of bids the estimated cost may be given in a lump sum (L. S.) instead of the quantity. On such individual items a contractor may bid as much as 25 per cent over the estimate but a still higher bid would disqualify him

Lump sum items may include land clearing, furnishing water wagons, pile drivers, and other limited use equipment, and maintenance and protection of road traffic.

Unclassified excavation is usually on a price-per-cubic-yard basis. The proposal contains an estimate of the number of yards, say 130,000. If the contractor figures he can dig, move, and compact the soil at a profit for 60¢ a yard, he will bid $78,000. If measurements of the completed job show he actually moved 150,000 yards he will be paid $90,000; if only 110,000 he will receive only $66,000.

Most items that can be measured are listed in this way, in terms of cubic yards (C.Y.), linear feet (L.F.), tons, barrels, acres, or square yards (S.Y.) In each case the contractor will be paid on the basis of his unit price times the quantity that he actually moved, provided, or processed.

This arrangement allows flexibility in making changes while the work is in progress. Correction of errors in original figuring and changes to meet field conditions can be made without re-negotiation of the contract.

After a price is set opposite each item on the proposal, the prices are totaled to obtain the bid price for the whole job. This is the figure that usually decides who gets the contract.

Additional Information. Engineering study of a job is likely to include borings to determine subsurface conditions, particularly the nature and bearing qualities of the soil, and the presence of ground water or rock. The results of such borings may or may not be shown on the plans, although they are taken into account in estimates of cost.

A contractor who does not find borings in the proposal, or who wants more detailed information than that supplied, can usually look at the drillers' logs and other first hand information at the proper Public Works office.

There he may also find detailed topographic maps of the work area, cross sections from which cut and fill were taken, notes about the presence of unsuitable soil, and other useful data.

Weighted Bids. A contractor who does not bid according to his estimate is said to weight his bid. He may increase or decrease his total bid according to the amount of competition and his need for the work, but weighting usually means juggling item prices to obtain special advantages.

A contractor may estimate $10,000 for clearing. If not restrained by the form of proposal, he might bid $50,000 for clearing, and lower various other prices to make up for the $40,000. In this way he would get some extra money early in the job, when he might expect to need it most.

He might estimate dirt excavation at 50¢ a yard and rock at $2.50. The proposal could give quantities at 90,000 yards of dirt and 10,000 of rock, on which he would normally bid $45,000 plus $25,000, a total of $70,000. But if his personal knowledge of the route indicated that it was more likely there would be 70,000 yards of dirt and 30,000 of rock, he might weight his bid to charge more for rock and less for dirt.

He might bid the dirt at 40¢ and the rock at $3.40, for the same total of $70,000. If his judgment were correct about the quantity of rock he would be paid $140,000, as against the $110,000 he would have for the same work if he had bid according to his estimate.

Public works departments do not like weighted bids, although some of them are quite tolerant of high land clearing charges. A weighted bid of the type just described can add greatly to the cost of a job, and place the state engineer in an embarrassing position. An entire bid may be rejected if any item is too far out of line, so the contractor cannot afford to let his weighting become obvious.

Re-Negotiation. A contract may have a

Fig. 11-25. Simple arrow diagram

provision for negotiating a new price on any item whose quantity is so far away from the estimate that a problem is created.

In the excavation job above only 500 yards of rock might be found. The contractor's cost per yard on such a quantity would be much greater than on the 10,000 yards in the proposal, so that he might obtain an extra payment per yard. The State would still save money because of the higher proportion of low priced dirt in the cut.

On the other hand, in a job estimated to have only 200 yards of rock the contractor might figure $5.00 or even $10.00 a yard. If 10,000 yards of rock were found, the State would try to reduce the price to one more reasonable for large scale work.

A too-severe specification may be modified either officially or unofficially. This is often necessary for the contractor's survival.

CRITICAL PATH SCHEDULING

Critical path scheduling is a relatively new tool for project planning, scheduling, and cost study. It was developed by J. E. Kelley, Jr. and Morgan B. Walker, and is often called the Kelley-Walker Critical Path scheduling method. This article is based on

information supplied by Mr. H. S. Dhillon of the Perini Corporation.

The U. S. Navy has a parallel system, known as PERT (Performance Evaluation Reporting Technique) that also uses a network model, but differs in many respects.

These systems permit visualizing projects, study, and working out sequences, time, and costs more readily than is possible with bar graphs.

Most projects include one or more jobs or job sequences that must be completed before another phase of the work can be begun. For example, one sequence may be first clearing then trenching a culvert site, another procuring and bringing in specially designed pipe. Completion of these two sequences is necessary before pipe can be laid.

If one job or one sequence takes longer than the others leading to the same result, its time determines the time for achieving that result, whether it is starting the next phase or finishing the project.

Because of its important effect on work scheduling, the operation or sequence taking the longest time is called the critical path.

Vocabulary. Critical path scheduling has been set up in a somewhat formal manner as to vocabulary and format, to enable its users to understand each other, and to permit solving its more intricate problems by means of computers.

For purposes of this work, the following words are limited to the meanings listed for them:

Chain: a sequence of jobs following each other

Crash: speedup or rush work

Duration: the time required by a job

Event: the start or finish of a job or jobs

Float: time available for a job, minus job duration

Job: one small activity or single class of work

A. LANDSCAPE GRADING   C. SPREAD TOPSOIL
B. LAY ELECTRICAL CONDUIT   D. MAKE ELECTRICAL CONNECTIONS

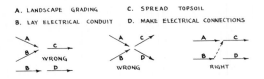

Fig. 11-26. Dummy arrow

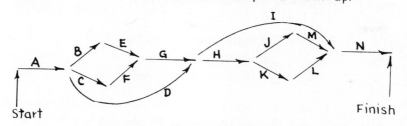

A. Decide to change.   F. Remove hub cap.   K. Lower car.
B. Take out jack.   G. Remove wheel.   L. Replace jack in trunk.
C. Take out wrench.   H. Replace wheel.
D. Take out spare.   I. Flat in trunk.   M. Put wrench in trunk
E. Raise car.   J. Replace hub cap. N. Clean up.

Start   Finish

Fig. 11-27. Three men changing a tire

Lead time: time between getting a job and starting work

Path: same as chain

Project: a group or series of related jobs

Arrow Diagram. Critical path schedules are worked out in arrow diagrams, the simplest form of which is shown in Figure 11-25. Each arrow represents a job or activity. It may be labeled by description as in this illustration, by code letters, or by an event numbering system. The first arrow is usually for lead time, getting ready to start work. An arrow may be added for cleanup.

Arrows are made in any convenient length, and may be straight or curved. They indicate only the sequence of the jobs in the pattern of a project, and have no scale.

Three questions should be asked and answered about each arrow:

1. What immediately precedes this job?
2. What immediately follows this job?
3. What can be concurrent with this job?

The arrow diagram must be worked out logically and thoroughly in regard to sequence and interdependence of jobs. Omission of any item gives a false picture and may lead to mistakes in scheduling. On the other hand, the simple act of placing each activity in a frame of reference with other jobs helps in building up an intimate knowledge of the project.

Dummy arrows are dash or dotted lines instead of solid. They are used where two jobs, say light landscape grading and laying of electrical conduit, must both precede spreading topsoil; but only the conduit is necessary to start making electrical connections to lamp poles.

In the left diagram in Figure 11-26 the B arrow for conduit is shown twice. The center diagram gives the false impression that grading is necessary to make electrical hookup. At the right, the dummy arrow correctly shows that S (topsoiling) cannot start until B is completed, without breaking the single line sequence from B to D.

Dummy arrows are also used to avoid duplication in event-numbering of solid arrows, as we will see below.

Detail. An arrow diagram may be drawn with as much or little detail as is desired. Figure 11-27 shows a possible arrangement of jobs when three men change a tire on a car. This could be made much more detailed perhaps showing the removal of each lug nut. On the other hand, in the office of a concern operating 30 or 40 cars the whole operation might be shown only as a single arrow.

If the tire were changed by one man the arrow diagram would be linear, that is, it would have only a single chain of jobs.

A big project is likely to be represented by a rather simple master diagram for general

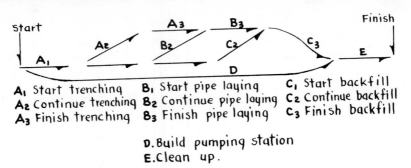

A₁ Start trenching   B₁ Start pipe laying   C₁ Start backfill
A₂ Continue trenching B₂ Continue pipe laying C₂ Continue backfill
A₃ Finish trenching   B₃ Finish pipe laying   C₃ Finish backfill

D. Build pumping station
E. Clean up.

Fig. 11-28. Overlapping jobs

reference, with separate detailed break-downs of its parts for close study and the use of various departments. A single arrow on the master sheet may be a 50 arrow diagram in the structures department analysis.

The expansion of a job into a diagram must be done with as much care as the construction of the original master diagram, to avoid omissions and wrong sequences.

Overlapping Jobs. It is usual for construction projects to have overlapping sequences. Brush clearing comes long before laying pavement, yet the two operations may go on at the same time in different parts of a highway section.

As an example, let us take laying a pipeline. On a simplified diagram this may be broken down into three activities that must be done in succession; trenching, laying pipe, and backfilling. A work section of a pipeline may be many miles long, but it may be possible to start each job as soon as a few hundred feet of the previous job are completed.

Each of these jobs may be considered to be done in three sections; the initial, continuing, and finish. Initial work must be completed before the next job can start, while the other two run at the same time in different areas. Figure 11-28 shows the arrangement of arrows to indicate this situation. Note that all of them are the same length, although initial work may take a day or less, while continuing work may go on for months.

The pipeline may require a pumping station. Building it would be a very different type of work that might be subcontracted, or diagrammed separately. However, the line is not usable without it, so it should be

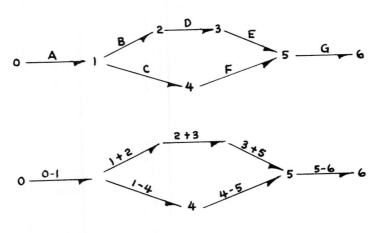

Fig. 11-29. Event numbers

represented in this master diagram as the lower arrow, D.

Events. The start and finish of every job is called an event. Therefore every arrow begins at an event and ends at one. These events are numbered, starting with one or zero at the beginning of the project, and continuing through an unbroken sequence of numbers to the finish. However, because of concurrent or parallel job chains, the numbers are not necessarily in sequence in any one chain. The only absolute rule in assigning these numbers is that the number at the head of an arrow must always be larger than that at its tail.

Event numbers are used to identify the arrows between them. In Figure 11-29, A is 0-1, B is 1-2, C is 1-4, and so forth. Since diagrams often include sufficient arrows to use the alphabet many times over, identification by event numbers is more practical than letter codes. It is also necessary when problems are to be handled by a computer.

Sometimes two or more jobs will begin and end at the same event. A borrow pit may require clearing, testing, and measuring before digging starts. In Figure 11-30 the three arrows B, C, and D in the top diagram would each be designated 1-2. This duplication is avoided by introducing a dummy, as shown in the two lower illustrations. The dummy may be either before or after the arrow. The junction between the arrow and the dummy is given an event number.

The rest of the illustrations in this section will use letter codes instead of event numbers, to avoid confusion with other figures. A real working diagram is drawn on a scale large enough to put in all necessary figures without crowding them.

Duration. When an arrow diagram has been completed and checked, the time that the job is expected to take is written under each arrow. This may be in hours, days, months, or any appropriate unit, but the same measurement must be used all the way through a diagram. If days are used, they

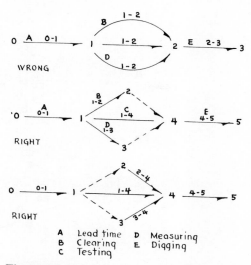

A Lead time D Measuring
B Clearing E Digging
C Testing

Fig. 11-30. Dummy arrows for identification

must be working rather than calendar days, to avoid confusing calculations with holidays and weekends.

The duration assigned is first a normal or average time, taken from experience, job studies, or an estimator's figures. In an ordinary construction project it would involve one-shift operation and use of equipment on hand or readily obtained, without either rush pressure or deliberate waste of time.

Dummy arrows always have a zero duration, as they are only symbols to show connection between jobs.

The top diagram in Figure 11-31 shows duration times.

Event Times. An event time is the sum of the duration of the jobs that precede the event, and represents the time that will elapse between the start of the project and that event. If two or more chains or sequences of jobs are needed to make the event possible, its time is determined by the slowest path.

The earliest event time, abbreviated EET, or sometimes as e.t., is defined as the earliest finish of the event by the slowest path. It is indicated on arrow diagrams by a number inside a square, and is worked out for

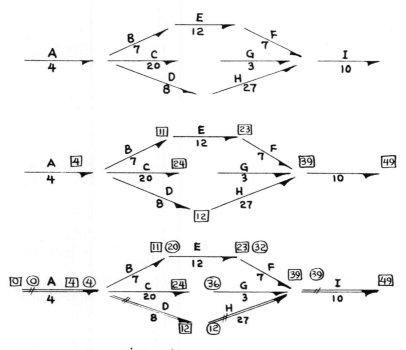

Duration figures below arrows
Earliest event times in squares
Latest event times in circles.

Fig. 11-31. Earliest and latest event times

each event in a diagram, as in the middle drawing.

Working out the EET is simply a matter of addition until a junction of two arrow heads is reached. Here both preceding chains are figured, and the larger number is used. In the illustration, the three chains leading up to be beginning of I add up to 30, 27, and 39, so 39 is used.

The earliest event time calculation shows how long the job will take under ordinary conditions.

The latest event time, abbreviated LET or l.t., is the latest time at which an event can be finished without delaying completion of the project. It is found by working back from the earliest event time for completion, along the slowest path. It is written in a

circle, alongside the square containing the EET, as shown in the bottom diagram of 11-31.

This is simple subtraction, working backward from the finish, except at a junction of arrow tails, where the smaller number is used. In the illustration for finish event of Job A, the three chains show LETs of 13, 16, and 4, so 4 is selected and written down.

The latest event times provide a quick method for determining what chain of jobs sets the time for the whole project, and is therefore its critical path.

Critical Path. The critical path is a sequence made up of one or more jobs or job series, whose duration is the determining factor in the length of the whole project. In order to be critical a job or series must conform to all of the following requirements:

1. EET and LET must be equal at the start

2. EET and LET must be equal at the finish

3. The time available for the job must be equal to its duration.

The time available for each job is found by subtracting the starting EET from the finish LET.

In Figure 11-31 the critical path is ADHI. F is not a critical job because the starting figures, 23 and 32, are different. G begins with an EET of 24 and a LET of 36. One non-critical job prevents a series from being critical.

As critical jobs are identified they are marked with a double hash stripe. When the critical path is worked out it may be emphasized by making the arrows heavier or with double lines, or by color.

There may be two or more critical paths, in which case each of them is marked.

The critical path ADHI determines the overall time of 34 days. Any efforts to save time and shorten the job should be first concentrated on it.

Float. The spare time in the quicker jobs, series, or paths is called float. In Figure 11-32 the critical path is ADEF, with a total

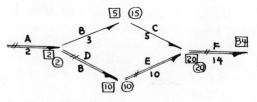

Fig. 11-32. Critical path

duration of 34. The alternate path, ABCF, would allow completion in 24 days if jobs D and E were not needed or could be speeded up. This also means that B could be started 10 days after the completion of A, without delaying project completion. This 10 days is float.

From a scheduling standpoint, float time may be regarded as waste time. It indicates a possibility of idle time for the men and equipment doing the jobs, and it presents a problem in utilizing them to speed up critical work.

For example, our illustration might represent grading for a highway, with a small cut, B clearing and C digging, and a larger one, D clearing and E digging. The original durations might have been assigned on the basis of an equal force in each cut. By taking men and equipment from the small job and assigning them to the critical path, the two cuts might be done in 13 days each, permitting bluetop work, F, 5 days earlier.

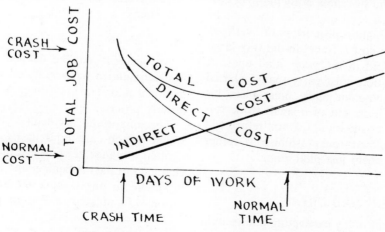

Fig. 11-33. Speedup and cost relationship

This is one way of using float time. It often happens that the same crews cannot work on the two jobs. Then an effort is made to shorten the critical path durations in other ways, to squeeze some or all of the float out of the faster series.

Crashing. Rushing a job through by an intensified effort that involves a substantial increase in costs is called crashing. Some jobs can be shortened by a big percentage at moderate cost, others respond poorly to unlimited extra expenditure.

A usual relationship between job duration and job cost is shown by the lower curve in Figure 11-33. Extending time beyond normal duration saves little money, and pouring in money after a saturation point saves very little time. The greatest gains in time for extra dollars spent occurs in the first few days saved, and the smallest gains per dollar are found near the minimum time end of the curve.

This curve is for direct costs only. Overhead or indirect costs are likely to be about the same per day whether the job is crashed or allowed to sleep. The straight line shows these, and the upper curve shows a total cost against project duration. The low point in this upper curve occurs at the most efficient time for the job.

The shape and pitch of the direct cost curve, and the steepness of the indirect cost line vary with each contractor and job. In general, the highest proportion of overhead cost to direct cost is found in the very large companies and in the very small ones. A big organization must carry many salaried people, a one-man outfit must meet his daily living costs out of a small work volume. In either situation, the best return will often be obtained from crashing jobs, rather than letting them just plod along.

CAUSES OF FAILURE

Every year many excavating and general contractors fail, or sustain losses that force

them to operate on a reduced scale, or give up. Most of the failures arise from one or more of the following causes:

Unforeseen price rises.
Abnormal labor cost.
Abnormal equipment breakage.
Death or disability of owner or key men.
Fire not adequately insured.
Liability or property damage not adequately insured.
Poor accident record.
Failure of subcontractors.
Adverse weather.
Unforeseen subsurface difficulties.
Faulty credit judgment.
Sudden restriction or withdrawal of credit.
Unavailability of materials.
Taking on too much work for financial resources.
Taking on too much work for adequate supervision.
Speculation.
Diversion of funds to non-business use.
Embezzlement by employees.

Some of these subjects have been discussed previously, others are of a general business nature and are too complex for discussion here.

Two subjects of particular importance to the excavator, however, are accidents and insurance.

ACCIDENTS

An accident may be defined as an unforeseen sudden happening, or as an unintentional and damaging interruption in an orderly process.

The important accidents are those in which persons are injured. However, this is often a matter of chance rather than the character of the happening, and an accident in which no one is hurt should be taken seriously, and steps taken to prevent its recurrence.

Employees should be protected by workmen's compensation insurance. This coverage is usually required by state law, but in any case is a MUST for any employer who is interested in the welfare of his employees, and in his own. Non-employees and property of others should be protected by liability and property damage insurance, lack of which can wipe out a prosperous business overnight. A contractor can protect his own equipment and property with fire and damage insurance.

However, the possession of full insurance does not justify the slightest negligence in regard to accident prevention. For one thing, the best insurance will only pay the more obvious costs. In small accidents that are most common, indirect uninsurable costs may run five times as high as the payments under compensation.

Some of these expenses and losses are:

1. Increase in insurance rates.
2. Payment to injured employee of wages for period too short for compensation.
3. Loss of time of other employees who stop work at the time of the accident and because of it.
4. Time spent by foremen and supervisors in assisting injured man, investigating the cause, selecting and briefing or training another man for the job, and preparing accident reports and attending hearings.
5. Slowdown of job, with possible failure to finish by deadline.
6. Paying full wages to employees who return to work before being capable of performing full duties.
7. Loss of chance for profit on man and his machine.
8. Lowering of morale of other men on the job.
9. Possible interference with work methods by public officials.
10. Unfavorable newspaper and other publicity.

Prevention. The first rule in accident prevention is to use common sense—in laying out a job, assigning machines and personnel to their duties, providing adequate supervision without fussiness, and in setting up sensible and reasonable safety rules.

Too many safety rules may be worse than none. Everyone of us has a limit in the amount of good advice we can absorb, and the limit is often painfully low. It is better to take a few important points at a time, and hammer them home, than to prepare long lists that will neither be read nor remembered.

Rules should be reasonably close to prevailing practice whenever possible, and should aim at greater rather than at lesser evils. A city whose streets are littered with newspapers and garbage should not start a cleanup campaign by arresting men for dropping cigarette butts. (But this is what the biggest city of them all has done.)

Enforcement of safety rules should not be so strict as to cause men to fail to report for first aid for minor accidents, or to lie about the way in which they occurred.

Excellent posters and leaflets can be obtained from insurance companies and safety councils, and when used in moderation bring very good results. Only those that have some bearing on the work should be selected.

Workers' suggestions, both on safety and other matters, should be encouraged and acted upon.

A worker's skill should not be taken for granted. In an emergency an unfamiliar machine might trick an experienced operator into the wrong move. Judgment should be used in giving out ticklish assignments. Training and refresher programs should be given periodically and whenever needed, and reference material on proper operation and procedures should be available.

Good housekeeping is important. Piles of junk, material, litter, boards with projecting nails, carelessly piled bags of material or heavy parts, and accumulations of grease and dirt cause accidents directly, and also indirectly by encouraging sloppy work attitudes.

Crowding causes accidents. On a rush job a boss tends to jam as many machines and men into the work area as it will take without bulging. That may mean collisions, and collisions lose time. A man can dig a ditch faster alone than with a helper who hits him on the head with a pick.

Piling Materials. High piles are dangerous piles, except loose material lying at its angle of repose. High piled bags or boxes may look all right, but only until something happens, and that doesn't have to be an earthquake. A truck running into or hooking a pile, or material leaking from a bottom bag, may bring a whole stack down, and the higher they are, the harder they fall.

Piles of bags or pieces of any height should be crosspiled, braced at the ends, and stepped back from the bottom.

In excavating, even a shallow ditch can injure a man seriously by caving, and deep ones are killers. High vertical faces around a cellar excavation might stay up, but it is safer not to trust them. Shore them up, and make sure the shoring is strong enough. Don't just guess, have it designed and inspected by an experienced and careful man.

Barricades. It is not only the workmen who must be kept out of accidents, but also the public. There are sidewalk superintendents who like to watch the work, and are apt to be foolish enough to fall into it if they have the chance. If there is an attractive danger spot, like a cellar excavation in a city, they must be fenced out. Such a fence should be strong, and at least seven feet high.

The fence or barricade must be secure itself, so that it will not fall into the excavation, or be left partly in space by a slide. It should have windows or peep holes in it. They build good will for the contractor, and make spectators less likely to move into the very dangerous truck drives that penetrate the fencing.

Barricades, signs, and flares can hardly be overdone on roadways. Any excavation that extends into a road, and particularly into a high speed highway, is just asking for trouble. And it is not enough to mark it so well that only one in a thousand would fail to notice it—10,000 cars might pass while it is open. And the police, the lawyers, and the newspapers will not be interested in the 9999 who didn't crack up in it. Just in the one who did.

Insect Stings. Clearing and excavating brings men into painful contact with hornets, yellow jackets, and other stinging insects so often that it is one of the special risks of the business.

While in most cases no serious injury results, such stings can be more dangerous than is commonly realized, and they cause a number of deaths every year.

They respond excellently to proper and early treatment.

There are three dangers:

Allergy to the injected poison, which will cause exaggerated reactions, and if very severe may result in shock or death from a single sting.

Stings close to the eyes or other vulnerable parts, that may disable a normally sensitive person.

Multiple stings from a swarm of insects may produce serious poisoning.

Most trouble comes from unexpected contacts. Preliminary scouting of an area on foot may reveal the location of nests, particularly of hornets on branches.

When possible, such nests should be destroyed in advance of the work. This can be done at night with little danger, as the insects are then sluggish and nearly blind. Also, as they are all nested, a 100% kill may be effected.

Ground nests are eliminated by pouring ¼ or ½ cup of insecticide down the hole, then tamping dirt in the top.

Paper hornet nests should be wrapped in wire screening, mosquito netting, or cloth, cut off the branch, and burned on a hot fire or kept under water for at least 48 hours.

The man doing this job can be protected by heavy clothing, gauntlet leather gloves, a hat or helmet, and a head-protecting mosquito net. The last item is the most important, as face stings are painful and dangerous.

If it is necessary to work among ground nests that have not been treated, they should be completely destroyed by pushing out or deep burial on the first approach. The insects are then disorganized and less likely to attack, particularly if the machine is kept in motion.

Minimum protection for operators in a danger area is a head net.

A man known to be particularly sensitive to stings should be kept on safer work until he can be desensitized to the poison by a series of shots.

Treatment. Treatment consists of stopping the swelling, slowing absorption of poison into the system, and stimulation to help to overcome its effects.

Three minims (a minim is 1/15 of a cubic centimeter) of epinephrine, divided among two or more shallow injections at the edge of the swelling, will constrict the blood vessels, stop enlargement of the swelling, and wall off the poison. This treatment should be a routine precaution for any sting near the eyes.

A dose of the same size injected in the upper arm rallies the system for defense. If no "lift" is felt the arm injection can be repeated in ten minutes, or sooner if the patient is unconscious.

These injections are made much more effective by addition of equal amounts of Chlor-Trimeton (strong solution) or some other injectible antihistamine to epinephrine before injection.

Ordinarily, injections can be made only by a doctor or a nurse. Sometimes it is possible to obtain bee sting kits including automatic injectors, for lay use in emergencies before medical help can be obtained.

A cortisone-like drug such as Prednisone (USV Pharmaceutical Corp.) may prevent or reduce allergic reaction to stings. A usual dose is two 5-milligram tablets immediately, followed by one every four hours if needed.

It may be supplemented by one or two 4-mg tablets of an anti-histamine such as Chlor-Trimeton (Schering Corp.).

These drugs require a doctor's prescription, so must be obtained in advance, and kept available for an emergency.

Stings left in the wound should be removed promptly, before swelling is severe, by pushing sideward with a needle or pin. Tweezers may squeeze more poison out of the stinger into the victim.

If the sting is on a hand, any rings or slip-over wrist watch bands must be removed at once. A dentist's drill is the ideal tool for cutting a ring after it is surrounded by the swelling.

The most vital factor in treatment is quick action. Every minute of delay increases the extent of the injury, and danger of shock. Even single stings in sensitive people, and multiple stings in anyone, should have prompt attention.

In the absence of other remedies, absorption of the toxin may be slowed by an ice pack on the stings, and/or a tourniquet above them. Danger of shock may be reduced by strong black coffee, taken by mouth if the patient is conscious, rectally if he is not.

Surface applications of mud or ointments may relieve pain, but have little or no effect on swelling or systemic reactions. Use of such remedies should not be discouraged, however, as they satisfy the man's desire to "do something."

INSURANCE

Every contractor needs insurance.

The only questions are what kinds and how much.

There are two types of insurance. One protects property owned by the insured, so he is paid if it is damaged or lost. The other protects him against claims for damage to other people because of his negligence. They are both important, but the second much more so than the first.

Much of the insurance protection a contractor needs is required by the majority of business men, but there are special angles.

To the layman, insurance policies are complicated and confusing. There are many kinds of coverage, some of them overlapping; and many circumstances that affect each type. It is important to go to a good broker or agent who can explain in detail the purpose of each policy and what it covers, and even more important, what it does not cover.

Self-protection. To protect his own property, a contractor should have fire insurance on his buildings and their contents, and separate all risk "floater" insurance on his equipment. Cars and trucks may be covered under the floater, or under separate motor vehicle policies for fire, theft, collision, and other damages.

The building insurance is made more complete by extended coverage, added at moderate additional cost, that protects against damage from wind, storm, hail, aircraft, vehicles, smoke, and certain other causes. Vandalism, earthquake, and some other coverages may need special endorsements on the policy. It should be remembered that these, and flood damage, are not included in extended coverage.

A good tools and equipment floater policy will protect a contractor against most damages to the machines—fire, theft, overturning, tornado, upset, and collapse of bridges. But riot, vandalism, malicious mischief (increasingly important), and "loss while waterborne" are probably included only on payment of an extra premium.

Such a policy may list all pieces of equipment covered, or list the large units and lump the smaller ones. Another method is to declare a gross value for all the machinery, and pay a premium on that. If equipment is listed individually, there is usually automatic coverage of new machines for a short period after purchase.

Compensation. Workmen's compensation insurance, required of employers by law in practically all states, and by common sense and self interest in all of them, pays medical expenses, part wages (as disability benefits) and damages to employees injured on the job. Usually there is a period of time, such as a week, in which compensation pays no wages unless the disability extends over a longer period. There may also be gradations from partial to full compensation for time lost, as the no-work period lengthens.

Premiums are based on the type of work and the amount of the payroll. Rates and requirements differ in various states, and a contractor working across state lines must take care to see that he is covered on both sides.

Liability and Property Damage. Liability insurance pays for injuries to people caused by acts of negligence for which the insured is liable. Property damage pays for similar injury to property.

A contractor is neither a business man nor a good citizen if he does not keep himself well insured for injuries and damage to others. His equipment and the nature of his work both make it likely that claims will be brought against him. He cannot afford to be put in bankruptcy by an operator's carelessness, nor should he risk causing damages for which he could not settle.

All too many contractors, and other business men also, think they are completely insured until an accident shows a hole in their coverage. This section will point out a few of the pitfalls, but the best precaution is to be friendly with a good insurance man, and talk to him freely about jobs and work methods.

Most liability policies have a minimum coverage of $5000 for injury to one person, and $10,000 for injury to two or more in the same accident. The policy covers each of a series of accidents in the same amounts, until it expires or is cancelled. Property damage minimum may be $1000.

In this time of fantastically high awards of damages, these minimum coverages are much too low. Amounts can be increased for a comparatively small cost, and it is usually sound policy to carry $100,000 to 300,000 liability, and 25,000 property damage.

In addition to the face amount of insurance, the company pays for investigation and for legal and trial costs, bonding fees, and release of attachments, which may add up to substantial costs.

Exact coverages of policies vary from company to company and state to state, so the following discussion is only a general guide to. what might be included.

First there is motor vehicle insurance, on personal and business cars, pickups, trucks, trailers, and on equipment that travels under its own power or is towed on public roads. This includes wheel tractors, graders, and self-powered scrapers.

Rates on trucks increase with their gross weight. At this writing, rates on wheel tractors and other heavy, slow moving equipment are prohibitively high. Arrangements can sometimes be made for coverage on job-to-job moves under the general contractors' liability. Careful investigation should be made of this point.

Towing a trailer of any kind may invalidate car or truck insurance, unless provided for in the policy, or the trailer is separately insured. If such towing of an uninsured trailer is rarely done, the company insuring the vehicle should be willing to issue a special endorsement or binder to cover the combination for a specific trip or time period, at little or no cost.

If there are a number of motor vehicles, economies may be affected by insuring them together in a fleet policy, and by keeping some of them on low mileage and therefore low rate local errands.

Contractors' Liability. There are a number of classifications of liability risks for the contractor that can be insured separately. It is good business to lump as many as possible in a comprehensive policy, to avoid extra payments on overlapping coverage, and to avoid confusion.

A comprehensive policy may cover:

Ownership, use, and operation of buildings and premises
Construction machinery, as above
Elevators
Completed work (Products) having defects causing injury or damage
Teams
All contractual work of kinds specified in the policy
Operations of subcontractors, except in maintenance of insured's property

It probably will not cover:

Dogs, animals, boats, aircraft, or vehicles
Blasting
Damage to subsurface pipes, conduits, and wires
Collapse of structures caused by excavation or underpinning work
Tunneling and bridge construction
Obligations assumed for others
Damage to rented or controlled equipment

The first exclusion in the above list is made because these risks should be covered by other types of policy. The next four are high-rate risks, and losses incurred under them can be more justly paid under special endorsements or other policies, by those who do such work, than by the larger number of contractors who do not.

These risks can be covered for specific jobs, usually only after inspection by the company so that it can see what it is letting itself in for, and set the premium accordingly. It is to the contractor's interest to have such inspections made to obtain the necessary coverage; not only for his own protection, but because it is only the most experienced of supervisors who will not benefit from talking over a job with a good inspector.

Employers often feel that inspectors are a threat and a nuisance, but they perform invaluable services both as safety engineers and job consultants. Contractors who will listen to their discussions of methods used on other jobs will often find that they will save more than the cost of the premiums charged and the safety procedures required.

"Obligations assumed for others" is a tricky one that has caused many painful surprises. It is a too-common practice for an owner to write up a work contract specifying that the contractor assumes all liability for everything that happens on the premises while he is working on them. This may extend the contractor's risks far beyond the premium he pays for his own activities. It is much better for the owner to take out an owner's risk policy for work in progress, and ideal if he can place it in the same company that insures the contractor.

If this is not possible, the contractor can show his contract to his own company, and pay an extra premium for an endorsement to cover any obligations he has assumed under it.

If such precautions are not taken, the results of the owner passing his responsibilities to the contractor may be disastrous to them both, as neither of them are insured for the owner's risks, and both are responsible for them.

"Damage to rented or controlled equip-ment" is another joker on which many a contractor has tumbled, although the amounts involved are usually modest. Liability policies are designed to protect against claims for others. If a contractor hires a machine, it is his for the period of use, and is no longer entitled to be the subject of a claim against him.

Coverage to protect such equipment can be obtained by endorsement of the liability policy. The extra premium is usually based on the rental cost.

Rates. Insurance is priced so that each class of risk will bring in enough money in premiums to pay sales, administrative, and legal costs, the claims that have to be paid, and to leave a surplus for reserves, and dividends to stockholders or policy holders.

An increase in losses automatically results in an increase in rates, although this effect may be delayed. The increase may be applied generally to all those having the particular type of insurance, or specifically to those whose accidents have piled up the claims.

Most insurance is written on one or more basic rates covering a general class of risk, with upward or downward revision depending on local conditions, and experience with a particular risk or a particular customer.

Fire insurance premiums are affected by how likely the property is to take fire, how readily and completely it will burn, and the availability of fire fighting equipment and water. A substantial drop in rates can sometimes be obtained by building alterations, digging a pond or providing access to it, or having a branch fire station established in the area.

Premiums are usually quoted on a basis of price of $1000 of coverage.

Contractors' liability and property damage rates are extremely variable. They are based first on experience with a particular type of work, so that blasting will have a higher rate than landscaping. Again, coverage for blasting in the country may be at nominal cost, where in a city it might be as high as 50% of the payroll.

A small contractor may be just carried at the average of the industry. A larger operator will be assigned an experience rating, based on the number of accidents he has had, and how expensive they have been. This rating may make his insurance more or less expensive than that of his competitors, and may thus affect his position in competitive bidding.

If the record is extremely bad, the rate may go so high as to make it difficult or impossible to stay in business. Companies might also refuse to write any insurance for him.

The premium for compensation insurance basically consists of a percentage of the payroll expressed in terms of dollars per $100 of wages. At the start of the policy term, the company and the insured define the risks that are to be covered, estimate the payroll for six months, and set the premium on the basis of the estimate. Then every six months the company makes an inspection of the insured's books, or perhaps only of his payroll tax returns, and an additional amount is charged or a credit issued for any difference from the estimated charge.

If a contractor has a number of different activities, and does not keep separate payroll records for them, he will be charged the rate of the most expensive coverage for all of them. It is therefore to his interest to keep the different classifications at least roughly divided.

Liability insurance may be assessed according to the payroll, or by the value of the work done during the period. Here also a separation should be made between jobs carrying different rates.

The contractor must pay liability premiums on all work done for him by subcontractors and by hired machinery unless he obtains and shows to his company certificates of insurance coverage from the subcontractors.

BONDS

The excavating contractor shares with other forms of business the danger of serious loss through dishonesty of an employee, or employees. For a contractor, the loss is as likely to be in property taken or sold "over the fence" as it is money.

Fidelity bonds of various types are available for protection against losses of this nature.

Construction contract bonds are required of contractors performing work for Federal, state, and local governments. There is an increasing use of them in contracts with private owners.

A bond is a three party agreement, made by the contractor and the bonding or surety company to protect the owner. It usually covers all obligations that the contractor assumes on the job, including completing the work to specification, and paying subcontractors and employees so that no liens or actions can be brought by them against the owner.

Three types of bonds may be involved. The first, the bid bond, accompanies a bid or proposal on a job, and guarantees that if the bidder is given the job, he will enter into a formal contract to complete it, and will supply bonds to complete the contract.

The bond supplied for the work itself is made up of two bonds, which are separate, but seldom if ever written separately. One is a Performance Bond, covering fulfillment of the contract, the other a Labor and Material Payment Bond, guaranteeing payment to personnel, suppliers, and subcontractors.

These last two are drawn separately so that no question of priority can arise when claims are presented by both the owner and those who have supplied services and materials. In the early days of bonding, the government had to be paid or satisfied first, and the others got what was left. This meant at least long delays, and in cases where the bond was too small, losses for the small claimants.

In order to obtain a bond, a contractor must convince the company that he is competent to do the job, and financially able to carry it. He pays the premium, usually not over 1% of the contract price, figures it as part of his cost, and passes it on to the owner in his bid or estimate.

Substantial all-around benefits are sometimes obtained from writing of construction bonds. The owner can let the contract to the lowest bidder without having to inquire into the question of whether he can complete it, as the bonding company guarantees performance. The contractor may save money by driving hard bargains with subcontractors who cut their figures a little closer because they know they will be paid.

If a contractor fails to complete the job or to pay the subcontractors, the bonding company takes over, lets a new contract to finish, and pays up the bills. Quite often, the new contract will be let to the contractor who defaulted, as his equipment is on the job.

The contractor is legally obligated to repay to the surety company everything that it has spent to finish his work. The company makes a more cooperative and intelligent creditor than a combination of an enraged owner and starving subcontractors, and in most cases the contractor is able to work his way out of his difficulties, and avoid a failure that might have been inevitable without the protection of the bond.

Unfortunately, there is another side to the picture. Many contractors who are thoroughly competent and reliable and have adequate resources for a job cannot get a bond to cover it. Potential low bidders may thus be weeded out, and work concentrated in the hands of a favored clique.

Inability to get a bond may result from a poor background, lack of resources, too many jobs already in progress, or other reasonable causes. All too often, however, it is the result of cloudy judgment or caprice.

Bonding companies usually know very little about construction work or the ability of men to do it. Yet they are acquiring the power to decide which contractors can work and which must starve, and to meddle with work and accounting methods. They work together so closely that a contractor rejected by one of them has very little chance of obtaining coverage from another.

Requirement of a bond in private work is therefore often productive of injustice and hardship and may result in substantially higher costs. The owner can usually obtain adequate protection by exercising good judgment in awarding the contract, and by the customary withholding of a percentage until the job is finished.

Courtesy of Contractors and Engineers

Bench cut on the New York Thruway, 1953

PART TWO

THE MACHINES

CHAPTER TWELVE

BASIC INFORMATION

In studying the construction of machinery, it is necessary, or at least convenient, to have an understanding of its basic principles. This chapter therefore includes a brief review of a few of the laws of physics involved in machinery construction and operation.

There are a number of parts or assemblies such as gears, brakes, and clutches that are common to machines of widely different types. These are described together for convenience of reference, and to avoid complicating the descriptions in the following chapters.

PRINCIPLES OF LEVERAGE

Speed and Force. An engine usually turns its crankshaft too fast and with too little force (torque) for the machine it powers. This speed can be reduced and the force increased in the same proportion by the use of gears, levers, pulleys, and other devices. Explanation of how this is done can best be started by referring back to some basic laws of physics.

Conservation of Energy. The first law is that of conservation of energy—that energy cannot be created or destroyed (with possible exceptions in the subatomic scale), but only changed from one form to another. In machinery, speed and force are applied by the engine. Some of this will be transformed by friction into heat and wasted, but most of the rest may be converted from speed to power, or power to speed, and back again.

Levers. The oldest and simplest converter is the lever which may be divided into three types. Figure 12-1 (A) shows a first class lever, consisting of a crowbar with its business end under a rock, a stone a foot away serving as a pivot or fulcrum, and a man pushing down on the handle end, three feet from the fulcrum. The man is three times as far from the fulcrum as the rock is, so that if he pushes the bar down three inches, the end under the rock will rise only one inch. This loss of distance (distance is a function of speed) is balanced by an increase in force—the 100 pounds pressure he puts on the bar becomes a lift of 300 pounds under the rock.

(B) shows the same example except that the fulcrum has been moved to a spot six inches from the rock, increasing the fulcrum-to-handle distance to three and a half feet, or 42 inches. The handle end of the lever is now seven times as long as the working end; it will have to be pushed down seven inches to raise the rock one inch, but the up push against the rock will be 700 pounds.

If the rock is too heavy to be lifted, and the fulcrum stone is pushed down into the ground by the force applied, the bar becomes a second class lever, the end under the rock becoming a fulcrum and the lower stone becoming the load.

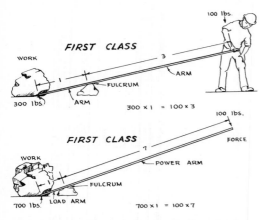

Fig. 12-1. First and second class levers

The same distance and force ratios apply. It is apparent that the difference between first and second class may be just which of the two points is weaker.

Increase of force in a lever or other device is called the mechanical advantage, abbreviated M.A.

A third class lever has the force applied between the fulcrum and the work so that it can be used only to increase speed or distance at the expense of force. Figure 12-2 shows a man lifting a board twenty feet long so that one end will rest on a wall top ten feet above him. He places one end under a weight, making it the fulcrum, and picks it up five feet from that end. As the other end is twenty feet from the fulcrum, it will move four times as fast and as far so that he need raise the board only two and a half feet to get its end to the wall top. The 100 pound lift he is exerting is

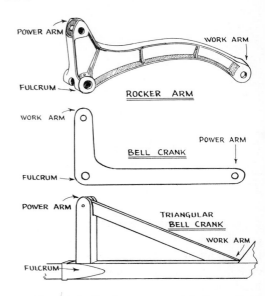

Fig. 12-3. Rocker arm and bell crank

reduced to 25 pounds at the upper end, but since in this case the board is both the lever and the load the lift need not be calculated at the end, but at the center of gravity ten feet from the fulcrum. His effective lift is thus reduced only to 50 pounds. For a heavier board, he should move further from the fulcrum where he would have to move the board a greater distance but with less force. If he picks it up more than ten feet from the fulcrum,

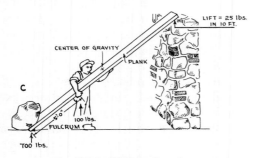

Fig. 12-2. Third class lever

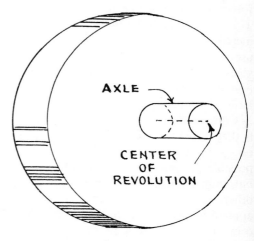

Fig. 12-4. Wheel and axle

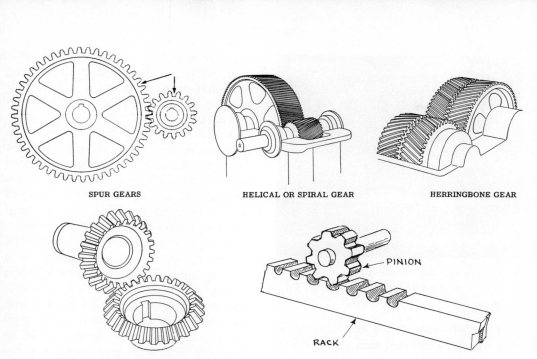

SPUR GEARS   HELICAL OR SPIRAL GEAR   HERRINGBONE GEAR

PINION

RACK

PAIR OF STRAIGHT TOOTH BEVEL GEARS
Courtesy of Power Transmission Council, Inc.

Fig. 12-5. Gears

the system becomes a second class lever in relation to the center of gravity, but remains a third class one with respect to the board end.

All parts of a lever move in curves rather than straight lines. The curves are arcs of circles centering at the fulcrum.

A lever must be rigid for the distance-force ratios to apply accurately. These are always figured in straight lines from the point where force is applied to the fulcrum, and from the fulcrum to the work-point. The lever itself may be straight, angled, curved or offset.

If a lever has a sharp angle at the fulcrum, as in Figure 12-3, it is called a bell crank. The space between the arms may be braced for greater rigidity, producing the triangular bell crank.

Wheel and Axle. A wheel and axle, Figure 12-4, is a specialized type of lever. For example, consider a wheel three feet in diameter rigidly fastened to a six inch axle, revolving at ten revolutions per minute (r.p.m.). Any point on it will describe a circle every six seconds, and the further the point is from the center the larger the circle will be, and the faster the point must move to get around it in six seconds. A point on the axle surface moves nineteen inches with each turn, and on the outer edge 113 inches. As with all mechanical devices with a fixed amount of power, greater speed means less force. This wheel may be considered to be a number of third class levers, with the center of rotation the fulcrum, the axle surface the power point, and the wheel perimeter the work.

If power is applied to the axle and work is done at the outer edge of the wheel, the result is increase of speed and decrease of force in a six to one ratio. If the power is applied to the wheel rim, the wheel acts as a number of second class levers, with the fulcrum again at the center of rotation but the work at the axle surface; power will be multiplied by six and speed (distance) cut by the same amount.

12-3

Courtesy of Unit Crane & Shovel Corp.

Fig. 12-6. Worm gears

GEARS

Types of Gears. If the axle is locked to two wheels of different size, it will deliver different proportions of power and speed to their circumferences because of the different leverage ratios. If the wheel edges are notched into teeth, which can be meshed with other toothed wheels, they are called gears and have almost unlimited capacity for changing the power-speed relationships between the input (driving) and output (driven) shaft.

Figures 12-5 and 12-6 illustrate some of the more common types of gears which may be roughly described as follows:

A spur gear has a rim so notched that the teeth are nearly the same size as the space between them, and the tooth edges are parallel to the axle. This is a simple, rugged type of gear that tends to develop noise and backlash more readily than the next two to be described.

A helical gear is wider with teeth set diagonally. The teeth are usually comparatively small in cross-section, and may be curved. It is usually quieter and smoother in operation than the spur but exerts considerable side thrust.

Gear teeth are often not flat on their contact surfaces, but have a definite bulge or crown that reduces friction.

A herringbone gear is a double helical, with teeth on the two sides sloping oppositely to each other. It has the advantages of the single helical and has no side thrust.

A rack or straight gear is a bar with teeth cut in one side. It is used where the forces are too heavy to depend on friction between a wheel and a track, or where an exact relationship between turning and moving must be preserved. A very large circular gear may be called a rack, also.

A worm is a cylinder with spiral teeth cut in it, and has a resemblance to a coarsely threaded bolt. It turns a worm gear or wheel gear, which is a type of spur gear with teeth cut in a curve to mesh with the worm. This accomplishes a very great speed reduction. It is usually made irreversible so that the gear cannot turn the worm, or semi-irreversible so a light automatic brake can prevent it from doing so.

Hypoid gears are bevels with curved teeth, and are between worm and bevel gears in use.

A sprocket gear usually has tapered or pointed teeth with rounded hollows between them. It is used in connection with roller chain and must be carefully matched with it for tooth spacing and hollow shape.

When two gears meshed together are of

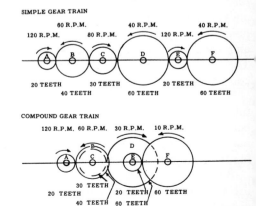

Courtesy of Power Transmission Council, Inc.

Fig. 12-7. Gear Trains

very different sizes, the smaller may be called a pinion and the larger a bull, ring, or rack gear. A gear which is meshed with others but performs no function except direction change, or adjustment, is an idler.

Meshed Gears. Figure 12-7 shows simple meshed gears. Any two that mesh with each other must rotate in opposite directions. If three are meshed in series, the two outer ones rotate the same way. The center counter-rotating gear is a reversing idler.

The speed ratio of two gears is proportional to the number of their teeth.

The rotation of a gear is described as clockwise, if from the point of view of the observer, its teeth turn in the same direction as a clock's hands. Counterclockwise is opposite rotation.

Clockwise rotation viewed from the front would be counterclockwise from the rear, so direction of view must be made clear.

In a simple set, the speed ratio of any two gears is proportional to the number of their teeth, regardless of whether they may be connected through larger or smaller gears.

(B) illustrates a compound gear set in which two gears of different sizes are locked on the same axle. Such sets permit great changes in ratio in a comparatively small space.

When it is important that gears be quiet they should be "matched," that is, ground or run together by the manufacturer so that the teeth match each other perfectly. This reduces noise, friction, and wear.

If a gear set of any type is taken apart, the gears should be marked so that they can be meshed together exactly as before, so that the advantages of their wearing in will not be lost.

Bevel Gear Set. Bevel gears can carry rotation around a right angle and change the speed-power ratio at the same time. An important additional use is to reverse direction of rotation of a shaft.

Figure 12-8 shows a set of three bevel

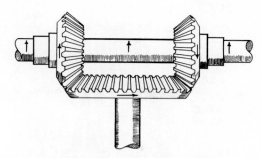

Fig. 12-8. Reversing set of bevel gears

gears. The two upper gears are the same size and revolve freely on a rotating horizontal shaft. The lower gear is larger and rotates more slowly on an axis at right angles to the horizontal shaft.

In the figure, the left hand gear is shown revolving in the same direction as the shaft, so that its lower teeth are moving toward the observer. The lower gear is thus caused to rotate from left to right, and turns the right hand gear so that it turns oppositely from the shaft, and the lower teeth move away from the observer. Two right angle turns have been made in the gear set, and the two upper gears, meshed with each other through the lower gear, turn in opposite directions at the same speed on the same shaft.

If the horizontal shaft were fitted with two clutches, by means of which the two upper gears could be separately locked to the shaft, engagement of the left hand clutch would rotate the gears as described above. Disengagement of this clutch, and engagement of the other, would cause the gears to rotate in the opposite direction. The lower gear, and any machinery driven by it, can therefore be rotated in either direction by the engagement of the proper clutch, or left idle by disengagement of both of them.

Differentials. A more complex set of bevel gears is the differential used in cars, trucks, and other machines, and illustrated in Figure 12-9 and in 18-13 and 18-14. It consists of a ring gear driven by a pinion

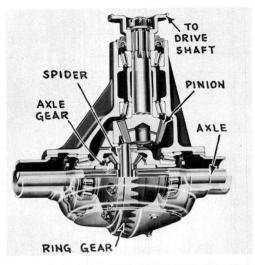

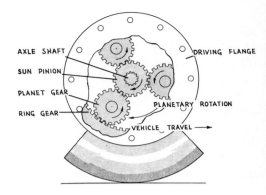

Courtesy of Euclid

Fig. 12-10. Planetary final drive

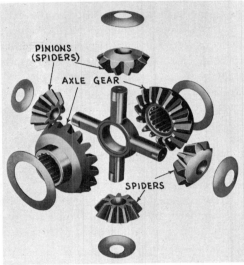

Courtesy of Timken-Detroit Axle Co.

Fig. 12-9. Differential

or a reducing gear set, and a pair of horizontal shafts (axles), driven by the ring gear through a set of six bevel gears. Two of these bevels are splined to the inner ends of the axles; the other four, called spiders, are mounted in opposed pairs in a case attached to the ring gear. The spiders rotate with the ring gear and turn the axle gears, the whole system revolving on the axis of the axles. If the load on the two axles is the same, the spiders do not revolve in-

dividually, but only with the case around the axles.

If both axles should be locked, none of these gears could turn. If one is locked each opposed pair of spider gears, in addition to revolving around the main axis, will revolve in opposite directions around their own axis (which turns end over end with the ring gear), rolling on the locked axle gear. The other axle continues to be driven by the rotation of the case, and is also turned by the turning of the spider gears, so that its speed is doubled. If one axle is slowed, the other will be speeded up proportionately.

Spur gears are sometimes used instead of bevels in a differential. The last Oliver crawler tractors had this design.

A differential is essential for delivering power to a pair of wheels which may have to revolve at different speeds, as in an automobile going around a curve, where the outside wheels go further than the inside ones. It is also used to steer machines and change gear ratios by proper application of brakes. A disadvantage is that it will deliver most of the power to the axle or unit which is disconnected or running free, as to a car wheel on a piece of ice which will spin while the other wheel on dry pavement is not turned at all.

There are many no-spin differentials in

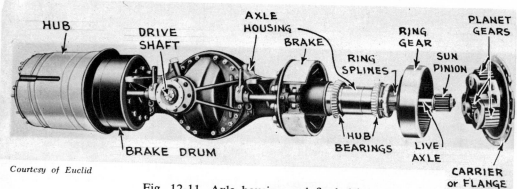

Courtesy of Euclid

Fig. 12-11. Axle housing and final drive

which this difficulty is overcome by the use of automatically locking spiders, or of worms and worm gears instead of bevel spiders, in which differential action in rounding corners is permitted, but which automatically lock against spinning. See Figure 17-13.

Planetary Gears. A planetary or sun gear

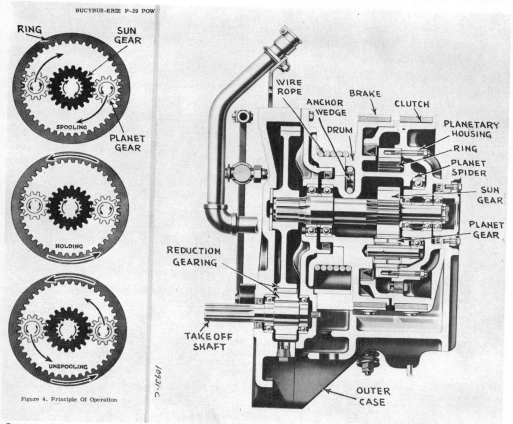

Courtesy of Bucyrus-Erie Company

Fig. 12-12. Planetary drive winch

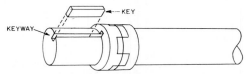

Fig. 12-13A. Keyed shaft

set, Figure 12-10, includes a central sun gear or pinion splined or keyed to a shaft or axle, two or more planet gears, and a stationary ring gear with internal teeth. The planet gears turn on shafts fastened to a carrier or driving flange. For reduction in speed and increase in torque, power is supplied to the axle and work is performed by members attached to or driven by the flange.

The planets are caused to rotate by the turning of the sun gear. This rotation forces them to walk around the toothed track formed by the ring gear, pulling the carrier or spider around with them in the same direction the axle rotates.

The torque multiplication is found by dividing the number of teeth in the ring by the number of teeth in the sun gear and adding one. For example, a ring of 30 teeth and a sun pinion of 10 teeth would produce a torque multiplication of four.

The final drive shown in Fig. 12-11 has a

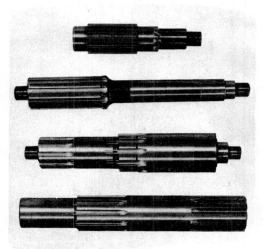

Courtesy of Unit Crane & Shovel Corp.

Fig. 12-13B. Splined shafts

sun pinion driven by the live axle, a ring gear splined to the end of the axle housing, and a flange which is bolted to the wheel hub.

A planetary gear set with a fixed ring such as this provides a compact and sturdy means of multiplying shaft torque. If the ring and the spider can be either held or allowed to rotate, brake and clutch action can be obtained.

The International power control winch, Figure 12-12, has a cable drum that is turned by a planet spider which can be stopped by an external band brake anchored to the outer case. The ring gear is bolted to the planetary housing which is held (or released) by a similar brake that is called a clutch because it is used as one.

The sun gear is driven through reducing gears by the tractor's power take-off shaft, and causes the planets to revolve on their pivots. If the ring gear is held by the "clutch," the planets travel around inside it, rotating the spider and the drum so as to wind in the attached cable. If the clutch band is released and the brake applied, the spider is held and the rotation of the planets turns the ring gear, which is a non-working part and imposes no load.

When both clutch and brake bands are released, pull on the line revolves the cable drum and unwinds the cable, with spider, planets, and ring gear turning freely. Control linkages are so arranged that the brake and clutch cannot be applied at the same time.

A similar arrangement could be used to produce a two-speed transmission. If both the spider and the ring were connected through gears or chains to a single output shaft, a change of gear ratio or of direction of rotation could be obtained by directing power one way or the other by means of the two brakes.

Power shift and automatic transmissions usually include planetary gear sets, controlled by disc clutches and/or brakes.

SHAFTS

Keys and Splines. Gears may be connected to the source of power or to the work by means of shafts or axles, or they may revolve freely on shafts that simply hold them in proper position.

A rigid connection can be made to the shaft by means of keys or splines. If keyed, a square sided slot called a keyway is cut in both shaft and gear hub. The key, a square strip or specially shaped piece of very hard steel, is inserted in the shaft slot and the gear forced over it. The key should be a tight fit in both slots so that the gear and the shaft cannot twist independently of each other.

A splined shaft, Figure 12-13B, has a number of grooves similar to keyways cut at evenly spaced intervals around it. Matching slots are cut in the gear hub, and it is slid or pressed on to the splines in the shaft. If it is a sliding fit, its position on the shaft may be controlled by a clutch-type yoke or collar. If a tight or press fit, it is held in place by friction, by set screws, snap rings, thrust washers, or other means. A splined connection is much stronger than a keyed one.

Universal Joints. Shafts carrying power from one unit to another seldom can stay on a single axis of revolution determined by the bearings in both units. Even if the original installation is exactly in line, wear of bearings, bending of frames, or settling of foundations are apt to cause misalignment. Often the units are deliberately installed out of line for convenience, or their position in regard to each other varies.

Even a very slight misalignment will cause a shaft to impose excessive strain and wear on its bearings, as exaggerated in Figure 12-14(A), and may also cause it to whip or break. It is therefore standard practice to place one or more (usually two) flexible couplings or universal joints in shafts connecting separate units, to per-

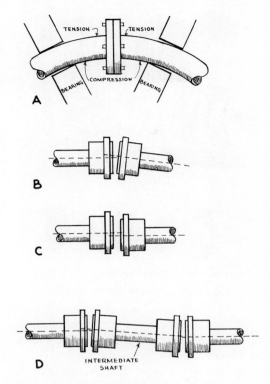

Fig. 12-14. Shaft misalignment

mit the shafts to adjust to differences in line.

Misalignment may be angular, as in (A) and (B), or parallel as in (C). If parallel, it is usual to handle it as two angles at the ends of an intermediate hub or shaft, as in (D). A simple type of flexible coupling which carries light loads through angles is shown in Figure 12-15. This also absorbs vibration and shock.

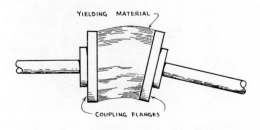

Fig. 12-15. Simple flexible coupling

Courtesy of Morse Chain Company

Fig. 12-16. Roller chain coupling

If the differences are slight, roller chain or silent chain couplings such as are shown in Figure 12-16 can be used.

A Morflex rubber biscuit coupling is shown in Figure 12-17. Two two-bolt flanges are placed facing each other. Each is bolted to two of a set of four rubber biscuits pressed into a steel holder. Misalignment of the shafts attached to the flanges is absorbed by flexing of the rubber.

The rubber will also absorb a certain amount of thrust or pull from the shafts, permitting their effective length to change.

An increase in angle to about 7° can be accommodated by mounting two couplings together, as in (C).

A Spicer universal joint is illustrated in Fig. 12-18B. It consists of a journal cross, two yokes, and four needle bearings. The driving yoke holds two opposite arms of the cross, and the driven yoke the other pair.

Each pair of arms rotates on the same axis as the yoke which holds it, and the two axes intersect at the center of the cross. The twist caused by revolving in two planes is absorbed by oscillation of the arms inside the yoke bearings.

Change in length of shaft is permitted by a sliding spline (slip joint) in one of the yokes.

The universal joint yokes or splines are

Courtesy of Morse Chain Company

Fig. 12-17. Rubber biscuit coupling

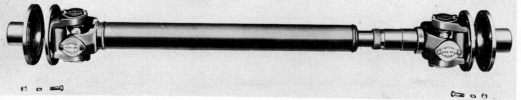

Fig. 12-18A. Shaft and universal joints

attached to the shafts by bolted flanges or welded stubs.

Journal cross universals are usually made for normal angles up to 8° or 10°, and momentary angles up to 15° to 20°. Higher angles require special construction.

Universal joints are generally used in pairs. This allows them to accommodate sideward as well as angular displacement, and it makes it possible to cancel out any pulsation caused in the driven shaft by joint action.

The constant velocity joint shown in Figure 12-19 uses rolling balls between the yokes to provide a constant-distance fulcrum and non pulsating drive. It can be used in most universal joint applications, but is particularly adapted to sharp angles, and places where only one joint can be used and a non-pulsating drive is desired. Front axles of four-wheel drive vehicles require these characteristics.

BUSHINGS AND BEARINGS

Where a gear or wheel turns upon a shaft, or the shaft turns within its supports, provision must be made to reduce friction and wear.

Bushing (Sleeve Bearing). There is no definite distinction between sleeve or solid bearings and bushings. They are sleeves or hollow cylinders installed between a shaft and its supports, or between a gear, pulley, or wheel and the shaft on which it turns.

A bushing is usually made of some metal alloy that is softer than steel, can retain an oil film that will prevent direct metal to metal friction, and is easily worked. Bronze, brass, babbit and oilite are some examples.

When such bushings are kept free of dirt and are properly lubricated they reduce friction and absorb most of the wear caused by the rotation of the shaft or wheel. A bushing may be replaced several times before the more expensive shaft has to be repaired, and they protect the supports (often called pillow blocks) for the life of the machine.

There are also hard steel bushings whose only function is to protect the support from damage by the shaft. They are usually found where shafts do not make complete revolutions (hinge pins of a front loader, for example), or where they turn very slowly. They may or may not be lubricated, and usually do not have seals against dirt.

Fig. 12-18B. Universal joint

Courtesy of Bendix Products Division, Bendix Aviation Corp.

Fig. 12-19. Constant velocity joint

Bushings in pillow blocks are usually split, that is, cut lengthwise into two pieces. This permits easy installation, as they do not have to be slid along the shaft to get in place. On some old machines you may find solid bushings that require stripping gears off a shaft to remove them. Labor can sometimes be saved by cutting them off, and replacing them with split bushings.

A bushing is always fastened to the outer member, whether it is stationary or rotating. It is easier to lock them against turning on the outside, and the inside turns a little more slowly than the outside, because of its smaller diameter.

Main and rod bearings in an engine have split bushings called bearing inserts.

If a bushing is in sight, or can be readily seen after guards are removed, wear can be observed. If the space between it and the shaft changes with differences in load or of direction of rotation, or if grease squeezes out and is drawn back in as it works, there is looseness that may or may not be serious.

Play in a bushing may also be detected by prying the shaft with a bar.

The friction area in a bushing is large

enough to convert an important amount of power into damaging heat if lubrication is not adequate. Heat is quickly followed by scoring (deep scratching) of both the bushing and the shaft. A worn bushing will not hold the shaft in alignment, and more trouble may follow.

Most bushings have a drilled passage to a grease fitting or oil line, and grooves that distribute lubricant through the full width. There is a slow motion of the lubricant toward the edges of the bushing, where it is usually lost. Supply must be fast enough or frequent enough to maintain a film of oil over the whole friction surface.

Anti-Friction Bearings. Anti-friction bearings are used instead of bushings at most rotary friction points where exact alignment and low friction are required, and the shape of the shaft permits their use.

The two principal types are ball bearings and roller bearings.

Ball bearings consist of two rings of hard steel, one of which fits closely on the shaft and has its outer surface machined into a smooth track; the other fits closely into the casing or hub and has a machined inner surface. Between these rings, known as the inner and outer rings or races, is a circle of hard steel balls held in place by a light cage. See Figure 12-21.

The inner race turns with the shaft, the outer race stays stationary with the casing, and the balls keep these races at an exact distance from each other and revolve so that all friction is changed from the sliding to the rolling type. Pushing a heavy piece of furniture, first without casters then with them, will give a somewhat exaggerated idea of the difference. The space around the

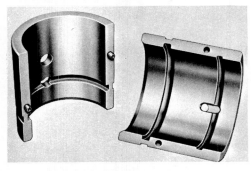

Courtesy of Caterpillar Tractor Company

Fig. 12-20. Split bushings (bearing inserts)

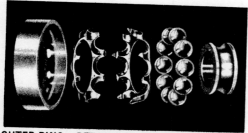

OUTER RING • RETAINER • BALLS • INNER RING

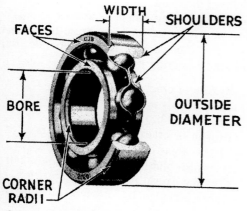

Courtesy of Ahlberg Bearing Company

Fig. 12-21. Ball bearing construction

A needle bearing is a special type of roller bearing that is used where space is limited and where rotation is partial, as in a universal joint. The rollers are very thin and turn directly on the shaft, as there is no inner race. The outer race is called the shell. See Figure 12-21.

Value. Bushings and bearings serve several functions. They reduce friction, provide for lubrication, and are replaceable individually when worn at less expense than replacing a whole shaft or casting.

For most uses anti-friction bearings are superior. They have less friction, can be more effectively lubricated, have a longer life, and keep parts in more exact alignment. However, they are more expensive, are subject to rust damage in idle machines, require more space than bushings, and since they cannot be split, they are used only where they can be slid onto a shaft.

Cleaning. Used anti-friction bearings must be cleaned before their condition can be judged. They are usually washed out in a

balls is filled with grease or oil which is either sealed in or renewed periodically.

Roller Bearing. Here the balls are replaced by steel cylinders which roll between two races in the same manner. An important variation is the tapered roller bearing illustrated in Fig. 12-22. The races are tapered, the inner one being called a cone, the outer a cup. This shape enables them to resist side thrust as well as normal loads, and since they are used in pairs with the small ends of the cone facing, tightening a nut behind one of them will force the cones nearer the cups, thus compensating for any wear in the rollers or races.

Tapered construction is also found in ball bearings, but less commonly. Two or more rows of balls may be used in either flat or tapered construction to increase load capacity without increase in diameter.

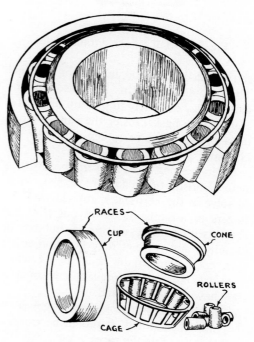

Fig. 12-22. Tapered roller bearing

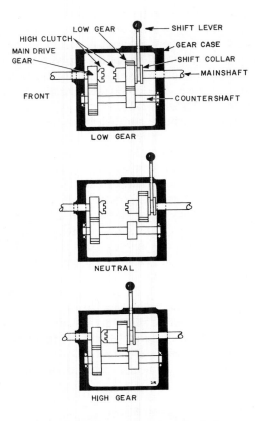

Fig. 12-23A. Two-speed transmission

face of a bearing, or corrosion on non-working surfaces, will probably not interfere with the usefulness of a bearing. Slight roughness or tendency to stick may only mean that it needs to be re-cleaned. A good test for real roughness is to press the two races toward each other while turning them or to lay the bearing flat on a clean surface and rotate it with heavy pressure from the palm of the hand.

The following defects should be looked for, and if found they are cause for scrapping the affected part if it is separate, or the whole bearing if it is not.

Broken or cracked rings
Dented seals or shields
Cracked or broken separators
Broken or cracked balls or rollers
Flaked areas on working surfaces
Brownish-blue or blue-black color indicating overheating
Races indented by pressure of balls or rollers

Looseness may be a characteristic of that model bearing, or may indicate extensive wear. Compare with a new bearing, if possible, after packing the old one with thin grease.

Ordering. Bearings need not be bought from equipment manufacturers, as they almost always buy them from others. Most anti-friction bearings can be obtained from bearing agencies or auto parts or contractors' supply houses, often at a substantial discount.

Each bearing carries a code number and a manufacturer's name, usually on one side of the rings. If the cup and cone are sold separately, each will have a number.

The code number tells a parts man everything about it—its dimensions, the load it can carry, and the way it is lubricated. If he carries the same make or one with the same code he can get it directly from stock. If not, he can look it up in a book and find its number in the make he has.

can or pan of kerosene, fuel oil, or gasoline. They should not rest on the bottom of the pail, as they are likely to pick up dirt there. Hang a wire basket inside the pail, or block up a piece of coarse screening above the bottom.

Open bearings can be scrubbed with a natural bristle brush (synthetic fibers may dissolve in the kerosene), rubbed with the fingers or a cloth, or blown out (but not spun) with compressed air. A number of cleanings and rinsings may be required. It is a good plan to dip them in oil when they are finished, to prevent corrosion if they have to stand around.

They should be packed with grease if they are to be stored. It should be worked well in by turning the races.

Inspection. Tarnish or stain on any sur-

There are several systems of code numbers used by different manufacturers.

TRANSMISSION

A transmission, gear set, or gear box is a set of gears and shafts which provide a change in the speed-power ratio. It may be a single speed or gear reduction type in which a small gear driven by the engine meshes with a larger gear that turns the working parts; or it may be a selective or sliding gear type with several ratios.

Two Speed. Figure 12-23A shows the simplest selective type, an auxiliary transmission having two forward speeds. The engine-driven jackshaft carries a main drive gear which meshes with a larger gear splined to the front of the countershaft. The countershaft also carries a smaller fixed gear which may mesh with a sliding gear on the main shaft. The main shaft is a continuation of the jackshaft and may revolve separately from it in low position (shown), or be connected to it in high by a jaw clutch, a device described later under CLUTCHES.

A fixed jaw is on the jackshaft, and a sliding jaw on splines on the main shaft. The sliding jaw is expanded into a gear which may be meshed with the countershaft gear. This combination gear and clutch jaw is controlled by a fork and shifting collar which can slide it forward to connect the clutch and disconnect the gear, to the center so neither will be engaged,

and to the rear to engage the gear only.

For power takeoffs, see page 15-4.

When the jaw clutch is engaged, power is transmitted straight through the transmission as if on a single shaft and the countershaft turns without doing any work. This is called direct drive or high gear. When clutch and gear are both disconnected, the transmission is in neutral and no power goes through it. When the gears are meshed the drive goes from jackshaft to countershaft and from countershaft to main shaft, involving two pairs of gears which can be made in ratios to give any desired gain in power through reduction in speed.

Figure 12-23B shows a three speed forward and one reverse speed transmission. The countershaft gear is enlarged to a cluster gear consisting of three gears made in one piece, the forward one being the largest and the rear the smallest. The center section is meshed with a gear on the idler shaft.

A combination sliding jaw and gear, similar to that described in the two speed transmission, slides forward along the main shaft to mesh with the jackshaft jaw for high gear, and back to mesh with the large part of the cluster gear for second. Another sliding gear on the main shaft is moved forward to mesh with the small member of the cluster gear for first or low gear, and backward to mesh with a gear on the idler shaft for reverse.

The jackshaft and mainshaft revolve in the same direction whenever they are connected directly through the two countershaft gears, regardless of their ratios. Putting an extra (idler) gear in the series reverses the rotation of the mainshaft. See Figure 12-7 for this effect.

Compound Gearing. The larger trucks and tractors often have two transmissions in series, the engine power going first through one, then through the other. The smaller one is called an auxiliary. In tractors it may have one high and one low gear, or a for-

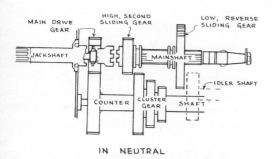

Fig. 12-23B. Three-speed and reverse transmission

ward and a reverse. In trucks it is likely to have from two to four forward speeds.

Trucks may have two speed axles. This means that there is a high-low shift in the differential, that has the same effect as an auxiliary two speed transmission.

The total number of gear speeds may be found by multiplying those in the two transmissions. For example, a jeep has a main transmission with three forward and one reverse, and a high-low auxiliary. This gives it six forward and two reverse speeds.

The two transmissions may be in separate cases with separate levers, or be both in one case, with one or two shift sticks. One stick usually has an interlock that permits starting only when it is in neutral.

Shift. Shifting gears is made possible by a friction clutch between the engine and the transmission. When under load, meshed gears stick together because of pressure against the teeth. When in neutral, a pair of gears cannot be engaged unless the two sets of teeth to be meshed are moving at about the same speed. To shift with the machinery stationary and the engine going, the engine clutch is disengaged, the jackshaft allowed time to slow or to stop its spinning, and the desired shift made. If the teeth will not mesh, the clutch is engaged slightly to turn the jackshaft to another position.

To shift while the machinery is moving, the engine clutch is disengaged to take the load off the gears, and the sliding gear is shifted into neutral. If the shift is from low to high, the jackshaft will be turning too rapidly for quiet engagement. Since it is no longer being turned by the engine it will soon slow to proper speed and can then be engaged. If the engine throttle is closed during shifting, this waiting period can be reduced by engaging and disengaging the clutch (double-clutching) while in neutral, as the engine loses speed more rapidly than the free spinning shaft.

If the shift is from high to low, the

jackshaft will be moving too slowly. After shifting into neutral, the clutch should be reengaged, the engine speeded up, the clutch disengaged, and the shift completed.

A clutch brake may be installed on the jackshaft which will stop it from spinning when the clutch is fully out. This makes shifting easier from low to high, but might make it more difficult from high to low.

Constant Mesh. In a constant-mesh type of transmission, the main shaft gear would be permanently engaged with the countershaft gear, but would spin freely on the main shaft unless keyed to it by a sliding jaw. This shifts more quietly because of special tooth design, and the slow speed of the gear hub as compared with its teeth.

A synchromesh transmission is a constant mesh type in which leather collars on the two jaws touch before the teeth do, providing just enough friction to slow or speed the jackshaft so that the jaws will synchronize and mesh quietly.

Power Shift. Transmissions that can be shifted while transmitting full engine power to the wheels are called power shift or shift-on-the-go units.

Power shift transmissions generally use sets of multiple disc clutches to control constant mesh gearing of either planetary or countershaft design.

Shifting is usually done by hydraulic pressure. Action may be anything from cushioned jerks to smooth changes, with smoothness becoming more common in newer designs, which may be called "soft shift." Gear clashing cannot occur.

Power shift units are usually teamed with torque converters, which effectively absorb drive train shock loads caused by changes in gear ratios. The transmissions may be designed specifically for cycling applications (loader or dozer) or for hauling (truck or scraper).

Cyclic Transmission. The Allison cycling transmission in Figure 12-24A is a twin turbine, drop-type planetary model

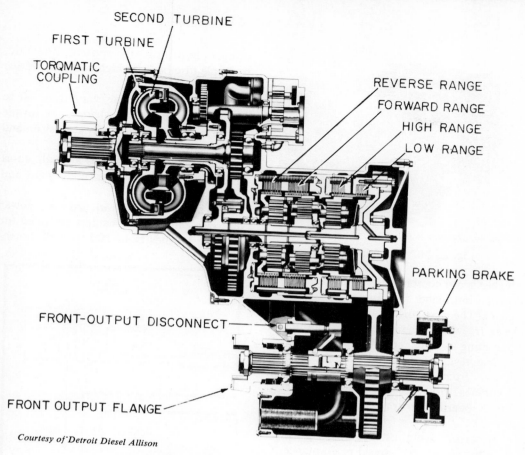

Fig. 12-24A. Cyclic transmission

with four forward and two reverse speeds.

Two speeds are provided automatically by the twin turbine torque converter, in which each turbine is connected to its separate output gear set, Figure 12-24B.

These gear sets are connected to the range gearing, which consists of a reverse and a low planetary gear set, plus a high gear clutch.

Range shifting is done by means of a control valve that directs hydraulic pressure to multiple disc clutches, which hold or release parts of the gear train to accomplish speed changes.

Each of the two turbines in the converter turns a different combining gear, to drive the forward-reverse range gears.

When vehicle motion is started, oil flow within the converter causes the first turbine

to turn, driving a low speed combining gear, to deliver high torque to the range gears.

As vehicle speed increases, the second turbine assumes the drive and, through its

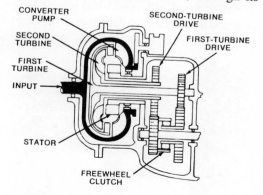

Fig. 12-24B. Twin turbine torque converter

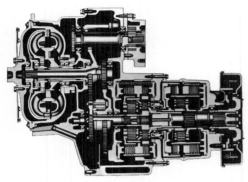

Courtesy of Detroit Diesel Allison

Fig. 12-24C. Short drop model

higher speed combining gear, delivers lower torque but higher speed drive to the range gears. The first turbine and its combining gear freewheel when the second turbine is operating at higher speeds.

The result is automatic 2-speed performance from the torque converter, which combines with two speeds in the forward range gearing to provide four forward speeds.

Reverse, having just one range, provides only the automatic 2-speed performance.

This twin-turbine transmission is available in several configurations, various capacity ranges, and with many options. These include long drop and short drop, indicating greater or less distance downward from input centerline to output.

Range speed in forward and reverse may be the same, but more often reverse is faster (2.66 to 1.96). Some models include an

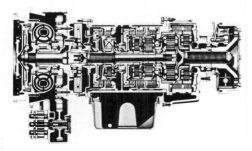

Courtesy of Detroit Diesel Allison

Fig. 12-24D. Dual-Path transmisison

integral hydraulically-applied, dynamic, mulitiple disc brake, which provides service braking for the vehicle.

Horsepower capacities vary from 70 to 235, with appropriate changes in torque converter, transmission physical size, and clutch plate areas.

Figure 12-24C illustrates a small, short drop model with equal ranges forward and reverse.

Hauler Transmission. Allison power shift transmissions for haulers are rated for use with engines from 100 to 1000 horse-

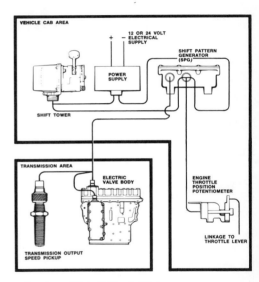

Courtesy of Detroit Diesel Allison

Fig. 12-24E. Automatic shift control

power. Generally, they have six speeds forward with one reverse, and consist of a torque converter, lock-up clutch, hydraulic retarder, splitter gearing, and range gearing.

The Dual-Path model is the largest and most recent of these. It divides power flow through the gearing to permit a more compact, lightweight design.

In first gear, torque from the converter is transmitted by the main shaft to a heavy duty first-gear combining planetary gear set which is connected to the output shaft. In each succeeding gear (second through sixth)

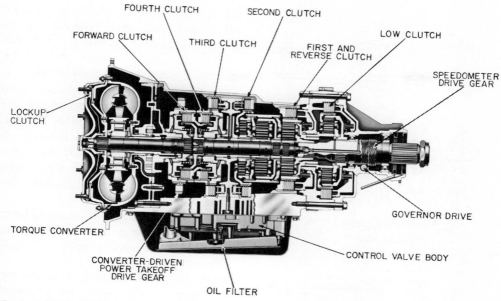

FOURTH CLUTCH   SECOND CLUTCH

FORWARD CLUTCH   THIRD CLUTCH   LOW CLUTCH

FIRST AND
REVERSE CLUTCH

SPEEDOMETER
DRIVE GEAR

LOCKUP
CLUTCH

GOVERNOR DRIVE

TORQUE CONVERTER

CONTROL VALVE BODY

CONVERTER-DRIVEN
POWER TAKEOFF
DRIVE GEAR

OIL FILTER

Courtesy of Detroit Diesel Allison

Fig. 12-24F. Automatic shift transmission

the torque converter is divided between the main shaft and a planetary gear section.

As the transmission is upshifted for increased vehicle speed, an increased percentage of power is directed through the planetaries along a hollow outer shaft, and a lessening amount is carried by the main or inner shaft. This "dual path" flow of power is combined by the first-gear planetary gear set.

This unit has an option of automatic electric shift.

Automatic Shift. Transmissions that change gear ratio without action by the operator, while transmitting full engine power, are termed automatic. They generally use sets of multiple disc clutches to control constant mesh gearing.

Automatic shifting is effected by hydraulic pressure or electrical signals that are proportional to vehicle speed, but are modulated or biased by a signal of throttle position. See Figure 12-24E.

These transmissions are teamed with torque converters, whose shock-absorbing and torque-multiplying properties are important for their success.

Many automatics are basic four speed forward with one reverse planetary types. There are also five and six range models.

A deep ratio low gear package may be built into a four-range model, to eliminate need for an auxiliary transmission in applications which have an extra heavy gross load.

CHAIN DRIVES

Roller Chains. Most gears transmit their power through direct meshing of teeth, but sprocket gears drive through roller chains.

Roller chains are used in revolving shovels in the track drive, crowd mechanism, and deck machinery, in ditching machines in track and digging drives, in grader final drives, and in many other places. It is usually the best drive to carry heavy torque at low speeds between parallel shafts that are at considerable or variable distances from each other.

Both offset and "standard" types are

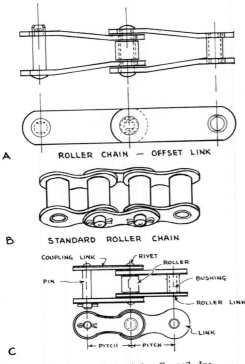

A   ROLLER CHAIN — OFFSET LINK

B   STANDARD ROLLER CHAIN

C

Courtesy of Power Transmission Council, Inc.

Fig. 12-25. Roller chain

used. Figure 12-25(A) shows offset construction.

The chain is a series of identical, tapered links, the narrow end of each fitting into the wide end of the next. The side plates are permanently fastened together by a hollow sleeve at the narrow end. Outside this sleeve is a roller sleeve which turns freely on the fixed inner sleeve. The side plates are fastened at the wide end of the link by a removable pin, which also passes

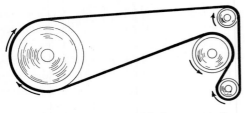

Fig. 12-26A. Rotation of belt-connected pulleys

through the inner sleeve of the small end of the next link, thus fastening the two together. Links can be strung together this way to any desired length, and the two end links joined to close the chain.

Standard construction is shown in (B) and (C). There are two types of link, each with straight, parallel side plates. The wide ones have removable pins fastening the plates together at each end; the narrow links are fastened together with sleeves (bushings) inside rollers. These links alternate in the chain.

Removable pins may be replaced by rivets, except for one or more left for opening the chain. Rollers are sometimes omitted.

Sprocket teeth must be the right size and shape to fit the chain or excessive wear will occur. As the chain goes around the sprocket it bends, hinge fashion, at each link end. The sleeve turns on the pin for the hinge action, and the roller contacts the sprocket teeth and guides the link into place, absorbing most of the scraping contact by a rolling motion. A small sprocket causes sharper bending and more wear in a chain than a large one. Short links bend less at each hinge and therefore can be used on smaller sprockets without damage.

If the chain has an even number of pitches (spaces between pins), the sprockets should have an odd number, and vice versa. This prevents a single link from contacting the same tooth each time, and favors more even wear and less vibration. It is called hunting tooth construction.

Sprockets connected by one side of a chain or belt revolve in the same direction. If connected on reverse sides, they revolve oppositely. See Figure 12-26A.

Drive between parallel shafts may be reversed by transfer between a set of meshed gears and a chain-and-sprockets set, as in the next illustration. One or the other (not both) of the sets is engaged with the driving shaft by a clutch, jaw, or tooth.

A chain or belt system needs a movable

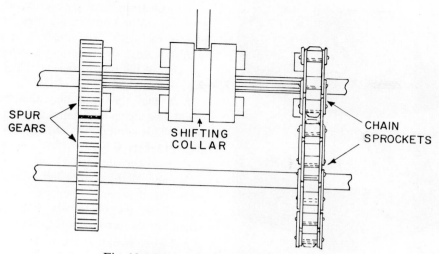

Fig. 12-26B. Chain-to-gear reversing shift

sprocket or pulley to permit adjusting tension.

Crawler tracks, described in Chapters 13 and 15, are special types of roller chain. Their load-carrying shoes may be bolted to chain links, or be links themselves.

Silent Chain. "Silent" chain, Figure 12-28, is made from a series of flat metal links which are tooth-shaped at each end. The ends are pierced for long cross pins. A center guide groove in the sprocket teeth, and retainer plates in the center links, may be used to hold the chain securely on the sprocket.

This chain is quieter and has less vibration than roller chain and can be run at higher speed. Its links are small in proportion to its strength so that it can be used on smaller diameter sprockets. It tends to cushion shocks and even out irregularities. It is more expensive.

Split pins or rockers are made in two longitudinal pieces, each of which is fastened rigidly in one of the overlapping sets of links. The two pin pieces rock on

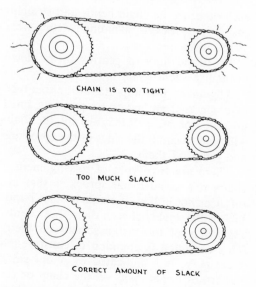

CHAIN IS TOO TIGHT

TOO MUCH SLACK

CORRECT AMOUNT OF SLACK

Courtesy of Power Transmission Council, Inc.

Fig. 12-27. Roller chain tension

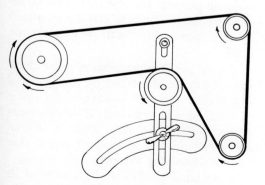

Fig. 12-26C. Adjustment pulley

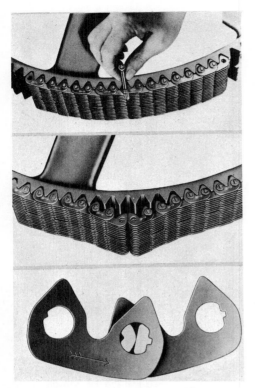

Fig. 12-28. Silent chain

each other as the chain bends and straightens at the sprockets, avoiding twisting wear in the links.

The usual limit of gearing up or down through a silent chain is 6 to 1.

Alignment. It is important for either type of chain that sprockets be properly lined up with each other. If they are not, the chain will tend to climb the sides of the sprockets and jump off them.

There are two types of misalignment — shafts not parallel and sprockets that are offset or not in line. If shafts are not parallel it is the angle between them that counts, regardless of the distance between them. This condition is common in construction machinery. It is indicated by a one-sided pattern of bright spots on the chain or the sprocket. Fig. 12-29A illustrates sprockets that are offset out of line. This condition

may result from poor installation, from the sprocket shifting on the shaft, or from end play in the shaft. It is checked by putting a straight edge or stretching a string along one side of the two sprockets. They should line up exactly.

If a shaft has end play the sprockets might line up at one time and not at another. This condition is checked by prying the sprockets sideward, first one way and then the other, with a bar.

Wear. Both roller and silent chains stretch as they wear. The case is often small enough so the chain will slap against it when it becomes loose. This slapping is a danger signal. It means that it is time to adjust the chain, or if there is no adjustment or the adjustment is fully extended, that the chain must be rebuilt or replaced. Otherwise the chain will probably start to jump, that is, to allow one of the sprockets to turn inside it. This will cause damage to the chain and the sprocket, and will prevent dependable delivery of power.

Chains that go around only two sprockets are adjusted by moving one of the sprockets. In engine drives it is usually the engine that is moved by loosening lock bolts, sliding it by means of crowbars or threaded adjusters, and re-locking.

If there is a third sprocket, adjustment is made by moving it inward or outward. A roller chain adjusting sprocket or idler may be either inside or outside the chain, while an adjuster for standard silent chain must be inside. Tension may be regulated by a threaded adjustment, by a spring, or both.

Repair and Replacement. A roller chain can usually be shortened by taking out a link in offset types. In standard chain a pair of links must be removed. If this shortens it too much, a half link, which is just an offset link used in a standard chain, may be substituted for the pair.

A stretched chain may have most of its life behind it. The proportion between its

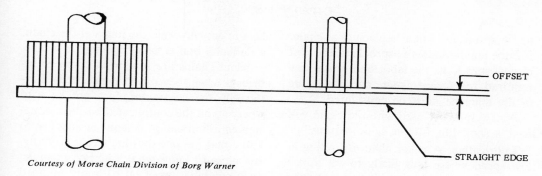

OFFSET

STRAIGHT EDGE

Courtesy of Morse Chain Division of Borg Warner

Fig. 12-29A. Checking sprockets for offset

length and the amount of stretch is important. If it is very long with short links a trifling amount of wear in each pin and bushing would add up to slack enough to justify removal of a pair of links. But that much stretch in a short chain, particularly if the links are long, would show the chain to be about worn out.

Each link gets a little longer as the chain stretches, so that they no longer mesh accurately with the sprocket. Even if proper tension is restored to a badly worn chain by removal of links the sprocket will be subjected to excessive wear, and jumping may occur.

Roller chains occasionally break links. This may be the result of a defective link, a shock overload, natural wear, or wear from rubbing against some stationary object. A few spare links should be kept on hand, as their help can put a machine back in service in a few minutes, that otherwise might be deadlined a week waiting for the parts. Links can be obtained from the equipment distributor, or from a sales agency for a chain manufacturer.

Broken links are often a symptom of severe wear. If they are badly worn, a new chain is probably needed.

An old chain with a few new links is likely to vibrate and pulsate except at low speeds. This would be unlikely to be noticed in crawler drive chains, it might or might not in a shovel crowd, but would be almost sure to show up in a shovel's transmission case.

It usually does not pay to try to prolong the life of a roller drive chain by repinning and rebushing, as the cost is likely to be as great as that of a new chain.

A silent chain can generally be restored to full usefulness by repinning if its stretch is less than 1.5 per cent. This job must be done by the chain dealer, as it requires special tools and equipment.

Sprocket teeth wear from contact with the chain. By the time the chain is worn out the teeth have had a definite change of shape, and will not fit a new chain properly. As a result, a new chain installed on old sprockets may have only half the life of the original. It is therefore a good idea to re-

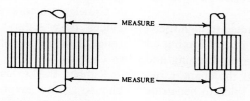

MEASURE

MEASURE

Courtesy of Morse Chain Division of Borg Warner

Fig. 12-29B. Checking shafts for parallel

12-21

place sprockets when replacing a chain. But the condition of the sprockets, the cost of new sprockets, the labor and down time required to replace them, and the value of the machine must be considered.

A new sprocket used with old chain will have a short life, for the same reasons.

Lubrication. A roller chain operating at low speed where it is likely to get mixed up with dirt should not be lubricated. The grease and oil will simply pick up sand and clay and make them into a grinding compound. Such a chain should be run dry, and occasionally taken off and washed in kerosene or fuel oil.

A shovel crowd chain moves too fast to operate dry. It does not get down in the dirt, but on many jobs it gets a steady sprinkling of dust. It is not practical to enclose it in a case. It is lubricated by painting with oil or thin grease, anywhere from twice a day to once a week.

Engine and transmission drives run in covered cases and are lubricated by partial submersion in an oil bath. There may also be a spray of pumped oil. Silent chains are always run in covered cases with lubricant, as they would be quickly ruined by exposure to dirt and dust.

The oil that does the work on a roller chain is not what is visible on the surface, but that which is between the sleeves and the rollers, and the rollers and the pins. Since it has to soak in, a thin oil is usually best, but good penetrative qualities are sometimes obtained with thicker oils and light greases. Be careful about this, however, as it is no use keeping the outside of the chain looking well lubricated if its insides are dry.

Roller chains are not suited to high speed work because there is no way of forcing oil continuously between the pin-bushing and bushing-roller surfaces. If the chain is run too fast, or the oil is too thick to penetrate or too thin to lubricate, sliding friction between pin and bushing may create enough heat to actually weld the joints of the chain, a difficulty that is called galling.

Silent chains are somewhat tricky to lubricate when they run at high speeds. Fortunately the contractor does not have to worry about the complicated business of designing a lubrication system—he just has to know that the one that is there is working. But he must know what it is. For example, on a combination of oil bath and pumped spray, lubrication is not adequate if the pump fails and the chain is being lubricated by the bath only.

If redesigning equipment will result in speeding up a chain, a manufacturer's engineer should check to see if lubrication is adequate for the higher speed.

Silent chain lubricant is good quality non-detergent petroleum base oil. It may have inhibitors against foam and rust, and to improve film strength.

Heavy oils and greases are not used except under exceptional conditions. Multiple viscosity oils, gear oils, and extreme pressure compounds are not used at all.

Automotive crankcase oils classified as AP1 Service ML are satisfactory.

Viscosity should be 20 for outside temperature up to 40° F., 30 from 40 to 100°, and viscosity 40 from there up.

Tracks. The crawler track is a specialized type of roller chain. There are two principal types.

Linked shoe tracks, described in Chapter 13, are made by fastening shoes or pads into an endless belt by means of hinge pins. They are used in the majority of machines that do not depend on thrust against the ground for most of their working force.

Crawler tractors and some other machines have separate shoes bolted onto roller chains. This construction is described in Chapter 15.

Rubber tracks, used on light fast machines, are an adaptation of the belt construction to be described below. A wide rubber band, reinforced by the steel cables, is stretched around the track wheels. The bull wheel may drive it by meshing the teeth with metal plates vulcanized in the rubber, or by friction of grooves against V shaped ridges in the rubber.

BELT DRIVES

Flexible belts made of fabric and rubber, or more rarely of leather, are widely used to convey power between parallel shafts. Such drives are easy to design and are usually cheap to make. They absorb shocks, are cheaper, and are easier to install and service than chains, but will not last as long, nor carry as heavy loads, and cannot be used where exact timing is required.

A belt will stretch under a heavy pull, and resume almost its original size when released. This characteristic makes a belt drive a shock absorbing one that gives some protection to the machine. An overloaded belt will usually slip, acting as a safety clutch to prevent damage. However, this slippage makes it impractical to use belts for most heavy power applications.

Pulleys. A belt runs around two or more pulleys. Those in contact with the inside of the belt will all turn in one direction. Any pulley in contact with the outside of the belt will revolve in the opposite direction, in the same manner as with roller chain.

One pulley in any belt system should be adjustable, Fig. 12-26C, as belts wear and stretch, and must be tightened by moving the pulleys apart. If the pulley can be moved rapidly and locked in tight and loose positions it can be used as a clutch. If it is loose, the drive pulley will turn inside the belt without moving it, but when tightened friction will cause the two to turn together.

If two pulleys in a belt system are different sizes, the larger one will turn more slowly. The driven shaft can be made to turn faster or slower than the driving or engine shaft by proper selection of pulley sizes.

A belt drive can be made to provide sev-

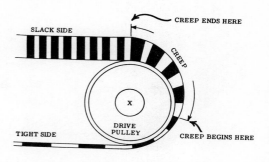

Fig. 12-30. Change in belt shape

eral speed ratios. Several pulley sizes are mounted side by side on the same shaft, and the belt is shifted from one set to another as required. Or a V-pulley may have a movable side which when slid inward will cause the belt to ride higher in a larger circle.

A belt changes shape as it goes around a pulley, as it is thicker on the slack side than on the loaded side. See Figure 12-30.

Flat Belts. Flat belts are of two types: the endless, which is made in a closed circle; and cut, which is fastened into the circular form by a pin passed through wire loops crimped into its ends.

A center bulge in the flat pulley helps to keep the belt on. Centrifugal force tends to throw the belt outward so that it tends to move to its line of greatest tension across the pulley, and stay there. A bulge puts more stress on the center of the belt than on the sides. Unless the pulleys are in good alignment with each other—their axes of revolution parallel to each other, and perpendicular to the direction of travel of the belt—the belt will climb the tighter side, and will probably slip off under heavy load. Flat belt require more tension than V-belts, to avoid slippage.

Flat belts are cheaper than V belts and are usually longer lived because their shape reduces internal friction, but they need much more attention. They are preferred when the work is done by a different machine than that which supplies the power,

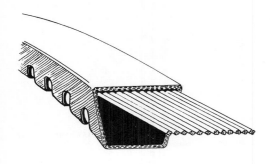

Fig. 12-31. Reinforced lugged belt

as, for a pulley on a tractor driving a cement mixer, a buzz saw, or a pump.

Conveyors. Conveyor belts are wide flat belts that transport dirt or other material. They will be discussed in Chapter 14.

V-Belts. The V-belt is the standard means of driving permanently mounted light load accessories, such as fans, generators, and water pumps.

V pulleys have such high steep flanges that the belt cannot climb up them readily. The driving friction is entirely on the sides of the belt, and is very effective because of the wedge action of the belt taper.

V-belts are kept small in cross section to minimize internal friction and heating. When too much power must be transmitted for the strength of one belt, two or more are used, working in parallel grooves in the same pulleys. Such belts must be the same size and dimensions or the tighter one will carry too much of the load, with resultant strain, slippage, and rapid wear. Since belts stretch in service, this means that if one belt breaks the whole set must be replaced.

Lugged belts have a series of slots in the inner surface. This gives them greater flexibility, better grip on the pulleys, and reduces fatigue so that they last longer. They are particularly good for small pulley heavy duty work. They are recommended whenever a chronic slippage problem is encountered.

A belt drive may slip because of a slippery glaze on belt and pulley surfaces. This can sometimes be removed by wiping with gasoline, or dusting with fullers earth.

Fit. V-belts are made in several side slopes. If one flares out more sharply than its pulley groove, contact will be made only at the upper belt corners. This gives it a poor grip, causing damaging distortion of the belt structure, excessive wear on the corners, and damage to the pulley groove.

If the belt does not flare out as much as the pulley groove it will ride hard on the bottom corners, with about the same results.

If the groove is too wide or too shallow the belt will ride on the bottom, so that it will lose the wedging effect of the sides. Such a belt will require excessive and damaging tension to reduce or prevent slippage.

If the belt is so wide at the top that it rides high in the groove it will tend to turn over, and the cover will wear quickly along its top contact.

A rough or irregular surface in the pulley groove will give the belt a poor grip and is likely to cause rapid wear. Smoothing can often be done with a fine file, grinding stone, or emery paper.

Projections on the pulley surface cut into a belt very rapidly, because it creeps and changes shape as it goes around the pulley, causing it to slide along the surface.

Installation. To install a V-belt, move the adjusting pulley inward as far as it will go. Put the belt in the groove around the least accessible pulley or pulleys, or the hardest one to turn. Then try to slide the belt onto the last pulley.

If it will not slip on, pull or push it on as far as possible, then turn the pulley to wind on the installed part. If for any reason the pulley cannot be turned the belt may be pried onto it with a screwdriver, taking care not to damage the fabric.

When the belt is in the grooves of all pulleys, the adjustment pulley is moved outward until the belt is under light tension, and is then locked in place.

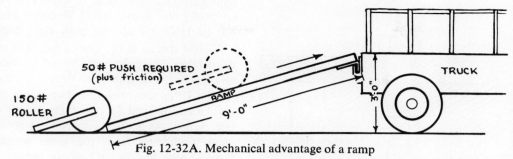

Fig. 12-32A. Mechanical advantage of a ramp

Correct tension for light service allows pushing the center of a section of belt inward about an inch for each foot of unsupported span between the two pulleys. For heavy service such as a traction drive in a small machine, the yield should be only half an inch to a foot of span.

An engine with a crankshaft driven front mounted power takeoff may require a major disassembly job to install an ordinary fan belt, or it may have a spring loaded jaw that can be pushed forward or back to open the shaft just far enough to slip a fan belt through its opening.

It is sometimes possible to obtain belts with a coupling that can be opened for installation, and closed after putting around the shaft.

INCLINED PLANES

Ramps. The inclined plane is applied generally in reducing force needed to lift heavy weights. Figure 12-32A shows a 150 pound roller that is to be placed in a truck body three feet above the ground. Two men could lift it onto the truck, but one can push it more easily up the ramp shown. A long ramp has less of an upgrade than a short one for the same height and will require less force to move an object up it, but the push must be exerted for a longer time. Friction may be very important in determining whether an inclined plane should be used, as a sliding load on a rough ramp might cause more friction resistance than its own weight.

The mechanical advantage (MA) of an inclined plane (ignoring friction) is found by dividing its length by its height. In the illustration it is three, which means that the force requirement of rolling up it is one third of that needed to lift it vertically.

Another use for ramps is to keep the object being raised in a good position. A heavy crate could be pulled onto this truck by means of a winch on the body. Although there might be plenty of power to lift the crate straight up, without the ramp it would catch under the rear of the truck body and could not be loaded without the use of other devices.

Threads. The house jack shown in Figure 12-32B is an application of inclined

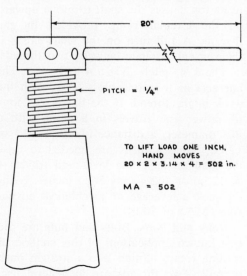

Fig. 12-32B. Screw-type house jack

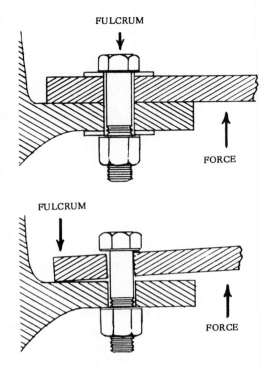

FULCRUM

FORCE

FULCRUM

FORCE

Fig. 12-33. Leverage against loose bolt

plane mechanics. The 2″ center shaft of the jack carries a spiral thread which fits within a similar thread in the jack body. Turning the shaft by means of a bar through a hole in its top causes the shaft to move upward on the circular ramp provided by the threads. A full turn on the shaft raises it the vertical distance between two threads.

The 1/4″ pitch gives a 4 to 1 ratio. The bar acts as a second class lever, with the work at the thread. If the bar is 20″ long its power end travels in a circle with a 40″ diameter, a distance of 40 × 3.14 or 125.6. Since 4 turns are needed to raise the load an inch, the lever end travels 4 × 125.6 or 502.4 (say 502) inches, giving the jack a leverage or mechanical advantage (MA) of 502.

Bolts and Nuts. Bolts and nuts are the best known applications of this method of turning rotary motion into a straight push or pull. Two threads are in general use, known as coarse (or more specifically, as

Unified National Coarse, abbreviated NC or UNC) and fine (Unified National Fine, NF or UNF for short), both right hand.

Coarse thread is the most used. It is cut deeper into the bolt, each thread is wider and heavier, and it is more steeply pitched so that it tightens with only two-thirds the revolutions needed for the same length of fine thread. As a result, it does not get as much tightening from the same wrench force as with fine, but it tightens faster.

Coarse threads, when compared with fine, are easier to start without cross threading, particularly in hard-to-reach areas, and are less likely to strip or to be ruined by rust or careless handling. Fine theads will develop greater bolt tension for the same wrench pull, they are more suitable for adjustments, take less away from bolt strength, they are somewhat less likely to work loose, and their heads and nuts are narrower so that they require less space.

A fine thread bolt is about 10 per cent stronger than a coarse thread of the same size and material. However, the thread itself is much weaker.

In general, a coarse thread will have enough strength to allow breakage of the bolt by turning the nut, but a fine thread will strip before the bolt breaks.

Three grades of steel are in common use in bolts in heavy equipment. They are rated by minimum tensile strength (resistance to permanent stretching) in pounds per square inch, abbreviated psi.

Grade 1 is a low carbon steel rated at 55,000 psi. It is used in machine bolts that have square heads, square nuts, coarse threads, and a black finish. Their use is limited because of low strength and the inconvenience of square nuts in places that are hard to get at.

Grade 2 is a somewhat stronger low carbon steel rated at 64,000 psi. It is used in bolts and nuts of natural steel color (bright finish), with six sided heads and nuts, in both fine and coarse threads.

Grade 5 is a medium carbon steel strengthened by tempering to a rating of 105,000. Finish is black, and threads are usually coarse. It is identified by three radial lines on the bolt head, and may have a manufacturer's symbol too. Nuts are higher than standard.

The use of these Grade 5 bolts cannot be too strongly recommended. Their extra strength is worth many times their slight extra cost.

Bolts must be tight to hold. Figure 12-33 shows how looseness may increase the leverage against them. Looseness also encourages nuts to loosen further.

Other Threads. Pipe thread is a different coarse thread, used only on pipes and fittings.

Thin walled tubing usually has a fine thread. Pipe and tubing thread will not mesh with each other, and the tubing will be damaged if an attempt is made to force them.

Wedges. The wedge is an allied device that is capable of building up enormous pressure at an angle to the direction of the force, and which can convert hammer blows into steady pressures. A wedge hammered into a log to split it is a familiar example. Friction losses are very large in most wedging work.

Cutting Edges and Teeth. The cutting edges of most excavating machines are wedge shaped, at least when new. This shape is strong and has high penetrative and disruptive ability.

A wedge tooth entering material which resists equally on both faces will tend to move in a direction midway between the slopes of its faces. If resistance is unequal the wedge will tend to slide on the face contacting the greatest resistance. Figure 12-34 shows the actions of a wedge tooth on various slopes while being moved horizontally. Held at the pitch (A) it will penetrate steep slopes; at (B) it will cut into any kind of slope but won't dig down; whereas

held as in (C) it will penetrate upslope or level. It will be noted that it is the angle between the under side of the tooth and the ground that determines its penetration. (D) shows the same angle as (C), but the point and part of the under slope of the tooth are worn away, destroying its "suck" and therefore its penetration.

Sufficient down pressure will make any of these pitches or conditions dig down to some extent, but this is an expensive and inefficient substitute for properly positioned and sharpened teeth. However, in hard materials, weight or down pressure greatly assists the digging by forcing the point of the wedge into the soil and keeping it there.

Separated teeth cut into hard earth and rock much better than a continuous cutting edge because they concentrate the force in small areas, allow the material to be displaced to each side as well as above and below, and usually so shatter or weaken material between them that the lip at the tooth bases can easily penetrate it.

A moldboard is a continuation of the upper surface of the tooth or cutting edge, and its shape is important to the efficiency of many machines. In a farmer's plow it

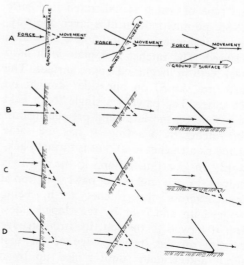

Fig. 12-34. Penetration of a wedge

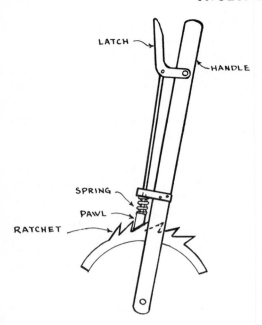

Fig. 12-35. Pawl and ratchet

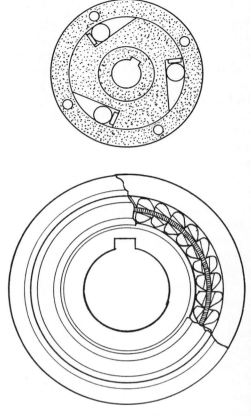

Fig. 12-36. Free-wheeling units

turns the slice of earth on its side or over and presses it into place. In a grader or dozer it rolls the earth upward until it falls forward, lessening friction in the load being pushed, and facilitating drifting of the material sideward if the blade is angled.

Ratchet and Pawl. The ratchet and pawl, one type of which is illustrated in Figure 12-35, permits motion of a lever one way, and resists reverse motion unless a catch is released. The ratchet is a plate with hard steel teeth in its curved upper surface. The teeth are sloped on one side, straight on the other. The pawl is a single tooth held in a slide on the lever and pressed against the ratchet by a spring. It has the same shape as a ratchet tooth but is turned backward. In the illustration, if the lever is moved to the right the pawl will move freely across the ratchet teeth, being wedged up and over by the sloped sides. But if moved to the left, its straight edge catches on the straight edge of the first tooth and remains until pulled up by a latch.

Over-running Clutch. An over-running clutch, or free-wheeling unit, may be a circular ratchet, or the types illustrated in Figure 12-36. In the first example, cylinders of graduated sizes lie in tapered hollows between inner and outer hubs. Rotation of the outer hub clockwise (or the inner one counter-clockwise) will move the cylinders into the small end of the slot, where they will jam the two parts together so that they will rotate as one. Reversing the rotation of either shaft will move the cylinders to the large end of the slot, where they can revolve freely without preventing the inner and outer hubs from rotating independently.

In the other type shown, rotation in one direction will cause the cams to lock the two hubs, and in the other will release them.

Figure 14-23 shows a circular ratchet with an automatic device that prevents unnecessary tooth contact.

CONTROLS

Levers. Clutches and brakes may be operated by foot pedals or hand levers The pedal is pushed down to release the clutch or apply the brake, and is returned to position by springs that may be in the unit, in the linkage, or in both.

Hand levers may be pushed or pulled. A single lever may control two clutches (example, the swing lever on a shovel) and move both ways from a center neutral or released position. On single controls, the linkage can sometimes be changed to reverse the direction of operation.

A lever is ordinarily returned to neutral by springs and/or compression of lining and parts in the unit. It may be held engaged by a hand operated pawl and ratchet, by a ball or wheel pushed into a socket by a spring (ball detent), or more commonly by locking over center.

Over center locking may be obtained in different ways in various linkages. A simple type is shown in Fig. 12-37A, in which a sliding collar causes two arms attached to it to move outward. Maximum pressure against the arms, and through them against the friction lining of the attached shoes, is obtained when the arms are in line with each other. Moving the collar further, against the stop, relaxes the pressure slightly. The collar is now held between the stop and the pressure required to move it back across center, and it will stay put until force is applied to move it.

Rod Linkage. Connections with mechanical controls may be of lever, clevis and rod construction, or stiff wire in a flexible casing.

An example of a pedal and rod linkage is shown in Figure 12-38I. This can be designed to carry pull (and somewhat less push) at a distance and to the side. It provides for remote control, leverage, adjustment, and change of direction.

This type of system is often difficult to design, but once built is sturdy and durable. Difficulties include friction in numerous pivot points, and the fact that slight wear in each of many spot adds up to lost motion that may make control difficult. Careless reassembly after an overhaul, particularly failure to get arms nearly perpendicular to the direction of pull, may make the system ineffective. A loose or bent support will allow the cross shaft to twist and bind.

It often happens that a machine has clutches and brakes which work easily and smoothly when new, but which gradually degenerate so as to require excessive effort. Relining, turning down of drums, or routine overhaul may fail to restore their efficiency. In such cases, the trouble may be in some adjustment which is repeatedly made wrong. More often, it is lost motion in the linkage. A tiny looseness at each clevis and pin, inside the clutch, in its connections with the pedal or lever, and weakening of arms so that they twist may be sufficient to destroy the delicate balance which is necessary for proper functioning. Complete rebuilding or replacement of the linkage may restore efficiency, at less expense than frequent stops for adjustments, and shutdowns for relining.

Pivots should be oiled daily or weekly, depending on amount of use. Cotter pins should be inspected and replaced if wearing or broken. Arm clamp bolts must be

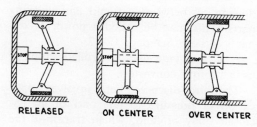

RELEASED   **ON CENTER**   **OVER CENTER**

Fig. 12-37A. Locking over center

kept tight to prevent sliding off or loss of keys.

Adjustments are made by taking the pin out of a threaded clevis, and turning it in or out along the rod, or by loosening and tapping off a splined arm and replacing it in a different position.

Flexible Wire. This connection consists of a flexible conduit held by clamps, and a stiff steel wire inside it that is pulled and pushed a short distance by a small finger grip. This device is simple, neat, and cheap to design and install, and works well when new. Unless it is exceptionally well made, rust and dirt make it harder to use, and the wire will eventually break where it is clamped to the mechanism, or perhaps the handle will pull off.

Its useful life will be prolonged by an occasional application of light lubricating oil to the sliding part of the handle, and to the entire outside of the conduit. The business end of the wire should move in and out as straight as possible, as bending will ultimately break it.

If the end of the wire breaks off it may be possible to shorten the conduit so that enough wire will be exposed for refastening. Otherwise the wire or the whole unit must be replaced.

The conduit should be anchored at enough points to prevent it from bending as the wire is pulled or pushed. Such motion causes waste travel distance of the control, and may prevent it from working dependably, or at all.

Hydraulic. There are at least three types of hydraulic control. One has valve-directed flowing oil between a pump and working parts. This may be called a power or dynamic system. See page 12-72.

A second type uses controlled slippage, as in a torque converter.

The third, discussed here, involves application of pressure to one part of an enclosed body of fluid, which transmits the pressure to other units in the system, mov-

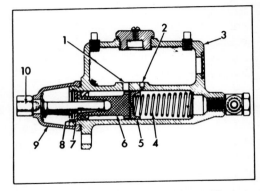

1. Intake port	6. Piston assembly
2. By-pass port	7. Piston stop washer
3. Body	8. Piston stop wire
4. Spring	9. Boot
5. Cup	10. Connecting link

Courtesy of Harnischfeger Corporation

Fig. 12-38A. Master cylinder (compensator)

ing them against working resistance and springs.

The most important and best known application of this static type of control is in automotive hydraulic brakes. For convenience, it will be called a hydraulic brake system here, although it has many other uses in machines.

A hydraulic brake system includes a master cylinder with a built-in reservoir, that is sometimes called a compensator; lines, and wheel or acting cylinders.

Pressure on a pedal or pull on a lever moves a piston in the master cylinder that forces fluid through the lines to the wheel cylinders, where the pressure forces other pistons to move and apply brakes or do other work. When pressure is released springs move the wheel pistons back to their original positions, forcing the fluid back into the master cylinder.

Hydraulic brake systems are nearly frictionless, and usually require less force to operate than mechanical systems do. Equal pressure is supplied to each wheel cylinder, but differences in action may be caused by design differences in wheel cylinder sizes or brake linings, or accidentally by different, dirty, oily, or wet linings.

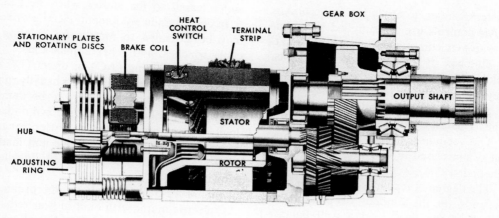

STATIONARY PLATES AND ROTATING DISCS
BRAKE COIL
HEAT CONTROL SWITCH
TERMINAL STRIP
GEAR BOX
OUTPUT SHAFT
STATOR
HUB
ADJUSTING RING
ROTOR

Courtesy of WABCO

Fig. 12-38B. Electric drive for winch

Maintenance consists chiefly of keeping the system full of fluid and free of air. Low fluid shows up by free play at the top of the pedal, and a corresponding lowering of the point at which the brake becomes effective.

Air may enter because of low fluid, leaks, or disconnecting lines during repairs. It makes the pedal feel spongy and the brakes become unresponsive and may not work at all.

Air is removed by bleeding. This is done in the field by filling the reservoir, putting gentle pressure on the pedal, and opening bleed screws in the wheel cylinders, one at a time. Fluid is allowed to flow out of the bleed opening until it is free of bubbles. The screw is re-tightened, and another opened in its turn. Several fillings of the reservoir may be necessary.

The wasted fluid may be caught in a container by means of a special bleeder tube, soaked up with rags, or allowed to drip, depending on conditions.

Leakage of fluid onto linings destroys their smooth-friction quality.

Brake systems use rubber in their sliding and sealing parts. It may swell and rot upon contact with any petroleum product, which includes most "hydraulic fluid". They can use only genuine brake fluid, whose main ingredient is glycerin.

Hydraulic brakes may have a vacuum booster. This mechanism is described in Chapter 18. It must be bled in addition to the wheel cylinders, and may have as many as three bleed connections.

Leaking lines must be replaced. Leaking cylinders can often be rebuilt.

Power Controls. Clutches, brakes, and other machinery controls can be operated by hand or foot pressure, or partly or wholly by power. In general, small and light machines will respond easily to manual controls, while heavy ones often need some sort of booster action.

The power may be mechanical, vacuum, air, hydraulic, or electric.

Mechanical boosters include brakes which are not strong enough to stop the drum against which they work, and which will be shifted by contact with the drum so as to pull or push on a linkage that will apply a larger brake.

Vacuum employs the intake suction of a gasoline engine, or a separate vacuum pump, to pull a piston in a cylinder. The movement of this piston may operate a brake through a mechanical linkage, or combine with pressure from a hydraulic master cylinder, to apply hydraulic brakes.

Air controls use air supplied by an engine driven compressor to operate brakes or clutches by moving short mechanical linkages, expanding flexible tubes so that they press against the bands, shoes or discs, or pushing pistons in hydraulic brake systems. The last is called "air over hydraulic."

The vacuum-hydraulic and air-mechanical systems used in trucks are described in Chapter 18.

The degree of response from an air brake depends on the amount of pressure admitted to the lines, and a well balanced valve is required for smooth action.

Hydraulic power systems are described later in this chapter, and in discussions of bulldozers, graders, and heavy trucks.

Electric controls operate by supplying current through a rheostat or other regulator to electric motors or magnets attached to mechanical linkages. Increase of current increases the force of application.

Power controls often eliminate a complicated mass of mechanical linkage which is hard to service. They permit using brakes or clutches of adequate size and rugged design without much consideration of the effort required to apply them.

However, in general they do not give as precise control, nor the "feel" of the load, as well as first class manual applications.

A power system of either the booster or the full acting type should be almost effortless to operate. However, in some of them valves are so poorly balanced, or such heavy springs are used on the levers, that effort expended is much greater than on a comparable manual unit.

ELECTRIC DRIVE

Accessories. A familiar example of electric drive is a generator on an automobile engine supplying electricity to a heater motor or a windshield wiper. Another example is the electric starter. In this case the generator's output must be stored in a battery,

and released to the starter when needed.

In construction equipment, electric drive may be extended to motors for steering, cable winches, and other applications.

Either alternating (AC) or direct (DC) current may be used. AC motors usually run at a set speed, and are likely to be damaged by low voltage or lugging down under load. DC motors will run at a wide range of speeds, depending on voltage and load, and are not easily damaged by lugging down.

Figure 12-38B shows a weatherproofed AC electric motor with built-in gear reduction designed for winches that control cable scrapers and dozers. It is equipped with a spring loaded brake that automatically locks the unit whenever current is cut off.

This motor is said to be able to accelerate from a standing start to its full speed of 180 revolutions per minute in 1/5 second, and to reverse for full power in the opposite direction in 1/4 second.

Such motors are installed at the point where power is applied, and substitute electric wiring for drive shafts and cable lines.

Diesel-Electric. In this type of drive, most or all of the engine power is used to turn a generator that supplies electricity to motors that do the heavy work of the machine.

This type of drive is found in medium large shovels and in dredges. It is standard in most railroad locomotives, and is being introduced in big scrapers, off the road trucks, and rollers.

Full Electric. Power for electric motors may be obtained from a ground cable or trolley wires connected to a stationary source of current. This may be a "high line" or regular industrial supply system, or a diesel generator unit set up for the job.

Full electric drive is simple and quiet, and avoids fueling problems. It is necessary where the power requirement of machines exceeds that of diesel engines they can carry, and is desirable under many other conditions.

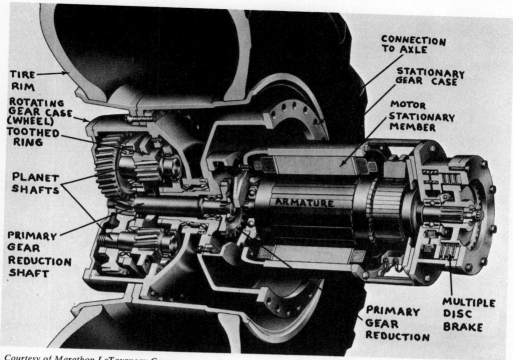

Courtesy of Marathon LeTourneau Company

Fig. 12-38C. Electric drive wheel

A machine may be made so that it can operate on either its own diesel-electric system or on outside power.

Electric Wheel. Diesel-electric may be applied to haulers by the electric wheel, a self contained unit that includes a DC motor of 40 to 400 horsepower, reduction gearing, a wide base rim, and a tubeless tire. Figure 12-38C shows a cutaway view of one.

The stationary member or gear box bolts to the outer end of a dead axle. The pole ring of the motor is bolted to it. The rotating armature is geared down to turn the primary reduction shaft. This drives a compound set of planetary gears in a stationary carrier. The large driven planet gears turn short shafts carrying smaller inner gears that are meshed with an internal toothed ring that is bolted to the rotating gear case that makes up the wheel of the machine.

The multiple disc brake is mounted on an inner extension of the armature shaft. It is applied by springs, and released by an electromagnet. It is applied automatically whenever there is no current in the brake circuit, either because the controller is in the off position, or because of a general power failure.

The motor also provides regenerative electric braking that is sufficient to slow the machine on down grades and stop it in normal operation. The disc brakes are used for sudden stops and for parking.

Power is usually provided by a diesel driven generator mounted on the machine. It may also be supplied by overhead trolley wires. A single machine may be designed to use current from either source.

Current is controlled by hand switches and rheostats. Since the motors are reversible the same peformance is obtained backward and forward, and the operator's station may swivel to face either way.

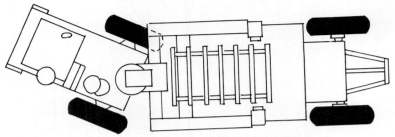

Fig. 12-38D. A scraper's articulation is forward

The electric wheel is a recent development. At present it is finding its principal construction applications in large all wheel drive scrapers, rollers, and trucks. Further references will be found in the sections about these machines.

WHEEL STEERING

Wheeled equipment uses many different steering systems. They include manual, booster, hydraulic, and others.

The manual type that was standard on all cars until a few years ago, transmits power from wheel through a worm gear to the steering linkage. Its operation is too familiar to justify description here.

Power Booster. This mechanism is similar to manual steering but it is geared higher so as to involve less spinning of the wheel, and includes in the linkage a hydraulic piston and valve. It is described in Chapter 18.

Operation is similar to power steering in a car. The maximum effort exerted to turn the front wheels under the worst conditions is the same as that required while rolling on a hard roadway. In each case it need be just sufficient to compress a spring.

If the front wheels are jammed so that they will not turn, a relief valve will open in the case. This may make a noise and/or cause the wheel to pulsate. A properly designed booster will not exert enough pressure to damage the linkage, but an overstrength one might, so it is best not to try to force it.

The pump is usually mounted on the en-

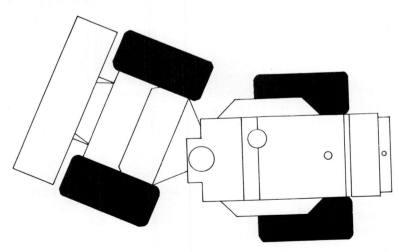

Fig. 12-38E. A tractor's articulation is centered

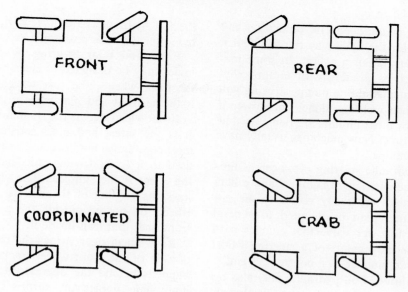

Fig. 12-38F. Front-and-rear steering

gine and driven by a belt. The most common difficulty is a loose belt, which will cause heavy, sluggish steering.

Steering may also be stiff when the oil is cold. The booster does not work at all if the engine is not running.

Brake-Assisted. If a brake is applied so as to slow one side of a vehicle and not the other, the vehicle will turn toward the braked side. One-side braking is the only steering means in most crawlers, and in a few skid-steer wheel tractors.

Most rear-drive wheel tractors have a combination, with sharp automotive-type angling of the front wheels, and individual braking on the rears. Most steering is done with the front wheels only. But when pulling heavy loads, and/or making very sharp turns, the front wheels may skid ahead instead of turning the tractor. This tendency can be corrected by applying the rear brake on the inside of the turn.

Full Hydraulic. This type of steering turns the front and/or rear wheels directly or through simple linkages by means of one or more hydraulic rams. Control is by a

valve operated by a lever, or less often by a wheel.

In this arrangement the position of the steered wheels depends directly on the position of a piston in a ram, as the height of a hydraulic dozer blade depends on its ram pistons. There is no direct relationship between the position of the control lever and that of the wheels.

For example, in making a left turn the lever is moved to the left. The wheels will keep swinging left as long as the lever is left, until they reach their stops. If the lever is moved far over they will turn at maximum speed (this may be disturbingly fast), if it is moved slightly they will turn slowly; but in each case they will go all the way.

When the wheels have reached the desired angle the steering lever is moved back to center or neutral, holding them at the angle. When the turn is completed the lever must be moved to the right and held until the wheels are straight.

It will be noted that this is very different from the automotive steering to which most of us are conditioned. Its difficulties are increased for the novice by the fact that it may be very rapid in action. In some machines such as graders with multiple hydraulic systems on one pump, steering will be fast when working alone, but slower if blade lift or other circuits are being used at the same time. New machines usually have separate pumps.

Full hydraulic steering of wheels is limited to machines such as graders and rollers whose wheels are easily visible to the operator. But it can be adapted to general use by a follower valve that enables it to follow the movements of a steering wheel in the same manner as a booster system.

Articulated. Full hydraulic may also be used to turn a whole section of a machine on a massive, vertical hinge. One or two double acting cylinders are employed. This construction, called articulated steering, has been standard in self powered overhung (four wheel) scrapers since they first appeared. It has also become the preferred construction in large four wheel drive tractors.

In the scraper, Figure 12-38D, the hinge, (kingpin) is near the front of the machine, and it is always the front or tractor part that swings when steering, with a result close to (although not exactly like) turning the front wheels.

In the tractor loader, Figure 12-38E, the hinge is midway between the axles, so both parts share equally in the swinging. This action produces the effect of the four wheel coordinated steering described below. Front and rear wheels run in each others' tracks, backward and forward, and the whole weight of the tractor lines up efficiently behind any pushed load.

Electric. Some scrapers have used electric motors, operating through reduction gearing, to control turning around the kingpin.

There were two fingertip switches, one for right and one for left. Operation followed the full hydraulic pattern, requiring return by power to center position.

Front and Rear Steering. Wheeled vehicles may be designed to steer by angling the front wheels, the rear wheels, and/or both the front and rear wheels.

Front wheel steer is the standard method. It is the safest and most satisfactory for most conditions, both because operators are used to it from driving cars, and because the effect is the simplest. The vehicle follows the angling of the wheels. The rear wheels do not go outside the path of the front ones, but trail inside it.

Rear wheel steer swings the rear wheels outside of the front wheel tracks. As this happens behind the operator it may give him some unpleasant surprises in close quarters. The principal advantage is greater effectiveness in handling off-center loads at either the front or rear, and in preventing "drift" down a side slope. It was widely used in front loaders in preference to front steer, as it kept machine weight squarely behind the bucket on turns, and kept the front wheels tracking inside the rears while backing from a bank, close to a truck. But it has been replaced by articulation.

A vehicle with both front and rear steer can use it in two separate ways. In coordinated steering the front wheels are turned one way and the rears to the same angle in the opposite direction. The wheels will then track, that is, the trailing wheels will always move in the same tracks as the leading wheels, whether the machine is moving forward or backward. This feature lessens rolling resistance in soft ground, as one pair prepares a pathway for the other.

Coordinated steering also provides maximum control of direction under load, enabling the machine to keep on a straight course in spite of side forces, and to force loads to change direction. It also permits accurate cutting along curved lines, and short turns in proportion to the maximum angle of the wheels.

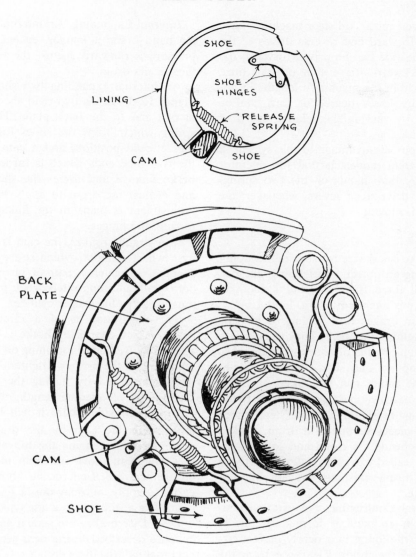

Fig. 12-38G. Two-shoe brakes

In all wheel crab steering, both sets of wheels are turned in the same direction. If both sets are turned at the same angle, the machine will move in a straight line at an angle to its center line. If it carries a straight dozer blade, this will meet dirt at an angle and sidecast in the manner of an angle dozer. Each wheel makes its own track, giving maximum flotation.

Special results can be obtained with either coordinated or crab steering by using different angles of turn on independently controlled front and rear wheels.

Skid Steer. Skid steering is effected by separately controlled power trains in the two sides of a four wheel drive tractor.

In some models, the mechanism is similar to that in a crawler tractor. One side is driven while the other is slowed or stopped, causing a turn toward the slow side.

However, most skid steer machines have separate forward and reverse drives to the sides. One side can be driven forward, the other backward, for a spin or spot turn, usually within the machine's length. In addition, any lesser degree of turn can be obtained by varying the relative speed of the two sides.

A serious disadvantage to this system in many current models is that the operator must keep both hands on the two spring-centered drive-steer levers whenever the machine is moving.

BRAKES

A brake is a device for slowing, stopping, or holding an object. A friction brake performs all three functions, but a tooth or jaw brake is intended only to hold.

Friction brakes turn vehicle or machinery momentum into heat. They usually consist of metal bands or shoes with composition lining on one side, with linkage by which the lining can be forced against a smooth, narrow, cylindrical drum. Bands are ribbons of slightly flexible steel; shoes are rigid pieces shaped to a drum. Brakes that are outside the drum and squeeze inward are called external contracting; those that are inside pushing out are internal expanding.

External Contracting. An external contracting brake band is formed so that it will hold the lining in a nearly perfect circle around the drum. Clearance from the drum is regulated by an adjusting bolt which pulls the ends of the band together, and by brackets and set screws which prevent the band from bulging at any point. The brake is applied by pulling the band ends together by means of a lever in the brake linkage, and released by allowing a spring to push them apart. The band is flexible enough to pull against the drum with fairly even pressure all the way around, although the pull is at the ends only.

Internal Expanding. An internal expanding band brake is similar, except that the pressure is outward against the inner surface of the drum.

An internal expanding two shoe brake, Figure 12-38G, has two rigid shoes hinged at one end to the back plate. There is a spring which holds the shoes together in the released position, and a cam between the two shoe ends which is turned by the brake linkage and forces the shoes apart and against the drum to apply the brake. Adjustment is made in the linkage or by moving the hinges.

In hydraulic brakes, the cam is replaced by a wheel cylinder, which is a cylindrical chamber in which a pair of pistons move out equally and oppositely when pushed by pressure from the master cylinder.

The illustration is of a direct acting brake. A self-energizing brake is designed so that the friction of the lining on the drum tends to pull the brake on tighter. This may be accomplished by having the shoes or their lining of unequal length, the longer being on the side where it will be pulled away from the cam by the drag of the drum; or by arranging the hinges so that the shoes can move on them into closer engagement. Self-energizing brakes are easier to apply, but may not be as precisely controlled as the direct acting type.

Disc. Disc brakes are similar to the disc clutches described in the next section, except that half the discs do not rotate.

The WABCO disc brake construction shown in Figure 12-38H has a set of alternately placed lined and unlined steel discs. The lined discs are splined to the brake hub and turn with it. The unlined discs are splined on the outside to a stationary housing.

The discs are held apart by springs while released, and are applied by air pressure that forces them together. Braking is obtained from the friction between the rotating and non-rotating surfaces, and is pro-

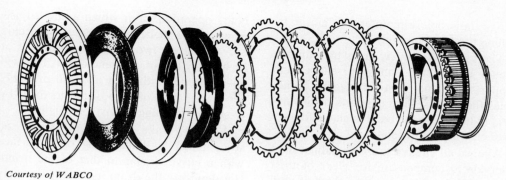

Courtesy of WABCO

Fig. 12-38H. Disc brake, exploded view

portional to the force with which they are squeezed together.

There are also single disc and shoe-type disc brakes. Application may be mechanical, hydraulic, or air.

Adjustments. There are at least four types of adjustments that may be made on brakes. Examples of these are shown in Figure 12-38I. First, pedal height may be adjusted in the brake lever for convenience of the operator. This particular adjustment has no effect on brake performance unless a low position should make the pedal rod strike the floorboard.

Then there is the clevis adjustment at the left end of the operating rod. Shortening the rod will tighten the brake, lengthening it will loosen the brake. In some machines this adjustment might be for pedal position only.

The most used adjustment on this type of band brake is the brake adjuster bolt and nut. Tightening the nut shortens the working part of the rod, pulling the band closer to the drum, and reducing free movement of the linkage necessary to make contact. In this example the adjustment affects only one end of the band directly, but it

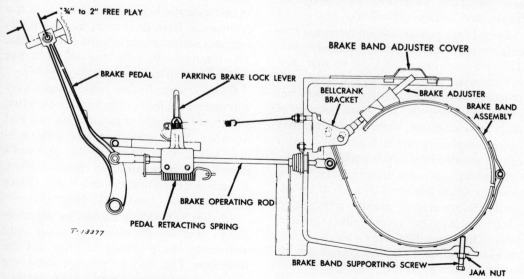

Courtesy of Fiat-Allis Construction Machinery, Inc.

Fig. 12-38I. Steering brake linkage

brings the whole circumference closer by pulling the band around.

The supporting screw helps to bring the anchor end of the band into shape, and supports the center against sagging. Sometimes there are several supports of this nature, to help keep the band in shape to have even clearance all around.

Almost all brakes must be adjusted at intervals to compensate for wear of their lining. The brake is usually tightened until it drags, then backed off until it is just freed from the drag. This provides the smallest clearance between lining and drum that will not allow destructive and power absorbing friction while the brake is released. If there are two or more major adjustments first complete one, do the other, then recheck the first.

Adjustments in some brakes are automatically tightened if the linkage moves past a set point in applying the brakes when the machine is moving backward.

Brake adjustments may be made difficult or unsatisfactory by broken, cracked, or uneven lining, loose pieces between the lining and the drum, warped or bent shoes or bands, and linkage difficulties.

Some brake pedals develop increased leverage as they near the bottom of their travel. They are sometimes left loose so that they will be easier to apply. This is a dangerous practice, as heating, reduced friction between the lining and the drum because of seepage of lubricant, water, or glazing, may cause complete failure to hold. This danger can be avoided by a tighter pedal.

Manufacturer's instructions vary, but a drum to lining gap of .015 to .030 by adjusting to no-drag is usually satisfactory. Clearance may be several times greater, close to the control linkage on a band brake, but elsewhere the need for a greater gap indicates something wrong.

A high spot that drags all the way around as the drum revolves is a defect of lining, shoe, or band. It may work out in a few hours operation, after which a new adjustment can be made. If the dragging occurs only at one or two spots the trouble is at least partly an out-of-round drum. This will not get better with wear and will probably get worse. It is often possible to fix it by re-cutting (turning down) on a lathe.

A bent brake band can usually be straightened by hammering on a surface with a curve similar to that of the drum. But don't use the drum itself, as it is likely to be brittle. Bent shoes usually have to be discarded.

The wise contractor keeps a spare lined band or shoe of each size for quick repairs.

Allowing For Heat. An external contracting brake that squeezes the lining inward upon a drum when the brake is applied, should be adjusted at normal operating temperature. This is often pretty hot. If it is set up correctly when cold the expansion of the drum as it heats will reduce clearance and is likely to cause drag, resulting in more heat and more expansion.

An internal expanding brake, in which the lining is pushed outward against the surface of a drum, is usually adjusted cold. In this case the heating in use expands the drum away from the brake, and a hot adjustment would result in a drag while cold. This is not as serious as drag when hot, but it should be avoided whenever possible.

Chatter. Some brakes (and clutches too) will chatter under certain conditions, often when only partially engaged. This is a real nuisance, often interfering with slow, delicate work. It is hard on the operator's nerves, and even harder on the equipment. It causes crystallization and early failure of shafts, cases, and brake parts.

Chatter is basically caused by defective design, but usually does not occur until something else is wrong to bring out the weakness. This may be gummy lining, the wrong type of lining, a bent band or warped shoe, out-of-round drum, wrong hook-up, or looseness in the linkage.

It may be very difficult to determine just what is causing a chatter, and how to stop it. The best advisor is a manufacturer's service man. Sometimes he can provide a package of replacement parts designed to cure the trouble, or can recommend a harder lining, a softer lining, or perhaps a different length of lining on the band.

Linings. Most brake linings are of the molded type. They are made principally of asbestos, but may also include metals including zinc, tin, lead, copper, and organic compounds such as natural or synthetic rubber, resins, drying oils, and ground cashew nut shells. These mixtures are carefully made up to meet the use of the particular lining, and are hardened in molds by heat and pressure. Such linings are brittle, and it is difficult to adapt them to drums that are more than 10 per cent larger or smaller than those for which they are shaped.

Woven lining is a fabric of asbestos and metal threads, together with some of the materials listed above. It is quite flexible, and may be bought by the roll, and cut to proper lengths as needed. It will adapt to different diameter drums and will not break while being handled. But it is much more easily damaged by contact with oil or brake fluid, as these materials soak right into it. It is not strong enough for the hard service often imposed on molded linings, but it can sometimes be used for a short period while waiting for the right lining.

Brake linings with similar appearance may vary greatly in performance. They are made to produce different coefficients of friction with drums. This "coefficient" is a measure of the amount of drag or stopping effect at a given pressure. A lining with a high coefficient will make a brake easier to apply and more likely to grab or jerk.

A low friction lining may require several times the pedal pressure or level pull needed by a high friction lining for the same stopping effect. For this reason linings

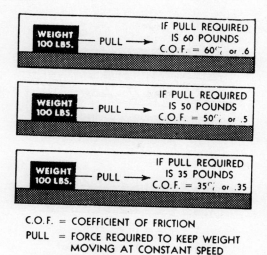

C.O.F. = COEFFICIENT OF FRICTION
PULL = FORCE REQUIRED TO KEEP WEIGHT MOVING AT CONSTANT SPEED

Courtesy of American Brakeblok

Fig. 12-38J. Coefficient of friction

should be bought from the equipment distributor, or from a supply house that knows enough to supply the correct type.

Variability in friction coefficient among different linings offers a way to cure some deficiencies in brakes. A grabbing brake might be smoothed down by using a lower friction lining, while an inadequate or heavy-pressure brake might be improved by a higher friction lining.

Most linings are fastened to shoes or bands by means of brass or copper rivets. Holes the size of the rivet shanks are drilled through the lining, in position to correspond exactly with the holes through the band or shoe. A larger hole, the size of the rivet head, is drilled about two-thirds the depth of the lining from the face or friction surface, the rivet placed in it, and expanded at the back of the band.

Lining must be replaced as soon as it is worn down to the rivet heads. Rivets are harder than lining and wear down more slowly, cutting grooves in the drum. New lining working against a grooved drum will not make full contact with it, the brake will not work properly, and the roughness will cut the new lining.

Bonded or rivetless lining is glued under

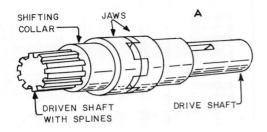

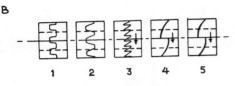

Fig. 12-39A. Jaw clutches

heat and pressure to brake shoes. When such a lining wears out the shoe is replaced with another that has been lined at the factory or in a shop. Bonding permits more economical use of lining, as it can be worn down nearer the shoes than a riveted lining before it must be discarded. Old shoes may be turned in for credit unless they are damaged.

Condition. Most linings are made to work dry, and they will hold only when they are clean and dry. If oil, grease, brake fluid, or other liquids get on them their friction characteristics will change. Usually they stop working, but occasionally friction will be increased so that they will grab and chatter with unpleasant effects on the operator's nerves and the machine's performance.
If you act promptly you can often save the lining by washing it in naphtha, benzene, or white gasoline, either on the machine or after taking it off. You can use carbon tetrachloride if there is plenty of ventilation. Don't use leaded gasoline, kerosene or fuel oil, as they leave harmful deposits.

Grease and other substances may leave a hard glazed or gummy deposit on brake lining that is difficult or impossible to remove with solvents. Its effect may be that the operator will have to push or pull harder to get the original braking result, it may

prevent the brake from working sufficiently with any pressure, or it may grab and slip alternately causing the brake to chatter.

Such a deposit may be removed by taking off the band or shoes and scrubbing the lining with a wire brush and solvent, or using a file, or it may be cut by using fullers earth. This is a finely powdered clay that can be obtained in drug stores and chemical supply houses. Use a rubber bulb syringe (another drug store item) to blow it into the brake while the drum is turning. The operator can do this whenever he feels the brake starting to misbehave, and he may keep the lining usable until it wears out.

If neither solvent and wire brushing nor fullers earth will make the brake work properly it should be relined.

Wet Linings. Some linings are made to work in oil, which cleans and cools them, but their use is more common in clutches than in brakes. It is aggravating to see how well these hold when oil-soaked, where a dry lining is made useless by a few drops.

These "wet" linings wear out very quickly if used dry.

CLUTCH

A clutch is a device by which two shafts turning on the same axis—that is, in line with each other—can be connected and disconnected. Clutch-like action can also be obtained by the use of a movable pulley in a belt system, by engaging and disengaging gears, and in other ways.

Jaw Clutch. Figure 12-39A shows jaw clutch constructions. Each unit consists of two toothed rings, called jaws, with the teeth facing each other. One jaw is keyed onto the drive shaft, the other is movably splined to the driven shaft. A circular groove is cut in the back of the driven jaw. In its hollow a brass ring or shifting collar rides, held from rotating by a yoke, which is connected to the clutch control linkage. It pulls the jaw back to disengage the clutch, or pushes it forward to engage its teeth with those of the lower jaw. When engaged, the two shafts will rotate as one, being bound together through the clutch teeth and the key and splines. When disengaged, one may rotate without affecting the other.

This type of clutch has the virtue of positive action as it cannot slip, although it might disengage itself or break. It is cheap to build, occupies little space, particularly as one jaw might consist merely of tooth sockets cut in the hub of a gear serving other purposes. It will operate dry or wet.

However, it is rough and inconvenient in action, hard to engage at any time, and impossible to disengage under load. If engaged at speed it may give damaging shocks to shafts and gears. If both jaws are stationary tooth may strike against tooth, so that one jaw must be turned until tooth is against hole to enable them to mesh.

In (B), (1) is the same as (A). (2) can be engaged at low speeds. It will show a tendency to slip out of engagement when overloaded, particularly as it becomes worn. It is the only design shown which can be disengaged under load.

The other three are one way clutches which will be disengaged automatically by reversing rotation. They can be engaged at moderate differences in speed, although (3) may break tooth points. (5) is the strongest construction.

For best results, a friction clutch should be placed between a jaw clutch and the power source.

Single Plate Clutch. The dry disc or single plate clutch is the friction type most used in cars, trucks, and many other machines. If an engine clutch it is built into the flywheel; if used elsewhere it has a similar housing.

Figure 12-39B shows a typical specimen. The flywheel turns with the engine crankshaft, and at its center is the pilot bearing in which the front of the jackshaft rests. The purpose of the clutch is to connect and

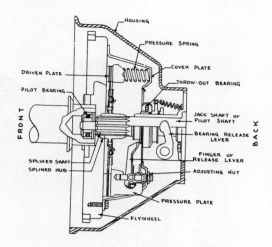

Fig. 12-39B. Single plate clutch

disconnect these two shafts which are in line with each other (have the same center line of rotation).

The pressure plate has a machined front surface and is mounted on bolts threaded into a cover plate bolted to the outer part of the flywheel. It can slide backward and forward on these bolts. Each bolt carries a coil spring which pushes the pressure plate forward.

The clutch driven disc, or plate, is between a machined rear surface of the flywheel and the pressure plate, but does not extend out from the hub as far as the bolts and springs. It has friction lining on both its front and rear, and is movably splined to the jackshaft.

The springs push the pressure plate against the disc, pushing it forward against the flywheel plate. The disc is squeezed between two plates which are turning with the engine, and the friction between the machined surfaces and its lining is sufficient to turn the disc, the jackshaft, and any load driven by it, up to the power of the engine.

The clutch is disengaged by pulling the pressure plate back against the springs by means of three levers, called fingers. These levers are pivoted on rear extensions of the

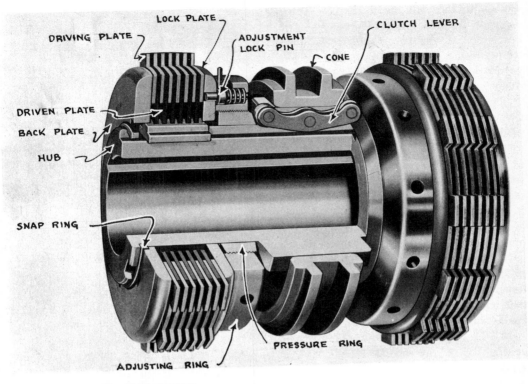

Courtesy of Twin Disc Clutch Company

Fig. 12-40. Multiple disc clutch, reversing control type

flywheel and are attached to brackets on the back of the pressure plate. The inner ends of the fingers are held in a ring attached to the clutch throwout bearing, which is moved backward and forward along the jackshaft by the clutch linkage. When the clutch pedal is depressed or the hand lever moved forward, the ring moves the fingers and the pressure plate back, disengaging the clutch. The pressure plate springs reengage the clutch when it is released.

Adjustment to compensate for lining wear may be made by threaded clevises in the pedal linkage, by screwing the throwout collar backward or forward on threads in the throwout bearing, or by set screws in the clutch fingers. When the lining wears so far that slippage occurs after these are properly adjusted, the disc must be removed and relined or replaced.

This clutch is standard for use in engine flywheels and finds many other applications. It is usually smooth in operation and long wearing. However, except in special constructions, it should not be slipped for extended periods because of excessive wear and possible heat warping or scoring of the plates.

Double Plate. A double plate clutch has two discs splined to the jackshaft. They are separated by an unlined plate driven by the flywheel. Extra friction surface is obtained without increasing diameter.

Heavy Duty. There are two types of special construction used for disc clutches in extra heavy service, as in tractors that carry shovel dozers or rippers.

Caterpillar uses a double plate clutch that operates in oil that is kept in circulation by a pump. Linings are a special type that grip when oil soaked, and the circulation carries away the heat generated by slippage.

Dry clutches may carry ceramic discs bonded to the plate instead of regular lining. The ceramic is made of sintered metal and clay or similar substances. It is highly resistant to both heat and wear, and will outlast regular lining several times under extreme conditions. In any construction, generous design size will tend to provide long life.

Multiple Disc. The multiple disc clutch consists of an engine-driven splined hub, a driving drum or spider with a splined inner surface; and a number of driven discs splined to the shaft and free of the drum, alternating with driving discs splined to the drum and rotating freely on the shaft. Also splined to the hub on each side of the discs are the fixed back plate and the movable pressure plate which have polished surfaces toward the discs. The clutch is engaged by moving the collar against the pressure plate, causing it to press the discs together so that friction between them transmits power from shaft to drum. To release, the collar is slid away from the plate and springs force the discs apart.

The discs may be all metal, or alternately metal surface and friction lined. Wear is taken up by means of a threaded collar.

The multiple disc clutch is as smooth in action as the single plate, is somewhat longer but much smaller in diameter for the same holding power, and it may be run dry or in oil. It is somewhat more complicated and expensive.

The unit shown in Figure 12-40 includes two multiple disc clutches, and is used both to disconnect power and to reverse its direction. The control lever has FORWARD, NEUTRAL (release), and REVERSE positions.

Shoe Clutch. Heavy duty equipment often uses internal expanding band or shoe clutches. They are similar to band or shoe brakes, except that the bands and linkage revolve, and therefore have to be operated through a throwout collar. This type of clutch is most used under severe conditions where it must operate partially engaged at

Fig. 12-41. Internal expanding band clutch

times, and is subject to rapid changes in load. See Figure 12-41.

Linkage. Clutches may be operated by a foot pedal, as in cars and trucks. In this case foot pressure releases the clutch, and it is re-engaged by the pressure plate springs and usually by a spring on the clutch pedal also, after foot pressure is removed. Clutches may also be operated by hand levers, as in shovels and most crawler tractors, in which case the pressure plate springs are comparatively weak. The lever is pulled back so that it locks over center and holds the clutch in engagement until it is pushed forward, where it is held in the released position by a light spring.

Adjustment. A newly adjusted clutch not equipped with a booster will require a moderate to heavy pull to lock over the hand lever. As the clutch is used day after day the pull needed will gradually diminish, perhaps to such an extent that it will not lock.

This easing results from wear of the facings, and too-easy operation usually indicates a clutch that is either slipping or will slip under severe strain. The operator may adjust such a clutch himself, or notify the foreman or service department.

When it is absolutely necessary to operate with a loose and slipping clutch, the lowest gear should be used, and heavy loads avoided. A properly adjusted clutch suffers wear of the facings while being engaged, disengaged, or deliberately slipped; one that slips wears under continuous load also, and will not last long.

Whenever the time and know-how are available, adjustment should be made in the clutch itself (fingers or collar) rather than in the linkage. This is usually done through a hole in the top or side of the housing, and involves loosening a lock, turning the collar, and re-locking.

Adjustment is usually made on the basis of free play and "feel" in the clutch lever or pedal. When it is moved from fully engaged position there should be a free travel space, 1/4 to one inch on pedals, one to three inches on levers, before the resistance of the pressure plate springs is felt.

Failure to leave this free play will cause excessive wear to the clutch collar and fingers, and may cause slippage as well. It has the same effect as riding the foot on a clutch pedal.

A clutch hand lever should lock over center in fully engaged position. Tightening the clutch increases the amount of pull to required to lock it, and over-tightening may prevent it from locking.

Instruction books usually specify the number of pounds pull that indicate a correctly adjusted clutch, and this may vary from 25 to 60 pounds, although 40 is quite usual. An experienced operator can judge the adjustment on his own machine by feel.

In shops pull is measured by a regular spring testing scale. If this is not available, the clutch can be pushed into engagement with a parcel post or bathroom scale, and the maximum reading noted. After a little practice the pull can be estimated without measurement.

A loose clutch will sometimes require a very heavy lever pull because of a worn throwout bearing. A clutch that is both loose when engaged and dragging when disengaged usually has a broken or separated finger.

The important thing about adjusting a clutch is that it must not drag when released nor slip when engaged.

Dragging. Drag is tested by disengaging the clutch, then moving the gear lever slowly from neutral toward any gear position. If it grinds continuously when the teeth are held in light contact, there is a drag.

A drag that occurs only when the engine is cold is not serious. This condition may be chronic in oil bath clutches. At operating temperatures, however, a drag is an extreme nuisance to the operator, and indicates that the clutch has too little clearance, so that loosening or repair is needed.

Slipping. A slippage test is made by holding the machine with its brakes or against an obstacle, and then engaging the clutch slowly in high gear. If the engine does not stall, and there is no fluid drive or safety slip clutch, the clutch is too loose.

If lever pull or pedal position indicate correct adjustment, slipping shows that servicing is needed.

If oil or grease gets on the facings of a dry clutch, it can be washed off by plugging the housing, partly filling it with gasoline (white gasoline or naphtha is best) and operating it for a while. If there is no housing, the cleaner is pumped in with an oil can.

Slipping may also be caused by glazing of the lining. Band and shoe clutches often respond well to blowing in fullers earth (from the drug store), but disc clutches may require relining.

Worn out lining is the most usual cause of slipping after adjustment, but worn or damaged collar or fingers, or wrongly hooked up linkage must be considered as possibilities.

Safety Clutches. There are cushion clutches which will slip a certain distance under a shock load before re-establishing a

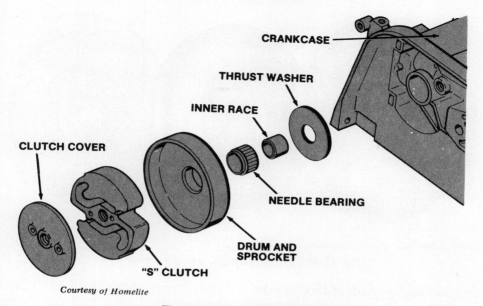

CRANKCASE

THRUST WASHER

INNER RACE

CLUTCH COVER

NEEDLE BEARING

"S" CLUTCH

DRUM AND
SPROCKET

Courtesy of Homelite

Fig. 12-42. Centrifugal clutch

solid connection, and safety clutches which will slip rather than transmit enough strain to break parts. These units save damage to machinery and cables from sudden increase of load or hitting obstacles.

Friction clutches on excavators and cranes are sometimes adjusted so that they will not carry the full engine power to the load, either to cushion shocks, to prevent overloading of the boom or cables, or as a safety precaution against picking up a tipping load. Except in light work, this is likely to result in excessive slippage, heating, and wear. In addition, it usually requires too-frequent adjustments, to keep on the hair line between dangerous slippage and solid engagement.

Centrifugal. A centrifugal or automatic clutch is usually a small friction clutch whose linkage is operated by a pair of weights flexibly mounted on the drive shaft.

Rapid rotation of this shaft will cause the weights to move outward by centrifugal force, moving the linkage to engage the clutch. Slowing the shaft will allow them to move back to center, disengaging it.

A well-known application is in the chain saw. The clutch is fully engaged at full throttle cutting speeds, and disengages when the engine is cut back to low speed. This avoids the hazard of motion in the cutting chain when it is not in use.

FLUID DRIVES

One of the most important advances in the design of construction machinery during recent years has been the development of fluid drives or couplings. These usually transmit engine power through fluid, without mechanical linkage, but may drive through gears in such a way that hydraulically controlled slippage occurs and similar cushioning is achieved.

The fluid drives discussed in this section are the hydraulic coupling and the torque converter. In both of these, the driving and driven members are inside the same case, and have the same axis of rotation.

Pump-and-motor and hydrostatic drives, which are quite different, will be covered in a later section in this chapter.

Most fluid drives are designed for use

Courtesy of Twin Disc Company

Fig. 12-43. Fluid coupling, exploded view

with engines having speeds of 1800 or more r.p.m. If used with lower speed engines, a larger size must be used in proportion to power. At engine cranking speeds they transmit no torque, but in cold weather they may constitute a drag which will in-

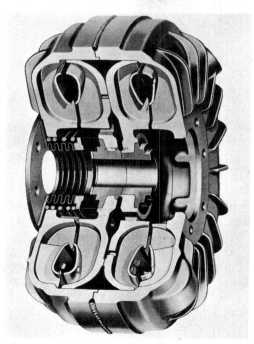

Courtesy of Twin Disc Clutch Company

Fig. 12-44. Fluid coupling, cutaway

crease starting difficulties unless a conventional engine clutch is used also. At engine idling speeds they transmit enough power to turn the output shaft against light loads so that the power train tends to "creep."

An engine protected by a fluid coupling cannot be stalled by a load, but it may stall while idling because of defects in ignition and carburetion.

Because their effects are achieved through slippage, fluid couplings may increase fuel consumption by an amount which varies with the type and design of the unit, and the class of service. Some models may be provided with lock-up clutches which will permit changing to a mechanical connection under conditions where cushioning is not required and slippage is undesirable.

The slippage generates heat. In torque converters this is usually dissipated through an outside radiator and cooling system. In most other installations the heat is radiated from the outside of the case. Cooling problems become more acute when the coupling is small in proportion to the power of the engine, when heavy continuous loads are carried and when air temperatures are high. Capacity of a unit can often be increased by an auxiliary cooling system.

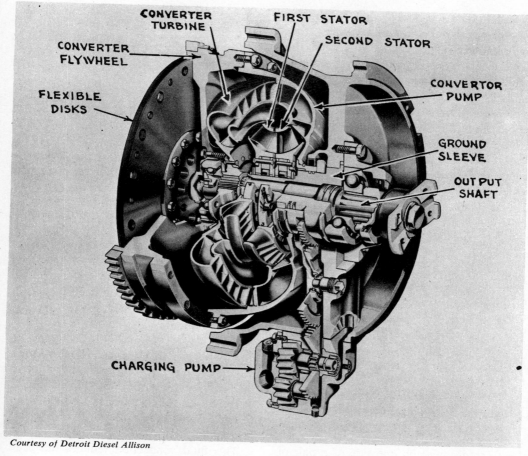

Courtesy of Detroit Diesel Allison

Fig. 12-45A. Single stage torque converter

Means may be provided to drain the fluid into a reservoir (dump the clutch) so that no power will be transmitted through it.

The smooth application of fluid power improves starting traction of wheels against dirt or pavement, and of pulleys on belts. Engine lugging ability is increased because engine speed can be maintained in spite of slowing of the power train. Torque converters have this effect and they also multiply the force exerted by the engine.

In many industrial applications there is a friction clutch also, giving a marked advantage for precise control.

Properly engineered fluid couplings reduce maintenance and repair costs, increase production, and relax the operators, advantages that more than offset increased fuel consumption in many types of service.

Hydraulic Coupling. The hydraulic or fluid clutch or coupling consists of an oil chamber which contains a set of pump vanes driven by the engine, and a turbine set connected to the driven machinery. These vane sets, somewhat resembling half grapefruits with the meat removed, are set close together, flat sides facing, and they turn on the same axis.

The Twin Disc coupling, Figures 12-42 and 12-43, uses two impellers and two turbines, thus providing increased capacity in proportion to diameter.

Rotation of the pump vanes by the engine-driven input shaft causes the oil to spin, and the oil rotates the turbine in the same direction as the pump. The action is similar to the rotating of sugar in the bottom of a cup when the liquid is stirred at the top. The high rotation speed and close clearances permit transmission of very heavy loads.

The force exerted against the turbine is slight at engine idle, and increases with speed until the two members turn almost as a unit. Load on the turbine or output shaft will cause increased slippage so that it will turn more slowly while the engine continues to work at full speed and power.

A fluid clutch is run about 85 per cent full of oil. The proper level is indicated by the location of the filler plug. If it is on the rim of the coupling, the coupling is turned until a plate reading "Top for Filling" is at the top. If the filler is on the side it is turned into its highest possible position, and fluid poured in until it runs out of a check hole.

The air space in the coupling provides for expansion of fluid as it heats in service. Provision for further expansion may be made in the shaft seal by a spring and bellows arrangement on one side of a pair of

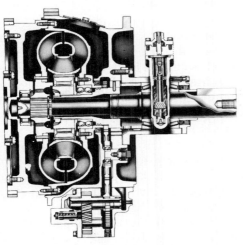

Courtesy of Detroit Diesel Allison

Fig. 12-45B. Twin turbine converter

matched seal rings in a positive type seal. High pressure will compress the bellows spring and allow air or fluid to escape between the rings.

Couplings may have spider drive. A ring gear with internal rubber or metal teeth is bolted to the flywheel, and a smaller ring with matching external teeth is bolted to the coupling. Such a connection can operate at a slight angle, so that installation need not be perfectly accurate. As a rule, however, problems of alignment are very critical.

Sometimes it is necessary to use a coupling that is oversize for the engine. Its capacity can be cut down by putting blade blocks in the runner. These are small steel stampings that block the outer spaces between the blades, reducing the volume of fluid circulation and the coupler capacity.

The coupling usually has no cooling system, although one or both sides may be fitted with vanes to increase air circulation around it. Overheating usually indicates overworking, either because the coupling is too small or the operator is loading it too heavily. Prolonged idling will also cause overheating.

This fluid connection reduces the peak power, or perhaps the shock power, that an engine can transmit to its work. This is because the momentum of a solidly connected engine can exert a momentary force much greater than its sustained power, and this force may permit the machine to accomplish work beyond its rated capacity. However, the use of this force may damage the machine out of proportion to the extra production, and its elimination may be considered either good or bad.

The fluid coupling is little used in construction machinery, having been almost entirely replaced by torque converters.

TORQUE CONVERTER

A torque converter is a fluid coupling so constructed that the flow caused in the oil by slipping under load is converted into additional force acting on the turbine vanes.

Single Stage. The Allison converter, Fig. 12-45A, is driven by flexible discs and a flywheel. The converter pump is bolted to the flywheel, and together they form an oil-tight housing within which the turbine and stator (reactor) operates. The turbine is splined to the output shaft at the front. The rear of this shaft is surrounded by a stationary casing called the ground sleeve.

When the pump is turned by the engine it picks up oil near the center and throws it outward and backward against the turbine vanes at a predetermined angle. The turbine has sharply curved blades which catch the speeding oil from the pump and extract maximum force from it by changing the direction of flow. This causes the turbine to rotate and the oil to move from the outer circumference to the center of the turbine, which it leaves traveling in the opposite direction from the rotation of the pump and turbine. Outlet passages are smaller than the inlet so that when the turbine is moving at low speed under load the oil emerges with greater velocity than it enters. It is still capable of exerting force, but will tend to exert it against the pump unless its direction is changed.

The stator is mounted on the ground sleeve and includes a free wheeling unit which allows it to spin freely in the direction of pump rotation, but locks when backward pressure is applied. When the stator is struck by oil speeding back from the turbine, it locks against the ground sleeve and provides curved passages that change the direction of oil flow so that it enters the pump hub moving in the same direction the pump is rotating. Its velocity is added to the velocity of the oil developed in the pump so that the total force of the oil leaving the pump is correspondingly greater. This regenerative action is the key to the torque multiplication developed in the converter, and increases automatically with slippage caused by increase in load on the output shaft.

When slippage decreases so that oil leaves the turbine vanes slowly, the stator turns in the direction of pump rotation, and the converter acts as a fluid coupling.

Torque converters can be supplied with a friction clutch on the input from the engine, or a rear disconnect adaptor. On some models, a hydraulic lockup clutch may be provided that can lock the flywheel and turbine together for direct or positive drive. A free wheel or over-speed lockup option prevents the turbine from turning faster than the engine.

The uses of these and other special arrangements are discussed later.

A total of five converter series are available from this manufacturer, with torque

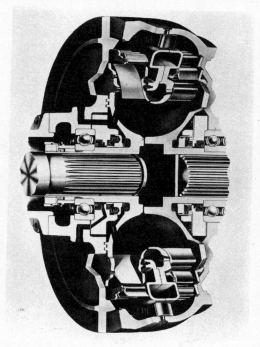

Courtesy of Twin Disc Clutch Company

Fig. 12-46. Three stage torque converter, cutaway

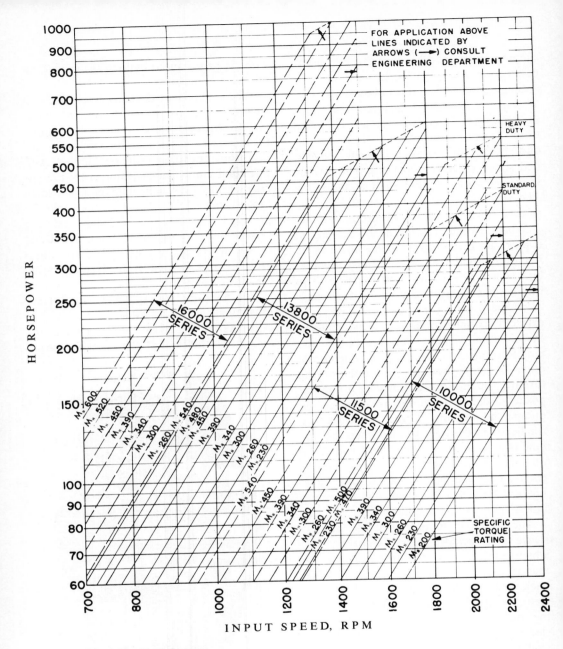

Courtesy of Twin Disc Clutch Company

Fig. 12-47A. Converter application chart

ratios ranging from 2.5 to 1 up to 4.0 to 1. Capacities cover engines from 60 to 600 horsepower.

Twin Turbine. A twin turbine converter, Figure 12-45B, has a single pump with two turbines, one inside the other. Each tur-

bine drives a different gear train in a transmission. If load on the drive shaft is high, as in starting to move a load, the first turbine turns, driving a low-gear system. If the load is reduced, higher velocity oil reaches the second turbine and turns it, along with

Courtesy of Twin Disc Clutch Company

Fig. 12-47B. Three stage torque converter, exploded view

its higher speed gear connections.

This results in automatic two speed action, which is usually supplemented by an additional choice of ranges in the transmission. See also page 12-17.

Three Stage. The Twin Disc, Figures 12-46 to 12-48A, is a 3-stage converter with fixed reactors. The fluid leaving the pump or impeller impinges on the first stage vanes in the rim of the turbine is diverted by a set of reactor vanes and passages in the case to exert a push on the second rim set of turbine vanes.

From these it passes through a second reactor and spends its remaining force on an inner set of turbine vanes. Most of it then goes to the pump, but a small part is diverted through the cooling system.

The multiple stages permit a torque multiplication as high as 6 to 1, which can be limited to a lesser amount. The unit may be used in bulldozers, trucks, and other heavy equipment with only a one or a two speed transmission.

There are four series of Twin Disc three stage converters, to accommodate engines from 60 to 1000 horsepower. Maximum torque ratios range from 2.0 to 1 to 6.0 to 1. Figure 12-47A provides information on their speed and power capacities.

Each series has a number of specific torque ratings, for a total of 30 capacities

available in the complete line, providing a close match to the requirements of various engines. Rating within a series is varied at the factory by a choice of impellers, or

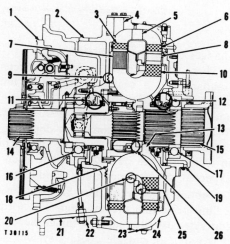

TORQUE CONVERTER
(CROSS-SECTION)

1–Clutch assembly. 2–Inlet from fluid cooler. 3–Turbine blade (first stage). 4–Outlet to fluid cooler. 5–First stage reactor blade (converter housing). 6–Turbine blade (second stage). 7–Impeller blade. 8–Second stage reactor blade (converter housing). 9–Labyrinth seal (impeller to clutch housing). 10–Turbine blade (third stage). 11–Seal assembly (impeller). 12–Seal assembly (turbine). 13–Turbine assembly. 14–Input shaft. 15–Output shaft. 16–Converter front bearing. 17–Converter rear bearing. 18–Impeller assembly. 19–Rear seal drain hole. 20–Labyrinth seal (impeller to turbine). 21–Clutch housing. 22–Front seal drain. 23–Drain plug. 24–Converter housing. 25–Labyrinth seal (turbine to reactor ring). 26–Freewheel assembly.

Courtesy of Caterpillar Tractor Company

Fig. 12-48A. Cross section of a 3-stage converter

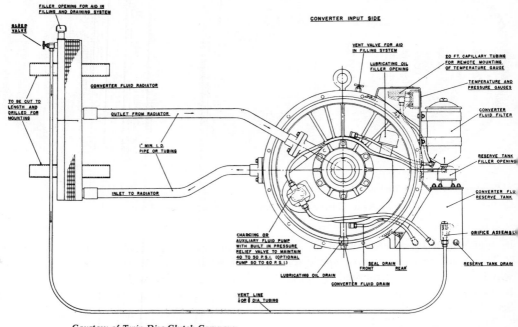

Courtesy of Twin Disc Clutch Company

Fig. 12-48B. Cooling circulation, three stage converter

changes in third stage turbine blades.

There may also be a choice between a standard and a heavy duty type within a series, with differences in shafts, bearings, and details of design.

The choice of a converter depends on engine characteristics, the type of machine in which it is to be installed, and even on expected machine operating conditions.

Cooling and Circulation. Slippage inside a fluid drive unit creates heat. The torque converter usually makes so much heat that it requires an outside cooling system, just as an engine does. The type of system varies with the make, model, and type of machine. Figure 12-48B is a three stage converter with a stationary housing and a cooling radiator. Oil pressure in the converter is higher at the rim than near the hub, so circulation is obtained by tapping the line to the radiator on the rim and the return line near the hub.

Cooling circulation in a single stage converter with a rotating housing depends on

the charging pump keeping pressure in the converter. Some of this fluid enters the pipe to the radiator or heat exchanger, and cooled oil goes into a sump to be picked up again by the pump.

Proper temperature is very important to a converter. Good operating range is between 180 and 220° F., with absolute minimum 160° and maximum for steady operation 250°. Operators should be reminded that they must watch their gauges, and should stop and report if they indicate trouble.

The basic circulation in the converter is from impeller to turbine to stator to impeller, over and over. In addition there is the circulation for cooling, mentioned above, and the bleed-and-charging circuit.

In the three stage unit, the charging pump draws fluid from a reserve tank, that may be the machine's diesel fuel tank, and pushes it through a filter and into the converter. A pressure relief valve built into the pump regulates the amount of pressure it can

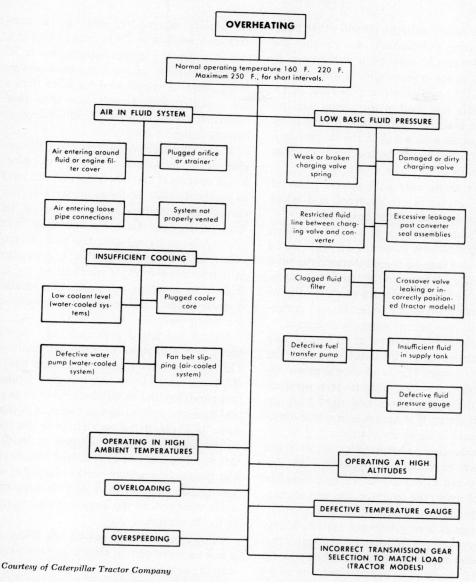

Fig. 12-48C. Causes of converter overheating

build up in the converter. This pressure varies from unit to unit, but it is very important that it be correct for each one. Low pressure systems, used chiefly in shovels, may require only 20 or 30 pounds, where a tractor converter may need from 45 up to 75 pounds.

The high point of the fluid system, located at the top of the radiator in our illustration, has a bleeder opening and tube into the

reserve tank, which excess oil enters through an orifice (small, precision opening) that will not allow passage of as much fluid as the charging pump can deliver. This restricted opening enables the charging pump to keep the pressure in the converter and cooling system as high as the setting of its relief valve.

Air. Torque converters can function properly only if they are filled solidly with fluid. Air mixed with the oil will cause poor performance, overheating, and possibly serious damage. It may enter the system if the fluid in the reserve tank is low enough to permit the charging pump to suck air (a moderately low level might permit sucking air if the machine is working on very steep slopes), or there is a leak in the suction line.

Suction leaks may be too small to show up by outward leakage of oil.

Of course air enters the system when oil or filter elements are changed and when the lines are opened for any reason.

Moderate quantities of air can be taken off by the bleeder system, as it will rise to the top and then be forced through the tube and orifice by pressure of liquid behind it. Partly opening the bleeder screw on top of the converter housing while the unit is running will also permit some air to escape.

Check for the presence of air by stopping and starting the engine, while watching the pressure gauge. The hand should move immediately from operating to zero pressure when the engine stops and right back up when it starts. If it hesitates in either direction, there is air in the system.

Overheating. Overheating is a major problem in converter operation. Its frequency and extent is affected by design, the type of work, the operator, the air temperature, and the condition of the unit.

Design is a very intricate matter. Manufacturers will not guarantee a converter that is used with any engine other than that for which it is suited, or one in which any major parts have been changed to obtain different performance. This is not your problem—if you can't make a unit give reasonably good performance, send for a factory man to take it from there.

A converter is likely to overheat if the work is very heavy. We will consider the ifs, ands, and buts of this statement just below, but the fact remains that if other conditions are the same, heavier work will generally mean more heat. Cooling systems are usually designed to take care of continuous operation at 70 per cent efficiency, that is, when the output horsepower is 70 per cent or more of the input horsepower. If the machine is operated for more than a few minutes with greater loss than this, particularly at or near stall conditions, overheating is apt to occur.

If the converter is undersize for the engine, or for the normal work of the machine, it will operate at low efficiency and will tend to overheat easily. So far as the work is concerned, the operator can usually relieve the load on the converter by operating in a lower gear, or by reducing the load. Whenever you get complaints that a converter in a tractor or truck is overheating try to find out by watching or by questions whether correct gears and sensible work methods are being used.

Use of too low a gear and high engine speed may also result in some overheating. This type of converter will also heat badly if left connected to an idling engine. Under this condition the input shaft turns at several hundred revolutions per minute, while there is likely to be enough drag on the output shaft to keep it stopped or hardly turning. The lost power is turned into heat, and the cooling system works sluggishly at low speeds because there is little pressure difference between the inner and outer parts of the converter.

A converter should not be left idling. If it has an input clutch, release it. Otherwise stop the engine.

Air in the converter will cause overheating. This problem we discussed earlier. Too much or too little oil is another possibility that is easy to check.

The trouble may be in the cooling circuit rather than in the converter or its work. A radiator may have its air circulation stopped by accumulation of leaves or other trash, the fan may not operate because of a broken or slipping belt, or the fluid passages in a radiator or heat exchanger may be partly blocked. The outside work is easily done, but clogged tubes need special solvents and probably some replacements.

Selection of Ratios. The torque converter automatically proportions the amount of torque delivered to the output shaft to the load through infinite gradations. However, it does not necessarily do this according to the desires of the machine operator, or even for maximum efficiency.

The accuracy with which a converter selects ratios depends on its internal design and the conditions of its use. Its efficiency usually drops when the speed of the output shaft is less than one half that of the input shaft. For these reasons it is often desirable or necessary to use a transmission in conjunction with it. This may have standard construction, or be automatic or semi-automatic in action. A converter with a narrower range of action and snappier response can then be used, and a much wider range of ratios supplied.

Cleanliness. Torque converters are precision made and depend on very close tolerances. Some parts are lapped smooth with irregularities allowed only up to twelve millionths of an inch. Oil in these units circulates at high velocity, and any foreign material it carries will wear down the edges and pit the hollows in vanes, changing the effective shape. This is in addition to bearing and seal damage always caused by dirty oil.

A converter for a medium size off the road truck may cost $2,000.00, and single parts for it bring several hundred dollars. It is certainly worth the trouble of seeing that every drop of oil that goes into it is clean, and that its filters are inspected and cleaned and replaced regularly.

Recommended interval for changing converter oil is every 1,000 hours or every three months. Since construction equipment has an average use of around 1,000 hours a year, the actual interval in your shop is something to be discussed with the equipment dealer. If the unit uses fuel oil from the engine supply tank changing is not necessary.

The three stage converter has four drain points. They are the bottom of the converter housing, the reserve tank, the filter, and the radiator. Air vents in the top must be opened to insure rapid and complete draining. The converter should be warm or hot so that the fluid will flow freely.

The filter element is replaced, and its case cleaned and rinsed, whenever the oil is changed.

To refill, pour fluid in the radiator until it comes out the converter housing vent. Close the vent, add fluid to the filler opening, and replace the radiator cap. Fill the reserve tank an inch below the full mark.

Then start the engine and run it at half throttle. Check fittings for leaks, and check the gauge for proper operating pressure. Keep checking the reserve tank dip stick, and add fluid if the level goes down.

Many torque converters are lubricated by the converter oil, as this is pumped through its bearings as part of the circulation system. The three stage type, however, is separately lubricated by dip and splash from a lubricating oil reservoir, or by grease from outside fittings.

This separation of fluid and lubricant requires the use of seals to keep each substance in its place. These are in double sets, with a drain between them that will take any leaking fluid into the reserve tank. This makes the unit safe against damage from diluted lubricant, that might occur if only a single internal seal separated them.

Converter Clutches. For efficient use in construction equipment it is usually desirable to supplement a torque converter with one or more clutches or clutch-like devices.

An input clutch is placed between the engine and the converter. It can be released to protect the converter against being overheated by an idling engine, and to prevent "creep" of the output shaft and the machine.

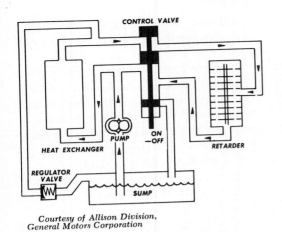

Fig. 12-48D. Hydraulic retarder

It is used chiefly in equipment that does not have a gearshift, or that does not require gear shifting on the move.

An output clutch is a heavier and more expensive unit, as it has to carry the multiplied torque. It is needed for mechanical shifting with the machine moving, as it relieves the jackshaft of the inertia of the converter. It will stop creep, but may not prevent overheating when idling.

A lock-up clutch locks the input and output shafts together, so that the converter is put out of operation. It prevents slippage and the resulting loss of efficiency during operations that do not require either torque multiplication or fluid cushioning. For example, an off the road hauler might need the converter in climbing out of a pit, but would perform better without it on a flat haul at the top. This device may also be called a direct drive, and may operate automatically.

The so-called free wheel produces the opposite effect from free wheeling in a car. It is an over running clutch that locks the shafts together whenever the output tries to move faster than the input. This permits use of the braking effect of engine compression in going down grades. Converter drag also helps in slowing the machine. This device, or a lock-up, are necessary in any downhill hauling operations.

Another advantage of the lock-up clutch and the free wheel is that either will make it possible to turn and start the engine by pushing or towing the machine. This is often very important.

Hydraulic Retarder. Hydraulic retarders are used to reduce speed of trucks and other haulers on down grades, and to slow them down preparatory to stopping with the friction service brakes.

A retarder consists of a paddle or vane-type rotor turned by the drive shaft or the converter output shaft, a fixed casing or stator fitted with vanes, and an oil circulation system.

When the casing is empty there is practically no friction. When oil is admitted it sets up a drag or resistance that slows the shaft, turning its energy into heat. The degree of retarding or braking depends on the amount of oil in the unit.

Oil may be supplied and cooled by a torque converter system, or by being forced out of a hydraulic cylinder and through a heat exchanger. The surplus heat is useful in keeping the converter or engine warm on long down grades.

A hydraulic retarder greatly increases the safety of hauler operation in hilly country, and usually permits higher average speed as well.

Output Shaft Governor. The hydraulic torque converter has opened up interesting new fields of governor applications. This is because torque converters sometimes produce excessive output shaft and vehicle speeds under light loads, as compared to the speeds under full rated loads.

It is usually desirable to maintain, within limits, a constant output shaft speed in spite of changes in the load placed on the converter. To obtain this control, it is necessary that the governing device be driven by the converter output shaft, so that it will respond to variations in output shaft speed rather than engine speed.

A dozer, for example, will need full power while cutting hard material, and little power while spreading it. If equipped with a torque converter and a conventional engine governor, it would move slowly in the cut at full throttle, then speed up as the load diminished until it would move too rapidly for accurate grading.

If engine speed were controlled from the output shaft only, the speed of the dozer would be the same through cut and fill, except when working beyond capacity. However, engine speed might be increased too much under heavy load conditions, unless an engine governor were used also.

The output shaft governor installation therefore consists of two separate governors, one driven by the engine and one driven by the converter output shaft, either of which may override to reduce fuel to the engine. Speed may be under the control of either, depending upon the load and speed settings.

At output speeds below the set value, the shaft governor will permit the engine to run at full speed, but the engine governor will hold it down to whatever maximum it is set for. When shaft speed reaches the set value the shaft governor will close the throttle sufficiently to prevent speeding, even if engine rpm is less than that allowed by the engine governor.

A single operating lever may be used to control both engine and output shaft speed settings, a control lever may be used for each governor speed setting, or one may be fixed and the other variable.

A single lever might be set up to change engine speed setting from idle to maximum for the first part of the lever travel, and output shaft speed setting over the last segment of travel, where its movement would then be used to vary road, hoisting or other speed at full engine rpm, and any single setting would maintain a fixed road speed regardless of conditions, up to the capacity of the machine.

AIR MOTORS

Rotating motors powered by compressed air are used in wagon drill feeds, in hoists, and in many other machines which operate near a supply of compressed air in conditions where exhaust fumes would be a problem, and where a light motor developing high torque is required.

A stalled air motor will resume turning if the load is reduced within its power range, without needing to be re-started. It will not be damaged or rendered inefficient by lugging down. It is lubricated by oil mixed with the incoming air.

Both piston and vane types are used. An Ingersoll Rand four cylinder piston (radial) motor is shown in the cutaway, Figure 12-49. All cylinders operate off one throw of a counterbalanced crankshaft. A rotary valve releases compressed air into each cylinder as its piston passes dead center on the upstroke. The air expands, driving the piston down, and is exhausted on the next upstroke.

Air motors are supplied in both reversible and non-reversible models. Speed is controlled by an air throttle. Gearing reduces speed and increases torque.

The vane motor, Figure 12-50, has a

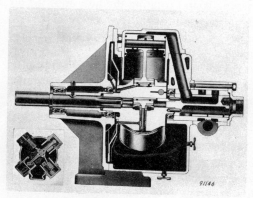

Courtesy of Ingersoll-Rand Company

Fig. 12-49. Radial air motor

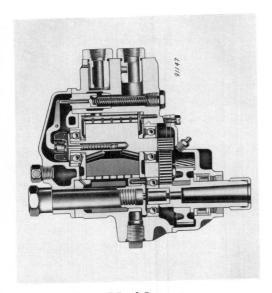

Courtesy of Ingersoll-Rand Company

Fig. 12-50. Vane air motor

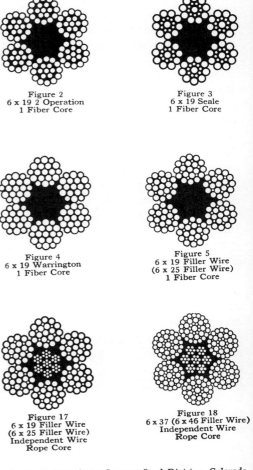

Figure 2
6 x 19 2 Operation
1 Fiber Core

Figure 3
6 x 19 Seale
1 Fiber Core

Figure 4
6 x 19 Warrington
1 Fiber Core

Figure 5
6 x 19 Filler Wire
(6 x 25 Filler Wire)
1 Fiber Core

Figure 17
6 x 19 Filler Wire
(6 x 25 Filler Wire)
Independent Wire
Rope Core

Figure 18
6 x 37 (6 x 46 Filler Wire)
Independent Wire
Rope Core

Courtesy of Wickwire-Spencer Steel Division, Colorado Fuel & Iron Corporation

Fig. 12-51. Wire rope cross sections

rotating cylindrical hub set off-center inside a casing cylinder. Flat vanes are set in longitudinal slots in the hub. Air pressure in the hub and centrifugal force move them outward into light contact with the cylinder wall.

The curved space lying between the rotor and the more distant section of the casing has an inlet at one end and an outlet at the other. Compressed air entering through the inlet pushes against the nearest vane, moving it and rotating the hub until the air can escape through the outlet. The five vanes provide for a smooth continuous rotation.

In reversible models the inlet and outlet connections can be switched to produce operation in the opposite direction.

CABLE (WIRE ROPE)

Cable, or wire rope, is one of the most important materials or parts used in excavation machinery. There are many types for different uses, but most of them are made up of carbon steel wires wound into strands, and strands wound with each other to make cable. The strands are wound

around a center or core, which may be an additional strand, a miniature cable, or a rope made of sisal or manila. The wire core is stronger and more resistant to crushing, but is less flexible and resilient than the hemp.

A cable is designated by its size, by the grade of steel wire used in it, as to whether it is preformed, by its lay, the number of strands not including the core, and the number of wires in each strand.

A widely used construction is the 6 x 19 that is, six strands of nineteen wires each.

REGULAR LAY ROPE

A regular lay rope is one in which the direction of lay of the strands in the rope is opposite to the direction of lay of the wires in the strands.

LANG LAY ROPE

A lang lay rope is one in which the direction of lay of the strands in the rope is the same as the direction of lay of the wires in the strands.

RIGHT LAY ROPE

A right lay rope is one in which the path of the strands in the rope is from left to right in a direction away from the observer. A right lay rope may either be regular lay or lang lay.

LEFT LAY ROPE

A left lay rope is one in which the path of the strands in the rope is from right to left in a direction away from the observer. A left lay rope may be either regular lay or lang lay.

Courtesy of Wickwire-Spencer Steel Division, Colorado Fuel & Iron Corporation

Fig. 12-52. Wire rope lays

The wires may be all of one size, or of two or more sizes. Additional wires, to a total of 25 per strand, may be added without changing the 6 x 19 designation. See Figure 12-51. Each construction has a name, often the name of the designer of the particular type. Variations in flexibility and in resistance to crushing and to abrasion are obtained. Small wires are desirable when the cable is subjected to sharp bending; large outer wires when it may be rubbed and chafed.

Lay. The lay of a wire rope is the direction of twist of the wires in the strands and

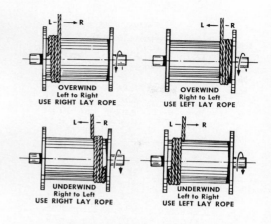

Fig. 12-53. Winding on drum

the strands in the cable. Four standard lays are illustrated in Figure 12-52. The right and left designations indicate the direction the strand takes in crossing the top of the cable as it winds away from the observer. In regular lay the wires in the strands are twisted in the opposite direction from the strands in the cable. In Lang lay the wires and the strands both have the same twist. In practice, the difference is that the Lang lay has better fatigue resistance because of the flatter exposure of the wire, but it has a tendency to untwist unless both ends are held.

Under conditions where coarser outer wires are needed to obtain resistance to abrasion, Lang lay may be used in order to regain some of the bending fatigue resistance lost by using the thicker wire.

In the field, this is commonly interpreted to mean that Lang lay has inherently better abrasion resistance than regular lay, but when the same size outer wires are used, the difference is negligible except under certain special conditions.

Right lay is the usual construction and is recommended for overwinding on a drum when the anchor is on the left, and for underwinding when the anchor is on the right. Left lay is preferable for overwinding from the right, or underwinding from the left. This is because the cable, when re-

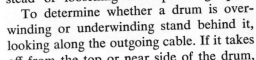

Rope Diameter Inches	Approximate Weight lbs./ft.	ACTUAL ROPE STRENGTH	
		Improved Plow Steel Tons of 2000 lbs.	Plow Steel Tons of 2000 lbs.
¼	0.10	2.74	2.39
⁵⁄₁₆	.16	4.26	3.71
⅜	.23	6.10	5.31
⁷⁄₁₆	.31	8.27	7.19
½	.40	10.7	9.35
⁹⁄₁₆	.51	13.5	11.8
⅝	.63	16.7	14.5
¾	.90	23.8	20.7
⅞	1.23	32.2	28.0
1	1.60	41.8	36.4
1⅛	2.03	52.6	45.7
1¼	2.50	64.6	56.2
1⅜	3.03	77.7	67.5
1½	3.60	92.0	80.0
1⅝	4.23	107.0	93.4
1¾	4.90	124.	108.
1⅞	5.63	141.	123.
2	6.40	160.	139.
2⅛	7.23	179.	156.
2¼	8.10	200.	174.
2½	10.00	244.	212.
2¾	12.10	292.	254.

NOTE:

For breaking strength of galvanized ropes, deduct 10% from strengths shown. For Wire Strand Cores and Independent Wire Rope Cores add 7½% to the listed strengths and 10% to the weights.

Courtesy of Wire Rope Institute

Fig. 12-54. Strength and weight table, 6x19 fiber core wire rope

lieved from strain, tends to twist slightly as if to unwind its strands, and if used as advised above, this twisting will cause the wraps on the drum to hug each other, instead of loosening and spreading apart.

To determine whether a drum is overwinding or underwinding stand behind it, looking along the outgoing cable. If it takes off from the top or near side of the drum, it is over; if the bottom or far side, under. See Figure 12-53.

Grades. Several grades of steel are used in wire rope. The two most suited for excavation machinery are plow steel and improved plow steel. The improved variety is about 15 percent stronger than the plow steel, which in turn is considerably stronger than the mild plow and traction steels used in elevator cable, stationary guy ropes, and highway guards. Weight and strength data are given in Figure 12-54.

There is also a higher grade, extra improved plow steel, which is 15 per cent stronger than improved plow.

Wire. Cable wire is stiff and springy. In ordinary non-preformed construction, if a wire is cut or broken the ends will straighten and project from the rope surface at an angle. These ends cause extra wear to sheaves, drums, and other wraps of cable, and will cut unprotected hands.

If such a cable is cut or broken, the wires and strands will untwist for several feet or yards on each side, unless bound (seized) or clamped.

Preforming. Preformed cable is made of wires which are shaped so that they lie

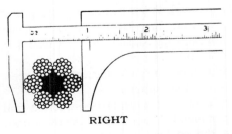

RIGHT

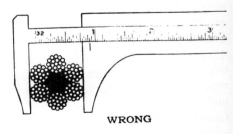

WRONG

Wire Rope is usually manufactured slightly larger than the nominal diameter. The diameter of a new rope may exceed the nominal diameter by the amounts as shown in the United States Federal Specification for Wire Rope.

Courtesy of Wickwire-Spencer Steel Division, Colorado Fuel & Iron Corporation

Fig. 12-55. Measuring wire rope

  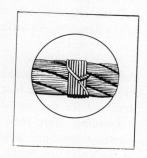

Courtesy of Wickwire-Spencer Steel Division, Colorado Fuel & Iron Corporation

Fig. 12-56. Seizing wire rope

naturally in their positions in the strand and the rope. They show little tendency to stand out from the surface or to unravel when cut.

Preformed cable is safer to handle than the straight wire type and is more resistant to fatigue caused by working over small sheaves, or around sharp angles. It is recommended for use in most excavating machines, and is replacing the older type.

Seizing. Wire rope is manufactured in long pieces that are cut into shipping or working lengths by the manufacturer, dealer, or user.

Cable formed from straight wire should be firmly bound (seized) with soft iron wire in two to four places on each side, before cutting. This process is illustrated in Figure 12-56. A tighter wrap can be obtained by holding the loose end of the wire under some tension and twisting it onto the rope with bar, as in 12-57.

Large cables require more and tighter seizing than small ones.

Seizing wire should be thin enough so that the cable end will go through a socket or clamp easily. If difficulty is experienced, the cable and binding can be flattened with a heavy hammer.

Courtesy of Wickwire-Spencer Steel Division, Colorado Fuel & Iron Corporation

Fig. 12-57. Tight-wrapping seizing wire

Preformed cable is usually bound only when it is to be placed in a poured socket, or is to be stored or roughly handled before use.

Cutting. The best tool for cutting is an oxy-acetylene torch, as it is quick and easy and will weld the ends, prevent raveling, and reduce strain on the seizing. Next choice is a regular cable cutter, which is a concave chisel in a vertical guide. The cable is placed in a groove under the edge and the chisel struck several times with a heavy hammer. An ordinary cold chisel may also be used.

Standard bolt cutters will be damaged by the hard wires. A hacksaw is rather tedious to use. It will work best if fitted with blades designed for use on hard steel.

Bending. Wire rope is very flexible, but repeated bending causes the metal to lose its resiliency and break, usually first in individual wires, producing a weak spot that finally breaks under strain. Sharp bends, as around small drums and sheaves, are more damaging than gradual bends around large ones. Large cables, and cables made up of coarse wires, are most damaged by bending. Reverse bends break down the wires faster than two or more bends in the same direction. The damage from bending is largely fatigue in the metal, but is increased by wear and nicks on the wires from friction between them.

Wire rope manufacturers recommend that drum or sheave diameter be 45 times the diameter of the cable to minimize bending stresses. Unfortunately, it is impractical to build excavating machines up to these standards, and ratios vary from thirty down to ten or below. This results in a generally short life for cable, and in some cases it is more economical to use a small rope which is not strong enough for shock loads, rather than the recommended size which is so damaged by bending over small sheaves that it fails quickly.

Crushing. Wire rope is damaged when

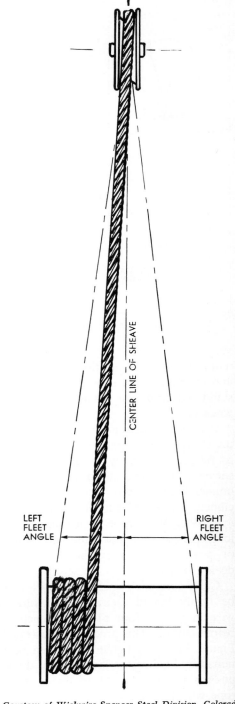

Courtesy of Wickwire-Spencer Steel Division, Colorado Fuel & Iron Corporation

Fig. 12-58. Fleet angle

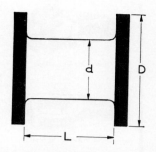

TO FIND THE WIRE ROPE CAPACITY OF A DRUM

Length of wire rope in feet that a drum will hold = $(D^2 - d^2)$ LX

D = Diameter of flange in inches
d = Diameter of drum in inches
L = Inside width of drum in inches
X = Rope Factor—See Table below

Rope Factors

1/8 inch	4.250	1 inch	.0655
1/4 inch	1.018	1 1/8 inch	.0516
3/8 inch	.466	1 1/4 inch	.0418
7/16 inch	.341	1 3/8 inch	.0347
1/2 inch	.262	1 1/2 inch	.0292
9/16 inch	.2064	1 5/8 inch	.0248
5/8 inch	.168	1 3/4 inch	.0214
11/16 inch	.138	1 7/8 inch	.0186
3/4 inch	.116	2 inch	.0164
13/16 inch	.099	2 1/4 inch	.0129
7/8 inch	.085	2 1/2 inch	.0105

Fig. 12-59. Rope capacity of winch drum

wound unevenly on a drum and placed under strain. The wraps of cable crush and kink each other. In many machines the working end of the rope does not stay in line with the drum so that a device of pulleys or rollers, called a fairlead, must be placed between the drum and the work to line the cable up for smooth winding on the drum.

Fleet angle, Figure 12-58, is the maximum angle made by a rope with a line perpendicular with the drum. It should not be more than 1½ degrees if the drum is smooth, or 2 degrees if it is grooved. It can be reduced by increasing the distance between the drum and the guide sheave.

Kinking. Cable is severely damaged by kinking which occurs when it is pulled tight while it has a loop or twist in it. To avoid this damage it should always be unwound, not lifted, from a reel or coil. A reel should be set so that it can revolve as the cable is pulled off it. A coil should be untied, the outside end laid on the ground, and the coil rolled away, leaving the cable extended behind it.

Sometimes a coil will get tangled while rolling. In this case the ends should be pulled through the tangle when necessary and walked around to unwind where possible and laid out fairly straight on the ground. One end should then be twisted oppositely from the obvious kinks while jerking it up and down so that the reverse twist will get to them. If the cable is short enough so that the far end can be flipped free of the ground, this alone may straighten it without hand twisting.

Lubrication. Wire rope needs a lubricant to minimize friction between the wires. The manufacturer lubricates it thoroughly and for short cables in continuous use this is sufficient. However, if it is to lie idle for any length of time it should be oiled or greased to avoid rust damage. Also, if it wears shiny in use it should be greased for the sake of the sheaves as much as that of the cable.

Safety Factor. The safety factor in rope is the ratio between the pull required to break a new cable and the load it carries in service. Safety factors of three, five, or better in relation to normal pull, are recommended. A shovel with a five ton pull on a drag cable should use one with a tested strength of fifteen to twenty-five tons. This safety factor allows a margin for shock loads, which may be several times normal load for a moment, and permits the rope to continue working after some of its wires are weakened and broken.

Highest safety factors are required for suspended loads over or near people or valuable property.

Cable life is prolonged and required margin for shock loads is reduced in excavator service by the use of a torque converter in the power train, in spite of the increase in torque at the drum shafts.

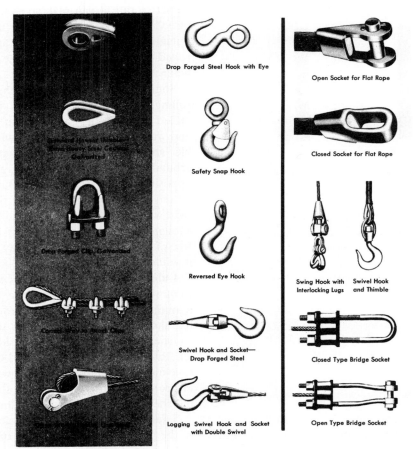

Courtesy of Wire Rope Institute

Fig. 12-60. Cable fittings

Capacity. Drum capacity may be calculated according to Figure 12-59.

CABLE SYSTEMS

Fittings and Anchors. In general, cable ends require some sort of clamp or fitting to attach them to the power source or work. A variety of these is shown in Figures 12-60 and 12-61.

The connection between the wire rope and the fitting may be secured by clamps, wedges, or fillers.

The standard clip, or cable clamp, consists of a U-bolt, a saddle, and two nuts. The cable is doubled over on itself, and the two thicknesses squeezed between the U and the saddle by tightening nuts. The grooved inner surface of the saddle has a better grip than the U, and is therefore used on the live or working end of the rope.

There are also heavy duty types of clip that use two identical saddles with rough inside surfaces, and can be put on from either side.

Two or more clips are used, the number increasing with the diameter of the cable. They afford a good grip, but are tedious to install and remove and occupy too much space for some uses.

The wedge socket jams the cable in too-small grooves in the outer surface of the wedge and in the inner suface of the socket.

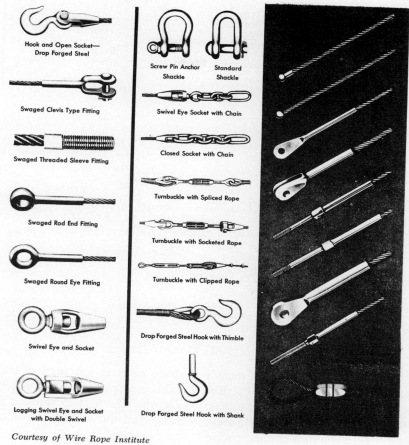

Hook and Open Socket—
Drop Forged Steel

Swaged Clevis Type Fitting

Swaged Threaded Sleeve Fitting

Swaged Rod End Fitting

Swaged Round Eye Fitting

Swivel Eye and Socket

Logging Swivel Eye and Socket
with Double Swivel

Screw Pin Anchor
Shackle

Standard
Shackle

Swivel Eye Socket with Chain

Closed Socket with Chain

Turnbuckle with Spliced Rope

Turnbuckle with Socketed Rope

Turnbuckle with Clipped Rope

Drop Forged Steel Hook with Thimble

Drop Forged Steel Hook with Shank

Courtesy of Wire Rope Institute

Fig. 12-61. More cable fittings

The cable is wrapped around the wedge which is tapped into the socket. Pull on the cable pulls the wedge farther, tightening the connection.

This device is best suited to excavator and other cables whose ends are more or less fixed, and which have to be removed at intervals. It is too bulky for many applications.

A poured (filled) fitting consists of a conical socket attached to an eye, loop, hook, or other device. It is installed by putting the end of the rope through the small end of the socket, fraying it out like a brush, removing the hemp center, if any, and lubricant or other foreign matter.

Molten zinc is then poured in the socket. It hardens, holding the individual wires in their expanded position so that they cannot be pulled through the small end.

Babbit and lead are too soft for use in these fittings unless the load is to be extremely light.

Drums. Power is usually imparted to a cable by anchoring it to a spool-type drum, rotating the drum so that the cable is wound in around it. This pulls the cable and the cable pulls anything to which it is attached.

The surface onto which the cable winds is called the lagging. There are three main types: smooth, cut with grooves to guide the cable into position, and tapered or coni-

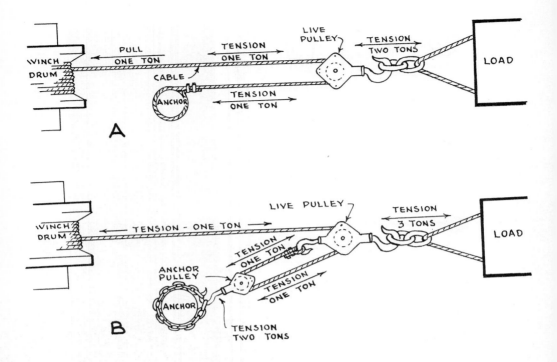

Fig. 12-62. Multiple lines

cal with the small diameter at the anchor side. The tendency of the cable to slide down the taper keeps the wraps tight against each other. This tapered lagging should not be used if there is sufficient rope to wind in a second layer.

A drum with cylindrical lagging, whether grooved or not, will pull in line at uniform speed until the lagging is entirely covered by the rope. The line will then wind in on top of itself, the first layer acting as lagging to receive the second. Because of the increase in diameter, the second layer will be pulled in somewhat more rapidly and with less power than the first.

A tapered lagging will pull each successive wrap with slightly more speed and less power than the previous one.

If a cable is attached to a reversible drum, that is, one which can be rotated by power in both directions, it is usually not anchored to it, but is wrapped around it two or more times. Whichever way the

drum turns it will wind in one section of cable and pay the other out in the same amount. The friction between the wraps of cable and the lagging provides the necessary traction.

Sheaves. The direction of cable pull may be changed by running it over a pulley or sheave (pronounced shiv). This is a steel wheel revolving on or with an axle, and having a groove in its rim to hold the particular size of cable to be used. Too large a groove will not support the cable properly and puts extra flexing strain on it; too small a groove will pinch and wear it.

Sheave grooves for $\frac{3}{8}$ to $\frac{3}{4}$ ropes should be $\frac{1}{32}$ to $\frac{3}{64}$ inches wider than the cable.

Multiple Lines. A pulley or set of pulleys may be used to change the power-speed ratio of a cable system. Figure 12-62 (A) shows a drum and a line attached to it passed around a pulley in a snatch block and brought back to an anchor near the drum. The drum is pulling in the rope with a force of one ton. This force, less friction losses in the block, is exerted on the anchor. There is tension of one ton on the line between drum and pulley, and on the line between pulley and anchor also. Since both pulls are exerted on the pulley, the pull against it is two tons. If it should yield to this force and move inward, it would move only one foot for each two feet of cable wound on the drum. The two feet is pulled out of the drum-to-pulley section, but as the block moves, one foot from the other section crosses it toward the drum, so the drum-to-pulley section is shortened by only one foot. It will be noted that the ratio checks —the force is doubled and the distance (speed) halved. Actually, there will be a loss of five percent or more in the pulley, and a slightly less than two-to-one advantage if the drum line and the anchor line are wider apart than the width of the sheave. The mechanical advantage of this system declines as the angle between the two cables increases and it becomes zero at 114°.

(B) shows the same drum and pulley with an additional block attached to the anchor. The rope is now run around the first pulley, around the anchor sheave, and anchored to a hook on the first sheave. If the drum winds in three feet of cable with a force of one ton, the first pulley will be subjected to a force of three tons, one ton for each line, and in response to this force it may move one foot. The pull on the anchor pulley is two tons, and whichever of the three units is least able to take the pull to which it is subjected will move inward.

The drum is said to be attached to a one part line, the anchor to a two part line, and the first pulley to a three part line. A four part line could be rigged by putting an additional pulley with the first one, and reeving the cable around that and back to the anchor, by which means a four ton pull (less friction) could be obtained.

The number of parts of line may be increased up to the space and pulleys available. The force exerted on any part of the system will be the force at the drum multiplied by the number of lines; and the distance any member will move is the length of cable wound in, divided by the number of lines acting on the member.

CHAIN

Chain is used in work where the line is subject to scraping, kinking, twisting, and other abuse. It is heavier and bulkier than cable of similar strength. Except in small sizes it is not used on drums or sheaves.

It is not damaged by sharp bending, is resistant to abrasion, and can be easily attached, detached, lengthened, shortened, and repaired.

Chain is made up of a series of interlocked links. Each of these may be a single piece of rod bent to shape and welded, or two half links butt-welded together. Chain size designation indicates the approximate diameter of the rod stock used. For example, $\frac{3}{8}''$ chain is composed of either $\frac{3}{8}''$ or $1\frac{3}{32}''$ round rod.

The strength of chain is affected by the size and shape of the links. Short links are stronger but do not move on each other as readily as larger ones, are more inclined to kink, and are harder to repair.

The most important strength factor, how-

Courtesy of S. G. Taylor Chain Company

Fig. 12-63. Logging chain

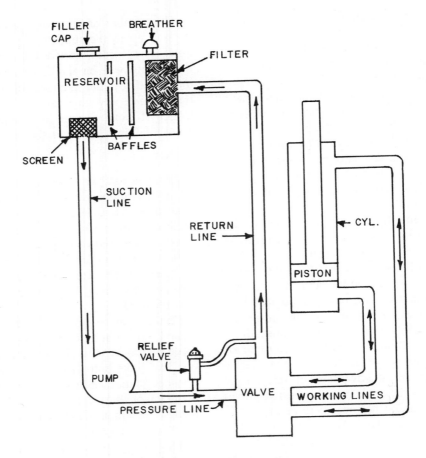

Fig. 12-64A. Hydraulic system

ever, is the quality of the metal used. A good alloy steel chain is more than three times as strong as the same size and weight of standard hardware store quality.

Chain and associated tackle are discussed in Chapters 1 and 21.

HYDRAULICS

Hydraulic systems of the dynamic or flowing type are of great and increasing importance in construction machinery, as well as in other industrial applications.

They depend on the fact that liquids cannot be compressed except under extreme laboratory conditions, so that pressure exerted on any part of a confined fluid will be exerted by the fluid everywhere on the confining surface.

In industrial hydraulics, the pressure is produced by flow from a pump driven by a rotating shaft. The flow is confined by conduits, casings, and hoses; directed by valves, and put to work by cylinders or motors.

Lines carrying fluid from the pump to working parts may be very short, very long, or anything in between. Power loss in the lines is slight if they are large enough for the flow. There is no need for alignment among the separate parts of the system. This is an important advantage over mechanical drive.

Fluid. Hydraulic fluid for flowing systems is a petroleum base oil, similar to but not identical with lubricating oil for engines. It must be distinguished from the glycerin base hydraulic fluid used in brakes and other static systems. The two liquids are absolutely incompatible, and cannot be substituted for, nor mixed with, each other.

Viscosity is measured in the same way as motor oil, with 5W very thin and 40 fairly heavy or thick. Too-thin oil tends to leak back through pumps and out through seals. If too thick, it consumes extra power and makes operation sluggish.

Viscosity index (VI) is a measure of a fluid's change in thickness as it heats during work and cools when idle. If the changes are slight, the VI is high, and the oil is good. If the index is low, it may be possible to raise it by adding a substance called a viscosity index improver.

A good fluid should lubricate the system. It should be able to penetrate fine spaces, resist being wiped off completely by pressure, keep friction to a minimum, and act as a cooling medium. It may contain an extreme-pressure additive to assure these characteristics.

The oil should resist breakdown by oxidation, and by reaction with parts of the system. It must be carefully refined to obtain this characteristic, and have one or more special chemicals added. Such special oil will also keep damage to metal from rusting or corrosion (eating away) to a minimum.

To retain its stability and neutrality the hydraulic fluid must be kept filtered to remove water, and particles of every kind.

Good hydraulic fluid may dissolve small quantities of air without harm, but it must not carry it in the form of foam or bubbles. These will compress in working parts, causing mushy, unreliable performance. Good oil resists foaming, but here again an additive may be needed.

The fluid should remain separate from any water that enters the system, so that it can be filtered out. Combining in the form of an emulsion allows the water to stay, and cause oxidation, deterioration of fluid characteristics, and deficient lubrication.

Pressure. The potential pressure at a working part is the same as that at the pump, as long as the line between them is open and adequate. But if the rate of flow is reduced below demand by a line that is too small or a fluid that is too thick, pressure will fall at the work end, unless (or until) movement of the working parts is slowed enough to let the fluid catch up with them.

Pressure inside the system is measured in pounds per square inch (psi). Changes in force and speed, equivalent to mechanical leverage advantages, are made by varying the relative areas of pistons and other movable parts.

Hydraulic systems have been operating at increasing pressures for many years. They now seem to range from 1500 to 4500 psi.

High pressure increases efficiency and promptness of response, and reduces bulk and weight. It intensifies problems of leakage, breakage, and wear.

Efficiency. Hydraulic systems, if properly designed and in good condition, may be highly efficient but are subject to many variables.

There are always some losses in friction, and in most systems there is some slippage and internal leakage. Movement of fluid, and these losses, create heat which absorbs power. The heat is often sufficient to require removal by a cooling system.

System efficiency may be as low as 70 per cent, but is often between 85 and 95 per cent.

Leverage, Hydraulic Jack. The hydraulic jack cross section in Figure 12-64B shows the way in which difference in area provides leverage.

Its rigid casing includes a large cylinder containing a piston which raises the car or other load, a small cylinder containing a piston, pump, and a reservoir.

When the pump handle is raised, it draws the small piston to the top of its cylinder, permitting oil from the reservoir to flow under it. The piston is pushed down, first closing the hole that admitted the oil, then pushing the oil through the check valve into the bottom of the large cylinder. This check valve consists of a ball spring held against a carefully machined seat. Pressure from the pump forces the ball back against the spring, opening the passage and allowing oil to pass through. The spring and pressure from the large cylinder force the ball back against its seat when the handle is raised so that no oil can pass back through it. Each stroke of the pump therefore adds liquid and builds up pressure in the large cylinder, thus raising the piston and whatever load may be on it.

The operator of the jack may apply fifty pounds pressure to the handle. If the leverage of the handle is six to one, three hundred pounds force is exerted on the pump piston which has an area of one quarter of a square inch. Pump pressure is therefore twelve hundred pounds per square inch, which pressure passes practically undiminished through the check valve, passage, and cylinder to the bottom of the large piston which has an area of five square inches. The fluid presses against this whole area, so the total pressure on the piston is six thousand pounds, or three tons. If the man applies a force of a hundred pounds to the handle, the lift will be increased to six tons.

The pump cylinder's effective length is one inch, so each stroke will deliver one quarter of a cubic inch of oil. In the lift this is spread out over an area of five square inches and will raise the piston one twentieth of an inch. Inside this particular jack the mechanical advantage (MA), or change in force-distance, is twenty to one, the force being multiplied by twenty and the distance

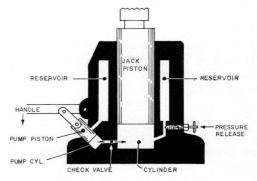

Fig. 12-64B. Hydraulic jack

moved divided by the same amount. Including the leverage of the handle, the jack has a mechanical advantage of one hundred twenty, with a friction loss of less than one percent. This ratio could be increased up to any force desired simply by reducing the size of the small piston or increasing that of the large one.

However, the useful power of any jack is limited by the strength of the casing and other parts.

The jack is lowered by opening a return passage by means of a thumbscrew. The weight of the piston and its load then pushes the fluid in the cylinder back into the reservoir.

PUMPS

The pump in the jack is a one cylinder reciprocating type, hand operated. But most pumps in excavators are driven by rotary shafts.

There are three principal types of rotary pump, knows as gear, vane, and piston. Each is made in several designs.

Pumps may also be classified as fixed displacement and variable displacement. The displacement is the amount of liquid it moves in one revolution or cycle. If it is fixed, the output should always be the same at any one speed of rotation. But output can of course be changed by varying the speed of revolution.

The variable type can be adjusted inter-

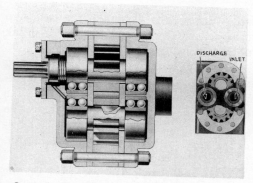

Fig. 12-65. Gear-type hydraulic pump

nally to change or (usually) to stop its output, without changing its speed. The hydrostatic piston pump usually has this feature, and is also reversible.

A pump may have a built-in pressure limiting device, or such a relief valve may be a separate unit in the system.

The pump may be permanently attached so that it works whenever the engine is running, as for example a tractor front-end pump; whenever the engine clutch is engaged, as in a rear-mounted tractor pump; or be connected specially each time it is to be used, as in a truck dump hoist. In any case, the pressure developed is utilized through one or more valves, usually of the spool (sliding core) type.

Even when pumps are new they are likely to allow a slight movement of oil from the pressure side back toward the inlet. Such losses occur along the side walls and between meshed teeth, around the ends of vanes, and, to a limited extent, past pistons. The rated capacity of pumps takes such losses into account.

As the pump wears the losses will increase, although not necessarily at a steady rate. The result will be a loss in volume that will be greatest at high pressures, and a drop in the maximum pressure it can develop. The pressure capacity loss may not be noticed until it drops below the setting of the relief valve. Volume decline may result in

a proportionately slower movement of light loads.

A pump that is small for its job lacks reserve capacity and will show results of wear much sooner than a bigger pump with the same work. Also, what seems to be a rather sudden pump failure may be just the final stage of a long slow decline.

A worn pump will give much better performance with thick oil than with thin oil. One of its symptoms is brisk performance only until the fluid is well warmed up.

The indications of a pump that is defective because of either wear or breakage are: slower than normal hoist or other movement, and less than normal power; both conditions being more noticeable when the hydraulic oil is hot than when it is cold.

Gear Pump. The gear pump, Figure 12-65, consists of two accurately meshed spur gears between side plates, turning in a chamber shaped so that the teeth are in close contact with a housing, for about half the circumference of each gear.

Oil enters the inlet side by gravity and/or atmospheric pressure, is caught in the hollows between the teeth in both gears, and is carried to the other side of the housing. Meshing of the teeth in the center prevents it from returning between the gears, so it is forced into the outlet passage. A continuous flow is generated.

Very close clearances are required, to prevent back leakage. The side plates are sometimes movable, and are then kept in contact with the gears by conducting pressure oil behind them.

This is the simplest and most economical of the rotary pumps, and is perhaps the most widely used. But it is not adapted to extreme high pressure.

A rotary may be the only pump in systems up to about 2000 pounds pressure, or a charging pump in a higher pressure system.

Vane. Figure 12-66A shows a vane pump of the unbalanced type. This is not widely used, but is easier to explain than

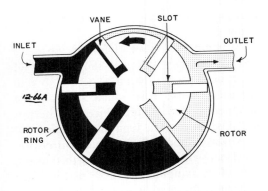

Fig. 12-66A. Unbalanced vane hydraulic
pump

the balanced model in the next illustration.

A cylindrical, engine-driven rotor turns off-center inside a circular rotor ring or case. This rotor has a number of radial slots that hold flat vanes, in a sliding fit. When it turns, the vanes are pushed outward by centrifugal force until they reach the ring, against which they then fit closely. They are also a sliding fit against side plates.

The vanes divide the crescent-shaped area between the rotor and the case into compartments which expand and shrink alternately as the rotor turns.

Fluid feeds into the inlet at the point where the chambers start to expand, creating a suction, and is discharged into the outlet as they contract. Beyond there, the chamber shrinks to the dimension of a lubricating film, so that pressure air cannot follow the vanes back to the inlet.

This pump does most of its work on one side, resulting in off-center loads that tend to cause rapid wear. For this reason, most vane pumps are the balanced type, Figure 12-66B, with two port sets on opposite sides. The ring is oval instead of circular, with a centered rotor.

The same pumping action is accomplished in half a revolution instead of a full revolution.

The balanced vane pump is somewhat more expensive and complicated than the gear pump, may develop somewhat higher pressure, and is probably second to it in popularity with manufacturers.

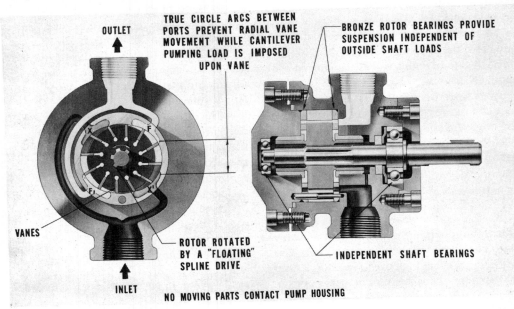

Courtesy of Vickers Incorporated

Fig. 12-66B. Balanced vane hydraulic pump

Part-swing vane pumps, also called vane cylinders because they are often quite long in proportion to diameter, have a fixed vane or block that prevents full revolution, and ports arranged for a back and forth (instead of around and around) motion.

Piston Pump (Hydrostatic). The pump described under this heading has as its full name: Inline Axial Piston Pump. The Inline is because there is another type that has a bent (out of line) axis. The Axial refers to the fact that piston movement is parallel to the axis of revolution (there are also radial piston pumps).

"Hydrostatic" is the popular name for these pumps, the almost-identical motors that are driven by them, and the drive systems in which they are used.

The term is inappropriate. Strictly speaking, hydrostatic means hydraulic pressure without motion, as in the push of water or wet soil against bracing in an excavation. However, industry has accepted the term for these systems, and it will be used here.

The axial piston or hydrostatic pump, Figure 12-66C to E, consists of a number (five or more) of pistons in a rotating cylinder block. The pistons are parallel to the axis of revolution, and have their heads in sliding contact with a non-rotating, tilted plate, known mostly as a swashplate but occasionally as a cam.

The casing has two ports, which carry low pressure fluid from a supply or charging pump, which goes through a check valve, an open tube, and an outlet to the working parts.

In one pumping position, the swashplate is tipped away from the upper port, which becomes an inlet. Charging pressure pushes pistons deep into their cylinders, until their ends press against the plate.

As the block turns, the inlet port closes, and the plate's slope forces the pistons back into their cylinders, putting a high wedging pressure against the fluid in them. At about half a turn from the inlet, this fluid is forced

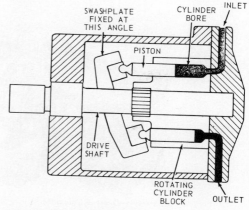

Courtesy of Deere & Company

Fig. 12-66C. Hydrostatic pump, fixed displacement

through the outlet. It goes through a motor circuit, then returns to the pump inlet.

The pistons are closely fitted in the cylinders. No fluid except a film for lubrication should get past them. For any one angle of the swashplate, the displacement (the amount of fluid pumped with each revolution of the shaft) and the potential pressure are the same, whether the engine is idling or turning at maximum speed.

In a fixed-displacement hydrostatic pump the swashplate angle cannot be changed.

Gear and vane pumps also have fixed displacement, but it is usually not so exact, because of leakage from the pressure to the inlet side. Such leakage is increased by thin oil and by pump wear, and is proportionately greater at low speeds than at high ones.

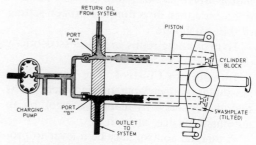

Courtesy of Deere & Company

Fig. 12-66D. Hydrostatic pump in operation

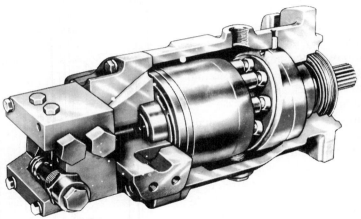

Fig. 12-66E. Hydrostatic (inline axial piston) pump

Variable Displacement. The displacement and pressure of the hydrostatic pump depend on the angle of the swashplate. If its slant is increased, the stroke of the pistons is lengthened, increasing the volume of fluid handled. The steeper angle reduces the strength of the wedging effect, so that less pressure can be devolped.

Flattening the angle of the plate has the reverse effect, reducing volume and increasing pressure.

If the plate is flat, at right angles to the axis of rotation, the pistons will not move in the cylinders, no fluid will move, no pressure will be produced, and no power will be consumed.

If the plate angle is reversed, the pump will take in oil at the port which was the outlet, and discharge it at the former inlet. This reverses flow in the system, so that a motor driven by this pump can be caused to rotate in either direction, by proper setting of the swashplate.

Hydrostatic circuits are discussed briefly on page 12-85.

Swashplate Control. The angle of the swashplate in variable displacement units may be regulated by manual control, by an automatic device, or both.

A manual control may be a lever to provide a neutral center, then gradually increasing speed with lever movement in either direction away from neutral. In a travel drive, moving the lever forward would cause forward travel; to the rear would reverse the machine.

In this arrangement, the control lever provides the effect of both a gearshift and a clutch, supplying infinitely graduated changes in speed-power ratio and a disconnect feature.

One automatic control in an excavator depends on engine speed. If engine rpm drops below its throttle setting because of hard digging slowing the bucket, a flow valve in a control circuit reduces the swashplate angle, thus decreasing the rate of flow to the bucket cylinder, increasing the pressure in it, and relieving load on the engine so that it can resume speed.

This effect is somewhat similar to that obtained by shifting a transmission into a lower gear, but adjustments can be made smoothly under power to any of an infinite number of ratios.

VALVES

Check Valve. The check valve is a simple and important device that allows passage of fluid in one direction only. A small ball check is used in the jack in Figure 12-64B, and greater detail is shown in 12-66E.

The principle is simple. A fluid passage is made with a machined collar. A solid plunger, which may be a ball or a plug, is shaped to a leak-tight fit against the collar, and is held against it by a spring. Pressure from behind, that is, in the same direction as spring pressure, holds it more firmly on its seat, keeping the passage blocked. Pressure from the other side pushes the plunger back by compressing the spring, so that fluid can flow past it.

Relief Valve. Most hydraulic systems, and most engine lubricating systems also, depend heavily on a relief valve for proper operation. The pump is built to deliver a much higher pressure than the system can use so that it will have reserve capacity to compensate for damage or wear. The relief valve is set for the pressure that the system should use, and spills any excess into a bypass back to the reservoir.

The pressure relief valve in Figure 12-67A is similar to the check valve in the hydraulic jack, except that it is on a larger scale, the spring is quite heavy, and the spring tension, and therefore the opening pressure, is adjustable.

Most relief valves make a rattling or squealing sound while they are open, so that you can make a rough check on their performance by operating the machine. Their normal behavior differs in various machines, but they should always open when the hydraulic mechanism is stalled by a load. If pump pressure and volume are very high compared with system requirements, the valve may open whenever movement is slowed by a load.

If the valve opens at too low a pressure, either because of a weakened spring or an improper adjustment, the hydraulic part of the machine will operate sluggishly and with less than its normal power. If the valve is set for too high a pressure action will be brisk and snappy, and usually jerky, power will be greater than normal, and bursting of hoses is likely to be frequent.

Many relief valves are adjustable by taking off a cover nut and turning an adjusting

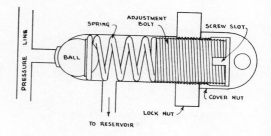

Fig. 12-67A. Pressure relief valve

screw that regulates spring tension. The tighter the spring the greater the pressure required to open the valve. Adjustment should not usually be changed unless a pressure gauge can be connected into the line. However, a rough judgment of pressure can be made by an experienced operator, on the basis of machine performance.

A relief valve may be in a "capsule" that can be removed from the machine complete, for shop checking and adjustment.

The relief is most often in the line from the pump to the valve(s), where it protects the whole system. But such valves may also be between the valve and cylinders, for protection in case of a collision that might put excessive pressure in a blocked line in holding position.

A cylinder may also have a bypass valve (opening) at one or both ends to allow relief of pressure while the piston is held at the end of its stroke.

Thermal Relief. If a machine is parked in the hot sun with lines blocked in HOLD position, the fluid in one or more of its cylinders might expand enough from heat to burst the cylinder or a hose. A thermal relief valve, installed in each line from the control valve to the cylinders, will be opened

Fig. 12-67B. Check valve

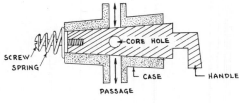

Fig. 12-68A. Simple rotary valve

by the heat, and allow enough fluid to escape to relieve the pressure.

This fluid is voided outside of the system, but its quantity is so small and activity so rare that it does not pay to provide for salvaging it.

Such heat damage may also be prevented in many systems by leaving the control valve in an open-line or FLOAT position while the machine is parked.

Control. Control valves for high pressure systems usually have a movable core with slots or holes that can be brought into and out of register with slots or holes in a case or body. The matching surfaces of the core and body are machined to a close sliding fit that prevents leakage between ports.

A rotary valve, in which the core is rotated inside the case (usually 60 degrees or

less) is good for low and moderate pressures, but in high pressure systems it is difficult to balance so that it will not stick when pressure is higher on one side than on the other.

The quarter-turn petcock, Figure 12-68A, is a simple, familiar type of rotary valve. It consists of a tapered bore with a hole drilled through it, held tightly in a similarly tapered case by a spring. The case has a hole through it, one end of which is attached to a tube or a tank. The core may be turned by means of a handle so that its hole will coincide with the casing hole and fluid can flow through it; or so that its hole is at right angles to that in the casing and its solid part blocks the passage and prevents fluid from passing through.

Figure 12-68B shows operating positions of a schematic four-way spool (sliding) control valve, of a type used in bulldozers. The case and the core are each drilled or cut away for a number of passages, which are opened into each other, or closed, as the spool slides in the case.

The passages in the casing are connected with four outside pipes—one bringing pressure from the pump, two leading to the working part of the system, and one emptying into the reservoir. The two working positions of the valve admit pressure from

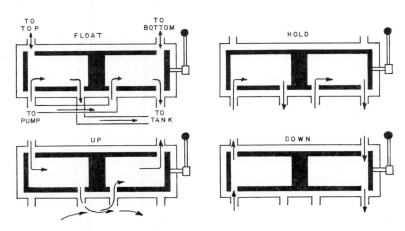

Fig. 12-68B. Schematic control valve

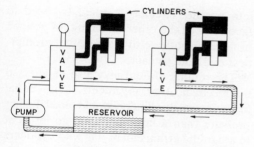

Fig. 12-69A. Series circuit

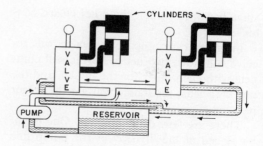

Fig. 12-69B. Parallel circuits

the pump and direct it to one or the other of the working lines, and connect the other working line to the reservoir. The "hold" position closes both working lines and opens the pump line to the reservoir. The "float" position opens the pressure and the working lines to the reservoir. In the hold and float positions the pump develops only sufficient pressure to overcome line friction on the way to the reservoir. Its pressure in the working position is determined first by the degree of resistance, second by the setting of the relief valve, and third by the ability of the pump.

The free flow of fluid when the valve is in neutral or HOLD characterizes this as an open center system. It would also be possible to block flow in neutral, and shut off the pump with an automatic device whenever flow stopped. It would then be a closed center system.

This four way valve might also be used in a dozer hoist, or a loader with a separate dump valve. Most excavator circuits have three way valves with no FLOAT position.

Two or more spool valves are often mounted side by side with interconnecting passages, or are built into a single casing.

A control lever may be mounted directly on the spool shaft, may be connected to it by rods or cables, or may power-operate it through air, hydraulic, or electric means. Use of power controls is increasing, because of the high pressure and volume to be regulated in modern systems.

Sequences. When two or more working circuits are powered by one pump, they must share its flow, which tends to always go to the area of least resistance. The designer must figure out where the power is needed in what proportions, and include appropriate devices to distribute it.

There are two basic ways to route the power flow; series and parallel. They are shown in Figures 12-69A and 12-69B for systems in which two circuits might operate together or separately.

In a simple series, fluid goes through the two valves in succession, and then returns to the tank. If the first circuit is used, it takes all the flow, but its exhaust goes through the second valve. If that is put in working position, resistance to its piston will cause back pressure on the first circuit. Pressure will become balanced in the two circuits, and the one with least resistance will get most or all of the fluid, "robbing" the other.

If only the second valve is used, it will get full flow through the open center neutral position of the first.

If the two valves were in series on the pressure line, but had separate outlets to the return, the second circuit could work only when the first was idle.

In a parallel system, the two valves have separate pressure and exhaust systems, but are designed so that putting one valve in working position will block the open center in the other also, or so that in neutral one valve has open center, the other closed center.

In either case, one active valve will receive full flow, two active valves will compete with each other for the intake, and the one with least resistance will get the larger part, or all.

There are flow divider valves. One might be placed at the division of the pump line

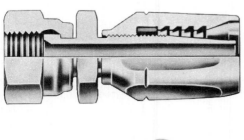

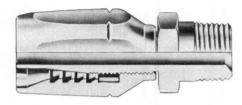

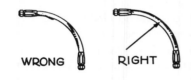

WRONG   RIGHT

Fig. 12-70A. Pressure hose and fittings

into parallel circuits. It could be fixed to favor one or the other in timing, in volume of fluid, or both; or to assure equal promptness and fluid to each.

The routing of fluid may also be affected by mechanical arrangements. With equal loads, a small cylinder would need higher pressure than a large one, and so lose out in a parallel arrangement. Or with cylinders of equal size, leverage might be made more favorable for one than the other, so it could operate on lower pressure.

Providing a separate pump for each circuit is costly, but under a wide range of conditions it justifies itself by superior performance.

Separate circuits may also include provision for reinforcing each other. A button, lever, pedal, or automatic device may cut off one and direct its power flow into another for temporary increase in its speed.

These are just some of the simpler factors in layout of a hydraulic system. And they must be manipulated according to the requirements of the machine, which usually must cope with a number of different situations.

LINES AND FITTINGS

Lines. Hydraulic pressure systems may use plumbing pipe (but not galvanized, it flakes) for rigid lines, but tubing is preferable. It is usually stronger, more resistant to vibration, and less likely to leak at connections. It can also be bent at the factory to fit the machine.

The wall strength of any conduit, whether rigid or flexbile, must be increased with increase in diameter, due to enlargement of area exposed to internal pressure.

A rigid line must be firmly clamped close to any connection to a flexible line, so that its movements will not have sufficient leverage to strain and eventually break the pipe at the clamp.

Most flexible lines in construction equipment are high pressure type hose woven of

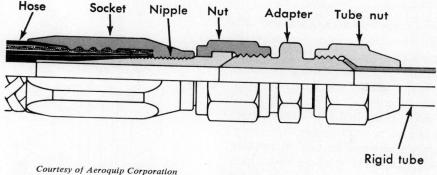

Courtesy of Aeroquip Corporation

Fig. 12-70B. Swivel fitting

fabric, rubber and metal. Their bursting strength is far greater than the pressure they are required to hold, but they become weakened by repeated bending or rubbing, so that they eventually blow out.

Useful life varies with the quality and pressure resistance of the hose, its flexibility, the pressure in it, the tightness of its curves, the sharpness and frequency of bending, and the way it is installed. A wrong-length hose can change a good arrangement into a bad one.

Rubbing can often be corrected by loosening and retightening in a slightly different position, or by tying with cord or tape. Damage may be prevented by a protective covering of tape, wire, or strap iron, or with a sleeve of radiator hose.

A hose should be set in a smooth curve, not being so long as to bulge outside it, nor so short as to make an angle at the fitting. If the angle changes as the machine works, adjust length for the most usual position.

A hose shortens as pressure inside it increases, so that a correct curve during installation may mean a tight bend under pressure, and most hose damage occurs under pressure.

Hose should be marked with a dash line or some other pattern by which you can tell whether you have it straight. Be sure to check this as you finish an installation, as any twist will damage a hose that is under pressure.

Each flexible hose is fastened to a threaded metal fitting at each end. Some fittings are permanently fastened, so that they have to be scrapped when a hose breaks. The majority, however, are reusable fittings that can be used with hose after hose.

Theoretically, the reusable fittings allow stocking just a roll of hose in each size needed and acquiring a few special tools for efficient assembly. But it is wiser to keep a spare hose, complete with couplings, for each place with a history of frequent breakage. A change can then be made in a few minutes, and the broken one fixed later. And a mechanic is more apt to do a good job making up a hose assembly in the shop than in the field.

Another advantage of the complete hose with fittings is that it will take care of trouble with a broken or stripped fitting also.

Fittings may have either coarse pipe thread or fine tubing thread. Never try to force mis-matched threads together, as they will strip.

Bolted flanges may be used instead of threaded fittings in many places in very high pressure systems.

A swivel fitting, Figure 12-70A top, and 12-70B, has a threaded collar which turns freely on the fitting when it is loose, but pulls into a rigid non-leaking connection when tightened. A pipe union is another example of this type of fitting.

In a hydraulic system, each hose and

each tube or pipe that both begins and ends in a solid connection, must include a swivel. This permits installation and tightening at the fixed end or ends, then using the swivel for the final assembly and tightening without twisting the line.

The swivel is the first joint to take apart, and the last one to put together.

Two wrenches must be used in working on swivels, to avoid twisting the lines. Open end or adjustable wrenches are best, but if a stillson is used carefully it should not damage the fittings.

Rotary Joint. It is occasionally necessary to make hydraulic connections between units that have rotary motion in relation to each other. A common example is an excavator whose rotating or swinging upper section contains pumps and valves controlling hydraulic motors that drive the tracks.

A rotary joint, also called swing joint and swivel joint, takes care of such situations. In Figure 12-70C there are two sleeves, centered on the axis of rotation of the unit, and machined for exact fit where their surfaces are in contact. The outer one, called a casing, is fastened to the lower unit (carbody). The inner piece or core rotates with the upper deck.

Horizontal ring-shaped channels are cut into the inner surface of the casing. They are sealed from each other by the close fit of the surfaces between them, or by O-rings.

For each hydraulic line, a conduit and passage from a valve in the revolving unit leads down the inside of the core, then through a port opening into a ring channel in the case. A port in the case opens into this ring, and conducts the pressure fluid to the hydraulic motor or other working part in the carbody. Fluid moves freely through this connection in any swing position, and in either direction.

Any reasonable number of lines can be connected, by providing the necessary number of passages, rings, and ports.

The joint has non-rigid attachments to

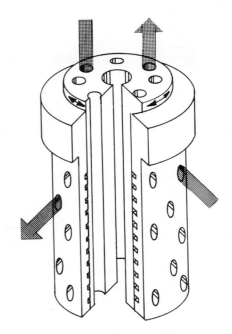

Fig. 12-70C. Swing joint

the excavator, to prevent distortion and damage from digging strains.

Reservoir. The reservoir of a hydraulic system serves basically as a supply tank, feeding fluid into the pump as needed, and receiving low pressure oil exhausted or by-

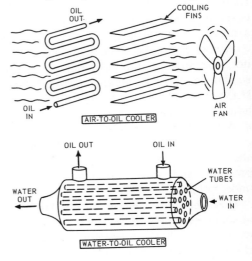

Courtesy of Deere & Company

Fig. 12-70D. Oil coolers

passed by working parts. It must hold sufficient oil to take care of pump demand in spite of tipping of the machine, surges in the system, normal leakage, and spills from broken lines.

Sometimes a capacity of hundreds of gallons is required, and since excavators often have little spare space, location of the tank may be difficult. It may be beside a tractor seat, in loader columns, in an enlarged transmission case, or almost anywhere.

The reservoir normally includes a filler cap and drain plug, an outlet to the pump and an inlet from discharge lines, an air breather (unless in a sealed system), a baffle or baffles to reduce sloshing and to prevent inlet oil from going directly out of the outlet, an oil level gauge, an outlet screen, and perhaps one or more filters.

Another function is to cool oil. Radiation from the sides of the tank may be enough for a system with moderate pressure and activity, but a separate cooler is needed in many high pressure systems.

Open or Closed. A hydraulic system may be more or less open to the air at the reservoir, or it may be sealed. If open, there is usually some sort of filter in the vent, to reduce intake of dirt with air.

The quantity of fluid in active use in a system may vary considerably. If there are one-way cylinders (now quite rare) they use oil when extended, return it when retracted. A suddenly lowered bucket with a load may push fluid into the tank from the bottoms of the cylinders, before it can be drawn back to fill the tops.

In an open system with a properly filled tank, the resulting surges that raise and lower level simply push air out or draw it back in. In a closed system, they build up pressure or pull it down to vacuum. A large air capacity may be needed to cushion these effects.

An over-full open tank may squirt oil out the vent, sometimes giving the operator a disagreeable hot shower. If underfilled, the

system may suck air, causing chattering, poor response, and other troubles.

The sealed construction is best suited to the modern ultra high pressure systems, because of their increased sensitivity to dirt damage, and possible danger in too-forceful sprays of oil.

Coolers (Heat Exchangers). Hydraulic oil, and seals and other parts, tend to deteriorate if system temperature rises over 180 to 200° F.

If radiation of heat from the reservoir and exposed parts is not sufficient protection, a cooling system must be installed.

A widely used system is to place an oil-cooling radiator in front of the engine water-cooling radiator, with alterations in fan size and other engine cooling details.

Oil is circulated through this cooler at a pressure of about 100 psi, which might be obtained by a constriction in the return line to the reservoir, or from a special pump. There is usually a bypass that will cut out the cooler if it becomes plugged, or if conditions make its use unnecessary.

There are also heavier and more complex systems of circulating oil through pipes in a chamber filled with circulating water.

CYLINDERS AND MOTORS

Pumps convert mechanical energy into fluid energy. Cylinders and motors turn the fluid energy into work. Cylinders push and pull in straight lines or arcs; motors rotate.

Cylinders of all types are often called rams, but this term is now most often limited to a rodless one-way cylinder.

Cylinder, Single-Acting. The single-acting or one-way cylinder may be a ram type, Figure 12-70E, or a piston-and-rod design, 12-70F.

A steel plunger or piston slides in a cylinder, both machined to be a close but not a wiping fit. The piston has one or more rings to seal the contact with the cylinder, so that only a thin film for lubrication can get past.

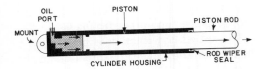

Fig. 12-70E. Hydraulic ram

The cylinder has a single port at its attachment end, through which pressure fluid enters to push against the face of the piston to move it, the rod if any, and any attached load, away from the cylinder.

When the oil pressure is released, the weight of the load, or sometimes a spring, moves the piston back to its original position, forcing oil out of the cylinder back to a reservoir. There is also a hold position, in which oil is trapped in the cylinder and will keep the piston and load in position.

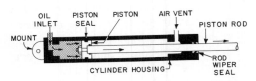

Fig. 12-70F. Hydraulic cylinder, one-way

If the rod is smaller than the piston, the opposite cylinder end should be dry except for a lubricating film on its walls. There must be an air vent, to prevent building up of pressure against the piston. This vent should also allow escape of any oil that leaks past the piston seal.

The rod or ram has an oil wiper seal where it enters the cylinder block.

The single-acting cylinder has comparatively limited use in excavating machinery,

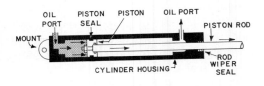

Fig. 12-70G. Hydraulic cylinder, two-way

as two-way power is usually advantageous. Exceptions include the lighter dump truck hoists (where two-way would be better), three-point tractor hitches, part-swing devices where two rams push against each other, and other specialties.

The plunger type of one-way cylinder has the advantages of a thicker and stronger (full piston diameter) rod, and no need for an air vent. But it does not stay in alignment as well, is more subject to side wear, and the piston may be pushed out of the cylinder if not restrained by the mechanical linkage. Pushing or blowing out occurs as a result of poor design, wear or breakage of a part, or carelessness during overhaul.

Cylinder, Double-Acting. The double-acting or two-way cylinder is the work horse of excavators, for functions that can

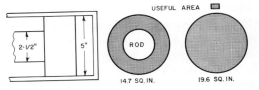

Fig. 12-70H. Effective piston areas

be performed by pushing and pulling. See Figure 12-70G.

The mechanism is similar to the one-way type with piston and rod, except that an oil port and line is substituted for the air vent, and a high pressure seal replaces the rod wiper.

A control valve directs pressure oil into either end of the cylinder while it permits flow out of the other end. The piston moves away from the pressure.

The valve may also block both ports, locking the piston in place, or open a passage between the two ends, allowing the piston to move freely under load, or float.

Many cylinders have a snubbing device, to prevent the piston from moving at full speed and pressure against a cylinder end, with resultant shock to the system. The

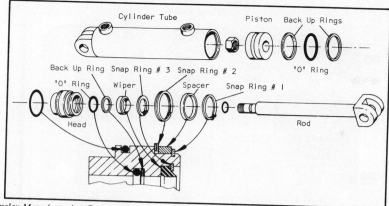

Courtesy of Insley Manufacturing Corporation

Fig. 12-71A. Hydraulic cylinder, disassembled

snubber may be a tapered or double port which is almost closed by the piston as it nears the end of its stroke. The reduced opening restricts outflow of fluid, creating back pressure to oppose and slow the piston.

Part of the port might be exposed behind the piston, relieving pressure into the exhaust passage. This also serves to reduce strain on the pump, if the valve is not shifted back into neutral when the piston has reached the end of its stroke.

Most double-acting cylinders are unbalanced, meaning that they have a stronger push (expansion) from the piston's open side than pull (contraction) from pressure on the rod side, because of difference in area exposed to fluid pressure.

A cylinder with a 5-inch piston attached to a 2½-inch rod, operating on 2000 pounds per square inch pressure, would have a push of about 39,000 pounds and a pull of 29,-000. A 3-inch rod would reduce pull to about 25,000 pounds.

When volume of flow through lines is a limiting factor, the pull is faster than the push for light loads.

Mounting. A hydraulic piston rod is strong only for straight push and pull, and has little resistance to bending stresses from the side. The cylinder and the outer end of the rod are usually mounted hinge-fashion,

and movements in the plane of these hinges impose no bending stresses.

If there is any possibility of side strains bending or twisting the working parts out of this plane, one or both ends of the unit must have a universal or spherical mounting, to permit sideward pivoting to align with the changed direction of the stress.

Hydraulic Motors. The hydraulic motor is a hydraulic pump in reverse. The pump is driven by a rotating shaft and produces fluid power. The motor uses that power to rotate a shaft.

Their use is a paying proposition because when two shafts are in different places, fluid lines may be an efficient and convenient way to carry power from one to the other. Such transfer may include a change of direction,

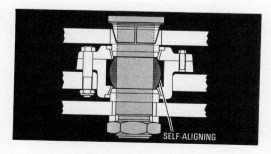

Courtesy of Fiat-Allis Construction Machinery, Inc.

Fig. 12-71B. Spherical hinge mounting

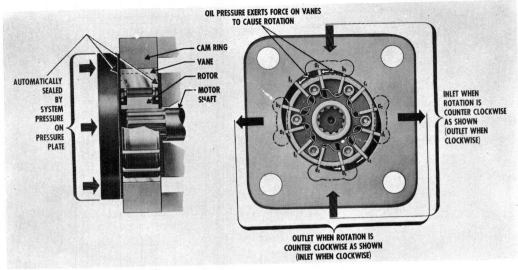

Courtesy of Vickers Incorporated

Fig. 12-71C. Hydraulic motor details

a reversal of rotation, and/or shift in speed-power ratio.

Motors of either gear or piston type may be almost identical in appearance and description to pumps of the same type. However, they are seldom interchangeable unless specifically designed for dual use. They differ in details of fluid flow, and in points of stress, and may use different types of bearings and seals.

A vane motor differs from a vane pump in having springs or clips to hold the vanes out against the housing when at rest, as it

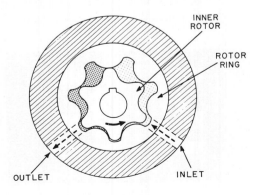

Courtesy of Deere & Company

Fig. 12-71D. Internal gear motor

cannot depend on centrifugal force (as the motor does) until the fluid starts to turn it.

Figure 12-71D is a cross section of a rotor or internal gear motor. Not actually gears, the moving parts are called the rotor and the rotor ring. The rotor is driven inside the rotor ring, which is held by the rigid case.

The rotor is mounted eccentric to the ring. The ring has one more lobe than the rotor, so that only one lobe is in full engagement at any time. This causes the rotor's lobes to slide over the outer lobes, making a seal.

In operation, pressure fluid enters the motor, striking the rotor and rotor ring lobes, forcing both to rotate. As they rotate, a seal is formed, then broken, as each inner lobe engages a cavity in the outer ring. The fluid is discharged at low pressure at the outlet port in the case.

Hydraulic motors may be used for almost any kind of rotary drive, and can be designed to slow or stop their output shafts when input stops. However, because of the possibility of slight internal leakage under sustained static pressure, they cannot be depended upon to keep a load locked in place.

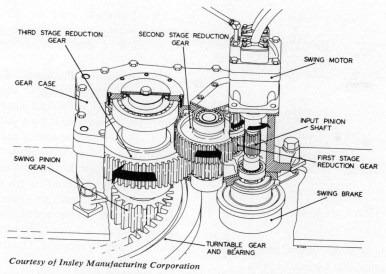

Courtesy of Insley Manufacturing Corporation

Fig. 12-71E. Hydraulic motor with reduction gearing

If locking is necessary, it is usual to put a friction brake or clamp in the mechanical part of the system. This is held engaged by springs when the motor is idle, and released by hydraulic pressure when it is working.

The efficient speed for a motor is often much higher than the desired speed for the driven unit. Motor shafts may then be part of a final drive type of gear box, as in Figure 12-71E. The gear set can then include the holding brake mentioned above.

Hydrostatic Circuit. A hydrostatic motor is very similar to the inline axial piston pump described earlier. High pressure oil from the pump enters cylinders on the high side of an angled swashplate. The pistons slide down the slope, causing its cylinder block to revolve, turning an attached output shaft.

The exhaust line from the motor is the inlet line of the pump.

It is usual to have an adjustable swashplate on the pump and a fixed one in the motor. However, when two motors with different requirements are driven from one pump, they may have the adjustable members.

A hydrostatic circuit is frequently a closed loop, involving one pump (with charging pump), one motor, and a pair of connecting lines. The pump and motor may be built into a single block with drilled passages for fluid, or be at considerable distances with connection by piping and/or flexible lines. One reservoir and filter may serve one circuit, or several.

In-block construction may serve simply as a transmission with variable speed and direction, or provide change in direction (angle drive) also. Separated units also allow replacement of more or less compli-

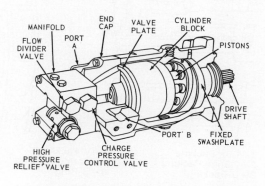

Courtesy of Deere & Company

Fig. 12-71F. Fixed displacement piston motor

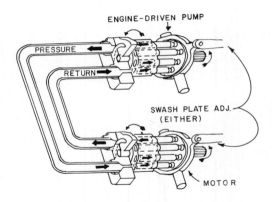

Fig. 12-71G. Hydrostatic circuit

cated layouts of shafts, gears, and/or chains with more adaptable hydraulic lines.

SEALS AND FILTERS

The proper functioning of a hydraulic system is completely dependent on having effective seals, that will keep fluid and pressure in and dirt out. Design, manufacture and installation of such seals becomes increasingly critical as pressures increase.

There are two distinct types. Static seals, which include gaskets, seal fixed parts together. Dynamic seals are for parts that move against each other. The motion may be rotation of shafts, or sliding of pistons and rods.

Compression Packing. Compression packing consists of string, or shaped inserts, that are wound or placed around a shaft, and tightened against it by a packing nut or gland. The sealing material may be cotton string, graphite and asbestos compounds, plastic, rubber, or flexible metals.

These seals will hold only a few hundred pounds pressure, and are found mostly on old equipment. Leaks are stopped by tightening the nut, but not tight enough to prevent lubrication seepage. They will not hold if the shaft is scored.

U- and V-Packing. U- and V-packing, often called chevron packing, is made of various materials compressed and molded into rings that are concave in one direction and convex in the other. The rings may be intact, for slide-on installation, or cut for installation from the side. Several of them are placed around a shaft with their open or concave sides toward the pressure, so that it will tend to expand them, forcing the

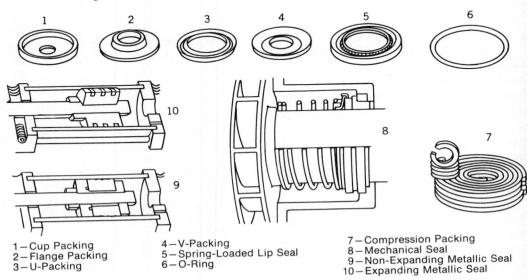

1 — Cup Packing
2 — Flange Packing
3 — U-Packing
4 — V-Packing
5 — Spring-Loaded Lip Seal
6 — O-Ring
7 — Compression Packing
8 — Mechanical Seal
9 — Non-Expanding Metallic Seal
10 — Expanding Metallic Seal

Courtesy of Deere & Company

Fig. 12-71H. Types of hydraulic seals

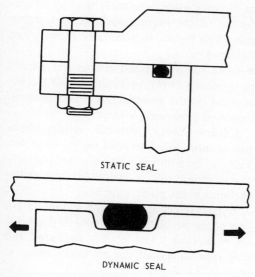

STATIC SEAL

DYNAMIC SEAL

Courtesy of Deere & Company

Fig. 12-71I. Static and dynamic

inner lip against the shaft, and the outer lip against the case.

They are rather lightly compressed by being held in a gland or case, with a flange or plate tightened by bolts. The bolts are often secured against loosening by a wire passed through drilled holes in the heads.

These packings are satisfactory for low to medium high pressure.

Similar packing rings made from more flexible material may be combined with circular springs to hold them firmly against the shaft. Such a spring-and-seal units is often supplied in a metal case that is forced into a bored hole in the case, and is used singly.

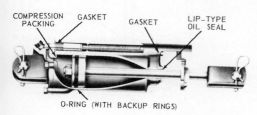

COMPRESSION PACKING   GASKET   GASKET   LIP-TYPE OIL SEAL

O-RING (WITH BACKUP RINGS)

Courtesy of Deere & Company

Fig. 12-71J. Seals in hydraulic cylinder

Spring loaded seals are more commonly used in lubrication systems than in hydraulics.

O-Ring. The O-ring is probably the most popular seal for sliding shafts such as piston rods, and is important for static use as a gasket.

It is composed of a slender ring of synthetic, oil-immune rubber, which is usually quite soft and flexible. It is used only between smooth, carefully shaped metal surfaces, as it does not have the bulk to take care of irregularities, and will wear rapidly or tear immediately on sliding parts that are

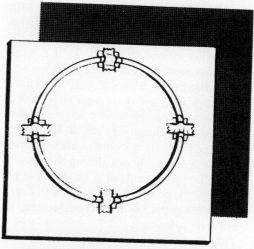

Fig. 12-71K. O-ring prepared for storage

rough, jointed or angled. It is also worn rapidly by rotating shafts.

The O-ring is compressed moderately, about ten per cent, in a specially shaped groove between the two parts. For gasket use it may be reinforced by a back-up ring of stronger material.

An O-ring must be the right size, both in ring diameter and in gauge, to do its work. It must also be in perfect condition. Tiny nicks or scrapes, loss of shape, or hardening from heat exposure, is likely to make it useless.

There is therefore a problem of keeping

all O-rings that may be needed for service in stock, and an at least equal problem of keeping them in useful condition.

An O-ring is best fastened to a piece of cardboard, after being arranged in its proper round shape. Fastening is usually with Scotch tape, but it must not touch the rubber. Put a little piece of paper over the ring, then apply the tape, as in Figure 12-71K. Additional cardboard, or at least a bit of paper, should then be put over it.

Mechanical Seal. The mechanical or positive seal is usually a pair of hard washers, lapped to make perfect rotating fit with each other, pressed together by diaphragms or springs. These are used chiefly for rotary seals at low speeds and low to moderate pressure. They are excellent to protect lubricated units, such as track wheels, that work in dirt and mud. See Figure 15-11.

Metallic Seals. Metallic seals for pistons and piston rods are quite similar to piston rings in engines. However, engine piston rings are expanding type; cut at one point, and shaped to push outward against the cylinder walls for a smooth sliding fit.

Metallic piston seals may be of the same type, or may be non-expanding. The latter need extreme precision in manufacture, to avoid serious leakage.

Piston rod seals may be fitted into the cylinder housing, and contract around the rod.

In general, metallic seals are subject to more leakage than other types. This may not matter on pistons, but for rods may call for an additional wiper seal.

Gaskets. Gaskets are flat sheets, formed to fit between flat pieces of metal, and cut away wherever necessary to allow for passages, bolt holes, or moving parts. Shapes for hydraulic components tend to be fairly simple. Gasketed parts are usually bolted together.

Gaskets are made from a number of quite different materials, which may be metallic, non-metallic, or both. They must be yielding enough to mold into all irregularities of the mating surfaces, but strong enough to resist penetration or blowing out sideward.

Bolts for a gasketed connection should be tightened firmly, up to the torque recommended by the manufacturer; and then retightened after some hours of use.

Filters. Every hydraulic system needs one or more filters to remove fine particles. This is in addition to the comparatively coarse screen that is customarily put at the entrance to the pump inlet line.

Bypass filters are arranged so that only a part of the circulating fluid goes through one on each cycle. The greatly preferred full flow type filters all the oil every time, at least until it clogs.

There are surface filters made of very fine wire mesh, porous paper in various shapes, or even of stacked discs of metal or paper. The surface may be made very large in a small container by elaborate folding.

Depth filters are packed with absorbent material such as cotton waste. Since oil can take many different paths through them, they tend to pick up particles at all levels, instead of just on the surface.

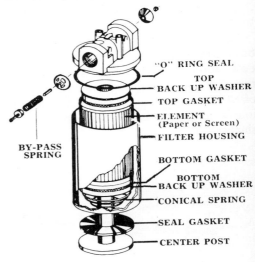

"O" RING SEAL
TOP
BACK UP WASHER
TOP GASKET
ELEMENT (Paper or Screen)
FILTER HOUSING
BY-PASS SPRING
BOTTOM GASKET
BOTTOM BACK UP WASHER
CONICAL SPRING
SEAL GASKET
CENTER POST

Courtesy of Drott Manufacturing Corporation

Fig. 12-71L. Hydraulic filter

Adsorbent filters are built like absorbent ones, but are chemically treated to attract and remove certain contaminants, which may be not only particles, but liquids and sludge from deteriorated oil.

A current standard for many filters is removal of particles down to 10 microns in diameter. This is about 1/4,000,000th of an inch, and is the average size of ordinary portland cement particles. A few systems trap much finer bits, some down to 2 microns.

Any filter may remove many particles finer than its rating indicates, when two or more try to go through a hole at once, and as the filter fills up with use.

Even a new filter impedes oil flow a little, so that pressure is slightly higher on the up-stream side. As it works, holes clog and the pressure difference increases until the filter is a serious obstacle to flow. A bypass should then open, filtration will cease, and a warning signal of some kind should appear or be heard.

There may be filters on both inlet and return lines, or on one. They may be built into a reservoir or pump body, or be just plugged into lines at covenient spots.

MAINTENANCE

Hydraulic systems are rugged and may operate for long periods with little attention. However, it is urgent that they do receive the servicing they require, as failures can be very expensive.

The first and most urgent requirement is clean oil. Filters must be the proper type and size, with cleaning or changing as often as necessary. Oil should be changed according to manufacturer's directions. This may be once or twice a year. The system should be operated first to warm it up. It may require filling and short period operation with flushing oil.

Leaks should be fixed promptly. They make the machine dirty and hard to service, and any hole that lets oil out might let dirt in, in spite of the great pressure difference.

All hoses should be inspected periodically, and sharp bends and rubbing areas corrected. Where rubbing is chronic, protection should be arranged.

LUBRICANTS

A fundamental necessity for machine operation is proper lubrication. The lubricant provides a slippery film between surfaces rubbing, turning, or scraping on each other. This film greatly reduces friction and the wasted power, wear, and heating that friction causes. Lubricant may also serve as a cooling medium and as a barrier or cleaner to keep abrasive material from getting or remaining between moving parts.

Oil and Grease. Lubricants are called oils or greases. Oils are fluid and vary from the extreme thinness of penetrating oil to the slow flowing transmission oils, which are more often called greases. The term grease may be said to include these thick oils, but more specifically means the semi-solid and solid mixtures of oil with special soaps or fillers which give the combination the qualities of body (flow resistance), adhesiveness, pressure endurance, water resistance, and melting point on the basis of which greases are selected.

Dip Lubrication. Transmissions and other gear boxes usually are partly filled with oil or fluid grease. Some of the gears are partly immersed in this lubricant and carry it on their teeth to the higher gears with which they are meshed. Other gears, bearings, and splines are lubricated by splash, by gravity flow of oil carried to higher points, or both.

The dip method is best suited to heavy lubricants which cling to parts enough so that adequate quantities will be picked up and transferred to higher levels. Rotation should be slow and construction simple enough so that local hot spots will not result from uneven distribution of the lubricant.

When engines are lubricated in this manner the crankshaft usually has projections which dip into the oil and splash it around so that it reaches all surfaces requiring it.

An engine or a gear box may be lubricated partly by dip and splash, and partly by pumped oil.

Pump Systems. A pump may pick up oil from a reservoir, usually the crankcase or oil pan under the engine, and force it through the crankshaft and camshaft in drilled passages which have openings in each bearing. The amount of oil which escapes at each point is regulated partly by the size of the outlet, but chiefly by the closeness of bearing fit. Connecting rods may also be drilled to carry oil to wrist pins. Parts not reached directly by pumped oil, such as cylinder walls, are lubricated by an oil mist generated by leakage out of the bearings, and by dipping and splashing as well. All oil returns to the reservoir to be picked up again by the pump.

A weakness of many of these systems is that the pump moves only sufficient volume for normal requirements. If bearings wear or the oil becomes too thin, through error in selection or because of dilution or too much heat, an excessive amount of oil will escape at the bearings. This will lower the oil pressure and make it likely that the last bearings in the series will receive too little lubricant, with resultant damage. When the engine is idling, pressure may also be inadequate to reach all bearings. Another disadvantage of low oil pressure is that it may materially reduce the volume and effectiveness of the oil mist.

These weaknesses should be avoided by using an oversize oil pump with a capacity in excess of any probable need. A pressure relief valve that will spill the excess back into the crankcase can then keep the oil pressure at a constant level in spite of thin lubricant, low speed, or loose bearings.

Dirty Oil. It is difficult or impossible to keep foreign materials out of oil. Dirt can enter an engine through outside contamination of oil in cans or funnels, by the oil dip stick that often is so located that it is very difficult to avoid touching it to dirty parts when checking oil level, through an inadequately protected or improperly serviced air intake and then past the piston rings, or through an improperly protected crankcase breather. (When an engine pulls, it tends to build up pressure in the crankcase, when it decelerates or holds back a load by compression a vacuum may develop which will suck air in. These effects are very slight in a new engine and increase with wear of piston rings and cylinders.) Carbon may work down from the combustion chamber and metal particles may appear from anywhere. A machine whose engine pan gets in the dirt, such as a tractor, may take in some of it through holes in the oil pan or past a defective seal on the rear main bearing.

Pump systems may be protected by filters. There is usually a screen at the pump intake, but this is a comparatively coarse mesh which is useless against the finer particles that cause most of the extra wear. Of more importance are the line filters which contain replaceable elements of fiber, cloth, or paper, or permanent ones of closely spaced metal discs or porous stone.

The difficulty with many filter systems is that the filter is in the return line from working parts to reservoir. Also, they are often bypass designs, that filter only part of the oil flow.

As we have seen, all serious sources of dirt in engine oil put it in the crankcase first. A particle may be put through the system several times before it happens to get in the filter. If it escapes through a bearing it may re-circulate dozens of times, each time taking a little metal out of the bearing, the shaft which rides in it, or both. Small particles, fine sand size and smaller, are likely to stay active much longer than coarse ones. If highly abrasive, like sharp silica particles, a fraction of a teaspoonful may cut a big engine to pieces before it is filtered out. More damage may be done in a few hours or even minutes than in years of normal operation.

The answer to this is the full flow filter, in the line from the pump to the working parts, which is large enough to easily filter all the oil going into the engine passages. The pressure gauge, if tapped into the line after the filter, will give warning of any clogging sufficient to reduce oil flow.

Dirt may get in a gear box through defective seals on shafts, from cans or funnels, from dirt dropped in while removing filler plugs, metal grindings, and suction caused by temperature changes. Thick oil and leisurely turning of the parts allow most of this material to settle down into a sump that should be provided just above the drain plug, where it will be largely drained out while changing oil. Some particles, however, will remain in circulation, damaging parts with every passage through them.

Since oil is changed in these units at long intervals, and breather plugs, if any, are small and easily serviced, the most serious contamination comes from metal filings. These are produced very slowly if the unit is in good condition, and more rapidly as bearings wear and shafts and gears get out of line. A large quantity can be permanently taken out of circulation by using magnetic drain and check-level plugs, which will hold them until removed for cleaning.

The only cure for dirt in a dip or splash system is to change the oil. This should be done when the unit has just been operating long and fast enough to warm the oil and pick up the dirt. It is a good plan to follow draining by putting in a thinner oil, running long enough to give it a chance to wash all parts, then draining that.

Diesel Lube Oil. Diesel engines tend to produce sludge and varnish-depositing compounds as by-products of combustion. Special heavy duty oils have been devel-oping that contain detergents tnat keep these substances in suspension, rather than making harmful deposits. It is absolutely necessary that these be used instead of ordinary motor oil.

Such oils can also be used to advantage in gasoline engines. On the first filling they may pick up so much accumulated sludge and varnish as to need to be changed very quickly, but this cleanout is very beneficial to the engine.

Some of the heavy duty motor oils sold at premium prices are suitable for use in diesels, but the engine manufacturer should always be consulted before trying any brand.

Injection Systems. The solidified lubricants are usually pumped into bearings and other parts requiring them at intervals calculated so that there will always be sufficient grease sticking to the rubbing surfaces. The pumping may be done by hand guns, air guns, or mechanical injectors, which will be described in Chapter 21. The bearing, or a tube that opens into it, has a grease fitting that matches the nozzle of the gun and permits grease at pressures as high as 10,000 pounds per square inch to pass from the gun into the bearing.

Some anti-friction bearings are greased and sealed at the factory and require no further lubrication at any time. More often bearings are greased periodically, the intervals depending on the speed, the load, and the danger of dirt getting into them. Bushings may require very frequent lubrication, often every four or five hours of operation.

Dirt. If sand or other abrasive particles are allowed to mix with grease, they form a grinding compound that will cut away the hardest metal rapidly. The rate of wear increases with speed and with load.

Hinges which must work in the dirt, such as unsealed track pins and bushings, last longer if run dry than if lubricated.

Rotating parts require lubrication unless both speed and load are very low, so that seals must be used or sufficient volume of lubricant supplied to keep dirt flushed out.

Seals. Grease or oil seals may serve to keep lubricant in, or dirt out, or both. A simple type is a curved leather ring held in the case by a metal retainer, and pressed against the shaft by a circular spring. Excess grease in the bearing will force its way out between the shaft and the leather, carrying with it any dirt which might be working its way in. If grease pressure were applied at its other side, it would first press the leather tightly against the shaft, increasing its effectiveness, but might eventually burst and destroy it.

Other grease seals of felt, leather, rubber, or metal are made so that pressure from either side will be resisted to the point of destruction. Caution should therefore be used in forcing grease into any fitting where resistant seals might be present.

Asphalt-Base Lubricants. Exposed gears on revolving shovels, various other types of open gearing, and sometimes wire ropes, may be lubricated with an asphalt derivative, known under various trade names, such as Crater Compound (Texaco) and Mobilcote S (Mobil).

In its natural state it is too hard for use in any grease gun or dispenser. It is heated, then poured in a thin stream on revolving gears, or painted on stationary ones. It may be heated in the original container, and poured into a number of small cans, one of which is kept hot on the exhaust pipe or manifold.

It can be obtained mixed with a volatile solvent, so that it can be applied without heating, and hardens on the gears as the solvent evaporates.

The most convenient arrangement, and the most expensive, is to get the thin solution in an aerosol can, so that it can be sprayed on by pushing a button.

Small amounts of any type should be applied often, as most of a heavy application runs or works off in a few minutes. The surplus builds up hard deposits underneath that may interfere with the gear or with other machinery below. Such accumulations are very difficult to remove, particularly when combined with dirt.

Asphalt-base lubricant cannot be removed from skin or clothing by ordinary cleansers. However, it is readily softened by lubricating oil and can then be removed by wiping or washing.

CABS

Machines may be fitted with cabs to protect the machinery, the operator, or both. Protection for the engine only is called a hood.

Revolving shovels, highway-type trucks, and some other machines carry cabs as standard equipment.

Cabs or canopies may be fitted on tractors, rollers, graders, and most bare machines that are large enough to carry them. They are used to protect the operator against rain, sun, cold, or wind.

Rain protection requires a roof, a glass windshield with a wiper, and walls, windows, or curtains at the sides and rear. Means should be provided to roll, swing, or fold all the vertical panels out of the way or to remove them when they are not needed.

A sun canopy consists of a roof on four posts and one or more movable pieces which can be set to block off a low sun. Walls of any type will cut off air circulation, and glass or plastic will admit radiant heat as well.

A large umbrella fastened to the back of the seat by a strong swivel joint provides almost ideal sun protection as it can be tilted toward the sun and does not cut off any breeze.

An increasing number of cabs are entirely enclosed, with glass windshield, and glass windows, some of which can be raised and lowered. This construction makes it possible to provide effective heating in cold

Fig. 12-71M. Protective cab

weather, and air conditioning where heat is excessive.

The entire cab should be readily removable in case it becomes uncomfortable or interferes with operation or repairs.

A cab will increase operator comfort under many conditions, and often permit work in weather too severe for an open machine. It is essential for snow plowing and will pay handsome dividends in work when weather is very hot, cold, or rainy.

A cab will interfere more or less with the operator's view of his surroundings so that it may reduce accuracy of work or cause an accident hazard. It may make it difficult to enter and leave the seat.

ROPS (Rollover Protection Standards) now make it obligatory to equip a wide variety of machines with cabs including members (usually tubing) strong enough to protect the operator if the machine overturns.

POWER PLANTS

The majority of excavating and hauling machines use gasoline or diesel engines. These are called internal combustion engines because fuel is burned inside the same unit that turns the shaft. They were given the name to distinguish them from steam engines, which burn fuel to make steam and then pipe the steam to an engine that converts it to usable power.

The latest entry into this field is the gas turbine, which so far is used only in very large off the road trucks. It has the advantages of simplicity, light weight, and high power. But it is efficient only at full speed, and may be a fuel hog at even 50 per cent throttling.

Gasoline and diesel engines have many things in common. They burn a mixture of air and fuel, turning the heat of their explosive combination into pressure against a piston on a crank that turns a shaft. They must use clean air and clean fuel, keep a film of oil on all moving parts, should be kept at an even temperature by a cooling system, and usually have a throttle to regulate speed. Industrial engines such as are used in excavators have a governor that automatically opens and closes the throttle to maintain proper speed.

The exhaust from both engine types is extremely noisy, and requires muffling to avoid operator ear damage, and to keep job noise down to endurable levels.

Air Filter. Dust must be filtered out of air taken in by an engine to prevent it from wearing moving parts by scratching and grinding them and from building up gummy deposits by combining with the lubricating oil. Excavating machines work on dusty jobs so often that it is particularly important that they have good filters that are properly cared for.

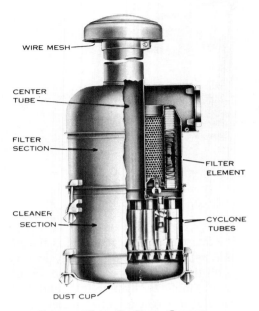

WIRE MESH

CENTER TUBE

FILTER SECTION

FILTER ELEMENT

CLEANER SECTION

CYCLONE TUBES

DUST CUP

Courtesy of Caterpillar Tractor Company

Fig. 12-72. Dry-type air filter

Equipment made before 1960 is usually equipped with both dry and oil bath cleaners. A dry pre-cleaner at the top (inlet) of a vertical intake pipe has vanes that give incoming air a rotary motion that throws heavy particles out of the air stream into a hollow case with a sight window. When dirt is halfway up the glass the operator stops the engine, unscrews a wing nut, removes first a cap and then the case, which he turns upside down to shake the dirt out of it. The case is put back carefully, with care that the gasket between the cap and the case is seated properly.

At the bottom of the intake pipe the air stream turns sharply upward, blowing across the surface of an oil pool and through a maze of oil-covered wire, which remove most of the remaining dirt. The reservoir should be cleaned and refilled and the filter mesh cleaned and re-oiled frequently.

A low oil level reduces the velocity of air, so that more dust particles are left for the filter. If the oil is high, too much of it may be picked up by the air stream, and

some of it might reach the engine. This oil will burn in diesel cylinders, and may cause runaway speed not controlled by the governor, that may damage or destroy the engine. It can be stopped by shutting off the air supply, or by putting the machine in high gear and stalling it with a load.

New equipment now is likely to have a Donaldson dry cleaner, shown in Figure 12-72. The pre-cleaner is replaced by simple wire mesh, that serves to prevent entrance of coarse particles. A straight intake pipe brings air to a chamber in the top of the cleaner section.

The air enters the top of the outer chambers of the 32 to 42 nylon cyclone tubes. Vanes cause it to spin. At the bottom the air makes a sharp U-turn to go up through the center tube into the filter chamber. Dirt, thrown out of the air by centrifugal force and the sharp turn, drops out of the bottom of the tubes into the dust cap. This section takes out about 95 per cent of the dirt.

The air moves from the outside of the filter chamber inward through a resin impregnated pleated cellulose element that removes the remainder of foreign particles from the air. Filter efficiency ranges from 99.8 to 100 per cent.

The dust cap is removed, emptied, and replaced daily, or as necessary. It is very important that no oil be put in it. The cyclone tubes need no routine attention, as they are self-cleaning.

The filter may be serviced by using a jet of clean dry air at a pressure of 40 pounds or less. The air is directed first against the inside of the element, then the outside, alternating until it is clean. Clean water may be used instead of air, at the same pressure. The air or water is directed along the complete length of each pleat.

Oil or sooty deposits may be removed by washing with warm water and a non-sudsing household detergent.

Excessive engine exhaust smoke and/or

loss of power may indicate that the filter needs servicing. If the trouble continues after cleaning, and is not due to other causes, the filter element must be replaced. Its life may be as short as 125 hours or as long as 3,000 hours. Life is affected by the quantity and nature of the dust and other pollutants, and by care or lack of it in emptying and cleaning.

Engines with more than one intake pipe will have a complete air cleaner for each.

Altitude. High altitudes reduce the power of all internal combustion engines.

This is because the air becomes less dense as height above sea level increases. The engine therefore draws in fewer molecules with each stroke, develops lower compression, and has less oxygen to combine with fuel. Two cycle engines suffer a smaller decline up to about 10,000 feet, because their blower feed packs more air in.

Installation of a supercharger on an engine to be used at high altitudes may restore or increase its sea level power. However, the same engine with the same supercharger would show increased power if taken down to sea level.

Many diesels now have turbochargers (turbo-superchargers) as standard equipment. The units are expensive, but they increase engine horsepower and altitude adaptability substantially, without increase in engine size and weight.

Engines to be used entirely at high altitudes should be specially constructed and adjusted.

Block and head. An internal combustion engine has one or more cylinders in which a piston moves up and down or, in the rare horizontal engine, back and forth. The small space between the piston and cylinder wall is sealed by rings that are set in grooves in the piston and push out against the wall with gentle pressure. The upper rings are shaped to prevent pressure in the combustion chamber from blowing

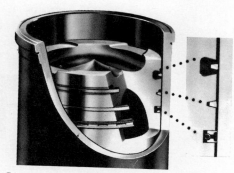

Courtesy of International Harvester Company

Fig. 12-73. Piston rings

by; the lower one distributes oil on the wall, and keeps it from working up into the combustion chamber. See Figure 12-73.

The cylinder is kept cool by circulation of water through spaces (jackets) surrounding it. This water is pumped through a cooling radiator and returned to the jackets. Air cooled engines have fins on the outside of the cylinders that radiate heat into a stream of air.

Industrial engines usually have replaceable cylinder liners, such as those in Figure 12-74. These can be replaced when worn,

Courtesy of International Harvester Company

Fig. 12-74. Cylinder liners

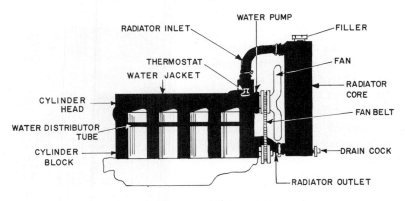

Fig. 12-75A. Water cooling system

so that it is not necessary to re-bore the cylinder and fit it with oversize pistons.

The cylinder casting is called the block. Another casting, the head, is bolted to the top of it. The two pieces have flat machined surfaces facing, with a gasket or seal rings to insure an air and water tight fit.

Lubrication is usually by pressure from a pump to all bearings, and by splash and mist to cylinder walls.

Each piston is attached by a wrist pin (hinge pin) to a connecting rod. This is attached to the crankshaft by a lubricated bearing (a babbit bushing, called a rod bearing). Up and down motion of the piston, powered by fuel-air explosions above it, rotates the crankshaft.

This shaft is usually connected to the working parts of the machine by a friction clutch or torque converter, and a transmission.

If the stroke—the distance the piston moves up and down—is short, the piston speed at any given engine speed will be less than if the stroke is long. The compression ratio need not be affected, as a smaller space above the piston will compensate for its smaller displacement.

Four cycle engines have intake and exhaust valves opened and closed by the camshaft, which is driven by the crankshaft through timing gears or chain.

Water Cooling. Water is forced by a centrifugal pump throughout the block and head, where it takes heat from walls of cylinders and combustion chambers. It then flows through small tubes in a radiator, where it delivers the heat to air that is pulled or pushed past the outside of the tubes by a fan.

The cooled water flows into the pump, and returns to the engine.

A thermostat, usually located in the upper radiator connection, closes to prevent or reduce circulation whenever the engine is too cold for efficient operation.

The fan is mounted on and driven by the engine. In vehicles and haulers, it pulls air from front to rear. In tractors, it may pull, push, or be reversible.

The air stream of a pull fan is aided by the wind created by vehicle motion forward. The pusher is helpful in keeping dust away from the operator, and is very suitable for loaders. Reversing may help to free the radiator from accumulations of dust and fine trash.

Cooling water should contain additives to reduce corrosion of metal parts. It requires addition of ethylene glycol anti-freeze wherever there is a possibility of freezing. Alcohol is not suitable because of its low boiling point.

Air Cooling. Engines may also be cooled directly by flow of air around the outside of the cylinders, which are formed with fins to

Cooling System Capacity in Quarts	QUARTS OF ANTI-FREEZE REQUIRED										
	3	4	5	6	7	8	9	10	11	12	13
5	−62°										
6	−34°										
7	−17°	−54°									
8	−7°	−34°	−69°								
9	0°	−21°	−50°								
10	4°	−12°	−34°	−62°							
11	8°	−6°	−23°	−47°							
12	10°	0°	−15°	−34°	−57°						
13		3°	−9°	−25°	−45°	−66°					
14		6°	−5°	−17°	−34°	−54°					
15		8°	0°	−12°	−26°	−43°	−62°				
16		10°	2°	−7°	−19°	−34°	−52°				
17			5°	−4°	−14°	−27°	−42°	−58°			
18			7°	0°	−10°	−21°	−34°	−50°	−65°		
19			9°	2°	−7°	−16°	−28°	−42°	−56°		
20			10°	4°	−3°	−12°	−22°	−34°	−48°	−62°	
21				6°	0°	−9°	−17°	−28°	−41°	−54°	−68°
22		33% MINIMUM FOR CORROSION PROTECTION		8°	2°	−6°	−14°	−23°	−34°	−47°	−59°
23				9°	4°	−3°	−10°	−19°	−29°	−40°	−52°
24				10°	5°	0°	−7°	−15°	−24°	−34°	−46°

Per Cent Anti-Freeze	Protects to	Boiling point increase over water
60	−62°F	19°F
50	−34°F	14°F
40	−12°F	10°F
33	0°F	8°F

Fig. 12-75B. Protection chart for ethylene glycol antifreeze

increase their heat-radiating surfaces.

The air is forced through passages by a blower-type fan, with multiple blades turning in a case. The passages are designed to distribute the air stream equably, so that all parts of each cylinder are kept at a similar temperature.

In air cooling, there is a much greater temperature difference between the metal parts and the cooling medium, than in water cooling. The engine may be run much hotter, with temperature limited chiefly by the necessity of maintaining an oil film on the cylinder walls.

Air cooling uses a smaller volume of air than liquid cooling, eliminates the need of filling and maintaining a cooling system, has fewer parts and connections, and eliminates danger of (and precautions against) freezing and boiling.

However, there are problems of even heat distribution, regulation of engine temperature, and cylinder distortion, which are harder to solve than with conventional water cooling.

GOVERNOR

Most heavy equipment is required to operate at a fairly steady speed regardless of load changes that would tend to slow the engine or speed it up. Engine speed is kept at the proper level by a governor. There are three kinds in common use.

Courtesy of Deutz Diesel Corporation

Fig. 12-75C. Air cooled diesel

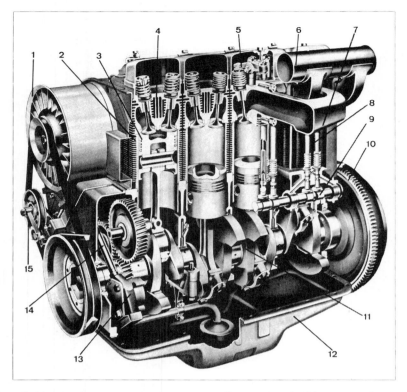

1) cooling air blower
2) cylinder liner with cooling fins
3) piston with combustion chamber (direct injection)
4) light-metal cylinder head with inlet and exhaust valves
5) rocker arm
6) air intake manifold
7) exhaust manifold
8) pushrod with cover tube
9) camshaft
10) starter ring gear on flywheel
11) crankshaft with counterweights
12) oil sump
13) lube oil pump
14) timing gears
15) V-belt idler

Courtesy of Deutz Diesel Corporation

Fig. 12-75D. Cutaway view of air cooled diesel

Mechanical. In a mechanical governor there are weights that revolve with an engine driven shaft. They are held in toward it by a spring or springs, and moved away from it by centrifugal force which increases with speed. When the shaft turns rapidly the weights move out, when it slows they move inward. An arm controlled by the position of the weights regulates engine speed by controlling fuel injection in diesels or the air valve in gasoline engines.

For simplicity, we will discuss only the application to diesels.

When the engine is required to move a heavy load, it slows down, causing the weights to drop inward and increase the fuel supply, as an automobile driver pushes the accelerator to keep speed up a hill.

If the load is reduced and the engine starts to race, the higher speed moves the weights outward, reducing the fuel supply and keeping engine slowed to proper speed

The operator can usually change the speed setting of the governor anywhere in its range from idle to full speed by moving a hand throttle, which changes the tension of the governor spring. Some machines also have a foot accelerator or a foot decelerator which can over ride the hand lever setting.

A mechanical governor cannot keep the engine running at exactly the same speed with changes in load. An increase in load causes the steady speed of the engine to reduce slightly; and loss of load will increase the steady speed slightly. This is because a return to the original steady speed while carrying a load would reduce the fuel so that the load could not be carried, and speed would fall off again.

The change in steady speed with load is called speed droop. It may amount to 5 to 10 per cent of engine speed. Adjustments can often be made to regulate the amount of droop. If it is reduced too far the governor will "hunt" or cause the engine to speed up and slow down alternately, an effect which is greatly increased by delay in response of the engine to changes in fuel supply.

Hydraulic. Hydraulic governors are operated by revolving weights and springs in the same manner, but the linkage controls a hydraulic valve in a pressure system, which in turn controls the fuel supply. This makes it possible to obtain very prompt response, and to reduce or eliminate speed droop.

Velocity. Velocity-type governors for gasoline engines may be built into the carburetor or may be a self-contained unit installed between the carburetor and intake manifold, Figure 12-72D. Operation depends on the velocity of air-fuel mixture impinging on a throttle plate normally held open by a spring. When the accelerating engine reaches the desired maximum speed, the air velocity will overcome the spring, causing the throttle plate to close, thus

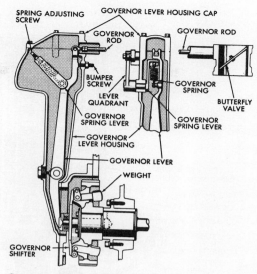

Courtesy of Waukesha Motor Co.

Fig. 12-75E. Governor details

maintaining the engine at that speed. When the engine speed decreases, the spring will pull the throttle plate open.

This type of governor is not as quick to respond as a mechanical type, allowing the speed to drop more under full load. Velocity governors are less expensive to purchase and service, and operate very satisfactorily where immediate speed response is not absolutely essential or where

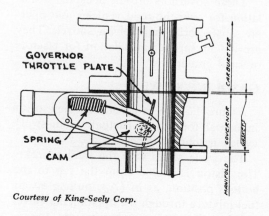

Courtesy of King-Seely Corp.

Fig. 12-75F. Velocity-type governor

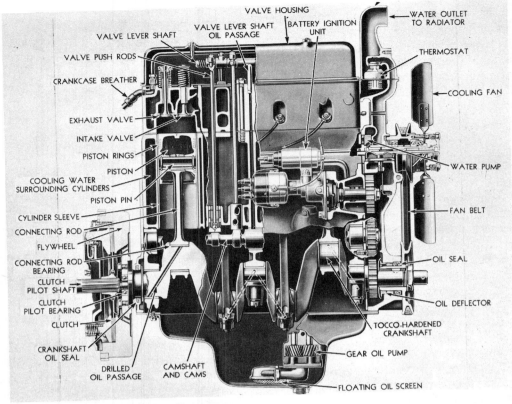

VALVE HOUSING
VALVE LEVER SHAFT
VALVE LEVER SHAFT OIL PASSAGE
BATTERY IGNITION UNIT
WATER OUTLET TO RADIATOR
VALVE LEVER SHAFT
VALVE PUSH RODS
THERMOSTAT
CRANKCASE BREATHER
COOLING FAN
EXHAUST VALVE
INTAKE VALVE
PISTON RINGS
PISTON
COOLING WATER SURROUNDING CYLINDERS
PISTON PIN
WATER PUMP
CYLINDER SLEEVE
CONNECTING ROD
FLYWHEEL
CONNECTING ROD BEARING
CLUTCH PILOT SHAFT
CLUTCH PILOT BEARING
CLUTCH
FAN BELT
OIL SEAL
OIL DEFLECTOR
TOCCO-HARDENED CRANKSHAFT
CRANKSHAFT OIL SEAL
DRILLED OIL PASSAGE
CAMSHAFT AND CAMS
GEAR OIL PUMP
FLOATING OIL SCREEN

Courtesy of International Harvester Company

Fig. 12-76A. Cutaway view of industrial gasoline engine

the engine is geared down so far that this characteristic does not show in performance.

These governors are the accepted installation for over-the-road vehicles and operate satisfactorily on revolving shovels. They are not the proper installation for tractors or road graders.

GASOLINE POWER

Figure 12-76A is a cutaway view of an industrial four cycle (actually four-stroke cycle) gasoline engine.

The cycle starts with the intake stroke. The piston is pulled from the top to the bottom position, as in (A), pulling an air fuel mixture through the open intake valve.

When the piston reaches bottom, the intake valve closes, and the piston moves up,

squeezing the air mixture into the combustion chamber. This movement is called the compression stroke. The amount that the air is compressed is called the compression ratio. If this ratio is 7 to 1 it means that the charge in the cylinder is compressed to $\frac{1}{7}$ of its former volume.

As the piston nears the top of the compression stroke, a spark jumps across the electrodes of a spark plug and sets fire to the air-gasoline mixture, so that it explodes and drives the piston down on the power stroke. The timing of the spark is regulated by a rotating contact in the distributor, that is also driven by the timing chain or gears.

As the piston reaches the bottom and starts up the exhaust valve opens, so that the gases resulting from the explosion can be pushed out into the exhaust passages.

This is the exhaust stroke, and it finishes the cycle. The next move of the piston pulls in a fresh charge of air and gasoline, which is where we started.

Carburetor. Air and gasoline are mixed in the carburetor, shown schematically in Figure 12-75. This contains a pool of gasoline supplied by a low pressure fuel pump, and kept at a constant level by a float that shuts it off when too high, and lets it in when low. As air is pulled through the carburetor by the intake strokes of the piston, it picks up a spray of gasoline through metering jets. If the jets pass a lot of gasoline, the mixture is called rich; if only a little, it is lean.

There are two valves to control the air stream through the carburetor. One, the throttle, is between it and the engine. When it is closed, it allows just enough air-fuel mixture through to allow the engine to idle. When it is open, the engine pulls in all it can hold. In any position except wide open, the throttle causes a partial vacuum in the manifold, as the piston tries to pull more air out of it than the throttle lets in.

The other valve, called the choke, is between the carburetor and the open air. When that is closed, the vacuum pulls on the gasoline supply so that raw gasoline is pulled into the engine, and such a high vacuum is caused that much of it changes to vapor, even if it is cold. This rich mixture of gasoline helps to start a cold engine and keep it running, but is too much in a warm engine.

Whenever the throttle is open to allow more air-fuel mixture to get into the cylinder, the explosion becomes more powerful and tends to make the piston move faster and turn the crankshaft with more power.

The gasoline engine requires a fuel that turns easily into an explosive vapor, and an electric spark to set off each power explosion.

In some engines, the gasoline is injected

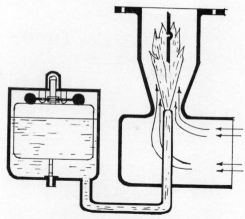

Fig. 12-76B. Simplified carburetor jet

into the cylinder in about the same manner as in a diesel, so that the carburetor is eliminated.

Compression. The compression ratio is important to the efficiency and performance of the engine. When the molecules of air and gasoline are squeezed tightly together, they will deliver a higher proportion of their explosive energy to the piston, and waste less in heating the chamber, than when compression is low. However, many practical problems are created by high compression. Industrial engines usually have a ratio between 7.0 and 8.0 to 1. Some automobile engines have still higher compression ratios.

One limit to compression is that air is heated when it is compressed, and a very high ratio will make the mixture hot enough to explode without waiting for a spark. This heat is what makes a diesel go, as even its relatively inert fuel burns automatically when compression reaches 15.0 to 1.

Of more immediate importance in gasoline combustion is uneven burning. As shown in Figure 12-77, if a highly compressed gas-air mixture is ignited, the burning of the part near the spark plug will cause expansion and so compress the part of the charge not yet reached by the flame front that it may detonate. A detonation is

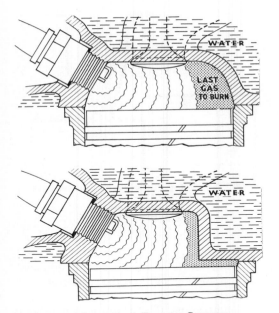

Courtesy of International Harvester Company

Fig. 12-77. Combustion chambers

a nearly instantaneous explosion. It produces a sharp blow on the top of the piston, causing the "ping" so familiar to the motorist.

This difficulty is met by changing the shape of the combustion chamber, to produce more even burning, to cool the section where detonation would occur, or to cause the detonation to occur in a side chamber so that the edge is taken off it by the time it reaches the piston; by locating the spark in the center of the combustion chamber, by putting lead or other anti-detonation substances in the gasoline, by reducing the compression ratio, or adjusting the distributor so that the spark fires later. This last may result in still-burning gas going out the exhaust, wasting fuel and power and creating excessive heat.

As engine speed increases, it is able to take in less of a charge on each stroke, as the incoming air is slowed by friction in the cleaner, carburetor, and manifold passages; and the interval during which the intake valve is open becomes shorter.

Up to a rather high limit, the faster an engine goes the more horsepower it produces, but after a certain speed, usually somewhere near half to two-thirds of top speed, its torque or actual twisting power decreases, as shown in Figure 12-78, because of the smaller charges in the cylinders, and the briefer time during which the explosion can affect the piston.

A "souped-up" engine usually has bigger air passages, and valves that open further, so that it can fill its cylinders at higher speed.

An industrial gasoline engine is more heavily built than a corresponding automotive model, as its operating and load conditions are more severe.

Fuel. The standard fuel for industrial gasoline engines is of course gasoline.

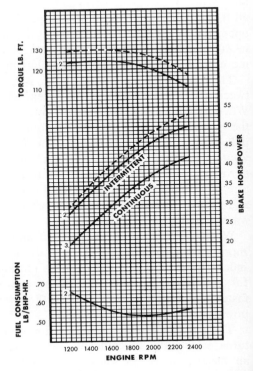

Fig. 12-78A. Performance curves, gasoline engine

This is usually the "regular" gasoline supplied for automobiles. It includes a small quantity of tetraethyl of lead, a poisonous compound that reduces tendency to knock, and/or other chemicals to improve performance; together with a dye to give warning of the poison, or to identify make or quality.

Gasoline is classified by octane rating. Octane is an excellent fuel in anti-knock characteristics, and is given a 100% rating. Heptane is a poor anti-knock fuel and is given a 0% rating. The octane number of a gasoline is the octane percentage in a mixture of octane and heptane which it matches in antiknock value.

Commercial range now is approximately 85 to 90 for regular gasoline with additives, 91 to 95 for premium quality, and 100 to

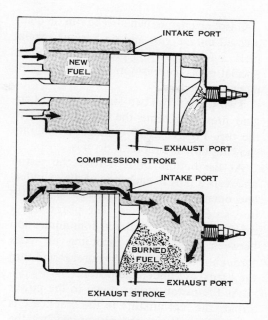

Fig. 12-79. Two-stroke cycle, gasoline

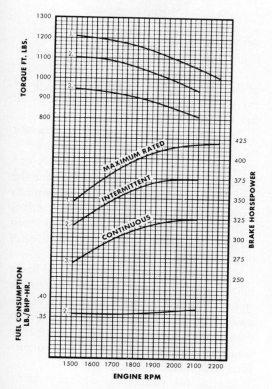

Fig. 12-78B. Performance curves, diesel engine

115 is expected for aviation gasoline.

The treated and dyed gasolines are perfectly satisfactory in industrial engines that are used regularly. However, in machines that are seldom used, such as a compressor owned by a contractor who seldom blasts, or a pump with an owner who is lucky enough to do most of his work dry, gasoline evaporates in the carburetor and lines, leaving a gummy deposit that may interfere with starting and operation, and is a nuisance to remove. However, white gasoline may be very difficult to obtain.

The chief objection to "doped" gasolines on the job is that they are poisonous, and may be very irritating to the skin, so they should not be used to clean parts, and require special care in handling. The lead may enter the body through the skin or lungs and cause cumulative poisoning, similar to painter's colic.

Leaded or doped gasoline cannot be used in engines equipped with the catalytic type of combination muffler and exhaust scrubber.

Gasoline engines can also be run on compressed butane gas, and on various special fuels available in oil production centers. Special tanks, carburetors and other equipment are needed.

Two Cycle. The two-stroke-cycle gas engine fires each time the piston comes to the top. Figure 12-79 contains a diagram showing how it is able to do this.

At the bottom of its stroke the piston uncovers a port in each side of the cylinder, one opening into the crankcase, the other into an exhaust passage. The piston has a ridge that prevents direct communication between these ports. Incoming air goes through a conventional carburetor and into the crankcase under the piston.

As the piston moves down in the power stroke, the fuel mixture in the crankcase is compressed. When the piston reaches the bottom of the stroke, it opens both ports. The compressed charge enters through one opening while the burned gases go out the other. The piston goes up on its compression stroke, a spark plug fires the charge near the top, and the cycle is repeated.

The intake of air through the crankcase prevents keeping a reservoir of oil such as provides lubrication for four cycle engines. Oil is therefore mixed with gasoline, usually in the proportion of one pint of lubricating oil to four gallons of gasoline. The oil mist that results from vaporizing the gasoline lubricates all parts.

At the same crankshaft speed, a two cycle will fire twice as often as a four cycle. It does not put out double the power, size for size, however, as there is a little lost motion covering and uncovering the ports.

The two cycle gasoline engine is used chiefly in small, high speed units, and is of little importance in excavating.

DIESEL ENGINES

In a diesel engine plain air is drawn in on the intake stroke. It is compressed so tightly that it becomes very hot, and a spray

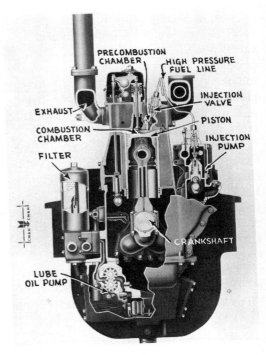

Courtesy of Caterpillar Tractor Co.

Fig. 12-80. Cutaway view of diesel engine

of fuel oil injected in it near the top of the stroke is ignited just by contact with the heated air.

The diesel therefore needs no carburetor or ignition system. But it does need a method of metering the fuel correctly and injecting it at just the right time. Since the fuel is a non-volatile liquid, precautions must be taken to make sure that it mixes with the air thoroughly enough to burn cleanly and completely in the very brief time available.

A diesel must be more strongly and finely built than a gasoline engine, because of the strains imposed by the higher compression and temperatures. For this reason diesels cost about twice as much as gas engines in small models, but this difference diminishes with increase in size.

Special oil containing detergent compounds must be used to prevent deposits of sludge and varnish from forming. This diesel lube oil can also be used to advantage in gasoline engine crankcases, as both a lubricant and a cleaner.

Diesel is more economical than gasoline for several reasons. One, possibly a temporary one, is that the fuel is cheaper. Another is that the higher compression makes it much more efficient. The fuel has a higher heat value. A greater proportion of the available power in the fuel is set to work turning the crankshaft. This means less fuel to buy and to handle. Fire danger is greatly reduced.

Diesel engines used in excavators have a size range from approximately 60 to 1,600 horsepower. They are available in four cycle and two cycle construction.

Four Cycle. In the four cycle (four stroke cycle) diesel, Figure 12-81, a downward stroke of the piston with the intake valve open fills the cylinder with air. The upward stroke against the closed valve compresses it, injection and burning of the fuel occurs at the top, the burning drives the piston down on the power stroke, and the next upward stroke pushes the products of combustion out past the open exhaust valve. See Figure 12-81.

While the piston and valve action are the same as in the four stroke gas engine, the power principle is quite different. The air drawn into the cylinder is just air, and its flow is not regulated by a butterfly valve as in a gas engine.

Compression is very high, the minimum ratio being about 16 to 1, and the highest at present is over 20 to 1. Diesel (non-volatile) fuel is injected into the combustion chamber when the piston is at or near the top of the compression stroke, and is ignited by the heat of compression. Engine speed is regulated by controlling the amount of fuel injected on each stroke. Pressures and temperatures in the cylinder are much higher than in gas engines.

Two Cycle. The two cycle (two-stroke cycle) diesel, Figure 12-82, has one power

stroke for each two strokes of the piston, that is, every down stroke produces power. A blower in the intake passage, Figure 12-83, pushes air at low pressure into a chamber in the block that opens into the cylinder through a ring of holes or ports that are covered by the piston except at the bottom of its stroke.

The power stroke starts by injection of fuel into hot compressed air when the piston is at the top of its stroke. This burns and expands, forcing the piston down. As the piston nears the bottom the exhaust

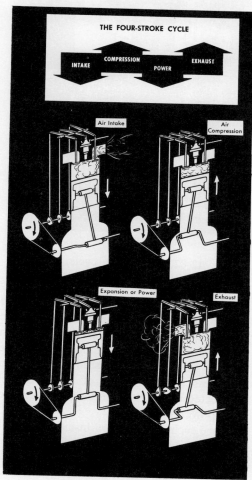

Courtesy of Detroit Diesel Allison

Fig. 12-81. Four-stroke cycle, diesel

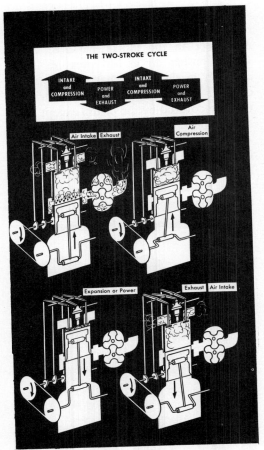

Courtesy of Detroit Diesel Allison

Fig. 12-82. Two-stroke cycle, diesel

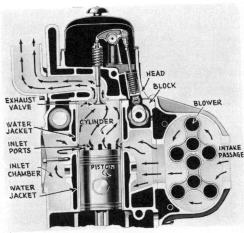

Courtesy of Detroit Diesel Allison

Fig. 12-83. Cutaway view, two-stroke cycle diesel

Fuel. Ignition quality of diesel fuels is rated by cetane numbers, which may be between 35 and 60. Higher numbers indicate easier starting.

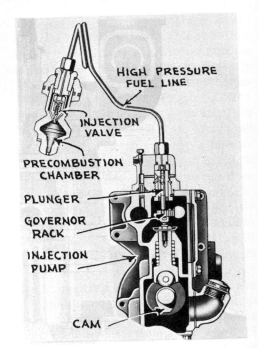

Courtesy of Caterpillar Tractor Co.

Fig. 12-84. Diesel injection pump and line

valve opens and the burned gas starts to escape. Further movement of the piston uncovers the intake ports, so that an inrush of clean air forces the remainder of the burned gases out. The piston then rises to compress its new filling of air, and fuel is injected at the right moment to start the next power stroke.

Since each down stroke, instead of every other down stroke, is a power stroke, as few as two cylinders can be used without excessive roughness.

Most two-cycles use a poppet intake valve, but some are "loop scavenged" and have no moving valves.

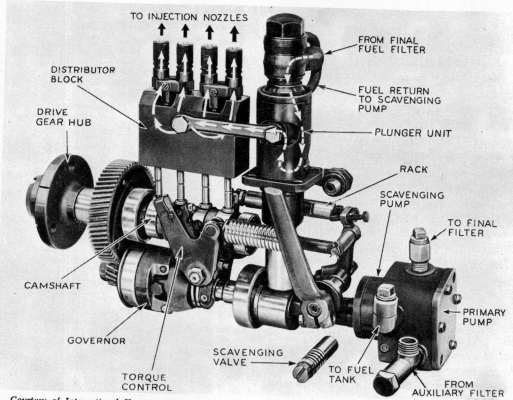

TO INJECTION NOZZLES

DISTRIBUTOR BLOCK

DRIVE GEAR HUB

CAMSHAFT

GOVERNOR

TORQUE CONTROL

SCAVENGING VALVE

TO FUEL TANK

FROM FINAL FUEL FILTER

FUEL RETURN TO SCAVENGING PUMP

PLUNGER UNIT

RACK

SCAVENGING PUMP

TO FINAL FILTER

PRIMARY PUMP

FROM AUXILIARY FILTER

Courtesy of International Harvester Company

Fig. 12-85. Injection pump and distributor

Most diesels in construction equipment use #2 diesel oil. This is a little heavier in body and less flammable than kerosene, evaporates very slowly, and has a slight lubricating value.

Some engines may permit substitution of #2 furnace oil, used widely for domestic heating. Diesel oil cetane number is between 45 and 47, as against 37 to 40 for heating oil. Costs of these fuels were discussed in Chapter Eleven.

Some machines have heated fuel systems so that they can burn much heavier grades, up to bunker C residues. Others need a #1 diesel oil that is similar to kerosene.

Many diesels cannot use kerosene or gasoline because they depend on fuel for lubrication of moving parts in the fuel system. Any of them may have trouble with gasoline due to pre-ignition, or vapor lock in the fuel lines.

In any grade of fuel the most important requirement is cleanliness. Many fuels contain sulphur and other corrosive chemicals in sufficient quantity to damage pumps and injectors. Any of them will have more or less foreign matter that absolutely must be strained out, as the close fits in a diesel fuel system will not tolerate any solids. It is customary to have both primary and secondary filters, and often a final one at each injector.

Second grade fuel usually increases down time and maintenance expense.

Fuel Supply. In order to get fuel into each cylinder in the right quantity and the right time, four separate functions must be performed. The fuel must be measured, directed to the proper cylinder, timed to reach it at the right time, and put under sufficient pressure to enter it in a vaporizing spray. As might be expected, there is a considerable variety of methods used in different makes.

Distributor Pump. A number of engines combine all four functions in a high pressure pump.

The Caterpillar, Figure 12-84, uses a low pressure (15 p.s.i.) gear type transfer pump to move fuel from the tank lines through a pair of filters to the high pressure injection pumps. There is an individual pump for each cylinder. These are set in line· in a single case. The pump plungers are raised by cams on a shaft, and lowered by coil springs. The length of stroke is the same at all times.

As the plunger rises it first cuts off an inlet port, then drives the fuel above it through a check valve, a section of high pressure tubing, an injection valve, and into the precombustion chamber. There is a spiral groove in the plunger, that is open to the high pressure fuel above it, and that at some point in the stroke will open into the low pressure inlet port, so that pressure will be relieved and injection will cease.

The point in the upward travel of the plunger at which the groove will contact the port is regulated by turning the plunger by means of a rack (straight gear) bar controlled by the governor and throttle. Horizontal movement of this bar turns all plungers equally, so that the same amount of fuel is injected into each cylinder.

The tubing connecting the pump and the

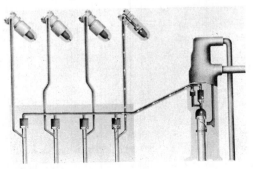

Courtesy of International Harvester Company

Fig. 12-86. Fuel distribution

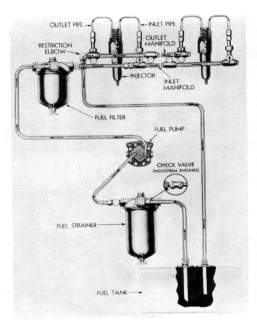

Courtesy of Detroit Diesel Allison

Fig. 12-87. Unit injector system

injection valve is strongly built and carefully designed to withstand high pressure and shock. Each one of a set is the same length, any differences in space to be crossed being compensated by special bending.

The injection valve is held against its seat by a spring. The high pressure impulse from the pump lifts its plunger and allows the fuel to pass it and spray into the precombustion chamber through a single opening.

Pump and Distributor. The pump and distributor mechanism may be separate, as in the four cylinder International shown in Figures 12-85, 12-86. The single plunger pump puts the low pressure fuel from the transfer pump under high pressure, and also meters the amount by a spiral relief groove and a governor-controlled mechanism for turning the plunger.

The high pressure jets of fuel from this pump enter the distributor passage. Four poppet valves, lifted by cams, direct the fuel through lines to the injectors in accordance

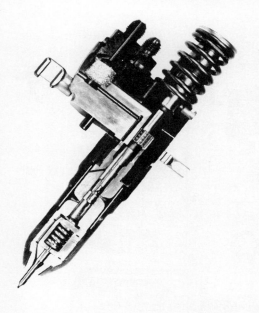

Courtesy of Detroit Diesel Allison

Fig. 12-88. Unit injector

with engine firing order. One of these will be open for each push of the high pressure pump, and the measured fuel will pass it into high pressure tubing that will conduct it to an injector. There it will unseat a valve and spray into the pre-combustion chamber.

Pre-Combustion Chamber. A pre-combustion chamber is a small chamber connected by an open passage to the main combustion chamber. Fuel is injected into it and is ignited by compression heat. There is not enough air to burn it fully. The burning mixture expands and rushes into the main chamber, where it mixes with more air to burn completely.

This arrangement makes possible use of relatively low pressures in the fuel injection system, promotes efficient performance on standard fuels under a wide range of load and speed conditions, permits low idling speed, and reduces exhaust pollutants.

The prompt and thorough mixing of fuel and air are very important, and there

are a number of designs using different shapes of chambers and piston tops.

Unit Injector. In the unit injector system, the whole job—pressure, timing, distribution, and measurement, is handled by the injectors.

The General Motors diesel, Figures 12-87 and 12-88 has a low pressure (about 45 p.s.i.) fuel pump, located between the primary and secondary filters, that delivers fuel directly to the injectors in the head. These are actuated by the main camshaft acting through rocker arms. The throttle and control rack regulates the amount of fuel dispensed by each by turning the injector plunger, the camshaft determines the distribution and the timing, and the pressure is produced by a plunger in the injector. This pressure can be much higher than can be carried by outside tubing, and makes possible injecting the oil as a very fine spray directly into the main combustion chamber.

Pressure-Time. The Cummins PT injection system regulates amount of fuel supplied to the cylinder by varying the pressure in the supply to the injectors.

The fuel flow is from the tank through a medium pressure transfer pump, a pressure

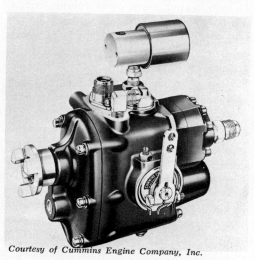

Courtesy of Cummins Engine Company, Inc.

Fig. 12-89. Pressure-Time pump

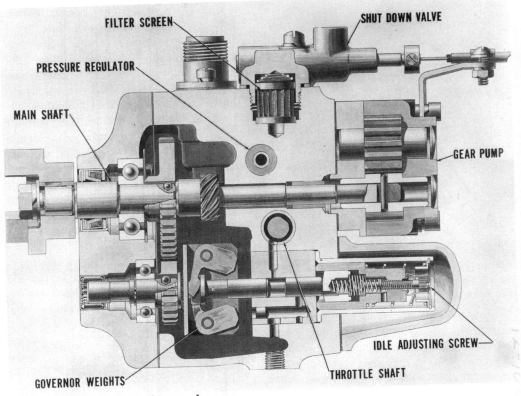

FILTER SCREEN

SHUT DOWN VALVE

PRESSURE REGULATOR

MAIN SHAFT

GEAR PUMP

IDLE ADJUSTING SCREW

GOVERNOR WEIGHTS

THROTTLE SHAFT

Courtesy of Cummins Engine Company, Inc.

Fig. 12-90. Pressure-Time pump, cutaway view

regulating valve, a throttle shaft, a governor plunger, a shut down valve, tubing, injector passages, and drain tubing back to the tank. Fuel enters the injector cups on the up stroke of the plunger, in an amount determined by its pressure.

Figures 12-89 and 12-90 show the pump and regulating mechanism in assembled and cutaway views. The throttle shaft regulates the flow of fuel between the pressure regulator and the governor. When idling it cuts off the fuel flow through the main passage and forces it to pass through an idling hole to the governor idling port. The governor controls the fuel pressure at idling speed and when the engine reaches maximum speed, the governor shuts off the operating fuel supply. It is interesting to note that it does this directly by a sliding plunger in a fuel passage, and that it is not connected to the throttle by any linkage.

The shut-down valve is used for stopping the engine, and should remain closed until the engine is to be re-started. This shut down may be either a manual or electric solenoid type.

The injectors, Figures 12-91 and 12-92 have no throttle connection, and meter the fuel according to its pressure, which is responsive to the position of the throttle.

Starting. Diesel engines are started by cranking in much the manner of gasoline engines. Procedures are described later in this chapter.

There is usually an electric starting motor of the type used for automobile engines, but of much heavier construction. The battery or battery set may be 12 to 48 volts.

A heavy diesel may be started by a compressed air motor, supplied from the machine's air brake system. This does not have to rest until it runs out of air. It is important that there be no leaks that might empty the system overnight.

An air starter needs a large reservoir (air receiver), and a connection by which air can be supplied in emergencies by another machine or system.

Smoke. The diesel is normally a clean burning engine, as the cylinder is charged with more than enough air to burn the maximum amount of fuel injected. It has a big advantage over gas engines in that the exhaust is almost free from carbon monoxide. However, it does exhaust some bad smelling irritating and moderately toxic gases, that prohibit its use in poorly ventilated places unless an efficient exhaust scrubber is attached.

In view of its generally clean-burning characteristics, it is unfortunate that so many diesel trucks trail clouds of black smoke behind them, to the annoyance of everyone on or near the highway. This nuisance is caused by injecting more fuel than the engine is designed to use, so that the mixture is too rich.

Excavating machines and off-the-road haulers almost never show black exhaust smoke, while all too many highway trucks put on a good imitation of a coal burning locomotive firing up. Smoking is an indica-

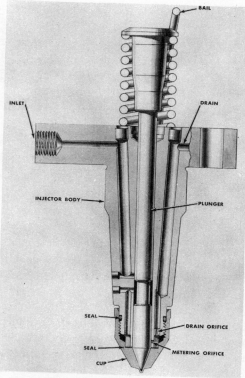

Courtesy of Cummins Engine Company, Inc.

Fig. 12-92. Injector, cutaway view

tion of fuel being wasted, oil being contaminated with sludge, and exhaust valves and mufflers being damaged by contact with still-burning gases. For this reason, any alert foreman or operator will send for a service man if he sees a dirty exhaust.

Some of the smoky exhausts seen on highways indicate defective parts or adjustments which will be fixed at the next service stop. Most of them, however, are a result of a driver or mechanic "souping up" or "hot-rodding" the engine by tampering with injectors or pumps, or replacing them with oversize ones designed for different service. Some increase in acceleration or power may be obtained, as the hydrogen in the fuel gets more BTUs out of the available oxygen than the carbon can, but the wasted fuel, damage to the engine, and nuisance to the public far outweigh this advantage.

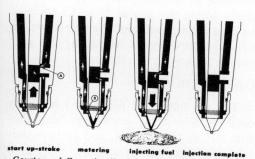

start up-stroke   metering   injecting fuel   injection complete
Courtesy of Cummins Engine Company, Inc.

Fig. 12-91. Fuel injection cycle

Fig. 12-93. "Souped-up" diesel on the highway

If a diesel has any tendency to smoke, it will do so at wide open throttle, and particularly when lugged down below its normal operating speed. Since the same quantity of air is drawn into the cylinder regardless of throttle setting, it follows that the mixture becomes richer as the throttle is opened and more fuel is injected. At below-normal speeds, the slower piston stroke allows a considerable part of the heat of compression to be absorbed by the cylinder walls, and the resulting lower temperature flame does not use as much of the available oxygen as at higher speeds.

Superchargers. A supercharger is a blower or pump that forces air into the intake of an engine at higher pressure than atmosphere. As a result, more oxygen is packed into a cylinder, so that more fuel can be burned and more power is produced by the same size engine, or equal power by a smaller engine.

A supercharger will overcome friction of air passages so that cylinders can fill completely at higher speeds than when atmospheric pressure alone pushes the air. It can also compensate for the reduced density of air at high altitudes.

The turbosupercharger, Figure 12-102, includes a set of turbine vanes driven by the high velocity exhaust gases, that spin compressor or blower vanes that force air into the engine.

The exhaust gases are channeled through nozzles in which their velocity is built up to exert maximum force on the turbine vanes, and the impulses from individual cylinders are blended together into a smooth stream. This process creates some backpressure, but less than buildup of intake pressure.

The turbosupercharger adds to engine efficiency by converting waste energy in the exhaust gas to intake pressure that increases compression, cleans out burned gas completely and may even convert the intake stroke into an air pressure power stroke.

The turbocharger may rotate at 80,000 rpm or more. One side is kept hot by exhaust gas right out of the engine, the other is cooled by incoming air. Proper care includes idling the engine 5 to 15 minutes between a hard pull and shutting down, to equalize temperatures. Neglecting this precaution is likely to ruin the unit.

GAS TURBINE

The gas turbine is presently not widely used in construction machinery, being limited to very large trucks, and perhaps mobile electric generating sets.

The basic unit consists of a compressor, a combustion chamber, and a turbine. Compressed air is mixed with burning fuel in the chamber, goes through nozzles at high velocity, and strikes turbine vanes, causing the turbine to revolve. The turbine shaft drives the compressor (internal work), and an output shaft (net work).

Speed is generally inversely proportional to size. A 10,000 horsepower unit might turn at 3,000 rpm, where a 1,000 hp might rev up to 20,000. In trucks with 2 or 3,000 horsepower, rpm might be about 14,000.

Conversion of fuel energy to rotation is highly efficient, 80 to 90 per cent. However, more than half this power is consumed in driving the compressor. Net efficiency is usually less than in a diesel at full speed, and very much less when throttled down. However, they are competitive in larger sizes, from 1,500 horsepower up.

Gas turbines are remarkably light and compact in proportion to power produced, and are considered to be simple and easy to maintain. They will burn any non-corrosive liquid fuel. Diesel fuel oil or kerosene is commonly used.

Drive to working parts may be through planetary reducing gears, an alternator to produce AC current, a rectifier, and DC wheel motors.

EXHAUST CONDITIONERS

Gasoline exhaust contains substantial quantities of carbon monoxide, an odorless, poison gas, together with small quantities of other chemicals. Diesel exhaust has little carbon monoxide, but a wealth of irritating gases of varying degrees of toxicity. These include hydrocarbons, aldehydes and oxides of nitrogen.

Because diesel-contaminated air usually becomes 'intensely irritating before it is a serious health danger, diesels are safer than gasoline engines wherever ample natural ventilation is lacking. No matter how efficient protective devices may be, there is always a chance that they will fail unexpectedly.

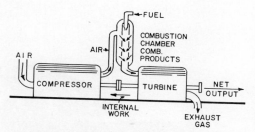

Fig. 12-94. Schematic of gas turbine

The amount of irritating and poisonous chemicals exhausted can be kept at a minimum by a proper air-fuel mixture in the cylinder. In a diesel 15 to 22 parts of air to one of fuel is the best range. If there is too much air, as when the engine is idling or running at low throttle, there may be excessive production of irritants. If the mixture is too rich, there are many products of incomplete combustion, and particles of solid carbon as well.

The noxious exhaust gases are soluble in water, and combine with it to produce acids. In the presence of heat and a catalytic agent they will combine with oxygen

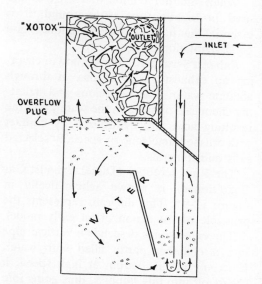

Fig. 12-95. Exhaust scrubber, schematic diagram

Courtesy of The Ruth Company

Fig. 12-96. Exhaust conditioner mounted on tractor

to form carbon dioxide and water. The carbon dioxide has a suffocating rather than a poisoning effect, and is comparatively easy to control by ventilation.

When internal combustion engines are used in buildings or underground, it is essential that both exhaust conditioners and good ventilation be provided.

Exhaust Scrubber. One method of cleaning up exhaust gas is to pass it through alkaline water. The poisonous and irritating chemicals dissolve in the water, and the resulting acids are neutralized by contact with crushed limestone or specially selected salts and minerals.

The Ruth scrubber (Diesel Exhaust Gas Conditioner) is shown schematically in Figure 12-95. This diagram represents the principle of the action of an early model.

The exhaust gas descends a vertical pipe into a chamber partly filled with water. Emerging at the bottom at high speed, it shears off into tiny bubbles, thus going into intimate contact with the water. Some of these bubbles rise to the surface immedi-

ately, others are carried around in the circulation one or more times.

Above the water surface is a bed of XOTOX, an alkaline mineral compound supplied for these units. The violent movement of the exhaust gas carries considerable water up into the bed, where it dissolves enough of the mineral to neutralize the acids dissolved out of the gas. The XOTOX serves not only to keep the water alkaline, but separates water from the exhaust

Courtesy of Oxy-Catalyst, Incorporated

Fig. 12-97. Catalytic Exhaust muffler

Courtesy of Oxy-Catalyst, Incorporated

Fig. 12-98. Parts of Catalytic Exhaust

stream, and acts directly on remaining gases.

A venturi tube is built on the outlet to mix the exhaust with five times its volume of air, so that any remaining deleterious substances will be thinned out before being breathed.

The scrubber is contained in a case that is mounted in any convenient place on the machine. Crawler tractors usually carry it on the rear, as in Figure 12-96. An oversize pipe conducts the exhaust from the manifold to the unit. A check valve is used to prevent water from being sucked back into the engine if it should rotate backward, as in slipping down a hill after stalling. However, since this might not work, the operator must be careful to declutch immediately if he stalls.

The conditioner must be carefully chosen to fit the particular engine and its load requirements, to provide sufficient capacity and minimum backpressure. The standard backpressure for these units is one and one-half inches of mercury (about three-quarters of a pound). Power loss is about one-half of one percent.

Catalytic Exhaust. The OCM Catalytic Exhaust, Figures 12-97 and 12-98, is a replacement muffler containing sticks of ceramic porcelain coated with a platinum compound. When heated to 500 degrees Fahrenheit, this catalytic film will cause the carbon monoxide and hydrocarbons to combine with the fresh air supply drawn through the venturi, producing water and carbon dioxide. The heat of the exhaust, assisted by that generated by this secondary combustion, keeps the catalyst tubes hot and the reaction going.

A pyrometer in the exhaust tube indicates whether the unit is working at correct temperature, and may also give information about the fuel mixture in the engine.

It takes about five minutes of full load operation to create sufficient heat to start the reaction. With gasoline engines, it will continue even at idling speed, but diesels must be kept at about 60 percent load to maintain it.

The catalytic units are said to be heat proof and shock proof. After 2000 to 2500 hours of operation they should be exchanged for fresh units, so that they can be re-vitalized by the factory. No daily maintenance is required, and installation and replacement costs are said to be low.

These conditioners are used successfully with white gasoline, diesel fuel, and LP (propane) gas. However, they are not satisfactory with leaded gasoline, so that a supply of untreated fuel must be assured before they are installed.

This is the type of emission-control muffler that is standard equipment in most 1976 model automobiles, and which is able to take the place of many of the annoying and uneconomic devices installed in earlier models to reduce pollution.

There are several designs, some using pellets or perforated shapes instead of sticks or rods.

The principal problem is avoiding leaded gasoline. There is a question about whether one tank of it will put the catalytic muffler out of action, but it is certain that repeated use will. Control is much more difficult with machines in the field than it is with cars on the road.

FUEL CONSUMPTION

Manufacturers often show fuel consumption per engine brake horsepower as an extra line on torque and horsepower graphs, as in Figures 12-78A and B. The line is usually a flat curve, with highest consumption rate at the lowest revolutions per minute, and the lowest rate near the highest rpm.

Unfortunately, consumption is usually given in pounds. This may be satisfactory to engineers, but it is a nuisance to contractors, who must measure and pay for their fuel in gallons. To convert to gallons, divide pounds of gasoline by 6.0, and pounds of diesel fuel by 7.0, 7.1, or 7.3. There is a difference of opinion among authorities as to whether the 7.1 or 7.3 figure is right, and for most purposes 7.0 is close enough, and it is a lot easier to use.

Load. Fuel curves on engine graphs are for full load operation. Few excavators or haulers work continuously at full load, and their engines could not maintain performance if they did. The shovel may coast on its swing and has little load while lowering its bucket, the dozer backs up without load, and the truck and scraper return empty and may descend grades. Average load may run from less than 1/4 to over 3/4 of maximum. An engine running without load uses about 30 per cent as much fuel as it would under full load at the same speed. Any reduction in load results in saving fuel, so that in most operations fuel consumption is well below rated levels.

Work in hard digging or hauling on soft ground or at high speed will usually consume more fuel than more favorable conditions. But there are exceptions. A shovel scraping at a hard bank may use less fuel per hour (although more per yard) than when in easy digging that permits deep penetration and heavy bucket loads.

Efficiency. Many engines do not operate anywhere near their top efficiency. Dirty air cleaners, wear and faulty adjustments in the engine and its fuel system, compression leakage past worn or stuck valves or rings, and the drag of constantly engaged accessories such as generators and hydraulic pumps all add to fuel consumption. A new

Fuel	Type of Service		
	Light	Average	Heavy
Diesel	.02 to .03	.04	.07
Gasoline	.06 to .07	.08	.10

Fig. 12-99. Fuel consumption, gallons per horsepower hour

or newly rebuilt engine that has been properly broken in will use less fuel than its full load rating on practically any job, but as it approaches the end of its useful life, or is allowed to deteriorate, it may use more at part load than it is supposed to at full load.

Engines that are governed or altered to produce less than their rated horsepower will use fuel in proportion to the power that they are permitted to develop, rather than to their rating.

High altitude and high temperature increase fuel consumption unless special adjustments are made.

The table in Figure 12-99 indicates fuel consumption that may normally be expected under three classes of conditions. "Light" is for equipment in excellent condition with light loads, "heavy" covers old engines and equipment in difficult working conditions, while "average" represents the ordinary run of equipment on jobs.

For fast, rough figuring, use 80 per cent of rated full load fuel consumption as a basis. But be sure to change pounds of fuel to gallons, to avoid major confusion.

The contractor who has records of fuel consumption should use his own figures, as they are likely to be much more accurate for him than average figures such as those above.

POWER

Power may be measured in terms of force—push, pull, twist, or lift, as a tractor may have a drawbar pull of 30,000 pounds.

It may be measured in terms of work, as if the tractor pulled a 30,000 pound load

six feet, it would have done work amounting to 6 feet × 30,000 pounds, or 180,000 foot-pounds.

Small loads and small distances may be measured in inch-pounds. That is, if a 20 pound weight were lifted 7 inches, the work would amount to 140 inch-pounds. Dividing by 12 will convert inch-pounds to foot-pounds.

The general formula is:

Work = Force × Distance

The third measurement is in terms of work and time. If the tractor pulled 30,000 pounds 100 feet in one minute, it would have a work rate or production of 30,000 pounds × 100 feet or 3,000,000 foot-pounds a minute.

Torque. Torque is defined as the turning power of a shaft. It is the measure of twisting power.

Engine torque is rated in terms of pounds-feet, which is the amount of force exerted at a distance of one foot from the center of the crankshaft.

A dynamometer may be used to measure engine torque. A simple type called the Prony brake is shown in Figure 12-100. A brake acts on a drum that revolves with the flywheel. The brake carries a force arm that has a scale at its outer end. Engine speed is measured in revolutions per minute (rpm) by a tachometer.

When the brake is applied, its drag against the revolving drum turns it and the

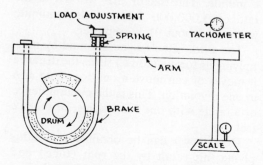

Fig. 12-100. Prony brake or dynamometer

arm, and the force is registered by the scale.

Whenever the brake holds engine speed to a lower level than its throttle setting calls for, the scale indicates the amount of force the engine is exerting at that speed against the resistance or load of the brake.

If the force arm were one foot long the scale reading would indicate the torque. If it were two feet long its reading would have to be multiplied by two to obtain the torque.

With a two foot arm, the throttle wide open, and the brake holding the engine to a speed of 1,000 rpm, the scale might register 401 pounds. Multiplying by two, we would have a torque figure of 802 pound-feet (pounds for short) at that speed.

The brake can be used to hold the engine at a number of test speeds, perhaps at 200 rpm intervals throughout its speed range. The result for one diesel engine is shown in Figure 12-101.

If the test is to be of value to the people who will use the results in estimating, the engine should be fully equipped with water pump, fan, generator, governor, and muffler. It will carry these accessories when it is installed in equipment, and each of them takes some power.

Most engine testing is now done on dynamometers, that use an electrical generator (dynamo) to apply the load, and measure power output by current flow. One horsepower equals .746 (say ¾) of a kilowatt.

Horsepower. Horsepower is a work-time measurement, equalling 33,000 foot pounds a minute. The tractor that worked at the rate of 3,000,000 foot-pounds a minute was using about 91 horsepower.

The work of an engine is found by multiplying its torque (force) times the distance it moves or can move a one foot long arm in one revolution. This is the circumference of a circle with one foot radius, or 6.2832 feet.

Horsepower can be obtained by multiplying this result by the rpm (time) and dividing by 33,000. The basic formula is:

$$HP = \frac{Torque \times rpm \times 6.2832}{33,000}.$$

However, since the two figures, 6.2832 and 33,000, are always present, the formula can be simplified by dividing one into the other. The formula will then be:

$$HP = \frac{Torque \times rpm}{5252}$$

Applying this formula to the torque readings, we obtain the horsepower ratings in the second column of Figure 12-101. A comparison of the two sets of figures shows that horsepower continues to rise with increasing speed long after torque is dropping.

Torque and horsepower figures such as these are usually posted to graph paper and used to plot curves, such as those in Figure 12-78, A and B.

When an engine is slowed down by load (lugs down) it increases the force it is putting into the work until speed drops below the peak of the torque curve. While working in the range between full speed and maximum torque it is said to be lugged down.

The ability of an engine to lug down is very important in certain types of work, particularly for excavators. However, operating an engine at a lugged down speed for more than a few seconds is inefficient, and puts heavy strain on the engine. The load should be reduced or the gear changed to allow it to regain full speed.

If torque readings are honestly taken, the calculated horsepower will be correct. But advertised horsepower is full of pitfalls. It may be that the peak advertised horsepower will occur at engine speeds that are seldom or never reached in operation, as for example an automobile rated at 120 miles per hour. Torque readings are often taken without a generator and sometimes even with no cooling system or with higher-than-normal air pressure.

Horsepower Ratings. There are a number of different kinds of horsepower ratings. Knowing what they are sometimes makes it easier to understand specifications.

Brake Horsepower is engine horsepower measured as described above, or in any way that would give similar results.

Net horsepower is brake horsepower with the drag of accessories deducted.

Flywheel horsepower is engine horsepower with or without accessories.

Drawbar horsepower is engine horsepower less friction losses in the power train.

Maximum horsepower is the most that an engine can develop for five consecutive minutes.

Intermittent horsepower is power that can be developed for changing loads such as are provided by construction equipment.

Continuous horsepower is the rating for long life performance under steady loads such as deep well irrigation pumping.

A.M.A. Horsepower. This is an arbitrary rating used in taxing motor vehicles. It may also be called taxable horsepower. It depends only on bore (cylinder diameter) and the number of cylinders, and has no constant relationship to either torque or brake horsepower. The formula is:

$$\text{A.M.A. hp} = \frac{\text{Bore}^2 \times \text{No. of cylinders}}{2.5}$$

Displacement. A manufacturer's specification sheet should state the cubic inches displacement. This is the sum of the piston areas times the length of the stroke, both in inches.

An efficient internal combustion engine should develop at least 5/8 (0.625) pounds-feet of torque for each cubic inch of piston displacement. This relationship varies, but can be used as a general check. To find the approximate torque, multiply the displacement by 5 and divide the result by 8. That is,

$$\text{Torque} = \frac{5 \times \text{Displacement}}{8}$$

Engine RPM	Measured Torque, Pounds-Feet	Brake Horsepower
800	725	114.2
1000	802	152.7
1200	825	188.5
1400	815	217.7
1600	790	240.7
1800	756	259.0
2000	710	270.00

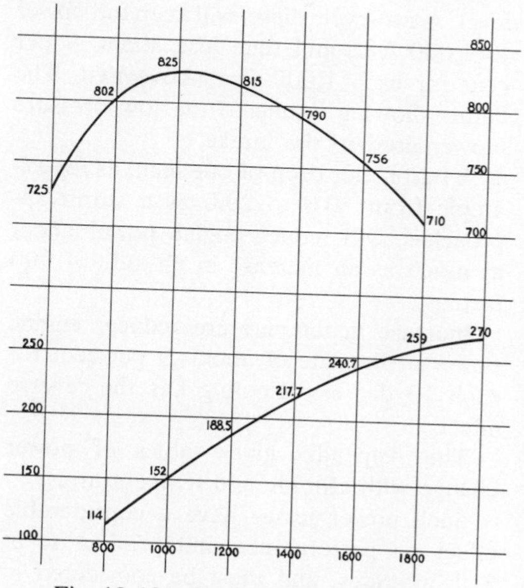

Fig. 12-101. Engine speed, torque, and horsepower

Altitude and Temperature. Air becomes thinner as altitude increase or its temperature is raised. The thinner air contains less oxygen to combine with fuel, and provides less substance for compression.

As a result, engine power diminishes with increases in altitude and/or temperature. Standard tests are made at a sea level barometer pressure at 29.92 and a temperature of 60 degrees Fahrenheit. If test conditions differ, the report should specify what they were.

In general, increase in altitude will reduce the power of a four cycle engine about

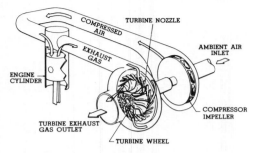

Courtesy of Cummins Company

Fig. 12-102. Turbocharger

3 per cent for each 1,000 feet above sea level. A two-cycle diesel will keep full power to 1,000 feet, and then lose about .9 per cent for each 1,000 feet above that. The better showing is due to the low pressure blower used in the intake.

A barometer drop of one inch, as for example from 30.0 to 29.0 as a storm approaches, will reduce engine power about as much as an increase in altitude of 900 feet.

Increase in temperature reduces engine power at the rate of about .9 per cent for each 10 degrees. Cooling has the reverse effect.

The Appendix gives tables of power change with altitude and temperature.

Such power losses have a considerable effect on performance and production at high altitudes, and must be considered in estimates. For example, a loaded truck might be able to climb a 10 per cent grade in San Francisco in third speed, but could manage only 8.5 per cent in that gear in Denver. If it went up to work on a pass in the Rockies at 12,000 feet altitude the grade ability would be reduced to 6.4 per cent. Lower gears would have to be used for all climbs.

Fuel consumption increases as power is lost with altitude, as the thinner air does not provide an efficient mixture. Special carburetor, injector, or pump adjustments may be made to save some of the fuel, but they cannot restore the power.

Supercharging. Superchargers, usually of the turbo or exhaust driven type, are widely used to provide more than normal power in proportion to displacement, and/or to retain normal power in engines used at high altitudes.

Most supercharged engines will also lose power if altitude is increased, or gain it if brought nearer sea level. But such losses are unlikely to equal the additional power obtained from supercharging, and the charger can usually be adjusted to restore the altitude losses.

HAULER POWER TRAIN

In most construction machinery engine power is delivered to working parts through gears, chains, belts, convertors and/or pulleys and cables that multiply torque and reduce speed.

Gearing. Except for considerable losses in friction and slippage that will be discussed later, such stepping up of power or torque is always exactly proportional to the loss in speed. If an engine produces 300 foot pounds of torque at 1,800 revolutions per minute, a 3 to 1 gear set will convert this to 900 pounds of torque at 600 rpm. If gearing is 25 to 1, output will be 7,500 pounds torque at 72 rpm.

A standard truck transmission may drive through two sets of gears (mainshaft to countershaft, and back to mainshaft) in all speeds except one, which is direct through the mainshaft. An auxiliary transmission may have two additional gear sets. A rear end has one set of gears if it is single reduction, two if it is double reduction. It may have two speeds.

An off the road truck or scraper may have these gears, and another reduction set in the final drives in the hubs.

The total torque multiplication, or total gear reduction (TGR) is found by multiplying all gear reductions being used in the power train by each other. If reductions in low gear were 6 to 1 in the transmission, 4 to 1 in the rear end and 6 to 1 in the final drive, then total reduction would be $6 \times 4 \times 6 = 144$. It is usual to simplify figuring by combining the rear end and final drive ratios into an axle ratio, in this case 24 to 1. The total reduction would then be found by multiplying this by the transmission ratio, $6 \times 24 = 144$. In direct drive in the transmission the total reduction would be 24.

Step-up gearing to increase speed at the expense of power may be used in the highest transmission gear or gears. Such gears are called overdrives. If the truck above used an overdrive with a ration of .7 to 1, the total reduction would be $.7 \times 24 = 16.8$.

Converter. Torque multiplication in a torque converter is infinitely variable from 1 to 1 up to its capacity, that may be anywhere from 2.1 to 1 to 6 to 1. A hauler similar to the above, equipped with a converter with a maximum ratio of 2.7 to 1 and a low transmission gear of 4.0 to 1 would have a total reduction of $2.7 \times 4 \times 24 = 259.2$. This would be at stall speed, and the reduction would automatically decrease as it gained speed, up to a minimum reduction at no load in low gear of $1 \times 4 \times 24 = 96$. In direct gear maximum reduction would be $2.7 \times 24 = 64.8$, and reduction at light load $1 \times 24 = 24$.

Mechanical Efficiency (ME). The mechanical efficiency of a mechanism is the percentage of the power put into it that it delivers to its work. If a hauler engine has 300 pounds of torque (T) at 1,800 rpm and has a total gear reduction (TGR) of 100 to 1, it should deliver 30,000 pounds-feet of torque to the rear hubs. If the hub torque (HT) is found by testing to be 24,000 pounds, the friction loss is 6,000 pounds. To find the percentage of engine power we use the formula:

$$ME = \frac{HT}{T \times TGR}$$

we have:

$$ME = \frac{24,000}{300 \times 100}$$

$$= .8 \text{ or } 80 \text{ per cent}$$

The loss of torque in a mechanical power train varies with the number and type of gear sets, the design, the perfection of finish, the condition of the unit, and the amount of torque carried. It may range from 2 to 5 per cent for each gear set.

Losses are greatest in the lowest gear, in which the greatest number of reductions are used, and in which the greatest strain is placed on the drive line. Friction is increased by high pressures on gear teeth and bearings, and by shafts changing shape under twist. These effects are greatly increased as equipment wears.

A rule of thumb figure for the overall efficiency of an off the road hauler is 80 per cent for new equipment. An old and badly maintained machine may go down to 70 or even 60 per cent efficiency. Highway trucks have higher efficiency than off the road models. Tandem drive consumes extra power.

Torque converter machines usually have lower mechanical efficiency because of slippage. An average figure may be taken as 75 per cent.

Mechanical efficiency should not be confused with work or job efficiency. It is simply a measurement of power lost between the engine and the rear axle.

Loaded Tire Radius (LR). The loaded tire radius is the distance from the center of the wheel hub to the ground when the tire is carrying its full rated load. It is the length of the power arm through which the torque at the hub acts to drive the machine. Lengthening this radius by putting on bigger tires increases vehicle speed and decreases power.

Figure 12-103 gives the static loaded radius for some tire sizes. This radius is increased somewhat by centrifugal force when the truck is in motion, and is greater

Rock Grip Excavator H.D. Tire
Static Loaded Radius

Size	Plies	Radius	Size	Plies	Radius
7.50-20	12	17.6	13.00-24	18	23.7
8.25-20	12	18.1	14.00-24	20	25.2
9.00-20	12	19.1	16.00-24	20,24	27.0
10.00-20	14	19.7	16.00-25	24	27.0
11-22.5	14	19.8	18.00-24	20	29.4
10.00-24	14	21.7	18.00-25	16,20	29.4
11.00-20	14	20.5	18.00-32	24	33.4
12-22.5	14	20.5	18.00-33	32	33.4
11.00-22	14	21.5	21.00-24	20,24	31.2
12-24.5	14	21.5	21.00-25	16,24	31.2
11.00-24	14	22.5	24.00-25	24	33.6
12.00-20	16	21.0	24.00-29	24,36	35.7
12.00-24	16	23.0	27.00-33	30,36	40.1

Courtesy of Firestone Tire & Rubber Company

Fig. 12-103. Loaded radius of some tire sizes

if load is less than that for which the tire is rated.

Loaded radius is reduced by low tire pressure and by tread wear. A big tire may have a change of 3½ inches from new tread down to fabric. It can be measured on the vehicle by parking it on hard, level footing, placing a carpenter's level at the center of the hub, leveling it, and measuring from its other end to the ground.

The loaded radius is always less than the unloaded radius, which is one half the overall diameter. The difference is about half an inch in a 6.00-16 and more than 5 inches in a 30.00-33.

The variables in loaded tire radius alone are enough to make it important to field check the calculated performance of haulers.

Rim Pull (Tractive Effort). Rim pull is the torque that a machine is capable of exerting at the contact of its drive tires with the ground; that is, at the distance from the axle represented by the loaded radius of the drive tires.

The term is not a very clear one, as the rim is actually the outer part of the metal wheel, and except in this particular use it does not mean the tread surface of the tire. Tractive effort is a term that may be used instead of rim pull. It comes closer to expressing the meaning, as this is the effort that the machine can exert to move itself. However, the term rim pull is in general use and has been used for a long time, it is not as easily confused with other terms having to do with traction, and it is two syllables shorter. It will therefore be used here.

Rim pull is stated in foot pounds unless some other measurement is specified. It is different in each gearshift position, being greatest in the lowest and slowest gears.

Figuring Rim Pull. Rim pull is figured as the hub torque (HbT) divided by the loaded tire radius (LR) in feet. The hub torque is the engine torque (EnT) times the total gear reduction (TGR) times the mechanical efficiency (ME). Combining these, we have

$$RmP = \frac{EnT \times TGR \times ME}{LR, \text{ feet}}$$

Since tire radius is always given in inches it may be less confusing to change the formula to:

$$RmP = \frac{EnT \times TGR \times ME \times 12}{LR, \text{ inches}}$$

It often happens that the estimator does not know either the gear reduction or the loaded radius. But even one page flyers will usually give him the engine horsepower and the speed in each gear. He can use them to get an approximate value for rim pull by using the arbitrary formula:

$$RmP = \frac{300 \times \text{Engine Horsepower}}{\text{Maximum Speed, mph}}$$

Figure 12-105 gives power and speed data for a simplified 300 horsepower hauler that will be used as the basis of sample calculations.

TRAVEL RESISTANCE

Rim pull must overcome travel resistance before it can move a machine, and in order to keep it moving.

Travel resistance is rolling resistance, plus or minus any hindrance or help from grades. It is figured as a percentage of machine gross weight, without reference to engine power.

Rolling Resistance. Rolling resistance is the power consumed by the ground surface and the tire as the vehicle moves. It is expressed in pounds of pull per ton of vehicle weight, or as a percentage obtained by dividing the pounds of pull per ton by 2,000.

On a paved level road rolling resistance

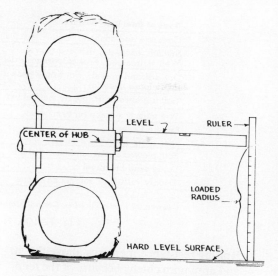

Fig. 12-104. Measuring loaded radius

usually is 40 to 60 pounds per ton or 2 to 3 per cent of vehicle weight. It includes power absorbed in flexing of tires and by friction in wheel bearings and between the tire tread and the road.

Rolling resistance may be as high as 600 pounds per ton. Figure 12-106 shows the amount that may be expected on various surfaces. Such figures are approximations, but serve as a convenient basis for working out power requirements and speed of haulers under various conditions.

Gear	Total Gear Reduction	Speed, mph	Pounds Rim pull, full governed speed*	Rim Pull, pounds per ton** Vehicle Weight	
				Loaded	Empty
1st	180	2.5	36,000	720	1,440
2nd	90	5.0	18,000	360	720
3rd	60	7.5	12,000	240	480
4th	36	12.5	7,200	145	290
5th	22.5	20.0	4,500	90	180
6th	15	30.0	3,000	60	120

*Maximum rim pull with the engine lugged down may be 10 to 15% greater
**These two columns not ordinarily supplied in specifications, but are easily obtained by dividing rim pull by vehicle weights.

Fig. 12-105. Power in gears, simplified hauler

Type of Haul Road Surface	Pounds per Ton of Gross Vehicle	Per cent of Gross Vehicle Weight
Concrete and asphalt...............	30 lbs.	1.5%
Smooth, hard, dry dirt and gravel. Well maintained. Free of loose material....	40 lbs.	2%
Dry dirt and gravel. Not firmly packed. Some loose material....................	60 lbs.	3%
Soft unplowed dirt, poorly maintained ...	80 lbs.	4%
Wet muddy surface on firm base.........	80 lbs.	4%
Snow—Packed......................	50 lbs.	2.5%
4" Loose......................	90 lbs.	4.5%
Soft, plowed dirt or unpacked dirt fills.....	160 lbs.	8%
Loose sand or gravel.................	200 lbs.	10%
Deeply rutted, or soft spongy base.......	320 lbs.	16%

Courtesy of Euclid Division, General Motors Corporation

Fig. 12-106. Rolling resistance

Rolling resistance increases above the 40 pound minimum chiefly because of the tendency of tires either to sink into a yielding surface or depress the general area on which they travel, so that it is as if they were constantly climbing a grade to carry the truck along on a level. There is also a factor of roughness of surface.

On rough ground large diameter tires will meet less rolling resistance than small ones. This is because the big tire bridges over hollows and absorbs humps, and the higher center of revolution tends to roll it over humps instead of trying to push it through them.

Hard tires show least friction and rolling resistance on hard surfaces, and the most sinking and highest resistance on soft surfaces.

Grade Resistance. The extra power required to overcome gravity in moving a machine up a grade is called grade resistance. It amounts to about 20 pounds per ton, or 1 per cent of loaded weight, for each per cent of grade. That is, a 5 per cent grade would offer a resistance of 100 pounds for every ton of vehicle weight.

The estimator adds grade resistance to rolling resistance for uphill work, and subtracts it for downhill hauls. If assistance from a favorable grade is greater than the rolling resistance the machine should be able to coast down. It is then time to consider the power of its brakes or retarders, rather than that of the engine and the gear reductions.

The grade at which a machine starts to roll freely provides a rough field check on the accuracy of estimates of rolling resistance.

Figure 12-107 gives the travel resistance for combinations of rolling resistance from 40 to 200 pounds with various grades up to 25 per cent.

Grade Ability. Travel resistance is compared with pull in various gear speeds to determine which one will be used.

Vehicle weight is taken from its specifications, or by weighing in the field. The figure is rounded off to the nearest 1,000 pounds or the nearest ton for convenience in arithmetic. This weight, multiplied by the proper figure selected from Figure 12-107, will give the pounds resistance that must be overcome for that grade and ground condition. It is compared with the gear speed and pull chart, as in Figure 12-105, and a gear with power greater than the resistance is selected. Haul time is figured on the basis of speed in that gear.

For example, our sample hauler with a gross weight of 50 tons must climb a 7 per cent grade on a haul road with a rolling resistance of 4 per cent (80 pounds to the

Rolling Resistance, Pounds per ton	Grade per cent																
	-10	-6	-4	-2	-1	0	+1	+2	+3	+4	+5	+6	+8	+10	+15	+20	+25
0	-200	-120	-80	-40	-20	0	20	40	60	80	100	120	160	200	300	400	500
30	-170	- 90	-50	-10	10	30	50	70	90	110	130	150	190	230	330	430	530
40	-160	-80	-40	–	20	40	60	80	100	120	140	160	200	240	340	440	540
60	-140	-60	-20	20	40	60	80	100	120	140	160	180	220	260	360	460	560
80	-120	-40	–	40	60	80	100	120	140	160	180	200	240	280	380	480	580
100	-100	-20	20	60	80	100	120	140	160	180	200	220	260	300	400	500	600
120	-80	–	40	80	100	120	140	160	180	200	220	240	280	320	420	520	620
150	-50	30	70	110	130	150	170	190	210	230	250	270	310	350	450	550	650
200	–	80	120	160	180	200	220	240	260	280	300	320	360	400	500	600	700

Fig. 12-107. Travel resistance

ton). From Figure 12-107 we take 220 for the combined or travel resistance. Then:

Required Rim Pull $= 50 \times 220 = 11,000$

According to Figure 12-105, this truck has a rim pull of 12,000 pounds in 3rd gear, and only 7,200 pounds in 4th. Third gear is selected. The 1,000 pound margin gives security against minor errors in measuring grade or selecting the soil factor.

Maximum speed in 3rd is 7.5 miles per hour. Whether it will attain this speed depends on its start up the grade and its length, as it does not have enough surplus power for rapid acceleration.

Traction. Tire treads must be able to grip the ground surface to use the power that is applied to them. Their ability to do so is called tractive efficiency, which is discussed in Chapter 3.

A machine cannot use more power than its traction permits. If travel resistance is greater than traction, wheels will spin without moving the machine.

Acceleration. In any haul or haul section where the equipment has to start from a stop, or enters the section at a speed lower than it can attain, the average speed is affected by the time required to accelerate.

Acceleration is rapid when the rim pull is high in proportion to the travel (rolling plus grade) resistance. Since the rim pull is less in the higher gears, more acceleration time is needed in them, particularly when a gear is used that is just able to pull the load.

The rule for computing acceleration is: for each pound of rim pull per gross ton of machine and load that is in excess of that required to move the machine, an acceleration of .01 miles per hour in each second can be expected.

That is, if a machine can exert a rim pull of 300 pounds for each ton of its weight, and travel resistance is 80 pounds, the excess or accelerating rim pull (AccRmP) is 220 pounds.

The acceleration formula is:

Accel, mph per second $=$ AccRmP $\times$ .01

This does not work well for the lowest gears where there is a very great amount of excess rim pull. Acceleration in them is very rapid, but not as fast as is required by the formula. This difference arises partly from sluggishness within the engine, and partly from other causes.

Length of Section, Feet	Starting From Stop	Headway when entering
0–250	.25 to .4	.5
250–500	.4 to .5	.6
500–1,000	.5 to .6	.7
1,000–2,000	.6 to .7	.8
2,000–5,000	.7 to .8	.8

Fig. 12-108. Factors, maximum to average speed

Results will be more accurate if arbitrary reductions are made in high excess rim pulls· per ton, as follows:

Over 300 pounds, subtract 75%
From 200 to 299 pounds, subtract 50%
From 100 to 199 pounds, subtract 25%

In a hauler with 640 pounds per ton of excess rim pull in low gear, this would work out:

Accelerating rim pull		640
Subtract 75% of 340	255	
50% of 100	50	
25% of 100	25	330
Corrected accelerating pull		310
Acceleration, mph per second		3.1

Starting from a dead stop with a load may take another 2 to 4 seconds. Shifts with sliding gears average almost 4 seconds, synchromesh shifts 1 to 2 seconds, and power shifts no time. Start and shift time are added to acceleration time.

If this machine has a top speed of 2.5 mph in low, it will take 2.5/3.1, or .8 seconds, plus perhaps 2 seconds starting time, to reach full speed in this gear. In a high gear with only 10 pounds excess rim pull accelerating time would be 75 seconds.

Production tables for scrapers often figure in acceleration and deceleration as part of "fixed" time.

Accelerating time may also be figured by multiplying the maximum speed by the appropriate factor from Figure 12-108. But much more accurate results may usually be obtained by working out figures for the hauler and the job conditions.

A hauler may require 1,800 feet or more to reach full speed in a gear in which it has very little excess rim pull.

TIRES

Importance. Tires make up a substantial fraction, usually from 1/10 to 1/4 at list prices, of the purchase price of rubber mounted machinery. They usually wear out or break up more rapidly than the rest of the machine, so that they require replacement one or more times during its life. In addition, they need repairs.

Repair and replacement costs are greatly affected by care and conditions of use. Proper tire selection and good maintenance and work practices often add substantially to profits.

Good tire management includes:

1. Buying tires that have the proper size, strength, tread and speed rating for the job.

2. Using tires only for purposes, loads, and speeds for which they are designed.

3. Maintaining tires at proper pressure and tires and machines in good condition.

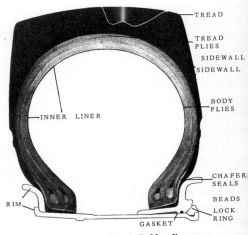

Courtesy of Firestone Tire & Rubber Company

Fig. 12-109. Cross section of tire

12-126

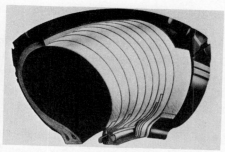

Courtesy of Goodyear Tire & Rubber Company

Fig. 12-110. Cutaway view of tire

4. Keeping excavation areas, haul roads, and fills smooth and free of spillage.

Construction. All tires used in construction equipment are of the pneumatic type, in which air or fluid is held under pressure inside a flexible, ring-shaped container.

There are many different sizes and types of tire, but they agree in the principles of construction and parts shown in Figure 12-109 and 12-110.

The typical tire is made up of the following:

1. BEADS. The tire beads are bundles of strong steel wire that prevent any change of shape that would interfere with fit on the rim.

2. CASING, CORD BODY, OR CARCASS. The fabric body of the tire, that gives it strength to hold the internal pressure that supports the load.

3. PLIES. The plies are the individual fabric layers in the cord body. They are usually woven of cotton, rayon, or nylon cords, or steel wires, surrounded by rubber and looped around the beads.

4. TREAD. The part of the tire that contacts the road. It must provide traction, long wear, cushioning qualities, and cut resistance.

5. SIDEWALLS. The sides of the tire, between the tread and the bead. Technically, may refer only to the protective rubber outside coating on the sides.

6. TREAD PLIES OR BREAKER. Part plies that are only under the tread, and that serve to resist and distribute road shocks, and to resist penetration by sharp objects.

7. CHAFER PLIES. Part plies that protect the tire from rim damage near the bead.

In addition, a tubeless tire has an air tight inner layer on the casing that seals in the pressure. The rim must also be air tight.

Courtesy of Goodyear Tire & Rubber Company

Fig. 12-111. Tire delivery is often urgent

Other tires have a soft inner tube to contain the air. This is protected from possible damage from the rim by a cloth flap.

Fabric. Cotton used to be the standard material for tire casings, but it has been largely superseded by rayon and nylon.

These two synthetics are competing vigorously for the market. Right now (1975) nylon seems to be in the lead for high quality tires. It is less subject to heat damage and is slower to blow than other fabrics.

Steel wire has increasing application in heavy duty tires. Heat conduction is excellent, so that the tire does not over-

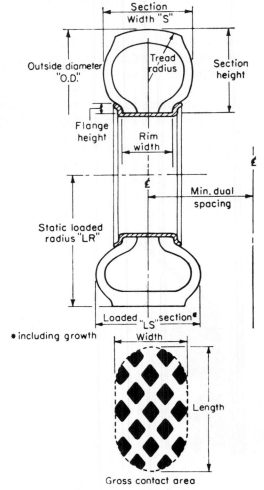

* including growth

Courtesy of Goodyear Tire & Rubber Company
Fig. 12-112. Tire dimensions

heat readily, and great strength to resist impacts and cuts.

Weight. Tire weights range from about 40 pounds for the smallest truck tire, up to about 1700 for a 27.00-33 off the road unit. Tube type tires, add about 10 per cent for tube and flap.

Manufacturers do not list tire weights, but they may be figured from the excise tax charged. The 1975 rate is 5¢ a pound for off-highway tires, 10¢ for on-highway.

Measurements. Figure 12-112 gives accepted names for the various tire dimensions. Tires are identified as to size by the approximate width of a section of inflated tire that is not under load (section S in the diagram) and by the diameter of the rim where the tire bead rests on it. A tire with a width of 10 inches on a 20 inch rim would be called a 10.00-20 tire.

Most tires are nearly round in cross section, so that this tire would have an outside diameter slightly larger than the rim size plus twice the width, or 20 + (2 × 10) = 40 inches. This size relationship does not exist in wide base tires.

All tires in sizes up to 11.00 width, and tube type tires of standard cross section in any size, have the width figure shown with two decimal places. For example, your car might have a 6.70-15 tire, meaning a width of 6.7 inches and a 15 inch rim.

Tubeless tires from 7.00 through 11.00 inches width for highway use are marked with a whole number for width, and a number ending with .5 for rim diameter. For example, a 10 inch width tire on a 22.5 inch rim would be marked 10-22.5.

This tire would be interchangeable for size and load capacity with a tubed tire of size 9.00-20. On all sizes from here up thru 11:00 you add one to the tire width and 2½ to the rim diameter of a tubed tire to find the equivalent tubeless size. These tubeless tires are mounted on rims with a 15 degree slope outward from the center, that makes them 2½ inches larger at the bead than flat rims of the same center size. See Figure 12-113.

In widths of 12.00 and above both tubed and tubeless tires use 5 per cent slope rims, and carry the same markings.

Wide base tires, made only in large sizes, are wider than they are high, and can be used only on special rims. They are marked with widths always ending in .5, and rim sizes are in odd numbers. For example, 26.5-25, and 37.5-33.

Wide base tires look much bigger and stronger in their measurements than they really are. You subtract from 4.5 to 6.5 from their width to find the equivalent size and load capacity in conventional tires.

The greater tread width and lower pressure of wide base tires give them much greater ground contact area, and therefore better traction and flotation on soft ground than standard types.

Ply Rating. The strength of tire bodies formerly varied directly with the number of plies or fabric layers, so that the load a tire should carry might be figured according to the number of plies. Modern practice is to use cords of different strength and type, such as cotton, rayon, nylon, or steel, and to use different weights and spacings of cord in the fabric.

As a result, the number of plies no longer has much meaning by itself. However, since plies were a convenient way of rating tire strengths, the term has been retained as "ply rating." This term indicates the strength of the casing in relation to plies of standard strength. The amount of load recommended by the Tire and Rim Association for each ply rating is included in the specification table in Figure 12-114. When there are two or more ply ratings in one size, the higher number can take higher pressure and carry heavier loads.

At the same air pressure, the recommended load is usually the same for a tire size, regardless of the ply rating. The difference is that the heavier construction can

take more air pressure, and the greater pressure will carry a heavier load.

However, the heavier tire is more resistant to carcass damage under lighter pressures and loads. Where damage from bruising is an important factor, the heavier tire is likely to be the better investment even if its extra load capacity is not needed.

Load. Load is probably the most critical factor in tire costs. Load capacity increases with the size of the tire and with its ply rating. Both size and plies are expensive.

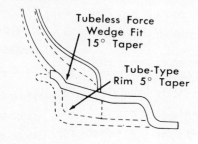

Tubeless Force
Wedge Fit
15° Taper

Tube-Type
Rim 5° Taper

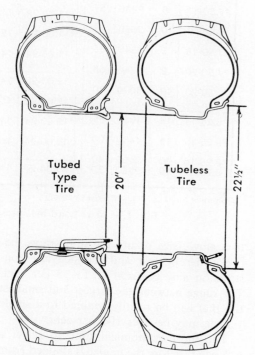

Tubed
Type
Tire

20"

Tubeless
Tire

22½"

Courtesy of Firestone Tire & Rubber Company

Fig. 12-113. Rim shapes for tubed and tubeless tires

Size of Tire	Ply Rating	Max. Load Lbs.	Infl. Press. Lbs. per Sq. In.
*6.00-16	6	1255	45
*6.50-16	6	1420	45
*6.50-20	8	2180	65
*7.00-15	6	1520	45
*7.00-15	8	1800	60
*7.00-16	6	1580	45
7.00-17	6	1740	45
*7.00-17	8	2060	65
*7.00-18	8	2140	60
*7.00-20	8	2310	60
7.00-20	10	2630	75
*7.50-15	8	2060	60
*7.50-15	10	†3310	80
*7.50-15	12	†3660	95
*7.50-16	6	1815	45
*7.50-16	8	2140	60
*7.50-17	8	2440	65
7.50-17	10	2650	75
7.50-18	8	2540	65
*7.50-20	8	2740	65
*7.50-20	10	3090	80
*8.25-15	12	†4060	85
*8.25-15	14	†4450	100
*8.25-16	10	2600	60
8.25-17	10	2980	70

Size of Tire	Ply Rating	Max. Load Lbs.	Infl. Press. Lbs. per Sq. In.
*8.25-20	10	3330	70
8.25-20	12	3730	85
*9.00-15	12	†4680	80
9.00-18	10	3690	70
*9.00-20	10	3960	70
9.00-20	12	4480	85
*10.00-15	14	†5480	85
*10.00-20	12	4580	75
10.00-20	14	5210	90
*10.00-22	12	4880	75
*11.00-20	12	5150	75
11.00-20	14	5730	90
*11.00-22	12	5480	75
11.00-22	14	6100	90
*11.00-24	12	5810	75
11.00-24	14	6260	85
12.00-20	14	6020	80
12.00-20	16	6450	90
*12.00-24	14	6780	80
12.00-24	16	7280	90
13.00-20	16	7150	80

Size Tubeless Tire	Ply Rating	Max. Load Pounds	Infl. Press. P.S.I.
6.00-16	6	1255	45
6.50-16	6	1420	45
*7-17.5	6	1520	45
7-22.5	6	1870	50
*7-22.5	8	2180	65
*8-17.5	6	1735	45
8-17.5	8	2060	60
8-19.5	6	2090	50
*8-19.5	8	2440	65
*8-22.5	8	2740	65
*8-22.5	10	3090	80
9-22.5	12	3730	85
*10-22.5	10	3960	70
10-22.5	12	4480	85
*11-22.5	12	4580	75
*11-24.5	12	4880	75
*12-22.5	12	5150	75
12-22.5	14	5730	90
*12-24.5	12	5480	75

Courtesy of Firestone Tire & Rubber Company

Fig. 12-114A. Load and pressure ratings, transport truck tires

In a single size of tire the price range between the lightest and the heaviest construction may be from 10 to 35 per cent. The range between the largest and smallest tire that can be readily adapted to a single vehicle may be over 100 per cent.

The most economical time to select a tire size is before the hauler is shipped from the factory. This is particularly the case if optional tires are enough larger in size to require special rims, wheels, and/or a change in axle ratio.

Tire Sizing. Haulers are usually factory-equipped with tires that are barely adequate for the vehicles' recommended loads. Offering larger tires as standard equipment would increase the advertised price of the unit, putting it at a competitive disadvantage with others offering minimum equipment. If the hauler is to be used only for rated or smaller

Tire Size	TIRE LOADS AT VARIOUS INFLATION PRESSURES											
	25	30	35	40	45	50	55	60	65	70	75	80
7.00-20	......	1810	1980	2140	2300	2440	2580	2710	2850[10]	2980	3100	3210[12]
7.50-20 / 8-22.5	......	2030	2220	2400	2570	2740	2900	3040	3190[10]	3330	3470[12]	
8.25-20 / 9-22.5	......	2390	2620	2820	3030	3220	3400	3580[10]	3750	3920	4070[12]	
9.00-20 / 10-22.5	......	2840	3100	3360	3590	3830	4050[10]	4250	4460[12]			
10.00-20 / 11-22.5	......	3200	3500	3780	4050	4310	4560	4780[12]	5020	5250[14]		
10.00-22 / 11-24.5	......	3400	3720	4030	4310	4590	4850	5100[12]	5350	5590[14]		
10.00-24	......	3620	3960	4280	4580	4870	5160	5420[12]	5680	5940[14]		
11.00-20 / 12-22.5	......	3540	3870	4190	4480	4770	5050[12]	5300	5560[14]			
11.00-22 / 12-24.5	......	3760	4110	4440	4760	5060	5360[12]	5620	5910[14]			
11.00-24	......	3990	4360	4720	5060	5380	5700[12]	5980	6270[14]			
12.00-20/21	......	4020	4390	4750	5080	5400	5740	6030[14]	6320	6600[16]		
12.00-24/25	......	4520	4930	5330	5720	6080	6450	6780[14]	7110	7430[16]		
13.00-24/25	......	5320	5820	6280	6740	7170	7590	7980	8370	8750[18]		
14.00-20/21	5040	5620	6140	6630	7100[12]	7550	8000	8400[16]	8820	9200	9580[20]	
14.00-24/25	5630	6270	6850	7400	7920	8430	8920	9400[16]	9830	10280	10690[20]	11100
16.00-20/21	6670	7430	8120	8780	9410[16]	10010	10580	11130[20]	......			
16.00-24/25	7370	8200	8970	9700	10400[16]	11070	11710	12300[20]	12900	13490[24]	14020	14560[28]
18.00-24/25	9580	10670[12]	11670	12620[16]	13520	14380[20]	15230	16000[24]	16780	17500[28]		
18.00-32/33	11350	12640	13820	14930	16000	17020	18010	18930[24]	19850	20720	21580	22400[32]
★18.00-49	......	......	......	19600	21000	22350	23650	24850	26100	27300	28350	29400[32]
21.00-24/25	12250	13640[16]	14910	16120[20]	17280	18370[24]	19450	20440[28]				
★21.00-35	......	......		19650	21050	22400	23750	24950	26150	27350	28450	29500[36]
24.00-25	15130	16860	18420[18]	19920	21340[24]	22700	24020[30]	......	......	......		
24.00-29	16330	18200	19890	21500	23020[24]	24500	25930[30]	27250	28590[36]			
24.00-32/33	17570	19600	21400	23140	24800[24]	26380	27900	29330	30760[36]			
27.00-33	22700	25280	27620	29890[24]	32030	34060[30]	36020	37890[36]	......			
30.00-33	27790	30950	33830[28]	36590	39200[34]	41700[40]	......					

★Subject to TRA approval.

The small index numbers are Ply Ratings, and denote maximum recommended load for that Ply Rating. For example: with 7.50-20 tire, or 8-22.5 corresponding tubeless size, 3190[10] is maximum load for 10-ply rating, and 3470[12] maximum for 12-ply rating.

Courtesy of Rubber Manufacturers Association

Fig. 12-114B. Load and pressure ratings, earthmover tires

loads, such tires may be satisfactory, but they leave no margin of safety.

Occasionally heavy trucks are offered with tires that are definitely too small. It is worth while to look up the truck's loaded weight and weight distribution in its specifications, and its tires in Figure 12-114 or similar tables, to see whether they will carry the load.

For example, an off the road truck with a loaded or gross vehicle weight of 90,000 pounds, with 64,000 pounds on the rear axle is equipped with four 18.00-25, 24 ply drive tires. They are rated at 16,000 pounds at 30 miles an hour, 17,920 pounds at 10 miles an hour. Maximum vehicle speed is 31 mph.

At full rated speed the drive tires have a combined capacity of 64,000 pounds, exactly equal to the rated load. There is no

Fig. 12-115. Off-the-road tire treads

margin of capacity for accidental or deliberate overload, shifting of load to the rear while climbing steep grades or for weight shift on curves. To keep any margin of safety the owner would have to underload or keep speed down to 20 mph or below.

For average work involving moderate to high speeds, full loads, and some road roughness, it would probably be wise to specify 28 ply tires in the same size, with a load capacity of 17,500 each, or a margin of 1,500 pounds per tire.

If still greater capacity were desired, different wheels, rims, and rear end gearing could be used, and 24-ply 21.00-25 tires used.

A contractor might make either of these extra investments to reduce tire damage (he would be more than repaid if he were saved one ruined tire), or to permit overloading the truck without tire damage.

Tread. Figure 12-115 shows a few of the tread designs that are available for hauler tires.

In general, high, widely separated diagonal cleats are used for drive tires on soft earth; wide closely spaced bars or lugs for rock, button or diamond treads for free rolling tires, or drivers on sand or dry humus; straight ribs for grader and farm tractor front wheels, notched ribs for highway travel, and smooth treads for compactors and loose sand.

Diagonal thin cleats tend to break up on rock, and the wide spaces between them do not have enough tread to protect the body against bruising. Rock treads tend to fill up on soft earth or mud. Any design using cross or diagonal cleats will wear irregularly on pavements, and will probably be very noisy. Designs that lack cleats or lugs may not have sufficient traction in soft ground.

Many tread designs are intended to serve reasonably well under two or more different conditions, rather than very well under one.

Some jobs require a particular type of tread for efficient operation, others can get along on almost anything.

Rims. Small truck tires are mounted on drop center rims that may be part of the wheel, as in a car, or demountable by removing lug nuts.

Tires larger than 12.00 size are mounted on three to five piece rims held together with a lock ring. This ring can be very dangerous if a tire is inflated before it is properly secured, as it can blow out with explosive force.

Tapered rims are best for low pressure tires, as the tire bead is wedged against the sloping floor of the rim, reducing the tendency of the wheel to revolve inside the tire.

Two tires with different rim sizes may be interchangeable on a wheel without affecting the speed or pull of the machine. This is because rims may be flat, or have various degrees of taper. Two rims that fit the same wheel may have quite different diameters at the bead, where measurement is made. Greater rim diameter may be compensated by lower tire height, so that overall diameter remains the same.

TIRE WEAR

A tire parts with a little of its surface on every revolution, partly because it changes shape with a scuffing effect as it contacts the ground. This changing of shape is called deflection. It is increased if the pressure is decreased (less air to hold the tire stiff) or if the load is increased (the extra weight pushes down harder on the tire).

The amount of wear that will result from any given amount of deflection depends largely upon how abrasive or sharp the road or ground surface is. Surfaces such as new brush-finish concrete, sharp gravel, and crushed or blasted rock take tread off rapidly, while old concrete, blacktop, fine sand, and clay are less wearing.

Wet rubber cuts more easily than dry rubber, so damage from cuts and scratches is likely to be increased by rain. But sometimes water acts as a protective lubricant and reduces wear.

Spinning the wheels will take tread off, sometimes as fast as a buffing machine would. When the tread shows long scratches running part or all the way around the tire, the wheel has been spun. Using brakes increases tread wear, and the harder they are applied, the faster the tread comes off. If the wheel locks and slides a flat spot may be made on the tread, that will tend to get worse with ordinary use. A high spot on a brake will concentrate wear on one spot also.

Alignment. If an axle sags tires will wear on the inner edge. If a wheel leans outward the wear will be on the outer edge of the tread. Excessive toe-in or toe-out will produce feathered edge wear on one side.

Section of overinflated tire showing how center of tread is subjected to excessive wear.

Underinflated tires flex at every turn of wheel, resulting in high internal heat, fabric breaks.

Correctly inflated tires permit all of tread to contact road, are not soft enough to flex excessively.

Fig. 12-116. Effects of inflation on tire shape

Alignment is usually more troublesome in highway vehicles than in off the road equipment.

Cupping. Tires often wear irregularly in a pattern of alternating hollows and ridges, an effect that is called cupping. This may be caused by wheel alignment, but in earthmover tires it is usually caused by driving on paved roads.

Widely separated diagonal cleats are particularly subject to this difficulty, but any coarse tread not including at least two around-the-tire ribs is likely to cup.

Cupping occurs even at very low speeds, but it becomes more severe as speed increases. The tread may become so rough that the tire must be buffed down, retreaded, or scrapped.

Inflation. If a tire has too much pressure for the weight it is carrying it will round out so that only the center of the tread will be on the road. If it has too little pressure the center will fold up and in and it will ride on the edges. And whatever it rides on it wears. If there is center wear, the tire has traveled a long way with too much air; if it is worn on both edges it has probably been too soft or overloaded too long.

Air pressure in a tire should be sufficient to carry the load without either of these effects. The load carried determines the correct pressure for each type of service. The table in Fig. 12-114B shows that with each increase in pressure the allowable load is increased, until the maximum pressure for the tire is reached. If the load exceeds that amount, a tire with more plies or a bigger size must be used. This table should be read both ways, as it also means that with decrease in load, tire pressure should be decreased.

Dump truck and scraper tires are inflated according to the weight they carry when the vehicle is loaded.

When pressure is higher than that permitted by tire design, it tends to put too much stress on the cords. This tension makes the tire easier to break and puncture. For an example of this effect, push a pencil point lightly against a toy balloon that is only partly inflated, and then against the same balloon when it is blown up tight.

Tire pressure must be checked when the tire is cold, before it has been run. This is because the heat generated in it by use increases the air pressure. If it has the right air pressure when it is hot, it will be too soft when it cools off. A tire is built to stand the extra pressure that is caused by heat as long as it is not carrying more than its proper load.

Heat. An underinflated or overloaded tire flexes too much, causing internal shear and eventual separation in the fabric. This mechanical effect is damaging enough, but the heat that it creates is far more destructive.

Bending of any material creates heat. For example, bend a wire coat hanger rapidly back and forth a few times, then cautiously touch the spot where it bends. It may be hot enough to burn your finger.

Rubber is a poor conductor, so the heat created by flexing builds up in the tire. If it passes 250° F. the rubber will lose strength and some fibers will be weakened. At 280°, the temperature at which many tires are cured and retreaded, plies may separate, and the tire may blow out.

Overloads. There are two types of overloading. One is carrying a load that is within the rated tire capacity, but that is too much for the pressure in the tire. This is really underinflation, and has the same effects.

The other type of overload is placing more weight on the tire than it can carry at its maximum recommended pressure. Such overloading usually results from heaping loads and heavy materials. It produces the results of underinflation in a much more severe form.

Speed. Heat is generated in each part of the tire as it flexes in passing under the wheel, and it is radiated constantly from the tire surfaces and through the wheel. Speeding up will increase the generation of heat but will not affect the rate at which it is radiated away.

A tire with a 12 foot circumference will turn about 37 times a minute at five miles per hour, and 220 times at 30 miles per hour. Six times as much heat will be produced at the higher speed.

This tire might run moderately underinflated or with an overload at five miles per hour with little damage, as heat could be radiated away as fast as it was produced. But at thirty miles per hour, a high speed for earthmover tires, extreme overheating and a blowout would be likely.

Also, injuries from impact are more likely and more severe as speed increases. The force of the blow increases greatly with speed.

Retreading. Retreading is used here as a general term covering both retreading and recapping. Strictly speaking, a tire is retreaded by buffing the worn tread down close to the fabric, and vulcanizing on a new tread with part sidewalls that extend halfway to the bead. Recapping, a slightly cheaper and sometimes less satisfactory operation, replaces the tread only. However, most construction men use the terms interchangeably.

Retreading costs about half as much as a new tire, and is subject to the same fleet discounts. There are few places that can retread the very large tires, so they may involve extra cost for freight, and considerable lost time.

Some contractors and many mines find that they can retread most of their tires again and again, perhaps as many as four times, and that this is the most efficient and economical way to buy rubber. Others report that they have had no success whatever with retreading, and have adopted a policy of using only new tires. The in-between majority do some retreading or recapping and a lot of replacing.

If tread wears rapidly because of abrasive soil, steep grades, and/or sharp turns, and there is little rock, overloading, or abuse, it is likely that repeated retreading will be economical. But on rock jobs, particularly with overloads and/or high speed, the chances are that by the time a tire has worn off its tread it will have suffered such carcass damage that it should be scrapped. Overloads, heating, or recent bruising may weaken the body without visible damage, so that the tire might be thought suitable for retreading, but would fail soon after.

The advisability of retreading is something that each contractor must work out for himself, and he may find that the answer is not just the same on any two jobs or any two types or sizes of equipment.

Retread life may be estimated at anywhere from 50 to 100 per cent of new tire life. At 50 per cent it theoretically represents no saving, but may be used because of a shortage of cash, or to keep down investment in a unit that may be sold or scrapped.

A tire that is to be run to destruction can be used about 20 per cent longer than one that is to be retreaded, as it is not necessary to protect the fabric from wear spots. This should be considered in figuring costs.

TIRE LIFE

Differences of Opinion. There are sharp differences in the published literature about tire life. Manufacturers of earth moving equipment assume it at 5,000 hours. Most estimating books accept these figures. On the other hand, tire manufacturers state that such tire performance is most unusual, and that 1,000 to 3,000 hours is a more reasonable estimate.

The attitude of each side is understandable. Equipment men like to emphasize low cost of operation, and are therefore naturally optimistic about tire performance.

Tire men are also proud of the performance of their product. But tires are subject to severe conditions and abuse that are beyond their control, and many tires break up long before the end of their natural life. A tire manufacturer who brags about a 5,000 hour life is likely to be asked for a new set of tires, free, if they wear out in 3,000 miles. On the other hand, if he can show that life should have been only 2,000 miles under the conditions of use, he can probably sell the customer another set.

It is likely that the tire industry is much closer to an accurate estimate of tire performance than the equipment people are. However, there is enough evidence on both sides to justify their claims.

Wear Factors. Several tire companies have published tables of factors by which probable tire life can be calculated. Figure 12-117 is based in part on such tables, and in part on additional information obtained from contractors and mines.

Maximum normal life of off the road tires under favorable conditions is assumed to be 6,000 hours or 60,000 miles, with the hourly basis being more convenient for us to use.

Each of the groups of operating conditions has a series of decimal fractions indicating the part of full tire life that can be expected under each condition.

For example, in Group F, trailing tires that neither propel nor steer the vehicle, and therefore have a minimum of wear, are assigned a factor of 1.0, indicating that in this position full life may be expected.

In the same group, front wheels that have tread rubber scuffed off as they resist sideward push as the machine is steered are given a factor of .9, meaning that life should be 9/10 of that of a trailing tire under the same conditions. Scraper driving wheels, that operate under severe conditions including spinning, are given a factor of .6

Neutral conditions are given a rating of 1.0, and disappear in the arithmetic. Factors larger than one show better than ordinary conditions and increased life.

To find the effect of wheel position on tire service we multiply the proper factor by the maximum tire life of 6,000 hours. We find that if all other conditions are entirely favorable a tire is likely to run 6,000 hours on a trailing wheel, 5,400 on a front wheel, and 3,600 on a scraper drive wheel.

It is of course unlikely that all other conditions are excellent. A factor is selected from each group, representing as closely as possible the conditions on the job. These factors are multiplied by each other, and their product is multiplied by 6,000 to give the average life in hours of tires on the job being studied.

These figures assume that a tire's life is finished when the tread is worn off. If the casing is good it can be retreaded, but it then starts a new life on a different cost basis. The saving is taken into account as a reduction in replacement cost.

FACTORS IN EARTHMOVER TIRE LIFE
To be multiplied by 6,000 hours or 60,000 miles

Group A – MAINTENANCE
INCLUDES INFLATION

Excellent	1.1
Average	1.0
Poor	.7
Very bad	.4

Group B – MAXIMUM SPEEDS

10 miles per hour	1.2
20 miles per hour	1.0
30 miles per hour	.8
40 miles per hour	.5

Group C – CURVES

None	1.1
Moderate	1.0
Severe, single wheels	.8
Severe, dual wheels	.7
Severe, tandem wheels	.6

Group D – SURFACE

Snow, packed, no road exposed	3.0
Earth	
Hard packed earth	1.0
Soft earth or sand, maintained	1.0
Gravel road, well maintained	.9
Soft earth, some rock	.8
Mud, ordinary	.8
Gravel road, poorly maintained	.7
Mud, abrasive or with rock	.5
Blasted rock	
Soft coal	.9
Soft shale or limestone	.7
Granite, gneiss, trap, basalt, hard shale or limestone	.6
Slate or schist	.4
Lava, hard surface	.3
Obsidian, volcanic glass, flint	.1
Blacktop	
Clean, wet	1.4
Cold weather	1.2
Hot weather, 75 to 100° F.	.8
Very hot, over 100° F.	.5

Group E – LOADS

Recommended by Tire and Rim Assn.	1.0
50% underload	1.2
20% underload	1.1
10% overload	1.0
20% overload	.8
40% overload	.5

Group F – WHEEL POSITION

Trailing	1.0
Front (non-driving)	.9
Driving	
Rear dump	.8
Rear dump tandem	.7
Bottom dump	.7
Scraper, self propelled	.6

Group G – GRADES, DRIVE TIRES ONLY

Level	1.0
Firm surface	
6% maximum	.9
10% maximum	.8
15% maximum	.7
25% maximum	.4
Loose or slippery surface	
6% maximum	.6
10% maximum	.6
15% maximum	.4

Group H – MISCELLANEOUS CONDITIONS AND COMBINATIONS

Favorable or counteracting	1.5
None	1.0
Unfavorable	.8
Very unfavorable	.6

Fig. 12-117. Factor table, earthmover tire life

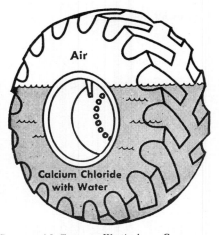

Fig. 12-118. Hydroflated tire

Miscellaneous Conditions and Combinations might include extreme heat as a factor by itself, or in combination with speed and/or overloads, some rock on road in combination with speed and overloads, cowboy behavior of operators, frequent ripping of scraper tires by dozer blades, or favorable factors such as low friction road surface of borax or serpentine.

If tread wear is extremely rapid, the tire life may depend entirely on its rate. On the other hand, fast driving on coarse rock with overloads might land every tire on the scrap pile with thick tread still on it. A ruinous factor like blasted volcanic glass determines tire life by itself, without multiplying by other factors.

Therefore, while this table of wear factors is a helpful guide, it cannot stand by itself. It should be used by men who are familiar with tires and with the conditions under which they operate.

TIRE BALLAST

Liquid Ballast (Hydroflation). Drive tires may be 75 to 100 per cent filled with fluid (hydroflated) to obtain additional weight, in order to increase traction and drawbar pull.

The most effective fluid is a solution of 5 pounds of commercial grade calcium chloride ($CaCl_2$) in each gallon of water. This solution weighs 10.6 pounds to the gallon. Freezing point is 52 degrees Fahrenheit below zero. Added weight for 75 per cent fill ranges from 395 pounds in a 15½-38 farm tractor tire to 5,060 in a 33.5 x 33.

Plain water may be used where freezing is not a problem, but it weighs only 8.3 pounds to the gallon. Some contractors use a 3 pound solution with a weight of 10.1 pounds to the gallon and a freezing point of three below zero.

Fluid may be put in a tire by gravity flow from an elevated tank, but is much more efficient to use a special pump. A number of models are available, for volumes from 240 to 600 gallons per hour.
per hour.

If a tire is to be hydroflated, the load is taken off it by jacking or by removing it from the machine. It is kept in a vertical position, with the air-water type valve at the top. An adapter is used to permit fluid to be pumped in and air vented at the same time.

The adapter is screwed onto the dust cover threads of the special valve. The adapter then removes the valve core housing. Tires equipped with jumbo type valve stems require a special adapter.

When fluid rises over the stem it spills through the vent, and filling is stopped. This provides the 75 per cent fill usually recommended for construction equipment tires.

Some manufacturers recommend 90 to 100 per cent filling, but this makes the tire too rigid for riding comfort.
to normal pressure. Pressure should be checked after a few hours operation. Checking is done with the valve in its highest position. Corrosion resistant gauges are needed to resist attack by the calcium chloride salt.

Heat is produced as calcium chloride dissolves in water. If water is poured on the chemical it will boil and splatter danger-

Tire Size	Gallons Water	*3½ lbs. Calcium Chloride per Gal. Water			†5 lbs. Calcium Chloride per Gal. Water		
		Gals. Water	Lbs. CaCl₂	Total Wt.	Gals. Water	Lbs. CaCl₂	Total Wt.
8.25-20	11	9	33	112	9	44	118
9.00-20	15	13	45	152	12	60	161
10.00-20	19	16	58	193	15	77	204
11.00-20	20	17	60	203	16	80	215
12.00-20	27	23	82	275	22	109	290
12.00-24	30	26	91	264	24	121	322
13.00-24	37	32	112	376	30	149	398
14.00-20	41	35	124	417	33	165	441
14.00-24	44	38	134	448	35	177	473
16.00-20	49	42	149	499	39	197	527
16.00-24	64	55	194	650	52	258	689
18.00-25	83	71	252	845	67	334	892
21.00-25	123	105	372	1250	99	496	1322
21.00-29	158	135	480	1610	127	637	1700
24.00-25	152	130	461	1549	123	612	1635
24.00-29	177	152	537	1800	143	714	1905
27.00-33	350	300	1060	3561	282	1411	3765
30.00-33	380	325	1151	3870	306	1531	4085
23.5-25	113	97	342	1147	78	456	1103
26.5-25	165	143	500	1678	133	665	1774
29.5-25	210	182	636	1817	169	846	2257
29.5-29	235	203	712	2033	189	947	2526
33.5-33	353	305	1070	3055	285	1423	3795

* 3½-lb. solution provides anti-freeze protection to —15°F, slush-free.
† 5-lb. solution provides anti-freeze protection to —53°F, slush-free.

Courtesy of Firestone Tire & Rubber Company

Fig. 12-119. Hydroflation table

ously, so the chemical should always be added to the water.

This reaction heats water so that it will dissolve more than the normal amount of 5 pounds to the gallon. However, when a stronger solution cools off in the tire it will crystallize, putting the tire out of balance and possibly damaging it. A hydrometer test should be made of a solution before it is put in tires.

Liquid tire ballast is particularly useful in tractor, tractor-loader, and grader drive wheels, as these machines usually have more power than they have traction. It is sometimes used in both drive and free rolling tires both for traction and to lower the center of gravity in rubber mounted cranes, loaders, fork-lifts, trenchers, and drills.

This type of ballast serves to improve traction and stability, reduces tire wear re-

Fig. 12-120. Traction (skid) chains

sulting from slippage, reduces the bounce of big tires, and is said to increase puncture resistance. These advantages far outweigh the moderate cost, and the messiness that may accompany tire repair.

Fig. 12-121. Ring cross links

Led Ballast. Tire ballast made of finely powdered minerals may be used to obtain greater weight and more effective reduction of bounce than is provided by calcium chloride. "Led" Ballast, Inc., now makes three grades, ranging from slightly heavier than 75 per cent loading with CaCl solution to more than double its weight.

Tires must be fitted with special large valves and stems. The mineral dust is blown from a storage hopper into the tire under 4 to 8 pounds pressure. The air space in the tire is filled 80 to 97 per cent when the ballast has settled, and 100 per cent when it is installed and when it is fluffed up by traveling. It is said to take 15 to 20 minutes to fill a large tire, and 30 to 45 minutes to empty it with the same apparatus. Final inflation is by compressed air to normal pressure and periodic checks are made in the ordinary manner.

TIRE CHAINS

Traction (Skid) Chains. Small and medium size earthmover tires, and highway tires of any size, may be equipped with automotive type tire chains, provided there is enough clearance to accommodate them. Each chain usually consists of a pair of circular chains smaller than tread diameter, resting against the sidewalls and connected across the tread by a number of short cross chains.

The side chains have locking clamps, which are opened for installation and removal. It is advisable to stretch spring tighteners to keep tension in the outer side chains.

Dual wheels may have double width cross chains that extend across both tires, or single chains that are fitted to the outer ones only.

These chains are essential equipment for most two-wheel drive vehicles or machines that work in snow or on ice. They may be valuable for occasional mud conditions, but tend to increase their severity by cutting and churning the ground. They increase danger of getting stuck in sand.

All-wheel drive machines function well in snow and mud without chains, but may need them on ice.

This type of chain has a very long life if used on dirt, ice, snow, or loose rock; or on pavement that is largely covered with snow or ice. But they wear badly on bare pavement, with the rate of wear of both chain and pavement increasing rapidly with speed. They cause no significant tire damage.

Installation. In well-equipped shops, chains may be installed by lifting the wheel

used horizontally. Cross chains are caught on tread lugs to hold them while manipulating other crosses.

The clamps can be inserted in any one of several side chain links. The chain on the inner side of the tire, which is harder to reach, is hooked up first. There should be little difficulty, as slack can be obtained by pulling on the cross chains.

The cross chains are then pulled to center them on the tire, and to space them more or less evenly. The outer clamp is then in-

Courtesy of Chain Systems

Fig. 12-122. Tire protection chains on loader

clear of the ground, placing the opened chain on top, and linking it at or near the bottom. If no adequate hoist or jack is available, the chain is laid on the ground and the machine moved until the wheel is resting on the cross chains, a foot or two from one end. The other end is lifted and dragged over the top of the tire, and brought down to be connected.

Big chains are heavy, and fenders or other structures are often in the way of lifting devices. They may be worked up by hand or with a bar, or with a chain hoist

serted in the furthest link it can reach, closed, and fastened. If it cannot be made to reach, the inner chain may be loosened. Links can be added if necessary.

The two sides should be about the same length, but a difference of a link or two may not matter. If spring tighteners are used, the inner side chain may be tighter than the outer.

Chains should be snug on the tire, but not tight. It should be possible to move the crosses slightly by hand. They should not be loose enough to allow either side chain to ride up on the tread, nor should the cross chains strike against fenders or other machine parts.

Chains may loosen substantially in the first few turns of the wheel. They should be checked, and tightened if necessary.

Protection Chains. Drive tires on loaders and scrapers working in abrasive blasted rock may have a very short life, due chiefly to grinding and gouging the tread as wheels spin. There may also be many premature casing failures because of damage from sharp rocks and momentary overloads.

Tread wear and casing damage may both be greatly reduced by covering the tire with chain mesh.

The Erlau chain consists of forged links connected by welded rings, usually in a diamond pattern. It covers the entire tread and most (or sometimes all) of the sidewalls of the tires. If kept quite snug, with three inches or less of slack at the bottom of a large tire, it will not allow the tire to turn inside it.

The mesh protects the tread from most direct abrasion, and the casing from most causes of puncture. It affords considerable protection against bruising. Tread wear under the chain is stated to be about 10 per cent every thousand hours, and the life of the chain under ordinary rugged conditions to be 5000 or 7000 hours. Chains cost about 25 per cent more than the tires they protect, but will still show substantial savings under tire-destroying conditions.

Use of these chains permits the operator to ignore possibility of tire damage, and to apply the full power of the machine to its work. On the other hand, there are ground conditions under which the chains may reduce traction seriously.

Steel Treads. Another approach to loader tire protection is found in the Caterpillar Dystred Cushion Track Loader. Specially shaped tires carry roller chains in grooves on each side of the smooth tread. Track pads, usually with two grousers, are bolted

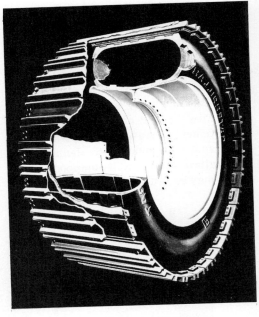

Courtesy of Caterpillar Tractor Company

Fig. 12-123. Dystred steel-treaded tire

to the chains, and extend a short distance beyond the tire on both sides. Inflating the tire, which is called a driver, makes a snug fit (adjustable by changing inflation pressure) that permits the rubber to transmit full power to the steel.

Single grouser shoes, with or without anti-side-slip side bars, are also available.

This construction provides complete protection for tire treads, but not for sidewalls

OPERATOR RESPONSIBILITY

The operator of excavation equipment must be a responsible person, for at least two reasons.

This machinery is heavy and powerful. It may be capable of enormous destruction to property, and injury and death to people (including the operator) if it is operated improperly or carelessly, or if it goes out of control because of mechanical failure.

This equipment is both very expensive and extremely vulnerable to damage aris ing from improper maintenance. In order to do its work it must be lubricated according to schedules that vary widely from machine to machine. There are many parts that deteriorate rapidly if used when worn or out of adjustment, and are likely to cause rapidly increasing damage to other parts if their repair is postponed. Mechanical neglect greatly increases danger of accidents.

The operator should check his machine before, during, and after work, and correct or report improper conditions. Such checking always includes level of lubricant, hydraulic oil, and coolant, and indications of unusual wear.

PRE-START CHECK

On arrival on the job, the operator should walk around his machine, looking at it carefully. The parts deserving particular attention vary with the type of machine, the work it is doing, and often with the service history of this particular unit.

Tires. If it has tires, proper pressure is of the first importance. It should take only a couple of minutes to check the whole set with a gauge, but an experienced operator may be able to check them adequately by eye.

Tires should be inspected for cuts, tears,

and worn tread. A pattern of scratches and grooves on drive tires indicates wasteful spinning. If the tread is smooth, traction will be poor except in sand, and the tire will soon be worn past the possibility of retreading.

Tracks. Track tension can be checked only if the machine is reasonably clean, and was moving forward when stopped for parking. Then there should be a slight sag on each side of the carrier (upper) roller. Too tight a track will wear rapidly, while if too loose it will also wear rapidly and may come off.

Bright patches of scraped metal usually indicate that something is loose or out of line, although work in loose gravel may produce a similiar effect. Prompt inspection by a mechanic is called for.

Low cleats and bent shoes on a tractor mean poor traction, and reduced safety on side slopes, but this is a long term situation.

A flat spot or spots on the carrier roller shows that it does not turn freely. If it is jammed by surface debris, free it. Otherwise, it is damaged inside, and must be repaired or replaced promptly.

Working Tool. The bucket, blade, edge, tooth or other part that is forced through dirt or rock during excavation is subject to both wear and breakage.

Fig. 12-124. This track should be reported

Wear is usually gradual, but may be quite rapid in abrasive material such as sharp gravel or sandstone. It usually makes the cutting edge thicker and duller, so more power is needed to do less work. It may change a straight edge into scallops.

Its most serious effect is working through a replaceable edge to the structure behind it, which may be difficult to repair. A worn backing plate cannot properly support a new cutting edge, unless built back to shape by tedious welding.

The operator should be alert to report excessive wear.

Breakage of cutting parts may occur without warning, but it is often preceded by cracking. Detection of cracks usually makes it possible to patch-weld and reinforce the unit, to postpone or avoid expensive repair. A break usually tears and bends the metal so as to make rebuilding difficult.

Loader and hoe buckets are particularly vulnerable to breakage following cracking.

Linkage. Most excavating tools are hinged to their power units. Linkage may be very simple, as in a direct lift dozer blade or ripper, or quite complex, as in a front loader.

In any case, each hinge is a point where lubrication is probably required, and where a pin might loosen and come out. Loosening of a pin is rare, but can be very damaging, so it pays to look at each one to see that it is properly in place.

Looseness due to wear of the pin and bushing is most readily detected during operation, but it can usually be studied in the parked machine by prying with a crowbar.

Steering linkage acting on the wheels is subject to damage from collision with rocks or other objects. Steering action in an articulated machine might be completely blocked by a careless mechanic leaving a safety brace attached.

Leaks. Surfaces of the machine, and the

Fig. 12-125. Leakage

ground under it, should be inspected for evidence of leakage of lubricating or hydraulic oil, or coolant.

Very slight oil or antifreeze leakage traps dust, making a dirty or greasy spot or smear around the leak. If more severe, a channel will be cut through the dirt, and an oily or moist spot will form on the ground under it.

The seriousness of leakage is variable. Any opening that will let oil get out may permit dirt to get in. In spite of the efficiency of modern filters, this is a danger. It may show that a seal is beginning to fail, or worse, that a case has cracked. Or it may be normal outflow from a pressure-greased fitting. Each leak must be judged individually.

In the hydraulic system, leaks are usually at joints. If the joint is tight, the leak probably indicates deterioration of the hose inside the joint clamp, and may (or may not) be the warning of a coming blowout at that point, or elsewhere in that hose.

If the hose itself starts to exude oil, it should be replaced immediately.

Radiator. A number of fluid levels should be checked at the beginning of each shift.

The radiator should be filled to within a few inches of the filler cap. You check it by unscrewing the cap (always with a cold or at least not-hot engine) and looking. The cooling system is sealed, so any loss of fluid is cause for alarm.

Leakage may occur externally through hoses or their connections, a worn water pump, or a punctured radiator. Such leaks should be visible.

More serious loss is through the cylinder head gasket into the cylinders. Then the coolant might work its way down into the crankcase oil, or be blown out the exhaust.

Worse yet, leakage of hot gas from the cylinder may heat the coolant to boiling.

The fluid is usually a mixture of water and ethylene glycol antifreeze, possibly with other additives. In summer it can probably be brought up to level with plain water, but in freezing weather, losses must be replaced with antifreeze mixture.

The air passages through the radiator may become plugged by trash, such as leaves, straw, and seeds. This is particularly likely during mowing or clearing operations. Interference with cooling increases rapidly with the area affected.

Removal may be easy, if the radiator is exposed, or the guard is a swing-away type. Then a brush will take off surface layers. If there is no access, or the trash is deeply embedded, use an air jet from the rear to

Fig. 12-126. Remember the dip sticks

blow them out. Washing out with a high pressure water spray may do it, if air is not available.

Reservoirs. Almost any piece of equipment has one or more fluid reservoirs whose level is checked by a dip stick (or more rarely, by a sight gauge), usually while the engine is shut off.

There is always a dip stick for the engine. The operator must be sure to use it, and to add oil if necessary, before starting. In addition, there may be dip sticks for a torque converter, the transmission, and other drive units. It is necessary to know where each of these is, and to remember to check it. Availability of the proper fluid may be a problem.

If the level is unusually low in any of these, it should be rechecked during the shift. Many of the sticks have an additional marking for proper level with the engine running.

Some reservoirs or units are checked with a dip stick only when the engine is running. If so, the dash should carry a plate or decal with instructions.

The tank for the hydraulic system may have a sight gauge. If not, a stick, or a look through the filler cap opening, is necessary.

Standard transmissions, differentials and final drives often are checked by removing a screw filler plug. Lubricant should be up to its level, or slightly above. A daily check is not customary for these, unless there are signs of leakage.

If there is an air pressure tank, open the drain cock (or drain cocks) in the bottom to let out any fluid, and re-tighten.

Check the fuel level, unless there is a gauge on the dash. If drain cocks are provided for removal of water, use them.

Air Cleaner. The air precleaner usually has an easily removable container (dust cup) for coarse trash. This may have a sight gauge on the dash, with a red strip to warn that it is full. This may not register unless it is tapped.

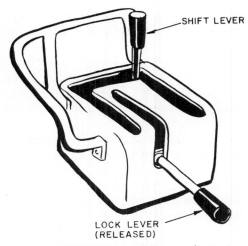

SHIFT LEVER

LOCK LEVER
(RELEASED)

Fig. 12-127. This machine won't start

If the indicator shows that trash is present, open and dump the container. If there is no indicator, open it to check.

STARTING THE ENGINE

Interlocks. Most new machines, and many old ones, have one or more safety devices, called interlocks, that prevent the starter from operating if controls are not in proper position.

To avoid delay and embarrassment (not to mention possible accident), it is wise to follow all instructions in the book and/or on the dash, before and during starting. The following list may not be complete.

If there is a seat belt, fasten it.

If the brakes can be locked down, press and lock them.

Put all gear shift levers in neutral. With power shift, there is usually an additional shift lock lever that must be engaged.

Put all equipment control levers in neutral, HOLD, or FLOAT.

Connect the battery to the electrical system. The disconnect, if there is one, may be a lever or handle in the battery case, or on or under the dash, or a key operated switch on the dash..

The starter switch itself may be a push button or pedal in old equipment. Now it is usually a turn switch, with or without a key. It may have three positions — off, run, and start.

Diesel. Most excavation equipment is diesel-powered.

The diesel is stopped by simply shutting off the fuel. It is started by opening the fuel passage and cranking with the starter. In cool or cold weather, starting aids are needed.

The fuel shutoff may be a separate lever or a dashboard knob that is pushed in to run, and pulled out to stop. Or it may be a position on the hand throttle or foot accelerator. When either of these speed controls is moved from a speed position to low idle, an additional pull past a stop puts it in shutoff position, where it will be held by a detent.

To start in warm weather above 60°F (16°C), put the separate lever in ON position, or move the throttle out of the detent into idle position, and turn the switch to START. Release switch when the engine starts.

A diesel in good condition should start within five or ten seconds in warm weather. If it doesn't start within fifteen, throttle position and fuel supply should be checked.

A starter should not be used continuously for more than thirty seconds, after which it needs about two minutes to cool off.

Cold Weather. Many diesels will start on a warm weather basis down to about 40° F (5° C), but this varies with make, model, and individual engines, and cannot be depended upon.

There are three principal methods for helping a cold diesel to start — ether, preheater, and glow plug.

Ether is sprayed sparingly from an aerosol can into the intake of the air cleaner while the engine is being cranked. Do not use it either before cranking or after starting. This volatile, highly flammable fluid ignites readily in the cylinders. Some machines have built-in ether dispensers, which may involve supplying it in can or capsules. The aerosol is generally more convenient and dependable.

Ether is effective all the way from moderate temperatures down to extreme cold.

Ether is very poisonous, flammable, and explosive. It must not be used in an unventilated room, or near heat, flame, or sparks. It is likely to explode in the manifold if used for starting at the same time as a preheater.

A preheater is often supplied for cold weather diesel starting. This may resemble a miniature furnace-type oil burner inside the intake passage. There is a spark plug, a nozzle, and a dashboard hand pump. The spark is switched on, fuel is forced out the nozzle and is ignited and the engine is turned with the starter. The burner heats the air going into the cylinders. It consumes some oxygen, but usually not enough to interfere with combustion in the cylinders. The principal drawback is that mechanical difficulties, particularly shorting out the plug with fuel oil, and failure of the pump to operate, may prevent it from working when most needed.

Some preheater effect can be obtained by removing the pre-cleaner and directing the flame of a blowtorch down the intake pipe while cranking. The long distance to the cylinders causes the loss of a large part of this heat, but effects are usually good. (Note—don't ever do this with a gasoline engine, as it not only is dangerous, but it wets the spark plugs with condensation.)

A glow plug is a small electric heater in a precombustion chamber or intake passage. At moderate temperatures it is switched on for a minute before using the starter, when below freezing for two minutes. It may be used alone, or in addition to ether injection.

Pony Engine. Many old diesels are started by a pony (auxiliary) gasoline engine. It is set beside the diesel, and turns it through a clutch and gears. The clutch is

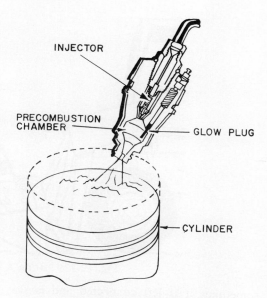

Fig. 12-128. Glow plug

disengaged, the gas engine started by hand cranking or preferably with an electric starter, and run until it is warm. Cooling systems may be connected so that it warms the diesel at the same time.

The drive gears are then meshed and the clutch engaged, so that the diesel is turned over by the gas engine. It is customary to have a "start" position on the throttle, that will admit air but will not supply fuel. This permits smooth warming up without danger of running fast enough to damage the kicker engine. Finally the throttle is moved to operating position, and the clutch and/or gears disengaged.

This reduces the problem of starting the diesel to one of starting the kicker engine, but unfortunately this is often difficult. One precaution is to stop it by shutting off the fuel, rather than the ignition, so that gasoline will not be left in the carburetor and lines where it will evaporate and leave residues of lead, dyes, and other foreign materials. White gas, if obtainable. is to be greatly preferred even with this precaution.

Controls for use of the auxiliary engine may be complicated and confusing, so that

Fig. 12-129. Bring a booster battery

sometimes it is left connected and is severely damaged by over-speeding when the diesel is revved up. Newer models have an automatic kick-out that prevents this from happening.

Gasoline. A gasoline engine is stopped by turning off the ignition, usually with a key. If it is warm or hot, it is started by turning on the ignition and using the starter, which is often on the same switch.

An exception to this statement is that a very hot engine may boil the gasoline in the line to the carburetor, creating a won't-start condition called vapor lock. Sometimes you can cool the line by wrapping it in a damp cloth, but usually you just wait.

If the engine is cool or cold it will need to be choked. A choke is a valve that cuts off almost all the air flowing into the carburetor. It is usually operated by a dashboard knob that you pull for choking (closing the choke). Cranking the engine then creates a vacuum in the carburetor, pulling extra gasoline through the jets. Partial choking creates a very rich fuel-air mixture, full choke (choke closed) draws liquid gasoline also.

Engines are quite individual in their response to choking, but in general, the colder it is, the more they need.

When the engine starts, the choke should be partly opened immediately, but partial choking may be necessary for several minutes. Not enough choke causes stalling and/ or misfiring or hesitating during acceleration. Too much choke causes rolling, or pulsation in speed, and may stall the engine if not corrected. Over-choking also may dilute lubricating oil and foul spark plugs.

If the engine does not start with the starter turning briskly and the choke full out, it may be over-choked. Many units can go directly from a not-enough to a too-much choke state without catching, and sometimes without even kicking. You test for this by opening (pushing) the choke and continuing to crank. A wide open throttle is helpful in getting rid of extra gas, but starting is better when it is partly closed.

If the engine has an automatic choke you simply open the throttle half way, use the starter, and hope for the best. You can usually check the action of the automatic by taking off the air cleaner and looking into the carburetor intake. Full choking can be obtained by entirely blocking the opening with your hand.

Extra gasoline may be supplied by pumping the accelerator.

Poor starting usually means that an ignition tuneup is needed.

Oil Drag. There are four factors that make cold weather starting difficult—the extra heat required to raise fuel-air mixtures to the ignition point, the slower vaporization of fuel (particularly important with gasoline), the drag of thick cold oil on all parts, and the lowered efficiency of cold batteries.

The oil drag can be very serious. In general, thinner oils should be used in cold than in warm weather, both to reduce drag and to supply better lubrication. If conditions are severe, or particular machines are hard-starting, it is a good plan to put a small quantity of gasoline, one or two cups, in the crankcase oil filter tube just before shutting down. The engine should be turned over afterward just enough to mix the oil and gas in all its parts.

This will thin the oil so that drag will be greatly reduced the next morning. As soon as the engine warms up, the gasoline will evaporate rapidly, returning the oil to its proper viscosity. The gas vapor will escape through the crankcase breather pipe, and if it has a filter element in it, danger of fire from this source is negligible.

Dry Fuel. A basic all-year precaution for good starting is to keep fuel oil and gasoline dry. Moisture gets in them from condensation in the tank above the fuel level, when the machine cools at night, and it may be present in the fuel supply.

Condensation can be prevented by filling the tank at the end of the work shift.

Water in the fuel can be made harmless by adding a small quantity of methanol (wood alcohol) daily. This prevents freezing, and enables water to mix with fuel so it seeps through filters harmlessly.

Booster Cables. If the battery is too weak for starting, you may be able to use another battery, or set of batteries, of the same rated voltage, either on the ground or in another machine.

Fig. 12-130. Grease pump

Heavy insulated copper cables with spring clamps on each end are used. One is usually marked in red, the other in green. Hook up negative to negative, and positive to positive, putting the clamps on the battery posts or cable clamps. It may be necessary to twist or wiggle them to get a good connection.

If the auxiliary battery has a good charge, the starter should work normally. Remove the cables when the engine has started.

Do not hold a clamp in position while the starter is being used. A cable with a defect inside the handle might generate enough heat from the current to melt the insulation, and inflict a severe burn.

Batteries generate explosive hydrogen gas, which might be ignited by a spark. The filler caps of each battery should therefore be in place, and covered snugly with a dry cloth, before putting on the clamps.

This hazard could be reduced by putting flame stops of fine mesh wire screen in the filler cap vents at the factory.

Push-Start. Many engines start best if turned by rolling, pushing, or pulling the machine rather than by using a starter. It is a very good plan to leave equipment in such position that it can be readily towed if necessary, as this may save hours of monkeying around with adjustments that might better be left for a slack time. Coasting down hill is a good means of starting, and a crawler machine may be backed up a steep pile, and started next day on a run of a few feet.

Unfortunately, this can seldom be done with newer machines, as torque converters have too much slippage unless equipped with lockup clutches, and power-shifts are likely to be damaged.

Warming Up. After starting, a cold engine should be run at one-third to one-half speed for three to five minutes, with no load or light load. The primary purpose is to get the lubricating oil warmed and thinned, and distributed to all parts needing its protection. If there is a turbocharger (see discussion under STOPPING), it is particularly important that it turn slowly until warmed and lubricated.

The dash gauges should be watched. The temperature indicator should gradually climb into proper operating range. Oil pressure should be correct or a little high within a few seconds. The ammeter should show a high charging rate for a few minutes.

Equipment having air brakes should not be moved, nor should the brakes be unlocked, until pressure reaches normal range, usually 100 pounds.

Running the engine will turn the hydraulic pump or pumps, and circulate fluid through the valve and reservoir. After a couple of minutes, all valves should be manipulated to move hydraulic-powered attachments through their full range of action, without load. This serves to warm the fluid in their lines, cylinders, and motors, and is an important check on their proper operation.

The steering should be checked twice through its full range, for the same reason.

None of these warm-up precautions are needed in re-starting after stopping for

Fig. 12-131. These tracks won't freeze down

servicing, lunch, or job delays, provided that operating parts are still hot to the touch. If they are merely warm, a minute or two of running may be advisable.

Toward the end of the warming up period, make any dip stick checks which are supposed to be done with the engine idling. These may include engine, torque converter and transmission levels.

Many machines have points that require once-a-shift greasing, and these are likely to be the operator's responsibility. These can be attended to at this time also. Usually, you pump in enough fresh grease with a hand gun or bucket pump to make some old grease come out. If there is resistance, check to make sure grease is supposed to come out, before forcing it.

STOPPING

Idling. It is a general rule that an engine that has been working under load should run half speed with no load for five minutes, then idle a half minute before being shut off. This permits gradual readjustment of temperature between hot spots and cooler sections. It is particularly important for a turbocharger.

In short stops, as for lunch, some manufacturers recommend allowing the engine to idle. Others never want it idled for more than a minute or two. This is be-

Fig. 12-132. Use ladders and handholds

cause some engines are adapted to idling, others are not. Consult the instruction book or the foreman.

The transmission should be put in neutral whenever the engine is run without the machine working.

Turbocharger. A turbocharger is a double blower on a single shaft. One side is rotated by exhaust gas, the other crams fresh air through the intake passages into the cylinders. With more air, the cylinder can burn more fuel, and power is increased without adding to engine size.

The blowers may turn at 100,000 or more revolutions a minute, while the diesel is revved up to 2,400 or less. The exhaust side is very hot, the other side is cool. It is lubricated by pressure from the engine oil pump.

When the engine stops, the flow of oil to the turbocharger bearings stops immediately. If it continues spinning at high speed for only a few seconds, the bearings are likely to burn out. The hot side may tend to cook the cool side.

When the engine runs at part speed without load, the turbocharger turns at moderate speed without picking up much heat. Idling slows it down still more, and prepares it for a no-damage stop.

Just one quick engine shutoff from full speed at full load may put the turbo-

charger out of action. And it is very expensive to repair or replace.

The charger may also be damaged by a quick speedup of a cold engine, before lubricating oil reaches it. It may spin in the wind, so the intake pipe must be plugged or covered if the machine is transported, to prevent this unlubricated rotation.

Parking. The machine should be parked on reasonably level ground. Mud spots, or areas subject to flooding, should be avoided.

In freezing weather, tracks must be cleaned, and kept from extensive ground contact by being driven onto planks, logs, or stones.

Most parts supported by hydraulics or by cables, such as buckets, blades and bowls, should be lowered to the ground, and controls left in the FLOAT or HOLD position. Shovel and crane booms, however, are usually left up.

The transmission (or transmissions) should be placed in neutral, and locked if power shift.

Shut off engine by turning the key, pulling the shutoff knob, and/or moving the throttle below its idle position.

Disconnect electrical system, if equipment permits.

Turn seat upside down, or cover it, to keep it dry, if it is exposed to weather.

If there is a cab, put up windows, close doors, and lock it. If the machine may stand in hot sun, leave windows open a crack at the top.

SAFETY

To operate safely, a man should be alert, observant, reasonably cautious, and willing to invest a little extra time and trouble in doing things right.

One elementary precaution is to use handholds, steps, ladders, or whatever helps are provided for getting in and out of the seat, and reaching service and inspection areas. An agile man can skip many of these, but he should not. Surfaces are often slippery with water, ice, or oil; and long steps and reaches make sliding more likely.

Warning devices should be kept in working order. These include intermittent blowing of a horn whenever reverse gear is engaged, flashing "lollipop" lights when working along roads, and bright head and rear lights for night operation.

Restricted vision is a problem that increases with machine size. Before resuming work after parking, an operator must walk around his machine to make sure no one is eating or dozing in its shade. When moving, he must keep aware of blind spots.

Rollover protection structures are important for safety, and cabs may add greatly to comfort. But both of them restrict vision even under good circumstances, and call for increased alertness. A cab with misted windows is very dangerous, and they must be wiped off as many times as needed, before and during operation. All glass must be kept clean.

Attachments should be on the ground when you leave the machine. An opened connection or an accidentally tripped valve may otherwise lead to disaster. However, it is customary to leave crane booms up when they are cable-supported.

After shutting off the engine, move all operating levers back and forth, to make sure you haven't left anything up that ought to be down.

These are only samples of various safety factors that must be kept in mind by the careful operator.

CHAPTER THIRTEEN

REVOLVING SHOVELS

BACKGROUND

Revolving shovels were the first important power excavators. Part-swing, steam-powered dipper models mounted on railroad cars or barges were in use over 130 years ago.

Other models—clamshell, dragline, and hoe, were gradually developed. In the first 20 years of this century, steam power was largely replaced by internal combustion engines and electric motors.

Also, full-swing replaced part-swing, and crawler self-propelled mountings became standard. More recently, rubber-tired carriers have become common.

Until around 1950 practically all these machines manipulated their buckets by means of cables (wire ropes), winding onto mechanically or electrically driven drums. Swing and travel depended on gear arrangements. Controls were mechanical, air, hand-or-foot-pressure hydraulic, or combinations of these systems.

Now, in 1975, production of cable and mechanical drives in backhoes has been largely replaced by hydraulics. Hydraulic-operated cranes are standard among smaller machines, and the hydraulic clamshell has gained acceptance. Also, the front end tractor loader (hydraulic) is competing successfully with the dipper shovel in many kinds of work.

As a result, the importance of the cable-operated excavators has declined greatly, particularly among small machines. However, they are dominant as draglines, clamshells, and big cranes and dipper models, and are still very widely owned and used in all sizes and types.

Therefore, the detailed discussion of their characteristics prepared for previous editions has been kept almost intact here.

Description of the hydraulic machines follows the cable units.

BASIC SHOVEL, CABLE-OPERATED

Structure. A shovel has three structural divisions. Anatomically, the top or revolving unit is the head and torso, the mounting or travel unit is the legs, and the various attachments are the arms and hands. A revolving and a travel unit together make up a "basic shovel."

There are five attachments (also called rigs or fronts) which have primary importance. These are known best under the names dipper shovel (or dipper stick), backhoe, dragline, clamshell, and crane. They are shown in Figure 13-1.

A basic shovel has three sets of machinery. One of them is made up of deck-mounted drums which are fitted with spools for cable or with sprockets for chain, and which are rotated and stopped by clutches and brakes. These control the in-and-out and up-and-down movement of the bucket for digging and dumping.

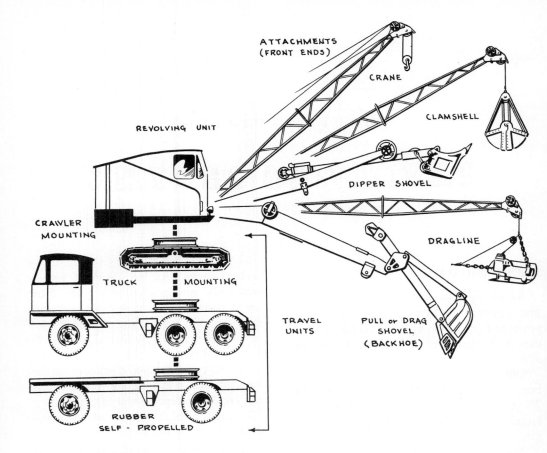

ATTACHMENTS
(FRONT ENDS)

CRANE

CLAMSHELL

REVOLVING UNIT

DIPPER SHOVEL

CRAWLER
MOUNTING

DRAGLINE

TRUCK MOUNTING

TRAVEL
UNITS

PULL or DRAG
SHOVEL
(BACKHOE)

RUBBER
SELF - PROPELLED

Courtesy of The Thew Shovel Company

Fig. 13-1. Shovel rigs and mountings

The second set rotates (swings) the deck, upper machinery, and front end around a hollow center pin. The upper unit is supported by rollers or balls on a circular track or turntable, and is rotated by a vertical pinion gear that meshes with bullgear teeth in the turntable. It "walks" around the bullgear, swinging the shovel as it does so. The gear is controlled by a reversing clutch (a unit consisting of two friction clutches and a set of bevel gears), which can turn it in either direction.

This mechanism is an important factor in the efficiency and adaptability of the shovel. It enables it to face in any direction for digging and dumping, and to move loads quickly anywhere within its reach. Comparatively few wearing parts are in-

volved, none of them work in the dirt, and friction losses are slight.

The third power train provides means to walk or propel the shovel. A vertical shaft extends downward through the deck and the hollow center pin to drive a horizontal axle, which has a clutch and a brake on each side. These may be of either jaw (tooth) or friction construction. A pair of sprockets on the ends of the outer axles drive the tracks through roller chains.

The propel mechanism is controlled by a reversing friction clutch set, usually the same one that controls the swing. A few machines have an independent travel shaft and clutches. Truck mountings depend entirely on the truck for travel.

When one set of friction clutches controls two or more functions, jaw clutches

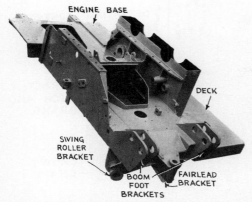

Courtesy of Bucyrus-Erie Company

Fig. 13-2. Deck frame, upper side

are shifted to connect the power train that is to be used, and to disconnect the other.

Small shovels use gasoline, diesel or electric engines, with mechanical drive to all moving parts. Large shovels may use diesels, with mechanical or electric drive; or several electric motors supplied from "high lines" through a cable.

The engine is fitted with a disc clutch, so that it can be cut off from the machinery. It may also have a torque converter, that serves to cushion it against shocks, to reduce lugging down, and to give additional power in hard digging.

There may be a two-speed engine transmission, chiefly to allow faster travel.

Size. There have been light shovels with ¼ yard buckets, and ½ yard was a very useful and popular size. Now the smallest is ¾ yard. It weighs over 20 tons. The largest models, used in strip mining, have dipper buckets up to 140 or more yards, or draglines over 200 yards, and weigh thousands of tons.

Small models usually (but not always) have faster cycle times than larger ones, but they cannot dig as hard material, nor move it as far.

REVOLVING UNIT

The revolving unit, or superstructure, is built on and around a heavy steel deck or bed plate. This carries the machinery frame, which may be welded to it, as in the Bucyrus-Erie in Figure 13-2, or bolted on. The lighter deck section under the operator's station, shown in 13-3, and the cab are bolted to the bed plate.

The deck is supported by the swing rollers which rest on the turntable of the travel unit. It carries the engine, transmission, and operating machinery.

A diesel (or, less commonly, gasoline) engine is mounted across the back. Drive from its clutch, converter, or transmission shaft is usually by silent (roller) chain in an oil-bath case, but may be by shaft and helical gear sets. In either case, it has a reduction ratio that makes the deck machinery turn more slowly than the engine.

In Figure 13-4, the large chain-driven sprocket turns a transmission or jackshaft (not shown), which is parallel with the engine crankshaft. A small gear on this shaft turns the large gear above it, which upper gear is splined to the rear drum

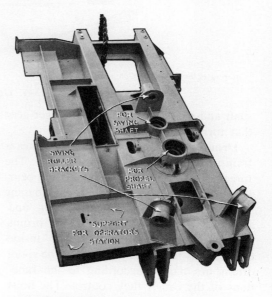

Courtesy of Bucyrus-Erie Company

Fig. 13-3. Deck, under side

13-3

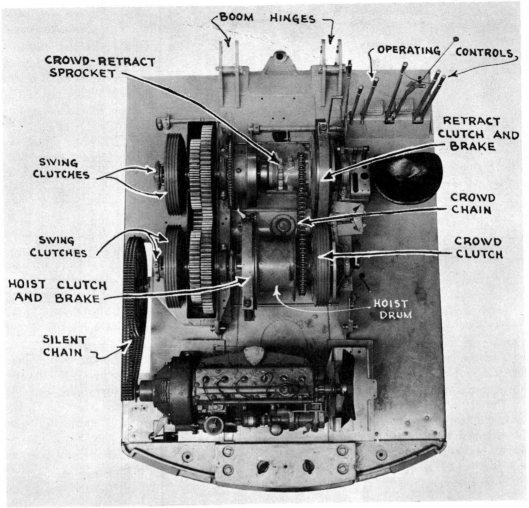

Fig. 13-4. Deck machinery

shaft and meshes with a slightly larger gear that turns the front drum shaft.

Drums. Figure 13-5 shows the power train involved in each drum operation. The three spool drums are mounted on bearings so that they are not affected by the rotation of the shafts within them, and are rigidly attached to brake-clutch drums similarly mounted. The clutches inside these drums turn with the shafts. The spool drums are named according to the functions of the cables they control. The hoist drum, on the rear shaft, is controlled by one internal expanding clutch and one external contracting brake. When the clutch is engaged the drum turns with the shaft and winds in the hoist cable attached to it. When the clutch is disengaged the drum may rotate freely, and any pull on the hoist cable will cause it to unwind unless the brake is applied to stop it.

The boom drum is also controlled by a clutch driven by the front drum shaft and by a brake. This brake automatically locks the boom drum whenever the control lever is released, and releases when the clutch is engaged. When both clutch and brake are released the drum is prevented from turning faster than the front drum shaft by a ratchet in the boom hoist unit.

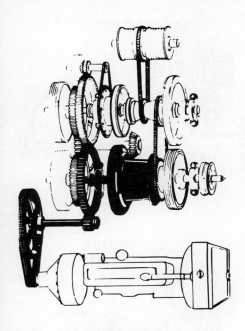

HOIST

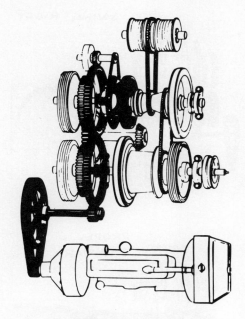

BOOM HOIST

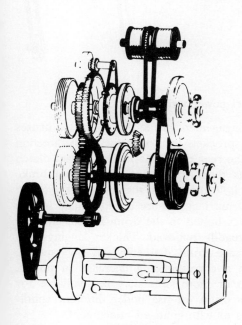

CROWD

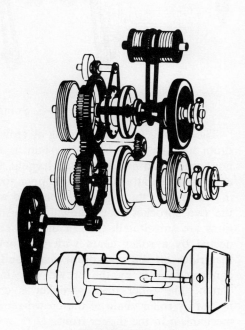

RETRACT

Courtesy of Bucyrus-Erie Company

Fig. 13-5. Drum power trains

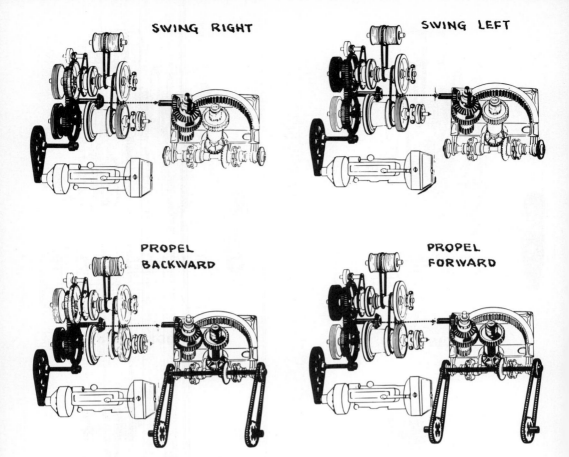

SWING RIGHT   SWING LEFT

PROPEL BACKWARD   PROPEL FORWARD

Fig. 13-6. Swing and propel power trains

The digging (also called drag or crowd) drum has the same control system as the hoist drum and has a grooved lagging, except when the attachment is a dipper stick, in which case the lagging is removed and the two sprocket gears shown in illustrations are substituted. One of these is connected by a roller chain with a sprocket that rides on the rear drum shaft, and is connected to it by the crowd (digging) clutch. The other front sprocket is connected by roller chain to the crowd-retract mechanism in the dipper front.

The two drum shafts rotate in opposite directions since their connection is a pair of spur gears. The front and rear sprockets, however, rotate in the same direction because they are chain-connected.

Engagement of the digging clutch causes the sprockets to turn in the same direction as the rear drum shaft. The retract clutch turns them oppositely. The crowd brake will hold them from turning either way.

Drum laggings are usually made in two halves bolted to the drum frame, and can be readily removed. Thicker laggings will increase drum diameter and line speeds, with a corresponding decrease in power. Two or more sizes may be available for each drum for use under different conditions or with changed attachments.

Laggings are frequently grooved in order to guide the cable into its proper place on the drum, and to support it against strain tending to flatten it. Smooth tapered lagging causes oncoming cable to slide toward the

Courtesy of Koehring Company

Fig. 13-7. Deck machinery, tandem drums

smaller anchor end so that it lies directly against the previous wrap.

Swing and Propel. Figures 13-4 and 13-6 show the swing and upper walking mechanism. At the left end of each drum

shaft is a swing clutch of the internal expanding type, the bands keyed to the shaft, and the drum separated from it by a bearing. A large spur gear is keyed to each clutch drum hub so as to revolve on

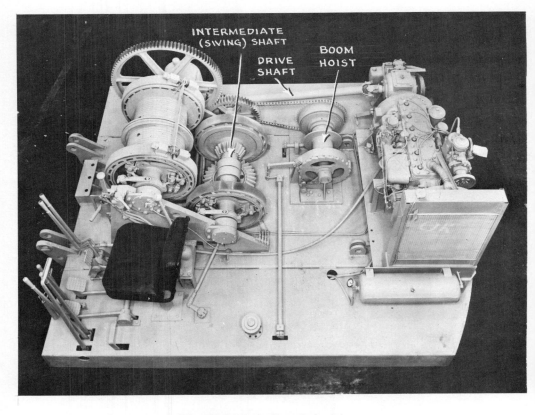

Fig. 13-8. Propeller shaft drive

the same bearing as the drum. These spur gears mesh with a smaller gear between and below them, to which is fastened a short horizontal shaft which terminates in a bevel gear. This in turn meshes with the bevel part of a combined bevel and spur gear, which turns on a bearing on the vertical swing shaft. The spur section meshes with another spur gear, which turns on a bearing on the vertical propel shaft.

Each of these vertical shafts has a jaw splined to it, which can be moved down to mesh with jaw teeth in the gears or raised until clear of them. When engaged, these jaw clutches will transmit rotary motion of the gears to the shafts.

At the lower end of the vertical swing shaft is a pinion (small spur) gear which meshes with the large internal toothed bull gear in the travel unit, Figures 13-21 and

13-23, which gear is centered on the vertical propel shaft. The propel shaft terminates in a bevel gear that meshes with a bevel (ring) gear on the live axle, which drives the tracks through chains.

Controls are so arranged that the operator can engage either swing clutch, but not both, and either jaw clutch, but not both. Figure 13-6, first position, shows the rear swing clutch and the swing jaw clutch engaged. The rear drum shaft, through the swing clutch, turns the horizontal propel and swing shaft, which turns the combination gear, the engaged swing jaw, and the vertical swing shaft. This causes the swing pinion to revolve and roll along the inner surface of the bull gear, pulling the revolving unit around it. A heavy collar and bearing transmit this pull to the shovel deck.

Since the two drum shafts revolve in op-

Fig. 13-9. Independent swing and propel

posite directions, release of the rear swing clutch and engagement of the front one will cause this power train to revolve in the opposite direction, reversing the direction of the shovel's swing. If both swing clutches are disengaged, the swing-propel mechanism will not move, unless outside forces acting on the shovel itself cause it to swing and rotate the pinion and all connected gears and shafts.

If the swing jaw clutch is disengaged, and the propel jaw clutch engaged, the use of the swing clutches will turn the live axle and cause the shovel to move forward or backward.

If both jaw clutches are disengaged, the engagement of swing clutches will spin the horizontal propel and swing shaft without performing any work.

Reversing or Swing Shaft. Another deck machinery arrangement is shown in Figure 13-7. In this old model Koehring shovel, the drive chain turns the swing shaft (reversing shaft), which also serves as a transmission shaft, driving the intermediate and drum shafts through a small gear on its opposite end.

Near the center of the swing shaft are two bevel gears, controlled by the two swing clutches. The bevel gears turn a combination bevel and spur gear horizontally, its direction of rotation depending on which clutch is engaged. This turns two spur gears, which may be connected and disconnected from the vertical swing and propel shafts by a pair of jaw clutches.

The intermediate shaft carries the boom hoist drum, clutch, and brake, and the retract sprocket controlled by the same clutch through a sliding jaw clutch.

Parallel Drums. The drum shaft is driven from the intermediate shaft by spur gears and rotates in the opposite direction. It carries both the hoist and digging drums. When a dipper stick is attached, the lagging is removed from the digging drum and two sprockets are substituted (see inset). One of these is connected by a roller chain to the retract sprocket, while the other transmits the crowd-retract power to the stick through another chain.

When the retract clutch is engaged, it acts to turn the sprockets and pull (retract) the bucket back out of the digging. When the crowd clutch is engaged, it rotates the sprockets the opposite way and pushes (crowds) the bucket into the digging. These two clutches are controlled by one lever

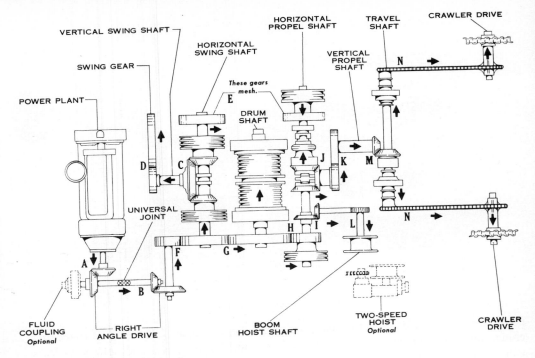

VERTICAL SWING SHAFT
SWING GEAR
POWER PLANT
HORIZONTAL SWING SHAFT
These gears mesh.
E
DRUM SHAFT
HORIZONTAL PROPEL SHAFT
TRAVEL SHAFT
CRAWLER DRIVE
VERTICAL PROPEL SHAFT
N
D
C
UNIVERSAL JOINT
J
K
M
A
F
G
H
I
L
N
B
TWO-SPEED HOIST
Optional
FLUID COUPLING
Optional
RIGHT ANGLE DRIVE
BOOM HOIST SHAFT
CRAWLER DRIVE

Courtesy of Gar Wood Industries, Inc.

Fig. 13-10. Deck layout

and cannot both be engaged at the same time. The digging (crowd) brake will hold the sprockets against movement in either direction.

In Figure 13-8 a right angle drive shaft transmits engine power to the intermediate or swing shaft close behind the drum shaft.

Propel Shaft. A layout that has two reversing shafts shown in Figure 13-9. The swing shaft is used only for swinging. The travel shaft walks the shovel, hoists the boom, and, with a dipper attachment, crowds the bucket. It is connected to these three functions by jaw clutches.

Figure 13-10 shows the deck layout of a Gar Wood with separate propel and swing.

If a live boom is not required, independent swing allows any rig but a dipper to walk without any shifting of jaw clutches, thus saving most of the time in move-ups. With any rig, it will allow simultaneous control of walk and swing, which saves time and effort in restricted spaces.

Some large draglines are equipped with a walking shaft that is independent of all other functions. Such machines cannot be readily converted to shovel use.

A-Frame and Gantry. The deck A-frame consists of a triangular frame of angles or tubing on the deck on each side of the machinery, and a cross bar slightly above the cab roof which carries sheaves for the boom support line. See Figure 13-13.

Larger shovels require higher supports for the boom cables. When the A-frame projects high above the roof, as in Figure 13-11, it is called a gantry. Gantries, and some A-frames, are of folding or collapsible construction so that they can be lowered by hand or power for shorter booms, or to go under obstructions.

Boom Foot. The boom foot is a pair of hinge brackets on the reinforced front edge of the deck. Booms are attached by horizontal pins running through these brackets so that they can pivot up and down, but are held from side movement.

Fig. 13-11. Gantry boom cable support

Cab. The cab is a shell of sheet metal and glass, designed to protect the machinery and the operator from the weather. One type of construction is shown in Figure 13-12. The window at the operator's station can be slid out of the way or removed. Numerous doors provide cable working space, access to machinery, and permit the operator to see through the cab in spots.

The circle vision cab, 13-13, allows the operator to see at a distance in all directions with doors closed.

Controls. The operator's seat may be on either side of the deck, forward of the center. Levers and pedals controlling all functions of the machine are near it. A representative set of controls is shown in Figure 13-14.

Fig. 13-12. Shovel cab

Fig. 13-13. Circle vision cab

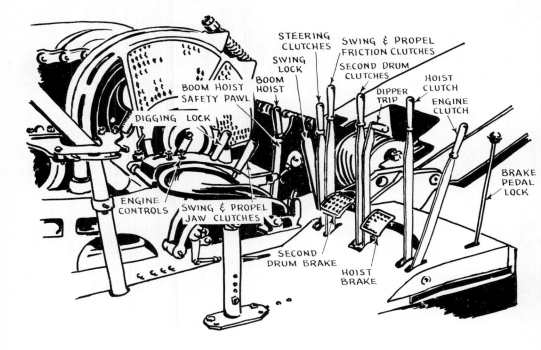

Fig. 13-14. Operator's control station, Bucyrus-Erie 22B

In this machine, the hoist lever engages the hoist clutch when pushed forward. The lever is held back by a light spring, but when pushed forward beyond the point necessary to engage the clutch, it locks over center, and requires a pull to disengage it. Use of this lever rotates the hoist or front drum so as to wind in the cable attached to it.

The digging, drag, or crowd lever when pushed forward engages the clutch of the same name, winding in cable that forces the bucket into material being dug. If used with a dipper stick it can be pulled backward from center also, engaging the retract clutch. These movements are made against spring tension, and both clutches lock over center.

The boom hoist lever, when pulled back, engages the boom clutch, causing the boom cable to wind in and raise the boom. In center position, it releases this clutch and allows an automatic brake to lock the boom drum. When pushed forward, it releases the brake and allows the weight on the cable to turn the boom drum as fast as the drum shaft turns. There is also a ratchet which can be set to prevent the boom cable from paying out at all.

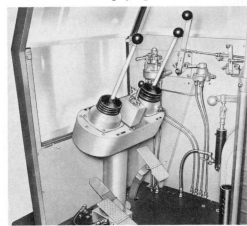

Courtesy of Thew Shovel Company

Fig. 13-15. Control station, Lorain 85A

Fig. 13-16. Combination swing and hook rollers

The swing lever, when pushed forward, engages the front swing clutch and causes the shovel to swing left or, in walking position, causes it to move forward. When pulled back, this lever engages the other swing clutch, with reversed results.

The swing lock, when pulled back, engages a toothed lock with the bull gear and prevents the shovel from swinging. The swing-propel shift lever, when forward, engages the jaw clutch for swinging. When back, it connects the walking jaw. In the middle both are disengaged.

The hoist brake locks or slows the rotation of the hoist drum, the drag brake does the same for the drag drum. These brakes are held in released position by springs, but can be locked down by engaging with a ratchet. Pushing on the top of the pedal locks it, on the bottom releases it.

The throttle regulates the speed of the engine by changing the governor setting. The steering lever is used only when walking. Forward turns to the right, backward to the left, center straight ahead, in forward motion.

The engine clutch is used to disconnect the machinery from the engine, for safety when adjusting or lubricating with the engine running, and for starting when cold.

The digging lock will lock the shovel tracks when pushed forward, will release them when pulled back, and in the center will allow the shovel to move forward only.

Controls are different in each make of shovel. Many use vacuum, air, or hydraulic controls, with shorter operating levers. Steering brakes are generally used to hold for digging also, and may be applied by hand or automatically. The direction of throw of operating levers can often be reversed.

A horn or a siren can and should be installed so that the operator can signal without interrupting work. It is handy in spotting and dismissing trucks, and for warnings and communications.

Lorain two-lever "Joy-Stick" controls are shown in Figure 13-15.

Swing Rollers. The whole weight of the revolving unit and attachment rests on rollers. These are tapered so that they can follow the circular turntable without any part dragging. Bushings or anti-friction bearings are used.

The swing roller may run between two tracks, as in Figure 13-16, so that it will also serve to hold the deck from tipping. Take up for wear is made by moving the rollers in from the bracket.

In 13-17 and 13-18 the swing rollers carry the load only, and hook rollers stabilize the machine. Adjustment is made in the hook roller brackets.

The deck may be held down without hook rollers by the use of a very heavy

Courtesy of Harnischfeger Corporation

Fig. 13-17. Separate swing and hook rollers

center pin, or pintle. This construction is not as efficient and adjustment is more difficult.

The upper weight may also be carried on race rollers. The many small rollers, shown in Figure 13-19, move on a circular track, and under a similar track fastened to the bottom of the deck. The roller axles are held between inner and outer retaining rings. The axles carry no weight except that of the rings, and serve only to keep the rollers at proper spacing.

The revolving unit is stabilized and held centered by hook rollers, the race flanges, and the center shaft.

Courtesy of Insley Manufacturing Corporation

Fig. 13-18. Hook roller

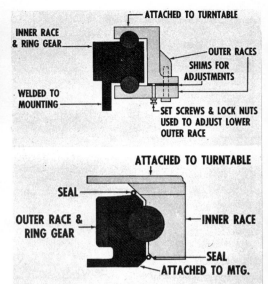

Courtesy of Thew Shovel Company

Fig. 13-18A. Ball race designs

Shear Balls. The rotating support may also be made of a number of hard steel balls and spacers in a circular channel between two races, one of which is attached to the base and the other to the revolving superstructure. The races overlap the balls in such a way that they cannot move up, down, or sideward relative to each other.

Figure 13-18A shows cross sections of two of the many possible constructions. The upper picture is of a two-row or two-ball arrangement. The outer race is composed of two rings bolted together. This permits assembly with the balls in place, taking apart for inspection, and adjustment for wear.

The lower cross section, and Figure 13-18B show a single row of balls with each race forged in one piece. This is assembled without the balls, that are then fed with their separators through a hole in the inner race. The hole is then plugged. Dirt is kept out of this oversize bearing, and lubricant retained in it, by upper and lower Neoprene seals.

This construction is stronger than with

a two-piece race, and the single row of balls permits more accurate machining and less overall height than either two-row or conventional roller construction. The larger Thew-Lorain shovels and cranes use a single row of 69 chrome steel balls three inches in diameter.

Counterweight. The weight of the attachment and its load are balanced by putting the center of rotation in the front center of the deck, so that the engine and most of the deck machinery are behind it. In addition, counterweight may be placed in the rear of the revolving unit in pockets behind the engine, or in a band around the rear of the deck, outside of the cab. The exposed counterweight makes an effective bumper and guard for the rear of the machine, but adds to its length, or tail swing, which is a disadvantage in close quarters.

Counterweight can be added or taken away to suit different applications.

Lubrication. Most of the deck machinery may be enclosed in oil tight cases, and lubricated by gears dipping oil out of one or more reservoirs and carrying it to distribution points, from which it flows

Courtesy of Harnischfeger Corporation

Fig. 13-19. Bearing-type swing rollers

over the other gears and bearings and back to the reservoirs. The oil level should be checked often through inspection plugs.

The turntable gear is greased with asphalt base lubricant. This may be thinned and pumped through a deck-mounted dispenser, or thinned or heated and poured through a hole in the deck, while revolving the upper frame. The gear should be checked at least once a day, and greased whenever bright spots appear on the teeth.

Pressure gun fittings throughout the shovel are greased in accordance with a schedule supplied with the machine. In general, bushings require more frequent attention than anti-friction bearings, some of

Courtesy of Thew Shovel Company

Fig. 13-18B. Ball race turntable

which may be sealed so as not to require lubrication. The crowd chains for the dipper stick should be kept painted with oil or light grease, but the tracks and drive chains should be left dry.

If the deck machinery is not enclosed, the gears must be kept coated with asphalt-base lubricant. Small, frequent applications are recommended, as any surplus works down to the floor, where it may combine with dirt to build up hard deposits which will obstruct the horizontal gears. If it gets in the underbody, it may cause jaw clutches to stick, and spoil traction brake linings.

Live and Dead Booms. The shovel described had a live boom, which, by releasing the fixed rachet pawl, could be

placed under the complete control of one lever which could be used in conjunction with other operating levers.

Another construction frequently found is to have the boom drum operated by the swing clutches, or some other clutch used in the regular routine of digging. These must be disconnected from their normal function by a jaw clutch, and connected to the boom drum by another jaw clutch or a sliding gear. If the swing clutches are used for swinging, walking, and boom control, two jaw clutches must be disengaged at all times. A separate brake, and a pawl and rachet are connected to the boom drum and must be released to lower it.

Such a dead boom is satisfactory under open, regular digging conditions where changes in boom angle are needed infrequently. It seldom hampers the efficiency of a hoe but greatly increases the difficulty of clamshell work.

There are also live booms which are controlled through worm gearing and which move quite slowly. These are designed primarily for precision crane work rather than digging.

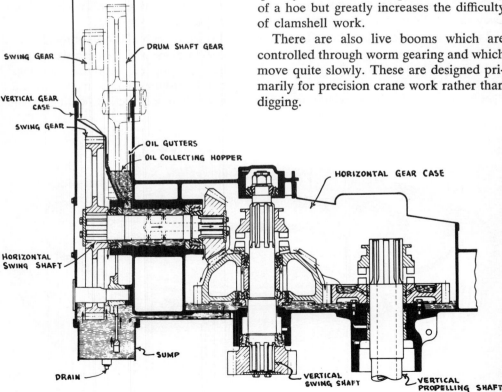

Courtesy of Bucyrus-Erie Company

Fig. 13-20. Lubrication of enclosed transmission

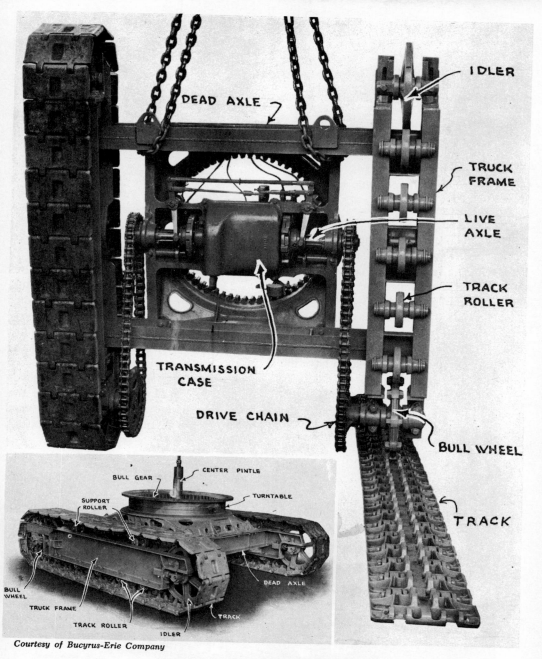

Courtesy of Bucyrus-Erie Company

Fig. 13-21. Lower frame and tracks

CRAWLER TRAVEL UNIT

Many shovels may be purchased with three different types of travel unit—crawler (track or cat mounting), truck mounting with two engines, and rubber-self-propelled, as shown in Figure 13-1. The crawler is the most widely used and will be described first.

General Construction. The crawler chassis, shown in Figure 13-21, is made up of a turntable welded to a frame consisting of

Courtesy of Harnischfeger Corporation

Fig. 13-22. Lower frame with tractor-type crawlers

Courtesy of Bucyrus-Erie Company

Fig. 13-23. Live axle details

two heavy I beams, called axles, that connect two heavy truck frames which rest on the track wheels and are surrounded by the track. The track usually consists of flat shoes hinged and pinned together at their ends, as shown in Figure 13-28A.

P&H employs tractor-type crawler belts with bolted-on grouser cleat or flat type shoes. Grousers are shown in Figure 13-22.

The travel unit is considered to have the bull wheels in the rear and the idlers in the front, regardless of whether this corresponds to the front and rear position of the revolving unit.

Live Axle. The 3-piece live axle, Figures 13-23 and 24, runs across the upper frame, parallel to the frame axles. Near its center is a bevel gear which is driven by the vertical propel shaft. On each side of this are steering clutches, with the inner jaws fixed to the center section of the axle.

The outer jaws slide outward along splines on the outer axles to disengage from the inner jaws. If they are moved farther than necessary to disengage, they are locked

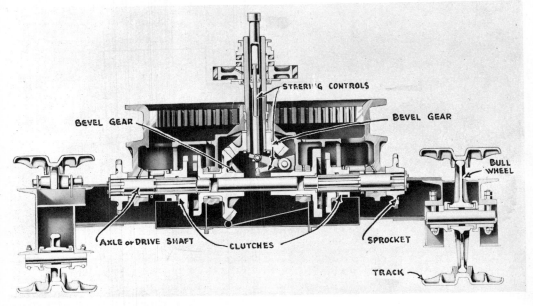

Courtesy of Bucyrus-Erie Company

Fig. 13-24. Propel power train in base

Courtesy of Unit Crane & Shovel Corporation

Fig. 13-25. Friction steering brakes

by engaging a fixed tooth. At the outer ends of the live axle are small diameter, large-toothed sprocket gears, which are linked by heavy roller chains to large sprockets attached to the bull wheels, which in turn engage with the track shoes.

In the Unit design shown in 13-25 the steering brakes are friction bands which are applied automatically by springs whenever the propel shaft is idle. Applying travel power releases them. Separate steering jaw clutches are engaged or disengaged as needed to apply power to the track.

Other machines use friction steering brakes which are manually controlled only.

Steering. If both axle jaw (steering) clutches are engaged, rotation of the vertical propel shaft will turn both tracks equally. If the right clutch is disengaged, the left track only will drive. The effect on the shovel's direction of travel will be determined by the steepness or roughness of the footing, and friction in the right track. Figure 13-26. On firm level ground the shovel will move straight forward, or a little to the right, on steep upgrades or on rough ground, it may turn quite sharply

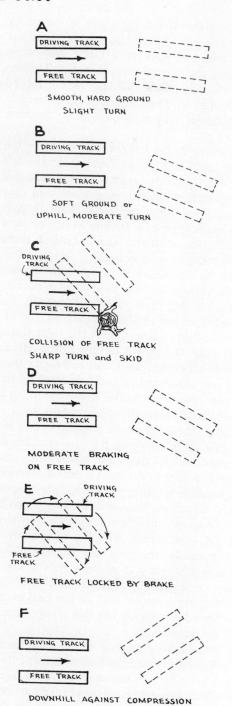

Fig. 13-26. Steering with clutches and friction brakes

to the right. If the right axle is locked by the brake, the right track will stop, and the left one will walk in a circle around it, turning the shovel very abruptly. Turning in soft material is difficult, as the locked track digs in. On slippery surfaces one or both tracks may skid, preventing accurate steering.

Turning to the left involves engaging the right hand clutch and disengaging the left one, and perhaps locking the left side if a sharp turn is desired. A gradual turn may be made by use of the clutch alone, by partial application of a friction brake, or by a succession of short, sharp braked turns with straight runs between.

If the machine is moving down a grade steep enough so that its engine is holding it back rather than pushing it forward, the steering action of the clutches will be reversed; that is, disengaging the right clutch will cause the shovel to turn to the left, as the disconnected track can roll faster than the one connected to and held back by the engine. When the brake is used, steering is about normal regardless of grade.

The shovel should be stopped before engaging or disengaging the jaw clutches or brakes. While they are transmitting the engine power to the tracks, they are almost impossible to disengage; and when apart they might be damaged by being forced into engagement while the jaws are turning at different speeds. The shovel is stopped by partially engaging the friction clutch that would move it in the opposite direction or by using friction brakes. In an emergency a ratchet lock or jaw brakes can be used for stopping but this is very jarring to the machinery.

The steering method is the same going backward (toward the bull wheels), except that for a turn in the same compass direction the opposite clutch and brake are used.

Digging Lock. Details of a ratchet digging lock are shown in Figure 13-27. The teeth of the inner fixed jaws of the steering clutches project farther from the axle than the outer jaw teeth. Above each of these inner jaws is a pair of pawls that may be held clear of the projecting teeth, or dropped individually or together to engage them. These pawls are cut so that one of each pair will allow rotation only in a clockwise direction, the other counter clockwise.

If both are engaged the axle cannot turn; if both are disengaged, it can turn in either direction. If one only is engaged, the axle can revolve away from it only. As long as the steering clutches are engaged the rotation of the inner axle controls the traveling of the shovel, so these pawls can be used to lock the shovel in place, or to hold it against digging strains from either direction.

Swing Lock. It is usually necessary to lock the shovel from swinging before it can be walked safely. This is partly because the power is applied to the vertical propel shaft from above the turntable, and carried to the live axle below it. This power should turn the shaft and axle, but if the resistance of the axle is sufficient, the shaft will remain stationary and the revolving unit rotate around it. In easy walking, the swinging tendency will be slight, in soft or uphill travel the power will all go into swinging unless checked.

The swing is usually locked by means of

Courtesy of Bucyrus-Erie Company

Fig. 13-27. Ratchet digging lock

a toothed bar anchored to the bottom of the deck, which can be engaged with the turntable gear by means of the swing lock control lever. A friction swing brake is used in some machines.

If the shovel is equipped with separate reversing clutches for propel and swing, the deck and boom can be kept in line or swung to avoid obstructions while walking by use of the swing clutches. However, it is usual for the operator to lock the swing for extended moves in the open, so that he will not have to keep alert to vary the engagement of the swing clutch if the propelling torque changes.

Steering Controls. The steering clutches and brakes may be operated by a single lever, which engages both clutches when in middle position, moves forward to disengage the right clutch, and backward to disengage the left, and engages the respective brakes by further movement. The control linkages for these, and for traction ratchets, are in the hollow center of the vertical propel shaft, as that is the only place where they would not be sheared off as the shovel revolves.

If the steering brakes are used to hold the shovel while digging, separate levers are used for right and left hand steering brake and clutch. The brake is applied and the clutch released at one end of the lever travel, and the brake released and the clutch engaged at the other end.

Tracks. Figure 13-28A shows part of a track that is used, with variations, on most shovels and other machines that do not depend on the tractive pull of their crawlers for their ability to work.

It is made up of a number of identical shoes cut and drilled at their ends so that they can be fastened together by pins. Wedge shaped projections are cast into the upper surfaces of the shoes to provide a grip for the sprocket or bull wheel, and to keep the track centered on the idler and the rollers.

Fig. 13-28A. Shovel track

Track pads are available in several widths. Wide tracks hold the machine up better on soft ground, but they somewhat increase the effort of steering, and are more subject to severe twisting stresses on uneven ground. They are heavier and more expensive, and cause complications by increasing the over-all width of the machine.

Extra long tracks increase stability, and are desirable for long boomed machines, and those which must handle heavy loads.

Flotation can be greatly increased by placing supporting platforms on the ground.

Track Behavior. Refer to Figure 13-28B. If the shovel moves forward, the bull wheel will turn so that the teeth will mesh with the pads under and ahead of it, picking them up one by one and passing them around its rear and then forward over its top. As the bull wheel moves forward on

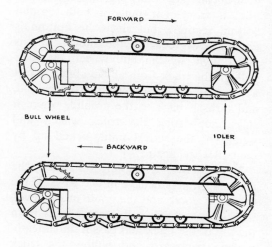

Fig. 13-28B. Track behavior

Fig. 13-29A. Big crawlers

the track, it pushes the truck frame and the idler ahead. The idler, pressing against the track in front of it, pulls on both top and bottom sections, but since the bottom is pinned down beneath the shovel, and is pulled tight by the bull wheel, the idler pulls in the track from above about as fast as the bull wheel supplies it from behind, and turns it back to the ground, where it stays, supporting the truck wheels and weight of the shovel until picked up by the bull wheel again.

If the shovel is walking backward, the mechanism may behave in two ways. If the footing is good, and the grade nearly level, the bull wheel will pull in some slack from the top of the track and pass it underneath, moving backward and pulling the truck frame and idler after it. The track it has

turned underneath supports the shovel and is then carried up by friction with the idler.

If the footing or grade is such that there is insufficient traction in the pads directly underneath the bull wheel to move the shovel, they will be skidded forward against the rear truck roller, the bull wheel will pull the upper track tight, and it will pull the idler backward. The idler moves truck frame, shovel, and bull wheel backward.

There are several reasons why a shovel or other tracked vehicle should be walked forward when possible. In forward motion the track is laid smoothly along the ground and its slack hangs harmlessly in gradual curves in the top section. The heavy strain is on the short piece from the bottom of the bull wheel to the first truck wheel supporting a good share of the shovel's weight. In

Fig. 13-29B. 140-yard stripping shovel

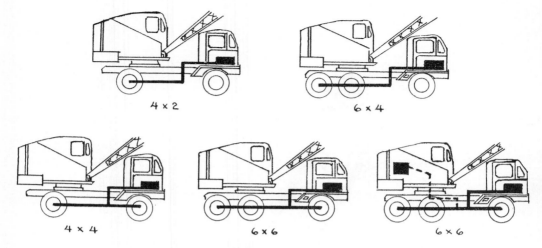

Fig. 13-30. Power trains in truck mountings

reverse movement, the track may be kinked in one or more damaging angles which the truck rollers must climb over or push down, and the traction stress is on the whole distance from the top of the bull wheel to the front slope of the idler, causing excessive wear and danger of breakage.

Track Maintenance. The usual track roller in a shovel revolves on a fixed axle, with hard steel bushings to take the wear. Grease is pumped to them through the axle and outside fittings.

No seals are used, and it is important to grease them daily, to keep dirt from working in and caking with old grease. Lubrication is needed even when the machine stands in one place to dig all day, as rocking with digging motions can work grease out and dirt in.

If neglected (and even sometimes in spite of conscientious servicing) lubricant and dirt may cake in the passages so that even extra high pressure guns will not force lubricant through.

It is a lot of work to remove and disassemble a roller to clean it out. A very easy cure, that is almost 100 percent effective, is to remove the grease fitting, insert a blasting cap deep in the passage, and fire it. It

will blow the passages clean, and will not damage the hard steel parts.

All precautions discussed in Chapter 9 should be followed in handling the caps. Some object should be leaned against the roller to stop fragments of the cap shell from flying.

This procedure may damage brass bushings, and is sure to ruin antifriction bearings and grease seals. It should NEVER be used except in bushed rollers or idlers without grease retainers.

Both the tracks and the drive chains wear, and need periodic adjustment. Bull wheels and idlers are mounted so they can slide backward and forward on the truck frame fastenings, and the position of each is fixed by heavy bolts. Lengthening the front bolts will force the idler forward and tighten the track only, lengthening the rear bolts will force the bull wheel and sprocket backward and tighten both the track and the drive chain. Care must be taken not to turn a wheel sideward by unequal adjustment of paired bolts, as it will then tend to walk out of the track.

If the track is allowed to become loose, it may come off when working or making turns on uneven ground. Slack also allows

Fig. 13-31. Michigan Truck Shovel

the shovel to rock back and forth under digging pulls, wearing the track parts, and makes proper control difficult when climbing ramps onto trailers. A too tight track may be badly strained or break on rocks, and wear rapidly because of tension on the hinge pins.

Maximum crawler speeds range from about three quarters of a mile an hour for medium large machines, to four or more for small ones.

Multiple Crawlers. Very large stripping shovels may weigh several thousand tons. They usually travel and work on the surface of coal seams that have limited resistance to crushing and shearing.

Fig. 13-32. 8 x 4 crane carrier

13-25

Courtesy of Clark Equipment Company, Construction Machinery Division

Fig. 13-33. Manual-type outriggers

They are supported on four crawler units, one at each corner of the lower frame. Each unit includes a pair of crawler frames and tracks, and an AC drive motor.

Universal mountings, that allow oscillation of individual tracks and swiveling of the whole unit for steering, connect the crawler units to massive hydraulic jacks that support the lower frame, and automatically keep it level on uneven ground.

Travel and steering controls are located on the lower frame.

RUBBER MOUNTINGS

Two-Engine Truck Mounting. The revolving unit may be carried on a turntable fastened to a truck chassis. The truck engine is then used for traveling, and the shovel engine for digging. Ordinarily, the shovel controls for walking, steering, and traction braking are disconnected or missing. Occasionally, however, arrangements may be made to connect the shovel engine to the[1] wheels through the truck transfer case, and to steer and brake the truck from the shovel cab. This mechanism is used for short moves at low speeds in the working area.

Any standard truck chassis of sufficient rated capacity can be used to carry a shovel, although considerable extra bracing is required, and better service should be obtained from a chassis specially engineered for the shovel. Tandem drive gives best support, and all wheel drive is advisable in work where mud or sand may be encountered.

The truck should have the right hand side of the cab cut away, to make it possible to carry the boom forward at a low angle. The frame and the rear axle should

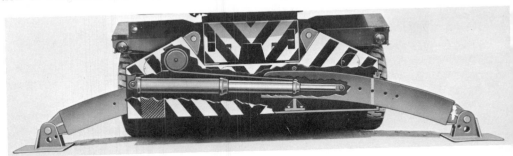

Courtesy of The Thew Shovel Company

Fig. 13-34A. Cutaway of hydraulic outrigger

be a single rigid unit while digging. Springs may be eliminated, locked during work, or replaced by stabilizer bars in the tandem construction.

The truck mounted shovel can ordinarily swing in a full circle, but with most attachments can work through only 270° because of interference of the cab and the truck front.

Figure 13-30 indicates various power trains used in the trucks.

Figure 13-31 shows a complete Michigan shovel, and Figure 13-32 side and bottom views of a Lorain 6 x 6 chassis.

Half tracks are sometimes used as carriers, for severe conditions. A front mounted winch is helpful in getting any unit out of bad spots.

Outriggers. Outriggers may be used to increase stability. These may be beams which can be slid or folded out of the bumpers, usually only at the rear, and which are supported by blocks or jacks. They provide a much larger and more rigid base than the tires. When they are used, lifting capacity is greater than that of a crawler of the same size, particularly when working off the back so that the truck engine acts as a counterweight.

Platforms may be put under the blocks to give support when working on soft ground, and the jacks may be used to lift the wheels out of mud or sand pockets.

There are also power outriggers, as in Figure 13-34. A curved arm with a hinged ground shoe is moved outward and downward, or inward and upward, by a hydraulic two way telescoping ram. The set of four can be used to steady, level, and even to lift the machine.

These outriggers are individually controlled from the truck cab. Automatic locking wedges relieve the hydraulic rams of strain during work, and prevent settlement that might be allowed by fluid leakage.

Outriggers are necessary for precision crane work and for using long booms.

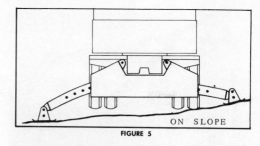

FIGURE 5 / ON SLOPE

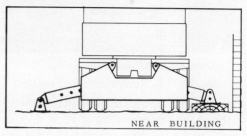

NEAR BUILDING

Fig. 13-34B. Using hydraulic outriggers

Uses of Truck Mounting. The advantage of truck mounting is its capacity for rapid and inexpensive movement from one job to another. With many models, the shovel can be placed and locked in traveling position in less than a minute, and then moved along roads at 20 to 35 miles an hour.

This is a pleasant contrast to the slow and laborious job of loading and securing a crawler type on a trailer, and unloading it at its destination.

On the job, however, the truck-shovel suffers from lack of maneuverability. Instead of turning in nearly its own length as the crawler, it must have considerable space in which to turn or side-step.

Its most important weakness for excavation work is the ease with which it can get stuck. Even with all wheel drive, its ability to get in and out of soft spots is greatly inferior to that of a crawler, and with rear drive only, constant care must be exercised to keep it out of trouble in soft ground and during rains. This disability has little effect on its usefulness when equipped as a crane, as most of its work is then on pavement.

An occasional problem is danger of tire

Fig. 13-35A. Rubber-mounted backhoe

damage. It is unwise to work in areas where garbage or scrap iron have been dumped, in blasted hard rock, or recently cut over land.

Self-Propelled. The rubber mounted, self-propelled shovel is a third type. It is somewhat more recent in origin and is in effect a compromise between the crawler and truck mountings. It can utilize any front end attachment.

Figures 13-35 and 13-36 show one of these machines, a Link-Belt speeder UC 68. The revolving upper is similar to those found on crawler mountings, except for different controls for steering, brakes, and stabilizers, which in this machine are power-hydraulic.

The travel unit has the same ring gear as a crawler and is similar in physical dimensions. It is constructed as a short-chassis, 2 or 4 wheel drive carrier, with dual wheels front and rear. There are no springs. The front axle is connected to the frame by an oscillating mechanism that can be leveled and locked by power hydraulic stabilizers. The rear axle is bolted solidly to the frame.

Two travel speeds are available in both directions. Torque converter drive gives a speed range from 0 to 5 miles an hour. Other machines of this type may have up to six travel gears and speeds up to 15 miles per hour.

Maneuverability on the job is subject to the same limitations as the truck except that the short wheel base, and in some

Fig. 13-35B. Undercarriage of self propelled wheel mounting

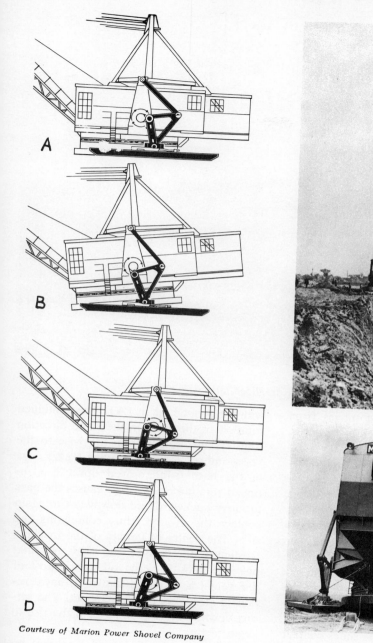

A

B

C

D

Courtesy of Marion Power Shovel Company

Fig. 13-37. Walking dragline

models four-wheel steering, allows it in tighter places. It is permitted to travel on highways, but for long trips it is better carried on a trailer.

WALKING DRAGLINES

In both sections of this book, men-tion is frequently made of crawler ma-chines walking, but this is with apologies to the true walking draglines (walkers), as shown in Figures 13-37 and 13-38.

These machines are made in sizes of five yards and up, and are so slow and bulky that they must be taken apart for

Courtesy of Page Engineering Company

Fig. 13-38. Walking draglines stripping coal

shipment by freight in most moves from one job to another. On the job, however, they are easily maneuvered, and have extremely small ground pressure.

They have a working base which consists of a wide, flattened cylinder called a tub, which rests on the ground and carries the turntable on its upper side. The walking mechanism is operated by cams or cranks, one on each end of a large diameter multiple-part shaft mounted across and extending beyond the sides of the upper frame and the tub. They are connected to the walking frames and the walking shoes. When the shaft is rotated the shoes contact the ground, raise the machine, and support its weight as a movement or step is made. The rear lifts clear, and the front drags.

When the dragline is in normal operation, it sets on its base and the shoes are elevated as shown in (D). To move, the walking shaft is rotated clockwise, thus turning the crank from its upright position and bringing the walking shoes in contact

with the ground, as in (A). The continued rotation of the shaft in the same direction transfers the weight from the base to the shoes and results in the leading (rear) edge of the base being moved upward. Continued rotation of the shaft causes the base to move backward and downward to again contact the ground, thus completing the step. The center of gravity of the machine is slightly forward of the shaft, so that only the back is raised off clear. Most of the weight is taken off the front, but it retains a steadying ground contact. The setting down is more gradual than the rise to avoid jarring.

The process is very similar to that by which a man on his hands and knees progresses by resting his weight on his hands and swinging his body forward with feet dragging, then rests on his knees and toes while he moves his hands forward again.

Steering is done by lifting the shoes and turning the revolving unit until its rear faces the desired direction, then stepping

BOOM POINT SHEAVES

STICK (HANDLE)

BOOM LINE

HOIST LINE

DUMP LINE

SHIPPER SHAFT

SADDLE BLOCK

HOIST LINE

BOOM

Courtesy of Bucyrus-Erie Company

Fig. 13-39. Dipper shovel attachment

off as described. This is probably the only steering device in heavy equipment which takes negligible power, does not tear up the ground and permits turns of any desired sharpness under any footing conditions.

This easy steering and the light ground pressure made possible by the large area of the base are the special advantages of this mounting. They enable it to work in places dangerous or impossible to crawler machines because of difficult terrain, soft footing, or nearness to loose down slopes.

Courtesy of Harnischfeger Corporation

Fig. 13-40. Chain crowd dipper

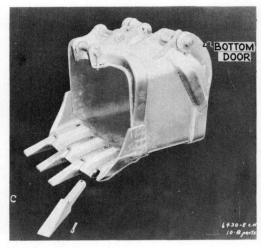

Courtesy of Bucyrus-Erie Company

Fig. 13-42. Detachable tooth

The disadvantages are that they are slow, with top speed around one fifth of a mile an hour, and are very wide, thirty feet or more. They must be disassembled to ship from one job to another, and are at a disadvantage in the bottom of a pit, both because of their width, and the fact that they must walk away from their boom, which might be blocked by a hill.

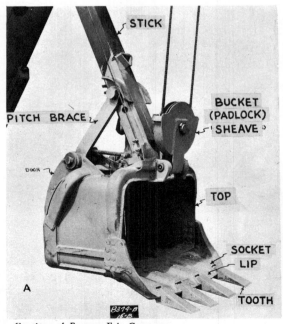

Courtesy of Bucyrus-Erie Company

Fig. 13-41. Dipper bucket

Fig. 13-43. Bucket capacity, 140 yards

DIPPER SHOVEL ATTACHMENT

The original shovel attachment, or front end, is the dipper. It is commonly called shovel front and often just shovel. Several constructions will be described.

General Construction. Reference is made to Figure 13-39. The boom is a massive beam, hinged to the deck at the boom foot, and extending diagonally upward. It is supported by a four-part boom line. It may contain a center slot, in which a saddle block and shipper drum are pivoted on the shipper shaft, or it may carry a divided stick in side blocks as in 13-40.

The dipper stick, or handle, slides back and forth through the saddle block and pivots on the shipper shaft. The upper or inner end of the stick is enlarged into a bumper which prevents it from sliding

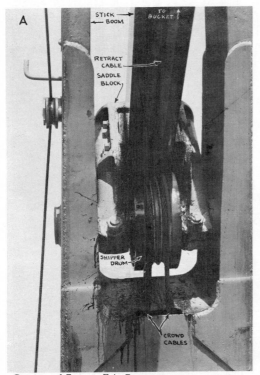

Courtesy of Bucyrus-Erie Company

Fig. 13-44. Front view of shipper drum

through the block. The lower end carries the bucket.

Bucket. A typical bucket is shown in Figure 13-41. It is a welded steel box open at the top, and closed on the bottom by a hinged door. Digging is done by the front top edge, which is reinforced by a lip. The lip contains tapered sockets to hold a set of four teeth. The reinforcing ridges running down from the sockets are called tooth bases.

Construction is heavy. The whole bucket, or at least the lip and other parts subject to severe wear, is of alloy steel.

Teeth are usually of manganese steel. They have tapered shanks to fit into the sockets. The fastening device indicated in 13-42 is simply a cotter pin through the shank which prevents the tooth from falling out. Digging stresses force it into the socket, so the pin is not under strain. The tooth

is removed by taking out the pin, striking the point with a hammer and pulling out by hand. If it is jammed, a hard steel wedge is put behind it and hammered down.

A number of other fastening devices are used. All will permit rather easy installation, and quick removal if not jammed, or cemented in place by soil particles. The teeth may have detachable points.

The teeth take the brunt of the digging, and each one should be strong enough to take the entire power of the shovel. When they wear dull, they may be reversed to partly restore cutting ability. Eventually they must be built back to size and edge by welding on caps, or building up with rod; or be replaced with new ones. It is good practice to keep a spare set of teeth on hand.

The digging ability of the shovel and its fuel economy are both diminished, and maintenance costs are increased, by using dull teeth.

The door is hinged high on the back of the bucket to allow it to swing wide when it opens. A sliding latch fastens it to the bucket front. It is opened by a sharp pull on a light cable. The trip mechanism will be described below.

The top of the bucket is pinned to the end of the stick, and it is supported near the door hinge by two pitch braces bolted to the stick. The braces can be moved up and down along the stick to change the digging angle of the teeth.

The padlock sheave is hinged to the bucket just above the stick, and carries the hoist line. Larger buckets may use a hinged yoke to fasten the sheave to the bucket. This construction is shown in Figure 13-43.

Stick and Boom Drums. The one piece Bucyrus-Erie stick rests on the shipper drum, and is held in alignment by wear plates in the saddle block. It moves freely back and forth in response to cable pull at its ends.

The shipper drum, Figures 13-44 and

C

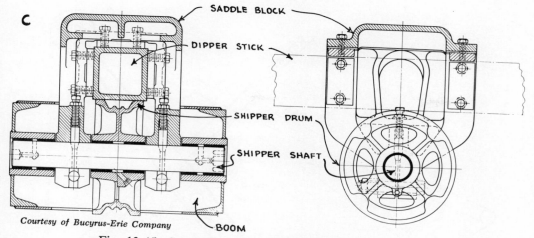

SADDLE BLOCK

DIPPER STICK

SHIPPER DRUM

SHIPPER SHAFT

BOOM

Courtesy of Bucyrus-Erie Company

Fig. 13-45. Cross section of shipper drum and saddle block

13-45, supports the stick on a pair of flat rims. Between the rims are three grooves. The outer ones carry the double crowd cable, and the center one holds the retract cable. The drum revolves on the same

axle (shipper shaft) that serves as a pivot for the saddle block.

The boom foot drum, 13-46, revolves on the center line of the boom foot pins, so that raising and lowering of the boom

CROWD-RETRACT CHAIN

BOOM FOOT DRUM

RETRACT LINE

CROWD LINE

BOOM

DECK

BOOM HINGE PIN

Courtesy of Bucyrus-Erie Company

Fig. 13-46. Front view, boom foot drum

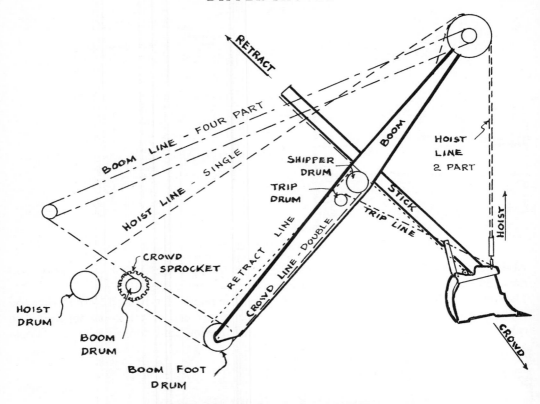

Fig. 13-47. Dipper shovel reeving, cable crowd

does not change its position relative to either the shipper drum or the deck machinery. It includes two grooved cable drums, and a center sprocket. It is turned in either direction by the crowd-retract chain from the front drum on the deck, and controls the crowd and retract cables.

The chain is kept at proper tension by an idler sprocket with a threaded adjustment.

Reeving. Figure 13-47 shows the reeving for this dipper attachment. The boom line runs from the front of the boom drum, up and back over a sheave on the left side of the A frame top, forward to a small sheave on the left side of the boom point, back to a horizontal sheave hinged to the A frame cross member, forward to the top of the right hand small sheave on the boom point, and back to an anchor on the A frame.

The hoist cable takes off from the top of the hoist drum, runs upward and forward over the large left hand sheave on the boom point, around the padlock or bucket sheave on the top front of the bucket, and over the large right hand boom point sheave to an anchor on the boom.

The crowd cable is anchored and wrapped on the outer side of the left hand boom foot drum, runs from the bottom of this through the hollow center of the boom to the bottom of the left groove in the shipper sheave, back along the stick to the end, around a curved holding plate and forward along the stick to the shipper sheave, around its right groove and to the bottom of the outer side of the right boom foot drum, where it is wrapped and anchored.

The retract cable is anchored to the inner side of the right hand boom foot drum, takes off from the top, runs over the

CABLE INSTALLATION

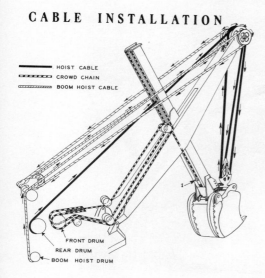

HOIST CABLE
CROWD CHAIN
BOOM HOIST CABLE

FRONT DRUM
REAR DRUM
BOOM HOIST DRUM

Courtesy of Unit Crane & Shovel Corp.

Fig. 13-48. Chain crowd reeving

boom to the shipper sheave, around the center groove and forward along the bottom of the stick to an eye bolt anchor that has a threaded adjustment by which the cable can be tightened.

The boom foot drum can be rotated in either direction by the opposite acting crowd and retract clutches. If turned so as to wind in the crowd cable, it will pull on the back of the stick, and force it to slide forward or downward through the saddle block. This will pull the retract cable, which unwinds from the drum at the same speed the crowd winds onto it, so that the two cables remain tight and balanced against each other. If looseness develops, it may be removed by tightening the eye bolt anchor of the retract.

When the foot drum rotation is reversed, the retract will wind onto the drum, pulling the stick back through the saddle block and unwinding the crowd cable.

The crowd normally pushes the bucket into the digging and the retract pulls it out. The crowd is therefore geared down for greater power, and a double cable is used for greater strength. Retract may be

fifty or sixty percent faster than the crowd.

Chain Crowd. The Unit shovel uses a roller chain for the combined functions of the chain and cables described above. The front drum is replaced by a triple sprocket, and the shipper shaft has three idler sprockets. Reeving is shown in Figure 13-48.

An endless three-part chain is used. The single retract runs from a reversible front drum, under the upper boom idler, and over the shipper sprocket to an anchor near the bucket. This is linked between the

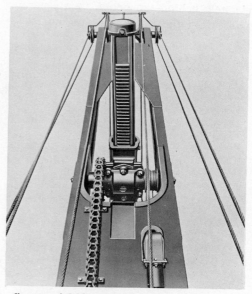

Courtesy of Baldwin-Lima-Hamilton Corp.

Fig. 13-49. Chain crowd details

13-37

Fig. 13-50. Crowd chain

halves of the double crowd chain, which goes under the lower boom sprocket and the shipper sprocket to an anchor on the back of the stick.

Chain crowd may also use a crowd re-tract chain to drive a boom foot sprocket, or a sprocket set mounted on the deck immediately behind the boom, then separate chains from the boom foot to the stick.

The stick may also be moved through the saddle block as in the Lima and Gar Wood machines in Figures 13-49 and 13-50. The shipper drum becomes a spur gear (rack pinion) which meshes with rack teeth cut in the bottom of the stick. This shipper gear is turned by a chain running from the boom foot sprocket. Rub plates in the saddle block hold the teeth in engagement.

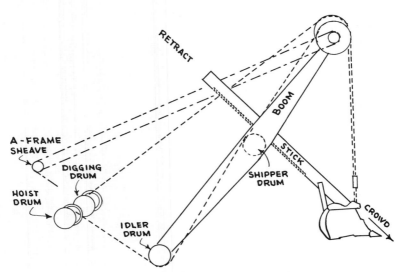

Fig. 13-51. Automatic crowd reeving

13-38

If the top of the shipper gear is rotated forward, the stick is crowded; if it moves back, the stick is retracted.

Chain crowd saves the expense and time loss of broken crowd cable, and suffers less damage from bending around small sheaves than cable of the same strength. However, it is more noisy, an extra item to lubricate, and costs more to replace when it eventually wears out.

Automatic Crowd. Both the cable and chain crowds described are the independent or positive type. Another crowd-retract mechanism, called automatic or gravity crowd, is illustrated in Figure 13-51. This is used on shovels which are not equipped with a reversible digging drum.

The dipper stick has teeth on the lower surface engaging with a spur gear keyed to the shipper shaft. The shipper drum, which may be spool or spiral in construction, is fastened to the shaft also.

A cable is reeved from the top of the digging drum, across the left boom point sheave, around the bucket sheave, over the right boom point and back to the top of the shipper drum. It is wrapped around this, and runs down along the boom, around an idler drum, and is then wrapped and anchored on the hoist drum.

If the hoist clutch is engaged and the digging brake applied, the bucket will be lifted. The cable being reeled in will turn the shipper drum, shaft, and spur gear, and retract the stick. If the digging brake is released, the weight of the bucket will strip the cable off the digging drum faster than it is reeled in on the hoist drum, and the bucket will drop but the retract motion will continue. If the brake is partially released the stick will still retract, and the bucket may be raised, held, or lowered by varying the brake pressure.

If the hoist and digging clutches are engaged at the same time, the bucket will hoist at twice normal speed, as the cable is being reeled in at both ends, and retract

action of the dipper stick will continue.

If the digging clutch is engaged, and the hoist brake held, the bucket will be raised without affecting the in-and-out position of the dipper stick. If the hoist brake is released, the bucket will drop, pulling the cable off the hoist drum and turning the shipper drum so as to crowd the stick forward. If the brake is partly engaged, the bucket will be lowered, held, or raised, depending on how hard it is applied. The stick will be crowded at the

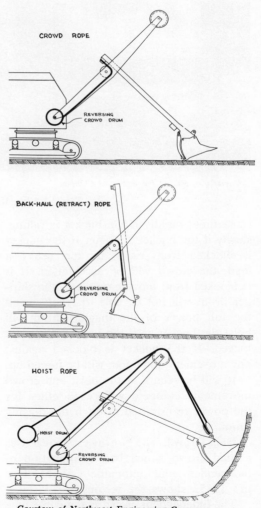

Courtesy of Northwest Engineering Company

Fig. 13-52. Dual crowd

13-39

Fig. 13-53A. Knee action crowd

same time; rapidly if the bucket is falling, slowly if the bucket is rising. If the bucket is blocked from rising by too heavy a load, the crowd will be accelerated. If it is blocked from moving outward, the shipper drum will not be able to revolve, and the full power of the digging drum will hoist the bucket as if the hoist brake were locked. If the bucket is blocked against any movement, the boom will be forced up.

It will be seen that the hoist and crowd movements compete with each other for the power transmitted through the digging drum.

Dual Crowd. The Northwest shovel uses a dual crowd which is a combination of independent and automatic systems. Figure 13-52 shows the reeving.

The independent crowd is the cable type described earlier. The hoist cable installa-

tion is also conventional, except that it is anchored on the crowd and backhaul drum at the base of the boom, and wrapped around it in the same direction as the retract cable.

Tension on the hoist cable, from drum pull and bucket weight, adds force to the crowd. When the hoist drum is stationary, crowding lowers the bucket and retracting raises it.

Big Stripping Shovel. The big Marion strippers use a special crowding mechanism, shown in Figure 13-53. The stick is in two pieces, hinged to each other and to a stiff leg pivoted on the deck behind the boom. The boom has a wide center opening, allowing the front stick to work through it without touching.

The rear stick or crowding handle has rack teeth, and is crowded and retracted by

Fig. 13-54. Stripping for coal

shipper shaft pinion gears driven by electric motors mounted in the gantry.

The boom carries hoisting strains only, and therefore can be lighter in construction than if it had to line up and crowd the stick as well.

The Bucyrus-Erie is the biggest of these machines at present. It is all electric, and uses the Ward Leonard controls described in the next section. Utility power is supplied at 7200 volts through a tail or trail cable 5 inches in diameter, and 5000 to 15,000 feet long. The alternating current motors that drive the main generators supplying all electricity total 9000 horsepower. There are eight hoist motors of 625 horsepower each. It uses about as much electricity as the homes and industries in a town of 15,000 people.

The boom has a fixed angle of 50 degrees. Its maximum digging radius is 218 feet, its maximum dumping reach 200. Maximum dumping height is 140 feet.

The dipper capacity is 140 yards. Boom is 200 feet long. The boom foot is about 50 feet above the ground, crawler bearing pressure is 60.5 pounds per square inch. The roller circle diameter is 54 feet.

The dipper handle and dipper are made of alloy steel, largely T-1. The high strength and light weight of this steel has reduced the dead weight of structural members so as to make possible a substantial increase in the capacity of the dipper.

An electric elevator with push button controls operates inside the hollow center pin. It has stops in the lower frame, and on the deck at the level of the operator's cab. It is rated for three people or 1000 pounds.

The machine as shipped from the factory weighs about 8000 tons, to which the user adds another 2000 tons of ballast. It can sidecast over 7000 bank yards per hour, or 5,000,000 yards a month.

Bucket Trip. The dump line for the bucket door may be operated by mechanical or electric power or by hand. Hand operation is not satisfactory except for very light duty service, as it does not permit using

Fig. 13-55. Mechanical dipper trip

13-41

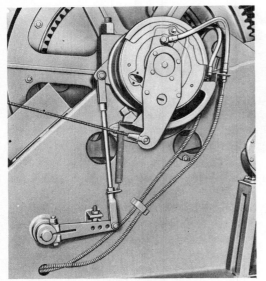

Fig. 13-56. Hydraulic control dipper trip

a strong latch spring to overcome friction and jamming.

The trips in Figures 13-55 and 13-56 are mounted on the end of a live shaft in the deck machinery. A drum rides freely on the shaft and can be connected to it by a brake-type clutch or a ratchet which is held out of engagement by a spring.

Fig. 13-57. Electric dipper trip

The drum contains a spring which keeps a tension on the dump line sufficient to pull in slack when the bucket is retracted. Engaging the clutch or pawl reels in the line with a snap sufficient to pull the latch and allow the door to open. Since the line is always tight, a quick light blow on the lever will make the trip operate fully.

The line runs up the side of the boom and over a sheave on or close to the shipper shaft.

The electric trip, Figure 13-57, consists of a small starter type electric motor equipped with an eccentric sheave. Actuating the sheave through a push-button switch controlling the motor produces a pull on the cable to release the latch.

The unit may be mounted on the lower boom, as shown, near the shipper shaft, or on the shovel deck.

ELECTRIC SHOVELS

Electrics and Diesel-Electrics. All-electric shovels and draglines normally operate on three-phase alternating current (AC) furnished at between 2000 and 6000 volts by utilities (high lines) or by generator sets assembled on the job.

Electric units are used to serve in the place of the internal combustion engine, and of the clutches, brakes, mechanical controls, and to eliminate some of the gears, chains, and cables, so that construction is quite different from that of the gas or diesel powered rig.

A diesel-electric shovel carries a diesel engine and a generator to supply current to electric motors and controls that perform all operating functions. A part electric machine has diesel power, with mechanical drive to some power trains, and electric to others. Electric swing clutches may be used in a machine that is otherwise all-mechanical.

In excavators used by contractors, electric and diesel-electric drives begin to appear in the three and four yard class, and are standard in very large machines. Even small shovels and cranes may be electric if they work in yards or underground, and do little or no moving around.

In many ways electricity is an ideal power for excavators. Since it allows packing a lot of power in a small place, motors can be placed so as to eliminate complex gear trains and chain drives; maintenance and fueling of an internal combustion engine are eliminated, exhaust gas is not a problem indoors or underground, and operation is smooth and quiet. Controls are easy to operate.

But first cost is higher, particularly in small units, and in any size the necessity of being near a power source and of having a cable connection can hamper activities severely. There are also technical problems involved in conveying the heavy current into the revolving frame, and in controlling and distributing it.

The diesel-electric is independent of outside power, and has the smoothness and some of the simplicity of electric drive. Engine fueling, maintenance, and noise are about the same as with mechanical drive.

Controllers. The main operations of the electric excavator depend on three master controllers at the operator's station. Two of them, the hoist and crowd-retract in the dipper shovel, or the hoist and drag in the dragline, have hand levers. The swing is operated by a pair of foot pedals.

By means of these, circuits carrying light current can be opened, closed, or "throttled" to regulate flow of current to a desired amount.

There are several types of controllers now used in excavators, of which three will be described.

One, the rheostat or resistance type, includes a number of circuits of different resistance, connecting a power source and the generator fields. A rotating switch provides means to close any one of these to permit current to flow through it. The

contacts are arranged so that in moving from the OFF position, the controller closes the circuit of maximum resistance first, and those of diminishing resistance in succession, until the final full flow connection is reached.

Since this current is used only for the field excitation to be discussed below, it need not be heavy enough to cause serious heating in the resistance units, or to require large contact areas.

The power applied usually cannot be increased smoothly, but rises in a series of steps as successive contacts are reached.

The impedance type control, Figure 13-58, uses a magnetic resistance to "throttle down," redirect, or cut off the flow of current. There are no contact points, and the transition from zero to maximum power is smooth.

The eight yard P&H Model 1800 has a stepless electronic tube control for all operating motions. This makes it possible to use a really minute current, about 1 watt. Response is said to be very rapid and sensitive, so that coordination is improved and cycle time shortened.

Courtesy of Harnischfeger Corporation

Fig. 13-58. Impedance-type control

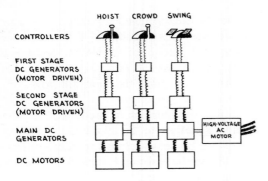

Fig. 13-59. Block diagram of Double Ward-Leonard system

Ward-Leonard System. Most electric excavators use a drive and control system known as the Ward-Leonard or Double Ward-Leonard, that makes it possible to control the heavy working current by varying a light current to generator fields.

Figure 13-59 indicates schematically the principle of operation, without in any way representing the actual mechanical and electrical layout.

One or more AC motors (one shown) turn a set of three generators, that supply direct (DC) current to the hoist, crowd, and swing motors. The output of each generator, and therefore the speed and power of the motor it powers, is regulated by the flow of exciting current through its field coils.

This current, in turn, is regulated by the position of the controller lever at the operator's station. Thus an easily handled low-voltage small-amperage current is stepped up to regulate the high voltage and very heavy amperage needed to provide the power required at the hoist drum and at the crowd and swing pinions.

In the Double Ward-Leonard system shown in the diagram there is a set of intermediate or second stage generators, whose field coils are excited by current from the first stage, and whose output excites the main generator fields.

This refinement permits both reducing the current required at the controller, and providing ample capacity at the main generator.

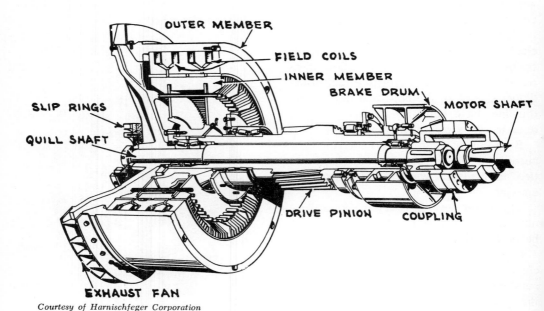

Courtesy of Harnischfeger Corporation

Fig. 13-60A. Magnetorque unit

The main AC motor turns at full speed whenever it is switched on, but current consumption is light when the generators are not charging and the DC motors are idle.

The swing and crowd-retract generators have means to reverse the direction of current flow. The crowd motor is reversible, the swing may use either one or two reversible motors geared to individual swing pinions.

The hoist motor is also reversible, so that it lowers by power at a controlled speed, instead of allowing the bucket to drop by gravity. A mechanical clutch (friction or pneumatic) is usually put between it and the drum, adjusted so that there will be slippage to absorb shock if the dipper stops suddenly.

A gearshift, or moving of jaw clutches, permits the hoist motor to be used to propel the shovel.

DC is usually selected for motor drive because of its ability to develop high torque when lugged down by heavy loads, and to use variable voltage without damage.

Magnetorque. Another system is the Magnetorque electro-magnetic clutch, Figure 13-60. At present this is used chiefly in the hoist of all-electrics, as shown, and in the swing of large diesel shovels.

The Magnetorque unit consists of inner and outer rotating members, fabricated of high strength alloy steel, concentric with one another and separated by an air gap. The outer member carries a sealed electrical winding and is rotated in one direction continuously by a quill shaft turned by a high voltage AC motor.

This electric clutch transmits power without friction or vibration. Decrease in speed of the inner member automatically results in increase in the torque being transmitted by the unit. The operator can alter the output characteristic merely by changing the position of the controller.

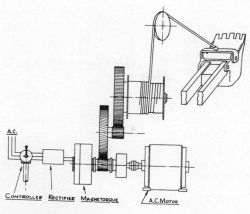

Courtesy of Harnischfeger Corporation

Fig. 13-60B. Magnetorque hoist drive

When the controller is in the OFF position no current flows in the coils of the outer member; hence the inner member is free to rotate in the opposite direction. This allows the dipper to be lowered rapidly by gravity.

QUARRY SHOVEL

The P&H electric shovels of 4 to 8 yards capacity, such as the machine in Figure 13-61, are representative of a class of excavator that is widely used in pits, quarries, and mines; usually in loading blasted rock and ore into hauling units. They are also found in heavy rock cuts for highways where fragmentation is poor and speed is essential.

Power Line. The power line is carried overhead, often on portable or temporary poles, to the immediate vicinity of the shovel; then along the ground in a flexible heavily insulated tail cable to a terminal box in the lower frame.

A ring-type high voltage collector is located at the base of the vertical propel shaft. This consists of three pairs of rings in an oil filled moisture proof case. One ring of each pair is stationary and connected to one of the three input wires; the other revolves with the upper frame and is

Courtesy of Harnischfeger Corporation

Fig. 13-61. Quarry shovel

connected to one of three lines going up through a protecting sleeve inside the hollow propel shaft and to the control panel. The rings keep constant contact in any swing position of the shovel.

Power requirement of the 4 yard sho⟨v⟩ is about 225 KVA (190 kilowatts) fr⟨om⟩ a high line, or 300 KVA (255 kilowatt⟨s⟩ from a diesel generating set.

Control Panel. The control panel c⟨o⟩

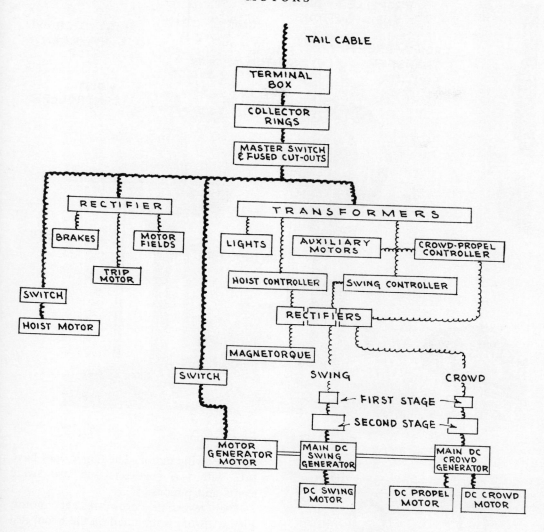

Fig. 13-62. Block diagram of electrical layout

tains the master switches and fuses both for the incoming AC current and for the DC circuits from main generators to motors; selenium rectifiers that convert AC current to DC to operate brakes, dipper trip motor, and main motor fields, and low voltage AC from the controllers into DC for generator field excitation.

There are also transformers to step down AC voltage for use in the controllers, lights, auxiliary motors, and the crowd-propel circuits.

The approximate path of current through various parts is indicated in Figure 13-62 and the operator's controls in Figure 13-63.

Motors. The identical crowd and swing DC generators are mounted on the extended shaft of the main motor, Figure 13-64. They are driven at constant speed, and output is varied from zero to maximum by the current supplied to their field coils by a modified Ward-Leonard hookup known as impedance control.

The smaller models have a single re-

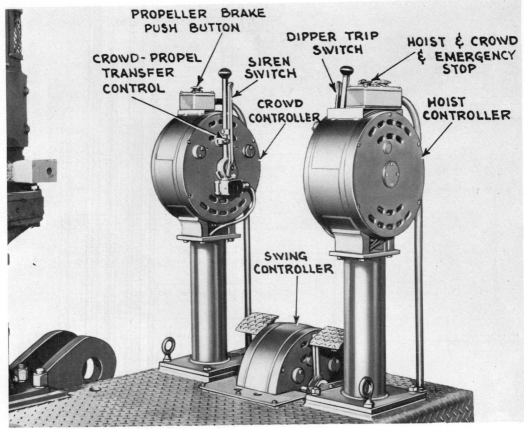

PROPELLER BRAKE
PUSH BUTTON

CROWD-PROPEL
TRANSFER
CONTROL

SIREN
SWITCH

DIPPER TRIP
SWITCH

HOIST & CROWD
& EMERGENCY
STOP

CROWD
CONTROLLER

HOIST
CONTROLLER

SWING
CONTROLLER

Courtesy of Harnischfeger Corporation

Fig. 13-63A. Operator's station

Courtesy of Harnischfeger Corporation

Fig. 13-63B. Operator's switch panel

versible swing motor. The larger ones have two, with two gear sets and two swing shafts and pinions.

The reversible crowd-retract motor, Figure 13-65, is mounted on the boom just below the shipper shaft. Crowding or retracting of the straddle-type sticks is accomplished through a worm, worm wheel, and a set of spur gears, the combination of which link the crowd motor to the gear racks on the under side of the dipper sticks. The motor is cooled by filtered air from a blower.

The hoist motor and Magnetorque drive were described earlier.

Brakes. Each motor or gear train is equipped with a powerful brake that is held out of engagement by an electromagnet, so that if power is cut off, either de-

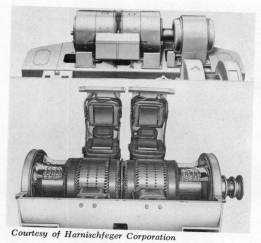

Courtesy of Harnischfeger Corporation

Fig. 13-64. Crowd and swing generators

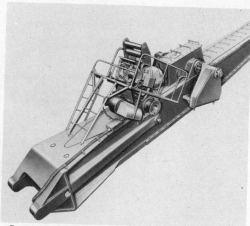

Courtesy of Harnischfeger Corporation

Fig. 13-65. Crowd motor

liberately or accidentally, the unit "freezes." Current is shut off and brakes are applied by push buttons at the operator's station.

The push button control is used to hold the bucket and sometimes the swing while waiting for arrival of trucks, or for other developments. The automatic application protects the machine and its surroundings against damage in the event of power failure.

Ventilation. A blower in the top of the cab draws in filtered air, as shown by the flow arrows in Figure 13-66, and maintains the cab above atmospheric pressure. Some air leaks out, but most of it is drawn out by blowers, one to each motor, that prevent overheating by keeping a current of air moving through them.

The main filter keeps the air clean so that abrasive dust is not blown into the motors.

There is a door between the operator's station and the main cab, so that he can use or avoid the air flow, as he wishes. If this door is open, his windows must be closed.

Propel. Current from the crowd generator can be diverted from the crowd motor to the propel motor by flipping a switch on the crowd lever (controller handle).

No gears or clutches are involved in this instantaneous one-finger shift. The automatic brake locks the crowd while current is used for propelling.

The propel motor is mounted vertically on the right side of the deck, and is permanently geared to the vertical propel shaft. This drives a horizontal shaft, which turns the live axles extending across the rear of the travel unit. This arrangement is shown in Figure 13-67.

One axle is permanently geared to the horizontal propel shaft; the other has a hydraulically operated jaw clutch and jaw brake for steering. The horizontal shaft has a spring-loaded friction brake that can be released hydraulically.

The steering clutch and brake are shifted hydraulically, under electric control. The current for this function originates at the crowd controller at the operator's station, and is carried into the lower unit by collector rings outside the base of the center pintle.

The propel shaft brake holds the machine while digging, and at any time that it is not disengaged by current activating the propel motor. The fact that the power train is permanently geared to one axle prevents any possibility of a runaway—a

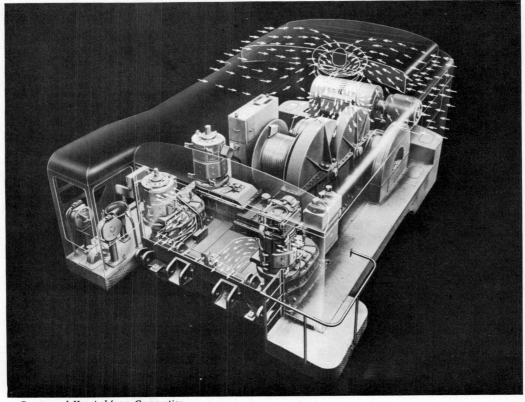

Courtesy of Harnischfeger Corporation

Fig. 13-66. Ventilation system

good consideration in an excavator of such weight, 160 to 260 tons.

The unit can turn to only one side by power. To make a turn in the opposite direction, it is steered while walking backward until heading in the desired direction. The steering control is then released, and the shovel walked straight forward.

The axles carry spur gears on their outer ends, instead of the usual chain sprockets. These are meshed with larger gears on the bull wheel axles.

Tracks. The tracks, Figure 13-68, are made of shoes fastened to each other by hinge pins. The inner surface is formed into a raised track to carry the track rollers. Pins are long, and project at each side so that the drive sprockets mesh with them, rather than with shoe lugs.

Since there is a fixed distance gear drive rather than a roller chain, the rear sprockets are not adjustable. Track wear is taken up only by moving the idler forward.

Boom. The boom feet rest in sockets on the front of the deck. The boom is held in line by two pairs of spring-cushioned sway braces. It is not pinned to the sockets, but held in them by the braces, that permit slight rolling of the boom under side stresses.

Standard equipment does not include a boom hoist drum. The boom is supported by a two part line anchored on the gantry. It can be changed by hoisting the bucket until the stick pulls the boom up enough to take the weight off the line.

These machines are used under such unchanging conditions—usually loading

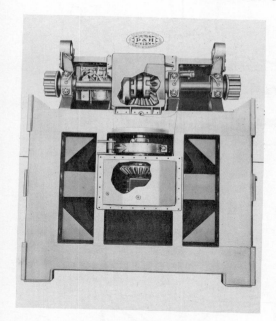

Fig. 13-67. Propel shaft and axles

blasted rock off a flat floor—that there is seldom need for changing boom height. For special purposes, an electric-powered live boom hoist can be supplied.

DIGGING—INDEPENDENT CROWD

Some operating positions of a shovel are shown in Figure 13-69.

The shovel is swung to face the section to be dug, with the hoist and crowd clutches disengaged and the brakes set. The stick is about horizontal with its center forward of the shipper drum. The hoist brake is released, the bucket falls in an arc, and the back of the stick rises, all motion being centered on the shipper shaft. The momentum of the bucket carries it back of the line of the shipper shaft and application of the hoist brake stops it before it hits the underside of the boom.

The controls cannot hold the bucket in this position and it will tend to swing down and forward by gravity. But as the hoist brake is applied, the crowd brake is re-

leased and the stick slides downward through the saddle block until the bucket teeth rest on the ground, as in (C).

It is now desired to move the bucket forward horizontally, to level irregularities on the ground, then to continue the level grade into the bank. This is done by releasing the hoist brake and engaging the hoist clutch, which pulls the bucket forward and downward in an arc around the shipper shaft. The crowd brake is released, and the retract clutch partially engaged so that it will pull the bucket up to compensate for the downward path of its arc, until the stick is vertical.

From here forward the hoist tends to pull the bucket upward. The bucket can now be kept to its level grade by releasing the retract clutch and partially engaging the crowd brake, so that the stick can slide down through the saddle block just enough to compensate for its upward curve from the hoist line. If the digging is hard, the crowd brake may be released and the crowd clutch engaged sufficiently to force the bucket against the dirt, but not enough to cause the boom to be raised by the crowding force.

As the teeth penetrate the bank, (E), the bucket fills. When it is full the crowd brake and clutch are disengaged, and the retract clutch is engaged so that the bucket is pulled back out of the bank

Fig. 13-68. Bull wheel and track

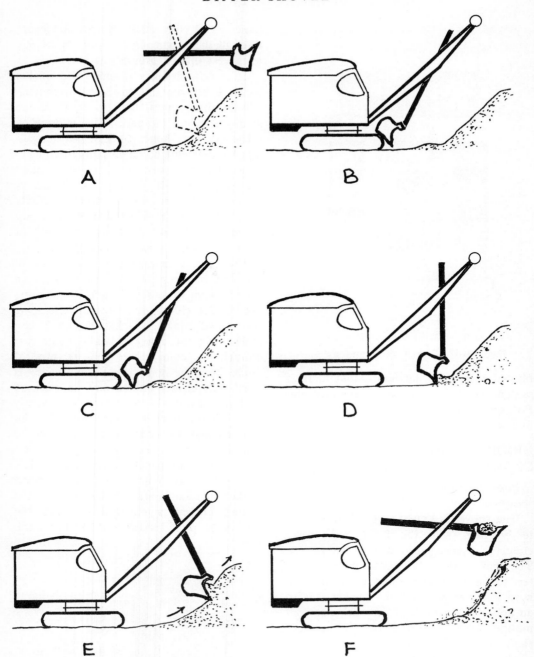

Fig. 13-69. Digging a bank

while still hoisting, and the shovel is now ready to swing.

To summarize, the bucket was swung back into digging position by first releasing the hoist brake, then the crowd brake. For the digging, the hoist clutch was continuously engaged, while the depth and angle of digging were controlled by the crowd-retract mechanism, which would also allow digging up or down slopes or even vertical steps.

If the bucket threatened to strike the bank or the ground while being swung back, the retract could be used to lift it without interrupting its backward motion. It is generally not necessary to start the digging back at the tracks, and the bucket can be dropped at any point by releasing the crowd brake and retract clutch. The usual procedure is to drop it at the toe of the bank.

Dumping. This shovel swings to the left when the swing lever is pushed forward. Well before the end of the swing, this lever should be pulled into neutral, the swing continuing from momentum until stopped by pulling the lever back to engage the other swing clutch just enough to act as a brake. If the swing is to the right the lever is moved in reverse direction.

During the swing the position of the bucket is adjusted by means of the hoist and crowd-retract controls so that it will be in correct position to dump as soon as it is over the truck.

If the truck is facing directly away from the shovel, the placement of the load in the front or rear of the body is regulated by the crowd-retract, placement from side to side by the swing, and the height above the body by the hoist; except that crowding beyond the boom point also raises the bucket if the hoist brake is held.

When the bucket is in correct position, the door in its bottom is unlatched by means of the dipper trip and swings down and back by gravity, allowing the contents to fall. If dirt sticks it can usually be dislodged by jerking with the crowd-retract lever, alternately engaging and disengaging both clutches. The load may be dumped in a heap or spread by moving the bucket as it dumps. The door is slammed shut by a quick downward or backward motion of the bucket. With the door closed the bucket can be used to tidy up the load on the truck, pushing and pulling the humps into the hollows.

Walking. As soon as the shovel has dug away so much of the bank that it does not reach enough to easily fill the bucket, the operator locks down the operating (hoist and crowd) brakes, engages the swing lock, disengages the swing jaw clutch and engages the walking jaw clutch by moving the swing-walk lever, then engages the forward swing friction clutch by pushing the lever forward, so that the machine moves forward. In his new position he releases the swing lever, disengages the walking jaw clutch, engages the swing jaw clutch, releases the swing lock and the operating brake locks, and is ready to work.

The shovel may be held in position against digging push by the ratchet lock on the live axle, which can be set to permit forward movement but hold against rolling back. If there is no ratchet it is necessary to set the traction brakes to dig and to release them to move. This might be done either manually or automatically, depending on the make and model of the machine.

Leveling. If the space into which the shovel is advancing is too rough to afford good footing it may be leveled by cutting off the humps with forward motion of the bucket, or by dragging the bucket backwards. It can be allowed to rest on the ground and be pulled back with the retract, releasing the hoist brake enough so that the bottom just brushes the loose material as it swings back.

A pit may also be leveled behind the shovel for the benefit of the trucks by holding the bucket in light contact with the ground and swinging the shovel. This will cause dirt to be pushed across the bucket marks smoothing them over, will pull wastage around the sides of a windrow back into digging reach, and will move loose rocks out of the way. It is a practice which is universally condemned by shovel manufacturers, as it imposes severe twisting strains on the stick, saddle block, and boom pins, shortening their life, or sometimes causing complete collapse. It will be seen that if the boom and stick are swinging with the momentum of their own weight and that of the shovel, and the bucket catches on an obstacle, the twist is severe.

However, no amount of disapproval will stop operators from using this technique as it is often the easiest way to finish work, and sometimes the only way. The twist may always do some damage, but if the front end is strongly built and the operator skillful enough to avoid contact with too-resistant material, the advantages will far outweigh the extra wear. No operator should do this, however, until he has acquired very precise control over the particular machine.

The shovel can also be swung while digging in order to square off a bank or boost a boulder or stump to the side. The same criticisms apply as to swing-scraping.

The drive chains of a dipper shovel should be kept to the rear for both traveling and digging. In this position, power applied to the bull wheels is exerted directly on the lower tracks where they are in contact with the ground, and there is minimum rocking back and forth under digging thrust. In addition, they are less likely to get in trouble with stones and loose dirt.

Pull shovels and draglines work with the chains toward the digging.

Tangled Cable. If the hoist clutch and brake are released when the bucket is in the air, it will fall, unwinding the hoist cable from the drum rapidly. When the bucket strikes the ground it will stop pulling on the cable but the drum will continue to revolve with its own momentum, pushing the cable. This will result in the cable leaving the drum and looping itself all over the machinery, or the drum revolving inside the wraps of cable, thus loosening all of them back to the anchor, or in both. If the hoist clutch is engaged and the cable wound back onto the drum the wraps will cross each other and tangle. If a heavy load is picked up by the cable in this condition it will be strained, deformed, or crushed. In addition to having its life shortened it will then be difficult to spool onto the drum smoothly unless under enough load to pull its kinks straight as they come in.

The easiest way to straighten out this mess is to manipulate the bucket so that it will pull most of the cable away from the drum. It is first necessary to engage the hoist clutch and reel in the disordered cable, then hoist the bucket, dumping it first if loaded. The bucket may then be crowded all the way out and rested on the ground, dropped in a hole, or pulled back close to the tracks. There will now be only a few wraps of cable on the drum, which can be straightened out with the gloved hand, or gently with a screwdriver and a ball pean hammer. The engine clutch should be disconnected before handling the cable as an inequality in the hoist clutch might cause it to tighten the cable with a jerk even while disengaged. Leather palm gloves should be worn because even a preformed cable may have projecting broken wires that cut like needles.

If it is preferred to straighten the cable without raising the bucket, the engine clutch and the hoist clutch and brake should be disengaged and the cable pulled off the drum by hand until it is possible to straighten out the wraps still on it. The engine and hoist clutches should then be engaged with the engine idling, and the cable guided cautiously into correct position by hand as it is reeled in. It is sometimes easier if the cable is stripped from the drum by pulling it from the live side of the bucket sheave, as the boom point sheave will then line it up for re-winding.

If the cable has been loaded at all while tangled, it should be put under heavy tension while fully extended to straighten out kinks.

Tangling the cable in this manner is one of the principal bugaboos for beginners. It can be avoided almost entirely by riding the brakes, that is, by keeping a light pressure on the foot brakes at all times—not enough to stop the drum from being turned, but sufficient to stop free spinning. This is bad operation as it gives the shovel a constant drag to overcome, increases lining wear, and heats the drums. However, none of these drawbacks are serious enough to counterbalance the wasted time and labor, the loss of confidence, and the damage to the cable resulting from tangles. When the new operator develops good coordination with the controls, he can then train himself to release the brakes fully instead of partially.

The crowd retract cables should not tangle unless loose. They are straightened by prying with a screwdriver, and working the stick back and forth.

Cable Breaks. Shovel cables are subject to shock, heavy loads, sharp bending, rapid motion, exposure to weather, and drag cables may be exposed to friction with earth and rock. They may last a few hours or for years, but sooner or later they will break unless replaced. If the break occurs during work, time is lost until another cable can be obtained and installed. If the break occurs at the wrong time, it may cause injury or death to personnel and damage to property.

A cable that is abused, is too small for its work, or is of poor quality or defective may break suddenly before showing any signs of wear. Usually, however, a wire rope will not break until weakened and inspection will show thinned and broken wires on its surface. Such a cable may last a long time but it is good operation to change it during maintenance time in order to eliminate work stoppage and possible damage when it fails.

Cables often break one strand at a time, giving the operator warning before parting completely. However, the fact that fifty cables may break in this manner is no proof that the fifty-first will not snap suddenly, either under load or reaction from load.

Torque converter drive prolongs rope life and reduces likelihood of sudden failure.

Replacing Cable. For safety's sake, the engine should be stopped before a cable is replaced. Some clutches may jerk a line even when they are disengaged, or a helper or a bystander might accidentally lean on a lever at just the wrong time. Drums may be turned and cable reeled in during work by using the starter, or by starting the engine, running it while necessary, and then shutting it off again.

A line that is being replaced is usually partially or wholly in place. If the machine is an unfamiliar one, it is of great impor-

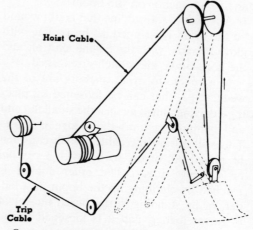

Hoist Cable

Trip Cable

Courtesy of Thew Shovel Company

Fig. 13-70A. Dipper hoist and trip lines

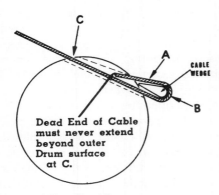

Fig. 13-70B. Wedge anchor to drum

tance to memorize its exact layout. If a service manual is on hand, the diagram should be checked to make sure it is both right and understandable. If there is no diagram, a rough sketch may be made as a memory help.

It is very embarrassing to pull a cable out and then not remember how it was reeved. Fortunately, shovels are less complicated this way than scrapers.

A typical hoist line reeving is shown in Figure 13-70A. The line runs from the hoist drum on the shovel deck over a boom point sheave, around the padlock or bucket sheave, over the other boom point sheave, and to an anchor on the boom.

Any part of the old line that is still wound on the drum is stripped off by pulling with gloved hands, with the engine shut off and the hoist brake and clutch disengaged.

Most cable to drum anchors are of the wedge type, that can be loosened by placing a driving punch against the small flat end and hammering until the wedge is loose. This will release the cable, which can then be pulled out of the socket with a pry bar or a screw driver. The other anchor is reached by climbing up the boom or a ladder, or by standing in the raised bucket of a tractor loader, and it is loosened in the same manner.

The old line should be rolled into a coil or coils and tied securely, even if it is to be thrown away. An old cable lying around loose is a nuisance and a danger anywhere—on the job, on a truck going to the dump, and in the dump also. But it is easily handled when coiled and fastened.

The new cable, cut to proper length ahead of time, or by measuring the old line, is drawn over the proper boom point sheave and down to the hoist drum. The end is pushed through the small end of the anchor socket and bent back into a loop, in which the wedge is placed. The slack is pulled back through the small end by hand, drawing the wedge into the socket. It is forced in further by a couple of hammer blows, just enough to hold it in place until tension on the cable seats it firmly.

The drum is turned to reel in the line until the other end is near the bucket sheave. That end is then reeved through this sheave, over the empty boom point sheave, and is anchored to the boom. Putting the cable on the drum and then drawing it off in this manner is a precaution against kinking it.

With the line fastened at both ends, the engine is started and idled, and the hoist clutch is engaged to wind the slack cable onto the drum. It should wrap smoothly, with each loop lying against the previous one. A man working alone will guide it from near the drum with a gloved hand; a helper can pull the slack down to the bucket sheave and keep light tension on the line from there. Some resistance or back pull helps to make it wind snugly.

The anchor clamps are tightened by running the empty bucket down and out until the cable is stripped from the drum, and stopping it abruptly with the brake. This should be done two or three times. It is important to keep the cable from running slack for the first hour or two of operation, as it is most likely to cross loops on the drum before it is broken in.

Badly worn or damaged cables should be replaced in non-working time before they break. Down time on a shovel loading a string of trucks costs much more than the cable and the labor of changing it.

Sometimes a cable is cut to the wrong length by mistake, or it is convenient to use a pre-cut line of slightly different length. One that is too long may be hard to fasten securely to the drum, as the extra wraps make it hard to put enough tension on the anchor to seat the wedge. If the drum is exactly the right size to hold the correct length of line, the extra will start a second layer, causing severe wear at the point where it snaps over.

If the drum is small enough so that there is overlap with standard length line, or wide enough to take the extra on the first layer, this causes no problem.

A hoist line that is too short may limit bucket movement, and is likely to cause a damaging jolt by the anchor stopping it if it is run out too far. An instruction book may specify a cable that is too short for comfortable operation.

The hoist is the most frequently broken line on a dipper shovel. Under severe conditions it may break several times a week, but this much trouble usually indicates overloading or other abuse. A too-large cable will be damaged by pinching in sheaves, one that is too small breaks from lack of strength.

Procedure is similar for a boom cable, except that the boom must be rested on the bucket before taking off the old cable, and reeving between the A-frame or gantry and the boom point is usually a four-part line that requires more care getting on right.

In order to change a crowd or retract cable the turnbuckle fastening the retract to the bucket must be loosened. The new cable or cables are cut carefully to proper length, and installed to be snug with the turnbuckle loose. Tightening will put it under proper tension.

Safety Precautions. An operator should lower the bucket to the ground before leaving the machine as the brakes may lose their grip as they cool and let the bucket down, gradually or with a rush. There is also the possibility of a would-be operator accidentally releasing the hoist brake while an admiring friend stands under the bucket.

Workmen and spectators should always keep beyond the furthest reach of the shovel, if possible, but they seldom do. A point to remember when in danger of being hit by the bucket is that a shovel is weak on the infighting, and that it may be safest to run toward it where the slower speed of its motion makes it easier to dodge and less damaging if it does connect.

The engine clutch should always be disengaged during maintenance work, except when it is necessary to move parts for access.

Booming Up and Down. The boom is normally held in a fixed position during digging. However, lowering the boom enables the bucket to dig deeper and further away and raising it permits a higher dump point. If desired, the boom latch may be locked out of engagement, the boom lowered to dig, raised during the swing to dump, and lowered again on the swing back. This uses extra power to lift the boom and may slow the digging

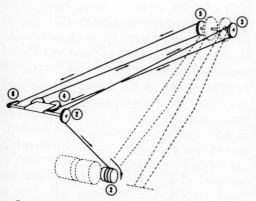

Courtesy of Thew Shovel Company

Fig. 13-70C. Dipper boom reeving

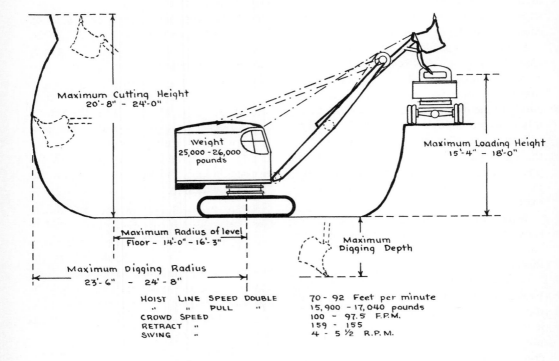

Fig. 13-70D. Weight, range, and speed of half yard dipper

cycle, but often simplifies the work. It is only possible with a live boom.

Booms should always be lowered slowly and stopped with the brake, not the ratchet.

Applications. The dipper shovel is a good machine for truck loading. It is accurate because of the three-way control of the bucket and the ability to dump in any position. It is fast because only a small part of its weight—the stick and bucket —are involved in the digging motion, and because of its ability to dig in any spot it can reach with minimum waste motion approaching and leaving the cut.

The dipper is also the attachment for hard, heavy digging—hardpan, boulders, ledge, and blasted rock. The weight of the boom, backed to a varying extent by the weight of the shovel itself, holds the bucket to its work. The effective cutting angle of the teeth in a bank, aided by the

variable direction of pressure provided by the hoist and crowd, enable it to cut highly resistant material, and to break up cracked and fissured rock formations. The shape of the bucket enables it to pick up objects much larger than itself without chaining. Larger size shovels are increasingly effective at this type of work.

However, it is necessary that the shovel be able to move into its work constantly in order to retain its effectiveness. Solid rock projecting from the pit floor will prevent getting at the bank unless ramps are built over the rock, or bulldozers used to push the bank to the shovel. In digging below the tracks, the downslope should not be much over 20 per cent if the machine is to follow the work. If it is to stand at the edge and dig, it can cut down steeply five to seven feet, but is limited as to width of cut, and unless digging against a bank may push away about

Fig. 13-71. Long boom for stripping overburden

as much material as it picks up. Narrow ditches can be dug effectively only in ground free from large boulders which would tear up the ditch sides, and firm enough so that the shovel can be worked straddling its ditch, with or without platforms.

Reach and Speed Data. Figure 13-70 indicates the working range and speed of half yard dipper stick shovels. These are taken from the published specifications of several standard makes. For simplicity, the differences caused by boom angle have been omitted, and each distance is the maximum that can be reached from the most favorable angle. Live boom machines can operate according to these figures for depth of digging, reach, and dumping height, by raising or lowering the boom during the swing. Other machines can be adjusted for maximum performance for one or two, but not for all

of the positions shown in the illustration.

A shovel does not work efficiently at its maximum reach as it loses in lift, in ability to fill the bucket in a short sweep, in tooth angle, and in accuracy of control. Good digging practice is to keep the shovel close to the work and trucks down at its level for easy, accurate loading. Special situations often arise, however, where every inch of reach must be utilized.

Swing speeds are quite variable, ranging from four to five and a half revolutions per minute. This difference is not very important, as the gain in speed usually means a loss of power. A slow swinging shovel may swing from digging to close dumping position more quickly than a fast swinging one of the same power because of more rapid acceleration, but will be slower on a long swing.

Special dipper shovels, called stripping

Fig. 13-72. Loading out of a bank

shovels, are made with booms and sticks proportionately much longer than those shown. Counterweight and power may be increased, or bucket size reduced to compensate.

Operating Suggestions. The following is a list of suggestions for efficient dipper stick operation:

1. Keep bucket teeth sharp and built up to proper size.
2. Don't crowd bucket so deep into the bank that it slows the hoist.
3. Pull bucket out of the bank as soon as it is full.
4. Keep close to the bank unless it is likely to slide.
5. If bank is very hard, or material needs mixing, make passes in it with bucket door open.
6. Place shovel so that the drive chains are away from the digging.
7. Spot trucks as close to digging as safety permits to save swing time.
8. Spot trucks on both sides if possible to save waiting time while trucks move.
9. Spot trucks so that they line up with arc of swing, or back them directly toward the shovel.
10. Don't swing over truck cab if you can avoid it.
11. Move up while waiting for trucks.

STICK SHEAVES

HOIST CABLE
(3 or 4 part)

JACK BOOM

BOOM LINE

DRAG CABLE
(SINGLE)

NORTHWEST

STICK

PADLOCK SHEAVE

DRAG CABLE
(two part)

BUCKET

BOOM

J. O. JOHNSON
APPLETON, WISCONSIN

Courtesy of Northwest Engineering Company

Fig. 13-73. Cable hoe

12. In mud, load driest material in bottom of truck, sticky stuff on top, for easier dumping.

13. Don't let tracks sink deep in ground—bury mud with dry fill or use poles or platforms.

14. Work smoothly—slamming and jerking are hard on machine and operator.

PULL SHOVEL OR HOE

The dipper shovel is primarily designed to dig at its own level. The pull shovel, Figure 13-73, also known as a hoe, backhoe, dragshovel and ditching shovel, is at its best digging below its own level.

Most hoes are now made with hydraulic operation, page 13-101, with more efficient "wrist-action" control of the bucket. A few cable hoes have this improved bucket.

But existing full-cable-operation hoes will continue to do good work for years, and deserve a careful description here.

Booms. The hoe attachment has a jack boom, a boom, and a stick. The two booms may be hinged on the same or separate pins in the boom foot, or one may be hinged to the foot, and support the other. Some makes use a convertible boom that can be used with either a dipper or a hoe by changing sheaves.

The jack boom supports the hoist sheaves. The main boom holds the stick hinge at its upper end. It either carries a sheave on each side, or a differential drum to guide the two-part drag cable. It may be straight, or "gooseneck" to permit pulling the bucket higher.

The stick is hinged to a double sheave at its upper end, and is attached to the bucket at the bottom.

Bucket. Figure 13-74 shows a general purpose bucket. Its open or front end faces toward the shovel when the stick is vertical. The bottom has a lip and teeth similar to those in the dipper bucket. It usually carries a pair of side cutters. Two types are shown.

Courtesy of Bucyrus-Erie Company

Fig. 13-74. Hoe bucket

The bucket is wider in front than in the rear for ease of dumping. The cutters are wider than the bucket and protect it against side wear and jamming. The width of cut is determined by the distance between the outside edges of the cutters.

Widths of half yard buckets vary from twenty-one to forty inches. Narrow ones are for bucket width trenches, wide ones for cellars and pit digging, and intermediate widths for general purpose work.

Narrow buckets do not fill as well as wide ones, especially in shallow work, and they do not dump as cleanly.

This bucket is fastened to the stick at the front with a pin, and at the back by a pair of angle irons bolted to the stick. There are several holes through which this bolt may be placed so that the angle of the bucket may be adjusted. A high fastening points the teeth down for better penetration at depths; a low fastening enables the load to be swung and dumped with less fall-off. Between the rear arms there is usually a back plate to catch spillage, with a narrow gap between it and the bucket back to allow water to drain off.

Brackets on the upper front part of the bucket provide fastening for short chains that are attached to the padlock sheave

case. This sheave must be heavily protected as it is frequently dragged through dirt and rocks.

The standard or solid bottom bucket can be dumped only by overturning.

Drop bottom hoe buckets, tripped in the same way as dipper buckets, are seen occasionally. The door is the bottom and includes the lip and teeth. It is hinged at the front and latched at the back.

A dump bucket enables the operator to dump in one spot without the excessive hoisting discussed below. However, it does not overcome the tendency to spill while preparing to dump, and it is necessary to dump it at a greater height to allow room for the door to swing.

The most efficient backhoe bucket is the wrist action type, with digging pitch and dumping controlled by a hydraulic ram. These are standard on the hydraulic machines, and are standard or optional on some cable rigs.

Reeving. Figure 13-75 shows a diagram for reeving a hoe. The jack boom line is the same pattern as the dipper's boom line, but it is much shorter.

The hoist line may be three part for

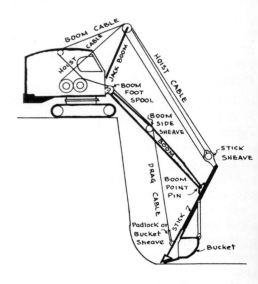

Fig. 13-75. Hoe reeving

ordinary digging, or four part for hard or heavy work. The cable goes from the drum over a large jack boom sheave, through a stick sheave, around the other jack sheave, and may then be anchored on the stick sheave case, or reeved through it and back to an anchor on the jack boom.

The drag cable usually runs from the drag (digging) drum, over the boom foot spool to the left side sheave on the boom, through the bucket or padlock sheave, over the right boom sheave, and to an anchor on the boom.

Some shovels use a differential drum, Figure 13-76, mounted on the top of the boom. This has a center section which is twice the diameter of the two side drums. The inner drag cable is powered by the drag drum on the deck and anchored on the center differential drum. The outer drag cable is threaded through the center drum, wrapped on the outer drums, and is anchored on the bucket chains.

Winding in the inner cable turns the differential drum and winds the outer cable onto the side drums which, being half the diameter of the center drum, exert twice the pull.

This construction avoids using a padlock sheave that must be dragged in the dirt. However, the inner cable is confusing to replace or adjust, and failure to center the outer cable exactly may result in one side taking most of the pull, with resulting breakage.

The jack boom is pulled forward by the weight of the boom resting on the hoist line, and held back by the boom cable. It is usually in a nearly vertical position, a little forward for deep digging, a little back for scraping down high banks.

Hoist and Drag. The functions of hoist and drag in a back hoe or pull shovel are inter-related in a very complex manner.

If the hoist clutch is engaged the stick sheaves will be pulled toward the top of the jack boom, the stick will pivot on

the boom end, and the bucket will move out as in Figure 13-77 (A). If the drag brake is held, or the bucket encounters resistance so that it cannot move out, the stick will be held against pivoting, and stick and boom will be pulled toward the jack boom as a unit. The boom cannot move in a straight line but pivots on its foot pins so as to rise as it comes back as in (B).

If the hoist brake is locked and the drag slipped, the bucket will move out and somewhat down, (C).

If the hoist brake is slipped and the drag brake is held, the boom and stick will swing down in an arc, as in (D). If the drag clutch is engaged while slipping the hoist brake, the bucket will move in horizontally, as in (E).

If the hoist brake is locked and the drag clutch engaged, the bucket will move in, but the stick, pivoting on the boom, will move the stick sheave out, tightening the hoist cable and causing boom and bucket to rise, as in (F).

Starting Cut. The bucket can be moved out horizontally from any raised position by releasing or slipping the drag brake so that the bucket swings out, and engaging the hoist clutch enough to prevent hitting the ground, Figure 13-78 (A). When it is almost fully extended or is at the desired distance, the drag brake is applied and the hoist drum is released. The boom, stick, and bucket will fall together striking the teeth on the ground with a pickaxe effect, (B). For soft digging the drop is cushioned with the hoist brake. In penetrating a hard surface the brake is applied

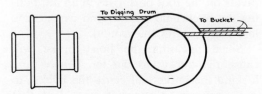

Fig. 13-76. Differential drag drum

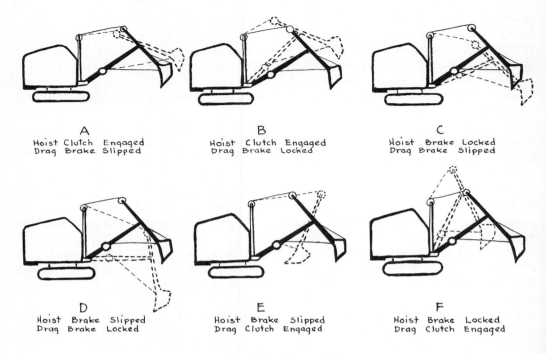

A
Hoist Clutch Engaged
Drag Brake Slipped

B
Hoist Clutch Engaged
Drag Brake Locked

C
Hoist Brake Locked
Drag Brake Slipped

D
Hoist Brake Slipped
Drag Brake Locked

E
Hoist Brake Slipped
Drag Clutch Engaged

F
Hoist Brake Locked
Drag Clutch Engaged

Fig. 13-77. Bucket positions

just as or after it strikes, in time to prevent cable from spinning off the drum.

If the drag clutch is then engaged and the hoist released, the bucket will move toward the shovel with its teeth in digging position. If the ground is normal the bucket will rapidly cut its way down until full, then resist any further pull.

If in the (B) position the drag clutch is engaged and the hoist brake locked, the bucket will move in causing the stick to pivot on the boom end. The stick sheave, held from pivoting forward by the hoist cable, acts as a fulcrum and the stick, now a second class lever, forces the boom point toward the shovel. The boom pivots on the foot pins, rises in an arc pulling the bucket up with it, and causes it to lose contact with the ground.

Since releasing the hoist brake will cause the bucket to dig too deeply, and locking the brake will cause the bucket to rise above its digging, it is evident that keeping the bucket at the desired depth

while being pulled in may be accomplished by either partial or intermittent use of the hoist brake. A smooth, sensitive brake can be partially engaged; a jumpy one is applied in a series of light jerks.

The bucket first acts somewhat as a dozer blade and pushes dirt (C), then curves under it and fills, and as it passes under the boom point the teeth start to point upward (D), and come out of the ground. The hoist clutch is engaged and drag continued until the bucket sheave guard nearly touches the boom, (E), when the drag clutch is released and the drag brake is applied with hoist continuing during the swing.

Dumping. When dumping position is reached, usually a 35° to 90° swing for ditching, or 45° to 180° for loading trucks, the drag brake is released and the bucket swings outward, being prevented from grounding by holding the hoist brake, or engaging the hoist clutch. Material on the teeth and lip falls first, then that in the body of the bucket, and last, dirt that is lightly stuck to the sides and back. More adhesive stuff may be shaken out by applying and releasing the drag brake in jerks during the dump.

The dump may be spread out over a distance of some twelve feet but is likely to be concentrated in a shorter distance. Loose, dry material dumps near the machine, and compact, sticky dirt at the furthest point where the bucket bottom is vertical or overhanging. The whole contents of the bucket may be dumped in the nearer area by only partially releasing the drag brake, so as to keep the bucket over one spot as the hoist pulls it up and overturns it, position (H).

After dumping, the bucket is dropped again at the beginning of the ditch and pulled in digging another full length slice, and this maneuver is repeated until the desired depth is obtained. The digging may also be done by cutting the bucket in

deeply after the first penetration, and lifting when filled, leaving the remainder of the slice for the next bite. If the spoil is to be dumped immediately alongside the ditch, spillage is unimportant, and the bucket can be hoisted and dumped without pulling it close to the boom.

Push Back. If a large amount of earth is to be dug, it is good practice to push the spoil pile back while dumping on it. This may be done by releasing the drag brake before hoisting far enough to clear the pile. The loaded bucket swings down and out by gravity, striking a blow against the heap which, if correctly aimed, will knock the top off. The momentum of the bucket plus the weight of the boom against the stick and the continuing hoist pull give a follow-through which should push the dirt out beyond dumping range, making room for the bucket contents in the pile. This operation should be started while the pile is low and the blow has

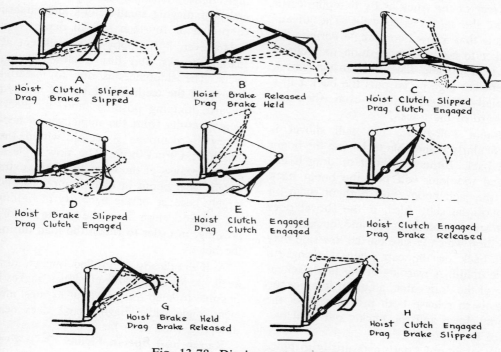

Fig. 13-78. Digging and dumping

maximum force with gravity assisting.

This technique may also be used to smooth over loose ground on which the shovel must walk, to boost boulders out of the way, and to dispose of piles left at the near end of trenches. The novice should use it with caution as mistakes in aim or timing will cause cable tangles.

Loading Trucks. The most convenient position for a truck to take to be loaded by a backhoe is very close to it to reduce spilling, and backed toward it to allow dumping in the full length of the body. The danger in this position is that a broken drag cable or other accident may permit the bucket to swing outward, taking the truck cab with it. For this reason trucks are often loaded from the side.

A pull shovel loads trucks more easily if it is on a higher level, so it is good practice to have the trucks in the pit or to build a dirt platform for the shovel. This last may be done by utilizing the spill at the rear of the truck or by dumping earth where the shovel is to walk. The advantage is that a low boom angle permits the bucket to be held in position to retain a full load at some distance from the shovel, whereas a high boom puts the bucket in a partially dumped position even when just forward of the tracks.

Boom Fall-Back. The pull shovel is particularly vulnerable to having the boom fall back. Either the hoist or drag if engaged too long will pull the boom back on the jack boom, and both of them on top of the cab. Also, if the shovel does not stand level a boom and bucket in a position which is safe on the low side may fall back on the cab when it swings to face uphill. A warning of this is a slackening of the drag cable, a cure if it is not too late is to release both brakes and let the boom down and the bucket out until balance is restored.

A structural safety measure is to hinge a piece of pipe to the A frame, and another to the jack boom, of such sizes that one will slide inside the other. A stop pin may be put in either to prevent them from telescoping together far enough to allow the jack boom to touch the cab, without interfering with normal use. If the A frame is a folding type, it might require bracing.

Applications. The pull shovel shares with the dipper stick the advantages of a rigidly attached bucket, with resultant control of digging, and ability to hold the bucket teeth to their work. In addition it is able to work below the grade on which it stands, keeping itself and sometimes attendant trucks out of the mud, rock, and ramp difficulties which often hamper digging from the floor.

The pull shovel is adept at stripping top soil, making shallow cuts and removing windrows without losing material at the sides, at removing overburden from rock prior to quarrying, at scraping down high banks, and particularly at its specialty of digging small cellars and ditches to twelve or fourteen feet in depth.

However, it is sloppy and inefficient at loading trucks and has a slow digging cycle. The sloppiness is due to material falling off the teeth as the bucket is lifted toward dumping position, and the inefficiency results from the maneuvering necessary to complete a dump within the length of a truck body. The slow cycle is partly because of the carrying of the extra weight of a boom in digging motions and largely because of the necessity of pulling the bucket close into the boom before hoisting, in order to retain the load during the hoist.

A hydraulically controlled bucket, now available for some cable machines, makes it possible to keep the bucket at its most efficient angle while digging, and does neat dumping, close in or far out.

Range and Speed. Figure 13-79 gives data for half yard hoes. The boom is al-

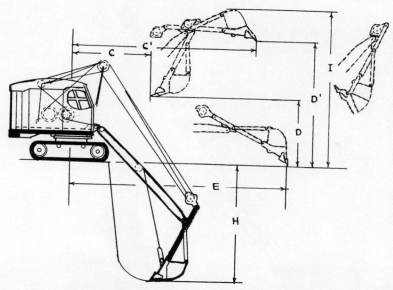

	MIN.	MAX.			MIN.	MAX.	
BOOM LENGTH	14'-0"	17'-0"	HOIST LINE SPEED - SINGLE	180	184	F.P.M.
STICK	6'-11"	7'-9"	HOIST LINE PULL - SINGLE	8,000	8,500	lbs.
H DEPTH	12'-0"	16'-0"	HOIST LINE NO. PARTS	2	4	
C RADIUS - BEGIN DUMP	9'-0"	9'-0"	DIGGING LINE SPEED - 2 PART	62	90	F.P.M.
C' RADIUS - END DUMP	19'-6"	22'-6"	DIGGING LINE PULL - 2 PART	12,800	32,000	
D CLEARANCE - BEGIN DUMP	8'-9"	9'-6"	WEIGHT	25,000	30,700	
D' CLEARANCE - END DUMP	14'-0	16'-4"				
E DIGGING RADIUS	22'-9"	27'-9"				
I MAX. HEIGHT	19'-6"	22'-0"				

Fig. 13-79. Range and speed, half yard hoe

ways live and the angle of the jack boom has little effect on the digging, so that the performance shown can be attained by either live or dead boom units.

The bucket fills best well below the tracks, but it loses penetration near maximum depth. This is because the teeth cannot strike the ground at a steep enough angle, and the direction of pull is more upward than inward.

Hoes are not usually made with bucket capacities greater than two yards.

Production. There are so many variables in hoe operation that a table of output would be of little use. Yardage may run from fifty to eighty percent of that moved by a dipper stick, under conditions favorable to both these types of attachment.

Suggestions for operation. The following suggestions for pull shovel operation may be of use:

1. Keep bucket teeth sharp and built up to proper size.
2. Line shovel up to cut outer edges of excavation in a straight line.
3. In deep digging, keep a fairly straight face and keep shovel back from it as far as possile. Don't work out on a "peninsula" which may cave.
4. Keep knocking the spoil pile back as you dump on it. You may need the space.

Courtesy of Northwest Engineering Company

Fig. 13-80. Dragline

5. Don't dig yourself into a trap—the spoil pile may slump across your way out.

6. Don't let the drag slide the shovel into trouble when the bucket hooks something solid.

7. Don't work with the drag cable in the dirt. If a pile builds up in front of the tracks, back away and knock it off.

8. Try to keep the shovel higher than trucks it is loading.

9. Don't pull boom back on top of shovel with drag or hoist.

10. Keep boom low while walking, or while swinging uphill.

11. Don't walk, or swing uphill, when drag cable is slack.

DRAGLINE

A dragline attachment is shown in Figure 13-80. It has a long light crane boom,

with a fairlead set at its foot, and a bucket attached to the machine only by cables.

Boom. The boom is of lattice construction. It may be of welded steel angles, angle corners and tubular braces, or all tubular. Very long booms sometimes use aluminum sections near the tip.

Each boom is made up of at least two sections, tapering from their bolted center connection toward each end. The bottom is reinforced to hinge on the boom foot pins, and the top to hold the point sheaves.

Additional sections, usually in lengths of five or ten feet, can be placed between the upper and lower sections to obtain extra reach or dumping height.

If the boom is intended for dragline work only, it carries one large sheave on the point, but if it is to be used for a clamshell also, it has two. A smaller sheave is carried on each side for the boom support line.

Fairlead. The fairlead is a device

13-68

mounted on the boom or boom foot, which lines up the drag cable to spool smoothly onto the drum, even when the bucket is far off to the side.

The model shown in Figure 13-81 has a pair of pulleys mounted in a frame that swivels around a guide tube or throat. Rub plates or rollers are set forward of these pulleys, and pressure of the cable against them swings the fairlead into proper position to receive it between the pulleys. The throat lines it up so that it will feed through them properly when spooling off the drum.

Another common type, 13-82, has the front pulleys mounted in a frame on a vertical hinge, and a pair of horizontal sheaves in a frame behind them.

Bucket. Figure 13-83 is a picture of an Erie general purpose bucket, which is somewhat similar to a pull shovel bucket without stick fastenings and side cutters. A pair of drag chains are attached to the front of the bucket, through brackets by which the pull point may be moved up or down. The upper position is used for deep or hard digging as it pulls the teeth into a steeper angle.

The drag chains converge in a drag yoke to which the drag cable is fastened. The hoist (bail) chains are attached to pivot (trunnion) pins toward the rear of the bucket sides, rise vertically to a spreader bar, then converge to fasten to the dump sheave housing, which in turn is fastened to the end of the hoist cable.

Dragline buckets are made in various weights: light ones for digging soft earth, and rehandling stockpiles of material; medium weight for general work, and heavy and extra heavy for deep and rocky digging. A light bucket means less weight to be lifted each cycle; a heavy one has better penetration and wear resistance. Light buckets may sometimes be obtained with a toothless cutting edge which is excellent for stripping soft topsoil, grading, and cleaning up.

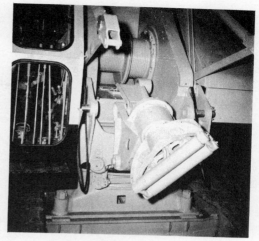

Courtesy of Koehring Company

Fig. 13-81. Fairlead with roller guard

Perforated or sieve buckets are standard buckets with a number of holes cut in the back and sides. These are useful in wet digging as water is pushed through the holes by incoming dirt and any remainder drains out while the bucket is being lifted. Water can be almost entirely manipulated out of a standard bucket if it is possible to take a deep bite so that a massive chunk of earth will push it out, particularly if the bucket can be pulled up a steep bank to

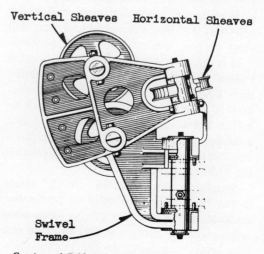

Courtesy of Baldwin-Lima-Hamilton Corp.

Fig. 13-82. Four-sheave fairlead

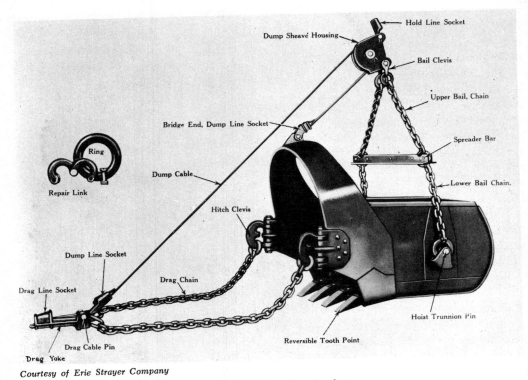

Labels in figure:
Hold Line Socket
Dump Sheave Housing
Bail Clevis
Upper Bail, Chain
Bridge End, Dump Line Socket
Spreader Bar
Ring
Dump Cable
Lower Bail Chain
Repair Link
Hitch Clevis
Dump Line Socket
Drag Chain
Drag Line Socket
Drag Cable Pin
Hoist Trunnion Pin
Reversible Tooth Point
Drag Yoke

Fig. 13-83. Dragline bucket

dump any remaining water. However, if it is not possible to get a good bite the perforations are necessary to avoid profitless carrying of water and sloppy spoil piles or loads. Very thin mud or fine dry soil may be lost through the holes, but most digging, wet or dry, can be handled.

It is questionable whether the perforated bucket dumps sticky mud more readily than a solid one, as the reduction in suction is counterbalanced by mud sticking in the holes. The slat bucket, also called a basket, is used for solid, sticky materials that come up in chunks, and is not suitable for ordinary excavation.

Some operators weld ⅜" or ½" chain in the rear corners of solid or perforated buckets, as the slapping of the loose ends helps to dump sticky soil and to clean out thin layers remaining on the bucket sides and bottom and in corners after dumping.

The effectiveness of penetration of a dragline bucket decreases with depth below the machine, as the drag cable then pulls in a more upward direction, raising the teeth out of the soil. This can be compensated for in part by reversing or sharpening the teeth; by using a longer boom which, by permitting digging farther from the shovel, decreases the upward angle of the drag cable; or by fastening the drag cable higher on the bucket. Larger and heavier buckets dig much better at the same depth and distance.

Choice of bucket size is determined by the materials to be handled and the length of the boom. For example, a ¾ yard machine usually has a 40-foot boom and uses a ¾ yard general purpose bucket. However, if the material to be dug is very heavy or tends to come up in amounts more than the bucket capacity, if a longer boom is used without extra counterweight, or digging is so hard or abrasive that a

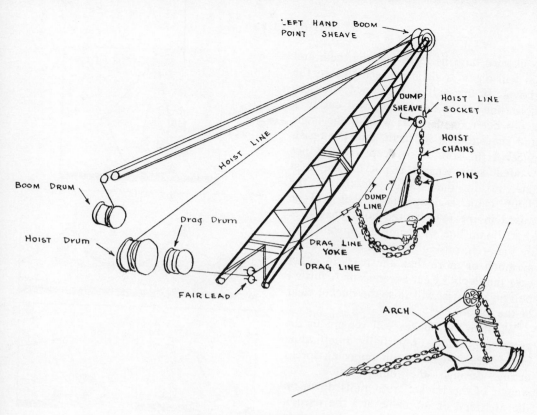

Fig. 13-84. Dragline reeving

heavy duty bucket is needed, ⅝ yard capacity should be more satisfactory. The same machine might use a ⅞ or one yard light bucket on a standard boom in handling coal or dry humus.

A plank fastened across the mouth of a bucket will convert it into a backfiller.

Reeving. The dump cable runs from the top of the bucket arch over the dump sheave and forward to the drag yoke.

The boom line is a standard four-part rigging, similar to the dipper boom support, except that a longer cable is needed. The hoist line runs from the hoist drum over a large boom point sheave and down to the dump sheave case. The drag cable runs from the drag (digging) drum through the fairlead to the drag yoke. See Figure 13-84.

Some machines have a light multiple line from the boom hoist drum to a hanging padlock sheave set, and a heavier two part line from there to the boom point. See Figure 13-103.

This shortens the inner cable, which is most subject to wear. One inner line can be used with different boom lengths.

Bucket Action. If the bucket is lifted with the hoist while the drag cable is slack, it will hang in fully dumped position. If tension is then put on the drag cable, it will pull on the dump cable before the slack is out of the drag chains. The dump cable will pull the front of the bucket up, toward the dump sheave, 13-84. Releasing the drag cable will allow the dump cable to run back over the sheave, and the bucket will return to dump position, pivoting on the hoist chain pins.

If the bucket is then lowered to the ground, it will turn to a horizontal position, or will rest on its teeth and arch, depending on its balance. A pull on the drag cable will now tip the bucket forward or backward onto its teeth and the teeth and lip will dig in as it is dragged toward the shovel. If the pull is continued with the hoist line slack, the bucket will cut to a depth determined by its weight, the angle and sharpness of its teeth, and the resistance of the soil. If so deep a cut is not wanted some tension is put on the hoist line, raising the bucket slightly. If the dump cable is long the bucket will be raised in the rear by the hoist chains. A short dump line will cause an upward pull on the arch, raising the front of the bucket as much or more than the rear. In either case the depth of cut is reduced. A further pull on the hoist will raise the bucket clear of the ground.

Whether the bucket will remain in the carrying position, or partially dump while being raised, depends on opposing forces acting through the dump cable. The weight of the bucket front pulls down on the arch end of the cable and the tension between the hoist and drag cables tends to stretch its drag-yoke-to-dump-sheave section and pull the bucket up. In effect, the dump cable must pinch the other two cables together in order to obtain slack to drop the front of the bucket. This pinching requires comparatively little force when the angle between the cables is small and becomes increasingly difficult as the angle flattens out. Also, if the dump cable is long it will not have to pull as strongly on the two cables to obtain slack as if it is short. The action of the bucket will therefore depend on the angle between the hoist and drag cables, the length of the dump cable, and the weight and distribution of the bucket load.

A wide angle between hoist and drag cables can be had when picking the bucket out of the soil, either by pulling the bucket close to the shovel, or by keeping the boom at a low angle. Bringing the bucket all the way in usually wastes time and causes wear on bucket and chains which

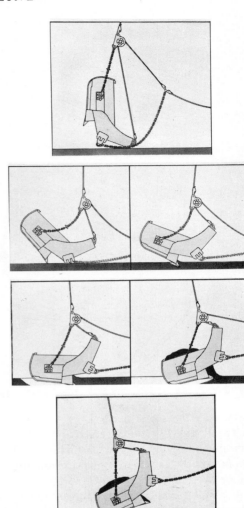

Fig. 13-85. Bucket action

might be avoided if it were picked up as soon as full. A low boom has a tendency to tip the shovel when heavily loaded, and often cannot be used because of obstructions or height of dump. A short dump cable makes it difficult to dump except directly under the boom point. If a live boom is used it is possible to dig low and dump high, but this takes extra time and work.

The technique used will depend on the

job and on the operator's preference. There is generally at least one good method of handling any situation.

Digging. In an ordinary dragline digging cycle the bucket is not thrown or cast. It is lowered into the pit with both lines taut, the hoist brake being almost wholly released, then re-applied smoothly as the bucket is about to strike the ground, and the drag brake is released enough to allow the bucket to drop straight instead of following an arc centering on the fairlead.

When the bucket rests on the ground the hoist cable is slackened slightly and the drag clutch engaged. The drag cable pulls the bucket, with the teeth digging in and cutting a slice of dirt which piles inside the bucket. If the hoist brake is locked, the bucket will move up in an arc

centering on the boom point, and on level ground may pivot so the teeth dig more sharply but no longer have the full weight of the bucket to force them in.

Ordinarily, the hoist brake is released enough to let the bucket cut level or follow the pit contour. If the pit slopes up toward the shovel, which is the most favorable digging condition, it may be necessary to partially engage the hoist clutch to avoid digging in, or to prevent the hoist line from becoming too slack and allowing the chains and dump sheave to slump into the bucket.

Hoisting. When the bucket is filled the hoist clutch is fully engaged and the bucket is lifted clear of the ground. If the bucket has a tendency to dump, the drag may be left engaged until the angle between drag and hoist cables is sufficiently wide to hold it. The drag clutch is then released and the hoist is continued, with the drag brake applied just enough to allow the hoist to pull the bucket forward and upward under the boom, without slackening the lines enough to dump it. If the drag brake is applied too tightly the bucket may hit the boom.

The swing is started as soon as the bucket is clear of the ground. This is the period of heaviest load in the dragline cycle, as the machine is simultaneously lifting a load, pulling against the drag brake with the hoist clutch, and swinging. If overloads are picked up in the bucket they may so slow the line and swing speeds that less dirt will be moved per hour than if smaller bites are taken.

Dumping. Hoisting is discontinued as soon as the bucket is high enough to be dumped. When the swing is completed, the drag brake is released, partially or wholly, and the bucket swings out and dumps. A long dump cable will permit normal dumping inside the boom point, a short one under it only. Raising the boom will bring the dump point closer,

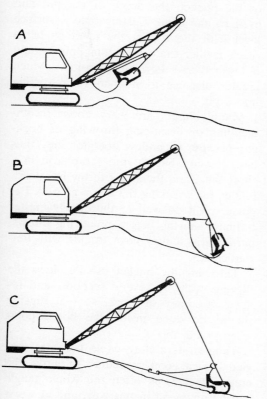

Fig. 13-86. Casting the bucket

lowering it puts it at a greater distance.

It is poor operation to raise the bucket higher than necessary before dumping. Clearance should be allowed so as not to strike trucks or other receptacles, but piles can be barely cleared. This saves time on a short swing and fuel and wear and tear under any circumstances.

Casting. If the bucket is to be thrown or cast it is pulled close in by the drag cable during the swing to the pit, with the hoist line held at a length that will keep the bucket above any obstructions during the cast, as in Figure 13-86. When the swing is completed, the drag brake and clutch are released and the bucket swings outward like a pendulum. When it is just short of the furthest point of this swing, the hoist brake is released and the bucket falls downward and outward (B). It is checked by gradual application of both brakes just before hitting the ground, as otherwise it might be damaged, over-turned, or tangled in its chains. It should then rest on the ground in the same position as if lowered and responds to drag pull similarly, except that if the pit floor is level, the hoist cable slackens as the bucket approaches the boom point and tightens again after it has passed under it.

A swing throw may also be used. During the swing from dump to pit, the drag cable is left slack and the hoist held so that the bucket is high enough to clear obstructions. The centrifugal force of the swing will pull the bucket outward, and the hoist brake should be released at the proper point so that the bucket will land where it is wanted. The swing should be checked as soon as the hoist brake is released, and hoist and drag brakes applied gently as or just before the bucket strikes. This swing throw requires much more expert operation than the other type and may damage the shovel seriously if improperly done.

Another technique is to make a pendulum throw while the shovel is swinging so that the centrifugal force adds to the outward sweep of the bucket.

Throwing to dump is a similar process, but usually the height of the piles prevents the use of a long enough hoist line to obtain much distance. The weight of the load, however, causes the bucket to cast farther than it could on the same length line if empty. A combination pendulum and swing throw is most effective, particularly if the shovel is dumping at 180° from the pit and can revolve in a full circle so that it can dump without stopping.

If the load sticks in a bucket which is thrown to dump, it may overturn the shovel.

The distance the bucket can be thrown is affected by the skill of the operator, the length and angle of the boom, the weight of the bucket, the depth of the pit, and even by the wind. Casting onto a surface level with the tracks, the teeth ought to reach farther than the boom point would if lowered to a horizontal position. Much greater distance is attained when casting into a pit.

Manufacturers are wary about recommending or discussing throwing of buckets, because a careless operator may thus bang a good bucket into scrap iron in a short time, and the swing throw has possibilities for wrecking the boom as well. Beginners will do well to thoroughly master ordinary digging before practicing throws, particularly if they have trouble with tangled cables.

Throwing the bucket usually slows the digging cycle by several seconds, and reduces accuracy of work, so that digging should be done inside the boom point when practicable.

Tangles. If a drag cable becomes tangled it is a good procedure to throw the empty bucket, as often the whole tangle can be unwound in this way, and at worst very few wraps will be left on the drum to be straightened. If the drag cable becomes bent or kinked so that it does not spool on smoothly, the bucket should be dragged in all the way a few times with a load to straighten it. If the bucket is swinging in the air the drag should be wound in while it is swinging out and holding the cable under tension, and held while it is swinging in and the cable is slack. If the cable is running smoothly, this precaution need not be taken.

If the hoist cable becomes tangled the bucket should be rested on the ground, the hoist brake released, and the machine swung. When the hoist cable is unwound past the tangles or to the anchor, the hoist clutch should be engaged and the swing clutch disengaged. The cable should then reel back properly, unless crushed or kinked, in which case tension should be put on it as it is wound in, by partially engaging the swing clutch to keep the boom from swinging back easily over the bucket.

The cables should not be allowed to run all the way out in a throw, for if the momentum of the bucket is stopped abruptly by the cable anchor the anchor may tear out of the drum. Manufacturers often supply and specify a drag cable which is so short that it is unsafe for casting.

Novice Operators. To the novice, the dragline is apt to seem a very loose, rough, and contrary machine. Special provisions have to be made in every movement to keep the bucket from jerking and swinging, but these soon become automatic and very precise control can be obtained eventually. The beginner will obtain the best results by keeping the bucket fairly close in, except when actually digging or dumping, by making slow starts and slow stops when swinging, and dragging the brakes slightly to avoid spinning out the cables.

Bucket Wander. As the bucket is pulled

in, it is liable to be deflected by irregularities and find a path considerably to one side of a direct line. Also, the shovel may swing by gravity during the haul-in if it is not standing level. In either case the drag cable will not be directly under the boom. The fairlead sheaves will put it in correct alignment for the drum, but if the angle is sharper than the fairlead pivot can meet, the cable will be dragged across the fairlead guard plate and will wear the cable, and wear or possibly tear off the guard. The shovel boom should be kept in line with the cable by the use of the swing lever.

Boom Twist. As a loaded bucket is lifted the boom will tend to swing over it. However, if the boom is held to one side by the swing clutches during hoisting, or if a heavy, oversize load such as a stump is partially lifted, then dragged along the ground by the swing, a powerful twisting force is applied to the boom.

In order to give a long reach without prohibitive weight, a dragline boom is of light skeleton construction that is not intended to withstand heavy side and twisting strains. Sometimes a boom so strained will collapse, but more often will twist slightly, bending some of the cross braces particularly on the lower side. Once twisted, normal loads may increase the damage and failure will follow if the boom is not straightened.

A good ironworker is needed to properly repair such a boom, but if the bent members are angle irons it may be kept in service quite a while by straightening them with a jack. A stout plank is placed inside the boom across the corner angles as a support for an automobile jack that can push the pieces straight.

This type of repair should not be attempted on tubular members as they lose their strength if flattened.

Applications. The dragline does not have the positive digging force of the dipper and the pull shovel, as the bucket is not weighted nor held in alignment by rigid structures, and can therefore bounce, tip over forward, or drift sideward when it encounters hard material. This weakness increases with depth, and is particularly noticeable in small size machines, as the weight and bulk of a large bucket are sufficient to give considerable stability and penetration.

The dragline experiences its greatest difficulty in cutting down and is able to continue a deep cut opened by other machines or by blasting in much harder material than it could dig from the surface.

The outstanding advantage of a dragline over most other rigs is long reach for both digging and dumping, plus the ability to dig below the tracks. It has the further good point of a high cycle speed, being second only to the dipper stick in this re-

gard. It will be preferred to the dipper stick for truck loading where the earth is not too tough, and where the original grade is better than the new, because of water, mud rock outcrops, steep ramps, or other problems within the excavation. Because of its greater reach and output, it will be preferred to the pull shovel in any situation where it is capable of digging the soil effectively, where precisely cut vertical sides are not required, and where there is room for it to swing.

The dragline is the only practical attachment for extensive digging in mud, as its reach enables it to handle a wide area from a single stand and the sliding motion of the bucket avoids trouble with suction. It is also the best machine for many stripping operations in which the spoil is high-piled away from the pit, but is rivaled by long boomed dipper stripping shovels if

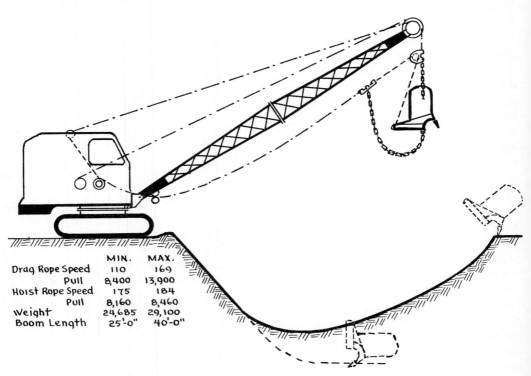

	MIN.	MAX.
Drag Rope Speed	110	169
Pull	8,400	13,900
Hoist Rope Speed	175	184
Pull	8,160	8,460
Weight	24,685	29,100
Boom Length	25'-0"	40'-0"

Fig. 13-87. Speed and pull, half yard dragline

the spoil is to be moved across a narrow pit as in some strip mining.

Draglines with skillful operators do an excellent job of topsoil stripping, grading, and spreading piles of earth, but except under wet conditions they do not usually do as well as bulldozers of comparable value.

Working Data. Figure 13-87 gives working diagrams and data for half-yard draglines with thirty foot booms. Shorter or longer booms may be mounted. A long boom increases reach and depth, but may require additional counterweight, reduces the safe load, slows the swing, and gets into difficulties with obstructions.

As in previous tabulations, the data are averaged from that published for several different makes.

Production. In easy digging, dragline production should be between sixty and ninety percent of that of a dipper shovel of the same size if trucks are loaded on the same level. If trucks are in the pit output may range from seventy to a hundred percent.

In hard digging, dragline yardage falls off much more rapidly than shovel yardage because of weaker penetration. High banks also slow them when loading on their own level or side casting, because of the extra time required for hoisting.

In side casting production on a volume-distance basis may be much higher than that of a dipper.

Operation Suggestions. The following are suggestions for dragline operation:

1. Keep bucket teeth sharp and built up to proper size.
2. Keep dump cable short so load can be picked up well out from the shovel.
3. Dig in layers, not ditches.
4. Keep digging surface sloped up toward shovel.
5. Don't drag in so far that dirt or rocks pile up in a ridge.

6. Keep drag cable from working in dirt.
7. Pick up bucket as soon as it is full.
8. Do not pull drag yoke into fairlead.
9. Inspect bucket chains frequently, especially at ends, and have them built up or replaced if worn thin.
10. Inspect fairlead frequently—worn bushings or spacers may let sheaves wear and cut cable.
11. Salvage short pieces of hoist or drag cable for dump cable. If dump anchors are too small, hammer the cable flat or change the anchors.
12. Do not guide bucket by swinging the boom while digging—you may twist the boom.
13. Do not swing until bucket or load is clear of the ground—you may twist the boom.
14. For heavy loads use high boom and swing slowly.
15. If the shovel starts to tip dangerously, drop the bucket. Don't try to dump it as its outward swing increases the tip.
16. Don't pick up overloads in bucket if they slow the hoist and swing, or if they tip the shovel.
17. Don't slap bucket against boom while hoisting.
18. Work with machine level. If you can't, load trucks at either the top or bottom of the swing. Loading is sloppy at the side as the counterweight swings the boom uphill as the load is dropped.
19. Don't work with a cable that is cross wound on the drum.
20. Don't travel uphill with a high boom.

CLAMSHELL

The clamshell front is a jack of all trades. It can do most of the jobs the other rigs will do, although less efficiently, and in addition has its own specialties of digging deep, narrow, straight-sided excava-

Courtesy of Gar Wood Industries, Inc.

Fig. 13-88. Clamshell

tions, and neat rehandling of materials.

Figure 13-88 shows this rig. It consists of a boom similar to that used for the dragline, with two boom line sheaves and two operating sheaves at the point, and a bucket hung from the operating sheaves by two cables.

Buckets. A number of different types of clamshell buckets are available. Figure 13-89 shows a center pull construction, and 13-90 an Erie lever arm type. Each consists of two jaws hinged on a movable bar, the main shaft, and hinged on their outer ends to brackets extending to an upper bar, the head shaft parallel to the mainshaft. The head shaft is supported by the head, or head beam, and the hoist cable. In 13-89, sheaves are fastened to each shaft and the digging cable is reeved through them and finally anchored to the headbeam. In 13-90, the lower sheaves are on the lever arm, which is an upward ex-

tension of one jaw. If the digging clutch is engaged, the cable will pull the two sets of sheaves together, powerfully raising the

Courtesy of The Wellman Engineering Company

Fig. 13-89. Center pull bucket

center hinges of the jaws. Since the corner arms are fixed and will not allow the outer ends of the jaws to rise, the jaws pivot on their outer hinges and rotate inward and upward until they meet. The bucket is now closed, and will pick up its load if raised. If the digging line is released while the bucket is held in the air by the hoist or holding line, gravity will cause the mainshaft to move down, pushing the jaws downward and outward and dumping the load.

These buckets, and certain other designs, can be obtained with various types of teeth with flat or curved lips, with toothless lips, with and without side cutters, with ballast plates for weight, and in various weights, widths, and shapes for special conditions.

There is not space in this volume to give proper consideration to all these variations. The items of importance in the selection of a bucket are weight, digging characteristics, and quality of construction. The weight is selected according to lifting capacity of the shovel, which varies with its size and weight, the length and angle of the boom, and the amount of counterweight used. The shovel manufacturer or dealer can supply

a table of lifting capacities, and the weight of the bucket and load should not exceed their recommendation. See Figure 13-101.

Figure 13-91 shows the ways in which bucket capacity is measured. Line of plate is the usual rating.

The more resistant the earth to be dug, the heavier the bucket must be in relation to its capacity. A good, extra heavy duty clamshell will dig almost anything except solid rock, but its closing action when fully reeved is very slow, and its massive weight reduces the payload. A light bucket, suitable for handling soft or loose dirt, will have faster action and a minmum of dead weight, but it will not penetrate hard materials, and will suffer damage if repeatedly banged and scraped on them. For miscellaneous work including both hard and soft, a medium weight, general purpose bucket is usually selected. Some of these have provision for adding or subtracting plates to change weight and strength.

A shovel can usually just handle a medium bucket of its own rated capacity, that is, a ¾ yard machine uses a ¾ yard bucket. In heavy duty work, the next

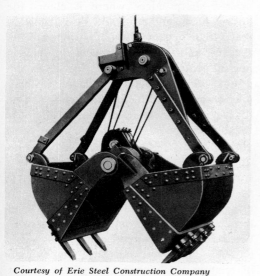

Courtesy of Erie Steel Construction Company

Fig. 13-90. Lever arm bucket

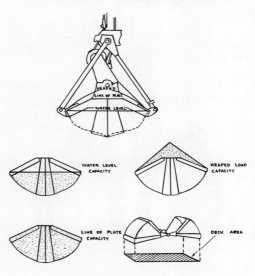

Fig. 13-91. Bucket ratings

Fig. 13-93. Scheidt grapple

smaller size is used, and for very light re-handling, a size larger. A small bucket must be used in mud as suction holds it, greatly increasing the force required to start it upward.

In general, teeth should be easily de-tachable, upper and lower sheaves should be shielded against dirt and should turn on high-grade bushings or anti-friction bearings; the closing line should be guarded against sharp edges where it enters the bucket head, and it should be possible to reeve the bucket without using all the sheaves, and without throwing the digging line off center.

For other uses than excavation there may be additional factors to consider. One of the commonest uses for clamshells is rehandling loose material such as coal, gravel, cinders, sand, etc., either in piles or in transfer to or from barges, freight cars, trucks, hoppers, and other receptacles. In this work, the bucket should be able to remove practically all the loose material from a hard surface, and should therefore be wide and equipped with a straight, toothless lip. If the dumping point is high in relation to the boom length, a bucket with minimum top-to-bottom measurement

Fig. 13-92. Orange peel bucket and tong grapple

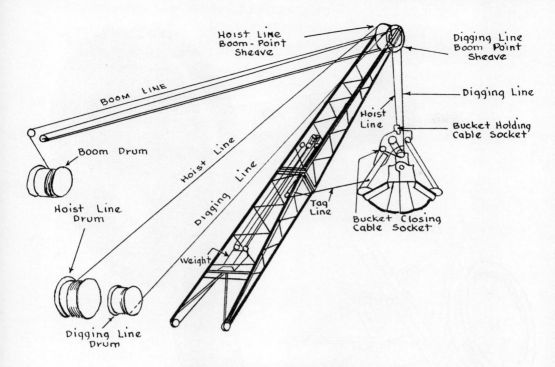

Fig. 13-94. Clamshell reeving

must be used for maximum efficiency.

Clamshells may be used to handle large rocks, cordwood, and other bulky objects, as the pinching effect of the jaws gives them an excellent grip. There are special buckets made for these jobs which may properly be considered multiple jaw clamshells.

Grapples. When a clamshell has more than two jaws, it is called a grapple, star, or orange peel bucket. The jaws may be made to fit accurately to each other when closed, Figure 13-92, upper, so either loose dirt or large objects can be lifted; or as independent tongs that will hold big pieces only, as in the lower cut. Both show a four-jaw Owen that is constructed so that each of these jaws can work independently, making it possible for each of the four points to get a firm grip on an irregular or off-center object.

The multi-tine grapple in Figure 13-93 is made with loose joints, so that the tines are free to feel their way as they close.

Loads can be gripped by their sides as well as by the points.

Reeving. Figure 13-94 indicates the reeving system. The hoist (holding) and digging (closing) lines are carried over the center pair of boom point sheaves and descend directly to the bucket. The hoist line is anchored to the head beam; the closing line goes through a guide in the head beam to the sheave sets. The bucket need not be fully reeved, as the full power of all the lines is not always needed. The cable may be anchored after rounding fewer sheaves for quicker but less powerful closing.

Tagline. The tagline is a light cable running from the boom to the bucket that serves to prevent the bucket from twisting or spinning in the air. It is kept on a light tension by a weight sliding on a track inside the boom, as in Figure 13-94, or by a spring-loaded drum, such as the electric dipper trip previously described, or the boom-mounted spring-wound Rud-O-

Courtesy of McCaffrey-Ruddock Tagline Corp.

Fig. 13-95. Drum type tagline

Matic tagline drum shown in Figure 13-95.

The cable is ordinarily fastened to two corners of the bucket by chains. If both are on one jaw, the hinge line of the jaw will be at right angles to the cable. If one corner of each jaw is held, the opening will be in line with the cable. The bucket can also be held in a diagonal position by attaching to only one corner.

The tagline pull is usually light enough so that a man can guide the bucket by hand or with a stick as it is lowered to place it exactly as desired. This is done when it is necessary to cut in a position other than that in which the tagline holds it.

The same company builds a tagline with an extra drum for electric cable, for use with a magnet. Contact with the electric source is through brushes. The electric line is set slightly slack, the tagline taut.

There is also a magnet reel that has electric line only, which is taut at all times.

Digging. In digging, the bucket is placed over the work by swinging the boom, and either moving the shovel or raising or lowering the boom to obtain correct distance. The digging (drag, closing, or crowd) brake is released, causing the bucket to open, and the hoist brake is then released, allowing the bucket to contact the ground on its teeth. If the ground is soft, the hoist brake will be only partially released, or reapplied just before the bucket lands. If it is hard the bucket will be allowed to fall freely, so that its weight will drive it into the ground for a good bite. In either case both brakes must be applied as soon as it has hit to avoid unspooling of the cable.

The digging clutch is then engaged to pull the jaws together. They first push the dirt inward, then curve and close under it. If the material is not too resistant and the bucket has proper weight a full or heaping load will be gathered. The digging line will lift the bucket as soon as the jaws are tightly closed. The hoist clutch should then be engaged so that the hoist (holding) cable will not become slack while the bucket is being raised.

Dumping. The swing is started as soon as the bucket is clear of obstructions. The distance of the bucket from the machine may have to be adjusted by raising or lowering the boom if precise dumping is being done. This, of course, is practical only if the shovel is equipped with a live boom.

When the bucket is properly positioned, the digging brake is released, the jaws open by their own weight and that of the load, and the earth is dumped. The bucket is swung back to the pit in the open position and lowered or dropped for another bite.

Centrifugal force of the swing puts the bucket out beyond the boom point, offering some increase of reach for both digging and dumping. Stopping the swing will allow the bucket to swing inside the boom point for close digging.

Chopping. If the bucket does not fill at the first closing it may be opened, hoisted, and dropped again, in which case the earth scraped together the second time will be added to that loosened the first time. This process may be repeated until a full bucket is obtained.

Signals. Clamshell work is often done at such depths that the operator cannot see the digging point. In this case it is best to have a man placed so that he can see the work and signal the operator. This is essential if there are men working in the pit where they might be hit by the bucket.

Hand Labor. If the bucket will not dig satisfactorily, either because it is too light in construction or slides off slopes, hand labor or explosives may be employed to loosen the dirt and the clamshell used to lift it out afterward.

Deep Digging. In deep ditches or shafts, it is difficult to keep the walls perpendicular, particularly if the earth is stony, as with each bite the bucket is edged a little away from the wall, thus causing the pit to grow narrower with depth. This tendency may be combatted by swinging the bucket against the wall as it drops, by hand trimming, or by making the top of the shaft enough oversize so minimum width will still be had at the bottom. It is helpful to equip the bucket with side cutters or corner teeth for this work.

Applications. A clamshell has the unique advantages of being able to stand on either the new grade or the old and to excavate at its own level, to a depth limited only by the length of cable its drums will carry, and at a height limited only by its boom length. It is the best of the rigs for handling bulky objects such as boulders, stumps, and logs, as chaining is usually unnecessary. It is the best tool for piling and burning loose brush and trees. It does not push or pull loose material to other positions as it digs. There are very few excavating jobs which cannot be done with a clamshell, and it is an excellent utility and odd job rig.

But it is slow. It ordinarily moves fewer yards per hour than any other rig which can do the particular job. This arises from the time consumed in closing the bucket and from the fact that the operator has minimum control over the position of the bucket, which is always directly under the boom point unless swung elsewhere. The bucket can be moved toward or away from the shovel by raising or lowering the

Fig. 13-96. Crane

boom, around it by swinging the shovel, outwardly by centrifugal force, and inward as a pendulum so that there is no point in its digging a range it cannot reach, but careful operation is required to get it where it belongs.

A number of different buckets are required for best handling of a variety of digging, and there is often loss of efficiency in the use of an unsuitable or compromise type of bucket.

A clamshell has the same reach as a dragline, except that the bucket cannot be cast as effectively. Since most digging is done directly under the boom point, maximum depth in "diggable" soil is determined by the amount of cable the drums will carry. Dumping height is controlled by the height of the bucket, which is variable; no diagram of working ranges is needed.

Operating Suggestions. Suggestions for clamshell operation are:

1. Keep bucket teeth sharp and built up to size.
2. Don't use more parts of line in the bucket than you need.
3. Be sure footing is solid.
4. Don't travel with a high boom—a bump may tip it back on the cab.
5. Don't swing uphill with a high boom.
6. Keep back from the edge of deep, wide cuts.
7. If machine tips dangerously, release both holding and closing lines.
8. Don't hit boom with bucket.

CRANE

The crane, Figure 13-96, may consist simply of a digging or hoist line run over a sheave on a lattice boom point, and down to a hook or clamp. More precise control may be obtained by attaching the hook to a sheave block, and reeving a line of two or more parts between it and the boom.

If the shovel does not have a live boom, the operating line not used for the hook can be reeved as a boom hoist line, and the regular boom drum left idle.

A small machine can carry a very long crane boom if the angle is kept high and the load is light.

A jib boom can be added to the regular boom to obtain greater reach and hold bulky loads away from the danger of swinging into the boom. The jib boom is hinged to the boom point and may be set straight or angled down. Its support cable is anchored on the boom. See Figure 13-97.

Lifting capacity of a jib boom is usually less than half that of the crane itself. It leaves the point sheaves free so that another line for heavier lifting may be reeved in the conventional manner.

A crane designed for precision work such as setting steel usually has a foot accelerator. Enough lines are put on the hook block so that the engine can handle full loads at close to idling speed. Loads are lifted, maneuvered in tight places, and set at low throttle, and the accelerator is used to speed them up while in the clear. Still slower hoist speed may be obtained by slipping the clutch.

Loads may be lowered by slipping the hoist brake.

In cranes equipped with a worm or other slow moving boom hoist, the boom may be raised or lowered by engine power to give more precise control than is afforded by the hoist clutch and brake. This will bring the load in toward the shovel as it is raised, and cause it to move further out as it is lowered.

Fluid Couplings. Smooth operation is aided greatly by installing a fluid clutch or torque converter between the engine and deck machinery. These units allow the foot accelerator to exert complete and smooth control over raising, holding, and

Courtesy of Koehring Company

Fig. 13-97. Jib boom

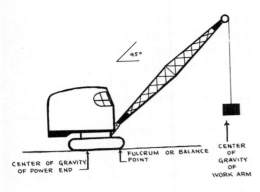

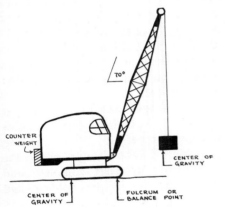

Fig. 13-98. Crane balance on level

lowering. If the engine idles fluid slippage will allow the load to descend. Proper acceleration will hold it stationary and further speed-up will raise it.

If the engine is set at sufficient speed to raise the load and the hoist clutch is engaged, the load may be held by applying the brake and raised by releasing it.

Some mechanical clutches—either engine or hoist—can be slipped to obtain somewhat the same effects, but delicate adjustments are required, wear is rapid, and heat damage is likely.

Applications. Cranes lift by means of chains, slings, ropes, or tongs which grip the load and are caught in the hook. Other attachments can be used in the same manner by passing a chain around some part of the bucket or stick, or catching it in a specially made hook or bracket, and using

the chain hook to pick up the load.

The crane is more efficient than other rigs for most hoisting work as it does not carry the dead weight of a bucket and other digging parts, slower and smoother lifting can be achieved by increasing the number of lines, and the operator has a better view of the hook and the load.

There are a few jobs where a digging attachment is more convenient, as the bucket can be used to move or overturn the load to make it easier to get a chain around it. Usually, however, the bucket is more of a nuisance than a help. This is particularly so when working space is limited.

Work can sometimes be done best by reeving the crane as a dragline, attaching the hook to the ends of the hoist and drag cables and leaving off the bucket. This permits in-and-out control of position without adjustment of the boom angle. However, the drag cable makes it unsafe for the ground crew to work between the load and the shovel.

Cranes can carry loads from place to place Those mounted on rubber tires do this better than crawlers. When mobility is more important than height of lift, a shovel dozer may be preferred for this type of work, particularly in tight quarters.

Stability. The lifting ability of a crane does not depend primarily on engine power but on balance. The balance point or line is the end or edge of the tracks, tires, or outriggers toward the load. The weight of the shovel itself tends to hold it flat on the ground, and that of the load, most of the boom, and the force exerted while raising the load all tend to tip it. Weight on either side becomes more effective as its distance from the balance point increases. The entire machine functions as a first class lever with the fulcrum at the balance line. Changes in balance are shown in Figures 13-98 and 13-99.

Since the ground contact of most cranes is longer than it is wide, they have greater stability to front and rear than off the sides.

The balance of a crane can be improved by adding counterweight (but not enough to encourage straining its structure), by raising the boom so that the load is brought in closer, or by using extra long tracks or outriggers.

Putting the crane on a slope will shift its center of gravity downhill, increase the effective boom height uphill, and decrease it downhill. More weight is placed on the downhill edge so that sinking into soft ground at that end may make the machine slope greater than that of the ground. Slanted ground will therefore increase stability against uphill loads, and seriously diminish it downhill.

Crawler cranes are rated to lift 75 per cent of their tipping load in the least stable direction at their maximum boom angle (usually 75 to 80 degrees above horizontal) when standing on hard level ground. Tipping is said to occur when shovel weight no longer rests on the outer track rollers, even if the track itself is still on the ground. Draglines and clamshells are usually rated at 67 per cent of tipping load because their burden may be increased by suction or by digging beyond the boom point, and they often work on less stable footing than do cranes.

Rubber tired cranes may be rated at 85 per cent of tipping load as they have more reserve stability at the tipping point. Figure 13-100 and 13-101 give range and capacity data for a P&H 555A-TC 35 ton truck crane. Note the importance of outriggers and boom angle.

High Boom Hazards. A crane can lift maximum loads only if the boom is high enough to place them close in. It can handle lesser loads with greatest safety and convenience if they are close enough so that there is no question about stability. The strain on the very vital boom hoist line and compressive forces on the boom

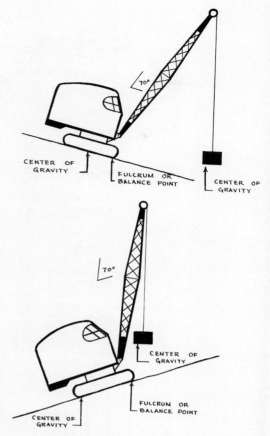

Fig. 13-99. Crane balance on slope

itself are greatest with a low angle and least with a high one.

It follows that it is customary to operate cranes with their booms held high. This practice involves two special dangers—the boom falling back on the cab, or overturning the machine by abrupt swinging.

A boom may fall back on a shovel if it is raised past its center of gravity; or as a result of swinging it from a downhill to an uphill direction, walking the machine onto an upgrade, pitching as it walks or swings on uneven ground, or recoil from a dropped or released load. See Figure 13-102.

The primary responsibility in avoiding this disaster—and that is what it is—is with the operator. There are two safety

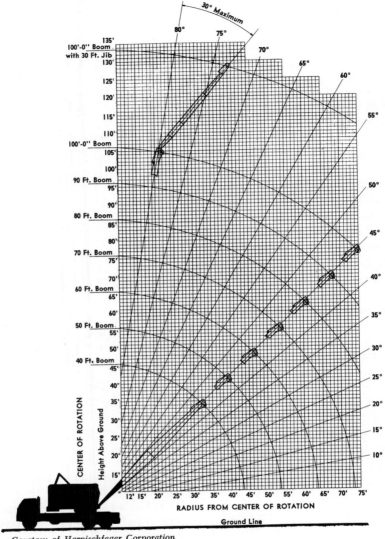

Courtesy of Harnischfeger Corporation

Fig. 13-100. Range diagram, truck crane

devices, however, which will afford some protection against his mistakes.

The boom hoist clutch may be equipped with an automatic release which disengages it when the boom reaches its maximum safe angle. This will provide only against pulling the boom over by power when the crane is level or working downhill.

The boom may be connected to the A-frame by a telescoping stop, consisting of a pair of rods or tubes sliding inside tubes. One member is hinged to the boom, the other to the frame. When the boom is lowered this device extends automatically. When it is raised the parts telescope until they hit stops at maximum height which prevents the boom from moving higher. See Figure 13-103.

This device will prevent the boom from falling back under most conditions. However, if a careless operator continues to

OPERATING RADIUS	TYPE OF SERVICE	40 FT. BOOM			60 FT. BOOM			80 FT. BOOM			100 FT. BOOM			Max. Jib Capacities With Outriggers
		Without Outriggers Over Side	Without Outriggers Over Rear	With Outriggers	Without Outriggers Over Side	Without Outriggers Over Rear	With Outriggers	Without Outriggers Over Side	Without Outriggers Over Rear	With Outriggers	Without Outriggers Over Side	Without Outriggers Over Rear	With Outriggers	
12'-0"	Crane	40500	52760	70000	.....	.....	.....	.....	.....	.....	.....	.....	.....	.....
15'-0"	Crane	30900	38480	70000	30100	37680	69600	.....	.....	.....	.....	.....	.....	.....
20'-0"	Crane	23600	26750	43900	22800	25950	43100	22000	25150	42300	21200	24350	41500	.....
25'-0"	Crane	17600	20200	31400	16800	19400	30600	16000	18600	29800	15200	17800	29000	.....
30'-0"	Crane	13100	15700	23200	12300	14900	22400	11500	14100	21600	10700	13300	20800	.....
35'-0"	Crane	10150	12750	18200	9350	11950	17400	8550	11150	16600	7750	10350	15800	10000
40'-0"	Crane	8400	10850	14500	7600	10050	13700	6800	9250	12900	6000	8450	12100	9000
45'-0"	Crane	.....	.....	.....	6200	8400	11500	5400	7600	10700	4600	6800	9900	8000
50'-0"	Crane	.....	.....	.....	5350	7250	10200	4550	6450	9400	3750	5650	8600	7000
55'-0"	Crane	.....	.....	.....	4670	6310	8830	3870	5510	8030	3070	4710	7230	5700
60'-0"	Crane	.....	.....	.....	4050	5560	7800	3250	4760	7000	2450	3960	6200	4700
65'-0"	Crane	.....	.....	.....	.....	.....	.....	2740	4140	6140	1940	3340	5340	3700
70'-0"	Crane	.....	.....	.....	.....	.....	.....	2300	3610	5430	1500	2810	4630	2800
75'-0"	Crane	.....	.....	.....	.....	.....	.....	2000	3200	4800	1200	2400	4000	2000

Radius is horizontal distance from center of rotation to hook.

All crane capacities based on 85% tipping load and were derived from tests made with the machine mounted on a P&H crane carrier with dual rear axle and pneumatic tires, fulcrum point of outriggers 7 ft. from center of truck and with the machine standing on a firm level uniformly supporting surface, using a Baldwin Southwark Load Cell and Brown Electronik Indicator for load measurement. Capacities without outriggers depend upon proper inflation, capacity and condition of tires.

Capacities shown include weight of hook, block, chains, etc.

Backstops are recommended for crane booms exceeding 50 ft. in length.

*Maximum boom length for use without outriggers, 70 ft.

For clamshell and magnet ratings deduct 20% from crane ratings. Limit on clamshell rating is 10,000 lbs. Combined weights (bucket or magnet, etc., plus contents) should not exceed clamshell capacities.

Maximum length of boom for clamshell operation is 60 ft.

Single part hoist line for loads up to 12,000 lbs.

Two part hoist line for loads up to 24,000 lbs.

Three part hoist line for loads up to 36,000 lbs.

Four part hoist line for loads up to 48,000 lbs.

Six part hoist line for loads over 48,000 lbs.

Above jib capacities for 30 ft. jib on all booms up to 100 ft. maximum, using outriggers.

Deduct 1,500 lbs. from main hook capacities when boom is equipped with jib.

Courtesy of Harnischfeger Corporation

Fig. 13-101. Capacities, 35 ton crane

hoist after meeting the stops he is likely to bend or collapse the boom. Also, if a very heavy load is set down while the boom is against the stops, the release of tension in the boom cable will shorten it, bending the boom back against the stops. Cushion stops, Figure 13-104, reduce this danger.

When a low boom is swung or stopped abruptly, as is common practice in excavator operation, it tends to twist the shovel chiefly in a horizontal plane and will do no damage unless the footing is slippery or unstable. But if the boom and load are high the force approaches the vertical and tends to overturn the shovel. If it is already operating near its lifting capacity, a rough start or stop to a swing may set it on its side, or crumple the boom before the operator realizes what is happening. Increase in boom length in proportion to crane weight increases the danger.

Also, a fast swing will cause the load to move outward by centrifugal force, increasing its leverage and the danger of tipping. An attempt to stop may make matters worse. Lowering or dropping such a load while coasting on the swing is the best procedure, but circumstances may make this dangerous or impossible.

A crane carrying a maximum load should not be backed away from it, but can be moved toward it as the reaction to

Courtesy of Albert Roth, Jr.

Fig. 13-102. A high boom and a dropped load

propelling torque shifts weight away from the direction of travel.

Danger of hitting the boom with the load increases with boom angle. A load will swing inward because of swinging and stopping forces, because of lifting clear when outside the boom point, or when it is being transported on rough ground.

Overhead Obstructions. The operator must be alert to the possibility of overhead obstructions. These are usually trees or wires. Electric lines are by far the most dangerous.

Contact may be made while working, or by wobbling into them while walking on rough ground. Unless hit by a falling wire, the operator is usually safe in his cab, even when the machine is in contact with high voltage, but he will not be comfortable. If he leaves the machine he must jump clear so as not to touch any part of it and the ground at the same time.

Avoiding Other Accidents. A crane operator usually has a much more responsible job than an excavator operator. He customarily works with a ground man or crew who will be endangered by his mistakes and he is liable to be near buildings or other valuable property. The loads he handles may be both expensive and fragile.

Most accidents are caused by falling or swinging loads. Slings or other fastenings can break or slip off. The crane operator should be fussy about both their strength and their manner of fastening.

He should inspect his cables frequently and change them as soon as they are under suspicion. He should be particularly careful about the boom cable, as a falling boom is many times more destructive than a load alone.

When a load is swung it is difficult for either the operator or ground crew to judge exactly the path that will be taken, particularly if the object is irregular in shape. Its distance in or out can be affected by centrifugal force, by soft or uneven ground under the machine, and by settling of the boom against a defective brake. Therefore, constant care is needed in close work to make sure that the load

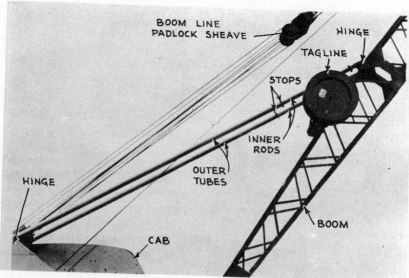

Fig. 13-103. Mechanical boom stop

does not hit workmen, or knock objects over on them.

A variation of the falling load is one that is dropped or set down in a hurry to avoid tipping. Such action can have serious or even fatal consequences, but even so may be less hazardous than overturning the crane onto men or equipment working near it.

Questionable loads should be tested before being lifted and swung. This can be done by lifting barely off the ground with the boom in the least stable position—over the side or downhill—with the boom lowered to give a radius about 25 per cent greater than that which will be used in handling. Another procedure is to lift the load, then boom out (down) until the machine starts to tip. A safe lifting radius is about 80 per cent of tipping radius for crawler cranes, and 90 per cent for rubber tired machines.

Breaker Balls. Secondary breaking of oversize rock in mines, quarries, and other excavations may be done by breaker balls, also known as skullcrackers. These tools are also used for demolishing buildings and

breaking and/or compacting metal scrap.

Balls weigh from a half ton to three tons, sometimes more. Figure 13-106 shows a construction with a replaceable U-hook that is fastened in sockets in the ball by poured lead. They may be made of a hard semi-steel, or of a tough nickel alloy. The alloy lasts longer, but it costs more. Its use is

Fig. 13-104. Cushion boom stop

Fig. 13-106. Breaker balls

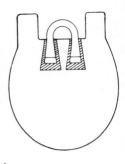

Fig. 13-105. Hand signals for crane operation

usually justified when rock is abrasive.

A swivel connection between the hoist line and the ball will increase the life of the cable and the hook, by eliminating twisting.

A breaker ball is lifted by a crane, positioned over the boulder, and dropped by releasing the hoist brake. The brake is reapplied just before the ball hits, to avoid spinning out the cable and to keep the ball

from traveling. A number of blows may be required to break a single boulder.

The drop may be from the crane boom tip, 30 or more feet in the air, or from a height of 5 or 10 feet, depending on the breaking characteristics of the stone. High drops wear the ball faster, and increase the nuisance of flying rock particles.

Breaking oversize with a ball is less noisy and dangerous than blasting. However, the operator must be protected by bullet proof glass or other shielding, and personnel must be kept out of the area in which chips are likely to fly.

There are also spherical drop balls that are lifted by an electromagnet and dropped by cutting off the current. These are used mostly in scrap yards, where magnets are needed anyhow, and where the problem of the ball rolling out of reach is less than in a pit.

HYDRAULIC CRANE

The hydraulic crane might more properly be called hydraulic-powered or hydraulic-operated. In it, the lattice boom is replaced by a hollow, telescoping box usually with rectangular or trapezoidal cross section. It is supported by hydraulic cylinders.

Carriers or travel units may be very similar to those that carry cable cranes. At this time, the most popular seems to be a four-wheel-drive-and-steer, but specialized truck carriers are also common.

There are usually four stabilizers or outriggers, one at each corner. They are under

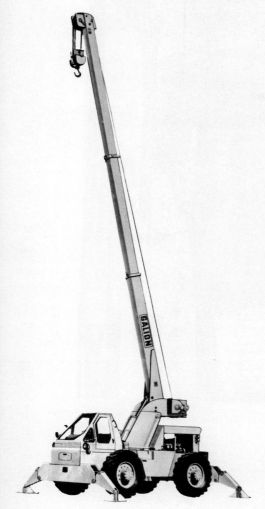

Courtesy of Jeffrey Galion Inc.

Fig. 13-107A. Hydraulic crane

center of rotation, across it from the boom tip and the load, to serve as a counterbalance.

A hydraulic winch may be mounted on the back of the boom, or on the turret. It powers the hoist line for the crane hook. There may be a second, auxiliary winch and line.

The operator's station and cab is usually beside the turret on the turntable, but may be in the carrier truck cab. Then a swivel seat permits the operator to face either set of controls.

The boom has internal hydraulic cylinders, and sometimes cables and sheaves also, by which it can be extended or retracted by substantial amounts. There may

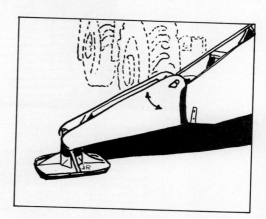

full individual hydraulic control, for both sideward movement and down pressure. If used, they are supposed to be forced down until they support the full weight of the crane.

A bubble indicator at the control station guides the operator in leveling the machine with the stabilizers.

The revolving unit (house) is mostly a turntable or turret that supports the boom. The turret hinges that carry the boom are often a substantial distance back of the

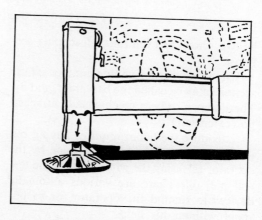

Fig. 13-107B. Two types of outrigger

1. Kruger Load Moment System Control Console
2. 360° Position Positive Swing Lock
3. Swing Foot Brake
4. Swing Set Brake Control Lever
5. Swing Control Lever
6. Swing Control Pedal
7. Outer Mid Section Telescope Control Lever
8. Mid Section Telescope Control Lever
9. Inner Mid Telescope Control Lever
10. Boom Elevation Control Pedal
11. Boom Elevation Control Lever
12. Aux. Hoist Control Lever
13. Foot Throttle
14. Main Hoist Control Lever
15. Hand Throttle
16. Outrigger Control Panel
17. Electronic Boom Angle Indicator

* Main Hoist Control Lever is Equipped with Controlled Free Fall Mechanism(Optional)

Courtesy of Grove Manufacturing Company

Fig. 13-107C. Controls, hydraulic crane

be two, three, or even four telescoping sections.

Three is the most common number. These sections may be called the base, mid, and fly. There may also be a mechanically attached jib (extension).

Extending the boom requires paying out the hoist cable. This is usually done by the operator releasing the winch so that the line will pull off. There are various automatic devices to remind the operator, or to unspool without his help.

In general, the operator should keep the boom as short as job requirements permit, to maintain stability and to limit problems in avoiding obstacles.

The boom can be lengthened at any time by operating boom extension levers and releasing the hoist line. If there is a suspended load, the release must be partial, so the line will pay out just enough to compensate for what is taken in by the boom, and neither lift nor lower it. In shortening the boom, the cable down to the load will lengthen, so you have to reel it in to compensate.

13-94

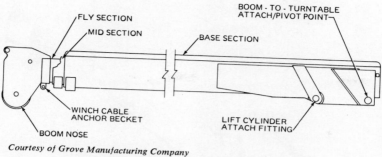

FLY SECTION

MID SECTION

BOOM - TO - TURNTABLE
ATTACH/PIVOT POINT

BASE SECTION

WINCH CABLE
ANCHOR BECKET

BOOM NOSE

LIFT CYLINDER
ATTACH FITTING

Courtesy of Grove Manufacturing Company

Fig. 13-107D. Telescoping boom, diagram

It is much better to adjust boom length between loads whenever possible.

Each movable section should be extended by about the same amount as the other(s).

Boom bounce may be your principal problem in adjusting to a hydraulic crane. The boom does not have the steadying effect of a separate line of supporting cable. It therefore has an increased tendency to whip up and down, immediately after raising, lowering, or setting down a load.

Bounce is most likely to occur with a fully extended boom at a low angle, but it is not limited to this situation. Your chief defense against it is smooth starting and stopping of vertical motions of either the boom or the hoist line. The design of the valves is very important to proper control.

Posted Information. There should be a chart posted in the cab, giving the safe lifting load of the crane at various distances, and/or at different boom angles.

There may be an indicator that shows the boom angle. In cable machines, there is occasionally an instrument that measures the load on the boom support cables, and gives an alarm if some set limit is exceeded.

In hydraulic cranes, an electronic device may measure stress in the boom, and sound a buzzer and light a warning signal, if safe loading is exceeded.

Usually, however, the operator depends largely on his own judgment, in keeping load and distance within safe limits.

PARTIAL SWING

In partial swing shovels, the engine and sometimes the deck machinery stay stationary while the boom and bucket swing on a small turntable.

Early shovels with railroad mountings had this construction. It is still found in railroad cranes, in dredges, and tractor-mounted hoes. Excavators carried on trucks

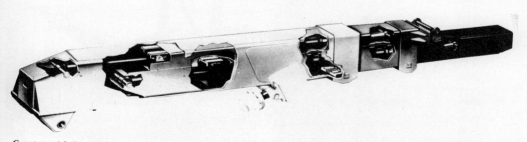

Courtesy of Jeffrey Galion Inc.

Fig. 13-107E. Telescoping boom, cutaway

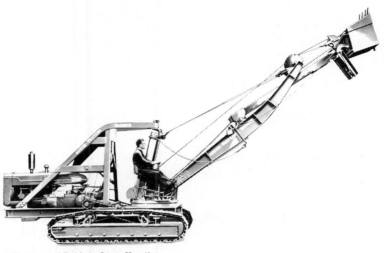

Courtesy of Baldwin-Lima-Hamilton

Fig. 13-108. Part swing shovel (obsolete)

are usually full-swinging, but may not be able to work full-circle because of the truck cab.

The partial swing, more often but less accurately called the three-quarter swing, cable-operated shovel in small sizes has become a part of history, but a picture of one is included for old times sake. A description will be found in earlier editions of this book.

Dredges. Part swing cable shovels in large sizes are still used as dredges. They are mounted on barges or ships, for underwater work in which bottom hardness, spoil disposal, or other conditions are unsuitable for the hydraulic dredges described in the next chapter.

Figure 13-108A shows a large dipper dredge and 13-108B a grapple. Loading is done on barges moored at either side, or on shore if it is close enough. Occasionally the spoil is dumped back in the water at the side.

Barges are towed to the job and maneuvered on it by winch lines attached to anchors. Pointed steel pipes, called spuds,

Courtesy of Bucyrus-Erie Company

Fig. 13-108A. Dipper dredge

Fig. 13-108B. Grapple dredge

are driven into the bottom and used as braces against digging thrust, wind, and waves.

Dipper dredges are adapted to deepening harbors and channels where the space is too restricted or the bottom too hard or coarse for hydraulic dredges. They are also used for digging canals through swamps where there is too much standing water to permit the use of land machinery.

Grapple dredges are best suited to work at depths beyond the economic limit of other types, and for cleaning up loose rock. Hydraulic dredges will be discussed in the next chapter.

HYDRAULIC FULL-REVOLVING EXCAVATORS

Like its cable-operated counterpart, the full-revolving hydraulic excavator may be said to be made up of three structures: the revolving deck unit, the travel base, and the attachment.

In hydraulic machines these distinctions remain clear only while considering the frames, arms, and other mechanical parts. The hydraulic system is operative in both deck and attachment, and often in the base as well.

This discussion will first cover the mechanical arrangements, including the placement of the principal hydraulic units; and will then deal with the generation and distribution of the hydraulic power.

REVOLVING UNIT

The revolving unit consists essentially of a heavy rectangular steel deck, formed and reinforced to carry engine, pumps, attachment, controls, and cab; and to rest and revolve on a turntable. Some parts are heavy steel plate and specially shaped ribs and bases, others may be open girders.

The center of rotation is usually forward of the center, and sometimes far forward. The area of greatest strain, and therefore of heaviest construction, includes this spot, the mounting hinges for the boom just forward of it, and the hinges for the hoist cylinder(s) on the front edge of the deck.

The forward location of the rotation center places the major part of the deck weight at the rear, where it serves to counterbalance the weight and pull of the hoe. This

13-97

Courtesy of Koehring Company

Fig. 13-109A. Hydraulic backhoe, crawler-mounted

effect is usually increased by placing the engine and pumps across the rear.

Counterweight. In addition to counterbalancing by arrangement of parts, there is usually a massive counterweight of shaped iron attached to the rear of the deck outside the cab. This may be very heavy, up to almost one-fifth of total excavator weight.

The swing axis is centered in the travel unit, so that the front of the deck is set back

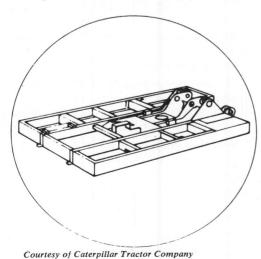

Courtesy of Caterpillar Tractor Company

Fig. 13-109B. Revolving deck

Courtesy of Insley Manufacturing Corporation

Fig. 13-109C. Counterweight hoist

from the front of the tracks, and the rear edge overhangs. This overhang is increased by the counterweight.

There is even greater overhang at the side, when the top is swung 90 degrees. This overhang or tail swing must be allowed for in placing the excavator for work.

It is sometimes desirable to remove counterweight, to suit a lighter attachment, to reduce tail swing, or to lighten the machine to move it across a bridge of doubtful strength, or to avoid overloading a trailer.

Most of the larger machines provide standard or optional means to handle counterweight by engine power. There may be a pair of hydraulic cylinders mounted on the back, with connections to a valve and pump. Or pulleys may carry an over-cab cable connection to the boom.

Engine. Standard power is diesel, from less than 100 horsepower to over 300. While bigger machines have bigger engines, there does not seem to be a standard ratio between power and either machine weight or bucket capacity.

There is a friction clutch to permit disconnecting the engine from the pumps, dur-

Courtesy of Hein-Werner Corporation

Fig. 13-110A. Triple pump set

ing starting or servicing. There is a clutch lever at the engine, and perhaps one at the control station also.

Drives. One or more hydraulic pumps may be driven directly by a shaft behind the clutch, in line with the engine crankshaft. But it is perhaps more usual for the pumps to be offset from this shaft, in a group of two or more, with drive through gears.

Most of these machines have hydraulic drive to all functions. Cylinders operate parts directly, or through simple levers. Motors drive through reduction gears, sometimes

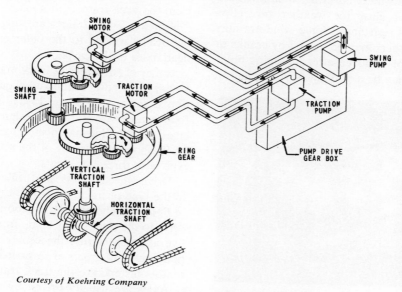

Courtesy of Koehring Company

Fig. 13-110B. Swing and traction circuits

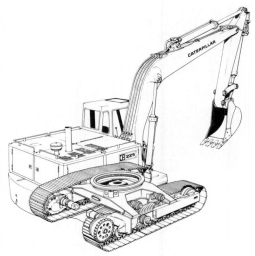

Courtesy of Caterpillar Tractor Company

Fig. 13-111. Carbody

through more or less extensive mechanical arrangements also.

Figure 13-110B shows hydraulic-mechanical arrangements for swing and propel in the Kormatic system.

Cab. The cab may be almost identical with that for a cable shovel. It is always of the circle vision type, with the roof higher than the machinery covering. Windows have shatterproof glass for weather protection and vandal resistance. Some windows may be fixed, others may be moved by cranks or on slides, or removed after loosening bolts.

There may be a skylight or roof window,

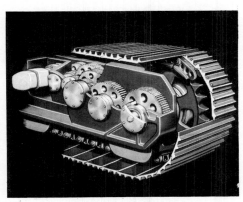

Courtesy of Caterpillar Tractor Company

Fig. 13-112. Gear-driven track

which is helpful in watching out for wires and tree branches.

Standard cab location is on the right front corner of the deck, beside the boom. However, it may be on the left side. Controls for all phases of machine operation, plus gauges and warning lights, are grouped here.

Controls will be discussed later, in connection with hydraulic circuits and operation.

TRAVEL UNIT

Most full-revolving hydraulic excavators seem to be advertised and sold with crawler mountings, so they will be the subject of the discussion below.

However, they are also well adapted to truck carriers and self-propelled wheel mountings, without major differences from those described earlier for cable rigs.

Carbody. The carbody is a massive frame that includes the turntable, and the dead axles or cross members that transmit its weight to the track frames.

It carries the large ring gear that engages with the swing pinion extending downward from the deck frame.

Tracks. Track frames are single or double beams welded to the outer ends of the dead axles in the carbody.

Either of two types of track may be used. One is the traditional linked-shoe construction described on Page 13-21. The other follows the crawler tractor design of a roller chain with bolted-on shoes, Pages 15-6 to 15-9. Shoes are usually of the semi-grouser type with three low cleats.

Drive wheels, idlers, and rollers conform to the type of track.

Propel. The propel, traction, or travel drive to the tracks may come from a pair of live axles set across the center of the carbody, or from a pair of reversible hydraulic motors fastened to the track frames.

Axles are usually driven by a hydraulic motor on the deck, through reduction gears,

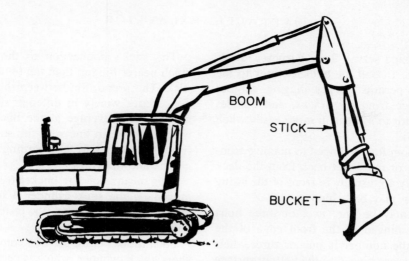

Fig. 13-113A. Hydraulic backhoe, diagram

a vertical shaft, and bevel gears. However, a hydraulic motor may be located in the axle housing, and turn them through a reduction-type transmission.

In the axle mechanism, a pair of brakes and jaw clutches, similar to those on page 13-18, provide for steering while traveling, and for holding the machine while working. Control is usually hydraulic. Friction brakes may be used instead of jaws.

The outer ends of the axles usually carry small sprockets, from which roller chains carry power to large sprockets on the bull (drive) wheel axles. Another arrangement uses a multiple gear train along the track frame.

Hydraulic motors, usually of the hydrostatic (axial piston) type, can be mounted in any convenient location on the inner surface of the track frames, and connected to the bull wheel axle by sets of reduction gears.

Direct-mounted hydraulic motors permit independent track movement on the two sides. This makes possible counter-rotation of the tracks for the spin turns discussed in Chapter 12. This feature makes it easy to turn in the length of the machine with little ground disturbance, and to maneuver accurately in very restricted spaces.

Turntable. Turntable and roller designs shown on pages 13-13 to 13-15 are used in hydraulic shovels also.

Poclain uses an arrangement in which rollers, tipped alternately to carry loads from two directions are substituted for balls in a race.

HOE (BACKHOE) ATTACHMENT

Most hydraulic excavators are designed for hoe use only. The hoe might therefore be considered as an integral part of the machine.

However, it can be taken off and put back on, there are often options in stick length, and it may carry working tools other than hoe buckets.

Present capacity of standard buckets is ½ to over 4 yards.

Construction. All hydraulic hoe attachments are made of three strong structural members, the boom, the stick, and the bucket. These are hinged to each other, and the boom is hinged to the excavator deck.

Movement at each of these hinges is controlled by two-way hydraulic cylinders.

Boom. The boom is always of the bent or gooseneck type, concave toward the ground. It usually has one bend or angle, but may have two.

This shape serves three purposes. It allows space to pull the bucket closer to the machine, permits deeper digging without interference from the tracks, and enables the operator to see past it more easily when it is raised.

The boom foot is hinged to massive trunnions two or more feet back from the deck edge. They are usually in front of the swing center, but may be behind it.

There may be one, two, or three hoist cylinders, hinged to the front edge of the deck. If the number is one or three, their piston rods are hinged to the bottom surface of the boom. If there are two, they are usually fastened to the sides.

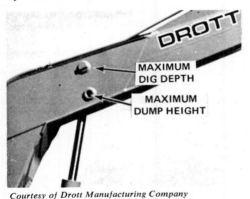

Courtesy of Drott Manufacturing Company

Fig. 13-113B. Depth-height adjustment

If there are two points of attachment to the boom, the upper one is used for maximum digging depth, the lower for maximum dump height.

The outer end of the boom is usually prolonged into a two-piece bracket, in which the stick is held by a heavy hinge pin or pins. The stick cylinder is mounted on the boom top.

Stick. The stick, dipper stick, or arm is hinged to the end of the boom, and is connected to the stick cylinder rod at its upper (back) end, and to the bucket and bucket dump arms at the bootom or front. It is usually one-piece, but may extend-retract by telescoping.

The stick's connection to the boom is much nearer the top than the bottom of the stick. The proportion between the two sections varies widely in different makes and models. The leverage at the bucket teeth, when they are in line with the stick, varies from 1 to 4 to 1 to 8, according to photographs and drawings.

This means that the bucket teeth will be moved by the stick cylinder 4 to 8 times faster and farther than its piston moves, with ¼ to ⅛ the force.

Such ratios permit use of a comparatively short, thick cylinder, which is necessary because of space limitations, and advantageous from a service standpoint.

Some machines provide two places for the boom-to-stick hinge. The one that is closer to the bucket will supply more power for hard digging; the other will provide more speed in easier work

The motions of the stick are variously described. Extending the cylinder forces the bucket in toward the machine, crowding it into the digging. This may be called pulling, in-pulling, digging, crowding, or just "in".

Contracting the cylinder forces the bucket outward, a motion called extending, reaching, or "out".

The bucket cylinder is hinged to the top or front side of the stick. Its piston rod is fastened to bucket dump arms, one pair of which is hinged to the stick.

Bucket Mounting. The bucket, which will be described below, is connected to the lower end of the stick by a hinge pin, and to a triangular set of paired dump arms. The other two angles of the arms are hinged to the bucket cylinder rod and to the stick. See Figure 13-114.

The dump arm mechanism is necessary because the bucket has such an extended arc of rotary movement around the stick hinge (from 136 to over 173 degrees, varying with the machine and the attachment points used), that a piston rod could not follow it. Dump arms supply the required around-a-curve reach, and prevent the cylinder from being pulled in against the stick as it extends.

When the cylinder is extended the bucket teeth move inward in a curling or digging motion. When it is retracted, the bucket opens or extends.

Several sets of holes may be provided, so that bucket action can be changed by moving hinge pins. The choice is between a combination of greater speed and range of movement in bucket control, or slower motion and greater digging and breakout force. Selection depends on the work being done and the operator's preference, and is likely to be changed only under unusual conditions.

Usually, the pair of arm-connection holes in the bucket that is nearest the teeth provides the greatest power.

Buckets. The bucket is somewhat similar to those of cable hoes described earlier. It is sometimes called a dipper or a tool.

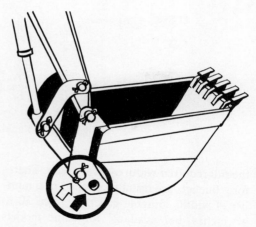

Fig. 13-114. Power-speed adjustment

Differences include location of attachment points, a smoother curve between floor and back, and sometimes lighter construction.

The primary use of a backhoe is digging ditches, although it is also well adapted to digging cellars and general excavation. For efficiency in ditches, the bucket should cut

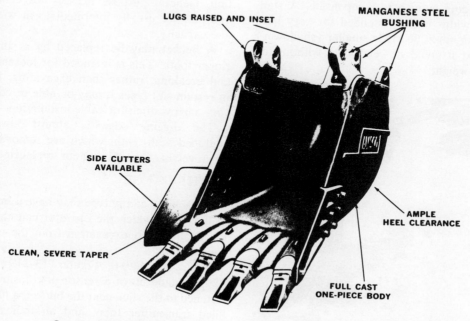

LUGS RAISED AND INSET

MANGANESE STEEL BUSHING

SIDE CUTTERS AVAILABLE

AMPLE HEEL CLEARANCE

CLEAN, SEVERE TAPER

FULL CAST ONE-PIECE BODY

Courtesy of Amsco Division of Abex Corporation

Fig. 13-115. Backhoe bucket details

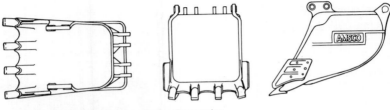

Courtesy of Amsco Division of Abex Corporation

Fig. 13-116. Backhoe bucket, three views

the full required width on every pass. Therefore, buckets are usually supplied in a number of widths, ranging generally from 30 to 48 inches, but available for some models down to 24 inches and up to 59 or more.

A single machine may also be fitted with a wide range in bucket size, with the largest for a single machine sometimes being more than 2½ times the capacity of the smallest.

Narrow buckets tend to be deep in proportion to width, and may fill poorly in chunky or rocky digging. If width is the same, reducing depth from the front edge reduces capacity, but may increase efficiency in loading enough to compensate. A standard width bucket intended for very hard digging might be made smaller (shallower) so that it could be reinforced without too great weight.

Courtesy of Construction Technology, Inc.

Fig. 13-117. Hoe-mounted ripper

The bucket is usually slightly wider at the open or front end, to reduce friction at the sides, and to allow for easier dumping. Additional clearance from trench walls may be obtained, and bucket cutting width increased by 2 to 8 or more inches, by installing sidecutters. They may be fixed-width or adjustable, smooth-edged or toothed.

Sidecutters are useful in accommodating a bucket to a wider trench, cramming more dirt into a narrow bucket, reducing drag in sticky soil, and reducing wear on the front edges.

Wide buckets may have poor penetration. General purpose buckets for cellars and pits are usually intermediate in width and capacity.

A bucket may be replaced by a single ripper tooth. This is intended for loosening and breaking, rather than excavating, but in certain soil types it may be able to cut a very narrow ditch for cable installation.

The digging edge is almost always equipped with teeth, which are removable for reversing, sharpening, or replacement.

DEPTH GAUGE

A backhoe of any type may have a built-in device by which the operator can check the depth of his excavation from the control station.

The ACCU-DEPTH, Figures 13-119 and 120, is made up of a sensor or transmitter fastened to the stick near the bucket, a fluid-filled transmitter tube, and an indicating head or gauge.

To set it for a reading, the operator

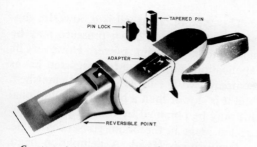

Courtesy of Amsco Division of Abex Corporation

Fig. 13-118. Detachable tooth

places the bucket on a reference point. He depresses a valve button on the head, then rotates its dial until ZERO lines up with a pointer. He then places the bucket, in the same position, on the bottom of the trench, and opens the indicator valve again. The pointer swings to indicate the vertical distance (depth) of the second position (ditch bottom) below the reference level.

The fluid in the system is supported by atmospheric pressure. Its weight at the sensor increases in direct proportion to the vertical distance between sensor and head, creating a vacuum whose variation moves the pointer. Maximum working depth is 28 to 30 feet, but within this range, error is said to be less than one-half per cent.

The reference point may be the ground surface beside the ditch, a section of completed bottom, a grade stake, or a stringline.

Other methods of checking depth are discussed under OPERATION.

OTHER ATTACHMENTS

Clamshell. Many hydraulic excavators

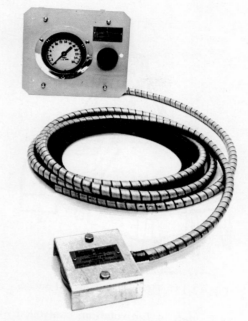

Courtesy of Equipment Supply Div., and AMTEK/ U.S. Gauge

Fig. 13-120. ACCU-DEPTH unit

may be fitted with a clamshell bucket, which replaces the hoe bucket or immobilizes it. The clam jaws are opened and closed by a two-way cylinder. This may be mounted horizontally across them, or be part of an attaching stick vertically above them. Closing power is supplied from the hoe bucket circuit.

The clam is usually attached to the end of the hoe stick by a universal mounting, which allows it to hang straight regardless of stick position or machine slope. There

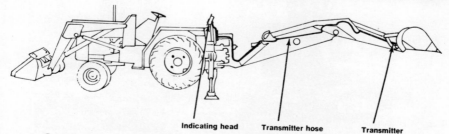

Indicating head   Transmitter hose   Transmitter

Courtesy of Equipment Supply Div., and AMTEK/U.S. Gauge

Fig. 13-119. ACCU-DEPTH installation

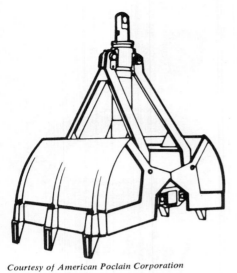

Courtesy of American Poclain Corporation

Fig. 13-121. Hydraulic clamshell bucket

may be part or full rotation, controlled by the operator.

Attachment may be direct to the mounting, or through one or more vertical extension rods. Extensions provide greater digging depth but less dumping height. Maximum length is limited by dumping requirements, or by clearance needed to lift the bucket out of the ground.

This type of clamshell is a powerful digging tool, in which down pressure may be added to jaw closing power. However, its applications are limited by restricted lift as compared with a cable operated clamshell, and it is slower than a hoe. Some additional lift may be obtained by using a power-controlled hinge in the boom gooseneck angle.

Demolition Head. A demolition head, operated either by the excavator's hydraulics, or by a separate air compressor, may be substituted for the bucket. This may be used for breaking boulders, weak bedrock, pavement, and structures, and driving trench bank supports. These devices are described in Chapter 19.

Even with a standard bucket, a big hydraulic hoe is a powerful demolition tool, because of its fairly long, high reach, with ability to both push and pull forcefully.

Dozer Blade. There may be an option of a frame-mounted bulldozer blade, for use in clearing the shovel's path of loose rocks, and occasionally in cutting and smoothing. It also affords protection against boulders pulled inward by the bucket.

Push arms are hinged to the track frames. Lift cylinders are controlled from the cab.

Courtesy of Liebherr-America

Fig. 13-122. Clamshell on double-hinged boom

13-106

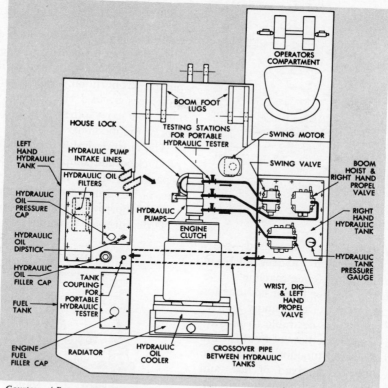

Fig. 13-123. Deck layout

HYDRAULIC SYSTEM

Excavator hydraulics are usually the highest pressure types of the systems discussed in the article starting on pages 12-68.

Pumps and Circuits. Most of the engine power is usually delivered to a set of hydraulic pumps, which provide pressure-flow of fluid to power all (or almost all) functions of the machine.

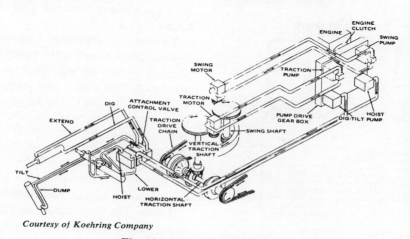

Fig. 13-124. Hydraulic circuits

HOIST
OPERATION
SECTION
DIG
OPERATION
SECTION
DIPPER
OPERATION
SECTION
CONTROL VALVE ASSEMBLY

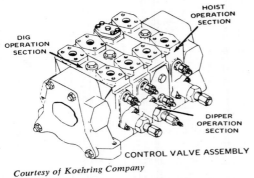

Courtesy of Koehring Company

Fig. 13-125. Valve bank

Three types of pump are used—gear, vane, and axial piston (hydrostatic)—in various combinations. Some machines have only two pumps, one for power and one for cooling—and others have as many as five.

Multiple pumps are desirable to provide a fixed and dependable flow of fluid for the use of each of several functions, so that one cannot be robbed (deprived of flow and pressure) by the opening of another circuit.

On the other hand, a pump is expensive, it consumes some power (and therefore fuel) when it is idling, and may be inefficient when run at part capacity. It is therefore economically desirable to provide it with as much useful work as it can handle.

These opposed factors are met in a variety of ways, including use of variable-output pumps, proportioning valves, and design of sequences to feed circuits that are important, and starve others that are less important at the moment.

The most desirable arrangement for both production and simplicity of operation is a separate circuit for each function that takes care of its normal needs, plus means to temporarily borrow fluid from another circuit.

Differences are too numerous and changeable to justify detailed discussion here. But the operator who is changing from one machine to another should be prepared to find differences in response to controls, particularly when two or more are used at the same time.

WRIST & DIG
CONTROL VALVE
SERIES CIRCUIT
FROM PUMP
48 GPM
TO TANK
100 GPM
71
GPM
71
GPM
WRIST
CYLINDER
DIG
CYLINDER
Faster dumping of
the load increases operating speeds saving time and money.
A special circuiting design recycles the oil
expelled from the wrist cylinder into the
dig cylinder, increasing the speed of
the oil flow. Working speed of the cylinders
is increased giving the 20-H Series
Three an accelerated dumping action.

Courtesy of Bucyrus-Erie Company

Fig. 13-126A. Integrated circuits

LINES TO
BOOM, STICK, BUCKET
PILOT
VALVE
MAIN
VALVE BANK

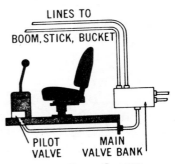

Courtesy of Caterpillar Tractor Company

Fig. 13-126B. Hydraulic over hydraulic

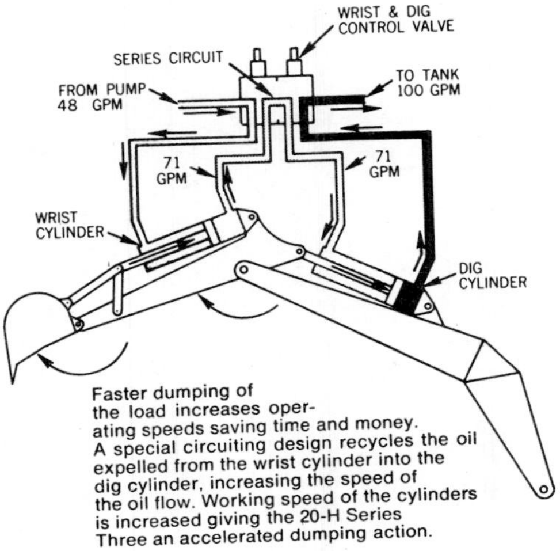

Layout. A revolving excavator has a rather complex hydraulic system, but fortunately it has enough space in which to arrange it. The engine and pumps are usually across the back; the swing motor and its vertical shaft must be above the big swing gear in the carbody. If the control valves are directly operated, they must be in the cab.

Otherwise, components can be arranged for efficiency, access, and convenience.

Lines. The basic piping on the deck and along the boom and stick is apt to be high strength tubing, factory-formed into the

lengths and curves needed. Where parts move in relation to each other during work, lines connecting them must be flexible hose. Hose may be used in other parts of the piping, for ease in installation, or to avoid stresses if parts shift slightly in position.

Stick and bucket lines must allow for very sharp bends. The length needed at sharp angles forms loops as the bends are straightened or reversed.

If hydraulic motors or other units are in the travel unit, a swing joint is installed at the center of revolution. This device, described in Chapter 12, provides for flow of hydraulic fluid regardless of the rotation of the upper unit. It requires a flexible or cushion mounting for one or both sleeves, to avoid distortion when digging strains cause the deck to rock slightly on the turntable.

Controls. A bewildering variety of operator's controls is offered. They vary in arrangement of levers and pedals, and in the means by which they operate the valves.

Levers are usually self-centering, that is, they return to neutral or HOLD position automatically when released. A propel stick may have a detent to hold it in full-on position. Brakes may have finger-released latches to hold them in ON positions.

Simple pedals are spring-returned to OFF or neutral, like an automobile's accelerator or foot brake. Swing and traction brake pedals may have hold-down latches.

Rocker pedals are hinged at the center. Pressing down on the front opens a valve to cause a member to move. Pressing the back moves the same part in the opposite direction. A spring returns it to neutral from either position.

Cylinders are either naturally self-locking when flow of fluid is blocked, or can be made so by an automatic check valve. Hydraulic motors may or may not have a braking effect when shut off, but they cannot be depended upon to hold parts against creep. Positive braking may be supplied by a fric-

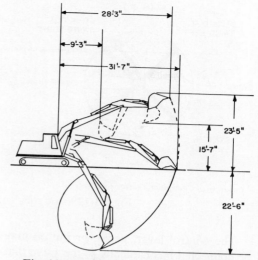

Fig. 13-127. Digging and dumping range

tion brake. This may be operated by a pedal. If automatic, it is engaged by spring pressure, and released by hydraulic pressure when the motor is used.

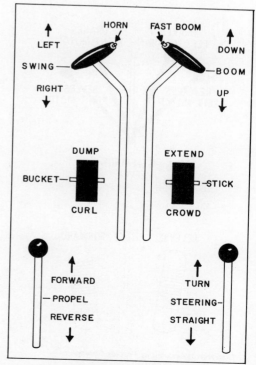

Fig. 13-128. Typical hoe controls

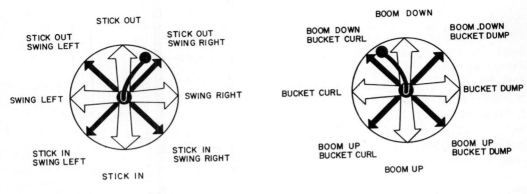

Courtesy of Caterpillar Tractor Company

Fig. 13-129. Joy stick controls, hydraulic excavator

The control lever, pedal, or button may be connected mechanically to the valve that it operates, by cable or rod-and-clevis linkage.

More often, a fluid system is used. In small machines, this may be static like a hydraulic brake system, with or without power boosters. Larger machines may use

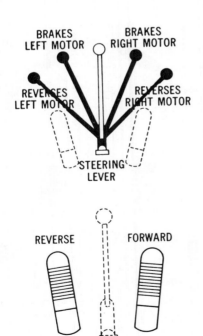

Courtesy of Caterpillar Tractor Company

Fig. 13-130. Travel and steering

power from a constantly running hydraulic pump, or air, or combinations of static hydraulic with air or pumped hydraulic. Different methods may be used in one machine.

Direct connection is simple and prompt. If valves are excellently balanced and leverage is adequate, this is a highly satisfactory arrangement.

Power controls should make the fineness of balance of the operating valves less important, but their own smoothness may be questionable. However, a power system usually means less operator effort, and less lever or pedal motion for a given effect.

OPERATION

Some aspects of hoe operation were discussed in Chapters 4 and 5, in connection with cellar excavations and ditches.

Operation of a full revolving hydraulic excavator or hoe resembles operation of a cable hoe in many ways, but this machine is under much more complete control because of two-way power, and "wrist-action" flexibility in the digging and dumping angle of the bucket.

Figure 13-127 shows the digging and dumping range of a typical machine. Part of its abilities must be used with discretion, as it is capable of digging a hole under itself and falling into it.

Effect of Controls. In the control se shown in Figure 13-128, the boom hois lever actuates the cylinder(s) that support the boom. Moving the lever forward forces the boom down, pulling it back raises the boom.

It may be unwise to move this lever all the way forward when the boom is raised, as in some machines it will allow the boom (and the whole attachment) to fall with dangerous speed and force.

There may be a finger button or some other device to divert fluid from the swing circuit to the boom circuit, to raise it more rapidly from a deep excavation.

Tilting the stick (crowd) pedal forward retracts the stick cylinder, pulling the top of the stick back and swinging the bucket forward. Back-tilting the pedal reverses these motions, crowding the bucket inward toward the cab.

Since the stick is fastened to the end of the boom, its position in space at any time depends on both the boom and the stick cylinders.

The bucket pedal, when tilted forward, causes the bucket cutting edge to move out, opening and dumping the bucket. When tilted back, it brings the edge inward, closing or curling the bucket.

The swing lever rotates the upper works to the left (counterclockwise) when pushed forward, and to the right when pulled back. In most machines, an automatic brake stops the swing smoothly when the lever is centered. There should be a lock-down pedal brake also, to lock the swing during travel.

In this example, the travel or propel lever moves the machine in the same direction it is moved. The right hand steering lever controls the clutch and brake for a right turn, the left lever is for the left controls.

All of these controls are (or should be) the metering type that can be partially opened to obtain fractional speed, for performing delicate operations slowly without reducing engine speed. But accuracy of response of part-open valves is quite variable. Also, all levers except possibly the travel

control are spring loaded to return to neutral (center), and automatically lock their unit when centered by holding fluid in cylinders. In motors, there is usually a brake that is applied by springs in neutral, and released automatically by fluid pressure.

Other Systems. Probably no two makes of excavator have exactly the same arrangements of levers and pedals. The type that is most conspicuously different from the example described is the so called joysticks for attachment control, and separate hydraulic drive to the two tracks.

One joystick controls both the stick and the swing. The stick valve is operated by the front-to-back motion of the stick, the swing by its side-to-side movement. The other stick moves forward and back to move the boom in the same directions, and moves left to curl the bucket and right to dump it. The two motions do not interfere with each other, and the valves operate in the same manner as when responding to individual levers or pedals.

Hydrostatic propel can be controlled in several different ways. In this example there are two foot pedals. Pushing the left one, accelerator fashion, moves the machine back; the other one moves it forward. Speed is regulated by pressure on the pedal, from zero in full-up position to full speed when depressed to the floor.

A single steering lever is moved to the right for right turns, to the left for left. There are three stages. In the first, nearest center, one track is driven faster than the other. The next locks one track while drive (regulated by the foot pedal) is continued

Fig. 13-131. Bucket tucked in

to the other. The third stage, reached by pushing the lever past a bumper spring, causes the tracks to rotate oppositely, for a spin or spot turn.

Attachment Motions. Assume that the excavator is parked with bucket drawn in under the boom, as in Figure 13-131, and that it is properly located on the centerline of a proposed ditch.

The bucket is freed from the ground by raising the boom, and moved toward digging position by extending the stick. The two movements are made at the same time. Until the stick reaches a vertical position, its arc of swing is down, and must be compensated by continued raising of the boom.

When the stick has passed the vertical position, its continued motion will take it in an upward arc. If it is to be kept near the ground, the boom is now lowered as the stick is moved outward.

When the stick is extended ¾ or ⅞ of maximum reach, the bucket is opened until the floor is about vertical (teeth straight down). It is then lowered (not dropped) to the ground by pushing the boom lever forward.

The stick may now be crowded inward, while the bucket is rotated gradually toward a half-closed or digging position. At the same time the boom is raised slowly, so that the downward arc of stick motion will not bury the bucket too deeply.

The bucket curl is adjusted for best penetration and filling, which is likely to be at an angle of 15 to 45 degrees to the surface being penetrated. Except when finishing off, the bucket should usually be filled in as short a digging pass as is feasible, without slowing the engine or dragging the shovel. Long passes take longer, and usually wear the bucket more.

Digging angle may be varied during digging. There should always be at least a slight angle, so that the bucket floor will not be dragged on undug dirt. Such drag-

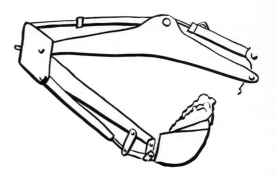

Fig. 13-132. Bringing up a full bucket

ging consumes power and causes excessive wear. Small changes in angle help the teeth in "hunting" for easiest penetration.

If the digging pass continues under the boom point, the bucket path will curve upward, and the boom will require lowering instead of raising if it is to be kept level.

Down pressure can be applied to the bucket to improve penetration, by not raising the boom, or by pushing it down, depending on bucket position.

When the bucket is full, it should be closed or curled, the crowd motion should then be stopped, and the bucket lifted out of the ditch by pulling back the boom hoist lever.

As soon as it is clear of the ground, it is swung to the side, and dumped by pushing the bucket lever. Generally, height is figured so that the teeth will just clear the pile. A higher position wastes time in excess hoisting, a lower one exposes bucket and stick to impact and twist.

However, if the pile will be big enough to threaten to slop into the excavation, the dump may be made in the pile, so that opening bucket and extending the stick will push soil away, to make room for more. The swing should be fully stopped before entering the pile, to avoid twisting strains.

The bucket is swung back to the excavation as soon as it is dumped and clear of the pile. During the swing it is brought

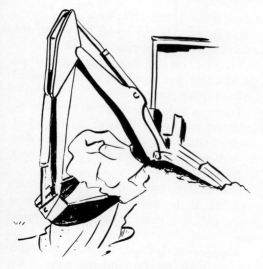

Fig. 13-133. Dragging up a boulder

back to digging position, both in degree of curl and in-and-out position on the stick control. It is lowered to the ground, and pulled in to refill.

To avoid building a pile of earth at the near edge of the excavation, always rotate the bucket toward a closed position, before bringing it out of the ground.

Finishing. The ditch is dug in a series of slices or chunks, down to the bottom. Depth may be measured by the operator or a helper, but if the trench is being dug to a fixed depth below ground surface, it is more convenient to put a mark on the stick that, when it is vertical and the bucket is flat on the ground, will indicate the correct depth.

The far wall can be trimmed by forcing down the bucket with the floor vertical, keeping it vertical as it goes down by curling out just enough to compensate for inward curve of its descent. When the bottom is reached, the bucket may be pulled in with a slight lift until clear of the back wall, then closed into its load.

Pipe may be set in the trench under laser control, close to the digging. A mark may be then made on the back of the bucket, to line up with the beam at correct depth. This provides absolutely accurate control.

The ditch floor is trimmed by keeping the bucket curled so that its bottom is either at right angles to the floor, or flat on it, and crowding it in with the stick while moving the boom up and then down to keep the bucket path level. Curl will need to be adjusted as the stick swings on its hinge.

These two bucket positions share the advantage of having no suck (tendency to dig in). Teeth-down leaves soil looser and may be harder to keep level.

To trim the near wall (this is usually not necessary), pull the bucket to it along the ditch bottom, with its floor flat. Curl it up into the bank, then raise it with the boom hoist.

The boom will tend to pull it away, so crowd the stick in just enough to compensate. This will increase the curl of the bucket, so keep extending (flattening) it to keep the teeth in cutting position.

After practice, you can balance these three movements so that the bucket will come straight up on the desired line, at a good cutting angle.

Boulders. To dislodge a boulder, dig down on both its far side and its near side. Then place the bucket just beyond it, with the floor sloping at a steep angle. Apply boom down pressure and move the bucket lever back and forth rapidly.

When the teeth have penetrated to the under side, crowd and curl the bucket. Crowding keeps it close to the rock, and curling exerts great breakout force.

A boulder too large to lift in the bucket may be rolled out by ramping the near side, then backing away, pulling and rolling it up the slope.

Stalling. The bucket may be stalled by digging resistance, either alone or in combination with a full bucket. Immediate action must be taken, either to reduce the resistance or to cut off the power. Other-

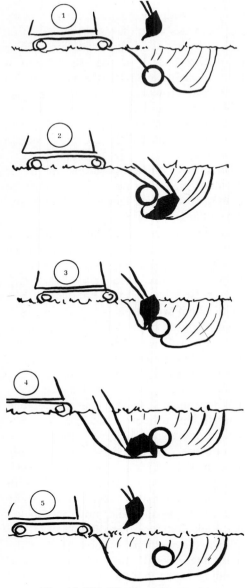

Fig. 13-134. Undercutting a pipe

wise, oil will overheat and parts will be strained.

Another effect of too much resistance is a tendency to drag the machine toward the bucket. This problem will be discussed in connection with tractor mounted hoes, in which it is more serious.

Chopping. Small quantities or layers of frozen ground, soft rock, or other resistant material, may be loosened by lifting the attachment and dropping it, teeth down, a technique known as chopping. It is hard on the machine, and should be done only when absolutely necessary.

To avoid serious damage, be sure that no cylinders are at the ends of their strokes. Lack of cushioning by the hydraulic system in such positions causes more severe shock to the parts.

Combining crowd and down pressure with wiggling of the bucket control often gives more effective penetration than chopping does.

Undercutting Pipe. To undercut a pipe, dig on both the far and the near sides, keeping at least a few inches away from it. Go a foot or preferably two feet below the pipe, then undercut with a flat bucket from the far side, and a widely opened one from the near side.

Temporary Support. When making a wide excavation in sandy soil, or above a soft, muddy subsoil, it might be advisable to rest the front of the tracks on a platform, plank or pole, and/or to avoid digging close to the tracks.

Applications. The hydraulic full revolving backhoe has all the advantages listed for the cable hoe on Page 13-66. In addition, it can usually dig deeper, and is neat and efficient at truck loading. Moreover, it is at less disadvantage in digging material at or above its own level.

Its typical cycle, digging at moderate depths and loading trucks (or piling) at its own level, is slower than that of a dipper shovel working in a bank, but is usually faster than a front loader. It varies from 20 to 40 seconds or more.

The cycle may be shortened by at least 25 per cent if trucks can be spotted in the pit for loading. Time in raising the bucket to the surface is saved, and swing can usually be made shorter.

In addition to typical backhoe operations

such as ditching, and digging cellars and sunken pits, these machines are valuable for mud excavation. They may use platform supports, described in Chapter 3, or depend on their own tracks, either standard or extra wide, for support.

In such work, they have a great advantage in their ability to apply force to the bucket to pull, push, or raise themselves out of trouble.

TRACTOR-MOUNTED BACKHOE

A hydraulic hoe attachment similar to that described for revolving excavators can be mounted on the back of a wheel tractor, a crawler tractor, a drag trencher, or a truck. A two wheel drive tractor equipped with a front loader is the standard type of carrier.

It is usual to install permanent mounting

Courtesy of Liebherr-America

Fig. 13-137A. Demonstrating down pressure

brackets on the tractor, then use pins, hooks, and/or tie bars for more or less quick installation and removal of the hoe unit. The quick-detachable model in Figure 13-138A is designed for four wheel drive units, and includes a seat. On two wheel

Courtesy of American Poclain Corporation

Fig. 13-137B. Wide-track hoe working in mud

Courtesy of Ware Machine Works

Fig. 13-138A. Quick-detachable hoe

drives, the tractor seat may swivel to give access to the hoe.

There are also integral units, in which the tractor has mountings for both hoe and loader built into its own frame.

These units are almost always smaller and lighter than full-revolving rigs, but they are nevertheless powerful and capable machines.

Boom, stick and bucket are similar to those on the larger machines, but there are many differences in detail.

Swing. The boom and associated digging parts are mounted on a small turntable that swings through only about half a circle, 175 to 210 degrees on different models. Full swing is not practical because the tractor is in the way.

The pivot is usually a pair of pins, upper

Courtesy of Ford Motor Company

Fig. 13-138B. Integral frame for tractor, hoe, and loader

and lower. Power is from a pair of hydraulic cylinders based on the non-revolving frame. The rods may be connected to brackets, or to opposite ends of a roller chain meshing with a sprocket on the rotating section. A part-swing vane motor may replace the pistons.

Swing is usually very fast, as there is no heavy deck full of machinery to move. Acceleration and deceleration are rapid, allowing 90 degree swings to be completed in as little as three seconds. Action is likely to be jerky.

Part swing causes little difficulty in straight open work, like digging a ditch across a field. But it creates serious problems with complicated jobs.

Stabilizers. A pair of heavy stabilizer arms or outriggers are hinged to the sides of the turntable base. They are raised and lowered by two way cylinders.

Cleated shoes on their outer ends are forced down against or into the ground while digging, to increase stability against tipping, and against being dragged by the bucket.

Hydraulics. The pump and reservoir are usually in the tractor. The most convenient arrangement is to have one pump for the hoe and another for the loader. It is more usual to have one pump, and a diversion valve to route the flow to whichever unit is in use.

A hoe and loader are not used for digging at the same time, but it is frequently necessary to make position adjustments in one while working with the other.

A pair of hoses with quick-detachable couplings brings the fluid to the valve bank in the hoe.

There are six operating valves, and usually six levers. But four functions may be performed by two "joy stick" levers, that move to the sides as well as back and forth.

Figure 13-139 shows a standard type of control with a usual but not necessarily standard arrangement of functions. There are also universal controls, whose levers actuate cables that can be switched from one valve to another, so that the operator can make up his own pattern.

Pressure in different machines is variable, and is not always stated in specifications. Many models are in the 2,000 to 2,500 psi range, with flow 15 to 35 gallons per minute.

Buckets. Buckets are similar in construction and linkage to those of the big machines, but differ in size. The smallest standard models seem to be 12 inches wide, with a capacity of 3 cubic feet; the largest 36 inches wide with ½ yard capacity. In addition, there are specialty buckets on miniature hoes as narrow as 6 inches.

Even a standard size tractor hoe may replace the bucket with a ripper tooth as narrow as 3 inches. This is primarily designed for penetrating very hard, frozen, or bouldery soil, but it may dig a slot for cable or conduit installation.

Applications. The tractor-mounted hoe is a small, powerful, and fairly economical package. It can be driven between jobs, or carried easily on a light trailer.

As a ditch-digger, it can work in places that are difficult or impossible for larger machines, cross lawns without damage (except under unusually soft conditions), make narrow ditches, and show a high production rate.

It does not trench as fast as the continuous type ditchers discussed in the next

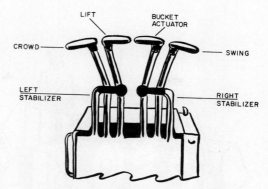

Fig. 13-139. Tractor hoe controls

chapter, but it can handle special situations that are difficult for them, work in rocky soil, and handle oversize pieces.

It is also a handy utility tool. It can dig out stumps and boulders so big that it cannot possibly lift them, load trucks with either the hoe or the loader end in emergencies, and serve as a light-duty (but very jerky) crane.

OPERATION

The boom, stick, and bucket are similar in action to full-swing attachments.

Differences in operation include managing the tractor as a travel unit and counterweight, use of stabilizers, arranging work so that it can be done with limited swing,

Courtesy of Ware Machine Works

Fig. 13-140. Universal controls

Fig. 13-141A. A dumped loader bucket helps stability

and doing heavy digging with a light machine.

Positioning for Digging. The tractor is driven to the work spot, and maneuvered so that it is centered on the centerline of a ditch, the rear wheels toward the starting point, and about ¾ of the hoe's maximum reach from it.

The operator centers the steering wheel, puts the tractor in neutral and locks its brakes. If one hydraulic pump supplies both loader and hoe, he sets a diversion valve to deliver flow to the hoe. If there are separate circuits, he lowers the loader bucket to the ground.

The bucket will hold the tractor most effectively if it is put in fully dumped position, and forced down against the ground.

He then flips the seat over or swings it around, and sits rearward on the tractor, facing the hoe and its controls. In the terminology apparently accepted by the trade, he is still facing forward, and hoe motions are described accordingly.

Stabilizers. The two outside controls are used to push the stabilizers outward and downward, pressing them firmly against

the ground so as to take just a little weight off the tractor's tires.

If the ground is uneven, one is extended further than the other, to make similar pressure on each side. If the ground slopes to the side, the low one is pushed down harder, to reduce or eliminate tractor tip.

Occasionally, you may wish to put a block under the stabilizer on the low side, or level the ground by some superficial digging.

The standard stabilizer shoe has a ridge or cleat on its under surface, to penetrate and grip the ground. This is likely to tear up a lawn, not so much from original penetration, as from dragging and twisting during digging. Very rarely, you may prefer to try digging without setting stabilizers, for this reason.

On soft ground, a plank may be placed to support the stabilizers.

Digging. Digging motions are the same as those described earlier for the full revolving excavator. However, this machine is probably very much lighter in proportion to digging power, and is more likely to be dragged into the work.

Fig. 13-141B. Extra support for soft ground

This dragging occurs as you pull or crowd the bucket inward against digging resistance. It is strongest at ground level, and reduces somewhat with depth.

Shifting of the machine shortens the digging strip it can work, often puts it out of line, makes the stabilizers tear up the ground, and may get the tractor in the ditch. It is the operator's responsibilty to make it a rare occurrence.

Cutting a thin slice generates less pull than a thick one, but it takes longer and wears the bucket more. Rapid, small changes in bucket angle as it is brought in lessens resistance.

Curling or closing the bucket provides powerful digging and breakout force at the bucket teeth, with little or no pull on the tractor.

Starting a deep cut, then curling the bucket as soon as resistance builds up, will usually provide good digging with little dragging.

In general, you try to dig for production, but lighten the cut every time you feel a drag-the-tractor force developing. The reaction is a split-second one, which is difficult at first, but which soon becomes practically automatic.

Moving. To move when a section of digging is finished, the stabilizers (and the loader bucket if it is down) are raised. The bucket is placed on the ditch bottom with its floor at a slight angle, with the stick almost vertical, but leaning slightly toward the tractor.

Pushing the hoist and crowd levers slowly forward will now lift the back of the tractor off the ground, and roll it forward on its front wheels. If it does not go far enough, the manuever is repeated. Then the stabilizers are put down, and digging resumed.

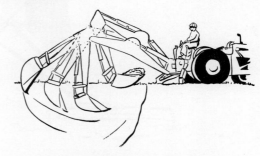

Fig. 13-142A. Digging motion

Fig. 13-142B. The hoe moves the tractor

The tractor may be moved to the side in somewhat the same manner. The bucket is curled halfway, placed in the ditch a little forward (away from the tractor) of the boom point, and forced down by the boom cylinder until the rear wheels are clear of the ground. The hoe is then swung until the tractor is in the desired position. Then it is lowered, and the stabilizers set for digging.

Stability. The total weight of a tractor-hoe, including its loader, is much smaller in proportion to digging power than that of a full revolving machine. Stability is therefore more of a problem.

Down pressure on the bucket tends to lift the rear wheels off the ground. Upward pull lifts the front wheels. Rearward (toward the tractor) pull may either dig the dirt or drag the tractor.

Tractor weight must be increased to its maximum by putting water and calcium chloride, or perhaps mineral dust, in all four tires. The loader bucket may be filled with dirt, or piled with rocks or other heavy objects. It may be held in the air for counterbalance (and for convenience in moving), or rested on the ground for resistance to dragging.

Stabilizers should be put down far enough to take a little of the weight off the rear tires, as will be indicated by a slight change in shape at the bottom. Tires must never be lifted clear of the ground, nor even so that most of the weight is off them, except one side in leveling on a slope. On ordinary ground, a ridge on each stabilizer pad sinks all the way in.

Rear wheels are locked by the tractor brakes.

Even with these precautions, weight and grip on the ground are small in proportion to the digging forces — crowd and curl — which may be more than six tons. The operator must limit the force he exerts to

Fig. 13-143. Leveling with stabilizers and block

ability of the machine to keep its position.

Digging without pulling the tractor may be the hardest thing for the beginner to learn, as was discussed earlier, but it ordinarily offers few problems to the experienced operator. It is largely a matter of learning manipulation of the bucket to obtain penetration without excessive pull.

When the tractor is dragged toward the digging, it is usually necessary to re-position it with the bucket. If the movement has not been straight, it may be necessary to drive the tractor to get it in proper alignment.

Stability against tipping is ordinarily not a major problem in digging. But if the tractor is not level, or is being used to lift heavy objects with a chain, this must be considered. It usually has least stability when the load is at a 60 degree angle from center, on either side if level, on the low side if not level. Tipping tendency is reduced by bringing the load in toward the tractor, or stopped by setting it down.

Slopes. A side slope affects stability of a hoe, making digging motions difficult to control, and produces a ditch that is out of plumb (side not vertical).

A slight slope may be ignored, or corrected with the stabilizers by putting pressure on the downhill one until it levels the tractor, or adding a block, Figure 32.

For a steeper slope it is advisable to make a cut or shallow trench for the uphill stabilizer and wheel, using the spoil to fill for the downhill side if necessary. This cut may be made by the loader bucket, preferably pushing downhill, or by the hoe.

A shallow leveling cut may be made by the hoe on fairly steep side grades, as little material is dug and accuracy is not important. This may extend full length of the hillside, or just be scooped out for digging position.

When the tractor is tipped, it is much more stable dumping uphill from the ditch than downhill, and this side is generally

Fig. 13-144. Side hill trenching

(but not always) favorable for backfilling.

If there is a choice between trenching up a hill or down it, work with the tractor heading downhill. This reduces the principal problem of dragging the tractor toward the digging, and is much more stable against tipping when dumping the bucket.

An exception to this advice is that in wet ground it may be desirable to dig uphill to avoid ponding of water in the digging area.

Close Work. Tractor mounting is better than full revolving for working close to buildings or trees, as it is not necessary to allow space for a tailswing. One wheel can be rubbed against the obstruction, if necessary.

It is sometimes possible to get closer to a wall by backing the machine to it at an angle for short cuts. The front wheels may be turned parallel to the wall, so that the

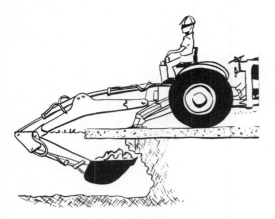

Fig. 13-145. Undercutting pavement

tractor can be boosted by the bucket from one digging position to another.

It is possible to undercut curbs and pavements for several feet, Figure 13-145. Increased under-reach may be had by digging deeper than the required bottom.

Undercutting to Side. Most hoes must ditch one-half of the tractor width away from obstructions on the ground. The bottom of a ditch can be brought closer to the obstruction (a house foundation, for example) by tilting the tractor away from the wall.

This is done by down pressure on the stabilizer on the wall side. Its lift can be increased by placing blocks under it.

The ditch will then slant, with the bottom

Fig. 13-146. Tilting the tractor to cut close

toward the wall. Pipe or conduit must be placed quickly, and without getting in the ditch, as the overhanging wall may be expected to fall.

AUTOMATIC DIG CYCLE

Some backhoe models may be factory-equipped with an automatic mechanism, introduced into a six-lever control bank. When actuated by holding down a foot pedal and pulling the crowd lever, it regulates crowd, curl, and hoist actions so as to

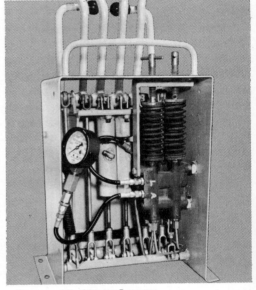

Courtesy of Ford Motor Company

Fig 13-147. Automatic digging controls

make a digging pass that will fill the bucket. Manual operation controls lowering into contact with the ground, hoisting out of the excavation, swinging, and dumping.

The device is said to be an all-hydraulic computer unit. It senses the resistance, or lack of it, at the bucket teeth, and makes appropriate adjustments to change the depth of bite. There are manually set pressure regulation valves on top of the console, with which it can be adapted to the hardness or softness of the digging.

Fig. 13-148. Hydraulic dipper shovel

Automatic operation can be over-ridden manually at any time, without even releasing the foot pedal, just by pulling or pushing the appropriate levers. This action is taken to either avoid or remove obstructions. All actions return to normal when the foot pedal is released.

This type of control is especially valuable for the inexperienced operator, who can obtain full production much sooner than with regular controls. It relieves the seasoned operator of the necessity of concentrating on this part of the cycle.

SHOVEL FRONT

A few hydraulic excavators may be equipped with a shovel front. These machines are used in the same general way as the cable-operated dipper shovels, but are more flexible in operation.

Boom and stick are wider than in the hydraulic hoe. The stick is moved directly by a pair of cylinders mounted on the boom sides, with piston rods hinged to the stick sides.

The bucket is a multiple purpose or 4-in-one type, which is described later under LOADERS. The whole bucket is pivoted on the end of the stick, with cylinders to move and hold it in a wide range of digging angles. In addition, the front-and-bottom piece is hinged so that it can be swung forward away from the back plate by additional cylinders, opening the bottom for dumping without need for the space and time required to overturn it; or closing so

Fig. 13-149. Dumping position

Courtesy of The Warner & Swasey Co., Gradall Division

Fig. 13-150. Multi-purpose hydraulic

as to clamp big objects, clamshell fashion.

These machines have great flexibility in digging. They can cut a bank from the ground level up in the same manner as a tractor loader, and can also enter it just as efficiently at higher levels. Control over bucket angle, and its easily manipulated crowd-retract that does not require moving the whole machine, make it adept at selective digging, prying out and picking up boulders, and utilizing planes of weakness in rock.

The clam-type opening of the bucket allows very high dumping, as no allowance need be made for swing of bucket or dropping of a door. The load can be let out in controlled amounts to avoid shock to truck bodies.

GRADALL

Figures 13-150 to 13-155 show the Gradall, a special type of full hydraulic pull shovel. It resembles a conventional shovel only in the construction of the turntable and swing rollers, and in having the engine mounted on the rear of the deck.

The engine drives a three-unit tandem pump that supplies pressure through control valves actuated by three hand levers and two foot pedals, to hydraulic cylinders and motors which power all working functions.

The boom, Figure 13-150, is a hollow box girder, one section of which telescopes inside the other in response to the push and

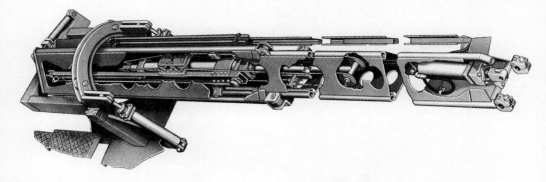

Courtesy of The Warner & Swasey Co., Gradall Division

Fig. 13-151. Cutaway of boom, old type

pull of a long two-way ram. It is mounted in a heavy steel cradle that serves for support, control, and counterweight. Cradle and boom are raised and lowered by two double-acting cylinders based on the platform. The boom can be lowered to a vertical position, for 25-foot digging depth, and raised to dump at over 18 feet. It can also be tilted (rotated on its long axis) 90° in either direction from the horizontal by a hydraulic motor.

Swing is controlled by hydraulic motors driving reduction gears.

Figure 13-154 shows the hydraulic controls. There are three hand levers and four foot pedals. The left hand lever raises the boom when it is pulled back, and lowers it if moved forward. The right hand lever retracts the boom when pulled back, and extends it when moved forward. These two important movements are quickly memoried because the boom moves in the same direction as the lever.

The center lever tilts or rotates the boom and bucket clockwise when moved back, counter clockwise when moved forward. The bucket is kept in horizontal position for most digging, so this control is used chiefly for finish grading, and occasionally for working out obstacles.

The left pair of pedals controls the angle that the bucket makes with the boom. De-

pressing the left pedal moves the bucket teeth outward to digging position, the right pedal brings them in to closed or tucked-in setting.

The right hand pedals control the swing, left for left and right for right.

All hydraulic valves are spring loaded so that they return to a neutral HOLD position as soon as they are released. Affected parts stay locked in position until controls are moved again.

Crawler travel is controlled by a pair of levers, one on each side of the seat. The tracks are independently driven by hydrostatic motors, with gear trains to the bull wheels, and can be counter-rotated for spin turns. They are locked for digging by a side-

Courtesy of Warner & Swasey Company

Fig. 13-152. Side sloping

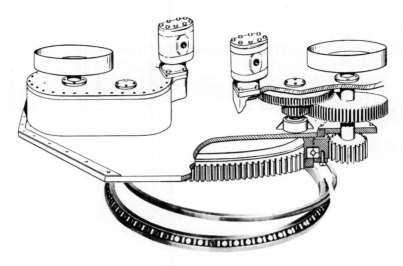

Courtesy of The Warner & Swasey Company

Fig. 13-153. Hydraulic motors and swing gears

to-side lever in front of the other controls.

A special truck carrier may be operated either from its own cab, or from the excavator control station.

Operation. Figure 13-115 shows a sequence of control settings for a digging cycle.

The normal digging procedure is to swing the machine and extend the boom until the bucket is over the spot where digging is to start, lower the bucket in digging position, pull it inward with the boom retract until it fills, rotate it to closed position, then lift the boom and swing it while extending or retracting as necessary to position the bucket over the dump spot. The bucket is then opened to dump, and the boom swung back to the digging.

The first few passes may be made by starting with the teeth at an angle of about 45 degrees with the ground, and flattening it as it is pulled in. If the slice is too thin a steeper angle may be used, if too thick a flatter one. It is good operation to fill the bucket in the shortest pull that does not cause the engine to lug down or the carrier to shift.

The bucket should be closed when full

and lifted out of the ground. Swing must not be started until it has lifted clear. If spoil is being dumped right alongside the ditch it may not be necessary to close the bucket fully, and swing will be very short. If dumping is into a truck it should be as near the ditch as possible, and swing may not be started until the bucket is high enough to clear the truck.

The ditch is deepened and cut toward the carrier by a series of similar passes. The far end can be trimmed to a vertical slope for a short distance down by opening the bucket wider, but if the ditch is deep there will be a curve at the bottom that cannot be reached from this position. However, the carrier can be moved backward far enough to cut a square end (if it is needed), as there is ample clearance between the nearest part of the ditch and the carrier.

The bottom of the ditch is finished by making the final pass with the bucket level, using the hydraulic down pressure of the boom. This cleans up soil spill, evens out irregularities and tooth marks, and leaves an excellent surface for pipe or granular material.

If a boulder or heavy root is encountered,

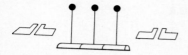

1. BOOM AT REST
CONTROLS IN NEUTRAL POSITION

Start the engine and let it run at idle for a few minutes. With engine at idle engage the clutch. Increase engine speed to full throttle.

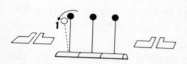

2. LIFT BOOM
FROM CARRIER BOOM REST*

Pull back on left lever to lift boom (1). When lever is released the boom will remain positioned.

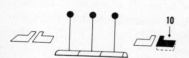

3. SWING BOOM
OVER PROPOSED EXCAVATION

Press outer right pedal to swing to right (10). Release pedal when boom is aligned with proposed trench.

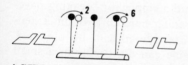

4. EXTEND AND LOWER BOOM TO START OF CUT

Push right lever forward to extend boom (6) and push left lever forward to lower boom (2). These two operations can be performed together.

Courtesy of The Warner & Swasey Company

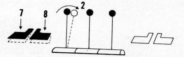

5. POSITION BUCKET TO START CUT

Press outer left pedal until bucket teeth are set for correct penetration (7). Push forward on left lever to lower boom for penetration (2). Press inner left pedal until bucket is set for desired grade (8).

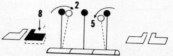

6. RETRACT BOOM, AND APPLY DOWN PRESSURE
KEEP BUCKET AT PROPER GRADE

Pull right lever back to retract boom (5) and push left lever forward to apply down pressure (2). Use inner left pedal to keep bucket from opening (8). Continue this operation until bucket is full.

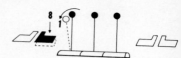

7. LIFT BOOM — AS BOOM RISES CLOSE BUCKET

For best production results, don't raise boom higher than necessary. Pull lever back (1) and as the boom rises, press inner left pedal (8). This closes the bucket and prevents excess spillage. When bucket is completely closed, release pedal.

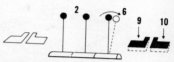

8. SWING OFF TO DUMP LOCATION
EXTEND BOOM IF REQUIRED

When bucket is clear of ditch press inner right pedal (9) (left swing) outer right pedal (10) (right swing). If required extend boom by pushing right lever forward (2). When bucket is over dump location release pedals and levers.

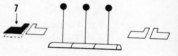

9. OPEN BUCKET FULLY TO DUMP

To release material press outer left pedal until bucket is fully open (7).

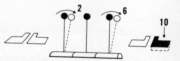

10. START NEXT CYCLE WITH ILLUSTRATION 3,
AND REPEAT

Press swing pedal as illustrated for proper alignment of boom and trench. Push right lever forward to extend boom to far end of first cut. Lower boom and repeat steps.

Fig. 13-154. Gradall operation

it may not be practical to remove it by direct pull of the boom retract. Much greater force can be applied by putting the bucket in a level or down-tilted position, getting the teeth under the obstacle, and closing the bucket. The floor will act as a fulcrum to aid the teeth in prying the object up, providing a powerful breakout force without dragging or tipping the machine.

If the ditch is dug slightly wider than the bucket, the bucket can be rotated when necessary to get under objects that cannot be gripped otherwise.

When a boulder is too large to be pried out in this manner, it is necessary to widen the ditch until it is freed. A ramp is dug from it to the ground surface, the carrier is moved ahead a short distance, and the retract used to drag or roll the stone to the surface, where it can be rolled to the side.

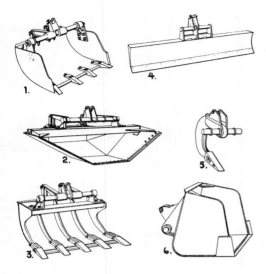

1. 36" Bucket
2. 72" Formed ditch digging bucket
3. 47" Pavement removal bucket
4. 8'0 Scraper blade
5. Single tooth ripper
6. 36" Push-type rehandling bucket

Courtesy of The Warner & Swasey Co., Gradall Division

Fig. 13-155. Buckets and attachments

Heavy lifts and any side pushing or pulling should be done with the boom retracted to obtain additional force and reduce strain.

Grading. These machines are particularly well adapted to finish grading. The retractable boom makes it possible to draw in a bucket or grading blade in a smooth line that is not disturbed by hinge action of a boom or the effect of bumps on wheels or tracks. Its reach, which ranges from 24 feet from center pin in the small machine with a standard boom to 39 feet in the large model with a boom extension, enables it to finish extensive areas that may be too soft or steep to carry machinery.

When doing light grading up or down a slope, the bucket or blade is kept in light contact with the ground as it is drawn in by the telescoping boom. When the ground is below the level of the boom pivot, retract-

ing tends to pull the bucket up, so that for a level grade the boom lowering valve must be kept partly open to compensate. If the ground slopes steeply upward the boom may have to be lowered as the bucket moves in, to keep a straight line contact.

The bucket or blade may be tilted to follow an uneven contour, or to work an even contour from a side angle.

Hydraulic control permits keeping the bucket at its most efficient digging angle at any point in its reach, and dumping just where desired. With the boom tilt it gives precise control, and allows exact cutting of floors and sloped or vertical walls in cellars and ditches. It is adept at shaping and fine grading roadsides under even difficult conditions, and can stay safely on the road as it works.

Pull type buckets are available in widths from 15" to 72", and push buckets, whose digging action resembles that of a dozer shovel bucket with remote control, in 36" and 60" width. There is an 8' scraper blade for grading where dirt does not have to be picked up, and a ripper for breaking pavement or loosening hard dirt. Boom extensions and offset booms can be obtained.

This machine can dig deeply in a small area, so that it can connect ditch sections from the side (see Page 5-3), and dig for septic tanks and manholes with little or no overcutting. It can push as well as pull and has the advantage of being able to reach under overhead structures, even through doors or windows, to pick up objects and bring them out horizontally. As long as loads and heights are within its range, it serves well as a crane.

No clutches or brakes are used by the operator. He does not need to take care to shift smoothly from clutch pull to brake holding and back, as the rams both pull and hold. Control valves may be held in any position intermediate between open and closed to regulate speed.

CHAPTER FOURTEEN

CONVEYOR MACHINERY

CABLE EXCAVATORS

Cable excavator is a general term which includes any cable-operated machine using an excavator bucket working between a head structure and a tail anchor spaced hundreds of feet apart. The head structure usually is a guyed mast but may be a movable self-supporting tower. The tail anchor usually is designed to move along an arc, the radius of which is determined by the length of the operating cables.

Traveling towers are used at both ends for levee construction and similar work where the cuts must be parallel instead of radial. Head and tail towers on such jobs move in unison.

SLACKLINE CABLEWAY EXCAVATORS

This type of excavator uses a bucket supported on a track cable which is slackened to lower the bucket into the digging, and tightened to hoist it.

Figure 14-1 shows typical installations. The main plant consists of an engine-driven drum hoist, a mast or tower, a hopper to receive the spoil, a set of cables which support and control a bucket, and an anchor.

Hoist. The hoist in Figure 14-2 consists of a two drum winch that resembles the deck machinery of a shovel in appearance. The front, inhaul, or drag drum has two speeds controlled by a pair of friction

clutches, and a holding brake. The second drum has one clutch and a brake, and holds the tension cable that regulates the height of the track cable.

Operating levers may be mounted on the side of the hoist bed, or at a distance. Remote control may be mechanical, hydraulic, air, or electric.

Engines are gasoline, diesel, or electric. Belt drive or steam may also be used.

The hoist must be far enough from the sheaves on the tower to permit the cable to feed smoothly and evenly onto the drums.

Head Mast. An engineered installation of a slackline cableway usually will have at the head end a tubular steel mast or a steel lattice mast, similar in design to the boom of a crane. The base should be arranged to allow some rocking in any direction to relieve tension caused by any unequal tightening of guys.

For simpler installations a trimmed tree can be used if it possesses the necessary strength and height.

A trussed mast of built-up timber or a tower of timber or steel also may be erected from plans supplied by manufacturers.

Whatever construction is used, the mast should be braced by four to eight guy cables extending from the top to ground anchors such as strong stumps or buried logs. Best results are obtained if the guys include turnbuckles or other takeups.

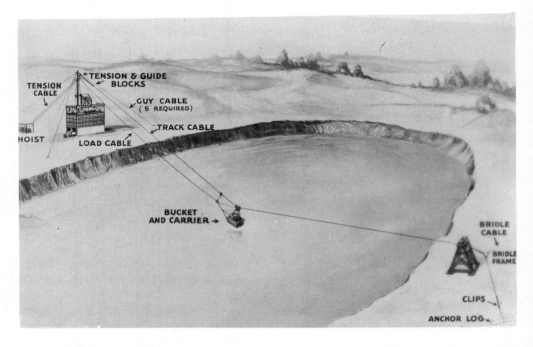

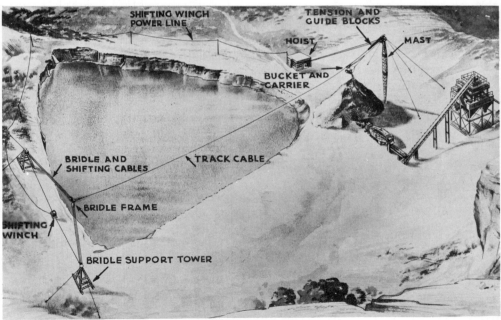

Courtesy of Sauerman Bros., Inc.

Fig. 14-1. Slackline cableways

The mast top should carry rings for mounting cable sheaves which will permit changes in direction of digging without twisting the mast.

If gravity bucket return is to be used the tower should be high enough to allow a track cable drop of fourteen feet in every hundred down to the anchor or tail tower.

Hopper. The hopper is a box into which the bucket dumps. It may be a funnel or chute to direct the spoil into a truck parked below or onto a conveyor belt, or it may feed directly into an elaborate washing, screening, or treatment plant.

It must be low enough so that the fully dumped bucket will not strike it, and just far enough away so that the dump stop will not interfere with the rigging at the mast. If it is placed further away it cannot be as high because of the down slope of the track cable.

An example of the spacing required is that a three quarter yard bucket should be dumped within twenty-four feet of the mast, and will require twenty-six feet of space below.

A hopper is not necessary for stockpiling. When dumping is to be done in the open the hoist should be elevated or placed well back to avoid danger of burial in the pile.

Bucket and Reeving. Figure 14-3 shows a dragline type bucket with carrier and chains, and the tower top and cable fastenings.

Carriers consist of parallel plates or frames which hold a set of sheave axles. Two to four sheaves ride on the track cable and support the carrier. The others carry chains which support the bucket and regulate its position.

The track cable is supported by the tension block. This is connected to the mast by a several-part line, one end of which is anchored to the second drum in the hoist. The number of parts of line varies with the size of the machine.

The drag cable runs from the front drum

Courtesy of Sauerman Bros., Inc.

Fig. 14-2. Two drum hoist

around the lower guide block to the dump block. A short single chain runs from the dump block frame to a clevis that connects it to two drag chains to the bucket, and to the arch chain that runs over a carrier sheave and down to the bucket arch.

The dump chain is anchored to the frame of the dump sheave block (traveler block) which rides on the track cable in front of the carrier. It is reeved around the dump block sheave and over a carrier sheave to a fastening on the back of the bucket. The bucket or support chain passes over a third carrier sheave and is anchored to the arch and the back of the bucket.

The dump button is clamped on the track cable near the tension block.

When pulled by the drag cable, the bucket normally assumes digging or carrying position, Figure 14-1(A), with the floor parallel to the track or tilted down in the back. The drag and arch chains pull the front of the bucket up. The weight of the rear of the bucket on the dump chain pulls the dump sheave close to the dump block.

Dumping. When the traveler block dump sheave is pulled into contact with the dump button, it stops. The dump block continues to move forward and away from the traveler block. This pulls the dump chain around its sheave so that it raises the back of the bucket and dumps it.

The dump button tends to move toward the mast as the track cable stretches. This change may be offset at first by leaving the track cable somewhat slack during the

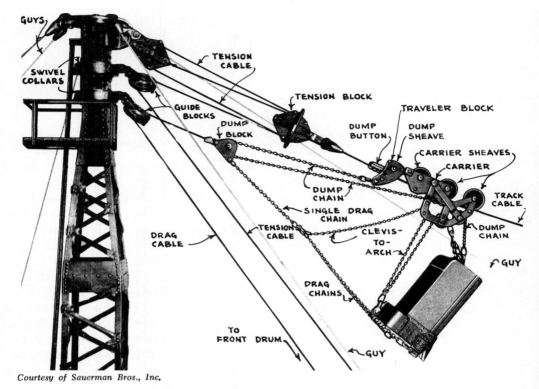

Courtesy of Sauerman Bros., Inc.

Fig. 14-3. Tower and bucket rigging

dump. If stretching continues the button is unclamped and moved or the track cable is shortened at the anchor.

Tail Anchor. The track cable may be anchored at the far side of the excavation in a number of ways. Choice is dictated by sites and materials available, and by the frequency with which the path of the bucket needs to be shifted. Deep digging in material which caves or flows into the bucket path permits extensive work in one position, while shallow cuts and earth which stands in steep slopes require frequent shifting. The anchored end of the cable should be several feet above the ground if digging is to be done near it.

The simplest anchorage is obtained by passing a chain or cable around a tree or tall stump, but this permits no side shifting. Figure 14-4 shows an arrangement al-

lowing for limited shifting. Two anchors, of any type, are connected by a slack bridle cable. A bridle hitch frame consisting of a pulley block with a fastening for the track cable is threaded onto the bridle. Clamps are placed on the bridle cable each side of the hitch to prevent it from shifting during digging.

The track cable is usually run over a sheave hung from a tail tower or A-frame, built of strong timbers bolted together and mounted on skids. This need only be high enough to keep the bucket clear of the ground at the closest point where it will dig.

The bucket path is changed by slackening the track cable, loosening the clamps on the bridle cable, and sliding the hitch to a new position. The clamps are reset and tightened, and the A-frame moved until its sheave is on a direct line between the head tower and the bridle hitch.

If the bridle cable is connected to a hand winch at each anchor the bridle can be shifted by taking in one end and paying out the other. A motorized winch controlled by the hoist operator can be used to shift the bridle frame, and its use results in easy and rapid shifting.

A heavy tractor or loaded truck that is too broken down for normal work may be used as a single anchor. The cable is anchored to a draw bar, pintle hook, or other fastening on the back or side of the machine and is supported by the tail tower sheave. To shift, the track cable is slackened and taken off the tractor which is maneuvered to tow the tower to its new position, then blocked in position to serve as an anchor again.

This is practical only when the ground is firm and fairly clear of obstructions. A crawler tractor can be used in spaces too restricted for trucks.

The track cable may be anchored to a tail tower or block of sufficient width and weight not to need anchors. This may be on tracks of either the railroad or crawler type, or on skids, and can be self-propelled or towed.

Operation. A typical slackline setup consists of a tower connected to a lower anchor by a heavy track cable, which can be tightened or slacked off by a winch operating through a lighter cable (the tension line) and pulleys. A dragline-type bucket is suspended from the track. It can be pulled to the tower by a two-speed winch, and tends to move toward the anchor by gravity. When pulled close to the tower it is dumped automatically into a hopper or onto a pile.

To start the digging cycle, the drag brake is released and the bucket allowed to coast out to the digging point. The fastest trip is obtained by holding the track cable at such tension that the bucket will

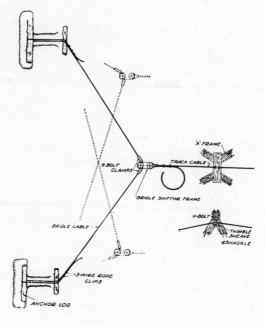

Courtesy of Sauerman Bros., Inc.

Fig. 14-4. Fixed tail anchor

just clear the ground or water as it reaches the digging point. This may require a tight track for distant digging or a slack one for working close to the mast.

When the bucket reaches the digging it is checked with the drag brake and the tension brake is released, allowing the track to sag and the bucket to rest on the ground. If the pit is under water the operator must get used to the feel of the controls and the angle at which the track cable enters the water when the bucket is making proper contact.

The low speed or drag clutch is engaged and the bucket pulled in. In good digging it will fill within a few bucket lengths. The tension clutch is then engaged so that the track tightens and lifts the bucket clear. The high speed drag or inhaul clutch is then used to pull the bucket in to the automatic dump over the hopper.

The maximum load of the cycle occurs when the bucket is being hoisted and

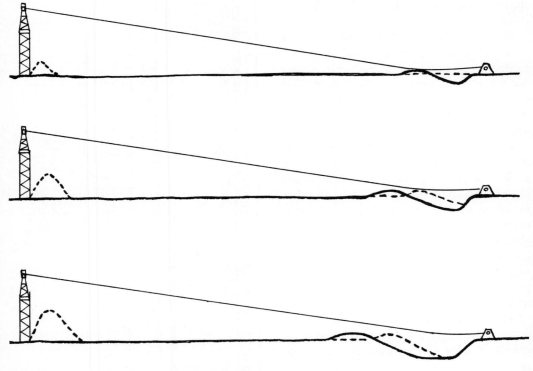

Fig. 14-5. Starting a slackline pit

pulled in at the same time. For this reason the high speed drag is usually not connected until the hoisting is complete.

When a pit or a line of digging is
opened the cut should be made at the far
end and worked toward the mast. This
provides a slope, Figure 14-5, which
allows the bucket to operate at maximum
efficiency. If excavation is selective, similar cuts and slopes can be developed in
each deposit.

In caving or flowing material, digging
can be continued on one line for a long
period. The tail anchor can be moved a
considerable distance each time it is reset.

If the material does not cave at all the
bucket may cut a slot wider than itself because of wabble, or exceptionally, cut so
narrow a slot that it cannot be deepened
because of jamming between the sides.
Tail anchor moves under such circumstances must be short, but should be suffi-

cient to place the new line of digging on
undisturbed ground in the outer area to
minimize trouble from skidding sideward
into the former cut. Ridges left standing
between the two cuts may be taken out by
moving the anchor halfway back after
completing the second one.

The digging ability of these buckets is
comparable to that of draglines and the
ability to handle hard formations increases
rapidly with size. The carrier prevents
them from bouncing and losing contact as
much as a dragline bucket, but the inability to vary the direction of digging makes
it more difficult to cope with boulders or
seams of harder material sloping across the
digging line.

Bottomless Buckets. A bottomless crescent bucket may also be used with this
type of slackline excavator. Such buckets
are generally used for open stockpiling,
but can dump in a hopper that is partly

Courtesy of Sauerman Bros., Inc.

Fig. 14-6. Crescent buckets

buried in the pile, or which has a hard surfaced ramp to carry the bucket to it.

The bucket is placed to dig in the same manner as the standard cableway bucket. However, having no bottom it cannot be raised from the ground without dumping its load, so it is dragged all the way in, usually with the low speed clutch.

Light and heavy buckets are shown in Figure 14-6. For slackline use the rear or backhaul chains are attached to a carrier block on the track cable.

When the bottomless bucket has a full load, the pressure of the dirt against the forward sloping top of the rear section raises (floats) the cutting edge out of the ground, so that there is little tendency to either dig in or overload, and resistance to pull is comparatively light.

The top or overhaul cable shown on the large bucket serves to keep the back down, resists overturning, and helps in righting the bucket if it does overturn.

DRAG SCRAPERS

The power drag scraper machine is the basic contrivance for utilizing a bottomless bucket to move earth. The bucket may vary in design, but is usually crescent shaped. No track cable is used and the bucket is not lifted off the ground by a normal cycle.

Characteristic drag scraper installations are shown in Figures 14-7 and 14-8. The head mast is placed at the low side of the pit, and the tail anchor moved around the high side. Initial digging is aided by gravity, but this advantage may be lost with increasing depth.

14-7

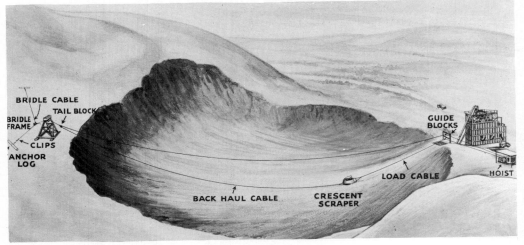

Courtesy of Sauerman Bros., Inc.

Fig. 14-7. Drag scraper pit, fixed tail anchor

Hoist. The hoist may have two or three drums, separately controlled by clutches and brakes. The front, drag, or inhaul drum is geared down for digging while the rear or backhaul drum, which returns the empty bucket, is two to three times faster. These two drums work oppositely, one reeling in while the other pays out. When a third drum is added, it is used to change and hold the position of the tail block.

Clutch and brake are interconnected so that the brake is released automatically when the clutch is engaged, and set when the clutch is released. The drag brake prevents excess slack from developing in the cable which is being overhauled (paid out).

Hopper. The hopper is set low and close to the mast where the pull on the bucket is high enough to prevent it from cutting in and catching the hopper edge. It is also good practice to protect the hopper by a sloping apron carrying metal sheets or rails. The hopper is equipped with bars, set parallel to the direction of digging movement, which support the bucket but allow the load to fall through. They may serve as a grizzly also, as oversize stones will rest on top and be pushed beyond the opening by the next load.

If an elevated hopper is required, the bucket is brought to it on an inclined chute. The hopper may be emptied by a conveyor belt, bucket conveyor, bulldozer, or trucks.

If the material is to be wasted or stock piled, a high mast is used and the pile built in front of it. Larger quantities can be dug from a mast setting if another machine moves spoil to the sides as it is brought in.

Cables. The inhaul cable runs from the front drum through the lower guide block, then out to the bucket drag chains. The outhaul line runs from the second drum through the top guide block, across the pit, around one or two anchor sheaves, and forward to chains on the rear bottom of the bucket.

A single piece of cable may be used for both inhaul and outhaul lines. It is passed over the bucket through sheave sets and is clamped to the bucket chains at front and rear. Its ends are anchored on the inhaul and outhaul drums.

Mast. Masts are similar to those used for slackline excavators except that since they are usually low they do not need to be so elaborate. Cross bars between two poles are frequently used.

Tail Fastenings. Three anchor fastenings

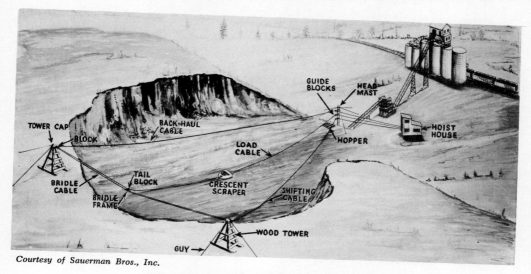

Fig. 14-8. Drag scraper, rapid shift anchor

are in general use for excavation. In Figures 14-7 and 14-9 a bridle cable is anchored at both ends, and a clamp or bridle frame can be attached anywhere along it. From the clamp a line, called the tail guy, is reeved over a thimble sheave fastened in an A-frame or tail tower and to the tail block. The outhaul line is reeved through the tail block sheave.

In 14-8 and 14-10 two wood towers are erected and fitted with metal caps, which serve as fastenings for guys, for the ends of the bridle cable, and for a guide block. The tail block is fastened directly to the bridle frame. The backhaul line is reeved through the guide block and tail block.

The position of the tail block is determined by the length of the shifting and holding cable. This can be regulated by loosening clamps and adjusting it when the bucket cables are slack, or by means of a hand winch. This arrangement makes possible a longer bridle cable, and enables the line of digging to be shifted without moving an A-frame.

The rapid shifting device in Figure 14-11 uses similar capped towers and bridle. The bridle frame rides on the bridle cable on two sheaves and is connected to the tail

block by a hinged arm, as in 14-12. The shifting cable runs from the third drum of the hoist through a mast guide block, then is anchored on the bridle frame.

When the shifting cable is reeled in, it pulls the bridle frame along the bridle cable. When it is released the normal tension on the backhaul cable pulls it in the

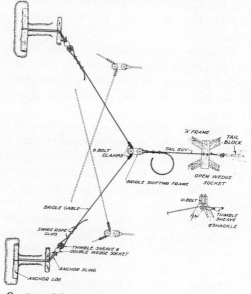

Fig. 14-9. Detail, fixed tail anchor

14-9

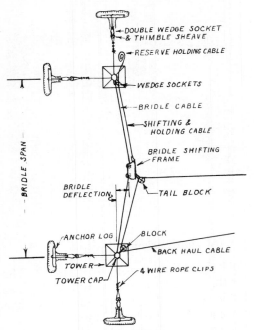

Fig. 14-10. Detail, manual shift anchor

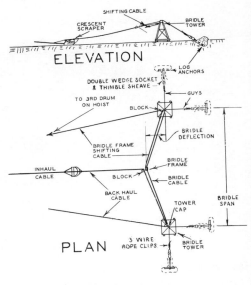

Fig. 14-11. Detail, power shift anchor

other direction. The clutch and brake on the third drum thus give full control over the position of the tail block between the towers.

Dragline Conversions. A dragline or crane can be used as a mobile hoist and tower for drag scraper or slackline operations. An application is shown in Figure 14-13.

The hoist line is strung over the boom point sheave to an anchor, and is used as a backhaul line or a combined tension and track cable. The drag cable, which may be used with a dragline fairlead or with a block fastened part way up the boom, provides the digging pull. The machine can usually handle a crescent bucket fifty percent bigger than the proper size dragline bucket.

For long spans or heavy loads the boom may be guyed for extra stability.

Cable capacity of the drums usually limits these machines to rather short spans. Maximum reach varies in different makes.

Operation. A drag scraper generally has a mast at the low side of the pit and a tail anchor which can be moved around the high side. The bottomless bucket is dragged by an inhaul line reeved through a pulley mounted in the mast to a winch drum. It is pulled back by a winch (backhaul) line reeved through an anchor pulley.

The bucket is moved out to digging position by engaging the backhaul clutch. This automatically releases the backhaul brake. The rear of the bucket is shaped so that it rides on the surface without cutting in.

To dig, the inhaul clutch is engaged, the bucket fills, then automatically rides up on its load so that it does little additional cutting. If it does continue to cut, and offers heavy resistance, it is probably not properly designed for the material and should be altered or replaced.

The bucket rests on the ground as it is pulled in. It is dumped by pulling it over a hopper, or by releasing the inhaul clutch and pulling it backward.

In deep soil which caves or flows into the bucket slot, the same line of digging

Courtesy of Sauerman Bros., Inc.

Fig. 14-12. Tail block

can be used for long periods. However, it is better practice to shift occasionally so as to lower wide sections fairly evenly.

If the soil stands up in banks, frequent shifts must be made as the existence of narrow, deep slots will make working difficult. If a rapid shifting device is used at the anchor end, the whole working width is taken down in layers, usually starting at the back and taking successive strips toward the hopper.

If the tail block is shifted by hand, each move should be long enough to enable the bucket to start a new path without sideslipping into the previous one.

When peak demand for spoil is in excess of average production, digging may be done at short range for quick returns and at the back of the pit for more leisurely stockpiling.

Uses. Both slacklines and drag scrapers are at their best in digging extensive areas of uniform material. Such conditions are often found in surface mining of sand, gravel, clay, or mine tailings. Large units may be used effectively in removing blasted rock, if breakage is good, and the bucket path can be kept parallel to the face.

Material may be dumped into conveyors, sorting and processing plants, or trucks, or stockpiled. Certain types of cut and fill jobs, such as building levees and embankments, can be handled by use of mobile head and tail towers working on parallel paths.

They find an important use in rehandling stockpiles. Material brought to a yard by trucks or railroad cars can be high piled and fed into hoppers as required. Drag buckets may be reversed on the cables so that inhaul and outhaul functions are exchanged to pile material and later reclaim it.

These machines are not suited to most types of selective digging, to soil containing boulders too large to go into the bucket, and to jobs where there is not sufficient yardage to justify the cost of setting up. In general, they are not good investments unless assured of enough work to pay for them, as they may be difficult to adapt to other jobs.

In suitable work, they are highly effi-

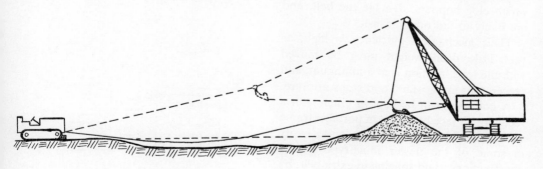

Fig. 14-13. Dragline conversion

cient because of the small number and light weight of moving parts.

A drag scraper installation is less costly than a slackline. It is equally satisfactory in many jobs, but when the haul is long, or the digging deep under water, the higher speed of the slackline makes it more desirable.

CONVEYORS

CONVEYOR BELTS

The belt conveyor is a transporting, elevating, or distributing machine made up of an endless flat belt which carries a load on its upper surface. It operates between a head and a tail pulley and is supported by idlers, that are supported in turn by a frame or by steel cables. Conveyors are made in small portable elevating units that are loaded with hand shovels, and as giants that carry millions of tons of earth for many miles.

As independent units, they are well suited for rapid transportation of loose material. They have less mobility and flexibility than trucks and scrapers, and are therefore used chiefly where large volumes of material are to be moved along one route. They are particularly applicable where the load must be lifted steeply, or carried across rough country where road construction would be difficult. They are desirable as feeders for processing plants because they provide an even and continuous flow. They simplify traffic problems where hauling space is restricted, as in tunnels and in busy pits. However, they are not adapted to hauling big chunks which clog hoppers, damage the belt, and are likely to fall off in transit.

Their mechanical efficiency is high, as very little dead weight must be moved with the load, friction is at a minimum, and power-consuming starts and stops are rare.

A conveyor can be kept in touch with a receding pit face by fitting in portable frame extensions (this can be done quite easily in some of the newer mine-type shuttle constructions) and splicing in extra belt, by installing a feeder conveyor between the hopper and the face, or by trucking from the excavator to the belt.

In addition to their use as independent and semi-independent belt haulers, conveyor belts are used as parts of loading, ditching, and processing machines.

The large permanent type of belt conveyor is almost unique among machines used in connection with excavation in that it is usually custom built and rules of design and construction are flexible, so that it can be "tailored" to the job, and to the customer's needs or whims; and there is no clear distinction among the functions of design, operation, and maintenance.

The belt itself is the most important single item. It is an endless flat strip of rubber-covered fabric laid up in plies, sometimes including steel cables. The fabric materials and number of plies determine belt strength. The rubber cover is not for strength, but protection from abrasion and weather. Its thickness and quality are varied for different types of service.

In an average installation, the belt is

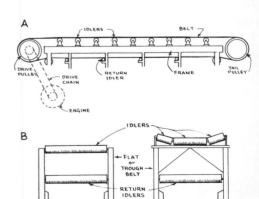

Fig. 14-14. Belt conveyor parts

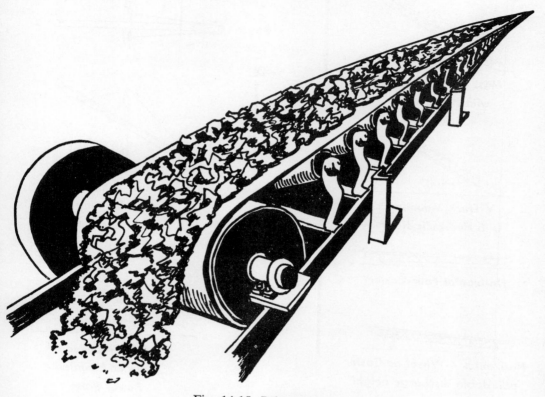

Fig. 14-15. Belt conveyor

about half of the original construction cost, and its repair and replacement is the biggest maintenance charge. Proper care of it is therefore very important.

The belt extends between a head pulley (which may be the drive pulley), and a tail or return pulley, and carries its load on its upper surface, usually toward the head pulley. Its upper strand is supported by idler sets whose three rollers are arranged to shape it into a trough, and the lower strand is supported at wider intervals by flat rollers called return idlers.

Figure 14-14 shows the parts of a simple conveyor and 14-15 one in action. There is a frame which keeps the working parts in position, an engine that turns a drive pulley that moves the belt by friction, a tail pulley to reverse the direction of belt travel, and idlers and return idlers.

Frame. Many frames are of the sectional type. A head and a tail piece must always be used, and as many intermediate sections as are required to obtain the desired length. Sections are bolted to each other.

Figure 14-16 shows several ways of mounting rigid frames. Wheel and caster carriers are suitable only for light machines. Pillar and floor are usually permanent, but may be made so that they can be disassembled for moving. Stationary installations usually include a catwalk from which the belt and rollers can be serviced.

Conveyors may also be carried on two strands of wire rope with movable head and tail sections, on skids, and in other ways.

Belt. The belt is an endless flat strip of rubber-covered cotton or rayon fabric laid up in plies. Very long or heavily loaded belts may be reinforced with steel cable.

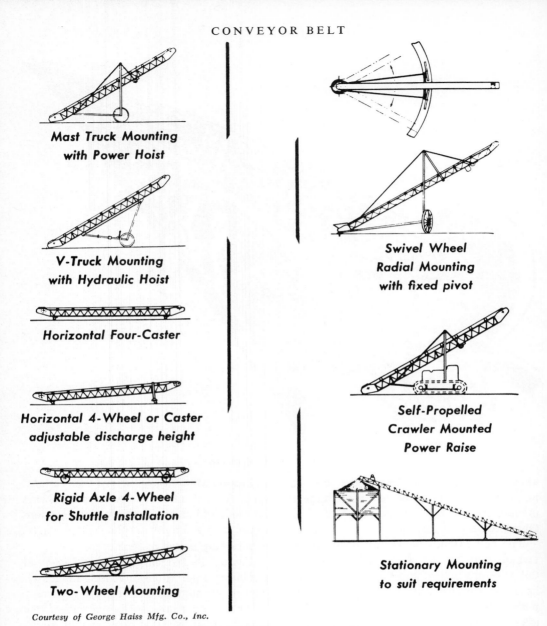

Mast Truck Mounting with Power Hoist

V-Truck Mounting with Hydraulic Hoist

Horizontal Four-Caster

Horizontal 4-Wheel or Caster adjustable discharge height

Rigid Axle 4-Wheel for Shuttle Installation

Two-Wheel Mounting

Swivel Wheel Radial Mounting with fixed pivot

Self-Propelled Crawler Mounted Power Raise

Stationary Mounting to suit requirements

Courtesy of George Haiss Mfg. Co., Inc.

Fig. 14-16. Conveyor mountings

The type of fabric, the number of plies, and any reinforcement which is present determine the strength of the belt. The rubber cover serves only to protect the fabric from abrasion and weather. Its thickness and quality are varied to suit different types of service. Several constructions are shown in Figure 14-17.

There is no definite limit to the length of a single belt. More friction surface for the drive and stronger belt construction are needed as the unit is made longer, the climb steeper, or the load heavier. Common practice limits the carrying distance to a quarter mile, but belts with a carry of a mile or more have been constructed.

Any distance can be crossed by a series of belts, each of which dumps on the next

STANDARD

Multiple layers of suitable duck of the same thickness and ply across the entire Belt. Standard is suitable for nearly all types of Conveyor service.

SHOCK PAD

Substantially the same as Standard but with a reinforced top *Cover* consisting of an abrasion-resistant tire-tread stock on the top surface, backed by a thick pad of resilient rubber. The pad yields to sudden, extreme impacts and pressures—protecting the cover from puncture or breakage and preventing rupture of the carcass.

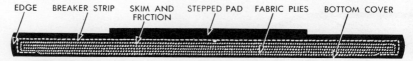

STEPPED PAD . . . Stacker-type Cover

Substantially the same as Standard construction except that it is moulded with the *Cover* having an additional thickness standing out in relief in the center Belt area. It is recommended when the loading of abrasive material is concentrated in the center of the Belt with only slight abrasion at the cover edges. Stepped Pad Belting is not recommended for two-pulley or internal drives.

STEPPED PLY

A smooth-top construction having a heavier cover in the center of the Belt than at the edges. This is accomplished by moving the middle portion of one or two of the top fabric plies to the sides and filling in the extra space with cover stock. A Stepped Ply Belt has more crosswise flexibility and troughs more easily. The extra thickness of cover stock gives longer life to the Belt under loading conditions where abrasion is concentrated in the center Belt area.

Courtesy of Hewitt-Robins Incorporated

Fig. 14-17. Belt cross sections

through a hopper or over a baffle plate. The length of each conveyor is figured between the centers of the head and tail pulleys.

Belts which move dirt or any loose material are usually run with a trough in the upper surface, which centers the load and reduces spill off the sides. Very short conveyors may carry a bigger load by using a flat surface with fixed side skirts.

Belts may be from eight inches to eight feet in width, but standard belts range from 12″ to 60″, with 30″ very common.

Drive. Power carries from the drive pulley to the belt by friction. If the resistance of the belt to moving is greater than the friction, the pulley will spin or slip inside the belt, with resultant loss of power, and wear on both surfaces. The amount of friction or traction is determined by the nature of the surfaces, the slack side tension on the belt, and the area of contact.

The load or tension on the carrying strand of the belt, which tends to cause slippage, is made up of the pull of gravity on this strand and its load, friction in idlers, pulleys, in the belt, and in its load; and the

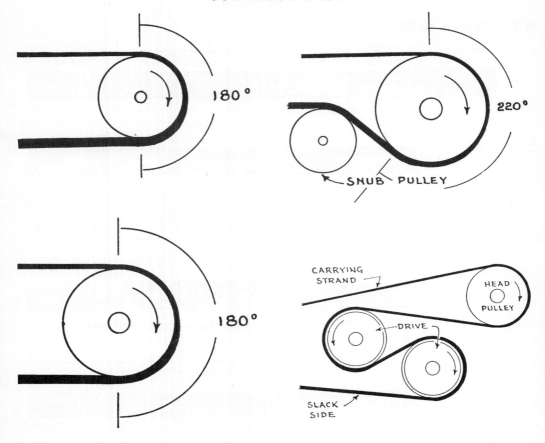

Fig. 14-18. Belt drives

inertia of the whole system when starting or accelerating.

The drive pulley surface may be bare metal, or be covered by smooth face or grooved rubber lagging. Such lagging may be bolted or vulcanized in place. It increases traction particularly when the belt is wet or frosty, and prevents pulley wear.

The belt is held in full contact with the pulley by its tension on the slack or low tension side. This tension is normally regulated by some form of gravity takeup, in which a hanging weight exerts a pull on the tail pulley, or a special takeup pulley, which moves outward if the belt slackens, and inward if it tightens. If the incline is steep, the weight of the slack side may maintain sufficient tension. Very short conveyors may

have threaded adjustments to move the tail pulley in or out.

The amount of drive traction which may be obtained by increasing tension is limited by sharply increased power requirement and shortened life of a too-tight belt.

The area of contact is determined by pulley diameter, the arc of contact, and the number of pulleys. A thicker pulley not only increases the contact area, but also reduces flexing strain on the belt. Its disadvantage is the cost of the pulley itself and of changes in frame and layout to accommodate it.

If an existing drive pulley is replaced by a larger one, the belt will then be driven at higher speed with less power, unless the gearing or the motor is changed. Putting on lagging has the same effect.

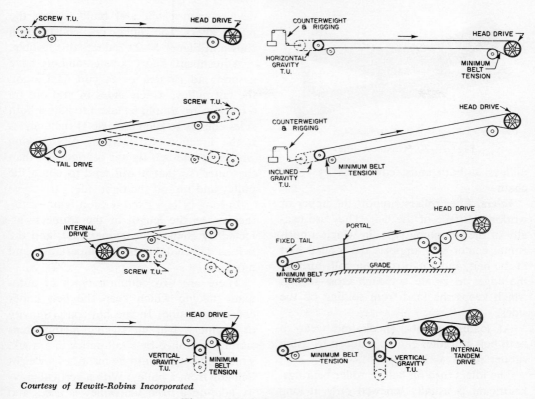

Courtesy of Hewitt-Robins Incorporated

Fig. 14-19. Drives and takeups

A belt whose strands are parallel will have 180° of contact with the drive pulley. This contact can be increased by a snubbing pulley on the slack side. Increasing the degree of wrap in this manner is the cheapest way to increase contact area.

Head pulley drives are adequate for short conveyors, and for those so steeply inclined that the weight of the return strand maintains a high tension on the slack side. For conditions requiring greater traction, tandem pulleys are used, Figure 14-18. Up to 440° of wrap can be obtained in this manner.

A belt which is pulling a load changes shape as it goes over a drive pulley. It is stretched thin where it first contacts it, and then fattens up as its tension is reduced. It moves fastest where it is thinnest, in the same manner that water is accelerated in going through a restricted place in its chan-

nel. The change in belt thickness and speed is quite small, but requires that the second of the tandem pulleys turn a bit slower than the first, if extra stress is to be avoided. The amount of difference varies with the load. Where electric drive is used, separate motors on the two pulleys can be made to automatically adjust their speeds to each other.

A possible disadvantage of tandem drive is that one pulley works on the load-carrying surface of the belt, which may be wet and slippery so as to afford a poor grip; or gritty so that excessive wear of both belt and pulley will occur. This may be avoided by using the head and lower pulley for drive, and the intermediate one as an idler.

A number of drive and takeup arrangements are shown in Figures 14-18 and 14-19.

Short reversible belts may have drive

Fig. 14-20. Cushion idler

pulleys at both ends, connected by roller chain.

Idlers. Idler rollers support the upper or working surface of the belt. They are usually of the troughing type, in which a horizontal center roll supports the loaded part of the belt, and a pair of outer rolls turn up the edges to create a trough cross-section which keeps the load from spilling off the sides.

At loading points, shock and wear to both belt and idler can be reduced by using rubber idlers as in Figure 14-20.

The idlers turn on ball or roller bearings. Lubricant is usually renewed only at long intervals, and may be sealed in.

In excavation work, the flat idler and belt are used only on skirted conveyors.

The return idlers support the unloaded lower stretch of the belt. They carry little weight and may be quite widely spaced. Construction and lubrication are similar.

If the belt carries sticky material, some of it may not be removed by the cleaning

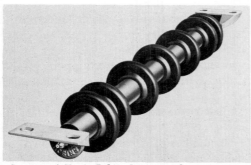

Fig. 14-21. Non-clog return idler

devices, and catch and build up on the return idlers. The rubber disc construction shown in Figure 14-21 reduces this trouble.

Adjustment. Short belts commonly have a screw adjustment or takeup (T.U.) on the tail pulley, which slides in and out on a track. Care should be taken to adjust both sides equally to keep the axle at right angles to the direction of belt travel. Any cocking will make the belt tighter on one side than the other, so that it will tend to climb the pulley and run off the tight side.

In long belts a fixed tension is not satisfactory, as the length of the frame is affected by changes in temperature, and the belt is affected by both temperature and moisture.

There are two common types of automatic takeup which keep the belt under constant tension. In the horizontal gravity or counterweight-and-rigging method tension is controlled by a tail pulley in a track, which pulls it outward by means of a weight hanging from a pulley. With a vertical gravity take-up a fixed tail pulley is used, and the weighted pulley hangs between two return idlers, preferably at or near the point of minimum belt tension. These constructions are indicated in Figure 14-19.

If the belt stretches too far to be adjusted a piece is cut out, the ends stapled or vulcanized together, and the adjustments reset.

Alignment. The belt is sensitive to tiny changes in frame and pulley alignment, which will cause it to wander out of a straight line. Internal changes in belt tension, or a splice which is beginning to pull apart, may have the same effect. Trouble with wandering can be greatly reduced by making both pulleys and frames wide enough so that the belt does not have to run absolutely straight to keep out of trouble. The wider construction is more expensive, of course, but through the years it should more than pay for itself in longer belt life, and in reduced checking and adjusting.

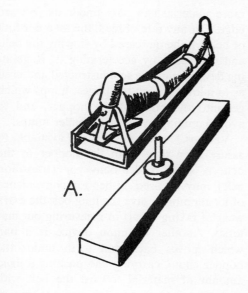

A.

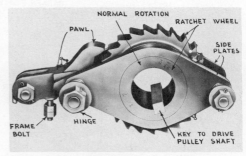

Courtesy of Hewitt-Robins Incorporated

Fig. 14-23. Holdback

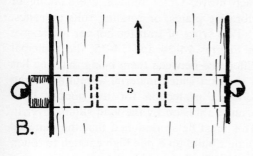

B.

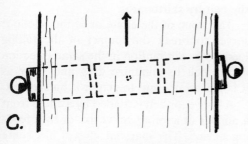

C.

Fig. 14-22. Training idler

If the framework is out of line, the carrying strand may still track well enough, being steered by the load seeking the idler troughs. The return strand, however, will find the shortest path, or allow itself to be influenced by sloping idlers, so that it will rub against stationary parts. Unfortunately, it is almost standard practice to carry the return strand inside the framework, where it is very difficult to see just what it is doing. It is much better to hang it below the frame members, where misbehavior can be readily observed and necessary corrections made before serious damage is done.

Troughed idlers will steer the belt if they are tipped in the direction it is moving. However, only a very slight tilt can be used, as if it is overdone it will set up a drag against the bottom of the belt which will wear it rapidly and consume extra power.

Self-aligning or training idlers are mounted on a center swivel, and have vertical spools set at each edge. If the belt rubs against a spool, it tilts and presses a lined brake shoe against the adjoining roller, slowing it, swinging the idler, and shifting the belt back toward center. This device creates very little drag, and if placed every 50 feet in place of regular idlers, will keep a belt in line under any ordinary conditions. Two placed at 30 foot intervals ahead of the tail pulley will line the belt up properly to go under the loading point.

Holdbacks. An inclined conveyor tends to run backward when power is cut off, because the weight of the load pulls the upper part of the belt downhill. This tendency can be overcome by means of a brake of sufficient size, but it is often more convenient to use a device which will automatically lock it against turning backward, without interfering with normal movement of the belt.

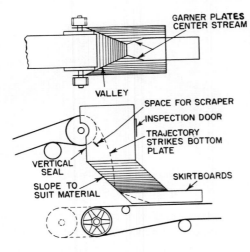

GARNER PLATES
CENTER STREAM

VALLEY

SPACE FOR SCRAPER

INSPECTION DOOR

TRAJECTORY
STRIKES BOTTOM
PLATE

VERTICAL
SEAL

SKIRTBOARDS

SLOPE TO
SUIT MATERIAL

Courtesy of Hewitt-Robins Incorporated

Fig. 14-24. Transfer hopper for non-abrasive material

The holdback shown in Figure 14-23 consists of a ratchet wheel keyed to the drive pulley shaft and held between a pair of side plates which are hinged to a toggle and tooth pawl anchored on the conveyor frame. When the belt and drive wheel are turning in the normal direction friction between the ratchet and the side plates lifts the unit on the hinge, holding the pawl clear of the ratchet. If the drive wheel starts to turn backwards, the hinge is pulled down and the pawl meshes with the ratchet.

This device cannot be used on a conveyor transporting down hill, as its load tends to move the belt forward.

Hoppers and Loading. Hoppers are of two principal types, the primary or loading hopper, and the transfer. The primary usually has enough storage capacity so that it can supply material steadily to the belt, although it is loaded intermittently by bucket or truck loads. It may have a screen or grizzly to keep out oversize pieces.

Loading hoppers must have a restricted chute capacity, or a gate or feeding device to regulate the rate of delivery to the belt. In some installations, increased flow is obtained by merely raising the chute, so that

more material can pass under its forward edge. For any rate of feed, the load per foot of belt will be increased by slowing the belt, and decreased by accelerating it. Regulating devices must be so constructed and used as to allow through the largest lumps which get into the body of the hopper, and to avoid jamming by several smaller lumps.

Various feeding devices are available to assure even flow from the hopper to the belt. A short feeder conveyor (transition belt) may be used, either for convenience of location or to save the long belt the extra wear of taking a part in measuring out material. Another common device is a pan which moves back and forth under the hopper chute, carrying and pushing a fixed amount of material toward the belt with each move. Other models use shaking or vibrating plates, or rotating rolls or vanes.

Two types of transfer hopper are shown in Figures 14-24 and 14-25, the principal difference between them being that one has a shelf on which material accumulates and forms a protection for the hopper back against abrasion by the load sliding over it. This must be so designed that the material cannot build up a pile high enough to touch the belt, as this would cause excessive wear. Material is guided to its proper position on the belt by skirtboards.

The slope of the chute or of the pile in it serves to break the impact of the falling load, to give it motion in the same direction as the belt to minimize drag and wear as it is picked up, and to center it on the belt.

A conveyor belt is loaded by a chute from a hopper or another conveyor, or by a feeding device. The material always has some vertical drop, and may be moving in a different direction and at a different speed than the belt. As a result there is impact to the belt in stopping its fall and giving it its proper speed and direction. Chutes and baffles should be so constructed that they convert as much as possible of vertical and cross movement into movement in the di-

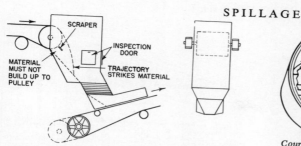

Courtesy of Hewitt-Robins Incorporated

Fig. 14-25. Transfer hopper for abrasive material

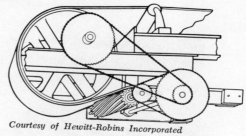

Courtesy of Hewitt-Robins Incorporated

Fig. 14-26. Belt cleaner

rection of belt travel, and so that they deposit their load squarely in the center of the belt. If the material is abrasive, the chute may be so constructed that a layer of the material piles up on its slope and protects it. When the main conveyor is a big one, it may pay to put in a short feeder conveyor (transition belt) to insure feed at the proper speed and in the right direction. This transfers wear from an expensive belt to an inexpensive one.

Coarse lumpy materials, particularly when they include sharp edges, require greater precautions at the loading point than fine or soft ones do. Impact damage to the belt can be reduced by using thicker rubber on both the top and bottom surfaces, and by using closely spaced rubber-cushioned idlers at the loading point.

A certain amount of belt wear from impact and abrasion occurs each time the belt goes over an idler, as it is a high point with a sag on each side. The slight but steady wear at these points can be reduced by minimizing the sag, which is affected by belt tension and idler spacing. Maximum smoothing effect can be obtained from any given number of idlers by using wide spacing near the head, where tension is greatest, and progressively narrower spacing toward the tail pulley.

A fact that is commonly overlooked is that any internal movement in the load on the belt is absorbing power. Such movement cannot be stopped, but it should be kept to a minimum.

If the load slides back on itself going up an incline, a lighter load may be more efficient. If it slides on the belt, a belt with a cleated surface may be required. Often a change in the moisture content of the earth will cause or cure this type of trouble.

Spillage. There is usually very little spillage off a belt which is wide enough for its load. However, many belts are overloaded in bulk if not in weight, and the jouncing over idlers, movements in the load, and changes in alignment result in pieces rolling and bouncing off the sides. If decking is placed under the full length of the belt, or sometimes under the ends and the loading point, this material will be kept from spilling onto the return belt. A clean belt and clean pulleys and idlers mean longer belt life.

It is important that a diagonal cleaner bar or plow be placed on the return strand just above the tail pulley, as a large sharp object might otherwise fall on the belt and catch between it and pulley, so as to puncture the belt or even cut it into strips.

Wet dirt and many other materials stick to belts and will build up to considerable amounts if permitted to do so. A scraper may be placed just after the discharge point, to clean the belt and drop the scrapings into the same receptacle as the main stream.

This cleaner may consist of a rubber or steel blade, or a series of them, or be a serrated rubber roll or bristle brush which revolves oppositely to the belt. Proper pressure against the belt and adjustment for wear are provided by counterweight or springs.

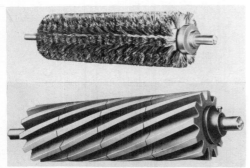

Courtesy of Hewitt-Robins Incorporated

Fig. 14-27. Brush and rubber roll cleaners

A water spray, with or without squee-gees, may be used alone or in conjunction with a mechanical cleaner.

Troughs and Skirts. Most belts that carry loose material—and this includes excavated soil and blasted rock—are troughed so that the center is lower than the sides. The side idlers are usually at a slope of 20 degrees. However, there is a trend toward steeper slopes, and 35 and 45 degrees may now be used to increase load and/or reduce spillage.

Skirts are used to prevent spillage off the sides of a belt at a loading point, and to increase the capacity of short belts. They usually consist of vertical or sloped faces of metal or wood, with a lower piece of flexible rubber which makes light contact with the belt. This rubber should be lowered as it wears, and replaced whenever neces-sary, to avoid damage from material wedg-ing under it.

Dual skirts have an inner board to form the load, and an outer member to prevent fines from seeping through.

Both troughs and skirts may be used at a loading point. The trough tends to keep ma-

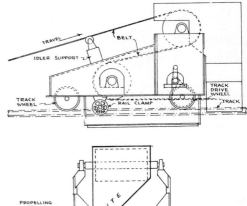

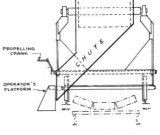

Fig. 14-29. Tripper

terial from reaching or pressing against the skirt and working under it. Trough slopes as steep as 50 degrees may be used.

Trippers. A conveyor belt may discharge into any of several bins or piles. For a dis-charge point not at the end of the belt frame a tripper may be used to dump the the load into a side-casting chute at any point. As shown in Figure 14-29 it consists primarily of a pair of straight idler pulleys, one of which turns the belt under and dumps it, and the second returns it to its original direction.

The model illustrated is moved along tracks by a hand crank, and is anchored in position by a clamp. Other types can be propelled by the belt or by a separate motor, and be under manual, automatic, or remote control.

Safety devices. A belt conveyor will run for long periods without attention. Idlers re-quire lubrication twice a year or less (if of the modern type with bearings and seals), a well founded and constructed frame will keep the parts in line, and properly designed and protected chutes may never clog. How-ever, sudden accidents happen which can

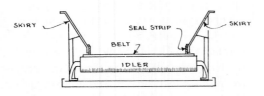

Fig. 14-28. Skirted belt

Fig. 14-30. Cleated belt

be very costly if their results are not controlled, and it is impossible to keep a man always on the alert for events which may not happen for years, or ever. Controls should therefore be installed which will automatically react to emergencies.

If the power fails on an inclined belt, it will tend to run backward, jamming its load into and around the loading chute. This movement can be prevented by a holdback ratchet on the head or drive pulley, which will allow free working motion, but immediately lock against backsliding.

If objects jam between the belt and a pulley, if discharged material backs up into the belt, or if the belt runs off its rollers the power requirement will be sharply increased. If drive is electric, an automatic overload switch in the line can be set to cut off the power, and prevent or limit the resulting damage.

It is also a good plan to use a motor which is not too big for its job. 200 horsepower can drag a belt into a lot more trouble than 100 can, without increasing current requirement as sharply. The oversize motor will also put greater stress on the belt when starting it with a load.

A moderate rise in power consumption without increase in load indicates increasing friction. It may be in dry or broken idlers, result from the belt rubbing the frame, or be caused by too much tension. A

Fig. 14-31. Incline bucket conveyor

record of current consumption will often show up such conditions before they would be observed otherwise.

Inclines. Belts are used for uphill, horizontal, and downhill (decline) transporting.

A belt will lift dry sand on an incline about 15°, and wet sand up to 20°. Some cohesive materials can be carried on grades as steep as 28° on standard belts. The limiting factors are slippage between the belt and the load, and sliding of the load on itself.

Rough surface belts may be used on somewhat steeper inclines if the load is thin. Belts with metal cleats, as shown in Figure 14-30, and sandwich or double belts, Figure 14-39, will carry loads up to 35 or 45 degrees. Belts or chains with attached buckets can lift at any slope or vertically.

Fig. 14-32. Cross-country conveyor

Capacities. Persons used to seeing dirt moved by the bucketful or truck load have difficulty appreciating the volume production represented by the thin ribbon of dirt on a belt.

Tables showing conveyor sizes, inclines, and recommended speed will be found in the Appendix. As one example, a 30″ belt carrying gravel weighing 3400# per yard, including occasional stones up to 14″ diameter at 450′ per minute will move about 520 yards an hour.

Output is usually figured in tons per hour, and conversion to yards made by means of average weight data. Production is determined by the width of the belt, the speed of the belt, and the height to which it is piled. Power needs are proportioned to tonnage and lift.

Those variables make design of a conveyor system a complex matter, with plenty of room for difference of opinion. In general, increasing belt width means more expensive construction throughout. Increase of speed beyond specifications shortens belt life, and may involve lost power through slippage between the belt and the load. Heavier loading may call for stronger frame and idler construction, and heavier belting. In addition, a problem of spilling off the sides may be encountered.

If capacity of an existing installation is to

Fig. 14-33A. Denny hill conveyor

Courtesy of Link-Belt Company

Fig. 14-33B. Digging in Denny Hill

be increased, the most economical, although often not the soundest, way is to increase power and speed and to maintain original load.

Recommended speeds for various materials range from 300 to 1200 feet per minute, with dirt, sand, and gravel in the higher brackets.

Conveyors many miles in length are often used in dam construction to bring material in from borrow pits. They can operate on grades much too steep for trucks, so that in rough country original construction may not be substantially more expensive than that of a haul road of equal capacity. Once built, it can show a much lower operating cost per yard than truck fleets.

Distribution of the fill may be handled by conveyors, trucks, or both.

An interesting application was in the removal of Denny Hill in Seattle—5,000,000 yards of earth had to be moved from the excavation through the business district of the city to barges. Operation of heavy trucks on this route would have been ex-

tremely objectionable, but the conveyor in the overhead structure shown in Figure 14-33A was able to transport the whole yardage with only slight inconvenience to the city.

Loading was done by dipper shovels directly into mobile hoppers, as shown in Figure 14-33B.

Shiftable Frames. A conveyor used in an open pit may require frequent sideward moves to keep near a bank that is being cut back. Dismantling a conventional unit and setting it up again is likely to be costly in labor and in lost production.

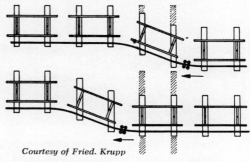

Courtesy of Fried. Krupp

Fig. 14-34A. Shiftable conveyor frames

Courtesy of Fried. Krupp

Fig. 14-34B. Side-shifting a conveyor

A German construction, shown in Figures 14-34, A and B, is to mount the entire conveyor system on ties or skids that rest on a strip of land graded as if for a road. Idlers supports are grouped on frames about fifteen feet long, that are not directly connected to each other. The frames are supported at each end by a pivot mounting on a cross tie or skid.

The ties are connected by one or two 90-pound railroad rails. A heavy four wheel drive tractor equipped with a side boom carries a roller clamp that holds the top of one of the rails. Shifting is done by lifting the clamp and rail slightly, and running the tractor parallel with the conveyor and at a slight distance from it. This will pull the rail and the whole conveyor three to six feet to the side. A return trip in reverse moves it the same distance again. The process is repeated until the conveyor reaches its new location. The rail is flexible enough for the bending involved.

The head and tail sections are usually mounted on skids or pontoons, and held by tightening anchor lines by hand winches. Sideward movement involves towing with the tractor, or pushing with a dozer. Very large drive stations may have pontoons that are equipped with hydraulic powered legs that can be used for tension adjustment and for side movement.

Ropebelt. The carrying run of a belt conveyor may be supported on flexible idler assemblies suspended from parallel tensioned wire ropes (cables). This system was first used underground in 1955, and is now finding many applications above ground as well.

A pair of 5/8 or 3/4 inch first quality steel cables are supported by stands of adjustable height at 10 to 20 foot intervals. The cables are anchored at the head and

Courtesy of Goodman Manufacturing Company

Fig. 14-35. Ropebelt conveyor

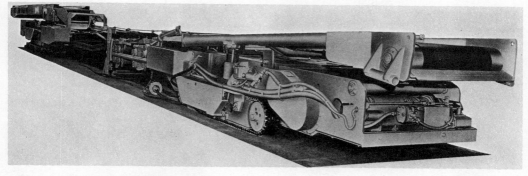

Courtesy of Goodman Manufacturing Company

Fig. 14-36. Extensible conveyor

tail conveyor units and at 200 to 300 foot intervals in between.

Idler sets made up of three idlers—one bottom, two sides—are connected to each other by roller chain hinges, and are clamped on the cables at intervals of five feet or more. Rigid spreader bars may be used between stands to maintain cable spacing. Return idlers are fastened to the stands, or suspended from the cables at required intervals.

The combination of flexible idlers and flexible side supports enables the idlers to move down or to change shape under impact from material on the belt, saving strain and wear. Deflection of the idler under load also forms a deeper trough that increases carrying capacity and reduces spill.

Frames, idlers, and spreaders can be easily shifted, added, or removed to meet changed conditions. Ropes are readily extended or shortened. The rope frame is therefore well suited to use in locations where changes are often necessary. The materials are lighter and much less bulky than standard conveyor frames, and are more easily stored while not in use.

Extensible Conveyor. An extensible con-

Courtesy of Link-Belt Company

Fig. 14-37A. Levee stacker

Fig. 14-37B. Coal-stripping stacker

veyor is usually a rope frame unit with mobile, crawler mounted head and tail units. The head or drive unit holds up to 200 feet of extra belt, moving back and forth over a series of takeup pulleys. The tail unit carries two reels of rope held by brakes.

In coal mining work the tail unit may follow a continuous miner, receiving its discharge directly or across a short loading bridge conveyor. It can also be loaded by shuttle cars or in any conventional manner.

During operation the tail unit travels as much as is necessary to keep in contact with the excavator, while the head unit is kept stationary at a point where it discharges into a main conveyor, haulers, or a processing plant.

Both the ropes and the belt pay out automatically as the distance between the head and tail is lengthened. It is only necessary to place extra idler frames to support the belt.

When the reserve belt has all been paid out into the working stretch, the conveyor is stopped and an extra piece of belt is spliced in. This should be long enough to cover the working distance and refill the storage rolls. The rope reels have sufficient capacity to take care of several belt extensions. Reserve belt 200 feet long will permit a conveyor advance of 100 feet.

If the conveyor discharges into hauling units the head section may be moved up instead of adding belt when the reserve is used up.

STACKER

A conveyor may end in a pivoted boom to allow placing material in separate piles or to facilitate building piles or embankments to a desired shape or size. The machine shown in Figure 14-37A is building a levee along the Mississippi River. The long conveyor in the background feeds it through a hopper. Such a conveyor might be loaded in a pit miles away, or by trucks at a nearby transfer point.

A mobile Link-Belt stacker with a 190′ boom is shown in 14-37B. This unit is self-propelled on wide gauge railroad-type tracks. It is side casting overburden across the working face of a pit, and is being loaded by a large dipper shovel in the background.

Stackers are usually custom-made. They can be stationary, towed, or self-propelled. Booms vary from 60 to 200 feet in length, belt width from 24″ to 54″, and capacity from 300 to over 1000 tons per hour.

They are frequently used in stacking of processed material in mine and factory yards. They also can load materials from a

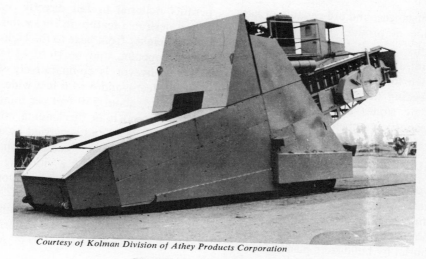

Courtesy of Kolman Division of Athey Products Corporation

Fig. 14-38. Portable belt loader

pit floor directly into trucks or trains on the high wall.

BELT LOADERS

A belt loader may be any of a number of machines that load haulers or sidecast by means of a conveyor belt. The material may be pushed or dumped into a receiving hopper by other machines, or dug by a knife, blade, plowshare, disc, or bucket wheel that is part of the loader and is moved by it.

Wheel and ladder ditchers conform to the belt loader definition, but the belt is such a small part of the operation, and the material is so seldom loaded, that they are not included.

In all of these machines (including the ditchers) the belt keeps the digging parts from being choked by accumulated cuttings, and is in itself the most economical type of elevator. As a result, these machines are a highly efficient type of loader under favor-

Courtesy of Kolman Division of Athey Products Corporation

Fig. 14-39. Moving into position

14-29

able circumstances, and are capable of very high volume of production in relation to power used.

In general, they do best in materials of a fairly even texture that can be cut readily by their feeding devices. They are inconvenienced by stones and lumps that are large in proportion to belt width.

PORTABLE BELT LOADERS

This type of belt loader does not propel or load itself, but can usually be moved around fairly readily. It consists of a heavy loading box or trap which is filled to the point of burial by bulldozers pushing material to it, a wide conveyor belt that carries the material to an elevated discharge point, an engine, and various accessories.

The trap must be very heavily built. Its opening may be either fixed or adjustable. If fixed, there is a reciprocating-plate feeder under it, to deliver a more or less steady flow of material to the belt. If adjustable, it permits material to fall directly onto the belt, a desirable feature in sticky soil.

Box height , from base to the high end of the trap opening, may be 10 to 18 feet.

The conveyor is comparatively short— 40 to 60 feet—and 4 to 6 feet wide. Speed is moderate, about 350 feet per minute.

The tail end is in the loading box. The discharge or head end is carried by a cantilever frame over the loading point. The head may be fitted with a vibrating grizzly or screen to carry oversize beyond the hauler.

The belt and any accessory units are driven through chains and intermediate shafts by a diesel engine of 100 to over 200 horsepower.

The whole machine is a single unit, supplied with a detachable pneumatic tired undercarriage, wheels, and a king pin attachment for use with a fifth wheel trailer hitch. In addition, the bottom of the trap is fitted with heavy skid plates, so that it can be dragged on the ground without damage.

Courtesy of Kolman Division of Athey Products Corporation

Fig. 14-40A. Working in bank

Placement. For operation, this belt loader is preferably placed in a slot at or below the bottom level of a high wide bank, at least 15 to 20 feet high, that is to be removed. If this is its first placement on the job, it will probably be backed in by the towing tractor; otherwise it is usually pushed in place by dozers.

The slot makes it possible to place the trap nearer the material to be loaded, and reduces filling around it.

The discharge end ordinarily extends over a truck roadway. Discharge clearance may be 10 to 12 feet. The road may be excavated if more is required.

Since a standard machine has a fixed discharge position, a truck that has been loaded must move away before an empty one can be placed. Since the belt must be idle during the exchange, the reduced spotting time in drive-through is much more efficient than backing in.

The trap should be below most of the dirt to be removed, as dozer production is much greater on a downhill push than on a level. It would be most economical to feed if placed in an evcavation below pit floor level, but depth is limited by the height of discharge needed for the haulers being used.

Feeding. Bulldozers push soil or broken rock into the trap. Their first loads are used to fill the slot outside it, and build a cushion of loose material on top, to reduce possible damage from (and to) dozer blades and tracks.

Crawler dozers with wide U-type or angling blades are the preferred digging-pushing units. They are started far enough back to fill their blades, then push the loads onto the trap. Gravity flow moves material into the opening as the belt or the plate feeder removes it.

Two or three big dozers are needed at the start at each location, and perhaps six at the finish. As cutting progresses, pushes become longer and take more time. The down grade toward the trap diminishes, or

may even be reversed to an uphill push, with serious reduction in size of loads.

This cycle, which can distort equipment balance on the whole project, can be reduced or eliminated by putting one or more dozers on long pushes right at the beginning.

When the dozer spread is not adequate to feed the loader, it can be worked more hours (yes, including overtime), bringing distant soil into stockpile near the trap when the belt is not working.

Rubber tired machines are generally not favored because of their greater travel distance in picking up loads. Their higher speed may outweigh this disadvantage on long pushes. Big wheel mounted front end loaders may also give a good account of themselves.

Materials. Belt loaders are used in about every class of volume digging, including sand, gumbo, glacial till, and blasted rock. This versatility is partly due to their large capacity, in production and in sizes of pieces that can be handled, which permits use of big machines to supply them.

Four, five, and 6-foot wide belts handle quite large pieces without difficulty, even

Courtesy of Kolman Division of Athey Products Corporation

Fig. 14-40B. Feeder trap

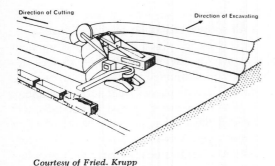

Courtesy of Fried. Krupp

Fig. 14-41. Wheel excavator in pit

the stickiest soils find it difficult to plug the large trap holes that go with such belts.

That is not to to say, however, that any one of these loaders will work well anywhere.

A plate feeder is desirable both for production and for the good of the belt in most materials, but if the soil is sticky it may be much better to drop it directly onto the belt.

Oversize. Loose rock in the bank may be in pieces that are too large for the trap, the belt, or the fill.

In general, pieces that will go through the trap opening will be carried by the belt, so the first two situations are a single problem.

Fortunately, any rock big enough to cause difficulty is likely to be noticeable in front of the blade, before reaching the trap. Operators can be instructed to separate oversize pieces from the feed before reaching the trap.

The trap may be protected by welding on heavy grizzly bars, but rocks retained on them may be difficult to reach to get them out of the way.

For protection of fill quality, a belt loader may be equipped (preferably at the factory) with a shaking or vibrating screen at the conveyor discharge. This can separate oversize from acceptable material, with minimum operating cost.

With a screen, there should be two discharge points. The soil will fall through the screen close to the conveyor tip; the rejects will be carried further, to the end of the screen. The two discharge streams must be sufficiently separated so that two trucks or other haulers can stand side by side, one under each discharge stream.

If the bank has been wisely selected, or the rock properly blasted, there should be much more undersize (throughs) than oversize. The truck receiving the latter might stand for considerable periods before being fully loaded. But this cost would be far less than any method of separating the same pieces at the fill.

Haulers. A belt loader can discharge into any type of hauler that has an open top. This includes trucks and trailers of rear dump, bottom dump, and side dump construction; either highway or off-highway models; and non-elevator scrapers.

The principal consideration is that the haulers should be big, the bigger the better. Time for spotting a truck, figured from stopping the conveyor to restarting it, may average 15 seconds. A 60-inch belt is rated at one cubic yard a second. Haulers carrying 45 yards (loose and heap) would keep the loader busy ¾ of the time, ones with 15 yard capacity would work it only half the time.

Belt loading is kind to haulers, as the material is poured into them, instead of being dumped in big chunks from a bucket. With portable loaders, one loading spot is likely to be used for concentrated traffic for several days, so that it is likely to be well prepared and maintained. Scrapers are spared the heavy stresses of being pushed through a digging run.

Yardage carried is comparable to that obtained from non-belt loaders. There are fewer large voids and heaps may be higher, but the soil is fluffed up so that it does not pack into the body as tightly. In scrapers, compaction from normal forced loading is lost, but heaps may be higher to compensate.

The belt loader may also discharge into any type of high capacity conveyor belt system for haulage.

Production. A 60-inch belt loader is rated at one yard per second, or 3600 yards per hour. A loss of ¼ or more due to hauler spotting may be assumed, reducing this to 2700 yph. Average production on a 45 minute hour basis would be almost 2000 yards, which is a lot of dirt.

Loaders with 48-inch belts are rated at 2000 to 2800 yards per hour, and 72-inch units at 4800. These figures are subject to the same markdowns for spotting time and 45 minute hours.

WHEEL EXCAVATOR WITH BOOM

Wheel excavators have been widely used in other countries, particularly Germany, for many years, but have been uncommon in the United States until recently. Most of them are big electric powered machines designed for large scale removal of soil overburden in strip and open pit mines.

Booms. The wheel excavator has two conveyor-equipped booms. The digging boom or ladder carries the cutting wheel, and the discharge boom or stacker disposes of the cuttings. The two booms are mounted on a revolving superstructure. They may be fixed in line with each other, or may have partial swing independent of each other. In-line booms balance each other, independent ladders and long stackers must be equipped with counterweight structures. The two booms usually have separate hoists.

The ladder may be raised and lowered on a fixed pivot, or it may have a slide on which it can be crowded into the digging or pulled back from it.

These machines may be designed for digging through a wide vertical range, or be usable only in a comparatively narrow band. Large units designed for a particular mine may be special for its needs. Combined boom lengths may be hundreds of feet.

Most wheels work only at and above their support level, but some are designed for deep cuts. Downward ladder angle is limited to about 20 degrees with ordinary conveyor

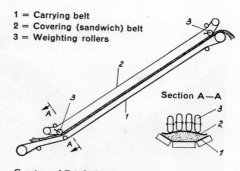

1 = Carrying belt
2 = Covering (sandwich) belt
3 = Weighting rollers

Section A—A

Courtesy of Fried. Krupp

Fig. 14-42. Sandwich belt

equipment, because of sliding of the load. A covered or sandwich belt, Figure 14-42, has a second conveyor belt run at the same speed upside down over the one carrying the load, with enough weighting rollers to keep it in contact. Such belts can be used at angles of 35 degrees or more.

Wheel. The digging wheel rotates on an axle at the tip of the ladder. It carries a number of toothed buckets, that will vary in size, number, and design for the material to be dug. The curved part may be made of link chain to reduce sticking.

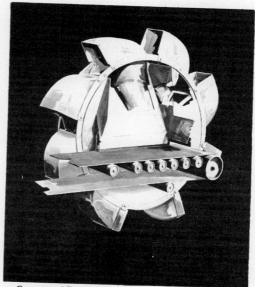

Courtesy of Fried. Krupp

Fig. 14-43A. Big digging wheel

Courtesy of Fried. Krupp

Fig. 14-43B. Big digging wheel

The buckets are open against the wheel, so that their loads spill as they pass over the top, Figure 14-43A. The material falls on a chute that slides it to a side conveyor belt. Chutes may have scrapers or rotating bars, or be rotating tapered ring design.

Wheel speed is limited by centrifugal force holding material from dumping, and by poor filling of buckets and increased tooth and lip wear at high speeds. Bucket lip speed may be from 200 to 1,000 feet per minute. Highest speeds are for firm, hard digging.

The smallest wheels are less than five feet in diameter. They are usually double, and discharge to a center conveyor. The largest wheel under erection is 71 feet in diameter at the bucket lips. It carries 18 buckets with a capacity of 8.2 yards each, and is turned by four 845 horsepower electric motors.

Production is figured by multiplying the revolutions per minute of the wheel times the number of buckets times the load in each bucket. Bucket load in bank yards may be half the capacity or less at high wheel

speeds or in poor digging conditions. It is therefore important to make field measurements of actual production. The largest machines may move 21,000 yards an hour.

Cutting is done largely with the sides of the buckets.

When working above track level the wheel rotates so that the far buckets come up as they fill. They are pushed into the digging by crowding the ladder or by forward travel. If work is done well below track level the direction of rotation and the position of the buckets are reversed, so that upward movement and digging are at the near part of the wheel, and the machine retracts or travels away from the cut to pull the buckets into the digging.

Discharge. The discharge or stacker boom may be long and high in order to pile soil at the far side of a wide, deep cut; it may be comparatively short and low for loading onto shiftable belt conveyors, or in to trucks or rail cars; or it may be a bridge to a separate unit that might be a stacker, car loader, or a mobile belt hopper. A single

14-34

Courtesy of Fried. Krupp

Fig. 14-44. Wheel loading a conveyor

machine may have both a long distance stacker for overburden and a short loader for coal or ore, for convenience in digging alternating layers.

The conveyor in a long stacker boom may be very high speed to keep weight to a minimum and to gain throw distance.

Hundred-Ton Model. A "small" Krupp wheel excavator is shown in Fig. 14-44. This 97 ton machine is mounted on a pair of shovel type crawlers. It carries a 14 foot wheel with 6 buckets of 6 cubic foot capacity each on a 31 foot ladder that is raised and lowered hydraulically. Cutting can be done up to 35 feet above track level but only 16 inches below. There is no crowding mechanism. The 66 foot stacker boom can discharge at 3 to 31 feet above ground level.

The ladder swings in a complete circle with the revolving frame, and is counterweighted by an engine or a tail counterweight on the frame. The stacker is set in a second counterweighted revolving structure that can be locked to swing with the ladder, or released and swung independently through a 180 degree earc. The independent swing makes it possible to hold it steady for loading trucks, hoppers, or cars during the swing of digging wheel, and to stockpile at an angle to the digging direction.

Maximum production in a 60 minute

hour would be 690 yards. Actual output in sand and gravel pits might be up to 520 yards an hour into a conveyor system, or about 20 per cent less into trucks.

In 1974 this unit was priced at $550,000 (rate of exchange $0.40 = 1 DM), set up on the job, in East Coast areas.

Independent Loading Unit. Fig. 14-45 shows a very large wheel excavator, digging a 100 foot bank and discharging spoil across a bridge conveyor into a separate mobile loading unit with a two train capacity. This machine can dig 100 feet above its tracks and 16 feet below. Its daily output is about 80,000 bank yards.

The excavator is mounted on three pairs of crawler tracks without leveling devices.

Courtesy of Fried. Krupp

Fig. 14-45. Excavator and car loader

Operation Without Crowd. A wheel excavator with a non-crowding boom uses its track propel mechanism for crowd-retract.

Soil banks are cut to a stable slope, usually between 45 and 70 degrees. The angle selected depends on the character of the soil, the length of time the bank will have to stand, and sometimes on the season. The high wall beside the excavator may be left with a more gradual slope than the bank in front of it.

In continuing an existing cut, as in Figure 14-41 and 14-46(A), the machine is positioned parallel to the high wall and far enough from it so that the wheel can just reach the top.

The machine is walked forward until the wheel cuts a thin slice as the boom is swung in an arc between the old high wall (pit face) and the new one. As the wheel reaches the end of its arc it is moved forward a few inches to reach material during the return swing. This operation is repeated without changing the boom height until the tracks reach the toe of the bank or the bottom of the boom touches it.

The excavator then travels backward until the wheel can be lowered to cut the next shelf. The boom is again swung in an arc as wide as the bank, with the propel crowding it forward at each end of the swing.

The arc of swing moves toward the pit each time the wheel is lowered. This is because the cut into the new high wall is shortened to preserve its slope, but the slope of the old wall lengthens the arc there.

The slice taken in this digging pattern is thickest in the direction of travel of the machine, and thins out toward the high wall and toward the pit side. At a uniform swing speed production is greatest where the slice is thickest. Steady output is obtained by increasing swing speed where the slice is thinner. This is done by the operator in small and medium machines, and automatically in big ones.

The slices are also thinnest at the bottom where the teeth enter the soil, and thickest at the top where they break out. With certain classes of hard material the shallow cutting angle may cause the teeth to ride the bottom without cutting it. This difficulty can be overcome by using a falling cut.

Figure 14-46 shows the falling cut method that may be used when the bank is hard soil or soft rock, where the weight of the wheel is used to improve penetration, and the teeth of the buckets enter the bank almost at right angles.

The top cut is started in the same manner, but forward propel is stopped when the wheel axle is directly above the edge of the bank. Further cuts are made by lowering and swinging the boom, at the same time moving the excavator back from the bank if necessary to keep the desired slope, which

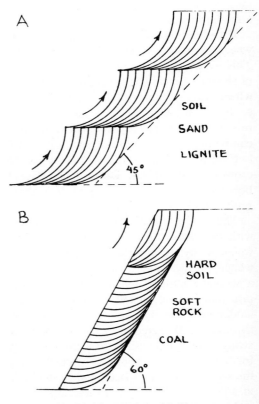

Fig. 14-46. Bench and falling cuts

Courtesy of Bucyrus-Erie Company

Fig. 14-47. Wheel excavator in coal mine

may be anywhere from 55 degrees to vertical.

When the bottom of the bank is reached, the boom is raised and the machine moved forward to start a new cut at the top.

Bucyrus-Erie. The Bucyrus-Ere Model 1054WX, Figure 14-47, is meant to dig dirt from 40 to 100 feet above the level at which it stands, and to dump it at a greater height 400 feet away.

The base and revolving frame are similar to those used in large stripping shovels. There are four pairs of crawlers, with hydraulic steering and leveling. Propel is possible only when the conveyors are either parallel or at right angles to the tracks, in order to assure stability. Travel speed is about 1/6 of a mile an hour.

The revolving frame has full rotation, with a speed of 1/6 to 1/10 of a revolution per minute. Side digging force on the buckets may be 27,000 pounds.

The digging ladder holding the wheel is 85 feet long, and has a 45 foot crowd-retract movement. The wheel is 24 feet in diameter over the bucket lips. The nine buckets have chain backs to minimize sticking of dirt.

Capacity is 1½ yards each, but average load is expected to be a yard, loose measure. Rotary digging force is 63,000 pounds at the bucket lips.

The wheel turns at 7 revolutions a minute. Average production is 2,000 yards an hour.

Two conveyor belts, each 54 inches wide, are used. The one in the ladder moves at speeds up to 900 feet per minute, the stacker unit at over 1,000.

The buckets spill by gravity as they come back over the top of the wheel. Dirt lands on an inclined metal plate that breaks its fall before it slides or is scraped onto the belt.

The machine is equipped with variable speed DC motors with Ward Leonard controls. Two 1,000 horsepower motor generator sets convert high line AC current into DC.

The digging and stacker booms are permanently fastened to swing together, so that each furnishes counterweight for the other, and the dump area is always directly opposite the digging area.

While digging the machine swings back

Courtesy of Barber-Greene

Fig. 14-48. Integral-wheel excavator, from bank

and forth in an arc, in somewhat the manner of a cutter dredge. Work is done within 30 degrees of a right angle to the length of the pit, in order to dump spoil at the maximum distance across the pit. The depth of cut is varied by the crowd, retract, hoist, and travel controls. A thin slice, usually about 6 inches deep, is taken.

OTHER WHEEL EXCAVATORS

Barber-Greene XL-50. This unit digs at its right side by means of a pair of 16-foot diameter wheels, each fitted with 12 buckets, for a total cut width of 10 feet. Rota-

Courtesy of Barber-Greene

Fig. 14-49A. Wheel from pit floor

tion is upward at the front. The cutting action holds the wheels down to grade, and breaks up and mixes the spoil, which falls on the first of a set of 2 or 3 belts arranged in series.

There are three crawler assemblies. The single front unit is at the left to clear the uncut bank. It has primary control of steering, and supports a hoist cylinder for front-to-rear grade adjustment. The right rear crawler, on the bank side, has a vertical cylinder for regulating cross slope. The left rear crawler is pivot-mounted directly to the frame.

The rear crawlers have individual hydrostatic drive, permitting any desired crowding speed, and assistance in steering when it is needed. There are two gear ratios, one for work and the other for travel up to 2.7 miles per hour. The wheels have two-speed mechanical drive.

The #1 conveyor is 4½ feet wide, runs on impact idlers, and receives material from the buckets and transfers it to the 5-foot wide, 31 foot long #2 belt. Both have hydrostatic drive. The #3 belt, which is optional, allows delivery to either of two haulers parked side by side, with instant shift from one to the other.

There are two angling blades used to clean the pit floor of loose material from the previous pass. One is ahead of the front crawler and casts to the left. Its windrow is caught by the wider center blade that feeds it to the buckets.

Preferred depth of cut is 8 feet, but deeper and shallower ones may be made. High banks may be taken in two or more lifts. The floor may be left flat, or sloped as desired. Production may be 5 yards per linear foot of travel, so even the largest trucks need not move far during loading.

The standard engine is 450 horsepower, with torque converter drive to pumps and the wheel transmission.

Easi-Grade. The Huron Easi-Grade has a full width digging wheel, and loads trucks ahead of it on the original grade, instead of to the side.

The digging drum is 8 feet in diameter and about 10 feet wide. It is made up of heavy parallel discs supporting a total of 18 backhoe buckets. These dig as the drum turns, dumping on conveyors that move the material to the side and then forward. Haulers move ahead of the excavator while being loaded.

The machine may be automatically controlled for line and depth of cut by a stringline. It is designed to make very tight curves, and to reduce a rough grade or untouched field to finish grade in one pass.

It has a 525 horsepower engine, and is rated for production up to 1000 yards an hour.

BLADE-FED BELTS

Elevating Graders. These machines dig with a plowshare or disc that cuts a slice of earth and slides it onto a conveyor belt, which elevates and dumps it. Models that are to be used for loading loose windrowed material may also have a set of chain driven paddles to move the dirt onto the conveyor.

The belt keeps the share from being choked with accumulated cuttings, and is in

Courtesy of Barber-Greene

Fig. 14-49B. Wheel marks in bank

itself the most economical type of elevator. As a result, these machines are the most efficient type of loader under favorable circumstances, and are capable of a very high volume of production. They cannot be used in rock, and are severely inconvenienced by boulders.

Autovator 1000. This very large machine cuts by means of a 16-foot high toothed knife cutter held vertically on the right side of the machine. The thickness of slice taken, to a maximum of 2 feet, is regulated by steering the whole machine.

There are two knives, one for working forward, the other for reverse.

Courtesy of Huron Manufacturing Corporation

Fig. 14-49C. Hoe buckets set in wheel

Fig. 14-49D. Autovator

Cut material falls onto a moldboard that guides it onto a 60-inch conveyor belt, that moves it upward and to the left for discharge into haulers.

There are eight crawler track assemblies, four of them with hydrostatic drive. There are automatic controls for grade and slope, with manual override. The two 425 horsepower engines drive a total of 7 hydraulic pumps, which provide power for all functions of the machine.

Grader Attachment. An elevating belt with a special cutter, usually a disc, may be mounted on a heavy grader, to replace the standard circle and blade.

The disc, which is of the type used in disc plows, can be raised and lowered independently of the conveyor. Disc and conveyor can be shifted to right or left as a unit to vary the cutting distance outside of the wheel line.

The disc is adjusted to secure the desired depth, and width is regulated either by side shift or by steering the machine. Digging is done alongside the machine rather than under it. Height of bank and width of slice taken vary with the material. Generally, a slice 20 to 30 inches high and 15 to 20 wide is good procedure.

Low gear is standard for digging to obtain the power and stability for heavy cuts, and for easy synchronization of speed with trucks being loaded. However, second is often used when cuts are light and the spoil is sidecast.

A more complex adaptation, Figure 14-51, enables a machine to make a quick, neat job of excavating a slot for road widening.

Fig. 14-50. Elevating belt on grader

PADDLE LOADERS

Athey. The Athey Force Feed Loader, Figure 14-52A, is a self-propelled, rubber-tired machine that is primarily designed for loading windrows of loose material, including dirt, gravel, broken pavements, snow, and leaves.

It consists of a greatly modified truck chassis carrying an inclined belt conveyor, a three piece funneling plowshare and mold-board, and chains fitted with paddles.

The plowshare wings rest on the ground and funnel loose material toward the center as the machine moves into it. The curved paddle wheels move down at the digging end, then back along the center moldboard, pushing the dirt onto the conveyor belt which elevates and dumps it. The mold-board is raised by a hydraulic ram and bell cranks, and lowered by gravity and hydraulic pressure. The conveyor height is regulated by a separate hydraulic control.

The paddle feeder can be lowered to almost touch the moldboard, or raised to any desired distance to afford better clearance. If it becomes choked by too much or too coarse material it automatically rises, compressing the springs above it. This feeder prevents clogging of the moldboard. Optional side augers may replace the moldboard wings.

Dumping is usually done immediately to the rear. Rear loading of trucks permits the entire operation of digging and hauling to be in one lane, minimizing traffic interference. There is an optional cross conveyor that permits loading or casting to either side if desired.

The loader is frequently teamed with a grader for road work. The grader windrows the surplus material and the loader pours it into trucks for disposal.

This machine can load about five yards a minute under favorable conditions. It can be used in aerating or mixing material, or feeding a portable crusher which is

Courtesy of Rivinius, Inc.

Fig. 14-51. Road-widening cut

towed behind it. With special side wings, it can dig stockpiles of loose material. It is not used for direct excavation.

Operation. The paddle loader is placed at the beginning of the windrow and a truck backed up to it. The transmission is put in one of the lower gears, the belt and feeder started, and the machine driven so that the windrow will be guided into the feeder paddles by the side wings.

The relative speeds of the machine and the feeder and conveyor depend on the transmission gear selected. The speed should be high enough to keep the paddles well filled, without choking them up. Changes are made in travel rather than feed and belt speeds.

Courtesy of Athey Products Corporation

Fig. 14-52A. Athey Force-Feed loader

As the machine moves, the truck must back up to keep under the conveyor. The truck should be directly behind the loader to center the load, and may have to move at a slightly different speed to distribute the load to front or rear.

Conveyor height should be sufficient to clear the tailboard of the truck when it is backed against the radiator guard. If the body is very long, additional height may be required to fill the front. Height should be kept to a minimum to avoid dust and spillage.

When the loader engine is running full speed the spoil is thrown off the belt in an arc. When throttled down it falls almost straight. In some models the belt is run only at high speed.

When the truck is filled both the driving and digging clutches are released. The truck is driven away, another is backed under the conveyor, and loading is resumed in the same manner.

These machines will pass a considerable amount of cobbles, sod, tin cans, and other trash, but the feeder mechanism may be badly jammed by wire or certain kinds of metal scrap.

Courtesy of George Haiss Mfg. Co., Inc.

Fig. 14-52B. Bucket loader, rear view

CHAIN BUCKET LOADERS

These machines are apparently no longer manufactured, but enough of them are on the job to justify a brief description.

They are intended primarily for rapid loading of stockpiled material. They mix as they load, so are good for combining different sizes of aggregate.

Topsoil may be loaded directly from the field, if not too hard to penetrate. Preliminary cultivation is not necessary, as lumps are broken and vegetation chopped and thoroughly mixed. The soil is fluffed up for easier handling and greater bulk.

These loaders are made up of a crawler or wheel mounting, a vane or auger feeding device, a bucket elevator, and a discharge chute or conveyor. A clean-up blade follows the feeder.

Operation. To dig a stockpile, a bucket loader is driven to its edge and the elevator lowered until the clean-up scraper rests on the ground or is slightly above it. The elevator clutch is engaged and the loader moved into the pile in low gear.

The truck should be under the discharge chute before the elevator is started. The truck driver should back it at such a speed that the body will always be under the chute, and the load will be distributed properly to front and rear. The loader operator can swing the chute by turning a hand wheel to compensate for slight errors in placing the truck.

Forward motion should be at such a speed that the buckets carry heaping loads but the material does not jam the feeders. If the pile is high it will be necessary to release the driving clutch occasionally and hold the machine with the brakes while it elevates the bulk of what is in the feeding mechanism. If the pile slides readily several truck loads may be obtained without moving.

The machine should not be allowed to dig deeply into a pile on one pass because

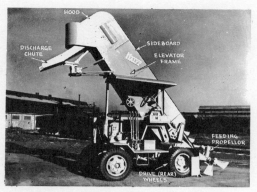

Courtesy of George Haiss Mfg. Co., Inc.
Fig. 14-52C. Bucket loader, side view

of danger of slides. These may go over the guards and bury the wheels or jam the tracks, or may make it difficult to escape because of side pressure against the guards. The pile is best dug back in a series of short cuts which may be parallel or fanned out.

A bank may be dug in the same manner. Even where the danger of slides is very slight no long single cuts should be made, as the machine might become wedged between the sides.

In bank digging care must be taken not to crowd in hard enough to slow the engine or strain the paddles or tail shaft. Cobbles do not interfere particularly with digging, but it is necessary to work around boulders. clean-up blade is rested on the ground, or forced into it, and the machine backed with the elevator going.

MACHINERY FOR UNDERGROUND

Continuous Miner. Coal, salt, potash, and other minerals are mined underground by machines that can cut them directly out of the solid vein, drop the material on a built-in chain conveyor, and load it into electric powered, rubber tired shuttle cars or continuous belt conveyors at the rear.

The Goodman boring type machine in Figures 14-53 and 14-54 has cutters

Courtesy of Goodman Manufacturing Company

Fig. 14-53. Boring-type continuous miner

mounted on two sets of rotating arms and on upper and lower chains that cut an oval shaped path 13 feet wide by 7 feet high. It travels on reversible crawler treads. It is electric-powered and hydraulic controlled.

The cutting heads are swung by steering the whole machine, and in addition can be tilted and retracted by special controls. The discharge end of the conveyor can be swung to either side.

The crawler tread gathering loader, Figues 14-55 and 14-56, is designed to load coal that has been blasted from the seam. As the entire machine moves forward the gathering arms on the loading head scoop up the coal and move it back to a built in chain conveyor.

The conveyor can be swung at an angle of 40 degrees to either side to permit loading wide areas with minimum maneuvering. The machine is built in models for work in low, intermediate, or high seams.

Mucking Machines. When there is sufficient space, almost any excavator can be used for underground loading of blasted rock, minerals, or ore, known under the general term "muck." However, there are a number of machines designed specially to work in small spaces, and to operate with air or electric power to minimize air pollution.

The Conway shovel, one model of which is shown in Figure 14-57, has a digging dipper or bucket hinged to a boom, which in turn is hinged to a turntable on the main

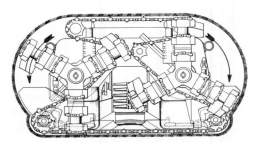

Courtesy of Goodman Manufacturing Company

Fig. 14-54. Cutting pattern

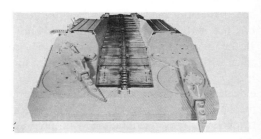

Courtesy of Goodman Manufacturing Company

Fig. 14-55. Digging pattern, loader

Courtesy of Goodman Manufacturing Company

Fig. 14-56. Swing motion loader

frame. A pair of chains extend from separately controlled drums to the dipper. A conveyor belt runs from the rear of the turntable up and back to a dumping point.

There are four loading motions, utilizing power from a single electric motor. They are tramming (crowd-retract motion of the whole machine), rooting (jiggling or rocking the bucket to assist penetration), swinging the boom and dipper, and hoisting.

Reeling in both bucket chains rotates the bucket backward on the boom hinge until it contacts stops. Further pull raises bucket and boom together until the load spills down the boom chute onto the conveyor, and is carried back to dump into a car coupled to the rear of the machine.

Reeling in one bucket chain will cause the bucket and boom to swing to that side.

This machine usually operates on railroad type tracks laid close to the face before the blast, so that much or all of the muck pile can be cleaned up before the tracks need be extended.

Courtesy of Goodman Manufacturing Company

Fig. 14-57. Conway shovel

Courtesy of Joy Manufacturing Company

Fig. 14-58. Conveyor-loader

The Joy 18-HR Loader, Figure 14-58, is essentially a chain flight conveyor mounted on crawlers, equipped with a gathering head which is pushed into the base of the pile.

Courtesy of The Eimco Corporation

Fig. 14-59. Downhill loading

The Eimco Rocker Shovels, which are illustrated in Chapter 16, have a front mounted bucket that is loaded by pushing into the pile, and lifted over the machine to dump into a car coupled to the rear.

Small models are rail mounted, and can swing the whole upper works to reach the sides. They can be used on steep down grades, as in Figure 14-59. A conveyor belt can be added for longer reach.

Any of these loaders can be used in conjunction with portable conveyor belts, which discharge far enough to the rear so that a whole train of cars can be run underneath the frame. When the front car is loaded, the locomotive moves ahead to put the next car under the discharge, until the last one. The loader and belt are then stopped, and the train pulls away. This setup is shown in Figure 14-60.

In addition to their mucking work, loaders may be used as locomotives in switching cars at the heading, and in moving the jumbo and other equipment.

Courtesy of The Eimco Corporation

Fig. 14-60. Use of portable conveyor in car loading

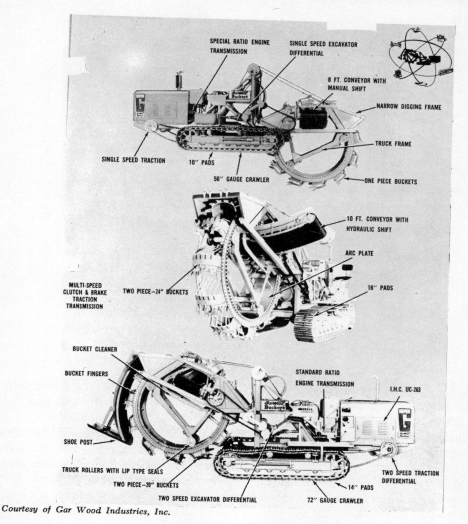

SPECIAL RATIO ENGINE TRANSMISSION
SINGLE SPEED EXCAVATOR DIFFERENTIAL
8 FT. CONVEYOR WITH MANUAL SHIFT
NARROW DIGGING FRAME
TRUCK FRAME
ONE PIECE BUCKETS
56" GAUGE CRAWLER
10" PADS
SINGLE SPEED TRACTION

10 FT. CONVEYOR WITH HYDRAULIC SHIFT
ARC PLATE
16" PADS
MULTI-SPEED CLUTCH & BRAKE TRACTION TRANSMISSION
TWO PIECE—24" BUCKETS

BUCKET CLEANER
BUCKET FINGERS
STANDARD RATIO ENGINE TRANSMISSION
I.H.C. UC-263
SHOE POST
TWO SPEED TRACTION DIFFERENTIAL
TRUCK ROLLERS WITH LIP TYPE SEALS
TWO PIECE—20" BUCKETS
TWO SPEED EXCAVATOR DIFFERENTIAL
14" PADS
72" GAUGE CRAWLER

Courtesy of Gar Wood Industries, Inc.

Fig. 14-61. Wheel ditcher

DITCHERS

WHEEL DITCHERS

Construction. The Buckeye Model 306, shown in Figure 14-61, is representative of this class of machine. It consists of a crawler mounting, an engine on a front over-hang frame, a cutting wheel, a conveyor belt, a shoe post, and the driving and control mechanisms.

The wheel, shown in more detail in Figure 14-62, is of channel construction with the buckets riveted or bolted to outside brackets. Both sides of the wheel are toothed and are driven by identical gears on the drive shaft (number one shaft). The wheel has no axle and is supported by idler wheels on an internal frame.

Various types of buckets are available. Solid bottom buckets can be used in any soil that is not sticky, and must be used when the dirt is loose and powdery. The slat type is suitable for any soil cohesive enough not to fall between the bars.

Spring loaded bars set on the frame at the dump point push the dirt out of slat buckets so that even the stickiest soils can

14-47

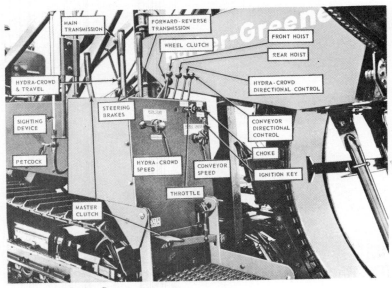

Courtesy of Barber-Greene

Fig. 14-62. Hydraulic ditcher controls

be dug. Large stones in the bucket will force the bars back against the springs.

The conveyor belt carries dirt dumped by the buckets to either side and leaves it in a windrow. The belt is reversible and can be shifted sideward.

The shoe post and crumber is hinged at the top to the wheel frame and held in position by two turnbuckles. It serves to support the back of the wheel, pushes spillage ahead of it until it is picked up again, and rounds and smooths the bottom.

The crumber may be expanded into a tile box such as that shown in Figure 5-46, or it may have a hydraulic lift.

The wheel frame is hinged on two shoes which slide in vertical tracks on a rear frame or mast. The shoes are lifted by lines across sheaves above the tracks. The back of the wheel frame or boom is raised by another pair of cables.

These cables may be operated by hydraulic cylinders expanding against sheave sets, or by self-locking winches.

Courtesy of Cleveland Trencher

Fig. 14-63A. Side to side shift

Courtesy of Cleveland Trencher

Fig. 14-63B. Wheel tilt

The transmission is compound, having one set of three speeds forward, and one of four forward and one reverse. This combination gives eight forward speeds for digging and four for traveling, with two digging reverse and one traveling reverse.

Gearing may be reduced to a two speed dig-or-travel shift, with hydraulic drive for infinitely variable ratios.

The drive shaft goes forward from the transmission to a differential and axle set in front of the engine. The bull wheels, at the front of the tracks, are driven through roller chains. Steering is by means of axle brakes.

A shaft which turns at engine speed extends to the rear of the transmission and powers the wheel drive and hoist. The wheel has two speeds forward and one reverse, and a disc clutch, which are independent of the track drive or hoist gearing.

The digging wheel drive is carried backward through a drive shaft to a differential and axle set, carrying sprockets on the ends. These are connected by roller chains to the number one or wheel drive shaft. This shaft is cut in the center, and carries a driven sprocket with wheel drive gear on each side. The differential assures equal driving force to the two sides of the wheel.

The conveyor is usually driven by variable speed hydraulic motors, at one or both ends. But power may be supplied through a two speed chain or shaft from the digging wheel drive train.

The operator's station is usually at the side, over the track. Figure 14-62 shows details of one in which most controls are hydraulic.

Wheel ditchers may dig from 4 to over 8 feet deep, with bucket speeds 120 to 300 feet a minute. Ground speed may be 2 to 20 feet a minute.

Some wheels can be side-shifted to work near obstructions, and/or tilted to make vertical ditches on side slopes.

Operation. The buckets cut the earth and

Courtesy of Cleveland Trencher

Fig. 14-63C. Crumber with hydraulic lift

carry it forward to the top where they dump most of it on a conveyor which carries it to the side and piles it.

In starting a cut, as the wheel is lowered to the ground the buckets will start to dig. The machine is stationary. Enough weight should be allowed to rest on the buckets so that they will fill heaping without gouging deep enough to slow the wheel.

If no shoe is used, the wheel can be lowered to cut bottom grade before walking the machine. The ditch will have a rounded start, as in Figure 14-63D (A). Position should be such that the center of the wheel is over the starting point of the full depth ditch.

If a shoe is used, the rear hoist is pulled high and the wheel lowered in the position shown in (B). When the buckets cut so

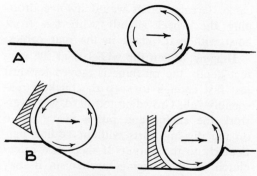

Fig. 14-63D. Starting wheel cut

far that the shoe rests on the ground, the ditcher is moved forward. The rear cables are left slack so that the rear weight rests on the shoe and only enough tension is kept on the front cable to prevent the buckets from biting too hard.

The wheel will cut to an increasing depth as the supporting shoe is pulled into the ditch, and the front lift cable is unreeled to allow the wheel to sink.

When bottom grade or the limit of depth of cut is almost reached, the wheel is held from further down-cutting by holding the front cables, while continuing motion. The shoe will move down the ramp until it levels out. The rest of the cutting at that depth is done with rear cables slack and front ones tight, unless it is the full depth to which the machine will dig in which case both cables can be slack.

Control of the machine is complicated by the fact that the digging center is behind the tilting and turning center of the tractor. If the tractor starts up a grade or over a bump, a shoeless wheel will dig down and will have to be raised to keep level. If supported by a shoe, it will not cut down and may be raised slightly.

In starting down a grade the tractor will pitch forward. The wheel will tend to rise out of the ground, and either the front or rear cable should be slacked to drop it to grade. If a shoe is used a slight pitch might not require any attention as the back will lower itself on the slack cable. Greater pitch would involve dropping the wheel slightly with the front hoist with extra slack in the rear cables.

Digging. When the wheel is at the correct depth, the machine is moved forward just fast enough to keep the buckets reasonably full. Crowding too hard will overwork the engine and put a heavy strain on the digging parts without adding particularly to the output. If there is a safety clutch on the digging mechanism it may slip excessively and heat under overload.

Selection of the right combination of gears for various types of digging must be based on experience. If either the machine or the soil conditions are unfamiliar, it is good policy to err on the side of underfilling rather than overcrowding.

Soft rock usually responds best to a high wheel speed with very slow walking speed. If dirt is very soft it can sometimes be crowded so that material in excess of the bucket capacity piles on each side of the ditch without damage.

Obstructions. Where boulders, heavy roots, or pipe lines are liable to be met, both walking and wheel speeds should be low and the engine should be throttled down at critical points.

Soft boulders will be cut through by the teeth. Hard ones may be pulled up to the surface. This is less likely to happen if the boulder is deeply buried, because the deeper boulder not only is held down by a greater weight of dirt, but the direction of the tooth contact tends to force it forward rather than up.

The wheel will usually ride over a deep stone it cannot move, so that it will be lifted above grade. If a big stone is near the top it may stop the forward motion of the machine, in which case power should be cut off promptly.

If a boulder is pulled to the surface, it is liable to land in a very inconvenient spot, forward of the wheel and between the tracks. It may be necessary to lift the wheel into transporting position, walk forward until clear of the rock, push it out of the way, and back up until the wheel can be lowered to the ditch bottom. If the boulder is too large for the wheel to clear, the wheel drive clutch may be released so that the wheel can turn as it crosses it.

It is sometimes easier to cut and repair tile lines than to work over them. Where they are expected, tile of proper sizes and joint fillers should be kept on the job.

Where a pipe must be left intact digging should be done cautiously, and the wheel lifted to inspect the digging face whenever contact is suspected.

When the wheel is lifted above grade to clear any obstruction, it may be worked back to grade at the other side in the same manner as the cut is started.

Turning. Turning must be done with great caution while digging, as the whole machine revolves around the center of one track, which causes the wheel to move sideward in the earth. If the buckets are equipped with long side teeth or side cutting bars are used, and the earth is reasonably soft, a gradual turn can be made without damage. However, too sharp a turn may bend the wheel frame or the wheel itself, or pull the wheel frame off the vertical track.

If a shoe is used, it further limits the ability to follow a curve, and if very long, such as a tile laying shoe, may prevent any significant turning unless it is loose enough, or hinged, so that it can trail instead of being swung wide.

Line and grade. In most ditching work, it is important to both keep the machine accurately in line and working at the proper depth, although the operator cannot see the bottom.

The ditch is first surveyed and the course and the depth of cut ascertained. A line is then established at a fixed and constant distance above the bottom grade, and offset from the center beyond the track line of the ditcher.

The line should be established at a height which will put it at least a few inches above the ground at all points, stakes driven along it, and the exact height marked with nails and a string.

A rigid bar is fastened to the front of the power unit of the ditcher with one end over the string when the ditching wheel is centered on the ditch line. A plumb bob or other weight is fastened to the bar so that it will hang directly over the string.

The operator can then keep the machine on correct line by keeping the plumb bob just over the string. If the ground is irregular, the cord holding the plumb bob can be run through eyes or pulleys so that the operator can reach an end of it to raise and lower it when necessary.

The same apparatus fastened on the side beam of the wheel can be used with a fixed length of string to determine both line and depth.

Targets. Targets, as in Figure 14-64A (A), can be used instead of the string. Each consists of a light pipe that can be pushed or driven into the ground, and a red and white strip on a collar which slides up and down the pipe and can be locked to it by a clamp screw.

The pipe stakes are set sufficiently far

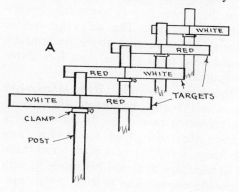

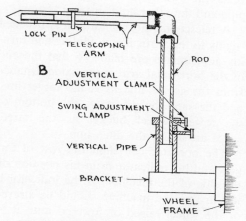

Fig. 14-64A. Targets and sight arm

Courtesy of Fuller Form, Inc.

Fig. 14-64B. Trapezoidal ditch

from the center line to avoid interference with the machine. The strips are set at some arbitrary distance above grade which should keep them at least four feet above the ground surface.

The device shown in (B) is attached to the wheel frame, preferably forward of the center of revolution. It consists of a bracket that will offset the device from the frame, a rigid vertical pipe, a rod made of smaller pipe or round bar inside it, with a double clamp, one part of which prevents the rod from sliding down in the pipe and the other holds it from turning.

A telescoping horizontal arm is fastened rigidly to the vertical rod. Sliding and turning adjustments are made so that the end of the horizontal arm is the same distance from the center line as is the target line, and the same height above the bottom of the lowest wheel bucket as the target strips are above the ditch line.

The operator can sight the row of targets and his indicator from his position on the side of the machine. If the indicator is not in line the machine should be steered or the wheel raised or lowered until it is. When a target is reached the swivel clamp

is loosened and the indicator swung back so as not to hit it.

A slight tilt of the wheel forward or back will not affect the accuracy of this gauge noticeably, but if the wheel is tilted

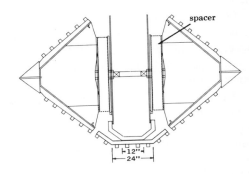

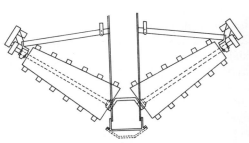

Courtesy of Fuller Form, Inc.

Fig. 14-64C. Sloper cones

Courtesy of The Parsons Company

Fig. 14-65. Ladder ditcher

considerably the indicator will be moved downward so that it will have to be held under the target line for accuracy.

Targets are best suited to work in straight lines on constant gradients. If changes in direction or grade occur, a tar-

get must be at each point of change.

Both laser and automatic following of stringline controls are used in trenching. These are discussed in Chapters 3 and 19.

Trapezoidal Ditchers. Permanent irrigation ditches usually have a flat bottom and sloped sides. A wheel ditcher may be fitted with side cutters, called cones, that enable the wheel to produce this shape in one pass.

Two types of cones are shown in Figure 14-64C. One is driven directly by the wheel through its axle. These cannot be adjusted for different slopes, but they can be set out with spacers for a wider ditch.

The others, (B), are turned by hydraulic motors. Their vertical angle can be increased or decreased by adjusting support rod length.

Both types move the cuttings down the slope to the wheel buckets, which carry them to the conveyor for discharge to the side.

In soft, caving soil, cones may be replaced by sloper bars, which are rigid blades held in side frames. They slice the dirt, which falls down to the buckets.

The bars are simpler and cheaper, but are too hard to pull in many soils.

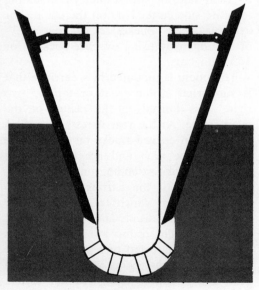

Courtesy of Cleveland Trencher

Fig. 14-66. Sloper bars

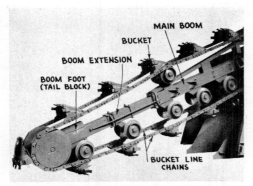

Fig. 14-67. Boom details

LADDER DITCHERS

Construction. The Parsons Model 250 Trenchliner, shown in Figures 14-65 to 14-69, is representative of this type of machine. Digging is done by a number of small buckets carried between two pairs of chains traveling around a boom. Mounting is usually on tracks, as shown, but wheels are satisfactory for small machines or light work.

The main frame consists of a flat bed and an arched superstructure. The tracks are suspended from the main frame at two points in the rear and a single pivotal point in front which allows for oscillation when traveling over uneven ground. The engine is supported at the front, and projects forward from the crawler to provide a count-

Fig. 14-68. Head shaft

erbalance for the weight and digging pull of the ladder.

The digging boom (ladder or elevator) includes the main or upper boom, the boom extension, the boom foot or tailblock assembly, the buckets, and the bucket chains. The boom extension telescopes into the main boom and can be secured in the desired position by clamp bolts. The boom foot contains a threaded adjustment for regulating the tension of the bucket chain.

The boom chains are made up of side bars, center bars, and attaching bars. The buckets attach to the latter by a single pin on each side. The chain parts are linked together with slotted pins which require no cotter keys or bolts to secure. The chains are driven by split-type sprockets clamped to the head shaft.

Buckets include a carrying portion and a lip which contains tapered tooth sockets. They are available in several cutting widths, and the cut of a particular size set of buckets can be increased by the use of flare-type sockets which permit setting teeth up to three inches further out than normal on each side, or still further by mounting separate side cutter bars on the chain. For example, a 24" bucket set can dig a 27" or 30" ditch by using oversize sockets, or 36" with bars.

The boom is mounted in a carriage that is supported at the front or top by two rigid arms hinged to the center of the main frame. At the rear it rests on rollers that ride on curved tracks on the outside of the frame arch and are kept on the tracks by hooks extending under them.

The front of the carriage carries two sheaves through which two-part lines are reeved to winch drums and anchors on the main frame. The drums are mounted on a single cross shaft and turn together. Drive is through a self-locking worm and worm gear and a double-acting friction clutch.

Reeling in the cable pulls the carriage forward and the curved track causes it to rise. When digging at maximum depth (twelve and a half feet for this model) the boom is about 35° from vertical. As it rises it tilts toward a horizontal position. If there is much shallow ditch it may be desirable to shorten the boom to keep a steeper digging face.

The boom can be shifted laterally across the carriage by loosening clamp bolts and pulling with a ratchet chain hoist. It can be brought to within 12″ of the outside of the machine on either side. Off-center work is called off-set digging. During the adjustment the drive sprockets and hubs slide along the square head shaft.

The conveyor is curved, skirted, revers-

this point prevent mud from sticking in the buckets.

Power Trains. The machinery involves six power trains. There is a high traction drive for traveling, rated in miles per hour, and low traction, rated in inches per minute, for moving it while digging. The excavating buckets and their chains are turned for digging by the excavator or bucket line drive. The conveyor has two drives, both reversible. One sideshifts it to regulate the discharge point, the other turns the belt to carry away the spoil dumped on it by the buckets.

Engine power first goes through a foot-operated clutch, then the engine transmis-

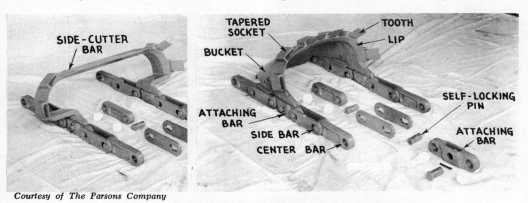

Courtesy of The Parsons Company

Fig. 14-69. Bucket, side cutter, and chain section

ible, and so mounted that it can be shifted to discharge at various distances on either side of the machine. It is kept in alignment by a V-strip molded in the inside of the belt, which fits in a center groove in the idlers. At maximum extension the discharge end is high enough to load trucks.

Small machines may use sloping metal slide plates instead of a conveyor to place the spoil beyond the ditch edge.

Digging is generally done on the lower or forward side of the boom. The loaded buckets dump by both centrifugal force and gravity. A combination hood and chute prevents scattering and guides the dirt to the conveyor. Spring cushioned cleaners at

sion with one reverse and three forward speeds. This is used to select speeds for high speed traction and for rotation of the bucket line and conveyor belt. The constant-speed countershaft drives the boom hoist drums through a roller chain and a reversing clutch of the multiple disc type.

The transmission main shaft is connected through a double universal joint to the main gear case, which directs power to all the other functions of the machine. The traction drive is to a shaft equipped with a pair of friction clutches and brakes, then through roller chains to the crawlers. The high traction drive is direct, but low traction goes through the reduction type low

Courtesy of The Parsons Company

Fig. 14-70. Ditching with a ladder

traction transmission on top of the case, which has one reverse and five forward speeds, then back through the case to the axle. In combination with the engine transmission this provides fifteen selections of low speeds for various digging conditions in addition to the three walking speeds of high traction.

The excavator drive is through a denture clutch and roller chain to the driven sprocket on the ladder head shaft. This is equipped with a safety release band clutch which is adjusted to carry normal loads, and to slip if the buckets hook into something solid.

Applications. The ladder ditcher can dig much deeper than the wheel type, is more maneuverable, does less overdigging on angles and curves, and needs less hand assistance at ditch ends and obstructions. It is not as well suited to hard digging.

Attachment of tile-laying shoes is not recommended by manufacturers, although contractors have successfully used shoes of their own design at depths of twenty feet in irrigation subdrainage work.

Courtesy of Vermeer Manufacturing Company

Fig. 14-71A. Drag trencher, with hoe and backfiller

Operation. This ditcher can start an almost vertical rear face by holding the machine with the steering brakes, engaging the bucket and conveyor drives, and lowering the boom. When full depth is reached, the machine is driven forward at such speed that the buckets will be kept full without straining or gouging.

Digging can also be done in reverse, in which case the dirt is cut off the rear wall and carried under the tail pulley. The buckets will not load as well and a considerable amount of spillage will be left on the trench floor. The elevator must be centered so that the tracks can straddle the ditch. It may be necessary to put planks across the trench, between the elevator and the tracks, to prevent caving.

Reverse is generally used only in undercutting pipes, curb stones, or other obstructions. The ditch is dug forward until the obstacle is reached. The elevator is raised to clear, the ditch dug a little past it, the elevator lowered, and cutting done in reverse right up to it. The slanting slice of earth under a pipe which cannot be reached may fall into reach of the buckets or be knocked down by hand.

DRAG TRENCHERS

The class of ditchers which pulls or drags cuttings to the surface, rather than lifting them in buckets, may conveniently be called drag trenchers. Many of them are small, three tons or less. Some of them carry other related equipment such as a cable-laying plow, a cable reel, or a backfill blade and/or a backhoe on the other end of the tractor.

Depending on size and arrangements, the operator may stand beside the trencher and steer it with handlebars, sit on it side saddle, or have a regular tractor-style seat and control station.

In the digging, the machine should be propelled forward fast enough to keep the cutters loaded with soil, but not so fast as

Courtesy of Vermeer Manufacturing Company

Fig. 14-71B. Drag trenching

to slow or stall them. Since the resistance of soil may change frequently, it is necessary to keep alert, and be ready to make adjustments often.

Usually the cutters are run at a uniform rate, and the ground speed adjusted as necessary. This is particularly the case with machines having hydrostatic drive to the wheels.

When the cutters are moving rapidly, they throw cuttings out of the trench with considerable force. Most of the pieces strike baffles and fall harmlessly to the sides, but they can be a nuisance and a danger to a person close to the discharge point.

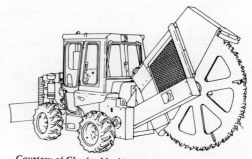

Courtesy of Charles Machine Works

Fig. 14-71C. Carbide-toothed wheel

They are usually moved back from the edge by rotation of a pair of auger flights, to prevent sliding back into the slot.

Cutters are usually carried on a roller chain. For cutting slots in rock or pavement, however, a solid wheel with carbide teeth may be used. Over-crowding can be very expensive with these, as it can cause breakage of a number of teeth in one turn of the wheel.

PUMP EXCAVATORS

HYDRAULIC DREDGES

Hydraulic dredges consist of floating pumps, which suck in mixtures of soil and water and deliver them to a disposal point through a pipe line. They can excavate underwater soil or waterside banks, transport spoil through pipes and rough grade it on the fills. A single machine, with proper accessories, may perform all three functions, or it may dig without transporting, or a dredge-type pump may transport from a hog box without digging the soil itself.

General Construction. The heart of the hydraulic dredge is the main centrifugal pump with its suction and discharge lines. Except for hog box work, the pump, power plant, and accessories are mounted in a floating hull. The suction line is carried in a live boom called the ladder. Its position and the working movements of the hull are controlled by winches and spuds on the hull, and by anchors on the bottom or on shore.

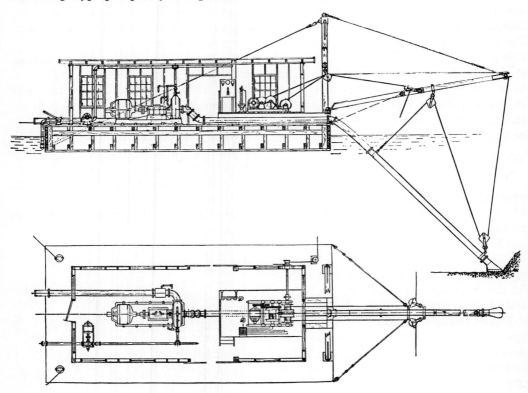

Courtesy of Morris Machine Works

Fig. 14-72. Suction dredge diagram

Courtesy of Morris Machine Works

Fig. 14-73. Suction heads

Plain suction dredges, such as the Morris shown in the diagram in Figure 14-72 erode the bank by the force of water flowing into the suction head, Figure 14-73. High pressure jets of water may be set around the rim of the head to loosen the soil, as in (B). These are suitable for dredging loose free flowing material such as sand and some gravel deposits. They should not be used where there are many pieces too large to go through the pump, or where the bottom must be cut to a definite grade.

As a rule, suction dredges are moved around the job by winch lines to bottom or shore anchors, and they dig a succession of overlapping pits.

Cutter dredges, shown in Figures 14-74 and 14-75 loosen material by means of revolving blades or chains. Position control is through two spuds on which the machine pivots, and a pair of swing anchors. While digging, the cutter is kept in almost continuous horizontal motion back and forth through an arc.

Dredges are usually rated as to size by

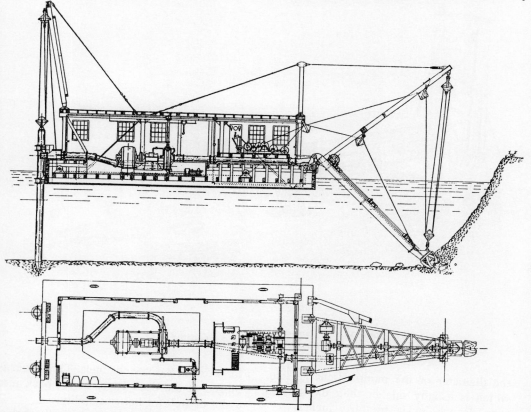

Courtesy of Morris Machine Works

Fig. 14-74. Cutter dredge diagram

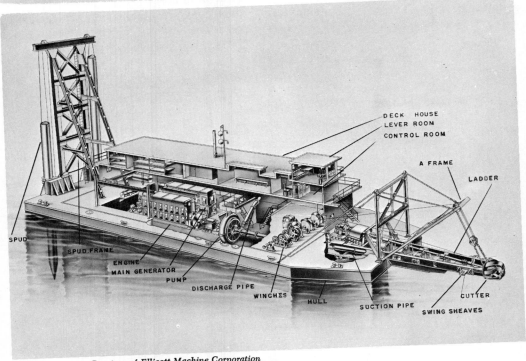

Courtesy of Ellicott Machine Corporation
Fig. 14-75. Cutter dredges

the diameter of the pump discharge line, which is usually smaller than the suction pipe. Range is six to thirty inches.

Hull. The standard hull shape is rectangular, but the bow (front) may be tapered for work in narrow channels. The ladder is hinged at the deck edge, or set back into a slot or well for better stability.

Modern practice is to make hulls of steel divided into compartments. The steel is de-

14-58

sirable to withstand the severe vibration of the dredging. The compartments prevent sinking from a single leak, and make it possible to trim (level) the dredge by selective pumping or filling.

The barge may be made of a number of pontoons bolted together. This construction is used when it is likely to be necessary to dismantle it for moves between jobs, or when it must be shipped overland to its first job.

The main machinery should be mounted in a single pontoon or section when possible, to avoid delay and damage during taking apart and reassembling.

Hull size is determined first by the weight of machinery to be floated, and secondarily by the need for supporting and counterbalancing the overhang of the ladder.

Cross section is varied according to special requirements. Dredging shallow channels requires light draft, so that length and width must be increased to provide support at the surface. Narrow channels call for reduced width with greater depth or length.

Large dredges may contain hoppers for storage of spoil, and may be equipped with screening, crushing, and washing plants for processing it.

Seagoing dredges are often of the hopper type, and may have a tapered hull propelled by their own engines. Other dredges are towed from job to job, or taken apart and shipped.

Power Plants. Power may be diesel, electric, or steam. The dredge in Figure 14-75A has two diesel engines. One drives the pump, the other turns the main generator that supplies current to run the cutter, winches, and accessories.

Two engines are used so that the pump can work steadily at full capacity, without being affected by the variable requirements of the other equipment.

All-electric drive is frequently used in dredges that work within reach of power lines. They may also be used with a shore generating set. Current is brought from shore by special submarine cable, or supported on the discharge pipe floats.

Electric power makes a cleaner and quieter dredge and eliminates the fueling problem. However, caring for the power lines may be difficult because of danger from high voltage, movements of the dredge, and change in length of the discharge line.

Steam power is used, but to a decreasing extent. Each unit may be driven by a separate steam engine, or only the pump may have steam drive, while a steam powered generator supplies current to individual electric motors.

The eight-inch Dragon dredge in Figure 14-75B has a single diesel engine that drives the dredging pump and also a hydraulic pump. The winches and cutter are operated by hydraulic motors.

Pumps. The main or suction pump is of the centrifugal type, with wide clearances which permit pumping a high proportion of solids and passing of large pieces. Construction is varied according to the depth of digging, the length and rise of the discharge line, and the coarseness or abrasive qualities of the spoil. The pump must be able to maintain a suction velocity of at least twelve feet per second.

These units are designed so that most of the wear is taken by replaceable liners or tips. White iron, and manganese and nickel alloys are extensively used for these parts. Shafting is extra heavy to minimize shock damage from jamming on stones or logs.

Because of the clearances required efficiency is usually only about fifty to sixty-five percent, and that is in a rather narrow speed range. If the engine has a different speed, pulleys or gears must be used to adapt it. Belt drives are often used because of their shock absorbing nature. Power requirement is greatest when pumping heavy volume against a low head.

The pump may be mounted partly or

—Artist's view of the cutter of a hydraulic pipe line dredge.

—A spiral cutter with seven blades.

—For special work, cutters may be provided with teeth of the so-called "shovel" type welded to the blades.

Courtesy of Ellicott Machine Corporation

Fig. 14-76. Cutter heads

wholly below outside water level, so that it will prime by gravity. If it will not, a valve may be closed in the discharge line and air pulled out by means of a vacuum pump or exhaust blower. Other units prime by raising the suction line and filling it by means of an auxiliary pump.

The auxiliary pump, usually a centrifugal, is used for wetting stuffing boxes and water lubricated bearings, priming, flushing decks, and miscellaneous jobs. If a high pressure type, it can supply water for jetting deposits around a plain suction line.

Suction Boom. A plain suction dredge boom may consist of a pipe, supported by cables attached to a jack boom, which is raised and lowered by a winch line. Rubber hose of special construction is used at the upper end where it meets the hull, and

heavy metal pipe for the balance. A suction hood, Figure 14-73 (A), is flared out and equipped with guards to keep out oversize rock. If agitation is used, as in (B), the opening is smaller and the high pressure jets are arranged around its rim. The pressure water line is fastened to the outside of the suction line.

A lattice boom or ladder may also be used to support the pipes.

Ladder. A cutter must be supported by a ladder boom strong enough to carry its weight, withstand side stresses, and to keep the drive shaft in alignment. The jack boom is fixed, and a multiple winch line run from it to the ladder point. The swinging lines are run from winches around swing sheaves near the cutter end, and outward to side anchors.

The open end of the suction line is carried in the suction head on the cutter end. This head contains the end bushing for the drive shaft and the supporting flange for the cutter.

Digging Depth. The downward angle of the ladder during digging varies from a few degrees to a standard maximum of forty-five. The greatest depth to which a dredge will dig is therefore about seven tenths of the length of the ladder. Large machines ordinarily carry longer ladders than small ones.

Increasing the digging depth of a dredge adds greatly to its cost. The longer ladder demands additional buoyancy at the front to support it and weight at the back to counterbalance it, which means a substantial increase in the size and weight of the hull. The pump may also need re-design and additional power to develop the added suction needed.

A dredge with an oversize ladder, hull, and pump engine with a special pump will have only about the output of a standard dredge of the same size, but the cost might be greater than that of a larger dredge with standard digging depth.

Under very special conditions, ladder angles up to 60° are sometimes used for increased depth.

Plain suction lines or ladders are much lighter and can be extended downward at less cost. A cutter machine under necessity of digging a loose deposit below its reach may have the cutter head removed and an extension suction pipe installed.

Depth of cut can sometimes be increased by partially draining the pond in which the dredge works or by diverting the waste water away from it, so that the digging will serve to pump out water and lower its surface level.

Forty feet is the deepest cut ordinarily recommended, but with special constructions and under favorable conditions depths down to one hundred feet below water surface have been excavated. Fifty feet usually makes an adequate channel for ocean-going vessels.

Cutter Heads. The cutter head, Figure 14-76, is driven through a shaft from a power plant mounted on the deck or on the upper part of the ladder. The lower end of the shaft is supported in a long bushing which is lubricated and flushed by clean water supplied by the auxiliary pump. The shaft may be fitted with a weak section that is easy to replace, or with a shear pin, for protection of the machinery if the cutter jams.

Cutter drives are occasionally reversible to aid in cleaning off vegetation caught in the blades.

The cutter head is fastened by a key and nut and is sometimes steadied by a backing flange. It encloses the suction head and serves to slice, chop, and stir up material so that it is easily picked up by the water flowing between the blades into the pipe.

There are several types of cutter heads. Each of them is made in a variety of sizes, and many in several weight classes as well. One-piece construction is the cheapest, but when worn will have to be built up by welding, or scrapped.

Others use detachable blades or renewable blade edges. Teeth can be welded on the blades or sockets welded to carry renewable teeth. The proper number, shape, and protection of blades for various types of digging will have an important effect on dredge output and cutter life.

Standard rotation is counter clockwise. Cutting is done chiefly on the right (starboard) swing where the blades move upward against the bank.

Large dredges with toothed cutters can dig cemented gravel, hard clay, and soft rock.

When in doubt about the proper cutter head it is usually best to take a heavy type, as reduction in maintenance and down time usually more than offsets the higher price.

Courtesy of Ellicott Machine Corporation

Fig. 14-77. Five drum dredge winch

In hard, abrasive digging it is good practice to keep a complete spare cutter head, as replacing wearing parts may take considerable time, particularly if the head has been sprung or damaged by rough usage.

A traveling chain cutter ladder consists of a double roller chain carrying bottomless buckets or cutter bars. It runs on the top and bottom of the ladder, around the lower end, and is driven from an upper sprocket. It is chiefly used in deposits full of boulders, which it carries back from the digging face where they will not interfere with cutting.

Winches. Plain suction dredges require a power winch to raise and lower the suction line and two to four power or hand winches for anchor lines and miscellaneous lifting.

Courtesy of Ellicott Machine Corporation

Fig. 14-78. Spud wells—and dredge factory

14-62

Fig. 14-79. Spuds and shop carrier

Cutter dredges need at least five drums under individual clutch and brake control. Two control the swinging of the dredge, one the raising and lowering of the ladder, and two the spuds. Additional power or hand winches may be provided for handling barges, shore lines, and materials.

A five drum winch set is shown in Figure 14-77. This is of the layshaft design, the drums being carried on bearings on fixed shafts. Each drum flange carries a spur pinion that meshes with a spur gear that turns freely on the drive or jackshaft, running parallel to the drum shafts. The drive gear can be rotated by engaging it to its shaft by a friction clutch, or stopped by a band brake. This construction provides full control of the drum without subjecting its bearings to strain when it is stationary under load.

Control levers may be at the winch, as shown, or grouped in a pilot house. Mechanical, hydraulic, or air controls are used.

Spuds. The spud wells, Figure 14-78 (A), are two pairs of guide collars mounted on the rear, stern, or spud end of the dredge hull. They are of two piece construction so that they can be opened for convenient installation of the spuds. They are heavily made and securely fastened to reinforcements on the hull as they have to withstand the reaction from the swinging of the dredge.

The spuds, 14-79, are heavy steel tubes, pointed at one end and provided with a lifting band and sheave block at the other. Hoist lines from the winch pass over sheaves in the spud gantry, around the spud sheaves, and are anchored on the tower.

Spuds must be long enough to reach the bottom without passing through the upper spud well. They are used to hold the dredge against digging resistance, wind, and tide and to advance it while working.

One, called the working spud, usually nearest to the discharge line, is down during digging, and provides stability against thrust and a pivot on which the dredge swings. The other, the walking spud, is used as a pivot in moving the dredge.

Canal Dredge. Figure 14-80 shows a

Fig. 14-80. Canal dredge

dredge made specially for construction and maintenance of narrow, shallow canals where spoil can be jetted over a bank or pumped through a short discharge line.

During digging this dredge is held in position by four spuds. The ladder and cutter move independently of the hull. Swing, hoist, and down pressure are hydraulic.

Advancing into the cut requires raising the front spuds, and rocking the rear ones alternately with a walking motion.

Sizes range from 6 to 12 inches, cutting widths in single passes are 12 to 20 feet, and maximum depth from 8 to 12 feet.

Bumboats and Anchors. When working in narrow channels dredges may be manipulated by lines to objects on shore or to anchors dropped in dug holes. Access to the shore may be by the discharge pipe, catwalks, or a rowboat. Floating pipe may be handled by a shore crane.

Under most working conditions, however, a small barge or boat with a hoist adequate to handle anchors and pipes, and an engine to move it around, are required. Sometimes the hoist will be on a barge which is moved by a small motorboat, but it is more efficient to combine the two.

A heavily made boat on the style of a large flat bottom rowboat with an outboard motor for propulsion and a hand winch for the hoist will be adequate for small and many medium dredges. Larger units may require a working deck, inboard power plant, and a power winch.

Anchors are usually of the fluke type, weighing three to four hundred pounds for small dredges. When used for swinging, they are placed beside the ladder, and well off to each side, making certain the cutters will not foul their lines at the ends of the swing. When the dredge has moved forward so that the lines pitch back too much for efficiency, they are picked up and moved forward. If the dredge path is curved the outside anchor is advanced farthest.

Courtesy of Morris Machine Works

Fig. 14-81. Float and discharge line section

Floating Discharge Line. If discharge is into a barge or hopper, a very short pipe with an open flap valve is used.

If the spoil is to be used as fill, heavy pipe takes it to the rear (spud end or stern) of the dredge. A lighter pipe on floats carries it to land where a still lighter shore pipe takes it to the discharge point.

Floating pipe is made of high carbon or alloy steel, generally in twenty to forty foot lengths that have raised rims at the ends. Pieces are connected by short pieces of rubber hose with screw clamps. This arrangement provides flexibility to allow for motion from wind, waves, and advancing of the dredge, combined with great resistance to pulling apart.

Steel ball joints with rubber flanges, or flexible rubber hose, are used to connect to the dredge and shore pipes. The ball joints are used in pairs.

A turn is generally right angled through an elbow with a smooth, continuous curve. It is customary to place one or more turns directly behind the dredge to give the line flexibility to follow easily for at least one pipe length, and to absorb the twisting movement of the dredge.

Floats may be composed of oil drums with a metal or steel frame, of steel tanks as in Figure 14-81 or small wooden barges. Each float normally carries one pipe section. An A-frame or cherry picker hoist with a hand winch may be built into the float nearest to shore to partly support that pipe so as to reduce the vertical angle with the shore pipe. Another hoist may be kept at the point behind the dredge where

A. DREDGE CAPACITY

PIPE SIZE OF DREDGE	CU. YDS. AT 10%	CU. YDS. AT 20%
6"	30	60
8"	60	120
10"	90	180
12"	125	250
16"	225	450
18"	285	570
20"	350	700
24"	500	1000
28"	700	1400
30"	800	1600

B. MAXIMUM LENGTH OF LINE

SIZE OF DREDGE	HEAVY MATERIAL	LIGHT MATERIAL
6"	800 Ft.	1500 Ft.
8"	1000 Ft.	1800 Ft.
10"	1400 Ft.	2500 Ft.
12"	1800 Ft.	4000 Ft.
16"	3500 Ft.	6000 Ft.

Courtesy of Ellicott Machine Corporation

Fig. 14-82. Dredge output tables

the line is opened to install extra sections.

The floats may have a light hand railing for use in walking along the pipe. Telephone and electric cables may be carried beside the pipe, or wires on short poles above the rail.

The pipe wears fastest on the bottom, and its life can be prolonged by rotating the straight pieces one third of a turn occasionally.

"Floating" pipe is sometimes laid on the bottom, underwater, to leave a channel open for shipping.

Shore Lines. Fixed shore lines leading to a permanent discharge point such as a gravel plant are customarily made of heavy pipe with bolted or threaded flanges.

Shore pipe of the standard movable type is of light welded construction, with one end expanded and the other tapered so that adjoining pieces make a sliding fit. Metal is high carbon or alloy steel, 10 to 16 gauge.

Each end carries a pair of shallow hooks to permit fastening the connection with wire or weaving wire rope along a number of sections.

The tapered joints are somewhat flexible so that the pipe can be laid and used slightly out of line. This causes additional back pressure and wear, but is often unavoidable near the discharge end.

The stream may be allowed to spurt directly out of the last pipe, or may be broken up and spread out by various types of baffles or Y- or T-ends which reduce pitting and make a wider fill.

Window valves are adjustable openings in the bottom of sections of pipe. They are used to distribute fill along the line of a pipe instead of concentrating it at the end, in order to produce a smoother grade.

Shore pipe ordinarily lies on the ground or fill surface, but is carried on a trestle of some sort if window valves are used, if a deep fill is to be made from a single discharge point, or if the discharge point is changed by removing pipe sections rather than adding them.

Y-shaped diversion valves operated by a hand crank are used to change the flow from one branch pipe to another.

Output. The output of a dredge, Figure 14-82, Table A, depends on the capacity

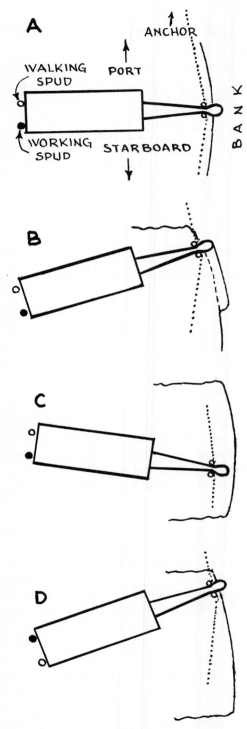

Fig. 14-83. "Walking" a dredge

of the pump, the depth of digging, the height of discharge, the line friction, and the percentage of solids. Pump capacity is expressed in terms of yards per hour of solids. Ability to maintain production against a high head requires increase in power over that needed for a low head.

The vacuum required to raise the load depends on depth and percentage of solids. Aside from friction, it must be able to lift the excess of weight in the suction pipe over the weight of the same volume of plain water, which is the underwater weight of the solids.

The height of discharge and the friction head combine to produce back pressure on the pump. Pumps may be designed to work against either high or low heads.

Recommended maximum lengths of discharge line for nominal heights are given in Table B. Much longer lines may be used, but with diminishing output or increased power requirement.

If the required line length of lift is beyond the capacity of the dredge pump, a booster pump of the same characteristics should be installed in the discharge line.

The most critical factor in output of any particular dredge is the proportion of solids carried. Except under conditions of overload, there is little difference in volume or cost between pumping low and high percentages of solids. Dredge output is usually calculated on a basis of ten percent, but if twenty percent solids can be carried, production is doubled.

Proportion of solids can be increased first by the proper use of the correct type of cutter, and secondly by increasing velocity of flow by the elimination of sharp bends, reducing the size of the discharge pipe or speeding up the pump.

The limit on percentage of solids is the plugging point of the pipe. This will be reached more quickly at low velocities, and with smaller loads of coarse material such as gravel, than with fine soil such as silt, or light stuff such as humus. Plugs are expensive to remove and must be avoided.

OPERATION

Advancing. The dredge is advanced into the digging by the pull of the swing winches and the pivot action of the spuds. One of these, called the working spud, is down while the dredge digs; the other is used for advancing. The swing lines go around sheaves near the end of the suction ladder to anchors placed at each side.

In Figure 14-83, the dredge is headed directly toward the bank with the working spud down. Pull on the left (port) swing line, as in (B), moves the cutter in an arc to the left, pivoting the dredge on the working spud. Pull to the right (starboard), (C), pivots the dredge to the right.

These movements are made with the pump operating and the cutter head revolving. Firm material is taken in slices from the top down by lowering the ladder at the end of the swing. Flowing or sliding soil may be dug at the toe only.

When the material within reach of the cutter has been dug to bottom grade, the ladder is lifted until the cutter is above the top of the bank and the dredge swung to starboard, in the (C) position. The walking spud is dropped, the working spud raised, and the dredge pulled to the left, as in (D). This causes the dredge to pivot on the walking spud, moving the working spud forward a distance determined by the extent of the side swing in (C) and (D).

The working spud is then dropped, the walking spud raised, and digging resumed.

Spuds must penetrate the bottom sufficiently for a firm grip. If the bottom is hard extra heavy spuds may be required, and it may be necessary to raise and drop one of them several times in one place in order to sink it.

Anchors must be raised and moved forward occasionally to avoid back pull from the swing lines.

Cutting. The standard dredge does its hard digging while swinging toward starboard, with the cutter blades slicing up into the bank. The swing back to port is usually made with the cutter at the same level, cleaning up loosened material.

The revolving of the cutter causes it to act as a driving wheel, pushing the ladder to the left. This push will be weak in loose sand and strong in hard formations. On the right swing the winch must overcome this resistance in addition to crowding the blades into the digging.

When swinging to the left, the cutter may pull the ladder too rapidly for effective cleaning and cause the winch line to run slack, and foul on the cutter. In this case the right hand winch brake is applied enough to hold the cutter back to line speed.

Dry banks are undermined by cutting below the water level so that they slide or cave into the water where they can be picked up by the suction. The height that can be safely reduced this way depends on how readily it slides and the size of the unit. If too much material comes down at one time, it may bury the ladder or damage the dredge.

It is usual for a dredge to cut to bottom grade at each stand, regardless of the depth of material to be moved. Moving the dredge with the attendant work of shifting anchors and extending discharge lines is laborious.

The percentage of solids moved is regulated by the size of the slice taken with each pass and the speed with which it is cut. The slice can be enlarged by advancing the dredge further toward the bank or by lowering the ladder. Cutting speed depends primarily on the hardness of the material and is controlled by the speed of the swing line, which can be reduced from maximum by throttling the engine, shifting to a lower gear, or slipping or intermittently disengaging the clutch. If the winch cannot pull at full speed production may be increased by turning the cutter faster.

The dirt cut or stirred up by the blades falls inside the cutter head or is carried around the outside. Water pulled in from the top and the left side by suction of the pump further breaks up the pieces and carries them up the suction pump.

A skillful operator will keep the intake of solids as high as their nature and the velocity of his discharge line permit without plugging. But he should not allow the line to plug as cleaning it out is liable to be a long and tedious job involving down time that will far outweigh the gain of extra few percent of solids that caused it. A blocked line is particularly serious in freezing weather.

A vacuum indicator connected to the suction line indicates roughly the percentage of solids since these are heavier than water and a higher vacuum is needed to lift them. A plugged intake will cause vacuum to rise sharply. This reading may also be affected by the angle of the ladder, being somewhat lower when taking off the top of a high bank than when working on the toe.

A pressure gauge attached to the discharge pipe indicates the head against which the pump is working. This will include the lift from the pump to the discharge opening, and the friction against the pipe walls, which is ordinarily a function of the length of the pipe and the velocity of the stream. If the line threatens to plug discharge pressure will rise abruptly as the stream backs up behind the obstruction.

A threatened plug may be cleared out by alternately speeding and slowing the pump, or raising or backing off the cutter so that clean water or a leaner mixture will be pumped. When working with heavy loads it may be good practice to run the pump at somewhat less than full speed so that extra force will be available to break through obstructions.

If the pipe does clog, the plug may be located by observing the small leaks usually found in shore pipe connections, or by tapping the tops of pipes with a light hammer. They will have different tones when full of water under pressure, or mud, and when partly empty.

Whenever a part of the pipe is to be cut off from the dredge to be taken apart, or when the dredge is to be shut down for more than a few minutes for any reason, plain water should be pumped until the discharge line is clean. Mud in a pipe will make it difficult or impossible to handle, and may dry out and cake in an unused line so that water will not move it again.

If a single discharge opening is used the dredge pump must be stopped whenever a new section is to be added.

The dredge operator is usually unable to see much of the discharge area, but he must fit his actions to its needs. Some quick signaling system must be used for efficient operation. Flags, telephone, and radio are suitable for different conditions.

A dredge is generally run on a two or three shift basis, because of its high original cost, and desire to spread its overhead over as many work hours as possible. It will often be run through lunch periods by a split crew.

Clearing. Vegetation should be cleared away from the path of a dredge as thoroughly as possible. Cattails, reeds, and other swamp growth may wind around the cutter head, causing constant delays in removing them. Stumps and logs may jam or break the cutter, or block the pump or line.

When the cut is across dry land, brush, trees, and stumps should be completely removed and heavy sods plowed and disked. If it is in tree or stump filled water,

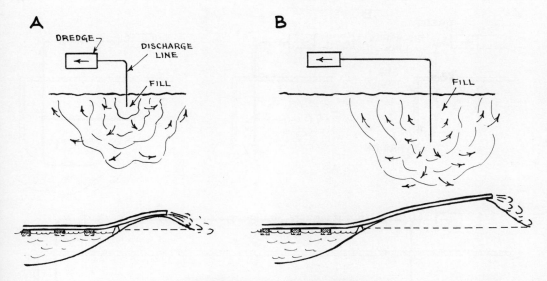

Fig. 14-84. Open hydraulic fill

it should be preceded by a snag boat, generally a grapple dredge. This will raise the stumps and logs, blasting them when necessary, and swing them outside of the cutting line or load them in barges for removal.

Discharge Pipe Patterns. Dredge discharge lines can be rather readily extended at either end but are difficult to move. Floating pipe is unwieldy to manage in long towed sections, particularly in a wind, and it is a slow job to disassemble and relocate it piece by piece. Shore pipe must always be taken apart for moving. It should be flushed out with clean water first to lighten it.

As the dredge advances, the discharge pipe must be lengthened. This is done by disconnecting at a joint immediately behind the hull, placing another section by means of a light floating crane, and refastening.

In view of the difficulty of moving and the desirability of keeping the number of curves to a minimum, patterns involving a few right angle turns and considerable extra footage of pipe are in general use. These are shown in Figures 14-84 and 14-85.

Wherever possible, two or more discharge openings with control by diversion valves are used. This system permits adding pipe without losing work time by stopping the dredge.

For simplicity, a balanced amount of cut and fill is assumed in these diagrams, so that the fill advances about the same distance as the dredge moves. This favorable condition will be found when the bottom is used merely as a borrow pit so that depth or width of cut can be varied to keep pace with the fill, and when the disposal area is large or deep enough to easily accommodate the spoil from specified channel work.

If a natural balance does not exist the depth of the fill may be varied to provide for disposal of greater or less yardage. This is a matter that should be carefully calculated in advance, so that seawalls or berms (earth dikes) may be built to correct height. Maximum depth of a single layer of hydraulic fill is limited by the pressure exerted on the dikes.

If the spoil is coarse or mixed, additional layers may be added by using the technique to be described for building

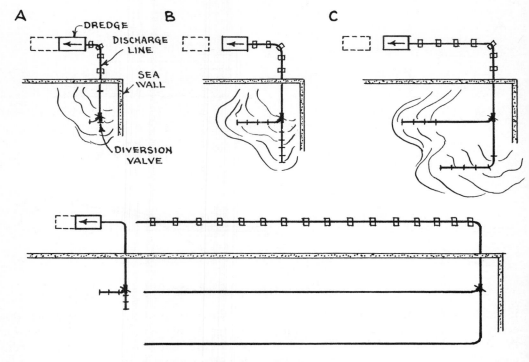

Fig. 14-85. Discharge pipe pattern

dams. The extra height of the upper layers may put additional load on the pump, which would have to be speeded up to maintain full output. However, the lift may be more than balanced by the use of lines shorter than those needed to spread the same material over a larger area.

Handling Pipe. The method of handling discharge pipe depends largely on the firmness of the fill surface. On quick draining sand or gravel a pile of pipes may be pushed up to the discharge end by a bulldozer, which can also place each pipe in a position to be engaged by hand. Shovel dozers, or cranes or cherry pickers mounted on crawlers or half tracks, can also be used for handling.

On softer fills, pipe may be dragged in place by winch lines or handled by a dragline on platforms. More often, men in hip boots roll pipes along the discharge line to its end and place them by hand. If footing is very bad it may be necessary to

lay boards on each side of the line to support the men. Supporting boards may also be placed under the pipes.

Fills are likely to be somewhat firmer in the immediate vicinity of the discharge than further away, because coarser particles tend to settle out first.

Pipe joints are often rather out of line when connection is first made. They are straightened out two or more lengths back from the discharge opening, usually by prying into better position and blocking up. A straight line reduces friction on the load and wear on the pipe.

Leaks at joints are quite common. They can sometimes be stopped by tightening the connection or prying the pipes into better position. They can be checked by forcing in the thin ends of shingles. Those on the bottom block themselves with spoil without requiring attention unless the pipe is on a trestle.

In general, pipes lying on the ground

do not need to be hooked together as the friction in the joint and with the ground holds them together. However, on curves and at changes in grade fastening is necessary.

Individual joints are fastened by winding wire around the pipe hooks, which are placed in line with each other, and twisting it tight with a short stick which is left caught against the pipe. On curves only the outside hooks need fastening.

If the line is on a trestle each joint must be fastened on both sides.

A number of joints may be fastened at a time with a cable. It is looped and clamped, or fastened in any manner to both ends of the section to be strengthened, put under the hooks on one side of the first joint, over the top of the pipe to be caught, under the opposite side on the next, and on alternate sides to the other anchor. The cable becomes tighter with each joint, and if the correct slack has been allowed, can be slipped over the last few joints only by means of a bar. Another method is to leave it loose enough to go over all hooks easily and then use a load binder to tighten it.

Light cable, three eighths or half inch, is most satisfactory, but heavy, worn out swing lines are often used.

HYDRAULIC FILLS

Hydraulic fills may be built by a dredge as wasting operations to dispose of unwanted spoil from channel dredging, they may be a secondary but valuable part of the work, or the dredging may be done merely to get the fill.

Open Fill. The simplest type of open hydraulic fill is shown in Figure 14-85. Spoil is being wasted on an area large enough so that the water deposits practically all its burden when allowed to flow freely away from the pipe. A small percentage of the soil is fines which remain in suspension and drain back into the dredging area.

The end of the discharge pipe is supported on a low, light horse, or a log, or any object that will keep it above the fill. It may or may not have baffles to spread the stream. The stream will first erode slightly where it strikes the ground, but deposit the heaviest particles immediately around it and lighter ones farther away. In

Courtesy of Ellicott Machine Corporation

Fig. 14-86. Filling behind a wall

this manner a cone will be built up. Average slopes may be as steep as 15 percent for chunky material or as flat as 3 percent for silt. A high content of solids will give a steeper slope than a light burden of the same material. The action of the flowing water will sort and pack in the pieces.

When the fill has built up to grade or to a point where it threatens to interfere with the stream, another horse is set up, a pipe length is brought forward, the stream is diverted to another line, or a signal is sent to the dredge operator to stop the pump, and the pipes are connected.

The surface may be a slope up from the start of the fill, or level. In either case it should be made as smooth as possible to simplify keeping pipe gradient and to save future grading work. If the material is so coarse that it tends to build up a series of cones, the pipe may be set on a trestle and window pipes used. If the openings can be closed a series of them can be used with open ports only in the last one or where needed. If the ports are simply slots the window pipe must be taken off and replaced at the end of the line when it is lengthened.

Settling Basins. If the proportion of suspended particles in the runoff water is large enough to constitute a serious waste or to cause undesirable silting where it is discharged, or if the edge of a fill is to be steeper than the natural slope the water must be retained or ponded in the fill area long enough to permit settlement. This is done by construction of dikes across the natural direction of flow. These may be permanent seawalls or berms thrown up by a dragline or dozer.

Berms must be of ample height and heavy enough not to be pushed out by pressure, even when softened by soaking. Dirt must be thoroughly temped or puddled around the spillway.

Spillways are usually of wood.

The amount of settlement obtained is determined by the turbulence in the settling pool, the length of time the water remains in it, and the fineness of the particles. The first two factors are largely fixed by the proportion between the area and depth of the pool, and the rate of inflow. As the fill progresses the area will be reduced. This may be counteracted for a while by raising the water level by putting more boards in the outlet gate, but ultimately additional land will have to be inundated if settlement is to be obtained.

Where the bottom being dredged includes known areas of fine and coarse soils the fines may be worked first while a large settling area is available, and coarse deposits used to complete the fill.

Dams. The velocity of the soil-water mixture discharged from a dredge pipe drops rapidly after it hits the ground because it spreads over a widening area and the water soaks into the fill. The stream may immediately lose its ability to carry cobbles, drop the gravel a few feet farther on, then the sand. Varying quantities of silt and clay may be carried after the water comes to rest in a settling basin.

This results in a gradation of material, from coarse at the pipe through finer sizes with increasing distance. The separation is not clean, as some fines are trapped near the pipe and some coarse particles are rolled or carried by occasional currents well into the edges of the cone.

An earth dam must have the impenetrability of clay and the slump resistance of gravel. One way to meet these requirements is the construction shown in Figure 14-87. A water resistant core of clay, silt, and fine sand is supported and protected by gravel sides.

This cross section can be obtained by the proper selection and placing of hydraulic fill. The bank material must have a higher proportion of fines than needed in the dam structure to allow for the amounts left in the coarser material

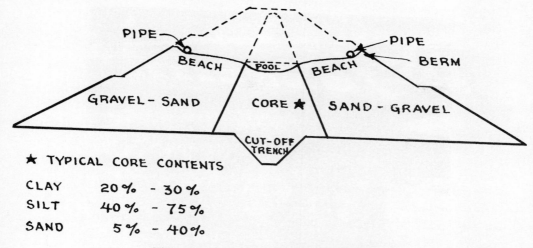

★ TYPICAL CORE CONTENTS

CLAY   20% - 30%

SILT   40% - 75%

SAND   5% - 40%

Fig. 14-87. Cross section of earth dam

and that carried away in the waste water.

The dam is built up in a series of low fills. Berms are built by dragline or dozer along the outer edges of the fill and discharge lines run inside them. The spoil builds up in a slope or beach from the outer line down to a central settling basin called the core pool. Coarse particles remaining in the stream are ordinarily deposited immediately upon entering this calm water. Occasionally, concentrated runs of water may build sand or gravel tongues out into the pool. This possibility is guarded against by anchoring beams or other floats just offshore where they will break up such formations.

The core pool water is drained into a pipe line or flume which is usually on one of the outer slopes, and is built up as the dam level rises. The width of the core is regulated by that of the pool.

CLEANUP DREDGE

The Mud Cat, Figures 14-88 and 14-89,

Courtesy of Mud Cat Division

Fig. 14-88. Clean-up dredge

14-73

Courtesy of Mud Cat Division

Fig. 14-89. Suction head auger and mud shield

is made in two models designed especially for cleaning mud deposits and weed growth out of shallow ponds and canals.

Its carrier is a specially designed barge, 30 or 39 feet long and 8 wide, which draws only 21 to 27 inches of water. A pair of side pontoons, connected by a floor and end bulkheads, are filled with polyurethane foam. The space between them is occupied by a 175 horsepower diesel engine, a dredge pump with an 8-inch inlet and a 6-inch discharge, and accessories, controls, and a cab.

The suction boom has a hydraulic hoist. It carries an 8-foot wide auger driven by a hydraulic motor. The auger is partly enclosed while working by a mud shield. Single layers up to 18 inches thick may be removed at one pass. Maximum working depth is 10½ feet for the smaller machine, 15 feet for the larger.

The dredge is propelled by a capstan winch and a 5/16-inch diameter cable. The cable is placed ahead of time, and anchored. After each pass it is moved 6 to 8 feet and re-anchored. This arrangement minimizes the equipment needed on the dredge, and can assure complete coverage of the area.

The suction line and the mud shield together assure that most of the material loosened by the auger or the shield will be drawn up into the dredge, so that little turbidity (which may be injurious to fish and other pond life) is caused, in comparison with draglines and other excavators.

The auger is able to chop up water weeds and reasonable amounts of trash. If jammed, it can be lifted to the surface and cleaned. The shield tends to support it against digging too deeply into the bottom.

The pump has a capacity of 2000 gallons per minute against a 155 foot head, and can transport a half mile or more without a booster. Removal of solid material may be at the rate of 50 to 120 yards an hour.

SPOIL SEPARATION

In dredging cleanout of ponds in built-up areas, two critical problems may be getting enough water to supply the pump and float the carrier, and obtaining adequate sites for settling basins.

Both problems may be reduced by use of mechanical and/or chemical means to accelerate separation of water and solids.

Water Supply. An enclosed body of water may be sucked dry by a dredge, unless it is large or is fed by generous stream flow, or its water is returned to it after separation of the solids. This subject is discussed on page 6-40.

Re-use or clarification of water may be desirable or required in order to avoid damage to drainageways or other property from overflow of a settling pond.

Processing that accelerates separation of water and solids favors both efficient recycling of the water, and disposal of any surplus.

Cyclone. The cyclone, Figure 14-90, is a machine that is widely used in mineral processing plants to de-water finely ground

Courtesy of Dorr-Oliver

Fig. 14-90. Cyclone, cutaway

material during or after processing. It has no moving parts, and is operated by the force of the water-solids mixtures that are pumped into it.

In the design shown, mixed water and solids enter a cylindrical chamber at a tangent. Pressure of about 22 pounds per square inch is sufficient to give the fluid high rotary speed. Centrifugal force presses solids against the outer wall, along which they move in a downward spiral, through a conical section, to an underflow or discharge gate. Water, with any solids too fine or too light to separate from it, overflows through a pipe at the top.

Adjustments for sharpness of separation

and/or thickness of underflow mud may be made either manually or automatically during operation. Efficiency in separating fine or light particles diminishes as machine size increases.

Figure 14-91 shows data on four standard models. It will be noticed that capacity is small in relation to dredge output. For example, it would take four 18-inch cyclones to process spoil from an 8-inch dredge with a 2000 gallon per minute flow.

Their separation range of 42 to 55 microns would enable them to remove all gravel and sand, a substantial fraction of silt, and some clay, with a variable proportion of organic material. Unfortunately, they tend to leave in the water precisely those materials that take longest to settle, and which demand the largest quiet water area for separation.

However, under many conditions they remove most of the bulk, outside disposal of which can make it possible to use a settlement basin for a much longer time.

Solids from the underflow usually contain enough water to make a flowing mud, which may have to be channelled, dragged, or pumped to a drying area, or hauled away in tight-bodied dump trucks.

Drying Equipment. Water may be almost entirely removed from mixtures with almost any soil by centrifuges or by vacuum separators. However, these machines have a rather small output, may be tricky to operate, have a high power requirement, and are very expensive.

There seems to be little prospect that they can be successfully applied to dredging operations in the near future.

Clotting. Organic mud, with or without included fine mineral particles, takes days and sometimes weeks to settle out of water naturally. But addition of certain chemicals, called polymers, causes such material to form into relatively large clots which sink promptly, leaving clear water in 20 to 30 minutes.

Their use is experimental at this time. Excellent results appear to follow injecting the polymer into the dredge outlet pipe about 20 feet short of its discharge point.

HYDRAULICKING

Hydraulicking may be said to be a dry land dredging operation. Banks are washed down by high pressure water jets and the mixed spoil and water moved by gravity, pumps, or both, to a fill. It is used in building hydraulic fill dams when sufficient material cannot be obtained by dredges, for breaking down banks, cleaning rock for blasting, and for various special conditions.

Plant. Hydraulicking equipment consists of water pumps, suction and discharge lines, monitors, and a flume or sluiceway to carry the mixed water and soil.

The centrifugal water pumps may be high pressure jet types, or medium pressure set in series with one pump's discharge supplying the intake of another. Usual pressure is between 80 and 150 pounds, but 275 may be used.

Steel or iron pipe is used except near the monitors, where flexible hose is required.

Monitors, also called giants, are large, high pressure nozzles mounted in swivels on heavy skid frames. Small sizes may be aimed by hand, large ones have a movable baffle which utilizes the power of the stream of water to turn the nozzle.

The sluiceway may be a gully of controlled width in the pit floor, a flume of wood or metal, or dredge discharge pipe. A gully should have a slope of at least 4 percent, but flumes can be somewhat flatter. A narrow deep waterway will give maximum velocity and transport the greatest bulk of solids.

The high pressure water may do the transporting or additional water may be supplied by low pressure pumps or gravity flow.

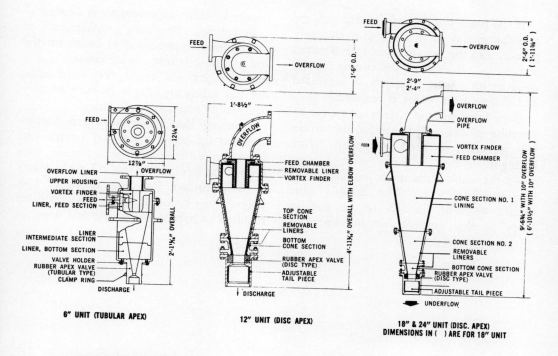

Courtesy of Dorr Oliver

Fig. 14-91. Cyclone details

OPERATING DATA

DIAMETER OF UNIT* INCHES	CAPACITY GPM	NORMAL OPERATING PRESSURE PSI	SEPARATION RANGE** MICRONS
6″	47-110		
12″	185-390	22	28-40
18″	520-1100	22	36-46
24″	710-1520	22	42-55
		22	50-68

*36″ and 48″ units are available on special request.
**Separation range is based on 2.7 specific gravity solids at a feed concentration below 25% by weight.

The sluiceway runs into a distributing pipe which takes it to the fill. This pipe should start high enough on the slope to give sufficient head to push the spoil through it.

Hog Boxes. If the grade is not sufficient to carry the spoil from the pit to the fill, it is poured into a hog box, which is a concrete trough from which a dredge-type pump moves it through a pipe line to a fill.

When gravity feed is not possible and the hydraulic fill method is required on the dam, dirt may be dumped through grizzlies into the hog box by trucks, washed into a sump by monitors, then pumped to the fill.

Operating a Monitor. The primary requirement of dam hydraulicking is the correct selection of material. Definite proportions of fines, sand, and gravel are usually required, and these are often not found in a single bank.

Several monitors may have to work in different sections, or in different pits, and increase or decrease their part of the output on the direction of the job inspectors.

If the bank will cave if undermined, the monitor should be as close to it as safety permits and the stream should be directed at the base. If most of the cutting can be done at an angle, efficiency is increased and danger from slides is reduced.

If the bank will not cave it is best cut by standing at such a distance that the stream will just start to bend downward before hitting it.

As the bank recedes the monitor must be moved forward. The flexible hose should be long enough to be laid in loops so that considerable movement is required before an extension is needed.

Sometimes draglines, bulldozers, or other machines are used to break up the soil and push it into the sluiceway.

Water jets will erode very hard soils, but not efficiently. Light blasting of clay will greatly increase production.

Cleaning Rock. Fire hose and nozzles supplied by small centrifugal pumps can often be used to advantage in cleaning thin layers and pockets of dirt off rock which is to be drilled. Pressures of 20 to 100 pounds may be used. Several men will be required to handle one nozzle and hose.

CHAPTER FIFTEEN

TRACTORS AND BULLDOZERS

TRACTORS

CRAWLER TRACTORS

Industrial crawler (track laying) tractors have a weight range from 3 or 4 tons to over 50 tons, without dozers or other equipment. Maximum horsepower is over 500.

Figure 15-1 shows a typical direct drive crawler tractor. Figure 15-2, also a cutaway, is a view of an older machine with the parts labeled. Each is made up of a center section or chassis that contains the engine, transmission, and steering units; and two track frames that supply traction and support.

Engine. The engine is usually diesel, although in the smaller sizes, it may be gasoline fueled. Power is rated first as net engine horsepower, meaning the net horsepower at the flywheel with the engine driving all accessories normal to tractor operation. Some manufacturers occasionally stray from this standard, and advertise stripped or gross horsepower without accessories.

The second standard of measurement is drawbar horsepower. This is a smaller figure that represents usable power at the drawbar under certain set conditions, after deducting losses in friction and slippage. This is the usual rating for tractors with direct mechanical drive.

Power may also be measured in terms of

pounds drawbar pull, a factor that is limited by traction and that may be increased by mounting a bulldozer or any extra weight. Crawlers with power shift and/or torque converters are generally rated in pounds of drawbar pull at a given speed.

Clutch. Crawlers with mechanical drive have an engine or flywheel clutch with one or two discs, that is used to cut off the drive train when stopping the machine or shifting gears. When the machine is to be used to carry a loader or for any other very heavy service, the discs may be set with long wearing, heat resistant ceramic discs; or be kept in a circulating bath of oil that reduces wear and takes away heat.

The clutch is usually controlled by a hand lever at the operator's left. It is moved forward to disengage the clutch, and pulled back to engage it. It has an over center locking device to hold it engaged.

Large machines may have a hydraulic booster to reduce operator effort and assure sufficient engagement pressure.

Torque Converter. Most large crawlers now have a torque converter instead of, or in addition to, an engine clutch. This device, described in Chapter 12, provides shock-absorbing slippage between engine and tracks, multiplies torque so that fewer transmission gear ratios are needed, and

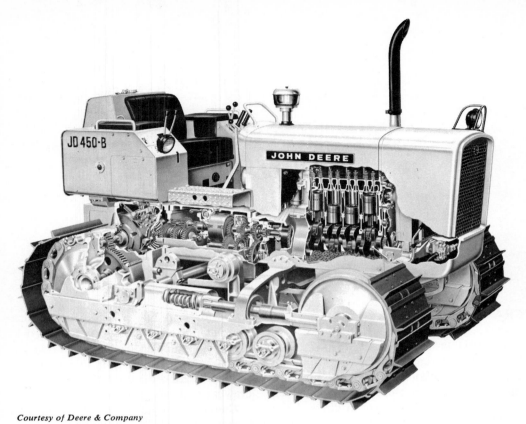

Fig. 15-1. Crawler tractor

allows on-the-move power shifting among those ratios that are used.

Center Frame. Rigidity of the center section is obtained by making the steering clutch and transmission housings in one heavy casting or weldment with internal braces, together with heavy construction of all cases forward to the radiator. In addition, a pair of heavy side beams may run forward from the transmission case, supporting the engine, the crankcase guard, and the radiator base.

A front pull hook may be bolted to the crankcase guard, which is a plate protecting the bottom of the engine. Use of this hook for heavy pulls is a greater strain on the tractor than use of the drawbar. In old tractors without side beams, extreme loads on the hook may break the flywheel housing and spring fastenings, and pull the engine out of the tractor.

Transmission—Manual Shift. This type of transmission, used in machines with friction clutches, is compact, with very heavy construction. Shifting is often done by sliding spur gears, but newer machines may have constant mesh helical gears with synchronized shifting.

The number of gear speeds varies in different models and may be from two to eight forward speeds, and one to six reverse speeds.

When more than one reverse speed is present the transmission is often compound, with two sets of shiftable gears in series, operated by two levers. In such a unit, one lever may be used for selecting most of the gears, while another will shift only between a high and a low range, or between forward and reverse.

When one lever takes care of only the shift between forward and reverse, it may

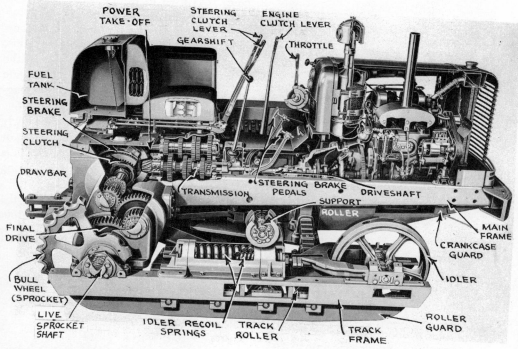

POWER TAKE-OFF

STEERING CLUTCH LEVER

ENGINE CLUTCH LEVER

GEARSHIFT

THROTTLE

FUEL TANK

STEERING BRAKE

STEERING CLUTCH

DRAWBAR

TRANSMISSION

STEERING BRAKE PEDALS

DRIVESHAFT

SUPPORT ROLLER

FINAL DRIVE

MAIN FRAME

CRANKCASE GUARD

IDLER

BULL WHEEL (SPROCKET)

LIVE SPROCKET SHAFT

IDLER RECOIL SPRINGS

TRACK ROLLER

TRACK FRAME

ROLLER GUARD

Courtesy of Fiat-Allis Construction Machinery, Inc.

Fig. 15-2. Crawler tractor, labeled

be called the reversing lever. It may provide two way movement in all gears, but the higher gears may be locked out of reverse.

A reversing lever usually provides a faster shift than a lever that controls speed change also, as it has straight line motion, and the gears are usually synchromesh. However, the engine clutch must be disengaged to shift it, and it is not as fast or easy as a reversing clutch or other power shift devices.

The engine clutch must be disengaged for any conventional gear shift. There may be a lock that will prevent any gear from disengaging unless the clutch is fully released.

There is a universal joint in the shaft from the clutch to the transmission to prevent damage if these two units should get slightly out of alignment.

The drive shaft from the rear of the trans-

mission ends immediately in a bevel gear that drives the live axle ring gear.

The transmission and bevel gears operate in a sealed box that contains sufficient oil

Courtesy of Caterpillar Tractor Company

Fig. 15-3. Underside of tractor

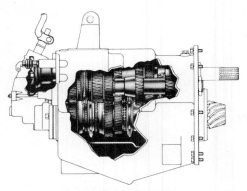

Courtesy of Caterpillar Tractor Company

Fig. 15-4A. Manual shift transmission

or fluid grease to lubricate all the gears and bearings by dip, splash, or pressure spray.

Transmission—Power Shift. Tractors equipped with torque converters usually have power shift (shift-on-the-go) transmissions with two or more speeds in each direction.

The shifting is done by pairs of friction clutches connecting and disconnecting gears in the drive line, or by brakes controlling planetary gear sets. The shifting lever may control these by mechanical linkage, or hydraulic means.

Power Takeoff. The standard power takeoff is a connection that will turn a shaft inserted through the rear wall of the gear

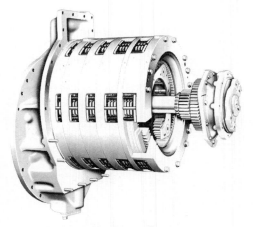

Courtesy of Caterpillar Tractor Company

Fig. 15-4B. Power shift transmission

case. It is used to power accessories such as cable control units, a winch, or a hydraulic pump. If the transmission is compound the takeoff may have two gear ratios, otherwise its speed is controlled only by the engine. It is engaged by releasing the engine clutch, meshing a sliding jaw, and re-engaging the clutch.

The takeoff usually turns more slowly than the engine. It operates in neutral or any gear, but not when the engine clutch is disengaged. It is not affected by the steering clutches.

Lack of power in the takeoff when the engine clutch is released, or when the output shaft of a torque converter is slowed by

Courtesy of Caterpillar Tractor Company

Fig. 15-4C. Final drive

heavy load, results in inefficiency in many operations. There are therefore an increasing number of constant running "live" power takeoffs, that can be operated whenever the engine is running, regardless of clutch position or converter action.

Such a takeoff may be driven by a quill or inner shaft inside the clutch and transmission main shafts, that is not affected by clutch action. Or it may be at the front, driven from an extension of the engine crankshaft, or on the engine and driven from the timing gears or the flywheel.

Live takeoffs may have their own clutches or disconnect devices, but often do not.

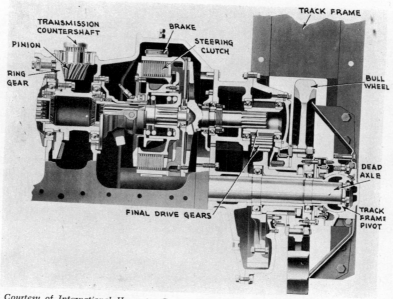

Courtesy of International Harvester Company

Fig. 15-5. Rear axle and final drive

Rear Drive Assembly. The rear drive or live axle assembly for clutch and brake steering is shown in the cut-away drawings in Figures 15-2 and 15-5.

At each side of the ring gear, the axle extends through a section of the case which is kept free of grease by means of seals on the axle and by a drain hole in the bottom. The holes must be plugged while working in mud or water. In each of these compartments is a multiple disc clutch and a band brake. This pair of clutches and brakes are used for steering in the same manner as those on the shovel live axles but, being always of the friction type, they are smooth in operation.

Next to the steering clutch compartment on each side is the final drive. This is a dip lubricated gear set of either single or double reduction construction. The large outer gear is attached to a short axle which turns the bull wheel (sprocket) that drives the track. Grease seals are used where the axle enters the final drive case, and where the bull wheel drive leaves it.

Dead Axle. The dead axle, or pivot shaft, is a hinge pin which runs across the back of the tractor, through or forward of the final drive cases. It ties the track frames and center section together but allows them to oscillate vertically. In addition, it usually serves as an axle for the bull wheel.

This dead axle may be a continuous shaft, or two shafts in line with each other, with their inner ends anchored to the transmission case.

Bull Wheel. The bull wheel is a big sprocket of very heavy construction. The wheel itself is usually a flat disc, widening to shallow teeth at the rim. The hollows mesh with the track pin bushings, providing positive drive to the track. The number of hollows may not divide evenly into the number of track pins, so that the bushings always contact different hollows on successive trips around. This hunting tooth design equalizes wear.

The wheel and sprocket teeth may be one piece, or the teeth may be bolted on in 3-hollow sections. This provides for replace-

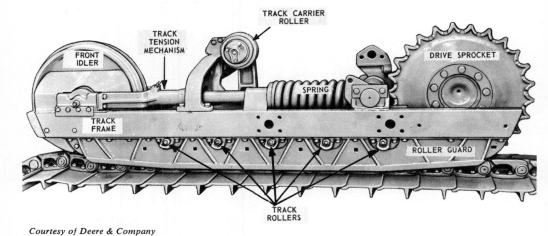

Courtesy of Deere & Company

Fig. 15-6A. Track frame assembly

ment without the major job of removing bull wheels. It also allows right-for-left exchanges of segments, if wear is one-sided.

Tracks. The track frame and mechanism is similar in general appearance to that described in Chapter 13.

The live axles turn large toothed wheels, called drive sprockets or bull wheels, that

Courtesy of Caterpillar Tractor Company

Fig. 15-6B. Sectional bull wheel

are at the rear of the truck frames (track frames). The frames rest on the small truck or track rollers. The idlers, which are smoothed flanged wheels similar in size to the drive sprockets, are mounted on spring cushioned yokes at the front of the frames. One or two small support rollers are mounted above each frame, except in very small machines, to prevent excessive sagging in the upper track section.

The track itself consists of a true roller chain and bolted-on shoes. The parts of a track chain are shown in Figures 15-6A, 6C, 6D, and 15-7. Each link pair is fastened together with a bushing at one end. A pin goes through the bushing and holds the overlapping ends of the next pair of links. The track is assembled on a hydraulic press, that is able to force the slightly oversize pins and bushings into the links so forcefully that they very seldom work apart in service. The pin turns easily in the bushing, providing the necessary hinge action.

This track can be opened (broken) by driving out any one of the pins. However, all the pins except one may be too tight in their links to make this practical in the field. If this is the case, the single master pin is very slightly smaller, so that it can

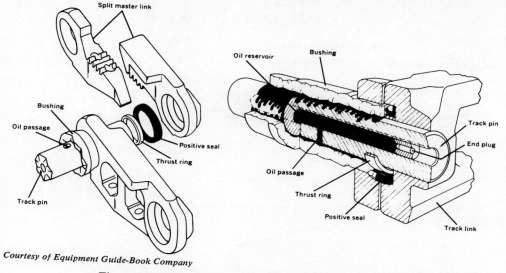

Courtesy of Equipment Guide-Book Company

Fig. 15-6C. Split master link, and lubricated track section

be driven out with a sledgehammer and drive pin. It is usually longer than the others, so that it can be recognized. Track pin steel is so hard that it is brittle, and the eyes should be protected against flying chips while hammering.

Another type of track chain uses longer bushings that project into the outer links, so that driving out the pin does not separate the track. In this construction, one short bushing is used between spacing washers around the master pin, so that the track can be opened at this point and at no other.

It is a common custom to cut one-piece master links with a torch when the track must be opened in the field, then replacing them with new ones when re-assembling. The split master link may be taken apart by loosening bolts.

Some tracks have seals at the ends of the bushings, to prevent or reduce entrance of dirt that grinds away their parts.

The inner surface of the rail or track chain is shaped to provide a smooth track to support the rollers and idler. The outer surface is flattened and provided with bolt holes—two to a side link—for fastening on the shoes or pads.

The bushings provide a grip for the bull

wheel. Usually the wheel has an odd number of teeth and the track an even number of bushings, or vice versa, so that the same bushing and socket do not connect twice in succession. This hunting tooth design distributes wear evenly.

Certain types of stiff mud, and wet snow, may build up in the track and in the sprocket hollows so that the sprocket will spin, usually with abrupt and damaging stops and starts, and will probably make the track over-tight at the same time. This condition may make work difficult or impossible, as repeated hand cleaning may be required. Ice and mud shoes usually have openings in the center that permit the

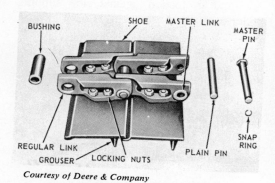

Courtesy of Deere & Company

Fig. 15-6D. Track links and shoes

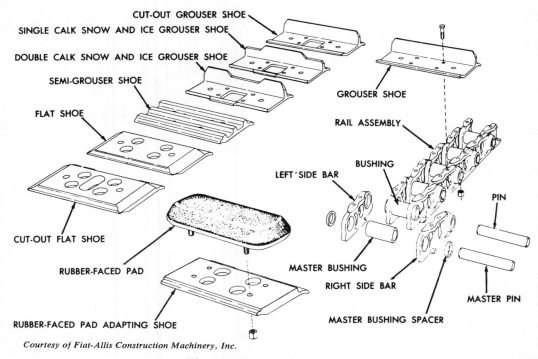

CUT-OUT GROUSER SHOE

SINGLE CALK SNOW AND ICE GROUSER SHOE

DOUBLE CALK SNOW AND ICE GROUSER SHOE

SEMI-GROUSER SHOE

FLAT SHOE

CUT-OUT FLAT SHOE

RUBBER-FACED PAD

RUBBER-FACED PAD ADAPTING SHOE

GROUSER SHOE

RAIL ASSEMBLY

BUSHING

LEFT SIDE BAR

PIN

MASTER BUSHING

RIGHT SIDE BAR

MASTER PIN

MASTER BUSHING SPACER

Courtesy of Fiat-Allis Construction Machinery, Inc.

Fig. 15-7. Track links and shoes

sprocket teeth to force the snow out through them, leaving the inner parts comparatively free. There are also cutaway sprockets that allow mud to squeeze through openings in the tooth bottoms.

There are a great many types of track shoes, a few of which are shown in Figures 15-7 to 9. The standard construction is a flat plate with a single high cleat or grouser across it. This affords good traction and protection against side slipping under most conditions, but will not grip on

Courtesy of International Harvester Company

Fig. 15-8. Ice shoes

ice or frozen ground, and tears up surfaces on which the machine works. If the tracks spin, each grouser acts like a bucket on a ditching machine, taking dirt from beneath and piling it at the rear. As a result, the machine may dig itself down into trouble very rapidly on soft ground when heavily loaded. In addition, such piles make backing up in the same path a very rough trip.

Flat shoes are used on machines that usually carry rather than push, and those that work in unpaved yards that would be cut up by the cleats. They have been used on bulldozers and shovel dozers, but usually do not give enough traction and permit a dangerous amount of sideslipping.

Rubber-face pads are used by crawler tractors that work inside buildings and on paved roads. Traction is better than flat shoes on hard surfaces, and scuffing and scarring are reduced to a minimum. They are usually not rugged enough for heavy pushing.

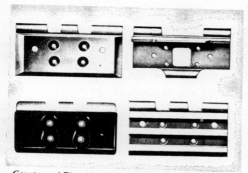

Courtesy of Fiat-Allis Construction Machinery, Inc.

Fig. 15-9. Special track shoes

Courtesy of Fiat-Allis Construction Machinery, Inc.

Fig. 15-10. Single flange track roller

Semi-grousers are flat shoes that carry two to three low cleats. They are the most satisfactory equipment for front loaders, as they do not dig up the ground in spinning and turning as much as full grousers, without reducing traction and stability as severely as flat shoes.

Snow and ice plates usually feature cutout or skeleton shoes, and high cleats.

Conversion attachments are also obtainable. Metal or rubber pads can be bolted over grousers to protect streets, and grousers can be bolted on flat or semi-grouser shoes. Most track shoes have six bolt holes—four to fasten the shoe to the track, and two for attaching extra pieces.

Pads of most of these types can also be obtained in various widths. Wide ones provide flotation for work on soft ground and give maximum traction; narrow ones provide easier steering and a slippage factor which reduces damage from overload and shocks.

The tracks are usually set quite close to the center section. In special wide gauge models, however, they are set farther away. The additional spacing may vary from a few inches to a foot or more. This change requires special final drive cases and shafts, and changes in braces and springs.

A wide track machine has much better stability on side slopes, and in carrying high loads, it fouls less with mud, and it

is possible to mount wide shoes on it for swamp work. It is heavier, is generally more clumsy to maneuver in restricted spaces, but can turn with a load with less difficulty than a standard model, and does not get stuck as readily.

Track Wheels. The rollers and idlers are flanged in order to keep the track in line. The idler customarily has a wide center flange that fits between the track links. The track (truck) and support rollers have outer flanges that are on each side of the track rail. They may also have an inner flange. On the bottom, it is usual to alternate single and double flanged rollers.

Rollers and idlers revolve on fixed axles. They may have tapered roller bearings, or solid sleeve types made of bronze or spe-

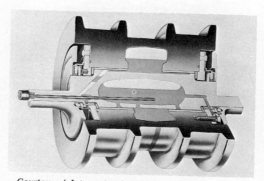

Courtesy of International Harvester Company

Fig. 15-11. Double flange track roller

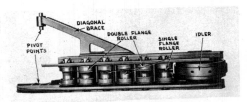

Courtesy of Fiat-Allis Construction Machinery, Inc.

Fig. 15-12. Track frame and rollers

cial metals. Good seals, to keep lubricant in and dirt out, are very important. The most successful type, the "positive" seal, consists of two finely machined rings, one attached to the axle and one to the hub, which are pressed against each other by springs. These rings fit so perfectly that neither dirt nor grease can get past them, and are hard enough to outlast other wearing parts.

Figures 15-10 and 15-11 show track rollers of single and double flange types, with roller and solid bearings. Figure 15-12 is a set that includes single and double flange construction mounted in the track frame.

Worn rollers are built up by welding on special rotating jigs.

Adjustment. The big idler is held in position by a sliding yoke backed by a spring, or by a cylinder of compressed nitrogen, which pushes it forward in order to keep tension in the track. If the track collides

with something, or an object is caught between the track and the idler, or sand or snow builds up on the sprocket or rails, the idler can move back by compressing the spring, thus absorbing the shock, or relieving the tension on the track.

Adjustment for track wear is made by turning the adjustment bolt (screw) that is threaded into the idler yoke and turns freely on the spring. Screwing it into the yoke pulls the idler back and loosens the track; screwing it out of the yoke pushes the idler forward and tightens the track. See Figure 15.13. This adjustment does not affect spring tension.

Threaded track adjustments are usually very troublesome. Packing in of fine dust,

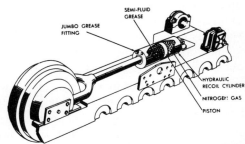

Courtesy of TEREX *Division, General Motors Corporation*

Fig. 15-14. Hydraulic track adjuster

broken down lubricant, and rust makes them very difficult to turn, and they often are hard to get at and provide little grip for a wrench. It is usual to have to employ penetrating oil, heat, and six foot pipe extensions on a wrench to get them to turn.

Hydraulic track adjusters usually consist of a piston in a tight cylinder that has a grease fitting. The piston rod is an extension of the idler yoke. Pumping grease into the cylinder pushes the idler forward and tightens the track, bleeding grease through the fitting or a relief valve allows the idler to move back and release the track.

Hydraulic adjusters are very easy to use, but any leakage in them is likely to put the tractor out of action.

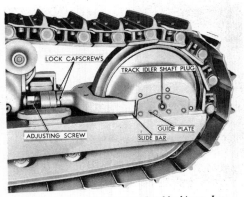

Courtesy of Fiat-Allis Construction Machinery, Inc.

Fig. 15-13. Mechanical track adjuster

The tractor should be moved backward and forward a few times during adjustment to equalize upper and lower tension.

A properly adjusted track should sag a little at the top when the machine moves forward.

Track Frame Fastenings. In many crawlers, the track frames pivot individually on the line of dead axle that runs across the rear of the tractor, usually between the bull wheel centers. The frame is extended straight back for an outer pivot point, and a diagonal brace extends inward to hinge near the center line of the tractor.

The front of the tractor center section may be supported on the side frames in a number of ways. It is desirable that the tracks be allowed to oscillate — that is, to have independent up and down movement at the front — as this improves riding comfort, accuracy of dozer work, and traction on uneven ground.

There are two types of oscillating connections—spring and stabilizer bar. The spring is usually a heavy leaf type, whose ends rest on the track frames, and whose center supports the central frame. This system is losing popularity because of the trend toward placing dozer hoists on the hood or radiator. Springs tend to compress and allow the center frame to move down when the dozer blade is lifted, and to move back up when it is dropped, interfering with accuracy of control.

The stabilizer is a rigid bar hinged to the track frames at its ends, and to the center frame. It allows oscillation without spring action.

Crawler tractors that are to carry loaders usually have a cross beam that is rigidly fastened to the track frames and the center frame, preventing any oscillation. This construction improves stability against tipping, and makes it possible to base the loader solidly on both track frames.

Drawbar. The drawbar is a heavy steel tow bar, fastened under the center of the

tractor, extending backward across a support bracket, and projecting to the rear. It can swing horizontally and is held in the desired position by a pin or bolt through the bracket or by a pair of bolts in a clamp. It is desirable that the anchor of the drawbar be as low and as far forward as the construction of the tractor will permit for best distribution of stress.

The one shown in Figure 15-15 has a safety catch to hold both the swing and the load pins.

The biggest tractors use a non-swinging drawbar rigidly attached to the frame.

Courtesy of Fiat-Allis Construction Machinery, Inc.

Fig. 15-15. Drawbar, with pin lock

Controls. The standard operating controls for a small gear shift tractor, Figure 15-16, consist of the main or flywheel clutch lever, which is held in the forward disengaged position by a spring, and locks over center in engaged position when pulled all the way back; the right and left steering clutch levers which are held in forward engaged position by springs, are released by pulling back, and do not lock over center; the gear shift lever, the right and left brake pedals which are depressed against springs to engage the brakes, and

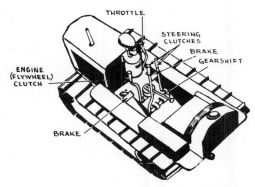

Fig. 15-16. Operating controls, small tractor

which may be locked down by moving the lock levers forward while the brakes are engaged.

Larger machines may have hydraulic boosters on some or all of the controls to offset the greater effort required by large clutches and brakes.

A decelerator pedal permits the operator to leave his hand throttle at full speed position, and still slow the engine as much as he wishes for accurate work, safety or comfort crossing rough ground, or noise reduction when receiving instructions. Pressing the pedal slows the engine, releasing allows it to return to the speed required by the throttle setting.

The decelerator is used in crawlers be-

Courtesy of Fiat-Allis Construction Machinery, Inc.

Fig. 15-17A. Controls, gear drive

cause their low ground speed makes it safe and usual to operate them with the hand throttle set at full working engine speed, which has to be cut back to idling only occasionally.

Wheel tractors doing the same work have accelerators instead. An operator must develop discrimination when changing from one to the other, in regard to when the pedal should be depressed.

OPERATION—MECHANICAL DRIVE

Starting the Engine. Instructions for starting diesel engines will be found in Chapter 12. The special problem in a tractor (as well as in many other machines) is no-start interlocks.

Most newer tractors have a live starter circuit only when the gearshift is in neutral. If there are two shift levers, either one may be the key.

There may also be a switch in the battery case that must be in connected position to allow the starter to turn.

Allow the engine to warm up at part speed for two or three minutes before operating. You can spend this time checking the tractor for condition and signs of abnormal wear or behavior.

Starting the Tractor. A conventional crawler tractor, Fig. 15-16, may be operated as follows: Release the engine clutch by moving it forward, shift into the gear to be used, set the throttle for the desired engine speed, pull back on the clutch lever until the machine starts to move or the engine starts to work, then pull slowly until the machine is moving and the clutch is carrying the full engine power without slipping, then pull it back rapidly until it locks over center. Starting with heavy loads, or in high gears, requires a longer slow pull than loads easily moved. The pull should be slow enough to avoid jerking or jumping the tractor, but any hesitation beyond this minimum will increase wear.

Some engines will operate efficiently

only if the throttle is wide open or nearly so. Others will perform reasonably well at any speed. On the whole, fast engine and tractor speeds are suited to rough, heavy work and fast traveling on smooth ground, and slow to moderate speeds for precise work, moving in close quarters and walking on rough ground.

The top speed of most crawlers is from four to seven miles an hour on a level, but this is in a high gear which has little reserve power for climbing hills or pulling loads. It is expensive to operate at top speed, as wear on the track chains is greater per mile of travel at high speeds, but the time saved or work performed will often more than justify the cost.

Gear Selection. For precise work, the lowest gear is desirable, not only because the slow pace gives more time to steer and make adjustments, but because the bulldozer blade or other unit will move faster in proportion to the speed of the tractor. This is because the blade speed is in proportion to the speed of the engine, while the tractor speed is varied by the gear ratio. The blade can therefore cut a grade more accurately in low gear than in second, even if the tractor speed is the same.

For less exacting work the gears should be the highest that can be used without lugging the engine below its governed speed, and which will give the amount of control needed.

Gear shifting is slow and cumbersome in some tractors because of the spur gearing used, the hand throttle setting which makes double clutching impractical, and the fact that friction in the tracks slows it so rapidly that considerable operator skill is required to complete a shift before the machine stops. The gear used throughout is therefore usually low enough to start the load, although a higher gear might move it once it is underway.

An increasing number of machines use constant mesh gears which shift more eas-

ily when standing, and which can be shifted on the move.

Speed may be increased or decreased by opening or closing the throttle, or by choice of gears. A decelerator pedal will slow the engine below throttle setting, when pushed down against a spring.

Reversing. To change direction of travel of a standard crawler, first bring it to a full stop. With the clutch released it will stop by itself in a few feet, except when going down a grade. It may be stopped in inches by its load, or by applying a brake.

In an ordinary transmission, you move the shift lever through neutral into the desired gear, then re-engage the clutch.

If the tractor has a reversing transmission, there will be two levers, one to control direction only. This has three positions, forward, neutral, and reverse. It may enable you to use all forward speeds in reverse, or it may have an interlock to keep it out of reverse position in the higher gears. Each gear is usually somewhat faster in reverse than forward.

The reversing transmission may use sliding spur gears, which are usually somewhat

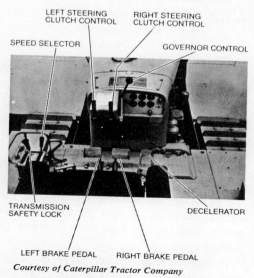

Courtesy of Caterpillar Tractor Company

Fig. 15-17B. Controls, converter drive

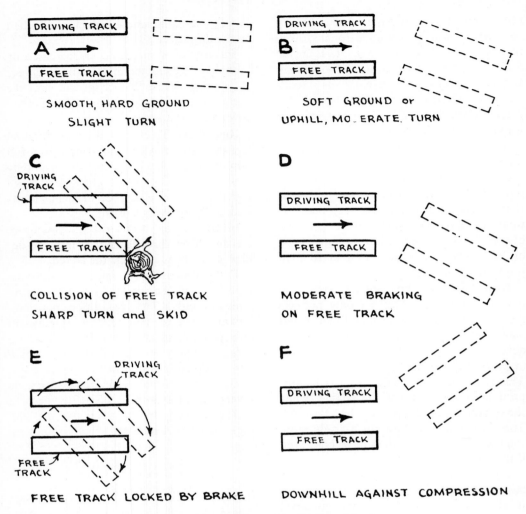

Fig. 15-17C. Steering with clutch and brake

clumsy or even difficult to shift, but should have helical gears with a sliding hub, which are faster and easier; or a pair of friction clutches.

Steering. Five factors are involved in steering a conventional crawler tractor— the effect of the steering clutches, of the steering brakes, of the footing, of the grade, and of the load being pulled or pushed.

If one steering clutch lever is pulled back while the tractor is moving forward, the track on that side will run idle and the other will continue to drive. The amount of turn in an unloaded tractor will depend on the resistance of the idle track to moving forward. Its resistance is made up of internal friction, which by itself will produce only very gradual change in directions and ground drag. As a result, on a smooth hard surface the machine will turn very gradually. In soft ground it will turn more sharply. If the non-driving track collides with a stump it will swing abruptly. See Figure 15-17C.

Rolling resistance in the free track can

be increased to any desired degree by applying its steering brake. If the track is locked, and traction is good, the machine will spin around it. Brake application can produce any degree of turn between this and the non-braked gradual change in direction. Braked turns are seldom smooth, as direction is usually changed in a series of steps.

When moving uphill, the free track is held back by gravity, and the weight of the machine also increases the tendency to turn, as shown in (B). Other conditions being equal, less braking will therefore be required for steering when going uphill than on the level.

If the ground is soft, both tracks will dig in. The driving track will tend to spin and dig; the braked track to push earth at the front in the direction of turn, and oppositely at the rear. At best turning consumes some power, and under soft conditions the resistance built up by the displaced earth may be great enough to prevent the turn and will put severe strain on the tracks. Under such conditions turns are best made in a series of short jerks, to give the tracks a chance to work away from the dirt ridges before turning further.

Since the machine drives from one track while being steered, only half of its normal traction is available. This complicates maneuvering in soft ground and with a load.

A heavy load, either on the drawbar or in front of a blade, has somewhat the effect of an upgrade in supplying a turning force when drive is from one side only; the load itself resists being turned. In general, a loaded tractor will steer more readily than a free one in response to a steering clutch only, and with greater difficulty in response to clutch and brake. Very sharp turns are usually difficult or impossible but the location and character of the load influence this.

Even with all clutches engaged, a tractor with a heavy offcenter load will tend to turn toward the loaded side.

A tractor so loaded that it is using nearly its full traction in straight going will have difficulty in turning, as the driving track will be inclined to spin. It may be necessary to relieve the load by backing in order to change direction. This is usually fairly easy with a pushed load but difficult with a pulled one.

Steering by means of clutches alone works backwards in going down a hill steep enough to make the tractor try to roll faster than the engine drives it. The effect of releasing a steering clutch in this case is to allow its track to roll free, and move faster than the other track, which is braked by the engine, so that the machine will turn away from the released clutch. Firm application of that brake will cause the machine to turn toward it in the normal manner. If precise steering is required, the brake should be applied before the clutch is released.

Steering in reverse is the same process as steering forward.

Use of either steering brake while its steering clutch is engaged will slow, stop, or hold the machine, but has no steering effect, as while the clutches are engaged the axle shaft acts as a solid unit from track to track.

There are crawler tractors which differ from the type described in important respects.

OPERATION—WITH TORQUE CONVERTER

Figure 15-17B shows a typical set of controls for a crawler tractor with torque converter and power shift.

In this example, steering clutches and brakes are similar in location and action to those on a mechanical drive machine.

Shift. This gear shift (speed selector) lever moves in a U-shaped slot, opening to the rear. The cross slot is neutral, the

Courtesy of Caterpillar Tractor Company

Fig. 15-17D. U-shift

right side is forward, the left side reverse.

There are two speeds in each direction. In other machines there may be three, or even four. Positions are in line, so that the lever moves through low in moving between neutral and high.

Notches on the side bars, and/or detents under the lever, should hold it in position, and provide the operator with feel for it.

An additional lever, on the side of the U, is moved forward to lock the shift lever in neutral. This lock should be set whenever the operator gets off the machine. An interlock prevents use of the starter when this lever is in operating position.

To start work, you first start the engine. Set the hand throttle to desired engine speed (indicated on the tachometer on the dash), which in most jobs is between ¾ and full rpm.

Unlock the speed shift, and move the shift lever to the desired direction and speed. This can be done at full throttle in most tractors. If this is undesirable, either because of instructions to the contrary, or because the machine is going through a jerky-

shift phase, slow the engine with the decelerator pedal during the shift.

Steering is done in the same manner as in a gear drive model.

Varying Speed. A torque converter causes the relative speeds of tractor and engine to vary widely. Under full throttle a heavy load may slow the machine to a creep or even stop, and with a light load or no load it will reach its maximum speed in the gear range being used.

The loss of speed involves an increase in pulling or pushing power, because of the torque multiplication explained in Chapter 12. It saves the operator the responsibility of shifting to a lower gear, lightening the load, or even slipping the clutch, which he would do in a gear drive model. Under most conditions, this slowing is an efficient and desirable response to increase in load.

Speeding up under light load is also efficient, but it is not always as welcome, as it may make it difficult or impossible to do precise work with a blade, bucket or implement. Fortunately, it can be controlled with the decelerator pedal, which will slow both engine and tractor to the desired rate of motion.

A machine might have an output shaft governor on the torque converter. This will keep the tractor, or a mounted unit such as a winch, at a steady speed regardless of load, as long as its requirement does not exceed engine power.

For this, there are two throttles, one controlling a standard engine governor, the other the output shaft governor. It is usual to keep the engine throttle wide open, and use the output governor to regulate speed of operation.

Proportioning Power. The proportioning of engine power between the tractor and its attachment, which is usually hydraulic-powered, is largely governed by the slippage factor in the torque converter.

Slowing the engine permits increased slippage in the converter, so that at idling speed

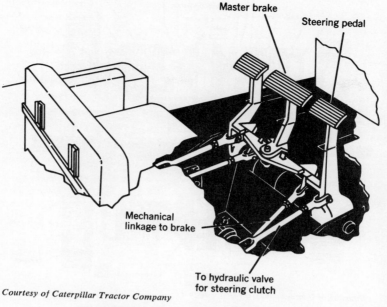

Master brake

Steering pedal

Mechanical
linkage to brake

To hydraulic valve
for steering clutch

Courtesy of Caterpillar Tractor Company

Fig. 15-18. Combination pedal controls

little or no power is transmitted. But the parts driven by the hydraulic pump will be slowed only in almost direct proportion to the engine, and will keep nearly full power.

As a result, decelerating will usually stop tractor motion while allowing attachment action to continue at reduced speed. The uses of this will be discussed in Chapter 16, under loader operation.

DIFFERENT CONTROL SYSTEMS

Control levers and pedals may be arranged in a number of different ways in various makes and models, as in the combination levers below.

There are also several basically different control systems, which call for changes in operation.

Combination Controls. Small tractors, and those with power boosters, may have a steering clutch lever or pedal also operate the steering brake. The first half or two-thirds of its travel disengages the clutch, then further movement contacts and applies the brake linkage.

This simplifies operation, but in many

machines makes it impossible to use the brakes to slow the tractor without sacrificing the braking effect of the engine.

This disadvantage may be overcome by adding another pedal, which applies both brakes for stopping or holding the machine, without affecting the clutches. It may have a lock-down attachment for parking.

Reversing Clutch. Forward-reverse shifting may be done by means of a pair of multiple disc clutches, which is the smoothest and most convenient method. In the past, there have been a few tractors that used such a reversing clutch as the main or engine clutch.

Figure 15-19 is a view of one possible layout. The engine crankshaft turns the main clutch shaft through a flexible connector. The shaft drives a clutch such as that shown in Figure 12-40.

A shift yoke between the two sets of discs can engage either set to the hub. One clutch spider will engage a countershaft gear for forward, the other an idler gear to turn the countershaft oppositely.

In power shift machines, the same result

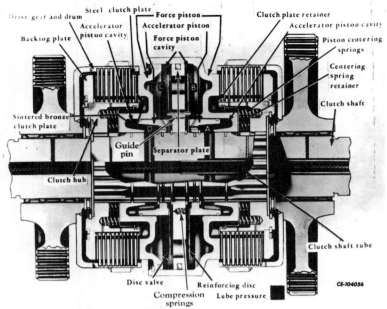

Courtesy of International Harvester Company

Fig. 15-19. Reversing clutch

is accomplished in much less space inside the main transmission, using multiple disc clutches and either planetary or countershaft gears.

From the operator's viewpoint, almost any clutch shift is to be preferred to a gear shift. For full usefulness, however, it should be able to be moved directly (although softly) from one direction to the other, without throttling down or waiting for the machine to stop.

Since this requires substantial extra smoothness or strength (and cost) in the clutch, many machines require stopping the tractor, and/or slowing the engine, before reversing.

Reversing clutches of various types are used in shovels, rollers, motor boats, and many other machines, and are made in capacities up to 1000 horsepower.

Planetary Steering. The International TD-25 uses a pair of two-speed planetary transmissions instead of steering clutches. This arrangement, shown in Figure 15-20, makes it possible to shift into a lower gear ratio without clutching, and to drive the·

tracks at different speeds so as to turn under full power.

The planetaries are controlled by brakes on three discs on each side. Application is by fingertip levers and hydraulic power. When the levers are forward the axles are in high drive. When they are in the middle gear ratio is increased about twenty-five percent. When fully back, the outer axles are disconnected from the drive and brakes are applied.

If the left lever is forward, and the right lever in the middle, the left track will move faster and the machine will turn to the right with power on both tracks. If the right lever is pulled back farther, the machine will turn with a sharpness depending on the amount the brake is applied, in the same way as a standard crawler. The left planetary may be in either high or low.

Left turns are made by reversed positions of the levers.

A foot pedal applies both disc brakes mechanically for downhill or for emergency operation and for use as a parking brake when the engine is not running.

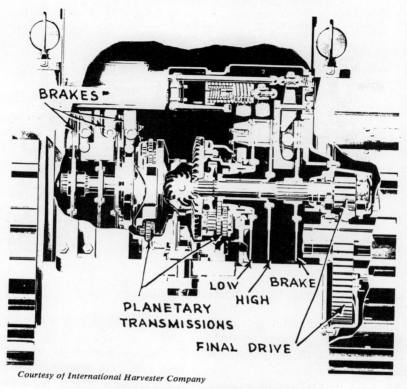

Courtesy of International Harvester Company

Fig. 15-20. Planet power steering

In gear drive models, the engine clutch is controlled by a hand lever. There are two gear shift levers, one for forward-reverse, the other for four speeds.

The high-low planetary ranges operated by the clutch levers combines with these gears to provide eight speeds, both forward and reverse.

In torque converter models, the power shift transmission has two speeds each way.

Differential Steering. The Oliver line of crawler tractors, now no longer in production, provide another type of live axle and steering control, shown in Figures 15-21 and 15-22. Drive is through a special spur gear differential. There is a pair of steering brakes, but the steering clutches are absent.

The spider shafts are extended beyond the differential to planet gears meshed with sun gears which rotate freely on the axle shafts. The sun gears are rigidly attached to brake drums.

If either brake is applied, the planets will start to walk around the attached sun gear, and the resulting rotation of their shafts and spiders will turn the nearer axle gear oppositely from the rotation of the whole assembly. This axle gear will continue to rotate in the same direction, but its individual movement will cause it to turn more slowly than the assembly, the amount of difference being determined by the force with which the brake is applied.

The spiders are also meshed with the spiders of the other axle gear, and will turn these so that it is accelerated by the same amount that the first one is slowed. The track on the braked side will therefore be driven at a lower speed than the other and the tractor will turn to that side.

This effect was formerly obtained by a

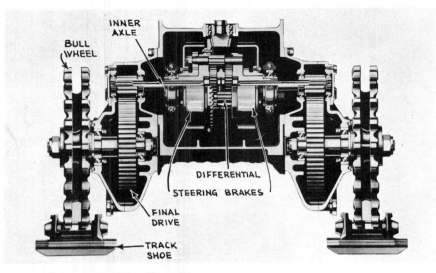

Courtesy of The Oliver Corporation

Fig. 15-21. Differential steering mechanism

brake on a drum fastened to the axle shaft itself. The planetary gearing provides greater braking effect with less effort and lining wear.

The differential, brakes, and final drives are sealed in a single case that is partially filled with oil. Brake linings are special type for this service.

The sharpness of the turn depends on the effectiveness of the brake. A short turn may be made, but usually it is not possible to make it spin on a locked track. This

Courtesy of The Oliver Corporation

Fig. 15-22. Details of steering differential

steering method is not affected by downgrades.

This mechanism is somewhat simpler than that first described and has several advantages. Turning is more smoothly done, tendency to bog down while turning on soft ground is much less because both tracks are being turned at all times, trouble with lubricant leaking from transmission or final drives onto steering clutches and brakes is eliminated, and it is not necessary to plug drain holes in the steering housings before working in mud or water.

There are also several disadvantages. It will not turn as sharply, which hampers it in close quarters. Also, in a turn, the speed of the free track is increased in proportion to the slowing of the braked one and its power is accordingly decreased. This increase of speed is equivalent to shifting into higher gear, and occurs automatically at a time when the tractor must overcome turning resistance, and needs more power rather than less.

It is therefore frequently necessary to shift into a lower gear in the transmission in order to make the turn, a complication which arises less often with steering

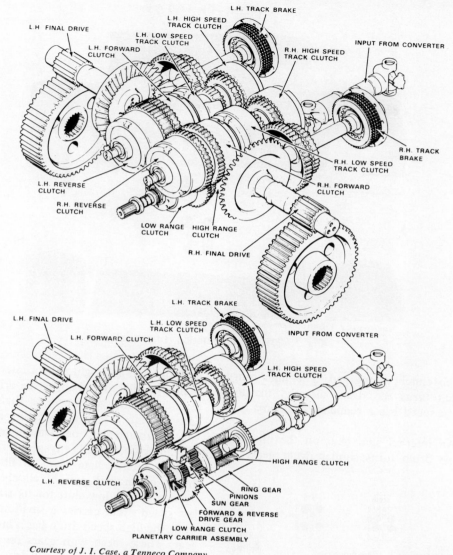

L.H. FINAL DRIVE

L.H. FORWARD
CLUTCH

L.H. LOW SPEED
TRACK CLUTCH

L.H. HIGH SPEED
TRACK CLUTCH

L.H. TRACK BRAKE

R.H. HIGH SPEED
TRACK CLUTCH

INPUT FROM CONVERTER

R.H. TRACK
BRAKE

R.H. LOW SPEED
TRACK CLUTCH

L.H. REVERSE
CLUTCH

R.H. REVERSE
CLUTCH

LOW RANGE
CLUTCH

HIGH RANGE
CLUTCH

R.H. FORWARD
CLUTCH

R.H. FINAL DRIVE

L.H. FINAL DRIVE

L.H. FORWARD CLUTCH

L.H. TRACK BRAKE

L.H. LOW SPEED
TRACK CLUTCH

INPUT FROM CONVERTER

L.H. HIGH SPEED
TRACK CLUTCH

L.H. REVERSE CLUTCH

HIGH RANGE CLUTCH

RING GEAR
PINIONS
SUN GEAR
FORWARD & REVERSE
DRIVE GEAR
LOW RANGE CLUTCH
PLANETARY CARRIER ASSEMBLY

Courtesy of J. I. Case, a Tenneco Company

Fig. 15-23A. Independent track drive

clutches. It is sometimes necessary to use a brake to prevent one track from spinning in slippery going.

Independent Track Control. A crawler can have independent drive to its two tracks, providing not only high-low but also forward-reverse shift for each.

Such a mechanism can add greatly to maneuverability, and is an advantage in close quarters and on soft ground. But it necessarily adds to complexity and cost.

The Case 1150B and 1450 Crawlers have a modulated clutch-shifted transmission. High-low range clutches affect both tracks. Each track drive has its own high-low and forward-reverse shifts. See Figure 15-23A.

A drive shaft from the engine through a torque converter drives a pair of hydraulically controlled multiple disc clutches inside two drums, which will rotate at shaft speed or at a lower speed, depending on

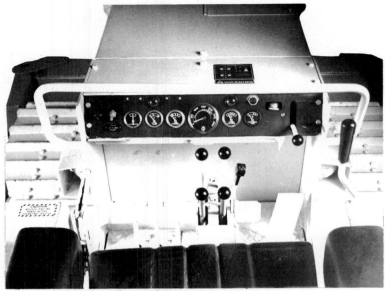

Courtesy of J. I. Case, a Tenneco Company

Fig. 15-23B. Control station, 1450 tractor

which clutch is engaged. The high range clutch turns the drum directly; the low range speed has a reducing planetary gear set.

An internal spur gear on the high-low range drum turns similar drum gears on

Fig. 15-23C. Independent-track shift patterns

Courtesy of J. I. Case, a Tenneco Company

each of a left and right pair of intermediate shafts, parallel to and above the drive shaft assembly. These intermediate shaft drums each contain a pair of reversing clutches, which turn the shaft one way or the other, depending on which is engaged.

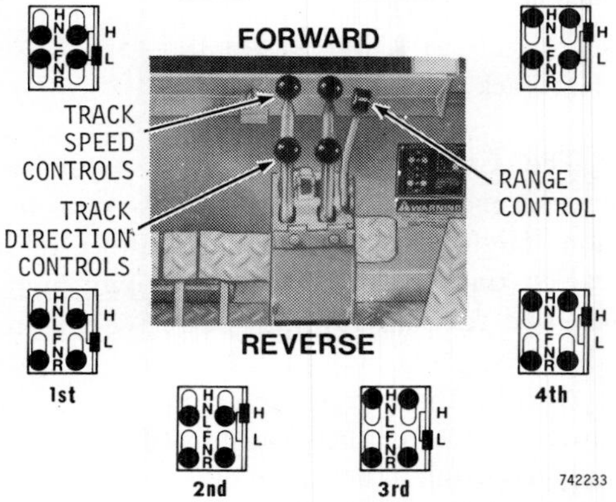

Each intermediate shaft is fitted with drums of another pair of clutches which provide a high-low shift for its track. The low speed drum carries a small spur gear, while the high speed drum has a larger spur gear. These mesh with gears on an outer shaft, which drives the final drive through a hypoid pinion and ring gear set. This shaft carries the multiple disc power track brake.

Controls are shown in Figure 15-23B and C. The hand controls consist of a set of five levers. The right hand lever controls the high-low range, the upper or forward pair controls the individual high-low track speeds, and the lower two control the individual forward-reverse clutches. These levers may be shifted on the go at any engine speed, individually or simultaneously.

The left and right steering brake pedals

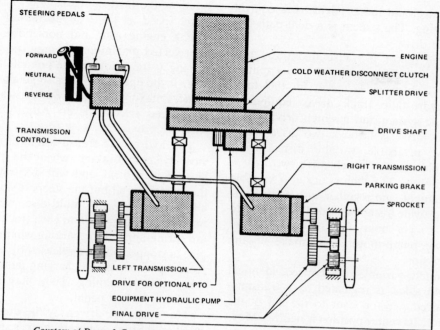

Courtesy of Deere & Company

Fig. 15-24. Hydrostatic track drive

power-operate the two brakes separately. Partial application releases the track clutch, further movement applies the power disc brake. A pedal between the steering brakes, called a power transfer pedal, acts in the same manner on both brakes through an equalizing device. The power transfer pedal provides fine metering of clutches and brakes when precise maneuverability is required in delicate work or tight quarters. The machine can be "inched" forward or reverse while maintaining full engine RPM for hydraulic functions.

A "manual" foot pedal operated by the right foot applies both brakes simultaneously without disengaging the clutches. This allows the operator to utilize both the brakes and engine compression to slow or stop the tractor.

This transmission provides the operator with means to supplement the torque converter with a choice of four gear ratios, two in the main drive, the other two in the track drives, in both forward and reverse. All shifts may be done at full throttle and machine speed.

Steering may be done by any one of four methods. By releasing the track clutch on one side while the other side continues to drive, a gradual turn can be made. A brake or pivot turn is made by depressing one of the steering brakes, the degree of turn varying with firmness of brake application and traction for the other track.

A power turn is made by shifting one track into high track speed, the other in low track speed. The machine will turn toward the low speed track, with power on both tracks, as described earlier for planetary steering.

For a counter-rotation turn, one track is driven forward while the other turns backward. The machine can then make a complete turn in its own length, making it highly maneuverable.

Hydrostatic. A line of John Deere crawlers now being introduced has a hydrostatic system for both fully automatic transmission control and independent track control for steering. The system is a dual-path hydrostatic drive.

Engine power is transmitted through a cold weather disconnect clutch to a splitter drive, which provides power for the dual-path hydrostatic track drive, equipment hydraulic system and a winch drive.

Track drive is through two drive shafts and two reversible, variable displacement hydrostatic pumps which drive two hydrostatic motors. The motors are also variable displacement, and together with pump variation, provide a stepless speed range of 0 to 6.5 miles per hour, forward and reverse. The two pump-motor pairs make up the transmission.

Infinitely variable travel speed, identical for both sides, is regulated by a transmission speed selector lever to the left of the operator. Its center position is neutral. Moving the lever forward causes the tractor to move forward with travel speed increasing with distance from neutral. Reverse is selected by moving the lever rearward from neutral with the travel speed increasing with the distance from neutral. As with conventional machines slow speeds yield maximum push and faster speeds yield less push.

The transmission is fully automatic. The speed selector lever is set at the maximum travel speed desired and the automatic control regulates the speed, reducing travel speed as the load increases and increasing travel speed as the load decreases, maintaining engine rpm and horsepower near the selected maximum, keeping the tractor moving at the quickest rate possible for the load. If the operator wishes to go slower he reduces maximum speed with the transmission speed selector lever. Adjustable detents are provided.

The hydrostatic transmissions also provide dynamic braking when the selector lever is in neutral, and will stop and hold the tractor on all but the steepest slopes. In addition, there are multi-disc wet type brakes in the gear train to each track. These are spring-applied for parking when the engine is stopped and are released by pressure from the transmission charging pump when the engine is running. They may also be applied by a foot pedal.

The tracks are driven by the dual transmission through separate gear reductions. There are no steering clutches or brakes. Steering is accomplished by independently varying the speed and direction of the right and left transmission. The optional steering controls are either two hand levers or two foot pedals, one for the right transmission and one for the left. Since the transmissions are infinitely variable in forward and reverse, steering is stepless enabling live power turns of any degree of sharpness as well as counter rotation. Both give added productivity in some operating conditions. Counter rotation is obtained by moving the steering control past the neutral position. Increased resistance is felt on the control when in the counter rotating position.

Air Motor Drive. The air or electric powered Eimco 630, a small crawler tractor used chiefly in tunnel work, has a separate reversible motor for each track. Power is obtained through a flexible hose or cable, and drive is through reduction gearing to the sprockets. Either track can be turned forward or backward or stopped without affecting the other track.

These machines are well adapted to work

Courtesy of Eimco Division, Envirotech Corporation

Fig. 15-25A. Crawler tractor with air power

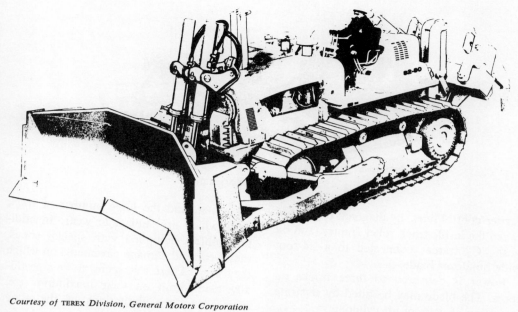

Courtesy of TEREX *Division, General Motors Corporation*

Fig. 15-25B. Twin engine tractor with U-blade

loaders or dozers, and to carry drills and other equipment, underground where space is limited and exhaust gas is undesirable. Their range is limited to areas where air or electricity can be supplied to them, and where the lines do not interfere with work too seriously.

This type of track drive, with air motors, is used in most crawler mounted drills.

Twin Engines. The Terex 82-80, shown in Figure 15-25B carrying a U-blade dozer, has two 220-horsepower diesel engines mounted side by side. Each of them drives one track, through its individual torque converter, power shift transmission, and planetary final drive.

The two sides of the tractor are built into separate frames that are allowed limited oscillation around a heavy dead axle. A front roller bracket prevents the halves from separating. Radiators are at the rear.

Steering can be done in several ways. The relative speeds of the engines can be changed by throttle opening or closing, one side may be put in a lower gear than the other for a power turn, the two sides can be driven in opposite directions for a spin turn, and conventional steering may be done with one side in neutral with its brake applied.

Dual Tractors. A pair of tractors may be linked together by bolted-in-place fastenings, with control by one operator.

The Caterpillar DD9G, Figure 15-26A, has a tandem, front-to-rear swivel fastening with a shock absorbing weight transfer cylinder. The operator's station on the front tractor controls the dual unit, or the front tractor when separated, by an air system. The rear tractor can be operated independently by manual controls.

This unit is designed primarily for pushing scrapers. The linkage and one-man control provide tandem power and traction without the lost time and fine work required to coordinate two separate tractors.

The D9Gs can also be paired side by side, as in (B), with similar one-station air control of both. They are connected at the rear by a 16″ diameter steel tube which

Fig. 15-26A. Dual tractors, tandem

carries control lines, by diagonal braces between the inside track roller frames, and by push "C" frames connected to a 24-foot wide bulldozer blade.

Power is delivered to three tracks on turns. The blade may be tilted by separate operation of pairs of lift cylinders.

WHEEL TRACTORS

Crawler tractors are the most compact, powerful, all-purpose pulling and pushing machines that have been developed. However, they are handicapped by low operating speed, and the large number of track parts which are subject to wear. In addition, unless equipped with special shoes, they wear and damage pavements on which they work. There also seems to be a gradually rising limit on their maximum size, imposed by the limitations of the alloy steels commercially available, and problems of transportation between jobs.

Crawler construction is disproportionately expensive in small models, and rate of wear may be very high in any size.

For these reasons, the crawler tractor, and other track-mounted machines, are

Fig. 15-26B. Dual tractors, side by side

Courtesy of Clark Equipment Company

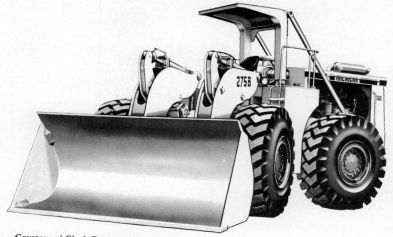

Fig. 15-26C. Articulated 4-wheel drive tractor, with loader

often replaced by equipment mounted on rubber tires. This may be described as wheel-mounted, rubber-mounted, or pneumatic tired.

There are two principal types of wheel tractor. The basic or original style drives from large rear wheels only, and has much smaller front tires. The other drives through all four wheels, which are of equal size.

The second type is of greater importance in earthmoving, and will be discussed first.

FOUR WHEEL DRIVE

The four wheel drive tractor occupies a position between the crawler and the two wheel drive. Traction is not usually as good as with tracks and grousers, but it is sufficient for most needs.

Such a machine needs more power and weight than a crawler to do the same class of work; it has great advantages over the crawler in speed and ability to use and

Courtesy of Clark Equipment Company

Fig. 15-26D. Rear-steer tractor and power train

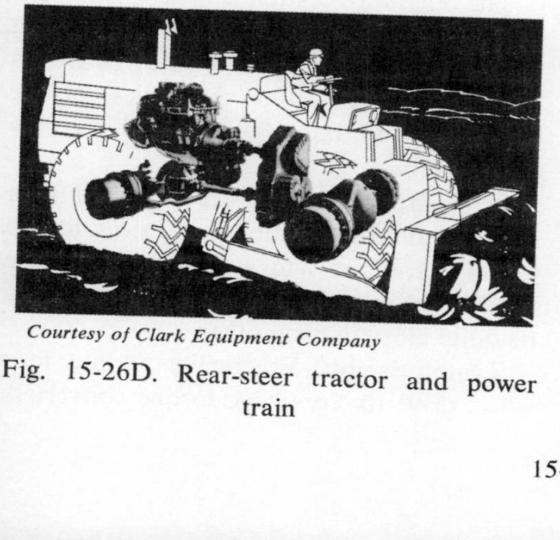

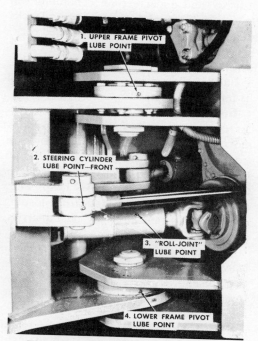

1. UPPER FRAME PIVOT LUBE POINT
2. STEERING CYLINDER LUBE POINT—FRONT
3. "ROLL-JOINT" LUBE POINT
4. LOWER FRAME PIVOT LUBE POINT

Fig. 15-26E. Center articulation parts

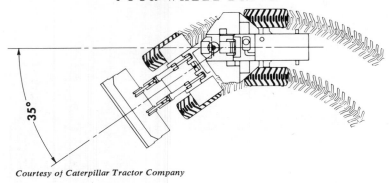

35°

Fig. 15-27. Articulation, full turn

work on highways; and it is more stable and comfortable on rough, hard terrain. It is not well adapted to mud work because of higher ground pressure and spinning of wheels.

Four wheel drive is rather ineffective at dealing with peak loads such as are found in dislodging stumps and boulders, but very good when a heavy load can be picked up gradually without loss of speed, as in dozer or scraper work in smooth material.

Big tires and heavy tracks are both expensive, and under certain conditions will have similar maintenance costs. Tracks wear less than rubber on sharp edged rock, but are rapidly cut to pieces by silica sand which does not damage tires.

Further discussion of these tractors will be found in Chapter 16.

Importance. The four wheel drive tractor has become a substantial factor in earthmoving. Its most successful application is in front end loaders, with bucket capacities up to 26 yards (a top of 12 to 15 yards is usual) with a moderate, SAE heap. They are the standard machines for truck loading in most yards, and in many pits, quarries, and highway cuts.

Most of these units are in the medium or heavyweight class, ranging from 70 to over 600 horsepower. Weight with loaders varies from 6 to over 75 tons, and larger models are being tested.

Four wheel drive dozers have also found wide acceptance in scraper pushing, light

to medium grading, pit cleanup work, and stockpiling. In addition, the front end loaders do dozer work in their spare time.

There are also lighter skid-steer units. At least one of these weighs less than a ton, complete with loader. They are so different from the big ones that they will be discussed separately, in a later section.

Construction. For some years, most of these machines had rigid frames and rear wheel steering, as in Figure 15-26D. Current models have the center-articulated construction shown in the other illustrations, and described in Chapter 12.

The engine is usually at the rear, with considerable overhang behind the axle to increase its value as counterweight for the bucket or blade. Loader columns and arm pivots are behind the front axle; dozer columns somewhat further forward.

The front drive shaft has a pair of sharp-angle universal joints for pivoting.

The operator's station is usually out front on rear steer machines, and over the pivot (on an extension of either frame) in articulated models. In either case, view of the bucket is good, but spill over its back may be dangerous.

Tires are large, and are weighted with calcium chloride solution or mineral dust to increase traction and stability. Wheelbase usually is short, so front and rear tires may be quite close to each other.

Transmission. Practically all big four wheel drive tractors have torque converters and power shift. Many of them have the twin turbine converter that teams a two-speed manual shift with a two-speed automatic shift, giving four speeds forward with two reverse.

Smaller models, 3 tons or less, are more likely to have hydrostatic or reversing clutch drives that are separate to the two sides. These will be described in the next section.

Depending on instructions with the tractor, shifts may be made under any conditions or at any throttle position, or be limited to closed throttle. Some models require that the machine be at a standstill before

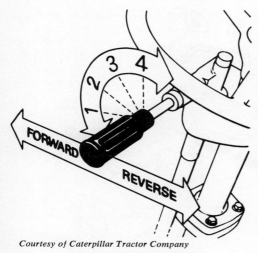

Courtesy of Caterpillar Tractor Company

Fig. 15-28. Twist shift

making a shift between forward and reverse. Instructions MUST be followed, if unnecessary and expensive breakdowns are to be avoided.

There may be a disconnect lever, to cut out drive to one axle. It is comfortable and economical to use two wheel drive for light loads and traveling.

Few, if any, of these transmissions are designed to transmit power from the wheels to the engine. The tractors should not be either pushed or pulled to start, nor towed from one place to another, unless the instruction book states specifically how it may

be done. Removal of both drive shafts is usually required for towing on the road.

Lubrication is usually supplied by a pump driven by the input shaft, which does not turn unless the engine is running. Towing will work the transmission without lubricating it, unless the transmission is disconnected from the drive wheels by a disconnect lever for one end, and drive shaft removal from the other.

Shift among the four or more speed ranges may be in a straight line, with a jog or obstacle at netural; or a U-pattern with forward on one side, reverse on the other, and the cross slot neutral.

Another shifting arrangement, Figure 15-28, utilizes a short lever under the left side of the steering wheel. It is pushed forward for FORWARD and pulled back for REVERSE. Speed ranges are selected by twisting a handgrip on the lever. Detents mark each position.

Brakes and Clutch. Brakes are on four wheels, and may be of any design. Vacuum-hydraulic is popular in smaller models, and air-over-hydraulic in big ones.

The torque converter may include a friction input clutch. This is usually released automatically by light pressure on a brake pedal. It disconnects the drive line to the transmission and wheels, without affecting the power takeoffs and pumps.

This effect is usually optional with the operator. There may be two brake pedals, only one of which operates the disconnect. Or there may be one pedal, with a separate lever whose position determines whether or not the brake will actuate the clutch.

The brake-operated clutch is particularly useful in loaders, and will be discussed in Chapter 16.

An input clutch may also be programmed to release automatically during any shift between reverse and forward.

Accelerator. Wheel tractors usually have a stay-in-position hand throttle and a foot accelerator. The throttle may be set at low or moderate speed (low idle or fast idle), and full engine speed for heavy digging or fast travel is then obtained by pressing the accelerator.

This follows the automobile and truck pattern, and is opposite from crawlers doing the same work. They ordinarily operate on a high speed throttle setting, with a decelerator pedal to slow them down.

The difference recognizes the speeds of the machines. Wheel jobs may travel up to 20 or even 40 miles an hour, so are safer if they slow down automatically on releasing an accelerator.

angles, requiring caution on side slopes or in loader operation.

The next most popular, but on-the-decline system, rear steer, causes the rear of the tractor to swing outside of the track of the front wheels, when on a curved path. This is the same problem as with a car or truck while backing up. It must be kept in mind whenever working close to obstructions or other machines.

The M-R-S four wheel drives have both front and rear steering, under separate control. See page 12-35A.

Skid steer is discussed in the next sec-

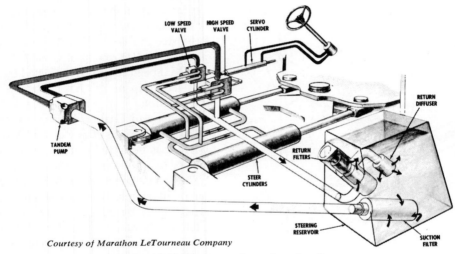

Courtesy of Marathon LeTourneau Company

Fig. 15-29. Two-speed steering circuits

Steering. There are several different steering systems, all power operated.

Most new four wheel drive tractors in medium and large sizes have articulated or pivot steering, with a kingpin or hinge halfway between the axles.

Power is by hydraulic rams, but there is usually a follower valve that makes them respond to the wheel in normal automotive fashion. One set of wheels always follows the other directly, so there is no problem of watching out for side swing of the wheels while turning.

There is loss of stability at sharp steering

tion in this chapter. Front steer is unusual.

Whatever system is used, control over direction with heavy pushed or pulled loads will often be inferior to that obtained with crawler machines. It is hard to turn with a heavy centered load, either pulled or pushed, and difficult to avoid turning if the load is off center. This is largely due to a failure of traction at the most heavily loaded wheels, and will vary widely with the footing.

If a dozer has a tilting blade, put it down on the side toward which you want to turn. A slight lift on the bucket will make a loader more steerable.

Traction. Most four wheel drives have more than enough power to spin the wheels readily. Such spinning is a normal part of operation, although it should be kept to a minimum in abrasive or rocky soils.

Both torque converter and wheel spin prevent you from applying shock loads to the work (and to the tractor). To some extent, you make up for this by utilizing momentum. Speed, even in the lowest range, gives the weight of the machine a substantial digging power, as it takes a lot of resistance to slow and stop it.

Calculate your depth of cut with a bucket to get maximum effect in the few seconds when momentum is added to tractive power. But keep in a low enough gear to carry through with straight digging.

In bulldozing, you will usually do best by cutting thin slices over a long enough strip to build up the load gradually, rather than tearing out a quick chunk as you would with a crawler. Under ordinary circumstances, if the wheels are spinning steadily, you are wasting too much power.

SKID STEER

There are many light four wheel drive tractors, with weight ranging from less than one ton up to nearly 6 tons (including a built-in front loader), with independently controlled drive to the two sides. All wheels are fixed in position, so that the power trains must serve to steer as well as to propel. See also page 16-10.

Several types of power train are used. These include hydrostatic drive, hydraulic drive, chains, and belts. Final drive is by a chain or chains from a central sprocket to the wheel hubs. Non-hydraulic drives include two pairs of reversing friction clutches.

Travel control is by means of two operating levers, one at each side of the driver's station. They are spring loaded to return to a central or neutral position, where hydraulic resistance acts as a service brake. This arrangement forces the operator to keep

Courtesy of Melroe Division, Clark Equipment Company

Fig. 15-30. Skid steer tractor-loader

both hands on the levers whenever the machine is moving.

Pushing a lever forward causes the wheels on that side to rotate to move the machine forward, pulling it back reverses the movement. When both levers are moved the same amount in the same direction, the machine

Fig. 15-31. Skid steer turn

Fig. 15-32. Two wheel drive tractor

will move in a straight line. If one lever is moved more than the other, the machine will usually turn toward the side of lesser movement.

If the levers are moved oppositely, one side will propel forward and the other backward, so that the machine will spin horizontally. In general, a machine with a loaded bucket will tend to slide the back wheels, with it empty it will slide the front wheels. The actual movement is rather complicated, but the result is a U-turn in little more than overall machine length.

Small size and sharp turning enable these machines to work in very restricted areas. The drive to all wheels gives them good pushing power in proportion to weight.

TWO WHEEL DRIVE

Most wheel tractors used in farming and for light construction work have two wheel drive. The rear or drive wheels and tires are very much larger than the front ones, which serve for support and steering. Heavy duty rear tires are needed in almost any construction work; while heavy front ones may

be needed only with front loader equipment.

Industrial. Models used in construction are called industrial, to differentiate them from farm tractors, which are usually not as strongly built. Figure 15-32 shows a standard type, although doing a farm job.

The gasoline or diesel engine at the front drives a disc clutch, a heavy duty reduction type transmission, a differential, and a pair of live axles which turn a pair of disc wheels equipped with large, low pressure tires. The radiator base, engine, flywheel housing, transmission case, and rear axle housing are heavily built and make up the frame of the tractor. At the front, two small wheels and tires support an I beam front axle, which is hinged to the radiator base plate. The axle will normally oscillate on rough ground without imparting twisting strains to the tractor. If a loader is to be carried, the axle may be rigidly fastened to the plate for greater stability.

The rear hubs may be designed to each carry one or two wheels. The use of dual wheels increases stability, traction, and weight, but makes the machine clumsy by

Courtesy of The Oliver Corporation

Fig. 15-33A. Old wheel tractor, cutaway

adding greatly to the width. This disadvantage, and the extra stress they cause in the drive line by eliminating wheel spin, limit their use to special conditions.

A number of different transmission designs are used. There may be 5 to 10 forward speeds, controlled by one lever or by two. There are usually 2 reverse speeds, sometimes 4. If there is a reversing gear or clutch, all forward speeds may be usuable in reverse also.

Older machines are usually equipped with hard-to-shift spur gears, and most newer ones have easily shifted constant mesh.

A power takeoff is standard equipment. It turns a shaft to the rear of the axle housing, which can be used to power attached implements. It can be engaged and disengaged by moving a lever with the clutch released.

It may be driven from the transmission gears, in which case it is controlled by the

Courtesy of Ford Motor Company

Fig. 15-33B. Tractor drive train

main clutch to the drive wheels. In modern machines, however, it is more often driven by a separate shaft from the engine through this clutch, and controlled by its own friction clutch. The two clutches are operated by the same pedal. Pushing it half way down releases the propelling clutch only. Further movement cuts off the power to the takeoff also.

If there is a three point hitch, its pump is controlled by the takeoff clutch, but it is not disconnected by disengaging the power takeoff lever.

Torque converters and power shift are unusual in the models that carry loaders or hoes, or handle 3-point equipment, but they are almost standard in hauler models for scrapers.

Steering is primarily by automotive mechanical linkage to the front wheels. They angle very sharply, so turning radius is short. A hydraulic booster is standard on large machines, and optional on small ones. It is a necessity with front loader equipment.

With manual steering, the wheel must be held firmly on rough ground, as otherwise a bump may cause it to spin with sufficient force to break a wrist. This danger is probably negligible with power steering, but the precaution is still a good one.

Brakes are on the rear wheels, and should be separately controlled by a pair of foot pedals, side by side. A single brake may be applied on the side toward which a turn is being made. The tractor tends to pivot on the braked wheel, thus making it easier for the angled front wheels to pull it around. This one-sided braking often makes it possible to turn sharply with loads that would cause unassisted front wheels to skid uselessly.

If one wheel spins on poor footing, and there is no differential lock, application of the brake on the spinning wheel will force the other one to turn.

The two brakes are applied together to stop. It is usually possible to lock the two

Courtesy of Ford Motor Company

Fig. 15-34. Tractor controls

pedals together, to simplify braking during travel without load.

Controls. Controls vary in details and placement. Those shown in Figure 15-34 are fairly representative.

The steering wheel is in top center, the clutch is on the floor at the front left, brake pedals front right, and the accelerator behind the brakes.

This machine has three gear shift levers. The tall center stick manipulates the main transmission through three speeds, the small right hand one controls the high-low shift (and the starter interlock), and one high on the left is for forward and reverse. Other makes or models may have only one gear lever, or two.

Another lever engages the power takeoff, which is under clutch control when connected. In this illustration a lever engages the differential lock when pulled back. In many machines this is done with a foot pedal. The lock holds itself in place while it is needed, and slips out automatically when traction is restored.

Engine speed is regulated by a governor, which is controlled by a lever on the steering column or dash, which holds any position in which it is put. The automotive type

Courtesy of Caterpillar Tractor Company

Fig. 15-35. Tractor for heavy hauling

accelerator pedal provides for increasing speed above throttle setting. There may be a duplicate accelerator on the left.

Top speed of utility industrial tractors may be 12 to 20 miles per hour, with 15 or less being usual. Haulers are much faster, sometimes to almost 40 mph.

Hauler Models. Scrapers and wagons may be pulled by two wheel drive four-wheel tractors with built-in kingpins for attachment. A big machine of this type is shown in Figure 15-35.

It is becoming more usual to make such tractors with the two drive wheels only, and the engine overhung to the front. This type will be described in Chapter 17, in connection with the scrapers to which such units are more or less permanently connected.

Traction. The principal weakness of two wheel drive is that it has poor traction on soft and slippery footing, particularly in reverse. This problem is reduced by putting the maximum possible share of the weight of the tractor and its attachments on the drive wheels, and by use of proper tires

(usually high-cleated ones, in the largest recommended size.)

Any rear drive unit has better traction moving forward than backward. This is largely due to reaction from turning the drive axle, as the whole machine tends to rotate in an opposite direction from the axle.

Referring to Figure 15-36, it will be seen that if the tractor is driven forward, the front of the drive wheels turn downward, and the equal force of reaction from the force applied tends to lift the front wheels up and back. The weight is thus concentrated on the drive wheels.

When the tractor is moved backward, the reaction from turning the axle forces the front wheels down, increasing their load, decreasing that on the drive wheels, and reducing traction.

If the tractor is moved up a hill, this shift in weight will be affected by a shift in the center of gravity. If the hill is very steep, there is danger that a tractor climbing it forward will overturn when the power needed to lift the front wheels becomes less than that needed to turn the rear wheels.

These effects will be more noticeable when the center of gravity is near the axle, or when the tire is large so that its contact with the ground is distant from the axle. Since wheel tractors normally have short wheel bases and big tires, it follows that reduction in traction while backing will be more severe than in cars or trucks.

Weight and Counterweight. With two wheel drive, traction problems are affected by (even determined by) weight distribution.

Liquid or powder in the drive tires increases their weight and traction. Wheel weights do the same. Balance is not greatly affected, and the proportionate loss of traction in reverse remains.

An attachment projecting to the rear of the tractor is 40 to 60 per cent more effective than wheel weight in improving traction. This is because it tends to raise the front of the tractor, with the drive axle as a pivot or fulcrum, and thus adds some of the front weight to the rear wheels.

A tractor carrying a couple of tons of rear-mounted hoe has traction so improved that the forward-reverse problem can usually be ignored, and very heavy front loads can be handled. Loader performance becomes excellent, with improvements in traction, balance, and extra weight (momen-

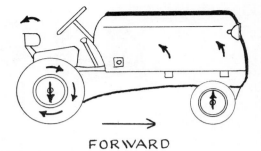

FORWARD

BACKWARD

Fig. 15-36. Driving torque and traction

tum) driving the bucket deep in the pile.

A big rotary mower is an effective although lesser counterweight, and may be less in the way. Mounted rippers or plows, or a drawbar-carried drum of sand or set of metal weights, are useful but lack most of the leverage advantage.

But a heavy weight attachment projecting to the rear may cause the tractor to become too light in the front for effective steering

Fig. 15-37. A mower is good counterweight

Courtesy of Schramm, Inc.

Fig. 15-38. Tractor-compressor

when going up a steep hill, and may even give it a tendency to turn over backward.

Steering is then done by cautious application of a wheel brake. Overturning may be prevented by moving slowly and smoothly, holding the attachment low for possible support, or backing up if there is sufficient traction.

Uses in Construction. The two wheel drive tractor is the cheapest type of mobile power unit and attachment carrier that a contractor can buy. It is also reasonably fast, highly maneuverable, comparatively non-damaging to its footing, and relatively easy to hook up to attachments. It is often driven on roads between jobs.

As a result, its uses are too numerous to be detailed here. The two most important are the carrying and powering of backhoes and front loaders. Rear and side attachments include mowers and light grader blades.

The backhoes, which were discussed in Chapter 13, are used or are at least usable on almost every construction job. They also improve the traction of the tractors that carry them.

A front loader adds greatly to the usefulness of a tractor. Except in a few models, its efficient work with two wheel drive is limited to soft digging, and rehandling of loose material, but it does this work well. It can protect the tractor and clear the way for it.

The loader also serves as a motorized wheelbarrow, a work platform of adjustable height, a jack or hoist for heavy objects and disabled equipment, and as a counterweight for a hoe.

The Schramm Pneumatractor has a compressor built into the engine block. This is a valuable extra that does not interfere with its capacity for all ordinary tractor work appropriate to its several model sizes and varied attachments.

THREE POINT HITCH

Most farm tractors, and many industrial models, have a built-in mechanism to provide semi-automatic depth, draft, and position control to implements mounted at the rear. The device is called a three point hitch, or sometimes a hydraulic drawbar.

The original construction had a single lever moving along a curved guide or quadrant, which provided depth control responsive to the load on the implement. Now there may be an additional quadrant lever that controls a fixed position—that is, height of lift or depth of penetration—relative to the tractor, without being influenced by load. A single quadrant lever may have both functions, being put in one or the other phase by a selector.

There may also be an adjustment to increase or decrease speed of response.

Both mechanisms differ from most hydraulic controls in that they regulate position rather than movement of the lift arms. An ordinary hoist valve such as is found in a bulldozer or a dump body, has UP, DOWN, and NEUTRAL or HOLD positions. In UP it will cause the ram or rams to raise the load until it meets the end of its travel, in DOWN it will allow it to lower or force it down as far as it will go, and in HOLD it will lock it in any position.

In three point hitch controls there are no specific UP or DOWN lever positions. The top or forward end of the quadrant usually means DOWN, and the bottom or rear, UP, but any intermediate position may result in raising, lowering, or holding the attached implement, depending on its position at the time, its weight, and the working resistance it is encountering.

In position control, and in depth control while conditions remain unchanged, a lever position corresponds to a specific implement position. The quadrant may have a movable stop, which can be locked to mark a desired setting, to which the lever can be returned accurately without looking at it.

Hoist and Linkage. The hoist is operated by a hydraulic pump, valve, and cylinder unit inside the casing of the rear axle and transmission. The pump may be driven by an ordinary power takeoff, but more often (and preferably) is separately driven so that it turns whenever the engine is running. However, its power is usually cut off by disengaging the power takeoff clutch.

The cylinder turns a cross shaft which raises a pair of lift arms, one on each side of the differential case. Rods extend from these down to the lower links hinged to the axle case.

The lower links are fitted at their ends with ferrules which slide over pins rigidly

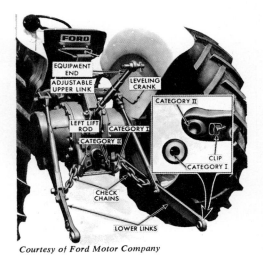

Courtesy of Ford Motor Company

Fig. 15-39. Three point hitch

fastened to the implement, and are locked by inserting lock (linch) pins in holes. The top link is an adjustable length bar that can be disconnected at either end. If the top link is not fastened, raising the bottom merely causes the implement to pivot backward without rising off the ground.

The linkage is made in two strength classes. Category I is standard equipment. Category II, usually offered at extra cost, is designed for heavy attachments and more rugged service.

The length of the top link determines the front-to-rear pitch of the implement. This may be very important in adjusting the suck of a grader blade or a plow, and may affect mower performance also. Shortening the link rotates the implement forward, increasing the suck of a plow, but decreasing that of a blade. If it is too long it will reverse these effects.

Too much suck causes diving and pitching, too little means no penetration.

Some links have hole-and-pin adjustment, but threads are easier to use and permit more accurate adjustment. If adjustments are required frequently or on-the-move, this top link may be replaced by a hydraulic cylinder.

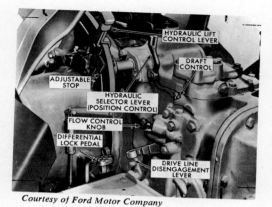

Courtesy of Ford Motor Company

Fig. 15-40. Controls, 3-point hitch, British model

One of the lift rods that connect the hoist lift arms to the lower links is equipped with a threaded length adjustment operated by a crank. This permits raising or lowering one side of the hoist for convenience in hitching, and to tilt an implement relative to the tractor. If the tool is a grader blade both lift rods may be adjustable to make possible a steeper tilt.

The hitch is limited in side to side swing by a pair of chains anchored on the power takeoff cover plate. If this plate is installed upside down the chains may be out of place and the lift will not work properly.

Some manufacturers offer an optional extra of brace bars (stabilizer kit, Figure 15-41) that will hold the attachment rigidly in line. Their use causes damage to some implements.

Depth Control. Semi-automatic depth control is ordinarily used for hitch-mounted implements that have edges or points that work below the ground surface. These include light grader blades, light rippers (subsoil plows), and moldboard and disc plows.

If the tractor is standing still, the hydraulic pump is operating, and the implement is resting on the ground; moving the depth control lever to the top of its quadrant will cause the implement to rise until the ram reaches its limit of travel. If the lever is not

quite to the top, the flow of oil may or may not be cut off by the movement of the ram before it reaches its limit, depending on the design of the particular unit.

If the valve lever is moved part way down the implement will fall and rest on the ground. If the tractor is moved forward the implement's teeth or share will penetrate the ground and encounter resistance that will rotate it forward so that it pushes against the top link. When this pressure is greater than that allowed by the position of the valve, it will open a valve port that will admit fluid under pressure to the ram, raising the implement until its pressure on the top link is in balance.

If the top link pressure is less than that required by the valve lever position, an exhaust port will open to drain fluid out of the ram, allowing the implement to sink and the draft pressure on the top link to increase to the proper amount.

If the front wheels of the tractor should start up a ridge, the tractor will assume a climbing position which tends to drop the implement deeper into the ground. This change in angle causes it to exert additional pressure on the top link so that fluid is supplied to the ram at higher pressure, thus raising the plow.

As the rear wheels cross the ridge, the tractor will nose down and tend to pull the teeth out of the ground. This will relieve pressure and perhaps pull against the top link, opening the drain port and allowing the plow to sink until balance is re-established. In this way the plow is kept at an approximately constant depth regardless of pitching of the tractor.

The regulating pressure on the top link also acts to push the rear axle and wheels down, adding to traction. However, if the plow strikes something solid enough to stall the tractor, the shock load against the top link will drive it past the valve port opening position to another port that will drain the ram, taking the weight of the implement off the tractor wheels so that they can spin instead of stalling the engine.

This control arrangement maintains a fairly constant load on the tractor, prevents damage from shock, and keeps the implement near proper depth on rough ground. But it tends to vary working depth with soil resistance, skimping hard areas and over-penetrating in soft ones. The operator can counteract this behavior by adjusting the control lever for greater or less depth (draft resistance) as necessary. There may also be means to easily adjust the valve for greater or less responsiveness.

The one-way cylinder does not provide down pressure, so there is no way to force the implement down to cut soil that is too hard for it to penetrate.

Position Control. Position control may be provided by the same lever after setting a selector, or by a second quadrant and lever alongside the depth control. Automatic response to changes in resistance are eliminated. The relationship between lever position and implement height is fixed, unless the ground or other surfaces raise the implement above its setting.

This control is used for attachments that work on or above the ground, not in it.

If the lever is set fully forward, the hoist does not operate, and the mower rests on the ground by its own weight, usually with additional rear support from a small caster wheel, skid shoes, or a height regulating bar. It will stay in contact regardless of any pitching of the tractor crossing ridges or obstructions.

In this position, a mower may tend to dig into the ground, cut too close, and scalp high spots. This can be prevented by raising its front slightly with the control. However, it will then be supported, and will lift clear of the ground as the tractor crosses a ridge, unless the operator lowers it.

The unused lever may have to be left in full UP position, to allow the other to function properly.

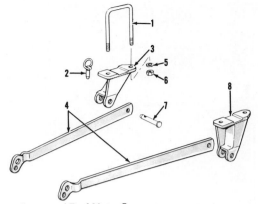

Courtesy of Ford Motor Company

Fig. 15-41. Stabilizer kit

Hooking Up. An implement is attached by backing the tractor to line up its hitch with the pins, raising or lowering the lower links until they are at the right height. The lower links have considerable free motion from side to side. One of them is slipped onto the corresponding pin, and the lock pin dropped and locked into place. The other link can now be made to line up by the adjustment crank, by raising or lowering, by moving the tractor, or moving the implement until the link ferrule can be slipped over the pin and locked.

The top link may originally be fastened to either the implement or the tractor, in either case it pivots freely up and down, but it may be either too short or too long to connect. If it is too long the lower links are raised so that the implement tips back and increases the space. If it is too short, the control lever is put at the bottom of the quadrant. If this does not bring it into position, the operator may stand on the links to push them down, raise the back of the implement, or tow it forward until it tilts into correct position. The ferrule is then slipped over the pin and locked. This may be done from the ground or the tractor seat.

The length of an adjustable link may be changed to attach it, then re-adjusted for operation.

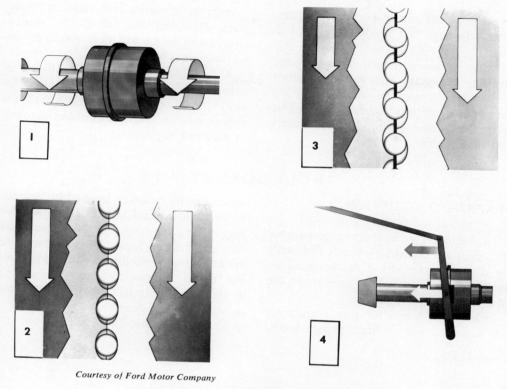

Courtesy of Ford Motor Company

Fig. 15-42. Load Monitor sensing units

Attaching is easy if the implement is light enough to be moved easily, or if the operator has been skillful or lucky in getting into just the right position relative to a heavy one. In general if a heavy tool is to be attached, the operator should have at least a strong back, and preferably a helper or a crowbar.

Detaching is easy. With the control lever down there should be little tension on the links so that they can be readily slid or hammered off the pins. A narrow unit, such as a one bottom plow or a subsoiler, may fall over unless supported by blocks, and must be handled carefully.

When hitching or unhitching indoors—even in a garage with the door open—the engine should be stopped when it is not being used to move the tractor or to lift the links, as even a moderate dose of exhaust gas may cause a headache that will take the fun out of working, and a heavy exposure can easily be fatal to the operator.

Advertising to the contrary, this type of linkage is usually more difficult to attach and detach than the old fashioned tongue and drawbar requiring only one pin. However, once attached it is so much easier to transport and use, particularly in small scale work, that it is worth many times the extra trouble.

An additional advantage is that an implement carried on the hitch increased rear wheel traction, and acts as counterweight for a front loader's load.

Load Monitor. Ford's Load Monitor provides the three point hitch with an additional type of semi-automatic control.

A torsion spring in a two-piece cylindrical steel coupling is connected to the gear box output shaft and rear axle input gear. Drive is through ball bearings in spherical ramps.

Higher or lower torque in the tractor drive line causes the balls to move against pressure of disc springs, and the coupling parts to twist slightly on each other. This motion moves an outer housing that triggers a hydraulic valve.

Increase in torque will raise an implement, decreasing the pull on the tractor. Decrease in torque will allow return to preset position.

A 5-position selector allows for use of either of the standard 3-point controls, or any of three Load Monitor variations. These include remote control of cylinders on towed or mounted implements.

BULLDOZERS

GENERAL DESCRIPTION

The bulldozer is the short range member of a class of excavators which ordinarily dig dirt and transport it to the dumping point, and often grade both the cut and the fill in the same operation.

Bulldozers (dozers for short) are tractors equipped with a front pusher blade,

which can be raised or lowered by hydraulic or cable control and which is used for digging and pushing. Angling dozers are bulldozers which have blades that may be set at angles to cast dirt to either side while the tractor moves forward. When their blades are set straight, they do the same work as straight dozers.

Crawler tractors, known more familiarly

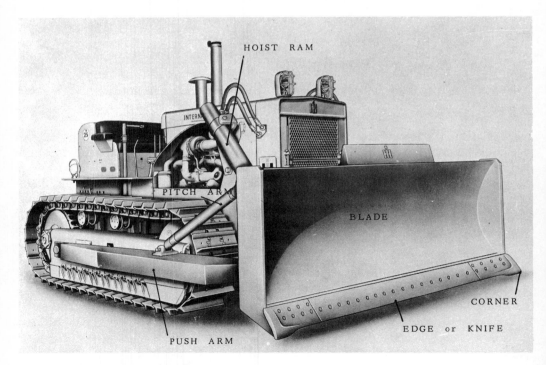

HOIST RAM

PITCH ARM

BLADE

CORNER

EDGE or KNIFE

PUSH ARM

Courtesy of International Harvester Company

Fig. 15-43. Direct lift hydraulic bulldozer, old model

Courtesy of TEREX *Division, General Motors Corporation*

Fig. 15-44A. Single ram hoist

as cats (when capitalized "Cat" is a trade-mark of the Caterpillar Tractor Co.), carry most of the bulldozers intended for excavation or heavy pushing, but an increasing number are mounted on special four wheel drive rubber tired tractors. Dozers mounted on two wheel drive tractors, trucks, and graders are used chiefly for spreading and backfilling loose material.

Blade. The blade is a massive structure that has a rectangular base and back. The leading edge of the base is a flat blade or knife of tough, hard steel which projects ahead of and below the rest of the blade. The front of the blade is called the mold-board and is concave and sloped back.

As this blade is pushed into the ground, the knife cuts and breaks up the dirt, which is then pushed up the curve of the moldboard until it falls forward. Material being pushed ahead of the blade is thus kept more or less in rotary movement, which tends to even up the load and offer less friction and a larger load than would be obtained with a flat vertical moldboard. The weight of the dirt first helps penetration, and then, as a full load is obtained, it pushes against the upper curve of the mold-board and "floats" the blade so that it cuts less deeply or not at all.

The edge is usually in three pieces, a wide center and two corners, which are bolted to the blade. They are reversible so that they can be removed and turned upside down when worn. The corners wear

fastest and are sometimes built up by welding until the whole edge is so worn as to need reversing or replacing.

The blade may be equipped with shoes located just behind the blade. They may be set below the cutting level so that the weight of the blade will rest on them, and it will not cut into pavement or other surface which is compact enough to support the shoes. At least three settings are usually provided—low, so that the cutting edge will not touch the pavement; level, so that it will scrape the pavement but will not cut it; and high, where the shoes are out of

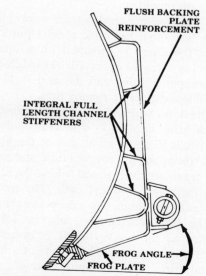

Courtesy of TEREX *Division, General Motors Corporation*

Fig. 15-44B. Cross section of dozer blade

the way and the blade operates normally. These shoes are absent on most dozers.

Push arms should be attached to the blade near its outside edges, as the greatest strains normally occur at the corners. An exception to this is scraper pushing, where the center is subject to violent collision contacts again and again.

A blade that is to be used for scraper pushing is likely to cave in the center unless it is heavily reinforced. An extra plate or cup may be welded on the outside, or the inside may be braced with channels and plate steel.

Push Arms. Push arms are heavy, hollow beams extending from a hinged connection with the tractor to the bottom of the blade. Most dozers have arms that are mounted on the outside of the track frames, but some have inside arms.

Outside arms are easier to design and to fasten to the tractor, as they need not get involved in the narrow clearances between tracks and center frame. Since blades are usually only slightly wider than the push arms, a wider blade and valuable extra clearance for the tracks in the slot cut by the blade are provided.

Inside arms are used when a track-width blade is wanted for some special purpose, when the dozer is mounted on a loader frame, or when the dozer will be used chiefly for scraper pushing and the narrow blade offers less danger of ripping tires.

The dozer with inside arms and a tractor width blade is often at a severe disadvantage. It is likely to climb out of the blade path when pushing around a turn, and rocks and debris are likely to roll behind the blade into the track or wheel path. The tractor can easily get jammed in a deep narrow cut.

The front of the arm may be solidly welded to the blade, but it is becoming more usual to connect by means of a horizontal hinge, that permits changing the pitch and tilt of the blade.

Diagonal horizontal braces may extend

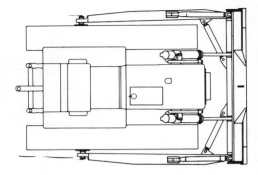

Fig. 15-45. Top view of bulldozer

from the inside of the push arms to the center of the blade.

Pitch (or tip) is the angle that the cutting edge or the blade itself makes with the ground. It is changed by rotating the blade around hinges on the front of the push arms.

Tilt is the side to side angle of the blade with the horizontal.

Pitch Arms (Pitch Braces). These are diagonal members between the push arms and the blade top. They brace the blade against loads above the line of the push arms, and may provide means for regulating its pitch and tilt.

Pitch arms may be rigid one-piece arms welded at both ends, in which case they serve only to brace the blade. Or they may be in two coupled pieces, hinged to the push arm and to the blade, and provided with a length-adjustment. This may be threaded, pin-and-hole, or a hydraulic cylinder.

Lengthening both pitch arms will tip the blade forward, shortening it will bring it back. The effect of such adjustments varies with digging conditions and the shape of the blade and the cutting edge. Usually, tipping forward will increase the suck of the edge and improve penetration in hard soils. It will also increase the upward pressure against the blade of loosened material coming up from the edge. This may be sufficient to float or carry the blade so that it will not be able to cut.

Since the blade is usually carrying a full

load by the time there is enough upward pressure to float it, this does little harm. It is a convenience in pushing loose loads, as it automatically keeps the blade full.

Tipping the blade back reduces penetration, and may prevent the edge from cutting down except under heavy pressure. Such a setting is good in cleaning loose material off a firm level surface, or in cutting off humps or knobs without regarding the whole surface.

Blade pitch adjustments are seldom made or needed with hydraulic lift bulldozers, but are often important in cable lift blades, because of their weaker penetration.

Lengthening one pitch arm will tilt the blade so that the corner opposite the long arm will be low. This will be discussed later.

Mountings. The direct mounted two-ram hoist shown in Figure 15-43 is now the most popular type for medium and large machines. The rams are carried in swivel brackets on the sides of a heavily reinforced hood and/or radiator guard structure. The piston rods are pinned to the back of the blade.

Direct mounting is best suited to tractors that have stabilizer bars or rigid frames instead of front springs. Otherwise the tractor springs will get involved in the stresses of lifting and lowering the blade. The engine will be pushed up by use of down pressure,

Courtesy of Gar Wood Industries, Inc.

Fig. 15-46. Front hydraulic unit

and pulled down when the blade is lifted; the weight of tracks cannot usually be put on the blade by down pressure, grading hard ground is less accurate, and the springs make it difficult to raise the tracks out of mud by pushing the blade down.

The single ram hoist substitutes a single thicker central ram for the two side rams. This arrangement allows all twisting strains to be absorbed by the blade and arms. It is better suited to pusher work than to general bulldozing.

Direct lifts may also be based on columns between the tracks and the hood.

Figure 15-46 shows an assembly that includes the radiator guard, a front mounted hydraulic pump, a valve controlled by mechanical linkage from the seat, hood mounting braces which contain the hydraulic oil reservoir, the hoist rams and their swivels, and the tubing and lines.

The track frame mounting shown in Figure 15-47 is going out of style, but is still used on small and medium machines. A frame or cradle is placed on each track frame, supporting a nearly horizontal two way short thick ram.

The cylinder piston rod is hinged to the upper end of a triangular bell crank. This rotates on its rear angle, and lifts the push arm by a rod at the front.

Track mountings place the weight as well as the push on the track frames, and permit spring mounting of the central frame without interference with dozer action. The leverage through the bell crank permits use of a shorter and thicker ram than direct lift, with a corresponding reduction in hydraulic service problems.

The track frame mounting has lost out largely because the hood mounting is simpler, with fewer wearing parts.

Hydraulic Lift. Most dozers now use a hydraulic system to raise and lower the blade. It may also provide from-the-seat adjustment of pitch, tilt, and even angle. Such a system includes a pump, a con-

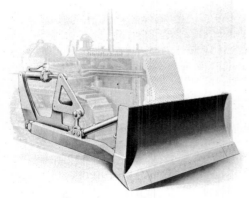

Courtesy of Caterpillar Tractor Company

Fig. 15-47. Outside lift arm bulldozer

trol valve, cylinders, a reservoir, lines, and filters. General design is discussed in Chapter 12.

The hydraulic pump turns whenever the engine does. It is usually either gear or vane construction. Pressure may be 1200 to over 2000 pounds per square inch.

Cable lift will be described later.

Blade Tilt. When ground that is being dug is harder on one side than on the other, a bulldozer blade will tend to cut down on the soft side, particularly if it is a plastic soil that pulls it down. On a side hill the blade will tend to hang a little low on the down hill side because of shift in its center of gravity, and will cut low on that side.

This difficulty is partially met by making the push arms and blade into a rigid unit that resists distortion. This is seldom wholly effective and as the machine ages flexibility will develop.

There are a number of operational techniques which will produce flat grades under adverse conditions, or convert flats into side slopes where desired, but these usually involve extra skill, time, and work.

Mechanical Tilt. Some blades can be tilted mechanically to cut deeper on one side than on the other. Pitch braces, if adjustable, can be used for this purpose by making one longer than the other. The blade will be low on the side of the long

brace. The twist of this adjustment is taken up by connections at the front and rear of the push arms.

Some are made in two pieces. One may slide inside the other, and be secured at any of several lengths by a heavy bolt or lockpin through matching holes. Or the sections may be threaded together, with one being turned on a swivel.

A mechanically tilted blade may rotate on a horizontal center pin, and be locked to the push arms by removable pins or wedges in curved tracks. Adjustment is usually made by removing the locks, then lowering the intended high side onto a support. A crowbar may be needed to get alignment for re-locking.

Mechanical tilting involves stopping the

Courtesy of Fiat-Allis Construction Machinery, Inc.

Fig. 15-48A. Hydraulic tilt

Courtesy of Clark Equipment Company

Fig. 15-48B. Hydraulic tilt at push arm hinge

dozer and working on the blade for several minutes. It is usually not practical to make these adjustments frequently, or for working down small areas. If not used occasionally, the parts may rust or stick so as to make changing tilt quite difficult.

Power Tilt. The nuisance of mechanical adjustments of blade tilt can be avoided by putting a two-way hydraulic ram in one of the pitch braces, as in Figure 15-48A, or in one or both of the push arm hinges, Figure 15-48B, and controlling tilt by a valve at the operator's station. This permits rapid and effortless adjustment during operation.

Power tilting adds greatly to the ability

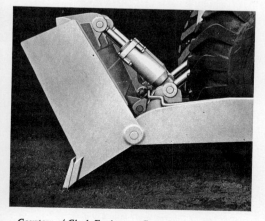

Courtesy of Clark Equipment Company

Fig. 15-48C. Hydraulic blade pitch control

of the machine to cut hard ground, to cut grades accurately up to walls or other obstructions, to crown roads, and to do accurate grading of slopes and curves.

If both pitch arms are adjustable, they can be used together to adjust blade pitch, in order to regulate penetration and pushing behavior. See Figure 15-48C.

Control Valve. The control valve may be a reciprocating or (rarely) a rotary type, operated by a hand lever. The operating position at the front is FLOAT, in which all the valve chambers are open (see the diagram, Figure 12-68B, so the pressure line empties into the exhaust line, and the ports to the fronts and backs of the rams are open so that no pressure is exerted on the bell cranks, and the blade rests by its own weight on the ground.

The next position of the valve control lever is DOWN, in which the pressure line feeds into the tops (backs) of the cylinders, and the bottoms (fronts) drain into the exhaust, causing the pistons to force the blade down to the limit of its travel, which may be five to twenty inches below the line of the tracks. If the blade meets resistance, the front of the tractor will be raised off the ground. HOLD position, sometimes called neutral, closes the ports to the rams and opens the pressure line into the exhaust. This locks the blade in position without loading the pump. The rear position is UP, in which the pressure is directed into the bottom cylinder sections, while the upper sections empty, so that the pistons pull their rods to raise the blade.

The piping may be standard plumbing or special tubing, except where the line is expected to bend with the movements of the machine. Flexible high pressure hose, woven of metal, fabric, and neoprene rubber, is used in such places.

Use of FLOAT. Most bulldozer control valves include a FLOAT position. FLOAT is ordinarily used to smooth areas by back-dragging, and in pushing dirt on pavements

Fig. 15-49A. Cable lift bulldozer

or other surfaces which are hard enough to support the blade, but which will be damaged if it is forced down. If the moldboard is so shaped and tilted that a full load tends to lift the edge to the surface, transporting can be done in FLOAT.

If the blade will not float, much of this work requires a high degree of operator skill and concentration, with results generally inferior to those obtained automatically by floating.

Replacement of a non-floating with a floating valve is likely to be expensive. However, it usually is possible to install a pipe connecting the lines which carry pressure to the backs and fronts of the rams. A high pressure quarter turn valve in this pipe will permit the blade to float when it is open, and to operate normally under control of the main valve when closed.

CABLE BULLDOZERS

Figure 15-49A is a sample of this type of dozer with its hoist mounted on the radiator guard. The push arms are hinged to brackets bolted to the outside of the track frames and to the bottom of the blade.

Power Control Units. The cable control

unit may be mounted on the radiator guard and driven from the engine crankshaft, or be on the rear of the transmission where it is driven by the power takeoff. Front units

Fig. 15-49B. Rear power control unit (PCU)

15-48

Courtesy of Fiat-Allis Construction Machinery, Inc.

Fig. 15-49C. Pusher plate

have one drum while rear ones may have one or two, or even three.

Each cable drum is controlled by a clutch and a brake applied by a single lever beside the operator. In the front unit, the cable is reeved upward to hoist sheaves on the gooseneck or radiator guard, and then in a multiple line around sheaves on the top or rear of the blade at the center. In the rear unit, the cable is led around sheaves to a pipe or guard channel which brings it to sheaves on the side of the radiator, from which it goes to the hoist sheaves.

A two-drum rear control unit is shown in Figure 15-49B. A cutaway view of another model will be found in Chapter 17.

Blade. The blade is similar to the hydraulic powered type, except that the cutting edge is sometimes advanced at a sharper angle, and may project farther below the bottom of the blade for better penetration.

Control Lever. The control lever has three positions, DOWN, HOLD, and UP. In DOWN, the brake and clutch on the cable control unit are released, the cable is slack, and the blade rests by its own weight on the ground. In HOLD, the cable drum is locked by the brake. If the cable is slack,

the blade can rest on the ground or dig; if it is tight, the blade is held against lowering. In UP, the brake is released and the clutch engaged so that the cable is spooled onto the drum and the blade raised.

PUSHERS

Pushing scrapers is an important use for heavy tractors equipped with dozer blades or special devices.

An ordinary dozer working as a pusher is likely to cave in the center of its blade unless it is reinforced. Reinforcement may be a plate or a cup welded outside, that will also serve to hold the scraper bumper from sideslipping, or it may be internal bracing.

Fixed plates, as shown in Figure 15-49C, are cheaper than dozers and are usually very rugged. However, they may not line up with bumpers on different models of scrapers, and they may not keep proper contact when pitching on rough ground. Similar plates may be mounted on angle dozer C-frames, where they are under as effective control as a blade, and do not cause as much hazard of ripping tires with their corners.

If two pushers are used in tandem behind one scraper, at least one of them should have a rear push bar or plate, with a frame that will carry the thrust directly to the front blade or plate.

If the tractor has conventional dozer push arms hinged to the outside of the track frames, the rear pusher should also be fastened to the track frames, to carry force directly. The rear pusher thrust will then not be carried by the central frame and gear cases.

If the front pusher is mounted on the central frame the rear pusher is mounted on the steering clutch case, so that push is taken through the tractor in a direct line.

For efficient operation a pusher should make rapid contact with the scraper, that may be either stationary or moving. Any miscalculation of speed and/or distance

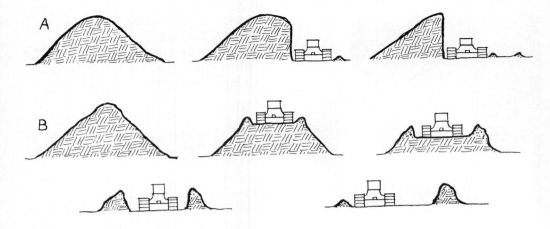

Fig. 15-50. Taking down a pile

may result in a collision that will damage both machines if repeated too often.

This situation is even more serious when tandem pushers are used. Doubling up the machines triples the likelihood of miscalculation and rough engagement. The front push blade or block must carry the power of two tractors, and therefore has less reserve strength.

Cushion. Use of spring cushioned blades and/or rear plates will permit much faster engagement and make any damage unlikely. A blade may be hinged to a special push frame at the bottom, with rubber discs or compressed nitrogen to absorb shock in the middle, where scraper contact is made.

A rear cushion plate is valuable in a center mounting in protecting the gear cases from shock blows.

BULLDOZER OPERATION

References. This section assumes familiarity with tractors. It is recommended that the sections on starting engines and on general operation procedures in Chapter 12, and the material on tractor operation earlier in this chapter, be reviewed.

Also, the operator's instruction manual for the particular machine should be studied, if it is available. It should provide control diagrams, operating instructions, and necessary information on lubrication and checking, which are part of the operator's responsibility on most jobs.

Digging. A tractor bulldozer of any type is worked by moving the tractor forward, or less commonly, backward, and raising and lowering the blade to contact material to cut, spread, or transport it.

As a dozer moves forward and digs, some of the soil cut by the blade will pile up in front of the blade and move with it, and some of it will drift off the sides, forming ridges or windrows. Resistance to the machine's movement is made up of the power absorbed in cutting and breaking up soil, and in friction in the loosened dirt. If the blade is lowered, more work will be done and resistance will increase, as a thick slice requires more digging power than a thin one, and the total amount of dirt resisting the blade is increased. If the blade is raised, the slice will thin or disappear, and the amount of earth being pushed will decrease, so both work and resistance are reduced.

In heavy digging, efficient operation involves pushing the most dirt without losing too much speed by engine slowing, slippage in the torque converter, or by spinning

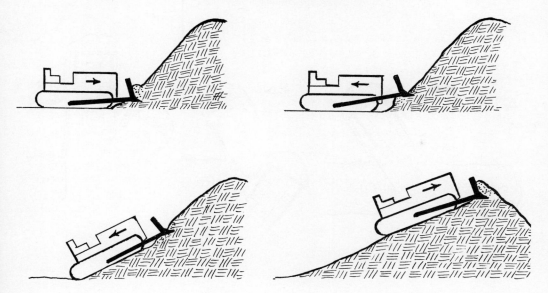

Fig. 15-51. Starting a slot

the tracks. The operator starts cutting a slice which should give this result, and if the machine slows, he raises the blade slightly; if it is not working to capacity, he lowers it. The upward blade movement should be made as gradually as possible to avoid leaving a bump in the path of the tracks.

If the blade is set to cut an even depth, digging resistance will remain about the same through the pass, but that of the loosened and transported dirt will increase steadily. This increasing resistance does not slow the machine at first, as it causes the governor to open the throttle to maintain tractor speed. Once the engine is wide open, further increase will slow it, so that the blade should be raised gradually to the surface of the ground where it can push the loose dirt without digging more. Sometimes the blade is "pumped" during this lift by being dropped and raised quickly, cutting out a bit of extra dirt each time it is lowered.

The dozer digs and transports much more effectively downhill than on a level or uphill, and work should be arranged to work down a grade when it can be done.

Breaking Piles. A pile of dirt may be knocked down by walking into it with the blade at the desired grade, after which it may be spread or piled elsewhere. If the heap is too large or hard for the machine to take at one pass, or if it is to be spread in more than one direction, the first pass may be made to cut away part of the pile to grade, Figure 15-50 (A), or to cut the top of it part way down, as in (B). If the second method is to be used but the dozer cannot move the part it can reach, a ramp may be made by loosening the soil by pushing then backblading with down pressure, Figure 15-51, so that the blade can contact the heap at a higher level.

If the pile is very large or hard in proportion to the power of the dozer, the side cut should be repeated from different angles, Figure 15-52 (A), in order to shorten the cut required for each pass. If the digging leaves a high face that might fall on the machine, the dozer should be turned toward it occasionally and driven into it with the blade held high. This should cause the bank to fall or slide with-

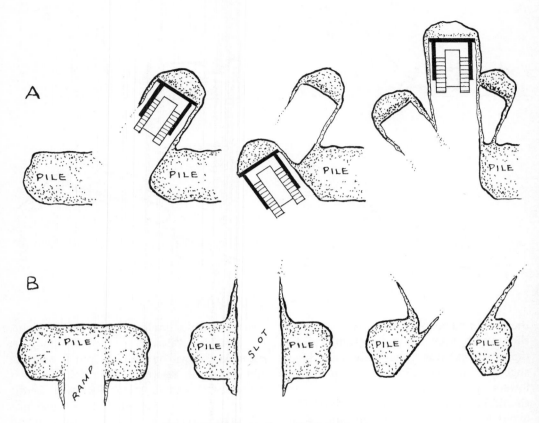

Fig. 15-52. Spreading pile from center

out burying the side of the dozer in it.

If the sides are not accessible, a center cut may be made by first ramping up and cutting a slot down to grade, widening it to both sides. This slot, and any cut more than a few inches in depth, should be made somewhat wider than the blade to avoid jamming the dozer between the walls. This may happen in very narrow cuts through rocks or roots in the sides being turned by grazing contact with the blade so that they project, or by creeping or falling in of the sides. Dozers whose blades are track width or very little wider are particularly subject to getting jammed in this manner.

A narrow cut also does not leave room to maneuver to get at a rock or other obstacle encountered in the floor.

Suck. Plastic (rubbery silt or clay) soils will pull the blade down as it is pushed through them, dragging down the front of the tractor at the same time. An adjustable blade may be set for less penetration, and overdigging may sometimes be avoided by making a number of very thin slices. More often digging is done in the regular way and gouges made below grade are refilled with loose material.

Dozers which have the cutting edge set at a pronounced forward angle and set well below the blade for a maximum suck; those with the hoist on the hood of a tractor with front springs; and any with spring cushions on the lift rods are particularly cranky in such soil.

Road Cuts. A cut which is to have steep sloped sides, as for a highway, should be

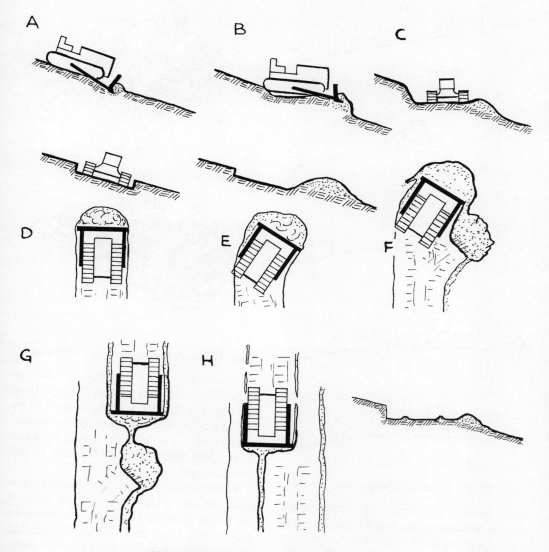

Fig. 15-53. Notching a slope from above

started full width, and necessary measurements made to assure cutting the slope correctly, as it may be difficult or impossible to get the machinery up on it afterward.

Side Slopes. A bulldozer not having a tilting control for the blade will cut deeper on the side which is downhill or in softer soil. On a slope, this tendency may be overcome in several ways. A shelf may be built by pushing downhill, as in Figure 15-53 (A)-(C), so the dozer can start its side cut level or tipped oppositely to the slope. Or the dozer may cut and turn downhill, raising the blade, as in (D) to (F), thus cutting a more or less level shelf which can then be enlarged or graded off as in (G) and (H).

Shallow stripping may be done by starting at the top of the slope so that in each

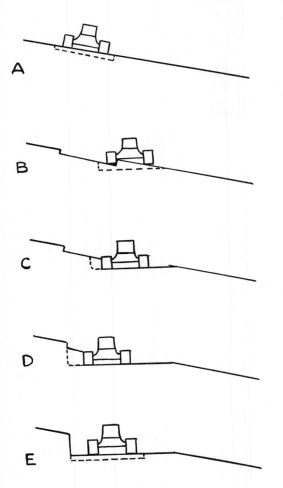

Fig. 15-55. Avoiding wastage on long push

Fig. 15-54. Notching a slope from the side

pass except the first, the upper track can be walked in the hollow made on the previous pass. This works best in pushing two ways from a central point. Stripping downhill is more efficient but it is not always possible.

Figure 15-54 shows a series of steps by which a short wide shelf may be cut in a hillside, working from the side only. The material is piled in the background.

Uneven cutting may often be corrected by taking advantage of soil windrows, rocks, or other high spots to tilt the tractor up on the side where it cuts too deep. In narrow cuts, the tractor may be backed

up on one side of the cut opposite the high or hard spots, or may turn so as to cut from the opposite direction. If no natural helps are available, the tractor may be walked onto boards or other lifts placed by hand at the low side, to start the cut at the desired slope.

Transporting. While transporting material the dirt which flows off the sides of the blade must be checked or replaced to keep a full load. It may best be checked by moving repeatedly in the same path, so that the ridges built up in earlier passes prevent dirt from leaving the blade. It may be replaced by digging down through the length of the push, just enough to replace the wastage, or by centering on a windrow left by a previous pass, which will replenish losses from the sides.

If the ground is smooth, the blade is held to just touch it or dig slightly. If the ground is uneven, an effort is made to cut enough in the humps to replace the dirt

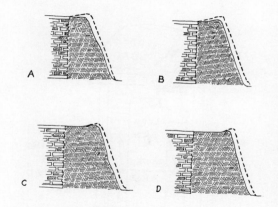

Fig. 15-57. Shaping a deep fill

Fig. 15-56. Piling with a dozer

that is lost in the hollows or drifts off at the sides.

Wastage through side spill may be reduced by having two or more dozers work side by side, with the blades touching, so that little or no material can be lost between them.

Piling. Figure 15-56 shows two common ways of piling dirt with a dozer. (A) is somewhat more efficient as it does not involve pushing so much dirt up, only to have it slide down again at the back, but the difference would be important only if very large amounts of material were involved. Often a pile started according to

method (A) is continued in the style of (B), because of the pile building back too near the digging.

If the earth is being pushed off a bank, the blade should be slowly lifted before reaching the edge. The loose fill sinks under the weight of the machine. The extra height built at the edge supplies a safety factor, and should be enough to keep it level after compaction.

Each push to the edge should be on a slightly different path, if space allows, in order to distribute its packing action.

Spreading. When spreading material, the blade is held somewhat above the original surface so that dirt can slip under it in a smooth layer on which the tractor can walk. A thin layer may be spread to the desired grade, but a thick layer should be built higher to allow for compaction. If there is not enough dirt ahead of the blade to reach the end of the area to be covered, it saves time to stop pushing as soon as the load is light and go back for more. The next bladeful will be pushed through the spot and can easily take the remnants of the first load with it.

It is best to vary the path used in spreading, as it is easier to keep track of the grade if no heavy windrows are built up.

Turning. A bulldozer has difficulty turning while pushing a heavy load. More

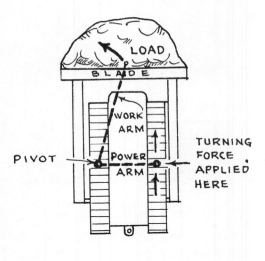

TURN TO LEFT

Fig. 15-58. Leverage in swinging a load

power is needed to swing the load than to push it straight ahead. As shown in Figure 15-58, action is that of a bell crank lever pivoting on the braked track. Turns are easier with a wide gauge machine, as the power arm is lengthened, and harder with an angledozer or other dozer with blade carried a distance ahead of the tractor, as that lengthens the work arm, reducing the leverage.

A clutch steering machine loses half its potential traction on a turn, and a differential steering job moves into a higher gear ratio, so that, in addition to an increase in load, traction or power is usually cut down. It is therefore necessary to lift the blade somewhat on a turn and to let part of a full load escape, to be picked up on another pass.

It is often easier to break the curve up into two or more straight lines separated by angles; to pile the soil at the angle points, and to make a separate process of pushing it along the next line. The heaps at the angle points are best moved frequently instead of being allowed to pile up until difficult to dig.

Scalloping. Accurate grading is difficult on rough ground, or in rocky soil, as any pitching of the tractor is exaggerated in the movement of the blade. The blade control is not fast enough in most machines to be kept level while the tractor oscillates. If it is allowed to dig in on a drop, the material is apt to be left in a pile just beyond, and the tractor, in walking into the hole and up on the heap, will pitch even more sharply so that a series of scallops are made. Once this process starts, only expert operators can level out without going back to the beginning and doing it over.

This scalloping is the bane of beginners and is liable to be troublesome until a feeling for the balance of the machine is acquired. An experienced operator can tell —usually without being conscious that he does so—when the machine is about to either rear up or pitch down, and start moving the blade to compensate. If the change promises to be very abrupt, he can slow the tractor by cutting the throttle or slipping the clutch and using brakes, if necessary, to get more time to change the height of the blade.

A difficulty with hydraulic dozers, and a few cable jobs, is that the drop below the line of the tracks is limited and usually is not enough to enable it to work over a peak and start a grade down the other side; or to make a sharp down cut from a level grade, except by preparing a heap or ramp to back up on, to tilt the dozer down.

Backblading. After an area has been graded it may look a bit rough because of small windrows of loose dirt, grouser prints, and piles left where the dozer turned. These may be smoothed down by backing over the area with the blade floating. It acts as a drag, smoothing off humps and filling hollows, but does not move enough dirt to change the grade. This is called "Okie dozing."

Soil that has been pushed into places

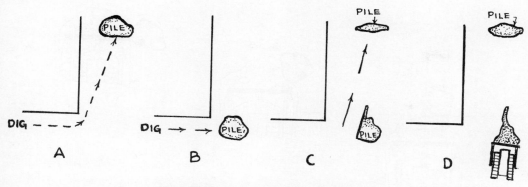

Fig. 15-59. Pushing around a corner

where the dozer cannot get behind it can be dragged in the same manner. Hard soil or a large quantity of loose dirt is better pulled with down pressure.

Rocks—in Cuts. Rocks of flat or irregular shape may catch under the blade knife and be pushed along with it, increasing the resistance or floating it out of the ground. The blade may sometimes be freed by shaking it up and down, but it is often necessary to stop and back in order to get behind and under it.

Firmly imbedded rocks even of rather small sizes may roll or slide the blade up so that a bump is left. It may be necessary to back up, then move forward forcing the blade into the ground deeply enough to get a grip on the stone that will roll or push it out. It may then be pushed ahead a few feet, with the blade high enough to let dirt slide under it, and that dirt backbladed into the gouge.

In digging out large rocks and stumps the blade action usually is a combination of push and lift. The push is faster and more powerful but objects requiring chiefly upward motion may be handled by slipping the flywheel clutch, or releasing one steering clutch in low gear, so the blade is pushed against the object without jamming it, and lifting at the same time. This technique should be used sparingly

because of excessive wear on the clutch.

In torque converter-equipped machines, the converter usually takes care of the drive slippage without resort to clutches.

Drive may be weakened in proportion to hoist by slowing the engine. No extra wear is involved, but full hoist speed is not maintained.

The brake-actuated input clutch mentioned earlier keeps full hoist speed with total loss of push.

Rocks—in Fills. Coarse material, such as rocks, lumps of sod, and other debris, made grading difficult or impossible. If a high fill is being made or a hole is dug to bury the trash it can be worked over the front of the fill and buried. As the blade is raised in spreading dirt the coarse material has a tendency to stay ahead of it to the last, although some will slip under and sometimes force the blade up so that the grade is lost. Such pieces can be moved along a bit further with the next pass, and eventually gotten over the edge, although it is often quicker to get off the tractor and move such small, difficult pieces by hand.

If the stones are not to be buried but are to be left at the side for other disposal, they can often be worked over without making special passes for the purpose. Referring to Figure 15-60 (A), it will be seen that if a dozer pushes a bladeful of dirt

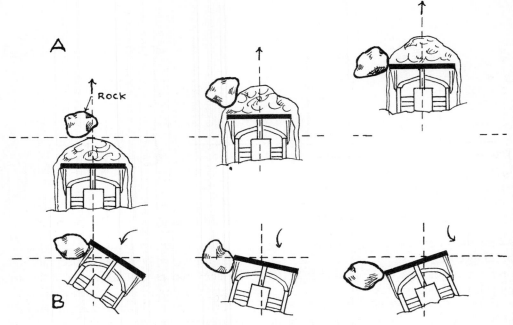

Fig. 15-60. Side-shifting a boulder

into a loose rock, the rock will tend to drift off to the nearest side. If, during a series of pushes, each one is aimed to slide the rock in the same direction, it may be moved out of the area without any direct contact with the blade.

While the dozer is backing up, it can move a rock sideward by the maneuver shown in (B). When the side of the blade touches the rock, the steering clutch on the opposite side from the stone is released and the brake applied hard. The dozer will spin on the braked track, the side of the blade will push the rock, and it may move several feet. The rear of the track is sometimes used in the same manner while going forward to move a rock a short distance when restricted space makes it difficult to get at it otherwise.

Very heavy rocks, or those embedded in soil, should not be moved in this manner because of excessive strain on the tracks.

Pitching. Tracked vehicles pitch badly in walking over ridges, stones, or poles. Figure 15-61 shows a series of positions assumed by a dozer walking over a small log. After overbalancing, the machine generally falls with a crash which is very damaging to both tractor and operator. Such a log or bump should be pushed out of the way or avoided if possible. Crossing, if necessary, should be done slowly and at an angle, so that one side of the machine crosses the top and starts down while the other is still climbing. This slows the fall, and avoids any danger of turning over backwards. If the bump is a soft ridge, turning the tractor sharply while crossing it will cut it down.

When a bulldozer is digging at capacity, the tracks often spin a little, then grip and move, and then spin again. Each time they spin, they build up piles of dirt at the back. If the machine backs up in the same track, these piles will have the effect of the log in the illustration, causing the back to rear up and then fall. This jolting can be avoided

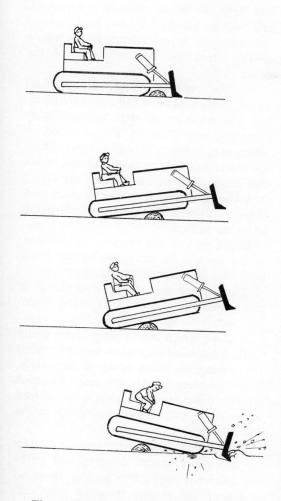

Fig. 15-61. Wrong way to cross a log

by keeping a wide enough work strip so that it is not necessary to use exactly the same path backward as forward. A tractor is not much affected if only one track crosses a bump.

Tracks will carry the machine over a narrow ditch without any pitching, if the movement of the tracks does not break down the banks. A pair of quite frail boards will often be sufficient to prevent caving as they will protect the bank against the backward push of the grousers. Crossing of a ditch wider than one quarter to one third of the length of the track on the ground is liable to be unsafe unless it is quite shallow.

Rear Power Control. Old dozers may have pumps or winches driven from a rear power take-off, so that they do not operate when the flywheel clutch is released. This lowers the efficiency of the machine in hard or rocky digging, in uprooting stumps, in handling of bulky objects, and in many jobs where it may be desirable to stop the machine momentarily to raise or lower the blade, and then continue. The operator of a rear pump job will de-clutch, shift into neutral, adjust the blade, shift into gear, and engage the clutch. With a front pump the clutch is merely disengaged and re-engaged.

The effect of a front pump may be obtained by stopping the tractor by releasing both steering clutches instead of the flywheel clutch. This involves greater driver effort, and may result in higher maintenance because of the greater cost of the steering clutches.

Gears. Bulldozing may be done in low or second gear. Low is less taxing for the machine in heavy pushing, and easier for the operator in precise work. Second is considerably faster, and in clean material good loads can be moved and smooth grades maintained. However, the higher gear makes it harder to cope with stones, hard spots, and other difficulties. Loose gravel may float a loaded blade so that it keeps a good grade automatically.

The amount of traction often determines whether a particular job can be done in second. A machine with narrow tracks, flat shoes, or worn grousers, or a standard machine on loose soil, spins the tracks easily so that when too deep a bite is taken or an obstacle is hit the load or shock to the engine is cushioned by slippage in the tracks, giving the operator time to raise the blade or disengage the clutch before stalling, even in a high gear.

Dozing Cycle. Most bulldozer digging is

done in shuttle fashion with the machine facing in one direction through the dig, push, spread, and return parts of the cycle. This is because the distances covered are usually quite short and turns, particularly in soft dirt, take time and spoil the grade, so that it is quicker and easier to back to the cut than to make two turns in order to use a higher gear. On pushes of one hundred feet or longer the turns may be better unless the machine has a fast reverse.

Many tractors in the medium and small class have only one reverse, while heavy ones may have from two to six reverse speeds. A bulldozer needs several reverse speeds almost as much as it does forward ones. Backing up is the unloaded part of the cycle where lack of work should allow high speed, but if a single reverse is used it must be powerful enough to climb steep grades and pull out of mudholes, and so cannot be fast.

The backup part of the cycle may be put to work by using back-ripper teeth on the blade to loosen soil for the next push.

Speed in reverse may also be limited by the quality of grading done during the push. In making heavy cuts it may not be efficient to take time and skimp loads to make a level floor with each pass, particularly if the soil is coarse. Gouges and bumps may be left by the blade and humps made by spinning tracks. The result is that a slow return may be made to avoid pitching even when a higher gear is available.

Hill Work. Dozers may be used on moderate side slopes and wide track models on steep ones up to 30° or more. However, they are quite likely to upset unless care is used. A machine which appears to have an ample margin of safety may be suddenly flipped on its side by running over a stone with the higher track, at the same moment that the lower track enters a hollow or soft ground. This is less apt to occur if the machine is pushing rather than walking, as it will then be moving slowly,

will have the blade close to the ground, and will be steadied by the load. It also obtains some support from the windrow spilling from the downhill side of the blade.

Working on frozen slopes is hazardous as the grousers may act like skates and allow the machine to slide uncontrollably downhill, regardless of the direction in which it is facing or trying to move. Sharp ice cleats will hold in such conditions but dirt grinds their points off very rapidly.

A similar danger is encountered on rock slopes, particularly shale with beds parallel to the surface.

Slopes on soft fills are very treacherous, as the tip will be increased by the lower track sinking deeper than the upper one.

If a machine starts to roll over slowly, it can sometimes be saved by turning downhill and lowering the blade.

A slope which is too steep to be safely worked sideward may sometimes be graded by running the dozer along it diagonally. If it is too steep for this, soil may be pushed straight down from the high spots, moved along the bottom, then pushed up to the low places.

Dozers can safely negotiate very steep up and down grades. Digging and pushing efficiency are much greater downhill and taper off to zero on steep upgrades. Steering is apt to be tricky on steep slopes, whether up or down, because of track slippage and shift in the center of gravity. Very steep grades of 25° or more should be climbed forward rather than in reverse, because of better balance and traction.

Cutting should be done downhill whenever possible and in very hard ground it may be advisable to dig it downhill, even if the spoil must then be pushed up the same hill for disposal.

The engine oil pressure gauge should be watched closely on steep work as some engines do not get proper lubrication when tilted steeply, especially at compound angles, and a low oil level which still gives adequate lubrication on a level may leave the pump dry on either up or down grades.

Where a run includes a down slope the operator may push several loads to the top, then push most of the resulting pile down in a single pass.

Unless the ground is loose a dozer can push much more of it than it can cut and move while cutting. Here again it may be good technique to drop one or more loads at the end of the cut, pushing the final load all the way through, along with the bulk of dirt piled previously.

Cable Operation. There are a number of differences in operation and performance between the hydraulic and cable controls. Cable usually moves the blade somewhat faster, the lift having a speed of a foot a second or better, and the drop is that of a free falling body. If a slower lift is desired, it is usually made in a series of short jerks, although some cable controls allow partial engagement of the clutch for a slow smooth rise.

The blade may be lowered by allowing it to fall, but it is desirable to check it with the brake just before it touches the ground to ease the shock, and it is necessary to have the brake applied just after it hits so that the drum will not continue to revolve, unspooling and snarling the cable. A smoother technique is to let the brake drag slightly as the blade goes down, or to alternately release and apply it (pump it down).

The depth of cut is regulated by the cable, but control is not as complete as in the hydraulics. The blade cannot be forced down by the weight of the tractor, so it tends to ride up over hard surfaces. If it uses a sharper cutting front, the blade is more apt to be sucked down into plastic soils. If a tractor has springs, a blade with too much suction will pull it down by compressing the tractor springs, and so will cut too deep unless carefully con-

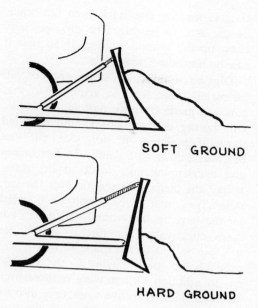

SOFT GROUND

HARD GROUND

Fig. 15-62. Dozer blade pitch

trolled. Engine torque partly counteracts the pull-down.

Digging may be done with a few inches of slack in the cable allowing the blade to find its own depth, or with a taut cable. Grading is done with a tight cable. It is bad operation to let the cable get very slack as it snarls and knots, shortening its life, the excess may catch on objects, and lifting of the blade is delayed until the slack is reeled in.

Cutting Hard Ground. If the blade refuses to cut down, it should usually be tipped forward by means of the pitch rods. This helps it to cut when it is empty, but a load can float it out of the ground more readily. If it will not cut into humps or a bank, which is less usual, it should be tipped back. For general pushing the blade should be centered or back, as it rolls the material most effectively in that position, reducing friction. See Figure 15-62.

If the blade will not cut after adjusting the pitch, a limited amount of ground can sometimes be cut up by spinning the tractor, first on one track then the other, on the area to be dug. The grousers may chew up the ground sufficiently so that it can be bladed off readily.

Digging in hard or stony ground will be easier if the work is arranged so that cuts are made by only part of the blade. One corner of the blade should be tilted down if possible. If not, one cut can be ground down to a depth of a few inches and overlapping cuts made at the sides. If half the blade is in the air over the cut, weight per inch of edge will be doubled for the part on the ground and it will cut more effectively. The dirt it gets under will add weight to the blade so that it will probably do some digging in the original cut in addition to removing a substantial bit at the side. Effective work may also be done by working outward from a cut, herringbone fashion.

If any considerable quantity of hard dirt must be bulldozed, it should be loosened by back ripper teeth, by a separate ripper, or dug with hydraulic controlled dozer shovels or dozers.

Because of the usually unlimited drop of the blade below the tracks, cable dozers are able to cross sharp ridge tops without losing blade contact with the ground and are therefore preferred in cutting roads or firebreaks across rough country.

Comparison, Cable and Hydraulic. New cable dozers are now quite rare. However, there are large numbers still in service, so the following discussion has been kept as a matter of general interest.

Cable bulldozers have a number of work differences from hydraulics. They have faster lift and drop, and start moving with a yank which is very effective at breaking out. The blade can be dropped far below the tracks and usually has a high lift also, but it cannot be forced down into the digging by the weight of the tractor. The sharper digging angle often given to the edge and lower moldboard to com-

pensate for this makes dumping loads difficult on steep upgrades, as the dirt rests on the raised blade.

On the maintenance side, cables need periodic replacement. The cost of the cable is not large but the nuisance of having it on the job and installing it during work time is a factor to consider. More lubrication points are present, and sheave bushings and clutch and brake linings need occasional replacement. The dozer is useless for lifting the machine out of mud.

Hydraulic systems have no fast wearing part to correspond to the cable, but packing glands need regular inspection, with adjustment and replacement; other leaks in the system must be watched for; breakage of flexible hose or pipes can put it out of action, and pumps, rams and valves need occasional replacement or repair. Oil must be added to the system rather frequently and changed occasionally. Operation is slowed by cold.

The decision about which type to use depends on the work and on the size of the machine. If the digging is such that a cable blade has difficulty cutting it and a hydraulic digs it easily, a hydraulic should be used. If it is so hard that the hydraulic has difficulty, it should be broken up with rippers, after which either can handle it easily. Stumps in soft ground respond best to the yank of the cable, but in hard ground down pressure is needed to get a standard blade under them.

Cable blades give much better penetration in large sizes than in small. The smallest dozer blade is about five feet wide and weighs two or three hundred pounds, while the largest mounted on a crawler tractor is about sixteen feet wide and weighs several tons. It will be seen that the weight on each foot of edge is much greater in large machines than in small, and it is this which largely determines whether a particular blade edge will pene-

Courtesy of Balderson Inc.

Fig. 15-63. U-blade handling garbage

trate. The largest machines have a penetration sufficient for most soils, and although hydraulic dozers of the same size can put several times as much weight on the knife, this extra is less often needed.

Hydraulic blades are therefore usually preferred in medium and small dozers, and cable in large units. This is not a hard and fast rule, as even a large dozer purchased for work in hard ground, in soft mud, or at miscellaneous work should be hydraulic, where a small one to be used only in pushing loose material might as well be cable.

Output. Dozer output falls off in almost direct ratio with increase in distance. Use of scrapers, or shovel and trucks, should be considered for pushes of over 100 feet.

If space is ample, the large dozer will move dirt at lower cost per yard than a small one, and its advantage is increased in hard or rocky digging. In land clearing, the large machine is even more economical.

In restricted quarters, as in landscaping, backfilling trenches in narrow spaces, and working inside buildings small dozers may show larger production and much lower cost than large ones.

Dozer output varies more than that of any other excavator. Production tables are included in the Appendix, but the figures should not be used in estimating without careful checking.

SPECIAL CONSTRUCTIONS

U-Blade. The U-Blade, Figs. 15-25B and 63, has the sides advanced further than the center. This makes it possible to transport a larger load by reducing side spill. The pointed corners assist penetration in hard soil and under stumps and boulders. It functions well in rough pioneering on side hills, and is ideal for handling coal, garbage, and other loose materials.

The edges will cut deeper than the center if the blade is below track level.

Bowldozer. The Bowldozer, Figure 15-64, is the highest capacity dozer blade. It utilizes a standard dozer mounting with tilt control.

The rear wall is similar to a regular blade, except that it has a scraper-type edge with an advanced center section. The sideboards extend forward from 4½ to 7 feet, and are held rigid by a cross beam connecting the lower front corners. This includes another cutting edge, set at a sharp angle.

This unit is bottomless. It is raised and backed at the end of its push. It can handle most types of material, but may be damaged in rough digging. It is specially adapted to light material, such as coal, or high volume work such as supplying a belt loader.

Angle Dozers. Angle dozers, known under various trade names such as Angledozer and Gradebuilder, are bulldozers with

15-63

Fig. 15-64. Bowldozer

blades that can be angled to left or right, in addition to a center straight across position. They may have either hydraulic or mechanical hoists.

The main frame or C-frame consists of the push arms and a V or U-shaped front connection between them. The blade is fastened to this by a center vertical hinge pin, or by a pair of pins.

The outer ends of the blade are hinged to adjustable landside arms, that connect each of them to any one of three brackets on the push arm. When both landside arms are in center brackets the blade is straight across, while if one arm is back and the other forward, the blade is angled to cast away from the forward fastening.

Adjustment of angle may be made by

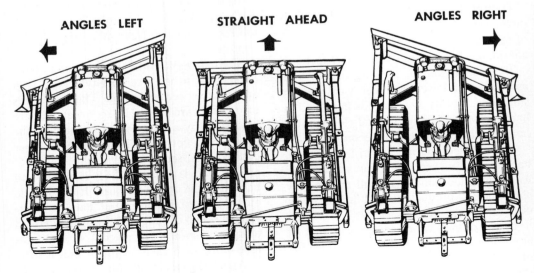

Fig. 15-65. Angling dozer

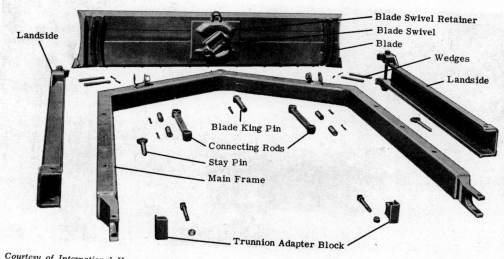

Landside — Blade Swivel Retainer — Blade Swivel — Blade — Wedges — Landside — Blade King Pin — Connecting Rods — Stay Pin — Main Frame — Trunnion Adapter Block

Courtesy of International Harvester Company

Fig. 15-66. Bullgrader parts

hand, with the hoist holding the blade clear of the ground. If it is too heavy or stiff for this, the lock pins are removed from the brackets, and one side of the blade is walked slowly into a bank. This will angle the blade. Pins are placed as soon as the landsides and the brackets line up.

The blade may be tilted by any one of the means used for standard blades, or by rotating it by hand on a horizontal center hinge, and using pin fastenings between the push arms and a curved track on the back of the blade.

The blade is wider than a straight dozer model to enable it to cut a full width path when angled. The bottom of the moldboard and the edge are flared out at the lower corners so as to make a full, sharp edged cut when side casting. This causes some undercutting of edges in straight work.

The moldboard may be lower than a straight blade and curved in more deeply, from bottom to top. This increases the rotary movement in dirt being pushed, and aids in drifting it toward the side.

When the angled blade is tilted so that the heel (rear corner) of the blade is down,

the heel may be used in dislodging difficult objects and in digging ditches narrower than the blade.

The angled positions add a sideward movement of the load to the forward motion which can be utilized in leveling and changing slopes whose slant is across the path of the machine, in crowning roads, in side casting earth back into small trenches, and in making shallow ditches. It tends to cause the tractor to turn oppositely to the side drift of the dirt.

Courtesy of J. I. Case, a Tenneco Company

Fig. 15-67. Hydraulic angle and tilt

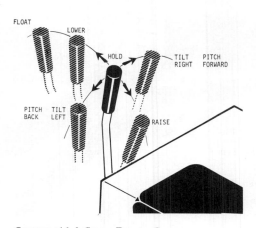

FLOAT
LOWER
HOLD
TILT RIGHT
PITCH FORWARD
PITCH BACK
TILT LEFT
RAISE

Courtesy of J. I. Case, a Tenneco Company

Fig. 15-68. Hoist-tilt-pitch control

The angling blade is desirable in pioneering roads across rough country, and is superior to the straight blade in light trench backfilling and some other jobs. Its drawbacks are somewhat greater weight, cost, and upkeep; clumsiness in restricted spaces, difficulty in turning with a load, and looseness in the joints. Many of its functions can be served equally well by a power tilting dozer.

Hydraulic Angle-Tilt-Pitch Dozer. The most versatile dozers perform all blade adjustments by hydraulic power, with on-the-move adjustments by valve controls within reach of the operator.

An example is shown in Figures 15-67 and 68. The main dozer frame is a C-frame, conventionally hinged near the tractor dead axles, and fastened to the center of the back of the blade by a universal bearing.

The C-frame is raised and lowered by a pair of cylinders mounted to the trunnions on the front wrapper. The blade is conventional, being about as high as a straight dozer.

Angling is effected by a pair of horizontal hydraulic cylinders mounted above the C-frame, and a control lever at the right side of the operator's compartment.

The angle cylinders are connected to a pair of struts extending forward to the top and near-bottom of the blade. The upper struts are hydraulic cylinders that lengthen or shorten to regulate tilt and pitch of the blade.

Blade lift and tilt are controlled by a single hand lever to the left of the angling lever. Pitch is controlled by the same lever in conjunction with a foot pedal. The lever is a standard 4-position control, with FLOAT at the front and RAISE at the rear. The lever is pushed sideward to tilt the blade.

If the valve foot pedal is depressed while simultaneously pushing the hand lever sideward, blade pitch is obtained rather than blade tilt. Maximum pitch can only be attained when the tilt adjustment is centered. A separate hand lever controls the angling function of the blade.

Clearing Blades. There are a number of special blades and attachments used chiefly or entirely in land clearing. These will be described in Chapter 21. They include stumpers, rake and rock blades, and tree dozers.

CHAPTER SIXTEEN

TRACTOR LOADERS

FRONT LOADER

The most advanced development of the bulldozer is the front end tractor shovel. This machine may be called a shovel dozer, dozer shovel, tractor loader, end loader, front loader, or just loader. It is used for digging, loading, rough grading, and limited hauling.

A typical front loader, Figures 16-1 and 3A, includes a support frame on a tractor, a hydraulic system, a pair of push or lift arms (the boom) hinged to the top of the support frame, a tractor-width bucket hinged to the front of the arms, and a pair of dump arms hinged both to the push arms and to the bucket. Non-tractor movements are under control of two pairs of hydraulic cylinders.

The front loader may be carried by any type of tractor. Crawler and four wheel drive tractors are used for heavy service, and two wheel drive for lighter work.

CRAWLER MOUNTING

Crawler tractors that carry loaders are usually specially designed for them, and differ from standard models. Tracks are wide gauge, and are made extra length, with an additional track roller on each side. The idler and sometimes the front roller are of extra heavy duty design. Width is necessary for stability against side tipping when carrying high loads. The long tracks move the center of balance forward so that heavier loads may be broken out and carried. The heavier idler and roller construction is required by the heavy front loads.

Tractors redesigned to carry loaders do not have springs. Most of them have a rigid connection between the track frames and main frame at the front, thus improving stability at the expense of some operator comfort and grading control.

If drive is mechanical, the engine clutch must be rugged, as a loader is very hard on it. It is likely to have ceramic discs instead of lining, or operate in a circulating and cooling oil bath.

However, most new loader tractors have torque converters, teamed with power shift transmissions. This construction avoids the problem of slipping a clutch, and improves lugging qualities in the bank. Power shift improves flexibility and shortens cycles.

Hydrostatic drive may replace the converter and transmission.

Loader Frame. The frame is composed of a massive weldment fastened to the track frames and/or the central frame. It carries the pivot or hinge pins for the push and dump arms and their hydraulic rams, and transmits the weight, thrust and twisting strains of the loader to the tractor.

In the past, the fastenings of the loader frame to the tractor have been a weak point

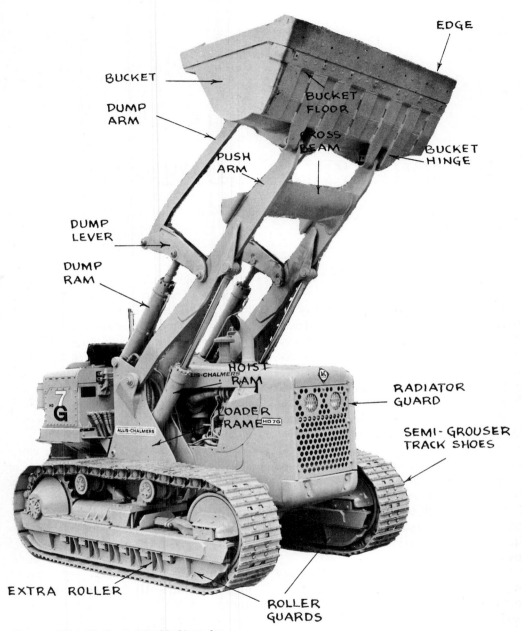

BUCKET

DUMP ARM

PUSH ARM

DUMP LEVER

DUMP RAM

EDGE

BUCKET FLOOR

CROSS BEAM

BUCKET HINGE

HOIST RAM

LOADER FRAME

RADIATOR GUARD

SEMI-GROUSER TRACK SHOES

EXTRA ROLLER

ROLLER GUARDS

Courtesy of Fiat-Allis Construction Machinery, Inc.

Fig. 16-1. Crawler mounted front loader, early model

of the machine. This difficulty has been largely overcome by wide basing and heavy pinning, Figure 16-2A, or by making a one-piece frame for both the tractor and the loader, Figure 16-2B.

In either case, the frame connections or

stress points should be inspected periodically, and tightened or welded as required.

Arms. The push or lift arms are hinged to the top of the columns or tower on the frame. They extend forward to hinges near the bottom of the bucket. A cross beam braces them near the front. The arms-and-brace assembly may be called a boom.

Lift arms are raised and lowered by two way cylinders in the bases of the columns.

In principle, the dump arms are a connection between the back of the bucket, above the lift arm hinges, and the columns. This connection includes hydraulic cylinders, whose lengthening dumps the bucket, and whose shortening rolls it back.

The actual mechanism is more complicated, to supply changes of leverage, stability, and/or a mechanical (parallelogram) linkage to automatically level the bucket floor as it is lifted. This last function may be taken over by hydraulic valve arrangements.

Bucket. The bucket is of simple box construction. A heavy cutting edge of tempered steel runs along the front and part way up the sides. The upper back curves forward. It is placed as close to the tractor as possible, for stability.

Buckets are usually the same width as the outside of the tracks. Some early models had narrow buckets, and light material buckets may be wider. Present size range for standard weights ranges from 1 to 4½ yards for crawlers. Wheel tractors have both smaller and larger sizes, from 5 cubic feet to over 20 yards.

Buckets are made in different sizes and weights for various types of material and work conditions. Light material buckets for handling humus, sawdust, or snow may be from 40 to 100 per cent larger than standard buckets. Rock buckets are heavily reinforced. Slat buckets are used for handling loose rock or wood. They allow unwanted dirt to fall away through the slots.

Teeth are standard equipment on rock buckets, and optional on standard weight. They help greatly in hard digging, and in handling rock, stumps, and brush, but they interfere more or less with grading. Their cost is partly offset by the protection they give to the cutting edge.

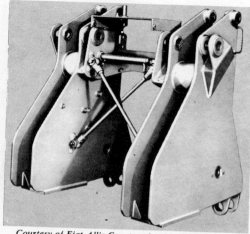

Courtesy of Fiat-Allis Construction Machinery, Inc.

Fig. 16-2A. Heavy loader frame

Design must be a compromise. The bucket should be strong enough to take any punishment the tractor can give it, but light enough to raise a big load without overbalancing the tractor, and without absorbing too much of its lifting power.

Bucket Action. The standard bucket has three working motions. It is raised and lowered by two way rams controlling it through the push arms, it is tilted or rolled between carrying and dumping position by the dump rams and linkage, and it is crowded and retracted by the forward and reverse travel of the tractor.

Dumping height is the elevation above ground level of the lip of the bucket in

Courtesy of Caterpillar Tractor Company

Fig. 16-2B. Integral loader-tractor frame

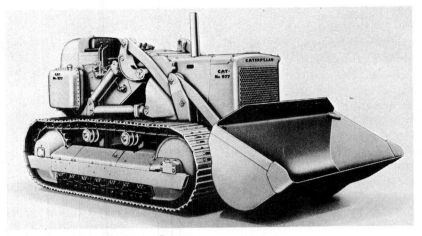

Courtesy of Caterpillar Tractor Company

Fig. 16-3A. Rolled back bucket

dumped position. It may be several feet below the height of the lip in carrying position, and 1½ to 2½ feet below the bucket hinges. Maximum dump height varies from 7 to 10 feet, being greatest in the larger and newer machines.

Manufacturers seldom supply figures on hoist speed. The bucket may raise at the rate of .9 to 1.5 feet per second, and lower at about twice that speed.

The bucket will usually tilt more than 100 degrees between dump and full back positions. At maximum height dumping slope of the bottom is 45 to 50 degrees. At ground level a bucket may be kept fully dumped for float-grading while moving forward.

Courtesy of Fiat-Allis Construction Machinery, Inc.

Fig. 16-3B. Slat loader bucket

A bucket is said to be rolled, tilted, or curled when the floor is tilted so as to retain a load. Some older machines could not tilt back from a flat position at ground level.

When a bucket is rolled back during penetration into soil, it pivots on its heel (rear of the floor) as a fulcrum, usually developing much more breakout force than can be provided by the hoist. It does this without pulling down the tractor.

Rollback is also useful in slicing upward in hard or heavy banks. It makes it easier to pick up heavy, heaped, or sloppy loads, and oversize objects. They can then be carried at a safe level, two or three feet above the ground.

Rolling back improves balance slightly by moving the load toward the tractor.

There should be an indicator on a loader arm to show the tilt of the bucket, as this is usually difficult to observe directly. Its usefulness may be improved by a paint mark, a strip of bright tape, or a weld tack at level-on-the-ground position.

Automatic control of rollback will be discussed under OPERATION.

Rolling back is useful in slicing upward in hard or heavy banks, and in picking up and carrying heavy, heaped, or sloppy loads, and oversize objects. It improves balance

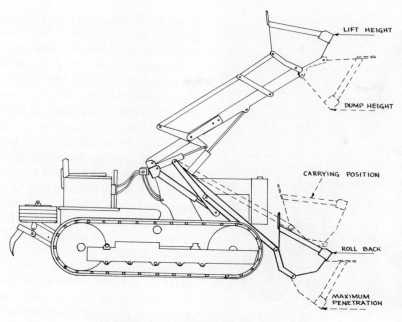

Fig. 16-4. Range of bucket positions

by moving the load toward the tractor.

Side Dump. The bucket shown in Figure 16-5 can be dumped forward in the usual manner, and in addition can be dumped to the left side by means of a cross ram and a separate control valve. The left side is prolonged into a chute to provide dumping clearance.

This bucket, available on both crawler and wheel mounted loaders, permits the machine to dig windrowed material along a roadside and load it into trucks standing in one lane of the pavement, without any turning of the tractor.

It may also be used for pit loading, dumping either to the side or ahead, depending on conditions. Digging thrust is slightly off center, and the extended edge is a disadvantage in hard digging, but on many jobs neither of these differences is noticeable.

Hydraulic System. Loader hydraulic pumps are usually mounted at the front of the engine for direct drive by a crankshaft extension. Capacities are large, varying from 15 gallons per minute upward.

The reservoir, holding up to 60 gallons of oil, is equipped with filters to remove outside dirt and products of wear. It may be a closed system, or an open one with a filtered air vent. It is important to keep oil at the proper level. Too little will allow the system to suck air and perform jerkily, too much may cause squirting out of the vent or building up of damaging pressure when lowering a loaded bucket.

Hoist and dump are controlled by two valves at the right of the cockpit. The hoist valve has four positions, FLOAT at the front, then DOWN pressure, HOLD, and UP at the back. Figure 16-6B shows four arrangements used for the dump valve. In each case it has three positions, a center position to lock it against change in tilt, one to dump it, and one to rotate it back.

The Fiat-Allis units and a number of other machines have a dump valve handle that moves the same way as the top of the bucket—forward to dump, back for roll-back.

In the Caterpillar the handle moves op-

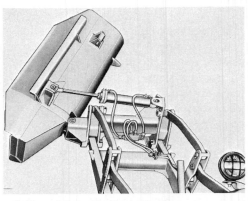

Courtesy of Caterpillar Tractor Company

Fig. 16-5. Side dump loader

positely, for convenience in controlling raising and dumping at the same time. Drott and others have a combination control, in which moving the hoist lever to the right (outward) dumps the bucket, and moving it left (inward) rolls it back.

The control bank may carry an additional valve and lever, for use in operating a multipurpose bucket, a side dump bucket, or a rear mounted ripper.

Sometimes there is a diversion valve used

to shut off the loader system, and divert pressure to other valves for control of a ripper, scraper, or other rear unit.

Counterweight. Present models are stable enough for almost any job if operated with reasonable care. However, a contractor owning an older, less stable loader, or wishing to handle very heavy loads, or to work frequently on rough ground, may obtain additional stability by weighting the back.

On a small machine, 500 to 1500 pounds will permit lifting and carrying heavier (perhaps too heavy?) loads, and will improve traction when carrying a load. Larger machines can carry proportionately greater rear weights.

Counterweight may be a dead weight of metal or concrete attached to the back of the transmission case by bolts. It is most effective if it projects well behind the tractor, but is then most in the way.

The weight may also be a working part, such as a rear mounted ripper, a power

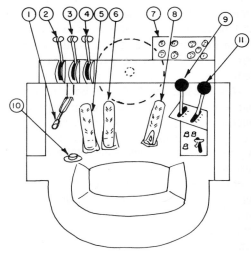

1.	Parking brake	6.	Brake pedal
2.	Hand throttle	7.	Instrument panel
3.	Range lever	8.	Accelerator
4.	Directional lever	9.	Bucket control
5.	Brake pedal with	10.	Horn
	drive cutout	11.	Boom control

Fig. 16-6A. Controls, four wheel drive

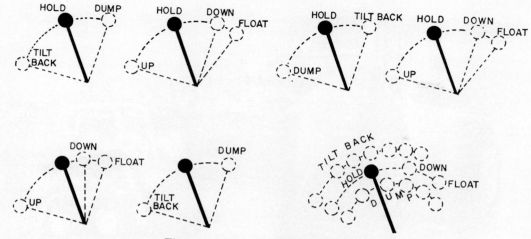

Fig. 16-6B. Loader control patterns

control unit, or a towing winch. The ripper enables the machine to break up soil in advance of grading or loading. However, the 7 to 30 inch layer of soil that can be reached by a ripper is not deep enough for efficient loading.

A counterweighted machine may be more difficult to steer with a load, as keeping the back down increases the track contact with the ground.

Track Shoes. Most crawler loaders have the semigrouser (3-cleat) shoes described on page 15-9. Some have 2-cleat types.

When shoe surfaces are flat, the tracks will spin rather readily on many footings, giving the effect of a slipping clutch. This tendency to spin cushions all parts of the tractor against shock loads, but it often interferes with steering and traction, and prevents the full power of the machine from being applied to its work. Semi-grouser and other special, semi-flat shoes give better traction but will still spin rather freely under slippery conditions. Full grousers grip well and aid in digging but make the machine very touchy and apt to stall, build up ridges behind the tracks, and subject machine and bucket to shocks and overloads that may shorten bucket life materially.

The chief objection to the use of grousers on a loader is that they tear up the ground when it turns, causing it to work itself down into holes that make work slow and sometimes dangerous. Flat shoes do this damage more slowly, or not at all if the ground is firm. Dirt loosened in this manner is easily smoothed off but will dig up again on the next turn. Under such conditions truck positions may have to be changed frequently to keep the loader on good footing.

Spinning grousers will also build up piles behind the tracks, which may upset the machine when it backs up with a load.

As semi-grouser shoes wear, the heads of the attaching bolts become exposed and wear off, and the edges of the shoes are weakened so that they may bend. Their life can be prolonged by welding an alloy strip along the center bar, or on all three bars. A tractor patched in this way will have good traction, but may give a rough ride on hard ground.

Flat or semi-grouser shoes are not suitable for work in mud, as they become so slippery as to give little traction, and will not grip boards or other objects placed under them to assist in climbing out of holes. They also permit the machine to

Fig. 16-7A. Wheel loaders, 1200 and 37 horsepower

slide downhill when working on a side slope.

For special conditions, grousers may be bolted on the regular shoes, or grouser shoes may be substituted for them. Grousers may be bolted on only six or eight evenly spaced track shoes on each side. This provides more traction in mud than a full set of grousers, but is limited to very soft ground, as it is too rough elsewhere.

WHEEL MOUNTINGS

Wheel mounted front loaders have the same basic principles as those on crawlers, but there are a number of differences in size and structure. Uses are strongly affected by type of carrier.

The most important units for heavy work are mounted on four wheel drive tractors. Smaller models are on skid steer or conven-

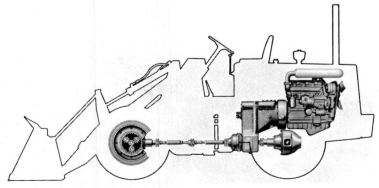

Fig. 16-7B. Power train, articulated loader

Courtesy of Ford Motor Company

Fig. 16-8A. Skid steer loader

tional two wheel drive machines. These carriers were described in Chapter 15.

Dump rams and arms may be hinged on the columns above or below the lift arms, as in the crawler machines. But they may also be on the same hinge pins, or based on the lift arms themselves. In the two last cases, the bucket remains at a fixed angle with the lift arms as they are raised or lowered, unless changed hydraulically by either manual or automatic controls.

Four Wheel Drive. Most of the medium and large size loaders on rubber are on articulated, four wheel drive tractors. Loader frame columns are forward of the swivel. Lift arms are therefore shorter than on the crawlers, and rise to a steeper angle for the same bucket height.

Most of these machines are equipped with torque converters and easy shift or power shift transmissions. It is important that the forward-reverse shift be of the clutch or power shift type, to avoid delays in meshing gears or synchronizing clutch and gears.

Operators sit high and forward, where they have a good view of the bucket, but are exposed to danger of objects falling off its back. The engine is in the rear.

Turning radius is much longer than that of crawlers, so more space is required to maneuver from bank to truck and back. Faster travel and usually faster shifting usually keep cycle time as short as that of a conventional crawler.

Wheel mounted loaders started out as re-handling machines, but are now rough and tough enough to handle almost any digging. They give best results in sand, gravel, and common earth, but also do well in hardpan and blasted rock.

In common with crawler mounted loaders, they can handle large boulders readily by carrying or pushing, can keep their pit area cleaned and leveled, and can be easily transferred from digging to dozer or tractor work.

They have the outstanding advantage of quick and easy moving from one part of a job to another. They can also be driven from one job to another, but most of them are so wide as to require special permits,

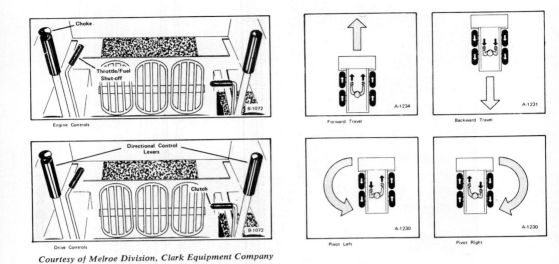

Courtesy of Melroe Division, Clark Equipment Company

Fig. 16-8B. Engine and steering controls, skid steer

and slow enough (30 mph or less) so that trailers may be preferred on long hauls.

Very large tires are generally used. They serve to provide excellent flotation, permitting work on most footings. Ground pressure is still much higher than with crawlers, but the packing effect of the tires and the more gradual turns make it possible to work easily on sandy ground that would tear up under crawlers, and cause excessive track wear. But slippery surfaces may cause loss of both traction and steering accuracy.

Skid Steer. Skid steer loaders are small and very compact. Weights range from less than a ton to almost 6 tons, with bucket capacities from 5 cubic feet up to 1⅜ yards (2 yards for light material). Dumping clearance is 6 to 8 feet. Drive is usually hydrostatic, but may be by belts or chains.

The lift arms are pivoted to triangular columns behind the rear wheels, and extend alongside the tractor for its full length. The cab includes steel mesh sides, to protect against operator entanglement with the arms. The entrance is at the front, across steps provided on the bucket.

Travel is controlled by a pair of hand levers, each of which causes its side of the machine to move forward or back, as the

lever is moved. The loading part is operated by rocker foot pedals, Figure 16-8C. The center pedal is installed for auxiliary hydraulically controlled equipment.

Quick coupling devices for changing buckets or other front end implements are available, and may be standard.

Skid steer enables these units to turn in approximately their own lengths. Combined with short wheelbase, this enables them to

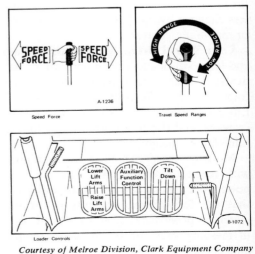

Courtesy of Melroe Division, Clark Equipment Company

Fig. 16-8C. Travel and loader controls

work comfortably in areas so restricted that many standard loaders could not even enter.

Two Wheel Drive. Two wheel drive loaders depend on good traction to dig competently. A large machine, equipped with a backhoe or a heavy rear implement, can give an excellent account of itself in medium digging, if the ground is neither slippery or loose sand. But poor ground or really hard digging make its work difficult and uneconomical.

These machines are excellent utility units, for general helping and cleanup work. In yards, they can load crushed rock and other stockpiles as well as four wheel drives.

Design is usually simple. Side frames are fastened to the rear axle housing and the sides of the radiator base, and columns, cross braced to each other, are erected on the frames. Lift arms hinged on the column tops extend forward of the radiator guard to the bottom hinge of the bucket. These are between the hood and the front wheels when lowered. Double-acting hoist rams are hinged on the column bases and push upward and forward against the arms.

The dump rams are hinged to the lift arms. Their piston rods are extended into dump arms that hinge to the upper back of the bucket. The pump is driven from the front of the engine. A single two-section control valve regulates movements of the hoist and dump rams. The standard bucket is box type, straight-edged with an option of teeth, and as wide as the rear wheel tread. Other widths are usually available.

DIGGING

Bucket Angle. Most loader digging is started with the floor flat, or tilted to a slight downward angle. This position gives maximum penetration into banks and high spots, and cuts a smooth path on which the tracks can follow.

Buckets that do not roll back are flat when the dump rams are fully retracted. For them, this is the best angle both for digging and for raising and carrying a load.

A roll back bucket is fitted with an indicator by which the operator can tell its angle of tilt. He may adjust it to flat position by reading the indicator, or by observing the bucket directly.

Cutting Down. For cutting down into a level surface, tip the bucket downward ten to thirty degrees. When it has penetrated to a depth of two to six inches it should be turned up to a flat or almost flat position, while the forward motion of the tractor is continued until the bucket is filled.

This tilt sequence combines good original penetration, sturdiest position of the bucket for most of the pass, and a powerful prying effect during the change in angle. Under some soil conditions continual minor adjustment of the angle while digging helps penetration.

Courtesy of Deere & Company

Fig. 16-9. Two wheel drive loader

The flat position is best for pushing a quantity of loose dirt, but the bucket should be turned down steeply for spreading and grading it so that dirt will flow freely off the floor into holes, and so it will not be pulled down by sticky soil. Care should be taken not to hook into solid obstructions at a steep angle, as the bucket is then in its weakest position, and leverage against the dump mechanism is at its maximum.

The depth of cut may be regulated either by the hoist or the dump rams. For any given position of the push arms the edge will be highest when flat and two or more feet lower when fully dumped.

Penetration. The loader bucket has much weaker penetration in proportion to size than a dipper stick, because it is larger in relation to the power and weight of the machine, has a wider edge in the digging, and may lack teeth. Another digging difficulty is that the hoist is very slow in proportion to the speed of the tractor, so that the bucket tends to get under more dirt than it can break loose and lift. The effect is of a dipper stick shovel with no retract and with the ratio between crowd and hoist fixed so that the bucket always tends to crowd too deeply into a bank.

Digging Banks — Gear Drive Crawler. A standard method is to use low gear to force the flat bucket into the bank toe at ground level. When resistance slows the tractor, the bucket is rolled back gradually and hoisted, while crowding with the tractor is continued.

If the tractor is slowed toward stalling before the bucket is full, it is stopped by disengaging the clutch, or slowed by alternately engaging and disengaging it. Clutch use may not be necessary if the tracks spin freely enough to allow the engine to maintain lugging speed.

Rolling back the bucket as it rises in the bank increases cutting efficiency by aligning the edge with its upward movement, and

Fig. 16-10. Starting a shallow cut

by retracting it for a thinner slice. Its own suction, and crowding by the tractor, tend to make the cut thicker.

The proper balance among these forces varies with the machine, the bank, and the position and momentum of the bucket. It is the operator's responsibilty to so balance them that he will get a good bucket load in minimum time.

If the bank is hard, a thin slice may be most productive. If it is vertical, no crowding may be needed after initial penetration. If it slopes back, more crowding and less rollback are needed.

In general, a nearly vertical cut face in a bank not more than a foot or two higher than the lift arm hinges, is most efficient for fast loading.

If the bank is higher than this, and overhangs, the upper part should be nudged with the full bucket occasionally to bring it

Fig. 16-11A. Bucket movement in bank

down. But if the overhang is substantial, dig somewhere else.

If the loader has a self-leveling device, it may be necessary to cut it off, or to override it, to keep proper tilt in the bank. If it

does not have it, you must remember not to give the bucket a high lift in full curl-back position, as this is likely to spill part of the load over the back - and probably on you.

If the bank is only two or three feet high, it may be dug by keeping the bucket at final grade and running beside the bank, cutting into it as much as possible without slowing the tractor excessively. The cutting will be done by the side and floor of the bucket which is in the bank. The soil will roll along the bucket and fill it, although the load will be heavier on the bank side.

If the machine is driven head-on into the bank and gets under too great a load, the bucket may remain stuck and the back of the tractor rise. Curl the bucket more, or shift into reverse and back a little. The tractor should then settle down, and the bucket come up with a good load.

Digging Banks — Torque Converter. If the tractor has a torque converter, resistance in the bank will affect tractor speed much more than hoist speed. Even with the tracks or wheels stalled, the engine loader's and the hydraulic pump may maintain good speed.

This has the favorable effect of slowing crowd speed relative to hoist. However, the crowding force is increased by multiplica-

Courtesy of Caterpillar Tractor Company

Fig. 16-11B. A big load in a low bank

Courtesy of Deere & Company

Fig. 16-12. Nudging down a high bank

tion in the converter, so the bucket is still likely to be crowded into more dirt than it can lift.

Converters cannot be operated at full throttle under stall or near-stall conditions for more than a few seconds without excessive heat buildup and strain.

If the throttle is cut back to idling speed,

the torque converter will exert almost no drive force. The loader pump will be slowed, but it will continue to .exert full pressure to lift the bucket at a slower speed. This combination favors breaking out an overloaded bucket. Part throttle weakens drive without stopping it.

The wheel mounted loader may be op-

Fig. 16-13. A good load to carry

erated with the hand throttle closed, so that engine governor setting is controlled by a foot accelerator. The operator cuts engine speed by lifting his foot.

Many four wheel drive loaders have an engine (input) clutch in addition to the torque converter. This clutch is power-released when you press one of the brake pedals lightly, or touch a separate pedal. This permits disconnecting the drive (crowd) without slowing the engine or the hoist.

The crawler usually works with the hand throttle wide open. A decelerator (opposite of accelerator) pedal is depressed to slow it down. This pedal you push when you start to get stuck in the bank.

The separate responses of push and lift to load and throttle are the chief features that distinguish torque converter operation. On a strange or old machine, freedom from fear of stalling or of damaging the clutch are also important.

Payload. The amount picked up in the bucket varies with the type of bucket, the power behind it, the traction, the nature of the material, and the skill of the operator.

For each yard of bucket capacity, you might pick up anything from one-half to one and a half yards, and should average seven-eighths of a yard in medium digging. A bucket must be rolled back to carry its maximum load.

When dumping close to the digging point, as in side casting, or loading a properly placed truck, keeping up a fast cycle is usually more important than getting maximum loads with each pass. As distance to the dump point increases, capacity loads become more important than the time taken to get them.

If the load must be carried over rough ground or backward up a slope, it should be limited to the weight the machine can carry easily without tipping.

If the bucket does not fill sufficiently the tractor should be backed, the bucket lowered to floor level, and another pass

Fig. 16-14A. Bucket, and chain grabs

made. If the load is one sided the second cut should be made at an angle to the bank so that the empty side will penetrate first.

It is often difficult for the operator to judge the amount in the bucket unless the digging is easy enough to permit dirt to be forced over the top. It is a good idea to cut a row of holes, about an inch and a half in diameter, in the upper back of the bucket, Figure 16-14A, through which the operator can see whether it is filled. Spillage through such holes is negligible in ordinary digging.

This figure also shows the construction

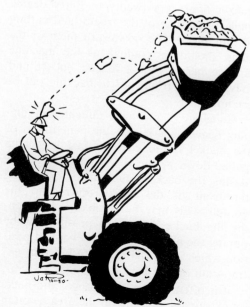

Fig. 16-14B. Too much tip-back

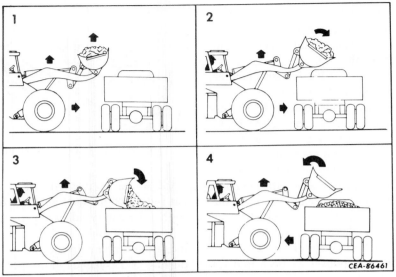

Courtesy of International Harvester Company

Fig. 16-15. Loading a truck

of grab brackets for convenient attachment of chains.

Ramping Down. If the digging is downward, as in cutting a cellar or a ramp, hard material may require pitching the bucket floor at a twenty to thirty degree angle to the line of the tracks, and cutting in thin slices. The downward pitch of the ramp should be gradual, as the machine is nose heavy with a loaded bucket and may tip over frontward in backing out of the hole.

Gouging. A common difficulty in digging heavy soils is that the penetration of the bucket is too good so that it will be pulled down by the slice it has dug, either raising the back of the tractor or pulling the front of the tracks down into the ground. This may be combatted by keeping the bucket as flat as possible, or by tipping the floor into a nearly vertical position which cures the difficulty but puts extra strain on the bucket.

It is often best to let it gouge, and make an extra pass to grade off the area as often as necessary.

Transporting. If the ground between digging and dumping is hard and smooth, backing, turning, and dumping can all be done at speed with safety. If the ground is rough the machine must move slowly, as going over a bump or ridge with a heavy load may cause it to fall forward, the bucket dropping to the ground and the operator's seat rising into the air. If part of the load dumps, the tractor will settle back. In crossing rough ground the operator should be alert to lower or dump the bucket if an upset starts. Lowering has the effect of taking the bucket weight off the tractor long enough to enable it to recover its balance.

Dropping a loaded bucket and stopping it abruptly in the air may burst a hoist ram hose if it does not overbalance the machine.

With crawlers, you cross ridges at such an angle that one track will be part way across before the other reaches it. If the ridge is soft it may be possible to cut a more level path through it by turning sharply while on it.

Loader Size, Cubic Yards	Loading Cycle Element	Well-Blasted Rock Average Cycle Element Time	Poorly-Blasted Rock Average Cycle Element Time	Common Earth Average Cycle Element Time
8 or less	Bucket Load	10		6
	Maneuver Loaded	10		9
	Dump	5	No Data	3
	Maneuver Empty	9		9
	Total Cycle Time	34		27
8 to 10	Bucket Load	8		5
	Maneuver Loaded	13		11
	Dump	5	No Data	4
	Maneuver Empty	11		10
	Total Cycle Time	37		30
10 or more	Bucket Load	10	15	8
	Maneuver Loaded	11	15	12
	Dump	5	6	6
	Maneuver Empty	9	12	11
	Total Cycle Time	35	48	37

Courtesy of U.S. Department of Transportation

Fig. 16-16. Cycle time, big front loaders

If the bucket is carried two to four feet above the ground the consequences of overbalancing are unlikely to be serious. If it is high the weight is not quite so far forward, so tipping is less likely; but if it occurs, it will give the operator a worse toss. A high held bucket also involves the danger of less likely but far more serious side tipping.

The bucket should not be given a high lift if there are rocks or lumps projecting over the back, as they might fall and injure the operator or the tractor. Normally such things fall to the ground, or on the front of the tractor without serious damage, but they might roll down the arms or bounce back off the hood. This danger is particularly to be guarded against when the operator's station is at the front of the tractor.

TRUCK LOADING

This discussion is based chiefly on popular-size loaders with buckets from one to 2½ yards capacity. Operations are described with general reference to direct drive, manually controlled machines, for the sake of simplicity.

Converters and automatic controls simplify this operation without changing it basically.

Procedures for the actual digging in the bank are the same as those already described.

Maneuvering. The normal cycle for truck loading is made up of digging in the bank, backing out and making a part turn toward the truck, then moving forward to the side of the truck while completing the turn and raising the bucket to clear the truck body.

The bucket should be high enough so that the downward movement of the bucket lip during dumping will not cause it to strike the truck, and it is good practice to have it high enough to clear the side to avoid accident while backing. The hoist is usually completed before the truck is reached. The control is then moved to HOLD and the tractor moved so that the bucket is over the truck body.

A good procedure is to time the lift so

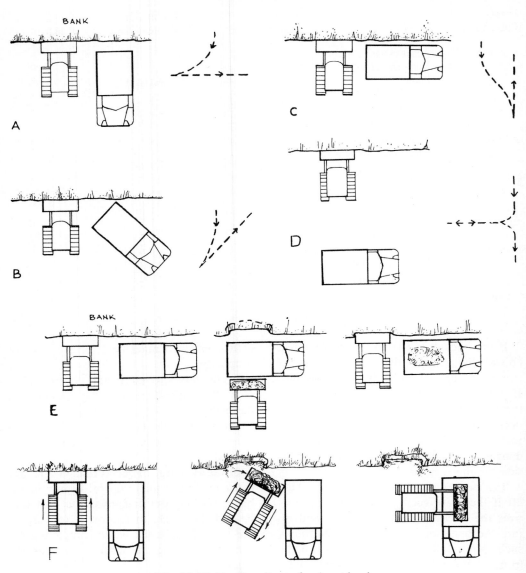

Fig. 16-17. Truck patterns for front loader

that the bucket will just safely clear the truck as it is moved over it and the control can be left in UP during the dump. When the dump valve is opened it "robs" the hoist valve so that lifting stops, to recommence immediately when the dump valve is closed. This "live" bucket is more easily controlled over the truck body than when the hoist is in HOLD.

In either case, the tractor is walked forward until the bucket is as far over the body as desired or until the radiator guard touches the truck body or tire. The main clutch is released (or, with non-clutch converter drive, the throttle closed), the tractor held with a brake, and the control moved forward to dump the load. The first bucket or two are best dumped slowly to reduce shock to the truck. If the soil is sticky the bucket may be shaken by banging against the dump stops by moving the dump valve lever rapidly back and forth.

The tractor is placed in reverse and backed away, the bucket lip being raised to clear the body if necessary. When clear of the truck, the machine is stopped, put in low or second, and headed toward the bank, the bucket being put in digging position and lowered during the return trip.

Front loader cycle time varies all the way from 20 to 50 seconds. Average with small machines is about 25 seconds, with big ones 35 or more. But in rough digging a big machine may have the shorter cycle.

Easy digging, good truck spotting, and power shift favor fast cycles.

Figure 16-16 shows the result of a study of cycle time in big wheel loaders.

Spotting Trucks. There are many possible patterns of digging and dumping, some of which are shown in Figure 16-8. (A) is the most used method, in which the side of the truck is at right angles to the face of the bank. This involves a quarter turn twice in each digging cycle. The turning stress is high.

In (B) the truck is parked at an angle of about 45° to the bank, so the loader's turns are only half as sharp. In this position some loads can be swung from the bank onto the truck with a very short backward movement, thus increasing loading speed. This is the best system for tracks, but because of driver resistance and indifferent supervision, it is little used.

In (C) the truck is parked parallel with the bank and the digging is done just behind it. This involves about the same amount of turning as (B) with a greater amount of walking.

In (D) the truck is parallel to the bank but at a distance from it, so that the tractor must make a 180 degree turn each time to put the load in the truck, and 180 de-

grees to head back to the digging. More walking is required than in other methods. This is the slowest and most unsatisfactory of the arrangements suggested, but is often used in muddy or sandy pits where trucks are restricted to certain drives.

In (E) the truck is again parallel to the bank, but it comes to the loader for each bucketful. The machine fills its bucket from the bank and backs straight away. The truck backs in front of it, the tractor advances and dumps into the truck, the truck moves forward, the loader takes another bite, backs up, and waits for the truck to come back.

This method involves no turning, is fast and efficient if properly coordinated, and keeps wear on the tractor to a minimum. It is particularly valuable when the soil is loose and very abrasive, so that it would wear the tracks rapidly if it were worked into them by steering. However, the method is so unpopular with truck drivers that it is seldom used.

(F) shows the effect of counter-rotating tracks (or wheels) in shortening travel.

For wheel loaders, extra travel distance from bank to truck, up to 50 to 90 feet, may not consume time because of more maneuver space and use of higher gears.

Loader output can usually be increased substantially by having a spotter place the trucks, or by training the drivers to be alert to the needs of the machine so that they will not only take a convenient position, but be ready to move if the machine works away from them. This last is particularly important where the digging is shallow and truck bodies large.

Dumping in Body. The width of standard buckets varies with machine width, usually from about 5½ to 9 feet, but a few up to 16 feet. Truck bodies (including trailers and off the road models) vary from 7 to over 25 feet.

Dumping height of the bucket, measured

Courtesy of Caterpillar Tractor Company

Fig. 16-18. Filling a long body

from the ground to the lip of the bucket, held at maximum height with the floor inclined downward at 45 degrees, varies from 8 to 12 feet, with a few higher. Highway dumpers have side heights from 5 to 10 feet, off-the-roads may be over 15 feet.

Ordinarily, loaders are matched in size to the trucks that they fill. However, it is often necessary to cope with a mis-matched set.

If the loader is too big for the truck, care must be taken to pour the load gently into the body, and allow excess to spill off the back.

A well matched body can be fully loaded by dumping in the center only. A long body is loaded by dumping alternately in the front and rear from the side, although there is a tendency to pile up too much in the center and skimp the corners.

Such a body may also be loaded from the side at the front, and finished by filling from the rear, over the tailgate.

For convenience and efficiency in side loading, the sideboard of the truck should be a foot or two lower than the lip of the bucket in full-lift dumped position. This makes it easy to place the load in the center or even in the far side of the body.

With increase in lift the bucket moves back toward the tractor and loses reach, so that high sides may require extra work moving the load across the body, or loading from both sides.

Many tractor loaders now have sufficient dumping height and reach to load the largest bodies readily without the special procedures described below.

High Trucks. If a body too high for convenience is loaded, a heaping pile of soil is built up on one side. This heap may be moved toward the other side by pushing it with a loaded bucket held with its floor parallel with the ground and just clearing the sideboards. The bucket is then dumped and its load pushed over by the next bucketful, this process being repeated until the body is full.

Dirt may also be worked over by putting the bucket in fully dumped position, dropping it just inside the body and rotating it part way toward flat position.

Very high trucks may be loaded by using both sides. The bucket must be rotated to flat position after each dump in order to be pulled back over the side.

If many big trucks are to be filled it may save time to dig slots two to three feet deep into which they can back while the tractor operates on the higher level. If the pit floor must not be torn up, a few bucketsful will build a ramp up which the tractor can walk to get an easy working height. The ramp should be made so that the machine is not tipped up steeply while dumping, as this reduces reach and increases the effort of holding position.

If there is uneven distribution of earth front and rear while loading, it may be corrected by cutting into the bank at an angle so that the side of the bucket which will dump over the low spot will get the heaviest part of the load. Dirt already dumped can sometimes be rearranged with the bucket in dumped position by pushing, pulling, or turning.

Distances. The loading is fastest if the truck is so close to the digging that the machine has just comfortable room to turn, although the hoist is not fast enough to lift the load to the required height in a very short travel distance. With steadily diminishing returns on a yards-per-hour basis the dozer shovel can load at considerable distances. When a pit is too wet or sandy for truck operation it can dig in the pit, carry material onto firm ground, and there put it in the truck.

Output. The old rule of thumb is that a front loader will load trucks at an average rate of 50 bank yards an hour for each yard of bucket struck capacity, half the rate of a dipper shovel.

Under favorable conditions this rate may be doubled on tracks, and tripled for wheel mountings. Figure 16-16 shows theoretical

Courtesy of Deere & Company

Fig. 16-19. Hoe weight helps loader digging

no-delay possibilities for large and very large wheel machines, based on Federal Highway Administration Report No. FHWA-RDDP-PC-520, "Production Efficiency Study on Large-Capacity, Rubber-Tired Front-End Loaders."

Conditions get unfavorable much more easily for loaders than for shovels, as their production diminishes more rapidly as the digging becomes harder or coarser, or the footing gets softer. On the other hand, an alert foreman and a good operator can often step up loader production way above average, simply by using sound procedures that are generally ignored.

When compared to a shovel of similar rate of production, the loader has the advantages of moving around more readily, cleaning up the pit floor and moving boulders without assistance, picking up bigger rocks without chaining, and use as a dozer while not loading. The dipper can handle harder digging, work on softer floors, and has lower repair and maintenance costs because its tracks and rollers have much less use.

OPERATION — TWO WHEEL DRIVE

A front loader that is mounted on a rear drive tractor reduces its traction when it is carrying a load. The front axle acts as a balance point, and any weight ahead of it will counterbalance the weight on the rear wheels, sometimes to the extent of raising them off the ground. Any reduction in weight on the driving wheels

Courtesy of Fiat-Allis Construction Machinery, Inc.

Fig. 16-20A. Load and carry, short haul

reduces their traction, which in any case is not as good backing as going forward because of reaction from driving torque.

As a result of these factors, the machines cannot carry good loads on loose, slippery, or soft ground, especially in reverse. This difficulty may be reduced by heavy counterweights on the rear wheels or on the back of the tractor, by using dual wheels, or by attachment of a rear mounted tool such as a ripper, a winch, or a scraper blade.

The rear tires should be filled with a water solution of calcium chloride. This is necessary for efficiency in any tractor work, and is the first step in counter-weighting a loader.

If the load is carried very high, more of the weight will be on the rear wheels although danger of side tipping is increased. Extra traction may be obtained for a moment by lifting the load high and letting it drop. While it is falling the tractor is almost free of its weight, and the

wheels may grip enough to get the machine moving. Sometimes the machine will be able to drag the loaded bucket backward on the ground or on skids.

Another consideration is that the front wheels normally carry less than half the tractor weight, but when the load is heavy they carry most of the tractor and all of the load. This results in hard turning, particularly with large machines, in which power steering is a necessity. Front tires must be the heaviest offered with the tractor.

The two wheel drive depends on its momentum to drive the bucket deeply enough into the pile to pick up a good load. The speed required may be greater than would be safe for turning and dumping, and it is therefore desirable to have a foot accelerator by which the operator can speed it up as it approaches the pile without changing the hand throttle adjustment.

The pit must be arranged so that the machine will not have to carry a load while it backs uphill or across rough or

Fig. 16-20B. Load and carry, long haul

sandy ground, as it is likely to lose time or be unable to work due to poor traction.

Direct digging of hard or firm soil can be made much easier by breaking up the ground with a ripper or subsoil plow, and cutting only to the depth the tool penetrates. Such a machine may be mounted on the rear of the tractor and serve for counterweight, or be a separate unit attached and pulled by the drawbar and detached during digging.

LOADER HAULING

The larger rubber tired loaders may carry loads in complete-operation digging, hauling, and spreading, over distances up to a quarter mile.

Bucket capacity is smaller that in scrapers or trucks of comparable price, travel is slower, and the tipping influence of a loaded bucket limits them to well graded routes.

They have an advantage over scrapers in being able to dig into banks from the floor of a pit or a cut, and in the ease with which they can build steep sided stock piles or dump over banks or into hoppers, without help of other equipment.

A loader can supply a low or medium bin, saving either the expense of a truck ramp and hopper, or rehandling by clamshell. If the edge is a little too high, the loader can use a steep ramp of dirt or aggregate.

If hauling is done by trucks, the loader (or an equivalent machine) is needed to fill them. A comparison graph such as that in Figure 16-21 can be worked out, to

Fig. 16-20C. Ramp up to a hopper

determine the relative cost per yard of hauling by truck or by loader. If truck hauling requires a spreading or piling machine at the dump, its cost must be included.

The cost is not the whole story, however. The work may require more material than the loader alone can deliver, and there could be many reasons why the acquisition of an additional loader would not be justified.

Job simplification may also be important. Where one unit can do the work of three, problems of scheduling are cut by at least two-thirds. This factor may be particularly important in a small operation. The loader might be able to do this hauling, and also take care of other work, so that trucks could be eliminated without adding other costs.

Carry or Push. If material is to be moved a short distance, 200 feet or less, and the quantity is large, it may be economical to push rather than carry it. A flat bucket floating on the ground, moving between windrows of spill, will move two to three times as much dirt or rock as it will carry. Loads may be even better if the push is down hill.

Disadvantages include probable use of a lower gear, poor loads until windrows build up, rocks rolling behind the bucket into the tires, possible damage to the push route, and time spent cleaning it afterward.

Pushing is standard procedure in supplying belt loaders, even over quite long distances.

Underground. The load and carry method is firmly established in subsurface mining, where it may be called load-haul-dump transport.

The Eimco 920 LHD, Figure 16-22A, is one of the largest machines of this type. It weighs about 43 tons, is 10 feet wide and 38 long, is rated to carry 16½ tons in its bucket (8 to 17 yard buckets for various weights of material), has almost 400 diesel horsepower, and is only 6½ feet high with the bucket in low carrying position. A water bath conditioner reduces exhaust fumes to a permissible level.

Operation is usually shuttle fashion, without turns. The machine self-loads at the digging face, rolls back the bucket, backs past a chute or hoist, and usually makes a small forward turning movement to dump

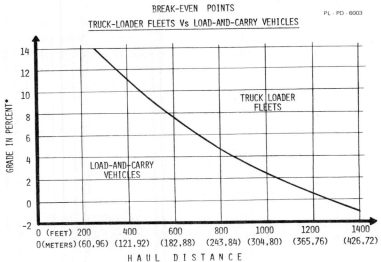

BREAK-EVEN POINTS
TRUCK-LOADER FLEETS Vs LOAD-AND-CARRY VEHICLES

PL - PD - 6003

*NOTE: PRODUCTION CALCULATIONS BASED ON 3% ROLLING RESISTANCE (SMOOTH, DRY EARTH).

Courtesy of International Harvester Company

Fig. 16-21. Load and carry, efficiency curve

Courtesy of Eimco Division, Envirotech Corporation

Fig. 16-22A. Load-haul-dump unit, large

into it. It then straightens out, and goes back to the heading.

Maximum speed is 15 miles an hour, but underground haulage roads usually demand slower movement. It has four wheel drive with articulated construction, and can climb a 40 per cent grade with a load.

Dumping height is the same as machine height, 6½ feet. An ejector bucket can dump a couple of feet higher.

The 911 LHD is the smallest in the line, with 4-foot width and a height of only 44 inches. It may have either diesel or electric power, and can work in a tunnel as small as 6 x 6 feet.

HANDLING OVERSIZE

The floor of a truck body should be protected by a layer of dirt or other cushioning material, before placing big rocks on it.

Courtesy of Eimco Division, Envirotech Corporation

Fig. 16-22B. Load-haul-dump unit, small

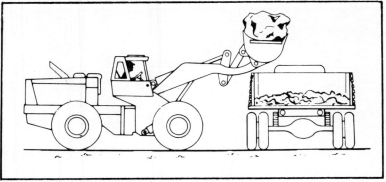

Courtesy of International Harvester Company

Fig. 16-23. A dirt cushion protects the truck

This is particularly important if they are to be dumped rather than lowered.

Loader buckets are not well adapted to picking up large loose objects on the ground. They tend to push ahead of the edge unless it is dug into the ground under them, and the overhang of the back makes it hard to balance anything bulky on the floor.

Roll back buckets can pick up many objects by getting the edge under them, then tilting back the bucket until they fall in. Larger pieces, such as stumps or boulders, may be crowded against a steep bank which will prevent them from falling out while the bucket is rolled back enough so that the object will slide in. Two loaders can lift the object between them until it settles into one of the buckets.

If the object is too large to be picked up in this way or there is nothing to crowd against, it can be maneuvered onto the floor and held from falling out by a chain to the top back of the bucket, as in Figure 16-24 (A). This method is especially effective with loose stumps resting upright.

A load usually settles back as it is raised or tilted back so that it is easy to unhitch before dumping. If it settles forward and puts so much tension on the chain that it cannot be released, its overhang may be lowered so as to be supported by the truck sideboard or the pile. The chain may then be removed and the load boosted into position.

The bucket may also be lowered over a loose stump in dumped position, and attached by a chain or chain and tongs, as in (B). The bucket is curled, prying the stump up to a forward carrying position, as in (C).

For efficient odd job use, or for crane work, it is almost essential to have a chain grab bracket, or a hook or hooks, fastened onto the top or back of the bucket. For use in dumped position, a second bracket on the rear beam is desirable.

A chain should not be anchored on a push arm and passed over the top of the bucket, as these change their relative positions during a lift, and the chain is likely to be stretched and broken.

Crane. The front loader may also be used handily as a crane by means of a chain fastened to the back of the bucket and dropped across the edge. Height of the load may be regulated both by the hoist and dump controls.

If the lift is not high enough to put the load on a truck, a small log or a couple of rocks may be placed where the tracks will climb them as the machine approaches the truck, which will cause the tractor to rear up and increase the load height.

The crane attachment for use in place

16-26

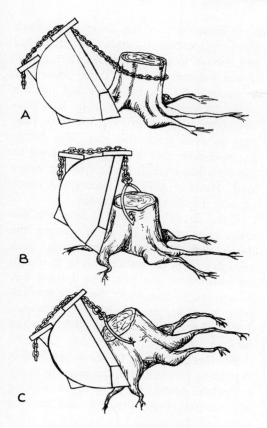

Fig. 16-24. Picking up stumps

hand. This "motorized wheelbarrow" work may be a profitable sideline when grading around buildings under construction.

BULLDOZER WORK

A regular wide bulldozer blade, Figure 16-26A, may be substituted for the bucket. Its use is advisable if the machine is to be used as a pusher for scrapers, or if it is desired to equip it with grousers and do heavy digging in rocky soil. The blade is more rugged than the bucket. A wide blade enables the machine to turn in its path without climbing onto the edges of the cut. However, the standard width bucket will do most bulldozing jobs better than a bulldozer, and is immediately available for loading or other special work.

This blade is somewhat more effective than on a regular bulldozer because the digging angle can be changed with the dump control to meet any change in the digging, and because the blade can be dumped at the top of steep piles, where part of the load has a tendency to ride back down the pile on a standard blade.

of the bucket has a somewhat higher lift and is more convenient to use than the chain and bucket, but for incidental work it does not justify its cost or the time required to change over. See Figure 16-25.

If the bucket is placed in a fully dumped position, a beam or log may be wedged or lashed to a dump arm so that it lies on the bucket and extends forward from it. This will be raised by lifting the push arms or flattening the bucket, and will give a very high lift to light loads.

Valve action is not smooth enough for precision crane work, but the mobility of the machine and its ability to work on soft footing and in restricted spaces qualify it for many hoisting and moving jobs.

The loader may also be used to carry or raise material piled in the bucket by

Courtesy of Fiat-Allis Construction Machinery, Inc.

Fig. 16-25. Crane attachment

Courtesy of Fiat-Allis Construction Machinery, Inc.

Fig. 16-26A. Dozer blade attachment

In addition, it can be floated forward in full dump position as well as backward to smooth over loose soil.

Bucket Manipulation. In dozer work, the bucket is used most efficiently by cutting with it in a digging position and spreading with it more or less dumped. In the latter position the bucket floor will have a tilt smiliar to that of a blade. Its advantages over a blade are the greatly increased cutting efficiency due to the knifelike edge; its ability to push bigger loads of soil because a good pushing load is contained in the bucket where it causes so little load or friction that nearly a full blade load can be pushed ahead of the bucket; faster grading work because of greater transporting capacity and ability to carry dirt to hollows without disturbing surfaces in between; the facility with which material can be backed out of banks, and the exactness with which it can be deposited where needed.

The transition from forward digging to spreading is most easily made by putting the lift lever at UP, and dumping the bucket a little at a time by rapid back and forth movements of the dump lever. Each dumping movement should lower the edge just enough to compensate for the amount it was hoisted while the dump lever was in neutral. As the hoist will not operate while the dump ram is moving, it is possible after practice to keep the edge running an even grade this way. The dumping operation may be stopped as soon as the floor has enough tilt to spill the entire load. Hoisting may be interrupted occasionally if a thin layer is wanted.

If the dirt is to be placed in a pile, the same method may be used, starting the dump with the edge kept down soon enough so that the bucket will be in dumping position at the pile. The bucket can be allowed to rise while the shift into reverse is made; or can be left stationary during the shift and lifted slightly while backing out.

Another system is to leave the bucket in pushing position until the place to dump is reached, putting the hoist in UP, shifting into reverse, then dumping and backing away.

If a high pile is being built, either method may be used, simply lifting the bucket to conform to the slope of the pile while climbing it. It is a good plan to knock the top off the pile occasionally, by forcing the bucket down in dumped position in the loose dirt a few feet short of the top. Several yards may be pushed over at a time this way, shortening the climb to dump the next few buckets.

In backfilling against a wall, you may fill the bucket, put it in a flat position, and push a large amount of material ahead of it. Let this fall into the hole, then go back with the full bucket for more. This avoids danger of hitting the wall with the upper part of the bucket.

Reach-Down. A loader has greater effective reach below itself than any bulldozer. In the vertical position the bucket edge will reach more than two feet below the tracks, and it will cut a sheer wall to that depth. Because of their wide bases no dozer blades can be forced straight down or reach such a depth in a short distance, even when capable of such a drop.

This depth ordinarily cannot be reached in one cut, but by several with the blade at a successively steeper angle, or the push arms lower each time. In itself it is only occasionally useful in ditching or squaring off sides of shallow excavations, as the tractor cannot follow the bucket over such an edge without special precautions.

However, this ability to reach down enables the dozer shovel to cross sharp ridges without the bucket losing contact with the ground, a matter of great importance in clearing work. The bucket may also be used to ease the machine down lesser drops by holding it a foot or so above the bottom and walking very slowly off the edge. When the tractor over-balances it will be supported by the bucket, which can be gradually hoisted, thus lowering the front of the tracks to the bottom. This process may crush the bank enough so that the machine can then be walked off it, or it may be necessary to put small blocks to reduce the drop of the back.

Backdragging. The bucket is efficient at backdragging because the floor, when tilted straight down, can penetrate fairly hard material vertically and the clear space between the bucket and the tracks is sufficient to accommodate a lot of dirt. When backdragging in order to smooth loose dirt, the operator has a choice of tilting it to full dump, where it will cut, through intermediate positions where it will not cut and will pull decreasing quantities of earth, to flat where it will move hardly any. In the flat or near flat positions, down pressure can be applied so that loose dirt may be compacted.

General Considerations. But it can be seen that the operator usually needs two con-

Courtesy of International Harvester Company

Fig. 16-26B. Backfilling against a wall

trols more or less continuously, as against one in a bulldozer, in order to bulldoze with a bucket. Some operators may dislike the machine for this reason.

Another drawback is that location of the push arms between the tracks and the narrow path cut by the bucket enable stones and dirt to roll between the bucket and the track, making it necessary to lift the bucket and back far enough to get behind them in order to get through or to keep a grade. Also, the top of the bucket may be in the way of pushing close to a building.

A more serious consideration is the possibility of breaking up the bucket. It is not practical to make it as heavy or as strong as a bulldozer blade, and maintenance costs in heavy digging, particularly among rocks, may be higher. This is especially true if the operator does not keep the bucket at a proper angle when digging or if full grouser track shoes are used.

Courtesy of International Harvester Company

Fig. 16-27. Backdragging with bucket

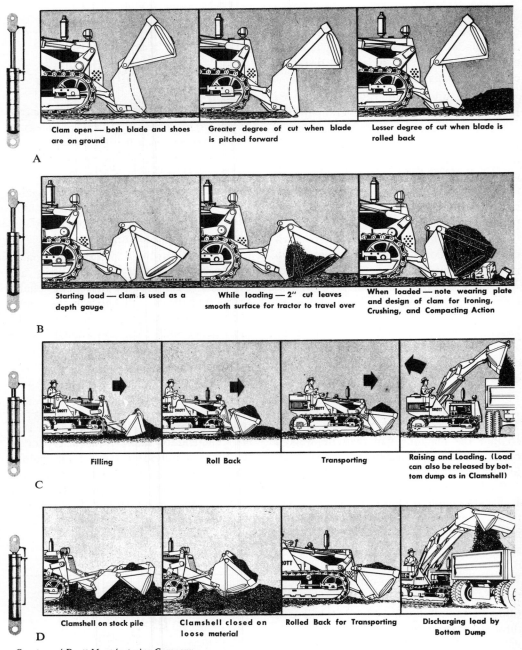

A

Clam open — both blade and shoes are on ground

Greater degree of cut when blade is pitched forward

Lesser degree of cut when blade is rolled back

B

Starting load — clam is used as a depth gauge

While loading — 2" cut leaves smooth surface for tractor to travel over

When loaded — note wearing plate and design of clam for Ironing, Crushing, and Compacting Action

C

Filling

Roll Back

Transporting

Raising and Loading. (Load can also be released by bottom dump as in Clamshell)

D

Clamshell on stock pile

Clamshell closed on loose material

Rolled Back for Transporting

Discharging load by Bottom Dump

Courtesy of Drott Manufacturing Company

Fig. 16-28A. Uses for multipurpose (4-in-1) bucket

The use of the heavily built rock bucket for hard digging will materially reduce maintenance costs, but it necessitates the lifting of several hundred pounds of extra weight when loading.

The dumping rams may be used to

push or pull the tractor when lack of traction or breakdown prevent it from walking in the regular way. The bucket is pushed down in firm ground or on a mat of poles or planks. Dumping the bucket will pull the tractor forward, closing it push back.

This control over the bucket edge can also be used in pulling stones back out of water or marsh, and in reducing the length of the machine to fit it in a short space.

In general bulldozing work where traction is good, a shovel dozer bucket should produce from 25 to 50 percent more work than a blade. It shows its greatest advantage in hard ground and among obstructions.

MULTI-PURPOSE (4-IN-1) BUCKET

This special bucket has a back that is similar to a dozer blade, and a separate floor attached to upper hinges by the bucket sides. A pair of double acting rams on the rear of the bucket can lift this clamshell-fashion, or clamp it firmly against the bottom of the blade. It is equipped with a cutting edge at the base of the back, and another on the bucket lip.

Figure 16-28A shows some ways to use its special features. When the floor is raised all the way, the back of the bucket can be used as an ordinary dozer blade. If it is lifted slightly, it acts as a float or depth gauge to regulate depth of cut, and as a bowl for holding the cuttings.

If fully back, operation is the same as with a one-piece bucket. In addition it can be used clamshell fashion for picking up loose material without pushing it around.

It can also grip and raise substantial pieces, such as tree trunks and boulders. Such things should be held at the center whenever possible, to avoid twisting the loader frame. Such clamped loads can be released only by opening the clam.

Loads lying in the bucket may be dumped through the bottom in this way, or by tilting the whole bucket forward, in the conventional way.

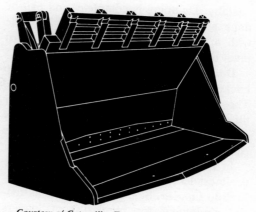

Courtesy of Caterpillar Tractor Company

Fig. 16-28B. Ejector bucket

Bottom dump permits loading very high trucks, as the bucket lip does not swing down when discharging the load. However, it tends to fill only the near side.

OTHER SPECIAL BUCKETS

Ejector Bucket. A rock ejector bucket has a back wall that slides forward to completely clear the bucket of its load. Its movement is controlled by the tilt mechanism through a mechanical linkage. In any position less than full height, it advances automatically as the bucket floor is tilted downward.

At full height, however, the bucket floor stays level when the tilt lever is moved to DUMP, so that the back moves forward hori-

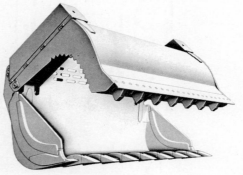

Courtesy of Caterpillar Tractor Company

Fig. 16-28C. Demolition bucket

zontally. Dumping is complete, and can be done at a greater height than is possible with a standard bucket, whose edge must pivot downward.

Demolition. The demolition or clamp bucket resembles a clam-shell on its side. The whole bucket responds to the regular tilt control. The upper jaw is independently operated by a pair of hydraulic cylinders concealed and protected in its side arms.

In a 5½ yard bucket, the clamping force is almost 13 tons with the jaws half open.

Side Loader. Side dump buckets as special equipment for regular front loaders were mentioned earlier. Such buckets are standard equipment on a number of small to medium crawler-mounted loaders designed for underground use.

In narrow tunnels, it is usually necessary for a mechanical loader to hoist the load overhead, and dump it behind onto a belt or into a hauler. A somewhat wider tunnel can have a belt, tracks, or a narrow roadway, with space for a loader to operate alongside it, shuttle fashion, on one or both sides.

A side loader can dig straight-on at the face, back alongside the dump point, and spill the load into it without turns. It then moves (trams) forward to refill.

Dumping is faster than with an overhead, and less roof clearance is needed. All cars of a train (instead of just the nearest one) can be loaded without shuffling.

Side loaders are usually 4½ and 8½ tons in weight, with buckets from 15 to 35

Courtesy of Eimco Division, Envirotech Corporation

Fig. 16-28D. Side loader

Courtesy of Eimco Division, Envirotech Corporation

Fig. 16-29A. Overhead shovel

cubic feet capacity. Width varies from 3 to 5 feet. Power is usually air or electric, but may be diesel.

OVERHEAD SHOVEL

The overhead shovel digs at the front in the same manner as a front end loader. The filled bucket is then lifted entirely over the tractor, and dumped behind it.

These units have declined in popularity, and are now little used above ground. However, in spite of increasing competition from side dump and load-and-carry models, they are extensively used in mines and tunnels.

Eimco. The bucket is mounted on a rocker frame that raises it as it is pulled back by a pair of leaf chains that wind on rear winch drums. After dumping it returns to digging position by gravity.

The hoist is driven by a live power take-off. It is two speed, and can be shifted under load.

The standard bucket is 1½ yards, apparently undersize for the machine's 150 horsepower and 17 tons. However, it is designed particularly for heavy rock digging

in mines, where pay dirt may be very heavy in proportion to bulk, and digging is hard. Larger buckets with different shapes are available for softer formations.

The loader can be obtained in three different loading heights, and can be converted from one to another in the field. The high discharge can load trucks as high as eleven feet, and standard high railroad gondolas. It requires a headroom of seventeen—too high for many underground jobs. The low discharge will load a 7 foot truck,

Courtesy of Eimco Division, Envirotech Corporation

Fig. 16-29B. RockerShovel

Courtesy of The Eimco Corporation

Fig. 16-29C. Loading in a tunnel

and has a maximum height of only twelve.

This company also makes air and electric RockerShovels for use in tunnels and mines where space is more restricted. The Model 21, Figure 16-29B, runs on the same rails that carry the cars it loads. It can dig straight ahead, or swivel about 30 degrees to either side. The loaded bucket is passed over the top of the machine, and dumped into a car directly behind. The operator

stands on a side platform. A great many models are made to suit particular requirements.

The controls of the Model 21 are shown in 16-29C. The operator stands on the step plate and holds the two rubber-covered handles. Moving the left (front) one forward moves the machine forward, moving this same lever back—backs the machine. The rear curved lever can be moved forward to raise the bucket, back to lower it, and to center neutral to lock it in any position.

The loading cycle is only six seconds, so this small machine can move a lot of muck in a shift.

The Model 40H, Figure 16-30, dumps its bucket onto a conveyor belt that carries the spoil well back from the machine. The front lever controls locomotion (crowd-retract) when moved forward and backward, and swings the bucket and deck when moved sideward. A small knee operated lever controls the conveyor motor.

A crawler mounted model with a separate reversible air motor in each track is shown in Figure 16-31.

Loading overhead. The standard overhead shovel or loader digs at the front, lifts the bucket, and dumps it to the rear.

Courtesy of The Eimco Corporation

Fig. 16-30. RockerShovel with conveyor

There are also other types, one of which can dig at either end but dumps only at the front.

The digging pattern of these machines is the shuttle type, Figure 16-32, with only sufficient turning to avoid cutting pockets in the bank, so that the pit floor is easily kept smooth. The machine, with bucket on the ground, is almost twenty feet long, so trucks should be parked at least thirty feet from the point of digging. Trucks should be parallel to the bank so that loading can be done from the side as dumping range may be too short to reach the front from over the tailgate.

If the walking distance is short, the length of the cycle is determined by the time it takes the bucket to move from dig to dump and back, as the tractor will have to be stopped, usually after it has reached the truck, for completion of the dump; and again to await return of the bucket to digging position.

If the distance is long, the bucket is checked at some point where it will tend to balance the tractor while it walks to the dumping or digging point. This reduces stresses on the front rollers. The

Courtesy of The Eimco Corporation

Fig. 16-31. RockerShovel on crawlers

distance which the bucket must be carried while stationary is waste motion and slows the cycle.

This type of machine is ordinarily backed until contact is made with the truck to be loaded. The operator cannot see into the truck unless the sides are quite low, but the loader can dump up to 9 feet.

Applications. The overhead loader can usually show a larger output than a front bucket of the same size, it can work on softer ground, and it does not grind up the tracks and the pit floor by constant turning.

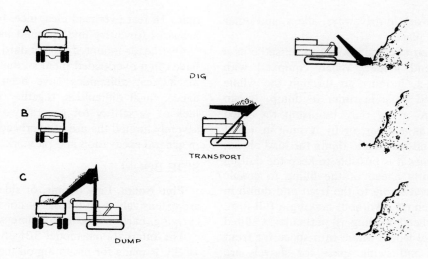

Fig. 16-32. Overhead loading

Courtesy of International Harvester Company

Fig. 16-33. Tractor with side boom

It can work in driveways, alleys, and other narrow places.

But an overhead is not as versatile as a front end loader. When equipped with a bucket, it must go through the whole overhead cycle in order to dump. It can therefore grade only by using the filled bucket as a dozer, or by turning in a half circle and backing to dump the load ahead. Sometimes it is possible to keep the cutting sufficiently ahead of the filling to enable the shovel to dig to the front and dump in holes behind it without making a full turn.

Overhead loading is particularly suited to tunnel work, where turn space for front loaders and swing space for shovels are usually lacking. The largest machines re-quire 18 feet overhead clearance, but there are sizes for every underground use.

Overheads mounted on standard tractors have often overloaded them so that serious mechanical difficulties have been experienced. Such difficulties, together with the lack of versatility for general work, have severely limited the use of overhead shovels in general excavation and pit work.

SIDE BOOM

Pipe Boom. Pipe booms, or side cranes, are extensively used for heavy hoisting and carrying, particularly in pipe line work.

The International model SBI-25, Figure 16-33, is made for mounting on the TD-25 crawler tractor, and is rated at 130,000

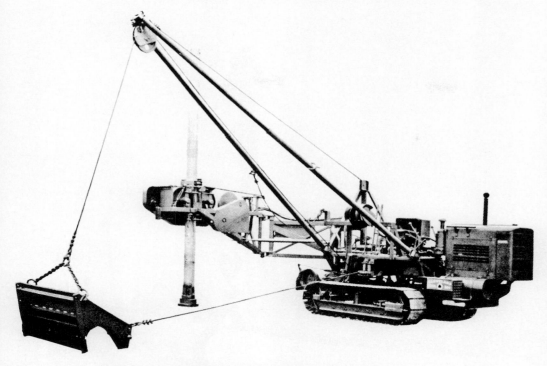

Courtesy of The Cleveland Trencher Company

Fig. 16-34. Side crane and backfiller

pounds tipping load at four foot overhang.

A power takeoff drives two side-mounted drums that are controlled through separate clutches and brakes. One of these controls the boom height, the other a load line which crosses a boom point sheave and is reeved into a multiple part hoist block line.

A heavy counterweight is hinged to the winch side of the tractor. It hangs close to the tracks for traveling and while handling light loads, and can be extended by hydraulic rams to counterbalance heavy loads.

Backfiller. The Cleveland 80W Backfiller, Figure 16-34, is a tractor unit designed specifically to operate a backfill bucket from a side boom. It has 48 speeds, both forward and reverse, with a range from a foot a minute to over 4 miles an hour. This is to enable it to move steadily as it backfills.

The bucket can be easily removed and the boom used as a side crane for placing

pipe and hoisting burdens. The boom is telescopic (manual adjustment), and is raised and lowered by a hydraulic motor operating through worm gearing, and can be held by a hand brake.

A backfiller can easily refill ditches on ground too soft to allow a dozer to work. If necessary, it can work on platforms, although since it has no means of swinging them to the front, they are best made up of logs or timbers that can be handled separately.

Position of the attaching cables can be reversed, so that the board will push instead of pull. A horizontal boom converts the hoist into a drag-out line.

This machine can be fitted with the hydraulic tamping unit shown, to compact backfill as soon as it is placed. It has a 30-inch stroke, and strikes over 40 blows a minute.

Fig. 16-35. Side cranes at work

CHAPTER SEVENTEEN

SCRAPERS

PLACE IN EARTHMOVING

The bottom-dump scrapers discussed in this chapter are also known as carrying scrapers or pans, and by various trade names. They are highly mobile excavators with a centrally located bowl that digs, carries, and spreads loads. There is a wide range of types and sizes. Struck capacity is usually between 7 and 40 yards, but there are larger and smaller units also.

The typical modern scraper is a self-powered, rubber-tired unit. There may be either two or three axles. Controls are usually hydraulic, but may be cable, or combination cable-hydraulic.

Standard or conventional models have power and traction sufficient for most hauling needs, but require help of pusher tractors or other machines in order to dig efficiently.

Self-loading scrapers which ordinarily do not need pusher help, may have two engines with separate drive axles for extra power, or an elevator that reduces loading resistance.

There are also full trailer scrapers, usually with cable control, which are towed by a drawbar attached to a separate tractor. These were once the dominant type, and are still useful in special situations. They are self-loading only if the tractor is large in proportion to scraper capacity.

Scrapers are of primary importance in earthmoving. They are the standard tool for alternating cuts and fills, under the wide range of conditions where the cut is firm enough to support them, and the soil (including rock that is soft, ripped or blasted) is digable.

The scraper digs, hauls, and spreads in a single cycle. It works in thin layers in the cut and on the fill, without limit as to the number of layers, so that its efficiency is not particularly affected by depth of cut or height of fill. Its use causes considerable compaction of fills, and favors proper use of rollers.

In various models, its economical haul distance ranges from less than one hundred feet to several miles. Where conditions are favorable, it can move earth at lower cost per cubic yard than any other type of earthmover, except for the gigantic shovels, draglines, and wheels used for sidecasting in strip mines.

It is not only an excellent machine for bulk earthmoving, but a precision finishing tool as well. The cutting edge is carried between front and rear wheels, so that it is unaffected by pitching, and the operator can control its position very accurately. If job conditions give him time, he can cut or fill accurately to grade, and when space is wide enough for maneuvering, can build crowns and slopes as well.

Fig. 17-1A. Two-axle scraper

TRACTOR-SCRAPER

The standard or conventional scraper of the year 1975 is a single-engine self-powered but not self-loading machine. It is made in two distinct sections, tractor and scraper, which are connected by a kingpin swivel and hydraulic lines. In steering, the two parts pivot on this swivel.

SCRAPER

The scraper has three basic operating parts, the bowl, the apron, and the ejector or tailgate. In addition, it includes the gooseneck and yoke, and the scraper (trailer) wheels.

Gooseneck. In front, the gooseneck or yoke has a vertical swivel connection with the tractor which is usually in two parts with two pins, upper and lower. It permits turns of 85 to 90 degrees each side of center. In addition, there is a longitudinal horizontal hinge that permits the two sections to tip independently from side to side, through a limited angle. In three-axle models, there is also a horiziontal cross hinge to keep all wheels on the ground on vertical curves.

Behind the swivel, it arches up to allow space for the tractor wheels to roll under it on turns, then widens into a very massive crossbeam, and is finally a pair of side arms extending backward and somewhat downward to trunnion fastenings on the sides of the scraper bowl.

The gooseneck carries the steering cylinders, the lift cyclinder and lever arm for the apron, and a pair of hoist cylinders for the bowl. All of these may have two-way action, or be one-way with return by gravity, springs, or counter-acting cylinder.

Bowl. The bowl, Figure 17-3, is the principal member and carries the cutting edge. It is substantially a box with rigid sides, with the apron forming a movable front, and the ejector a movable back. Extensions of the sides converge behind the rear axle, forming a case for the ejector cylinder, and support for a bumper by which the machine may be pushed.

The bowl is supported at the rear by the rear or scraper axle, at the center by trunnions on the ends of the yoke arms, and at the front by a pair of hydraulic cylinders suspended from the yoke.

The pull of the tractor is applied through the yoke. Most of it is transmitted by the trunnions at the bowl center, but a variable amount comes through the lift cylinders, depending on their position.

The floor is cut off forward of the center line, and fitted with a cutting edge, often called a knife. This edge is usually very hard steel plate in three pieces, a wide center one and narrower ends, fastened with plow bolts with smooth sides up. The sections can be removed, inverted and re-installed when worn on one side.

For most work, the center piece is set further forward than the sides. It is mounted back flush with the ends only when the job

Fig. 17-1B. Three-axle scraper

is grading, working light cuts, or in sand.

Teeth may be bolted to the center section, to improve penetration in difficult ground. They interfere with dumping and spreading, and are laborious to install and remove, particularly on a worn edge. They are used more often on elevating scrapers than on standard models.

The fronts of the bowl sides, at the bottom, usually have bolted-on wear plates, called side cutters. These receive less wear than the bottom edge, but eventually need replacement.

Apron. The apron forms the forward side and a variable amount of the bottom of the scraper assembly. When in a down or closed position it rests against the scraper bowl at the cutting edge. When lifted it moves upward and forward (again, in varying proportions) far enough to leave the whole front of the bowl open.

It is lifted, lowered, and in some models clamped down forcibly, by a hydraulic cylinder which usually is linked near the base of a lever hinged to the yoke.

Since apron movement calls for travel through a considerable distance without (usually) the need for much brute force, this third-class lever arrangement is efficient.

When the scraper is digging (loading) the apron is held at a moderate distance forward of and/or above the cutting edge. Dug material both moves backward into the bowl and forward into the apron. In sand, the apron may be kept in a dragging position for compaction, in chunky or boulder-

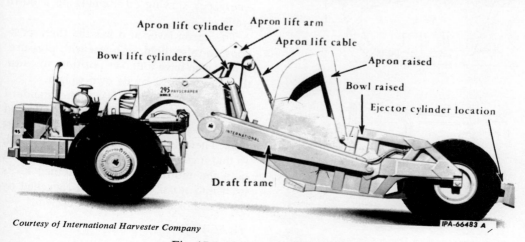

Courtesy of International Harvester Company

IPA-66483 A

Fig. 17-2. Names of scraper parts

17-3

filled soil it must be kept up and forward to be out of the way.

Occasionally, down pressure is used on the apron to clamp bulky objects against the bowl, in order to carry them in a half-loaded position.

There may be a mechanical adjustment in the apron lift, to obtain a larger opening at the expense of tight closing. The higher position may be useful in loading big pieces.

Ejector (Tailgate). The ejector is the rear wall of the bowl. It is usually a sliding or bulldozer type, that moves forward hori-

zontally, forcing the load out of the bowl, over the cutting edge. It is supported by rollers riding on the floor, and on tracks welded to the sides of the bowl.

A less common type of ejector includes both the rear wall and the bowl floor, and moves upward and forward on a hinge at the back of the cutting edge, dumping the load truck-style. See Figure 17-23.

Power is usually a two-way hydraulic cylinder inside the rear pusher block (bumper) frame. Machines of the larger sizes may have two cylinders, and they may be telescoping in design, to increase the length of push in proportion to casing length.

Controls. A typical arrangement of controls for scraper movements is shown in Figure 17-4. The levers operate independently unless some optional combination control is used.

The bowl lever has three standard positions, RAISE, HOLD, and LOWER (DOWN). LOWER is with pressure, so that on hard soil continued movement will raise the scraper, with weight resting on the cutting edge. An additional position may be standard on some machines and optional on others, QUICK DROP.

This last position is useful in starting a cut at an earlier-than-intended spot, and for emergency slowing or stopping on a down grade.

The apron lever also has the three positions, to raise, hold, or lower with pressure. It may also have a FLOAT position, to cause the apron to rest by its own weight.

The ejector control has three standard positions, FORWARD (EJECT), HOLD, and RETURN. There may be a FAST RETURN also. There is often an automatic control, so that it can be held in RETURN by a detent until the return is complete. It then kicks itself into HOLD. This cycle is cancelled by the operator moving any control.

TRACTOR

The tractor includes the engine, drive

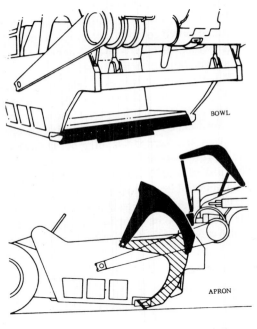

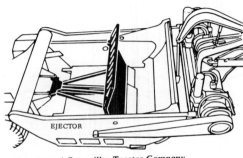

Courtesy of Caterpillar Tractor Company

Fig. 17-3A. Bowl edge, apron, and ejector

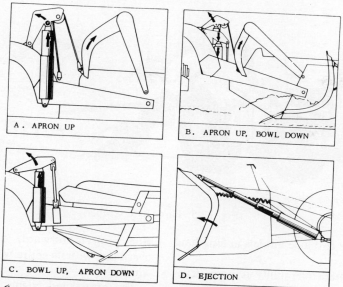

A. APRON UP
B. APRON UP, BOWL DOWN
C. BOWL UP, APRON DOWN
D. EJECTION

Courtesy of TEREX *Division, General Motors Corporation*

Fig. 17-3B. Operating positions

train and drive wheels, hydraulic pumps, and operator's station. It is more or less permanently attached to the scraper by a swivel or articulated joint.

The tractor itself may be a heavy, rear drive wheel tractor with two axles. Its combination with the scraper (which has an axle of its own) is called a three axle scraper. But the majority now in production have only one axle, and their combination is known as a two axle scraper.

These two types of tractor are very similar in engine, power train, controls, and other characteristics not related to steering.

Three-Axle. Figure 17-5 (A) shows a three axle scraper complete. The two-axle tractor part was shown alone in Figure 15-35.

A strong, nearly vertical kingpin is fastened to the tractor frame over the rear axle, on a hinge arrangement that permits tipping to either side and front to back. The scraper yoke surrounds this, and is aligned to it by upper and lower bearings.

Steering is done by turning the front wheels of the tractor while it is moving, with

assistance from individual control of rear wheel brakes.

The three axle scraper has the advantages of good stability, superior roadability at high speeds, both loaded and empty; and a reasonably comfortable ride. It is definitely

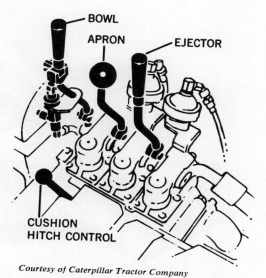

BOWL
APRON
EJECTOR
CUSHION
HITCH CONTROL

Courtesy of Caterpillar Tractor Company

Fig. 17-4. Scraper controls

Courtesy of Caterpillar Tractor Company

Fig. 17-5. Three-axle scraper

superior to the two axle models in long haul work.

However, maneuverability is poor, traction is weaker because weight resting on the front wheels is lost to the drive wheels, and design is less simple.

The industry appears to be expressing a strong preference for the two axle models.

Two-Axle (Overhung). A typical two axle scraper, Figure 17-6, is powered by a tractor having only a drive axle. Stability is supplied by the attached scraper. The engine projects forward, a position called overhung, with its entire weight on the drive wheels.

The articulated connection between the tractor and the scraper yoke varies in construction, but usually includes separate upper and lower hinges, and a pair of steering cylinders based on the yoke.

There may be a cushioning arrangement, such as that in Figure 17-7, to reduce a riding defect called loping. Oil displacement from this cylinder is regulated by a leveling

Courtesy of WABCO

Fig. 17-6A. Two-axle scraper

Courtesy of WABCO

Fig. 17-6B. Two-axle in tight turn

valve, and resisted by a nitrogen accumulator, so that a motion-dampening balance of pressure is obtained.

The articulated connection, and the high arch of the yoke, permit very sharp turns, up to 180° from one side to the other. As a result, the unit is highly maneuverable, and can often turn within its own length.

Drive Train. Practically all self-powered scrapers have torque converters and power shift. The shift may be partly automatic.

This discussion is focused on the two-wheel, overhung tractors.

The transmission is usually behind the drive axle, as in Figure 17-14. Input is by a long shaft from the converter to the top of the transmission; output is a shorter shaft from the bottom forward to the differential.

There may be anywhere from 4 to 10 forward speeds, and one or two reverse. There is a 9-speed unit that has 5 speed ranges, with a choice between converter drive or lockup-direct in the upper four. Lockup provides faster travel and less pulling power than converter drive in the same gear.

An 8-speed box has converter drive in the two lowest ranges, and automatic shift with direct drive in the top six. The automatic goes only up to the range set by the shift lever position.

Shift pattern may be straight line or U-shaped. If straight, the lever is in a slot with toothed edges, which indicate effective position for each range, and may make it necessary to move the lever only one position at a time.

With automatic shift, a hold pedal may be provided. When depressed, it keeps the transmission in the range being used at the time. The operator may use it to stay in a low ratio to increase hydraulic power by maintaining a high engine speed, or to keep in a low direct drive for safety on down grades.

A retarder in the transmission may be either optional or standard. It saves wear on

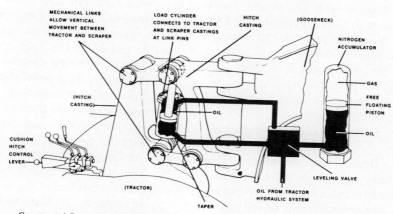

Courtesy of Caterpillar Tractor Company

Fig. 17-7. Cushion hitch for two-axle scraper

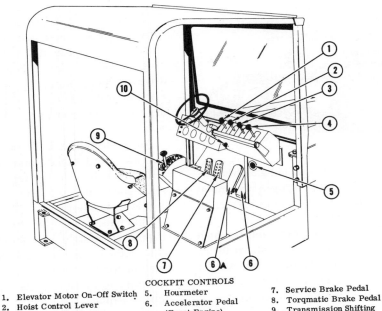

COCKPIT CONTROLS

1. Elevator Motor On-Off Switch
2. Hoist Control Lever
3. Tailgate Control Lever
4. Elevator Motor Reverse Switch (Optional)
5. Hourmeter
6. Accelerator Pedal (Front Engine)
6a. Accelerator Pedal (Front & Rear Engines)
7. Service Brake Pedal
8. Torqmatic Brake Pedal
9. Transmission Shifting Quadrant
10. Throttle Lock

Courtesy of WABCO

Fig. 17-8A. Scraper controls

brakes and prevents over-speeding down long grades. It is not for quick use, as it takes 2 to 6 seconds to become effective after it is engaged.

The differential may be a non-spin design, as in Figure 17-9, or a standard type that allows one wheel to spin. The standard unit may be fitted with a lock, applied by a foot pedal, that forces the two axles to turn together. The throttle should be cut to stop wheel spin before engaging it, and the machine should not be turned while it is in use. It releases automatically when traction becomes equalized.

Final drives are reduction gearing, usually planetary.

Brakes. Brakes may be of any kind suited for heavy service. Smaller units may have booster hydraulic, larger ones pressure-hydraulic, air-over-hydraulic, or full air.

A sequence valve may apply the rear (scraper) brakes a little before the front

ones to keep the rig running straight. In three-axle scrapers, this is a precaution against jacknifing.

In air systems, there is likely to be an emergency tank that provides power to automatically apply and lock the scraper brakes if air pressure falls below 55 pounds.

In three-axle models, there are no brakes on the tractor front wheels, and there are separate pedals for the right and left drive wheel brakes, allowing them to be used to assist steering. A third pedal allows application of all four brakes together.

Steering. Two-axle tractors (three-axle scrapers) have regular, automotive steering, hydraulic powered. The front wheels swivel on kingpins mounted at the ends of a rigid front axle, and keep a constant relationship to the position of the steering wheel.

Sharp turns, or any turns when there is a heavy load, are made easier by applying one brake to slow the drive wheel on the side

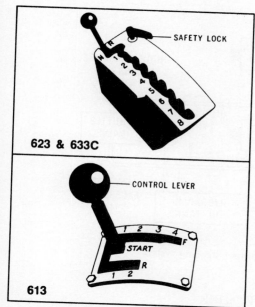

Courtesy of Caterpillar Tractor Company

Fig. 17-8B. Scraper shifts

toward which the tractor is being turned.

Attachment of the scraper makes the machine into a semi-trailer. This involves little change in forward driving, but backing requires special techniques. See Chapter 18.

Courtesy of WABCO

Fig. 17-8C. Scraper shift

In forward driving, the only important difference from the bare tractor is the need to take precautions against jacknifing. There is danger that if an abrupt stop is made, particularly with the front wheels turned, or when the front brakes are more effective than the rears, the scraper will swing forward beside the tractor, with total loss of control over direction of travel.

Two-axle scrapers steer somewhat in the way described earlier for articulated four

Courtesy of WABCO

Fig. 17-9. Non-spin differential

wheel drive tractors, but differences arise because the pivot point is much nearer to the front axle than to the rear. This makes the tractor unit do most of the swinging around the kingpin during steering, with a feeling similar to that of turning a tractor or a truck. The rear or scraper wheels do not follow the front wheels accurately on curves, but track inside them.

The steering can be used to walk the tractor out of soft spots, with or without

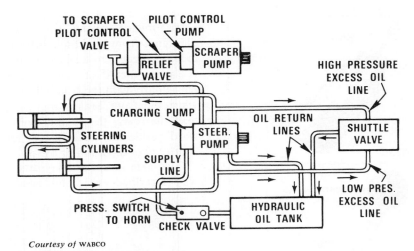

Courtesy of WABCO

Fig. 17-10. Schematic of scraper steering system

blocking. The diagrams in Figure 17-11A show a succession of moves which depend on the fact that when the tractor is turned while stationary one wheel moves forward, the other back, thus putting them both on a new footing. If necessary, the forward wheel can then be blocked and the machine steered in the opposite direction, so that the other wheel will move forward pulling the scraper after it. This technique may also be used without blocks while moving steadily forward. While standing still the strain is sufficient to make the steering system overheat. The technique should be used only in emergencies, and then slowly or intermittently.

PUSHER

A conventional scraper with only one pair of driving wheels does not have sufficient power or traction to load itself. It can usually pick up a few yards, then the weight of this partial load prevents entrance of more dirt. Two-engine and trailer scrapers do not have adequate self-loading ability for many conditions.

Additional power is usually supplied by one or more pusher tractors (or, more rarely, pull or snatch tractors), crawler or four wheel drive, which are discussed in Chapters 8 and 15. They supply the force needed to pick up a full load in an efficient time.

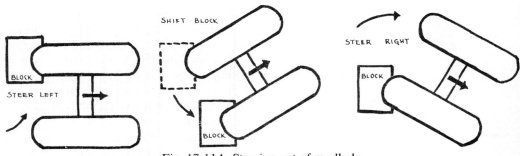

Fig. 17-11A. Steering out of mudhole

Fig. 17-11B. Push loading

Push-Pull. Scrapers may be used to assist each other in loading. The two are connected during loading by bumpers (push) or by cables, hooks, or brackets (pull). The power of the two machines is concentrated on loading one of them. When it is full, its bowl is lifted, and all power is diverted to the other.

The push-pull device in Figure 17-12 is fitted to a minimum of two tractors, preferably of the same model. Each carries a pusher block and a hook at the rear, and a cushion block and a loose-fitting, hydraulically controlled bail in front.

The front scraper starts loading; the second comes up behind it and establishes pushing contact between the bumpers, then lowers the bail onto the hook. It is a loose fit, so that exact lining up is not necessary. Loading of the front scraper is continued until the load appears adequate to the operator of the rear scraper, who then signals the front operator that he has dug enough.

The front scraper lifts its bowl, the rear one lowers it. The hitch opens so that the bumpers lose contact, and pull is exerted through the hook and bail. When the second load is complete, the bowl is lifted, and tension on the hook is decreased so that the bail can be lifted automatically. The two scrapers then operate independently to the fill, and hook up again on their return to the cut.

This arrangement has great advantages in job efficiency, but the total traction of a pair of conventional scrapers may not be sufficient for good loads under unfavorable conditions. However, the hitch can be used on the two-engine tractors described below, and should then provide excellent loading wherever tires can find a grip.

TWO-ENGINE SCRAPER

A scraper's power may be doubled, and

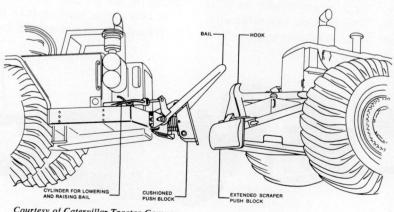

BAIL — HOOK

CYLINDER FOR LOWERING AND RAISING BAIL

CUSHIONED PUSH BLOCK

EXTENDED SCRAPER PUSH BLOCK

Courtesy of Caterpillar Tractor Company

Fig. 17-12. Push-pull connectors

Courtesy of TEREX *Division, General Motors Corporation*

Fig. 17-13A. Two-engine scraper

its traction often more than doubled, by mounting a second engine at the rear, above the ejector cylinder, so that it can drive the scraper wheels. The unit may be called two-engine, tandem-powered, or four wheel drive.

The engines must be compatible with each other in controls and performance, and equipped with torque converters and power shift transmissions. They are controlled from the operator's station in front, and under ordinary conditions are coordinated in rpm and transmission ratio.

The second engine and drive give better loading than push-pull, as all wheels are drive wheels, and power-weight ratio is much greater. The machine has substantial advantages in maneuverability and accelera-

tion, in ability to climb steep or slippery grades, and in independent (instead of paired) activity.

These two-engine scrapers can usually work alone in easy to medium digging. In hard digging they need, or at least benefit by, pusher help. When paired as push-pulls they can usually handle any type of scraper loading without outside help, except on slippery ground.

Even when pushers are used for loading, the second engine often pays its way in making it possible to pull loads up steep, loose, or muddy grades, to improve acceleration, and increase both loaded and empty speeds on soft ground and moderate grades.

A two-engine scraper may be extended by fastening a second scraper bowl and en-

Courtesy of TEREX *Division, General Motors Corporation*

Fig. 17-13B. Two-engine scraper drive

Courtesy of Deere & Company

Fig. 17-14. Elevating scraper, cutaway

gine on the rear of the first, so that there are two bowls powered by three engines, all still controlled by one operator.

ELEVATING SCRAPER

The elevating scraper is a truly self-loading machine. Pusher help is not required and is not even useful under ordinary conditions. If a pusher is used, it is usually a crawler on ground too slippery for tires, and it must be operated with care to avoid damage to the elevator.

Elevator. If these machines, the apron is replaced by an elevator made up of two roller chains carrying a number of crossbars called flights. Its foot is near the bowl cutting edge, it slopes back 40 to 45 degrees, and is somewhat higher than the sides of the bowl. It is driven through reduction gear boxes by one or sometimes two hydraulic (or more rarely, electric) motors on the upper sprockets, which are connected by a rigid cross shaft.

In digging, the elevator is rotated as shown in Figure 17-15. The flights may cut soil ahead of the cutting edge (A), or above it as in (B). The arrangement varies among different makes and models.

The backward motion of the flights (which is considered to be "forward" rotation) carries the cut soil into the bowl, along with that loosened by the edge.

The material falls or is thrown into the bowl as the loaded flights move upward. It cannot fall out because of the narrowness of the space between the edge and the elevator, and the continued motion of the flights. Angle and height are such that a full load, usually sloping from the elevator down to the tailgate, can be obtained.

The elevator is held in line at the bottom by a pair of brackets which may be spring loaded to permit the elevator to ride up and over boulders entering the bowl. There may be a mechanical adjustment of clearance, to suit different types of digging.

The elevator may have a single forward speed, usually between 225 and 290 feet per minute. Means to reverse it may be provided as standard or optional equipment. Some machines have several forward speeds, others provide infinitely variable forward

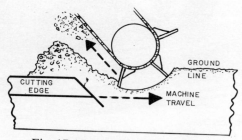

Fig. 17-15. Elevator cutting action

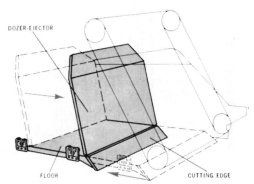

Courtesy of Caterpillar Tractor Company

Fig. 17-16. Ejector and floor movements

and reverse speeds with hydrostatic drive.

During loading, which is done in forward speed, the elevator may use half the power of the engine. An overload valve or switch will stop it if it jams.

Reverse is used for dislodging obstacles or sticky soil, and occasionally to prevent soil from entering the bowl during light grading.

The allocation of available power to the elevator or to the drive wheels is usually determined with efficiency by the torque converter delivering slower· motion with multiplied torque to the wheels, while a

power takeoff or generator delivers full speed full power to the elevator.

This difference enables the elevator to retain its speed while the forward motion slows in hard digging, to give it an opportunity to cut and/or scoop up the soil in thinner bites.

Bowl and Ejector. The general shape of the bowl is similar to that in a conventional scraper, but its structure must be different to allow for ejection.

It is not practical to lift the elevator up and forward in the manner of an apron, because of lack of space under the yoke. An opening for discharge of load is made by pulling the front half of the bowl floor to the rear. This dumps part of the material.

Then the rear wall of the bowl is moved forward, usually in conventional ejector fashion, pushing the rest of the load out the gap left by the backsliding floor.

During unloading, the machine is moving forward continually, and the discharged material is struck off by the cutting edge. The spread is usually quite uniform, as the flights break lumps and aerate soil as they load it.

The double action of ejection, first moving the floor and then the rear wall, is accomplished automatically in one position of

Courtesy of M-R-S Manufacturing Company

Fig. 17-17. Four wheel drive tractor and elevating scraper

the control lever. In some machines, one cylinder or a pair do both jobs through mechanical linkage, in others two or four cylinders operate separately in obedience to a sequence valve.

Ejection is sometimes started and stopped with an empty bowl, to move the edge back far enough to make it possible to pick up a boulder. The scraper is moved over it at full height until it is behind the elevator. The bowl is then lowered, and the object scooped up by the retracted edge, which is then slid forward to retain it.

The method described is only one of several used to get a normal load out the bottom. See Figures 17-16 and 17-28B.

Working Characteristics. The self-loading principle of these machines is that the elevator flights continuously remove the material that they dig, and that is dug by the bowl edge, so that it does not impose a back pressure or load on further digging.

The last yard in a load is therefore as easy to dig as the first, although it does take some extra power to hoist it to the top of the heap.

The problem of loading is reduced to one of cutting or chopping a slice of the ground. There is usually sufficient traction to permit this in fairly hard soil, although depth of cut may have to be reduced. Teeth bolted to the center section of the edge help digging, but are a nuisance in spreading.

Self-loading makes each scraper an independent unit, that can work alone or as a member of a large fleet, with equal efficiency. The cost, and the planning and supervision problems of pushers are eliminated. Coarse or hard soil is broken into small lumps or pulverized, reducing voids in the load and problems in the spreading of it.

There are drawbacks, of course. The elevator is expensive, and requires maintenance. It is heavy, and must be carried on the haul, with resulting loss of acceleration and of speed up grades or on soft footing. It cannot be used where there are large

rocks, and repairs are apt to be increased greatly by small ones, and even by hard soil.

The balance of advantages against disadvantages is definitely favorable, as is indicated by the increasing share of the scraper market being taken over by the elevating models.

Four Wheel Drive Tractor. The scraper in Figure 17-17 is powered by a tractor with four wheel drive and four wheel steer. Three-axle steadiness is provided, with increase rather than decrease in traction.

Two-Engine. The two-engine elevating scraper combines the two separate approaches to self loading.

Doubled power helps loading chiefly by shortening time by permitting a thicker slice. The front engine may put all or most of its power into driving the elevator, while the rear one supplies the push.

The principal two-engine advantage is increasing ability to work, or to work more rapidly, under unfavorable circumstances—slippery footing in the cut and on the haul, and/or steep grades and soft ground.

TOWED SCRAPER

The towed or full trailer scraper, pulled by a crawler tractor, was the dominant type for years. It has been replaced by the faster and more flexible self propelled machines, and is now of secondary importance.

However, it is still efficient under many conditions of short haul, difficult circumstances, and only-occasional need, and will continue to be used for years.

A description of these machines in earlier editions of this book is therefore being retained.

Carryall. Figures 17-18 to 17-22 show a trailer scraper formerly manufactured by LeTourneau-Westinghouse, now WABCO Construction & Mining Equipment Group.

The frame structure is that of a low-slung, pneumatic-tired wagon. The rear axle is rigid. The narrower front axle is fastened to the frame yoke or gooseneck

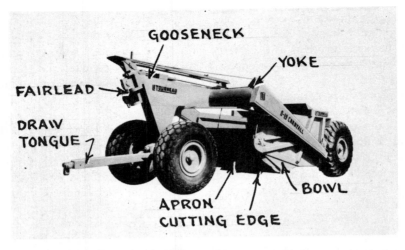

Fig. 17-18. Towed four wheel scraper

by a hinged kingpin which permits it to turn and to oscillate. The frame yoke is arched to allow the front wheels to pass under it so that very sharp turns may be made.

A tongue drawbar is rigidly fastened to the front axle and contains a swivel joint (universal forging), so that the scraper can overturn without damaging the tongue or affecting the towing tractor.

The bulk of the scraper consists of two parts—the front yoke which includes a cross beam and a pair of arms extending backward and downward from it; and the body which is hinged to the yoke arms, pivots on the rear axle, and includes the bowl.

Bowl. The bowl is substantially a box open at the top. The bottom and sides are part of the body. The front, or apron, is hinged near the center of the bowl so that it can be lifted away from the bottom. The front of the floor is equipped with a cutting edge or blade. The back, known as the tail gate or ejector, moves forward to the edge when pulled by a winch cable, and back

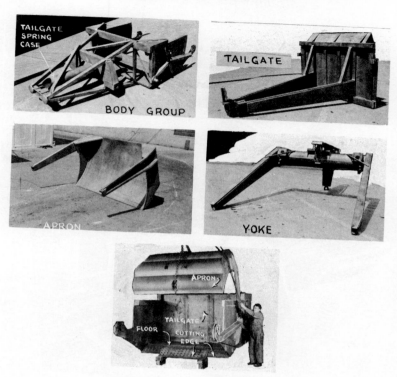

Fig. 17-19. Scraper parts

again from spring pressure, and keeps close sliding contact with the bottom and sides.

The cutting blade is composed of three pieces of wear-resistant steel bolted onto the bowl bottom. The center piece may be held forward of the side pieces for improved penetration or, more rarely, the pieces may be in one straight line for smooth finish grading.

The bowl is attached to the rear axle inside of the wheels, which are positioned so that they will roll in the path made by the bowl edge.

Control Units. The traveling, steering, and digging power of this machine is provided by the movements of the towing tractor. Individual motions of the bowl, apron, and ejector are powered by a two-drum power control unit mounted on the back of the tractor. When the scraper is not attached the drums may be used to operate other machines.

Figures 17-20 and 17-21 show a control unit. It consists of a pair of high speed winches driven from the power takeoff, and controlled separately by friction clutches

Fig. 17-20. Power control unit

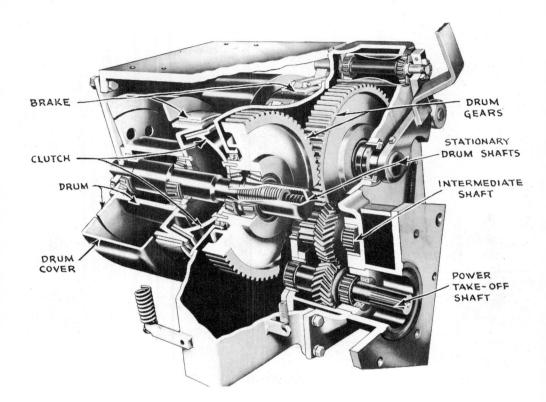

BRAKE

CLUTCH

DRUM

DRUM
COVER

DRUM
GEARS

STATIONARY
DRUM SHAFTS

INTERMEDIATE
SHAFT

POWER
TAKE-OFF
SHAFT

Fig. 17-21. Cutaway view of control unit

and brakes. One lever controls the clutch and brake for one drum, and has three positions—clutch engaged, brake off; clutch disengaged, brake on; and clutch and brake both released. The two control levers are behind and to the right of the tractor seat.

Each cable passes through a fairlead made of two sheaves which swivel on vertical hinges. These keep the cable lined up so that the winches can operate as effectively on turns as on straightaways. Additional fairleads are mounted on the scraper yoke to line the cable up with the scraper sheaves.

Reeving. Three functions of the machine are controlled by the two cable units—the lifting and lowering of the bowl for digging depth and carrying and spreading height, the lifting and lowering of the apron, and the forward and backward motion of the

tail gate. The cable threading or reeving is shown in Figure 17-22, but is somewhat different in other models.

The right hand (looking forward) control unit and lever operate the bowl hoist cable. It is reeved through a fairlead, then around a set of seven sheaves lying horizontally on the yoke top. Three of these sheaves revolve around a fixed axle at the front. The other four, called traveling sheaves, are mounted in a frame that can slide in a track. The frame carries an additional sheave at the rear. A heavier cable is threaded through this and around guide sheaves to anchors on the side beams of the bowl.

When the cable is reeled in by the control unit the traveling sheaves are pulled forward powerfully by an eight part line and raise the bowl by pulling the heavy

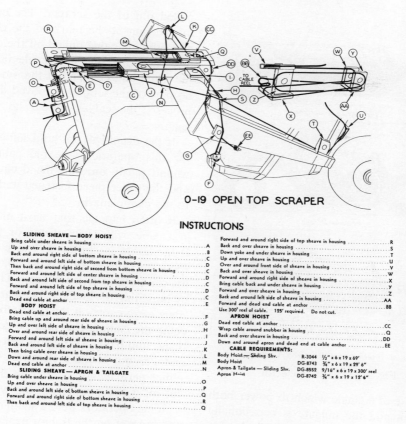

O-19 OPEN TOP SCRAPER

INSTRUCTIONS

SLIDING SHEAVE — BODY HOIST

Bring cable under sheave in housing	A
Up and over sheave in housing	B
Back and around right side of bottom sheave in housing	C
Forward and around left side of bottom sheave in housing	D
Then back and around right side of second from bottom sheave in housing	C
Forward and around left side of center sheave in housing	D
Back and around left side of second from top sheave in housing	C
Forward and around left side of top sheave in housing	D
Back and around right side of top sheave in housing	C
Dead end cable at anchor	E

BODY HOIST

Dead end cable at anchor	F
Bring cable up and around rear side of sheave in housing	G
Up and over left side of sheave in housing	H
Over and around rear side of sheave in housing	I
Forward and around left side of sheave in housing	J
Back and around left side of sheave in housing	K
Then bring cable over sheave in housing	L
Down and around rear side of sheave in housing	M
Dead end cable at anchor	N

SLIDING SHEAVE — APRON & TAILGATE

Bring cable under sheave in housing	O
Up and over sheave in housing	P
Back and around left side of bottom sheave in housing	Q
Forward and around right side of bottom sheave in housing	R
Then back and around left side of top sheave in housing	Q

Forward and around right side of top sheave in housing	R
Back and over sheave in housing	S
Down yoke and under sheave in housing	T
Up and over sheave in housing	U
Over and around front side of sheave in housing	V
Back and over sheave in housing	W
Forward and around right side of sheave in housing	X
Bring cable back and under sheave in housing	Y
Forward and over sheave in housing	Z
Back and around left side of sheave in housing	AA
Forward and dead end cable at anchor	BB

Use 300' reel of cable.   125' required.   Do not cut.

APRON HOIST

Dead end cable at anchor	CC
Wrap cable around snubber in housing	Q
Back and over sheave in housing	DD
Down and around apron and dead end at cable anchor	EE

CABLE REQUIREMENTS:

Body Hoist — Sliding Shv.	R-3044	½" x 6 x 19 x 69'
Body Hoist	DG-8743	¾" x 6 x 19 x 29' 6"
Apron & Tailgate — Sliding Shv.	DG-8552	9/16" x 6 x 19 x 300' reel
Apron Hoist	DG-8742	¾" x 6 x 19 x 12' 6"

Fig. 17-22. Cable threading diagram

cable. Brake application will hold it in any position. If the winch is released the weight of the bowl will pull the traveling sheaves back.

The other unit powers both the apron and the tailgate. The line goes through a set of stationary and traveling sheaves above the yoke. The traveling sheaves are fastened to a single heavy cable anchored well down on the front of the apron.

The winch line goes from the last stationary sheave of the apron set around guide sheaves to the rear of the tailgate thrust arm. Sheaves are fastened to the frame just behind the gate, and a six part line connects them with three sheaves fastened to the thrust arm or triangle which is

part of the gate. The gate is held to the rear by a heavy spring.

When this line is winched in, it tends to both raise the apron and to move the tailgate forward. The gate is held back by a spring, and usually by the dirt in the bowl, so that the apron rises as its traveling sheaves move forward. When it is fully up it hits stops so that the traveling sheaves are stalled by the apron cable, and cannot move farther. The line then moves the gate forward, causing it to eject the dirt from the bowl over the cutting edge and under the raised apron.

When the winch is released, the gate moves back under pressure from the spring and the apron descends by gravity.

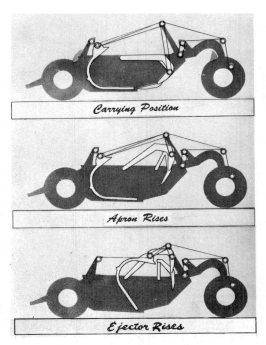

Carrying Position

Apron Rises

Ejector Rises

Courtesy of International Harvester Company

Fig. 17-23. Tilting ejector

The use of a single winch to power both the apron and the tailgate is occasionally awkward, but it is of course a considerable economy, particularly in towed scrapers, and has the added advantage of making it impossible for an amateur operator to attempt to eject a load without fully opening the apron—an error that would result in strained parts and broken cable.

The apron-tailgate line is anchored to the frame after being threaded through the third sheave on the thrust arm. It is then continued to a storage reel.

This reel or spool provides a reserve of cable that can be quickly installed. Due to the number of sheaves and the location of some of them, threading a cable is a difficult and confusing task. If an operator makes regular inspections of his cables he can usually find signs of serious wear before one breaks. He can then loosen the anchor clamp and wind the cable in on the winch drum until the old cable is entirely

pulled out of the scraper. The cable may then be cut at the mark of the clamp anchor, the old cable pulled off the drum and discarded, and the new cable fastened to the drum and clamped at the other end. If no part of the scraper is moved during this process, the cables should be the same length.

If the drum will not hold sufficient cable for this operation, the old cable can be wound partly on, cut, the drum stripped, and the outer piece fastened and reeled in to the clamp mark.

It often happens that most of the wear in a cable will occur at one or two spots. If the cable is shifted to move the worn spots into places of less strain, the cable life can be greatly lengthened. This is done by loosening the anchor clamp and winding some extra cable off the spool, then retightening the clamp. This may be done several times, and badly worn spots can sometimes be taken out of the system this way, being either permanently wound on the drum or cut off without replacing the whole cable.

Some operators weld a small cable cutter and bracket on the tractor near the control units. This is a valuable time saver when shifting or changing cables, as it cannot be mislaid.

Tilting Ejector. Some makes of scraper dump their loads by tilting the bowl floor and back upward and forward.

The tilting ejector is usually a curved piece that make up the back and most of the floor of the bowl. It is hinged to the bottom just behind the cutting edge. During the dump it is rotated forward on this hinge, scraping dirt from the sides of the bowl, and spilling the load over the front of the edge. The highest angle is steep for the floor and overhanging for the back. Return is started by spring action and finished by gravity.

Hydraulic. There are also towed scrapers with hydraulic control, and some that have combinations of cylinders and cables.

Fig. 17-24. Scraper digging action

Power is usually supplied by a pump on the tractor, with a valve that directs fluid through a set of hoses to cylinders on the scraper. These hoses have (or should have) quick-disconnect couplings on the rear of the tractor.

Location and function of cylinders is generally similar to those described for self-powered scrapers.

LOADING

Conventional. The conventional scraper needs the help of a pusher to obtain a worth-while load. However, to avoid confusion, the scraper operation will be decribed separately first. The cooperation of the pusher in the work will be described afterward.

For normal loading, the tailgate is always fully retracted. In rough or bouldery soil, the apron is held almost all the way up, in most digging it is just high enough not to drag, and in sand it may be allowed to drag for a compacting effect. Up to one-third of the load in the bowl may rest on the apron when it is kept fairly low.

The bowl is lowered so that its edge will cut a slice of ground. In hard ground, it might be forced down to lift the scraper wheels in the air.

Digging is done in low or second. The throttle is adjusted at the highest engine speed that will not cause the wheels to spin.

Depth of cut is regulated by raising or lowering the bowl. Except when working near final grade, the cut should be as deep as the machines can handle without spinning drive wheels. If speed is maintained, a deep cut fills the bowl fastest.

A spinning wheel causes expensive wear to its tire, and delivers less push than a non-spinning one.

There may be a differential lock pedal, which you push down if only one of a pair of wheels spins. They will then turn together. But release the accelerator a moment to stop spinning before using it. Don't steer while the differential is locked. The pedal will come back up when traction becomes equal on both sides.

The dirt piles up in the bowl, part of it falling forward on the apron. Incoming dirt must be forced to rise through an increasing depth in the bowl. Loading rate slows, and a point of refusal is likely to be reached where power is not sufficient to force any more dirt into the bowl, or when the soil will not take the thrust, so that it breaks up and is pushed ahead, or drifts off the sides.

If a full load has not been obtained, some may be added by pumping. The edge is raised sufficiently to decrease the draft and allow the tractor to regain its full speed, then dropped several inches below the normal cutting depth. Some soil will be forced into the bowl. The knife is lifted as the engine starts to labor and the process is repeated if desired.

The advantage obtained is partly the momentum of the accelerated scraper, and

Courtesy of TEREX *Division, General Motors Corporation*

Fig. 17-25. Two-engine scrapers loading coal at power plant

partly the ramrod effect of the thicker layer of dirt punching its way up through the load.

This punching effect is important. It will be found that clay and other heavy soils may be loaded most effectively in thin layers which reduce the cutting power necessary, without sacrificing too much thrust. Sand, however, must be taken in deeper cuts and requires pumping with a much smaller load. It is sometimes helpful to set the apron to drag on the surface of loose soil to compact it while loading.

When a satisfactory load has been obtained, the apron is lowered and the bowl edge raised an inch or two above the ground.

Courtesy of International Harvester Company

Fig. 17-26A. Pushing

This position is held for several feet before lifting to carrying position, to spread any loose material in front of the blade and leave the cut smooth.

Scraper cuts, loading and pushers were discussed in Chapter Eight, in connection with road building. There is some repetition in this section.

Pushers. Pushers are machines that supply extra power to a scraper during digging. For this minute or two, the machine needs several times as much power as it does at any other time. It is efficient to have one pusher provide it for several scrapers, rather than have it part of the scraper, and unused most of the time.

Pushers are usually tractors, either crawler or four wheel drive, equipped with dozer blades or pusher plates. Special pushing equipment may include a cushioned bumper, to reduce shock at unequal speeds. Two pushers may work together in tandem.

Because of their greater speed, rubber-tired pushers can service a maximum number of scrapers under average conditions. However, they do not exert as heavy a push in proportion to weight as a crawler can, and their usefulness declines rapidly if the cut becomes slippery.

On some jobs and under some conditions, the scraper moves into position to dig, then waits for a pusher. At other times, the scraper will start its run, loading a few yards before being pushed.

The pusher is driven up behind the scraper, stopped or moving, and contact made with its bumper as smoothly as possible. The two machines each exert as much power as possible without excessive spinning of wheels or tracks. Most of the digging power is supplied by the pusher, and the scraper engine may be cut back to half or two-thirds speed.

The push must be in a straight line, as a blade at an angle may cut scraper tires, or cause jacknifing or steering problems.

When a full load is obtained, the pusher drops back to service another machine, and the scraper goes into a higher gear and leaves. The pusher operator has a better view of the load, and may signal when it is ready.

In general, it does not pay to struggle to get a heaping load, as diminishing returns may make the last yard a very expensive one. This situation is discussed later, on page 17-36.

Push-Pull. Scrapers may take turns pushing each other. The most efficient arrangement is to equip them in pairs, each with a front bumper carrying a hook on its top, and a cushion rear bumper with a bail (loop) above it, controlled hydraulically. See Figure 17-12.

Elevating Scraper. The elevating scraper has chain-driven flights or paddles that remove soil from the cutting area and move (elevate) it into the bowl. The elevator replaces the apron.

The digging is shared between the flights and the bowl edge, in ways that differ among various machines, and may be subject to adjustment.

In general, the work of the elevator is increased with increase of digging speed and force, and with depth of cut. The elevator

is not as rugged as most parts of a scraper, and may be choked by too much dirt, or jammed by rocks or trash.

Elevator scrapers are designed for self-loading only. Pushers are likely to damage them.

The hydraulic or electric drive of the chain usually has overload protection, so the chain will stop moving if over-worked or jammed. Re-starting it can be a problem. Lifting it out of the ground is the first step. Then try it in reverse, if it has a reverse.

If there is no reverse, or if it is jammed, raise the bowl, put a log or other strong support under a flight on the forward (toward the tractor) side of the sprocket, and lower the bowl gently. This should turn the elevator backward to release the obstruction.

HAUL

The loaded scraper is driven to the fill area at the highest speed allowed by safety and reasonable comfort. Some of the driving problems are discussed later, under *RE-TURN*.

Scrapers can move loads successfully over soft and rough ground, but always at the expense of speed and maintenance expense. Yardage, both in the load and in the number of loads can be greatly increased by providing a smooth, hard surface for rubber-tired tractors, and a smooth, slightly soft surface for any track types. If a very large quantity is to be hauled, it pays to oil or blacktop the road.

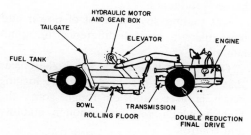

Fig. 17-26B. Elevating scraper

Courtesy of Caterpillar Tractor Company

Fig. 17-27A. Cross country haul road

This surface would rule use of crawlers.

In general practice, the haul route is dirt unless bad weather is expected. Then bank gravel, shell, or other material not seriously softened by water should be used. The road should be maintained by a grader, but if none is available it can be kept in shape at somewhat higher cost by the pans. In dry weather watering equipment will be required.

Upgrades slow the movement of loaded pans, the amount of loss depending largely on the ratio of the power of the tractor to the weight of the load. Two-engine rigs will get up the grades faster than single-engine ones of the same type. On severe grades level loads may make enough better time than heaped loads to give higher production.

Soft footing absorbs a lot of power and may make a critical difference in uphill hauling.

On graded haul roads the highest gear that will move the load without lugging down is used, and on rough going the highest which will give enough stability and control.

Bowl Position. A scraper bowl is car-ried high to avoid obstacles and low to keep the machine stable. The actual height will therefore vary with route conditions.

A good haul road maintained by graders should permit keeping the bowl within two or three inches of the ground. But such a road is usually good enough so that there is little danger of upsetting if it is held higher.

On a rough or soft road the bowl must be fairly high to avoid colliding and dragging, but the stability of a low bowl may be needed. Under these conditions, a number of up and down adjustments may be made during the haul.

The bowl may be deliberately dragged to slow the machine descending a steep hill, or in an emergency stop. But the wear on the edge makes this form of braking as expensive as it is inefficient.

The bowl should never be held fully up. In this position the hoist cylinders become rigid frame members, and must absorb damaging stresses on rough ground. A fully-up position may break the lines in a cable machine, for the same reason.

Shifting. The modern scraper shifts very easily. You just move a lever, and clutches inside the transmission make the shift. But you may have a problem wiggling the lever around various planned obstructions, and the shift may not be as smooth as you and the power train would like.

There may be as few as four or as many as nine forward ranges. In the upper group they may be in pairs that each have one set of gears, with drive through the torque converter in one range, and direct mechanical in the other. When you start up a grade or get into a soft spot, the converter will allow the road speed to drop but maintain engine speed. In direct, road speed will hold up better at first, but the engine will slow in proportion to any drop.

If the resistance continues or becomes heavier, direct will cause the engine to lug down so you have to shift, while the converter might keep going comfortably at reduced speed. In this last case, the next lower direct range might give better performance.

In some semi-automatics, the two lowest (digging) speeds are manual. From second you can move the lever direct to the highest gear you want to use, and the transmission will do its own figuring and shifting. You can override the automatic with the lever any time you want. There may be a pedal you can push down to hold it in whatever gear it is in, until the pedal is released.

Shifting is guided mostly by the feel of the machine and your estimate of conditions immediately ahead. But it may be controlled by dashboard gauges. You must manage to keep the engine within its efficient operating speed range. If it is on the fast side, shift into a higher gear. If it is near the bottom, shift into a lower gear.

If the torque converter fluid is too hot, use a lower gear.

Retarder. Many torque converters include a hydraulic retarder. This is a drag or resistance mechanism which runs free when empty, but when filled with oil it slows the drive shaft. Partial filling for partial effect may be possible.

The retarder is used to slow the machine

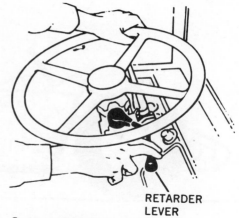

RETARDER
LEVER

Courtesy of Caterpillar Tractor Company

Fig. 17-27B. Retarder control

on down grades. You have to think in advance to take advantage of it, as it needs from two to six seconds to take in enough fluid to start working.

The energy of motion which it absorbs is changed into heat, which is dissipated by the converter cooling system. On a long descent it is useful in keeping the converter warm.

Brakes. Big scrapers usually have air brakes, smaller ones have hydraulic of either the automotive or the pressure-pumped type. All of these are absolutely dependent on tight, non-leaking systems. Most or all air brakes need 100 pounds pressure. If the gauge doesn't show that much, stop immediately and investigate.

There is usually an arrangement that puts on the rear (scraper) brakes sooner and/or harder than the front (tractor) set, to assure straight stops.

Air brakes should not be pumped (put on and off rapidly), as this exhausts the air supply.

DUMPING AND SPREADING

Dumping and spreading are one operation, as the scraper dumps only while moving forward, and the material falls under its edge and is automatically spread by it.

In dumping a conventional scraper, the

Fig. 17-28A. Scraper starting to dump

bowl is lowered until the edge will just allow a layer of dirt of the desired thickness to slide under it. The apron is then raised enough so that the falling dirt will supply the knife sufficiently to make a continuous smooth layer, without drop-

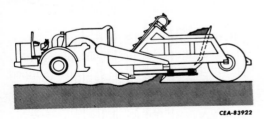

Move the ejector gate lever to "FORWARD" (toward operator) and hold there as necessary. (An "ON" and "OFF") operation of this lever will give the best results until the gate is fully forward.

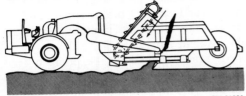

After the sliding bowl floor is all the way back, the ejector gate will move forward pushing material out of the bowl. (To remove sticky material from the bowl, place the elevator control lever in "UNLOAD".) (lever away from operator)

Courtesy of International Harvester Company

Fig. 17-28B. Dumping an elevating scraper

ping excess which would heap in front, or drift to the side to make windrows.

When the apron is fully raised the tailgate is moved forward gradually, pushing or dumping the dirt out of the bowl. Too fast a crowd will supply too much dirt to the knife, and place unnecessary strain on the cylinders and linkage of the ejector mechanism. If the dirt is sticky good results are obtained by advancing the gate a foot, allowing it to slide back six inches, and advancing it again when required. A lift gate may be banged against its stops to knock it clean. Dirt resting on the edge after the gate is fully forward may often be dumped by moving the gate back and forth.

When the dump is complete the gate is fully retracted, the apron dropped, and the bowl raised to carrying position.

Spreading is usually done in thin layers as it provides better compaction and eliminates or reduces the need for other grading equipment. It also keeps a more even grade on the dump, reducing the hazards of high speed work.

Dumping and spreading is ordinarily done heading away from the cut and at medium or ~~high~~ speed. In finishing a grade and under some other conditions a low

RIGHT WAY

WRONG WAY

**TO MAINTAIN
FILL SLOPE**

1 MAKE FILL HIGH ON THE OUTSIDE

2 THIS PREVENTS SCRAPER FROM SLIDING OVER SLOPE

3 ACCURATE SLOPES CAN THUS BE MAINTAINED TO DESIRED HEIGHTS ELIMINATING NECESSITY FOR HANDWORK

4 IF WET CONDITION PREVAILS ARRANGE FOR DRAINAGE TO PREVENT WATER POOLING IN CENTER OF FILL

**RESULT OF
INCORRECT METHOD**

1 SCRAPER WILL SLIDE OVER SIDE OF HILL

2 DAMAGE TO SLOPE WILL BE CAUSED

3 IMPOSSIBLE TO MAINTAIN ACCURATE DEGREE OF SLOPE, TENDENCY TO WORK AWAY FROM EDGE OF FILL

Courtesy of U.S. Army Engineers

Fig. 17-29A. Cross section of fill

gear may be used to give more exact control.

The fill should be started at its outer edges and kept built out to finish slope all the way up. This is important, as the type of pan ordinarily used has no way of dumping over edges and any patch fill made could not be readily compacted, so as to support the machinery building additional layers.

It is also desirable to have the fill slope up toward the edge, as this arrangement will place the larger part of the weight of both tractor and scraper on the side away from the slope, and reduce danger of caving. However, if rain is expected, the fill may be crowned enough to shed water at the end of the day.

The closeness with which a pan can approach a high edge is determined by the type of soil, the degree of compaction,

the slopes, the weight of the machines, and to some extent by the skill of the operator. In general, heavy soils which are not in a muddy condition are safer than sandy soils. Working right to the edge or overhanging it slightly with a tamping roller will give good compaction.

If conditions are such that the scrapers cannot safely go near enough to the edge to build a proper slope, they can drop the material a few feet in and a grader can distribute it.

The slope should be checked frequently for proper grade.

Elevating Scraper. The elevator usually stays in place during the dump. An opening for unloading is made by pulling back the knife and front half of the floor, sliding on guides and/or rollers under the rear half. The tailgate is then moved forward.

A substantial part of the load falls by

Courtesy of Young Corporation

Fig. 17-29B. Trailer scraper, current model

gravity through the floor opening. The gate and floor controls are interlocked, so that gate movement does not start until the floor is completely open. The control may be returned to HOLD for a few seconds after floor motion is complete, so as not to overload the opening.

RETURN

After completing the spread the scraper is turned, either on the fill or after going on to a turning point, and driven back to the cut at the highest safe speed.

Safe speed might be the maximum of the scraper at full governed engine speed, but there are usually limiting factors. The machine will be automatically slowed by upgrades and soft footing, and should be slowed for rough spots, curves, close approaches to other machines, downgrades and other hazardous conditions.

Accuracy of steering and efficiency of brakes must be taken into account. Two-axle machines may have a tendency to rythmic bouncing, called loping, which is most uncomfortable and may reduce control dangerously. Special cushion hitches may reduce this, but you should immobolize them during digging and spreading, and may forget to re-connect. Otherwise, a small increase or decrease in speed may steady it.

A loaded scraper (or truck) has the right of way over an empty unit.

TRAILER SCRAPER OPERATION

Connecting. Two-axle cable scrapers (sometimes called four-axle or full trailer) are usually stored with the bowl blocked up and the cables in place. To reconnect, the tractor is backed to within a foot or two of the tongue and each cable is threaded through its fairlead (in the model illustrated, around the bottom of the lower sheave and over the top of the upper) to an anchor on the control unit drum.

It is customary to attach the bowl line to the right drum and the apron and tail gate line to the left.

If the cable is not threaded through the scraper, an instruction book and/or a very experienced man are needed to get it through right. The sample reeving diagram in Figure 17-22 shows how complex this pattern is. Specific information is not good for any but a particular make, model, and year.

If the tongue is light or counterbalanced it can be raised by hand and the tractor backed into it. If it is too heavy to lift, a pulley or snubber should be fastened to one side, and one of the cables looped under this projection, or wedged into it. Reeling in the cable will then raise the tongue. The tongue may also be lifted with a jack or pry.

The tractor drawbar pin should be locked in place with a bolt or cotter pin.

Hydraulic two-axle scrapers depend on counterbalancing or jacking the tongue, or leaving it blocked at correct height when disconnecting. One-axle hydraulics allow manipulation of hitch height by applying down pressure to the bowl.

Cable operated one-axle scrapers should be carefully blocked at the correct height when they are disconnected.

Controls. All full trailer cable operated scrapers and many self-powered models are controlled by two levers behind and to the right of the operator. One or both of them has a horizontal handle projecting forward.

It is standard practice to operate the bowl hoist that controls digging depth and carrying height from the right hand lever and winch. Moving the control to the left engages the clutch and raises the bowl, center or neutral locks it in position, and moving it to the right releases it to lower.

Courtesy of Shunk Manufacturing Company

Fig. 17-29C. Changing scraper edge

The left lever and winch will then control both the apron and tailgate. Moving the lever to the right engages the clutch that first raises the apron until it hits its stops, then pulls the tailgate into dumping position.

Center position holds both units in place. Moving the lever to the left releases the winch, allowing the tailgate to first be moved back by springs or gravity, after which the apron will close.

The levers are spring loaded to return to

Courtesy of TEREX *Division, General Motors Corporation*

Fig. 17-29D. Two-engine scraper on steep shortcut

neutral automatically. Clutches and brakes may be so built that they should not be slipped, so that gradual changes are made in a series of light jerks.

Each control system is different. Also, a scraper may be hooked up to operate on opposite drums.

Hydraulic operated scrapers may use either two or three controls. The third valve and lever permits independent operation of the apron and the ejector.

Cable Tension. The cable should never be run slack. Hanging loops may run off sheaves and jam when pulled in, and the looseness causes damaging cross winding

Courtesy of Caterpillar Tractor Company

Fig. 17-30. Guard lowered for servicing

on the drum and lost motion and time when reeling in.

It is equally important not to operate a trailing scraper with either the bowl all the way up or the gate all the way forward. The distance from the winch fairleads to the scraper changes on rough ground and turns, and a line held by the winch brake at one end and bowl or gate stops at the other will be strained and probably broken.

Overturning. On side slopes always move slowly, keep the blade low, avoid going over rocks or bumps with the upper wheels or hollows or soft spots with the lower ones, and be extra careful when turning.

A full trailer pan should not be used on steep side slopes unless the operator is very skillful, or other equipment capable of righting it is on the job. The narrow front tread, oscillating front axle, and tendency to jackknife when backed up all make it liable to side tipping. It is also unsafe to use it close to the edge of poorly compacted fills.

If side hill work cannot be avoided, a scraper with minimum overhead structure should be used and the bowl should be carried low. Turns should be made uphill where possible and downhill turns should be made gradually, preferably with the apron fully down and the bowl scraping on the ground.

Danger of tipping is particularly severe if the soil under the upper side of the machine is harder than that under the downhill wheels or if the upper side is forced up by running over boulders.

Overturning a trailing scraper usually does little or no damage except in lost time, as the swivel in the draw tongue allows it to turn freely and the structure is sturdy enough to resist damage. A good operator, or a lucky one, may be able to right it by towing it over a hump or the edge of a hollow that will roll it back. If the cables are not too tangled the parts may sometimes be worked into a more suitable position before towing.

If it cannot be righted by towing it can be rolled by pulling a high part from the side, preferably by another machine. If none of sufficient power is around, the tractor must be disconnected in order to get beside it to right it.

A self powered scraper is less inclined to tip, but if it does it takes the operator with it, unless he is very quick at getting out.

Servicing. Servicing a towed cable scraper is dangerous, unless the operator or mechanic knows what he is doing, and takes proper precautions.

When changing blades or doing other

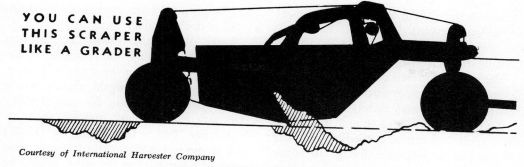

YOU CAN USE
THIS SCRAPER
LIKE A GRADER

Courtesy of International Harvester Company

Fig. 17-31. Grading with scraper

work under the scraper, both the bowl and the apron should be blocked, so that they cannot come down if the supporting cables break or are accidentally released.

If work is being done behind a forward-positioned or raised tailgate (ejector) it must be blocked from moving back.

The blocking must be heavy and positive, and should be checked by releasing the cable, then re-locking it.

The tractor should be immobilized by locking, or providing some well-understood indication at the controls that the scraper is being serviced.

The dumping mechanism is returned to loading position, or at least started back, by heavy springs or torque rods. These can be dangerous if their tension is accidentally released by work done on their cases.

Do not leave scraper parts in positions from which they would move if the control unit brakes were released.

Use gloves in handling cable.

Keep hands free of cable and sheaves while the rig is operating.

Do not use weak or frayed cable.

Do not drive at high speed on rough ground or turns.

Hydraulic. Hooking up a hydraulically controlled model is simplified by counter-balancing or jacking the tongue, or leaving it blocked at correct height when disconnecting. It may or may not be necessary to connect the hydraulic lines before the draw

tongue, to move it into a good position.

Disconnect couplings should be wrapped in plastic when they are opened, and checked carefully for cleanliness when re-connecting.

Hydraulic controls eliminate problems with cable, and provide down pressure on the bowl edge. General operation is similar to cable jobs.

GRADING

Scrapers are sometimes used to smooth and patch haul roads, and in grading and leveling areas where the quantity of dirt to be moved is not enough to fill the bowl, and where distances are too short for the loading-dumping cycle.

For grading, the center piece of the knife should be flush with the side pieces. Otherwise, finish work will be limited to the width of the center.

The location of the knife between front and rear axles provides good stability for grading, and controls are usually sufficiently sensitive for medium fine work. But there is no means to shift dirt from side to side, grader-fashion, so that crowning a road or shaping a slope is a laborious process.

On the other hand, the pan can move substantial amounts of material along a road, and can bring in borrow, or dispose of surplus off the road.

When cuts and fills are shallow and close-ly spaced, the usual working position is with the apron fully lifted, and the tailgate most of the way forward. It is bad to have it all the way forward, as it is then subject to twisting strains from the frame and the bowl.

The bowl is lowered or raised to give the desired depth of cut or thickness of fill, and acts much like a very steady dozer blade.

If a cut is deep or long enough to provide more spoil than can be carried in front, the tailgate can be retracted enough to admit some inside the bowl, and then used to push it out again when a fill is reached.

An elevating scraper must have the elevator operating in reverse, or the floor partly retracted, to avoid picking up loads. In either case, there is more than normal strain on the elevator flights, and it should not be used in this way to cut hard surfaces.

The self loading feature makes the elevating scraper ideal for grading work that requires a moderate amount of off-site borrow. It makes a good combination with a grader for this.

There is no definite line of separation between this light grading, and heavy cuts and fills.

STUMPS AND BOULDERS

Scrapers are not designed to dig or transport bulky objects, but they may be used for the purpose when more suitable equipment is not available.

Large objects can usually be dug and loaded rather readily after practice, unless the size is excessive or the shape ridiculous. But getting them out again is occasionally a real problem, calling for gate-jiggling, crowbars, and ingenuity, and perhaps finally a crane. Either the digging or ejecting of a stump or boulder might strain or bend parts of a scraper, so the work should be done carefully.

If the object is low enough for the tractor to walk over it, raise the apron to full height, drop the bowl two or three feet behind it, apply down pressure, and move forward. If the edge hooks into it, lift the bowl while inching forward. If it slips off, back for another try. You may have to try from other directions in order to get a grip.

If the piece is to be picked up and carried, the tailgate is kept in normal full-back position. But if you want to just loosen and shove it, keep the gate forward.

Often, or perhaps usually, the stump or boulder cannot be picked up on the same pass that loosened it, so you have to back off and take another bite. Try to capture it before it is entirely loosened, as you might then have trouble catching up with it.

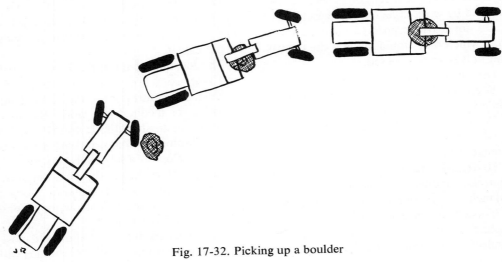

Fig. 17-32. Picking up a boulder

The apron can sometimes be used to push the object back into the bowl, or clamp it to the edge so that it can be carried.

If a stump is too high to walk across with the tractor, pass it very closely, and then steer sharply in front of it. This should bring the bowl into position to get its edge into or under it.

Scrapers, particularly of those models which have very high apron lifts, can pick up large stones or stumps. However, these sometimes turn or catch when in the bowl and become jammed when an effort is made to discharge them. They can be worked out by moving the gate back and forth and the bowl up and down; by digging enough dirt to shift its position; by putting the machine on a favorable slope or over a hole, prying with bars or saplings, lifting with a loader or a crane, or reducing the objects with chain saw, axes, chisels, sledges, or pneumatic drills, then dumping the pieces.

Oversize rock is liable to cause damage by denting, bending, or straining parts. Damage may also be done by accidental collision with boulders during ordinary digging or transporting.

JOB FACTORS

Operating Efficiency. The self powered scraper is more sensitive to weather delays than most excavators are. Rain quickly makes the surface of the cut too wet to use in the fill, stopping work. On resumption of work it may be necessary to waste the top layer, or stockpile it to dry.

Also, scraper drive tires quickly lose traction on rain-wet surfaces, particularly if they have been dusty.

A Bureau of Public Roads study covering 2660 hours of scraper operation in the southern states during a rainy year showed that these machines averaged a loss of 54 per cent of available working time because of rain, a substantially worse record than other equipment. Under the same conditions crawler-drawn scrapers lost only 25 per cent of their time because of weather.

Job Delays. In the same rainy year study, self powered scrapers showed a loss of about 2/5 of the time the weather allowed them to work, because of other delays.

These delays were divided about equally among the following five causes:

Waiting for opening of cuts

Waiting for pusher

Maintenance and repair

Lack of operators

Miscellaneous

The net result was that these scrapers worked about 35 minutes an hour during working periods, and 17 minutes an hour for the whole job.

Other studies, not quite so complete or carefully recorded, indicate somewhat better performance. In this discussion, an efficiency of 75 per cent or a 45-minute work hour will be assumed. However, each estimator must of course use his own judgment and figures.

Scrapers are not particularly subject to stoppage from breakdown of other equipment. A pusher can usually be replaced quickly by odd job dozers, scrapers can sometimes load each other, or they can operate for a while at part load. They can maintain their own haul road, at a price, and can continue to build a fill for quite a while in the absence of grading and compacting equipment, unless supervision is very exacting.

Ownership Expense. The self powered scraper, as a busy contractor's first line equipment, may be used from 600 to 1500 hours a year, with the probability that it will be less than 1,000.

For depreciation purposes it is considered to have a five year life. With reasonable care it may last much longer, many still-active units being over ten years old. On the other hand, rough treatment and lack of care can finish the useful life of a scraper in a year.

Insurance and property tax costs are average. Some of the smaller units may have occasional use on highways and require licensing, but most of them are too wide.

Pusher costs are those for dozers, medium to hard service.

Figure 17-33 gives a sample scraper cost setup including both fixed and operating expense. The costs are original ones from 1962, and are of course much too low. However, present costs are so changeable, and so variable across the country, that a new basis sound enough for use in an example could not be found.

This illustration should be used only as a framework, in which the contractor substitutes figures for his own conditions and costs of equipment.

Operating Expense. Self powered scrapers are expensive machines. Their work is usually hard and they get punishing treatment trying to keep up their output, particularly in being forced through resistant material by too-powerful pushers and in high speed travel on rough ground.

Fuel and Oil. Engines operate at full and nearly full load a greater part of the time than most excavators, as the highest possible gear is used most of the time, whether loaded or empty. Medium loading occurs in long level runs on firm roadways, light loads in down hill travel.

With equipment in excellent condition, fuel consumption is likely to average .05 to .06 gallons per rated brake horsepower per hour. In ordinary operating condition, figure .06 to .07. Chassis lubrication is usually once a shift of 8 or 10 hours. Oil changes vary with manufacturers' recommendations and dust conditions, but are average for construction machinery.

In lack of specific information, figure lubrication at one-third of fuel cost.

Maintenance. In cable operated scrapers the principal maintenance item is cable. Its life varies so much that it is impossible to set up any average or usual figure.

Bowl cutting edges need periodic reversal or replacement, always before they are so far gone that they let the bowl wear.

If the contractor has detailed cost records he can compute the cost of these items separately. Otherwise they are lumped together with repairs.

Scrapers are often parked outdoors for long periods between jobs. They are likely to suffer serious rust damage unless they are carefully protected by paint and grease.

Repairs. Repair costs increase with weight of load, length of struggle to get loaded, roughness and courseness of ground, grades, and the age of the machine. Severe damage may be done by boulders, ledge, and ripped rock, particularly with powerful pushers. Some makes and models are more subject to rock damage than others are.

For rule of thumb calculation in lack of specific records, a scraper's non-tire maintenance and repair will equal 80 per cent of its purchase price under ordinary operation for 5,000 hours. In light service repair costs may run only 50 per cent of purchase, in heavy service 150 per cent or more.

Tires. Tire repair, recapping, and replacement may make up one-third or more of the non-labor operating expense.

A particular problem with drive tires is that scraper low gears are faster than the low gear of crawler pusher tractors. If the scraper is kept at full throttle and the pusher is in low, considerable spinning of the drive wheels must occur. This is waste effort and waste rubber, as the scraper usually supplies a modest fraction of the loading power, and a spinning wheel has less push than one whose tread is engaging the ground.

Spinning is most apt to occur when loading is kept up too long. The scraper operator feels the struggle of the pusher and lack of load response, and has a tendency to tramp on the throttle in an effort to help.

Drive wheel spin in the cut may be reduced by driver education, running pushers in second gear, using rubber tired pushers,

SELF–PROPELLED SCRAPER
Struck Capacity 21 Yards
300 horsepower Empty Weight 30 tons

PURCHASE

Price at factory	$ 50,200	
Extras (cable reel, hardfacing, spare parts	500	
Sales tax, 3%	1,521	
Freight	907	
Purchase price, or total cost	$ 53,128	
Deduct, 4 tires @ $2,532 each, fleet price	10,128	
Net cost to be depreciated		$43,000
Repair factor 1/10,000 of net cost		$ 4.30
Average annual investment, 5 year basis, 6/10 net cost $ 25,800		

OWNERSHIP

Hourly depreciation, 6,000 hours total use, no salvage		$ 7.17

Carrying charges, per cent of average annual investment

Interest	0 to 11.3%	assume at	4.0%
Insurance	.5 to 2.5%	assume at	1.0%
Taxes	0 to 4.5%	assume at	3.0%
Storage	0 to 5.0%	assume at	.0%
Total			8.0%

Hourly carrying charges, .08 x $\dfrac{25,800 \text{ av. ann. inv.}}{1,200 \text{ work hours per year}}$ = $ 1.72

Total ownership costs per hour $ 8.89

OPERATING

		Hourly Cost	
Type of service, normal range	Light	Medium	Heavy
Original tire purchase, $10,128	$ 1.69	$ 1.69	$ 1.69
Tire replacements, 4 recaps @ $1,250	.83		
3 recaps, 1 new		1.05	
3 recaps, 4 new			2.31
Tire repairs, 15% of combined cost	.38	.41	.60
Mechanical repairs and maintenance,	3.87	4.73	6.67
Diesel fuel @18¢ a gallon,	2.70	3.24	3.78
Lubrication, 1/3 of fuel	.90	1.08	1.26
Total non-labor operating cost per hour	9.54	12.20	16.31
Ownership cost, above	8.89	8.89	8.89
Total non-labor cost	18.43	21.09	25.20
Operator's wages	5.00	5.00	5.00
Total hourly cost	23.43	26.09	30.20
Cost per minute, 45 minute hour	.52	.58	.67
Cost per 1,200 hour year	$28,116.00	$31,308.00	$36,240.00

Fig. 17-33. Scraper cost data, 1962 prices

increasing pusher power, or having scrapers equipped with torque converters.

Scraper drive tires start to slip and spin on loose or muddy haul routes on grades as low as 3 to 6 per cent. Under severe conditions of abrasive mud and steep grades a tire can be worn smooth in 500 hours of operation.

Labor. A scraper has one operator, and no helpers or hand laborers are ordinarily needed. Pusher expenses, including labor, are divided by the number of scrapers they service, and added to the cost of each machine.

Supervision. Policies on supervision vary. A grading foreman may be assigned to one cut or to one fill, or he may divide his time among a number of cuts and fills. In the absence of a foreman a cut may be supervised by the pusher operator and a fill by a grader operator.

Experienced scraper operators need little supervision. The critical items are proper loading, efficient haul speed, cutting side slopes correctly and finishing them before they get out of reach, keeping the pit floor reasonably smooth, and accurate finishing of the floor.

Operators may need a foreman to secure help or services for them when they are needed.

HOW BIG A LOAD?

Heaped capacity ratings of scrapers are for advertising purposes only. A rule of thumb is that even under good digging conditions with plenty of pusher power, a scraper should not be expected to pick up more than its rated struck capacity in bank yards. But it might be more profitable to leave the cut sooner with a yard or two less.

Rate of Loading. During a loading run, material first enters the empty scraper rapidly. As the load increases the rate of loading drops. When the weight of the dirt above the opening balances the digging force, the scraper does not pick up any more material.

The block graph in Figure 17-34A shows the amount that might be expected to be loaded during each 10-second time. It will be noted that this machine picked up 8 yards in the first 10 seconds, but only about 1/20th of a yard in the last 10. Loading cost of the last cubic foot is over 100 times as much as for the first one in the bowl.

Line 1 in Figure 17-34B shows the increase in average cost per yard of a load as the scraper keeps digging.

It is obvious that two minutes is too long to keep this scraper in the cut. The question

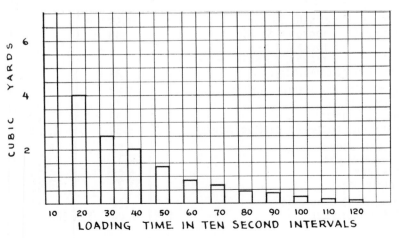

Fig. 17-34A. Scraper load increase in 10-second intervals

is, what is the most economical time to pull out? The answer depends partly on the hauling cost.

Haul Cost. Increase of load size decreases hauling cost per yard unless the speed and acceleration of the scraper are seriously reduced. Sometimes a moderate increase in load may reduce speed by forcing the machine into the next lower gear, but under slightly different conditions the same increase might have no effect.

Any increase in load will increase time required for acceleration and will have a small effect on dumping time.

In general, a light load can be hauled rapidly but the cost is divided among few yards, so the cost per yard is high. Larger loads move more slowly, but costs are shared by more yards, so are lower on a per yard basis.

Line 2 shows the cost per yard of hauling each of the 10-second loads 2500 feet, one-way. Note that the last yard and a half that added so much cost to loading do not lower hauling cost much.

Most Efficient Time. Line 3 shows the loading and hauling costs combined. It is a sagging curve, whose low section is the most economical time of loading For this particular problem it is 40 seconds, with 50 seconds so close that there is no prac-

tical difference. A range from 30 to 60 seconds might be tested in the field.

Maximum Production. If scrapers loaded themselves without help, the most efficient loading time would also give the greatest production. However, the pusher cost makes the most efficient time a little short of best production. Usually 10 to 20 seconds longer push will boost production, at a slightly higher cost.

Aiming for maximum production is the right idea where working time is limited. It may prove economical as well if the pusher has to wait for the hauling units.

When scrapers are waiting for the pusher it does not pay to keep them in the cut longer than the most efficient time.

LAND LEVELERS

These are towed scrapers with bottomless buckets which are chiefly used in leveling farm land, although they may be used profitably in other large scale grading operations. They include drag levels and the long-chassis land levelers.

Drag levels do the heavier part of the work, and the land levelers proper the final smoothing. The latter consist of a frame from 20 to 80 feet in length, which rests on wheels or skids on each end and carries

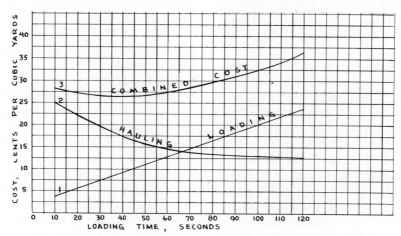

Fig. 17-34B. Cost per yard and loading time

Courtesy of Be-Ge Manufacturing Company

Fig. 17-35. Leveling drag scraper

a bottomless bucket in the middle. The bucket can be moved up or down by either hand wheels or hydraulic controls.

Some land levelers are made up of a drag scraper bolted into the long frame. Others are made so that they can be taken apart immediately behind the bucket and used as a drag scraper.

There are a number of makes and types of these machines, but a few examples of each will suffice to show the type of construction used.

Industrial Drag Scraper. The BE-GE TDS-18 model, shown in Figure 17-35, is eighteen feet wide and holds about nine yards. It is towed by tractors of 70 to 160 horsepower.

The full width bucket or bowl is shallow with no floor, and has a cutting knife or blade at the bottom of the back. It is rigidly fastened to a drawbar or tongue extending forward to the tractor and is hinged to a frame supported by a pair of wide steel wheels behind the bucket.

A two-way hydraulic ram is mounted on the top of the bucket. Its piston rod is hinged to an upward extension of the wheel frame. When the ram is extended the wheel frame rotates on the axles, raising the bowl through the hinge connection.

When the ram is retracted the bowl is lowered or forced down.

There are several working differences from a dozer that may be noted. Since the bucket is hung between the tractor and the rear wheels, it has little tendency to scallop and can run a good grade readily. No work is done ahead of the tractor so that an area cannot be worked unless the tractor can get through it or a dozer smooths its path. The blade is much wider in proportion to the power of the tractor and capacity is larger. Sharp sided cuts, ditches, and steep dump piles may be left as the tractor does not follow. It ordinarily does not dig hard soil as well.

As compared with a carrying scraper this machine grades a greater width with each pass, will penetrate more rapidly, and may be dumped completely while standing still to build dikes or to avoid bogging down. Its capacity is smaller and speed is lower because the load is dragged on the ground rather than carried.

The machine acts as a semi-trailer and after some practice can be backed readily. It is possible, therefore, to use it either in a shuttle or in any of the scraper patterns, the shuttle being favored for short hauls.

Landplane. The Marvin LANDPLANE,

Courtesy of Marvin Landplane Company

Fig. 17-36. Drag scraper, closeup

shown at work in Figure 17-37, has its bucket or blade centrally mounted under a long girder frame. Its front and rear sections (extensions) can be telescoped inward to reduce its wheelbase, for decreased pull and better handling in rough ground, and for convenience in moving between jobs.

Maximum frame length varies in different models from 35 to 90 feet, with bucket widths from 10 to 16 feet.

The drawbar, or tongue, is fastened to a hitch which allows for lateral movement in making 90 or 180 degree turns. The hitch does its pulling from diagonal braces fastened to the main frames close to the center of draft near the bucket, and extending downward to the front end of the machine.

Wheels may be all steel with flat rims, or equipped with single rubber pneumatic tires. They are mounted on forks set into fork heads on each corner, which allows all four wheels to caster so that they may roll freely in any direction.

These machines are intended to be used on land that is fairly level or that has been made so by scrapers of the carrier or drag type. Their principal feature is length. The longer the framework, the more accurate the leveling operation will be. The effective leveling length is the distance between the front and rear wheels. The bucket suspended midway between the wheels acts as a plane to shave off any high spots that appear in the field, and pushes and carries

Courtesy of Marvin Landplane Co.

Fig. 17-37. LANDPLANE

Courtesy of Be-Ge Manufacturing Company

Fig. 17-38. Scraper Plane

this dirt until a corresponding low place is found elsewhere in the field.

The bucket gives its load a rolling motion, pulverizing clods so that the fine dirt automatically filters out of the bottom of the blade into low places. The tractor operator sets the blade at the start of the operation to the point where there is enough dirt in the bucket to keep it from running empty during the operation. It is not necessary or desirable to raise and lower the bucket dozer- or grader-fashion but if it is found to be running empty the bucket may be lowered to gather more dirt from the high places in the field.

On extremely rough ground or during the first planing operation it may be necessary to work with the bucket blade higher than normal to work down the rough spots without making cuts so deep that they would stall the tractor. When the high places have been cut down the tractor operator can lower the bucket with the manual screws to the proper cutting depth.

Normally, the bucket blade is set to a depth that would be level with the bottom of the wheels, assuming that the wheels were resting on firm, unworked ground; however, it is necessary to compensate for the settling of wheels in soft ground by raising the bucket.

In very loose or sandy ground such as desert lands, the air pressure may be reduced in dual rubber tires to around 12 pounds, which will provide greater surface area and flotation to carry the weight of the machine. Under extreme sandy conditions oversized dual tires are used to provide the necessary flotation.

These machines will turn sharply in their own wheel base, particularly if the bucket is partly filled with dirt so that the machine will pivot on it. The field may be worked in strips or in any pattern that seems most desirable to efficiently level the area. The manufacturer recommends that the planing on farmed fields be done diagonally to the plowing or cultivating for the best results

Courtesy of Eversman Manufacturing Company

Fig. 17-39. Automatic land leveler

and easiest pulling. It is also recommended, where irrigation is practiced, that the last operation be done with the flow of water.

They are sometimes useful to contract levelers of desert land to level fields in both directions before the survey stakes are set, in order to move as much dirt as possible at the least amount of expense, and bring the field to a more uniform level before actually moving dirt to the proper grade. They are valuable for final smoothing after the stakes are pulled.

Correction leveling on an annual basis is advisable to restore smooth level surfaces on fields which are roughened by wind and water erosion, ditches, levees, furrows, and ground settling.

Other uses include leveling snow on airport runways in Greenland.

BE-GE Scraper Plane. This machine, Figures 17-38, has a flexible front suspension, a hydraulic dumped bucket, a front or main frame, center wheels, and a straight edge or tail boom.

The draw tongue supports the front of the front frame. It is in effect a beam extending from the tractor drawbar to the very short front axle of the machine. The frame is attached to the beam by a hinged kingpin which allows steering and oscillation.

The front frame widens and deepens toward the rear. It rests on an axle just behind the bucket that is carried by two pairs of tires. The front corners of the bucket or bowl are hinged to this frame. A vertical slot is cut in the back to allow space for the tail boom to move up and down while planing.

The tail boom rests on a hinge connection at the front of the main frame, and on a pair of casters supported by rear wheels. It consists of a single hollow steel beam.

It carries the dumping and depth regulating mechanism for the bowl. Dumping is done by means of a hydraulic ram controlled by a valve on the towing tractor.

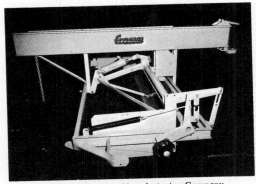

Fig. 17-40. Detail of automatic leveler

The ram will lift the bucket high enough to allow it to ride over obstructions without digging in, and lower it until the lift arm locks over center. The height at which the blade will lock is determined by the adjustment of a turnbuckle on the lift arm.

This arrangement prevents the bowl from slipping down due to hydraulic leaks or other defects, and allows it to be raised for dumping trash at edges of fields or to reduce the load in heavy cuts.

The mounting of the main frame on the drawbar reduces any raising and lowering of the blade caused by irregularities in the ground under the front wheels.

If front wheels cross a bump the front frame will be lifted about half as much as if it were carried directly by the axle.

The lift of the front frame has little direct effect on the blade because the frame pivots on the axle directly behind the bowl and is hinged to the bowl well forward of the blade.

The front of the rear frame is lifted about one third as much as the front of the front frame. These two reductions in lift result in a lift of the blade of only one fifth to one sixth the distance the front wheels rise, as contrasted with a blade rise of about one half the front wheel rise in the standard, rigid frame models.

The blade will be affected by pitching of the tractor as this raises or lowers the tongue. However, the average land being planed is well enough graded to minimize this effect. Also, the blade will be more affected by rising and falling of the rear wheels than a rigid model; but these wheels should ride in the smoothed track of the blade.

The machine can be shortened for towing on highways by driving out the hinge pin connecting the front and rear frames, disconnecting the hose from the lift ram, and putting blocks behind the rear wheels. The tractor is then backed. The two frames will telescope into a piggyback position where they can be locked with a pin. The machine can be extended by blocking the front of the rear wheels, removing the lock pin, and driving forward until working position is reached, then reconnecting.

Eversman. The Model 329 Automatic Land Leveler, Figures 17-39 and 17-40, has the center wheels mounted on a crank axle that is supported by the main frame and that supports the front of the bowl subframe. Linkage is arranged so that when the wheels move down in relation to the main frame the bowl is moved up, and when the wheels come up it is lowered.

When the center of the machine.is passing over a ridge these wheels ride up on it while the frame, supported by lower ground at the ends, will not. The crank axle will turn as the wheels move up, thereby lowering the blade to make a deeper cut. In crossing a hollow, the wheels drop relative to the main frame and the blade is lifted, thereby spreading a thicker layer of fill than a standard leveler would.

The bowl is raised to dump or to clear obstructions by means of a hydraulic ram. An indicator gauge tells the operator its position, and aids him in getting it back into automatic operation setting after use. The gauge pointer is not affected by bowl movements caused by the axle.

The ram is also used to relieve weight while making manual adjustments.

When the tractor rear wheels are elevated, the blade stays on grade and cuts off the ridge.

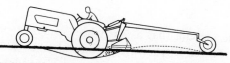

When the tractor rear wheels fall into a depression, the blade again maintains grade and fills the low spot.

Courtesy of Eversman Manufacturing Company

Fig. 17-41. Tractor mounted leveler

Working depth is set by a crank handle and screw at the hitch block connection between the draw tongue and the front frame member. The bowl should average a half load, filling in ridges and emptying in hollows. A high setting may be needed for the first working of uneven ground, or when the tractor has barely enough power to pull the leveler.

The rear of the bowl is supported by springs. An adjustment permits reducing tension when more weight is needed on heavy, cloddy ground, or increasing it to hold the bowl up in light, loose, or damp soil.

The forked rear section swivels on a vertical pin for ease in steering. It may be removed when the machine is being used chiefly as a dirt moving scraper. It, and the tongue and dolly, may be loaded on the main frame for transporting.

This machine is 32 feet long and 12 wide, is used chiefly in land smoothing and seed bed preparation, and requires a 3-plow tractor to pull it.

Eversman levelers are available in sizes up to 70-foot length with a 14 foot blade. The most popular unit (Model 4512-RT) is 45 feet long with a 12 foot blade.

Tractor Mounting. The Eversman TM leveler, Figure 17-41, is designed to mount on 3 point or 2 point tractor hitches. The cutting depth is regulated by a hydraulic cylinder, or by an optional screw crank arrangement.

Linkage is arranged so that the bowl cutting blade keeps its level relative to the tractor front wheels and the leveler rear wheels, as illustrated. Rise and fall of the tractor rear wheels does not effect the blade height. The blade will be raised if the front wheels climb a bump, but only a fraction of

Courtesy of Eversman Manufacturing Company

Fig. 17-42. Towed ditcher, working position

the amount that a rear wheel would lift it.

Towed Ditcher. Irrigation and other shallow ditches may be made and cleaned by the pull ditcher in Figure 17-42. It has two plow-like moldboards or wings flared out from a single share. It is raised and lowered by a hydraulic ram, or by an optional screw crank.

A pin and quadrant adjustment on the tow bar regulates the angle at which the wings enter the ground. When the bar is at the bottom of the quadrant their full length will enter the ground and a wide flat ditch is made. If the bar is higher, the ditcher will nose down and dig with only the front part, making a narrower, steeper sided cut.

Mounted ditchers are also made to fit most 3-point hitch tractors.

LIGHT SCRAPERS

There are a number of types of scrapers that are designed for use behind light tractors or animals. These may be carried on the tractor, on wheels, or dragged on the ground. Many of them are controlled through ropes and trip levers.

They are usually economical to buy and maintain and they work well in small

Courtesy of Eversman Manufacturing Company

Fig. 17-43. Mounted ditcher-plow

Courtesy of Digmor Equipment & Engineering Company

Fig. 17-44. Scara-Scraper

17-44

Courtesy of Miskin Scraper Works

Fig. 17-45. Closed bottom scraper

places. However, their capacity is too small for general work and they will not cut hard soil unless it is plowed or rooted. They are primarily farmers' tools and will be used by the contractor chiefly for spaces or quantities too small to justify the use of heavier equipment.

Scara-Scraper. These scrapers, designed for mounting on 3-point hitches, have an open-bottom box with two cutting edges at the rear, back to back. The front cross bar carries a set of scarifier teeth. This may be fixed in position, or controlled manually or hydraulically so that the teeth may be lifted when not needed.

In forward motion, the scarifier loosens soil and the forward knife cuts and piles it into the bowl, from which it is dumped by using the 3-point lift. In reverse, the rear edge acts as a dozer blade.

Miskin. A closed bottom scraper, Figures 17-45 and 17-46, is attached to a wheel tractor by a three point hitch, which adjusts its working or carrying level. Manipulation of tilt for digging or dumping is controlled by an auxiliary, two way cylinder.

The scraper bowl is hinged to the front bar and side plates. A two piece hinged rod runs from the top hitch connection to the rear of the bowl. A vertical cylinder based

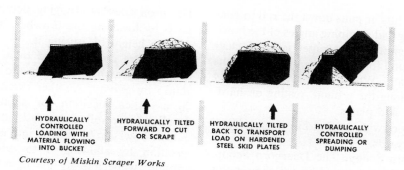

HYDRAULICALLY CONTROLLED LOADING WITH MATERIAL FLOWING INTO BUCKET

HYDRAULICALLY TILTED FORWARD TO CUT OR SCRAPE

HYDRAULICALLY TILTED BACK TO TRANSPORT LOAD ON HARDENED STEEL SKID PLATES

HYDRAULICALLY CONTROLLED SPREADING OR DUMPING

Courtesy of Miskin Scraper Works

Fig. 17-46. Loading and dumping closed bottom

Courtesy of Ford Motor Company

Fig. 17-47. Rear lift scoop

on the bar lifts or pulls down the rod to control digging and dumping angles of the bowl.

When empty, the scraper is lifted clear of the ground for transport, by the lift arms of the hitch. When loaded, it is usually dragged along the ground, resting on hardened steel skid plates.

An 8-foot wide bucket has a capacity of over two yards.

Dearborn Scoop. The Dearborn Scoop, Figure 17-50, is carried by the three-point hydraulic lift drawbar used on Ford and Ferguson tractors. Digging depth is regulated by lowering the drawbar, and transporting height by raising it.

Pulling a cord attached to a trip lever releases a catch and allows the bucket to dump by gravity. Another catch will hold in a vertical position for spreading and leveling loose soil, dozer fashion.

The bowl can be rotated into any desired position by allowing it to drag on the ground, while moving the tractor forward or backward.

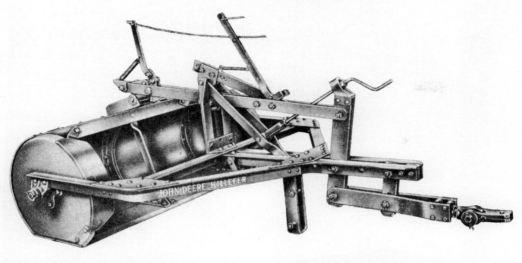

Courtesy of Deere & Company

Fig. 17-48. Revolving scraper

Capacity is about one third of a yard.

Rotary Drag Scraper. Although it is no longer manufactured, the machine shown in Figure 17-55 serves to indicate the construction of these units, which do not have wheels. The bucket edge is tipped down more or less for digging by a hand lever. When it is full, this model automatically rotates back to a carrying position. Other types are prevented from further digging by the load itself or are put in transport position by backing.

This scraper is dumped by pulling the latch lever which releases side catches and allows the bucket to rotate into an upside down position where it is supported by the wide flanges on the sides. The dirt being dumped passes under the edge, which spreads and smooths it.

The scraper may be dragged to the cut in the dumped position or backed. In reverse the catches do not hold and the bucket rolls freely. It is stopped in the digging position and pulled forward for another load.

Hand Scraper. The light scraper shown in Figure 17-49 is pulled by a line from any power source, and is controlled by one or more men at the handles. Raising the handles while it is moving causes it to dig and lowering them puts it in transport position. It is dragged back to the cut by hand.

These can be purchased at most large hardware stores and are very valuable for digging and smoothing mud and for cleaning small ponds and pits where there is no access for larger equipment, or where the job size does not justify bringing it in.

This may dump itself in hard digging by pivoting on the cutting edge, so the operator tries to prevent it from cutting in too deeply. He is likely to be flipped when it dumps, unless towing speed is kept very low.

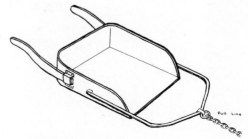

Fig. 17-49. Hand scraper

Courtesy of Marvin Landplane Company

Fig. 17-50. Final planing

CHAPTER EIGHTEEN

TRUCKS

DUMP TRUCKS

The dump truck is probably the most familiar of the machines used for excavation. However, its structure is rather complex and it is so important that a detailed description is considered to be in order.

It is composed of four major assemblies. The chassis includes the frame, bumper, springs, dead axles, wheels, and tires. The power train, which is supported by the chassis, consists of engine, clutch, transmission, drive shaft, differential, and live axles. The cab is the driver's compartment. The body assembly, which includes the carrying box, tailgate, cab shield, and the hydraulic system and controls, is an entirely separate unit, usually made by a different manufacturer and adaptable to different makes of truck.

The International in Figure 18-1 is not a current model, but still may be considered typical of the light medium trucks that carry three to five yard bodies, and it will be used for purposes of illustration.

LIGHT TRUCKS

Frame. The frame, shown in Figure 18-2, consists of two parallel pressed steel channels with cross braces, some of which serve as supports for the engine and transmission. The front cross member is extended to the sides and serves as a bumper.

The frame side members behind the cab have a 34″ spacing. The length from the back of the cab to the rear axle is five feet, and other chassis can be obtained in which this distance is extended to six feet, seven feet, or eight and a half feet. The width and lengths of this section are standardized for most makes of trucks for convenience in mounting bodies.

For dump use, the side members are cut immediately behind the rear cross member.

Pull hooks should be fastened to the top of the frame members just behind the front bumper, and on the rear cross member. A rear pintle or clevis hook is useful in towing other machines.

Springs. Springs are of the leaf type and are shown in Figures 18-3 and 18-4. They are fastened to the frame by two shackles, one of which is a single pin hinge, the other a U hinge that takes care of the increase in length of the spring as it is compressed.

The rear springs carry the largest part of the load and are proportionately heavy. A helper spring is placed just above the main spring under frame brackets which rest on its ends when the main spring is partly compressed. The helper adds sufficient strength to carry heavy loads without increasing stiffness under light loads.

Each spring is fastened to the axle by a pair of U bolts and by the spring center bolt, the head of which fits into a socket in

Fig. 18-1. Dump truck

the top of the axle. The braking power on all four wheels, and the driving power in the rear, are transmitted to the frame through the springs so that it is important to keep their fastenings very tight. If the U bolts are loose, the center bolt may shear and the axle move out of line.

The front axle is a drop center I beam, the rear a hollow casing which carries the differential and the axle shafts.

The front wheel hubs pivot on nearly vertical kingpins held by the ends of the axles. The steering mechanism is similar to that used in automobiles.

Brakes. The foot brakes are hydraulic with the type vacuum booster described below. Each brake shoe has a separate wheel cylinder. Some details of construction are shown in Figure 18-5.

The parking brake is a single mechanical

Fig. 18-2. Truck frame

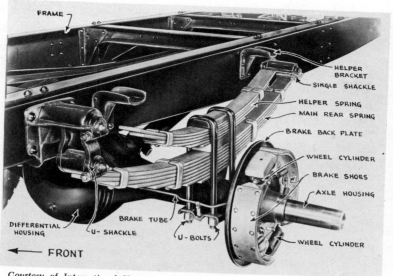

Courtesy of International Harvester Company

Fig. 18-3. Rear springs and axle housing

unit on the drive shaft behind the transmission. Its effective grip is multiplied by the rear end gearing. It is not designed to be used for stopping the truck.

Vacuum Brakes. The vacuum in an engine intake manifold can be used to apply brakes and do odd jobs around a machine. Sometimes it is given so much work that an auxiliary vacuum pump is installed to insure an ample and steady suction. Diesels do not create an intake vacuum.

It should be understood that vacuum is a minus quantity—an air pressure that is lower than that of the atmosphere—and that work is really done by pressure of atmospheric air trying to force its way into the vacuum. It is convenient to speak of a vacuum as having suction, but this suction is really pressure going the other way.

Vacuum brakes involve a cylinder or chamber with a movable piston or diaphragm. Atmospheric air on one side of

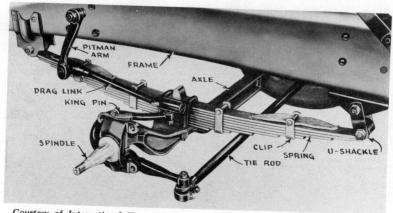

Courtesy of International Harvester Company

Fig. 18-4. Front spring and axle

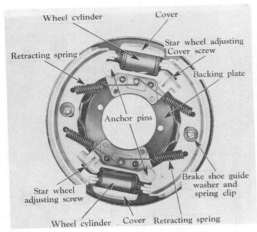

Courtesy of International Harvester Company

Fig. 18-5. Hydraulic brake

the piston pushes the piston into the vacuum on the other side of it to apply the brakes. The vacuum cylinder is usually used as a booster to make application of hydraulic brakes easier, but it may also supply all the braking effort through direct mechanical connections.

Figure 18-6 shows a cutaway view of a vacuum-hydraulic booster, and 18-7, a brake system. There are three principal parts in the booster—the control valve, the slave cylinder, and the vacuum cylinder. When the brake is off the vacuum piston is pushed full to the left by a spring, and has vacuum on both sides of it.

The control valve consists of four chambers which, from left to right in the illustration, contain hydraulic fluid, vacuum, a variable proportion of vacuum and air, and atmospheric air. The proportioning chamber is connected to the left side of the vacuum piston by tubing.

When the brake pedal is depressed hydraulic pressure from the master cylinder moves the control valve piston to the right,

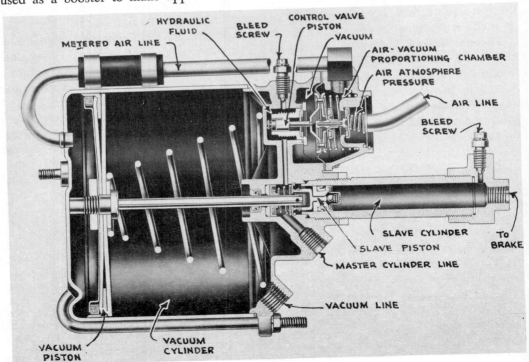

Courtesy of Bendix Products Division, Bendix Aviation Corporation

Fig. 18-6. Vacuum brake cylinder

closing the vacuum poppet valve, and then opening the atmosphere valve which allows air to flow from the air chamber to the proportioning chamber. The resulting increase in pressure (drop in vacuum) closes the atmosphere valve by its action on the diaphragm. Increased hydraulic pressure will reopen it. Decreased hydraulic pressure will allow it to remain closed and will open the vacuum valve that will drain air from the proportioning chamber away into the vacuum system until a new balance is reached. The amount of air pressure in the chamber and in the left end of the vacuum cylinder is therefore exactly regulated by the hydraulic pressure, which in turn depends on the amount of push exerted on the brake pedal.

Pressure from the brake cylinder also enters the left portion of the slave cylinder. It pushes the piston to the right, forcing the fluid beyond it into the brake lines. At the same time, air metered by the control valve pushes the vacuum piston to the right, adding its force to that of the slave piston. The brakes thus receive direct pressure from fluid and booster pressure from the vacuum mechanism. The initial movement of the vacuum piston closes a check valve in the

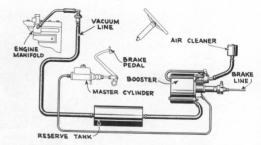

Courtesy of Bendix Products Division, Bendix Aviation Corporation

Fig. 18-7. Vacuum brake lines

slave cylinder piston so that higher pressures may be maintained in the lines on the brake system side of the piston.

The booster reduces the required pedal pressure from 30 to 70 percent, depending on its model, and the pedal ratio used.

If the vacuum system does not function, the vacuum piston will be held to the left by its spring, and if the brake is applied, fluid will flow directly from the master cylinder lines to the brake lines through the slave cylinder piston check valve which is open when the piston is in its leftward position.

When a trailer is used with the vehicle the layout in Figure 18-8 may be used. A supply vacuum line is run directly from the

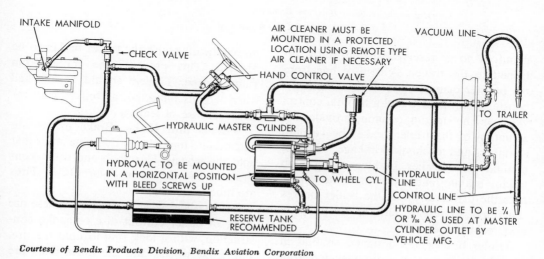

Courtesy of Bendix Products Division, Bendix Aviation Corporation

Fig. 18-8. Tractor-trailer brake lines

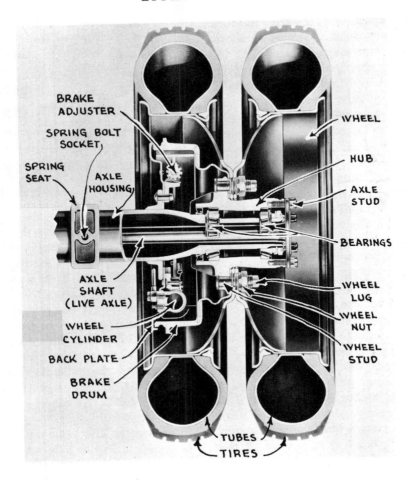

BRAKE
ADJUSTER

SPRING BOLT
SOCKET

SPRING
SEAT

AXLE
HOUSING

WHEEL

HUB

AXLE
STUD

BEARINGS

AXLE
SHAFT
(LIVE AXLE)

WHEEL
LUG

WHEEL
CYLINDER

WHEEL
NUT

BACK PLATE

WHEEL
STUD

BRAKE
DRUM

TUBES

TIRES

Courtesy of International Harvester Company

Fig. 18-9. Rear hub

manifold to a reserve vacuum tank attached to the trailer. A control line from the booster control valve or from both the booster control valve and a manual control valve on the steering column, provides for actuation of the trailer brakes by vacuum chambers and mechanical linkage. The first method gives trailer braking along with tractor braking and the manual valve permits the driver to apply the trailer brakes separately. A reserve tank in the trailer is also recommended.

Trailer brakes are designed so that in case the trailer becomes disconnected from the tractor the broken vacuum lines will automatically cause the trailer brakes to be applied. This prevents the trailer from running free. Trailers should be blocked when left disconnected, as brakes may release because of leakage.

Wheels and Hubs. Brake drums are anchored to the hubs by the same studs that hold the wheels. Rear hub and drum construction are shown in Figure 18-9. Tires are tubeless in new trucks.

Six identical wheels are used. These are steel, of either cut-out disc or spoke construction, with a lock ring to hold the tire. The front wheels and the inner rears are mounted with the convex side out, and the outer rears have that side in, in order to meet the hub.

The wheel stud is pressed through the hub from the back so that its head holds the drum. The inner wheel is fastened by five hollow lugs with inner and outer threads. The outer wheel is then mounted and fastened with large nuts which screw on to the outer threads of the lugs.

It is very important that both inner and outer fastenings be tight, as the driving and stopping power of the truck are transmitted through them. If any looseness develops they will wear and ultimately break.

Studs, lugs, and nuts on the right hand hubs have the usual right hand thread, but those on the left have a left thread. The part of the lug or nut nearest the rim moves faster than that near the center, and during an abrupt stop will tend to revolve on the stud in the direction the wheel turns. The thread arrangement causes this force to tighten the connection.

TIRES

Tire construction is described in Chapter 12. The table of wear factors given there does not apply to highway use, where no similar figures have been worked out.

Tires built for highway use wear faster with increasing speed, but the rate of wear does not increase as rapidly as with the heavier off the road construction. Forty to fifty miles an hour are considered moderate highway speeds.

Speed is chiefly damaging when combined with curves, over use of brakes, or rough or littered roads. Underinflation and overloading may cause severe damage.

Standard tires for all wheels are 7.50-20, 8-ply, with a rated capacity of 2740 pounds. Optional tires are offered up to 8.25-20, 10-ply, rated at 3330 pounds. For dump use, the largest tires are usually the best investment because of extra traction and resistance to abuse. Tires of the desired size should be supplied at the factory,

as it is more economical than to change over. In addition, a change of rear tire size necessitates changing the speedometer gear.

The front wheels usually carry much less weight than the rear wheels, and can often safely use smaller and lighter tires. However, when all tires are of the same size, only one spare is needed, and the life of rear tires can be prolonged by a rotation program that puts them on the front part of the time. Large front tires may rub on the frame when steered sharply.

From a safety standpoint, the front wheels should have good tires as a front blowout on a loaded truck may put it out of control; while a blown-out rear will be carried temporarily by the tire next to it.

It is usually not practical to carry a spare tire on a dump truck unless the body is specially made to accommodate it. It cannot be mounted on the side of the body because of the eight foot width limit. It can be placed on a reinforced cab shield, but its weight is too great for one man to handle it there. However, it is sometimes possible to make a rack for it under the right side of the body.

Generally, if an empty truck has a rear flat, it can return to its base. If the trip is long, it is advisable to remove the flat tire. A front flat can be removed, and one of the outer rears substituted for it. If the truck is loaded, however, it should be parked and a tire brought to it.

Outer rear tires may be changed without a jack by running the inner wheel up on a block so that the outer one will be held clear of the pavement.

Tires are expensive and are quickly destroyed by neglect or abuse. Proper inflation, which can be determined only by a pressure gauge, is extremely important when heavy loads are carried. If the tire bulges prominently at the bottom when resting on a smooth surface, it is either soft or overloaded, and running it will develop destructive heat, weakening the fabric.

Measuring with Endless Tape

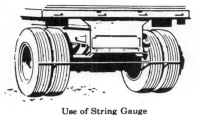

Use of String Gauge

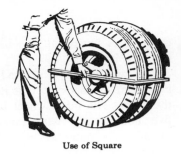

Use of Square

Courtesy of Rubber Manufacturers Association

Fig. 18-10. Matching dual tires

Dual Tires. Dual tires are two tires mounted on two wheels that are bolted to the same hub. All highway dump trucks and most off the road rear dumps use dual drive tires. Dual front tires are very rare.

The two tires of the set work as a unit. They must be the same size and ply rating, have nearly the same amount of tread, and carry the same pressure. Otherwise the larger or harder tire will carry more than its share of the load, and is likely to be damaged.

New tires of the same size but of different makes may differ in outside diameter. Tires of the same size and make but of different ply ratings are likely to differ either in the outside diameter or the loaded radius, or both.

Scuffing. The two tires of a dual pair do not travel the same distance on a curve. If a truck equipped with dual 8.25-20 tires spaced 11 inches on centers makes a U-turn between curbs 60 feet apart, the outer outside tire would travel 94.2 feet, the inside one of the pair 91.4, a difference of 2.8 feet. The difference in travel distance in the inner pair is similar.

Travel distance differences between tires that are locked together causes slippage and scuffing. Because of leverage and road crown the outside tires slip more than the inside, so they wear faster. As they get smaller they carry less of the vehicle weight, are less firmly pressed against the road surface, and do a larger share of the slipping.

The more heavy and powerful the truck,

the bigger the tire, the wider the spacing, and the more pronounced the wear.

It is important to rotate the tires before the inside tires become overloaded and the difference becomes too great for proper matching. The maximum permissible differences, measured in inches, are:

Tire Size	Diameter	Circumference
8.25 and smaller	1/4	3/4
9.00 and larger	1/2	1½

Where differences exist that are within these limits, the truck owner has a choice between two recommended practices. He may put the two larger tires on the outside wheels to conform to the crown of the road, or he may put both of them on the right side, and let the differential take care of the differences.

The second method should also be used for temporary use of tires with greater differences than are allowable for continuous operation.

Measurement. Sets of duals should be checked for size differences at least every thousand miles, after making sure that air pressure of the four tires is exactly the same. Replacement tires should be measured and compared before they are put on the truck.

Figure 18-11 shows three ways of checking sizes. In (A) the circumference is measured with a tape, either before mounting or after jacking up the wheel. In (B) a straight edge or a taut string checks both pairs at once. In (C) a "square" made of two 1 x 2" wood strips rigidly fastened to make an exact right angle can be laid along the side of the outside tire and across the treads.

Rotation. Systematic rotation of tires will prevent damaging size differences from developing. The simplest system is to move the right tires to the left side and the left tires to the right side, putting inside tires on the outside and outsides on the inside.

Tandems. Tandem drives, to be described later, have two sets of axles, one in front of the other, each equipped with dual tires.

Scuffing is much greater with this arrangement. The outer tires wear more than the inner, and all tires are dragged sideward on turns. The side drag is hardest on the rear set.

Many tandems do not have any differential or power divider between axles, so that all eight tires rotate at the same speed. In this case it is very important that the tires be matched so that the average diameter of those on one axle is within 1/4" of the average diameter on the other axle.

Front Tires. Truck front tires should last much longer than the rears, as they do not transmit driving force and carry much lighter loads. However, they are subject to excessive wear from being run out of line.

Front wheels do not roll exactly parallel to each other, as steering is more stable if they toe in slightly (fronts of the tires closer together than the rears), have a slight camber (bottoms of the tires closer together than the tops), and a slight caster, or backward tilt of the kingpin.

Too much toe-in shows a feathered edge at the inside of the tread design, while toe-out feathers the outside edge. Too much camber concentrates smooth wear on the outside of the tire, negative or reversed camber wears the insides and makes it look as if the axle is sagging. Wrong caster may result in cupping wear.

Misalignment of front wheels may result from poor adjustment, but more often it is

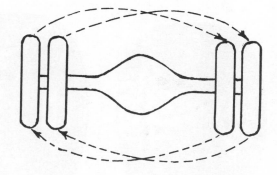

Fig. 18-11. Rotating position of dual tires

the result of bending caused by glancing blows against curbs or banks. Severe misalignment may be felt by the driver, but if it is slight it will show up only in tire wear. The two precautions are careful driving and frequent tire inspection.

Wobble. Rapid tire wear, with or without cupping, may be caused by wheel wobble. The trouble may be a bent wheel, in which case only the tire is damaged.

A loose or broken bearing will allow a wheel to wobble. Unless repair is made the wheel may come off. Defective bearings are a common cause of rear axle breakage.

POWER TRAIN

Engine. Trucks in this class usually have gasoline engines, 225 to 440 cubic inches, with compression ratios that permit using regular gasoline, and that can be converted to burn LPG (liquified petroleum gas).

Diesel engines are also available in several sizes.

Clutch. The standard clutch is a dry single plate type located in the engine flywheel housing. It is held in engagement by springs, and released by pushing a foot pedal.

Transmission. The transmission may have either four or five forward speeds, with one reverse. The five speed may have direct drive in either fourth or fifth. All gears except first are synchromesh.

Four wheel drive may be provided by

SUPER MILEAGE LUG   TRANSPORT   SUPER TRANSPORT

ALL TRACTION   SUPER ALL TRACTION   TRANSPORT '100'

Courtesy of Firestone Tire & Rubber Company

Fig. 18-12. Truck tire treads

factory installation of a two speed transfer case behind the transmission, from which power may go to the rear driveshaft only, or to both front and rear shafts, and a live front axle.

The power takeoff is in the left side of the transmission, and, for dump use, drives a shaft to the body hoist pump. The takeoff is operated by engaging its drive gear in the transmission. The clutch must be de-

pressed while engaging it, but, except when under load, it can be disconnected without clutching.

The drive shaft from the rear of the transmission supports a drum for the mechanical parking brake, and continues through a bearing in the center frame cross member. Behind this the shaft is fitted with two universal joints, one of which includes a sliding connection. This makes it possible

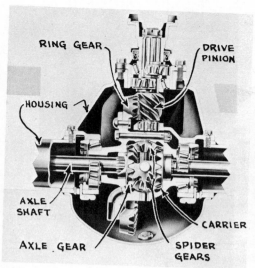

Fig. 18-13. Differential

for the axle to move up and down relative to the frame without damage to the shaft.

Automatic Transmission. Optional automatic transmissions are similar to those described in Chapters 12 and 17.

The friction clutch is replaced by a torque converter. The Allison transmission has four or five forward ranges and one reverse. Gears are constantly engaged, and power is directed through them by hydraulically shifted disk clutches.

Five speeds are used when a very low or creeper gear is required, as in dump or transit mix work.

In lower ranges, the converter acts to give variable speed-power ratios within the range; upper gears have an automatic lockup for non-slip drive.

In operation, the driver uses the shift lever to select the highest range he wishes to use. Shifting will then be automatic up to and down from that range. The lever can be moved at any time.

Automatic shifting is regulated by speed, torque, and throttle position.

Differential. The standard differential, shown in Figure 18-13, uses hypoid drive and is similar to that in a car except for

heavier construction. Different sets of pinion and ring gears may be obtained to vary gear reduction.

A two speed, electric shift axle, shown in Figure 18-14, is available as optional equipment. In high range, the ring gear turns the carrier and spider gears directly. In low, it works through the sun gear set shown to the right of the spiders.

The smaller gears turn freely on the cross shaft, which is attached to the ring gear. A two-sided clutch jaw is splined to the center of the shaft. It is controlled by a vacuum diaphragm or by an air chamber on the outside of the housing. When its teeth lock with the teeth of either gear, power is transmitted to that gear and through the meshing large gear to the differential. It will not stay in neutral position.

Two speed axles are desirable for most dump work.

Final Drive. The live axles are splined into the axle gears in the differential and are bolted to the hubs, so that they revolve on the bearings supporting these parts and carry no weight.

The bolts which fasten the axle shaft flange to the hub must take the full drive

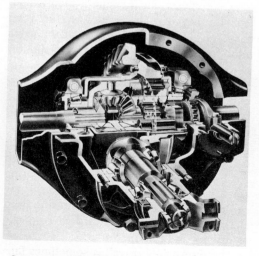

Fig. 18-14. Two speed differential

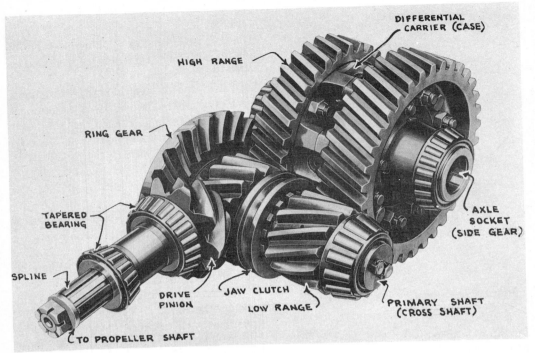

DIFFERENTIAL CARRIER (CASE)

HIGH RANGE

RING GEAR

TAPERED BEARING

SPLINE

DRIVE PINION

JAW CLUTCH

LOW RANGE

AXLE SOCKET (SIDE GEAR)

PRIMARY SHAFT (CROSS SHAFT)

TO PROPELLER SHAFT

Courtesy of The Timken-Detroit Axle Company

Fig. 18-15. Double reduction two speed differential

torque of the truck, and must be kept very tight to avoid shearing.

The power is transmitted from the hub through studs and friction contact to the wheels. The wheel and tire act as a unit as long as sufficient air pressure is maintained. Friction between the tires and the road converts the rotary movement of the axles into straight forward or backward motion of the axle housings and the truck.

This horizontal push of the tires is transmitted through the hub bearings to the axle housings, and through the rear springs and their front shackles to the frame.

The rear springs have four functions: to carry the weight of the truck, to absorb road shocks, to steady the axle housing against twisting in reaction to the turning of the wheels, and to keep the housing from shifting forward or backward in response to driving and braking forces.

The term "final drive" is sometimes limited to reducing gear sets that carry power from axles to hubs.

Cab. The cab, shown in Figure 18-16, resembles a passenger car coupe body except that it is cut off immediately behind the driver's seat.

The clutch, brake, and accelerator pedals are identical in location and function with those in standard passenger cars. The gearshift lever may be on the floor or on the steering column.

Rear vision mirrors, projecting beyond the body line, are required on both sides. There must also be blinker-light turn signals, backup light, and possibly an automatic backup horn signal.

Such items as heater, windshield wiper and washer, adjustable seat, seat belts, and trim are similar to or identical with automotive equipment.

Cab over Engine. In this truck construction, which is usually known by the abbreviation C.O.E., the frame is shortened and the cab mounted over the rear of the engine.

Advantages are shorter turning radius,

better load distribution, less overall length, and improved vision from the cab. Disadvantages are awkwardness in servicing the engine, extra climb up into the cab, and difficulty in keeping the cab cool.

BODY

A dump body unit consists of the box or body proper, the tailgate, body hardware such as chains and pins, and optional equipment such as cab guards. The hoist, which is often sold as a separate unit, includes a subframe, pump, valve, cylinder, and the controls.

A very wide range of body and hoist constructions is available for every truck. These units are usually not made by the truck manufacturer.

Figures 18-17 and 18-18 show a medium duty body made by Gar Wood. A hoist subframe extending backward from the cab is bolted onto the truck frame. This is attached at the rear by heavy hinges to the body frame which consists of two beams that rest on the subframe, and cross pieces to support the floor and sides.

The sides are sheet metal reinforced by flanges at top and bottom, and V-type or pyramidal side braces, and are welded to the flanged front wall. At the rear, heavy corner posts and the rear frame member combine to make a structure rigid enough to resist bending outward.

The front and rear corners have slots or gusset pockets in which sideboards can be placed. These usually consist of planks 1½″ thick, which may be as high as desired. They may be used to increase the capacity of the body or to prevent spillage off the sides.

Body capacity is figured level with the sides and is stamped on a plate on a front corner. Sideboard capacity and heap must be calculated. See pages 2-51 and 52.

The double-acting tailgate is somewhat higher than the sides, and usually has offset hinges at the top to increase clearance for

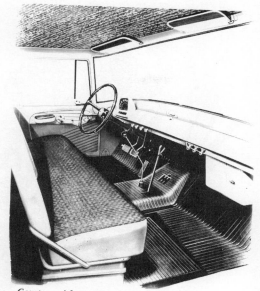

Courtesy of International Harvester Company

Fig. 18-16A. Truck cab

dumping bulky objects and to make closing more positive. It is made of steel plate with box reinforcing.

The upper hinges are equipped with removable pins. The lower hinge pin is a fixed part of the gate, but the hinge itself can be opened by means of a lever at the left front corner of the body, within reach of the driver.

If the body is flat on the subframe and the latch is open, the gate will hang in a closed position with the lower hinge pins lying in the hollow of the latches. When the tailgate lever is moved up it moves the latches upward and forward against the post, locking the bottom of the gate tight to the body.

The body and the gate are held in this position for loading and transporting. To dump, the gate lever is pushed down and the body raised in the front, pivoting on the rear hinges. Its own weight and the pressure of the load sliding against it cause the gate to swing outward on the upper hinges. When the load is fully dumped, the body is lowered, the gate swings into closed posi-

Courtesy of International Harvester Company

Fig. 18-16B. Cab over engine chassis

tion, then is clamped there by pulling the lever up.

Each rear corner post contains two keyhole slots of such size that a chain fastened to the top of the gate can pass through the upper part of the hole freely, but any link dropped in the lower slot will be caught.

The tailgate can be held at any desired angle by unfastening it at the top hinges, keeping it clamped at the lower ones, and passing the chains through the upper slots. Adjustment is made by raising the tailgate

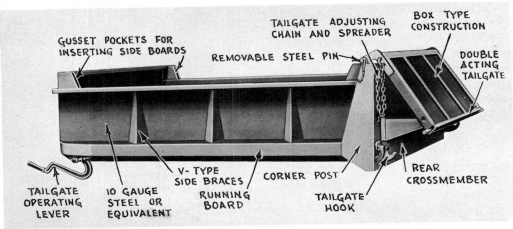

Courtesy of Gar Wood Industries, Inc.

Fig. 18-17. Dump body

by hand to take its weight off the chain, then lifting the chain out of the slot and moving it in or out to the desired length. The other chain is then adjusted to as near the same tension as possible.

The tailgate may also be fastened at the top hinges, released at the bottom, and restricted in opening by the chains. Each is passed down the back of the gate through a hole in its slide flange, and forward to the lower slot. The gate opening is restricted in order to spread thin, even sheets of free flowing material.

The cab guard, which is an optional extra, is a sheet of reinforced steel, curving or angling upward and forward over the cab. This is almost a necessity for trucks that are to be loaded overhead. Holes should be cut in it to allow the driver to see into the body through the rear window of the cab.

A heavy duty body of similar size is shown in Figure 18-19. Heavier plate and reinforcing is used. Separate chains are used for adjusting the tailgate, and holding it for spreading. The tailgate hook is hinged above the tailgate pins instead of below.

Courtesy of Gar Wood Industries, Inc.

Fig. 18-18. Body frame

Except in very light service, a heavy duty body will usually save enough maintenance to repay the additional cost.

Hoist. A direct type hoist is shown in Figure 18-20. It consists of a hydraulic pump, a valve, and a cylinder.

The pump is driven from a transmission power takeoff through shafts and universal joints, and works only when both the engine clutch and the takeoff gear are engaged.

The valve in many hoists is built into the pump body and has three positions, UP, HOLD, and DOWN. It is controlled from the cab, either through a floor lever or a knob and wire on the dash. The lever is more sturdy but takes up floor space, inter-

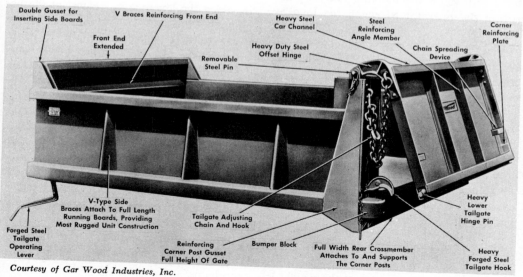

Courtesy of Gar Wood Industries, Inc.

Fig. 18-19. Heavy duty body

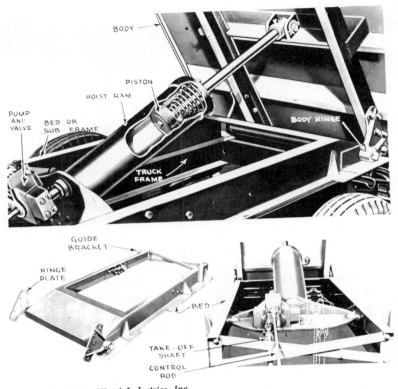

Fig. 18-20. Simple hoist

feres with insulating the cab, and may not permit the operator to watch the load as carefully.

The single-acting ram is bolted to a cross member which is hinged to the body sub-frame, and the piston rod is hinged to a crossbeam of the body. A spring may be placed between the piston and the ram head to cushion the piston when forced to the limit of its travel, and to help to start the body down when pressure is released.

When the body is down, the ram slopes up a little from the horizontal. When it is expanded, it pushes both back and up. The body hinge pins are made strong enough to resist the backward pressure so that the body is forced up. Leverage is lowest and load greatest at the start of the dump. As the body rises a large part of the weight is transferred from the ram to the rear hinges.

A number of hoist linkages are offered in which the leverage is greatest at the beginning of the dump so that the body moves slowly at first and more rapidly as it approaches the top. See Figure 18-21.

The Gar Wood cam and roller hoist, Figures 18-22 and 18-23, has a wedging action. The ram is approximately horizontal and its piston rod is fastened to an axle carrying two sets of rollers. The inner ones run on a pair of straight tracks between the sub-frame beams, and the outer ones run under curved tracks or cams fastened to the body frame. When the double sets of rollers are forced backward between the lower tracks and the cams, the cams are wedged upward. The pitch of the cams causes this action to start gradually and to accelerate as the body rises.

Since there is no solid connection be-

Courtesy of Dodge Brothers Corporation

Fig. 18-21. Leverage hoist

Courtesy of Gar Wood Industries, Inc.

Fig. 18-22. Cam and roller hoist

tween the piston rod and the body, a chain with a shock-absorbing spring is hooked to the body and the sub-frame to prevent the body from going too far when dumping downhill, or when the load catches in the back.

It is usually possible to get several different size hoists for one body. If the largest one is used it will probably have enough strength not to suffer from any size load that will be put upon it and it will not work hard on normal loads. This will reduce strain on all its parts and give them a longer life. The extra cost, in proportion to the whole cost of a truck, is not large.

Detachable Body. Hand-loading stand-

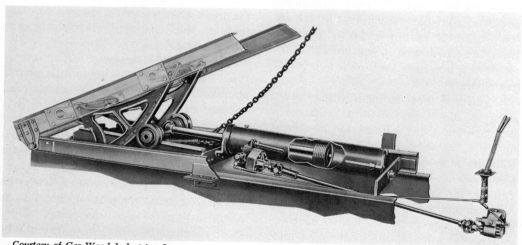

Courtesy of Gar Wood Industries, Inc.

Fig. 18-23. Cam and roller cutaway

COLLECT **LOAD** **LOAD**

HAUL **DUMP**

Courtesy of The Heil Company

Fig. 18-24. Detachable body, tip-dump type

ard truck bodies may be quite inefficient. Time and energy are consumed in getting the material up into the body, and the truck may have to stand for long periods in order to get a load.

Hand loading may be made easier by using a low truck and body, by filling from the back with the tailgate down, or by installing hinged or removable side panels for side loading. The reduction in height is limited, however, and idle time is not affected.

Several manufacturers build detachable bodies or containers which can be left on the ground for loading and lifted onto the truck by a hydraulic or mechanical hoist. One truck may take care of a number of containers.

Containers may be built with sloping sides so that they will nest inside each other, and up to six empties may be carried on the truck at one time. Some are made with the rear open to facilitate rolling in heavy objects, or in a number of special designs for carrying garbage or liquids. Some dump by upsetting, others through hinged bottoms or backs. Bodies have capacities from 1½ to 40 yards or 15 tons.

The Heil Load Lugger hoist, Figure 18-24, is a pair of bell cranks controlled by two two-way hydraulic rams and a three-

position valve. The containers are lifted by a pair of chains fastened at the middle to the hoist, and having locking rings which engage knobs in the body sides.

The underbody on which containers rest while being carried is a flat deck. Bodies are wedged against side sway and held from slipping forward or back by the hoist chains. A hook at the rear of the deck can be raised by a lever in the cab so as to catch the front of a container being lowered and dump it.

A jackleg or support under the rear of the truck chassis can be lowered to carry the extra load imposed when a full container is lifted.

The Dempster Dumpster, shown in Figure 18-25, uses a drop-bottom detachable container which is lifted up and moved forward onto an inclined rest where it is locked into position for transporting.

Normal use of these trucks involves leaving containers in convenient locations for hand loading. The truck comes to the various parts of the job as required, leaves an empty container, then picks up the full one and takes it to the unloading point.

The hoist and flat body base may also be used without containers for handling and transporting heavy objects such as concrete pipe or machine parts.

HEAVY TRUCKS

Large trucks for use on highways are very similar in design to lighter ones, except that all parts must be stronger and heavier. Either gasoline or diesel engines can be obtained. Air brakes are standard. The transmission contains more speeds and an auxiliary high-low box may be provided. Standard, rock, or quarry type bodies may be used.

Power Steering. Power steering is standard equipment on heavy trucks, and is optional on all sizes.

A typical unit, the Vickers hydraulic steering booster is shown in Figure 18-26. It consists of a two-way ram with built-in valving and is powered by an engine-driven hydraulic pump. The ram is bolted to the back of the drag link and the piston rod is fastened to the frame by a ball joint. The pitman arm of the steering gear is clamped in the control valve between the ram and the drag link, and can move slightly either way against light springs before being stopped by the valve case.

If the steering wheel is turned so as to move the pitman arm to the right it moves the valve spool with it, opening a passage for oil under pressure from the pump to the rod end (right side in the illustration) of the piston. The piston rod is anchored

Courtesy of Dempster Brothers Inc.

Fig. 18-25. Drop-bottom detachable container

to the frame, so this causes the ram, valve casing, and drag link to move to the right

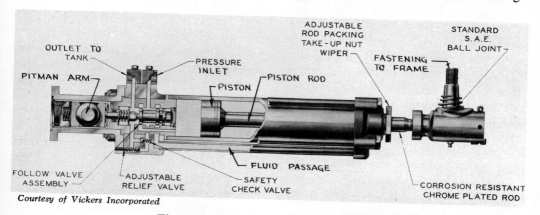

Courtesy of Vickers Incorporated

Fig. 18-26. Hydraulic steering booster

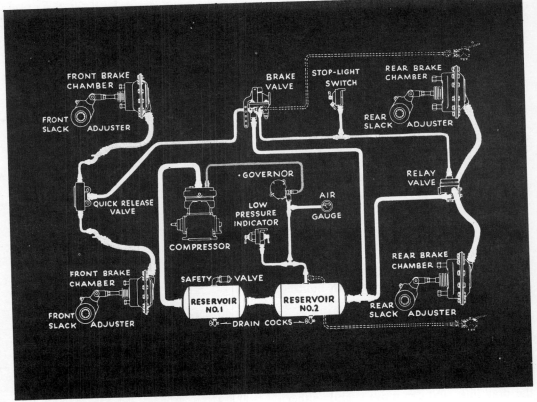

Fig. 18-27. Air brake piping diagram

until the valve casing catches up with the spool and closes the passage.

If the wheel is turned the other way the pitman moves the valve spool forward to open a passage to the front of the piston. The ram will then move forward until the passage is closed.

If there should be no pressure in the system a check valve opens which permits fluid to move from one side of the piston to the other. Pressure of the pitman arm will then move the valve case and ram in either direction mechanically, and will operate the drag link as if the ram were not present.

Road shocks are transmitted from the drag link through the ram to the truck frame without affecting the steering wheel.

This type of control may also be used in graders, tractors, and other machines; but usually not in rollers or scrapers.

Air Brakes. Brakes on heavy trucks and wheel tractors are usually applied by compressed air. Figure 18-27 contains a diagram of a system that is widely used in heavy trucks.

The air is supplied by a compressor constantly driven by the engine, which unloads, or stops pumping, when full pressure —usually 100 pounds—is reached in the reservoir or receiver and resumes when it falls to 90 pounds.

A valve operated by a foot treadle or conventional brake pedal allows exact control of air pressure in lines leading to the brake chambers, where it acts against diaphragms that move rods and levers called slack adjusters which apply the brakes. The front brake lines are provided with a quick release valve which drains them rapidly when released to prevent any lag that might interfere with steering or vehicle balance. The rear brakes, and trailer brakes if used, may have a relay

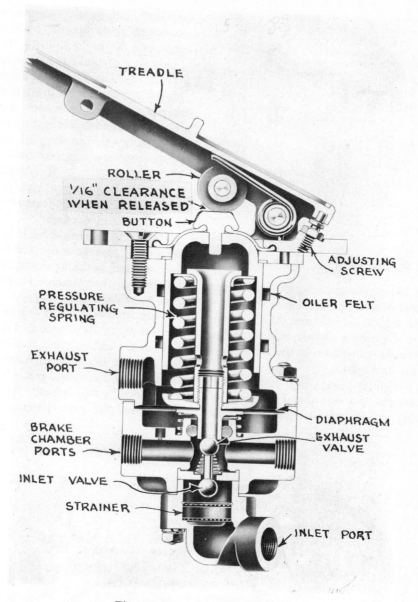

Fig. 18-28. Air brake valve

valve that feeds air direct from the reservoir into the lines.

The brake valve is shown in Figure 18-28 in applied position. In applying, the treadle compresses the pressure regulating spring which pushes down on a diaphragm. This pushes the exhaust valve seat down on its ball valve, closing it, and with slight further movement pushes the inlet valve down off its seat, opening it.

This admits compressed air to the brake line passage and to the bottom of the diaphragm. When its pressure against the diaphragm is sufficient to lift it against the mechanical pressure on the spring, it will move upward to close the inlet passage so

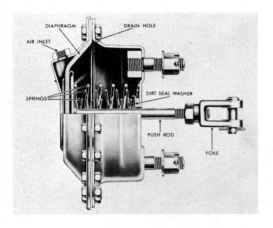

Fig. 18-29. Brake chamber

air, and immediately responds to any movement of the treadle by increasing or decreasing the pressure in the brake lines.

One type of brake chamber is shown in Figures 18-29 and 18-30. It contains a diaphragm attached to a push rod which is moved in a direction to apply the brake by air pressure, and is released by coil springs. The force applied to the push rod depends on the pressure supplied from the brake valve, and the size of the diaphragm area against which it pushes. The force applied to the brake is further affected by the length of the slack adjuster lever.

The quick release valve, Figure 18-31, has a cross passage connected to the brake lines, an inlet and an exhaust passage, and a diaphragm valve which is spring loaded so as to seal the inlet passage.

that the brake passage is sealed against the movement of air in or out.

If the treadle is released, the air will push the diaphragm higher, unseating the exhaust valve and allowing the air to flow out of the brake lines.

This system is always in a state of balance between the downward pressure of the spring and the upward pressure of the

When the brakes are applied inlet air pressure forces the diaphragm down so that it closes the exhaust port, and allows air to flow into the brake lines. When the brakes are being held inlet and brake line pressure is balanced, the exhaust port remains closed, and the inlet is closed by the rim

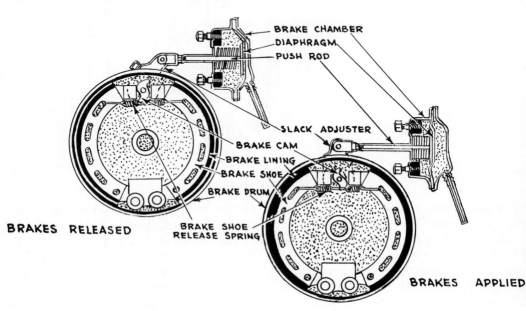

Fig. 18-30. Brake shoe action

of the diaphragm. When the brakes are re-leased pressure drops in the inlet, and pressure in the brake lines forces the diaphragm body upward so that the air can flow from the brake lines through the exhaust port.

This allows the brakes to release more rapidly than if the escaping air had to follow the pressure line back into the brake valve.

The relay valve, Figure 18-32, provides a mechanism by which air in a direct line from the reservoir can be admitted to the rear brake lines at brake valve pressure. Air from the brake valve acts on a diaphragm to open the supply valve, admit reservoir air to the brake lines and to the bottom of the diaphragm. The diaphragm also seals a circular exhaust slot.

When pressure is balanced, the supply valve is closed and the diaphragm held in a horizontal position by the supply valve spring. If brake valve pressure is reduced, air from the brake lines will lift the edges of the diaphragm and escape into the exhaust port until balance is restored.

This relay acts as a booster in obtaining prompt braking action and as a quick release valve as well. The bleeder passage allows a small amount of air to move

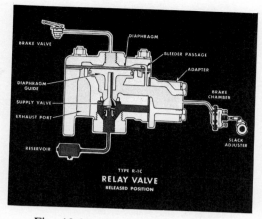

Fig. 18-32. Air brake relay valve

through the diaphragm to assure positive balancing of pressures.

When a truck is equipped with air brakes, it is absolutely essential that it should not be operated when there is not enough pressure to apply them. Warning of low pressure may be given by a buzzer or other noise maker, but this may not be heard over engine and road noise.

The visible warning shown in Figure 18-33 is simple and effective. It hangs directly in the driver's line of vision except when pushed out of the way by air pressure.

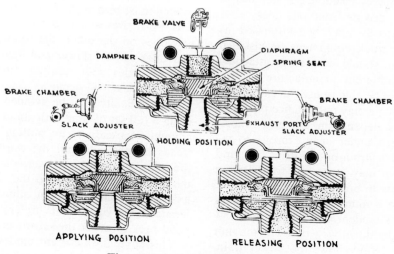

Fig. 18-31. Quick release valve

Courtesy of Monroe Standard, Inc.

Fig. 18-33. Low pressure warning signal

TANDEM DRIVE

The load carrying capacity of a 2-axle truck of any size can be increased by installing an extra axle in the rear. For dump use this should be a driving axle. A truck so equipped may be called a six wheeler, a ten wheeler, or a tandem. The double axle unit itself may be called a tandem or a bogie.

Tandem drive permits carrying much heavier loads in proportion to tire size and axle strength by distributing both weight and driving strains over twice as many units. It improves traction but not nearly as much as a driving front axle does.

Tandem drives may be purchased as part of new truck equipment or installed later. Caution should be used in attaching a bogie of different manufacture, as the truck may not be able to use the extra traction or the bogie may not balance properly. Some trucks depend on the rear wheels spinning when under heavy torque to relieve excessive strain on the drive line, and extra traction may cause the transmission or other parts to break.

There are a number of different makes and types of tandem drives. Figure 18-34 is a side and a detail view of a Timken unit and Figure 18-35 shows the power train.

Two identical axle housings are held parallel to each other by eight torque rods. The upper four rods are hinged to the axle housings and to the side frame brackets, the lower set to the bottom of the housings and to the bottom of a heavy cross member rigidly fastened to the frame. These rods hold the axles against twisting and absorb power and braking thrusts.

The weight of the truck is carried on two leaf springs, fastened on an oscillating collar on the lower frame cross member. The ends of the springs are held on the tops of the axle housings by spring pads and rebound brackets which allow the axles to move forward and back and to tip, without putting extra strain on the spring.

If the load on the truck is increased, the spring will flex and the frame move down relative to the axles. As the frame settles, the axles will first be forced apart by the torque rods, and then pulled together. In and out movement of the axles also occurs as the tires hit bumps in the road.

The propeller shaft crosses the top of both differentials, driving them through either worm gearing or hypoids with double reduction gears. The shaft between the two axles is fitted with a pair of universal joints and a slide coupling. No differential is used between the axles in the model illustrated, so they must turn at the same speed.

If worn tires are used on one axle and new ones on the other, the rotation of the new ones will be slower and fight between the two axles with resultant tire scuffing and gear wear will occur.

This difficulty may be avoided by either using a full set of tires with about the same amount of tread, or by putting the good treads on one side and the worn ones on the other so that the average diameters of the tires on each axle are the same. The axle differentials can then adjust the differences. The right side is preferable for the good tires to offset the crown of the road.

This company's current models have an inter-axle differential to equalize such dif-

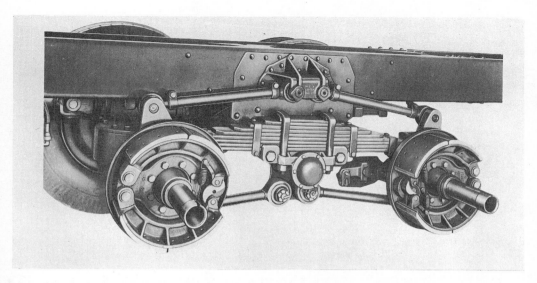

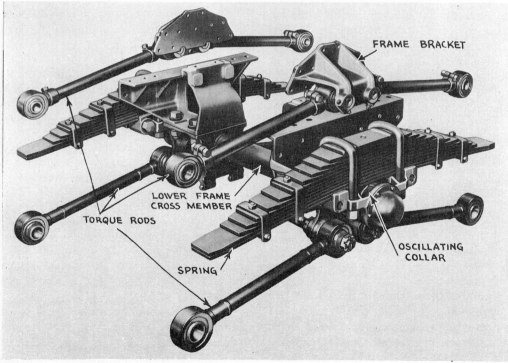

FRAME BRACKET

LOWER FRAME
CROSS MEMBER

TORQUE RODS

OSCILLATING
COLLAR

SPRING

Courtesy of The Timken-Detroit Axle Company

Fig. 18-34. Tandem drive—springs and rods

ferences. It can be locked to provide positive traction to both axles.

The Mack bogie or tandem, shown in

Figure 18-49, uses a parallelogram of springs, instead of torque rods. The springs are anchored to the axles in rubber Shock

Courtesy of The Timken-Detroit Axle Company

Fig. 18-35. Tandem drive—power train

Insulators (pillow blocks) enclosed in steel boxes under pre-load. The upper and lower boxes are carried on trunnions, so that any twisting of the double pairs of springs is avoided.

The propeller shaft is fitted with an inter-axle power divider which is just forward of the front bogie axle. From it a hollow shaft turns the front axle pinion. A smaller shaft inside it drives the rear axle. Two universals and a slip joint are used.

The two axles may also be driven by separate shafts originating in the transmission or transfer case. The rear shaft is carried in a pillow block and bearing on top of the front housing. A differential or divider may or may not be used with this construction.

In any of these tandem drives, some tire scuffing will occur because neither axle has any provision for steering. As the truck turns the rear bogie wheels have a tendency to slide toward the outside of the curve, and the front set toward the inside.

If a bogie is improperly designed, it may tend to nose down in front when stopping and rear up when accelerating with resultant bouncing and excessive wear on parts.

Six wheelers are more difficult to steer than standard trucks on slippery surfaces because of the resistance of the extra axle to sideward movement.

In three-axle bogies the additional axle is usually at the front and is non-driving. It improves load carrying capacity, but increases tire scuffing and steering problems.

There may be an arrangement by which the non-driving wheel set can be lifted clear of the road whenever loads are not heavy enough to require its help.

Highway Limitations. Vehicles to be used on highways must not be more than eight feet wide, and they are limited in length, gross weight, and weight on any one axle. These restrictions vary in different states, but a limit of nine tons for the load on any one axle is usual.

Dual drive tires are always used, with either single or tandem axles. The size and weight bearing capacity of tires is limited by width restriction and the space required by the frame.

As a result of both weight and width regulations, the highway dump truck is usually limited to a maximum carrying capacity of 10 to 15 yards, and the highway semitrailer to about 22 yards.

Courtesy of Mack Motor Truck Corporation

Fig. 18-36. Double-spring tandem drive

Uses. The highway dumper usually has a combined weight of chassis and body equal to only ¾ of its rated payload, as against a usually higher ratio for off-highway haulers in comparable sizes. Highway trucks may go over 70 miles an hour, nearly double the speed of other haulers.

The result is that the highway truck is a highly efficient hauler when operating conditions are suitable. It should not be used constantly in rough or soft pits or on poorly maintained haul roads, or on excessively steep grades, or under big shovels loading coarse rock.

However, the highway dumper is the preferred hauler for jobs where distances are long and roads well maintained, and loading and ground conditions in the pit and on the fill are not too rough.

It is necessary to the contractor whose hauling must be done on streets and highways. This includes most excavating for foundations, city and suburban roads, supplying sand, gravel, fill, and topsoil, and most other small to medium size work. On major highway construction it is usually possible to obtain permits to operate off highway vehicles over short sections of public road, but the contractor is likely to have to rebuild them after he is through using them.

OFF THE ROAD

Trucks that are built to operate in mines or pits, or in other types of excavation in which the use of public roads is not required, are not subject to any legal restrictions in size or weight.

Off-highway rear dump trucks may be 9 to 26 feet wide, 22 to 50 feet long, 11 to 18 feet high, with loading height (body sides) between 8 and 17 feet. Capacity range is from 13 to 210 tons, with body ratings up to 112 yards struck. Larger models are being tested.

Courtesy of TEREX *Division, General Motors Corporation*

Fig. 18-37. "Small" off the road truck

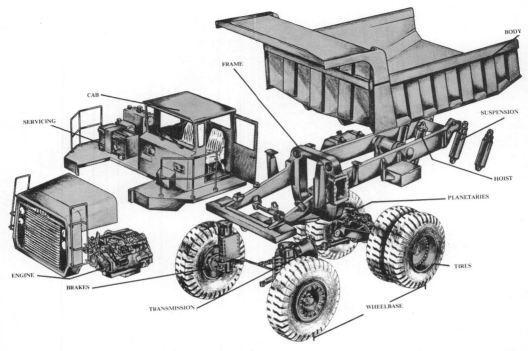

Courtesy of TEREX *Division, General Motors Corporation*

Fig. 18-38. Off the road truck, exploded view

Empty weight (chassis and body) may equal the payload capacity, or be as low as 4/5 of it.

Construction is heavier than in highway trucks, in order to stand up under conditions of rough footing, heavy loads, and short hauls. Substantial amounts of high-strength steel may be used.

Top speed is usually 30 to 35 miles an hour, with some models going over 40. Road conditions and tire wear limit practical speed.

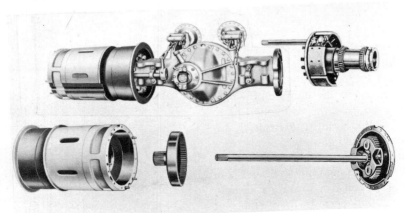

Courtesy of Euclid Division, General Motors Corporation

Fig. 18-39. Rear axle and final drive

Fig. 18-40. A very big truck

Construction. Figures 18-37 to 18-44 are representative of this type of truck. Components tend to be massive and comparatively simple.

Road shocks may be absorbed by conventional leaf springs, coil springs, nitrogen-and-oil (air over hydraulic) cylinders, rubber discs, oscillating bars, and various combinations of these. Older models may have no springs or cushions, except rubber shock pads.

Horsepower range is from about 160 up to 2000, with 4 to 12 diesel cylinders. Turbine engines may be used in large sizes.

Most of these haulers have torque converters and power shift transmissions, with hydraulic retarders either standard or optional. A few of the smaller ones have mechanical drive, and some of the largest use a generator and electric wheel motors.

Differentials may be either standard or limited action. They are usually single reduction, but may be double. Further gear reduction is obtained through planetary final drives in the wheel hubs.

There are usually two axles, with drive through dual wheel sets on the rear. There are also tandems and all wheel drives.

Tires are among the largest made, up to 36:00-51, with the same size front and rear.

Steering is usually automotive type, with front wheels swiveling on a rigid axle, but there are also articulated models.

Body. Bodies are heavy duty rock or quarry type. The sides may flare out to make a larger loading target. The floor is usually a single heavily reinforced plate, but in older models may be double. Air ducts may be provided for internal heating of both floor and sides with exhaust air, to prevent loads from freezing down during sub-zero hauls.

There is no tailgate. The body floor may have a continuous upward slope from front to back to retain the load, or it may be flat or nearly so, with an upturned chute in the rear.

Standard hoist construction is a pair of direct acting three-stage telescoping cylinders, with power up and part way down. Power-down permits raising the body to a very steep dumping angle, 55 degrees or more, then pulling it back until gravity can lower it the rest of the way.

Fig. 18-41. Articulated truck frame

Courtesy of Euclid Division

Fig. 18-42. Rock body

Most materials can be dumped readily, even when backing up a grade.

Shuttle Truck. The Koehring Dumptor, Figures 18-45 to 18-49, is an off-the-road truck specialized for short hauls. Its chassis is similar to that of a wheel tractor, having a short wheel base, big rear and small front tires, solid connection between frame and rear axle, and an oscillating front axle with a center cushion spring. Body capacity is ten cubic yards struck.

The Dumptor has a two speed, constant-mesh transmission with torque-converter drive and power steering. Top speed in either direction is about 21 miles an hour.

Brakes are four wheel air operated. Parking brake is spring set, air released.

Dual controls with a pivoting seat enable

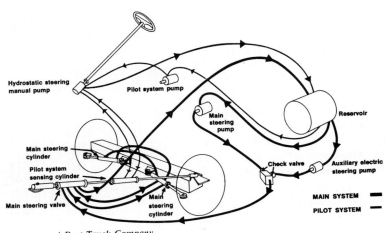

Courtesy of Dart Truck Company

Fig. 18-43. Two-stage hydraulic steering

Fig. 18-44. Telescoping hoist cylinders

Courtesy of Koehring Company

Fig. 18-45. Gravity dump body

the operator to always face the direction of travel. To change direction, the operator swings around on the seat and takes over the other set of controls. If a cab is used, it has full windshields both front and rear.

The Dumptor body is designed so that when loaded it is out of balance and will dump by gravity when the operator releases a by-pass for the hydraulic fluid in the body hoist cylinders. For controlled dumping a regular hydraulic valve admits fluid to the cylinders to raise and hold the body at proper height for spreading or dumping into hoppers. The body is always returned to carrying position by hydraulic power.

As with some conventional off the road trucks, the rear of the body can be used as a bulldozer when in the dumped position to spread piles enough to make it possible to cross them with the next load. However, it cannot cut down to the level of the wheels. Any dump graded entirely by this means will have an ascending grade.

Courtesy of Koehring Company

Fig. 18-46. Shuttle-type dump truck

Courtesy of Koehring Company

Fig. 18-47. Kick-out pan

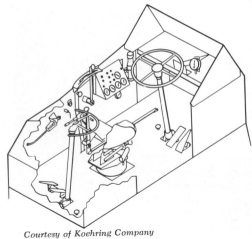

Courtesy of Koehring Company

Fig. 18-48. Dual controls

This machine has a very short wheelbase and has a turning radius of 24½ feet in either direction. However, it is designed primarily to shuttle from a shovel to a dump without turning. Time can be saved on short hauls by eliminating turns, by high speed travel both ways, and by the fast dump.

Due to its no-turn operation, the Dumptor is well adapted to tunnel and mine work where turn space is seldom available. Its use can also simplify layout and traffic problems in a pit where space is restricted, ground is too soft for convenient turning, or traffic is very heavy.

The Model 60 Dumptor has a five yard body with a latch type gravity dump, with-

out hydraulics. Angle of dump is limited by spring cushioned chains. It is returned to carrying position by moving forward and stopping sharply, or by backing suddenly. If this body is loaded too much to the front to dump by gravity, it can be tipped by driving backward and stopping quickly.

ELECTRIC DRIVE

Some of the biggest off the road trucks have electric drive. A diesel or a turbine engine drives a generator, which supplies current to electric drive motors. These may be located in each wheel, or in differentials between pairs of wheels.

The motors use direct current (DC) on

Courtesy of Koehring Company

Fig. 18-49. Shuttle pattern

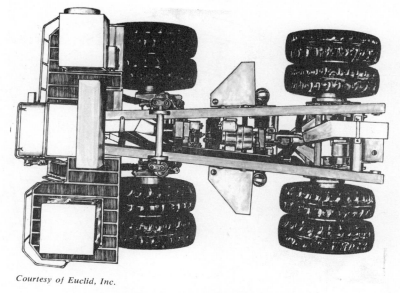

Courtesy of Euclid, Inc.

Fig. 18-50. Turboelectric truck chassis

account of its superior lugging performance. The generator may produce DC directly, or it may be an alternator whose AC output is converted to DC in a rectifier.

The motors operate through reducing gears, as in Figure 12-38C, and are reversible.

Electric drive eliminates all gear shifts and speed ranges, automatic or otherwise. There is smooth, stepless transition from zero to maximum speed, which is determined by the amount of reduction in the motor gearing, and by grade and rolling resistance. It may be 24 to 36 miles an hour.

Drive motors also supply dynamic braking, to slow the hauler on down grades and before stops.

Turboelectric. The Euclid R-210 Turbo electric truck is powered by an Avco Lycoming gas turbine engine which provides 2250 shaft horsepower on a standard day. Because the engine is so compact—only 38.3 cubic feet in gross volume—it can fit between the truck frame box rails.

The engine drives a General Electric traction alternator at 2100 rpm through a planetary reduction gear having a 6.2 re-

duction. The engine, reduction gear and generator are close coupled, i.e. mounted to each other, and are mounted slightly aft of the center of the truck. AC output from the alternator is rectified to DC which powers the four wheel motors, one motor at each corner of the truck.

These motors are electrically converted to generators during dynamic braking, their output being dissipated in heat by a fan cooled resistance grid. Friction brakes are also provided to stop the vehicle at low speeds and in an emergency, from high speeds. A control box provides for forward, neutral, reverse, power and braking.

This machine is a prototype, now being tested. Changes in design may be made when it goes into production.

TRACTION

Soft Footing. Truck efficiency drops rapidly on soft ground. The wheels sink in so that they must constantly climb a slope to move the truck horizontally. As the front wheels sink they have an increasing tendency to push in the manner of a sled runner instead of rolling up and out, so that the

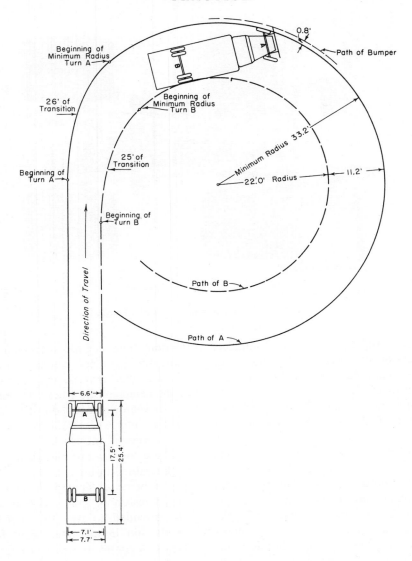

Courtesy of Department of Public Works, State of California

Fig. 18-51. Right turn, two-axle truck

force required to move them is greatly increased.

If the power required to roll a driving wheel up the slope in front of it is greater than that needed to spin the wheel in the hole, it will spin and usually dig the hole deeper and polish its slope. The ordinary differential will deliver its full drive power to the spinning wheel so that the standard truck will not be able to move even if only one side lacks traction.

Another difficulty is that the front wheels, if turned to climb out of a rut, are liable to stop revolving and skid sideward so that the truck cannot be steered.

When the truck is reversed, the reaction

Courtesy of Marmon-Harrington Company, Inc.

Fig. 18-52. Four wheel drive conversion parts

to the twist of the driving axles tends to lift the back of the truck and to push down the front, thus reducing the weight on the driving wheels and their traction.

Similar difficulties are encountered in sand which allows all wheels to sink somewhat and, although not slippery, will first compress under the push of a driving wheel, then shear, allowing the wheel to spin a part turn carrying the sand with it, then compress to hold it momentarily before shearing again. This produces a suc-

cession of shocks to the power train, while digging the wheel in deeper each time it moves.

These factors severely limit the use of trucks in many circumstances and may be combatted in a number of ways.

In muddy pits, the use of big tires is advisable because of their larger area which increases contact and reduces sinking. It may be said that a tire will sink in mud until a large enough area is in contact to carry its weight. If the tire is small it will

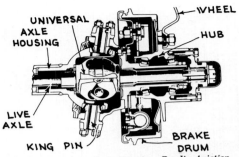

UNIVERSAL
AXLE
HOUSING

WHEEL

HUB

LIVE
AXLE

KING PIN

BRAKE
DRUM

Courtesy of Bendix Products Division, Bendix Aviation Corporation

Fig. 18-53. Front hub, four wheel drive truck

sink much farther to develop enough contact; therefore, it will have a higher slope to climb to get out of the hole and have more side friction to hold it in.

Tires with coarse block treads afford better grip on most muds but are at a disadvantage on hard slippery surfaces and in sand. Tire chains are helpful in mud and on slippery surfaces, but not in sand. Block treads and chains on front wheels help in steering out of ruts.

Tandem drives are helpful, particularly if tires are kept as large as those which would be used with a single axle. Limited action differentials keep all wheels driving instead of wasting power in spinning one wheel or pair of wheels.

Live Front Axle. The most effective way to combat bad footing is to supply driving power to the front axle as well as to the rear. The increase in traction will vary between fifty and two hundred percent, depending on the number of rear driving wheels, ground conditions, load distribution, and steepness of climb.

The advantages obtained are conversion of the front wheels from a dead obstacle to a driving force, ability to pull around a turn instead of having only a straight forward push, tendency of front wheels to climb out of a rut instead of sliding in it, and good traction in reverse.

The two principal problems connected with the all-wheel drive are driving the front wheels through sharp turning angles, and making allowance for the fact that the front wheels go farther than the rear wheels on curves, both backward and forward.

Figure 18-51 shows why the front wheels have longer travel. The rear wheels take a smaller circle on a curve so that they need turn fewer times in the same truck travel distance. On loose surfaces this difference adjusts itself, but means should be provided to allow the front and rear driveshafts to revolve at different speeds on hard pavements in order to prevent excessive wearing of tires and strain on shafts, pressure on gear teeth, and waste of power.

Figure 18-52 shows parts required to convert a standard truck to four wheel drive.

An auxiliary two-speed transmission and transfer case is placed immediately behind the regular transmission, and drives the front and rear propeller shafts.

The front shaft is attached to the transfer case through a jaw clutch. When this is engaged, the front shaft is driven at the same speed as the regular or rear propeller shaft. When it is released, drive is to rear wheels only.

The truck or other vehicle is customarily kept in two wheel drive. The front shaft is engaged whenever extra traction is needed for either moving or stopping.

The front wheels are driven through constant velocity universal joints protected by ball sleeves. Figure 18-53 is a detail of a hub.

Drive to four wheels greatly increases efficiency of brakes on slippery surfaces. Front wheels do not lock as readily, so that steering control is better maintained.

Warn Hubs. The front hubs may be fitted with free wheeling attachments, so that the axles will turn the wheels forward but not backward, and the wheels will not turn the axles when the vehicle is moving forward.

This allows the front hubs, axles, differential, and drive shaft to stop whenever

Courtesy of The Four Wheel Drive Auto Co.

Fig. 18-54. Power train, FWD

drive is disconnected in the transfer case, reducing wear and noise.

But in four wheel drive, hubs in free wheeling will not transmit power in reverse, nor permit the steadying effect of front drive on braking.

When hubs are locked, a vehicle can be shifted in and out of four wheel drive instantly at any speed, which is a safety factor in intermittently slippery conditions. With hubs free wheeling, it is necessary to stop to shift into four wheel drive.

The hubs are easily shifted into a solid connection by turning their caps with the fingers. This adjustment can be made readily if it is necessary to engage them to back out of a mud hole. Control from the driver's seat may be available.

It is common practice to keep the hubs in free wheeling position except when bad footing is expected. Sometimes they are locked for all winter driving.

FWD. The Four Wheel Drive Auto Company's line of four and six wheel drive trucks uses a differential in the transfer case that can be locked for positive drive by a hand lever. The differential compensates for all speed differences between the front and rear axles. The front drive is permanently connected.

Drive from the transmission to the transfer case is by means of a wide silent chain in some models, and by helical gears in others.

This power train, shown in Figures 18-54A and 54B, is available only in trucks made by this manufacturer.

Positive Drive. All the systems described thus far have the weakness that if one front and one rear wheel have no traction, they will spin and the truck will be unable to move.

The Walter overcomes this difficulty by the use of a set of three self-locking differentials employing worm gears instead of

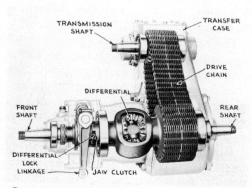

Courtesy of The Four Wheel Drive Auto Co.

Fig. 18-55. Chain drive

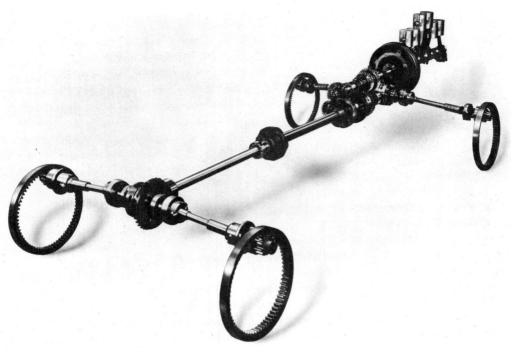

Courtesy of Walter Motor Truck Co.

Fig. 18-56. Walter power train

bevel spiders. These will permit sufficient differential action for steering but will lock against spinning.

Figure 18-56 shows the power train. The engine is located ahead of the front wheels and is mounted as a unit with a six forward, two reverse speed compound transmission. This unit includes the center differential and the front axle differential also.

The rear differential has rubber mountings suspended from the frame, so that it is also not subject to moving with the wheels and does not form part of the unsprung weight. All wheels are carried on beam type dead axles with attachment to the chassis through conventional springs.

The live axles each include a pair of universal joints, and carry brake drums close to their differentials. An additional pair of brakes is located in the rear hubs.

The live axles end in small spur gears which turn large internal tooth ring gears in the hubs.

This construction permits the use of lighter axles than could be used if the speed reduction were in the differentials only, or if the axles carried the weight of the truck.

6 x 6. So far, the all-wheel drive systems discussed have been of the two axle, or 4 x 4 type. Any of these may be expanded to a three axle, or 6 x 6, drive by substituting a bogie for the rear axle and extending and reinforcing the frame. The extra axle improves flotation, traction, and carrying capacity, but increases cost and makes the truck less maneuverable.

General Considerations. All-wheel drive is of great value wherever mud, sand, or snow is encountered. Trucks so equipped can keep a job going under conditions that would be impossible to two wheel or tandem drive trucks.

On highways, the extra traction is de-

Courtesy of American Steel Foundries

Fig. 18-57. Fifth wheel

sirable for snowplowing and operating under-frame or towed graders, and sometimes in towing trailers.

However, the traction and flotation obtained cannot be compared with that obtained by crawler equipment. Mud must have a fairly solid bottom near the surface for any truck to operate, and very slippery surfaces require chains no matter how many wheels are driving.

All-wheel drive has several disadvantages. It is expensive, particularly when it is installed as an accessory rather than in the original assembly line. The extra cost may be anywhere from twenty-five to one hundred percent of the price of the standard truck.

In return for this, the truck can carry heavier loads under much worse conditions than before conversion, so that the cost is justified in many lines of work.

The most serious operating fault is steering. A standard truck can turn the front wheels at an angle of 35° or better, while the various front drive hubs are limited to 28° to 30°. It follows that their turning radius is much longer, giving a small truck the clumsiness of a large one. This may not be particularly important on large scale open work, but for the contractor working in woods, around residences, and in storage yards it is a serious consideration.

The all-wheel drive vehicle is high, particularly if a conversion from a standard make. This feature is useful in rough ground but detracts from the appearance of the truck and makes it harder to load with small machines or by hand.

Maintenance costs are somewhat higher, but the extra parts decrease strain and wear on the standard ones. If a truck is to be consistently given heavy overloads the front drive may partially pay for itself in reduced maintenance on the rear drive.

The extra machinery is usually noisy, more so while it is working that if it is disconnected from the engine.

There is a question whether mechanical front wheel drive absorbs power and causes driver fatigue, or the reverse. Manufacturers of four wheel drive equipment claim that they have proved that its use increases average road speed, does not reduce power, and prevents fatigue. Most drivers feel strongly the opposite way on each of these counts. However, if the front drive can be disconnected when not needed, these bad effects are largely cancelled out.

It would seem that the ideal mechanical hookup would be one which would give the driver the triple option of direct drive to the rear wheels only, for easy going; a four wheel drive with differentials for heavy loads on good footing, and a positive drive to all

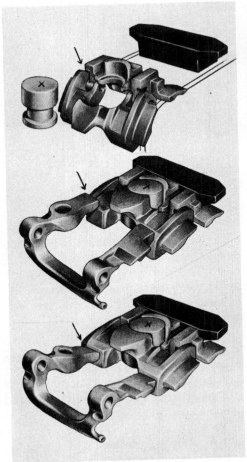

Fig. 18-58. Fifth wheel latching action

wheels when needed for sand, mud, or ice.

TRAILERS AND WAGONS

Dump bodies may be mounted on semi-trailers and trailers, with either standard or special constructions.

A semi-trailer is a frame which has supporting axles and wheels at the rear and rests on the prime mover or tractor at the front. Tractors for highway type semi-trailers are equipped with a connecting device called a fifth wheel, one variety of which is shown in Figures 18-57 and 18-58. To connect, the trailer is held at a proper height by jacks or wheeled standards, the wheels blocked, and the tractor backed into it. The upper fifth wheel, or trailer hitch, is hit by the tractor fifth wheel and wedged slightly up until its knob reaches the socket, into which it slips and locks automatically. The pin and socket take the pull and thrust of towing and stopping, while the fifth wheel surface carries the weight. The knob can be released from the hitch when not under load by moving a hand lever.

Off-the-road semi's may use a kingpin arrangement such as that described in the preceding chapter.

Full trailers, often called wagons, are equipped with supporting axles and wheels at both ends, so that none of their weight rests on the prime mover. See Figure 18-59. The front axle may swivel or may have steering linkage. It is fastened to a drawbar that is attached to the prime mover and which carries a fifth wheel or other support for the trailer frame.

Tractors may be short coupled trucks or have more of the characteristics of large wheel tractors, there being no sharp distinction between the types. The two wheel or four wheel tractors used with self-powered pans may be used with semi-trailers.

Trailers of either semi or full construction must be equipped with brakes. These are operated by air, vacuum, fluid, or electricity from the tractor, and are usually fixed to go on automatically if the connection is broken. This is to prevent the trailer from rolling free if it breaks away from the tractor. Brakes may be synchronized with the tractor system, controlled separately, or both.

Trailer lights are controlled from the tractor cab. Current is conducted along a multiple wire cable with plug-ins for both tractor and trailer slotted so that connection cannot be made incorrectly.

Rear dump bodies are usually installed

Courtesy of Fruehauf Trailer Co.

Fig. 18-59. Full trailer chassis

on semi- rather than full trailers, and may be dumped hydraulically as in Figure 18-60, or by cables. The hydraulic dump conforms to the systems already described but may be much larger and heavier. Bodies may be very long and require level ground and good overhead clearance. The pump and valve may be in the tractor and connected to the hoist ram or rams by pipes and flexible tubing.

A semi-trailer can be operated in more restricted areas than a truck of the same capacity. Since the drive wheels are not at the rear, it can be backed somewhat

Courtesy of Gar Wood Industries, Inc.

Fig. 18-60. Dump bodies on semi-trailers

Courtesy of Challenge-Cook Bros.

Fig. 18-61. Bottom dump

closer to the edges of fills and farther onto soft ground than a truck. However, traction is not as good and more driving skill is required.

BOTTOM AND SIDE DUMPS

Bottom dumpers are usually semitrailers, occasionally full trailers. They are almost always too wide for highway use. Tapered bodies are shaped to dump through a full length bottom opening controlled by one or two pairs of clamshell doors.

The doors are operated by either hydraulic or air cylinders, to slide or pivot sideward and upward along the outside of the body.

Dumping is done with the unit moving forward, without any particular limit as to travel speed. The load is deposited in a windrow. Its cross section is determined by the width to which the doors are opened, and the speed of movement. Maximum height is limited by the rear clearance of the body, but this may increase as the rear wheels ride up on the dumped material.

Body walls are much lower than those of rear dumps of comparable size, which makes loading easier. Dumping on the move saves time, and windrows are easier to spread than full piles.

Disadvantages are inability to dump off edges or close against obstructions. Receiving hoppers must be made specially long

with drive-over provisions to accommodate them.

Capacities range from about 20 tons up to 120, with the smaller sizes becoming rare.

Side Dump. Side dumps are used in building out edges of fills which are long enough to allow a broadside approach. They are lower and more stable with a raised body than a rear dump, and can be unloaded at higher speeds.

Figure 18-63 shows Gar Wood two-way

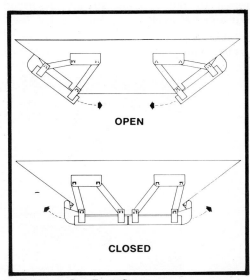

Courtesy of Dart Truck Company

Fig. 18-62. Bottom door action

Courtesy of Gar Wood Industries, Inc.

Fig. 18-63. Side dump bodies

side dump trailer-mounted bodies. They are fastened to their sub-frames by hinge latches at both sides. Hydraulic rams are mounted vertically at the front and rear.

The body is dumped by unlatching it at one side and extending the rams. The free side rises, pivoting on the other side hinge. The lower side gate moves outward and downward automatically, making a chute to carry the spoil away from the wheels.

Courtesy of Gar Wood Industries, Inc.

Fig. 18-64. One-side dumper

Figure 18-64 shows a one way side dump mounted on a truck.

The Athey side dumps, Figures 18-65 and 18-66, are mounted on semi-trailers designed for the Caterpillar 630 and 660 tractors. They can be equipped with sloped sides for spill-over dumping, or vertical automatic-opening side gates, and can be arranged to dump to either side.

The hoist is a three-sleeve telescoping ram. Spring loaded plungers on the frame side reduce dumping shocks, and assist body return. Center of gravity is low to permit dumping on the move without the special stabilizing arrangements that were used in earlier models.

Side dump bodies may be made without hoists and dumped into a hopper by lifting one side with a crane.

Track Trailers. Some construction jobs are too soft for rubber tired wagons, and others are in such abrasive material that tires are very short lived. Under such circumstances dump bodies mounted on tracks may be used behind crawler tractors.

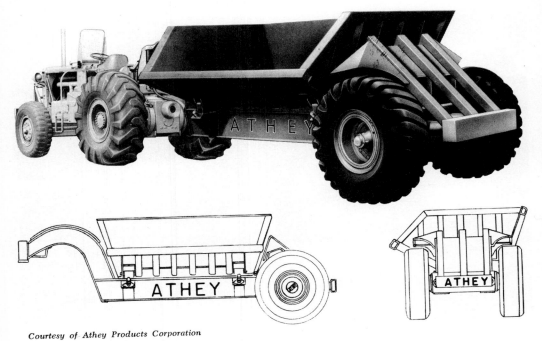

Courtesy of Athey Products Corporation

Fig. 18-65. Quarry-type side dump

Figure 18-67 shows an Athey two-way dump quarry trailer. The dump mechanism has locking hinges on both sides and a single telescoping ram. The sides are flared and overhanging. Box bodies with automatic side gates and bottom dump models are also available.

The tracks are of a special design with short, wide pads and self-cleaning pins. Each track uses two dual idlers. Truck frames are hinged at their centers to the trailer frame.

The tracks are short to minimize turning resistance. However, they provide much more ground contact than the largest of tires.

Two of these units may be pulled by a tractor of sufficient size.

TRUCK SIZES

A desired hauling capacity can be obtained by a few large trucks, or a larger number of small ones. Truck fleets may contain only one size unit or a variety.

A big truck should move dirt more cheaply than a small one of the same speed, particularly on long hauls. Its purchase and maintenance costs are usually lower on a per yard basis, and there is a definite saving in driver's wages. Insurance premiums and registration costs are variable.

A big body presents a larger target for loading by a big shovel or dragline, and may increase its production as much as twenty percent in better cycle time and reduced spillage. However, a high sided body will make loading by a small shovel or dozer shovel difficult or impossible.

The large truck requires substantially more room to maneuver and if it does not have it, may waste so much time getting placed to load or dump that its production will be smaller and more costly than that of a small truck.

One 50-yard truck in a fleet may smash up a haul road that would last indefinitely under 20-yard units. Damage is particularly likely when the road surface, whether

Fig. 18-66. Ready to dump

of gravel, blacktop, or concrete, rests on soil which supports moderate loads but either flows or compresses under great weights. Costly additional fill or surfacing may be required for protection. This problem is affected both by ground pressure of the individual tires and by the total weight of the whole machine.

Fleets made up of big trucks do not have traffic problems as frequently as more numerous small ones, and in general offer fewer problems to the dispatcher; but breakdown of one unit will cripple the fleet more seriously.

On very short hauls where space is so restricted that two trucks will lose time

Fig. 18-67. Track trailer

waiting for each other, a single large truck may allow greater shovel production than two smaller units having a greater total capacity.

Overloads. Trucks are generally capable of carrying much heavier loads than are recommended by their manufacturers. However, overloading uses up the reserve of strength that is built into its parts, and causes accelerated wear and more frequent breakdowns. Brakes and the holding power of engine compression will not be as effective or safe under additional tonnage. Overloads will slow acceleration, reduce pulling power, and lower speeds on rough ground so that under severe conditions they will decrease production.

Conditions favoring overloads include smooth roads, hard ground in loading and dumping areas, level or easy grades, long hauls, and careful operators. Rough or soft surfaces and steep grades will slow and strain a heavily loaded vehicle, and rough or careless operation will damage it far more than if it carried fewer tons.

The contractor wishing to do heavy hauling can either purchase a truck of proper size, or build up a lighter one to the same capacity. The first course should prove more economical if the machine is to have fairly continuous use. However, if the necessary capital is not available, or the investment is not justified by the work in sight, the use of the lighter unit may be advisable.

If the decision is to overload a light truck, and it is bought new, it should be equipped with the largest and heaviest tires the manufacturer offers and have a two-speed, heavy-duty axle. If springs are not adequate extra leaves should be added.

The frame should be reinforced. The point of greatest stress is between the body and the cab. The usual flat reinforcing fish-plate may not be satisfactory because it reinforces against vertical stresses only, and twist is an important factor in failure of frame members. Use of angles or channels produces better results. A channel inserted and welded in the frame channel is the preferred method. Fishplating should extend along at least two feet of frame, preferably three or more, and should be securely welded top and bottom. Bolting or riveting weakens both the frame and the plate.

The rear cross member may require strengthening also. Any type of fishplating, or welding a heavy pipe between the frame members immediately ahead of it, should give sufficient support.

Reinforcement is most easily installed before the body is mounted, and most effective before the frame has been strained by carrying heavy loads.

Bodies can be built up with wood sideboards to fifty percent or more over capacity. For very bulky loads, a bigger body or metal sideboards may prove more satisfactory. A shallow body with large floor area will carry a larger heap in proportion to level capacity than a short deep one.

The hoist should be the largest capacity that will fit the body. Overloaded hoists need frequent repairs which are much more costly than the extra price of the bigger hoist.

An expensive preparation for overload is installation of tandem drive, together with lengthening and reinforcement of the frame. Some makes of trucks give excellent service with tandems while others have frequent transmission and clutch trouble.

If soft or slippery ground is an important factor conversion to all-wheel drive will increase the carrying capacity, but only at considerable expense. Wider rims and oversize tires are often installed and frames reinforced in connection with this change.

DRIVING A DUMP TRUCK

Operating a conventional highway-type dump truck is somewhat like driving a car. However, the truck is much heavier and

Courtesy of Athey Products Corporation

Fig. 18-68. Standing clear while being loaded

more bulky, has very poor visibility to the rear, and usually requires special methods of shifting gears and climbing and descending hills, because of greater weight in proportion to engine power and braking capacity.

In addition, checking the truck, spotting it to load and dump, and operation of the body are special techniques not related to any experience gained driving a car.

Checking. It is particularly important that tire inflation and condition be checked at least daily, as the tires are the most vulnerable and most expensive wearing parts of a truck. Running them soft with a load may ruin them in a few minutes.

If the truck has air brakes, it should not be moved until the dash gauge shows safe operating pressure.

Direction signals and stop lights must be in working condition for driving on highways or on heavily traveled haul roads.

Stones stuck between dual tires should be removed immediately as they may be thrown behind the truck with deadly effect. A stone can usually be pried out with a bar or board, but it is occasionally necessary to let the air out of one tire to free it.

Each load should be checked for rocks or other objects which might fall off. These may be moved to safer positions or be pushed off in the pit.

Loading. Truck drivers may add sub-

Courtesy of Koehring Company

Fig. 18-69. Back toward a hoe for loading

stantially to shovel production by getting promptly in proper position. If a spotter is present or the shovel operator is giving signals, directions should be followed exactly regardless of the driver's ideas of where the truck should be. If a spot log is used the rear wheels should be backed squarely against it.

If the shovel has not moved a truck will generally be spotted in the same place as the last one loaded. Following tire tracks or observing the location of spilled material makes it easy to get in the same position.

In general, a truck hauling from a revolving shovel is placed to require the shortest practical swing from the digging. It should be facing directly away from the shovel so that the shovel can reach from back to front of the body by use of the crowd mechanism, or at a right angle so that the swing controls can be used to distribute the spoil along the length of the body.

When large dipper shovels are loading rock the truck is sometimes placed facing

the shovel. The open door of the bucket tends to prevent rock from rolling toward the shovel and therefore serves to protect the cab and the operator.

The distance from a dipper shovel should be such that the center of the body is in the middle of the crowd arc of the bucket. With draglines and clamshells it should be directly under the boom point.

A hoe or a dragline loads most efficiently when it is on the bank and the trucks are on the floor of the pit. For draglines and hydraulic hoes, the difference in cycle time is the number of seconds needed to hoist the load to the higher level. For the cable hoe, there is also time consumed by awkwardness in dumping the bucket above ground level.

If you must be on the same level as a cable hoe, back in very close to it.

When being filled by a front end loader, the truck should usually be backed against the bank at an angle of forty-five degrees so that the tractor need make only one eighth of a turn to dump. The truck should be close and the loader walking time should not be greater than the time

Courtesy of Caterpillar Tractor Company

Fig. 18-70A. Good truck placement

required to lift the bucket. Wheel tractors require more space to maneuver than crawlers.

It is often necessary to move a truck while it is being filled. This is particularly true with belt or bucket loaders, which may themselves be moving, or if stationary, may not distribute the load throughout the body.

The driver can remain in the cab during loading, but a broken shovel cable may allow the truck to be struck by a falling boom or an uncontrolled bucket. The size of the shovel and the amount of cab protection largely determine the amount of danger. A cable hoe loading from behind the truck is particularly hazardous.

Gear Shift. A small but increasing number of trucks have automatic or semi-automatic transmissions, over which the driver has little control. Others have constant-mesh gears that shift easily.

This section is limited to the majority now running that have sliding spur gears in the transmission. In order to avoid clashing when shifting it is necessary that the two gears to be meshed have about the same tooth speed. Shifting arrangements are quite variable. Figure 18-71B shows patterns which are widely used.

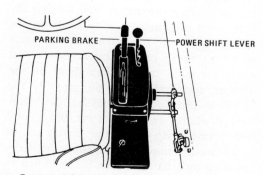

Courtesy of Caterpillar Tractor Company

Fig. 18-70B. Power shift control

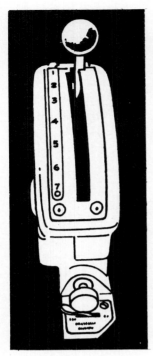

Fig. 18-71A. Automatic shift control

When the truck is standing, the clutch is pushed down and after a slight pause the gear lever is moved to the first or low gear position. The pause gives the jackshaft, extending from the clutch into the transmission, a chance to slow down or stop so that it can be meshed.

After the truck is moving, the shift to second is started by releasing the accelerator, depressing the clutch, and moving the gear lever into neutral. The teeth on the sliding gear will be moving at the same speed as those on the countershaft first gear, but more rapidly than those on the smaller countershaft second gear. If the clutch is held down the jackshaft will slow until its sliding gear can be meshed quietly into second. However, the engine slows down much more rapidly than the free-spinning jackshaft, so that if the clutch is engaged again after the shift to neutral and then depressed, the jackshaft will be slowed so that the shift can be completed immediately upon releasing the clutch the second time.

This operation is known as double-clutching and is used with any shift to a higher gear. It does not work if the engine does not throttle back to idling speed.

If a truck has a properly adjusted clutch brake it will slow the jackshaft without the necessity of double-clutching.

When a shift is made from a higher to a lower gear, as in climbing a hill, double clutching is required even if there is a clutch brake. In this case the jackshaft gear will be turning too slowly to engage with the next lower gear, therefore the clutch is pushed down, the gears put in neutral, the engine speeded up, the clutch let up and pushed down, and the shift completed.

Shifting up is mostly a matter of timing, but shifting down requires both correct engine speed and timing and is somewhat more difficult to master. It should usually be done quickly before valuable momentum is lost by the truck.

If the truck has an auxiliary transmission it is usually of the two speed, sliding gear type. Shifting is done in the same manner as in the main transmission. However, the jackshaft has much more inertia because the main jackshaft and transmission gears are spinning with it, so if a bad shift is made, the danger of damage to the teeth is greater.

If a transmission is of the synchromesh type, double clutching is not necessary unless the synchronizing mechanism becomes ineffective.

Automatic and semi-automatic transmissions should be operated in accordance with manufacturers' instructions. Jerky shifts should be promptly reported and serviced, as they are hard on the power train and on the driver's disposition.

Axle Shift. Many trucks are equipped with two-speed axles with power shifting.

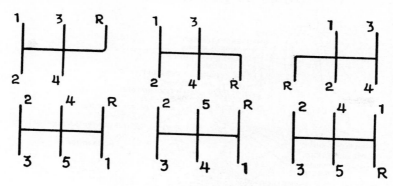

Fig. 18-71B. Manual gear shifts

These are controlled by a valve or switch on the dashboard or the gearshift lever. The power may be vacuum, air, or electricity.

These shifts are of the pre-selective type. A truck in low axle ratio may have the control moved to high axle, but as long as the engine is driving the truck the pressure on the teeth will prevent the gears from disengaging. If the accelerator is released, the tension will be relieved, and the power mechanism will move the shifting collar or other device out of the low axle position and push it toward engagement with the high axle. There will be a discrepancy in speed, but the driveshaft loses speed rapidly, and the shift should be completed automatically when speeds are synchronized.

In the shift from high to low, the hand control is shifted, and the accelerator is released and then pushed down.

When going downhill against the engine the shift is made by pre-selecting the ratio and opening the throttle briefly to reverse the torque.

A "split shift" is shifting transmission gears and axle ratio at the same time. This is done by pre-selecting the axle ratio and depressing the clutch and shifting gears in the usual manner. The axle shift will take place while the clutch is out or immediately after re-engagement.

Hills. A truck with a gross weight of 16,000 pounds may use the same engine as a car weighing 3,500. Brakes, although larger, are not increased in size in proportion to the extra weight which must be controlled.

Hills, therefore, present two problems to trucks—of pulling up and of getting down safely. It is sometimes possible to rush small hills and get over them in high, and to descend moderate grades in high gear with use of the brakes.

Generally, however, the gear shift must be used. When the truck slows below the efficient torque point of the engine (usually between one third and two thirds of top speed in the gear being used), a shift should be made to the next lower gear. If the shift is delayed too long the next lower may not be able to pull so that another shift down must be made immediately.

If the truck is overloaded or the grade very steep the truck may not be able to start moving from a dead stop, so that unless gears are shifted properly it may become necessary to back the truck to the bottom again or to get help. Under such circumstances the amateur truck driver may do well to put it in low before he starts up.

Failure to shift properly going downhill may have much more serious consequences. If too high a gear is used the brakes may become overworked so that the drums heat and expand away from

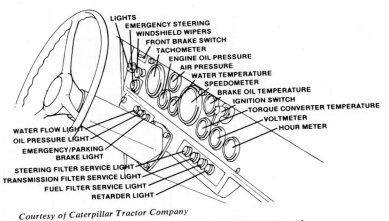

LIGHTS
EMERGENCY STEERING
WINDSHIELD WIPERS
FRONT BRAKE SWITCH
TACHOMETER
ENGINE OIL PRESSURE
AIR PRESSURE
WATER TEMPERATURE
SPEEDOMETER
BRAKE OIL TEMPERATURE
IGNITION SWITCH
TORQUE CONVERTER TEMPERATURE
VOLTMETER
HOUR METER

WATER FLOW LIGHT
OIL PRESSURE LIGHT
EMERGENCY/PARKING
BRAKE LIGHT
STEERING FILTER SERVICE LIGHT
TRANSMISSION FILTER SERVICE LIGHT
FUEL FILTER SERVICE LIGHT
RETARDER LIGHT

Courtesy of Caterpillar Tractor Company

Fig. 18-71C. Controls, off the road truck

them. If the brakes are loose, the pedal may strike the floor or require "pumping," before it can make an effective contact. If they are tight, there is still the danger that the curve of the shoes will no longer fit the expanded drums, so that not enough lining will be in contact to stop the truck. Heat glaze may destroy the effectiveness of the lining.

If a shift down is attempted when the truck is moving too fast, it will be difficult to complete both because of the speed and its mental effect, so the gears may remain in neutral. A truck going downhill out of gear is even more out of control than when in high gear.

Proper procedure is to slow to a near stop before starting down a grade and to shift into a gear low enough to hold the truck to a safe speed with only occasional braking. If the speed still increases too much, a near or full stop should be made before a shift into a lower gear.

Downhill speed should be kept within the normal range of the gear being used. If the engine is allowed to "wind up" past its maximum operating speed severe damage may result. If the truck is allowed to coast rapidly in a low gear with the clutch out, the clutch may be wrecked by centrifugal force.

Hill problems are directly related to load. An empty truck needs only a little more care than a passenger car, while a severely overloaded one may be unsafe even with the most careful handling.

Curves. Curves must be taken more slowly with a load than without. The truck should be slowed before entering the curve and accelerated or held at an even speed while in it.

When rounding a curve the centrifugal force tends to pull the truck toward the outside, and this force is resisted by the friction of the tires on the road. If traction is good the truck will tilt, compressing the outside springs and tires. Increase in load will both raise the center of gravity and increase the force tending to tip or slide the truck.

The force required to turn the truck is exerted by the front wheels, which slow the truck and cause it to nose down very slightly. The outside wheel does more than half of the turning work because the tip of the truck adds to the weight it is carrying. This wheel is therefore under the greatest strain of any part of the truck.

When brakes are applied, the forward momentum of the truck is resisted by the wheels. This transfers weight forward, compressing the front springs and relieving

Courtesy of Caterpillar Tractor Company

Fig. 18-71D. Hauling out of a mine

the rears. This increases the burden of the front wheels, particularly the outside one which may be already overloaded.

Compression of the outside front spring by both turning and braking forces throws the truck off balance and makes it difficult to control. The overloaded wheel may roll its tire, hop, or skid.

Accelerating the engine causes the rear wheels to push forward against the inertia of the truck, which tends to make the truck nose up, transferring some weight from the front wheels to the rear ones, and improving stability.

If the road is banked, the centrifugal force acts to pull the truck to the pavement so that stability is improved. Reverse banking, as in a left curve on a crowned road, will cause it to pull the truck away from the road.

Skids. In normal operation the direction in which a wheel and tire are rolling controls the course of the vehicle and considerable resistance exists against forces tending to slide the wheel sideward. This

resistance decreases if the wheel is turning either faster or slower than required by the speed of the truck. If the wheel is locked it moves sideward almost as readily as along its plane of rotation.

If the front wheels are locked while rounding a curve they will no longer turn the truck which then moves straight ahead. This type of skid most often occurs when brakes are applied too hard, but on very slippery surfaces the wheels may lock when turned sharply without brakes. When the brake is released the wheel may revolve again, or may remain locked until turned in the direction of the skid so that its friction with the road tends to revolve rather than to twist it sideward or bend it.

When the front wheels are again revolving they will substantially control the direction of the movement of the front of the truck until again locked by brakes or too sharp a turn.

If the rear wheels lock on a curve and the front ones do not, centrifugal force causes the rear of the truck to slide out-

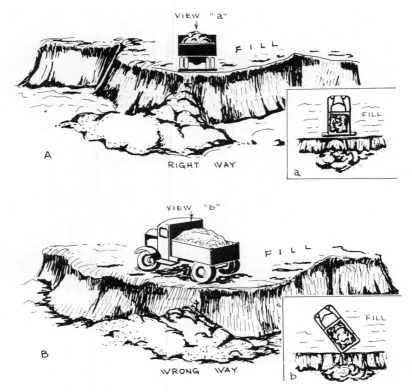

Fig. 18-72A. At the dump

ward, pivoting on the front axle. The brakes should be immediately released and the front wheels straightened so that the side pull of the turn will be stopped.

If the rear wheels have skidded too far sideward to be started rolling in this manner, the tendency of the truck to spin may be checked by steering sharply in the direction of skid. The front wheels will travel in the same direction as the back is sliding, and the truck will move diagonally sideward. The front wheels will tend to get in the line of skid, and the rear wheels to start turning. If the engine stalls it should be started immediately.

Should the rear wheels slide due to spinning, traction can usually be restored by releasing the accelerator and pressing it down gradually.

Sometimes it is more important to avoid

collision with a person or a particular object than to straighten out the skid. Also, once a truck has started to "pinwheel," straightening it out is a matter of skill and luck rather than rules. The safest system is to avoid skidding by driving slowly and cautiously when roads are slippery.

Backing. Most fills using rear dump trucks require turning and backing to the edge of the fill. If possible, a fill should be arranged so that the turn is made near the dump spot, the turn spot should be wide enough so that reverse gear need be used only once, turns in reverse should be toward the driver's left, and the truck should be level or facing uphill while dumping.

If the truck has a two speed axle or an auxiliary transmission, two reverse speeds are available. Low should be used when the load is heavy, the ground soft or rough, or when complicated steering is required.

Vision to the rear is blocked off by the body. The cab guard may be solid so as to render the rear window of the cab useless. If there is no guard, or it has eyeholes, view of the ground will still be blocked by the load or the tailgate.

A view to the left can be obtained by watching the mirror, or better by leaning out of the cab past the side of the body. Unless the body is very narrow or the cab very wide, it may be necessary to open the door and stand on the left hand running board. The gas and brake are controlled by the right foot and the steering done by both hands or the right hand alone. To stop, the driver swings back onto the seat so that he can use both feet for clutch and brake.

The driver can see to the rear of the right side of the truck in a right hand mirror, or directly by partly raising the body, to look between it and the frame.

Under many circumstances, such a view to the right is not sufficient for accuracy. The route must then be plotted ahead of time either before or while backing, so that steering can be done in reference to the left wheels with enough space left to keep the right ones out of trouble.

Whenever a truck (or a car) is turned while backing, the front and rear wheels will go in opposite directions from the centerline of the machine. It follows that a truck cannot get around curves in a driveway unless the drive is wider than the truck. The additional space needed depends on the length of the truck, the sharpness of the curve, and the skill of the driver. In backing, the front wheels tend to go off the drive on the outside, while in forward motion the rear wheels tend to run off on the inside.

Since the front wheels are steered oppositely to the direction of turn, a vehicle

Fig. 18-72B. Dumping a load

cannot back away from a wall which its side touches.

Because of restricted vision to the rear and the weight of a truck, the driver must take every precaution to avoid the possibility of children, spectators, or workmen getting behind the blind side of the truck while backing.

Dumping. When dumping off the edge of a fill, the driver should back so that both rear wheels will be the same distance from it, Figure 18-72A (A), rather than at an angle to it, as in (B). If one wheel sinks in deeper than the other, it may not be possible to either dump the truck safely or to pull out with the load.

Safe distance from the edge is determined by circumstances and the judgment of the driver. If the fill is shallow or of firm material or the truck has tandem or all wheel drive, a very close approach is

Fig. 18-72C. Dumping on a side hill

possible. Certain rear dump semi-trailers can be backed over the edge. On the other hand, it may be necessary to keep six or more feet back if the fill is soft, slippery, sandy, or otherwise treacherous. Very high fills should always be treated with respect.

Another factor is the means used to spread. If one or more large bulldozers are spreading and easily keeping ahead of the loads, the truck need not risk a close approach, but the danger is reduced as it can be pulled out readily if stuck. If spreading is done by hand or by overworked machines, dumping over the edge will speed the job but it will be less convenient to extract the truck if stuck.

When the truck is in position to dump the tailgate latch is released, the transmission power take-off engaged, with use of the clutch and a hand control, and the hoist valve placed in UP position. The engine is accelerated to a moderate speed but not raced.

Most hoists are constructed so that they can be left in the UP position to hold the body at its high position. However, if this appears to strain the mechanism the valve may be moved to an intermediate HOLD position as soon as the body is at its steepest angle.

As the body rises, the load slides backward along the floor and under the tailgate, Figure 18-72B (A). Unless right at

the edge of a bank the load will tend to pile up until it blocks the gate, as in (B). The truck is then placed in low and moved forward without disturbing the hoist controls until there is space for the remainder of the load to slide out.

If part of the load sticks in the body the truck may be backed into the pile to shake it loose. This may have to be done several times if the load is sticky or the dumping is uphill. The clutch should be released just before the wheels hit the pile to avoid shock to the power train. If the truck rolls away from the pile on the rebound the shaking out can be done without shifting out of reverse, otherwise low gear is used to drive forward a few feet and reverse to hit the pile.

If any considerable part of the load cannot be shaken out, it must be dislodged with a hand shovel or other tool. Most of the body can be reached from the pile and tailgate but the upper end may require climbing up on the body running boards. Sticking can be reduced if the shovel operator will put the driest material he has in the bottom of the truck so that it will slide out carrying the wetter dirt.

If the load includes rocks, stumps, or mats of trash large enough to have difficulty going under the tailgate, the gate should be left latched at the bottom and unfastened at the top so that it can drop down and the load slide over it. However, it is not safe to back against the pile to shake out the load with the gate down as it will be bent or knocked off. If much bulky material is to be carried, it is good practice to take the gate off and use higher sideboards so that a full load can be piled toward the front.

If the tailgate becomes twisted it may jam the upper hinge pins so that they cannot be removed. In this case, the bottom hinges should be released. This will relieve tension on the upper pins which can be removed while the gate is held in position by hand. The bottom is then relatched and the gate is lowered.

Loads put more strain on the hoist when heaped in front than when evenly distributed.

If dumping is done with the gate hinged at the top in the usual manner, and the load jams instead of sliding under it, the upper part of the body usually empties and the load piles up at the back. Conventional hoists have one-way rams which serve only to lift the body and depend on gravity to lower it. However, when the load is all in the back the body is out of balance and will remain in the dumped position until it is emptied.

The weight of dirt makes it difficult to remove the obstruction or the tailgate. Sometimes the body can be lowered by driving forward and applying the brakes sharply, with the control in LOWER. Other times it is possible to pull it down by a winch cable. Most often, however, hand tools are used to dig the dirt off the top and loosen it from the bottom until the object can be pried, cut, or smashed into a size or position which will enable it to slide through.

A stuck load may overbalance the truck so that the front wheels will rise several feet off the ground. This usually does no particular harm, but gives the driver quite a surprise.

A truck in this position will usually settle down if towed in the conventional manner by front bumper pull hooks. If it does not, the line should be attached to a front corner of the body. If the load is dislodged the truck may right itself suddenly.

The low lift also makes it difficult or impossible to dump uphill, so the truck should be on a level or faced uphill (so the load will slide downhill) whenever it can be arranged.

Trouble in dumping also comes from defects in the hoist. If it acts normally except that it will not lift to full height, the trouble is apt to be lack of oil. In this case, the body should be blocked up and oil poured into the system. A quart or two is often sufficient.

If the lift is slow and tends to creep back down if the engine is idled, the hoist control lever may have slipped. Sometimes it can be made to work properly in a different position, and at other times the linkage must be adjusted.

A loaded body should not be raised into dumping position unless the rear wheels are nearly level. Any slant will cause the center of gravity of the body to shift downhill as it is raised. The twisting force exerted may break the body hinges, tear the sub-frame off the chassis, or overturn the truck.

When the dump is completed the body is lowered, and the latch closed. This is often forgotten, with resulting embarrassment in the pit when part of the next load spills through. If the gate is sprung out of shape, or material sticks in its way, it may not latch on the first attempt.

Spreading. The load may be spread over a considerable space instead of being dumped in a pile. The procedure is to start the dump in the manner described, and when the body is high enough to begin to spill, put the truck in gear and drive forward while continuing the hoist. If low gear is used and the hoist is fast the spread will be short and heavy. If the hoist is

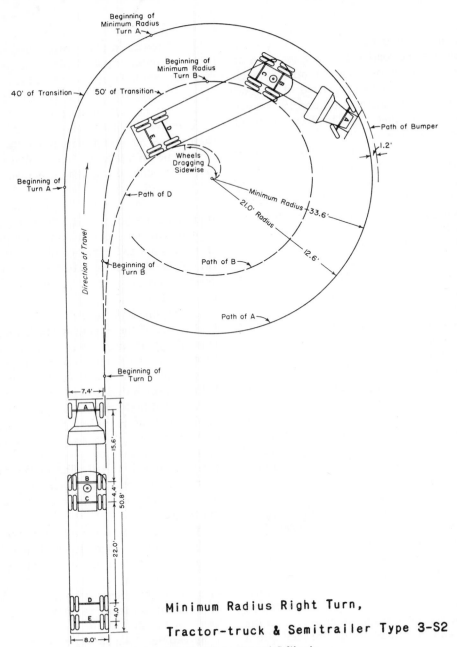

Minimum Radius Right Turn,
Tractor-truck & Semitrailer Type 3-S2

Courtesy of Department of Public Works, State of California

Fig. 18-73. Tire tracks of semitrailer

slow, or is so constructed that it can be slowed by opening the valve only partially, the material will be spread thinner.

Use of higher gears for the truck will also thin the spread. If the hoist cannot be set partially open and if the highest gear that will move the truck is not fast enough for the spread desired, the hoist valve can be opened and closed as the truck is driven forward.

This type of dumping is liable to result in a series of piles rather than a smooth sheet. If the load will flow freely and contains no lumps a much smoother spread can be obtained by chaining the bottom of the tailgate to the body so that it will open only a few inches.

When the body is raised so that the load starts to pour out under the gate in a sheet, the truck is driven forward. The body is hoisted just enough to keep the opening fully supplied, as raising it to full height might cause part of the load to slide over the gate or might overbalance the truck.

With any one material, the thickness of spread will be determined by the gate opening and the speed of the truck. A larger opening will increase the thickness, a higher speed decrease it.

Surface Dumping. Dumping on the surface of a fill is usually arranged so that it can be done without backing. Rear dump trucks may be used, although their height in full dump position makes it advisable to keep on smooth ground and move slowly. They may spread the load or dump in piles.

Such surface building is usually done by scrapers, or by bottom or side dump trucks or wagons.

The bottom dump wagon moves rapidly while dumping. The thickness of the spread is determined by the relation between the speed and the width of the door opening. The maximum height of the windrow is determined by the clearance of the wagon.

The wagon may be followed immediately by a dozer which spreads the material in a single pass, or a number of windrows may be built both behind and beside the front one. The latter way is economical of dozers but is liable to slow down the wagons and might cause upsets because of the necessity of traveling on the rough fill.

A side dump may be used on the surface of the fill in the same manner as a bottom dump wagon, or dump off an edge which is long enough to permit it to get broadside to it.

TRAILER OPERATION

Semitrailer operation differs from truck driving in backing, and in precautions needed when steering sharply, and when stopping.

Very sharp turns may be made in proportion to the length of the whole unit, particularly when the tractor is a short coupled, off-the-road type, and the trailer yoke is high enough to allow space for the tractor rear tires under it. However, turns must be made carefully to avoid overrunning and jackknifing.

In a solid truck only the front wheels are turned and the momentum of the truck pushes against them at their center of revolution. The light weight and fairly positive control of the wheels, impossibility of turning the wheels beyond their stops, and the low line of push make this a rather safe and stable arrangement.

A sharp fast turn with a semi is dangerous. When the tractor is turned the weight of the trailer pushing straight ahead tends to skid the driving wheels to the outside of the turn, increasing its sharpness, so that the unit may jackknife with the two parts, still connected, trying to move in different directions. This may do no more than cause a tangle that will be difficult to straighten without a tow, but it can result in serious damage, overturning both units and tearing the hitch apart.

Jackknifing will also occur as a result of a fast stop in which the tractor brakes are stronger or find better traction than the trailer brakes.

In rounding a curve the tractor rear wheels will track toward the inside of the curve from the front, and the trailer wheels will track still further in. Swinging turns wide enough to keep the trailer clear and avoiding stops sudden enough to cause jackknifing are special steering requirements for forward driving.

Jackknifing is a common occurrence when backing a semi, but is then frustrating rather than destructive.

When a semi is backed around a curve the trailer wheels will move toward the same side as the front wheels, and any difference in direction between the two sections increases as the tractor is backed straight. Jackknifing is more apt to occur if the trailer turns sharply and the tractor slowly, rather than when the reverse is true.

If the swivel point between tractor and trailer is any noticeable distance behind the tractor rear axle, the swivel moves in an arc as the tractor turns and accentuates the steering effect on the trailer.

In backing down a hill the trailer brakes should be released and those of the tractor applied sufficiently to create a drag. The trailer has more tendency to roll straight when pulling than when being pushed.

Full trailers are much more difficult to control than semi's. Most drivers cannot back them more than a few feet.

In general, full trailers should not be used where backing is required. Emergency measures for getting one out of a blind street include unhooking so as to tow it backward.

MACHINERY TRAILERS

CONSTRUCTION

Most excavating machines are not designed to travel under their own power on highways, particularly for long moves, and are carried from job to job on machinery trailers. These are manufactured in a wide variety of sizes and types, of which only a few can be described here.

A semi-trailer has a rigid drawbar supported by the towing truck. This may carry the direct weight of more than half the trailer and load if the wheels are in the rear, or only serve to stabilize against rocking if the axle is under the middle of the load. Full trailers are usually supported in front by a swiveling axle and connected to the truck by a comparatively light hinged draw tongue.

Any trailer can and should be equipped with brakes and lights operated from the truck or tractor cab. Even a small one can produce an unpleasant mess by jackknifing during an abrupt stop, and the side shift of its weight may put the truck out of control.

A trailer must have a deck or carrying space large enough to support or hold the machine to be carried, it must be strong

Courtesy of Rogers Brothers Corporation

Fig. 18-74. Deck trailer

enough for the job, and must provide for loading and unloading with a minimum of difficulty and danger. It should be pulled by a truck or tractor amply powerful to climb any grade it may meet, and have brakes adequate for stopping it going down.

In general, a trailer should be purchased which is of sufficient size for the biggest and heaviest machine to be regularly carried. For occasional use most trailers will take considerable overloads, particularly if oversize or extra-ply tires are installed, but frequent overloading is almost sure to result in sagging beams which are very expensive to repair. Braking effectiveness may be dangerously reduced by too much weight.

A limiting factor in trailer size is often the power of the tractor or truck that is available. It is a far too common practice to use small units to haul disproportionate loads with resultant rapid wear and breakage in the power train, embarrassing failures to get up hills when the engine is a little below par, and the danger of disastrous runaways on down grades. Big rugged trailers are heavy, particularly when in the full trailer deck construction, and their own extra weight may make a critical difference in the ability of tractors to handle the load.

Another factor is the size and weight of ramps. In deck jobs height increases as the span from rear axle to front support gets longer and as load capacity increases. Larger tires also raise the deck. Higher decks require longer and heavier ramps or more dependence on loading from a bank. A crane or loader, or two to four men, may be required to place and remove ramps to a high trailer.

Maneuverability may be critical. From this standpoint semitrailers are much to be preferred. When in small sizes an experienced operator can spot them almost as readily as he can the truck. Large ones may find themselves in difficulties in country roads and residential districts.

1. Bearing plates   4. Walking beam
2. Bronze bushings   5. Wheel bearings
3. Lube reservoir   6. Brake shoe
        7. Detachable drums

Courtesy of Rogers Brothers Corporation

Fig. 18-75A. Walking beam suspension

A small trailer is much cheaper than a big one to buy, maintain, license, and insure, it gets around more easily, does not require as much towing power and weight, and can be more readily equipped with a low deck, or the means for reducing or eliminating need for ramps. Under some circumstances it may be good practice to own one of them and to hire a big one for infrequent use. At other times, the small unit may pay its way by saving a big one from being worn out running around carrying small machines.

Deck Trailer. Figure 18-74 shows a medium weight trailer in both level deck and

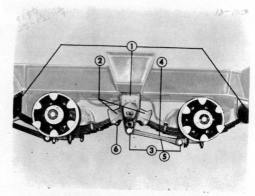

1. Steel beams   4. Torque arms
2. Oscillation   5. Adjustable arm
3. Rubber bushings   6. Wear pads

Courtesy of Rogers Brothers Corporation

Fig. 18-75B. Spring suspension

Courtesy of Rogers Brothers Corporation

Fig. 18-76. Tandem wheel oscillation

drop deck models. Capacity range is 15 to 40 tons.

The trailer proper consists of a girder frame supporting a flat wood deck, usually hardwood planks two inches thick. Standard width is 8 feet. It is supported at the rear by four dual wheel sets (8 tires), and at the front by a welded gooseneck.

When the unit is used as a semitrailer, the gooseneck is attached to the "fifth wheel" of a tractor truck. For full trailer operation, it rests on a dolly.

The wheels may be mounted on a pair of walking beams, left and right, to allow front to rear oscillation to maintain road contact and reduce jolting. The tops of the tires are even with the top of the deck on level ground. Cutaway deck sections permit them to go higher as they oscillate.

In another design, there are two axles in tandem, with leaf springs and torque arms.

Wheels are fifteen inch. Tire sizes are 8:25 up to 11:00, 10 to 16 ply. Level deck height may be 28 to 36 inches, differences depending mostly on tire size. Drop decks are 6 to 9 inches lower.

Air or vacuum brakes are used on all four hubs. These are controlled from the truck cab through piping and flexible hoses

Courtesy of Rogers Brothers Corporation

Fig. 18-77. Trailer sub-assemblies

with snap-on couplings. They are designed so that if the hoses become disconnected, the trailer brakes will lock on automatically to prevent a runaway. They can be released by bleeding the air or vacuum tank, or may release very gradually by leakage.

The dolly consists of a towing platform, spring-mounted on a single axle. Two perforated discs are used in the swivel connection to the gooseneck to hold grease, to increase bearing surface, and to eliminate sway or whip.

The draw tongue is hinged to the base of the platform. A pair of breakaway chains are attached to the frame, and can be hooked to the back of the truck to hold the trailer if the tongue should become disconnected.

If the trailer is to be used both with and without a dolly, the dolly can be equipped with a tractor-type fifth wheel (bottom right in illustration) instead of the standard turntable. The tongue is then fitted with a bracket so that it can be made to hold the dolly upright during the change.

Dollies are also available with two rocking axles, each carrying two or four tires.

Level deck construction is shorter, provides more unencumbered deck space, and is somewhat easier to load and unload from the back. The drop deck lowers the overall height of the loaded machine six inches or more and is easier to load from the side, but reduces road clearance.

The sloping deck section behind the wheels, called a beavertail, may be standard or optional equipment. It reduces ramp requirements for loading from the rear. It has a bracket on its rear face, usually full width, to hold metal angles on the ends of ramps. If there is no tail, a similar bracket is built into a rear bumper.

Heavier trailers may have three or more axles. The model shown in Figure 18-78 is convertible between two and three. The rear and the frame and deck above it are hinged to the main frame. In working position the

Courtesy of Rogers Brothers Corporation

Fig. 18-78A. FLIP-UP third axle

rear axle carries its full share of the load. When not needed, it can be unlocked, then swung up, pivoting on top pins, so that it lies upside down on the deck. It can be disconnected and removed by pulling the pins and uncoupling the brake and light lines.

Reduction of overall length in this way reduces the need for or cost of over-length permits, makes the unit more maneuverable, and reduces tire wear.

Detachable Gooseneck. Some meduim to heavy deck trailers are made so that the

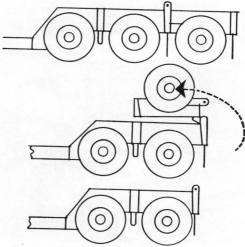

Courtesy of Rogers Brothers Corporation

Fig. 18-78B. FLIP-UP axle changes

Courtesy of Transport Trailers

Fig. 18-79A. Detachable gooseneck

gooseneck can be removed, leaving the front of the frame resting on the pavement. Machinery can then be loaded readily using very short ramps (usually hinged to the deck) or small blocks.

The gooseneck and frame are held in alignment by removable pins or other safety locks. With these released, and brake and electric lines opened, the two units are lowered to the ground by a hydraulic jack in the gooseneck, or by a line from a winch mounted on the tractor. The gooseneck is then detached from the frame, and carried or dragged a short distance by the truck.

Ramps are flipped over to rest on the ground, and the machine driven up onto the

Courtesy of Talbert Trailers

Fig. 18-79B. Extra-heavy trailer with multiple dollies

Courtesy of Rogers Brothers Corporation

Fig. 18-80. Versatile gooseneck

trailer. The gooseneck is backed into place, attached, lifted, and locked, lines are reconnected, and the ramps are folded back onto the deck.

The Rogers Croucher, Figure 18-80, also uses the hydraulic control in the gooseneck to lower the front of the trailer to get a high load under an obstruction, or to raise it to increase clearance over ground rises. For tire changing, a support may be placed under the center of the trailer, and the wheels raised clear of the ground by down pressure on the gooseneck.

Girder or I-Beam Trailers. It is often not practical to transport shovels on deck trailers because the combined height of trailer and shovel is too great to go under

bridges on the route. This difficulty may be reduced by the use of a girder type trailer which supports the frame of the shovel and allows the tracks to hang within a few inches of the ground.

Girder construction also eliminates the necessity of using long ramps, which may be so heavy as to require a crew of men to handle them, and which are dangerous to use when slippery.

The tracks of a shovel on a girder trailer hang very low and may rest on the road on sharp humps or hill crests. If the tracks are not locked, they will reduce resulting drag by turning. If the machine hangs up badly it can be walked forward while the trailer is towed at the same speed until

Courtesy of Talbert Trailers

Fig. 18-81. Girder trailer

Courtesy of Alfred Stauffer Machine Works

Fig. 18-82. Low bed trailer

the trailer frame will raise the tracks clear of the ground again.

Loose tracks can be prevented from hanging below the rollers by raising the upper section with a jack on the truck frame, or by working pieces of wood in between the track and the bull wheel by turning it.

Occasionally the tracks must be removed to obtain sufficient clearance from the ground when the machine is carried in the conventional manner.

Low Bed Trailers. Figure 18-82 shows two sizes of the Stauffer low bed semi-trailer which is used to carry machines weighing ten tons or less. In order to keep the overall width down to eight feet the deck must be limited to a width of six feet four inches.

In the lighter model the wheel stub axles and brake back plates are welded directly to the deck rails. The tandem wheels in the larger size are mounted on rocker arms allowing front and rear oscillation.

The draw tongue is adjustable for height,

and is fastened to the truck by a vertical pin through a bracket on the rear cross-member. Connection is made by jacking the trailer and tongue to the required height, backing the truck so that upper and lower holes in its tow bracket line with the tongue hole, and then placing and locking the pin. The pin is a loose fit in the hole so that the trailer can tip somewhat without binding on it.

A built-in hydraulic jack can be obtained as optional equipment. Ordinarily the trailer is blocked up when left so that it need not be jacked to reconnect.

Deck height is fourteen inches. Length is fourteen to sixteen feet. Load is about 60% on the pintle hook of the towing truck.

Ramps are not required for loading crawler machines and very short ones for wheeled vehicles. A pair of blocks placed behind the trailer will cause tracks to rear up so as to pass over the rear edge of the deck in loading, and will ease them down in unloading.

Courtesy of Transport Trailers

Fig. 18-83A. Loading a tilt trailer

Tilt Trailer. In a tilt trailer the deck is hinged to a subframe that connects the gooseneck or drawbar to the axles. The deck can be tilted so that the back edge rests on the ground. A machine being loaded can walk directly up the sloped deck as if it were a ramp. As its center of gravity passes the hinge, the deck automatically goes back to level position.

To load one of the simplest and smallest of these units, the trailer is attached to the truck, the deck lock is released, and the front of the deck is pulled up by hand until the rear rests on the ground. The machine is then driven from the ground directly up the sloping deck. When the machine has moved past the hinges, the deck will swing into carrying position and will lock automatically.

The machine must be moved very slowly as it starts to overbalance the deck, which is liable to come down so rapidly as to strike a heavy blow on the lower frame, and to subject the towing truck to shock.

Damage to the truck can be prevented by supporting the tongue on blocks or a standard during loading.

If blocks are jammed under the tongue, moving the machine to the back of the trailer, with the deck locked, will release them. For transportation, the machine may be moved well forward so that the deck will remain level even if the lock fails.

Most modern tilt trailers are equipped with a hydraulic ram or other snubbing device, so that the deck will move smoothly from loading to travel position. Heavy units may start the tilting motion by hydraulic power. Otherwise, they may be tilted by standing on the rear.

Large tilt trailers (maximum capacity about 25 tons) may have regular gooseneck attachment to a dolly or fifth wheel. Smaller units have a rigid two bar attachment to a special bracket on the rear of dump or other truck. The bar usually includes a jack for raising it to the truck, and for supporting it during parking and loading.

There may be a full width solid deck over the wheels, requiring a deck height of 34 to 41 inches. Slope in loading position may be 15 to 18°, too steep for a weak machine or a slippery deck. This angle can be reduced

Courtesy of Transport Trailers

Fig. 18-83B. Narrow tilt deck

by putting blocks under the edge before lowering it, and putting blocks or very short ramps behind it.

Another construction uses a narrower deck, a little over 6 feet wide, on a low frame between the wheels. This reduces the loading angle, and lowers the center of gravity during transport, but restricts the trailer to carrying small machines.

Tagalong. The Rogers Tagalong semitrailer, which is no longer made, has a deck height of 24 inches. Four full-size truck tires on a fixed axle are used in the rear, and loading is done at the front or sides.

This trailer is designed for use with dump trucks. Its tongue is fastened to a vertical swivel pin welded into the truck frame behind the rear axle. It is raised to connect with the pin by the mechanism shown in Figure 18-84. A cable attached to the underside of the dump body is looped around a pulley on the trailer tongue, then anchored on the truck frame. Raising the body draws the tongue up, after which it is fastened with a horizontal connecting pin. The trailer is disconnected by raising the body until the cable takes the weight off the pin, withdrawing the pin, and lowering the body. The cable is then pulled off the tongue so that the truck can be driven away.

When not in use, the cable lies under the body, and occasionally may be in the way of the body beams or dump mechanism. A light cable on a spring wound reel may be installed by the owner to take up undesirable slack.

Loading is done with blocks, or with a pair of short ramps, with the tongue resting on the ground. When the machine is positioned on the trailer, the truck is backed into position and the tongue is picked up by the cable and pinned. If the body hoist will not pick up the loaded trailer, jacks must be used.

For any machine maneuverable enough to turn on the deck, it is more convenient to load from the side with the trailer attached to the truck. The deck is still low enough to make it possible to use the short ramps or blocks. This trailer will tip seriously as the machine is turned unless blocked up under the front corners.

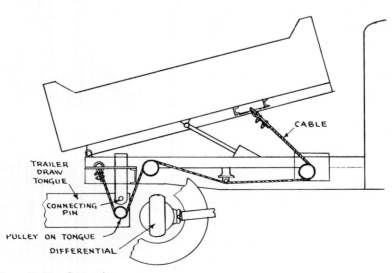

Fig. 18-84. Trailer lift by body cable

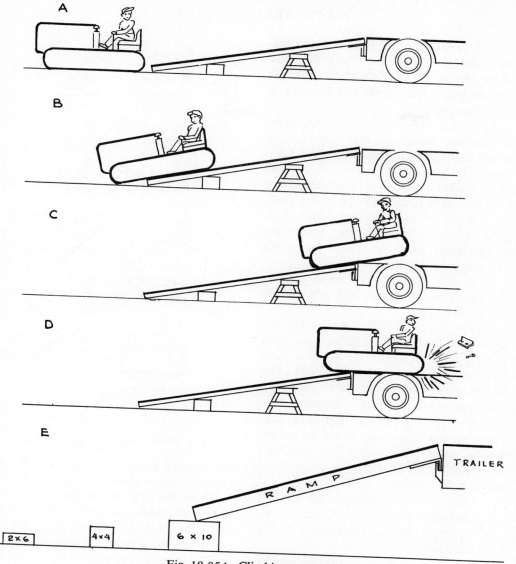

Fig. 18-85A. Climbing a ramp

LOADING

Rear Loading. It is standard practice to drive self-propelled machinery onto deck trailers over the back with the aid of ramps, blocks, or banks, or combinations of them.

Ramps. Ramps may be made of planks of oak or other strong wood, two to five inches thick (three is usual), and ten to sixteen inches wide. A metal angle is fas-

tened under one end of the plank, and rests in an angle on a shelf on the rear of the trailer.

Ramps may also be made of steel box or channel construction, or of wood reinforced with steel plates or angles.

They are ordinarily so heavy that it takes at least two men to handle them, so that it is often advantageous to place them with the machine that is to be loaded. However, they are usually not strong enough to take the weight of a heavy machine unless supported between the ends. Short pieces of heavy timbers, railroad ties, or heavy saw horses can be used to block them up.

In loading crawler machines, the point of greatest strain is near the ground where the tracks are forced upward into a climb. In unloading, bracing is particularly needed where the machine strikes the ramp as it pitches off the trailer deck. In addition to these special points, it is good policy to have blocking under one or two more points to reduce the length of unsupported spans.

Thorough blocking will make possible the use of lighter ramps and add materially to their life.

Ramps supplied with a trailer are designed to load most machines on level ground. Length should be at least three feet for each foot of trailer height. The incline should be considerably less than the machine is capable of climbing.

Rubber-tired machines impose a concentrated but fairly even strain on ramps. Rollers must have very slight gradients and need help from dirt piles or banks to get on a deck trailer.

Crawlers may be loaded either forward or backward, but are under better control when backing uphill, particularly if the tracks are loose.

Ramps are set to line with the tracks or wheels of the machine, which is placed so that it can move up them without turning. If the ramps are wider than the tracks, they are set so that the operator can see an edge and steer accordingly. If they are narrow so that he cannot watch them, a competent man should stand on the trailer or the ground and guide the operator by signals.

The machine should be in its lowest gear and should move slowly, but with sufficient throttle to avoid any chance of stalling. It should not be steered sharply

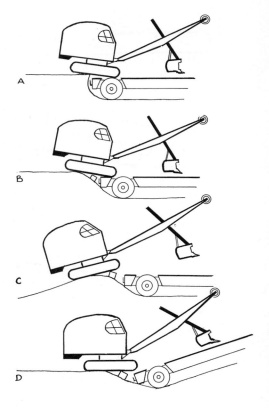

Fig. 18-85B. Loading from a bank

nor stopped except in emergency. A loading sequence is shown in Figure 18-85A.

When the top is reached, the upper end of the tracks will move upward above the tires and deck, and fall when the machine overbalances. A jarring fall can be avoided by moving very slowly. A sloping extension on the rear will provide an intermediate grade between the ramp and deck which reduces the abruptness of the fall.

If the ramp is short it can be extended by blocking, as in (E).

Wet ramps are dangerous and should be sprinkled with sharp sand. Steel surfaces may be too slick to climb even when sanded. Machines should never be loaded up ramps when there is snow or ice on the ramps, the deck, the tracks, or the ground where the machine is standing.

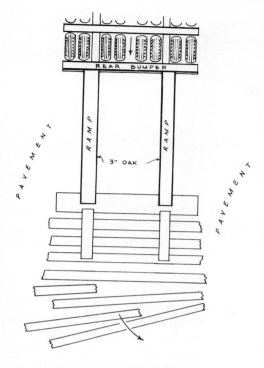

Fig. 18-86. Protecting pavement

Since the machine must walk over the tires after leaving the ramps, tracks should be inspected and any sharp pieces of metal removed. Grousers will not damage tires, but some varieties of ice cleat might when new.

If the trailer can be backed against a bank, or into a deep gutter, it may be possible to load it without ramps, and perhaps even without blocks. Figure 18-85B shows four situations in which ramps are not needed. Lower banks or shallower gutters may be useful in making it possible to ramp up at an easier gradient than from a level.

These trailers are also used extensively for carrying bulky objects such as tanks, machine parts, small buildings, and pipe.

Picking up and removing "dead" loads, including disabled machinery, is greatly simplified by a rear mounted winch on the tractor.

Pieces which are wider than the trailer

may be jacked and blocked up, the trailer backed under them, and the blocks knocked out.

Protection of Pavements. When crawler machines are loaded or walked on paved streets, the road surface must be protected. Damage can be caused by side scuffing during turns, and by digging in of grousers.

Crushing is particularly likely on oiled gravel when the subgrade is wet, or on any blacktop in hot weather. It is guarded against by laying planks or boards as in Figure 18-86, to spread weight over a wider area. Note the extra boards where the front of the tracks tends to dig in just below the ramps.

In many localities, it is unlawful to walk crawler machines on streets but nothing is said if no visible signs of the move are left. Bruising of pavements, and scuffing on turns, can be prevented by laying thin boards under the tracks. A single thin board, narrower than the track, may protect hard pavements, while softer ones may need wide planks, or narrow planks in pairs.

If the walk is long, the same boards can be used repeatedly.

Digging in of grousers may be prevented by walking on boards in the same manner.

Side Loading. It is often more convenient to load from the side. Normal curb and sidewalk heights above the gutter permit use of shorter and lighter ramps. Drop deck models are lower at the side than at the back. In addition, the machine may have to enter or leave the job through a driveway or other narrow access. By side loading, it can be walked directly into the drive, where rear ramps will put it on the street pavement and necessitate a right-angle turn, often under tight conditions. Also, it is frequently possible to pull up alongside a roadside bank where backing into it would not be practical.

Trailers are not ordinarily equipped with ramp-supporting rails on the sides, but they can be welded on easily. There is usually enough recess in the girders to do this without increasing width.

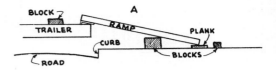

Ramps can be laid on the edge of the deck, Figure 18-87 (A), but this is not recommended since they may be kicked off by the movements of the machine. Ramps can be supported on blocks, (B), or blocks used alone, (C).

Unless it is blocked up the trailer will tip sideward under the weight of the machine as it climbs on. This simplifies loading from blocks or a low bank, and may make it possible for a large tractor to walk on from ground level, pulling the trailer down with its grousers. But tipping may make unblocked loading dangerous or impossible if the deck is slippery with rain.

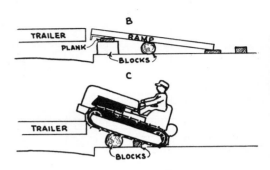

Fig. 18-87. Side loading

The machine can be carried crosswise or turned on the deck. If the machine is large in proportion to the deck space, turning may be difficult or impossible. Machines narrower than the deck can turn rather readily unless the planking is wet and slippery.

The front corners of the deck should be blocked up before turning, but the blocks must not be so high that the deck will rest on them when loaded.

Transporting. The amount of load on rear trailer tires, and on those of the dolly or tractor, can be adjusted by placing the machine toward the front or rear of the deck. Placement will also be affected by the need to rest the bucket or blade, and the direction in which the boom is to be carried.

No part of the load should rest over the tires unless high enough to allow them to oscillate without hitting.

A dipper stick bucket may be tucked under the boom on the deck, rested on the gooseneck or between the rear tires, or, if the tractor is a dump truck, in its body. Long booms are usually faced to the rear and carried low, with the bucket or hook resting on the deck. It may be

necessary to have the operator stay with a large shovel in transit to raise, lower, or swing the boom to avoid obstacles.

When the machine is correctly positioned, it is left in gear with the clutch engaged and the brakes locked. The ramps and blocks are loaded onto the deck and the machine secured against sliding by blocks or chains, or both.

Chains may be fastened to the four corners of the machine, and to the "D" loops on the trailer sides. At least two of these chains should be tightened and locked with load binders. Sometimes a machine can be lashed by passing a chain from a loop through or over it to a loop on the other side, and tightening it.

Blocks should be placed at each end of both tracks. The ramp blocks or the ramps themselves can be used, but triangular blocks or ones curved to fit the track or tire are much more effective. These may be wedged in place, or nailed down lightly.

Many machines are moved without any chaining or blocking, particularly on short trips. Crawler tractors tend to stay put very well. Shovels, because of their height, are less stable. Rubber-tired vehicles tend to bounce and creep.

Courtesy of Huber Corporation

Fig. 18-88. Portable roller hitched to truck

Blocking is recommended for all machine transportation, since a minor accident may become a serious one if it slides a heavy unit off a trailer. In some places, truckers are required to block and chain trailer loads.

In most states, truck and trailer widths are limited to eight feet. Wider vehicles must obtain special permits, which may be limited as to route and time of day, and which may also be void while streets are wet or icy. In practice, many localities do not enforce width restrictions for local moves. However, if an accident occurs to an over-width vehicle lacking a permit, the operator may find himself in a difficult position.

Drivers of full trailers should be sure of their route so as not to get into blind streets. The average driver cannot back such a rig any distance, or turn it around in restricted quarters. It may be necessary to disconnect the towing truck, turn that around, and tow the trailer out backwards. One or two men might be able to steer the trailer by manipulating the drawbar. If the load is a revolving shovel, it can pick up the tongue with a chain and steer by swinging. A jeep or wheel tractor following the trailer can sometimes manipulate the drawbar successfully.

Another method of getting out of a blind street is to unload the machine and drag the empty trailer sideward until it faces the exit.

Trailer routings must also be planned carefully to avoid underpasses too low to get through, and bridges too weak to cross over. Long low trailers can also get hung up crossing banked highways or railroad tracks and on sharp hilltops on country roads.

Trucks. Machinery may also be transported on trucks with either flat bed or dump bodies. This gives better maneuverability but involves the use of longer and heavier ramps or dependence on banks. It has the disadvantages of labor for handling the ramps, danger in using them, and of greater height of the machine when loaded.

When a platform body is used, it is helpful to extend and curve it down at the rear. This decreases the ramp length or bank height required, and increases stability by reducing the angle between deck and ramp.

A dump body can be extended by making braces to support the tailgate. These braces should be made to slide or swing out of the way when not in use, to allow hanging the tailgate down if desired. Tailgate clamps cannot support machines safely.

Hoist Hitches. Rubber tired machines are often transported by raising either the front or the rear and attaching it by a swivel to the rear of a truck or trailer.

If the front wheels are on the road they must be securely fastened to roll straight.

Serious damage may be done by towing with the rears on the road and the engine stopped. Tractor transmissions may depend on turning of the clutch (input) shaft for lubrication, and this does not turn with the rear wheels in neutral.

Towing speed may be so much faster than that normal to the tractor that power train parts may burn out or even fly apart.

Machines with automatic transmissions must not be towed with drive wheels on the ground unless the drive shaft is removed.

Cleated tires wear irregularly on hard pavements, and may overheat and blow at ordinary highway speeds.

The usual way to make connection is to place a pair of inclined blocks behind the truck, drive the machine up on them until the hitch can be locked. The truck is then backed a few inches and the blocks are picked up. Unhooking is done by pulling the machine up on blocks.

The hitch may not allow a sharp angle between truck and tow, so caution must be used in making turns and in backing.

A small steel wheel roller may be equipped to convert into a semitrailer by hitching a drawbar to a truck's pintle hook, then forcing down a pair of rubber tired wheels to raise the steel rolls off the ground.

CHAPTER NINETEEN

GRADING AND COMPACTING MACHINERY

GRADERS

MOTOR GRADER

The grader, motor grader, or motor patrol, is a machine used principally in shaping and finishing, rather than in digging or transporting. It is available in sizes ranging from 30 to over 600 horsepower, in a variety of constructions.

Except for a few light models that are rear-mounted tractor accessories, a grader consists of a wide, controllable blade (moldboard) mounted at the center of a long-wheelbase, rubber tired prime mover. The blade is usually held at an angle to travel direction. It will cut, fill, and sidecast. These functions can be performed separately or at the same time, in any combination.

A grader may be used as a prime mover or carrier for a number of accessories. These include snow plows and side wings, belt loaders, arm-mounted clearing tools, bank slopers, trailer drags, and disc harrows, in addition to scarifiers or rippers.

HEAVY GRADER

Most heavy graders are tandem drive units of the type shown in Figure 19-1. The diesel engine, with 125 to 340 horsepower, is mounted at the rear, and drives four single wheels in tandem pairs through gears and chains. The frame connection to the front axle is long and high, to allow space for carrying and manipulating the moldboard or blade under it.

The central location of the blade, and the long wheelbase of the machine, provide natural stability for the cutting edge. Smooth-acting, multi-directional control provides precision finishing ability.

Weight range is from 13 to 30 tons. Lighter and heavier grading machines will be discussed in other sections.

Clutch. In many types of grader operation, the engine clutch is used frequently under heavy load. It is required to be always smooth in operation. Oil cooked discs are commonly used.

A clutch may be retained even when the machine is equipped with torque converter.

Transmission. Graders must use a drop type transmission, with two or more sets of parallel shafts, as the drive axle is under the engine.

These units need slow, powerful gearing for heavy or precise work, moderate speeds for lighter or less fussy jobs, and also travel

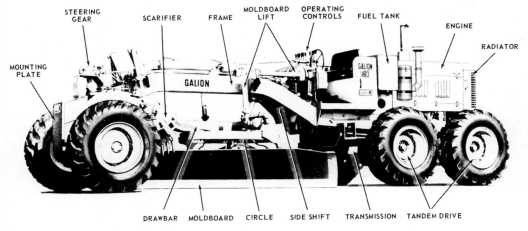

STEERING GEAR · SCARIFIER · FRAME · MOLDBOARD LIFT · OPERATING CONTROLS · FUEL TANK · ENGINE · RADIATOR

MOUNTING PLATE

GALION

DRAWBAR · MOLDBOARD · CIRCLE · SIDE SHIFT · TRANSMISSION · TANDEM DRIVE

Courtesy of Galion Manufacturing Company

Fig. 19-1. Heavy grader

speeds up to 20 to 35 miles per hour. There must also be a choice of reverse speeds, as some heavy work may be done backing up, but most reversing is for return travel only, and should be brisk.

There may be 6 to 9 forward speeds, and 2 to 9 reverse, in either direct drive or power shift. There may be options of up to 3 additional speeds. Most of the transmissions are compound, with shifting done by one, two, or three levers.

Figure 19-2 shows a direct drive, easy shift transmission with helical gears, with 8 speeds forward and 4 in reverse, with an op-

tion of three additional very low (creeper) speeds. Figure 19-3 is a torque converter and power shift unit, with four speeds each way, which is optional on the same machine.

A 9-speed compound power shift pattern is shown in Figure 19-4.

If there is a converter, it is usually single stage. It may be equipped with an output shaft governor, to keep ground speed uniform in spite of changes in load. There may be an input clutch in the converter, or a modulating pedal to momentarily disengage the transmission (and power to the wheels) while moving the blade. The clutch is also used for small, exact machine movements.

Hydrostatic Drive. The CMI tandem graders use an all-hydraulic drive system that replaces clutch, torque converter, transmission, and a parallel shafting. There is a fixed displacement piston pump driven by the engine, which supplies power to a pair of motors geared to a 2-speed rear axle. One of the motors is fixed displacement, the other variable, with a series hook-up that provides the operator with an infinite choice of power-speed ratios.

Control is by a hand lever engine throttle, and a rocker foot pedal regulates direction, drive ratio, and ground speed.

The axle is shifted by a hand lever. It

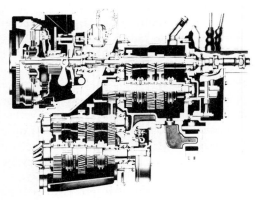

Courtesy of WABCO

Fig. 19-2. Transmission, mechanical drive

provides a WORK RANGE and a TRAVEL RANGE.

Power Train. Figure 19-5 is a top rear view of the rest of a power train in a typical transmission-equipped grader. A bevel gear on the transmission output shaft is meshed with a bevel on an intermediate shaft, which turns the inner, live, or drive axles through spur gears. There is usually no differential.

The live axles each have a pair of sprockets near the outer ends, from which separate roller chains turn outer or wheel axles to front and rear. There may be a set of planetary reducing gears (final drive) inside the drive sprockets. The wheel hubs are keyed to the outer axles. Gear drive may be used instead of chains.

The tandem drive case carries the outer axles, supports the weight of the machine, and provides protection and lubrication. It is pivoted on the inner axle housing, to permit both wheels to follow ground contour.

Tires. Tires are special heavy duty design for grader work. They are preferably the same size all around, but front ones may be smaller and/or lighter. Traction tread is usual, but the fronts may be ribbed.

Size range is from 13.00-24, 10 ply, up to 26.5 x 25, 16 ply.

Brakes. There are usually brakes on each of the four drive wheels, but none on the fronts. They may be either shoe or mul-

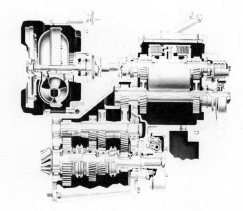

Courtesy of WABCO

Fig. 19-3. Transmission, with torque converter

tiple disc design, and are applied by booster-assisted hydraulic pressure. Or there may be a single disc brake in the transmission, acting through the four wheels.

Older machines may have brakes on two wheels only. These may be inadequate for job-to-job speeds, so that extra precautions in down-shifting or even lowering the blade to drag may be needed.

The parking brake is usually on the lower transmission shaft, on the end opposite the drive pinion, and is operated by a hand lever.

Throttle. Basic governed engine speed is set by a hand throttle, which may be a lever on the console.

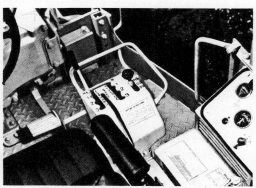

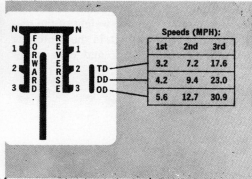

	Speeds (MPH):		
	1st	2nd	3rd
TD	3.2	7.2	17.6
DD	4.2	9.4	23.0
OD	5.6	12.7	30.9

Courtesy of Caterpillar Tractor Company

Fig. 19-4. Power shift

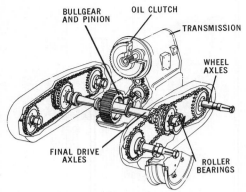

Courtesy of Caterpillar Tractor Company

Fig. 19-5. Grader power train

There is usually an accelerator for speeding the engine beyond throttle setting, and a decelerator for slowing it. These may be combined in one rocker pedal, providing acceleration for toe pressure and deceleration for heel pressure.

Frame and Front Axle. The rear of the frame is a pair of beams that support the engine and the power train. Forward of the operator's station these slope upward and converge into a single beam of reinforced box cross section. This may slope down to the front axle hinge, or end in a column above it.

The front of the frame may be a mounting plate for front end accessories, and/or project forward far enough to serve as a narrow bumper.

The front axle is compound. The lower section usually carries the weight of the machine, may be arched high in the center for clearance over windrows, and ends below

Courtesy of WABCO

Fig. 19-6. Grader frame

the wheel spindles. It oscillates on a central pin, and is hinged to wheel spindle brackets just below the kingpins.

The upper section, of lighter construction, is substantially a straight bar (lean bar) hinged to the tops of the hub brackets. It can be moved from side to side by a hydraulic cylinder, or by mechanical means. When moved off-center, it causes the front wheels to lean sideward. The angle may be as much as 16 degrees.

Wheels are leaned to increase their resistance to sliding sideward on the ground because of load on the blade, or steering stresses.

Courtesy of Fiat-Allis Construction Machinery, Inc.

Fig. 19-7 Leaning front wheels

Steering may be mechanical with a booster, or by a hydraulic cylinder located behind or above the axle, and operating through a more or less conventional linkage. The cylinder control is equipped with a follower valve that keeps a direct relationship with the position of the steering wheel. Special arrangements may be made to avoid interference between leaning and steering.

In the CMI double parallelogram axle, both sections are hinged at the center. The weight of the chassis is carried on a pair of hydraulic cylinders running diagonally down to the lower section. Extending them raises the frame, together with the circle and blade, retracting them lowers these parts. See Figure 19-12.

This permits adjusting the whole blade upward and downward without disturbing its cross slope, or using the individual lift cylinders. It simplifies the use of the automatic grade control to be discussed later.

BLADE AND MOUNTING

Blade. The blade is the principal working part of a grader. Most of it is a curved piece of steel called a moldboard. The distinction is chiefly that "moldboard" emphasizes its function in causing dirt to roll and mix as it is moved.

Standard blade (moldboard) lengths in tandem graders are 12 and 14 feet, with some 10 footers in small machines and 16's in large ones. Two-foot extensions are sometimes added for light work or long side reach. Height is usually 24 inches, but may be as much as 31.

The cross section is such that dirt being pushed has a rotary movement, rising at the bottom and falling forward at the top. This characteristic, combined with the usual sideward drift caused by working it at an angle,

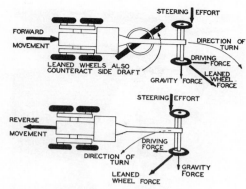

Courtesy of Caterpillar Tractor Co.

Fig. 19-8. Leaning front wheels action

makes spillage over the top very unusual.

The bottom is fitted with a removable and reversible wearing edge, with separate pieces (end bits) at the corners.

The blade is supported and held in position by a pair of heavy curved brackets, called circle knees. They are attached to the underside of a rotatable ring, called a circle.

Tilt (Pitch). The blade may be fastened to the knees by a hinge bar at the bottom, and a curved slide at the top. This construc-

Courtesy of Caterpillar Tractor Company

Fig. 19-9A. Grader moldboard, from rear

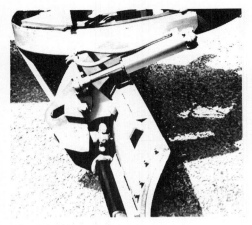

Courtesy of WABCO

Fig. 19-9B. Moldboard with power tilt

tion permits changing the tilt (forward lean) of the moldboard.

The adjustment may be manual, with clamp nuts to be loosened and then re-tightened, as in Figure 19-9 (A), or by hydraulic cylinders controlled from the operator's station, as in (B).

The blade is usually kept near the center of its tilt adjustment. Effects of changing it are discussed later, under Operation.

Blade Side Shift. Many grader operations can be made more efficient by extending the blade a greater than normal distance to the side. This move can usually be made with the circle controls, but they disturb the side-to-side slope. It is often better or easier to shift the blade itself, by sliding it along the knees. For extreme reach, both the circle and the blade adjustments may be used.

For manual blade shift, clamp nuts or lugs are loosened, one end of the blade is placed in the ground or against an obstruction, and the circle or the grader moved so as to push against the blocked end. The blade should slide. When it has reached its desired position, the clamps are tightened or the lugs are reset.

With power sideshift, the move is made by holding the blade clear of the ground and obstructions, and sliding it sideward by means of the control valve for its hydraulic cylinder.

Drawbar and Circle. The drawbar is a V-shaped (or sometimes T-shaped) connection between the front of the grader frame and the circle. It is rigid with the circle, and fastened to the frame by a ball and socket, or a universal fitting, that allows limited angular movement from side to side and up and down.

The drawbar carries the full horizontal

Courtesy of WABCO

Fig. 19-10A. Circle

Courtesy of Caterpillar Tractor Company

Fig. 19-10B. Bottom of circle

Fig. 19-11. Circle lift, direct hydraulic

load on the blade, as other connections provide only vertical and side support.

The circle is a toothed ring that is rotated in or on a supporting frame by the circle turn (circle drive or circle reverse) mechanism. Pads and shims provide for low-friction movement, and for adjustment of clearances.

The circle is turned by a spur pinion gear meshing with teeth cut all around it, usually in the inside. The pinion may be turned by a shaft driven either by the engine power takeoff or by a hydraulic motor, or by a direct-mounted hydraulic motor and reducing gears. In each case, movement is controlled by a lever in the control console, at the operator's station.

It is important that the blade angle, which is controlled by circle rotation, be capable of exact adjustment, and that it remain in place during work, without drifting or creeping.

Means to keep it locked against hydraulic creep include drive through an irreversible worm gear, automatic clamping brake on the shaft, or clamping the circle itself. The clamps are applied by springs, and released by hydraulic pressure whenever the control lever is used.

Rotation in full size graders is usually continuous in both directions. However, a full circle can be turned only when the blade is in such a position that it will not strike the ground or any part of the machine as it is turned. Some machines have very limited clearances, in others they are generous. Scarifier teeth must usually be removed.

In general, for rotation into reverse blading position, the blade should be level, a few inches above the ground, and approximately centered.

Lift. The blade is lifted and lowered through the circle by a pair of arms or cylinders, supported by the frame and fastened to the sides of the drawbar at the rear. These are separately controlled, but are influenced by each other.

Direct lift by a pair of two way cylinders, Figure 19-11, is superficially the simplest method. But cylinder rods are subject to change of position caused by seepage of fluid past the piston, or in the valve. Such creep can be prevented, or compensated, but corrective devices involve loss of simplicity.

The most usual lift is a pair of arms hinged to shaft-mounted levers or cranks, Figure 19-10A. The shafts are rotated by mechanical drive, or by hydraulic motors with locking devices similar to those used for the circle reverse.

Range of possible vertical motion in a

Courtesy of CMI *Corporation*

Fig. 19-12. Axle lift for blade

crank-controlled shaft is limited by crank length, which in turn is restricted by available space. It may therefore be necessary to extend or shorten the shaft with a telescopic adjustment for extreme blade positions. This is done by taking weight and pressure off the shaft, pulling a pin or loosening a clamp, sliding the parts to their new position by power, and re-locking.

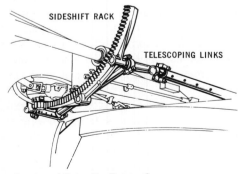

Courtesy of Caterpillar Tractor Company

Fig. 19-13. Mechanical circle sideshift

Hydraulic cylinders can usually be made long enough to take care of full range of position without manual adjustments.

The CMI graders have a pair of hydraulic lift cylinders for conventional control of the two sides of the blade. In addition, a pair of cylinders based on the front axle raise and lower the whole front of the machine, moving the circle and blade a somewhat smaller vertical distance.

This front lift has no effect on the cross slope of the blade. It makes possible automatic operation in which possible conflict between grade and slope controls is avoided.

Circle Side Shift. The circle side shift, or lateral shift, swings the circle and blade to the side, usually to the right.

When the move is mechanically powered, a curved track is mounted on the underside of the frame. A long curved gear section slides on this, being moved by a pinion gear based on the frame. An arm is fastened by ball and socket joints to the right side of the gear section and the left side of the drawbar, for reach out to the right.

Turning the pinion for right shift moves the gear outward and upward to the right, pulling the drawbar and circle with it. For left shift, the arm is fastened to the left side of the gear and the right side of the drawbar, and the pinion rotated oppositely.

Sideshift to the right raises the blade increasingly on the right side and lowers it somewhat on the left. The lift levers must be manipulated if it is to be kept level.

A hydraulic cylinder mounted on one side of the frame with its piston connected to the opposite side of the drawbar can be used to shift the circle.

Blade lift cylinders may be based on a hydraulic saddle or cross piece on the grader frame. This may be designed to rotate on the frame when locks are released. When free, it can be turned by action of one lift cylinder, then locked in a side position.

The principal use of circle or lateral sideshift is to position, or start to position, the

Courtesy of WABCO

Fig. 19-14A. Blade sideshift

Courtesy of Fiat-Allis Construction Machinery, Inc.

Fig. 19-14B. High reach

blade at a steep or vertical angle offside, for slope work. On many machines, manual adjustment of telescoping arms is required also.

A combination of circle and blade sideshifts will put the blade a long way out, to work in spots too soft to support the grader.

CONTROLS

The control levers for the tools or working parts of the grader are usually arranged in a more or less straight line on a console on the dash.

There are five basic controls that are always present: the two blade lifts (right and left), circle turn, circle shift, and wheel lean. Common additional controls are for scarifier, and blade shift. A dozer or a snow plow might have another lever, or use the scarifier control. A rear ripper will have one or two levers.

Location of controls is not completely standardized. Figure 19-15A shows the control station of a Galion 160, and 15B indicates location of the major control levers in four other machines. Individual graders may have lever locations switched to suit operator preferences.

All these levers are two-way, operating in forward and back positions, and holding in the center. There are no FLOAT settings.

Mechanical—Old Type. For many years, the standard controls in graders were all-mechanical. Many thousands of these machines are now in use, and presumably will be for years to come, so a description of

a typical control box in a Caterpillar 12 is provided here.

The engine clutch shaft and the upper transmission main shaft are hollow. An inner (quill) shaft, driven at all times by the flywheel, extends through this hollow to drive a propeller shaft ending in a worm gear under the dash. This meshes with a gear on a vertical shaft, shown in Figure 19-16.

At the top of the vertical shaft is a bevel gear in mesh with two bevel gears turning freely on a horizontal reversing shaft. Spur gears cut on the backs of these two bevels are meshed in series with spur gears on five similar shafts.

The inner side of each gear is cut as a clutch jaw with sloping teeth.

The center of each of the reversing shafts is splined and carries a center collar with sloping jaw teeth cut on each side. This collar is slotted to hold a shifting fork connected to an overhead sliding bar, which is held in a center position by light springs. The fork is constructed to drag on the collar when it is centered to prevent the shaft from turning.

The toothed collar may be slid either forward or backward on the splines so that its teeth can engage one or the other of the gears. Since these revolve in opposite directions, the reversing shaft can be rotated in either direction by engaging the collar with the proper gear.

The beveled jaw teeth allow easy mesh-

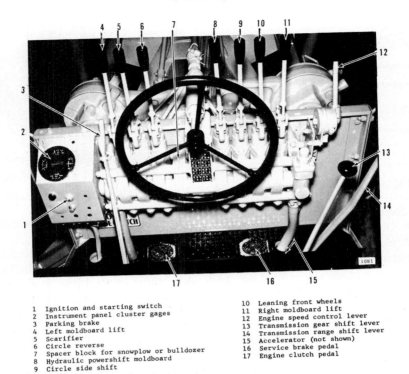

1 Ignition and starting switch
2 Instrument panel cluster gages
3 Parking brake
4 Left moldboard lift
5 Scarifier
6 Circle reverse
7 Spacer block for snowplow or bulldozer
8 Hydraulic powershift moldboard
9 Circle side shift

10 Leaning front wheels
11 Right moldboard lift
12 Engine speed control lever
13 Transmission gear shift lever
14 Transmission range shift lever
15 Accelerator (not shown)
16 Service brake pedal
17 Engine clutch pedal

Courtesy of Galion Manufacturing Company

Fig. 19-15A. Grader controls

ing and fit snugly without backlash when fully engaged. They also exert a thrust that tends to push them out of engagement. Under normal conditions, a light hand pressure on the control lever will keep them interlocked, but increasing loads will add to the tendency to kick out. If the driven unit jams or reaches the end of its travel, the jaw clutch will forcibly disengage.

This arrangement enables the operator to feel the working conditions and guards the mechanism against overload. There is also a shear pin at the base of the control column.

Worn teeth and/or a loose shaft will cause the mechanism to kick out under normal loads, unless held in with considerable force. This condition is fatiguing to the operator, and repairs should be made as soon as possible.

The two lift shafts are connected directly to gear sets on the sides of the dash. The other four are extended into propeller shafts leading forward to their gear boxes.

A similar mechanism used in medium-light graders is shown in Figure 19-17A. Drag-creating devices to prevent creep are shown in 17B.

Hydraulic-over-Mechanical. The larger mechanically operated machines now have a changed construction, Figure 19-18, in which the toothed collar is replaced by a pair of small friction clutches. These are engaged and disengaged hydraulically by movement of a valve operated by the control lever.

Hydraulic. Many graders use all-hydraulic control for some or all of their functions. The three-position valves are similar to those used in other hydraulic equipment, but must be of the metering type to allow gradual and/or partial engagement when-

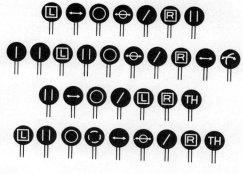

BLADE LIFT, LEFT
BLADE LIFT, RIGHT
MOLDBOARD SIDE SHIFT
CIRCLE REVERSE
CIRCLE SIDE SHIFT
LEAN WHEELS

SCARIFIER
RIPPER
MOLDBOARD TILT
THROTTLE
DOZER

Fig. 19-15B. Four grader control sets

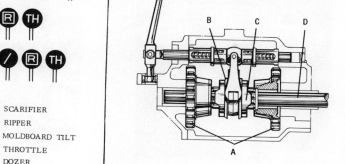

Courtesy of Caterpillar Tractor Company

Fig. 19-17A. Mechanical controls, new style

ever necessary for slow and precise movement.

Precautions against creep in these systems include irreversible worm gearing, and clamps that hold shafts from turning or piston rods from sliding, when valves are in neutral. They are pressure-released whenever valves are engaged to move the parts.

OPTIONAL EQUIPMENT

Scarifier. The scarifier is a set of teeth used for breaking up surfaces too hard to be readily penetrated by the blade. The teeth,

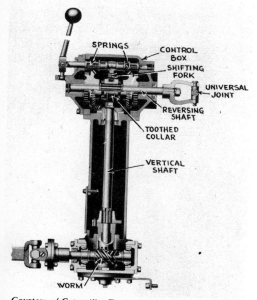

SPRINGS
CONTROL BOX
SHIFTING FORK
UNIVERSAL JOINT
REVERSING SHAFT
TOOTHED COLLAR
VERTICAL SHAFT
WORM

Courtesy of Caterpillar Tractor Company

Fig. 19-16. Control box, old style

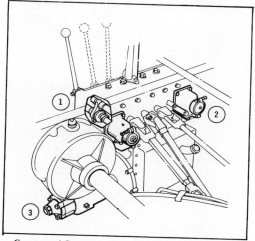

Courtesy of Caterpillar Tractor Company

Fig. 19-17B. Anti-creep devices

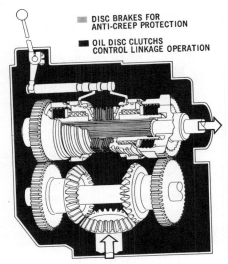

DISC BRAKES FOR
ANTI-CREEP PROTECTION

OIL DISC CLUTCHS
CONTROL LINKAGE OPERATION

Courtesy of Caterpillar Tractor Company

Fig. 19-18. Hydraulic-over-mechanical control

Courtesy of Galion Manufacturing Company

Fig. 19-20. Standard V-scarifier

consisting of rather slender shanks with replaceable tips, are set in a bar with a flattened V-shape, that is narrower than the grader; or in a wider straight bar. It is pulled by a pair of beams hinged to the bottom front of the frame, and is usually raised and lowered by a hydraulic cylinder. Angle of penetration may be adjustable from the operator's station.

The shanks are wedged or clipped in place, and may be readily adjusted for height, or removed. A full set, often eleven in number, is used for shallow penetration and light work, and in material than crumbles into fine pieces. For deeper penetration,

or slabby material, every other tooth or two out of three may be removed.

Individual teeth are not designed to take the full push of the grader, and caution must be used in hard material and among rocks to avoid bending or breaking them.

Scarifiers used to be almost standard equipment but are rather less used now. One can speed up grader work, and greatly reduce wear on the blade. It may be used in the same pass with the blading, or separate passes may be made for loosening and for shaping.

It is usually necessary to remove the shanks, or set them high in the bar, in order

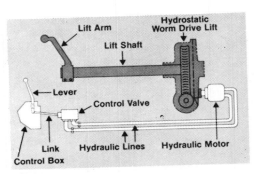

Courtesy of WABCO

Fig. 19-19. Hydraulic lock

Courtesy of CMI *Corporation*

Fig. 19-21. Circle-mounted scarifier

Fig. 19-22. Rear-mounted ripper

Fig. 19-23. Dozer attachment

to reverse the blade. They may interfere with cutting flat bottom ditches and handling high windrows.

Circle Mounting. In the CMI Autograde line, the scarifier is mounted behind the blade, on a pair of arms extending backward and downward from the circle. It rotates with the circle, and is lifted, lowered, and sloped by the circle lift arms. A separate control tips it to put the teeth down to work, or up to be out of the way.

This unit can be automatically controlled for depth and slope in the same manner as the blade.

Ripper-Scarifier. This unit, Figure 19-22, which mounts on the back of a grader, may be fitted with 11 teeth for scarifying depths up to 9 inches, or with 5 heavier ripper teeth for penetration to 14 inches.

It is raised and forced down by a hydraulic cylinder. The lift frame is of the parallelogram type that keeps the teeth at the same vertical angle, whether up or down.

A rear scarifier or ripper will handle much heavier work than a standard front mounted one, it can process a strip the full width of the machine, and it is not in the way of the blade. However, it adds to machine length so that it is awkward in close quarters, usually produces coarser and more difficult-to-grade pieces, and cannot break ground on the same trip that it is bladed.

Front Attachments. The front of many grader frames can be built out to form a narrow vertical bumper, forward of the tires. This may have limited use for protection in straight-on collisions, and for occasional pushing of other vehicles.

It may carry a dozer blade, raised and lowered by the scarifier cylinder or by separate controls, which can be convenient in rough work, particularly in leveling piles.

A snow plow, preferably of the V-type, is often fitted to the heavier machines, to provide possible work in the winter months. Chains should easily be put on the drive wheels for snow plowing.

Blade Stabilizer. The stabilizer is a shoe or skid plate that can be rested on or pressed against the ground behind the blade.

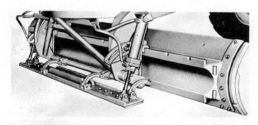

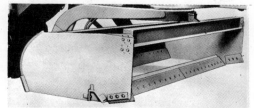

Fig. 19-24. Blade stabilizer and bowl

Courtesy of Deere & Company

Fig. 19-25. Tight turn with articulated grader

In high speed, shallow work a grader blade may develop various kinds of vibration, including a long period rhythmic bounce or lope, or a short period chatter. These cause a wavy, rough, or washboard surface on the work, discomfort to the operator, and extra machine wear.

The stabilizer, which is under separate hydraulic control, can be put in a dragging position where it will steady the blade and dampen or eliminate vibration. It can also be used as a support shoe to limit the cutting depth of the blade.

Blade Bowl. A blade bowl consists of two sideboards, a spreader bar, and a cutting edge. It is attached at the ends of a standard moldboard, and is supported from the circle by an adjusting bolt. The edge is ordinarily a little lower than the moldboard edge.

This attachment converts the blade into a bottomless scraper bowl. The leading edge cuts and shatters material, which is held by the sides and caught and moved forward by the blade. When a depression is reached, dirt falls into it and is smoothed off by the moldboard edge.

The bowl is used only with the blade set

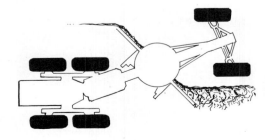

Fig. 19-26. Crab steering with articulation

Courtesy of Clark Equipment, Austin-Western Division

Fig. 19-27A. 4 x 4 grader

Courtesy of Clark Equipment Company, Austin-Western Division

Fig. 19-27B. 6 x 6 grader

straight across, and where somewhat more dirt must be moved than is normal to grader operation.

ARTICULATED (FRAME) STEERING

A few graders have frames that are hinged just forward of the engine, with pivoting controlled by a pair of hydraulic cylinders. This permits the front frame section, which carries the circle and blade, to be turned at an angle of 20 degrees or more to the tandem power section, as in Figure 19-25. In addition, normal sharp angle steering of the front wheels is retained.

With two points of turn, these machines can make a U-turn in less than one and a half times overall length, a remarkable feat for a tandem grader. A differential may be required between the drive axles to minimize scuffing.

Independent steering of wheels and frame may provide operational advantages. As indicated in Figure 19-26, the front end can be offset to the side by steering front and rear in the same direction (crab steering), so that its wheels will not have to tangle with the windrow or other obstruction, while the tandem rear is kept squarely behind the

blade. Front wheels can still be leaned or turned as necessary to resist any remaining side thrust.

The double-articulated Ray-Go Giant will be described later.

ALL WHEEL DRIVE AND STEER

Austin-Western. The Austin-Western power graders, Figures 19-27A to 19-28C, drive and steer on all four or six wheels.

The engine is rear-mounted in the standard manner. A torque converter with an input clutch delivers power to a 4-speed power shift transmission with a separate forward-reverse lever, and a high-low range auxiliary, providing 8 speeds in both directions. A transfer case drives front and rear propeller shafts.

The front shaft is in an under-the-frame arch, with multiple universal joints to enable it to conform to its curve.

Both front and rear axles are double reduction type. Front differential is standard, no-spin rear differential is standard on 4-wheel models.

The current machines are produced in two types. The Pacer models have a pair of rear wheels, the Super have tandem wheels

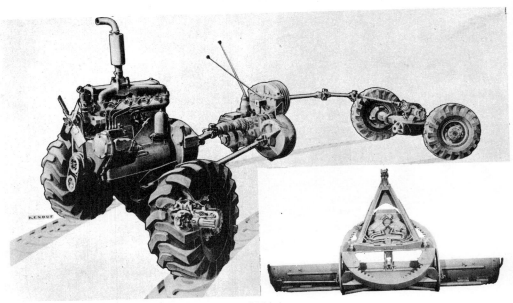

Courtesy of Clark Equipment Company, Austin-Western Division

Fig. 19-28A. Four wheel drive and steer

with enclosed chain drive. Overall dimensions of the machines are similar, but the tandems add 2,800 to 3,500 pounds to machine weight.

Steering. The rear wheels are steered hydraulically up to 15 degrees turn to either side, with a lever control. Single wheels pivot individually, tandems as a four-wheels-and-frame unit, 5th wheel fashion.

Front wheels have hydraulic steering up to 25 degrees either way. A steering wheel and follower valve give automotive type response.

Brakes. The disc service brake is located on the transmission output shaft, next to the transfer case. The parking and emergency brake, also disc type, is on the rear gear carrier. There are no brakes on the wheels.

The service brake is operated by either of two pedals. The right one operates the brake only, the left one also automatically releases the converter clutch.

Blade Mechanism. The circle is of the internal tooth type, with a V drawbar. Circle lift is by means of a pair of two-way cylinders, mounted in swivel cradles on the

Courtesy of Clark Equipment Company, Austin-Western Division

Fig. 19-28B. Front drive axle

frame. The piston rods are connected directly to the sides of the drawbar by ball joints.

The circle reverse, operated by a hydraulic motor, turns a self-locking worm gear set. There is full rotation both ways, without removing scarifier teeth.

There is hydraulic circle side shift and blade side shift, with optional power tilt blade. The 13-foot moldboard has a triple radius curve, with curvature increasing from top to bottom. All positions can be reached without mechanical adjustments.

Applications. As in other wheel-mounted equipment, all-wheel drive make the full weight of the machine available for traction, where rear drive can only utilize the weight on the drive wheels. In conventional graders, this is about 70 per cent.

As a finishing tool, the grader should not often be called upon to work in mud. But it does have to work in loose soil or sand, and it may be urgently needed to skim slippery films off haul roads. A grader may also be exposed to poor footing on turnarounds, between work strips, and in traps such as loosely filled ditches. In any of these cases, the reserve traction of all-wheel drive is valuable.

The pulling ability of powered front wheels improves ability to resist the side thrust of heavy blade loads, whether on straightaways or turns.

Extra traction provides for more effective

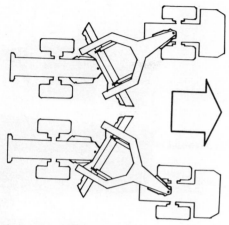

Courtesy of RayGo, Inc.

Fig. 19-29A. Double articulation

handling of accessories such as scarifier, dozer blade, and snow plow.

Applications of all-wheel steer will be discussed under Heavy Grader Operation.

Double Articulation. The huge two-engine, four-wheel drive Ray Go Giant, Figures 19-29A to 29C, has a frame that is hinged just forward of the rear engine, and is also pivoted to the front engine-and-wheels unit. The two swivels are under separate control.

The blade, which may be 20 or 24 feet wide, is permanently set straight across the frame. It can be raised and lowered with independent control of the two sides, but it has no circle mechanism.

Courtesy of Clark Equipment, Austin-Western Division

Fig. 19-28C. Circle and drawbar

Courtesy of RayGo, Inc.

Fig. 19-29B. Giant grader in open pit mine

Courtesy of RayGo, Inc.

Fig. 19-29C. Offset axles for sidecasting

Sidecasting blade angles are obtained by crab steering, as illustrated.

The front 318 horsepower engine supplies all power to the hydraulic system, in addition to driving the front wheels. The same-size rear engine just turns the rear wheels. Both have torque converters and air shifted transmissions with three speeds forward and reverse.

This machine is designed primarily for heavy, rough work on the largest scale, as in open pit mining, dam building, and major highway construction. Its jobs include maintaining (and building) haul roads, spreading and grading fill, and cleaning off blast litter.

AUTOMATIC CONTROLS

Automatic Blade Control. Several automatic systems are available for control of

Courtesy of Caterpillar Tractor Company

Fig. 19-30A. Stringline for grade control

Courtesy of Caterpillar Tractor Company

Fig. 19-30B. Grade control wheel

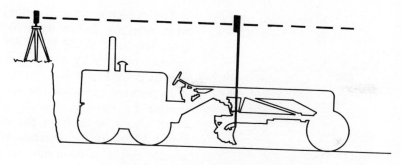

Fig. 19-31. Laser grade control

hydraulically hoisted grader blades. Slope regulation is the most widely used.

With this, there is a pendulum device mounted on the grader frame, that remains perfectly vertical in a right-to-left or cross plane, regardless of movements of the grader. It is coordinated with transistor-computer circuits.

The operator sets a dial at the cross slope he wants, which is usually from job specifications, but may be his own estimate. He then operates the grader, keeping the leading edge of the blade at the desired grade.

The instrument, operating on battery power and through solenoid valves, raises and lowers the heel of the blade to keep the correct slope. Corrections are made at the rate of 10 or more a second, so a high degree of accuracy can be obtained.

It is usually necessary to keep the blade at a predetermined angle to have the slope the same as that set on the dial. A sharper angle reduces the effective width, and therefore changes the amount of slope.

Some controls include a sensor on the blade circle, which corrects for changes in angle, and allows more flexibility.

The next step up is adding a sensor to keep the leading edge of the blade on grade. This is carried by an arm projecting beyond the side of the machine. It may terminate in a finger (sensing fork) guided by a string-line, or a small wheel rolling on pavement or on a previous pass by the grader.

The grade and the slope corrections are usually separate, with grade controlling the leading edge, and slope the trailing one.

Such controls are by no means a sleeping license for the operator. He has to steer, work the throttle (so far), and be sure he doesn't run into anything. If a too-heavy cut lugs down the grader, he must override the control and raise the blade.

Laser. An almost-final step in automation is grade control by laser. It is adapted only to graders that already have slope control.

The background of this system is a laser, a beam of intense (usually red), highly concentrated light, that can maintain an almost pencil-thin beam for many hundreds of feet.

Lasers, and many of their uses, are described in Chapter 2. They are becoming increasingly important in many phases of construction work.

For mobile machines, the laser shines down on a prism, which casts a basically horizontal beam. This can be adjusted to job grades, slope up or slope down. The prism revolves rapidly, casting the beam over a circular pattern.

A mast is fastened to the leading edge of the blade, and kept vertical by a pendulum and hydraulic mini-cylinders. It carries a set of laser-sensitive cells that activate solenoids in the lift valve to lift or lower the blade whenever necessary.

In the CMI Autograde graders, grade corrections are made by raising or lowering

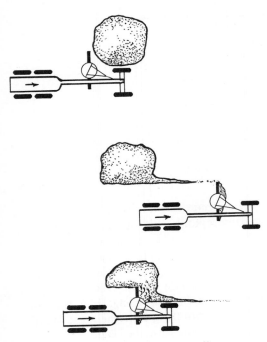

Fig. 19-32. Spreading a pile

The grader is normally moved forward while working. It can work in reverse by using the back of the blade for spreading or smoothing; or by reversing the blade so that its cutting edge faces to the rear.

In forward operation the grader is kept in line by steering, and by leaning the front wheels away from the direction of thrust or toward the direction of turn. The wheels may also be leaned to avoid rubbing vertical banks.

Adjustments. The blade is controlled in a number of ways. The ends can be raised or lowered independently of each other, or together. It may be positioned across the line of travel, parallel to it, or at any angle. It can be shifted to the side and into a vertical position by power. There are also mechanical adjustments for extending its range.

The lift arm adjustments are made by removing a pin holding an outer and an inner section together, raising or lowering the lift crank until the set of holes nearest the desired length are in line, reinserting the pin, and locking it.

Manual side shift adjustments are most conveniently made by two men. They are seldom required in open work. The results achieved by the extra reach can often be obtained by better operating technique or by adding a blade extension. Most work can be done with a standard blade centered on the beams, and the lift arms adjusted to within one hole of center.

The blade is ordinarily kept near the center of the tipping adjustment so that the top of the blade is directly over the edge. Increasing the lean forward decreases cutting ability and causes the blade to ride over its load rather than to push it. It diminishes the likelihood of catching on solid obstructions and may be used for rapid, light planing of rather regular surfaces and for mixing operations. When leaned back the blade cuts readily but tends to let the load ride over its top, and

the machine at the front axle, which was described earlier. This avoids problems in coordinating two sets of commands to the blade lift cylinders.

At present, automatic operation is for uniform grades, and cannot be used in vertical curves, where an incline changes or merges into a level stretch. Steering, and adjustment of blade angle and tilt, are still up to the operator.

Load. Automatic grade regulation devices usually know nothing about load. Any one of them will stall a machine if a cut to grade is too heavy for it, or will try to fill a hollow without enough material in front of the blade. They are therefore limited to finishing sections that are already almost on grade, and must be closely supervised by the operator and the foreman.

HEAVY GRADER OPERATION

General. The following discussion refers to the tandem rear drive grader with manual control, unless otherwise indicated.

to dig into obstructions. In machines not carrying a scarifier this tilt may be used to cut hard surfaces.

Bulldozing. A grader blade may be used as a bulldozer to a limited extent, often in spreading piles of loose material. If there is space to work beside the pile the blade should be extended well to the side, and the pile reduced in a series of cuts, as shown in Figure 19-32.

If there is not room enough to do this and the piles are not too high, the front wheels may be driven over them. The front axle will push the top off and the blade will cut as much more as power permits.

The blade should be kept well below its highest position so that if the machine gets hung up on it and loses traction it will be possible to raise it to restore weight to the wheels.

Piles to be distributed by a grader should be spread as much as possible while being dumped.

The grader can also be used for light cut and fill work in building and regrading roads.

The load to be pushed is limited by the power and traction of the machine and will usually be much less than by a crawler of the same weight, although it will be moved faster. The blade itself is quite low, but being more concave than the dozer blade imparts a more pronounced rolling action to the load so that a large quantity can be pushed without spilling over the top.

If the blade is lowered on the right only the other lift arm will remain stationary and the circle and blade will pivot around it, causing the left end of the blade to rise about one quarter of the distance the right side lowers. If the left corner is to be held in position it must be lowered intermittently while the right side is lowered steadily. This effect occurs on either side.

Keeping track of both ends of the blade is necessary and may be the hardest technique for the beginner to learn.

If the blade is raised to its full height and the controls left engaged, the lift cranks will continue to rotate inward until stopped by the frame. The blade will be lowered during the movement from top center to the frame, although the controls are in the raise position. If the controls are moved to DOWN the blade will rise until the cranks are past center after which it will respond normally.

If much dozer work is to be done, it may be advisable to install a front dozer blade.

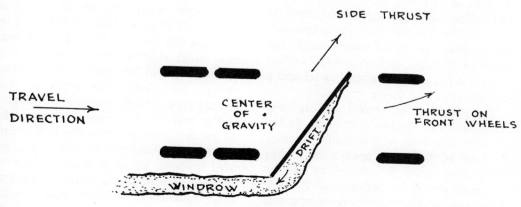

Fig. 19-33. Side casting

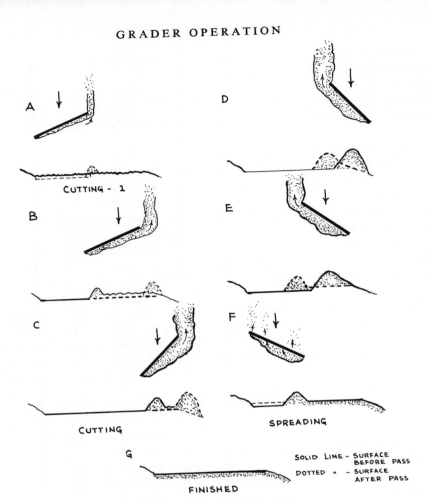

Fig. 19-34. Refinishing a flat section

Side Casting. When the blade is set at an angle, the load pushed ahead of it tends to drift off to one side, Figure 19-33. The rolling action caused by the curve of the moldboard assists this side movement. As the blade is angled more sharply the speed of the side drift increases, so that the dirt is not carried forward as far and a deeper cut can be made.

The sideward movement of the load exerts a thrust against the blade in the opposite direction, which tends to swing the front of the grader toward the leading edge. This thrust is handled by leaning the front wheels to pull against the side draft, and steering enough to compensate for any side slipping which occurs in spite

of setting the front wheels to lean.

The most usual way in which to describe blade setting is to say that a blade set straight across, Figure 19-32, is at zero and all other settings are described by their angular distance from that position. Most road shaping and maintenance is done at a 25° to 30° angle, with straighter settings for distributing windrows and sharper ones for hard cuts and ditching.

The angle of the blade is regulated by the circle reverse control. The mechanism is self-locking and can be turned any desired amount. In some makes it can be adjusted only while the blade is empty or doing light work, in others while pushing a heavy load.

Side shifting the circle from center changes the slope of the blade, requiring compensating lift adjustments. Side shifting the blade itself does not affect slope.

Planing. If the blade is set at an angle it can be used to plane off irregular surfaces by lowering them sufficiently so that enough material will be cut off the humps to fill the hollows. Enough extra material should be cut to keep a partial load in front of the blade. The forward and sideward movement of the loosened dirt serves to distribute it effectively. If a windrow is left at the trailing edge of the blade it is picked up on the next pass. On the final pass a lighter cut is made and the trailing edge of the blade is lifted enough to allow the surplus material to go under rather than around it, to avoid leaving a ridge.

This type of light planing will produce a smooth surface under favorable conditions, but the fill in the hollows is liable to settle or be compressed below the cut sections. Also the blade may chatter in a very shallow cut, particularly if the mechanism is loose or worn.

A more thorough method is shown in diagrams in Figure 19-34. A series of cuts are made across the area to a depth sufficient to reach the bottoms of the holes, or at least to two inches. The large windrow of loosened dirt is then spread back evenly over the area.

It is easier to get a smooth surface with this method because of the advantage of working loosened material and the more uniform distribution with a full blade. The surface will tend to remain smooth after settling or rolling.

It is desirable to vary the blade angle during this work, the first cuts being taken with a straighter blade than the later ones, and the first spreading pass having a sharp angle which will be reduced on each following pass as the size of the windrow is reduced.

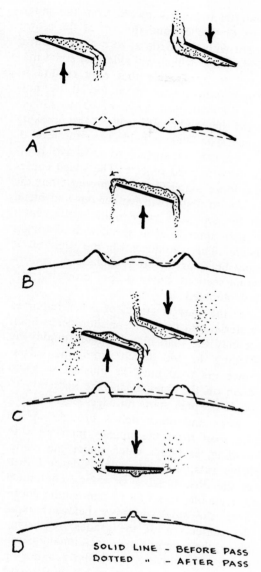

A

B

C

D

SOLID LINE - BEFORE PASS
DOTTED " - AFTER PASS

Fig. 19-35. Crowning a road

Windrows should not be piled in front of the rear wheels as they will interfere considerably with grading accuracy and traction.

Crowning. When the piece to be smoothed is a dirt or gravel road, it is generally crowned so that water will flow off to the sides. Figure 19-35 shows a sequence of passes in a crowning operation. Road

19-23

material is bladed inward from the shoulders or ditches, and the top of the crown is cut with the blade at zero angle, or at a slight angle which will side cast some material to either side that may require it. The windrows are then spread by putting the blade at an angle of 10° to 25° toward the center, and using a fast working speed. The blade is held above the level of the undisturbed surface so as to avoid collision with solid objects. The speed causes the loose material to be thrown from the blade, so that it will feather out and blend at the top. Any ridge built in the center is then spread out at speed with a straight blade. This should finish the job, but it may be desirable to backblade or rework some sections where proper crowning was not obtained.

If the road is gravel or other imported material and the ditch or shoulder is loam or clay the road may be made muddy by blading in too much from the edges. Since the road must be finished so as to drain off the sides, it may be necessary to blade off high spots on edges outward.

Sods and other debris brought up on the road from the ditches interfere with grading, as lumps catch under the blade, leave ruts, and block the sideward drift of the dirt. If labor or raking equipment is available, it is best to remove the debris from the side windrows before making the center cut. If the grader is working alone, the operator may occasionally climb down to remove a particularly annoying piece, but most of it will be left on the road to dry out and be broken up by traffic.

Stones are a more serious nuisance both in making smooth grades difficult or impossible and in causing damage to machine and operator during cutting. If the cut is shallow or the road dirt compacts readily it is possible to blade loose stones out of the road while grading it.

Large rocks or those firmly embedded in road or shoulders are dangerous. If the blade hooks into one too solid to move, the grader will jump sideward or stop abruptly. This imposes severe shock on the blade, circle, and power train and is liable to throw and injure the operator. Danger of serious consequences increases greatly with higher speeds, so that all cutting should be done in low gear and often at part throttle where buried rocks or roots are expected.

Work Patterns. The following discussion will concern work done by single graders. However, it is often efficient to have two or more of the machines working together, each performing one step in the work sequence. This speeds the job, produces better results than working small sections with individual machines, and reduces or eliminates blocking of roads with windrows.

Road building and grading may be done on three general patterns. One is to work the two sides alternately, turning at the end of the strip. Another is to do one side at a time, working in both directions by means of a reversible blade. The third is doing one side at a time with the reverse trips non-working or utilized for light work with the back of the blade.

The pattern used is determined by the length of the strip being worked, the turning space and footing, and the reversibility of the blade. The machine does its best work going forward, and this increased efficiency must be balanced against the time, labor, and risks involved in turning. In a long run even difficult turning conditions may take an insignificant part of the working time, where easy turns may not be justified on a short run. A general rule is that a grader should not be turned if the strip is less than 1,000 feet long.

Turns. In turning, the front wheels are leaned all the way over in the direction the front of the frame is to turn, and left in this position for both forward and reverse

Fig. 19-36. Turning on rough ground

movements until the turn is completed. If ditches or rough ground must be used the machine should be backed into or across them. The oscillating tandem drive will readily climb ditches or obstructions which would be difficult to cross with front wheels.

Turning may be a serious problem. Because of lack of a differential the rear wheels drive straight forward and the tandem arrangement gives them considerable resistance to sideward movement. Weight on the front wheels is light and even when leaning properly the tires may not have enough traction to turn the machine on loose material, but will skid and slide sideways.

If the front tires do grip enough for a sharp turn (minimum radius is about 36 feet), the rear wheels on the inner side of the curve must spin or those on the outside drag enough to compensate for the different distances they travel. Turning a grader sharply therefore results in scuffing and gouging which may be sufficient to damage soft surfaces.

It is therefore often necessary to make a gradual turn either by swinging in a wide circle or by jockeying backward and forward in order to avoid damaging or loosening the turn spot. It is sometimes possible to do most of the turning on ungraded areas where no damage will be done.

Other factors affecting the choice between turning and working in reverse may be idle travel distance between the end of the work and the first possible turning area, interference with traffic while turning, and chances of getting stuck.

Tandem drive affords powerful traction, but a careless operator may hang the blade or circle up on ridges or rocks and the front wheels particularly may sink into mud with the same result. Recently filled ditches which have become water-soaked make grader-traps. Repeated sharp turns on sandy soil may loosen it enough so that the rear wheels will spin in it.

If the lifts are in good condition, many graders can be unstuck by putting blocks under the blade and applying down pressure. This will normally lift the front wheels off the ground so that they can be shored up and the blade then lifted clear. If the pressure is applied to the rear corner of a sharply angled blade, it may lift the rear wheels on one side. The scarifier will lift the front only.

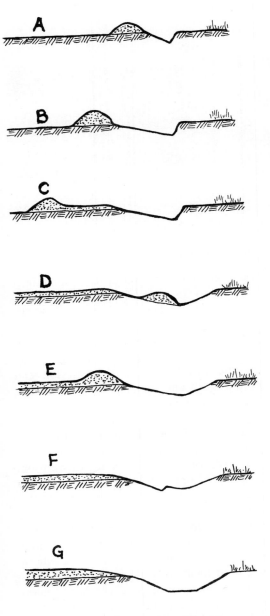

Fig. 19-37. Ditching for fill

Large front tires—the same size as on the drive wheels—give better front end flotation, are more effective in climbing ditches and banks and in holding the machine to its course on slopes, turns, and loose ground.

Reverse Blading. When the blade is turned so that it will cut at an angle in reverse gear, any heavy load will tend to push the rear wheels sideward, as they can neither be steered or leaned to resist the thrust. Reverse work should be arranged so that these wheels can be kept against a bank that will prevent them from sideslipping. Therefore, when doing ditch work in both directions, after the first cut made going forward, the cuts from the bank are made in reverse and the spoil moved away and spread on the forward trips.

Older and smaller models of graders may not be able to turn the blade to cut in reverse. Many of the newer ones cannot be reversed unless teeth are removed from the scarifier, and if the soil were hard enough to require loosening, the reverse position of the blade would not be available until the loosening was finished. While the change can be made quickly in some machines, in others maneuvering the blade into reverse position may take more time than would be justified to work a short trip.

The back of the blade is also effective for smoothing loose material, and can sidecast limited quantities of it. It can be safely used in fast reverse as it will ride over obstructions rather than digging into them. During the cutting and transporting passes back blading is not effective, but it is useful in finishing.

Road Building. A grader, without assistance from other machines or hand work, can shape up a road across a field by digging a pair of parallel ditches and using the soil to build the road crown. However, sod can make the finishing operation tedious and unsatisfactory as it tends to ball up under the blade and catch and pull out of loose surfaces. For this reason, the strip should be thoroughly disked before grading is commenced.

The outer ditch lines are marked by

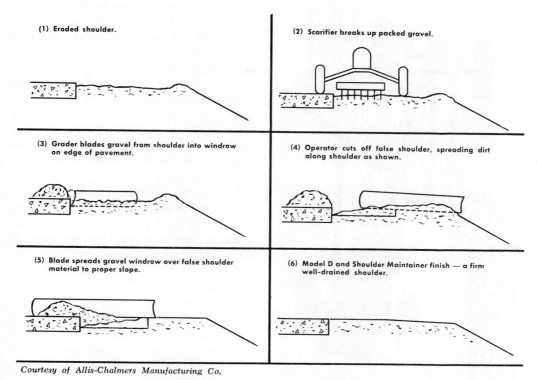

(1) Eroded shoulder.

(2) Scarifier breaks up packed gravel.

(3) Grader blades gravel from shoulder into windrow on edge of pavement.

(4) Operator cuts off false shoulder, spreading dirt along shoulder as shown.

(5) Blade spreads gravel windrow over false shoulder material to proper slope.

(6) Model D and Shoulder Maintainer finish — a firm well-drained shoulder.

Courtesy of Allis-Chalmers Manufacturing Co.

Fig. 19-38. Re-shaping a road shoulder

stakes or by the edge of the disked strip. The first cut on each side is made about two feet inside the edge, Figure 19-37 (A). The blade is held at a very sharp angle, perhaps fifty or sixty degrees, with the leading edge just outside the wheel track and the windrow rolled off under the grader. The cut is a light one made primarily to mark the edge of the work and to hold the wheels against side slipping.

The next cut is made at about a 25° angle, as in (B), casting the spoil beyond the inner wheels. If the windrow is large enough it is spread toward the center, as in (C). Otherwise, additional ditch cuts are made until sufficient material is piled to justify a spreading pass.

Ditch cuts, alternated with casting or spreading, are continued until the ditch is the proper depth. The outer slope is

then cut, as in (D), and the spoil moved up the inner slope, as in (E), from where it can be spread over the road.

The other side is done in the same manner and the fills are blended at the top. Ditch cuts, except the first one or two, can be made either in forward or reverse. Light casting and spreading can be done in either direction but forward is more efficient if the windrow is heavy.

Manufacturers recommend doing forward ditch cuts and other heavy grader work in second gear at a speed of three or four miles per hour. Blading windrows and similar handling of loose soil can often be done in third gear at speeds up to six miles. However, when there is loose rock, lower speeds will pay off in improved quality of work. In the presence of buried obstructions heavy enough to stall the grader, very slow movement may be re-

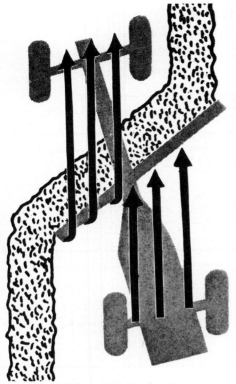

Fig. 19-39. Balancing load with four-wheel
steer

quired for protection of both the operator
and the machine.

If a wide bottom ditch is required the

further operations shown in (F) and (G)
are undertaken. Slices are cut from the
inside slope of the ditch, cutting down to
the level of the original bottom and leaving a ridge. This is cut out by running the
grader with its outer wheels in the original
bottom, setting the blade with its leading
edge even with the outer tires and at a
sharp angle so that it will cut only the
width of the desired flat bottom. The
spoil is piled on the bottom of the inner
slope from where it is cast onto the road
and spread by following passes.

The number and sequence of passes
are affected by the depth of the ditch, the
width of the road, and the resistance of
the soil.

Figure 19-38 indicates a series of steps
that can be taken to restore a shoulder on
a paved road.

All-Wheel Drive and Steer. Traction on
all wheels makes possible pushing larger
loads with the same weight grader, and
working in mud or sand too soft for conventional drive.

All-wheel steering allows the machine to
follow tight curves accurately, as the rear
wheels can be turned so that they will follow in the front wheel tracks instead of
trailing off on the inside of the curve. It
makes possible the same effectiveness in

Fig. 19-40. Work with offset axles

blading in reverse and forward. The turning radius is reduced somewhat, and ability to turn with backing in cramped quarters greatly improved. Scuffing is reduced.

In working, the front and rear axles can be offset from each other to either side, as in Figure 19-39, by turning both sets of wheels in the same direction. This steadies the grader against side thrust, and enables it to increase its effective side reach by running a front wheel up on a bank, or in a soft gutter, while the rear wheels stay on more level or firm ground. It also allows side shifting of the trailing wheels to avoid running on a windrow, or making tire marks on a finished surface, as in 19-40A.

Under certain conditions, the 4 x 4 grader may show a tendency to oscillate vertically, so that a smooth finish is difficult to obtain. Also, operators unused to four wheel steer may have difficulty keeping the rear wheels in place against side thrust.

An optional 6 x 6 construction avoids these difficulties. With this, operation is similar to that of tandem drive machines, while the rear wheels are kept straight. Rear- and all-wheel steer are still instantly available when required.

LIGHT GRADERS

The graders discussed so far have been of the high powered, heavy duty type, or lighter models patterned directly after the big units.

There are also lighter machines, of less than 100 horsepower, that are simplified in structure. The most conspicuous difference is that the blades cannot rotate in a full circle. Circle sideshift may be limited or lacking.

Fiat-Allis. The Fiat-Allis 60 horsepower 5-ton Model M 65 has standard tandem drive, no differential, with the engine over the drive wheels. The transmission is 4-speed mechanical, with a reversing lever. Extra creeper (very slow) gears are optional.

The drawbar is T-shaped, bracketed to the circle, and pulled from the front of the frame by a ball and socket connection.

The blade is 10 feet long, and is supported by a ¾-turntable resting on the circle. It is rotated through 139 degrees by a hydraulic motor and self-locking reduction gears.

Blade lift is by two direct acting cylinders

Courtesy of Fiat-Allis Construction Machinery, Inc.

Fig. 19-41. Medium-light grader

Fig. 19-42. Light grader

cradled on frame brackets. Circle sideshift is manual, with pin-and-hole adjustments of a telescoping arm. Blade side shift may be either manual or hydraulic.

Steering and leaning front wheels are hydraulically controlled. A scarifier can be mounted behind the blade, or at the rear.

Huber Maintainer. In these 3½ to 5 ton machines tandem drive is replaced by a pair of large single drive wheels, which reduces machine length substantially.

A 59 horsepower diesel or a 65 horsepower gasoline engine, with no clutch, drives

Fig. 19-43. Drag leveler

the rear axle through a hydrostatic transmission with two speed ranges, 0 to 8.8 miles per hour in low and 0 to 20.5 in high. The rear hubs have planetary reduction final drives.

There is a differential, and separately operated rear wheel brakes may be used to assist steering. The front wheels do not lean.

The 9 or 10-foot blade swivels on a pusher A-frame, and can be swung through a 38 degree arc (maximum angle 19) by a hydraulic cylinder. It has power sideshift.

Optional equipment includes a front mounted bulldozer blade or a loader, a snow plow, or a scarifier. A drag or berm leveler, Figure 19-43, may be mounted on the rear, to level off ridges left by the blades. It may be followed by a small broom to clean off a paved surface.

The side dozer, next illustration, is a 4-foot blade mounted on a telescopic bar that replaces the blade. Its principal use is removing material between fence posts along highways, which otherwise would require hand work.

Short wheel base and maneuverability make this a very useful machine in close quarters.

Comparison with Heavy Graders. Light graders do not provide the full range of blade positions, blade angle usually cannot be changed while working, and total weight and power are not only much less, but are even smaller in relation to blade width.

This disproportion is necessary because chassis width cannot be reduced far without loss of stability, and a moldboard should be long enough to reach past the wheel line even when sharply angled.

The result is that full width cuts must be shallow, except in loose material, and the operator should not expect to load the blade as full or move it as fast as with the heavier machines.

These graders cannot slope banks too steep for them to walk on, dig flat bottom ditches narrower than the tread, cut in reverse, nor move heavy windrows in one pass.

On the other hand, they can work in narrow crooked roads and driveways, in buildings, and among obstructions where the large machines would be at a disadvan-

Courtesy of Huber Corporation

Fig. 19-44. Side dozer

tage. They can often do maintenance and light construction jobs more economically.

Drawbar Grader. Light grader blades can be readily mounted on the back of wheel tractors equipped with a three point hitch, described in Chapter 15.

A simple model is shown in Figure 19-45. The blade can be locked straight across, at seven different angles to either side, or reversed, by pulling a lockpin and rotating it by hand. It can be given a moderate cross

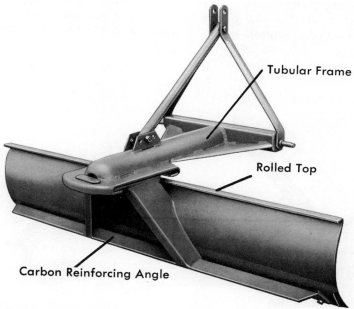

Tubular Frame

Rolled Top

Carbon Reinforcing Angle

Courtesy of Arps Corporation

Fig. 19-45. Light blade for a 3-point hitch

Fig. 19-46. Rake blade

blade to obtain greater reach on one side.

As the blade is entirely behind the tractor, it lacks the stability of the under-frame mountings, tending to dig when the tractor starts to climb, and to rise when the tractor starts to tip forward. This is partly compensated by the use of an automatic leveling device in the drawbar control valve. When this is used, the position of the hoist is controlled partly by a hand lever and partly by the pull or pressure exerted by the grader on the upper link. Under favorable circumstances, an even grade may be obtained by setting the hand lever and letting it take care of itself. For more exacting work, the automatic mechanism can be cut out and direct control used.

There are a number of useful accessories, such as rollers or shoes behind the blade to provide more stability and/or to regulate cutting depth, box sides to increase dirt-dragging capacity, and ditching points to help penetration of the leading edge.

Rake Blade. A light grading blade may be made up of a set of curved teeth fastened to a crossbar frame, as in Figure 19-46. This model has alloy steel teeth 28 inches long, an inch wide, and spaced an inch apart.

slope by raising or lowering one of the lower hitch arms. Suck is adjusted by shortening or lengthening the upper arm.

More elaborate models have a greater choice of working angles, built-in adjustments for slope, and means to offset the

Fig. 19-47. Under-truck grader

The caster wheels are removable for doing rough work. The cross frame just behind the tractor wheels can be fitted with ripper teeth. There is a low grader blade that can be folded down across the bottoms of the teeth when rake action is not desired. A three point hitch provides for lift and adjustments.

The principal use of the rake is to sift stones out from gravel on roads or from soil being shaped and planted. The stones are sidecast into a windrow, along with other oversize objects, while loose soil shifts between the teeth.

Stones may be sidecast off the area by a succession of passes, pulled into piles with the rake set straight across, or picked up by a loader or by hand.

Rakes are also used to spread crushed stone and blacktop, to remove trash, break up lumps, distribute soil in seedbed perparation, and to cover seed.

UNDER-TRUCK GRADERS

Hydraulic controlled grader blades can be mounted under truck frames, forward of the rear wheels. Four-wheel drive trucks are best adapted to their use.

The operator cannot watch this blade and the road at the same time so he must learn to operate by feel. Ordinarily, this type of unit is used for planing and reworking roads which are fairly well shaped.

A vertical blade has best penetration, but can be used only at low speeds and with caution as it may hook into obstructions. As the blade is tilted forward, penetration decreases and safe speed increases. Horizontal position is used for carrying the blade when not in use.

Figure 19-47 shows the St. Paul Hydraulic Truck Patrol. It uses the dump body pump, with separate levers to regulate the grader attachment.

A pair of double-acting rams mounted on the truck frame raise and lower the blade through bell cranks. Hydraulic or mechanical rotation of the circle are optional.

The moldboard is stabilized against chattering by a levelizer shoe that rides on the ground behind the leading corner of the blade. When not in use the blade is raised vertically.

TOWED GRADERS

Towed graders depend on a tractor or truck for movement. The grader controls

Courtesy of York Modern Corporation

Fig. 19-48. Towed grader, with scarifier and rake

Courtesy of York Modern Corporation

Fig. 19-49. Blade and rake action

may be hydraulic, mechanical, or manual. Those with power controls may carry a small gasoline engine, or be operated by a valve and tubing from the tractor. The tractor operator may run the grader also, but it is more usual to have two operators.

Steering, or tracking, is controlled by angling the draw tongue and shifting the frame along the rear axle.

Towed graders are made in quite small sizes, and are sometimes equipped with stone rakes, smoothing drags, and other accessories. Sometimes the front axle is removed, and the gooseneck mounted directly on a hydraulic lift tractor drawbar. This change makes it possible to back the machine as a semitrailer—an important convenience in re-doing short pieces and in repeated scraping of hard spots. The weight transfer improves wheel tractor traction.

The 5-ton full trailer machine in Figure 19-48 has full hydraulic controls, powered by a rear mounted engine and pump. It carries a scarifier, a grader blade, and a two-section rake. These units may be used individually, or in any combination.

Operation. In conventional pull grader work, two operators are employed and proper coordination between them is very important. The tractor driver is responsible for keeping in correct line of work; varying speed to suit working conditions, and turning, slowing, or stopping to avoid hitting obstructions.

If the leading edge of the blade hooks into something solid, the operator's platform is thrown sideward with a force depending on the tractor speed. This may throw or injure the operator unless he has a secure grip, or jumps off.

If the obstruction is visible the grader operator may avoid it by side shifting the

Courtesy of WABCO

Fig. 19-50. The first leaning wheel grader

Courtesy of WABCO

Fig. 19-51. Early grader opening a ditch

Courtesy of CMI *Corporation*

Fig. 19-52. Preparing a subgrade

drawbar if this control is available, or by raising the blade to pass over it. Either process may be too slow to avoid a collision.

Arrangements may be made to signal by means of a gong or horn, or a cord fastened to the tractor operator's arm. A line can also be run from the grader, around a pulley on the tractor dash, and to the master clutch hand lever (if there is one), to enable the grader operator to stop the tractor in an emergency.

Towed graders can do most of the jobs handled by motor patrols, but not as efficiently. They are lacking in flexibility, and in continuous work show higher operating costs when two operators are used. They find their best use where the need for a grader is not frequent enough to justify the cost of buying a power grader, and in working shoulders and slopes too soft for driving wheels.

A towed grader can also be used to advantage behind a motor grader to avoid blocking roads and to eliminate the need for return trips. When a ditch is being cleaned out toward the road, the resulting windrow may be a serious obstacle to traffic. If the patrol tows a grader behind it, offset toward the center of the road, it can spread the windrow in the same trip.

If a tractor operator works the towed grader controls also, he must alternate between looking ahead for steering and behind for grading. Either the quality or the speed of work usually suffers as a result, but it is still a satisfactory practice under many conditions.

AUTOMATIC LANE PROCESSORS

There are quite different automatic leveling, spreading, and trimming machines that are used in final preparation of highway lanes for paving. These represent the most elaborate and efficient development of the grading machine. They are also readily changed over to paving machines, for either concrete or asphalt.

Autograde. The CMI Dual Lane AUTO-GRADE Trimmer-Spreader, Figures 19-52 to 62, is the original and apparently the most successful type and will be used for description.

The basic machine consists of two augers and two strikeoff blades under a wide, short frame that is supported and propelled by four crawler track units, at the ends of horizontal legs.

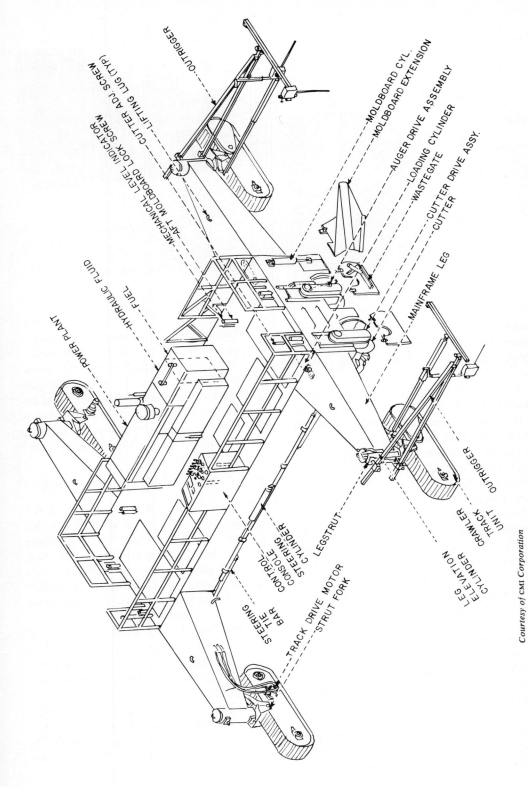

OUTRIGGER

MOLDBOARD CYL.
MOLDBOARD EXTENSION
AUGER DRIVE ASSEMBLY
LOADING CYLINDER
WASTEGATE
CUTTER DRIVE ASSY.
CUTTER DRIVE ASSY.

MAINFRAME LEG

LIFTING LUG (TYP.)
CUTTER A.D.I. SCREW
MECHANICAL LEVEL INDICATOR
AFT MOLDBOARD LEVEL LOCK

POWER PLANT
HYDRAULIC FLUID
FUEL

OUTRIGGER
TRACK UNIT
CRAWLER

ELEVATION
ELEVATOR CYLINDER

LEGSTRUT
STEERING CYLINDER
CONSOLE
CONTROLE
TIE BAR
STEERING

TRACK DRIVE MOTOR
STRUT FORK

Fig. 19-53. Automatic 2-lane processor

Courtesy of CMI Corporation

19-36

The augers and blades are under both hydraulic and mechanical control from the operator's station. Leveling of the machine by means of hydraulic jacks in the leg supports, and steering, are under both manual and automatic control.

References used for automatic control are stringlines and/or completed grades or pavements.

FRAME

Mainframe. The mainframe of this model is a box-like unit, about 28 feet wide and 10 feet long.

Its flat deck supports the power unit, which is made up of a diesel engine and a set of hydraulic pumps; the operator's station with its control console or panel, and a two-section tank containing fuel and hydraulic fluid.

The augers and blades, called collectively the subframe components, are held within and beneath the deck by slide plates or bars, and are supported by hydraulic cylinders and mechanical stops.

The flooring is partly an open grating that permits the operator to watch the behavior of the units beneath him.

A waist-high railing surrounds the deck. An opening and climbing rungs are provided at each side. A cab is optional.

Legs. This size machine is supported by four legs, extending forward and backward to crawler drive units. Single lane models may have three legs.

Each leg is normally attached to the mainframe by a pair of horizontal hinge pins at the bottom, and the piston rod of a hydraulic cylinder near the top. When in working position the cylinder is retracted to clamp the leg firmly in place.

Extending the cylinder forces the leg to pivot downward. Control is by two manual valves, one for front and one for rear. This action raises the frame for loading on a trailer, or for convenience in inspection and repair under it.

Each leg is attached at its outer end, through a steering strut fork, to a vertical hydraulic cylinder whose piston rod is supported by a bracket on the crawler frame. This cylinder raises or lowers the leg to establish and maintain the level of the mainframe.

A light outrigger frame may be mounted on the outer side of each leg, to support a sensing unit for automatic controls.

Crawler Units. The crawler units have tractor-type frames and tracks, with the 3-cleat pads used on loaders. Any of several track lengths, up to 10 feet, and widths up to 2 feet, are chosen according to machine size and expected firmness of ground.

When steered, the crawler frames swivel in a limited arc around the centerlines of the leg elevation cylinders.

Drive is by means of hydrostatic (axial cylinder) motors, usually mounted on the inboard sides of the frames, with speed-reducing roller chains to sprockets on short axles that turn the bull wheels and tracks.

The rear crawler motors have fixed displacement, the front ones variable displacement. Drive is to all four during work, but may be shifted to the rears only when traveling.

Steering. The crawlers are swiveled for steering, with front and rear sets under separate control.

The same mechanism is used front and rear. A two way cylinder is mounted horizontally across the outside of the mainframe. The piston rod is anchored to the frame. The cylinder is fastened to a bracket-supported slide bar.

In Figure 19-53, expanding the cylinder moves the bar to the left, retracting it moves it right. The bar is connected at both ends to link rods that move wide, rigid steering arms under the legs. The arms swivel the crawlers through yokes.

Steering is usually controlled automatically by front and rear sensors and a stringline while working. During travel, and work

Courtesy of CMI *Corporation*

Fig. 19-54. Crawler unit

They are made in a number of pitches, from 10 to 40 degrees.

There is an option on some machines of leg connections whose angle can be changed by hydraulic jacks controlled from the operator's station.

SUBFRAME COMPONENTS

The work of the Autograde is done chiefly by two full-width, horizontal, 30-inch diameter augers, backed by two high, full-width blades or moldboards. These units are known collectively as subframe components.

Each of them is divided into right and left sections, with a hydraulic lift cylinder at each outer end, and one in the center. Mechanical depth indicators extend from near the lift points up through the deck.

without a line, it is operated by a pair of switches on the console.

Turning front and rear tracks to the same angle in opposite directions enables the machine to follow curves accurately. Turning them equally in the same direction sideshifts it without change in centerline direction. Intermediate positions make possible very delicate maneuvers.

Changing Tread Width. On straight going, the outside edges of the tracks are normally in line with the sides of the mainframe. However, it is sometimes desirable to have the tread narrower or wider than the working parts of the machine.

Right and left hand legs with their track assemblies may be exchanged. This brings each track inward about two feet. Drive motors are then on the outboard side of the legs.

Wedges may be installed horizontally between frame and legs. If their wide parts face outward, the legs will be angled inward to narrow the tread. If wide parts are inward, the legs will be spread apart and tread will be widened.

The wedges are special parts, as they must maintain the frame-to-leg rigidity, which is necessary for control of grade.

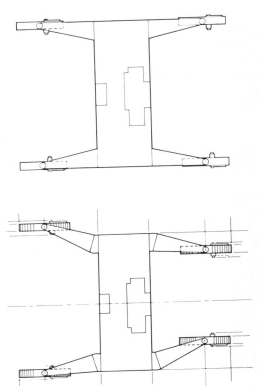

Courtesy of CMI *Corporation*

Fig. 19-55. Leg arrangements

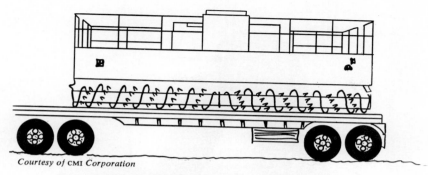

Fig. 19-56A. Mainframe on a trailer, showing cutter

Cutter. The front auger is called a cutter. It takes care of initial digging, pulverizing, and spreading operations.

There are two 14-foot, 12-inch diameter tubes. They carry one welded-on spiral of helicoid flighting, and a spiral of digging teeth. The tooth shanks are welded into the tube, and have replaceable carbide points.

Spirals are usually oppositely pitched on the two halves, so that if they are turned in the same direction, all material will be moved outward to both sides, or inward. If rotated in opposite directions, material will be shifted to one side.

The front moldboard is close behind the cutter, and holds material in contact with it until it passes under the edge, or off to the sides, assuring a constant supply for filling hollows.

Discharge to the sides is controlled by the cutter doors. These are hinged plates which are raised manually to allow free flow of material, or lowered for various degrees of restriction, and secured in position by pins.

The axle ends are mounted in slide plates, one for both inner ends, and one for each outer end. Each plate is lifted and lowered by a two way hydraulic cylinder.

Each outer plate carries a hydraulic motor and reducing gear box on its inner side. A shaft through the plate and a roller chain

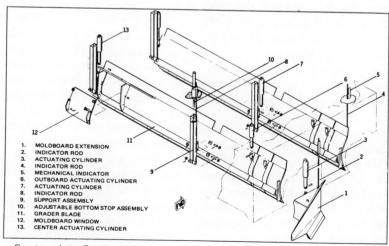

1. MOLDBOARD EXTENSION
2. INDICATOR ROD
3. ACTUATING CYLINDER
4. INDICATOR ROD
5. MECHANICAL INDICATOR
6. OUTBOARD ACTUATING CYLINDER
7. ACTUATING CYLINDER
8. INDICATOR ROD
9. SUPPORT ASSEMBLY
10. ADJUSTABLE BOTTOM STOP ASSEMBLY
11. GRADER BLADE
12. MOLDBOARD WINDOW
13. CENTER ACTUATING CYLINDER

Fig. 19-56B. Moldboards, right side

Courtesy of CMI *Corporation*

Fig. 19-57. Base reclaimer

carry the drive down to the axle. Since motor and axle both move with the plate, rotation is not affected by lifting or lowering.

Auger. The rear auger has only smooth helicoid flights, with no teeth. The flighting is installed in short hinged sections, wrapped and bolted on an inner cylinder.

Otherwise, general construction is similar to that of the cutter. It is split into right and left sections with separate drive, with lift cylinders at the center and sides. Drive motors and mechanism are identical.

The flow of material out the sides is controlled by waste-gates. These are similar to the cutter doors, but are hydraulically adjusted from the operator's station.

Moldboards. The front and rear moldboards are generally similar in construction. There are separate right and left halves, with a shared lift mechanism in the center and separate lifts at the sides. They are about 43 inches high, only slightly concave to the front. Wearing edges are standard grader blades, bolted in place.

Each lift includes a hydraulic cylinder, and a slide bar that moves up and down inside a square channel secured to the mainframe. This mechanism braces the moldboard against working stresses.

Each of the rear moldboard sections has a large, square opening or window near the inner end. This is normally closed by a removable section, which is kept on hangers on the back of the moldboard when the window is open.

These windows are opened when surplus material is to be left in a windrow in the center of the work strip for later removal, or is to be fed into a reclaimer which will be described below. When windows are open, the auger sections are rotated to move material toward the center. Some adjustment of flighting sections may be advisable for maximum efficiency.

There are three adjustable bottom stops for the rear moldboard — one center and two sides. The adjusting screws are contained in pipes that extend above the deck.

The stops are set when the machine is originally lined up with the stringline at the beginning of a work run, and the moldboard normally rests on them. They permit the

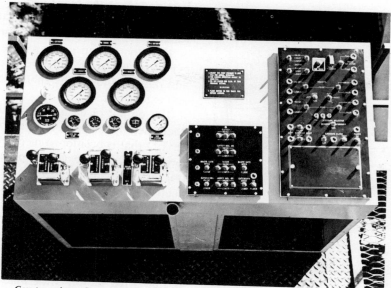

Courtesy of CMI *Corporation*

Fig. 19-58. Control console

operator to raise the moldboard hydrauli-cally if necessary, and then return it to its exact setting without re-calibration.

Moldboard Extensions. Rear mold-board extensions may be carried on mount-ing brackets outside the frame, and raised and lowered by hydraulic cylinders con-trolled from the operator's panel.

A pair of short extensions are standard equipment. Their use is required when so much surplus material is being discharged to the side that it tends to slump back into paving area behind the machine.

Longer extensions have additional sup-port from auxiliary arms. They are used chiefly in spreading base material, and may add 50 per cent or more to the effective width of the machine in light trimming operations.

Mechanical Indicators. Each of the subframe components have three (center and two outside ends) mechanical indica-tors for elevation. These are directly fas-tened rods that extend up through slots in the mainframe floor, where they are easily observed by the operator.

Calibration is both ways from zero, through plus and minus numbers, represent-ing inches.

Side extensions are also equipped with mechanical indicators for visual control.

Each indicator has a movable pointer, which the operator can set to remind him of the correct level.

Base Reclaimer. The subgrade or base reclaimer is an accessory mounted across the rear Autograde legs. It picks up surplus material discharged through windows in the rear moldboard on a conveyor belt.

This belt discharges onto another belt that can direct the material anywhere within a 180 degree arc. It is high enough to load following trucks over the cab.

The belts are driven and controlled hy-drostatically by two extra pumps mounted on the forward part of the Autograde en-gine. Swing and height of the second con-veyor are adjusted by hydraulic cylinders connected to the constant pressure circuit.

Controls are at the operator's station on the Autograde.

Receiving Hopper. When the machine

Fig. 19-59A. Stringline and sensor

is used to spread aggregate or other material, it is helpful to have the correct amount constantly available.

A receiving hopper may be installed between the front legs. It is equipped with rollers, so that it can push a truck while it is dumping. It has hydraulically adjusted gates, controlled from the Autograde, which meter out and strike off the desired amount of material and leave it in a neat windrow.

HYDRAULIC SYSTEM

Pumps. The hydraulics are powered by six pumps. There are five axial-piston hydrostatic pumps driven from the rear of the engine, and one pressure compensated pump at the front. These, with the electrical system, use the full power of the engine, as there are no mechanical drives.

All of the hydrostatics are driven by a gear box mounted at the rear of the engine. Each of them is a cam-regulated, reversible, variable displacement unit operating in a closed loop with its motor or motors. However, a single 35-gallon reservoir, a cooler, and a 3-filter system is used by all of them.

Four pumps supply fluid separately to the

four motors at the outer ends of the cutter and auger sections. The fifth usually takes care of all four of the travel motors.

The front pump is for all the hydraulic cylinders that steer and level the machine, and raise and lower the subframe members and extensions.

The output of each hydrostatic pump is controlled by the pitch of a cam (see Chapter 12) that is positioned mechanically by a lever on the console. Center position is neutral or non-pumping. Moving away from neutral increases speed, crossing neutral reverses direction.

Additional hydrostatic pumps may be mounted, to power attached equipment such as a belt reclaimer.

Track Drive. Each track has an axial piston drive motor, with fluid circulation through high pressure lines in the leg. All motors have fixed displacement, so that torque and speed is regulated by the pressure and volume received from the pump.

Cylinders and Valves. All cylinders are double-acting. Wherever necessary, they are linked with slide bars or other devices that assure straight-line motion of the parts they control.

Valves in the steering and leveling circuits are electrically controlled, and respond either to automatic sensors actuated by stringlines, or to manual control switches on the panel.

The rest of the cylinders, for subframe and outboard units, have electrically actuated valves controlled by panel switches only.

Gauges. Each hydraulic system has a pressure gauge. These are located with the engine gauges on the left side of the console panel. Control levers are below them, and each is aligned with its own gauge.

A single gauge reports fluid temperature at the outlet to the reservoir. During normal operation, the temperature should be 180 degrees F., and it should never be permitted to go over 200 degrees, where thermal decomposition of the oil becomes likely.

MANUAL CONTROLS

The operator's control console, Figure 19-58, has three panels or groups. On the left, below the gauges, are what might be called rotation control levers, two for the left and right cutter sections, two for the auger, one for the tracks, and one for the engine throttle.

Except for the throttle, these levers are neutral when centered, increase speed as they move away from neutral, and change direction of drive when crossing neutral.

The square center panel controls the height of the working units relative to the frame. Rubber-covered toggle switches actuate valves in solenoids that raise and lower cutter, auger, and moldboard sections, and the waste gates and any extensions.

The right hand panel has the same type of switches to raise, level, and lower the whole machine, for either work or for loading on a trailer. The working-leveling section is labeled TRACERS. It also controls the track drive selection, steering, and the master electrical switch.

Leveling and steering switches have signal lights that indicate when a circuit is active. Their principal value is keeping the operator informed about the actions of the automatic controls.

AUTOMATIC CONTROLS

The standard automatic control system is made up of stringlines (one or two), tracers in contact with the strings, sensors that convert tracer movements into electrical impulses, and solenoid valves that direct flow of hydraulic fluid in response to these impulses.

A tracer and a sensor together make up a sensor unit. There are six of them, four for level and elevation, and two for steering. The six are mounted on four outriggers, attached to the outer sides of the legs.

Outriggers are obtained in different lengths, for reaching two to 20 feet.

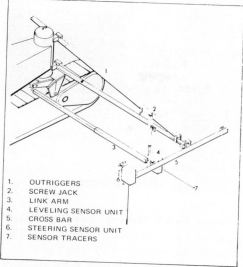

1. OUTRIGGERS
2. SCREW JACK
3. LINK ARM
4. LEVELING SENSOR UNIT
5. CROSS BAR
6. STEERING SENSOR UNIT
7. SENSOR TRACERS

Courtesy of CMI *Corporation*

Fig. 19-59B. Outrigger and sensor unit

Stringline. The general principles of setting up a stringline were discussed in Chapter Two. It is a continuous line of taut string, or occasionally wire, that is exactly parallel horizontally and vertically with an edge of a proposed grade or surface.

For satisfactory results with the Dual Lane Autograde, the string should be between 18 and 30 inches above desired grade, and 5 inches above obstructions.

Distance between a pair of stringlines may be 32 to 64 feet, depending on the outrigger equipment. If the machine has a cross slope system, one line, on either side, is sufficient.

Since automatic controls cause a machine to follow a stringline exactly, it is absolutely essential that it be accurate.

Tracer. A tracer or wand is usually an aluminum rod, extending about 18 inches from the sensor. An extension across the sensor carries a counterweight that presses the tracer lightly upward against the string. This keeps it in contact, and tends to counteract any slight sag of the line.

Steering wands are balanced to press lightly outward.

Courtesy of CMI *Corporation*

Fig. 19-60. Ski tracer unit

Other Grade Guides. Leveling may be regulated by the surface of an existing grade or pavement, instead of by a stringline.

One method is to attach a ski frame to the outrigger, to slide along the grade to be matched. A string is stretched in the frame, and the wand is placed in contact with it, Figure 19-60. As the ski moves along the surface, the frame and string are raised or lowered by any change, and the information is transmitted to the sensor.

Another method is to attach a small wheel directly to the tracer, and allow it to roll on the finished surface.

Sensor Box. The tracer and its counterweight extension are both secured by set screws to a rotary switch shaft. A shock absorbing spring protects the shaft from damage from tracer collisions.

If the stringline raises a tracer, this indicates that the machine is too low at that point. The tracer rotates the switch, making a contact that sends current to the RAISE side of a solenoid valve, and also lights a signal on the operator's console.

The valve directs pressure fluid to the lower side of the leg cylinder, the leg and sensor are raised, lift on the wand ceases, the switch rotates back to null (neutral), the valve closes, and the signal goes out.

If the wand dipped while following a stringline, a too-high grade would be indicated. The switch would be turned in the reverse direction, opening the LOWER side

of the valve, and lighting a different signal.

Signal lights tell the operator what the controls are doing, which is important if any malfunction is suspected. The lights are also important while adjusting controls at the beginning of a job.

The steering sensors respond to movement toward or away from the inside of the string, opening and closing the proper side of the steering valve, and showing signals.

The automatic changes are usually each too small to be felt, and are positively indicated only by the signals.

A horn blows to warn the operator if any tracer jumps or otherwise loses contact with the string.

Over-Riding. The operator can use the toggle switches on the control panel to override the automatic action, and lift, lower, or turn the machine if he wishes. As soon as he releases a manual control, the automatic returns the machine to its pre-set condition.

Cross Slope Control. New machines may be equipped with an electronic cross slope system that makes it possible to operate automatically and accurately with one stringline.

The operator can set a transverse slope from zero to 20 degrees, either left or right. As with grader cross slope devices, one side is kept at a grade reference, and the system will keep the other side raising or lowering as necessary to stay in correct relationship.

The system includes a pair of aluminum tubes, one front, one rear, mounted across the ends of the legs. See Figure 19-61. Each contains a sensor that controls two valves. A control panel is mounted on a pedestal beside the console.

The valves are in parallel with the regular leg cylinder valves.

Hydraulic Sensor System. In a recent development, electrical operation of the automatic controls can be replaced by all-hydraulic.

Movements of the wand open and close ports in a very small and precise hydraulic valve. Fluid in a pressure line is blocked, or directed through either of two lines to small cylinders that operate the RAISE or LOWER (or LEFT or RIGHT) ports in the main control valve.

Subframe Components. The four subframe tools—cutter, auger, and two blades—are lifted, lowered, and leveled as a group by the automatic leg elevation mechanism. But they remain largely under individual manual control.

The rear moldboard is set with threaded bottom stops, so that it cannot be dropped below grade. But it can be raised to clear an obstacle or for some other reason, but then should be immediately returned to bottom.

The other three units are not usually limited, as they have less (or no) effect on final grade, unless manipulated so as to starve the rear moldboard of material.

The cutter should be adjusted upward if tooth marks show behind the machine, or downward if the rear auger or moldboard has to grind through hard, unbroken material.

OPERATING CHARACTERISTICS

Starting a Grade. The relationship between stringline setting and the grade and centerline varies from job to job and section to section. The grade changes in successive cutting and spreading passes in the same strip. It is therefore necessary to set the automatic controls for each work run.

First a section large enough for the Autograde is graded to a reasonably smooth finish and approximate grade. This may be done by other equipment, or by the machine itself under manual control. Stringline(s) are set or re-checked carefully.

The Autograde is then operated manually to place it in precise level and side-to-side position. The rear moldboard is set so that its bottom edge is exactly at grade, and preferably flush with the bottom of the side frame.

Threaded bottom stops are set manually from the deck to hold it from going lower.

The sensors are adjusted, one at a time, to be in a null or neutral position when in proper contact with the string. The machine is then operated for a short stretch, the grade it produced checked by instrument, and any necessary adjustments made.

The finished strip should be checked by engineers at frequent intervals during work.

Courtesy of CMI *Corporation*

Fig. 19-61. Making a second pass

Courtesy of CMI *Corporation*

Fig. 19-62A. Straddling a trailer

Rough Ground. These machines are usually operated on finished or semi-finished surfaces, where corrections to final grade are relatively small.

However, it may also be operated on irregular or rough ground that is free of large stones and vegetation. But heavy work may interfere with accuracy, so that it must be brought back for a second or finishing pass.

If the rough work includes filling hollows, a second pass must be made after compaction.

Whenever a second trip will be needed, the first grade should be left high so that the machine will have something to work with when it comes back.

Compaction. Autogrades have no provision for compacting the grade that they produce. If it is made by cuts and fills, the fills are subject to loss of volume under the weight of following machines, pavement, and traffic.

Fill depth should therefore be kept to a minimum. This may be done by pregrading with a preliminary pass, or working the strip with graders or scrapers. The surface is then compacted, preferably with rubber tired rollers, before the final grading.

Compaction problems may be reduced or eliminated by making the rough or preliminary grade high enough so that the final work is entirely cuts, deep in high spots and shallow in lows.

All-cut grading calls for disposal of large amounts of surplus material. Extension moldboards may push it well back, but unless used for shoulder or other-lane fill, it must be removed eventually.

A base reclaimer may be used to pick up material wasted through rear moldboard windows, to load into following trucks, or windrow it well off to the side.

Superelevations and Crowns. Superelevated (banked) curves have a straight but tipped cross section. This may be indicated by two stringlines. If these are properly set, the machine will automatically take care of transitions from flat cross section to banked slope and back again.

However, if the straight sections are crowned, careful operator attention is required to set the crown originally, and to make proper transitions from crown to superelevation and back to crown.

In the original setup, the outer edges of the rear moldboard are set at zero on the indicators, and the center is raised hydraulically by the amount indicated on the plans.

In making a transition to a curve, the op-

Courtesy of CMI *Corporation*

Fig. 19-62B. Removing legs

erator must have the engineers' instructions about the stations at which it begins and ends, and divide that part of the run into a number of small bits of lowering of the center, to make a smooth and precise adjustment.

As the curve ends, the crown is gradually restored by lifting the center.

If operation is on one stringline with a cross slope system, the operator must also keep track of variations in transverse slope as the roadway enters and emerges from the most steeply banked stretch.

LOADING ON TRAILER

The Autograde is placed on a trailer by raising it on its legs, backing the trailer under it, then raising the legs.

First select a reasonably firm and level spot, where the low bed trailer can be backed under the machine from the side. Put all rotary drives in OFF position. Raise the subframe components all the way. Adjust the constant pressure pump to 1800 to 2000 pounds.

Using the four LOADING switches in the upper left corner of the right hand column, lower all the legs, taking care to keep the frame fairly level as the legs push it up. When raising is finished, put the loading switches in neutral.

The trailer needs a clear deck space 10 feet wide and more than 28 feet long, and must be accurately aligned under the machine. When it is in place, 2 x 6 timbers are set under the ends of the mainframe, and it is lowered onto them by raising the legs.

Reduce the hydraulic pressure to 1000 to 1100. Lower the cutter and auger until they rest lightly on the trailer bed.

If there is sufficient clearance, the trailer might be able to carry the machine to another work site on the job with its legs extended. If this is not possible, or if the rig must use a highway, the legs must be removed.

This job is done with the help of a crane or other hoist that can handle more than 7000 pounds. With the weight of a leg and crawler unit supported by the hoist, two bottom pins and locks are taken from the bottom of the leg, and the steering and loading ram connections are opened. The leg is then removed, and placed on a second trailer. This is done with each of them.

At the new work area, the machine is reassembled, and lifts itself off the trailer by reversing the loading process. It is usual to make a thorough inspection, and lubricate all points, before putting it back to work.

COMPACTORS

COMPACTION

Compaction is an absolute requirement in at least some part of most earthmoving projects, and is usually worth its cost in filling and backfilling even if it is not required. Its underlying principles were discussed in Chapter 3.

Soil may be compacted naturally (settled) by time and weather. If it is porous, settlement may be speeded by soaking it and allowing it to dry, a method called puddling. But Nature is very slow, and neither process can be depended upon to produce the high densities required in modern construction.

This discussion will be limited to mechanical means of obtaining high density compaction promptly. There are three basic methods: rolling, vibrating, and hammering. The first two are most widely used, and are often combined. Adequate rolling can sometimes be obtained by systematic routing of hauling equipment.

Compaction Effects. The primary effect of a roller is usually to compress material under it by dead or static weight. There is also a kneading effect which is usually small under smooth steel rolls, and more important with tamping and rubber-tired rollers.

A vibrator shakes soil particles individually, causing a tendency to move into closer contact with each other, and to displace excess water. Loose soil such as sand or clean gravel may be sufficiently compacted by vibration alone, but it is usually desirable and often essential to have weight (sometimes great weight) on the vibrating surface.

Most vibrators used in compaction of earth or asphalt are rollers, but a few are flat plates. Tubular internal vibrators, such as are used in placing concrete, may be useful in stabilizing mud pockets in granular or mixed granular soil.

Hammers are made in a number of different designs, including slow and heavy weight droppers and jump rammers that crush particles together, to air and hydraulic hammers with such rapid strokes that part of their work is done by vibration. They are

Courtesy of Galion Manufacturing Company

Fig. 19-63. Three-wheel roller

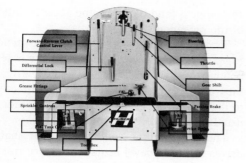

Courtesy of Huber Corporation

Fig. 19-64. Three-wheel roller, rear

Courtesy of Koehring Road Division

Fig. 19-65. Gears in drive train

used chiefly in trenches, and in places where obstructions prevent use of rollers.

STEEL WHEEL ROLLERS

Smooth steel rolls are used in consolidation of most blacktop surfaces, and in rolling gravel roads and road bases, and some subgrades. They produce a smooth, solid surface under favorable conditions, but may fail to compact hollows narrower than the roll, and do not compact deeply in proportion to their weight.

Standard models of smooth steel rollers are three-wheel and tandem. There are also three-wheel tandems, and models that have one smooth roll, with drive from tires or a tamping or grid roll.

Any steel drum that is not a drive wheel can be equipped with a vibrator. This is done most often with combination or tandem models. The mechanism and its effects will be discussed in a later section.

Three-Wheel. The standard three-wheel roller has a pair of large drive rolls in the rear, and a smaller but wider two-piece steering roll in the front. Weights are usually between 5 and 15 tons, but both smaller and larger machines may be available. The drive wheels carry about 70 per cent of the weight.

Weight can usually be increased substantially by filling the rolls with water or damp sand ballast, or sometimes by adding wheel weights. If the descriptive rating for a machine has two figures, as 10-14 ton, the first represents empty tonnage, the second the total of empty weight and maximum ballast.

Compression produced is measured in pounds per lineal inch of roll width. It is much greater in the rear rolls. The difference in compression and in diameter between front and rear produces some kneading effect that assists compaction.

Figure 19-63 shows one of these ma-

TORQUE CONVERTER   TAIL SHAFT GOVERNOR   2-SPEED TRANSMISSION

Courtesy of Huber Corporation

Fig. 19-66A. Roller drive train assembly

Courtesy of Huber Corporation

Fig. 19-66B. Drive roll

chines, with operator station on top. Figure 19-64 details controls in a machine in which the operator sits or stands at the rear.

Clutch and Transmission. A roller drive train must be designed so that starting, stopping, and reversal of direction can be done easily and with great smoothness. Rough or sudden response to any of these actions tends to cause shifting of material and over-compaction under the drive rolls. On fresh, hot blacktop (asphaltic concrete) this is likely to do permanent damage to the

Courtesy of Huber Corporation

Fig. 19-67. Guide roll

smoothness (and therefore the acceptability) of the pavement.

Two types of power train are shown in Figures 19-65 and 66. A diesel engine is set behind the front roll. In mechanical drive, the transmission jackshaft is coupled directly to the rear of the crankshaft. It is now more usual to have a torque converter here.

The jackshaft or converter output shaft drives a pair of bevel gears which turn freely on a reversing shaft, in opposite directions. A pair of disc clutches can be manipulated by a control lever to engage either one (but not both) to the shaft, or to release both of them. This last position, NEUTRAL, has the effect of an engine clutch for mechanical drive, or an output clutch with a converter.

The reversing shaft will not revolve when the clutches are disengaged, and will rotate according to the direction of the controlling bevel gear when one of them is engaged.

If a mechanical drive arrangement is used, two spur gears of unequal size slide on splines on the center of the reversing shaft and mesh with larger gears at the center of the intermediate shaft. Engagement of one or the other of these gears yields the primary shift between high and low, while the clutches provide forward and reverse shifting.

The intermediate shaft to the rear carries an outer pair of wide spur gears of unequal sizes, which are in constant mesh with two large differential housing gears. Either of these intermediate shaft outer gears may be engaged to the shaft by sliding jaws, and drive to the differential will be through whichever one is so engaged. This provides the secondary low—high shift, which, compounded with the primary shift and reversing clutches, gives four forward and four reverse speeds.

If there is a converter, the intermediate shaft is omitted. The sliding gears on the reversing shaft drive the large gears directly, providing a single two-speed shift between working and traveling speeds.

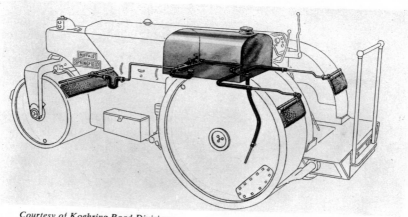

Courtesy of Koehring Road Division

Fig. 19-68. Watering system

The reversing clutches serve as a master clutch and must both be disengaged in order to shift gears.

A non-spin bevel gear heavy-duty differential carries the power to a pair of axles carrying spur-pinion gears at their outer ends. These mesh with large ring gears bolted to the inside of the driving rolls.

Hydrostatic Drive. In hydrostatic drive, a variable-displacement, reversible piston pump is mounted on the engine output shaft. It drives a similar, but usually fixed displacement, motor geared to a simplified transmission, which may be one-speed or two-speed.

This mechanism, which was discussed in Chapter 12, provides smooth reversing of direction and a stepless change of speed ratios, from zero to maximum speed at the throttle setting, by movement of a single lever. It also provides dynamic braking when in neutral.

Guide Roll and Steering. The front or guide roll is made up of two sections which turn independently. It is connected by a yoke and a horizontal hinge to an overhead kingpin. The weight of the frame is carried on roller bearings on the kingpin shoulder, and the kingpin itself is kept in line by tapered roller bearings. Steering is

by means of a two-way hydraulic ram, acting against a lever clamped to the kingpin. Extending the ram turns the steering roll to the left, retracting it, to the right.

In many machines steered with a bar or tiller, control is by a 3-position valve. Angle of turn continues increasing as long as the valve is held open, until stops are reached. When the valve is put in neutral, the wheel will be locked in its position, whether straight ahead or turned.

This steering mechanism may be equipped with a follower valve, that synchronizes the position of the roll with the position of a steering wheel, to produce automotive-type steering response.

The rolls are equipped with scraper blades, two to a wheel, which are held in light contact with the surface by springs unless locked out of contact. These are essential to prevent material from sticking to the roll and building up so as to spoil the smooth surface.

The roller may be equipped with a sprinkling system. This consists of a tank, valves, and piping to the wheels. At each outlet the water trickles over a cocoa mat, a fabric of wood fibers that distributes the water as a film of moisture over the entire roll surface as it turns against it. Moist rolls limits the depth to which the unit can be depressed. Depth of penetration is also regulated by setting the teeth up or down in their clamps, and by holding the wheels above the ground.

A roller has rather weak traction, so in order to do effective scarifying, it is usually necessary to put teeth or lugs in the driving wheels, in tapered sockets provided for the purpose. These sockets are normally filled with plugs flush with the roll. Use of spike teeth not only gives good traction, but damages the surface so that the scarifier has an easier job breaking it.

Courtesy of Koehring Road Division

Fig. 19-69. Hydraulic-lift scarifier

are required for certain types of black top work. The mats are swung out of contact when not in use.

Scarifier. A scarifier is a common accessory. Its base frame is bolted to the rear of the roller. The teeth are mounted in individual arms, fastened to a shaft which is raised or lowered mechanically or by a two-way ram hinged to the carrying frame. A pair of wheels which ride on the surface

TANDEM AND PORTABLE ROLLERS

Tandems. Tandem rollers are particularly adapted to compacting pavements, but they are found rather often on fills, particularly on light work. The size range is from about a ton and a half to twenty tons.

They usually have two rolls but may have three. Drive and guide rolls are the same width, but drive rolls are larger, carry 60 per cent of roller weight, and have about twice as much compression as the guides.

Courtesy of Koehring Road Division

Fig. 19-70. Tandem roller

Courtesy of Koehring Road Division

Fig. 19-71. Right-angle roll drive

Engines are center mounted, and may be either parallel to, or at right angles to, the direction of travel.

Steering is usually done by swiveling the front or guide roll by means of a hydraulic cylinder. At least one model also has an articulated frame swivel under independent control. This permits increasing the effective width of the roller by crab steering, with a single instead of a double rolling for each pass of the machine.

In gear drive, a reversing clutch is teamed with four speeds. Converter models have two speeds, hydrostatics either two or one. All speeds are reversible. Upper gear(s) are for travel only.

The single drive roll is fitted with a large bull gear. This may be driven at right angles through a small pinion, or by parallel shafts and reducing gears.

The brake is usually inside the transmission. In hydrostatics, most stopping is done automatically by the drive mechanism.

The frame is outside the rolls so that the machine cannot work against vertical walls or other obstructions high enough to hit the frame. Vertical clearance is 17 to 20 inches. Projection is greater on the drive side.

The segmented guide roll, Figure 19-72, provides greater compaction density. The solid compression roll largely irons out its prints.

Three-Axle. The three-axle tandem is similar in general construction to the two-axle, but has an extended gooseneck, which rests on a walking beam, or on hydraulically controlled kingpins. The beam or kingpins

Courtesy of Koehring Road Division

Fig. 19-72. Segmented guide roll

Courtesy of Koehring Road Division

Fig. 19-73A. Three-axle tandem

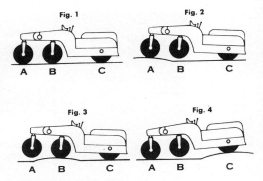

Walking beam locked

are supported in turn by two guide rolls in tandem.

The hydraulic steering synchronizes the turning of the rolls, so that scuffing is avoided.

The gooseneck-to-rolls mechanism can be locked to concentrate the front weight on either roll which is crossing a hump; unlocked so that weight will be distributed equally between them, regardless of ground shape, so that vertical curves can be followed, or partially locked so that the front roll can rise out of line but the middle one cannot, for maximum compaction on the high spots only.

Portables. There are a number of light tandem rollers, 6 tons or smaller, that are "portable", that is, they can be towed as semitrailers. They have a pair of rubber tired wheels that are held up out of the way during work, but can be lowered to support them when towed behind a truck.

In the simpler models, the wheels are moved from upper to lower positions manually, after driving one roll up a ramp or bank. In others, they are shifted hydraulically.

In either case, the wheels carry part of the roller weight when they are lowered, causing the guide roll (which is the lighter end) to rise off the ground.

Before lowering the wheels hydraulically, a towbar at the drive roll end is attached to the pintle (tow) hook of a truck. It is forced down hydraulically to lift the drive roll, so that the roller is supported on the rubber

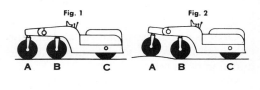

Walking beam semi-locked

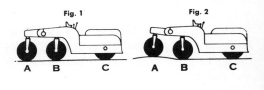

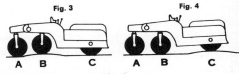

Walking beam unlocked

Courtesy of Koehring Road Division

Fig. 19-73B. Walking beam action

OPERATING

WHEELS SWUNG BACK

DRIVE ROLL RAISED

BOTH ROLLS RAISED

Courtesy of Koehring Road Division

Fig. 19-74. Portable roller

tires and the towbar. It can then be towed at moderate highway speeds.

Trench Rollers. Trench rollers are used for compaction of backfills in pavements and ditches, and for road widening when the strip is narrow.

The model shown in Figure 19-75 has a 16-inch wide compression roll and a 12-inch wide steering roll in tandem.

The leveling wheel, which may have either a rubber or an iron tire, is mounted on a ball crank controlled by a hydraulic ram. This wheel is raised or lowered until the frame and rolls are level or conform to road slope. At maximum height, it will allow rolling 16 to 23 inches below the leveling wheel.

OPERATION

Controls. So far as the controls are concerned roller operation is simple. In certain machines the reversing clutch lever is put in neutral, the shifting lever moved into the desired gear, and the clutch lever moved forward to roll forward, back to move backward. In some other rollers the main clutch is used for shifting and the reversing clutch for choosing direction and for starting and stopping.

With either torque converter or hydrostatic drive, there is only one gear for work-

Courtesy of Koehring Road Division

Fig. 19-75. Trench roller

ing speeds, and probably another for travel. The forward-reverse lever, the steering bar or wheel, the throttle, and sometimes the brake, are all you have to use.

Modern machines have hydraulic steering. If there is a steering wheel, control resembles that of an automobile.

Units steered by a tiller or steering bar are turned by moving the bar in the direction of turn until the front roll is at the angle desired. The bar is returned to center until the turn is completed, moved to the other side until the front roll is straightened out, and is then returned to center. Any adjustments in sharpness of turn may be made by moving the bar one way or the other.

The error to be avoided is leaving the bar in the side position as one would a car's steering wheel. This causes the ram to continue to move, making the angle sharper and sharper until the stops are reached.

Steering sharply is liable to cause scuffing and may result in severe damage to the surface when the roller is moving backward. A turned front roll then has a tendency to dig in on the edge turned toward the rear, and then tip so as to dig in further so that a rut is plowed in the surface. Work sequences should be arranged so that turns can be made while moving forward.

The differential lock is not used during compacting work. It is intended to give additional traction when scarifying, or walking over rough or steep ground.

The sprinkling system cannot be used in compaction of subgrades as dirt or gravel will stick to the wet rolls and come up in chunks, spoiling the surface. If water is needed for the compaction it should be supplied by water wagons sufficiently in advance so that it can dry slightly before the roller reaches it.

Stopping. Stopping and starting should be done gradually in order to avoid scuf-

fing of the surface. It is best to disengage the clutch before reaching the end of the run so that the roller can drift to a stop without using the brake or reversing the clutch. When it is stopped, the clutch for movement in the opposite direction is gradually engaged for a smooth start.

The roller should not be stopped repeatedly in the same place, particularly on blacktop, as this may cause creeping and formation of pockets.

Obstructions. Manholes or other obstructions in the road interfere with regular rolling patterns. The way they are handled will vary greatly with their height, construction, and location. In rolling up to an obstruction head-on, the clutch should be released or partially released before coming to it, so that the roll will just touch it without having momentum enough to break or climb it.

The pattern is rolled the first time as nearly normally as possible and curving passes made later. Spaces which the roller cannot reach without excessive maneuvering should be hand tamped or a smaller roller should be used in them.

If the obstruction is both low and strong it may be possible to ride the front roll over it and straddle it with the rear rolls.

Sequence. Rolling speeds are slow. One and a half to three miles an hour is usual.

The rear wheels of a roller do most of the compacting, particularly on the three wheel type. The smaller and lighter front roll serves to "work" and stabilize the soil.

In rolling deep loose material such as fill or gravel all passes in a series except the first should be overlapped at least half the width of the drive roll. Gradual extension of the rolled into the unrolled area makes possible greater concentration of weight on local ridges and high spots, and keeps the rolls running at a truer grade.

In rolling a graded area with a side slope, as a crowned or banked road, work

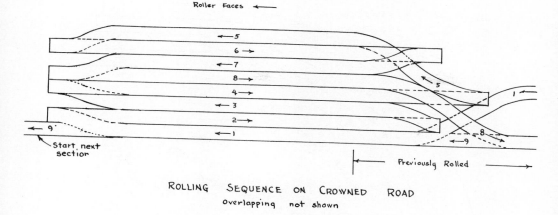

ROLLING SEQUENCE ON CROWNED ROAD
Overlapping not shown

Fig. 19-76. Rolling sequence, straight

should always be done from the bottom up. The lower edges of the rolls have a tendency to push downhill, which can be best resisted by compacted material. In working uphill the creep of soil away from the upper edge helps to preserve the slope.

A crowned road is rolled according to the pattern in Figure 19-76, starting at one edge and working up until the center is reached by the upper roll, then moving diagonally to the opposite side and working up from there. Each re-rolling is started at the bottom in the same manner.

It is efficient to roll sections as long as are ready for work at one time. Two procedures in general use are shown in Figure 19-77. In (A) a section is done completely at a time with ends overlapped a few feet, and in (B) a short section is started and lengthened steadily in one direction, following grading operations. When the original piece appears to have had sufficient rolling it is abandoned and the strip being rolled kept at one length afterward, being extended at one end and shrinking at the other.

Banked or super-elevated curves are

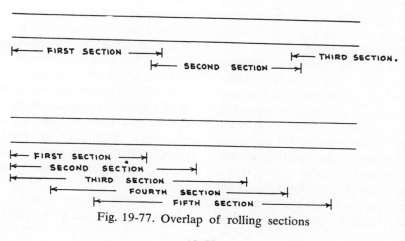

Fig. 19-77. Overlap of rolling sections

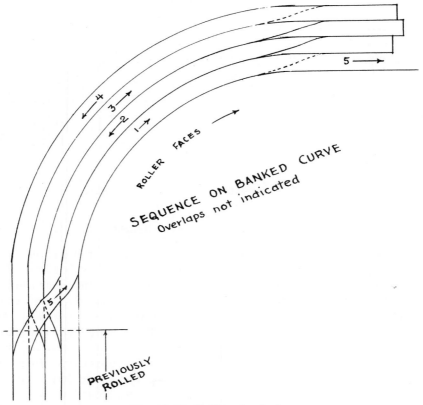

Fig. 19-78. Rolling banked curve

rolled from the inner edge to the outer edge, Figure 19-78. The transition from crown to bank is made by a diagonal from center to low side. From bank to crown the move is from either edge straight into adjoining low side. The meeting of these two types of grade is a convenient place to end a rolling section if the continuous advance system is not being used.

Rolling should be continued until no advantage is noted from successive passes. Presence of too much water in the subgrade may make its compaction impossible, but long rolling will at least bring much of the water to the top where it can evaporate more readily. The waterlogged condition results in a rubbery action of the ground, in which it goes down under the rolls and springs back into nearly its original condition when they have passed.

This condition may not be apparent at the start of the work, as the larger air spaces in the unconsolidated soil may be adequate to hold the water. As these spaces are reduced, however, the water is forced out of them and becomes a lubricant between all the particles.

Scarifying. The scarifier is used principally in ripping up old blacktop or oiled surfaces, but can also be used to pull out rocks or to loosen hard soils for a grader.

A roller does not have much traction on hard surfaces, so under many conditions it does not have enough pull to scarify effectively. This may be partially compensated by working downhill, by shallow cutting, and by reducing the number of teeth.

If a heavy pull is required, the drive rolls may be fitted with spikes. Tapered

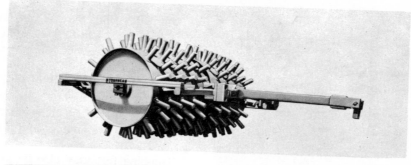

Courtesy of WABCO

Fig. 19-79. Tamping rollers

plugs are driven out of the rolls and points substituted. This is a laborious operation unless the rolls are new or spikes have been installed frequently. With worn rolls, there may also be some difficulty matching the plugs into the holes when replacing them, so that the smooth surface will not be spoiled by pits or bumps.

The actual scarifying is done by locking the roller differential, lowering the teeth into the ground or pavement by hydraulic control, and driving forward in low gear. At the end of the pass the roller is turned if the strip is long, or backed to the beginning if it is short.

If a pavement breaks into pieces too large to pass between the teeth they will become plugged and the scarified material will pile up ahead of them, as in front of a dozer blade. When this occurs the scarifier should be lifted and forward motion continued until the teeth have passed over the pile. They are then positioned so

that they just clear the pavement. The roller is backed and the pile pushed backward by the teeth until it is cleared off the unbroken pavement.

The teeth are then lowered into the pavement at the nearest point where full depth penetration was obtained, and the scarifying resumed. This cleaning off may be done as often as it is necessary.

Blocking by slabs can be reduced by taking out some of the teeth.

OTHER DRUM ROLLERS

Tamping Rollers. Towed tamping or sheepsfoot rollers were for a long time the standard tool for compacting fills. They consist of steel drums fitted with projecting "feet" and towed by means of box frames. On a soft fill layer, the roll will compact the surface somewhat while the feet compress the base with greater force. As the soil becomes packed, the feet do not penetrate as far, and first lift the roll clear and finally

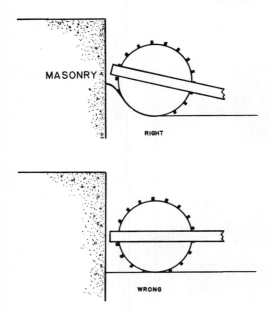

Fig. 19-80. **Compacting against a wall**

walk themselves almost out of the ground.

Feet are usually from seven to nine and a half inches long. Two types are used. The original sheepsfoot has an enlarged, off-center sole and a straight shank. The tapered foot diminishes from base to sole and may be round, square, or angular. The sheepsfoot is easier to pull, but is liable to tear up the ground when backed, and compacts only below the sole. The tapered foot kneads and compresses soil laterally for its full length, and works as effectively backward as forward.

Two or more units can be combined in multiple frames. These are hinged to oscillating bars in the rear, and to tow bars in front. The towing tongue should be spring-cushioned to minimize shock to tractor and roller.

Cleaner bars are usually put on the rear of the frame to remove dirt caught between the teeth. If the roller is to be used extensively in reverse on sticky soils it should have front cleaners also. Clogging destroys the effectiveness of the machine.

Individual rolls may be from four to six feet wide, and three to five feet in diameter, not including the feet. They can be filled with ballast of water or sand, and may carry sand boxes in addition. Foot pressures range from about 150 to 750 pounds per square inch, depending on the size of the unit, and the amount and kind of ballast used.

Tamping feet may also be mounted on multiple rings or narrow wheels, with a rear mounted weight box. The rings can turn independently, reducing or eliminating scuffing and the extra resistance it causes.

These rollers may be fitted with vibrators.

In compacting dirt against a masonry wall, the roller can operate parallel to the wall if there is room. If not, it can be backed in and the fill kept built up to conform to the roller's curve, as in Figure 19-80. This will reduce hand tamping requirements to a minimum.

A bulldozer may tow a tamping roller while spreading fill so that both grading and compaction can be done by one operator.

Self-Propel. Most tamping rollers are now self-propelled. They may be specially designed machines, or four- or two-wheel drive tractors with special wheels. These machines are usually center-articulated, with arrangements for rear wheel feet to step accurately between prints of the leading wheels, in forward or reverse movement.

Traction with four-wheel drive is remarkable, and it is often profitable for the machine to carry a dozer blade, for use in rough grading while compacting.

Self-propelled tamping rollers may be capable of speeds up to 20 miles an hour or more, forward and reverse. But long feet tend to kick out and loosen soil at over 6 miles, and thus limit working speed.

Shorter feet, or the pads described below, allow higher speed during compaction.

There are many tamping roller combinations. A single machine, usually articulated, may have a pair of rubber drive wheels and

Fig. 19-81. Pad rolls

Fig. 19-82A. Ridged rolls

a tamping drum, a tamper and a smooth wheel, or a tamper and a vibrating roll. Usually, each has its own special uses, with an additional range of general applications.

Pads and Crushers. Steel drums may be obtained with a variety of surfaces, some of which are shown in accompanying illustrations.

The pad roller carries wider, closer-spaced feet than a tamping roller, for more concentrated compaction effect to a lesser depth. It has increased effectiveness in loose soil.

Feet may be combined into parallel ridges around the drum, with height varying according to a pattern. Used on self-powered, center-articulated tractors, these are designed so that the rear ridges track between the front ones, even while turning. They are effective crushers of miscellaneous material, including soft rock, old pavement, garbage, and vegetation.

For most types of work, these rollers must have cleaner bars, preferably with replaceable tips. They cut out sticky dirt or trash in the ring grooves between lines of teeth, or ridges.

They do not reach material between teeth or pads in a row, but this usually squeezes into the cleaned groove with the next revolution of the drum.

RUBBER TIRE ROLLERS

The rubber or pneumatic tire roller substitutes a number of wheels and tires for

Fig. 19-82B. GRID roller

Fig. 19-83. Rubber tired roller

the steel drums discussed so far. They add to their downward pressure a kneading effect, as material is pressed toward spaces between tires.

Wheels are mounted in pairs that can oscillate, or singly with spring action, so that tires can move down into soft spots that would be bridged by a drum. There may be means to lock out individual movement for special work.

The typical rubber tire roller is a weight box mounted on two rows of tires in tandem. The rear row may have an even number, the front an odd number of wheels. They are aligned so that the rear tires more than cover the spaces left between front tire tracks.

Ballast in the boxes is usually dry, and almost anything will do. Sand, gravel, stone, and/or chunks of metal are used.

These rollers may be trailers. More often, they are self-propelled with (usually) torque converter, reversing clutch, and two to four speeds each way. Final drive may be by

Fig. 19-84. Wheel action in rubber tired roller

Fig. 19-85. Drive train

roller chains to pairs of rear wheels. Top speed may be 15, or over 20, miles per hour. Weights range from about 5 to 35 tons.

Rubber (pneumatic) tire rollers are successful in most types of compaction work, from raw fills to asphalt surfaces. On pavements, they apply compactive pressures similar to those from heavily loaded trucks. A final rolling with steel wheels removes any tire marks, if done promptly.

The tires, which are usually smooth tread, may be fitted with scrapers for sticky soil, and with mats and a watering system for blacktop work.

Heavy Compactors. There are larger rollers, made as trailers only, in which there is a single row of two or four tires across the center of the weight box. Stability is provided by the hitch to the towing tractor. There may be a vibratory attachment.

With ballast, these weigh from 50 to 200 tons. They are used for compaction of earth fills in layers up to two feet thick, and even thicker layers of broken rock.

These big jobs are not nearly as popular as the two-axle types. They are criticized as being slow and cumbersome, and prone to getting stuck, or tipping. Weak fills may be damaged by shearing, due to extreme weight. But under a number of conditions, they are highly efficient tools.

VIBRATORY ROLLERS

Vibration. Vibration of a tool in contact with the ground produces a rapid series of impacts that develop pressure waves which penetrate the soil, setting its particles in motion. If the tool exerts static pressure, the combined effect will be to rearrange the particles, and force them into a compact structure with a minimum of voids.

Vibration is usually produced by rapid rotation of an axle carrying off-center weights, called eccentrics. It is transmitted to the vibrator parts through ordinary bearings, and insulated from non-vibrating parts

Fig. 19-86. Fifty ton Rollopactor

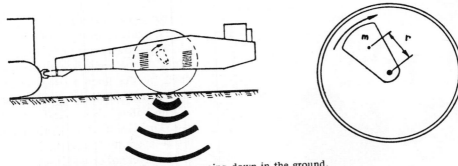

Figure 1. A vibrating roller develops pressure waves going down in the ground.

Courtesy of Vibro-Plus Products

Fig. 19-87. Vibration, and vibrator

of the machine by pads or hangers of flexible material, called isolators. Power from a fixed source to a vibrating member is usually carried by flexible belts, or hydraulic hose.

Measures of vibratory motion are frequency and amplitude. Frequency is cycles per minute, and depends on the rotation speed of the eccentric axle. There may or may not be means to change frequency during operation.

Frequency available in rollers ranges from a few hundred vibrations per minute (vpm) to 2400 or more, but in any one machine, the range is usually much smaller.

Amplitude is the distance of movement of the vibrating surface from a central position. It is half the total movement. It varies with the elastic and damping properties of the soil, but may be given a nominal rating, neglecting soil effect.

In some machines, amplitude may be adjusted by transfer of fluid between rotating

chambers, one of which adds to off-center weight, while the other counterbalances it.

In general, increase in ground speed calls for faster vibration. Otherwise, there is a great deal of contradictory evidence about the relative advantages of various speeds.

Most soils have a resonant frequency at which they vibrate most readily. A vibrator in contact wtih the soil will develop its greatest amplitude at that frequency. Some authorities consider that best compaction is obtained at resonant frequency, while others consider it less important.

Dynamic force is the centrifugal force produced by the eccentrics. In a general way, it represents the power behind the vibratory stroke.

Vibration in Rollers. A roll vibrates if it is mounted on a vibrating axle. Figure 19-88B shows a one-roll trailer with a 20 horsepower diesel engine, which is shear-mounted to isolate it from any vibration of the frame.

Drive is through a flexible coupling and

Courtesy of Tampo Manufacturing Company

Fig. 19-88A. Vibrating (eccentric) axle

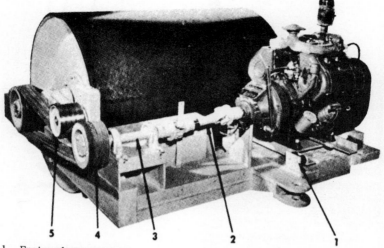

1. Engine shear-mounts.
2. Flexible coupling
3. Belt drive shaft
4. Dampened drive belt
5. Belt pre-load

Courtesy of Tampo Manufacturing Company

Fig. 19-88B. Towed vibrating roller

a flat belt to the eccentric axle. The axle turns in end bearings supported in vibration-damped brackets on the frame, and in another set of bearings on which the roll revolves.

The roll vibrates with the eccentric, and may be called the vibrating mass. The frame and the engine have little vibration, but their weight is essential to the effectiveness of the roll. They may be called the surcharge.

An eccentric may also be turned by a hydraulic motor and V-belt, substituting flexible hose for belts and couplings, and providing means for control of frequency independent of engine or travel speed, and for reversing rotation of the eccentrics. This reversal may be made automatic when direction of travel is changed, to reduce sinking and scuffing.

Many vibratory rollers are of articulated construction, with an engine and control section driving by means of a pair of large wheels, which are usually rubber-tired, but may be steel drums, smooth or with tamping

or pad feet. The other section is a drum roller, smooth or otherwise, with a vibratory axle turned by a hydraulic motor.

There is also a double-drum, double-articulated machine for asphalt, with engine, controls, and drive wheels in the center section. The manufacturer recommends that on breakdown work, the lead (leading or front) roll be static, the rear vibrating; that both vibrate during compaction, and that only the lead should vibrate on the finish pass. The smooth drive tires carrying the center section are not rated for any compaction effect.

Vibrating rollers usually give best results when moving slowly, from 1½ to 3 miles per hour. They will provide some compaction at higher speeds, but must usually do the same area a greater number of times.

Vibration greatly increases roller effectiveness on loose, granular soils, but makes littel difference on silt and clay. The rule of thumb is: the larger the grain, the greater the effect. Comparisons between vibratory

Fig. 19-89. Vibrating roll, non-vibrating drive wheels

and static rolling are therefore of little value unless the soil type is specified.

Under favorable soil conditions, a vibrating roller may produce compaction equivalent to that of a static roller 2 to 4 times as heavy. Some of them are therefore very small, even so small that the operator walks alongside instead of riding.

Reversing. Stopping, then moving in the opposite direction, may be a critical operation with any roller, particularly on plastic material such as hot blacktop or damp clay.

The roll stands longer in the stop-spot than any other, and may sink in. The danger is increased by stopping resistance and starting torque, and can be avoided only by smooth and prompt operation of controls.

The problem is even more acute with a vibrating roller. The vibration is usually not affected by travel speed. It keeps working the roll downward during the pause, and may create an embarrassing hole if not properly controlled.

One machine reduces this problem by

Fig. 19-90. Double vibrating drums, and double articulation

automatically reversing the eccentric shaft when travel direction is reversed. This brings the eccentric axle to a momentary stop as the roll stops.

TAMPERS AND VIBRATORS

It is often not practical to use rollers against walls, between manholes, in ditch bottoms, and many other places where space is restricted.

Soil in such spots is compacted by tamping. Considerable benefit can be obtained by walking or stamping down two or three inch layers, or pounding with sledges or other tools.

Hand Tampers. Special hand tampers usually consist of a square or rectangular

Courtesy of Koehring Road Division

Fig. 19-91. Pad roller

steel plate, with a handle fastened at right angles to the center of one side. It is lifted and dropped or pounded down on the soil as many times as are required. Best results

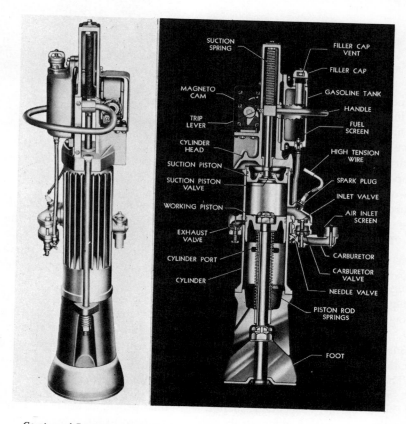

Courtesy of Barco Manufacturing Co.

Fig. 19-92. Barco rammer

Courtesy of Master Division, Koehring Company

Fig. 19-93. Vibrating tamper

Courtesy of Jackson Vibrators, Inc.

Fig. 19-94. Vibrating compactor

are obtained if the dirt is hand shoveled to the spot in thin layers during the tamping.

The tamping foot may have other shapes for particular work such as packing material in narrow crevices, and under ties or other supports.

This work is slow and laborious, and results vary with the skill and energy with which the tool is handled.

Barco Rammer. Mechanical tampers, which may be operated by gasoline, air, or electricity, may operate by jump pounding, by leaving the shoe on the ground and hammering it, by vibration, by combined hammering and vibration, or by weight dropping.

The Barco Pegson rammer, Figure 19-92, includes a single cylinder, two-cycle engine. The working piston is attached to the foot. Fuel mixture is drawn into the cylinder by a spring suspended suction piston. At the top of the suction stroke, a magneto cam is tripped, firing the spark plug. The force of the explosion drives down against the foot which is resting on the ground. Reaction from the downward blow lifts the whole machine a foot or more into the air. It falls by gravity, and is fired again when the suction piston reaches the top of its stroke.

The machine weighs about two hundred pounds, and can be operated at a rate of 50 to 60 jumps per minute. Tilting it slightly in the direction of the desired motion while working will cause it to walk in that direction.

This machine is no longer made, but similar principles are used in current makes. It is more usual, however, to have a small two cycle engine of standard design operate the jump mechanism through a crank.

Paving Breakers. Paving breakers, which will be described in the next chapter, can be equipped with tamping feet. The foot normally stays on the ground and is hammered by the piston and anvil block.

The blow is quicker and lighter than that delivered by the jump type. Considerable vibration is set up which tends to rearrange and pack the soil particles, and adds to the effectiveness of the hammering.

Depth of layers compacted varies from four inches up, depending on soil type, the weight of the breaker, and the size of the foot.

Vibrating Tamper. The master backfill tamper, Figure 19-93, has a gasoline engine that operates a reciprocating type vibrator, with frequencies between 6000 and 9000 cycles per minute. It has a very short stroke.

The machines weigh from 75 to 100 pounds. Working depths depend largely on soil characteristics. Compaction may be obtained for some distance around the foot, by vibratory settling. The machine tends to move in the direction of lean.

The machine can also be equipped with edged tools for cutting pavement or hard soil.

Traveling Vibrator. This flat vibrating unit, Figure 19-94, is primarily intended for compaction of blacktop, loose granular soils, clean gravel, and crusher rock.

These machines move themselves along the working surface at speeds up to 30 feet a minute, when hand operated. They can also be mounted in gangs held in a lift frame on a tractor, which is also equipped with a generator to provide current for them.

Vibratory Probe. Sandy soil may be compacted to depths up to 60 feet by means of an open pipe clamped in a pile-driving vibrator. This is sunk and withdrawn vertically at calculated intervals, usually 4 to 10 feet, in a square grid pattern over the area to be stabilized.

Vibration packs the soil particles closely together, and forces excess water out of the top of the formation. If the soil type is suitable, compaction can be accomplished under water.

Weight-Dropping Tamper. The unit in Fig. 19-96A, which is no longer manufactured, is a self-propelled weight dropper which is effective in breaking pavement, and in tamping backfill in trenches.

The weight which may be 400 or 600 pounds is suspended in a tower by a cable over a head sheave. The sides of this weight fit into vertical guide tracks, and a fitting on the bottom provides for the attachment of cutting or tamping tools.

A one-way ram is mounted on the side of the tower. Its piston rod carries a double sheave. The hoist line for the weight is reeved around these pulleys and a pair on the top of the tower. Extending the ram pulls the cable so that it raises the weight; releasing the ram allows the weight to drop. The single line attached to the weight moves four times as fast and as far as the piston rod sheaves. Impact can be controlled between 100 and 4000 or 5000 foot pounds, depending on the weight used. Stroke is about 25 per minute, maximum compaction depth 14 feet.

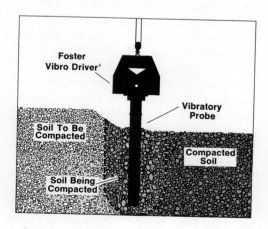

Courtesy of L. B. Foster Company

Fig. 19-95. Vibratory probe

TOWER

HOIST RAM

HAMMER

30" TAMPER

TOWER-TILT ADJUSTMENT

Fig. 19-96A. Weight-droping tamper

The machine is six feet wide and can straddle ditches up to four feet. The tower and front of the chassis can be shifted hydraulically (sliding tower control) to place the weight over any part of the four foot width, without moving the wheels. The tower can be tilted back hydraulically to reduce overall height while traveling, and can be adjusted to a vertical position when the machine is headed up or down hill.

A hydraulic clutch is used to give an infinite variety of creep speeds for working. Maximum road speed is 18 miles per hour. Machine weight is about 5000 pounds.

Shallow to medium depth trenches are completely backfilled and pounded down. More fill may be added and compacted if necessary to bring it up to the bottom of the pavement. Deep trenches may be filled in two or more lifts.

Power-Down Hammer. Newer portable hammers may use hydraulic pressure to supply power for the down stroke, as well as for the lift. The CMI Champion in Figure 19-96B may operate on the rear of either a truck or a tractor.

The truck is much faster and more convenient moving between jobs, while the tractor is more maneuverable.

This model has a downstroke force that is adjustable by the operator to an impact force from 80 to 16,000 pounds. Cycle can be varied from zero up to 110 strokes per minute.

As with the slower and less forceful weight droppers, this tool is used to compact and stabilize deep, narrow fills, as in trenches. It can be fitted with tools for breaking pavement (illustrated) or boulders, or for cutting reinforcing bars.

PILES

Piles are relatively long, thin pieces of strong material that are driven (hammered) or vibrated into the ground to provide structural support. The unmodified word "pile" usually means a bearing or foundation pile whose strength is required in a nearly vertical plane to carry a load, usually a structure.

Sheet piles, or sheet piling, are used chiefly to resist horizontal movement of earth and/or water.

Bearing piles may be classified as end bearing or friction. If end bearing, they are designed to have their lower ends on or in a strong formation, and act primarily as columns between that base and the structure.

A friction pile transmits the load by friction to the soil surrounding it, and in conservative design, is likely to carry the full load by its sides, and none at the bottom.

The standard materials for piles are wood (timber), concrete, and steel.

BEARING PILES

Timber (Wood). Wood is the original pile material, and supported prehistoric buildings and bridges. It may still be the most widely used, and under many conditions it is the most economical type.

Almost any tree that grows a straight trunk may be used as piles, but preference is given to a few varieties, such as oak and

Courtesy of CMI *Champion*

Fig. 19-96B. Tractor hammer

southern yellow pine in the East, and Douglas fir in the West. Strength, soundness, and rot-resistance are desirable qualities.

Most piles for permanent structures are now pressure treated with creosote or other preservatives. This is a protection against marine worms below salt water level, and against a variety of destructive organisms above ground water everywhere. It reduces the importance of the durability of the native wood.

A pile should be as straight as possible, as bends create problems in driving and in load bearing. If a straight line between the centers of the ends is not entirely within the wood, the pile is not usable for most purposes.

Wood has the largest surface area in proportion to weight of any pile material, making it most suitable for friction bearing. It is easy to handle, as it can be picked up in any convenient way without damage, and it is resistant to all kinds of mild abuse.

However, it must be driven with caution, to avoid splitting and brushing out of the top. Splicing one onto another is difficult and unsatisfactory. Wood is subject to destruction by fire while in storage, and in its above-ground sections when part of a structure. Even treated wood is shorter lived than either steel or concrete under most conditions.

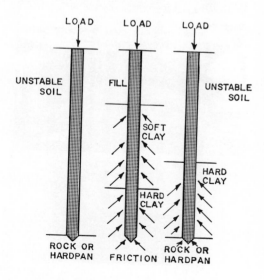

Fig. 19-97A. Load bearing piles

Courtesy of Intrafix

Fig. 19-98. Cutting pile tops

Concrete. Concrete piles differ greatly among themselves in the way they are made, and in shape. They may be cast in a yard, and either prestressed or posttensioned (compressed longitudinally by tension in reinforcing rods or cables) to increase strength for weight, and to reduce danger of breakage. They still must be handled with extreme care, and they are very heavy.

Most of them are cast in place, that is, made in the opening in the ground that they will occupy when part of a structure. The first installation step is to drive a cylindrical or tapered steel shell to full depth. This is usually a rather thin-walled tube, given strength by inserting a steel filler called a mandrel. The mandrel is withdrawn as soon as driving is complete.

The shell may then be filled with concrete, and left to form part of the finished pile. Or an inner shell or liner may be placed to serve as a form, and the original or outer shell withdrawn after concrete is placed. Sometimes the original shell is filled with concrete, and then withdrawn from it before it has set.

A shell that is left in place, or that is withdrawn while the concrete is still plastic, maintains the pressure of the pile against the surrounding soil, which may be important to its load bearing capacity.

Concrete piles are the longest lasting type under most conditions.

With prefabricated sections, new ones can be secured to those in the ground by welding reinforcements, and/or fastening concrete to concrete with epoxy mortar, so great depths can be reached.

Steel. The most common cross section shapes used for steel piles are H and pipe. The H is shaped like an I-beam, but has a web (H cross-bar) that is shorter (lower) in proportion to the flanges, and it is heavy steel in all parts. Size ranges from 8 x 8 inches to 14 x 14.

The pipe pile is usually filled with concrete, but sometimes the filler is sand. If it is small, between 6 and 18 inches in diameter, the bottom may be closed, preferably with a flat plate, but sometimes with a point. Larger sizes are driven with the bottom open, and dirt is cleaned out afterward with high pressure water and/or air.

If diameter is over 36 inches (or even 30), a pipe may be considered a caisson.

Steel piles may also be fabricated in a number of different shapes, for special conditions, or to use available material such as railroad track.

Steel handles fairly well, takes very hard driving, and will carry heavy loads when driven to rock or into resistant material. The main drawbacks are low skin friction in proportion to weight, and poor resistance to rust damage in above-water sections.

Steel sections can be welded together readily, making a high strength joint. It is also easily cut when too high after driving.

Composite. A composite pile is one that has its lower part made of one material and the upper part of another. The most common type has wood below, and concrete from a few feet below water level to the top. The object is to avoid wood in the area where it is apt to decay.

A composite of steel and concrete usually has steel at the bottom, for the same reason. Occasionally, it is made the other way for structural reasons.

The joint between two materials may be a technical problem. If any bending stresses are expected, it is made inside a steel collar.

Cutting Piles. Piles are usually driven until some specified resistance is met. They are then often too high for the requirements of a structure, or are too irregular to use, so that some or all must be cut off.

Steel is cut readily with an ordinary oxy-acetylene torch. Reinforced concrete is more difficult. A carbide saw, or a jackhammer and a cutting torch, may be used.

Either type of pile can be burnt off quickly with an Oxy-Gun melting torch. This uses rods or tubes made up of metal

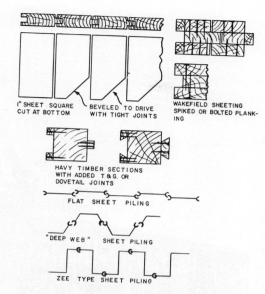

Fig. 19-99. Sheet pile cross sections

wires enclosed in iron. Oxygen is fed through the tube, and its tip is ignited. Flow of oxygen causes the rod to burn slowly at the tip with intense heat, up to 6600 degrees F. This is sufficiently hot to penetrate concrete quickly by melting it.

The pile is usually burned through in just enough places to cut all the reinforcing. This weakens the concrete, which can then be broken off with a sledgehammer, or by nudging with some machine.

SHEET PILES

Wood. Wood sheet piling may be made

Fig. 19-100. There are many interlock designs

Courtesy of Vulcan Iron Works, Inc.

Fig. 19-101. Pile driver on trestle

of planks, either square edge or tongue and groove. Another type is made of three layers of square edge planks laminated and spiked together, with the center layer offset to produce tongue and groove effect at the edge. This may be called Wakefield sheeting.

The first of a set of tongue and groove piles is flat on the bottom, and is driven straight down. The rest of a set are cut diagonally, Figure 19-99, with the point on the groove side. They are driven on the tongue side of those in place, so that the shape of the bottom will crowd each one into the necessary close contact with its neighbor.

Wood piling will not take hard driving, and may be difficult to keep interlocked.

Steel. Steel sheet piling has the advantage of an interlock arrangement that can hold and seal both in compression and in tension. This makes it easier to keep it together while driving, and assures that once driven, it will not be pulled apart by any ordinary forces.

There are three principal cross sections, as illustrated: flat, deep web, and Z. The second is most widely used. The corrugated effect obtained for the finished wall increases its strength greatly over the flat design, and it is easier to use than the even more rugged Z (zee) type. The arches are usually faced in opposite directions alternately, but if a narrow wall is required, they may be faced one way, with loss of stiffness.

These piles can be obtained in lengths up to 70 feet, but because of difficulties in handling, it may be preferred to get shorter lengths, and weld on additional pieces as needed.

Ordinary mild or structural steel is standard, but alloy piles can be obtained. Their extra cost is often more than covered by savings in bracing.

King piles are standard sheet piles to which an H or an I-beam has been welded, to provide stiffening.

Sheet piles may have one or two holes for handling at one end. These should be welded or plugged if they are driven to near water level. Handling is a delicate job, as they are easily bent, and once bent, may be difficult or impossible to keep locked during driving.

Setting. A guide structure is necessary for lining up the piles and starting the driving. This may be the top level of permanent bracing, or a temporary structure. It is desirable to place all the piles for a wall section, or for the whole perimeter of a coffer dam, before any serious driving is done. But a few, say one in ten or one in five, may be driven a few feet for temporary anchorage.

Wind pressure is a major problem, which tends to increase with length of wall. High temporary braces, or wires run from anchors to the handling holes, or both, may be necessary to avoid blow-over after setup and before driving. Even after the bottoms are securely embedded, a pile wall might be bent over by high wind.

Setting should start at a corner. Each pile must interlock with the previous one, and they must be vertical.

Driving. Driving is done a few feet at a time, working from one side to the other, or around a cofferdam. Corner piles and king piles are most rigid, and should be driven a little ahead of others, which can adjust to small misalignment more readily. A light, fast hammer is best for starting. A vibratory hammer may do the whole job in cohesionless soils.

Interlocks may be greased to reduce friction, or left dry to avoid trouble with soil-

Courtesy of Vulcan Iron Works, Inc.

Fig. 19-102. Drop hammer

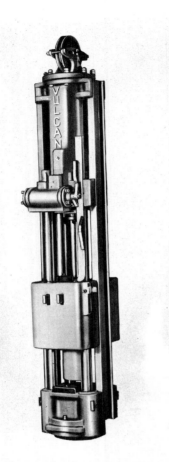

Courtesy of Vulcan Iron Works, Inc.

Fig. 19-103. Single-acting hammer

grease mixtures if in sandy formations.

Sheet piling is driven down far enough to prevent soil and/or water from entering an excavation. If bottom is firm, a few feet below the floor is enough, but if it is very soft, a considerable extra depth will be needed. A trench needs less over-driving than a wide area.

A newly driven cell of sheet piling will be apt to leak badly until water can be lowered enough to cause outside pressure to force the interlocks into sealed position.

Finishing. When the job is finished, sheet piling may be cut off flush with the ground or foundation line and left, or it may

be pulled out for re-use. Extractors are crane-held hammers or vibrators with gripping devices. In general, the crane should exert only moderate upward pull, allowing the extractor to do most of the work.

PILE DRIVERS

A pile driver is a machine that hammers piles, poles, or sections made of wood, steel, or concrete into the ground. Piles are used to support buildings on soft ground, to make retaining walls, and to keep soil and water out of excavations.

The most widely used pile driver, Figure 19-101, is a revolving shovel that has a crane boom and a lead (or leader) which serves as a guide for a weight that is dropped or hammered against the pile. There are also big specialized machines which are moved around the job on skids or tracks, and homemade contrivances made up of a tripod, a leader, and a weight, with power provided by travel of a vehicle or animal.

The spotter is an approximately horizontal connection between the shovel deck and the leader, hinged at both ends. It may be one-piece or adjustable for length.

Drop Hammer. The drop hammer is a weight that is hoisted along the leader guides by a cable, and released to fall by gravity onto the pile. If hoisting is done by a winch on the shovel deck, as in Figure 19-102, its brake and clutch are released for the drop. If it is done by towing, or by a winch that will not unspool freely, nippers are used. These grip the weight when lowered onto it, and release it when they contact a tripper which can be set at any point on the leader.

Steam Pile Hammer. In these devices, the hammer rests on the pile or pile cap throughout the operation, and the blow is struck by a ram that moves up and down on guides within the hammer. In a single-acting hammer the ram is raised by steam or compressed air and dropped by gravity;

in a double-acting type it moves by power both ways.

Steam is generated by a boiler mounted on the shovel deck. Many very old steam shovels are still in active service because their single power unit provides for all crane controls, and for the hammer also. (Many construction men wish that some company would build a modern steam-powered crane.)

Gasoline or diesel powered cranes may mount an auxiliary boiler and fittings on a rear extension of the deck. Others obtain steam through hose from stationary boilers.

Compressed air can be used in many steam hammers without changing over any parts. The compressor may be mounted on the shovel or be on the ground. Recommended capacity varies from 50 to 1250 cubic feet per minute, depending on the hammer size.

Diesel. The diesel pile hammer strikes the anvil with the combined forces of gravity fall of its ram-piston, and the firing of a fuel-air mixture.

Figure 19-105 shows the cycle of a single-acting diesel. With the hammer in a vertical position in the leader, and its anvil and drive cap resting on the top of a pile, the piston is raised by a load line, until disconnected by an automatic trip mechanism.

The piston then falls, freely at first. It actuates a cam on a fuel pump, which delivers a measured amount of diesel oil which falls into a cup in the top of the anvil. The falling piston next covers the cylinder's exhaust ports, and begins compression and heating of air trapped in the bottom.

A ball point on the piston mates perfectly with the cup in the anvil, so that on impact it sprays the fuel in the cup into a ring of hot, high-pressure air between the rims of the piston and anvil, where it ignites.

The resulting explosion drives the piston upward, and the anvil and pile downward. The piston rises until stopped by gravity, then falls to repeat the cycle. Action is

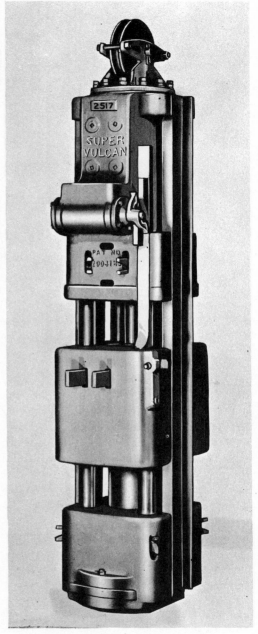

Courtesy of Vulcan Iron Works, Inc.

Fig. 19-104. Double-acting hammer

stopped by pulling a rope that disengages the fuel pump cam.

The pile is driven first by the gravity fall

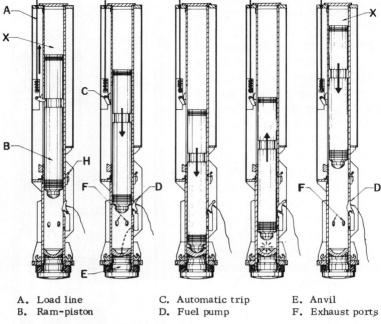

A. Load line   C. Automatic trip   E. Anvil
B. Ram-piston   D. Fuel pump   F. Exhaust ports

Courtesy of MKT *Division, Koehring Company*

Fig. 19-105. Diesel hammer action

of the ram-piston against the anvil, then by the explosion of fuel. This acts against the pile and the piston, with the force tending to take the line of least resistance.

If the pile drives easily, it will absorb most of the force by moving downward, and the piston will be raised only a short distance. Greater pile resistance will mean less movement, diverting explosive energy to throwing the piston higher. This will result in a longer fall and greater impact on the next stroke.

In this way, the diesel hammer automatically increases its impact as driving becomes harder.

The hammer must encounter a certain amount of resistance in order to fire, so it may not operate satisfactorily with light piles in soft ground, or with mis-aligned piles that cushion the blow by whipping. A dynamic weight may be put between the anvil and the drive cap to increase impact resistance.

The diesel hammer has the advantage of not requiring outside help except for the lifting crane. Neither a steam generator nor an air compressor is needed.

Bases and Heads. A standard or solid base for a steam hammer has a conical recess that holds the pile so that it cannot slip out of line, and a center hole through which the ram strikes the top of the pile. Heads are available for steel, pipe, wood, and concrete piling which fit this recess on the top, the pile on the bottom, and provide for directing and spreading the blow so as to avoid damage to the pile top. Cushioning may be included. A steel disc may be fastened to the top of a wooden pile.

Another form of base is provided for soft wooden piles in which a flat steel plate inside the base rests on the pile.

Extractor. A pile extractor, Figure 19-106, consists of a steam or air hammer which strikes its blow on the up stroke, and is equipped with clamps or other devices by which it can hold the top of the pile or sheeting that must be pulled. A crane keeps a steady lift on the extractor which grips the pile and hammers it upward.

The vibratory units to be described below are highly effective extractors.

Hammer Weight. The striking energy of the hammer or ram depends upon the weight of the moving part and the velocity at which it strikes the pile. The weight, however, is the more important of the two factors in securing penetration. A fast blow tends to cause shattering, where a slower one of the same total energy will have more driving power. This is in part due to the inertia and resilience of the pile. Under various conditions, blows struck at the rate of 80 to 150 per minute on a short stroke have an efficient velocity.

The blow should be heavy enough to drive the pile with a minimum number of contacts. This is particularly important when the driving must be to grade, rather than until a specified resistance is encountered.

It is usually recommended that the weight of the ram be equal to the weight of the pile, and in no case should it be less than one quarter of the pile's weight.

A much lighter hammer may be adequate if water jets are used ahead of the pile to erode a path for it.

Operation. The crane is walked to the approximate working position. The hammer is raised by the hoist line high enough in the leader to clear the pile. The pile is raised by the other winch line until vertical and hanging in the leader, then lowered at the point where it is to be driven and lined up.

Most piles are driven vertically. In such cases the boom point is moved by walking, swinging, and raising or lowering until the lead and pile hang straight to the driving point. If the lead is solidly connected to the deck, boom angle is fixed and cannot be changed while lining up.

For oblique (batter) driving, the boom can be raised until its point is inside the spotter hinge so that the lead and pile will lean toward the shovel and will be driven away from it. Lowering the boom will line up the pile to be driven under the crane. An adjustable spotter that will move the foot of the leader toward or away from the deck simplifies these adjustments and makes a wider angle of operation possible.

The drop hammer is lifted by the hoist line to the desired height when the hoist drum is released to allow the hammer to fall freely. Just as it hits the pile, the clutch

Courtesy of Vulcan Iron Works, Inc.

Fig. 19-106. Pile extractor

Courtesy of Vulcan Iron Works

Fig. 19-107. Portable pile hammer

is engaged for the next lift, sufficient interval being allowed for the follow-through of the blow, but not enough to allow the line to spin out.

The steam hammer is rested on the pile and the steam or air line opened to activate it. The hoist line is held slightly slack and is allowed to play out freely as the hammer works down.

The first few blows are struck lightly to get the pile started down smoothly. If at any time during the driving the top of the pile should show signs of failure, the velocity of the blows should be reduced. This is done by reducing the lift of a drop hammer, or throttling down a steam hammer. If much steam hammer work is to be done at lower velocity, the pressure at the boiler or receiver may be reduced.

If a steam hammer is shut down long enough to cool, it should be started slowly to work water out of the cylinder.

It is usual for the resistance of the pile to increase with successive strokes, both because of increased friction and penetrating more and denser layers. The point of refusal is where it cannot be driven further.

The safe bearing load of a pile is calculated from the weight and velocity of the striking hammer and the "average set" or penetration obtained from the last few blows.

Portable Pile Hammer. Figure 19-107 shows a 5000-pound pile hammer, with striking parts weighing 900 pounds, that can be used with light hoisting equipment and self propelled drills. It requires up to 125 cfm of compressed air.

The leader often consists of a drill feed mast, or a short slide bracket allowing only limited movement that can be mounted on the bucket arms of a front or swing loader. With such mountings the hammer can drive short piles, sheeting, fence posts, and heavy pipe.

A tool base may be substituted for the driving base, for attachment of demolition tools for breaking concrete, masonry, or boulders; or tampers for compacting fill in trenches and against buildings.

Vibratory Driving. Piles may be driven and extracted by vibration instead of hammering.

One design of vibratory driver/extractor is shown in Figure 19-108. This unit is suspended from a crane without leads, and clamped to the top of the pile.

The vibrating unit includes two (or sometimes four) eccentrics mounted on shafts that are belt or chain driven by electric or hydraulic motors. The motors and much

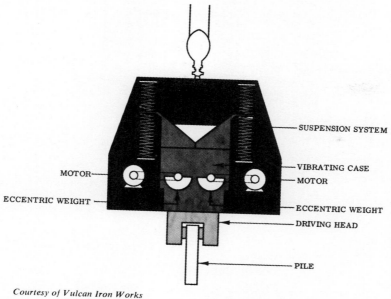

SUSPENSION SYSTEM

VIBRATING CASE
MOTOR
MOTOR

ECCENTRIC WEIGHT
ECCENTRIC WEIGHT
DRIVING HEAD

PILE

Courtesy of Vulcan Iron Works

Fig. 19-108. Vibratory driver-extractor

of the weight of the unit are insulated from vibration by a suspension system.

The eccentrics are equal in weight and operate oppositely and symmetrically, generating vibration in a vertical plane only. Vibration is transmitted to the pile by a hydraulically actuated clamp in the driving head.

Vibration of the pile breaks the grip of soil on its sides and permits it to flow under its tip, while the combined weight of the pile and the driving unit move it down. There is no hammering, so the noise of driving is only that made by its power source and the crane engine, usually both diesels.

Three types of vibratory designs are used. They differ chiefly in frequency, and are rated as high, 3500 to 10,000 vibrations per minute, medium at 1500 to 3500, and low (the example illustrated), 400 to 1500.

A number of factors enter into a choice. Low and medium frequency cause only up

and down motion of the pile, high frequency (sonic) causes it to expand and contract horizontally. Power requirement increases as frequency becomes higher.

Vibratory drivers are most effective in granular soils, but may do well in fine soils that are wet. Thick piles (unless hollow and open) are more difficult than thin ones, because of longer travel distance of particles vibrating from under the tip.

Whenever the vibrators are applicable, penetration speed is usually much greater than with impact driving. Pile tops are almost never damaged.

The vibratory driver may be used for extraction without change or adjustment, and is almost always highly effective. A moderate upward pull is supplied by the crane. Vibration loosens soil grip on the pile, or causes the soil to shear a few inches out from it.

CHAPTER TWENTY

COMPRESSORS AND DRILLS

RECIPROCATING COMPRESSORS

Light and medium drilling and breaking tools are usually powered by air supplied by portable compressors that are driven by standard types of industrial engines. These machines draw in atmospheric air, compress it, and deliver it through hose to the tools. The compressed air provides a power medium somewhat similar to steam in performance, with the important difference that it can be used cold.

Types of Construction. Until recently all compressors used in excavation work were of the reciprocating type. These machines have cylinders, each of which takes in air during a suction stroke, and discharges it through a check valve at higher pressure during a compression stroke. These may be in the same block as the engine cylinders, but more often are in a separate block with a disk clutch connection between crankshafts.

The diagram in Figure 20-1 shows the air flow in a single stage compressor. The inlet passage receives atmospheric air through a cleaner. The discharge is at full working pressure (usually 100 pounds), and opens into a line to a storage tank (air receiver) from which it is piped to the tools.

The two stage machine, Figure 20-2, has one or more primary or low pressure cylinders which draw in atmospheric air, compress it to about 30 pounds, pass it through a cooling radiator (intercooler) to a secondary or high pressure cylinder, where pressure is increased to 100 pounds and discharge is to the receiver.

The newer compressors are rotary vane or screw-type air pumps, with either one or two stages. Most portable compressors now being manufactured are rotaries.

In all constructions, it is important that the incoming air be filtered to prevent dust from causing excessive wear, fouling of moving parts, and choking of passages. Intake may be through one or several passages or manifolds.

Rating. Compressors are rated according to the number of cubic feet of atmospheric air they take in each minute, when working at maximum governed speed, with a specified pressure, usually 100 pounds, at the discharge. This rating is abbreviated to CFM or cfm. The "actual CFM" is this intake measurement, less any losses by leakage in the machine.

Actual CFM, or actual air delivery, is the volume of compressed air coming from the receiver outlet, transformed back to standard atmospheric pressure at sea level.

Some garage and industrial compressors are rated according to piston displacement per minute at full speed. This figure will always be larger than the CFM in the same compressor, as air left in the space above the piston on the compression stroke ex-

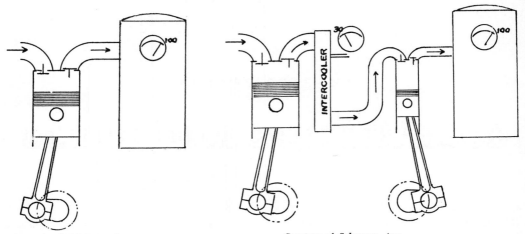

Fig. 20-1. Single stage compression

Fig. 20-2. Two stage compression

pands and partly fills the cylinder on the suction stroke so that the full displacement cannot be filled with new air.

The relationship between CFM and piston displacement is called volumetric efficiency. This figure may be between 65 and 80 percent, and depends largely on the air space above the piston and the increase in pressure attained in the cylinder. In two stage machines, it is calculated on the basis of the displacement in the primary cylinders only.

Pressure Control. Air-operated auto-matic controls are used to keep the pressure in the receiver within certain limits, usually 90 to 100 pounds. In addition, the engine must have a mechanical governor to prevent it from racing when air pressure is low.

During operation the engine and compressor continue turning at all times. When the receiver pressure reaches maximum, the compressor intake valves are held opened so that air that enters on the suction stroke is forced out again on the compression stroke. In this condition

Fig. 20-3. Single stage compressor

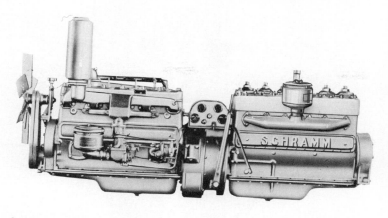

Courtesy of Schramm, Inc.

Fig. 20-4. Engine-compressor unit

the compressor is said to be unloaded. The engine is throttled down at the same time the unloading takes place. In rotaries, the throttling of the engine automatically unloads the compressor.

Heat. Compression of air produces heat. The working parts of the compressor tend to become very hot and must be cooled by fins, water jackets, or an oil bath.

The ability of air to hold moisture is decreased by pressure and increased by heat. As the hot compressed air enters the receiver these effects are often balanced, but as the air cools in the receiver or in the lines extensive condensation may occur, particularly on damp days.

Compressed air loses some pressure as it

cools so that the compressor must pump air to compensate for shrinkage, as well as to supply the tools.

Single Stage. The Schramm compressor, shown in Figures 20-3 and 20-4, is an example of single stage construction.

The compressor proper consists of four or six cylinders in line in a single block. Standard industrial engine types of pistons, rods, crankshaft, bearings, and cooling system are used, and lubrication is by force feed through passages in the shaft.

The intake valve, Figure 20-5, is a poppet type which opens during each downstroke of the piston. The push rod consists of two opposed pistons, sliding in an open cylinder. These pistons are nor-

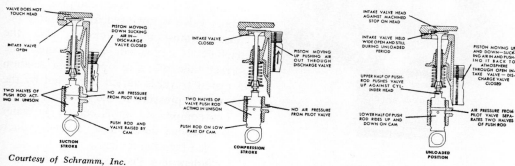

Courtesy of Schramm, Inc.

Fig. 20-5. Unloading-type poppet valve

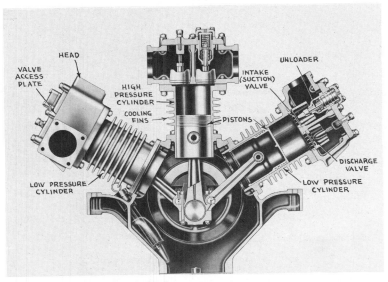

Courtesy of Worthington Corporation

Fig. 20-6. Two stage compressor

mally in contact with each other and transmit the thrust of the cam to the valve stem.

When air pressure in the receiver reaches its maximum, air passes through a pilot valve and tubing into the center of the push rod cylinder, forcing the upper piston upward, unseating the valve, and holding it against the cylinder head. The lower piston rides up and down on the

cam, working only against the air cushion above it.

When air pressure drops sufficiently in the receiver, air is bled from the unloader lines, and the two pistons resume contact and operate as a push rod.

In some models, the intake valves are in the head, and the upper push rod piston is in contact with a rocker arm.

The exact timing and wide opening of the poppet valve favors complete filling of the cylinder, permitting high volumetric efficiency.

The automatic, check type exhaust valve is similar to those to be described in connection with two stage compressors. A poppet valve is not used for discharge because it must remain closed while the machine is unloaded, and because the correct time of valve opening in relation to piston travel is varied somewhat by the amount of pressure in the receiver.

The compressed air is cooled by water jackets around the exhaust passages, which are called the aftercooler.

Cylinders and Pistons. Compressor pistons may have either flat or convex tops.

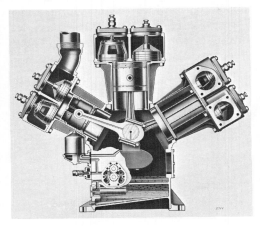

Courtesy of Ingersoll-Rand Company

Fig. 20-7. Two stage compressor

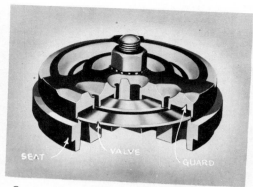

Fig. 20-8. Cutaway view of check valve

Single stage compressors ordinarily have a number of identical cylinders and pistons in line in a water cooled block. In two stage units, the low pressure stage requires more cylinder capacity than the high pressure, so that if a single cylinder of each type is used, the primary is much larger. Two or three first stage may be used for each second stage cylinder so that all cylinders can be of similar or identical size.

Such cylinders are arranged radially, as shown in Figures 20-6 and 20-7, so as to operate off one throw of the crankshaft. Large models may have two or more sets, each on a separate throw.

Two stage cylinders may be either air or water cooled. Air cooling employs the thin cylinder walls and outside fins illustrated. The intercooler fan keeps air moving over the fins.

Connecting rods may be fastened to the crankshaft with individual bearings, as in 20-7, or the rod for the second stage piston may have a heavier bearing and carry the primary rods pinned to its sides, as in 20-6.

Heads and Valves. On two stage machines, each head contains a discharge valve and passage, and usually an inlet valve and passage also. The primary stage inlet valves are fitted with an unloading device.

Valves are of the automatic check type. Suction valves are similar or identical to discharge valves, but are inverted to act oppositely.

Check type valves, shown in Figures 20-7 to 20-10, vary in construction in dif-

Fig. 20-9. Inlet valve parts

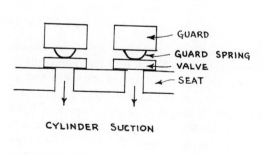

CYLINDER SUCTION

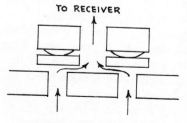

CYLINDER PRESSURE

Fig. 20-10. Exhaust valve action

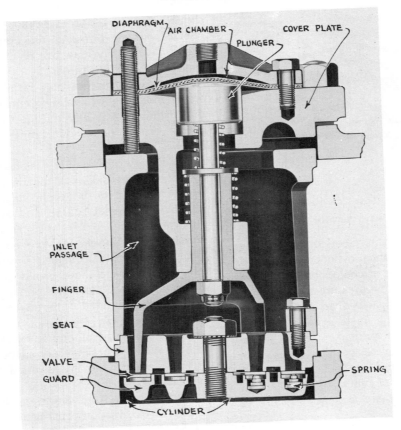

Courtesy of Gardner-Denver Company

Fig. 20-11. Inlet valve unloader

ferent compressors, but agree in having fixed relatively heavy seat and guard plates with sets of perforations that are not in line with each other. The valve proper, also known as the disc or the strip, is a light, movable piece or pieces between the two plates, held against the seat by springs. This part must be light to avoid impeding air flow and hammering on the seat.

The direction of air flow is through the seat, around the disc or strip, and then through the guard, also known as the bumper. A suction valve is placed with the seat plate on top, opening into the inlet passage. A discharge valve has the seat on the bottom, opening into the cylinder.

The seat has a finely machined surface, against which the valve is pressed by light springs that may be of strip, coil, or conical construction. Air moving through the seat in response to the stroke of the piston compresses the springs, unseating the valve, so that it can pass around it, as in Figure 20-10.

Air pressure on the opposite side will hold the valve firmly on the seat so that it will not be able to pass by.

Suction valves are usually equipped with an unloading mechanism, one type of which is shown in Figure 20-11. The unloader is a plunger which has a number of fingers attached to its skirt. These extend into the ports in the valve seat. Springs hold the plunger at the top of its travel while the compressor is working.

When the compressor unloads (stops

compressing), air is admitted to the chamber above a diaphragm, forcing the diaphragm and plunger down until the fingers press the valve away from the seat. Air drawn in on the suction stroke is now discharged back through the inlet valve on the compression stroke, and no work is done. When air pressure above the diaphragm is released, springs push the fingers out of engagement.

This type of unloader may also be worked by air pressure against a piston attached to the plunger, rather than against a diaphragm.

Valves are usually accessible for service by removal of a cover plate, without need for disturbing other parts of the machine.

Intercooler. The intercooler, used only in two stage machines, is a tubular cooling radiator connecting the first stage discharge passages with the second stage inlet. A fan mounted on the rear of the compressor provides air circulation both around these tubes and the compressor cylinders. Intercooler efficiency varies, but it is expected to reduce the temperature of the contained air to within a few degrees of that of the atmosphere.

Intercooler pressure has a natural relationship to the receiver pressure, generally in the neighborhood of 30 to 100. If intercooler pressure rises disproportionately, leakage in high pressure suction valves is often indicated.

The intercooler is fitted with a spring loaded safety valve set for about 35 pounds. This can be tripped by hand, and should be blown daily to avoid the accumulation of material which might clog it.

A drain valve in the bottom should also be opened daily, or oftener, under pressure, to blow out accumulations of water or oil.

Intercoolers are sometimes fitted with an unloader or automatic drain, which consists of a blowoff valve that is opened by air pressure in the control tubing.

Air Receiver. The air receiver is a cylindrical tank with convex ends, usually mounted horizontally on the rear of the frame. It must conform with various Federal and state safety regulations and National Board certification is normally stamped on it.

The receiver acts as a small reservoir between the compressor and the tools, which reduces the frequency of unloading in light use. It serves to separate moisture and oil from the air and provides a place for draining them.

It is equipped with an inlet from the compressor, and with one or more outgoing air lines with shut-off valves. A smaller air line goes to the pressure gauge and the automatic controls.

The following fittings are required for safe and convenient operation:

Safety valve of the spring type that will open when air pressure rises above the highest operating pressure. This must have the capacity to let air out faster than the compressor can put it in. It should have a hand trip mechanism by which it can be

Compressor Size	50	105	210	315	420
Safety Valve	1"	1 1/4"	1 1/2"	2"	3"
Pressure Gauge	3 1/2"	3 1/2"	3 1/2"	6"	6"
Drain Valve	1/2"	1/2"	1/2"	1/2"	1/2"
Fusible Plug	3/8"	3/8"	3/8"	3/8"	3/8"
Air Line Valve	3/4"	1"	1 1/2"	2"	3"

Courtesy of Schramm, Inc.

Fig. 20-12. Minimum size of fittings

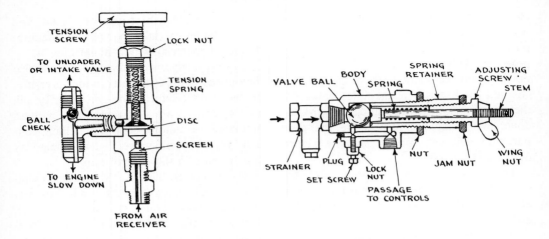

Fig. 20-13. Pilot or trigger valve

opened and any sediment blown out daily.

A gauge indicating pressure in the receiver. It is the standard practice to use a 200# face, so that the needle will be vertical at normal high pressure of 100 pounds.

A drain cock or valve at the low point in the receiver. This is used for draining or blowing out water, oil, and sediment.

A fusible plug, which will melt if the air gets hot enough to be near the flash point of lubricating oil vapor.

The table in Figure 20-12 gives recom-

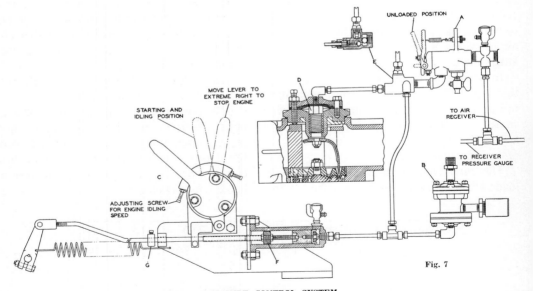

Fig. 7

PRESSURE CONTROL SYSTEM

A — Pilot Valve   C — Throttle Control
B — Intercooler Unloader   D — Inlet Valve Unloader
                    E — Delay Valve

Courtesy of The Jaeger Machine Company

Fig. 20-14. Pressure control system

Fig. 20-15. Pneumastat fuel saver

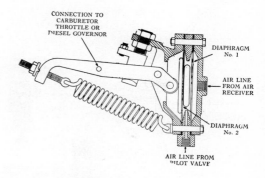

CONNECTION TO CARBURETOR THROTTLE OR DIESEL GOVERNOR

DIAPHRAGM No. 1

AIR LINE FROM AIR RECEIVER

DIAPHRAGM No. 2

AIR LINE FROM PILOT VALVE

Fig. 20-16. Cross section of Pneumastat

mended safe sizes for these attachments.

Air Pressure Controls. Automatic controls are necessary to keep sufficient pressure in the receiver to operate the tools and to prevent building up of excessive pressure.

Air from the receiver is admitted to the control system by the pilot or trigger valve, two types of which are shown in Figure 20-13. Receiver air contacts a disc or a ball held on a finely machined seat by a spring. When pressure becomes high enough to force it off its seat, a much larger area can be reached by the compressed air so that the push against the spring is greatly increased, and the disc or ball is snapped back far enough to allow air to escape quickly into a side passage to the controls.

When pressure drops sufficiently, the disc will be pushed down by the spring. When it passes the side passage, air from the passage presses against its upper surface, equalizing the receiver pressure below, and allows the spring to push the disc back on its seat. The air in the passage then leaks out through the top of the valve.

A pressure control system is shown in Figure 20-14. The air passage from the pilot valve is divided. Air directed to the unloading mechanism in the inlet valves goes through a ball check, which allows free passage to the unloaders but blocks

air moving from the unloaders to the pilot valve, allowing it to leak slowly through a small drilled passage.

Air from the pilot valve is also piped to the throttle control valve, and to the intercooler unloader, if one is used. Air entering the throttle valve pushes a plunger which closes the throttle against tension of a spring. Release of air pressure allows the spring to reopen the throttle.

When the pilot valve opens in response to high receiver pressure, the three controls operate simultaneously—the compressor and intercooler unload, and the engine cuts back to idling speed.

When the pilot valve closes and releases the pressure in the controls, the throttle opens and the intercooler unloader closes immediately, but the compressor does not load up until the air in its unloaders has leaked back through the small hole in the check valve. This delay gives the engine a chance to gain speed before starting to pump air.

In the system of control described, the engine throttle is either fully open or in idling position, and is liable to move frequently from one to the other. This wastes fuel because of the non-productive time while unloaded and the extra consumption during acceleration. The change of speed and load also increases wear.

Most manufacturers offer devices that

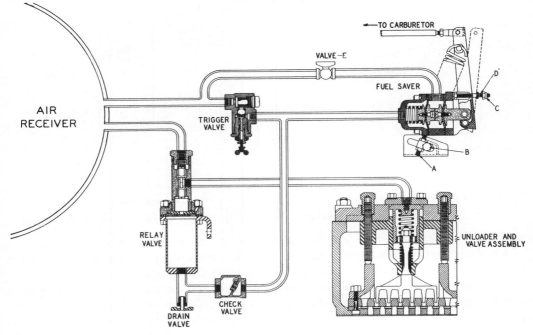

Courtesy of Worthington Corporation

Fig. 20-17. Air control system

will proportion engine speed to receiver air pressure within the limits of the on-and-off system already described. Various constructions are employed, but in principle they consist of a diaphragm or piston acted upon directly by air from the receiver, which pushes against the spring held throttle lever, so that increased pressure will partly close the throttle, and falling pressure open it. Adjustment to the desired pressure-speed relationship may be made by adjusting the tension of the spring.

Figures 20-15 and 20-16 show the Pneumastat, made by Schramm, which controls both the on-and-off and the gradual change mechanism.

In some makes, a shutoff valve is provided for the regulator's receiver air line so that the machine can operate on the on-and-off basis only. The pressure should be low when this is closed, unless it is possible to bleed the line.

These "fuel savers" may show substantial economies when the average consumption of the tools used is below the capacity of the compressor, as they will permit continuous operation on part throttle. However, when the air demand is intermittent, or as great or greater than the compressor output, they are of little use.

Relay Valve. In order to deliver full air pressure promptly to the unloaders, without increasing the size of the trigger valve, some manufacturers use a relay valve. Figure 20-17 shows a Worthington control system that includes both a fuel saver and a relay valve.

When the compressor is ready to unload, air from the trigger valve passes through the check valve into the relay. There it pushes upward against a diaphragm, unseating a plunger which opens a passage from the receiver to the unloader tubing.

When the machine is ready to load up, the air drains back through the check valve, the plunger reseats and cuts off air coming from the receiver, and the air in the tubing leaks out.

Gauges and Controls. The instrument panel should carry the indicators for the compressor oil pressure and receiver air pressure, in addition to the usual engine gauges. A two stage compressor should have a gauge for intercooler pressure also.

The clutch control is usually an over-center lever, but in one make it is a hydraulic jack which is pumped to disengage the clutch and released to engage it. Auxiliary levers are provided for hand operation of the throttle and the unloading mechanism.

Capacities. Air requirement of tools varies with weight, model, and condition. In general, a 60 or 85 cfm compressor can supply two light hand drills or paving breakers, or one of medium weight. A 105 or 125 can take care of a heavy hand drill, or two or more light ones. A wagon drill requires a 210 or 315, a light crawler drill a 600.

When tools are used intermittently, a compressor may be able to take care of a larger number than is indicated by its capacity, but it will do so at the risk of reduced pressure and delays.

Mountings. All makes of portable compressors may be had in a variety of mountings. Some of these are shown in Figure 20-18.

Rubber mounted compressors with a capacity of 60 CFM and smaller can be moved short distances by hand; 105's may be towed by a car or shifted by hand on level pavements. Larger models should be handled with a truck or tractor.

Tractor-Compressors. Many small contractors have only occasional use for a compressor, with the result that the machine stands idle so much of the time that it cannot make an adequate return on investment, and tends to deteriorate rapidly because of excessive idle time, usually exposed to the weather. However, this occasional use may be sufficiently important to justify the ownership of the machine.

Courtesy of Worthington Corporation

Fig. 20-18. Compressor mountings

To meet this problem, and to provide easier maneuvering of machines on the job, several manufacturers build units which serve both as tractors and compressors.

The Schramm Pneumatractors are diesel wheel tractors whose power plants are combination engine-compressors. The 3 or 4 power cylinders are in the same block with

Courtesy of Schramm, Inc.

Fig. 20-19. Pneumatractor and Skyworker

an equal number of compressor cylinders. They all share both the cooling and the lubrication systems.

Air compression may be stopped during tractor operation by unloading the intake valves.

The Skyworker attachment shown in the illustration is a pair of 13-foot hinged booms which carry a basket, air hose, and tools. It can be used to place a worker high in the air and in a wide radius around the tractor, to facilitate drilling, scaling, and other work in hard-to-reach places. It is steadied by hydraulic outriggers.

The tractor can be equipped with all standard accessories and with other special drill mountings.

They are made in three sizes, the largest with a 200 cfm compressor, the smallest with 125 cfm with 41 horsepower. All models have mechanical drive, with four speeds forward, and either one reverse or a hydraulic reverser for all speeds.

The Le Roi Tractair, shown in Figure 20-20A, has a four cylinder engine of 42 horsepower, and a two cylinder 125 cfm single stage compressor in the same block. The air cylinders are larger than the power ones. The compressor may be unloaded by either air or mechanical controls.

Courtesy of Le Roi Company

Fig. 20-20A. Tractair

Operation. Before starting a reciprocating type compressor, routine check should be made of oil level in engine and compressor, fuel, and cooling system. The receiver drain cock should be opened and the clutch disengaged.

After starting, the engine is run for several minutes to warm, then the clutch is engaged. Soon after it turns the compressor without choking or laboring the drain cock may be closed.

If the engine is not thoroughly warm or is in poor condition, it may not be able to build up full receiver pressure immediately. If it threatens to stall, the clutch should be disengaged or the compressor unloaded by a hand control, and extra warm-up time allowed.

During work the drain in the bottom of the receiver should be opened occasionally to get rid of water from condensation. In hot damp weather this can amount to sev-

Courtesy of Schramm, Inc.

Fig. 20-21. Crawler-type tractor-compressor

Courtesy of Le Roi Division

Fig. 20-20B. Small rotary compressor

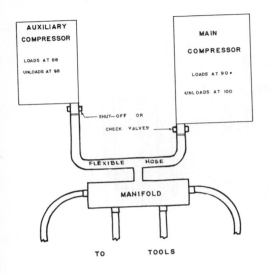

Fig. 20-22. Piping for main and auxiliary compressors

est oil that will give proper lubrication, and which will not be pumped into the cylinders in excessive quantities.

Thorough cleaning of incoming air is essential.

If exhaust valves leak some of the hot compressed air which has just been forced out of the cylinder will come back in and be recompressed, resulting in very high pressure locally, and excessive temperatures which will boil off the lubricant.

Under these conditions enough oil may be present in the air to cause explosions, most often in the cylinder (dieseling) or the passages, but occasionally in the receiver or even in the tools.

This danger is avoided by replacing piston rings in the compressor as soon as they allow pumping of oil, keeping valves free of carbon and dirt, and replacing broken or worn parts.

Multiple Units. Air demand too large for a single compressor may be supplied by two or more of any size discharging into a single supply system.

The hookup can be arranged so that one machine does most of the work, or so that two or more work equally.

The first method is most satisfactory when one machine can supply the normal load but cannot keep up with peak demand. The compressor which is to carry the full burden has its pilot valve set for normal pressures, probably loading at 90 pounds and unloading at 100. The auxiliary has both settings about two pounds lower. Discharge lines from both receivers should be flexible hose leading into a common manifold, from which air is taken by the tool lines, as illustrated in Figure 20-22.

When the machines are started, both will run until a pressure of 98 pounds is built up in the receivers and manifold. The auxiliary will then cut out and will be unloaded as long as the main machine can keep pressure above 88 pounds.

eral gallons a day, and it is important to keep it down.

The safety valves on the receiver and the intercooler should be tripped by hand at least once a day to blow out any carbon or sludge deposits which might prevent them from working in an emergency.

The filters in the air intake for both engine and compressor, and any in the control system, should be cleaned as often as necessary.

Once started, the operation of the compressor is completely automatic, except for the occasional attentions listed above. The oil and air pressure gauge should be checked occasionally.

A spare valve assembly should be kept on hand.

Carbon and Explosions. Air temperatures are high enough to cause some vaporization of lubricating oil. The non-volatile residue, in combination with any dirt in the air, is liable to build up hard or gummy deposits that will interfere with valve action, and sometimes will choke air passages. Since thin oils leave less residue than heavy ones, it is advisable to use the light-

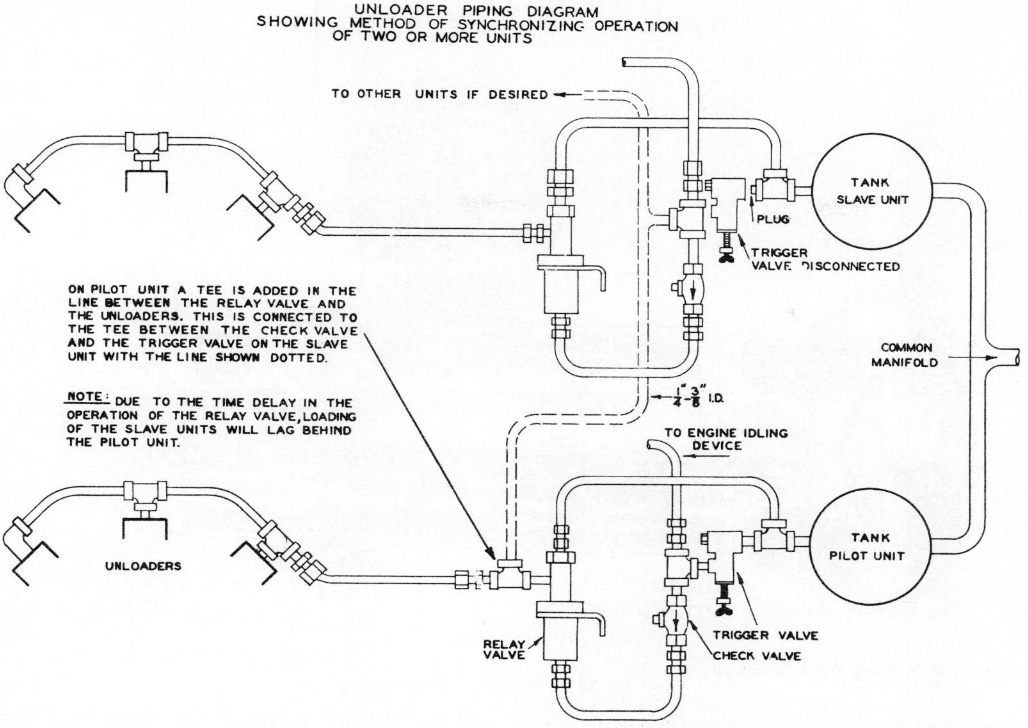

TO OTHER UNITS IF DESIRED

TANK
SLAVE UNIT

PLUG

TRIGGER
VALVE DISCONNECTED

ON PILOT UNIT A TEE IS ADDED IN THE
LINE BETWEEN THE RELAY VALVE AND
THE UNLOADERS. THIS IS CONNECTED TO
THE TEE BETWEEN THE CHECK VALVE
AND THE TRIGGER VALVE ON THE SLAVE
UNIT WITH THE LINE SHOWN DOTTED.

NOTE: DUE TO THE TIME DELAY IN THE
OPERATION OF THE RELAY VALVE, LOADING
OF THE SLAVE UNITS WILL LAG BEHIND
THE PILOT UNIT.

COMMON
MANIFOLD

$1\frac{3}{8}$" I.D.

TO ENGINE IDLING
DEVICE

TANK
PILOT UNIT

UNLOADERS

RELAY
VALVE

TRIGGER VALVE

CHECK VALVE

Courtesy of Worthington Corporation

Fig. 20-23. Piping for pilot and slave compressors

If pressure falls below that, the auxiliary will load, and will assist until pressure is restored to 98 pounds.

If the work is to be shared equally by two or more machines, one pilot valve should operate all of the air controls. This is most conveniently done in systems including relay valves, as shown in Figure 20-23. The method would be similar if no relay valves were used, except that the air admitted through the one trigger valve would have to operate all controls directly, causing a lag in the "slave" units.

If air consumption falls so that all machines on a manifold are not needed, the surplus ones may be cut out by stopping their engines.

ROTARY COMPRESSION

The Ingersoll-Rand Gyro-Flo compressor, which is shown in Figures 20-24A and 20-24B, operates on the principle of a vane-type air motor. A cylindrical rotor is mounted off-center inside a larger cylindrical casing. Sliding vanes are fitted into lengthwise slots in the rotor. Centrifugal force causes them to keep in contact with the casing wall whenever the rotor is turning rapidly.

The vanes divide the space between the rotor and the casing into a series of compartments. Because of the off-center mounting, these are very small as they pass one side of the casing, and large on the opposite side.

Slots in the casing allow air to enter the compartments as they expand. As they contract, the air is compressed and is forced out an exhaust passage just before the point of closest contact between rotor and casing.

The compressor shaft carries two rotors in tandem. The larger one compresses atmospheric air to about 30 pounds, and

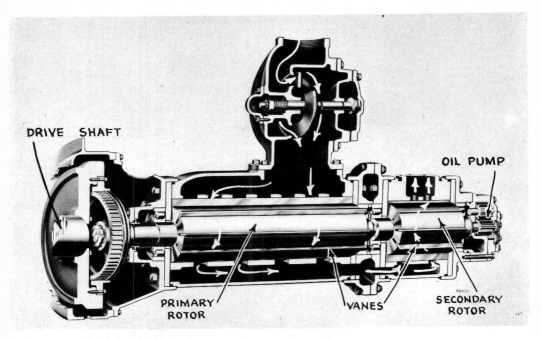

DRIVE SHAFT

OIL PUMP

PRIMARY ROTOR

VANES

SECONDARY ROTOR

Courtesy of Ingersoll-Rand Company

Fig. 20-24A. Rotary compressor

discharges it through a separating wall into the smaller secondary unit, where pressure is raised to 100 or more pounds.

Vanes may be made of special steel, aluminum, or laminated plastic that is not hard enough to score the casing. Air slippage around them is largely prevented by a bath of lubricating oil that is sprayed into the compartments by jets.

Most of this oil is carried out with the air and is salvaged in a large separator. The bulk of it drops out when the air stream loses velocity and the balance is removed by filters. The oil is cooled by flow through a radiator, unless diverted by a thermostatic valve. It is filtered before returning to the bearings and compartments of the compressor.

In addition to serving as a lubricant and sealer, the oil absorbs a large part of the heat of compression so that no intercooler or other air cooling device is used.

The flow of air is steady enough so that no receiver is required, the air passing directly from the separator into the lines. Automatic controls proportion the air output to the demand by throttling the engine and also the compressor air intake.

Standard pressure is 100 pounds at full speed. As the engine is throttled down, pressure rises to a maximum of 110 pounds.

When the engine is started the compressor is automatically unloaded by the vanes remaining deep in the rotor slots through the full revolution. It is therefore not necessary to disconnect the compressor to start the engine, and no clutch is provided.

When the engine idles there is not enough centrifugal force to keep the vanes in contact with the casing, so the compressor is automatically unloaded by leakage around them.

Single Stage. The Worthington line of QT Monorotor compressors are located inside a tank. This serves as air receiver, oil storage sump, and noise reducer. It replaces most of the air plumbing and many of the oil lines ordinarily needed, so that layout can be simplified.

This machine is equipped with a disc type

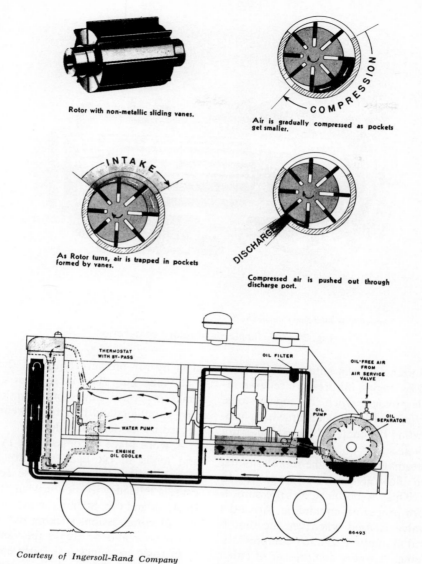

Rotor with non-metallic sliding vanes.

COMPRESSION

Air is gradually compressed as pockets get smaller.

INTAKE

As Rotor turns, air is trapped in pockets formed by vanes.

DISCHARGE

Compressed air is pushed out through discharge port.

THERMOSTAT WITH BY-PASS

OIL FILTER

OIL-FREE AIR FROM AIR SERVICE VALVE

OIL PUMP

OIL SEPARATOR

WATER PUMP

ENGINE OIL COOLER

86493

Courtesy of Ingersoll-Rand Company

Fig. 20-24B. Rotary compressor details

clutch between the engine and compressor, to help with cold starting and to make it possible to operate the engine alone for adjustment or other reasons. However, the clutch is not supposed to be moved into engagement while the engine is running, as the impact might break compressor vanes.

The automatic air control system works first on the engine throttle and then on the compressor air intake. It works gradually instead of in the off-on manner of some of the controls described earlier.

When final discharge pressure is reached, the engine speed control starts to slow the engine. If pressure continues to rise the suction will start to close. The air intake will be entirely closed at about 10 per cent above set working pressure.

A temperature switch shorts out the engine ignition if discharge air pressure

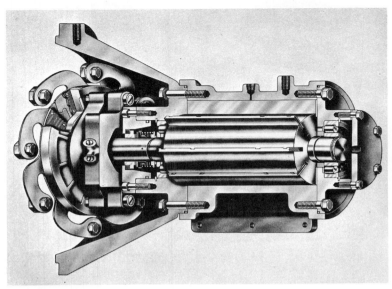

Courtesy of Worthington Corporation

Fig. 20-25A. Single stage rotary compressor

becomes excessive. An automatic unloader operated by engine oil pressure relieves receiver air pressure when the engine stops. Another control in the starter wiring prevents starting the engine if there is pressure in the air receiver.

An automatic valve in the receiver outlet stops air flow if pressure drops to 45 pounds. This is because air pressure is needed for proper oil circulation. An outlet check valve is used whenever the unit is connected in multiple to other compressors.

Servicing. A rotary compressor of either type requires little attention, except to make sure it has clean air and clean oil of the type and viscosity recommended by the manufacturer. The air filter must be serviced frequently, condensation water must be drained out of the oil filter daily, the oil filter must be in excellent condition, and oil must be changed when it starts to break down.

The immediate result of allowing the oil to become dirty is excessive wear on the vanes, also called blades. They are pressed against the back or trailing edge of the rotor slots by air pressure ahead of them, and each vane may slide in and out a million times in twelve hours of operation. It is easy to understand that rapid wear will be caused by lubrication failure.

A worn blade becomes loose in its slot, and tends to tilt back. It also develops grooves and ridges. Both these conditions cause a tendency to stick instead of moving freely in and out. Sticking in the slot causes loss of compression, sticking out jams the compressor and/or breaks the vane.

In some machines it is fairly easy to remove vanes for inspection; in others it is difficult. Contractors who are very careful about keeping air and oil clean may get more than 6,000 hours from a set, while others who are less careful may have to replace them every 1,500 hours or run serious risk of breakage.

If a vane does break, it is usually necessary to disassemble the compressor sufficiently to make sure that no broken pieces are left in it, and flush the oil reservoir and air receiver and service the oil filters.

If the compressor starts to use more oil

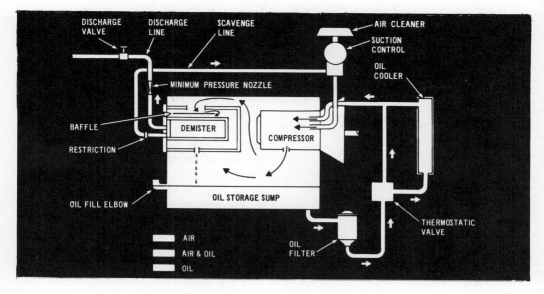

Courtesy of Worthington Corporation

Fig. 20-25B. Oil flow, rotary compressor

than is usual, it may be that the final separator filter element is torn or matted, or that the return lines to the oil tank are blocked. Inspection and repair should be made immediately, as the oil circulates very rapidly and a large amount may be wasted if even a small percentage of flow is allowed to go out with the air.

Rotascrew. The Rotascrew compressor, Figure 20-26, uses a pair of cylinders with matching threads to compress air in a one stage operation. There are no sliding vanes or metal to metal contact. Oil provides sealing and cooling.

AIR LINES AND ACCESSORIES

Metal Pipe. If the compressor cannot be brought close to the work, ordinary plumbers' pipe may be used to carry the air. The pipe should be at least as large as the receiver discharge line, and preferably should be a size larger.

Angles in the line should be kept to a minimum. If the line is reduced in size, reducing couplings instead of couplings with bushings should be used for smoother air

flow. If air may be required later between the compressor and the end of the line, tees with plugs may be used instead of couplings between lengths of pipe, so that additional outlets can be added without opening and reconnecting the line.

The drop in pressure in an air line depends on:

1. The size and length of the pipe
2. Pipe bends, and type of fittings used

Courtesy of Gardner-Denver Company

Fig. 20-26. Rotascrew threaded cylinders

TABLE I *
AIR PRESSURE DROP
per Hundred Feet of Straight Pipe

Air Flow C.F.M.	Diameter of Air Line						
	3/4"	1"	1 1/4"	1 1/2"	2"	2 1/2"	3"
50	2.51	.7	.16	.07	—	—	—
105	—	3.1	.71	.31	.09	—	—
210	—	—	2.82	1.26	.34	.13	—
315	—	—	—	2.73	.76	.29	—
420	—	—	—	—	1.35	.53	.17

TABLE II **
EQUIVALENTS OF FITTINGS *in feet of straight pipe*

Type Fitting	Diameter of Pipe Fitting						
	3/4"	1"	1 1/4"	1 1/2"	2"	2 1/2"	3"
Elbow or Tee	1.2	1.6	2.2	2.6	3.6	4.4	5.7
Globe Valve	3.5	4.7	6.5	7.8	10.6	13.1	17.1

* Adapted from Schramm Data Sheet C-50 B
** Copied from Schramm Data Sheet C-50 B

Courtesy of Schramm, Inc.

Fig. 20-27. Air pressure drop data

3. The volume of air flowing
4. The pressure of air as it enters the line

The table in Figure 20-27 gives the drop of pressure per 100 feet of straight pipe line, and the equivalent in feet of straight pipe of 90° turns (elbows or tees), and of shut off valves. If the tools use less than full compressor output, the drop will be proportional to their consumption rather than to the capacity of the compressor.

Hose and Fittings. Connection between the receiver or metal lines and the tools is made by flexible hose or tubing of rubber and fabric. The rubber should be neoprene or of some other oil resistant type.

Most hose have between three and seven plies, or one to three braid layers, and may be of either wrapped or molded construction. The molded type with rayon braid fabric is lighter and more flexible for its size and strength. Heavy duty hose with tough covers are required for mining, for quarrying rock which breaks with sharp edges, and where the hose is subject to abuse from machines, tools, or rock falls. For less severe conditions the lighter hose is both more economical and easier to handle.

Hose should not be used after the inner tubing starts to deteriorate as broken pieces will clog filters or valves in the tools. Such pieces are often so small as to be mistaken for carbon.

Hose are fastened to each other and to other units by threaded, quarter turn, or snap-on couplings.

The threaded connections are best suited to connections which are seldom changed

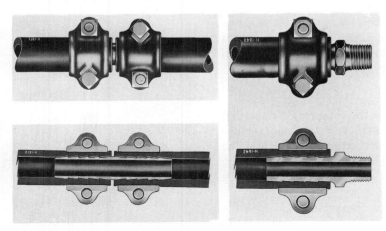

Courtesy of Worthington Corporation

Fig. 20-28. Air line connectors

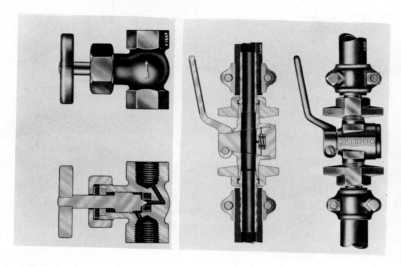

Courtesy of Worthington Corporation

Fig. 20-29. Air line valves

during the work. They take more time to assemble and disassemble than the other types, but usually require less servicing. No gaskets are needed.

The quarter turn or quick detachable are usually obtainable only for medium and small hose. The connectors in any one make are all the same size, so different size hose can be connected without the use of bushings. Both surfaces and the gasket should be kept out of the dirt when apart and should be cleaned before coupling. Spare washers should be kept, as these may be lost or damaged.

Air pressure should be turned off and the line bled before opening either type of connection, for ease of operation, because of danger from the hose whipping, and possible damage to eyes from blowing dirt.

The snap-on couplings are assembled by pulling back a sliding collar on the socket, inserting the plug on the other hose piece, and releasing the collar. They are opened by pulling back the collar and pulling the plug out. The socket automatically shuts off the air when the plug is out, and should be installed on the side toward the receiver.

At present, these are available only in small sizes. They are quick and convenient to use. The sockets should be wrapped in a cloth or otherwise protected from dirt if they are to be left open on the ground.

Reels. Hose can be wrapped around the compressor or coiled in tool boxes when not in use. However, keeping it on a reel is good policy as it keeps it away from contact with sharp or heavy objects and avoids kinking.

A dead reel, similar to the type used for garden hose, may be used for storage. If the inside end of the hose is left projecting near the axle, it may be connected to the receiver so that the necessity of removing it entirely from the reel is avoided.

A more convenient device is a live reel permanently mounted on the compressor. Air from the receiver is admitted to the axle through a rotating, pressure-sealed connection, and from the axle to the hose by an ordinary coupling.

The outer end of the hose should be tied to the reel when the compressor is moved to avoid unspooling.

Manifolds. When several working lines

Courtesy of Worthington Corporation
Fig. 20-30. Line manifold

Courtesy of Schramm, Inc.
Fig. 20-31. Tank manifold

are to be taken off one receiver or supply line, a manifold or "pig," such as those shown in Figures 20-30 and 20-31, is used.

Oilers. Most air hammers are equipped with oiling systems supplied by small reservoirs in the tool itself. However, these are often neglected, need frequent attention, and may not function satisfactorily in a worn tool.

A line oiler, illustrated in Figure 20-32, is a reservoir which feeds into the air stream through a needle valve. The oil is blown into a fine spray which is carried into the tool to keep it lubricated. Oil will feed only under pressure.

Oilers are built into track drills and other heavy air-operated equipment. Otherwise they should be as close to the tool as possible, as the air and oil may separate almost completely in ten or fifteen feet. However, sufficient distance must be allowed for easy manipulation of the tools, and some lubrication will be afforded at long distances by the air stream pushing condensed oil along the inside of the hose.

Some line oilers will operate in any position while others must be upright. They should always be placed in the line so that the air moves in the direction of the arrow

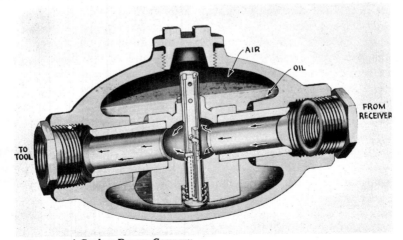

Courtesy of Gardner-Denver Company
Fig. 20-32. Line oiler

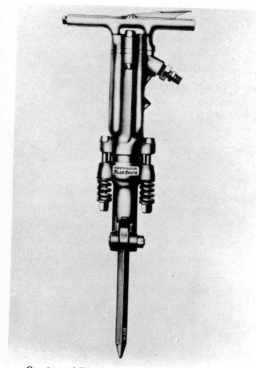

Fig. 20-33. Paving breaker

on the case. Hose between the oiler and the tool must be oil resistant.

HAND BREAKERS

Most of the light, hand-held breakers,

sheeting drivers, and tampers are air hammers that do not rotate the steel. An anvil block may be used to transmit the blow of the piston to the steel.

The Worthington WB-81 paving breaker, Figures 20-33 and 20-34, is a characteristic construction. The tools shown in Figure 20-38 will adapt the machine to most kinds of work. Special shapes can be made by blacksmiths when required.

Gasoline Hammer. There are also paving breakers powered directly by gasoline engines, as shown in Figure 20-36. This unit, which is made in 55 and 65 pound models, has a two cycle engine of about 2 horsepower, with a centrifugal clutch. It strikes 1250 to 1350 blows per minute, with an impact of 20 to 40 foot pounds.

These breakers often can be converted to shallow depth hand drills by changing the bottom assembly. In the drill, power from the percussion action supplies both rotation and a stream of blow air.

Hydraulic. Breakers may also be hydraulically operated. Figure 20-37 shows one that is powered by a loader, by means of hoses to tees inserted in its bucket lift and lower lines. In absence of other equipment, it may be run by an engine driven pump on a small trailer.

Requirement in two sizes ranges from 4

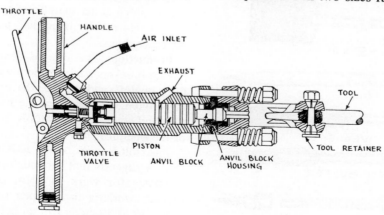

Fig. 20-34. Cross section of paving breaker

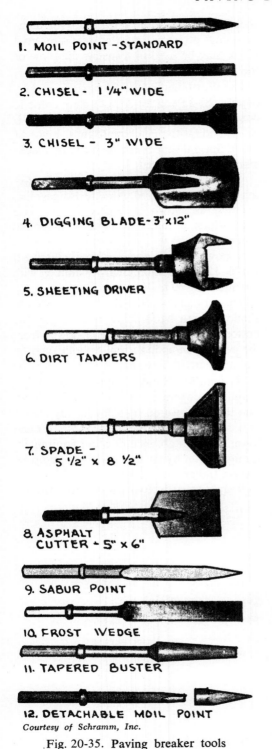

1. MOIL POINT - STANDARD

2. CHISEL - 1 1/4" WIDE

3. CHISEL - 3" WIDE

4. DIGGING BLADE - 3" x 12"

5. SHEETING DRIVER

6. DIRT TAMPERS

7. SPADE - 5 1/2" x 8 1/2"

8. ASPHALT CUTTER - 5" x 6"

9. SABUR POINT

10. FROST WEDGE

11. TAPERED BUSTER

12. DETACHABLE MOIL POINT

Courtesy of Schramm, Inc.

Fig. 20-35. Paving breaker tools

Courtesy of Wacker Corporation

Fig. 20-36. Gasoline powered hammer

to 7 gallons a minute, at 800 pounds pressure. Since machines with such hydraulic systems are on most jobs, hooking up is usually much easier than bringing in a compressor. No muffler is needed, as it has no external exhaust.

Production. The rate of demolition by paving breakers varies widely with the nature, quality, and area of the material being broken, the weight of the hammer, the type and condition of its tools, and the skill of the operator.

Ingersoll-Rand's book, ROCK DRILL DATA suggests that for rough preliminary estimates for work in plain concrete, a breaker working in an open street might break 20 to 30 cubic feet an hour, while in narrow cuts for trenches it might process only 3 or 4 cubic feet. Heavy machine foundations might yield at the rate of 5 to 10 cubic feet an hour, and ordinary floors and slabs at 12 to 15 feet.

Reinforcing in the concrete would reduce production by about one-half.

Hydraulic Splitters. Hydraulic splitters apply the plug and feathers mechanism to very heavy work.

The splitter in Figure 20-38 has a pair of hard shims (feathers) at the base of a two way hydraulic cylinder. They are rounded on the outside and slightly enlarged at the bottom to fit a straight drilled hole $1\frac{3}{16}$ to $1\frac{5}{8}$ inches diameter. The inner, facing sides are flat with a gradual taper, and are smoothly finished.

The plug is a slender wedge, also very hard, that is forced between the feathers by the hydraulic piston. An enormous side pressure, up to 400 tons, is exerted against the sides of the hole.

This model is supplied with 7,000 pound hydraulic pressure from a separate engine-driven pump, with hose connections. Another make uses a hand pump on the cylinder.

Almost any sound rock can be cracked by proper use of these devices. Boulders usually break up, but for solid rock excavation, a backhoe-mounted demolition hammer may be used to loosen and reduce the pieces before loading them.

The outward force should be exerted at right angles to the proposed crack. As in blasting, there must be a free face toward

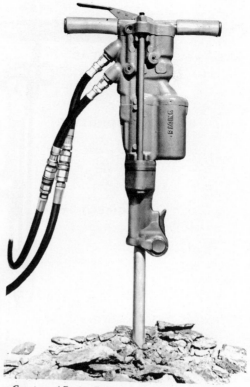

Courtesy of Racine Federated Industries

Fig. 20-37. Hydraulic hammer

which at least one side can move, for the pressure to be fully effective.

The primary use for splitters is rock breakage in areas where explosives would be very dangerous to people or property, or where the noise of blasting could not be tolerated.

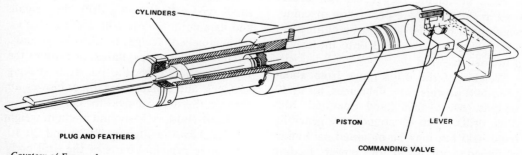

CYLINDERS

PLUG AND FEATHERS

PISTON

LEVER

COMMANDING VALVE

Courtesy of Emaco, Inc.

Fig. 20-38. Hydraulic rock splitter

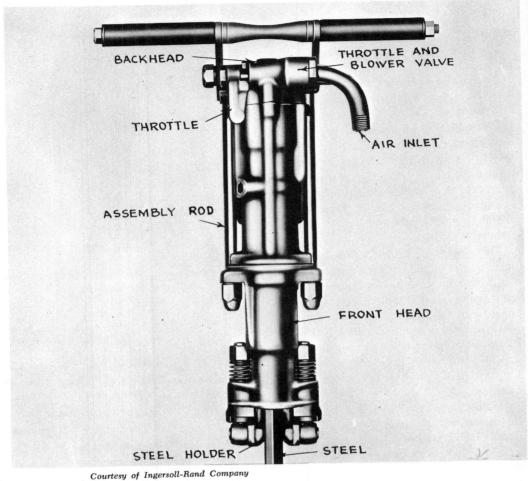

BACKHEAD

THROTTLE AND BLOWER VALVE

THROTTLE

AIR INLET

ASSEMBLY ROD

FRONT HEAD

STEEL HOLDER — STEEL

Courtesy of Ingersoll-Rand Company

Fig. 20-39. Jackhammer

PERCUSSION ROCK DRILLS

Most pneumatic rock drills (air drills) used in excavation are percussion machines which both hammer and turn a cutting bit, which is usually threaded onto the end of a hollow steel rod.

Weights of the drills range from about 10 to 500 pounds. The light ones, up to 30 pounds, are always hand operated. Medium drills, up to 80 pounds, are generally hand held but may be mounted on frames equipped with hand or power feeding mechanisms. Heavier models are almost always frame mounted and power fed.

Hand rock drills are called sinkers, hammers, and jackhammers (jack hammers). They are air operated, and except for a small type called a plug drill, they rotate and hammer a hollow drill steel to which a bit is attached. Weight is 15 to 70 pounds.

Hammering chips, flakes, or crushes the rock, rotation gives the bit fresh striking surfaces, and exhaust and direct blower air through the steel removes the cuttings.

Hand-Held. A 55-pound medium weight drill is shown in Figures 20-39 and 20-40. The three principal parts of the drill body are the upper end or back head, the cylinder, and the front head. These are

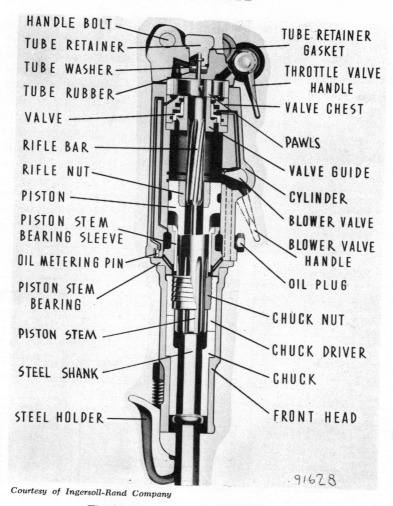

HANDLE BOLT
TUBE RETAINER
TUBE WASHER
TUBE RUBBER
VALVE
RIFLE BAR
RIFLE NUT
PISTON
PISTON STEM BEARING SLEEVE
OIL METERING PIN
PISTON STEM BEARING
PISTON STEM
STEEL SHANK
STEEL HOLDER

TUBE RETAINER GASKET
THROTTLE VALVE HANDLE
VALVE CHEST
PAWLS
VALVE GUIDE
CYLINDER
BLOWER VALVE
BLOWER VALVE HANDLE
OIL PLUG
CHUCK NUT
CHUCK DRIVER
CHUCK
FRONT HEAD

91628

Courtesy of Ingersoll-Rand Company

Fig. 20-40. Jackhammer cross section

machined to fit each other accurately without gaskets, and are held together by a pair of alloy steel bolts, called assembly rods.

Air reaches the drill through flexible hose, then a curved metal tube with a swivel connection to the back head, and the throttle valve. The throttle can be set in closed, wide open, and several intermediate positions so that the speed of drill action can be regulated.

The piston is moved rapidly up and down in the cylinder by compressed air. On the down stroke its stem strikes the upper end of the drill steel. It is kept in alignment by the cylinder bore, rifle bar, the stem bearing, and the chuck.

The rifle bar is splined and slightly twisted. Four pawls attached to its upper end allow it to rotate in the direction of the twist. The rifle nut, in the top of the piston, has matching splines.

As the piston moves up, the rifle nut splines exert a twisting force on both the rifle bar and the piston. The rifle bar is held by the ratchet, so the piston turns. On the return stroke, the rifle bar turns, having less rotation resistance than the piston, which drives straight down. Each up and down cycle of the piston therefore rotates it an amount determined by the twist in the rifle bar and the spacing of the ratchet teeth.

Rifle bars with matching chuck nuts can be obtained in several pitches, and sometimes with a reverse twist which will give rotation on the down stroke.

The lower end of the piston, called the stem, is splined into the chuck driver that is threaded into the chuck, which is made in upper and lower jaws. The lower jaw

collar and the piston near the bottom of its stroke. As the piston completes its down stroke, it drives the steel down (or the drill up), until the collar reaches the steel puller. The piston is then forced up by air entering the bottom of the cylinder through the automatic valve located around the upper part of the rifle bar, and the weight of the drill causes it to slide down along the shank until the chuck rests on the collar again.

It is important that the shank be the correct length, as the drill will not operate properly on a longer or shorter stroke than

Drifter	Type of Rotation	Bore	Weight	Chuck (Std.)	Air Consumption	Hole Sizes
			ENGLISH			
D300A	Rifle Bar	3''	65 lb.	7/8'' x 4 1/4'' Hex	220 cfm	1 1/4''-1 3/4''
DC35	Rifle Bar	3 1/2''	111 lb.	1 1/4'' Round	270 cfm	1 1/2''-2''
URD350	Air Motor	3 1/2''	156 lb.	1 1/4'' Round	355 cfm	1 1/2''-2''
X71WD	Rifle Bar	4''	164 lb.	1 1/4'' Round	271 cfm	1 3/4''-3 1/2''
D40	Rifle Bar	4''	156 lb.	1 1/4'' Round	328 cfm	1 3/4''-3 1/2''
D475A	Rifle Bar	4 3/4''	239 lb.	1 1/2'' Round	425 cfm	1 3/4''-4 1/2''
URD475U	Air Motor	4 3/4''	335 lb.	1 1/2'' Round	520 cfm	2 1/2''-6''
URD550	Air Motor	5 1/2''	460 lb.	2'' Round	800 cfm	
			METRIC			
D300A	Rifle Bar	76.2 mm	29.6 Kg	22 mm x 108 mm Hex	6.2 M³/min.	32.0-45 mm
DC35	Rifle Bar	85.8 mm	50.4 Kg	32 mm Round	7.6 M³/min.	38.0-51 mm
URD350	Air Motor	85.8 mm	71.0 Kg	32 mm Round	10.0 M³/min.	38.0-51 mm
X71WD	Rifle Bar	101.6 mm	74.4 Kg	32 mm Round	7.7 M³/min.	44.5-89 mm
D40	Rifle Bar	101.6 mm	70.8 Kg	32 mm Round	9.3 M³/min.	44.5-76 mm
D475A	Rifle Bar	121.0 mm	109.0 Kg	38 mm Round	12.0 M³/min.	44.5-104 mm
URD475U	Air Motor	121.0 mm	152.0 Kg	38 mm Round	14.7 M³/min.	44.5-104 mm
URD550	Air Motor	139.7 mm	209.0 Kg	51 mm Round	22.6 M³/min.	63.5-152.4 mm

Courtesy of Ingersoll-Rand Company

Fig. 20-41. Drifter specifications

has a socket into which the shank of the steel fits. Rotation is carried from the piston stem through the chuck parts into the steel.

Chucks may be obtained to fit the hexagon shank shown, octagonal designs, or round steel with lugs. Steel diameters range from 7/8 to 1 1/2 inches. One inch is generally standard for medium drills.

The steel puller is a clamp that holds the steel in position during work. It is released by foot pressure on the projecting lever.

The steel is shown in its highest position, with the chuck resting on the steel

that for which it is designed. Standard length is three and a quarter inches from the collar to the end.

From the throttle, air goes through a drilled passage in the back head, past the pawls on the rifle bar, into the valve, which directs it alternately to the top and bottom of the cylinder. Air exhausted from the top of the cylinder passes into the open air while the hammer is working. Exhaust from below the piston goes through a drilled passage into the piston stem bearing, and downward along the chuck driver splines into the space below the piston stem. Here it enters the hole in the center of the steel, which takes it to the bit where it serves to blow rock chips out of the hole during drilling.

In deep holes or soft rock the amount of air provided by this puff blowing may not keep the hole clean. All the air that comes through the throttle can be turned down the steel by moving the blower or throttle valve to the blow position.

A bypass in the back head can be connected to divert part of the air into the steel while leaving the bulk of it to run the hammer. This provides more cleaning in proportion to drilling.

Lubrication is provided by a reservoir around the piston stem bearing, which is filled through a plug in the side. This oil is forced by air pressure to points where it directly lubricates some surfaces, and it is carried by the air stream to others. However, it is safer to use a line oiler.

Drifters. Percussion drills that weigh 65 pounds or more are usually supplied with supports, and with automatic devices to feed them into the work. They are called drifters or drifter drills, that is, drills intended for mounting on "drifter" or power feeds, such as are carried by crawler drills or jumbos.

Drifters are rated both by cylinder bore and weight. Figure 20-41 gives specifications for the current Ingersoll-Rand line. The smallest has a 3-inch bore and weighs 65 pounds, the largest a 5½-inch bore and a weight of 460 pounds.

Recommended hole diameter is usually smaller than cylinder bore.

Rotation may be powered by rifle bar, as described for the smaller drill, rifle bar with double sets of pawls to permit reversing, or by a reversible air motor turning the chuck through reduction gears. Rifle bar rotation is simple, and economical of air, but it lacks flexibility.

Reverse rotation is used in unscrewing (breaking) steels and bits, and in pulling back out of holes.

The drill assembly includes a striking bar, which is a short piece of hollow steel. The upper end receives the blows of the piston; the lower end is threaded for coupling to the drill steel. It is held in line by a chuck bushing and other parts, and in place by a threaded fronthead or chuck housing cap, which usually has lugs to allow loosening it by hitting it with a hammer or other tool.

Fit must be good. Play to the side, caused by an undersize striking bar or by worn drive ring or chuck bushing, will cause extensive damage.

Blow air is carried by an air tube from the backhead through the center of the piston (in the rifle bar, if there is one) and into the striking bar. The tube is a sliding fit in the bar. If it is loose it will leak high pressure air and cause "short stroking". If too tight, it will probably break.

Some blow air, both from the high pressure line and from the exhaust, feeds through the tube during drilling. Full air flow is turned into it in BLOW position of the control valve.

The throttle controls for drilling, blowing, and rotation may be mounted on the backhead, but are usually at an operator's station where other movements of the machine may be controlled also.

Down Hole. A down the hole drill is a simple, heavy pneumatic unit in which the piston blow is delivered against the bit shank, without any steel or rod between. The drill is very slender in proportion to its weight and strength. It has no rotation mechanism in itself. Air is exhausted through the bit, and carries chips to the surface, around the outside of the rods.

Diameter of piston hammers varies from 2½ to 5 inches, and weighs from 13 to 175 pounds. Strokes may be as short as 7/8 inches, or as long as 5. Blows per minute range from 500 to 2,700 in standard units, and to 3,800 in high pressure designs.

The theoretical limit on frequency of strokes is about 6,000 per minute. Lubri-

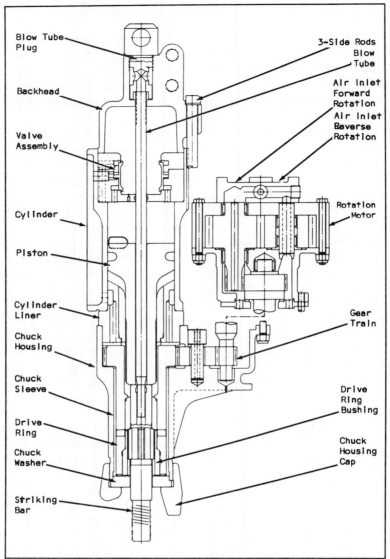

Blow Tube Plug

Backhead

Valve Assembly

Cylinder

Piston

Cylinder Liner

Chuck Housing

Chuck Sleeve

Drive Ring

Chuck Washer

Striking Bar

3-Side Rods Blow Tube

Air Inlet Forward Rotation

Air Inlet Reverse Rotation

Rotation Motor

Gear Train

Drive Ring Bushing

Chuck Housing Cap

Courtesy of Worthington-CEI

Fig. 20-42. Drifter, with parts labeled

cation is supplied by heat resistant drill oil added to the air from the compressor.

The drill rod above the drill supplies air, keeps it in contact with the rock at proper pressure, and rotates it. The rotation, usually at the rate of 15 or 25 revolutions per minute, is supplied by an air or hydraulic motor that has the position on the chain or other feed that is usually occupied by a drill.

A carbide insert bit is held in the drill chuck by a split collar and a nut. In taking out the bit, the nut is loosened or "broken" by a hydraulic wrench, and then spun off by the rotary mechanism.

Down hole drills are made to operate at the usual construction machinery air pressure of 100 pounds. High pressure models may run at 250 up to 500 pounds. There is a tendency to increase pressures for greater efficiency.

A pressure of 100 pounds at the receiver usually means about 90 pounds at the drill. Back pressure from the restricted openings in the bit and from the rising column of air and chips around the rods is likely to be about 20 pounds, leaving an effective pressure of only 70 pounds to operate the drill. If receiver pressure were 300 pounds, losses would be the same 30 pounds or less, so the efficiency would be much greater. Higher pressure is usually accompanied by decrease in stroke and increase in frequency and drilling rate.

As with all "souping up" operations, the possible damages must be weighed against the increase in production.

The down hole drill is used for making holes of 5 to 7½ inch diameter in hard and medium rock. Since none of its striking force is absorbed by the rod (steel for big and deep working drills is usually called rod or pipe), its working depth is limited only by the ability of the air stream to keep cuttings blown out of the hole, and the capacity of the rotary bearings to carry the weight of a long string of rods.

Down pressure or pulldown on these drills is much less than with rotaries. It seldom exceeds 2,500 pounds even in very hard rock.

The exhaust air rising from the bit should have a velocity of almost 3,000 feet a minute to clean the hole. In rapid drilling with coarse chips 5,000 feet may be needed.

When the hole is large in proportion to rod diameter, air velocity will be low because it has a wide passageway. It can be speeded up by using an auxiliary compressor, decreasing the size of the bit, or using bigger rods.

A principal disadvantage is the danger of losing a whole drill as a result of a rock fall, or formation of mud collars. For this reason, use may not be prudent in badly fractured formations, or in wet shale or other muddy strata.

Volume needed for blowing only can be

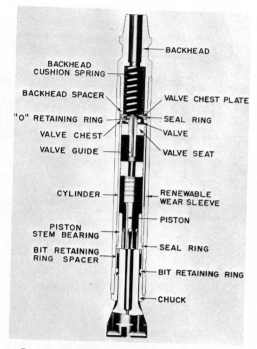

Courtesy of Ingersoll-Rand Company

Fig. 20-43. Down hole drill assembly

worked out approximately by using the formula:

$$\text{(hole diam.)}^2 - \text{(rod diam.)}^2 \times 16.4 = cfm$$

STEELS, RODS, AND PIPES

Steels and rods are the connectors between non-downhole percussion drills and their bits. If only one piece is needed it is usually called a steel; if two or more pieces are fastened together, they may be called either sectional steel or rods.

Rotary drill rods, when made with tapered threads, are usually called stems or pipes, sometimes rods, but almost never steels.

In moderate to deep drilling the set of connectors and bit is called the string. In percussion work it carries hammering and rotation with light to moderate down pressure, in rotaries it is subjected to rotation and moderate to very heavy pressure. For

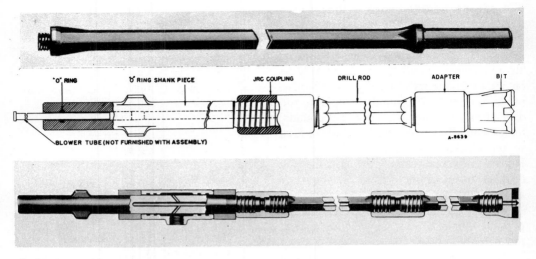

Courtesy of Ingersoll-Rand Company

Fig. 20-44. Drill steel and sectional steel

both, it carries pressure air to blow cuttings out of the hole. In larger rotaries air may be replaced by fluid.

Steels. Hollow drill steels for hand and machine-mounted percussion drills are made in outside diameters of ⅞ to 2 inches. There are standard lengths of one to 10, 12, 14, and 20 feet. Many odd lengths are produced by cutting down during repair work.

Cross section may be hexagonal, octagonal, or round, or the whole steel may have a continuous, rounded thread.

One end may be forged into a shank, which fits in the chuck of a hand drill. For drifters, the shank is usually a short, separate piece, called a striking bar. The oppo-

site end of shank pieces, and both ends of most other steels, carry a male thread which may or not be backed by a raised shoulder. Such threads are coarse, left handed, and made in several non-matching types.

A steel may have one tapered end to carry a tapered bit, or be forged into a cutting bit. This last construction was once standard, but is now very rare.

Sections of steel (sectional steel) are fastened together by couplings, which are short, strong tubes with internal threads. Bits have female threads also, for direct fastening to steels.

Most of the hammering impact shock is carried between rod ends, which butt against each other inside the coupling, or from rod

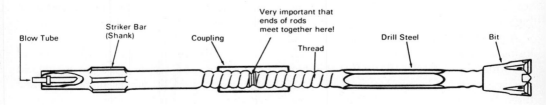

*Courtesy of Worthington-*CEI

Fig. 20-45A. Drill steel assembly

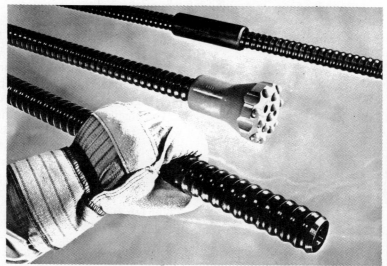

Fig. 20-45B. Spiral steel

end to bit socket (bottom). If there are shoulders on the steel threads, they carry shock through the coupling, or the shoulder of the bit.

If a connection is loose, or ends or shoulders are not properly shaped, some or all of the shock will be carried by the threads, with resulting rapid deterioration.

Threads and tapers are designed to be tightened by the rotation of the drill, to jam tightly enough so that reverse twisting will not loosen them, but to loosen when struck sharp blows when free of pressure or twist.

Threads need to be well lubricated by special (heat resistant) grease that prevents direct metal to metal contact. Without this, full tightening may be difficult, and impact may create tiny weld points that both prevent natural tightening while drilling, and make loosening excessively difficult.

Threads and couplings deteriorate in service, and require every-use inspection and frequent repair. Forging machines are available that will make (upset) new threaded ends, with shoulders when required, or reshape old ones. Continuously threaded rod is cut to a new flat face, but must have the rim chamfered for the final 1/8 inch, to about 30 degrees.

Every steel must have a hollow center to carry air to the bit for hole cleaning and cooling. Care must be taken not to plug or narrow it with mud, trash, or repair mistakes.

Steels are long and thin, and not nearly as strong as they look. They are easily bent by accident, rough handling, and particularly by being blasted out of stuck holes. Their life in service may be limited by metal fatigue, and by cracks starting at dents and

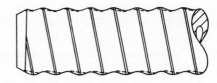

Cut it off before it's worn out

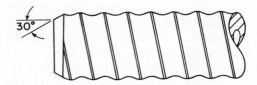

Chamfer carefully and evenly

Fig. 20-46. Trimming spiral steel

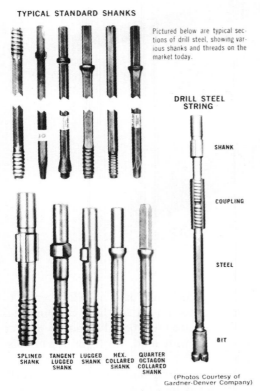

TYPICAL STANDARD SHANKS

Pictured below are typical sections of drill steel, showing various shanks and threads on the market today.

DRILL STEEL STRING

SHANK

COUPLING

STEEL

BIT

SPLINED SHANK — TANGENT LUGGED SHANK — LUGGED SHANK — HEX. COLLARED SHANK — QUARTER OCTAGON COLLARED SHANK

(Photos Courtesy of Gardner-Denver Company)

Courtesy of Timken Company, Rock Bit Division

Fig. 20-47. Drill steel shanks

scratches on the surface. In general, long steels and thin steels are most often and most severely damaged.

A steel should NEVER be used as a crowbar or a lever, even if the load seems light. Tubes have almost no strength for such work.

A blacksmith shop, if available, can usually straighten and rethread steel for less than the cost of replacement, unless it is the carburized type.

A set of hand steels is a series in graduated lengths from starter to the longest piece required. For hand work, a two foot increase with each change is usual. The length difference between successive pieces is called the steel change.

Shanks. Three types of shank are: hexagonal, quarter octagonal (square, with

chanfered corners), and round lugged. The round shanks with two lugs are largely used with the heavier drills and thicker steels. Shank diameter is normally the same or only slightly smaller than steel diameter.

The shank collar serves to limit the upward penetration of the steel into the chuck, and enables the puller to hold the steel from sliding out of the drill.

Shank lengths, from the upper side of the collar to the end, are three and a quarter inches for small steels and four and a quarter for the larger sizes. It is important that this length should not vary more than one sixteenth inch, and that the shank end be kept straight across.

In drifter drills, the shank is replaced by the striking bar, which is basically a short shank with a threaded lower end, which is kept clamped in the drill.

Drill Pipe. Down hole and rotary drills use a heavier type of steel, called rod or pipe, that is designed to carry torque rather than impact. It uses A.P.I. (American Petroleum Institute) steeply tapered threads, male on one end, female on the other.

Standard lengths for deep drilling are 21, 29, and 43 feet. For mobile rigs they are made up to suit tower height and steel change, in lengths of 20, 25, or more feet.

Metal. There are four principal types of steel used in making drill steels (or rods): plain carbon, high carbon, alloy, and carburized.

Plain carbon steel (.80 to .85 per cent carbon) is the oldest type, and is still standard for hand drills and other light hammers. It is the least expensive, usually has shoulders behind the threads, and can be rethreaded and repaired in a small shop. Outside diameters range from 7/8 to 1 1/4 inches.

High carbon (.95 to 1.05) steel is used in the same sizes, but without shoulders, for somewhat heavier service with light drifters. Skill and good heat treating equipment are needed for proper repair, unless this con-

Courtesy of Ingersoll-Rand Company

Fig.20-48A. Detachable bit, and shoulder threads

PERCUSSION BITS

Design. Percussion rock bits chip or crush rock with hammer blows. Rotation changes the area of impact with each blow. Compressed air, or air and water, supplied through the center passage in the steel, and through one or more holes in the bit, blows the chipped and powdered rock away from the bit and out the drill hole.

Bits are usually made of deep hardened alloy steel. They may have inserts of silicon carbide at impact areas. Most of them can be restored by grinding when worn.

Left hand female threads, or smooth off-center taper, are provided to match one or more styles of steel.

The work of smashing the rock is usually done by four, but occasionally two or six, ridges that are radial to a center hole, with their faces at right angles to the centerline of the steel.

Two ridges (four wings or points) at right angles to each other are called a four point cross, the most common construction. If not at right angles, it is an X bit. A six wing (six point) design is called a rose bit.

The wings may be replaced by round

sists only of cutting the damaged tip off a long thread.

Various steel alloys such as manganese-chrome-molybdenum are made into 1¼ and 1½ inch diameters, which are harder, up to 45 Rockwell C scale, and highly fatigue-resistant. They can be rethreaded in the field with carbide tools, without heat treatment. Other repairs require special shops.

Carburized steel is low carbon, but surface treated for great hardness, both inside and outside, after forming and threading. It is expensive but high-performance and long lasting. It cannot be rethreaded or repaired in the field.

Outside diameters range from one to 2 inches, for use with the heaviest drills.

Courtesy of Ingersoll-Rand Company

Fig. 20-48B. Multiple use steel bit

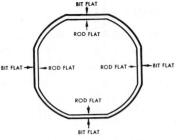

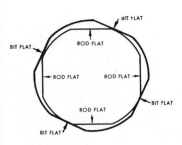

FLATS of the Liddicoat bit and drill steel connection are shown here enlarged for illustration. Flats of the bit line up directly with the flats of the drill steel when connecting. With the first machine blow the bit is seated and camming action results.

CAMMING ACTION is shown above in the sectioned drill steel and bit. The bit tightly grips the drill steel when the rod is seated. Tight stretching of the bit skirt on the drill steel cylinder results from the rotation and impact of the machine.

Courtesy of Western Rock Bit Manufacturing Company

Fig. 20-49. Bit-to-steel connection

buttons of carbide set in an almost continuous, slightly convex surface.

The hollows between the wings are flutes. The width across the face of a pair of wings (the width of the bit) is the gauge or size.

The bit is widest at the face, and tapers to the back. The taper is called the gauge angle. It is usually three degrees in carbide bits.

Standard taper of the wings, that is, the angle btween the two side slopes, is 110 degrees.

Bits are selected according to hole size (the same as bit gauge), size and type of

Courtesy of Ingersoll-Rand Company

Fig. 20-50A. Carbide-insert bits

thread, type of rock, and hole footage to be expected.

Air Holes. The number and placement of air holes in percussion bits is quite variable.

Small bits for light drills may have a single hole in the center of the face. This is the most effective position for instant removal of cuttings, but it is vulnerable to plugging. Downward movement presses it into a core that is not cut directly. This may be forced into the hole in soft or tough material.

Most bits have this center hole. In addition, there may be one to five holes in the volutes, or recesses between the wings. In general, bits designed for soft and/or wet rock have more holes than those meant for hard, dry formations.

A venturi bit has one or more holes to direct air back up the hole at an angle. Reverse air holes tend to break up mud collars, a problem to be discussed later, either while they are forming or when the bit is being pulled out.

Smooth-face button bits may have a front hole somewhat off center, and other holes in volutes that are cut into the bit to provide an escape route for the cuttings.

Carbide Inserts. The drilling speed and service life of bits can be greatly improved by carbide inserts.

Courtesy of Timken Company, Rock Bit Division

Fig. 20-50B. Button and venturi bits

The material used is sintered tungsten carbide, a mixture of 9 to 12 per cent cobalt (CO) and 91 to 88 per cent tungsten (WC), with a hardness of 91 to 87 on the Rockwell A scale. Fine powders of these materials are mixed in whatever proportion is desired, and cemented into a solid by heat and pressure.

Maximum hardness is obtained with maximum tungsten, minimum cobalt, and finest grain size. Increase in cobalt and grain size reduces hardness but increases toughness. These variations are within a narrow range, but may be important in performance.

The harder inserts will give maximum footage between regrinds, but may chip or break under certain conditions. The softer ones will wear more rapidly (but far, far more slowly than steel), but are less susceptible to breakage.

The Timken insert bits are made in three grades — G or hard, J or intermediate, and H or tough. New work is usually started with the J, and a change in either direction made if indicated. Under many ground conditions, G will be used with light drills and H with heavy ones.

Excessive rotation speed and heavy down pressure is likely to break or chip inserts of any grade.

A carbide bit will cut faster than an all

steel bit of similar design, and may drill a hundred times as much footage before requiring reconditioning or replacement.

Carbide bits are expensive, so that their use is not practical unless loss in holes, or by carelessness or theft, can be controlled.

Under average drilling conditions, however, they will usually prove more economical than steel bits.

Their use is particularly indicated in deep holes where their negligible taper saves the waste of drilling oversize at the top; in hard and abrasive rock where taper, lost time, and expense of bit replacement are important; and where transportation and reconditioning of bits is difficult.

In fissured rock, the large investment in carbide bits may be too risky, because of danger of sticking.

Any pieces of carbide loose in a hole, as a result of chipping or falling out of a bit, are extremely destructive to the rest of the bit, and to other bits. If they cannot be blown or flushed out of the hole, it should be abandoned, and a new hole drilled alongside.

Bit Action. With the down stroke of the drill piston, each edge of the bit is driven into the rock like a cold chisel hit with a hammer, a tiny fraction of an inch from where it hit with the previous stroke.

If the bit is sharp and rotation rapid

enough, it will tend to cut off a wedge-shaped slice. If it is dull, it is more likely to crush the edge to powder to a shallower depth, a much less efficient operation.

For one quality of rock frequency of stroke, and size and sharpness of bit, the thickness of the slice will depend on the speed of rotation. The depth of cut is governed by the force of impact, plus down pressure, through the drill string.

Most drills have a fixed blow frequency at proper air pressure and full throttle. Rifle bar drills have a fixed rotation rate, so with them the operator has only control over down pressure. This is light in hand held hammers, but may be regulated up to a ton or more in drifters and downholes. Too little pressure may reduce depth of chipping and therefore footage (rate of drilling). Too much down pressure prevents the bit from getting back up on top of the rock for a clean stroke, and may reduce footage even more. In addition, it subjects the whole mechanism to damaging strain.

The depth of penetration by each blow is usually very slight. The wings follow each other closely, overlapping their work, and cut a round hole with smooth sides.

However, in soft rock the bit may penetrate so rapidly so that each wing cuts a spiral thread in the wall, which is not entirely removed by the following wing. This process, which is called rifling, results in a threaded, undersize hole, from which it may be difficult or impossible to remove the bit.

Rifling may be reduced or eliminated by supporting the drill to reduce down pressure, or by using dull bits, rose bits, or X bits.

A dull bit spreads blows over too large a surface, and does no chipping and very little crushing. In addition, it sets up a powerful dragging effect, which results in over-tightening thread connections, and gently increased strain on the drill string and the rotation mechanism.

Efficient drilling is absolutely dependent on removal of rock chips and dust as fast as they are formed. This important subject will be discussed under HOLE CLEANING.

Reconditioning. There are a few throw-away bits that are designed to be used only until worn, and then discarded. These include steel bits with surface hardening only, and bits with shallow carbide inserts or buttons.

One-use bits may have a pilot consisting of advanced cutting edges near the center. These facilitate starting a hole, cut more rapidly in some formations, and avoid the nuisance of rehandling.

Most bits, however, are designed to be reconditioned several times.

As a bit is used it wears along the cutting edges and on the sides of the wings. The side wear is the more serious as it soon reverses the taper so that the edge becomes narrower than the part behind it. This causes the bit to drive into a hole that is too small for it, so that it sticks. The degree of wear which causes sticking varies with the type of rock.

Worn bits of the multi-use type can be

Courtesy of Western Rock Bit Manufacturing Company

Fig. 20-51. Liddicoat one-use bit

reconditioned in special machines. Generally it is not economical to attempt to sharpen them by hand.

A deep hardened bit can usually be ground several times. It is cut back from the sides until the reverse taper is removed, and on the face to restore the edge.

A shallow-hardened bit will be softened by deep grinding but may be hot milled. It is heated, reshaped, and rehardened in a special forging machine. Many deep hardened bits will also give better service if hot milled than if ground.

The gauge size is reduced by reconditioning.

Carbide bits may be reconditioned by hand or machine grinding, using a silicon carbide wheel.

In soft, abrasive rock it is important to use bits with pronounced taper to avoid excessive sticking and possible frequent loss. In fissured rock where sticking does not depend on bit design, the large investment in carbide bits is a risky proposition, and is not recommended.

Under average drilling conditions, however, they will usually prove more economical than steel bits.

Their use is particularly indicated in deep holes where their negligible taper saves the waste of drilling oversize at the top; in hard and abrasive rock where taper, lost time, and expense of bit replacement are important; and where transportation and reconditioning of bits is difficult.

Carbide bits may be reconditioned by hand or machine grinding, using a silicon carbide wheel.

Hole Taper. As a bit wears, its hole becomes smaller. In soft rock wear may be so slight that the same bit can be used on several steels in succession. In hard or abrasive rock it may wear badly in a few inches. Holes taper smoothly with bit wear and in steps at bit changes.

Whenever a bit is changed in a hole the new bit must be smaller to avoid sticking.

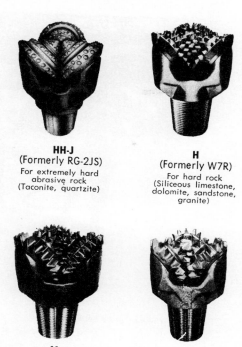

HH-J
(Formerly RG-2JS)
For extremely hard
abrasive rock
(Taconite, quartzite)

H
(Formerly W7R)
For hard rock
(Siliceous limestone,
dolomite, sandstone,
granite)

M
(Formerly OW)
For medium rock
(Limestone, sandstone,
sandy shales)

S
(Formerly OSC-1G)
For soft formations
(Calcite, shale, clay)

Courtesy of Hughes Tool Company

Fig. 20-52. Rota-Blast bits

If all new bits are used the reduction must be an eighth inch. Used bits can be compared by setting their edges together, and very small reductions can often be made.

The starting bit must be large enough to allow for the taper and still have a big enough hole at the bottom for the quantity of explosive needed.

ROTARY BITS AND DRIVE

The rotary bit cuts, chips, and/or grinds, without hammering. In construction work, it is usually carried by mobile tower drills, described in a following section. A rotating table or head above ground supplies rotation. This motion, down pressure, and cleaning air, are carried by hollow drill pipe with threaded connections.

Courtesy of Le Roi Division

Fig. 20-53A. Effect of dust collector

The bits are usually of the roller or tri-cone type discussed under DEEP ROTARY DRILLING, with special adaptions to being cleaned by air instead of mud. Rotation of the bit against the bottom causes the cones to rotate, setting up a chipping and crushing action, with a variable amount of scraping and abrading.

In general, large teeth with wide spacing in each row are used for soft formation, and tooth size is progressively reduced for harder drilling. Teeth are steel, often surfaced with tungsten carbide or other hard facing, except that full carbide inserts are used for the hardest and most abrasive formations.

In standard bits the air stream comes from the bits behind the cones, tending to clean them before reaching the bottom and blowing out the cuttings there. This construction is best for formations that are moist enough to cause cuttings to pack between the teeth, and is satisfactory for all non-abrasive rocks.

The jet bit for very abrasive rock (the carbide insert HH-J in the illustration is one) directs the air streams directly on the bottom. This avoids sandblasting the teeth and cones with fine particles sucked into the air stream. In either construction part of the stream of clean air may go through the bearings to cool and lubricate them.

The blast hole rotary drive is usually a rotary head, fastened through a chuck to the top of the drill rod. It is turned by an air, hydraulic, or electric motor with variable speeds from zero up to 150 or more revolutions per minute. The head is raised and lowered with the drill string by the feed mechanism, which is described in a later section.

Rigs built primarily for down hole drilling can usually be changed over to use rotary bits when ground is soft enough to make this advisable. However, rotation may not be as fast or down pressure as great as in machines that are built primarily for rotary drilling.

Rotaries are the fastest drills in soft rock, compete closely with down hole types in medium cutting, and are usually not economical in very hard formations. These statements are general, however, and there are exceptions.

Bit footage may be as little as five feet, but usual expectation is from several hundred to several thousand. Life is affected by the drillability of the formation, the design of the bit, and the manner of use and abuse.

HOLE CLEANING

As the bit reduces rock to chips, sand, and dust, these particles must be removed promptly, or they will form a layer that will prevent the bit from striking the rock.

The basic method of removal is by a current of compressed air, entering the hollow steel at the drill, and emerging from the bit in holes at the front or bottom (one) and sides (none to four). Some of the air is exhausted from below the piston (puff blowing) at moderate pressure, and part is blow air admitted directly from the high pressure chamber supplying the drill, by putting the throttle in BLOW position.

In a properly balanced rig, the air that cleans cuttings from the bottom of the hole has sufficient volume and velocity to carry them up the hole and out the top. There the chips and coarse sand will pile in a ring around the hole, fine sand will go a little further, and dust will drift with the wind.

Chips and sand must occasionally be pushed back or removed from the edge of a deep hole, so as not to slide back in. This is a minor nuisance. The material itself may be useful as stemming for explosives, grit on slippery surfaces, and perhaps in other ways.

Dust. The dust, however, is a major nuisance. It cannot be tolerated underground. Even in the open, it is unhealthy to inhale for both men and engines. It contaminates exposed lubricant and lubricated surfaces, often turning them into grinding tools. And lately, its suppression has become required by law.

Courtesy of The Spencer Turbine Co.

Fig. 20-53B. Vacuum dust collector

There are three principal methods to control dust in air exhausted from drill holes. They are: vacuum separation, wet or damp (detergent) drilling, and foam.

Vacuum (Filter) Collectors. A hood may be placed over the drill hole, with a loosely gasketed fit around the rod, and connected by flexible tubing to a container. A vacuum apparatus pulls the air out of this through dry filters and/or cyclonic separators leaving dust and chips behind. The clean air is discharged to the atmosphere, and the ground rock is removed as often as necessary.

There may be considerable leakage of outside air into the hood, particularly on rough ground. This is not serious, as even a weak vacuum will remove the dust completely, leaving harmless coarse particles behind. The hood may depend principally on the drill rod for support on steep slopes.

The collector is usually built into a mobile tower drill and often into crawler drills. Skid or wheel mounted units are used for smaller crawlers, and for hand drills. One collector may have several intake hoses to service a number of holes at the same time. Power may be gasoline or electric.

There are also venturi systems for hand drills whose vacuum is created by the flow of compressed air.

A separate vacuum apparatus is suited to quarries with fairly regular work surfaces and drilling patterns, but may be an almost hopeless nuisance in rough and pioneer work. Its great advantage is that it permits bit and hole cleaning to be done efficiently by dry air.

Detergent System. Dust may be suppressed by dampening it at the bit.

Water in a tank is mixed with a small amount of detergent, and fed slowly into the hole cleaning air. The detergent breaks the surface tension of the water, enabling it to coat dust particles thoroughly, so that they stick together in little balls that are much too heavy to float in the air.

Unfortunately, the dampening slows the escape of the material from the face, so part of it is still there for the next blow. This regrinding absorbs power, weakens impact on the rock, and wears the bit. The loss in footage in damp drilling as compared with dry drilling may be 10 per cent in hard rock.

The detergent is necessary for efficiency. Without it, much more water is needed (wet drilling), and action is less uniform. The extra water turns part of the dust to mud, which may build up mud collars that make pulling the bit difficult.

In winter the detergent-water needs antifreeze. Alcohol (methanol) is standard for this. Fuel oil, which may be kept in suspension by the detergent, is sometimes used.

Detergent-dampening is the most widely used dust suppression system at present. Any crawler drill can easily carry a tank of 30 to 50 gallons capacity, enough for a full shift of drilling. Skid, trailer, or even wheelbarrow mounted tanks can be used for smaller units. Aside from refilling, little maintenance is required. The principal disadvantage is reduced production, with increased bit wear.

Foam. A foam system uses the same equipment as the detergent method, with change only in chemicals.

The detergent in the water is replaced by a special foaming agent, plus some diesel fuel. When drawn into the air stream, this generates a stiff foam of fine air bubbles which carries dust and chips out of the hole in a dampened and harmless condition.

The bubbles hold and buoy up rock particles of all sizes, so that they do not tend to sink or fall down against the current. Much less air is therefore required to clean the hole, so that full-blow air may be needed seldom or not at all. The saving in air may be 150 to 350 cfm, reducing compressor requirement and cost.

There is a loss in penetration and increase in bit wear compared to dry drilling, but the loss is probably less than with detergent. On the other hand, the chemical is more expensive and may not be obtained as readily.

NOISE SUPPRESSION

Percussion drilling is an inherently noisy operation. Aside from the compressor, which was, considered separately, the drill admits and ejects high pressure air at a rapid frequency, while it also hammers metal against metal and metal against rock. Exhaust from air motors may add to the din.

There seems to be little hope of convert-

Courtesy of Worthington-CEI

Fig. 20-54. Detergent tank

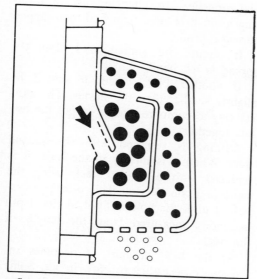

Fig. 20-55A. Muffler for exhaust air

ing this into a quiet operation, but there are methods of lowering the noise level noticeably.

One means is to install mufflers, usually called silencers, on direct exhausts of pressure air from the drill piston, and from any air motors. These are usually simple devices, involving abrupt changes in direction of air flow. They are designed to cause minimum back pressure. On small drills it is a major problem to keep them small and light enough, and sufficiently out of the way, so as not to interfere seriously with the use of the tool.

Exhaust air coming out of the hole has greatly lowered velocity, and is not an important noise problem in itself.

The blow of drill piston on striker bar or shank is the most difficult noise to reduce. Some improvement can be obtained by reducing the resonance of the parts involved, as by a vibration-absorbing clamp. But the only way to substantial reduction would be to encase the entire drill in a sound-absorbent structure. This would create extreme problems for the operator and service man.

Bit-on-rock noise is diminished by depth, and is further reduced by foam dust suppression, as the bubbles are an effective absorber of noise.

A downhole drill becomes much quieter as soon as it enters the hole, and might be considered on noise-sensitive jobs for just that reason.

HAND DRILLING

Starting. The rock drill can be carried by hand or truck to the work area and connected to a flexible air line from the compressor. A short starter steel, a foot to thirty inches in length, is fitted with a bit and clamped in the chuck. This bit should be the largest size which is to be used in the hole.

The area surrounding the hole should be free of dirt and litter that would interfere with starting the hole or might slide into it during drilling.

If the rock surface is horizontal, the drill is held vertically and the throttle partly opened. If the bit wanders on the surface instead of cutting it, it can be held

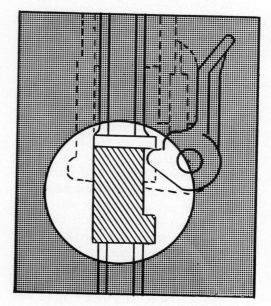

Fig. 20-55B. Steel silencer

in position with a foot pressed against the lower end of the steel. If the rock slopes the steel should be held at a right angle to the slope instead of straight up and down until it has dug a pit that will hold it.

When the bit has cut in enough so that it is not likely to bounce out the hole is said to be collared. The throttle can then be fully opened. The drill can be simply supported against leaning over, or if it tends to bounce, it can be pushed down with one or both hands, or (over manufacturers' protests) by hanging a leg over it.

In soft rock it is not advisable to push down on the drill as it increases any tendency toward rifling or binding. It is sometimes necessary to pull up slightly to avoid these difficulties. In hard rock extra weight is helpful, particularly if the drill is light.

While the hole is shallow the automatic puff blowing should keep it clean. However, it is good practice to turn the valve to blow occasionally. The amount of extra chips and dust raised will indicate whether the puff blowing has been doing the job.

Dust around an air drill is a damaging nuisance, an its suppression is required on an increasing number of jobs. Methods are discussed under HOLE CLEANING. They make checking bit action more difficult.

Deepening. When the hole is cut so the drill is close to the ground, the throttle should be closed and the drill lifted until the bit and steel are out of the hole. The steel is taken out and another steel about two feet longer is locked in the chuck. If it is not much worn, the starting bit may be put on the second steel.

The bit is detached by holding the steel firmly and striking the bit upon the wings so as to unscrew it. The steel may be held under the foot. Some operators prefer to use wrenches—a pair of Stillsons, or special wrenches—for changing bits. In installing a bit, it is hand tightened then tapped to seat it firmly.

It should be remembered that the thread is left hand.

If another bit is used, it should be slightly smaller than the starter. The easiest way to compare sizes is to put the two bits together.

Drilling is resumed with the second steel. When it is mostly in the hole, it is replaced with a longer steel, using the same or a smaller bit. This process is repeated until the hole is finished. Its depth is readily measured from the length of the steels.

As the hole deepens, it becomes increasingly necessary to blow it out. Some hammers can be adjusted to send down an extra stream of air while drilling, which reduces the need to stop drilling to blow.

If air and dust stop coming out of the hole it should be blown immediately. If it will not blow the steel should be pulled out. The trouble is usually a plugged bit. The cuttings can be removed with a thin punch, a nail, or an ice pick.

An attempt to drill without a flow of air will overheat and spoil the bit and probably cause a collar of compacted cuttings to form just over it, which will make it very difficult to pull.

Whenever possible, holes should be drilled vertically. This puts the weight of the hammer on the bit instead of on the operator and simplifies keeping the hole straight.

Jamming. There are several conditions under which getting the steel and bit out of the hole are a greater problem than the drilling. No drill runner is so good that he never sticks a steel.

Generally, the principal cause of sticking is the use of worn bits. In most ground the bit gives a warning that it is reaching a danger point before it jams. This may be slower penetration, production of fine rock dust instead of chips, twisting of the drill in the hands, or failure of the steel to rotate. Condition of the hole may be roughly checked by raising the drill a few inches occasionally to see if it is free.

In hard or brittle rock a bit may function satisfactorily which is so worn that it would stick immediately in a soft formation. However, worn bits are also dull bits and drilling is slowed by their use. It is also poor economy to use a worn bit because excessive wear may make it impractical to recondition it, particularly by grinding.

In soft rock a bit may rifle the hole. This trouble may be expected whenever the drilling rate is very rapid. New, sharp bits are worse offenders than worn ones. Precautions which can be taken include use of special X or rose bits, using a lighter drill or partly supporting the weight of a heavy one, grinding back the cutting edges of the bit, or reconditioning bits by grinding the sides to restore taper without sharpening the edges.

Seams of dirt or finely disintegrated rock in a formation will cause plugging of the air hole. At the same time there is rapid penetration and dirt falls on the top of the bit, where it may compact into a hard, tight collar.

When the speed of penetration increases suddenly drilling should be stopped immediately and the hole blown. Drilling can then be resumed on part throttle with frequent, thorough blowing. If air stops coming out of the hole the steel should be pulled and the hole in the bit opened.

If the ground is composed of sloping layers of rock of varying hardness, or seams or cracks slope across the line of drilling, the hole will tend to drift down the slope. The curved hole that results will bind the steel.

Drilling on a dip at right angles to the seams should prevent this trouble. Often an intermediate direction will be satisfactory. In vertical drilling, curving can often be prevented by using light pressure and sharp bits and arranging that the drill gets maximum air pressure. This may involve cleaning the drill air filter, shortening and straightening the line from the compressor, or temporarily shutting down other units using air from the same source.

Pulling Steels. In taking a steel over three feet long out of a hole, the usual procedure is to unclamp the steel puller, set the hammer aside, and lift the steel by hand. However, if the steel is stuck, the hammer should be left attached and lifted with the throttle alternately in drilling and blowing position. The heavy vibration and rotation of the steel and the air pressure help to break it free.

Two men—one on each handle—can lift the drill much more effectively than one.

If the sticking is caused by spiraling (rifling) it may be possible to extract the steel by turning it against the direction of rotation while lifting. If two men are working, one can lift the drill with throttle closed while the other turns the steel and drill with a wrench. If the operator is alone he should remove the drill, place the wrench just under the shank collar, and pull the wrench upward while turning it.

This may cause the steel to unscrew from the bit. If the hole is full depth a steel bit can be sacrificed economically. If it is not full depth, leaving the bit in will make it necessary to drill another hole. If the bit is carbide, every effort should be made to salvage it by direct pull.

Usually steel and bit can be raised together by jacking. No really suitable jack is generally available, but a chain jack or a regular bumper jack with a chain grab fastened to the lifthook can be used. The jack is placed on the edge of the hole, the end of a light chain given a double turn around the steel and hooked, and the chain pulled tight and caught in the jack

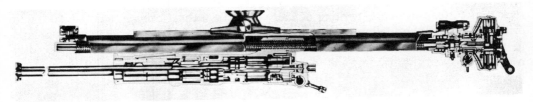

Courtesy of Joy Manufacturing Company

Fig. 20-56A. Screw-type drifter feed

head. The best grip is just under the collar, but a small wood wedge will generally enable it to hold anywhere on the steel.

This jack does not pull straight up and jack and steel will tip as the steel rises. This may cause bending of the steel or the jack if lifting is continued too long on one hold. It is a good plan to release the chain and fasten it again further down before bending is severe.

A chain fastened to wagon drill feeds, loader buckets, bulldozer blades, cable control units, or to any type of shovel or crane may be used for steel pulling.

If all methods of extraction fail a parallel hole is drilled nearby and the steel picked out of the loosened rock after the blast. It is generally bent by the explosion, and unless it lands on top, may be further twisted by a shovel.

Care of the Drill. A drill is built of special steels and is very finely machined throughout. It therefore costs many times as much per pound as a shovel or dozer, and it deserves much better care than it usually gets.

The oil reservoir is built into the bottom of the cylinder and may require filling from one to four times a day, depending on the make and condition of the drill. This oil is forced or sucked into lubricating passages, and into the air stream during work. The oil mist in the exhaust air gives an indication of the amount being used.

A line oiler should be used instead of or in addition to the reservoir for best results, as it has a larger capacity and is more reliable.

The frequency with which either a reservoir or an oiler must be checked is best determined by experience. If exhaust air ceases to carry oil, the cause should be found immediately.

Air drills have very slight clearances and have a severe temperature drop. Air often enters them at a temperature of a hundred degrees or more, and is exhausted close to freezing.

In addition, water condensing from the air or leaking from a water tube may wash oil off the parts which need it. Special oils designed for these conditions will give better service than standard lubricants. However, any oil is better than no oil.

Every possible precaution should be taken to keep the oil clean. This is a difficult matter when the air is full of rock dust, but it is important. Dirt may clog oil or air passages or score sliding parts.

Use of an inlet air filter may save trouble with scoring or clogging, but it should be cleaned frequently. Most filters are so small that a little debris can reduce the air passage enough to starve the drill.

A drill needs periodic cleaning, the interval depending on the quality of air it receives. An oil-pumping compressor, very hot air, or deteriorated hose may foul it in a day or less. With clean air it may function well for weeks.

A quick cleanout can be done by disconnecting the air hose, pouring a cup of kerosene or fuel oil into the drill, reconnecting the air, and running the drill idle for about a minute. The hose is again disconnected, a cup of rock drill oil

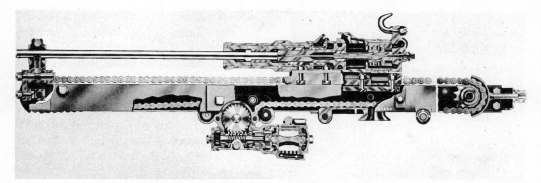

Fig. 20-56B. Drifter and chain feed

poured in, and the drill operated again.

If it is very dirty, several doses of a mixture of three parts kerosene and one part oil can be used.

At much longer intervals, it is good practice to take the drill to the shop (drill doctor), to be disassembled, thoroughly cleaned, and checked for worn parts.

POWER FEEDS

Drills that weigh 75 pounds and more are usually supplied with supports and with automatic feeds.

Air Leg. The air leg used in stopers provides both support and automatic feed for light drifter drills. It is used almost entirely in underground work. The drill is supported by an air cylinder, whose piston ends in a stinger point or some other device to hold its position on rock. Air is admitted to the leg cylinder by a control valve, causing it to expand sufficiently to keep the bit in contact with the rock as it cuts into it.

Screw Feed. Heavier machines use a feeding arrangement that is separate from the support. The complete feed unit may be called a power feed, mast, carriage, or tower. It is used both for feeding the bit into the hole, and in pulling it back out.

A screw power feed, used chiefly for drifter drills used underground, is shown in Figure 20-56A. A vane type air motor turns a threaded rod inside a shell or casing. The

drill mounting, called a cone, is threaded onto the rod, and moves back and forth along a slot in the shell as the rod pushes or pulls it by turning.

The pressure on the bit is regulated by

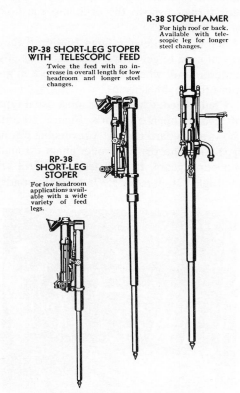

R-38 STOPEHAMER
For high roof or back. Available with telescopic leg for longer steel changes.

RP-38 SHORT-LEG STOPER WITH TELESCOPIC FEED
Twice the feed with no increase in overall length for low headroom and longer steel changes.

RP-38 SHORT-LEG STOPER
For low headroom applications available with a wide variety of feed legs.

Fig. 20-57A. Stopers

Courtesy of Ingersoll-Rand Company

Fig. 20-57B Jackleg

changing the amount of air reaching the feed motor. More air, more pressure.

There are also hand screw feeds. The air motor is replaced by a crank that the operator turns to move the drill into the hole or to retract it.

Screw feeds develop high friction if they are allowed to get dirty. Their use is chiefly underground, where dust is kept to a minimum for health reasons.

Chain Feed. Chain is the most usual type of feed in open air work. Figure 20-56B shows a standard design. A reversible piston type air motor drives a roller chain through reduction gears. It both feeds the drill bit

"Light" stopers weigh 75 to 100 pounds,

Courtesy of Gardner-Denver Company

Fig. 20-57C. Column drill support

into the rock and, when reversed, pulls it out of the hole.

The chain goes around idler sprockets at the top and bottom of the mast. The drill is fastened to the chain, and moves along the shell on lubricated guides.

Hydraulic. Hydraulic feed may be a hydraulic motor operating in the same manner as an air motor to power a chain feed, or a two way hydraulic ram that may either work through a chain or a direct fastening to the drill.

MOUNTINGS

Stoper. In the standard stoper, Figure 20-57A, the leg and the drill make up a single straight line unit. The operator places bit and steel in the drill, places the pointed bottom of the leg, called the stinger, on the floor or in a wall crevice, and opens the valve to extend the feed leg until the bit is held against the rock in drilling position. During drilling air is fed to the leg cylinder so as to keep an even pressure on the bit.

When the steel is all in the hole or the leg is fully extended, the leg is retracted, pulling the steel out of the hole, and a longer steel is substituted.

Stoper operation is tricky. The point may slip out of position and even jump onto a foot (safety shoes are a MUST for stoper operators), the moderately heavy machine must be supported until the hole is collared, and collaring may prove difficult because of the strain of keeping the stinger down and handling the drill.

Some deluxe or safety stopers have a double stinger so that they can lock at both ends to support the entire weight during collaring.

The stoper is intended primarily for overhead drilling, as in other positions its weight is more difficult to handle. Holes should be inclined away from the vertical, and the operator should stay on the uphill side, out from under the hole.

heavy ones up to 130. Cylinder bores are

2¾ or 3 inches.

Jackdrill. The air feed leg shown in Figure 20-57B is hinged to the drill. This arrangement permits convenient use in horizontal and angle drilling, as well as vertical.

This unit may be set up with the leg and drill almost in line, so that the feed supplies both support or push, or at a sharp angle where the leg supports and the operator pushes. The feed tends to push the steel against the top of the hole, but a slight down pressure on the drill will keep it in line.

The operator should keep a firm grip on the handle of an air leg machine, as if the steel sticks it may whip out of his hand and cause injury.

Column. A screw or chain feed drifter drill in underground use may be supported by a pipe column. The column, of 3 or 4 inch diameter, usually extends from floor to ceiling of a tunnel, but it may be placed directly against the rock or timbering, or have foot and head blocks. Columns are usually tightened by means of threaded sections. Pneumatic columns are extended by air pressure in the manner of an air leg, and may carry a stinger point on each end.

The drifter is supported by a saddle on the column itself, or on a side arm. It can swivel vertically and horizontally. The saddle can be moved along the arm, and the arm can be moved along the column.

Column mountings are somewhat clumsy to use. They are being replaced by air legs in light work, and by jumbos in big jobs.

Wagon Drill. The wagon drill has been made obsolete by the more efficient crawler drill, and is no longer manufactured. However, there are still many of them at work.

The standard wagon drill carries a drifter drill and mast hinged to a U-bar, that is hinged in turn to a three wheel chassis that can be towed or pushed from place to place. Pointed rods called pegs or steady points can be set against the ground to hold it on grades and during drilling. See Figure 20-58A.

The mast can be pivoted on the U-bar and the U-bar can be raised and lowered by a hand crank or a hydraulic jack. The combination of these two adjustments provides an almost unlimited range of drilling position and directions in the center plane of the machine as shown in illustration. The mast may also have a slide to allow it to be moved along its clamp to the U-bar, and a swivel that will permit limited side tilting to compensate for the frame being off level on slopes.

The **single front wheel swivels, and** is attached to a drawbar. The heavier rear wheels have two positions, for travel as shown, and at right angles to permit accurate sideward movement for line drilling of closely spaced holes.

The drill has a standard feed travel on the mast of eight or ten feet, so that six or eight foot steel changes can be used. Feeds for longer steel changes are optional.

Steel diameters are commonly 1¼ to 1½ inches. A two-arm steel centralizer at the

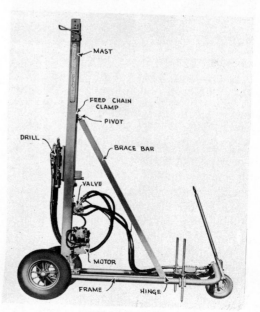

Courtesy of Gardner-Denver Company

Fig. 20-58A. Wagon drill

Courtesy of Gardner-Denver Company

Fig. 20-58B. Twindril

bottom of the mast can be latched on the steel to hold it in line while starting (collaring) the hole, and swung out of the way when it is not needed. A loop at the top of the mast simplifies handling and vertical stacking of the steels.

An air powered winch may be installed to enable the machine to move itself or a compressor for short distances.

Twindril. This unit, Fig. 20-58B, is essentially a pair of wagon drill masts, each having its own drill and feed mechanism, that are clamped to a welded frame adapted for carrying by a tractor boom or crane. This is used in rock trenching for pipelines. The steels are held three to five feet apart. Nar-

rower trenches can be drilled by holding the frame at an angle to the centerline of less than 90°.

BOOMS

Drill Booms and Jibs. Air drills and feeds of the type carried by wagon and crawler rigs may be mounted on hydraulically controlled booms. Such booms can be mounted in any desired number from one up on any type of base large and strong enough to carry them. See Figure 20-58C for typical construction.

The boom is attached to its support by a universal or two-hinge bracket. A hoist ram raises and lowers the boom, and a side ram swings it to right and left. The two controls can be used together to place it in a wide range of positions.

The drill tower or mast is fastened to the outer end of the boom by a double hinge clamp connection called a cone. There may be a hydraulic ram that can move the mast back and forth along this clamp to adjust the reach, and there may be hydraulic means to lower (dump) and tilt (roll) it.

Booms are often mounted in pairs on old 20-ton tractors such as the D8, HD19, and TD24 that are no longer suitable for heavy work. A compressor may be mounted on the same machine.

Jumbos. Modern rock tunneling involves drilling a number of holes in the face at the same time. The drills and their operators are supported on a movable platform, called a jumbo. These devices are made in great variety, including single or double platforms that merely support drill runners and hand drills, arrangements of columns and swivels to support either jack hammers or automatic feed drills, hand jacked carriers, or full hydraulic control as the Hydra-Boom machine, which is shown in Figure 20-58D.

Where muck is hauled away from the face by rail cars the jumbo is usually rail mounted. It may use the same tracks, or

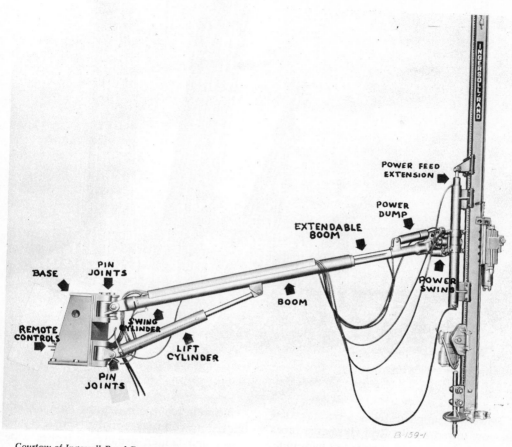

Courtesy of Ingersoll-Rand Company

Fig. 20-58C. Hydra-boom

much wider ones that permit it to straddle the train without blocking them. It may also carry a crane for transferring cars.

Rubber tired or crawler mounted jumbos are now widely used, particularly where haulage is by truck.

CRAWLER DRILL

This section deals with the lighter models of crawler drill, which have both a mast and a boom. This is the dominant type of drill rig for hole diameters of 2½ to 4½ inches, up to 80 or perhaps 100 feet deep. Some models can drill larger holes, and most of them can go deeper, but at lower efficiency

than the big mobile tower or hinged-mast machines.

In smaller holes, and depths of 20 feet or less, they compete with hand-held drills.

General Description. A crawler drill has a pair of tracks, a body and turret mounted between them, a boom that is based on the turret and can be raised, lowered, and swung in a part circle, a mast on a universal mounting at the boom point, and a percussion drifter drill that can be moved along the mast by a power feed. Some manual adjustments (pin changes) may or may not be necessary to obtain the full range of drilling positions.

Fig. 20-59A. Track-mounted drills, range of sizes

Power is supplied by a separate compressor, towed by the drill rig. Some functions are air-operated, others are hydraulic, with a pump driven by an air motor.

Fig. 20-58D. Hydra-Boom jumbo

Travel. The tracks are the type used on crawler tractors, except that they are narrower and lighter, usually 10 inches wide. Single grouser shoes are standard.

Travel or propel is called tramming, a term derived from mining. Each track is driven independently, forward or backward, by an air motor. Maneuverability is excellent.

Brakes usually go on automatically when tramming controls are put in neutral. These brakes can (and must) be locked out of operation by a manual control if the machine is to be towed.

Travel speeds are slow, 2 or 2½ miles an hour. Speed while being towed must not exceed five.

Before moving, the mast foot must be lifted sufficiently to clear the highest spots on the ground to be crossed. If the move is longer than between adjacent holes, it is

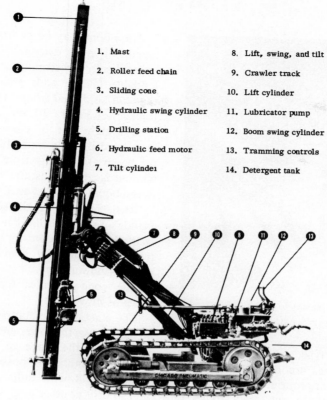

1. Mast
2. Roller feed chain
3. Sliding cone
4. Hydraulic swing cylinder
5. Drilling station
6. Hydraulic feed motor
7. Tilt cylinder

8. Lift, swing, and tilt
9. Crawler track
10. Lift cylinder
11. Lubricator pump
12. Boom swing cylinder
13. Tramming controls
14. Detergent tank

Courtesy of Chicago Pneumatic Tool Company

Fig. 20-59B. Crawler drill, labeled

good practice to lower the boom, and tilt the mast back into a carrying position. The drifter should be directly over the turret.

The track frames are pivoted to the body at the rear, and support it on a walking beam (hinged cross member) near their centers. The mechanism may provide for extremes of oscillation, up to 24 degrees difference in vertical angle between the tracks.

Tramming motors may each be 10 to 16 horsepower, and are usually the piston type. They are located in the body, and turn the track bull wheels through reducing final

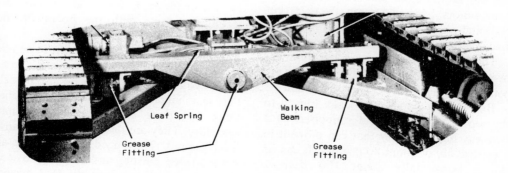

Leaf Spring

Walking Beam

Grease Fitting

Grease Fitting

Courtesy of Worthington-CEI

Fig. 20-59C. Oscillating track mechanism

Fig. 20-59D. Crawler drill turret

drives, all-gear type, or gear and chain.

These machines are designed to tow a compressor of appropriate size (600 to 900 cfm) except on extremely rough or steep ground. For frequent use under such conditions, an air powered winch may be installed in the rear. The crawler can then go ahead, by either track drive or winch power, until the connecting hose is fully extended. It can then stop, be braced, and winch the compressor to itself.

Body. The body, supported by the rear axle and the walking beam, contains or carries the air manifold and lubricator, tramming motors, hydraulic pump, the detergent tank and air and hydraulic controls for all machine functions except drifter movements.

The body serves as the connection between the tracks, and as the support for the pedestal or turret that carries the boom.

Boom. The boom is a beam, hinged to the top of the turret, and supported by a two-way hoist cylinder that can move it in a vertical arc of about 80 degrees. It can be swung horizontally about 100 degrees, either by another cylinder, or by rotation of the turret.

The boom may be one piece, with fixed length, or made extendable by having two hydraulically telescoping sections. The extendable feature adds greatly to the area that can be drilled from one position of the tracks. See Figure 20-59E.

A straight line of closely spaced holes may be drilled by extending or retracting the boom as it swings from one to another, without need to adjust the vertical angle between boom and mast each time.

Mast. The mast is the support and slide (guide) for the drill. It is attached to the right side of the boom tip by a multi-directional, hydraulically controlled connector called a power cone.

The cone provides for three separate types of movement.

The mast may be dumped, that is, swung so as to move the lower end or foot forward or back in the plane of the boom. This movement makes it possible to keep the mast vertical with different boom angles, and to drill at other angles within an arc of about 180 degrees relative to the boom.

The dump angle may be used together with boom angles to provide for drilling in almost any direction, including vertically upward.

The mast may be rolled, that is, tilted to right or left. The arc varies from 80 to 180 degrees in different models.

It can also slide several feet on the cone, so that its foot can be extended or retracted (raised or lowered, if vertical) without affecting its angle with the boom or the ground.

Mast height is usually sufficient to handle 10 or 12 foot steel changes, and starter steels two feet longer. There are also models that can handle 20-foot steels two inches thick, but these high, heavy masts limit flexibility. They are desirable only on fairly level ground. They need hydraulic or air powered steel hoisting and handling devices.

Feed. The mast carries the feed mechanism, which supports, crowds, and retracts the drill. Power is usually air, but may be hydraulic.

The basic chain feed mechanism was described earlier.

Feed pressure and pull may be varied from zero up to 3,000 or 4,000 pounds. Speed may be 100 feet per minute.

Drifter. The drifter (drill), also described earlier, may have either rifle bar or independent rotation. There is a range of sizes, usually from 4½ inch bore and 290 pounds up to 6 inch bore and 460 pounds. Smaller and larger sizes may be used.

The drill is slidably fastened to the mast, and clamped to the feed chain, by a bolted-on backing plate called a slab back.

If a downhole drill is used, the drifter is replaced by, or converted to, a rotary unit that turns the rod, and also transmits the lowering and raising movements of the feed, but does not hammer.

Air System. Compressor size is determined by the needs of the drifter, including the rotation motor if it has one. The tramming motors and the hydraulic system (except feed, if it is hydraulic) are shut off during drilling. Their requirements, when in use, are much smaller than those of the drilling.

The compressor is usually 600 to 900 cfm rating at 100 pounds, but one machine requires a 1,200 size.

Recommended main or bull hose is two inch diameter, 100 feet long, with surface reinforcement against scuffing. Larger hose may be too stiff for good maneuverability. Length must be sufficient to allow a good working area for the drill rig from one stand of the compressor, and to allow it to be parked away from the dust if drilling is dry and open.

Air enters through a main shutoff valve, a manifold, and a lubricator. From there it feeds the valves and motors for travel and the hydraulic pump, and a main supply hose runs forward to the mast-and-drill control station.

It is essential that the compressed air be clean. It is filtered at the compressor, and with tight hoses, there is no problem. However, any hose that is opened should be blown out before replacement. This particularly applies to the bull hose between the compressor and the drill rig. This often gets in complications that may be solved by disconnecting, and it is right down in the dirt.

Always disconnect at the crawler drill end of the hose, and always blow it out thoroughly before reconnecting.

Most of these machines are meant to operate at 100 pounds of air, and do so badly at less than 90 that they should be stopped if pressure falls that far. The problem might be using too small a compressor, or either engine or air trouble in one of proper size.

Hydraulic System. Hydraulics are used instead of air in all position adjustment functions because air, being compressible, does not hold position against changes in load.

A small hydraulic pump, protected by a relief valve, is driven by an air motor. The hydraulics usually control setup functions after the machine is in position, and before drilling starts. The air is therefore shut off and the pump stopped during travel, parking, and drilling.

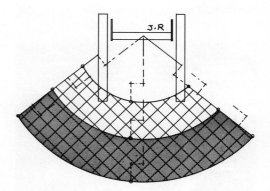

Fig. 20-59E. Shaded area shows extra range of extensible boom

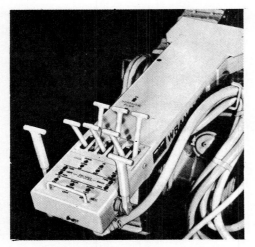

Courtesy of Worthington-CEI

Fig. 20-60A. Tramming and hydraulic controls

A typical control bank, Figure 20-60A, is located on an arm at the left rear corner of the machine. It has all the hydraulic valve levers, plus air controls for tracks and the pump. This bank may be on a swivel, so the operator can have access to it while walking behind the machine during travel, or when standing beside it while positioning the mast.

Standard two-way cylinders may be used for boom rotation and lift. The power cone at the boom tip may include a pair of actuators, each made up to two opposed cylinders with one piston rod. The central part

Courtesy of Worthington-CEI

Fig. 20-60B. Swivel panel for drilling controls

of this rod carries a rack gear that meshes with a pinion. The pinion shaft actuates the swiveling of the mast. One controls the dump action, the other the side tilt.

Drilling Controls. The control panel in Figure 20-60B has four rotary valves, each of a metering type to allow fine adjustment of opening.

The two left hand valves are one way, on or off. DRILL controls the percussion or hammering action, BLOW the volume of cleaning air.

The two right handles govern reversible action—FEED down or up, and rotation for drilling (FWD), or reversing (REV) for bit or steel removal, or getting out of bad holes.

If the drifter had a rifle bar, the ROTATE handle would control direction only, not speed.

CRAWLER DRILL OPERATION

Travel. Crawler drills are moved between jobs on trailers, or if distances are short, may be towed very slowly. On the job they move themselves, towing the compressor that provides their power.

For self-propelling (tramming) the operator walks behind or beside the machine, within reach of the track controls. There are usually two levers, one for each side. Moving a lever forward turns that track forward; backward moves it back. An automatic brake goes on when the lever is centered.

Steering is done by varying the relative speed and/or forward-reverse direction of the two tracks. When they go in opposite directions, the machine will spin-turn in its own length. Very precise maneuvering and spotting is possible.

The rig usually pulls the compressor with it by a pintle hook or drawbar connection on any but the shortest moves. Then, and when moving independently, the operator has a major responsibility to avoid damaging the hose by stretching, kinking, or walking over it.

Tramming is done with the mast above

any possible obstructions, and preferably folded back to a near-horizontal position, for maximum stability.

Setting Up. To set up for a single vertical drill hole, choose the most convenient and stable position for the machine. If ground conditions permit, try to drill in the area between the line of tracks and just forward of them.

Raise the boom to an angle of about 25 or 30 degrees, and set the mast vertically. Position the foot of the mast so the hole marker lines up with the steel centralizer, by swinging the boom, and extending or retracting it (if it is extendible), or by raising or lowering it. Recheck the mast for vertical, then slide it down until the foot rests solidly on the ground.

The crawlers give the machine a good grip on most types of ground, their brakes lock automatically, and it is heavy enough to stay in place. But if there is any question of its sliding even a fraction of an inch, block it with wood or rocks.

Once the boom and mast are set for drilling, shut off the air motor that drives the hydraulic pump, unless feed is hydraulic.

Starting the Hole. Move the drifter (drill) to the top of the mast with the feed motor.

Make sure that all threads are the same —steels, couplers, and striking bar.

Grease both threads on the first length of drill steel. Screw a good bit onto one end and a coupling on the other, finger tight. Remember, these are left hand threads. Swing the centralizer arms open, stand the steel (or rod) on its bit in the centralizer.

Move the drill down with the feed so that the striking bar threads engage the rod coupling. With the rotation motor, turn the bar counterclockwise until it is firmly seated in the coupling. Be sure NOT to start hammering.

Lower the feed until the bit rests firmly on the ground. Close and latch the centralizer. With the drifter's rotation motor,

turn the striking bar counterclockwise until the rod and bit turn with it, and threads appear firmly seated.

Start part throttle hammer action of the drill, continue rotation at moderate speed, and feed it down slowly. When the bit has penetrated any overburden and has made a good socket (collared the hole) in rock, open the centralizer arms wide.

Drilling. The interaction of bit and rock, and some of the effects of varying pressure and feed, were discussed under **Bit Action.**

Rotation should be fairly rapid when penetrating overburden or other loose material, but should be slow to moderate in rock. Fast rotation keeps giving the edges thicker slices of rock than they can break efficiently,

*Courtesy of Worthington-*CEI

Fig. 20-61A. Centralizer

therefore slowing penetration while using more air. It is also likely to cause the drill string to whip and vibrate, and break prematurely.

Pressure should hold the bit firmly against the bottom, but pressure in excess of that does not allow proper rebound from the strokes. This hampers the bit in the same manner that a low ceiling interferes with using a sledgehammer.

Insufficient pressure, which usually means partly supporting the weight of the string on the feed, prevents the bit from following

REACH OUT   REACH UP

REACH OVER   REACH IN

Courtesy of Gardner-Denver Company

Fig. 20-61B. Drilling positions, crawler drill

through enough on its strokes, reducing penetration. It also loosens threads and heats up the bottom of the drifter.

A good stream of blow air is essential. It is the most important part of penetrating overburden or soft rock, as these can readily plug up the holes in the bit. After this, there is nothing to do but pull it out for cleaning —promptly.

There should be a continuous stream of chips, sand, and dust coming out of the hole. If you are drilling with detergent, the dust should be in little balls, but it should still come out. The volume and coarseness of this material gives you a reading on bit action.

The positive indication of progress, of course, is how fast the rod goes down into the hole.

Adding Steel. If the hole is to be deeper than the length of the first rod, you have to add another. This is done when the striking bar coupling has almost reached the centralizer.

You start by checking the hole. Turn off the hammer, continue rotation slowly, and raise and lower the bit a foot or so to make sure it is not binding.

Then turn off the rotation, lower the bit to the bottom, and alternately start and stop the hammer action of the drifter a few times, to loosen the coupling threads. Lifting the string so that it bounces may help. The ringing note of the steel changes when loosening occurs.

It is important to keep rotation shut off while hammering to loosen.

Operate the drill rotation motor in reverse, to turn the striking bar clockwise and unscrew it from the coupling. Grease the striking bar threads, and raise the drill to the top of the mast.

Grease the threads on another section of steel, turn a coupling onto one end by hand, and raise that end to line up with the striking bar. Fit the bottom of the steel into the coupling on the piece that is in the hole, and turn to catch the threads. Lower the drifter so that the bar threads enter the upper coupling.

Tighten the threads by turning the rotation counterclockwise slowly. Blow air and rotate some more, then start the hammer action after both threads are thoroughly seated.

This operation is repeated for each steel added.

Withdrawing Steel. When the hole has been drilled to required depth, turn blow air on full, rotating the string slowly without hammering, until air coming out of the hole is clean.

Stop the rotation and blowing, and loosen the threads as described under **Adding Steel.** Stop the hammer action, and raise the string by running the drifter up to the top of the mast.

Then close and latch the centralizer arms, then lower the string until the coupling below the top steel section rests on the centralizer, so that it supports the weight of the string.

It is absolutely essential that the centralizer be latched, and that the coupling be above it. Otherwise, the whole string might be lost in the hole.

Rotate the drill slowly in reverse (clockwise) and raise it until the striking bar thread is loosened from its coupling. Stop the rotation, and unscrew the steel by hand from its bottom coupling, which is resting on the centralizer. Remove the steel section.

Lower the drifter until the striking bar thread engages the coupling on the centralizer. Rotate the bar to thread it loosely

Courtesy of Gardner-Denver Company

Fig. 20-61C. Pulling steel

into the coupling, sufficiently to support it. Stop rotation and raise the string to take its weight off the centralizer.

Open the centralizer arms. Run the drifter up the mast until the next coupling is above the centralizer. Close and latch the centralizer, and lower the string until the coupling rests on the centralizer. Then repeat steps described above, for each section of steel in the hole.

The bit is usually removed after the steel has been laid aside. It should be inspected for wear, and for chipping or loss of inserts. If unscrewed over the hole, the hole should be covered by a board.

The steels should be laid out in a definite pattern. Their positions should be changed in the string on the next hole, so that throughout the job wear and fatigue will be

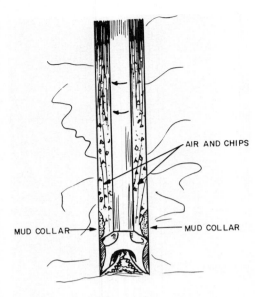

AIR AND CHIPS

MUD COLLAR

MUD COLLAR

Fig. 20-62A. Mud collar

about equal throughout the set being used.

Problems. The drill string does not always come out readily. Occasionally, it does not come out until a loader picks it out of the blasted rock.

If the ground is wet, or if water is being added to blow air to suppress dust, the drill chips and dust may combine with the water to build a mud collar above the bit. Sometimes it is thick enough to just allow space for the rod, but a thickness of 1/16 of an inch may be enough to cause trouble.

The bit must have a taper so that it can follow the cutting edges without dragging and binding on the sides of the hole. If the hole becomes partly clogged, this taper causes a wedging action that compresses and often hardens mud to increase its resistance, as the bit is pulled up.

The operator can usually get through this resistance by some combination of force (the feed can usually lift a ton or more), rotation, and hammering, accompanied by maximum blowing of air. Only experience can fill in the details of just how to do it.

The most important advice is not to get

stuck in the first place. Most times, the collar results from improper drilling. The most common error is not supplying enough blow air in proportion to rate of penetration, so that an overburdened air stream allows accumulation just above the bit.

Another cause of sticking, most common in big holes, is pieces of rock falling out of the wall onto the top of the bit, jamming it. The operator should not blame himself for this, but he struggles with it in much the same way.

The stream of air and cuttings out of the hole should be checked frequently, as changes in its character usually call for changes in technique. Finer grain and/or reduced output means harder rock or a dull or broken bit, calling (perhaps) for another bit, or greater pressure. Coarser grain and/or increased penetration rate mean softer rock, with possible need for faster rotation and/or decreased pressure to reduce danger of a mud collar.

Chips of carbide detached from a bit will grind up the rest of the bit, or any other bit. Repeated blowing might get them out, or the hole may be abandoned.

Angle Drilling. Setup procedures for angle drilling are the same in principle as for vertical holes, but are harder to visualize. Arranging for the first hole may take extra time and figuring.

If the direction of the hole is above horizontal, the rods will have a tendency to slide out of the hole. The coupling is therefore kept on the opposite side of the centralizer from the drill, while adding or withdrawing steels.

If the angle is nearly horizontal, the blow air may travel along the top, leaving excessive drilling debris on the bottom. It is a good plan to pull the steel back toward the mouth of the hole occasionally, with full blow air, to drag this material toward the opening.

Alignment. It is necessary that the hole and the line of motion of the drill be in

ROCK DRILL OIL (Typical Spec.)	Below 30°F. Grade #10	30-80°F. Grade #30	Above 80°F. Grade #40-50
	Light	Medium	Heavy
Viscosity (S.S.U. @ 100°F.)	175-225	500-850	1500-2000
Flashpoint (C.O.C.)	350°F. Min.	380°F. Min.	395 Min.
Carbon Residue (Conradson)	0.20% Max.	0.30% Max.	.040% Max.
Film Strength (P.S.I.)	12,000 Min.	12,000 Min.	12,000 Min.
Falex Wear Test	3000# Min.	3370# Min.	——
Steam Emulsion No.	1200 Min.	1200 Min.	1200 Min.
Pour Point	10,000 SSU Max. @ Min. Operating Temp.		

Courtesy of Worthington-CEI

Fig. 20-62B. Typical specifications for rock drill oil

close alignment. Careful attention to instructions on starting a hole should assure a correct beginning. But if tracks should move, the boom sink, or the mast tilt during drilling, alignment will be lost and problems will arise.

Drilling out of alignment puts severe strain on the drill string, and on the mast and its positioning mechanism. Drilling force will be lost in flexing of rods, and in friction against the sides of the hole. The life of both rods and couplings will be shortened by bending stresses. Failure may occur immediately, or some time later.

When misalignment is detected promptly, the operator may have a choice of methods to correct it. He may decide to re-position the mast by using the boom and cone controls, or he may move the whole machine.

However, if the hole is already crooked, it will probably be necessary to abandon it, and drill another hole close alongside, with greater care.

Lubrication. Chassis lubrication is conventional. There are a number of high pressure grease fittings and a few gear case reservoirs that need daily or weekly attention, according to type of use and a schedule supplied with the machine.

The drifter and other air-operated units are lubricated by oil in the air stream. A special rock drill oil (see a typical specification in Figure 20-62B) is kept in a 5-gallon tank, from where it feeds through a regulating needle valve into the manifold, where it is picked up by the air stream.

One recommendation for lubricator adjustment is to turn the needle valve handle to closed, then open it two turns. Then open the main air hose at the drifter, let air blow until it shows an oil mist, then reconnect.

Visible mist is the standard and dependable indicator of the presence of oil in the air. But it does not tell how much, except to the expert. Correct proportion is indicated in many of these machines by a slow seepage of oil out of the bottom of the chuck housing and down the striker bar while the drill is operating.

If this area is dry, there is not enough oil and the lubricator valve should be opened further. Oil running down calls for cutting down the feed.

The oil tank must be full at the start of the shift, and it should be checked at least once during the shift. Many expensive components can be ruined by working briefly without oil.

MOBILE TOWER DRILLS

Larger drill rigs usually have a higher mast or carriage, often called a tower, hinged directly to the travel base, with hydraulic cylinders to raise it to working position or to lower it for travel.

Most units drill only with the tower vertically fixed to the carrier, but means are sometimes provided to compensate for uneven ground, or to drill at angles.

Tower height may be 30 to 60 feet, or more, to provide feed length or stroke of 20 to 55 feet.

The tower is usually at the rear of the carrier, which then becomes the "front" of

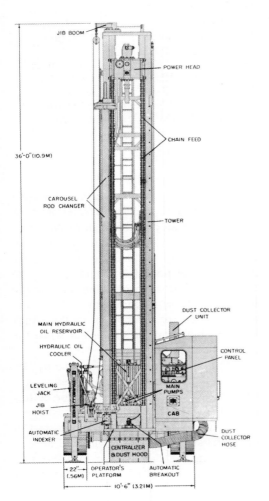

JIB BOOM

POWER HEAD

CHAIN FEED

36'-0"(10.9M)

CAROUSEL
ROD CHANGER

TOWER

DUST COLLECTOR
UNIT

MAIN HYDRAULIC
OIL RESERVOIR

HYDRAULIC OIL
COOLER

CONTROL
PANEL

LEVELING
JACK

MAIN
PUMPS

JIB
HOIST

CAB

AUTOMATIC
INDEXER

DUST
COLLECTOR
HOSE

CENTRALIZER
& DUST HOOD

22"
(.56M)

OPERATOR'S
PLATFORM

AUTOMATIC
BREAKOUT

10'-6" (3.21M)

Courtesy of Ingersoll-Rand Company

Fig. 20-63A. Mobile tower drill

the whole unit. This is the simplest construction, permits accurate spotting, and enables the whole machine to keep further away from dangerous edges. Central mounting permits using a greater proportion of machine weight for down pressure, and is used only in very heavy units.

The mounting may be on a special carrier with individually controlled shovel-type tracks, on a big crawler tractor, or on a heavy duty truck similar to a crane carrier. When crawler tracks are driven by separately reversible air or hydraulic motors, it

can maneuver accurately in limited space. However, it is not suited to rough or steep ground. Truck mounting is preferred where job locations are changed frequently, as in water well drilling.

The machine is leveled up before drilling by means of three hydraulic jacks separately controlled from the operator's station. A bubble or pendulum level on the tower shows when it is vertical. Check valves hold pressure in the jacks if the hydraulic system fails. The hydraulic pump may be driven mechanically or by an air motor.

Drill rod is ordinarily rotated from the top by an air, hydraulic, or electric motor, geared down. There may be two or three speeds, or infinitely variable rates from zero to 80, 100, or higher rpm.

The rotary head must include a swivel head through which compressed air enters the rod. Oil field units may use the mechanically driven fixed height rotary table described later under deep rotary drilling.

The feed mechanism may be an air motor and chain, a two way hydraulic ram connected directly or operating a chain through a pulley, or a cable and drum crowd-retract with electric drive for hoisting and lowering and hydraulic motor drive for feed control.

The feed keeps the bit in contact with the rock and provides proper drilling pressure (pulldown). This may be as much as 25 tons in these mobile machines, and much more with stationary deep well rigs. The feed is reversed to pull rods and bit out of the hole when it is finished or when trouble develops.

The tower includes a rack on which extra drill stems may be stored, with hydraulically controlled devices for handling them, in adding to the drill string or in pulling it out.

Air requirement is 250 to 1300 cfm, mostly for blow air. The compressor is usually mounted on the tractor or truck that carries the drill, making a complete self-

ROTARY OR PERCUSSION
(Cost Comparisons)

Type of Rock	—Medium (3.0 to 4.0 on Moh's Scale)
Daily Production	—5,000 Tons
Hole Size	—4" Diameter
Hole Depth	—30 Feet
Hole Spacing	—8' x 10'

Equipment Used	—Rotary: 1 unit in 15,000 lb. Pulldown Class, 250 CFM Compressor, Tungsten Carbide Drag Bit
	—Percussion: 2 "Air Tracs"® w/600 CFM Compressors, 4½" Bore Drill w/independent rotation, 1½" Steel

Cost Factor	Cost Per Foot	
	Rotary	Percussion
Capital Equipment (5 yr. Amortization)	$ .030	$ .070
Drill Steel, Couplings and Shanks	None	.051
Kelly Bar and Subs	.016	None
Bits	.075	.015
Bit Grinding	.003	.004
Rock Drill Oil and Coupling Grease	.002	.008
Compressor, "Air Trac" & Drill Maint.	None	.056
Rotary Drill Maint.	.028	None
Fuel	.011	.056
Labor	.020	.040
Total Cost Per Foot of Hole	.185	.300
Cost per Ton	.030	.048

Type of Rock	—Hard (4.0 to 5.5 on Moh's Scale)
Daily Tonnage	—3,500
Hole Size	—4½" Diameter
Hole Depth	—30 Feet
Hole Spacing	—9' x 12'

Equipment Used	—Rotary: 2 units in 15,000 lb. Pulldown Class, 250 CFM Compressor, Roller Cone Bits
	—Percussion: Heavy "Air Trac,"® 900 CFM Compressor, 5" Bore Independent Rotation Drill, 1¾" Hex. Steel

Cost Factor	Cost Per Foot	
	Rotary	Percussion
Capital Equipment (5 yr. Amortization)	$ .120	$ .110
Drill Steel, Couplings and Shanks	None	.102
Kelly Bar and Subs	.020	None
Bits	.340	.101
Bit Grinding	None	.001
Rock Drill Oil and Coupling Grease	.002	.018
Compressor, "Air Trac" & Drill Maint.	None	.046
Rotary Drills Maint.	.026	None
Fuel	.045	.067
Labor	.080	.040
Total Cost per Foot of Hole	.633	.485
Cost Per Ton	.072	.055

Courtesy of Gardner-Denver Company

Fig. 20-63B. Costs — rotary and percussion drilling

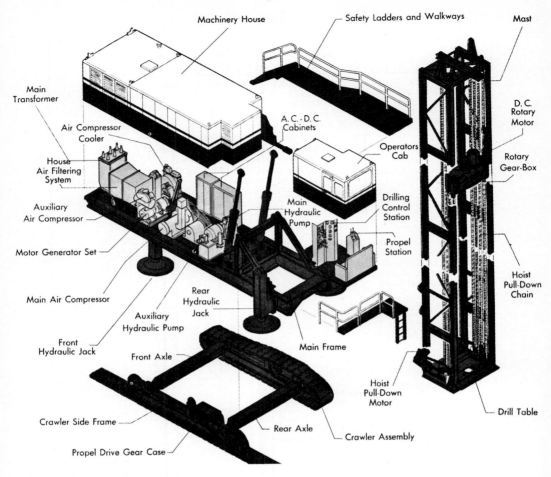

Labels on figure:

Machinery House
Safety Ladders and Walkways
Mast
Main Transformer
D. C. Rotary Motor
Air Compressor Cooler
A. C.-D. C. Cabinets
Operators Cab
Rotary Gear-Box
House Air Filtering System
Drilling Control Station
Auxiliary Air Compressor
Main Hydraulic Pump
Motor Generator Set
Propel Station
Main Air Compressor
Rear Hydraulic Jack
Hoist Pull-Down Chain
Auxiliary Hydraulic Pump
Front Hydraulic Jack
Front Axle
Main Frame
Crawler Side Frame
Hoist Pull-Down Motor
Drill Table
Propel Drive Gear Case
Rear Axle
Crawler Assembly

Courtesy of Marion Power Shovel Company

Fig. 20-63C. Large tower drill, exploded view

contained unit. Small or special models may tow a compressor, or use piped air from a stationary or distant source.

Power may be supplied by the carrier engine through a power takeoff, by a separate diesel or gasoline engine, or by electric motors.

The compressor or compressors may be the only power source, with all drill functions operated by air motors. Some machines are powered by the carrier engine through a power takeoff, or by a separate diesel engine, with various combinations of air, hydraulic, and mechanical drive.

One large unit with electric drive uses 50 horsepower for travel, 125 for air com-

pression, 15 for a hydraulic pump, and 10 for air compression. Another uses two 125 horsepower motors for air, and two 25s for hoist and rotary drive.

These rigs may be equipped with down hole drills for holes up to 9 inch diameter, and with rotary bits for diameters up to 15 inches. Augers are sometimes used. Models designed primarily for down hole equipment may have slower rotation and less down pressure than those meant for rotary work. The down hole drill may require more air capacity and higher pressure than the rotary. However, most machines are convertible to some extent between types.

Conventional drifter drills of 4½ or 5¼

inch size may be used for drilling holes up to 4 inches in the conventional manner. Rotation may be provided by either the head or the drill. Depth is limited by the absorption of hammer impact by long drill steel. Efficiency starts to decline at 30 feet in hard rock, and at somewhat greater depths in soft material. They are sometimes used down to 100 feet or more.

Down hole and rotary drills are limited in depth by the ability of the rotary table bearings to carry the weight of long rods, and by the ability of the air stream to keep the cuttings cleaned out of the hole. Down holes are intended to work down to about 200 feet, and have drilled to 800 feet or more. An extra compressor may be hooked in to help clean the hole at great depths.

If a rotary drill is used, and an optional mud circulating system substituted for air blast cleaning, there is no exact limit to depth. However, the special rigs to be described later under the heading DEEP ROTARY DRILLING are more efficient for very deep holes.

The maximum practical depth of drilling is different with each make and model, and often must be learned by experience. In general, the biggest machines with the most elaborate controls are able to maintain efficiency to the greatest depths, but not always.

JET PIERCING

The Linde JPM 3 Piercing Drill uses a high velocity, high temperature (4,300 F.) flame to drill certain hard rocks that are rich in silica. The principal use of these machines at present is in drilling taconite, a hard, tough, low-grade iron ore. It may also be used in granite, sandstone, and cherty varieties of rock.

Jet piercing penetrates rock by flaking or spalling of the rock surface. Sudden expansion of the surface layer in contact with the flame breaks it off from the cooler rock behind it, and the particles are carried away by steam and exhaust gases.

Courtesy of Linde Company

Fig. 20-64. Jet piercing burner

Rotary. The burner is operated from a mobile tower drill on crawlers, usually with electric power obtained through a trailing cable. It is leveled by hydraulic jacks, which may also be used to tilt it for angle drilling. Tower height is 60 feet.

The 55 foot blowpipe or kelly, that serves as the only drill rod, is fastened at its upper end to a swing joint, that in turn is hung from a cable that is allowed to pay out as the drill penetrates. The blowpipe is turned either by a rotary table, or by a 4-speed rotary head. High pressure oxygen, fuel oil, and water are supplied through the swing joint to separate tubes in the blowpipe.

Burner. The burner is at the lower end of the blowpipe. It is similar in design to some military rockets. Fuel oil is atomized by the oxygen blast, and the two combine

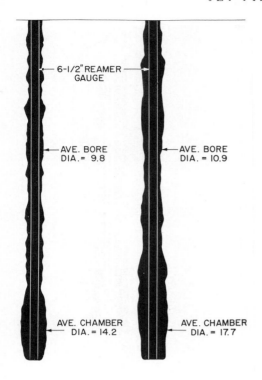

Courtesy of Linde Company

Fig. 20-66. Cross section of jet holes

in a combustion chamber, from which three flames issue at a downward angle. Combustion gases attain a supersonic velocity of 6,000 feet per second, or 4,000 miles an hour. The burner is protected from contact with the hole bottom or sides by a reaming collar, whose 6½ inch outside diameter determines the minimum hole size.

A current of water cools the burner and the collar, and is converted to steam that assists the combustion product gases in blowing the flaked rock to the surface. A short casing is sometimes used at the hole collar. A rotor draws fumes and steam from the hole top and exhausts them through a 20 inch stack.

The readiness with which the rock spalls may vary every few inches in bedded rock, to make a succession of wide and narrow openings. In ordinary drilling the size may vary from the 6½ inch minimum up to 12 inches or more, with an average diameter of 10 or 11 inches.

Speed of penetration is variable. The operator must pay out line to keep the burner within a few inches of the face, but should not allow the collar to rest on it. He has various semi-automatic controls and indicators to help him regulate the rate of feed.

Oxygen and fuel may be supplied by fixed lines, but it is usually more efficient to use truck or crawler mounted supply tanks stationed near the drill, that are in turn supplied by mobile units. The burner has an hourly consumption of 10,000 cubic feet of oxygen, 40 gallons of fuel oil, and 1,000 gallons of water. New models now under test may increase oxygen consumption to 12 or 15,000 feet, and other requirements proportionately.

Penetration rates in various materials as published by the manufacturer are shown in Figure 20-65. Burning (drilling) time may be between 55 and 75 per cent of working time.

Jet piercing has its greatest competitive advantage in hard solid formations that produce excessive wear on drill bits.

Chambering. Flame is the only drilling method that can easily produce blast holes that are larger at the bottom than the top. Such chambers are made by moving the burner up and down in the lower part of the hole after it has reached bottom. Each trip removes an additional layer of rock and enlarges the hole.

Chambering can be used to increase the concentration of explosives at the bottom of the hole where they are most effective. The need for drilling below grade and the danger of leaving unblasted ribs are reduced.

Hole Measurement. The irregular size of holes makes it desirable to measure them before loading, so that the concentration of explosives can be varied. This is done by a special three-finger inside caliper called a Contourometer. The instrument is lowered into the hole and an electrical pickup registers its reading on the dial of a meter at the collar of the hole as it is raised.

Suspension Piercing. A burner of slightly different design, providing a single downward flame, is used in non-rotating jet piercers. The carrier is usually a churn drill modified to provide a very short stroke. This type is limited to materials that spall readily, and have few cracks or seams. Under favorable conditions it can sink holes as deep as 180 feet.

This same type of burner can be used with a swinging blowpipe to cut channels in granite for taking out blocks for dimension stone.

AUGER DRILLS

Augers are rotary drills that have a helix or screw thread on drill rods that are called augers or flights. Drilling resistance and/or control of the rate of feed prevents the thread from penetrating in proportion to its turning speed, so that material cut by the bit is gripped by the threads and forced out of the hole by a screw conveyor action.

The flights are made in sections proportional to the feed length of the drill unit. They are connected to each other by bolting or pinning.

Boring Head. The boring heads are bits of the drag or fishtail type that cut by rotary scraping. There is a great variety in design and material to meet different conditions. There is usually an advanced center or pilot cutter. The teeth, called fingers, are generally detachable. They may be set in a separate head or in the leading edge of the auger flight. Cutting edges are of steel hardened by various processes, or of tungsten carbide. Worn steel teeth may be built up with borium or carbide hardfacing.

The head should be slightly larger than the auger flights, so that they will not bind in the hole.

Capacity. Augers are made in standard

Courtesy of Peterson Engineering Company

Fig. 20-67. Auger boring heads

Courtesy of Salem Tool Company

Fig. 20-68. Vertical auger drill

skill of the operator. A big machine may put a small bit down in firm soil so rapidly that most of the working time is spent latching on additional augers.

Uses. Augers are primarily earth drills, but are also adapted to penetrating soft to medium rock such as shale, soft limestone, and sandstone.

Augers are widely used for soil testing, prospecting for minerals, blast holes, drainage holes, putting pipes and conduits under fills, pavements or obstructions, placing deep footings, setting fence posts and utility poles, mining, and many other purposes.

Construction. Units for vertical auger drilling are somewhat similar to blast hole rotaries, in that they have a rotating head that grips the top of the rod, and a power feed mechanism. However, the feed or carriage stroke is usually quite short, and the rotation mechanical instead of air or hydraulic. Wagon or crawler drills may be used by substituting an air driven rotary head for the drill.

The 54 horsepower truck mounted McCarthy Model 106-24 in Figure 20-69A is designed to handle augers of 24 to 8 inch diameters to depths of 30 to 100 feet. It has a feed stroke of 8 feet and uses 6 foot sections. Maximum carriage speed in feet per minute is 40 downward and 34 up. Down pressure is about 7 tons, and lift capacity 8 tons.

Rotary drive is mechanical with 8 ratios, ranging from 13 to 270 revolutions per minute at full engine speed. Levelling jacks may be mechanical or hydraulic. Maximum torque at the bit is about 3,000 pounds.

In horizontal units the drill frame makes a bed or cradle for supporting the auger as it is fed into the hole.

Figure 20-69B shows a light auger designed for mounting on a jeep or light truck. It is driven by the carrier's power takeoff.

The rotary table is fixed near the top of the mast, and is driven by a vertical shaft and a 3-speed transmission. A square drive

diameters from 2 inches to 6 feet. Some of them are easily operated by one man, others require rigs weighing 35 tons or more. Vertical holes are usually limited to a depth of 100 feet or less, while horizontal holes may extend over 250 feet.

Most auger drills will handle a variety of sizes. With any one machine, increasing the size of the hole decreases the speed of penetration and maximum depth obtainable, and limits the hardness of material that can be drilled.

The rule of thumb is that maximum depth is in inverse proportion to auger diameter. That is, if a machine will drill down 50 feet with a 12 inch bore, it should not be expected to go much deeper than 25 feet with a 24 inch auger.

Capacity diminishes in regard to both diameter and depth of hole when the going gets tough.

Drilling rates vary with the power of the unit, the hole size, the material, and the

Courtesy of Paris Manufacturing Company

Fig. 20-69. Jeep auger

stem or kelly can slide up and down through a square socket in the table. The kelly is supported by bearings in a cross head or frame that is raised and lowered for a 6 foot stroke by a pair of hydraulic rams. The rams are fully extended at the start of the digging stroke, and the top auger flight is coupled to the bottom of the kelly.

Augers are carried in a rack on the front bumper.

Vertical Drilling. The machine is carefully leveled by means of its jacks. If the ground is soft, planks or blocks should be used to spread the weight, as any movement of the carrier during drilling is likely to damage or break the augers.

The drill chuck is raised to the top of its stroke. A cutting head is fastened to an auger, and the other end of the auger is fastened in the chuck.

The cutting head is lowered to the ground. A guide or centralizer may be latched around the auger just above it, the drill clutch or hydraulic valve is engaged to rotate the auger, which is then pushed into the ground with the feed control.

As the bit cuts the ground the spoil is brought to the surface on the auger thread.

Courtesy of Acker Drill Company

Fig. 20-70A. Vertical auger with tower

Most of it is thrown off by centrifugal force and makes a circular pile around the hole. A shield may be used to prevent it from building so close to the pile that it will tend to fall in when the auger is removed. Cuttings may be removed with a hand shovel from time to time.

When the auger is mostly underground, the rotation is stopped, the guide removed, the auger unlatched, the chuck raised all the way. Another auger is attached to the one in the ground and to the chuck, then drilling is resumed. The guide is used only for the first length.

When full depth is reached, the feed is stopped and the auger revolved until the hole is clean of cuttings. The chuck is raised all the way, bringing the top section clear out of the ground and showing the next one. An auger fork or some other device is caught around the lower auger to prevent it from sliding down.

The upper section is then detached at both ends and set aside. The chuck is lowered, fastened to the top of the next auger and raised. Sections are removed one by one in this manner.

Some drills have an auxiliary tower with a cable hoist that permits raising several sections at a time.

The head should be inspected for wear or breakage of the bits after every hole. If the bits are not sharp and full size, they will cut an undersize hole, causing excessive wear on the head and auger flights, and causing the augers to bind and possibly break off.

The positions of the auger sections should be changed with each hole. The section immediately behind the head wears more rapidly, and if used in this place repeatedly, it will wear to a smaller gauge than the others and will pack cuttings against the wall instead of bringing them up.

Bits and wearing parts of augers can be built up by hardfacing with borium or carbide rod.

Horizontal Auger Drills. The Parmanco drill shown in Figure 20-71 is used largely in drilling horizontal blast holes in strip mine overburden. It has a truck rear axle and wheels, and can be engine driven through the drill transmission by shifting jaw clutches. It is steered by the draw tongue on the small front wheels, or can be towed by other vehicles by the same tongue.

Feed is by means of a cradle which moves forward and back by power.

When the machine is in position, jacks are used to set it at the proper height and angle for the hole. The drive wheels may be removed if necessary for a very low hole.

The feed carriage is moved all the way back, and an auger and head placed so that the secondary cutters (interruptions) in the

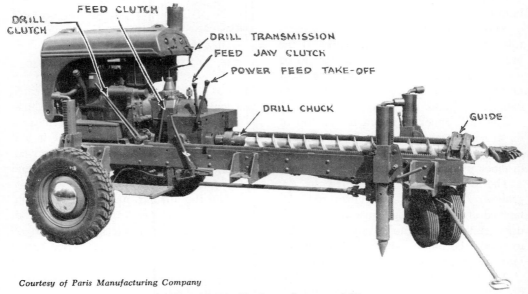

Courtesy of Paris Manufacturing Company

Fig. 20-71. Horizontal auger drill

Courtesy of The Salem Tool Company

Fig. 20-72. McCarthy coal recovery drill

auger flight are in front of the guide. The feed is then moved forward and the auger attached to the chuck. The feed is moved forward until the drill head makes contact.

The drill transmission is placed in first speed and its clutch engaged. When the head has penetrated two or three feet the guide is removed.

When the drill cradle reaches the forward end of its track the chuck is discon-

nected from the auger and the cradle is moved to the rear.

Another auger section is then latched to the rear of the front auger and inserted in the chuck. Except for the first few feet, which are drilled in low gear, any one of four transmission speeds may be used.

When the hole is finished the drill is put in third or fourth speed and the auger turned for two or three minutes. It is some-

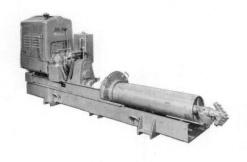

Courtesy of Salem Tool Company

Fig. 20-73A. Trench drill at work

times kept turning while being pulled out also.

Augers are withdrawn one section at a time, uncoupled, and the cradle moved forward to engage the next one. The positions of the auger sections should be changed with each hole.

Coal Recovery Auger. The McCarthy coal recovery drill, Figure 20-72, is a horizontal auger that uses sections from 16 to 54 inches in diameter, and up to 24 feet long. Penetration may be 60 to 250 feet.

The compound drilling head, with inner and outer cutters, is designed to remove soft coal in lumps, with minimum fine breakage. The cuttings are carried back by the auger flights to a conveyor belt that loads into trucks.

Sections are handled by a crane that may be hydraulic or hand powered. The ma-

chine moves itself hydraulically from hole to hole on mobile skids.

This drill is used chiefly for mining coal from the edges of strip pits. Standard practice is to cut parallel holes, as shown, leaving thin pillars or ribs between them to support the overburden. Sometimes a single hole is cut for a center line tunnel, and coal from each side is drilled and blasted into it.

Such auger drilling may be a final operation, where the coal seam has become too thin or the overburden too heavy to justify further stripping. On other jobs parallel slots or keyways are cut at intervals, and the coal seam between them is removed by augers.

Trench Drill. The trench drill is a horizontal auger specially constructed to bore under roads, driveways, and other surface features that would be damaged by trenching. It can bore under tree roots or through them.

These machines are usually skid mounted, and are lowered into working position by a sideboom or a crane. Power may be a gasoline engine or a hydraulic motor driven from a separate pump or machine at the surface. Various model sizes handle auger diameters from 3 to 36 inches. Practical length of hole is about 250 feet, so that it is possible to drill under an entire city block by working from both ends.

Very accurate leveling and aiming are required for successful drilling of long holes. Even if this is done properly, changes in soil or contact with underground obstructions may put the augers out of line. A tendency to go off course can be corrected by means of bars or wood wedges from a pit dug in its path.

In firm soil the hole can be drilled bare. When the auger comes out at the other side, a line is attached to it and is pulled through as the auger is retracted and dismantled. This line can then be used to pull a pipe or conduit through the hole.

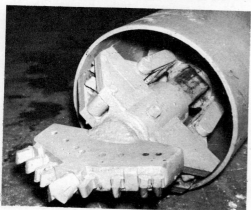

Courtesy of Salem Tool Company

Fig. 20-73B. Collapsible boring head

If the dirt is likely to cave or the pipe is too large to pull, it is pushed through while the hole is drilled. The boring head is slightly larger than the outside of the pipe, and the auger a free fit inside it. The pipe is advanced by a pusher plate and adapter ring. Its front edge stays slightly behind the bit. The cuttings are fed through the pipe by the auger flights. Auger and casing sections are added at the same time.

A collapsible boring head should be used, so that the augers can be pulled if necessary without bringing the pipe out also. Such heads are expanded by drilling pressure and fold back inside the casing when they are retracted.

Back Boring. An auger drill may make a hole up to 50 per cent greater diameter than its normal capacity by using a back boring head to ream out the hole. The hole is made in the normal manner through an embankment, until the bit projects on the far side. The bit is removed, and a larger back boring one installed. Rotation in the drilling direction is used, with the feed pulling instead of pushing. Cuttings fall into the auger flights for removal. Each time the feed reaches the end of its stroke an auger section is removed.

Obstructions. An auger may be stopped by a hard rock seam or a boulder. This may sometimes be penetrated by changing to a carbide tipped bit designed for harder material, or by drilling a small hole first and then enlarging it.

At other times it is necessary to blast. The augers are withdrawn, and a charge of the fastest acting explosive available—high strength gelatin dynamite or a slurry — is primed, lowered or pushed to the obstruc-

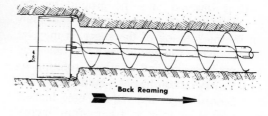

Back Reaming

Courtesy of Salem Tool Company

Fig. 20-73C. Back boring

Courtesy of Salem Tool Company

Fig. 20-73D. Back boring head

tion, and fired electrically or with Prima-cord. The charge may be a pound for a small hole, or several pounds in a larger one.

The obstruction might be shattered into drillability by one blast, or several, or it may be necessary to abandon the hole, depending on conditions. The explosion sometimes loosens the formation so that the auger gets squeezed and caught when drilling is resumed in a horizontal hole.

Post Hole Digger. The Danuser Post Hole Digger, Figure 20-73E, is an attachment that mounts on a tractor 3-point hydraulic lift, and is driven by a propeller shaft from the power takeoff.

The auger can pivot on the support arms so that it will drill straight down even if the tractor is headed up or down hill. It is adjusted for side tilt by the leveling arm in the drawbar.

It is lowered into the ground by the drawbar control. Spoil is raised and spun off by the rotation of the auger. It may be necessary to raise and lower a few times, or to stop lowering and continue rotation to avoid carrying too heavy a load.

Standard auger diameter is nine inches, but widths of four to twenty-four inches are obtainable. Except for the smallest sizes they have removable cutting edges. The auger is attached to the drill shaft by a soft bolt that will shear if the auger jams.

This machine is used chiefly for making holes in which to set posts for fences and for planting seedlings. It can also be used for test holes to find depth and quality of topsoil or other shallow surface layers.

PNEUMATIC UNDERGROUND PIERCERS

An air powered mole, Figure 20-73F, may be used instead of an auger for small underground horizontal bores.

This is usually a cylinder, 4 to 5 feet long and 3½ to 5 inches in diameter, tapered at the front and attached to a one-inch air hose at the back. Weight may be 60 to 190 pounds.

It contains a valve and a two-way cylinder chamber. Whenever it has air pressure, a piston moves back and forth in the cylinder, striking heavy blows against an anvil that is rigidly fastened into the tapered nose. Striking force may be 150 or more foot pounds, with blow frequency up to 400 a minute.

When the unit is embedded in soil, each blow tends to drive it forward. The tapered nose makes an opening, displacing material to make a hard packed tube. Slipping back

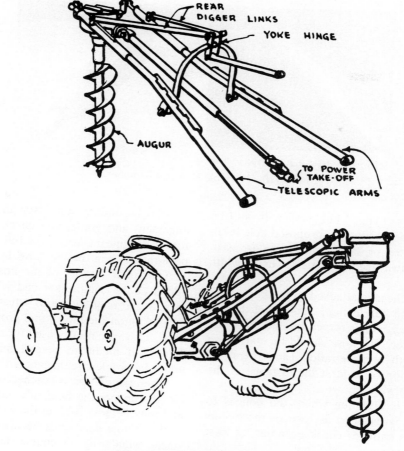

Fig. 20-73E. Post hole digger

between strokes is prevented by friction, and by roughening the cone with spiral threads.

Working air pressure is about 90 pounds. Requirement is 60 to 150 cfm in various models. Lubrication is provided by a line oiler near the compressor, or by occasionally opening the air hose and pouring in a small quantity of A transmission fluid.

It is recommended that the hole be at least three feet below the surface in ordinary soils, and four feet in sand or gravel. Otherwise, displaced dirt may tend to move chiefly toward the surface, causing the hole to incline upward.

Some units have a reversible valve. It is shifted by disconnecting the hose at a coupling, and twisting it (and the rear of the mole) counterclockwise a prescribed number of times. Air hammering will then bring the tool out. The hose must be pulled by hand to avoid binding.

If there is no reverse, there should be a bracket on the unit for attachment of a light retrieval cable, which follows it into the hole and can be used to pull it back if necessary.

An oversize hole may be made by adding a collar. A coupling may be fastened to the front to enable the tool to drive pipe or conduit ahead of it. Such pipe must be fitted with a tapered nose cone.

Operation. A starting pit is dug, similar

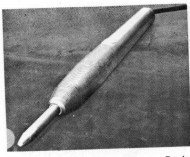

Courtesy of Allied Steel and Tractor Products

Fig. 20-73F. Hole Hog and launcher

to that for an auger, but smaller. It must be slightly deeper than the proposed hole level. If a starter or launching frame is used, the pit must accommodate its extra width also.

An air hose is attached to the back, provided with lubrication, and coupled to a compressor. Air is turned on and off at the compressor. The hose should be marked so that length of penetration (and therefore, the location or at least the distance of the tool) is always known.

A pit is also dug where the hole is to finish. It is usual to dig one or more slots across its path for check purposes, at least until accustomed to the tool.

Place the mole in its pit and against the face to be drilled. Level and align it carefully, by instrument or otherwise. Open the air valve part way, and push on the cylinder or the hose to help it get in. When about one-third is underground, stop and re-check alignment and grade, and check again before it disappears. When the full length has penetrated, open the air valve fully.

Additional lengths of air hose may be added at the compressor end when necessary. Speed may be one to 4 feet a minute, depending on the model and ground conditions. These are mostly used for rather short holes, but might go several hundred feet under favorable circumstances.

When the exit pit is reached, shut off the air, disconnect the hose, lift out the mole, and pull the hose back to the compressor.

If the hole dead ends because of an obstacle, wandering off course, or because a blind hole is wanted, the tool must be brought back by reversing it, or by pulling a retrieval cable or the hose. If, for any reason, it cannot be moved either backward or forward, it must be dug up.

Courtesy of Allied Steel and Tractor Products

Fig. 20-73G. Starting and finishing a hole

Courtesy of Bucyrus-Erie Company

Fig. 20-74A. Spudding drill, truck mounted

CHURN DRILLS (SPUDDING DRILLS)

Until a few years ago the churn drill was the basic tool for putting down holes in the four to twelve inch diameter range. It lost its lead first in the oil fields, where it was largely replaced by the faster deep drilling rotaries to be described in the next section. It has now been displaced in mining and construction by mobile tower machines with rotary or down hole drills.

The churn drill has many things in its favor. It is rugged and dependable, the principle of operation is simple, and with suitable adaptions in the drill string and winches, can drill any class of material from dirt to taconite, to over a mile in depth.

But it makes a hole at only two to ten feet an hour, including bailing and changing time. Competitive machines can make holes at five to ten times its rate. As a result, labor costs per foot are high, and several times as many machines and shifts may be needed to do a job with a churn.

For example, a quarry that used three churn drills three shifts a day, seven days a week, has replaced them with one down hole unit working a 40-hour week. If increased stone production should cause the blasters to catch up with the new drill, it could easily pull away again by working overtime, while the churns would need help from another unit. Cost comparisons also favor downholes and rotaries.

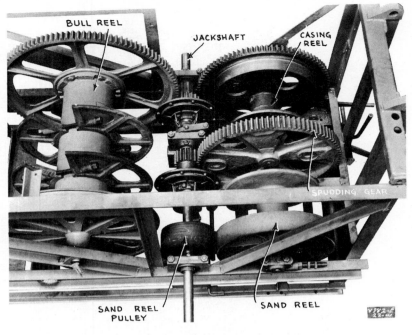

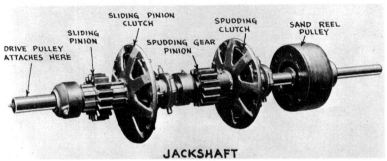

Courtesy of Bucyrus-Erie Company

Fig. 20-74B. Deck machinery

Churns are still important in the water well drilling field, but are losing popularity. Here their principal disadvantages are the long period during which they must be on the job, and noise and heavy pounding of their work. Water supply for drilling may also be a laborious nuisance.

Although churn drills are declining in use, there are still thousands of them at work all over the world, and they will be an important drilling tool for many years. The following description of typical designs, and the technique of drilling with them, is therefore being retained.

Well and blast hole models are usually limited to depths of 1500 feet or less by cable and lifting capacity, and design.

A spudder drills by raising and dropping a chisel bit in a hole partly filled with water. The cuttings mix with the water to form a mud or "sand" that is bailed out at intervals. The bit is suspended or "swung" on a cable, and may be weighted with sections called drill stems, and other units. The assembly of units attached to the bit is called the string of tools. Water well drilling requires maximum mobility, including extensive use of public roads and maneuvering into tight places. Truck bed mounting is usually preferred, although some operators prefer to use a semi-trailer so that one truck can take care of two or more rigs, or so that the truck can be detached and used for commuting and for carrying water and other supplies.

Blast hole work ordinarily requires frequent moves in a limited area, often on steep slopes and rough ground. Self-powered crawlers are therefore standard equipment. However, in many quarries, which have relatively flat and smooth drilling areas, trucks, or trailers with either pneumatic or steel tires will prove more economical.

In the oil fields, churn drills or spudders have been developed for top-to-bottom drilling to 6000 feet, and servicing to 7000 feet. The largest of these models handles 6000 pounds of tools and with an auxiliary A-frame can take static casing loads up to 160,000 pounds. Such machines are apt to be skid or trailer mounted.

Well Drills. Well drills are frequently used in blast hole work on small jobs when surface conditions are good, and as auxiliaries to speed up a drilling program.

The Bucyrus-Erie 22-W well drill, shown in Figure 20-72, is representative of a class of light truck or trailer mounted rigs. It is designed to drill holes of four to ten inches diameter, to depths of 350 to 1000 feet, and to handle (swing) 1500 to 1800 pounds of tools.

The frame is a box of steel sections that can be bolted directly to a truck or trailer frame or to skids. The derrick or tower is a lattice boom in two telescoping sections. It is hinged to the upper machinery frame and is lowered to a horizontal position for moving. It is erected by a power winch which first pulls the foot of the tower against the rear of the deck, then extends the upper section. It is secured by jamming the slide with chocking irons and fastening the tubular bracing shown.

Courtesy of Bucyrus-Erie Company

Fig. 20-74C. Cushioned derrick sheave

Fig. 20-74B shows the deck machinery. An engine, mounted across the front of the frame, drives the jackshaft through a flat belt on an external pulley (not shown). A belt is used because it absorbs the drilling shocks which otherwise would be transmitted to the engine.

The jackshaft carries two pinion gears driven through friction clutches, and the sand reel pulley which is keyed to the shaft. One pinion slides on a splined hub so that it can turn either the bull reel or the casing reel, or run free in neutral. The other pinion drives the spudding or crank gear which gives the up and down motion to the tools.

The bull reel, or spudding drum, lowers the tools into the hole, keeps them in proper cutting position, and lifts them out. It is operated by a jackshaft clutch and a large band brake. When the jackshaft pinion is shifted out of engagement the reel is controlled only by the brake.

Moveable flanges divide the drum into working and storage sections. Only one layer of cable should be kept on the working section to avoid crushing. Blast hole work rarely requires much extra cable, but in deep well work most of the cable is kept in the storage sections.

Courtesy of Bucyrus-Erie Company

Fig. 20-75. Blast hole drill (without derrick)

The casing or calf reel, also driven by the bull reel pinion, is optional equipment. It is used to lift and place casings which are pipe sections used to line the hole to prevent the sides from falling into it. Casings are usually not required in rock. This winch, which controls a line over the right hand derrick sheave, is also useful for miscellaneous lifting jobs.

The sand reel is a high speed winch used in bailing the hole. It includes a cast iron wheel which is mounted between the lined pulley on the jackshaft, and a brake shoe. It is mounted on an eccentric and can be tipped by a hand lever so as to contact either the driving pulley or the brake. In the middle position it turns free.

The spudding gear is located at the end of its axle and is driven by the center jackshaft pinion. It carries a crank pin connected by a bar or pitman arm to the spudding beam that is hinged to a crossbar near the front of the frame. The rotation of the spudding gear raises and lowers the pitman arm and the rear of the beam. The length of this stroke may be adjusted by placing the crank pin near to the gear axle for a short stroke, or far from it for a long one. Three holes are provided for this adjustment, with stroke lengths of 24″, 30″, and 36″

The spudding beam is furnished with two sheaves. The heel sheave is mounted on the shaft on which the beam is hinged. The beam sheave is at the rear. The sheaves may be of all steel construction, or have steel hubs and rims connected by rubber cushion rings which absorb drilling shock.

The bull reel cable is reeved behind and over the heel sheave, under the beam sheave, over the main derrick sheave, and down to the tools.

The main derrick sheave is cushioned by

a set of alternating rubber and steel washers, Figure 20-74, which compress under load, allowing the sheave to move down a short distance on tracks. When the load is released they thrów the sheave back to its top position.

The spudding action is as follows: The bull reel cable suspends the tools in the hole with the bit edge close to the bottom. When the spudding beam is pushed up by the crank, the cable is able to take a shorter path between the heel and the derrick sheaves. The resulting slack allows the tools to drop. When the slack is used their momentum pulls hard on the cable, compressing the cushion and pulling down the derrick sheave. While this is down the bit strikes the bottom of the hole. This relieves the weight on the cushion so that it is able to snap the sheave upward at the moment when the bit has a tendency to bounce, and when the spudding arm has started its down stroke, pulling the cable back across the derrick sheave and raising the tools. This cycle is repeated 50 to 60 times a minute.

Blast Hole Drill. The Bucyrus-Erie 22-T blast hole drill, Figure 20-75, is

Courtesy of Bucyrus-Erie Company

Fig. 20-76. Tool guide (centralizer)

crawler mounted and uses a box section derrick. The truck frames are connected by a rigid rear (drilling end) axle and an os-

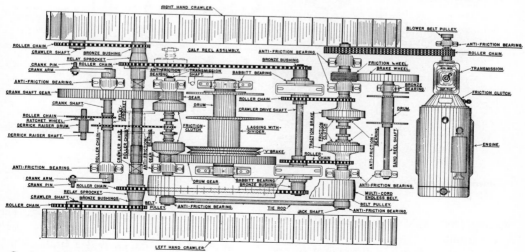

Courtesy of Stardrill-Keystone

Fig. 20-77. Deck layout

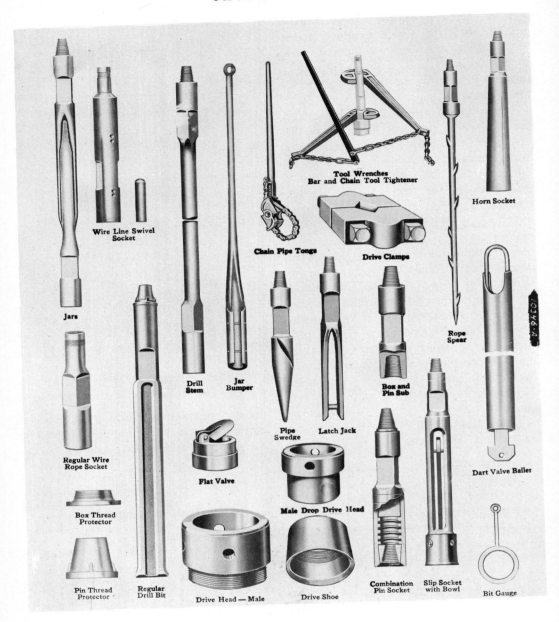

Wire Line Swivel Socket

Jars

Regular Wire Rope Socket

Box Thread Protector

Pin Thread Protector

Regular Drill Bit

Drill Stem

Jar Bumper

Flat Valve

Drive Head — Male

Chain Pipe Tongs

Tool Wrenches
Bar and Chain Tool Tightener

Drive Clamps

Pipe Swedge

Latch Jack

Male Drop Drive Head

Drive Shoe

Box and Pin Sub

Combination Pin Socket

Horn Socket

Rope Spear

Slip Socket with Bowl

Dart Valve Bailer

Bit Gauge

Courtesy of Bucyrus-Erie Company

Fig. 20-78. Spudding drill tools

cillating front axle that is hinged to the center frame.

The tracks are driven through a reversing transmission on the end of the jackshaft. This turns the walking shaft, under the bull reel, through a roller chain. The walking shaft has two pinion gears, each controlled by a clutch and a brake which drive the tracks separately through intermediate gears and chains.

The tubular tool guide, Figure 20-76, is latched together on the tool string when starting a hole, to keep it in alignment. It is separated and folded back as soon as the hole is deep enough to serve as a guide.

The Keystone 51 blast hole drill, Figure 20-77, may be mounted on a self-powered crawler, a tandem truck, or a half track.

The engine drives a parallel jackshaft through a roller chain. This drives the crawlers through reduction chains, controlled by paired friction clutches and brakes. The jackshaft also turns the sand wheel through a friction pulley and the transmission shaft through a flat belt.

The transmission shaft drives the bull reel and the spudding or crank shaft gear through friction clutches and spur gears. The calf reel and clutch can be mounted on the end of the transmission shaft.

Two pitman arms are hinged to cranks at the end of the crank shaft and operate a double spudding beam.

The crown pulley has steel springs under it, and a hydraulic recoil cylinder over it to give a snap-back action to the stroke.

Tools. A set of drilling, bailing, and fishing tools is shown in Figure 20-78. These are connected to each other by A.P.I. (American Petroleum Institute) tapered threads. The male end is called pin thread; the female, box thread. They should be covered with protecting caps during handling and storage. Each tool has one or two squared or half squared sections to afford a grip for a wrench.

If the drill line is steel cable its end is threaded through the head of a swivel socket and sweated into a ferrule, which is then drawn inside the socket and supports it. The socket can be released by unscrewing a bushing in the top and drawing the ferrule through the enlarged hole.

This connection allows the socket to rotate in either direction without twisting the line. A non-swiveling socket is also available.

The jars may be fastened to the bottom of the socket by screwing together and tightening hard. They consist of an upper and lower section interlocked in the same manner as two chain links. In normal drilling the line is not slack enough to allow the pieces to telescope together. However, if the bit gets stuck, the line can be held so that the upper piece will drop on the down stroke and be pulled with a hammerlike blow on the up stroke, thus driving the stuck tools up. The free motion in this device is about five inches, but wear may increase it to as much as 18 inches before it must be rebuilt or discarded.

One or more drill stems may be fastened below the jars. Their function is to increase the weight and length of the tool string.

The bit is the cutting tool at the bottom of the string. It will be discussed below.

The dart valve bailer is a hollow cylinder up to 25 feet long. The flat plunger shown projecting from the bottom is connected by a short vertical rod to a ball valve, which in its closed position rests on a fitted seat by gravity. When the bailer is lowered to the bottom of the hole the dart is pushed up, opening the valve and allowing the cylinder to fill. The plunger shoulders prevent the cylinder from plugging by resting on the bottom of the hole.

Drive clamps are used to hammer down casings. They are fastened near the top of a squared section of any string tool except the bit. The top of the casing pipe is protected by a drop head or a threaded drive head or drive shoe.

Fishing tools are used in seizing and pulling up lost tools or other objects. Tools may be lost through line or tool breakage, accidental unscrewing of threaded connections, or jamming. Wrenches or other objects may be dropped in the hole.

Drill Bits. Figure 20-79 shows the face of a hard rock bit, and Figure 20-80 explains various terms used in describing it.

A drill bit has four important functions.

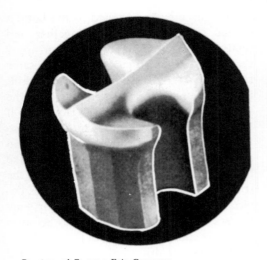

Courtesy of Bucyrus-Erie Company

Fig. 20-79. Spudding bit

It must penetrate the rock, grind it into fine enough grains to mix readily with water, keep the water and cuttings mixed, and keep the hole reamed out to full size and round cross-section. In order to perform these tasks, it must be properly shaped when it is placed on the string, and hard and tough enough to stay that way for a reasonable time.

A worn or damaged bit can be reshaped and hardened by forging and hammering. A forge may be set up beside the drill, using a hand bellows or one powered by the drill engine, and the shaping done with hand tools. Or the bit may be sent to a shop with a special furnace and a semi-automatic ram and hammer shaper.

A shop can do a better job, particularly with large bits, both in shaping and tempering. It often happens, however, that a drilling crew can do satisfactory work in spare time while the drill is running smoothly and thus save considerable time and investment.

A rig should be provided with sufficient spare bits so that it is never necessary to use a dull one.

Setting Up. The manner of moving and setting up varies with the type of mounting.

In each case, however, after being placed in position to drill, the machine is leveled by means of three or four jacks; one or two at the front and two at the rear corners. The jacks may be screw-type house jacks or built-in hydraulics with either hand or engine pumps. Planks and timbers are placed under them when necessary to give more bearing surface. If the site is on a grade blocking may have to be built up several feet.

A test of leveling work is to have the spudding cable and tools hang so that they are exactly centered in the tool guides.

The socket is attached to the end of the drilling line, raised, and the rest of the string of tools are attached one by one, or the whole string is assembled on the ground and then attached to the socket to be raised as a unit.

Drilling. The bull reel is turned until the tools hang free with the bit just above the ground. The spudding or crank clutch is engaged, starting the tools in an up and down motion. They are raised by engaging the bull reel clutch or lowered by releasing the bull wheel brake until the bit strikes a strong blow at the ground, without developing slack in the jars or the cable. The bit may be steadied by mechanical guides or by hand until it has penetrated far enough to be kept in line by the sides of the hole.

As the bit penetrates, the bull wheel brake is partially released. This may be done by hand or by the vernier, which permits different settings on the brake control lever to leave just sufficient drag on the brake so that it will turn to pay out line enough to keep proper contact, without allowing it to become slack. This is possible because if the bit goes too far before it strikes, its momentum is stopped by the line rather than by the ground, resulting in an increase in line tension which will turn the drum against a moderate drag. The amount of drag determines the closeness

FIG. 1

The Angle of Clearance is the taper on the outside or reaming edges which comes in contact with walls of the drill hole. Drawing shows drill bit having wide angle of clearance and no wearing surface. Arrow points to angle of clearance.

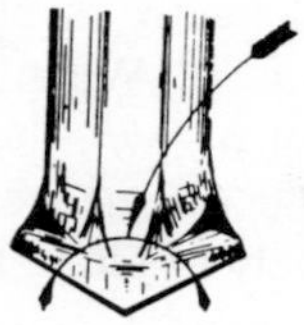

FIG. 2

The angle of penetration is the bevel on the cutting edge which penetrates or breaks up the material in the bottom of the drill hole. Drawing shows drill bit having a penetrating angle of 120°. Arrow points to the angle of penetration.

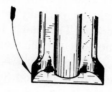

FIG. 3

The Wearing Surface is the area which has no clearance and is in actual contact with the wall of the drill hole. This drill bit has a large area of wearing surface and no angle of clearance. Arrow points to wearing surface.

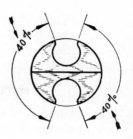

FIG. 4

The Reaming Edge is the outside edge of the bit and is measured as a part of the full circumference. Drawing shows drill bit having 80 per cent reaming edge—40 per cent reaming edge on either side of the water courses.

FIG. 5

The Crushing Face is the area of surface on the bottom of the bit and is compared by measuring its percentage of the total area of the drill hole. Arrow points to crushing face — the shaded portion of the drawing. Water courses shown in white.

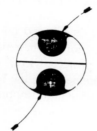

FIG. 6

The Water Course is that portion of the hole which is not filled by the bit and through which the water or cuttings must pass when the bit is moved up or down in the hole. Arrows point to water courses, crosshatched in black in this drawing.

FIG. 7

The Contour of Penetrating Edge on a drill bit may be concaved, straight or convex, and the degree of contour is measured by the change in angle from a square line across the bottom of the bit. The drawing shows a concaved penetrating edge.

FIG. 8

The Cross Section of a drill bit is the size of the body at a point back from the end when it is not upset. Arrow points to cross section. The drawing shows the cross section of a drill bit back a few inches from the cutting edge.

Courtesy of Bucyrus-Erie Company

Fig. 20-80. Bit sections and names

with which the bit will follow the receding bottom.

Best drilling results are obtained when the bit is pulled back sharply as soon as it has hit. The effect on most rock is more shattering if there is no follow-through. Snappy return can be obtained through use of manila rope; or elastic cushions under the derrick sheave.

As soon as the hole is a few inches deep

water is poured in from a hose or can. The most convenient water supply is a piped pressure system, but very often it must be brought to the job in drums, dipped out in cans, and poured into the hole. Consumption may be several hundred gallons a day, and work will be slowed if there is not enough to keep the sand (chippings) from forming a paste.

The tools are pulled out of the hole periodically. The bailer, attached to the sand line, is then placed in the hole and lowered to the bottom by wholly or partially releasing the sand reel brake. The bailer is left in the bottom a moment to fill, then lifted clear of the hole, swung to the side, and lowered. The valve will be opened by the ground, and the contents will run out. The dumping place must have drainage away from the hole in some direction where minimum damage will be done by the sand flowing away from it.

More water is put in the hole before drilling is resumed.

While the tools are out for the bailing operation the drill bit is inspected. As in the case of the hammer and wagon drill bits the wear on the sides is most serious, as if it tapers in instead of flaring out at the bottom it may wedge into a too-small hole and get stuck.

If the bit is worn or broken it is removed from the string and a spare bit substituted.

Side Drift. When drilling through rock with sloping joints or bedding planes the bit may have a tendency to drift down the slope. This can be combatted by using only sharp bits, perhaps having a sharper angle of penetration than is standard, and braking the bull reel so that the blow is quicker and lighter than usual.

A very hard sloping layer may throw the bit off line in spite of these precautions. In such a case, exploding 10 or more sticks of dynamite on the bottom should "burn" a pit in it so the drill can bite.

Running Casing. If the material being drilled is unconsolidated earth or rock that is eroded by water or shatters extensively when pounded, it is necessary to protect the hole with casing pipe so that it will not cave in. When the hole has penetrated a foot or two, the tools are pulled up and a pipe of such size that the bit can move freely inside it, and of a length short enough to handle conveniently, is placed in the hole and the bit lowered inside it. The pipe is held upright by hand or with ropes and the drilling resumed. In many soils the churning of the bit and water will undercut the pipe so that it will sink in of its own weight until side friction becomes great enough to support it.

After that it is driven in. The hole is drilled a distance below it, which depends on the character of the soil. The tools are raised until a square section can be reached. A drive clamp is bolted on the square and the line payed out until the clamp is in a position to hammer the pipe, whose top is protected by a cap. The casing is then hammered to the bottom of the hole by the raising and dropping of the tools and drive clamp; the digging line being payed out to keep contact in the same manner as when drilling.

When the casing reaches bottom, the drive clamp is removed and the hole is

ANCHORED REEL

Fig. 20-80B. Hand spudding drill

drilled another few feet before the clamp is replaced and the casing driven farther.

When a casing section is about to go underground, the tools are pulled up and another section is threaded on its top.

Home Made Spudder. It is sometimes necessary to drill holes in out of the way locations where it is not practical to bring in a power drill suited for the job. Examples would be putting down a well for water supply, making pilot holes for hand driving of piles, or blowing out a piece of pond bottom for drainage.

A setup using a jeep or truck wheel for power is shown in Figure 20-80B. The wheel is jacked up, and fitted with projecting lug or pulley. It may be necessary to remove the wheel and fasten it to the lugs in the hub.

A tripod of heavy poles is erected over the spot to be drilled, the three pieces securely fastened at the top, and a pulley is hung from it. A drill bit is suspended from a rope passed over the pulley and under the wheel lug.

The bit may be a regular churn drill bit, or any piece of heavy metal of the same general shape and the right size. The bottom should be shaped into a dull edge slightly wider than the rest of the bit, and a means to fasten the rope to the top must be provided.

The other end of the rope is held by hand, or wound on a hand winch or any other reel that can be held by a brake.

Drilling is similar to regular churn drill practice. The man at the end of the rope pays out enough line so that the bit will strike the ground when the spudding lug is up. Another man will have to steady the bit until it is deep enough to take care of itself. The hole is kept partially full of water, and drilling must be stopped occasionally to bail it out with whatever apparatus is available. A tin can on a stick may be used.

A number of problems arise with makeshift apparatus such as this, that must be met by the ingenuity of the men in utilizing whatever materials they have. Hemp rope may fray badly where the spudding lug strikes it. A piece of rubber hose would protect it.

If no motor vehicle is available, two men may drill a well by pulling the bit up with the rope, and releasing it. This is hard, monotonous work. If the bit is heavy a pulley arrangement to provide a longer, lighter pull is helpful.

Casing or piles may be driven by using a log or a metal weight instead of the bit.

DEEP ROTARY DRILLS

The rotary drills to be discussed in this section are primarily used in boring deep holes for oil. They employ a toothed bit which is rotated against the bottom of the hole by a drilling string driven by a power plant. Mud or, occasionally, water or air is pumped through the hollow drilling string and bit and rises outside of them.

A diagram of a large stationary unit for deep work is shown in Figure 20-81A.

Derricks and Masts. The derrick shown is built on the job out of pre-cut and drilled angle iron on poured concrete footings. Other constructions include prefabricated sections of angle or tubing, which are assembled and bolted; complete derricks assembled on the ground and raised by winches; and hydraulic or cable controlled telescoping or folding masts, transported to the job on large trailers which are used as both support and substructure. The derricks are generally centered on the hole and are self-supporting, while the masts are on one side of the hole, lean over it, and are supported by guys.

The derrick or mast serves as a support for a pair of hoist blocks which handle the drill pipe and casing. Its strength must be sufficient to easily carry the weight of the longest string of pipe or casing that might be used on the job, and it is therefore de-

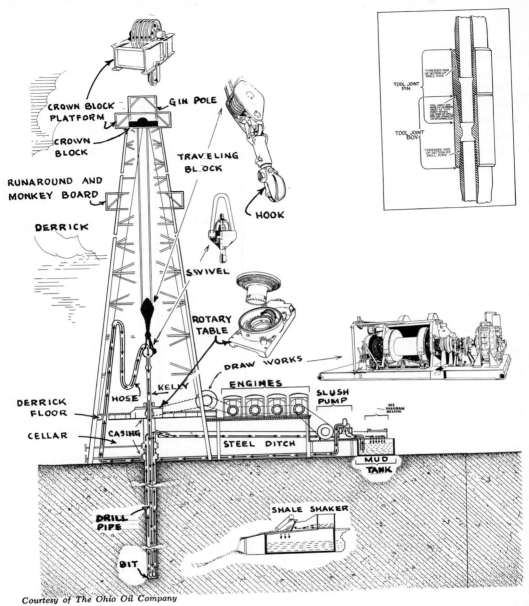

CROWN BLOCK PLATFORM

GIN POLE

CROWN BLOCK

TRAVELING BLOCK

RUNAROUND AND MONKEY BOARD

HOOK

DERRICK

SWIVEL

ROTARY TABLE

KELLY

DRAW WORKS

HOSE

ENGINES

SLUSH PUMP

DERRICK FLOOR

CELLAR

CASING

STEEL DITCH

SEE DIAGRAM BELOW

MUD TANK

SHALE SHAKER

DRILL PIPE

BIT

TOOL JOINT PIN

THREADED END OF SECTION OF DRILL PIPE

TOOL JOINTS ARE MADE UP WITH A SHOULDER AT THIS JOINT WHEN THE PINS AND BOXES ARE DISCONNECTED

TOOL JOINT BOX

THREADED END OF SECTION OF DRILL PIPE

Courtesy of The Ohio Oil Company

Fig. 20-81A. Structures for rotary drill

signed with reference to the estimated drilling depth.

Its height is determined by the length of pipe to be pulled as a unit. Economic drilling of very deep wells requires removing sections 120 feet or more in length, where shallow holes may justify handling 30 feet or less. Maximum derrick height at this writing is 192 feet.

Derrick structures and many other items of oil drilling equipment should be approved by the A.P.I. (American Petroleum Institute), which has set up standards for promotion of safety and efficiency.

Substructure. Substructures are massive supports and platforms of steel and wood which carry the power plants, drilling machinery, and pipe racks. They are usually separate from the derrick and can be removed without disturbing it.

The drilling floor is inside the derrick legs and high enough above the ground to provide easy access to the space under it, called the cellar.

Shallow holes cased with pipe protruding a few feet above the derrick floor are drilled opposite the draw works. One of these, called the "rat hole," supports the kelly and swivel when not in use. The other, called the "mouse hole," supports a single piece of drill pipe ready to be added to the drilling string.

When the surface soil is full of boulders, when sub-zero weather might freeze the kelly into the conventional rat hole, and under some other conditions, a mechanical rat hole is used. It is a 40 foot track, pivoted like a seesaw, which bears a dolly that supports and facilitates moving the kelly. It is placed at the corner of the derrick.

Power Plant and Controls. Power requirements are heavy and may exceed 1500 horsepower. Multiple engines are commonly used and may be high pressure steam, electric, or internal combustion with electric, mechanical, or fluid drive.

The type of internal combustion engine is usually decided by the fuel available locally. Standard diesels or, less often, gasoline engines are used, or adaptions of either type to use crude oil, natural gas, or butane.

During drilling the rotary machinery absorbs about one third of the power expended and the pump the balance. The hoisting mechanism requires as much power as the two combined. In small units one engine may drive the pump and a smaller one the rotary table, with provision made for hooking them both to the hoist winch.

Large plants may use as many as six engines, driving into a transmission box from which the power of any one engine, or any group of them, can be applied to

any function. Engines can be disconnected and stopped when not needed.

Engines driving the rotary are subject to severe shock and it is good practice to protect them with torque converters.

Draw Works. The draw works is the power distribution and control mechanism. It includes one or more variable speed transmissions; clutches and brakes operated mechanically, or by air or fluid; a heavy duty winch for the main hoist; one or more cat heads or light winches for handling pipe, miscellaneous lifting, and coupling and uncoupling pipe; and chain belt or shaft drives to the rotary table and the pump.

The instrument panel has engine gauges and indicators for the weight of the drilling string, rotary speed, and mud pressure.

Rotary Table. The rotary table consists of a stationary base with a machined cup, and a rotary, also called turntable or table, which rests on a wide ball bearing in the cup, and is turned by bevel gears.

The rotary is pierced by a large, square shouldered hole in which the two piece outer or table bushing rests. This has a square hole through it which holds the inner or kelly drive bushing. This is also of split construction, and has a square hole with smoothly machined sides.

This arrangement transmits rotary drive to the kelly, a square shaft which is free to move vertically in the inner bushing. The hole can be enlarged for passage of the drill string by lifting the inner bushing; and further widened to pass the bit by removing the outer one.

A soft rubber wiper collar is fastened to the lower side of the base, to clean mud off the drill string as it is raised.

Kelly. A simple drilling string consists of a kelly, lengths of drill pipe, joints, some heavier drill collars, and a bit. A reamer may be required under special conditions. They are all made of high grade alloy steels.

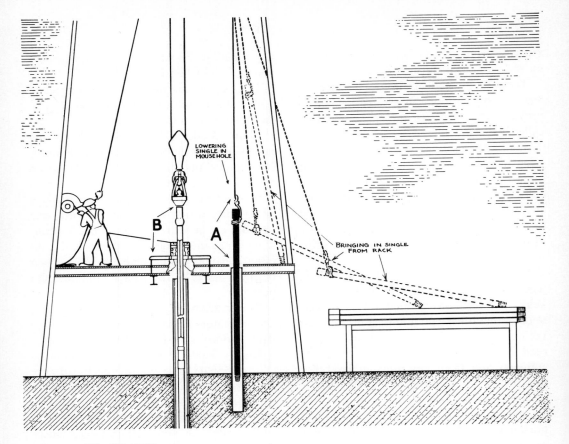

LOWERING
SINGLE IN
MOUSEHOLE

B

A

BRINGING IN SINGLE
FROM RACK

Courtesy of The Ohio Oil Company

Fig. 20-81B. Bringing drill pipe section from rack

The kelly or grief stem is a hollow square shaft (hexagonal or splined cross-sections are rare) which slides up and down in the inner table bushing. The ends are round and are upset, or hot hammered, into a heavier cross-section. Tapered A.P.I. threads are used, left hand on the top, right hand on the bottom. A recess is machined into the upset upper portion to provide a grip for the hoist swivel.

The lower end is subject to constant connecting and disconnecting of pipe so that threads wear rapidly. The kelly is therefore protected by a joint protector, called a sub, and this in turn is protected by another, the sub saver. These are more economical to repair or replace than the kelly.

A kelly should have a square section somewhat longer than the drill pipes which are to be used.

Drill Pipe. The drill pipe is made in standard lengths of 21, 29, and 43 feet, known respectively as range one, two, and three. Standard outside diameters are $4\frac{1}{2}$, $5\frac{9}{16}$, and $6\frac{5}{8}$ inches. In the field, lengths will be variable because of shortening due to repair work. Construction is seamless steel tubing. The ends are upset to give thicker walls. Taper right hand male threads are used on each end.

Protecting joints, used for attachment to other pipes or collars and for protection of the drill pipe threads, are heated before screwing on so as to form a very tight connection, and are sometimes welded in addition.

Pins and connectors have a larger out-

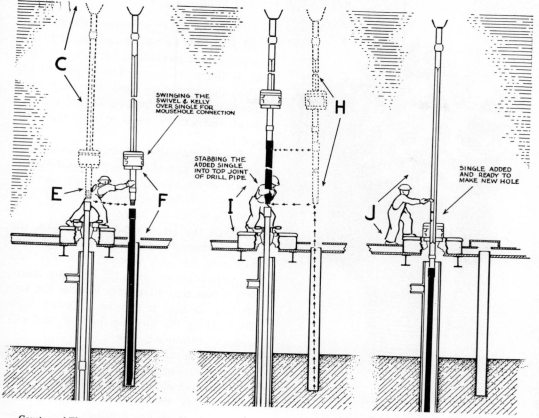

Fig. 20-81C. Adding drill pipe to string

side diameter than the pipe.

Drill Collars. The drill collars are heavy, thick-walled pipes of the same internal diameter as the drill pipes. They are used just above the bit, for a distance of 30 to 360 feet. Their total weight should be 50 to 100 percent greater than the weight to be carried on the bit.

They serve to concentrate the required weight at the bottom of the hole, so that the greater part of the drill string is under tension, holding the excess weight off the bit. This keeps the drill pipe hanging straight, discourages whipping, and greatly reduces metal fatigue.

In addition, they serve as a flywheel to keep the bit turning when it encounters ground that may stick or stall it. The momentum of the collars will absorb part or all of many shock overloads, protecting the drill pipe against twisting and the engine from stalling.

Bits. Rotary bits may be classified as drag bits and roller bits. The drag type may be subdivided into fishtails, which have metal or metal and carbide knives, and the diamond type, which have a shaped matrix studded with points of diamond, carbide, or hard steel. Roller bits have toothed cones or wheels that grind and chip as they rotate.

Both types may again be divided into types which drill a clean hole, and those which drill on the circumference only, and leave an uncut core to be pulled out for inspection.

Due to the complexity of the problems surrounding bit use, only a brief mention

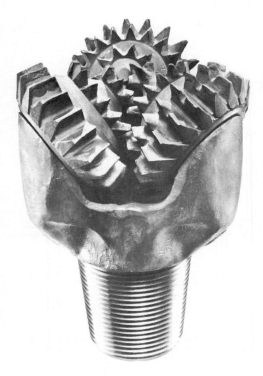

Fig. 20-82. Tri-Cone rotary bit

of some of their features is possible here.

In using any bit, it is essential that the cuttings be removed in order to expose fresh ground. This is done by pumping fluid or air through the center of the bit at sufficient velocity to enable it to carry the cuttings up the outside of the bit and drill string, and out of the hole. This same fluid should cool and lubricate the cutting parts and perhaps the bearings as well.

Fishtail bits are suitable for soft and fairly even formations. Two bladed ones give the best penetration but are very subject to stalling. When four blades are used, two should be somewhat shorter and of narrower gauge than the others. Hard surfacing is used in a thin layer on the cutting side so that a sharp edge will be preserved as it wears down.

Fluid from nozzles in the bit body jets along the cutting side of the knife.

The diamond-type is set with commercial diamonds or bits of carbide. It is suitable for hard formations and is less subject to stalling. It is expensive, particularly when diamonds are used, but may make much more hole footage before replacement and can be used at relatively high speeds.

The Hughes tri-cone bit, Figures 20-82 and 20-83, has three conical cutting elements which can rotate on roller and ball bearings on bearing pins machined integrally with the bit body.

When the bit is rotated in service, the cones rotate about their axes through contact with the rock being drilled. This sets up a twisting-tearing and chipping-crushing action that effectively loosens the formation so that it can be removed from the hole by the flushing fluid or mud circulating through the bit.

The teeth on the cones are hard faced to increase resistance to abrasive wear. The rows of teeth on the inner end of each cone interfit with the grooves on the mating cones thus providing a self-cleaning action.

In general, cone bits for soft or sticky formations have long, widely spaced teeth while those for drilling harder formations have shorter and more closely spaced teeth.

Bit sizes range from 3¾ inch to 26 inch; 7⅞ and 8¾ inch are the most popular.

Rotation speeds vary from 40 to 500 revolutions per minute, with most drilling being done between 100 and 350. In most formations, there is an optimum drilling speed for a particular bit beyond which faster turning will yield diminishing advantage, and ultimately a loss of footage. Speeds may be limited by the rotary power available or the capacity of the pump to clean out the cuttings. Also, a harmonic vibration may occur at certain speeds which interferes with proper drilling and shortens the life of the drill string.

The compressive strength and drillability of formations vary widely and prevent adoption of a universal drilling practice. The weight upon the bit, speed of rotation, and volume of flushing fluid circulated are important factors in obtaining maximum penetration rate.

The determination of drilling practice factors must be the result of actual field tests and experience.

Reamers. Reamers are placed directly behind the bit or at any point in the drill string. Sometimes two or more are used.

They serve to straighten a hole or to serve as a fulcrum in curving it, to restore the size when reduced by the movement of the walls, or to enlarge a hole. Enlargement might be desired in order to run oversize casing, or to make it easier to drill on a curve.

Mud. The mud or drilling fluid serves to remove cuttings, to lubricate the bit and the outside of the drill string, to hold down pressure zones of gas, oil, or water, to support the walls against caving and to seal them against seepage in either direction.

It may be wholly made up of water and dirt and pulverized rock obtained from the drilling, or may be largely composed of imported materials. Chemicals may be added to change its characteristics, and oil can be used instead of water.

Ordinarily, the solids should be clay or silt with no more than two to five percent sand. Imported materials include various clays of particular fineness and colloidal characteristics, of which bentonite may be the best known; and finely ground metallic salts which are used to increase weight.

Mud should be thin enough for the pump

Courtesy of Hughes Tool Company

Fig. 20-83. Rotary bit parts

to move it at the necessary velocity to keep the bit clean; but it should not have surplus water to be absorbed by rocks that might swell and narrow the hole. The deposit on the walls, known as filter cake or mud cake, should not build up to a sufficient thickness to impede pulling the bit or installing casing.

Getting the right quality mud and keeping it that way is a complex problem requiring both engineering knowledge and practical experience.

Circulating System. The mud follows the path of the arrows in Figure 20-81. It is drawn from a storage and treatment tank by a reciprocating pump called the mud or slush pump, which may develop pressure up to 2000 pounds and have a capacity up to 1000 gallons per minute. A standby pump of equal or smaller capacity, preferably with a separate power plant, is used in emergencies and to circulate mud when the drilling plant is shut down.

The high pressure fluid is delivered to the top of the kelly opening through a swivel. This swivel must support the weight of the drilling string, permit the kelly to turn freely, and prevent leakage. The weight is carried on anti-friction bearings and pump type packings are used.

The fluid then goes down the drill string, through the bit, and rises on the outside of the string and enters the surface casing. This is a pipe which is placed and cemented in the hole to hold surface soil against caving, to prevent high pressure blowouts around the hole, and to direct and control the flow of mud. In areas where high pressures are known to exist it may go down a thousand feet, but it is often run only to a footing in firm rock. Its inside diameter must be large enough to allow free passage of the size bit and reamer needed for the deep drilling.

This casing is extended several feet above ground. It is fitted with two shutoff valves by means of which the outflow of mud can be impeded or stopped so that high pressure can be held in the hole to control blowoffs.

Above the valves the flow pipe carries the spill-over of mud into the ditch, a steel trough which discharges onto the shale shaker, a fine mesh vibrating screen. The mud goes through this into the tank, and sand and chips are rejected into the shale pit.

Hoist. The hoist mechanism for the drilling string consists of a crown block in the derrick head, and a traveling block suspended from it by a multiple line powered by the main winch in the draw works. This line runs to a clamp anchor on the derrick and thence to a storage reel. At intervals, a short piece of cable is fed in from the reel and an equal length taken off at the drum to spread wear more evenly. Storage may also be on a standby or auxiliary winch.

The traveling block supports the hook, which is usually a set of clamp arms that supports the swivel, and the drill pipe elevator. The elevator is pushed to the side when the swivel is in use, but the swivel must be removed to use the elevator. The hoist machinery must be capable of pulling the whole drill string, which may weigh hundreds of tons, at a rapid rate, and be under very delicate control for feeding the bit into the ground.

In feeding, the weight on the bit is varied according to the hardness of the formation. Soft formations drill best at pressures of around two tons, while hard ones will require six tons or more. The operator must know the length and weight of his drill string, and subtract the suspended weight shown by the gauge to find the amount resting on the bit. The rate of feed is varied to keep pressure constant in any one condition, and to change it for a different cutting condition.

A great variety of semi-automatic and full automatic controls are available, and considerable differences of opinion exist about their relative value. They may depend for action on the suspended weight, the torque of the rotary, or on both.

Drilling. Drilling is commenced with an oversize bit fastened directly on the bottom of the kelly. Water or mud is pumped into the kelly and is discharged on the ground. If swivel construction permits, the hoist line may be slack to allow the weight of the traveling block to rest on the string to help penetration.

When the kelly has worked down through the bushing until its bumper almost touches, it is pulled out of the hole and the bit is unscrewed. A drill collar is pulled in from the rack by a light line and placed in the mouse hole. The kelly is hoisted together with the inner bushing, and the pin thread on its sub or sub saver is lined or stabbed into the box thread on the collar, and the two tightened by a spinning line or other tools. The kelly and collar are lifted, then lowered through the table and the bit attached. Drilling and pumping are resumed and the kelly again works down through the bushing to the upper end of the squared section.

The kelly is then lifted until the top of the drill collar protrudes above the turntable. Wedges, called slips, are placed between it and the outer bushing to prevent the collar from slipping down. The hoist line is slackened, the kelly disconnected and swung to the side to be coupled with another drill collar or a drill pipe which has been placed in the mouse hole. This is lifted and stabbed into the drill collar joint in the rotary. After tightening, the kelly is connected to its top, the string is lifted a few inches, the slips removed, the bit lowered to the bottom of the hole, and the drilling resumed.

An extra length of pipe is added in this manner whenever the working space (square shank) of the kelly is fed through the kelly bushing, until the proper depth

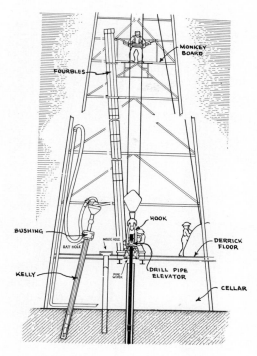

Courtesy of The Ohio Oil Company

Fig. 20-84. Pulling drill pipe—latching on

or formation is reached for setting the surface casing, or until the bit becomes dull or damaged and needs replacement. The whole string must then be pulled out of the hole.

Pulling Drill Pipe. Pipes may be removed one, two, three, or four at a time, depending on their length and the height of the derrick. Multiple sections are called doubles, thribles, or fourbles. Fourbles are shown in the diagrams in Figures 20-84 to 20-86.

The kelly and bushing are raised and the drill pipe secured with slips. The kelly is disconnected and parked in the rat hole, and the swivel unlatched from the hook, and left on the kelly.

The drill pipe elevator is latched on the pipe joint above the table, and the string is raised until the joint below the fourth pipe can be reached. The slips are locked, the joint broken (twisted open), and the four-

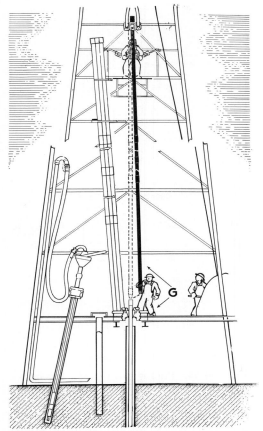

Fig. 20-85. Pulling drill pipe—hoisting

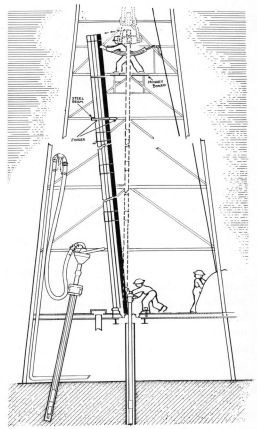

Fig. 20-86. Pulling drill pipe—stacking

ble lifted to clear. It is then pushed to the side of the derrick and leaned against a beam. It is moved by men on the derrick floor and by the derrick man on the monkey board near its top, while still suspended by the elevator.

Additional sections are removed in the same manner. The drill collars are larger than the pipe and do not have a projecting joint shoulder. A handling sub is threaded into them to afford a secure grip for the elevator.

The bit is ordinarily too large to be pulled through the opening left by the kelly bushing. The outer bushing is therefore split and removed to allow the bit through. As the bit offers no grip for a wrench, a special tool called the bit breaker is inserted in the table and holds the bit when it is lowered into it.

When the drilling string is to be put back in the hole it is refastened and lowered a section at a time, by reversing the processes by which it was raised and stacked.

Running Casing. The setup for setting (running) casing is shown in Figure 20-87. Casing is required to prevent the sides from caving or leaking, and to keep out high pressure water or gas. The casing is small enough to slide into the hole without pounding, and further drilling after it is installed requires the use of a smaller bit which will pass through it.

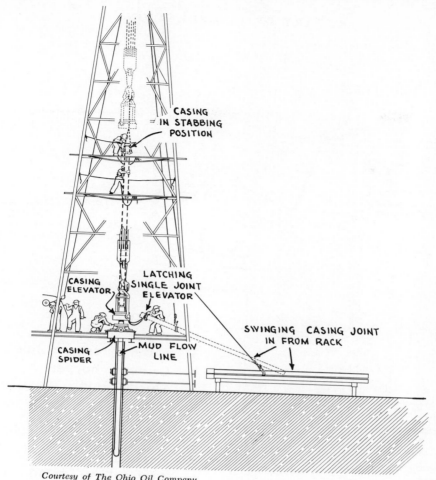

CASING
IN STABBING
POSITION

LATCHING
SINGLE JOINT
ELEVATOR

CASING
ELEVATOR

SWINGING CASING JOINT
IN FROM RACK

CASING
SPIDER

MUD FLOW
LINE

Courtesy of The Ohio Oil Company

Fig. 20-87. Running casing

For casing work the rotary table and base is replaced by a casing spider, which contains a tapered socket, with three wedges or slips for holding the pipe. The drill pipe elevator is replaced by the automatic grip casing elevator which contains a similar socket and wedges.

The single joint elevator, also operated by the main hoist, is used for bringing in casing sections from the rack.

The bottom casing pipe has a float shoe on the bottom and a float collar on the top. These are both check valves that will prevent mud from entering the pipe as it is lowered. The parts which are inside the line of the casing wall are of cement, plastic, rubber, or other readily drillable mate-

rials. The empty pipe is buoyed up by the air it holds, and is substantially lighter than if the mud were allowed to fill it.

The shoe and collar are threaded and welded to the bottom pipe, which is then picked up by the light or single joint elevator and lowered (stabbed) into the casing spider. The wedges are adjusted to hold it. Another pipe is then brought in by the single joint elevator, lowered into contact with the float collar, and the threads caught. The light elevator is then released and the automatic-grip lowered over it. The two sections are screwed together tightly, welded, and the elevator wedges engaged.

The pipe is raised a few inches to release

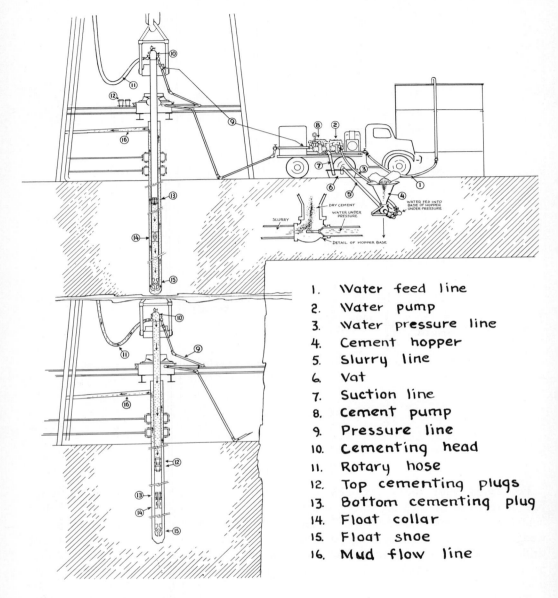

1. Water feed line
2. Water pump
3. Water pressure line
4. Cement hopper
5. Slurry line
6. Vat
7. Suction line
8. Cement pump
9. Pressure line
10. Cementing head
11. Rotary hose
12. Top cementing plugs
13. Bottom cementing plug
14. Float collar
15. Float shoe
16. Mud flow line

Courtesy of The Ohio Oil Company

Fig. 20-88. Cementing

the load on the spider wedges and to allow them to be moved out of contact. The pipe is then lowered until the elevator is a foot or two above the spider, the spider wedges are engaged, the elevator lowered slightly, and its wedges disengaged. The elevator is then raised and another pipe brought in by the single joint clamp to be attached.

This process is repeated until the string of casing has reached the bottom of the hole, except that after the first few joints no welding is done. This is required at the bottom for reinforcement against the twisting force of the bit while drilling.

If the hole is deep mud may have to be pumped into the casing to fill the lower part to brace it against crushing by hydrostatic pressure.

Cementing. The space between the casing and the wall of the hole should be filled with cement grout. This reinforces the casing, seals off circulation between horizons of oil or gas which may have been by-passed, and stops erosive water action. The cementing arrangement is shown in Figure 20-88.

The casing is held by the automatic grip elevator, and is frequently raised and lowered a few inches to secure even distribution. A cementing head, connected to the cement and mud pressure lines, is attached to the top of the casing.

The casing is filled with mud to provide a support for the descending column of cement. The head is then removed, the bottom cementing plug put in the pipe, and the head replaced. This plug has a valve in the center which prevents mud from working back into the cement.

The amount of cement grout or slurry required is figured in advance. Water and dry cement are mixed by high pressure jets or mechanically, and are pumped into the casing through the cementing head. The slurry pushes down the plug and the mud beneath it is forced to flow out of the casing and up around it into the mud lines and tank.

When the bottom cementing plug hits the float collar valve, the cement passes through it and the shoe, and flows upward around the outside of the casing.

When the calculated amount of cement has been pumped in, the cementing apparatus is shut down, the head removed, and one or two solid plugs put in the casing. The head is replaced and mud pumped in. This forces the plugs and the column of cement down, and causes continued upward flow of cement along the walls of the hole. When the plugs hit the bottom

cementing plug, they seal the hole through it and block the mud. Pressure inside the casing rises abruptly and the mud pump is shut down.

Pressure of 1000 to 1500 pounds is maintained in the casing for about seventy-two hours, by which time the cement should have set. In deep wells, it may be necessary to relieve some pressure occasionally to prevent it from becoming high enough to burst the casing.

Directional Drilling. Drilling is almost always started straight down, as any angle will cause difficulties with the derrick machinery. However, once underground, the hole may be changed in direction (offset), to reach areas not accessible from the surface, to correct a drift out of line, or to get around a bit lost in the bottom. Procedures are complex and delicate, and a technique can only be roughly indicated here.

The change in direction may be started by inserting a long thin wedge, edge up, at one side of the hole, that will divert the bit into the opposite wall. The angle at which the hole leaves the vertical is determined by the pitch of the wedge. It can then be maintained or increased by choice of tools and regulation of weight.

If a reamer is placed close behind the bit, rotation kept slow, and sufficient weight placed on the string to force it against the outside of the curve the reamer will act as a fulcrum to pry the bit toward the inside of the curve, thus increasing the angle. An eight to twelve foot drill collar is usually placed between the bit and the reamer, to avoid too abrupt deflection. Reducing weight, increasing the distance between bit and reamer, or speeding up rotation, will allow the hole to straighten.

Directional drilling is strongly affected by the dip and character of the formations being drilled, and by the type and sharpness of bit. Tools must be pulled out frequently, and the direction checked by means of

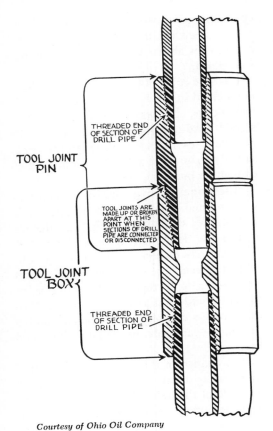

TOOL JOINT PIN

THREADED END OF SECTION OF DRILL PIPE

TOOL JOINTS ARE MADE UP OR BROKEN APART AT THIS POINT WHEN SECTIONS OF DRILL PIPE ARE CONNECTED OR DISCONNECTED

TOOL JOINT BOX

THREADED END OF SECTION OF DRILL PIPE

Courtesy of Ohio Oil Company

Fig. 20-89A. Tool joints

magnetic-photographic devices lowered into the hole. Dangers to be avoided include spiral wandering of the bit, and reverse turns.

In working through a curve, the drill string flexes as it turns in the hole in the manner of a speedometer cable in its casing. Curves must be wide enough so that it will not bind or be strained, and they must also permit driving casing through them. The usual maximum change in deflection allowed is four degrees per hundred feet,

CAISSON DRILLS

Caisson drills are specialized auger or rotary rigs that drill holes from 12 inches up to 10 feet in diameter, to depths up to 100 or 200 feet, either vertically or at a slant. They may be equipped with devices to bell out the bottoms to 3 to 5 times the hole diameter.

Their principal use is in making holes for poured concrete piers. They have largely replaced the driven pile in large areas where deep clay soils are found, and they are used in many other soil types as well.

They are not particularly suited for drilling in loose soil that tends to cave, nor for boulders or solid rock that is too hard to be dug with a pick. However, they can sink holes in loose ground when casing is used, and boulders may be broken by lowering a man in holes 2 or more feet in diameter, to hand-break or blast them.

For continuous hard digging, many of the rigs can use the rotary shaft cutters described in a later section, or rotating "bundles" of down hole drills.

In addition to pier work, caisson drills may be used for pre-boring for driven piles, sinking ventilation shafts, soil and mineral sampling, and digging for manholes, cesspools, and shallow wells.

When used for sampling, they can take big chunks of formation for thorough checking, and permit lowering a man to inspect the undisturbed formations in the wall.

Figure 20-89B shows a heavy duty foundation drilling machine on a special truck carrier, with the mast folded down into transport position. It has a rotary table at the foot of the mast, that is driven by a transmission with 4 forward and 2 reverse speeds. This receives engine drive through a pair of input clutches, one low speed for drilling, the other high speed for spinning off cuttings.

The kelly or drive stem telescopes—The inner stem is 4½ inches square, the outer one 7 inches. The two bars are linked by an automatic latch-release that permits application of both down pressure and hoist pull.

Figures 20-89D and F show a caisson drill unit mounted on a standard 1½ yard

Courtesy of Hugh B. Williams Mfg. Company

Fig. 20-89B. Caisson drill with tower folded down

crane. Its frame is supported by the boom foot and by tie rods to the roof structure. It carries its own engine, transmission, and rotary table. Torque converter drive is used, and manual or power shift transmission is optional. The kelly may be solid, telescoping, or "triplescoping."

Drilling may be done by augers with short flights that are lifted straight to the surface to unload by spinning, or by buckets that are lifted and swung for dumping to the side.

Buckets may be fabricated out of steel tubing, with convex hinged bottoms. Cutters and slots provide for filling the bucket by rotating it on the bottom.

Hole bottoms may be belled out by side cutters that are kept folded into the bucket walls while drilling hole, or by various types of undercutting reamers.

SHAFT SINKING

Rotary drills are being used to an increasing extent to sink ventilation and escape shafts into underground mines. The diameter of the shaft that can be drilled is now limited more by the sizes needed, than by limitations of the machinery or the technique.

Such holes are drilled by conventional heavy duty oil field rigs, with rotary tables altered to accommodate the huge bits and drill collars needed.

Most big hole work is done in two stages. First a pilot hole is drilled several diameters deeper than the intended bottom. The big bit has an extension or stinger that is guided by the pilot hole to keep exact alignment. Rock cutting is done by rows of rotary shaft cutters bolted to the body of the bit.

Big bits require very heavy weights to keep them cutting. A 90 inch ventilation shaft was drilled recently in New Mexico with three lead-filled collars 40 inches in diameter weighing 21 tons each above the

Courtesy of Calweld Inc.

Fig. 20-89C. Caisson drill buckets

Fig. 20-89D. Crane attachment and drilling bucket

bit. The bit carried a total weight of about 50 tons, after deducting bouyancy of the mud and tension support from above.

Rotation was 12 revolutions per minute, average penetration rate 2.15 feet an hour, with a maximum of 5 feet. Maximum mud circulation rate was 8,000 gallons a minute.

A drilled hole may be lined by lowering reinforced steel casing welding each joint before it goes in, or by pouring concrete from the top into forms that are lowered in stages to follow the bit.

Costs of drilling and digging shafts are somewhat similar. The drilled shaft has the advantages of speed, safety because no men need to work in it until it is finished and lined, minimum difficulty with water flow or caving because of mud pressure, simplicity of setting lining, and better chance of using the shaft without lining.

Fig. 20-89E. Belling reamers

In unlined ventilating shafts, the smooth-walled circular hole cut by machine may have the same air capacity as a blasted and dug rectangular hole with twice its cross section.

Shaft Sinking Machines. There are also drilling units that operate entirely inside the shaft. The design shown in Figure 20-89H cuts a 76 inch diameter hole by enlarging a 12¼ inch pilot hole that has been drilled into an opening of a mine beneath.

The machine is suspended by the rotary drilling rig used to put down the pilot hole. It is in three sections, with a rotary drive motor case at the top, a rotating cutter head at the bottom, and a works cylinder in the middle.

The machine is lifted out of the hole by the surface rig when inspection and servicing are needed.

Three 60 horsepower electric motors turn the cutter head at 6 rpm, through a central shaft and reducing gears. The head is set with rolling cutters of design suitable for the formation being drilled, and has a stinger projecting down into the pilot hole.

The cylinder is held in position against drilling thrust by hydraulic jacks powered by an electric pump. They push shoes outward against the shaft walls. The upper directional jacks keep the cylinder lined up with the pilot hole.

The drilling unit is lowered by the sur-

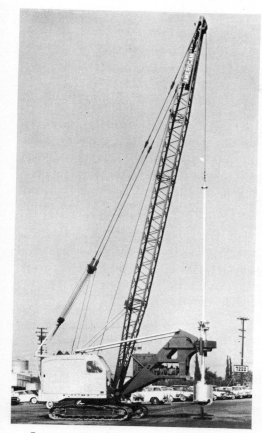

Courtesy of Calweld Inc.

Fig. 20-89F. Caisson-drilling crane attachment

face rig until the cutters rest on the hole bottom. The holding jacks are set and the directional jacks are adjusted. The cutter head is then rotated, and pushed down by sets of hydraulic jacks that can exert thrust up to 450,000 pounds.

A flush system blows cuttings down the pilot hole into the mine, where they are removed by mine haulage equipment, or blown up a nearby ventilating hole.

When the thrust jacks have reached the end of their stroke, rotation of the head is stopped, holding and directional jacks are released, the hoist line to the surface is slackened, and the cylinder is lowered by releasing the thrust jacks. The cylinder is then locked in position, and drilling is resumed.

Courtesy of Hugh B. Williams Mfg. Company

Fig. 80-89G. Shaft drill

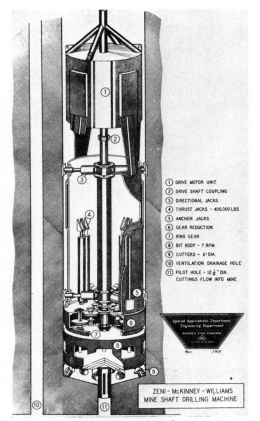

Fig. 20-89H. Shaft sinking machine

Cutters are available to permit drilling any formation from soft shale to granite and quartzite.

A limiting factor on use of this machine, and the raise borers described below, is that there must be access to the bottom of the proposed shaft, to provide for removal of cuttings. Raising them to the surface with air or mud is more difficult and costly when hole diameter is large.

Raise Borers. A machine working at the top can enlarge a pilot hole from the bottom. The unit shown in Figure 20-90 first cuts an 11-inch hole down to a lower level passage, using a 10-inch drill pipe in 5-foot sections that weigh 710 pounds each. Cuttings are washed up out of the hole with air, water, or drilling mud. Bits may be

steel tooth or carbide insert, depending on the formation.

When the drill breaks through into the passage, the bit is removed and replaced by a much larger back reamer, 4 to 12 feet in diameter. Steel or carbide insert teeth, or steel disc cutters, are mounted upon its upper surface.

The reamer is rotated and pulled, with a force that may be as high as 200 tons, and penetrates at rates from two to twenty feet an hour. Cuttings fall down the enlarged opening, and are removed by other equipment at the bottom.

The bored raise may be vertical, or pitched at any slope steep enough to drop the cuttings down away from the bit. Maximum depth may be 1000 to 2000 feet, depending on the machine and the job conditions.

TUNNELING MACHINES

An increasing share of tunneling work is being done by boring or tunneling machines, often called borers or moles. The problems that they encounter, or perhaps the circumstances under which they work, are discussed in Chapter 9.

The principal elements in a tunneling machine are usually a full size cutting head that rotates or oscillates to cut the ground, a complete or partial tubular casing or shield, a crowding or propelling mechanism with provision for steering, a muck disposal system, and electric and hydraulic power units. There are usually means to place tunnel lining and/or rock bolts.

Diameter may be as small as 6 feet, or as large as 37 feet, but both extremes are rare. At the moment, sizes from 10 to 25 feet appear to be the most common.

Machines are preferably assembled on the outside, at portals, to dig their own way underground. However, it may be possible to lower one in sections down a large shaft, and assemble it in a drill-and-blast stub tunnel.

Fig. 20-90. Drilling head of raise borer

Cutterhead. Cutterheads are made in a variety of designs, to suit different conditions, and to work out ideas of their engineers, and of the contractors who use them.

Most of them rotate on a central structural member. They are fabricated to hold a number of cutters. There are slots or windows to take cuttings through the head, away from the face, and buckets or scoops to gather and load them into a conveyor system.

The head may revolve as a single unit.

This results in difference in speed between cutters near the center and those at the rim. This may be compensated by variations in spacing, and in type of cutter. In other machines, the center is driven separately at higher rpm. It often has a forward extension or stinger that helps to prevent wandering off course.

Oscillating heads have a motion like a windshield wiper for each of a number of radial cutter arms. Their arcs overlap and distance of motion can be adjusted. They

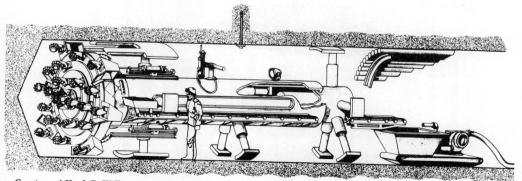

Fig. 20-91A. Tunneling machine

Fig. 20-91B. Starting a tunnel

are used chiefly in soft or variable ground.

Cutters. Cutters may be simple steel teeth set in rigid sockets, if the ground is soft. More often, they are cones, spools, or discs whose contact with the face causes them to rotate with a dragging, crushing,

or fracturing effect on the rock formation.

Their principle is generally similar to the rotary drill bits described earlier, but individual cones may be widely separated. Teeth may be smaller for harder formations. Carbide inserts or disc type cutters are used for both medium and hard rock.

Steel cutters primarily cut and chip, carbide inserts grind. Each cutter should make a set of grooves and ridges. The following cutter will break out or slice the ridges in making its own grooves. As the rock becomes harder, the cuts get shallower, and the efficiency of the operation declines.

Some rock is too hard for present day boring machines, but progress is being made in designing cutters to cope with it.

Shield. Except when rock is unusually stable, a borer is likely to be protected by a tubular shield or casing, or at least by a curved roof section. Shield length and characteristics are determined largely by the stability of the ground.

The shield may be in two longitudinal sections. The front one moves with the cutterhead. The rear one is anchored to the tunnel or its lining, and serves as a thrust block for hydraulic cylinders that move the

Fig. 20-91C. Revolving cutterhead

Courtesy of Calweld Division, Smith International

Fig. 20-91D. Oscillating cutter coming out of a bore

20-91B

Courtesy of Calweld Division, Smith International

Fig. 20-91E. Conveyor system

front section and cutterhead slowly forward into the digging.

In soft, caving, or flowing ground the cutterhead may operate inside the shield, whose top, and sometimes sides and bottom, penetrate the soil before the cutters reach it. In extreme conditions, the head may function more as a barrier against admitting too much material, than as a cutting tool. Flow of material into the machine may then be regulated by adjustable openings, called windows. Shields may also be operated without cutterheads.

In hard ground, the cutterhead is out in front of the shield, and cuts a circle large enough for it to follow without binding.

A rearward extension of the shield may be used for setting lining for the tunnel. A system suitable for very unstable ground is to fabricate the lining inside the shield from curved sections, building each ring onto a previously completed one. Such a lining is necessarily smaller than the tunnel bore. The annular space is filled by immediately pumping in fillers such as pea gravel and/or cement grout.

If the ground is judged to be stable enough to maintain itself without support for a few hours, the lining may be fabricated or expanded behind the shield, so that it may be full tunnel size and require less filler behind it.

On other jobs, roof bolts and wire mesh may be placed instead of lining. The rear section of the shield can have ports or slots through which holes and can be drilled and bolts inserted.

Tunnels that are left bolted or unsupported during excavation are often lined later as a separate project.

Courtesy of The Robbins Company

Fig. 20-91F. Cutting discs

Propel (Crowd). The tunneling machine is usually moved forward into the digging by bracing or locking the rear portion to the tunnel walls or lining, then moving the front portion forward by means of hydraulic cylinders. When these reach the end of their stroke, the front portion may be locked to the walls, and the rear drawn up to it with the same cylinders. The cycle is then repeated. The re-setting part takes from a half minute up to 3 minutes.

The stroke of the pistons may be anywhere from 2 to 6 feet in various models. Crowding force varies up to 8000 tons.

Power. Power is usually electric and hydraulic. The cutterhead is driven directly by one or more electric motors, with a total of 100 to over 1000 horsepower, depending on machine size and expected digging resistance.

Other functions are powered by hydraulics, with electric driven pump or pumps. Big machines carry these pumps with them, small ones usually tow them behind.

The electricity is preferably supplied by power lines from outside. If this is not practical, it is generated by a diesel-electric set towed or self-propelled behind the borer.

Muck Disposal. Cuttings are scooped up by buckets in the rim of the cutterhead, and carried up to dump on a conveyor belt. This carries them back through the machine, and a substantial distance to the rear. It usually dumps into rail cars, a string of which is pushed into the tunnel by a locomotive. As each car is loaded, the train moves the next one into position.

In dry ground or under any dusty conditions, there are water sprays at the cutter head, and at each transfer point on or off the belt, to keep down dust. In addition, the ventilation system is usually of the suction type, so that any escaped dust, fumes, or rock gas are continuously drawn out of the work area.

Muck cars may be hauled out of a portal, hoisted up a shaft by a crane, or dumped within reach of hoisting machinery. The spoil is usually then reloaded into surface hauling units for removal from the area.

SUBMARINE DRILLING

Barge. When underwater rock is to be blasted to improve a channel, it is usual to drill and load it from a specially designed barge. See Figure 20-91H.

The barge must be of sufficient size and buoyancy to carry equipment and supplies, with an ample margin of safety for wind, wave and tide action, and must not have so much draft that it will go aground in the work area. It is equipped with winches to provide for movement between holes, and four corner spuds to keep it stationary during drilling.

It is moved around in much the same manner as the dredges described in Chapter 14, except that it is not swung on a spud as a pivot. Winch lines are fastened to anchors in the water or on shore, and are pulled in and payed out as necessary to move to new positions. Winches and spud hoists are air powered. Spuds may be lifted by winch lines, or by air motor feeds that can also apply down pressure.

Courtesy of Hugh B. Williams Mfg. Company

Fig. 20-91G. Inside a drilled tunnel

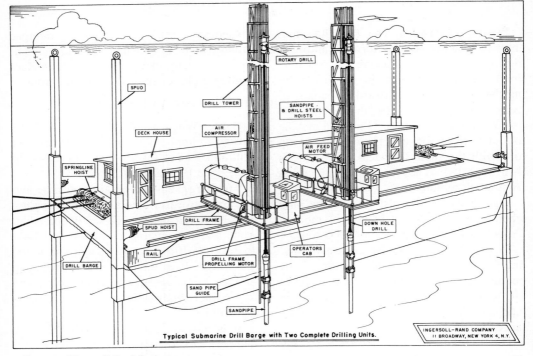

Typical Submarine Drill Barge with Two Complete Drilling Units.

INGERSOLL-RAND COMPANY
II BROADWAY, NEW YORK 4, N.Y.

Courtesy of Ingersoll-Rand Company

Fig. 20-91H. Typical submarine drilling barge with two drilling units

The compressor or compressors have gasoline or diesel engines, or electric motors supplied by a shore cable.

Drills. The drill is similar to the mobile tower units described earlier, and forms a complete unit assembled on a base that can be moved along the barge deck on rails, so as to drill two or more holes from a single barge position.

The tower is usually high enough to carry feed and drill pipe to reach the deepest hole depth required. This avoids lost time and possible mishaps during drill pipe changes or additions. Possibility of tide and weather changes make it important to complete a hole promptly once it is started.

A unit with a rotary table and a down hole drill is illustrated. Straight rotary drills are also used, choice depending on rock characteristics and the preference of the contractor.

Sand Pipe. The sand pipe is a temporary casing that extends from above the water surface down to the collar of the hole. Ordinary pipe or casing or a special telescoping construction may be used. The top is belled out into a funnel. Inside diameter is slightly greater than that of the bit.

The sand pipe keeps loose material from falling into the hole, centralizes the bit and drill rod, and keeps contact with the hole if it becomes necessary to pull the bit for replacement. It is lowered into place with a hoist line that is left slack, so that the pipe can follow the bit down through soil and into the initial overbreak in rock.

Drill cuttings are blown up the sand pipe outside the drill rod. It is sometimes slotted near the bottom to allow part of the cuttings to escape without being carried to the top.

The sand pipe is left in position when the hole is completed and the bit pulled. It may be used directly in loading explosives or a smaller loading tube may be used inside it. It is raised when loading is finished.

Courtesy of Joy Manufacturing Company

Fig. 20-911. Skid-mounted diamond drill

Holes. Drilling patterns are similar to those used on land, except that holes are made larger in proportion to the size of powder cartridges, and they are extended to greater depths below grade.

The larger size is to avoid possible difficulty with cartridges hanging up in these hard-to-get-at holes. The water around the charge prevents loss of explosive power that would occur during expansion in air-filled holes.

Holes are drilled deeper below grade to reduce danger of leaving ridges of unbroken rock that would make excavation difficult.

Marking such spots, then bringing the barge back for re-drilling and re-shooting, would be very expensive.

Blasting. Powder factors are extremely variable, ranging from one-half to five pounds per cubic yard of rock. Both the character of the rock and the fineness of fragmentation needed affect this. Very fine breakage is needed if a suction or cutter dredge is to do the cleaning up.

No stemming is needed. Loading may be done only to the height usual in stemmed holes, or right to the top.

Primers are placed in the bottom, or both in the bottom and the top. The wire or detonating cord leads to above the surface, where it is tied to a float. Lines are hooked up for firing from a small boat, after the barge has pulled up its spuds and moved to a safe distance.

Light blasting can be done a hole at a time without moving the barge.

DIAMOND DRILLS

Construction. Diamond drills are small portable rotary drills used for securing samples of rock or ore lying beneath the surface, for seismic prospecting, and for drilling blast holes.

The drilling unit may be gas, air, or electric powered. It is usually mounted on skids, but can be permanently attached to a truck or trailer.

Courtesy of Joy Manufacturing Company

Fig. 20-91. Angle drilling

Truck mounted machines may carry a folding derrick; others use a tripod assembled on the job from pipe sections, or a portable mast.

Figures 20-90 to 20-92 show several Joy diamond drills.

The engine is fitted with a disc clutch. It drives the rotary, or swivel head, and hoist winch through a variable speed transmission and separate clutches.

The swivel head is hinged on the rear frame. Its case can be rotated to drill in any direction in a plane perpendicular to the drive shaft, and can be unclamped and swung away from the machine for convenience in pulling rods.

It contains a bushing that is rotated by bevel gears. This is fitted with internal splines or hexagon sliding grip for a kelly, which has a lock clamp or chuck at the bottom for gripping drill rod.

Hydraulic feed employs two parallel double acting cylinders connected to the top of the kelly by a yoke. A three position valve allows any desired variation in forcing down, supporting, or raising the drill string. Fluid may be water from the drill pump, or oil from a standard hydraulic pump.

Mechanical feed requires a kelly with external threads, with two slots, which are engaged by the drive chuck. A threaded collar turned through a three speed reduction transmission from the drive shaft, meshes with the kelly's threads and raises and lowers it a predetermined amount for

Courtesy of Joy Manufacturing Company

Fig. 20-92. Pulling a drill string

each revolution of the kelly and the string.

The hoist winch is used for lifting and lowering drill rod, for dragging skid mounted machines on short moves, and for odd jobs. A cat head may be installed also for general utility work.

The drill string consists of a number of drill rods, a core barrel, one or more reamer shells, and a bit.

The rod is pipe with flush threaded connections that will slide through the kelly.

Core barrels, Figure 20-93, are the same outside diameter as the bit. They contain cylindrical sockets for holding a core, and a choker spring near the bottom for gripping it. The double tube type contains passages in the shell for bypassing drill water so as not to erode the core.

Reamer shells are cutters of the same type as the bit, and a few thousandths oversize, which follow it to assure a full size hole in spite of possible wear or damage to the bit itself.

Figure 20-94 shows some designs of reamer shells, and 20-95 of bits. Plug bits are used for drilling "solid" holes, for blasting, or sludge sampling. Casing bits cut a ring, leaving a core inside, which is seized in the barrel.

Bits and reamers cut by means of commercial diamonds set in a matrix of special composition, which resists both wear and shock. Pieces of carbide are occasionally substituted for diamonds. Fishtail and chopping bits may be used under special circumstances.

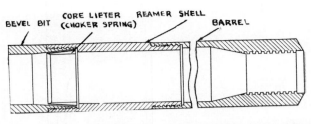

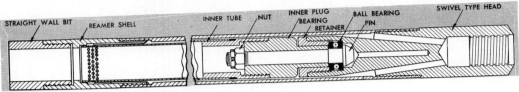

Courtesy of Joy Manufacturing Company

Fig. 20-93. Core barrels

The pump delivers water through a flexible hose to the top of the drill rod through a swivel. Water goes down through the drill string and bit, cools the bit, and washes out rock, dust, and chips, which it carries with it to the surface around the outside of the string.

Courtesy of Joy Manufacturing Company

Fig. 20-94. Reamer shells

The mud of water and grindings flows out of the top of the hole into a settling basin.

Solid Drilling. It is good practice to level the machine and essential to block it firmly in position before drilling. The tripod or derrick is erected so that the front of the hoist sheave is directly in line with the swivel head drive chuck and with the proposed hole. In angle drilling the tripod head may have to be far off center in order to line up.

Solid drilling is done with a bit, a reamer, and a string of rods. The first rod may be lifted by the hoist and lowered through the kelly; or the head may be swiveled and the rod slid in from the side. The reamer and the bit are threaded on to the bottom of the rod, then tightened. The water swivel is attached to the top.

The kelly is lifted to its highest position, the swivel head rotated into the direction of drilling, and the bit lowered until it touches the ground. The kelly chuck is clamped firmly.

The kelly, rod, and bit are rotated by engaging the rotary jaw clutch, then the engine clutch. The pump is started. The kelly is pushed down hydraulically by admitting fluid into the rams by hand valve, or automatically by screw feed. The bit should be kept in firm contact with the rock but heavy pressure should be avoided as it may break the diamonds or overheat the bit.

When the kelly reaches the bottom of its stroke the chuck is loosened, the kelly raised, the chuck tightened, and drilling resumed. Each new grip of the chuck is on a higher point on the rod so that it is fed into the ground in a series of strokes.

When the rod has been worked down so that it projects only a short distance above the kelly, power is cut off from the swivel head by opening the jaw clutch, water pumping is stopped, and the water hose swivel is removed and attached to another rod. This is raised by the hoist, lowered into contact with the threads on the first rod, and attached by the use of chain wrenches.

The winch clutch is released, the rotary clutch and the pump engaged, and drilling resumed. Additions of successive rods are made in the same manner.

If hoist height permits, two or three rods may be threaded together on the ground and raised and fastened into the string in a unit.

In surface drilling the feed is used to put pressure on the bit, but in deep work it must hold back much of the weight of the drill string to avoid excessive pressure.

Maximum depth is limited by the power of the machine to turn a long string and the capacity of the hoist to raise it. Depths of 50 feet to 15,000 feet may be obtained with different units and in various rock formations.

The settling basin is cleaned out at regular intervals. Samples of the grindings are taken and labeled according to the depth of drilling that produced them. These form an index of the type of formations encountered. Allowance must be made for the time it takes the mud to rise from the bit to the overflow, and the possibility of the water eroding soft strata and mixing

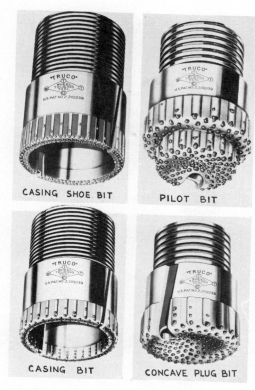

CASING SHOE BIT   PILOT BIT

CASING BIT   CONCAVE PLUG BIT

Courtesy of Joy Manufacturing Company

Fig. 20-95. Diamond bits

the product with the material being drilled.

If the rig is to be shut down for more than a few minutes, the hole should be thoroughly flushed to avoid separating and caking of mud during idle time.

In solid formations and under expert analysis, sludge or mud samples will give good data. However, the taking of intact cores of rock is generally a more reliable method, and under many conditions is the only way in which an accurate record may be obtained.

Core Drilling. Core drilling may be started at the surface or after solid drilling at any depth.

A barrel is substituted for the bottom rod. It is the same diameter and can be fed through the kelly in the same manner. A hollow casing bit is used instead of a plug or pilot bit.

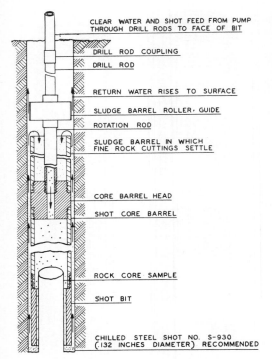

CLEAR WATER AND SHOT FEED FROM PUMP
THROUGH DRILL RODS TO FACE OF BIT

DRILL ROD COUPLING

DRILL ROD

RETURN WATER RISES TO SURFACE

SLUDGE BARREL ROLLER· GUIDE

ROTATION ROD

SLUDGE BARREL IN WHICH
FINE ROCK CUTTINGS SETTLE

CORE BARREL HEAD

SHOT CORE BARREL

ROCK CORE SAMPLE

SHOT BIT

CHILLED STEEL SHOT NO. S-930
(.132 INCHES DIAMETER) RECOMMENDED

Courtesy of Acker Drill Company

Fig. 20-96. Shot core hole

Courtesy of Acker Drill Company

Fig. 20-97. Trimming a bit

As the bit cuts it leaves a core standing in the center. When the core reaches the back of the barrel drilling is stopped and the bit raised slightly. In many formations, the core will be lifted by the choker spring (core lifter), and will break loose so that it can be pulled with the drill string to the surface.

If the core will not break, the bit should be turned slowly without advancing, and with little or no water circulation. The churning will dry out the mud between the core and the inner wall of the bit and cause it to heat and bind, twisting the core off.

If the core is of broken material the lifter may not be able to hold it intact. In this case, turning without water will build a mud cake inside the bit that will support it while it is raised.

The string is raised to take a core or to inspect or replace a bit by pulling up with

the hoist, and holding the string with a safety clamp laid across the top of the hole. The clamp has jaws which are released during hoisting, and set to prevent slipping down while the sections are removed.

The swivel is removed from the top and the first rod is unscrewed while still in the kelly. The swivel head is then swung aside. Rods are then removed one or more at a time, depending on the height of the hoist. In deep work most of the working time is spent pulling out and putting back strings so that a high lift that will handle several sections at once is a great time saver.

If the formations being tested are water soluble, a saturated solution of the same chemicals is used in the pump. Under some conditions, mud or liquids other than water may be used.

Grease is used on the drill rod in deep work to reduce wall friction, unless the nature of the testing forbids it.

SHOT CORING

Operation. Shot coring is accomplished by the rotation of a hollow steel cylinder or bit that is slotted and allows chilled steel shot to feed under its flat bottom area. During rotation the shot breaks into small sharp pieces that erode the rock, which is the softer material. A small amount of wear takes place at the face of the bit.

The drill mechanism, drill rods, water circulation, and core barrels are similar in a general way to those used for diamond drilling. However, it is considered economical to use shot when cores are larger than 6 inches in diameter. Cores up to 26 and 48 inch diameters may be obtained.

As the shot bit wears, the face becomes slightly rounded and allows the shot to feed from underneath it. After the bit becomes dulled or has lost its flat surface, it can be flattened out by beating with a hammer, filing, or grinding. The slot can be lengthened with a saw, so that considerable wear can take place before the bit is discarded.

The secret of success in operating shot core barrels is to put in the right quantity of shot during the boring. Too small additions starve the cutting progress and too much shot causes the bit to round off with little or no cutting progress. Water supply should be just enough to wash the cuttings away from the face of the bit to the top of the hole, where the water is settled and re-used if there is a scarcity of water.

Core Removal. As the shot bit cuts it leaves a core standing in the center of the core barrel. When the core reaches the head of the barrel the drilling is stopped, however, before completely stopping the operation the water flow may be reduced, to make a thick mud that will plug the core in the end of the core barrel. After disconnecting the drill rods at the surface, a small amount of grit is fed through them into the core barrel.

The grit assists in locking the core to the barrel. There may be a spring tine core lifter that fits into the bit and holds the core until it is removed at the surface.

A slight tap on the core usually breaks it at the bottom. After it is secured and broken off, the string of tools is raised to the surface and the core is removed from the barrel. Inspection is made and the bit is trimmed up for its next coring.

CHAPTER TWENTY-ONE

MISCELLANEOUS EQUIPMENT

ROOTERS AND RIPPERS

Towed Rooter. Towed rooters of the type shown in Figure 21-1 are used for loosening hard ground, breaking weak rock, ripping up pavements, laying underground cable, and a variety of other purposes. Most of them have been replaced in construction by tractor-mounted rippers, but a description of them may still be of interest.

This machine is very simple in design. Frame and draw tongue are a box beam, carrying sockets for three shanks or standards at the rear. The shanks are fastened in sockets by single pins. The tooth points are detachable. Weight may be as great as 17 tons.

A pair of triangular bell cranks, rigidly fastened to each other, are hinged to the frame and to the axles. A double sheave on a crossbeam connecting the top of the bell cranks is connected to another on the draw tongue by a four-part line from a control unit on the towing tractor.

When the cable is reeled in, the upper end of the triangle frame is pulled forward. The bottom rotates on the axles, and the back rises, bringing the frame and teeth up. Releasing the cable allows the frame to drop by its own weight. The front surfaces of the teeth and shanks are sloped so that the digging action tends to pull them deeper.

Towed rooters may also be operated hydraulically. Such units have down pressure so that the full weight of the unit may be placed on the teeth. However, this does not add nearly as much weight as it does in tractor mounted models.

These machines are quite subject to overturning while ripping ground that contains large boulders, or that breaks off in slabs, particularly if only the center tooth is being used. Another tractor can easily tip the rooter back on its wheels from the side, but the towing tractor must be disconnected to get at it, and then reconnected. It can sometimes be righted by turning the tractor and manipulating the hoist.

Turning while the teeth are in the ground is apt to break the shanks.

Mechanical Lift. Towed rippers can also be equipped with a mechanical or ground wheel lift, similar to that used on agricultural plows. Such machines are called panbreakers, subsoil plows, or subsoilers.

The lift mechanism usually consists of a hook or dog which can be caught on a rack or gear on the wheel so that rotation of the wheel raises the frame to transporting position. Releasing a catch will allow

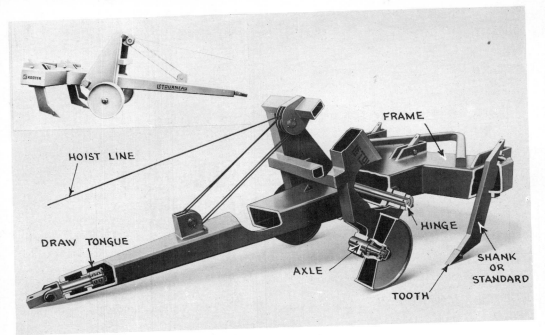

Courtesy of Le Tourneau-Westinghouse Company

Fig. 21-1. Rooter with cable control

it to drop. Working depth is regulated by a threaded crank. Control is by trip rope. Some models have a rope for raising and another for lowering, while others use successive pulls on the same rope for the two operations.

Some of these machines are built to carry only a single shank or standard. Others may carry one to three heavy standards, or up to five lighter ones (chisels). Maximum depth of penetration varies between 20 and 36 inches, and horsepower requirement from 20 to 90.

They are designed primarily for agricultural work. They are also useful to the excavating contractor, who can use them for breaking hard soils on jobs too small to justify bringing in a heavy duty rooter, for shallow ditching, and for subdrainage. Their cost is comparatively low, and since they do not rely on cable or hydraulic controls, they can be towed by any ma-

Courtesy of Pullman-Standard Car Manufacturing Company

Fig. 21-2. Rooter with hydraulic control

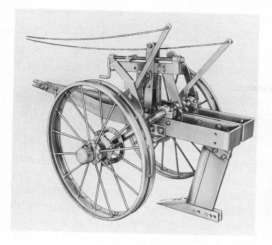

Courtesy of John Deere, Moline, Illinois

Fig. 21-3. Subsoil plow

Courtesy of John Deere, Moline, Illinois

Fig. 21-4. Breaking hardpan layer

chine with sufficient pull. However, they are not efficient in pavements or frost heavy enough to require upward breakage, as the lift does not operate except when the wheels are turning, and too much vertical resistance will cause them to skid.

The Panbreaker shown in Figures 21-3 and 21-4 is an example of this type of equipment.

This subsoiler can be equipped with a V-type plow for making ditches 16 to 26 inches deep. Drift wings can be used to spread and flatten the spoil.

Moles. A mole or mole-ball, Figure 21-5, is a ripper accessory that will open a drainage tunnel in a plastic soil. The type illustrated is a torpedo-shaped piece of iron, attached to the heel of the standard. As it is pulled through the ground, the mole presses the soil outward with great force, leaving an open tunnel with a firm lining. Seepage from the surface is aided by soil breakage.

Gradient is determined by surface slope and the route taken across it. It should be between six inches and two feet to a thousand feet, and in no case should be

steep enough to permit erosion in the tunnel. A piece of tile should be placed in each outlet to protect it from erosion or stoppage.

This device is effective only in soils which are damp and plastic enough to be molded, and not soft or loose enough to flow or cave into the passage. An uneven surface or stones in the soil make it difficult or impossible to run it at an even gradient.

Mole drains may not work at all, may stop up in a couple of weeks, or give satisfactory service for years. Occasionally their good effects are accomplished as

Courtesy of John Deere, Moline, Illinois

Fig. 21-5. Mole ball

Fig. 21-6A. Ripper, radial mounting

much or more by the incidental breaking of an impervious hardpan than by the drains themselves. The work is much more economical than tiling.

Fig. 21-6B. Ripper, parallel mounting

TRACTOR MOUNTED RIPPER

A ripper is a tooth, or a set of teeth, which is usually mounted in a frame hinged to the back of a crawler tractor, with hydraulic hoist. It is pulled through rock or hard soil to break and loosen it.

The mounting may be a radial type, in which the teeth move in an arc, so that they tend to point downward in carrying position, and swing forward as they are lowered. A parallelogram or double mounting has two pairs of hinge arms, upper and lower, that keep the shank(s) at a constant angle to the ground as it is raised or lowered.

In an adjustable parallelogram, the upper arms are replaced by a pair of hydraulic cylinders which can be used to regulate shank angle, either before or during ripping.

Adjustment is useful in selecting an efficient point angle for a particular formation. It allows using a steep angle for entry, and a flatter, point-saving one for ripping. Dull points may sometimes be sharpened during ripping by changing from a steeper to a flatter angle.

The hydraulic pump and controls are built into the tractor, which should carry a bulldozer. This should have hydraulic tilt, for efficiency in handling oversize pieces of rock.

Shanks and Teeth. A tooth is made up of a long shank, and the tooth proper, which may be called a point, a tip, or a cap.

A shank is usually fastened to a ripper bracket by two removable pins. If the bracket is open at the rear, teeth not being used may be buried outside the work area, pins driven out, and the tractor driven away. The procedure is reversed for re-installation. Burial avoids need for a hoist to handle the heavy shank, and the open end greatly simplifies lining up for the pins.

Shanks may be straight or curved. Straight ones are generally used for massive or blocky formations; curved ones for bedded or laminated rock or road pavement

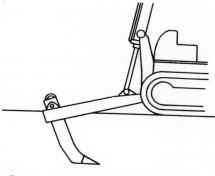
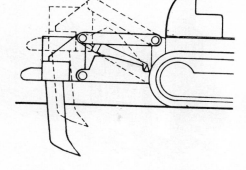

Courtesy of International Harvester Company

Fig. 21-6C. Radial and parallel mountings

that is further shattered by the lifting action of the bottom of the curve. Brackets may be loosely pinned to allow the tooth to swivel slightly to "hunt" weaknesses in rock, and to reduce side strain when the tractor is turned. They may allow for adjustment of depth of cut and/or angle of penetration.

Teeth are detachable, and may be reversible. They may be built up with hard face rod when worn, but this may spoil their shape and efficiency. It is better practice to weld on a forged cap.

Tooth points may have a service life of 30 minutes to 1,000 hours, most of the difference depending on the abrasive qualities of the rock. The operator should glance at them every time they come out of the ground. Dull teeth are inefficient, worn-out or broken ones allow destruction of the shank.

Tips may cost from $20 to $50 each, depending on size and material. The range in per-hour cost, from two cents to $100, is an estimator's nightmare. However, reasonably accurate predictions can usually be made on the basis of rock or soil type.

Most heavy duty rippers may be used with three teeth, with the center tooth only, or with the side teeth only. Use of all the teeth requires maximum power, works a strip somewhat wider than the ripper, and produces most thorough breakage where full penetration can be obtained. But it may

limit depth and thoroughness, and even make any penetration impossible. Trouble may occur with slabs and boulders pushing ahead of the shanks.

Ripping with only the outer teeth increases efficiency of penetration, reduces power required, and usually produces coarser pieces than with three teeth. Break-

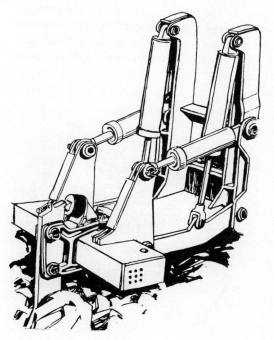

Courtesy of ATECO

Fig. 21-6D. Adjustable parallel ripper

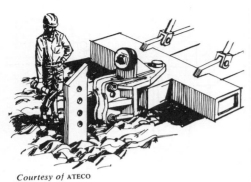

Fig. 21-6E. Open shank bracket

age may be poor or lacking in the center, so that overlapping passes may be required.

Most very heavy ripping work is done with a single center tooth. All the weight and pull available can be applied to the single tooth, and it can hunt and follow weaknesses in the rock more readily.

One mistake to be avoided is running the tractor for extended periods "on tiptoe". Down pressure on the ripper may raise the rear of the tractor, so that only the front of the tracks provides traction. This reduces production, wears the undercarriage and provides the operator with the roughest kind of a ride.

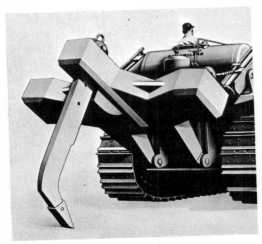

Fig. 21-7C. Coal ripper

Ripability. The factors involved in the resistance of rock to ripping are discussed on page 3-5.

Seismic testing is the recognized method of testing ripability. It depends on the relationship between the cohesiveness of rock and the speed of vibration through it.

A diagram of the method used is shown in Figure 21-7B.

Penetration. Maximum penetration of general purpose heavy rippers varies from about 15 to 30 inches, with 24 to 28 inches being the most common. Overall width between outer teeth may be 8 to 11 feet, to correspond with tractor width.

There are also special purpose rippers used in pipe line work, loosening coal, and subdrainage that may have single shanks permitting penetration as deep as eight feet.

Heavy duty rippers may be designed for operation by tandem tractors. The front machine carries and pulls the ripper; the second pushes it with a dozer blade held by a shelf or socket at the rear of the bracket or the shank. A pair of D9 tractors can put as much as 60,000 pounds of down pressure on a ripper tooth, making it possible to fracture very resistant rock.

The front tractor may do the whole job wherever it can produce good results alone, and be helped by the pusher only when necessary. The pusher blade may be allowed to float most of the time, with down pressure only if the ripper tries to ride up.

Depth. Except when breaking hard or tough ground (pavement or frost) over soft material, teeth usually should be near full depth, with the pull beam approximately horizontal.

Shallow ripping tends to be irregular in depth, and causes tooth wear out of proportion to production.

However, there are a number of conditions that may call for part depth ripping. Full depth may create too heavy a load, or may require so much down pressure that the tractor is on the front of the tracks only (on

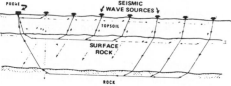

By dividing the distance in feet from the plate to the phone by the time lapse in seconds, the velocity in feet per second is obtained.

$$V \text{ (in FPS)} = \frac{D}{T} \quad \frac{\text{(Distance from phone to plate)}}{\text{(Time lapse in seconds)}}$$

Fig. 21-7B. Seismic recorder and readings

Courtesy of International Harvester Company

Fig. 21-7C. Tiptoe operation is poor operation

tiptoe), with loss of traction and production, with increase in wear. Or the rock may have a natural horizontal weakness part way down, that should be utilized.

In any of these cases, it may pay to make a shallow cut. But if better depth can be reached easily by removing a tooth, or teeth, they should probably come off.

If a ripper tooth can be kept at a constant depth, breakage along its path is uniform down to the tooth point. There may be additional loosening below the tooth, particularly when digging against sloping beds. The ground between teeth, or between successive passes with one tooth, may break to the full depth, part way down, or not at all. If the spacing is right for the depth and the formation, there should be good breakage to at least half of tooth depth in the spaces between them.

Many formations do not permit constant depth ripping. Hard spots, slabs, or comparatively small flat pieces riding under a tooth may cause it to come up a few inches

Courtesy of American Tractor Equipment Corporation

Fig. 21-8A. Ripper and pusher

or all the way to the top. Resistant rock may stop the tooth, and make it necessary to raise it before the tractor can resume forward movement.

It takes only one unbroken rock section every 20 to 30 feet to make scraper loading difficult, and one every 10 feet may make it impossible. The average depth of ripping is important in measuring production, but it is the depth of complete breakage that determines effectiveness.

Direction. Any direction is suitable in soil, frost, rock with bedding planes parallel to the surface, and any material without a definite structure.

On slopes it is safest to rip up or down hill, as large chunks might tend to lift and overturn the tractor. Difficult ground might be ripped downhill only.

When bedding planes or joint structure are at an angle with the surface, primary ripping is usually done against the grain, so that the slope of the beds tends to pull the tooth down. However, if this results in excessive pulling up of big slabs, another direction may be used.

If bedding planes are perpendicular to the surface, ripping should be across them. If parallel or nearly parallel, the tooth or teeth may tend to cut steep sided grooves with unbroken ribs between them.

Whenever it is practical, tractors doing double duty as scraper pushers and as rippers rip in the direction of scraper travel, to avoid turns.

Irregular depth and poor breakage can often be corrected by cross ripping; that is, ripping the same area again at right angles to the original direction. The tractor must walk on rocks turned up the first time, and may find the going very rough. It may be necessary to push some of the pieces off to the side or smash them before cross ripping.

Spacing. Wide spacing means fewer passes, and therefore increased production on an area basis. However, close spacing may be needed, either to obtain reasonably

Courtesy of ATECO

Fig. 21-8B. Ripping direction in sloping beds

even bottom breakage, or to produce pieces fine enough for removal by the equipment used.

With full penetration three feet between one-tooth passes is often satisfactory. Slab material may be loosened on spacings of 8 or even 10 feet, if the pieces can be handled. Crumbly material may call for wide spacing to reduce fines.

Slabs. Many formations, frozen ground, strong or rubbery rock with weak bedding planes, and concrete pavements, tend to break out in big slabs that are pushed ahead by the ripper shanks, so that they lift the ripper out of the ground and/or jam against the tractor. A number of small slabs may combine to block multiple teeth.

A big slab may make it necessary to lift the teeth high, drive forward to clear it, lower the teeth to the ground surface, push the slab back until the teeth reach broken ground, then force them down to resume ripping. If the teeth are simply raised to clear the slab and forced in again at the other side, an unbroken pinnacle of rock will be left.

Slabs may be broken by forcing the tooth or teeth down on them, by climbing them with the tractor and turning, or by a separate machine with a drop ball or pile driver.

Operation. Most ripping is done in low gear, at one to 1½ miles per hour. Higher

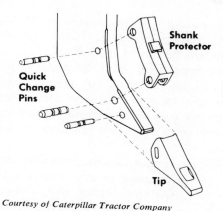

Quick
Change
Pins

Shank
Protector

Tip

Courtesy of Caterpillar Tractor Company

Fig. 21-8C. Ripper tooth details

speed tends to increase wear out of proportion added production.

A torque converter reduces repair cost, as it cuts down both track spin and shock to the gear train. However, it may also reduce production substantially, particularly in rough work, as shock breaks even more rock than it does tractor parts.

Ripper operation in hard rough rock is the toughest service ever required of a crawler tractor. Hoist pull, down pressure, and a portion of the drawbar pull are taken by the transmission case. An exceptionally solid rock mass may bring the tractor from walking speed to a dead stop in a fraction of a second. Tracks spin on smooth slabs, then are stopped abruptly, by a grouser catching an edge. The underframe and power train cases may be hammered and jammed by pieces of rock too high for clearance, turned up by the tracks, or forced forward by the ripper. Re-ripping an area may involve constant pitching over loose rock, even after bulldozing the biggest pieces to the side.

Damage to the tractor and ripper are greatly affected by operator skill and attitude. A rough operator on such a rough job can wreck a new crawler in a season or less. A very careful man might be able to obtain almost as much production, and keep the machine running for its normal span of

years, although repair costs will be high in this work. The careful man will last longer himself, as bouncing a big tractor around on rock is rough on an operator's insides.

Ripper Uses. The ripper has two principal uses in earthmoving. One is to make a digable soil easier to dig, the other is to compete with explosives in loosening otherwise undigable material. Secondary uses include laying underground cable, cutting tree roots, and rolling out boulders.

The first use is a very old one, being chiefly employed in aiding scrapers and dozers to dig hard soil. The machine often has the double job of ripping, and pushing scrapers or bulldozing.

The use of rippers in soil being dug by scrapers is discussed in Chapters 8 and 17. The particular problem is that a very fine or loose soil is harder to load than one that is a little too hard for easy cutting. Ripping of a borderline soil may have the effect of decreasing production, or of not increasing it enough to justify the expense of ripping. Fine breakage with three teeth may cause loads to be smaller than if one or two teeth are used. Many contractors consider that if three teeth can be pulled through a soil at full depth, it does not need ripping.

A bulldozer does not have this problem, as it will almost always do better in loose soil than in formations that are even moderately hard. The worst that ripping can do here is to be a waste of effort. It cannot harm the digging and it usually helps it.

Production. A ripper designed for mounting on a 20-ton crawler tractor usually has an overall width of 9 feet (3 yards) when carrying three teeth, and should loosen a 12-foot strip. Maximum penetration is 2 feet or more. In loosening soil or decomposed rock that is free from hard ledge or boulders, the ripper might be worked at full depth in low gear at a bit better than one mile an hour, say 30 linear yards a minute, less 5 for turn time, leaving a net of 25.

Multiplying this by the 4-yard strip width,

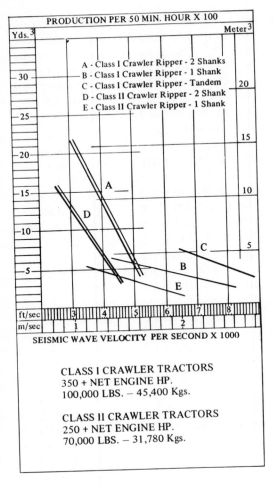

PRODUCTION PER 50 MIN. HOUR X 100

A - Class I Crawler Ripper - 2 Shanks
B - Class I Crawler Ripper - 1 Shank
C - Class I Crawler Ripper - Tandem
D - Class II Crawler Ripper - 2 Shank
E - Class II Crawler Ripper - 1 Shank

SEISMIC WAVE VELOCITY PER SECOND X 1000

CLASS I CRAWLER TRACTORS
350 + NET ENGINE HP.
100,000 LBS. – 45,400 Kgs.

CLASS II CRAWLER TRACTORS
250 + NET ENGINE HP.
70,000 LBS. – 31,780 Kgs.

Courtesy of ATECO

Fig. 21-8D. Ripper production graph

it covers 100 square yards a minute. At two foot depth, this would yield 66 bank yards a minute, or 3000 an hour, enough to keep a whole spread of scrapers in full time operation.

Even after deducting one-third for contingencies, there is a very respectable 2000 yards per hour. Potential production such as this is the reason that contractors try to assign ripping as just a part time job for pushers.

Difficult soil containing boulders or frost or interrupted by ledges, is loosened much more slowly. And all-rock ripping may go

at a rate as low as 100 yards an hour before it is given up. In very hard work the work hour may be figured at 20 minutes.

In ripper work most decreases in production caused by difficult ground conditions also increase machinery costs. There is probably no other earthmoving tool that shows such a wide range of output and unit cost.

Ripping is most apt to be profitable where the power is ample or even excessive for the job. A 10-ton tractor with a ripper might tear itself to pieces with little production in a formation that a 20-ton machine would walk through, loosening substantial yardage with only moderate strain on itself. This larger machine might in its turn be nearly helpless in a harder formation that would yield easily to tandem tractors or a single 30-ton unit. The heavier and higher powered units usually provide better and more uniform breakage than the overworked smaller ones.

Ripping Cost—Soil. For an example of the method of figuring ripping cost, we will use as an example a crawler tractor in the 30 ton class (Caterpillar D9, International TD30), carrying a heavy duty ripper with three teeth and a hydraulic dozer with tilting blade. At a purchase price of $140,000, an active life of 6000 hours, owning and operating costs might be $70.00 an hour. Divided by 3000 yards loosened, this comes to about 2⅓ cents a bank yard. On the actual job, the cost might be even less, or three times as much.

If the ripping were done as spare time work by a pusher, which was able to do it in addition to taking good care of its scrapers, the cost might sink to amortization and repairs on the ripper only. This could be a very small figure, or quite a large one if the ripper were to have only occasional use throughout its life.

For soil use only the contractor might prefer to get the back ripper teeth discussed below.

Ripping Cost — Rock. A thin, weak bedded shale might be ripped at little more cost than hard soil. However, a really tough rock might prove so resistant that only one tooth could be used, penetration would still be poor, and breakage so difficult and irregular that production would be cut down to 100 yards an hour.

A tractor ripping difficult rock is likely to be worn out within 3,000 hours if it keeps at it. Repair costs can easily be double those in ordinary heavy duty service. If the right ripper is bought for the job it should last at least as long as the tractor. Some heavy duty rippers outlast two or three of the tractors that carry them.

On this basis, depreciation and ownership expense would be up to $61.00 per hour (doubled depreciation) and operation up 50 per cent on repairs, to $43.00, a total of $104.00. Production of 100 yards would cost an uneconomic $1.04 an hour, but 200 or 300 yards would bring it down within talking distance of blasting.

Production may be substantially increased in many formations, and machine life lengthened, by light shake-up blasting before ripping. Such blasting might be done for $0.05 to $0.20 a yard.

If the ripper were strong enough, a pusher tractor with a down pressure blade could be used, doubling both pull power and penetration rate. Assuming that the rock had good breaking qualities and just needed more force, output might be increased to 400 or even 600 yards an hour.

The tractor carrying the ripper would be under less strain, but the ripper would be under much more. The net effect would usually be a substantial reduction in repairs, and increase in tractor life. But even if no savings were obtained in this way, the pusher would add only 50 per cent to costs, and would be paid for over and over by increased production.

These are only samples to show how costs might be worked out.

Other Costs. Any comparison of ripping and blasting costs must take into account the method by which the broken rock is to be moved.

In a side hill cut it may be possible to sidecast the ripped rock with a dozer at a fraction of the cost of using a shovel in blasted rock. This would be true even if ripper breakage were coarse and irregular.

A dozer, often the same one that carries the ripper, may be used economically for pushing ripped rock up to 200 or 300 feet along the road route to a fill. Production would be much better with fine breakage than with coarse.

For longer hauls the scraper is the preferred machine. If the material is suitable for scraper loading, it can probably move it for 10 to 20 cents a yard less than a shovel and trucks could. In the numerous formations that can be scraper-loaded after ripping, but not after blasting, the combined saving of ripping instead of blasting, and scraper instead of shovel, may be very substantial.

Where the rock cut is shallow, ripper-scraper combinations are even more desirable because of the relative inefficiency of both blasting work and shovels or loaders in low banks.

Fig. 21-9A. Back ripper teeth

RIPPING

BULLDOZING

RETRACTED

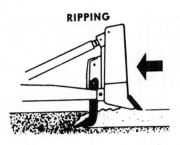

Courtesy of Preco, Incorporated

Fig. 21-9B. Back ripper action

But a scraper is not designed to handle coarse rock. Where ripper breakage is poor, scraper loading is likely to be slow, repair costs high, and useful life short. Scrapers may also be damaged at the dump by riding over rocks, and in struggles to eject over-size pieces.

On slopes, both fine and coarse ripped rock may be bulldozed downhill to belt loaders that put it into trucks.

If the rock cannot be ripped into sizes for scrapers or belts, and there are too many big pieces to make it economical to break them with a drop ball, pile driver, or explosives, it will probably be necessary to bulldoze the rippings into piles for a front loader or dipper shovel.

The loader can do its own dozing. Work conditions are apt to be very rough, production low, and costs high.

If ripped rock can be pushed off a nearby face the yardage from several layers can be heaped for efficient shovel loading. The cost of ripping can then be added to that for dozing, to get a figure to compare to that for drilling and blasting.

If there is no face to push to, or when repeated cutting of the top destroys it, the loose rock may be pushed up into windrows for the shovel. The shovel will not work efficiently, as side material will tend to slide away from it, and dozer help will be needed to keep the pile trimmed. For this condition the ripping method should be charged both

Courtesy of Deere & Company

Fig. 21-9C. Multiple-shank ripper on wheel tractor

with the piling expense, and the lowered shovel efficiency.

Back Rippers. Ripper teeth may be hinged to the back of a dozer blade in such a way that they float on the ground when the blade is moving forward, and dig in when the blade is dragged backward.

A back ripper makes it possible to use the nonworking part of a dozer or pusher cycle to loosen ground for the next push or the next scraper pass.

The rippers can be prevented from digging by raising the blade high enough to keep them off the ground, or by pulling them up in their brackets by hand, and inserting lock pins.

A hydraulic blade with down pressure will of course permit ripping harder formations than the now-rare cable blade. Costs of wear and breakage increase with the resistance of the ground.

Light Mounted Rippers. Figure 21-9C shows a ripper of lighter construction mounted on a heavy 2-wheel drive tractor equipped with a front loader. It is similar in design to the heavy units, except that it has four teeth and one ram.

This unit is primarily designed to loosen soils to be loaded by the same machine, but can do medium duty rooting on many projects. It also serves as counterweight.

The John Deere subsoil plow, 21-9D, is carried on a hydraulic lift drawbar. It is essentially an agricultural tool, but will break up quite hard soils, and can prove a valuable aid to a light loader or dozer. It is inexpensive, readily mounted and removed, and does not interfere with many other uses for the tractor.

CABLE LAYERS

Either mounted or towed rippers can be used to lay cable and flexible conduit underground, without ditching. The soil is preferably even-textured. It is essential that the ripper be able to penetrate it and hold a reasonably even depth.

Courtesy of Deere & Company

Fig. 21-9D. Subsoiler

The cable may be mounted on a reel carried on the tractor, or on the ripper frame; or may be laid on the ground and carried over or under the tractor by sheaves or tubes. One end is fed through a conduit or around sheaves to take it down the back of the single ripper shank, and curve backward near the heel of the tooth.

This end is anchored at the beginning of the work strip. The tractor moves forward, the ripper makes a slot in the ground, the cable (or conduit) slides through its guides to be left near the bottom of the slot. The

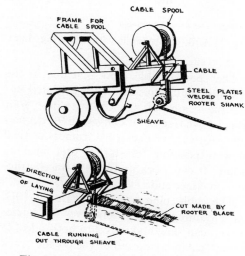

Fig. 21-10A. Cable-laying attachment

cut in the ground may heal itself, or be pressed together by one passage of the tires of a car or truck.

The ability of a cable-laying ripper or plow to penetrate ground, keep in line, and make minimum surface disturbance, is improved by high frequency vibration of the shank. However, the vibration may have an unfavorable effect on some types of cable.

The cable plow in Figure 21-10B has a vibrating shank (saber) attached to a cable static chute by vibration-absorbing hinges. It is mounted on a Category II 3-point hitch on a heavy wheel tractor with hydrostatic drive or creeper gears.

Courtesy of Arps Corporation

Fig. 21-10B. Cable layer with vibrating shank

The special cable plow machine in the next illustration has the saber mounted under the center of a short-wheelbase four wheel drive tractor. It can swing 90 degrees either forward or backward from its vertical plowing position, for starting and finishing a line. It can work up close to an obstruction on either end.

The cable/pipe plow in Figure 21-10D, for mounting on crawlers over 70 horsepower, can slide and swivel to both sides, to permit cable-laying either inside the track path, or at an offset to either side.

WINCHES

A winch is a steel spool or drum that is capable of powerful rotary movement, and that has means to attach a line (usually cable, officially called wire rope) that winds in as it turns.

The operating drums in cable-powered machines such as shovels and cranes answer to this description, but the term "winch" is usually limited to units that normally exert their power outside the machine that carries them.

An exception to the definition is the capstan or cathead winch, described later, that is not attached to the hemp rope to which it provides power.

A winch must have power to wind in its line, and means to release the line to be pulled to its next load. Pull out may be arranged by releasing the drum from the drive train, so that it will turn and unwind as the line is pulled; or by providing a reverse gear. If there is a reverse, it may be twice as fast as the inpull.

A winch is said to underwind when the cable is reeled on and off the bottom of the drum, and to overwind when it is at the top.

A few winches are hand operated, but most of them are powered by the engine of the carrier. Drive is usually mechanical, through a power takeoff, but may be hydraulic or electric.

Power. A winch is usually power-rated on the basis of the maximum pull it can exert on its line. Specifications should also give the speed of the line, in feet per minute (fpm). These figures vary with drum diameter, in inverse ratio, if drive is mechanical.

A small drum winds in less cable per revolution than a large one turned with the same power and speed. It therefore exerts more pull, but at a lower line speed.

The effective diameter of a drum increases with each full layer of line it wraps around itself, as additional line must wind

Courtesy of Vermeer Manufacturing Company

Fig. 21-10C. Center-shank cable layer

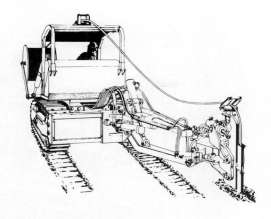

Courtesy of ATECO

Fig. 21-10D. Side-pivoting cable layer

on top of it. This means that pull will decrease as the drum fills, and line speed will increase.

A drum eight inches in diameter carrying ⅞ cable might have a line pull of 28,000 pounds bare, and about 16,000 on a fifth layer. But line speed would increase from 80 to 140 feet per minute. Additional effects of multiple layers are discussed later.

Line speed specifications vary with the tractor or other carrier. Pull is assumed to be fixed, and should not be exceeded even if power permits.

Torque Converter. If the winch is on a power shift tractor, its drive should be through the torque converter. This device confuses the speed-power relationship, as indicated by the graph in Figure 21-13.

When pull on the line is light, the full drum shows the relatively fast line speed that would be expected. But increasing load slows down a full drum much more than it does a bare drum, because of greater leverage against the converter. At peak load (30,000 pounds in this example) the line moves faster on the bare drum than on the full one, as reduced slippage in the converter provides that much faster drum speed.

Brakes. A full service winch should have a brake that can lock the drum against full load. It may be necessary to hold a sus-

pended or springy load when an engine stalls or direction of pull must be adjusted.

The brake may also be needed to hold the drum when the tractor is pulling the load.

The brake may be an external band, contracting on a drum on an intermediate or side shaft in the winch case. It may be applied manually, hydraulically, and/or automatically.

Courtesy of Hyster Company

Fig. 21-11. Towing winch

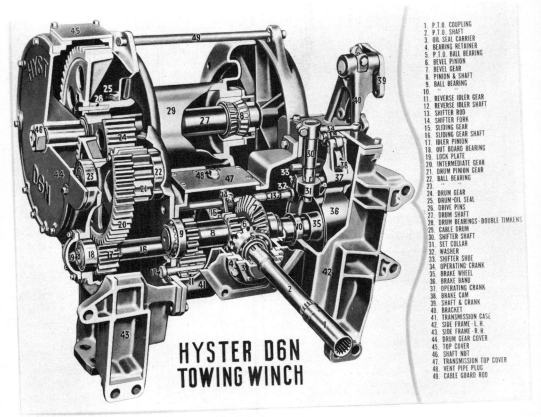

1. P.T.O. COUPLING
2. P.T.O. SHAFT
3. OIL SEAL CARRIER
4. BEARING RETAINER
5. P.T.O. BALL BEARING
6. BEVEL PINION
7. BEVEL GEAR
8. PINION & SHAFT
9. BALL BEARING
10. "   "
11. REVERSE IDLER GEAR
12. REVERSE IDLER SHAFT
13. SHIFTER ROD
14. SHIFTER FORK
15. SLIDING GEAR
16. SLIDING GEAR SHAFT
17. IDLER PINION
18. OUT BOARD BEARING
19. LOCK PLATE
20. INTERMEDIATE GEAR
21. DRUM PINION GEAR
22. BALL BEARING
23. "   "
24. DRUM GEAR
25. DRUM-OIL SEAL
26. DRIVE PINS
27. DRUM SHAFT
28. DRUM BEARINGS-DOUBLE TIMKENS
29. CABLE DRUM
30. SHIFTER SHAFT
31. SET COLLAR
32. WASHER
33. SHIFTER SHOE
34. OPERATING CRANK
35. BRAKE WHEEL
36. BRAKE BAND
37. OPERATING CRANK
38. BRAKE CAM
39. SHAFT & CRANK
40. BRACKET
41. TRANSMISSION CASE
42. SIDE FRAME - L. H.
43. SIDE FRAME - R. H.
44. DRUM GEAR COVER
45. TOP COVER
46. SHAFT NUT
47. TRANSMISSION TOP COVER
48. VENT PIPE PLUG
49. CABLE GUARD ROD

HYSTER D6N
TOWING WINCH

Courtesy of Hyster Company

Fig. 21-12. Winch gearing, mechanical shift (old model)

In automatic position, a pawl locks the brake whenever a line that is being wound in starts to move out and unwind.

In free spooling, when line is being pulled off the drum to extend it to a load, there is a danger that the drum will spin, that is, turn faster than the removal of the line, with resulting loosening and tangles. This is prevented by setting the brake to drag slightly, so that the drum will turn only the amount that the line pulls it.

Reversing clutches may be used as a brake by engaging both of them while a winch input clutch disconnects power.

A light winch may also not have a brake. In this the gearing is usually irreversible, and can be turned only by the takeoff shaft. The drum is connected to it by a jaw clutch.

If the clutch is engaged and the power take-off disengaged, the drum will be locked. In spooling out, spinning can be avoided by holding a foot against the drum, leaning a log against it, or snubbing it with a wood wedge.

Towing Winch. The towing or logging winch is usually mounted on the rear of a crawler tractor, and driven by its power takeoff. Standard rotation direction is over-winding.

If the takeoff is clutch-controlled, the winch may have a jaw or sliding gear to connect the drum to the power train, and depend on the tractor clutch for starting and stopping of rotation, and for shifting gears or jaw clutches. For both safety and con-venience, such a unit should have duplicate clutch and throttle controls that can be reached from behind the winch.

A live or constant running takeoff in a power shift tractor may drive the winch through a pair of multiple disc clutches on a reversing shaft, so that the drum may be power-rotated in either direction. Clutch control may be manual, or hydraulically actuated.

Most winches now sold domestically are power-shift models.

In either type, several stages of gear reduction are provided in the winch between takeoff drive and the drum shaft, to provide for slow and powerful rotation.

A towing winch drum has rounded flanges, which turn inside closely fitted and rounded casing parts. This construction prevents, or at least reduces, damage to line pulled at such an angle that it drags over these edges.

If the work is such that pulling at angles

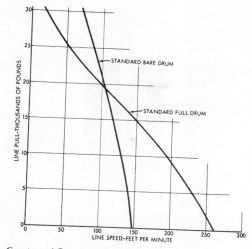

Courtesy of Carco Winch Products

Fig. 21-13. Line speed and pull with torque converter

is frequent, additional protection can be obtained from a fairlead. Three rollers line up the cable with the drum from the sides and top. But it still is harmful to reel it with

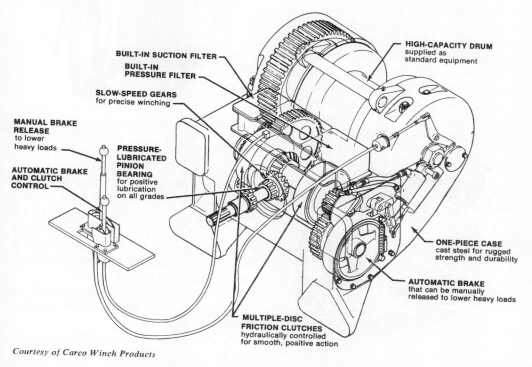

Courtesy of Carco Winch Products

Fig. 21-14. Winch with clutch shift

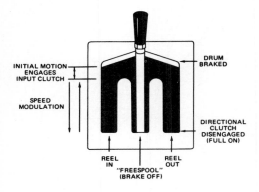

INITIAL MOTION
ENGAGES
INPUT CLUTCH

DRUM
BRAKED

SPEED
MODULATION

DIRECTIONAL
CLUTCH
DISENGAGED
(FULL ON)

REEL
IN

REEL
OUT

"FREESPOOL"
(BRAKE OFF)

Courtesy of Caterpillar Tractor Company

Fig. 21-15A. Single-lever winch control

The small side drum shown in this figure is a cathead, or capstan. It is optional factory equipment on most winches and is fastened to an extended intermediate or drum shaft. It is used with hemp rope for handling light loads. The rope is wrapped around it one or more turns, and driven by friction when the loose end of the rope is pulled tight. When tension on the end is relaxed, the spool turns inside the wraps without pulling them.

A donkey is adapted to both field and yard logging, and can operate a drag scraper or other digging equipment.

great force at sharp angles, because of bending and crushing stresses in the cable.

A towing winch line can be reeved over a boom to provide a tractor crane. However, only a fraction of its power can be used in such an application.

Extra Drums. A winch may have two or more drums, under separate control.

The donkey or yarding winch, shown in Figure 21-16, has two drums with separate clutches and brakes. Such units may have two to four speeds in one direction. A reverse can sometimes be obtained on special order. One drum is generally used for pulling in the load, and the other for pulling the cable out for the next load.

Courtesy of Caterpillar Tractor Company

Fig. 21-15C. Cable reel

Power Control Units. The light, fast power control winches used for operating cable dozers and scrapers can be used for light clearing and dragging work. The scraper type is particularly **convenient in** land clearing, because the fairleads insure proper spooling of cable.

Tractors with this equipment usually do heavy pulling by a chain or cable attached to the drawbar.

Wheel Tractor Mounting. Wheel tractors may be fitted with winches mounted either on the front or rear. Rear winches should

Courtesy of Deere & Company

Fig. 21-15B. Towing winch with fairlead

Courtesy of Hyster Company

Fig. 21-16. Yarding winch and cathead

underwind and their tractors should be anchored by a line from the front for heavy loads, to avoid danger of overturning backward.

In the past, wheel tractors have been used to carry extremely powerful winches, used for machinery rescue and land clearing, which are apparently no longer available.

The Evans winch shown in Figure 21-18 could get a slow 50-ton pull from the engine of a light wheel tractor. The winch case had a connection for an anchor line extending forward under the tractor.

Applications. Winch operation was discussed in Chapter 1.

A towing winch provides accessible storage for a long piece of cable, that enables the pull of the tractor to be readily exerted

Courtesy of Hyster Company

Fig. 21-17. Winch helping hoe up a grade

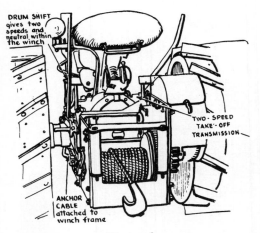

Courtesy of Al Evans Winches, Inc.

Fig. 21-18. 50-ton pull on a wheel tractor

at a distance, and it enables the tow line to be adjusted to the exact length required.

A long cable permits use of snatch blocks and anchors to build up the line pull to any power required. Snatch block applications and other winch uses will be discussed later.

The winch provides a pull equal to or greater than that of the tractor drawbar, and is largely independent of traction. In muddy or sandy conditions, or in working

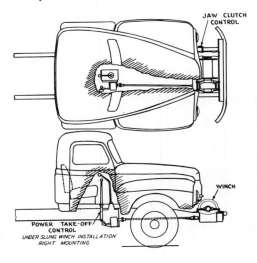

Courtesy of Sperry Vickers Tulsa Division

Fig. 21-19. Front winch on light truck

on roads or landscaped ground, much less disturbance is made by winch than track pull. If the tractor is dragged backward even with the brakes set it may be anchored by a line to a stump or tree or braced by placing a log behind it. If there is no objection to tearing up the ground, the tractor can try to pull the load with tracks spinning until a mound of earth is built up behind them. This will act as a block for the tractor when the winch is used.

The cable is pulled off the drum by hand to attach it to the load. If the drum is disconnected from the gearing so that the pull on the cable will turn it, the brake should be applied enough to prevent it from spinning free and tangling the cable when the pull stops. If two men are available, one can use the tractor engine and winch controls to turn the drum to unspool while the other pulls away the cable.

Large winches require heavy cable, and dragging hundreds of feet of it is laborious, even for several men. If many loads are to be brought out of one place it may be advisable to pull the line out by means of another winch with lighter cable reeved around a pulley behind the loads.

The second winch may be a towing winch, a cable control unit on another machine, or it may be a second drum on the same winch.

TRUCK MOUNTING

Winches in a range of sizes may be installed at the front or rear of the frames of trucks, pickups, and jeeps, or on platform bodies. Rear frame mountings may interfere with body type, or with access for loading, and are seldom used.

Mechanical drive is generally through a power takeoff on the transmission, and a propeller shaft with universal joints, but may be hydraulic from an engine pump.

A takeoff may be one speed in one direction, one speed each in forward or reverse, or two or more speeds in one or both di-

Courtesy of Ramsey Winch Manufacturing Company

Fig. 21-20. Winch on front of pickup truck

rections. Vehicles with four wheel drives operate the winch through the main transmission with its choice of speeds, with the auxiliary shift in neutral.

Ordinarily, the only control on the winch itself is a jaw clutch. The driver at the cab controls cannot watch the line, so that two men are required for efficient and safe operation. However, it is usually fairly easy to add clutch and throttle controls beside the winch, so that one man can operate it safely, although not always conveniently. It must be possible to lock the extra clutch lever in disengaged position.

Figure 21-19 contains a diagram of a Tulsa front mounted winch and controls. When the jaw clutch in the drum is engaged, the winch can be connected or disconnected at the power takeoff, and controlled through the engine clutch and throttle.

Truck winches may have either manual or automatic holding brakes. The drum is released or reverse gear is used to strip the cable. Recent model winches are equipped with drag brakes that prevent over-spooling and snarling of cable when the clutch is released for free spooling. In older models and in those without brakes, it may be necessary to snub the drum. A brake may be

Fig. 21-21. Truck cherry picker

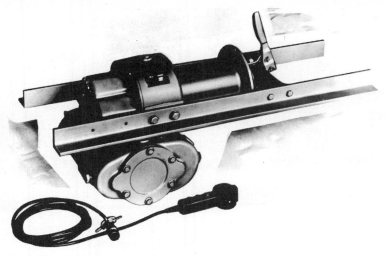

Fig. 21-22. Electric winch

applied lightly, or a foot placed on the drum flange.

A winch may be used to power a crane of the A-frame or cherry picker type.

Hydraulic. A winch of almost any kind, in any position, may be powered by a hydraulic pump and motor set, instead of a power takeoff shaft. Such a motor has a speed reducing gearbox, and may have a stepdown chain drive as well. If design is adequate, performance should be satisfactory.

Hydraulic drive puts some additional weight and bulk at the winch or close to it, but avoids the often-difficult problem of getting a shaft drive to it. Hydraulic hoses can reach almost anywhere, although care must be taken to wrap or cushion all points where they rub or even might rub against moving or vibrating parts.

Electric. Electric winches, powered by the battery of the carrying truck, tractor, or car, are usually made in small sizes, with pull of four tons or less.

These units are compact, offer minimum problems in installation, and may be economical in price. The principal problem is that the average good battery may provide peak power for only about a minute. Many winch jobs, even getting out of a ditch, take longer than that.

There is usually a hand crank that will turn the drum after the battery refuses to. It is often able to provide enough pull to complete a job, although the purpose is chiefly to relieve cable tension if power fails during a pull.

The engine should be running during a pull, fast enough to provide maximum help from the alternator. Also, the winch may run down the battery so that the starter would not work.

Booster cables may be used to provide extra winch action from another battery.

HAND WINCHES

Winches with hand-crank power may be used for occasional jobs that do not justify the cost of a power unit, and in places that are not accessible to machinery. They are made in two, five, and fifteen ton sizes. The two smaller ones are the most popular.

The Beebe winch, Figure 21-23, consists of a spool drum with a double reduction set of gears operated by a hand crank. The handle may be inserted in the small outer gear, as shown, for full power, or in the larger gear for faster cranking and lighter pull. Gear ratios are 24 to 1 and 4 to 1.

A spring held ratchet may be engaged to prevent slipping back while changing the handle or resting, and the load can also be held or controlled with a friction brake.

The frame can be bolted to a deck or skids or attached to a tail yoke for anchoring by a chain or cable.

The two ton winches of the same make are similar in construction, and also have two gear ratios, 22 to 1 and 4 to 1.

Either size may be carried by two men and readily set up in remote places. If the unit is clean and well lubricated it may be able to pull stumps or rescue machinery better than a power winch of the same capacity, because of sustained pull and ability to take up a little at a time.

Substituting a ratchet handle for the standard type will greatly reduce the labor of winding in.

It is often possible to rig a hand winch for power drive.

LAND CLEARING TACKLE

Tow Lines. Standard types of cable (wire rope) are used on winches. A 6 x 19 construction is most common. Independent wire core is recommended by manufacturers because of its greater strength, but many operators prefer fiber center because of its flexibility.

Wire core resists mashing but if deformed it will not straighten out well under load so that it may become difficult to spool on the drum or feed through tackle.

Use of heavy cable decreases the length that can be wound on the drum and increases the labor of handling. However, land clearing which includes stump pulling usually requires the heaviest cable recommended for the winch. If lighter lines are used the tractor should not be anchored and maximum pulls should be avoided.

Strength of cable is given in a table in Chapter 12.

Light and medium cables are ordinarily attached to a tail chain and round hook. The chain is usually made of stock of the same diameter as the cable, and may be

Courtesy of Beebe Bros.

Fig. 21-23. Hand winch

from eighteen inches to eight feet long. It should be made of high strength or alloy steel, to give it substantially greater strength than the wire rope.

If the chain is long enough to make a complete choker around most of the loads or anchors, it may end in a standard slip or round hook. If it is so short that it functions merely to relieve the cable end of twists and kinks, it should have a hook with a wide, rounded inner surface which will do minimum damage to the rope wires. It is

Courtesy of Warren Axe & Tool Company

Fig. 21-24. Cable-choker hooks

STRENGTH AND WEIGHT TABLE
CHAINS

Safe Working Load – Pounds *

SIZE	Approximate Weight per foot	Dredge Iron	Proof Coil	BBB Coil	High Test	Alloy
1/4	.8	—	900	1100	2500	2750
5/16	1.1	—	1400	1700	4000	—
3/8	1.6	—	1900	2300	5100	6600
7/16	2.1	—	2500	3100	6600	—
1/2	2.8	—	3300	4100	8200	11250
9/16	3.4	—	4100	5100	—	—
5/8	4.3	6930	5000	6300	11500	16500
3/4	6.2	10140	7100	8900	16200	23000
7/8	8.5	14000	9600	12000	—	28750
1	10.5	18600	12400	15500	—	38750
1 1/8	12.6	23400	15600	19500	—	44500
1 1/4	15.4	28800	19200	24000	—	57500

Safe working Load is about one fourth of average breaking strength.

Data supplied by S. G. Taylor Chain Company

Fig. 21-25. Chain weight and strength

also desirable to have the hook end turned in, to minimize catching on obstructions when pulled in empty. See Figure 21-24.

Winch lines of 7/8″ diameter and larger may be fastened directly to a hook. Attachment may be by cable clips, wedge and socket, or poured fitting. Cable clips are light, easily obtained and inexpensive, but are a nuisance to install, hard to remove, difficult to salvage, make a clumsy joint, and are apt to damage the cable.

Wedge clamps are comparatively easy to install and to take apart, can be used many times without replacement or repair, and do not damage cable. However, they are heavy and bulky and may be put out of action by loss of an improperly installed or tightened wedge.

Poured fittings are compact, light, and strong but require special materials and techniques for installation or removal.

A winch line may also be ended in a shackle, a cluster, a loop or a thimble, or a cable takeup with or without a swivel.

It is good practice to use separate attachment or choker lines. They are more readily repaired or replaced than the winch cable, and will serve to save it from the severe wear resulting from contact with the choker hook and load.

Chain. Towing chain is made in a variety of sizes, strengths, and designs, four of

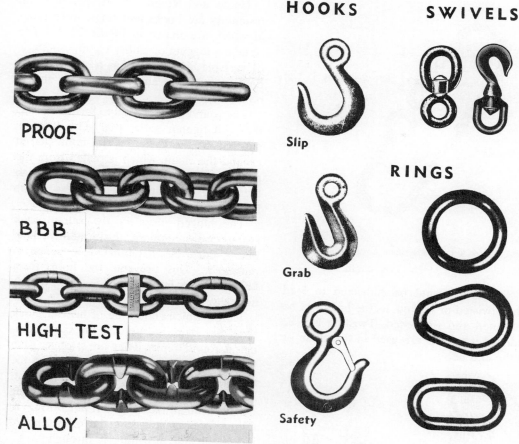

HOOKS

Slip

Grab

Safety

SWIVELS

RINGS

Courtesy of S. G. Taylor Chain Company

Fig. 21-26. Chain

Courtesy of The Thomas Laughlin Company

Fig. 21-27. Chain fastenings

which are shown in Figure 21-26. Size is designated by the approximate diameter of the round bar stock used in making the links, and ranges from ¼″ to 2″ or larger.

Strength is determined by the bar size, by the quality and treatment of the steel, and to a smaller extent by the shape of the links. The table in Figure 21-25 gives data for various types. The safe load shown is about one quarter of the average breaking strength.

Alloy steel chain is very expensive, but is desirable for land clearing work and load handling because of its light weight in proportion to strength. For example, ⅜″ alloy chain is thirty percent stronger than ⅝″

proof chain and weighs only two-fifths as much. In addition, the lighter chain gets a better grip on objects and can be passed through narrow spaces. Substantial savings in labor and machine time are realized from its use, but only in the hands of men who will not abuse it or lose it.

Alloy chain is almost standard for permanent attachment to winch cables. When used as a separate piece it should be dipped in red paint to distinguish it from less strong and valuable chain and to render loss less likely.

Short links are stronger than long ones but make a chain more balky and inclined to kink.

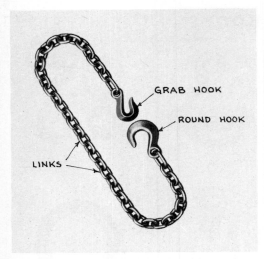

Courtesy of S. G. Taylor Chain Company

Fig. 21-28. Log chain

No chain should be subjected to load when twisted or kinked, as the links will be deformed and weakened. Twisting can be reduced if swivels are used in the chain or at the end hooks.

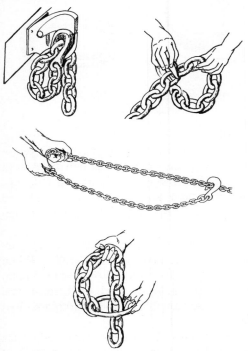

Fig. 21-29. Grab hook uses, and stump choker

Hooks and Rings. The most used chain fastenings are hooks and rings, some types of which are shown in Figure 21-27. The standard logging or utility chain, 21-28, is of variable length, and has a round (slip) hook on one end and a grab hook on the other.

The round hook is used to form a choker that will tighten under pull. The chain is put around the load, the slip hook placed around the chain, which is pulled from the grab end. The chain tends to slide through the hook, increasing the tightness of its hold in proportion to the force exerted.

The hinged piece in the safety hook swings back automatically to admit a line, but it must be manipulated in order to release it. This prevents the hook from falling off the line when it is slack, avoiding delay and damage.

Rings, which should be of heavier stock than the chain, are used in the same manner as round hooks. They are stronger than hooks and will not lose their hold when the chain is slack, but they are less convenient. A stump choker may be made by pulling chain through the ring but anchoring to a tree requires releasing the other end of the chain and threading it through.

"Rings" are made in a number of different shapes. The three illustrated are the most popular. If two are used on a chain, one should be small or narrow enough to slide through the other.

The grab hook is used to form a chain loop that will not tighten. It slides over any link in the proper size chain, as in Figure 21-29, but will not slide along the chain.

Grab hooks can be used to shorten chains by lengthening the loop or by blocking the chain so that it will not slide through a drawbar or other opening.

Shackles, Figure 21-30 (A), can be used in place of rings or hooks for many purposes, and are handy means of attaching lines to objects. They can be used for emergency chain repair.

Fig. 21-30. Shackles

Loadbinders, 21-31, are used chiefly in tightening chains. The binder is expanded, the slack pulled out of the chain, and the hooks attached to links each side of the slack. The lever is pulled to shorten the binder, and locks over center. This process may have to be repeated on successive links. The handle should be tied down when left as part of a load lashing, as otherwise any slackening of the chain will allow it to fall open.

Loadbinders which do not have springs can be used with a chain to move heavy objects.

Repairs. It is common practice to overload winch and land clearing chains to destruction, with resulting heavy costs in repair and replacement. This is in part because the extremely heavy and variable pulls required to uproot tough stumps or to free jammed logs call for a chain which in ordinary quality is too heavy to handle and to use.

Broken chains can be repaired by hot forging of new links or by using special repair or connecting links, two types of which are shown in Figure 21-32.

Such links are purchased assembled and separated by driving a chisel or a very sharp screwdriver between the pieces. This is most conveniently done in the shop.

If the links to be connected have pulled out of shape it may not be possible to get the repair pieces through them. Such a link can be opened up by placing it on a block with a hole in it, and driving a big punch through it.

A good repair link is somewhat stronger than a standard chain link.

Fig. 21-31. Load binders

Snatch Blocks. A snatch block is a pulley or set of pulleys on an axle which is held in a portable case that can be attached to one or two pull lines.

These blocks are generally employed to increase the force exerted by a line. They are used for hoisting, dragging heavy loads, changing direction of pull, and land clearing. The principles involved are described

MISSING LINK

Swing Link

Fig. 21-32. Repair links

Swivel Hook

Swivel Hook

Swivel Shackle

Single

Double

Triple

Courtesy of Boston & Lockport Block Co.

Fig. 21-33. Snatch blocks

in Chapter 12, and their use in land clearing is discussed in Chapter 1.

Figure 21-33 shows several types of block. Single sheave blocks should have latch fastenings that permit opening to insert or remove cable, but double and triple sheaves have solid cases.

The pulleys (sheaves) turn on a single axle. Self-lubricating bronze bushings are commonly used but bushings greased through the axle and sealed anti-friction bearings are also obtainable.

These general purpose blocks can be obtained with a loose hook, a swivel hook, or a rigid or swivel shackle. The shackle is the strongest unit in proportion to weight, but is also the least convenient to use. Swivels have the advantage of allowing the block to adjust itself automatically to line of pull so

that rubbing of the lines against the case is kept to a minimum.

The latched blocks are much more efficient than the solid ones. The average hook or other attachment is too large to pass between the pulley and a fixed case so that it must be removed from the line for reeving. If it will pass, it is still necessary to thread the end of the cable through the block and to pull it out the same way. This takes much more time than inserting and removing the cable from the side, particularly when a number of blocks and lines are used.

Fixed blocks are often left rigged during moving, and crossing and tangling of lines may result, particularly if they are not pulled snugly together.

Figures 21-34 and 21-35 illustrate some

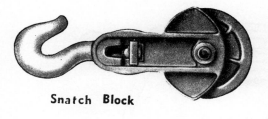

Snatch Block

Sling Block

Courtesy of Jacques Power Saw Company

Fig. 21-34. Heavy duty blocks

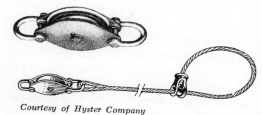

Courtesy of Hyster Company

Fig. 21-35. Power chokers

heavy duty blocks designed specially for land clearing. The sling block permits use of two parts of light choker line instead of a single heavy one.

When a choker cable or chain is permanently fastened to a block, the combination is called a power choker. One type shown is attached to the choker line by a shackle, and the other by a poured fitting. The shackle or loop on the opposite end makes it possible to attach an extra line, and to use the block as part of a direct line when the extra power of the pulley is not required.

Block and Fall. A block and fall is a set of light pulley blocks using fiber rope. In small sizes, pull is by hand. In larger units, several men, a light tractor, or a horse may provide power.

They range in size from very small, four pounds or less complete with ⅜-inch rope, to heavy duty models with one-inch rope.

In the smallest sizes they are used chiefly for tightening fence wire. In general, they are useful in countless situations where required pull is greater than can be exerted by a man, and machine power either cannot

be readily applied, or might be too strong, fast, or jerky for the job.

Nylon rope should be avoided, because it stretches, wasting space and energy.

The block and fall is now little used in land clearing, but is mentioned here because of its kinship with cable-reeved blocks.

The chain hoist is a specialized block and fall, rigged with chain and provided with an automatic lock. It is commonly used in repair shops for lifting engines and other heavy assemblies. In emergencies, it can often be taken down and used for horizontal pulling.

Root Hook. This is a very heavily constructed hook used for pulling small stumps, and individual roots on large ones. It is designed more for strength than for gripping power, and it is often necessary to notch a stump or dig behind a root, in order to give it a grip.

Tongs. Tongs are often used in place of chokers to grip loads that are to be hoisted or pulled. The type shown in Figure 21-36 has points which are pulled into the load and which tighten automatically. It is sometimes necessary to tap the points in order to obtain the first grip. They can often be shaken loose after the load is placed, without hand work.

These are used for picking up or drag-

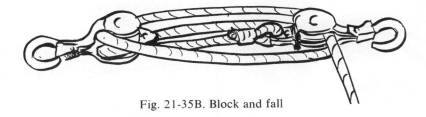

Fig. 21-35B. Block and fall

ging loose stumps, logs, and similar objects. They are seldom used for pulling "solid" stumps, as considerable weight would be required to obtain sufficient strength, and the points have a tendency to tear through wood under extreme stress.

They are of particular advantage in handling objects which are so placed or piled that it is difficult to get chains under them, or to remove chains after they are placed.

The Johnson grubbing tongs shown in Figure 21-37 grip with the inner surfaces of the arms. They are heavily built for rugged service in pulling stumps and trees up to ten or twelve inch diameters, and can also be used for hoisting. Under many conditions they are more readily attached and detached than chains, and are of particular advantage when the pulling is done by a shovel dozer or some other machine which can carry them to the point of attachment.

In heavy work, it is important to keep the line of pull straight through the tongs, as a side pull imposes excessive strain.

DOZER EQUIPMENT

The bulldozer is an important land clearing tool in itself. But its performance in this field can be extended and improved by replacing the standard blade with special blades and devices, the more important of which are described below.

Perhaps the most necessary special construction for clearing is protection devices, for both operator and machine. A strong cab, OSHA approved, should include an extra strong roof and verticals, and a rear protection of heavy screening. The radiator must have a strong, small-mesh guard. The engine must have side guards. Fine mesh cab screening for operator protection against insects may be needed.

The operator must be constantly alert for falling trees and branches, for high stumps, rocks, and pits concealed by vegetation, for poles thrusting toward him, and for buildup of dangerous tensions in pushed material.

Stumper. The stumper in Figure 21-38B

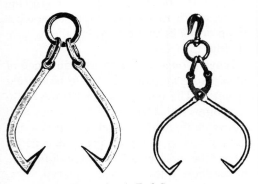

Courtesy of Warren Axe & Tool Company

Fig. 21-36. Grapple hook and skidding tongs

Courtesy of Re-Bo Manufacturing Company

Fig. 21-37. Grubbing tongs

Courtesy of Rome Industries

Fig. 21-38A. Cab guard

Courtesy of Fleco Corporation

Fig. 21-38B. Stumper

is designed to fit over and be pin-fastened to the center swivel of an angling dozer C frame, or to a special frame operated by a bulldozer hoist.

These tools are usually two and a half feet wide, and have a drop below the C frame of two or more feet. Construction is very massive. A serrated edge assists pene-

Courtesy of Fleco Corporation

Fig. 21-38C. Tree dozer

Fig. 21-38D. V-tree cutter

tration in earth or wood, and keeps it from skidding to the side.

The stumper is used for pushing over trees and stumps, driving under stumps to boost them out, and digging around them when necessary. It is also effective in dig-

ging out boulders, knocking dirt off loose stumps, digging up railroad ties, ripping up shale and old paving, and making shallow ditches.

It concentrates the full power of the tractor on a narrow front. It makes possible cutting and lifting underground roots with minimum soil disturbance, and without wasting power in unnecessary digging of a wide strip of soil. It is not subject to the twisting strains which shorten the life of full blades used for stumping.

It is not good at piling or transporting loose stumps, nor at backfilling holes or clearing brush. It should be teamed with a dozer or used alternately with a rake or blade on the same tractor.

Tree Pusher (Knockdown Beam). A tree pusher is usually a heavy tractor equipped with a dozer, stumping, or angle or V-blade, and a higher push frame with longer reach. The push frame is preferably under separate control, but may be linked to the blade so that it is raised and lowered by its controls. Sometimes a pusher is carried without a blade.

Trees are pushed by the upper frame. This may uproot them, or just put them un-

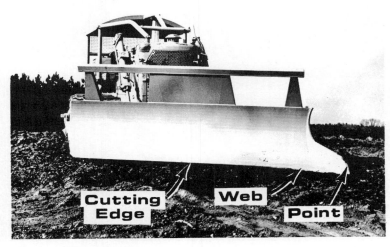

Fig. 21-38E. Angle-blade tree cutter

der heavy tension, so that the blade can drive under them readily to tear them out of the ground.

A V or angled blade can cast the tree to the side. Disposal is usually handled by other equipment.

V-Tree Cutter. A V-cutter for trees, Figure 21-38D, consists of a V-shaped, dozer-mounted blade fitted along the bottom with horizontal, scalloped cutting edges, and a center mounted splitting point or stinger. Floats keep the edges from digging in.

The V-cutter is designed to shear off all vegetation at ground level, whether it is large or small, and to cast the debris to each side, where it is left for other equipment.

Trees that are too large to be sliced from the side are rammed and split by the stinger. The machine might then be able to slice the halves without stopping, or might have to back up to make additional passes.

Operation is most efficient on even surfaced ground where the edges maintain good contact, where it is firm enough to hold the roots in position while the trunk is cut, and where there are no rocks hard or large enough to damage the cutting edges.

Under favorable conditions, cleared areas will be left entirely firm, and free of projecting stumps. The cut trees and brush may be pushed off by a dozer, preferably equipped with a rake blade.

Angle-Blade Cutter. This somewhat similar device has the stinger mounted on the forward end of a long blade set at a 30 degree angle. The full length cutting edge is straight.

This arrangement permits dropping and pushing all trunks to one side and doing after-cutting windrow piling. In very heavy work, there may be problems with off center loads.

The blade can be tilted downward to use the stinger as a stumper, or the whole blade in cutting shallow ditches.

Rake Blade. A rake blade may replace either a dozer blade or a front loader bucket, or be fastened by brackets and pins to a standard blade. It is made up of a set of tines that may be more or less vertical at top and center, and curve forward at the bottom. An upward extension is called a brush guard. There may be a solid center section to protect the radiator.

The tines may be operated below ground

Courtesy of Fleco Corporation

Fig. 21-38F. Rake blade

Courtesy of Fleco Corporation

Fig. 21-38G. Tree shear

level to bring up roots and boulders, without moving solid masses of soil. However, separation does not usually occur automatically, except perhaps in dry sand.

Wet, sticky, or lumpy dirt may build across the teeth, blocking the slots between them, either by itself or after becoming matted with brush or roots. Under such conditions, it takes patience, skill, and experience to take growth (and rocks) and leave the soil behind.

Rakes vary greatly in weight, strength, and tooth spacing, to suit various types of work. The heaviest ones, designed for grub-

bing out stumps, heavy roots, and boulders, must have tines so strong that any one of them can take the full push of the tractor without bending. Tooth spacing is usually wide in this type.

The lightest ones are intended for raking up loosened material on the surface, and for underground removal of light or weak root system. Tines are lighter and more closely spaced. They may be badly damaged by heavy work.

When grubbing under difficult conditions, the blade may be forced to its full working depth at the start of a push, then raised gradually as it picks up a load. Shaking the blade with the hoist encourages flow of dirt between the tines.

Brush piles that are dirty can often be cleaned by putting the tines under the edge, then hoisting and moving forward to roll the whole pile.

Heavy soils, such as clay and silt, are much more troublesome than sand, gravel, or loam. Almost any soil is easier to separate when dry than when wet.

Tree Shear. Tractor-mounted, hydraulic tree shears are presently able to cut hardwood trees up to 20-inch diameter, and softwood over 30 inches. They snip or shear the tree a few inches above the ground. Capacity is reduced if the wood is frozen.

The unit in Figure 21-38G is shown in open and closed positions. It is supported by a dozer (or loader) frame and lift. The left side is a massive fixed jaw or anvil, the right side is a hinged knife (cutter blade) that is moved by a powerful hydraulic cylinder.

The tractor is maneuvered until the tree trunk is between the knife and the anvil, which usually rests on the ground. The cutter is then forced through the wood. Unless badly out of balance, the tree falls across the anvil, away from the knife.

The shear may crush the bottom few inches of the trunk. A slice would have to

Fig. 21-38H. Accumulator to hold cut trees

be trimmed off saw timber, but not from a log to be pulped. No loss of timber is involved, as hand cutting is usually done much higher, and would leave more than that section on the stump.

Such a shear may be teamed with a handling device mounted on the same tractor. The accumulator in Figure 21-38H is a double clamp that can hold one large tree, or a number of small ones, both while being cut, and while moving on to the next tree. When the arms are full, the tree or the stack is laid (piled down) on the ground, for dragging (skidding) away, or to be picked up by a chipper.

In another construction, the shear is off-set to the side of the tractor, and a platform and grapple immediately in front of it clamp and carry the trees horizontally after they fall. This combination may be called a feller-buncher.

Root Plow. A root plow, or root cutter, is usually a horizontal knife, straight or V-shaped, supported by vertical standards or shanks at each side, and carried in a rear mounted frame hinged to the track frames of a big crawler tractor.

The model illustrated in Figure 21-39A has a cable lift, but hydraulic operation is optional. Working depth is 8 to 20 inches, depending on growth size, soil condition, and tractor power.

Angled fins slide roots up to the surface, but cleave through soil with little disturbance.

Most plants that are able to regenerate and sprout after being cut, depend on their ability to grow new shoots from a ring or crown of dormant buds near ground level. Undercutting with a root plow eliminates or tremendously reduces this source of regrowth.

Some sprouting may still occur from misses, and from the plants that can regenerate from fragments of roots or stems. But such sprouting is quite weak, compared to that produced by intact stumps or root systems. It is combatted first by planting grass

Courtesy of Fleco Corporation

Fig. 21-39A. Root plow

with a mechanical seeder mounted on the plow (grass competition kills many young plants of other species), then by mowing trouble areas or running the root plow through again.

A fringe benefit from root plowing is almost complete loosening of soil, which increases water absorption. This good result may be accomplished with very little, if any, increase in erosion, which is retarded by innumerable fragments of vegetation left to rot in the disturbed soil.

Root Rake. Roots and stumps cut loose by the plow may be brought to the surface and piled by either a rake blade or a root rake. They are most effective in sandy soil and with coarse pieces. Heavy, wet soil, and

Courtesy of Rome Industries

Fig. 21-39B. Root rake

fine matted roots, may cause almost constant clogging.

Towed Chain. Trees and stiff brush may be uprooted by a chain towed between two dozers of 180 horsepower or more. Chains must be very heavy, with bar diameter in the links 2 to 3 inches, and weight up to almost 90 pounds per foot. They should be about three times as long as the spacing between tractors, and 2½ times the height of the tallest tree.

One or more steel balls, from 3 to 10 feet in diameter, which may be hollow, filled, or solid may be fastened in the chain by universal connectors. They should be centered or equally distributed in the center half of the chain. Balls may serve to hold the chain up to slide over stumps and ridges, and to add momentum to overcome sudden resistance.

The chain should include several links, preferably one at each end and two in between, to prevent development of damaging twists. A supply of quick repair links should be available.

A third tractor with a pusher may be needed to follow the chain, and assist with stubborn trees, or lifting over obstacles.

Chaining is seldom used for construction,

Fig. 21-39C. Chopping brush with a disc harrow

being suited best to very large areas. Special problems are handling the chain between jobs, and danger of the towing tractors falling into pits or colliding with obstacles. Advance scouting on foot may be impractical in very thick growth.

It is often necessary to cover the area twice, in opposite directions, to complete the uprooting. But there is danger of loosened trees moving with the chain, and creating a massive, tangled pile.

CHOPPERS, SHREDDERS, AND CHIPPERS

The actual bulk of vegetation in any area is far less than it appears to be when it is standing, or even after it has been cut, uprooted, or piled. The difference is usually greater with brush and saplings than with trees with substantial trunks.

As a result, sufficient clearing for many purposes can often be accomplished by chopping vegetation into small pieces, without removing it. Such pieces may be partly or wholly buried by the chopping process, or left scattered on the surface.

When chopping involves partial burial, as by a disc harrow, the debris cannot be removed. Shredding and surface scattering by a rotary mower permits raking up for burning or removal, with increased difficulty in getting all the pieces. Chips are delivered through a spout, may be piled or scattered, but can conveniently be put directly in trucks for removal and possible use elsewhere.

Disc Harrow. Weeds, brush and small saplings may be knocked down, chopped or mangled, and partially buried by a heavy disc harrow.

These units usually have six or more con-

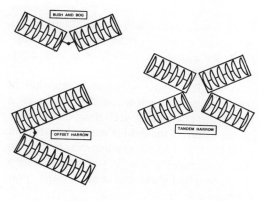

Fig. 21-39D. Disc arrangements

Fig. 21-39E. Rolling chopper

cave discs, two feet or more in diameter, with sharp, scalloped edges. They are mounted on an axle (or axles) whose angle to the direction of movement can be regulated. If it is parallel to the tractor's axle, the discs roll freely, if at an angle they cut down with a strong slicing effect, and mix and overturn the soil and vegetation.

At the best, the result will be well-chopped vegetation so well mixed with loosened earth that only a small part will be visible on the surface. The surface will usually have low parallel ridges, but might be almost smooth, and ready for planting with a cover crop, preferably some legume.

If the ground is too hard or rocky, or the vegetation is too coarse for the size harrow being used, there will be spots or areas where one or more of the functions—chopping, loosening, or burial—was not complete.

A single harrow sidecasts, somewhat in the manner of a plow, so that it leaves a furrow on one side of each pass, and a ridge on the other. If the field is worked in straight lines back and forth, the line where the two directions meet will be left as a trough or a ridge. Skipped spots will create rough spots for the same reason.

In a rough field these irregularities probably will not matter. In crop land, they can be corrected at the next plowing. But they should be kept in mind when planning the work.

A bog harrow has two axles, which angle to both cast outward, creating a center trough and side ridges. These are smoothed by overlapping passes by a half width.

In tandem construction, with two axles, one behind the other, the discs in the second set throw the opposite way from the first, thus eliminating troughing of the dirt except on turns.

Rolling Chopper. A rolling chopper is usually a big drum fitted with cutting blades, set in a tow frame with a draw tongue. It is towed behind a tractor, usually a crawler, of 60 horsepower or more. Weight can be adjusted by putting water in the drum, or draining it.

A fully ballasted drum may put 1½ to 2 tons weight on each foot of blade, providing a tremendous crushing and cutting force. Widths are available up to 16 feet. Smaller units may be towed in tandem, with different blade spacing on front and rear drums, or in triangular groups of three.

Choppers are usually pulled through standing trees and brush. The tractor should have a dozer blade or, for heavy growth, a tree pusher.

These units are generally used on jobs

where the chopped vegetation can be left on and in the ground to decay. If the purpose is to produce range land for grazing, grass seed may be broadcast directly behind the towing tractor, so that the chopper will mix in and cover some of it.

A variable amount of material is mixed into the soil and buried, but this is usually not considered an important part of the result.

If the downed vegetation is to be piled, this may be best done by rake blades working at right angles to the direction of chopper travel.

The rolling chopper is most effective on ground that is firm enough to support sticks under the cutter, but soft enough to allow the cutter itself to sink in. Quality of work increases with speed, so it is advantageous to tow with a tractor powerful enough to use a high gear. Small stones do not bother it, but large ones interfere with the work and damage the blades.

The size growth that can be handled is limited by the ability of the tractor to knock it down, and by the resistance of the wood. Results vary with roller weight, tree type, and ground conditions. A few of the larger trunks on a job may be left almost intact.

Rotary Mixer. The rotary mixer is a machine that is used chiefly in mixing and stabilizing road bases and surfaces, but is also an excellent tool for clearing and mulching brushland.

These mixers are made by several manufacturers, in a variety of sizes. They may be rear-mounted tractor accessories, or complete self-propelled machines.

A rotor assembly, which consists of a shaft, tine-holding plates, and tines is mounted across the direction of travel under a mixing chamber which controls the movement of materials so that two things are accomplished: first, a high percentage of the materials is deflected ahead of the rotors so that constant re-mixing cycles are established; second, the material is at the

final stage controlled in such a way that coarse and fines are mixed, blended, and placed so that aggregate segregation is entirely corrected. All sizes of the material it is working, from dust to the largest gravel, are uniformly distributed, the aggregate is keyed and interlocked, and the voids filled with fines which securely mortar-in the stone.

A rolled plate at the trailing edge of the mixing chamber acts as a strike-off to provide a smooth surface and to partially compact the mix.

Rotor speeds are controlled by a multi-speed transmission and by the engine throttle. Depth of cut is regulated by raising or lowering the rotor relative to mixing chamber. Adjustable springs carry part of the

Courtesy of American-Marietta

Fig. 21-39F. Rotary mixer

Courtesy of American-Marietta

Fig. 21-39G. Rotor assembly

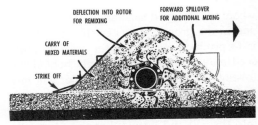

Courtesy of American-Marietta

Fig. 21-40. Rotary mixing action

rotor weight to permit it to ride over obstructions.

The tine-holding plates are driven by the shaft through individual friction clutches, designed to slip momentarily under shock loads. A variety of interchangeable tines can be obtained for different working conditions.

Smaller units of 3-foot working width may be purchased for mounting directly on most wheel tractors of 2-plow capacity or over, and still smaller sizes for garden tractors, but these are not adapted for industrial work. Principles of operation are similar but there are wide differences in ruggedness and in details of construction. All sizes are useful in landscape work.

They pulverize topsoil, mix it thoroughly with subsoil, sand, fertilizer, humus, or any other material desired, smooth it off, and leave it in ideal condition for planting. They will chop up sod and weeds and mix them with soil, increasing its bulk and making its appearance more attractive.

The large motorized and self-propelled mixers do a good job of clearing brush and palmetto, and will handle trees up to 2½″ to 3″ in diameter. Light brush is cut up and completely buried; larger branches and trunk sections will be partly or wholly buried if the rear section of the hood is left down, or scattered on the surface for removal if it is raised. It is unsafe to walk behind the machine when the plate is up as

Courtesy of Triumph Machinery Company

Fig. 21-41A. Front mounted shredder

objects are thrown out with great force.

Mobile Shredder. The Shred-King, Figure 21-41A, resembles the rotary mixers in having a full width horizontal rotor turning inside a protective case. This rotor is made up of the drive axle and 20 steel discs about 19 inches in diameter. The discs carry four shafts, 90 degrees apart, near their rims. A pair of heavy flail cutters is mounted between each pair of discs, on opposite shafts.

The cutters are stirrup-shaped, and can turn freely on the shafts. When the rotor is spinning, they are held out beyond the discs by centrifugal force. Top speed at the tips is about 113 miles an hour, so each cutter can deliver almost 2000 impacts per minute. Chips are thrown backward and downward.

If a cutter hits a rock, or any other material that it cannot cut, it swings back to a protected position between the discs, and then is swung out again centrifugally for another blow. Such material therefore does not break the cutter nor shock the drive train, but it can produce rapid cutter wear.

One model of this machine is designed to mount on a three point hitch, for power takeoff drive by a tractor of 80 horsepower or more. Another model, with its own diesel engine, can replace the bucket on a loader capable of handling its almost 3-ton weight. In any case, the tractor should have hydrostatic drive or creeper gears, for very slow crowding.

The frame is equipped with a pair of gathering arms to direct standing brush and saplings into the cutters. A cross bar between them bends the trunks slightly away from the machine as the cutters reach them. This serves to make them shred more readily because of tension, and to prevent the tops from falling over the tractor.

These machines are designed to shred brush and trees up to six inches in diameter, fallen trees and branches, scrap wood, and brush piles. In normal operation, all vegetation is cut smoothly to ground level and reduced to fine shreds, chips, and occasional

Courtesy of Triumph Machinery Company

Fig. 21-41B. Rear mounted shredder

sticks a foot or less in length. Fairly large stumps may be taken down to ground level by moving into them, or down onto them, very slowly.

The machine will work almost equally well going forward or backward. If the shredder is mounted at the front, or if the growth can be knocked down and driven over by the tractor, forward motion is most convenient. Heavier growth, and piles of branches and other debris, are usually done in reverse with a rear mounted unit.

Rotary Mower. Rotary mowers, which in small sizes are the homeowners' favorite for cutting the lawn, are also made in big, rugged models capable of mowing tall weeds, thick brush and small saplings.

Courtesy of Bush Hog Division of Allied Products

Fig. 21-41C. Rotary brush mower

Fig. 21-41D. Heavy mowing for a rotary

Light duty rotaries cut by means of a knife that rotates parallel to the ground. It is fastened to a vertical axle, which is often the extended end of the engine crankshaft.

If the knife collides with something it can't cut, it is likely to bend or break, and may bend the axle as well. Even if some slippage device is included in the axle fastening, it may not be adequate to save the blade.

A heavy duty rotary such as the Bush Hog, Figure 21-41B, has a flat disc blade holder or flywheel fastened to the mower axle. Two heavy blades or flails are fastened to it by hinge pins.

When the blade holder is turning at high speed, up to 776 revolutions a minute, centrifugal force causes the blades to extend straight out, in cutting position, even against considerable cutting resistance.

But if a blade strikes something it cannot cut, it simply folds back on its hinge. It extends again immediately, and yields again at the next collision.

This arrangement entirely prevents shock damage to the shaft or drive, and limits blade damage to dulling or chipping. It also allows continued effort to cut or grind through the obstruction.

Most of these mowers are designed for mounting on the rear of medium size wheel tractors with a three point rear hitch. This allows positive (but not rigid) in-line fastening, and hydraulic lift. Drive is from the tractor's power takeoff, through a shaft and universal joints. The shaft has a protective shield sleeve to reduce winding up of brush and vines.

In ordinary operation, the mower rests on the ground, supported at the rear by a swivel wheel that may be adjustable for height. In front it may rest on narrow skid shoes, and/or be supported by the hydraulic lift. Lower gears are used, the choice depending on the power available, and the heaviness of cut. The engine is run at ¾ to full throttle.

Quality of cutting is somewhat better going forward than backward, but the difference is usually not important. If the tractor has a reversing transmission, or a suitable reverse gear, mowing may be done neatly in straight adjoining lines, backward and for-

Courtesy of Mitts & Merrill, Incorporated

Fig. 21-41E. Brush chopper

ward. Otherwise, the area may be cut by working inward from all sides, with some re-mowing of skips in the corners.

If cutting becomes very heavy, the engine will lug down and efficiency will drop sharply. It may be restored by raising the mower, if close cutting is not required. Or the tractor may be put into a lower gear.

In general, a heavy duty rotary will chop up anything that its towing tractor can easily go over. For a 40 horsepower tractor and a six-foot mower, that would mean almost any thickness of brush, densely growing saplings up to 1½ inches diameter, and occasional trunks up to 3 inches.

Heavy stands often cannot be done in reverse, as they block the mower box. But for small pieces of cleanup at edges, or to chop piles of brush, the mower may be raised, backed into or over the material, and then lowered with the tractor standing still and the blades turning.

Most vegetation cut by this mower is left in rather short pieces, with considerable shredding, and it can usually be left on the ground to rot. However, a certain number of long pieces escape by lying flat on the ground. Sapling stumps may not be cut off flush, and might require repeat cuttings. Work is usually not as thorough and neat as the shredder's.

Vines may wind around the drive shaft, and have to be cut off with hand clippers or a heavy knife. This trouble is greatly reduced by making first passes in reverse.

Rocks damage the blades slightly, and repeated contacts will wear them away, or make them impossibly dull. The rocks themselves are often smashed.

The machine is dangerous, as it may throw rocks and other hard pieces with great force for more than 50 feet, chiefly to the rear. It should always be stopped if someone approaches it.

Rotary brush choppers are versatile and economical clearing tools, for work within their capacity. And they make excellent junior partners for bigger machinery, for controlling areas that have re-sprouted after clearing.

Courtesy of Morbark Industries

Fig. 21-41F. Self-feeding tree chipper

Brush Chipper. The brush chipper or grinder is the standard tool for processing hand-cut brush, saplings, and branches, wherever burning is impractical.

This type of vegetation is reduced to small chips by an engine-driven toothed cylinder or disc turning at high speed. The unit is usually mounted in a light trailer, and towed by a truck into which it can discharge the chips. It may be moved by a tractor on rough ground, or where complicated maneuvering is required.

The chips have only a tiny fraction of the bulk of the brush that produces them. Where brush must be hauled away, 10 to 20 truckloads may be reduced to one load of chips.

On rough ground, in or bordering on fields and woods, the chips can be scattered to save hauling away. Small quantities are inconspicuous, and their decay adds compost to the soil. But piles or thick layers produce barren areas that may persist for years.

Cut material is fed into the machine by hand, in single stems or in bundles. The roller or its feed mechanism pulls them in. Great care must be taken not to allow hands to follow the brush. More time is usually needed than would be required to throw the material on a fire.

The chipper can usually be kept close to the cutting or gathering, so less carrying time is needed than with fires. Problems of building and maintaining fires, preventing spreading, and putting them out when leaving the area, are all avoided.

Disadvantages include a high noise level, cost of purchase and maintenance, and consumption of fuel.

It is most efficient when kept close to the

Courtesy of Vermeer Manufacturing Company

Fig. 21-41G. Stump chipper

work, so that pieces can be ground as they are cut. If this is not practical, brush should be piled with all the butt ends facing the side where the chipper is to stand.

Tree Chipper. There are also chippers that can process full size trees, up to 20-inch diameter or more. The Model 75 in Figure 21-41F is mounted on a semitrailer with a fifth wheel for connection to a highway type tractor.

Its operator can pick up trees (or bunches of small trees) with a grapple, and place them on a chain conveyor and into the grip of vertical and horizontal compression rollers. These force it against a rotating 75-inch, 3-knife chipping disc.

The chips are screened to separate fines (mostly bark dust) and oversize pieces. A conveyor and chute take them from there, and can be adjusted to load them into trucks, pile them, or scatter them.

The compression rolls are powerful enough to pull most trees into the cutter complete, without need to trim any branches.

The chipper accepts only cut trees, or uprooted ones with the stump cut off and discarded. Stumps are almost sure to carry dirt to dull the knives, and are likely to include rocks to chip or break them.

In many areas, it is possible to sell chips to a paper mill, or to other wood product factories. Value depends on quality in relation to the processors' requirements, possible profit may depend largely on haul distance. But it often pays to sell chips for less than cost, when other disposal methods would be more expensive. And utilization is greatly preferable to wasting.

If there is no market for chips, they may be sprayed on slopes for erosion control, usually in conjunction with tree planting, or spread thinly over the ground for decomposition into humus. But thick layers and piles should be avoided except in wasteland, as they may make the covered areas barren for years.

Stump Chipper. It is often difficult or

Courtesy of Vermeer Manufacturing Company

Fig. 21-41H. Eliminating a stump

impossible to get to a stump with equipment big enough to pull or dig it out. Even more often, the tearing up of the area and creation of a big hole are unacceptable. Hand digging is less destructive, but it is laborious and prohibitively expensive, and leaves a hole to be filled.

A stump chipper of appropriate size can chew almost any stump to chips, with little ground disturbance outside of its cutting arc. The work is done by a cutting wheel equipped with teeth, preferably carbide. Working depth is 6 to 24 inches in various models.

Figure 21-41G shows one of the larger

Courtesy of Hyster Company

Fig. 21-41I. Logging arches

and more sophisticated machines. It has a 36-inch wheel carrying 48 teeth, some of which are on the rim, but most on both sides. This wheel can be lifted and lowered, and moved back and forth sideward continuously at a controllable rate during digging.

The unit is a trailer, attached to a light towing vehicle by a drawbar that can be swung from side to side by a hydraulic cylinder for accurate machine placement. Another cylinder extends and retracts the bar to move the unit and its wheel backward and forward during work.

The hole made by chipping a stump is much smaller than one made by digging or pulling, as its lower parts are left in place. Chips may be used for temporary backfill, but they are loose, barren, and shrink with decay. Soil, or a mixture of soil and chips, should be put in for a longer lasting repair.

It is economical to cut the stump off as close to the ground as is practical, before chipping it.

Stump chippers are used in land clearing, where stumps are to be removed after tree cutting, and in parks and built up areas. Teeth are likely to be extensively damaged by contact with buried rock, so the wheel should be operated at slow speed and with

great caution if its presence is suspected. Visibile rock should be removed if possible.

Log Movers. There is a great variety of specialized equipment designed to move tree trunks and shorter logs from the place they are cut to a nearby processing or transporting facility. These units are only occasionally of interest in construction projects, and will be very briefly considered.

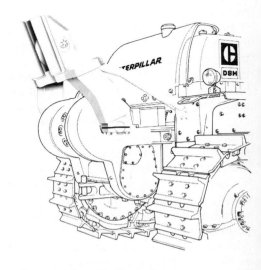

Courtesy of Caterpillar Tractor Company

Fig. 21-41J. Integral arch

A standard tool for through-the-woods hauling of long, heavy logs, singly or in bunches, is the logging arch or sulky, Figure 21-41I. This is a one axle trailer, on tires or tracks, with a straight, inclined beam from a drawbar connection with a tractor up to sheaves and a fairlead. The tractor (skidder), usually a crawler, is equipped with a logging winch.

The winch line ends in a massive hook. It is threaded through the fairlead. The log or logs are fastened individually at their butts (thick ends) by choker cables to the hook. The winch line is wound in until the near ends of the logs are at least clear of the ground, and as much higher as the tractor operator's ideas of balance and convenience require. The winch is then locked, and the logs are dragged with only their thinner and lighter ends resting on the ground, and the butts safe from digging in.

Figure 21-41J shows an integral arch, which is a fairlead mounted on top of a logging winch. The line is led from the front of the winch up through the fairlead and back to the log chokers. This device serves the important arch function of lifting the butt or near ends of logs off the ground. However, it imposes their possibly substantial weight on the rear of the tractor, which on long hauls or rough ground might be a disadvantage. On the other hand, it does not interfere with tractor maneuverability, and does not need to be hitched or unhitched.

A rear mounted towing grapple has a pair of pincer arms that are closed hydraulically on the butt ends of a log or pile of logs, then lifted to tow them. A universal mounting allows the grapple and its load to swing and tilt, to accommodate turns and rough ground. The hoist may be either a direct or a parallelogram type, similar to those in rippers.

A grapple does not require the fastening of individual choker lines to the logs, but it may require that they be pre-piled in position to be picked up.

Courtesy of Rome Industries

Fig. 21-41K. Towing grapple

The vertical and horizontal accumulators used for carrying sheared trees were mentioned earlier.

There are several clamp designs that replace loader buckets. They carry one or more logs, up to the limits of clamp size and loader stability, at right angles to travel direction. These need plenty of clear space, and there may be problems of catching ends against the ground, and of side tipping on slopes and rough ground.

CHAIN SAWS

A chain saw is a hand held power tool designed to cut and trim trees. Generally, it has a high speed, two cycle, rope-start gasoline engine.

There is an automatic centrifugal clutch. This turns a sprocket, usually by direct connection, but sometimes through a speed reducing gearbox.

The sprocket meshes with a toothed cutting chain that slides in a groove around a a flat bar. The bar is sometimes fitted with a very thin sprocket or wheel in its outer end.

The only operating control is the throttle. Cutting is done at top engine speed. The clutch disconnects the chain when the engine drops toward idling speed.

Power Head. A chain saw, minus its bar and chain, may be called a power head, saw body, or just body.

There is a case of sheet metal that en-

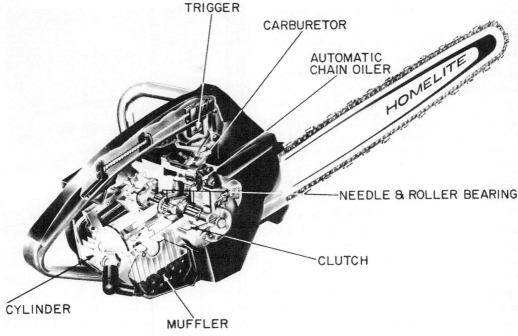

TRIGGER

CARBURETOR

AUTOMATIC
CHAIN OILER

NEEDLE & ROLLER BEARING

CLUTCH

CYLINDER

MUFFLER

Courtesy of Homelite, a Textron Division

Fig. 21-41L. Chain saw, cutaway

closes the engine and the reservoirs. In the Homelite XL2 in Figures 21-41L and 21-41M the engine is horizontal, with the cylinder at the rear, and the crankshaft across the center. Reservoirs for fuel and chain oil are at the front left.

The case also encloses the carburetor and ignition system, and carries the re-wind rope starter unit on the left side, opposite the sprocket.

There is a curved handle, running from front to back of the body. This carries the throttle trigger or triggers, and a throttle latch or hold-down, and is often called the throttle handle.

The handle bar is at right angles to the throttle handle, at the front of the saw.

The throttle handle should be held in the right hand, and is the principal holding member, while the handle bar is used to guide, tilt, and steady the saw. Both must be held at all times during operation.

The engine is a two cycle design turning

at high speed, 8000 rpm in this model. It requires a mixture of regular gasoline with oil for fuel. The type and proportion of oil will be discussed under lubrication.

To the left side of the bar, looking toward its tip or nose, the body protrudes in a slightly rounded structure called the bumper. In ordinary cutting this is rested against the near side of the log, and the saw is

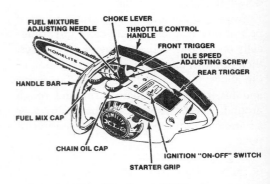

FUEL MIXTURE
ADJUSTING NEEDLE

CHOKE LEVER

THROTTLE CONTROL
HANDLE

FRONT TRIGGER

IDLE SPEED
ADJUSTING SCREW

REAR TRIGGER

HANDLE BAR

FUEL MIX CAP

CHAIN OIL CAP

IGNITION "ON-OFF" SWITCH

STARTER GRIP

Courtesy of Homelite, a Textron Division

Fig. 21-41M. Chain saw

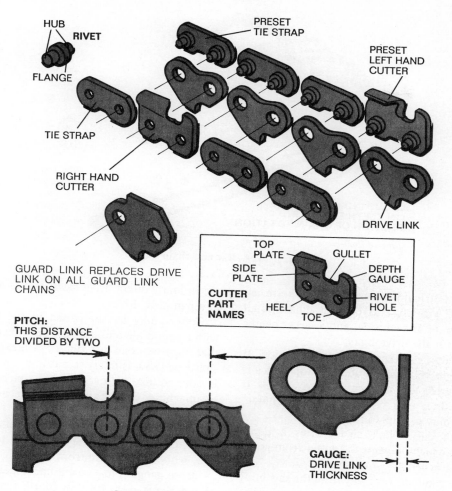

Courtesy of Omark Industries

Fig. 21-41N. Saw chain parts

pulled tightly against it by the pull of the cutting action.

The bumper is used as a fulcrum in pivoting the saw in a number of the cuts described in the next section. Its effectiveness is greatly increased by fastening a set of spike teeth to it. This can usually be obtained as standard or optional equipment.

Chain. A section and an exploded view of a typical chain are shown in Figure 21-41M. A chain is made in three interlocked narrow sections or strips. The center is a succession of identical drive links, which have bottom tangs for meshing with the drive sprocket, keeping alignment in the bar groove, and cleaning sawdust out of it.

Sides of the chain are called right and left, looking in the direction of motion.

The side strips are made up of cutters alternating with triple sets of tie straps. Right and left cutters are mirror images of each other, and are placed with the side plate out, and the top plate pointing across. The cutters are staggered in the assembled chain, so that right and left cutters alternate, but are separated by one tie strap. This arrangement may be interrupted where the chain ends are joined.

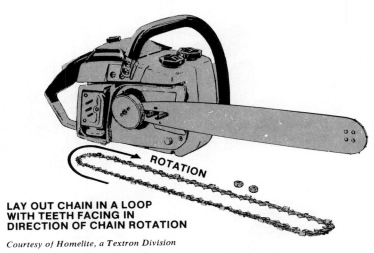

LAY OUT CHAIN IN A LOOP
WITH TEETH FACING IN
DIRECTION OF CHAIN ROTATION

Courtesy of Homelite, a Textron Division

Fig. 21-42A. Bar and chain

The left hand cutters and straps are assembled with preset rivets. These have small gauge hubs at each side, and a larger center flange for the drive links.

A cutter is L-shaped in cross section. If the angle between its side and top plates is 90 degrees or sharper, it is a chisel. If rounded, it is a chipper. Other manufacturers may use different terms.

Each cutter link also carries a depth gauge, which prevents the cutter from gouging in too deeply.

The chain moves away from the saw body along the top of the bar, and returns along the bottom. Most cutting is done at the bottom. Chain speed is about 3500 feet a minute in many direct drive saws, and is much lower in gear drives.

Cutters may be sharpened in the field, with a round file and a guide, but it is a job that must be done just right. An inexpertly sharpened saw is likely to cut in a curve. It is safer to take it to a shop, for sharpening by machine, until experienced instruction can be obtained.

One design of chain can be sharpened by a grinding wheel specially attached to the saw, without removing from the bar.

Depth gauges must be cut back occasionally, perhaps every third or fourth sharpening. They must be .020 to .030 below the top point of the cutters, depending on the chain type.

Saw Bar. The saw bar is a flat piece of steel with rounded ends, with a narrow groove to accommodate the chain drivers. Length may be ten inches to over 3½ feet, in various models. It is clamped into the saw body, in line with the sprocket, and is drilled or cut to allow for clamping and adjustment.

It is possible to obtain a bar having either a wheel or a sprocket in the outer end (nose), to reduce friction as the chain is pulled around it. These items must be extremely thin.

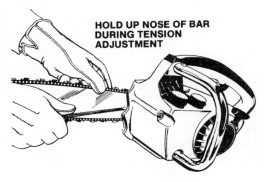

HOLD UP NOSE OF BAR
DURING TENSION
ADJUSTMENT

Courtesy of Homelite, a Textron Division

Fig. 21-42B. Chain adjustment

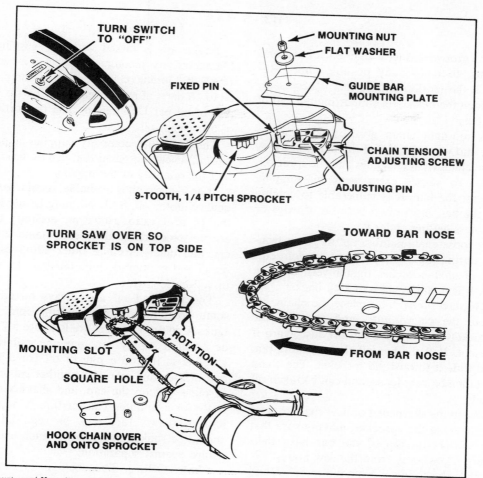

TURN SWITCH TO "OFF"

MOUNTING NUT

FLAT WASHER

FIXED PIN

GUIDE BAR MOUNTING PLATE

CHAIN TENSION ADJUSTING SCREW

ADJUSTING PIN

9-TOOTH, 1/4 PITCH SPROCKET

TURN SAW OVER SO SPROCKET IS ON TOP SIDE

TOWARD BAR NOSE

MOUNTING SLOT

ROTATION

SQUARE HOLE

FROM BAR NOSE

HOOK CHAIN OVER AND ONTO SPROCKET

Courtesy of Homelite, a Textron Division

Fig. 21-42C. Chain installation

Chain Adjustment. The chain must be loose enough to move freely around the bar, and tight enough so that the drive tangs cannot jump out (or be flipped out) of the groove. The chain changes in length, as it is longer when it is hot than when it is cold, and it becomes longer as it wears.

Tension is adjusted by moving the bar outward, away from the drive sprocket, to tighten; and inward to loosen. This bar is moved by means of a pin or dog on an adjustment screw in the saw body or drive case, which is parallel to the bar and is turned by a screwdriver, clockwise for tightening. The pin fits into a square or round hole in the bar.

This bar is held in alignment by the adjustment pin, and by a fixed pin that engages a long, straight slot. It is clamped in place by one or two bolts.

Instructions vary with different saws and chains. Lacking information to the contrary, adjust a cold chain so that it is nearly snug in the groove all around, and a warm chain so that the tangs at the bottom center hang about halfway out of the groove. Let a hot chain cool.

Adjustment is made by loosening the clamp bolts, supporting the tip of the bar so that it will not slump downward, and turning the screw. When proper tension has apparently been obtained, pull the chain out of the groove and let it snap back at several points, to relieve any possible kinks. Recheck the tension and correct if necessary. Tighten the bolts, still supporting the end of the bar.

A too-tight chain absorbs power, runs hot, and wears itself and the bar rapidly. A loose chain chews its tangs and the sprocket, and also causes extra wear by slapping against the bar. It is vulnerable to coming off the bar, most often because of twigs or splinters getting under it.

A loose chain is more apt to get in trouble during horizontal cuts, or random cutting of small branches, than in vertical down cut. It is a major cause of kick-backs.

Replacement. To replace the chain, remove the clamp nuts, and take off the sprocket housing, and the sawdust guard if present. Pull the bar off the adjusting pin, and slide it toward the sprocket. The chain will then be very loose, and can be taken off easily.

Wrap the sharpened or new chain around the bar and the sprocket, making sure that the cutters on top of the bar have their sharp edges away from the saw body.

Slide the bar away from the sprocket, and fit it onto the adjusting pin. It is often necessary to move the pin by turning the screw, before it will engage. Put back the guard and the housing, being careful that the bar does not come off the adjuster while you do it. Turn down the clamp nuts with your fingers only, or very lightly with a wrench.

Adjust chain tension as described above, being sure to snap it to relieve kinks. Tighten the clamp nuts firmly.

A new chain should have extra lubrication, usually from an oil can, before it is run on the saw. The first half hour of work should be light cuts, while the chain and bar are wearing in. Tension is adjusted as required.

Engine Lubrication. The standard chain saw requires regular lubrication only for the engine and the chain.

Engine oil is mixed with the gasoline. Manufacturers recommend special 2-cycle (SAE 30) motor oil in the ratio of one part oil to 16 parts of regular gasoline. This is ½ pint of oil per U.S. gallon, or a proportion of 6 per cent oil.

The Homelite company supplies a premium 2-cycle motor oil that can be used in half the quantity of the regular.

If no 2-cycle oil is available, regular good quality motor oil, SAE 30, may be used in the 16 to 1 ratio. However, ordinary oils are not recommended by saw manufacturers, and detergents and other additives in diesel and multigrade lube oils tend to foul up a 2-cycle engine.

The fuel and oil are always mixed in some container, before pouring into the tank. A common procedure is to use a 2½- or 2-gallon supply can, put one gallon of gasoline in it, add a pint of oil (or a half pint of premium oil), add another gallon of gasoline, put on the cap, and shake vigorously for a minute. The mixture is poured into the saw's fuel tank as required. It should be shaken again before each use, to assure against separation.

Chain Oiling. The chain is lubricated from a reservoir, which should be filled each time the machine is refueled. There are special bar and chain oils which are formulated to flow even in cold weather, and to cling to the chain.

If these are not available, any clean motor oil of SAE 30 weight may be used. In cold weather, weight 20 or 10 may be better, or 30 cut with one part of kerosene or fuel oil to 4 of oil. Multigrade oil may be used.

In most saws the oil is transferred to the chain by a tiny pump. This may be engine driven if lubrication is automatic, or be operated by a plunger that is pressed with the right thumb at intervals, if manual.

There should be sufficient oil fed to the chain to keep it damp around the connecting links, and to allow a very fine spray to be thrown off the front at full speed. Such throwoff may be much less when special chain oil is used.

More oil is needed in heavy cutting than in light work. The hand pump allows the operator to increase the amount, but it also makes it possible for him to under-lubricate, or even not-lubricate.

A few machines have both automatic chain oiling and a hand pump. With this arrangement, the operator has the option of pumping more oil when he thinks it desirable, without having full responsibility.

With manual oiling, a stroke may be given at intervals that vary between 10 seconds and one minute of actual cutting time. If a little oil is left each time the fuel tank goes dry, the chain is probably being properly lubricated.

Running without oil, or with too little, will wear out both the chain and the bar groove, and may cause the chain to break.

Indications of a dry chain are a dry appearance, too-quick stopping when the throttle is closed, and sometimes thin smoke, tightening, and engine stalling.

The bearing in the clutch hub usually has enough lubricant to last between overhauls. A nose sprocket or wheel should be lubricated with a special miniature grease gun every time the fuel tank is filled, or at least once a day. This should be done when it is warm, so that dirt will come out with the old grease. Lubrication at the end of the day fills the chamber and prevents condensation of moisture in it.

Controls. The saw is usually equipped with a throttle trigger, a latch (for starting) that holds the throttle partly open, perhaps a chain oil pump handle, an ignition switch, and a choke. Some models have two throttle triggers for convenience in changing hand holds on the throttle bar.

There may be a lock to prevent accidental opening of the throttle. If located on the top of the throttle bar, it is kept disengaged by pressure of the palm of the hand while gripping the bar.

PULL STARTER STRAIGHT UP TO CRANK ENGINE

CHAIN MUST BE IN THE CLEAR

Courtesy of Homelite, a Textron Division

Fig. 21-42D. Starting

CHAIN SAW OPERATION

Chain saw work is also discussed in Chapter One.

Starting. Since tank capacity is very small, in the interest of light weight, it is customary to fill with oil-gas for the engine and oil for the chain before each use.

Even if reservoirs were left filled after the last saw work, they should be checked, in case of leakage.

Usual procedure is to flip on the ignition, latch the throttle in starting position, and pull out the choke. Details vary among different makes and models.

The starter is a re-wind pull rope. You hold the saw down firmly with one hand, and brace it with a foot or knee; grasp the starter handle in the other hand, pull a few inches until you feel the starter engage, then pull briskly to give the engine a fast spin. Do not pull all the way against the end stop, as this may damage the starter or the rope. Keep hold of handle, relaxing enough to let the starter springs pull the handle back down. Then pull again.

Repeat the pull several times, until the engine fires. It may simply kick without taking hold. This is usually a signal that it has

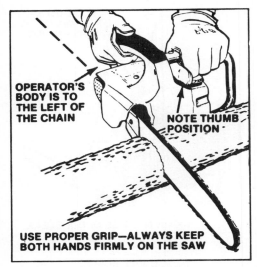

OPERATOR'S BODY IS TO THE LEFT OF THE CHAIN

NOTE THUMB POSITION

USE PROPER GRIP—ALWAYS KEEP BOTH HANDS FIRMLY ON THE SAW

Courtesy of Homelite, a Textron Division

Fig. 21-42E. Operating position

had enough choke. Push the choke button in (or half in). The engine should start on the next pull. If not, re-choke and pull some more.

It may or may not need half choke for a few seconds to keep running. Let it run on the latched throttle briefly, then release the latch by squeezing the throttle, so the engine can idle. It may or may not need more warmup time before working.

Cutting. Hold the saw with the left hand by the cross bar (handle bar), and with the right hand by the throttle or side bar. The throttle, and the chain oil pump if there is one, should be within reach of your right fingers.

To cut a log lying on the ground, the bar is lowered onto the cutting line, at right angles (or any desired angle) to the log, with the bumper just touching its near side. The throttle is opened just before the cutters reach the log.

Enough down pressure should be applied to cause a moderate to rapid cut, but not enough to make the engine slow down or the clutch slip. A substantial stream of coarse sawdust or chips should spray out of the cut.

Moderate down pressure can be exerted and satisfactory cuts made by pressing straight down. However, it is much more efficient and more usual to pivot the saw in the cut, holding it with the cross bar, and raising the rear of the throttle bar so that it pivots on the bumper, forcing the bar down by leverage.

The bottom of the cut may be kept horizontal, or shaped as desired, by making a moderate rocker cut, then drawing the saw back to slice off the ridge left on the near side, then pivoting it in again, as in Figure 21-42F. Cutting can then be resumed on the far side.

Varying the angle of approach shortens the line along which cutting is taking place, and makes it easier for the cutters to bite in. It is helpful under almost any conditions, and is necessary with big logs or dull chains. But a dull chain should be sharpened or changed.

Pressure required and cutting rate obtained vary widely with conditions. The most important factor is chain sharpness. A very dull chain might not cut through a log, but still slice a two-inch stick easily. A moderately dull chain may cut a stick rapidly but a log slowly or not at all. Fine sawdust shows the chain is too dull for its job, or, more rarely, that its depth gauges need filing.

Procedures in cutting a standing tree were discussed in Chapter 1.

Hard wood such as oak requires a sharper saw and cuts more slowly than a soft wood. Cutting with the grain (rip sawing) is slower and more difficult than across the grain, and requires more than normal lubrication. Low cuts in stumps are slowed by very dense wood and cross-graining.

Direct drive saws, which are the most common type, have very high speed chains and require (and will accept) very little pressure. The old gear drive saws usually needed heavy pressure for heavy cuts.

Ground Damage. If a log that is being

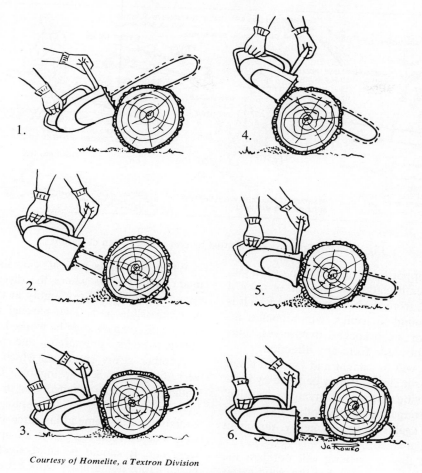

Courtesy of Homelite, a Textron Division

Fig. 21-42F. Working through a thick log

cut is on the ground, or close to it; it is difficult but important to prevent the chain from touching it. Even fine, soft dirt causes rapid wear, multiplied by the fact that it is drawn into the cut in the wood. A stone can dull and damage, and perhaps ruin, a chain in a fraction of a second.

If the cut must be completed in place it is done by using only the nose (boring), pointing it downward, using light pressure, and being alert to pull it back as it goes through.

If a number of cuts are being made in one trunk, it may be necessary to complete only one of them. The cut-off log can then be rolled over, with levers if necessary, to put the uncut sections on top of the rest of the cuts. They can then be cut easily and safely.

Stresses. Cut or fallen trees and their branches are usually under bending stress of some kind. An important exception is straight trunks lying on smooth ground. When a stressed piece of wood, a long log for example, is being cut, the two pieces will tend to move at the point of weakness being developed. This movement may tend to close the cut, to widen it, or to twist the two pieces around it.

If the log is supported at the ends, it will tend to sag in the middle and close a cut made in the top, and open one in the bot-

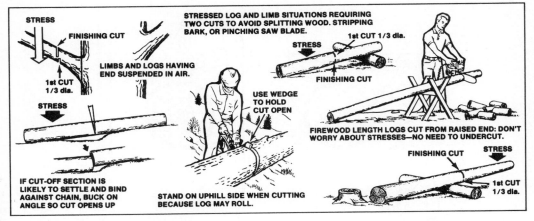

Courtesy of Homelite, a Textron Division

Fig. 21-42G. Stress in wood must be considered when cutting it

tom. If supports are on one side of the cut, it will open if made from the top or close if it is made from the bottom. Twist, which is usually found in trunks with branches attached, or in branches themselves, may close a cut made from any direction.

A saw will be caught and jammed if left in a cut as it closes. It may be freed by jacking or prying up the log, driving in a wedge, or by cutting it out with another tool. If the pressure against it is very heavy, the chain may be damaged.

A cut that is opening will not jam the saw, and the tension makes the wood easier to cut. But the weight of the free section may split the other part, and its direction of fall might be hazardous.

When both top and bottom are accessible, it is good practice to cut the log ¼ to ⅓ through on the side that is expected to close, then complete the cut from the other side. This avoids working in wood that is under heavy stress, and prevents splitting.

Courtesy of Homelite, a Textron Division

Fig. 21-42H. Springpole

When twisting is present, cuts should be made first at spots where the direction of stress is simple, and preferably at the thinnest accessible part. Two parallel cuts, an inch or three apart, may be worked in alternately, to reduce pressure on one spot.

Another stress situation, which can be very dangerous, is a springpole. This is a sapling bent under a felled tree or wedged between obstacles. This sapling might straighten with tremendous force if the log were cut or removed, or if its upper end were cut. It should be snipped off at the base, before doing other work around it.

Undercutting. Undercutting means using the upper side of the bar for cutting upward. It is standard procedure where a top cut would bind, and there is space to get the bar underneath the log.

Its principal drawback is that the resistance of the wood to the saw tends to force it against the operator, who must brace himself to withstand a moderate force. Any relaxation lets the chain push itself and the saw out of the cut. This usually results in a kickback, which will be discussed below.

A secondary problem is that it is difficult to use a pivoting movement for pressure. With gravity against the saw too, it takes more effort (usually) to cut up than to cut down.

Courtesy of Homelite, a Textron Division

Fig. 21-42I. Boring with the nose of the bar

Boring. Boring is cutting with the nose of the bar. The principal problem is keeping the nose down and forward, where it can't climb and kick.

Boring is necessary where a cut, usually an undercut but not always, must be made in a place where there is no room for the saw outside the wood to be cut. So it is forced—bored—straight in. If an undercut, it is essential that there be enough uncut wood above the saw to keep it down, and enough forward pressure exerted to prevent it from backing out from under.

Kickback. In normal cutting position the resistance of wood to the cutters simply pulls the saw bumper more tightly against the log. In undercutting it is a straight horizontal force that can be resisted by straight pressure.

But when the force is at the tip or nose, or the upper side, leverage is completely against the operator. If the nose is not held by an obstruction, it will flip upward at nearly the speed of the chain, about 60 feet a second, with enough force to swing it through a considerable and possibly dan-

gerous arc. This may be a truly shocking experience when it is unexpected.

Figure 21-42J shows three conditions under which kickback often occurs. Compare the right hand picture with the left hand picture in the previous illustration. The angle of approach made the difference between a successful cut and a shock. In the other two examples, the nose extended further than expected, and touched an unnoticed obstacle.

There is a special chain construction (guard link) which is supposed to reduce kicking. But avoiding it is chiefly a matter of the operator being vigilant and using good judgment.

BRUSH BURNER

Under many circumstances, burning is still the preferred and often the only practical way to dispose of cut or uprooted vegetation.

Trained crews with adequate equipment can usually burn clearing debris under almost any conditions, wet or dry. But the time (including lost time) and labor in-

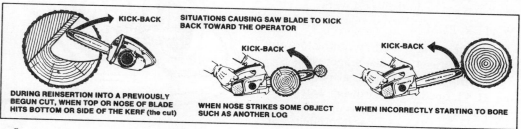

Courtesy of Homelite, a Textron Division

Fig. 21-42J. Kickback situations

Fig. 21-42K. Brush burner

volved in getting new fires started under difficulties, may be excessive.

Substantial help may be obtained from brush burners, such as those in Figure 21-42K. These trailer-mounted devices have a small (3 to 12 horsepower) engine, an airplane type propellor, and a fuel oil sprayer.

The air blast, delivered off the back of the trailer, may be enough to invigorate a fire, and to push it into the material to be burned, which may be brush, logs, or stumps. With help of a mist spray of fuel oil, injected into the air stream as needed, it also provides whatever additional heat is required.

However, the basic principles of properly starting and maintaining fires, as given in Chapter 1, should be understood if these machines are to do their best work.

WATER PUMPS

The pumps most used by contractors are of the centrifugal and diaphragm types.

Centrifugal. The centrifugal pump, Figure 21-43, operates by throwing water outward from a center by means of rapidly revolving vanes. The vacuum thus created at the center is constantly filled by water "sucked" (forced by atmospheric pres-

sure) into it through the inlet passage. The velocity of the water thrown off the tips of the vanes creates pressure which drives the water out of the pump through the outlet passage.

Figures 21-44 to 21-47 show the construction of Marlow centrifugals. The water enters the system through a strainer at the bottom of a hose fastened to the inlet. A foot valve may be installed above the

Fig. 21-43A. Centrifugal water pump

Fi.g 21-43B. Pump supported by safety raft

strainer to prevent the hose from draining when the pump is shut down.

The inlet hose is made of heavy plies of stiffened fabric and rubber which will not flatten under atmospheric pressure. It is usually made in ten to twenty foot sections.

Each section ends in a metal coupling. Four inch diameter and smaller lines usually have pipe thread couplings, while larger sizes have bolted flanges.

The upper end of the hose may be led directly into the pump body or through an

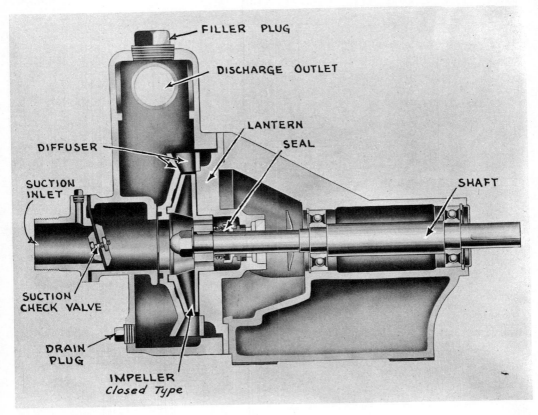

Courtesy of Marlow Pumps

Fig. 21-44. Cross section of centrifugal pump

elbow, usually with an angle of 30° to 45°, which can be set in different positions by unbolting and rotating.

The hose-to-pump connections each include a small threaded and plugged hole. A vacuum gauge should be put in the inlet. This can be loosened or removed to admit air in order to drain a hose not equipped with a good foot valve, or to allow the pump to run without pulling water. The outlet fitting will take a pressure gauge.

The connector is fastened to the pump with a gasket, and hose flanges are fastened to each other in the same manner. Any threaded connections are smeared with sealing compound before connecting. Even very small leaks on the inlet side will substantially reduce the amount of water handled by the pump, and may stop it from working at all.

The pump body is a rigid casing which serves as a support for the pumping mechanism and as a tank to supply priming water.

The inlet check valve serves to prevent water in the tank from flowing back down in the inlet hose when the pump is stopped, will stop the outlet hose from draining if its end is under water, and prevents siphoning of water out of a higher discharge point.

The diffuser or liner forms the inner shell of the pump. It provides an inlet passage by which water reaches the center of the impeller vanes, a plate which limits leakage of water or air past the back of the vanes, and peeler passages which take the water from the tips of the vanes and convert its

Fig. 21-45. Cutaway pump, diffuser type

velocity into pressure. It is the most widely used device to make a centrifugal pump self-priming.

The impeller consists of vanes curved back from the direction of rotation so as to minimize shock and turbulence where they

Fig. 21-47. Diffuser priming action

hit the water at the center, and so as to give maximum outward velocity at the tips. They add a substantial wedging action to the centrifugal force.

Three types of impeller are in general use—the open, semi-open, and closed. Effi-

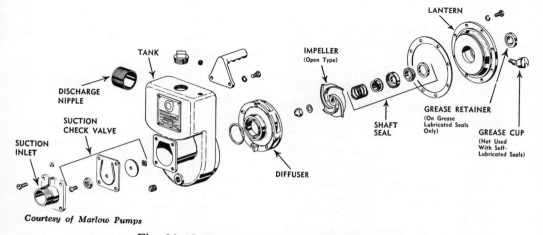

Fig. 21-46. Exploded view of centrifugal pump

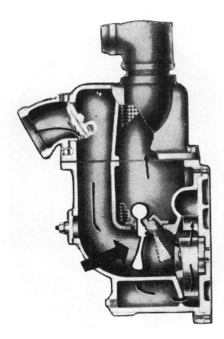

Courtesy of Marlow Pumps

Fig. 21-48. Recirculation priming valve

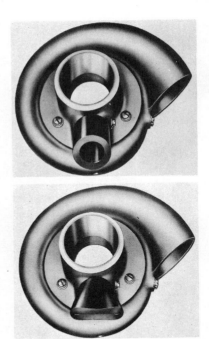

Courtesy of Homelite Corporation

Fig. 21-49. Rubber tube recirculation control

ciency, maximum pressure, and ability to handle air are ordinarily greatest with the closed types, while the open ones are best able to pass solids.

The impeller is backed by the lantern which closes the front of the casing, carries the drive shaft bearing a seal, and, with open impellers, prevents leakage past the front of the vanes.

In some makes, the diffuser and lantern are equipped with renewable wear plates facing the impeller.

The discharge or outlet tee is bolted to the top of the body and can be turned. The filler plug is removed in order to pour in water for priming.

The outlet hose is of lighter construction than the inlet and will flatten out when empty. Couplings are identical.

The drive shaft extends forward from the impeller which is splined or keyed to it, and through the lantern to a direct connection with the engine drive shaft. Portable pumps usually have combustion engines,

while fixed ones have electric power; but there are many exceptions to this generality. Electricity is more satisfactory where power is available because of ease of arranging for automatic operation and avoidance of the nuisance of fueling, lubricating, and adjusting the engine.

The drive shaft is usually on anti-friction bearings, although bushings may be used. The rear bearing may be lubricated by the water in the pump, by an outside grease fitting, or be sealed.

The water seal may be a standard packing box in which a number of packing rings are compressed by a collar nut, or it may consist of metal rings and springs. Its most important function is to keep air out while the pump is running, but it must also prevent loss of prime through leakage of water when it is stopped.

Priming. A self-priming centrifugal pump must be filled with water (primed) in order to function. The tank should contain enough water to cover the impeller. Water acts as a sealer of the clearance spaces next to the impeller, mixes with air to give it enough weight to be affected by centrifugal force, and serves as a check valve to prevent air on the discharge side of the pump from working back to the inlet. In most machines it provides lubrication for the shaft bushings and seals.

Contractors' centrifugals are of the so-called self-priming type. The pump will not prime itself—that is, draw water into itself—when dry, but once it is filled with water it will remain full and prime and reprime itself through any number of starts and stops. And as long as water is in the tank, and the pump is turning, it tends to move air and water from the inlet to the discharge.

If the inlet hose fills with air instead of water, the pump is said to have lost its prime.

There are two general methods of circulating the priming water so as to pump air. The diffuser method depends on gravity forcing the tank water between the outer or discharge tips of the vanes, without its being able to reach the center because of their rotation. The water at the outside and the air at the inside mix where they make contact, and air bubbles are thrown from the impeller tips into the diffuser passages, from which they rise to the surface of the tank and escape through the discharge pipe, while the water continues to press against the vanes. This process is illustrated in Figure 21-47.

As air is removed in this manner, a vacuum is created in the inlet pipe that causes water to rise in it until it enters the pump. The priming process is then complete. Water is forced through the pump hundreds of times faster than air, and vacuum increases sharply.

The recirculation method is to open a passage by which water can flow from the tank, which is on the outlet side, into the inlet passage. This mixes with air as it is caught by the impeller. On the discharge side the air bubbles to the surface and the water returns to the inlet.

The priming passage may have a shutoff with a manual control on the outside of the body; an automatic valve, 21-48, which is closed by the rush of water when priming is completed, or a soft rubber tube, 21-49, which is closed by the high vacuum created when water reaches the impeller.

Efficiency of priming at any one height is affected by factors of design too complex for consideration here, by the clearance between the impeller and the plates to the front and rear, by the clearance between the tips of the impeller vanes and the tips of the "peelers," by leakage of air into the inlet, and by any back pressure from the discharge line. Since the movement of air is quite slow, a very small leak may delay or prevent building up sufficient vacuum.

The height of the pump above water is an important consideration, as higher lifts require higher vacuums. A new pump may be able to pull water up 28 feet (although volume handled drops rapidly with increase in lift), but a badly worn unit may have difficulty raising it five. Another factor is that under high vacuum warm or hot water may evaporate so rapidly that the pump will be kept busy pumping water vapor.

Even a badly worn pump can draw water to a considerable height, once the air is exhausted. In emergencies, defective pumps may be primed by forcing water into the inlet hose from another pump, or siphoning water from a higher point through the discharge line, with the pump stopped. When the inlet is filled, the pump is started, and the direction of flow reversed.

In general, it is cheaper to overhaul a pump when it first shows signs of weakness, than to delay or lose jobs through its inefficiency.

Fire, well point, and other special centrifugal pumps may have an auxiliary vacuum pump which can be engaged by a

Courtesy of Homelite, a Textron Division

Fig. 21-50. Hand-carry pump

gear, which will pull air out of the main pump to prime it quickly.

Portables. In addition to the heavy wheel-mounted pumps, there are light hand-carry models (45 to 105 pounds) in 1½, 2, and 3 inch sizes, that are made of aluminum and magnesium alloys, and driven by high speed two cycle engines. One is shown in Figure 21-50. Electric drive is also available.

Their output is standard for their hose size, and their extreme portability makes them invaluable where access is difficult, and when jobs are small and scattered. They are handy where frequent changes of position must be made to follow a receding water level, and for keeping down inflow during work.

Several may be used in one excavation if there is too much water for one. They are ordinarily used on jobs needing intermittent rather than steady round-the-clock use, but contractors often keep them running for days at a stretch. An auxiliary gas supply can be rigged up to feed automatically into the pump tank, by a float valve in the tank, and a line to a drum or other container.

As with all two cycle gasoline engines, it is absolutely essential that lubricating oil be mixed with the gasoline.

Jetting Pump. Jetting pumps are usually centrifugal pumps of heavy construction, so designed that they can deliver water at very high pressure but which have comparatively weak suction.

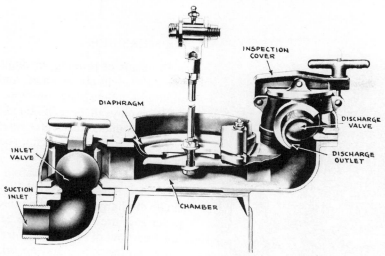

Courtesy of Marlow Pumps

Fig. 21-51. Diaphragm pump cross section

They may be of the self-priming construction described or may use an auxiliary vacuum pump for priming.

Their special uses include sinking foundation piles by water pressure, hydraulic excavating and flushing, making test holes, fire fighting, and dewatering where the discharge point is too high for the ordinary pump.

Well Point Pumps. These are centrifugal pumps which are able to handle water with a large proportion of air because of special internal design, or by means of auxiliary vacuum pumps.

They are used for removing underground water from the soil in excavation areas, and were discussed in Chapter 5.

One pump will be supplied from numerous points or inlets. These consist of pipes with gauze or slotted screens of fine mesh which are driven or jetted into the ground. The well points are connected with the surface by pipes which are connected to a main or header pipe to the pump through hose or pipes.

Diaphragm Pump. Figures 21-51 and 21-52 show a Marlow "Mud Hog":

It consists chiefly of a movable diaphragm in a closed chamber between two check valves. The diaphragm moves vertically. On the up stroke it causes a suction which makes the inlet ball rise off its seat and pulls the contents of the inlet pipe. On the down stroke, the inlet valve is seated and the outlet valve forced open, discharging part of the contents of the chamber into the outlet pipe.

The action of this pump is positive so that it will handle any material which will

Courtesy of Marlow Pumps

Fig. 21-52. Diaphragm pump

Courtesy of The Jaeger Machine Co.

Fig. 21-53. Eccentric drive diaphragm

flow through the pipe. It is not subject to air lock, but because of its elasticity air moves through it slowly, particularly on high lifts.

It can be slowed or put out of action by trash lodging in the valves. For this reason, the valve chambers are made quickly accessible for easy removal of foreign matter.

The maximum lift is somewhat less than that of the centrifugal pump, because some pressure in the intake line is required to open the inlet valve, and power is not provided for high outlet pressure. On high lifts it is good practice to set the pump about halfway between inlet and discharge levels so as to equalize the resistance to the up and down strokes.

A Jaeger diaphragm pump is illustrated in Figure 21-53. The shaft is driven by an eccentric instead of a rocker arm. The suction valve is a flap type, and the discharge valve is a ball.

Suction lift is similar to that of a centrifugal but the outlet (force) lift is less.

SUBMERSIBLES

Problems of priming, of diminished output with height above water, and of impossibility of pulling water up more than 25 or 28 feet, may all be eliminated by placing the pump under the water, and providing power from above through a shaft, hydraulic lines, or electric wires.

Shaft Drive. The Crisafulli pump, not illustrated, has a two-sided impeller and a single discharge line. Both sides of the impeller case are open near the center, allowing free flow of water (and trash) into it. There are no valves and no priming mechanism.

This pump is ordinarily built into a long, light, two-wheel trailer. A driveshaft extends forward from the pump. It may be connected to the power takeoff of the towing tractor, or to an engine or electric motor on the trailer tongue.

A butyl rubber discharge hose is bolted onto the pump, which is then backed into the water. The tractor or motor remains above water.

The pump body fills with water by gravity, so there is no priming problem. As no power is consumed in suction lift, a large volume of water can be moved in proportion to power input.

Ordinary amounts and types of trash go through the impellers, so inlet screens, with their cleaning problems, are usually not needed. If required for special conditions, they can be very coarse and therefore comparatively trouble free.

Electrics. Small submersible electric powered pumps are commonly used for household water supply. They may be used in dewatering construction sites by suspending in drilled holes, if rate of water inflow is not large.

For heavier service, a caisson pump, see Figure 21-54, may be suspended vertically in the hole by means of a cable. The short

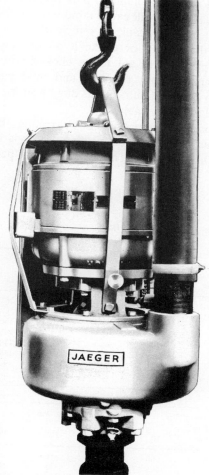

Courtesy of The Jaeger Machine Co.

Fig. 21-54. Caisson pump

suction line is attached to the bottom center of the pump. The outlet line must reach a disposal point, and, if digging goes on, must have additional sections added. This can also be operated horizontally for ordinary pumping.

PUMP OPERATION

If the water is small in volume or contains a heavy load of mud or other solids, a diaphragm pump is preferred. For larger quantities of water a centrifugal pump is needed. Both light hand-carry and heavier

wheel models of centrifugals are available.

Most satisfactory results are obtained when the capacity of the pump or pumps is substantially greater than the inflow, particularly when there is a large volume of standing water.

Setting Up. The pump should be level and placed as near water level as possible, as centrifugals can push more strongly than they can pull. The table of output, Figure 21-55 shows the loss in volume which occurs as a pump is raised above the water. The capacity of a pump is greatest when total lift is low.

The pump should be supported on a platform, or on boards as in Figure 21-43. Otherwise vibration, and softening or washing of the ground, may cause it to settle off level, or even fall into the pit.

A strainer should always be used when there is a possibility of sucking up stones or other objects which might damage or clog the line; and it is a good practice to use one whenever possible.

Use of a foot valve is optional. If the hose diameter is small, the lift low, the pumping steady, and the inlet opening unlikely to be exposed, it is not needed. On high lifts, particularly where the pump is worn and the inlet valve defective, a foot valve is very desirable.

Foot valves are not particularly dependable, being more subject to jamming by trash or mud than parts of the pump proper. In addition, the end of the inlet hose is frequently horizontal or nearly so, and some valves do not function well in this position.

Inlet Protection. The inlet should be well below the surface of the water—about six times the inside diameter of the hose when possible. This is frequently not practical, particularly when the place must be pumped dry.

At lesser depths a whirlpool may form over the inlet, and air enter through its center in sufficient quantity to form an air lock and cause the pump to lose its prime. Such a vortex will not form if the end of the hose is vertical, and is most likely to occur if it is horizontal.

The whirlpool may be prevented by digging a sump pit to lower the inlet; by floating a square or round piece of wood two or more feet in diameter over the inlet, or by bolting a roof over the strainer.

The inlet should not be allowed to rest in soft mud. It is liable to pick up abrasive material which will cause rapid wear of the pump; and may sink in sufficiently to reduce the intake, or choke it off entirely. It is particularly apt to be choked when the pump is not running, and mud can slump into the hollow it has dug by sucking in soil with the water.

This can be prevented by resting the strainer in a wood box, or on a wide board. A light metal plate may be fastened per-

STANDARD TABLES FOR SELF-PRIMING CENTRIFUGAL PUMPS

The following tables give capacities in gallons per minute

Model 5-M, 1-1/2 inch

Total Head Including Friction	Height of Pump Above Water In Feet			
	10	15	20	25
15 feet	85	-	-	-
20	84	68	-	-
25	82	67	-	-
30	79	66	49	35
40	71	60	46	33
50	59	52	41	28
60	42	40	32	22
70	22	22	20	12

Model 10-M, 2 inch

Total Head Including Friction	Height of Pump Above Water In Feet			
	10	15	20	25
25 feet	166	-	-	-
30	165	140	110	-
40	158	140	110	75
50	145	130	106	70
60	126	117	97	68
70	102	100	85	60
80	74	74	68	48
90	40	40	40	32

Model 17-M, 3 inch

Total Head Including Friction	Height of Pump Above Water In Feet			
	10	15	20	25
25 feet	284	-	-	-
30	278	245	203	158
40	260	239	198	155
50	236	224	187	150
60	204	200	172	138
70	164	164	149	120
80	122	122	118	98
90	77	77	77	70
95	54	54	54	33

Model 20-M, 3 inch

Total Head Including Friction	Height of Pump Above Water In Feet			
	10	15	20	25
30 feet	333	280	235	165
40	315	270	230	162
50	290	255	220	154
60	255	235	205	143
70	212	209	184	130
80	165	165	157	114
90	116	116	116	94
100	60	60	60	60

Model 30-M, 4 inch

Total Head Including Friction	Height of Pump Above Water In Feet			
	10	15	20	25
30 feet	500	435	350	250
40	495	430	345	250
50	475	415	340	245
60	450	400	325	240
70	415	370	300	230
80	355	325	270	210
90	250	240	215	175
100	100	100	100	100
105	20	20	20	20

Model 40-M, 4 inch

Total Head Including Friction	Height of Pump Above Water In Feet			
	10	15	20	25
25 feet	665	-	-	-
30	660	575	475	355
40	645	565	465	350
50	620	545	455	345
60	585	510	435	335
70	535	475	410	315
80	465	410	365	280
90	375	325	300	220
100	250	215	195	145
110	65	60	50	40

Model 90-M, 6 inch

Total Head Including Friction	Height of Pump Above Water In Feet			
	10	15	20	25
25 feet	1500	-	-	
30	1480	1280	1050	790
40	1430	1230	1020	780
50	1350	1160	970	735
60	1225	1050	900	690
70	1050	900	775	610
80	800	680	600	490
90	450	400	365	300
100	100	100	100	100

Model 125-M, 8 inch

Total Head Including Friction	Height of Pump Above Water In Feet			
	10	15	20	25
25 feet	2100	1850	1570	-
30	2060	1820	1560	1200
40	1960	1740	1520	1170
50	1800	1620	1450	1140
60	1640	1500	1360	1090
70	1460	1340	1250	1015
80	1250	1170	1110	950
90	1020	980	940	840
100	800	760	710	680
110	570	540	500	470
120	275	245	240	240

Model 200-M, 10 inch

Total Head Including Friction	Height of Pump Above Water In Feet			
	10	15	20	25
20 feet	3350	3000	-	-
30	3000	2800	2500	1550
40	2500	2500	2250	1500
50	2000	2000	2000	1350
60	1300	1300	1300	1150
70	500	500	500	500

Courtesy of *Contractors' Pump Bureau.*

Fig. 21-55. Pump capacities

manently to the under side of the strainer.

When leaves, grass, or other vegetable debris are in the water, a strainer may become plugged every few seconds. The best cure for this is to make an additional strainer of wire of sufficiently fine mesh to catch the bulk of the trash, with a large enough area so that the water will move through it at too low a velocity to hold the trash against it.

Even if too small for self-cleaning, such an additional screen will clog less often, and be more easily cleaned, than the standard strainer. Two constructions were illustrated in Chapter 5.

A riskier but less laborious method is to remove the hose strainer, and allow the trash to go through the pump.

Air in Inlet. A contractor's centrifugal pump is built to handle solid water, and pumps air with difficulty. When the pump and inlet hose are empty, the pump must be filled with water (primed) before starting, and will then work for a while slowly drawing the air out of the inlet pipe. When the water is sucked into the pump its efficiency rises abruptly. If air is permitted to enter the intake, the pump will lose its grip, and have to again slowly work up enough vacuum to lift the water into it. This process is fairly quick in new pumps, but in worn ones is often a slow process if the lift is at all high.

Very often failure of a pump to develop sufficient vacuum to prime itself is due to air leaks in the inlet line. These are most often past the gaskets at hose couplings, but may be at any spot on the inlet side. This includes the hoses themselves; gasket between the inlet elbow and the body, cracks in the inlet or body, threaded fittings, loose or worn shaft packings, or lack of grease in a shaft bushing:

Such leaks are hard to detect, as, if the pump is worn or the lift is high, a very small leak will do the damage. Wet clay should be smeared over all suspicious spots on the hose, couplings, gaskets, or casing parts. This makes a good temporary seal for small leaks; and will show up larger ones by disappearing into them. Leaks along the shaft will generally drip when the pump is stopped.

Gauges. The pump casing should be checked for cracks. Gaskets and connector threads should be smeared with pipe dope or Permatex #2. Installation of a vacuum gauge on the inlet side is an excellent idea, as it will immediately indicate the nature of the trouble. Low vacuum indicates leakage, a worn pump, or lack of prime water, and high vacuum a plugged inlet—usually the strainer, but sometimes a stuck valve. A fairly normal vacuum, with little or no production, indicates back pressure from too high an outlet or a plugged line.

A vacuum gauge may be calibrated in pounds or inches. If pounds, it measures the difference between air pressure inside and outside of the inlet chamber in pounds per square inch. If inches, it indicates the height of a column of mercury which could be supported by this difference in pressure.

Figure 21-56 contains data for conversion of inches of vacuum into the number of feet that water should rise in the inlet hose, and for some other calculations.

A primed pump starting with an empty hose will increase its vacuum slowly, until the water rises into the pump and displaces the air. The increased efficiency of pumping water instead of air will cause the vacuum to jump to a higher figure than that needed to pull the water, after which it should drop to normal and remain steady as long as conditions are unchanged. As the water level drops as a result of pumping, the vacuum increases proportionately.

If the hose is full, vacuum will rise very rapidly to full pumping level.

Outlet or Discharge Line. If the outlet hose is not long enough to reach a disposal point, a level or down hill extension may be made by means of a wood flume, a ditch, or parallel dikes.

If the discharge hose end is under water, the pump is stopped, and the inlet and foot valves leak or are lacking, water will siphon from the higher level through the pump and out the inlet. On a high lift, the force may be sufficient to turn the

TABLE CONVERTING INCHES VACUUM INTO FEET SUCTION

To convert inches vacuum into feet, multiply by 1.13.

Inch Vac.	Feet	Inch Vac.	Feet	Inch Vac.	Feet
1	1.13	11	12.47	21	23.81
2	2.27	12	13.61	22	24.95
3	3.41	13	14.74	23	26.08
4	4.54	14	15.88	24	27.22
5	5.67	15	17.01	25	28.35
6	6.80	16	18.14	26	29.48
7	7.94	17	19.28	27	30.62
8	9.07	18	20.41	28	31.75
9	10.21	19	21.55	29	32.89
10	11.34	20	22.68		

CONVERSION FACTORS FOR PRESSURE AND VACUUM

Feet head (water) $\times$ .433 = pounds pressure

Pounds pressure (water) $\times$ 2.31 = feet head

Feet head (brine, sp. gr. = 1.2) $\times$ .52 = pounds pressure

Pounds pressure (brine) $\times$ 1.92 = feet head

Feet head (gasoline, sp. gr. = .75) $\times$ .325 = pounds pressure

Pounds pressure (gasoline) $\times$ 3.08 = feet head

Inches of mercury $\times$ 1.132 = feet head of water

Feet suction lift of water $\times$ .882 = inches of mercury

Courtesy of Contractors' Pump Bureau

Fig. 21-56. Conversion tables

pump and engine backward. Otherwise the flow is leakage past the vanes.

If the discharge point is high, thirty or more feet above the pump, it may be advisable to install a check valve just above the pump, to relieve the pump of destructive water hammer when shut down.

If the outlet hose is long or crooked, pump capacity may be increased by using oversize hose. This reduces the velocity and friction of the water. Sharp turns or abrupt changes in size should be avoided through use of taper fittings and 30° or 45° elbows.

It is frequently desirable to reduce the output of the pump, as when the water level has been lowered, and pumping need only remove an inflow which is less than the pump capacity. Gasoline power permits throttling the engine down. If this does not afford sufficient reduction, and it is not convenient to stop and start the pump frequently, a gate valve may be put on the outlet connection, and output regulated by opening or closing it. This does not add materially to the strain on the pump, whereas throttling the intake line does.

Freezing. Special precautions are necessary in operating a pump in freezing weather.

Whenever the pump is to be shut down, the body and hoses should be drained.

The drain plug at the bottom of the case, and the filler plug, should be removed. If the end of the discharge hose is above the pump and underwater, it should be lifted out to avoid siphoning water down to the pump.

The discharge hose can be emptied of water by holding it so that there is a continuous slope to the ends. If this is not practical, water can be worked out of it by raising it at the high end, and walking toward the other end, holding it up in a loop high enough to force the water over other high points. If the line is metal, it should be opened at its low points, and blown out if air is available.

If the inlet hose is light, it can be disconnected, pulled out of the water, and drained. If it is heavy, and has no foot valve, it may drain itself if the pump check valve is not air tight. If it does not drain, an opening should be made in the metal at its upper end. A petcock may

be provided, or a vacuum gauge removed, or a new hole cut and threaded. The air admitted in this way will allow the water inside the hose to drop to the surrounding water level.

This opening must be sealed when the pump is used again.

If freezing is apt to be severe enough to form a plug of ice at water level in the hose, the opening at the top should be made large enough to permit pouring sufficient alcohol or fuel oil into the hose to protect it. If the hose is deep in the water, one dose should give protection for a long time if the pump is not used. Should the opening in the hose be close to the surface, alcohol may have to be renewed every few days.

If a foot valve is required for the pumping it will be necessary to lift the lower end of the hose out of the water to open the valve to drain it unless a "tickler" is installed. Any homemade device which will unseat the flap in the foot valve by pulling on a line or a rod will be adequate.

If the pump body does not drain completely it may be necessary to tip it to get all the water out, or to pour in enough alcohol to prevent the residue from freezing.

The use of salt or calcium chloride as an anti-freeze is not recommended, because of probable damage to the pump.

The engine should be protected by antifreeze in the regular manner.

In freezing weather, it is advisable to start the engine before priming the pump, although it should not be allowed to run dry longer than a minute or two. If the temperature is below 0° F., it may be advisable to use hot water or a weak antifreeze mixture for priming.

If the pump is frozen, the engine will not turn over. Filling the tank with hot water or with a strong anti-freeze solution should free it.

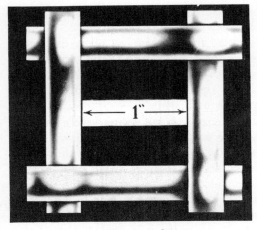

"Space"

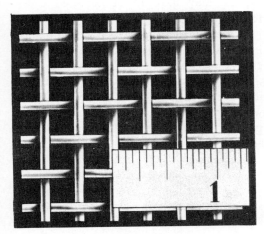

"Mesh"

Courtesy of Hewitt-Robbins Incorporated

Fig. 21-57. Wire cloth measurements

DIAMETER OF ROUND HOLE	SIZE OF SQUARE OPENING	DIAMETER OF ROUND HOLE	SIZE OF SQUARE OPENING
1/8	3/32	2 1/4	1 7/8
3/16	5/32	2 3/8	2
1/4	3/16	2 1/2	2 1/8
5/16	1/4	2 3/4	2 1/4
3/8	5/16	3	2 1/2
1/2	3/8	3 1/4	2 3/4
5/8	1/2	3 1/2	3
3/4	5/8	3 3/4	3 1/8
7/8	3/4	4	3 5/16
1	7/8	4 1/4	3 1/2
1 1/4	1	4 1/2	3 3/4
1 3/8	1 1/8	4 3/4	4
1 1/2	1 1/4	5	4 1/4
1 5/8	1 3/8	5 1/4	4 1/2
1 3/4	1 1/2	5 1/2	4 3/4
1 7/8	1 5/8	5 3/4	4 7/8
2	1 3/4	6	5
—	—	7	5 3/4
—	—	8	6 3/4

Courtesy of Hewitt-Robbins Incorporated

Fig. 21-58. Equivalent sizes, square and round

The outlet hose should be checked for ice before starting. Even if it has been drained, a small amount of water may collect in a low spot and glue the walls together. A hose blocked in any manner is liable to blow out under working pressure. Such ice can be readily broken by tapping the hose with a hammer or a block of wood.

Thawing an ice blocked inlet hose is a major operation, and methods will depend on the facilities on hand. A gasoline burn-ing heater equipped with a blower and a warm air duct is most effective.

Under extreme conditions, a pump may freeze up while running. Provided the inlet is deep in the water, this may be prevented by wrapping hoses and pump body with sacking or blankets, or by applying heat (engine exhaust, bonfire or large blowtorch) to the case. Flame should not be used if there are gasoline leaks anywhere, and caution should be used if the engine is oily or dirty.

GRIZZLIES AND SCREENS

Screens of various types are used to eliminate oversize stone from fill or crusher feeds, and to separate mixtures into uniform sizes.

In order for separation to take place, the raw material must be moved or shaken on the screen surface. Sticky or wet materials and fine openings require the maximum amount of movement.

Movement may be accomplished by gravity flow along an inclined screen, by rotating, shaking, or vibrating the screen, or less commonly, by raking devices. Combination of gravity and other methods is usual.

Screens may be made of welded bars, sheet steel or rubber with round, square, slot, or octagonal holes; or wire cloth. Cloth is the most popular material in a wide range of applications.

A cloth may be described in terms of "space" or "mesh." The difference is illustrated in Figure 21-57. Space is the actual dimension of the clear opening between adjacent parallel wires. Mesh is the number of openings per inch measured from center to center of parallel wires.

A square opening, whether in cloth or plate, will allow larger material through it than will pass a round hole of the same diameter. Figure 21-58 contains a table of equivalent sizes of square and round openings. As an example of its use, if a screen with round holes two inches in diameter is separating the aggregate properly, a square-hole cloth to replace it should have spacing of 1¾ inches.

Rectangular openings pass larger material than a square of their short dimension, but smaller than a square of the long dimension. Each shape rectangle will have different characteristics.

Screen cloth is composed of woven wire. Figure 21-59 shows the actual size (plus or minus an allowance for distortion in pho-

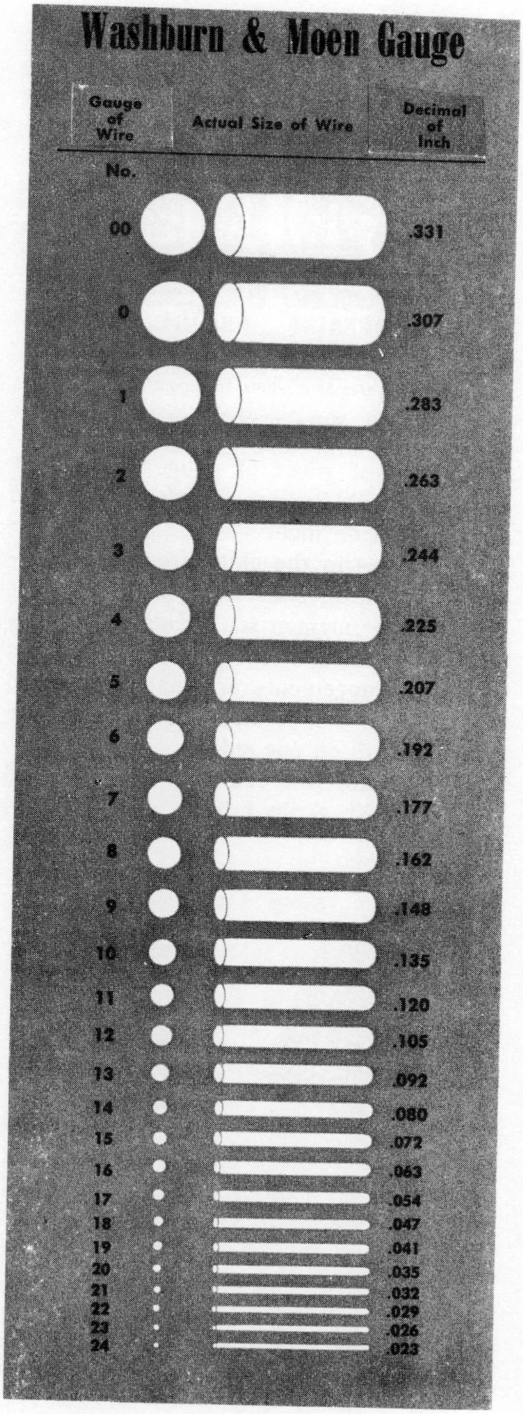

Washburn & Moen Gauge

Gauge of Wire No.	Actual Size of Wire	Decimal of Inch
00		.331
0		.307
1		.283
2		.263
3		.244
4		.225
5		.207
6		.192
7		.177
8		.162
9		.148
10		.135
11		.120
12		.105
13		.092
14		.080
15		.072
16		.063
17		.054
18		.047
19		.041
20		.035
21		.032
22		.029
23		.026
24		.023

Courtesy of Hewitt-Robbins Incorporated

Fig. 21-59. Wire size

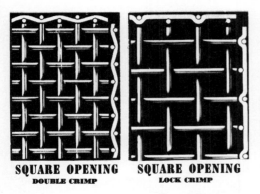

| SQUARE OPENING | SQUARE OPENING | RECTANGULAR | NON-BLIND |
| DOUBLE CRIMP | LOCK CRIMP | | |

Courtesy of Hewitt-Robbins Incorporated

Fig. 21-60. Wire cloth, square openings

Courtesy of Hewitt-Robbins Incorporated

Fig. 21-61. Wire cloth, rectangular openings

tography) of wires of gauge (diameter) from 00 to 24, together with their size in decimals of an inch.

The wires in the cloth are crimped together—that is, bent under extreme pressure to a permanent set. Crimping may affect only one set of wires, or both. Various special arrangements are made to obtain extra tightness of connection, or to have the cloth smooth on one side.

Manufactured screen cloth does not need to be welded, as the crimps hold the wires in place. Homemade cloth—usually made of bars for rough grizzly work—must be welded in order to keep in alignment.

Rectangular openings are used when a steep screen slope reduces the effective length of the squares. Non-blind cloth may have the longer wires of a smaller size than the cross ones, so that they vibrate and tend to dislodge and pass pieces sticking between them. "Blinding" is the partial or complete clogging of the screen with inert material.

Figures 21-60 and 21-61 show four of the many screen cloth constructions that are available. Each picture is triple, showing the surface and two cross sections.

Gravity. A grizzly is generally a primary screen with fairly coarse openings that rejects boulders or large fragments from run-of-pit material.

Portable gravity grizzlies are shown in Figure 21-62. (A) is a light unit which is placed on a truck body before loading. Gravel or soil is deposited on the top of the slope and pours and rolls down along the surface, the undersize falling through the spaces and the oversize rolling clear. After loading, the shovel picks up the grizzly and places it on the next truck. A helper is needed to steady it during placement.

When used for stockpiling, it can be bolted to the truck body, as in (B). It must

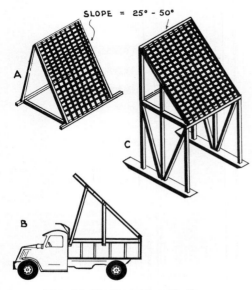

SLOPE = 25° - 50°

Fig. 21-62. Portable grizzlies

Fig. 21-63. Fixed grizzly

then not be high or heavy enough to over-balance the body in dumped position.

(C) is a similar unit built up on a skid frame. The truck backs under it to be loaded. It often becomes partly buried, and the frame should be strong enough to with-stand heavy pulls. It may be braced be-tween the skids also.

Figure 21-63 shows a fixed installation that may feed into trucks, a conveyor, a crusher, or a small stockpile.

These units may have screen surfaces of parallel steel bars or coarse mesh, sup-ported by heavy cross members. The bars may be made of triangular, square, dia-mond, or round cross-section, or of in-verted rails. Mesh construction works more accurately than parallel bars.

A major difficulty with this type of screen may be slightly oversize rocks stick-ing between the bars. This trouble is least with triangular pieces with a flat side up,

and greatest with round sections. A gradual increase in slot width toward the bottom is helpful.

Slot type surfaces pass oversize flat rocks readily. If the underbracing is placed at such close intervals that the slots are re-duced to squares, this difficulty is decreased. However, a wider bar spacing or a longer slide is then required to give the same ca-pacity.

For a given material and bar spacing the amount of material passing through the grizzly will be increased by lengthening the screen or by reducing the slope. A long screen occupies more space both vertically and horizontally. A too flat slope will allow material to rest on the surface and clog the slots. A too steep or short slope will reject undersize material.

Spacing may be from 2″ to 2′, depending on the type of service.

Loose dry dirt will separate satisfactorily

Fig. 21-64. Vibrating bar grizzly

at a grade of 30° to 40°. Wet gravel or damp loam requires slopes up to 50°. Wet sticky materials cannot be satisfactorily processed.

Increase in width of slots will usually make it possible to reduce the slope or the length. Grizzlies used only to reject occasional large boulders, using bar spacings of two feet, may be flat if some means of removing the boulders is provided.

Use of these units is restricted by inaccuracy of sizing, by their tendency to clog, or the space required. They are desirable in that they can be built cheaply on the job, often out of scrap material, and are easy to operate and repair.

The cantilever grizzly has rigid bars held only by two tie rods near the upper end. Stones are not blocked by cross supports, and the projecting bars vibrate under impact, thus speeding the flow of material and freeing wedged stones. This requires less slope than the solid grizzly, and can be used with narrower spacing. It is not as suitable for use with material containing heavy boulders.

The slope of a grizzly can be reduced and the accuracy of sizing and capacity increased by shaking or vibrating the screen. Best results are obtained by purchasing a made up unit, such as the Nordberg shown in Figure 21-64, in which the screen has a limited movement relative to the frame and is connected to a shaking or vibrating device. Performance of a standard grizzly may often be improved by fastening a vibrator tightly to a side or top frame.

Revolving. Revolving screens, such as the two Telsmith models shown in Figure 21-65, are usually cylinders of wire cloth or perforated plate set at a slope of five to seven degrees. They are rotated at a speed of fifteen to twenty rpm.

Material placed inside the upper end of the cylinder is rotated up the side of the screen until its weight overcomes adhesion and centrifugal force and it falls to the bottom. The upward movement is slightly toward the lower end and the fall is straight, so material tends to move from the higher to the lower end in a series of short steps.

The amount of rise is affected by the stickiness and speed of the particles. Sticky pieces on a fast turning screen will cling to it for full revolutions, clogging or "blinding" the holes, while hard smooth particles on a slow screen will appear to move straight down the bottom. In general, speed should be adjusted to carry the load less than one third way up the screen.

In the upper view, one cylinder is used with several grades of perforated cloth. The upper end has a fine mesh and the lower sections are successively coarser. As the load moves down, it loses first the finest particles, then successively larger grades. Stones too large for any of the mesh are dropped out the lower end. Screenings from each section are chuted separately.

A disadvantage of this construction is that the full load, including large pieces, must be handled by the finest and most delicate of the screens. Its life may be prolonged by setting reinforcing bars inside which will hold the larger stones from contact part of the time. This construction is chiefly used for coarse sizes.

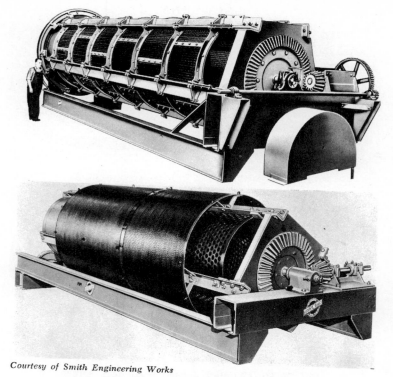

Fig. 21-65. Single and double rotary screens

Two concentric cylinders may be used. The first screens on the inner or main cylinder are coarser than the finest products required. Material passing through them is further classified by the finer mesh outer screens which do not have to carry any rough material. This type is well adapted to sand and gravel.

Screen frames may be made up with hexagonal or other polygonal cross-sections. These allow the use of flat sections of mesh or plate and agitate the load more.

Blinding. Blinding, or clogging of screen holes, may be a serious problem. It is most severe when the material is sticky, when most of the load is a size close to that of the screen opening; when the screen has insufficient slope, or wires that are rough with rust; or when it is overloaded.

Corrective steps include shaking or vibrating the screen, drying the material, changing particle size by adjusting the crusher, heating the screen wires or the aggre-gate, or tapping or scraping from below.

Shaking. Shaking screens are flat and rectangular, are suspended on loose or flexible attachments, and are shaken longitudinally by connecting rods or other eccentrics. Slope is usually around sixteen to eighteen degrees. The throw may be as much as a foot, but six inches is more usual. The number of strokes per minute is usually the quotient of six hundred divided by the length of throw in inches. This would be one hundred strokes for a six-inch throw.

Material dumped on the upper end works its way down, most of the motion being on the back or up stroke. Small particles fall through the screen and the rejects move off the end.

Shaking screens set up heavy vibrations. These are reduced by arranging them in oppositely moving pairs, which preferably should carry equal loads.

The frame or a counterweight may be hung to move oppositely from the screen.

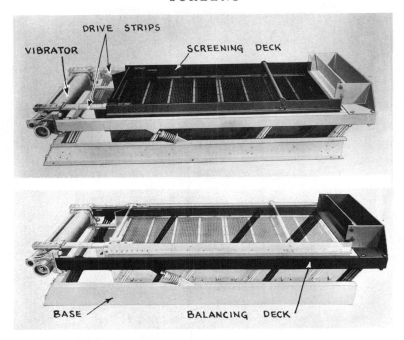

Fig. 21-66. Counterbalanced level deck screen

One shaking screen has a rotary movement on a ten-inch radius. Material moves chiefly on the down stroke.

Vibrating. Screens may be vibrated mechanically or electrically to agitate the material lying on them, to increase the rate of passage through the holes, to separate particles which are stuck together, and to move all the rejects off the discharge end promptly.

Mechanical vibrators include eccentric, cam, bumper, and unbalanced flywheel types. Electric vibrators are electromagnets. Motions imparted to the screen may be reciprocating or rotary. Frequency ranges from 600 to 3600 per minute, and

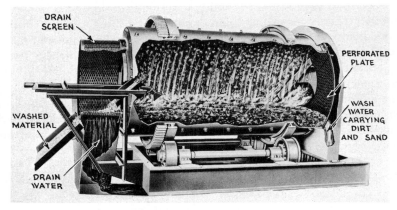

Fig. 21-67. Rotary washer

throw up to one half inch. Slopes vary from horizontal to 40 degrees. Standard widths are one to six feet, and lengths two and a half to sixteen feet.

Vibrating screens are usually mounted on springs or cushions, or suspended from cables. Solid contact would dampen the screen and vibrate the building.

Scalping. Scalping screens are coarse vibrating screens used instead of the smaller sizes of grizzly. Maximum openings are about eleven inches square, and slopes of twelve to eighteen degrees are standard.

Screen construction is very heavy as they are fed pit-run material that may include big pieces. Occasionally they are protected by widely spaced grizzly bars.

Deck. Deck screens consist of two or more vibrating screens placed one above the other. The lowest screen may be short, and if so, is called a half deck. They are usually vibrated through the frame so that motion is the same on each deck.

The top deck has the coarsest grid. Material passed through it is further separated into sizes by each succeeding layer. The rejects move down the screen surfaces to chutes or conveyors.

Courtesy of Smith Engineering Works

Fig. 21-68. Scrubber drum

The number of decks can often be changed in the field by adding or removing one or more screens.

Figure 21-66 shows a screen of the counterbalanced level type.

WASHERS

Aggregate (usable material) frequently comes to the plant mixed with clay or other impurities which have to be washed out. The amount and kind of impurity and the water supply largely determine the method.

Courtesy of The W. S. Tyler Company

Fig. 21-69. Spray washer

Fig. 21-70. Shovel and crusher capacities

Courtesy of Pit and Quarry Publications

TABLE I.—SIZES OF SHOVEL DIPPERS AND CORRESPONDING CRUSHERS

The dimensions of the largest stone that a given dipper will handle are those of a stone which will pass through the dipper. The crusher recommended is the smallest that will handle the stone without regard to capacity.

Rated Capacity of Dipper (cu. yd.)	Size Stone to Pass Dipper (in.)	Size of Jaw Crusher (in.)	Size of Gyratory (in.)
¾	32 by 35	36 by 48	13 or 16
1	33 by 38	36 by 48	16 or 20
1¼	33 by 40	42 by 48	20 or 26
1½	30 by 36	36 by 42	20 or 26
1¾	33 by 45	42 by 48	20 or 26
2	33 by 45	42 by 48	20 or 26
2½	36 by 48	48 by 60	36 or 42
3	40 by 48	48 by 60	42 or 48
3½	44 by 50	48 by 60	42 or 48
4	48 by 57	56 by 72	48 or 60
5	48 by 60	66 by 86	60 or 72

TABLE II.—SHOVEL CAPACITIES

Based on stone with a specific gravity of 2.6.

Size of Dipper		Swings per Minute	Tons Loaded per Hour, 70% Efficiency
Cu. yd.	Tons		
½	.45	2.4	46
¾	.675	2.36	67
1	.90	2.23	86
1¼	1.125	2.15	103
1½	1.35	2.10	120
1¾	1.575	2.00	136
2	1.80	1.97	150
2¼	2.025	1.91	165
2½	2.25	1.85	177
2¾	2.475	1.80	190
3	2.70	1.75	202
3¼	2.925	1.70	213
3½	3.150	1.65	222
3¾	3.375	1.61	233
4	3.600	1.57	240

TABLE III.—APPROXIMATE CAPACITIES OF JAW CRUSHERS

Material weighing 100 lbs. per cu. ft.
Discharge opening, closed, in inches.

Size (in.)	Smallest Discharge Opening (in.)	Capacity (tons per hr.)	Largest Discharge Opening (in.)	Capacity (tons per hr.)	Horsepower	Height (ft. and in.)
16 by 10	1½	15	4	45	15	2—4
24 by 15	2	30	5	80	35	3—6
36 by 24	3	75	6	160	75	4—11
42 by 40	4	130	8	250	125	7—9
48 by 36	5	175	8	275	150	7—6
48 by 42	5	175	8	275	150	8—6
60 by 48	5	240	9	450	200	9—2
84 by 56	8	350	12	600	200	10—7
84 by 60	8	350	12	600	250	10—7
84 by 66	8	400	12	600	250	11—8

TABLE IV.—APPROXIMATE CAPACITIES OF GYRATORY CRUSHERS

Material weighing 100 lbs. per cu. ft.
Discharge opening—open side

Size (in.)	Smallest Discharge Opening (in.)	Capacity (tons per hr.)	Largest Discharge Opening (in.)	Capacity (tons per hr.)	Horsepower	Height (ft. and in.)
2½	⅜	.5	½	.75	3	1—1
8	1	15	2	40	20	5—6
12	2	40	2¾	70	45	7—5
16	2½	100	4½	160	80	8—3
20	3	150	5	250	125	10—1
26	3½	225	6	400	150	11—0
30	4	250	6½	450	175	12—2
36	4½	370	7	600	225	14—0
42	5	420	7½	700	250	15—5
48	5½	750	9	1,200	300	18—6
54	6¼	900	9½	1,600	350	19—0
60	7	1,200	10	2,000	400	20—0
72	9	2,000	12	3,000	500	20—0

TABLE V.—APPROXIMATE CAPACITIES OF SINGLE-ROLL CRUSHERS

Material weighing 100 lbs. per cu. ft.

Size (in.)	Capacity in Net Tons per Hour			Horsepower	Height (ft. and in.)
	4 in.	6 in.	10 in.		
21 x 48	125–150	200–250	300–375	75–100	3–10
21 x 60	150–175	225–275	325–400	100–125	3–10
24 x 48	150–175	225–275	325–400	100–125	4– 5
24 x 60	175–200	250–300	350–425	125–150	4– 5
30 x 48	175–200	250–300	350–425	125–150	5– 0
30 x 60	200–225	275–325	375–450	150–175	5– 0
36 x 48	200–225	275–325	375–450	150–175	6– 0
36 x 60	225–250	300–350	400–475	175–200	6– 0
42 x 48	225–250	300–350	400–475	175–200	7– 0
42 x 60	250–275	325–375	425–500	200–225	7– 0

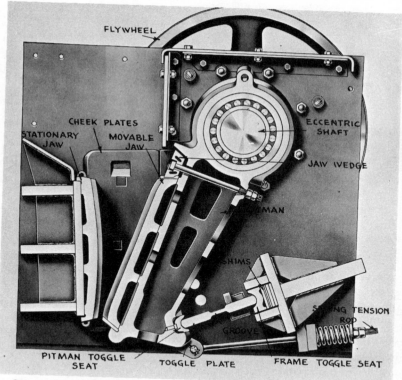

FLYWHEEL

CHEEK PLATES

STATIONARY JAW

MOVABLE JAW

ECCENTRIC SHAFT

JAW WEDGE

PITMAN

SHIMS

SPRING TENSION ROD

GROOVE

PITMAN TOGGLE SEAT

TOGGLE PLATE

FRAME TOGGLE SEAT

Courtesy Baldwin-Lima-Hamilton Corp.

Fig. 21-71. Jaw crusher

Two principal methods are immersion and spray. Either may be assisted by vibration, which tends to separate stuck particles.

Figures 21-67 and 21-68 show cutaway views of a Telsmith rotary scrubber. A Ty-rock water spray screen is illustrated in Figure 21-69.

ROCK CRUSHERS

Rock (or stone) crushers reduce rocks to smaller and more uniform sizes. They are chiefly used in connection with blasted rock, and for cobbles and boulders found in gravel deposits.

The reduction may be accomplished by pressure, impact, shearing, or by combinations. The crushers must be of heavy construction and surfaces in contact with the stone should be renewable plates of manganese or other special alloys. Flat or corrugated surfaces are used for different conditions.

Most crushers are provided with overload reliefs so that "tramp" iron (iron scrap) or other uncrushable material will not seriously damage the machine. These may be springs, shear pins, or easily accessible under-strength parts.

Crushers are roughly classified as primary and secondary but these types overlap. The primary work on run-of-pit material, have relatively large hopper openings and produce a coarse product. Secondary crushers will accept only small stone from the primaries, and turn out a finer and more uniform product. Both types are commonly protected against both oversize and undersize pieces by grizzlies and/or screens. Primary and secondary functions may be com-

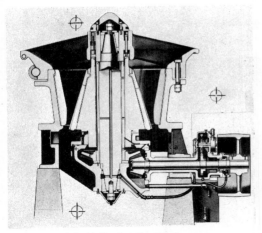

Courtesy of Smith Engineering Works

Fig. 21-72. Gyratory crusher, conical type.

bined in one crusher, or in two or three crushers in series.

The ratio of reduction of a crusher is the difference in size between the largest pieces it will process readily and the product it obtains from them.

Figure 21-70 contains tables in general use in the industry, showing the capacities

of shovels and certain crushers in size of pieces and in hourly production. Production figures should not be taken as a basis for estimates as improvement in machines, and differences in field conditions cause wide variation.

Jaw. Jaw crushers are simple and economical in construction and require minimum power.

The usual construction is called swing-jaw. In this, one jaw is movable and the other is fixed in position. In twin-jaw models both of them move.

Figure 21-71 shows a cross-section of an Austin Western. The crushing surfaces consist of two jaws which do not quite touch at the bottom, and which are widely separated at the top. Jaw faces may be flat or convex.

One jaw is fixed, the other is attached to the pitman or eccentric arm, which is hinged on a toggle and spring fulcrum. At the top the pitman is mounted on an eccentric shaft running between two heavy balance wheels. Rotation of the shaft causes

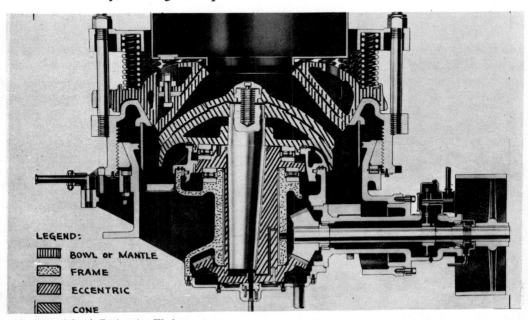

LEGEND:

⬚ BOWL or MANTLE

⬚ FRAME

⬚ ECCENTRIC

⬚ CONE

Courtesy of Smith Engineering Works

Fig. 21-73. Gyratory crusher, dome type

the pitman and jaw to gyrate, leaning first toward the fixed jaw, then away from it, with some vertical rubbing motion. The pitman acts as a second-class lever, with longest movement of ⅜″ to ½″ at the top, and greatest power toward the bottom.

The toggle bar runs the full length of the jaw. It is held in replaceable sockets by heavy springs. It is notched to produce a line of weakness at which it will break before other crusher parts if tramp iron (iron chunks) get between the jaws.

Adjustment for fineness of product is made by moving the toggle forward, usually by means of spacers placed behind its frame socket.

Rocks resting in the V between the jaws are crushed by pressure, then allowed to drop as the jaw moves back. This process is repeated until the rocks are reduced to pieces small enough to pass through the narrow space at the bottom of the jaws.

Jaw crushers are made in considerable variety of construction and in sizes up to a jaw opening of five and a half by seven feet. They are classified according to size of opening. For example, a 10″ x 30″ crusher has a ten inch opening at the top, and the jaws are 30″ long (or wide).

The maximum diameter round rock that will be accepted or "nipped" is about 80 percent of the width of the opening.

Gyratory. These machines have a conical or domed crushing member called the cone, head, or sphere, which moves in a small circle around a vertical axis, inside a fixed bowl or mantle. Three Telsmith constructions are shown.

The cone may be relatively stationary at the top and move at the bottom only, may gyrate equally at top and bottom, or be mounted so that the head can wobble as well as gyrate.

The crushing head is free to turn under thrust from material being crushed. Modern secondary units provide relief from break-

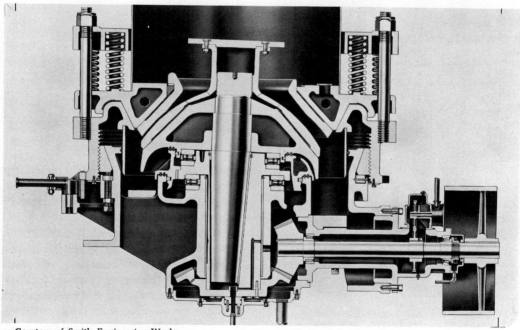

Courtesy of Smith Engineering Works

Fig. 21-74. Gyratory crusher, secondary type

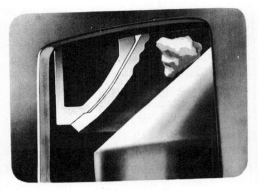

Position 1. With the head in the position of maximum opening, a large stone is just entering the crushing cavity.

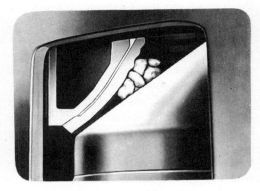

Position 4. The head has once more moved to the closed side and, receiving another impact, the particles are again reduced in size.

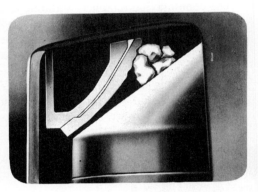

Position 2. As the head moves to the closed side, the stone receives its initial impact and is broken into several smaller particles.

Position 5. The particles again take a vertical path, spread across the head and advance farther into the crushing cavity.

Position 3. The broken particles fall vertically toward the head as it recedes from under the stone when moving to the open side.

Courtesy of Nordberg Manufacturing Co.

Fig. 21-75A. Gyratory crushing action

ing strains by heavy springs holding the mantle in place against normal loads. Fineness of product is adjusted by raising or lowering the mantle.

The Gyratory Breaker, Figure 21-72, is designed for primary crushing. It has an unobstructed feed opening and wide space between the upper crushing parts.

Figure 21-73, a style S Gyrasphere, is intended for coarse secondary crushing. 21-74, a style FC Gyrasphere, which is for fine secondary crushing, has a center plate that distributes the feed evenly to the full

Position 6. With another crushing impact, a further reduction in size takes place corresponding to the opening of the cavity at this point.

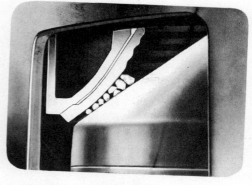

Position 8. Again a reduction takes place with those particles now in the parallel zone reduced to the size to which the crusher is set.

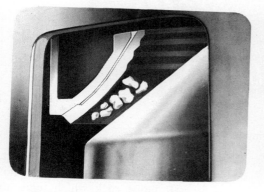

Position 7. The material has traveled farther on its downward path and is just entering the wide parallel zone at the bottom of the head.

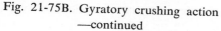

Courtesy of Nordberg Manufacturing Co.

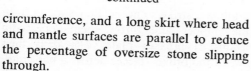

Fig. 21-75B. Gyratory crushing action
—continued

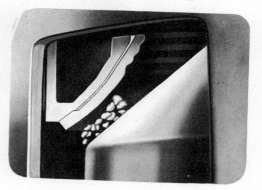

Position 9. Here the head is again at the open side with all the material now in the parallel zone. Note the large opening created for the discharge of the crushed fines.

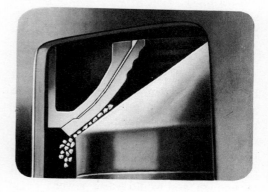

Position 10. All particles are now reduced in size, having received five impacts from the head in passing through the crushing cavity.

circumference, and a long skirt where head and mantle surfaces are parallel to reduce the percentage of oversize stone slipping through.

The crushing chamber is annular (ring-shaped), and wedge-shaped in cross-section. Rock fed into the top falls between cone and mantle and is crushed as the opening narrows with the movement of the cone. When it widens again the pieces fall farther, to be further crushed on its return. The action is somewhat similar to that of the jaw crusher, but the squeeze comes

HOPPER CHUTE

FLAIL

BREAKER PLATE

AXLE

ADJUSTMENT

GRATE

FINES

COARSE

Courtesy of Iowa Manufacturing Company

Fig. 21-76. Swing-hammer mill

from the side rather than the bottom and the curve of the chamber breaks up flat pieces.

Figure 21-75 shows a succession of steps in the reduction of a large stone in a Symons Cone Crusher. This machine features a more rapid gyration and a greater side travel between open and closed position than most crushers of this type, a construction that minimizes difficulty with clogging.

Cone speed and distance of travel must be carefully synchronized. A wide space allows pieces to fall more freely than a narrow one, and if coupled with slow movement would allow pieces to fall too far before the next impact. Fast gyration and short travel would not allow them to fall far enough, and would waste power.

Gyratory crushers are more expensive than jaw types, require more power, and need more vertical space for installation, but have greater production and will pro-

duce a finer and more uniform product.

Most jaw and cone crushers are not suited to soft moist rock which tends to cake under pressure, nor to rock mixed with clay.

Swing-Hammer Mills. These machines, one of which is illustrated in Figure 21-76, break by impact. Rapidly revolving flails, with tip speeds often over two miles a minute, strike stones as they slide in from the hopper and throw them repeatedly against a breaker plate. The flails then sweep the broken pieces across a bar grate, through which they fall if they are small enough. Any that are oversize are carried around to the breaker plate again for recrushing.

The grate openings may be all the same width, or be in several sets, with narrow openings near the breaker plate and progressively wider ones away from it. The latter arrangement permits the use of several hoppers under the grate, and the sepa-

Fig. 21-77. Traveling breaker plate

Fig. 21-78. Pusher feed

ration of crushed stone into different sizes.

These machines are subject to extreme peak loads when encountering large or hard rocks, and require heavy flywheels to maintain momentum.

Ring-hammer mills are of similar construction except that rings are used instead of hammers. These turn on impact and distribute wear over their whole circumference.

Hammer mills have the largest ratio of reduction of any type crusher, and in weak rock of favorable structure may reduce forty-eight inch cubes to one-inch cubes in one pass. They do primary crushing in soft and medium rock, secondary in any type, and are also garbage shredders.

Their product tends toward cubical shape more than that of the pressure type crushers. Fineness of crushing can be obtained by setting the breaker plate close to the hammers; but fineness of product is determined by the setting of the grate bars.

Some crushing is done against the bars, but this is kept to a minimum as they are not as sturdy as the breaker plates.

Sticky material may be fed to the hammers by an armored chain or traveling breaker plate, as in Figure 21-77, or a pusher, as in 21-78.

Roll Crusher. The single roll crusher

Fig. 21-79. Toothed roll crusher

Courtesy of Eagle Crusher Company

Fig. 21-80. Mobile crushers

consists of a toothed or fluted roll which revolves close to a crushing plate. The teeth or projections are called sluggers and act as sledges in breaking large stones. Smaller pieces are dragged between the roll and the plate and crushed by dragging pressure.

This does its best work on stratified or laminated rock that is not particularly abrasive. It passes clay or other sticky material with little difficulty. Product is usually coarse.

Double roll crushers consist of two power-driven rolls which rotate in opposite directions, with their top surfaces moving toward each other. Stone is pulled down between them by gravity and by friction with the roll surfaces. The rolls may be smooth, corrugated, or toothed. A Jeffrey toothed model is shown in Figure 21-79. Another make uses one set of axles for two stages of crushing, using different size rolls in parallel boxes.

The drive roll is fixed in position. The driven roll is mounted in a carriage pushed toward the drive roll by springs or cushions, and held by adjustable stops. Tramp iron will compress the springs, avoiding damage. Or drive may be through rubber tires, to protect by slippage.

These crushers are made in lab sizes, and with rolls up to 96 inch diameter and width (face) to 36. They are favored for finish crushing of most rock types.

In general, the ratio of reduction for feed materials over one inch in diameter is restricted to four to one, but smaller pieces may be reduced as much as ten to one.

The size of rock that can be processed depends on the angle of nip and friction between the stone and roll surfaces.

The angle of nip is found by drawing lines from the roll centers to the points of contact with the stone, and drawing tangents to those lines. The angle at which the tangents intersect is the nip. It should not be over 31° for general use of smooth rolls.

This angle is reduced by using smaller rock or larger rolls, or by pulling the rolls apart so that the product will be coarser. The friction will be affected by the hardness or slipperiness of the rock and the surface of the roll. A toothed, pitted, or corrugated surface will increase gripping power.

Roll surfaces tend to wear in grooves so that a side shifting adjustment is often provided to equalize wear. This may be manual or automatic.

Mountings. Heavy crushers are installed on concrete foundations. Smaller sizes may be set in masonry, in steel frames for semipermanent installations, or mounted on mobile chassis of either self-powered or trailer types.

The engine may be permanently hooked up to the crusher or be in a separate unit connected by a shaft or belting. Diesel, gasoline, or electric power is used.

Figure 21-80 shows two mobile mountings for Eagle light crushers.

PROCESSING PLANTS

Complete plants for the processing of blasted rock, or of pit-run gravel containing oversize stone, perform the function of taking in a coarse and variable material and turning out one or more classes of graded and uniform product of greater value.

Equipment includes a hopper into which spoil is dumped by shovel, conveyor or haul units, screens to separate pieces according to size both before and after crushing, from one to three crushers to reduce oversize pieces, and hoppers or chutes and conveyors for individual handling of different sizes separated by the screens. Provision must be made for removal or storage of the products.

The plant units should be protected by a grizzly (coarse primary screen) on the hopper, which would prevent rocks too large for the passages or primary crusher from entering them. Feed from the hopper to the primary crusher or screens may be by gravity, by apron conveyor, or belt conveyor. Such conveyors may be a few feet or several hundred feet long. They serve both to transport the material and to regulate the amount entering the crusher.

Bypassing and Reprocessing. A crusher is most efficient when all pieces in the feed are too large to pass through it without crushing. It is therefore desirable to separate and direct the raw material so that the fines will go around the crusher rather than through it.

The product of the primary crusher will have a number of pieces small enough to fall through the secondary crusher, and it is also usual to cause the fines which bypass the primary to mix with its product. A further separation is therefore required to avoid overloading the secondary with small sizes.

Any crusher will pass some oversize stone, particularly of material that tends to splinter into long thin pieces. Crushed ma-

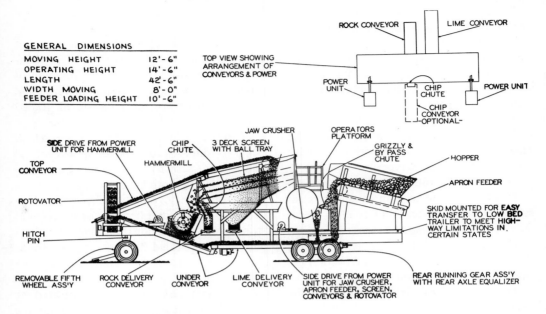

GENERAL DIMENSIONS

MOVING HEIGHT	12'-6"
OPERATING HEIGHT	14'-6"
LENGTH	42'-6"
WIDTH MOVING	8'-0"
FEEDER LOADING HEIGHT	10'-6"

Courtesy of Universal Engineering Corp.

Fig. 21-81. Limerok crushing and screening plant

terial is therefore returned to the feed stream from which the finished pieces will be screened out and delivered to output hoppers, intermediate sizes routed to the secondary crusher, and oversize sent back to the primary crusher.

This rehandling of material gives the plant an appearance of complexity and duplication of work. However, conveyors and screens have lower power and maintenance requirements than crushers and pay for themselves in saving crusher size and wear, in addition to making possible more uniform and valuable products.

The use of two or three crushers in series may be economical even where a single crusher could be obtained which would complete the reduction itself. The ratio of reduction—that is, comparative size of the feed and the product—determines both the power requirement and the output of a crusher. The table of single roll crusher capacity in Figure 21-70 indicates the increase in output obtained by increasing spacing to obtain a coarser aggregate. Also,

when the opening is large a greater percentage of the feed can be bypassed, again increasing the volume of unsorted material handled.

If a secondary crusher is used the primary may have quite a low reduction so that a comparatively small, light, and inexpensive crusher can be substituted for a heavy one. Also, if a third stage is used, the secondary unit will be relieved of part of its burden and will have to reduce that which it does get much less.

The size of the primary crusher in a stationary plant is determined largely by the production required, the size stone to be handled, and the cost.

Required production may be obtained by installing one crusher large enough to do the whole job, a series of crushers that will do it in stages, or two or more separate units each of which processes a part of the material.

Size. Usually, crushers are purchased to provide output required by estimated demand and excavators are obtained in sizes to keep them supplied. Occasionally, however, a crusher is selected to match the output of the excavators. The capacity tables can be used in calculating sizes required in any case, but because of the very wide variations in proportion of fines, and in production of both types of machine, results should be checked against field experience.

The primary crusher will not have to handle all material supplied to the plant as much of it will be undersize, and will be screened out for direct delivery to the secondary. The proportions will vary in different pits, in different sections of the same pit, and in the same section for successive blasts.

Large crusher openings reduce both blasting and loading costs. If big pieces can be processed, fragmentation need not be so fine so that holes can be more widely spaced or lightly loaded. Or drilling and loading can be standard and savings obtained in reduced secondary blasting. Shovels can be less selective about the size rocks they pick up.

Big shovels are more efficient than small ones in rock piles, and where sufficient output is required will give substantially lower digging costs. However, they pick up big rocks and should be teamed with crushers large enough to take anything that will go through the bucket. The tables indicate the size crushers required for various bucket capacities. This does not allow for oversize balanced on top of the dipper.

In large operations, several shovels and crushers will be used. Cost of plant may be reduced by including one very large crusher that will take rock rejected by grizzlies on smaller units. This arrangement will require either occasional or constant use of some machine capable of collecting the boulders and feeding them to the big crusher.

In general, rock which is too large to crush should not be hauled unless the plant is near an abandoned pit into which oversize can be rolled or pushed. Secondary blasting near the plant must be done under mats and is more expensive and risky than when done in the pit.

Mobile Crushing Plants. The Universal Limerok plant, shown in cross-section in Figure 21-81, is a mobile, single unit, double reduction rock crushing and sorting plant designed for use in limestone quarries.

The hopper may be protected by a bar grizzly to prevent the entrance of boulders too large for the primary crusher. Blasted rock may be dumped in the hopper by a shovel, conveyor, or trucks on a platform or bank.

An apron feeder carries the rock forward to fall on a grizzly which passes pieces small enough not to require primary crushing, and slides the larger ones into the 18″ x 24″ jaw crusher.

This crusher dumps on a conveyor belt which also carries the pieces that dropped through the sorting grizzly. Transfer is made to a second belt which runs up an incline and discharges into a rotary elevator made up of a series of buckets carried inside a revolving drum which is set across the frame. This "Rotovator" elevates the load and drops it on an upper conveyor which discharges onto a three deck screen.

Stones which fail to pass the coarse mesh screen on the top deck move forward into the secondary crusher, a hammer mill. Product of the hammer mill falls on the forward bottom belt which takes it back to the screen.

Pieces passing the upper deck, but failing to go through either the second or third deck, slide forward into a chute opening onto the rock delivery conveyor. Movable gates are arranged so that when smaller sizes are required, the rejects from the second, or from both the second and third decks, may be directed into the secondary crusher. An extra "chip" conveyor can be installed with an additional chute to allow side delivery of two sizes of crushings.

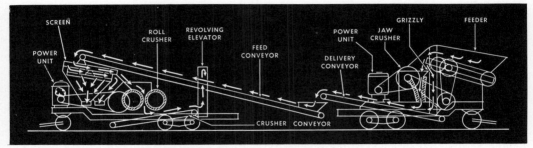

Courtesy of Gruendler Crusher & Pulverizer Co.

Fig. 21-82. Two-unit portable plant

Fine particles which go through all three screens fall into a hopper that feeds the lime delivery conveyor.

The side delivery conveyors for rock, lime, and chips fold up against the sides of the machine for transit. The hopper and primary crusher are mounted on a subframe and skids so that they can be removed for separate transportation, if necessary.

Two engines are used. These are mounted separately on rubber tires. They are positioned beside the main unit and drive through propeller shafts or belts. One operates the hammermill, the other the primary crusher and the rest of the machinery.

The use of two engines permits separate control of the speed of the secondary crusher. High speed produces a greater percentage of fines for agricultural limestone; low speed, pieces for road metal.

The Gruendler two-unit portable rock and gravel plant, Figure 21-82, has two separate trailer chassis, one of which carries the feed hopper and primary crusher, the other the screens, secondary

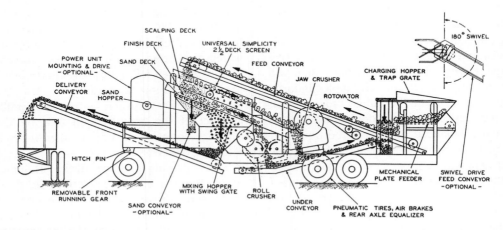

FLOW OF MATERIAL

From the charging hopper the pit material is uniformly fed by the mechanical feeder onto the feed conveyor which conveys it to the 2½-deck screen. Material retained on the top scalping deck is chuted to the primary jaw crusher. Material passing top deck and retained on second deck flows to the roll reduction crusher. Material passing second deck and retained on bottom half-deck sand screen flows to the mixing hopper. A swinging gate permits desired amount of fines to be mixed with the finished product. Material from both crushers falls onto the under conveyor where it is conveyed to the Rotovator which elevates it back onto the feed conveyor which again takes it to the screen. Finished material is delivered from the mixing hopper to truck by conveyor.

Courtesy of Universal Engineering Corp.

Fig. 21-83A. Material flow diagram

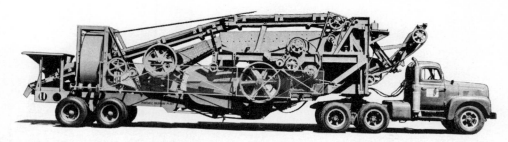

Courtesy of Diamond Iron Works Division of Goodman Manufacturing Company

Fig. 21-83B. Plant on a semitrailer

crusher, and output conveyors. The power unit for each section is mounted permanently on its chassis to facilitate moving.

The Universal 880 Gravelmaster, Figure 21-83A, serves to reduce and size run-of-pit gravel. In this the feed is taken from the hopper by conveyor belt to a two and a half deck gyrating screen, which sends rejects from the top deck into the primary jaw crusher, from the secondary deck into the secondary roll crusher, and from the half deck onto the delivery or output conveyor.

Material passing all the screens enters

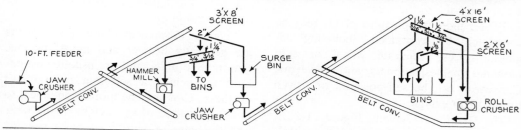

Flowsheet of material through plant from primary crusher to finished sizes

Courtesy of Baldwin-Lima-Hamilton Corporation

Fig. 21-84. Stationary crushing and processing plant

Courtesy of Barber-Greene Company

Fig. 21-85A. Asphalt mixing (blacktop) plant

the sand hopper. An adjustable gate permits adding any desired quantity of this sand to the gravel on the output conveyor. A sand conveyor can be installed to stockpile it or to side load it onto trucks.

The product of both crushers falls on a conveyor which takes it through the bucket drum and back to the screens.

Trailer mounted portable plants may be completely self-contained, including engines, or use separate wheel or skid mounted engines, hoppers, and conveyors. Conveyors can be self-powered or driven by the plant engine.

The Diamond 77 plant, Figure 21-83B, is mounted on a semitrailer frame, wheels and fifth wheel, for maximum maneuverability. The 2½ deck screen operates at an 18° angle, and can be lowered hydraulically to travel position.

Conveyors may discharge plant product into trucks, piles, or bins; or into a washing and screening plant on a separate trailer.

When maximum mobility is required, as in a unit which is loaded by a shovel in a low bank, use of a built-in power plant is desirable as it can be moved without delay for detaching parts and reassembling or lining up.

Fixed or Permanent Plants. Fixed plants are generally heavier and more spacious than mobile units, and are usually custombuilt to include a selection of standard crushers. They may be more economical to operate because of the weight of the units and the ease of getting at parts needing service. Moving is difficult and expensive.

Figure 21-84 shows a layout of an Austin-Western plant that includes several conveyor-connected buildings.

ASPHALT MIXING PLANTS

The asphalt plant is the key unit in most bituminous paving. It manufactures blacktop (officially asphalt concrete or bituminous concrete), and usually loads it into trucks that take it direct to the job. It may also store and dispense the fluid asphalts used in surface treatments and road mix.

Asphalt plants come in all sizes. There are permanent factories covering many acres, semi-permanent or knockdown plants that may be quite large but made of sub-assemblies that can be readily detached and moved, mobile plants mounted on one or two trailers that can follow the work, and even patching rigs that are towed behind a pickup and turn out batches as small as 4 cubic feet.

This section will describe briefly the process of manufacturing asphalt concrete in a large or medium-size plant. The product will be called blacktop most of the time, as this term is simpler and shorter than the official names, and is used almost exclusively by the men who work with it.

Blacktop (Asphaltic or Bituminous Concrete). Blacktop was described and discussed in Chapter 8. It will be considered from the manufacturing angle here.

It is a mixture of asphalt, which is usually in its basic form, called asphalt cement; coarse aggregate, which is usually crushed rock, but may be natural gravel; fine aggregate, which is sand, either natural or (more rarely) crusher-made; and a small amount of fines, sometimes called mineral fillers.

The asphalt functions chiefly as a binder or glue, the aggregate supplies strength, and the binder gives body to the asphalt.

Plant Components. Almost all plants capable of making high grade asphaltic concrete to meet state specifications are made up of a number of more or less standard components. They are designed and assembled so as to produce a required volume of material, accurately mixed in predetermined proportions at elevated temperature, very rapidly and uniformly.

The major components are systems for cold feed, heating and drying, screening, hot storage of aggregate and of asphalt, proportioning and mixing ingredients, and emission control.

While these basic components are similar in most plants, operating and quality controls offer many different levels of sophistication. Regulators for proportioning and mixing vary from a man operating hand levers to systems that are completely controlled by computers.

Cold Feed. The typical cold feed system includes a handling system for incoming aggregate, storage piles, a transfer or conveying system, and a set of hoppers called cold bins.

Supplies of aggregate may be unloaded from barges or rail cars by clamshells or conveyors, dumped from trucks and piled by dozers, or handled by any suitable bulk materials system.

Storage piles serve both as an hour-by-hour source of material for the plant, and as reserves for future needs. There are always at least two, one of fine aggregate (sand) and one of coarse aggregate (crushed rock). It is usual to keep the coarse material in two or more piles, each of a different size. The number of different-size piles is affected by the production rate of the plant, and details of its operation.

The aggregate supply piles must include all the sizes that will be used in the mixes. It is not essential (although it is helpful) for them to be in the same proportions in which they will be used, nor that they be completely separated from each other. They will be mingled in the dryer anyhow, and separated again by screening.

Where storage space is limited, piles may be separated and backed by masonry or metal walls, so that in effect they are in large open bins.

A large installation may have a tunnel

and conveyor under the storage area, with a remote controlled trap and gate under each pile or location. Feed may then be moved direct from the piles to the dryer, without use of cold bins.

If supply barges or other containers are unloaded by clamshell, it may also have the job of keeping the cold bins filled. It may do this directly from the barge, but more often all or most of the aggregate goes first into storage.

Rubber tired front loaders are used for longer distances between piles and bins. They need a ramp up to one side of the bins, for easy dumping.

Fig. 21-85B. Dryer

Level of material in cold bins is usually checked visually. The plant operator and a loader operator may be able to see part way into them. A clamshell man can check level by lowering the bucket until it rests on the aggregate, and gauging depth by the wraps of cable on the hoist drum. It is good policy to keep them filled to the top, where the level is easily observed.

These bins have controllable discharge arrangements. Sand may flow through a bottom chute, and rest on a variable speed conveyor belt, whose speed determines the rate of feed. Coarse aggregate is carried by a vibrating or reciprocating feeder. Both

types discharge onto a belt, which carries the combined feeds to the dryer.

Cold feeds are set according to the needs of the mix or mixes being made. If any sizes are over or under supplied to the hot bins, feed adjustments are made either manually or by remote control. The system may be stopped by switching off the feed motors.

Dryer. Specifications for blacktop should call for dry aggregate. In most parts of the country, aggregate is damp, with water content varying from 2 to 15 per cent. Under similar conditions, fine aggregate will hold more moisture than coarse. In systems design, a water content of 5 per cent is often assumed.

This excess moisture is removed in a dryer. This is a revolving drum fitted with internal slats, in which aggregate is continuously lifted, dropped, and tumbled in a blast of flame and hot air. This process serves to both drive off the water and heat the material.

The dryer drum is slightly tilted, so that the aggregate works its way by gravity from the high feed end to the lower discharge chute. Non-adjustable bars or paddles may be installed to either speed up or retard the flow.

Heat is supplied by a large burner, fueled with oil, gas, or a combination, and located at the discharge end. The draft it creates is supplemented by powerful air blowers, pressure at the burner end, and suction at the exhaust.

The output capacity of a dryer varies directly with its cross section area, that is in proportion to the square of its diameter. The relationship to length is less exact, but a long drum properly heated will outproduce a shorter drum of the same diameter.

Gas velocity—the speed of heated air through the drum—is an important factor, but increasing it beyond normal limits is expensive and aggravates the dust problem.

The percentage of moisture to be removed is a critical factor. Wet aggregate

Courtesy of Barber-Greene Company

Fig. 21-85C. Large portable plant

absorbs much more heat than dry. It may have to be fed into the drum at a lower rate to get proper results. If the dryer tends to be the plant bottleneck (it often is), this factor may be important enough to justify arranging at least partial cover for the storage piles.

Aggregate is discharged from the dryer through a chute into the base of a vertical bucket elevator (hot elevator) that lifts it to the top of the tower, from where it can move through the rest of its processing by gravity.

Dust. The particles or pieces in aggregate tend to wear as they slide or fall over each other in any sort of handling or processing. The reduction in particle size is usually not significant, but the dust created by the wear is a major problem. If retained, it makes the aggregate dirty; if removed, it becomes a control and sometimes a disposal problem.

Heat and tumbling in a dryer produce a lot of dust. The air stream blows it off the pieces, together with dust left from previous handling, in such quantities that its control was considered necessary even before air pollution was taken seriously.

Dust control is important enough to deserve a section of its own. It will be discussed under EMISSION CONTROL.

Fines. Some or all of the dust collected by dry control systems can be used in the product, as a certain percentage of fines (mineral filler) are required by blacktop specifications.

If the quantity is insufficient, or if wet collection makes the dust too difficult to use, fines are supplied by purchase of portland cement, fly ash, ground limestone, or similar materials. These may be purchased in bag or bulk form.

Fines must be handled in dust-retaining equipment, measured accurately, and added to the aggregate between the hot bins and the mixer.

TOWER

Screens. Aggregate lifted by the hot elevator is discharged onto a set of screens, which separate it into the sizes specified by the job mix formula.

The screening unit is usually a compact, vibrating type, with 2 to 3½ decks to obtain separation into four sizes. These might be ⅛ (sand), 5/16, 9/16, and 1 or 1⅛.

Each size is conveyed or chuted into a separate hot bin.

Hot Bins. The hot bins are hoppers or containers with bottom gates. Each holds a limited supply of one size of aggregate, heated in the dryer and only moderately cooled by hoisting and screening.

For full plant efficiency, hot bins should

Fig. 21-86. Mixing tower, cutaway

be insulated, but they usually are not. Lack of insulation permits cooling below formula temperature during a comparatively brief plant breakdown, or interruption in the flow of trucks. Small deficiencies may be tolerated, large ones require recycling the material into the dryer feed.

In a batch plant, a measured weight of aggregate is delivered from each bin to the the weigh hopper, in accordance with a job mix formula. If operation is manual, an operator opens the gates one at a time. A dial gauge shows the weight discharged. When it reaches the correct figure for that size, the operator closes the gate and opens the next until its proper weight is shown. When all bins have been tapped, the total aggregate

can be discharged into the pugmill for mixing.

In a continuous type plant, feeders under each bin are set to deliver measured quantities to a conveyor or elevator that takes the total discharge of all of them to the mixer.

Emptying of any bin will stop operation, so they must be equipped with indicators. The minimum is a low level warning; a preferred system is a set of three lights at the operator's station, showing constantly whether the level is low, middle, or high. Supply to a bin is adjusted by changing the proportion of its size in the cold feed.

Asphalt. The asphalt used in blacktop manufacture is the asphalt cement described in Chapter 8, which is semi-solid at ordinary temperatures. It must be heated in the storage tank in order to pump it through supply lines. The temperature is made high enough, 275 to 325 degrees Fahrenheit, to assure quick and thorough coating of aggregate in the mixer.

Heat must be applied with caution, as overheating cooks or cracks asphalt, destroying its value.

Any of four different heating systems may be used. The tank may contain coils, which are heated by circulating fluid from outside, either steam or hot oil. The tank may include immersion-type electric heaters, limited to moderate temperatures, which contact the asphalt directly. Oil burners with various safeguards may also be used for direct heating.

It is desirable to have two tanks, one for accepting fresh supplies, and one for feeding the plant. A smaller tank is easier to heat rapidly after a shutdown, and the feed tank is not then subject to sudden temperature drop if a large quantity of cooler asphalt is delivered. However, temperature drop may also be controlled by circulation arrangements.

Tanks, pumps, and lines to the mixer should all be heavily insulated, both for convenience in operation and for fuel economy.

The line to the pugmill is usually a loop in which hot liquid asphalt is circulated continuously by a pump while the plant is in operation, and through short shutdowns. This keeps asphalt at proper temperature instantly available for mixing, simply by opening a valve. When the plant is shut down for the night or for major repairs, the pump is reversed to empty the line before being stopped.

Pugmill. An asphalt-aggregate mixer seems to be always called a pugmill, whether in a batch or continuous mixer, or in a mobile mix-on-the-job unit.

A standard pugmill is a tub containing mixing paddles attached to a pair of axles

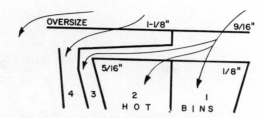

Fig. 21-87A. Typical screen arrangement

rotating in opposite directions. Paths of the tips overlap in the center. Sides and bottom are curved for close clearances.

Batch Mixing. Mixing is most efficient with a load that just allows the paddle tips to be exposed at the top of their circle. The space below a horizontal plane at this level, less bulk of shafts, paddles, and liners, is called the live zone, and a load may be stated in percent of live zone.

In rough calculations, the batch is assumed to weigh 100 pounds per cubic foot. The pugmill (and batch plant) may be rated on the maximum net capacity, expressed in pounds or in tons. Sizes of 1½ to 5 tons are usual. Batches may take up 45 to 95 per cent of live zone volume.

Weighed aggregate is fed from the hot bins in a stream into the mixing box, usually being mixed dry for a few seconds before being sprayed with a measured quantity of asphalt. Total mixing time is the number of seconds between cutting off the feed, and opening the discharge gate. This may be as little as 20 seconds, or as long as 75. Forty seconds is a usual automatic setting.

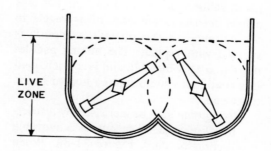

Fig. 21-87B. Pugmill cross section

Fig. 21-87C. Pugmill paddles

Mixing time is affected by many factors, including design, condition, and efficiency of the mixer, the mixture design, and the condition of the materials. The most important variable, however, is often the judgment of the engineer or the public authority who specifies the time. There is a general tendency to shorten the process, particularly the dry mixing part.

The batch may be discharged directly into a truck, which remains under the chute until it has obtained enough batches to make up a load. Or it may go into a hot storage bin, either directly or through a conveying system.

Continuous Flow. In a non-batching plant, calibrated feeders under the hot bins deliver measured amounts of aggregate to the entrance to the pugmill, where it is sprayed with asphalt. The mixing paddles move it as they blend it, and it is discharged in a continuous stream into a hopper at the other end.

Mixing time may be increased by restricting the discharge by means of a movable obstruction or dam, Figure 21-88. When this is raised, it increases the time the mate-

rial is in the mixer. The rate of flow remains the same, except for a brief period while the mixer fills to the higher level.

The continuous process allows setting the controls just once for a particular mix, at the proper rate of feed for each ingredient, and then continuing production at those settings until the job is completed or a change is made.

Direct Flame Process. In the direct flame or turbulent mass continuous process, not illustrated, hot bins are eliminated, and mixing is done in the dryer. Carefully proportioned cold aggregate is fed into a revolving drum dryer, the discharge end of which is a pugmill.

The asphalt, which is usually emulsion, but may be cutback or cement in some models, is added to the hot, dry aggregate in the oxygen-free atmosphere beyond the flame. Water in emulsion goes up the stack as steam, while volatiles from cutbacks are burned in the stack. The asphalt removes the fines from the air and returns them to the mix. Lack of oxygen prevents oxidizing of the asphalt.

The mix is discharged into an insulated

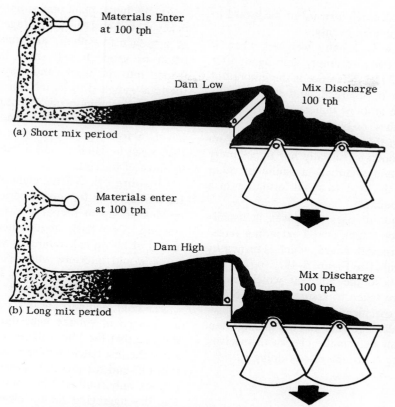

Fig. 21-88. Mixing time regulation in continuous flow

hopper, from which it is discharged into trucks.

These machines are also made in small portable patching models, with capacities as small as 3 tons per hour, with discharge onto trays from which the mix may be hand shoveled.

Hot Storage Bin. Either type of pugmill is capable of a steady rate of production over a period of hours, but usually is idle part of the time because there is no truck under the discharge gate. There may be delays in spotting a truck that is waiting to load, and/or a lapse of time between truck arrivals.

Even a high capacity plant may have a rate of production that is low for loading big trucks. A 3-ton mixer turning out a 2½ ton batch every minute would take 6 min-

utes to load a 15-ton truck. Smaller batches and bigger trucks take even longer.

When a line of trucks is waiting, which is usual the first thing in the morning and not uncommon at other times, slow loading is costly in truck time.

Irregularities can be smoothed out by a hot storage bin, in which hot mix can be stored as fast as the mixer can turn it out. An atmosphere of inert gas such as nitrogen prevents or slows oxidation, and insulation and a heating system prevents cooling. A large capacity gate permits loading a truck in seconds.

Trucking. Blacktop is usually delivered directly into trucks from the mixer discharge of a batch plant, or through a hopper with a control door in a continuous plant.

The truck may remain in one position

until filled, or move forward or backward to distribute the load evenly.

The truck is usually weighed when it comes into the yard empty, and again after it is loaded. The difference is the amount of material to be charged, and it should be close to the amount indicated by the number of batches, or by flow meters.

While on the scales, or waiting to get on them, the operator usually has opportunity to spread and secure a tarpaulin or other cover over the load, to avoid formation of a crust by cooling.

When not in use, the cover is usually wound up on a stick or rod extending across the top of the cab guard, which is turned by a hand crank. See Figure 8-49.

This is also the point at which testing samples are taken to assure that the black-top meets specifications. The principal test-ters are state engineers, but the management of the plant may make checks themselves, particularly if no one else is doing it.

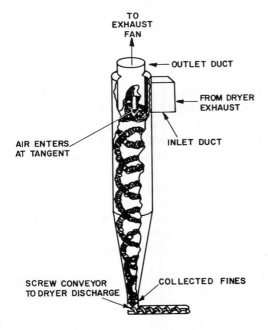

Fig. 21-89A. Cyclone, collecting dust

An automatic plant may supply a delivery ticket that shows the percentage by weight of aggregates, asphalt, and fines in each batch or drop. It will show the amount loaded into the truck, and may also show the total supplied to its job during the day.

A rather common misadventure is loading a truck whose tailgate has not been fastened. A front loader can close the gate so that it can be latched, and it can also clean up most of the spill.

Cleaning Up. When a plant is shut down for the night, or for longer periods, it must be cleaned up in preparation for the next start-up. Two basic requirements are removal of all asphalt remnants whose hardening would interfere with operation; and emptying of non-insulated bins. Fortunately, the two tasks can be combined.

The cold feed and the dryer are stopped or reduced in volume near the end of the shift, so that the bins will be partly empty. After the last batch is mixed, the asphalt is shut off, and the pugmill operated with aggregate only, until the hot bins are empty. The dry aggregate should clean off all asphalt on working surfaces, but they must be inspected to make sure that it has actually done so.

The aggregate is discharged from the mixer into a truck, and dumped on a clean surface in the yard, or in a special bin, to be reclaimed when work resumes. The small amount of asphalt coating it has acquired will burn off in the dryer.

Asphalt or blacktop deposits not reached by this operation may be removed with various combinations of shovels, scrapers (including loaders for spills on the ground), flame guns, brushes, and kerosene.

When a change is made from a base mix with large aggregate to a finer one, the first batch may be run through dry—that is, without asphalt—and routed back to the dryer feed. This reduces the chance of the fine mix being difficult to work because of leftover big pieces.

EMISSION CONTROL

Emission control units make up a substantial part of both the bulk and the cost of a blacktop plant. They are chiefly or entirely designed to remove dust that is created or loosened by handling and heating aggregates, and may be called either separators or collectors. The bulk of this is in the dryer exhaust, with much smaller amounts from processes in the tower.

Three principal methods are used. There is dry separation by centrifugal force in cyclones, or by filtering through fabric in a bag house. Wet separation dampens the dust particles so that they increase bulk by sticking together, then throws them by centrifugal force against wet surfaces.

It is usual to have two stages of separation. The first or primary step takes out coarse particles, usually with some finer ones; then secondary units take out the balance.

Fine particles are usually measured in microns. A micron is 1/1,000 of a millimeter, or 1/25,400 of an inch. They are then divided into three classes, ultra-microscopic (1/10 to 2 microns), microscopic (2 to 10 microns), and macroscopic (10 to above 400 microns). Portland cement particles average about 10 microns, smoke about 3/10 of one micron.

Fineness of required separation varies with laws in different localities, which tend to become more severe.

Centrifugal (Cyclone) Collectors. A cyclone, Figure 43-89A, is a combination cylinder and cone. The dirty (dust laden) air enters the upper cylinder on a tangent (at an edge) and is forced into rotary motion. It moves down into the cone, whose smaller diameter increases its rotational speed, then turns upward through a central exhaust opening.

The dust particles are heavier than the air, and are thrown against the curved wall of the chamber. They move downward

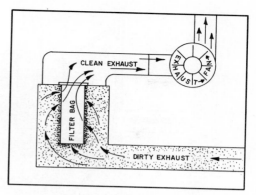

Fig. 21-89B. Single filter

along the wall with the spiraling of the air, and by gravity. When the air stream turns up to the exhaust, the particles slide down to accumulate in a trap at the bottom. They may be removed by a constantly running screw conveyor, or by intermittent opening of a dump valve.

Cyclones are usually in parallel sets, each receiving a fraction of the air to be cleaned.

They are most effective with large and heavy particles, and are therefore highly favored for primary separation. However, reducing the diameter of a cyclone improves its ability to trap smaller particles, so it can be used in secondary separation if there is tolerance for very fine particles. But such a unit would tend to choke up on coarse particles, unless protected by a primary separator.

A cyclone that removes 99 per cent of 30-micron particles might capture 60 per cent of the 10 size and 20 per cent or less of 5-micron. Or it might be designed to trap only 10 per cent of the 10-microns if the trapped dust were to be used as fine sand rather than as mineral filler, which could be obtained from the secondary stage.

Fabric Filters (Baghouse). Particles can be removed from air by passing it through a fabric filter fine enough to hold them. This principle is applied in many different ways, of which the most familiar is the vacuum cleaner.

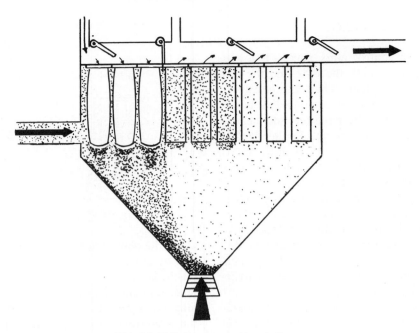

Fig. 21-89C. Baghouse circulation

The application that has been found most suitable to hot mix plants is called a baghouse. An airtight building is divided into two chambers or plenums. Dirty gas enters the lower chamber, passes through filters in the ceiling into the clean gas chamber, and then into the open air.

The most effective filter material at present is a felted fabric made of DuPont's Nomex, a high temperature nylon, which can stop particles smaller than one micron.

The fabric is usually shaped into bags which are fitted over cylindrical wire cages, which are placed to cover ports over the dirty gas chamber, with closed ends down. The exhaust enters with moderate pressure, air and gas flow through the fabric, leaving the dust on its outer surfaces.

Accumulation of particles will clog the filter rapidly. It is freed of them by reversing the air flow, to blow and knock them off. One method of doing this is to divide the clean air chamber into sections sealed from each other, each equipped with a damper to close its exhaust passage.

A blower raises the air pressure in these sections, one at a time in regular rotation, so that air flow through its bags is reversed, causing them to balloon out with a snap. This jolt combines with the reverse flow of air to dislodge practically all the dust. Pressure is then diverted from that upper chamber section to another, and filtering action is resumed.

The dislodged dust falls onto a smooth metal floor that slopes steeply from each side to a center trough, in which dust is removed by a screw conveyor.

The bag cleaning cycle may be anywhere from one to 8 or more seconds. It is adjusted by experiment to the type of dust, and other conditions. The capacity of the system allows for the number of bags normally out of service during each cleaning cycle.

Exhaust temperature when entering a baghouse is usually between 250 and 350° F. Nomex will be damaged if it rises over 450°. It should be protected by a thermostat that will shut off the dryer burner or

Courtesy of Barber-Greene Company

Fig. 21-90A. Paver, rear view

cause other protective action if temperature reaches 400°.

A low temperature must be avoided, as the exhaust is heavily charged with water vapor that might condense and clog the filters if the air is allowed to cool below its dew point.

Wet Collectors (Scrubbers). Scrubbers may use water in the form of a very fine spray or mist to wet particles. Finer particles require finer mist, and this has greater power demands. Mist may be generated by high pressure spray through nozzles, or air pressure drawing water through venturis.

Impingement type scrubbers do not need a fine water mist, but depend on particles striking a wet surface on which they are trapped, and from which they can be washed.

Wetted particles may be collected in cyclones similar in principal but not in detail to the dry ones discussed earlier. Wetting causes particles to stick together, increasing

their size so that they are more readily captured.

An important problem with wet systems is disposal of the mud, which is extremely fine grained, and may be acid because of sulfur in the burner fuel oil. Solids and water are partly separated in settling basins. The water may be re-used, the mud must be dredged out. It would require drying and grinding for use as mineral filler. It is undesirable land fill because of its fineness, which makes it unstable when either very wet or very dry.

Noise. Blacktop manufacture is a fairly noisy operation, and many plants in or adjoining residential areas have difficulties as a result.

The principal source of noise is the dryer, particularly the burner and the blowers. They can be muffled by jacketing and controlled air intake. Exhaust is already muffled by dust collectors.

Noise from falling and sliding stone in the dryers, in chutes and on screens may be

Fig. 21-90B. Paver, side view

reduced by changes in design detail, and by jacketing. Conveyors are quieter than bucket elevators.

PAVING MACHINE (FINISHER)

Most blacktop (bituminous concrete) is spread and graded by paving machines, also called pavers and finishers. These units are variable in size, in sophistication of controls, and in price; but the larger ones are quite similar to each other in basic design.

Each is made up of two basic sections, which are always used together, the tractor and the screed.

Tractor. The diesel-engined tractor may have either crawler tracks or wheels for support and propulsion. It carries a receiving hopper in front. The center of the hopper floor is part of a pair of conveyors which move material from the hopper through the tractor to two distributing screws at the rear.

The hopper is usually hinged each side of the conveyors, and the sides may be tilted to dump into the center. The hopper frame carries rollers, which enable the paver to push trucks by their rear tires as they dump. There may also be a pair of hydraulically controlled arms that swing roller tips into the hollows of the rear truck wheels, to hold the machines together on down grades.

The conveyors are usually the bar type. Material is pulled along a flat steel plate by cross bars attached to chains. The two conveyors occupy more than half the width of the machine. Their discharge to the screws is regulated by flow-control gates.

There are two variable-speed spreading screws or augers, each synchronized to the conveyor and gate on its side. They distribute the mix to the sides, and send it against a curved deflector plate and down to the pavement. Retaining plates prevent (usually) spillover at the ends.

The augers, and the chamber or box in which they turn, may have means to slide or telescope inward and outward hydraulically, to keep the outer edge lined up with pavement edge or markers. Extensions may increase working width substantially.

The ends of the spreading boxes are open at the top. Mix required for patching and trimming is taken out of them with hand shovels.

The hopper, conveyors, and screws are sprayed or painted with kerosene at the start of a shift, and when resuming work after a shutdown for which they have been emptied. This prevents sticking and building up on the cool metal. A few moments of contact with fresh, hot mix warm the parts so that further applications are not needed during work.

Tracks and Wheels. Crawler tracks are usually the full length of the tractor. Idlers and bull wheels may be set a little high, so that machine weight is carried only on the large track wheels, and the crawlers make their own ramps in moving from one level to another. Shoes may be flat, have several low cleats, or be fitted with rubber pads.

In wheel carriage, there are two large pneumatic-tired drive wheels at the rear, just forward of the screws. There are usually two pairs of small, hydraulically steered front wheels in tandem, under the hopper sides. Their tires may be solid rubber, with wide flat treads.

Drive is usually hydrostatic, to provide an infinite range of working speeds to suit varied conditions.

Controls. Controls may be manual, or largely automatic with manual override.

There may be a single control station, or a pair of them, one at each rear corner. There may be an additional set of controls (usually buttons) for the spreader side shifts, placed at the rear and accessible to a man standing on the screed platform.

If there is one operator for dual controls, he is usually at the side of the machine where problems are most likely to develop. In a routine large scale job, this would be where matching to a set of guide markers, or to a previously laid pavement strip. In resurfacing work, it might be an irregular outer edge of the old pavement.

If there are two operators, both may be at their stations, or one may elect to stay on the screed to manipulate sideshifts, or to keep closer watch on the appearance of the mat.

Screed. The screed is made up of the tamping bar, the screed plate, and their adjustments and controls. It follows closely behind the spreading screws. It completes the spreading action, and provides a smoothly finished surface and some compaction. Its

Courtesy of Barber-Greene Company

Fig. 21-90C. Front of paver, showing hopper

special function is to provide means to produce a level upper surface in spite of irregularities in the original grade.

The tamping bar strikes off the loose material left by the spreading screws, and presses it down with a beveled under surface. It is followed by the screed plate, usually adjusted to a slight climbing angle, which provides further compaction and smoothing, leaving the surface ready for a roller. It may vibrate, to provide additional compaction and a tighter surface.

Both the bar and the plate are hinged at the middle to provide for making either flat or crowned surfaces. There are delicate adjustments for regulating height, cross slope, and pitch.

Each half of the screed plate is equipped with a kerosene or propane burning heater. This must be used when starting up cold, to prevent sticking, but is shut off during work.

The top of the plates, or a platform built above or behind them, provides a walkway from one side of the machine to the other, and is often an observation post.

The screed assembly is attached to the tractor by a pair of long screed or pull arms that pivot on brackets on the forward part of the tractor track frames or side frames. The height of these pivots may be hydrauli-

Courtesy of Barber-Greene Company

Fig. 21-90D. Paver on crawlers

cally adjustable. The arms pull the screed, but do not support it, so that it rests on the mix that it is smoothing.

When the machine is in forward motion, the screed tends to seek a level at which the path of the bottom surface of the plate is parallel to the direction of pull. The exact level, and therefore the thickness of the material being laid, is affected by the climbing angle of the plate, the resistance of the mix to shaping, and the relative height of the pivots that pull the arms.

The pivots' location, and the length of the tractor, serve to provide gradual reactions to changes, and to dampen out possible pivot movements from irregularities in the footing. However, a pivot will rise or fall as the tractor encounters changes in level that are two-thirds its length or more. Such vertical motion will change the line of draft, and may cause thickening or thinning of the mat (layer) being put down, maintaining mat thickness but not accuracy of surface, or combinations of these defects.

With manual controls, it is the operator's responsibility to make adjustments to maintain both the specified surface grade and the required thickness of material. In general, the paver's built-in self-leveling character-

istic enables it to automatically smooth over or cancel out ½ to ⅔ of the longitudinal irregularities in the base with a binder course, and the same fraction of the remainder in placing a surface course over the binder. This is with careful but not necessarily expert operation.

However, correction of side slopes (wedging) and establishment of crowned or banked cross sections, is entirely the responsibility of the operator and the foreman.

Automatic Controls. Medium and large pavers may be fitted with automatic controls. These machines can follow a stringline, or a guide shoe or ski on an existing surface, and leave a mat with correct grade and cross slope, almost regardless of base irregularities. However, variable thicknesses must be placed in several layers, each thoroughly compacted, or the surface will sink in the deeper areas during rolling and use.

Automatic devices include hydraulic lifts on the pull pivots for the screed arms, keeping them in line in spite of any vertical movements of the tractor. In addition, the bar and screed controls will respond automatically to any change in their action, and to instructions to change cross section.

The automatics may include means to keep the paver working at an even speed, in spite of changes in load caused by pushing trucks, or going up or down hill. Since speed is one of the factors that affects screed plate resistance and level, this is a worthwhile refinement.

Accessories. Two or four corners of the machine may be fitted with extension brackets from which light chains may be hung. One or more of these may be used to keep close visual reference with a guide edge of line. The chain is hung so that it is directly above and close to the pavement or string.

A pail of kerosene may be hung from the side of the paver, at any convenient point, for dipping tools to prevent asphalt from sticking to them, or to soften or remove it after it has stuck.

The paver may be equipped with a side burner for cold-weather use in heating and softening hard bituminous material at longitudinal joints, against which fresh material is being laid, to assure a good bond. If there is no such heater, the old edge may be painted with emulsion.

Small Paver. There are a number of paving machines used for smaller jobs that are simpler than the machines just described.

A typical paver in this class does not have either a tractor or a conveyor. A truck loaded with blacktop mix is backed against rollers on the hopper frame. Side arms are swung to place hook rollers in the grooves in the rear truck wheels, between hub and rim. The truck tows the paver with these while it dumps into the hopper.

The hopper opens directly onto the surface being paved, through a control gate. The material is struck off by the rear plate of the hopper, and is spread, smoothed, and slightly compacted by an adjustable, full width, one piece screed.

As soon as a load is fully dumped, the paver operator swings the tow arms out of contact with the wheels. The truck drives away, and is replaced by another. For ordinary intervals, the hopper is left partly full while waiting, to prevent cooling.

These machines may have either wheels or tracks. They may be towed slowly for short moves. For longer trips, they are lifted onto a trailer by means of a permanently fastened set of chains.

Work Crew. It is theoretically possible for a full size paver on a routine job to place a long strip of paving with only an operator, or with automatic controls, with no operator. But in practice, from two to six men, in addition to one or two operators, are needed in order to keep the job and the machine functioning at full efficiency, and to obtain best-quality results.

Some details of hand-working blacktop are discussed in Chapter 8.

Truck drivers need, or at least appreciate,

Courtesy of Barber-Greene Company

Fig. 21-90E. Paver control station

guidance as they back the last few feet to the hopper. Hot blacktop does not always dump smoothly, and as the driver cannot see inside the body, he may need to be told whether to raise or hold, in order to avoid overfilling or starving the hopper.

Dumping is often incomplete, with some fraction left sticking to the body. The truck is driven away from the paver with the body raised, then shaken by driving alternately backward and forward a few feet, with abrupt stops. Any mix falling in the path of the paver tracks or wheels is moved with a hand shovel or other tool.

A fully dumped body on a large truck may put the cab guard over 20 feet above the pavement. At this height it is likely to rake off utility wires and tree limbs. The truck spotter must watch out for overhead obstacles, and warn drivers.

In making a longitudinal joint with a previously laid strip, the screed is overlapped two or three inches, and held a little high. The resulting step should be beveled off by pushing its edge back over the fresh mat with a lute.

Stopping and re-starting the paver leaves a cross ridge, which should be smoothed with a lute.

If the mix is laid at a temperature less

Fig. 21-90F. Beware of overhead obstacles

than 290 degrees Fahrenheit (some states specify 275°) the surface is likely to be pulled and stretched by the trailing edge of the screed. Pieces of aggregate, and occasionally of foreign material, make grooves from ½ inch to several feet long. While these marks disappear during rolling, if closely spaced they involve a deficiency of material, making it advisable to hand shovel a thin layer of mix over them.

Maximum hand work is usually required for a levelling course. It is essential that pot holes be filled and tamped ahead of the paver. This is best done several days beforehand, but it often is not, or is not done thoroughly.

When building up an irregular outer edge, the end of the spreader box may run out of material momentarily, leaving gaps that must be filled by hand. Sometimes excess mix is left from spillovers, and must be removed or flattened.

Feathering out at high spots often causes paver irregularities that must be corrected with a lute or shovel.

Paving intersections with streets or driveways can seldom be done economically by a paving machine, and is done by hand. If the area is large, the laborers are supplied

directly by truck, rather than by shoveling out of the paver boxes.

At the end of a shift, all mix must be removed from working surfaces, or they must be soaked with kerosene to soften it permanently.

ASPHALT DISTRIBUTOR

Asphalt in surface treatments is usually applied by a spray bar or a hand spray on an asphalt distributor. This machine is mounted on a truck or semitrailer chassis, and ordinarily includes a heated tank, a pump, a spray bar and circulating system, and a measuring wheel.

A distributor usually has both a driver and an operator. The driver of the truck is responsible for steering, speed, and immediate obedience to bell signals. The distributor itself is more complicated, and its operator is responsible for all its non-travel functions.

The machine is dangerous, and must be treated with great respect. Fluid temperatures range from 160 to 400° F., or higher. The pipes and hoses are usually too hot to touch; and accidentally released fluid inflicts immediate, severe burns. The cutbacks or fluid asphalts are often hotter than their

Courtesy of Seaman Company

Fig. 21-91A. Asphalt distributor

flash points, and may catch fire explosively from flame or spark.

Cooling asphalt sticks to everything, including clothes, skin, hair, work platforms, control levers, and hand grips. Freshly spread asphalt is almost as slippery as wet ice, and anyone who steps on it, or gets the drive wheels of any machine in it, is likely to be in trouble.

Most grades of asphalt cool or set into solids or semi-solids that can block pumps and passages, glue tools down or to each other, and make almost permanent markers for mistakes in spraying.

The safeguards built into the machines, careful training of operators and assistants, and care in operation combine to maintain a low accident rate in spite of the hazards.

Tank. The tanks is usually elliptical in cross section, being wider than it is high. The shell is of moderately heavy steel such as 10 gauge, with heavier ends, covered with two inches of glass fiber or molded asbestos insulation under protective metal. The interior has two or more baffle plates set across it to reduce surging from truck movements.

An 18- or 20-inch manhole near the center provides limited access for service, in-

spection, and stick-measurement of load.

A sump with a coarse strainer, and sometimes baffles to prevent vortex (whirlpool) formation, permits almost complete delivery of the load into the pump suction line.

There is usually a ball float that operates a gauge to indicate the quantity in the tank, and a warning signal that appears when the tank is 80 per cent full.

A thermometer of the bar type, calibrated from 50 to 600° F., is kept in a well and withdrawn for inspection. It is a nuisance to read. There should be an option of a much more convenient dial thermometer which is always in sight.

Tanks are made in a number of sizes. The range in standard distributors is 800 to 3000 gallons. Light models, called maintenance distributors, hold from 400 to 1000 gallons. Special models may hold as much as 5500 gallons.

Ratings are for cold (60 degree) asphalt, and include a 5 per cent allowance for expansion during heating.

There must be a vent to carry off pressure from vaporization of cutback solvents, and from heating expansion of fluid. This is usually a small pipe from the air space in the top of the tank, discharging under the ma-

Courtesy of Rosco Manufacturing Company

Fig. 21-92A. Distributor tank

chine. In addition, the manhole is fitted with a relief device, to open at a few pounds internal pressure.

Vents and overflows are dangerous. While heating, it is important to watch for heavy fumes, and to make sure that no flames or sparks are present or probable in the area. RC cutbacks are particularly hazardous, as their naphtha vapors are very explosive.

Tank Heaters. Means must be provided to heat asphalt in the distributor. Suppliers store it at lower temperature than is required in using it; and long job delays can cause further cooling and even congealing in the distributor tank.

A standard arrangement is to have two U-shaped tubes, 8 inches or so in diameter, in or near the bottom of the tank, with openings in the rear. Burners force flame and heated air into one end of each tube, to be exhausted into a stack at the other.

Asphalt in contact with the outside of the

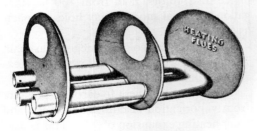

Courtesy of Rosco Manufacturing Company

Fig. 21-92B. Tank baffles and heat tubes

tubes is heated and tends to rise. It is also circulated by the bitumen pump, which should be operated during heating. Until the material is warm enough for pump circulation, you can obtain some mixing by going back and forth jerkily with the truck.

Temperature must be checked frequently, and overheating avoided. Burners should be shut off when the asphalt is a few degrees cooler than wanted, as the heat in the tubes will warm it some more. A full tank will stay hot for hours.

Too high temperature breaks down emulsion, boils off solvents, alters cutback mixtures, increases fire danger, and finally (at about 600° F.) causes the asphalt itself to crack and deteriorate.

The standard burner is a generating type which uses kerosene, supplied at low pressure by a fuel pump. It is started like an old fashioned blow torch, by lighting a wick in a puddle of fuel in a tray. This heats a burner coil above it, causing the kerosene in the coil to vaporize and emerge from a nozzle as a high pressure gas. Once lit, its flame keeps the coil hot. If it should go out, it should not be re-ignited until flues have had time to ventilate.

An optional arrangement is an atomizing oil burner similar to that used in a furnace, with either a hydraulic or electric motor. Another is use of liquefied petroleum gas (LPG).

The generating burner is the simplest, the cheapest, and the most dependable. Its only defects are that it must be started by hand, and warmed up for a couple of minutes. In general, these disadvantages are outweighed by the very important factor of reliability.

The same fuel pump may supply additional heaters. A portable heater may be attached to a flexible hose, and used for heating road patches, a clogged spray bar, or a locked pump.

Provision may be made for heating the pump with engine exhaust.

Power. Power for operating the bitumen

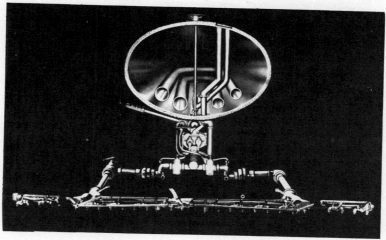

Courtesy of E. D. Etnyre & Company

Fig. 21-92C. Tank and spray bar

pump, the hydraulic system (if any), and accessories is provided by a gasoline engine or a hydrostatic motor.

On a full size distributor, the engine may be four cylinder, about 40 horsepower, with electric starter and standard accessories. Drive to the pump is through a friction clutch and a 2- or 4-speed transmission. A flexible coupling in the drive shaft allows for minor misalignment.

The engine, the pump, and the controls may be located either at the rear of the tank, or in front of it, behind the truck cab.

At the rear, both engine and operator are subjected to the fumes, heat, dust, and fire hazards associated with spraying the hot liquids. Communication with the driver is usually by bell signals only. Drive and circulation systems are likely to be short and efficient. The operator can watch the strip of oil being laid down, and take immediate steps to correct deficiencies.

In the front location, the engine and operator are relatively cool and clean. He can talk to the driver if they shout. But he can't watch the spray bar.

The truck engine provides power for hydrostatic drive. The pump is driven by a power takeoff on the transmission. The ma-

chine is much quieter, work associated with fueling and servicing a separate engine is avoided, and a fixed relationship exists between truck speed and pump output. However, problems arise from driving two systems through one clutch.

Operating characteristics of the different systems will be discussed later in this section.

Bitumen Pump. The main asphalt pump is usually a positive displacement gear type, of about 400 gallons per minute capacity. It may be mounted under the tank, behind it, or in front of it.

Courtesy of E. D. Etnyre & Company

Fig. 21-93. Bitumen pump

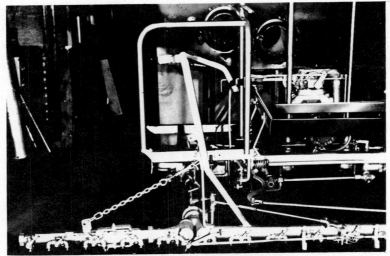

Courtesy of E. D. Etnyre & Company

Fig. 21-94A. Spray bar and piping

A tachometer in the truck cab registers pump output in gallons per minute.

The discharge port is usually at the bottom, so that the pump is self draining. This feature reduces problems with congealed asphalt not removed or sufficiently thinned by flushing. It is still a good idea to pour a quart or two of kerosene into the filler when shutting down.

If these precautions are not taken, or the tank cannot be drained, a residue of cold, thickened asphalt will "freeze" the pump so that it will not turn. Almost all asphalts and asphalt mixtures set up overnight, regardless of the weather.

At every cold start, the clutch should be engaged gradually, and disengaged immediately if the pump refuses to turn.

If it is locked, it must be freed by heating. The asphalt supplier usually has a steam line available, which will heat and open lines rapidly with reasonable safety. Some machines can use exhaust from the distributor engine to warm the pump. If neither of these aids is available, a portable heater (flame gun) is used.

Courtesy of E. D. Etnyre & Company

Fig. 21-94B. Nozzle

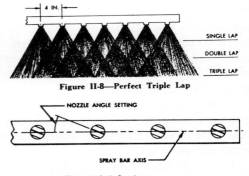

Courtesy of The Asphalt Institute

Fig. 21-94C. Nozzle settings

Courtesy of Seaman Company

Fig. 21-94D. Distributor at work

Spray Bar. The spray bar receives liquid asphalt through lines from the bitumen pump, and distributes it on the road surface through nozzles.

A circulating spray bar can be kept hot, and maintain constant pressure and fluid circulation through a pair of passages. Flow through the nozzles is controlled by a cross bar linked to valves in their stems. There may also be means to shut off nozzles individually.

The non-circulating bar, found chiefly in older machines, has a single passageway, with valveless nozzles screwed directly into its wall. Flow into the bar and through the nozzles is controlled by a valve or valves between the pump and the spray bar.

Either type has a basic or inner (central) section a little less than 8 feet wide, which may be in one or 2 sections. Extensions, usually one and 2 feet wide, may be added to either side, or to both, for a total width up to 24 or even 30 feet. Extensions may be partially supported by chains or springs.

Swivel joints are usually provided to allow extensions to swing back or forward if they hit something, and the inner bar may be swiveled to its supply pipes, and fitted with a center shear pin.

The bar is mounted in a cross frame be-

hind the truck wheels, usually close to the rear edge of the catwalk. The frame can be lifted vertically, or pivoted, to shift the bar between working position, 6 to 12 inches above the road, and a higher carrying position. In addition, there are fine adjustments to regulate working height. Other adjustments may conform the bar to flat or crowned road cross sections.

Position may be controlled mechanically by means of a hand lever, with the weight of the bar carried largely by counterbalance springs. Hydraulic controls are usually optional, and may be standard.

Nozzle. A nozzle is usually a rather simple threaded hollow plug, with a straight slot that shapes the emerging fluid into a fan shaped spray. Standard slot opening is ⅛ inch.

Nozzles are turned so that slots are at an angle of 15 to 30 degrees (usually 30) to the axis of the spray bar. The height of the bar is adjusted so that the fans of adjacent nozzles just meet (single lap, seldom used), overlap one half of their full width (double lap), or overlap two-thirds (triple lap). See Figure 21-94B. The disadvantage of the single lap is that a miscalculation, a fault in a nozzle, or a drop in pressure, would be apt to leave a bare streak or streaks.

Fig. 21-95A. Hand spray unit

To adjust for a double lap, close every other nozzle (if you can), and make a trial run at the pump pressure that is to be used. The spray fans should just meet on the pavement, making a continuous band of asphalt without overlap. If they overlap, lower the bar, if they don't meet, raise it, until the correct height is found. Mark it.

For a triple lap, increase double lap height by 50 percent. It may be tested by closing two out of three nozzles.

During a run, lightening of the load as the tank empties allows the truck springs to raise the frame and spray bar upward, spoiling the adjustment. A very few distributors have an automatic compensator that prevents this. In most machines, the operator may keep lowering the bar as its frame rises. But it is more usual to ignore this factor.

In the illustration, it may be noticed that the outer edges of the oiled strip are skimped in either overlap arrangement. This can be partly corrected by special outer-end nozzles that spray only straight down and inward, or by a flap or curtain that deflects asphalt from the edge inward.

In general, if liquid asphalt of proper viscosity is applied at a reasonable rate under proper temperature conditions, it will make a satisfactorily smooth layer even if nozzle adjustment and bar height are not exactly right. It is possible to take very fine adjustments too seriously.

Too high pressure and/or temperature in the system tends to atomize the fluid coming out of the nozzle, so that some of it drifts away as a fog. Too low pressure causes streaking and dribbling.

Excessive heat alone makes the fluid so thin that it spreads on the surface like paint or water, and cannot make a thick enough coating to hold aggregate properly.

Tool Box. A distributor should be equipped with a spray bar (tool) box, long enough for whatever spray bar extensions

Fig. 21-95B. Spraying by hand

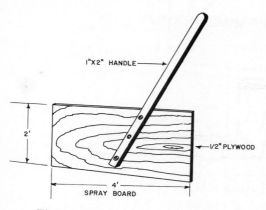

Fig. 21-95C. Spray (splash) board

Fig. 21-95D. Pour pot

are used. It carries these pieces, tools, and probably a few spare parts such as spray nozzles.

The box is covered. A shallow puddle of fuel oil or kerosene is kept in it. This splashes around when the machine is in motion, softening and dissolving asphalt on the sections and tools.

Hand Spray. A hand spray, sometimes called "the hose" comsists of one to three nozzles at the end of a long pipe fitted with two handles, one of which controls a turnoff valve. The pipe is connected to the distributor pressure system by a long, flexible, heat resistant hose.

One valve on the distributor directs liquid into the hose, another adjusts its pressure. The turnoff handle on the spray itself can also reduce pressure at the nozzles by partial closing.

The hand spray is useful and often essential for applying asphalt to areas than cannot be reached efficiently or at all by the fixed spray bar.

Operation is a hot job. It requires some experience to be able to apply an eveh coat, and some touching up may have to be done with a lute. Control is easiest with one nozzle, but work is faster with two.

After each use, the distributor valves should be adjusted to suck fluid out of the spray and hose, which must then have its hand valve open. It is then cleaned by suck-

ing up a small amount of kerosene through the nozzles.

The hose may or may not be disconnected between jobs, but it and the spray must be very carefully arranged on the distributor to be safe from damage and out of the way.

Any sprayer will splash some asphalt up onto vertical surfaces such as curbs, or along a level area past the point of effective application. This can be stopped or reduced by holding a splash board between the spray and the surface to be protected. An expert operator may control the spray with one hand and the board with the other, but it is usually much better to use two men.

Pour Pot. A pour pot, Figure 13-95D, is a specialized version of a watering can. The spout is shaped to spread its discharge into a fan, but pressure is so low that it does not spatter. This is the preferred tool for working narrow strips alongside curbs or other vertical objects that must not be marred.

It is also efficient in filling in small strips or areas too small to justify hooking up, or perhaps even carrying, a pressure hand spray.

Circulation. The basic circulation system includes the tank, the pump, the spray bar, and their connecting lines, Figure 19-96. With the circulating type bar shown, the asphalt liquid is in continuous motion through the system, keeping all parts close

MATERIAL FLOW DURING <u>SPRAY</u> OPERATION

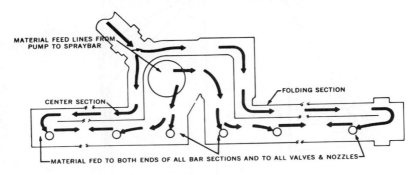

MATERIAL FLOW DURING <u>CIRCULATE</u> OPERATION

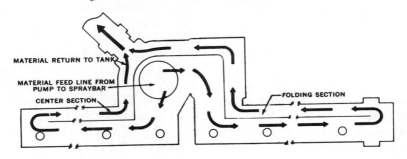

Courtesy of Rosco Manufacturing Company

Fig. 21-96. Circulation in spray bar

to tank temperature, preventing freezing and clogging in lines and bar, and overheating near the tank heaters.

Control valves are reversible, to allow sucking material out of the bar and lines, and returning it to the tank.

There is also a flush cycle, in which a few quarts of kerosene are poured or drawn into the system from the heater tank, and circulated to dissolve asphalt that might freeze parts or restrict passages at the next cold start. It is not necessary to clean out residues completely, but to partly remove and partly soften them so that they will not be troublesome.

There are usually provisions for pumping liquid from an outside source into the tank, from the tank into another container, and from one outside container to another, without putting it through the tank.

The pressure system to the spray bar includes two outside taps controlled by hand valves. One supplies the hose for the hand spray. The other is a tap for filling pots for hand pouring.

OPERATION

Temperature. The temperature of the asphalt in the distributor, at the time that it is being applied to the road, is of critical importance both in distributor performance and pavement quality.

Almost any asphalt, whether cement, cutback, or emulsion, becomes thinner as it is heated. In the distributor, thinner liquids pump and flow more easily, require less pressure to move through a nozzle, and are more inclined to float away as fog droplets after discharge. Behavior in fan spraying from nozzles changes, so that a liquid producing a smooth coat at one temperature may streak at another. If the quantity of application is regulated by pressure in the spraybar, the rate will increase as the liquid is thinned.

The temperature to be used on a job is usually specified, and unless there are strong indications to the contrary, heat should be regulated to provide it. Unfortunately, both distributor operators and personnel on the paving job are often careless or indifferent, so that application temperatures vary widely from desirable ones.

Asphalt and cutbacks are often deliberately overheated. This may be done to increase ease of pumping, but most often to put down a shallower-than-paid-for layer without detection. The lower viscosity permits a thinner layer to flow together smoothly, and would cause one of proper depth to run off the sides. A given stretch of road can then be done with less asphalt, but the resulting surface is of correspondingly lower quality.

This practice could be reduced by customers' insistence on having visible, display type thermometers on distributors making delivery to them.

Amount of Application. The amount of asphaltic material left on each square yard of surface is critical in regard to both the quality and the cost of the treatment. It is a function of the amount sprayed in gallons per minute and the speed at which the spray bar moves.

If the rate of application is fixed by adjustment of the flow of the pump, or by the pressure at the nozzles, the amount deposited will then depend entirely on the speed of the carrier truck. If it doubles its speed, it will halve the application.

On the other hand, at a given truck speed, the application can be increased by running the pump faster or increasing pressure, or reduced by slowing the pump or decreasing its pressure.

The job specification usually calls for a certain amount, say 0.5 gallons per square yard. A 12-foot bar would cover 4 square yards with every yard the truck moved. At 3½ miles an hour it would travel 102.5 yards (308.5 feet) per minute, and would need to pump out at the rate of 205 gallons per minute.

If the truck speeded to 4 miles an hour and continued to pump at the same rate, application would decrease to about 0.44 gallons per square yard. Slowing to 3 mph would increase application to 0.58 gallons.

A truck speedometer is not accurate or sensitive enough to cope with such need for exactness. The machine is therefore equipped with a measuring wheel. It is carried on a hinged bracket on the left side of the frame.

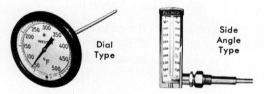

Courtesy of E. D. Etnyre & Company

Fig. 21-97A. Thermometers

When not in use it is held above the road, when needed it is lowered or pressed down against it.

A speedometer cable runs from its hub to a registering head (odometer) in the truck cab. This indicates vehicle speed in feet per minute, by a clock-type hand, and total distance traveled in feet in digits. There is usually an indicator hand that the driver can set at the speed that is to be used, after which he can concentrate on keeping the two hands together.

This assembly—wheel with frame, cable, and head—toegther make up a very accurate speedometer installation. However, to distinguish it from the ordinary speedometer, it is called a recording bitumeter.

Coordinating. There are at least three systems in use for coordinating rate of spray with rate of travel. For each of them, the

TACHOMETER HEAD

PNEUMATIC TACHOMETER WHEEL

Courtesy of Rosco Manufacturing Company

Fig. 21-97B. Tachometer head and wheel

amount of oil needed for each linear foot of road is first calculated. The formula is:

$$\text{gal. per linear foot} = \frac{\text{specified gal. per sq. yd.}}{9}$$
$\times$ spray bar width in feet.

A tachometer in the cab registers revolutions per minute of the pump. Since the pump is rated at a certain output per revo-lution, it is possible to have the tachometer dial show gallons per minute.

This gpm reading is divided by the spray bar width in feet and by truck speed in feet and then multiplied by 9 to give gallons per square yard. Either pump speed or truck speed may be adjusted to meet the required amount. This calculation is usually done with a cardboard slide rule supplied by the

Courtesy of Seaman Company

Fig. 21-97C. Trailer-mounted maintenance distributor

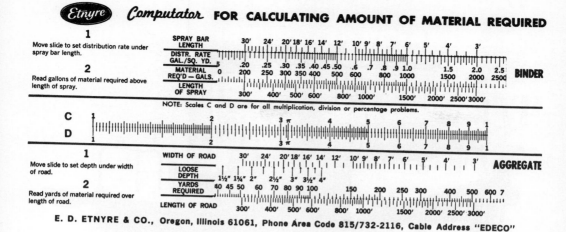

Etnyre *Computator* **FOR CALCULATING AMOUNT OF MATERIAL REQUIRED**

1
Move slide to set distribution rate under spray bar length.

2
Read gallons of material required above length of spray.

BINDER

NOTE: Scales C and D are for all multiplication, division or percentage problems.

1
Move slide to set depth under width of road.

2
Read yards of material required over length of road.

AGGREGATE

E. D. ETNYRE & CO., Oregon, Illinois 61061, Phone Area Code 815/732-2116, Cable Address "EDECO"

FOR DETERMINING DISTRIBUTOR AND PUMP SPEEDS

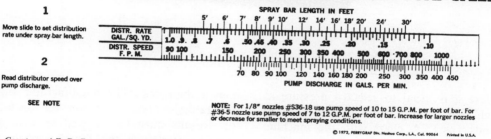

1
Move slide to set distribution rate under spray bar length.

2
Read distributor speed over pump discharge.

SEE NOTE

NOTE: For 1/8" nozzles #S36-18 use pump speed of 10 to 15 G.P.M. per foot of bar. For #36-5 nozzle use pump speed of 7 to 12 G.P.M. per foot of bar. Increase for larger nozzles or decrease for smaller to meet spraying conditions.

© 1972, PERRYGRAF Div. Nashua Corp., L.A., Cal. 90064   Printed in U.S.A.

Courtesy of E. D. Etnyre & Company

Fig. 21-98A. Cardboard slide rule for distributor calculations

distributor manufacturer, as in Figure 19-98A. To work it out without such a helper, start with a convenient truck speed, perhaps 300 feet per minute, and adjust pump speed to fit.

If the width of the spray bar were increased, it would be necessary to increase pump speed or slow down the truck. Leakage in the pump, from wear or malfunction, would decrease its output and require speeding it up to compensate (thus registering more than actual production on the tachometer), or slowing the truck.

Another system is the use of an adjustable pressure relief valve on the return line from the spray bar to the tank. The operator would then set the valve at the spraying pressure needed. The pump would supply more volume than was needed, and the ex-

cess would return to the tank. Pressure in the spray bar, and the amount put through each nozzle, would not be affected by bar width.

Courtesy of E. D. Etnyre & Company

Fig. 21-98B. Single-lever circulation control

21-121

Courtesy of E. D. Etnyre & Company

Fig. 21-98C. Distributor with hydrostatic pump

Here it is usual to select a pressure at which the nozzles perform well, perhaps 20 pounds (psi). A chart with the machine should give the gallons per minute per foot of spray bar at that pressure. Truck speed is calculated to spread that production over the proper area of road. Again, figuring is usually done with a slide rule, or amounts read out of a table.

The thickness of oil cover can be increased at any time by either increasing the pump pressure, or slowing the truck; or decreased by opposite actions.

This system is sensitive to changes in viscosity. At any one pressure, a thin liquid will go through the nozzles more rapidly than a thick one. The manufacturer supplies tables to make corrections for this factor. Results must be checked as soon as spraying starts, as the liquid asphalt is sometimes different from that which was ordered.

Hydrostatic. A distributor may have its bitumen pump powered by hydraulic drive from the truck transmission, instead of having an engine of its own. A takeoff shaft drives a variable displacement hydraulic pump, whose discharge turns a fixed displacement motor that drives the bitumen pump.

For this system to operate, the truck's power takeoff and engine clutch must be engaged, and the engine must be turning fast enough to provide enough power for the pump.

There is a recording bitumeter for truck speed; and a tachometer to show pump speed, with a fine vernier-type control for regulating the cam angle in the pump, and therefore its output. As explained in Chapter 12, cam adjustment provides stepless control from zero to maximum rate.

Courtesy of Rosco Manufacturing Company

Fig. 21-99A. Pressure-opened manhole cover

Fig. 21-99. Start of a spillover

The basic calculation is the same as the first method described under **Coordinating.** Pump output in gallons per minute is divided by spray bar width in feet, then multiplied by 9 to adjust (square) feet to yards. Truck speed is selected for convenience, and pump speed adjusted to fit, preferably figuring with a slide rule supplied by the manufacturer.

The difference from separate-engine systems is that once the correct ratio between a truck speed and a pump speed is set, speeding or slowing the truck will not change the rate of application. This is because the truck engine is driving both the rear axles and the pump shaft. However, the ratio may be changed at any time by adjusting the pump output with the vernier control.

The ability to maintain relationship between road speed and pump output automatically is a substantial advantage of hydraulic pump drive. This is accompanied by avoidance of the noise and complications of operating a separate distributor engine, and the convenience of having a hydraulic system to operate controls.

However, there are disadvantages, most of which seem to be associated with the present power takeoff system, which stops whenever the truck clutch is disengaged.

The bitumen pump, with any hydraulic controls, can operate whenever the truck is traveling, and when it is standing in neutral, but not during gear shifting or the first seconds of movement after a stop. Special operation techniques are needed to offset the resulting problems.

It is to be expected that future models will have live power takeoffs, with a two-step clutch or other arrangement to permit pump operation whenever the truck engine runs.

Measurement. The accuracy of figuring the application can and should be checked by measurement of the contents of the tank at the start and at the end of a run. The difference between the two figures should equal the square yards covered (distance in feet times spray bar width in feet divided by 9) times the gallons or fraction of gallon per square yard.

The principle is simple, but measurement may not be. If you have an indicator for asphalt level in the tank, it is usually at one end, and will not be accurate unless the truck is level, front to rear. There are large sections of the country where a dependably level piece of road is a rarity.

If your distributor is one of the few properly equipped with a pair of indicators, one front and one rear, you average their readings, even when the road looks level.

If the manhole is at or very close to the center of the tank, stick measurement can be accurate regardless of moderate slopes, if done carefully. The stick he held at right angles to the top or long axis of the tank, which on a grade it NOT straight up and down, and lowered until it just touches the top of the liquid. The reading is then taken off the stick and a marker on the manhole edge. The stick should be marked directly in gallons. Remembering, you are measuring what isn't there, and must subtract your figure from tank capacity to find how much is there. But if the tank was full at the beginning of the run, the stick reading can show the amount used directly.

Wipe off the bottom of the stick with a kerosene rag before putting it back in its holder.

Cleaning Up. The pump, bar, and circulating system should be cleaned out at the end of the shift, or whenever the machine will stand long enough to cool. This is easily done by first sucking back the asphalt in the lines and returning it to the tank, then pouring in a few quarts of kerosene and circulating them. The cleaner may be discharged through the bar nozzles if conditions permit, or pumped into the tank. Its quantity is too small to have a noticeable effect on the next load.

When spreading the last load of the day, the tank may be allowed to drain fully. If the load was a cutback at proper temperature, it will be left quite clean. Emulsions, however, leave heavy coatings.

A dirty tank is often cleaned at the asphalt plant by taking on a load of thin cutback, then returning it to the supply tank.

Asphalt may be removed from outside surfaces by various combinations of heat, solvents (kerosene or fuel oil, or much more expensive special cleansers), and scrapers. It is burned off hand tools in a fire cart.

Removal from the skin may be a two stage operation, first softening it with lubricating oil, then washing it off with kerosene.

Water. Water may be present in a distributor tank and lines as a residue from a load of emulsion, or as a result of condensation in an empty or partly empty tank. If very hot asphalt cement is taken on, an explosion might occur.

The usual problem, however, is frothing. The cement, or any cutback over the water's boiling point, will change it to steam. This is likely to form into very fine bubbles, expanding the asphalt into a froth that boils the top of the load out the overflow, and is likely to push up the manhole cover also.

Frothing also occurs as a load is heated, and often at temperature apparently far below boiling. A full load of cutback taken on

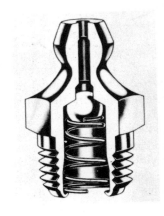

Courtesy of Stewart-Warner Corporation

Fig. 21-100. Hydraulic fittings

at 160° F. might boil over when brought up to 180. Circulation in the tank is slow, and the part of the liquid against the heater tubes is much hotter than the rest of it. The boiling starts there, and may start a complex reaction that will push an embarrassing number of solid gallons out the top. It is a poor way to start a job.

The extent of the mess is somewhat proportional to both the amount of water and the heat of the asphalt.

Water may be flushed out of the system by circulating a couple of gallons of kerosene in both the tank and the lines, and discharging it through the hose line or spray bar. If this is not done, circulate the hot material through the lines when the tank has just started to fill, and there is plenty of room for foaming without overflowing.

A special chemical, Dow-Corning DC-200, may be added to asphalt to reduce or prevent foaming.

LUBRICATION EQUIPMENT

High Pressure Fittings and Couplings. There are at least six different fittings which are attached to machinery to permit pumping grease. Of these, the most popular is the hydraulic fitting, Figure 21-100. The gun nozzle or coupling, 21-101, contains jaws which are wedged apart by forcing them over the top of the fitting, and which clamp below the bulge. High pressure developed in the gun pushes against the head of the fitting and tends to push the gun away, but it also presses the tops of the jaws outward, wedging the bottoms inward for a tight grip. If grease passes freely and pressure is light, the gun can be removed by direct pull. If pressure is high, it is taken off by tilting sideward until the nozzle touches the shoulder and pries the jaws off.

The ball and spring should prevent grease from leaking back from the bearing. They are lacking in some small fittings.

The Zerk, or push fitting, Figure 21-102,

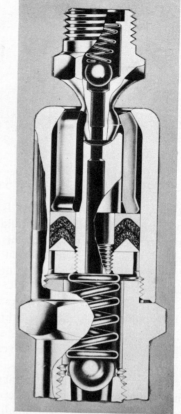

Courtesy of Stewart-Warner Corporation

Fig. 21-101. Hydraulic nozzle and fitting

provides no grip by which the nozzle can be held on. It is being superseded by the hydraulic fitting. Zerk guns will work on either fitting, but hydraulic guns will work only on hydraulic fittings.

These two types have the advantage of speed, with the hydraulic providing the

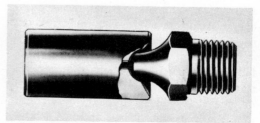

Courtesy of Stewart-Warner Corporation

Fig. 21-102. Zerk fitting and nozzle

Courtesy of Stewart-Warner Corporation

Fig. 21-103. Button head fitting and nozzle

more positive seal. If the machine is exposed to dirt or dust, the fitting should not be cleaned off after greasing, as the dab of grease sticking to it will prevent dirt from getting in the neck of the opening.

Standard threads are ⅛″ P. T. (pipe thread), ¼″ P. T. for oversize fittings, and ¼″-28 for small ones. A variety of threads are available for special uses. Many small fittings have no threads and are driven in. These may not stand high pressures.

Large (¼″ P. T.) hydraulic fittings are sometimes made of a small fitting screwed or driven into a larger base. The small part may come out when the coupling is pulled away. If much trouble is experienced, one piece units, or ⅛″ P. T. fittings in ⅛ to ¼″ bushings should be substituted.

Hydraulic fittings are made in a variety of angles and lengths. Special extensions can be made up with ⅛″ pipe material.

If grease leaks at the fitting-coupling connection, it usually indicates that one of these units, most often the fitting, has been scarred or damaged so that the coupling cannot seat properly. The defective part should be replaced.

The button head, Figure 21-103, has a flat surface across which the coupling slides. It can be wiped clean whether dry or greasy. It is close coupled, more rugged than the hydraulic fitting, and has a larger grease passage. It is made in standard and giant sizes, in a variety of threads.

The coupling may connect to it by pushing, pulling, or by sideward motion. Leakage of grease around it usually indicates that the neoprene face seal is damaged or worn. The coupling cannot be pulled off while the grease in it is under high pressure.

The pin type, 21-104, is the original Alemite fitting. The coupling locks on it with a quarter turn. This fitting projects farther than the others described, and is more subject to accident. Making connection is slower than with the hydraulic. It is no longer in wide use.

The Dot, 21-105, is made in standard and Mogul sizes. The gun is screwed on a full turn, usually forcing some grease into the fitting. Any additional amount required is then pumped in.

The flush type, 21-106, is used where the fitting cannot be allowed to project. It has limited use in earth moving equipment because of the difficulty of cleaning.

Courtesy of Stewart-Warner Corporation

Fig. 21-104. Alemite fitting and nozzle

Courtesy of Stewart-Warner Corporation

Fig. 21-105. Dot fitting and nozzle

Special Fittings. Fittings may be obtained with a reservoir that will store grease and release it slowly. If the bearing is close fitting, grease may be supplied only when it is in motion.

The mechanism is shown in Figure 21-107. Grease pumped through the hydraulic fitting is restricted by the enlarged lower end of a metering pin, and rises into the reservoir, pushing back the follower and compressing the spring. When the cup fills, the metering pin blocks the passage into the

reservoir. The fitting is lubricated directly during filling.

Pressure of the follower spring then feeds the grease gradually to the bearing around the metering pin.

Bearings and their casings and seals may be protected against too high pressures by the use of a fitting with a relief valve allowing excess lubricant to bleed out of the side or along a slot in the threads. A special bushing can be used with a standard fitting for the same effect.

Fittings are also available which will shut off the passageway after a certain quantity has been admitted, or a certain pressure reached. Water pump fittings may be threaded for a water-tight and dirt-proof cap.

Hand Guns. The majority of hand grease guns now in production are the lever type. Figure 21-108 shows a typical example in cross-section. The cylindrical barrel has a

Courtesy of Stewart-Warner Corporation

Fig. 21-106. Flush fitting and nozzle

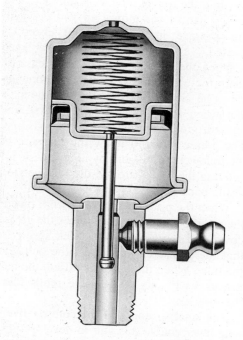

Courtesy of Stewart-Warner Corporation

Fig. 21-107. Reservoir fitting

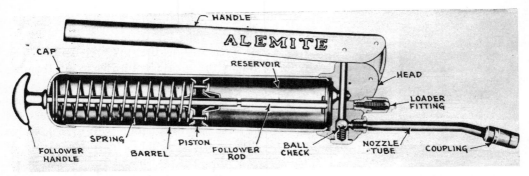

Fig. 21-108. Lever-type hand gun

smooth inside finish. The metal piston with leather seals is pushed toward the head by a light spring. The follower rod is used to pull the piston back against the spring when refilling by suction. The collar groove in the rod permits locking it in the back position.

The head contains a fitting through which grease can be pumped into the reservoir, and a passageway from the reservoir into the nozzle tube. A piston actuated by the hand lever moves up and down in this passage in which a ball check is located.

When the piston is pulled up, grease or air in the tube is prevented from following it by the check. This leaves a vacuum so that when the passage to the reservoir is opened, grease is sucked into the passage. The grease is urged in by the pressure of the follower spring, and by atmospheric air entering the barrel around the follower rod in the back cap.

When the piston is moved down it blocks the reservoir passage, then forces the grease down and compresses the check spring so that the grease can flow past the ball into the tube. When the piston is raised, the ball reseats itself and the passage refills from the reservoir.

The small piston and the comparatively long lever enable this gun to develop pressure up to 10,000 pounds per square inch. Most nozzles and fittings are designed to

take 20,000 pounds pressure. However, the seals and casings of the parts being lubricated will often bend or break at less pressure so caution must be used when forcing grease into them.

The gun may be filled in three ways. The easiest is to pump grease into it through the fitting in the head casing from a loader pump. Precautions should be taken against pumping in air with the grease.

If there is no loader fitting, or no loader pump is available, the head casting is unscrewed, the head of the barrel cleaned and pushed well down in the grease supply, and the follower arm drawn back slowly. If the grease is thin enough to flow, it will be sucked into the barrel. It may be necessary to move the gun around in the container to prevent air from entering. When the follower is fully back it is locked with a sideward motion, the head screwed on, and the follower released. It is good practice to keep the grease in a warm place to keep it soft enough to flow.

If the grease cannot be pumped or sucked into the gun, it may be put in with a small paddle and air kneaded out of it. It is difficult to avoid air pockets with this method.

Air in the grease may form a pocket in the passageways that will prevent the gun from working. The block may be temporary until the air is worked out, or permanent if the gun is worn enough to

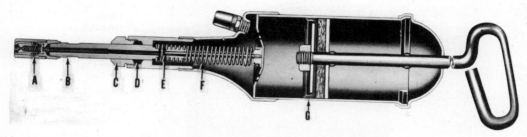

A. Nozzle or coupler

B. Stem

C. Slide cup

D. Packing

E. Piston

F. Main spring

G. Follower piston

Courtesy of Stewart-Warner Corporation

Fig. 21-109. Push-type hand gun

allow it to return beside the piston into the reservoir.

Air takes much longer to get through the head than the same bulk of grease. However, many times a gun is said to be air locked when the trouble is partly or wholly foreign matter which prevents the ball check from seating properly. This allows grease or air to be sucked into the cylinder from the outlet tube on the up stroke, and pushed back into it on the down stroke.

In either case the cure is to disassemble the unit, clean it, and pack it with fresh grease.

The outlet tube consists of a piece of 1/8" pipe, or of equivalent size flexible hose. Pipe thread is used. Any of the standard couplings can be attached to this.

Thick grease may not feed properly in this type of gun, as combined atmospheric and spring pressure may not be enough to make it flow. It may be persuaded to work by tapping or by heating. Best results may be obtained if the follower piston is so built that the rod can be locked to it (as

in this example) so that pressure can be applied while pumping.

A special gun may be obtained in which the grease is forced along the gun by twisting a threaded follower.

The push type gun, Figure 21-109, is convenient for narrow spaces and high spots. The newer models may be filled through the loader fitting shown, but for older types it is necessary to remove the back plate, pull out the follower, and pack it by hand. The amount of lubricant in the gun is indicated by the distance the follower handle projects.

The gun is held by the handle and pushed against the fitting. After the coupler grips, the end of the gun slides on a plunger which pushes the grease out of the stem into the fitting. Pressure is then released, the spring forces the gun back, the stem fills with grease, and it is ready for the next push. A ball check in the coupling prevents air from entering the stem.

Screw type guns are of simple construction. The follower rod is threaded and screwing it into the rear cap forces the

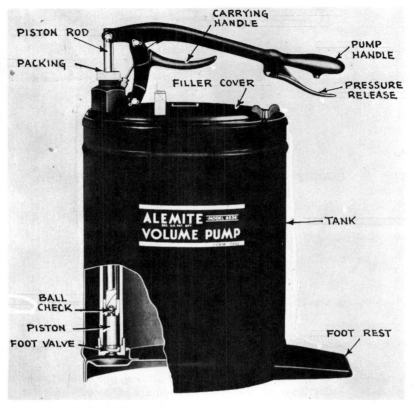

PISTON ROD

PACKING

CARRYING HANDLE

FILLER COVER

PUMP HANDLE

PRESSURE RELEASE

ALEMITE MODEL 6536 VOLUME PUMP

TANK

BALL CHECK

PISTON

FOOT VALVE

FOOT REST

Courtesy of Stewart-Warner Corporation

Fig. 21-110. Bucket pump

follower piston against the grease, which passes directly into the coupling.

Bucket Pumps. As used here, this term includes all large capacity hand-operated grease pumps. Only a few of many models available will be described. Pressure varies from a few pounds for transfer pumps up to 15,000 pounds in heavy-duty models.

Such pumps generally consist of a pail, a pail cover, a hand pump fastened in the cover, a flexible hose, and a coupling or nozzle.

An Alemite volume bucket pump is shown in Figure 21-110 with its parts labeled.

The pressure release lever opens a by-pass so that grease pumped is returned to the tank, and the hose can drain back into the tank also. This allows the release of any

pressure binding a coupling on a fitting, allows the handle to be pushed down to carrying position without forcing grease out of the coupling, and allows priming the pump after refilling, without forcing air into the hose.

Figure 21-111 shows a foot operated pump. This may be fitted at the factory with a small piston for high pressure, or a larger piston for medium pressure and volume delivery.

The bucket pump shown in Figure 21-112 includes a bracket for carrying a hand gun that may be mechanically loaded by applying the gun loader fitting to the loader assembly on the pump.

Figure 21-113 shows a pump assembly that clamps directly on a 25 pound grease bucket. These pumps operate at low pressure and usually handle only light-bodied or semi-fluid greases.

Somewhat similar portable transfer pumps with crank handles are used for dispensing gear oil and other fluid lubricants. The crank actuates a piston type pump. An open nozzle or trigger faucet is used on the delivery end of the hose.

Air-Operated Pumps. Wherever possible, grease is pumped by compressed air. Many pumps are intended for use in a standard 55 gallon (400 lb.) drum.

A single vertical rod connects the piston of a one cylinder, double acting air motor to the grease pump piston. This is hollow and carries a ball on a machined seat at the bottom. When the piston moves down, grease unseats the valve and fills the piston. When it moves up, the ball seats and the grease is forced up the space around the piston rod and out toward the nozzle.

The mechanical primer is a large plunger which moves grease toward the ball check with each up stroke.

When this unit is connected to the air

Courtesy of Stewart-Warner Corporation

Fig. 21-111. Foot lever bucket pump

line, it pumps until full pressure is built up in the grease line. This pressure then shuts off the air by closing the passage through the trigger valve. The unit is then said to be balanced. When grease pressure drops, air is admitted again and pumping is resumed.

Courtesy of Stewart-Warner Corporation

Fig. 21-112. Loader for hand gun

Courtesy of Stewart-Warner Corporation

Fig. 21-113. Pail cover pump

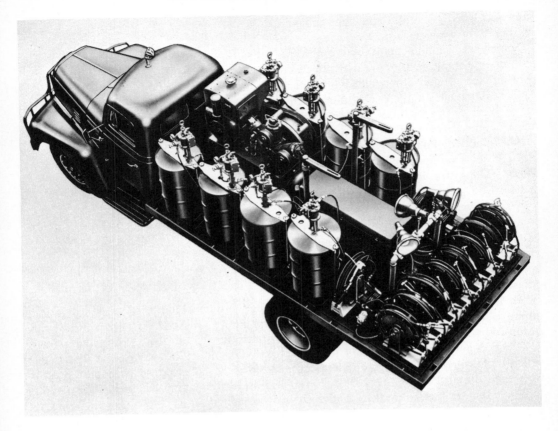

Courtesy of Stewart-Warner Corporation

Fig. 21-114. Grease truck

These pumps are available in high and low pressure constructions, and similar units dispense oil in large volume and at very low pressure.

Centralized Lubrication. In a complex machine many points requiring lubrication are difficult to reach with any type of grease coupling, and other fittings may be overlooked so that bearings run dry. Lubrication may be neglected because of the time required to take care of it, or because of difficulty with the grease gun. Whatever the cause, failure to supply each fitting with grease as often as it needs it is liable to result in excessive wear and costly breakdowns.

This danger can be avoided in part by the use of sealed cases containing oil or fluid grease which is automatically distributed through the moving parts by splash or pressure, and by the use of sealed bearings which contain sufficient grease for several years of operation.

Otherwise inaccessible points can often be connected by tubing to fittings in a convenient location. It is very important that only one point be greased by each fitting, as otherwise the lubricant follows the path of least resistance and may over-lubricate one point and not reach any others on the same line.

Points requiring greasing so frequently as to interfere with normal operation may be fitted with reservoir fittings.

There are also systems by which a single reservoir and pump of the hand, air, or

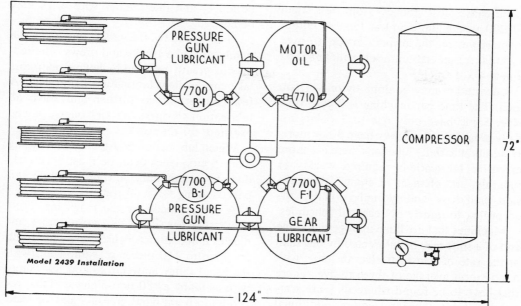

PRESSURE GUN LUBRICANT

MOTOR OIL

7700 B-1

7710

COMPRESSOR

7700 B-1

7700 F-1

PRESSURE GUN LUBRICANT

GEAR LUBRICANT

Model 2439 Installation

72"

124"

Courtesy of Stewart-Warner Corporation

Fig. 21-115. Grease truck layout

electric type supplies a large number of fittings through tubing and metering valves or injectors.

Such systems are generally applicable only to stationary machinery, deck parts of swing shovels, and other protected parts, because of the danger of damaging tubing. In appraising a centralized system, an important point to consider is whether the operator is given a warning if any point does not get grease because of blocking or leakage.

Grease Trucks. When a contractor has ten or more pieces of equipment in the field, he may consider the advisability of buying or building a mobile greasing unit to take care of their lubrication requirements. Such units are shown in Figures 21-114 and 21-115. These are only samples of possible arrangements. A wide variety of equipment and many layout plans are available.

The usual carrier is a flat bed truck with a capacity of 2 tons or more. A beefed up pickup may carry a small unit. A van type body may be used when the climate is cold

or rainy. Four wheel drive trucks are used for rugged conditions. Semitrailers are suited to part time use, as they can be parked and the tractor used for other jobs.

A grease truck may carry from 2 to 10 drums of grease and oil, each weighing about 400 pounds; a gasoline engine that drives an air compressor, an air pump in each active drum (spares may be carried), and a reel of hose for each lubricant.

The reels receive grease through their axles and feed it into a connection to the hose. They may be hand wound or equipped with pull back springs that allow the hose to be pulled out, then wind it back when it is released.

The nozzles may be plain or equipped with meters. Pressure in the hoses is automatically maintained by the air pumps.

The truck may carry drums of fuel, antifreeze, and/or water in addition to the lubricants. If it is not needed full time for greasing, it may be equipped with a small electric generator, drills, lights, vise, and tools, so that it can be used for field repairs.

Purchase cost ranges from less than $1,000 where the simplest type of equipment is put on the owner's truck, to over $15,000 for an elaborate setup mounted on a new 4 x 4 truck.

Lubrication may be done by one or two men. The time per machine is highly variable. Some have only one to 3 points for daily greasing, others may have 30 or more. In addition to daily lube jobs there are more extensive lubrications required weekly or monthly, and changes of engine, transmission, final drive, and hydraulic oil are made according to hourly or seasonal schedules.

Machines that are lined up together at the end of a shift may be lubricated by one man at the rate of two to twelve an hour, with a long term average of three to five. Those that have to be found where they are scattered around the job require additional travel and search time. If the system is to catch and stop machines at work, time will be longer and interference with work may be costly.

Time spent by the grease crew, or the number of lube trucks and crews required, may be greatly reduced by having operators take care of light lubrication themselves. A few contractors go further and leave all lubrication to operators and oilers, paying overtime for the work.

When lubrication must be done during work hours, trucks can be most effectively done by a stationary unit, or a mobile unit parked along the haul route.

A survey made by the magazine, CONSTRUCTION METHODS AND EQUIPMENT, indicates that contractors use one grease truck for 7 up to 110 pieces of equipment, with the usual range between 20 and 80. Some, with as many as 80 non-highway units, relied entirely on hand greasing and/or fixed grease racks.

APPENDIX

This section is made up of technical, statistical, and advisory material largely contributed by others, that has been reproduced without change or comment. Some of it is in partial or complete disagreement with material presented in other parts of this book.

Some of the production and cost information has been made up in 1962, while other parts date back as far as 1949. Much of it has been obtained in the field, but some of it is calculated from time and motion studies.

Working conditions vary so widely that averages such as may be found in these tables and examples cannot be safely applied to any specific job. Changes and refinements in equipment may have changed production characteristics by the time this is read.

Because of these factors, the figures serve for general information only, and should not be used directly in estimates of production or cost, nor in comparison of different makes and models of equipment.

Equipment specifications have been obtained from sources believed to be reliable, but their accuracy cannot be guaranteed. They are subject to change.

Machinery rental rates vary with locality, demand for equipment, and policies of owners. Rates are given to serve only as a guide to current practices in the United States and Canada, and not to recommend any particular arrangement or price basis.

The following is a guide to the major subjects in this section:

POWER SHOVEL YARDAGES ▶

CONDITIONS:

1. Cu. yds. bank measurement per hour.
2. Suitable depth of cut for maximum effect.
3. No delays.
4. 90° swing.
5. All materials loaded into hauling units.

● Grey figures denote optimum depth of cut.
● Black figures denote yards per hour.

SHOVEL DIPPER CAPACITY IN CU. YDS.

Class of Material	3/8	1/2	3/4	1	1 1/4	1 1/2	1 3/4	2	2 1/2
Moist Loam or Light Sandy Clay	3.8' / 85	4.6' / 115	5.3' / 165	6.0' / 205	6.5' / 250	7.0' / 285	7.4' / 320	7.8' / 355	8.4' / 405
Sand and Gravel	3.8' / 80	4.6' / 110	5.3' / 155	6.0' / 200	6.5' / 230	7.0' / 270	7.4' / 300	7.8' / 330	8.4 / 390
Good Common Earth	4.5' / 70	5.7' / 95	6.8' / 135	7.8' / 175	8.5' / 210	9.2' / 240	9.7' / 270	10.2' / 300	11.2' / 350
Clay, Hard, Tough	6.0' / 50	7.0' / 75	8.0' / 110	9.0' / 145	9.8' / 180	10.7' / 210	11.5' / 235	12.2' / 265	13.3' / 310
Rock, Well Blasted	40	60	95	125	155	180	205	230	275
Common, with Rocks and Roots	30	50	80	105	130	155	180	200	245
Clay, Wet and Sticky	6.0' / 25	7.0' / 40	8.0' / 70	9.0' / 95	9.8' / 120	10.7' / 145	11.5' / 165	12.2' / 185	13.3' / 230
Rock, Poorly Blasted	15	25	50	75	95	115	140	160	195

HAUL UNITS NEEDED TO SPOT UNDER SHOVEL PER HOUR IN MEDIUM DIGGING

Size excavator dipper	Minimum haul unit capacity at 4 times dipper size	Approx. shovel cycle time in seconds 90° swing . . . no delays loading on grade	Loading time for 4-dipper truck in sec.	To synchronize loading spot one truck every	No. spots by hauling units at shovel needed per hr.
3/8	1 1/2 yd.	19	76	1.26 min.	48 spots
1/2	2 yd.	19	76	1.26 min.	48 spots
3/4	3 yd.	20	80	1.33 min.	45 spots
1	4 yd.	21	84	1.4 min.	43 spots
1 1/4	5 yd.	21	84	1.4 min.	43 spots
1 1/2	6 yd.	23	92	1.53 min.	39 spots
2	8 yd.	25	100	1.66 min.	36 spots
2 1/2	10 yd.	26	104	1.73 min.	35 spots

SHORT BOOM DRAGLINE PERFORMANCE — CU. YDS.

CLASS OF MATERIAL	3/8	1/2	3/4	1	1 1/4	1 1/2	1 3/4	2	2 1/2
Light Moist Clay or Loam	5.0' / 70	5.5' / 95	6.0' / 130	6.6' / 160	7.0' / 195	7.4' / 220	7.7' / 245	8.0' / 265	8.5' / 305
Sand or Gravel	5.0' / 65	5.5' / 90	6.0' / 125	6.6' / 155	7.0' / 185	7.4' / 210	7.7' / 235	8.0' / 255	8.5' / 295
Good Common Earth	6.0' / 55	6.7' / 75	7.4' / 105	8.0' / 135	8.5' / 165	9.0' / 190	9.5' / 210	9.9' / 230	10.5' / 265
Clay, Hard, Tough	7.3' / 35	8.0' / 55	8.7' / 90	9.3' / 110	10.0' / 135	10.7' / 160	11.3' / 180	11.8' / 195	12.3' / 230
Clay, Wet, Sticky	7.3' / 20	8.0' / 30	8.7' / 55	9.3' / 75	10.0' / 95	10.7' / 110	11.3' / 130	11.8' / 145	12.3' / 175

NOTE: Top figure denotes optimum depth of cut. Bottom figure denotes yards per hour.

Courtesy of Power Crane and Shovel Association

Shovel and dragline performance, 60 minute hour

Theoretical Excavator Production in Loose Yards, for 60 minute hour without delays, for various bucket loads and cycle times

NOT FOR DIRECT USE IN ESTIMATING*

Cycle time, Seconds	12	13	14	15	16	17	18	19	20
Cycles per Hour	300	276.9	257.1	240	225	211.8	200	189.5	180
Bucket Load, Cubic Yards									
3/8	111.5	103.8	96.3	90	84.3	79.4	75	71.1	67.5
1/2	150	138.4	128.5	120	112.5	105.9	100	94.7	90
3/4	225	207.6	192.6	180	168.6	158.8	150	142.2	135
1	300	276.9	256.8	240	224.8	211.2	200	189.5	180
1-1/2	450	415.2	385.2	360	337.2	317.7	300	284.4	270
2	600	554	514	480	450	424	400	379	360
2-1/2	750	692	642	600	562	530	500	474	450
3	900	831	770	720	674	634	600	568	540
3-1/2	1050	970	899	840	786	740	700	663	630
4	1200	1108	1028	960	900	847	800	758	720
5	1500	1384	1284	1200	1125	1059	1000	947	900
6	1800	1662	1540	1440	1350	1271	1200	1137	1080

Cycle Time, Seconds	21	22	23	24	25	30	35	40	50
Cycles per Hour	171.4	163.6	156.5	150	145	120	102.9	90	72
Bucket Load, Cubic Yards									
3/8	65.5	61.3	58.7	56.2	54.3	45	38.6	33.5	27
1/2	85.7	81.8	78.2	75	72.5	60	51.4	45	36
3/4	128.5	122.7	117.3	112.5	108.7	90	77.2	67.5	54
1	171.4	163.6	156.5	150	145	120	102.9	90	72
1-1/2	256.9	245.4	234.7	225	217.5	180	154.3	135	108
2	343	327	313	300	290	240	205.4	180	144
2-1/2	429	409	391	375	362	300	257	225	180
3	514	491	470	450	435	360	309	270	216
3-1/2	600	573	548	525	507	420	360	315	252
4	686	655	616	600	580	480	412	360	288
5	858	818	782	750	724	600	515	450	360
6	1028	982	940	900	870	720	617	540	432

*See Figure 13-131

Excavator production before delays

Compacted Cubic Yards of Run-of-Bank Gravel = Square Yards x Depth in inches x .033

Square Yards	1"	2"	3"	4"	5"	6"	8"	10"
1	.033	.066	.099	.132	.165	.198	.264	.330
2	.066	.132	.198	.264	.330	.396	.528	.660
3	.099	.198	.297	.396	.495	.594	.792	.990
4	.132	.264	.396	.528	.660	.792	1.056	1.320
5	.165	.330	.495	.660	.825	.990	1.320	1.650
6	.198	.396	.594	.792	.990	1.188	1.584	1.980
7	.231	.412	.693	.924	1.155	1.386	1.848	2.310
8	.264	.528	.792	1.056	1.320	1.584	2.112	2.640
9	.297	.594	.891	1.188	1.485	1.782	2.376	2.970
10	.330	.660	.990	1.320	1.650	1.980	2.640	3.300
20	.660	1.32	1.98	2.64	3.30	3.96	5.28	6.60
30	.990	1.98	2.97	3.96	4.95	5.94	7.92	9.90
40	1.32	2.64	3.96	5.28	6.60	7.92	10.56	13.20
50	1.65	3.30	4.95	6.60	8.25	9.90	13.20	16.50
60	1.98	3.96	5.94	7.92	9.90	11.88	15.84	19.80
70	2.31	4.62	6.93	9.24	11.55	13.86	18.48	23.10
80	2.64	5.28	7.92	10.56	13.20	15.84	21.12	26.40
90	2.97	5.94	8.91	11.88	14.85	17.82	23.76	29.70
100	3.30	6.60	9.90	13.20	16.50	19.80	26.40	33.00
200	6.60	13.20	19.80	26.40	33.00	39.60	52.80	66.00
300	9.90	19.80	29.70	39.60	49.50	59.40	79.20	99.00
400	13.20	26.40	39.60	52.80	66.00	79.20	105.60	132.00
500	16.50	33.00	49.50	66.00	82.50	99.00	132.00	165.00
600	19.80	39.60	59.40	79.20	99.00	118.00	158.40	198.00
700	23.10	46.20	69.30	92.40	115.50	138.60	184.80	231.00
800	26.40	52.80	79.20	105.60	132.00	158.40	211.20	264.00
900	29.70	59.40	89.10	118.80	148.50	178.20	237.60	297.00
1000	33.00	66.00	99.00	132.00	165.00	198.00	264.00	330.00
2000	66.00	132.00	198.00	264.00	330.00	396.00	528.00	660.00
3000	99.00	198.00	297.00	396.00	495.00	594.00	792.00	990.00
4000	132.00	264.00	396.00	528.00	660.00	792.00	1056.00	1320.00
5000	165.00	330.00	495.00	660.00	825.00	990.00	1320.00	1650.00
6000	198.00	396.00	594.00	792.00	990.00	1180.00	1584.00	1980.00
7000	231.00	462.00	693.00	924.00	1155.00	1386.00	1848.00	2310.00
8000	264.00	528.0(	792.00	1056.00	1320.00	1584.00	2112.00	2640.00
9000	297.00	594.00	891.00	1188.00	1485.00	1782.00	2376.00	2970.00
10000	330.00	660.00	990.00	1320.00	1650.00	1980.00	2640.00	3300.00

Example: 546 square yards x 6 inches deep: 500 s.y.   99.000 cu. yd.
  40 s.y.   7.920 cu. yd.
  6 s.y.   1.188 cu. yd.
  108.108, say 108 cu. yd.

For crushed stone, add 10% to figures in table
For loose yards of gravel (truck measure) add 30% to figures in table.

Courtesy of Peckham Road Corporation

Compacted gravel requirement, various areas and depths

WEIGHT AND MEASURE TABLES

Measure of Length
1 Mile = 1760 Yds. = 5280 Ft. = 63,360 Inches
1 Mile = 8 Furlongs = 80 Chains
1 Furlong = 10 Chains = 220 Yds.
1 Chain = 4 Rods = 22 Yds. = 66 Ft. = 100 Links
1 Rod = 5.5 Yds. = 16.5 Ft.

Measure of Length—English to Metric
1 Mile = 1.609 Kilometer
1 Yard = 0.9144 Meter
1 Foot = 0.3048 Meter = 304.8 Millimeters
1 Inch = 2.54 Centimeters = 25.4 Millimeters

Measure of Length—Metric to English
1 Kilometer = 0.6214 Mile
1 Meter = 39.37 Inch = 3.2808 Ft. = 1.0936 Yd.
1 Centimeter = 0.3937 Inch
1 Millimeter = 0.03937 Inch

Square Measure
1 Sq. Mile = 640 Acres = 6400 Sq. Chains
1 Acre = 10 Sq. Chains = 4840 Sq. Yds. = 43,560 Sq. Ft.
1 Sq. Chain = 16 Sq. Rods = 484 Sq. Yds. = 4356 Sq. Ft.
1 Sq. Rod = 30.25 Sq. Yds. = 272.25 Sq. Ft. = 625 Sq. Links
1 Sq. Yd. = 9 Sq. Ft.
1 Sq. Ft. = 144 Sq. Inches
An Acre is equal to a Square 208.7 Feet per Side

Square Measure—English to Metric
1 Sq. Mile = 2.5899 Sq. Kilometers
1 Acre = 0.4047 Hectare = 40.47 Ares
1 Sq. Yard = 0.836 Sq. Meters
1 Sq. Foot = 0.0929 Sq. Meters = 929 Sq. Centimeters
1 Sq. Inch = 6.452 Sq. Centimeters = 645.2 Sq. Millimeters

Square Measure—Metric to English
1 Sq. Kilometer = 0.3861 Sq. Mile = 247.1 Acres
1 Hectare = 2.471 Acres = 107,640 Sq. Ft.
1 Are = 0.0247 Acre = 1076.4 Sq. Ft.
1 Sq. Meter = 10.764 Sq. Ft. = 1.196 Sq. Yd.
1 Sq. Centimeter = 0.155 Sq. Inch
1 Sq. Millimeter = 0.00155 Sq. Inch

Cubic Measure
1 Cubic Yd. = 27 Cu. Ft.
1 Cubic Ft. = 1728 Cu. Inches
1 Cord = 128 Cu. Ft.
1 Gallon = 0.1137 Cu. Ft. = 231 Cu. Inches
1 Cubic Ft. = 7.48 U. S. Gallons
1 U. S. Gallon = 0.83268 Imperial Gallon
1 Imperial Gallon = 1.2009 U. S. Gallons

Cubic Measure—English to Metric
1 Cubic Yd. = 0.7646 Cubic Meters
1 Cubic Ft. = 28.316 Liters
1 Cubic Inch = 16.38 Cubic Centimeters
1 U. S. Gallon = 3.785 Liters
1 U. S. Quart = 0.946 Liters
1 U. S. Pint = 0.473 Liters
1 Imperial Gallon = 4.542 Liters

Cubic Measure—Metric to English
1 Cubic Meter = 35.314 Cu. Ft. = 1.308 Cu. Yd. = 264.2 U. S. Gallons
1 Cubic Centimeter = 0.061 Cu. Inch
1 Liter = 0.0353 Cu. Ft. = 61.023 Cu. Inches
1 Liter = 0.2642 U. S. Gallon = 1.0567 U. S. Quart

Measures of Weight—English and Metric
1 Long Ton = 2240 Lbs. = 1016.05 Kilograms
1 Short Ton = 2000 Lbs. = 907.18 Kilograms
1 Metric Ton = 2204.6 Lbs.
1 Kilogram = 2.2046 Lbs.
1 Lb. = 0.45359 Kilograms

Specific Gravity—is a number indicating how many times a certain volume of material is heavier than an equal volume of water.

ENGLISH SYSTEM—If a material has a specific gravity of 2.7 for instance, multiply this by 62.4 lbs. (weight of 1 cu. ft.) of water to obtain the weight in lbs. per cu. ft. of the material in question.

METRIC SYSTEM—If a material has a specific gravity of 2.7 for instance, multiply this by 1000 kilograms (weight of 1 cu. meter of water) to obtain the weight in kilograms per cu. meter of the material in question.

Equivalents of Density—English and Metric
1 Lb. per Cu. Yd. = 0.5933 Kg. per Cu. Meter
1 Kg. per Cu. Meter = 1.6856 Lbs. per Cu. Yd.

Equivalents of Pressure—English and Metric
1 Lb. per Sq. Inch = 0.0703 Kg. per Sq. Centimeter
1 Kg. per Sq. Centimeter = 14.244 Lbs. per Sq. Inch

Weights of Diesel Fuel
1 U. S. Gallon = 7 lbs. average.
1 U. S. Gallon = 3.17 kilograms.

Reproduced from "Estimating Production and Costs."
Courtesy of Euclid Division of General Motors Corporation

Weight and measure tables

MATERIAL WEIGHTS

Material	Weight in Bank per Cubic Yard	Percent of Swell	Swell Factor	Loose Weight per Cubic Yard
Ashes, Hard Coal	700-1000 lbs.	8%	.93	650-930 lbs.
Ashes, Soft Coal with Clinkers	1000-1515 lbs.	8%	.93	930-1410 lbs.
Ashes, Soft Coal, Ordinary	1080-1215 lbs.	8%	.93	1000-1130 lbs.
Bauxite	2700-4325 lbs.	33%	.75	2020-3240 lbs.
Brick				2700 lbs.
Cement, Portland	94 lbs. per bag			
Cement, Portland	2970 lbs. (packed)	20%	.83	2450 lbs.
Coke, Lump, Loose				620-865 lbs.
Coke, Solvay, Egg, Chestnut or Pea				840 lbs.
Coke, Gas, Egg, Chestnut or Pea				785 lbs.
Coke, Gas Furnace				730 lbs.
Concrete	3240-4185 lbs.	40%	.72	2330-3000 lbs.
Concrete, Mix Wet				3500-3750 lbs.
Copper Ore	3800 lbs.	35%	.74	2800 lbs.
Gasoline, 56° Gaume	6.3 lbs. per gallon			
Granite	4500 lbs.	50 to 80%	.67 to .56	3000-2520 lbs.
Iron Ore, Hematite	6500-8700 lbs.		.45	3900 lbs.
Iron Ore, Limonite	6400 lbs.			
Iron Ore, Magnetite	8500 lbs.			
Kaolin	2800 lbs.	30%	.77	2160 lbs.
Lead Ore, Galina	12550 lbs.			
Lime				1400 lbs.
Limestone, Blasted	4200 lbs.	67 to 75%	.60 to .57	2400-2520 lbs.
Limestone, Loose, Crushed				2600-2700 lbs.
Limestone, Marble	4600 lbs.	67 to 75%	.60 to .57	2620-2760 lbs.
Mud, Dry (Close)	2160-2970 lbs.	20%	.83	1790-2460 lbs.
Mud, Wet (Moderately packed)	2970-3510 lbs.	20%	.83	2470-2910 lbs.
Oil, Crude	6.42 lbs. per gallon			
Phosphate Rock	5400 lbs.			
Sand, Dry	3250 lbs.	12%	.89	2900 lbs.
Sand, Wet	3600 lbs.	14%	.88	3200 lbs.
Sandstone	4140 lbs.	40 to 60%	.72 to .63	2980-2610 lbs.
Shale, Riprap	2800 lbs.	33%	.75	2100 lbs.
Slag, Sand	1670 lbs.	12%	.89	1485 lbs.
Slag, Solid	4320-4860 lbs.	33%	.75	3240-2640 lbs.
Slag, Crushed				1900 lbs
Slag, Furnace, Granulated	1600 lbs.	12%	.89	1430 lbs.
Slate	4590-4860 lbs.	30%	.77	3530-3740 lbs.
Trap Rock	5075 lbs.	50%	.67	3400 lbs.

Wood & Lumber

Beechwood	3250 lbs. per cord	Hemlock	2200 lbs. per cord
Chestnut	2350 lbs. per cord	Hickory	4500 lbs. per cord
Elm	2350 lbs. per cord	Pine, Norway or White	2000 lbs. per cord
		Poplar	2350 lbs. per cord

Reproduced from "Estimating Production and Costs."
Courtesy of Euclid Division of General Motors Corporation

Material weights

The COMPACTION TEST FOR OPTIMUM MOISTURE CONTENT is similar in purpose to the STANDARD PROCTOR TEST. The two tests differ in details of procedure as to the number of dirt layers and thickness of dirt, weight of the tamper used for compacting and the distance through which the tamper is moved.

TABLES OF USEFUL ENGINEERING DATA
MECHANICAL-ELECTRICAL EQUIVALENTS

Power

1 horsepower (hp	= 550 foot-pounds (ft.-lb.) per second (sec.)
	= 33,000 ft.-lb. per minute (min.)
	= 1,980,000 ft.-lbs. per hour (hr.)
	= .275 ft.-tons per sec.
	= 16.5 ft.-tons per min.
	= 990 ft.-tons per hr.
1 horsepower-second (hp-sec.)	= 550 ft.-lb.
	= .275 ft.-tons.
1 horsepower-minute (hp-min.)	= 33,000 ft.-lb.
	= 16.5 ft.-tons.
1 horsepower-hour (hp-hr.)	= 1,980,000 ft.-lb.
	= 990 ft.-tons
1 horsepower (hp)	= 746 watts (w)
	= .746 kilowatts (kw)

Energy

1 horsepower-hour	= 2544 BTU
	= .746 KW-hr.
1 Kilowatt-hour	= 3413 BTU

Pressure

1 lb. per sq. in.	= 2.0360" of mercury at 32° F.
	= 27.71" of water at 32° F.
	= 2.3091 ft. of water at 60° F.
	= 144 lb. per sq. ft.
1 in. of mercury	= .491 lb. per sq. in.
1 in. of water	= 5.2 lb. per sq. ft. = .0361 PSI.

MEASURE OF ANGLES

Degrees	Rise in Inches per ft.	Degrees	Rise in Inches per ft.	Degrees and Minutes	Per cent Rise in ft. per 100 ft.	Degrees and Minutes	Per cent rise in ft. per 100 ft.	Degrees and Minutes
1	.210	¼	1° 11′	1	34.4′	36	19° 48′
2	.419	½	2° 23′	2	1° 8.7′	37	20° 18′
3	.629	¾	3° 35′	3	1° 43.1′	38	20° 48′
4	.839		4° 46′	4	2° 17.5′	39	21° 18′
5	1.050	1¼	5° 56′	5	2° 51.8′	40	21° 48′
6	1.261	1½	7° 7′	6	3° 26.0′	41	22° 18′
7	1.473	1¾	8° 18′	7	4° 0.3′	42	22° 47′
8	1.686	2	9° 28′	8	4° 34.4′	43	23° 16′
9	1.901	2¼	10° 37′	9	5° 8.6′	44	23° 45′
10	2.116	2½	11° 46′	10	5° 42.6′	45	24° 14′
11	2.333	2¾	12° 54′	11	6° 16.6′	46	24° 42′
12	2.551	3	14° 2′	12	6° 50.6′	47	25° 10′
13	2.770	3¼	15° 9′	13	7° 24.4′	48	25° 38′
14	2.992	3½	16° 15′	14	7° 58.2′	49	26° .6′
15	3.215	3¾	17° 21′	15	8° 31.9′	50	26° 34′
16	3.441	4	18° 26′	16	9° 5.4′	51	27° 1′
17	3.669	4¼	19° 30′	17	9° 38.9′	52	27° 28′
18	3.900	4½	20° 33′	18	10° 12.2′	53	27° 55′
19	4.132	4¾	21° 36′	19	10° 45.5′	54	28° 22′
20	4.368	5	22° 37′	20	11° 18.6′	55	28° 49′
21	4.606	5¼	23° 38′	21	11° 51.6′	56	29° 15′
22	4.848	5½	24° 37′	22	12° 24.5′	57	29° 41′
23	5.094	5¾	25° 36′	23	12° 57.2′	58	30° 7′
24	5.313	6	26° 34′	24	13° 29.8′	59	30° 32′
25	5.596	6¼	27° 31′	25	14° 2.2′	60	30° 58′
26	5.853	6½	28° 27′	26	14° 34.5′	61	31° 23′
27	6.114	6¾	29° 22′	27	15° 6.6′	62	31° 48′
28	6.381	7	30° 16′	28	15° 38.5′	63	32° 13′
29	6.652	7¼	31° 8′	29	16° 10.3′	64	32° 37′
30	6.928	7½	32°	30	16° 42.0′	65	33° 1′
31	7.210	7¾	32° 51′	31	17° 13.4′	66	33° 25′
32	7.498	8	33° 41′	32	17° 44.7′	67	33° 49′
33	7.793	8¼	34° 30′	33	18° 15.8′	68	34° 13′
34	8.094	8½	35° 19′	34	18° 46.7′	69	34° 36′
35	8.403	8¾	36° 5′	35	19° 17.0′	70	35° 0′

USEFUL CONVERSION FACTORS FOR RAPID APPROXIMATION

Feet	X	.00019	= Miles
Links	X	.66	= feet
Feet	X	1.5	= links
Square inches	X	.007	= square feet
Square feet	X	.111	= square yards
Acres	X	4,840.	= square yards
Square Yards	X	.002066	= acres
Width in chains	X	8.	= acres per mile
Cubic feet	X	.04	= cu. yds. (Ap.)
Cubic inches	X	.00058	= cu. ft.
U. S. bu.	X	.046	= cu. yds.
U. S. bu.	X	1.244	= cu. ft.
U. S. bu.	X	2,150.42	= cu. in.
Cubic feet	X	.8036	= U. S. bu.
Cubic inches	X	.000466	= U. S. bu.
U. S. gals.	X	.13368	= Cu. ft.
U. S. gals.	X	231.	= cu. in.
Cubic feet	X	7.48	= U. S. gals.
Cubic inches	X	.004329	= U. S. gals.
Cylindrical feet	X	5.878	= U. S. gals.
Cylindrical in.	X	.0034	= U. S. gals.
Pounds	X	.009	= cwt. (112 lbs.)
Pounds	X	.00045	= tons (2,240 lbs.)

Example: Given seven acres of land. To find number of square yards multiply seven by 4,840. Answer: 33.880 square yards.

TABLE 13
PERCENT OF SEA LEVEL HORSEPOWER AVAILABLE FOR A FOUR CYCLE, GASOLINE OR DIESEL ENGINE FOR VARIOUS ALTITUDES

Altitude in feet	110	90	70	60	50	40	20	0	—20
0	95.4	97.1	99.1	100.0	100.8	101.8	103.9	106.2	108.5
1000	92.0	93.7	95.5	96.4	97.4	98.4	100.3	102.5	104.8
2000	88.7	90.4	92.1	93.0	93.8	94.8	96.8	98.8	101.0
3000	85.5	87.2	88.8	89.6	90.5	91.4	93.3	95.2	97.4
4000	82.5	84.0	85.6	86.5	87.3	88.2	89.9	91.8	93.8
5000	79.5	80.9	82.5	83.3	84.2	84.9	86.7	88.5	90.4
6000	76.7	78.1	79.5	80.3	81.1	82.0	83.6	85.3	87.2
7000	73.8	75.2	76.7	77.5	78.2	79.0	80.6	82.3	84.0
8000	71.2	72.5	73.9	74.6	75.4	76.2	77.6	79.3	81.1
9000	68.6	69.9	71.3	72.0	72.7	73.4	74.8	76.4	78.2
10000	66.2	67.5	68.7	69.3	70.7	70.7	72.2	73.7	75.3

TABLE 14
PERCENT OF SEA LEVEL HORSEPOWER AVAILABLE IN TRACTORS AT VARIOUS ALTITUDES POWERED BY G.M.C. TWO-CYCLE DIESEL ENGINE (APPROXIMATE ONLY)

Altitude in feet	Percent of Horsepower Available	Altitude in feet	Percent of Horsepower Available
0	100.0	6000	96.0
1000	100.0	7000	95.3
2000	99.1	8000	94.7
3000	98.2	9000	94.2
4000	97.5	10000	93.6
5000	95.8		

Reproduced from "Earth Moving and Construction Data." Courtesy of Allis-Chalmers Manufacturing Company

Miscellaneous information

RULES OF THUMB

The following "Rules of Thumb" are approximately only.

ROUND TRIP HAUL TIME IN MINUTES for

$$\text{tractor-scraper} = \frac{\text{one way haul in ft.}}{100} + 1\frac{1}{2}.$$

THE ESTIMATED HOURLY OPERATING AND OWNERSHIP COST for a crawler tractor is equal to the delivered price multiplied by .0003 (Does not include operators wages).

GRADE RESISTANCE is equal to twenty pounds per ton of tractor weight for each 1% of grade.

THE MAXIMUM POUNDS DRAWBAR PULL of a crawler tractor is equal to 90% of its weight.

REPAIRS AND REPAIR LABOR COSTS for a crawler tractor will amount to about 100% of the delivered price of the machine based on a 5-year life of 10,000 hours.

SHEEPSFOOT COMPACTION OF SUBGRADE— continue passes until tamper "Walks itself out."

TO CORRECT ENGINE HORSEPOWER RATING FOR ALTITUDE:

1. For a gasoline or four stroke cycle engine deduct 3% from 1000 ft. of altitude above sea level.

2. For a two stroke cycle engine deduct 1% for each 1000 feet above 1000 feet.

TO CORRECT ENGINE HORSEPOWER FOR TEMPERATURES:

1. Deduct 1% of rated power at 60°F. for each 10° temperature rise.

2. Add 1% of rated power at 60°F. for each 10° temperature drop.

MILES PER HOUR IN FEET PER MINUTE AND FEET PER SECOND

Miles Per Hour	Feet Per Minute	Feet Per Second
1	88	1.46
2	176	2.94
3	264	4.4
4	352	5.87
5	440	7.33
6	528	8.8
7	616	10.26
8	704	11.73
9	792	13.2
10	880	14.67
11	968	16.13
12	1,056	17.6
13	1,144	19.07
14	1,232	20.52
15	1,320	22.00
16	1,408	23.47
17	1,496	24.93
18	1,584	26.4
19	1,672	27.86
20	1,760	29.33
21	1,848	30.8
22	1,936	32.26
23	2,024	33.72
24	2,112	35.2
25	2,200	36.67
26	2,288	38.14
27	2,376	39.6
28	2,464	41.04
29	2,552	42.50
30	2,640	44.00

ANGLES OF SLOPES

Slopes ½ to 1		= 63° 30′
Slopes ¾ to 1		= 53° 00′
Slopes 1 to 1		= 45° 00′
Slopes 1¼ to 1		= 38° 40′
Slopes 1½ to 1		= 33° 42′
Slopes 1¾ to 1		= 29° 44′
Slopes 2 to 1		= 26° 35′
Slopes 3 to 1		= 18° 25′
Slopes 4 to 1		= 14° 2′

TABLE FOR CONVERTING PRESSURE PER SQUARE INCH INTO FEET HEAD OF WATER

Pounds per Sq. In.	Feet Head	Pounds per Sq. In.	Feet Head	Pounds per Sq. In.	Feet Head
1	2.31	45	103.90	140	323.26
2	4.62	50	115.45	150	346.34
3	6.93	55	126.99	160	369.44
4	9.24	60	138.54	170	392.53
5	11.54	65	150.08	180	415.62
6	13.85	70	161.63	190	438.71
7	16.16	75	173.17	200	461.80
8	18.47	80	184.72	225	519.52
9	20.78	85	196.26	250	577.25
10	23.09	90	207.81	275	634.97
15	34.63	95	219.35	300	692.70
20	46.18	100	230.90	325	750.42
25	57.72	110	253.99	350	808.15
30	69.27	120	277.08	375	865.87
35	80.81	125	288.62	400	923.60
40	92.36	130	300.17	500	1154.45

TABLE FOR CONVERTING FEET HEAD OF WATER INTO PRESSURE PER SQUARE INCH

Feet Head	Pounds per Sq. In.	Feet Head	Pounds per Sq. In.	Feet Head	Pounds per Sq. In.
1	.43	55	23.81	190	82.37
2	.87	60	25.98	200	86.60
3	1.30	65	28.14	225	97.42
4	1.73	70	30.31	250	108.25
5	2.17	75	32.47	275	119.07
6	2.60	80	34.64	300	129.90
7	3.03	85	36.80	325	140.72
8	3.46	90	38.97	350	151.55
9	3.90	95	41.13	375	162.37
10	4.33	100	43.30	400	173.20
15	6.50	110	47.63	500	216.50
20	8.66	120	51.96	600	259.80
25	10.83	130	57.29	700	303.10
30	12.99	140	60.62	800	346.40
35	15.16	150	64.95	900	389.70
40	17.32	160	69.28	1000	433.00
45	19.49	170	73.61	....	......
50	21.65	180	77.94	....	......

Reproduced from "Earth Moving and Construction Data."
Courtesy of Allis-Chalmers Manufacturing Company

Miscellaneous information (continued)

TRAVEL TIME IN MINUTES
TRAVEL DISTANCE IN FEET

Speed in Miles Per Hour	100	200	300	400	500	600	700	800	900	1000
1.0	1.136	2.275	3.405	4.540	5.67	6.82	7.95	9.08	10.22	11.36
2.0	.568	1.136	1.705	2.275	2.84	3.41	3.98	4.55	5.12	5.68
3.0	.379	.758	1.136	1.515	1.89	2.27	2.66	3.03	3.41	3.79
4.0	.284	.568	.853	1.136	1.42	1.70	1.99	2.27	2.56	2.84
5.0	.227	.454	.682	.910	1.14	1.36	1.59	1.82	2.04	2.27
6.0	.189	.378	.568	.758	.95	1.14	1.32	1.51	1.70	1.89
7.0	.162	.324	.486	.648	.81	.97	1.14	1.30	1.46	1.62
8.0	.142	.284	.427	.570	.71	.85	.99	1.14	1.28	1.42
9.0	.126	.252	.378	.505	.63	.76	.88	1.01	1.14	1.26
10.0	.114	.227	.341	.455	.57	.68	.80	.91	1.02	1.14
11.0	.103	.206	.310	.414	.52	.62	.72	.83	.93	1.03
12.0	.095	.189	.284	.379	.47	.57	.66	.76	.85	.95
13.0	.087	.174	.262	.349	.44	.52	.61	.70	.79	.87
14.0	.081	.162	.244	.325	.41	.49	.57	.65	.73	.81
15.0	.076	.152	.227	.303	.38	.46	.53	.61	.68	.76
17.5	.065	.129	.194	.259	.32	.39	.45	.52	.58	.65
20.0	.057	.113	.170	.227	.28	.34	.40	.45	.51	.57
22.5	.050	.101	.151	.202	.25	.30	.35	.40	.45	.50
25.0	.045	.090	.136	.181	.23	.27	.32	.36	.41	.45

EXAMPLE: To estimate time required to travel 550 feet at 6.0 MPH.

First establish time for 500 ft. at 6.0 MPH .... .95

50 ft.—½ of time shown for 100 ft. at 6.0 MPH .... .09

Enter 1.04 min. for travel time for 550 ft. at 6 MPH.   1.04 Minutes

WEIGHTS AND THICKNESSES OF VARIOUS PIPE USED FOR HIGHWAY DRAINAGE

MATERIAL	DIAMETER IN INCHES							
	12	15	18	24	30	36	42	48
Corrugated Metal Gage....	16	16	16	14	14	12	12	12
Wall thickness, in..........	1/16	1/16	1/16	5/64	5/64	7/64	7/64	7/64
Weight per ft. lb..........	10.5	12.9	15.3	25.2	30.9	51.0	59.5	68.0
Spiral Corr. Cast Iron Wall Thickness, in...........			5/16	3/8	7/16	7/16		
Weight per ft. lb..........			65	90	135	180		
Vitrified Tile D.S. Wall Thickness, in...........	1	1¼	1½	2	2½	2¾		
Weight per ft. lb..........	50	65	100	80	290	385		
Cast Iron Class "B" Wall Thickness, in...........	.62	.68	.75	.89	1.03	1.15	1.28	1.42
Weight per ft. lb..........	82	115	150	233.33	333.3	454	592	750
Reinforced Concrete Wall Thickness, in...........	2	2¼	2½	3	3½	4	4½	5
Weight per ft. lb..........	88	121	164	264	378	500	655	870

Reproduced from "Earth Moving and Construction Data."
Courtesy of Allis-Chalmers Manufacturing Company

Miscellaneous information (continued)

SHORING AND TIMBERING IN TRENCHES

A VISUAL AID TO ONTARIO REGULATION 419/73

MADE UNDER THE CONSTRUCTION SAFETY ACT, 1973

GENERAL NOTES

1. Drawings illustrate minimum shoring requirements. All cleating is not shown.

2. Spacing of timbers as shown is maximum spacing.

3. The shoring and timbering shall extend at least one foot above the top of unsloped trench walls.

4. The composition of materials used for shoring and timbering shall be No. 1 Structural Grade spruce.

5. Struts and wales shall be supported by cleats spiked to wales and sheathing respectively.

6. Posts shall be used between wales; and between the lowest wale and trench bottom.

7. A level area extending two feet from the edge of the trench shall be maintained free of equipment and materials.

8. Shoring and timbering for a trench more than twenty-five feet in depth or more than twelve feet in width shall be designed by a professional engineer.

9. Prefabricated support systems (trench boxes) shall be designed by a professional engineer.

10. For further information; and shoring and timbering details refer to The Construction Safety Act, 1973 and the regulations thereunder.

TABLE OF MINIMUM TRENCH SHORING

Depth of Trench	Spacing of Sheathing (2″ x 8″ except 3″ x 8″ where marked with asterisk*)				Size of Wales (spaced not more than 4 ft. apart vertically)		Size of Struts (spaced not more than 4 ft. apart vertically and 8 ft. horizontally)				
							Width of trench 6 ft. or less			Width of trench 6 ft. to 12 ft.	
	Soil Type				Soil Type		Soil Type			Soil Type	
Feet	1	2	3	4	1,2,3	4	1,2	3	4	1,2,3	4
4 - 10	4 ft.	4 ft.	close	close	6 x 6	8 x 8	4 x 4	6 x 6	6 x 6	8 x 8	8 x 8
10 - 15	4 ft.	4 ft.	close	close*	8 x 8	10 x 10	6 x 6	6 x 6	8 x 8	8 x 8	10 x 10
15 - 20	2 ft.	close	close	close*	8 x 8	12 x 12	6 x 6	6 x 6	10 x 10	8 x 8	12 x 12
20 - 25	close	close	close	close*	10 x 10	14 x 14	8 x 8	8 x 8	12 x 12	8 x 8	12 x 12

SOIL TYPES 1) Soil that is hard and solid.
2) Soil that may crack or crumble.
3) Soil that is loose, soft, sandy or that has been previously excavated.
4) Soil that is wet, muddy or will flow easily unless supported immediately after excavation.

NOTES a) All timber dimensions are full size (in inches).
b) All sheathing, Number 1 Grade spruce planks; all wales and struts, Number 1 Structural Grade spruce timbers.
c) All spacings are centre to centre.
d) "Close" means space between sheathing planks not to exceed 1/2 in.
e) Wales may be omitted in type 1 soil only, up to 10 ft. depth.

Courtesy of Construction Safety Association of Canada

Trench shoring standards

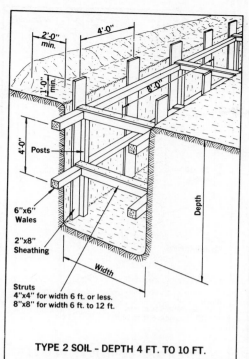

TYPE 2 SOIL – DEPTH 4 FT. TO 10 FT.

Labels on figure: 2'-0" min.; 4'-0"; 1'-0" min.; 8'-0"; 4'-0"; Posts; 6"x6" Wales; 2"x8" Sheathing; Depth; Width; Struts 4"x4" for width 6 ft. or less. 8"x8" for width 6 ft. to 12 ft.

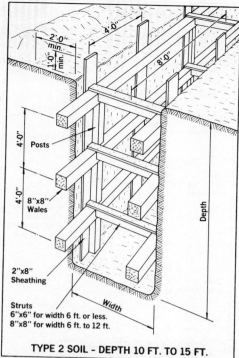

TYPE 2 SOIL – DEPTH 10 FT. TO 15 FT.

Labels on figure: 2'-0" min.; 4'-0"; 1'-0" min.; 8'-0"; 4'-0"; 4'-0"; Posts; 8"x8" Wales; 2"x8" Sheathing; Depth; Width; Struts 6"x6" for width 6 ft. or less. 8"x8" for width 6 ft. to 12 ft.

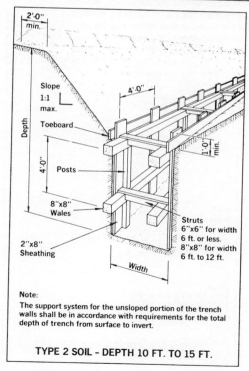

Note:

The support system for the unsloped portion of the trench walls shall be in accordance with requirements for the total depth of trench from surface to invert.

TYPE 2 SOIL – DEPTH 10 FT. TO 15 FT.

Labels on figure: 2'-0" min.; Slope 1:1 max.; Toeboard; Depth; 4'-0"; 4'-0"; 1'-0" min.; Posts; 8"x8" Wales; 2"x8" Sheathing; Width; Struts 6"x6" for width 6 ft. or less. 8"x8" for width 6 ft. to 12 ft.

Courtesy of Construction Safety Association of Canada

Trench shoring standards (continued)

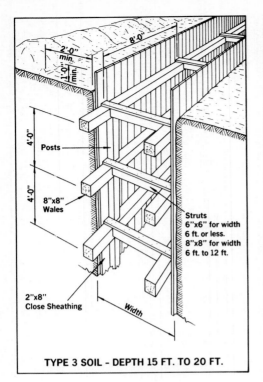

TYPE 3 SOIL – DEPTH 15 FT. TO 20 FT.

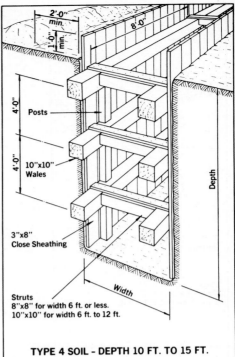

TYPE 4 SOIL – DEPTH 10 FT. TO 15 FT.

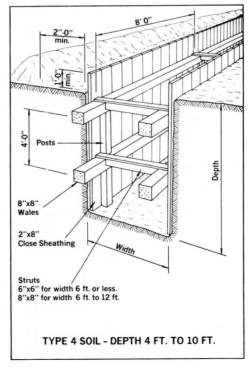

TYPE 4 SOIL – DEPTH 4 FT. TO 10 FT.

Courtesy of Construction Safety Association of Canada

Trench shoring standards (continued)

TIRES FOR EARTH MOVING, MINING AND LOGGING SERVICE FOR SHORT HAULS
Maximum Speed 30 Miles Per Hour

Tire Loads at Various Inflation Pressures

Tire Size	25	30	35	40	45	50	55	60	65	70	75	80	85	90
7.00-20		1810	1980	2140	2300	2440	2580	2710	2850[10]	2980	3100	3210[12]		
7.50-20 / 8.22-5		2030	2220	2400	2570	2740	2900	3040	3190[10]	3330	3470[12]			
8.25-20 / 9.22-5		2390	2620	2820	3030	3220	3400	3580[10]	3750	3920	4070[12]			
9.00-20 / 10.22-5		2840	3100	3360	3590	3830	4050[10]	4250	4460[12]					
10.00-20 / 11.22-5	2870	3200	3500	3780	4050	4310	4560	4780[12]	5020	5250[14]				
10.00-22 / 11.24-5	3040	3400	3720	4030	4310	4590	4850	5100[12]	5350	5590[16]				
10.00-24	3240	3620	3960	4280	4580	4870	5160	5420[12]	5680	5940				
11.00-20 / 12.22-5	3180	3540	3870	4190	4480	4470	5050[12]	5300	5560[14]					
11.00-22 / 12.24-5	3390	3760	4110	4440	4760	5060	5360[12]	5620	5910[14]					
11.00-24	3580	3990	4360	4720	5060	5380	5700[12]	5980	6270[16]					
12.00-20/21	3600	4020	4390	4750	5080		5740	6030[14]	6320	6600[16]				
12.00-24/25	4060	4520	4930	5330	5720	6080	6450	6780[14]	7110	7430[16]				
13.00-24/25	4760	5320	5820	6280	6740	7170	7590	7980	8370	8750[18]				
14.00-20/21	5040	5620	6140	6630	7100[12]	7550	8000	8400[16]	8820	9200	9580[20]			
14.00-24/25	5630	6270	6850	7400	7920	8430	8920	9400[16]	9830	10280	10690[20]	11100	11510	11900[24]
16.00-20/21	6670	7430	8120	8780	9410[16]	10010	10580	11130[20]						
16.00-24/25	7370	8200	8970		10400[16]	11070	11710		12900	13490[24]	14020	14560[28]		
18.00-24/25	9580	10670[12]	11670	12620[16]	13520	14380[20]	15230	16000[24]	16780	17500[28]				
18.00-32/33	11350	12640	13820	14930	16000	17020	18010	18930[24]	19850	20720	21580	22400[32]		
21.00-24/25	12250	13640[16]	14910	16120[20]	17280	18370[24]	19450	20440[28]						
24.00-25	15130	16860	18420[18]	19920	21340[24]	22700	24020[30]							
24.00-29	16330	18200	19890	21500	23020[24]	24500	25930[30]	27250	28590[36]					
24.00-32/33	17570	19600	21400	23140	24800[24]	26380	27900[30]	29330	30760[36]					
27.00-33	22700	25280	27620	29890[24]	32030	34060[30]	36020	37890[36]						
30.00-33	27790	30950	33830[28]	36590	39200[34]	41700[40]								

The small index numbers are Ply Ratings, and denote maximum recommended load for that Ply Rating. For example: with 7.50-20 tire, or 8.22-5 corresponding tubeless size, 3190[10] is maximum load for 10-ply rating, and 3470[12] maximum for 12-ply rating.

WIDE BASE TIRES FOR EARTH MOVING SERVICE

Tire Loads at Various Inflation Pressures
MAXIMUM SPEED 30 MILES PER HOUR

Tire Size	25	30	35	40	45	50	55
20.5-25	8630	9590[12]	10500	11350[14]	12170	12940[20]	
23.5-25	11150[12]	12440	13610[16]	14700[20]	15760	16760[24]	
26.5-25	13940[14]	15530	17000[20]	18350	19700	20970[24]	22100[28]
29.5-25	17190[16]	19160	20950[22]	22600	24210[28]		
29.5-29	18600[16]	20650	22620[22]	24450	26200[32]	27870	29490[34]
33.5-33	25850	28750	31420[26]	34000	36470[32]	38800	41050[34]
37.5-33	31450	35020[24]	38350	41400[28]	44400[36]		

Tire Loads at Various Inflation Pressures
MAXIMUM SPEED 10 MILES PER HOUR

Tire Size	25	30	35	40	45	50	55
20.5-25	9670	10740[12]	11760	12710[16]	13630	14490[20]	
23.5-25	12490[12]	13930	15240[16]	16460[20]	17630	18760[24]	
26.5-25	15610[14]	17390	19040[20]	20550	22070[24]	23480	24760[28]
29.5-25	19250[16]	21460	23470[22]	25310	27110[28]		
29.5-29	20830[16]	23130	25330[22]	27390	29340[28]	31200	33000[34]
33.5-33	28950	32200	35190[26]	38080	40850[32]	43460	45980[38]
37.5-33	35230	39230[24]	42950	46370[30]	49730[38]		

The small index numbers are Ply Ratings, and denote maximum recommended load for that Ply Rating. For example: with 20.5-25 tire, 9520[12] is maximum load for 12-ply rating, 11350[14] maximum for 16-ply rating and 12840[20] maximum for 20-ply rating.

EFFECT OF INFLATION AND LOAD ON TIRE MILEAGE

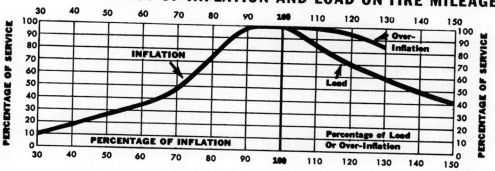

Courtesy of Rubber Manufacturers Association

Tire load and inflation table

General industry practices

The following is for information purposes only to explain the general industry practices of equipment rental, and is not to suggest or influence the rates or terms of rental of any item as these are matters which must be determined between the lessee and lessor. To avoid disagreements, the rental terms should be agreed upon prior to rental and, especially where larger equipment is involved, the terms should be spelled out in a written agreement signed by both the lessor and lessee.

Time basis of rates

It is the general practice in the industry to base rates upon one shift of 8 hours per day, 40 hours per week, or 176 hours per month of a 30 consecutive day period.

If the equipment is rented by the day, the rate for overtime is one-eighth of the daily rate for each hour in excess of eight. If it is rented by the week, the rate for overtime is 1/40 of the weekly rate for each hour in excess of 40. If it is rented by the month, the overtime rate is 1/176th of the monthly rate for each hour in excess of 176 hours in any one 30 consecutive day period.

Many distributors do not rent by the day or by the week, especially in the case of large equipment, even though daily and weekly rates appear in this compilation.

Costs of repairs

Tractor equipment: In the case of tractor equipment and/or rubber-tired hauling equipment, the difference between "normal" and "abnormal" wear and tear is not easily discernible and can be the cause of unpleasant relationships between the lessor and lessee. Most lessors, therefore, require the lessee, in their lease agreements, to bear all costs of repair to such equipment, regardless of the cause. Many distributors measure tread wear to determine charges to lessee for tire wear. Tractor equipment would include crawler as well as rubber-tired.

Non-tractor equipment: The lessor, under the lease agreement, usually bears the cost of repairs due to normal wear and tear on non-tractor equipment, and the lessee bears all other costs. **Cranes and shovels:** Rubber-tired or crawler mounted cranes or shovels would not be included in the category of either tractor equipment or rubber-tired hauling equipment and generally the lessor agrees to bear the cost of repairs due to normal wear and tear. In individual cases (especially where the equipment may be subjected to unusual abuse or excessive wear and tear), lessors insist on the lessee bearing all costs of repair on equipment regardless of its type or classification or charge a higher rental rate.

"Normal wear and tear"

"Normal wear and tear" is that which would be expected to result from the use of the equipment under normal circumstances, provided the equipment is properly maintained and serviced. This is a question of fact which must be settled by the interested parties.

Fuel and Lubricants

In all cases, the lessee bears costs of fuel and lubricants.

Operator

In no case does the average rental rate in this publication include the cost of operator.

Condition of equipment

The equipment rented is to be delivered to the lessee in good operating condition, and is to be returned to the lessor in the same condition as delivered, less normal wear, unless there is an agreement which may vary this general practice. Where excess cleaning or repairs are necessary, many lessors have made additional charges for this cleaning and repair.

Freight charges

The rates in the schedule are all f.o.b. the lessor's warehouse or shipping point. The lessee pays the freight or drayage charges from shipping point to destination and return. When loading, unloading, dismantling or assembling is necessary, the lessee pays this additional charge.

Rental period

On local rentals the rate starts when equipment leaves the lessor's warehouse and stops when it is returned to such warehouse. On out-of-town shipments, the rental starts on the date of the bill of lading of shipments and stops on the date of the return bill of lading.

Payment, taxes and insurance

Rentals are payable in advance and subject to the terms and conditions of the lessor's rental contract. No insurance, license, sale or use taxes are included in these rates.

Courtesy of Associated Equipment Distributors (AED)

Equipment rental information

Type of Equipment	$ Per Month	$ Per Week
Air Compressors, portable, all types		
Diesel, 125 cfm	270.00	91.00
Diesel, 210 cfm	402.00	135.00
Diesel, 365 cfm	670.00	232.00
Diesel, 600 cfm	952.00	336.00
Diesel, 1000 cfm	1570.00	552.00
Gas, 60 cfm	152.00	52.25
Gas, 125 cfm	247.00	85.50
Bulldozer, crawler, gear drive		
Diesel, 45 to 59 horsepower	1452.00	510.00
Diesel, 92 to 115 horsepower	2671.00	913.00
Bulldozer, crawler, power shift		
Diesel, 61 to 89 horsepower	1681.00	530.00
Diesel, 120 to 144 horsepower	3116.00	947.00
Diesel, 251 to 300 horsepower	5271.00	1590.00
Cranes, lifting, cable, crawler		
Diesel, 12½ ton capacity	2133.00	575.00
Diesel, 30 ton capacity	2905.00	927.00
Diesel, 60 ton capacity	4244.00	1546.00
Cranes, lifting, cable, truck		
Diesel, 12½ ton	1925.00	—
Diesel, 30 ton	2796.00	1038.00
Diesel, 60 ton	4266.00	1479.00
Cranes, lifting, hydraulic, telescoping boom		
Diesel, 12½ ton	1433.00	468.00
Diesel, 30 ton	3702.00	1223.00
Diesel, 75 ton	4973.00	1497.00
Drills, rock, crawler mounted		
4-inch drifter, straight boom	996.00	372.00
4-inch drifter, swing boom	1112.00	451.00
5-inch drifter, swing boom	1252.00	422.00
Drills, rock, hand held (jackhammers)		
Utility, up to 30 pounds	74.75	26.50
55 to 64 pounds	86.00	10.25
Front loader, crawler, power shift		
Diesel, 1 yard bucket	1255.00	420.00
Diesel, 2 yard bucket	2601.00	832.00
Diesel, 5 yard bucket	6621.00	—
Front loader, 4-wheel drive, power shift, articulated frame		
Diesel, 1¼ yard bucket	1298.00	440.00
Diesel, 2 yard bucket	1864.00	625.00
Diesel, 5 yard bucket	4020.00	1331.00
Front loader, 4-wheel drive, power shift, rigid frame		
Diesel, 1 yard bucket	968.00	326.00
Diesel, 2 yard bucket	1744.00	597.00

Type of Equipment	$ Per Month	$ Per Week
Grader, tandem drive, direct drive		
Diesel, 5 to 8 tons	953.00	313.00
Diesel, 13 to 14 tons	1896.00	611.00
Grader, tandem drive, torque converter		
Diesel, 11 to 13 tons	1778.00	602.00
Diesel, 15 to 22 tons	2898.00	873.00
Pumps, water, diaphragm		
Gas, 2 inch single	111.00	39.25
Gas, 4 inch double	277.00	94.00
Pumps, water, portable, centrifugal		
Diesel, 6 inch, 90M	451.00	156.00
Diesel, 10 inch, 200M	878.00	296.00
Gas, 2 inch, 7 to 10M	102.00	36.25
Gas, 4 inch, 30 to 40M	222.00	77.50
Rollers, pneumatic tired		
Diesel, 12-ton, 9-wheel	761.00	255.00
Diesel, 20-ton, 9-wheel	1588.00	538.00
Rollers, steel wheel, 2-axle tandem		
Gas, 2 to 3 ton	405.00	136.00
Gas, 5 to 8 ton	582.00	195.00
Rollers, 3 steel wheel		
Diesel, 8 to 10 ton	634.00	237.00
Diesel, 10 to 12 ton	776.00	252.00
Diesel, 12 to 14 ton	837.00	260.00
Rollers, vibratory, tandem		
Gas, 7 horsepower, 1400 lb	226.00	80.75
Gas, 12 horsepower, 4000 lb	463.00	159.00
Scraper, 2-wheel tractor, single diesel		
250 horsepower, 18 yard struck	3441.00	1150.00
450 horsepower, 25 yard struck	6364.00	—
Scraper, 2 wheel tractor, two diesels		
300 horsepower, 14 yard struck	5424.00	1715.00
550 horsepower, 18 yard struck	6990.00	2455.00
Shovels, backhoe, hydraulic, crawler		
Diesel, ½ yard bucket	1787.00	591.00
Diesel, ¾ yard bucket	2397.00	758.00
Diesel, 1 yard bucket	3046.00	959.00
Diesel, 2 yard bucket	5029.00	1661.00

Courtesy of Associated Equipment Distributors (AED)

Abbreviated equipment rental table

EXPLANATION OF TABULATED ITEMS

ALL PERCENTAGE RATES ARE OFFERED AS A GUIDE AND SHOULD BE REVIEWED BY THE INDIVIDUAL EQUIPMENT OWNER.

DEPRECIATION AND OBSOLESCENCE

The straight-line method of depreciation based on the useful life of the machine is best suited to a compilation of this type. A uniform percentage is charged off yearly during the economic life of each piece of equipment.

Obsolescence is an element whose effects are hidden but nonetheless real. When recently purchased equipment is outmoded by advanced models or new methods it is apparent that the purchaser's business potential is affected.

OVERHAUL, MAJOR REPAIRS, PAINTING

This section covers the cost of periodic major repair and reconditioning work, usually performed in the owner's shops. Broadly speaking it is the cost of offsetting the effects of the heavy wear-and-tear to which construction equipment is subjected, and of repairing damage which may be discovered only during a period of overhaul. The percentage factors listed for major repairs are for average working conditions and should be re-negotiated if abnormal conditions exist.

GENERAL OVERHEAD

This broad heading incorporates a number of items which are constant for all items of equipment and consequently have been grouped together. The total of these is 10% and they are divided into four classifications:—

INTEREST. An interest rate of 4% has been used in this schedule.

STORAGE. This expense consists of the cost of providing land and buildings for the storage of equipment when not in use, starting or running machines in storage, the maintenance and servicing of the property and any direct overhead involved. An average figure for this expense is 1%.

INSURANCE. An allowance of 1% is made to cover the cost of general policies covering equipment.

OVERHEAD. An allocation of 4% is made to cover the general expense of doing business. Overhead includes all business operating costs such as office space, supplies and office equipment, salaries and wages and other essentials for the conduct of a business.

TOTAL PERCENTAGE

This is the total of the percentages of the first three columns. Under average conditions this figure covers the cost of doing business and ensures funds to purchase new equipment.

AVERAGE USE MONTHS PER YEAR

To convert the total to a monthly rental percentage rate it is necessary to determine, for each piece of equipment, the average number of months of operation annually. Dividing the total percentage by the average use in months per year produces the monthly percentage rate. There is little need to stress the extent to which the number of working months is affected by the vagaries of demand, job nature, climate and other factors, and the fact that the monthly rental rate varies accordingly.

MONTHLY RENTAL RATE

The monthly rental percentage rate is based on a minimum period of not less than one month, and operational time of not more than 200 hours monthly. A month shall be considered from the date of commencement of the rental period up to, but not including, the same date in the next calendar month. After the minimum rental period has expired, the rent payable for a fraction of any succeeding month shall be the proportionate part of the monthly rental rate in dollars according to the number of calendar days in such fraction.

CURRENT REPLACEMENT VALUE

The current replacement value of new equipment including Provincial and Municipal sales taxes must be used as the basis for calcuating equipment rental rates, and not the original purchase price. The necessity for this is apparent when it is appreciated that changes in the purchase price of new equipment are normally accompanied by corresponding changes in the cost of all the factors which comprise the rental rate. When the monthly rental percentage rate is applied to current replacement value there is automatic compensation for such changes. This would not

Courtesy of Canadian Construction Association

Percentage basis for rental rates

be the case if the original purchase price were used. Consequently, all percentages in this schedule are expressed in terms of current replacement value.

Current Replacement Value ($) x Monthly Percentage (%) = Monthly Rental ($).

WEEKLY AND DAILY RATES

Regardless of the length of the rental period it is necessary for an owner to prepare equipment for delivery in good working order and, upon its return, to inspect and restore it as nearly as possible to its original condition. When the rental period is short, it is apparent that the cost of inspection and handling would be out of proportion to the earned rental if the monthly rate were used.

This condition is adjusted by the use of a weekly rate which is ⅓ of the monthly rate, and by a daily rate which is 1/12 of the monthly rate.

The weekly rental rate is based on operational time of not more than 60 hours in seven consecutive days, while the daily rate has a basis of not more than 10 working hours in a twenty-four hour period.

OVERTIME

Since rental rates are based on a maximum number of working hours in a given period it is necessary to charge for the additional wear-and-tear on equipment due to overtime operation. Wear-and-tear increases in proportion to the overtime while other factors may undergo little or no change.

Overtime should be charged by the hour at 66⅔% of the straight-time rate. For example, when equipment is rented on a monthly basis the hourly

cost to the lessee is $\dfrac{Monthly\ Rental\ in\ Dollars,}{200}$

and overtime hours should be charged at 66⅔% of this figure. Similarly, the overtime hourly rate on weekly rentals is 66⅔% of $\dfrac{Weekly\ Rental\ in\ Dollars}{60}$

and, for daily rentals, 66⅔% of $\dfrac{Daily\ Rental\ in\ Dollars.}{10}$

"ALL-FOUND" RENTALS

It is the practice of some owners to provide equipment on an "operated" basis. Under this arrangement it is the responsibility of the owner to pay all expenses in connection with normal operation of the equipment, with the usual exception that the lessee pays transportation costs.

The rates listed in this schedule are bare "machine" rates and will not be adequate under such conditions. A separate and additional rate must be calculated for those items which may be classified as job or operating costs. This rate, in dollars, is added to the bare "machine" rate in dollars to produce the required "all-found" rental.

Typical of the job costs which must be incorporated into an "all-found" rental are wages for operators and other labour; unemployment insurance; holiday pay allowance; workmen's compensation; public liability insurance; equipment insurance; fuel, lubricants and supplies; labour and material for field repairs; unloading; erecting; dismantling; loading; and an allowance for overhead.

RENTAL AGREEMENTS

While the terms of agreement may be subject to negotiation between lessor and lessee it may be of value to enumerate briefly those conditions which are commonly accepted.

Equipment to be rented is assumed to be in good mechanical condition, clean and with all standard attachments and accessories. Rental starts when the equipment leaves the lessor's shipping point and ends on and includes the date of actual delivery of the equipment to the Lessor or at any other equidistant point if instructions to do so are given by the Lessor. Rental is normally f.o.b. the lessor's premises and the lessee pays the cost of transportation both ways, plus unloading and loading costs at the job. Costs of erecting, dismantling and operating (including field repairs) are borne by the lessee. Rental is payable in advance by the month.

In the interests of uniformity it is strongly recommended that standard rental agreement forms be used, a copy of which is included on pages 63 to 65.

Courtesy of Canadian Construction Association

Percentage basis for rental rates (continued)

Description of Equipment	Depreciation, Obsolescence	Overhaul Major Repairs, Painting	Total, Including 10% Overhead	Average Use, Months per Year, Canada	Monthly Rental Rate, Per Cent
Air					
compressor, portable, gas or diesel, 125 to 1200 cfm	12	10	32	7	4.5
drill, crawler mounted	14	16	40	8	5.0
drill, hand, up to 60 pounds	30	20	60	6	10.0
pavement breakers, all weights	30	35	75	5	15.0
receivers and lubricators	13	9	32	4	8.0
Asphalt					
distributor, truck-mounted	18	12	40	5	8.0
paving machine	18	17	45	5	9.0
Backhoes, complete					
cable, crawler, 1 to 4 yard	9	11	30	6	5.0
hydraulic, crawler or truck, ½ to ⅞ yard	13	13	36	6	6.0
hydraulic, crawler, 1 to 2½ yard	9	11	30	6	5.0
Buckets					
clamshell, excavator, ½ to 4 yard	15	25	50	5	10.0
clamshell, rehandling, ½ to 5 yard	15	15	40	5	8.0
dragline, general purpose, ½ to 5 yard	15	15	40	5	8.0
Compaction equipment					
pneumatic tired roller, self-propelled	11	14	35	5	7.0
steel wheel roller, tandem or 3-wheel, 10 to 20 ton	9	11	30	5	6.0
vibrator, manually guided	22	28	60	5	12.0
vibratory rollers, self-propelled	11	19	40	5	8.0
Concrete					
curb and gutter machine	18	20	48	4	12.0
mix plant, portable, 4 to 10 cy	15	10	35	5	7.0
pumps	15	15	40	4	10.0
spreader-finisher, for forms	22	18	50	5	10.0
spreader-finisher, slipform	18	12	40	5	8.0
trucks, dump, with agitator	18	22	50	5	10.0
trucks, ready-mix, 6 to 15 cy	15	20	45	6	7.5
Cranes, lifting					
cable, 12 to 100 ton	9	11	30	6	5.0
hydraulic, self-propelled	9	11	30	6	5.0
Ditching machines, ladder or wheel	15	15	40	5	8.0
Draglines and clamshells, with bucket					
½ to 1¾ yard, crawler	13	10	36	6	6.0
2 to 4½ yard, crawler	9	11	30	6	5.0
Front loaders					
crawler or wheel, 1 to 6½ yard	15	17	42	7	6.0
four wheel drive, 8 to 12 yard	11	14	35	7	5.0
Graders, motor, 80 to 260 horsepower	11	14	35	7	5.0
scarifier, rear, hydraulic	15	15	40	4	10.0

Courtesy of Canadian Construction Association

Rental rates as percentage of replacement cost

CANADIAN RENTALS

Description of Equipment	Deprecia- tion, Ob- solescence	Overhaul Major Re- pairs, Painting	Total, Including 10% Over- head	Average Use, Months per Year, Canada	Monthly Rental Rate, Per Cent
Gravel and chip spreaders, self-propelled	15	15	40	5	8.0
Pile driving equipment					
air or steam hammers, single acting	15	15	40	5	8.0
air or steam hammers, double acting	22	18	50	5	10.0
diesel hammers	22	18	50	5	10.0
drop hammers, not including leads	10	10	30	3	10.0
electric vibratory, with generator	15	17	42	6	7.0
Pumps, centrifugal					
diesel, 4 to 10 inch	15	15	40	5	8.0
gasoline, 2 inch	45	25	80	4	20.0
gasoline, 4 to 8 inch	18	22	50	5	10.0
Scrapers					
standard, 1 or 2 engines, all sizes	15	17	42	7	6.0
elevating, all sizes	15	17	42	7	6.0
trailer, without tractor	9	11	30	5	6.0
Screening plants, portable	9	11	30	5	6.0
Shovels, cable dipper, crawler					
½ to ⅞ cy	13	13	36	6	6.0
1 to 4½ cy	9	11	30	6	5.0
Soil pulverizor, self-propelled	22	28	60	4	15.0
Stackers, belt conveyor	11	14	35	5	7.0
Tractors					
backhoe attachments	15	15	40	4	10.0
diesel crawler, with dozer, 40 to 185 horsepower	13	19	42	7	6.0
diesel crawler, with dozer, 186 to 440 horsepower	15	17	42	6	7.0
diesel or gasoline, industrial farm,					
2-wheel drive, with or without loader	15	17	42	7	6.0
diesel rubber tired, with dozer	15	17	42	6	7.0
Trailer, machinery (Float), with or without					
tractor, all sizes	15	15	40	5	8.0
Trucks					
dump or stake, diesel or gasoline, highway	18	21	49	7	7.0
dump, off highway, rack body	13	19	42	7	6.0
Dumpster, dual control	13	19	42	7	6.0
lubrivan, complete	18	21	49	7	7.0
pickup, two wheel drive	23	23	56	8	7.0
pickup, four wheel drive	30	40	80	8	10.0
water	18	17	45	6	7.0

Courtesy of Canadian Construction Association

Rental rates as percentage of replacement cost (continued)

ESTIMATED HOURLY OWNERSHIP AND OPERATING COST

Several examples of how to calculate the approximate hourly cost of owning and operating International and Hough Construction Equipment are shown below. *These are examples offered as a guide only.* The figures should be corrected to the actual job conditions, delivered price, local cost of fuel and lubricants, etc. No guarantee is made, nor implied, that actual job costs will not be higher or lower than those shown.

HOURLY OWNERSHIP COSTS		E-200 Elevating Pay scraper	E-270 Elevating Pay scraper	295B Pay scraper	PH-180 45 Ton Pay hauler	H-400B Pay loader	TD-25B PS Crawler Tractor w/Hydraulic Semi "U" Blade	D-500 Pay dozer
(A-1)	Depreciation	$ 3.137	$ 6.585	$ 9.051	$ 9.190	$ 7.250	$ 5.251	$10.290
(A-2)	Interest, Insurance, Taxes	1.404	3.001	4.214	4.153	5.000	2.100	4.690
	TOTAL HOURLY OWNERSHIP COSTS	$ 4.541	$ 9.586	$13.265	$13.343	$12.250	$ 7.351	$14.980
HOURLY OPERATING COSTS								
(B-1)	Fuel	$ .850	$ 1.530	$ 2.720	$ 2.040	$ 2.210	$ 1.445	$ 3.740
(B-2)	Engine Oil	.103	.120	.149	.157	.120	.116	.190
(B-3)	Transmission Oil	.028	.031	.038	.045	.016	.888	.010
(B-4)	Final Drive Lubricant	.007	.013	.020	.016	.020	.092	.018
(B-5)	Steering, Hydraulic Oil	.038	.026	.038	.024	.080	.016	.040
(B-6)	Chassis Lubricant	.020	.020	.030	.030	.023	.012	.020
(B-7)	Filter Costs	.230	.250	.290	.290	.127	.150	.250
(B-8)	Tire Costs	1.460	3.062	4.943	3.975	5.420	- 0 -	4.760
(B-9)	Repairs and Labor	2.823	5.926	8.146	7.352	4.350	4.726	6.170
	TOTAL HOURLY OPERATING COSTS	$ 5.559	$10.978	$16.374	$13.929	$12.366	$ 7.445	$15.198
TOTAL ESTIMATED HOURLY OWNERSHIP AND OPERATING COST		$10.100	$20.564	$29.639	$27.272	$24.616	$14.791	$30.178
(B-10)	Operator's Wages per Hour, Add	—	—	—	—	—	—	—

DATA USED TO FIGURE THE HOURLY COST IN THE PRECEEDING SEVEN EXAMPLES

	E-200	E-270	295B	PH-180 45 Ton	H-400B	Hyd. Dozer TD-25B & Semi "U"	D-500
F.O.B. Factory Price	$35,095	$75,034	$111,018	$103,830	$125,000	$51,300	$117,200
Freight ($2.00 per cwt)	530	980	1,486	1,272	2,500	1,210	2,860
Tires @ Fleet Purchase Price	3,726	9,186	14,830	11,924	16,277	52,510	14,285

A-1 — Depreciation (less tires), figured straight line method: 10,000 hours, 5 years.

A-2 — Interest, insurance and taxes: 4% × Delivered Price/1000. (Consumption rates given in this section under the headings B-1 through B-9)

B-1 — Fuel for average load: $0.17 per gallon, all machines listed. B-2 — Engine oil: $0.83 per gallon, all machines listed. B-3 — Transmission oil: $1.11 per gallon, all machines listed. B-4 — Final drive lubricant: $0.77 per gallon, all machines listed.

B-5 — Steering, hydraulic oil: $0.77 per gallon, all machines listed.

B-6 — Chassis Lubricant: $0.20 per pound.

B-7 — Filter costs: all filters for average conditions.

B-8 — Tire Costs: Replacement tire costs divided by estimated tire life give hourly tire costs. Replacement tire cost figured as 45% discount off delivered price for all IH rubber tired machines and 40% discount off delivered price for Hough machines. Estimated tire life taken as 3000 hours.

B-9 — Repairs and labor: Taken as 90% of the depreciation rate for the TD-25B, PS-295B, E-270 and the E-200; as 80% for PH-180 and D-500; and as 60% for the H-400B.

B-10—Operator's wage: In all cases this item was left as an "Add" item, hence the total estimated ownership and operating cost shown is minus the operator's wage.

IT IS NOT INTENDED THAT THESE COSTS SHOULD REPRESENT ANY GIVEN SET OF JOB CONDITIONS.

Courtesy of International Harvester Company

Estimating equipment cost

PRODUCTION OF POWER SHIFT CRAWLERS WITH SEMI-U BLADES
(3,000 lb. material — loose cubic yards)

Tractor Model	DOZING DISTANCE (IN FEET)					
	50	100	150	200	300	350
TD-25C	960	620	475	390	290	260
TD-25B	770	510	385	315	240	215
TD-20C	520	335	255	205	145	125
TD-15B	420	275	210	165	115	105

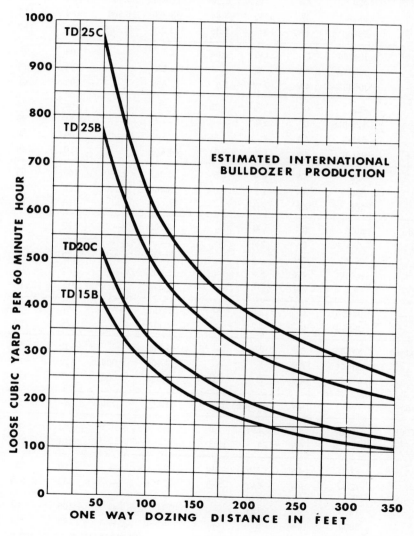

Courtesy of International Harvester Company

Dozer production graph

A-21

Make and model		Drive	Capacity, yards Heaped	Capacity, yards Struck	Capacity, lbs. Lift	Capacity, lbs. Carry	Maximum Dump Clearance Inches	Width, inches	Length, inches, Bucket on Ground
J. I. Case	350	C	¾	0.686	3,750	—	96	63	156
	1150B	C	1¾	1.52	9,200	—	104	81	189
	680E	RW	1¼	1.02	5,550	—	115	84	198
	W26B	4W	3	2.6	29,300	—	119	113	291
Caterpillar	931	C	1	0.87	9,100	3,185	96	70	153
	983	C	4½	3.74	50,240	17,584	145	114	267
	920	4W	1½	1.19	11,930	5,965	109	89	225
	980B	4W	4½	3.84	33,920	16,960	126	115	295
	992(g)	4W	10	8.36	93,970	46,985	171	157	436
John Deere	JD450C	C	1¼	1.04	—	—	103	72	164
	JD301A	RW	¾	—	2,500	—	94	75	182
	JD644B	4W	2	2.5	12,895	—	108	101	259
Fiat Allis	FL9	C	1⅝	1.39	—	—	105	74	180
	12G-B(a)	C	2¾	—	—	—	131	96	229
	545-B	4W	2	1.64	13,100	—	109	89	243
	945-B	4W	6½	5.5	44,620	—	130	129	336
Ford	335	RW	0.6	0.5	3,385	2,240	103	68	168
	550	RW	1	0.9	8,000	4,700	105	78	180
	A66	4W	2¾	—	23,000	—	110	96	259
International Harvester	500E	C	¾	—	4,500	—	99	52	155
	250C PS	C	2¾	—	—	—	129	96	225
	3200B	4W	—	—	1,600	1,250	89	43	116
	H-60E	4W	2	—	15,200	—	102	98	235
	580	4W	21	—	232,470	—	211	216	574
Komatsu	D315-15	C	1	—	3,530	—	94	66	173
	D95S-1	C	4.2	—	14,110	—	128	109	243
Massey Ferguson	MF500	C	2¼	—	11,000	—	101	82	180
	MF50A	RW	1	—	4,000	4,000	112	80	196
	MF88	4W	6	5.18	70,000	18,000	129	129	354
Melroe (Clark)	700	4W	0.34+	0.27+	1,250	1,250	87	53	111
Michigan (Clark)	35AWS	4W	1¼	1.0	—	—	103	83	202
	675	4W	24	19.5	—	—	215	215	607
Terex	72-31	4W	2½	2⅛	22,500	—	120	104	247
	72-71	4W	6½	5⅓	77,100	—	147	139	390
Yale	1500	4W	1½	—	13,000	—	106	84	227
	6000	4W	6	—	45,576	—	129	134	345

Courtesy of Construction Methods and Equipment

Sample specifications for front loaders

FRONT LOADERS

Make and Model		Tire Size, rear	Outside Turning Radius inches	Rated Horsepower and RPM		Number of Cylinders	Displacement, cu. in.	Max. speed mph		Steering	Operating Weight, Pounds
								Forward	Reverse		
Case	350	—	—	39	@ 2,000	4	188	4.9	4.4	—	11,000
	1150B	—	—	105	2,100	6	—	6.2	7.4	Hydr.	25,500
	680E	16.9 - 24	204	80	2,200	4	336	19.5	19.5	Front	14,700
	W26B	23.5 - 25	244	165	2,200	6	504	29.9	30.8	Artic.	30,245
Cat	931	—	—	62	2,400	4	318	6.9	3.2	—	15,300
	983	—	—	275	2,060	6	893	6.3	7.4	—	71,600
	920	15.5 - 25	205	65	2,400	4	318	15.0	6.6	Artic.	13,400
	980B	26.5 - 25	288	260	2,200	6	638	26.7	12.0	Artic.	49,700
	992(g)	38.00 - 39	350	550	2,000	12	1,786	24.9	27.1	Artic.	141,800
JD	450C	—	—	65	2,500	4	219	6.7	6.4	—	16,200
	301A	14.9 - 24	122	43	2,500	3	152	15.5	4.6	Front	6,120
	644B	20.5 - 25	185	145	2,200	6	531	25.8	10.2	Artic.	28,280
Fiat	FL9	—	—	87	2,100	4	329	5.7	6.5	—	24,120
	12G-B	—	—	195	2,000	6	516	4.8	5.4	—	46,800
	545-B	17.5 - 25	178	102	2,400	6	301	22.0	22.7	Artic.	19,135
	945-B	29.5 - 29	243	335	2,100	6	844	20.0	21.9	Artic.	63,890
Ford	335	12.4 - 28	117	41.4	2,000	3	175	21.0	18.4	Front	3,400
	550	16.9 - 24	—	56.0	2,200	3	201	19.3	19.1	Front	11,000
	A66	20.5 - 25	202	136	2,200	6	401	21.0	8.0	Artic.	28,400
IH	500E	—	—	44	2,500	3	155	5.9	5.9	—	12,350
	250C	—	—	190	2,400	8	573	5.3	5.9	—	45,512
	3200B	7.00 - 15	—	30	2,800	4	102	8.0	8.0	Skid	3,600
	H-60E	15.5 - 25	223	110	2,500	6	360	22.0	26.0	Artic.	20,800
	580	65.50 - 51	487	1,075	1,900	12	1,788	17.8	17.8	Artic.	272,000
Ktsu	D315-15	—	—	63	2,350	4	243	4.5	5.8	—	15,870
	D95S-1	—	—	240	1,850	4	783	6.3	7.2	—	62,130
MF	500	—	124	136	2,100	8	511	4.1	4.1	—	29,700
	50A	16.9 - 24	—	61	2,000	4	203	20.2	20.2	Front	7,175
	88	29.5 - 29	285	286	2,100	6	855	25.0	27.3	Artic.	58,815
Melroe	700	7.00 - 15	65	—	2,400	2	85	7.0	7.0	Skid	3,800
Mich.	35AWS	14.00 - 24	185	78	2,300	4	248	23.2	23.2	Fr. & Rr.	13,055
	675	50.5 - 51	521	635	2,100	12	1,710	16.3	16.3	Artic.	381,475
Terex	72-31	20.5 - 25	255	146	2,300	4	284	20.6	7.4	Artic.	28,100
	72-71	29.5 - 29	323	336	2,300	8	568	20.8	20.0	Artic.	76,250
Yale	1500	13.00 - 24	221	98	2,500	4	248	25.0	9.5	Artic.	16,000
	6000	29.5 - 29	294	309	2,100	6	855	20.9	20.9	Artic.	67,700

Courtesy of Construction Methods and Equipment

Sample specifications for front loaders (continued)

Make and model		Engine cylinders	Rated horsepower	Operating wt., less bucket, lbs.	Minimum width, ft.-in.	Overall length, transport ft.-in.	Height, transport, ft.-in.	Counter-weight, pounds
American	35	6	250	84,000	11 - 2	41 - 5	12 - 3	14,000
Bantam	C-366	6	182	57,080	10 - 2	36 - 7	12 - 6	7,000
Bucyrus-Erie	15H	4	95	25,000	8	26	9 - 8	2,650
	31H	6	208	63,600	11 - 4	34 - 9	12 - 10	13,500
	40H IPA	6	331	115,500	11 - 11	41 - 1	12 - 0	25,000
J. I. Case	980	6	180	43,460	9 - 10	33 - 7	10 - 5	6,600
Caterpillar	225	8	125	46,100	10	32 - 3	10 - 5	5,700
	245	6	325	123,100	12	43 - 3	15 - 1	17,000
Drott	35EC	6	131	34,450	8 - 0	26 - 6	11	6,600
	120	6	180	132,000	11 - 11	46 - 1	11 - 2	27,000
John Deere	JD690-B	8	456	38,300	7 - 11	29 - 6	10 - 11	4,500
Ford	H48	6	107	37,300	8 - 6	27	9 - 9	—
Harnischfeger	P&H H312	4	105	35,200	8	28 - 9	10	4,000
	P&H H2500	8	274	105,190	12	41 - 5	12	25,000
Hein-Werner	C-148	6	175	49,500	10 - 4	32 -6	10 - 3	6,700
Insley	H-560C	4	118	31,800	9 - 5	26 - 10	10 - 7	2,500
	H-2000	8	280	85,480	12	40 - 5	14 - 10	18,000
International Harvester	3964	6	100	31,000	8	27-5	10 - 10	3,960
JCB	807	6	110	36,940	9 - 5	29 - 2	11 - 2	5,400
Koehring	4660	6	190	61,950	10 - 9	35 - 8	10 - 10	13,400
	8660	8	325	116,000	11 - 11	43 - 8	12 - 8	26,800
	12660	24 (2)	760	255,700	17 - 10	59 - 4	15 - 5	30,800
Liebherr America	R901LCB	3	60	25,300	9	24	9 - 5	4,000
	R981	12 (2)	320	144,000	14 - 2	—	—	22,500
Lima	2505	6	208	70,000	11	38 - 1	11 - 1	10,000
Link-Belt Speeder (FMC)	LS 4500	6	208	71,950	10 - 6	33 - 10	10 - 10	9,100
Lorain	L48H	6	250	76,800	11	42 - 3	12 - 6	17,800
Poclain	TCS	4	69	27,720	8 - 10	25 - 7	10	2,530
	HC300L	8	280	106,450	11 - 11	40 - 9	13 - 9	18,700
Unit	H-202-C	4	152	43,140	8	31 - 7	10 - 11	—
Warner & Swasey	211	4	160	52,760	9 - 11	33 - 9	10 - 6	5,130
	1900	16 (2)	308	201,000	13 - 10	47 - 8	15 - 6	24,400

Courtesy of Construction Methods and Equipment

Sample specifications for hydraulic backhoes

Make and model		Struck capacity, cu. yd.	Bucket width, in.	Depth, ft.-in., for 8 ft. level bottom	Loading height, ft.-in.	Reach from center	Digging force, stick cyl., lbs.	Speed range, mph
American	35	to 2¼	35 - 45	26 - 1	21 - 9	40 - 1	115,600	0 - 1.4
Bantam	C-366	¾ - 1¼	30 - 60	22 - 9	19 - 10	35 - 3	—	1.2 - 2.3
Bucyrus-Erie	15H	to ⅝	19 - 36	16 - 6	16 - 2	25 - 9	—	1.2
	31H	to 1¾	42 - 48	18 - 3	18 - 6	31 - 3	—	1.8
	40H IPA	to 4	68 - 78	21	20 - 9	37	—	1.7
J. I. Case	980	⅝ - 1⅛	24 - 42	20 - 4	15 - 3	30 - 3	—	1.45
Caterpillar	225	7/16 - 1⅛	24 - 48	20 - 5	18 - 11	31 - 6	22,600	2.3
	245	1⅝ - 2⅝	36 - 60	31 - 7	26	46	44,900	1.9
Drott	35EC	½ - ⅞	24 - 60	20 - 8	22	29 - 8	68,025	1.3
	120	1¼	42	31	22 - 7	47 - 7	190,825	2.8
John Deere	JD690-B	3¼	60	—	15	30	16,000	1.7
Ford	H48	½ - ¾	24 - 60	19 - 7	15 - 1	29 - 8	14,100	0 - 1.4
Harnischfeger	P&H H312	⅜ - 1	24 - 40	18 - 10	16 - 3	28 - 10	—	1.1
	P&H H2500	1½ - 3	33 - 60	27 - 6	21	41 - 7	—	.66 - 1.5
Hein-Werner	C-148	½ - 1¼	18 - 60	23 - 10	16	33 - 9	86,590	0.9 - 1.5
Insley	H-560C	⅜ - 3¾	18 - 48	19 - 7	15 - 3	26 - 9	—	1.5
	H-2000	1½ - 2	30 - 55	27 - 5	17 - 7	39 - 10	—	.75 - 1.5
International Harvester	3964	¼ - ⅞	18 - 45	19 - 10	17 - 2	28 - 3	8,600	1.25
JCB	807	½ - 1	24 - 60	21	21 - 10	32 - 8	17,900	1.48
Koehring	4660	to 2	to 62	22 - 6	17 - 6	36 - 7	150,700	1.6
	8660	to 3½	to 74	29 - 6	20- 4	46	212,600	1.27
	12660	to 7	to 88	30 - 6	26 - 8	60 - 10	339,300	1.18
Liebherr America	R901LCB	—	18 - 40	14 - 1	14 - 9	25 - 5	4,200	1.4
	R981	—	32 - 75	36 - 6	27 - 0	52	—	0 - 1.4
Lima	2505	1½	31 - 56	23 - 10	20 - 6	36 - 7	125,665	.85 - 1.7
Link-Belt Speeder (FMC)	LS 4500	¾ - 1¼	33 - 45	21 - 2	15 - 9	33	81,201	1.0
Lorain	L48H-150	1½	39	25 - 6	14 - 6	39	38,200	1.0 to 2.0
Poclain	TCS	5/16 - ⅞	18 - 41	14 - 6	20 - 10	25	14,134	1.29
	HC300L	1¼ - 3	36 - 60	28 - 4	26 - 10	41 - 8	54,905	0.72
Unit	H-202-C	¾ - 1	24 - 72	21 - 4	23 - 4	30 - 2	—	1.6
Warner & Swasey	211	9/16 - 1¼	24 - 48	22	17	31 - 7	96,212	.63 - 1.6
	1900	3 - 4½	46 - 65	35 - 1	25 - 6	48 - 10	261,341	.47 - .92

Courtesy of Construction Methods and Equipment

Sample specifications for hydraulic backhoes (continued)

SOIL CLASSIFICATION

In order to describe soils, the Public Roads Administration has investigated various soil types which exhibit characteristic field behavior. On the basis of this study soils have been divided into eight distinct classes. These classifications are sufficiently detailed so that characteristics such as compressibility, elasticity, capillary action, cohesion, shrinkage and moisture content—all extremely vital considerations to a good subgrade—can, when considered with local climatic and usage conditions, give a good index to the adequacy of the soil for a desired purpose. These eight soil classifications are as follows:

A-1—Well graded material, coarse and fine, excellent binder. Highly stable under wheel loads irrespective of moisture conditions. Functions satisfactorily when surface treated or when used as a base for relatively thin wearing courses.

A-2—Coarse and fine materials, improper grading or inferior binder. Highly stable when fairly dry. Likely to soften at high water content caused either by rains or high capillary rise from saturated lower strata, when an impervious cover prevents evaporation from top layer, or to become loose and dusty in long continued dry weather.

A-3—Coarse material only, no binder. Lacks stability under wheel loads, but is unaffected by moisture conditions. Not likely to heave because of frost, nor to shrink or expand in appreciable amounts. Furnishes excellent support for flexible pavement of moderate thickness and for relatively thin rigid pavements.

A-4—Silt soils, without coarse material, and with no appreciable amount of sticky colloidal clay. Has a tendency to absorb water very readily in quantities sufficient to cause rapid loss of stability even when not manipulated. When dry or damp presents a firm riding surface which rebounds but very little upon the removal of load. Likely to cause cracking in rigid pavements as a result of frost heaving, and failure in flexible pavements because of low supporting value.

A-5—Similar to Group A-4 but have highly elastic supporting surfaces with appreciable rebound upon removal of load even when dry. Elastic properties interfere with proper compaction of macadams during construction and with retention of good bond afterwards.

A-6—Clay soils without coarse material. In stiff or soft plastic state absorb additional water only if manipulated. May then change to a liquid state and work up into the interstices of macadams or cause failure due to sliding in high fills. Furnish firm support essential in properly compacting macadams only at stiff consistency. Deformations occur slowly and removal of load causes very little rebound. Shrinkage properties combined with alternate wetting and drying under field conditions are likely to cause cracking in rigid pavements.

A-7—Similar to Group A-6 but at certain moisture contents deforms quickly under load and rebounds appreciably upon removing of load, as do subgrades of Group A-5. Alternate wetting and drying under field conditions leads to even more detrimental volume changes than in Group A-6 subgrades. May cause concrete pavements to crack before setting and to crack and fault afterwards. May contain lime or associated chemicals productive of flocculation in soils.

A-8—Very soft peat and muck incapable of supporting a road surface without being previously compacted.

To classify a given soil, a sample is run through a series of tests to determine into which of the above groups it most closely falls. The tests to determine its classification are as follows:

1. SIEVE ANALYSIS TEST—This test determines the per cent of total quantities that will pass through seven different size sieves. Certain further checks are made to determine the distribution of material passing through a No. 40 sieve.

2. MOISTURE EQUIVALENT TEST—This test determines the per cent of weight difference between a dry sample and a moist sample.

3. LIQUID LIMIT TEST—This test is defined as the per cent of moisture at which soil changes from a plastic to a liquid condition. The test is conducted by thoroughly mixing a sample with water, smoothing it out, marking a groove in the sample and then determining the number of controlled shocks necessary to close the groove. By repeated tests it is determined what moisture content will permit the groove to close with twenty-five shocks. This moisture content is the liquid limit.

4. PLASTIC LIMIT TEST—This is defined as the per cent of moisture at which the soil changes from a solid to a plastic condition. Test is conducted by moisting a sample and rolling it into a $\frac{1}{8}$″ diameter thread with the palm of the hand. The moisture content at the time the thread begins to crumble determines the Plastic Limit.

5. PLASTICITY INDEX—The numerical difference between the liquid limit and plastic limit.

6. SHRINKAGE TEST—This test determines the "Shrinkage Limit" and the "Shrinkage Ratio". Test is conducted by putting a sample in a test bowl, drying out, and noting volume change.

$$\text{The Shrinkage Limit} = (\% \text{ moisture content}) - \frac{(\text{Volume of dish} - \text{volume dry soil})}{\text{Weight dry soil}} (100)$$

$$\text{The Shrinkage Ratio} = \frac{\text{Weight of dry soil}}{\text{Volume of dry soil}}$$

7. FIELD MOISTURE CONTENT—Minimum moisture content, expressed as a percentage of the weight of the oven dried soil, at which a drop of water placed on a smoothed surface of the soil will not immediately be absorbed, but will instead spread out over the surface and give it a shiny appearance.

8. SOIL ACIDITY OR ALKALINITY—Determine pH value with colorometric test equipment. One purpose is that a lime content has certain beneficial characteristics.

Reproduced from "Earth Moving and Construction Data."
Courtesy of Allis-Chalmers Manufacturing Company

Soil classification

When the above tests have been made the results are compared by use of charts and the soil classed accordingly. Many soils will be border line cases as to classification.

Although the soil tests and resulting classification will usually give a good index to the behavior of a soil, it does not fill the need for practical soil classification terminology required by the engineer out on the job. Under field circumstances he may be able to test the soil only by visual examination. One of the common classification methods used by many engineers in the field is grouping soils by texture and structure. The terms are general and the range in any one group may be great.

These groups are as follows:

SANDY SOIL—Loose and granular soil, the individual grains of which can readily be seen or felt and may range from very fine sand to coarse sand.

CLAY SOILS—Clay soil is a fine-textured soil which forms hard lumps or clods when dry.

LOAM—A loam is a soil having a relatively even mixture of sand, silt and clay.

SANDY LOAM—A soil containing much sand but having sufficient silt and clay to render it coherent.

SILT LOAM—When this class of soil is dry and powdered, it is often called "rock flour." It is a soil having a moderate amount of fine sand and clay, over half the particles being of the size called "silt". The dry lumps are easily broken and then feel soft and floury.

CLAY LOAM—A fine-textured soil having a large percentage of clay. When dry, the clods are hard and difficult to break.

GRAVELLY OR STONY SOILS—All the above soils, if mixed with a considerable amount of pebbles, are classed as gravelly sand loams; sand clay loams; sandy clay soils, etc.

SOIL COMPACTION

The primary objective of compacting soil by sheepsfoot rollers, flat wheel rollers, pneumatic tired units, or other means is to obtain a soil of a specific density in order that it will carry specified loads without undue settlement. Much has been written on this subject, but soil types, equipment, operating conditions and moisture content are so variable that it is not practical to attempt to state definitely what work is required and what equipment is needed to get certain definite results from compaction.

The work necessary to get the desired compaction on a specific job should be determined by actual test on the job.

Soil settlement occurs under load for two reasons: (1) Air and water are expelled from the earth due to compression; and (2) The earth is forced out laterally into the surrounding soil.

Compaction operations attempt to do these things artificially by means of various types of rollers or tampers so that settlement after construction work is completed will be held to a minimum. To do this, two principles of action are involved.

1. It is necessary to place the earth in layers sufficiently thin to permit air and water to be expelled efficiently and easily. Some soils, depending upon their permeability, may be put down in thicker layers than others. For example, clay must be placed in thin layers whereas a sandy soil could be rolled in thick layers.

2. The second principle to consider is that the compression of soil particles requires movement of the individual particles in order to fit them together and fill in the voids. Before movement can take place friction must be reduced. Lubrication of the soil particles by means of moisture will help to overcome friction. Too little moisture will not materially reduce friction; too much moisture only means that the excess water must be expelled. There is, then, an optimum or ideal moisture content.

Tests have been developed for determining the adequacy of soil compaction. There is some difference in the exact procedure of tests as used by the Army Engineers and the various States, but the fundamental principles remain the same. The tests generally used are based on procedures established by the American Association of State Highway Officials (AASHO). Three main tests are used to test soil for proper compaction.

1. Moisture-content test.
2. Unit-weight determination or density test.
3. Compaction test for optimum-moisture content.

The MOISTURE CONTENT TEST (Similar to Public Roads Administration "Moisture Equivalent Test") is used to determine the ratio of the weight of the water contained in a given sample to the dry weight of the sample. The answer is expressed in per cent. The test is conducted by weighing a moist sample of earth, drying it in an oven, then noting the loss due to the water evaporation. The weight of water lost divided by the weight of the dry sample and multiplied by 100 equals the per cent of moisture content.

The UNIT WEIGHT determination is a test for determining the weight of a unit volume. The answer is expressed in pounds per cubic foot.

The COMPACTION TEST FOR OPTIMUM MOISTURE CONTENT (Modified ASSHO method) is an important test used to determine what quantity of moisture in earth will permit the greatest compaction. If too much water is present more work must be done to expel the excess water. If insufficient water is present the dirt will not compact easily. This test is made by compacting in a standard test machine a quantity of the sample dirt which has been thoroughly mixed with water. After compaction the weight per unit volume of the compacted material is determined. Next, samples of the compacted earth are taken and the moisture content is determined as in the MOISTURE CONTENT TEST discussed above. From this information the moisture content for a unit weight of dirt is now known. This same procedure is repeated on several samples with varying amounts of water added until the addition of more water does not give any weight increase for a given volume. The moisture content which results in the greatest weight per volume is the OPTIMUM MOISTURE CONTENT.

Reproduced from "Earth Moving and Construction Data." Courtesy of Allis-Chalmers Manufacturing Company

Soil compaction

Adopted by The Institute of Makers of Explosives
February 1, 1964

Definitions

1. The term "explosives" as used herein includes any or all of the following: dynamite, black blasting powder, pellet powder, blasting caps, electric blasting caps and detonating cord.
2. The term "electric blasting cap" as used herein includes both instantaneous electric blasting caps and all types of delay electric blasting caps.
3. The term "primer" as used herein means a cartridge of explosives in combination with a blasting cap or an electric blasting cap.

When Transporting Explosives

1. DO obey all federal, state and local laws and regulations.
2. DO see that any vehicle used to transport explosives is in proper working condition and equipped with a tight wooden or non-sparking metal floor with sides and ends high enough to prevent the explosives from falling off. The load in an open-bodied truck should be covered with a waterproof and fire-resistant tarpaulin, and the explosives should not be allowed to contact any source of heat such as an exhaust pipe. Wiring should be fully insulated so as to prevent short circuiting, and at least two fire extinguishers should be carried. The trucks should be plainly marked so as to give adequate warning to the public of the nature of the cargo.
3. DON'T permit metal, except approved metal truck bodies, to contact cases of explosives. Metal, flammable, or corrosive substances should not be transported with explosives.
4. DON'T allow smoking or unauthorized or unnecessary persons in the vehicle.
5. DO load and unload explosives carefully. Never throw explosives from the truck.
6. DO see that other explosives, including detonating cord, are separated from blasting caps and/or electric blasting caps where it is permitted to transport them in the same vehicle.
7. DON'T drive trucks containing explosives through cities, towns or villages, or park them near such places as restaurants, garages and filling stations, unless it cannot be avoided.
8. DO request that explosive deliveries be made at the magazine or in some other location well removed from populated areas.
9. DON'T fight fires after they have come in contact with explosives. Remove all personnel to a safe location and guard the area against intruders.

When Storing Explosives

10. DO store explosives in accordance with federal, state or local laws and regulations.
11. DO store explosives only in a magazine which is clean, dry, well ventilated, reasonably cool, properly located, substantially constructed, bullet and fire resistant, and securely locked.
12. DON'T store blasting caps or electric blasting caps in the same box, container or magazine with other explosives.
13. DON'T store explosives, fuse, or fuse lighters in a wet or damp place, or near oil, gasoline, cleaning solution or solvents, or near radiators, steam pipes, exhaust pipes, stoves, or other sources of heat.
14. DON'T store any sparking metal, or sparking metal tools in an explosives magazine.
15. DON'T smoke or have matches, or have any source of fire or flame in or near an explosives magazine.
16. DON'T allow leaves, grass, brush, or debris to accumulate within 25 feet of an explosives magazine.
17. DON'T shoot into explosives or allow the discharge of firearms in the vicinity of an explosives magazine.
18. DO consult the manufacturer if nitroglycerin from deteriorated explosives has leaked onto the floor of a magazine. The floor should be desensitized by washing thoroughly with an agent approved for that purpose.
19. DO locate explosives magazines in the most isolated places available. They should be separated from each other, and from inhabited buildings, highways, and railroads, by distances not less than those recommended in the "American Table of Distances."

When Using Explosives

20. DON'T use sparking metal tools to open kegs or wooden cases of explosives. Metallic slitters may be used for opening fiberboard cases, provided that the metallic slitter does not come in contact with the metallic fasteners of the case.
21. DON'T smoke or have matches or any source of fire or flame within 100 feet of an area in which explosives are being handled or used.
22. DON'T place explosives where they may be exposed to flame, excessive heat, sparks, or impact.
23. DO replace or close the cover of explosives cases or packages after using.
24. DON'T carry explosives in the pockets of your clothing or elsewhere on your person.
25. DON'T insert anything but fuse in the open end of a blasting cap.
26. DON'T strike, tamper with, or attempt to remove or investigate the contents of a blasting cap or an electric blasting cap, or try to pull the wires out of an electric blasting cap.
27. DON'T allow children or unauthorized or unnecessary persons to be present where explosives are being handled or used.
28. DON'T handle, use, or be near explosives during the approach or progress of any electrical storm. All persons should retire to a place of safety.
29. DON'T use explosives or accessory equipment that are obviously deteriorated or damaged.
30. DON'T attempt to reclaim or to use fuse, blasting caps, electric blasting caps, or any explosives that have been water soaked, even if they have dried out. Consult the manufacturer.

When Preparing The Primer

31. DON'T make up primers in a magazine, or near excessive quantities of explosives, or in excess of immediate needs.
32. DON'T force a blasting cap or an electric blasting cap into dynamite. Insert the cap into a hole made in the dynamite with a punch suitable for the purpose.
33. DO make up primers in accordance with proven and established methods. Make sure that the cap shell is completely encased in the dynamite or booster and so secured that in loading no tension will be placed on the wires or fuse at the point of entry into the cap. When side priming a heavy wall or heavy weight cartridge, wrap adhesive tape around the hole punched in the cartridge so that the cap cannot come out.

When Drilling And Loading

34. DO comply with applicable federal, state and local regulations relative to drilling and loading.
35. DO carefully examine the surface or face before drilling to determine the possible presence of unfired explosives. Never drill into explosives.
36. DO check the borehole carefully with a wooden tamping pole or a measuring tape to determine its condition before loading.
37. DO recognize the possibility of static electrical hazards from pneumatic loading and take adequate precautionary measures. If any doubt exists, consult your explosives supplier.
38. DON'T stack surplus explosives near working areas during loading.
39. DO cut from the spool the line of detonating cord extending into a borehole before loading the remainder of the charge.
40. DON'T load a borehole with explosives after springing (enlarging the hole with explosives) or upon completion of drilling without making certain that it is cool and that it does not contain any hot metal, or burning or smoldering material. Temperatures in excess of 150° F. are dangerous.
41. DON'T spring a borehole near another hole loaded with explosives.
42. DON'T force explosives into a borehole or through an obstruction in a borehole. Any such practice is particularly hazardous in dry holes and when the charge is primed.
43. DON'T slit, drop, deform or abuse the primer. DON'T drop a large size, heavy cartridge directly on the primer.
44. DO avoid placing any unnecessary part of the body over the borehole during loading.

Courtesy of E. I. du Pont de Nemours & Co. (Inc.)

"Don't" rules for explosives

45. DON'T load any boreholes near electric power lines unless the firing line, including the electric blasting cap wires, is so short that it cannot reach the power wires.

46. DON'T connect blasting caps, or electric blasting caps to detonating cord except by methods recommended by the manufacturer.

When Tamping

47. DON'T tamp dynamite that has been removed from the cartridge.

48. DON'T tamp with metallic devices of any kind, including the metal end of loading poles. Use wooden tamping tools with no exposed metal parts, except non-sparking metal connectors for jointed poles. Avoid violent tamping. Never tamp the primer.

49. DO confine the explosives in the borehole with sand, earth, clay, or other suitable incombustible stemming material.

50. DON'T kink or injure fuse, or electric blasting cap wires, when tamping.

When Shooting Electrically

51. DON'T uncoil the wires or use electric blasting caps during dust storms or near any other source of large charges of static electricity.

52. DON'T uncoil the wires or use electric blasting caps in the vicinity of radio-frequency transmitters, except at safe distances. Consult the manufacturer or the Institute of Makers of Explosives pamphlet on "Radio Frequency Hazards."

53. DO keep the firing circuit completely insulated from the ground or other conductors such as bare wires, rails, pipes, or other paths of stray currents.

54. DON'T have electric wires or cables of any kind near electric blasting caps or other explosives except at the time and for the purpose of firing the blast.

55. DO test all electric blasting caps, either singly or when connected in a series circuit, using only a blasting galvanometer specifically designed for the purpose.

56. DON'T use in the same circuit either electric blasting caps made by more than one manufacturer, or electric blasting caps of different style or function even if made by the same manufacturer, unless such use is approved by the manufacturer.

57. DON'T attempt to fire a single electric blasting cap or a circuit of electric blasting caps with less than the minimum current specified by the manufacturer.

58. DO be sure that all wire ends to be connected are bright and clean.

59. DO keep the electric cap wires or leading wires disconnected from the power source and short circuited until ready to fire.

When Shooting With Fuse

60. DO handle fuse carefully to avoid damaging the covering. In cold weather warm slightly before using to avoid cracking the waterproofing.

61. DON'T use short fuse. Know the burning speed of the fuse and make sure you have time to reach a place of safety after lighting. Never use less than two feet.

62. DON'T cut fuse until you are ready to insert it into a blasting cap. Cut off an inch or two to insure a dry end. Cut fuse squarely across with a clean sharp blade. Seat the fuse lightly against the cap charge and avoid twisting after it is in place.

63. DON'T crimp blasting caps by any means except a cap crimper designed for the purpose. Make certain that the cap is securely crimped to the fuse.

64. DO light fuse with a fuse lighter designed for the purpose. If a match is used the fuse should be slit at the end and the match head held in the slit against the powder core. Then scratch the match head with an abrasive surface to light the fuse.

65. DON'T light fuse until sufficient stemming has been placed over the explosive to prevent sparks or flying match heads from coming into contact with the explosive.

66. DON'T hold explosives in the hands when lighting fuse.

In Underground Work

67. DO use permissible explosives only in the manner specified by the United States Bureau of Mines.

68. DON'T take excessive quantities of explosives into a mine at any one time.

69. DON'T use black blasting powder or pellet powder with permissible explosives or other dynamite in the same borehole in a coal mine.

Before And After Firing

70. DON'T fire a blast without a positive signal from the one in charge, who has made certain that all surplus explosives are in a safe place, all persons and vehicles are at a safe distance or under sufficient cover, and that adequate warning has been given.

71. DON'T return to the area of any blast until the smoke and fumes from the blast have been dissipated.

72. DON'T attempt to investigate a misfire too soon. Follow recognized rules and regulations, or if no rules or regulations are in effect, wait at least one hour.

73. DON'T drill, bore, or pick out a charge of explosives that has misfired. Misfires should be handled only by or under the direction of a competent and experienced person.

Explosives Disposal

74. DON'T abandon any explosives.

75. DO dispose of or destroy explosives in strict accordance with approved methods. Consult the manufacturer or follow the Institute of Makers of Explosives pamphlet on destroying explosives.

76. DON'T leave explosives, empty cartridges, boxes, liners, or other materials used in the packing of explosives lying around where children or unauthorized persons or livestock can get at them.

77. DON'T allow any wood, paper, or any other materials employed in packing explosives to be burned in a stove, a fireplace, or other confined space, or to be used for any purpose. Such materials should be destroyed by burning at an isolated location out-of-doors and no person should be nearer than 100 feet after the burning has started.

ADDITIONAL "DO'S AND DON'TS" PARTICULARLY APPLICABLE TO SEISMIC PROSPECTING

1. DO post "Explosives" signs conspicuously near explosives magazines. These signs should be so placed that a bullet passing through them at right angles cannot strike a magazine.

2. DO provide separate storage compartments for the dynamite and electric blasting caps on the shooting truck, where it is permitted to transport them on the same vehicle. These compartments should be lined with some soft material such as wood or rubber. If detonating cord is to be used, it should be carried in the dynamite compartment.

3. DON'T make up more charges than needed to be loaded and fired in one shot.

4. DO place the cap near the top of a dynamite charge, either midway in the side of the top cartridge or in the top of the second cartridge. When side-priming, wrap adhesive tape around the hole punched in the cartridge so that the cap cannot come out.

5. DO use sufficient half-hitches with the cap wires to secure the electric blasting cap in the cartridge. One half-hitch may be inadequate to prevent tension on the wires from pulling the cap out of the cartridge.

6. DO make sure, particularly in dry holes, that the hole is cool and there are no hot pieces of drill steel in the hole. If in doubt, pour in enough water or dirt to cover the bottom of the hole or wait at least one hour before loading it. Temperatures in excess of 150°F. are dangerous.

7. DON'T drop into a borehole an explosive charge that contains an electric blasting cap. DON'T drop the next explosive unit if any must be placed on top of the one containing the cap.

8. DO make certain that the charge is securely placed at a safe depth in the hole. Use shot anchors if there is any chance that the charge may "float," such as in heavy drilling mud or in water containing marsh gas.

9. DO securely anchor any casing, if there is a possibility that it may blow out of the borehole.

10. DON'T approach any explosives that have been thrown out of the borehole until it is evident that they are not burning.

11. DON'T return to any borehole until the smoke and fumes from the blast have dissipated and it is certain that the borehole has ceased to "blow."

Courtesy of E. I. du Pont de Nemours & Co. (Inc.)

"Don't" rules for explosives (continued)

SUGGESTED SPEEDS WHICH ARE TODAY CONSIDERED GOOD PRACTICE FOR VARIOUS WIDTHS OF BELT HANDLING VARIOUS MATERIALS

KIND & CONDITION OF MATERIAL HANDLED	WIDTH OF BELT										
	14"	16"	18"	20"	24"	30"	36"	42"	48"	54"	60"
UNSIZED COAL, GRAVEL STONE, ASHES, ORE, OR SIMILAR MATERIAL	300'	300'	350'	350'	400'	450'	500'	550'	600'	600'	600'
SIZED COAL, COKE OR OTHER BREAKABLE MATERIAL	250'	250'	250'	300'	300'	350'	350'	400'	400'	400'	400'
GRAIN, WET OR DRY SAND	400'	400'	500'	600'	600'	700'	800'	800'	800'	800'	800'
CRUSHED COKE, CRUSHED SLAG OR OTHER FINE ABRASIVE MATERIAL .	250'	250'	300'	400'	400'	500'	500'	500'	500'	500'	500'
LARGE LUMP ORE, ROCK SLAG OR OTHER LARGE ABRASIVE MATERIAL . .	—	—	—	—	350'	350'	400'	400'	400'	400'	400'

TABLE 2		CAPACITIES OF TROUGHED BELT CONVEYORS BASED ON SPEED OF 100 F.P.M. FOR VARIOUS WEIGHTS OF MATERIAL									TABLE 2
BELT WIDTH	MAXIMUM LUMPS		CAPACITY PER HOUR AT SPEED OF 100 F.P.M.									
	SIZED	UNSIZED	30# cu. ft.	50# cu. ft.	90# cu. ft.	100# cu. ft.	125# cu. ft.	150# cu. ft.	160# cu. ft.	180# cu. ft.	200# cu. ft.
14"	2"	2½"	9T	15T	28T	31T	39T	46T	49T	56T	62T
16"	2½"	3"	13T	21T	38T	42T	52T	63T	67T	75T	83T
18"	3"	4"	16T	27T	48T	54T	67T	81T	86T	97T	107T
20"	3½"	5"	20T	33T	60T	67T	83T	100T	107T	120T	133T
24"	4½"	8"	30T	50T	90T	100T	125T	150T	160T	180T	200T
30"	7"	14"	47T	79T	142T	158T	197T	236T	252T	284T	315T
36"	9"	18"	70T	117T	210T	234T	292T	351T	374T	420T	467T
42"	11"	20"	100T	167T	300T	333T	417T	500T	534T	600T	667T
48"	14"	24"	138T	230T	414T	460T	575T	690T	736T	828T	920T
54"	15"	28"	178T	297T	534T	593T	741T	890T	948T	1070T	1190T
60"	16"	30"	222T	369T	664T	738T	922T	1110T	1180T	1330T	1480T

Courtesy of Hewitt-Robins Incorporated

Belt speed and capacity tables

For finding inclined length and degree of slope for Conveyors of known horizontal distance and vertical lift. The example indicated by dotted lines shows that a Conveyor with a horizontal distance of 85′ and a lift of 27½′ would have an inclined length of 89½′ and a slope of 18°

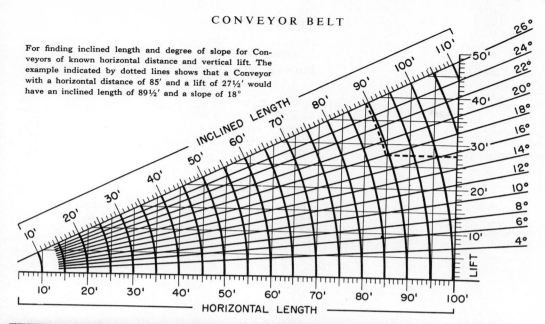

MAXIMUM SAFE INCLINATIONS OF TROUGHED BELT CONVEYORS FOR HANDLING VARIOUS BULK MATERIAL								
MATERIAL	ANGLE	RISE PER. 100 FT.	MATERIAL	ANGLE	RISE PER. 100 FT.	MATERIAL	ANGLE	RISE PER. 100 FT.
CEMENT—LOOSE	22°	40.4′	EARTH—LOOSE	20°	36.4′	PACKAGES— PAPER WRAP	16°	28.6′
CLAY—FINE DRY	23°	42.4′	GLASS—BATCH	21°	38.4′	ROCK—FINE CRUSHED	22°	40.4′
CLAY—WET LUMP	18°	32.5′	GRAIN	16°	28.6′	ROCK—MIXED	20°	36.4′
COAL—MINE RUN	18°	32.5′	GRAVEL—BANK RUN	18°	32.5′	ROCK—SIZED	18⁶	32.5′
COAL—SIZED	18°	32.5′	GRAVEL—SCREENED	15°	26.8′	SALT	20°	36.4′
COAL—BIT SLACK	23°	42.4′	GYPSUM—POWDERED	23°	42.4′	SAND—DRY	15°	26.8′
COAL—ANTHRACITE	16°	28.6′	LIME—POWDERED	23°	42.4′	SAND—DAMP	20°	36.4′
COKE—OVEN RUN	18°	32.5′	LIMESTONE	18°	32.5′	SAND—TEMP'R'D FOUNDRY	24°	44.5′
COKE—SIZED	18°	32.5′	ORE—FINE	22°	40.4′	SULPHUR POWDERED	23°	42.4′
COKE—BREEZE	20°	36.4′	ORE—CRUSHED	20°	36.4′	WOOD—CHIPS	27°	50.9′
CONCRETE—WET	15°	26.8′	ORE—SIZED	18°	32.5′			

The horsepower required at the prime mover to drive a Belt Conveyor is the sum of these integral parts:

1. Power to move the empty Belt over the Idlers.
2. Power to move the load horizontally.
3. Power required to lift or lower the load.
4. Power to turn the pulleys.
5. Power required by Trippers.
6. Drive losses.

This may be expressed in a basic formula:

L = Length of Conveyor (feet)
S = Belt Speed (fpm)
T = Capacity (short tph)
H = Height of Lift or Drop (feet)
F_P = Pulley Friction
C = Idler Friction Factor*
Q = Weight of Moving Parts Per Ft. of Conveyor*

$$\text{Hp at Motor Shaft} = \frac{.03\ LSCQ}{1000}_{\substack{Empty \\ Belt}} + \frac{CLT}{1000}_{\substack{Load \\ Horiz.}} + \frac{TH}{1000}_{\substack{Load \\ Lift}} + Fp + Trippers + Drive\ Losses$$

Courtesy of Hewitt-Robins Incorporated

Belt inclination and power requirement

METRIC SYSTEM

LENGTH

Unit		Metric Equivalent		U.S. Equivalent	
millimeter	(mm)	0.001	meter	0.03937	inch
centimeter	(cm)	0.01	meter	0.3937	inch
decimeter	(dm)	0.1	meter	3.937	inches
METER	(m)	1.0	meter	{ 39.37	inches
				3.28	feet
dekameter	(dkm)	10.0	meters	10.93	yards
hectometer	(hm)	100.0	meters	328.08	feet
kilometer	(km)	1000.0	meters	{ 3280.8	feet
				0.6214	mile

WEIGHT OR MASS

Unit		Metric Equivalent		U.S. Equivalent	
milligram	(mg)	0.001	gram	0.0154	grain
centigram	(cg)	0.01	gram	0.1543	grain
decigram	(dg)	0.1	gram	1.543	grains
GRAM	(g)	1.0	gram	15.43	grains
dekagram	(dkg)	10.0	grams	0.3527	ounce avoirdupois
hectogram	(hg)	100.0	grams	3.527	ounces avoirdupois
kilogram	(kg)	1000.0	grams	2.2	pounds avoirdupois

CAPACITY

Unit		Metric Equivalent		U.S. Equivalent	
milliliter	(ml)	0.001	liter	0.034	fluid ounce
centiliter	(cl)	0.01	liter	0.338	fluid ounce
deciliter	(dl)	0.1	liter	3.38	fluid ounces
LITER	(l)	1.0	liter	1.05	liquid quarts
dekaliter	(dkl)	10.0	liters	0.284	bushel
hectoliter	(hl)	100.0	liters	2.837	bushels
kiloliter	(kl)	1000.0	liters	264.18	gallons

AREA

Unit		Metric equivalent		U.S. Equivalent	
square millimeter	(mm²)	0.000001	square meter	0.00155	square inch
square centimeter	(cm²)	0.0001	square meter	0.155	square inch
square decimeter	(dm²)	0.01	square meter	15.5	square inches
square meter (also CENTARE)	(m²) (ca)	1.0	square meter	10.76	square feet
square dekameter (also ARE or AR)	(dkm²) (a)	100.0	square meters	0.0247	acre
square hectometer (also HECTARE)	(hm²) (ha)	10,000.00	square meters	2.47	acres
square kilometer	(km²)	1,000,000.00	square meters	0.386	square mile

VOLUME

Unit		Metric Equivalent		U.S. Equivalent	
cubic millimeter	(mm³)	0.001	cubic centimeter	0.016	minim
cubic centimeter	(cc, cm³)	0.001	cubic decimeter	0.061	cubic inch
cubic decimeter	(dm³)	0.001	cubic meter	61.023	cubic inches
cubic meter (also STERE)	(m³) (s)	1.0	cubic meter (or stere)	1.308	cubic yards
cubic dekameter	(dkm³)	1000.0	cubic meters	1307.943	cubic yards
cubic hectometer	(hm³)	1,000,000.0	cubic meters	1,307,942.8	cubic yards
cubic kilometer	(km³)	1,000,000,000.0	cubic meters	0.25	cubic mile

SYMBOLS

g	10^9 times (1,000,000,000): giga-	m	10^{-3} times (1/1000 or .001): milli-
m	10^6 times (1,000,000): mega-	μ	10^{-6} times (1/1,000,000 or .000001): micro-
k	10^3 times (1,000): kilo-	n	10^{-9} times (1/1,000,000,000): nano-
h	10^2 times (100): hecto-	$\mu\mu$	10^{-12} times (1/1,000,000,000,000): micromicro-
dk	10 times: deka-		
d	10^{-1} times (1/10 or .1): deci-	Å, λ	Angstrom unit
c	10^{-2} times (1/100 or .01): centi-	$\mu\mu$	micromicron
		μ	micron

TEMPERATURE CONVERSION

Fahrenheit and Celsius (Centigrade) temperatures may be converted into each other by use of either of the following formulas:

$$\text{Temperature, Fahrenheit} = \frac{9 \times \text{Temperature, Celsius}}{5} + 32$$

$$\text{Temperature, Celsius} = \frac{5 \times (\text{Temperature, Fahrenheit} - 32)}{9}$$

This conversion works out to whole numbers every 5 degrees on the Celsius scale and every 9 degrees on the Fahrenheit scale, as in the table below, where equivalent temperatures on the two scales are placed beside each other.

Fahrenheit	Celsius	Fahrenheit	Celsius	Fahrenheit	Celsius
−40	−40	122	50	302	150
−22	−30	140	60	320	160
0	−17.7	150	65.5	350	176.6
14	−10	176	80	356	180
32	0	200	93.3	392	200
50	10	212	100	400	204.4
68	20	230	110	428	220
86	30	250	121.1	450	231.1
100	37.7	275	135	464	240
104	40	300	148.9	482	250
				500	260

WIND CHILL INDEX

How you find Wind Chill Temperature: Locate actual temperature in top row and wind speed in left hand column. Equivalent (wind chill) temperature is found where these two intersect. Example: with temperature of 10° and wind speed of 10 mph, the Wind Chill Temperature is −9°.

(From the U.S. Weather Bureau)

Wind Speed (mph)	THERMOMETER READING						
	30	20	10	0	−10	−20	−30
	EQUIVALENT (WIND-CHILL) TEMPERATURE (F.)						
calm	30	20	10	0	−10	−20	−30
5	27	16	7	−6	−15	−26	−35
10	16	2	−9	−22	−31	−45	−58
15	11	−6	−18	−33	−45	−60	−70
20	3	−9	−24	−40	−52	−68	−81
25	0	−15	−29	−45	−58	−75	−89
30	−2	−18	−33	−49	−63	−78	−94
35	−4	−20	−35	−52	−67	−83	−98
40	−4	−22	−36	−54	−69	−87	−101

Date __October 5, 1970__

Name __Olson Excavating__   Address __Lyndhurst, Ohio__
Operation __Route 303 Relocation__ Production req'd __600 BCY/hr.__ Location __Hinckley__
TEREX Model __S-24__   Struck Capacity, cu. yds: __24__   Capacity lbs. __80,000__
Material __Sandy earth__   Bank Yd. __3200__ lbs. Swell Factor __0.85__ Loose Yd. __2720__ lbs.
Pay Load per Cycle; Loose Cu. Yds. _____ Bank Cu. Yds. __25__ lbs. __80,000__
Type of Loading Unit __Pushed by 82-80__ Bucket Size __-- __ No. of Passes to Load __--__
Loading Conditions __well maintained__ cut Loading Production _____ Tons or Bank Cu. Yds./Hr.
                                                                0.75 Min.

A. LOADING TIME __Average Conditions__

Loaded Haul—Total Length __2400__ Ft. Elevation __500__ Ft.

Road Section	Length in Ft.	Rolling Resist.	Per Cent Grade	Trans. Gear	Max. Speed	Speed Factor	Average Speed	Hauling Time in Min.
	200	5	0	4	14.3	.33	4.7	.48
	600	3	1	5	19.5	.65	12.7	.54
	1000	3	0	6	22.5	.75	16.9	.67
	400	3	8	2	7.5	1.20	9.0	.51
	200	5	0	4	14.3	.40	5.7	.40

                                         TOTAL HAULING TIME __2.60__ Min.

B.

Return Empty—Total Length __2400__ Ft.

Road Section	Length in Ft.	Rolling Resist.	Per Cent Grade	Trans. Gear	Max. Speed	Speed Factor	Average Speed	Return Time
	200	5	0	4	29.5	.40	11.8	.19
	400	3	-8	6	31.5	.72	22.7	.20
	1000	3	0	6	29.5	.81	23.9	.47
	600	3	-1	6	29.5	.77	22.7	.30
	200	5	0	4	29.5	.40	11.8	.19

                                         TOTAL RETURN TIME __1.35__ Min.

C.

D. Turning and Dumping—Conditions __Average__ Turning and Dumping Time __0.60__ Min.

E. Spot at Loading Machine—Conditions __Average__ Spotting Time __0.60__ Min.

                                                      5.90 Min.

F. TOTAL TIME PER COMPLETE HAULING CYCLE (A+B+C+D+E) _____

G. Average Trips per Hour $= \dfrac{\text{50 Min. Prod. Hr.}}{\text{(F) Total Cycle Time}} =$ __8.48__ Trips per Hour

H. Hourly Production = (G) Trips per Hour × Pay Load = __8.48 x 25 = 212__ Tons or Bank Yds. per Hour

J. Number of TEREX Req'd $= \dfrac{\text{Hourly Production Req'd}}{\text{(H) Bank Yds. or Tons per TEREX per Hour}} =$ __600/212 = 2.83 or 3__ TEREX

Fleet Production per Hour = J × H = __3 x 212 = 636__ Bank Yds. or Tons

Hauling Cost per Bank Yd. or Ton

     Hourly Cost of Owning and Operating _____ TEREX @ _____ each = _____
     Hourly Cost of Owning Spare TEREX _____ @ _____ each = _____

K. Hourly Cost for Fleet of _____ TEREX _____ Total _____

ESTIMATED HAULING COST PER YD. OR TON $= \dfrac{K}{\text{Fleet Production}} =$ _____

            or $\dfrac{K}{\text{Req'd Production}} =$ _____

*This is an important checking figure. Its relation to the rated capacity of the unit determines the need for top extensions or smaller body, should unusually light or heavy materials be handled.

Courtesy of Terex Division of General Motors Corp.

Sample estimate of hauler production

GLOSSARY

Abrasion. Wear by rubbing of coarse, hard, or sharp materials.

Abutment. The part of a bridge that supports the end of the span, and prevents the bank from sliding under it.

A foundation that carries gravity and also thrust loads.

Acre. Unit for measuring land, equal to 43,560 sq. ft.; or 4840 sq. yds.; or 160 sq. rds.

Adhesion. The soil quality of sticking to buckets, blades, and other parts of excavators.

After Cooler. Any device which will cool compressed air after it is fully compressed.

A-frame. An open structure tapering from a wide base to a load-bearing top.

Aggregate. Crushed rock or gravel screened to sizes for use in road surfaces, concrete, or bituminous mixes.

Air Receiver. The air storage tank on a compressor.

Air Waves. Air borne vibrations caused by explosions.

Alloy Steel. Steel compounded with other metals to improve its quality.

Ampere. The intensity of electric current produced by one volt acting through a resistance of one ohm.

Angle. The difference in direction of two lines which meet or tend to meet. Usually measured in degrees.

Angling Dozer (Angle dozer). A bulldozer with a blade which can be pivoted on a vertical center pin, so as to cast its load to either side.

Annular. Ring-shaped.

Anvil Block. In a paving breaker, a movable piece of steel between the air piston stem and the steel.

A.P.I. American Petroleum Institute.

A tapered thread used in drill strings and accessories.

Apron. The front gate of a scraper body.

A short ramp with a slight pitch.

Assembly Rod. An external bolt holding a machine together.

Atmospheric Pressure. Pressure of air enveloping the earth, averaged as 14.7 lbs. per sq. in. at sea level, or 29.92 inches of mercury as measured by a standard barometer.

Auger. A rotating drill having a screw thread that carries cuttings away from the face.

Auxiliary. A helper or standby engine or unit.

Avalanche Protector. Guard plates that prevent loose material from sliding into contact with the wheels or tracks of a digging machine.

Axis. A straight line around which a shaft or body revolves.

The centerline of a tunnel.

Axle, dead. A fixed shaft functioning as a hinge pin.

A fixed shaft or beam on which a wheel revolves.

Axle, live. A revolving horizontal shaft.

Babbitt. A soft antifriction metal composed of tin, antimony, and copper in varying proportions.

Backfurrow (Land). The first cut of a plow, from which the slice is laid on undisturbed soil.

Back Haul. A line which pulls a drag scraper bucket backward from the dump point to the digging.

Backfill. The material used in refilling a ditch or other excavation, or the process of such refilling.

Backfire. A fire started to burn against and cut off a spreading fire.

An explosion in the intake or exhaust passages of an engine.

Backhoe. A hoe or pull shovel.

Bail. A hinged loop used for lifting.

A hoist yoke or bracket.

Bailer. A hollow cylinder used for removing rock chips and water from churn drill holes.

Ballast. Heavy material, such as water, sand or iron, which has no function in a machine except increase of weight.

Ball Joint. A connection, consisting of a ball and socket, which will allow a limited hinge movement in any direction.

Bank. Specifically, a mass of soil rising above a digging or trucking level. Generally, any soil which is to be dug from its natural position.

Bank Gravel. A natural mixture of cobbles, gravel, sand, and fines.

Bank Measure. Volume of soil or rock in its original place in the ground.

Bank Yards. Yards of soil or rock measured in its original position, before digging.

Barrel. The water passage in a culvert.

Base Line (Traversing). The main traverse or surveyed line running through the site of proposed construction, from which property lines, street lines, buildings, etc., are located and plotted on the plan.

Batter. Inward slope from bottom to top of the face of a wall.

A pile driven at an angle to widen the area of support and to resist thrust.

Batter Boards. Horizontal boards placed to mark line and grade of a proposed building.

Battery. A storage battery or dry cell.

In blasting, often a blasting machine.

Bearing. A part in which a shaft or pivot revolves.

Bearing, anti-friction. A bearing consisting of an inner and outer ring, separated by balls or rollers held in position by a cage.

Bearing, needle. An anti-friction bearing using very small diameter rollers between wide faces.

Bearing, pilot. A small bearing that keeps the end of a shaft in line.

Bearing, solid. A one piece bushing.

Bearing, throwout. A bearing that permits a clutch throwout collar to slide along the clutch shaft without rotating with it.

Bed. A base for machinery.

Bedding. Ground or supports in which pipe is laid.

Bedding Plane. A separation or weakness between two layers of rock, caused by changes during the building up of the rock-forming material.

Bedrock. Solid rock, as distinguished from boulders.

Bell. An expanded part at one end of a pipe section, into which the next pipe fits.

Bell Crank. A lever whose two arms form an angle at the fulcrum, or a triangular plate hinged at one corner.

Belt Conveyor. An endless pulley-driven belt supported on rollers, which transports material placed on its upper surface.

Bench. A working level or step in a cut which is made in several layers.

Bench Mark. A point of known or assumed elevation used as a reference in determining and recording other elevations.

Bench Terrace. A more or less level step between steep risers, graded into a hillside.

Bends (Caisson disease). A cramping disease induced by too rapid decrease of air pressure after a stay in compressed atmosphere, as in a caisson.

Bent (Set). In tunnel timbering, two posts and a roof timber.

Berm. An artificial ridge of earth.

Bid. To make a price on anything; a proposition either verbal or written, for doing work and for supplying materials and/or equipment.

Binder. Fines which hold gravel together when it is dry.

A deposit check that makes a contract valid.

Bit. The part of a drill which cuts the rock or soil.

Bit, carbide. A bit having inserts of tungsten carbide.

Bit, chopping. A bit that is worked by raising and dropping.

Bit, coring. A bit that grinds the outside ring of the hole, leaving an inner core intact for sampling.

Bit, diamond. A rotary bit having diamonds set in its cutting surfaces.

Bit, drag. A diamond or fishtail bit.

A bit that cuts by rotation of fixed cutting edges or points.

Bit, fishtail. A rotary bit having cutting edges or knives.

Bit, multi-use. A bit that is sharpened for new service when worn.

Bit, plug. A diamond bit that grinds out the full width of the hole.

Bit, roller. A bit that contains cutting elements that are rotated inside it as it turns.

Bit, throwaway. A bit that is discarded when worn.

Bituminous. Containing asphalt or tar.

Black Powder. Gunpowder. A mixture of carbon, sodium or potassium nitrate, and sulphur.

Blade. Usually a part of an excavator which digs and pushes dirt but does not carry it.

Blanket. Soil or broken rock left or placed over a blast to confine or direct throw of fragments.

Blast. To loosen or move rock or dirt by means of explosives or an explosion.

Blast Hole. A vertical drill hole 4 or more inches in diameter, used for a charge of explosives.

Blasting Gelatin. A jelly-like high explosive made by dissolving nitrocotton in nitroglycerin.

Blasting Machine (Battery). A hand operated generator used to supply firing current to blasting circuits.

Blasting Mat. A steel blanket composed of woven cable or interlocked rings.

Bleed. To remove unwanted air or fluid from passages.

Blinding. Compacting soil immediately over a tile drain to reduce its tendency to move into the tile.

Clogging of a screen.

Block. A pulley and its case.

Block, crown. A sheave set suspended at the top of a derrick.

Block, snatch. A sheave in a case having a pull hook or ring.

Block, sling. A frame containing two sheaves mounted on parallel axles, so that they will line up when pulled from opposite directions.

Block, traveling. A frame for a sheave or a set of sheaves that slides in a track.

Blockholing. Blasting boulders by means of drilled holes.

Blue Tops. Grade stakes whose tops indicate finish grade level.

BM. Bench mark.

Body. The load carrying part of a truck or scraper.

Body, quarry. A dump body with sloped sides.

Body, rock. A dump body with oak planking set inside a double steel floor.

Bogie (Tandem) (Tandem drive unit). A two axle driving unit in a truck. Also called tandem drive unit or a tandem.

Boom. In a revolving shovel, a beam hinged to the deck front, supported by cables.

Any heavy beam which is hinged at one end and carries a weight-lifting device at the other.

Boom, crane. A long, light boom, usually of lattice construction.

Boom, jack. A boom whose function is to support sheaves that carry lines to a working boom.

Boom, lattice. A long, light shovel boom fabricated of criss-crossed steel or aluminum angles or tubing.

Boom, live. A shovel boom which can be lifted and lowered without interrupting the digging cycle.

Booster. An auxiliary device that increases force or pressure.

Booster Pump. A pump that operates in the discharge line of another pump, either to increase pressure, or to restore pressure lost by friction in the line or by lift.

Boring. Rotary drilling.

Borrow Pit. An excavation from which material is taken to a nearby job.

Boulder. A rock which is too heavy to be lifted readily by hand.

Bowl. The bucket or body of a carrying scraper.

Sometimes the moldboard or blade of a dozer.

Box. A transmission.

A dump body.

Box Girder. A hollow steel beam with a square or rectangular cross section.

Box Thread. The female side of A.P.I. tapered thread.

Brake. A device for slowing, stopping, and holding an object.

Brake, disc. A brake which utilizes friction between fixed and rotating discs, or between discs and shoes.

Brake Drum. A rotating cylinder with a machined inner or outer surface upon which a brake band or shoe presses.

Brake, friction. A brake operating by friction between two surfaces rotating or sliding on each other.

Brake Horsepower. The horsepower output of an engine or mechanical device. Measured at the flywheel or belt, usually by some form of mechanical brake.

Brake, self energizing. A brake that is applied partly by friction between its lining and the drum.

Brake, tooth (Jaw brake). A brake used to hold a shaft by means of a tooth or teeth engaging with fixed sockets. Not used for slowing or stopping.

Braze. To solder with brass or other hard alloys.

Break. To twist open or disconnect.

A short rest period.

Breast Board. A temporary barrier to prevent the digging face from caving or flowing into a tunnel.

Breast Timber. A leaning brace from the floor of an excavation to a wall support.

Bridge. In an electric blasting cap, the wire that is heated by electric current so as to ignite the charge.

Sometimes the shunt connection between the cap wires.

Bridle Cable. An anchor cable that is at right angles to the line of pull.

Bridle Hitch. A connection between a bridle cable and a cable or sheave block.

Brinell Test. A method of determining the hardness of metal by the indentation of a standard steel ball of known hardness under a definite load.

British Imperial Gallon. A fluid gallon equal to 1.2 U.S. gallons approximately; contains 277.42 cu. in. There are 6.23 such gallons per cu. ft.

Bucket. A part of an excavator which digs, lifts and carries dirt.

Bucket Loader. Usually a chain bucket loader, sometimes a tractor loader or shovel dozer.

Bucket Sheave (Padlock sheave). A pulley attached to a shovel bucket, through which the hoist or drag cable is reeved.

Bucket, slat. An openwork bucket made of bars instead of plates, used in digging sticky soil.

Bucking. Sawing a long log into shorter pieces.

Buffer. A pile of blasted rock left against or near a face to improve fragmentation and reduce scattering from the next blast.

A movable metal plate used in tunnels to limit scattering of blasted rock.

Bulkhead. A wall or partition erected to resist ground or water pressure.

Bull Clam. A bulldozer fitted with a curved bowl hinged to the top of the front of the blade.

Bull Gear. A toothed driving wheel which is the largest or strongest in the mechanism.

Bull wheel. A large driving wheel or sprocket.

Bulldozer. A tractor equipped with a front pusher blade.

A cleaning blade that follows the wheel or ladder of a ditching machine.

In a machine shop, a horizontal press.

Bullgrader. Trade name for an International (formerly Bucyrus-Erie) angling dozer.

Bumboat. A small boat equipped with a hoist and used for handling dredge lines and anchors.

Bumper (Guard). A slotted or perforated

plate that holds a check type air valve near its seat.

Burden. The distance from a drill hole to the face, or the volume of rock to be moved by the explosive in a drill hole.

Burn. To cut with a torch.

To pulverize with very heavy explosive charges.

Burn Cut. A narrow section of rock pulverized by exploding heavy charges in parallel holes.

Bushing. A metal cylinder between a shaft and a support or a wheel, that serves to reduce rotating friction and to protect the parts.

Bushing, split. A bushing made in two pieces, for ease of insertion and removal.

Butt Joint (Open joint). In pipe, flat ends that meet but do not overlap.

Cab Guard. On a dump truck, a heavy metal shield extending up from the front wall of the body and forward over the cab.

Cable. Rope made of steel wire.

Cable, backhaul. In a cable excavator, the line that pulls the bucket from the dumping point back to the digging.

Cable Control Unit. A high speed tractor winch having one to three drums under separate control. Used to operate dozers and towed equipment.

Cable, drag. In a dragline or hoe, the line that pulls the bucket toward the shovel.

Cable Excavator. A long range, cable-operated machine which works between a head mast and an anchor.

Cable, inhaul (Digging line). In a cable excavator, the line that pulls the bucket to dig and bring in soil.

Cage. A circular frame that limits the motion of balls or rollers in a bearing.

Cairn. A pile of stones used as a marker.

Caisson. A box or chamber used in contruction work under water.

Cam. A rotating or sliding piece, or a projection on a wheel, used to impart exactly timed motion to light parts.

Camber. Vertical convex curve in a culvert barrel.

Outward lean of the front wheels of a motor vehicle.

Cantilever. A lever-type beam that is held down at one end, supported near the middle, and supports a load on the other end.

Cap. A detonator, set off by electric current or a burning fuse.

A fitted or threaded piece to protect the top of a pile from damage while being driven.

A pipe plug with female threads.

The roof or top piece in a three piece timber set used for tunnel support.

Cap, delay. An electric blasting cap that explodes at a set interval after current goes through it.

Cap, millisecond delay (Short delay). A detonating cap that fires from 20 to 500 thousandths of a second after the firing current passes through it.

Capillary Attraction. The tendency of water to move into fine spaces, as between soil particles, regardless of gravity.

Capillary Movement. Movement of underground water in response to capillary attraction.

Capillary Water. Underground water held above the water table by capillary attraction.

Capstan (Cat head). A non-winding winch used with soft rope.

Carbide. Tungsten carbide, a very hard and abrasion-resistant compound used in drill bits and other tools.

Carbide Bit. A steel bit which contains inserts of tungsten carbide.

Carbon Steel. Usually a hardened steel not alloyed with other metals.

Carriage. A sliding or rolling base or supporting frame.

Carrier. A rotating or sliding mounting or case.

Carryall. Trade name for Le Tourneau-Westinghouse scrapers.

Cartridge. A wrapped stick of dynamite or other explosive.

Casing. A pipe lining for a drilled hole.

Casing Spider. A frame and wedge set that supports the top of a casing string while new sections are added.

Caster. A wheel mounted in a swivel frame so that it is steered automatically by movements of its load.

In an automotive vehicle, the toe-in of the front wheels.

Cat. A trade marked designation for any machine made by the Caterpillar Tractor Company. Widely used to indicate a crawler tractor or mounting of any make.

Cat Head. A capstan winch.

Catskinner. Operator of a crawler tractor.

Catwalk. A pathway, usually of wood or metal, that gives access to parts of large machines.

Cave (Caving). Collapse of an unstable bank.

Center of Gravity. That point in a body about which all the weights of all the various parts balance. It is found experimentally by balancing on a knife edge or a point.

The center of mass of a cut or a fill.

Center of Mass. In a cut or a fill, a cross section line that divides its bulk into halves.

Centerpin (Center pintle). In a revolving shovel, a fixed vertical shaft around which the shovel deck turns.

Centigrade. A temperature scale on which the freezing point of water at sea level atmospheric pressure is indicated as 0° and its boiling point as 100°. Degrees centigrade (°C) equals Fahrenheit (°F) minus 32 multiplied by $\frac{5}{9}$.

Centralizer. A device that lines up a drill steel or string between the mast and the hole.

Centrifugal Force. Outward force exerted by a body moving in a curved line. It is the force which tends to tip a car over in going around a curve.

Centripetal Force. The force or restriction exerted inward to keep a body moving in a curved line. The force which keeps a car from being thrown out of a curve about which it is moving is centripetal force.

Cetane Number. An indication of diesel fuel ignition quality. The cetane number of a fuel in the percentage by volume of cetane in a mixture of cetane and alpha-methylnapthlalene which matches the unknown fuel in ignition quality. American diesel oil usually varies from 30 to 60 cetane.

c.f.m. Cubic feet per minute. A standard capacity or performance measurement for compressors.

C-frame. An angling dozer lift and push frame.

Chain. A tow line or drive belt made of interlocked links.

A surveyor's steel tape measure.

Chain, breakaway (Safety chain). A chain that holds a tractor and a towed unit together if the regular fastening opens or breaks.

Chain Bucket Loader (Bucket loader). A mobile loader that uses a series of small buckets on a roller chain to elevate spoil to the dumping point.

Chain, leaf. A silent chain designed for low speed heavy duty work.

Chain, logging. A chain composed of links of round bar bieces curved and welded to interlock, with a grab hook at one end and a round hook at the other.

Chain, roller. Generally, any sprocket-driven chain made up of links connected by hinge pins and sleeves.

Specifically, a chain whose hinge sleeves are protected by an outer sleeve or roller that is free to turn.

Chain, silent. A roller-type chain in which the sprockets are engaged by projections on the link side bars.

Chain, stud type. A roller chain in which the inner (block) links are connected solidly by non-rotating bushings.

Chanfer (Chamfer). To bevel or slope an edge or corner.

Channel Terrace. A contour ridge built of soil moved from its uphill side, which serves to divert surface water from a field.

Check Dam. A dam that divides a drainageway into two sections with reduced slopes.

Check Valve. Any device which will allow fluid or air to pass through it in only one direction.

Cherry Picker. A small derrick made up of a sheave on an A-frame, a winch and winch line, and a hook. Usually mounted on a truck.

Chip Blasting. Shallow blasting of ledge rock.

Chipping. Loosening of shallow rock by light blasting or air hammers.

Chock. A block used under and against an object to prevent it from rolling or sliding.

Choker. A chain or cable so fastened that it tightens on its load as it is pulled.

Choker Hook (Round hook). A hook that can slide along a chain.

Chord. A straight line connecting two points on a curve.

Chuck. The part of a drill that rotates the steel.

A device that clamps a rod or shaft.

Churn Drill (Spudding or well drill). A machine that drills holes by dropping and raising a bit and drill string hung by a cable.

Circle. In a grader, the rotary table which supports the blade and regulates its angle.

Circle Reverse. The mechanism that changes the angle of a grader blade.

Clam. A clamshell bucket.

Clamshell. A shovel bucket with two jaws which clamp together by their own weight when it is lifted by the closing line.

A shovel equipped with a clamshell bucket.

Clay. A "heavy" soil composed of particles less than 1/256 mm in diameter.

Clean. Free of foreign material. In reference to sand or gravel, means lack of binder.

Cleavage Plane. Any uniform joint, crack, or change in quality of formation along which rock will break easily when dug or blasted.

Clevis. A shackle.

A split end of a rod, drilled for insertion of a pin through the two sections.

Clinometer. A hand instrument for measuring grades by sighting.

Clod Buster. A drag that follows a grading machine to break up lumps.

Closing Line (Digging line). The cable which closes the jaws of a clamshell bucket.

Cloth, wire. Screen composed of wire or rod woven and crimped into a square or rectangular pattern.

Clutch. A device which connects and disconnects two shafts which revolve in line with each other.

Clutch, automatic. A clutch whose engagement is controlled by centrifugal force, vacuum, or other power without attention by the operator.

Clutch Brake. A device to slow the jackshaft when a clutch is released, to permit more rapid gear shifting.

Clutch, centrifugal. A clutch that is kept in engagement only by centrifugal force, so that it automatically disconnects the power train when the engine idles.

Clutch, denture. A jaw clutch.

Clutch, disc. A coupling that can be engaged to transmit power through one or more discs squeezed between a back-plate and a movable pressure plate, and that can be disengaged by moving the plates apart.

Clutch, fluid. A fluid coupling other than a torque converter.

Clutch, jaw (Positive or denture clutch). A toothed hub and a sliding toothed collar that can be engaged to transmit power between two shafts having the same axis of revolution.

Clutch, lockup. A clutch that can be engaged to provide a non-slip mechanical drive through a fluid coupling.

Clutch, overrunning (Free wheeling unit). A coupling that transmits rotation in only one direction, and disconnects when the torque is reversed.

Clutch, slip (Safety clutch). A friction clutch that protects a mechanism by slipping under excessive load.

Clutch, wet (Oil clutch). A clutch that operates in an oil bath.

Cobble. Rounded stone with diameter of 4 to 12 inches.

Cocking. Tipping sideward.

Running off center.

Cockpit. The part of a tractor or grader containing the operator's seat and controls.

Cocoa Mat. A fabric of wood fibers used to distribute water evenly over a smooth surface.

Cofferdam. A set of temporary walls designed to keep soil and/or water from entering an excavation.

Cohesion. The soil quality of sticking together.

Conduit. A pipe or tile carrying water, wire, or pipes.

Collar. A sliding ring mounted on a shaft so that it does not revolve with it. Used in clutches and transmissions.

The open end of a drill hole.

Collaring. Starting a drill hole. When the hole is deep enough to hold the bit from slipping out of it, it is said to be collared.

Compaction. Reduction in bulk of fill by rolling, tamping or soaking.

Compacted Yards. Measurement of soil or rock after it has been placed and compacted in a fill.

Compensating Drive. In a four wheel drive truck, a free wheeling unit in the front propeller shaft that allows the front wheels to go farther than the rear on curves.

Compression. For steel wheel rollers, the compacting effect of the weight at the bottom of the roll, measured in pounds per linear inch of roll width.

Compression Ratio. The ratio of the volume of space above a piston at the bottom of its stroke to the volume above the piston at the top of its stroke.

Compression Roll (Drive roll). The drive wheel of a steel wheel roller.

Compressor. A machine which compresses air.

Concussion. Shock or sharp air waves caused by an explosion or heavy blow.

Cone of Depression. The dried up area of soil around a single underground suction point.

Contour Line. A level line crossing a slope.

Conveyor. A device that transports material by belts, cables, or chains.

Conveyor, apron. One or more endless chains carrying overlapping or interlocking plates that carry bulk materials on their upper surface.

Conveyor Belt. An endless belt of rubber-covered fabric that transports material on its upper surface.

Conveyor, decline. A conveyor that transports downhill.

Conveyor, feeder. A short conveyor belt that supplies material to a long belt.

Conveyor, screw. A revolving shaft fitted with auger-type flights that moves bulk materials through a trough or tube.

Corduroy. A road made of logs laid crosswise on the ground or on other logs.

Cordwood. Wood cut in 4-foot or shorter lengths to be used as fuel.

Core. A cylindrical piece of an underground formation cut and raised by a rotary drill with a hollow bit.

Core Barrel. A hollow cylinder containing a socket and choker springs for holding a section of drilled rock.

Core Drill. A rotary drill, usually a diamond drill, equipped with a hollow bit and a core lifter.

Corrosion. Wear or dissolving away through chemical action as by rusting, or acids.

Countershaft. A shaft which receives power from a parallel mainshaft, and transmits it to another part of the mainshaft or to working parts.

Counterweight. A "dead" or non working load attached to one end or side of a machine to balance weight carried on the opposite end.

A working part attached or positioned partly for the purpose of improving machine balance.

Coyote Holes. Horizontal tunnels in which explosives are packed for blasting a high rock face.

Cradle. A support bracket with a hinged connection to its load.

A carriage.

Crane. A mobile machine used for lifting and moving loads without use of a bucket.

Crankshaft. The engine shaft that converts the reciprocating motion and force of pistons and connecting rods to rotary motion and torque.

Crawler. One of a pair of roller chain tracks used to support and propel a machine, or any machine mounted on such tracks.

Creep. Very slow travel of a machine or a part.

Unwanted turning of a shaft due to drag in a fluid coupling or other disconnect device.

Crimp. A tight bend in metal made under pressure.

Cross Hair. A hair mounted horizontally in a

telescope so as to divide the field of view into halves.

Crosshead. A connection between a connecting rod and a piston rod which is guided so as to move in a straight line.

Cross Section. A profile taken at right angles to the centerline of a project.

Cross Section Paper. Paper ruled in squares for convenience in drawing and measuring.

Crowd. The process of forcing a bucket into the digging, or the mechanism which does the forcing. Used chiefly in reference to machines which dig by pushing away from themselves.

Crown. The elevation of a road center above its sides.

The curved roof of a tunnel.

Crown Fire. A fire burning in tree tops.

Crumber. A "bulldozer" blade that follows the wheel or ladder of a ditching machine to clean and shape the bottom.

Crusher. A machine which reduces rocks to smaller and more uniform sizes. *See also Jaw, Gyratory, and Hammer Mill.*

Culvert. A pipe or small bridge for drainage under a road or structure.

Curtain Drain (Intercepting drain). A drain that is placed between the water source and the area to be protected.

Curve, vertical. A change in gradient of the center line of a road or pipe.

Cut. To lower an existing grade.

An artificial depression.

To stop an engine, or throttle it to idling speed.

Cut and Cover. A work method which involves excavation in the open, and placing of a temporary roof over it to carry traffic during further work.

Cut, gross. The total amount of excavation in a road or a road section, without regard to fill requirements.

Cut, net. The amount of excavated material to be removed from a road section, after completing fills in that section.

Cutter (Cutter head). On a hydraulic dredge, a set of revolving blades at the end of the suction line.

Cutting. Excavating.

Lowering a grade.

Cycle, digging. Complete set of operations a machine performs before repeating them.

Cylinder. In hydraulic systems, a hollow cylinder of metal, containing a piston, piston rod and end seals, and fitted with a port or ports to allow entrance and exit of fluid.

Cylinder, slave. A small cylinder whose piston is moved by a piston rod controlled by a larger cylinder.

Dart Valve. A drain for a well bailer that opens automatically when rested on the ground.

Datum. Any level surface taken as a plane of reference from which to measure elevations.

Dead Axle. See Axle, dead.

Dead Furrow. The line in a field where two directions of plowing meet, and the slices are turned away from each other.

Deadheading. Traveling without load, except from the dumping area to the loading point.

Deck Screens. Two or more screens, usually of the vibrating type, placed one above the other.

Decking. Separating charges of explosives by inert material which prevents passing of concussion, and placing a primer in each charge.

Decompression. The process of reducing high air pressure gradually enough not to injure men who have been working in it.

Deflagration. To burn with sudden and startling combustion. Describes explosion of black powder, in contrast with more rapid detonation of dynamite.

Degree of Curve. The number of degrees at the center of a circle subtended by a chord of 100 ft. at its rim. Occasionally in highway surveying it is defined as the central angle subtended by an arc of 100 feet.

Delay. An electric blasting cap which explodes at a set interval after current is passed through it.

Delay, short period (Millisecond delay). An electric blasting cap that explodes 1/50 to ½ second after passage of an electric current.

Density. The ratio of the weight of a substance to its volume.

Derrick. Usually a non-mobile tower equipped with a hoist, but may be used as a synonym for crane.

Detail Drawing. A large scale drawing showing all small parts, details, dimensions, etc.

Detergent. A chemical compound that acts to clean surfaces and to keep foreign matter in solution or suspension.

Detonation. Practically instantaneous decomposition or combustion of an unstable compound, with tremendous increase in volume.

Detonator. A device to start an explosion, as a fuse or cap.

Dewatering. Removing water by pumping, drainage, or evaporation.

Diamond Drill. A light rotary drill, most often used for exploratory work and blast holes.

Diaphragm. A flexible partition between two chambers.

Dieseling. In a compressor, explosions of mixtures of air and lubricating oil in the compression chambers or other parts of the air system.

Differential. A device that drives two axles and allows them to turn at different speeds to adjust to varying resistance.

Differential, non-spin (Limited action differential). A differential that will turn both axles, even if one offers no resistance.

Differential, two speed. A differential having

a high-low gearshift between the drive shaft and the ring gear.

Diffuser. Inner shell and water passages of a centrifugal pump.

Digging Line. On a shovel, the cable which forces the bucket into the soil. Called crowd in a dipper shovel, drag in a pull shovel and dragline and closing line in a clamshell.

Dike. A long low dam.

A thin rock formation that cuts across the structure of surrounding rock.

Dimension Stone. Rock quarried in blocks to predetermined sizes, in such a manner as not to weaken or shatter it.

Dip. The slope of layers of soil or rock.

Dipper. A digging bucket rigidly attached to a stick or arm.

Dipper Stick. A name for the standard revolving shovel (dipper shovel), and for the straight shaft which connects the bucket with the boom.

Dipper Trip. A device that unlatches the door of a shovel bucket to dump the load.

Direction of Irrigation. Direction of flow of irrigation water. Usually at right angles to the supply ditch or pipe.

Ditch. Generally, a long narrow excavation.

In rotary drilling, a trough carrying mud to a screen.

Ditcher, ladder. A machine that digs trenches by means of buckets mounted on a pair of chains traveling on the exterior of a boom.

Ditcher, wheel. A machine that digs trenches by rotation of a wheel fitted with toothed buckets.

Diversion Valve. A valve which permits flow to be directed into any one of two or more pipes.

Dog. A heavy duty latch.

Dolly. A unit consisting of draw tongue, an axle with wheels, and a turntable platform to support a trailer gooseneck.

A small wheeled carriage designed to support heavy machines.

Donkey. A winch with two drums which are controlled separately by clutches and brakes.

Dope. A viscous liquid put on pipe threads to make a tight joint.

Double. In rotary drilling, two pieces of drill rod left fastened together during raising and lowering.

Double-Clutching. Disengaging and engaging the clutch twice during a single gear shift, in order to synchronize gear speeds.

Downstream Face. The dry side of a dam.

Dozer. Abbreviation for bulldozer or shovel dozer.

Dozer Shovel (Shovel dozer). A tractor equipped with a front-mounted bucket that can be used for pushing, digging, and truck loading.

Draft. Resistance to movement of a towed load.

Drag. Pulling a bucket into the digging, or the mechanism by which the pulling is done or controlled.

Drag Brake. On a revolving shovel, the brake which stops and holds the drag (digging) drum.

Drag Scraper. A digging and transporting device consisting of a bottomless bucket working between a mast and an anchor.

A towed bottomless scraper used for land leveling. Called "leveling drag scraper" to distinguish from cable type.

Dragline. A revolving shovel which carries a bucket attached only by cables, and digs by pulling the bucket toward itself.

Dragshovel (Hoe, Backhoe, or Pullshovel). A shovel equipped with a jack boom, a live boom, a hinged stick and a rigidly attached bucket, that digs by pulling toward itself.

Drain, intercepting (Curtain drain). A drain that intercepts and diverts ground water before it reaches the area to be protected.

Drainage Head. The furthest or highest spot in a drainage area.

Draw. A small valley or a gully.

Drawbar. In a tractor, a fixed or hinged bar extending to the rear, used as a fastening for lines and towed machines or loads.

In a grader, the connection between the circle and the front of the frame.

Drawbar Horsepower. A tractor's flywheel horsepower minus friction and slippage losses in the drive mechanism and the tracks or tires.

Drawbar Pull. The pull a tractor can exert on a load attached to the drawbar. Depends on power, weight, and traction.

Draw Knife. A curved, two handled knife used in digging clay.

Draw Pin. A removable pin that attaches a load to a drawbar.

Drawpoint. A spot where gravity fed ore from a higher level is loaded into hauling units.

Draw Tongue. A bar hinged to a towed machine, fitted with some device for attaching it to a tractor.

Draw Works. The power distribution and control machinery of a rotary drill.

Dredge. To dig under water.

A machine that digs under water.

Drift. A small nearly horizontal tunnel.

Drifter. An air drill mounted on a column or cross bar, and used for horizontal drilling underground.

Drill, auger. See Auger.

Drill Bit. See Bit.

Drill, blast hole. A machine capable of drilling holes 4 inches or more in diameter to a depth of 100 or more feet.

Drill, churn (Spudding drill). A drill that cuts its hole by raising and dropping a chisel bit.

Drill Collar. Thick walled drill pipe used immediately above a rotary bit to provide extra weight.

Drill. core. A drill that cuts around a cylinder

of rock or soil, and lifts it to the surface for inspection.

Drill, diamond. A rotary drill that uses a diamond-studded bit.

Drill Doctor. A mechanic or shop that sharpens and services drill bits, tools and steels.

Drill, percussion. A drill that hammers and rotates a steel and bit.

Sometimes limited to large blast hole drills of the percussion type.

Drill Pipe. The sections of a rotary drilling string connecting the kelly with the bit or collars.

Drill, quarry. A blast hole drill.

Drill Steel. Hollow steel connecting a percussion drill with the bit.

Drill String. In rotary drills, all revolving parts below the ground.

In churn drills, the tools hanging from the drilling cable.

Drill, well. A churn drill, mounted on a truck.

Drilling, core. Exploratory drilling that includes cutting cylinders of rock or soil and bringing them to the surface for inspection.

Drilling, directional (Offset drilling). Curving a rotary drill hole to avoid obstacles or to reach side areas.

Drilling, solid. In diamond drilling, using a bit that grinds the whole face, without preserving a core for sampling.

Drive. To dig or make a tunnel.

To hammer down piling.

Drive Clamp. A collar fitted on a churn drill string to enable it to be used as a hammer to drive casing pipe.

Drive, positive. A driving connection to two or more wheels or shafts that will turn them at approximately the same relative speeds under any conditions.

Drop Hammer. A pile driving hammer that is lifted by a cable and that obtains striking power by falling freely.

Drum. A rotating cylinder with side flanges, used for winding in and releasing cable.

Drum, spudding. In a churn drill, the winch that controls the drilling line.

Dry Well. A deep hole, covered, and usually lined or filled with rocks, that holds drainage water until it soaks into the ground.

Dynamic. Forces tending to produce motion.

Dynamic Balance. A condition of rest created by equal strength of forces tending to move in opposite directions.

Dynamite. A mixture of an explosive or explosives with relatively inert material.

Dynamite, straight. A dynamite in which nitroglycerin is the principal or only explosive.

Earth Drill. An auger.

Eccentric. A wheel or cam with an off-center axis of revolution.

Ejector. A cleanout device, usually a sliding plate.

Elevating Grader. See Belt Loader.

Elevation (Surveying). The height of a point above a plane of reference.

Elevator. A cage hoist.

A machine that raises material on a belt or a chain of small buckets.

Embankment. A fill whose top is higher than the adjoining surface.

Erosion. Wear caused by moving water or wind.

Excavation, unclassified. Excavation paid for at a fixed price per yard, regardless of whether it is earth or rock.

Exploit. Excavate in such a manner as to utilize material in a particular vein or layer, and waste or avoid surrounding material.

Explosive. A chemical compound that can decompose quickly and violently.

Explosive, high. A material that detonates, that is, explodes almost instantaneously.

Face. The more or less vertical surface of rock exposed by blasting or excavating, or the cutting end of a drill hole.

An edge of rock used as a starting point in figuring drilling and blasting.

The width of a roll crusher.

Factor of Safety. The ratio of the ultimate strength of the material to the allowable or working stress.

Fairlead. A device which lines up cable so that it will wind smoothly onto a drum.

False Set (Horsehead). A temporary support for forepoles used in driving a tunnel in soft ground.

Fast Powder. Dynamites or other explosives having a high speed detonation.

Faulting. In geology, the movement which produces relative displacement along a fracture in rock.

Feather. To blend the edge of new material smoothly into the old surface.

Feed. A mechanism which pushes a drill into its work.

The process of supplying material to a conveying or processing unit.

Feeder. A pushing device or short belt that supplies material to a crusher or a conveyor.

Feed Travel. The distance a drilling machine moves the steel shank in traveling from top to bottom of its feeding range.

Ferrule. A short unthreaded tube or bushing shrunk or soldered onto a tube or line.

Fifth Wheel. The weight-bearing swivel connection between highway-type tractors and semi-trailers.

An unnecessary machine or person working on a job.

Fill. An earth or broken rock structure or embankment.

Soil or loose rock used to raise a grade.

Soil that has no value except bulk.

Fill, net. In sidehill work, the yardage of fill required at any station, less the yards of material obtained from the cut at that station.

Fill, net corrected. Net fill after making allowance for shrinkage during compaction.

Filter Bed. A fill of pervious soil that provides a site for a septic field.

Filter Cake (Mud cake). A deposit of mud on the walls of a drill hole.

Final Drive. A set of reduction gearing close to or inside of a drive wheel.

Fines. Clay or silt particles in soil.

Finish Grade. The final grade required by specifications.

Fishing. The operation of recovering an object left or dropped in a drill hole.

Fitting, poured. A wire rope attachment fastened to it by separating the wires, expanding them in a conical socket, and filling it with molten zinc.

Fitting, wedge socket. A wire rope attachment in which the rope lies in a too-small groove between a wedge and a housing, so that pull on the rope tightens the wedge.

Flail. A hammer hinged to an axle so that it can be used to break or crush material.

Flame Gun. A large blowtorch using kerosene for fuel.

Flange. A ridge that prevents a sliding motion. A rib or rim for strength or for attachments.

Fleet Angle. The maximum angle between a rope and a line perpendicular to the drum on which it winds.

Flight. The screw thread (helix) of an auger.

Float. In reference to a dozer blade—to rest by its own weight, or to be held from digging by upward pressure of a load of dirt against its moldboard.

Flotation. Separation of minerals by floating the lighter ones in a fluid.

The weight supporting ability of a tire, crawler track, or platform on soft ground.

Flow Gradient. A drainageway slope determined by the elevation and distance of the inlet and outlet, and by required volume and velocity.

Fluid Clutch. A hydraulic coupling which does not increase torque.

Fluid Drive. A connection between two shafts that transmits torque through a fluid.

Flume. An artificial channel, often elevated above the ground, used to carry fast flowing water.

Follower. A piston that maintains a light pressure against a variable amount of fluid in a container.

Foot. In tamping rollers, one of a number of projections from a cylindrical drum.

Foot Pins. The hinge which attaches the boom to a revolving shovel.

Foot-Pound. Unit of work equal to the force in pounds multiplied by the distance in feet through which it acts. When a 1 pound force is

exerted through a 1 foot distance, 1 foot pound of work is done.

Foot Valve. A check valve in the inlet end of a pump suction hose.

Footing (Foot wall). A sill under a foundation.

Ground, in relation to its load bearing and friction qualities.

Ford. A place where a road crosses a stream under water.

Forepole. A plank driven ahead of a tunnel face to support the roof or wall during excavation.

Fork. A two-pronged rod or yoke used to slide shifting collars along their shafts.

Fork Head. A wheel-guiding frame with a swivel connection to the machine or vehicle that rests on it. (A caster frame.)

Foul Air Duct. A suction line in a tunnel ventilation system.

Four by Four (4 × 4). A vehicle with four wheels or sets of wheels, all engine driven.

Four-Part Line. A single rope or cable reeved around pulleys so that four strands connect the fixed and the movable units.

Fourble. In rotary drilling, a unit of four drill pipes left coupled together.

French Drain (Rubble or stone drain). A covered ditch containing a layer of fitted or loose stone or other pervious material.

Friction. Resistance to motion when one body is sliding or tending to slide over another.

Front. The working attachment of a shovel, as dragline, hoe, or dipper stick.

Front End Loader. A tractor loader with a bucket which operates entirely at the front end of the tractor.

Frost. Frozen soil.

Frost Line. The greatest depth to which ground may be expected to freeze.

Fulcrum. A pivot for a lever.

Full Trailer. A towed vehicle whose weight rests entirely on its own wheels or crawlers.

Fumes. Usually smoke from an explosion.

Fumes, excellent. Fumes that contain a minimum of toxic and irritating chemicals.

Fumes, poor. Toxic or irritating chemicals produced by an explosion.

Fuse. A thin core of black powder surrounded by wrappings, which, when lit at one end, will burn to the other at a fixed speed.

Fuse, detonating. A string-like core of PETN, a high explosive, contained within a waterproof reinforced sheath. "Primacord" is the best known brand.

Gantry. An overhead structure that supports machines or operating parts.

An upward extension of a shovel revolving frame that holds the boom line sheaves.

Gauge (Gage). Thickness of wire or sheet metal.

Spacing of tracks or wheels.

Gear. A toothed wheel, cone, or bar.

Gear, bevel. A gear made of teeth cut in the surface of a truncated cone.

Gear, bull. A gear or sprocket that is much larger than the others in the same power train.

Gear, cluster. Two or more gears of different sizes made in one solid piece.

Gear, helical. A gear with straight or curved teeth cut at an angle of less than 90° to the direction of rotation.

Gear, herringbone. A gear with V-teeth.

Gear, idler. A gear meshed with two other gears that does not transmit power to its shaft. Used to reverse direction of rotation in a transmission.

Gear, pinion. A drive gear that is smaller than the gear it turns.

Gear, planetary set. A gear set consisting of an inner (sun) gear, an outer ring with internal teeth, and two or more small (planet) gears meshed with both the sun and the ring.

Gear, rack. A toothed bar.

Gear, sprocket. A gear that meshes with roller or silent chain.

Gelatin, blasting. A high explosive made by dissolving nitrocotton in nitroglycerin. It is the strongest and highest velocity commercial explosive.

General Drawing. A drawing showing elevation plan, and cross section of the structure, also the borings for substructure and the main dimensions, etc.

Giant (Monitor). In hydraulicking, a large high pressure nozzle mounted in a swivel on a skid frame.

Glory Hole. A vertical pit, material from which is fed by gravity to hauling units in a shaft under the pit bottom.

Gooseneck. An arched connection, usually between a tractor and a trailer.

Grade. Usually the elevation of a real or planned surface or structure. Also means surface slope.

Grader. A machine with a centrally located blade that can be angled to cast to either side, with independent hoist control on each side.

Grade Stake. A stake indicating the amount of cut or fill required to bring the ground to a specified level.

Gradient. Slope along a specific route, as of a road surface, channel or pipe.

Grapple. A clamshell-type bucket having three or more jaws.

Gravel. Rock fragments from 2 mm to 64 mm (.08 to 2.5 inches) in diameter. Or a mixture of such gravel with sand, cobbles, boulders, and not over 15 percent of fines.

Grease. Thick oil.

A solid or semi-solid mixture of oil with soap or other fillers.

Grid. A set of surveyor's closely spaced reference lines laid out at right angles, with elevations taken at line intersections.

Grief Stem. A kelly.

Grizzly. A coarse screen used to remove oversize pieces from earth or blasted rock.

A gate or closure on a chute. (May be spelled "grizzlie.")

Ground Pressure. The weight of a machine divided by the area in square inches of the ground directly supporting it.

Ground Waves. Vibrations of soil or rock.

Grouser. A ridge or cleat across a track shoe, which improves its grip on the ground.

Grout. A cementing or sealing mixture of cement and water, to which sand, sawdust, or other fillers may be added.

Grubbing. Digging out roots.

Guage Size. The width of a drill bit along the cutting edge.

Guard (Bumper). In a compressor check valve a backing or retaining plate for the movable part.

Gudgeon. A reinforced bushing or a thrust absorbing block.

Guy. A line that steadies a high piece or structure by pull against an off-center load or other guys.

Gypsy Spool (Cat head). A capstan winch.

Gyratory Crusher. A crusher having a central conical member with an eccentric motion in a circular chamber tapering from a wide top opening.

Half Track. A heavy truck with high speed crawler track drive in the rear and driving wheels in front.

Hammer Mill (Hammermill). A rock crusher or a shredder employing hammers or flails on a rapidly rotating axle.

Handle (Stick). In a dipper shovel or hoe, the arm that connects the bucket with the boom.

Hand Level. A sighting level that does not have a tripod, base, or telescope.

Hardpan. Hard, tight soil.

A hard layer that may form just below plow depth on cultivated land.

Harrow. An agricultural tool that loosens and works the ground surface.

Haul. Average Haul—The average distance a grading material is moved from cut to fill.

Haul Distance—Is the distance measured along the center line or most direct practical route between the center of the mass of excavation and the center of mass of the fill as finally placed. It is the distance material is moved.

Haul, station yards of—Equals the number of cubic yards multiplied by the number of 100 ft. stations through which it is moved.

Haul, free—Is the distance every cubic yard is entitled to be moved without an additional charge for haul.

Haul, over-—Is the distance in excess of that

given as the stated haul distance to haul excavated material.

Haulageway. A main tunnel connecting underground excavation areas with an exit.

Haulaway. An excavation method which involves hauling the spoil away from the hole.

Haunch. In pipe, the sides of the lower third of the circumference.

Head. Height of water above a specified point. The back-pressure against a pump from a high outlet.

Heading. In a tunnel, a digging face and its work area.

Head Mast. In a cable excavator, the tower that carries the working lines.

Headwall (Sidewall). A culvert sidewall. Sometimes only the upstream wall.

Heap. The soil carried above the sides of a body or bucket.

Heel. A floor brace or socket for wall-bracing timbers.

The trailing edge of an angled blade.

Heeling In. Temporary planting of trees and shrubs.

Helical. Spiral.

H. I. Height of instrument.

High Line. A high tension electric line.

Electric power supplied by a utility.

High Wall. A face which is being excavated, as distinguished from spoil piles.

Undisturbed soil or rock bordering a cut.

Hinge. A connection which allows swinging motion in one plane.

Hitch. A horizontal shelf along the side of a rock tunnel, that supports roof timbers.

A connection between two machines.

Hoe (Backhoe, pullshovel). A shovel that digs by pulling a boom-and-stick-mounted bucket toward itself.

Hog Box. A concrete box in which water and dirt are mixed to be pumped to a fill.

Hoist. The mechanism by which a bucket or blade is lifted, or the process of lifting it.

Hood. A casing on the end of a suction line that causes it to pick up material from the bottom only.

A curved baffle that prevents scattering and separation of material discharged by a conveyor belt.

Hook, cable. A round hook with a wide beveled face.

Hook, grab. A chain hook that will slide over any one link, but will not slide along the chain.

Hook, pintle. A towing bracket having a fixed lower part, and a hinged upper one, which when locked together make a round opening that can hold a tow ring.

Hook, round (Slip hook). A hook that has smooth inner surface, and will slide along a chain.

Hook, safety (Lockon hook). A round hook with a hinged piece across the opening, that

allows a line to enter it readily, but requires special manipulation to remove it.

Hook, swivel. A hook with a swivel connection to its base or eye.

Hopper. A storage bin or a funnel that is loaded from the top, and discharges through a door or chute in the bottom.

Horizon. A horizontal layer.

Horse. A saw horse or other simple frame or support.

Horsehead (False set). A temporary support for forepoles used in tunneling soft ground.

HP (hp). Horsepower.

Horsepower. A measurement of power that includes the factors of force and speed.

The force required to lift 33,000 pounds one foot in one minute.

Horsepower, drawbar. Horsepower available to move a tractor and its load, after deducting losses in the power train.

Horsepower, Indicated. The horsepower developed in the cylinders. Determined by use of an indicator gauge. Does not include engine friction losses.

Horsepower, rated. Theoretical horsepower of an engine based on dimensions and speed.

Power of an engine according to a particular standard.

Horsepower, shaft (Flywheel or belt horsepower). Actual horsepower produced by the engine, after deducting the drag of accessories.

Holdback. An automatic safety device that prevents a conveyor belt from running backward.

Holding Line. The hoist cable for a clamshell bucket.

Hot Mill. To heat metal, then shape it.

Housing. A heavy case or enclosure for rotating parts.

Hub. The strengthened inner part or mounting of a wheel or gear.

Hull. The substructure and deck of a ship or dredge.

Humus. Decayed organic matter.

A dark fluffy swamp soil composed chiefly of decayed vegetation, that is also called peat.

Hunting Tooth. A sprocket and roller chain combination in which one has an odd number of contacts and the other an even number, so that no tooth will contact the same pin twice in succession.

Hydraulic Dredge. A floating pump that sucks up a mixture of water and soil, and usually discharges it on land through pipes.

Hydraulic Fill. Fill moved and placed by running water.

Hydraulic Gradient. The slope of the surface of open or underground water.

Hydraulicking. Excavating on dry land by means of water jets.

Hydrometer. A device (usually a float in a glass tube) for measuring the specific gravity of fluids.

Hydrostatic. Relating to pressure or equilibrium of fluids.

Hygroscopic. Water absorbed from the atmosphere.

Hypoid. A pinion-and-ring gear set transmitting rotation through a right angle by means of teeth having structure intermediate between a bevel and a worm set.

I.C.C. Interstate Commerce Commission.

Idler. A wheel or gear which changes the direction of rotation of shafts, or the direction of movement of a chain or belt.

Impeller. A rotary pump member using centrifugal force to discharge a fluid into outlet passages.

Impervious. Resistant to movement of water.

Inclined Plane. A slope used to change the direction and speed-power ratio of a force.

Inertia. The property of matter by which it will remain at rest, or in uniform motion in a straight line, unless acted upon by an external force.

Inhaul. The line or mechanism by which a cable excavator bucket is pulled toward the dump point.

Injector. In a diesel engine, the unit that sprays fuel into the combustion chamber.

Instrument. A telescopic level, such as a transit or a builders' level.

Intercepting Drain. Curtain drain.

Intercooler. A radiator in which air is cooled while moving from low pressure to high pressure cylinders of a two stage compressor.

Intermediate Shaft. A shaft which is driven by one shaft, and drives another.

Interruptions. Secondary cutters in auger drills.

Invert. The inside bottom of a pipe or tunnel.

Jack. A mechanical or hydraulic lifting device. A hydraulic ram or cylinder.

Jack Boom. A boom which supports sheaves between the hoist drum and the main boom in a pull shovel or a dredge.

Jack Hammer. An air drill that hammers and rotates a hollow steel and a bit, and that can be operated by one man.

Jackknife. A tractor and trailer assuming such an angle to each other that the tractor cannot move forward.

Jackleg. An outrigger post.

Jackshaft. A short drive shaft, usually connecting a clutch and transmission.

Jars. A tool in the churn drill string which contains slack to allow hammering upward to free a stuck bit.

Jaw. In a clutch, one of a pair of toothed rings, the teeth of which face each other.

In a crusher, one of a pair of nearly flat faces separated by a wedge-shaped opening.

Jaw Clutch. A clutch consisting of two toothed jaws, one of which slides along its shaft to engage or disengage from the other.

Jaw Crusher. A fixed and a movable jaw widely spaced at the top and close at the bottom, with means to move one jaw toward and away from the other.

Jetting. Drilling with high pressure water or air jets.

Jetty. A long fill or structure extending into water from the shore, that serves to change the direction or velocity of water flow.

Jib Boom. An extension piece hinged to the upper end of a crane boom.

Jig. A guide used in shaping pieces of wood or metal.

Journal. That part of a rotating shaft or axle which turns in a load-supporting bearing.

Jumbo. A number of drills mounted on a mobile carriage, and used in tunnels.

Kelly. A square or fluted pipe which is turned by a drill rotary table, while it is free to move up and down in the table. Also called grief stem.

Key. A hard steel strip inserted in matching grooves (keyways) in a shaft and a hub to make them turn as a unit.

Keyhole Slot. A slot enlarged at one end to allow entrance of a chain or bolt that can then be held by the narrow end.

Keyway. A square edged lengthwise slot in a shaft or hub.

Kill. Cut off electric current from a circuit. Stop an engine.

Kilowatt. An electrical unit of work or power. Equal to 1000 watts, 1.34 horsepower, and 1.18 KVA.

Kingpin (King pin). A vertical swivel or hinge pin, usually supported at both top and bottom.

Knife. The dirt cutting edge of a digging machine.

KVA (Kilovolt-ampere) Approximately $8\frac{9}{100}$ of a kilowatt.

Lacing. Small boards or patches that prevent dirt from entering an excavation through spaces between sheeting or lagging planks.

Ladder. The digging boom assembly in a hydraulic dredge or chain-and-buckets ditcher.

Ladder Ditcher. A machine that digs ditches by means of buckets in a chain that travels around a boom.

Lag. Delay in one action following another. To install lagging, or increase the diameter of a drum.

Lagging. The surface or contact area of a drum or flat pulley, especially a detachable surface or one of special composition.

In a tunnel, planking placed against the dirt or rock walls and ceiling, outside the ribs.

Boards fastened to the back of a shovel for blast protection.

Lagging, split. Drum lagging made in two

pieces to allow changing it without dismantling the drum.

Land. A backfurrow.

Land Leveler. A towed scraper with a bottomless bucket centrally mounted in a long frame. Used chiefly in agricultural grading.

Land Tile. Porous clay pipe with open (butt) joints.

Laminated. In thin parallel layers.

Lantern. In a centrifugal pump, a hollow casing on the engine side of the pump body.

Lapped. Overlapped and fitted together.

Lay. The direction of twist in wires and strands in wire rope.

Lay, lang. A wire rope construction in which the wires are twisted in the strands in the same direction as the strands are twisted in the rope.

Lay, regular. A wire rope construction in which the direction of twist of the wires in the strands is opposite to that of the strands in the rope.

Layshaft. A fixed shaft supporting revolving drums.

Lead (Leader). In a pile driver, the usually vertical hanging beam that guides the hammer and the pile.

Lead Wires. In blasting, the heavy wires that connect the firing current source or switch with the connecting or cap wires.

Leg. A side post in tunnel timbering.

A wire or connector in one side of an electrical circuit.

Level. To make level or to cause to conform to a specified grade.

Any instrument that can be used to indicate a horizontal line or plane.

Leveling Rod (Surveying). A telescoping rod marked in feet and fractions of feet, and fitted with a movable target or sighting disc.

Lever. A bar that pivots so that force applied at one part can work at another, usually with a change in the force-distance ratio.

Lever, first class. A bar having a fulcrum (pivot point) between the points where force is applied and where it is exerted.

Lever, second class. A lever whose force is exerted between the fulcrum and the point where it is applied.

Lever, third class. A lever to which force is applied between the fulcrum and the work point.

Lift. A step or bench in a multiple layer excavation.

Line. A cable, rope, chain, or other flexible device for transmitting pull.

To line pieces up in order to couple them together.

Line, drilling. In a churn drill, the cable that supports and manipulates the tools.

Line Oiler. An oil reservoir and metering device placed in a compressed air line to lubricate air tools.

Line, spinning. A line wrapped around a threaded pipe, so that a pull will rotate the pipe to fasten or unfasten it from another.

Lip. The cutting edge of a bucket. Applied chiefly to edges including tooth sockets.

Liquid Limit. Minimum moisture content which will cause soil to flow if jarred slightly.

Load. To place explosives in a hole.

To transfer material to a hauling unit or hopper.

Load Binder. A lever that pulls two grab hooks together, and holds them by locking over center.

Load, deck. Charges of dynamite spaced well apart in a borehole, and fired by separate primers or by detonating cord.

Load Factor. Average load carried by an engine, machine, or plant, expressed as a percentage of its maximum capacity.

Loader, belt (Elevating grader). A machine whose forward motion cuts soil with a plowshare or disc and pushes it to a conveyor belt that elevates it to a dumping point.

Loader, bucket. A machine having a digging and gathering rotor, and a set of chain mounted buckets to elevate the material to a dumping point.

Loader, front end. A tractor loader that both digs and dumps in front.

Loader, paddle. A belt loader equipped with chain driven paddles that move loose material to the belt.

Loader, reversed. A front end loader mounted on a wheel tractor having the driving wheels in front and steering at the rear.

Loader, swing. A tractor loader that digs in front, and can swing the bucket to dump to the side of the tractor.

Loader, tower. A front end loader whose bucket is lifted along tracks on a more or less vertical tower.

Loader, tractor. A tractor equipped with a digging bucket that can dump into hauling equipment.

Loam. A soft, easily worked soil containing sand, silt, and clay.

Lock. In a compressed air system, a chamber that can be opened to pressure air at one end, and to atmospheric air at the other.

Logging Tongs. Tongs with end hooks that dig in when the tongs are pulled.

Loose Yards. Measurement of soil or rock after it has been loosened by digging or blasting.

Low Bed. A machinery trailer with a low deck.

Lug Down. To slow down an engine by increasing its load beyond its capacity.

MA (Mechanical advantage). Increase in force obtained at the expense of speed or distance.

Machined. A smooth surface finish on metal. Shaped by cutting or grinding.

Magazine. A structure or container in which explosives are stored.

Manifold. A chamber or tube having a number of inlets and one outlet, or one inlet and several outlets.

Mass Diagram. A plotting of cumulative cuts and fills used for engineering computation of highway jobs.

Mass Profile. A road profile showing cut and fill in cubic yards.

Mass Shooting. Simultaneous exploding of charges in all of a large number of holes, as contrasted with firing in sequence with delay caps.

Mast. A tower or vertical beam carrying one or more load lines at its top.

Mastic. A soft sealing material.

Mat. A heavy, flexible fabric of woven wire rope or chain used to confine blasts.

A wood platform used in sets to support machinery on soft ground.

Mechanical Efficiency. As applied to engines it is the ratio of the useful horsepower available at the flywheel or power takeoff to the horsepower developed in the engine cylinders, expressed in percent.

Mesh. In wire screen, the number of openings per lineal inch.

Metering Pin. A valve plunger that controls the rate of flow of a liquid or a gas.

Millwright. A mechanic specializing in installation of heavy machinery in permanent plants.

Military Crest. A ridge that interrupts the view between a valley and a hilltop.

Millisecond Delay (Short period delay). A type of delay cap with a definite but extremely short interval between passing of current and explosion.

Mining. Usually removal of soil or rock having value because of its chemical composition.

Misfire. Failure of all or part of an explosive charge to go off.

Mixed Face. In tunneling, digging in dirt and rock in the same heading at the same time.

mm. Millimeter.

Mole (Mole ball). An egg-shaped device pulled behind the tooth of a subsoil plow to open drainage passages.

Moldboard. A curved surface of a plow, dozer, or grader blade, or other dirt mover, which gives dirt moving over it a rotary, spiral, or twisting movement.

Monitor (Giant). In hydraulicking, a high pressure nozzle mounted in a swivel on a skid frame.

Mouse Hole. In a rotary drill substructure, a socket that holds a single piece of drill pipe ready to be added to the string.

Muck. Mud rich in humus.

Finely blasted rock, particularly from underground.

Mud. Generally any soil containing enough water to make it soft.

In rotary drilling, a mixture of water with fine drill cuttings and added material, which is pumped through the drill string to clean the hole and cool the bit.

Mudcapping. Blasting boulders or other rock by means of explosive laid on the surface and covered with mud.

Multi Use Bit. A detachable drill bit that can be sharpened and reshaped when worn.

Multiple Lines. A single line reeved around two or more sheaves so as to increase pull at the expense of speed.

Net Cut. In sidehill work, the cut required less the fill required at a particular station or part of a road.

Net Fill. The fill required, less the cut required, at a particular station or part of a road.

New York Rod. A leveling rod marked with narrow lines, ruler-fashion.

Nip. The seizing of stone between the jaws or rolls of a crusher.

Nip, angle of. In a roll crusher, the angle between tangents to the roll surfaces at the widest point at which they will grip a stone.

Nipple. A short piece of pipe with male threads on each end.

Nipple, close. A nipple so short that its two sets of threads meet in the middle.

Nitroglycerin. A powerful liquid explosive that is dangerously unstable unless combined with other materials.

Normal Haul. A haul whose cost is included in the cost of excavation, so that no separate charge is made for it.

Octane Number. Percent of iso-octane by volume in a mixture of iso-octane and normal heptane that has the same anti-knock character in a standard variable compression Cooperative Fuel Research test engine as the fuel under test. Octane has anti-knock characteristics. A mixture having 75% octane and 25% heptane is said to have an octane rating of 75.

Off-Set Digging. In a ladder ditcher, digging with the boom not centered in the machine.

OHM. Unit of electrical resistance to current flow. It is equal to a fall in potential of 1 volt when a current of 1 amp. flows.

Oil. Any fluid lubricant.

Any liquid petroleum derivative that is less volatile than gasoline.

One on Two (One to two). A slope in which the elevation rises one foot in two horizontal feet.

One Part Line. A single strand of rope or cable.

Open-Cut. A method of excavation in which the working area is kept open to the sky. Used to distinguish from cut-and-cover and underground work.

Optimum. Best.

Ore. Rock or earth containing workable quantities of a mineral or minerals of commercial value.

Oscillation. Independent movement through a limited range, usually on a hinge.

Outrigger. An outward extension of a frame which is supported by a jack or block. Used to increase stability.

Overbreak. Moving or loosening of rock as a result of a blast, beyond the intended line of cut.

Overburden. Soil or rock lying on top of a pay formation.

Overhang. Projecting parts of a face or bank.

Overhaul. In many highway contracts, a movement of dirt far enough so that payment, in addition to excavation pay, is made for its haulage.

Overhead Shovel. A tractor loader which digs at one end, swings the bucket overhead, and dumps at the other end.

Overtopping. Flow of water over the top of a dam or embankment.

Overwinding. A rope or cable wound and attached so that it stretches from the top of a drum to the load.

Pad (Shoe or plate). Ground contact part of a crawler-type track.

Pan. A carrying scraper.

Parallel. An arrangement of electric blasting caps in which the firing current passes through all of them at the same time.

Parallel Series. Two or more series of electric blasting caps arranged in parallel.

Part Swing Shovel. A shovel in which the upper works can rotate through only part of a circle.

Parts of Line. Separate strands of the same rope or cable used to connect two sets of sheaves.

Pass. A working trip or passage of an excavating or grading machine.

Paving Breaker. An air hammer which does not rotate its steel.

Pawl. A tooth or set of teeth designed to lock against a ratchet.

Pay Formation. A layer or deposit of soil or rock whose value is sufficient to justify excavation.

Peat (Humus). A soft light swamp soil consisting mostly of decayed vegetation.

Peg Point (Steady point). A pointed bar in a slide clamp. Used to brace a machine during work.

Pellet Powder. Black powder made up into hollow cartridges.

Peeler. One of a set of blades that pick up and channel water moved outward by the impeller of a centrifugal pump.

Perched Water Table. Underground water lying over dry soil, and sealed from it by an impervious layer.

Permissible. Low-flame explosive used in gassy and dusty coal mines.

Petcock. A small drain valve.

pH. Percentage of free hydrogen ions. A measurement of soil acidity. pH 7 is neutral, smaller readings increasingly acid.

Philadelphia Rod. A leveling rod in which the hundredths of feet, or eighths of inches, are marked by alternate bars of color the width of the measurement.

Pig. An air manifold having a number of pipes which distribute compressed air coming through a single large line.

Pilot Valve. In a compressor, an automatic valve which regulates air pressure.

Pillow Block. A metal-cased rubber block that allows limited motion to a support or thrust member.

Pin, track. A hinge pin connecting two sections or shoes of a crawler track.

Pin, master. The only pin in an integrated crawler track that will open the track when driven out.

Pin, taper. A straight-sided pin that is smaller at one end than at the other.

Pin Thread. The male side of A.P.I. tapered thread.

Pintle. A vertical pin fastened at the bottom that serves as a center of rotation.

Pintle Hook. A towing device consisting of a fixed lower jaw, a hinged and lockable upper jaw, and a socket between them to hold a tow ring.

Pioneering. The first working over of rough or overgrown areas.

Pioneer Road. A primitive, temporary road built along the route of a job, to provide means for moving equipment and men.

Piston Displacement. The amount of air displaced by moving all pistons of an engine or compressor from the bottom to the top of their stroke.

Piston, free running. A piston not connected with a rod, that does its work by hammer-like blows.

Piston, slave. A small piston having a fixed connection with a larger one.

Piston Speed. Total feet of travel of a piston in one minute.

Pi (π) A number, approximately 3.1416 or $3\frac{1}{7}$, which when multiplied by the diameter of a circle, will give the circumference.

Pit. Any mine, quarry, or excavation area worked by the open-cut method to obtain material of value.

Pit, dig-down (Sunken pit). A pit that is below the surrounding area on all sides.

Pitman Arm. An arm having a limited movement around a pivot.

Pivot. A non-rotating axle or hinge pin.

Pitch. The slope of a surface or tooth relative to its direction of movement.

In a roller or silent chain, the space between pins, measured center to center.

Pitch Arms (Pitch braces) (Pitch rods). Rods, usually adjustable, which determine the digging angle of a blade or bucket.

Pivot Tube. A hollow hinge pin.

Pivot Shaft. A tractor dead axle, or any fixed shaft which acts as a hinge pin.

Planimeter. A device that measures an area on a map when run around its edges.

Plate, pressure. A flywheel-driven plate that can be slid along a clutch shaft to squeeze a lined plate against the flywheel.

Platform. A wood mat used in sets to support machinery on soft ground. Also called a pontoon.

An operator's station on a large machine, particularly on rollers.

Plastic Limit. The minimum amount of water in terms of percent of oven-dry weight of soil that will make the soil plastic.

Plastic Soil. A soil that can be rolled into ⅛" diameter strings without crumbling.

A soft, rubbery soil.

Plenum. Use of compressed air to hold soil from slumping into an excavation.

Plug. A stoppage in the discharge line of a dredge, or in an underground drain.

Plug and Feathers. A set of two half-round pieces of hard steel and a gradual-taper wedge, used for splitting drilled boulders.

Plug, magnetic. A drain or inspection plug magnetized for the purpose of attracting and holding iron or steel particles in lubricant.

Plumb Bob. A pointed weight hung from a string. Used for vertical alignment.

Plumbers' Dope. A soft sealing compound for pipe threads.

Ply. One of several layers of fabric or of other strength-contributing material.

Pneumatic. Powered or inflated by compressed air.

Point, well. A pipe having a fine mesh screen and a drive point at the bottom. Used for pumping out ground water.

Pond. A small lake.

In dredge work, an area where discharge water is held long enough to allow fine soil particles to settle.

Pontoon. A float supporting part of a structure, such as a bridge.

A wood platform used to support machinery on soft ground.

Poppet Valve. A valve shaped like a mushroom, resting on a circular seat, and opened by raising the stem. Standard automotive equipment.

Port. Left side of a ship or boat.

Portal. A nearly level opening into a tunnel.

Pot Hole. A small steep-sided hole, usually with underground drainage.

Poured Fitting. A connecting device which is fastened to the end of a cable (wire rope) by inserting the cable end in a funnel shaped socket, separating the wires and filling the socket with molten zinc.

Power Arm. The part of a lever between the fulcrum and the point where force is applied to the lever.

Power Control Unit. One or more winches mounted on a tractor and used to manipulate parts of bulldozers, scrapers, or other machines.

Power-Divider. A non-spin differential.

Power Takeoff. A place in a transmission or engine to which a shaft can be so attached as to drive an outside mechanism.

Power Train. All moving parts connecting an engine with the point or point where work is accomplished.

Powder. Black powder or gunpowder.

General term for explosives including dynamite, but excluding caps.

Powder, black. A mixture consisting mostly of carbon, sodium or potassium nitrate, and sulphur, used as an explosive.

Preform. In wire rope, to shape the wires so that they will lie in place.

Pre-Selective. An arrangement by which a gear lever can be moved, but the resulting speed shift will not take place until the clutch or the throttle is manipulated.

Pressure Plate. In a clutch, a plate driven by the flywheel or rotating housing, which can be slid toward the flywheel to engage the lined disc or discs between them.

Primacord. Trade marked name for a detonating fuse.

Prime. To provide means to start a process, as to supply sufficient water to a pump to enable it to start pumping. In blasting, to place a detonator in a cartridge or charge of explosive.

Primer. Usually the combination of a dynamite cartridge and a detonating cap.

Prime Mover. A tractor or other vehicle used to pull other machines.

Primary Excavation. Digging in undisturbed soil, as distinguished from rehandling stockpiles.

Profile. A charted line indicating grades and distances, and usually depth of cut and height of fill for excavation and grading work. It is commonly taken along the centerline.

Projected Pipe. A pipe laid on the surface before building a fill that buries it.

Propogation. Spread of an explosion through separated charges by concussion waves in water or mud.

Propel Shaft. In a revolving shovel, a shaft which transmits engine power to the walking mechanism.

Propeller Shaft. Usually a main drive shaft fitted with universal joints.

Prospecting, seismic. Underground exploration conducted by measuring vibrations caused by explosions set off in drill holes.

Protractor. A device for measuring angles on drawings.

P.S.I. (psi). Pressure in pounds per square inch.

P.T. Pipe thread.

Puddle. To compact loose soil by soaking it and allowing it to dry.

Puff Blowing. Blowing chips out of a hole by means of exhaust air from the drill.

Pull. To loosen the rock around the bottom of a hole by blasting. Usually used with a negative to describe a blast which did not shatter rock to the desired depth.

Pulley. A wheel that carries a cable or belt on part of its surface.

Pull Shovel (Dragshovel or hoe). A shovel with a hinge-and-stick mounted bucket that digs while being pulled inward.

Pump, centrifugal. A pump that moves water by centrifugal force developed by rapid rotation of an impeller.

Pump, diaphragm. A pump that moves water by reciprocating motion of a diaphragm in a chamber having inlet and outlet check valves.

Pump, jetting. A water pump that develops very high discharge pressure.

Pump, mud (Slush pump). The circulating pump that supplies fluid to a rotary drill.

Pump, well point. A centrifugal pump that can handle considerable quantities of air, and is used for removing underground water to dry up an excavation.

Pumping. Mechanical transfer of fluids.
Alternately raising and lowering a digging edge to increase the volume of dirt being transported.

Pulpwood. Wood to be used in making paper.

Pusher. A tractor that pushes a scraper to help it pick up a load.

Quadrant. A quarter of the circumference of a circle.
A curved guide for a lever.
A curved scale for measuring angles.

Quarry. A rock pit.
An open cut mine in rock chosen for physical rather than chemical characteristics.

Quarter Octagonal. A square shaft with corners cut back.

Quicksand. Fine sand or silt that is prevented from settling firmly together by upward movement of ground water.
Any wet inorganic soil so unsubstantial that it will not support any load.

Quill Shaft. A light drive shaft inside a heavier one, and turning independently of it.

Rat Hole. In a rotary drill substructure, a socket that supports the kelly and swivel when they are not in use.

Races. The inner and outer rings of a ball or roller bearing.

Radial. Lines converging at a single center.

Radius. Horizontal distance from the center of rotation of a crane to its hoisting hook.

Rake Blade. A dozer blade or attachment made of spaced tines.

Rake, brush. A rake blade having a high top and light construction.

Rake, rock. A heavy duty rake blade.

Rail. A piece of railroad type track.
The chain or inner surface of a crawler track.

Raise. A shaft being dug upward from a tunnel.

Ram. A hydraulic cylinder.
The moving weight in a pile driving hammer.

Ram, one way or single acting. A hydraulic cylinder in which fluid is supplied to one end so that the piston can be moved only one way by power.

Ram, two way or double acting. A hydraulic cylinder in which fluid can be supplied to either end, so the piston can be moved by power in two directions.

Ramp. An incline connecting two levels.

Range Pole. A pole marked in alternate red and white bands one foot high.

Ratchet. A set of teeth which are vertical on one side and sloped on the other, which will hold a pawl moving in one direction, but allow it to move in another.

Ratio of Reduction. The relationship between the maximum size of the stone which will enter a crusher, and the size of its product.

Reamer. A cutting device that enlarges or straightens a hole.

Reamer Shell. A cutter just above a diamond bit, used to assure a full-size hole.

Rearing. Rising of the front of a tractor when pulling a heavy load.

Receiver. The air tank or reservoir on a compressor.

Reciprocating. Having a straight back-and-forth or up-and-down motion.

Reduction, double. Two sets of gears in series that both reduce speed and increase power.

Reduction, single. A gear set that causes one shaft to turn another at reduced speed.

Reclaiming. Digging from stockpiles.
Reprocessing previously rejected material.

Reel. A revolving rack used for storage of hose and cable.
In a churn drill, the winches are usually called reels.

Reel, bull (Spudding reel). The churn drill winch that lifts and lowers the drill string.

Reel, calf (Casing reel). The churn drill winch used for handling casing and for odd jobs.

Reel, dead. A storage reel.

Reel, live. A reel that supplies air, water, or electricity to the inner end of the hose or wire wound on it.

Reel, sand. In a churn drill, the high speed winch that lifts the bailing cylinder.

Reeving. Threading or placement of a working line.

Refusal. The depth beyond which a pile cannot be driven.

Relay. A valve or switch that amplifies or restores original strength to an air, hydraulic, or electrical impulse.

Relief Holes. Holes drilled closely along a line, which are not loaded, and which serve to weaken the rock so that it breaks on that line.

Relief Valve. A valve which will allow air or fluid to escape if its pressure becomes higher than the valve setting.

Retaining Wall. A wall separating two levels.

Retract. The mechanism by which a dipper shovel bucket is pulled back out of the digging.

Reverse Bend. To bend a line over a drum or a sheave, and then in the opposite direction over another sheave.

Reversing Clutch. A forward-and-reverse transmission which is shifted by a pair of friction clutches.

Revetment. A wall sloped back sharply from its base.

A masonry or steel facing for a bank.

Revolving Shovel. A digging machine in which the upper works can revolve independently of the supporting unit.

Rheostat. A device that regulates flow of electricity by varying the amount of resistance in the circuit.

Rib. A ridge projecting above grade in the floor of a blasted area.

Ridge Terrace. A ridge built along a contour line of a slope to pond rainwater above it.

Rifling. Forming a spiral thread on the wall of a drill hole, which makes it difficult to pull out the bit.

Rifle Bar. A cylinder with curved splines.

Rifle Nut. A splined nut that slides back and forth on a rifle bar.

Rig. A general term, denoting any machine. More specifically, the front or attachment of a revolving shovel.

Riprap. Heavy stones placed at water's edge to protect soil from waves or current.

Riparian Rights. Rights of a land owner to water on or bordering his property, including right to prevent diversion or misuse of upstream water.

Ripper. A towed machine equipped with teeth, used primarily for loosening hard soil and soft rock.

Road Metal. Crushed stone used in road surfaces.

Hard pavement.

Roadster. Low priced model of a scraper or a truck.

Rob. To remove part of an installation for use elsewhere.

To take out supporting pillars or walls of pay rock in a mine.

Rock. The hard, firm, and stable parts of earth's crust.

Any material which requires blasting before it can be dug by available equipment.

Rocker Arm. A lever resting on a curved base so that the position of its fulcrum moves as its angle changes.

A bell crank with the fulcrum at the bottom.

Rocking. Pushing a resistant object repeatedly, and backing or rolling back between pushes to allow it to reach or cross its original position.

Rod Stock. Round steel rod.

Roll. The wheel of a roller.

Roll, compression. The drive wheel of a roller.

Roll, guide. The front or steering wheel of a roller.

Roller, hook. In a revolving shovel, a roller attached by a bracket to the revolving section, and contacting the lower face of a circular track on the travel unit.

Roller, support. In a crawler machine, a roller that supports the slack upper part of the track.

Roller, swing. In a revolving shovel, one of several tapered wheels that roll on a circular turntable and support the upper works.

Roller, track. In a crawler machine, the small wheels that rest on the track and carry most of the weight of the machine.

Roller, truck. A track roller.

Root Buttress. A root that is above ground where it joins the trunk.

Root Hook. A very heavy hook designed to catch and tear out big roots when it is dragged along the ground.

Rooter. A heavy duty ripper.

Rotary Table. The part of a rotary drill which turns the kelly and drill string.

Rotation Firing. Crushing a small piece of rock with a first explosion, and timing other holes to throw their burdens toward the space made by that and other preceding explosions. Or row shooting.

Rotary (Rotary table). In a rotary drill, the unit that turns the kelly and drill string.

Rotary Tiller. A machine that loosens and mixes soil and vegetation by means of a high speed rotor equipped with tines.

Rotor. Any unit that does its work in a machine by spinning, and does not drive other parts mechanically.

Round. A blast including a succession of delay shots.

Row Shooting. In a large blast, setting off the row of holes nearest the face first, and other rows behind it in succession.

R.P.M. or *rpm.* Revolutions per minute.

Rubble Drains. French drains.

Rule of Thumb. A statement or formula that is not exactly correct, but is accurate enough for use in rough figuring.

Run Levels. To survey an area or strip to determine elevations.

Running. Operating, particularly a drill.

Saddle Block. In a dipper shovel, the boom swivel block through which the stick slides when crowded or retracted.

Sand. A loose soil composed of particles between $\frac{1}{16}$ mm and 2 mm in diameter.

Rock chips and other waste produced by drilling action.

Sandhog. A man who works in compressed air.

Safety Clutch. A clutch that slips instead of transmitting loads beyond the capacity of the machine.

Safety Factor. The ratio between breakage resistance and load.

Scaling. Prying loose pieces of rock off a face or roof to avoid danger of their falling unexpectedly.

Scalping Screen. Usually a vibrating grizzly.

Scarifier. An accessory on a grader, roller, or other machine, used chiefly for shallow loosening of road surfaces.

Scavenge. To clean out thoroughly.

To pick up surplus fluid and return it to a circulating system.

Schematic. Showing principles of construction or operation, without accurate mechanical representation.

Scoria. Cinderlike lava filled with bubbles.

Brick-like material formed by volcanic heating of clay beds.

Scour. Erosion in a stream bed, particularly if caused or increased by channel changes.

Scraper (Carrying scraper) (Pan). A digging, hauling, and grading machine having a cutting edge, a carrying bowl, a movable front wall (apron) and a dumping or ejecting mechanism.

Scraper, bottom dump. A carrying scraper that dumps or ejects its load over the cutting edge.

Scraper, drag. A digging bucket operated on a cable between a mast and an anchor, that is not lifted off the ground during a normal cycle.

A two wheel tractor-towed scraper equipped with a bottomless bucket.

Scraper, rear dump. A two wheel scraper that dumps at the rear.

Scraper, self powered. A scraper built into a single unit with a tractor.

Scraper, two axle. A full trailer type carrying scraper.

Screen. A mesh or bar surface used for separating pieces or particles of different sizes.

A filter.

Screen, deck. Two or more screens placed one above the other for successive processing of the same run of material.

Screen, scalping. A coarse primary screen or grizzly.

Screen, shaking. A screen that is moved with a back-and-forth or rotary motion to move material along it and through it.

Screen, vibrating. A screen that is vibrated to move material along it and through it.

Seam. A layer of rock, coal, or ore.

Section. An area equal to 640 acres or 1 square mile.

A part of a work area or strip.

Seepage. Movement of water through soil without formation of definite channels.

Seize. To bind wire rope with soft wire, to prevent it from ravelling when cut.

Selective Digging. Separating two or more types of soil while digging them.

Semi-Grouser. A crawler track shoe with one or more low cleats.

Semi-Trailer. A towed vehicle whose front rests on the towing unit.

Serrated. (An edge) cut into a line of teeth.

Series. An arrangement of electric blasting caps in which all the firing current passes through each of them in a single circuit.

Set. The distance a pile penetrates with one blow from the driving hammer.

Set, timbering. A tunnel support consisting of a roof beam or arch, and two posts.

Sewer Tile. Glazed waterproof clay pipe with bell joints.

Shackle. A connecting device for lines and draw bars which consists of a U shaped section pierced for a cross bolt or a pin.

Shaft. A round bar that rotates or provides an axis of revolution.

A vertical or steeply inclined tunnel.

Shaft, cam (Camshaft). A shaft carrying cams which open and close valves.

Shaft, counter (Countershaft). A shaft that allows one end of a (main) shaft to drive the other through reduction gears.

Shaft, crank (Crankshaft). The main shaft of a piston-type engine, that converts reciprocating motion into rotation.

Shaft, idler. A shaft that carries a gear that reverses direction of rotation in a transmission.

Shaft, input. The shaft that delivers engine power to a transmission or clutch.

Shaft, jack (Jackshaft). A short driveshaft, usually connecting a clutch and a transmission.

Shaft, lay. A fixed shaft supporting rotating drums or gears.

Shaft, main (Mainshaft). The transmission shaft forming a continuation of the input shaft.

Shaft, output. A shaft that transmits power from a transmission or clutch.

Shaft, reversing. A shaft whose direction of rotation can be reversed by use of clutches or brakes.

Shaking Screen. A suspended screen which is moved with a back-and-forth or rotary motion with a throw of several inches or more.

Shale. A rock formed of consolidated mud.

Shale Pit. A dumping place for coarse material screened out of rotary drill mud.

Shale Shaker. A screen in the mud circulating system of a rotary drill.

Shank (Standard). The connecting bar between a ripper or scarifier tooth and the frame.

The part of drill steel that fits into the drill.

Sheave (pronounced "shiv"). A grooved wheel used to support cable or change its direction of travel.

Sheave, traveling. A sheave block that slides in a track.

Sheave, padlock. The bucket sheave on a dipper or hoe shovel.

A sheave set connecting inner and outer boom lines.

Sheave Block. A pulley, and a case provided with means to anchor it.

Sheepsfoot. A tamping roller with feet expanded at their outer tips.

Sheet Erosion. Lowering of land by nearly uniform removal of particles from its entire surface by flowing water.

Sheeting Jacks. Push-type turnbuckles, used to set ditch bracing.

Sheet Piling. Steel strips shaped to interlock with each other when driven into the ground.

Sheeting. $7/8''$ tongue and groove board.

Planks used in shoring and bracing.

Sheeting Driver. An air hammer attachment that fits on plank ends so that they can be driven without splintering.

Shift. A work period.

Shift, graveyard. Work "day" from midnight to 8:00 A.M.

Shift, swing. Work "day" from 4:00 P.M. to midnight.

Occasionally refers to the midnight to 8:00 A.M. shift.

Shipper Shaft. In a dipper shovel, the hinge on which the stick pivots when the bucket is hoisted.

Shoe. A ground plate forming a link of a track, or bolted to a track link.

A support for a bulldozer blade or other digging edge to prevent cutting down.

A cleanup device following the buckets of a ditching machine.

Shoe, tile. A box towed behind a ditching machine, in which tile can be laid on the ditch bottom.

Shot Rock. Blasted rock.

Shoot. Blast.

Shoring. Temporary bracing to hold the sides of an excavation from caving.

Shovel. A digging and loading machine or tool.

Shovel, dipper (Shovel) (Dipper stick). A revolving shovel that has a push type bucket rigidly fastened to a stick that slides on a pivot in the boom.

Shovel Dozer (Dozer shovel). A tractor equipped with a front-mounted bucket that can be used for pushing, digging, and truck loading.

Shovel, hoe (Dragshovel, pullshovel, ditching shovel, backhoe). A revolving shovel having a pull-type bucket rigidly attached to a stick hinged on the end of a live boom.

Shovel, hydraulic. A revolving shovel in which drums and cables are replaced by hydraulic rams and/or motors.

Shovel, part swing. A revolving shovel that cannot swing through a full circle.

Shovel, revolving. A digging machine that has the machinery deck and attachment on a vertical pivot, so that it can swing independently of its base.

Shoulder. The graded part of a road on each side of the pavement.

The side of a horizontal pipe, at the level of the center line.

Shrinkage. Loss of bulk of soil when compacted in a fill. Usually is computed on the basis of bank measure.

Shunt. A connection between the two wires of a blasting cap which prevents building up of opposed electric potential in them.

Shuttle. A back and forth motion of a machine which continues to face in one direction.

Sidecasting. Piling spoil alongside the excavation from which it is taken.

Side Hill. A slope that crosses the line of work.

Sidehill Cut. A long excavation in a slope that has a bank on one side, and is near original grade on the other.

Sidewalls. Walls, usually masonry, at each end of a culvert.

Silicosis. A lung disease caused chiefly by inhaling rock dust from air drills.

Silt. A soil composed of particles between 1/256 mm and 1/16 mm in diameter.

A heavy soil intermediate between clay and sand.

Silting. Filling with soil or mud deposited by water.

Silt Trap. A settling hole or basin that prevents water-borne soil from entering a pond or drainage system.

Siphon. A tube or pipe through which water flows over a high point by gravity.

Six by Six (6×6). A truck having drive to the front wheels and to tandem rear wheels.

Six Wheeler (Ten wheeler). A truck with two sets of rear axles.

Skewed. On a horizontal angle, or in an oblique course or direction.

Skip. A non-digging bucket or tray that hoists material.

Skirt (Skirt board). A vertical strip placed at the side of a conveyor belt to prevent spillage or to increase capacity.

Skiving. To dig in thin layers.

Skullcracker. A steel ball swung from a crane boom. Used for demolishing buildings and for breaking boulders.

Slab. The deck or floor of a concrete bridge.

Any horizontal section of masonry.

Slack Adjuster. In air brakes, the connection between the brake chamber and the brake cam.

Slackline (Slackline cableway). A cable excavator having a track cable which is loosened to lower the bucket, and tightened to raise it.

Slag. Refuse from steel-making.

Slave Unit. A machine which is controlled by or through another unit of the same type.

Sleeper. In corduroy roads, a cross log or timber supporting the stringers (longitudinal supports).

Slick Hole. A hole column loaded with explosive, without springing.

Slick Sheets. Thin steel plate spread on a tunnel floor before a blast, to make hand mucking easier.

Slide. A small landslide.

Slide Coupling. A slip joint.

Sling. A lifting hold consisting of two or more strands of chain or cable.

Sling Block. A frame in which two sheaves are mounted so as to receive lines from opposite directions.

Slat Bucket. A digging bucket of basket construction, used in handling sticky, chunky mud.

Slip Joint. A splined connection loose enough to allow its two parts to slide on each other to change shaft length.

Slow Powder. Black powder, often called gunpowder. Also, some of the slow acting dynamites.

Sludge Samples. Samples of mud from a rotary drill, or sand from a churn drill, used to obtain information about the formation being drilled.

Slugger. A tooth on a roll-type rock crusher.

Sluice. A steep, narrow waterway.

Slurry. Cement grout.

Slush Pump. The mud pump for a rotary drill.

Slusher. A mobile drag scraper with a metal slide to elevate the bucket to dump point.

Smart Aleck. A limit switch that cuts off power if a machine part is moved beyond its safe range.

Smoother Bar. A drag that breaks up lumps behind a leveling machine.

Snag Boat. A boat equipped with a hoist and grapple for clearing obstacles from the path of a dredge.

Snake Hole. A hole driven into a toe for blasting, with or without vertical holes.

Snakeholing. Drilling under a rock or face in order to blast it.

Snaking. Towing a load with a long cable.

Inserting a tow or hoist line under an object without moving the object.

Snatch Block. A pulley in a case which can be easily fastened to lines or objects by means of a hook, ring, or shackle.

Soil. The loose surface material of the earth's crust.

Soil, heavy. A fine grained soil, made up largely of clay or silt.

Solid Loading. Filling a drill hole with all the explosive which can be crammed into it, except for stemming space at top.

Space. In a screen, the actual dimension of the clear opening between adjacent parallel wires or bars.

Spaced Loading. Loading so that cartridges or groups of cartridges are separated by open spacers which do not prevent the concussion from one charge from reaching the next.

Spacing. The distance between drill holes along a line parallel to the face.

Spall. To break off from a surface in sheets or pieces.

Specific Gravity (Solids or liquids). The ratio of the mass of a body to an equal volume of water.

Spider Gear (Carrier pinion). A differential gear which rotates on its shaft in a rotating case.

Spile (Forepole). A plank driven ahead of a tunnel face for roof support.

Spillway. An overflow channel for a pond or a terrace channel.

Spinning Line. A chain or rope used as a wrench in attaching and detaching drill pipe sections.

Spiraling. Rifling.

A drill hole twisting into a spiral around its intended center line.

Spiral Cleaner. A device for removing dirt from a conveyor belt.

Spirit Level. A glass tube containing fluid and an air bubble.

Spline. A set of parallel grooves running lengthwise of a shaft.

Split Sprocket. A two piece sprocket that can be assembled on a shaft without removing the shaft bearings.

Spoil. Dirt or rock which has been removed from its original location.

Spool. The movable part of a slide-type hydraulic valve.

To wind in a winch cable.

Spot (Trucks). To direct to the exact loading or dumping place.

Spot Log. A log or marker placed to show a truck driver the spot where he should stop to be loaded.

Spotter. In truck use, the man who directs the driver into loading or dumping position.

In a pile driver, the horizontal connection between the machinery deck and the lead (pile guide).

Spring, helper. On a truck rear axle, an upper spring that carries no weight until the regular spring changes shape under load.

Springing. Enlarging the bottom of a drill hole by exploding a small charge in it.

Spring Line. The meeting of the roof arch and the sides of a tunnel.

Spring Loaded. Held in contact or engagement by springs.

Sprocket. A gear that meshes with a chain or a crawler track.

Spuds. On a dredge, steel tubes pointed at the bottom and provided with lifting tackle at the top which are used to hold and to move the dredge.

Spud Well. On a dredge, a pair of guide collars for a spud.

Spudding Drill (Churn drill). A drill that makes hole by lifting and dropping a chisel bit.

Spur. A rock ridge projecting from a side wall after inadequate blasting.

Spur Valley. A short branch valley.

Squib. A detonator consisting of a firing device, and a chemical that will burn with a flash which will ignite black powder.

Stab. In adding to a drill string, the action of lining up and catching the threads of the loose piece.

Stabilize. To make soil firm and to prevent it from moving.

Stadia. Measurement of distance by proportion to the space on a vertical rod seen between upper and lower instrument cross hairs. Usual proportion is one vertical to 100 horizontal.

Stake, side. On a road job, a stake on the line of the outer edge of the proposed pavement.

Any stake not on the center line.

Stake, slope. A stake marking the line where a cut or fill meets the original grade.

Stacker. A large mobile elevating belt.

Starboard. Right side of a boat.

Starter. In drilling, a short steel used to start a drill hole.

Static Balance. A condition of rest created by inertia (dead weight) sufficient to oppose outside forces.

Static Load. A load that is at rest and exerts downward pressure only.

Station. Any one of a series of stakes or points indicating distance from a point of beginning or reference.

Station, minus. Stakes or points on the far side of the zero point from which a job was originally laid out.

Stator (Reactor). In a torque converter, a set of fixed vanes that change the direction of flow of fluid entering the pump or the next stage turbine.

Steady Point (Peg point). A pointed steel bar which can be locked in a clamp, and is used to brace a drill frame against the ground.

Steel. In air hammers, the hollow or solid steel bar which connects the hammer with the cutting tool.

Steel, alloy. Steel compounded with other metals to increase strength, wearing or rust resistance, or to obtain other desired qualities.

Steel Centralizer. On a wagon drill, a guide to hold the starting steel in proper alignment.

Steel Changes. The difference in length between successive steels used in drilling one hole.

Steel Puller. A hinged clamp on the bottom of a hand drill.

Steering Brake. A brake which slows or stops one side of a tractor.

Steering Clutch. A clutch which can disconnect power from one side of a tractor.

Stemming. Dirt or other inert material placed in parts of a drill hole instead of explosives.

Stick or Handle. In a dipper shovel or pull shovel, a rigid bar hinged to the boom and fastened to the bucket.

Stockpile. Material dug and piled for future use.

Stone. Rock.

Stone Boat. A flat steel sled with an up-curved front.

Stope. An underground excavation that is made in a series of steps or benches.

Street Ell. A pipe elbow with male threads on one end, female on the other.

Strength. In an explosive, the energy content in relation to its weight.

Stress. The force per unit area. When the force is one of compression it is known as "pressure." It is an internal force which resists an external force.

String (of tools). In a churn drill, the tools suspended on the drilling cable.

String Loading. Filling a drill hole with cartridges smaller in diameter than the hole, without slitting or tamping them.

Stringer. A beam running lengthwise of a bridge or wood road.

String Level. A spirit level equipped with prongs so that it can be hung from a string.

Strip. Remove overburden or thin layers of pay material.

Stripping. Removal of a surface layer or deposit, usually for the purpose of excavating other material under it.

Stripping Shovel. A shovel with a specially long boom and stick that enable it to reach farther and pile higher.

Stoper. A hand-size air drill mounted on a column or other support.

Strut. An inside brace.

Stud. A bolt having one end firmly anchored.

Stuffing Box. A space around a shaft filled with soft packing to prevent fluids or gases from leaking along it.

Stumper. A narrow heavy dozer attachment used in pushing out stumps.

Sub (Joint protector). A threaded thread protector used with drill pipe.

Subgrade. The surface produced by grading native earth, or cheap imported materials, which serves as a base for a more expensive paving.

Subsoil Plow (Pan breaker). A one-tooth ripper designed for agricultural work.

Sub Saver. A protector for the thread protector on the kelly of a rotary drill.

Suck. The shape of the bottom of a cutting

edge or tooth which tends to pull it into the ground as it is moved.

Suction. Atmospheric pressure pushing against a partial vacuum.

The "pull" of a pump.

Adhesion of a mass of mud to the underside of an object being lifted out of it.

Sump. A low spot to which water is drained, and from which it is removed by a pump.

Sun Gear. The central gear in a planetary set.

Sun Gears. A planetary gear set consisting of a central gear, an internal-tooth ring gear, and two or more planet gears meshed with both of them.

Supercharger. A blower that increases the intake pressure of an engine.

Surge Bin. A compartment for temporary storage, which will allow converting a variable rate of supply into a steady flow of the same average amount.

Surveying. To find and record elevations, locations, and directions, by means of instruments.

Sweat. To unite two closely fitting pieces by enlarging the outer one by heat.

Swell (Growth). Increase of bulk in soil or rock when it is dug or blasted.

Swing. In revolving shovels, to rotate the shovel on its base.

In churn drills, to operate a string of tools.

Swing Angle. The distance in degrees which a shovel must swing between digging and dumping points.

Switchback. A hairpin curve.

Swivel Head. In a diamond drill, the mechanism that rotates the kelly and drill string.

Synchromesh. A silent-shift transmission construction, in which hub speeds are synchronized before engagement by contact of leather cones.

Tagline. A line from a crane boom to a clamshell bucket that holds the bucket from spinning out of position.

Tail. The rear of a shovel deck.

The anchor end of a cable excavator.

Tail Anchor. The anchor for a track cable, or the turn point for a backhaul line in a cable excavator.

Tailblock. The boom foot and idler sprocket assembly on a ladder ditcher.

Tailboard. Tailgate.

Tailgate. The hinged rear wall of a dump truck body.

The hinged or sliding rear wall of a scraper bowl.

Tailings. Second grade or waste material separated from pay material during screening or processing.

Tail Swing. The clearance required by the rear of a revolving shovel.

Talus. Loose rock or gravel formed by disintegration of a steep rock slope.

Tamp. Pound or press soil to compact it.

Tamper. A tool for compacting soil in spots not accessible to rollers.

Tamping Roller. One or more steel drums, fitted with projecting feet, and towed by means of a box frame.

Tandem. A double-axle drive unit for a truck or grader. (A bogie).

A pair in which one part follows the other.

Tandem Drive. A three-axle vehicle having two driving axles.

Tangent. A line that touches a circle and is perpendicular to its radius at the point of contact.

Taproot. A big root that grows downward from the base of a tree.

Target Rod. A leveling rod.

Tee. A pipe fitting that has two threaded openings in line, and a third at right angles to them.

Telescope. To slide one piece inside another.

Terrace. A ridge, a ridge and hollow, or a flat bench built along a ground contour.

Terrain. Ground surface.

Ten Wheeler (Six wheeler). A truck with tandem rear axles.

Three Part Line. A single strand of rope or cable doubled back around two sheaves so that three parts of it pull a load together.

Thrible. Three sections of drill pipe handled as a unit.

Thorough Cut. Through cut.

Through Cut. An excavation between parallel banks that begins and ends at original grade.

Throw. The longest straight distance moved in the stroke or circle of a reciprocating or rotary part.

Scattering of blast fragments.

Throwout Bearing. A bearing, sliding on a clutch jackshaft, that carries the engage-and-disengage mechanism.

Thrust Arm. A cable-controlled bar that can slide by power in two directions.

Thrust Washer. A washer that holds a rotating part from sideward movement in its bearings.

Tight. Soil or rock formations lacking veins of weakness.

Blasts or blast holes around which rock cannot break away freely.

Tile. Pipe made of baked clay.

Tile, land. Short pieces of porous pipe with butt (open) joints, used for underground drainage.

Tile, sewer. Glazed clay pipe with bell joints.

Tile Shoe (Tile box). A device that permits laying tile directly behind a ditcher.

Tilth. Soil condition in relation to lump or particle size.

Timber. Wood beams larger than 4 x 6.

Trees.

Timbering. Wood bracing in a tunnel or excavation.

Toe. The projection of the bottom of a face beyond the top.

Tongs. A pair of curved arms pivoted to each other, scissor fashion, so that a pull on a ring or chain connecting the short ends will cause the long ends to close to grip an object between them.

Tongue. Drawbar of a towed vehicle.

Tooth Base. The inner part of a two piece tooth on a digging bucket.

Occasionally, the socket in which a tooth fits.

Topographic Map. A map indicating surface elevation and slope.

Topsoil. The topmost layer of soil. Usually refers to soil containing humus which is capable of supporting a good plant growth.

Topping. Fine material forming a surface layer or dressing for a road or grade.

Torque. The twisting force exerted by or on a shaft, without reference to the speed of the shaft.

Torque Converter. A hydraulic coupling which utilizes slippage to multiply torque.

Torque Rod. A bar having the function of resisting or absorbing twisting strains.

Track. A crawler track.

A railroad-type track.

Track, crawler. One of a pair of roller chains used to support and propel a machine. It has an upper surface which provides a track to carry the wheels of the machine, and a lower surface providing continuous ground contact.

Tilting Dozer. A bulldozer whose blade can be pivoted on a horizontal center pin to cut low on either side.

Track Frame (Truck frame). In a crawler mounting, a side frame to which the track roller and idler are attached.

Track Roller. In a crawler machine, the small wheels which are under the track frame and which rest on the track.

Traction. The total amount of driving push of a vehicle on a given surface.

Tractive Efficiency. A measure of the proportion of the weight resting on tracks or drive wheels which can be converted into vehicle movement.

Tractor. A motor vehicle on tracks or wheels used for towing or operating vehicles or equipment.

Tractor Loader (Tractor shovel or shovel dozer). A tractor equipped with a bucket which can be used to dig, and to elevate to dump at truck height.

Trailer (Full trailer). A towed carrier which rests on its own wheels both front and rear.

Trailer, semi (Semitrailer). A towed carrier that rests on the tractor in front, and on its own wheels in the rear.

Tramp Iron. Scrap metal entering a crusher.

Transfer Case. In an all-wheel drive vehicle, a transmission or gear set that provides drive to the front shaft.

Transfer Point. Turning point.

Transit. A surveying instrument that can measure both vertical and horizontal angles.

Transmission. A gear set that permits change in speed-power ratio and/or direction of rotation.

Transmission, clutch-shifted. A constant-mesh transmission in which power is directed through gear trains by engagement of friction clutches.

Transmission, compound. A gear set in which power can be transmitted through two sets of reduction gears in succession.

Transmission, reduction-type. A transmission whose output shaft (usually the countershaft) always turns more slowly than the input shaft.

Transmission, reversing. A transmission that has only a forward and reverse shift.

Transite. Cement-asbestos pipe.

Transition Belt (Feeder conveyor). A short belt carrying material from a loading point to a main conveyor belt.

Tread. The ground contact surface on a tire or track shoe.

Occasionally, a high-friction lagging on a belt pulley.

Treadle. A foot pedal hinged to the floor at one end.

Trench. A ditch.

Trestle. A bridge, usually of timber or steel, that has a number of closely spaced supports between the abutments.

Trickle Drain. A pond overflow pipe set vertically with its open top level with the water surface.

Trim Holes (Relief holes). Unloaded drill holes closely spaced along a line to limit the breakage of a blast.

Trip. A release catch.

Tripper. A double pulley that turns a short section of a conveyor belt upside down in order to dump its load into a side chute.

Tripod. A three-legged support for a surveying instrument.

Troughing. Making repeated dozer pushes in one track, so that ridges of spilled material hold dirt in front of the blade.

Truck, bottom dump. (Dump wagon). A trailer or semitrailer that dumps bulk material by opening doors in the floor of the body.

Truck, dump. A truck or semitrailer that carries a box body with a mechanism for discharging its load.

Truck, platform (Rack body truck). A truck having a flat open body.

Truck, rear dump (End dump). A truck or semitrailer that has a box body that can be raised at the front so the load will slide out the rear.

Truck frame. Track frame.

Trunnion (Walking beam or bar). An oscillating bar which allows changes in angle between a unit fastened to its center, and another attached to both ends.

A heavy horizontal hinge.

Trussed. Braced by an assembly of members into a rigid unit.

T.U. Takeup. A mechanism for adjusting belt or chain tension.

Tub. The base of a walking dragline.

Turbine. A rotary engine driven by pressure of liquid or gas against its vanes.

Turn Angles. To measure the angle between directions with a surveying instrument.

Turning Point (Transfer point). A point whose elevation is taken from two or more instrument positions to determine their height in relation to each other.

Turntable. A base that supports a part and allows it to rotate or swing.

In a shovel, the upper part of the travel unit.

Two Part Line. A single strand of rope or cable doubled back around a sheave so that two parts of it pull a load together.

Universal Joint. A connection between two shafts that allows them to turn or swivel at an angle.

Upset. To enlarge an end of a bar by shortening it.

Vein. A layer, seam, or narrow irregular body of material different from surrounding formations.

Venturi. A pressure jet that draws in and mixes air.

Vernier. A device permitting finer measurement or control than standard markings or adjustments.

In a spudding drill, a brake adjustment that permits the line to pay out automatically as the hole deepens.

Vertical Curve. The meeting of different gradients in a road or pipe.

Vertical Drains. Usually columns of sand used to vent water squeezed out of humus by weight of fill.

Viaduct. A bridge, usually of concrete, that is supported on piers between its abutments.

Vibrating Screen. A screen which is vibrated to separate and move pieces resting on it.

Viscosity. The resistance of a fluid to flow. A liquid with a high viscosity rating will resist flow more readily than will a liquid with a low viscosity. The Society of Automotive Engineers (S.A.E.) has developed a series of viscosity numbers for indicating viscosities of lubricating oils.

Vitrify. Glaze during heat treatment.

Volt. The electromotive force which will cause a current of one ampere to flow through a resistance of one ohm.

Voltage (Potential). Electromotive force.

Volumetric Efficiency. In compressors, the relationship between c.f.m. and piston displacement.

Wadding. Paper or cloth placed over explosive in a hole.

Wagon. A trailer with a dump body.

Wagon Drill. A wheeled frame holding a pneumatic drill and a mechanism for feeding it into the rock and retracting it.

Walker. A walking dragline.

Walking Beam (Trunnion). A rigid member whose ends rest on supports that may move up and down independently, and whose center is hinged to the load it carries.

Walking Bar. A trunnion or walking beam.

Walking Dragline. A dragline shovel which drags itself along the ground by means of side mounted shoes.

Wash Boring. A test hole from which samples are brought up mixed with water.

Waste. Digging, hauling and dumping of valueless material to get it out of the way; or the valueless material itself.

Watershed. Area which drains into or past a point.

Water Table. The surface of underground, gravity-controlled water.

Watt. The power of a current of one ampere flowing across a potential difference of one volt.

Wedge. A piece that tapers from a thick end to a chisel point.

Weld. To build up or fasten together metals by bonding on molten metal.

Well. A slot in the front of a hydraulic dredge hull in which the digging ladder pivots.

A wall around a tree trunk that protects it from fill.

Well Drill. A churn drill used for water wells. It usually has a limited depth capacity and a truck or trailer mounting.

Well Point. A pipe fitted with a driving point and a fine mesh screen, used to remove underground water.

A complete set of equipment for drying up ground, including well points, connecting pipes and a pump.

Weldment. A base or frame made of pieces welded together, as contrasted with a one piece casting or a bolted or riveted assembly.

Wetting Agent. A chemical that reduces the surface tension of water so that it soaks into porous material more readily. Example—synthetic soap powder.

Whaler. A horizontal beam in a bracing structure.

Wheel, track (Truck wheel). One of a set of small flanged steel wheels resting on a crawler track and supporting a track frame.

Wheel, bull. A driving sprocket for a crawler track.

Wheel Ditcher. A wheel equipped with digging buckets, carried and controlled by a tractor unit.

Winch. A drum that can be rotated so as to exert a strong pull while winding in a line.

Winch, capstan (Cat head). A revolving spool

that exerts a pull by friction with one or more loops of fiber rope.

Winch, donkey (Yarder). A two drum towing winch.

Winch, oil field. An extremely powerful low speed winch on a crawler tractor.

Winch, power control (Power control unit). A high speed tractor mounted winch with one to three drums. Used chiefly for operation of bulldozers, scrapers, and rooters.

Winch, towing (Logging winch). A heavy duty winch mounted on the rear of a crawler tractor.

Window Pipe. A dredge discharge pipe with one or more openings in the bottom.

Windrow. A ridge of loose dirt.

Wing. Projection on an air drill bit.

Wing Wall. A wall that guides a stream into a bridge opening or culvert barrel.

Work Arm. The part of a lever between the fulcrum and the working end.

Working Cycle. A complete set of operations. In an excavator, it usually includes loading, moving, dumping, and returning to the loading point.

Working Drawing. Any drawing showing sufficient detail so that whatever is shown can be built without other drawings or instructions.

Worm. A gear formed of a cylinder with spiral threads cut in its surface.

Worm Wheel. A modified spur gear with curved teeth that meshes with a worm.

Wrist Action. In a bucket, the ability to change its digging or dumping angle by power.

INDEX

Listings in this index are arranged according to the word by word method, so that all entries beginning with one word are completed before starting those that begin with a different word or a changed form of the same word. For example, "tail swing" precedes "tailings."

Abbreviations and sets of initials are treated as single words, so that "CFM" is found after "caving." When there is a question about whether a compound word such as "backhoe" or "centerline" is one word or two, it has been divided and treated as two. However, when a term has grown so firmly together that its meaning changes when separated, as "dragline," it is placed as a single word.

Occasional small liberties have been taken with alphabetical sequence in order to keep a related series together.

Arrangement of double headings is not consistent, as it has followed the assumed interest and phraseology of those using the index. Most men think of a dipper shovel as a shovel, so the principal listing is given under "Shovel, dipper." On the other hand, the hoe shovel is considered primarily a hoe, so is detailed under "hoe." Cross references are supplied. Short listings are often duplicated, long ones may be abbreviated under the secondary heading.

Many of the listings are over-long, some of them containing dozens of page references. This is because of the size of the book, and impossibility of making clear cut or useful divisions in the subjects. In such cases, the chapter numbers can serve as a guide to the type of reference. For example, the listing for "Stability, machine" reads in part: 13-74,**86 to 89,** 16-5,6,34. The "13-" indicates Chapter 13, which deals with revolving shovels. It may therefore be assumed that pages 86 to 89 of that chapter refer to revolving shovel stability, while the pages whose numbers are preceded by "16-" deal with tractor loaders, the subject of Chapter 16.

Boldface type is used to emphasize the more important references in a group.

For convenience in reference, the chapters are listed here:

PART ONE

Chapter
1 Land Clearing (and fire fighting)
2 Surveys and Measurements, Lasers
3 Soil and Mud (and stuck machinery)
4 Cellars
5 Ditching and Dewatering (ditches, drains, pipes, pumping)
6 Ponds
7 Landscaping, and Agricultural Grading
8 Roads, including scraper work, figuring yardage, and paving
9 Blasting and Tunneling

This Index covers the text, the illustrations, and the Appendix. Negative references are included if they convey information of interest.

Much time and effort has been devoted
to making this index as complete and useful
as possible. However, in a work of this
size, it is not possible to entirely avoid
omissions and errors, and the lack of a
reference does not necessarily mean that
the desired information is not in the book.
It may also be looked for under related
subjects (if not under Steering, look under
Turn, if not there, try Truck) or in the
detailed Table of Contents in the front of
the book.

INDEX

*In hyphenated numbers the first part indicates the chapter number, the other the page number. Following simple numbers refer to additional pages in the same chapter. For example, the entry "9-31, **42**, 12-6 to 8, 10" refers to pages 31 and 42 in Chapter 9, and pages 6, 7, 8, and 10 in Chapter 12, with page 9-42 the most important reference.*

Index references cover the text and illustrations, and the Appendix, but not the Glossary.

INDEX